THE EUROPEAN GARDEN FLORA

The European Garden Flora is the definitive manual for the accurate identification of cultivated ornamental flowering plants. Designed to meet the highest scientific standards, the vocabulary has nevertheless been kept as uncomplicated as possible so that the work is fully accessible to the informed gardener as well as to the professional botanist. This new edition has been thoroughly reorganised and revised, bringing it into line with modern taxonomic knowledge. Although European in name, the Flora covers plants cultivated in most areas of the United States and Canada as well as in non-tropical parts of Asia and Australasia.

Volume I contains accounts of all the Monocotyledons, which includes those groups known informally as the 'petaloid monocotyledons' (the Liliaceae and Amaryllidaceae in the first edition, divided here among 17 families), the grasses and sedges (Gramineae and Cyperaceae), the aroids (Araceae) and the large and diverse Orchidaceae.

James Cullen has been a professional plant taxonomist for over 50 years, working particularly on the classification and identification of plants in cultivation (especially Rhododendron) at Liverpool and Edinburgh Universities, at the Royal Botanic Garden Edinburgh, and in Cambridge. With the late Dr S. M. Walters, he was the initiator of the first edition of The European Garden Flora and is responsible for two spin-offs, The Orchid Book (1992) and Manual of North European Garden Plants (2001).

Sabina Knees is a taxonomist at the Royal Botanic Garden Edinburgh and although now working on plants of the Middle East, particularly the Flora of the Arabian Peninsula and Socotra, she spent over 20 years working as a horticultural taxonomist for the Royal Horticultural Society and the Royal Botanic Garden Edinburgh and is a founder member of the Horticultural Taxonomy Group (HORTAX). She was editor of The New Plantsman for seven years and worked initially as a research associate and then as a member of the editorial committee on the first edition of The European Garden Flora.

Suzanne Cubey has worked at the Royal Botanic Garden Edinburgh (RBGE) since 1987 originally as a researcher, and then later becoming the Assistant Secretary on the Editorial Board for the first edition of The European Garden Flora. Since 2005 her main role has been as Assistant Herbarium Curator with particular responsibility for the Cultivated Plants, where she curates the cultivated specimens in the RBGE herbarium and manages the vouchering of research material from the living collections.

THE EUROPEAN GARDEN FLORA
FLOWERING PLANTS

A manual for the identification of plants cultivated in Europe, both out-of-doors and under glass

VOLUME I

Angiospermae – Monocotyledons

Second edition

edited by

James Cullen, Sabina G. Knees, H. Suzanne Cubey

assisted by

J.M.H. Shaw, P. Harrold, L. Banfield,
A. Laporte-Bisquit, M.F. Gardner, S. Neale,
G.D. Rowley, N. Zantout & C.D. Brickell

sponsored by

The Stanley Smith (UK) Horticultural Trust
The Royal Botanic Garden Edinburgh
Cambridge University Botanic Garden

CAMBRIDGE
UNIVERSITY PRESS

CAMBRIDGE UNIVERSITY PRESS
Cambridge, New York, Melbourne, Madrid, Cape Town,
Singapore, São Paulo, Delhi, Tokyo, Mexico City

Cambridge University Press
The Edinburgh Building, Cambridge CB2 8RU, UK

Published in the United States of America
by Cambridge University Press, New York

www.cambridge.org
Information on this title: www.cambridge.org/9780521761475

Second edition first published 2011
First edition published 1984–2000

Printed in the United Kingdom at the University Press, Cambridge

A catalogue record for this publication is available from the British Library

Library of Congress Cataloging-in-Publication Data

The European garden flora, flowering plants: a manual for the identification of plants cultivated in Europe, both out-of-doors and under glass/edited by James Cullen, Sabina G. Knees, H. Suzanne Cubey; assisted by J.M.H. Shaw ... [et al.]. – 2nd ed.
5 v.; cm.
Includes index.
Contents: Vol. 1. Monocotyledons: Alismataceae to Orchidaceae – Vol. 2. Dicotyledons: Casuarinaceae to Cruciferae – Vol. 3. Dicotyledons: Resedaceae to Cyrillaceae – Vol. 4. Dicotyledons: Aquifoliaceae to Hydrophyllaceae – Vol. 5. Dicotyledons: Boraginaceae to Compositae.
ISBN 978-0-521-76167-3 (set) – ISBN 978-0-521-76147-5 (Vol. 1) – ISBN 978-0-521-76151-2 (Vol. 2) – ISBN 978-0-521-76155-0 (Vol. 3) – ISBN 978-0-521-76160-4 (Vol. 4) – ISBN 978-0-521-76164-2 (Vol. 5)
1. Angiosperms–Europe–Identification. 2. Monocotyledons–Europe–Identification. 3. Dicotyledons–Europe–Identification. 4. Plants, Ornamental–Europe–Identification. I. Cullen, J. (James) II. Knees, Sabina. III. Cubey, H. Suzanne, 1964- IV. Shaw, J. M. H.
QK495.A1E97 2011
635.9094–dc23 2011021496

Volume I ISBN 978-0-521-76147-5 Hardback

CONTENTS

MAPS AND FIGURES

Figures

CONTRIBUTORS TO THE FIRST EDITION

European Advisers

Mrs E.M. Allen
Professor C.D.K. Cook
Professor H. Ern
Dr H. Heine
Sr A. Pañella
Professor P. Wendelbo

Contributors

J. Akeroyd
J.C.M. Alexander
D. Allison
S. Andrews
G. Argent
K.B. Ashburner
T. Ayres
M-J. Balkwill
P.G. Barnes
P. Berry
S.A. Bird
J.J. Bos
A. Brady
W. Brandenburg
C.D. Brickell
M.I.H. Brooker
N. Brown
S.T. Buczacki
R.K. Brummitt
G.S. Bunting
E.J. Campbell
D.C.G. Cann
S. Carter
D.F. Chamberlain
M. Cheek
J.Y. Clark
W.D. Clayton
J. Compton
B. Conn
A.J. Coombes
C.J. Couper
E.J. Cowley
C.J. Cox
T.B. Croat
J. Cullen
S.J.M. Droop
J. Dransfield
W.W. Eddie
J. Edmondson
U. Eggli
H. Ern
A.R. Ferguson
D. Fräenz
C. Fraile
D. Frodin
J. Fryer
T.W.J. Gadella
A. Galloway
M.F. Gardner
C. Gorman
R. Gornall
Z. Gowler
V. Graham
M.L. Grant
P.S. Green
C. Grey-Wilson
N. Groendijk-Wilders
R.F.L. Hamilton
E.H. Hamlet
C.J. Harrison
E. Haston
F. Hibberd
P. Hoch
J. Howe
G. Hull
D.R. Hunt
J.A. Le Huquet
R. Hyam
B. Hylmö
C. Innes
E. van Jaarsveld
N. Jacobsen
A. James
V. Jamnicky
M. Jebb
S.J.L. Johnstone
L. Jones
P.C. de Jong
P.-M. Jörgensen
S.L. Jury
J. Kendall
C.J. King
S.G. Knees
J.M. Lamond
D.J. Leedy
J.M. Lees
A.C. Leslie
B.E. Leuenberger
D.Z. Li
H. McAllister
B. MacBryde
D. McClintock
F. McIntosh
D. McKean
M.T. Madison
V. Malécot
J. Mann Taylor
J.B. Martinez-Laborde
B. Mathew
V.A. Matthews
H.S. Maxwell
R.D. Meikle
M. Mendum
D. Middleton
R.R. Mill
D.M. Miller
R.J. Mitchell
C.M. Mitchem
B.D. Morley
E.C. Nelson
D. Nicol-Brown
D.H. Nicolson
W.P. Oates
U. Oster
J. Parnell
M. Peña Chocarro
D.M. Percy
F.H. Perring
D. Philcox
D. Pigott
D.A.H. Rae
J.A. Ratter
S.S. Renner
S. Renvoize
A.J. Richards
J.E. Richardson
I.B.K. Richardson
M. de Ridder
E.M. Rix
N.K.B. Robson
G.D. Rowley
A. Rutherford
M.J.P. Scannell
P. Sell
J.M.H. Shaw
A.J. Silverside
N. Sinclair
R.M. Smith
S.A. Spongberg
L. Springate
C.A. Stace
W.T. Stearn
D.C. Stuart
G.J. Swales

D.M. Synnott
N.P. Taylor
M.C. Tebbitt
S.A. Thompson
S. Thornton-Wood
J. Tun-Garrido
T. Upson
R. Vickery
F.T. de Vries
K.S. Walter
S.M. Walters
T. Wangdi
J.D. Wann
M.C. Warwick
M.F. Watson
D.A. Webb
C. Whitefoord
C.M. Whitehouse
A.C. Whiteley
D.O. Wijnands
J.R.I. Wood
P.J.B. Woods
P.F. Yeo

PREFACE TO THE SECOND EDITION

Work on the first edition of this Flora was begun in 1976, and publication of 6 volumes followed reasonably regularly thereafter (vol. II, 1984; vol. I, 1986; vol. III, 1989; vol. IV, 1995; vol. V, 1997 and vol. VI, 2000). During the whole period of production, the work of 175 authors and contributors was supervised by an editorial committee, chaired from 1977 to 1989 by the late Dr S.M. Walters and by myself from 1989 to 2000. The membership of the committee varied, but included (at different times) Dr J.C.M. Alexander, the late Mr A. Brady, Mr C.D. Brickell, Dr J.R. Edmondson, the late Mr P.S. Green, Professor V.H. Heywood, Professor P.-M. Jørgensen, Dr S.L. Jury, Ms S.G. Knees, Dr A. Leslie, the late Mr J. Lewis, Ms V.A. Matthews, Ms H.S. Maxwell, Mrs D.M. Miller, Dr E.C. Nelson, Dr N.K.B. Robson, the late Professor D.A. Webb, the late Professor D.O. Wijnands and the late Dr P.F. Yeo. Over the whole period the project was sponsored by the Royal Horticultural Society of London, the Royal Botanic Garden Edinburgh (where the secretariat was based) and the Stanley Smith Horticultural Trust. Financial and other support was provided by the Cory Foundation of Cambridge University Botanic Garden, the Coke Trust, the Cruden Foundation, the Edinburgh Botanic Garden (Sibbald) Trust, the Humphrey Whitbread Trust, the John S. Cohen Foundation, the Lok Wan Tho Memorial Fund, the Royal Botanic Garden Edinburgh (especially staff time and office space, logistic and IT support), the Royal Horticultural Society, the Stanley Smith Horticultural Trust (the major provider of funds over the whole period), the Will Charitable Trust, the William Adlington Cadbury Charitable Trust and the Wolfson Industrial Fellowship. More detail on the history of the project can be found in an article in *The Plantsman* (**7**: pp. 164–7, 2000).

Following completion of the work on volume VI in 1999, Ms S.G. Knees was appointed to the Stanley Smith Horticultural Fellowship at the Royal Botanic Garden Edinburgh, and for 3 years worked on the correction, updating and improvement of the text (in both paper and computerised form); this work now provides the foundation for the new second edition, which has been edited by myself, Ms S.G. Knees and Ms H.S. Cubey (née Maxwell), assisted by Mr J.M.H. Shaw (Nottingham), Ms L. Banfield (RBG Edinburgh), Mr P. Harrold (RBG Edinburgh), Ms A. Laporte-Bisquit (RBG Edinburgh), Mr M.F. Gardner (RBG Edinburgh), Ms S. Neale (RBG Edinburgh), Mr G.D. Rowley (Reading), Miss N. Zantout (RBG Edinburgh and Bremen) and Mr C.D. Brickell (Pulborough).

We are particularly grateful to the Royal Botanic Garden Edinburgh for allowing Ms Knees and Ms Cubey time from their official duties at the garden to complete the necessary changes to, and organisation of, the text. We are also grateful to all those contributors who have revised the accounts written for the first edition.

The current edition contains many corrections, updates and improvements to the orginal text. The major change is the omission of the ferns and their allies (Pteridophyta) and the Gymnosperms. These groups, considered as cultivated plants, do not fit well with a Flora-type treatment, being mainly cultivars of relatively few genera and species; there are many, more appropriate, guides to them available. Because of this omission, it has been thought sensible to combine the original volumes I & II into a single large volume (I) containing all the Monocotyledons; volumes II–V (equivalent to III, IV, V & VI of edition 1) contain the Dicotyledons. Contributions that have been only lightly revised are presented here under the name of the original author. Those that have been more heavily altered are under the name of the original author plus that of the editor.

Because the original was first conceived during the 1970s, long before the current and continuing storm of changes to family and generic placements caused by the introduction of cladistic and molecular taxonomic methods, the text was organised and families and genera were recognised according to the system presented in H. Melchior's edition (edn 12) of *Syllabus der Pflanzenfamilien* (1964). Use of this system is no longer tenable, but the editors were loath to accept any of the current labile and uncertainly documented systems (e.g. those developed by the Angiosperm Phylogeny Group) as these are more likely to confuse users of this book rather than help them. Instead, for the recognition of genera and families, we have chosen to follow R.K. Brummitt's *Vascular Plants, Families and Genera* (1992), which provides a published and easily available listing of all the families and genera recognised at Kew in the 1990s. Because Brummitt's list is in alphabetical order, the order of families and genera used here is rather arbitrary, but retains essentially that of the genera found in edition 1, thus maintaining some degree of continuity with the original publication.

Many other changes in detail have been made to the text, most of which do not require mention here. However, it is important to note that, though we have tried to list (under the heading 'Literature') taxonomically useful, up-to-date literature for all genera and families, it does not necessarily follow that the text of the Flora will contain any or all of the changes proposed in this literature. Further relevant literature continues to appear on a regular basis and interested readers are referred to *Kew Record*, or the extremely valuable *ePIC* plant information database available on the website of the Royal Botanic Gardens Kew (www.kew.org.uk).

Each volume contains its own index, and a consolidated index to accepted Families and Genera will be found in volume V.

James Cullen
Director, Stanley Smith (UK) Horticultural Trust
Cambridge, April, 2009.

PREFACE TO THE FIRST EDITION

The ideas that led to the production of *The European Garden Flora* derived from two independent sources, and were developed during the late 1960s and early 1970s. One of us (SMW), then Director of the University of Cambridge Botanic Garden and heavily involved in the completion of *Flora Europaea*, had been thinking of how botanic gardens might make some contribution to the problems of horticultural taxonomy and classification. He had indeed discussed these matters with the British amateur botanist David McClintock, who was himself very anxious to further research towards the production of a Flora for gardens. The other (JC), then Assistant Regius Keeper at the Royal Botanic Garden Edinburgh, formerly Assistant Director at Liverpool University Botanic Gardens, Ness, had independently developed similar ideas and had produced some samples to show how such a Flora might be prepared.

These two sources came together at the 1976 Scientific Review Group (of which SMW was a member) of the work of the Royal Botanic Garden Edinburgh. Discussion between us soon revealed our common concerns. The need for such a Flora and the recommendation that the RBG Edinburgh should be actively involved in its production were written into the Review Group report, to become part of the official policy of the Garden.

A meeting of the British botanists and horticulturists who might wish to be involved in such a production was called by us in Autumn 1976 and held at the offices of the Royal Horticultural Society in London. It was agreed that such a Flora should be produced, that its content should be mainly those species, subspecies and (botanical) varieties that were grown in Europe (both outdoors and under glass), that it should be organised from RBG Edinburgh, that it should use a minimum of taxonomic jargon, and that it should seek contributors from all parts of the world. The group at this first meeting constituted itself into the Editorial Committee (which was later strengthened by the inclusion of members from other countries in Europe) to oversee the project, and to organise and administer the grants necessary to see the production through. A twenty-year timescale was envisaged.

For the first two years, the preparation of lists of species that could or should be included in the projected Flora was the main concern. For this purpose, a vacation student (Ms Margaret McDonald, the first of many) was employed by RBG Edinburgh to make an index from all the available nurserymen's catalogues. This early predecessor of *The Plant Finder* became the basis for the content of the Flora. The employment and training of vacation students has been a constant feature of the development of the project.

In 1979 the Editorial Committee approached the Stanley Smith Horticultural Trust for a grant to employ a Research Associate to work full-time at Edinburgh. The application was successful, and in October 1979, Dr Crinan Alexander began work in this capacity. He continued, supported by grants from other donors, until he was succeeded by Sabina G. Knees in 1987. Since 1994 various other workers have been employed on an ad hoc basis; they are individually acknowledged in volumes IV, V and VI.

The first volume of the Flora (volume II) was published in 1984; subsequent volumes followed in 1986, 1989, 1995 and 1997; this, the final volume, will be published in 2000. If we take the originally envisaged twenty-year span as beginning with the first appointment of the Research Associate, then the production has slipped by only a single year.

Full acknowledgement of the contributions of the various authors (now over 200), editors and collaborating institutions has been given in the individual volumes. Here we would like to add some more general acknowledgements. Firstly, to the Royal Botanic Garden Edinburgh, which has fully honoured its commitment to the 1976 Review Group by providing staff-time, secretariat and logistical support (telephones, stationery, computers, etc.) over the whole 24-year period. Similarly, though on a smaller scale, the University of Cambridge Botanic Garden has provided background support. The Royal Horticultural Society has provided funds and facilities over a long period, especially since 1987. Several grant-giving bodies have been associated with the project; again, these are acknowledged in the various volumes, but a special acknowledgment is due to the Stanley Smith Horticultural Trust (now the Stanley Smith (UK) Horticultural Trust) and its first Director, Sir George Taylor FRS, for its continued support, but especially for its faith in the project at the beginning. Finally, all the volumes have been published by Cambridge University Press; we are grateful to them for persisting with the project over such a long period and especially to the two editors who have seen the process through – Martin G. Walters and Dr Maria Murphy.

James Cullen
S. Max Walters
Cambridge, September 1999

ACKNOWLEDGEMENTS

In addition to the contributors mentioned above, the editors are grateful to the following for editorial support: Bridie Andrews, Jane Bulleid, Christine Couper, Rosanna Colangeli, M.J. Howard, Linda McGee, Valerie Muirhead and Moira Watson. Specialist taxonomic advice was given by Wilhelm Barthlott, David Chamberlain, Ulf Molau and Paddy Woods. Futher horticultural advice was provided by J. Bogner, L. Buchan, R.U. Cranston, P. Cribb, J.D. Donald, R. Kerby, L.A. Lauener, Mrs E. Molesworth-Allan, S. Mayo, A. Paxton, R.L. Shaw, Dr. D. Wills, and Luwasa (Hydroculture) Ltd. Many other individuals provided information over the years and their help is also acknowledged here.

Special thanks are due to Sophie Durlacher, who undertook the arduous task of project managing this second edition.

We are grateful to J.C.M Alexander, Diane Allison, Peter Barnes, Anke Berg, Elaine Campbell, Eleanor Catherine, Mark Coode, Urs Eggli, M. Flanagan, Christine Grey-Wilson, Geoffrey Herklotts, Frances Hibberd, Michael Hickey, Christopher Hogg, Julia Howe, Sarah Johnstone, Christabel King, Rosemary King, Debbie Maizells, Victoria A. Matthews, Suzanne Maxwell, Gillian Meadows, Mary Mendum, Diane Nicol-Brown, Susan Oldfield, Christina Oliver, Donald Pigott, Rodella Purves, Sally Rae, Gordon Rowley, Alison Rutherford, Julian Shaw, Niki Simpson, Rosemary M. Smith, Mark Tebbitt, Margaret Tebbs, Vicky Thomas, Femke de Vries, Maureen Warwick and Rosemary Wise for the preparation of illustrations and also to Edinburgh University Press for allowing us to re-use illustrations from volumes 6 (1978) and 7 (1982) of P.H. Davies (ed.) *Flora of Turkey.*

INTRODUCTION

Amenity horticulture (gardening, landscaping, etc.) touches human life at many points. It is a major leisure activity for very large numbers of people, and is a very important means of improving the environment. The industry that has grown up to support this activity (the nursery trade, landscape architecture and management, public parks, etc.) is a large one, employing a considerable number of people. It is clearly important that the basic material of all this activity, i.e. plants, should be readily identifiable, so that both suppliers and users can have confidence that the material they buy and sell is what it purports to be.

The problems of identifying plants in cultivation are many and various, and derive from several sources, which may be summarised as follows:

(a) Plants in cultivation have originated in all parts of the world, many of them from areas whose wild flora is not well known or documented. Many have been introduced, lost and reintroduced under different names.

(b) Plants in gardens are growing under conditions to which they are not necessarily well adapted, and may therefore show morphological and physiological differences from the original wild stocks.

(c) All plants that become established in cultivation have gone through a process of selection, some of it conscious (selection of the 'best' variants, etc.), some of it unconscious (by methods of cultivation and particularly propagation), so that, again, the populations of a species in cultivation may differ significantly from the wild populations.

(d) Many garden plants have been 'improved' by hybridisation (deliberate or accidental), and so, again, differ from the original stocks.

(e) Finally, and perhaps most importantly, the scientific study of plant classification (taxonomy) has concentrated mainly on wild plants, largely ignoring material in gardens.

Nevertheless, the classification of garden plants has a long and distinguished history. Many of the herbals of pre-Linnaean times (i.e. before 1753) consist partly or largely of descriptions of plants in gardens, and this tradition continued, and perhaps reached its peak, in the late eighteenth and early nineteenth centuries – the period following the publication of Linnaeus' major works, when exploration of the world was at its height. This is the period that saw the founding of *Curtis' Botanical Magazine* (1787) and the publication of J.C. Loudon's *Encyclopaedia of plants* (1829, and many subsequent editions).

The further development of plant taxonomy, from about the middle of the nineteenth century to the present, has seen an increasing divergence between garden and scientific taxonomy, leading on the one hand to such works as the *The Royal Horticultural Society's dictionary of gardening* (1951 and reprints), based on G. Nicholson's *Illustrated dictionary of gardening* (1884–1888), *The new Royal Horticultural Society's dictionary of gardening* (1992), and the very numerous popular, usually illustrated, works on garden plants available today, and, on the other hand, to the Floras, Revisions and Monographs of scientific taxonomy.

Despite this divergence, a number of plant taxonomists realised the importance of the classification and identification of cultivated plants, and produced works of considerable scientific value. Foremost among these stands L.H. Bailey, editor of *The standard cyclopedia of horticulture* (1900, with several subsequent reprints and editions), author of *Manual of cultivated plants* (1924, edn 2, 1949) and founder of the journals *Gentes Herbarium* and *Baileya*. Other important workers in this field are T. Rumpler (*Vilmorin's Blumengärtnerei*, 1879), L. Dippel (*Handbuch der Laubholzkunde*, 1889–93), A. Voss & A. Siebert (*Vilmorin's Blumengärtnerei*, edn 3, 1894–6), C.K. Schneider (*Illustriertes Handbuch der Laubholzkunde*, 1904–12), A. Rehder (*Manual of cultivated trees and shrubs*, 1927, edn 2, 1947), J.W.C. Kirk (*A British garden flora*, 1927), F. Enke (*Parey's Blumengärtnerei*, edn 2, 1958), B.K. Boom (*Flora Cultuurgewassen*, 1959 and proceeding), V.A. Avrorin & M.V. Baranova (*Decorativn'ie Travyanist'ie Rasteniya Dlya Otkritogo Grunta SSSR*, 1977) and R. Mansfeld (*Verzeichnis landwirtschaftlicher and gärtnerischer Kulturpflanzen*, edn 2, 1986).

The present Flora, which, of necessity, is based on original taxonomic studies by many workers, attempts to provide a scientifically accurate and up-to-date means for the identification of plants cultivated for amenity in Europe (i.e. it does not include crops, whether horticultural or agricultural, or garden weeds), and to provide what are currently thought to be their correct names, together with sufficient synonymy to make sense of catalogues and other horticultural works. The needs of informed amateur gardeners have been borne in mind at all stages of the work, and it is hoped that the Flora will meet their needs just as much as it meets the needs of professional taxonomists. The details of the format and use of the Flora are explained in section 2 below (p. xvi).

In writing the work, the Editorial Committee has been fully aware of the difficulties involved. Some of these have been outlined above; others derive from the fact that herbarium material of cultivated plants is scanty and usually poorly annotated, so that material of many species is not available for checking the use of names, or for comparative purposes. Because of these facts, attention has been drawn to numerous problems which cannot be solved but can only be adverted to. The solution of such problems requires much more taxonomic work.

The form in which contributions appear is the responsibility of the Editorial Committee. The vocabulary and the technicalities of plant description are therefore not necessarily those used by the contributors.

1. SELECTION OF SPECIES

The problem of determining which species are in cultivation is complex and difficult, and has no complete and final answer. Many species, for instance, are grown in botanic gardens but not elsewhere; others, particularly orchids, succulents and some alpines, are to be found in the collections of specialists but are not available generally. Yet others have been in cultivation in the past but are now lost, or perhaps linger in a few collections, unrecorded and unpropagated. Further problems arise from the fact that the identification of plants in collections is not always as good as it might be, and some less well-known species probably appear in published lists under the names of other, well-known species (and vice versa).

The Flora attempts to cover all those species that are likely to be found in general collections (i.e. excluding botanic gardens and specialist collections) in Europe, whether they are grown outdoors or under glass. In order to produce a working list of such species, a compilation of all European nursery catalogues available to us was made in 1978 by Margaret McDonald, a vacation student working at the Royal Botanic Garden, Edinburgh. Since then, numerous additions have been made. This list (known as the 'Commercial List'), which includes well over 12,000 specific names, forms the basis of the species included here. Since 1987 the annual production by the Hardy Plant Society (later, the Royal Horticultural Society) of the *Plant Finder* has made the process of ascertaining which plants are on sale very much simpler. Similar publications on the continent of Europe – Erhardt, A. & W., *Pflanzen–Einkaufsführer* (1990), Pereire, A. & Bonduel, P., *Où trouver vos Plantes* (1992) and van der Laar, H.J., *Naamlijst van Houtige Gewassen* (1985–89) – have made the process of scanning catalogues no longer necessary. Another work of great benefit to cultivated plant taxonomy is *Index Hortensis* by Piers Trehane; volume I covering herbaceous perennials was published in 1989; other groups will be covered in subsequent volumes. This work gives the nomenclaturally correct names and synonyms for plants in cultivation and also has extensive lists of cultivars. In addition to the 'Commercial List', *The Plant Finder* and *Index Hortensis*, several works on the flora of gardens have been consulted, and the species covered by them have been carefully considered for inclusion. These works are: *The Royal Horticultural Society's dictionary of gardening*, edn 2 (1956, supplement revised 1969); *The new RHS dictionary of gardening* (1992); Encke, F. (ed.), *Parey's Blumengärtnerei* (1956); Boom, B.K., *Flora der Cultuurgewassen van Nederland* (1959 and proceeding); Bean, W. J., *Trees and shrubs hardy in the British Isles* (edn 8, 1970–81); Krüssmann, G., *Handbuch der Laubgeholze* (edn 2, 1976–78); *Manual of cultivated broad-leaved trees and shrubs* (English edition, translated by M.E. Epp, G.S. Daniels, editor, 1986); Encke, F., Buchheim, G. & Seybold, S. (eds), *Zander's Handworterbuch der Pflanzennamen* (edn 14, 1993), and, since 1986, Mansfeld, R., *Verzeichnis landwirtschaftlicher und gärtnerischer Kulturpflanzen* (edn 2, 1986) and Jelitto L. & Schacht, W., *Die Freiland Schmuckstauden* (1963); Jelitto, L., Schacht, W., & Fessler, A. (eds), (edn 3, 1985), *Hardy herbaceous perennials revised*, English edition (1989); and Brickell, C.D. (ed.), *The Royal Horticultural Society A–Z encyclopedia of garden plants* (2003). Most of the names included in these works are covered by the present Flora, though some have been rejected as referring to plants no longer in general cultivation.

As well as the works cited above, several relating to plants in cultivation in North America have also been consulted: Rehder, A., *Manual of cultivated trees and shrubs* (edn 2, 1947) and *Bibliography of cultivated trees and shrubs* (1949); Bailey, L.H., *Manual of cultivated plants* (edn 2, 1949); *Hortus Third* (edited by the staff of the L.H. Bailey Hortorium, Cornell University, 1976).

The contributors have also drawn on their own experience, as well as that of the family editors, European advisers and other experts, in deciding which species should be included.

Most species have a full entry, being keyed, numbered and described as set out under section 2c below (p. xvii). A few, less commonly cultivated, species (additional species) are not keyed or numbered individually, but are described briefly under the species to which they are most likely to key out in the formal key.

2. USE OF THE FLORA

a. *The taxonomic system followed in the Flora*. Plants are described in this work in a taxonomic order, so that similar genera and species occur close to each other, rendering comparison of descriptions easier than in a work where the entries are alphabetical. The families (and higher groups) follows R. K. Brummitt's *Vascular Plants*: Families and Genera (1992).

The order of the species within each genus has been a matter for the individual author's discretion. In general, however, some established revision of the genus has been followed, or, if no such revision exists, the author's own views on similarity and relationships have governed the order used.

b. *Nomenclature.* The arguments for using Latin names for plants in popular as well as scientific works are often stated and widely accepted, particularly for Floras such as this, which cover an area in which several languages are spoken. Latin names have therefore been used at every taxonomic level. A concise outline of the taxonomic hierarchy and how it is used can be found in C. Jeffrey's *An introduction to plant taxonomy* (1968, edn 2, 1982). Because of the difficulties of providing vernacular names in all the necessary languages (not to say dialects), they have not been included. S. Priszter's *Trees and shrubs of Europe, a dictionary in eight languages* (1983) is a useful source for the vernacular names of woody plants.

Many horticultural reference works omit the authority that should follow every Latin plant name. Knowledge of this authority prevents confusion between specific names that may have been used more than once within the same genus, and makes it possible to find the original description of the species (using *Index Kewensis*, which lists the original references for all Latin names for higher plants published since 1753). In this Flora authorities are therefore given for all names at or below genus level. These are unabbreviated to avoid the obscure contractions that mystify the lay reader, and on occasion, the professional botanist. In most cases we have not thought it necessary to include the initials or qualifying words and letters that often accompany author names, e.g. A. Richard, Reichenbach filius, fil. or f. (the exceptions involve a few, very common surnames).

In scientific taxonomic literature, the authority for a plant name sometimes consists of two names joined together by *ex* or *in*. Such formulae have not been used here: the authority has been shortened in accordance with *The international code of botanical nomenclature*: e. g. *Capparis lasiantha* R. Brown ex de Candolle becomes *Capparis lasiantha* de Candolle; *Viburnum ternatum* Rehder in Sargent becomes *Viburnum ternatum* Rehder. The abbreviations *hort.* and *auct.*, which sometimes stand in place of the authority after Latin names, have not been used in this work as they are often obscure or misleading. The situations described by them can be clearly and unambiguously covered by the terms *invalid*, *misapplied* or *Anon. Invalid* implies that the name in question has not been validly published in accordance with the Code of Nomenclature, and therefore cannot be accepted. *Misapplied* refers to names which have been applied to the wrong species in gardens or in literature. *Anon.* is used with validly published names for which there is no apparent author.

Gardeners and horticulturists complain bitterly when long-used and well-loved names are replaced by unfamiliar ones. These changes are unavoidable if *The international code of botanical nomenclature* is adhered to. Taxonomic research will doubtless continue to unearth earlier names and will also continue to realign or split up existing groups, as relationships are further investigated. However, the previously accepted names are not lost; in this work they appear as synonyms, given in brackets after the currently accepted name; they are also included in the index. Dates of publication are not given either for accepted names or for synonyms.

c. *Descriptions and terminology.* Families, genera and species included in the Flora are mostly provided with full-length descriptions. Shorter, diagnostic descriptions are, however, used for genera or species that differ in only a few characters from others already fully described, e.g.:

3. P. vulgaris Linnaeus. Like *P. officinalis* but leaves lanceolate and corolla red …

This implies that the description of *P. vulgaris* is generally similar to that of *P. officinalis* except in the characters mentioned; it should not be assumed that plants of the two species will necessarily look very like each other. Additional species (see p. xvi), subspecies, varieties, formae and cultivars (see p. xviii) are described very briefly and diagnostically.

Unqualified measurements always refer to length (though 'long' is sometimes added in cases where confusion might arise); similarly, two measurements separated by a multiplication sign indicate length and breadth respectively.

The terminology has been simplied as far as is consistent with accuracy. The technical terms that, inevitably, have had to be used are explained in the glossary (p. 611). Technical terms restricted to particular families or genera are defined in the observations following the family or genus description, and are also referred to in the glossary.

d. *Informal keys.* For larger genera and families an informal key is often given; this will not necessarily enable the user to identify precisely every species included, but will provide a guide to the occurrence of the more easily recognised characters. A selection of these characters is given, each of which is followed by the entry-numbers of those species which show that character. In some cases, where only a few species of a genus show a particular character, the alternative states are not specified, e.g.:

Leaves. Leathery: **18**, **19**.

This means that only species **18** and **19** in the particular genus have leathery leaves; the other species may have leaves of various textures, but they are not leathery.

e. *Formal keys.* For every family containing more than one genus, and for every genus containing more than one full-entry species, a dichotomous key is provided. This form of key, in which a series of decisions must be made between pairs of contrasting characteristics, should lead the user step by step to an entry number followed by the name of a species. The reader should then check this identification with the description of that species; in some cases, other less commonly cultivated species may be mentioned under the description of the full-entry species, when the brief descriptions of these should also be scanned, so that a final identification can be made. A key to all the families of the Monocotyledons to be included in the Flora is provided (p. 3).

f. *Horticultural information.* Notes on the cultural requirements and methods of propagation are generally included in the observations on each genus; more rarely, such information is given in the observations under the family description. These notes are generally brief and very generalised, and merely provide guidance. Reference to general works on gardening is necessary for more detailed information.

g. *Citation of literature.* References to taxonomic books, articles and registration lists are cited for each family and genus, as appropriate. No abbreviations are used in these citations (though very long titles have been shortened). The citation of a particular book or article does not necessarily imply that it has been used in the preparation of the account of the particular genus or family in this work.

h. *Citation of illustrations.* Where possible, references to good illustrations are given for each species, subspecies or variety; the names under which they were originally published (which may be different from those used here) are not normally given. The illustrations may be coloured or black and white, and may be drawings, paintings or photographs. Up to five illustrations per species have been given, and an attempt has been made to choose from widely available, modern works. Where no illustrations are cited, they either do not exist, as far as we know, or those that do are considered to be of doubtful accuracy or of very restricted availability.

There are considerable difficulties with the citation of references to illustrations, particularly with regard to their dates of publication. These difficulties arise from two main sources. 1. Many illustrated plant books of the 19th and early 20th centuries were published in fascicles, as illustrations became available, out of proper order. This means, however, that the dates of publication of individual illustrations are often more or less unknown, and, anyway, these dates do not correspond to the date of the final binding. 2. Many popular illustrated books published since the 1950s have gone through many reprints, each of which bears its own date, so that the same book may well be cited with different dates in different places in this work. We have tried to cite meaningful dates wherever this was possible, but the user should understand that the date given here for a particular volume may not correspond with the date printed in the copy available to him.

In searching for illustrations, use has been made of *Index Londinensis* (1929–31, supplement 1941), R.T. Isaacson's *Flowering plant index of illustration and information* (1979) and an extensive index compiled over the last 30 years at the Royal Botanic Garden, Edinburgh.

Several pages of figures of diagnostic plant parts are included with various groups in the Flora, and should be particularly helpful when plants are being identified by means of the keys. Some of these are original, others have either been redrawn from various sources or are silhouettes of photocopies of leaves.

i. *Geographical distribution.* The wild distribution, as far as it can be ascertained, is given in italics at the end of the description of each species, subspecies or variety. The choice and spelling of place names in general follows *The Times Atlas*, Comprehensive edition (1994 reprint), except:

(1) Well-established English forms of names have been used in preference to less familiar vernacular names, e.g. Crete instead of Kriti, Naples instead of Napoli, Borneo instead of Kalimantan;

(2) New names or spellings will be adopted as soon as they appear in readily available works of reference.

(3) Because of a lack of precise information, certain areas, now politically divided, are referred to under their older names (e.g. 'former Jugoslavia', 'former USSR', etc.).

j. *Hardiness* (see map, p. xix). For every species a hardiness code is given. This gives a tentative indication of the lowest temperatures that the species can withstand:

G2 – needs a heated glasshouse even in south Europe.
G1 – needs a cool glasshouse even in south Europe.
H5 – hardy in favourable areas: withstands 0 to −5 °C minimum.
H4 – hardy in mild areas: withstands −5 to −10 °C minimum.
H3 – hardy in cool areas: withstands −10 to −15 °C minimum.
H2 – hardy almost everywhere: withstands −15 to −20 °C minimum.
H1 – hardy everywhere: withstands −20 °C and below.

The map of mean January minima (p. xix) shows the isotherms corresponding to these codes.

k. *Flowering time.* The terms spring, summer, autumn and winter have been used as a guide to flowering times in cultivation in Europe. It is not possible to be more specific when dealing with an area extending from northern Scandinavia to the Mediterranean. In cases where plants do not flower in cultivation, or flower rarely, or whose time of flowering is not recorded, no flowering time is given.

l. *Subspecies, varieties and cultivars.* Subspecies and varieties are included, where appropriate. This is done in various ways, depending on the number of such groups; all are self-explanatory.

No attempt has been made to describe the range of cultivars of any species, either partially or comprehensively. The former is scarcely worth doing, the latter virtually impossible. Reference to individual, commonly grown cultivars is, however, made in various ways:

(1) If a registration list of cultivars exists, it is cited in the 'Literature' paragraph (see section 2g) following the description of the genus.

(2) If a particular cultivar is very widely grown, it may be referred to, either in the description of the species to which it belongs (or most resembles), or in the observations to that species.

(3) If, in a particular species, cultivars are numerous and fall into reasonably distinct groups based on variation in some obvious character, then these groups may be referred to, together with an example of each, in the observations to the species.

m. *Hybrids.* Many hybrids between species (interspecific hybrids) and some between genera (intergeneric hybrids) are in cultivation, and some of them are widely grown. Commonly cultivated interspecific hybrids are, where possible, included as though they were species. Their names, however, include a multiplication sign indicating their hybrid origin; the names of the parents (when known or presumed) are also given. Other hybrids that are less frequently grown are mentioned in the observations to the parent species they most resemble. In some genera, where the number of hybrids is very large, only a small selection of those most commonly grown is mentioned.

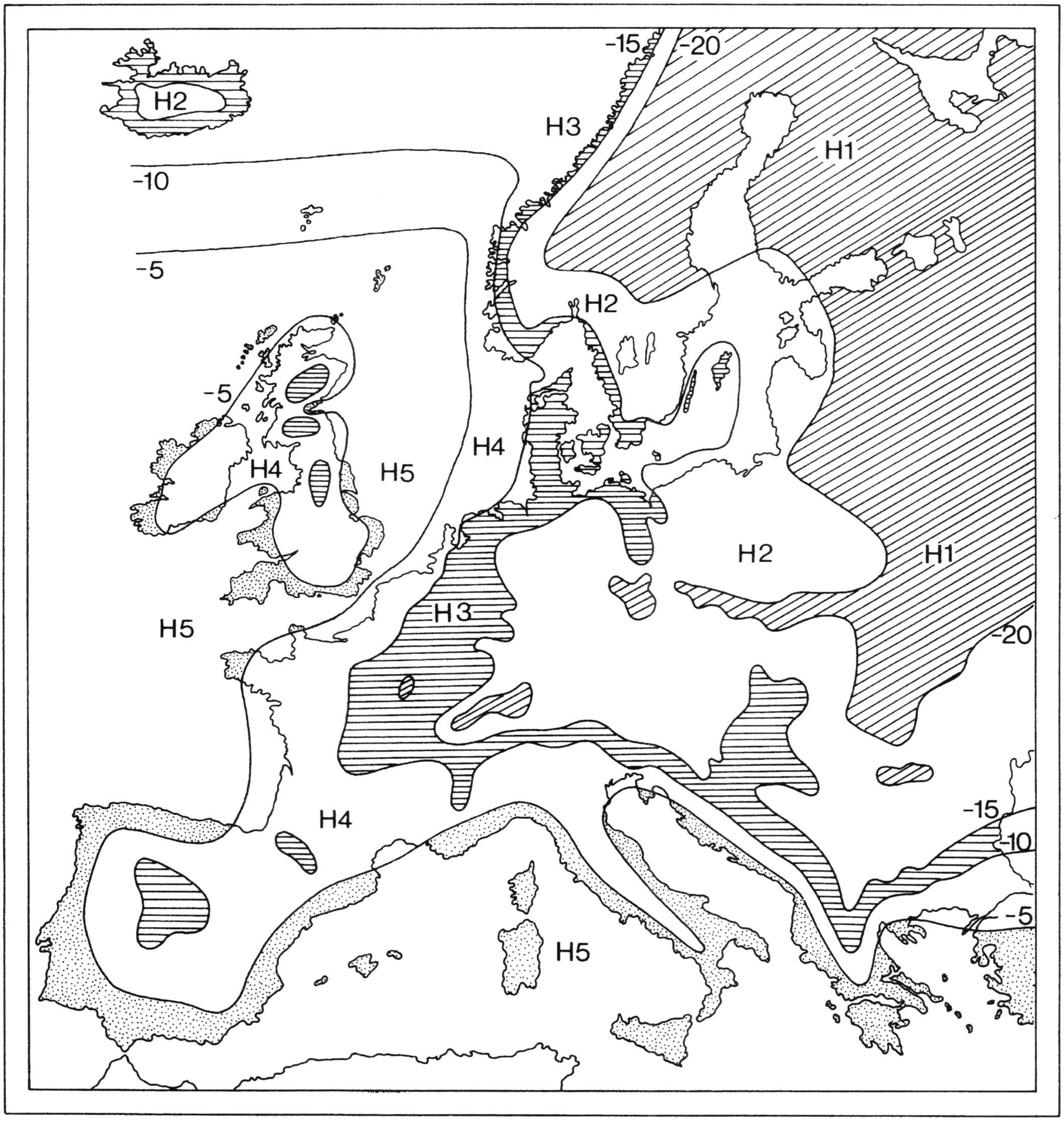

Map 1. Mean minimum January isotherms for Europe (hardiness codes). (After Krüssmann, *Handbuch der Laubgehölze*, 1960 and *Mitteilungen der Deutsche Dendrokigische Gesellschaft* **75**: 1983.)

MONOCOTYLEDONS

KEY TO CLASSES

1a. Cotyledons usually 2, lateral; leaves usually net-veined, with or without stipules, alternate, opposite or whorled; flowers with parts in 2s, 4s or 5s, or parts numerous; primary root system (taproot) usually persistent, branched **Dicotyledons**

b. Cotyledon 1, terminal; leaves usually with parallel veins, sometimes these connected by cross-veinlets; leaves without stipules, opposite only in some aquatic plants; flowers usually with parts in 3s; mature root system wholly adventitious **Monocotyledons**

MONOCOTYLEDONS

KEY TO FAMILIES

1a. Ovary superior or flowers completely without perianth (including all aquatics with totally submerged flowers) **Group I**

b. Ovary partly or fully inferior (if plants aquatic then flowers borne above water-level) **Group II**

Group I

1a. Trees, shrubs or prickly scramblers with large, pleated, usually palmately or pinnately divided leaves; flowers more or less stalkless in fleshy spikes or panicles which often have large, sometimes woody, basal bracts (spathes) **46. Palmae**

b. Combination of characters not as above 2

2a. Totally submerged aquatic plants of fresh or saline water 3 **6. Potamogetonaceae**

b. Terrestrial or epiphytic plants, if aquatic then not totally submerged, occasionally entirely floating 3

3a. Small floating aquatic plants not differentiated into stem and leaves **49. Lemnaceae**

b. Plants aquatic or not, clearly differentiated into stem and leaves (the latter rarely scale-like) 4

4a. Perianth hyaline or papery, translucent or brown, or reduced to bristles, hairs, narrow scales or totally absent 5

b. Perianth well-developed, though sometimes small, not hyaline or papery, usually coloured, sometimes greenish 12

5a. Flowers in small, 2-sided or cylindric spikelets provided with overlapping bracts (spikelets sometimes 1-flowered) 6

b. Flowers arranged in heads, superposed spikes, racemes, panicles or cymes, never in spikelets as above 7

6a. Leaves alternate in 2 ranks on a stem which is usually hollow and with cylindric internodes; leaf-sheath usually with free margins, at least in the upper part; flowers arranged in 2-sided spikelets (sometimes 1-flowered) each usually subtended by 2 sterile bracts (glumes); each flower usually enclosed by bracts - a lower lemma and an upper palea (sometimes absent); perianth of 2 or 3 concealed scales (lodicules), more rarely of 6 scales or absent; styles generally 2, feathery **45. Gramineae**

b. Leaves usually arranged on 3 sides of the cylindric or more usually 3-angled stems which usually have solid internodes; young leaf-sheaths closed, though sometimes splitting later; flowers arranged in 2-sided or cylindric spikelets, often with a 2-keeled or 2-lobed glume at the base; each flower subtended by a glume only; perianth of several bristles, hairs, scales or absent; style 1 with 2 or 3 papillose stigmas **53. Cyperaceae**

7a. Dioecious trees or shrubs often supported by stilt-roots, with stiffly leathery, sharply toothed leaves; fruits compound, often woody **50. Pandanaceae**

b. Combination of characters not as above 8

8a. Inflorescence a simple, fleshy spike (spadix) of inconspicuous flowers, subtended by or rarely joined to a large bract which encloses it in bud (spathe); leaves often net-veined and/or lobed (plant rarely a small evergreen floating aquatic) 9

b. Inflorescence various but not as above; leaves not as above 10

9a. Leaves iris-like, parallel-veined, fragrant when crushed; spadix protruding obliquely sideways from what appears to be a normal foliage leaf **48. Acoraceae**

b. Leaves various, but not as above, usually net-veined and/or lobed; spike and spadix various, not as above **47. Araceae**

10a. Flowers bisexual; perianth-segments 6, hyaline or brownish; ovary with 3-many ovules **42. Juncaceae**

b. Flowers unisexual; perianth-segments a few threads or scales; ovary with 1 ovule 11

11a. Flowers in 2 superposed, elongate, brownish or silvery spikes; ovary borne on a stalk which has hair-like branches **52.Typhaceae**

b. Flowers in spherical heads; ovary not stalked **51. Sparganiaceae**

12a. Carpels free, or very slightly united at the base only, or ovary apparently of a single carpel with a single style 13

b. Carpels united for most of their length, though the styles may be free 17

13a. Inflorescence a spike, sometimes bifid; perianth-segments 1–4 14

b. Inflorescence not a spike; perianth-segments 6 15

14a. Stamens 6 or more; carpels 3–6; perianth-segments 1–3, petal-like **5. Aponogetonaceae**

b. Stamens 4; carpels 4; perianth-segments 4, not petal-like **6. Potamogetonaceae**

15a. Ovules few, borne basally in each free carpel **1. Alismataceae**

b. Ovules many, borne parietally on the walls of the free carpels 16

16a. Leaves linear; latex absent; all perianth-segments petal-like **2. Butomaceae**

b. Leaves with stalk and expanded blade; latex present; perianth differentiated into 3 green sepals and 3 coloured petals **3. Limnocharitaceae**

17a. Perianth differentiated into greenish calyx and petaloid corolla 18

b. Perianth-segments all similar, not differentiated into calyx and corolla or absent 20

18a. Flowers solitary or in umbels; leaves broad, opposite or in a single whorl near the top of the flowering stem **22. Trilliaceae**

b. Flowers in spikes, heads, cymes or panicles; leaves not as above 19

19a. Stamens 6 or 3–5 with 1–3 staminodes; anthers basifixed; leaves usually borne on the stems, often with closed sheaths, never grey with scales; bracts neither overlapping nor conspicuously toothed **44. Commelinaceae**

b. Stamens 6, staminodes 0; anthers dorsifixed; leaves mostly in basal rosettes, often rigid and spiny-margined, when on the stems then usually grey with scales; bracts usually overlapping and conspicuously toothed **43. Bromeliaceae**

20a. Plants obviously woody, at least at the bases of the stems, rarely composed of rosettes of fleshy leaves borne on woody rhizomes which are often concealed 21

b. Plants entirely herbaceous 28

21a. Leaves reduced to scales, photosynthesis being carried out by flattened or divided stems (cladodes); fruit a berry 22

b. Leaves well-developed; fruit usually a capsule, more rarely a berry 23

22a. Flowers borne directly on the cladodes **24. Ruscaceae**

b. Flowers borne in inflorescences of various kinds directly on the stems, among the cladodes **23. Asparagaceae**

23a. Leaves succulent 24

b. Leaves leathery or parchment-like or herbaceous but not succulent 25

24a. Flowers bell-shaped or almost spherical; perianth-segments united only at the extreme base **27. Agavaceae**

b. Flowers tubular, at least the 3 outer perianth-segments united for more than half their length, the 3 inner segments united to the outer to varying degrees **14. Aloaceae**

25a. Leaves clearly stalked with 3–9 main veins arising at the top of the petiole clearly connected by cross-veinlets; climbers, rarely shrubs or creeping subshrubs 26

b. Leaves not stalked, veins not as above, cross-veinlets obscure; shrubs or small trees 27

26a. Leaf-stalks bearing 2 tendrils; flowers inconspicuous; placentation axile **26. Smilacaceae**

b. Leaf-stalks without tendrils; flowers very conspicuous, usually large; placentation often parietal **25. Philesiaceae**

27a. Flowers all bisexual, stalks not jointed; fruit a fleshy berry; ovules 1 in each cell of the ovary **30 Dracaenaceae**

b. Flowers mostly unisexual, stalks jointed near their bases; ovules 2 in each cell of the ovary; fruit dry, indehiscent **29. Nolinaceae**

28a. Leaf-stalks bearing 2 tendrils; flowers foetid **26. Smilacaceae**

b. Combination of characters not as above 29

29a. Leaves erect and blade-like, generally forming a tuft arising from a rhizome, at ground level, more rarely borne on a stem **30. Dracaenaceae**

b. Combination of characters not as above 30

30a. Leaves reduced to small, scale-like spines, photosynthesis carried out by flattened or much-divided cladodes **23. Asparagaceae**

b. Leaves well-developed, cladodes absent 31

31a. Flowers solitary or in umbels; leaves broad, opposite or in a single whorl near the top of the flowering stem **22. Trilliaceae**

b. Combination of characters not as above 32

32a. Flowers in umbels (rarely solitary, when subtended by 1 or more spathes) **19. Alliaceae**

b. Flowers not in umbels, when solitary not subtended by spathes 33

33a. Plants emergent aquatics, either rooted in the substrate or free-floating **40. Pontederiaceae**

b. Plants terrestrial 34

34a. Placentation parietal or basal **21. Asteliaceae**

b. Placentation axile 35

35a. Leaves equitant, distichous, forming fans; inflorescence a panicle **31. Phormiaceae**

b. Combination of characters not as above 36

36a. Nectar secreted from glands on the outside of the ovary, generally in the grooves between the carpels 37

b. Nectar secreted from glands on the perianth or rarely on the bases of the staminal filaments, rarely nectar absent 46

37a. Plants with bulbs **18. Hyacinthaceae**

b. Plants without bulbs, usually with rhizomes 38

38a. Leaves reduced to basal sheaths, without blades; perianth blue, flowers in heads **10. Aphyllanthaceae**

b. Leaves with well developed blades; perianth and flowers not as above 39

39a. Leaves all basal, inflorescences borne on scapes (occasionally the scapes may bear small bracts) 40

b. At least some foliage leaves clearly borne on the flowering stems 43

40a. Leaves fleshy and succulent; perianth-segments of the outer whorl united for part of their length, those of the inner whorl variably joined to the outer, all red, orange or yellow, tips often greenish **14. Aloaceae**

b. Leaves not fleshy and succulent; perianth not as above 41

41a. Leaves with a broadened blade and narrow, petiole-like base; flowers in as 1-sided raceme; perianth-segments fused below into a narrow tube which broadens upwards **11. Hostaceae**

b. Leaves without broadened blades and petiole-like bases; flowers in racemes, spikes or panicles, not as above; perianth-segments free or joined into a tube which is not narrow below and broadened above 42

42a. Perianth-segments united into a distinct tube below; stamens declinate **12. Hemerocallidaceae**

b. Perianth-segments free or very slightly united at the base; stamens not declinate **9. Anthericaceae**

43a. Ovules 1–many, axile from the upper inner angle of each cell of the ovary; sap often orange-red; perianth persistent in fruit, usually hairy outside **32. Haemodoraceae**

b. Combination of characters not as above 44

44a. Fruit a berry; perianth urceolate to bell-shaped or almost tubular, segments all united **20. Convallariaceae**

b. Fruit a capsule; perianth various, not usually as above 45

45a. Carpels usually free above, styles usually 3; perianth-segments free **7. Melanthiaceae**

b. Carpels united, style 1; perianth-segments free or united **8. Asphodelaceae**

46a. Style 1 sometimes divided above, or stigmas sessile; nectaries usually on the perianth-segments, rarely absent; plants usually with bulbs **16. Liliaceae**
b. Styles usually 3, free to the base or almost so, more rarely 1; nectaries on the perianth-segments, on the bases of the staminal filaments or absent 47
47a. Style 1, ovary stalked; perianth-segments united; seeds hairy **13. Blandfordiaceae**
b. Styles 3; ovary not stalked; perianth-segments free or if united then ovary borne below soil-level, in the corm; seeds not hairy 48
48a. Plants with corms; nectaries on the perianth or on the staminal filaments **15. Colchicaceae**
b. Plants with rhizomes; nectaries absent **7. Melanthiaceae**

Group II

1a. Ovary not completely inferior, generally about half above the point of attachment of the perianth and androecium 2
b. Ovary totally inferior 4
2a. Anthers opening by pores **35. Tecophilaeaceae**
b. Anthers opening by slits 3
3a. Styles 3; nectar secreted by glands on the perianth-segments **7. Melanthiaceae**
b. Style 1; nectar secreted by glands on the the ovary, situated in the grooves between the carpels **20. Convallariaceae**
4a. Perianth actinomorphic or weakly zygomorphic; stamens 6, 4, 3 or rarely many, very rarely 5, 2 or 1 5
b. Perianth strongly zygomorphic or asymmetric; stamens 5, 2 or 1 (very rarely 6), petaloid staminodes often present; if the perianth is weakly zygomorphic, then stamen 1, united with the style to form a column 18
5a. Ovules 1–many, axile from the upper, inner angle of each cell of the ovary; perianth persistent in fruit, hairy outside; sap often orange **32. Haemodoraceae**
b. Combination of characters not as above 6
6a. Unisexual climbers with heart-shaped or very divided leaves; rootstock tuberous or woody **39. Dioscoreaceae**
b. Combination of characters not as above 7
7a. Rooted or floating aquatics; stamens 2–12; ovules distributed all over the carpel walls (placentation diffuse-parietal) **4. Hydrocharitaceae**
b. Terrestrial or marsh plants, or epiphytes; stamens 3 or 6, rarely many; placentation axile or parietal (when ovules restricted to a few well-defined placentas) 8
8a. Stamens 3 **41. Iridaceae**
b. Stamens usually 6 or 3 + 3 staminodes, rarely many 9
9a. Placentation parietal; flowers in an umbel with the inner bracts long, thread-like and hanging **38. Taccaceae**
b. Placentation usually axile; bracts and inflorescence not as above 10
10a. Perianth consisting of an outer calyx-like whorl and an inner corolla-like whorl; bracts usually overlapping and conspicuously coloured **43. Bromeliaceae**
b. Perianth not differentiated as above (though the segments of the 2 whorls may differ slightly); bracts not as above 11
11a. Leaves long-persistent, evergreen, usually borne in a rosette, terminating in a spine or cylindrical body, more rarely not in a rosette but on creeping, slightly woody stems 12
b. Leaves and stems not as above 14
12a. Plant a subshrub with creeping, woody stems **37. Velloziaceae**
b. Plant forming a loose to dense rosette, stems scarcely apparent 13
13a. Leaves each terminating in a spine; flowers up to 6 cm long **27. Agavaceae**
b. Leaves each terminating in a cylindrical body; flowers 10–15 cm long **28. Doryanthaceae**
14a. Flowers in a spike; leaves fleshy, often spotted with brown **27. Agavaceae**
b. Flowers in heads, umbels or solitary 15
15a. Leaves hairy, pleated or with prominent veins **36. Hypoxidaceae**
b. Leaves generally not as above 16
16a. Leaves reversed by an 180° twist in the narrow base; inflorescence a terminal umbel without spathes, with lateral flowering branches below it; seeds pale **17. Alstroemeriaceae**
b. Leaves not reversed; flowers in a terminal umbel or solitary, subtended and enclosed in bud by 1–4 spathes; seeds black 17
17a. All leaves basal **34. Amaryllidaceae**
b. At least some leaves borne on the flowering stems **33. Ixioliriaceae**
18a. Fertile stamens 6; perianth-segments all similar, united below into a curved and unevenly swollen tube; stem below ground, fleshy **27. Agavaceae**
b. Stamens 5, 2 or 1, very rarely 6; staminodes, which may be petal-like often present; perianth-segments usually differing among themselves; fleshy underground stems rare 19
19a. Fertile stamens 2 or 1, united with the style to form a column; pollen usually borne in coherent masses (pollinia); leaf-veins, when visible, all parallel to margins **60. Orchidaceae**
b. Fertile stamens 5 or 1, rarely 6, not united with the style; pollen granular; leaf with a distinct midrib more or less parallel to the margins, the secondary veins parallel to each other, running at an angle from the midrib to the margins 20
20a. Fertile stamens 5 or rarely 6 21
b. Fertile stamen 1, petal-like staminodes 5 22
21a. Leaves and bracts spirally arranged; flowers unisexual; fruit a banana **54. Musaceae**
b. Leaves and bracts in 2 ranks; flowers bisexual; fruit not a banana **55. Strelitziaceae**
22a. Fertile stamen with a thread-like filament and wider anther of 2 pollen-bearing lobes, not petal-like 23
b. Fertile stamen in part petal-like, with only 1 pollen-bearing anther-lobe 24
23a. Leaves spirally arranged; lower apparent perianth-segment (the lip) formed from 5 united staminodes **57. Costaceae**
b. Leaves distichous; lower apparent perianth-segment (lip) formed from 2 joined staminodes **56. Zingiberaceae**
24a. Leaf-stalk with a swollen band (pulvinus) at the junction with the blade; ovary smooth, with 1–3 ovules **59. Marantaceae**
b. Leaf-stalk without a pulvinus; ovary usually warty, with numerous ovules **58. Cannaceae**

1. ALISMATACEAE

Aquatic herbs, often scapose, without latex. Leaves alternate, often broad; ptyxis supervolute. Inflorescence usually much-branched, rarely a single umbel. Flowers unisexual or bisexual, actinomorphic. Perianth 2-whorled: sepals 3, petals 3. Stamens 6–many. Ovary of 6–many free carpels, superior; ovules 1–2 per carpel, basal or marginal. Fruit a group of achenes.

Thirteen genera and about 90 species of water plants. Six genera are native to Europe, 4 to North America. Species of 5 genera are grown as ornamentals.

Literature: Buchenau, F., Alismataceae, Das Pflanzenreich **16**: 1–66 (1903); Charlton, W.A., Studies in the Alismataceae: 4. Developmental morphology of Ranalisma humile and comparisons with two members of the Butomaceae, Hydrocleis nymphoides and Butomus umbellatus, *Canadian Journal of Botany* **51**(5): 899–910 (1973).

Leaves. Sagittate: **1**.
Flowers. Some unisexual: **1**; all bisexual: **2, 3, 4, 5**.
Stamens. Six: **3, 4, 5**; more than six: **1, 2**.
Carpels. Spirally arranged: **1, 2, 3**; in a single, sometimes irregular, whorl: **4, 5**.
Style. Apical: **1–4**; lateral: **5**.
Achenes. Laterally compressed: **1, 5**; ovoid: **3, 4**; winged: **1**; ribbed: **2, 3, 4**.

1a. Leaves sagittate **1. Sagittaria**
b. Leaves not sagittate 2
2a. Some flowers unisexual; achenes winged **1. Sagittaria**
b. Flowers all bisexual; achenes ribbed or furrowed but not winged 3
3a. Stamens 9 or more **2. Echinodorus**
b. Stamens 6 4
4a. Achenes with 3 ribs on the back and a double inner rib, spirally arranged in a spherical head **3. Baldellia**
b. Achenes otherwise ribbed, in a single, sometimes irregular whorl 5
5a. Achenes 6–15, oblong-ovoid, in a hemispherical head; style apical **4. Luronium**
b. Achenes 11–28, laterally compressed, in a more or less flat head; style lateral **5. Alisma**

1. SAGITTARIA Linnaeus

C.J. King

Rootstock often having stolons and tubers. Leaves aerial, floating or submerged. Some flowers unisexual, usually in whorls of 3, forming racemes or panicles with female or bisexual flowers at the base and male flowers above, or occasionally with the flowers all male or all female. Petals white. Stamens usually numerous. Carpels numerous, each with 1 ovule, spirally arranged on the receptacle, developing into a head of laterally compressed, winged, beaked achenes.

A genus of about 20 species, mostly from America. Several species are in cultivation as waterside plants or in aquaria. They need a moist or wet loamy soil and are readily increased by division, seed or underground tubers when produced.

Literature: Bogin, C., Revision of the genus Sagittaria (Alismataceae), *Memoirs of the New York Botanical Garden* **9**: 179–233 (1955); Rataj, K., Revision of the genus Sagittaria, *Annotationes Zoologicae et Botanicae (Bratislava)* **76**: 1–31 (1972) & **78**: 1–61 (1972).

Stolons. Present: **2, 3, 5, 6, 7**.
Leaf-blades. Sagittate: **1, 2, 3**.
Flowers. 1.2 cm or less in diameter: **5**; 1.5–2.5 cm in diameter: **2, 6**; **3** cm or more in diameter: **1, 3, 4, 7**.
Petals. With a purple blotch at base: **1, 2**.
Stamens. Eleven or fewer: **1, 6**; twelve to nineteen: **1, 4, 5**; twenty or more: **2, 3, 4, 5, 7**.
Filaments. Hairy: **4, 5**; dilated at base: **5, 6**.

1a. Aerial leaves sagittate 2
b. Aerial leaves not sagittate 4
2a. Stolons absent; petals usually with a purple blotch at base **1. montevidensis**
b. Stolons present; petals with or without a blotch 3
3a. Flowers to 2.5 cm in diameter; petals usually with a purple blotch at base **2. sagittifolia**
b. Flowers 3–4 cm in diameter; petals without a blotch **3. latifolia**
4a. Stolons absent; filaments hairy **4. lancifolia**
b. Stolons present; filaments hairy or not 5
5a. Flowers to 1.2 cm in diameter; filaments hairy **5. graminea**
b. Flowers larger than 1.2 cm in diameter; filaments hairless 6
6a. Flowers 1.5–2 cm in diameter; stamens 7–10 **6. subulata**
b. Flowers 3–4 cm in diameter; stamens 20–30 **7. macrophylla**

1. S. montevidensis Chamisso & Schlechtendal. Illustration: *Botanical Magazine*, 6755 (1884).

Rootstock without stolons, leaf-stalks to c. 20 cm, blades to 15 × 10 cm, broadly to narrowly sagittate, submerged leaves to 40 × 1 cm, linear. Stem 60–100 cm. Flowers c. 5 cm in diameter, in 5–9 whorls. Petals usually each with a dark purple blotch bordered with yellow at the base. Stamens 9–15; filaments rough with short, stiff hairs. *S America*. G2. Summer.

Suitable for the tropical aquarium.

2. S. sagittifolia Linnaeus. Illustration: Hegi, Illustriertes Flora von Mittel-Europa **1**: t. 20 (1935); Ross-Craig, Drawings of British plants **31**: t. 10 (1973); Phillips, Wild flowers of Britain, 133 (1977); Grey-Wilson & Mathew, Bulbs, t. 1 (1981).

Rootstock with stolons and tubers, which are 1–1.5 cm. Leaf-stalks to 45 cm, blades to 25 × 22 cm, broadly to narrowly sagittate, submerged leaves to 80 × 2 cm, linear. Stem to 1 m. Flowers to 2.5 cm in diameter, in 3–5 whorls. Petals usually with a purple blotch at the base. Stamens 20–25; filaments linear, hairless. *Europe, Asia*. H2. Summer.

The asiatic form with pure white petals is sometimes referred to as var. **leucopetala** Miquel; a double form of this and the typical variety bear the cultivar name 'Flore Pleno'.

3. S. latifolia Willdenow (*S. gracilis* Pursh; *S. hastata* Pursh; *S. simplex* Pursh). Illustration: *Addisonia* **2**: t. 54 (1917); Hess et al., Flora der Schweiz **1**: 212 (1967); Hay & Beckett, Reader's Digest encyclopaedia of garden plants and flowers, 636 (1978); Brickell (ed.), RHS A–Z encyclopedia of garden plants, 944 (2003).

Rootstock with stolons and large tubers, leaf-stalks to 40 cm, blades 10–30 × 8–20 cm, broadly to narrowly sagittate, submerged juvenile leaves linear. Stem to 1.2 m. Flowers 3–4 cm in diameter, in 4–8 whorls. Stamens 20–40; filaments linear, hairless. *N America.* H2. Summer.

The most variable species in the genus with numerous names for the various wild variants. A double-flowered variant, 'Flora Pleno', is found in cultivation.

4. S. lancifolia Linnaeus (*S. falcata* Pursh). Illustration: *Botanical Magazine*, 1792 (1816); *Addisonia* **12**: t. 406 (1927).

Rootstock without stolons, leaf-stalks to 1 m, blades 15–45 cm, narrowly to broadly lanceolate, the midrib and 2 outer veins arising from the base, the other veins arising higher up the midrib. Stem to 2 m. Flowers 3–5 cm in diameter, in 5–12 whorls. Stamens 12–30; filaments linear, hairy. *SE USA, West Indies, C & northern S America.* H5. Summer.

5. S. graminea Michaux. Illustration: Das Pflanzenreich **16**: 55 (1903); Rickett, Wild flowers of the United States **2**: t. 16 (1967).

Rootstock with creeping stolons, leaves long-stalked, blades 5–20 × 1.5–6 cm, linear to narrowly elliptic, the veins all arising from the base, or entirely linear if submerged. Stem to 50 cm or more. Flowers 1.2 cm in diameter, in 5–7 whorls, stamens 12–20; filaments dilated at the base, hairy. *E USA.* H3. Summer.

A variant with the submerged leaves terete instead of ribbon-like is known as var. **teres** (Watson) Bogin or is sometimes treated as a separate species, *S. teres* Watson.

6. S. subulata (Linnaeus) Buchenau (*S. filiformis* Smith; *S. natans* Michaux not Pallas). Illustration: Muenscher, Aquatic plants of the United States, 96 (1944).

Rootstock with stolons. Leaves very variable, mostly linear, to *c.* 100 × 1.5 cm, submerged, rarely terminating in a floating blade, 5 × 2.5 cm. Stem to 30 cm. Flowers 1.5–2 cm in diameter, in 1–3 whorls. Stamens 7–10; filaments dilated at the base, hairless. *E USA.* H3. Summer.

7. S. macrophylla Zuccarini

Rootstock with stolons. Leaf-stalks to 50 cm, blades to 25 × 10 cm, lanceolate to ovate, sometimes truncate or hastate at base, submerged leaves to 40 × 1.5–4 cm, linear. Stem to 1 m. Flowers 3–4 cm in diameter, in 3–5 whorls. Stamens 20–30; filaments linear to spindle-shaped, hairless. *Mexico.* G1. Summer.

Suitable for tropical and subtropical aquaria.

2. ECHINODORUS Engelmann

C.J. King

Rootstock sometimes bearing stolons, leaves aerial, floating or submerged. Flowers bisexual, arranged in several simple whorls or the inflorescence compound with branches bearing whorls of flowers. Petals white. Stamens 6–30. Carpels several to many, spirally arranged on the receptacle, developing into a head of ribbed, beaked achenes.

A genus of about 47 species from the warmer regions of America. Several species are cultivated in heated aquaria. They require a loamy soil containing peat and sand, and are propagated by division or seed.

Literature: Matias, L.Q., O genero Echinodorus (Alismataceae) no dominio da Caatinga brasileira, *Rodriguesia* **58**(4): 743–774 (2007).

Stolons. Present: **1, 2**.
Leaf-blades. Broadly ovate: **3**.
Whorls of flowers. One: **1, 2**; two to eight: **3, 4, 5**; ten or more: **6**.
Flowers. Less than 1 cm in diameter: **1, 2**; more than 2 cm in diameter: **5**.
Flower-stalks. To 1 cm: **2, 4, 6**; 1–2.5 cm: **1, 5**; more than 2.5 cm: **3**.
Stamens. Nine: **1, 2**; ten to twelve: **3, 4, 5**; more than twelve: **3, 5, 6**.
Achenes. 1 mm: **1**; 1.5–3 mm: **2, 3, 5, 6**; more than 3 mm: **3, 4**.

1a. Stolons present 2
b. Stolons absent 3
2a. Leaf-blades not more than 3 cm **1. tenellus**
b. Leaf-blades 4–8 cm **2. magdalenensis**
3a. Leaf-blades broadly ovate **3. cordifolius**
b. Leaf-blades linear to lanceolate or elliptic 4
4a. All main veins arising from base of leaf-blade 5
b. Some main veins arising higher up the midrib **4. brevipedicellatus**
5a. Leaf-blades linear-lanceolate to elliptic, more than 3 times as long as broad **5. paniculatus**
b. Leaf-blades elliptic to oblong, less than 3 times as long as broad **6. longistylis**

1. E. tenellus (Martius) Buchenau. Illustration: Muenscher, Aquatic plants of the United States, 83 (1944); de Wit, Aquarium plants, 151 (1964).

Rootstock bearing stolons, mat-forming. Leaf-stalks to 5 cm, blades 2–3 × 1.2 cm, linear-lanceolate to oblong, submerged leaves to 10 cm × 2 mm, linear. Stem to 10 cm. Flowers to 1 cm in diameter, 3–6 in a single whorl. Flower-stalks 1.2–2.5 cm. Stamens 9. Achenes 1 mm, each with a short, hooked beak. *C & E USA, West Indies, C & S America.* G2. Summer.

2. E. magdalenensis Fassett. Illustration: de Wit, Aquarium plants, 144 (1964).

Rootstock with many stolons, leaf-stalks 5–10 cm, sometimes longer, blades 4–8 × 1–4 cm, lanceolate to narrowly ovate. Stem to 30 cm. Flowers *c.* 5 mm in diameter, 4–7 in a single whorl. Flower-stalks 1 cm. Stamens 9. Achenes 1.5–2 mm, with a short beak. *Colombia.* G2. Summer.

3. E. cordifolius (Linnaeus) Grisebach (*E. radicans* (Nuttall) Engelmann). Illustration: Muenscher, Aquatic plants of the United States, 83 (1944); Rickett, Wild flowers of the United States **6**: t. 15 (1973).

Leaf-stalks 5–15 cm, blades 10–20 × 6–10 cm, broadly ovate, base cordate to truncate. Stem to 1 m or more. Flowers 1.2–1.8 cm in diameter, in 2–6 whorls, 5–15 flowers in each whorl. Flower-stalks to 6 cm. Stamens 12–20. Achenes 1.8–3.5 mm, with a short beak. *S, C & E USA, Mexico.* G1. Summer.

4. E. brevipedicellatus (Kuntze) Buchenau. Illustration: de Wit, Aquarium plants, 138 (1964).

Leaf-stalks 5–15 cm, blades 20–30 × 2–4 cm, narrowly lanceolate, some main veins arising from the base, others arising higher up the midrib. Stem to 1 m. Flowers 1–2 cm in diameter, in 4–6 whorls, 3–8 flowers in each whorl. Flower-stalks 1 cm. Stamens 12. Achenes 3–4 mm. *Brazil.* G2. Summer.

5. E. paniculatus Micheli.

Leaf-stalks 10–30 cm, winged, blades 25–45 × 3–8 cm, linear-lanceolate to elliptic, sometimes sickle-shaped, main veins all arising from the base. Stem to 2 m. Flowers 1.2–2.5 cm in diameter, in 5–8 whorls, 6–8 flowers in each whorl. Flower-stalks 1.5 cm. Stamens 10–24. Achenes 2 mm, each with a short beak. *Brazil & Venezuela to Paraguay.* G2. Summer.

6. E. longistylis Buchenau.

Leaf-stalks to 50 cm, blades 15–25 × 10 cm, elliptic to oblong, main veins all arising from the base. Stem to 1 m. Flowers 1–2 cm in diameter, in 10–15 whorls, 6–9 flowers in each whorl. Flower-stalks 1 cm. Stamens numerous. Achenes 2.5 mm, each with a long beak. *Brazil.* G2. Summer.

3. BALDELLIA Parlatore

C.J. King

Rootstock sometimes bearing stolons. Leaves elliptic to lanceolate or linear-lanceolate. Flowers bisexual, in 1–3 whorls in umbels or racemes. Stamens 6. Carpels numerous, free, each with 1 ovule, spirally arranged in a spherical head. Achenes ovoid, longitudinally 5-ribbed (3 ribs on the back and an inner double rib), each with a short apical beak.

A genus of 2 species from Europe and N Africa; the more widespread species is cultivated in an unheated aquarium or in a bog garden.

1. B. ranunculoides (Linnaeus) Parlatore (*Alisma ranunculoides* Linnaeus; *Echinodorus ranunculoides* (Linnaeus) Engelmann). Illustration: Bonnier, Flore complète **10**: t. 570 (1929); Ross-Craig, Drawings of British plants **31**: t. 8 (1973); Grey-Wilson & Mathew, Bulbs, t. 1 (1981).

Leaf-stalks to 25 cm, blades to 10 × 1.5 cm or sometimes larger, decurrent, submerged leaves linear. Stem to 30 cm. Flowers 1–1.5 cm in diameter. Petals white or pale pink with a yellow blotch at the base. Achenes 2–3.5 mm. *Europe northwards to S Scandinavia, N Africa,* H2. Summer.

4. LURONIUM Rafinesque

C.J. King

Stems to 50 cm or more, floating or creeping and rooting at the nodes. Floating leaves long-stalked, blades to 4 × 1.5 cm, elliptic to ovate, submerged leaves linear. Flowers bisexual, 1.2–1.5 cm in diameter, long-stalked. Petals white with a yellow blotch at the base. Stamens 6. Carpels 6–15 in an irregular whorl, free, each with 1 ovule. Achenes oblong-ovoid, 2.5 mm, with 12–15 longitudinal ribs and a short, apical beak.

A genus of 1 species, formerly known as *Elisma* Buchenau, cultivated in unheated aquaria. It requires a loamy soil containing peat and sand, and is propagated by division.

1. L. natans (Linnaeus) Rafinesque (*Alisma natans* Linnaeus; *Elisma natans* (Linnaeus) Buchenau). Illustration: Hegi, Illustrierte Flora von Mittel-Europa **1**: t. 19 (1935); Ross-Craig, Drawings of British plants **31**: t. 9 (1973); Fitter, The wild flowers of Britain and northern Europe, 259 (1974); Grey-Wilson & Mathew, Bulbs, t. 1 (1981).

Europe. H2. Summer.

5. ALISMA Linnaeus

C.J. King

Leaves aerial, floating or submerged. Flowers bisexual, in panicles or occasionally in racemes or umbels. Stamens 6. Carpels 11–28 in a single whorl, free, each with 1 ovule. Achenes laterally compressed, each with a short beak.

A genus of 9 species, almost cosmopolitan, but indigenous in the northern hemisphere. A few species are cultivated as waterside plants, and some are suitable for unheated aquaria. They will grow in any good wet soil, but prefer one that contains abundant organic matter. Propagation is by division of the rootstock or by seed.

Literature: Hendricks, A.J., A revision of the genus Alisma (Dill.) L., *American Midland Naturalist* **58**: 470–93 (1957); Bjorkqvist, I., Studies in Alisma L., *Opera Botanica* **17**: 1–128 (1967) & **19**: 1–138 (1968).

1a. Styles shorter than the ovaries, recurved **1. gramineum**
b. Styles as long as or longer than the ovaries, more or less erect 2
2a. Leaves lanceolate to elliptic, cuneate at base; petals purplish pink **2. lanceolatum**
b. Leaves ovate or elliptic-ovate to lanceolate, usually cordate or truncate at base; petals white or purplish white **3. plantago-aquatica**

1. A. gramineum Lejeune (*A. plantago-aquatica* Linnaeus subsp. *graminifolium* (Steudel) Hegi). Illustration: Bonnier, Flore complète **10**: t. 571 (1929); Grey-Wilson & Mathew, Bulbs, t. 1 (1981).

Aerial leaves long-stalked, blades to 10 × 4.5 cm, elliptic or narrowly oblong, submerged leaves to 100 × 1.5 cm, linear. Stem to 1 m. Flowers about 6 mm in diameter. Petals white or purplish white. Anthers almost circular. Styles shorter than the ovaries, recurved. Achenes 2–2.75 mm. *Europe, Asia, N Africa.* H2. Summer.

A variant suitable for the aquarium, with entirely submerged leaves (forma **submersum** Gluck), is maintained by using sand at the base instead of soil.

2. A. lanceolatum Withering. Illustration: Bonnier, Flore complète **10**: t. 571 (1929); Ross-Craig, Drawings of British plants **31**: t. 7 (1973).

Leaves aerial, long-stalked, blades to 27 × 7.5 cm, lanceolate to elliptic, tapered at base. Stem 30–45 cm. Flowers 6–9 mm in diameter. Petals purplish pink. Anthers elliptic. Styles as long as or longer than the ovaries, more or less erect. Achenes 2– 3 mm. *Europe, Asia, N Africa.* H2. Summer.

3. A. plantago-aquatica Linnaeus. Illustration: Bonnier, Flore complète **10**: t. 570 (1929); Ross-Craig, Drawings of British plants **31**: t. 6 (1973); Grey-Wilson & Mathew, Bulbs, t. 1 (1981); Brickell (ed.), RHS A–Z encyclopedia of garden plants, 95 (2003).

Leaves usually aerial, long-stalked, blades to 30 × 12 cm, ovate or elliptic-ovate to lanceolate, mostly more or less cordate or truncate at base, but sometimes tapered. Stem 20–100 cm. Flowers to *c.* 1 cm in diameter. Petals white or purplish white. Anthers elliptic. Styles as long as or longer than the ovaries, more or less erect. Fruiting heads to 6 mm in diameter. Achenes usually 2–3 mm. *N temperate regions.* H2. Summer.

Two varieties are occasionally cultivated: var. **americanum** Schultes & Schultes (*A. triviale* Pursh) from N America and temperate E Asia, with white petals and fruiting heads 4–4.5 mm in diameter; and var. **parviflorum** (Pursh) Torrey (*A. parviflorum* Pursh; *A. subcordatum* Rafinesque) from eastern N America, with flowers 3–3.5 mm in diameter and fruiting heads *c.* 3.5 mm in diameter.

2. BUTOMACEAE

Aquatic herbs without latex. Leaves basal, not clearly divided into blade and stalk; ptyxis supervolute. Flowers solitary or in umbels, subtended by bracts, bisexual, actinomorphic. Perianth 1-whorled; segments 6, all petal-like. Stamens 1–many. Ovary of 6–many free carpels, superior; ovules many in each carpel, placentation diffuse-parietal. Fruit a follicle or group of follicles.

A single genus, native to Europe, with a single species, often cultivated as an ornamental aquatic.

1. BUTOMUS Linnaeus
A. Brady
Description as for family.

A genus of a single species from temperate regions of Europe and Asia. It is suitable for cultivation in bog gardens, watercourses with still or slow-moving water, and at pool margins. Propagation is by division of the rootstock or by seeds.

1. B. umbellatus Linnaeus. Illustration: *English Botany*, t. 1441 (1805); *Botanical Magazine*, 9125 (1926); Keble Martin, The concise British flora in colour, 79 (1965); Cook, Water plants of the world, 166 (1974).

Leaves 50–100 cm × 3–10 mm, apex acuminate. Flowering stem to 1 m. Sepals and petals 1–1.5 cm; sepals greenish outside, sepals inside and petals pinkish white with darker pink veins. *Eurasia*. H1. Summer.

3. LIMNOCHARITACEAE

Aquatic herbs with latex. Leaves on the stems clearly divided into a blade and stalk. Flowers solitary or in umbels subtended by bracts, bisexual, actinomorphic. Perianth of 2 whorls: sepals 3, petals 3, persistent. Stamens 1–many. Ovary of 6–many free carpels, superior; ovules many in each carpel, placentation diffuse parietal. Fruit a follicle or group of follicles.

Three genera, of which species of 2 cultivated as ornamental aquatics.

1a. Leaves floating, stalks with transverse septa **1. Hydrocleis**
b. Leaves usually erect, above the water, stalks without transverse septa **2. Limnocharis**

1. HYDROCLEIS Richard
E.C. Nelson
Annual or perennial herbs, with stolons. Leaves basal or in groups along the stems; stalks with transverse septa. Flowers in axillary clusters or in terminal umbels or solitary. Petals yellow. Stamens 6–many; staminodes present or absent. Carpels 3 or 6, united at the base.

A genus of about 9 species from S America. The plants may be free-floating or sometimes rooted in shallow water. Only 1 species is grown in aquaria in Europe. It will thrive in a rich loam in about 20 cm of water; it is not frost-hardy, but may survive out of doors in mild winters. Propagation is by cuttings.

1. H. nymphoides (Willdenow) Buchenau. Illustration: *Botanical Magazine*, 3248 (1833); Cook, Water plants of the world, 298 (1974); Brickell (ed.), RHS A–Z encyclopedia of garden plants, 555 (2003).

Perennial herb with stolons, sometimes rooting in shallow water. Leaves 5–7 cm in diameter, oval with cordate bases, thick, floating, deep green. Flowers yellow, bowl-shaped, to 5 cm in diameter. *Tropical S America*. H5–G1. Summer.

2. LIMNOCHARIS Humboldt & Bonpland
E.C. Nelson
Annual or perennial herbs without stolons. Leaves in rosettes; stalks triangular in cross-section, without transverse septa. Umbel with 2–15 flowers. Petals yellow. Stamens numerous, surrounded by staminodes. Carpels numerous, laterally flattened, closely adpressed and apparently forming a single ovary.

The 2 species are natives of the warmer regions of America, but are now naturalised elsewhere. They are tender, so should be grown indoors in aquaria or tubs, or in shallow pools. Propagation is by seed, offsets or suckers. Only 1 species is well known in European gardens.

1. L. flava (Linnaeus) Buchenau. Illustration: Cook, Water plants of the world, 299 (1974).

Perennial with erect stems. Leaves to 60 cm, blades *c.* 20 cm, ovate with cordate bases. Umbels with 6–12 flowers. Petals pale yellow bordered with white, to 2.5 cm in diameter. *West Indies, C & tropical S America*. G2. Summer.

4. HYDROCHARITACEAE

Aquatics, usually with at least the flowers emerging from the water. Leaves alternate, variable, usually with distinct stalk and blade. Flowers usually borne in a bifid spathe or between 2 opposite bracts, rarely solitary, unisexual or bisexual, actinomorphic. Perianth in 2 whorls: sepals 3, petals 3. Stamens 1–many. Ovary of usually 3–6 united carpels, inferior; ovules many, diffuse parietal. Fruit usually a capsule, rarely berry-like.

There are 15 genera and about 100 species of submerged, floating or emergent aquatics. Ten genera occur wild in Europe, and 10 in North America. Species of 9 genera are grown as ornamentals or aquarium-oxygenating plants.

Habit. Free-floating: **5, 6, 9**.
Leaves. All basal: **4, 6, 7, 8, 9**; arranged along the stem: **1, 2, 3, 5**. Linear: **1, 2, 3, 5, 9**. Rigid: **8**. Stalked: **4, 6, 7**. Spirally arranged: **5**.
Flowers. Conspicuous: **1, 4, 5, 6, 7, 8**. Petals white: **1, 4, 7, 8**; petals pinkish: **5**.

1a. Leaves with distinct stalks, in basal rosettes 2
b. Leaves arranged along the stem, or, if in basal rosettes, then stalkless 4
2a. Plants rooted in the substrate, without stolons; spathes tubular, ribbed or winged **7. Ottelia**
b. Plants floating, with stolons; spathes not ribbed or winged 3
3a. Leaves with spongy tissue beneath; petals absent, or, when present, at most 1.5 times as long as sepals **6. Limnobium**
b. Leaves without spongy tissue beneath; petals present and more than 1.5 times as long as the sepals **4. Hydrocharis**
4a. Leaves in a basal rosette 5
b. Leaves distributed along the stems 6
5a. Leaves rigid, margins with prominent, spiny teeth; plant aloe-like **8. Stratiotes**
b. Leaves not rigid but ribbon-like, margins without prominent, spiny teeth **9. Vallisneria**
6a. All leaves spirally arranged **5. Lagarosiphon**
b. Leaves opposite or whorled, at least in the upper part of the stem 7
7a. Flowers conspicuous, held above the water surface, with nectaries **1. Egeria**

b. Flowers minute, submerged or borne at the water surface, without nectaries 8
8a. Leaves in whorls of 3–8, with fringed nodal scales **3. Hydrilla**
b. Leaves in whorls of 3–4, with inconspicuous, unfringed nodal scales **2. Elodea**

1. EGERIA Planchon
E.C. Nelson
Dioecious or monoecious, aquatic, perennial herbs. Leaves submerged, whorled, stalkless, linear. Flowers conspicuous, with nectaries, held above the water surface. Male spathes 2–4-flowered; petals of male flowers much longer than sepals, stamens 9. Ovary stalkless, styles 3, free, bifid or occasionally 3-fid.

A genus of 2 species from warm temperate regions of S America, now naturalised in Europe and elsewhere. Both are cultivated in ponds and aquaria, and are sometimes listed in *Elodea*, but differ in their flowers: in *Egeria* these are conspicuous and held above the water surface. Propagation is by cuttings.

1a. Leaves to 4 cm; plants robust **1. densa**
b. Leaves at most 2.5 cm; plants very slender **2. najas**

1. E. densa Planchon (*Elodea densa* (Planchon) Caspary; *Elodea canadensis* Michaux var. *gigantea* Bailey). Illustration: Cook, Water plants of the world, 257 (1974); Brickell (ed.), RHS A–Z encyclopedia of garden plants, 401 (2003).

Stems stout, to 1 m. Leaves crowded in whorls of 3–5, to 4 cm × 5 mm, minutely toothed. Sepals 3–4 mm; petals 8–11 mm, obovate, white. *S America.* H1. Summer.

A good aquarium plant, an excellent oxygenator.

2. E. najas Planchon (*Elodea najas* (Planchon) Caspary).

Stems very slender. Upper leaves in whorls of 4–6, linear, to 2.5 cm × 1.5 mm, sharply toothed. Petals *c.* 7 mm, white. *S America.* H5. Summer.

Not as commonly cultivated as *E. densa.*

2. ELODEA Michaux
E.C. Nelson
Dioecious or monoecious, aquatic perennial herbs, often with stolons. Leaves submerged, whorled or sometimes opposite, stalkless, usually linear, with 2 minute scales at each node. Flowers inconspicuous, without nectaries, borne and pollinated at the water surface. Spathes axillary, stalkless, with 1 or rarely 3 male flowers, or 1 female flower, or rarely 1 bisexual flower. Sepals 3; petals 3. Stamens 3–9. Ovary stalkless; styles 3, free, bifid or occasionally entire.

A genus with poorly understood limits. Some plants commonly included in it belong to other genera (e.g. *Egeria, Hydrilla*). There are perhaps 15 species, all native to America, but now widely naturalised in other countries, where they are sometimes serious weeds of aquatic habitats. There are 3 species that are widely cultivated in ponds and aquaria; they are easily grown and propagated by cuttings or stolons.

1a. Leaves less than 2 mm broad; sepals of female flowers less than 2 mm **3. nuttallii**
b. Leaves usually at least 2 mm broad; sepals more than 2 mm 2
2a. Leaves *c.* 10 × 2.5 mm, rounded at the apex **1. canadensis**
b. Leaves *c.* 2.5 cm × 2.5 mm, tapered to acute at the apex **2. callitrichoides**

1. E. canadensis Michaux. Illustration: Cook, Water plants of the world, 257 (1974).

Leaves usually *c.* 10 × 2 mm, but the largest may be up to 17 × 4 mm, in whorls of 3, or the lower opposite, all linear to oblong, minutely toothed, rounded at the apex. Sepals of female flowers 2–2.7 mm; petals white or pale purple. *N America.* H1. Summer.

A common aquarium plant, now widely naturalised in Europe, having been introduced into Ireland in 1834.

2. E. callitrichoides (Richard) Caspary (*E. ernstae* St John).

Leaves to 2.5 cm × 2.5 mm, in whorls of 3 or the lowest opposite, all gradually tapering to an acute apex, minutely toothed. Sepals of female flowers 3–3.5 mm; petals white. *S America.* H4. Summer.

Like the other species, this is naturalised in Europe. It is not as hardy as *E. canadensis.* For the use of the name, see Cook, *Watsonia* **15**: 117 (1984).

3. E. nuttallii (Planchon) St John.

Leaves to 1.5 cm, less than 2 mm broad, in whorls of 3 or occasionally 4, or the lowest opposite, all linear, tapering to an acute apex, minutely toothed. Sepals of female flowers 1–1.8 mm. Male flowers breaking free and floating to the water surface. *N America.* H1. Summer.

This plant is sometimes confused with *Hydrilla verticillata* and the name *E. nuttallii* has been misapplied to *Hydrilla.*

3. HYDRILLA Richard
M.J.P. Scannell
Dioecious, submerged, branched herbs with stolons, overwintering as turions. Leaves 1–4 cm × *c.* 2.5 mm, in whorls of 3–8, or the lowest opposite, lanceolate, each with a single midvein, stalkless, minutely toothed, with 2 brown, fringed scales, *c.* 0.5 mm, in each axil. Flowers borne and pollinated at the water surface, solitary, emerging from a tubular spathe. Sepals 3; petals 3. Male flowers breaking free and floating on the water surface at maturity; stamens 3. Female flowers stalkless, with an elongate perianth-tube (to 4.5 cm); sepals 1.5–3 × 0.3–0.5 mm; petals smaller; staminodes 3; ovary red; styles not apparent; stigmas 3 (rarely to 5). Fruit cylindric.

A misunderstood genus of a single species, widely distributed in isolated temperate and tropical habitats; sometimes cultivated in fresh-water aquaria or ponds. Propagation is by cuttings or by turions.

Literature: Scannell, M.J.P. & Webb, D.A., The identity of the Renvyle Hydrilla, *Irish Naturalists' Journal* **18**: 327–30 (1976).

1. H. verticillata (Linnaeus) Royle (*Elodea nuttallii* misapplied). Illustration: *Irish Naturalists' Journal* **18**: t. 1–2 & f. 1–5 (1976).

Europe (British Isles, Germany to Former USSR), E Asia, E Africa, Australia. H2. Summer.

European plants are hardy and can be easily cultivated. The plant is a good oxygenator.

4. HYDROCHARIS Linnaeus
E.C. Nelson
Aquatic perennial herbs. Leaves stalked, all basal. Flowers unisexual, held above the water. Male spathes stalked, with 1–4 flowers. Female spathes stalkless, each 1-flowered; petals much longer than sepals. Stamens 9–12, the outer usually reduced to staminodes. Styles 6, bifid, free.

A genus of 2 species, of which only 1 is widely cultivated in Europe. Propagation is by cuttings, stolons or seeds. The plants will grow floating in deep water, but will sometimes become rooted in mud in shallow water. They are suitable for still-water habitats, ponds and aquaria.

1. H. morsus-ranae Linnaeus.
Illustration: *English Botany*, 1444 (1805); *Botanical Magazine*, 3248 (1833); Keble Martin, The concise British flora in colour, 79 (1965); Cook, Water plants of the world, 261 (1974).

Leaves circular to kidney-shaped, entire, cordate at base. Petals *c.* 1 cm, obovate, white, each with a yellow spot at the base. *Europe, W Asia, N Africa.* H1. Summer.

5. LAGAROSIPHON Harvey
E.C. Nelson

Dioecious, submerged plants. Leaves stalkless, each with 2 minute scales at the node. Flowers submerged until open, pollinated at the water surface. Sepals 3, petals 3. Male spathe with many flowers; stamens 3, staminodes 3. Female spathe usually with 1 flower; ovary stalkless, styles 3, free, bifid.

A genus of about 15 species, native in tropical and subtropical regions of Africa, Madagascar and India. Only 1 species is cultivated in Europe, where it has escaped from cultivation and become naturalised. It will grow easily in aquaria and ponds in milder regions, but thrives best when the water temperature is 12–18 °C throughout the year. Propagation is by cuttings.

1. L. major (Ridley) Moss (*L. muscoides* Harvey; *Elodea crispa* misapplied).
Illustration: Cook, Water plants of the world, 259 (1974); Brickell (ed.), RHS A-Z encyclopedia of garden plants, 605 (2003).

Stems to 60 cm, sometimes branched. Leaves dark green, linear, to 3 cm × 3 mm, densely arranged at the tips of the branches, reflexed, minutely toothed. Perianth-tube to 3.5 cm; sepals and petals tinted pink. *Southern Africa.* H4. Summer.

The plant resembles a large *Elodea* and is sometimes incorrectly called *E. crispa*. It is a good oxygenator, but requires plenty of light.

6. LIMNOBIUM Linnaeus
E.C. Nelson

Floating perennial herbs, with stolons. Leaves all basal, with distinct stalks; blade with a layer of spongy tissue beneath. Flowers unisexual, held above the water surface. Sepals 3, petals 3. Male spathes with 3–6 flowers; stamens 6–12, filaments united. Female spathes with 1 flower; ovary 6–9-celled, stigmas as many as cells, bifid. Fruit many-seeded.

A genus of about 3 species, found in the warmer parts of America, but introduced and naturalised elsewhere. The plants are generally free-floating, but may occasionally root in shallow water. Propagation is by stolons. The genus is not common in cultivation, and only 1 species is seen in European gardens.

1. L. laevigatum (Willdenow) Heine (*L. stoloniferum* (Meyer) Grisebach).
Illustration: Cook, Water plants of the world, 261 (1974).

Leaf blade to 3 cm, elliptic to ovate, wavy, pale green; stalk to 5 cm. Sepals oblong; petals only slightly longer than the sepals and narrower, linear. *West Indies, C & tropical S America.* G1. Summer.

This species needs plenty of light, but not direct sunlight. It grows best in shallow water over a peaty substrate.

7. OTTELIA Persoon
E.C. Nelson

Monoecious or dioecious aquatic herbs. Leaves submerged, basal, stalked. Flowers held above the water surface. Spathes stalked, tubular, ribbed and winged; male spathes with many flowers, the female (and occasional bisexual spathes) with 1 flower. Petals larger than sepals. Stamens 6–20, anthers opening laterally. Ovary stalkless; styles 3–15, bifid, free.

A genus of about 40 species from warmer regions of the world, mainly Africa and E Asia. Only 1 species is widely cultivated in Europe, and it is naturalised in N Italy. Propagation is by division or seed.

1. O. alismoides (Linnaeus) Persoon.
Illustration: *Botanical Magazine*, 1201 (1809); Cook, Water plants of the world, 265 (1974).

Perennial with tuberous rootstock. Leaves varying greatly in size, generally submerged but sometimes partly emergent; juvenile leaves linear or with lanceolate blades; older leaves with broad, elliptic blades to 20 cm, thin, translucent pale green, margins wavy; stalk to 50 cm, triangular in cross-section. Petals 2–3 cm, obovate, white, each with a yellow basal spot. Stamens 6–9. Styles 6–9. *E Asia, NE Africa, Australia.* H5. Summer.

This species prefers warm water (22–28 °C) that is clear and not alkaline; it should be planted in a rich compost.

8. STRATIOTES Linnaeus
E.C. Nelson

Dioecious, submerged, aloe-like perennial herbs, with stolons. Leaves all basal, stalkless. Flowers held above the water. Male spathes with several flowers, female spathes with 1 flower. Petals larger than the sepals. Stamens 12, surrounded by numerous staminodes. Ovary stalkless, styles 6, bifid, free.

A genus of a single species that resembles an aloe in form. The plants are submerged, but rise to the surface at flowering time. It is easily cultivated in outdoor ponds and aquaria and is propagated very easily by means of the stolons. It can become a weed in suitable situations.

1. S. aloides Linnaeus. Illustration: *English Botany*, 379 (1791); *Botanical Magazine*, 1285 (1810); Keble Martin, A concise British flora in colour, 79 (1965); Cook, Water plants of the world, 266 (1974).

Leaves *c.* 50 × 2 cm, in a rosette, linear-lanceolate, brittle, tapered to the apex, with numerous marginal teeth. Petals white, 1.5–2 cm, broader than long. *Eurasia.* H1. Summer.

9. VALLISNERIA Linnaeus
E.C. Nelson

Dioecious, submerged plants, with stolons. Leaves all basal, alternate, stalkless. Flowers pollinated at the water surface. Sepals 3; petals 3. Male spathes almost stalkless or stalked, many-flowered; stamens 2–3. Female spathes on long stalks, 1-flowered; ovary linear, styles 3, bifid, free.

A genus of about 10 species distributed throughout the world apart from the coldest regions. They grow in still or running water, and reproduce by means of stolons; some cause serious problems in irrigation canals. Only 3 species are cultivated in Europe and these are occasionally naturalised. Propagation is by stolons or offsets. In aquaria the plants will root in sand, and a layer of loam beneath the sand will improve their growth.

1a. Plants becoming dormant in winter, even in aquaria **1. americana**
b. Plants growing throughout the year 2
2a. Stalks of female spathes coiled after flowering **3. spiralis**
b. Stalks of female spathes straight and remaining so **2. gigantea**

1. V. americana Michaux.

Leaves to 2 m × 2 cm, linear. Male spathes to 1.5 cm. Female spathes on stout stalks which curve or spiral in fruit. Fruit curved. *E USA.* H4. Summer.

2. V. gigantea Graebner.

Leaves to 1 m × 2 cm, linear, with 5–9 conspicuous veins and longitudinal brown or blackish stripes, indistinctly toothed. Female spathes on straight stalks. Fruits striped with brown or black. *S Asia to Australia.* H5–G1. Summer.

Although it will survive outdoors in the mildest parts of Europe, this is generally grown in indoor aquaria. It is adaptable, and can survive water temperatures of over 30 °C. It is a good oxygenator.

3. V. spiralis Linnaeus. Illustration: *Botanical Magazine*, 4894 (1856), 6912 (1886); Cook, Water plants of the world, 268 (1974).

Leaves linear, ribbon-like, to 1 cm broad, with minute teeth towards the apex. Male spathes on short stalks, longer than those of *V. americana.* Female spathes linear, their stalks becoming spirally coiled in fruit. Fruit straight. *S Europe, W Asia.* H5. Summer.

A common aquarium plant that grows continually and is a good oxygenator. Several cultivars are available, including 'Torta', which has shorter and broader, spirally coiled leaves, and which may be grown outdoors in warm areas.

5. APONOGETONACEAE

Aquatics. Leaves alternate, long-stalked, with sheathing bases and expanded blades. Inflorescence a simple or forked spike. Flowers usually bisexual and actinomorphic. Perianth 1-whorled; segments 1–3 or rarely absent; when only 1 segment, this bract-like. Stamens 6 or more. Ovary of 3–6 free carpels, superior; ovules few in each carpel, basal. Fruit a group of follicles.

A single genus with about 44 species, a few of which are cultivated as ornamental aquatics.

1. APONOGETON Linnaeus
E.C. Nelson
A genus of about 20 species, formerly known as *Ouvirandra* Thouars. Only 2 species are well known in cultivation in Europe. Propagation is by division, offsets or seeds.

1a. Leaves entire **1. distachyos**
b. Leaves reduced to a lace-like structure **2. madagascariensis**

1. A. distachyos Linnaeus. Illustration: *The Botanist's Repository*, 290 (1803); *Botanical Magazine*, 1293 (1810); Brickell (ed.), RHS A–Z encyclopedia of garden plants, 127 (2003).

Leaves floating, to 25 × 7 cm, elliptic to oblong. Spike with 2 branches, each bearing *c.* 10 fragrant, white flowers. Perianth-segment 1, ovate, 1–2 cm. Stamens 6–18. *South Africa.* H4. Summer.

Widely cultivated in ponds out of doors in the British Isles and in south western Europe, where it is sometimes naturalised. It will grow in water to 60 cm deep. There are several cultivars listed, including 'Roseus', which has pink flowers.

2. A. madagascariensis (Mirbel) van Brugg (*A. fenestralis* (Poiret) Hooker). Illustration: *Botanical Magazine*, 4894 (1856).

Leaves submerged, to 20 × 12 cm, elliptic or oblong, reduced to a lace-like structure formed from the persistent network of veins. Spikes branched; flowers numerous, white; perianth-segments small, scale-like. Stamens 6. *Madagascar.* G2. Summer.

A botanical curiosity on account of its strange leaves. It is difficult to grow, as the water must be kept clean and free from sediment, must be maintained at *c.* 20 °C, and should be deep and continuously circulating. Shade is also required.

6. POTAMOGETONACEAE

Aquatics. Leaves alternate, long-stalked, with sheathing bases and expanded blades. Inflorescence a simple or forked spike. Flowers usually bisexual and actinomorphic. Perianth 1-whorled: segments 1–3 or rarely absent; when only 1 segment, this bract-like. Stamens 6 or more. Ovary of 3–6 free carpels, superior; ovules few in each carpel, basal. Fruit a group of follicles.

A single genus with about 44 species, a few of which are cultivated as ornamental aquatics.

1. POTAMOGETON Linnaeus
A. Brady
Rhizomes branching, sometimes producing turions. Leaves alternate, submerged or partly floating; submerged leaves thin, translucent, linear; floating leaves leathery, opaque. Stipules present, free or attached to the leaf base. Flowers wind- or water-pollinated. Fruit a drupe.

A genus of about 100 species distributed throughout the world; only 2 are widely cultivated in aquaria and ponds as oxygenating plants. They reproduce by rhizomes or turions.

1a. Margins of submerged leaves markedly wavy when mature **1. crispus**
b. Margins of submerged leaves flat **2. natans**

1. P. crispus Linnaeus. Illustration: Keble Martin, The concise British Flora in colour, 90 (1965); Ross-Craig, Drawings of British plants **31**: t. 36 (1973); Cook, Water plants of the world, 495 (1974).

Branches to 100 cm. Submerged leaves stalkless, membranous, *c.* 4 cm × 8–15 mm, almost translucent, margins wavy when mature. Floating leaves stalked, pointed, leathery. Stipules small, membranous, evanescent. Spike 1–2 cm, loose, on a stalk *c.* 2.5 cm. Fruit 4–5 mm, with a long, curved beak that equals the rest of the fruit in length. Turions 1–5 cm with thick, horny, spiky-bordered leaves. *Most of the Old World to Australia and New Zealand.* H1. Summer.

2. P. natans Linnaeus. Illustration: Butcher, A new illustrated British flora, 597 (1961); Keble Martin, The concise British flora in colour, 89 (1965); Ross-Craig, Drawings of British plants **31**: t. 16 (1973).

Leafy stem more than 100 cm. Submerged leaves to 20 × 3 cm, stalked, lanceolate, obtuse at apex, margins flat; these leaves fall quickly. Floating leaves to 12.5 × 7 cm, ovate to elliptic or lanceolate, rounded to almost cordate at the base, opaque; stalk often longer than the blade. Stipules large, often fibrous, persistent. Spike 3–8 cm, dense, axillary, cylindric; stalk 5–12 cm. Fruit 3–5 mm, with a short, straight beak, green, ovoid, compressed. *Eurasia.* H1. Summer.

7. MELANTHIACEAE

Plants terrestrial, herbaceous, with rhizomes. Leaves well developed, borne on the stems. Flowers in racemes or panicles, if solitary then without spathes, bisexual, actinomorphic. Perianth of 6 segments, nectaries absent or on perianth-segments. Stamens 6. Ovary of 3 free or partly united carpels, usually superior, rarely half-inferior, not stalked; ovules numerous, placentation axile; styles 3. Fruit a capsule. Seeds not hairy.

There are 22 genera with about 80 species; 4 are found in Europe, 14 in North America, and several are cultivated. Includes *Nartheciaceae*.

1a. Style 1, sometimes 3-branched at the apex, or style absent, stigmas united borne directly on the ovary 2
b. Styles 3 (rarely 4), free, separate and distinct, or stigmas 3, separate, borne directly on the ovary 4
2a. Perianth-segments united at the base for more than one-tenth of their total length **12. Aletris**
b. Perianth-segments free at the base, or united for less than one-tenth of their total length 3
3a. Leaves evergreen; seeds with linear or thread-like appendages at both ends; filaments not woolly **6. Heloniopsis**
b. Leaves deciduous; seeds without such appendages; filaments woolly **2. Narthecium**
4a. Plants obviously hairy 5
b. Plants not at all hairy 7
5a. Perianth-segments conspicuously clawed **10. Melanthium**
b. Perianth-segments not conspicuously clawed 6
6a. Leaves broadly ovate to elliptic with many conspicuous veins, narrowed towards the stalk-like, sheathing bases **11. Veratrum**
b. Leaves narrow, without conspicuous veins, not narrowing towards the sheathing bases **9. Zigadenus**
7a. Plants with a bulb; perianth-lobes linear, curled **8. Stenanthium**
b. Plants with rhizomes or fleshy roots; perianth-lobes not as above 8
8a. Perianth-segments very unequal, the 2 or 3 lower ones small and thread-like **7. Chionographis**
b. Perianth-segments all more or less equal 9
9a. Leaves narrowed towards the base, shortly stalked 10
b. Leaves not narrowed towards the base, not stalked 11
10a. Flowers unisexual **5. Chamaelirium**
b. Flowers bisexual **4. Helonias**
11a. Stem-leaves numerous; flower-stalks very long **3. Xerophyllum**
b. Stem-leaves few; flower-stalks very short or absent **1. Tofieldia**

1. TOFIELDIA Hudson
J.C.M. Alexander

Perennial herbs with rhizomes. Leaves narrowly sword-shaped, densely tufted in 2 opposite ranks. Flowering stems erect, with or without leaves. Flowers small, in terminal racemes or heads. Flower-stalks short, bearing a bract at the base and sometimes a bracteole just below the flower. Perianth persistent; segments 6, free, narrow, ascending or spreading. Stamens free; filaments hairless; anthers broadly ovate, basifixed. Ovary superior; carpels 3, free above, gradually narrowed into 3 short persistent styles. Fruit a spherical to ellipsoid capsule; seeds very small, numerous, ellipsoid to narrowly oblong.

A north temperate genus of about 18 species, not regarded as ornamental but sometimes grown in gardens devoted to native or local floras. Only the 2 European species are described here. They perform best in acid, peaty soils in cool, moist situations such as the banks of ponds or streams. Propagation can be effected by division of the rhizome or from seed. *T. pusilla* is notoriously difficult to establish and seed is best sown in soil taken from its wild habitat; it must be moved without disturbing the roots.

1a. Bract at base of flower-stalk unlobed; bracteole just below flower 3-lobed **1. calyculata**
b. Bract at base of flower-stalk 3-lobed; bracteole absent **2. pusilla**

1. T. calyculata (Linnaeus) Wahlenberg (*T. palustris* Hudson, in part; *T. alpina Smith*, not Sternberg & Hoppe; *T. rubra* Braun; *T. glacialis* Gaudin). Illustration: Bonnier, Flore complète **10**: pl. 573 (1929); *Journal of the Linnean Society, Botany*, **53**: 197 (1947); Fitter et al., Wild flowers of Britain and northern Europe, 261 (1974).

Leaves 1.5–15 cm, with 4–10 veins. Stem 5–35 cm, usually bearing 1–3 leaves. Flowers yellowish, rarely reddish, up to 30 in a 2–6 cm raceme, or rarely in a 3-flowered head *c.* 5 mm long. Bract at base of flower-stalk unlobed; bracteole just below flower 3-lobed. Perianth-segments 2–3.5 mm. Capsule 3.5 mm. *Gotland (Sweden) & Estonia south to the Pyrenees, Yugoslavia & the Ukraine.* H1. Summer.

2. T. pusilla (Michaux) Persoon (*T. palustris* Hudson in part; *T. borealis* (Wahlenberg) Wahlenberg; *T. alpina* Sternberg & Hoppe not Smith). Illustration: *Journal of the Linnean Society, Botany*, **53**: 197 (1947); Ross-Craig, Drawings of British Plants **29**: pl. 32 (1972); Fitter et al., Wild flowers of Britain and northern Europe, 261 (1974).

Leaves 1–8 cm, with 3–7 veins. Stem 1.5–25 cm, leafless. Flowers white or greenish, in a 3–20 mm head. Bract at base of flower-stalk 3-lobed; bracteole absent. Perianth-segments 1.5–2.5 mm. Capsule 2.5–3 mm. *N Europe, Alps, W Carpathians & C Urals.* H2. Summer.

2. NARTHECIUM Hudson
J. Cullen

Perennial herbs with rhizomes and fibrous roots. Leaves basal and also borne on the stem. Flowers in a usually dense raceme; bracts about as long as the flower-stalks. Perianth-segments 6, free, spreading in flower, erect and persistent in fruit. Stamens 6, versatile, opening to the outside of the flower; filaments woolly. Ovary superior, 3-celled, 3-sided, ovules numerous in each cell; style simple. Fruit a capsule splitting between the septa. Seeds numerous, tailed.

A genus of about 8 species from north temperate regions, of which 1 is occasionally grown in places where the soil is boggy and very acidic.

1. N. ossifragum (Linnaeus) Hudson. Illustration: Keble-Martin, The concise British flora in colour, t. 86 (1965); Everard & Morley, Wild flowers of the world, t. 24 (1970).

Leaves 4–30 cm × 2–5 mm, rigid, often curved. Stems to 40 cm. Perianth-segments 6–8 mm, linear-lanceolate, yellow. Anthers orange. Stem, perianth and ovary becoming deep orange after flowering. *W Europe, from Scandinavia to N Spain and Portugal.* H3. Summer.

3. XEROPHYLLUM Michaux
J.C.M. Alexander

Perennial, deciduous herbs, with stout, woody, stem-like rhizomes. Leaves numerous, densely tufted, glaucous on backs, mostly basal, linear, slightly broader at base and finely tapered at tip; margins very hard and rough or finely toothed. Flowering stems stout, unbranched, erect, bearing smaller and shorter stem-leaves with membranous margins. Flowers white or yellowish white, numerous in long, dense terminal racemes. Perianth-segments 6, free, spreading, ovate, in 2 similar whorls, persistent. Stamens 6, free. Ovary superior; styles 3, thread-like, erect, becoming curled. Fruit a 3-lobed, 3-celled capsule with 2–4 wedge shaped seeds per cell.

A genus of 2 or 3 species from N America, which perform best if given plenty of space in a damp, peaty soil. They should be left undisturbed for as long as possible and are easily propagated by seed or division of the rhizome.

1a. Basal leaves to 45 cm × 2.5 mm; flowers *c.* 9 mm across; stamens shorter than perianth-segments **1. asphodeloides**

b. Basal leaves to 90 cm × 6 mm; flowers *c.* 15 mm across; stamens as long as or longer than perianth-segments **2. tenax**

1. X. asphodeloides (Linnaeus) Nuttall (*X. setifolium* Michaux). Illustration: *Botanical Magazine*, 748 (1804); *Gardener's Chronicle* **13**: 433 (1880); Rickett, Wild flowers of the United States **2**: pl. 39 (1967).

To 1.5 m. Basal leaves to 45 cm × 2.5 mm. Raceme to 30 cm; flowers yellowish white, *c.* 9 mm across, fragrant. Perianth-segments *c.* 6 mm; stamens shorter than perianth-segments. *Eastern N America.* H4. Summer.

2. X. tenax (Pursh) Nuttall (*X. douglasii* Watson; *X. setifolium* misapplied). Illustration: *Botanical Magazine*, 8964 (1922); RHS Dictionary of gardening, 2294 (1956); *Journal of the Royal Horticultural Society* **91**: pl. 7 (1966); Rickett, Wild flowers of the United States **5**: pl. 41 (1971).

To 1.8 m. Basal leaves to 90 cm × 6 mm. Raceme to 60 cm; flowers white to cream-coloured, *c.* 15 mm across. Perianth-segments 8–9 mm; stamens as long as or longer than perianth-segments. *Western N America.* H4. Summer.

4. HELONIAS Linnaeus

S.G. Knees

Evergreen perennial producing leaves and inflorescences from a horizontally branched, tuberous rhizome. Leaves 15–45 × 3–5 cm, entire, shortly stalked, lanceolate, in basal rosettes. Racemes dense, spike-like, conical, 2–3 cm long, borne terminally. Scapes 4–6 cm × 8–10 mm, stout, hollow, with bracts. Flowers small (*c.* 5 mm), 25–30, bisexual, star-shaped, sweetly scented, subtended by bracts. Perianth of 6 persistent, pink segments. Stamens 6, with long filaments and projecting, blue-grey anthers. Ovary superior, 3-celled; styles 3, divided to the base. Fruit a 3-celled, dehiscent capsule, 3–4 mm. Seeds many, with white appendages.

A genus of a single species, grown for its spring flowering, pink, orchid-like inflorescences, distinguished from the closely related *Heloniopsis* by its dense spikes of smaller flowers. It is best suited to a moisture-retentive, but humus-rich soil, reflecting its native habitat in swamp. Showing a preference for shaded conditions, it succeeds well in the woodland or bog garden. Propagation is by seed or division.

1. H. bullata Linnaeus. Illustration: *Botanical Magazine*, 747 (1804); *The Botanical Cabinet* **10**: t. 961 (1824); Britton & Brown, Flora of the northern United States and Canada, edn 1 **1**: 402 (1896), edn 2 **1**: 488 (1913); Hunt (ed.), Marshall Cavendish encyclopedia of gardening **26**: 877 (1968).

Eastern N America. H4. Spring.

5. CHAMAELIRION Willdenow

A.C. Leslie

Dioecious herb with a bitter, tuberous rootstock. Male plants 30–70 cm, female to 1.2 m, both erect, unbranched, hairless. Lower leaves in a loose rosette, 5–20 × 1–4 cm, obovate to spathulate, stalked; stem-leaves smaller, the uppermost linear to lanceolate, stalkless. Inflorescence a dense, cylindric raceme, the male 4–12 cm, often drooping at the tip at first, the female more slender, to 30 cm, erect. Flower-stalks to 5 mm, without bracts. Flowers *c.* 7 mm wide; perianth-segments 3 mm, linear-spathulate, white. Stamens with almost spherical anthers. Female flowers usually with 6 staminodes. Ovary superior, oblong or obovoid, with 3 short styles; stigmas decurrent. Fruit a capsule. Seeds numerous, linear-oblong, broadly winged.

A genus of a single species, uncommon in cultivation. The flowers become yellowish when dried. It requires a rich, damp soil and is shade-tolerant.

1. C. luteum (Linnaeus) Gray. Illustration: *Botanical Magazine*, 1062 (1808); Gleason, Illustrated flora of the north-eastern United States and adjacent Canada **1**: 405 (1952).

Eastern N America. H2. Summer.

6. HELONIOPSIS Gray

P.G. Barnes

Perennial evergreen herbs with short rhizomes. Leaves basal, scape arising from the centre of the rosette, erect, unbranched, with reduced, bract-like, lanceolate leaves crowded below. Racemes few-flowered, often condensed and umbel-like. Flowers pendent. Perianth-segments 6, spreading, equal. Stamens 6, filaments subulate. Ovary superior, 3-celled; style simple, with a capitate stigma. Fruit a 3-lobed capsule. Seeds numerous, linear, each with a thread-like appendage at each end.

Three or four species in Japan, Korea and Taiwan. They are easily cultivated in humus-rich soil in light shade and sometimes form colonies by means of viviparous leaf-tips.

1. H. orientalis (Thunberg) Tanaka (*H. japonica* Maximowicz). Illustration: *Botanical Magazine*, 6986 (1888); Brickell (ed.), RHS A–Z encyclopedia of garden plants, 522 (2003).

Leaves oblanceolate, acute, sometimes viviparous at the apex, gradually narrowed to the base. Flower-stalks 1–1.5 cm, at first curved, elongating and becoming erect in fruit. Perianth-segments 1–1.5 cm, narrowly spathulate or oblong, obtuse, pink or violet, persistent in fruit and becoming greenish. Filaments subulate, *c.* 1.5 cm, anthers bluish violet. Fruit 1.2–1.5 cm. *Japan, Korea, Former USSR (Sakhalin).* H3. Spring.

Cultivated material is mostly var. **orientalis**; var. **breviscapa** (Maximowicz) Ohwi differs only in the smaller, pale pink or whitish flowers.

7. CHIONOGRAPHIS Maximowicz

J.C.M. Alexander

Hairless herbaceous perennials with short, thick rhizomes. Basal leaves elliptic to ovate or oblanceolate, often with long stalks; stem-leaves smaller and stalkless. Flowers many, in a terminal spike, white, bisexual, bilaterally symmetric. Perianth-segments 3–6, unequal; the upper 3 or 4 petal-like, linear to thread-like; the lower 2 or 3 very small or absent. Stamens 6, filaments very short or absent. Ovary superior, 3-celled; styles 3, free. Fruit a capsule with 2 seeds per cell.

A genus of about 7 species from China, Japan & S Korea grown for their delicate brush-like flower-spikes. They do well in moist, humus-rich loam in partial shade, such as found in woodland gardens. Some winter protection may be needed. They are easily divided in early spring or just after flowering and can also be grown from seed.

1. C. japonica Maximowicz. Illustration: *Botanical Magazine*, 6510 (1880); RHS Dictionary of gardening, 463 (1951); Hutchinson, The families of flowering plants, 595 (1959).

Rhizome oblong. Basal leaves several in a loose rosette, to 8 × 3 cm, narrowly ovate to oblong, entire, wavy or irregularly toothed, with short or long stalks; stem-leaves to 4 cm, erect, linear to linear-lanceolate. Flowering stem 15–45 cm, erect. Flowers to 1.8 cm across in a spike 5–20 cm long. *Japan & Korea.* H3. Spring-early summer.

8. STENANTHIUM (Gray) Kunth
J.C.M. Alexander
Perennial herbs with small bulbs. Basal leaves 2–4, long and narrow, bright green, arched above, hairless, acute. Flowering stem slender, erect; stem-leaves few, bract-like above. Flowers bisexual or unisexual, in dense or loose, terminal racemes or panicles. Perianth with a tube below, fused to the base of the ovary, lobes 6, linear, curled. Stamens 6, short, borne at base of perianth-lobes. Ovary superior; styles 3, free. Fruit a 3-beaked capsule with many narrow, oblong seeds.

A genus of about 5 species from N America and Sakhalin which generally perform well in a dry, sunny, well-drained position on slightly acid, humus-rich loam. *S. occidentale* (and its close relative *S. sachalinensis* Schmidt) are rather different from the other species and are sometimes regarded as the separate genus *Stenanthella* Rydberg. They prefer cooler, moister conditions with some lime in the soil.

1a. Flowers star-like, white to pale greenish or purplish in a dense panicle; perianth-lobes fused at base only **1. gramineum**
b. Flowers bell-shaped, brownish or greenish purple in a loose raceme; perianth-lobes *c.* half fused **2. occidentale**

1. S. gramineum (Ker Gawler) Morong (*S. angustifolium* Kunth). Illustration: *Botanical Magazine*, 1599 (1813); Rickett, Wild flowers of the United States **2**: pl. 37 (1967).

Bulb oblong. Basal leaves 4, linear, 30–40 × up to 1.8 cm, keeled and channelled, spreading. Flowering stem to 1.5 m, bearing a dense panicle to 60 cm long, with ascending, spreading or drooping branches; branches and upper part of stem spike-like. Flowers many, variable in size on a single plant, 1.2–1.8 cm across, white to greenish or purplish, fragrant, star-like, lower ones often male only. Perianth-lobes *c.* 6 mm, linear-lanceolate, acuminate, shortly fused at base. Stamens very short. Style *c.* half filament length; stigmatic branches very short, spreading. Fruit hanging. *SE United States.* H3. Late summer.

Var. **robustum** (Watson) Fernald (*S. robustum* Watson). Illustration: Rickett, Wild flowers of the United States **2**: pl. 37 (1967); Everett, Encyclopedia of horticulture, 3226 (1982). Leaves to 2.5 cm wide. Flowering stem to 1.8 m, bearing a denser panicle with ascending or spreading branches. Fruit often erect. *SE United States.* H3. Summer.

2. S. occidentale Gray. Illustration: Rickett, Wild flowers of the United States **5**: pl. 41 (1971); Clark, Wild flowers of the Pacific northwest, 62 (1976).

Bulb ovoid. Basal leaves 2–4, linear to very narrowly oblanceolate, 15–30 × *c.* 1.8 cm, slightly keeled, ascending. Flowering stem to 60 cm, bearing a loose raceme to 20 cm long, rarely with a few short basal branches. Flowers few, bell-shaped, 1.2–1.8 cm long, brownish or greenish purple, slightly fragrant, hanging. Perianth-lobes linear, fused for *c.* half their length. Stamens *c.* half segment length. Stigmatic branches long and slender, erect. *Western N America.* H2. Summer.

9. ZIGADENUS Michaux
S.J.M. Droop
Perennial herbs with rhizomes or bulbs. Leaves usually narrow, linear, hairless. Stems simple, leafy. Flowers in terminal panicles or racemes, greenish to yellowish white. Perianth-segments 6, free, spreading, persistent, ovate to lanceolate, each with 1 or 2 greenish glands near the base. Ovary superior or partly inferior, 3-celled with several ovules in each cell; styles 3, persistent. Fruit a 3-lobed capsule that splits between the septa to the base.

A genus of about 18 species distributed across N America and with 1 species in N Asia. A few species are cultivated; they are easily grown in most garden soils. Propagation is by division of the rootstock or by seed. The genus name is sometimes wrongly spelled 'Zygadenus'. The plants are poisonous to animals.

1a. At least the lower one-fifth to one-third of the ovary inferior; gland at the base of each perianth-segment reversed heart-shaped or divided into 2 equal, round glands 2
b. Ovary completely superior; gland at the base of each perianth-segment ovate, its outer margin drawn out into long, green lines 4
2a. Each perianth-segment with 2 distinct glands at the base; plant with a rhizome **3. glaberrimus**
b. Each perianth-segment with a single gland at the base; plant with a bulb 3
3a. Inflorescence usually a simple raceme, rarely branched below, but not branched throughout; capsule twice as long as the persistent perianth-segments **1. elegans**
b. Inflorescence usually a panicle, branched throughout; capsule only slightly longer than the persistent perianth-segments **2. glaucus**
4a. Perianth-segments more than 1 cm, stamens about half as long; panicle more or less corymbose, the lower flower-stalks much longer than the upper **4. fremontii**
b. Perianth-segments mostly less than 8 mm, stamens as long as them or longer; panicle not corymbose 5
5a. Perianth-segments 3–6 mm, distinctly clawed at the base; leaves not sickle-shaped **6. venenosus**
b. Perianth-segments 6–8 mm, not clawed; leaves sickle-shaped **5. nuttallii**

1. Z. elegans Pursh. Illustration: *Botanical Magazine*, 1680 (1814) – as *Helonias glaberrima*; Hitchcock et al., Vascular plants of the Pacific northwest **1**: 816 (1969).

Plant with a bulb. Stem to *c.* 80 cm. Leaves mostly basal, to 1 cm broad, sharply pointed. Inflorescence usually a simple, loose raceme, sometimes branched at the base. Upper bracts with scarious margins.

Perianth-segments 8–12 mm, each with a heart-shaped gland near the base. Ovary with its lower fifth to third inferior. Capsule conical, about twice as long as the persistent perianth-segments. *Western N America.* H4. Summer.

2. Z. glaucus Nuttall. Illustration: Rickett, Wild flowers of the United States **2**: t. 13 (1967).

Like *Z. elegans*, but leaves blunt, bracts fleshy throughout, inflorescence a panicle, the capsule only slightly exceeding the persistent perianth-segments. *Eastern N America.* H3. Summer.

The differences between this and *Z. elegans* are not always clear-cut.

3. Z. glaberrimus Michaux. Illustration: *Botanical Magazine*, 1703 (1815) – as *Helonias bracteata*; Rickett, Wild flowers of the United States **2**: t. 13 (1967).

Plant with a rhizome. Stems to 1.2 m, leafy. Inflorescence a loosely pyramidal panicle. Perianth-segments 1–1.5 cm, clawed at the base, each with 2 distinct glands at the base. Ovary with its lower fifth to third inferior. Capsule usually shorter than the persistent perianth-segments. *Southeast N America.* H4. Summer.

4. Z. fremontii (Torrey) Watson. Illustration: *Botanical Magazine*, n.s., 361 (1962); Rickett, Wild flowers of the United States **5**: t. 8 (1971); Brickell (ed.), RHS A–Z encyclopedia of garden plants, 1098 (2003).

Plant with a bulb. Basal leaves to 3 cm wide. Stem smooth, leafy. Inflorescence racemose or paniculate, often corymbose, the lower flower-stalks much longer than the upper. Perianth-segments 1–1.6 cm, each with a single gland near the base, the margins toothed towards the apex. Stamens about half as long as the perianth-segments. Ovary completely superior. Capsule 1.5–3 cm, oblong. *Western N America.* H4. Summer.

5. Z. nuttallii Gray. Illustration: Rickett, Wild flowers of the United States **6**: t. 8 (1973).

Plant with a bulb. Stems to 75 cm, leafy. Leaves mostly basal, *c.* 1 cm wide, keeled and sickle-shaped. Inflorescence usually a raceme, sometimes branched below. Perianth-segments 6–8 mm, not clawed, with a single gland at the base, the margins toothed towards the apex. Stamens about equal to the perianth-segments. Ovary completely superior. Capsule ovoid to ellipsoid, 3–4 times longer than the persistent perianth-segments. *Western N America.* H4. Summer.

6. Z. venenosus Watson. Illustration: Rickett, Wild flowers of the United States **5**: t. 8 (1971).

Plant with a bulb. Stem to 60 (rarely 70) cm. Leaves mostly basal, stem leaves reduced. Inflorescence usually a raceme, occasionally branched below. Perianth-segments to 3–6 mm, clawed, those of the outer whorl slightly shorter than those of the inner, each with a single gland at the base, margins toothed towards the apex. Stamens exceeding the perianth-segments. Ovary completely superior. Capsule 8–15 mm, cylindric. *Western N America.* H4. Summer.

10. MELANTHIUM Linnaeus

J. Cullen

Perennial herbs with rhizomes and fibrous roots. Leaves basal and on the stem. Inflorescence a panicle with its axis, branches and the backs of the flowers roughly hairy. Flowers bisexual and male in the same inflorescence. Perianth of 6, free, widely spreading segments, each with a conspicuous, narrow claw and an expanded blade that has 2 glands (appearing as dark spots) at the base. Stamens 6, borne on the claws of the perianth-segments. Ovary superior, cylindric, 3-celled with several ovules in each cell; styles 3, free, divergent. Fruit an inflated capsule, 3-lobed in section, 3-beaked at the apex. Seeds flat, broadly winged.

A genus of about 5 species from N America. One or two are occasionally grown as ornamentals. Cultivation as for *Veratrum*.

1. M. virginicum Linnaeus. Illustration: *Botanical Magazine*, 985 (1807); House, Wild flowers, t. 10 (1961); Rickett, Wild flowers of the United States **1**: t. 8 (1966).

Stem to 1.7 m. Leaves linear, folded but not obviously tapering towards the base, 1.5–3 cm broad. Perianth-segments at first cream, later green, the blade broadly oblong to ovate, flat, blunt, 2–3 times longer than the claw. *Eastern USA, from New York to Texas, inland to Indiana.* H1. Summer.

M. hybridum Walter (*M. latifolium* Desrousseaux). Illustration: Rickett, Wild flowers of the United States **1**: t. 8 (1966). More slender, leaves tapering to the base, perianth-segments with an almost circular blade that has crisped-undulate margins, and is about as long as the claw. *Eastern USA, from Connecticut to Georgia.* H1. Summer.

11. VERATRUM Linnaeus

S.J.M. Droop

Robust perennial herbs with rhizomes. Stems simple, leafy, usually hairless below, downy or coarsely hairy above. Leaves alternate, broadly ovate to elliptic, many-veined, pleated, more or less erect and overlapping, narrowed into long, sheathing bases. Flowers numerous in a terminal panicle or raceme, usually bisexual, occasionally male only. Perianth-segments 6, free, widely spreading, elliptic or lanceolate, persistent in fruit, white, green, reddish brown or almost black. Stamens 6, free. Ovary superior, 3-celled, ovules numerous. Fruit a 3-celled capsule containing numerous seeds.

A genus of about 45 species widely distributed in the northern hemisphere. Only a few are grown; they do best in moist, rich soils and are propagated by division of the rhizome.

Literature: Loesener, O., Die Gattung *Veratrum*, Feddes Repertorium **24**: 61–72 (1927), **25**: 1–10 (1928).

1a. Flowers reddish brown to black; leaves hairless **2. nigrum**
b. Flowers white to green; leaves shortly downy beneath **1. album**

1. V. album Linnaeus (*Helonias viridis* misapplied). Illustration: *Botanical Magazine*, 1096 (1808); Polunin, Flowers of Europe, t. 162 (1969); Brickell (ed.), RHS A–Z encyclopedia of garden plants, 1068 (2003).

Plant to 2 m. Leaves hairless above, hairy beneath, the lower elliptic to broadly elliptic, the upper progressively narrower. Flowers in a much-branched panicle with erect to drooping branches; flower-stalks 2–3 mm. Perianth-segments to 1.8 cm, white to pale green inside, sparsely to densely hairy outside, margins wavy to irregularly toothed. Capsule hairless to sparsely downy. *North temperate regions.* H1. Summer.

Variable and sometimes the N American species split off as below:

V. californicum Durand. Illustration: Rickett, Wild flowers of the United States **5**: t. 9 (1971). Leaves oblong to ovate, loosely hairy beneath, the hairs mostly near the margins; lower panicle branches not drooping; perianth-segments white to pale green, very sparsely hairy to hairless outside, margins entire. *Northwest USA.* H5. Summer.

V. viride Aiton. Illustration: Rickett, Wild flowers of the United States **5**: t. 9 (1971). Leaves oblong to ovate, loosely hairy beneath, panicle with distinctly drooping lower branches, perianth-segments pale to bright green, hairy outside at least on the midrib and margins, the margins of at least some irregularly toothed. *N America.* H5. Summer.

2. V. nigrum Linnaeus. Illustration: *Botanical Magazine,* 963 (1806); Polunin, Flowers of Europe, t. 162 (1969); Brickell (ed.), RHS A–Z encyclopedia of garden plants, 1068 (2003).

Plant to 1.3 m. Leaves hairless, the lower to 35 cm, broadly elliptic, the upper progressively smaller and narrower. Flowers in a raceme or sparsely branched panicle. Perianth-segments usually less than 7 mm, reddish brown to black, hairless or sparsely hairy outside. Capsule hairless. *Eurasia.* H1. Summer.

V. maackii Regel. Similar, but with smaller, narrower, lanceolate leaves. *Temperate Asia.* H1.

12. ALETRIS Linnaeus

J. Cullen

Perennial herbs with rhizomes and fibrous roots. Leaves mostly in a basal rosette. Flowering stem erect, bearing a terminal, spike-like raceme. Flowers almost stalkless, held ascending-erect. Perianth united into a tube for most of its length, the outside of the tube warty and 6-ridged. Stamens 6. Ovary superior, 3-celled, each cell with several ovules. Fruit a beaked capsule, surrounded by the persistent but dried and shrunken perianth.

A genus of about 10 species from eastern N America and E Asia. Only 2 species, both from N America, are grown. They are easily cultivated in a sunny position, and can be propagated by seed or division.

1a. Perianth tubular, to 1.2 cm, white **1. farinosa**
b. Perianth bell-shaped, to 6 mm, yellow **2. aurea**

1. A. farinosa Linnaeus. Illustration: *Botanical Magazine,* 1418 (1811); Rickett, Wild flowers of the United States **2**: t.12 (1967).

Leaves to 20 × 1.5 cm. Stem to 1 m. Perianth to 1.2 cm, tubular, white, very warty. Anthers orange-red, exposed at the mouth of the perianth-tube. Beak of capsule exposed in fruit. *E USA.* H3. Summer.

2. A. aurea Walter. Illustration: Rickett, Wild flowers of the United States **2**: t. 12 (1967).

Leaves to 8 cm, yellowish green. Stem to 70 cm. Perianth yellow, bell-shaped, to 6 mm. Beak of capsule not exposed in fruit. *SE USA.* H3. Summer.

8. ASPHODELACEAE

Plant terrestrial, herbaceous, with rhizomes. Leaves borne on the flowering stem, not reduced to sheaths. Flowers in racemes or panicles, without spathes when solitary, bisexual, usually actinomorphic. Perianth of 6 segments, free or united. Stamens 6. Ovary of 3 united carpels, superior, ovules many, axile; style 1; nectar secreted on the ovary. Fruit a capsule.

Twelve genera and about 150 species. Five genera are native to Europe, and a few are cultivated.

1a. Perianth-segments united at the base for more than one-tenth of their total length **8. Kniphofia**
b. Perianth-segments free at the base or united for less than one-tenth of their total length 2
2a. Leaves borne on the flowering stem **2. Asphodeline**
b. Leaves all basal 3
3a. At least 3 of the filaments hairy 4
b. None of the filaments hairy 5
4a. Perianth whitish inside, purplish outside **7. Simethis**
b. Perianth yellowish or brownish outside **6. Bulbine**
5a. Ovules 2 in each cell of the ovary 6
b. Ovules 4–8 in each cell of the ovary 7
6a. Perianth yellow **5. Bulbinella**
b. Perianth white or purplish **1. Asphodelus**
7a. Anthers dorsifixed; raceme 1-sided **3. Paradisea**
b. Anthers basifixed; raceme not 1-sided **4. Eremurus**

1. ASPHODELUS Linnaeus

V.A. Matthews

Annuals, or perennials with rhizomes. Leaves basal, linear, with a membranous sheathing base. Inflorescence a dense raceme or panicle. Bracts membranous. Flower-stalks jointed. Perianth-segments free or united at the extreme base, 1-veined, spreading. Stamens shorter than perianth-segments, free. Anthers dorsifixed, filaments hairless. Ovary superior, 3-celled, with 2 ovules in each cell. Fruit a capsule. Seeds winged, 6.

A genus of about 12 species growing from the Mediterranean to the Himalaya. They will grow in a sunny position in any well-drained soil and are best propagated by division.

Leaves. More or less round, hollow: **1**.
Flowering stem. Absent or very short: **5**; up to 1 m: **1, 2, 3**; more than 1 m: **3, 4**; hollow: **1**; solid: **2, 3, 4**.
Bracts. Pale green: **3**; brown: **2**.
Perianth-segments. 5–14 mm: **1, 4**; 1.5–2 cm: **2, 3**; 3.5–4 cm: **5**.

1a. Flowering stem absent or very short; perianth-segments 3.5–4 cm **5. acaulis**
b. Flowering stem present; perianth-segments 5–20 mm 2
2a. Perianth-segments 5–14 mm 3
b. Perianth-segments 1.5–2 cm 4
3a. Flowering stem 15–70 cm, hollow; leaves more or less round **1. fistulosus**
b. Flowering stem 1–2 m, solid; leaves flat **4. aestivus**
4a. Inflorescence unbranched or with a few short branches; leaves without a keel beneath; capsule 1.6–2 cm, if less then bracts dark brown **2. albus**
b. Inflorescence much-branched with the side-branches almost as long as the terminal; leaves with a shallow keel beneath; capsule 8–14 mm; bracts whitish to pale green **3. ramosus**

1. A. fistulosus Linnaeus (*A. tenuifolius* Cavanilles). Illustration: Reichenbach, Icones florae Germanicae et Helveticae **10**: pl. 513 (1848); Polunin & Huxley, Flowers of the Mediterranean, 238 (1965).

Annual or short-lived perennial. Leaves 5–35 cm, more or less round, hollow. Flowering stem 15–70 cm, hollow, branched or unbranched. Bracts whitish. Perianth-segments 5–12 mm, white or pale pink with a pinkish or brownish midvein. Capsule 5–7 mm. *Mediterranean area, SW Europe, SW Asia, India.* H4. Summer.

2. A. albus Miller (*A. delphinensis* Grenier & Godron). Illustration: Taylor, Wild flowers of Spain and Portugal, 95 (1972);

Grey-Wilson & Blamey, Alpine flowers of Britain and Europe, 269 (1979); Polunin, Flowers of Greece and the Balkans, pl. 55 (1980); Brickell (ed.), RHS A–Z encyclopedia of garden plants, 147 (2003).

Perennial. Leaves 15–60 cm, flat. Flowering stem 30–100 cm, solid, usually unbranched or with a few short branches. Bracts membranous, whitish or dark brown. Perianth-segments 1.5–2 cm, white or pale pink with a darker midvein. Capsule 8–20 mm. *S & C Europe.* H4. Spring–summer.

3. A. ramosus Linnaeus (*A. cerasiferus* Gay). Illustration: Reichenbach, Icones florae Germanicae et Helveticae **10**: pl. 514 (1848).

Perennial. Leaves 15–40 cm, flat, with a shallow keel beneath. Flowering stem 40–150 cm, solid, usually much-branched with the side branches almost as long as the terminal. Bracts whitish to pale green. Perianth-segments 1.5–2 cm, white with a reddish brown midvein. Capsule 8–14 mm. *S Europe, N Africa.* H4. Summer.

4. A. aestivus Brotero (*A. microcarpus* Salzmann & Viviani). Illustration: Polunin, Flowers of Europe, pl. 163 (1969); Megaw, Wild flowers of Cyprus, pl. 30 (1973); Nehmeh, Wild flowers of Lebanon, pl. 21, 22 (1978).

Perennial. Leaves 25–45 cm, flat. Flowering stem 1–2 m, solid, branched, with short side branches. Bracts whitish. Perianth-segments 1–1.4 cm, white with a pink or brownish midvein. Capsule 5–7 mm. *Canary Islands, through S Europe and N Africa to Turkey.* H4. Spring–early summer.

5. A. acaulis Desfontaines. Illustration: *Botanical Magazine*, 7004 (1888); *Bulletin of the Alpine Garden Society* **34**: 172 (1966); *RHS Lily Year Book* 1967: 24 (1966).

Perennial. Leaves 15–30 cm, flat. Flowering stem absent or very short. Bracts whitish. Perianth-segments 3.5–4 cm, white or pink with a greenish midvein. *Algeria, Morocco.* H5. Spring.

2. ASPHODELINE Reichenbach

V.A. Matthews

Biennial or perennial herbs with clusters of fleshy roots. Leaves numerous, linear, with wide, membranous sheathing bases and often rough margins. Inflorescence branched or unbranched with the flowers in racemes. Bracts membranous. Flower-stalks jointed. Flowers more or less bilaterally symmetric. Perianth united at the base, each lobe with 3 central veins (superficially appearing as 1 vein), the outer 3 lobes narrower than the inner. Stamens unequal; stamens and style curved downwards; anthers dorsifixed. Ovary superior, 3-celled. Fruit a capsule with 6 unwinged, sharply angled seeds.

A genus with 19 species occurring from the Mediterranean area to the Caucasus. The plants should be grown in a sunny position. Propagation is by division.

1a. Flowers white **3. taurica**
b. Flowers yellow 2
2a. Stem leafy up to the raceme; capsule 1–1.5 cm **1. lutea**
b. Leaves confined to the lower half of the stem; capsule 6–9 mm **2. liburnica**

1. A. lutea (Linnaeus) Reichenbach (*Asphodelus luteus* Linnaeus). Illustration: Polunin, Flowers of Europe, pl. 162 (1969); Huxley & Taylor, Flowers of Greece, 335 (1977); Brickell (ed.), RHS A–Z encyclopedia of garden plants, 147 (2003).

Perennial. Stems to 1 m, leafy up to the raceme. Leaves with smooth margins, sometimes rough towards the apex. Inflorescence unbranched. Bracts longer than the fruit-stalks. Perianth-lobes 2–3 cm, yellow. Capsule 1–1.5 cm, ovoid to spherical. *Mediterranean area.* H4. Spring–summer.

2. A. liburnica (Scopoli) Reichenbach. Illustration: Polunin, Flowers of Greece and the Balkans, 1594 (1980).

Perennial. Stems to 1 m, leafy in the lower half. Inflorescence usually unbranched. Bracts shorter than the fruit-stalks. Perianth-lobes 2.3–3 cm, yellow. Capsule 7–8 mm, spherical. *S Europe, Turkey.* H4. Summer.

A. brevicaulis (Bertoloni) Baker. Stems only to 50 cm. Perianth-lobes 1.5–2 cm, pale yellow. *SW Asia.* H4. Spring–summer.

3. A. taurica (Pallas) Kunth. Illustration: *Gardener's Chronicle* **21**: 175 (1897); Flora SSSR **4**: pl. 3, 1 (1935).

Perennial. Stems to 1 m, leafy up to the raceme. Bracts longer than the fruit-stalks. Perianth-lobes 1.2–1.7 cm, white. Capsule 8–12 mm, ovoid to oblong. *E Mediterranean area, Former USSR (Caucasus).* H4. Summer.

A. tenuior (Fischer) Ledebour. Stem 15–40 cm, leafy in the lower half or all leaves basal, leaf sheaths with ciliate margins. Inflorescence branched or unbranched. Perianth-lobes 1.7–2 cm, whitish, sometimes tinged with pink. Capsule 6–8 mm, spherical to obovoid. *Turkey, Iran, Former USSR (Caucasus).* H4. Summer.

A. damascena (Boissier) Baker (*A. balansae* Baker; *A. isthmocarpa* Baker). Biennial. Stem to 1.5 m, leafy to slightly above the middle, or all leaves basal; leaf sheaths not ciliate. Inflorescence branched or unbranched. Perianth-lobes 1.5–2.5 cm, white to pale pink. Capsule pear-shaped to cylindric. *SW Asia.* H4. Summer.

3. PARADISEA Mazzucato

V.A. Matthews

Similar to *Anthericum* but flowers always borne in a raceme. Perianth trumpet-shaped, the segments with a claw. Style and stamens curved upwards and anthers dorsifixed.

There are 2 species, native to the mountains of S Europe. Cultivation as for *Anthericum* (p. 25). Propagation is by division of the rootstock, or by seed.

1. P. liliastrum (Linnaeus) Bertoloni (*Anthericum liliastrum* Linnaeus). Illustration: Robert, Alpine flowers, pl. 8 (1938); Uberto, Mountain flowers, 45 (1978); Grey-Wilson & Blamey, Alpine flowers of Britain and Europe, 265 (1979); Grey-Wilson & Mathew, Bulbs, pl. 12 (1981).

Stem 30–60 cm. Leaves 4–7, 12–25 cm. Raceme with 3–10 flowers, usually rather 1-sided. Perianth-segments 3–5 cm, white often tipped with green. Capsule 1.3–1.5 cm. *Mountains of S Europe.* H3. Summer.

'Major' has larger flowers and is a more robust plant. Illustration: Brickell (ed.), RHS A–Z encyclopedia of garden plants, 769 (2003).

4. EREMURUS Bieberstein

V.A. Matthews

Perennials with thick rhizomes, the rhizome neck bearing fibrous or membranous remains of old leaves. Leaves basal, forming a tuft or rosette, usually narrow and often keeled beneath. Flowers many, borne in a raceme on an unbranched stem. Bracts membranous. Perianth-segments 6, almost free, with 1, 3 or 5 dark central veins, the inner segments often broader than the outer. Anthers basifixed. Ovary superior; style 1. Fruit a 3-celled, usually spherical, capsule. Seeds usually winged.

A genus of 40–50 species native to W and C Asia. They grow best in a rich well-drained soil in a sunny position. Propagation is by division or seed.

Literature: Wendelbo, P., On the genus Eremurus in south-west Asia, *Acta Universitatis Bergensis. Series Mathematica Rerumque Naturalum* **5** (1964); Fedtschenko, Eremurus, Kritische Übersicht über die Gattung, Cramer plant monograph reprints **3** (1968); Wendelbo, P. & Furse, P., Eremurus of south-west Asia, *RHS Lily Year Book* **1969**: 56–69 (1968).

Leaves. With rough margin: **1, 3, 4, 5, 6, 7, 8**; with smooth margin: **1, 2, 4, 7, 8**.
Scape. Hairless: **1, 2, 3, 5, 6, 7, 8**; downy, at least at the base: **2, 3, 4, 8**.
Bracts. Entirely hairless: **3, 4**.
Outer perianth-segments. With 1 vein: **3, 4, 5, 6, 7, 8**; with 3 or 5 veins: **1, 2**; bright yellow: **3**; yellowish green: **1, 2, 4**; white: **4, 5, 6, 7, 8**; pink or pinkish brown: **2, 4, 5, 6, 8**.
Flower length. 8–11 mm: **1, 2, 3, 4, 8**; 1.2–1.6 cm: **3, 4, 8**; 1.7–2.5 cm: **4, 5, 6, 7**.
Stamens. Shorter than perianth: **6, 8**; equal to perianth: **2, 4, 5, 7**; longer than perianth: **1, 2, 3, 4, 7**.
Capsule length. 5–14 mm: **1, 2, 3, 4, 7, 8**; 1.5–3.5 cm: **2, 5, 6, 7, 8**.

1a. Flowers bright yellow **3. stenophyllus**
b. Flowers white, pink, brown, pale yellow or yellowish green 2
2a. Outer perianth-segments with 3 or 5 veins 3
b. Outer perianth-segments with 1 vein 4
3a. Perianth-segments pale yellow, often flushed with green **1. spectabilis**
b. Perianth-segments whitish, yellowish green or pinkish brown **2. turkestanicus**
4a. Stamens more or less equal to or longer than perianth-segments 5
b. Stamens shorter than perianth-segments 8
5a. Perianth-segments 1–1.64 cm **4. olgae**
b. Perianth-segments 1.5–2.5 cm 6
6a. Scape hairless for entire length 7
b. Scape downy, at least at the base **4. olgae**
7a. Perianth-segments white, sometimes yellow at the base; flower-stalks up to 3 cm **7. himalaicus**
b. Perianth-segments pink, rarely white; flower-stalks 3–4 cm **5. robustus**
8a. Scape 35–100 cm; perianth-segments 1–1.5 cm **8. bucharicus**
b. Scape 90–250 cm; perianth-segments 1.8–2.5 cm **6. aitchisonii**

1. E. spectabilis Bieberstein. Illustration: *Botanical Magazine*, 4870 (1855); *RHS Lily Year Book* 1969: 16 (1968).

Leaves narrowly strap-shaped, to 4.5 cm broad, hairless, margins rough or smooth. Scape to 1.5 m, hairless. Raceme to 80 cm. Bracts linear, ciliate. Flower-stalks *c.* 1 cm, recurving in fruit. Perianth-segments 9–10 mm, pale yellow flushed with green, with 3 veins. Stamens *c.* 2 × longer than perianth; anthers brick red, filaments orange at the base. Capsule *c.* 1.2 cm, transversely ribbed. Seeds with very narrow wings. *SW Asia, Former USSR (Soviet Central Asia).* H4. Late spring–summer.

2. E. turkestanicus Regel. Illustration: Regel, Turkestan flora, t. 2 (1876); Fedtschenko, Eremurus, Kritische Übersicht über die Gattung, Cramer plant monograph reprints **3**: t. II (1968).

Leaves broadly linear, hairless, margins smooth. Scape to 1 m, hairless or finely downy. Raceme 40–60 cm. Bracts narrowly linear, hairy and ciliate. Flower-stalks *c.* 1 cm. Perianth-segments *c.* 1 cm, curling inwards with age, the outer greenish yellow with a whitish margin and a central brownish stripe and with 5 veins, the inner white with a yellowish green stripe and with 3 veins. Stamens equal in length to the perianth, projecting as the perianth-segments curl inwards. Capsule smooth 5–14 mm. *Former USSR (Soviet Central Asia).* H4. Summer.

E. comosus Fedtschenko. Illustration: Fedtschenko, Eremurus, Kritische Übersicht über die Gattung, Cramer plant monograph reprints **3**: t. VII (1968). Differs in having downy leaves, broadly lanceolate to ovate bracts, flower-stalks 2–3 cm and pinkish brown perianth-segments. *Former USSR (Soviet Central Asia), Afghanistan, W Pakistan.* H4. Summer.

3. E. stenophyllus (Boissier & Buhse) Baker (*E. bungei* Baker; *E. aurantiacus* Baker). Illustration: *Botanical Magazine*, 7113 (1890); Morley & Everard, Wild flowers of the world, pl. 50 (1970); Huxley (ed), Garden perennials and water plants, 108 (1971); *Gartenpraxis* 1981(8): 360.

Leaves linear, hairless or finely downy, margins rough. Scape 30–150 cm, hairless or finely downy. Raceme densely-flowered. Bracts narrowly linear, hairless or ciliate. Flower-stalks to 2 cm. Perianth-segments 9–12 mm, bright yellow, turning orange, then brownish with age, with 1 vein. Stamens longer than the perianth, with orange anthers. Capsule 5–8 mm, wrinkled. Seeds with broad wings. *Former USSR (Soviet Central Asia), Afghanistan, Iran, W Pakistan.* H4. Summer.

Plants may be divided into 2 subspecies: subsp. **stenophyllus** (*E. bungei* Baker), which has hairless leaves and scapes; and subsp. **aurantiacus** (Baker) Wendelbo (*E. aurantiacus* Baker), in which the scape and leaves are finely downy.

Hybrids between *E. stenophyllus* and *E. olgae* are offered by the trade under names **E. × isabellinus** Vilmorin, **E. × warei** Mottet, **E. × shelfordii** Hort., E. 'Shelford Hybrids' and E. 'Highdown Hybrids'. They have flowers that vary from white to yellow, orange, bronze or pink.

E. × tubergenii van Tubergen is a hybrid between *E. stenophyllus* and *E. himalaicus*, and has pale yellow flowers borne on a scape up to 240 cm in height.

4. E. olgae Regel. Illustration: *Gartenflora* **30**: t. 1048 (1881); *RHS Lily Year Book* 1969: 15 (1968); Morley & Everard, Wild flowers of the world, pl. 50 (1970); Rix & Phillips, The bulb book, 148–9 (1981).

Leaves narrowly linear, to 1.5 cm broad, hairless, margin rough or smooth. Scape 60–160 cm, downy at the base. Bracts narrowly linear, hairless. Flower-stalks 1.5–4 cm. Perianth-segments 1.2–1.6 cm, pale pink, or sometimes white with a yellow base, with 1 vein. Stamens equal in length to the perianth. Capsule 1–1.2 cm. Seeds with narrow wings. *N Iran, N Afghanistan, Former USSR (Soviet Central Asia).* H4. Late summer–autumn.

E. sogdianus (Regel) Bentham & Hooker. Illustration: Fedtschenko, Eremurus, Kritische Übersicht über die Gattung, Cramer plant monograph reprints **3**: t. V (1968). Differs from *E. algae* in having hairy bracts, longer flower-stalks (4–7.5 cm), flowers with the outer segments yellowish green with 1 vein and the inner segments white with 3 green veins, the stamens longer than the perianth and the capsule only 6–7 mm. *Afghanistan, Former USSR (Soviet Central Asia).* H4. Summer.

E. kaufmannii Regel. Illustration: *Journal of the Royal Horticultural Society* **93**: 58 (1968); *RHS Lily Year Book* 1969: 15 (1968). Leaves to 3.5 cm broad, downy, margins smooth. Scape downy. Bracts narrowly ovate-triangular, hairy. Flower-stalks 1–2 cm. Perianth-segments 1.5–2.2 cm, white with a greenish yellow base and 1 vein. Stamens more or less equal in length to perianth-segments. *Afghanistan, Former USSR (Soviet Central Asia).* H4. Summer.

5. E. robustus (Regel) Regel. Illustration: *Gartenflora* **22**: t. 769 (1873); *Botanical Magazine*, 6726 (1883); Huxley (ed.), Garden perennials and water plants, 107 (1971); Petrová, Flowering bulbs, 171 (1975).

Leaves strap-shaped, up to 4 cm broad, hairless, margins rough. Scape to 3 m, hairless. Raceme to 1 m, bearing up to 800 flowers. Bracts linear, widening at the base, densely hairy. Flower-stalks 3–4 cm. Perianth-segments 1.8–2.1 cm, pink, occasionally white, with 1 vein. Stamens equal in length to the perianth. Capsule *c.* 2 cm. *Afghanistan, Former USSR (Soviet Central Asia).* H4. Summer.

6. E. aitchisonii Baker (*E. elwesii* Micheli; *E. robustus* (Regel) Regel var. *elwesii* (Micheli) Leichtlin). Illustration: *Journal of the Linnean Society, Botany* **19**: t. 27 (1882); *Revue Horticole* **69**: 280 (1897); *RHS Lily Year Book* 1969: 15 (1968).

Leaves narrowly lanceolate, to 8 cm broad, hairless, margins rough. Scape 90–250 cm, hairless. Raceme loosely flowered, *c.* 30 cm. Bracts linear, widening at the base, ciliate. Flower-stalks 2–3.5 cm. Perianth-segments 1.8–2.5 cm, pale pink, occasionally white, with 1 vein. Stamens a little shorter than the perianth. Capsule 1.5–2 cm. *E Afghanistan, Former USSR (Soviet Central Asia).* H4. Summer.

7. E. himalaicus Baker. Illustration: *Botanical Magazine*, 7076 (1889); Blatter, Beautiful flowers of Kashmir **2**: pl. 60 (1928); *Journal of the Royal Horticultural Society* **73**: 43 (1948); *Gartenpraxis* 1981 (8): 361.

Leaves strap-shaped, to 4 cm, broad, hairless, margins rough or smooth. Scape to 2.5 m, hairless. Raceme densely-flowered, to 90 cm. Bracts narrowly linear, ciliate. Flower-stalks to 3 cm. Perianth-segments 1.7–2 cm, white, with 1 vein. Stamens equal in length to the perianth. Capsule 1.2–1.4 cm, wrinkled. *Afghanistan, NW Himalaya.* H4. Late spring–summer.

E. lactiflorus Fedtschenko. Illustration: Fedtschenko, Eremurus, Kritische Übersicht über die Gattung, Cramer plant monograph reprints **3**: t. XVI (1968); Rix & Phillips, The bulb book, 148 (1981). A smaller plant, rarely exceeding 1 m, with a loosely flowered raceme. The white perianth-segments are yellow at the base and the stamens are slightly longer than the perianth. The capsule is *c.* 3.5 cm, smooth. *Former USSR (Soviet Central Asia).* H4. Summer.

8. E. bucharicus Regel. Illustration: *Gartenflora* **39**: t. 1315 (1890), **44**: t. 1420 (1895); Fedtschenko, Eremurus, Kritische Übersicht über die Gattung, Cramer plant monograph reprints **3**: t. XVIII (1968).

Leaves linear, up to 5 mm broad, hairless or somewhat downy, margins rough or smooth. Scape 35–100 cm, hairless, or downy at the base. Raceme loosely flowered. Bracts narrowly linear, widening at the base, sparsely ciliate. Flower-stalks 2–3 cm. Perianth-segments 1.3–1.5 cm, white or pale pink with 1 vein. Stamens shorter than the perianth-segments. Capsule 1–1.8 cm. Seeds with broad wings at the top. *Afghanistan, Former USSR (Soviet Central Asia).* H4. Summer.

E. anisopterus (Karelin & Kirilow) Regel (*E. korolkowii* Regel). Illustration: Regel, Turkestan flora, t. 1 (1876); Fedtschenko, Eremurus, Kritische Übersicht über die Gattung, Cramer plant monograph reprints **3**: t. XVII (1968). Differs mainly in having the perianth-segments *c.* 1 cm, concave, the scape always hairless and the leaves usually with smooth margins. *Iran, Former USSR (Soviet Central Asia).* H4. Summer.

5. BULBINELLA Kunth

J. Cullen

Roots fleshy. Plants stemless, leaves in rosettes, somewhat fleshy. Scape well-developed, overtopping the leaves. Raceme with many flowers, terminal. Perianth of 6 free segments, persistent in fruit. Stamens 6, all the filaments hairless. Ovary superior, 3-celled, with 2 ovules in each cell. Fruit a capsule opening between the septa.

A genus of about 20 species from Africa and New Zealand. Only 2 species, both from New Zealand, are in general cultivation. One of these has functionally unisexual flowers, the males and females being borne on separate plants. Cultivation as for *Bulbine.*

1a. Flowers bisexual; ovary and fruit borne on a short but distinct stalk **2. hookeri**

b. Flowers functionally unisexual (i.e. the anthers small and without pollen, or the ovary small and without ovules), plants dioecious; ovary and capsule stalkless **1. rossii**

1. B. rossii (J.D. Hooker) Cheeseman.

Plant large, robust. Leaves oblong or oblong-lanceolate, to 30 × 6 cm, tapering, the tips recurved and obtuse to acute. Scape exceeding the leaves. Raceme cylindric with many closely packed flowers. Flowers yellow, functionally unisexual, 1–1.4 cm in diameter. Perianth-segments spreading in male flowers, more erect in the female flowers, remaining erect and becoming hardened as the capsule develops. Ovary and capsule stalkless. *New Zealand.* H5–G1.

2. B. hookeri (J.D. Hooker) Cheeseman. Illustration: Salmon, Field guide to the alpine plants of New Zealand, 107 (1968); Brickell (ed.), RHS A–Z encyclopedia of garden plants, 200 (2003).

Plant to 1 m (usually much less in cultivation). Leaves to 30 × 3 cm, erect. Scape exceeding the leaves. Raceme cylindric with many flowers, which are bisexual and yellow. Perianth spreading, to 1.4 cm in diameter, shrivelling and hanging below the capsule as it develops. Ovary and capsule borne on a short stalk. *New Zealand.* H5–G1.

6. BULBINE Wolf

J. Cullen

Herbaceous perennials with somewhat to very swollen, tuber- or bulb-like bases, usually stemless. Leaves in rosettes, fleshy, terete or flat, occasionally subterranean. Scape well-developed. Flowers few to many in terminal racemes. Perianth of 6 free, yellow or brownish, spreading segments. Stamens 6, free, the filaments of all, or of 3 only, with a patch of hairs. Ovary superior, 3-celled with several ovules in each cell. Fruit a capsule containing few seeds.

A genus of about 30 species, mostly from Africa, a few from Australia. The plants are easily cultivated, though may require protection in winter in colder areas. Propagation is by seed, though *B. caulescens* can be propagated by cuttings.

Leaves. Only 2, subterranean except for the apex: **4**; several, aerial, terete: **1**; several, aerial, flat: **2, 3, 5**.
Stem. Present: **1**; absent: **2, 3, 4, 5**.
Filaments. All with hairy patches: **1, 2, 3, 4**; some hairless: **5**.

1a. 3 filaments hairy, 3 hairless **5. semibarbata**
b. All 6 filaments hairy 2
2a. Leaves 2, mostly subterranean, their apices at soil level, shallowly concave, translucent **4. mesembrianthoides**
b. Leaves several, most of their length above ground, the apex not as above 3
3a. Plant with a branched, woody stem above ground, its branches ending in rosettes of terete leaves **1. caulescens**
b. Plant stemless, the flat leaves borne on a swollen, tuber- or bulb-like base 4
4a. Leaves 4–6 mm broad near the base, tapering gradually to a fine, acute apex **2. alooides**
b. Leaves to 2 mm broad, parallel-sided for most of their length, the apex rounded **3. bulbosa**

1. B. caulescens Linnaeus (*B. fruticosa* Anon.; *B. frutescens* Anon.). Illustration: Eliovson, Wild flowers of southern Africa, pl. xviii, 18 (1980).

Plant with short (to 25 cm), often branched, creeping, slightly woody stems. Scapes to 25 cm. Leaves several in rosettes at the ends of the branches, terete, somewhat glaucous. Raceme loose. Perianth brownish yellow, segments to 5 mm. Each filament with a patch of hairs above the middle. *South Africa (Cape Province).* H5. Summer.

2. B. alooides (Linnaeus) Willdenow. Illustration: *Botanical Magazine*, 1317 (1810).

Plant stemless, the rosettes of leaves borne on a swollen, tuber-like base. Leaves several, flat, lanceolate, 5–25 cm × 4–6 mm, tapering to an acute apex. Scape 12–45 cm. Raceme loose, with many flowers. Perianth yellow, segments to 10 mm. Each filament with a dense patch of hairs from just above the middle to immediately beneath the anther. *South Africa.* H5. Spring.

3. B. bulbosa (R. Brown) Haworth. Illustration: *Botanical Magazine*, 3017 (1830), 3129 (1832), as *B. semibarbata.*

Plant stemless, the rosettes of leaves borne on a swollen, tuber-like base. Leaves several, flat, linear, to 35 cm × 2 mm, apex rounded. Scape 20–70 cm. Raceme loose, with many flowers. Perianth yellow, segments 8–18 mm. Each filament with a patch of hairs above the middle. *E Australia.* H5–G1. Spring.

4. B. mesembrianthoides Haworth. Illustration: *Hooker's Icones Plantarum* **28**: t. 2528 (1897); *Flowering plants of Africa* **10**: t. 377 (1930).

Plant stemless, with tuberous roots. Leaves 2, very fleshy, unequal, borne below ground with only the shallowly concave, translucent apex exposed. Scape to 10 cm. Raceme with up to 8 flowers, loose. Perianth bright yellow, segments to 7 mm. Each filament with hairs over most of its length. *South Africa (Cape Province).* H5–G1.

A remarkable example of a 'window plant', in which the chlorophyll-containing parts are subterranean; light is transmitted to them through the translucent, exposed leaf apex.

5. B. semibarbata (R. Brown) Willdenow (*B. annua* misapplied). Illustration: Rotherham et al., Plants of western New South Wales and southern Queensland, 130 (1975); Cunningham et al., Plants of western New South Wales, 183 (1981).

Plant stemless, the base scarcely swollen. Leaves, flat, linear or lanceolate, acute, 8–19 cm × 1–5 mm. Scape to 35 cm. Raceme with 3–many flowers. Perianth yellow, segments to 6 mm. Three of the filaments bear patches of hairs just above the middle, the other 3 are hairless. *Australia.* H5–G1. Spring–summer.

7. SIMETHIS Kunth

V.A. Matthews

Roots fleshy, from a short rhizome. Flowering stem 12–40 cm, hairless. Leaves mostly basal, 15–60 cm × 2.5–7 mm, linear with sheathing bases. Flowers borne in a terminal panicle, on non-jointed stalks. Perianth-segments free, 8–11 mm, white above, purplish beneath. Filaments white, hairy; anthers yellow, dorsifixed. Ovary superior, ovules 2 per cell; style single, straight. Fruit a capsule. Seeds 3–6, black, with a whitish swelling at one end.

A genus containing one species mainly occurring in the west Mediterranean area, which is occasionally cultivated. It grows best in peaty soil in a semi-shaded position. Propagation is by seed.

1. S. planifolia (Linnaeus) Grenier (*S. bicolor* (Desfontaines) Kunth). Illustration: Ross-Craig, Drawings of British plants, **29**: pl. 17 (1972); Fitter et al., Wild flowers of Britain and northern Europe, 261 (1974); Rose, The wild flower key, 409 (1981).

W & S Europe, N Africa. H3. Summer.

8. KNIPHOFIA Moench

S.J.M. Droop

Plants with thick rhizomes, rarely with aerial stems. Leaves usually basal, in several ranks, linear, tapering gradually to the apex, usually keeled beneath, margins smooth or very finely toothed. Scape simple, erect, with a few small, sterile bracts beneath the raceme. Raceme usually dense, flowers usually opening from the bottom upwards, occasionally from the top downwards, spreading or drooping, white or red, the buds nearly always red. Bracts scarious, exceeding the flower-stalks. Perianth with a cylindric or narrowly bell-shaped, often curved, tube to 4.5 cm and 6 small lobes. Stamens 6, often becoming spirally twisted and withdrawn after shedding pollen, filaments attached to the perianth near the base. Ovary superior, 3-celled with numerous ovules in each cell; style usually projecting. Capsule spherical or ovoid, sometimes 3-angled.

A genus of about 70 species distributed mainly in eastern and southern Africa, with 1 species in Madagascar and 1 in southern Arabia. About 11 species are found in cultivation, along with many hybrids and cultivars (mainly derived from *K. uvaria*). They are relatively hardy in the more sheltered parts of Europe, where they can be grown out of doors in a light soil in a sunny position, with plenty of water during spring and summer. Some protection against frost, usually by means of leaf litter or straw, is helpful even in mild areas. Propagation is by division of the rhizomes or by seed.

Literature: Codd, L.E., The South African species of Kniphofia (Liliaceae), *Bothalia* **9**: 363–511 (1968); Marais, W., The tropical species of Kniphofia (Liliaceae), *Kew Bulletin* **28**: 465–83 (1973).

1a. Leaves 8–12 cm broad, not keeled **5. northiae**
b. Leaves to 8 cm broad, keeled beneath 2
2a. Plants with a distinct simple or branched stem to 60 cm; leaves very glaucous **4. caulescens**

b. Plants without a stem; leaves not or slightly glaucous 3
3a. Inflorescence opening from the top downwards **10. pumila**
b. Inflorescence opening from the bottom upwards 4
4a. Open perianth white, spreading; buds ascending **1. gracilis**
b. Open perianth cream to red or orange, or, if white, then perianth pendulous and buds spreading 5
5a. Bracts acute, obtuse, rounded 6
b. Bracts acuminate, sharply pointed or drawn out into a tail 7
6a. Stamens not or scarcely projecting at flowering, withdrawn later **8. uvaria**
b. Stamens projecting at flowering, up to 5 mm beyond perianth, usually withdrawn later **1. gracilis**
7a. Stamens projecting by more than 5 mm at flowering and not usually withdrawn later 8
b. Stamens not projecting or projecting by less than 5 mm at flowering, usually withdrawn later 10
8a. Bract apices acute **11. foliosa**
b. Bract apices acuminate 9
9a. Leaf margins and keel conspicuously finely toothed **3. ensifolia**
b. Leaf margins and keel smooth **10. pumila**
10a. Perianth constricted above the ovary 11
b. Perianth not constricted above the ovary 12
11a. Bract apices long-acuminate; inflorescence rather loose to loose **9. thomsonii** var. **snowdenii**
b. Bract apices sharply rounded; inflorescence dense **8. uvaria**
12a. Inflorescence loose (axis visible between the individual flowers), usually 8–25 cm long **2. rufa**
b. Inflorescence denser (axis not visible between the flowers), less than 8 cm long 13
13a. Buds redder than the open flowers **6. galpinii**
b. Buds and open flowers dull red to orange-yellow **7. triangularis**

1. K. gracilis Baker (*K. modesta* in the sense of Baker; *K. sparsa* N.E. Brown). Illustration: *Botanical Magazine*, 7293 (1893); *Bothalia* **9**: t. 7 (1968).

Leaves usually less than 2 cm wide, channelled above, keeled beneath, margin smooth or sparingly toothed towards the apex. Scape 25–80 cm, usually overtopping the recurved leaves. Raceme oblong to narrowly cylindric, dense to loose, 9–3 5 cm; buds ascending, exceeded by the bracts; flowers spreading or deflexed. Bracts ovate to lanceolate, acute, obtuse or rounded. Perianth almost cylindric to narrowly funnel-shaped, to 2.2 cm, constricted above the ovary. Stamens projecting for up to 5 mm at flowering, usually withdrawn later. *South Africa (E Cape Province, Natal).*

Usually with yellow or orange buds and creamy yellow flowers and ovate, obtuse or rounded bracts. The variant with pinkish buds, whitish flowers and lanceolate bracts was named *K. sparsa*, but this intergrades with typical *K. gracilis.*

2. K. rufa Baker (*K. natalensis* Baker var. *angustifolia* Baker). Illustration: *Botanical Magazine*, 7706 (1900); *Bothalia* **9**: t. 2a (1968).

Leaves soft, at first erect, later arching over, usually less than 1 cm broad, dull green to glaucous, keeled beneath; margins and keel remotely and finely toothed towards the apex. Scape slender, 40–65 cm. Raceme loose, cylindric, 8–25 cm; buds cream or dull yellow to orange-red, flowers pendulous, white to cream, yellow or dull red. Bracts lanceolate to linear-lanceolate, acute to acuminate. Perianth more or less cylindric, not constricted above the ovary, tube to 3.3 cm. Stamens not or slightly projecting at flowering, later withdrawn. *South Africa (Natal).* H5. Summer.

3. K. ensifolia Baker (*K. pumila* Baker not (Aiton) Kunth). Illustration: *Botanical Magazine*, 7644 (1899); *Flowering Plants of South Africa* **22**: t. 866 (1942); *Bothalia* **9**: t. 9 (1968).

Leaves glaucous, usually less than 4 cm broad, V-shaped in section; margins and keel usually conspicuously but finely toothed. Scape 60–80 cm. Raceme very dense, cylindric, 9–20 cm; buds spreading, greenish white or dull to bright red, flowers pendulous, greenish white to cream. Bracts linear-lanceolate, acute to acuminate. Perianth narrowly funnel-shaped. Stamens projecting more than 5 mm at flowering. *South Africa (Cape Province, Transvaal, Orange Free State).* H5. Summer.

Only subsp. **ensifolia** (described above) is in cultivation.

4. K. caulescens Baker. Illustration: *Botanical Magazine*, 5946 (1872); Brickell (ed.), RHS A–Z encyclopedia of garden plants, 599 (2003).

Plants with stems to 60 cm crowned with a large rosette of leaves. Leaves glaucous, fleshy, V-shaped in section, to 8 cm broad. Scape stout, 30–60 cm. Raceme oblong to cylindric, very dense; buds dull orange to red, flowers pale greenish yellow to creamy yellow. Bracts linear-oblong to linear, acute to acuminate. Perianth slightly constricted above the ovary. Stamens conspicuously projecting at flowering. *South Africa (Cape Province, Natal, Orange Free State).* H5. Summer.

5. K. northiae Baker. Illustration: *Botanical Magazine*, 7412 (1895).

Plants usually with stems 20–170 cm, with a rosette of leaves at the stem apex. Leaves 8–12 cm broad, leathery, recurved, not keeled beneath, channelled above, margins conspicuously finely toothed. Scape 20–30 cm, up to 2.5 cm in diameter. Raceme cylindric to ovoid, very dense, 10–22 cm, buds and flowers more or less spreading, buds pinkish or orange-red, flowers whitish or yellow. Bracts oblong, acute to obtuse. Perianth more or less cylindric, to 3.2 cm, slightly constricted above the ovary. Stamens conspicuously projecting at flowering. *South Africa (Cape Province, Natal, Orange Free State), Lesotho.* H5. Summer.

6. K. galpinii Baker. Illustration: *Flowering Plants of South Africa* **20**: t. 783 (1940); *Bothalia* **9**: t. 14 (1968).

Leaves narrow, grass-like, usually less than 6 mm broad, dull green, markedly veined, triangular in section, margins smooth or minutely toothed towards the apex. Scape 30–60 cm. Raceme almost spherical to ovoid, dense, 5–8 cm; buds and flowers pendulous, buds bright red to orange-red, flowers yellow to orange. Bracts linear-lanceolate, acuminate or tailed. Perianth to 3 cm, not constricted above the ovary. Stamens not or scarcely projecting at flowering, later withdrawn. *South Africa (Transvaal Natal), Swaziland.* H5. Summer.

7. K. triangularis Kunth (*K. macowanii* Baker; *K. nelsonii* Masters). Illustration: *Botanical Magazine*, 6167 (1875); *Flowering plants of Africa* **30**: t. 1184 (1954–5); *Bothalia* **9**: t. 15 (1968); Brickell (ed.), RHS A–Z encyclopedia of garden plants, 601 (2003).

Leaves soft, somewhat fibrous, to 8 mm broad, triangular in section, margin sparingly to conspicuously finely toothed. Scape 30–60 cm. Raceme more or less cylindric or ovoid, pyramidal at the apex, dense, 5–8 cm; buds and flowers pendulous, all dull red to orange-yellow. Bracts ovate-lanceolate to lanceolate, acute to long-acuminate. Perianth to 4 cm, not constricted above the ovary, expanded abruptly at the mouth. Stamens projecting for up to 3 mm at flowering, later withdrawn. *South Africa (Cape Province, Natal, Orange Free State), Lesotho.* H5. Summer.

Subsp. **triangularis** is described above; other subspecies are known in the wild but are not cultivated.

8. K. uvaria (Linnaeus) Hooker (*K. alooides* Moench; *K. burchellii* (Lindley) Kunth; *Tritoma uvaria* (Linnaeus) Ker Gawler). Illustration: *Botanical Magazine*, 758 (1804); *Flowering plants of Africa* **33**: t. 1289 (1959); *Bothalia* **9**: t. 24 (1968).

Plants to 1.2 m. Leaves stiffly erect to arched and spreading, dull green to glaucous, tough, to 2 cm wide, keeled beneath, margins and keel smooth or very sparingly finely toothed towards the apex. Scape to 1 m. Raceme oblong to almost spherical, dense, 4.5–11 cm; buds spreading, bright red to greenish, tinged with red, flowers becoming pendulous, orange-yellow to greenish yellow. Bracts ovate, rounded, acuminate or obtuse. Perianth to 4 cm, slightly constricted above the ovary. Stamens not or slightly projecting at flowering, withdrawn later. *South Africa (Cape Province).* H4. Summer-autumn.

Long cultivated, the parent of many hybrids; it has also given rise to numerous cultivars. All of these are very difficult to distinguish.

K. rooperi (Moore) Lemoine. Illustration: *Bothalia* **9**: t. 23 & 72 (1968). Leaves clear green, usually more than 2 cm wide. *South Africa (Cape Province, Natal).* H5. Summer-autumn.

9. K. thomsonii Baker var. **snowdenii** (Wright) Marais (*K. snowdenii* Wright). Illustration: *Botanical Magazine*, 8867 (1920).

Leaves soft, somewhat glaucous, less than 4 cm wide, keeled but not V-shaped in section, the upper surface almost flat, margins and keel smooth. Scape 60–300 cm. Inflorescence rather loose, to 40 cm. Flowers yellow, orange or red. Bracts ovate, long-acuminate, green, fleshy and keeled. Perianth to 3.5 cm, inflated above the ovary, much constricted above, curved, gradually widening towards the mouth. Stamens not or scarcely projecting at flowering, later withdrawn. *Kenya, Uganda.* G1. Summer.

10. K. pumila (Aiton) Kunth (*K. leichtlinii* Baker; *Tritoma pumila* (Aiton) Ker Gawler). Illustration: *Botanical Magazine*, 764 (1804), 6716 (1883).

Leaves to 2 cm broad, slightly glaucous, prominently keeled, margins smooth or minutely papillose, keel smooth. Scape 50–150 cm. Raceme 4–12 cm, very dense, oblong-cylindric, flowers opening from the top downwards or the bottom upwards, yellow to red. Bracts lanceolate to ovate-lanceolate, acuminate. Perianth to 3 cm, bell-shaped, slightly constricted above the ovary. Stamens much projecting at flowerspaning, the filaments yellow to red. *Ethiopia, Zaire, Sudan & Uganda.* G2. Summer.

K. leichtlinii is the name given to the variant, known only in cultivation, in which the flowers open from the top of the raceme downwards; otherwise it is essentially the same as *K. pumila*.

11. K. foliosa Hochstetter (*K. quartiniana* Richard). Illustration: *Botanical Magazine*, 6742 (1884).

Leaves to 7 cm broad, keeled, the upper surface almost flat, margins finely toothed, the keel smooth below, finely toothed towards the apex. Scape 30–120 cm. Raceme 20–40 cm, very dense, cylindric; flowers yellow, orange or red. Bracts oblong or ovate-lanceolate, acuminate or, rarely, the upper bracts rounded. Perianth to 3 cm, constricted above the ovary, narrowly funnel-shaped. Stamens conspicuously projecting at flowering. *Ethiopia.* G2. Summer.

9. ANTHERICACEAE

Plants terrestrial, usually with rhizomes. Leaves mostly basal, without broadened blades and stalk-like bases, not fleshy and succulent, nor reduced to sheaths. Flowers in racemes, spikes or panicles, bisexual, usually actinomorphic. Perianth of 6 segments, free or very slightly united at the base, all similar. Stamens 6, not declinate. Ovary of 3 united, superior carpels, nectar secreted on the ovary; placentation axile. Fruit a capsule.

There are 9 genera with about 230 species. One genus is native to Europe and 2 to North America; a few are cultivated for ornament.

1a. Ovary below ground, hidden in the corm or bulb **8. Leucocrinum**
b. Ovary exposed, above ground 2
2a. Leaves borne on the flowering stem; flower solitary **1. Herpolirion**
b. Leaves all at the base of the flowering stem; flowers not usually solitary 3
3a. Perianth-segments free at the base, or united for less than one-tenth of their total length 4
b. Perianth-segments united for at least one-tenth of their total length 7
4a. Anthers with tails at their bases that are papillose, hairy or warty, sometimes joined to the filaments for all or part of their length **6. Arthropodium**
b. Anthers without tails; filaments sometimes hairy 5
5a. Outer perianth-segments narrower than the fringed, inner segments **5. Thysanotus**
b. All perianth-segments similar, none fringed 6
6a. Capsule conspicuously 3-angled; raceme or panicle often producing plantlets **4. Chlorophytum**
b. Capsules smooth; inflorescence not producing plantlets **2. Anthericum**
7a. Scape flattened; perianth white **3. Comespermum**
b. Scape terete; perianth blue **7. Pasithea**

1. HERPOLIRION J.D. Hooker
E.H. Hamlet

Rootstock a creeping, wiry rhizome. Leaves 6–8, in 2 ranks, 2–6 cm × 1.5–4 mm, bluish green, strongly keeled towards the base. Flowering stem short, erect, covered with the sheathing bases of the leaves, bearing a solitary terminal flower. Perianth-segments free, 1.1–1.5 cm × *c.* 5 mm, pale blue to white, or cream, linear to narrowly obovate, spreading from near the middle, the inner 3 slightly narrower. Filaments finely downy, except at the apex; anthers almost basifixed, 2–3 mm long. Ovary superior, more or less spherical with many ovules per cell; style 1, cylindric, narrowing to small stigma. Capsule with 1 or 2 seeds per cell. Seeds black, more or less flattened.

A genus of one species from SE Australia and New Zealand. In cultivation it prefers a moist soil and needs protection in winter. Propagation is by seed or division of the rhizome.

1. H. novae-zelandiae J.D. Hooker. Illustration: Salmon, Field guide to the alpine plants of New Zealand, 170 (1968); Cochrane et al., Flowers and plants of Victoria, 167 (1973); Moore & Unwin, The Oxford book of New Zealand plants, 184 (1978); Costin et al., Kosciusko alpine flora, pl. 137 (1979).

SE Australia, New Zealand. H5. Summer.

2. ANTHERICUM Linnaeus

V.A. Matthews

Perennial hairless herbs with rhizomes and somewhat fleshy roots. Leaves basal, linear with a membranous sheathing base. Flowers in a loose raceme or panicle. Bracts usually small and membranous. Flower-stalks usually jointed. Perianth-segments spreading, free or shortly joined at the extreme base, each with 3 central veins. Stamens equal, straight; filaments hairless; anthers basifixed. Ovary superior, 3-celled, with 4–8 ovules in each cell. Fruit a smooth capsule.

A genus of about 300 species, occurring mainly in tropical and southern Africa and Madagascar, but also present in Europe, America and E Asia. Only 2 are in cultivation. They are best grown in a well-drained soil, in a sunny position. Propagation is by division of the rootstock, or by seed.

1a. Flowers usually borne in a raceme; perianth-segments 1.5–2 cm; style curved upwards; capsule 8–10 mm with a pointed apex **1. liliago**

b. Flowers borne in a panicle; perianth segments 8–14 mm; style straight; capsule *c.* 5 mm with a blunt apex **2. ramosum**

1. A. liliago Linnaeus (*A. algeriense* invalid). Illustration: Felsko, Portraits of wild flowers, pl. 121 (1959); Grey-Wilson & Blamey, Alpine flowers of Britain and Europe, 265 (1979); Grey-Wilson & Mathew, Bulbs, pl. 12 (1981); Brickell (ed.), RHS A–Z encyclopedia of garden plants, 123 (2003).

Stem 20–70 cm. Leaves 12–40 cm. Raceme with 6–10 flowers, rarely shortly branched at the base. Perianth-segments 1.5–2 cm, white, narrowly elliptic. Stamens 9–13 mm. Style curved upwards. Capsule 8–10 mm, egg-shaped with a pointed apex. *Europe, Turkey.* H4. Late spring–summer.

2. A. ramosum Linnaeus. Illustration: Huxley, Mountain flowers, 893 (1967); Grey-Wilson & Blamey, Alpine flowers of Britain and Europe, 265 (1979); Polunin, Flowers of Greece and the Balkans, pl. 57 (1980).

Differs from *A. liliago* in having the flowers in a panicle, shorter perianth-segments (8–14 mm), a straight style and the capsule only *c.* 5 mm with a blunt apex. *Europe, Turkey.* H4. Summer.

3. COMOSPERMUM Rauschert

V.A. Matthews

Rootstock a short rhizome. Leaves evergreen, basal, in 2 ranks, narrowly strap-shaped, curved, 10–50 cm. Flowering stem 15–40 cm, flattened, with a slightly winged margin, bearing a many-flowered panicle. Flowers white tinged with lilac, to pale pink, of 2 forms that occur on different plants: one form has the stamens projecting, the other has the stamens equal in length to the perianth-segments. Flower-stalks jointed. Perianth united at the base. Anthers dorsifixed. Ovary superior, ovules 2 per cell; style 1, with a slightly thickened stigma. Seeds 3-angled, brown with a tuft of white hairs at one end.

A genus of 1 species from Japan. In cultivation it is tolerant of shade, but not very free-flowering. Propagation is by seed, or by division of the rhizome in spring. Treated as *Alectorurus* in European Garden Flora, edn 1.

1. C. yedoensis (Franchet & Savatier) Rauschert (*Alectorurus yedoensis* (Franchet & Savatier) Makino). Illustration: *Botanical Magazine*, 8336 (1910); *Bulletin of the Alpine Garden Society* **8**: 5 (1940); Makino, Makino's new illustrated flora of Japan, 834 (1963); Kitamura et al., Coloured illustrations of herbaceous plants of Japan (Monocotyledoneae), pl. 35, 231 (1981).

Japan. H3. Summer.

4. CHLOROPHYTUM Ker Gawler

S.G. Knees

Evergreen perennials with rhizomes and fibrous or fleshy roots. Leaves all basal; stalkless and linear, or stalked with ovate or lanceolate blades. Scape bearing a raceme or panicle of bisexual, stalked flowers. Perianth star-shaped, of 6 free segments, each with 3–7 veins, white, greenish or yellowish. Stamens with basifixed anthers opening inwardly. Ovary superior, spherical to ovoid, 3-celled, with several ovules per cell; style thread-like, stigma capitate. Fruit a deeply 3-angled capsule containing black, round, slightly flattened seeds.

A genus of 215 species native of tropical America, Africa, India and Australia, but only 1 is widely cultivated in Europe. Propagation is by seed, offsets or division in early spring. In most areas greenhouse conditions are necessary; a rich, sandy loam provides an ideal growing medium for the majority of species.

Literature: Dress, W.J., Chlorophytum in cultivation, *Baileya* **9**: 29–50 (1961).

1. C. comosum (Thunberg) Jacques. Illustration: Graf, Exotica, edn 3, 1068 (1963).

Leaves linear to linear-lanceolate, 20–45 cm × 5–20 mm. Inflorescence branched or simple, 30–80 cm. Flowers 1–3 per node on jointed stalks. Perianth-segments ovate, acute, to 8 mm, white. Sometimes rosettes of leaves, often with roots, also arise from several of the inflorescence nodes. Capsule greenish, *c.* 9 mm, containing 3–5 seeds per cell. *South Africa.* G1. Flowering sporadically throughout the year.

In cultivation this species is most commonly represented by its variegated cultivars: 'Mandaianum' has a central creamy-white band along the leaf, while 'Variegatum' has white or creamy-white margins. The name *C. capense* (Linnaeus) Voss is often wrongly applied to the species.

5. THYSANOTUS R. Brown

V.A. Matthews

Rootstock fibrous, rhizomatous or tuberous. Leaves basal, linear, with membranous, sheathing wings at the base. Flowers in panicles or umbels. Perianth-segments free, the outer much narrower than the inner, which (in cultivated species) are fringed. Stamens 3 or 6; anthers basifixed or almost so. Ovary superior; style 1, unlobed. Capsule enclosed within the persistent perianth-segments. Seeds black, with an aril.

There are 47 species, the majority of which come from Australia; 2 species extend into SE Asia. Two species occur uncommonly in cultivation; other species have been introduced from time to time but always die out quickly. *Thysanotus* species are not frost-hardy and need plenty of sun to flower well. They will grow in any well-drained soil and water should be more or less withheld after flowering. Propagation is by seed.

Literature: Brittan, N.H., Revision of the genus Thysanotus R. Br. (Liliaceae), *Brunonia* **4**: 67–181 (1981).

1. T. tuberosus R. Brown. Illustration: *Edwards's Botanical Register* **8**: pl. 655 (1822); Galbraith, Field guide to the wild flowers of south-east Australia, pl. 1 6 (1977); *Australian Plants* **12**(94): front cover (1983).

Roots tuberous. Leaves 5–15, 20–30 cm, linear, almost terete towards the tip, hairless. Flowering stem 20–60 cm bearing a panicle of 1–5-flowered umbels. Perianth-segments 7–20 mm, purple. Outer segments lanceolate, with 3 or 4 veins; inner segments ovate, fringed. Stamens 6, unequal to almost equal. Style straight or curved. *E Australia.* H5–G1. Summer.

T. multiflorus R. Brown (*T. proliferus* Lindley; *T. multiflorus* var. *prolifer* (Lindley) Bentham). Illustration: *Edwards's Botanical Register* **24**: pl. 8 (1838); *Australian Plants* **12**(94): 77 (1983). Differs mainly in that the flowering stem usually bears a single, 6–20-flowered umbel of blue-violet flowers with only 3 stamens, twisted to one side. *W Australia.* H5–G1. Summer.

6. ARTHROPODIUM R. Brown

J. Cullen

Perennial herbs with rhizomes, usually forming tufts; roots fibrous, but often accompanied by some that are swollen and tuber-like. Leaves in a basal rosette, evergreen or deciduous. Inflorescence a terminal panicle with few or many branches; flowers often 2 or 3 in the axil of each bract. Flower-stalks conspicuously jointed at or about the middle. Perianth of 6 free, similar, spreading or slightly reflexed segments, persisting in fruit. Stamens 6, the anthers bearing hairy 'tails' attached to the filament for all or most of their lengths. Ovary superior, 3-celled, ovules several; style 1. Fruit a capsule containing several black seeds, opening between the septa.

A genus of about 12 species from Australia, New Zealand, New Guinea, New Caledonia and Madagascar, of which 4 are occasionally seen in gardens. They will grow in any well-drained garden soil, and can be propagated by seed or by division.

1a. Leaves evergreen, more than 3 cm wide; flowers 2 cm or more in diameter **1. cirratum**
b. Leaves deciduous, mostly less than 1 cm wide, occasionally to 3 cm wide; flowers to 1.5 cm in diameter 2
2a. Flowers deep purple, borne singly in the bract axils **4. minus**
b. Flowers white or mauve, at least the lower bracts with 2 or 3 flowers in their axils 3
3a. Perianth-segments to 5 mm, white; anther tails white **2. candidum**
b. Perianth-segments 6–10 mm, mauve or white, mauve in bud; anther tails yellow **3. milleflorum**

1. A. cirratum (Forster) R. Brown. Illustration: *Botanical Magazine*, 2350 (1822); *Baileya* **1**: 43 (1953).

Roots thick. Leaves evergreen, 30–60 × 3–10 cm, fleshy, green above, whitish green or glaucous beneath. Scape robust, to 1 m, bearing a panicle with numerous, rather short, lateral branches that are spreading or ascending. Flowers 2 or 3 together in the axils of conspicuous bracts. Perianth white, 2–4 cm in diameter. Anther tails free and curled at their lower ends, purplish at the top, white in the middle, yellow at their free ends. *New Zealand.* G1.

Variable in size of all its parts. The name was spelled as above by Forster, but it is often found in the literature as *A. cirrhatum*.

2. A. candidum Raoul (*A. reflexum* Colenso). Illustration: Salmon, Field guide to the alpine plants of New Zealand, 197 (1968).

Rhizome short, roots swollen. Leaves dying down in winter, 10–30 cm × 3–10 mm. Scape slender, bearing a few, erect branches; at least the lower bracts with 2 or 3 flowers in their axils. Flower-stalks 8–20 mm. Perianth to 1 cm in diameter, white. Anther tails white, joined to the filament for the whole of their lengths. *New Zealand.* G1.

3. A. milleflorum (de Candolle) MacBryde (*A. paniculatum* R. Brown; *A. pendulum* invalid). Illustration: *Botanical Magazine*, 1421 (1812); Rotherham et al., Flowers and plants of New South Wales and southern Queensland, 35 (1975); Elliott & Jones, Encyclopaedia of Australian plants **2**: 235 (1982); Brickell (ed.), RHS A–Z encyclopedia of garden plants, 123 (2003).

Rhizome short, most roots fibrous. Leaves dying down in winter, to 30 × 1–3 cm. Scape to 1 m, bearing few to many erect or ascending branches. Flower-stalks 8–20 mm, usually 2 or 3 together in the axil of each bract. Perianth mauve in bud, white or mauve when open, segments 6–10 mm. Anther tails yellow, attached to the filament for all of their length, or the lower part free and curled. *E Australia (Queensland to Tasmania and S Australia).* G1.

4. A. minus R. Brown. Illustration: Cunningham et al., Plants of western New South Wales, 181 (1981).

Rhizome short, some roots swollen and tuberous. Leaves dying down in winter, to 20 cm × 2–4 mm. Scape to 30 cm, bearing few, erect branches. Flowers fragrant, solitary in the bract axils; flower-stalks to 8 mm. Perianth deep purple, to 1 cm in diameter. Anther tails purple, attached to the filament for the whole of their lengths. *Australia.* G1.

7. PASITHEA D. Don

V.A. Matthews

Rootstock a short, erect rhizome with swollen roots. Leaves mostly basal, in 2 ranks, linear, to 25 cm. Flowers borne in a terminal panicle to 60 cm. Perianth united at the base, blue, each lobe with a darker central stripe; flowers *c.* 2.5 cm in diameter. Stamens 6, 3 shorter than the others; anthers versatile. Ovary superior, with 4 ovules per cell; style 1, stigma 3-lobed. Capsule spherical.

This genus contains a single species from SW South America. Cultivation as for *Camassia* (p. 114). Propagation by seed.

1. P. caerulea (Ruiz & Pavon) D. Don. Illustration: *Botanical Magazine*, 7249 (1892); RHS Dictionary of gardening, 1489 (1951).

Chile, Peru. H5–G1. Summer.

8. LEUCOCRINUM A. Gray

V.A. Matthews

Rootstock a cluster of short, fleshy roots. Leaves narrowly linear, forming a tuft, 10–20 cm × 2–6 mm, surrounded at the base by membranous bracts. Stem absent to very short, below ground level, flowers borne on subterranean, 5–30 mm stalks, appearing at ground level, white, fragrant. Perianth joined into a slender tube, 5–12 cm, the free lobes 1.4–2 cm, narrowly oblong. Stamens 6, borne near the top of the perianth-tube; anthers 4–6 mm. Ovary subterranean, with several black, angled seeds in each cell; style 1, slender, stigma slightly 3-lobed.

A genus with one species native to western USA. Plants grow best in a sunny position but need a soil that never dries out completely. Propagation is by seed.

1. L. montanum A. Gray. Illustration: *Bulletin of the Alpine Garden Society* **39**: 125 (1971); Rickett, Wild flowers of the United States **5**: pl. 18 (1971); Niehaus & Ripper, Field guide to Pacific states wildflowers, 9 (1976); Cronquist et al., Intermountain Flora **6**: 83 (1977).

Western USA. H4. Spring–early summer.

10. APHYLLANTHACEAE

Plants terrestrial, with rhizomes. Leaves without blades, reduced to basal sheaths. Flowers in heads, bisexual, actinomorphic. Perianth of 6 free segments. Stamens 6. Ovary of 3 united, superior carpels, placentation axile; nectar secreted on the ovary. Fruit a capsule.

A single genus, native to Europe and sometimes cultivated.

1. APHYLLANTHES Linnaeus
D.A. Webb
Perennial with a dense, fibrous rootstock. Leaves reduced to reddish brown sheaths, which surround the lower part of the numerous slender, wiry, glaucous stems 15–40 cm high. Flowers terminal, solitary or in compact groups of 2 or 3, subtended by about 5 reddish brown, scarious bracts. Perianth-segments free, erect in the lower half, spreading above. Anthers dorsifixed, introrse. Ovary superior; style 1; stigma 3-lobed. Fruit a 3-seeded capsule enclosed by the bracts.

A genus of a single species with very distinctive habit. It requires a sunny position in well-drained soil. In less sunny areas, even if the winter is mild, it is best grown in a frame or a cool house.

1. A. monspeliensis Linnaeus. Illustration: Coste, Flore de la France **3**: 349 (1906); Bonnier, Flore complète **10**: t. 593 (1929); Reisigl & Danetsch, Flore méditerranéenne, 101 (1979).

Flowers 2 cm across, varying from fairly deep to very pale blue. *SW Europe and Morocco.* H5. Early summer.

11. HOSTACEAE

Plant terrestrial, entirely herbaceous, usually with rhizomes. Leaves all basal, with a broadened blade and petiole-like basal part, not fleshy and succulent. Flowers in a 1-sided raceme, somewhat zygomorphic. Perianth of 6 segments, united below into a narrow tube that broadens upwards. Stamens 6. Ovary of 3 united carpels, superior, placentation axile, nectar secreted on the ovary. Fruit a capsule.

A single genus with about 26 species, very widely cultivated, sometimes found in gardening literature under the synonym Funkiaceae.

1. HOSTA Trattinick
J.C.M. Alexander
Perennial, clump-forming herbs with short, thick, fleshy rhizomes bearing dense clusters of roots. Leaves mostly basal. Leaf-stalks fleshy, erect or ascending, channelled, often winged, sometimes purple-spotted. Leaf-blades elliptic or oblong to ovate or almost circular, acuminate, acute or obtuse; with tapered, rounded, truncate or cordate bases, often decurrent, membranous to leathery, flat or with marginal undulations or wrinkles, yellowish or mid-green to dark green (with many yellow and variegated varieties and cultivars), dull, shiny or glaucous, with 2–14 pairs of side veins. Leaf-surface usually raised between the veins and sometimes also puckered, with wrinkles at right-angles to the veins. Flowering stems fleshy, erect or ascending, to *c.* 1.75 m, almost always unbranched, occasionally with a few leaves or large leaf-like bracts below; upper bracts erect or spreading, often channelled, persistent or withering at flowering. Flowers few to many in a loose to dense head, bell- or trumpet-shaped, white or various shades of bluish purple; perianth fused below into a narrow tube, which broadens upwards; perianth-lobes 6, spreading or recurved. Stamens 6, not or slightly projecting, free from perianth-tube (except in *H. plantaginea*), curved upward just below anthers; anthers versatile, white, yellow or purplish. Ovary superior, 3-celled; style projecting beyond stamens, curved upwards. Seeds many, membranous, winged.

A genus variously regarded as having 20–50 species, known as *Funkia* Sprengel until early this century, mostly native to Japan, with a few representatives in Korea and China. Cultivated for centuries in Japan, they were first imported into Europe early in the nineteenth century, since when a vast and confusing array of varieties and cultivars has arisen from sports or from spontaneous or deliberate hybridisation. Many of these cannot be assigned to any known wild species. Revisions of the Asiatic species are of very limited use when attempting to put names on plants growing in Europe. The account that follows is not a revision, it merely attempts to account for and describe most of what is grown in Europe and keeps as separate entities several species that might best be combined.

Hostas do particularly well if grown in partial shade where the roots can get down to a steady supply of water beside a stream or pond, but will also perform well in an open herbaceous border. They flower from early summer onwards and are easily propagated by dividing the clump with a spade. This can be done by cutting out a segment without digging up the clump. Though many clones are sterile, others reproduce freely by seed; however, cross-pollination is rife and offspring will very often not be true to the female parent. Horticultural (and taxonomic) interests will best be served by not growing hostas from seed.

Literature: Grey, C.H., Hardy bulbs **3**: 292–305 (1938); Maekawa, F., The genus Hosta, *Journal of the Faculty of Science, Tokyo (Botany)* **5**: 317–425 (1940); Hylander, N., The genus Hosta in Swedish gardens, *Acta Horti Bergiani* **16**: 331–420 (1954); Hylander, N., The genus Hosta, *Journal of the Royal Horticultural Society* **85**: 356–66 (1960); Thomas, G.S., Notes on the genus Hosta, *Bulletin of the Hardy Plant Society* **2**: 92–101 (1960), photograph captions all transposed; Hensen, K.J.W., De in Nederland gekweekte Hostas, *Mededelingen Directeur Tuinbouw* **26**: 725–53 (1963); Brickell, C.D., Notes from Wisley, *Journal of the Royal Horticultural Society* **93**: 365–72 (1968); Hylander, N., RHS Dictionary of Gardening, Supplement, 347–50 (1969); Fujita, N., The genus Hosta in Japan, *American Hosta Society Bulletin* **10**: 14–41 (1979), a very stilted translation from *Acta Phytotaxonomica et Geobotanica* **27**: 66–96 (1976); Grenfell, D., A survey of Hosta and its availability in commerce, *The Plantsman* **3**: 20–44 (1981).

The key that follows is based almost entirely on vegetative characters because of the relative uniformity of the flowers, and will thus be useful even when plants are not flowering. Some of the characters used in the key and descriptions need clarification.

Leaf Dimensions. These have generally been taken from the largest leaves on mature plants in late summer. In species with clearly defined blade and stalk, measurements are not a problem; however, many have tapered, decurrent blades that merge with the winged stalks. In some of these, the sides of the base of the blade are curled forward across the top of the stalk; this point where the sides approach each other is taken to be the base of the blade (Figure 1(8), p. 29). For flat, tapering leaves, the position of the base has to be decided arbitrarily.

Leaf surfaces. Many species have a grey or bluish bloom on one or both leaf-surfaces; they are referred to as *glaucous.* The description of a leaf as dull implies that it is not glaucous.

Number of pairs of side-veins. This has been used extensively, and there can be uncertainty as to how many of the smaller side-veins should be included. In this account, only those veins cutting a line running forwards from the widest part of the leaf have been counted (Figure 1(9), p. 29).

1a. Leaves not variegated 2
b. Leaves variegated 52
2a. Leaves glaucous or bloomed below 3
b. Leaves dull or shiny below, not glaucous or bloomed 24
3a. Leaves yellow to bright greenish yellow at least in spring and early summer, not densely white-bloomed below 4
b. Leaves mid-green to dark or slightly yellowish green, sometimes densely white-bloomed below 6
4a. Leaves becoming pale to mid-green in summer; blades 19 × 12 cm or more, with 8 or 9 pairs of side-veins; leaf-stalks 30 cm or more **7. fortunei** var. **albopicta** forma **aurea**
b. Leaves remaining yellow; blades 15 × 11 cm or less, with 7 pairs of side-veins or fewer; leaf-stalks less than 25 cm 5
5a. Leaves ovate, with rounded, truncate or cordate bases, slightly leathery, with 6 or 7 pairs of side-veins **24. nakaiana**, see **'Golden Nakaiana'**
b. Leaves narrowly lanceolate to elliptic, tapered at base, membranous, with *c.* 3 pairs of side-veins **16. lancifolia**, see **'Wogon Gold'**
6a. Leaves with 7 pairs of side-veins or fewer 7
b. Leaves with 8 pairs of side-veins or more 9
7a. Leaves lanceolate, glaucous above and below **11. × tardiana**
b. Leaves ovate, dull mid-green to dark green above 8
8a. Leaves mid-green, acuminate; leaf-stalks spotted **24. nakaiana**
b. Leaves dark green, very shortly acuminate; leaf-stalks not spotted **24. nakaiana**, see **'Pastures New'**
9a. Leaves slightly shiny yellowish green above, densely white-bloomed below **8. hypoleuca**
b. Leaves dull or glaucous and mid-green to dark green above, if slightly yellowish then not densely white-bloomed below 10
10a. Leaves very glaucous above 11
b. Leaves not or slightly glaucous above 18
11a. Leaves not puckered, though often raised between the veins 12
b. Leaves puckered, with wrinkles at right-angles to the veins 14
12a. Leaf-blades to *c.* 26 × 17.5 cm, erect or steeply ascending, only slightly arched at the tips **2. nigrescens**, see **'Krossa Regal'**
b. Leaf-blades to *c.* 15 × 12 cm, spreading, horizontal or strongly arched 13
13a. Leaf-stalks spotted; blades with 8 or 9 pairs of side-veins, not cupped **11. × tardiana**, see **'Halcyon'**
b. Leaf-stalks not spotted; blades with 10–12 pairs of side-veins, cupped **5. tokudama**, see **'Buckshaw Blue'**
14a. Leaves ovate to oblong; length usually at least 10 cm more than width **6. sieboldiana**, see **'Snowden'**
b. Leaves broadly ovate; length usually 5 cm more than width or less 15
15a. Leaf-blades with a very thin, white 'pencil-line' along margin **7. fortunei** var. **hyacinthina**
b. Leaf-blades without such a line 16
16a. Leaf-blades 25 × 15 cm or more, not cupped, floppy, greyish green, with 12–14 pairs of side-veins **6. sieboldiana**
b. Leaf-blades 20 × 15 cm or less, cupped, stiff, rather bluish, with 10–12 pairs of side-veins 17
17a. Leaf-blades almost circular, strongly puckered, bluish green **5. tokudama**
b. Leaf-blades broadly ovate, weakly puckered, very blue **5. tokudama**, see **'Buckshaw Blue'**
18a. Young leaves yellow or bright greenish yellow with a dark green margin and cross-veins, later entirely green, though margin remaining darker **7. fortunei** var. **albopicta**
b. Leaves not variegated when young, lacking a darker margin at any time 19
19a. Young leaves yellow or bright greenish yellow, later pale green **7. fortunei** var. **albopicta** forma **aurea**
b. Leaves mid-green to dark or slightly yellowish green 20
20a. Mature leaves slightly glaucous above, on and between the veins 21
b. Mature leaves not glaucous above or only glaucous in vein-depressions 22
21a. Leaves not puckered, with a very thin, white 'pencil-line' along margin **7. fortunei** var. **hyacinthina**
b. Leaves strongly puckered, without such a line **7. fortunei** var. **rugosa**
22a. Flowering-head just above leaf-mound **7. fortunei** var. **stenantha**
b. Flowering-head well above leaf-mound 23
23a. Leaf-blades lustrous dark green above, very broadly ovate to almost circular, with *c.* 8 pairs of side-veins **7. fortunei** var. **obscura**
b. Leaf-blades dull mid-green or slightly yellowish green, ovate to elliptic, with 10–13 pairs of side-veins **4. elata**
24a. Leaves yellow **16. lancifolia**, see **'Wogon Gold'**
b. Leaves green or yellowish green 25
25a. Leaf-blades 7 cm long or less 26
b. Leaf-blades 8 cm long or more 28
26a. Leaf-blades more than 3 times longer than broad, gradually tapered at base, often slightly twisted **12. tortifrons**
b. Leaf-blades *c.* 2 times longer than broad, rounded or shortly tapered at base, not twisted 27
27a. Leaves dark green; margin sometimes finely wrinkled; back of leaf-stalk spotted throughout **14. gracillima**

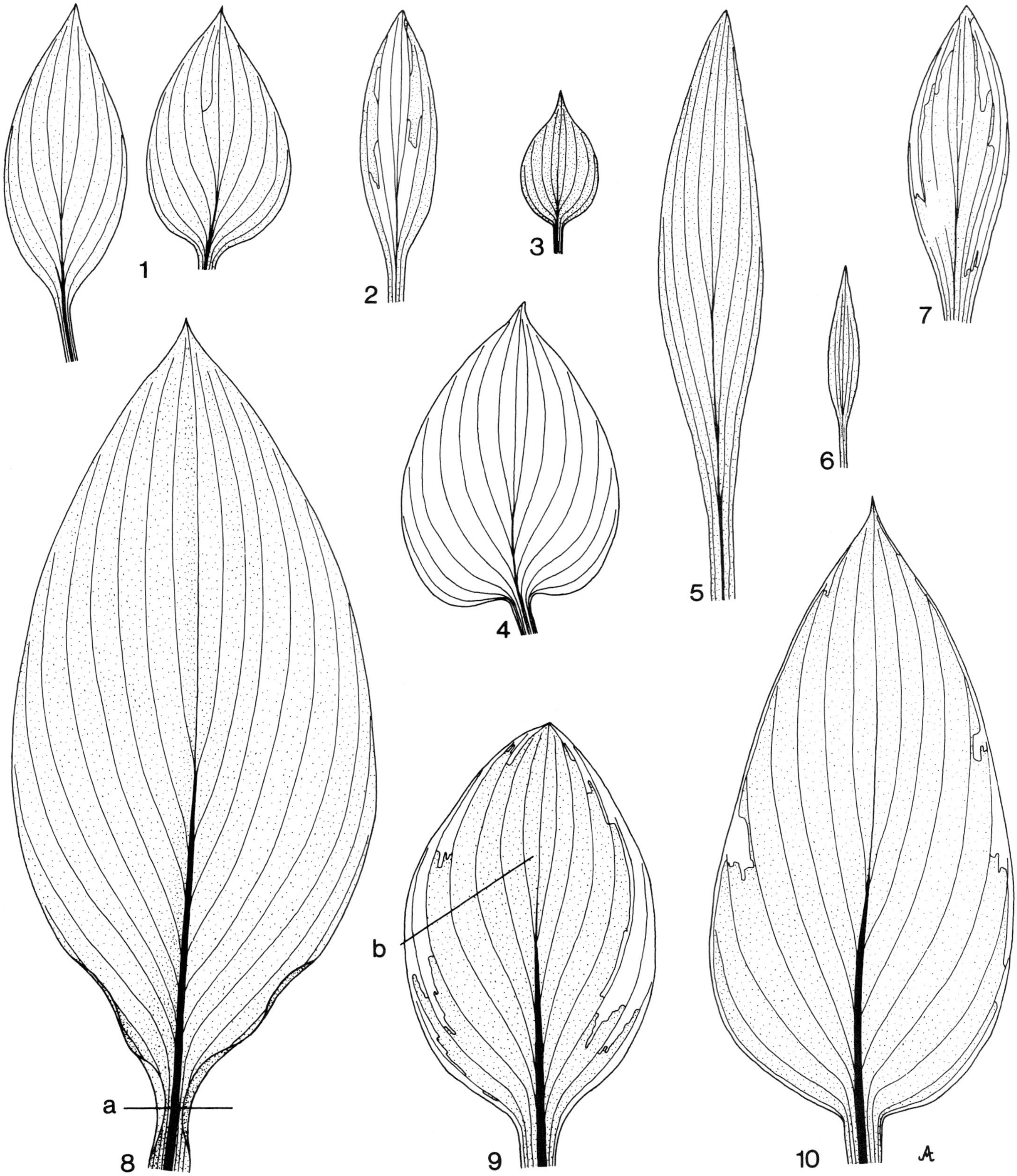

Figure 1. Leaf-blades of *Hosta* species and cultivars (× 0.5). 1, *H. minor.* 2, *H. lancifolia* 'Kabitan'. 3, *H. gracillima.* 4, *H.* 'Golden Nakaiana'. 5, *H. longissima.* 6, *H. tortifrons.* 7, *H. helonioides* forma *albopicta.* 8, *H. rectifolia* (a, point taken as base of leaf-blade). 9, *H. decorata* (to show number of side-veins cut by a line (b) running forwards from the widest point of the leaf: in this case, 6). 10, *H.* 'North Hills'.

b. Leaves mid-green; margin with 1 or 2 undulations; back of leaf-stalk spotted only in lower half **23. venusta**

28a. Leaf-blades with 5 or fewer pairs of side-veins 29

b. Leaf-blades with 6 or more pairs of side-veins 35

29a. Leaf-blades less than 2 times longer than broad 30

b. Leaf-blades 2 or more times longer than broad 31

30a. Leaf-blades *c.* 16 × 11 cm; flowers up to 25 per head **20. decorata** var. **normalis**

b. Leaf-blades *c.* 8.5 × 5.5 cm; flowers up to 10 per head **22. minor**

31a. Leaves shiny dark green above **16. lancifolia**

b. Leaves dull mid-green, yellowish or dark green above 32

32a. Leaves membranous 33

b. Leaves leathery 34

33a. Leaves erect or ascending **18. helonioides**

b. Leaves spreading or arched **15. sieboldii**

34a. Leaf-tips acuminate, stalks strongly spotted **10. tardiflora**

b. Leaf-tips acute, stalks not spotted **17. longissima**

35a. Leaves pale to bright yellowish green 36

b. Leaves mid-green to dark or slightly yellowish green 37

36a. Leaves elliptic; flowers very pale bluish purple **1. plantaginea**, see **'Honey Bells'**

b. Leaves ovate; flowers white **1. plantaginea**

37a. Leaves 13 cm wide or more 38

b. Leaves 12 cm wide or less 44

38a. Leaves dark green 39

b. Leaves mid-green or slightly yellowish green 41

39a. Leaves membranous **1. plantaginea**, see **'Royal Standard'**

b. Leaves slightly leathery 40

40a. Leaf-blades broadly elliptic to almost circular, abruptly acuminate, lustrous dark green above, not rolled at base **21. ventricosa**

b. Leaf-blades elliptic to ovate, acuminate, slightly shiny dark green above, sides rolled forwards at base **2. nigrescens**

41a. Leaves with 10–13 pairs of side-veins **4. elata**

b. Leaves with 9 or fewer pairs of side-veins 42

42a. Leaves broadly ovate, rounded to cordate at base **1. plantaginea**, see **'Royal Standard'**

b. Leaves lanceolate or narrowly ovate to elliptic, shortly tapered at base 43

43a. Flowering stems bearing large leaf-like bracts; leaves spreading **26. undulata** var. **erromena**

b. Flowering stems lacking large leaf-like bracts; leaves erect or steeply ascending **19. rectifolia**, see **'Tall Boy'**

44a. Leaves very shiny above **16. lancifolia**

b. Leaves dull or slightly shiny above 45

45a. Leaf-bases tapered 46

b. Leaf-bases rounded, truncate or cordate 48

46a. Leaves broadly ovate to broadly elliptic, *c.* 1.5 times longer than broad **20. decorata** var. **normalis**

b. Leaves lanceolate, narrowly ovate or narrowly elliptic, *c.* 2 times longer than broad 47

47a. Leaf-blades erect or steeply ascending **19. rectifolia**

b. Leaf-blades spreading or arched **26. undulata** var. **erromena**

48a. Leaves membranous 49

b. Leaves leathery 51

49a. Leaf-blades less than 6 cm wide **22. minor**

b. Leaf-blades more than 8 cm wide 50

50a. Leaf-tips obtuse **20. decorata** var. **normalis**

b. Leaf-tips acuminate **9. longipes**

51a. Leaves shiny below, rounded to truncate and rolled forwards at base **13. rupifraga**

b. Leaves dull below, cordate and not rolled forwards at base **25. capitata**

52a. Paler colour on leaf margin 53

b. Paler colour in middle of leaf-blade, or leaf mottled 63

53a. Leaves with 4 pairs of side-veins or fewer 54

b. Leaves with 5 pairs of side-veins or more 57

54a. Leaf-blade plus stalk less than 18 cm; leaf-tips 45° or less **15. sieboldii**, see **'Ginko Craig'** & **'Louisa'**

b. Leaf-blade plus stalk more than 20 cm; leaf-tips more than 45° 55

55a. Leaf-margins pale yellow, more than 8 mm wide in places; leaves erect or ascending **18. helonioides** forma **albopicta**

b. Leaf-margins white to cream-coloured, less than 6 mm wide; leaves spreading or arched 56

56a. Leaf-stalks 20 cm or more, less than 1 cm wide **15. sieboldii**

b. Leaf-stalks 15 cm or less, more than 1 cm wide **15. sieboldii**, see **'Emerald Isle'**

57a. Leaves glaucous below 58

b. Leaves dull or shiny below 59

58a. Leaves with 10 pairs of side-veins or more **6. sieboldiana**, see **'Aureo Marginata'**

b. Leaves with *c.* 8 pairs of side-veins **7. fortunei** var. **obscura** forma **marginata**

59a. Leaves strongly curled and often twisted at tip **3. crispula**

b. Leaves flat or with shallow undulations 60

60a. Leaf-tips obtuse **20. decorata**

b. Leaf-tips acute or acuminate 61

61a. Leaves less than 10 cm wide **26. undulata** var. **albo-marginata**

b. Leaves more than 10 cm wide 62

62a. Leaves lanceolate, dull above, dull to slightly shiny below; pale margin less than 1 cm wide **7. fortunei**, see **'North Hills'**

b. Leaves broadly ovate, lustrous above, very shiny below; pale margin more than 1 cm wide in places **21. ventricosa**, see **'Variegata'**

63a. Leaf-bases gradually tapered 64

b. Leaf-bases truncate, rounded or cordate 65

64a. Leaves with *c.* 3 pairs of side-veins, 4 cm wide or less, yellow in middle **16. lancifolia**, see **'Kabitan'**

b. Leaves with 6 or more pairs of side-veins, 6 cm wide or more, white in middle **26. undulata**

65a. Leaves predominantly yellow; green margin less than 1 cm wide **7. fortunei**, see **'Gold Leaf'** & **'Gold Standard'**

b. Green margin to 2 cm wide in places or leaves green with paler mottling 66

66a. Leaves cupped, abruptly pointed, glaucous above and below **5. tokudama**, see **'Variegata'**

b. Leaves not cupped, slightly acuminate, dull above, shiny or slightly glaucous below 67

67a. Leaves yellow or bright greenish yellow with green margins when young, later entirely green, though margin remaining darker, slightly glaucous below, puckered, rounded at base **7. fortunei** var. **albopicta**

b. Leaves green, mottled with pale yellowish green when young, later almost entirely green, shiny below, not puckered, truncate or slightly cordate at base **21. ventricosa**, see '**Aureomaculata**'

1. H. plantaginea (Lamarck) Ascherson (*Hemerocallis japonica* Redouté, not Thunberg; *Hemerocallis alba* Andrews; *Funkia subcordata* Sprengel; *Hosta* 'Thomas Hogg' misapplied). Illustration: *Botanical Magazine*, 1433 (1812); *Journal of the Faculty of Science, Tokyo (Botany)* **5**: 343 (1940); *Acta Horti Bergiani* **16**: pl. 24 (1954).

Leaf-stalks 15–28 cm, deeply channelled, winged, often with overlapping edges in upper part. Leaf-blades 14–22 × 9.5–17 cm, ovate, shortly acuminate at tip, rounded to slightly cordate at base, decurrent, membranous, flat or with several small undulations, bright yellowish green, slightly shiny above, very shiny below, with 6–9 pairs of side-veins. Flowering stems to 65 cm, mostly erect, yellowish green; lower bracts leaf-like, upper bracts 3–8.5 cm, greenish white. Flowers *c.* 11.5 cm, white, fragrant; anthers yellow; filaments fused to perianth-tube at base. *China.* H2. Late summer–autumn.

Plants with larger flowers and broader perianth-segments have been called var. **grandiflora** (Siebold & Zuccarini) Ascherson & Graebner (*H. grandiflora* Siebold & Zuccarini). Illustration: *Flore des serres*, t. 158, 159 (1846); Thomas, Plants for ground cover, f. 34, 35 (1970). They are variously described as having leaves broader than, or narrower than, the species.

'**Honey Bells**' (*H. plantaginea* × *?H. lancifolia)* has longer, pale green, elliptic, scarcely acuminate leaves with 6 or 7 pairs of side-veins and very pale, bluish purple, fragrant flowers. '**Royal Standard**' (*H. plantaginea* × *?)* has broadly ovate, mid-green to dark-green leaves, and fragrant, pure white flowers. These cultivars flower a little earlier than *H. plantaginea.*

2. H. nigrescens (Makino) Maekawa (*H. sieboldiana* (Hooker) Engler var. *nigrescens* Makino). Illustration: *Journal of Japanese Botany* **13**: 902 (1937); *Journal of the Faculty of Science, Tokyo (Botany)* **5**: 353, 354 (1940).

Young plant with dense greyish bloom that later vanishes. Leaf-stalks 20–50 cm, channelled, winged in upper half, purple-spotted. Leaf-blades 20–29.4 × 13–21 cm, ovate to elliptic, acuminate at tip, tapered to rounded or cordate at base, sides rolled forward at base forming a funnel, decurrent, slightly leathery, flat or with a few shallow undulations, dark green and slightly shiny above, mid-green and very shiny below, with 8–13 pairs of side-veins. Flowering stems to 142 cm, arched, bearing 2–4 spreading, sterile bracts below, and 30–40 flowers. Flowers 4.5–5 cm, white flushed with violet. *Japan.* H2. Summer.

Forma **elatior** Maekawa is larger in all its parts: flowering stems to 170 cm; flowers *c.* 6.5 cm. Illustration: *Journal of the Faculty of Science, Tokyo (Botany)* **5**: 323 (1940).

'**Krossa Regal**' (*H. nigrescens* forma *elatior* misapplied) is a statuesque plant with erect or steeply ascending, slightly arched leaves. The whole plant is densely bluish-glaucous when mature. Illustration: *Journal of the Royal Horticultural Society* **107**: 197 (1982).

3. H. crispula Maekawa (*H. latifolia* Matsumura var. *albimarginata* Wehrhahn; *H. undulata* var. *albomarginata* misapplied; *H.* 'Thomas Hogg' misapplied; *H. albomarginata* misapplied). Illustration: *Acta Horti Bergiani* **16**: pl. 13 (1954); *Journal of the Royal Horticultural Society* **85**: f. 104 (1960) & **93**: f. 187 (1968); Thomas, Plants for ground cover, pl. XIIB & XIIA (1970).

Leaf-stalks to 45 cm, channelled, winged in upper half. Leaf-blades 13–27 × 6–10.5 cm, lanceolate to narrowly elliptic, attenuate at tip, tapered, decurrent and often rolled at base, slightly leathery, strongly curled and often twisted at tip, mid-green to dark green with an irregular white margin to *c.* 2 cm wide, dull above, shiny below, with 7–9 pairs of side-veins. Flowering stems 50–85 cm bearing 30–40 flowers, without leaf-like bracts below; bracts 1.3–2 cm, pale, spreading. Flowers *c.* 4 cm, pale violet; anthers yellow, becoming purple. *Unknown in wild.* H2. Summer.

4. H. elata Hylander (*H. montana* invalid; *H. fortunei* (Baker) Bailey var. *gigantea* Bailey; *H. fortunei* (Baker) Bailey 'Robusta'). Illustration: *Gentes Herbarum* **2**: 134 (1930); *Journal of the Faculty of Science, Tokyo (Botany)* **5**: 361 (1940); *Acta Horti Bergiani* **16**: pl. 6, 10–12 (1954); *Journal of the Royal Horticultural Society* **81**: f. 130 (1956).

Leaf-stalks 15–47 cm, channelled, winged in upper half. Leaf-blades 18–32 × 10–24 cm, broadly ovate, rarely almost elliptic, acuminate at tip, rounded to cordate at base, decurrent, slightly leathery, with several shallow undulations, mid-green or slightly yellowish green, slightly glaucous, though becoming dull, especially above, with 10–13 pairs of side-veins. Flowering stems to *c.* 90 cm, rarely with a leaf-like bract below; bracts pale yellowish green. Flowers 4.5–6 cm, pale bluish violet; anthers pale yellow. *Japan.* H2. Early summer.

Sometimes considered part of *H. sieboldiana* though much less glaucous and somewhat yellower in the leaf and with a much longer flowering stem. It has been suggested that some clones grown as *H. montana* are distinct enough to be maintained as a separate species. This seems unlikely as Hylander published *H. elata* because he discovered that the name *H. montana* was invalid; he was not combining *H. montana* with anything else.

Species 4–6 are sometimes considered to be a single species, *H. sieboldiana.*

5. H. tokudama Maekawa (*H. sieboldiana* Engler var. *fortunei* Voss not Ascherson; *H. sieboldiana* Engler var. *glauca* Makino & Matsumura; *H. glauca* misapplied). Illustration: *Journal of the Faculty of Science, Tokyo (Botany)* **5**: 367 (1940); *Journal of the Royal Horticultural Society* **93**: f. 184 (1968).

Leaf-stalks to 35 cm, shallowly channelled, slightly glaucous. Leaf-blades to *c.* 20 × 15 cm, broadly ovate to almost circular, cupped, abruptly acuminate, cordate at base, not decurrent, stiff and leathery, puckered, dark bluish green, densely glaucous below, less so above, with 10–12 pairs of side-veins. Flowering stem 30–45 cm, barely exceeding leaves, glaucous; bracts stiff, purplish. Flowers 4–4.5 cm, very pale purple; anthers yellowish. *Japan.* H2. Summer.

Sometimes considered part of *H. sieboldiana* though generally compacter with more cupped leaves; probably close to *H. sieboldiana* var. *elegans.*

'**Variegata**' (*H. tokudama* Maekawa forma *aureonebulosa* Maekawa) has breaks of pale or yellowish green in the middle of the leaf-blades. This name is often misapplied to another unnamed cultivar, which has pale or yellowish green margins and resembles a small form of *H. sieboldiana* 'Aureo Marginata' (*H.* 'Frances Williams').

'**Buckshaw Blue**' (*H. tokudama* × ?) has narrower, intensely glaucous, more deeply cupped leaves.

6. H. sieboldiana Engler (*H. glauca* (Siebold) Stearn; *H. caerulea* misapplied; *H. sieboldii* misapplied). Illustration: *Botanical Magazine*, 3663 (1838); *Gentes Herbarum* **2**: 131 & 132 (1930); *Journal of the Faculty of Science, Tokyo (Botany)* **5**: 369 & 370 (1940); *Acta Horti Bergiani* **16**: pl. 1, 2 & 10 (1954).

Leaf-stalks to 70 cm, deeply channelled, not winged, glaucous. Leaf-blades 25–36 × 14.5–32 cm, ovate (some very broadly), acuminate, rounded to cordate at base, shortly decurrent, leathery, puckered, margin with many shallow or a few deeper undulations, mid-green and glaucous above, paler, greyish green and more densely glaucous below, with 12–14 pairs of side-veins. Flowering stems to 50 cm, arched above, about equalling leaves; bracts 2.5–6.5 cm, pale yellowish green, lanceolate, persistent. Flowers dense, 5–5.7 cm, white to pale greyish purple; anthers pale yellow. *Japan*. H2. Early summer.

Var. **elegans** Hylander has rounder, more abruptly tipped, greyer leaves with deeper puckering, similar to *H. tokudama* though less cupped. Illustration: *Acta Horti Bergiani* **16**: pl. 3 & 4 (1954); *Journal of the Royal Horticultural Society* **85**: f. 103 (1960); Thomas, Plants for ground cover, pl. XIIA & XXXII & f. 32 (1970).

'**Aureo Marginata**' (*H.* 'Frances Williams'; *H.* 'Gold Edge', not *H.* 'Gold Edger') has an irregular broad margin up to 4 cm wide, of cream, pale yellow or yellowish green to the leaves. Illustration: Thomas, Plants for ground cover, f. 33 (1970); *Journal of the Royal Horticultural Society* **107**: 197 (1982).

'**Snowden**' (*H. sieboldiana* Engler × *H. fortunei* (Baker) Bailey var. *albopicta* (Miquel) Hylander forma *aurea* (Wehrhahn) Hylander) has mid-green to dark green, ovate to oblong, puckered leaves, 28.5–35 × 17.6–20 cm, with 10 or 11 pairs of side-veins and rounded bases which are more decurrent than in *H. sieboldiana*.

7. H. fortunei (Baker) Bailey (*H. sieboldiana* Engler var. *fortunei* Ascherson & Graebner).

Leaf-stalks channelled and winged, often glaucous. Leaf-blades lanceolate to broadly ovate, gradually or abruptly acuminate, rounded to cordate at base, decurrent, membranous to slightly leathery, flat or with a few shallow undulations, sometimes puckered, mid-green to dark green, glaucous below and sometimes above, often variegated, with 7–10, occasionally to 13, pairs of side-veins. Some clones are bright yellow to pale yellowish green in spring but become darker green later in the season. Flowering stem just exceeding or well clear of the leaves, usually with 1 or 2 leaf-like bracts below and many densely packed flowers; bracts purplish, often glaucous, boat-shaped. Flowers pale violet, rarely reddish, with narrow lobes; anthers bluish. *Origin uncertain*. H2. Summer.

This name has long been applied to a group of similar cultivars and varieties, most of which are probably the result of hybridisation in cultivation both in Japan and Europe, with *H. sieboldiana* being a likely parent. *H. fortunei* is therefore a 'species of convenience' (Brickell, 1968) rather than a true botanical species. Among the most popular cultivars and varieties are:

Var. **albopicta** (Miquel) Hylander (*H. fortunei* 'Aureomaculata'; *H. lancifolia* (Thunberg) Engler var. *aureomaculata* Wehrhahn). Illustration: *Acta Horti Bergiani* **16**: pl. 6 & 8 (1954); Thomas, Plants for ground cover, f. XIA (1970). Leaves in spring yellow to pale greenish yellow with irregular mid-green to dark green margins to 2 cm, and dark cross-veins, becoming wholly green later in the season, though margin remaining slightly darker. Leaf-blades 17–20 × 10–12 cm, ovate, slightly acuminate, rounded at base, slightly leathery, with a few shallow undulations, dull above, slightly glaucous below, with 8 or 9 pairs of side-veins. Some recently developed seedlings or budsports such as '**Gold Leaf**' and '**Gold Standard**', which also have dark margins, keep their intense yellow colour to the end of the season.

Var. **albopicta** (Miquel) Hylander **forma aurea** (Wehrhahn) Hylander (*H. fortunei* 'Albopicta Aurea'; *H. fortunei* 'Aurea'; *H. lancifolia* (Thunberg) Engler var. *aurea* Wehrhahn). Illustration: Thomas, Plants for ground cover, f. XIB (1970). Similar to var. *albopicta* but lacking the darker green margin to the leaf and becoming pale green later in the season. Leaf-stalks to 45 cm; leaf-blades 19–26 × 12–18 cm, lanceolate to ovate or narrowly elliptic, acute, very slightly acuminate, rounded or tapered at base, membranous, slightly puckered, dull or slightly glaucous above, more strongly glaucous below.

Var. **hyacinthina** Hylander (*H. fortunei* var. *stenantha* misapplied; *H. glauca* misapplied). Illustration: *Acta Horti Bergiani* **16**: pl. 5, 9 & 10 (1954); Thomas, Plants for ground cover, f. 32 (1970); Thomas, The art of planting, f. 50 (1984). Leaves not variegated, very glaucous below, less so above, with a very narrow 'pencil-line' of white along the margin, appearing bluish to greyish green *en masse*. Leaf-blades 18.5–22 × 14–19 cm, broadly ovate, shortly acuminate, rounded to cordate at base, slightly leathery, flat or with 1 or 2 shallow undulations, slightly puckered, with 9 or 10 pairs of side-veins.

'**North Hills**' has dark green leaves with a narrow white margin to *c.* 8 mm wide. Leaf-blades 20–25 × 10–14 cm, lanceolate to oblong, shortly acuminate, rounded to slightly cordate at base, dull to slightly shiny on both surfaces, with 7 or 8 pairs of side-veins (Figure 1(10), p. 29).

Var. **obscura** Hylander. Illustration: *Acta Horti Bergiani* **16**: pl. 7 (1954). Leaves not variegated, dark green and lustrous above, paler and slightly glaucous below. Leaf-blades 20–23 × 14–17 cm, very broadly ovate to almost circular, acuminate or with a short abrupt tip, cordate at base, leathery, with *c.* 8 pairs of side-veins. Flowering stems much taller than leaf-mound.

Var. **obscura** Hylander forma **marginata** Hylander (*H. fortunei* 'Obscura-Marginata'; *H. fortunei* 'Aureo-marginata'). Illustration: Thomas, Plants for ground cover, f. 31 (1970); Thomas, The art of planting, f. 43 (1984). Differs from var. *obscura* in having leaves with irregular yellow margins, becoming paler later in the season. A form with pure white margins is known as '**Marginato-Alba**'.

Var. **rugosa** Hylander. Illustration: *Acta Horti Bergiani* **16**: pl. 5 (1954). Leaves strongly puckered between the veins, slightly glaucous above, more densely glaucous below. Leaf-blades to 30 × 23 cm, ovate, shortly acuminate, cordate at base, with up to 10 shallow undulations, leathery, with 8–13 pairs of side-veins.

Var. **stenantha** Hylander (*H. fortunei* var. *hyacinthina* misapplied). Illustration: *Acta Horti Bergiani* **16**: pl. 7 & 8 (1954).

Similar to var. *rugosa* but leaves less puckered. Often confused with var. *hyacinthine*, from which it differs in its less rigid leaves, which are dull or glaucous only in vein-depressions, and lacking the marginal 'pencil-line'. In full sun the leaves can become rather yellowish green. Leaf-blades to 35 × 25 cm, broadly ovate, acuminate, cordate at base, slightly leathery, with 6–8 shallow undulations and 9–11 pairs of side-veins. Flowering stems barely exceeding leaf-mound.

8. H. hypoleuca Murata. Illustration: *Acta Phytotaxonomica et Geobotanica* **19**: 67 (1962).

Leaf-stalks to 35 cm, deeply channelled, glaucous, purple-spotted below. Leaf-blades 20–45 × 10–30 cm, broadly ovate, obtuse or very slightly acuminate, cordate at base, not or slightly decurrent, leathery, with up to 20 small marginal wrinkles, yellowish green and slightly shiny above, very densely white-bloomed below, with 9–12 pairs of side-veins. Flowering stems not exceeding leaves, 25–35 cm, arched or bent sideways, bearing 10–18 flowers; bracts 1.4–2 cm, ovate to oblong, very pale. Flowers 3.5–4.5 cm, white; anthers *c.* 3 mm, yellow to purple. *Japan.* H2. Summer.

9. H. longipes (Franchet & Savatier) Bailey (*H. sieboldiana* var. *longipes* (Franchet & Savatier) Matsumura; *H. minor* misapplied). Illustration: *Gentes Herbarum* **2**: 138 (1930); *Journal of Japanese Botany* **13**: 901 (1937); *Journal of the Faculty of Science, Tokyo (Botany)* **5**: 386 & 387 (1940).

Leaf-stalks to 16 cm, broadly channelled, winged in upper half, purple-spotted. Leaf-blades to 14 × 8–10 cm, ovate to elliptic, shortly acuminate, truncate or slightly cordate at base, decurrent, membranous, flat or with a few undulations, mid-green to dark green, dull above, shiny below, with 6–8 pairs of side-veins. Flowering stems to *c.* 30 cm, usually much exceeding leaves, often branched, purple-spotted; bracts membranous, pale purple. Flowers *c.* 4 cm, pale violet; anthers *c.* 2.5 cm, bluish purple. *Japan.* H2. Late summer–autumn.

10. H. tardiflora (Irving) Stearn (*Funkia tardiflora* Irving; *H. lancifolia* (Thunberg) Engler var. *tardiflora* (Irving) Bailey; *H. sparsa* Nakai). Illustration: *Acta Horti Bergiani* **16**: pl. 20 (1954); *Botanical Magazine*, n.s., 204 (1953); *Journal of the Royal Horticultural Society* **93**: f. 186 (1968).

Leaf-stalks 12–27 cm, very shallowly and broadly channelled, not winged, strongly purple-spotted. Leaf-blades 10–15 × 2.8–6.5 cm, lanceolate to narrowly elliptic, acuminate, tapered at base, not decurrent, firm and leathery, flat or with 1–3 very shallow undulations, dull mid-green to yellowish green above, dull to slightly shiny mid-green below, with 4 or 5 pairs of side-veins. Flowering stems to *c.* 35 cm, ascending at *c.* 45°, bearing up to 50 flowers; bracts 1–1.5 cm, ovate to oblong, pale purple. Flowers 4–4.5 cm, pale bluish purple; anthers 3.5–4 mm, greyish purple. *Known only in cultivation.* H2. Autumn.

11. H. × tardiana Anon; *H. tardiflora* (Irving) Stearn × *H. sieboldiana* Engler var. *elegans* Hylander. Illustration: *Journal of the Royal Horticultural Society* **107**: 198 (1982).

Differs from *H. tardiflora* in its thicker, often broader, very glaucous leaves, which have 6–9 pairs of side-veins and very tough stalks. Flowers pale violet. Differs from *H. sieboldiana* var. *elegans* in its much smaller, narrower, unpuckered leaves. *Garden origin.* H2. Late summer.

This cross has given rise to many named and unnamed clones. Among the bluest and most popular are '**Halcyon**' and '**Harmony**'.

12. H. tortifrons Maekawa (*H. lancifolia* (Thunberg) Engler var. *tortifrons* invalid). Figure 1(6), p. 29. Illustration: *Journal of the Faculty of Science, Tokyo (Botany)* **5**: 323 (1940).

Leaf-stalks less than 5 cm, almost flat, slightly spotted. Leaf-blades to *c.* 7 × 1.3 cm, oblong to very narrowly elliptic, tapered at base and tip, slightly decurrent, leathery, often twisted through *c.* 90°, slightly shiny dark green above, shiny mid-green below, with 2 or 3 pairs of very indistinct side-veins. Flowers not seen, and not described by Maekawa. *Japan.* H2.

13. H. rupifraga Nakai (*H. longipes* (Franchet & Savatier) Bailey var. *latifolia* Maekawa). Illustration: *Journal of the Faculty of Science, Tokyo (Botany)* **5**: 390 (1940); *Botanical Magazine, Tokyo* **44**: 28 (1930).

Leaf-stalks to 12(23) cm, shallowly channelled, winged above, purple-spotted below. Leaf-blades 9–13(16) × 4–8.5(12) cm, ovate, acuminate, rounded to truncate and funnel-shaped at base, shortly decurrent, leathery, with a few marginal undulations, mid-green to dark green and dull above, slightly paler and shiny below, with 6–8 pairs of side-veins. Flowering stems to *c.* 50 cm, bearing many flowers; bracts pale purple. Flowers 3.2–4.1 cm, very pale violet. *Japan.* H2. Late summer.

One clone in cultivation under this name has much larger leaves than those described in previous publications (including the original description). These sizes are given in brackets. The name is often wrongly spelled 'ruprifraga'.

14. H. gracillima Maekawa (*H. longipes* (Franchet & Savatier) Bailey var. *gracillima* (Maekawa) Fujita). Figure 1(3), p. 29. Illustration: *Iconographia Plantarum Asiae Orientalis* **1**(4): pl. 33 (1936); *Journal of the Faculty of Science, Tokyo (Botany)* **5**: 396 (1940).

Leaf-stalks 2–8 cm, very narrow, shallowly channelled, slightly winged above, spotted on back from bottom to top. Leaf-blades 2.5–6.2 × 1.1–3 cm, lanceolate to ovate, acute or slightly acuminate, rounded to slightly tapered at base, slightly decurrent, membranous to slightly leathery, usually with several small marginal wrinkles, rarely flat, dark green, dull, above, slightly shiny below, with *c.* 3 pairs of side-veins. Flowering stems erect or spreading, to 25 cm, bearing *c.* 10 flowers; bracts 1–2 cm, narrowly triangular. Flowers 3–3.5 cm, pale pinkish purple. *Japan.* H2. Late summer–early autumn.

This name is widely misapplied to several clones with a range of leaf-shapes and sizes, particularly to a probably unnamed clone with narrow, wrinkled, almost oblong leaf-blades to *c.* 18 cm, and leaf-stalks to 15 cm. This is probably due to the name 'gracillima' being loosely applied to many dwarf or slender hybrids and sports.

15. H. sieboldii (Paxton) Ingram (*Hemerocallis sieboldii* Paxton; *Hosta lancifolia* (Thunberg) Engler var. *albomarginata* (Hooker) Stearn; *H. albomarginata* (Hooker) Ohwi; *H. sieboldiana* misapplied). Illustration: *Botanical Magazine*, 3657 (1838); *Journal of the Royal Horticultural Society* **85**: f. 103 (1960); Thomas, Plants for ground cover, f. 29 (1970).

Leaf-stalks to 26 cm, shallowly channelled, broadly winged, often spotted. Leaf-blades to 15 × 6 cm, lanceolate to elliptic, acute or obtuse, not acuminate, tapering at base, decurrent, membranous, flat or with a few broad undulations, mid-green (yellowish in full sun), dull above, shiny below, usually with a narrow cream-coloured or very pale yellow margin (many clones have plain green leaves), with *c.* 4 pairs of side-veins. Flowering stems erect, slender, to *c.* 30 cm, exceeding leaves, with a few small leaf-like bracts below; upper bracts green, concave. Flowers 5–6 cm, bluish purple or white; anthers yellow. *Japan.* H2. Summer.

Plants with white flowers and plain leaves are known as **H. sieboldii 'Alba'** (*H. albomarginata* (Hooker) Ohwi var. *alba* (Robinson) Hylander). Illustration: *Acta Horti Bergiani* **16**: pl. 19 (1954). All combinations of white or purple flowers with plain or variegated leaves are found in cultivation. The variegated form has not been found in the wild. Several clones similar to *H. sieboldii*, though generally smaller, are becoming popular; among these are **'Louisa'**, **'Ginko Craig'** and **'Emerald Isle'**.

Species 15–19 are sometimes considered to be a single species, *H. sieboldii*.

16. H. lancifolia (Thunberg) Engler (*H. japonica* (Thunberg) Ascherson & Graebner not Redouté; *Hemerocallis lancifolia* Thunberg; *Hosta lancifolia* (Thunberg) Engler var. *thunbergii* Stearn). Illustration: *Acta Horti Bergiani* **16**: pl. 17 (1954); *Journal of the Royal Horticultural Society* **85**: f. 102 (1960); Thomas, Plants for ground cover, f. 28 (1970).

Leaf-stalks to 31 cm, broadly channelled, winged above, purple-spotted. Leaf-blades 10–17 × 4.7–8.5 cm, lanceolate to elliptic, acute, slightly acuminate, shortly tapered at base, shortly decurrent, membranous to slightly leathery, flat or with a few undulations, shiny dark green above, very shiny and slightly paler below, with 5 or 6 pairs of side-veins. Flowering stems to 45 cm, erect, well exceeding leaf-mound, with 2 or 3 sterile bracts and 20–30 flowers; upper bracts 1.5–3 cm, shiny green. Flowers *c.* 5.3 cm, strong bluish purple; anthers bluish purple. *Japan.* H2. Late summer–autumn.

Clones with slenderer, more acuminate leaves have been called var. *gracilis* Anon.

'Kabitan' (*H. lancifolia* forma *kabitan* Maekawa). Illustration: *Journal of the Faculty of Science, Tokyo (Botany)* **5**: 401 (1940). This is smaller in all its parts and has leaves that are yellow to pale yellowish green with an irregular dark green margin. Leaf-stalks to 10 cm; leaf-blades 8–10.5 × 2.5–3.5 cm, elliptic, acute, tapered at base, membranous, with a few strong undulations, dull above, shiny below, with *c.* 3 pairs of side-veins. (Figure 1(2), p. 29).

'Wogon Gold' with entirely golden yellow, broader, elliptic, flimsy leaves and broadly winged leaf-stalks may be a 'Kabitan' hybrid.

17. H. longissima (Honda) Honda (*H. lancifolia* (Thunberg) Engler var. *longifolia* Honda; *H. lanceolata* invalid). Figure 1(5), p. 29. Illustration: *Acta Horti Bergiani* **16**: pl. 21 (1954).

Leaf-stalks to 22 cm, shallowly channelled, narrowly winged, not spotted. Leaf-blades to *c.* 20 × 4 cm, narrowly elliptic to oblong, acute, gradually tapered at base, narrowly decurrent, slightly leathery, flat, mid-green to dark green, dull above, slightly shiny below, with 3 or 4 pairs of side-veins. Flowering stems to 55 cm, well exceeding leaves, erect, slender, few-flowered; bracts shiny, green, concave. Flowers *c.* 3.3 cm, pale reddish. *Japan.* H2. Summer.

Var. **brevifolia** Maekawa (*Funkia lancifolia* (Thunberg) Engler var. *angustifolia* Gaerdtner) is described as having narrower leaf-blades to *c.* 19 × 3 cm, and smaller flowers, but may not be distinct.

18. H. helonioides Maekawa. Illustration: *Journal of Japanese Botany* **13**: 896 (1937); *Journal of the Faculty of Science, Tokyo (Botany)* **5**: 402 (1940).

Leaf-stalks *c.* 16 cm, channelled, broadly winged. Leaf-blades to *c.* 12 × 5 cm, elliptic, acute to obtuse, tapered at base, strongly decurrent, mid-green, dull above, slightly shiny below, thin and membranous, flat or with a few undulations, with 3 or 4 pairs of side-veins. Flowering stems well exceeding leaves, purple-spotted below; lower bracts clasping stem. Flowers *c.* 5 cm, pale bluish purple; anthers *c.* 3.2 mm, blue. *Japan.* H2. Late summer.

Most commonly seen in cultivation as forma **albopicta** Maekawa, with irregular pale yellow leaf-margins to 1.5 cm wide. (Figure 1(7), p. 29).

19. H. rectifolia Nakai (*H. longipes* misapplied). Figure 1(8), p. 29. Illustration: *Journal of the Faculty of Science, Tokyo (Botany)* **5**: 411 (1940); *Acta Horti Bergiani* **16**: pl. 21 (1954); *Botanical Magazine*, n.s., 138 (1951).

Leaf-stalks to *c.* 50 cm, shallowly channelled, broadly winged, unspotted. Leaf-blades erect or steeply ascending, to *c.* 25 × 12 cm, narrowly ovate to elliptic, shortly acuminate, tapered at base, decurrent, slightly leathery, dull mid-green above and below, with 6–9 pairs of side-veins. Flowering stems to *c.* 60 cm, sometimes spotted, bearing 15–35 flowers; bracts green, concave. Flowers to *c.* 4.8 cm, violet; anthers *c.* 3 mm, yellow. *Japan & Sakhalin.* H2. Summer–early autumn.

Many of the clumps seen in British gardens are somewhat mixed, containing some plants with much broader, cordate leaves. These do not conform to the original description of *H. rectifolia*. Some clones have fewer pairs of side-veins.

'Tall Boy' (*H. rectifolia* × ?) is larger in all its parts, with leaves to *c.* 15 cm wide and 8 or 9 pairs of side-veins; flowering stems to *c.* 80 cm. Illustration: Thomas, Perennial garden plants, pl. 140 (1976).

20. H. decorata Bailey (*H. decorata* var. *marginata* Stearn; *H.* 'Thomas Hogg' misapplied). Figure 1(9), p. 29. Illustration: *Gardeners' Chronicle* **90**: 88 (1931); *Botanical Magazine*, 9395 (1935).

Leaf-stalks to 30 cm, broadly channelled, winged, spotted at base. Leaf-blades to 16 × 11 cm, broadly ovate to broadly elliptic, obtuse, not acuminate, rounded to tapered at base, decurrent, membranous to slightly leathery, flat or with a few shallow undulations, mid-green to dark green with an irregular white to cream-coloured margin to *c.* 1.5 cm, dull above, slightly shiny below, with 5–7 pairs of side-veins. Flowering stems to *c.* 60 cm, occasionally with 1 or 2 leaf-like bracts below, bearing up to 25 flowers. Flowers 4–5 cm, dark lilac, paler at base; anthers *c.* 4 mm, yellow. *Known only in cultivation.* H2. Late summer.

Var. **normalis** Stearn has plain green leaves that are sometimes slightly acuminate. Illustration: *Journal of the Faculty of Science, Tokyo (Botany)* **5**: 418 (1940).

21. H. ventricosa Stearn (*Funkia ovata* Sprengel; *H. caerulea* (Trattinick) Andrews, not Jacquin). Illustration: *Botanical Magazine*, 894 (1806); *Gentes Herbarum* **2**: 128 (1930); Parey's Blumengärtnerei **2**: 241 (1958).

Leaf-stalks to 40 cm, broadly channelled, winged, spotted below. Leaf-blades to 25 × 18 cm, broadly ovate to elliptic or circular, abruptly acuminate, rounded to truncate or slightly cordate at base, decurrent, leathery, flat or with several shallow undulations, dull dark green with a metallic lustre above, very shiny mid-green to dark green below, with 7–9 pairs of side-veins, and conspicuous small transverse veins. Flowering stems to *c.* 90 cm, bearing 20–30 flowers; bracts 1.6–1.8 cm. Flowers 5.5–5.7 cm, bell-shaped, bluish purple; anthers *c.* 5 mm, purple-spotted. *China.* H2. Summer.

'**Aureomaculata**' has leaves variegated with patches of yellow or yellowish green that become darker and less distinct later in the season.

'**Variegata**' ('Aureomarginata') has broad, irregular, cream-coloured to yellow margins. Illustration: *Journal of the Royal Horticultural Society* **93**: f. 185 (1968); Thomas, Plants for ground cover, f. 34 (1970).

22. H. minor (Baker) Nakai (*Funkia ovata* Sprengel var. *minor* Baker). Figure 1(1), p. 29. Illustration: *Gentes Herbarum* **2**: 140 (1930); *Journal of the Faculty of Science, Tokyo (Botany)* **5**: 419 (1940).

Leaf-stalk to *c.* 8 cm, broadly channelled, winged. Leaf-blades *c.* 8.5 × 5.5 cm, broadly ovate, acute to obtuse, very slightly acuminate, rounded to truncate at base, decurrent, membranous, with very shallow undulations or none, dark green, slightly lustrous, with 5 or 6 pairs of side-veins. Flowering stems to *c.* 60 cm, bearing up to 10 flowers. Flowers 4.5–5 cm, bluish purple. *Japan & Korea.* H2.

Forma **alba** (Nakai) Maekawa has white flowers.

The name *H. minor* is much misused in gardens and reference works for diminutive varieties of several species. True *H. minor* is probably not commonly grown, and was originally considered a variety of *H. ventricosa.*

23. H. venusta Maekawa. Illustration: *Journal of the Faculty of Science, Tokyo (Botany)* **5**: 415 (1940); *Botanical Magazine*, n.s. 499 (1967).

Leaf-stalks to *c.* 6.5 cm, slender, channelled, winged above, spotted with pinkish purple below. Leaf-blades to 6.5 × 2.7 cm, lanceolate to narrowly elliptic, acute, slightly acuminate, rounded or tapered at base, decurrent, membranous to slightly leathery, with 1 or 2 broad, shallow undulations, mid-green, dull above, shiny below, with 4–6 pairs of side-veins. Flowering stems 10–35 cm, erect, with 3–8 flowers; bracts to 1 cm, ovate to elliptic. Flowers 2.5–3.2 cm, bluish purple; anthers *c.* 2.7 mm, greyish purple. *Quelpaert Island (Korea) & Japan.* H2. Summer.

24. H. nakaiana Maekawa. Illustration: *Journal of Japanese Botany* **11**: 688 (1935); *Journal of the Faculty of Science, Tokyo (Botany)* **5**: 419 (1940), flowers only.

Leaf-stalks to *c.* 10(21) cm, deeply channelled, winged, spotted below. Leaf-blades 6–7.5(14) × 2.5–4.5(8) cm, broadly ovate, acuminate to attenuate, truncate at base, decurrent, slightly leathery, with several small marginal undulations, mid-green, dull above, slightly glaucous below, with 5–7 pairs of side-veins. Flowering stem to 45 cm, much exceeding leaves, erect or curved; bracts 1.4–1.7 cm, overlapping, elliptic to circular. Flowers in dense almost spherical heads, 4.6–5.2 cm, pale bluish purple; anthers white, 3–3.2 mm. *Japan & Korea.* H2. Summer.

At least one clone in cultivation under this name has leaves considerably larger than those described in previous publications. These measurements are given in brackets.

'**Golden Nakaiana**' (*H. nakaiana* × ?) has broader, more leathery, slightly cordate leaves that are bright greenish yellow throughout the season, and have a rough colourless margin. (Figure 1(4), p. 29).

'**Pastures New**' (*H. nakaiana* × ?) has small, flat, rather bluish, slightly acuminate, leathery leaves to *c.* 10 × 6 cm.

25. H. capitata (Koidzumi) Nakai (*H. coerulea* (Trattinick) Andrews var. *capitata* Koidzumi).

Similar to *H. nakaiana* though generally larger, with abruptly acuminate leaves that are cordate rather than truncate at the base, and have 7–9 pairs of side-veins. *Japan.* H4. Early summer.

26. H. undulata (Otto & Dietrich) Bailey.

Var. **undulata** (*H. japonica* (Thunberg) Ascherson & Graebner forma *undulata* (Otto & Dietrich) Voss 'Albo-Variegata'; *H. lancifolia* (Thunberg) Engler var. *undulata* Bailey 'Medio-Variegata'). Illustration: *Gentes Herbarum* **2**: 136 (1930); *Journal of the Royal Horticultural Society* **85**: f. 105 (1960), **107**: 197 (1982). Leaf-stalks to 25 cm, deeply channelled, winged, sometimes faintly spotted below, striped. Leaf-blades to 17 × 6.8 cm, lanceolate to elliptic, acute, slightly acuminate, tapered at base, decurrent, membranous to slightly leathery, with a few strong undulations, tip sometimes twisted, mid-green with a large irregular area of white or very pale yellow in the middle, dull above, very shiny below, with 6 or 7 pairs of side-veins. Flowering stems to 1 m, bearing large, leaf-like bracts below, and up to 10 flowers above; upper bracts 2–2.6 mm, ovate to elliptic, variegated. Flowers 5.5–6 cm, pale bluish purple; anthers purple. *Known only in cultivation.* H2. Summer.

Var. **univittata** (Miquel) Hylander. Illustration: *Acta Horti Bergiani* **16**: pl. 15 (1954); Thomas, Plants for ground cover, f. 27 (1970). Very similar though generally more vigorous. The pale area is confined to the centre of the leaf and ends in a narrow stripe towards the leaf-tip. Plants often sport back to var. *undulata.*

Var. **erromena** (Stearn) Maekawa (*H. erromena* Stearn; *H. japonica* (Thunberg) Ascherson & Graebner var. *fortis* Bailey). Illustration: *Journal of the Faculty of Science, Tokyo (Botany)* **5**: 422 & 423 (1940); *Acta Horti Bergiani* **16**: pl. 14 & 16 (1954). Differs from var. *undulata* in its larger, plain mid-green leaves. Leaf-stalks to 50 cm. Leaf-blades to 22.5 × 13 cm, mid-green, with 8 or 9 pairs of side-veins. Flowering stems to 1.5 m.

Var. **albo-marginata** Maekawa (*H.* 'Thomas Hogg'). Illustration: *Journal of the Faculty of Science, Tokyo (Botany)* **5**: 423 (1940); Thomas, Plants for ground cover, pl. XIB (1970). Leaves broader than in var. *undulata,* flat or with a few very shallow undulations, dark green with a narrow cream-coloured to pale yellow margin to *c.* 1 cm wide, dull above, shiny below.

H. undulata, like *H. fortunei,* is a group of related, largely sterile clones, probably hybrid in origin, with *H. decorata* being a possible parent.

The name *H. japonica* is still sometimes encountered in gardens, catalogues and reference works. Its correct application was unclear from the time it was first published, and it was never unequivocally linked to a particular species.

12. HEMEROCALLIDACEAE

Plant terrestrial, entirely herbaceous, usually with a rhizome. Leaves all basal, without broadened blades and petiole-like bases, not fleshy and succulent, nor reduced to basal sheaths. Flowers in racemes, spikes or panicles, bisexual, actinomorphic or slightly zygomorphic. Perianth of 6 segments united into a tube below, funnel-shaped. Stamens 6, declinate. Ovary of 3 united carpels, superior, nectar secreted on the ovary; ovules many, placentation axile. Fruit a capsule.

As treated here, a family of a single genus with 15–20 species, native to Europe, many species and hybrids cultivated for ornament. Genera here treated as Phormiaceae (p. 194) are sometimes combined with the Hemerocallidaceae.

1. HEMEROCALLIS Linnaeus
E. Campbell & J. Cullen
Herbaceous perennials with rhizomes, roots fibrous, often swollen and fleshy at their ends. Leaves in 2 ranks, deciduous or persisting well into winter, linear, tapered gradually to the apex, usually recurved, flat or folded at the base. Flowers in a raceme or panicle borne on a scape, bracts small or large. Perianth united into a tube at the base, more or less funnel-shaped or openly funnel-shaped, 6-lobed, yellow or orange, rarely reddish. Stamens 6, deflexed downwards, attached to the top of the perianth-tube. Ovary superior, 3-celled, with many ovules, style thread-like. Fruit a 3-angled or 3-winged capsule.

A genus of about 15 species from E Asia, cultivated in China and Japan for several centuries. Its classification is difficult, as many of the plants in cultivation (as well as plants naturalised from gardens in China and Japan) are selections or hybrids of unknown parentage. Hybridisation has continued and many cultivars are available (see the list cited below). Twelve species are included here, and many of these are probably involved in the parentage of both old and new cultivated hybrids.

Literature: Bailey, L.H., Hemerocallis: the day lilies, *Gentes Herbarum* **2**: 143–56 (1930); Stout, A.B., Daylilies (1934); Stuntz, M.F. et al., Hemerocallis Checklist, 1893–1957 (1957).

Leaves. Persisting well into winter: **9**; erect or spreading, not recurved or bent over in the upper half: **7**.
Scape. Branched above: **3, 4, 5, 6, 7, 8, 9, 10**; not branched above: **1, 2**.
Bracts. Large and cup-shaped: **1, 2**; small, not cup-shaped: **3, 4, 5, 6, 7, 8, 9, 10**.
Flowering time. Spring: **2, 3, 4, 7**; summer: **1, 5, 6, 8, 9, 10**.
Flowers. Yellow: **1, 3, 4, 5, 6**; orange, brownish orange or reddish: **2, 7, 8, 9, 10**; more than 20 in each inflorescence: **5, 8**; less than 20 in each inflorescence: **1, 2, 3, 4, 6, 7, 9, 10**; fragrant: **4, 5, 6, 7**; not fragrant: **1, 2, 3, 7, 8, 9, 10**.

1a. Scape not forked or branched above, the flowers apparently stalkless and close together, their stalks hidden by the large, enfolding, cup-shaped bracts **2**
b. Scape forked and sometimes further branched above, or bearing only a single flower, the stalks clearly visible, not hidden by the bracts, which are small and not cup-shaped **3**
2a. Flowers 8–10 cm; inner perianth-lobes 2–2.5 cm wide **1. middendorffii**
b. Flowers 5–7 cm; inner perianth-lobes less than 2 cm wide **2. dumortieri**
3a. Flowers yellow **4**
b. Flowers orange, brownish orange or reddish **7**
4a. Tube *c.* one-quarter of the total length of the perianth; spring-flowering **5**
b. Tube of perianth longer, about one-third to half of the total length; summer-flowering **6**
5a. Leaves 30–45 cm × 5–9 mm; flowers not fragrant **3. minor**
b. Leaves 50–65 × 1–1.4 cm; flowers fragrant **4. lilio-asphodelus**
6a. Flowers opening at night; leaves robust, 1–1.6 cm broad; raceme with 20–65 flowers **5. citrina**
b. Flowers opening in the morning; leaves 5–10 mm broad; raceme with 3–15 flowers **6. thunbergii**
7a. Leaves rather stiff, erect or arching but not bent over in the upper half, to 30 cm; scape not or scarcely exceeding leaves **7. forrestii**
b. Leaves flaccid, arching and bent over in the upper half, 30 cm or more; scape clearly exceeding the leaves **8**
8a. Scape with many long branches bearing 20–75 flowers, glaucous above **8. multiflora**
b. Scape forked, but without other long branches, bearing up to 20 flowers, not glaucous above **9**
9a. Leaves persisting well into winter; flowers orange, sometimes flushed with red-purple but without bands or stripes of a darker colour **9. aurantiaca**
b. Leaves dying down in autumn; flowers brownish orange or rarely reddish, usually with bands or stripes of darker colour **10. fulva**

1. H. middendorffii Trautvetter & Meyer. Illustration: *Addisonia* **14**: t. 463 (1929).

Roots mostly fibrous, some slightly swollen. Leaves recurved, soft, to 30 × 1–2.5 cm. Scape erect, exceeding the leaves. Flowers few, in a compressed raceme, the flower-stalks hidden by the cup-shaped, enfolding bracts. Flowers 8–10 cm, yellow, inner perianth-lobes 2–2.5 cm broad; tube to 2 cm. *Japan, Former E USSR, Korea, N China.* H2. Summer.

2. H. dumortieri Morren. Illustration: *Addisonia* **14**: t. 462 (1929); Bloom, Perennials for your garden, 80 (1975).

Roots mostly swollen. Leaves recurved, to 35 × 1.5 cm. Scapes shorter than or slightly longer than the leaves. Flowers few in a compressed raceme, the flower-stalks hidden by cup-shaped bracts. Flowers 5–7 cm, orange-yellow, the inner perianth-lobes less than 2 cm broad; tube *c.* 1 cm. *Japan, Former E USSR, Korea.* H2. Spring.

3. H. minor Miller (*H. graminea* Andrews; *H. graminifolia* Schlechtendahl). Illustration: *Botanical Magazine*, 873 (1805); *Addisonia* **14**: t. 458 (1929).

Roots mostly fibrous, rarely a few swollen. Leaves recurved, 30–45 cm × 5–9 mm. Scapes exceeding the leaves, branched above at least once, erect, raceme with 2–5 flowers. Flowers 5–7 cm, yellow; tube about one-quarter of the total length, lobes narrow. *Japan, China, Former E USSR.* H2. Spring.

4. H. lilio-asphodelus Linnaeus (*H. flava* Linnaeus). Illustration: *Botanical Magazine*, 19 (1788); *Addisonia* **14**: t. 456 (1929); Brickell (ed.), RHS A–Z encyclopedia of garden plants, 524 (2003).

Rhizome spreading. Roots mostly swollen. Leaves recurved, 50–65 × 1–1.4 cm. Scapes ascending, exceeding the leaves, branched above, raceme with 8–12 flowers. Flowers fragrant, 7–8 cm, yellow; tube about one-quarter of the total length. *E Asia.* H3. Spring.

'Major', with larger flowers, darker leaves and wavy perianth-lobes, reflexed at their tips, is widely grown.

5. H. citrina Baroni. Illustration: *Addisonia* **15**: t. 482 (1930); Brickell (ed.), RHS A–Z encyclopedia of garden plants, 523 (2003).

Roots very fleshy. Leaves recurved, 70–80 × 1–1.6 cm. Scapes erect, exceeding the leaves, branched above, raceme with 20–65 flowers. Flowers fragrant, opening at night, 9–12 cm, lemon yellow; tube about one-third of the total length, lobes narrow. *China.* H3. Summer.

'Baroni' is often referred to *H. citrina* but correctly covers a hybrid between it and *H. thunbergii;* it is like *H. citrina* but has broader perianth-lobes.

6. H. thunbergii Baker. Illustration: *Addisonia* **14**: t. 459 (1929); Kitamura et al., Herbaceous plants of Japan **3**: 241 (1978).

Roots slender, cylindric, sometimes slightly swollen. Leaves recurved, 30–60 cm × 5–8 (rarely to 10) mm. Scapes erect, branched above, exceeding the leaves, raceme with 3–15 flowers. Flowers fragrant, opening in the morning, 9–11 cm, yellow; tube about one-third of the total length, lobes narrow. *China, Korea.* H2. Summer.

There is a related species, *H. vespertina* Hara, in Japan; this is often included in *H. thunbergii* or *H. citrina,* but has broader perianth-lobes than either. It is uncertain whether or not this is in cultivation in Europe.

7. H. forrestii Diels.

Roots cylindric but markedly swollen at their ends. Leaves erect or ascending, not recurved, to 30 × 1.5 cm. Scapes erect, not or scarcely exceeding the leaves, branched, raceme with 4–8 flowers. Flowers *c.* 7 cm, orange; tube less than one-quarter of the total length. *China (mainly the SW).* H5. Spring.

H. plicata Stapf. Very similar, but leaves narrower and raceme with usually 8 or more flowers. The leaves are reputedly folded at the base, but this requires confirmation. *SW China.* H5. Spring.

H. nana Smith & Forrest. Also similar, but the whole plant much smaller, leaves spreading, raceme with usually 1 (rarely as many as 3) flowers, which are fragrant. *SW China.* H5. Spring.

8. H. multiflora Stout. Illustration: *Addisonia* **14**: t. 464 (1929).

Roots fleshy. Leaves recurved, 60 × 1 cm or more. Scapes erect, exceeding the leaves, glaucous above, richly branched with 20–75 widely spaced flowers. Flowers 7–8 cm, orange; tube about one-quarter of the total length. *China (Honan).* H4. Summer.

9. H. aurantiaca Baker. Illustration: *The Garden* (1895); *Addisonia* **14**: t. 461 (1929); Kitamura et al., Herbaceous Plants of Japan **3**: 244 (1978).

Rhizomes spreading, roots swollen. Leaves persisting well into winter, recurved, 60–80 × 1.5–2.5 cm, glaucous. Scape exceeding the leaves, forked but not extensively branched above, raceme with up to 20 flowers, closely grouped. Flowers 10–13 cm, orange, sometimes flushed with red-purple but without bands or stripes of a darker colour; tube much less than one-quarter of the total length. *China.* H3. Summer.

10. H. fulva Linnaeus. Illustration: *Botanical Magazine*, 64 (1798); *Addisonia* **15**: t. 483, 484 (1930).

Rhizomes spreading. Roots fleshy at their ends. Leaves recurved, 30–90 × 1–2.5 cm. Scapes exceeding the leaves, erect, forked and slightly further branched above, raceme with up to 20 flowers. Flower 7–10 cm, brownish orange or rarely reddish, usually with zones or stripes of darker colour; tube one-quarter to half the total length. *Origin uncertain, perhaps China or Japan, naturalised there and elsewhere.* H2. Summer.

This is the most widely cultivated and variable of the species. Var. **longituba** Miquel (*H. longituba* (Miquel) Maximowicz) has narrower leaves and the perianth-tubes about one-third of the total length. Many cultivars are grown, including 'Rosea' with reddish purple flowers, and 'Kwanso' which has double flowers.

13. BLANDFORDIACEAE

Plants terrestrial, herbaceous. Leaves well developed, mainly basal. Flowers in racemes, bisexual, actinomorphic. Perianth of 6 segments, nectar secreted on the perianth. Stamens 6. Ovary of 3 united carpels, superior, stalked, placentation axile; style 1. Fruit a capsule. Seeds hairy.

A single genus with 4 species, occasionally cultivated.

1. BLANDFORDIA J.E. Smith

S.G. Knees

Deciduous, hairless perennial herbs with fibrous roots and often with rhizomes. Leaves sheathing, linear, triangular in cross-section, often 2-ranked. Perianth tubular to bell-shaped with 6 equal lobes, varying in colour from red to orange and yellow. Scapes solitary, bearing 3–20 bisexual flowers with bracts, in terminal racemes. Stamens 6, included in perianth and attached at or below the middle of the tube. Ovary superior, 3-celled with many ovules in each cell; style 1. Fruit a stalked, 3-celled, ovoid, capsule. Seeds many.

A genus of 4 species, all Australasian in distribution. None is widely cultivated, and one, *B. punicea,* is reputedly difficult to grow. A sandy loam should provide an ideal growing medium. Although readily grown as pot plants, care should be taken to ensure that they do not dry out. This can be facilitated by planting in partial shade. Propagation is usually from seed, but large plants can be divided in early spring.

1a. Leaf-margins smooth **1. cunninghamii**
b. Leaf-margins toothed 2
2a. Flowers 4–6 cm **2. grandiflora**
b. Flowers usually less than 4 cm 3
3a. Leaves 8–12 mm wide; raceme with 15–25 flowers **4. punicea**
b. Leaves 1–5 mm wide; raceme with 3–15 flowers **3. nobilis**

1. B. cunninghamii Lindley. Illustration: *Botanical Magazine*, 6209 (1876).

Leaves with smooth margins, 35–70 cm × 4–9 mm. Scape to 85 cm, bearing 5–20 flowers in compact racemes; flower-stalks 1.5–5 cm, equalling bracts. Perianth 4–5 cm, tubular, brownish red with yellow, obtuse lobes *c.* 3.8 mm long. Stamens attached below middle of perianth-tube. *Australia (New South Wales).* H5. Spring.

2. B. grandiflora R. Brown (*B. flammea* Hooker). Illustration: *Botanical Magazine*, 5374 (1868); Elliot & Jones, Encyclopaedia of Australian plants, 325 (1982); RHS dictionary of gardening, edn 1, 287 (1956); Brickell (ed.), RHS A–Z encyclopedia of garden plants, 183 (2003).

Leaves linear, bearing small rounded teeth, 60–80 cm × 1–4 mm. Scape to 60 cm, bearing 7–10 flowers. Perianth tubular to bell-shaped, 4–6 cm, varying in colour from red with yellow apical lobes to pure yellow. *Australia (Queensland and New South Wales).* H5. Spring–early summer.

3. B. nobilis Smith. Illustration: *Botanical Magazine*, 2003 (1818); Elliot & Jones, Encyclopaedia of Australian plants, 326 (1982); *Edwards's Botanical Register* **4**: t. 286 (1818).

Leaves linear, finely toothed on margins, 40–80 cm × 1–5 mm, crowded at the base.

Scape 40–70 cm, bearing narrowly tubular flowers, 2–3 cm × 8–12 mm, brick red or orange, often streaked with yellow, 3–15 per raceme. Fruits narrowly ovoid, 4–5 cm. *Australia (New South Wales).* H5. Spring–early summer.

4. B. punicea (Labillardière) Sweet (*B. marginata* Herbert). Illustration: *Edwards's Botanical Register* **31**: t. 18 (1845); Elliot & Jones, Encyclopaedia of Australian plants, 326 (1982).

Leaves strap-like, leathery, finely toothed, 70–90 cm × 8–12 mm, coarsely sheathed at base, with prominent midrib. Scape to 80 cm, stout, bearing 3–4 bracts and a loose raceme of 15–25 flowers in almost opposite pairs; flower-stalks 2–5 cm. Perianth 3–4 cm, narrowly bell-shaped, pinkish scarlet, with orange-yellow tips. Stigmas slightly projecting, stamens not projecting. Fruits 8–10 × 3.5–4.5 cm. *Australia (Tasmania).* H5. Summer.

14. ALOACEAE

Plants terrestrial, herbaceous, with rhizomes. Leaves all basal, fleshy and succulent. Flowers in racemes, bisexual, mostly actinomorphic. Perianth of 6 segments in 2 whorls, those of the outer whorl united for part of their length, those of the inner whorl variably joined to the outer, or of 6 free or united segments. Stamens 6. Ovary of 3 united carpels, placentation axile; nectar secreted on the ovary. Fruit a capsule.

A family of 7 genera and about 560 species. Many are cultivated as ornamental succulents.

1a. Perianth segments entirely free 2
b. Perianth segments united, usually at the base 4
2a. Apex of perianth bilaterally symmetric, 2-lipped **1. Haworthia**
b. Apex of perianth radially symmetric, not 2-lipped 3
3a. Perianth white or yellowish green; flowers borne all round the raceme-axis **2. Astroloba**
b. Perianth orange-red; raceme 1-sided **3. Poellnitzia**
4a. Perianth-tube cylindric or swollen evenly at the base **5. Aloe**
b. Perianth-tube swollen 1-sidedly at the base **4. Gasteria**

1. HAWORTHIA Duval

V.A. Matthews

Plants stemless or with a short stem. Leaves numerous, in a rosette or densely clothing the stem, and overlapping, succulent, variable in shape, smooth or with tubercles or ridges; margin often with teeth, often with a keel (sometimes 2 keels in broad leaves), sometimes with a recurved flattened end-area (retused), or with a translucent tip (window). Inflorescence a raceme or panicle. Flowers bilaterally symmetric, tubular, the perianth-segments free at the base, the outer sometimes adherent to the inner above, white or greenish, each segment with a green, brown or pink central stripe. Anthers dorsifixed. Ovary superior, 3-celled. Fruit a capsule, splitting between the septa.

A genus of *c.* 70 species native to South Africa. The plants are not frost-hardy and require protection. They grow best in a well-drained medium and should not be over-watered; protection from strong sun is advisable for most species. Propagation is best from offsets; because the species tend to cross, seed may produce hybrid plants.

Literature: Bayer, M.B., The new Haworthia handbook (1982); Pilbeam, J., Haworthia and Astroloba, a collector's guide (1983); Scott, C.L., The genus Haworthia, a taxonomic revision (1985) – not seen.

1a. Perianth triangular or rounded-triangular at the base; flowers borne in a raceme; style upcurved (Subgenus *Haworthia*) 2
b. Perianth hexangular or rounded-hexangular at the base; flowers borne in a simple or branched inflorescence; style straight 17
2a. Leaves truncate 3
b. Leaves not truncate 4
3a. Leaves arranged in 2 opposite rows, leaf-end rounded-oblong **1. truncata**
b. Leaves arranged spirally, leaf-end more or less irregularly circular **2. maughanii**
4a. Leaves with a recurved flattened end-area (retused), or a convex upper surface above the middle 5
b. Leaves lacking a recurved, flattened end-area, or with a convex upper surface at the middle 6
5a. Leaves usually smooth **3. retusa**
b. Leaves rough, with tubercles or papillae, or if smooth then leaves not acuminate **5. mirabilis**
6a. Leaves lacking marginal teeth 7
b. Leaves with marginal teeth or bristles, sometimes rather sparse, sometimes only at the base 8
7a. Leaves obovoid, sometimes attenuate at the tip, upper surface usually convex, terminal awn 1–5 mm **6. cymbiformis**
b. Leaves long-triangular, upper surface flat to concave, if convex then with a terminal awn 10 mm or more, or leaves strongly incurved **4. turgida**
8a. Leaves bearing at least some tubercles or raised spots **9. angustifolia**
b. Leaves lacking tubercles 9
9a. Leaf margin bearing usually dense, long, often wispy bristles that often overlap and interlace between the leaves 10
b. Leaf margin bearing short or rather stout bristles or teeth, sometimes minute 11
10a. Leaves green **11. arachnoidea**
b. Leaves blue-green or grey-green **10. translucens**
11a. Leaves of 1 colour, or spotted, never with lines 12
b. Leaves with longitudinal lines or a net-like pattern, sometimes also with flecks 13
12a. Upper surface of leaf smooth **13. variegata**
b. Upper surface of leaf with scattered translucent teeth **12. herbacea**
13a. Leaves light to dark green, often turning red, orange, purple or brown in full sun 14
b. Leaves blue-green, grey-green, yellow-green or brownish 16
14a. Leaves with longitudinal lines, and with obvious translucent flecks or clear areas towards the tip **6. cymbiformis**
b. Leaves with a net-like pattern, obvious flecks absent 15
15a. Leaves gradually tapered to a long tip **7. batesiana**
b. Leaves almost spherical to shortly pointed **8. reticulata**
16a. Leaves spreading **4. turgida**
b. Leaves erect or incurved **10. translucens**
17a. Perianth-tube curved, hexangular at the base, narrowing gradually to the junction with the flower-stalk; inflorescence simple or branched (Subgenus *Hexangulares*) 18

b. Perianth-tube straight, rounded-hexangular at the base, narrowing suddenly to the junction with the flower-stalk; inflorescence branched (Subgenus *Robustipedunculares*) 33
18a. Plants with columnar leafy stems 19
b. Plants with leaves in flattish, stemless rosettes 25
19a. Leaves in 3 vertical or sometimes spiral rows **14. viscosa**
b. Leaves in more than 3 rows 20
20a. Leaves with tubercles confined to the margin and keel **18. armstrongii**
b. Leaves lacking tubercles, or with tubercles at least on the lower surface 21
21a. Leaves 4–8 cm 22
b. Leaves 1.5–3 cm 24
22a. Leaves with rounded tubercles, or occasionally tubercles absent 23
b. Leaves with flattened tubercles **16. reinwardtii**
23a. Leaves, green **15. coarctata**
b. Leaves, brownish green or yellowish green **20. fasciata**
24a. Leaves grey-green or blue-green, with 5–7 darker ridges or stripes **17. glauca**
b. Leaves green or grey-green, lacking ridges or stripes, but sometimes with a net-like pattern **16. reinwardtii**
25a. Leaf margin with teeth, or lower surface of leaf with green tubercles or ridges 26
b. Leaf margins untoothed, although often tuberculate; lower surface of leaf with white tubercles or ridges, or smooth 30
26a. Leaves 10–25 cm, margin never toothed 27
b. Leaves 2–10 cm, if longer then margin toothed 29
27a. Leaf with irregularly arranged tubercles **24. limifolia**
b. Leaf with tubercles in patches, or in rows that may be indistinct 28
28a. Leaves rough, bright or bluish green, 10–15 cm, the tubercles in distinct rows **21. glabrata**
b. Leaves almost smooth, shiny dark green, 15–25 cm, the tubercles in patches and often in indistinct longitudinal rows **23. longiana**
29a. Leaves pale or yellowish green, with tubercles in rows, margin lacking teeth **25. ubomboensis**
b. Leaves dark green, the tubercles scattered or merging together into ridges, margin with teeth **24. limifolia**
30a. Upper surface of leaves covered in dense, tiny white tubercles, or with white ridges 31
b. Upper surface of leaves smooth or with sparse tubercles 32
31a. Upper surface of leaves tuberculate **22. radula**
b. Upper surface of leaves with ridges **24. limifolia**
32a. Leaves bright or dark green, upper surface with sparse tubercles **19. attenuata**
b. Leaves light green, grey-green, yellow-green or brownish, rarely dark green, upper surface smooth **20. fasciata**
33a. Leaves dark grey-green, blue-green or yellow-green **26. minima**
b. Leaves dark to bright green or brownish green **27. pumila**

Subgenus **Haworthia**. Inflorescence simple. Perianth-tube curved, triangular or rounded-triangular at the base. Style upcurved.

1. H. truncata Schoenland. Illustration: van Laren, Succulents, 16 (1935); Rauh, Schöne Kakteen und andere Sukkulenten, 246 (1978); Rowley, Illustrated encyclopedia of succulents, 165 (1978); Bayer, The new Haworthia handbook, 39 (1982).

Rosettes solitary or forming clusters, with 7–13 leaves. Leaves *c.* 2 cm, in 2 ranks, more or less erect, oblong, truncate at the top, which is translucent at first, both surfaces with tubercles. *South Africa.* G1. Spring–summer.

Von Poellnitz splits this species into 3 forms, using the thickness of the rounded-oblong truncated leaf-ends: forma **truncata** has leaf-ends 6–8 mm thick, forma **tenuis** von Poellnitz has leaf-ends only 3–5 mm thick, and in forma **crassa** von Poellnitz the leaf-ends are 9–11 mm thick, often with white radiating lines. These forms are not recognised by Bayer.

2. H. maughanii von Poellnitz. Illustration: Rauh, Schöne Kakteen und andere Sukkulenten, 247 (1978); Rowley, Illustrated encyclopedia of succulents, 165 (1978); Bayer, The new Haworthia handbook, 24 (1982); Pilbeam, Haworthia and Astroloba, 94–5 (1983).

Rosette usually solitary. Leaves *c.* 2.5 cm, spirally arranged, more or less erect, cylindric, truncate at the top giving an irregularly circular leaf-end 1–1.2 cm across. *South Africa.* G1. Spring–summer.

A hybrid between *H. maughanii* and *H. truncata* is available commercially.

3. H. retusa (Linnaeus) Duval (*H. fouchei* von Poellnitz; *H. geraldii* C.L. Scott, *H. longebracteata* G.G. Smith; *H. retusa* var. *multilineata* G.G. Smith). Illustration: *Botanical Magazine*, 455 (1799); van Laren, Succulents, 4 (1935); Bayer, The new Haworthia handbook, fig. 33a (1982); *British Cactus and Succulent Journal* **1**: 32, 33 (1983).

Rosette stemless, 7–10 cm in diameter, offsetting from the base, with 15–20 leaves. Leaves 3–8 cm, recurved, ovate-triangular, end-area more or less tapered-triangular and translucent, with 5–8 paler lines; margin with sparse teeth, terminal bristle 3–6 mm. *South Africa.* G1. Spring.

Var. **acuminata** Bayer – illustration: Bayer, The new Haworthia handbook, fig. 33b, (1982) – has leaves that are long-pointed with slightly roughened end-areas. It often appears in cultivation under the names *H. retusa* var. *densiflora* G. G. Smith, *H. magnified* or *H. paradoxa*. Var. **dekenahii** (G.G. Smith) Bayer – illustration: Bayer, The new Haworthia handbook, fig. 33c, (1982) – has silver-spotted leaves.

H. comptoniana G.G. Smith. Illustration: *Excelsa* **4**: 29 (1974); Bayer, The new Haworthia handbook, 9 (1982); Pilbeam, Haworthia and Astroloba, 56–7 (1983). Leaves *c.* 12, glossy with white flecks, and 5–7 white lines connected to make a net-like pattern. *South Africa.* G1. Spring.

H. heidelbergensis G.G. Smith. Illustration: Bayer, The new Haworthia handbook, 18 (1982); Pilbeam, Haworthia and Astroloba, 81 (1983). Rosette to 7 cm in diameter with *c.* 35 leaves. Leaves acuminate, end-areas with 3 or 4 lines. *South Africa.* G1. Spring.

4. H. turgida Haworth (*H. laetivirens* Haworth; *H. caespitosa* von Poellnitz; *H. turgida* var. *pallidifolia* G.G. Smith; *H. turgida* var. *suberecta* von Poellnitz). Illustration: Bayer, The new Haworthia handbook, 40 (1982); Pilbeam, Haworthia and Astroloba, 132–3 (1983).

Rosette stemless, 5–6.5 cm in diameter, offsetting from the base. Leaves 1–4 cm, usually grey-green, long-triangular with 3–7 longitudinal lines and sometimes with whitish flecks; margin and keel smooth or margin with teeth. *South Africa.* G1. Spring

Var. *erecta* invalid, of nurserymen's catalogues, is probably var. *suberecta*, which is not distinguished here.

H. aristata Haworth (*H. unicolor* von Poellnitz; *H. venteri* von Poellnitz). Illustration: *National Cactus and Succulent Society Journal* **35**: 11–12 (1980); Bayer, The new Haworthia handbook, fig. 41a (1982); Pilbeam, Haworthia and Astroloba, 39 (1983). Rosette *c.* 4 cm in diameter. Leaves 4–6 cm, green to brownish, margin and terminal bristle smooth. *South Africa.* G1. Spring.

Var. **helmiae** (von Poellnitz) Pilbeam (*H. unicolor* von Poellnitz var. *helmiae* (von Poellnitz) Bayer). Illustration: Bayer, The new Haworthia handbook, fig. 41b (1982). Rosette more compact, and the leaf margin and terminal bristle with teeth.

H. rycroftiana Bayer. Illustration: Bayer, The new Haworthia handbook, 34 (1982). Rosette 5–8 cm in diameter with up to 50 leaves. Leaves pale green with an inconspicuous net-pattern; margin and keel smooth. *South Africa.* G1. Spring. Often grown under the names *H. aristata* or *H. unicolor.*

H. zantneriana von Poellnitz. Illustration: Bayer, The new Haworthia handbook, 45 (1982); Pilbeam, Haworthia and Astroloba, 144 (1983). Leaves dull grey-brown, often turning lilac in full sun, with irregular translucent blotches and longitudinal stripes; margin and keel smooth, margin pale. *South Africa.* G1. Spring.

H. habdomadis var. **inconfluens**, and var. **morrisiae** may both key out here (see under 10. *H. translucens*). They differ in having bright green, yellow-green or pale brownish green leaves.

5. H. mirabilis Haworth (*H. mirabilis* forma *beukmannii* (von Poellnitz) Pilbeam; *H. nitidula* von Poellnitz; *H. triebneriana* von Poellnitz, including var. *napierensis* Triebner & von Poellnitz, var. *rubrodentata* Triebner & von Poellnitz and var. *sublineata* von Poellnitz). Illustration: *Botanical Magazine*, 1354 (1811); Das Pflanzenreich **33**: 100 (1908); Bayer, The new Haworthia handbook, fig. 25a (1982); Pilbeam, Haworthia and Astroloba, pl. 5, 97–8 (1983).

Rosette stemless, 5–6 cm in diameter. Leaves to 8 cm, lanceolate, recurved, end-area somewhat translucent, with a prominent central line extending almost to the tip and 2 shorter lines, lower surface with tubercles; margin with white teeth, terminal bristle to 1 cm. *South Africa.* G1. Spring.

Subsp. **badia** (von Poellnitz) Bayer. Illustration: *Cactus and Succulent Journal (U.S.)* **48**: 105 (1976); Bayer, The new Haworthia handbook, fig. 25b (1982); Pilbeam, Haworthia and Astroloba, pl. 5, 98 (1983). Leaves strongly recurved, usually reddish brown, not tuberculate beneath, margin smooth or slightly rough.

Subsp. **mundula** (G.G. Smith) Bayer. Illustration: Bayer, The new Haworthia handbook, fig. 25c (1982); Pilbeam, Haworthia and Astroloba, pl. 5, 99 (1983). Leaves strongly recurved, short, thick and blunter than the other subspecies, not tuberculate beneath; margin with minute greenish white teeth.

H. emelyae von Poellnitz. Illustration: *National Cactus and Succulent Society Journal* **34**: 28–31 (1979); Court, Succulent flora of southern Africa, 160 (1981); Bayer, The new Haworthia handbook, fig. 14a (1982); Pilbeam, Haworthia and Astroloba, pl. 3, 69–70 (1983). Rosette *c.* 4 cm in diameter. Leaves 3–4 cm, recurved, end-areas usually marked with brown, lilac or pink in full sun, tuberculate; margin and keel with tiny sparse teeth. *South Africa.* G1. Spring.

Var. **multiflora** Bayer. Illustration: Bayer, The new Haworthia handbook, fig. 14b (1982); Pilbeam, Haworthia and Astroloba, 71 (1983). Leaves to 4.5 cm, almost erect, narrow, sharply pointed, scarcely retused, upper surface with longitudinal lines towards the leaf-tip; margin and keel with obvious teeth.

H. magnifica von Poellnitz. *South Africa.* G1. Spring.

Var. **magnifica.** Illustration: Bayer, The new Haworthia handbook, fig. 22a (1982); Pilbeam, Haworthia and Astroloba, pl. 4, 89 (1983). Leaves *c.* 3.5 cm, dark green, stiff, triangular-elliptic, acuminate with a long terminal bristle, end-area triangular with 4 or 5 longitudinal lines, almost smooth or sometimes tuberculate; margin and keel with whitish teeth. Flowers with green lines and a green throat.

Var. **atrofusca** (G.G.Smith) Bayer. Illustration: Bayer, The new Haworthia handbook, fig. 22b (1982); Pilbeam, Haworthia and Astroloba, 90 (1983). Leaves *c.* 4 cm, dark blackish green in full sun, upper surface with small green tubercles, end-area rather swollen, with 3–5 indistinct longitudinal lines; margin and tip of keel with small, transparent teeth.

Var. **major** (G.G. Smith) Bayer. Illustration: Bayer, The new Haworthia handbook, fig. 22c (1982); Pilbeam, Haworthia and Astroloba, pl. 4, 90 (1983). Rosettes larger than in var. *magnified* and forming clusters reluctantly, leaves colouring to brown or pink in full sun, with many tubercles, each with a small bristle.

Var. **maraisii** (von Poellnitz) Bayer (*H. maraisii* von Poellnitz; *H. schuldtiana* von Poellnitz). Illustration: Bayer, The new Haworthia handbook, fig. 22d (1982); Pilbeam, Haworthia and Astroloba, pl. 4, 90 (1983); *British Cactus and Succulent Journal* **1**: fig. 7, 32 (1983). Rosettes readily forming clusters. Leaves *c.* 3.5 cm, with irregular tubercles, end-area with 3 longitudinal lines; margin with prominent teeth.

Var. **meiringii** Bayer. Illustration: Bayer, The new Haworthia handbook, fig. 22e (1982); Pilbeam, Haworthia and Astroloba, pl. 5, 90 (1983). Leaves erect, green, end-area poorly defined, both surfaces with bristly tubercles; margin and keel with pronounced teeth.

Var. **notabilis** (von Poellnitz) Bayer. Illustration: Bayer, The new Haworthia handbook, fig. 22f (1982); Pilbeam, Haworthia and Astroloba, 91 (1983). Leaves spreading-erect, green becoming brown-purple in full sun, end-area not markedly separate, both surfaces with bristly tubercles; margin and keel toothed. Flower usually with a yellow throat.

Var. **paradoxa** (von Poellnitz) Bayer (*H. paradoxa* von Poellnitz). Illustration: Bayer, The new Haworthia handbook, fig. 22g (1982); Pilbeam, Haworthia and Astroloba, 91 (1983). Leaves *c.* 3 cm, dark green, similar to var. *magnifica* but with irregular lines of tubercles and pale longitudinal lines on the end-area, and pale tubercles and transparent flecks on the lower surface towards the tip. Terminal bristle *c.* 1 mm.

H. parksiana von Poellnitz. Illustration: Bayer, The new Haworthia handbook, 28 (1982); Pilbeam, Haworthia and Astroloba, 103 (1983). Rosette 2–3 cm in diameter. Leaves dark green to almost black, slightly to strongly recurved, covered with minute tubercles the same colour as the leaf. *South Africa.* G1. Spring.

H. pygmaea von Poellnitz. Illustration: *Excelsa* **5**: 82 (1975); Bayer, The new Haworthia handbook, 31 (1982); Pilbeam, Haworthia and Astroloba, 106–7 (1983). Rosette to 3 cm in diameter. Leaves

2.5–3 cm, strongly recurved, more or less papillose, end-area rounded-triangular with 4 or 5 pale longitudinal lines; margin smooth. *South Africa.* G1. Spring.

Pilbeam recognises two forms: forma **major** Pilbeam in which the rosettes are up to 6 cm in diameter, and forma **crystallina** Pilbeam in which the leaves have heavily papillose end-areas.

H. springbokvlakensis C.L. Scott. Illustration: *Journal of South African Botany* **36**: 289 (1970); *Excelsa* **4**: 29 (1974), **5**: 82 (1975); Bayer, The new Haworthia handbook, 37 (1982); Pilbeam, Haworthia and Astroloba, 124 (1983). Rosette *c.* 7 cm in diameter. Leaves 3.5–4 cm, often greyish pink, very swollen, with minute tubercles the same colour as the leaf, end-area with long and short branched lines; margin with minute teeth. *South Africa.* G1. Spring.

6. H. cymbiformis (Haworth) Duval (*H. obtusa* Haworth; *H. lepida* G.G. Smith; *H. planifolia* Haworth var. *exulata* von Poellnitz; *H. planifolia* Haworth var. *setulifera* von Poellnitz). Illustration: *Botanical Magazine*, 802 (1805); van Laren, Succulents, 22 (1935); Bayer, The new Haworthia handbook, fig. 11a (1982); Pilbeam, Haworthia and Astroloba, pl. 1, 61–5 (1983).

Rosette stemless, 3–10 cm in diameter. Leaves 2–4 cm, green, turning yellow or orange-red in full sun, obovoid, upper surface flat to somewhat convex, with longitudinal lines and often with translucent flecks towards the leaf-tip, lower surface convex usually with a prominent keel towards the tip; margin and keel smooth, terminal bristle 1–5 mm. *South Africa.* G1. Spring.

There is a variegated form in cultivation.

Var. **incurvula** (von Poellnitz) Bayer (*H. incurvula* von Poellnitz). Illustration: Marloth, Flora of South Africa **4**: pl. 223 (1915); Bayer, The new Haworthia handbook, fig. 11c (1982); Pilbeam, Haworthia and Astroloba, pl. 3, 66 (1983). Leaves narrow, incurved, with rounded, translucent teeth.

Var. **transiens** (von Poellnitz) Bayer (*H. planifolia* Haworth var. *translucens* Triebner & von Poellnitz). Illustration: Bayer, The new Haworthia handbook, fig. 11d (1982); Pilbeam, Haworthia and Astroloba, 66 (1983). Leaves 1.5–2.5 cm, translucent for entire length with 8–10 longitudinal stripes; margin with minute teeth. The leaves turn brownish red in full sun.

Var. **umbraticola** (von Poellnitz) Bayer (*H. umbraticola* von Poellnitz). Illustration: Bayer, The new Haworthia handbook, fig. 11e (1982); Pilbeam, Haworthia and Astroloba, 66 (1983). The ends of the leaves have large clear areas separated by dark green lines.

Forma **gracilidelineata** (von Poellnitz) Pilbeam. Illustration: Pilbeam, Haworthia and Astroloba, 63 (1983). Rosettes to 3 cm across, in dense clusters. Leaves incurved, almost completely translucent.

Forma **ramosa** (G.G. Smith) Bayer (*H. ramosa* G.G. Smith). Illustration: Bayer, The new Haworthia handbook, fig. 11b (1982); Pilbeam, Haworthia and Astroloba, 65 (1983). Rosettes smaller than other forms or varieties, with elongated stems.

7. H. batesiana Uitewaal. Illustration: Bayer, The new Haworthia handbook, 5 (1982); Pilbeam, Haworthia and Astroloba, 45–6 (1983).

Rosettes stemless, 4–5 cm in diameter, readily clustering to form dense cushions. Leaves 2–2.5 cm, green, ovate-lanceolate tapering to a long wispy tip, indistinctly tessellated beneath the surface; margin and keel with minute teeth. *South Africa.* G1. Spring.

H. marumiana Uitewaal. Illustration: Bayer, The new Haworthia handbook, 23 (1982); Pilbeam, Haworthia and Astroloba, 95 (1983). Leaves 1.2–2 cm, green, turning purple in full sun, terminal bristle *c.* 2 mm, both surfaces tessellated, smooth or with bristles that arise from translucent flecks. *South Africa.* G1. Spring.

8. H. reticulata Haworth (*H. haageana* von Poellnitz; *H. haageana* von Poellnitz var. *subreticulata* von Poellnitz; *H. reticulata* Haworth var. *acuminata* von Poellnitz; *H. reticulata* Haworth var. *subregularis* (Baker) Pilbeam; *H. subregularis* Baker; *H. hurlingii* von Poellnitz var. *ambigua* Triebner & von Poellnitz). Illustration: *Botanical Magizine*, 1314 (1810); *National Cactus and Succulent Society Journal* **27**: 10–11 (1972); Bayer, The new Haworthia handbook, fig. 32a (1982); Pilbeam, Haworthia and Astroloba, 114–15 (1983).

Rosettes stemless, 4–7 cm in diameter, eventually forming large clumps. Leaves 2–3.5 cm, green, turning orange or red in full sun, narrowly triangular-ovate, both surfaces with a net-like pattern, pointed, with a terminal bristle to 1 cm; margin minutely toothed. *South Africa.* G1. Spring.

Var. **hurlingii** (von Poellnitz) Bayer (*H. hurlingii* von Poellnitz). Illustration: *Flowering Plants of Africa* **29**: pl. 1149 (1953); Bayer, The new Haworthia handbook, fig. 32b (1982); Pilbeam, Haworthia and Astroloba, pl. 7, 115 (1983). Rosette 2–3 cm across. Leaves to 2 cm, relatively fatter and less pointed than in var. *reticulata*, margin and keel with tiny teeth towards the base.

9. H. angustifolia Haworth (*H. monticola* Fourcade; *H. angustifolia* var. *albanensis* (Schoenland) von Poellnitz; *H. angustifolia* var. *subfalcata* von Poellnitz). Illustration: Bayer, The new Haworthia handbook, fig. 1a (1982); Pilbeam, Haworthia and Astroloba, 35 (1983).

Rosettes stemless, 2–6 cm in diameter, clustering readily. Leaves 2–10 cm, dull or dark green, lanceolate, tapered, tip with a whitish bristle, upper surface with a prominent central line and 5–7 indistinct longitudinal lines, rough from a covering of tiny tubercles; margin with tiny teeth. *South Africa.* G1. Spring.

H. emelyae may key out here (see under 5. *H. mirabilis*). It differs in the margin and keel having obvious teeth. *H. marumiana* may key out here (see under 7. *H. batesiana*). It differs in the leaves being 2 cm or less, with a tessellated pattern.

10. H. translucens Haworth (*H. gracilis* von Poellnitz; *H. translucens* var. *deliculata* (Berger) von Poellnitz). Illustration: *Botanical Magazine*, 1417 (1811); Bayer, The new Haworthia handbook, fig. 38a (1982); Pilbeam, Haworthia and Astroloba, 129 (1983).

Rosette stemless, 4–6 cm in diameter, offsetting from the base. Leaves *c.* 3 cm, erect to slightly incurved, grey-green, becoming lilac-grey in full sun, translucent with green net-like lines, narrowly lanceolate, tapering to a fine bristle; margin and keel with translucent teeth. *South Africa.* G1. Spring.

Subsp. **tenera** (von Poellnitz) Bayer (*H. tenera* von Poellnitz). Illustration: Bayer, The new Haworthia handbook, fig. 38b (1982); Pilbeam, Haworthia and Astroloba, 131 (1983). Rosette 1–3 cm in diameter with strongly incurved leaves.

H. bolusii Baker. Illustration: *Desert Plant Life* **7**: 119 (1935); Bayer, The new Haworthia handbook, fig. 7a (1982); Pilbeam, Haworthia and Astroloba, 48 (1983). Rosette stemless, *c.* 5 cm across, not readily offsetting. Leaves 2–4 cm, blue-green or grey-green, incurved, with large translucent areas at the tips; margin and keel with teeth 5–6 mm. *South Africa.* G1. Spring.

Var. **blackbeardiana** (von Poellnitz) Bayer (*H. blackbeardiana* von Poellnitz). Illustration: Court, Succulent flora of southern Africa, 162 (1981); Bayer, The new Haworthia handbook, fig. 7b (1982); Pilbeam, Haworthia and Astroloba, pl. 2, 49 (1983). Rosette to 15 cm in diameter.

H. cooperi Baker (*H. vittata* Baker; *H. pilifera* Baker; *H. cooperi* forma *pilifera* (Baker) Pilbeam; *H. obtusa* Haworth var. *pilifera* (Baker) Uitewaal). Illustration: *Cactus and Succulent Journal (*US) **53**: 70–71 (1981); Bayer, The new Haworthia handbook, fig. 10a (1982). Pilbeam, Haworthia and Astroloba, pl. 3, 57–9 (1983). Rosette to 9 cm across. Leaves 2–2.5 cm, blue-green or grey-green with translucent tips, round-ended or with a terminal bristle to 6 mm, both surfaces convex with dark longitudinal lines; margin and keel with translucent teeth. *South Africa.* G1. Spring.

Var. **leightonii** (G.G. Smith) Bayer (*H. leightonii* G.G. Smith). Illustration: Bayer, The new Haworthia handbook, fig. 10b (1982). The longitudinal lines towards the end of the leaf are reddish. This variety often appears in nurserymen's lists as *leightoniae.*

H. decipiens von Poellnitz. Illustration: Bayer, The new Haworthia handbook, fig. 12 (1982); Pilbeam, Haworthia and Astroloba, pl. 3, 67–8 (1983). Rosette *c.* 7 cm or more in diameter. Leaves 2–4 cm, spreading with incurved tips, grey-green with darker tessellation beneath the surface, turning yellow or red in full sun, terminal bristle *c.* 1 cm; margin with curved white teeth 2–2.5 mm, keel smooth. *South Africa.* G1. Spring.

H. habdomadis von Poellnitz. Illustration: Bayer, The new Haworthia handbook, fig. 17a (1982); Pilbeam, Haworthia and Astroloba, pl. 3, 78 (1983). Leaves 2.5–3.5 cm, pale greyish green to brownish green with darker tessellation, terminal bristle 3–4 mm; margin and keel with translucent teeth. Wide leaves may have 2 keels. *South Africa.* G1. Spring.

Var. **inconfluens** (von Poellnitz) Bayer (*H. altilinea* Haworth var. *limpida* (Haworth) von Poellnitz; forma *inconfluens* von Poellnitz). Illustration: Court, Succulent flora of southern Africa, 164 (1981); Bayer, The new Haworthia handbook, fig. 17b (1982); Pilbeam, Haworthia and Astroloba, 79 (1983); *British Cactus and Succulent Journal* **1**: 32, fig. 6 (1983). Leaves pale brownish green, strongly incurved, margin and keel usually untoothed, terminal bristle *c.* 6 mm.

Var. **morrisiae** (von Poellnitz) Bayer. Illustration: Court, Succulent flora of southern Africa, 164 (1981); Bayer The new Haworthia handbook, fig. 17c (1982); Pilbeam, Haworthia and Astroloba, 80 (1983). Leaves 2–5 cm, bright or yellowish green with a brown tip; margin and keel untoothed, sometimes with tiny teeth on the margin of young leaves.

11. H. arachnoidea (Linnaeus) Duval (*H. setata* Haworth). Illustration: van Laren, Succulents, 3 (1935); Rauh, Schöne Kakteen und andere Sukkulenten, 244 (1978); Bayer, The new Haworthia handbook, 2 (1982); Pilbeam, Haworthia and Astroloba, 37 (1983).

Rosette stemless, 3–4.5 cm in diameter or more, eventually offsetting to form clumps. Leaves *c.* 1.5 cm, lanceolate, opaque green with darker longitudinal lines, terminal bristle 6–7 mm; margin and keel with whitish bristles about 1.5 mm apart. *South Africa.* G1. Spring.

H. aranea (Berger) Bayer (*H. bolusii* Baker var. *aranea* Berger). Illustration: Engler, Das Pflanzenreich **33**: 114 (1908); Bayer, The new Haworthia handbook, 3 (1982); Pilbeam, Haworthia and Astroloba, 38 (1983). Rosette to 15 cm in diameter, flattish with a depressed centre. Leaves uniformly green, margin and keel with fine, dense bristles less than 1.5 mm apart. *South Africa.* G1. Spring.

H. semiviva (von Poellnitz) Bayer (*H. bolusii* Baker var. *semiviva* von Poellnitz). Illustration: *Cactus and Succulent Journal (U.S.)* **48**: 256 (1976); Bayer, The new Haworthia handbook, 35 (1982); Pilbeam, Haworthia and Astroloba, pl. 8, 122 (1983). Rosette 7–8 cm across. Leaves green, incurved, the tip translucent and flattened, drying up in the resting period, margin and keel with long, translucent bristles. *South Africa.* G1. Spring.

H. aristata may key out here (see under 4. *H. turgida*). It differs in the margin having relatively sparse teeth and the terminal bristle being toothed.

12. H. herbacea (Miller) Stearn (*H. atrovirens* (de Candolle) Haworth; *H. submaculata* von Poellnitz; *H. luteorosea* Uitewaal). Illustration: van Laren, Succulents, 14 (1935); *National Cactus and Succulent Society Journal* **27**: 51–52 (1972); Bayer, The new Haworthia handbook, 19 (1982); Pilbeam, Haworthia and Astroloba, pl. 4, 81(1983).

Rosette stemless, 2–8 cm in diameter, eventually offsetting to form small clumps. Leaves 1–4 cm, pale to dark green, erect to incurved, broadly lanceolate, both surfaces convex, with translucent flecks especially towards the leaf-tip, often each fleck giving rise to a tooth; margin and keel with translucent teeth. *South Africa.* G1. Spring.

H. maculata (von Poellnitz) Bayer. Illustration: Bayer, The new Haworthia handbook, 21 (1982); Pilbeam, Haworthia and Astroloba, pl. 4, 89 (1983). Differs from *H. herbacea* in its fewer and more turgid leaves, which turn violet in full sun and are spreading to somewhat recurved. *South Africa.* G1. Spring.

13. H. variegata L. Bolus. Illustration: Bayer, The new Haworthia handbook, 42 (1982); Pilbeam, Haworthia and Astroloba, 134 (1983).

Rosette stemless, 3–4 cm in diameter, with 20–30 leaves. Leaves 5–6 cm, dark green or brownish green with pale spots, linear-lanceolate, long-tapering; margin and keel with sharp, translucent teeth. *South Africa.* G1. Spring.

H. chloracantha Haworth. Illustration: Engler, Das Pflanzenreich **33**: 110 (1908); Bayer, The new Haworthia handbook, fig. 8a (1982); Pilbeam, Haworthia and Astroloba, 51 (1983). Rosette *c.* 6 cm. Leaves 3.5–4 cm, pale to yellowish green; margin and keel with small teeth the same colour as the leaf or darker. *South Africa.* G1. Spring.

Var. **denticulifera** (von Poellnitz) Bayer (*H. angustifolia* Haworth var. *liliputana* Uitewaal). Illustration: Bayer, The new Haworthia handbook, fig. 8b (1982); Pilbeam, Haworthia and Astroloba, 51 (1983). Rosette smaller, leaves dark green or purplish green.

Var. **subglauca** von Poellnitz. Illustration: Bayer, The new Haworthia handbook, fig. 8c (1982); Pilbeam, Haworthia and Astroloba, 51 (1983). Rosette more robust, leaves blue-green.

Subgenus **Hexangulares** Bayer. Inflorescence simple or branched. Perianth-tube curved, hexangular at the base, narrowing gradually to the junction with the flower-stalk. Style straight.

14. H. viscosa (Linnaeus) Haworth (*H. asperiuscula* Haworth; *H. beanii* G.G. Smith; *H. viscosa* var. *concinna* (Haworth)

Baker; *H. viscosa* var. *caespitosa* von Poellnitz). Illustration: van Laren, Succulents, fig. 8, 12 (1935); Barkhuizen, Succulents of southern Africa, 191 (1978); *National Cactus and Succulent Society Journal* **36**: 99–100 (1981); Bayer, The new Haworthia handbook, 62 (1982).

Rosette columnar, 7–20 cm tall, offsetting at the base. Leaves to 4 cm, dark green, in 3 straight or spiral rows, erect to projecting, triangular-lanceolate, both surfaces covered in small tubercles. *South Africa.* G1. Summer.

Many variants of *H. viscosa* have been given names and nurserymen's catalogues often list several varieties (var. *pseudotortuosa* (Salm-Dyck) Baker with spirally arranged leaves; var. *torquata* (Haworth) Baker with long, twisted leaves; and var. *subobtusa* von Poellnitz with shorter blunt leaves) but Bayer maintains that the variation in leaf size and arrangement is influenced by the environment, and so they are not recognised here.

H. nigra (Haworth) Baker (*H. nigra* var. *schmidtiana* (von Poellnitz) Uitewaal). Illustration: *Cactus and Succulent Journal (U.S.)* **48**: 105 (1976); Court, Succulent flora of southern Africa, 170 (1981); Bayer, The new Haworthia handbook, 55 (1982); Pilbeam, Haworthia and Astroloba, 100–1 (1983). Rosette 6–12 cm tall. Leaves to 7 cm, dark green, turning almost black in full sun, triangular-ovate, both surfaces with large, dense tubercles. *South Africa.* G1. Summer.

As in *H. viscosa,* many varieties and forms have been named and some are offered commercially, but they are not generally considered to be worthy of recognition.

15. H. coarctata Haworth (*H. fulva* G.G. Smith; *H. musculina* G.G. Smith; *H. greenii* Baker; *H. chalwinii* Marloth & Berger; *H. reinwardtii* (Salm-Dyck) Haworth var. *conspicua* von Poellnitz; *H. reinwardtii* var. *fallax* von Poellnitz; *H. reinwardtii* var. *committeesensis* G.G. Smith). Illustration: Marloth, Flora of South Africa **4**: pl. 22a (1915); *Desert Plant Life* **7**: 108 (1935), **18**: 68 (1946); Bayer, The new Haworthia handbook, fig. 48a–b (1982); Pilbeam, Haworthia and Astroloba, pl. 2, 53–5 (1983).

Rosette columnar, 5–20 cm tall, offsetting from the base. Leaves 4–6 cm, triangular-lanceolate, tips incurved, shining green, turning red in full sun, with greenish white rounded tubercles in longitudinal or sometimes transverse rows, density of tubercles very variable. *South Africa.* G1. Summer.

Var. **tenuis** (G.G. Smith) Bayer (*H. reinwardtii* (Salm-Dyck) Haworth var. *tenuis* G.G. Smith; *H. reinwardtii* var. *riebeckensis* G.G. Smith). Illustration: Bayer, The new Haworthia handbook, fig. 48c (1982). Pilbeam, Haworthia and Astroloba, pl. 2, 55 (1983). Rosettes to 45 cm, usually trailing and rooting where they touch the soil, leaves to *c.* 3 cm.

Subsp. **adelaidensis** (von Poellnitz) Bayer (*H. reinwardtii* (Salm-Dyck) Haworth var. *adelaidensis* von Poellnitz; *H. reinwardtii* var. *bellula* G.G. Smith). Illustration: Bayer, The new Haworthia handbook, fig. 48d (1982); Pilbeam, Haworthia and Astroloba, pl. 2, 50 (1983). Rosette to 13 cm with narrow leaves.

16. H. reinwardtii (Salm-Dyck) Haworth (*H. reinwardtii* var. *major* Baker; *H. reinwardtii* var. *pulchra* von Poellnitz; *H. reinwardtii* var. *archibaldiae* von Poellnitz; *H. reinwardtii* var. *peddiensis* G.G. Smith). Illustration: van Laren, Succulents, 13 (1935); *Flowering Plants of Africa* **29**: pl. 1148 (1953); Bayer, The new Haworthia handbook, fig. 57a (1982); Pilbeam, Haworthia and Astroloba, 108–9 (1983).

Very similar to *H. coarctata,* differing mainly in having flattened tubercles on the leaves, the leaves themselves being more densely arranged on the stem and less thick. *South Africa.* G1. Summer.

Forma **chalumnensis** (G.G. Smith) Bayer (*H. reinwardtii* var. *chalumnensis* G.G. Smith). Illustration: Court, Succulent flora of southern Africa, 167 (1981); Bayer, The new Haworthia handbook, fig. 57b (1982); Pilbeam, Haworthia and Astroloba, 109, 112 (1983). Rosette to 13 cm or more × 6–7 cm with the leaves tending to stand out from the stem. Leaves becoming red-purple in full sun, and with prominent transverse bands of almost confluent white tubercles on the back.

Forma **kaffirdriftensis** (G.G. Smith) Bayer. Illustration: Bayer, The new Haworthia handbook, fig. 57c (1982); Pilbeam, Haworthia and Astroloba, 109 (1983). Rosette to 12 × 3.5 cm with incurved leaves. Leaves becoming dark green to red-purple in full sun and with longitudinal rows of clear white, very large tubercles on the back which tend to coalesce into vertical ridges.

Forma **olivacea** (G.G. Smith) Bayer. Illustration: Bayer, The new Haworthia handbook, fig. 57d (1982); Pilbeam, Haworthia and Astroloba, 110 (1983). Rosettes *c.* 8 cm tall with rather narrow leaves. Leaves olive-green becoming orange-brown in full sun, with greenish white tubercles.

Forma **zebrina** (G.G. Smith) Bayer. Illustration: Bayer, The new Haworthia handbook, fig. 57e (1982); Pilbeam, Haworthia and Astroloba, 110–11 (1983). The incurved leaves have transverse bands of coalescing white tubercles on the back.

Var. **brevicula** G.G. Smith. Illustration: Bayer, The new Haworthia handbook, fig. 57f (1982); Pilbeam, Haworthia and Astroloba, 111 (1983). A smaller and more compact plant with rosettes to 8 × 2–3 cm. The backs of the dark green leaves bear bright white scattered tubercles.

The name *H. reinwardtii* var. *minor* Baker appears in catalogues, but the original description of this plant is inadequate and the variety cannot be assigned either to *H. reinwardtii* or *H. coarctata. H. venosa* subsp. *granulata* may key out here (see under 24. *H. limifolia*). It differs in the leaf surface having a faint net-pattern and grey-green tubercles.

17. H. glauca Baker. Illustration: Baker, The new Haworthia handbook, fig. 51a (1982); Pilbeam, Haworthia and Astroloba, 75 (1983).

Rosette columnar, 5–12 cm tall, *c.* 5 cm wide, offsetting from the base. Leaves 2–2.5 cm, grey-green turning reddish in full sun, narrowly ovate to lanceolate, somewhat incurved, with 5–7 darker longitudinal ridges on the back. *South Africa.* G1. Summer.

Var. **herrei** (von Poellnitz) Bayer (*H. herrei* von Poellnitz; *H. eilyae* von Poellnitz; *H. jacobseniana* von Poellnitz; *H. jonesiae* von Poellnitz). Illustration: Bayer, The new Haworthia handbook, fig. 51b (1982); Pilbeam, Haworthia and Astroloba, 76–7 (1983). Rosettes to 20 cm or more, usually trailing when they exceed 10 cm. Leaves blue-green, lacking tubercles or with irregular rows of whitish tubercles on the back.

18. H. armstrongii von Poellnitz (*H. glauca* Baker var. *herrei* (von Poellnitz) Bayer forma *armstrongii* (von Poellnitz) Bayer). Illustration: Pilbeam, Haworthia and Astroloba, 40 (1983).

Rosettes columnar, to 15 cm tall, 3–4 cm wide, offsetting from the base and lower part of the stem. Leaves 3–4 cm, dark green, grey-green or brownish green, triangular-lanceolate, with whitish tubercles confined to the margin and keel. *South Africa.* G1. Summer.

19. H. attenuata Haworth (*H.fasciata* (Willdenow) Haworth var. *caespitosa* Berger; *H. attenuata* var. *caespitosa* (Berger) Farden; *H. attenuata* var. *argyrostigma* (Baker) Berger; *H. attenuata* var. *deltoidea* Farden; *H. attentuata* var. *linearis* Farden; *H. attenuata* var. *minissima* Farden; *H. attenuata* var. *odonoghueana* Farden; *H. attenuata* var. *uitewaaliana* Farden). Illustration: *Botanical Magazine*, 1345 (1811); Rauh, Schöne Kakteen und andere Sukkulenten, 243 (1978); Bayer, The new Haworthia handbook, fig. 46a (1982).

Rosette stemless or almost so. Leaves 3–6.5 cm, dark green, erect to spreading-recurved, lanceolate, lower surface with sparse to dense white tubercles, usually unevenly distributed, upper surface with fewer tubercles. *South Africa.* G1. Summer.

Forma **britteniana** (von Poellnitz) Bayer. Illustration: Bayer, The new Haworthia handbook, fig. 46b (1982). Leaves with many large, white, separate tubercles.

Forma **clariperla** (Haworth) Bayer. Illustration: Bayer, The new Haworthia handbook, fig. 46c (1982). Leaves with large white tubercles merging together in horizontal bands on the lower surface.

Pilbeam's concept of the forms of *H. attenuata* does not agree with that of Bayer, and is not followed here. There is a plant in cultivation under the name H. 'Kuentzii' which seems to be a hybrid of *H. attenuata.* Illustration: Pilbeam, Haworthia and Astroloba, 83, (1983). The rosettes eventually tend to become shortly columnar and the leaves are shorter than those of *H. attenuata*, with tubercles that are separate or partly merged together.

20. H. fasciata (Willdenow) Haworth (*H. fasciata* forma *sparsa* von Poellnitz; *H. fasciata* var. *subconfluens* von Poellnitz; *H. fasciata* forma *vanstaadensis* von Poellnitz). Illustration: van Laren, Succulents, 2 (1935); Rowley, Illustrated encyclopedia of succulents, 165 (1978): Bayer, The new Haworthia handbook, fig. 49a (1982); Pilbeam, Haworthia and Astroloba, 71–2 (1983).

Rosette stemless, 5–7 cm in diameter. Leaves 3–5.5 cm, grey-green, incurved, lanceolate, upper surface smooth, lower surface with white tubercles that merge together into horizontal bands. *South Africa.* G1. Summer.

Forma **browniana** (von Poellnitz) Bayer (*H. browniana* von Poellnitz). Illustration: Bayer, The new Haworthia handbook, fig. 49b (1982); Pilbeam, Haworthia and Astroloba, 71–2 (1983). Rosette with a tendency to become columnar, to 12 cm. Leaves brownish green or yellowish green, longer than in forma *fasciata*.

Var. *staadensis* sometimes appears in catalogues and is probably an error for forma *vanstaadensis*, a synonym of *H. fasciata.*

H. starkiana von Poellnitz. Illustration: Barkhuizen, Succulents of southern Africa, 189 (1978); Court, Succulent flora of southern Africa, 169 (1981); Bayer, The new Haworthia handbook, fig. 60a (1982); Pilbeam, Haworthia and Astroloba, 125 (1983). Rosette stemless or more or less so. Leaves 4–5 cm, spreading, smooth, yellowish green, margin slightly thickened. *South Africa.* G1. Summer.

Var. **lateganiae** (von Poellnitz) Bayer. Illustration: Bayer, The new Haworthia handbook, fig. 60b (1982); Pilbeam, Haworthia and Astroloba, 125 (1983). Leaves dull green, longer (to 10 cm) and narrower.

21. H. glabrata (Salm-Dyck) Baker. Illustration: Bayer, The new Haworthia handbook, 50 (1982); Pilbeam, Haworthia and Astroloba, 74 (1983).

Rosette stemless, to 18 cm across. Leaves 10–15 cm, often spreading-recurving, lanceolate, bright green or bluish green with rows of many tiny tubercles of the same colour. *Not known in the wild.* G1. Summer.

22. H. radula (Jacquin) Haworth. Illustration: Engler, Das Pflanzenreich **33**: 91 (1908); *National Cactus and Succulent Society Journal* **23**: 92 (1968); Bayer, The new Haworthia handbook, 56 (1982); Pilbeam, Haworthia and Astroloba, 107 (1983).

Rosette stemless, offsetting readily. Leaves 6–8 cm, pointing in all directions, green, lanceolate, densely covered with tiny, white, evenly distributed tubercles. *South Africa.* G1. Summer.

23. H. longiana von Poellnitz. Illustration: *Excelsa* **4**: 29 (1974); Court, Succulent flora of southern Africa, 168 (1981); Bayer, The new Haworthia handbook, 54 (1982); Pilbeam, Haworthia and Astroloba, 88 (1983).

Rosette stemless, offsetting from the base. Leaves 15–25 cm, rigid, often dying back from the tips, lanceolate, shiny dark green with small tubercles of the same colour, in patches and often forming indistinct longitudinal rows. *South Africa.* G1. Summer.

24. H. limifolia Marloth (*H. limifolia* var. *stolonifera* Resende). Illustration: *National Cactus and Succulent Society Journal* **22**: 80 (1967); Bayer, The new Haworthia handbook, fig. 53a (1982); Pilbeam, Haworthia and Astroloba, 84 (1983).

Rosette stemless, 8–10 cm in diameter, with stolons that produce offsets up to 10 cm or more from the parent rosette. Leaves to 7.5 cm, borne in a tight spiral, spreading, dark green, broadly triangular-ovate, both surfaces with 15–20 transverse ridges the same colour as the leaf. *South Africa.* G1. Summer.

Var. **gigantea** Bayer. Illustration: *Journal of South African Botany* **28**: pl. ix (1962); Bayer, The new Haworthia handbook, fig. 53b (1982); Pilbeam, Haworthia and Astroloba, pl. 4, 86 (1983). Rosettes to 23 cm across. Leaves 7–11 cm, with small pale green tubercles, usually arranged irregularly.

Var. **striata** Pilbeam. Illustration: *Journal of South African Botany* **28**: pl. fig. (1962); Pilbeam, Haworthia and Astroloba, pl. 1, 85 (1983). Similar to var. *limifolia*, but the transverse ridges on the leaves are white.

H. venosa (Lamarck) Haworth. Illustration: *Botanical Magazine*, 1353 (1811); Bayer, The new Haworthia handbook, fig. 61a (1982); Pilbeam, Haworthia and Astroloba, 135 (1983). Rosettes not producing stolons. Leaves 2–12 cm, recurved; upper surface with 5 or 6 longitudinal lines, usually with net-like connexions; lower surface rough with irregular tubercles, margins with small white teeth. *South Africa.* G1. Summer.

Subsp. **granulata** (Marloth) Bayer (*H.* granulata Marloth). Illustration: *Cactus and Succulent Journal (U.S.)* **50**: 74, 76 (1978); Bayer, The new Haworthia handbook, fig. 61b (1982); Pilbeam, Haworthia and Astroloba, 136 (1983). Rosette columnar, to 9 × 4 cm. Leaves 2–3 cm, lower surface with many green tubercles more or less in transverse rows.

Subsp. **tessellata** (Haworth) Bayer (*H. tessellata* Haworth; *H. recurva* (Haworth) Haworth). Illustration: Marloth,

Flora of South Africa **4**: pl. 22c (1915); van Laren, Succulents, 6 (1935); Bayer, The new Haworthia handbook, fig. 61c (1982); Pilbeam, Haworthia and Astroloba, 136–8 (1983). Leaves 3–5 cm, broadly triangular-ovate, strongly recurved, upper surface with a strong, net-like pattern, lower surface with tiny tubercles, margin with recurved white teeth.

25. H. ubomboensis Verdoorn (*H. limifolia* Marloth var. *ubomboensis* (Verdoorn) G.G. Smith). Illustration: *Flowering plants of South Africa* **21**: pl. 818 (1941); Bayer, The new Haworthia handbook, fig. 53c (1982); *British Cactus and Succulent Journal* **1**: 32, fig. 4 (1983); Pilbeam, Haworthia and Astroloba, pl. 8, 133 (1983).

Rosette stemless, with offsets produced at the ends of stolons. Leaves 3–6.5 cm, pale or yellowish green, becoming lilac-green in full sun, broadly lanceolate, margin hard and horny, lower surface with sparse green tubercles in rows. *South Africa.* G1. Summer.

Subgenus **Robustipedunculares** Bayer.
Inflorescence branched, perianth-tube straight, rounded-hexagonal, narrowing suddenly to the junction with the flower-stalk. Style straight.

26. H. minima (Aiton) Haworth. Illustration: *Excelsa* **4**: 29 (1974); *Cactus and Succulent Journal (U.S.)* **53**: 227 (1981); Bayer, The new Haworthia handbook, 66 (1982); Pilbeam, Haworthia and Astroloba, pl. 5, 96 (1983).

Rosette stemless. Leaves blue-green or grey-green, triangular-ovate, with dense, prominent white tubercles on both sides. *South Africa.* G1. Summer.

H. marginata (Lamarck) Stearn (*H. albicans* Haworth; *H. uitewaaliana* von Poellnitz). Illustration: Engler, Das Pflanzenreich **33**: 95 (1908); Barkhuizen, Succulents of southern Africa, 187 (1978); Bayer, The new Haworthia handbook, 65 (1982); Pilbeam, Haworthia and Astroloba, pl. 5, 92–3 (1983). Leaves yellow-green, smooth or with occasional tubercles beneath. *South Africa.* G1. Summer.

27. H. pumila (Linnaeus) Duval (*H. margaritifera* (Linnaeus) Haworth; *H. papillosa* (Salm-Dyck) Haworth; *H. semiglabrata* Haworth). Illustration: Rauh, Schöne Kakteen und andere Sukkulenten, 245 (1978); Bayer, The new Haworthia handbook, 68 (1982); Pilbeam, Haworthia and Astroloba, pl. 6, 106 (1983); *Cactus and Succulent Journal (U.S.)* **55**: 164 (1983).

Rosette stemless, to 15 cm across and 30 cm tall. Leaves 7–8 cm, triangular-ovate to broadly lanceolate, dark brownish-green, both surfaces with large white tubercles in indistinct rows. *South Africa.* G1. Summer.

H. kingiana von Poellnitz. Illustration: *Excelsa* **4**: 29 (1974); Barkhuizen, Succulents of southern Africa, 185 (1978); Bayer, The new Haworthia handbook, 64 (1982); Pilbeam, Haworthia and Astroloba, 83 (1983). Rosette to 7 cm across and 8 cm tall. Leaves bright green, with rather indistinct, flattened tubercles, occasionally lacking tubercles. *South Africa.* G1. Summer.

The following names are of doubtful application, but are found either in catalogues, or are widely used in the labelling of plants. **H. antilinea** Haworth: the original description cannot be applied to any known species and, in view of its mottled history, the name is rejected as being confused. **H. asperula** Haworth: this name has been applied at various times to *H. pygmaea*, *H. magnified*, *H. magnifica* var. *maraisii*, *H. pubescens* Bayer, *H. emelyae* and *H. mutica* Haworth; it is now such a source of confusion that it should not be used. **H. cassytha** Baker: plants bearing this name are offered for sale and appear to be near to *H. viscosa*. However, Baker himself was unsure if the plant that he originally described from material at Kew was in fact a *Haworthia*. The name therefore cannot be validly applied to any known species. **H. columnaris** Baker: Baker considered this species to be related to *H. affinis* Baker and *H. bilineata* Baker; both these names have now been discarded due to confusion. Later workers related *H. columnaris* to *H. cooperi*, but its identity is still a mystery, so the name cannot be used. **H. cuspidata** Haworth: the name was applied originally by Haworth to a plant related to *H. mucronata* Haworth. It is now often used erroneously for a hybrid of garden origin between *H. cymbiformis* and *H. retusa*, Bayer recommends that the name should be discarded. **H. ferox** von Poellnitz: Bayer considers that the plant originally described under this name is probably *Aloe humilis* (Linnaeus) Miller. Plants sold by nurserymen as *H. ferox* usually turn out to be *H. herbacea*, *H. magnifica* var. *paradoxa*, robust forms of *H. turgida*, or *H. aristata* var. *helmiae*. **H. integra** von Poellnitz: this species was allied originally to *H. reticulata* and *H. cymbiformis*. Plants cultivated under this name differ from these two species in their longer, narrower leaves with longitudinal lines and a terminal white bristle; they cannot be assigned to any named species. **H. janseana** Uitewaal: the origin of this plant is unknown; it is thought that it may be a garden hybrid involving *H. turgida*. Plants sold or grown under this name often turn out to be *H. angustifolia*. **H. mucronata** Haworth: it is not known where this plant was originally collected and there are no original specimens. So much confusion surrounds the name that Bayer has suggested that it should not be used. **H. otzenii** G.G. Smith: although the place where this plant was first collected is known, there are no original specimens, and plants subsequently collected there do not fit Smith's description. Nevertheless, Bayer regards it as a synonym of *H. mutica* Haworth. Plants in cultivation under this name appear to be hybrids of *H. mirabilis*. **H. subfasciata** (Salm-Dyck) Baker: this name is sometimes applied erroneously to plants in cultivation, but it is not known to which plant the name should properly apply.

The following names apply to garden hybrids: **H. × icosiphylla** Baker (possibly *H. tortuosa* (see below) × *H. glabrata*); **H. kewensis** von Poellnitz; **H. rigida** (Lamarck) Haworth (possibly *H. tortuosa* (see below) is involved); **H. rugosa** (Salm-Dyck) Baker (possibly *H. attenuata* × *H. radula*); **H. ryderiana** von Poellnitz; **H. sampaiana** Resende; **H. subattenuata** (Salm-Dyck) Baker (possibly *H. pumila* × *H. attenuata*, most commonly seen in its variegated form); **H. subulata** (Salm-Dyck) Baker; **H. tisleyi** Baker (possibly *H. tortuosa* (see below) or *H. glabrata* is involved); **H. tortuosa** Haworth (very similar to and probably a hybrid of *H. viscosa*; it is commonly cultivated, as are its varieties, and there is also a variegated form).

2. ASTROLOBA Uitewaal

V.A. Matthews

Similar to *Haworthia* but with radially symmetric, usually white or yellow-green flowers, and spirally arranged, thick, ovoid-triangular, pointed leaves crowded onto elongated stems.

A genus of about 10 species native to South Africa. Cultivation as for *Haworthia* (p. 38).

Literature: Pilbeam, J., Haworthia and Astroloba (1983).

1a. Leaves bearing tubercles on the lower surface **3. aspera**
b. Leaves lacking tubercles, except sometimes on the keel or margin 2
2a. Leaves spreading more or less horizontally **1. foliolosa**
b. Leaves borne at an angle of *c.* 45° **2. pentagona**

1. A. foliolosa (Haworth) Uitewaal (*Apicra foliolosa* (Haworth) Willdenow). Illustration: *Botanical Magazine*, 1352 (1811); Jacobsen, Lexicon of succulent plants, pl. 22 (1974); Pilbeam, Haworthia and Astroloba, 151 (1983).

Stems to 20 cm or more, eventually sprawling, 4–5 cm wide, producing offsets from the base, leaves in 5 ranks, dark green, smooth, spreading, upper surface flat to convex, lower surface with a keel, the margin with minute tubercles. Flowers yellow-green. *South Africa (Cape Province).* G1. Summer.

A. deltoidea (J.D. Hooker) Uitewaal – illustration: *Botanical Magazine*, 6071 (1873); Jacobsen, Lexicon of succulent plants, pl. 22, (1983) – is a commonly cultivated species, which is so similar to *A. foliolosa* that it cannot be separated satisfactorily. Recent unpublished work has shown that it should probably be regarded as synonymous.

A. congesta (Salm-Dyck) Uitewaal (*Apicra congesta* (Salm-Dyck) Baker). Illustration: Pilbeam, Haworthia and Astroloba, 150 (1983). Stems *c.* 7 cm wide, leaves tending to become concave above. *South Africa (Cape Province).* G1. Summer. Recent unpublished work suggests that it may be best to consider *A. congesta* as a subspecies of *A. foliolosa.*

A. herrei Uitewaal. Illustration: Jacobsen, Lexicon of succulent plants, pl. 22 (1974); Pilbeam, Haworthia and Astroloba, 152 (1983). Stems branching from the base and along the stem. Leaves grey-green, with darker longitudinal lines below the surface. *South Africa (Cape Province).* G1. Summer.

2. A. pentagona (Aiton) Uitewaal. Illustration: Jacobsen, Lexicon of succulent plants, pl. 22 (1974); Pilbeam, Haworthia and Astroloba, 153 (1983).

Stems to 20 cm or more × 3–4 cm, erect, offsetting from the base, leaves in 5 ranks, green, becoming reddish in full sun, borne at an angle of *c.* 45°, both surfaces smooth, upper surface flat or convex, lower surface with 1 or 2 keels, keels and margins minutely tubercled. Flowers greenish. *South Africa (Cape Province).* G1. Summer.

A. dodsoniana Uitewaal. Illustration: Pilbeam, Haworthia and Astroloba, 151 (1983). Stems to 25 cm or more × 4–6 cm, with blue-green or grey-green leaves, appearing whitish when grown in full sun, upper surface somewhat concave. Flowers white, striped with green. *South Africa (Cape Province).* G1. Summer.

A. spiralis (Linnaeus) Uitewaal (*Apicra spiralis* (Linnaeus) Baker). Illustration: *Botanical Magazine*, 1455 (1812); Pilbeam, Haworthia and Astroloba, 153–4 (1983). Stems *c.* 3 cm wide, eventually sprawling, with grey-green or blue-green leaves which have a finely roughened keel and margin. *South Africa (Cape Province).* G1. Summer.

3. A. aspera (Haworth) Uitewaal (*Apicra aspera* (Haworth) Willdenow). Illustration: Pilbeam, Haworthia and Astroloba, 148 (1983).

Stems to 20 × 2–3 cm, sprawling, producing offsets from the base and along the stem. Leaves in 3 or 4 ranks, green, turning reddish in full sun; upper surface smooth, flat to slightly convex; lower surface keeled, with irregular lines of small green tubercles. Flowers pinkish white. *South Africa (Cape Province).* G1. Summer.

Var. **major** Haworth. Illustration: Pilbeam, Haworthia and Astroloba, 149 (1983). Stems more upright, 5–6 cm wide.

The hybrid genus × **Astroworthia** Rowley contains hybrids between *Astroloba aspera* and *Haworthia pumila* and occurs naturally in the wild in South Africa. × **Astroworthia bicarinata** (Haworth) Rowley (*Astroloba bicarinata* (Haworth) Uitewaal) – illustration: Jacobsen, Lexicon of succulent plants, pl. 22 (1974); Pilbeam, Haworthia and Astroloba, 149 (1983) – has stems to *c.* 20 cm and leaves which are obliquely curved towards the tip, with 1 or 2 keels beneath and green tubercles. The keel and margin are minutely toothed.

The plant that is generally seen in cultivation under the name *Astroloba skinneri* (Berger) Uitewaal, has now been proved to belong to × *Astroworthia* and has been given the name × **Astroworthia bicarinata** nothomorph **skinneri** (Berger) Rowley (illustration: Jacobsen, Lexicon of succulent plants, pl. 22 fig. 5, 1974). Pilbeam suggests that its status should be raised to × *Astroworthia skinneri* but has not published this name correctly. The stems grow 10–15 cm tall and are 7–8 cm wide at the base, with leaves which usually have 2 paler green keels beneath and have paler green tubercles and margin. The name *Astroloba bulbulata* (Jacquin) Uitewaal appears to be a synonym of this hybrid plant.

3. POELLNITZIA Uitewaal

V.A. Matthews

Differs from *Astroloba* in having larger, orange-red flowers that tend to be borne on only 1 side of the raceme. Stems to 30 × 7 cm, with ovoid-triangular leaves in 5 spiral ranks. Flowers to 2.5 cm. Perianth-segments incurved at the tips; margins of the outer segments fused to the midribs of the inner ones.

A genus of 1 species from South Africa. Cultivation as for *Haworthia* (p. 38).

1. P. rubriflora (L. Bolus) Uitewaal (*Apicra rubriflora* L. Bolus; *Aloe rubriflora* (L. Bolus) Rowley). Illustration: Pilbeam, Haworthia and Astroloba, 147, 154 (1983).

South Africa (Cape Province). G1. Summer.

4. GASTERIA Duval

V.A. Matthews

Succulent perennials, usually stemless. Leaves in rosettes, arranged spirally or in 2 ranks, fleshy, flat or somewhat keeled, with spots or tubercles, margin entire. Flowers borne in loose racemes or panicles, red or pinkish, often green at the apex. Perianth with a tubular, curved, inflated tube, formed by the joining of the outer lobes. Stamens dorsifixed. Ovary superior, 3-celled; style 1. Fruit a 3-celled capsule opening between the septa and containing flat, winged seeds.

About 50 species, native to South Africa. They will grow in any well-drained soil in a cool greenhouse, and need semi-shady conditions. Propagation is by seed or leaf cuttings; the plants form clumps of rosettes that can be separated to produce more plants.

1a. Leaves 3–5.5 cm 2
b. Leaves 6 cm or more 3
2a. Leaves linear, 1–1.5 cm wide at the base **14. liliputana**
b. Leaves tongue-shaped to triangular-ovate, 2–3.5 cm wide at the base **12. armstrongii**

3a. Mature rosettes with leaves in 2 ranks or almost so, occasionally the ranks arranged spirally 4
b. Mature rosettes with all leaves arranged spirally, or in several rows 18
4a. Leaves with green spots 5
b. Leaves with white or greenish white spots or tubercles 6
5a. Stem 1–1.5 cm; leaves 1.8–2 cm wide at the base **11. caespitosa**
b. Stem 15–25 cm; leaves 3–4 cm wide at the base **10. marmorata**
6a. Leaves 25–35 cm 7
b. Leaves to 25 cm 8
7a. Leaves 1.3–2.5 cm wide at the base **9. pulchra**
b. Leaves 4–5 cm wide at the base **7. picta**
8a. Stem 10–30 cm 9
b. Stem absent or very short 11
9a. Leaves *c.* 34, 4.5–5 cm wide at the base **8. maculata**
b. Leaves 12–20, 2–3 cm wide at the base 10
10a. Leaves with white spots arranged in transverse bands **9. pulchra**
b. Leaves with scattered white spots **6. planifolia**
11a. Leaves 1.3–2 cm wide at the base **1. verrucosa**
b. Leaves 2–5 cm wide at the base 12
12a. Leaves with scattered spots or tubercles 13
b. Leaves with spots or tubercles arranged in transverse rows, or more or less so 15
13a. Leaf margin with tubercles 14
b. Leaf margin lacking tubercles **5. nigricans**
14a. Leaves 4–10 **1. verrucosa**
b. Leaves 12–16 **13. excavata**
15a. Flowers to 2.2 cm **5. nigricans**
b. Flowers 2.4 cm or more 16
16a. Leaves 4–6, 18–20 cm; flowers *c.* 3.5 cm **4. neliana**
b. Leaves 8–12, 20–25 cm; flowers *c.* 2.5 cm 17
17a. Leaves with a straight margin **2. angulata**
b. Leaves somewhat wavy along the margin **3. disticha**
18a. Leaves 6–22 cm 19
b. Leaves 25–35 cm 28
19a. Leaves 1–1.5 cm wide at the base **14. liliputana**
b. Leaves 2–6.5 cm wide at the base 20
20a. Leaves with white tubercles on the upper surface 21
b. Leaves with white or greenish white spots or blotches on the upper surface 22
21a. Leaves 2.5–3 cm wide at the base **17. subcarinata**
b. Leaves 3.5–5 cm wide at the base **16. carinata**
22a. Leaves 25 or more **8. maculata**
b. Leaves 8–20 23
23a. Leaves convex on the upper surface **5. nigricans**
b. Leaves concave on the upper surface 24
24a. Leaves 8–10 cm **15. humilis**
b. Leaves 10–22 cm 25
25a. Leaves 10–15 cm 26
b. Leaves 15–22 cm 27
26a. Young rosettes with leaves in 2 ranks **20. schweickerdtiana**
b. Young rosettes with leaves in a spiral **19. obtusa**
27a. Leaves 8–12 **20. schweickerdtiana**
b. Leaves 12–15 **18. nitida**
28a. Leaves curved, acute **21. acinacifolia**
b. Leaves straight, blunt **22. candicans**

1. G. verrucosa (Miller) Duval. Illustration: Step, Favourite flowers of garden and greenhouse **4**: pl. 270 (1897); Das Pflanzenreich **33**: 42, 43 (1908); Lamb & Lamb, Illustrated reference on cacti and succulents **1**: 228 (1955); Jacobsen, Lexicon of succulent plants, pl. 87, 4 (1974).

Rosettes with a very short stem. Leaves 6–10, in 2 ranks; 10–22 × 1.3–2 cm, lanceolate, shortly acuminate at the apex, grooved above, convex beneath, dull green with scattered white tubercles; margin thickened, tuberculate. Flowering stems to 60 cm; flowers 1.8–2.5 cm. *South Africa.* G1.

G. prolifera Lemaire. Illustration: Riha & Subik, Illustrated encyclopedia of cacti and other succulents, 369 (1981). Differs in the rosettes being stemless, and having leaves 20–30 × 1.3–1.5 cm, spotted with white. *South Africa.* G1.

G. pillansii Kensit. The rosettes are stemless and the obtuse, mucronate leaves are 4–4.5 cm wide at the base, spotted with white. *South Africa.* G1.

G. × repens Haworth. Leaves 6–8 cm with small white tubercles in transverse rows. A hybrid between *G. verrucosa* and *G. carinata.*

2. G. angulata (Willdenow) Haworth.

Rosettes stemless. Leaves 8–10, in 2 ranks; 20–25 × *c.* 5 cm, grooved above, convex beneath, green with small white spots in transverse rows. Flowering stems to 90 cm; flowers *c.* 2.5 cm. *South Africa.* G1. Summer.

3. G. disticha (Linnaeus) Haworth (*G. lingua* (Thunberg) Berger). Illustration: Das Pflanzenreich **33**: 44 (1908); Jacobsen, Lexicon of succulent plants, pl. 86, 6 (1974).

Rosettes stemless. Leaves 10–12, in 2 ranks; 20–25 × 3–5 cm, tongue-shaped, green with white spots in transverse rows; margin wavy. Flowering stems to 90 cm; flowers *c.* 2.5 cm. *South Africa.* G1. Spring.

4. G. neliana von Poellnitz. Illustration: Lamb & Lamb, Illustrated reference on cacti and succulents **2**: 497 (1959).

Rosettes stemless. Leaves 4–6, in 2 ranks; 18–20 × 4–4.5 cm, tongue-shaped, pointed at the tip, both surfaces slightly convex, emerald green with whitish green blotches in more or less transverse rows; margin minutely toothed. Flowers *c.* 3.5 cm. *South Africa.* G1.

5. G. nigricans (Haworth) Duval. Illustration: Das Pflanzenreich **33**: 45 (1908).

Rosettes with a very short stem. Leaves 10–20, in 2 ranks or somewhat spiraled; 6–20 × 4–5 cm, tongue-shaped, pointed at the tip, both surfaces convex, dark green with white or greenish white, often rather indistinct, spots in irregular transverse rows; margin horny. Flowering stems to 90 cm; flowers 1.8–2.2 cm. *South Africa.* G1.

G. brevifolia Haworth. Illustration: Hay et al., Dictionary of indoor plants in colour, 250 (1974). Rosettes stemless, leaves with white spots that merge into transverse bands, margin minutely toothed. *South Africa.* G1. Summer.

G. triebneriana von Poellnitz. Illustration: *The Cactus Journal* **7**: 27 (1939). Leaves fewer, with whitish spots that may be scattered or coalesced into indistinct transverse bands. Older leaves much curved inwards, the upper side with a distinct central rib. Flowers 2–2.2 cm. *South Africa.* G1.

6. G. planifolia (Baker) Baker. Illustration: Das Pflanzenreich **33**: 47 (1908).

Rosettes columnar with a stem 15–25 cm. Leaves 12–20, in 2 ranks; 15–20 × 2–2.5 cm, lanceolate, glossy, dark green with scattered, confluent white spots. Flowering stems to 180 cm; flowers *c.* 1.8 cm. *South Africa.* G1.

G. bicolor Haworth. Illustration: Das Pflanzenreich **33**: 46 (1908). Rosettes only 10–15 cm, leaves dull light green with scattered white spots, especially at the base. *South Africa.* G1.

7. G. picta Haworth.

Rosettes becoming short-stemmed with age. Leaves 12–20, in 2 spiral ranks; 25–35 × 4–5 cm, triangular-lanceolate, slightly convex above, glossy blackish green, with confluent white spots in transverse rows. Flowering stems to 90 cm; flowers *c.* 1.8 cm. *South Africa.* G1. Summer.

8. G. maculata (Thunberg) Haworth. Illustration: Rowley, Illustrated encyclopedia of succulents, 163 (1978); Riha & Subik, Illustrated encyclopedia of cacti and other succulents, 367 (1981).

Rosettes columnar with a stem 20–30 cm. Leaves *c.* 34, in 2 ranks or arranged spirally; 15–20 × 4.5–5 cm, tongue-shaped, with 2 keels beneath, glossy dark green with white spots that are in rows towards the tip. Flowering stems to 120 cm; flowers 1.8–2 cm. *South Africa.* G1. Summer.

9. G. pulchra (Aiton) Haworth. Illustration: *Addisonia* **23**: pl. 752 (1955); Rauh, Die Grossartige Welt der Sukkulenten, t. 68 (1967).

Rosettes columnar with a stem 15–30 cm. Leaves 15–20, more or less in 2 ranks; 20–30 × *c.* 2.5 cm, linear, often with 2 keels beneath, shiny grey-green with confluent white spots in transverse rows; margins horny, white. Flowering stems to 90 cm; flowers *c.* 1.8 cm. *South Africa.* G1. Spring–summer.

G. prolifera, which differs in its stemless rosettes, and leaves only 1.3–1.5 cm wide at the base with scattered spots, may key out here (see under 1. *G. verrucosa*

10. G. marmorata Baker. Illustration: Jacobsen, Lexicon of succulent plants, pl. 87, 2 (1974).

Rosettes columnar with a stem 15–25 cm. Leaves 20–30, in 2 ranks; 13–15 × 3–4 cm, lanceolate, blunt, dark green mottled with darker green. Flowering stems to 75 cm; flowers *c.* 1.8 cm. *South Africa.* G1.

11. G. caespitosa von Poellnitz.

Rosettes with a short stem 1–1.5 cm. Leaves in 2 ranks; 10–14 × 1.8–2 cm, lanceolate, green spotted with light green above, the spots coalescing into indefinite transverse bands beneath. Flowers 1.9–2 cm. *South Africa.* G1.

12. G. armstrongii Schönland. Illustration: Lamb & Lamb, Illustrated reference on cacti and succulents **3**: 98 (1963); Court, Succulent flora of southern Africa, 175 (1981); Riha & Subik, Illustrated encyclopedia of cacti and other succulents, 368 (1981).

Rosettes stemless, or with a short stem to 4 cm. Leaves few, in 2 ranks when young, spirally arranged in older rosettes; 3–5.5 × 2–3.5 cm, tongue-shaped to triangular-ovate, apex pointed, green with white tubercles in tranverse rows. Flowers *c.* 2 cm. *South Africa.* G1.

G. batesiana Rowley. Illustration: *National Cactus and Succulent Society Journal* **10**: 2 (1955); Rauh, Die Grossartige Welt der Sukkulenten, t. 67 (1967); Jacobsen, Lexicon of succulent plants, pl. 81 4 (1974); Barkhuizen, Succulents of southern Africa, 179 (1978). Rosettes stemless, with spirally arranged leaves. Leaves blackish green with large green and small white tubercles in irregular transverse rows, especially in the lower half. Flowers *c.* 3.5 cm. *South Africa.* G1.

G. stayneri von Poellnitz. Illustration: *The Cactus Journal* **7**: 18 (1938). Leaves dark green with many dark green tubercles when young, becoming spotted with greenish white when mature. *South Africa.* G1.

13. G. excavata (Willdenow) Haworth. Illustration: Das Pflanzenreich **33**: 48 (1908).

Rosettes stemless. Leaves 12–16, in 2 spiral ranks; 10–15 × 2.5–3.5 cm, lanceolate, pale green with small whitish spots; margin tuberculate. Flowering stems to 60 cm; flowers *c.* 2.5 cm. *South Africa.* G1. Summer.

G. pseudonigricans (Salm-Dyck) Haworth (*G. subnigricans* Haworth). Leaves in 2 ranks; 15–20 × 3.5–4 cm, glossy green with indistinct white spots. Flowering stems 90–100 cm; flowers 1.8–2 cm. *South Africa.* G1.

14. G. liliputana von Poellnitz. Illustration: *Flowering Plants of South Africa* **9**: pl. 360 (1929); Jacobsen, Lexicon of succulent plants, pl. 87, 3 (1974); Barkhuizen, Succulents of southern Africa, 180 (1978).

Rosettes with short stems. Leaves arranged in a spiral; 3.5–6.5 × 1–1.5 cm, lanceolate, dark shiny green with white spots that often coalesce into transverse rows; margin with tiny teeth. Flowering stems to 10 cm; flowers *c.* 1.5 cm. *South Africa.* G1.

15. G. humilis von Poellnitz.

Rosettes stemless. Leaves 8–12, arranged in a spiral; 8–10 × 2–4 cm, long-triangular, concave above, glossy dark green with whitish spots that may be scattered or coalescent into transverse rows; margin tuberculate. Flowering stems to 60 cm; flowers *c.* 2.3 cm. *South Africa.* G1.

16. G. carinata (Miller) Duval. Illustration: *Botanical Magazine*, 1331 (1811); Das Pflanzenreich **33**: 50 (1908); *Flowering plants of South Africa* **8**: pl. 291 (1928).

Rosettes stemless. Leaves 4–8, arranged in a spiral; 12–15 × 3.5–5 cm, triangular-lanceolate, narrowed to a spine at the tip, concave above, dull green with white tubercles that may be scattered or coalescent into transverse rows. Flowering stems to 90 cm; flowers 2.5–4 cm. *South Africa.* G1. Summer.

17. G. subcarinata (Salm-Dyck) Haworth.

Rosette with a short stem. Leaves 10–15, in several rows; 10–15 × 2.5–3 cm, triangular-lanceolate, concave above, green with flat, white tubercles; margin minutely toothed. *South Africa.* G1.

Possibly a hybrid between *G. carinata* and *G. disticha.*

18. G. nitida (Salm-Dyck) Haworth. Illustration: Marloth, Flora of South Africa **4**: pl. 21 (1915).

Rosette with a short stem. Leaves 12–15, arranged in a spiral; 15–22 × *c.* 6.5 cm, broadly triangular-lanceolate, glossy green with whitish spots that coalesce into indistinct transverse rows on the lower surface. Flowering stems to 90 cm; flowers 2.2–2.5 cm. *South Africa.* G1. Summer.

19. G. obtusa (Salm-Dyck) Haworth.

Leaves 11–13, arranged in a spiral; 10–15 × *c.* 3.5 cm, triangular, glossy green with indistinct spots in transverse rows; margin minutely toothed. Flowers *c.* 2.5 cm. *South Africa.* G1.

20. G. schweickerdtiana von Poellnitz.

Rosettes stemless or with a very short stem. Leaves 8–12, in 2 ranks in young plants, later arranged in a spiral; 14–18 × 3.5–4 cm, concave above, convex beneath, light glossy green with white spots that are usually scattered on the upper surface, coalescent into transverse rows beneath. *South Africa.* G1.

G. trigona Haworth. Leaves spirally arranged in many rows; 3–3.5 cm wide, with white spots in transverse bands. *South Africa.* G1.

21. G. acinacifolia (Jacquin) Haworth. Illustration: Das Pflanzenreich **33**: 52 (A–F), (1908); Rauh, Die Grossartige Welt der Sukkulenten, t. 67 (1967); Olmos, Los Cactus y las otras plantas succulentas, 91 (1977); Court, Succulent Flora of southern Africa, 175 (1981).

Rosettes stemless. Leaves arranged in many rows; to 35 × 5 cm, lanceolate, concave above, dark green with small greenish white spots in transverse rows; lower part with tubercles. Flowering stems to 120 cm; flowers to 5 cm. *South Africa.* G1. Summer.

22. G. candicans Haworth.

Rosettes stemless. Leaves 12–20, arranged in many row;. 25–30 × *c.* 7 cm, broadly triangular-lanceolate, concave above, glossy green with tiny spots in indistinct transverse rows; margin white. Flowers 4–5 cm. *South Africa.* G1. Summer.

Names of doubtful application. The names *G. nova* and *G. variegata* have been found in nurserymen's catalogues; it is not known to which species these names apply.

5. ALOE Linnaeus

J. Cullen, edited by S.G. Knees

Plants herbaceous or, more frequently, woody, succulent, with or without distinct stems, sometimes with trunks and forming small trees. Leaves usually in distinct rosettes either at ground level or at the ends of the branches, more rarely dispersed along the branches, fleshy, often tough and leathery, and usually with soft, firm or sharp spines, at least along the margins. Inflorescence a raceme or panicle, one to several from each rosette; flowers stalked, each subtended by a bract, the inflorescence-stalk often bearing sterile bracts. Perianth of 6 lobes in 2 whorls, those of the outer whorl free or united for part of their length, those of the inner whorl variably joined to those of the outer; perianth-segments yellow, orange or red, their tips often greenish. Stamens 6, the anthers usually projecting from the perianth one after another. Ovary superior, 3-celled, with many ovules in each cell; style and stigma usually projecting from the perianth. Fruit a papery or woody capsule with many seeds.

A genus of about 360 species, mostly from southern Africa, but also occurring in Madagascar, the Canary Islands, east Africa and the southern part of the Arabian peninsula; several species are naturalised in southern Europe. Its classification is difficult, made so by the occurrence of natural hybrids, and the fact that material in cultivation may often differ morphologically from the wild stock. Reynolds (Aloes of South Africa, edn 4, 104, 1974) goes so far as to say: 'In the wild state, and in their various geographical situations, plants vary considerably in size, length and width of leaf, leaf markings, racemes, length of pedicels and flowers, while plants grown in gardens and greenhouses under conditions very different from their natural habitats, frequently modify out of almost all recognition.' In spite of the last part of this sentence, an attempt is made here to provide means of identification for all the species likely to be found in general cultivation; the existence of hybrids and the difficulties mentioned by Reynolds must, however, be borne in mind when attempts at identification are made. The elaborate hierarchy of Sections and Series proposed by Reynolds (references below) is not used here, as its value for identification seems limited. Various members of the Agavaceae (p. 174–187) may be incorrectly keyed out to this genus; all of these have leaves with hard, spiny tips, a feature uncommon in the cultivated Aloes. Some hybrids have been raised between species of *Gasteria* (p. 46) and *Aloe*: these are known as × *Gastrolea* Walther.

In the descriptions, the leaves must be considered to be acute unless the opposite is stated; the measurements of bracts and flower-stalks refer to those of the lowest open flowers in the raceme.

Aloes are easily grown in well-drained, rich soils; watering should be reduced in the resting season and resumed when new growth appears; most species can tolerate or benefit from full exposure to sunlight. Propagation is mainly by seed.

Literature: Berger, A., Das Pflanzenreich **33**: 159–330 (1908); Reynolds, G.W., The Aloes of tropical Africa and Madagascar (1966); Reynolds, G.W., The Aloes of South Africa, edn 4 (1974).

1a. Plants without distinct stems or trunks, or stems not exceeding 50 cm — 2
b. Plants with obvious and distinct stems or trunks that exceed 50 cm — 23
2a. Outer perianth-lobes free for more than half of their length — 3
b. Outer perianth-lobes united for more than half of their length — 14
3a. Perianth 3.5 cm or more — 4
b. Perianth less than 3.4 cm — 9
4a. Leaves 30–60 × 10–15 cm — 5
b. Leaves to 20 × 4 cm — 6
5a. Racemes inclined almost to the horizontal, the flowers borne on the upper side; flower-stalks *c.* 8 mm — **29. ortholopha**
b. Racemes erect or spreading, the flowers borne all round the axis; flower-stalks 3–4 cm — **12. glauca**
6a. Leaves broadly triangular, *c.* 3 × longer than the basal breadth; plant producing numerous offsets from the base — **11. brevifolia**
b. Leaves lanceolate, usually much more than 3 × longer than the basal breadth; plant not producing offsets — 7
7a. Spines on the margins of mature leaves black, *c.* 1 cm — **10. melanacantha**
b. Spines on the margins of mature leaves white, 2–3 mm or absent — 8
8a. Leaves at most 2 cm broad, tuberculate or prickly on the upper and/or lower surfaces; flower-stalks 2.5–3.5 cm — **8. humilis**
b. Leaves 2–4 cm broad, neither tuberculate nor prickly; flower-stalks up to 1.8 cm — **9. krapohliana**
9a. Leaves 30–50 × 5–10 cm; flower-stalks less than 5 mm or more than 2.5 cm — 10
b. Leaves 7–13 cm × 8–20 mm; flower-stalks more than 5 mm and less than 2.5 cm — 12
10a. Flower-stalks 2.5–3 cm, leaves with H-shaped spots — **27. microstigma**
b. Flower-stalks to 5 mm; leaves unspotted or with obscure lines — 11
11a. Inflorescence usually branched; flowers loose, their perianths clearly exceeding the bracts — **17. vera**

b. Inflorescence simple; flowers very dense and touching, bracts exceeding the perianths **5. broomii**
12a. Stem present, 10–20 cm; perianth orange at the base, shading to yellow at the mouth, the tips of the lobes green **3. bakeri**
b. Stem absent; perianth orange-pink 13
13a. Leaves to 1 cm broad, both surfaces rough and with round or lens-shaped pale green spots; perianth to 1.5 cm **1. bellatula**
b. Leaves 1.5–2 cm broad, both surfaces smooth and with pale, H-shaped spots; perianth *c.* 2.5 cm **2. rauhii**
14a. Leaves copiously spotted on at least the upper surface, the spots sometimes forming the bases of spines 15
b. Leaves unspotted or very rarely sparsely spotted, without spines on the surface 20
15a. Leaves with white spots forming the tuberculate bases of spines; perianth *c.* 5.5 cm, upcurved in the upper third; style projecting *c.* 2.5 cm **4. longistyla**
b. Leaves with spots that do not form the bases of spines; perianth not more than 4.5 cm, not upcurved in the upper third; style not projecting, or projecting at most for 5 mm 16
16a. Leaves 100–150 in each rosette, each leaf tapering to a long, dry 'awn' **6. aristata**
b. Leaves to 25 in each rosette, acute or acuminate, but not as above 17
17a. Racemes crowded at the ends of the inflorescence branches; flower-stalks 3.5–4.5 cm; leaves 8–12 cm broad **13. saponaria**
b. Racemes elongate along the inflorescence branches; flower-stalks 4–15 mm; leaves 4–7 cm broad 18
18a. Flower-stalks 1–1.5 cm **15. grandidentata**
b. Flower-stalks 4–7 mm; 19
19a. Leaves borne in rosettes; flower-stalks 6–7 mm **14. zebrina**
b. Leaves arranged in 3 ranks; flower-stalks 4–6 mm **39. variegata**
20a. Leaf margin entire, untoothed, forming a pink or red zone 2–3 mm wide **16. striata**
b. Leaf margin toothed (sometimes distantly so), not as above 21
21a. Leaves 2–3 cm broad; inflorescence simple, unbranched **7. virens**
b. Leaves 7.5–12 cm broad; inflorescence branched 22
22a. Perianth bright red, ultimately fading yellow; leaves to 35 × 7.5 cm **18. perryi**
b. Perianth orange or yellow; leaves 50–60 × 8–12 cm **19. camperi**
23a. Leaves borne strictly in 2 ranks, their margins entire and their apices rounded **38. plicatilis**
b. Leaves in more than 2 ranks, their margins usually toothed and their apices usually acute 24
24a. Plant tree-like with a richly, apparently dichotomously branched crown **37. dichotoma**
b. Plant not as above 25
25a. Outer perianth-lobes united for half or more of their length 26
b. Outer perianth-lobes free for half or more of their length 30
26a. Leaves 40–65 × 10–20 cm, forming a distinct, tight rosette, the stem scarcely visible between the leaves 27
b. Leaves 10–15 × 1–2.5 cm, forming an open rosette dispersed along the stem, which is clearly visible 29
27a. Leaf margin entire, untoothed, forming a pink or red zone 2–3 mm wide **16. striata**
b. Leaf margin distinctly toothed, not as above 28
28a. Flowers distinctly curved; leaves dull green to almost glaucous **31. africana**
b. Flowers straight; leaves reddish or purple near the base, shading to pink near the apex, the whole with a violet bloom **30. rubroviolacea**
29a. Leaf-base conspicuously auricled and ciliate; flowers *c.* 3 cm **21. ciliaris**
b. Leaf-base not auricled and ciliate; flowers 1.1–1.4 cm **20. tenuior**
30a. Flower-stalks 3–4.5 cm 31
b. Flower-stalks 1–6 mm 35
31a. Leaves 15–25 cm, marginal teeth golden yellow, brown or black; bracts 8–10 mm 32
b. Leaves 30 cm or more, marginal teeth white or reddish brown; bracts 1.5–3 cm 33
32a. Leaves not, or very sparingly, white-spotted **24. mitriformis**
b. Leaves conspicuously and densely white-spotted, especially on the lower surface **25. distans**
33a. Marginal teeth of leaves hard and sharp; leaves 10–15 cm broad; bracts *c.* 3 cm **12. glauca**
b. Marginal teeth of leaves firm but not hard and sharp, leaves 5–10 cm broad; bracts to 2 cm 34
34a. Leaves widely spreading, deflexed towards their apices; perianth orange-red **28. arborescens**
b. Leaves spreading at the extreme base, then curving upwards; perianth red to pink **26. succotrina**
35a. Leaves to 25 × 2.5 cm, with conspicuously striped sheaths **22. striatula**
b. Leaves 50–160 × 7–25 cm, sheaths not conspicuously striped 36
36a. Racemes held horizontally or obliquely, the flowers borne on 1 side of the axis **36. marlothii**
b. Racemes erect or ascending, the flowers borne all round the axis 37
37a. Rosette loose, open, the leaves spread along the stem, which is clearly visible between them **23. divaricata**
b. Rosette tight, dense, the stem scarely visible between the leaves 38
38a. Raceme 15–25 cm; flower-stalks 1–2 mm 39
b. Raceme 50–80 cm; flower-stalks 4–6 mm 40
39a. Marginal teeth of the leaves 5–6 mm; perianth reddish orange **35. excelsa**
b. Marginal teeth of the leaves *c.* 2 mm; perianth pale yellow **34. thraskii**
40a. Leaves spreading, flat or only slightly channelled; inner perianth-lobes with brown tips **32. ferox**
b. Leaves much recurved, deeply channelled; inner perianth-lobes without brown tips **33. candelabrum**

1. A. bellatula Reynolds. Illustration: Reynolds, Aloes of tropical Africa, t. 89 & 407–9 (1966).

Plants small, stemless, suckering and forming dense tufts. Leaves *c.* 16 in a basal rosette; 10–13 cm × 9–10 mm, linear but tapering, both surfaces roughened with papilla-like points, dark green, copiously spotted on both surfaces with pale, lens-shaped spots; margins with soft, triangular teeth. Raceme to 60 cm, rarely branched; flower-stalks *c.* 1.2 cm, bracts 4–6 mm. Perianth *c.* 1.3 cm, cylindric to bell-shaped, pale orange-pink, outer lobes free for little more than half their length. Anthers and style scarcely projecting. *Madagascar.* H5–G1.

2. A. rauhii Reynolds. Illustration: Reynolds, Aloes of tropical Africa, 422–3 (1966); *Cactus & Succulent Journal of America* **51**: 49 (1979).

Stem absent or very short. Rosettes *c.* 10 cm in diameter, with about 20 leaves. Leaves 7–10 × 1.5–2 cm, triangular-lanceolate, greyish brown or brownish with numerous H-shaped spots on both surfaces; margins with soft, white teeth. Raceme to 30 cm, rarely branched; flower-stalks *c.* 1 cm, bracts 4–5 mm. Perianth *c.* 2.5 cm, constricted above the ovary then cylindric, orange-pink, outer lobes free to the base. Anthers and style projecting *c.* 1 mm. *Madagascar.* H5–G1.

3. A. bakeri Scott-Elliott. Illustration: Reynolds, Aloes of tropical Africa, t. 91 & 424–6 (1966).

Stem 10–20 cm, branched, producing offsets at the base. Rosettes with about 12 leaves. Leaves *c.* 7 cm × 8 mm, lanceolate, green with a reddish tinge, usually unspotted, rarely with a few pale green spots; margins with firm, white teeth. Racemes to 30 cm, simple; flower-stalks 1–1.2 cm, bracts *c.* 3 mm. Perianth *c.* 2.3 cm, cylindric, orange shading to yellowish at the mouth, outer lobes free to the base. Anthers and style projecting *c.* 1 mm. *Madagascar.* H5–G1.

4. A. longistyla Baker. Illustration: Jeppe, South African Aloes, 35 (1969); Reynolds, Aloes of South Africa, edn 4, t. 6 & 142–3 (1974).

Plant stemless. Rosettes with 20–30 leaves. Leaves 12–15 × *c.* 3 cm, arching-erect, incurved, glaucous, both surfaces with soft or firm spines that arise from white, tubercular bases; margins with similar spines. Inflorescence to 20 cm, simple; flower-stalks 6–8 mm, bracts 2.5–3 cm. Perianth *c.* 5.5 cm, cylindric, upcurved in the upper third, pink to red, outer lobes united for more than half their length. Anthers and style projecting 2–2.5 cm. *South Africa (Cape Province).* H5–G1.

5. A. broomii Schönland. Illustration: Jeppe, South African Aloes, 54 (1969); Reynolds, Aloes of South Africa, edn 4, 147–9 (1974); *Flowering plants of South Africa* **16**: t. 605 (1936).

Plant stemless or stem very short. Rosette to 1 m in diameter, with numerous leaves. Leaves about 30 × 10 cm, broadly ovate-lanceolate, terminating in a sharp spine, unspotted but with obscure lines; margins with pale, triangular teeth. Inflorescence a very dense, simple raceme to 1.5 m; flower-stalks 1–2 mm, bracts exceeding the perianths. Perianth 2–2.5 cm, cylindric, slightly curved, pale yellow, outer lobes free for more than half their length. Anthers and style projecting 1.2–1.5 cm. *South Africa (Cape Province).* H5–G1.

Easily recognised by its tight, dense, cylindric racemes, the flowers with short stalks and long bracts.

6. A. aristata Haworth. Illustration: Jeppe, South African Aloes, 15 (1969); Reynolds, Aloes of South Africa, edn 4, 157–9 (1974); Brickell (ed.), RHS A–Z encyclopedia of garden plants, 100 (2003).

Plants small, stemless, forming compact groups of rosettes. Leaves 100–150 in each rosette; 8–10 × 1–1.5 cm, incurved, narrowly lanceolate, tapering to a long, dry 'awn', green with white spots on both surfaces; margins with soft, white teeth. Inflorescence with 2–6 branches, to 50 cm; flower-stalks to 3.5 cm, bracts 1–1.5 cm. Perianth *c.* 4 cm, cylindric, curved, swollen around the ovary, red, the outer lobes united for most of their length. Anthers and style projecting 1–2 mm. *South Africa.* H5–G1. Summer.

7. A. virens Haworth. Illustration: *Botanical Magazine*, 1355 (1811).

Plant stemless, rosettes producing offshoots and forming groups. Leaves numerous in each rosette; *c.* 20 × 2–3 cm, narrowly lanceolate, acuminate, marked with lines and faint spots; margins with firm, white teeth. Inflorescence simple, to 60 cm, most of its axis without bracts; flower-stalks *c.* 3 cm, bracts *c.* 1.5 cm. Perianth *c.* 4 cm, cylindric, broadening upwards, red, the outer lobes united for more than half their length. Stamens and style scarcely projecting. *Origin uncertain.* H5–G1. Spring–summer.

This species has been in cultivation for almost 200 years, but it is not definitely known in the wild. Reynolds (Aloes of South Africa, edn 4, 173, 1974) suggests that it may be a hybrid, with *A. humilis* as one of the parents.

8. A. humilis (Linnaeus) Miller. Illustration: Jeppe, South African Aloes, 14 (1969); Reynolds, Aloes of South Africa, edn 4, 166–8 (1974).

Plants stemless, forming dense groups. Leaves 20–30 in each rosette; *c.* 10 × 1.2–1.8 cm, erect or incurved, ovate-lanceolate, glaucous, both surfaces usually tuberculate and with some soft, white spines; margins with soft, white teeth. Inflorescence simple, to 35 cm; flower-stalks 2.5–3.5 cm, bracts 2.5–3.5 cm. Perianth cylindric but slightly swollen in the middle, 3.5–4.2 cm, almost straight, orange-red or orange, the outer lobes free for most of their length. Anthers and style scarcely projecting. *South Africa (Cape Province).* H5–G1. Spring.

A variable species; var. **echinata** (Willdenow) Baker, which has leaves with soft prickles (rather than tubercles) on the upper surface, is widely grown.

9. A. krapohliana Marloth. Illustration: Jeppe, South African Aloes, 34 (1969); Reynolds, Aloes of South Africa, edn 4, t. 8 & 170–1 (1974).

Plants stemless, or stem to 20 cm, solitary. Leaves 20–30 in each rosette, incurved; to 20 × 2–4 cm, glaucous, smooth, the lower surface sometimes with differently coloured bands; margins with soft, white teeth. Inflorescence usually simple, to 40 cm; flower-stalks to 1.8 cm, bracts somewhat shorter. Perianth cylindric, orange-red, *c.* 3.5 cm, the outer lobes free to the base. Anthers and style projecting 2–3 mm. *South Africa.* H5–G1.

10. A. melanacantha Berger. Illustration: *Flowering plants of South Africa* **11**: t. 433 (1931); Jeppe, South African Aloes, 25 (1969); Reynolds, Aloes of South Africa, edn 4, t. 9 & 172–3 (1974).

Plants with short stems, ultimately to 50 cm. Rosettes to 30 cm in diameter. Leaves numerous, to 20 × 4 cm, incurved, triangular-lanceolate, dull green, the upper surface smooth, the lower with a few black prickles; margins with hard, sharp, black spines. Inflorescence simple, to 1 m; flower-stalks *c.* 1.5 cm, bracts to 2.5 cm. Perianth *c.* 4.5 cm, cylindric, orange-red becoming yellow, the outer lobes free almost to the base. Anthers and style projecting *c.* 5 mm. *South Africa, South West Africa.* H5—G1.

Easily recognised by the sharp, black spines on the leaf margins.

11. A. brevifolia Miller. Illustration: Jeppe, South African Aloes, 13 (1969); Reynolds, Aloes of South Africa, edn 4, 177–9 (1974).

Plants stemless with many offsets from the base, forming large groups. Rosettes 7–8 cm in diameter, with 30–40 leaves. Leaves *c.* 6 × 2 cm, broadly triangular-lanceolate, glaucous, the upper surface without spots or spines, the lower with a few soft spines; margins with firm, white teeth. Inflorescence simple, to 40 cm; flower-stalks *c.* 1.5 cm, bracts *c.* 1.5 cm. Perianth *c.* 3.8 cm, cylindric, orange-yellow, the outer lobes free to the base. Anthers and style projecting 4–5 mm. *South Africa (Cape Province).* H5–G1.

12. A. glauca Miller. Illustration: Jeppe, South African Aloes, 54 (1969); Reynolds, Aloes of South Africa, edn 4, 197–200 (1974).

Stems variable in height, to 60 cm or rarely more. Rosette open, with many, spreading leaves. Leaves to 40 × 15 cm, broadly lanceolate, glaucous, lined on both surfaces; margins with dark brown, hard, sharp teeth. Inflorescence simple, to 1 m; flower-stalks to 4 cm, bracts 2.5–3 cm. Perianth *c.* 3.5 cm, cylindric, pink-orange fading yellow, outer lobes free to the base. Anthers and style projecting 2–3 mm. *South Africa.* H5–G1.

13. A. saponaria (Aiton) Haworth. Illustration: Lamb & Lamb, The illustrated reference to cacti and other succulents **2**: 443 (1959); Jeppe, South African Aloes, 67 (1969); Reynolds, Aloes of South Africa, edn 4, t. 12 & 230–3 (1974); Brickell (ed.), RHS A–Z encyclopedia of garden plants, 101 (2003).

Plant stemless or with a stem to 50 cm. Rosettes open, with 12–20 spreading leaves. Leaves 25–30 × 8–12 cm, narrowly or broadly lanceolate, the upper surface pale to dark green with numerous, dull, white, oblong spots arranged in irregular, transverse bars, lower surface paler and not spotted; margins with hard, sharp, triangular teeth. Inflorescence branched, 40–60 cm, flowers congested at the ends of the branches; flower-stalks 3.5–4.5 cm, bracts 1.7–2.2 cm. Perianth conspicuously constricted above the ovary, cylindric above, 3.5–4.5 cm, yellow to orange, outer lobes united for more than half their length. Anthers and style projecting 1–5 mm. *South Africa.* H5–G1. Summer.

14. A. zebrina Baker. Illustration: Jeppe, South African Aloes, 87 (1969); Reynolds, Aloes of South Africa, edn 4, 302–3 (1974).

Plants stemless or with very short stems, suckering freely and forming groups. Rosettes with 15–25 leaves. Leaves 15–30 × 6–7 cm, linear-lanceolate, tapering from about the middle, dark green with a glaucous bloom and with large, whitish spots arranged in irregular transverse bands above; margins with hard, sharp, brownish teeth. Inflorescence to 1.5 m, branched; flower-stalks 6–7 mm, bracts 1.2–1.5 cm. Perianth 3–3.5 cm, conspicuously and abruptly narrowed above the ovary, then cylindric and curved downwards, dull red, the outer lobes united for most of their length. Anthers and style not or scarcely projecting. *Southern Africa, from Angola to Mozambique southwards.* H5–G1.

A very variable species.

15. A. grandidentata Salm-Dyck. Illustration: Jeppe, South African Aloes, 63 (1969); Reynolds, Aloes of South Africa, edn 4, 305–7 (1974).

Like *A. zebrine* but leaves 10–15 cm, the apex dried and twisted; flower-stalks 1–1.5 cm; perianth 2.8–3 cm, cylindric to club-shaped, not conspicuously and abruptly constricted above the ovary. *South Africa.* H5–G1. Summer.

16. A. striata Haworth. Illustration: Lamb & Lamb, The illustrated reference to cacti and other succulents **1**: 167 (1955); Jeppe, South African Aloes, 63 (1969); Reynolds, Aloes of South Africa, edn 4, t. 20 & 316–17 (1974); Brickell (ed.), RHS A–Z encyclopedia of garden plants, 101 (2003).

Plants prostrate, usually stemless, more rarely with a stem to 1 m. Leaves 12–20 in a dense rosette, 50 × 20 cm, lanceolate, spreading or incurved, glaucous, sometimes tinged with red, unspotted; margins entire, forming a pink or red zone 2–3 mm wide. Inflorescence branched, to 1 m; flower-stalks 1.5–2.5 cm, bracts *c.* 5 mm. Perianth 2.5–3 cm, constricted abruptly above the ovary, then cylindric and curving downwards, orange-red, outer lobes united for most of their length. Anthers and style projecting 1–2 mm. *South Africa.* H5–G1. Spring.

17. A. vera (Linnaeus) Burmann (*A. perfoliata* var. *vera* Linnaeus; *A. barbadensis* Miller). Illustration: Reynolds, Aloes of tropical Africa, 150–1 (1966); Brickell (ed.), RHS A–Z encyclopedia of garden plants, 101 (2003).

Plants stemless or with very short stems, suckering freely. Leaves *c.* 16 in each rosette, 40–50 × 6–7 cm, lanceolate, erect-spreading and compact, grey-green or reddish, striped, unspotted; margins with firm, pale teeth. Inflorescence usually branched, to 1 m; flower-stalks *c.* 5 mm, bracts *c.* 10 mm. Perianth 2.8–3 cm, cylindric or slightly swollen below, yellow, outer lobes free for more than half their length. Anthers and style projecting 3–5 mm. *Probably Yemen, widely cultivated and escaped elsewhere.* H5–G1. Winter–spring.

This species, which has a long history in cultivation, was introduced to Britain from Barbados; hence the inappropriate species name given by Miller.

18. A. perryi Baker. Illustration: *Botanical Magazine*, 6596 (1881).

Stem to 30 cm. Leaves 12–20 in a rosette, *c.* 35 × 7.5 cm, lanceolate, spreading, glaucous and reddish-tinged; margins with small, pale brown teeth. Inflorescence to 60 cm, branched; flower-stalks *c.* 8 mm, bracts 4–6 mm. Perianth 2–2.5 cm, cylindric, bright red turning yellow, outer lobes united for more than half their length. Anthers and style slightly projecting. *Yemen (Socotra).* G1–2.

19. A. camperi Schweinfurth (*A. eru* Berger). Illustration: Lamb & Lamb, The illustrated reference to cacti and other succulents **2**: 436 (1959); Reynolds, Aloes of tropical Africa, t. 45 & 216–17 (1966).

Stems to 50 cm. Leaves 12–16 in each rosette, 50–60 × 8–12 cm, lanceolate, spreading and arching downwards, channelled, dark green, unspotted or occasionally with a few pale spots; margins reddish with hard, sharp, brownish red teeth. Inflorescence to 1 m, branched; flower-stalks 1.2–1.8 cm, bracts *c.* 2 mm. Perianth 2–2.2 cm, cylindric, orange-yellow, outer lobes united for more than half their length. Anthers and style projecting 3–4 mm. *Ethiopia.* G1–2. Spring.

Material in cultivation as *A. abyssinica* Lamarck (a name of doubtful application) probably belongs to this species or a hybrid of it.

20. A. tenuior Haworth. Illustration: Jeppe, South African Aloes, 111 (1969); Reynolds, Aloes of South Africa, edn 4, 374–5 (1974).

Stem slender but fleshy, 1–3 m. Leaves in a very loose rosette, the stem clearly visible between them, 10–15 × 1–1.5 cm, linear-lanceolate, glaucous, unspotted, not auriculate at the base; margins with very small, white teeth. Inflorescence usually simple; flower-stalks 3–5 mm, bracts 4–6 mm. Perianth 1–1.4 cm, tubular or slightly bell-shaped, yellow or red, outer lobes united for more than half of their length. Anthers and style projecting 4–6 mm. *South Africa.* H5–G1. Winter.

A variable species; Reynolds (pp. 347–52) divides it into 3 varieties but it is uncertain which of these is in cultivation.

21. A. ciliaris Haworth. Illustration: Jeppe, South African Aloes, 110 (1969); Reynolds, Aloes of South Africa, edn 4, t. 29 & 380 (1974); Brickell (ed.), RHS A–Z encyclopedia of garden plants, 101 (2003).

Plants with somewhat fleshy and scrambling stems to 5 m. Leaves dispersed along the stem, 10–15 × 1.5–2.5 cm, linear-lanceolate, tapering, the base broadly auricled, the auricles ciliate; margins with firm, white teeth. Inflorescence simple; flower-stalks *c.* 5 mm, bracts *c.* 3 mm. Perianth *c.* 3 cm, cylindric, orange-red, the outer lobes united for most of their length. Anthers and style projecting 2–4 mm. *South Africa (Cape Province).* H5–G1. Winter.

22. A. striatula Haworth. Illustration: Jeppe, South African Aloes, 114 (1969); Reynolds, Aloes of South Africa, edn 4, t. 31 & 392–7 (1974).

Stem to 1.75 m. Leaves dispersed along the stem, *c.* 25 × 2.5 cm, linear-lanceolate, tapering, green; margins with firm, white teeth, the base not auriculate but the sheath prominently striped. Inflorescence simple; flower-stalks 3–5 mm, bracts 1.5–2.5 mm. Perianth 4–4.5 cm, cylindric, slightly curved, yellowish or reddish-orange, the outer lobes free almost to the base. Anthers and style projecting 4–7 mm. *South Africa.* H5–G1.

23. A. divaricata Berger. Illustration: Reynolds, Aloes of tropical Africa, t. 105 & 530–2 (1966).

Stems 2–3 m. Leaves dispersed along the stems for the upper 50–100 cm, to 65 × 7 cm, broadly lanceolate, dull grey-green tinged with red, the base not auriculate, the sheath not striped; margins with sharp teeth. Inflorescence branched, to 1 m; flower-stalks *c.* 6 mm, bracts *c.* 4 mm. Perianth 2.8–3 cm, constricted sharply above the ovary, then cylindric, curved, red, the outer lobes free to the base (they may cohere in the lower half but are not united). Anthers and style projecting 2–4 mm. *Madagascar.* G2.

24. A. mitriformis Miller (*A. perfoliata* misapplied). Illustration: *Botanical Magazine*, 1270 (1810); Jeppe, South African Aloes, 20 (1969); Reynolds, Aloes of South Africa, edn 4, 406–7 (1974); *Il Giardino Fiorito* **47**: 624 (1981).

Stems 1–2 m, mostly sprawling on the ground. Leaves many in a rosette, to 20 × 15 cm, ovate-lanceolate, spreading or incurved, glaucous; margins with hard, sharp, golden yellow, brown or black teeth. Inflorescence branched, to 60 cm; flower-stalks 4–4.5 cm, bracts *c.* 10 mm. Perianth 4–4.5 cm, cylindric, dull orange-red, the outer lobes free to the base (they may cohere to some extent in the lower half, but are not united). Anthers and style projecting at most 3 mm. *South Africa (Cape Province).* H5–G1. Summer.

25. A. distans Haworth. Illustration: Lamb & Lamb, The illustrated reference to cacti and other succulents **1**: 165 (1955); Jeppe, South African Aloes, 21 (1969); Reynolds, Aloes of South Africa, edn 4, 414–15 (1974); Brickell (ed.), RHS A–Z encyclopedia of garden plants, 101 (2003).

Stems to 3 m, sprawling on the ground. Leaves in elongate rosettes, to 15 × 7 cm, broadly lanceolate or lanceolate, glaucous, with whitish spots towards the base; margins with golden yellow teeth. Inflorescence branched, to 60 cm; flower-stalks 3–4 cm, bracts *c.* 8 mm. Perianth *c.* 4 cm, cylindric, slightly curved, orange-red, the outer lobes free to the base. Anthers and style little projecting. *South Africa (Cape Province).* H5–G1. Summer.

A. stans Berger is a name of doubtful application; the species was placed close to *A. distans* by Berger, and specimens in cultivation under the former name may well be the latter or a hybrid of it.

26. A. succotrina Lamarck. Illustration: *Botanical Magazine*, 472 (1800); Jeppe, South African Aloes, 50 (1969); Reynolds, Aloes of South Africa, edn 4, 433–4 (1974).

Stems usually to 2 m. Leaves numerous in a dense rosette, to 50 × 7–10 cm, curved upwards or erect, broadly lanceolate, dull to greyish green, sometimes with a few whitish spots; the margins dull with firm, white teeth. Inflorescence usually simple, to 1 m; flower-stalks *c.* 3 cm, bracts to 2 cm. Perianth *c.* 4 cm, cylindric, red or pink, glossy, the outer lobes free to the base. Anthers and style projecting 3–5 mm. *South Africa (Cape Province).* H5–G1. Winter.

This species was once thought to have originated in Socotra (hence the species name); it is not, however, found there, being restricted to the Cape peninsula in South Africa.

27. A. microstigma Salm-Dyck. Illustration: Jeppe, South African Aloes, 33 (1969); Reynolds, Aloes of South Africa, edn 4, 436–7 (1974).

Stems to 50 cm. Leaves numerous in a dense rosette, to 30 × 6.5 cm, spreading and arching upwards, broadly lanceolate, green, usually with H-shaped, whitish spots; margins with hard, sharp teeth. Inflorescence simple, to 80 cm; flower-stalks 2.5–3 cm, bracts 1.2–1.5 cm. Perianth 2.5–3 cm, cylindric, orange fading to yellow, the outer lobes free to the base. Anthers and style projecting 1–3 mm. *South Africa, South West Africa.* H5–G1.

28. A. arborescens Miller. Illustration: *Botanical Magazine*, 8663 (1916); Jeppe, South African Aloes, 48 (1969); Reynolds, Aloes of South Africa, edn 4, 450–2 (1974); *Il Giardino Fiorito* **47**: 624 (1981).

Stems 2–3 m. Leaves numerous in each rosette, 50–60 × 5–7 cm, widely spreading and curved downwards towards their apices, lanceolate, dull green; margins with firm, pale teeth. Inflorescence usually simple; flower-stalks 3.5–4 cm, bracts 1.5–2 cm. Perianth *c.* 4 cm, cylindric, slightly constricted above the ovary, orange-red, outer lobes free to the base. Anthers and style projecting *c.* 5 mm. *South Africa north to Malawi.* H5–G1. Winter–spring.

A. × principis (Haworth) Stearn is the hybrid between this species and *A. ferox* (no. 32). It is intermediate between its parents, flowering somewhat later than *A. arborescens.*

29. A. ortholopha Christian & Milne-Redhead. Illustration: *Flowering Plants of South Africa* **23**: t. 882 (1943); Reynolds, Aloes of tropical Africa, t. 51 & 237 (1966).

Stems solitary, less than 50 cm. Leaves 30 or more in a rosette, to 50 × 12–14 cm, spreading and curving upwards, broadly lanceolate, dull grey-green tinged with pink; margins with sharp teeth. Inflorescence branched, its stalk arching so that the branches are almost horizontal, the flowers borne on the upper sides of the axes; flower-stalks *c.* 8 mm, bracts 1–1.5 cm. Perianth *c.* 4 cm, cylindric, somewhat swollen in the middle, orange-red to deep red, the outer lobes free for most of their length. Anthers and style projecting 1–1.5 cm. *Zimbabwe.* H5–G1.

30. A. rubroviolacea Schweinfurth. Illustration: *Botanical Magazine*, 7882 (1903).

Stems to 1 m. Leaves numerous in each rosette, to 60 × 10–11 cm, dense, spreading and recurved, broadly lanceolate, reddish or purplish at the base, shading to pink towards the apex; margins with small, hooked teeth. Inflorescence to 1 m or more, branched with erect branches; flower-stalks 3–4 mm, bracts 2–2.5 cm. Perianth 3–4 cm, cylindric, bright red, the outer lobes united for half their length. Anthers and style projecting 1–1.5 cm. *Saudi Arabia, Yemen*. G1.

31. A. africana Miller. Illustration: *Flowering Plants of South Africa* **9**: t. 333 (1925); Jeppe, South African Aloes, 41 (1969); Reynolds, Aloes of South Africa, edn 4, 503–4 (1974).

Stems to 2 m. Leaves *c.* 30 in a rosette, to 65 × 12 cm, widely spreading, broadly lanceolate, dull green to glaucous, with a few spines on both surfaces; margins with sharp, reddish brown teeth. Inflorescence to 80 cm, branched; flower-stalks 5–6 mm, bracts 1–1.2 cm. Perianth to 5.5 cm, cylindric, conspicuously curved, yellow to orange-yellow, the outer lobes united for more than half their length. Anthers and style projecting 1.5–2 cm. *South Africa (Cape Province)*. H5–G1. Spring.

32. A. ferox Miller. Illustration: *Botanical Magazine*, 1975 (1818); Lamb & Lamb, The illustrated reference to cacti and other succulents **1**: 166 (1955); Jeppe, South African Aloes, 40 (1969); Reynolds, Aloes of South Africa, edn 4, 509–13 (1974).

Stems to 3 or rarely 5 m. Leaves 50–60 in a dense rosette, to 100 × 15 cm, spreading, lanceolate, flat or slightly channelled, dull green with a reddish tinge, the surface usually spiny; margins with stout, reddish brown spines. Inflorescence branched, the individual racemes 50–80 cm; flower-stalks 4–6 mm, bracts 8–10 mm. Perianth *c.* 3.3 cm, cylindric, straight, orange-red, the outer lobes free for most of their length, inner lobes with brown tips. Anthers and style projecting 2–2.5 cm. *South Africa*. H5–G1.

A variable species, particularly in the size and spininess of the leaves.

33. A. candelabrum Berger. Illustration: *Flowering Plants of South Africa* **24**: t. 945 (1944); Jeppe, South African Aloes, 39 (1969); Reynolds, Aloes of South Africa, edn 4, 520–2 (1974).

Like *A. ferox*, but leaves deeply channelled and recurved, perianth pink to orange-red, the inner lobes without brown tips. *South Africa (Natal)*. H5–G1.

34. A. thraskii Baker. Illustration: Jeppe, South African Aloes, 42 (1969); Reynolds, Aloes of South Africa, edn 4, t. 54 & 531 (1974).

Stems to 2 m. Leaves numerous in dense rosettes, to 1.6 m × 22 cm, recurved and channelled, lanceolate, dull green or glaucous, sometimes with a few spines on the surface; marginal spines reddish, *c.* 2 mm. Inflorescence branched, individual racemes to 25 cm; flower-stalks 1–2 mm, bracts *c.* 9 mm. Perianth 2.5–3 cm, cylindric, pale yellow, the outer lobes free for most of their length. Anthers and style projecting 1.5–2 cm. *South Africa (Natal)*. H5–G1.

35. A. excelsa Berger. Illustration: *Flowering Plants of South Africa* **2**: t. 62 (1922); Reynolds, Aloes of tropical Africa, t. 72 & 310–11 (1966).

Stem to 4 m. Leaves *c.* 30 in a rosette, to 80 × 15 cm, spreading or reflexed, broadly lanceolate, channelled, dull green, usually with some spines on the lower surface; marginal spines sharp, 5–6 mm, reddish brown. Inflorescence to 1 m, branched; flower-stalks *c.* 1 mm, bracts 4–6 mm. Perianth *c.* 3 cm, cylindric, reddish orange, the outer lobes free for most of their length. Anthers and style projecting 8–10 mm. *Malawi, Zimbabwe*. G1–2.

36. A. marlothii Berger. Illustration: *Flowering Plants of South Africa* **6**: t. 171 (1925); Jeppe, South African Aloes, 37 (1969); Reynolds, Aloes of South Africa, edn 4, t. 56 & 535–9 (1974); *Il Giardino Fiorito* **47**: 624 (1981).

Stems 2–6 m. Leaves 40–50 in a rosette, 1–1.5 m × 20–25 cm, spreading, broadly lanceolate, grey-green or glaucous, the surfaces usually spiny; margins with stout, sharp, reddish brown spines. Inflorescence branched, the branches held almost horizontally, the flowers borne on their upper sides; flower-stalks *c.* 5 mm, bracts 8–9 mm. Perianth cylindric, 3–3.5 cm, orange to yellowish orange, the outer lobes free for most of their length. Anthers and style projecting 1.5–2 cm. *Botswana*. H5–G1.

37. A. dichotoma Masson. Illustration: Jeppe, South African Aloes, 57 (1969); Reynolds, Aloes of South Africa, edn 4, 551–5 (1974).

Plant tree-like, with a single trunk and a richly and apparently dichotomously branched crown, total height 6–9 m. Leaves in rosettes at the ends of the branches, 25–35 × 4–6 cm, spreading, linear-lanceolate, glaucous; margins with brownish yellow teeth. Racemes *c.* 15 cm; flower-stalks 5–10 mm, bracts 5–7 mm. Perianth *c.* 3.3 cm, swollen near the base then gradually tapered to the mouth, bright yellow, the outer lobes free for most of their length. Anthers and style projecting 1.2–1.5 cm. *South Africa, South West Africa*. H5–G1.

A. bainesii Dyer. Crown less branched, leaves 60–90 cm, dull green, perianth pink, 3.3–3.7 cm. *South Africa*. H5–G1.

38. A. plicatilis (Linnaeus) Miller. Illustration: *Botanical Magazine*, 457 (1799); Lamb & Lamb, The illustrated reference to cacti and other succulents **2**: 441 (1959); Jeppe, South African Aloes, 67 (1969); Reynolds, Aloes of South Africa, edn 4, 570–1 (1974).

Stem to 5 m. Leaves borne in dense rosettes, on 2 sides of the stem and forming a flattened 'fan', *c.* 30 × 4 cm, linear or oblong-linear, glaucous, flat; margins entire, the apex very rounded. Inflorescence simple, to 50 cm; flower-stalks *c.* 1 cm, bracts *c.* 8 mm. Perianth 4.5–5.5 cm, cylindric, orange-red, the outer lobes united for more than half their length. Anthers and style projecting 1–5 mm. *South Africa (Cape Province)*. H5–G1.

Very distinct with its fans of glaucous, tongue-shaped, entire, blunt leaves.

39. A. variegata Linnaeus. Illustration: Brickell (ed.), RHS A–Z encyclopedia of garden plants, 101 (2003).

Plant stemless or stems very short, to *c.* 15 cm, with 12 or more leaves stacked in three groups or sometimes spirally arranged. Leaves 10–15 × 3–6 cm, fleshy, rigid, double triangular, thornless or with small spines on the whitish edged margins, white speckled markings in uneven rows on a dark green and purple background. Flowers borne in several simple racemes reaching 20–25 cm; flower-stalks 4–6 mm. Perianth 3–4 cm, outer lobes united for more than half their length, slightly constricted above the ovary, orange-pink to dull reddish pink. *South Africa, Namibia*. H5–G1.

Very distinct with its leaf arrangement and attractive white markings.

15. COLCHICACEAE

Plants terrestrial, herbaceous, with corms. Leaves well developed. Flowers in racemes or without spathes if solitary, actinomorphic. Perianth of 6, free or united segments; if free, then often connivent. Stamens 6, nectaries on the staminal filaments. Ovary of 3 united carpels, superior, not stalked, below ground level in corm; placentation axile; styles 3. Fruit a capsule. Seeds not hairy.

There are 8 genera and about 140 species; three genera are cultivated. Several genera often found in this family have been removed here to the Liliaceae (see p. 62).

1a. Style 1, branched above **2. Bulbocodium**
b. Styles 3, free 2
2a. Perianth with a tube below **1. Colchicum**
b. Perianth segments free to the base, though superficially appearing to be united **3. Merendera**

1. COLCHICUM Linnaeus
C.D. Brickell revised by C.D. Brickell
Perennial stemless herbs with corms that occasionally have long, horizontal, rhizome-like outgrowths; tunics membranous, papery or leathery, frequently extended into a tubular, persistent false-stem or neck. Developing leaves and flowers enclosed within a membranous sheath. Leaves basal, partially developed at flowering or developing after flowering (occasionally developing as the flowers fade). Flowers solitary or in clusters, each subtended by a small bract and very shortly stalked, the stalks elongating as the capsules ripen. Perianth bell-shaped, funnel-shaped or star-shaped, purple, pink or white, sometimes tessellated; with a tube at the base and 6 lobes in 2 equal or almost equal whorls, bases of the lobes occasionally with auricles, the throat of the perianth sometimes ridged on either side of the filament-bases to form a short 'filament-channel', the ridges hairless or downy. Stamens borne near the bases of the perianth-lobes, in 1 or 2 series; filaments slender, sometimes thickened at the base; anthers dorsifixed and versatile, or basifixed and rigid. Ovary subterranean; styles 3, free stigmas point-like or unilaterally decurrent along the style. Capsule 3-celled, splitting along the septa. Seeds numerous, spherical or almost so.

A genus of about 45 species extending from E Europe and N Africa to W Asia, through Afghanistan to N India and W China (Xizang). The large-flowered species whose leaves develop after flowering thrive in any sunny, well-drained site, with the exception of *C. variegatum*, which, like the majority of those with leaves partially developed at flowering, is better grown in a bulb frame or alpine house.

Literature: Stefanoff, B., Monographie der Gattung Colchicum, *Sbornik Bulgariskata Akademiya na Naukita* **22**: 1–100 (1926); Bowles, E.A., A handbook of Crocus and Colchicum for gardeners, edn 2 (1952); Feinbrun, N., The genus Colchicum of Palestine and neighbouring countries, *Palestine Journal of Botany, Jerusalem series* 6, **2**: 71–95 (1953); Feinbrun, N., Chromosome numbers and evolution in the genus Colchicum, *Evolution* **12**: 173–88 (1958); Burtt, B.L., Meikle, R.D. & Furse, J.P.W., Colchicum and Merendera, *RHS Lily Year Book* **31**: 90–103 (1968). Brickell, C.D., Colchicum L., Flora Europaea **5**: 21–25 (1980); Brickell, C.D., Colchicum L., Davis, P.H. (ed.), Flora of Turkey and the East Aegean Islands **8**: 329–351 (1984); Persoon, K., Colchicum L., Guner, A., Ekim, T. & Baser, K.H.C. (eds.), Flora of Turkey and the East Aegean Islands **11**: 246–265 (2001); Persoon, K., Nomenclatural synopsis of the genus Colchicum (Colchicaceae) with some new species and combinations, *Botanische Jahrbücher für Systematik* **127**(2):165–242 (2007).

1a. Anthers basifixed, rigid 2
b. Anthers dorsifixed, versatile 3
2a. Flowers white with pale to deep red-purple stripes **2. kesselringii**
b. Flowers pale to deep yellow, sometimes purple-brown at base **1. luteum**
3a. Leaves partly developed at flowering or appearing immediately after flowering; flowering in autumn, winter or spring 4
b. Leaves undeveloped at flowering and appearing from December to April; flowering from August to November 13
4a. Corms with more or less horizontal, underground stolons 5
b. Corms erect, without underground stolons 6
5a. Leaves 2 or 3, narrowly linear to linear-lanceolate, 7–15 cm × 2–15 mm at maturity; flowers star-shaped to narrowly funnel-shaped; anthers black or purplish black **3. psaridis**
b. Leaves 3, narrowly lanceolate to very narrowly lanceolate, 20–32 × 2.4–4.5 cm at maturity; flowers bell-shaped or funnel-shaped; anthers yellow **4. baytopiorum**
6a. Corm-tunics membranous, frequently short-lived; leaves 3 (rarely 4) 7
b. Corm-tunics leathery to papery, persistent; leaves 2–12 8
7a. Leaves narrowly to very narrowly lanceolate, recurving, 20–32 × 2.4–4.5 cm at maturity; anthers yellow; flowering from October to November **4. baytopiorum**
b. Leaves linear-lanceolate, erect-spreading, 11–15 cm × 5–11 mm at maturity; anthers purplish black or purplish green; flowering from February to June **5. triphyllum**
8a. Corm-tunics strongly leathery, with distinct longitudinal corrugations; filaments hairy at least near the base **6. burttii**
b. Corm-tunics papery, somewhat leathery or leathery, lacking distinct longitudinal corrugations; filaments hairless 9
9a. Flowering from September to December 10
b. Flowering from December to June 12
10a. Leaves 2 (rarely 3), 2–18 mm wide at maturity; corm-tunics leathery **7. cupanii**
b. Leaves 3–12, 1–5 mm wide at maturity; corm-tunics papery to somewhat leathery 11
11a. Anthers yellow; perianth-lobes bright purplish pink, 1.5–3 cm × 2–9 mm **8. stevenii**
b. Anthers purplish black to dark brown or greyish brown; perianth-lobes white to pale rosy lilac, 1–2 cm × 1.5–4 mm **9. pusillum**
12a. Leaves hairless, strap-shaped to very narrowly linear-lanceolate, usually 2–3.5 cm wide at maturity **10. szovitsii**
b. Leaves ciliate, the upper surface partially or entirely covered with short hairs, narrowly linear-lanceolate, 1–2 cm wide at maturity **11. hungaricum**

13a. Flowers lightly to strongly tessellated 14
b. Flowers not tessellated 20
14a. Stigmas point-like or decurrent for not more than 0.5 mm; flowers usually 3–10, rarely fewer or more, lightly tessellated **20. cilicicum**
b. Stigmas decurrent for at least 1.5 mm; flowers up to 8, lightly to strongly tessellated 15
15a. Perianth-lobes spreading; pollen grey-green; leaves strongly pleated longitudinally, ovate to elliptic-ovate, 24–42 × 11–15.5 cm **25. macrophyllum**
b. Perianth-lobes spreading or forming a bell- or funnel-shape; pollen yellow; leaves never strongly pleated longitudinally, strap-shaped to narrowly lanceolate, never more than 35 × 7 cm 16
16a. Leaves 3–7, spreading, margins usually wavy; perianth-lobes hairless, spreading or forming a bell- or funnel-shape 17
b. Leaves 4–11, usually 5–9, more or less erect, margins not wavy, but sometimes twisted towards the apex; flowers broadly bell- to funnel-shaped; perianth-lobes downy along the ridges of the filament-channels 18
17a. Flowers lightly tessellated, funnel-shaped to bell-shaped; anthers yellow **27. lingulatum**
b. Flowers strongly tessellated, perianth-lobes spreading; anthers purplish black or purplish brown **24. variegatum**
18a. Flowers strongly tessellated, broadly bell-shaped; anthers purplish black or purplish brown **26. bivonae**
b. Flowers lightly tessellated, bell-shaped to funnel-shaped; anthers yellow 19
19a. Flowers narrowly bell-shaped, purplish pink to white; leaves 3–5, linear-lanceolate to broadly lanceolate, 14–35 × 1–7 cm, hairless, green **21. autumnale**
b. Flowers bell- to funnel-shaped, deep red-purple, occasionally paler; leaves 5–9, strap-shaped to narrowly lanceolate, 9–19 × 1–3.5 cm, ciliate (sometimes obscurely so), somewhat glaucous **28. turcicum**
20a. Stigmatic surface decurrent for at least 1.5–2 mm 21
b. Stigmatic surface point-like or shortly decurrent, usually to not more than 1 mm (if to 1.5 mm, then anthers purple or purplish brown) 27
21a. Anthers (undehisced) purple or purplish brown **23. bornmuelleri**
b. Anthers yellow 22
22a. Leaves 3–9, strap-shaped to oblong, occasionally narrowly lanceolate, usually not more than 7.5–19 × 1–3.5 cm 23
b. Leaves 3–5, elliptic, oblong-elliptic or oblong-lanceolate (occasionally the innermost leaf strap-shaped), 10–35 × 3–9.5 cm 24
23a. Leaves 3–7, spreading, margins hairless, usually wavy; flowers 1–3 (rarely 4), pale to deep purplish pink, perianth-lobes hairless **27. lingulatum**
b. Leaves 5–9, more or less erect, often twisted near the apex, ciliate (sometimes obscurely so), not wavy; flowers 3–8 (rarely 1), deep red-purple, occasionally paler, perianth-lobes downy along the ridges of the filament-channels **28. turcicum**
24a. Perianth-lobes narrowly oblong-elliptic, hairless; leaves elliptic; corm-tunic with a very long, persistent, strongly fibrous neck to 25 cm, obscuring the membranous sheath **18. balansae**
b. Perianth-lobes oblanceolate, oblong-elliptic to elliptic, downy along the ridges of the filament-channels; neck of corm to 14 cm, not strongly fibrous; membranous sheath usually projecting well beyond the neck 25
25a. Flowers funnel-shaped; leaves 10–12 (rarely to 16) × 3–5 cm **19. kotschyi**
b. Flowers bell-shaped, sometimes narrowly; leaves 18–35 × 2–9.5 cm (rarely less) 26
26a. Anthers 5–8 mm; flowers narrowly bell-shaped; leaves linear-lanceolate to broadly lanceolate **21. autumnale**
b. Anthers 1–1.2 cm; flowers bell-shaped; leaves narrowly elliptic to oblong-lanceolate **22. speciosum**
27a. Corms with more or less horizontal, underground stolons **12. boissieri**
b. Corms erect, without underground stolons 28
28a. Perianth-lobes lilac-purple to deep rosy purple, usually 5–7.5 × 1.1–2.5 cm; leaves very narrowly elliptic to narrowly elliptic-lanceolate 29
b. Perianth-lobes white to purplish pink, usually 1.5–4.5 cm × 2–12 mm; leaves linear, linear-oblanceolate, very narrowly linear-lanceolate or strap-shaped 30
29a. Anthers yellow; stigmatic surface point-like or shortly decurrent to 0.5 mm; flowers usually 3–25 **20. cilicicum**
b. Anthers purple or purple-brown; stigmatic surface 0.5–1.5 mm; flowers usually 1–3, rarely to 6 **23. bornmuelleri**
30a. Leaves 4–10, usually 6–8, linear to narrowly linear, 7–10 cm × 1–4 mm; flowers funnel- or bell-shaped **16. parlatoris**
b. Leaves usually 2–6, strap-shaped to very narrowly linear-lanceolate; if linear, then leaves only 2 or 3, more than 10 cm and flower funnel-shaped, becoming star-shaped 31
31a. Flowers narrowly bell-shaped to funnel-shaped; anthers 2–3 mm; leaves 2–5, strap-shaped to linear-lanceolate, usually 8–15 cm **13. alpinum**
b. Flowers funnel-shaped to star-shaped; anthers 3–8 mm; leaves 2–8, strap-shaped or linear to very narrowly linear-oblanceolate, usually 12–30 cm 32
32a. Corm-tunics membranous, reddish brown, neck often only slightly developed; leaves 2 or 3, linear to linear-oblanceolate; flowers 1–2, rarely to 5 **14. micranthum**
b. Corm-tunics leathery or almost so, dark blackish brown, neck persistent; leaves 3–8, strap-shaped to very narrowly lanceolate; flowers 1–12 33
33a. Flowers 1–3, rarely to 6; perianth-lobes narrowly oblanceolate to linear-elliptic, usually 1.5–2.5 cm × 2–6 mm; filaments 5–8 mm; anthers 3–4 mm **15. umbrosum**
b. Flowers 2–6 or more; perianth-lobes narrowly oblong-lanceolate, 2.8–4.5 cm × 4–11 mm; filaments 1.5–2 cm; anthers 6–8 mm **17. decaisnei**

1. C. luteum Baker. Illustration: Rix & Phillips, The bulb book, 33 (1981); Brickell (ed.), RHS A–Z encyclopedia of garden plants, 294 (2003).

Corm 1.5–3 × 1–2.5 cm, oblong-ovoid; tunic dark chestnut brown, papery, produced into a short neck. Leaves 2–5, linear-lanceolate, more or less erect and 1–3 cm at flowering, 10–30 cm × 6–20 mm at maturity, hairless. Flowers 1–4, funnel-shaped to narrowly bell-shaped. Perianth-lobes pale to deep yellow, narrowly oblong-lanceolate, 1.5–2.5 cm × 2–6 mm, tube sometimes purple-brown. Filaments 3–4 mm, hairless; anthers yellow, 6–13 mm, basifixed; pollen yellow. Styles straight, stigmas point-like. Capsule ovoid, 2–3 cm. *Former USSR (Soviet Central Asia), Afghanistan, N India, SW China (Xizang).* H3. Spring–summer.

2. C. kesselringii Regel (*C. regelii* Stefanoff; *C. crociflorum* Regel). Illustration: Rix & Phillips, The bulb book 32 (1981); Brickell (ed.), RHS A–Z encyclopedia of garden plants, 294 (2003).

Corm 1–3 × 1–2 cm, oblong-ovoid; tunic dark brown, leathery, produced into a short neck. Leaves 2–7, linear-lanceolate, more or less erect and 1–2 cm at flowering, 7–10 cm × 3–10 mm at maturity, hairless or margins slightly rough. Flowers 1–4, narrowly bell-shaped to funnel-shaped. Perianth-lobes with pale to deep red-purple central stripes, narrowly linear-lanceolate to narrowly elliptic, 1.5–3 cm × 2–7 mm; tube red-purple. Filaments 3–4 mm, hairless; anthers yellow, 8–10 mm, basifixed; pollen yellow. Styles straight, stigmas point-like. *Former USSR (Soviet Central Asia) & N Afghanistan.* H3. Early spring–summer.

3. C. psaridis Halacsy. Illustration: Grey-Wilson & Mathew, Bulbs, pl. 10 (1981).

Corm more or less horizontal with runners, 2–5 cm × 4–12 mm, irregular in shape, sometimes swollen; tunic pale brown, papery or membranous. Leaves 2 or 3, narrowly linear to narrowly linear-lanceolate, usually partly developed at flowering, erect or almost so, extending to 7–9 cm at flowering, 7–15 cm × 2–15 mm at maturity, margin hairless or partly ciliate. Flowers 1–3, star-shaped to narrowly funnel-shaped. Perianth-lobes white to pinkish purple, very narrowly elliptic or narrowly elliptic-oblong, 1.1–2.7 cm × 2–6 mm. Filaments 4–8 mm, hairless; anthers black or purplish black, 2–3 mm, dorsifixed; pollen yellow. Styles straight, stigmas point-like. Capsule oblong-ovoid, 1.5 cm. *S Greece.* H4. Autumn–winter.

4. C. baytopiorum Brickell. Illustration: Mathew & Baytop. The bulbous plants of Turkey, 68 (1984).

Corm 2.5–3.5 × 1.5–2.5 cm, narrowly ovoid to almost spherical, sometimes more or less horizontal with runners, 4–6 × 1.3–2 cm, irregular in shape; tunic reddish brown, membranous to papery, frequently short-lived. Leaves 3, narrowly to very narrowly lanceolate, more or less erect at first, then recurving, extending to 1–8 cm at flowering, 20–32 × 2.4–4.5 cm at maturity, hairless. Flowers 1–3, rarely to 5, bell-shaped or funnel-shaped. Perianth-lobes bright purplish pink, elliptic or oblong-elliptic to oblanceolate, 2.2–4.2 cm × 5–11 mm. Filaments 1–1.4 cm, hairless; anthers yellow, 5–7 mm, dorsifixed; pollen lemon yellow. Styles straight, stigmas point-like. Capsule narrowly ellipsoid, 2–2.5 cm. *W Turkey.* H4. Autumn.

5. C. triphyllum Kunze (*C. bulbocodioides* Bieberstein not Brotero; *C. biebersteinii* Rouy; *C. catacuzenium* Stefanoff; *C. ancyrense* Burtt; *C. montanum* misapplied). Illustration: *Botanical Magazine*, 9652 (1943); Grey-Wilson & Mathew, Bulbs, pl. 10 (1981); Mathew & Baytop, The bulbous plants of Turkey, 66 (1984).

Corm 1.5–2.5 × 1–1.5 cm, oblong-ovoid; tunic chestnut brown, membranous, short-lived, lacking an extended neck. Leaves 3, linear-lanceolate, erect-spreading, extending to 2–9 cm at flowering, 11–15 cm × 5–11 mm at maturity, margins rough (sometimes obscurely so) or smooth. Flowers 1–6, bell-shaped to funnel-shaped. Perianth-lobes purplish pink or white flushed with purplish pink, narrowly elliptic to oblanceolate, 1.5–3 cm × 5–12 mm, occasionally with thread-like outgrowths or auricles at the base. Filaments 7–9 mm, hairless; anthers purplish black or purplish green, 2.5–3.5 mm, dorsifixed; pollen yellow. Styles straight, stigmas point-like. Capsule ovoid to oblong-ovoid, 2–3 cm. *NW Africa, Spain, Greece, Turkey, Former USSR (S Russia).* H3. Early–late spring.

6. C burttii Meikle. Illustration: *Botanical Magazine*, n.s., 735 (1977); Mathew & Baytop, The bulbous plants of Turkey, 63 (1984).

Corm 3–5 × 1.5–2 cm, narrowly ovoid to almost spherical; tunic blackish brown, very leathery, with prominent longitudinal corrugations, apex fringed with rigid fibres extending for 1–5 cm. Leaves 2–4, narrowly linear to linear-lanceolate, erect-spreading and extending for 1–4 cm at flowering, recurved and 10–15 cm × 8–10 mm at maturity, shortly hairy at least on the margins. Flowers 1–4, funnel-shaped, becoming star-shaped. Perianth-lobes white or pale purplish pink, narrowly oblanceolate, 1.5–4 cm × 3–8 mm. Filaments 8–10 mm, thinly hairy at least near the base; anthers dark purplish black or black, 2–3 mm, dorsifixed; pollen yellow. Styles straight, stigmas point-like. Capsule (immature) narrowly ellipsoid, *c.* 1 cm, roughly hairy at the apex. *W Turkey.* H4. Early spring.

7. C. cupanii Gussone (*C. glossophyllum* Heldreich). Illustration: Mathew, Dwarf bulbs, pl. 21 (1973); Grey-Wilson & Mathew, Bulbs, pl. 11 (1981); Rix & Phillips, The bulb book, 184, 185 (1981).

Corm 1–2 × 1–1.5 cm, ovoid; tunic dark brown, leathery, with a short neck. Leaves 2, very rarely 3, linear to linear-lanceolate, to 10 cm at flowering, 7–15 cm × 2–18 mm at maturity, hairless or rarely ciliate at the base. Flowers 1–12, funnel-shaped to star-shaped. Perianth-lobes pale to deep purplish pink, narrowly elliptic, 1.8–2.5 cm × 3–5 mm. Filaments 6–10 mm, hairless; anthers purplish black, 2–3 mm, dorsifixed; pollen yellow. Styles straight, stigma point-like. Capsule oblong-ovoid, 1–1.3 cm. *N Africa, France, Italy, Greece, Crete.* H3. Autumn–early winter.

8. C. stevenii Kunth. Illustration: *Botanical Magazine*, 8025 (1905).

Corm 1.6–3 × 1–1.5 cm, ovoid to almost spherical; tunic dull, dark brown, papery to almost leathery, produced into a neck 1–6 cm long. Leaves 4–12, more or less erect, extending for 1–12 cm at flowering, recurved and often wavy, 8–18 cm × 1–5 mm at maturity, hairless or ciliate. Flowers 1–10, funnel-shaped. Perianth-lobes bright purplish pink, oblong-elliptic to oblanceolate, 1.5–3 cm × 2–9 mm. Filaments 8–12 mm, hairless; anthers yellow. Styles straight, stigmas point-like. Capsule oblong-ovoid, 1–1.5 cm. *Cyprus, Turkey, W Syria.* H4. Autumn–early winter.

Similar plants from Greece and the islands have been described as *C. peloponnesiacum* Rechinger & Davis and *C. andrium* Rechinger & Davis. Their status is uncertain.

9. C. pusillum Sieber (*C. montanum* Linnaeus var. *pusillum* (Sieber) Fiori). Illustration: Rix & Phillips, The bulb book, 177 (1981).

Corm 1.5–2.5 cm × 9–17 mm, ovoid to almost spherical; tunics dull, dark brown, papery, the neck 1–6 cm. Leaves 3–8, narrowly linear, more or less erect and extending for 1–4 cm at flowering, recurved and 8–11 cm × 1–5 mm at maturity, hairless or sometimes ciliate. Flowers 1–6, funnel-shaped, opening star-shaped. Perianth-lobes white to pale rosy lilac, narrowly oblanceolate to narrowly oblong-oblanceolate, 1–2 cm × 1.5–4 mm. Filaments 5–8 mm, hairless; anthers purplish black, brownish black or grey-brown, 1.5–4.5 mm, dorsifixed; pollen yellow. Styles straight, stigma point-like. Capsule ovoid, 1–1.3 cm. *Greece, Crete, Cyprus.* H4. Autumn.

10. C. szovitsii Fischer & Meyer (*C. szovitsii* var. *nivale* Boissier & Huet; *C. bifolium* Freyn & Sintenis; *C. szovitsii* var. *bifolium* (Freyn & Sintenis) Bordzilowski; *C. armenum* Fedtschenko; *C. nivale* Stefanoff). Illustration: *Botanical Magazine*, 8040 (1905); Rix & Phillips, The bulb book, 32 (1981); Mathew & Baytop, The bulbous plants of Turkey, 65 (1984).

Corm 1.5–4 × 1–3 cm, ovoid; tunic blackish brown, papery to almost leathery, neck absent or vestigial. Leaves 2 or 3, strap-shaped or very narrowly linear-lanceolate, more or less erect, extending for 2–12 cm at flowering, 12–25 × 1.2–3.5 cm at maturity, hairless. Flowers 1–7, ovoid-bell-shaped or sometimes bell-shaped or narrowly so. Perianth-lobes deep to pale purplish pink or white strongly suffused with purplish pink, oblanceolate to very narrowly elliptic, 2.1–3.5 cm × 4–10 mm, occasionally with basal auricles. Filaments 7–11 mm, hairless; anthers purplish black or greenish black, 2–5 mm, dorsifixed; pollen yellow. Styles straight, stigmas point-like. Capsule ellipsoid to spherical, 3.5–4 cm. *Turkey, Iran, Former USSR (Caucasus).* H3. Early spring–summer.

11. C. hungaricum Janka (*C. doerfleri* Halacsy). Illustration: Grey-Wilson & Mathew, Bulbs, pl. 11 (1981); Rix & Phillips, The bulb book, 32 (1981).

Corm 2–3 × 1–2 cm, oblong-ovoid; tunic dark brown, papery to almost leathery, with a short neck. Leaves 2, very rarely 3, narrowly linear-lanceolate, more or less erect, 3–10 cm at flowering, up to 20 × 1–2 cm at maturity, ciliate; upper surface sometimes partially or entirely covered with short hairs. Flowers 1–8, bell-shaped to funnel-shaped. Perianth-lobes purplish pink to white, elliptic-lanceolate or narrowly elliptic, 2–3 cm × 6–7 mm. Filaments 7–10 mm, hairless; anthers purplish black, 2–3.5 mm, dorsifixed; pollen yellow or orange. Styles straight or slightly curved at the apex, stigmas point-like. Capsule almost spherical, *c.* 1 cm. *Hungary, Jugoslavia, Albania, Bulgaria, Greece.* H3. Late winter–early spring.

White variants are referred to as forma **albiflorum** Maly.

12. C. boissieri Orphanides (*C. procurrens* Baker). Illustration: Grey-Wilson & Mathew, Bulbs, pl. 10 (1981); Rix & Phillips, The bulb book, 175 (1981); Mathew & Baytop, The bulbous plants of Turkey, 70, 77 (1984) – in the latter as *C. sieheanum*.

Corm with horizontal runners, 2.5–3 cm × 3–13 mm, irregular, often with tooth-like projections, occasionally irregularly ovoid or oblong-ovoid and more or less vertical; tunic pale reddish brown, membranous. Leaves 2 or 3, developing after flowering, linear, erect or almost so, 11–22 cm × 2–8 mm, margins hairless or partly ciliate. Flowers 1 or 2, bell-shaped to narrowly funnel-shaped. Perianth-lobes bright rosy lilac, very narrowly elliptic to narrowly elliptic-obovate, 2.5–5 cm × 5–15 mm. Filaments 1–2 cm, hairless; anthers yellow, 5–10 mm, dorsifixed; pollen yellow. Styles straight or slightly curved at the apex, stigmas point-like. Capsule ellipsoid, *c.* 2 cm. *S Greece, W Turkey.* H3. Autumn.

13. C. alpinum de Candolle. Illustration: Grey-Wilson & Mathew, Bulbs, pl. 9 (1981).

Corm 1–2.5 × 1–1.5 cm, almost spherical to ovoid; tunic dark reddish brown, membranous or somewhat leathery, neck 1–2 cm. Leaves 2 or 3, developing after flowering, strap-shaped to linear-lanceolate, 8–15 cm × 2–14 mm, hairless. Flowers 1 or 2, narrowly bell-shaped to funnel-shaped. Perianth-lobes purplish pink, occasionally white, narrowly oblong-elliptic, 1.7–3 cm × 4–10 mm. Filaments 2–9 mm, hairless; anthers yellow, 2–3 mm, dorsifixed; pollen yellow. Styles straight, stigmas point-like. Capsule oblong-ellipsoid, 1.5–2 cm. *France, Switzerland, Italy, Corsica, Sardinia.* H2. Late summer–autumn.

C. arenarium Waldstein & Kitaibel. Leaves 3–5, somewhat longer and broader, flowers slightly larger. *Eastern C Europe.* H2. Summer–autumn.

C. corsicum Baker. Illustration: Rix & Phillips. The bulb book, 175 (1981). Leaves 3 or 4, shorter and broader, stigmas shortly decurrent. *Corsica.* H2. Summer–autumn.

14. C. micranthum Boissier. Illustration: Grey-Wilson & Mathew, Bulbs, pl. 10 (1981); Mathew & Baytop, The bulbous plants of Turkey, 75 (1984).

Corm 2.5–3 × 1.3–2.7 cm, almost spherical to ovoid; tunic reddish brown, membranous, neck 1–4 cm but often only slightly developed. Leaves 2 or 3, developing after flowering, linear to very narrowly linear-oblanceolate, almost erect, 10–20 cm × 3–10 mm, hairless. Flowers 1 or 2, narrowly bell-shaped or funnel-shaped. Perianth-lobes white to pale purplish pink, narrowly oblanceolate, oblanceolate or narrowly linear-elliptic, 1.8–4 cm × 3–11 mm. Filaments 6–10 mm, hairless; anthers yellow, 3–5 mm, dorsifixed; pollen yellow. Styles straight, shortly curved and slightly swollen at the apex, stigmas decurrent for 1 mm. Capsule narrowly ellipsoid to ovoid, 1.8–2.5 cm. *NW Turkey.* H4. Autumn.

15. C. umbrosum Steven (*C. arenarium* Waldstein & Kitaibel var. *umbrosum* Ker Gawler). Illustration: Darnell, Hardy and half-hardy plants **1**: 139 (1930); Flora SSSR **4**: t. 2, fig. 8 (1935); Mathew & Baytop, The bulbous plants of Turkey, 80 (1984).

Corm 1.5–3 × 1.2–2.5 cm, ovoid to almost spherical; tunic dark blackish brown, usually leathery, usually with a persistent neck to 7 cm. Leaves 3–5, developing after flowering, strap-shaped to very narrowly lanceolate, more or less erect, 8–17 × 1–2.7 cm, hairless. Flowers 1–6, funnel-shaped at first, star-shaped when fully open. Perianth-lobes white to purplish pink, narrowly oblanceolate to linear-elliptic, 1.5–3 cm × 2–6 mm. Filaments 5–8 mm, hairless; anthers yellow, 3–4 mm, dorsifixed; pollen yellow. Styles straight or slightly curved and swollen at the apex, stigmas decurrent for 0.5–0.75 mm. Capsule oblong-ellipsoid, 2–3 cm. *Romania, Former USSR (Crimea), N Turkey.* H3. Late summer–autumn.

16. C. parlatoris Orphanides. Illustration: Rix & Phillips, The bulb book, 184 (1981).

Corm 2–4 × 1.5–2 cm, ovoid; tunic dark brown, leathery, with a neck 3–4 cm. Leaves 4–10, usually developing after flowering, linear to narrowly linear, 7–10 cm × 1–4 mm, hairless. Flowers 1 or 2, narrowly bell-shaped. Perianth-lobes purplish pink, narrowly elliptic,

8–50 × 4–12 mm. Filaments 1–1.2 cm, hairless; anthers yellow, 4–6 mm, dorsifixed; pollen yellow. Styles straight, stigmas point-like. Capsule ovoid-oblong, *c.* 1 cm. *S Greece.* H4. Late summer–autumn.

17. C. decaisnei Boissier. Illustration: *Botanical Magazine*, 6901 (1886).

Corm 3–6 × 2.5–4 cm, ovoid; tunic dark blackish brown, papery to almost leathery, apex produced into a persistent neck to 9 cm. Leaves 3–8, developing after flowering, strap-shaped, erect-spreading, 10–30 × 1.2–4.5 cm, hairless, ciliate or thinly hairy. Flowers 2–12, narrowly funnel-shaped to star-shaped. Perianth-lobes white to pale purplish pink, narrowly oblong-lanceolate, 2.8–4.5 cm × 4–11 mm. Filaments 1.5–2 cm, hairless; anthers yellow, 6–8 mm, dorsifixed, with membranous longitudinal margins; pollen yellow. Styles erect or slightly curved at the apex, stigmas point-like or obliquely decurrent for 0.5 mm. Capsule ellipsoid, 1.5–3 cm. *Turkey, W Syria; N Israel, Lebanon.* H4. Autumn–winter.

18. C. balansae Planchon (*C. candidum* Boissier). Illustration: Mathew & Baytop, The bulbous plants of Turkey, 67 (1984).

Corm 4–5 × 2–3 cm, ovoid; tunic dull, dark blackish brown, membranous to almost leathery, apex produced into a very long, thick, persistent, strongly fibrous neck to 25 cm. Leaves 4 or 5, developing after flowering, elliptic to oblong-elliptic (the outer) to strap-shaped (the inner), erect-spreading, 18–24 × 4–7.5 cm. Flowers 3–11, funnel-shaped. Perianth-lobes white to purplish pink, narrowly oblong-elliptic, 4.5–7.5 cm × 4–13 mm. Filaments 1–1.9 cm, hairless; anthers yellow, 1–1.7 cm, dorsifixed, with membranous longitudinal margins; pollen yellow. Styles curved at the apex, stigmas decurrent for 1.5–3 mm. Capsule oblong-ovoid, 3–3.5 cm. *Turkey, Greece (Rhodes).* H4. Late summer–autumn.

19. C. kotschyi Boissier (*C. candidum* Boissier var. *hirtiflorum* Boissier). Illustration: *Botanical Magazine*, n.s., 520 (1967); Wendelbo, Tulips and irises of Iran, 13, t. 2 (1977); Mathew & Baytop, The bulbous plants of Turkey, 73 (1984); Brickell (ed.), RHS A–Z encyclopedia of garden plants, 294 (2003).

Corm 3–5 × 2–3 cm, ovoid; tunic dark brown, membranous to almost leathery, apex produced into a persistent neck 3–14 cm. Leaves 3–5, developing after flowering, elliptic-lanceolate to oblong-elliptic, erect-spreading, 10–16 × 3–5 cm, hairless, margins slightly wavy. Flowers 3–12, funnel-shaped. Perianth-lobes white to purplish pink, narrowly oblanceolate to oblong-elliptic, 2.3–5.5 cm × 4–12 mm, downy along the bases of the ridges of the filament-channels. Filaments 7–12 mm, hairless; anthers yellow, 6–12 mm, dorsifixed, with membranous longitudinal margins; pollen yellow. Styles curved at the apex, stigmas decurrent for 2–4 mm. Capsule oblong-ovoid, 3–4.5 cm. *Turkey, Iran, Iraq.* H3. Late summer–autumn.

20. C. cilicicum (Boissier) Dammer (*C.* × *byzantinum* Ker Gawler var. *cilicicum* Boissier; *C. decaisnei* misapplied; *C. byzantinum* misapplied). Illustration: *Botanical Magazine*, 9135 (1928); Bowles, A handbook of Crocus and Colchicum, opposite 156 (1952); Rix & Phillips, The bulb book, 175 (1981); Mathew & Baytop, The bulbous plants of Turkey, 72 (1984).

Corm 4–6 × 3–4.5 cm, almost spherical to ovoid; tunic dull, dark brown, papery to almost leathery, apex produced into a persistent neck 5–17 cm. Leaves 4 or 5, developed after flowering, very narrowly elliptic to narrowly elliptic-lanceolate, more or less erect, 30–40 × 4–11.5 cm, hairless. Flowers 3–25, funnel-shaped to bell-shaped. Perianth-lobes pale lilac-purple to deep rosy purple, occasionally lightly tessellated, oblanceolate to elliptic, 4–7.5 × 1.2–2.5 cm, downy along the ridges of the filament-channels. Filaments 2.5–3.5 cm; anthers bright yellow, 6–10 mm, dorsifixed; pollen bright yellow. Styles straight or slightly curved at the apex, often with a dull crimson tip, very frequently projecting from the perianth, stigmas point-like or decurrent for not more than 0.5 mm. Capsule ellipsoid to obovoid, 3–4 cm. *Turkey, Syria, Lebanon.* H2. Autumn.

A deeper-coloured clone is offered as 'Purpureum'.

C. × **byzantinum** Ker Gawler. Illustration: *Botanical Magazine*, 1122 (1808). Probably a hybrid derived from *C. cilicicum.* The foliage does not appear until the spring, unlike that of *C. cilicicum* in which the leaves begin growth soon after flowering; the leaves are also broader and more strongly ribbed than those of *C. cilicicum.*

21. C. autumnale Linnaeus. Illustration: Grey-Wilson & Mathew, Bulbs, pl. 9 (1981); Rix & Phillips, The bulb book, 178 (1981).

Corm 2.5–6 × 2–4 cm, ovoid to almost spherical; tunic dark brown, membranous to leathery, neck 2–4 cm. Leaves 3–5, occasionally more, developing after flowering, linear-lanceolate to broadly lanceolate, erect or erect-spreading, 14–35 × 1–7 cm, hairless. Flowers 1–6, narrowly bell-shaped to bell-shaped. Perianth-lobes purplish pink, occasionally tessellated, to white, narrowly elliptic to oblong-elliptic, 4–6 × 1–1.5 cm. Filaments 1–1.6 cm, hairless; anthers yellow, 5–8 mm, dorsifixed; pollen yellow. Styles curved at the apex, stigmas decurrent for 3–5 mm. Capsule oblong-ovoid, 2–6 cm. *S, W & C Europe to Former USSR (Belorussiya & Ukraine).* H1. Late summer–autumn.

Very variable; several closely related taxa are cultivated:

C. tenorii Hort. non Parlatore. Flowers slightly tessellated, anthers yellow, crooks of the stigmas purplish. *Italy.*

C. lusitanicum Brotero. Flowers lightly tessellated, anthers pale or deep purplish black. *SW Europe, N Africa.*

C. longifolium Castagne (*C. neapolitanum* Tenore *var longifolium* (Castagne) Rouy). Leaves shorter, linear-lanceolate; flowers slightly smaller. *S Europe.*

C. parnassicum Sartori, Orphanides & Heldreich. Corm-tunics membranous, leaves arched. *Greece.*

Plants cultivated as *C. laetum*, with numerous star-shaped, pale purplish pink flowers, are akin to *C. byzantinum* (above). True *C. laetum* Steven is similar to *C. autumnale* but has 1–3 smaller flowers and shortly decurrent stigmas.

Cultivars of *C. autumnale* available include the double white 'Alboplenum' and the double lilac-pink 'Pleniflorum'.

22. C. speciosum Steven. Illustration: *Botanical Magazine*, 6078 (1874); Wendelbo, Tulips and irises of Iran, 11, t. 1 (1977); Mathew & Baytop, The bulbous plants of Turkey, 78 (1984).

Corm 5–8 × 2.5–4 cm, oblong-ovoid; tunic dull mid-brown, papery to almost leathery, apex produced into a persistent neck to 12 cm. Leaves 3–5, developing after flowering, narrowly elliptic to oblong-lanceolate, more or less erect, 18–25 × 5.5–9.5 cm, hairless. Flowers 1–3, narrowly bell-shaped to bell-shaped. Perianth-tube green or white flushed with purple; lobes pale to deep rosy purple, sometimes with a white throat, occasionally white throughout, oblanceolate to oblong-oblanceolate or elliptic, 4.5–8 × 1–2.7 cm, downy along the ridges of the filament-channels. Filaments 1–1.8 cm, hairless; anthers orange-yellow, 1–1.2 cm, dorsifixed; pollen deep yellow. Styles curved but not, or only very slightly, swollen at the apex; stigmas decurrent for 2–4 mm. Capsule ellipsoid, 4–5 cm. *N Turkey, Iran, Former USSR (Caucasus).* H2. Autumn.

Variants of *C. speciosum* with green perianth-tubes and conspicuously white-throated flowers and yellow anthers are frequently, but incorrectly, grown as *C. bornmuelleri,* which has purple-brown anthers. White variants of *C. speciosum* occur scattered through populations of purple-flowered plants in the wild, and are available in cultivation. The plant grown under the name *C. giganteum* Arnott (*C. illyricum superbum* invalid) is very closely related to *C. speciosum* but has broadly funnel-shaped, not bell-shaped, flowers.

23. C. bornmuelleri Freyn. Illustration: *Gardeners' Chronicle* **52**: 157 (1912).

Corm 2.5–4.5 × 2.5–4 cm, ovoid to almost spherical; tunic dull mid-brown, papery to membranous, apex produced into a neck 1–8 cm. Leaves 3 or 4, developing after flowering, very narrowly elliptic, more or less erect, 17–25 × 2.6–4.5 cm, hairless. Flowers 1–6, bell-shaped, sometimes narrowly so. Perianth-lobes rosy purple (throat usually white), oblanceolate to narrowly elliptic, 4.5–7 × 1.1–2.4 cm, downy along the ridges of the filament-channels. Filaments 1.3–2.8 cm, hairless; anthers purple or purple-brown, 8–12 mm, dorsifixed; pollen yellow. Styles slightly curved, stigmatic surface 0.5–1.5 mm, confined to the swollen apex of the style. Capsule narrowly ovoid or narrowly ellipsoid, 4–5 cm. *Turkey.* Late summer–autumn.

Probably a variant of *C. speciosum* Steven.

24. C. variegatum Linnaeus (*C. parkinsonii* Hooker; *C. variegatum* var. *desii* Pampanini; *C.* × *agrippinum* misapplied). Illustration: Grey-Wilson & Mathew, Bulbs, pl. 9 (1981); Mathew & Baytop, The bulbous plants of Turkey, 81 (1984).

Corm 2–5 × 2–3.5 cm, ovoid to almost spherical; tunic dark brown, more or less leathery, apex produced into a persistent neck 5–15 cm. Leaves 3 or 4, developing after flowering, linear-lanceolate or strap-shaped, 9–15 cm × 7–25 mm, hairless, margins cartilaginous, wavy. Flowers 1 or 2. Perianth-lobes spreading, deep red- or violet-purple, occasionally paler or white at the base, strongly tessellated, lanceolate to elliptic or oblanceolate, 2–2.7 cm × 5–25 mm, frequently slightly twisted near the obtuse or acute apex, hairless. Filaments 1.5–4 cm, hairless; anthers purplish black or purplish brown, 4–9 mm, dorsifixed; pollen yellow. Styles straight, occasionally slightly curved and swollen at the apex, stigmas decurrent for 1.5–2 mm. Capsule oblong-ovoid, 2 cm. *Greece, SW Turkey.* H4. Autumn.

C. agrippinum Baker is probably a hybrid between *C. variegatum* and *C. autumnale.* It resembles *C. variegatum* closely, but is distinguished by its upright flowers and the narrower-based, less tessellated and wavy perianth-lobes. It increases quickly and is a very easy garden plant.

25. C. macrophyllum Burtt (*C. latifolium* misapplied? *C. latifolium* Sibthorp & Smith var. *longistylum* Pampanini). Illustration: Goulimis, Wild flowers of Greece, 143 (1968); Rix & Phillips, The bulb book, 177 (1981); Grey-Wilson & Mathew, Bulbs, pl. 9 (1981); Mathew & Baytop, The bulbous plants of Turkey, 76 (1984).

Corm 5–7 × 4–6 cm, ovoid to almost spherical; tunic dull, dark brown, leathery, apex produced into a persistent, somewhat fibrous neck to 24 cm. Leaves 3 or 4, developing after flowering, ovate to elliptic-ovate, erect-spreading, 24–42 × 11–15.5 cm, almost acute to bluntly acuminate, strongly pleated, hairless. Flowers 1–5. Perianth-lobes spreading, lilac-purple to rosy purple, tessellated, often paler or white in the throat, elliptic or oblong-elliptic, 4.5–7 × 1.5–3 cm, hairless. Filaments 2–2.5 cm, hairless; anthers purple, 8–12 mm, dorsifixed; pollen grey-green, drying paler. Styles shortly curved at the apex, stigmas decurrent for 1.5–2.5 mm. Capsule ovoid, 4–5 cm. *Greece (Crete, Rhodes), SW Turkey.* H3. Autumn.

26. C. bivonae Gussone (*C. latifolium* Sibthorp & Smith; *C. bowlesianum* Burtt; *C. sibthorpii* misapplied). Illustration: Goulimis, Wild flowers of Greece, 141 (1968); Polunin, Flowers of Greece and the Balkans, pl. 55 (1980); Rix & Phillips, The bulb book, 180 (1981); Brickell (ed.), RHS A–Z encyclopedia of garden plants, 294 (2003).

Corms 2.5–6 × 2.5–4 cm, ovoid to almost spherical; tunics mid-brown, papery to almost leathery, apex produced into a short neck 1–2 cm. Leaves 4–11, developing after flowering, strap-shaped or linear-lanceolate, more or less erect, 12–30 × 1–4.5 cm, hairless. Flowers 1–6, bell-shaped, sometimes broadly so. Perianth-lobes rosy purple, strongly tessellated, sometimes white at the base, narrowly to broadly elliptic, occasionally obovate-elliptic, 4–8.5 cm × 8–35 mm, downy along the ridges of the filament-channels. Filaments 1–2.5 cm, hairless; anthers purplish black or purplish brown, 7–12 × 2–3 mm, dorsifixed; pollen bright yellow. Styles curved and slightly swollen at the apex, stigmas decurrent for 2–4 mm. Capsule oblong-ellipsoid, 3.5–4 cm. *S Europe from Corsica and Sardinia to W Turkey.* H3. Late summer–autumn.

A parent of many garden hybrids with *C. speciosum, C. bornmuelleri* and *C. autumnale.*

27. C. lingulatum Boissier (*C. sibthorpii* misapplied). Illustration: Grey-Wilson & Mathew, Bulbs, pl. 10 (1981); Mathew & Baytop, The bulbous plants of Turkey, 74 (1984).

Corm 3–6 × 2–3.5 cm, ovoid to almost spherical; tunic blackish brown, more or less leathery, apex produced into a persistent neck to 8 cm. Leaves 3–7, developing after flowering, oblong to strap-shaped, spreading, hairless, margins cartilaginous and usually wavy. Flowers funnel-shaped to bell-shaped. Perianth-lobes pale to deep purplish pink, sometimes lightly tessellated, narrowly oblanceolate to narrowly oblong-elliptic, 2.5–5 cm × 2–12 mm, hairless. Filaments 6–14 mm, hairless; anthers yellow, 4–8 mm, dorsifixed; pollen yellow. Styles curved at the apex, stigmas decurrent for 2–3 mm. Capsule oblong-ovoid, 1.5–2.5 cm. *Greece, NW Turkey.* H4. Late summer–autumn.

C. chalcedonicum Aznavour. Foliage similar, flowers deep rosy purple, tessellated. *NW Turkey.*

There are records of *C. lingulatum* from Greece with slightly tessellated flowers, so the two species may not be distinct.

28. C. turcicum Janka. Illustration: Polunin, Flowers of Greece and the Balkans, pl. 55 (1980); Grey-Wilson & Mathew, Bulbs, pl. 10 (1981).

Corm 3–5 × 2.5–4 cm, almost spherical to ovoid; tunic dark blackish brown, leathery, with an extended, persistent neck to 11 cm. Leaves 5–9, developing after flowering, strap-shaped to narrowly lanceolate, more or less erect and often twisted at the apex, 9–19 × 1–3.5 cm, somewhat glaucous, margins thinly cartilaginous, ciliate (sometimes obscurely so). Flowers 1–8, bell-shaped to funnel-shaped. Perianth-lobes red-purple, occasionally paler, sometimes lightly tessellated, elliptic to narrowly obovate, 3–6 cm × 3.5–13 mm, downy along the ridges of the filament-channels. Filaments 1–2.5 cm, hairless; anthers yellow (? sometimes purplish), 5–8 mm, dorsifixed; pollen yellow. Styles straight, curved at the apex, stigmas decurrent for 3–4 mm. Capsule ovoid to oblong-ovoid, 2–3 cm. *Balkans, NW Turkey.* H3. Late summer–autumn.

The closely related **C. atropurpureum** Stearn, with small, dark magenta-red flowers, is occasionally cultivated.

2. BULBOCODIUM Linnaeus
J.C.M. Alexander
Perennials with corms. Leaves linear to linear-lanceolate, obtuse. Perianth of 6 free segments, each divided into a blade and a long, narrow claw; claws held together by teeth or auricles at base of blade, thus forming a tube. Stamens 6, with slender filaments borne at base of blade. Ovary superior, but below ground, 3-celled, containing many ovules. Style 3-fid above, undivided below. Seeds spherical.

A genus of 2 species from S & E Europe, sometimes considered part of *Colchicum* but differing in the undivided lower style, and distinctly clawed perianth-segments. Cultivation conditions are similar to those for *Colchicum* (see p. 55).

1. B. vernum Linnaeus (*Colchicum bulbocodium* Ker Gawler; *C. vernum* (Linnaeus) Ker Gawler). Illustration: *Botanical Magazine*, 153 (1791); Polunin, Flowers of Europe, t. 163 (1969); Polunin & Smythies, Flowers of SW Europe, 389 (1973); Rix & Phillips, The bulb book, 111 (1981).

Corm oblong, to 3 cm across; scales almost black. Leaves 3 or 4, appearing with the flowers but reaching full size later, to 15 × 1.5 cm, with sheathing base and spreading blade. Flowers solitary (rarely 2 or 3), borne close to the ground, 4–8.5 cm, pinkish purple, rarely white. Blade linear-lanceolate, acute, toothed at base. Capsule 1.5–2 cm. *Pyrenees, SW & SC Alps.* H3. Winter–spring.

3. MERENDERA Ramond
C.D. Brickell
Perennial, stemless herbs. Corms oblong-ovoid, occasionally narrow and more or less horizontal, enclosed by a membranous or leathery tunic which is usually extended into a short neck. Developing leaves and inflorescences enclosed within a membranous, cylindric sheath. Leaves basal, partially developed at flowering or developing after flowering. Flowers purple to pink or white, solitary or in clusters. Perianth-segments free, each with a long, narrow claw and a broader, linear or narrowly elliptic to narrowly obovate blade, with or without 2 auricles at the base. Stamens borne close to the base of the blade; anthers versatile or basifixed, pollen yellow. Ovary subterranean; styles 3, free to the base; stigmas point-like. Capsule opening along the septa with 3 abruptly pointed flaps, maturing at or just above ground level by the elongation of the flower-stalk. Seeds numerous, spherical or almost so.

A genus of about 10 species extending from S Europe, N Africa and W Asia to Afghanistan, with outliers in Israel, Jordan and Ethiopia. It is very closely related to *Colchicum* (and included in it by some authors), but differs in the free perianth-segments. *M. montana* may be grown successfully in the open in a well-drained, sunny site. The remaining species are best cultivated as bulb-frame or alpine house plants in sharply drained, but fairly rich compost. Now included in *Colchicum* by some authors.

Literature: Stefanoff, B., Monographic der Gattung Colchicum, *Sbornik Bulgariskata Akademiya na Naukita* **22**: 1–100 (1926).

1a. Anthers basifixed 2
b. Anthers versatile 4
2a. Leaves partially developed at flowering; anthers greenish yellow; flowering in spring **5. robusta**
b. Leaves developing after flowering; anthers yellow; flowering in autumn 3
3a. Leaves 3 or 4, 4–10 mm wide; anthers usually 8–12 mm **1. montana**
b. Leaves 5–15, 1–3 mm wide; anthers usually 5–8 mm **2. filifolia**
4a. Plant with narrow, more or less horizontal corms **3. sobolifera**
b. Plant with erect corms 5
5a. Leaves 3, lanceolate to narrowly lanceolate, to 17 × 3 cm at maturity; corm-tunics brown, membranous; anthers yellow **4. kurdica**
b. Leaves 2–6, linear to linear-lanceolate, to 18 cm × 3–8 (rarely to 18) mm; corm-tunics black or blackish brown, leathery; anthers black, purple or brown 6
6a. Leaves 2–4, hairless; flowers 1–3; blade of perianth-segment oblanceolate to narrowly oblanceolate, 2–3 cm × 4–9 mm, auricles usually present **6. trigyna**
b. Leaves 2–6, ciliate or with margins roughened towards the base; flowers 2–6; blade of perianth-segment linear to narrowly elliptic, 1.5–2.7 cm × 2–5 mm, auricles usually absent **7. attica**

1. M. montana (Linnaeus) Lange (*M. bulbocodium* Ramond; *M. pyrenaica* (Pourret) Fournet). Illustration: Rix & Phillips, The bulb book, 176f (1981); Grey-Wilson & Mathew, Bulbs, pl. 11 (1981); Brickell (ed.), RHS A–Z encyclopedia of garden plants, 688 (2003).

Corm oblong-ovoid, 2–3 × 1.5–3 cm; tunic black or dark brown, leathery, neck 1–2 cm. Leaves 3 or 4, developing after flowering or as the flowers fade, narrowly linear, to 22 cm × 4–10 mm at maturity. Flowers 1 or 2, purple to red-purple. Blade of perianth-segment narrowly elliptic to narrowly oblong-oblanceolate, 2–6.5 cm × 3–11 mm. Filaments 4–6 mm; anthers yellow, 5.5–12 mm, basifixed. Capsule oblong-ellipsoid, 1–2.5 cm. *Iberian peninsula, C Pyrenees.* H2. Late summer–autumn.

2. M. filifolia Cambessedes (*Colchicum filifolium* (Cambessedes) Stefanoff). Illustration: Mathew, Dwarf bulbs, 38 (1973).

Corm oblong-ovoid, 1.5–2 × 1–2 cm; tunic black or dark brown, leathery, neck 1–5 cm. Leaves 5–15, developing after flowering or as the flowers fade, thread-like, to 15 cm × 1–3 mm at maturity. Flowers solitary, purple to red-purple. Blade of perianth-segment narrowly elliptic to narrowly oblong-elliptic, 2–4 cm × 2–8 mm. Filaments 4–7 mm; anthers yellow, 5–10 mm, basifixed. Capsule oblong-ellipsoid, 8–12 mm. *SW Europe, N Africa.* H5. Autumn.

3. M. sobolifera Fischer & Meyer (*Colchicum soboliferum* (Fischer & Meyer) Stefanoff). Illustration: *Botanical Magazine*, 9576 (1939); Polunin, Flowers of Greece and the Balkans, 497, 5 (1980).

Corms narrow, horizontal, 2–8 cm long; tunic brown, membranous. Leaves 3, linear to strap-shaped, 2–10 cm at flowering, 10–21 cm × 3–13 mm at maturity. Flowers 1 or 2, white, pink or pale purple. Blade of perianth-segment linear to narrowly elliptic, 2–4 cm × 2.5–8 mm, frequently with basal, usually linear, auricles. Filaments to 7 mm; anthers blackish violet, 1.5–3.5 cm, versatile. Capsule oblong-ovoid, 1.5–2 cm. *E Europe to Iran & Former USSR (Turkestan)*. H3. Spring–early summer.

4. M. kurdica Bornmüller (*Colchicum kurdicum* (Bornmüller) Stefanoff).

Corm oblong-ovoid, 2–4 × 1.5–2 cm; tunic brown, membranous, neck 1–3 cm. Leaves 3, lanceolate to narrowly lanceolate, 4–12 cm at flowering, to 17 × 3 cm at maturity. Flowers 1 or 2, pale to deep purple. Blade of perianth-segment oblanceolate to narrowly oblanceolate, 3–3.5 cm × 5–10 mm, with auricles at the base. Filaments 8–9 mm; anthers yellow, 3–4 mm, versatile. Capsule ellipsoid, 2–3 cm. *SE Turkey, N Iraq*. H5. Spring.

5. M. robusta Bunge (*M. persica* Boissier; *M. aitchisonii* J.D. Hooker; *Colchicum robustum* (Bunge) Stefanoff). Illustration: *Botanical Magazine*, 6012 (1873).

Corm oblong-ovoid, 3–6 × 2–3 cm; tunic dark brown, leathery, neck 1–3 cm. Leaves 3–6, partially developed at flowering, linear to linear-lanceolate, to 5 cm at flowering, to 25 × 2.5 cm at maturity. Flowers 1–4, deep pink to white. Blade of perianth-segment oblong-elliptic to linear-elliptic, 1.8–4 cm × 2–9 mm. Filaments 8–10 mm; anthers greenish yellow, 8–12 mm, basifixed. Capsule oblong-ovoid, 1–3 cm. *Iran, Afghanistan, Former USSR (Turkmeniya) & N India*. H4. Spring.

M. hissarica Regel. Illustration: Rix & Phillips, The bulb book, 32 (1981). Differs in having basifixed anthers, solitary white flowers and 2 leaves per corm. *Former USSR (Soviet Central Asia), Afghanistan*. H4. Spring.

6. M. trigyna (Adam) Stapf (*Bulbocodium trigynum* Adam not Janka; *M. caucasica* Bieberstein; *Colchicum caucasicum* (Bieberstein) Sprengel; *C. trigynum* (Steven ex Adam not Janka) Stearn; *M. eichleri* Boissier). Illustration: *Botanical Magazine*, 3690 (1839); Rix & Phillips, The bulb book, 32 (1981); Mathew & Baytop, The bulbous plants of Turkey, 99 (1984).

Corm oblong-ovoid, 1.5–3.5 × 1–2 cm; tunic black or blackish brown, leathery, neck 2–3 cm. Leaves 2–4, linear to linear-lanceolate, 1–4 cm at flowering, to 17 cm × 3–18 mm at maturity, acute, green, hairless. Flowers 1–3, white to purplish pink. Blade of perianth-segment oblanceolate to narrowly oblanceolate, 2–3 cm × 4–9 mm, base usually with auricles. Filaments 1.5–4 mm; anthers black, purplish black, greenish black or olive brown, 2–4 mm, versatile. Capsule ellipsoid to ovoid, 1.5–3 cm. *Turkey, Iran, Former USSR (Caucasus)*. H2. Spring–early summer.

7. M. attica (Tommasini) Tommasini (*Colchicum atticum* Spruner in Tommasini; *M. rhodopea* Velenovsky; *M. brandigiana* Markgraf). Illustration: Stojanoff & Stefanoff, Flora Bulgarica, 220, 243 (1925); Grey-Wilson & Mathew, Bulbs, pl. 11 (1981).

Corm oblong-ovoid, 1.5–3 × 1–2 cm; tunic black or blackish brown, leathery, neck 1–2 cm. Leaves 2–6, linear to linear-lanceolate, 3–5 cm at flowering, to 18 cm × 3–8 mm at maturity. Flowers 2–6, white to purplish pink. Blade of perianth-segment linear to narrowly elliptic, 1.5–2.7 cm × 2–5 mm, base not usually with auricles. Filaments to 8 mm; anthers black, deep purplish black or greenish black, 2–4 mm, versatile. Capsule ellipsoid to ovoid, 1.5–2.5 cm. *Balkan peninsula, Turkey*. H2. Late winter–spring.

M. androcymbioides Valdes (*Colchicum androcymbioides* (B. Valdes) K. Persoon). Resembles *M. attica* but anthers yellow, basifixed, leaves slightly broader. *SW Spain*. H5.

16. LILIACEAE

Plants with bulbs, rhizomes or corms, terrestrial, herbaceous. Leaves generally borne on the stems. Flowers in racemes or panicles, without spathes when solitary, usually actinomorphic. Perianth of 6 free or united segments, well-developed; nectaries on the perianth-segments, rarely absent. Stamens 6. Ovary of 3 united carpels, superior. Ovules many, axile. Style 1 sometimes divided above or stigmas sessile. Fruit a capsule.

In the restricted usage employed here, there are 17 genera and about 415 species. Five genera are native to Europe, 6 to North America. The assignment of genera to these smaller families is still somewhat controversial; here, 3 genera more often included in the Colchicaceae (*Gloriosa, Littonia* and *Sandersonia*) are put with the reduced Liliaceae, as are two others whose family placement is very controversial, *Tricyrtis* and *Uvularia*. All of these genera have conspicuous nectaries on the perianth, and an overall similarity to genera always retained in the Liliaceae. *Calochortus* is sometimes separated off as the Calochortaceae, *Tricyrtis* as the Tricyrtidaceae, and *Uvularia* as the Uvulariaceae.

1a. Perianth urn-shaped, the segments united except at the extreme tips **3. Sandersonia**
b. Perianth not as above, the segments free to the base or almost so 2
2a. Plants with bulbs or corms 3
b. Plants with rhizomes or fibrous roots 12
3a. Perianth segments without nectaries at or near the base 4
b. Perianth segments with nectaries at the base 5
4a. Flowers in panicles **4. Gagea**
b. Flowers solitary or in racemes **8. Tulipa**
5a. Style absent, stigmas borne directly on the ovary **6. Calochortus**
b. Style present 6
6a. Anthers basifixed 7
b. Anthers dorsifixed, often versatile 9
7a. Perianth-segments reflexed from the base or very near it **8. Erythronium**
b. Perianth-segments not reflexed from the base, reflexed, if at all, at their tips only 8
8a. Perianth-segments in 2 distinct whorls; bulb-tunics usually absent, thin and white if present **9. Fritillaria**
b. Perianth-segments not in 2 distinct whorls; bulb-tunics always brown **5. Lloydia**
9a. Perianth-whorls differing markedly in size, colour or markings; filaments swollen at their bases; nectary usually formed from a number of plates arranged in a fan shape **12. Nomocharis**

b. Perianth-whorls similar or segments of the inner whorl slightly narrower than those of the outer; nectary various, not as above 10

10a. Bulb persistent after flowering, over-wintering; leaves various but neither long-stalked and heart-shaped nor linear and very rarely produced in the winter **10. Lilium**

b Bulb dying after flowering, the plant persisting by offsets; leaves long-stalked and heart-shaped or linear and produced in winter 11

11a. Leaves long-stalked and heart-shaped **13. Cardiocrinum**

b. Leaves linear and produced in winter **11. Notholirion**

12a. Leaves with tendril-like tips; plants usually climbing or scrambling (if not climbing then style bent sharply to one side at its base) 13

b. Leaves without tendril-like tips; plants not climbing nor scrambling; style never bent sharply to one side at the base 14

13a. Perianth bell-shaped; style not bent abruptly to one side at its base **2. Littonia**

b. Perianth widely open, the segments spreading widely or reflexed; style bent sharply to one side at its base **1. Gloriosa**

14a. Perianth pure yellow, not spotted; flowers pendent **14. Uvularia**

b. Perianth spotted; flowers erect **15. Tricyrtis**

1. GLORIOSA Linnaeus

V.A. Matthews

Rootstock a tuberous rhizome. Stems often climbing, to 2.5 m. Leaves alternate, opposite or whorled, stalkless, often with a tendril at the tip. Flowers borne in the axils of the upper leaves, pendulous. Perianth-segments free, reflexed, yellow to red or purplish or variously bicolored, margins wavy or not. Stamens spreading, shorter than the perianth. Ovary superior, 3-celled. Style bent at a right angle at the base, with 3 branches. Fruit a capsule. All parts of the plant are poisonous.

A genus of 1 species, although in the past about 30 species have been recognised. Recent work by D.V. Field maintains that the various forms cannot be distinguished in the wild, and they are best regarded as forms of *G. superba*. He recommends that they should be given cultivar names for horticultural purposes but so far this has not been done. The main variation is in the height of the stem, possession of leaf tendrils, colour and width of perianth-segments and their degree of crisping or waving, and the width of the leaves. Flower colour is further complicated because it may change as the flower ages.

Gloriosa is generally grown in a cool greenhouse, although, in warm areas, plants may be put outdoors during the summer. Any well-drained soil is suitable. Propagation is by offsets, division of the rhizome, or by seed.

Literature: Field, D.V., The genus Gloriosa, *Lilies and other Liliaceae* **1973**: 93–5 (1972).

1. G. superba Linnaeus (*G. simplex* Linnaeus; *G. virescens* Sims; *G. carsonii* Baker; *G. abyssinica* Richard; *G. minor* Rendle; *G. lutea* Anon.; *G. rothschildiana* O'Brien; *G. verschuurii* Hoog). Illustration: *Journal of the Royal Horticultural Society* **94**: pl. 1 (1969); Perry, Flowers of the world, 176 (1972); Hay et al., Dictionary of indoor plants in colour, pl. 258 (1974); Herklots, Flowering tropical climbers, 127 (1976).

Tropical Africa, tropical Asia. G1. Summer–autumn.

2. LITTONIA Hooker

D.A. Webb

Perennial herbaceous climbers with tubers. Leaves stalkless, often whorled below, usually opposite or alternate above, with the apex drawn out into a tendril. Flowers nodding, single in the leaf-axils, or sometimes with the stalk arising beside or just below the leaf. Perianth bell-shaped, the lobes not opening widely but free almost to the base. Anthers basifixed, opening outwards. Ovary superior. Style simple, straight, continuing the axis of the flower. Stigma of 3 linear lobes. Fruit a many-seeded capsule.

A genus of about 8 species, ranging from South Africa to Senegal and Arabia, best grown in a peaty loam and dried off during the dormant season. A conservatory plant in all but the hottest regions of Europe.

1. L. modesta Hooker. Illustration: *Botanical Magazine*, 4723 (1853); Jeppe, Natal wild flowers, pl. 3 (1975); Brickell (ed.), RHS A–Z encyclopedia of garden plants, 642 (2003).

Tuber 3 cm across. Stems slender, unbranched, to 1.2 m. Lower leaves ovate-lanceolate, tapering from a broad base to the tendril; upper leaves usually narrower. Flower-stalks short. Perianth orange-yellow, the lobes lanceolate, acuminate. *South Africa (Transvaal, Natal, Orange Free State).* H5–G1. Early summer.

Var. **keitii** Leichtlin, with a branched stem and more numerous flowers, makes a more decorative plant than var. *modesta*.

3. SANDERSONIA Hooker

V.A. Matthews

Rootstock a tuberous rhizome. Stem up to 75 cm, bearing scattered, stalkless leaves up to 10 cm and usually with a tendril at the tip. Flowers solitary in the axils of the upper leaves, pendulous on 2–3 cm stalks. Perianth 2–2.5 cm, orange, spherical-urn-shaped, the perianth-lobes fused except at the tips, which curl outwards, the base of the perianth with 6 short nectar-bearing spurs. Stamens *c.* one-third as long as perianth, borne at the base of the lobes; anthers opening by slits. Ovary superior. Style with 3 branches. Fruit a capsule.

A genus with 1 species, native to South Africa. Cultivation as for *Gloriosa* (see above).

1. S. aurantiaca Hooker. Illustration: *Botanical Magazine*, 4716 (1853); *Flore des serres* **9**: t. 862 (1853–4); Jeppe, Natal wild flowers, pl. 3 (1975); Mathew, The larger bulbs, 127 (1978).

SE South Africa. G2–H5. Summer.

4. GAGEA Salisbury

E.M. Rix

Perennial herbs with small bulbs that are sometimes closely entwined with persistent, thickened roots. Basal leaves 1 or 2, arising from the bulb, linear or linear-lanceolate, hollow, solid or flat. Flowers usually yellow, rarely white, either in an umbel subtended by an opposite pair of leaf-like spathes on an otherwise leafless stem, or in a few-flowered panicle with at least 1 leaf on the stem below the point of branching. Perianth-segments free, equal, without nectaries. Anthers basifixed. Ovary superior. Style 1. Seeds flat or pear-shaped.

A genus of about 50 species, mainly found in C Asia. Most species have small, yellow, star-shaped flowers and are not considered worthy of cultivation. The Mediterranean species are grown in shallow, heavy soil, wet in spring, dry in the summer; the alpine and woodland species need cool, humus-rich soil.

Literature: Uphof, J.C.T., A review of the genus Gagea Salisb., *Plant Life* **14**:

124–32 (1958), **15**: 151–61 (1959) & **16**: 163–76 (1960).

Basal leaves. Flat: **1, 2, 6**; thread-like: **5**; hollow: **3**; solid, sometimes channelled: **2, 4, 5**.
Inflorescence. An umbel: **1, 2, 3, 4**; a panicle: **5, 6**.
Perianth. Yellow: **1, 2, 3, 4, 5**; white: **6**.

1a. Flowers white **6. graeca**
b. Flowers yellow 2
2a. Inflorescence a panicle, with a leaf on the stem below the first branch **5. peduncularis**
b. Inflorescence an umbel with no stem-leaves below the point of branching 3
3a. Basal leaves hollow **3. fistulosa**
b. Basal leaves flat or solid, sometimes channeled 4
4a. Basal leaves linear-lanceolate, flat, 7–15 mm wide **1. lutea**
b. Basal leaves linear, flat or channelled, 1–4 mm wide 5
5a. Basal leaf 1 per bulb **2. pratensis**
b. Basal leaves 2 per bulb **4. villosa**

1. G. lutea (Linnaeus) Ker Gawler (*G. silvatica* (Persoon) Loudon. Illustration: Reichenbach, Icones florae Germanicae et Helveticae **10**: 477, f. 1045 (1848); Ross-Craig, Drawings of British plants **29**: pl. 28 (1972).

New bulb formed within the old. Basal leaf 1 per bulb, linear-lanceolate, flat, 7–15 mm wide. Inflorescence an umbel of 1–7 flowers. Spathes ciliate. Flower-stalks hairy or not. Perianth-segments 1.5–1.8 cm, greenish yellow. Seeds pear-shaped. *Europe, from N England and Russia southwards to Spain, Greece and the Caucasus.* H1. Spring.

2. G. pratensis (Persoon) Dumortier (*G. stenopetala* Reichenbach). Illustration: Reichenbach, Icones florae Germanicae et Helveticae **10**: 474, f. 1038 (1848).

New bulb formed below old, outside the tunic, sometimes with a second new bulb present. Basal leaf 1 per bulb, linear, flat or channelled, 2–4 mm wide. Inflorescence an umbel of 1–4 flowers. Spathes lanceolate, sometimes ciliate. Flower-stalks 1–2 cm, hairless. Perianth-segments 9–16 mm, to 2.2 cm in fruit, yellow inside, green outside. Seeds pear-shaped. *Europe, from Denmark and Finland southwards to Spain, Turkey and the Crimea.* H1. Spring.

3. G. fistulosa Ker Gawler (*G. liotardii* (Sternberg) Schultes & Schultes; *G. anisanthos* Koch). Illustration: Grey-Wilson & Mathew, Bulbs, pl. 16 (1981).

New bulb formed beside old. Basal leaves 1, or 2 and unequal, per bulb, hollow, 6–20 × 3–5 mm when flattened. Inflorescence an umbel of 1–6 flowers, or replaced by bulbils. Spathes unequal, lanceolate. Flower-stalks 1.5–5 cm, usually hairy. Perianth-segments 1–1.8 cm, yellow, curling and elongating after flowering. Seeds pear-shaped. *S Europe, Greece, Turkey & Iran.* H1. Spring.

4. G. villosa (Reichenbach) Duby (*G. arvensis* (Persoon) Dumortier). Illustration: Reichenbach, Icones florae Germanicae et Helveticae **10**: 479, f. 1049 (1848).

New bulb formed beside old. Basal leaves 2 per bulb, linear, solid, D-shaped in section, 1–2.5 mm wide. Inflorescence more or less umbellate with 1–15 flowers. Spathes without bulbils in their axils. Flower-stalks usually hairy. Perianth-segments 7–10 mm, to 1.8 cm in fruit, yellow. Seeds pear-shaped. *Europe, from the Netherlands and Finland southwards to Spain, N Africa and Iran.* H1. Spring.

This species is a common weed of cultivation in the Mediterranean area; bunches of very small bulbils are often formed at ground-level, but not at the bases of the flower-stalks, as in the rather similar *G. granatelli* Parlatore.

5. G. peduncularis (Presl & Presl) Pascher. Illustration: Rix & Phillips, The bulb book, 44, 45 (1981).

New bulb formed beside old. Basal leaves 2 per bulb, linear, solid, 1–2 mm wide, exceeding the inflorescence. Inflorescence a panicle with 1–7 flowers. Stem-leaf 1, lanceolate, acuminate, 2–5 mm wide. Flower-stalks to 6.5 cm, usually 3–4 cm, densely woolly above, elongating during flowering. Perianth-segments 8–16 mm, to 1.8 cm in fruit, yellow. Seeds pear-shaped. *S Europe, N Africa & Turkey.* H2. Spring.

One of the most showy species and very common in the eastern Mediterranean area.

G. bohemica (Zauschner) Schultes & Schultes (*G. saxatilis* (Mertens & Koch) Schultes & Schultes). Illustration: Reichenbach, Icones florae Germanicae et Helveticae **10**: 480, f. 1052 (1848). Basal leaves thread-like, flower-stalks not elongating during flowering. *From C England and Germany south to Portugal & Israel.* H2. Spring.

6. G. graeca (Linnaeus) Terraciano (*Lloydia graeca* (Linnaeus) Kunth). Illustration: Polunin & Huxley, Flowers of the Mediterranean, pl. 239 (1965).

New bulb formed within the old; bulbs usually crowded together, surrounded with thickened roots. Basal leaves 2 per bulb, linear, flat, 1–2 mm wide, hairless. Inflorescence a panicle, usually with 1–5 flowers. Stem-leaves usually 3, alternate. Flowers pendent in bud, never opening flat. Perianth-segments 7–16 mm, oblanceolate, obtuse, white, with 3 purplish veins. Seeds flat. *Greece, Turkey.* H3. Spring.

5. LLOYDIA Reichenbach

V.A. Matthews

Perennial hairless herbs possessing bulbs with a brown, fibrous membranous tunic. Leaves basal and on the stem, linear. Flowers 1 or 2 at the top of the stem. Perianth-segments free, each usually with a gland at the base. Stamens borne at the base of the perianth-segments; anthers basifixed. Ovary superior, 3-celled. Fruit a capsule with 3 ribs.

About 12 species native to the temperate northern hemisphere. They grow best in a well-drained, peaty soil in a shady position. Propagation is by seed.

1. L. serotina (Linnaeus) Reichenbach. Illustration: Polunin, Flowers of Europe, pl. 164 (1969); Everard & Morley, Wild flowers of the world, pl. 7 (1970); Fitter et al., Wild flowers of Britain and northern Europe, 261 (1974); Grey-Wilson & Mathew, Bulbs, pl. 16 (1981).

Stem 5–15 cm. Leaves 7–20 cm, basal, usually 2, usually exceeding stem; stem-leaves 2–5. Flowers 1–1.5 cm, erect, white with reddish or purplish veins. *Temperate northern hemisphere.* H2. Late spring–summer.

L. longiscapa Hooker. Illustration: *Hooker's Icones Plantarum* **9**: pl. 834 (1852). Flowers larger, usually pendulous, each perianth-segment with a brownish purple or orange blotch at the base. *Himalaya, W China.* H2. Summer.

6. CALOCHORTUS Pursh

S.J.M. Droop

Perennial herbs with bulbs. Bulbs ovoid, with membranous or fibrous coats. Stems leafy or sometimes scape-like, often branched, frequently bearing bulbils. Leaves usually linear, the basal one large and conspicuous at flowering, or withering before flowering. Inflorescence cymose or more or less umbellate. Flowers conspicuous, often marked with colour contrasts. Perianth of 2 distinct whorls. Sepals usually lanceolate, hairless. Petals obovate to oblanceolate, usually more or less hairy on the inner surface and with a depression or gland near the base. Stamens 6, anthers oblong to linear, attached to the bases of the perianth-segments, the filaments broadened at the base. Ovary superior, 3-celled with numerous ovules in each cell, triangular or 3-winged in section. Fruit spherical to linear, 3-angled or 3-winged in section, erect or drooping, containing many seeds.

A genus of about 57 species from western N & C America, mainly from California. Probably about half the species have been in cultivation from time to time, but only a few are commonly grown. Cultivation is quite difficult in most of Europe, as the plants are easily killed by excessive moisture. They may be grown in a cool house or frame, in deep rich soil in a sunny position. Propagation is by seed, bulblets or bulbils.

Literature: Ownbey, M., A monograph of the genus Calochortus, *Annals of the Missouri Botanical Garden* **27**: 371–560 (1940).

Bulb-coats. Membranous: **1, 2, 3, 4, 5, 6, 7, 8, 9, 10, 11, 12, 13, 14, 15, 16, 17, 18, 19**; thickly fibrous: **20, 21**.
Stem. Usually bearing bulbils: **9, 12, 13, 14, 15, 16**; sometimes bearing bulbils: **11, 17, 18, 19, 21**; never bearing bulbils: **1, 2, 3, 4, 5, 6, 7, 8, 10, 20**.
Basal leaf. Large and conspicuous at flowering: **1, 2, 3, 4, 5, 6, 7, 8, 9, 10, 21**; withered before flowering: **11, 12, 13, 14, 15, 16, 17, 18, 19, 20**.
Stem leaves. Tightly curled under at the tip: **16**.
Flowers. Spherical to narrowly bell-shaped: **1, 2, 3, 4**; bell-shaped: **5, 6, 7, 8, 9, 10, 11, 12, 13, 14, 15, 16, 17, 18, 19, 20, 21**; drooping: **1, 2, 3, 4, 21**; erect or spreading: **2, 5, 6, 7, 8, 9, 10, 11, 12, 13, 14, 15, 16, 17, 18, 19, 20, 21**.
Sepals. Usually exceeding petals: **3, 16**; usually about equalling the petals: **5, 20, 21**; usually shorter than petals: **1, 2, 4, 6, 7, 8, 9, 10, 11, 12, 13, 14, 15, 17, 18, 19**; tightly curled under at the tip: **10, 11, 12, 13, 14, 15**; sparsely hairy on the lower surface: **21**.
Petals. Fringed: **1, 2, 3, 4, 5, 6, 7, 8, 20, 21**; not fringed: **9, 10, 11, 12, 13, 14, 15, 16, 17, 18, 19, 20, 21**. Hairs on the upper surface club-shaped: **5, 17, 18**; hairs on upper surface branched and glandular: **19**.
Gland (see figures 37 & 38, p. 573 & 578). Traversed by several membranes: **1, 2**; densely covered with simple or branched hair-like processes: **11, 12, 13, 14, 15, 16, 17, 18, 19**; naked or very nearly so: **3, 4, 5, 6, 7, 8, 9, 10, 11, 20, 21**; naked but covered from below by a single, fringed membrane: **5, 6, 7, 8, 9, 10**; not associated with any membrane: **3, 4, 11, 12, 13, 14, 15, 19, 20, 21**; completely surrounded by a ring of hairs: **19, 20, 21**; completely surrounded by a membrane: **19, 20**.
Anthers. Usually longer than filaments: **7, 8, 16, 17, 19**; usually shorter than filaments: **1, 3, 4, 6, 9, 11, 15, 21**; about equalling filaments: **2, 5, 10, 12, 13, 14, 18, 20**.
Fruit. Drooping: **1, 2, 3, 4, 5, 6, 7, 8, 9**; erect: **10, 11, 12, 13, 14, 15, 16, 17, 18, 19, 20, 21**; spherical to oblong, 3-winged: **1, 2, 3, 4, 5, 6, 7, 8, 9, 10**; linear to linear-oblong, 3-angled: **11, 12, 13, 14, 15, 16, 17, 18, 19, 20, 21**.

1a. Fruit more or less spherical to oblong, 3-winged, usually less than 3 times longer than broad and drooping; inflorescence more or less umbellate 2
b. Fruit linear to linear-oblong, 3-angled, more than 3 times as long as broad, erect; inflorescence cymose 11
2a. Flowers spherical to narrowly bell-shaped, usually drooping 3
b. Flowers openly bell-shaped, erect or spreading 6
3a. Flowers white to deep purplish pink; gland with several associated membranes 4
b. Flowers yellow; gland not associated with membranes 5
4a. Flowers white or tinged with pink, always drooping; lowest gland membrane less than two-thirds of the width of the petal at the level of the gland **1. albus**
b. Flowers deep purplish pink, sometimes erect or spreading; lowest gland membrane as wide as the petal at gland level **2. amoenus**
5a. Petals 1.6–2 cm, inner face hairless or with a few thick hairs near the gland, margin densely fringed, the fringe 1–1.5 mm deep **3. amabilis**
b. Petals 2.5–3.3 cm, inner face hairy to the tip, margin less densely fringed, fringe 1–2 mm deep **4. pulchellus**
6a. Flowers yellow; hairs on surface of petal club-shaped **5. monophyllus**
b. Flowers white to greenish, bluish or pale purple; hairs on surface of petal, if any, not club-shaped 7
7a. Fruit erect; petals hairless except for a single gland membrane and occasionally a few hairs near it **10. nudus**
b. Fruit nodding; petals with hairs associated with the gland and also near it 8
8a. Stem barely reaching above the surface of the ground, with many bulbils **9. uniflorus**
b. Stem more than 3 cm, without bulbils 9
9a. Stem usually branched, if unbranched then with a single stem leaf; anthers shorter than filaments **6. tolmiei**
b. Stem unbranched and without leaves; anthers slightly longer than filaments 10
10a. Anthers oblong, acute; petals 8–12 mm **7. coeruleus**
b. Anthers lanceolate, tapering to a fine point; petals 1.2–2.5 cm **8. elegans**
11a. Gland densely covered with short, thick, simple or branched processes, or, if naked, then not surrounded by a ring of hairs or a membrane; bulb-coats membranous 12
b. Gland naked or nearly so, but surrounded by a ring of hairs which are joined at the base or not; bulb-coats thickly fibrous-netted 21
12a. Gland not surrounded by a fringed membrane, or, if the outermost hairs are more or less joined at the base to form a membrane, then hairs on the petals gland-tipped 13
b. Gland obviously surrounded by a fringed membrane; hairs on petals sometimes club-shaped but never gland-tipped 18
13a. Petal hairs gland-tipped; anthers longer than filaments **19. gunnisonii**
b. Petal hairs not gland-tipped; anthers equalling or shorter than filaments 14
14a. Gland broader than long, not isodiametric 15
b. Gland more or less isodiametric, circular or diamond-shaped 17

15a. Gland more or less oblong, slightly arched above; petals yellow or yellow with reddish brown markings **15. luteus**
b. Gland distinctly arched above, more or less linear; petals with a reddish brown marking surrounded by yellow, never uniformly yellow 16
16a. Gland like an inverted V, without a notch at the apex; petals with a reddish brown mark surrounded by a bright yellow zone **13. superbus**
b. Gland more or less doubly crescent-shaped; petals with a pale yellow zone **14. vestae**
17a. Gland circular, less than one gland diameter from base of petal, sometimes naked; inflorescence distinctly cymose **11. splendens**
b. Gland more or less circular or diamond-shaped, usually more than 2 gland diameters from the base of the petal, never naked; inflorescence umbellate **12. venustus**
18a. Gland sagittate at base; petals acute to acuminate, purple, each with a median longitudinal green stripe; leaves tightly curled under at the tip **16. macrocarpus**
b. Combination of characters not as above 19
19a. Flowers vermilion to orange **17. kennedyi**
b. Flowers yellow 20
20a. Stem 10–50 cm; petals without zig-zag marking above the gland; anthers usually longer than filament **17. kennedyi**
b. Stem 50–100 cm; petals with zig-zag marking above the gland; anthers about as long as filaments **18. clavatus**
21a. Flowers drooping, usually yellow, often purplish; basal leaf conspicuous at flowering; hairs on petals purple; anthers oblong, much shorter than filaments **21. barbatus**
b. Flowers erect, orange-yellow; basal leaf usually withering before flowering; hairs on petals yellow; anthers linear, about as long as filaments **20. weedii**

Section **Calochortus.** Bulb-coats membranous. Stem usually without bulbils. Basal leaf conspicuous at flowering, tapering at both ends. Flowers spherical to openly bell-shaped. Ovary 3-winged, contracted to a short style and persistent, 3-fid stigma. Fruit spherical or oblong, 3-winged, drooping (erect in *C. nudus*).

1. C. albus Bentham. Figure 2(4), p. 67. Illustration: Rickett, Wild flowers of the United States **5**: t. 5 (1971); Brickell (ed.), RHS A–Z encyclopedia of garden plants, 213 (2003).

Stem stout, often branched, 20–80 cm. Basal leaf 30–70 cm, present at flowering. Flowers spherical to spherical-bell-shaped, drooping, white to pink. Sepals 1–1.5 cm, ovate to lanceolate, acuminate. Petals 1.5–3 cm, elliptic-obovate, acute to obtuse, fringed, sparsely covered above the gland with long, slender hairs. Gland depressed, traversed by *c.* 5 deeply fringed membranes, extending for one-third to two-thirds of the width of the petal at gland level. Anthers oblong, obtuse or acute, usually shorter than filaments. *USA (California).* H4. Spring–summer.

2. C. amoenus Greene (*C. albus* Bentham var. *amoenus* (Greene) Purdy & Bailey). Illustration: Rickett, Wild flowers of the United States **5**: t. 5 (1971).

Like *C. albus* but stems slender, to 50 cm; basal leaf 20–50 cm; flowers narrowly bell-shaped to spherical-bell-shaped, erect or drooping, dark purplish pink; sepals lanceolate; gland broad, arched, slightly depressed, extending right across the base of the petal; anthers obtuse, about as long as filaments. *USA (California).* H5. Late spring.

3. C. amabilis Purdy (*C. pulchellus* Bentham var. *amabilis* (Purdy) Jepson; *C. pulchellus* misapplied). Figure 3(17), p. 68. Illustration: Rickett, Wild flowers of the United States **5**: t. 4 (1971); Brickell (ed.), RHS A–Z encyclopedia of garden plants, 214 (2003).

Stem erect, often branched, 10–30 cm. Basal leaf 20–50 cm, exceeding the stem. Flowers spherical or spherical-bell-shaped, drooping, deep yellow. Sepals more or less spreading, usually exceeding the petals, ovate to lanceolate, acute to acuminate. Petals 1.6–2 cm, diamond-shaped, clawed, apex sharply rounded, densely fringed to 1–1.5 mm, the inner face hairless or with a few thick hairs near the gland. Gland deeply depressed, arched and bounded above by a transverse line of hairs 2–3 mm long, pointing into the centre of the depression. Anthers oblong, obtuse, usually shorter than the filaments. *USA (California).* H5. Late spring.

4. C. pulchellus Bentham (*Cyclobothra pulchella* Anon.). Figure 3(14), p. 68. Illustration: *Gardeners' Chronicle* 765 (1883).

Like *C. amabilis* but sepals not usually exceeding petals, petals 2.5–3.3 cm, less deep, yellow and hairy within almost to the tip; margin more deeply but less densely fringed. *USA (California).* H5. Late spring.

5. C. monophyllus (Lindley) Lemaire (*C. benthamii* Baker). Figure 3(15), p. 68. Illustration: Rickett, Wild flowers of the United States **5**: t. 4 (1971).

Stem erect, usually branched, 8–20 cm. Basal leaf 10–30 cm, greatly exceeding the stem. Flowers openly bell-shaped, usually erect, yellow. Sepals 1.6–2 cm, elliptic, acuminate, about as long as petals. Petals obovate, densely covered above the gland with club-shaped hairs, and sparsely fringed with similar hairs. Gland naked, spreading across the entire base of the petal, bounded below by a deeply fringed membrane and above by a few short hairs, hairs and membrane densely and shortly papillose. Anthers lanceolate, abruptly pointed, about as long as filaments. *USA (California).* H5. Late spring.

6. C. tolmiei Hooker & Arnott (*C. elegans* J.D. Hooker not Pursh; *C. maweanus* Baker; *C. coeruleus* (Kellogg) Watson var. *maweanus* (Baker) Jepson). Figure 2(8), p. 67 & 32(21), p. 578. Illustration: *The Garden* **87**: 474 (1923).

Stem erect, usually branched, 5–30 cm, with 1 stem leaf. Basal leaf 10–40 cm, usually exceeding the stem. Inflorescence more or less umbellate. Flowers 1–10, openly bell-shaped, erect or spreading, white or cream tinged with pink or purple. Sepals 8–20 mm, oblong-lanceolate, acute to acuminate. Petals 1.2–2.5 cm, obovate, hairy above the gland, sparsely fringed. Gland transverse, arched above, naked, bounded below by an irregularly toothed to fringed membrane, and above by a band of short, thick processes, these and the fringes of the membrane papillose. Anthers lanceolate, acute to abruptly pointed, shorter than filaments. *USA (Oregon & California).* H5. Late spring–early summer.

7. C. coeruleus (Kellogg) Watson. Figures 31(5) & 32(16), p. 573 & 578. Illustration: Rickett, Wild flowers of the United States **5**: t. 5 (1971).

Like *C. tolmiei* but stem unbranched, scape-like, 3–15 cm; flowers bluish, sepals

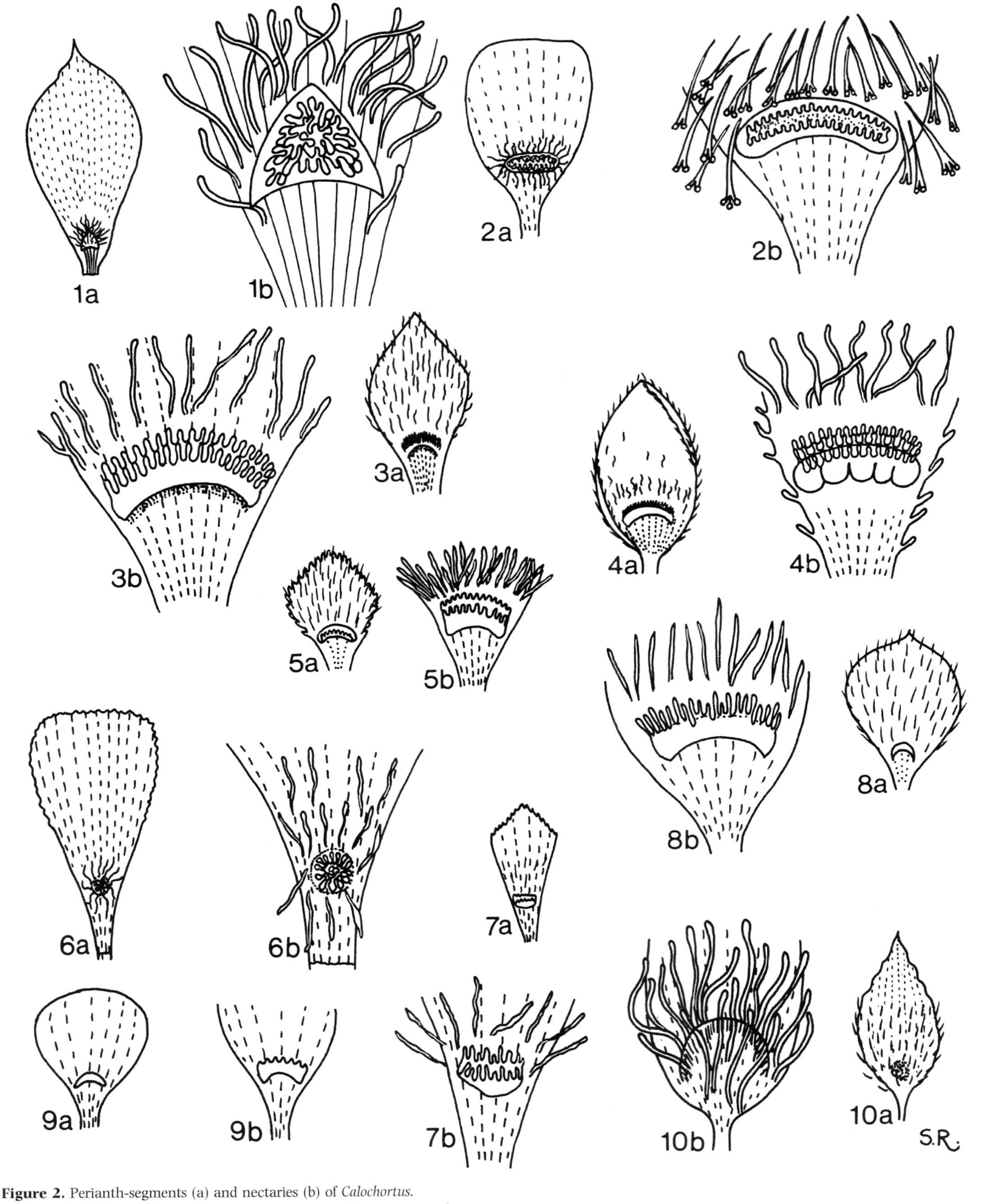

Figure 2. Perianth-segments (a) and nectaries (b) of *Calochortus*.
1, *C. macrocarpus*. 2, *C. gunnisonii*. 3, *C. elegans*. 4, *C. albus*. 5, *C. coeruleus*.
6, *C. kennedyi*. 7, *C. uniflorus*. 8, *C. tolmiei*. 9, *C. nudus*. 10, *C. barbatus*.

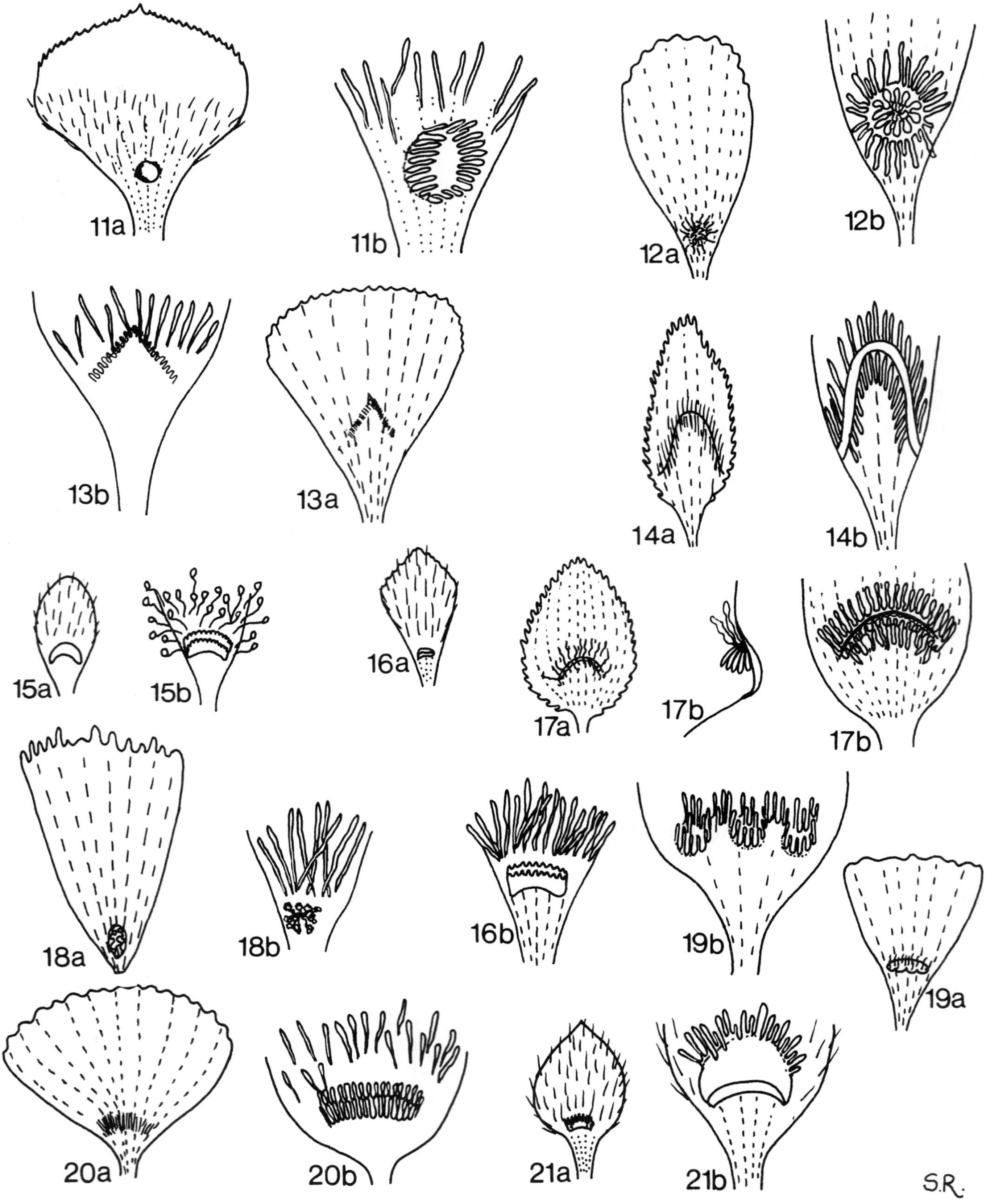

Figure 3. Perianth-segments (a) and nectaries (b) of *Calochortus* (continued). 11, *C. weedii.* 12, *C. venustus.* 13, *C. superbus.* 14, *C. pulchellus.* 15, *C. monophyllus.* 16, *C. coeruleus.* 17, *C. amabilis* (nectary in side and face view). 18, *C. splendens.* 19, *C. vestae.* 20, *C. luteus.* 21, *C. tolmiei.*

8–10 mm, petals 8–12 mm, broadly obovate; anthers oblong, acute, usually slightly longer than filaments. *USA (California).* H5. Summer.

8. C. elegans Pursh. Figure 2(3), p. 67. Illustration: Rickett, Wild flowers of the United States **5**: t. 5 (1971).

Like *C. tolmiei* but stem unbranched and scape-like; flowers greenish white, often with a purple crescent on each segment; gland membrane deeply fringed; anthers finely pointed, usually longer than filaments. *USA (C Idaho and adjacent Washington & Oregon).* H5. Late spring–early summer.

9. C. uniflorus Hooker & Arnott (*C. lilacinus* Kellogg). Figure 2(7), p. 67. Illustration: Rickett, Wild flowers of the United States **5**: t. 5 (1971).

Stem short, simple, barely reaching above ground surface, with many bulbils. Basal leaf 10–40 cm, much exceeding the inflorescence. Inflorescence more or less umbellate. Flowers 1–5, openly bell-shaped, erect on very long stalks, pale purple. Sepals 1.2–1.6 cm, elliptic-lanceolate, acuminate. Petals 1.5–2.8 cm, rounded and irregularly toothed above, with a few hairs near the gland. Gland naked, acute below, truncate above, bounded below by a broad, ascending, fringed membrane and above by a band of slender processes. Anthers oblong, obtuse or acute, shorter than the filaments. *USA (Oregon & California).* H5. Late spring.

10. C. nudus Watson. Figure 2(9), p. 67. Illustration: Rickett, Wild flowers of the United States **5**: t. 6 (1971).

Stem erect, scape-like, 5–15 cm. Basal leaf 10–20 cm, usually shorter than inflorescence. Flowers 1–6, openly bell-shaped, erect, white to pale purple. Sepals 1–1.6 cm, elliptic-lanceolate, acuminate. Petals 1.4–2 cm, rounded or acute and irregularly finely toothed, not fringed, hairless except occasionally for very few hairs near the gland. Gland naked, transverse, shallowly arched above, bounded below by an ascending, fringed membrane, the fringe minutely papillose. Anthers linear-oblong, obtuse to acute, about as long as filaments, pale blue at first. *USA (California).* H5. Summer.

Section **Mariposa** Wood. Bulb-coats membranous. Stems often with bulbils in the axils of the stem leaves. Leaves linear, tapering, the basal one withering before flowering. Flowers bell-shaped, erect. Ovary linear, tapering to a persistent, 3-fid stigma. Fruit linear-oblong to linear, 3-angled, erect.

11. C. splendens Bentham. Figure 3(18), p. 68. Illustration: Rickett, Wild flowers of the United States **5**: t. 6 (1971).

Stem erect, branched, rarely with bulbils, 20–60 cm. Inflorescence cymose. Flowers 1–4, bell-shaped, erect, pale purple, often with a darker purple spot on each segment. Sepals 2–2.5 cm, ovate-lanceolate, acuminate. Petals 2.5–5 cm, obovate, rounded and irregularly finely toothed above, sparsely covered with silky hairs from above the gland to about the middle. Gland small, circular, very near the base of the petal, naked or more usually covered with thick, branched processes. Anthers linear-oblong, obtuse to pointed, blue to purple, shorter than filaments. *USA (California), Mexico (Baja California).* H5. Late spring–early summer.

12. C. venustus Bentham. Figure 3(12), p. 68. Illustration: Rickett, Wild flowers of the United States **5**: t. 6 (1971); Brickell (ed.), RHS A–Z encyclopedia of garden plants, 214 (2003).

Stem erect, usually branched, with bulbils, 20–60 cm. Inflorescence more or less umbellate. Flowers 1–3, bell-shaped, erect, white to yellow, purple or dark red, each petal with a conspicuous, median, dark red mark, often with a second, paler mark above the first. Sepals 2–3.5 cm, linear-lanceolate, acuminate, curled under at the tip. Petals 2–4.5 cm, obovate, rounded above, sparsely hairy near the gland with long, slender hairs. Gland more or less circular or diamond-shaped, covered with simple, hair-like appendages. Anthers linear-oblong, acute to obtuse, about as long as filaments. *USA (California).* H5. Summer.

13. C. superbus Howell (*C. luteus* Lindley var. *citrinus* Watson; *C. luteus* Lindley var. *oculatus* Purdy & Bailey not Watson). Figure 3(13), p. 68. Illustration: Rickett, Wild flowers of the United States **5**: t. 6 (1971); Brickell (ed.), RHS A–Z encyclopedia of garden plants, 214 (2003).

Like *C. venustus* but flowers white to yellowish or pale purple, petals streaked with purple below and usually with a median reddish brown mark surrounded by a zone of bright yellow; sepals lanceolate, tapering, usually not curled under at the tip; petals rounded to acute, with a few short hairs near the gland; gland more or less inverted V-shaped; anthers linear-lanceolate to linear-oblong. *USA (California).* H5. Late spring–early summer.

14. C. vestae Purdy (*C. venustus var.* vestae (Purdy) Wilson; *C. luteus* Lindley var. *oculatus* Watson). Figure 3(19), p. 68. Illustration: Brickell (ed.), RHS A–Z encyclopedia of garden plants, 214 (2003).

Like *C. venustus* but flowers white to purplish, petals streaked below with red or purplish and usually with a median reddish brown mark surrounded by a zone of pale yellow; sepals lanceolate, tapering, usually not curled under at the tip; gland transverse, more or less doubly crescent-shaped, covered with short, thick processes. *USA (California).* H5. Late spring–early summer.

15. C. luteus Lindley (*C. citrinus* Anon., not Baker). Figure 3(20), p. 68. Illustration: Rickett, Wild flowers of the United States **5**: t. 5 (1971); Brickell (ed.), RHS A–Z encyclopedia of garden plants, 214 (2003).

Like *C. venustus* but flowers 1–4, dark yellow, the petals usually with reddish brown radiating lines and often also with a reddish brown median mark; sepals sometimes curled under at tip; gland transverse, more or less crescent-shaped, covered with short, thick processes. Anthers shorter than filaments. *USA (California).* H5. Summer.

16. C. macrocarpus Douglas (*C. cyaneus* Nelson). Figure 2(1), p. 67. Illustration: Rickett, Wild flowers of the United States **6**: t. 4 (1973).

Stem erect, often with bulbils, 20–50 cm. Leaves becoming tightly curled under at the tip. Inflorescence more or less umbellate. Flowers 1–3, bell-shaped, erect, purple, each petal with a median longitudinal green stripe and sometimes a dark purple band above the gland. Sepals 3.5–6 cm, narrowly lanceolate, tapering, hairless, usually exceeding the petals. Petals 3.5–6 cm, oblanceolate or broader, acute to acuminate with a few slender hairs above the gland. Gland more or less sagittate, bound by a fringed membrane and covered with slender, usually branched process. Anthers linear, obtuse, longer than the filaments. *Western N America (British Columbia south to Nevada and California).* H4. Summer.

17. C. kennedyi Porter. Figure 2(6), p. 67. Illustration: Rickett, Wild flowers of the United States **4**: t. 4 (1970).

Stem erect, usually simple, rarely with bulbils, 10–50 cm. Inflorescence more or less umbellate. Flowers 1–6, bell-shaped, erect, usually vermilion to orange, more rarely (var. **munzii** Jepson) yellow. Sepals 1.5–2.5 cm, ovate-lanceolate, acute. Petals 2.5–5 cm, obovate, usually rounded above, with a few club-shaped hairs near the gland. Gland circular, covered with simple or branched processes and surrounded by a fringed membrane. Anthers narrowly lanceolate, acute or obtuse, usually longer than filaments. *USA (Arizona, California).* H5. Summer.

18. C. clavatus Watson. Illustration: Rickett, Wild flowers of the United States **5**: t. 4 (1971).

Stem zig-zag, simple or branched, rarely with bulbils, 50–100 cm. Inflorescence more or less umbellate. Flowers 1–6, bell-shaped, erect, yellow, sometimes with red-brown markings on the sepals and a zig-zag line above the gland on the petals. Sepals 2.5–3.5 cm, ovate-lanceolate, acute. Petals 3.5–5 cm, broadly obovate, usually rounded above, with club-shaped hairs above the gland. Gland circular, depressed, densely covered with short, much-branched processes, and surrounded by a fringed membrane. Anthers oblong, obtuse or acute, reddish or purplish brown, about as long as filaments. *USA (California).* H5. Summer.

19. C. gunnisonii Watson. Figure 2(2), p. 67. Illustration: Rickett, Wild flowers of the United States **4**: t. 3 (1970).

Stem erect, simple, rarely with bulbils, 20–60 cm. Inflorescence more or less umbellate. Flowers 1–3, bell-shaped, erect, white to purple, greenish within, often with a transverse purple band above the gland and a purple spot on the claw of each petal and sepal. Sepals 2.5–3 cm, lanceolate, acute. Petals 3–4 cm, obovate, obtuse and rounded above, near the gland densely covered with branched, gland-tipped hairs. Gland transverse, more or less oblong, arched above, densely covered with branched processes, the outermost of which are sometimes joined at the base to form a membrane. Anthers lanceolate, acute to abruptly pointed, longer than filaments. *USA (Dakota, Montana, south to New Mexico).* H5. Summer.

Section **Cyclobothra** (D. Don) Baker. Bulb-coats thickly fibrous-netted. Stems usually branched, rarely with bulbils in the axils of the upper leaves; basal leaf conspicuous, withering before flowering or not. Flowers usually 2, narrowly to broadly bell-shaped, erect or drooping. Ovary linear, not winged, tapering to a persistent, 3-fid stigma. Fruit linear, 3-angled, erect.

20. C. weedii Wood (*C. luteus* Lindley var. *weedii* (Wood) Baker; *C. citrinus* Baker). Figure 3(11), p. 68. Illustration: Rickett, Wild flowers of the United States **4**: t. 2 (1970).

Stem erect, without bulbils, 30–60 cm. Basal leaf usually withering before flowering. Flowers 2, broadly bell-shaped, erect, orange-yellow, flecked and often margined with red-brown. Sepals 2–3 cm, ovate-lanceolate, tapering. Petals 2.5–3 cm, broadly obovate, rounded to acute, the margin minutely toothed to fringed, hairy over most of the inner surface with long, yellow hairs. Gland circular, nearly naked, bound by a dense ring of hairs that are sometimes united below to form a more or less continuous membrane. Anthers linear, acute to shortly acuminate, about as long as filaments. *USA (California), Mexico (Baja California).* H5. Summer.

21. C. barbatus (Humboldt, Bonpland & Kunth) Painter (*C. flavus* Schultes). Figure 2(10), p. 67. Illustration: Mathew, Dwarf bulbs, t. 17 (1973) – as *C. luteus.*

Stem erect, usually branched, rarely with bulbils, 30–60 cm. Flowers usually 2, openly bell-shaped, drooping, yellow, or sometimes the petals and frequently the sepals purplish. Sepals 1–2 cm, elliptic-lanceolate, acute or obtuse, sparsely hairy within, often with a gland like those on the petals. Petals 1.5–2 cm, obovate, obtuse or acuminate, entire or fringed laterally, densely hairy with tapering, purple hairs. Gland more or less circular, naked, bound by a ring of small hairs. Anthers oblong, shortly pointed, shorter than the filaments. *Mexico.* H5. Autumn.

7. ERYTHRONIUM Linnaeus

V.A. Matthews & J.M. Lees

Perennial herbs with membranous-coated corms. Leaves usually 2, basal or almost so, often mottled. Flowers nodding, solitary or several in a raceme, rarely in an umbel whose stem is below ground-level. Perianth-segments free, usually recurved, one or both whorls usually with 2 or 4 inflated appendages near the base. Stamens shorter than the perianth-segments; anthers basifixed. Stigma entire to 3-lobed. Ovary superior. Fruit a 3-celled capsule.

A genus of about 25 species from temperate N America with 1 species in Eurasia. The plants grow best in semi-shade in a soil that has plenty of peat or leaf-mould and does not dry out in summer. They are especially useful for planting beneath trees.

Literature: Applegate, E.I., The genus Erythronium: a taxonomic and distributional study of the western North American species, *Madroño* **3**: 58–113 (1935); Parks, C.R. & Hardin, J.W., Yellow Erythroniums of the eastern United States, *Brittonia* **15**: 245–59 (1963).

Leaves. Mottled: **1, 4, 5, 6, 7, 8, 9, 10, 11, 12**; not mottled: **2, 3, 11**.
Perianth-segments. White, cream or pale yellow: **1, 3, 4, 5, 8, 9, 10, 11**; golden yellow: **2, 3, 12**; pink to purple: **1, 6, 7, 8, 11**; inner segments with basal appendages: **1, 2, 3, 5, 6, 7, 8, 9, 10, 11, 12**; inner segments lacking basal appendages: **4, 7, 11**.
Anthers. White: **3, 4, 5, 8, 9, 10, 11**; yellow: **1, 2, 3, 7, 8, 11, 12**; dark red: **3**; blue, purple or brown: **1, 6, 12**.
Stigma. Entire or almost so: **2, 4, 5, 6, 7**; lobed: **1, 3, 7, 8, 9, 10, 11, 12**.

1a. Leaves mottled 2
b. Leaves not mottled 13
2a. Flowers golden yellow **12. americanum**
b. Flowers white, cream, pale yellow, pink, lavender or purple 3
3a. Anthers blue, purple or brown 4
b. Anthers white or yellow 5
4a. Stigma deeply lobed **1. dens-canis**
b. Stigma entire or almost so **6. hendersonii**
5a. Flowers pink 6
b. Flowers white, cream or pale yellow 8
6a. Flowers pale pink, tips of perianth-segments twisted **8. oregonum**
b. Flowers bright or rose pink, tips of perianth-segments not twisted 7
7a. Perianth-segments with a band of brown or purplish markings towards the base **1. dens-canis**
b. Perianth-segments lacking a brown or purplish band **7. revolutum**
8a. Stigma entire or almost so 9
b. Stigma 3-lobed 10
9a. Inner perianth-segments with basal appendages **5. citrinum**
b. Inner perianth-segments lacking basal appendages **4. howellii**
10a. Anthers white, filaments not broadened at the base 11

b. Anthers yellow, or if white then filaments broadened at the base 12
11a. Leaves 10–15 cm; flower with a greenish yellow centre that is surrounded by a band of yellow or brown; stigma with short, erect or spreading lobes **9. californicum**
b. Leaves 3.5–10 cm; flower with a greenish yellow centre that is not surrounded by a band of yellow or brown; stigma with long, thread-like, recurved lobes **10. multiscapoideum**
12a. Filaments broadened at the base; perianth-segments with basal appendages **8. oregonum**
b. Filaments not broadened at the base; perianth-segments lacking basal appendages, or if appendages present then stigma lobes erect to spreading and perianth-segments not twisted at the tips **11. albidum**
13a. Inner perianth-segments lacking basal appendages **11. albidum**
b. Inner perianth-segments with basal appendages 14
14a. Stigma entire or almost so **2. tuolumnense**
b. Stigma deeply 3-lobed **3. grandiflorum**

1. E. dens-canis Linnaeus. Illustration: Synge, Collins guide to bulbs, edn 1, pl. 11, f. 9 (1961); Lippert, Fotoatlas der Alpenblumen, f. 12 (1981); Grey-Wilson & Mathew, Bulbs, pl. 4 (1981); Rix & Phillips, The bulb book, 100, 101 (1981).

Leaves 10–15 cm, oblong, mottled. Flower-stem 15–30 cm, bearing a solitary flower. Perianth-segments *c.* 2.5 cm, rose-pink to purple, sometimes white, with a pale yellow or brownish base inside, above which is a band of brown or purplish markings. Inner segments with basal appendages. Anthers blue to purple. Stigma deeply lobed. *Eurasia.* H2. Spring.

There are several named cultivars based on differences in flower colour, e.g. 'Album' and 'Snowflake' with white flowers, and 'Purpureum' with purple-violet flowers.

Var. **japonicum** (Decaisne) Baker (*E. japonicum* Decaisne). Illustration: Kitamura et al., Coloured illustrations of herbaceous plants of Japan (Monocotyledoneae) **3**: pl. 31 (1978); Rix & Phillips, The bulb book, 100 (1981). Perianth-segments 5–6 cm, violet-purple with a dark, often 3-lobed mark at the base inside. *Japan, Korea.*

Var. **sibiricum** Fischer & Meyer. Perianth-segments 2.5–5.5 cm; anthers yellow. *Former USSR (Siberia), Mongolia.*

2. E. tuolumnense Applegate. Illustration: Synge, Collins guide to bulbs, edn 1, pl. 11, f. 7 (1961); *Journal of the Royal Horticultural Society* **89**: f. 106 (1964); Rickett, Wild flowers of the United States **5**: pl. 11 (1971); Rix & Phillips, The bulb book, 100 (1981).

Leaves 20–30 cm, lanceolate to broadly oblanceolate, yellow-green, not mottled. Flower-stem to 30 cm, bearing 1–several flowers. Perianth-segments *c.* 3 cm, golden yellow with a greenish base and 1 or 2 pairs of basal appendages. Anthers yellow. Stigma entire or almost so. *USA (C California).* H3. Spring.

The often-grown cultivars 'Pagoda' and 'Kondo' are hybrids between *E. tuolumnense* and a species with mottled leaves.

E. klamathense Applegate. Illustration: Rickett, Wild flowers of the United States **5**: pl. 12 (1971). Flowers usually solitary, sometimes a few. Perianth-segments 2–2.5 cm, white or cream, yellow in the lower third. *W USA (SW Oregon, NW California).* H3. Spring.

3. E. grandiflorum Pursh (*E. grandiflorum* subsp. *candidum* Piper, subsp. *chrysandrum* Applegate and var. *pallidum* St. John; *E. giganteum* misapplied). Illustration: Synge, Collins guide to bulbs, edn 1, pl. 11, f. 4 (1961); Rickett, Wild flowers of the United States **5**: pl. 11 (1971); Clark, Wild flowers of the Pacific northwest, 50 (1976); Rix & Phillips, The bulb book, 101 (1981).

Leaves 10–20 cm, oblong-elliptic, not mottled. Flower-stem 30–60 cm, bearing 1–several flowers. Perianth-segments 2–3.5 cm, golden yellow with a paler base and well-developed basal appendages. Anthers dark red, or sometimes yellow or white. Stigma with long, recurved lobes. *Western N America from British Columbia S to Oregon and Utah.* H3. Spring–summer.

Several cultivars are commercially available, such as 'Album', which has white flowers with yellow centres, and 'Robustum', which is a variant with taller stems and larger leaves.

E. montanum Watson. Illustration: Hitchcock et al., Vascular plants of the Pacific northwest **1**: 789 (1969); Clark, Wild flowers of the Pacific northwest, 35 (1976); Rix & Phillips, The bulb book, 99 (1981). Leaf-blade abruptly narrowed at the base. Flowers white with a yellow or orange centre. Anthers always yellow. *Western N America (British Columbia, Oregon, Washington).* H3. Spring–summer.

4. E. howellii Watson. Illustration: Abrams, Illustrated flora of the Pacific states **1**: 430 (1923); Synge, Collins guide to bulbs, edn 1, pl. 11, f. 10 (1961).

Leaves to 15 cm, lanceolate to lanceolate-oblong, mottled. Flower-stem to 20 cm, bearing 1–4 flowers. Perianth-segments *c.* 2.5 cm, white or pale yellow, with an orange base or yellow bars inside, lacking basal appendages. Anthers white. Stigma almost entire. *W USA (S Oregon, NW California).* H3. Spring.

5. E. citrinum Watson. Illustration: Synge, Collins guide to bulbs, edn 1, pl. 11, f. 8 (1961); Niehaus & Ripper, Field guide to Pacific states wildflowers, 11 (1976).

Leaves 10–15 cm, lanceolate to oblong, mottled, margin crisped. Flower-stem to 30 cm, bearing 1–3 flowers. Perianth-segments 2.5–4 cm, creamy white to pale yellow with a greenish yellow base inside. Inner segments with basal appendages. Anthers white. Stigma entire or slightly lobed. *W USA (SW Oregon, NW California).* H3. Spring.

6. E. hendersonii Watson. Illustration: *Botanical Magazine*, 7017 (1888); Synge, Collins guide to bulbs, edn 1, pl. 11, f. 6 (1961); Rickett, Wild flowers of the United States **5**: pl. 12 (1971); Rix & Phillips, The bulb book, 99 (1981).

Leaves 10–20 cm, lanceolate-oblong, mottled, margin crisped. Flower-stem to 30 cm, bearing 1–10 flowers. Perianth-segments 2.5–4 cm, lavender, with a purple base inside, above which is an indistinct yellowish band. Inner segments with basal appendages. Filaments purple, anthers brownish or bluish. Stigma entire or almost so. *W USA (SW Oregon, NW California).* H3. Spring–summer.

7. E. revolutum Smith (*E. revolution* var. *smithii* (Hooker) Orcutt and var. *johnsonii* (Bolander) Purdy). Illustration: Synge, Collins guide to bulbs, edn 1, pl. 11, f. 3 (1961); Rickett, Wild flowers of the United States **5**: pl. 12 (1971); Clark, Wildflowers of the Pacific northwest, 46 (1976); Rix & Phillips, The bulb book, 96–7 (1981).

Leaves 15–20 cm, lanceolate-ovate, mottled, margin crisped. Flower-stem 15–40 cm, bearing 1–several flowers. Perianth-segments 3.5–4.5 cm, rose-pink with 1 or 2 yellow bands at the base inside, and with basal appendages.

Filaments broadened at the base, anthers yellow. Stigma with recurved lobes. *Western N America from British Columbia to N California*. H3. Spring–summer.

The cultivar 'Pink Beauty' is thought to be either a robust variant of *E. revolutum* or a hybrid.

E. propullans Gray. Illustration: *American Naturalist* **5**: 299 (1871); Britton & Brown, Illustrated flora of the northern United States and Canada **1**: 421 (1896). Leaves 5–10 cm; flowers solitary, perianth-segments 1–1.4 cm, not or scarcely recurved, bright pink with a yellow base and no basal appendages; stigma entire. *USA (Minnesota)*. H3.

8. E. oregonum Applegate (*E. giganteum* misapplied). Illustration: *Botanical Magazine*, 5714 (1868); Synge, Collins guide to bulbs, edn 1, pl. 11, f. 1 (1961); Rickett, Wild flowers of the United States **5**: pl. 11 (1971); Rix & Phillips, The bulb book, 98–9 (1981).

Leaves 12–15 cm, lanceolate to oblong-lanceolate, mottled. Flower-stem 15–30 cm, bearing 1–6 flowers. Perianth-segments 3.5–5 cm, twisted at the tips, white to pink, tinged with brown or reddish at the base outside, and with a yellow base inside, above which are zig-zag marks of orange-brown. Basal appendages present. Filaments broadened at the base, anthers yellow, or white (subsp. *leucandrum* (Applegate) Applegate). Stigma with long, recurved lobes. *Western N America from British Columbia to Oregon*. H3. Spring.

'White Beauty', which is frequently offered by nurserymen, may be a variant of of *E. oregonum* or a hybrid.

9. E. californicum Purdy. Illustration: Abrams, Illustrated flora of the Pacific states **1**: 428 (1923); Rickett, Wild flowers of the United States **5**: pl. 11 (1971); Rix & Phillips, The bulb book, 98–9 (1981); Brickell (ed.), RHS A–Z encyclopedia of garden plants, 428 (2003).

Leaves 10–15 cm, narrowly oblong to ovate-oblong, strongly mottled. Flower-stalk 10–35 cm, bearing 1–3 flowers. Perianth-segments 2.5–3.5 cm, white to cream with a greenish yellow base inside, above which is a band of yellow, orange or brown. Basal appendages present. Anthers white. Stigma with short, erect or spreading lobes. *W USA (N California)*. H3. Spring.

10. E. multiscapoideum (Kellogg) Nelson & Kellogg (*E. purdyi* Anon.; *E. hartwegii* Watson). Illustration: *Botanical Magazine*, 7583 (1898); Niehaus & Ripper, Field guide to Pacific states wildflowers, 11 (1976).

Leaves 3.5–10 cm, oblanceolate, mottled. Flower-stems to 20 cm, apparently bearing solitary flowers (the flowers are borne in an umbel whose stem is below ground-level). Perianth-segments 2.5–4 cm, white with a pale greenish yellow base; inside, the inner segments with small basal appendages. Anthers white. Stigma with long, recurved lobes. *W USA (N California)*. H3. Spring.

11. E. albidum Nuttall (*E. mesochoreum* Knerr). Illustration: *Addisonia* **18**: pl. 593 (1934); Rickett, Wild flowers of the United States **2**: pl. 10 (1967); Dean et al., Wildflowers of Alabama, 23 (1983).

Leaves to 15 cm, elliptic, green or mottled. Flower-stalk to 30 cm, bearing a solitary flower. Perianth-segments 2.5–5 cm, white with a pale bluish or pinkish tinge outside and a yellow base inside; basal appendages absent. Anthers yellow. Stigma with spreading lobes. *Eastern C USA and adjacent S Canada*. H3. Spring–summer.

E. helenae Applegate. Illustration: Synge, Collins guide to bulbs, edn 1, pl. 11, f. 5 (1961); *Botanical Magazine*, n.s., 834 (1981); Rix & Phillips, The bulb book, 98 (1981). Leaves always strongly mottled. Flowers 1–4, white with a yellow centre, segments with basal appendages. *W USA (N California)*. H3. Spring.

E. purpurascens Watson. Illustration: Niehaus & Ripper, Field guide to Pacific states wildflowers, 11 (1976). Leaves narrowly lanceolate to lanceolate, bright yellow-green. Flowers to 8, perianth-segments 1–1.5 cm, white or cream with a yellow base and turning pink or purplish with age. Anthers white. *W USA (California)*. H3. Spring.

12. E. americanum Ker Gawler. Illustration: *Botanical Magazine*, 1113 (1808); Peterson & McKenny, Field guide to wildflowers of north-eastern and north-central North America, 103 (1968); *Journal of the Royal Horticultural Society* **102**: 18 (1977); Rix & Phillips, The bulb book, 97 (1981).

Leaves to 15 cm, elliptic, mottled. Flower-stalks to 30 cm, bearing a solitary flower. Perianth-segments to 5 cm, golden yellow, often spotted with brownish purple at the base; inside, the inner segments with small basal appendages. Anthers yellow or brown. Stigma-lobes short. *Eastern N America*. H3. Spring–summer.

8. TULIPA Linnaeus
V.A. Matthews & C. Grey-Wilson

Bulbous perennials. Bulb-tunics of various textures, hairless inside or lined with hairs. Leaves few, rather fleshy, alternate, decreasing in size up the stem. Flowers usually solitary, rarely to 12. Perianth-segments free. Nectaries absent. Filaments broadened towards the base, hairy or not. Anthers basifixed, dehiscing inwards. Ovary superior. Style very short or absent. Stigma 3-lobed. Fruit a spherical or ellipsoid capsule containing numerous, flat seeds.

A genus of *c.* 100 species from the north temperate Old World, especially Soviet Central Asia. Most species require a sunny position where the bulbs can be baked during the summer. In cooler, wetter areas it is advisable to lift the bulbs after the leaves have died down, store them and replant in the autumn.

Literature: Hall, A.D., The genus Tulipa (1940); Classified list and international register of tulip names (revised 1971); Botschantzeva, Z.P. (translated and edited by H.Q.Varekamp), Tulips (1982); Stork, A. L., Tulipes sauvages et cultivées, Série documentaire **13** des Conservatoire et Jardin botaniques de Genève (1984); Arnold, P., Tulips (1992); Zhao, Y.Z., A new species of the genus Tulipa (Liliaceae) from China, *Novon* **13**(2): 277–8 (2003).

1a. Main colour of perianth-segments pink, mauve or purple **Group A**
b. Main colour of perianth-segments not pink, mauve or purple 2
2a. Main colour of perianth-segments white or cream **Group B**
b. Main colour of perianth-segments not white or cream 3
3a. Main colour of perianth-segments yellow or greenish **Group C**
b. Main colour of perianth-segments orange, brownish red, red or crimson-red **Group D**

Group A (Perianth-segments pink, mauve or purple)

1a. Filaments hairless 2
b. Filaments hairy 3
2a. Anthers twisted **25. kaufmanniana**
b. Anthers straight **13. gesneriana,** including **platystigma**

3a. Stem 15–45 cm; bud nodding; leaves green **4. saxatilis**
b. Stem 5–20 cm; bud erect; leaves somewhat glaucous **5. humilis**

Group B (Perianth-segments white or cream)

1a. Anthers twisted **25. kaufmanniana**
b. Anthers straight 2
2a. Perianth-segments with a basal blotch 3
b. Perianth-segments lacking a basal blotch 9
3a. Filaments hairy 4
b. Filaments hairless 7
4a. Leaves glaucous 5
b. Leaves green **8. tarda**
5a. Stem hairy **7. turkestanica**
b. Stem hairless 6
6a. Perianth-segments with a yellow basal blotch **6. biflora**
b. Perianth-segments with a bluish basal blotch **5. humilis**
7a. Perianth-segments with a deep red, purple, reddish purple or purplish brown basal blotch 8
b. Perianth-segments with a yellow basal blotch **9. clusiana**
8a. Outer perianth-segments 1.5–3.2 cm; 2 or 3 bracts present below flowers **32. edulis**
b. Outer perianth-segments 3–6.5 cm; flowers lacking bracts **9. clusiana**
9a. Filaments hairy, yellow **1. sylvestris**
b. Filaments hairless, blue-black **13. gesneriana**, see **marjolettii**

Group C (Perianth-segments yellow or greenish)

1a. Filaments hairy 2
b. Filaments hairless 4
2a. Visible stem less than 10 cm **2. urumiensis**
b. Visible stem 10 cm or more 3
3a. Bud nodding **1. sylvestris**
b. Bud erect **3. orphanidea**
4a. Perianth-segments greenish **13. gesneriana**, see **viridiflora**
b. Perianth-segments yellow, sometimes striped or marked with green or another colour 5
5a. Perianth-segments pure yellow (excluding basal blotch if present) 6
b. Perianth-segments yellow, flushed or striped with another colour 11
6a. Filaments yellow, at least at the base 7
b. Filaments purple or black 9
7a. Anthers olive-green **12. urumoffii**
b. Anthers yellow or blackish 8
8a. Stem to 30 cm **10. linifolia**
b. Stem 30–60 cm **13. gesneriana**, including **retroflexa**
9a. Leaves streaked or mottled with reddish or purplish brown **23. greigii**
b. Leaves not so marked 10
10a. Stem to 25 cm; leaf margin wavy **29. armena**
b. Stem 30–60 cm; leaf margin not wavy **13. gesneriana**
11a. Perianth-segments with a red margin **13. gesneriana**, see **grengiolensis**
b. Perianth-segments variously marked, but not with a red margin 12
12a Perianth-segments with a basal blotch or V-shaped mark 13
b. Perianth-segments lacking a basal blotch or V-shaped mark 16
13a. Leaves streaked and mottled with reddish or purplish brown **23. greigii**
b. Leaves not so marked 14
14a. Perianth-segments with a black, purple or olive-green basal blotch **13. gesneriana**
b. Perianth-segments with a yellow or orange basal blotch, or red V-shaped mark 15
15a. Stem 3–10 cm, hairless **9. clusiana**
b. Stem 10–35 cm, downy **25. kaufmanniana**
16a. Perianth-segments 7.5–13 cm, long-pointed, twisted **13. gesneriana**, see **acuminata**
b. Perianth-segments 1.5–7.5 cm, neither long-pointed nor twisted 17
17a. Bud nodding **27. kolpakowskiana**
b. Bud erect 18
18a. Inside of bulb-tunics more or less hairless, or with a few short hairs towards the apex **13. gesneriana**, including **galatica**
b. Inside of bulb-tunics with short hairs at least at the base and apex, sometimes with a tuft of hairs protruding from the neck 19
19a. Leaf margin wavy **29. armena**
b. Leaf margin not wavy **9. clusiana**

Group D (Perianth-segments orange, brownish red, red or crimson-red)

1a. Perianth-segments with a basal blotch or V-shaped mark 2
b. Perianth-segments lacking a basal blotch or V-shaped mark 58
2a. Basal blotch or mark yellow or red 3
b. Basal blotch black, dark blue, dark purple, brownish or olive-green, bordered or not with a contrasting colour 10
3a. Stem smooth, hairless 4
b. Stem rough, bristly or downy at least in upper part 5
4a. Leaf margin wavy **29. armena**
b. Leaf margin not wavy **13. gesneriana**, including **fulgens, mauritiana**
5a. Anthers twisted **25. kaufmanniana**
b. Anthers straight 6
6a. Outer perianth-segments to 5.8 cm 7
b. Outer perianth-segments 6–10.5 cm 8
7a. Stem 7.5–25 cm; perianth-segments 1–3 cm wide **29. armena**
b. Stem 20–55 cm; perianth-segments *c.* 4 cm wide **22. carinata**
8a. Leaf margin wavy towards the apex **22. carinata**
b. Leaf margin not wavy 9
9a. Filaments brown or yellow **30. vvedenskyi**
b. Filaments purple **13. gesneriana**, see **elegans**
10a. Basal blotch with a contrasting border 11
b. Basal blotch lacking a border 35
11a. Leaves streaked and mottled with reddish or purplish brown 12
b. Leaves not so marked 13
12a. Anthers black **23. greigii**
b. Anthers violet or yellow **24. micheliana**
13a. Border of basal blotch white **10. linifolia**
b. Border of basal blotch not white 14
14a. Leaf margin wavy, at least towards the apex 15
b. Leaf margin not wavy 25
15a. Filaments yellow, reddish or brownish 16
b. Filaments black or dark blue, sometimes yellow or white at the apex 17
16a. Bud erect; outer perianth-segments 2.5–5.8 cm; anthers 5.5–9 mm **29. armena**
b. Bud nodding; outer perianth-segments 5–8 cm; anthers *c.* 12 mm **28. ostrowskiana**
17a. Leaves downy, margin ciliate or not 18
b. Leaves hairless, margin ciliate or not 22
18a. Filaments white at the apex; pollen reddish purple **18. lanata**

b. Filaments not white at the apex; pollen yellow, brown, pink, purple or green 19
19a. Inside of bulb-tunics hairy at the base, sparsely hairy elsewhere **14. undulatifolia**
b. Inside of bulb-tunics woolly or adpressed-hairy 20
20a. Leaf margin wavy only towards the apex **22. carinata**
b. Leaf margin wavy from the base to the apex 21
21a. Stem to 25 cm; anthers yellow or blackish **29. armena***
b. Stem 30–60 cm; anthers purple **20. tubergeniana**
22a. Anthers purple **10. linifolia**
b. Anthers black, blue, yellow or cream 23
23a. Filaments blue, sometimes yellow at the apex **16. agenensis**
b. Filaments black or yellow 24
24a. Inside of bulb-tunics lined with woolly matted hairs **16. agenensis**
b. Inside of bulb-tunics lined with short or adpressed hairs **29. armena**
25a. Filaments yellow, at least at the base 26
b. Filaments black, dark blue, purplish, reddish or olive-green, sometimes yellow at the apex 29
26a. Stem 7.5–15 cm; anthers yellow **29. armena**
b. Stem 15–60 cm; anthers usually dark purple or olive-green 27
27a. Leaves hairless although cilate near the apex; inside of bulb-tunics sparsely hairy or hairless 28
b. Leaves downy on upper surface; inside of bulb-tunics densely hairy **21. fosteriana**
28a. Stem 15–30 cm; anthers greenish; leaves glaucous **12. urumoffii**
b. Stem 30–60 cm; anthers yellow or purple; leaves green **13. gesneriana**
29a. Perianth-segments orange with a brownish green basal blotch **15. praecox**
b. Perianth-segments orange, red or crimson; basal blotch black, dark blue, dark purple or dark olive-green 30
30a. Filaments purplish or greenish, sometimes yellow at the apex 31
b. Filaments black, dark blue or reddish 32
31a. Filaments hairless **13. gesneriana**
b. Filaments hairy **3. orphanideam**
32a. Stem slightly downy **21. fosteriana**
b. Stem hairless 33
33a. Outside of perianth-segments yellowish or paler than the inside 34
b. Outside of perianth-segments neither yellowish nor paler **17. hoogiana**
34a. Leaves green; bulb-tunics lined with densely matted woolly hairs **16. agenensis**
b. Leaves glaucous; bulb-tunics lined with long hairs, or with short hairs at the base and top, hairs never matted or woolly **29. armena**
35a. Anthers twisted **25. kaufmanniana**
b. Anthers straight 36
36a. Filaments yellow, sometimes black or green at the apex, or red with a black base 37
b. Filaments black, purple, greenish or brown 41
37a. Stem hairless 38
b. Stem downy or bristly 39
38a. Filaments red with a black base **11. montana**
b. Filaments yellow, sometimes black at the apex **29. armena**
39a. Stem downy **19. ingens**
b. Stem bristly, or at least with rough hairs above 40
40a. Perianth-segments 2–6 cm **29. armena**
b. Perianth-segments more than 6 cm **30. vvedenskyi**
41a. Outer perianth-segments 8 cm or more 42
b. Outer perianth-segments to 8 cm 46
42a. Leaves streaked with reddish or purplish brown **24. micheliana**
b. Leaves not streaked 43
43a. Stem downy, at least above; leaf margin wavy at least towards the apex 44
b. Stem bristly; leaf margin not wavy **30. vvedenskyi**
44a. Inner perianth-segments more or less rounded at the apex; basal blotch of outer perianth-segments *c.* 5.5 × 3 cm **20. tubergeniana**
b. Inner perianth-segments markedly pointed at the apex; basal blotch of outer perianth-segments 8–45 × 6–10 mm 45
45a. Outer perianth-segments 4–9 cm; inside of bulb-tunics woolly **22. carinata**
b. Outer perianth-segments 8–15 cm; inside of bulb-tunics sparsely hairy **19. ingens**
46a. Filaments hairy **3. orphanidea**
b. Filaments hairless 47
47a. Stem to 25 cm 48
b. Stem 30–60 cm 54
48a. Leaf margin wavy or crisped, at least towards the apex 49
b. Leaf margin neither wavy nor crisped 53
49a. Leaves downy on upper surface, sometimes sparsely so 50
b. Leaves hairless on upper surface **29. armena**
50a. Leaves streaked with reddish or purplish brown **24. micheliana**
b. Leaves not streaked 51
51a. Anthers to 9 mm **29. armena**
b. Anthers 10 mm or more 52
52a. Inside of bulb-tunics woolly **22. carinata**
b. Inside of bulb-tunics sparsely hairy, more densely hairy at the base **14. undulatifolia**
53a. Perianth-segments 2–6 cm **29. armena**
b. Perianth-segments more than 6 cm **30. vvedenskyi**
54a. Leaf margin wavy or crisped, at least towards the apex 55
b. Leaf margin neither wavy nor crisped 57
55a. Leaves streaked with reddish or purplish brown **24. micheliana**
b. Leaves not streaked 56
56a. Inside of bulb-tunics woolly **22. carinata**
b. Inside of bulb-tunics sparsely hairy, more densely hairy at the base **14. undulatifolia**
57a. Outer perianth-segments to 10.5 cm, with a hairy midrib on the back; filaments yellow or brown **30. vvedenskyi**
b. Outer perianth-segments 4–7.5 cm, the midrib not hairy; filaments purple or black **13. gesneriana,** including **didieri**
58a. Perianth-segments red, brownish red or orange on both sides 59
b. Perianth-segments red or orange, marked or flushed with another colour, or outer segments yellowish buff on the outside 65
59a. Stem hairless 60
b. Stem downy 62
60a. Leaf margin at least slightly wavy; inside of bulb-tunics lined with adpressed hairs or densely woolly **29. armena**
b. Leaf margin not wavy; inside of bulb-tunics hairless or with a few silky hairs 61

61a. Stem 15–30 cm; filaments yellow at the base, olive-green at the apex **12. urumoffii**
b. Stem 30–60 cm; filaments purple, white or wholly yellow **13. gesneriana,** including **acuminata**
62a. Anthers twisted **25. kaufmanniana**
b. Anthers straight 63
63a. Flowers 1–5; filaments red, shading to yellow at the base **26. praestans**
b. Flowers solitary, rarely 2; filaments yellow, purple, grey or black 64
64a. Leaf margins wavy **29. armena**
b. Leaf margins not wavy **13. gesneriana**
65a. Perianth-segments 7.5–13 cm, long-pointed, twisted **13. gesneriana**, see **acuminata**
b. Perianth-segments to 7.5 cm, neither long-pointed nor twisted 66
66a. Outside of outer perianth-segments differing in colour from inner ones; filaments red **31. sprengeri**
b. Outside of outer perianth-segments not differing in colour from inner ones; filaments black, purple or yellow 67
67a. Leaf margins wavy **29. armena**
b. Leaf margins not wavy **13. gesneriana**

1. T. sylvestris Linnaeus (*T. australis* Link; *T. celsiana* de Candolle). Illustration: Hall, The genus Tulipa, pl. 3 (1940); Grey-Wilson & Mathew, Bulbs, pl. 5 (1981); Rix & Phillips, The bulb book, 122 (1981).

Inside of bulb-tunics with short hairs at the base and top. Stem 10–45 cm, hairless. Leaves 2–4, hairless. Bud nodding. Flowers usually solitary, sometimes 2, often fragrant; perianth-segments yellow, occasionally cream; outer segments 2–6.5 cm × 4–25 mm, usually tinged with green or red on the back; inner segments 2–7 cm × 6–25 mm. Filaments hairy, yellow; anthers and pollen yellow. *Origin unknown: naturalised from Europe and N Africa to the Middle East and Siberia.* H3. Spring.

In the wild, *T. sylvestris*, *T. australis* and *T. celsiana* cannot be distinguished satisfactorily from one another, although plants named *T. sylvestris* tend to be lowland plants of more northerly latitudes and plants named *T. australis* and *T. celsiana* tend to occur further south and at higher altitudes. Plants in cultivation represent only parts of the range of variation found in the wild; as offered commercially they show the following characteristics:

T. sylvestris has a stout stem, at least 2.5 mm across, and the outer perianth-segments are 3.5–6.5 cm, often tinged with green on the back. The filaments are 9–14 mm and the anthers 4–9 mm.

T. australis (*T. sylvestris* ssp. *australis* (Link) Pampanini) has a more slender stem, not more than 2 mm across, and the outer perianth-segments are 2–3.5 cm, tinged with red on the back. The filaments are 5–8 mm and the anthers 2.5–4 mm.

T. celsiana tends to have a shorter stem (to 15 cm) but the flowers are similar to those of *T. australis.* The leaves differ in that they tend to twist and lie flat on the ground, rather than being erect-spreading.

T. celsiana is usually late-flowering, often 2–3 weeks after *T. sylvestris* and *T. australis.*

2. T. urumiensis Stapf. Illustration: Synge, Collins guide to bulbs, edn 2, pl. 29, f. 18 (1971); *Bulletin of the Alpine Garden Society* **39**: 265 (1971); *Journal of the Royal Horticultural Society* **100**: 504 (1975); Rix & Phillips, The bulb book, 124 (1981).

Inside of bulb-tunics sparsely hairy towards the base. Stem 10–20 cm, mostly below ground, hairless. Leaves 2–4, usually in a flat rosette, green or slightly glaucous, hairless. Bud nodding. Flowers 1 or 2; perianth-segments to 4 × 1 cm, yellow, flushed with lilac or reddish brown on the back. Filaments hairy; anthers and pollen yellow. *NW Iran.* H4. Spring.

T. dasystemon (Regel) Regel. Illustration: Rix & Phillips, The bulb book, 117 (1981). Differs in the solitary flower having perianth-segments 1.5–3 cm × 4–7 mm, yellow with a central band of brownish or greenish. *Former USSR (Soviet Central Asia: Pamir Alai, Tien Shan).* H4. Spring.

3. T. orphanidea Heldreich (*T. hageri* Heldreich; *T. whittallii* Hall). Illustration: Hay & Synge, Dictionary of garden plants in colour, f. 907 (1969); Synge, Collins guide to bulbs, edn 2, pl. 29, f. 2, 6 (1971); Grey-Wilson & Mathew, Bulbs, pl. 5 (1981); Rix & Phillips, The bulb book, 124 (1981).

Inside of bulb-tunics with some hairs at the base, denser hairs at the top. Stem 10–35 cm, hairless or downy. Leaves 2–7, green, hairless, margin often reddish purple. Bud erect. Flowers 1–4; perianth-segments 3–6 cm, pale to bright orange or brownish red, rarely yellow, flushed with red, 80 basal blotch blackish, sometimes with a yellow margin; the outer segments buff and usually tinged with green and sometimes purple. Filaments hairy, purplish or greenish; anthers 5–12 mm, dark green or brown; pollen brownish, olive-green or yellow. *Bulgaria, Greece, Crete, W Turkey.* H4. Spring.

In addition to *T. orphanidea*, *T. hageri* and *T. whittallii* are listed in commercial catalogues. The plants grown in gardens can usually be roughly separated, but in the wild the three species cannot be distinguished. The clones grown in gardens show the following features: **T. orphanidea** has dull orange-brown flowers tinged with green and purple outside; **T. orphanidea** 'Flava' has yellow flowers flushed with red; **T. hageri** has dull red flowers, tinged with green outside; **T. hageri** var. **nitens** Anon., has orange-scarlet flowers and rather glaucous leaves; **T. hageri** 'Splendens' has flowers that are brownish red inside and crimson-scarlet outside; **T. whittallii** is taller than the other species (usually 30–35 cm) and the flowers are bright bronzy orange.

T. orphanidea is closely related to *T. sylvestris* and may, in fact, be a hybrid that has *T. sylvestris* as one of the parents. *T. orphanidea* 'Flava' is almost indistinguishable from *T. sylvestris.*

T. kurdica Wendelbo. Illustration: Rix & Phillips, The bulb book, 118 (1981). Differs in usually having a shorter stem (6–15 cm), and the solitary flower with bright brick-red to orange-red perianth-segments with a greenish black basal blotch. Anthers 1.3–1.4 cm. *NE Iraq.* H4.

T. primulina may key out here (see under 6. *T. biflora*). Differs in having grey-green leaves and 1 or 2 flowers, which often droop slightly. Perianth-segments whitish to creamy yellow, outer segments flushed with pink or green. Anthers yellow.

4. T. saxatilis Sprengel (*T. bakeri* Hall). Illustration: *Journal of the Royal Horticultural Society* **85**: pl. 118 (1960); Grey-Wilson & Mathew, Bulbs, pl. 5 (1981); Rix & Phillips, The bulb book, 126 (1981); Mathew & Baytop, The bulbous plants of Turkey, pl. 120 (1984).

Inside of bulb-tunics densely hairy at the base and the top. Stem 15–45 cm, hairless. Leaves 2–4, shiny green, hairless. Bud nodding. Flowers 1–4, fragrant; perianth-segments pink to lilac-purple, basal blotch yellow bordered with white; outer segments 3.8–5 × 1–2 cm; inner segments 4–5.5 × 1.5–3 cm. Filaments hairy, yellow; anthers yellow, purple or brown; pollen yellow. *Crete, W Turkey.* H4. Spring.

The clones introduced into cultivation differ from one another in that *'T. saxatilis'* has clear lilac-pink flowers and the flowers of *'T. bakeri'* are a deeper, more purplish pink colour.

5. T. humilis Herbert (*T. violacea* Boissier & Buhse; *T. violacea* var. *pallida* Bornmüller; *T. pulchella* (Regel) Baker; *T. pulchella* var. *albocaerulea-occulata* van Tubergen; *T. aucheriana* Baker). Illustration: Hall, The genus Tulipa, pl. 6–9 (1940); Hay & Synge, Dictionary of garden plants in colour, f. 906 (1969); Synge, Collins guide to bulbs, edn 2, pl. 29, f. 17, 19, 21 (1971); Rix & Phillips, The bulb book, 118, 125, 126 (1981).

Inside of bulb-tunics with some hairs near the base and top. Stem 5–20 cm, hairless. Leaves 2–5, usually in a basal rosette, somewhat glaucous, hairless. Bud erect. Flowers usually solitary, occasionally to 3; perianth-segments pale pink to purplish pink or magenta, rarely whitish flushed with pink, basal blotch yellow, olive-green or blue-black often bordered with white or yellow; outer segments 2.3–4.7 cm × 5–18 mm, sometimes tinged with greyish green on the back; inner segments 2.5–5 cm × 9–20 mm. Filaments hairy, yellow or sometimes dark bluish purple; anthers yellow, brown, purple or blackish; pollen yellow, bluish or greenish. *SE Turkey, N & W Iran, N Iraq, Former USSR (Azerbaijan).* H4. Late winter–spring.

T. humilis, T. violacea, T. pulchella and *T. aucheriana* are commonly offered by nurserymen as separate species. As grown in gardens, they represent only part of the variation displayed in the wild where they cannot be separated. Commercially available plants usually have the following characteristics: **T. humilis** has pinkish magenta flowers with a yellow centre and yellow filaments; **T. violacea** has violet-pink flowers with a yellow or blue-black centre and deep purplish filaments; **T. pulchella** has purplish flowers with a yellow edged blue-black centre and yellow filaments; **T. aucheriana** has pink flowers with a yellow centre and yellow filaments and tends to flower later than the other 'species'. Plants sold as *T. pulchella* var. *albocaerulea-occulata* or *T. violacea* var. *pallida* have whitish flowers with a dark blue centre and usually flower rather early.

6. T. biflora Pallas (*T. polychroma* Stapf). Illustration: Synge, Collins guide to bulbs, edn 2, pl. 29, f. 16 (1971); Lodewijk, The book of tulips, 107 (1979); Grey-Wilson & Mathew, Bulbs, pl. 5 (1981); Rix & Phillips, The bulb book, 119 (1981).

Inside of bulb-tunics hairy, sometimes densely so. Stem 5–10 cm, sometimes more, hairless. Leaves 2, grey-green, hairless, sometimes with a ciliate margin. Bud erect. Flowers 1–5, scented; perianth-segments white, flushed with greenish grey or greenish pink on the back and yellow at the base; outer segments 1.5–3.5 cm; inner segments 2–3 cm. Filaments hairy, yellow; anthers yellow, often dark purple or blackish at the tip; pollen yellow. *E Turkey, Iran, Afghanistan, Former SW USSR.* H4. Late winter–spring.

T. primulina Baker. Illustration: *Botanical Magazine*, 6786 (1884); Hall, The genus Tulipa, pl. 1 (1940); Synge, Collins guide to bulbs, edn 2, pl. 29, f. 8 (1971); Rix & Phillips, The bulb book, 124 (1981). Differs in having 1 or 2 flowers, which are often slightly drooping; perianth-segments whitish to creamy yellow, deeper yellow at the base inside; outer segments *c.* 2 cm broad, flushed with pink or greenish on the back. *Algeria.* H4. Spring.

7. T. turkestanica Regel. Illustration: Lodewijk, The book of tulips, 107 (1979); Rix & Phillips, The bulb book, 117, 118 (1981); Botschantzeva, Tulips, pl. 39 (1982); Brickell (ed.), RHS A–Z encyclopedia of garden plants, 1060 (2003).

Similar to *T. biflora* but with a hairy stem, 10–30 cm, and 2–4 leaves. There may be up to 12 slightly smaller flowers on a stem, which usually smell somewhat unpleasant, and have a yellow or orange centre. The anthers are yellow with a purplish tip, or completely purple or brown. *Former USSR (Soviet Central Asia), NW China.* H4. Late winter–spring.

Very close to *T. biflora;* further study may show that these two species are the same.

8. T. tarda Stapf (*T. dasystemon* misapplied). Illustration: Hall, The genus Tulipa, pl. 13 (1940); Hay & Synge, Dictionary of garden plants in colour, f. 904 (1969); Hay & Beckett (eds), Hay & Beckett (eds), Readers' Digest encyclopaedia of garden plants and flowers, 721 (1971); Rix & Phillips, The bulb book, 126 (1981).

Inside of bulb-tunics hairless or with sparse hairs at the base and the top. Stem to 15 cm, hairless. Leaves 3–7, shiny green, hairless, margin often ciliate. Buds erect. Flowers 4–6; perianth-segments 3–4 × 1.2–2 cm, white, tinged with greenish and sometimes red on the back, and yellow in the lower half inside. Filaments hairy, yellow; anthers and pollen yellow. *Former USSR (Soviet Central Asia: Tien Shan).* H4. Spring.

T. cretica Boissier & Heldreich. Illustration: Hall, The genus Tulipa, pl. 9 (1940); Grey-Wilson & Mathew, Bulbs, pl. 5 (1981); Rix & Phillips, The bulb book, 122 (1981). Differs in having only 2 or 3 leaves and 1–3 flowers. Perianth-segments 1.5–3 cm × 4–11 mm. *Crete.* H4. Spring.

9. T. clusiana de Candolle (*T. aitchisonii* Hall). Illustration: Hall, The genus Tulipa, pl. 14, 17 (1940); Hay & Synge, Dictionary of garden plants in colour, f. 892 (1969); Synge, Collins guide to bulbs, edn 2, pl. 29, f. 1 (1971); Wendelbo, Tulips and irises of Iran, 35 (1977).

Bulb-tunics with hairs protruding from the apex. Stem 3.5–30 cm, hairless. Leaves 2–5, *c.* 1 cm broad, glaucous, margin sometimes wavy. Bud erect. Flowers 1 or 2; perianth-segments white, the outer deep pink on the back except at the edge, basal blotch purple or crimson; outer segments 3–6.5 cm; inner segments 2–5 cm. Filaments hairless, purple; anthers and pollen purple. *NW Pakistan, N India to Iran; naturalised in S Europe and Turkey.* H4. Spring.

Var. **stellata** (Hooker) Regel (*T. stellata* Hooker). Illustration: Hall, The genus Tulipa, pl. 16 (1940); Synge, Collins guide to bulbs, edn 2, pl. 29, f. 3 (1971). The basal blotch of the perianth-segments and the stamens are yellow.

Var. **chrysantha** (Hall) Sealy (*T. stellata* Hooker var. *chrysantha* Hall; *T. aitchisonii* Hall subsp. *cashmeriana* Hall). Illustration: Hall, The genus Tulipa, pl. 15, 16 (1940); *Botanical Magazine*, n.s., 13 (1948); Synge, Collins guide to bulbs, edn 2, pl. 29, f. 11, 14 (1971); Hay & Beckett (eds), Readers' Digest encyclopaedia of garden plants and flowers, 720 (1971). Flowers 1–3; perianth-segments yellow, except usually for red or brownish purple coloration on the back of the outer segments; stamens yellow.

There is some controversy about the relationship of var. *stellata* and var. *chrysantha* to *T. clusiana*. Further work may show that it would be better to raise these varieties to subspecific or even specific level.

T. altaica, which has solitary yellow flowers tinged with red or green on the outside and leaves *c.* 3 cm broad, may key out here (see under 29. *T. armena*).

10. T. linifolia Regel (*T. batalinii* Regel; *T. maximowiczii* Regel). Illustration: Dykes, Notes on tulip species, pl. 19 (1930); Hall, The genus Tulipa, pl. 18 (1940); Synge, Collins guide to bulbs, edn 2, pl. 29, f. 9, 12 (1971); Rix & Phillips, The bulb book, 123 (1981).

Inside of bulb-tunics hairy towards the top. Stem 10–30 cm, hairless. Leaves 3–9, usually up to 10 mm wide, greyish green, hairless, margin often wavy and often ciliate, usually red. Bud erect. Flower solitary; perianth-segments 2–6 × 1.2–3.5 cm, red with a blackish purple basal blotch often bordered with white, cream or yellow, or pale yellow with a deep yellow or brownish basal blotch. Filaments hairless, black or yellow; anthers and pollen dark purple or yellow. *N Iran, Afghanistan, Former USSR (Soviet Central Asia: Uzbekistan).* H4. Spring.

A variable species, represented in cultivation by clones that can be distinguished to some extent. As grown in gardens, **T. linifolia** has bright red flowers with a blackish purple centre bordered with cream or yellow, and black filaments with dark purple anthers containing purple pollen; **T. batalinii** has flowers of pale yellow with a deep yellow or brownish centre and yellow stamens and pollen; **T. maximowiczii** has red flowers but the black centre has a white border, the black filaments are white at the apex and the purplish or yellow anthers contain purple pollen.

Hybrids have been produced between *T. linifolia* and *T. batalinii* that have apricot or bronze flowers, e.g. 'Bronze Charm', 'Bright Gem' and 'Apricot Jewel'.

T. julia Koch. Illustration: Rix & Phillips, The bulb book, 123 (1981); Botschantzeva, Tulips, pl. 4 (1982); Mathew & Baytop, The bulbous plants of Turkey, pl. 112 (1984). Bulb-tunics lined with dense matted hairs; leaves 1–3 cm wide; perianth-segments red to orange-red, pinkish or orange on the back with a yellow-bordered black basal blotch. *Former USSR (Transcaucasia), NW Iran, E Turkey.* H4. Spring.

T. albertii, which differs in having red, orange or yellow perianth-segments, always with a yellow-bordered black basal blotch, and segments 4–6 cm wide, may key out here (see under 29. *T. armena*).

11. T. montana Lindley (*T. wilsoniana* Hoog). Illustration: Dykes, Notes on tulip species, pl. 17, 18 (1930); Hall, The genus Tulipa, pl. 19 (1940); *Botanical Magazine*, n.s., 433 (1963); Botschantzeva, Tulips, pl. 20 (1982).

Inside of bulb-tunics hairy towards the top and with hairs protruding from the apex. Stem 20–30 cm, hairless. Leaves 3–6, glaucous, hairless, margin of basal leaves wavy, often red. Bud erect. Flower solitary; perianth-segments 4–5.5 × 2–2.5 cm, red with a greenish black basal blotch. Filaments hairless, red, black towards the base; anthers and pollen yellow. N Iran, *Former USSR (Soviet Central Asia: Turkmeniya).* H4. Spring.

12. T. urumoffii Hayek (*T. rhodopea* (Velenovsky) Velenovsky). Illustration: Hall, The genus Tulipa, pl. 22 (1940); Grey-Wilson & Mathew, Bulbs, pl. 6 (1981).

Inside of bulb-tunics lined with sparse silky hairs. Stem 15–30 cm, hairless. Leaves 3–5, glaucous, hairless, margin ciliate towards the apex. Bud erect. Flowers usually solitary, sometimes to 3; perianth-segments 3.5–6 × 1.7–2.5 cm, yellow or red to brownish red, basal blotch absent or black with a yellow border. Filaments hairless, yellow shading to greenish at the apex; anthers greenish; pollen pale brownish or greenish. *S Bulgaria.* H4. Spring.

13. T. gesneriana Linnaeus (*T. billetiana* Jordan; *T. bonarotiana* Reboul; *T. connivens* Levier; *T. etrusca* Levier; *T. passeriniana* Levier; *T. pubescens* Willdenow; *T. scabriscapa* Fox-Strangways; *T. sommieri* Levier; *T. spathulata* Bertoloni; *T. strangulata* Reboul; *T. variopicta* Reboul). Illustration: *Revue Horticole* 1890: 475 (1890); *The Garden* **44**: 534 (1893); Grey-Wilson & Mathew, Bulbs, pl. 6 (1981).

Inside of bulb-tunics without hairs or with a few hairs towards the top. Stem 30–60 cm, hairless or finely downy. Leaves 2–7, green, hairless, margin ciliate towards the apex. Bud erect. Flower solitary; perianth-segments 4–8.2 cm, inner broader than outer, red, orange, yellow or purplish, sometimes 'broken' (see below), with or without a yellow or blackish basal blotch that may have a yellow border. Filaments hairless, yellow or purple; anthers purple or yellow. *Origin unknown, possibly SW & SC Asia; naturalised in SW Europe.* H4. Spring.

A 'broken' flower is one in which the normal self-colour is variegated with another colour. The 'breaking' is caused by a virus. 'Broken' flowers were very common and new 'breaks' much sought-after during the period of tulipomania in Holland in the seventeenth century and in Britain from *c.* 1835 to the end of the nineteenth century.

T. gesneriana is an extremely complex species from which most of the garden cultivars have been derived. From the fifteenth century onwards there was a great influx of tulips from western and central Asia and some of the selections from these eventually became naturalised in Europe: they were known as the Neo-tulipae. Several of these selections were given species names and still appear in nurserymen's catalogues, but cannot really be separated satisfactorily from the mass of material included under *T. gesneriana*. Some of these, as sold commercially and grown in gardens, are described below.

T. acuminata Hornemann (*T. cornuta* Delile). Illustration: *Revue Horticole* 476 (1890); Everett, The New York Botanical Gardens illustrated encyclopedia of horticulture **10**: 3426 (1982); Mathew, P. J. Redouté: Lilies and related flowers, 103 (1982); Brickell (ed.), RHS A–Z encyclopedia of garden plants, 1054 (2003). Stem 30–45 cm. Perianth-segments 7.5–13 cm, long-pointed and tapering to a twisted point, pale red or yellowish, usually tinged with green or red. Filaments yellow or white; anthers reddish brown.

T. didieri Jordan. Stem 30–40 cm. Perianth-segments red, tinged with crimson on the back, basal blotch dark purple, margin irregular, creamy white or yellow, segment tips reflexed. Anthers black, occasionally yellow.

T. elegans Baker. Illustration: *The Garden* **32**: 514 (1887). Stem to 45 cm, downy. Perianth-segments red with a yellow basal blotch. Filaments and anthers purple.

T. fulgens Baker differs from *T. elegans* in its hairless stem and yellow anthers.

T. galatica Freyn. Illustration: Hall, The genus Tulipa, pl. 21 (1940). Stem *c.* 5 cm.

Perianth-segments pale yellow tinged with olive-green on the back and with an olive-green basal blotch. Filaments yellow.

T. grengiolensis Thommen. Stem 25–40 cm. Perianth-segments pale yellow, margined with red.

T. marjolettii Perrier & Songeon. Stem 40–50 cm. Perianth-segments creamy-white, edged with deep pink and tinged with purple on the back. Filaments blue-black, anthers yellow.

T. mauritiana Jordan (*T. mauriana* Jordan & Fournier). Stem 30–40 cm. Perianth-segments scarlet, yellow at the base. Filaments yellow, anthers purple.

T. platystigma Jordan. Stem to 55 cm. Perianth-segments rose-pink, blue at the base and suffused with orange towards the margins.

T. retroflexa Baker. Illustration: *The Garden* **32**: 514 (1887) & **85**: 241 (1921). Perianth-segments 7.5–10 cm, bright yellow, long-pointed, reflexing from the middle as the flower opens. Filaments yellow.

T. viridiflora Anon. Illustration: *The Garden* **32**: 514 (1887). Stem to 60 cm. Perianth-segments pale green, yellowish or whitish at the edge, acuminate.

T. cypria Turrill. Illustration: *Botanical Magazine*, 9363 (1934); Megaw, Wild flowers of Cyprus, pl. 33 (1973). Differs in the bulb-tunics being densely woolly inside and stem only to 35 cm. Perianth-segments dark red with a blue-black basal blotch usually bordered with yellow, outside green at the base. *Cyprus.* H4. Spring.

T. lehmanniana Bunge. Illustration: Botschantzeva, Tulips, pl. 21, 22 (1982). Differs in the bulb having a long neck and the tunics being densely woolly inside and the stem only to 25 cm. Leaf margin wavy, not ciliate. Bud nodding. Perianth-segments yellow, orange or red, flushed with orange-red or reddish brown, basal blotch black, olive-green or violet. *Former USSR (Soviet Central Asia), Afghanistan, NE Iran.* H4. Spring.

T. stapfi Turrill. Illustration: *Journal of the Royal Horticultural Society* **82**: pl. 45 (1957) & **88**: pl. 198 (1963). Differs in the bulb-tunics being densely woolly inside and the stem up to 30 cm. Leaves glaucous. Perianth-segments red with a yellow-bordered purplish basal blotch. *W Iran, N Iraq.* H4. Spring.

T. tschimganica may key out here (see under 25. *T. kaufmanniana*). Differs in having a stem to 25 cm and glaucous leaves. Perianth-segments yellow with red V-shaped marks towards the base inside, the outer with red midribs on the back, or segments entirely red, or red with a yellow base. Filaments yellow with a brownish tip.

T. albertii, which has stems to 20 cm and red, orange or yellow flowers with a black, yellow-bordered basal blotch, may key out here (see under 29. *T. armena*).

14. T. undulatifolia Boissier (*T. baeotica* Boissier & Heldreich; *T. eichleri* Regel). Illustration: *Journal of the Royal Horticultural Society* **88**: pl. 158 (1963); Hay & Synge, Dictionary of garden plants in colour, f. 893 (1969); Grey-Wilson & Mathew, Bulbs, pl. 5 (1981); Rix & Phillips, The bulb book, 126 (1981).

Inside of bulb-tunics hairy at the base, sparsely hairy above. Stem 15–50 cm, usually minutely downy. Leaves 3 or 4, glaucous, usually minutely downy on the upper surface, margin wavy or crisped, ciliate. Bud erect. Flower solitary; perianth-segments orange-red to red, paler red, bluish red or buff on the back, downy, basal blotch dark olive-green or black, often with a yellow border; outer segments 3–7.5 × 1.5–2.5 cm; inner segments slightly longer. Filaments hairless, black; anthers dark purple; pollen dark green or dirty yellow. *Jugoslavia, Greece, W Turkey, Iran, Former USSR (W Soviet Central Asia).* H4. Spring.

5. T. praecox Tenore. Illustration: Dykes, Notes on tulip species, pl. 29 (1930); Hall, The genus Tulipa, pl. 25, 27 (1940); Grey-Wilson & Mathew, Bulbs, pl. 6 (1981); Mathew & Baytop, The bulbous plants of Turkey, pl. 114 (1984).

Inside of bulb-tunics densely woolly. Stem 20–65 cm, hairless or slightly downy. Leaves 3–5, glaucous, hairless. Bud erect. Flowers solitary; perianth-segments orange tinged with green on the back, basal blotch brownish green with a yellow border; outer segments 4–10 × 2–5 cm; inner segments 3.5–7.5 × 1.5–3.5 cm, with a yellowish central stripe. Filaments hairless, black or dark green; anthers black or dark green; pollen dark green or yellow. *Origin unknown; naturalised in S Europe & W Turkey.* H4. Spring.

16. T. agenensis de Candolle (*T. oculis-solis* St. Amans; *T. lortetii* Jordan). Illustration: Hall, The genus Tulipa, pl. 23 (1940); Megaw, Wild flowers of Cyprus, pl. 32 (1973); Grey-Wilson & Mathew, Bulbs, pl. 6 (1981); Mathew, P.J. Redouté: Lilies and related flowers, 107 (1982).

Inside of bulb-tunics lined with dense matted woolly hairs. Stem to 20 cm, hairless. Leaves 3–5, green, margins flat or somewhat wavy, often ciliate. Bud erect. Flower solitary; perianth-segments dull red inside, lighter red or yellowish on the back, with a black basal blotch bordered with yellow; outer segments 4.2–8.5 × 1.8–3 cm; inner segments 3.7–7 × 1.5–2.3 cm. Filaments 8–12 mm, hairless, black or dark blue; anthers black; pollen yellow. *W & S Turkey, NW Iran; naturalised in S France & Italy.* H4. Spring.

T. sharonensis Dinsmore. Differs in the perianth-segments having a very large, deep olive-green, yellow-bordered basal blotch covering almost the basal half of the segment; outer segments *c.* 3 × 1 cm. *Israel.* H4. Spring.

T. systola Stapf. Illustration: *Journal of the Royal Horticultural Society* **88**: pl. 198 (1963) & **100**: 257 (1975); Rix & Phillips, The bulb book, 125 (1981). Differs in having the red perianth-segments slightly greyish on the back. Filaments 6–7 mm, dark blue with a yellow apex. *Iran.* H4. Spring.

17. T. hoogiana Fedtschenko. Illustration: Gardeners' Chronicle **48**: 53 (1910); Hall, The genus Tulipa, pl. 29 (1940); Rix & Phillips, The bulb book, 125 (1981).

Inside of bulb-tunics lined with thick matted woolly hairs. Stem to 45 cm, hairless. Leaves 3–5, greyish green, hairless, margin long-ciliate. Flower solitary, at first strongly scented; perianth-segments 7.5–10 cm, orange-red, basal blotch black or dark olive-green with a yellow border. Filaments black or reddish; anthers black, brown, pinkish or yellow; pollen brown, pinkish or yellow. *Former USSR (Soviet Central Asia: mountains of Turkmeniya).* H4. Spring.

T. aleppensis Regel. Illustration: Hall, The genus Tulipa, pl. 26, 27 (1940). Differs in having stems 10–25 cm, and perianth-segments 5–5.5 cm, red-crimson. *Syria.* H4. Spring.

T. kuschkensis Sealy. Illustration: *Botanical Magazine*, 9370 (1934); *Gartenflora* **84**: 295 (1935); Hall, The genus Tulipa, pl. 28 (1940); Botschantzeva, Tulips, pl. 2 (1982). Differs in having scentless flowers and outer perianth-segments whose margins roll back. *Former USSR (Soviet Central Asia: Turkmeniya), NW Afghanistan.* H4.

18. T. lanata Regel. Illustration: *Botanical Magazine*, 9151 (1926); Hall, The genus

Tulipa, pl. 29 (1940); Botschantzeva, Tulips, pl. 5 (1982).

Inside of bulb-tunics thickly lined with straight hairs. Stem 15–60 cm, downy. Leaves 4 or 5, glaucous, downy, margin wavy, red, ciliate. Bud erect. Flower solitary; perianth-segments 6–12 × 3–6 cm, red, somewhat silvery on the back and downy on the upper half, basal blotch olive-green with a yellow border. Filaments hairless, black, white at the tip; anthers and pollen reddish purple. *NE Iran, Former USSR (Soviet Central Asia: Turkmeniya), Afghanistan.* H4. Spring.

Very closely related to the three following species; further research may show that they all belong to a single species.

19. T. ingens Hoog. Illustration: *Gardeners' Chronicle* **32**: 14 (1902); Dykes, Notes on tulip species, pl. 51 (1930); Hall, The genus Tulipa, pl. 30 (1940); Botschantzeva, Tulips, pl. 6 (1982).

Inside of bulb-tunics sparsely hairy. Stem to 25 cm, downy, often reddish or purplish in upper part. Leaves 3–5, glaucous, downy, margin wavy and ciliate. Flower solitary; perianth-segments 8–15 cm, red with a black basal blotch. Filaments hairless, blackish or dirty yellow; anthers and pollen dark purple. *Former USSR (Soviet Central Asia: Pamir Alai).* H4. Spring.

Very close to *T. lanata.*

20. T. tubergeniana Hoog. Illustration: *Gardeners' Chronicle* **35**: 358 (1904); *Flora and Sylva* **3**: 94 (1905); Hall, The genus Tulipa, pl. 30 (1940); Botschantzeva, Tulips, pl. 1 (1982).

Inside of bulb-tunics thickly hairy. Stem 30–60 cm, downy. Leaves 3–5, glaucous, downy, margin wavy and ciliate. Flower solitary; perianth-segments to 10 × 3.5 cm, red, basal blotch olive-green or blackish, with or without a narrow yellow border. Filaments hairless, black; anthers dark purple; pollen brownish purple or dirty yellow. *Former USSR (Soviet Central Asia).* H4. Spring.

Very close to *T. lanata.*

21. T. fosteriana Irving. Illustration: Hall, The genus Tulipa, pl. 32 (1940); Botschantzeva, Tulips, pl. 7 (1982).

Inside of bulb-tunics densely hairy. Stem 20–45 cm, slightly downy. Leaves 3–6, glaucous, downy on upper surface, margin ciliate. Bud erect. Flower solitary, slightly fragrant; perianth-segments 4.5–8 cm, occasionally to 12.5 cm, bright red, basal blotch purplish black with a yellow border. Filaments hairless, black or yellow; anthers purplish black; pollen purplish brown or yellow. *Former USSR (Soviet Central Asia).* H4. Spring.

This species is closely related to *T. lanata.* A number of named cultivars is available as well as hybrids with *T. greigii.*

22. T. carinata Vvedensky. Illustration: Botschantzeva, Tulips, pl. 8 (1982).

Inside of bulb-tunics sparsely to densely woolly. Stem 20–55 cm, downy in upper part. Leaves 3 or 4, greyish green, downy, keeled below, margin wavy towards the apex, ciliate. Flower solitary; perianth-segments red, 4–9 × *c.* 4 cm, with the basal blotch yellow, black, or black bordered with yellow. Filaments hairless, blackish, *c.* 1 cm; anthers violet or yellow, *c.* 1.8 cm; pollen yellow, pinkish or buff. *Former USSR (Soviet Central Asia—Pamir Alai).* H4. Spring.

23. T. greigii Regel. Illustration: *Botanical Magazine*, 6177 (1875); *Journal of the Royal Horticultural Society* **97**(4): front cover (1972); Witham Fogg, Bulbs, 42 (1980); Rix & Phillips, The bulb book, 124 (1981).

Inside of bulb-tunics sparsely, or sometimes densely, hairy towards the top. Stem 20–45 cm, downy. Leaves 3–5, streaked and mottled with reddish or purplish brown, margin sometimes wavy, upper leaves often downy with ciliate margins. Bud erect. Flower solitary; perianth-segments red or yellow, basal blotch blackish with a yellow border; outer segments 3–10 cm; inner segments 4–11 cm. Filaments hairless, black; anthers black; pollen yellow. *Former USSR (Soviet Central Asia: Tadjikistan).* H4. Spring.

Numerous named cultivars of *T. greigii* are available, as well as hybrids between *T. greigii* and *T. fosteriana.*

24. T. micheliana Hoog. Illustration: *Gardeners' Chronicle* **31**: 353 (1902); *Revue Horticole* 206 (1903); *Journal of the Royal Horticultural Society* **90**: pl. 233 (1965); Botschantzeva, Tulips, pl. 18 (1982).

Inside of bulb-tunics lined with adpressed hairs that are denser at the base and the top. Stem 25–35 cm, sometimes to 60 cm, downy. Leaves 3–5, glaucous, streaked with reddish or purplish brown, more or less downy, margin ciliate and usually wavy. Bud erect. Flower solitary; perianth-segments 5.5–10 cm, bright red to crimson, the long basal blotch purplish black with a white or yellow border that may be very narrow or absent. Filaments hairless, black; anthers violet or yellow; pollen violet, yellow or brown. *NE Iran, Former USSR (Soviet Central Asia: Turkmeniya).* H4. Spring.

25. T. kaufmanniana Regel. Illustration: Hall, The genus Tulipa, pl. 34 (1940); Hay & Synge, Dictionary of garden plants in colour, f. 897 (1969); *Journal of the Royal Horticultural Society* **100**: 504 (1975); Rix & Phillips, The bulb book, 119, 121 (1981).

Inside of bulb-tunics lined with adpressed hairs, denser at the base and top. Stem 10–35 cm, slightly downy. Leaves 3–5, often in a basal rosette, greyish green, hairless, margin minutely ciliate. Bud erect. Flowers 1–5, often fragrant; perianth-segments 3–10 × 1–5 cm, cream or yellow flushed with pink or greyish green on the back, or pink, orange or red, often with a basal blotch of a different colour. Filaments hairless, yellow; anthers twisted, yellow; pollen yellow. *Former USSR (Central Asia: Tadjikistan, Kirgizia).* H4. Late winter–spring.

A large range of cultivars is available, as well as hybrids with *T. fosteriana* and *T. greigii.*

T. dubia Vvedensky. Illustration: Rix & Phillips, The bulb book, 121 (1981). Differs in the leaves being more or less downy, with a wavy margin. Flower solitary; perianth-segments 2–4 × *c.* 1.5 cm, yellow, streaked on the back with bluish pink, basal blotch orange. Anthers not twisted. *Former USSR (Soviet Central Asia: Tien Shan).* H4.

T. tschimganica Botschantzeva. Illustration: Rix & Phillips, The bulb book, 120, 121 (1981). Differs in having downy leaves and solitary flowers. Perianth-segments yellow with red V-shaped marks towards the base inside, the outer with red midribs on the back, or segments entirely red, or red with a yellow base. Anthers not twisted. *Former USSR (Soviet Central Asia: Tien Shan).* H4. Spring.

T. subpraestans Vvedensky. Illustration: Botschantzeva, Tulips, pl. 11 (1982). Differs in the bulb-tunics being hairless or almost so inside, and flowers usually solitary. Perianth-segments red, lacking a basal blotch. Filaments red; anthers yellow or black, twisted. *Former USSR (Soviet Central Asia: Pamir Alai).* H4. Spring.

26. T. praestans Hoog. Illustration: Hall, The genus Tulipa, pl. 35 (1940); Hay & Synge, Dictionary of garden plants in colour, f. 901 (1969); Rix & Phillips, The bulb book, 118, 125 (1981); Botschantzeva, Tulips, pl. 10 (1982).

Inside of bulb-tunics sparsely hairy at the top. Stem 10–45 cm, minutely downy. Leaves 3–6, greyish green, downy, margin ciliate. Flowers 1–5; perianth-segments 5.5–6.5 × *c.* 3 cm, scarlet-orange. Filaments hairless, red shading to yellow at the base; anthers yellow or purplish; pollen yellow, greyish brown or reddish. *Former USSR (Soviet Central Asia—Pamir AM).* H4. Spring.

The following cultivars are more frequently offered in the trade than is the species itself: 'Tubergen's Variety' (illustration: *Botanical Magazine*, 7920, 1903) has 2–5 flowers, often yellow at the base, and increases very freely; 'Fuselier' (illustration: Lodewijk, The book of tulips, 107, 1979; Rix & Phillips, The bulb book, 122, 1981) produces 3–5 flowers and blooms a little earlier than the species itself.

27. T. kolpakowskiana Regel. Illustration: Hay & Synge, Dictionary of garden plants in colour, f. 899 (1969); Synge, Collins guide to bulbs, edn 2, pl. 29, f. 13 (1971); *Journal of the Royal Horticultural Society* **100**: 505 (1975); Rix & Phillips, The bulb book, 118, 121 (1981).

Inside of bulb-tunics lined with adpressed hairs. Stem 10–30 cm, hairless. Leaves 2–4, greyish green, hairless, margin wavy and ciliate. Bud nodding. Flowers usually solitary, sometimes 2–4; perianth-segments 3.7–5.5 × 1.5–2 cm, yellow, marked with crimson, orange or olive-green on the back. Filaments hairless, yellow; anthers yellow. *Former USSR (Soviet Central Asia), Afghanistan.* H4. Spring.

T. heterophylla (Regel) Baker (*Eduardoregelia heterophylla* (Regel) Popov). Illustration: Rix & Phillips, The bulb book, 117 (1981). Differs in having the bulb-tunics hairless inside and the stem 5–15 cm. Flower solitary; perianth-segments 1.5–3 cm, yellow, marked with dull violet or green on the back. *Former USSR (Soviet Central Asia: Tien Shan).* H4. Spring.

T. iliensis Regel. Illustration: *The Garden* **17**: 311 (1880). Differs in having a downy stem and 1–5 flowers. Perianth-segments 1.5–3.5 cm × 4–10 mm, yellow, the outer segments tinged with red or olive-green on the back. *Former USSR (Soviet Central Asia).* H4. Spring.

T. tetraphylla Regel. Illustration: *Gartenflora* **28**: t. 964 (1879). Differs in having 1–4 flowers; outer perianth-segments 2–4 cm, yellow tinged with violet and green on the back. *Former USSR (Soviet Central Asia: Tien Shan).* H4.

28. T. ostrowskiana Regel. Illustration: *Gartenflora* **33**: t. 1144 (1884); *The Garden* **45**: 486 (1894); Botschantzeva, Tulips, pl. 27 (1982).

Inside of bulb-tunics densely hairy especially at the base and top. Stem 20–35 cm, hairless. Leaves 2–4, slightly glaucous, hairless, margin wavy and sometimes sparsely ciliate. Bud nodding. Flower solitary; perianth-segments 5–8 × 1.8–3 cm, red, basal blotch dark olive-green with an irregular yellow border; outer segments with a slight purplish sheen on the back. Filaments yellow, brown or reddish; anthers yellow or black; pollen yellow, black, brown or reddish. *Former USSR (Soviet Central Asia: Tien Shan).* H4. Spring.

Closely related to *T. kolpakowskiana;* further research may show that these two species should be united.

29. T. armena Boissier (*T. suaveolens* Roth). Illustration: Hall, The genus Tulipa, pl. 20 (1940); Rix & Phillips, The bulb book, 124 (1981).

Inside of bulb-tunics lined with long hairs, or with short hairs at the base and the top. Stem 7.5–25 cm, hairless or with rough hairs at least in the upper part. Leaves 3–6, glaucous, hairless, or downy on the upper surface, usually recurved, margin wavy and often ciliate. Bud erect. Flower solitary, often fragrant; perianth-segments red to crimson, often lighter red or yellowish on the back, with a black or dark blue, or sometimes yellow or greenish basal blotch that may be bordered with yellow. Sometimes the segments are yellow, or partly yellow and partly red, with or without the basal blotch. Outer segments 2.3–5.8 cm × 9–27 mm; inner segments 2–4.5 cm × 7–20 mm. Filaments hairless, black or yellow; anthers yellow or blackish. *Turkey, NW Iran, Iraq, Former USSR (Transcaucasia).* H4. Spring.

There is much confusion about *T. schrenkii* Regel, and where it stands in relation to other species. In the wild it is extremely variable in flower colour, white, yellow, pink and red flowers being known. In gardens it is usually represented by a variant in which the perianth-segments are red, edged with deep yellow or orange. Recent publications (Hortus Third, 1976, and Stork, 1984) as well as Hall (1940) equate *T. schrenkii* and *T. suaveolens*, and Mathew & Baytop in The bulbous plants of Turkey (103, 1984) have reduced *T. suaveolens* to a synonym of the variable *T. armena*. It would therefore seem reasonable to assume that *T. schrenkii* should be sunk into *T. armena.*

T. albertii Regel. Illustration: *Revue Horticole* **53**: 430 (1881); *Botanical Magazine*, 6761 (1884); Botschantzeva, Tulips, pl. 15 (1982). Differs in having red, orange or yellow perianth-segments, always with a yellow-bordered black basal blotch, and segments 4–6 cm wide. *Former USSR (Soviet Central Asia).* H4. Spring.

T. altaica Sprengel. Illustration: Hall, The genus Tulipa, pl. 37 (1940). Differs in having yellow perianth-segments without a basal blotch and filaments always yellow. *Former USSR (Soviet Central Asia to W Siberia).* H4. Spring.

T. borszczowii Regel. Illustration: *Gartenflora* **33**: t. 1175 (1884); Botschantzeva, Tulips, frontispiece (1982). Differs in having orange-red perianth-segments, or the inner segments deep yellow, or the outer segments red with a yellow margin. The filaments are grey or dark purple. *Former USSR (Soviet Central Asia: Kazakstan), ?SW Iran.* H4. Spring.

T. butkovii Botschantzeva. Illustration: Botschantzeva, Tulips, pl. 12 (1982). Differs in having the bulb-tunics lined with short hairs and a downy stem. The perianth-segments are brick-red with a dark red basal blotch. *Former USSR (Soviet Central Asia: Tien Shan).* H4. Spring.

T. ferganica Vvedensky. Illustration: Rix & Phillips, The bulb book, 121 (1981). Differs in having downy leaves and 1 or 2 flowers. Perianth-segments yellow, flushed with bluish pink or pinkish brown on the back. Filaments deep yellow. *Former USSR (Soviet Central Asia: Pamir Alai, Tien Shan).* H4. Spring.

Closely related to and possibly the same as *T. altaica.*

T. goulimyi Sealy & Turrill. Illustration: Goulandris, Wild flowers of Greece, 159 (1968). Differs in the bulb-tunics being densely woolly inside and the bright orange to brownish red perianth-segments lacking a basal blotch. Filaments hairy. *S Greece.* H4. Spring.

T. julia may key out here (see under 10. *T. linifolia*). Differs in the red or orange-red perianth-segments being pinkish or orange on the back.

T. lehmanniana may key out here (see under 13. *T. gesneriana*). Differs in the bud nodding and the leaf margin being hairless. Perianth-segments yellow, orange or red, flushed with orange-red or reddish brown, basal blotch black, olive-green or violet. The bulb has a long neck.

30. T. vvedenskyi Botschantzeva. Illustration: *Journal of the Royal Horticultural Society* **91**: pl. 162 (1966); Rix & Phillips, The bulb book, 124 (1981); Botschantzeva, Tulips, pl. 13, 14 (1982).

Inside of bulb-tunics sparsely lined with hairs, more densely so at the top. Stem 15–45 cm, bristly, glaucous, sometimes tinged with purple. Leaves 4 or 5, glaucous, often downy, margin slightly ciliate. Flower solitary; perianth-segments red with a black or yellow basal blotch; outer segments to 10.5 × 6.2 cm; inner segments to 1.5 × 5.2 cm. Filaments hairless, brown or yellow; anthers violet or yellow; pollen brownish, violet or yellow. *Former USSR (Soviet Central Asia: Tien Shan)*. H4. Spring.

T. cypria Turrill may key out here (see under 13. *T. gesneriana*). Differs in the bulb-tunics being densely woolly inside, the stem hairless and the blue-black blotch at the base of the dark red perianth-segments being bordered with yellow. The filaments are purple.

T. tschimganica may key out here (see under 25. *T. kaufmanniana*). Differs in having a downy stem and yellow perianth-segments with red V-shaped marks towards the base inside, the outer with red midribs on the back, or segments entirely red, or red with a yellow base.

31. T. sprengeri Baker. Illustration: *Gartenflora* **44**: t. 1411 (1895); Synge, Collins guide to bulbs, edn 2, pl. xxiii (1971); Rix & Phillips, The bulb book, 134 (1981); Mathew & Baytop, The bulbous plants of Turkey, pl. 108 (1984).

Inside of bulb-tunics hairless or more or less so. Stem 30–45 cm, hairless. Leaves 5 or 6, green, hairless, Bud erect. Flower solitary; perianth-segments orange-red to red; outer segments 4.5–6.5 × 1.5–2 cm, yellowish buff on the back; inner segments slightly larger. Filaments hairless, red, 3 long and 3 short; anthers and pollen yellow. *Turkey*. H4. Late spring–early summer.

32. T. edulis (Miquel) Baker (*Amana edulis* (Miquel) Honda). Illustration: Hall, The genus Tulipa, pl. 40 (1940); *Botanical Magazine*, n.s., 293 (1956); Mathew, Dwarf bulbs, pl. 82, f. 57 (1973); Rix & Phillips, The bulb book, 117 (1981).

Inside of bulb-tunics densely woolly. Stem 5–15 cm, hairless, bearing 2 or 3 linear bracts below the flowers. Leaves *c.* 6, 15–25 cm × 5–10 mm. Flowers 1 or 2; perianth-segments 1.5–3.2 cm × 4–6 mm, white with reddish brown or purplish veins on the back, basal blotch deep purple with a yellow border. Filaments hairless. *S Japan, NE China, Korea.* H5–G1. Late winter–spring.

Var. **latifolia** Makino (*Amana latifolia* (Makino) Honda; *Tulipa latifolia* (Makino) Makino) differs in the leaves being shorter and generally broader, 10–15 cm × 7–15 mm.

9. FRITILLARIA Linnaeus

E.M. Rix

Spring ephemerals, dormant in summer. Bulbs of 2 or more thick, closely wrapped scales, usually spherical or spindle-shaped, rarely with a thin, white, papery tunic, or bulbs of thick, separate scales, often surrounded at the base with numerous, white, rice-like bulbils. Basal leaves (on non-flowering shoots) 1, lanceolate or rarely linear. Stem-leaves usually alternate, more rarely opposite or whorled. Bracts similar to the leaves but smaller, alternate or whorled. Flowers usually nodding, bell-shaped to tubular, rarely saucer-shaped or conical. Perianth-segments free, in 2 whorls, those of the inner whorl wider than those of the outer, each with a usually conspicuous nectary. Stamens more or less equal to the style; anthers more or less basifixed, rarely versatile. Style 1, entire or 3-fid. Capsule erect (nodding in *F. japonica*), flat-topped, sometimes 6-winged, the style falling. Seeds flat (spherical in *F. japonica*).

A genus of about 100 species found throughout the temperate regions of the northern hemisphere (excluding eastern North America), with most species in Mediterranean area, the mountains of SW Asia and in California. Most species succeed in a well-drained soil kept dry in summer, but some species will tolerate damp in summer and a few (e.g. *F. camschatcensis* and *F. cirrhosa*) require it.

Literature: Beck, C., Fritillaries (1953); Furse, P., Fritillaries in Britain, *RHS Lily Year Book*, **22**: 92–100 (1959); Macfarlane, R.F., Fritillaria in California, *Lilies, 1975*, 53–66 (1975); Rix, E.M., Fritillaries in Iran, *Iranian Journal of Botany* **1**: 75–95 (1977); Turrill, W.B. & Sealy, J. R., Fritillaries, *Hooker's Icones Plantarum* **39**: parts 1 & 2 (1980).

1a. Style papillose, sometimes minutely so, or only in part 2
b. Style smooth, except for the stigmatic surface 15
2a. Style undivided or 3-lobed with lobes less than 1 mm 3
b. Style 3-fid with branches more than 1 mm 8
3a. Flowers pink; nectaries visible as bulges on the outside of the perianth; bracts paired **6. stenanthera**
b. Flowers not pink; nectaries not visible as bulges on the outside of the perianth; bracts solitary or in a group of 3 4
4a. Bracts in a group of 3; flowers green, unmarked **34. alfredae**
b. Bract solitary; flowers not green and unmarked 5
5a. Flowers pure yellow, unmarked **31. carica**
b. Flowers purplish, greyish or striped outside 6
6a. Leaves linear **28. elwesii**
b. Leaves lanceolate 7
7a. Leaves glaucous; style 1–2 mm in diameter **32. pinardii**
b. Leaves shining green; style 3–4 mm in diameter **33. uva-vulpis**
8a. Flowers in an umbel surmounted by a tuft of leaves; nectary circular, white, at the base of each perianth-segment 9
b. Flowers usually solitary or in a raceme; nectary not white and at the base of each segment 10
9a. Flowers orange or red, sometimes deep yellow in cultivated plants; nectary *c.* 5 mm in diameter **1. imperialis**
b. Flowers greenish yellow; nectary *c.* 2 mm in diameter **2. raddeana**
10a. Leaves linear 11
b. Leaves lanceolate 13
11a. Flowers narrowly bell-shaped **28. elwesii**
b. Flowers broadly bell-shaped 12
12a. Flowers purple or pink, tessellated with purple, or pure white **7. meleagris**
b. Flowers greenish, heavily tessellated with purplish brown or black **27. montana**
13a. Flowers greenish, tessellated with purplish or brown **21. hermonis**

b. Flowers purplish at base, yellowish at apex, rarely all yellow 14
14a. Bracts paired; flowers usually in a long raceme **26. reuteri**
b. Bracts not paired; flowers usually solitary **23. davisii**
15a. Style undivided, or lobed at apex for less than 1 mm 16
b. Style 3-fid, the branches more than 1 mm 22
16a. Flowers commonly more than 4 in a raceme 17
b. Flowers solitary or rarely more than 4 in an umbel 20
17a. Bracts in pairs; flowers whitish green **5. bucharica**
b. Bracts solitary or absent; flowers not whitish green 18
18a. Flowers pink; style lobed at the apex **42. pluriflora**
b. Flowers black to brown, green or yellowish; style not lobed 19
19a. Flowers not flared at the mouth; nectary triangular or rectangular **3. persica**
b. Flowers flared at the mouth; nectary linear **4. sewerzowii**
20a. Flowers purplish brown inside and out **32. pinardii**
b. Flowers yellow to green inside 21
21a. Flowers yellow outside; bulb surrounded at the base with more than 10 bulbils **41. pudica**
b. Flower green or purplish outside; bulbils, when present, fewer than 10 **35. bithynica**
22a. Leaves mostly in whorls of 3 or more 23
b. Leaves mostly alternate or opposite, only the bracts sometimes in a whorl of 3 28
23a. Bulb surrounded at the base with more than 10 bulbils 24
b. Bulb not surrounded at the base with more than 10 bulbils 26
24a. Flowers black; nectaries inconspicuous **38. camschatcensis**
b. Flowers not black; nectaries conspicuous 25
25a. Flowers orange or red **46. recurva**
b. Flowers brownish and green **45. affinis**
26a. Flowers whitish, lightly tessellated, or pinkish **11. thunbergii**
b. Flowers brown and green, tessellated, sometimes pale yellowish green 27
27a. Leaves linear in 3–5 whorls of 2 or 3; bracts coiled, tendril-like **13. cirrhosa**
b. Leaves lanceolate in 6 or 7 whorls of 4 or 5; bracts not tendril-like **12. roylei**
28a. Bracts in a group of 3 29
b. Bracts solitary or alternate 36
29a. Flowers tessellated 30
b. Flowers not tessellated 35
30a. Leaves linear, less than 10 mm wide 31
b. Leaves lanceolate, more than 10 mm wide 33
31a. Plant very small, with about 5 leaves; flowers whitish; nectary linear **37. japonica**
b. Plant tall with 7–10 leaves; flowers green or brown; nectary ovate 32
32a. Flowers mainly green; nectary less than 10 mm long **15. involucrata**
b. Flowers mainly brown; nectary more than 10 mm long **17. messanensis**
33a. Flowers tessellated, purple on a white ground **8. latifolia**
b. Flowers greenish or yellowish 34
34a. Flowers greenish yellow, lightly tessellated with red-purple; nectary 2 mm long **10. pallidiflora**
b. Flowers green, tessellated, sometimes heavily, with brown to purple; nectary 4–5 mm long **22. graeca**
35a. Flowers broadly bell-shaped, green marked with red-brown; capsule winged **16. pontica**
b. Flowers narrowly bell-shaped, red-brown or sometimes purplish outside; capsule not winged **32. pinardii**
36a. Nectary at the base of the perianth-segment; flower narrowly bell-shaped or conical, rarely saucer-shaped or flat 37
b. Nectary at the angle of the bell, above the base of the perianth-segment; flowers broadly bell-shaped 42
37a. Flowers pale pink, sometimes saucer-shaped **36. alburyana**
b. Flowers not pink, narrowly bell-shaped or conical 38
38a. Flowers pure yellow **30. conica**
b. Flowers not yellow 39
39a. Flowers brownish; leaves linear **28. elwesii**
b. Flowers blackish or whitish; leaves lanceolate 40
40a. Flowers white or pale green **40. liliacea**
b. Flowers blackish 41
41a. Leaves glaucous, scattered up the stem **29. tuntasia***
b. Leaves shining green, predominantly basal **39. biflora**
42a. Nectary linear to linear-lanceolate or oblong, more than 4 times as long as wide 43
b. Nectary ovate to lanceolate, less than 4 times as long as wide 49
43a. Leaves linear to linear-lanceolate **18. lusitanica**
b. Leaves lanceolate to ovate 44
44a. Flowers purple at the base, yellow at the apex **25. michailovskyi**
b. Flowers green, tessellated with brown or black, rarely yellowish 45
45a. Nectary linear, extending from the corner of the bell to near the apex of the segment **24. crassifolia**
b. Nectary linear-lanceolate or oblong, less than half as long as the apical portion of the segment 46
46a. Leaves usually glaucous; perianth-segments striped **22. graeca**
b. Leaves usually shining green; perianth-segments not striped 47
47a. Flowers yellow, lightly tessellated **9. aurea**
b. Flowers greenish or brownish purple 48
48a. Flowers greenish, mottled and veined; style *c.* 1.5 cm **43. purdyi**
b. Flowers purplish brown, heavily tessellated; style 7–9 mm **23. davisii**
49a. Flowers green, not tessellated but sometimes shaded with brown 50
b. Flowers tessellated, not usually green 52
50a. Leaves lanceolate to ovate, clasping the stem **20. gussichae***
b. Leaves linear to linear-lanceolate, not clasping the stem 51
51a. Perianth-segments strongly recurved at the apex, the inner much wider than the outer **19. acmopetala**
b. Perianth-segments not recurved at the apex, the inner slightly wider than the outer **17. messanensis**
52a. Flowers yellow, usually lightly tessellated 53
b. Flowers purplish or blackish, heavily tessellated 55
53a. Flowers pale greenish yellow, usually more than 2 per stem; capsule winged **10. pallidiflora**
b. Flowers pure yellow, sometimes greenish outside, rarely more than 1 per stem; capsule not winged 54
54a. Style-branches 2–5 mm, not recurved **9. aurea**
b. Style-branches 6–9 mm, very slender, recurved **44. glauca**

55a. Flowers very broadly bell-shaped, with angular shoulders to the bell, ground colour white, tessellated with purple **8. latifolia**
b. Flowers broadly bell-shaped, with rounded shoulders to the bell; ground colour greenish yellow, tessellated with brown or black **14. pyrenaica**

1. F. imperialis Linnaeus. Illustration: *Journal of the Royal Horticultural Society* **88**: f. 65 (1963); Wendelbo, Tulips and irises of Iran, f. 23 (1977); Brickell (ed.), RHS A–Z encyclopedia of garden plants, 457 (2003).

Bulb a compressed sphere to 10 cm in diameter with a pungent, foxy smell. Stem 50–150 cm. Leaves lanceolate, in 3 or 4 whorls of 4–8. Bracts 10–20 in a close group above the flowers. Flowers 3–5 in an umbel, broadly bell-shaped, red or orange (to yellow in cultivated clones). Perianth-segments 4–5.5 cm, each with a large, circular nectary *c.* 5 mm in diameter at its base. Style 3–4.5 cm, papillose towards the apex, 3-fid, the branches 1–4 mm. Capsule winged. *S Turkey eastwards to NW India (Kashmir).* H1. Spring.

Several cultivars are grown, differing in size and flower colour.

F. eduardii Regel. Illustration: Rix & Phillips, The bulb book, 89 (1981). Differs in its more conical flowers, smaller nectaries and scentless bulb. *USSR (Soviet Tadzhikistan).* H1. Spring.

2. F. raddeana Regel (*F. askabadensis* Micheli). Illustration: *Journal of the Royal Horticultural Society* **90**: f. 210, 215 (1965); Rix & Phillips, The bulb book, 79 (1981).

Like *F. imperialis* but with 6–20 smaller, pale yellow, conical flowers. Perianth-segments 3–3.7 cm, with nectaries *c.* 2 mm in diameter. *NW Iran, Former USSR (SW Soviet Turkmenia).* H2. Spring.

F. imperialis var. **chitralensis** Anon., has 1–4 flowers of deeper yellow, but is otherwise similar to *F. raddeana.*

3. F. persica Linnaeus (*F. libanotica* (Boissier) Baker; *F. eggeri* Bornmüller). Illustration: Wendelbo, Tulips and irises of Iran, f. 24 (1977); Rix & Phillips, The bulb book, 88 (1981) – 'Adiyaman'.

Bulb spindle-shaped, to 6 cm high, of closely fitting scales. Stem 20–150 cm. Leaves lanceolate, all alternate. Bracts 1, at the base of each flower-stalk, or absent. Flowers 7–20 or more in a raceme, narrowly bell-shaped, greyish, purplish, blackish or greenish. Perianth-segments 1.5–2 cm, all alike, each with a small, triangular to rectangular nectary *c.* 2 mm long above the base. Style 6–8 mm, smooth, slender. Capsule narrowly winged. *S Turkey to W Iran and south to Jordan and Israel.* H2. Spring.

'Adiyaman' is the finest and most commonly grown variant.

4. F. sewerzowii Regel (*Korolkowia sewerzovii* (Regel) Regel). Illustration: Rix & Phillips, The bulb book, 85 (1981); Brickell (ed.), RHS A–Z encyclopedia of garden plants, 458 (2003).

Bulb a compressed sphere to 6 cm in diameter, of 1 or 2 closely fitting scales. Stem 15–20 cm. Leaves broadly lanceolate, alternate or the lowest opposite. Bracts to 7 × 2 cm, 1 at the base of each flower-stalk. Flowers 4–12 (rarely 1) in a long raceme, narrowly bell-shaped, widely flared at the mouth, greenish or purplish outside, yellowish to brick-red inside. Perianth-segments 2.5–3.5 cm, all alike, each with a grooved, linear nectary 1–1.5 cm long. Anthers versatile, fixed *c.* 2 mm above the base. Style 1.5–2 cm, smooth, slender, undivided. Capsule winged. *Former USSR (Soviet Central Asia), NW China.* H1. Spring.

5. F. bucharica Regel. Illustration: Rix & Phillips, The bulb book, 86, 87 (1981).

Bulb spherical, to 4 cm in diameter, of 2 large scales. Stem 10–35 cm, papillose. Leaves lanceolate to ovate, the lower in 2 pairs, the rest alternate. Bracts 2 at the base of each flower-stalk. Flowers 1–10, usually 4–7 in a raceme, cup-shaped, horizontal or pendulous, whitish with green veins. Perianth-segments 1.4–2 cm, lanceolate, each with a deeply indented nectary about 3 mm long. Style 3–4 mm, smooth, slender. Capsule winged. *Former USSR (Soviet Central Asia), N Afghanistan.* H1. Spring.

6. F. stenanthera (Regel) Regel. Illustration: Rix & Phillips, The bulb book, 55 (1981).

Like *F. bucharica* but usually shorter, with the lowest leaves ovate, opposite, the rest much narrower. Flowers pinkish, conical, widely flared at the mouth, the outer perianth-segments larger than the inner, the nectaries visible as bulges on the outside of the perianth. Style *c.* 5 mm, papillose below, slender. Capsule with 6 horn-like wings on the corners. *Former USSR (Soviet Central Asia).* H1. Spring.

7. F. meleagris Linnaeus. Illustration: Rix & Phillips, The bulb book, 77 (1981); *Hooker's Icones Plantarum* **39**: t. 3801 (1980); Brickell (ed.), RHS A–Z encyclopedia of garden plants, 458 (2003).

Bulb spherical, to 2.5 cm in diameter, of 2 large scales. Stem 12–30 cm. Leaves 4–6, linear, alternate. Flower 1, very broadly bell-shaped, purple or pinkish, tessellated all over, to pure white. Perianth-segments 3–4.5 cm, the inner slightly wider than the outer, all acute, each with a linear (7–10 mm long), green nectary at the angle of the bell. Style 1.3–1.6 cm, papillose, 3-fid, the branches 2–5 mm. Capsule not winged. *Europe, from S England and Sweden (naturalised) to N Yugoslavia, Romania and W Russia.* H1. Spring.

Various cultivars, some with greenish flowers, have been named, but are now seldom offered. The white-flowered plants appear to be most common in England, rare in southern Europe.

8. F. latifolia Willdenow (*F. nobilis* Baker). Illustration: Mathew & Baytop, The bulbous plants of Turkey, t. 89 (1984); Brickell (ed.), RHS A–Z encyclopedia of garden plants, 458 (2003).

Bulb spherical, to 2.5 cm in diameter. Stem 4–35 cm. Leaves 5–9, usually 6 or 7, ovate to lanceolate, alternate, shining green. Flower 1, very broadly bell-shaped, dark reddish purple, tessellated. Perianth-segments 3.5–5 cm, obtuse or rounded, incurved at the apex, each with a narrowly ovate nectary, 3–4 mm long. Style 1–1.5 cm, smooth, 3-fid, the branches 2–5 mm. Capsule not winged. *NE Turkey, USSR (Caucasus); ?NW Iran.* H1. Spring.

F. tubiformis Grenier & Godron (*F. delphinensis* Grenier & Godron). Illustration: Rix & Phillips, The bulb book, 86 (1981). Like *F. latifolia,* but with narrower, glaucous leaves, flowers greyish outside and style 1.2–1.3 cm, branches *c.* 1 mm. *France, Italy (SW Alps).* H3. Spring.

9. F. aurea Schott (*F. bornmulleri* Haussknecht). Illustration: Rix & Phillips, The bulb book, 81 (1981).

Like *F. latifolia* but with usually smaller, yellow flowers. Perianth-segments 2–4.8 cm, with diamond-shaped nectaries *c.* 2 mm across. *Turkey.* H1. Spring.

F. collina Adams (*F. lutea* Bieberstein). Illustration: Rix & Phillips, The bulb book, 85 (1981). Leaves narrower, flowers with acute perianth-segments, the inner fringed on the margin, nectaries deeply indented.

Former USSR (S Russia, Georgia). H1. Spring.

F. tubiformis Grenier & Godron subsp. **moggridgei** (Boissier & Reuter) Rix. Flowers greenish yellow, perianth-segments rounded with linear-lanceolate nectaries 4–5 mm long, style 1.3–1.7 cm. *France, Italy (SW Alps).* H3. Spring.

These are not merely semi-albinos of *F. latifolia* and *F. tubiformis* (see also under 8, *F. latifolia*); they occur in separate populations and differ in minor characters. Pure white albinos of *F. tubiformis* are recorded, though very rare.

10. F. pallidiflora Schrenk. Illustration: *Journal of the Royal Horticultural Society* **85**: f. 127 (1960); Brickell (ed.), RHS A–Z encyclopedia of garden plants, 458 (2003).

Bulb spherical to spindle-shaped, to 5 cm in diameter. Stem 10–80 cm. Leaves broadly lanceolate, opposite or alternate, glaucous. Flowers 1–6 (rarely to 12), very broadly bell-shaped, pale greenish yellow, faintly tessellated with reddish brown. Perianth-segments 2.5–4.5 cm, rounded, each with an ovate, deeply indented nectary *c.* 2 mm long at the angle of the bell. Style 1.4–1.7 cm, smooth, 3-fid, the branches 2–4 mm. Capsule with 6 wide wings. *Former USSR (E Siberia), NW China.* H2. Spring.

11. F. thunbergii Miquel (*F. verticillata* var. *thunbergii* (Miquel) Baker). Illustration: Rix & Phillips, The bulb book, 87, 91 (1981); Brickell (ed.), RHS A–Z encyclopedia of garden plants, 458 (2003).

Bulb spindle-shaped, to 4 cm in diameter. Stem 30–80 cm. Leaves many, linear, the lowest opposite, the rest in whorls or alternate; bracts coiled and tendril-like. Flowers 2–6 (rarely 1), broadly bell- or cup-shaped, whitish, faintly tessellated or veined with green. Perianth-segments 2.3–3.5 cm, obtuse, each with a linear-lanceolate nectary, 3–4 mm long, *c.* 2 mm above the base. Style *c.* 10 mm, smooth, 3-fid, the branches 2 mm. Capsule winged. *?C China; widely cultivated in China and Japan.* H1. Spring.

F. verticillata Willdenow. Illustration: *Botanical Magazine*, 3083 (1831) – as *F. leucantha*; *Hooker's Icones Plantarum* **39**: t. 3827 (1980). With fewer, larger flowers, which are usually unmarked, and fewer leaves. Nectaries deeply indented. *NW China, Former USSR (E Siberia).* H1. Spring.

F. walujewii Regel. Illustration: Rix & Phillips, The bulb book, 87 (1981). Flowers reddish inside, reddish or white outside, nectary larger, lanceolate. *Former USSR (Soviet Tadzhikistan), NW China.* H1. Spring.

12. F. roylei J.D. Hooker. Illustration: *Hooker's Icones Plantarum*, t. 860 (1852).

Bulb to 3 cm in diameter, spindle-shaped. Stem 20–60 cm. Leaves many, lanceolate, in 6 or 7 whorls of 4 or 5. Flowers 1–4, broadly bell-shaped, yellowish green, spotted and streaked with dull purple. Perianth-segments 3.5–5 cm, rounded, each with a large, ovate nectary at the angle of the bell. Style *c.* 1.2 cm, 3-fid, the branches *c.* 5 mm. Capsule narrowly winged. *N India (Kashmir, Punjab).* H1. Late spring.

13. F. cirrhosa D. Don. Illustration: *Botanical Magazine*, n.s., 255 (1955); Brickell (ed.), RHS A–Z encyclopedia of garden plants, 457 (2003).

Like *F. roylei* but with fewer, usually linear leaves in 3–5 whorls of 2 or 3; the upper 3 bracts tendril-like. Flowers purplish, greenish or whitish, variously marked. Perianth-segments acute, nectary at the angle of the bell, smaller than in *F. roylei.* Capsule winged. *Himalaya, China.* H1. Late spring.

14. F. pyrenaica Linnaeus (*F. nigra* Miller). Illustration: *Hooker's Icones Plantarum* **39**: t. 3817 (1980); Rix & Phillips, The bulb book, 91, 93 (1981).

Bulbs spherical, to 3 cm in diameter. Stem 15–30 cm. Leaves 7–10, lanceolate to linear-lanceolate, alternate. Flowers solitary (rarely 2 together), broadly bell-shaped, outside blackish purple to brownish, heavily tessellated; inside yellowish green, the basal half tessellated with brown (sometimes pure yellow—var. *lutescens* Baker). Perianth-segments 2.5–3.5 cm, the inner obtuse, recurved at the apex, each with a triangular to ovate-lanceolate nectary at the angle of the bell. Style 8–9 mm, smooth, 3-fid, the branches 2–4 mm. Capsule not winged. *SC France, NW Spain.* H1. Spring.

15. F. involucrata Allioni. Illustration: *Hooker's Icones Plantarum* **39**: t. 3807 (1980); Rix & Phillips, The bulb book, 92 (1981).

Bulbs spherical, to 3 cm in diameter. Stem 15–25 cm. Leaves 7–10, linear-lanceolate to linear, opposite or almost so; bracts in a whorl of 3. Flower solitary, broadly bell-shaped, green, variously tessellated with purplish brown, or not tessellated. Perianth-segments 2.5–4 cm, all obtuse and not recurved at the apex, each with an ovate, blackish nectary at the angle of the bell. Style 1.2–1.5 cm, smooth, 3-fid, the branches 5–7 mm. Capsule not winged. *SE France, NW Italy.* H1. Spring.

16. F. pontica Wahlenberg. Illustration: *Botanical Magazine*, 8865 (1920); Rix & Phillips, The bulb book, 92 (1981).

Bulb spherical, to 3 cm in diameter, often with an antler-like projection when dormant. Stem 15–45 cm. Leaves about 8, lanceolate to linear-lanceolate, opposite or almost so; bracts in a whorl of 3, close to the flower. Flowers solitary or rarely 2 together, broadly bell-shaped, green, marbled with reddish brown or unmarbled, not tessellated. Perianth-segments 2.4–4.5 cm, the inner obtuse, somewhat recurved at the apex, each with a round, blackish nectary at the angle of the bell. Style 1.2–1.5 cm, smooth, 3-fid, the branches 5–7 mm. Capsule with 6 narrow wings on the angles. *S Albania, S Yugoslavia, N Greece, NW Turkey.* H1. Spring.

17. F. messanensis Rafinesque (*F. oranensis* Pomel). Illustration: Rix & Phillips, The bulb book, 91 (1981).

Bulb spherical, to 4 cm in diameter. Stem 15–45 cm. Leaves 7–10, linear, the lowest usually opposite; bracts in a whorl of 3. Flowers solitary or rarely up to 3 together, broadly bell-shaped, green, tessellated with reddish brown. Perianth-segments 2.2–4.2 cm, the inner pointed and recurved at the apex, each with a large, ovate-lanceolate, green nectary near the base. Style *c.* 10 mm, smooth, 3-fid, the branches 5–6 mm. Capsule not winged. *N Africa, Sicily, S Italy, Crete, Greece.* H1. Spring.

Subsp. **gracilis** (Ebel) Rix (*F. gracilis* (Ebel) Ascherson & Graebner). Illustration: Rix & Phillips, The bulb book, 90 (1981). Bracts usually alternate, perianth untessellated, the segments not recurved. *Yugoslavia, Albania.* H1. Spring.

Var. **atlantica** Maire (*F. messanensis* forma *oranensis* misapplied) has very glaucous, shorter leaves. *Morocco (Atlas mountains).* H1. Spring.

18. F. lusitanica Wikström (*F. hispanica* Boissier & Reuter). Illustration: Rix & Phillips, The bulb book, 73 (1981).

Bulb to 2.5 cm in diameter. Stem 10–50 cm. Leaves usually 6–9, linear to narrowly lanceolate, alternate; bracts alternate. Flowers 1–3, broadly bell-shaped, green to brown outside, yellowish inside, marked or tessellated with brown.

Perianth-segments 2–4 cm, the inner usually acute, sometimes recurved at the apex, each with a linear to linear-lanceolate nectary, 1–1.2 cm long, 3–4 mm above the base. Style 8–11 mm, smooth, 3-fid, the branches 3–4 mm. Capsule not winged. *C & S Spain, Portugal.* H2. Spring.

The fritillaries of the Iberian peninsula (except for the north where *F. pyrenaica* is found) do not seem to be satisfactorily classifiable. They range from the coastal variants from Portugal (illustration cited above) to high alpine plants from the Sierra Nevada, which have shorter stems and much longer leaves and flowers, but all have alternate, narrow leaves and long, narrow nectaries.

19. F. acmopetala Boissier. Illustration: *Botanical Magazine*, 9148 (1928); Rix & Phillips, The bulb book, 92 (1981); Brickell (ed.), RHS A–Z encyclopedia of garden plants, 457 (2003).

Bulb to 3 cm in diameter, often with bulblets. Stem 14–45 cm. Leaves 7–11, linear, alternate. Flowers solitary or rarely 3 together, broadly bell-shaped, green, marked but not tessellated with reddish brown. Perianth-segments 2.5–4 cm, the inner pointed, recurved at the apex, each with a large ovate to ovate-lanceolate, green or blackish nectary *c.* 5 mm above the base. Style 8–12 mm, smooth, 3-fid, the branches 3–5 mm. Capsule not winged. *S Turkey, Syria, Lebanon, Cyprus.* H2. Spring.

20. F. gussichae (Degen & Doerfler) Rix (*F. graeca* Boissier & Spruner var. *gussichae* Degen & Doerfler). Illustration: *Hooker's Icones Plantarum* **39**: t. 3822 (1980); Rix & Phillips, The bulb book, 92 (1981).

Bulb to 25 cm in diameter. Stem 20–30 cm. Leaves 5–8, ovate to lanceolate, alternate, stem-clasping, very glaucous. Flowers 1–3 in a long raceme, broadly bell-shaped, green, marked but not tessellated with reddish brown. Perianth-segments *c.* 3 cm, obtuse, not recurved at the apex, each with an ovate, green nectary *c.* 6 mm above the base. Style *c.* 9 mm, smooth, 3-fid, the branches 4 mm. Capsule narrowly winged. *Yugoslavia, N Greece, Bulgaria.* H2. Spring.

F. olivieri Baker. Illustration: *Journal of the Royal Horticultural Society* **88**: f. 70 (1980); *Hooker's Icones Plantarum* **39**: t. 3818 (1980). Leaves usually green, narrowly lanceolate, capsule unwinged. *Iran.* H2. Spring.

21. F. hermonis Fenzl.

Bulb to 2 cm in diameter, often with bulblets or stolons. Stem 8–35 cm. Leaves 5 or 6 (rarely to 9), lanceolate or oblong, alternate. Flowers 1 or 2, broadly bell-shaped, green, faintly tessellated with brown or purple. Perianth-segments 2.5–3.5 cm, the inner obtuse, each with an ovate, green or blackish nectary at the angle of the bell. Style 8–12 mm, papillose, 3-fid for 3– 4 mm. Capsule not winged. *S Turkey to Lebanon.* H1. Spring.

Subsp. **amana** Rix (*F. crassifolia* misapplied). Illustration: *Kew Bulletin* **29**: 649 (1974). This is the commonly cultivated variant, described above. Subsp. *hermonis*, which is found only on Mt. Hermon, is not in cultivation.

22. F. graeca Boissier & Spruner. Illustration: Rix & Phillips, The bulb book, 83, 90 (1981); Brickell (ed.), RHS A–Z encyclopedia of garden plants, 457 (2003).

Bulbs spherical or spindle-shaped, to 2.5 cm in diameter. Stem 6–20 cm. Leaves 5–12, usually 7–10, alternate or the lowest opposite, glaucous; bracts alternate. Flowers 1–3, broadly bell-shaped, green, tessellated, often heavily, with black or brown, usually with a clear, green, central stripe to each segment. Perianth-segments 1.8–2.4 cm, the inner obtuse, each with a lanceolate nectary at the angle of the bell. Style 7–10 mm, smooth, 3-fid, the branches 3–6 mm. Capsule not winged. *S Greece.* H1. Spring.

Subsp. **thessala** (Boissier) Rix. Illustration: *Hooker's Icones Plantarum* **39**: t. 3806 (1980) – as *F. pontica* var. *ionica*; Rix & Phillips, The bulb book, 90 (1981). Generally larger, not glaucous and has the lower leaves opposite, the intermediate leaves opposite or alternate, the bracts in a whorl of 3; the perianth-segments are 2.8–3.8 cm, green, usually lightly tessellated. *S Albania, Yugoslavia, NW Greece.*

Subsp. *thessala* is clearly distinguished from *F. pontica* (p. 84) by its unwinged capsule, tessellated flowers and longer flower-stalks.

23. F. davisii Turrill. Illustration: *Hooker's Icones Plantarum*, t. 3427 (1943); Rix & Phillips, The bulb book, 83 (1981).

Like *F. graeca* but the leaves shining green, the lowest wider, spreading. Flowers without a clear, green stripe on each segment. Style 7–9 mm, 3-fid, branches 3–7 mm. *S Greece.* H3. Spring.

F. rhodocanakis Orphanides. Flowers dark purple, not tessellated, each segment with a yellow apex, style often papillose. *S Greece (Idhra).* H3. Spring.

Intermediates between this and *F. davisii* are found on the Greek mainland close to Idhra.

24. F. crassifolia Boissier & Reuter.

Like *F. graeca* but leaves always alternate, nectary linear, 8–12 × 1–2 mm, forming a raised ridge from the angle of the bell to near the apex of the segment, flowers often yellowish or greenish with smaller brownish tessellations and less distinct green stripes. *SE Turkey.* H1. Spring. Subsp. *crassifolia* is rarely cultivated.

Subsp. **kurdica** (Boissier & Noë) Rix (*F. karadaghensis* Turrill) is described above. Illustration: Rix & Phillips, The bulb book, 82 (1981).

25. F. michailovskyi Fomin. Illustration: *RHS Lily Year Book* **33**: f. 25 (1969); Rix & Phillips, The bulb book, 82 (1981); *Gartenpraxis* **11**: t. 4 (1982).

Bulb to 2.5 cm in diameter. Stem 6–24 cm. Leaves 5–9, lanceolate, alternate or the lowest opposite or almost so. Flowers 1–4, more or less in an umbel, broadly bell-shaped, purplish brown or greenish outside, the apical one-quarter or one-third bright yellow. Perianth-segments 2–3.2 cm, the inner obtuse, each with a linear nectary from the angle of the bell to near the apex of the segment. Style 7–9 mm, smooth, 3-fid, the branches 2–3 mm. Capsule not winged, the dried perianth remaining attached to it. *NE Turkey.* H1. Spring.

Hybridises very easily with *F. crassifolia* subsp. *kurdica;* the hybrids have the flowers shading to greenish yellow at the apex and not with a clearly defined zone of yellow.

26. F. reuteri Boissier. Illustration: *Journal of the Royal Horticultural Society* **94**: f. 61–2 (1969); *Botanical Magazine*, n.s., 658 (1973).

Like *F. michailovskyi* in flower colour, but usually taller (15–25 cm), with up to 8 cup-shaped flowers, with perianth-segments 1.5–2.5 cm; style minutely papillose. *C Iran.* H1. Spring.

27. F. montana Hoppe (*F. racemosa* Ker Gawler; *F. orientalis* misapplied; *F. tenella* misapplied; *F. nigra* misapplied). Illustration: *Hooker's Icones Plantarum* **39**: t. 3803 (1980); Rix & Phillips, The bulb book, 86 (1981).

Bulb to 2.5 cm in diameter. Stem 16–40 cm. Leaves 8–20, linear, the lowest usually opposite, the rest alternate or in whorls of 3; bracts not coiled at the tips. Flowers 1–3, broadly bell-shaped, green, heavily tessellated with blackish purple or brown. Perianth-segments 1.8–3.2 cm, not recurved at the apex, each with a linear nectary 1–1.5 cm long, borne *c.* 5 mm above the base. Style 8–10 mm, papillose, 3-fid, the branches 2–7 mm. Capsule not winged. *S France, Italy, Yugoslavia, N Greece.* H1. Spring.

F. ruthenica Wikström. Illustration: *Hooker's Icones Plantarum* **39**: t. 3829 (1980). Bracts coiled; capsule winged. *Former USSR (European Russia, W Siberia).* H1. Spring.

F. orientalis Adam (*F. tenella* Bieberstein). Very like *F. montana* in leaf arrangement, but similar to *F. meleagris* (p. 83) in flower. *Former USSR (Caucasus).*

28. F. elwesii Boissier. Illustration: *Botanical Magazine*, 6321 (1877) – as *F. acmopetala*; *Hooker's Icones Plantarum* **39**: t. 3820 (1980).

Bulb to 3 cm in diameter. Stem 15–55 cm. Leaves 4–8, linear, alternate. Flowers 1–4, narrowly bell-shaped, brownish purple with clear, green stripes. Perianth-segments 2–3 cm, the inner obtuse or abruptly pointed, each with a small (2–3 mm), lanceolate nectary at the base. Style 7–11 mm, very papillose, stout, 3-fid, the branches 1–3.5 mm. Capsule not winged. *S Turkey.* H2. Spring.

F. latakiensis Rix. Style slender, smooth or minutely papillose. *S Turkey, Syria.*

Both *F. latakiensis* and *F. elwesii* are variable, and probably originated by hybridisation between *F. acmopetala* (p. 85) and *F. assyriaca.*

F. assyriaca Baker. Illustration: Rix & Phillips, The bulb book, 78 (1981). Plant much shorter at flowering, stem elongating in fruit, flowers 1.2–2.5 cm, greyish outside and not clearly striped, the style always undivided. *C Turkey.* H2.

29. F. tuntasia Heldreich. Illustration: Rix & Phillips, The bulb book, 86 (1981).

Bulb to 2.5 cm in diameter. Stem 10–35 cm. Leaves 8–25, narrowly lanceolate, alternate. Flowers 1–6, conical to bell-shaped, blackish purple and glaucous outside. Perianth-segments 1.5–2.5 cm, each with a linear nectary *c.* 4 mm long at the base. Style *c.* 10 mm, smooth, 3-fid, the branches 1–1.5 mm. Capsule not winged. *Greece (Kithnos, Serifos).* H2. Spring.

F. obliqua Ker Gawler. Plant shorter, with fewer leaves, flowers slightly larger (perianth-segments 2–3 cm). *S Greece (Attica).* H2. Spring.

30. F. conica Boissier. Illustration: Rix & Phillips, The bulb book, 80 (1981).

Bulb to 2 cm in diameter. Stem 7–25 cm. Leaves 5–7, shining green, the lowest broadly lanceolate, more or less opposite, the rest alternate. Flowers 1 or 2, conical or bell-shaped, yellow, not tessellated. Perianth-segments 1.2–2 cm, each with a greenish nectary *c.* 1 mm at the base. Style 7–9 mm, smooth, 3-fid, the branches 2–4 mm. Capsule not winged. *S Greece.* H2. Spring.

31. F. carica Rix (*F. sibthorpiana* misapplied). Illustration: *Botanical Magazine*, n.s., 129 (1950); Rix & Phillips, The bulb book, 80 (1981).

Bulb to 3 cm in diameter. Stem 5–15 cm. Leaves 6–8, lanceolate, all alternate, glaucous. Flowers 1–3, narrowly bell-shaped, yellow. Perianth-segments 1.3–2 cm, obtuse, each with a small (2–4 mm), blackish, narrowly ovate nectary at the base. Style 6–10 mm, papillose, stout, entire or 3-lobed at the apex, the lobes to 1 mm. Capsule not winged. *SW Turkey.* H2. Spring.

F. sibthorpiana (Smith) Baker. Leaves 2 or 3, ovate-lanceolate. *Greece (Symi), SW Turkey.* H2. Spring.

F. forbesii Baker. Leaves 5–10, linear or linear-lanceolate. *SW Turkey.* H2. Spring.

32. F. pinardii Boissier (*F. syriaca* Hayek & Siehe). Illustration: *Botanical Magazine*, n.s., 227 (1954); Rix & Phillips, The bulb book, 80, 81 (1981).

Bulb to 3 cm in diameter, often with bulblets. Stem 6–20 cm. Leaves 3–8, lanceolate, alternate, glaucous. Flowers 1 or 2 (rarely to 4 together), narrowly bell-shaped, greyish or purplish with a glaucous bloom outside, orange, yellow or greenish inside, not tessellated. Perianth-segments 1.5–2.5 cm, the inner obtuse, each with a small (3–5 mm), linear-lanceolate nectary at the base. Style 7–10 mm, papillose, slender, less than 3 mm in diameter, undivided or 3-fid at the apex, the branches 1–2 mm. Capsule not winged. *Turkey.* H1. Spring.

F. armena Boissier. Flowers dark purplish brown inside and outside. *NE Turkey.*

F. caucasica Adams. Flowers dark purplish brown inside and outside, style 9–17 mm, smooth, undivided. *NE Turkey, adjacent Caucasus.*

F. minuta Boissier & Noë (*F. carduchorum* Rix). Flowers brownish orange, style smooth, deeply 3-fid. *SE Turkey.*

F. drenovskii Degen & Stojanoff. Illustration: Rix & Phillips, The bulb book, 92 (1981). Flowers purple-brown or greenish inside, sometimes greyish outside, leaves scattered, narrowly lanceolate. *NE Greece, SW Bulgaria.*

33. F. uva-vulpis Rix (*F. assyriaca* misapplied). Illustration: *Iranian Journal of Botany* **1**: t. 5 (1977); *Hooker's Icones Plantarum* **39**: t. 3831 (1980); Rix & Phillips, The bulb book, 80 (1981).

Bulb to 3 cm in diameter, usually with a few large bulblets. Stem 10–35 cm. Leaves 3–5, lanceolate, alternate, shining green. Flowers solitary (rarely 2 together), narrowly bell-shaped, rounded, outside purplish grey, glaucous, inside yellowish. Perianth-segments 2–2.8 cm, the inner obtuse, each with a large, ovate nectary at the base. Style 5–7 mm, papillose, stout (3–4 mm in diameter), undivided. Capsule not winged. *SE Turkey, N Iraq, W Iran.* H1. Spring.

34. F. alfredae Post.

Bulb to 3 cm in diameter. Stem 10–36 cm. Leaves 9–11, the lowest oblanceolate to ovate, often opposite; bracts usually in a whorl of 3. Flowers 1–3, narrowly bell-shaped, green, glaucous outside, yellowish green inside, unmarked. Perianth-segments 1.2–3 cm, the inner spathulate, each with an ovate, green nectary, *c.* 5 × 2 mm, at the base. Style 5–10 mm, papillose, stout (to 3 mm in diameter), undivided or 3-lobed. Capsule broadly 6-winged. *S Turkey.* H2. Spring.

Subsp. **glaucoviridis** (Turrill) Rix (Illustration: Botanical Magazine, 9462, 1936; *Hooker's Icones Plantarum* **39**: t. 3830, 1980) is described above; subsp. *alfredae* from the Lebanon is a smaller, more slender plant, and is not cultivated.

35. F. bithynica Baker (*F. dasyphylla* Baker; *F. schliemannii* Sintenis; *F. citrina* Baker; *F. pineticola* Schwarz). Illustration: *Botanical Magazine*, 6321 (1877).

Bulb to 2 cm in diameter. Stem 7–20 cm. Leaves 5–12, usually 6–8, the lowest oblanceolate to ovate, usually opposite; bracts usually in a whorl of 3. Flowers 1 or

2 (rarely to 4), narrowly bell-shaped, outside glaucous or yellowish green, rarely with purple markings, inside greenish yellow. Perianth-segments 1.7–2.7 cm, the inner obovate, cuneate, each with a lanceolate nectary, *c.* 3 × 1 mm, at the base. Style 7–10 mm, smooth, slender, undivided. Capsule usually winged. *W Turkey, Greek Islands (Samos, Khios).* H2. Spring.

F. ehrhartii Boissier & Orphanides. Illustration: Rix & Phillips, The bulb book, 84 (1981). Flowers purple outside, apex yellow; capsule always unwinged. *Greece (Aegean area).* H2. Spring.

36. F. alburyana Rix. Illustration: Rix & Phillips, The bulb book, 83 (1981); *Gartenpraxis*, 1982(1) cover.

Bulb to 2 cm in diameter, often with bulblets. Stem 4–10 cm. Leaves 3 or 4, lanceolate, alternate. Flowers 1 or 2, upright, horizontal or drooping, cup-shaped to almost flat, pale pink, tessellated. Perianth-segments 2–3 cm, each with an elliptic, green nectary, *c.* 1 mm, at the base. Anthers semi-versatile. Style 9–15 mm, slender, smooth, 3-fid, the branches 1–2 mm. Capsule not winged. *NE Turkey.* H1. Spring.

37. F. japonica Miquel.

Bulb to 1 cm in diameter. Leaves 5–7, linear-lanceolate, the lower opposite; bracts in a whorl of 3. Flowers solitary (rarely 2 together), broadly bell-shaped, drooping, whitish, lightly marked and tessellated with brown and purple. Perianth-segments 2–2.5 cm (fringed on the margin in var. **koidzumiana** Nakai), each with a linear nectary, 5–8 mm long, from the angle of the bell to near the apex. Style 8 mm, 3-fid, the branches to 5 mm. Capsule drooping, not winged. *Japan.* H2. Spring.

38. F. camschatcensis (Linnaeus) Ker Gawler. Illustration: *Hooker's Icones Plantarum* **39**: t. 3844 (1980); Rix & Phillips, The bulb book, 159 (1981).

Bulb to 2.5 cm in diameter, of numerous closely-packed scales, with many small bulblets around the base. Stem 15–75 cm. Leaves many, all lanceolate, in 1–3 whorls of 5–7 below, alternate above. Flowers 1–8, broadly bell-shaped or cup-shaped, purple-brown to black, rarely green or yellowish. Perianth-segments 2–3 cm, acute, each with *c.* 12 raised lines inside and a rather inconspicuous, narrowly oblong nectary. Anthers versatile. Style 8–10 mm, smooth, deeply 3-fid, the branches 6–8 mm, recurved. Capsule angled but not winged. *Japan, E Siberia, N America from Alaska to Washington.* H1. Late spring.

39. F. biflora Lindley. Illustration: *Lilies, 1975*, 10 (1975); *Hooker's Icones Plantarum* **39**: t. 3843 (1980).

Bulb to 2 cm in diameter, of about 3, thick, loose scales. Stem 15–40 cm. Leaves mostly basal, ovate-lanceolate, alternate or opposite, shining green. Flowers 1–6 (rarely to 12), bell-shaped, blackish brown to purple, suffused with green. Perianth-segments 2–3.5 cm, each with raised, longitudinal lines inside, and an inconspicuous linear nectary. Anthers versatile. Style *c.* 9 mm, deeply 3-fid, the branches *c.* 5 mm, recurved. Capsule 6-angled, not winged. *USA (California).* H3. Spring.

40. F. liliacea Lindley. Illustration: *Botanical Magazine*, 9541 (1939); *Hooker's Icones Plantarum* **39**: t. 3842 (1980); Rix & Phillips, The bulb book, 94 (1981).

Bulbs to 2.5 cm in diameter, of 4–6 fleshy, ovate, loose scales. Stem 15–35 cm. Leaves 2–20, mostly basal, oblong-lanceolate or obovate, shining green, the lower opposite; bracts alternate. Flowers 2–4 (rarely to 8), openly conical to bell-shaped, white with a greenish or yellow throat. Perianth-segments 1.2–2.5 cm, each with a small, oblong, greenish yellow, sometimes purple-spotted nectary. Style *c.* 8 mm, smooth, 3-fid to more than halfway, the branches 4–5 mm, recurved. Capsule angled but not winged. *USA (California).* H4. Spring.

F. agrestis Greene. Illustration: Rix & Phillips, The bulb book, 94 (1981). Leaves often glaucous, flowers dull yellow-green, usually spotted with purple-brown inside, nectaries conspicuous. *USA (California).* H3. Spring

F. grayana Reichenbach & Baker (*F. roderickii* Knight). Illustration: Rix & Phillips, The bulb book, 93 (1981). Leaves narrower, oblanceolate; flowers greenish to reddish brown, with a cream or green patch at the apex outside; perianth-segments with dark brown raised lines and each with an inconspicuous, linear nectary almost as long as the segment. *USA (California).* H3. Spring.

41. F. pudica (Pursh) Sprengel. Illustration: *Botanical Magazine*, 9617 (1942); *Hooker's Icones Plantarum* **39**: t. 3840 (1980); Rix & Phillips, The bulb book, 94 (1981); Brickell (ed.), RHS A–Z encyclopedia of garden plants, 458 (2003).

Bulb a thick disc of 2–4 scales, surrounded at the base by numerous bulblets. Stem 7–30 cm. Leaves 2–7, linear to narrowly lanceolate, the lowest usually opposite, the rest alternate. Flowers 1 or 2 (rarely to 6), narrowly bell-shaped, yellow to orange-yellow. Perianth-segments 1–2.5 cm, each with a small dark nectary at the base. Style 8 mm, smooth, slender, undivided. Capsule not winged. *Western N America.* H1. Spring.

42. F. pluriflora Torrey. Illustration: *Botanical Magazine*, 7631 (1898); Rix & Phillips, The bulb book, 95 (1981).

Bulb of up to 8 ovate, fleshy scales. Stem 18–40 cm. Leaves 6–13, mostly basal, oblong-lanceolate. Flowers 1–4 (rarely to 12), drooping or facing outwards, pink to pinkish purple, unmarked. Perianth-segments 2–3.5 cm, each with a linear nectary running for two-thirds of the length of the segment. Style 1–1.5 cm, entire or 3-lobed at the apex. Capsule 3-ridged, not winged. *USA (California).* H3. Spring.

F. striata Eastwood. Illustration: Rix & Phillips, The bulb book, 95 (1981). Flowers sweetly scented, pale, perianth-segments recurved at their tips, each with a circular nectary at the base. *W USA.*

43. F. purdyi Eastwood. Illustration: Rix & Phillips, The bulb book, 94 (1981).

Bulb of 3 or 4 ovoid, fleshy scales, 8–15 mm long. Stem 10–40 cm. Leaves mainly basal, forming a rosette, elliptic to oblong-lanceolate, margins usually wavy. Flowers 1–4 (rarely to 14), bell-shaped, drooping, pale greenish or whitish heavily mottled and veined with purplish brown. Perianth-segments *c.* 1.5 cm, each with an oblong nectary, 2 × 10 mm. Style *c.* 1.5 cm, 3-fid, the branches to 1 cm, spreading and recurved. Capsule 6-angled, not winged. *USA (California).* H2. Spring.

44. F. glauca Greene. Illustration: *Hooker's Icones Plantarum* **39**: t. 3840 (1980) – poor specimen.

Bulb 5–10 mm in diameter, of 2 or 3 large scales and a few smaller ones at the base. Stem 12–18 cm. Leaves 2–4, mainly basal, alternate, recurved, glaucous. Flowers solitary (rarely up to 4 together), broadly bell-shaped, drooping, yellow mottled with brown or rarely all brown. Perianth-segments 1.5–2 cm, each with a lanceolate or elliptic nectary 5–6 mm long.

Style to 1.2 cm, 3-fid, the branches 6–9 mm, recurved. Capsule 6-angled, not winged. *USA (California, Oregon).* H2. Spring.

45. F. affinis (Schultes) Sealy (*F. lanceolata* invalid). Illustration: *Hooker's Icones Plantarum* **39**: t. 3847 (1980); Rix & Phillips, The bulb book, 94 (1981).

Bulb to 2 cm in diameter with numerous bulblets around its base. Stem 15–120 cm. Leaves in 1–3 whorls of 3–5, linear-lanceolate to ovate; bracts 1 per flower. Flowers 1–4 (rarely to 12) in a raceme, broadly bell-shaped, drooping, greenish yellow to purple, tessellated. Perianth-segments 2–4 cm, each with a prominent, triangular to ovate-lanceolate nectary 1–2 cm long. Style 3-fid, the branches recurved. Capsule 6-winged. *Western N America.* H2. Spring.

F. atropurpurea Nuttall. Illustration: Rix & Phillips, The bulb book, 95 (1981). Leaves linear, flowers smaller, more open, perianth-segments 1–2 cm, each with an indistinct, circular nectary near the base. *Western N America.* H1. Spring.

46. F. recurva Bentham. Illustration: *Botanical Magazine*, 9353 (1934); Rix, Growing bulbs, pl. 4 (1983).

Bulb 2–3 cm in diameter, a fleshy disc studded with bulblets around the base. Stem 20–90 cm. Leaves in 3–5 whorls of 3–5, linear to linear-lanceolate, often glaucous. Flowers 3–6 (rarely to 12), borne in a raceme, narrowly bell-shaped, orange-scarlet to red, tessellated with yellow. Perianth-segments 2–3.5 cm, recurved at the tip, each with a lanceolate nectary 5–6 mm long near the base. Style to 1.3 cm, 3-fid, the branches to 3 mm, not recurved. Capsule 6-winged. *USA (S Oregon, California).* H2. Spring.

F. gentneri Gilkey. Illustration: Rix & Phillips, The bulb book, 94 (1981). Perianth-segments 3.5–4 cm, not recurved at their tips, style more deeply divided with recurved branches. *USA (S Oregon).* H2. Spring.

10. LILIUM Linnaeus

V.A. Matthews

Rootstock a bulb composed of fleshy overlapping scales, sometimes stoloniferous. Stem erect, unbranched, leafy. Leaves scattered or in whorls, usually linear or lanceolate, sometimes with bulbils in the axils. Flowers in a terminal raceme or umbel, or sometimes solitary, funnel-shaped, cup-shaped, bowl-shaped or bell-shaped, with the perianth-segments usually spreading or reflexed to a greater or lesser extent, or of turk's-cap type (see Figure 4, p. 89). Perianth-segments free, each with a nectar-bearing furrow or gland at the base, papillose or sometimes hairy at the apex. Stamens borne at the base of the perianth-segments, usually free from one another; anthers versatile. Ovary superior; style 1. Fruit a capsule with 3 cells. Seeds many, flat, 2 rows in each cell.

About 100 species native to the temperate northern hemisphere.

Many species produce annual roots from the underground part of the stem that is above the bulb; these species are referred to as 'stem-rooting' in the descriptions, and should be planted more deeply (at *c.* 20 cm) than the others. In general, lilies grow best in a well-drained soil in semi-shade or sun, never in a very hot, dry position. There are exceptions e.g. *L. catesbaei* and *L. grayi*, which prefer damp, but not stagnant conditions. Most species will grow in acid or alkaline soils but again there are exceptions, such as *L. henryi*, which requires lime, and *L. speciosum*, *L. superbum* and *L. auratum*, which will not tolerate it. It is best to consult the literature for the requirements of particular species.

Lilies are especially susceptible to aphid-borne virus diseases. So far there is no cure, and affected plants should be discarded. Propagation is by seed, bulb scales, or, in those species that produce them, stem bulbils.

Literature: Elwes, H.J., A Monograph of the genus Lilium (1877–80); Grove, A. & Cotton, A.D., Supplement to Elwes' monograph of the genus Lilium, parts 1–7 (1933–40); Turrill, W.R., Supplement to Elwes' Monograph of the genus Lilium, parts 8 & 9 (1960–62); Woodcock, H.B.D. & Stearn, W.T., Lilies of the world (1950); Synge, P.M., Lilies (1980); A.C. Leslie, The International Lily Register *(1982)*, and supplements. Matthews, V.A. The International Lily Register *(2007)*.

1a. Basic colour of perianth-segments yellowish green 2
b. Basic colour of perianth-segments not yellowish green 6
2a. Leaves with minutely ciliate margin 3
b. Leaves with hairless margin 4
3a. Perianth-segments 4–6.5 cm, with dark purple spots and lines inside, and purplish warts; flowers of turk's-cap type **23. pyrenaicum**
b. Perianth-segments 6–15 cm, unmarked or variously blotched or suffused with purplish red inside, and lacking warts; flowers funnel-shaped **14. primulinum**
4a. Perianth-segments 2.5–5 cm; flowers of turk's-cap type **41. kelleyanum**
b. Perianth-segments 6–15 cm; flowers funnel-shaped 5
5a. Leaves with 5 or 7 veins; filaments and stigma green; perianth-segments thin with deep longitudinal ribs and transverse corrugations **13. nepalense**
b. Leaves with 1 or 3 veins; filaments and stigma purple or with purple spots; perianth-segments thick, lacking longitudinal ribs and corrugations **14. primulinum**
6a. Flowers of turk's-cap type 7
b. Flowers not of turk's-cap type 56
7a. Basic colour of perianth-segments white, greenish white, pink, crimson or purple 8
b. Basic colour of perianth-segments cream, yellow, orange or red 15
8a. At least some leaves in whorls 9
b. All leaves scattered 10
9a. Perianth-segments 3–5.5 cm **25. martagon**
b. Perianth-segments 6–10 cm, if less then upper leaves scattered **45. michauxii**
10a. Perianth-segments with wavy margins; pollen brown or dark red **37. speciosum**
b. Perianth-segments without wavy margins; pollen orange, yellow or lilac 11
11a. Perianth-segments white or greenish white, sometimes flushed with purple, or turning red or purple with age 12
b. Perianth-segments pink, lilac, purple or crimson-maroon 14
12a. Anthers yellow or orange-red; nectary-furrows usually green, warty **31. duchartrei**
b. Anthers purplish or grey; nectary-furrows purple or brown, not warty 13
13a. Leaves up to 6 mm broad with 1 or 3 veins that are minutely hairy beneath; perianth-segments white with purple spots **32. taliense**
b. Leaves 2–3.5 cm broad with 5 or 7 veins that are hairless beneath; perianth-segments greenish white, often reddish inside towards the base **13. nepalense**

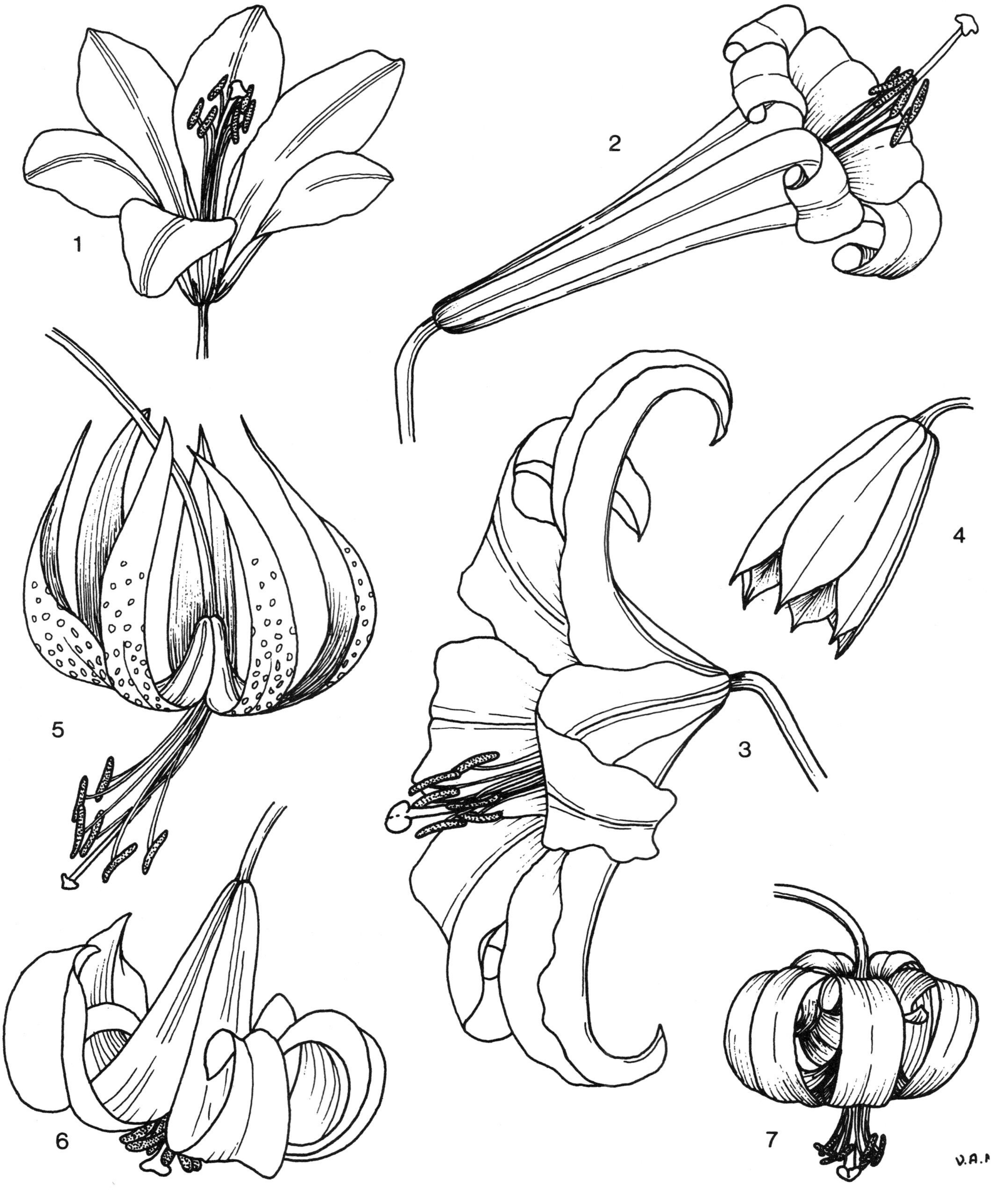

Figure 4. Types of *Lilium* flowers. 1, cup-shaped. 2, narrowly funnel-shaped. 3, bowl-shaped. 4, bell-shaped. 5–7, turk's-cap types.

14a. Nectary-furrows hairless **34. wardii**
b. Nectary-furrows bordered with warts or hairs **33. lankongense**
15a. At least some leaves in whorls 16
b. All leaves scattered 28
16a. Basic colour of perianth-segments yellow 17
b. Basic colour of perianth-segments orange-yellow, orange or red 20
17a. Perianth-segments 2.5–5 cm 18
b. Perianth-segments 6.5–10 cm 19
18a. Leaves oblanceolate; anthers orange **40. columbianum**
b. Leaves lanceolate to broadly elliptic; anthers brown **41. kelleyanum**
19a. Leaves with veins hairless beneath; anthers orange **40. columbianum**
b. Leaves with veins downy beneath; anthers purplish or reddish brown **39. humboldtii**
20a. Perianth-segments with some spots encircled with yellow **42. pardalinum**
b. Perianth-segments with spots not encircled with yellow 21
21a. Stem usually downy; leaves with veins downy beneath **39. humboldtii**
b. Stem hairless; leaves with veins hairless or minutely roughened beneath 22
22a. Perianth-segments 2.5–5.5 cm 23
b. Perianth-segments 6–10 cm 25
23a. Anthers 2–2.2 cm **28. hansonii**
b. Anthers 5–13 mm 24
24a. Anthers orange **40. columbianum**
b. Anthers purplish, brown or reddish brown, if orange then stem roots present **41. kelleyanum**
25a. Anthers 1.6–2 cm, red **44. superbum**
b. Anthers 6–11 mm, orange, purplish brown or red 26
26a. Leaves lanceolate to broadly elliptic **43. michiganense**
b. Leaves oblanceolate to obovate 27
27a. Stem roots present; flowers usually 1 or 2, occasionally up to 5; anthers purplish brown **45. michauxii**
b. Stem roots absent; flowers usually 6–10, occasionally up to 40; anthers orange **40. columbianum**
28a. Basic colour of perianth-segments cream, yellow or orange 29
b. Basic colour of perianth-segments red or orange-red 45
29a. Perianth-segments up to 5.5 cm 30
b. Perianth-segments 6 cm or more 34
30a. Anthers red **23. pyrenaicum**
b. Anthers brown, reddish brown or maroon 31
31a. Margin of leaves hairless **41. kelleyanum**
b. Margin of leaves ciliate or minutely rough 32
32a. Leaves with veins hairless beneath **23. pyrenaicum**
b. Leaves with veins downy beneath 33
33a. Stem hairless; leaves with 3–15 veins **23. pyrenaicum**
b. Stem downy; leaves usually with 3 veins **38. amabile**
34a. Leaves with veins hairless beneath 35
b. Leaves with veins minutely rough or downy beneath 38
35a. Perianth-segments recurved in the upper half to two-thirds **14. primulinum**
b. Perianth-segments strongly recurved 36
36a. Anthers 4–10 mm **23. pyrenaicum**
b. Anthers 1.2–1.7 cm 37
37a. Stem hairless; leaves with many veins, 2–3 cm broad **36. henryi**
b. Stem usually cobwebby-hairy; leaves with 5 or 7 veins, 1–2 cm broad **30. lancifolium**
38a. Basic colour of perianth-segments yellow-orange or orange 39
b. Basic colour of perianth-segments cream or yellow 41
39a. Anthers 1.2–1.7 cm, orange **36. henryi**
b. Anthers 4–10 mm, red 40
40a. Leaves linear; flowers 6–12; perianth-segments with reddish spots **22. × testaceum**
b. Leaves lanceolate; flowers usually 1–6; perianth-segments with brownish purple spots **23. pyrenaicum**
41a. Pollen red or reddish brown 42
b. Pollen yellow, orange or brown 43
42a. Perianth-segments 3–7 cm; anthers 4–10 mm **23. pyrenaicum**
b. Perianth-segments 6.5–12 cm; anthers 1–2.5 cm **29. leichtlinii**
43a. Perianth-segments bearing warts towards the base **36. henryi**
b. Perianth-segments without warts 44
44a. Leaves with 1 or 3 veins **14. primulinum**
b. Leaves with 9, 11 or 13 veins **26. monadelphum**
45a. Bulbils present in the leaf axils **30. lancifolium**
b. Bulbils absent 46
46a. Upper leaves adpressed to stem, lower leaves spreading **21. chalcedonicum**
b. Upper and lower leaves not changing in aspect 47
47a. Stem minutely rough or downy, sometimes only in the lower part, or hairy at the leaf axils 48
b. Stem hairless 51
48a. Perianth-segments 3–3.5 cm **27. pumilum**
b. Perianth-segments 5–8.5 cm 49
49a. Flowers with an unpleasant scent; nectary-furrows warty **38. amabile**
b. Flowers unscented; nectary-furrows hairy 50
50a. Flowers 1–6; stem with sparse hairs at the leaf axils, otherwise hairless **29. leichtlinii**
b. Flowers 5–40; stem downy, especially in the lower part **35. davidii**
51a. Perianth-segments bright red 52
b. Perianth-segments orange-red 55
52a. Leaves 1–3 mm broad 54
b. Leaves 4–17 mm broad 53
53a. Leaves downy on the veins beneath **23. pyrenaicum**
b. Leaves hairless on the veins beneath **24. pomponium**
54a. Perianth-segments 3–3.5 cm **27. pumilum**
b. Perianth-segments 4–7 cm **24. pomponium**
55a. Perianth-segments 3–7 cm; anthers 4–10 mm **23. pyrenaicum**
b. Perianth-segments 6.5–8.5 cm; anthers 1–1.7 cm **29. leichtlinii**
56a. Basic colour of perianth-segments white, greenish white, cream, pink, purple or purplish brown 57
b. Basic colour of perianth-segments yellow, orange, red or crimson 78
57a. At least some leaves in whorls 58
b. All leaves scattered 59
58a. Flowers nodding **51. mackliniae**
b. Flowers erect or borne horizontally **47. rubescens**
59a. Flowers open bowl-shaped, 20–30 cm in diameter; perianth-segments basically white with a yellow or red central band, spotted or not **5. auratum**
b. Flowers funnel-shaped, trumpet-shaped or bell-shaped, if bowl-shaped then less than 20 cm in diameter 60
60a. Perianth-segments up to 8 cm 61
b. Perianth-segments 9–30 cm 64

61a. Perianth-segments white, sometimes flushed with green outside **1. candidum**
b. Perianth-segments pink to purple or brownish purple, if white then flushed with pinkish purple 62
62a. Perianth-segments 6–8 cm **12. rubellum**
b. Perianth-segments 1–5.5 cm 63
63a. Nectary-furrows fringed **52. nanum**
b. Nectary-furrows hairless **51. mackliniae**
64a. Perianth-segments pure white 65
b. Perianth-segments cream, pink or purplish, if white then flushed with green, yellow, brownish purple or pinkish purple at least towards the base 66
65a. Filaments smooth **8. longiflorum**
b. Filaments papillose towards the base **3. formosanum**
66a. Pollen yellow 67
b. Pollen orange, reddish brown or brown 72
67a. Perianth-tube yellow inside, at least at the base 68
b. Perianth-tube not yellow inside 69
68a. Leaves with 1 vein that is minutely rough beneath; perianth-segments 12–15 cm, pinkish purple outside **2. regale**
b. Leaves with 3 or 5 veins that are hairless beneath; perianth-segments 15–30 cm, white or creamy outside, tinged with green or yellow **7. wallichianum**
69a. Perianth-tube greenish at the base inside **7. wallichianum**
b. Perianth-tube not greenish inside 70
70a. Nectary-furrows hairless **8. longiflorum**
b. Nectary-furrows bordered with hairs and/or warts 71
71a. Filaments.smooth **9. philippinense**
b. Filaments papillose towards the base **3. formosanum**
72a. Nectary-furrows hairless 73
b. Nectary-furrows bordered with hairs and/or warts 76
73a. Bulbils present in the leaf-axils **10. sulphureum**
b. Bulbils absent 74
74a. Perianth-segments pink inside and outside **11. japonicum**
b. Perianth-segments flushed with purplish red or brownish purple outside, white inside 75
75a. Leaves of flowering plants with 1 vein; pollen reddish brown to brown **6. leucanthum**
b. Leaves of flowering plants with 3 or 5 veins; pollen orange **8. longiflorum**
76a. Bulbils present in the leaf axils **5. sargentiae**
b. Bulbils absent 77
77a. Leaves 5–10 mm broad, margin recurved **3. formosanum**
b. Leaves 1–2.5 cm broad, margin not recurved **4. brownii**
78a. Flowers bell-shaped, funnel-shaped or trumpet-shaped, usually drooping or borne horizontally 79
b. Flowers cup-shaped or star-shaped, erect 83
79a. Nectary-furrows fringed **52. nanum**
b. Nectary-furrows hairless or with short hairs 80
80a. Veins of leaves minutely rough beneath **50. canadense**
b. Veins of leaves hairless beneath 81
81a. Anthers 3–4 mm **46. parvum**
b. Anthers 5–9 mm 82
82a. Leaves in whorls; perianth-segments 3–4.5 cm; pollen deep yellow **48. bolanderi**
b. Leaves usually scattered, sometimes 1 or more whorls present towards base of stem; perianth-segments 7–10 cm; pollen brownish red **49. parryi**
83a. At least some leaves in whorls **19. philadelphicum**
b. Leaves scattered 84
84a. Perianth-segments 3–4 cm; flowers star-shaped **20. concolor**
b. Perianth-segments 5–12 cm; flowers cup-shaped 85
85a. Stem hairless to sparsely hairy or rough **18. maculatum**
b. Stem woolly or cobwebby-hairy at least towards the top 86
86a. Perianth-segments with yellow towards the base, or completely yellow with black spots **17. dauricum**
b. Perianth-segments lacking yellow **16. bulbiferum**

1. L. candidum Linnaeus. Illustration: Elwes, Monograph of the genus Lilium, pl. 9 (1877); Hay & Synge, Dictionary of garden plants in colour, 101 (1969); Perry, Flowers of the world, 167 (1972); Grey-Wilson & Mathew, Bulbs, pl. 3 (1981).

Stem roots absent. Stem 90–200 cm, dark reddish purple, hairless. Basal leaves up to 22 × 5 cm with 3 or 5 veins, hairless. Stem leaves scattered, up to 7.5 × 1 cm, lanceolate, hairless. Flower-stalks erect or spreading. Flowers 5–20, broadly funnel-shaped, fragrant. Perianth-segments 5–8 × 1–4 cm, white, yellow at the base inside, obovate, recurved slightly at the tips. Anthers 5–18 mm; pollen yellow. *W & S Greece, S Yugoslavia, SW Turkey, Lebanon, Syria, Israel* H2. Summer.

L. candidum differs from almost all other species of *Lilium* in producing over-wintering basal leaves in the autumn. Fruits are not produced in cultivation.

Var. **salonikae** Stoker differs in being usually fertile, having a green stem and more widely spreading perianth-segments. *W Greece.* H2. Summer.

Var. **plenum** Weston has double flowers and is now extremely rare.

L. bakerianum Collett & Hemsley. Illustration: Grove & Cotton, Supplement to Elwes' monograph of the genus Lilium, pl. 7 (1935); Woodcock & Stearn, Lilies of the world, f. 26 (1950). Stem roots present. Stem usually shortly hairy or minutely rough below. Leaves linear to lanceolate with 1 or 3 veins, minutely rough on the upper surface, margin and veins beneath. Flowers 1–6, bell-shaped, nodding. Perianth-segments white, flushed with green outside and with reddish brown spots inside. *Nepal, N Burma, W China.* H3. Summer.

2. L. regale Wilson. Illustration: Woodcock & Stearn, Lilies of the world, f. 99 (1950); *Journal of the Royal Horticultural Society* **95**: pl. 264 (1970); Perry, Flowers of the world, 168 (1972); Synge, Lilies, pl. xv (1980).

Stem roots present. Stem 50–200 cm, grey-green, spotted and flushed with reddish purple, with tiny rough hairs, leafless at the base. Leaves scattered, 5–13 cm × 4–6 mm, linear with 1 vein, minutely rough on margin and vein beneath. Flower-stalks spreading or somewhat ascending, 1.8–13 cm. Flowers sometimes solitary, usually up to 25 or more, funnel-shaped, fragrant. Perianth-segments 12–15 cm, outer *c.* 2 cm wide, inner up to 3.7 cm wide, pinkish purple outside, white inside, recurved towards the tips; perianth-tube flushed with yellow inside. Anthers and pollen yellow. *W China (W Szechuan).* H2. Summer.

L. × imperiale Wilson (illustration: Hay & Beckett (eds), Readers' Digest encyclopaedia of garden plants and flowers, 402, 1971) is a hybrid between *L. regale* and *L. sargentiae.* Stems up to 120 cm, grey-green. Leaves hairless, with 1, 3 or 5 veins, the upper leaves with bulbils in their axils. Flowers similar to *L. regale* but the tips of the perianth-segments are broader and more recurved. Anthers orange-brown.

3. L. formosanum Wallace. Illustration: *Botanical Magazine*, 9205 (1930); Woodcock & Stearn, Lilies of the world, f. 47 (1950); Leeburn, Garden lilies, pl. 20 (1963); Synge, Lilies, pl. 9 (1980).

Stem roots present. Stem 30–150 cm, purplish brown especially towards the base, usually somewhat rough-hairy. Leaves scattered, often crowded near the base of the stem, 7.5–20 cm × 5–10 mm, linear to narrowly oblong with 3, 5 or 7 veins, hairless, margin recurved. Flower-stalks erect or ascending, 5–15 cm. Flowers usually 1 or 2, sometimes up to 10, narrowly funnel-shaped, fragrant, borne horizontally. Perianth-segments 12–20 cm, outer 2.5–3 cm wide, inner up to 5 cm wide, flushed with reddish purple outside, rarely white, white inside, recurved at tips. Nectary-furrows green, bordered with hairs and warts. Filaments papillose towards the base; anthers *c.* 9 mm, yellow to purplish; pollen brown or yellow. *Taiwan.* H5. Summer–autumn.

Var. **pricei** Stoker. Illustration: Woodcock & Stearn, Lilies of the world, f. 46 (1950). An earlier-flowering, shorter plant (stems usually under 60 cm) with solitary flowers that are more deeply flushed with reddish purple. It is generally hardier than var. *formosanum. Taiwan.* H3. Summer.

4. L. brownii Miellez. Illustration: *RHS Lily Year Book* **27**: f. 5 (1963); Leeburn, Garden lilies, pl. 17 (1963); Synge, Lilies, pl. vii (1980).

Stem roots present. Stem up to 1 m, flushed with dark purplish brown, hairless. Leaves scattered, 17–30 × 1–2.5 cm, lanceolate, with 5 or 7 veins, hairless. Flowers 1–4, narrowly funnel-shaped, slightly fragrant, borne horizontally or slightly drooping. Perianth-segments 12–15 × 2.5–6.5 cm, pinkish purple outside, sometimes tinged with green, white inside, recurved at the tips. Nectary-furrows bordered with hairs and warts. Anthers 1–1.8 cm, brown with reddish brown pollen. *China, Upper Burma.* H4. Summer.

Var. **australe** (Stapf) Stearn is taller (to 3 m), with narrower leaves, and flowers that are white or slightly flushed with green outside. *SE China, Hong Kong.*

Var. **viridulum** Baker (*L. brownii* var. *colchesteri* (van Houtte) Elwes) usually has a green stem and the flowers tinged outside with yellowish green and sometimes with pale pinkish purple, and yellow inside the tube fading to white with age. The flowers are strongly scented. *China.*

5. L. sargentiae Wilson. Illustration: Woodcock & Stearn, Lilies of the world, f. 104 (1950); Synge, Lilies, pl. xvii (1980).

Stem roots present. Stems up to 1.5 m, grey-green, spotted and tinged with pinkish purple, with tiny, sparse, rough hairs, leafless at the base. Leaves scattered, 10–20 cm × 5–20 mm, narrowly oblong with 3, 5 or 7 veins, hairless, bearing bulbils in their axils. Flower-stalks spreading or somewhat ascending, 3–10 cm. Flowers 2–5, funnel-shaped, fragrant, borne horizontally or slightly drooping. Perianth-segments 12–15 cm, outer 1.5–2 cm wide, inner 3.5–5 cm wide, pinkish or brownish purple or sometimes greenish outside, white inside, recurved at the tips; perianth-tube yellow inside. Nectary-furrows with sparse warts on the margins. Filaments somewhat warty-pubescent in lower one-third; anthers purplish with brown pollen. *W China (W Szechuan).* H4. Summer.

L. alexandrae may key out here (see under 10. *L. sulphureum*). It differs in having the flowers borne horizontally to almost erect, and has white perianth-segments flushed with green towards the base and sometimes tinged pinkish.

6. L. leucanthum (Baker) Baker var. centifolium (Stapf) Stearn (*L. centifolium* Stapf). Illustration: *Botanical Magazine*, 8960 (1923); Woodcock & Stearn, Lilies of the world, f. 72 (1950); *RHS Lily Year Book* **30**: f. 2 (1966); Synge, Lilies, pl. xii (1980).

Stem roots present. Stem 2–3 m, glaucous, densely leafy. Leaves scattered, 15–20 cm, linear to lanceolate with 1 vein (in non-flowering plants the leaves are shorter and have 3 veins), hairless. Flower-stalks horizontal to ascending, 6–12 cm. Flowers up to 18, funnel-shaped, fragrant, spreading or slightly drooping. Perianth-segments strongly recurved at the tips, outer 15–18 × *c.* 3 cm, inner 14–17 × 4.5–6 cm, flushed with purplish red outside and greenish towards the base, white inside; perianth-tube yellow inside. Nectary-furrows green, hairless. Filaments warty and downy towards the base; anthers 1.8–2 cm, reddish brown with reddish brown or dark brown pollen. *W China (W Kansu).* H4. Summer.

7. L. wallichianum Schultes & Schultes. Illustration: *Botanical Magazine*, 4561 (1851); Woodcock & Stearn, Lilies of the world, f. 117 (1950); Synge, Lilies, pl. 16 (1980).

Bulb stoloniferous. Stem roots present. Stem up to 2 m, green or tinged with purple, hairless. Leaves scattered, up to 25 × 1.2 cm, linear or lanceolate with 3 or 5 veins, hairless. Flowers 1–4, funnel-shaped, fragrant, borne horizontally. Perianth-segments 15–30 cm, white or creamy with a greenish or yellowish tinge outside towards the base, and yellow or greenish at the base inside, recurved at the tips. Stamens shorter than style; pollen yellow. *Himalaya.* H5. Summer–autumn.

Var. **neilgherrense** (Wight) Kara (*L. neilgherrense* Wight). Illustration: Elwes, Monograph of the genus Lilium, pl. 6 (1877); Woodcock & Stearn, Lilies of the world, f. 82 (1950). Usually a shorter plant (*c.* 90 cm) with broadly lanceolate leaves up to 12 × 1.5–3 cm. *S India.* H5. Summer–autumn.

8. L. longiflorum Thunberg. Illustration: Hay & Beckett (eds), Readers' Digest encyclopaedia of garden plants and flowers, 406 (1971); Brickell (ed.), RHS A–Z encyclopedia of garden plants, 631 (2003).

Stem roots present. Stem 30–100 cm, green, hairless. Leaves scattered, 10–20 cm × 5–15 mm, lanceolate to oblong-lanceolate with 3 or 5 veins. Flower-stalks spreading, up to 12 cm. Flowers 1–6, funnel-shaped, fragrant, borne horizontally. Perianth-segments 13–20 × *c.* 3.5 cm, white, slightly recurved at the tips. Nectary-furrows green, hairless. Anthers *c.* 9 mm, pollen yellow. *S Japan, Taiwan.* H4. Summer.

Var. **eximium** (Courtois) Baker. Flowers with a narrower tube and perianth-segments more strongly recurved.

Var. **takeshima** Duchartre. Stems tall, brownish purple. Flowers flushed with brownish purple outside. Pollen orange.

9. L. philippinense Baker. Illustration: *Botanical Magazine*, 6250 (1876); Elwes, Monograph of the genus Lilium, pl. 3 (1877).

Bulb stoloniferous. Stem roots present. Stem usually 30–45 cm, occasionally to

100 cm, green sometimes marked with purple towards the base, hairless. Leaves scattered, up to 15 cm × 5 mm, linear with 3 veins, hairless. Flowers usually 1 or 2, sometimes up to 6, trumpet-shaped, fragrant, borne horizontally. Perianth-segments 18–25 × 4–5 cm, white with a slight green flush and sometimes with purple streaking outside towards the base, spreading at the tips. Nectary-furrows with downy margins. Anthers and pollen yellow. *Philippine Islands.* H4. Summer.

L. nobilissimum (Makino) Makino. Illustration: Woodcock & Stearn, Lilies of the world, f. 87 (1950); *RHS Lily Year Book,* **30**: f. 27 (1966). Stem to 170 cm. Leaves with short stalks and bearing pale green bulbils in their axils. Flowers erect. Perianth-segments up to 15 cm. Anthers brown. *S Japan.* H5. Summer.

10. L. sulphureum Baker (*L. myriophyllum* Franchet). Illustration: *Botanical Magazine,* 8102 (1906); Grove & Cotton, Supplement to Elwes' monograph of the genus Lilium, pl. 21 (1939); Woodcock & Stearn, Lilies of the world, f. 82 (1950).

Stem roots present. Stem 1.5–2 m, green, marked with purple, minutely rough especially below, slightly ribbed. Leaves scattered, to 20 × 1.2–2 cm, shorter in lower part of stem, narrowly lanceolate with 3, 5 or 7 veins, hairless, bearing brown bulbils in their axils. Flower-stalks spreading or drooping, up to 20 cm. Flowers up to 15, trumpet-shaped, strongly fragrant, spreading or drooping. Perianth-segments 14–25 cm, outer up to 4 cm wide, inner up to 5.7 cm wide, cream to greenish white, flushed with pink outside, inside deep yellow in the tube, recurved at the tips. Nectary-furrows hairless. Anthers 1.8–2 cm, purplish brown with orange or brownish pollen. *W China (Yunnan), N Burma.* H4. Late summer.

L. alexandrae (Wallace) Coutts. Illustration: Grove & Cotton, Supplement to Elwes' monograph of the genus Lilium, pl. 15 (1936); *RHS Lily Year Book* **30**: f. 26 (1966); Synge, Lilies, pl. 4 (1980). Stem only to 1 m. Leaves lanceolate to narrowly ovate, shortly stalked, up to 4 cm wide. Flowers 1–5, bowl-shaped, horizontal or almost erect. Perianth-segments up to 17 cm, outer *c.* 3 cm wide, inner up to 4.5 cm wide, white flushed with green at the base inside and outside, sometimes tinged with pink. Pollen reddish brown. *Japan.* H5. Summer.

11. L. japonicum Houttuyn. Illustration: Woodcock & Stearn, Lilies of the world, f. 66 (1950); Hay & Synge, Dictionary of garden plants in colour, 102 (1969); Synge, Lilies, pl. x (1980).

Stem roots present. Stem 45–100 cm, hairless. Leaves scattered, 5–20 cm × 5–25 mm, narrowly lanceolate with 3 or 5 veins, hairless except for the minutely rough margin. Flowers 1–5, broadly funnel-shaped, fragrant, borne horizontally. Perianth-segments 9–17 × 2–3.5 cm, pink, oblanceolate to oblong. Nectary-furrows greenish, hairless. Anthers brown with orange-brown or reddish pollen. *Japan.* H4. Summer.

Var. **platyfolium** Anon. has wider leaves and is rather more vigorous.

12. L. rubellum Baker. Illustration: Woodcock & Stearn, Lilies of the world, f. 102 (1950); Hay & Synge, Dictionary of garden plants in colour, 103 (1969); Hay & Beckett (eds), Readers' Digest encyclopaedia of garden plants and flowers, 408 (1971); *Journal of the Royal Horticultural Society* **100**: f. 51 (1975).

Stem roots present. Stem 30–80 cm, green, marked with brownish purple, hairless. Leaves scattered, *c.* 10 × 4–5 cm, each with a short stalk, narrowly elliptic to narrowly ovate with 5 or 7 veins, hairless on veins beneath, margin minutely rough. Flowers up to 9, usually less, broadly funnel-shaped, fragrant, borne horizontally. Perianth-segments 6–8 × 1.7–2.5 cm, pink, unspotted or with maroon dots at the base, oblanceolate to oblong. Anthers yellow. *Japan.* H4. Summer.

13. L. nepalense Don. Illustration: Woodcock & Stearn, Lilies of the world, f. 84 (1950); Everard & Morley, Wild flowers of the world, 107 (1970); *Journal of the Royal Horticultural Society* **103**: 311 (1978); Synge, Lilies, pl. xiv (1980).

Bulb stoloniferous. Stem roots present. Stem 70–100 cm, pale green, hairless. Leaves scattered, 9–15 × 2–3.5 cm, lanceolate to oblong-lanceolate with 5 or 7 veins, hairless. Flower-stalks spreading to ascending up to 10 cm. Flowers 1–3, funnel-shaped, sometimes unpleasantly scented, more or less drooping. Perianth-segments up to 15 × 4 cm, greenish white to pale greenish yellow, reddish inside towards the base, deeply ribbed and with transverse corrugations outside, spreading in the upper two-thirds and recurved at the tips. Nectary-furrows brown or purple, hairless. Anthers purplish with yellow or yellow-brown pollen. *Bhutan, Nepal, N India (Kumaon).* H4. Summer.

Var. **concolor** Cotton. Illustration: Woodcock & Stearn, Lilies of the world, f. 85 (1950). Flowers entirely greenish yellow, and lacking the reddish throat. *Bhutan, ?Assam.*

14. L. primulinum Baker.

Bulb stoloniferous or not. Stem roots present. Stem to 2.4 m, brown, or green spotted with brown, hairless or minutely rough below. Leaves scattered, to 15 × 4 cm, lanceolate with 1 or 3 veins, margin hairless or minutely ciliate. Flower-stalks ascending or horizontal, decurved below flower, up to 15 cm. Flowers usually 2–8, occasionally to 18, funnel-shaped, usually fragrant, drooping. Perianth-segments 6–15 × 1.3–4.5 cm, yellow or greenish yellow, unmarked or variously blotched or suffused with deep purplish red inside, oblong-lanceolate, varying in the degree to which they are recurved. Nectary-furrows green, hairless. Pollen orange-brown. *W China (Yunnan), N Burma, Thailand.* H4. Summer–autumn.

Var. **primulinum** Illustration: *Botanical Magazine,* 7227 (1892). Flowers entirely yellow or greenish yellow with the perianth-segments recurved for about half their length. *N Burma.*

Var. **burmanicum** (Smith) Stearn. Illustration: Grove & Cotton, Supplement to Elwes' monograph of the genus Lilium, pl. 19 (1937); Woodcock & Stearn, Lilies of the world, f. 95 (1950). Stem 1.2–2.4 m, leaves to 4 cm wide, flowers purplish red in the throat with perianth-segments 6.5–15 cm, recurved in the upper third. *N Burma, Thailand.*

Var. **ochraceum** (Franchet) Stearn. Illustration: *RHS Lily Year Book* **30**: f. 12 (1966). Stem to 1.2 m, leaves to 2 cm wide, flowers purplish red in the throat with perianth-segments *c.* 6 cm, recurved in the upper two-thirds. *W China (Yunnan).*

15. L. auratum Lindley. Illustration: Elwes, Monograph of the genus Lilium, pl. 15 (1878); Perry, Flowers of the world, 169 (1972); Witham Fogg, Bulbs, 58 (1980); Rix & Phillips, The bulb book, 165 (1981).

Stem roots present. Stem 60–150 cm, occasionally to 225 cm, purplish green, hairless. Leaves scattered, 10–22 × 1.5–4 cm, narrowly or broadly lanceolate with a short stalk and 5 or 7 veins, hairless. Flowers usually up to 10, sometimes to 30, open bowl-shaped,

20–30 cm in diameter, fragrant, borne horizontally or slightly drooping. Perianth-segments 12–18 × 4–5 cm, white with a yellow or crimson central stripe, and usually yellow or crimson spots, slightly recurved at the tips and with scattered warts towards the base. Pollen brownish red. *Japan.* H4. Summer–autumn.

The following variants are in cultivation:

Var. **platyphyllum** Baker. Illustration: de Graaff & Hyams, Lilies, pl. 36 (1967); Synge, Lilies, pl. vi (1980). The leaves are especially broad and the large flowers have fewer spots than var. *auratum* and a central yellow band to each segment.

Var. **virginale** Duchartre. Illustration: de Graaff & Hyams, Lilies, pl. 37 (1967). Flowers white with yellow spots and bands.

Several cultivars have been selected and are commonly offered by nurserymen. *L. auratum* crossed with *L. speciosum* has produced **L. × parkmanii** Moore, the name given to a range of hybrids in which the perianth-segments are usually crimson or deep pink with a paler or white margin.

16. L. bulbiferum Linnaeus (*L. bulbiferum* var. *chaixii* (Maw) Stoker).

Stem roots present. Stem 40–150 cm, green, sometimes with purple spots and woolly above, ribbed. Leaves scattered, 5–15 × *c.* 1.5 cm, narrowly to broadly lanceolate with 3, 5 or 7 veins, margin ciliate, otherwise hairless, with or without bulbils in the axils. Flowers usually up to 5, sometimes up to 50, shallowly cup-shaped, erect. Perianth-segments 6–8.5 × 2–3 cm, bright orange, marked at the base with deeper orange and often orange-red towards the tips, warty and with maroon or blackish spots inside, narrowly ovate, narrowed to a claw at the base. Nectary-furrows greenish, hairy. Anthers 6–8 mm, brown with orange pollen. *C Europe south to N Spain, S Italy and N Yugoslavia.* H2. Summer.

Var. **bulbiferum.** Illustration: Elwes, Monograph of the genus Lilium, pl. 23 (1877); Grey-Wilson & Blamey, Alpine flowers of Britain and northern Europe, 271 (1979); Grey-Wilson & Mathew, Bulbs, pl. 3 (1981). Flowers redder and bulbils usually present.

Var. **croceum** (Chaix) Persoon. Illustration: Robert & Schroeter, Alpine flowers, pl. 6 (1938); Huxley, Mountain flowers, 17 (1967); Grey-Wilson & Blamey, Alpine flowers of Britain and northern Europe, 271 (1979); Synge, Lilies, pl. i (1980). This variety has orange flowers and usually lacks bulbils.

17. L. dauricum Ker Gawler. (*L. pensylvanicum* Ker Gawler). Illustration: *Botanical Magazine*, 872 (1805); Elwes, Monograph of the genus Lilium, pl. 21 (1877); Woodcock & Stearn, Lilies of the world, f. 40 (1950).

Bulb stoloniferous. Stem roots present. Stem to 75 cm, green spotted with reddish brown, ribbed and covered with white cobwebby hairs towards the top. Leaves scattered, 5–15 cm × 5–35 mm, linear to lanceolate with 3 or 5 veins, margin often cobwebby-hairy. Flowers to 6, open cup-shaped, erect. Perianth-segments 5–10 × 1.5–3.5 cm, red to orange-red, yellow towards the base, with warts and usually with brownish red spots, oblanceolate, slightly recurved at the tips, narrowed to a claw at the base. Nectary-furrows hairy. Anther 4–10 mm, pollen red. *NE Asia.* H1. Summer.

L. wilsonii Leichtlin. Illustration: Woodcock & Stearn, Lilies of the world, f. 120 (1950). Stem to 1 m, leaves hairless, usually with 5 or 7 veins. Perianth-segments 7.5–10 × 2.5–4 cm, reddish orange or apricot, with a yellow stripe from the centre of the segment to the base, densely dark-spotted. *Japan.* H1. Summer.

Very similar to and possibly only a variety of *L. dauricum*, 'Luteum' has yellow flowers spotted with black.

18. L. maculatum Thunberg (*L. elegans* Thunberg; *L. thunbergianum* Schultes & Schultes).

Stem roots present. Stem to 60 cm, ribbed, hairless to sparsely hairy or minutely rough. Leaves scattered, 4–10 × 1–2 cm, lanceolate to elliptic with 3, 5 or 7 veins, hairless. Flowers 1 or several, cup-shaped, erect. Perianth-segments 8–10 cm, yellow, orange or red with sparse or dense spots, clawed at the base. Nectary-furrows densely hairy. Anthers reddish. *Japan.* H2. Summer.

Thought by some botanists to be a hybrid between *L. dauricum* and *L. concolor.* A large number of variants is cultivated.

L. × hollandicum Woodcock & Stearn. Illustration: Hay & Beckett (eds), Readers' Digest encyclopaedia of garden plants and flowers, 399 (1971). This is the name given to hybrids between *L. maculatum* and *L. bulbiferum.* They usually grow 70 cm or more in height and generally have leaves with 3 veins. The flowers are red, orange or yellow. Many cultivars have been raised and selected.

L. catesbaei Walter. Illustration: *Botanical Magazine*, 259 (1794); Elwes, Monograph of the genus Lilium, pl. 25 (1877). Stem roots absent. Stem hairless. Flowers usually solitary, occasionally 2. Perianth-segments 5–12 × 1–2.5 cm, deep yellow, scarlet towards the recurved tip, blotched with maroon near the base, claw half the length of the segment, greenish outside. Pollen orange-red. *SE USA.* H5. Late summer.

19. L. philadelphicum Linnaeus. Illustration: Elwes, Monograph of the genus Lilium, pl. 17 (1877); Woodcock & Stearn, Lilies of the world, f. 91 (1950).

Stem roots present. Stem to 125 cm, pale green, hairless, usually leafless in the lower one-third. Leaves in whorls of 4–9, usually with some scattered between the whorls, 5–10 × 1.2–1.5 cm, oblanceolate with 3 or 5 veins, hairless, sometimes minutely rough on the margin. Flowers usually 1 or 2, occasionally to 5, cup-shaped, erect. Perianth-segments 4–7.5 × *c.* 2 cm, deep yellow to orange-crimson, usually yellow towards the base with purple or maroon spots, broadly oblanceolate, narrowed into a claw at the base. Nectary-furrows hairless. Anthers 6–9 mm; pollen red. *Eastern N America.* H4. Summer.

Var. **andinum** (Nuttall) Ker Gawler. Leaves mostly scattered and narrower (up to 5 mm wide). The distribution is more westerly.

L. tsingtauense Gilg. Illustration: Grove & Cotton, Supplement to Elwes' monograph of the genus Lilium, pl. 16 (1936); Woodcock & Stearn, Lilies of the world, f. 114 (1950); Synge, Lilies, pl. 15 (1980); Rix & Phillips, The bulb book, 162 (1981). Leaves in whorls of 5–14, the upper usually scattered, 4–12.5 × 1.5–3 cm, hairless. Flowers 1–5, often slightly bilaterally symmetric. Perianth-segments up to 5 × 2 cm, orange to orange-red with maroon spots, slightly warty. Anthers and pollen orange. *NE China, EC & S Korea.* H3. Summer.

20. L. concolor Salisbury.

Stem roots present. Stem 30–90 cm, green, variably flushed with purple, slightly hairy. Leaves scattered, 7–9 cm × 5–15 mm, linear to narrowly lanceolate with 3, 5 or 7 veins, slightly ciliate on margins and veins beneath. Flowers up to 10, star-shaped, fragrant, erect. Perianth-segments

3–4 cm × 7–10 mm, orange-red, unspotted, oblanceolate, slightly recurved at the tips. Anthers 7–9 mm, orange-red. *C China*. H4. Summer.

Var. **pulchellum** (Fischer) Regel. Illustration: *Botanical Magazine*, 6005 (1872); Elwes, Monograph of the genus Lilium, pl. 18 (1879); Woodcock & Stearn, Lilies of the world, f. 39 (1950). Stems green, hairless. Flower buds hairy and flowers spotted. *NE Asia*. H3. Summer.

21. L. chalcedonicum Linnaeus (*L. heldreichii* Freyn). Illustration: Elwes, Monograph of the genus Lilium, pl. 43 (1877); *RHS Lily Year Book* **31**: pl. 2 (1967); Synge, Lilies, pl. ii (1980); Grey-Wilson & Mathew, Bulbs, pl. 4 (1981).

Stem roots present. Stem 45–150 cm, hairless, occasionally rough-hairy below. Leaves scattered, lower 5–12 cm × 5–20 mm, lanceolate to obovate with 3 or 5 veins, margin ciliate and veins hairless or minutely rough beneath; there is an abrupt change between the lower spreading leaves and the smaller upper adpressed leaves. Flowers up to 12, turk's-cap type, faintly scented, nodding. Perianth-segments 5–7 cm, red or orange-red, unspotted, with warts inside towards the base. Nectary-furrows hairless. Anthers and pollen reddish. *Greece, S Albania*. H3. Summer.

Var. **maculatum** Stoker. Illustration: Woodcock & Stearn, Lilies of the world, f. 35 (1950). The flowers are spotted with deep purple or maroon.

22. L. × testaceum Lindley. Illustration: *RHS Lily Year Book* **22**: f. 7 (1958); Harrison, Know your lilies, 48 (1970); Rix & Phillips, The bulb book, 165 (1981).

Stem 100–150 cm, purplish and with a grey-green bloom, hairless. Leaves scattered, 5–10 cm, linear with ciliate margin and hairy on the veins beneath, often twisted. Flowers 6–12, turk's-cap type, fragrant, nodding. Perianth-segments *c*. 8 cm, tawny yellow to pale orange, usually with sparse reddish spots inside. Anthers red with orange pollen. *Garden origin*. H3. Summer.

A hybrid between *L. chalcedonicum* and *L. candidum*.

23. L. pyrenaicum Gouan.

Stem roots often present. Stem 15–135 cm, green, sometimes with sparse purple spots, hairless. Leaves scattered, 3–15 cm × 3–20 mm, linear to lanceolate or narrowly elliptic with 3–15 veins, margin very minutely cilate, veins hairless to densely downy beneath. Flowers to 12, turk's-cap type, faintly to strongly and unpleasantly scented, nodding. Perianth-segments 3–7 cm × 4–13 mm, yellow, orange or red with dark purplish or reddish brown spots and lines inside, especially towards the base, and usually with purplish warts. Filaments greenish or yellowish, smooth or papillose. Anthers 4–10 mm, with orange, red or brown pollen. *N Spain & Pyrenees, SE Alps, Balkans, NE Turkey, Former USSR (Georgia)*. H3. Summer.

1a. Leaf-veins hairless beneath 2
b. Leaf-veins with at least some hairs beneath 3
2a. Flowers spotted subsp. **pyrenaicum**
b. Flowers unspotted subsp. **carniolicum**
3a. Perianth-segments with warts towards the base inside; filaments smooth subsp. **carniolicum**
b. Perianth-segments lacking warts; filaments papillose subsp. **ponticum**

Subsp. **pyrenaicum.** Illustration: Woodcock & Stearn, Lilies of the world, f. 98 (1950); Synge, Lilies, pl. 3 (1980); Grey-Wilson & Mathew, Bulbs, pl. 3 (1981); Rix & Phillips, The bulb book, 167 (1981). Stem 30–135 cm, green with sparse purple spots. Leaves 7–15 cm × 3–20 mm with 3–15 veins that are hairless beneath. Flowers to 12. Perianth-segments 4–6.5 cm, yellow to greenish yellow with dark purple spots and lines inside and purplish warts. Filaments smooth. *N Spain, SW France*.

Forma **rubrum** Stoker (*L. pyrenaicum* 'Rubrum') has orange-red flowers. 'Aureum' with deep yellow flowers is sometimes offered commercially.

Subsp. **carniolicum** (W. Koch) Matthews. Stem to 120 cm, green. Leaves 3–11 cm × 4–17 mm, with 3–9 veins that are hairless or downy beneath. Flowers usually to 6, sometimes to 12. Perianth-segments 3–7 cm, yellow, orange or red, often spotted with brownish purple, and with purplish or, rarely, yellow warts. Filaments smooth. *SE Europe*.

Three varieties of subsp. *carniolicum* are in cultivation:

Var. **carniolicum** (*L. carniolicum* W. Koch). Illustration: Elwes, Monograph of the genus Lilium, pl. 45 (1877); *Journal of the Royal Horticultural Society* **103**: 209 (1978); Grey-Wilson & Mathew, Bulbs, pl. 4 (1981). Stem to 120 cm. Leaf-veins densely downy beneath. Flowers red or orange, spotted. *SE Alps (Austria & Italy), N Yugoslavia*.

Var. **albanicum** (Grisebach) Matthews (*L. albanicum* Grisebach; *L. carniolicum* W. Koch subsp. *albanicum* (Grisebach) Hayek). Illustration: *Botanical Magazine*, n.s., 839 (1982). Stem to 40 cm. Leaf-veins hairless beneath. Flowers yellow, unspotted. *SW Yugoslavia, Albania, NW Greece*.

Var. **jankae** (Kerner) Matthews (*L. jankae* Kerner; *L. carniolicum* W. Koch subsp. *jankae* (Kerner) Ascherson & Graebner). Stem to 80 cm. Leaf-veins downy beneath. Flowers yellow, spotted or unspotted. *NW Italy, NW Yugoslavia, W Bulgaria, C Romania*.

Subsp. **ponticum** (C. Koch) Matthews. Stem 15–90 cm, green. Leaves 3–8 cm × 8–20 mm, with 7–15 veins that are downy beneath. Flowers to 5, occasionally to 12. Perianth-segments 3.5–5.5 cm, deep yellow or deep orange with dense red-brown or purplish stripes and spots towards the base inside, but lacking warts. Filaments papillose. *NE Turkey, Former USSR (Georgia)*.

Var. **ponticum** (*L. ponticum* C. Koch; *L. georgicum* Mandenova; *L. carniolicum* W. Koch subsp. *ponticum* (C. Koch) Davis & Henderson). Illustration: *RHS Lily Year Book* **27**: f. 26, 27 (1963) & **33**: pl. 29e (1969). Stem usually stout. Perianth-segments deep yellow, 7–10 mm wide. *NE Turkey*.

Var. **artvinense** (Mischenko) Matthews (*L. artvinense* Mishchenko; *L. ponticum* C. Koch var. *artvinense* (Mishchenko) Davis & Henderson). Illustration: *RHS Lily Year Book* **33**: pl. 29f (1969). Stem more slender than in var. *ponticum* and with fewer, less crowded, leaves. Perianth-segments deep orange, 5–6 mm wide. *NE Turkey, Former USSR (Georgia)*.

L. ledebourii (Baker) Boissier. Illustration: *RHS Lily Year Book* **32**: pl. 5 (1968); Wendelbo, Tulips and irises of Iran, 32, 33 (1977). Flowers up to 5, fragrant, perianth-segments 6–7.5 cm, creamy-white, becoming greenish yellow, and dotted with purple or red towards the base. Stamens with the filaments joined together at base; anthers and pollen orange-red. Former *USSR (Azerbaidjan), NW Iran*. H3. Summer.

L. ciliatum Davis. Illustration: Woodcock & Stearn, Lilies of the world, f. 94 (1950) – as *L. ponticum*; *RHS Lily Year Book* **33**: pl. 29a & 29c (1969); Synge, Lilies, pl. 2 (1980). Stem

60–160 cm, sometimes with cobwebby hairs in the upper part. Leaves long-ciliate on the margin, veins hairless beneath. Flowers up to 20. Perianth-segments 4–5.5 cm, cream or pale yellow, marked with purplish brown towards the base, apex hairy. *NE Turkey*. H3. Summer.

L. callosum Siebold & Zuccarini. Illustration: Elwes, Monograph of the genus Lilium, pl. 41 (1877); Woodcock & Stearn, Lilies of the world, f. 30 (1950). Stem to 90 cm. Leaves 8–13 cm × 3–8 mm, linear with 3 or 5 veins, hairless. Flowers up to 10, scentless. Perianth-segments 3–4 cm, orange-red with faint black dots towards the base. *Former USSR (Amur region), N & C China, Korea, Japan, Taiwan*. H3. Late summer.

24. L. pomponium Linnaeus. Illustration: *Botanical Magazine*, 971 (1807); Woodcock & Stearn, Lilies of the world, f. 93 (1950); Grey-Wilson & Mathew, Bulbs, pl. 4 (1981); Rix & Phillips, The bulb book, 161 (1981).

Stem roots present. Stem 20–100 cm, green with purple spots below, hairless. Leaves scattered, 9–15 cm × 2–5 mm, linear with 1 or 3 veins, margin ciliate, otherwise hairless. Flowers usually 1–6, sometimes to 10, turk's-cap type, unpleasantly scented, nodding. Perianth-segments 4–7 cm, bright red with purplish or black spots and lines and purplish warts inside. Anthers 4–7 mm, sometimes to 1.4 cm, red. *Maritime Alps of France and Italy*. H3. Summer.

25. L. martagon Linnaeus. Illustration: Brickell (ed.), RHS A–Z encyclopedia of garden plants, 631 (2003)

Stem roots present. Stem 90–200 cm, purplish or green, hairy or hairless. Leaves in whorls of 6–10, up to 16 × 6.5 cm, oblanceolate with 7 or 9 veins, hairless or downy on the veins beneath. Flowers to 50, turk's-cap type, scented, nodding. Perianth-segments 3–4.5 cm × 6–10 mm, white, pink or dark purplish red to maroon, often spotted. Nectary-furrows with warty margins. Anthers orange-yellow or reddish purple. *Europe, Asia east to Mongolia*. H1. Summer.

1a. Flowers white; stem green — var. **album**
b. Flowers not white; stem purplish — 2
2a. Stem hairless or more or less so — 3
b. Stem hairy — 4
3a. Flowers maroon — var. **sanguineo-purpureum**
b. Flowers dull purplish pink — var. **martagon**
4a. Flowers maroon, unspotted — var. **cattaniae**
b. Flowers purplish pink to wine red, spotted — 5
5a. Buds hairy; leaves hairless on veins beneath — var. **pilosiusculum**
b. Buds not hairy; leaves downy on veins beneath — var. **hirsutum**

Var. **album** Weston. Illustration: de Graaff & Hyams, Lilies, pl. 1 (1967); *RHS Lily Year Book* **33**: pl. 8 (1969). Stem green and hairless, flowers white, unspotted.

Var. **cattaniae** Visiani (var. *dalmaticum* Elwes). Illustration: Elwes, Monograph of the genus Lilium, pl. 33 (1877); Synge, Lilies, pl. iii (1980); Grey-Wilson & Mathew, Bulbs, pl. 3 (1981). Stem purplish, hairy. Buds hairy, opening to dark maroon unspotted flowers.

Var. **hirsutum** Weston. Stem purplish, hairy. Leaves downy beneath. Flowers purplish pink, spotted.

Var. **martagon.** Illustration: Robert & Schroeter, Alpine flowers, pl. 5 (1938); Polunin, Flowers of Europe, pl. 167 (1969); Aichele, Wild flowers, 259 (1975); Grey-Wilson & Mathew, Bulbs, pl. 3 (1981). Stem purplish, stem and leaves hairless or almost so, flowers dull purplish pink, spotted.

Var. **pilosiusculum** Freyn. Stem purplish, cobwebby-hairy. Leaves narrow with hairs only on the margin. Bracts and buds hairy. Flowers wine red, with sparse spots.

Var. **sanguineo-purpureum** Beck. Stem purplish, hairless. Flowers dark maroon with heavy spotting.

The following varieties have been in cultivation and may still persist in some gardens: var. **albiflorum** Vukotinovic, with white flowers which are spotted with pink; var. **caucasicum** Mishchenko, which grows to 2.5 m and has lilac-pink flowers.

There are several hybrids and hybrid groups derived from *L. martagon* crossed with *L. hansonii*.

L. × marhan Baker (illustration: Hay & Beckett (eds), Readers' Digest encyclopaedia of garden plants and flowers, 399, 1971) was raised from *L. martagon* var. *album* and *L. hansonii* and has flowers with thick perianth-segments that are dull orange with reddish brown spots, especially towards the tips.

L. × dalhansonii Powell (illustration: *The Garden* **44**: t. 927, 1893) resulted from a cross between *L. martagon* var. *cattaniae* and *L. hansonii*. The abundant flowers are maroon, flushed with orange at the middle of the segments, and heavily spotted. The 'Backhouse Hybrids' have flowers varying from cream, yellow or pink to maroon, all with dark spots; a number of clones have been named.

L. polyphyllum, which differs in having less recurved perianth-segments, greenish white outside, inside creamy-white, with green or purplish streaks, and yellow anthers with orange pollen, may key out here (see under 31. *L. duchartrei*).

L. kelloggii Purdy. Illustration: Woodcock & Stearn, Lilies of the world, f. 67 (1950); Synge, Lilies, pl. 20 (1980); Rix & Phillips, The bulb book, 160 (1981). Stem roots absent. Stem glaucous. Leaves in whorls of 12–18, 6–10 × 1–2 cm, with 1 or 3 veins, hairless. Flowers to 20, very fragrant. Perianth-segments 3–5.5 cm, mauvish, pink or white, usually with a yellow band below the middle, spotted with purple in lower part. Anthers reddish to orange with orange pollen. *USA (S Oregon, NW California)*. H4. Summer.

26. L. monadelphum Bieberstein (*L. szovitsianum* Fischer & Avé-Lallement). Illustration: *RHS Lily Year Book* **27**: pl. 4 (1963); de Graaff & Hyams, Lilies, pl. 2 (1967); Synge, Lilies, pl. iv, v (1980); Rix & Phillips, The bulb book, 166, 167 (1981).

At least a few stem roots present. Stem 50–200 cm, hairless. Leaves scattered, 5–14 × 1–3 cm, lanceolate to oblanceolate or ovate with 9, 11 or 13 veins, margins ciliate, downy on veins beneath. Flowers usually 1–5, occasionally to 30, turk's-cap type, fragrant, nodding. Perianth-segments 6–10 × 1–2.2 cm, pale to mid-yellow, usually spotted with purple or maroon inside, flushed with purplish brown at the base outside, varying in the degree to which they are recurved. Stamens with the filaments joined at the base or free; anthers 8–13 mm, brown or maroon, pollen orange or yellow. *Former USSR (Caucasus), NE Turkey*. H3. Summer.

Var. **armenum** (Mishchenko) Davis & Henderson. Illustration: *RHS Lily Year Book* **33**: pl. 29b (1969). Inner perianth-segments acute, 1–1.6 cm wide. *Former USSR (Armenia), NE Turkey*.

There is a lily from the Caucasus that is occasionally grown under the names *L. kosa* or *L. kossii*. Both names are invalid because they have not been properly published. The lily was originally said to be closely related to *L. monadelphum*, and said to differ in having pink-tinged flowers, a feature which was observed in wild plants. However, plants grown in gardens have produced only pale yellow flowers, and do not appear to differ from *L. monadelphum*.

L. kesselringianum Mishchenko. Illustration: Woodcock & Stearn, Lilies of the world, f. 68 (1950); *RHS Lily Year Book* **33**: pl. 29d (1969). Flowers usually 1–3. Perianth-segments cream to pale yellow, spotted with purple or reddish brown towards the base, recurved in the upper two-thirds but not reflexed. Stamens with free filaments. *Former USSR (Georgia), NE Turkey.* H3. Summer.

27. L. pumilum de Candolle (*L. tenuifolium* Schrank). Illustration: Elwes, Monograph of the genus Lilium, pl. 42 (1877); Woodcock & Stearn, Lilies of the world, f. 97 (1950); Hay & Beckett (eds), Readers' Digest encyclopaedia of garden plants and flowers, 407 (1971).

Stem roots present. Stem 15–45 cm, occasionally to 90 cm, green, hairless or minutely rough, especially below. Leaves scattered, 3–10 cm × 1–3 mm, linear with 1 vein, hairless or slightly rough on margin and veins beneath. Flowers to 30, turk's-cap type, slightly fragrant, nodding, buds sometimes white-hairy. Perianth-segments 3–3.5 cm, scarlet, unspotted or with sparse black dots at the base. Nectary-furrows with a papillose border. Anthers 4–8 mm; pollen scarlet. *Former USSR (Siberia), Mongolia, N China, N Korea.* H2. Summer.

28. L. hansonii Moore. Illustration: Elwes, Monograph of the genus Lilium, pl. 34 (1877); Woodcock & Stearn, Lilies of the world, f. 49 (1950); Brickell (ed.), RHS A–Z encyclopedia of garden plants, 631 (2003).

Stem roots present. Stem 1–1.5 m, green. Leaves in whorls of 12–20, up to 18 × 4 cm, oblanceolate to elliptic with 3 or 5 veins. Flowers 3–12, turk's-cap type, fragrant, nodding. Perianth-segments very thick, 3–4 × *c.* 1.5 cm, deep orange-yellow spotted with purplish brown towards the base, recurved from about the middle. Nectary-furrows hairless. Anthers 2–2.2 cm, purplish with yellow pollen. *Former E USSR, Korea, Japan.* H3. Summer.

29. L. leichtlinii Hooker. Illustration: *Botanical Magazine*, 5673 (1867); Elwes, Monograph of the genus Lilium, pl. 39 (1877); Woodcock & Stearn, Lilies of the world, f. 71 (1950); Jefferson-Brown, Modern lilies, 48 (1965).

Bulb stoloniferous. Stem roots present. Stem up to 120 cm, green, hairless or with sparse white hairs especially at the leaf axils. Leaves scattered, up to 15 × 1.25 cm, linear to narrowly lanceolate with 1 or 3 veins, margin minutely rough and midrib slightly minutely rough beneath. Flowers 1–6, turk's-cap type, scentless, nodding; buds usually white-hairy. Perianth-segments 6.5–8.5 × 1.3–2.7 cm, yellow, spotted with purplish maroon. Nectary-furrows hairy. Anthers 1–1.7 cm, reddish brown; pollen reddish brown. *Japan.* H3. Summer.

Var. **maximowiczii** (Regel) Baker (*L. maximowiczii* Regel; *L. leichtlinii* var. *tigrinum* (Regel) Nicholson). Illustration: Elwes, Monograph of the genus Lilium, pl. 40 (1879). Stem to 2.5 m. Flowers reddish orange with purplish brown spots. *Japan, C Korea, Former E USSR.*

L. rhodopaeum Delipavlov. Illustration: *RHS Lily Year Book* **32**: pl. 6 (1968); Grey-Wilson & Mathew, Bulbs, pl. 4 (1981). Stem 80–100 cm. Leaves with ciliate margin, downy on the veins beneath. Perianth-segments 8–12 cm, yellow, unspotted, less recurved than in *L. leichtlinii.* Greece, Bulgaria. H4. Summer.

30. L. lancifolium Thunberg (*L. tigrinum* Ker Gawler). Illustration: Rix & Phillips, The bulb book, 164 (1981); Mathew, P.J. Redouté: Lilies and related flowers, 45 (1982); Brickell (ed.), RHS A–Z encyclopedia of garden plants, 631 (2003).

Stem roots present. Stem 60–150 cm, dark purple, usually with white cobwebby hairs. Leaves scattered, 12–20 × 1–2 cm, narrowly lanceolate with 5 or 7 veins, margins minutely rough, veins hairless beneath; axils with bulbils. Flowers to 40, turk's-cap type, scentless, nodding. Perianth-segments 7–10 × 1–2.5 cm, pinkish orange to orange-red with deep purple spots and purplish warts inside. Nectary-furrows bordered by warty hairy ridges. Anthers 1.2–1.4 cm, orange-red to purplish with reddish brown or purplish brown pollen. *Japan, Korea, E China.* H4. Summer–early autumn.

Var. **flaviflorum** Makino (*L. tigrinum* var. *flaviflorum* (Makino) Stearn). Flowers yellow.

Var. **fortunei** (Standish) Matthews (*L. tigrinum* var. *fortunei* Standish). Illustration: Synge, Lilies, pl. xix (1980). A larger plant with densely woolly stem and many orange-red flowers.

Var. **splendens** (van Houtte) Matthews (*L. tigrinum* var. *splendens* van Houtte). A vigorous plant with larger flowers of a brighter orange-red.

'Flore-pleno' is a selection with double flowers.

L. lancifolium sold by nurseries often turns out to be *L. speciosum*.

31. L. duchartrei Franchet (*L. farreri* Turrill; *L. duchartrei* var. *farreri* (Turrill) Grove). Illustration: *Botanical Magazine*, 8847 (1920); Woodcock & Stearn, Lilies of the world, f. 43 (1950); Synge, Lilies, pl. 8 (1980); Rix & Phillips, The bulb book, 162 (1981).

Bulb stoloniferous. Stem roots present. Stem 60–100 cm, green or flushed with purplish brown, slightly ribbed, hairless or slightly papillose; leaf axils often hairy. Leaves scattered, 5–10 × 1–1.5 cm, lanceolate with 3 or 5 veins, margin minutely rough and veins sometimes minutely rough beneath. Flowers to 12, turk's-cap type, fragrant, nodding. Perianth-segments up to 5.5 × 3 cm, white, often slightly flushed with purple outside and green or purplish at the base, streaked and spotted with purple inside. Nectary-furrows green with warty margins. Anthers 1–1.3 cm, yellow or orange, with orange pollen. *W China.* H4. Summer.

L. ledebourii, which differs in its narrower perianth-segments and orange-red anthers, may key out here (see under 23. *L. pyrenaicum*).

L. polyphyllum Don. Illustration: Elwes, Monograph of the genus Lilium, pl. 48 (1877); Woodcock & Stearn, Lilies of the world, f. 109 (1950); *Journal of the Royal Horticultural Society* **93**: pl. 126 (1968); Rix & Phillips, The bulb book, 158 (1981). Stem green, hairless. Leaves scattered or with the lower in a whorl, 5–12 cm × 3–15 mm, margin minutely rough. Flowers usually 1–6, sometimes to 30, very fragrant. Perianth-segments less recurved than in *L. martagon*, greenish white outside, inside creamy-white, with green or purplish streaks. Anthers 8–9 mm. *E Afghanistan, W Himalaya.* H4. Summer.

32. L. taliense Franchet. Illustration: Grove & Cotton, Supplement to Elwes' monograph of the genus Lilium, pl. 24 (1939); Woodcock & Stearn, Lilies of the world, f. 111 (1950); *RHS Lily Year Book* **27**: f. 8 (1963); Everard & Morley, Wild flowers of the world, 108 (1970).

Bulb shortly stoloniferous. Stem roots present. Stem to 140 cm, deep purple or green marked with purple, hairless or minutely rough towards the top. Leaves scattered, 7–13 cm × 4–6 mm, narrowly lanceolate with 1 or 3 veins, margin minutely rough, veins minutely rough beneath. Flowers to 12, turk's-cap type, fragrant, nodding. Perianth-segments 4–6 cm × 7–20 mm, white, spotted with purple except at the tips, strongly recurved. Nectary-furrows purple, hairless. Anthers 7–11 mm, mauve or greyish with yellow pollen. *W China.* H4. Summer.

33. L. lankongense Franchet. Illustration: Woodcock & Stearn, Lilies of the world, f. 69 (1950); de Graaff & Hyams, Lilies, pl. 5 (1967); *RHS Lily Year Book* **34**: front cover (1970); Synge, Lilies, pl. xi (1980).

Bulb stoloniferous. Stem roots present. Stem to 120 cm, green, minutely rough, slightly ribbed. Leaves scattered, 3–10 cm × 4–8 mm, lanceolate with 3, 5 or 7 veins, margin minutely rough, veins hairless beneath. Flowers to 15, turk's-cap type, fragrant, nodding. Perianth-segments 4–6.5 × 1–2.5 cm, pink sometimes flushed with mauve, spotted with reddish purple especially on the incurved margins and with a green median band, strongly recurved. Nectary-furrows green, bordered with white warts. Anthers 8–12 mm, purplish with orange-brown pollen. *W China (SE Xizang).* H4. Summer.

L. papilliferum Franchet. Illustration: de Graaff & Hyams, Lilies, pl. 4 (1967); Synge, Lilies, pl. 11 (1980). Stem minutely rough except near the base. Flowers 1–3, sometimes more. Perianth-segments 4–5 × *c.* 1 cm, deep crimson-maroon, becoming paler towards the base and bearing warts. Anthers brown. *W China (Yunnan).* H4. Late spring–late summer.

L. cernuum Komarov. Illustration: Woodcock & Stearn, Lilies of the world, f. 34 (1950); de Graaff & Hyams, Lilies, pl. 3 (1967); Synge, Lilies, pl. 6 (1980). Stem to 60 cm, green or marked with purplish brown, base and upper part more or less leafless. Leaves scattered, densely crowded in middle part of stem, 10–18 cm × 1–5 mm, linear with 1 or 3 veins, hairless. Perianth-segments 3.5–5 cm × 7–12 mm, lilac, pinkish purple or sometimes white, spotted with purple. Pollen lilac. *Korea north to Former E USSR.* H3. Summer.

34. L. wardii Stern. Illustration: Woodcock & Stearn, Lilies of the world, f. 118 (1950); Synge, Lilies, pl. xx (1980).

Bulb stoloniferous. Stem roots present. Stem to 1.5 m, dark green tinged with purple, hairless or minutely rough. Leaves scattered, 3–8 cm × 6–20 mm, narrowly lanceolate to elliptic with 3 veins, margin hairless or minutely rough, veins usually hairless beneath. Flowers up to 35, turk's-cap type, fragrant, nodding. Perianth-segments 5–6.5 cm × 7–15 mm, pink or pinkish purple spotted inside with purple and with a purple median line in the lower half. Nectary-furrows hairless. Anthers 10–11 mm, yellow or purple with orange pollen. *W China (SE Xizang).* H4. Summer.

35. L. davidii Elwes. Illustration: *RHS Lily Year Book* **18**: f. 17 (1954); Synge, Lilies, pl. 7 (1980).

Bulb stoloniferous or not. Stem roots present. Stem 1–1.4 m, green, usually marked with brownish purple, downy especially in lower part, usually with cobwebby hairs at the leaf axils. Leaves scattered, 6–10 cm × 2–4 mm, linear with 1 vein that is minutely rough beneath, margin usually inrolled and minutely rough. Flowers 5–20, turk's-cap type, unscented, nodding; buds usually hairy. Perianth-segments 5–8 cm × 8–25 mm, red or orange-red, densely spotted with deep purple raised spots except at the apex, warty in the lower part, strongly recurved. Nectary-furrows bordered with white hairs. Anthers 9–12 mm, purplish orange with orange or red pollen. *W China (Szechwan, Yunnan).* H4. Summer.

Var. **macranthum** Raffill. Stems taller (to 2 m) and flowers larger, borne 2 or 3 together on a stalk.

Var. **unicolor** (Hoog) Cotton. Stems shorter (to 1 m) and leaves longer and very crowded. Flowers paler, unspotted or with reddish or mauve spots.

Var. **willmottiae** (Wilson) Raffill. Illustration: Woodcock & Stearn, Lilies of the world, f. 41 (1950); Synge, Lilies, pl. viii (1980); Brickell (ed.), RHS A–Z encyclopedia of garden plants, 631 (2003). Bulb stoloniferous. Stem to 2 m, arching under the weight of the flowers. Leaves to 6 mm wide. Flowers to 40 per stem.

36. L. henryi Baker. Illustration: *Botanical Magazine*, 7177 (1891); Woodcock & Stearn, Lilies of the world, f. 60 (1950); Hay & Synge, Dictionary of garden plants in colour, 101 (1969); Synge, Lilies, pl. ix (1980).

Stem roots present. Stem 1–3 m, green with reddish purple markings, hairless. Leaves scattered, 8–15 × 2–3 cm, lanceolate with many veins, margin hairless, hairless or sparsely hairy on the veins beneath, sometimes with bulbils in the axils. Flowers usually 4–20, occasionally up to 70, 1 or 2 on each flower-stalk, turk's-cap type, nodding. Perianth-segments 6–8 × 1–2 cm, orange with dark spots towards the base and long warts, shortly clawed. Nectory-furrows green. Anthers 1.2–1.7 cm, orange. *C China.* H3. Summer.

Var. **citrinum** Anon. Flowers pale yellow, lightly spotted with brown or reddish brown. *L. henryi* crossed with *L. sargentiae* gave rise to **L. × aurelianense** Debras. The original cross is similar to *L. henryi*, but with the flowers borne almost horizontally, pale yellow to yellowish orange, unspotted and trumpet-shaped with the segments recurved at the tips. Later back-crosses show more variation. Illustration: Synge, Collins guide to bulbs, edn 2, pl. 19 (1961); Hay & Beckett (eds), Readers' Digest encyclopaedia of garden plants and flowers, 402 (1971).

37. L. speciosum Thunberg. Illustration: Elwes, Monograph of the genus Lilium, pl. 13 (1878); Everard & Morley, Wild flowers of the world, 107 (1970); Perry, Flowers of the world, 170 (1972).

Stem roots present. Stem 120–170 cm, green or flushed with purple, hairless. Leaves scattered, shortly stalked, up to 20 × 6 cm, broadly lanceolate to oblong-lanceolate with 7 or 9 veins, hairless. Flowers usually to 12, occasionally to 40, turk's-cap type, fragrant, nodding. Perianth-segments to 10 × 4.5 cm, white or pale pink tinged with deep pink towards the base and spotted with pink or crimson, warty in the lower part, strongly recurved, margins slightly wavy. Nectary-furrows green, hairless. Anthers 1.3–2.2 cm with brown or dark red pollen. *Japan, China, Taiwan.* H4. Late summer.

Var. **gloriosoides** Baker. The spots and warts are almost scarlet and the warts are confined to an area towards the base but are absent from the basal one-quarter of the segments.

Several cultivars that vary mainly in flower colour are commonly grown.

38. L. amabile Palibin. Illustration: Grove & Cotton, Supplement to Elwes' monograph of the genus Lilium, pl. 6

(1935); Woodcock & Stearn, Lilies of the world, f. 20 (1950).

Stem roots present. Stem 30–90 cm, downy, leafless in the lower part. Leaves scattered, 4–9 cm × 5–10 mm, lanceolate, usually with 3 veins, hairy. Flowers to 10, turk's-cap type, scent rather unpleasant, nodding. Perianth-segments 5–5.5 cm, red with black spots, strongly recurved. Nectary-furrows warty. Anthers 1.1–1.3 cm, brown with reddish pollen. *Korea.* H4. Summer.

Var. **luteum** Anon. Flowers yellow.

39. L. humboldtii Duchartre. Illustration: Elwes, Monograph of the genus Lilium, pl. 32 (1877); *RHS Lily Year Book* **33**: pl. 19 (1969).

Stem roots absent. Stem up to 225 cm, green often tinged with purple, often slightly downy. Leaves in whorls of 10–20, 9–12.5 × 1.5–3 cm, oblanceolate with 5 veins, margin wavy and downy. Flowers usually 10–15, occasionally to 80, turk's-cap type, nodding. Perianth-segments 6.5–10 × 1.5–2.5 cm, yellow to orange, spotted with maroon or purple towards the centre, strongly recurved. Anthers 1.2–1.3 cm, purplish with orange pollen. *USA (C California).* H4. Summer.

Var. **ocellatum** (Kellogg) Elwes (var. *magnificum* Purdy). Illustration: Woodcock & Stearn, Lilies of the world, f. 62 (1950); Synge, Lilies, pl. xxii (1980). Stem roots present. Perianth-segments have red tips and the spots are encircled with crimson.

L. humboldtii has been crossed with *L. pardalinum* to produce **L. × pardaboldtii** Woodcock & Courts. The stem reaches 150 cm and the leaves are in whorls; the turk's-cap type flowers are reddish orange, spotted with dark crimson.

The 'Bellingham Hybrids' were also developed from this cross.

L. iridollae Henry. Illustration: *Bartonia* **24**: pl. 1–5 (1946); Woodcock & Stearn, Lilies of the world, f. 64, 65 (1950); *RHS Lily Year Book* **17**: f. 4 (1953). Stem up to 175 cm. Leaves usually in whorls, downy on margin and veins beneath. Flowers to 8. Perianth-segments yellow, spotted with brown. Anthers reddish brown. *SE USA (S Alabama, N Florida).* H4. Late summer.

40. L. columbianum Baker. Illustration: Elwes, Monograph of the genus Lilium, pl. 31 (1877); Woodcock & Stearn, Lilies of the world, f. 38 (1950); Synge, Lilies, pl. 18 (1980); Rix & Phillips, The bulb book, 160 (1981).

Stem roots absent. Stem to 2.5 m, hairless. Lower leaves in whorls of 5–13, upper scattered, 5–14 × 1.2–4 cm, oblanceolate with 3 or 5 veins, hairless. Flowers usually 6–10, occasionally to 40, turk's-cap type, nodding. Perianth-segments 3.5–6.5 cm × 8–12 mm, yellow to orange-red with purplish red spots towards the base, strongly recurved. Anthers 6–11 mm, orange, with yellow to brownish pollen. *N America (Pacific coast from S Canada to N California).* H3. Summer.

L. wigginsii Beane & Vollmer. Stem roots present. Stem to 90 cm. Leaves scattered or in 2–4 whorls halfway up the stem, to 22 cm, narrowly lanceolate. Perianth-segments deep yellow, spotted with purple. Pollen yellow. *W USA (S Oregon, N California).* H4. Summer.

41. L. kelleyanum Lemmon (*L. nevadense* Eastwood; *L. nevadense* var. *shastense* Eastwood). Illustration: Woodcock & Stearn, Lilies of the world, f. 86 (1950); *RHS Lily Year Book* **24**: f. 10 (1960).

Bulb stoloniferous. Stem roots absent. Stem to 1 m, sometimes to 2 m. Leaves in 2 or 3 whorls of 7–16, scattered above and below, or all leaves scattered, 5–15 × 1.5–3.5 cm, lanceolate to broadly elliptic, hairless. Flowers up to 25, turk's-cap type, fragrant, nodding. Perianth-segments 2.5–5 × *c.* 1 cm, greenish yellow, yellow or orange, spotted with purplish red or brown towards the base, tips sometimes reddish. Anthers 5–7 mm, brown. *W USA (California, S Oregon).* H4. Summer.

L. occidentale Purdy. Illustration: Grove & Cotton, Supplement to Elwes' monograph of the genus Lilium, pl. 25 (1940); Woodcock & Stearn, Lilies of the world, f. 88 (1950); *RHS Lily Year Book* **31**: pl. 9 & f. 19 (1967); *Journal of the Royal Horticultural Society* **92**: pl. 245, 247 (1967). Stem to 2 m or more. Leaves in whorls, scattered above. Flowers usually 1–5, sometimes up to 20, unscented. Perianth-segments 3.5–6.25 cm, crimson, green or yellow at the base, or orange-red with an orange base and spotted with purplish brown. Anthers 7–8 mm, purple to reddish brown. *W USA (coastal S Oregon & N California).* H4. Summer.

L. medeoloides Gray. Illustration: Elwes, Monograph of the genus Lilium, pl. 35 (1879) – as *L. avenaceum*; Woodcock & Stearn, Lilies of the world, f. 80 (1950). Stem roots present. Stem to 75 cm, hollow, hairless or minutely rough below. Leaves in 1 or 2 whorls about halfway up the stem, with scattered leaves above. Flowers scentless. Perianth-segments apricot or orange-red, unspotted or with dark spots. Anthers purplish with orange-red pollen. *Former NE USSR, S Korea, China, Japan.* H3. Summer.

L. distichum Kamibayashi. Illustration: Woodcock & Stearn, Lilies of the world, f. 42 (1950). Stem roots present. Stem hollow, ribbed. Leaves in 1 or 2 whorls with upper leaves scattered. Flowers 3–8. Perianth-segments pale orange, usually with darker spots. Anthers 8–10 mm. *Former E USSR, NE China, N & C Korea.* H3. Summer.

L. pitkinense Beane & Vollmer. Illustration: Turrill, Supplement to Elwes' Monograph of the genus Lilium **9**: pl 3 (1962); *RHS Lily Year Book* **27**: pl. 6 (1963). Leaves linear. Flowers to 8. Perianth-segments orange-red, yellow towards the base with tiny maroon spots. Anthers brownish purple with reddish brown pollen. *W USA (California).* H4. Summer.

42. L. pardalinum Kellogg. Illustration: Elwes, Monograph of the genus Lilium, pl. 28, 29 (1877–9); Woodcock & Stearn, Lilies of the world, f. 108 (1950); Hay & Synge, Dictionary of garden plants in colour, 102 (1969), Witham Fogg, Bulbs, 48 (1980).

Bulb stoloniferous. Stem roots absent. Stem 2–3 m, green, hairless. Leaves in whorls of 8–19, with a few scattered above and below, to 20 × 5.5 cm, narrowly elliptic to oblanceolate with 3 veins, hairless. Flowers to 10, turk's-cap type, often fragrant, nodding. Perianth-segments 5–9 × 1.2–1.8 cm, dark red in upper half, orange to orange-red in lower half and heavily spotted with maroon, some spots encircled with yellow, strongly recurved. Anthers reddish or brown with orange pollen. *W USA (S Oregon, N California).* H4. Summer.

Var. **angustifolium** Kellogg (*L. roezlii* Regel). Leaves narrow.

Var. **giganteum** Woodcock & Coutts. Illustration: Synge, Lilies, pl. xxiii (1980). Tall-growing with up to 30 crimson and yellow, densely spotted flowers.

L. pardalinum has been crossed with *L. parryi* to produce **L. × burbankii** Anon. The leaves are whorled and the strongly reflexed perianth-segments are yellow with brown spots and often flushed red at the tips.

43. L. michiganense Farwell. Illustration: Woodcock & Stearn, Lilies of the world, f. 81 (1950); *RHS Lily Year Book* **19**: f. 13 (1955); Turrill, Supplement to Elwes' Monograph of the genus Lilium **8**: 15 (1960).

Bulb stoloniferous. Stem 60–150 cm, green, hairless. Leaves in whorls of 4–8, 9–12 cm, lanceolate to elliptic with 3 veins, usually minutely rough on margin and veins beneath. Flowers usually 3–6, sometimes to 25, turk's-cap type, nodding. Perianth-segments 6.5–8 cm × 7–18 mm, orange-red with a green zone at the base, spotted with deep red or purple, strongly recurved. Anthers 6–9 mm, orange with orange pollen. *Eastern N America.* H3. Summer.

L. vollmeri Eastwood. Illustration: Woodcock & Stearn, Lilies of the world, f. 115 (1950). Leaves scattered or with a few in whorls, hairless. Perianth-segments deep orange, tinged with red towards the tip at the margin, spotted with blackish maroon. *W USA (S Oregon, N California).* H4. Summer.

L. occidentale, which differs in having leaves with usually 7 veins, perianth-segments 3.5–6.25 cm, and purple anthers with orange-red pollen, may key out here (see under 41. *L. kelleyanum*).

44. L. superbum Linnaeus. Illustration: Elwes, Monograph of the genus Lilium, pl. 26 (1877); Woodcock & Stearn, Lilies of the world, f. 110 (1950); Hay & Beckett (eds), Readers' Digest encyclopaedia of garden plants and flowers, 408 (1971); Synge, Lilies, pl. xxv (1980).

Bulb stoloniferous. Stem roots present. Stem 1.5–3 m, green, usually flushed with purple, hairless. Leaves in whorls of 4–20, sometimes the upper ones scattered, 3.5–11 cm × 8–28 mm, lanceolate to elliptic with 3, 5 or 7 veins, sometimes minutely rough on margin and veins beneath. Flowers to 40, turk's-cap type, nodding. Perianth-segments 6–10 × 1–2 cm, orange flushed with red, especially at the tips, and spotted towards the green base with maroon. Nectary-furrows green, hairy. Anthers 1.6–2 cm, red with orange-brown pollen. *E USA.* H3. Summer.

45. L. michauxii Poiret. Illustration: *Edwards's Botanical Register* **7**: t. 580 (1821) – as *L. carolinianum*; Elwes, Monograph of the genus Lilium, pl. 25 (1877); Woodcock & Stearn, Lilies of the world, f. 79 (1950).

Bulb stoloniferous. Stem roots present. Stem to 120 cm, green or slightly spotted, hairless. Leaves in whorls of 3–7, upper ones usually scattered, up to 12 × 4 cm, oblanceolate to obovate, glaucous, hairless. Flowers usually 1 or 2, sometimes up to 5, turk's-cap type, scented, nodding. Perianth-segments up to 10 cm, orange-red to pale crimson, yellowish towards the base, spotted with purple or black. Anthers purplish brown with orange-red pollen. *SE USA.* H4. Summer–early autumn.

L. occidentale, which differs in having up to 20 flowers, with deep crimson perianth-segments 3–5–6.25 cm, may key out here (see under 41. *L. kelleyanum*).

46. L. parvum Kellogg. Illustration: *Botanical Magazine,* 6146 (1875) – as *L. canadense* var. *parvum*; Elwes, Monograph of the genus Lilium, pl. 30 (1877); Rix & Phillips, The bulb book, 160 (1981).

Bulb stoloniferous. Stem roots absent. Stem to 2 m. Leaves usually in whorls of 7–10, occasionally scattered, 5–12.5 cm × 5–30 mm, linear or lanceolate with 1 or 3 veins, hairless, margin occasionally minutely rough. Flowers to 30, open bell-shaped, borne horizontally or almost erect. Perianth-segments 3–4 cm, red or crimson, usually heavily spotted with dark red or purple, slightly recurved at the tips. *W USA (SE Oregon to C California).* H4. Summer.

Forma **crocatum** Stearn. Illustration: *RHS Lily Year Book* **31**: pl. 9 (1967). Flowers yellow or orange.

L. maritimum Kellogg. Illustration: Elwes, Monograph of the genus Lilium, pl. 12 (1879); Woodcock & Stearn, Lilies of the world, f. 75 (1950); Synge, Lilies, pl. 21 (1980). Stem 30–70 cm, occasionally to 200 cm. Leaves usually scattered, sometimes with 1 whorl near the middle of the stem. Flowers up to 12, nodding. Perianth-segments deep red or reddish orange, spotted inside with maroon, recurved at the tips. *W USA (coastal N California).* H4. Summer.

47. L. rubescens Watson. Illustration: Woodcock & Stearn, Lilies of the world, f. 101 (1950); *RHS Lily Year Book* **19**: f. 19 (1955); Synge, Lilies, pl. 22 (1980).

Bulb sometimes stoloniferous. Stem roots absent. Stem usually to 180 cm, sometimes to 300 cm, green suffused with purple, hairless. Leaves in whorls of 5–12, with some leaves scattered in lower part of stem, 5–12 × 1.5–3.5 cm, oblanceolate with 3 or 5 veins, hairless. Flower-stalks more or less erect, 3–6 cm. Flowers usually to 30, occasionally 100 or more, trumpet-shaped, fragrant, more or less erect. Perianth-segments 3.5–6.5 × 7–10 mm, at first white or whitish, changing to pinkish purple or purple, unspotted or sometimes with deep purple dots towards the base, spreading or recurved at the tips. Anthers and pollen yellow. *W USA (S Oregon, NW California).* H4. Summer.

L. washingtonianum Kellogg. Illustration: Elwes, Monograph of the genus Lilium, pl. 10 (1878); Woodcock & Stearn, Lilies of the world, f. 119 (1950); *RHS Lily Year Book* **33**: pl. 20 (1969). Stem to 2.5m. Leaves with wavy margins. Flowers broadly funnel-shaped, borne horizontally on almost erect stalks. Perianth-segments 8–10 × 1–2.5 cm, white or lilac with purplish dots towards the base, recurved towards the tips. Anthers pinkish. *W USA (S Oregon, NW California).* H4. Summer.

Var. **purpurascens** Stearn (var. *purpureum* Purdy). Illustration: *RHS Lily Year Book* **24**: pl. 3 (1960) & **31**: pl. 9 (1967); Synge, Lilies, pl. 23 (1980). Flowers more open, almost bowl-shaped and changing colour as they age from white to pink, finally becoming purplish.

48. L. bolanderi Watson. Illustration: Woodcock & Stearn, Lilies of the world, f. 28 (1950); *RHS Lily Year Book* **22**: f. 4 (1958); Synge, Lilies, pl. 17 (1980); Rix & Phillips, The bulb book, 160 (1981).

Stem roots absent. Stem to 140 cm, usually less, grey-green, sometimes marked with brown. Leaves in whorls of up to 12, 3–6 × 1–2.5 cm, oblanceolate with 1 or 3 veins, grey-green. Flower-stalks spreading or ascending, decurved below flower, up to 6.5 cm. Flowers to 18, funnel-shaped, scentless, usually somewhat drooping. Perianth-segments 3–4.5 × *c.* 1 cm, slightly recurved at the tips, dark crimson with darker red or purple spots, greenish at the base outside, the throat and tube yellowish spotted with purple or brown. Anthers purple with deep yellow pollen. *W USA (S Oregon, N California).* H4. Summer.

49. L. parryi Watson. Illustration: Botanical Magazine, 6650 (1882); Woodcock & Stearn, Lilies of the world, f. 89 (1950); Synge, Lilies, pl. xxiv (1980); Rix & Phillips, The bulb book, 162 (1981).

Bulb somewhat stoloniferous. Stem roots absent. Stem to 2 m, hairless. Leaves mostly scattered, sometimes with 1 or more whorls towards base of stem,

8–15 cm × 5–22.5 mm, linear-lanceolate to narrowly ovate with 3 veins, margins minutely rough. Flowers usually *c.* 12, occasionally up to 50, narrowly trumpet-shaped, fragrant, drooping or almost horizontal. Perianth-segments 7–10 cm × 8–12 mm, yellow, sometimes with deep red or brown dots near the base, narrowly oblanceolate with narrow claws, recurved at the tips. Anthers 7–9 mm, orange-brown with brownish red pollen. *SW USA (S California, S Arizona).* H4. Summer.

50. L. canadense Linnaeus (*L. canadense* var. *flavum* Pursh). Illustration: Elwes, Monograph of the genus Lilium, pl. 27 (1878); Synge, Lilies, pl. xxi (1950); Hay & Synge, Dictionary of garden plants in colour, 101 (1969); Brickell (ed.), RHS A–Z encyclopedia of garden plants, 630 (2003).

Bulb stoloniferous. At least some stem roots present. Stem to 1.5 m, green. Leaves mostly in whorls of 6–10 but some scattered, up to 15 × 2 cm, lanceolate to oblanceolate with 5 or 7 veins, minutely rough on margin and veins beneath. Flowers usually *c.* 10, sometimes up to 20, bell-shaped, nodding. Perianth-segments 5–7.5 × 1.2–2.5 cm, yellow, usually with maroon or purplish brown spots towards the base, tips recurved. Anthers 6–7 mm with yellow to reddish brown pollen. *Eastern N America.* H3. Summer.

Var. **coccineum** Pursh (var. *rubrum* Moore). Flowers red with a yellow throat.

Var. **editorum** Fernald. Flowers red with narrow perianth-segments (8–13 mm broad). Leaves broader than in var. *canadense*.

L. grayi Watson. Illustration: *Botanical Magazine*, 7234 (1892); Woodcock & Stearn, Lilies of the world, f. 48 (1950); *RHS Lily Year Book* **23**: pl. 5 (1959); Synge, Lilies, pl. 19 (1980). Stem roots absent. All leaves in whorls, 5–12 × 1.5–3 cm. Perianth-segments 4–4.5 × *c.*1 cm, deep red outside, inside paler and often yellow at the base, spotted with purple, tips not recurved. *E USA.* H3. Summer.

51. L. mackliniae Sealy. Illustration: Woodcock & Stearn, Lilies of the world, f. 74 (1950); Hay & Beckett (eds), Readers' Digest encyclopaedia of garden plants and flowers, 406 (1971); Synge, Lilies, pl. xiii (1980); Rix & Phillips, The bulb book, 165 (1981).

Bulb not stoloniferous. Stem roots present. Stem to 40 cm, green or tinged with brownish purple. Leaves scattered or in whorls near the stem apex, 3–6 cm × 4–10 mm, narrowly lanceolate to narrowly elliptic with usually 3 veins, hairy at the base. Flowers 1–6, broadly bell-shaped, more or less nodding. Perianth-segments 4.5–5 × 1.6–1.8 cm, purplish pink. Nectary-furrows green, hairless. Anthers purple with yellow, orange or brown pollen. *NE India (Manipur).* H4. Late spring–summer.

L. oxypetalum, which differs in having fringed nectary-furrows, may key out here (see under 52. *L. nanum*).

L. sherriffiae Stearn. Illustration: *Journal of the Royal Horticultural Society* **75**: f. 108 (1950); *Botanical Magazine*, n.s., 141 (1951); Synge, Lilies, pl. 13 (1980). Leaves scattered, linear to narrowly lanceolate. Flowers usually solitary, narrowly bell-shaped. Perianth-segments deep brownish purple outside, chequered with deep yellow and green inside towards the tips. Nectary-furrows with warty-crested margins. *Bhutan, Nepal.* H4. Summer.

L. henrici Franchet (*Nomocharis henrici* (Franchet) Wilson). Illustration: Woodcock & Stearn, Lilies of the world, f. 59 (1950); Turrill, Supplement to Elwes' Monograph of the genus Lilium **8**: 3 (1960); *RHS Lily Year Book* **27**: f. 51 (1963); Rix & Phillips, The bulb book, 162 (1981). Stem to 90 cm. Leaves 10–15 cm, rough on the margin. Perianth-segments whitish flushed with pinkish outside especially at the base; inner segments dark purple at the base with many longitudinal veins, slightly 3-lobed at the apex. Anthers yellow. *W China (Yunnan).* H4. Summer.

52. L. nanum Klotzsch (*Nomocharis nana* (Klotzsch) Wilson). Illustration: Turrill, Supplement to Elwes' Monograph of the genus Lilium **9**: 9 (1962).

Stem 6–45 cm, green, hairless. Leaves scattered, 4.5–12 cm × 2–5 mm, narrowly linear to linear with 3, 5 or 7 veins, margin minutely rough, hairless beneath. Flowers solitary, bell-shaped, fragrant, nodding. Perianth-segments 1–3.8 cm × 3–16 mm, pale purple to pink often with tiny reddish purple or brown dots, hairy or papillose towards the base inside. Nectary almost circular on the outer segments, more elongated on the inner. Anthers yellowish brown. *Himalaya, W China (SE Xizang, Yunnan).* H4. Summer.

Var. **flavidum** (Rendle) Sealy. Illustration: *Botanical Magazine*, n.s., 218 (1953). Flowers pale yellow

L. oxypetalum (Royle) Baker (*Nomocharis oxypetala* (Royle) Wilson). Illustration: Synge, Lilies, pl. 10 (1980). Stem to 25 cm. Leaves scattered, below the flower often in a whorl, to 7 × 1.2 cm. Flowers broadly bell-shaped. Perianth-segments 3.5–5.5 cm, yellow, with or without purplish spots towards the base; inner segments with fringed glands at the base. *NW Himalaya.* H4. Summer.

Var. **insigne** Sealy. Illustration: *Botanical Magazine*, n.s., 274 (1956). Flowers purple.

11. NOTHOLIRION Boissier

V.A. Matthews

Differs from *Lilium* in flowering once and then dying (the plant perennates by offsets) and having bulbs with few scales covered by a ribbed tunic; long basal leaves produced in autumn and winter; style with 3 narrow branches.

A genus of 6 species occurring from Afghanistan to W China. Most of the species like a cool position in peaty but not waterlogged soil; *N. thomsonianum* seems to require a sunnier place than the other species. They can also be grown in pots in an unheated greenhouse, especially in those districts where the leaves are subject to frost-damage.

Literature: Woodcock, H.B.D. & Stearn, W.T., Lilies of the world, 373–80 (1950).

1a. Perianth-segments with green tips; raceme with 10–40 flowers — 2
b. Perianth-segments without green tips; raceme with up to 6 flowers — **3. macrophyllum**
2a. Perianth-segments 4–6 mm wide — **1. thomsonianum**
b. Perianth-segments 8–18 mm wide — **2. bulbuliferum**

1. N. thomsonianum (Royle) Stapf. Illustration: *Botanical Magazine*, 4725 (1853) – as *Lilium roseum*; Grove & Cotton, Supplement to Elwes' monograph of the genus Lilium, t. 28 (1940); Woodcock & Stearn, Lilies of the world, f. 121 (1950); Rix & Phillips, The bulb book, 139 (1981).

Stem to 1 m. Raceme with 10–30 flowers. Flowers funnel-shaped, fragrant, borne horizontally. Perianth-segments 5–6.5 cm × 4–6 mm, pale pinkish lilac tipped with green, recurved at the tips. *Afghanistan, W Himalaya.* H4. Spring–summer.

2. N. bulbuliferum (Lingelsheim) Stearn (*N. hyacinthinum* (Wilson) Stapf). Illustration: Grove & Cotton, Supplement to Elwes' monograph of the genus Lilium, t. 29 (1940); Woodcock & Stearn, Lilies of the world, f. 124 (1950); *RHS Lily Year Book* **31**: pl. 7 (1967).

Stem 60–100 cm. Raceme with 10–30 flowers. Flowers funnel-shaped at the base, spreading wide towards the mouth, slightly fragrant, borne horizontally. Perianth-segments 2–4 cm × 8–18 mm, pale lilac tipped with green. *Nepal to W China.* H4. Summer.

N. campanulatum Cotton & Stearn. Illustration: Grove & Cotton, Supplement to Elwes' monograph of the genus Lilium, t. 30 (1940); Woodcock & Stearn, Lilies of the world, f. 123 (1950). Flowers 4–5 cm, scentless, drooping, deep crimson tipped with green. *N Burma, W China (Xizang, Yunnan).* H4. Summer.

3. N. macrophyllum (Don) Boissier. Illustration: Woodcock & Stearn, Lilies of the world, f. 125 (1950); *Journal of the Royal Horticultural Society* **94**: f. 101 (1969); Everard & Morley, Wild flowers of the world, pl. 108 (1970); Rix & Phillips, The bulb book, 163 (1981).

Stem 20–40 cm. Raceme bearing up to 6 flowers. Flowers broadly funnel-shaped, *c.* 5 cm across at the mouth, horizontal or slightly drooping. Perianth-segments 3–5 cm, pale lavender-pink spotted with purple inside and without green tips. *Nepal, Bhutan, N India, (Sikkim), W China (Xizang).* H4. Summer.

12. NOMOCHARIS Franchet

V.A. Matthews

Similar to *Lilium* but with the flowers opening flat or sometimes shallowly cup-shaped, and with the inner perianth-segments entire to fringed and bearing basal nectaries that usually have ridges of tissue arranged in a fan shape. Filaments swollen, with a needle-like appendage (aciculus) at the top.

A genus of 7 species from W China (including Xizang), Burma and N India where they are found in the alpine zone of high mountains. They grow best in a semi-shady position in which the soil never completely dries out. They resent root disturbance and attempts to move them are rarely successful. Propagation is by seed or bulb scales.

Literature: Sealy, J.R., A revision of the genus Nomocharis Franchet, *Botanical Journal of the Linnean Society* **87**: 285–323 (1983); Matthews, V., The genus Nomocharis, *The New Plantsman* **2**(4): 195–208 (1995).

1a. Leaves in whorls; filaments cylindric or broader at the top — 2
b. Leaves scattered; filaments subulate, often flattened below or on one side — 3
2a. Style 7–10 mm; perianth-segments with spots from base to apex, or if spotted only near the base then anthers 3–4 mm — **1. pardanthina**
b. Style 12–15 mm; perianth-segments spotted only near the base; anthers 4–6 mm — **2. farreri**
3a. Style 8–13 mm; flowers opening flat — **3. aperta**
b. Style 4–6 mm; flowers opening shallowly cup-shaped — **4. saluenensis**

1. N. pardanthina Franchet (*N. mairei* Léveillé; *N. leucantha* Balfour). Illustration: *RHS Lily Year Book* **10**: f. 23, 24, 26 (1946); Synge, Lilies, pl. xxix, (1980); *Botanical Journal of the Linnean Society* **87**: 293 (1983); Brickell (ed.), RHS A–Z encyclopedia of garden plants, 726 (2003).

Stem 25–90 cm. Leaves in whorls. Flowers 1–8, occasionally to 24, erect, horizontal or nodding, opening flat, 5–8.5 cm in diameter. Outer perianth-segments 2.5–4.5 × 1.5–2.5 cm, entire, white to pale pink with purple or maroon-crimson blotches, the base dark purple or dark maroon-crimson. Inner perianth-segments 2.6–4.5 × 2.4–3.6 cm, the upper half with fringed margins, white to pale pink with purple or maroon-crimson blotches that are deeper and more numerous than the blotches of the outer segments, the base dark purple. Nectaries with erect ridges of tissue in a fan shape on each side. Filaments 6–8 mm, cylindric or broader at the top, purplish, aciculus 1.5–3 mm; anthers 3–4 mm, crimson-maroon; pollen yellow or orange. Style 7–10 mm; style and ovary sometimes absent. *W China (Yunnan, Szechuan).* H3. Summer.

Forma **punctulata** Sealy (Illustration: *Gardeners' Chronicle* ser. 3, **59**: f. 138, 1916; *RHS Lily Year Book* **10**: f. 25, 1946) has perianth-segments with fine spots only towards the base, and the margins of the inner segments with tiny irregular teeth rather than fringed.

N. melegrina Franchet. Flowers nodding, 7–9.5 cm in diameter, white or white flushed with pink on the back, uniformly blotched with purple. Outer perianth-segments 4–5.5 cm, dark purple at the base. Inner perianth-segments 4–6 cm, margin entire or wavy to irregularly toothed or slightly fringed, dark purple at the base. Anthers 4–7 mm. Style and ovary sometimes absent. *W China (Yunnan, SE Xizang).* H3. Summer.

2. N. farreri (Evans) Harrow (*N. pardanthina* Franchet var. *farreri* Evans). Illustration: *Botanical Magazine*, 9557 (1939); Synge, Lilies, pl. xxviii (1980).

Stem 35–100 cm. Leaves in whorls. Flowers to 20, nodding at first, eventually horizontal or erect, opening flat, 5.5–9.5 cm in diameter. Outer perianth-segments 2.5–5.5 × 1–2.5 cm, entire, white to pink, blotched or spotted with purple or dark crimson for up to 1 cm at the base and with a larger basal blotch of the same colour. Inner perianth-segments 2.5–5.5 × 1.4–3.5 cm, entire or with small irregular teeth, white to pink, the base with a purplish crimson or reddish brown blotch, above which is a band of yellow-green, and then some reddish purple spots for up to 1 cm from the base. Nectaries with erect ridges of tissue in a fan shape on each side. Filaments 6–11 mm, cylindric or broader at the top, deep pink or purplish, aciculus 2–2.5 mm; anthers 4–6 mm; pollen yellow. Style 12–15 mm. *NE Burma.* H3. Summer.

There is a tendency for hybrids to occur when certain species of *Nomocharis* are grown together, and for them to combine the characters of the parents in varying degrees. Hybrids between *N. farreri* and *N. saluenensis* have been named **N.** × **notabilis** Sealy (illustration: *RHS Lily Year Book* **32**: pl. 3, 1968). Hybrids between *N. farreri* and *N. pardanthina* are known as **N.** × **finlayorum** Synge (illustration: *RHS Lily Year Book* **32**: pl. 4, 1968).

3. N. aperta (Franchet) Wilson. Illustration: *Botanical Magazine*, 9533 (1938); *RHS Lily Year Book* **10**: f. 16 (1946) & **29**: f. 37 (1965); Synge, Lilies, pl. xxvii (1980).

Stem 30–80 cm. Leaves scattered. Flowers 1–6, horizontal or nodding, opening flat, 5–10 cm in diameter. Outer perianth-segments 2.6–5 × 1–2.3 cm, entire, pale to deep pink, spotted with crimson in the lower part or all over, sometimes unblotched, the base with a dark purplish or crimson blotch, above which is a greenish or creamy area. Inner perianth-segments 2.5–4.1 × 1.3–3.2 cm, entire, similar in colour to the outer segments, but never unblotched. Nectaries with a fan-shaped swelling on each side, or more usually with a fleshy hump on one or both sides. Filaments 10–15 mm, subulate, plano-convex in section, purplish except for the yellow apex, aciculus 2–2.5 mm; anthers 3–6 mm; pollen yellow. Style 8–13 mm; style and ovary sometimes absent. *W China (Yunnan, Szechuan, SE Xizang), N Burma.* H3. Summer.

N. synaptica Sealy. Illustration: *Kew Bulletin* **1950**: 275 (1950). Flowers white, flushed with purplish pink and spotted all over with dark purple, the base of each perianth-segment dark purple with a narrow yellow border. Nectaries with erect ridges of tissue in a fan shape on each side. *India (Assam).* H3. Summer.

4. N. saluenensis Balfour. Illustration: *New Flora and Silva* **1**: f. 22 (1929); *Botanical Magazine*, 9296 (1933); *RHS Lily Year Book* **10**: f. 19 (1946); *Botanical Journal of the Linnean Society* **87**: 313 (1983).

Stem 20–85 cm. Leaves scattered. Flowers 2–5, horizontal or nodding, opening shallowly cup-shaped, 5–6.5 cm in diameter. Outer perianth-segments 3.3–4.3 × 1.3–2.3 cm, entire, white to pink, whitish or yellowish in lower part with a dark purple basal blotch, spotted with crimson near the base and sometimes more faintly so over the lower half. Inner perianth-segments 3.2–4.3 × 1.6–2.5 cm, entire, similar in colour to the outer segments. Nectaries with a fleshy hump on one or both sides. Filaments 8–11 mm, slightly flattened in lower part, purplish, aciculus 1.5–3 mm; anthers 7.5–8 mm, dark or blue-green; pollen yellow or greenish. Style 4–6 mm; ovary and style sometimes absent. *W China (Yunnan, SE Xizang), NE Burma.* H3. Summer.

13. CARDIOCRINUM (Endlicher) Lindley

V.A. Matthews

Differs from *Lilium* in the monocarpic bulbs that die after flowering (the plant perennates by offsets), the large, long-stalked, heart-shaped leaves and the capsule that splits into 3 valves, which are fringed with teeth.

A genus of 3 species from the Himalaya, China and Japan, formerly included in *Lilium*. The plants grow best in semi-shade, in a rich soil with plenty of leaf-mould. The bulbs should not be planted too deeply.

1a. Stem 1.5–4 m, leafy for whole length; flowers 15–20 cm **1. giganteum**
b. Stem 1.2–2 m, leafless in lower part; flowers 15 cm or less 2
2a. Leaf-blade longer than broad; raceme with 1–5 flowers **2. cathayanum**
b. Leaf-blade as broad as long; raceme with 4–10 or more flowers **3. cordatum**

1. C. giganteum (Wallich) Makino *(Lilium giganteum* Wallich*)*. Illustration: Brickell (ed.), RHS A–Z encyclopedia of garden plants, 231 (2003).

Stem 1.5–4 m. Basal leaves in a rosette, blade up to 45 × 40 cm, stem leaves decreasing in size up the stem, all dark shiny green above, paler beneath. Raceme with up to 25 flowers. Flowers 15–20 cm, regularly funnel-shaped, fragrant, pure white or with a green tinge, striped with reddish purple inside. Capsule 5–6.5 cm.

Var. **giganteum.** Illustration: Woodcock & Stearn, Lilies of the world, frontispiece & f. 54 (1950); Synge, Collins guide to bulbs, edn 1, pl. III (1961); Hay & Beckett (eds), Readers' Digest encyclopaedia of garden plants and flowers, 116 (1971); Rix & Phillips, The bulb book, 163 (1981). Stem 1.5–4 m, green, flowers white, only greenish outside when young, and the lowest flowers in the raceme opening first. *Himalaya, NW Burma, China (SE Xizang).* H3. Summer.

Var. **yunnanense** (Elwes) Stearn. Illustration: Synge, Lilies, pl. 26 (1980). Stem 1.5–2 m, deep purple, flowers often tinged with green and the flowers at the top of the raceme opening first. *W & C China.* H3. Summer.

2. C. cathayanum (Wilson) Stearn (*Lilium cathayanum* Wilson). Illustration: *Journal of the Royal Horticultural Society* **70**: f. 27, 28 (1945).

Stem to 1.5 m, green, often dark purple below. Leaves absent in lower part of stem, blade up to 20 × 13 cm, dark shiny green. Raceme with 1–5 flowers. Flowers 10–13 cm, irregularly funnel-shaped, white or greenish white, the lower segments marked with reddish brown inside and often with reddish spots at the tips of the perianth-segments. Capsule *c.* 5.5 cm. *E & C China.* H3. Summer.

3. C. cordatum (Thunberg) Makino (*Lilium cordifolium* Thunberg; *L. cordatum* (Thunberg) Koidzumi). Illustration: *Botanical Magazine*, 6337 (1878); Woodcock & Stearn, Lilies of the world, 371 (1950).

Stem 1.2–2 m, green. Basal leaves in a rosette, blade up to 30 × 30 cm, marked with reddish brown when young; stem leaves absent in lower part of stem. Raceme usually with 4–10 flowers, occasionally with up to 24. Flowers to 15 cm, irregularly funnel-shaped, fragrant, creamy white, the lower segments with a yellow blotch and reddish marks and spots. Capsule *c.* 5.5 cm. *Japan, Former USSR (Sakhalin Islands).* H3. Summer.

14. UVULARIA Linnaeus

V.A. Matthews

Rootstock a rhizome. Stem simple or branched, leafy. Leaves alternate, stalkless, sometimes perfoliate. Flowers drooping, yellow, narrowly bell-shaped, the perianth-segments free. Stamens much shorter than perianth-segments; anthers opening by slits. Ovary superior. Style with 3 branches. Fruit a 3-lobed or 3-winged capsule that splits between the septa.

A genus of 5 species native to eastern N America. The species with non-perfoliate leaves were formerly placed in the genera *Oakesia* Watson and *Oakesiella* Small. They are best grown in shade, in a well-drained soil. Propagation is by division.

1a. Leaves perfoliate 2
b. Leaves not perfoliate **3. sessilifolia**
2a. Leaves hairless beneath; perianth-segments 2–3.5 cm, pale yellow with tiny projecting glands on the inner surface **2. perfoliata**
b. Leaves downy beneath; perianth-segments 2.5–5 cm, bright yellow and smooth on the inner surface **1. grandiflora**

1. U. grandiflora Smith. Illustration: *Botanical Magazine*, 1112 (1808); Everard & Morley, Flowers of the world, 167 (1970); *Journal of the Royal Horticultural Society* **100**: 527 (1975); Klein, Gartenblumen **1**: Frühlingsblumen, 8 (1979).

Stem to 75 cm, hairless. Leaves perfoliate, 5–13 cm, oblong to narrowly ovate, downy beneath. Perianth-segments 2.5–5 cm, bright yellow, inner surface smooth. Stamens longer than the style. *Eastern N America.* H3. Spring–summer.

2. U. perfoliata Linnaeus. Illustration: Smith, Exotic botany **1**: t. 49 (1804); Rickett, Wild flowers of the United States **1**: pl. 27 (1965); Peterson & McKenny,

Field guide to wildflowers of north-eastern and north-central North America, 103 (1968).

Stem 20–60 cm, hairless. Leaves perfoliate, 5–9 cm, oblong to narrowly ovate, hairless, paler beneath. Perianth-segments 2–3.5 cm, pale yellow, inner surface with tiny projecting glands. Stamens shorter than the style. *Eastern N America.* H3. Spring–early summer.

3. U. sessilifolia Linnaeus (*Oakesiella sessilifolia* (Linnaeus) Small). Illustration: Smith, Exotic botany **1**: t. 52 (1804); *Botanical Magazine*, 1402 (1811); Rickett, Wild flowers of the United States **1**: pl. 27 (1965); Niering & Olmstead, Audubon Society field guide to North American wildflowers: Eastern region, pl. 318 (1979).

Stem to 40 cm, hairless. Leaves 3.5–7.5 cm, narrowly oblong, hairless, paler beneath, margin rough. Perianth-segments 1.5–3 cm, greenish yellow, inner surface smooth. Ovary and capsule borne on a stalk. *Eastern N America.* H3. Spring–summer.

U. caroliniana (Gmelin) Wilbur (*U. pudica* (Walker) Fernald; *U. puberula* Michaux). Illustration: *Loddiges' Botanical Cabinet* **13**: t. 1260 (1827). Stem with lines of downy hairs, leaves bright green on both surfaces and ovary and capsule not borne on a stalk. *SE USA.* H4. Spring–summer.

15. TRICYRTIS Wallich

P.F. Yeo

Perennial herbs. Stems (in cultivated species) 20–110 cm, simple or few-branched, leafy. Leaves 5–20 cm, alternate, sometimes in 2 rows, ovate to lanceolate, stalkless, lightly pleated, sometimes spotted with dark green. Flowers bisexual. Perianth-segments erect at base, spotted, 3–6 times as long as broad, the three outer each with a conspicuous, sometimes bilobed, spherical or ovoid nectarial pouch. Stamens curved outwards at the top. Ovary superior, 3-celled, narrowly oblong; style single, with 3 recurved bifid stigmatic branches. Perianth-segments (mainly on upper surfaces), stamen filaments and stigmatic branches usually with violet to reddish purple spots. Fruit a fusiform, furrowed capsule opening along the septa. Seeds flat.

A genus of 11–16 species from the mountains of Japan, China, Taiwan, Philippines and E Himalaya, often growing in woods. They are suitable for semi-shade or full sun and benefit from a moist and humus-rich soil. Autumn-flowering species need a warm position in northern countries to ensure flowering before the first frosts. The flowers are unusual in both form and colour.

Literature: Mathew, B., *The Plantsman* **6**: 193–225 (1985); Hong, W-P., Greenham, J.R., Jury, S.L. & Williams, C.A., Leaf flavonoid patterns in the genus Tricyrtis (Tricyrtidaceae senus stricto, Liliaceae sensu lato), *Botanical Journal of the Linnean Society*, **130**: 261–266. (1999).

Stems. Arching and drooping: **1, 4**; not more than 15 cm: **2**.
Flowers. Bell-shaped: **1**.
Ground-colour of flowers. Yellow: **1, 2, 3, 5**; white or pinkish: **6, 7, 8**.
Flowers. On separate stalks: **1, 2, 3, 4, 7**; in cymes: **5, 6, 7, 8**.

1a. Ground-colour of flowers yellow — 2
b. Ground-colour of flowers white or pinkish — 6
2a. Stems arching and drooping; perianth-segments not mucronate — 3
b. Stems erect; perianth-segments mucronate, conspicuously so in bud — 4
3a. Leaves cordate and embracing stem; flowers drooping, bell-shaped — **1. macrantha**
b. Leaves perfoliate; flowers ascending, funnel-shaped — **4. perfoliata**
4a. Flowers 2 to several in branched cymes — **5. latifolia**
b. Flowers 1 or few on their own stalks in the leaf-axils — 5
5a. Stems 5–15 cm; flower-stalk much shorter than flower — **2. nana**
b. Stems 20–50 cm; flower-stalks longer than or slightly shorter than flower — **3. flava**
6a. Perianth-segments not more than 2 cm, with the upper two-thirds spreading abruptly or recurved — **7. macropoda**
b. Perianth-segments 2.5–3 cm, diverging gradually — 7
7a. Plant tufted, thickly hairy; leaves pale green, not glossy, usually evenly tapered from near the base — **8. hirta**
b. Plant with stolons, moderately or sparsely hairy; leaves deep green, more or less glossy, ovate to elliptic — **6. stolonifera**

1. T. macrantha Maximowicz. Illustration: Makino, An illustrated flora of Japan, edn 2. 756 (1961); Kitamura et al., Coloured illustrations of herbaceous plants of Japan (Monocotyledoneae), t. 39, f. 247 (1970); *Journal of the Royal Horticultural Society* **107**: 228 (1982).

Stems arching and drooping, with coarse pale brown hairs (except in subsp. **macranthopsis** (Masamune) Kitamura). Leaves ovate-oblong to lanceolate, cordate, embracing the stem. Flowers 3–4 cm long, in upper axils, 2 or few on separate stalks, bell-shaped, drooping; ground-colour yellow; nectarial pouch ovoid, bent outwards. *Japan.* H2. Early autumn.

2. T. nana Yatabe. Illustration: Makino, An illustrated flora of Japan, edn 2, 755 (1961); Kitamura et al., Coloured illustrations of herbaceous plants of Japan (Monocotyledoneae), t. 39, f. 249 (1970).

Stems 5–15 cm, erect, shortly hairy. Leaves few, lanceolate to ovate, embracing the stem, heavily blotched. Flower-stalks much shorter than flower and fruit. Flowers few, solitary in leaf-axils, erect, funnel-shaped. Perianth-segments about 2 cm, with apical bristle to 2 mm; ground-colour yellow. Anthers 2 mm. *Japan*, H2.

3. T. flava Maximowicz. Illustration: Makino, An illustrated flora of Japan, edn 2, 755 (1961); Kitamura et al., Coloured illustrations of herbaceous plants of Japan (Monocotyledoneae), t. 39, f. 250, 251 (1970).

Stems 20–50 cm, erect, hairy. Leaves embracing the stem, blotched, the upper more or less broadly ovate. Flower-stalks longer than flower and fruit or only slightly shorter. Flowers 1 or few on separate stalks in leaf-axils, erect, funnel-shaped. Perianth-segments about 2.5 cm; apical bristle of outer segments 3–4 mm; ground-colour yellow. Anthers 2.5–3 mm. *Japan.* H2. Autumn.

Subsp. **ohsumiensis** (Masamune) Kitamura (illustration: Kitamura et al., Coloured illustrations of herbaceous plants of Japan (Monocotyledoneae), t. 39, f. 251, 1970) has perianth-segments 2.5–3.5 cm, with apical bristle not more than 2 mm and anthers about 4 mm. *Japan.* H2.

4. T. perfoliata Masamune. Illustration: Kitamura et al., Coloured illustrations of herbaceous plants of Japan (Monocotyledoneae), t. 39, f. 248 (1970); *The Plantsman* **6**: 204 (1985).

Stems 50–70 cm, arching and drooping, hairless. Leaves perfoliate. Flowers ascending, solitary in leaf-axils, funnel-shaped. Perianth-segments 2.5–3 cm, with bright yellow ground-colour. *Japan.* H2.

5. T. latifolia Maximowicz (*T. bakeri* Koidzumi; *T. macropoda* misapplied; *T. puberula* Nakai & Kitagawa). Illustration: *Botanical Magazine*, 6544 (1881); Kitamura et al., Coloured illustrations of herbaceous plants of Japan (Monocotyledoneae), t. 40, f. 252 (1970); *The Plantsman* **6**: 193 (1985).

Stems erect, hairy above. Leaves broadly ovate, cordate, thin, embracing the stem. Flowers 2–2.5 cm, erect, 2 or several in branched cymes, funnel-shaped, with bracts. Perianth-segments diverging from the middle; ground-colour yellow. *Japan, China.* H2. Summer.

6. T. stolonifera Matsumura (*T. formosana* Baker.). Illustration: *Botanical Magazine*, 8560 (1914); Brickell (ed.), RHS A–Z encyclopedia of garden plants, 1046 (2003).

Plant forming a dense, even stand of erect, somewhat hairy stems and spreading by stolons. Leaves more or less glossy, deep green with darker green spots, especially when young, the lower narrowed towards base, continued downwards for about 1–1.5 cm as a stem-clasping sheath; upper leaves ovate, cordate, embracing the stem. Flowers 2.5–3 cm, erect, several in a loosely branched terminal cyme, widely funnel-shaped; perianth-segments with a yellowish zone above the nectarial pouches, otherwise more or less brownish purple outside in bud and pinkish purple above the yellow zone when open, nearly white inside but densely purple-spotted. *Taiwan.* H3. Early autumn.

Some plants in cultivation in Britain under the names *T. stolonifera* and *T. formosana* have cordate-clasping, non-sheathing leaf-bases throughout. They have well-developed stolons but are probably hybrids between *T. formosana* and *T. hirta*. They vary in hairiness, leaf-colour and intensity of spotting of the flower.

7. T. macropoda Miquel. Illustration: Makino, An illustrated flora of Japan, edn 2, 754 (1961); Kitamura et al., Coloured illustrations of herbaceous plants of Japan (Monocotyledoneae), t. 40, f. 253 (1970).

Plants sometimes hairy. Stems erect. Leaves rounded or slightly cordate at base and forming a short, open sheath. Flowers 1.5–2 cm long, erect, several in terminal and axillary cymes that may be glandular-hairy; perianth-segments with the upper two-thirds reflexed; ground-colour white. *Japan, China.* H2. Early autumn.

T. maculata (D.Don) Macbride. Illustration: *Botanical Magazine*, 4955 (1855). Hardly different from *T. macropoda.* Cymes loose, repeatedly branched, with flowers held well above the leaves. Upper two-thirds of perianth-segments spreading. *C Nepal to Burma.* H2. Summer.

T. affinis Makino (*T. macropoda* Miquel subsp. *affinis* (Makino) Kitamura). Illustration: Makino, An illustrated flora of Japan, edn 2, 754 (1961); Kitamura et al., Coloured illustrations of herbaceous plants of Japan (Monocotyledoneae), t. 40, f. 254 (1970). Plant hairy. Flowers like those of *T. macropoda* but on separate stalks and with upper two-thirds of perianth-segments spreading, not reflexed. *Japan.* H2. Autumn.

8. T. hirta (Thunberg) W.J. Hooker. Illustration: *Botanical Magazine*, 5355 (1863); Makino, An illustrated flora of Japan, edn 2, 754 (1961); Kitamura et al., Coloured illustrations of herbaceous plants of Japan (Monocotyledoneae), t. 40, f. 255 (1970).

Plant tufted, without stolons, thickly hairy. Stems slightly divergent from the rootstock. Leaves cordate, the base forming a short open sheath, pale green, usually evenly tapered from near base. Flowers 2.5–3 cm long, erect, 1–5 in axils on separate stalks usually shorter than flowers, or the upper sometimes in cymes with short internodes, funnel-shaped; perianth-segments spreading or recurved only at the tips; ground-colour white, spots violet or purple, sometimes lacking. *Japan.* H2. Early autumn.

Sometimes cultivated and illustrated in books as *T. macropoda.*

17. ALSTROEMERIACEAE

Terrestrial plants, herbaceous. Leaves reversed by a twist at the base, not hairy or pleated. Flowers bisexual in an umbel without spathes, sometimes with additional flowers below the terminal umbel, weakly zygomorphic. Perianth of 6 segments that are generally similar, but often 2 or 3 of the inner whorl narrower and spotted. Stamens 6. Ovary of 3 united carpels, fully inferior; placentation axile. Fruit a capsule. Seeds pale.

There are 5 genera with 65 species. One genus is native to the southern parts of the USA, and two are widely cultivated.

1a. Perianth bilaterally symmetric; flowers erect; plants not climbing or scrambling **1. Alstroemeria**
b. Perianth radially symmetric; flowers drooping; plants climbing or scrambling **2. Bomarea**

1. ALSTROEMERIA Linnaeus
V.A. Matthews edited by S.G. Knees
Rootstock fibrous, with clusters of tubers or creeping rhizomes. Leaves alternate or scattered up the stem, larger on the sterile shoots, with short stalks that are twisted through 180°, many-veined. Flowers usually bilaterally symmetric, borne usually in a terminal, simple or compound umbel, rarely solitary. Perianth-segments free, clawed at the base, almost equal or the outer 3 broader and shorter than the inner. Stamens often unequal, usually curved downwards, then upwards at the tips; anthers basifixed. Ovary inferior, 3-celled with numerous ovules in each cell; style slender, curved downwards, with a 3-fid stigma.
Fruit a capsule containing numerous seeds.

A genus of about 50 species from S America. They will grow in any well-drained fertile soil, in semi-shade or sun. The more tender species should be given winter protection or the roots may be lifted and stored during the winter; alternatively they can be grown under glass.
Propagation is by seed or division of the rootstock.

Literature: Garaventa, A., El genero Alstroemeria en Chile, *Anales del Museo de Historia Natural de Valparaiso* **4**: 63–108 (1971); Muñoz Schick, M. & Moreira Muñoz, A., Alstroemerias de Chile (2003).

Leaf-margin. Hairy: **3**, **6**; hairless: **1**, **2**, **4**, **5**, **6**, **7**.
Umbel. Simple: **4**, **6**; compound: **1**, **2**, **3**, **5**, **6**, **7**.
Flowers solitary: **1**, **6**.
Basic colour of flowers. Whitish or white: **2**, **6**; yellow: **1**, **4**, **5**; orange: **1**, **2**, **3**, **4**; red: **2**, **3**, **5**, **4**, **6**; pink: **2**, **6**; lilac to purple: **2**, **7**.
Spots or stripes of perianth-segments. On all segments: **4**; on inner segments only: **1**, **2**, **3**, **5**, **6**, **7**; yellow: **2**, **6**; brown: **5**, **6**; purplish brown or reddish brown: **4**; purple: **2**, **3**, **4**, **7**; reddish purple: **3**, **6**; red: **1**, **2**.

1a. All perianth-segments spotted **4. psittacina**
b. Only inner perianth-segments spotted 2
2a. Flowers whitish or white, often flushed and splashed with mauve or pink, or mauve, purple, pink or crimson 3
b. Flowers yellow, orange, or red 5
3a. Flowers solitary or with simple umbels **6. pelegrina**
b. Flowers in compound umbels with 3–15 branched rays 4
4a. Leaves narrowly linear to linear-lanceolate or narrowly obovate; flowers white to lilac, pink or crimson, if purple then upper inner perianth-segments yellow **2. ligtu**
b. Leaves ovate-oblong; flowers violet; inner upper perianth-segments lacking any yellow, although sometimes orange at the base **7. violacea**
5a. Leaf margins hairy **3. haemantha**
b. Leaf margins hairless 6
6a. Inner perianth-segments with brown spots **5. brasiliensis**
b. Inner perianth-segments spotted or streaked with white, yellow, red or purple 7
7a. Flowers yellow to orange **1. aurea**
b. Flowers reddish **2. ligtu**

1. A. aurea Graham (*A. aurantiaca* D. Don; *A. aurea* Meyen). Illustration: *Edwards's Botanical Register* **22**: t. 1843 (1836); Hay & Synge, Dictionary of garden plants in colour, 120 (1969); Brickell (ed.), RHS A-Z-encyclopedia of garden plants, 103 (2003); Muñoz Schick, & Moreira Muñoz, Alstroemerias de Chile, 122–125 (2003).

Stems to 1 m. Leaves 7–10 cm, lanceolate, hairless. Umbel with 3–7 rays, each ray bearing 1–3 flowers. Perianth-segments 4–5 cm, orange; the outer segments tipped with green; the upper inner segments spotted and streaked with red, the lower inner segments spotted or unspotted. Stamens shorter than perianth-segments. *Chile.* H3. Summer.

Several cultivars are available with paler, deeper or larger flowers.

A. pygmaea Herbert differs in having its stem below the ground and a solitary flower growing from a tuft of grey-green leaves that are *c.* 2.5 cm. The perianth-segments are *c.* 5 cm, yellow with the inner ones spotted with red. *Argentina.* H5. Summer.

A. patagonica Philippi – illustration: Muñoz Schick, & Moreira Muñoz, Alstroemerias de Chile, 126, 127 (2003) – is another low growing, yellow-flowered species that is occasionally seen in cultivation.

2. A. ligtu Linnaeus (*A. chilensis* Lemaire). Illustration: Loddiges, Botanical cabinet **1**: t. 17 (1818); *Edwards's Botanical Register* **25**: pl. 13 (1839); Mathew & Redouté, Lilies and related flowers, 121 (1982); Muñoz Schick, & Moreira Muñoz, Alstroemerias de Chile, 112–117 (2003).

Stems 45–60 cm. Leaves 5–8 cm, narrowly lanceolate, hairless. Umbel with 3–8 rays, each ray bearing 2 or 3 flowers. Perianth-segments 4–5 cm, varying from whitish to lilac, pink or reddish; the upper inner segments spotted and streaked with white or yellow and red, or purple. Stamens shorter than perianth-segments. *Chile, Argentina.* H4. Summer.

The commonly grown 'Ligtu hybrids' were bred from *A. ligtu* var. *angustifolia* Anon. and *A. haemantha.* They show great variation in height and habit and have pink, orange or yellow flowers.

A. revoluta Ruiz & Pavon – illustration: Muñoz Schick, & Moreira Muñoz, Alstroemerias de Chile, 108, 109 (2003) – differs in having very narrowly linear leaves and purple or crimson flowers only 8–20 mm, with the upper inner perianth-segments yellow, spotted with purple. *Chile.* H5–G1. Summer.

3. A. haemantha Ruiz & Pavon. Illustration: Sweet, Ornamental Flower Garden **1**: t. 14 (1854).

Stems 60–90 cm. Leaves 7–15 cm, narrowly lanceolate, with hairs on the margin. Umbel with 3–15 rays, each ray bearing 1–4 flowers. Perianth-segments 4–5 cm, orange to deep orange-red; outer segments tipped with green; upper inner segments yellowish with purple stripes, lower inner segments similar in colour to outer segments, but striped with dark red. Stamens shorter than perianth-segments. *Chile.* H3. Summer.

Recent work suggests that this is another variant of *A. ligtu*, but as it is so distinct horticulturally it has been maintained here.

4. A. psittacina Lehmann. Illustration: *Edwards's Botanical Register* **18**: t. 1540 (1832); Sweet, Ornamental Flower Garden **3**: t. 183 (1854).

Stems up to 90 cm, spotted with pale purple. Leaves to 7.5 cm, lanceolate, hairless. Umbels with 4–6 unbranched rays. Perianth-segments 4–4.5 cm, deep red, tipped with green and spotted with reddish brown or purplish brown. Stamens more or less equal in length to perianth-segments. *N Brazil.* H4. Summer.

Plants sold under the name *A. pulchella* often turn out to be *A. psittacina* or *A. ligtu. A. pulchella* Linnaeus filius seems to be similar to *A. psittacina*, but the name is surrounded by confusion and work needs to be done on this problem.

A. versicolor Ruiz & Pavon – illustration: Muñoz Schick, & Moreira Muñoz, Alstroemerias de Chile, 110, 111 (2003) – differs in having shorter stems (20–60 cm) and umbels with 2–4 rays. The perianth-segments are yellow or orange, spotted with purple. *Chile.* H5–G1. Summer–autumn.

5. A. brasiliensis Sprengel.

Stems to 120 cm. Leaves 5–10 cm, lanceolate, hairless. Umbel with *c.* 5 rays, each ray bearing 1–3 flowers. Perianth-segments *c.* 4 cm, reddish yellow; inner segments with brown spots. Stamens shorter than perianth-segments. *Brazil.* H4. Summer–early autumn.

6. A. pelegrina Linnaeus. Illustration: Reader's Digest encyclopaedia of garden plants and flowers, 30 (1971); Brickell (ed.), RHS A–Z encyclopedia of garden plants, 104 (2003); Muñoz Schick, & Moreira Muñoz, Alstroemerias de Chile, 68 (2003).

Stems 30–60 cm. Leaves 5–8 cm, lanceolate, hairy or hairless. Flowers solitary or in umbels with 2 or 3 unbranched rays. Perianth-segments *c.* 5 cm, whitish flushed with mauve or pink, and with a darker central patch; the upper inner segments yellow at the base and spotted with brown or reddish purple. Stamens shorter than perianth-segments. *Peru.* H5–G1. Summer–autumn.

'**Alba**' (var. *alba* Anon.) – illustration: *Paxton's Magazine of Botany* **1**: 199 (1841) – has white flowers, flushed with green and the upper inner segments are spotted with green and often have a yellow blotch.

A. hookeri Loddiges. Illustration: Loddiges', Botanical Cabinet **13**: t. 1272 (1827); Maxwell, Flowers: a garden notebook, pl. 11 (1923); Muñoz Schick, & Moreira Muñoz, Alstroemerias de Chile, 102 (2003). The flowers are pink, with the upper inner segments blotched with yellow and spotted with reddish purple. *Chile.* H4. Summer.

7. A. violacea Philippi. Illustration: *Botanical Magazine*, n.s., 42 (1948); Muñoz Schick, & Moreira Muñoz, Alstroemerias de Chile, 26 (2003).

Stems 50–100 cm. Leaves 6–9 cm, ovate-oblong, hairless. Umbel with 3–6 rays, each ray bearing 3–5 flowers. Perianth-segments 3.5–5.5 cm, violet; the upper inner white below, spotted with purple, sometimes orange at the base. Stamens shorter than perianth-segments. *N Chile.* H4. Summer.

A. violacea has been crossed with *A. aurea* to produce 'Walter Fleming', which has deep yellow flowers flushed with pink and spotted with maroon.

2. BOMAREA Mirbel

V.A. Matthews

Herbs, usually climbing, growing from a rhizome with slender roots that bear tubers at their tips. Stem leafy. Leaves with short stalks that are twisted through 180°, many-veined. Flowers in terminal, simple or compound umbels. Perianth narrowly bell-shaped with free segments in 2 whorls that are usually rather dissimilar. Stamens borne at the bases of the segments and equal in length to them; anthers basifixed. Ovary superior, 3-celled; style slender, with 3 branches. Capsule 3-celled, with 6 ribs. Seeds orange or red, numerous.

About 120 New World species occurring mainly from Mexico to the Andes. The stems normally die back to ground level in the winter. They require a well-drained soil and almost no water in the resting season. They grow best with complete frost-protection, but in northern countries can be planted outdoors during the summer in a sheltered position. Propagation is by seed or division of the rhizome.

Perianth-segments. Equal or almost so: **1, 4, 5, 6, 7**; unequal: **2, 3**. Outer segments 1–1.5 cm: **7**; 2.5–3.5 cm: **1, 2, 3, 4, 6**; 5–6.5 cm: **3, 5, 6**. Outer segments crimson, pink or purplish: **3, 4, 5, 7**; scarlet, orange or yellow: **1, 2, 3, 6**. Inner segments spotted: **1, 2, 3, 4, 5, 6**; inner segments unspotted: **2, 7**.
Umbel. Simple: **1, 2, 3**; compound: **4, 5, 6, 7**.

1a. Umbel simple — 2
b. Umbel compound — 4
2a. Perianth-segments more or less equal in length — **1. multiflora**
b. Inner perianth-segments longer than outer — 3
3a. Inner perianth-segments 2.5–3.5 cm — **2. caldasii**
b. Inner perianth-segments 3.5–7.5 cm — **3. patacocensis**
4a. Flowers 1–1.5 cm, inner perianth-segments not spotted — **7. salsilla**
b. Flowers more than 1.5 cm; inner perianth-segments spotted — 5
5a. Outer perianth-segments orange-red — **6. shuttleworthii**
b. Outer perianth-segments crimson-purple or pink — 6
6a. Flowers 2.5–3 cm — **4. edulis**
b. Flowers 5.5–6.5 cm — **5. cardieri**

1. B. multiflora (Linnaeus filius) Mirbel. Illustration: *Journal of the Royal Horticultural Society* **41**: 449 (1892); Menninger, Flowering vines of the world, fig. 21 (1970); Harrison, Climbers and trailers, 23 (1973).

Stem downy. Leaves oblong, 7–10 cm. Flowers 20–40 in a simple umbel. Perianth-segments more or less equal, *c.* 2.5 cm. Outer segments tinged with red; inner segments reddish yellow with brown spots. *Colombia, Venezuela.* H5–G1.

B. oligantha Baker. Umbel with only 6–8 flowers that are up to 3.3 cm. *Ecuador.* H5–G1.

B. acutifolia, which has umbels with 13–20 flowers, 2.5–4 cm, may key out here (see entry under 6. *B. shuttleworthii*).

2. B. caldasii (Humboldt, Bonpland & Kunth) Ascherson & Graebner (*B. caldasiana* Herbert; *B. kalbreyeri* misapplied, not Baker). Illustration: *Gartenflora* **32**: 21 (1883); *Illustration Horticole* **32**: pl. 550 (1885); *Botanical Magazine*, n.s., 465 (1965); Harrison, Climbers and trailers, 21 (1973); Brickell (ed.), RHS A–Z encyclopedia of garden plants, 186 (2003).

Stem to 4 m. Leaves oblong, to 15 × 2.5 cm, almost hairless or downy on the veins beneath. Flowers 5–40 or more, in a simple umbel. Perianth-segments unequal; outer segments 2–2.5 cm, yellow to red or reddish brown; inner segments 2.5–3.5 cm, yellow to orange, unspotted or spotted with green, brown or red. *Andes of Colombia and Ecuador.* H5–G1.

3. B. patacocensis Herbert (*B. conferta* Bentham).

Stem to 5 m, downy. Leaves oblong-lanceolate, to 15 cm, downy beneath. Flowers 20–60 in a simple umbel. Perianth-segments unequal; outer segments *c.* 3.3 cm, orange or crimson; inner segments to 6.5 cm, yellow or crimson with brown or violet spots, orange at the tip. *Andes of Colombia and Ecuador.* H5–G1.

Plants grown under this name often turn out to be *B. racemosa.*

B. racemosa Killip. Stem densely red-hairy, outer segments 5–6.3 cm, scarlet; inner segments 6.3–7.5 cm, scarlet, spotted with brown, yellow at the base. *Colombia.* H5–G1.

B. frondea Masters. Illustration: *Gardeners' Chronicle* **17**: 669 (1882); *Journal of the Royal Horticultural Society* **41**: 444 (1892); *Botanical Magazine*, 7247 (1892). Outer segments yellow, inner segments yellow with dark red spots. *Colombia.* H5–G1.

4. B. edulis (Tussac) Herbert. Illustration: *Addisonia* **2**: pl. 65 (1917); Herklots, Flowering tropical climbers, 28 (1976).

Leaves lanceolate, to 12.5 × 2.5 cm, hairless or downy beneath. Flowers borne in a compound umbel. Perianth-segments more or less equal, 2.5–3 cm; outer segments pink; inner segments green or yellowish, spotted with purple or dark pink. *West Indies, C America south to Brazil & Peru.* G1.

5. B. cardieri Masters. Illustration: *Revue Horticole* **49**: 187 (1877); *Journal of the Royal Horticultural Society* **55**: 130 (1899); *Botanical Magazine*, 9601 (1940); Herklots, Flowering tropical climbers, 31 (1976).

Stem to 4 m, hairless. Leaves oblong-lanceolate, acuminate, to 18 × 7 cm. Flowers up to 40 in a compound umbel. Perianth-segments equal, 5.5–6.5 cm; outer segments pink, finely speckled inside with deep purple, and with a short, green, horn-like projection on the back just below the tip; inner segments pink or sometimes green, spotted with deep purple. *Colombia.* H5–G1.

6. B. shuttleworthii Masters. Illustration: *Gardeners' Chronicle* **17**: 85 (1882).

Leaves ovate-lanceolate, to 15 × 5 cm, hairless beneath. Flowers borne in a compound umbel. Perianth-segments equal, to 5 cm; outer segments orange-red with dark spots towards the tip; inner segments yellow at the base merging into green at the tip, with dark spots towards the tip and a red midrib. *Colombia.* H5–G1.

B. acutifolia (Link & Otto) Herbert. Illustration: *Botanical Magazine*, 3871 (1841), 6444 (1879); Herklots, Flowering tropical climbers, 26 (1976). Leaves downy beneath, flowers 2.5–4 cm, outer segments not spotted. *Mexico.* G1.

7. B. salsilla (Linnaeus) Herbert. Illustration: *Revue Horticole* series 4, **5**: 61 (1856); *Journal of the Royal Horticultural Society* **29**: 260 (1886).

Stem to 2 m, hairless. Leaves lanceolate to oblong. Flowers 6–18, borne in a compound umbel. Perianth-segments equal, 1–1.5 cm, crimson or purplish red, tinged with green at the tips. *Chile.* H5–G1. Summer.

18. HYACINTHACEAE

Plants terrestrial, herbaceous, with bulbs. Leaves basal, well developed. Flowers in racemes or panicles, when solitary without spathes, usually actinomorphic. Perianth of 6 free or united segments. Stamens 6. Ovary of 3 united carpels, superior; placentation axile; nectar secreted on the ovary. Fruit a capsule.

A large family of bulbous plants with 67 genera and about 950 species. Twelve genera are native to Europe, 9 to North America, and many are grown as ornamentals.

1a. Perianth-segments completely free to the base or united for less than one tenth of their length 2
b. Perianth-segments united for more than one tenth of their length 12
2a. Leaves borne on the flowering stem **3. Chlorogalum**
b. Leaves all basal 3
3a. Mature plant leafless consisting of a bulb and much-branched, climbing or scrambling stems **1. Bowiea**
b. Mature plants with normal leaves, not as above 4
4a. Flowering stem ending in a tuft of bracts above the raceme **10. Eucomis**
b. Flowering stem without a tuft of bracts above the raceme 5
5a. Inner perianth-segments erect, usually hooded or each with a swelling at the tip, outer segments spreading **14. Albuca**
b. All perianth-segments more or less similar, not as above 6
6a. Filaments flattened, usually broadest at the base, occasionally winged and toothed; flowers not blue **9. Ornithogalum**
b. Filaments thread-like, or, if flattened, then lanceolate to elliptic, narrow at the base; flowers often blue 7
7a. Perianth withering and falling after flowering **15. Urginea**
b. Perianth withering but not falling after flowering, persisting around or below the capsule 8
8a. Ovules 2 per cell; perianth-segments to 6 mm **2. Schoenolirion**
b. Ovules several per cell; perianth-segments more than 6 mm 9
9a. Each perianth-segment 3-veined **8. Camassia**
b. Each perianth-segment 1-veined 10
10a. Bracts 2 to each flower **6. Hyacinthoides**
b. Bracts 1 to each flower or bracts absent 11
11a. Perianth-segments spreading from the base; ovary not stalked **4. Scilla**
b. Perianth-segments curved to form a cup at the base, spreading from the top of the cup; ovary stalked **5. Ledebouria**
12a. Perianth-tube longer than the lobes 13
b. Perianth-tube shorter than to as long as the lobes 19
13a. Leaves 2, ovate, oblong or almost orbicular, spreading horizontally on the ground; scape absent or very short; perianth-lobes reflexed, almost as long as the tube **25. Massonia**
b. Leaves various, not as above; perianth-lobes much shorter than the tube, not usually reflexed 14
14a. Bracts obvious, as long as or almost as long as the flower-stalks 15
b. Bracts minute or absent 16
15a. Leaves to 6 mm wide; flowers blue to violet-blue or white **20. Brimeura**
b. Leaves much wider; flowers pink or pinkish purple, if whitish then spotted with red **23. Veltheimia**
16a. Perianth constricted at the mouth of the tube **21. Muscari**
b. Perianth not constricted at the mouth of the tube 17
17a. Capsule 3-angled **18. Bellevalia**
b. Capsule rounded, not angled 18
18a. Perianth 4—9 mm **19. Hyacinthella**
b. Perianth 1—3.5 cm **17. Hyacinthus**
19a. Flowers very unequal within the corymbose raceme, the outer with bilaterally symmetric perianths, the inner with almost radially symmetric perianths **26. Daubenya**
b. Flowers all similar within the raceme 20
20a. Outer perianth-lobes shorter than the inner **22. Lachenalia**
b. All perianth-lobes more or less of the same length 21
21a. Raceme ending in a tuft of bracts above the flowers **10. Eucomis**
b. Raceme not ending in a tuft of bracts above the flowers 22
22a. Perianth disc-shaped, the lobes spreading from a short obconical tube 23
b. Perianth cylindric or bell-shaped, the lobes spreading or not 24
23a. Filaments united into a corona with projecting lobes **16. Puschkinia**
b. Filaments flat, not united **7. Chionodoxa**
24a. Plant to 1 m; flowers 2.5 cm or more, white or greenish **24. Galtonia**
b. Plant much smaller; flowers less than 2.5 cm, variously coloured 25
25a. Perianth to 6 mm, greenish white; leaves spotted **12. Drimiopsis**
b. Perianth more than 1 cm, not greenish white; leaves not spotted 26
26a. Perianth blue or bluish purple **18. Bellevalia**
b. Perianth white, greenish or brownish orange, or pink 27
27a. Perianth-tube bell-shaped, lobes spreading or reflexed **11. Drimia**
b. Perianth-tube tubular to bell-shaped, the outer 3 lobes spreading, the inner 3 erect **13. Dipcadi**

1. BOWIEA J.D. Hooker
J. Cullen

Plants with green bulbs, composed of fleshy scales, which are mostly exposed above the soil. Leaves absent, apart from 1 or 2 small, quickly deciduous, scale-leaves produced on the bulb. Stem scrambling, richly branched to several orders of branching, green, photosynthetic. Flowers borne singly at the apices of the short, upper branches. Perianth-segments 6, free, green. Stamens 6, shorter than the perianth-segments. Ovary superior, 3-celled; style simple, 3-lobed at the apex. Fruit a capsule.

A genus of 2 species from southern Africa, 1 of them occasionally grown as a novelty (and sometimes also as a crop for the sake of the medicinally active chemicals it contains). It is easily grown in light, well-drained soil, but requires protection from frost throughout Europe.

Propagation is by seed or by offsets from the bulb.

1. B. volubilis J.D. Hooker. Illustration: *Botanical Magazine*, 5619 (1867); *Flowering plants of South Africa* **21**: t. 815 (1941); Brickell (ed.), RHS A–Z encyclopedia of garden plants, 189 (2003).

Bulb to 15 cm in diameter. Stem to 2 m or more. Perianth to 1.3 cm in diameter. *South Africa, South West Africa.* G1. Summer.

2. SCHOENOLIRION Durand

V.A. Matthews

Bulbs with membranous or fibrous tunics. Leaves basal, linear. Flowers in a raceme that is sometimes branched at the base. Flower-stalks jointed. Perianth-segments free, becoming papery with age. Stamens 6, joined to base of perianth-segments; anthers versatile. Ovary superior, with 2 ovules in each cell; style 1, short. Capsule ovoid, with black seeds.

A genus of 5 species found in SW and SE United States. Only 1 species is occasionally cultivated. It grows best in a sunny position, in a fairly moist soil. Propagation is by seed.

1. S. album Durand. Illustration: Abrams, Illustrated flora of the Pacific states **1**:412 (1923); Rickett, Wild flowers of the United States **5**: pl. 13 (1971); Niehaus & Ripper, Field guide to Pacific states wildflowers, 13 (1976).

Leaves 15–60 cm × 4–12 mm. Flowering stem 25–150 cm. Bracts 2–3 mm, papery. Flower-stalks *c.* 1 mm. Perianth-segments 3–6 mm, white tinged with green or pink. Stamens 3–5 mm. *W USA (N California, S Oregon).* H4. Summer.

3. CHLOROGALUM (Lindley) Kunth

V.A. Matthews

Bulbs with a fibrous or membranous tunic. Leaves mostly basal, linear; stem-leaves much reduced. Flowers borne on jointed stalks, in a terminal panicle. Perianth-segments free, with 3 central veins, twisting together over the ovary after pollination. Stamens with versatile anthers. Ovary superior; style 1, slightly 3-lobed at the apex. Capsule with 1 or 2 black seeds in each cell.

A genus of 5 species from west coastal USA, extending into N Mexico; 2 are sometimes cultivated. The flowers do not open until late afternoon, which is somewhat disadvantageous to the gardener. Cultivation as for *Camassia* (p. 114). Propagation is by seed.

Literature: Hoover, R.F., A monograph of the genus Chlorogalum, *Madroño* **5**: 137–47 (1940).

1a. Perianth-segments 1–2.5 cm; stamens *c.* two-thirds the length of the segments; leaves with wavy margins **1. pomeridianum**
b. Perianth-segments 8–12 mm; stamens more or less equal in length to segments; leaves not wavy **2. angustifolium**

1. C. pomeridianum (de Candolle) Kunth. Illustration: *Edwards's Botanical Register* **7**: pl. 564 (1821); Rickett, Wild flowers of the United States **4**: pl. 11 (1970); Niehaus & Ripper, Field guide to Pacific states wildflowers, **13** (1976); Spellenberg, Audubon Society field guide to North American wildflowers – western region, 79 (1979).

Leaves to 75 × 2.5 cm, margins wavy. Flowering stem 30–250 cm. Flower-stalks 5–35 mm. Perianth-segments 1–2.5 cm, white with a purple or green central stripe. Stamens about two-thirds the length of the segments. Style 1–1.5 cm. *W USA (S Oregon to N California).* H4. Summer.

2. C. angustifolium Kellogg. Illustration: *Proceedings of the California Academy of Natural Sciences* **2**: 105 (1863); Rickett, Wild flowers of the United States **5**: 48 (1971); Niehaus & Ripper, Field guide to Pacific states wildflowers, 13 (1976).

A smaller plant than *C. pomeridianum*. Leaves shorter and narrower with flat margins; flower-stalks *c.* 3 mm; perianth-segments 8–12 mm, white with a greenish yellow central stripe; stamens more or less equal to segments; style 3–4 mm. *W USA (N & C California).* H4. Summer.

4. SCILLA Linnaeus

S.G. Knees & P.G. Barnes

Perennial herbs with ovoid to spherical bulbs composed of numerous free scales that are progressively renewed annually. Roots annual or perennial. Leaves few to several, all basal, linear to elliptic, sometimes channelled, usually hairless, appearing before, with, or after the flowers. Scapes hairless, erect, usually 1–4 per bulb. Flowers few to many, in terminal racemes. Bracts absent or 1 subtending each flower. Perianth-segments 6, free, although occasionally close at the base giving the appearance of a short tube, blue, purple, pink or white, often each with a darker midrib. Filaments 6, free, equal, inserted at the base of the perianth; anthers dorsifixed. Ovary superior, almost spherical to ovoid, 3-celled, with 1–12 ovules per cell; style 1, straight; stigma small, truncate. Fruit a capsule; seeds spherical or oblong, sometimes angled, pale brown to black, each occasionally bearing an appendage.

A genus of up to 90 species occurring in Europe, Asia and southern Africa, valued for its ornamental, usually blue, flowers and suited to a variety of sites in the garden, particularly screes, rock gardens or raised beds. Some can be naturalised in grass or the wild garden and a few are best suited to lightly shaded positions. They are occasionally grown as specimen plants in the alpine house, whilst the more tender species require cool greenhouse conditions. Propagation is by offsets or by seed, which is freely produced.

Literature: Anderson, E.B. & Synge, P.M., Hardy Scillas, *RHS Lily Year Book* **24**: 27–30 (1961); Anderson, E.B. & Meikle, R.D., A Lily Group discussion of Scillas and Chionodoxas, *RHS Lily Year Book* **25**: 116–33 (1962); Speta, F., Die frühjahrsblühenden Scilla-Arten des östlichen Mittelmeerraumes, *Naturkundliches Jahrbuch der Stadt Linz* **25**: 19–198 (1980).

Perianth length. More than 1 cm: **3, 4, 6, 8, 9, 10, 15, 16**; less than 1 cm: **1, 2, 4, 5, 7, 11, 12, 13, 14, 15, 16, 17**.
Number of flowers. More than 15: **5, 7, 11, 14, 16, 17**; fewer than 25: **1, 2, 3, 4, 5, 6, 8, 9, 10, 12, 13, 15, 17**.
Flowering season. Spring: **1, 2, 3, 5, 6, 8, 9, 10, 12, 13**; late spring and summer: **4, 7, 11, 14, 15, 16**; autumn: **17**.

1a. Lower bracts more than 4 mm long 2
b. Lower bracts less than 4 mm long or absent 10
2a. Inflorescence with more than 20 flowers 3
b. Inflorescence with fewer than 20 flowers 6
3a. Leaves 9–15; lower bracts 3–8 cm **16. peruviana**
b. Leaves 3–8; bracts less than 2 cm 4
4a. Summer-flowering; bracts *c.* 1 cm 5
b. Spring-flowering; lower bracts 2–7 mm **5. persica**
5a. Flowers more than 40 in a long raceme; bracts longer than the flower-stalks **11. natalensis**
b. Flowers 15–35 in a conical raceme; bracts and flower-stalks similar in length **14. litardierei**

6a. Bracts equalling or longer than flower-stalks 7
b. Bracts much shorter than flower-stalks 8
7a. Bulb 3–5 cm, scales loose; leaves 1–3 cm wide; perianth-segments 9–12 mm **15. lilio-hyacinthus**
b. Bulb 1–2.5 cm, scales tight; leaves 2–5 mm wide; perianth-segments 5–8 mm **12. verna**
8a. Leaves usually 1–1.3 cm wide; perianth-segments less than 1 cm, light violet-blue **13. monophyllos**
b. Leaves 3–20 mm wide; perianth-segments more than 1 cm 9
9a. Perianth-segments 2–3 mm wide, pale blue, about as long as the flower-stalks **8. hohenackeri**
b. Perianth-segments 4–8 mm wide, whitish blue, half as long as the flower-stalks **9. mischtschenkoana**
10a. Perianth-segments less than 1 cm 11
b. Perianth-segments 9–25 mm 14
11a. Summer-or autumn-flowering; perianth-segments 3–5 mm 12
b. Spring-flowering; perianth-segments 5–10 mm 13
12a. Flowers more than 40; leaves 4–7 mm wide **7. scilloides**
b. Flowers 5–25; leaves 1–2 mm wide **17. autumnalis**
13a. Leaves usually 2, 3–15 mm wide; flowers 1–10, perianth-segments shorter than the lower flower-stalks **1. bifolia**
b. Leaves 5–7, 1.2–2.5 cm wide; flowers 7–20, perianth-segments similar in length to the lower flower-stalks **2. messeniaca**
14a. Flowers 1–4, pendent; perianth-segments usually more than 1.2 cm, exceeding the lower flower-stalks 15
b. Flowers 2–8, erect or horizontal; perianth-segments 9–12 mm, equalling or shorter than the flower-stalks 16
15a. Perianth-segments strongly recurved, pale blue **6. rosenii**
b. Perianth-segments straight, bright deep blue (occasionally pale blue or white) **3. siberica**
16a. Flowers mauve-blue; bracts absent **4. amoena**
b. Flowers whitish or pale blue; bracts 1–5 mm 17
17a. Perianth-segments 2.5–3.5 mm wide; flower-stalks 3–10 mm **10. puschkinioides**
b. Perianth-segments 4–8 mm wide; flower-stalks 1–2.5 cm **9. mischtschenkoana**

1. S. bifolia Linnaeus. Illustration: *Botanical Magazine*, 746 (1804); Polunin & Huxley, Flowers of the Mediterranean, pl. 240 (1965); Grey-Wilson & Mathew, Bulbs, pl. 17 (1981); Rix & Phillips, The bulb book, 40 (1981).

Bulb 5–25 mm in diameter, pinkish beneath the brown tunic. Leaves usually 2, occasionally 3–5, 5–20 cm × 3–15 mm, broadly linear, appearing with the flowers. Scape solitary, 5–30 cm, terete, partially sheathed at the base by the leaves. Raceme slightly 1-sided, with 1–10 flowers; bracts absent, or, if present, ovate-lanceolate, *c.* 1 mm; the lower stalks 1–4 cm, the upper less than 1 cm. Perianth-segments 5–10 × 1–3 mm, blue to purplish blue, ovate to elliptic, slightly hooded at the apex. Seeds almost spherical, *c.* 2 mm, brownish, each with an irregular, white appendage.
C & S Europe & Turkey. H4. Spring.

White- and pink-flowered variants are grown under the cultivar names 'Alba' and 'Rosea' (Illustration: *Botanical Magazine*, 746, 1804; Synge, Collins' guide to bulbs, pl. 24, 1961). A smaller variant, with fewer, violet flowers has been called **S. nivalis** Boissier.

2. S. messeniaca Boissier. Illustration: *Botanical Magazine*, 8035 (1905); Rix & Phillips, The bulb book, 41 (1981).

Bulb 2–3 cm in diameter, ovoid, tunic pale brown. Leaves 5–7, 15–25 × 1.2–2.5 cm, broadly linear, appearing with the flowers and sheathing the scapes at the base. Scapes 1 or 2, 5–15 cm, angled. Flowers 7–20 in a dense, conical raceme, 4–12 cm long; bracts *c.* 1 mm, linear, occasionally bifid; lower flower-stalks 4–8 mm, ascending or spreading. Perianth-segments 6–8 mm, lilac-blue, somewhat spreading, linear, apex obtuse. Stamens 4–6 mm, with pale blue filaments, anthers dark violet-blue. Ovary almost spherical, ovules 2 per cell. Capsule almost spherical, broadly 3-angled, *c.* 7 mm. *S Greece.* H4. Spring.

3. S. siberica Haworth. Illustration: *Botanical Magazine*, 1025 (1807); Synge, Collins' guide to bulbs, pl. 24 (1961); Hay & Synge, Dictionary of garden plants in colour, pl. 881 (1969); Rix & Phillips, The bulb book, 44 (1981).

Bulb 1.5–2 cm in diameter, ovoid, tunic dark purplish brown. Leaves 2–4, 10–15 cm × 5–15 mm, broadly linear, slightly hooded at the tips, shorter than the scape at flowering. Scapes 1–4, 10–20 cm, purplish above, finely ribbed. Flowers 1–4, occasionally more, broadly bowl-shaped, pendent, in loose racemes; bracts 1–2 mm, white; flower-stalks 8–12 mm, purple. Perianth-segments 1.2–1.6 cm × 4–6 mm, elliptic-oblong, obtuse, bright blue with a darker central stripe. Stamens 5–8 mm, filaments lanceolate, white below, blue above, anthers grey-blue. Ovary ovoid, pale green, ovules 2–12 per cell; style 4–5 mm, white. Capsule almost spherical, 8–10 mm in diameter, seeds spherical, pale brown, each with a long appendage. *Former S USSR; naturalised in C Europe.* H2. Spring.

Somewhat variable in colour: 'Spring Beauty' ('Atrocaerulea') is a clone with particularly bright, deep blue flowers (Illustration: *Addisonia* **16**: pl. 516, 1931; Rix & Phillips, The bulb book, 40, 43, 1981). The white-flowered cultivar 'Alba' is less vigorous.

4. S. amoena Linnaeus. Illustration: *Botanical Magazine*, 341 (1796); *Loddiges' Botanical Cabinet* **11**: t. 1015 (1825); Reichenbach, Icones florae Germanicae et Helveticae **4**: t. 464 f. 1014 (1847); Bonnier, Flore complète **10**: pl. 578 (1929).

Bulb 1.5–2 cm in diameter, ovoid, whitish beneath a dark, purplish brown tunic. Leaves 4–7, 15–22 × 1–2 cm, linear, often reddish near the base, usually appearing before the flowers. Scapes 1–5, 15–20 cm, angled. Inflorescence without bracts, with 3–6 flowers; flower-stalks 6–15 mm, the lower ones longer than the perianth; flowers erect. Perianth-segments 9–12 mm, mauve-blue with deeper blue midribs. Filaments lanceolate, anthers blue. Capsule spherical, containing 6–8 seeds per cell, each seed with a scarcely developed appendage. *Origin uncertain but naturalised in SE Europe.* H4. Spring–summer.

5. S. persica Haussknecht. Illustration: *Botanical Magazine*, n.s., 486 (1966); Mathew, Dwarf bulbs, 53 (1973); Rix & Phillips, The bulb book, 73 (1981).

Bulb 2.5–3 cm in diameter, broadly ovoid to almost spherical, outer tunic dark brown, inner violet-purple. Leaves 3–7, 30–45 × 1–1.5 cm, slightly keeled, linear with acute tips and rolled bases forming a tube, appearing with the flowers. Scapes 1–3, 20–40 cm. Flowers 20–80 in a compact, conical raceme 5–10 cm long, elongating to 20–25 cm in fruit; bracts 2–7 mm, fringed or 2-lobed above; flower-stalks 1.5–4 cm, thickening in fruit. Perianth-segments 7–8 mm, lanceolate to elliptic, slightly concave, spreading, bright blue. Ovary almost spherical, 3 mm in diameter, ovules 2 or 3 per cell. Seeds 4 × 2.5 mm, dark brown. *W Iran, N Iraq.* H4. Spring.

6. S. rosenii C. Koch. Illustration: Flora SSSR **4**: pl. 9 (1935); *Bulletin of the Alpine Garden Society* **30**: 126 (1962).

Bulb 1–2.5 cm in diameter, ovoid, tunic dark brownish violet. Leaves 2–3, 10–15 cm × 6–15 mm, linear, apex hooded, appearing with the flowers. Scapes 1–4, terete, 10–25 cm. Flowers usually 1 or 2, occasionally more, pendent, in short racemes; bracts 2–3 mm, 2-lobed; flower-stalks 1–1.5 cm, arching. Perianth-segments 1.5–2.5 cm × 4–8 mm, oblong, obtuse, becoming recurved, pale blue with a deeper median line on the outside, whitish near the base within. Stamens 8–12 mm, filaments slightly widened towards the base, white. Ovary obovoid, *c.* 5 mm; style longer than it. Seeds to 3 × 2 mm, pale brown, each with a white appendage. *Former USSR (Caucasus).* H3. Spring.

7. S. scilloides (Lindley) Druce (*S. chinensis* Bentham; *S. sinensis* (Loureiro) Merrill; *Barnardia scilloides* Lindley). Illustration: *Edwards's Botanical Register* **12**: t. 1029 (1826); *Botanical Magazine*, 3788 (1840); *Bulletin of the Alpine Garden Society* **20**: 180 (1952); Li, Flora of Taiwan **5**: t. 1285 (1978).

Bulb 1.5–2 cm in diameter, ovoid, tunic blackish. Leaves 2–7, 15–25 cm × 4–7 mm, linear, flaccid, appearing with the flowers, apex obtuse and slightly hooded. Scape 20–40 cm, slightly angled. Flowers 40–80 in a dense, oblong raceme, 7–12 cm long; bracts 1–2 mm, linear; flower-stalks 4–8 mm, ascending. Perianth-segments 3–4 mm, narrowly oblong, acute, spreading, mauve-pink. Filaments widened at their bases, finely hairy. Ovary ovoid, finely hairy on the angles, ovules few. Capsule *c.* 5 mm, obovoid-spherical. Seeds black. *China, Korea, Taiwan, Japan & Ryukyu islands.* H3–H5, depending on origin. Summer.

8. S. hohenackeri Fischer & Meyer. Illustration: *Botanical Magazine*, n.s., 453 (1964).

Bulb 1.5–2 cm in diameter, ovoid, tunic papery, greyish brown. Leaves 3–5, 10–25 cm × 3–10 mm, linear, V-shaped in cross-section, appearing before the flowers. Scape 5–20 cm. Flowers 4–12, in a loose raceme; bracts conspicuous, 5–6 mm, spurred below, fringed and 2-lobed above; flower-stalks 1–1.5 cm, spreading or recurved, equalling the perianths, but elongating to *c.* 2.5 cm in fruit. Perianth-segments 1.2–1.5 cm × 2–3 mm, oblanceolate, pale blue, recurved at maturity. Filaments pale blue, anthers greenish blue. Capsule 3-angled, pale brown, containing 2–4 glossy, black seeds per cell. *S Iran, Former USSR (Azerbaijan).* H3. Spring.

9. S. mischtschenkoana Grossheim (*S. tubergeniana* Stearn). Illustration: *Botanical Magazine*, n.s., 106 (1950); Hunt (ed.), Marshall Cavendish encyclopedia of gardening, 2003, pl. 1 (1967); Hay & Synge, Dictionary of garden plants in colour, 882 (1969); Rix & Phillips, The bulb book, 40 (1981).

Bulb 1.5–3 cm in diameter, ovoid or almost spherical, tunic papery, grey-brown. Leaves 3–5, 4–10 cm × 4–20 mm, linear to oblanceolate, appearing with the flowers. Scapes 1–3, 5–10 cm, terete. Flowers 2–6 in a loose raceme 6–12 cm long; bracts oblong, 1–5 mm; flower-stalks to 2.5 cm, ascending. Perianth-segments 1–1.2 cm × 4–8 mm, spreading, oblong-elliptic, obtuse, whitish blue with deeper blue mid-veins outside. Stamens 6–11 mm, slightly widened at the base, white, anthers blue. Ovary almost spherical, with usually 6 ovules per cell; style longer than the ovary. Capsule *c.* 9 mm in diameter, almost spherical. Seeds spherical, brown or black, each with a conspicuous, pale appendage. *Iran, Former USSR (Caucasus).* H4. Spring.

10. S. puschkinioides Regel. Illustration: *Gartenflora* **30**: t. 1050 (1881); Rix & Phillips, The bulb book, 43 (1981).

Bulb 1–2 cm in diameter, ovoid, tunic greyish. Leaves 2–5, 10–15 cm × 3–6 mm, narrowly oblanceolate or linear, obtuse, appearing with the flowers. Scape 10–15 cm, terete. Flowers 2–8 in a short raceme, stalks 3–10 mm, longer than the bracts. Perianth-segments 1–1.2 cm × 2.5–3.5 mm, oblong, acute, whitish or pale blue, each with a darker central line. Stamens 8–12 mm, filaments subulate, white, anthers blue. Ovary ovoid. Capsule flattened-spherical. *Former USSR (Soviet Central Asia).* H2. Spring.

11. S. natalensis Planchon. Illustration: *Flore des serres* **10**: t. 1043 (1855); *Botanical Magazine*, 5379 (1863); Eliovson, South African flowers for the garden, edn 2, 50 (1957); Fabian & Germishuizen, Transvaal wild flowers, pl. 11a (1982).

Bulb 7–10 cm in diameter, ovoid or almost spherical, tunic brown. Leaves 4–8, lanceolate, acuminate, to 20 cm at flowering, later extending to 30–60 × 7–10 cm. Scape 30–45 cm, terete. Flowers 50–100 in a raceme 15–30 cm long; bracts *c.* 1 cm, linear; flower-stalks 2–4 cm, ascending, blue. Perianth-segments 6–10 × 3–4 mm, elliptic-oblong, obtuse, spreading, light violet-blue or occasionally pink or white. Filaments widened towards the base, white. Ovary ovoid, usually white, ovules 10–12 per cell; style equalling the ovary. *E South Africa, Lesotho.* H5–G1. Summer.

12. S. verna Hudson. Illustration: Sowerby, English botany **1**: t. 23 (1790); Ross-Craig, Drawings of British plants **29**: pl. 23 (1972); Rix & Phillips, The bulb book, 47 (1981).

Bulb 1–2 cm in diameter, ovoid. Leaves 2–7, 3–20 cm × 2–5 mm, linear, slightly channelled, obtuse, appearing before the flowers. Scape 5–20 cm, terete. Flowers 2–12 in a short, corymbose raceme; bracts 5–15 mm, linear; lower flower-stalks 5–12 mm, ascending, elongating slightly in fruit. Perianth-segments 5–8 × 2–3 mm, narrowly oblong-ovate, light violet-blue. Stamens 2–4 mm, filaments lanceolate, white, anthers purplish blue. Ovary oblong, blue. Capsule almost spherical, seeds oblong, black, each with a small appendage. *W Europe.* H3. Spring.

S. ramburei Boissier. Illustration: Rix & Phillips, The bulb book, 73 (1981). Differs in its more robust habit, bulb 1.8–2.5 cm in diameter, broader leaves, 6–30 flowers and lower flower-stalks 1.5–5 cm. *Portugal, S Spain.* H3. Spring.

Sometimes regarded as a subspecies of *S. verna*.

13. S. monophyllos Link. Illustration: *Botanical Magazine*, 3023 (1830); Polunin & Smythies, Flowers of south-west Europe, pl. 59 f. 1636 (1973); Rix & Phillips, The bulb book, 73 (1981).

Bulb 1–2 cm in diameter, ovoid, tunic light brown. Leaf solitary, 10–25 × 1–3 cm, broadly lanceolate to elliptic, sheathing the scape at the base, appearing with or slightly before the flowers. Scape 5–15 cm. Flowers 3–15 in a loose raceme 5–10 cm long; bracts 4–7 mm, linear-lanceolate, acuminate; lower flower-stalks 1–1.5 cm, ascending. Perianth-segments 6–9 mm, elliptic, pale violet-blue. Stamens 3–5 mm, filaments linear, pale violet, anthers blue. Ovary almost spherical, ovules few per cell. *Spain, Portugal, NW Africa.* H4. Spring.

14. S. litardierei Breistroffer (*S. pratensis* Waldstein & Kitaibel; *S. amethystina* Visiani). Illustration: Reichenbach, Icones florae Germanicae et Helveticae **4**: t. 463 f. 1011 (1848); *RHS Lily Year Book* **24**: 6 (1961); Polunin, Flowers of Greece and the Balkans, pl. 58 (1980); Grey-Wilson & Mathew, Bulbs, pl. 17 (1981).

Bulb 1.5 cm in diameter, ovoid, tunic brown. Leaves 3–6, 25–30 cm × 4–8 mm, linear, appearing with the flowers. Scapes usually solitary, occasionally 2 or 3 together, 10–25 cm. Flowers 15–35 in narrow, conical racemes 5–15 cm long; bracts *c.* 1 cm, ovate; flower-stalks 8–12 mm, pale bluish violet. Perianth-segments 4–6 mm, ovate, pinkish or bluish violet. Stamens 3–5 mm, filaments pale blue, anthers deep violet-blue. Ovary with 3 or 4 ovules per cell. Capsule *c.* 4 mm in diameter. Seeds brown or black, to 3 × 1–2 mm. *Yugoslavia.* H4. Summer.

15. S. lilio-hyacinthus Linnaeus. Illustration: *Journal of the Royal Horticultural Society* **73**: 265 (1909); Bonnier, Flore complète **10**: pl. 578 (1929); Polunin & Smythies, Flowers of south-west Europe, 133, 16 (1973); Rix & Phillips, The bulb book, 68 (1981).

Bulb 3–5 cm in diameter, ovoid, with loose, overlapping, yellow scales, without a tunic. Leaves 6–10, 15–30 × 1–3 cm, glossy, oblanceolate, acute, appearing with the flowers. Scapes 2 or 3, 5–10 cm. Flowers 5–15 in a dense conical raceme 3–8 cm long; bracts 1–2.5 cm, ovate, papery, whitish; flower-stalks 8–12 mm, ascending, bluish violet. Perianth-segments 9–12 mm, ovate to elliptic, bright violet-blue, rarely white. Stamens 7–9 mm, filaments pale, 1–2 mm wide, lanceolate, anthers deep blue. Ovary almost spherical, with 6–10 ovules per cell. Capsule slightly 3-angled, seeds obovoid, blackish. *SW France, Spain.* H4. Summer.

16. S. peruviana Linnaeus. Illustration: Reichenbach, Icones florae Germanicae et Helveticae **4**: t. 465, f. 1017 (1848); Polunin & Huxley, Flowers of the Mediterranean, 243 (1965); Polunin, Flowers of Europe, pl. 168 (1969); Grey-Wilson & Mathew, Bulbs, pl. 18 (1981).

Bulb 6–8 cm in diameter, ovoid, tunic brown with woolly outer scales. Leaves 9–15, occasionally fewer, 40–60 × 1–4 cm, linear to lanceolate in a dense rosette, sometimes ciliate. Scapes solitary, rarely 2 together, 15–25 cm, stout. Flowers 40–100 in a broadly conical raceme 5–20 cm long; lower bracts 3–8 cm, subulate, papery; lower flower-stalks 3–5 cm, greenish, elongating in fruit to 6–10 cm. Perianth-segments 8–15 mm, elliptic, deep violet-blue or white. Stamens lanceolate-elliptic, pale blue; anthers yellow. Ovary almost spherical. Capsule ovoid, acuminate, with 3 or 4 seeds per cell. *SW Europe, NW Africa.* H4. Summer.

S. hughii Gussone is larger in all its parts, with leaves 4–6 cm wide and deep violet flowers. *Sicily.* H5. Summer.

17. S. autumnalis Linnaeus. Illustration: Reichenbach, Icones florae Germanicae et Helveticae **4**: t. 463 f. 1012 (1848); Polunin, Flowers of Europe, pl. 168 (1969); Grey-Wilson & Mathew, Bulbs, pl. 17 (1981).

Bulb 1.5–3 cm in diameter, ovoid, tunic brown or pinkish, occasionally with a few membranous scale-leaves. Leaves 5–12, 2–18 cm × 1–2 mm, linear, usually appearing after the flowers. Scapes 1–3, 5–30 cm. Raceme with 5–25 flowers, conical, without bracts, elongating in fruit; flower-stalks 3–10 mm, ascending. Perianth-segments 3–5 mm, spreading, lilac to pink. Capsules *c.* 4 mm, ovoid-spherical, obscurely 3-angled. Seeds 2 per cell, *c.* 2 × 1 mm, black, finely roughened. *S, W & C Europe, NW Africa to S Russia, Iran & Iraq.* H4. Summer–autumn.

5. LEDEBOURIA Roth

P.G. Barnes

Perennial herbs with bulbs. Leaves, all basal often spotted or striped. Inflorescence a slender, usually flexuous, raceme. Flowers small (*c.* 6 mm in the species described here), greenish or purplish, the segments free recurved. Filaments free to base. Ovary superior, broadly conical, abruptly flared at the base and stalked, with 2 basal ovules per cell. Seed very rarely produced.

A genus of 16 species in southern Africa and 1 (not cultivated) in India. Formerly a subgenus of *Scilla,* distinguished mainly by the stalked, conical ovary and usually spotted or striped leaves. Easily cultivated in a cool greenhouse for the attractive foliage. Propagation is by offsets.

Literature: Jessop, J.P., Studies in the bulbous Liliaceae: 1, *Journal of South African Botany* **36**(4): 233–66 (1970).

1a. Bulb ovoid, more or less above soil-level, narrowed into a neck; leaves narrowed to the stalk-like base, silvery above **2. socialis**
b. Bulb spherical, subterranean, without an elongated neck; leaves not silvery above 2
2a. Leaves narrowed to the base, green or spotted above **3. revoluta**
b. Leaves scarcely narrowed to the base, green or striped with dark brown **1. cooperi**

1. L. cooperi (J.D. Hooker) Jessop (*Scilla adlamii* Baker). Illustration: *Botanical Magazine,* 5580 (1866); *Bulletin of the Alpine garden Society* **42**: 341 (1974).

Bulb spherical. Leaves 1–5, 4–20 × 1–2 cm, narrowly oblong-lanceolate, scarcely narrowed to the base, both surfaces green or with brown longitudinal stripes. Perianth-segments pale purple or with a green keel. *South Africa.* H5–G1. Spring–summer.

Highly variable in the wild, but mainly represented in cultivation by a variant with stolons and heavily brown-striped leaves (*Scilla adlamii* Baker).

2. L. socialis (Baker) Jessop (*Scilla violacea* Hutchinson). Illustration: *Baileya* **18**: 52 (1971); Brickell (ed.), RHS A–Z encyclopedia of garden plants, 615 (2003).

Bulb 2–4 cm in diameter, ovoid, produced into a neck 1–2 cm long, often purple. Leaves 4–15 × 1–3 cm, lanceolate, narrowed to a channelled, stalk-like base 1–2 cm long; upper surface silvery with copious dark green blotches, lower surface green or purple. Perianth-segments pale purple with green keels. *South Africa.* H5–G1. Spring–summer.

Variable in leaf colour. The commonly cultivated variant has the leaf underside and the bulb deep purple, but plants lacking purple pigment in leaves and bulb are also grown.

3. L. revoluta (Linnaeus filius) Jessop (*Scilla lancifolia* (Jacquin) Baker). Illustration: Saunders et al. Refugium Botanicum **3**: t. 182 (1870); *Loddiges' Botanical Cabinet,* t. 1041 (1825).

Bulb 2–4 cm in diameter, spherical. Leaves 4–8, 4–15 × 1–3 cm, lanceolate to narrowly ovate; upper surface green, with or without darker spots. Perianth-segments greenish. *South Africa.* H5–G1. Spring–summer.

6. HYACINTHOIDES Medikus
A.C. Leslie
Perennial herbs. Bulbs with tubular, coalescent scales, completely renewed each year. Flowers in a raceme, each subtended by 2 bluish, linear-lanceolate bracts. Perianth-segments free to the base, erect or spreading, usually blue. Filaments free, attached to the perianth-segments; anthers versatile. Ovary superior, 3-celled; style simple; stigma capitate. Fruit a capsule. Seeds spherical, black.

A genus of 3–5 species in W Europe and N Africa. They are easily grown, and *H. non-scripta, H. hispanica* and hybrids between them will naturalise readily in deciduous woodland. Propagation is by offsets or seed.

Literature: Adolfi, K., Zure Unterscheidung von Hyacinthoides non-scripta (L.) Chouard und Hyacinthoides hispanica (Mill.) Rothm., *Gottinger Floristische Rundbriefe* **11**(2): 33–4 (1977); Quere-Boterenbrood, A.J., Over het vorkomen van Scilla non-scripta (L.) Hoffmannsegg & Link, S. hispanica Miller en hur hybrids in Nederland, *Gorteria* **12** (5): 91–104 (1984).

1a. Perianth-segments less than 1 cm **3. italica**
b. Perianth-segments more than 1 cm 2
2a. Raceme 1-sided, flowers pendent; anthers cream **1. non-scripta**
b. Racemes not 1-sided, flowers erect; anthers blue **2. hispanica**

1. H. non-scripta (Linnaeus) Rothmaler (*Endymion non-scriptus* (Linnaeus) Garcke; *Scilla nutans* Smith; *S. non-scripta* Hoffmannsegg & Link). Illustration: Coste, Flore de la France **3**: 323 (1906); Keble Martin, The concise British Flora in colour, pl. 85 (1965); Ross-Craig, Drawings of British plants **29**: pl. 4 (1972).

Leaves 20–45 cm × 7–15 mm, linear to linear-lanceolate. Flowering stem 20–50 cm. Racemes loose, 1-sided, drooping at the tip. Flowers erect in bud, finally pendent, scented. Perianth-segments 1.8–2 cm, oblong-lanceolate, erect below (so that the flower appears tubular), reflexed at the tips, violet blue or rarely pink or white. Outer filaments inserted at about the middle of the perianth, the inner inserted near the base; anthers cream. *W Europe.* H2. Spring.

2. H. hispanica (Miller) Rothmaler (*Endymion hispanicus* (Miller) Chouard; *E. patulus* Dumortier; *Scilla campanulata* Aiton). Illustration: Coste, Flore de la France **3**: 322 (1906); Bonnier, Flore complète **10**: pl. 588 (1929); Polunin & Smythies, Flowers of south-west Europe, 69 (1973).

Like *H. non-scripta*, but leaves often broader; racemes erect, not 1-sided; at least the upper flowers remaining erect or spreading, unscented; perianth-segments spreading (so that the flower is openly bell-shaped), not reflexed at their tips; filaments all inserted below the middle of the perianth; anthers blue. *SW Europe, N Africa.* H2. Spring.

Hybrids between *H. non-scripta* and *H. hispanica* are frequent in gardens and a complete range of intermediates can be found. Plants offered as *H. hispanica* are often hybrids.

3. H. italica (Linnaeus) Rothmaler (*Scilla italica* Linnaeus). Illustration: *Botanical Magazine*, 663 (1803); Coste, Flore de la France **3**: 321 (1906); Bonnier, Flore complète **10**: pl. 578 (1929).

Leaves 10–25 cm × 3–12 mm, linear to linear-lanceolate. Flowering stem 10–40 cm. Racemes dense, conical, erect, not 1-sided. Perianth-segments 5–7 mm, spreading widely, bluish violet (rarely white). All filaments inserted at base of perianth; anthers blue. *SE France, NW Italy, Spain, Portugal.* H2. Spring.

7. CHIONODOXA Boissier
E.M. Rix
Perennial herbs with small bulbs, tunics brown. Leaves usually 2. Scapes with 1–15 flowers in a loose raceme. Flowers blue or pinkish, often with a white central zone. Perianth with a short, obconical tube at the base, the lobes spreading. Stamens borne at the apex of the perianth-tube; filaments of unequal length, flattened; anthers dorsifixed. Ovary superior; style 1. Fruit an almost spherical capsule.

A genus of 6 species from western Turkey, Crete and Cyprus, grown for their very early, rich blue flowers. The genus is very close to *Scilla* (especially *S. bifolia*) and is sometimes united with it. Hybrids between the two are frequent; those between *C. forbesii* (and probably other species) and *Scilla bifolia* are grown as × **Chionoscilla allenii** Nicholson, and are intermediate between their parents, with very short perianth-tubes, a small, pale central zone to the flower and pale blue filaments.

Chionodoxas grow in well-drained but rich soil, and a position under deciduous shrubs is ideal.

Literature: Meikle, R.D., Chionodoxa luciliae, a taxonomic note, *Journal of the Royal Horticultural Society* **95**: 21–24 (1970); Speta, F., Über Chionodoxa Boiss., ihre Gliederung und Zugehörigkeit zu Scilla L., *Naturkundliches Jahrbuch der Stadt Linz* **21**: 9–79 (1976).

Flowers. 1 or 2, rarely to 4 on each scape: **3, 4, 5**, 4 or more on each scape: **1, 2**. Bright blue with pale centre: **2, 3**; bright blue without pale centre: **1, 4**; not bright blue: **5** (and cultivars of other species).
Perianth-lobes. More than 1.2 cm: **2, 3, 4**; less than 1.2 cm: **1, 2, 5**.

1a. Flowers bright blue, without a pale central zone 2
b. Flowers bright blue, lavender-blue or pink, with a pale central zone or entirely pale 3
2a. Scape with 2–4 flowers **4. lochiae**
b. Scape with 4–12 flowers **1. sardensis**
3a. Scape with 4–12 flowers **2. forbesii**
b. Scape usually with 1–3 flowers 4
4a. Perianth-lobes 1.2 cm or more **3. luciliae**
b. Perianth-lobes less than 1.2 cm **5. nana**

1. C. sardensis Whittall. Illustration: *Botanical Magazine*, n.s., 50 (1949); Rix & Phillips, The bulb book, 44 (1981).

Leaves erect or spreading. Scape usually 1 per bulb, to 40 cm, with 4–12 flowers. Flowers slightly pendent. Perianth bright blue, without a white central zone; tube 3–5 mm, almost spherical; lobes 8–10 × 2–4 mm. Filaments 2 mm and 3 mm, white. *W Turkey.* H2. Spring.

2. C. forbesii Baker (*C. luciliae* misapplied; *C. tmolusi* Whittall; *C. siehei* Stapf). Illustration: *Botanical Magazine*, 6433 (1879) – as *C. luciliae*, & 9068 (1925); Brickell (ed.), RHS A–Z encyclopedia of garden plants, 263 (2003).

Leaves 7–28 cm, erect or spreading. Scape usually 1 per bulb, to 30 cm, with 4–12 flowers. Flowers slightly pendent. Perianth rich blue with a white central zone, or pink with a white centre (in 'Pink Giant'); tube 3–5 mm, almost spherical; lobes 1–1.5 cm × 4–5 mm. Filaments 2 mm and 3 mm, white. *W Turkey.* H2. Spring.

3. C. luciliae Boissier (*C. gigantea* Whittall). Illustration: *Journal of the Royal Horticultural Society* **95**: f. 19, 20 (1970); Rix & Phillips, The bulb book, 40, 44 (1981); Brickell (ed.), RHS A–Z encyclopedia of garden plants, 263 (2003).

Leaves 7–20 cm, often recurved. Scape usually 1 per bulb, to 14 cm, with 1 or 2 flowers. Flowers erect. Perianth lavender-blue with a white central zone; tube 2.5–4 mm; lobes 1.2–2 cm × 3–8 mm. Filaments 2.5 mm and 3 mm, white. *W Turkey.* H2. Spring.

4. C. lochiae Meikle. Illustration: *Botanical Magazine*, n.s., 281 (1956).

Leaves 7–18 cm, erect. Scape usually 1 per bulb, to 18 cm, usually with 2–4 flowers. Flowers horizontal or pendent. Perianth bright blue without a white central zone; tube 5–7 mm; lobes 1.2–1.3 cm × 4–6 mm. Filaments 3 mm and 4 mm, white. *Cyprus.* H2. Spring.

5. C. nana (Schultes & Schultes) Boissier (*C. cretica* Boissier). Illustration: Rix & Phillips, The bulb book, fig. 42e (1981).

Leaves 8–18 cm, spreading. Scape usually 1 per bulb, to 20 cm, usually with 1–3 flowers. Flowers erect. Perianth blue with a white central zone; tube 3–5 mm, narrowly conical; lobes 6–11 × *c.* 3 mm. Filaments *c.* 3 mm, the outer longer and narrower, white. *Crete.* H2. Spring.

C. albescens (Speta) Rix. Flowers *c.* 10 mm, pale pinkish or lavender. *Crete.* H2. Spring.

8. CAMASSIA Lindley

C.J. King

Perennial herbs. Bulbs usually large, ovoid or spherical, with brown or black tunics. Leaves basal, linear, keeled, sheathing at the base, hairless, with entire margins. Scape terete, bearing a dense or loose raceme of flowers. Flower-stalks in axils of lanceolate-acuminate bracts. Perianth-segments spreading, persistent, blue, purple or white, free. Stamens 6, attached at the base of the segments; filaments thread-like; anthers versatile. Ovary 3-celled; style thread-like; stigma 3-lobed. Fruit a 3-lobed capsule, opening by 3 slits, with several black seeds in each cell.

A genus of 5 species from N America, and possibly 1 from S America. The bulbs do not increase much by division, but seed is freely produced and plants will reach flowering size in 3 or 4 years. They grow best in a rather moist border of good, well-drained loam, in full or at least half sun, and the bulbs should be left undisturbed. While camassias are generally quite hardy, they should be protected from severe frost in colder areas.

Literature: Gould, F.W., A systematic treatment of the genus Camassia Lindl., *American Midland Naturalist* **28**:712–42 (1942).

Bulbs. Clustered, and with an unpleasant smell: **5**.
Leaves. More than 7: **5**. Glaucous above: **1, 5**.
Flowers. With one perianth-segment deflexed: **1, 5**; regular: **1, 2, 3, 4**.
Perianth-segments. White: **1, 2, 4**. Twisting together as they wither: **1, 2, 3**; remaining separate: **1, 4, 5**.

1a. Perianth-segments twisting together around and above the capsule after flowering 2
b. Perianth-segments withering separately at base of capsule after flowering 4
2a. Flowers bilaterally symmetric, one perianth-segment deflexed; flower-stalks erect in fruit **1. quamash**
b. Flowers radially symmetric; flower-stalks not erect in fruit 3
3a. Capsule narrowly ovoid or oblong, dull green **2. leichtlinii**
b. Capsule broadly ovoid to nearly spherical, glossy green **3. howellii**
4a. Capsule ovoid or oblong 5
b. Capsule nearly spherical **4. scilloides**
5a. Bulbs clustered, with an unpleasant smell; leaves more than 7, 2–4 cm broad **5. cusickii**
b. Bulbs rarely clustered, without an unpleasant smell; leaves fewer than 7, 5–20 mm broad **1. quamash**

1. C. quamash (Pursh) Greene (*C. esculenta* Lindley not (Ker Gawler) Colville; *C. teapeae* St. John). Illustration: *Edwards's Botanical Register* **18**: t. 1486 (1832); *Flore des serres* **3**: t. 275 (1847); *Journal of the Royal Horticultural Society* **20**: t. 302 (1881); Rickett, Wild flowers of the United States **5**(1): t. 12 (1971), **6**(1): t. 11 (1973).

Bulb 2–5 × 1–3 cm, broadly ovoid to spherical, usually with black tunic. Leaves 5 or 6, 20–50 cm × 5–20 mm, green above and beneath, or glaucous above. Scape 20–80 cm, deep green. Raceme dense or loose, 5–30 cm. Bracts narrow, acuminate, shorter or longer than the flower-stalks. Flower-stalks 5–30 mm, spreading horizontally or curving upwards and becoming erect in fruit. Flowers either regular, or irregular with 1 perianth-segment deflexed. Perianth-segments 1–3.5 cm, linear-oblong, pale blue to deep violet-blue or white, either remaining separate as they wither, or twisting together over the young capsule. Stamens two-thirds as long as the segments; filaments pale blue or white; anthers yellow or blue. Ovary oblong-ovoid, bright green; style as long as or longer than the stamens, pale purple or white. *Western N America east to Montana and Utah.* H3. Late spring–early summer.

A very variable species, with up to 8 subspecies or varieties recognised by a combination of geographical locality and flower and foliage characters. The white form (Illustration: *Botanical Magazine*, 2774, 1827) has been called 'Flore Albo'.

2. C. leichtlinii (Baker) Watson.

Bulb 1.5–4 cm in diameter, broadly ovoid to spherical, with brown tunic. Leaves 4–6, 20–60 cm × 5–25 mm, bright green. Scape 20–130 cm, light green. Raceme loose, 10–30 cm. Bracts linear-lanceolate, *c.* 20 mm. Flower-stalks 1.5–5 cm, directed slightly upwards. Flowers regular. Perianth-segments 2–5 cm, oblong-lanceolate, creamy white or blue to violet, twisting together as they wither. Stamens half to three-quarters as long as the segments; filaments white or bluish; anthers yellow. Ovary narrowly ovoid to oblong, dull green; style rather shorter to slightly longer than the stamens, white. *Western N America.* H3. Late spring–early summer.

Subsp. **leichtlinii** (Illustration: *Botanical Magazine*, 6287, 1877; *Journal of the Royal Horticultural Society* **46**: t. 983, 1894) has creamy white flowers and is found wild only in Oregon.

Subsp. **suksdorfii** (Greenman) Gould (Illustration: Rickett, Wild flowers of the United States **5**(1): t. 12, 1971) has blue to violet flowers and extends from British Columbia to California.

Several of the colour forms have received cultivar names, and are perhaps the most desirable garden plants in the genus.

3. C. howellii Watson. Illustration: Abrams, Illustrated flora of the Pacific States **1**: f. 1019 (1940).

Bulb *c.* 1.5 cm in diameter, ovoid. Leaves 3–5, *c.* 30 cm × 5–10 mm. Scape 30–100 cm. Raceme 20–25 cm. Bracts linear, less than 1.2 cm. Flower-stalks

1.5–2.5 cm, spreading horizontally. Flowers regular. Perianth-segments 1–2 cm, oblong, pale blue or purple, twisting together as they wither. Stamens slightly shorter than the segments; filaments pale purple; anthers yellow. Ovary broadly ovoid to nearly spherical, glossy green; style about as long as stamens, pale purple. *W United States (SW Oregon).* H3. Late spring.

4. C. scilloides (Rafinesque) Cory (*C. esculenta* (Ker Gawler) Colville not Lindley; *C. fraseri* Torrey; *C. hyacinthina* (Rafinesque) Palmer & Steyermark). Illustration: *Botanical Magazine*, 1574 (1813); *Addisonia* **7**: t. 239 (1922); Rickett, Wild flowers of the United States **6**(1): t. 11 (1973).

Bulb 4 × 3 cm, ovoid, with an almost black outer tunic. Leaves 20–60 cm × 5–20 mm, bright green. Scape 20–80 cm. Raceme fairly loose, 8–10 cm. Bracts linear, 1–2 cm. Flower-stalks 5–30 mm, directed slightly upwards. Flowers regular. Perianth-segments 5–15 mm, narrowly oblong, blue, blue-violet or white, remaining separate as they wither. Stamens slightly shorter than the segments; filaments coloured as the perianth-segments; anthers yellow. Ovary nearly spherical, green; style as long as the segments, white or bluish. *C & E North America.* H3. Late spring–early summer.

5. C. cusickii Watson. Illustration: *Garden and Forest* **1**: f. 32 (1888); *Botanical Magazine*, n.s., 319 (1958).

Bulbs up to 9 × 5 cm, clustered, ovoid to almost spherical, with an unpleasant smell. Leaves 8–20, 40–80 × 2–4 cm, wavy, glaucous above, bright green beneath. Scape 60–80 cm, green. Raceme up to 40 cm. Bracts narrowly triangular-acuminate, the lowermost *c.* 8.8 cm, decreasing in size upwards to 1.5 cm, green and white at first, becoming membranous. Flower-stalks 1.1–2.3 cm, green, spreading at first, curving inwards towards the scape after flowering, and finally erect. Flowers irregular, with one perianth-segment deflexed. Perianth-segments 2.5–2.7 cm × 4–5 mm, linear-oblong, pale blue, remaining separate as they wither. Stamens shorter than perianth; filaments 1.5–1.6 cm, white; anthers yellow. Ovary oblong-ovoid, light green; style 1.5 cm, straight at first, becoming curved, white. *W United States (NE Oregon).* H3. Late spring–early summer.

9. ORNITHOGALUM

J. Cullen

Plants with bulbs that are usually subterranean with whitish or brownish papery tunics, more rarely partly exposed and green and fleshy. Leaves in a rosette, linear to lanceolate or obovate, margins ciliate or not, sometimes marked with a whitish line above. Raceme corymbose, pyramidal, almost spherical or cylindric, with 2-many flowers. Bracts usually conspicuous. Perianth-segments 6, equal, usually white, more rarely orange, yellow or red, each usually marked with a green stripe on the outside, free, usually widely spreading, more rarely erect. Stamens 6, filaments flattened, often broadened to the base, sometimes winged or abruptly widened towards the base. Ovary superior, cylindric to spherical, yellow, green or purplish black, 3-celled; ovules numerous; style terminal, long or short. Fruit a capsule with many seeds.

A genus of about 80 species with 2 main centres of distribution: southern Africa and the Mediterranean area. Species from both of these centres are grown. The South African species are rather tender and mostly require glasshouse protection (at least in northern Europe); in mild areas they can be stood outside in the summer. All can be grown in pots or in the open ground in well-drained soil on a sunny site. Several species reproduce by both seed and bulblets, and can become invasive.

Literature: Publications on the Mediterranean species are numerous and confusing: helpful accounts can be found in Flora Europaea **5**: 35–40 (1980), and in Davis, P.H. (ed), Flora of Turkey **8**: 227–45 (1984). The South African species are covered in Obermeyer, A.A., Ornithogalum: a revision of the South African species, *Bothalia* **12**: 323–76 (1978).

Bulb. Upper part exposed, with succulent, green tunics: **6**. Surrounded by numerous bulblets: **1, 6, 10, 13**.
Leaves. With a white line above: **10, 13, 14**. Margins ciliate: **3, 4**.
Raceme. Corymbose: **11, 12, 13, 14**; cylindric: **1, 3, 5, 6, 7, 8, 9, 10**; triangular to almost spherical: **2, 3, 4, 6**. With 2–20 flowers: **1, 10, 11, 12, 13, 14**; with 20–300 flowers: **1, 2, 3, 4, 5, 6, 7, 8, 9, 10**.
Perianth-segments. Without a green stripe outside: **1, 2, 3, 4, 5**. Orange, red or yellow: **3, 9**.
Ovary. Black, greenish black or purplish: **1, 2**.
Filaments. Some or all of them winged, the wings projecting as teeth beside the anther: **10**.

1a. Ovary black, greenish black or purplish 2
b. Ovary yellowish or green 3
2a. Perianth-segments 1.5–3.2 cm, erect in fruit **1. arabicum**
b. Perianth-segments 1–1.5 cm, reflexed in fruit **2. saundersiae**
3a. Filaments of at least 3 of the stamens winged for their whole length, the wings ending in teeth at either side of the anther; flowers drooping; perianth-segments not widely spreading **10. nutans**
b. Filaments various but not as above; flowers horizontal to erect; perianth-segments widely spreading 4
4a. Inflorescence a cylindric, triangular or almost spherical raceme with 20–300 flowers 5
b. Inflorescence a corymbose raceme with 2–20 flowers 11
5a. Perianth-segments without a green stripe outside, sometimes dark-coloured at the base inside 6
b. Perianth-segments with a conspicuous green stripe outside, not dark-coloured at the base inside 8
6a. Style very short or almost absent; perianth-segments yellow, red or orange, rarely white. **3. dubium**
b. Style as long as the ovary or longer; perianth-segments white or creamy white 7
7a. Filaments of the inner stamens with conspicuous, expanded bases that curve around the ovary; raceme pyramidal to almost spherical; scape 20–50 cm **4. thyrsoides**
b. Filaments of the inner stamens thread-like or with small broadened bases; raceme cylindric; scape 40–100 cm **5. conicum**
8a. Bulb with its upper part exposed, with succulent, green tunics, surrounded by numerous bulblets; bracts much exceeding the flowers **6. longibracteatum**
b. Bulb well below ground, with brownish or whitish tunics, without bulblets; bracts shorter than to a little longer than the flowers 9
9a. Perianth-segments pale yellow **9. pyrenaicum**
b. Perianth-segments white 10

10a. Ovary 2–2.5 mm; style usually shorter than the ovary, thickened and conical at its base **7. pyramidale**
b. Ovary 2.5–5 mm; style longer than the ovary, thread-like throughout its length **8. narbonense**
11a. Leaves broadest at ground level, tapering from there to an acute apex **11. montanum**
b. Leaves broadening above ground level, or equally broad for most of their length, tapering abruptly to a rounded apex 12
12a. Leaves usually 2–4, conspicuously broadened above, without a white line on the upper surface **12. oligophyllum**
b. Leaves usually more than 4, parallel-sided for most of their length, with a white line on the upper surface 13
13a. Fruit-stalks robust, 5–9 cm, rigid and horizontal; bulb surrounded by bulblets **13. umbellatum**
b. Fruit-stalks slender, 2–3.5 cm, weak, curved or ascending; bulb without bulblets **14. orthophyllum**

1. O. arabicum Linnaeus. Illustration: *Botanical Magazine*, 3179 (1832) – as *O. corymbosum*; Grey-Wilson & Mathew, Bulbs, t. 15 (1981).

Bulb below ground, surrounded by numerous bulblets. Scapes 30–80 cm. Leaves 5–8, more or less erect, dark green. Raceme cylindric with 6–25 flowers. Perianth-segments 1.5–3.2 cm, white or creamy white. Filaments broadening below. Ovary black or purplish black. Perianth-segments withering but not reflexing in fruit. *Mediterranean region.* H5–G1. Spring.

A species of uncertain origin, widely introduced in southern Europe from former extensive cultivation as a cut-flower crop.

2. O. saundersiae Baker. Illustration: Jeppe, Natal wild flowers, t. 6 (1975).

Bulb below ground. Scapes 30–100 cm. Leaves 6–8, erect, dark green and glossy above. Raceme triangular, with many flowers. Perianth-segments 1–1.5 cm, white or creamy white. Stamens with filaments conspicuously broadened below. Ovary black or greenish black. Perianth-segments persistent and reflexed in fruit. *South Africa (Transvaal, Natal, Swaziland, Zululand).* G1. Winter–spring.

3. O. dubium Houttuyn (*O. miniatum* Jacquin; *O. aureum* Curtis). Illustration: *Botanical Magazine*, 190 (1792); Marloth, The Flora of South Africa **4**: t. 28 (1915); *Bothalia* **12**: f. 8 (1978).

Bulb below ground. Leaves 3–8, yellowish green, margins ciliate. Raceme cylindric to almost spherical with 20 or more flowers. Perianth-segments 1–2 cm, orange, red, yellow or rarely white, often greenish or brownish at the base within. Stamens with filaments broadened below. Ovary yellowish green; style much shorter than the ovary. *South Africa (Cape Province).* G1. Winter–spring.

4. O. thyrsoides Jacquin. Illustration: Marloth, The Flora of South Africa **4**: t. 28 (1915); Hay & Synge, Dictionary of garden plants in colour, t. 870 (1969); Mason, Western Cape sandveld flowers, 35 (1972).

Bulb below ground. Scapes 20–50 cm. Leaves 6–12, ascending, with ciliate margins. Raceme triangular to almost spherical, with many flowers. Perianth-segments 1–2 cm, translucent white. Filaments of the inner stamens with broadly and abruptly expanded bases that curve around the ovary. Ovary yellowish green; style as long as the ovary. *South Africa (Cape Province).* G1. Winter.

5. O. conicum Jacquin (*O. lacteum* Jacquin). Illustration: *Botanical Magazine*, 1134 (1808).

Very similar to *O. thyrsoides*, but scapes 40–100 cm, leaves without ciliate margins, filaments only slightly expanded at their bases. *South Africa (Cape Province).* G1. Winter.

6. O. longibracteatum Jacquin (*O. caudatum* Aiton). Illustration: *Botanical Magazine*, 805 (1805).

Bulb with the upper part exposed and with a green, succulent tunic, surrounded by many bulblets. Scapes 1–1.5 m. Leaves 8–12, flaccid and succulent. Raceme triangular to cylindric with up to 300 flowers. Bracts conspicuous, much longer than the flowers. Perianth-segments to 9 mm, white with a green stripe outside. Filaments broadened towards the base. Ovary yellowish green. *South Africa (Cape Province, Natal).* G1. Flowering intermittent.

7. O. pyramidale Linnaeus.

Bulb below ground. Scapes 30–120 cm. Leaves 4–7, glossy green. Raceme cylindric with many flowers. Bracts shorter than the flower-stalks. Perianth-segments 1.1–1.5 cm, translucent white with a green stripe outside. Ovary 2–2.5 mm, green; style shorter than the ovary, thickened and conical at the base. *C Europe, Yugoslavia, Romania.* H4. Spring.

8. O. narbonense Linnaeus. Illustration: *Botanical Magazine*, 2510 (1824); Huxley & Taylor, Wild flowers of Greece and the Aegean, t. 364 (1977); Grey-Wilson & Mathew, Bulbs, t. 15 (1981); Brickell (ed.), RHS A–Z encyclopedia of garden plants, 748 (2003).

Very similar to *O. pyramidale* but ovary 2.5–5 mm, style longer than the ovary, thread-like throughout. *Mediterranean area, Former USSR (Caucasus), NW Iran.* H4. Spring.

9. O. pyrenaicum Linnaeus (*O. flavescens* Lamarck). Illustration: Grey-Wilson & Mathew, Bulbs, t. 15 (1981).

Very similar to *O. pyramidale* but generally smaller and more delicate, perianth-segments 9–13 mm, pale yellow with a narrow green stripe outside, ovary 2.4–3 mm, style longer than ovary. *Europe, Turkey, Former USSR (Caucasus).* H2. Spring.

10. O. nutans Linnaeus. Illustration: *Botanical Magazine*, 269 (1794); Polunin & Huxley, Flowers of the Mediterranean, t. 241 (1965); Hay & Synge, Dictionary of garden plants in colour, t. 869 (1969); Huxley & Taylor, Wild flowers of Greece and the Aegean, t. 361 (1977).

Bulb below ground, with numerous bulblets. Scapes to 60 cm. Leaves several, pale green, each with a white line on the upper surface. Raceme cylindric with 10–20 drooping flowers. Perianth-segments 2–3 cm, translucent white with a broad green stripe outside, not widely spreading. Filaments of at least the 3 inner stamens winged, the wings terminating in teeth at either side of the anther. Style longer than the green ovary. *Europe (naturalised in many areas), SW Asia.* H3. Spring.

11. O. montanum Cyrillo. Illustration: Polunin & Huxley, Flowers of the Mediterranean, t. 237 (1965); Grey-Wilson & Mathew, Bulbs, t. 15 (1981); Brickell (ed.), RHS A–Z encyclopedia of garden plants, 748 (2003).

Bulb below ground. Scapes to 20 cm. Leaves several, arching outwards, broadest at the ground level and tapering evenly to an acute apex. Raceme corymbose with 7–14 flowers. Perianth-segments 1.1–1.6 cm, white with a green stripe outside. Fruit-stalks erect-spreading, 4–7 cm. *S Europe from Italy eastwards, Turkey, Lebanon, Israel.* H4. Spring.

O. lanceolatum Labillardière. Very similar, but leaves broader and less acute, raceme stalkless, borne directly on the leaf rosette. *Turkey, Syria, Lebanon.* H4. Winter–spring.

12. O. oligophyllum Clarke (*O. balansae* Boissier). Illustration: Grey-Wilson & Mathew, Bulbs, t. 15 (1981); Brickell (ed.), RHS A–Z encyclopedia of garden plants, 749 (2003).

Bulb below ground. Scapes to 15 cm. Leaves usually 2–4, obovate, widening upwards then abruptly rounded to the apex. Raceme with 2–5 flowers, corymbose. Perianth-segments 1–1.6 cm, white with a green stripe outside. Fruit-stalks erect-spreading, 1–3 cm. *Balkans, Turkey, Former USSR (Georgia).* H4. Spring.

13. O. umbellatum Linnaeus. Illustration: Hay & Synge, Dictionary of garden plants in colour, t. 871 (1969); Grey-Wilson & Mathew, Bulbs, t. 15 (1981).

Bulb below ground, with numerous bulblets. Scapes 10–30 cm. Leaves several, linear, parallel-sided, tapering abruptly to the apex, with a white line on the upper surface. Raceme corymbose with 6–20 flowers. Perianth-segments 1.5–2.2 cm, white with a green stripe outside. Fruit-stalks horizontal, rigid, the lower 5–9 cm. *Most of Europe, N Africa, Turkey, Syria, Lebanon, Israel, Cyprus.* H2. Spring.

Can become an invasive weed, difficult to remove from rock gardens or screes.

14. O. orthophyllum Tenore (*O. tenuifolium* Gussone). Illustration: Matthews, Lilies of the field, opposite 25 (1968).

Similar to *O. umbellatum*, but bulb without bulblets, fruit-stalks spreading and curving upwards, the lower 2–3.5 cm. *S & C Europe to N Iran.* H3. Spring.

10. EUCOMIS L'Héritier

C.J. King

Perennial herbs. Bulbs usually large, ovoid, with shiny tunics. Leaves many, basal, strap-shaped to obovate, glossy, appearing with the flowers. Scape terete, bearing a simple, dense raceme of flowers, topped by a tuft of green bracts. Perianth fused at base, persistent; lobes 6–17 mm, spreading. Stamens 6, attached near base of lobes; filaments broadened and fused below into a shallow cup; anthers versatile. Ovary rounded to obovoid, 3-angled, with many ovules. Fruit a triangular capsule, inflated and membranous, or dry, hard and compact, opening by 3 slits. Seeds rounded or ovoid, black or brown.

About 10 species, ranging from south tropical Africa to South Africa. Most species are easily raised from seed and will flower in 2–5 years, or they can be increased by offsets. The plants require a sunny position in a rich, well-drained soil. They may be grown outdoors as border plants in favourable areas, but elsewhere they need the protection of a cool greenhouse. The single tropical species, *E. zambesiaca*, requires warm glasshouse conditions.

Literature: Reyneke, W.F., Three subspecies of Eucomis autumnalis, *Bothalia* **13**: 140–2 (1980), part of an unpublished MSc thesis, 'n Monografiese studie van die genus Eucomis L'Herit. in Suid-Afrika', submitted to the University of Pretoria (1972).

1a. Perianth-lobes green with distinct purple margin (except cultivar 'Alba') **1. bicolor**
b. Perianth-lobes entirely green 2
2a. Scape and leaves spotted or striped with purple; ovary purple **2. comosa**
b. Scape and leaves green; ovary green 3
3a. Scape not more than 30 cm; raceme not more than 15 cm **3. autumnalis**
b. Scape 40 cm or more; raceme 20 cm or more **4. pallidiflora**

1. E. bicolor Baker. Illustration: *Botanical Magazine*, 6816 (1885); Everard & Morley, Wild flowers of the world, t. 85 (1970); Brickell (ed.), RHS A–Z encyclopedia of garden plants, 434 (2003).

Bulb about 5 cm across. Leaves 30–50 × 8–10 cm, oblong with wavy margins. Scape cylindric, to 60 cm, sometimes spotted with dark purple. Raceme dense, many-flowered, 8–10 cm. Perianth pale green; lobes 1.2–1.5 cm, with a distinct purple margin. Ovary green. Bracts 30–40, ovate or oblong, acute, 5–8 cm, with crisped, sometimes purple, margins. *South Africa (Griqualand East, Natal).* G1. Late summer.

In cultivar 'Alba' the perianth is uniformly green.

2. E. comosa (Houttuyn) Wehrhahn (*E. punctata* L'Héritier). Illustration: *Botanical Magazine*, 913 (1806); *Flore des serres* **22**: t. 2307 (1877); Batten & Bokelmann, Wild flowers of the eastern Cape Province, t. 9(2) (1966); Brickell (ed.), RHS A–Z encyclopedia of garden plants, 434 (2003).

Bulb about 7 cm across. Leaves 30–77 × 3–8 cm, oblong-lanceolate with numerous small, dark purple spots near the base of the under-surface, and a slightly wavy margin. Scape cylindric, 45–70 cm, spotted with dark purple. Raceme dense, many-flowered, to 30 cm. Perianth pale green; lobes *c.* 1.2 cm. Ovary purple. Bracts 8–20, lanceolate, 2.5–8 cm, sometimes with purple margins. *South Africa (Eastern Cape Province, Natal).* G1. Late summer.

A variety where the spotting on the leaves and scape merges to form stripes is called var. **striata** (Houttuyn) Wehrhahn.

3. E. autumnalis (Miller) Chittenden.

Bulb 5–10 cm across. Leaves usually linear, sometimes lanceolate or ovate, with wavy margins. Scape cylindric or club-shaped, green. Raceme more or less dense, many-flowered, 5–15 cm. Perianth at first white or yellowish green, later green; lobes 6–17 mm. Ovary green. Bracts 10–45, oblong, ovate, or lanceolate, 2.5–8 cm, with crisped margins. G1. Late summer.

Subsp. **autumnalis** (*E. undulata* Aiton). Illustration: *Botanical Magazine*, 1083 (1808); *Flowering plants of South Africa* **6**: t. 220 (1926); Batten & Bokelmann, Wild flowers of the eastern Cape Province, t. 9(1) (1966); Graf, Tropica, 603 (1978). Leaves 15–55 × 6–13 cm, linear to ovate. Scape cylindric, sometimes slightly club-shaped, 6–30 cm. Raceme to 15 cm. Perianth-lobes 1–1.3 cm. Bracts 10–45, oblong, 3–7 cm. *South Africa (Eastern Cape Province, Transvaal), Rhodesia and Malawi.*

Subsp. **clavata** (Baker) Reyneke (*E. clavata* Baker; *E. robusta* Baker; *E. albomarginata* Barnes). Leaves 15–60 × 6–13 cm, lanceolate to ovate, sometimes linear. Scape usually club-shaped, 7–13 cm. Raceme 8–15 cm. Perianth-lobes 1.2–1.7 cm. Bracts 15–30, ovate, 5–8 cm. *South Africa (Transvaal, Orange Free State, Natal), Swaziland and Botswana.*

Subsp. **amaryllidifolia** (Baker) Reyneke (*E. amaryllidifolia* Baker). Leaves 13–50 × 1.5–4 cm, linear. Scape club-shaped, 6–23 cm. Raceme 5–8 cm. Perianth-lobes 6–8 mm. Bracts 13–20, lanceolate, 2.5–8 cm. *South Africa (Eastern Cape Province, Orange Free State).*

4. E. pallidiflora Baker (*E. punctata* var. *concolor* Baker). Illustration: *Botanical Magazine*, n.s., 561 (1970).

Bulb about 10 cm across. Leaves 50–70 × 10–12.5 cm, broadly sword-shaped with minutely crisped margins. Scape stout, cylindric, 45–75 cm, green. Raceme fairly dense, many-flowered, 24–43 cm. Bracts about 20, elliptic or lanceolate, acute, 3–4 cm, with minutely crisped margins. Perianth greenish white or green; lobes 1.2–1.6 cm. Ovary green. *South Africa (Natal?, Transvaal).* G1. Summer.

E. pole-evansii N.E. Brown. Illustration: *Flowering plants of South Africa* **24**: t. 953 (1944); Graf, Tropica, 603 (1978). Like *E. pallidiflora* in floral and other characters, differing principally in its larger leaves (75–120 × 10–17.5 cm), taller scape (90–180 cm), and larger bracts (5–10 cm). It may prove to be simply a large form of *E. pallidiflora* and therefore not worthy of separate specific rank. At present the range of variation within the 2 species is not known.

11. DRIMIA Willdenow

C.J. King

Perennial herbs. Bulbs with loose or tightly packed scales. Leaves 1–many, terete or flattened, usually appearing after the flowers. Inflorescence-stalk erect, terete, bearing a loose or dense raceme with usually many flowers. Flower-stalks each subtended by a membranous persistent bract. Flowers horizontal or pendent. Perianth white, greenish or greenish brown, usually with a short tube and reflexed lobes. Stamens 6, attached at the base of the perianth-tube. Ovary ovoid or ellipsoid, 3-celled. Fruit a capsule, opening by 3 slits. Seeds angled or flattened, black.

A genus with about 15 species in southern Africa, or, if enlarged to include *Urginea*, comprising about 120 species in Europe, Asia and Africa. Only one South African species, *D. haworthioides*, is normally found in cultivation in Europe, and this requires greenhouse protection. It will grow well in sandy loam and peat, but must be kept dry or almost dry during the summer resting period.

Literature: Jessop, J.P., The taxonomy of Drimia and certain allied genera, *Journal of South African Botany* **43**: 265–319 (1977).

1. D. haworthioides Baker. Illustration: Marloth, Flora of South Africa **4**: t. 24 (1915); Jacobsen, Handbook of Succulent Plants **1**: f. 370 & 371 (1960).

Bulb 2–3 cm high, with loose, stalked scales giving it a diameter of 5–6 cm. Leaves up to 10, 6–8 × 1–1.5 cm, oblong-lanceolate, usually ciliate. Inflorescence-stalk 20–40 cm, bearing 15–30 flowers. Bracts 1–4 mm. Flower-stalks 3–8 mm. Perianth 1–1.5 cm, greenish or greenish brown, fused for 2–4 mm into a basal tube, lobes reflexed. Stamens projecting. Capsule ellipsoid, 7–10 mm. *South Africa (Eastern Cape Province).* G1. Winter.

This species is occasionally cultivated, usually because of its peculiar bulb, which sits on the soil surface. When the leaf-blades wither, the leaf-stalks lengthen and become considerably enlarged at their tips, forming a group of purplish grey storage organs that may persist for several years. The plant may be increased readily by means of these swollen bulb-scales.

12. DRIMIOPSIS Lindley

J. Cullen

Herbaceous perennials. Bulb more or less spherical, pale, green where exposed. Leaves in a loose rosette, stalked or not, pale green, spotted with darker or brownish green above. Scape exceeding the leaves, bearing a many-flowered raceme; bracts absent. Perianth united into a tube below, the lobes little spreading, hooded at their apices. Stamens 6, filaments triangular, attached to the top of the perianth-tube. Ovary superior, 3-celled, ovules 2 in each cell. Fruit a few-seeded capsule.

A genus of about 7 species from tropical and southern Africa, of which 1 or 2 are occasionally grown. They require rather warm, moist conditions, but are not particularly ornamental. Propagation is by seed.

1. D. maculata Lindley & Paxton. Illustration: Saunders et al., Refugium Botanicum, t. 191 (1870); *Flowering plants of South Africa* **24**: t. 957 (1944).

Leaves to 10 × 5 cm, ovate or ovate-oblong, with a distinct stalk that is channelled and as long as, or longer than, the blade. Scape to 40 cm. Flowers at first white, later greenish. Perianth to 4 mm. *South Africa.* G1–2. Spring.

D. kirkii Baker. Illustration: *Botanical Magazine*, 6276 (1877). Plant larger, leaves oblong, tapering slightly to the base but not stalked; perianth *c.* 6 mm, persistently white. *Tanzania (Zanzibar).* G2.

13. DIPCADI Medikus

S.G. Knees

Hairless perennial herbs with bulbs. Leaves linear, often sheathing at the base. Flowers bisexual in loose racemes. Perianth tubular to bell-shaped, united in the lower third, the inner 3 lobes held erect, the outer 3 spreading. Stamens 6, included in perianth. Ovary superior; style 1. Fruit a 3-celled capsule.

A genus of 55 species, from South and tropical Africa, SW Europe and the East Indies. Only the European species is cultivated to any extent. Although easily grown in any light sandy soil, protection from winter wet may be necessary in the northern European countries. Propagation is by seed or offsets, which should be taken in spring.

1. D. serotinum (Linnaeus) Medikus. Illustration: *Botanical Magazine*, 859 (1805); Grey-Wilson & Mathew, Bulbs, t. 20 (1981); Polunin & Smythies, Flowers of south-west Europe, t. 59 (1973).

Herbaceous plant with scape 5–45 cm. Leaves deciduous, few, all basal, usually shorter than the scape, 5–20 × 1–2 cm, broadly V-shaped in cross-section. Racemes 1-sided, of 4–20 stalked flowers, subtended by scarious bracts. Perianth 1–1.6 cm, dull brownish orange. *SW Europe, from S Italy & France to Spain.* H5. Spring.

D. fulvum (Cavanilles) Webb & Berthelot. Illustration: *Botanical Magazine*, 1185 (1809). Similar but with brownish pink perianth-lobes. *S Spain, Morocco & Canary Islands.* H5. Autumn.

14. ALBUCA Linnaeus

E.H. Hamlet & V.A. Matthews

Bulbous perennials. Leaves basal, linear or lanceolate, usually hairless. Flowers in loose, terminal racemes. Perianth-segments free, often with a central band of a contrasting colour, the outer spreading, the inner erect, hooded at the tips. Outer whorl of stamens sometimes sterile. Ovary superior; style 1. Fruit a capsule; seeds black, flattened.

A genus of *c.* 30 species, from tropical and South Africa. The plants are not frost-hardy and need cool-house protection in the colder parts of Europe. Propagation is by seed.

Bulb. With fibres at the top: **9**.
Leaves. Downy: **4**. To 15 cm: **2, 4, 9**; 15–60 cm: **3, 5, 6, 7, 8, 9, 10**; 90–120 cm: **1**.
Perianth-segments. Outer white: **1, 2, 3**; green: **4, 10**; yellow: **5, 6, 7, 8, 9**. All striped: **1, 3, 5, 6, 7, 8, 9**; inner unstriped: **2, 4**; outer unstriped: **4, 10**; all unstriped: **4**.

Stamens. All fertile: **1, 2, 5, 6, 10**; outer sterile: **3, 4, 7, 8, 9.**

1a. Basic colour of perianth-segments white 2
b. Basic colour of perianth-segments yellow to green 4
2a. Leaves 45–120 cm; perianth-segments 1.5–4 cm, all with a central green or red-brown band 3
b. Leaves 7.5–15 cm; perianth-segments *c.* 1.2 cm, the inner ones lacking a central band **2. humilis**
3a. Perianth-segments 3–4 cm; outer stamens fertile **1. nelsonii**
b. Perianth-segments 1.5–2.5 cm; outer stamens sterile **3. altissima**
4a. Basic colour of outer perianth-segments green 5
b. Basic colour of outer perianth-segments yellow 6
5a. Leaves 4 or 5, 30–45 cm; inner perianth-segments white with a green stripe; outer stamens fertile; style cylindric **10. wakefieldii**
b. Leaves 10–20, 10–15 cm; inner perianth-segments green with yellow tips; outer stamens sterile; style 3-angled **4. namaquensis**
6a. Leaves 6–9; outer stamens fertile 7
b. Leaves 3–6; outer stamens sterile 8
7a. Perianth-segments 1.5–2 cm, with a central green band that remains green; flowers more or less pendulous, fragrant **6. fragrans**
b. Perianth-segments 2–3 cm, with a central green band that turns red-brown with age; flowers erect, not fragrant **5. aurea**
8a. Flowering stem 15–50 cm; leaves to 20 cm, if more then bulb with a tuft of fibres at the top 9
b. Flowering stem 40–100 cm; leaves 20–60 cm **8. major**
9a. Leaves lanceolate; central band of perianth-segments green **7. canadensis**
b. Leaves linear; central band of perianth-segments green, turning red-brown with age **9. cooperi**

1. A. nelsonii N.E. Brown. Illustration: *Botanical Magazine*, 6649 (1882); Brickell (ed.), RHS A–Z encyclopedia of garden plants, 93 (2003).

Leaves 4–6, 90–120 cm, lanceolate. Flowering stem 60–150 cm, bearing a raceme *c.* 30 cm long. Perianth-segments 3–4 cm, white with a green or red-brown central band. All stamens fertile. Style 3-angled. *South Africa.* H5. Early summer.

2. A. humilis Baker. Illustration: Brickell (ed.), RHS A–Z encyclopedia of garden plants, 93 (2003).

Leaves 1–3, 7.5–15 cm, narrowly linear. Flowering stem to 10 cm, bearing 1–3 flowers. Perianth-segments *c.* 1.2 cm, white; outer segments with a green central band; inner segments unstriped and often tipped with yellow. All stamens fertile. Style 3-angled. *South Africa.* H5.

3. A. altissima Dryander. Illustration: Le Roux & Schelpe, Namaqualand and Clanwilliam: South African wild flower guide **1**: 31 (1981).

Leaves 5 or 6, 45–60 cm, lanceolate, with recurved margins. Flowering stem 50–100 cm, bearing a raceme 30–45 cm long, with *c.* 50 flowers. Perianth-segments 1.5–2.5 cm, white, with a broad, green central band. Outer stamens sterile. Style 3-angled. *South Africa.* H5.

4. A. namaquensis Baker. Illustration: *Flowering plants of Africa* **32**: pl. 1255 (1957).

Leaves 10–20, 10–15 cm, almost cylindrical and very narrow, minutely downy. Flowering stem to 30 cm, bearing a raceme 10–20 cm long, with 6–10 flowers. Perianth-segments 1.5–2 cm, green, the inner with yellow tips. Outer stamens sterile. Style 3-angled. *South Africa.* H5.

5. A. aurea Jacquin.

Leaves 6–9, 45–60 cm, linear-lanceolate. Flowering stem 30–60 cm, bearing a raceme to 30 cm long. Flowers erect. Perianth-segments 2–3 cm, yellow, with a green central band that turns red-brown with age. All stamens fertile. Style 3-angled. *South Africa.* H5. Early spring.

6. A. fragrans Jacquin. Illustration: Kidd, Wild flowers of the Cape peninsula, pl. 74 (1973).

Simlar to *A. aurea* but flowers more or less pendulous, very fragrant with perianth-segments 1.5–2 cm, yellow, with a green central band that remains green. *South Africa.* H5.

7. A. canadensis (Linnaeus) Leighton (*A. minor* Linnaeus). Illustration: Levyns, Guide to the flora of the Cape peninsula, f. 25 (1966); Mason, Western Cape sandveld flowers, pl. 10 (1972); Kidd, Wild flowers of the Cape peninsula, pl. 53 (1973).

Leaves 3–6, *c.* 15 cm, lanceolate. Flowering stem 20–40 cm, bearing a raceme with up to 7 flowers. Perianth-segments 1.5–2 cm, yellow to greenish yellow with a broad, green central band. Outer stamens sterile. Style 3-angled. *South Africa.* H5. Late winter–spring.

8. A. major Linnaeus.

Similar to *A. canadensis* but leaves 20–60 cm, flowering stem 40–100 cm, raceme with 6–12 flowers, and perianth-segments with a green or brownish central band. *South Africa.* H5. Late winter–spring.

Possibly not distinct from *A. canadensis.*

9. A. cooperi Baker. Illustration: Kidd, Wild flowers of the Cape peninsula, pl. 66 (1973); Le Roux & Schelpe, Namaqualand and Clanwilliam: South African wild flower guide **1**: 31 (1981).

Bulb with a tuft of fibres at the top. Leaves 2–4, 15–30 cm, linear. Flowering stem 15–50 cm, bearing a raceme with 4–8 flowers. Perianth-segments 1.5–2.5 cm, yellow with a broad, green central band that turns red-brown with age. Outer stamens sterile. Style 3-angled. *South Africa.* H5. Spring.

10. A. wakefieldii Baker. Illustration: *Botanical Magazine*, 6429 (1879).

Leaves 4 or 5, 30–45 cm, linear. Flowering stem longer than the leaves, bearing a raceme 15–23 cm long, with 10–12 flowers. Perianth-segments *c.* 2.5 cm; outer segments green; inner segments white with a green central stripe. All stamens fertile. Style cylindric. *E Africa.* H5.

15. URGINEA Steinheil

P.G. Barnes

Perennial herbs. Bulbs with numerous free scales. Leaves narrow, deciduous. Flowers stalked, in long, erect racemes, the perianth falling after flowering, the stalks with small, often persistent, bracts. Perianth-segments 6, free, spreading. Filaments thread-like, sometimes dilated at the base, inserted at the base of the perianth-segments. Style equalling or exceeding the stamens. Fruit a 3-angled capsule, seeds many, flattened and winged.

A genus of about 100 species, mostly African, 3 occurring in the Mediterranean region.

Literature: Jury, S.L., Urginea undulata, *New Plantsman* **2**(3): 153 (1995).

1. U. maritima (Linnaeus) Baker. Illustration: *Botanical Magazine*, 918 (1806); Rix & Phillips, The bulb book, 173 (1981).

Bulb 5–15 cm diameter. Leaves 30–100 × 3–10 cm, narrowly lanceolate, somewhat glaucous and fleshy. Scape 50–150 cm, flowers very numerous, appearing after the leaves, in a dense raceme up to 40 cm long. Flower-stalks up to 3 cm long, spreading, exceeding the normally quickly deciduous, subulate bracts. Perianth-segments 6–8 mm, oblong, white with green or purple mid-veins. *Mediterranean coasts and Portugal.* H4. Summer.

Easily cultivated in a well-drained soil, but needing a sunny position and not free-flowering in northern Europe.

16. PUSCHKINIA Adams
E.M. Rix
Perennial herbs with small bulbs, tunics brown. Leaves usually 2. Scapes 5–20 cm, leafless, with 4–10 flowers in a loose raceme. Flowers almost stalkless or the lowest with stalks to 10 mm, pale blue with darker stripes, rarely white or greenish. Perianth 7–10 mm, with a short tube at the base, lobes erect or slightly spreading, not widely open; throat with a 6-lobed corona, its lobes alternating and projecting between the anthers. Anthers almost stalkless, borne on the corona, dorsifixed. Ovary superior; style 1. Fruit an almost spherical capsule.

A genus with a single, variable species, grown for its early, ice-blue flowers. It does well in any well-drained soil and benefits from being dry in summer.

1. P. scilloides Adams (*P. libanotica* Zuccarini; *P. hyacinthoides* Baker). Illustration: *Botanical Magazine*, 2244 (1821); Rix & Phillips, The bulb book, 42, 43 (1981); Brickell (ed.), RHS A–Z encyclopedia of garden plants, 869 (2003).

Former USSR (Caucasus), Turkey, N Iran, N Iraq & Lebanon. H1. Spring.

17. HYACINTHUS Linnaeus
P.F. Yeo
Perennial herbs with bulbs. Leaves present in spring, 2 or more on each flowering bulb, with smooth or rough margins. Flowers in a raceme arising between the leaves. Bracts small, 2-lobed. Flower-stalks much shorter than perianth. Flowers bisexual. Perianth with a narrow tube for about half to two-thirds of its length, its lobes divergent, spreading or recurved. Stamens attached to perianth at or below middle of tube, not exceeding perianth-lobes. Ovary superior, 3-celled, with 2-several ovules in each cell; style much shorter than perianth-tube. Capsule nearly spherical. Seeds black, wrinkled.

Three species from W and C Asia, all cultivated. Only *H. orientalis* is commonly grown and is one of the most important of the ornamental spring bulbous plants of Europe. The numerous cultivars are far more showy than the wild plant, and exhibit an extended range of colours. For bedding purposes or indoor display, new bulbs from the producers have to be used and are planted in autumn. For domestic decoration cold-treated bulbs are supplied, which flower earlier. For informal garden use, bulbs continue to give a display and to increase in number from year to year, but a rather rich soil is needed to sustain plants with well-formed racemes.

Literature: Bentzer, B. & von Bothmer, R., Cytology and morphology of the genus Hyacinthus L. s. str. (Liliaceae), *Botaniska Notiser* **127**: 297–301 (1974); Persson, K. & Wendelbo, P., The artificial hybrid Hyacinthus orientalis × transcaspicus (Liliaceae), *Botaniska Notiser* **132**: 207–9 (1979).

1a. Filaments shorter than the anthers **1. orientalis**
b. Filaments longer than the anthers 2
2a. Anthers held outside throat of flower; perianth-lobes shorter than tube **2. transcaspicus**
b. Anthers just reaching throat of flower; perianth-lobes longer than tube **3. litwinowii**

1. H. orientalis Linnaeus.
Plant to about 30 cm. Leaves 4–6, 15–35 cm × 5–40 mm, erect at first, channelled, bright green. Scape thick, hollow, collapsing in fruit. Raceme loose to dense, with 2–40 very fragrant flowers. Perianth 2–3.5 cm, pale to deep violet-blue, pale to deep pink, white or yellowish; tube constricted above ovary, as long as or longer than the lobes; lobes spreading, recurved at tips. Anthers longer than filaments, their tips well below throat of flower. Capsule about 1–1.5 cm, conical-spherical, fleshy. Seeds with an appendage. *C&S Turkey, NW Syria, Lebanon.* H1. Early spring.

Subsp. **orientalis**. Illustration: Reichenbach, Icones florae Germanicae et Helveticae **10**: t. 450, f. 1005 (1848); Huxley & Taylor, Wild flowers of Greece and the Aegean, t. 365 (1977); Everett, The New York Botanical Gardens illustrated encyclopedia of horticulture **5**: 1736 (1981). Leaves 4–5 mm (rarely to 11 mm) wide, linear. Flowers usually 2–12. Perianth 2–3 cm, pale violet-blue at base, nearly white above; lobes half to four-fifths as long as tube. *SC Turkey, NW Syria, Lebanon; naturalised in S Europe.*

Plants naturalised in Europe may have broader leaves and up to about 18 flowers.

Susp. **chionophilus** Wendelbo. Illustration: Rix & Phillips, The bulb book, 46 (1981); Mathew & Baytop, The bulbous plants of Turkey, t. 96 (1984). Leaves 1.2–1.5 cm (rarely to 3.6 cm) wide, linear-lanceolate, perianth-lobes about as long as tube, otherwise like subsp. *orientalis*. *C Turkey.*

The numerous cultivars of *H. orientalis* – illustration: *Botanical Magazine*, 937, 1806; Synge, Collins' guide to bulbs, t. 16, 1961; Rix & Phillips, The bulb book, 196, 1981 – have broad leaves (about 2–4 cm) and 20–40, more crowded, flowers in the raceme, and they show all the colours listed in the description of the species above. The perianth-lobes are usually shorter than the tube, as in subsp. *orientalis*.

2. H. transcaspicus Litwinow. (*H. kopetdaghi* Czerniakowska). Illustration: *Botaniska Notiser* **127**: 298 (1974); Wendelbo, Tulips and irises of Iran, 45 (1977).

Leaves 2 or 3, 7–16 cm × 4–15 mm, linear to lanceolate, channelled, green. Flowers 4–10, 1.3–1.5 cm, dark or medium violet-blue; tube cylindric; lobes shorter than tube, abruptly divergent or spreading. Filaments longer than anthers; anthers not at all enclosed by perianth, blue-black. Seeds without an appendage. *Former USSR (Turkmenistan: Kopet Dagh), NE Iran (E Elburz).* H1. Early spring.

3. H. litwinowii Czerniakowska. Illustration: Flora SSSR **4**: 306 (1935); *Botaniska Notiser* **127**: 298 (1974); Rix & Phillips, The bulb book, 46 (1981).

Plant 10–25 cm. Leaves 3 or 4, 15–17 × 1.5–5 cm, ovate-lanceolate to ovate, glaucous above, sometimes becoming spreading and wavy. Flowers 3–13, unscented. Perianth 1.8–2.5 cm, tube slender, constricted above ovary, dark greenish blue; lobes longer than tube, their basal parts continuing the tube upwards, their upper parts narrow, flared and recurved above the level of the anthers, very pale blue with deeper blue central band outside. Filaments longer than anthers. Seeds without an appendage.

Former USSR, E Iran (Kopet Dagh). H1. Early spring.

18. BELLEVALIA Lapeyrouse
P.F. Yeo

Perennial herbs with bulbs. Leaves 2–several on each flowering bulb, with margin cartilaginous and sometimes ciliate, appearing in autumn or spring. Flowers in a raceme arising between the leaves. Bracts usually minute, 2-lobed. Flowers bisexual, radially or slightly bilaterally symmetric, nodding or horizontal. Perianth with a tube for one-quarter of its length or usually more, violet-blue to brownish, forming a 6-lobed bell that is not constricted at the mouth though (rarely) the lobes close the mouth. When withered the perianth is sometimes carried on the fruit as a cap. Stamens attached to perianth at bases of lobes, rarely lower, not projecting beyond them; filaments shorter or longer than anthers, flattened, broadly to narrowly triangular, shortly united with each other at base. Ovary 3-celled, with 2–6 ovules in each cell; style 1. Fruit a capsule, obtuse at apex, the cells in cross-section forming prominent, acutely to acuminately angled lobes. Seeds rounded or occasionally pear-shaped, smooth, dull, black, but sometimes with a bluish bloom.

About 50 species in the Mediterranean and Black Sea regions, extending eastwards to NE Afghanistan and Tadjikistan (Former USSR). They are small- to medium-sized spring-flowering bulbs of varying appearance, some resembling *Muscari* or *Hyacinthella;* most of them are, or have been, cultivated for research purposes, but only a few have attractively coloured flowers. The limits of the genus have long been uncertain but the number and appearance of the chromosomes are decisive in the placement of borderline species. *Bellevalia* species may be grown in pots or planted in the open.

Literature: Feinbrun, N., A monographic study on the genus Bellevalia Lapeyr., 1: Caryological part, *Palestine Journal of Botany, Jerusalem Series* **1**: 42–54 (1938), 2: Taxonomical-geographical parts, ibid., 131–42 (1938) & 336–409 (1940); Freitag, H. & Wendelbo, P., The genus Bellevalia in Afghanistan, *Israel Journal of Botany* **19**: 220–4 (1970); Wendelbo, P., Notes on Hyacinthus and Bellevalia (Liliaceae) in Turkey and Iran, *Notes from the Royal Botanic Garden, Edinburgh* **38**: 423–34 (1980); von Bothmer, R. & Wendelbo, P., Cytological and morphological variation in Bellevalia, *Nordic Journal of Botany* **1**: 4–11 (1981).

Flower-stalks. Almost none at flowering time: **9**:
Raceme. Broadly conical, loose, flower-stalks when with young fruits 7 cm or more: **1**, **2**; cylindric or ovoid, loose, flower-stalks never more than 5 cm: **2**, **3**, **4**, **5**, **6**, **8**; narrowly conical, dense: **7**, **8**, **9**.
Leaf-margin. Ciliate: **1**, **2**, **4**, **7**, **9**; rough: **1**, **2**, **4**, **7**; hairless and smooth: **1**, **3**, **6**, **8**.
Flower-colour. White: **3**; pale to dark blue, not changing: **7**, **8**, **9**; bright violet-blue, becoming brownish: **6**; white to dull violet in bud, becoming brownish or purplish green: **1**, **2**, **4**, **5**.

1a. Flowering and fruiting parts of raceme loose, with the flower-stalks as long as or longer than the flowers and easily visible 2
b. Flowering and usually fruiting parts of raceme dense, with the flowers crowded, concealing the flower-stalks that are about as long as flowers or shorter 7
2a. Flower-stalks more or less straight, diverging from main axis of raceme above the horizontal and remaining the same from late bud stage to fruit; perianth white or greenish white **3. romana**
b. Flower-stalks with a different posture, or curved, or changing during the life of the flower; perianth at least partly pigmented from the time of opening onwards 3
3a. Flower-stalks greatly elongating during and after flowering, becoming 4–15 cm in fruit and making the raceme conical 4
b. Flower-stalks elongating only slightly during flowering and not afterwards, not exceeding 2 cm (except in **4. B. warburgii**); raceme becoming cylindric 5
4a. Longest flower-stalks 7–14 cm in fruit; leaves not glaucous above; flowers 9–13 mm **1. ciliata**
b. Longest flower-stalks 6 cm or less in fruit; leaves glaucous above; flowers 7–10 mm **2. saviczii**
5a. Flower-stalks in fruit 2.5–5 cm; flowers 1–1.3 cm **4. warburgii**
b. Flower-stalks in fruit not more than 1.5 cm; flowers 6–10 mm 6
6a. Flower-buds white or nearly so, changing to greenish brown when open; leaves usually prostrate and wavy, with rough or shortly ciliate margins **5. flexuosa**
b. Flower-buds bright violet-blue, changing to greenish brown when open; leaves channelled, with smooth margins **6. dubia**
7a. Flowers bright or pale blue or deep purplish blue with widely spreading lobes and exposed anthers **9. hyacinthoides**
b. Flowers blue-black; lobes not widely spreading; anthers not exposed 8
8a. Flowers erect in bud, nodding when open, not yellowish inside, bell-shaped **7. atroviolacca**
b. Flowers nodding in bud and when open, yellowish inside and on edges of lobes, irregularly urceolate **8. pycnantha**

1. B. ciliata (Cyrillus) Nees (*Hyacinthus ciliatus* Cyrillus). Illustration: *Edwards's Botanical Register* **5**: t. 394 (1819); Coste, Flore de 1a France **3**: 325 (1905); Polunin & Huxley, Flowers of the Mediterranean, t. 242 (1965).

Plant 20–50 cm. Leaves usually 4–6, to 30 × 3 cm, lanceolate, glossy, densely ciliate on margins. Racemes very loose, in flower 10–12 cm, broadly conical in fruit, normally to 15 × 20 cm. Flower-stalks erect in bud stage, spreading and becoming recurved during flowering, ultimately becoming straighter and horizontal. Flowers 9–11 mm, narrowly bell-shaped, with tube brown or lilac and the lobes greenish or white, becoming brownish or purplish. Capsule 1.5–2 cm × 8 mm, oblong with shortly tapered base and notched apex. *Aegean region, NW Africa.* H2. Spring.

B. longipes Post. Illustration: Rix & Phillips, The bulb book, 75 (1981). Differs from *B. ciliata* mainly in having 3 or 4 leaves, rough or smooth but hairless leaf-margins and even longer fruit-stalks, the raceme in fruit being up to 30 cm wide. Flower-stalks recurved, horizontal or ascending immediately after flowering. Flowers 9–13 mm. *N Syria, S Turkey, NW and W Iran, Former USSR (Transcaucasia).* H1. Spring.

2. B. saviczii Woronow.

Plant to 40 cm. Leaves 3–6, 15–25 cm, glaucous above, with shortly ciliate or rough margins, the outermost 5–20 mm wide. Racemes loose, in flower cylindric-oblong to ovate, in fruit broadly conical, then to 15 × 12 cm. Flower-stalks widely spreading during flowering, recurved afterwards. Flowers 7–10 mm, white in bud, bell-shaped when open and then greyish brown to pale greyish green. Capsule 1–1.2 cm, obovate-oblong, slightly notched at apex. *E Iran, Afghanistan and adjacent Former USSR.* H1. Spring.

3. B. romana (Linnaeus) Reichenbach (*Hyacinthus romanus* Linnaeus). Illustration: *Botanical Magazine*, 939 (1806); Reichenbach, Icones florae Germanicae et Helveticae **10**: t. 458, f. 1002, 1003 (1848); Grey-Wilson & Mathew, Bulbs, t. 19 (1981).

Plant to 30 cm. Leaves usually 3–5, emerging in autumn, reaching to top of racemes or beyond, channelled, with smooth margins, outermost to 1.5 cm wide. Racemes rather loose, in flower oblong or ovoid, to about 4 cm wide, in fruit cylindric, not widening. Flower-stalks 7–20 mm, diverging above the horizontal from late bud stage onwards, nearly straight. Flowers 8–10 mm, widely bell-shaped or broadly obconical, held in line with the stalk, white, becoming tinged with green, violet or brown, with conspicuous blue-black anthers. Capsule 1–1.5 cm, ellipsoid. *C & S Mediterranean area, SW France.* H2. Spring.

4. B. warburgii Feinbrun. Illustration: *Palestine Journal of Botany, Jerusalem Series* **1**: t. 16 (1940).

Plant to 60 cm. Leaves usually 3–6, not longer than scape, with shortly ciliate or rough margins, outermost to 3.5 cm wide. Racemes loose, soon becoming cylindric and longer than scape, 7–11 cm wide. Flower-stalks usually 1–1.5 cm in flower and to 5 cm in fruit, ascending in bud, becoming recurved in flower and young fruit, finally nearly straight. Flowers 1–1.3 cm, narrowly bell-shaped, white with green veins in bud, tube becoming purplish brown when open. Capsule about 1 cm, ovate in profile. *Israel.* H3. Spring.

5. B. flexuosa Boissier. Illustration: *Palestine Journal of Botany, Jerusalem Series* **1**: 371 & t. 17 (1940); Feinbrun & Zohary, Flora of the Land of Israel, Iconography, t. 12 (1940).

Plant to 50 cm. Leaves usually 4 or 5, usually longer than the scape, with shortly ciliate or rough margins, prostrate, wavy and up to 1.5 cm wide or, in shade, sometimes erect, not wavy and up to 3 cm wide. Racemes rather loose, soon becoming cylindric, about as long as scape, to 5 cm wide in fruit. Flower-stalks 8–20 mm, not lengthening after flowering, spreading or recurved in flower, ascending or horizontal in fruit. Flowers usually 7–8 mm, rarely to 10 mm, widely bell-shaped or broadly obconical, white in bud with green veins, sometimes tinged with lilac, tube becoming purplish brown when open. Capsule 8–10 mm, circular or ovate in profile. *Syria, Israel, Egypt.* H3. Early spring.

6. B. dubia (Gussone) Reichenbach (*Hyacinthus dubius* Gussone). Illustration: Huxley & Taylor, Flowers of Greece and the Aegean, t. 369 (1977); Grey-Wilson & Mathew, Bulbs, t. 19 (1981); Rix & Phillips, The bulb book, 75 (1981).

Plant to 40 cm. Leaves 2–5, longer than the scape, channelled, prostrate, with smooth margins, outermost to 1.3 cm wide. Racemes rather loose, narrowly cylindric, about 3 cm wide, not widening in fruit. Flower-stalks erect in young bud, in flower as long as perianth, soon spreading and then lightly recurved, more recurved after flowering but approximately horizontal in fruit and then about equalling the fruit in length. Flowers 6–8 mm, ellipsoid to weakly bell-shaped, bright violet-blue with green tips in bud, when open with tube gradually turning brown and lobes yellowish. Capsule 6–10 mm, obovate in profile. *Sicily to Balkan Peninsula and W Turkey.* H2. Spring.

B. dubia superficially resembles some species of *Muscari* allied to *M. comosum* (p. 125), but is easily distinguished by the unconstricted mouth of the perianth. *In* addition, the palest part of flowers that have turned brown is the apex, whereas in *Muscari* it is the base.

7. B. atroviolacea Regel. Illustration: Rix & Phillips, The bulb book, 47 (1981).

Plant 8–30 cm. Leaves 3–6, as long as or longer than scape, slightly channelled and slightly wavy, glaucous, with margins shortly ciliate or rough, the outermost to 2 cm wide. Racemes 3–6 cm in flower, densely conical, longer and looser in fruit. Flower-stalks at flowering time much shorter than perianth, mostly concealed, becoming 1–2 cm long in fruit. Flowers 8–9 mm, erect in bud, overlapping and sharply nodding when open, bell-shaped, uniformly blackish violet in all stages. *Former USSR (Turkestan & Tadjikistan), NE Afghanistan.* H1. Spring.

8. B. pycnantha (C. Koch) Losina-Losinskaya (*Muscari paradoxum* misapplied; *M. pycnantha* C. Koch). Illustration: *Botanical Magazine*, 6763 (1903); *RHS Lily Year Book* **30**: 113 (1967); Wendelbo, Tulips and irises of Iran, 47 (1977); Rix & Phillips, The bulb book, 72 (1981); Mathew & Baytop, The bulbous plants of Turkey, t. 61 (1984).

Plant to 40 cm. Leaves usually 3, much longer than scape, channelled, deep green, glaucous on upper surface, with smooth margins, outermost to about 2.5 cm wide. Racemes densely conical at first, with numerous overlapping flowers, later becoming ellipsoid or cylindric and loose. Flower-stalks about as long as flowers, concealed by the flowers in young racemes, to 1.2 cm in fruit. Flowers about 6 mm, nodding from late bud stage onwards, irregularly urceolate, angled in cross-section, with somewhat folded lobes, blue-black in all stages, greenish yellow on edges and inner surfaces of lobes. Capsule 8 mm, circular in profile with notched apex *Former USSR (Transcaucasia), E Turkey, N&W Iran, N Iraq.* H1. Spring.

9. B. hyacinthoides (Bertoloni) Persson & Wendelbo (*Hyacinthus spicatus* Sibthorp & Smith not Moench; *Strangweja spicata* (Boissier) Boissier). Illustration: *Botaniska Notiser* **132**: 66 (1979); Grey-Wilson & Mathew, Bulbs, t. 19 (1981); Rix & Phillips, The bulb book, 63 (1981); Brickell (ed.), RHS A–Z encyclopedia of garden plants, 172 (2003).

Plant to 15 cm. Leaves 4–8, emerging in autumn, much longer than the raceme, prostrate, slightly channelled, glaucous above, with shortly ciliate margins, to 5.5 mm wide. Racemes dense, ellipsoid to cylindric, with few (less than 20) crowded flowers. Flower-stalks not developed. Flowers 7–12 mm, horizontal or turned upwards, bell-shaped, with tube one-third to half as long as lobes, very pale blue with deeper blue to deep purplish blue veins. Filaments very short, with a prominent tooth on either side at base; anthers exposed. Capsule about 7 mm, wider than long in profile, notched. Seeds pear-shaped. *Greece.* H2. Spring.

19. HYACINTHELLA Schur

P.F. Yeo

Small perennial herbs with bulbs, leafless in summer and winter. Bulb-scales coated with powdery crystals. Leaves 2 or sometimes 3 on each (undivided) flowering bulb, the second (or third if there are 3) narrower than the first. Flowers in a raceme on a stalk arising between the leaves and overtopping them. Bracts minute. Flower-stalks not longer than flowers. Flowers not more than 1.2 cm, bisexual, radially symmetric or nearly so, more or less bell-shaped, blue to whitish or pinkish. Perianth-tube longer than the lobes, persistent in fruit. Stamens (in our species) attached just below the perianth-lobes and with filaments shorter than to slightly longer than the anthers. Ovary superior, 3-celled with 2–4 ovules in each cell; style simple. Fruit a rounded capsule to 5 mm wide with the cells rounded on the back. Seeds black; seed-coat wrinkled.

Sixteen species of E and SE Europe and W Asia, which have mostly been included in *Hyacinthus* or *Bellevalia* in the past. They have all been brought into cultivation for study at Göteborg, Sweden. They are spring-flowering bulbs, usually grown in pots under glass or in frames on account of their small size and inconspicuous flowers; their ornamental value is slight.

Literature: Feinbrun, N., Revision of the Genus Hyacinthella Schur, *Bulletin of the Research Council of Israel, Botany* **10D**: 324–47 (1961); Persson, K. & Wendelbo, P., Taxonomy and cytology of the genus Hyacinthella (Liliaceae-Scilloideae) with special reference to the species in SW Asia, Part 1, *Candollea* **36**: 513–41 (1981), Part 2, *Candollea* **37**: 157–75 (1982).

1a. Flowers 7–9 mm; flower-stalks less than 1 mm **4. nervosa**
b. Flowers 4–6 mm; flower-stalks 1 mm or more 2
2a. Leaves with distinct short or usually long hairs on margins and often also with hairs on surface **2. lineata**
b. Leaves without distinct hairs but often rough on margins 3
3a. Leaf-margins not rough **1. glabrescens**
b. Leaf-margins rough 4
4a. Leaves to 1.2 cm wide, erect, nearly flat, with a hooded tip **3. leucophaea**
b. Leaves to 6 mm wide, arching outwards, gutter-shaped **5. pallens**

1. H. glabrescens (Boissier) Persson & Wendelbo (*H. lineata* (Steudel) Chouard var. *glabrescens* (Boissier) Feinbrun). Illustration: *Bulletin of the Research Council of Israel, Botany*, **10D**: t. 3, f. 9 (1961); *Candollea* **36**:521 (1981).

Plant to *c.* 15 cm. Leaves hairless, with smooth margins, spreading. First leaf 8–20 mm wide, narrowly oblong-lanceolate to narrowly oblong-ovate to elliptic; second leaf half as wide as first or more. Inflorescence-stalk scarcely elongating in fruit. Flower-stalks 2–7 mm, scarcely elongating in fruit. Flowers 15–30, 4.5–6 mm, cylindric or slightly bell-shaped, medium to deep blue-violet, with lobes half as long as tube. *S Central Turkey*. H2. Spring.

2. H. lineata (Steudel) Chouard. Illustration: *Bulletin of the Research Council of Israel, Botany*, **10D**: t. 2, f. 5, 6 (1961); *Candollea* **36**: 527 (1981).

Plant to 14 cm in flower. Leaves with usually long hairs on margins and sometimes some on the surface near base, spreading, often twisted and often purplish at base. First leaf usually 5–15 mm wide, narrowly oblong-lanceolate to elliptic-ovate; second leaf more than half as wide as first. Inflorescence-stalk lengthening to 20–29 cm in fruit. Flower-stalks 2–6 mm, increasing to 4–12 mm in fruit. Flowers 6–25, usually 4.5–5.5 mm, weakly bell-shaped, deep blue to violet-blue, with lobes half as long as tube. *W Turkey*. H2. Spring.

H. hispida (Gay) Chouard. Illustration: Grey, Hardy bulbs **3**: 311 (1938); *Candollea* **36**: 519, f. 1a–c (1981). Leaves copiously stiff-hairy on margins and surfaces, erect or spreading, the first to 1.2 cm wide, the second less than half as wide as first. Flowers 5–6 mm in a very loose raceme, dark blue-violet, cylindric or weakly bell-shaped. *SC Turkey*. H2. Spring.

3. H. leucophaea (C. Koch) Schur. Illustration: Polunin, Flowers of Greece and the Balkans, t. 59, fig. 1643d (1980); Grey-Wilson & Mathew, Bulbs, t. 18 (1981) – as *H. dalmatica*; Rix & Phillips, The bulb book, 46 (1981).

Plant to 15 cm in flower. Leaves hairless but with rough margins, erect, nearly flat. First leaf to *c.* 1.2 cm wide, linear-lanceolate; second leaf more than half as wide as the first. Inflorescence-stalk lengthening to 25 cm in fruit. Flower-stalks 1–3 mm. Flowers up to 25, 4–5 mm, broadly cylindric or bell-shaped, pale or very pale blue, often green when old, with lobes one-third as long as tube. *E Europe*. H2. Spring.

The figure cited above from Rix & Phillips is stated in the text to be *H. atchleyi* (Jackson & Turrill) Feinbrun, but it does not show the dark blue flowers of that species.

4. H. nervosa (Bertoloni) Chouard. Illustration: Grey, Hardy bulbs **3**: 313 (1938); *Candollea* **36**: 535 (1981).

Plant to *c.* 17 cm in flower. Leaves hairless but with rough margins, erect. First leaf to 2 cm wide; second leaf half as wide as the first. Flower-stalk less than 1 mm. Flowers 10–25, 7–9 mm, urn-shaped or narrowly bell-shaped, pale blue or violet-blue, with lobes about half as long as tube. *SC Turkey to Jordan and eastwards to NE Syria & Iraq*. H3. Spring.

5. H. pallens Schur (*H. dalmatica* (Baker) Chouard). Illustration: *Bulletin of the Research Council of Israel, Botany* **10D**: f.1b & t. 1 f. 3 (1961); Polunin, Flowers of Greece and the Balkans, t. 61, f. 1643e (1980); Grey-Wilson & Mathew, Bulbs, t. 18 (1981) – as *H. leucophaea*; Rix & Phillips, The bulb book, 135 (1981).

Plant to 10 cm. Leaves hairless but with rough margins, arching and spreading outwards, channelled. First leaf to 6 mm wide; second leaf half as wide as first. Flower-stalks *c.* 2 mm. Flowers usually 6–20, 4–5 mm, rather narrowly bell-shaped, medium blue, with lobes *c.* half to two-thirds as long as tube. *Yugoslavia*. G2. Early spring.

20. BRIMEURA Salisbury

P.F. Yeo

Small hairless perennial herbs with bulbs, leafless in summer and winter (?). Leaves several on each flowering bulb. Flowers in a raceme arising between the leaves. Bracts about as long as flower-stalks or longer. Flowers bisexual, radially symmetric. Perianth with a long tube and 6 shorter lobes. Stamens concealed or exposed. Ovary superior, 3-celled with 2–4 ovules in each cell; style simple. Fruit a capsule, 5 mm long *(B. amethystina)*. Seeds black, shining, finely wrinkled *(B. amethystina)*.

Two highly dissimilar species, each with an interrupted distribution in S Europe, both cultivated. **B. fastigiata** (Viviani) Chouard, of slight horticultural value, is deceptively like a species of Scilla; *B. amethystina* is a delightful late spring dwarf bulb suitable for the rock garden, open border and positions under shrubs.

1. B. amethystina (Linnaeus) Chouard (*Hyacinthus amethystinus* Linnaeus). Illustration: *Edwards's Botanical Register* **5**: 398 (1819); Polunin & Smythies, Flowers of south-west Europe, t. 59, f. 1643 (1973); Grey-Wilson & Mathew, Bulbs, t. 19 (1981); Brickell (ed.), RHS A–Z encyclopedia of garden plants, 194 (2003).

Stem 10–25 cm. Leaves becoming slightly longer than stem, to 6 mm wide, channelled, bright green. Flower-stalks half to one and a quarter times as long as flower, longer in fruit. Flowers 5–15, *c.* 10 mm, horizontal or nodding and turned to one side, narrowly bell-shaped, bright blue or sometimes violet-blue or pure white. Perianth-tube cylindric or slightly constricted below throat; perianth-lobes one-third to half as long as tube. Stamens attached about halfway up the tube, small, not reaching the throat. *Pyrenees and NE Spain, NW Yugoslavia.* H3. Late spring–early summer.

21. MUSCARI Miller

D.C. Stuart

Herbaceous perennials with bulbs, with or without offsets. Leaves 1–8, basal. Scape present, flowers in racemes with minute bracts. Perianth united for most of its length into a cylindric, tubular, bell-shaped or urceolate tube, the 6 lobes small, blue, brownish, yellow, whitish or white, not changing as the flower dies, often of a different colour from the tube. Stamens 6; anthers not projecting from perianth-tube. The raceme often has sterile flowers near the apex, which may differ in colour or tone from the fertile, sometimes forming a tuft (comus). Ovary superior, 3-celled, with 2 ovules in each cell. Perianth carried up on the developing capsule and falling from there. Capsule strongly angled, with 2 seeds per cell. Seeds black, often shiny, often minutely wrinkled.

A genus of about 30 species from the Mediterranean area and SW Asia; only 12 of the species are in general cultivation. The genus is variable, and is sometimes split into 4 small genera: *Muscari* in the strict sense, *Leopoldia* Parlatore, *Botryanthus* Kunth and *Pseudomuscari* Garbari & Greuter. However, as each group contains at least 1 species that seems intermediate to those in the other groups, it seems sensible to retain the widest possible circumscription of *Muscari* here.

Chromosome morphology and number, floral, capsule and seed characters clearly separate *Muscari* in this wide sense from *Bellevalia* (p. 121) and *Hyacinthella* (p. 122). The commonly cultivated *M. comosum* and *M. armeniacum* both have wide geographical ranges, are very variable (with many synonyms, most of which are not cited here) and have given rise to monstrous forms which are frequently cultivated.

Literature: Schschian, A., Sistematika i geographiy kavkazskich vidov roda Muscari Miller, *Trudy Tbilisskii Botanicheskogo Instituta* **5**(10): 203–5 (1946); Stuart, D.C., Muscari and allied genera, *RHS Lily Year Book* **29**: 125–38 (1966); Speta, F., Uber die Abgrenzung und Gliederung der Gattung Muscari, und uber ihre Beziehungen zu anderen Vertreten der Hyacinthaceae, *Botanische Jahrbucher* **103**: 247–291 (1982).

Leaves. Single: **11**; two, broadened above: **7, 8**.
Fertile flowers. White: **6, 7, 8**; yellow: **2**; greenish: **1, 3**; buff or light brown: **4**; blackish blue: **9, 10, 11**; mid to light blue, sometimes purplish: **5, 6, 7, 8**.
Inflorescence. Branched or monstrous: cultivars of **4, 6**.
Perianth-lobes of fertile flowers. Of the same colour as the tube: **5, 10, 11**; of a different colour from the tube: **1, 2, 3, 4, 6, 7, 8, 9**.

1a. Raceme simple, most flowers fertile, sterile flowers at the apex 2
b. Inflorescence branched, most flowers reduced to coloured threads 12
2a. Perianth bell-shaped, not constricted towards the apex, blue or rarely white **5. azureum**
b. Perianth constricted towards the apex, blue, white, purplish, yellow, buff, brown or greenish 3
3a. Mature fertile flowers blue, rarely purplish or white 4
b. Mature fertile flowers ivory, yellow, greenish, buff or brown 9
4a. Perianth-lobes of the same colour as the tube 5
b. Perianth-lobes white or pale blue, the tube a different colour 6
5a. Leaf single, 1–3 cm broad, linear-oblanceolate **11. latifolium**
b. Leaves 2 or more, 5–15 mm broad, linear **10. commutatum**
6a. Leaves spathulate (sometimes narrowly so) 7
b. Leaves linear to linear-oblanceolate 8
7a. Raceme becoming loose in fruit; leaves not sickle-shaped, often ribbed **7. botryoides**
b. Raceme remaining dense in fruit; leaves sickle-shaped, not ribbed **8. aucheri**
8a. Perianth of fertile flowers blackish blue **9. neglectum**
b. Perianth of fertile flowers sky-blue, rarely purplish or white **6. armeniacum**
9a. Perianth yellow **2. macrocarpum**
b. Perianth buff, greenish or ivory 10
10a. Perianth buff, lobes cream or yellowish **4. comosum**
b. Perianth greenish to ivory, lobes blackish purple or brown 11
11a. Flowers narrowly urceolate, with prominent shoulders towards the apex, forming a brownish corona; sterile flowers absent or minute and violet **1. muscarimi**
b. Flowers oblong-urceolate, without a corona; sterile flowers conspicuous, pinkish violet **3. massayanum**
12a. Inflorescence blue **6. armeniacum** 'Blue Spike'
b. Inflorescence violet **4. comosum** 'Plumosum'

1. M. muscarimi Medikus (*M. moschatum* Willdenow). Illustration: *Botanical Magazine*, 734 (1804); Redouté, Liliacées **3**: t. 132 (1805).

Bulbs 2–4 cm across, roots thick. Leaves 3–6, 10–20 cm × 5–15 mm, pale greyish green. Scape erect to prostrate, shorter than the leaves. Fertile flowers 8–14 mm, narrowly urceolate, at first purple, becoming pale greenish to ivory, strongly contracted towards the apex, with expanded shoulders that form a distinct, brownish corona. The flowers smell strongly of musk. Sterile flowers rarely present, if so, small and purplish. *SW Turkey.* H5. Spring.

2. M. macrocarpum Sweet (*M. luteum* Todaro). Illustration: *Botanical Magazine*, 1565 (1813); Griffith, Collins' guide to alpines, t. 32 (1964); RHS Lily Year Book **29**: f. 29 (1966); Brickell (ed.), RHS A–Z encyclopedia of garden plants, 704 (2003).

Bulbs as in *M. muscarimi*. Leaves to 30 cm. Fertile flowers 8–12 mm long, oblong-urceolate, at first purplish, later bright yellow, shoulders expanded to form a deep brown or, rarely, yellow corona, smelling of bananas. Sterile flowers mostly absent, rarely few (when present, purplish). *Greek islands & W Turkey.* H5. Early spring.

3. M. massayanum Grunert. Illustration: *Journal of the Royal Horticultural Society* **107**: 335 (1982).

Bulbs 2–6 cm across, without offsets, tunics ivory. Leaves 2–4, to 25 × 1–2.5 cm, linear, sickle-shaped or sinuous, thick, glaucous, apex hooded. Scape stout, to

22 cm, mauvish. Raceme cylindric, dense. Fertile flowers to 1.1 cm, narrowly urceolate, at first pink to violet, becoming light greenish or yellowish brown, . Perianth-lobes dark brown to blackish. Stalks of the sterile flowers ascending to spreading, forming a dense, fleshy, pink or violet tuft. *E Turkey*. H5. Spring.

4. M. comosum (Linnaeus) Miller (*M. pinardii* Boissier). Illustration: Polunin, Flowers of Europe, t. 169 (1969); Rix & Phillips, The bulb book, 75 (1981).

Bulbs to 3.5 cm across, without offsets, tunics pink. Leaves 3–7, erect-spreading, generally shorter than the scape. Fertile flowers oblong-urceolate, pale brown, the shoulders rounded and darker brown, lobes cream or yellowish. Stalks of sterile flowers fleshy, ascending, bright violet, to 4 mm. Sterile flowers almost spherical or obovoid, rarely tubular, bright violet, many, forming a conspicuous tuft. *S & C Europe, N Africa, SW Asia*. H5. Spring.

In the wild the tuft of sterile flowers is very variable in development. In horticultural variants it is the main reason for growing the plant. In *M. comosum* 'Plumosum' (Illustration: Puttock, Bulbs and corms, 96, 1958) the branched inflorescence consists only of sterile flowers and their stalks, and is extremely showy.

5. M. azureum Fenzl (*Bellevalia azurea* (Fenzl) Boissier; *Hyacinthella azurea* (Fenzl) Chouard). Illustration: Rix & Phillips, The bulb book, 70 (1981); Mathew, Dwarf bulbs, 73 (1973) – as M. *azureum* 'Album'.

Leaves 2 or 3, 6–20 cm × 3–15 mm, erect or spreading, narrowly oblanceolate, often rather hooded at the apex, the upper surface pale and glaucous. Raceme dense, ovoid, 1.5–3 cm. Fertile flowers 4–5 mm, bell-shaped, not constricted, pale to bright blue, each lobe with a darker stripe, or rarely all white. Lobes of the same colour as the tube. Anthers blue. Sterile flowers few, smaller and paler than the fertile. *E Turkey*. H5. Spring.

An attractive species, widely grown. Pale blue variants are very similar to *M. coeleste* Fomin, a species from the Caucasus and E Turkey that is not yet in cultivation.

6. M. armeniacum Baker (*M. szovitsianum* Baker; *M. cyaneo-violaceum* Turrill). Illustration: *Botanical Magazine*, 6855 (1886); Rix & Phillips, The bulb book, 68, 71 (1981); Brickell (ed.), RHS A–Z encyclopedia of garden plants, 703 (2003).

Bulbs often with bulbils. Leaves 3–7 (usually 3–5), linear to linear-oblanceolate, upper surface sometimes pale. Inflorescence dense, flowers overlapping or contiguous. Fertile flowers 3.5–5.5 × 2.3–3.5 mm, obovoid to oblong-urceolate, bright sky-blue, sometimes with a weak or strong purplish cast, rarely all white. Lobes always paler than the tube or white. Sterile flowers few, smaller and paler than the fertile, rarely of the same colour. *SE Europe to the Caucasus*. H3. Spring.

An attractive, variable and prolific species, found in quantity both in gardens and in the wild. A rare, purple variant from Varna (Bulgaria) was at one time given specific rank as *M. cyaneo-violaceum*. The monstrous variant, now known as 'Blue Spike', is an ancient garden plant.

7. M. botryoides (Linnaeus) Baker. Illustration: *Journal of the Royal Horticultural Society* **26**: 136 (1884); *RHS Lily Year Book* **29**: f. 27 (1966); Polunin, Flowers of Europe, t. 170 (1969); Rix & Phillips, The bulb book, 68 (1981).

Leaves 2–4, 5–25 cm × 5–13 mm, erect, spathulate, rarely narrowly so, apex abruptly contracted, hooded or acuminate, upper surface often prominently ribbed. Raceme dense at first, generally becoming loose and cylindric. Fertile flowers 2.5–5 mm, almost spherical, sky-blue, rarely white; lobes white. *C & SE Europe*. H2. Spring–early summer.

8. M. aucheri (Boissier) Baker (*Botryanthus aucheri* Boissier; *M. tubergenianum* Turrill). Illustration: *Gartenflora* **77**: 56 (1928); *Die Gartenstauden* **1**: 165 (1929); Brickell (ed.), RHS A–Z encyclopedia of garden plants, 704 (2003).

Leaves 2 (rarely 3 or 4), 5–20 cm × 2–15 mm, erect-spreading, sickle-shaped, often symmetric, spathulate (sometimes narrowly so), upper surface pale green, glaucous, apex hooded. Raceme dense, ovoid or cylindric. Fertile flowers 3–5 × 2.3–5 mm, almost spherical or obovoid, bright sky-blue, rarely white, lobes paler blue or white. Sterile flowers larger or smaller than the fertile, paler, as many as them or fewer. *Turkey*. H3. Spring.

M. aucheri is most often cultivated in the attractive variant formerly known as *M. tubergenianum*, which has a prominent bunch of sterile flowers (paler than the fertile) in the upper part of the raceme.

9. M. neglectum Gussone (*M. racemosum* Lamarck & de Candolle; *M. atlanticum* Boissier & Reuter). Illustration: *Journal of the Royal Horticultural Society* **26**: 136 (1884); Polunin, Flowers of Europe, t. 170 (1969); Rix & Phillips, The bulb book, 71 (1981).

Bulbs 1–2.5 cm across, sometimes with offsets. Leaves 3–6, 6–40 cm × 2–8 mm, erect-spreading to prostrate, channelled to almost terete, bright green, sometimes reddish beneath. Raceme dense, becoming loose in fruit. Fertile flowers 1.5–3.5 × 3.5–7.5 mm, ovoid to oblong-urceolate, strongly contracted towards the apex, very dark to blackish blue, lobes white, recurved. Sterile flowers smaller and paler, rarely white. *Europe, N Africa, SW Asia*. H4. Early spring.

Quite an attractive garden plant, of variable size, the perianth-lobes making a strong contrast with the tube. Most cultivated variants increase readily by means of bulblets.

10. M. commutatum Gussone (*Botryanthus commutatus* (Gussone) Kunth). Illustration: Sweet, British Flower Garden, t. 369 (1833–1835); Polunin, Flowers of Europe, t. 169 (1969); Rix & Phillips, The bulb book, 71 (1981).

Leaves 2–5, erect-spreading, linear to linear-lanceolate. Raceme dense, ovoid, flowers often overlapping. Fertile flowers 3.5–7 mm, obovate-urceolate, strongly contracted and with prominent shoulders, deep violet-black, the lobes of the same colour, recurved. *Italy, Yugoslavia, Greece*. H5. Early spring.

11. M. latifolium Kirk. Illustration: *Botanical Magazine*, 7843 (1902); Mathew, Dwarf bulbs, 168 (1973).

Leaf single (rarely 2 together), 7–30 × 1–3 cm, erect, broadly oblanceolate, apex acuminate and often hooded. Raceme dense at first, becoming loose. Fertile flowers 5–6 mm, oblong-urceolate, strongly contracted towards the apex and with well-developed shoulders, deep violet-black, lobes of the same colour, recurved. *S & W Turkey*. H5. Spring.

A distinctive and handsome species.

22. LACHENALIA Murray

E. Campbell

Herbaceous perennials with bulbs. Leaves basal, usually 2 but occasionally up to 10, linear, usually with clasping bases, sometimes spotted or pustulate. Flowers in a raceme or spike, borne on a scape.

Perianth united into a tube at the base, bell-shaped, persistent, the 3 outer lobes usually with apical swellings and usually shorter than the 3 inner lobes, variously coloured. Stamens 6, attached to the top of the perianth-tube; anthers versatile, opening towards the centre of the flower. Ovary superior, 3-celled, with many ovules; style simple; stigma capitate. Fruit a 3-celled capsule opening between the septa.

A genus of about 50 species endemic to South Africa. Identification of the cultivated species is difficult as many of the plants grown are selections or hybrids. Fifteen species are included here; they are easily grown as pot-plants.

Literature: Ingram, J., Notes on the cultivated Liliaceae, 4: Lachenalia, *Baileya* **14**: 123–36 (1966).

Leaves. More than two: **1, 2**; one or two: **3, 4, 5, 6, 7, 8, 9, 10, 11, 12, 13, 14, 15**. Spotted or patched with different colours: **2, 7, 8, 9, 11, 12**; not spotted or patched: **1, 3, 4, 5, 6, 8, 9, 10, 13, 14, 15**. Pustulate: **3, 4, 5, 6**; not pustulate: **1, 2, 7, 8, 9, 10, 11, 12, 13, 14, 15**. With thickened margins: **5, 8, 9, 10**; without thickened margins: **1, 3, 4, 6, 7, 11, 12, 13, 14, 15**.

Scape. Spotted or patched with different colours: **1, 3, 5, 7, 8, 9, 11, 12, 15**; not spotted or patched: **2, 3,4, 6, 8, 10, 13, 14**.

Flowering time. January–May: **1, 2, 3, 4, 5, 6, 7, 8, 9, 10, 11, 14, 15**; August–September: **12, 13**.

Flowers. White or yellowish: **1, 2, 4, 5, 6, 8, 13, 14**; blue or bluish: **3, 7, 8, 9, 10**; red and/or yellow: **6, 11, 12, 15**. Fragrant: **3, 5, 6, 8, 9**; not fragrant: **1, 2, 4, 7, 10, 11, 12, 13, 14, 15**.

Outer perianth-lobes. More than three-quarters of the length of the inner lobes: **1, 2, 3, 4, 5, 7, 8, 9, 12, 14, 15**; up to three-quarters of the length of the inner lobes: **1, 6, 10, 11, 13**. With apical swellings: **1, 4, 5, 6, 7, 8, 11, 12, 14, 15**; without apical swellings: **2, 3, 9, 10, 13**.

Stamens. Projecting from the perianth: **6, 13**; not projecting: **1, 2, 3, 4, 5, 7, 8, 9, 10, 11, 12, 14, 15**.

Style. Projecting from the perianth: **2, 3, 6, 8, 12, 13, 15**; not projecting: **1, 4, 5, 7, 9, 10, 11, 14**.

1a. Leaves more than 2 — 2
b. Leaves 1 or 2 — 3
2a. Scape shorter than to as long as leaves; leaves 6–10; outer perianth-lobes with apical swellings — **1. contaminata**
b. Scape longer than leaves; leaves 4 or 5; outer perianth-lobes without apical swellings — **2. orthopetala**
3a. Leaves pustulate — 4
b. Leaves not pustulate — 7
4a. Flowers purplish blue — **3. purpureo-caerulea**
b. Flowers white or yellowish — 5
5a. Leaves narrow, less than 1.5 cm broad — **4. liliiflora**
b. Leaves wider, more than 2 cm broad — 6
6a. Leaves with thickened margins; neither stamens nor style projecting from the perianth — **5. pallida**
b. Leaves without thickened margins; stamens and style projecting — **6. pustulata**
7a. Leaf 1, less than 1 cm broad, with distinct, transverse, dark purple bands on the clasping base — **7. unifolia**
b. Leaves usually 2, more than 1 cm broad, without distinct, transverse bands on the clasping base — 8
8a. Leaves with thickened margin — 9
b. Leaves without thickened margin — 11
9a. Outer perianth-lobes with apical swellings; style projecting from the perianth — **8. glaucina**
b. Outer perianth-lobes without apical swellings; style not projecting — 10
10a. Flowers closely packed, the uppermost fertile — **9. orchioides**
b. Flowers loosely packed, the uppermost sterile and abortive — **10. mutabilis**
11a. Leaves with purple patches on the upper and/or lower surface — 12
b. Leaves without purple patches on either surface — 13
12a. Outer perianth-lobes much shorter than the inner; flowering in winter and spring — **11. aloides**
b. Outer perianth-lobes almost as long as the inner; flowering in autumn — **12. rubida**
13a. Flowers less than 1.5 cm; stamens longer than perianth — **13. unicolor**
b. Flowers more than 2 cm; stamens shorter than to as long as the perianth — 14
14a. Leaves becoming reflexed, with margins finely wavy; outer perianth-lobes with red apical swellings — **14. reflexa**
b. Leaves not becoming reflexed, margin not wavy; outer perianth-lobes with green apical swellings — **15. bulbifera**

1. L. contaminata Aiton (*L. angustifolia* Jacquin). Illustration: *Botanical Magazine*, 1401 (1811).

Leaves 6–10, to 20 cm × 3 mm, channelled, nearly erect, without spots. Scapes shorter than or just equalling leaves, spotted. Flowers to 5 mm, many, broadly bell-shaped; flower-stalks to 2 mm. Perianth-lobes all spreading, white; the inner tipped with greenish brown; the outer often flushed with red, bearing reddish apical swellings, and about three-quarters of the length of the inner. Stamens as long as the inner perianth-lobes. Style not projecting. *South Africa (Cape Province, Natal)*. H5–G1. Winter–spring.

2. L. orthopetala Jacquin. Illustration: *Baileya* **14**: f. 43 (1966).

Leaves 4–5, 10–15 cm × *c.* 6 mm, channelled, spotted. Scape to 25 cm, exceeding the leaves, flexuous, without spots. Flowers to 1.3 cm, numerous in a spike-like raceme, cylindric; flower-stalks *c.* 2 mm. Perianth white or tinged with red; the outer lobes without apical swellings and almost as long as the inner. Stamens as long as perianth. Style projecting. *South Africa*. H5–G1. Winter–spring.

3. L. purpureo-caerulea Jacquin. Illustration: *Botanical Magazine*, 745 (1804).

Leaves 2, 15–20 × *c.* 1 cm, pustulate, without spots. Scape to 23 cm, exceeding the leaves, patched with purple. Flowers to 1 cm, 30–40 in the raceme, bell-shaped, fragrant. Perianth-lobes all spreading, purplish blue; the outer without apical swellings and about four-fifths of the length of the inner. Stamens as long as perianth. Style projecting. *South Africa*. H5–G1. Spring.

4. L. liliiflora Jacquin.

Leaves 2, to 23 × 1 cm, pustulate, without spots. Scape to 30 cm, exceeding the leaves, without spots. Flowers to 2 cm, 12–20 in a spike-like raceme, cylindric; flower-stalks *c.* 4 mm. Perianth-lobes all spreading, white; the outer with green apical swellings, more than three-quarters of the length of the inner. Stamens shorter than the perianth. Style not projecting. *South Africa*. H5–G1. Winter–spring.

5. L. pallida Aiton (*L. lucida* Sims; *L. racemosa* Sims). Illustration: Saunders et al., Refugium Botanicum, t. 170 (1870).

Leaves 2, 15–23 × 2–3 cm, pustulate, without purple spots and with thickened margins. Scape to 25 cm, exceeding the leaves, minutely spotted. Flowers to 1.8 cm, 12–30 in a dense raceme, bell-shaped, fragrant; stalks very short. Perianth white, the outer lobes with green apical swellings and about four-fifths of the length of the inner. Stamens shorter than the perianth. Style not projecting. *South Africa.* H5–G1. Spring.

6. L. pustulata Jacquin. Illustration: *Botanical Magazine*, 817 (1805).

Leaves 2, to 28 × 2.5 cm, pustulate, without purple spots, margin not thickened. Scape to 26 cm, exceeding the leaves, without spots. Flowers *c.* 8 mm, many in a dense, spike-like raceme, bell-shaped, fragrant; stalks 2–5 mm. Perianth white or yellowish with spreading inner lobes; the outer lobes with pale green apical swellings and three-quarters of the length of the inner. Stamens and style projecting. *South Africa.* H5–G1. Winter–spring.

7. L. unifolia Jacquin. Illustration: *Botanical Magazine*, 766 (1804).

Leaf 1, 15–30 cm × 5–10 mm, patched with purple and with distinct, transverse purple banding on the clasping base. Scape to 32 cm, exceeding the leaves, spotted. Flowers to 1.5 cm, few to several in a loose raceme, bell-shaped; stalks to 1 cm. Perianth blue near the base, white often tinged with pink towards the apex; the inner lobes spreading; the outer with green apical swellings and five-sixths of the length of the inner. Stamens as long as perianth. Style not projecting. *South Africa.* H5–G1. Spring.

8. L. glaucina Jacquin (*L. sessiliflora* Andrews). Illustration: *Botanical Magazine*, 3552(1837).

Leaves 2, to 27 × 2.5 cm, with thickened margins, sometimes spotted with purple. Scape to 40 cm, exceeding the leaves, sometimes spotted. Flowers to 2 cm, many in a moderately dense spike, bell-shaped, fragrant; stalks very short or absent. Perianth iridescent, in various colours; usually the inner lobes pale purple at the base, white at the apex, slightly spreading, with wavy margins; the outer blue at the base, pink at the tip, with darker apical swellings, more than three-quarters of the length of the inner. Stamens as long as the perianth. Style projecting. *South Africa (Cape Province).* H5–G1. Spring.

9. L. orchioides Aiton (*L. pulchella* Kunth). Illustration: *Botanical Magazine*, 1269 (1810).

Leaves 2, to 28 × 2 cm, dark green, sometimes spotted with purple, with thickened margins. Scape to 35 cm, exceeding the leaves, sometimes spotted. Flowers to 1 cm, 10–50 in a dense spike, fragrant. Perianth iridescent, in various colours; usually the inner lobes cream tinged with pale purple, spreading and with wavy margins; the outer blue or blue-green, without apical swellings, four-fifths of the length of the inner. Stamens shorter than perianth. Style not projecting. *South Africa (Cape Province).* H5–G1. Winter–spring.

10. L. mutabilis Loddiges. Illustration: *Loddiges' Botanical Cabinet* **9**: t. 1076 (1825); *Botanical Magazine*, 9433 (1936).

Leaves 2, to 20 × 2 cm, bright green, unspotted, with thickened margins. Scape to 30 cm, exceeding the leaves, unspotted. Flowers to 1.2 cm, numerous in a loose spike, tubular; the uppermost flowers sterile and abortive, borne on long stalks; the fertile flowers more or less stalkless. Perianth with the inner lobes spreading, yellow-green at the base shading to reddish brown at the tips; the outer lobes blue to grey-green at the base, without apical swellings and three-quarters of the length of the inner. Stamens shorter than perianth. Style not projecting. *South Africa (Natal).* H5–G1. Winter–spring.

11. L. aloides (Linnaeus) Ascherson & Graebner (*Phormium aloides* Linnaeus; *L. tricolor* Jacquin). Illustration: *Botanical Magazine*, 82 (1789), 1704 (1815), 5992 (1872); Brickell (ed.), RHS A–Z encyclopedia of garden plants, 604 (2003).

Leaves 2, to 20 × 2 cm, spotted with purple. Scape to 29 cm, usually exceeding the leaves, spotted, arching or hanging. Flowers to 2.5 cm, 10–20 in a loose raceme, tubular, hanging; stalks to 5 mm. Perianth with the inner lobes yellow at the base shading to green with purple tips, all overlapping; the outer lobes bright red at the base, shading to yellow at the tips and with green apical swellings, half to two-thirds of the length of the inner. Stamens as long as perianth. Style not projecting. *South Africa (Cape Province, Natal).* H5–G1. Winter–spring.

This species is very variable in flower colour, and many selections and hybrids of it have been given specific names, e.g. *L. aurea* Lindley, *L. luteola* Jacquin, *L. nelsonii* Anon, and *L. quadricolor* Jacquin.

12. L. rubida Jacquin (*L. punctata* Jacquin). Illustration: *Botanical Magazine*, 993 (1807).

Leaves 2, to 14 × 2.5 cm, spotted or patched with purple on one or both sides. Scape to 26 cm, exceeding the leaves, spotted. Flowers to 3 cm, few to several in a loose raceme, tubular, hanging; stalks 2–4 mm. Perianth with inner lobes deep red at base shading to pale yellow at the tips, all overlapping; the outer lobes deep red with yellow-green apical swellings, seven-eighths of the length of the inner. Stamens as long as the perianth. Style projecting. *South Africa (Cape Province).* H5–G1. Autumn.

13. L. unicolor Jacquin. Illustration: *Botanical Magazine*, 1373 (1811).

Leaves 2, to 15 × 1.5 cm, without spots. Scape about equalling the leaves, without spots. Flowers to 8 mm, 30–60 held horizontally in a loose raceme, shortly cylindric; stalks 2–3 mm. Perianth white or tinged with pink; the inner lobes spreading; the outer without apical swellings, two-thirds of the length of the inner. *South Africa.* H5–G1. Autumn.

14. L. reflexa Thunberg (*L. pusilla* Kunth). Illustration: *Wiener Illustrierte Garten Zeitung* **17**: t. 418 (1892).

Leaves 2, to 15 × 3 cm, bright green, spreading just above the ground, then reflexed, with wavy margins, unspotted. Scape not exceeding the leaves, without spots. Flowers to 2.8 cm, few to many in a loose spike, bell-shaped, erect; stalks to 2 mm. Perianth white or yellowish green with a swollen tube; the inner lobes overlapping, each with a red median stripe; the outer lobes with red apical swellings, five-sixths of the length of the inner. Stamens as long as perianth. Style not projecting. *South Africa.* H5–G1. Winter–spring.

15. L. bulbifera (Cyrillo) Ascherson & Graebner (*Phormium bulbiferum* Cyrillo; *L. pendula* Aiton). Illustration: *Loddiges' Botanical Cabinet* **3**: t. 267 (1818); *Flowering Plants of South Africa* **4**: t. 158 (1924).

Leaves 2, to 30 × 4 cm, without spots. Scape shorter than leaves, faintly spotted. Flowers 3–4 cm, few to many in a loose raceme, tubular, hanging; stalks to 5 mm. Perianth with the inner lobes pale yellow with a purple and green apex and a red median stripe, all overlapping; the outer purplish red with green apical swellings, eight-ninths of the length of the inner. Stamens as long as the perianth. Style just slightly projecting. *South Africa*. H5–G1. Winter–spring.

23. VELTHEIMIA Gleditsch
C.J. King
Perennial herbs with ovoid bulbs. Leaves several, in a basal rosette, lanceolate, oblong or strap-shaped, rather thick, with wavy margins. Flowers in a dense raceme on a stout scape, nodding, bisexual. Flower-stalks short, subtended by bracts. Perianth tubular with tooth-like lobes, persistent. Stamens 6, attached at about the middle of the perianth-tube. Fruit a 3-celled, winged capsule. Seeds 2 in each cell, pear-shaped, black.

Two species from South Africa with a long history of nomenclatural confusion. Both have been cultivated in Europe for about 2 centuries, either in cool greenhouses or, in suitably mild areas, outdoors. They prefer a sunny position, a light, loamy soil, and good drainage.

Literature: Obermeyer, A.A., *Flowering Plants of Africa* **34**: t. 1356 (1961); Marais, W., Notes from fellows, *Journal of the Royal Horticultural Society* **97**: 483–4 (1972).

1a. Leaves glossy, dark green, strap-shaped, to 35 × 10 cm; perianth 3–4 cm **1. bracteata**
b. Leaves glaucous, lanceolate, to 30 × 4 cm; perianth 2–3 cm **2. capensis**

1. V. bracteata Baker (*V. capensis* misapplied; *V. viridifolia* Jacquin). Illustration: *Botanical Magazine*, 501 (1799) & n.s., 215 (1953); *Journal of the Royal Horticultural Society* **97**: f. 239 (1972); Brickell (ed.), RHS A–Z encyclopedia of garden plants, 1068 (2003).

Bulb broadly ovoid, about 7 × 6 cm, with fleshy scales. Leaves 8–12, to 35 × 10 cm, strap-shaped, oblong, shining green on both surfaces. Scape to 45 cm, dark purple, finely spotted with yellow, fleshy. Raceme 7.5–12.5 cm; flowers numerous, up to 60, pendent. Perianth 3–4 cm, pinkish purple with some pale yellow speckling. Fruit narrowly obovate in profile, conspicuously 3-winged. *South Africa (Eastern Cape Province)*. H5. Winter–early spring.

This is the more robust of the 2 species and the easier to cultivate. It stays in growth most of the year and therefore does not require a dormant period. A paler coloured variant with yellowish flowers tinted with red has been named 'Rosalba' (Illustration: *Journal of the Royal Horticultural Society* **97**: f. 240, 1972).

2. V. capensis (Linnaeus) de Candolle (*Aletris capensis* Linnaeus; *V. glauca* (Aiton) Jacquin; *V. roodeae* Phillips; *V. viridifolia* Jacquin). Illustration: *Botanical Magazine*, 1091 (1808), 3456 (1835) – as *V. glauca*; *Flowering plants of South Africa* **4**: t. 126 (1924) – as *V. roodeae*; *Flowering plants of Africa* **34**: t. 1356 (1961).

Bulb ovoid, up to 13 × 6 cm, with a flattened base, long neck, and thin scales. Leaves about 10, to 30 × 4 cm, lanceolate, glaucous green, shortly mucronate at the apex. Scape to 30 cm or more, glaucous, mottled with purple. Raceme 5–15 cm. Perianth 2–3 cm, varying from whitish spotted with red, to pink with a greenish tip or wine-red. Fruit a membranous, inflated, 3-furrowed capsule, rounded in profile. *South Africa (SW Cape Province)*. G1. Winter–early spring.

A more delicate-looking plant than *V. bracteata*, less easy to grow and propagate, and requiring a dormant period during the summer when the bulbs should be kept dry. The darker colour-form of this species has been named 'Rubescens'. A more compact plant, growing up to about 25 cm high with a sturdy stem and short, crisped-margined, greyish leaves, has been called **V. deasii** Barnes. But this is now considered to be merely a smaller form of *V. capensis* and, although horticulturally distinct, should be treated only as a variety or cultivar (Illustration: *Botanical Magazine*, 8931 (1938) – as *V. deasii*).

24. GALTONIA Decaisne
C.J. King
Perennial herbs, the bulbs with membranous tunics. Leaves few, basal, flaccid, flat. Scape terete, bearing a loose raceme of white or green flowers. Flower-stalks more or less spreading at flowering time, curving upwards in fruit, subtended by large bracts. Flowers bell-shaped, horizontal or slightly nodding. Perianth persistent, the tube slightly to much shorter than the obovate or oblong lobes. Stamens 6, attached just below the mouth or at about the middle of the perianth-tube, shorter than the lobes. Ovary oblong, 3-angled. Fruit a 3-celled capsule. Seeds numerous, angled, brownish or black.

Three species of medium to tall bulbous plants from the eastern region of South Africa. One of these, *G. candicans*, is grown much more frequently than the others, being considered the most attractive and also the hardiest. All species require a sunny position in a light, rich, well-drained soil, and once planted should not be frequently disturbed. Some protection during the winter may be necessary in colder areas. They are well adapted for growing in clumps in borders, but may also be potted up for the conservatory or greenhouse.

1a. Leaves to 5 cm broad, narrowing gradually to the apex; flowers usually white or with some green at the base 2
b. Leaves to 10 cm broad, the lower narrowing abruptly to form a broad apex with a distinct point; flowers pale green **1. viridiflora**
2a. Flowers *c.* 3 cm, perianth-lobes distinctly longer than the tube **2. candicans**
b. Flowers less than 3 cm, perianth-lobes about as long as the tube **3. princeps**

1. G. viridiflora Verdoorn. Illustration: *Flowering plants of Africa* **30**: t. 1188 (1955).

Leaves about 7, up to about 60 × 10 cm, forming a short column at the base, pale yellowish green, in parts thinly glaucous, at least the lower narrowing abruptly at the apex, mucronate. Scape to 1 m, bearing a 15–30-flowered raceme. Bracts to 3 cm, greenish, becoming papery, lanceolate, acuminate. Flower-stalks 2–5 cm, pale green. Perianth about 2–5 cm, pale green, the lobes longer than the tube, with whitish margins. Stamens all attached about the middle of the perianth-tube, or the 3 inner slightly higher. Capsule erect, 3-lobed, ovate in profile, 2.5 × 1.4 cm. Seeds brownish, weakly 3-angled. *South Africa (Orange Free State, Natal), Lesotho*. H4. Late summer.

2. G. candicans (Baker) Decaisne (*Hyacinthus candicans* Baker). Illustration: *Flore des serres*, t. 2172–3 (1875); Rix & Phillips, The bulb book, 169 (1981); Brickell (ed.), RHS A–Z encyclopedia of garden plants, 467 (2003).

Leaves 4–6, 50–100 × 5 cm, broadly lanceolate, slightly greyish green, fleshy. Scape stout, to 120 cm in cultivation, bearing up to 30 flowers. Bracts 3–4 × 1 cm, ovate-lanceolate, acute. Flower-stalks 2.5–6 cm, rigid, green. Perianth *c.* 3 cm, white, faintly tinged with green at the base of the tube, the lobes about twice as long as the tube. Stamens attached just below the mouth of the tube. Capsule erect, about 3 cm. Seeds black. *South Africa (Orange Free State, Natal), Lesotho.* H4. Late summer.

3. G. princeps (Baker) Decaisne (*Hyacinthus princeps* Baker). Illustration: *Botanical Magazine*, 8533 (1914).

Leaves 4–6, 40 × 4 cm, narrowly lanceolate, sheathing at the base. Scape up to 90 cm. Flowers fewer than in *G. candicans*. Bracts about 3 × 1 cm, ovate-lanceolate, acute, papery. Flower-stalks about 3 cm, lengthening to 8 cm in fruit, green. Perianth to 3 cm, tube green, lobes greenish white, only slightly longer than the tube. Stamens attached about the middle of the tube. Seeds black. *South Africa (Natal, Transkei).* H4. Late summer.

25. MASSONIA Houttuyn

C.J. King

Perennial herbs; the bulbs with whitish scales surrounded by brown membranous scales. Leaves 2, ovate, oblong or almost circular, spreading horizontally on the ground. Scape absent or very short. Inflorescence usually surrounded by large bracts. Flowers stalked, usually pleasantly scented. Perianth-lobes fused in the lower part, usually spreading or reflexed above. Stamens attached at the mouth of the perianth-tube. Ovary superior; style usually slightly longer than the stamens. Fruit a winged or deeply lobed capsule, opening by slits. Seeds numerous, black.

A genus now reduced to about 8 species, native to South Africa. The plants may be grown outside in an essentially frost-free, Mediterranean climate, and are suitable in those regions for sunny, well-drained positions in the rock garden. They can also be easily grown in sunny greenhouses, where the soil should be kept dry during the dormant period but moderately moist while they are in leaf. Propagation is normally by seed, plants reaching flowering size in about 3 years.

Literature: Jessop, J.P., Studies in the bulbous Liliaceae in South Africa: 6, The taxonomy of Massonia and allied genera, *Journal of South African Botany* **42**: 401–37 (1976).

1a. Anthers more than 2 mm; leaves usually hairless, or hairy only on the margins **1. depressa**
b. Anthers less than 2 mm or, if more, then the leaves hairy or papillose 2
2a. Leaves usually hairless, rarely papillose; perianth-lobes about half as long as the tube **2. jasminiflora**
b. Leaves usually hairy or papillose; perianth-lobes about as long as the tube 3
3a. Leaves papillose on upper surface **3. pustulata**
b. Leaves usually hairy, at least on the margins, but not papillose **4. echinata**

1. M. depressa Houttuyn (*M. brachypus* Baker; *M. latifolia* Linnaeus filius; *M. sanguineum* Jacquin). Illustration: *Botanical Magazine*, 848 (1805); *Flowering plants of South Africa* **2**: t. 46 (1922); Everett, The New York Botanical Gardens illustrated encyclopedia of horticulture **7**: 2154 (1981).

Bulb usually ovoid, 2–3.5 cm. Leaves 7–15 × 4–10 cm, circular to oblong, acute, hairless or with hairs only on the margins. Inflorescence capitate, with *c.* 20–30 flowers. Perianth green, yellow, white, cream, pink, red or brown, occasionally flecked with purple; tube 1–1.5 cm; lobes 8–10 mm. Filaments 1–1.6 cm, often cream, yellow or green, less often purplish; anthers 2.5–4 mm, yellow or purple. *South Africa (Cape Province).* H5. Late autumn–early winter.

2. M. jasminiflora Baker (*M. bowkeri* Baker). Illustration: *Botanical Magazine*, 7465 (1896); *Flowering plants of South Africa* **10**: t. 367 (1930).

Bulb ovoid to ellipsoid, 1–2 cm. Leaves 3–6 × 1.5–5 cm, ovate to broadly oblong, acute, hairless or occasionally with hairs only on the margins, rarely papillose. Inflorescence capitate, generally with fewer than 15 flowers. Perianth white or pink; tube 8–20 mm; lobes 4–8 mm. Filaments 2–4.5 mm, white or pink; anthers 1–1.5 mm, green, blue, dark purple or black. *South Africa (Cape Province, Orange Free State), Lesotho.* H5. Late autumn–early winter.

3. M. pustulata Jacquin. Illustration: *Botanical Magazine*, 642 (1803); Grey, Hardy bulbs **3**:428 (1938); *Flowering plants of South Africa* **23**: t. 915 (1943); *Journal of the Royal Horticultural Society* **109**: 418 (1984).

Bulb spherical or ovoid, 1–2.5 cm. Leaves 3–10 × 2–7 cm, ovate to oblong, acute, papillose on the upper surface, especially towards the apex, and with the margins often minutely hairy or minutely toothed. Inflorescence capitate, with 15–25 flowers. Perianth pink, white, yellow or greenish; tube 6–11 mm; lobes 4.5–10 mm. Filaments 4–12 mm, white; anthers 1–1.75 mm, yellow or reddish. *South Africa (Cape Province).* H5. Late autumn–early winter.

Closely allied to *M. echinata*, but distinguished by the relatively uniformly papillose upper surface of the leaves.

4. M. echinata Linnaeus filius (*M. amygdalina* Baker; *M. bolusiae* Barker; *M. longifolia* Jacquin var. *candida* Ker Gawler; *M. muricata* Ker Gawler; *M. scabra* Thunberg). Illustration: *Botanical Magazine*, 559 (1802); *Edwards's Botanical Register* **9**: t. 694 (1823); *Flowering plants of South Africa* **11**: t. 429 (1931).

Bulb usually ovoid, 1–2 cm. Leaves 2–8 × 1–4 cm, ovate to oblong, acute or obtuse, usually hairy, at least on the margins. Inflorescence capitate, with 5–20 flowers. Perianth yellow, white or pink; tube 5–7 mm; lobes 4–8 mm. Filaments 4–8 mm, white; anthers 0.5–1.25 mm, yellow or purple. *South Africa (Cape Province).* H5. Late autumn–early winter.

26. DAUBENYA Lindley

P.F. Yeo

Hairless, perennial, summer-dormant herbs with bulbs. Leaves 2, 5–15 × 3.5–7.5 cm, ovate or oblong, strongly parallel-veined. Inflorescence-stalk *c.* 4 cm, concealed in neck of bulb. Inflorescence a head-like corymb of *c.* 10 flowers, held close to the leaf-bases. Bracts to 3 × 2.2 cm, acute, overlapping, the inner smaller than the outer. Flowers erect, strongly unequal within the inflorescence. Outer flowers to 6.5 cm, with tube to 2.5 cm and *c.* 4 mm in diameter, and bilaterally symmetric limb to 4 cm, consisting of a large lower lip with 3 blunt, hooded lobes and a minute, 3-lobed upper lip; the lower lips encircle the corymb. Innermost flowers nearly radially symmetric with tubes similar to those of outer and lobes 2–6 mm. Transitional flowers also occur. Stamens inserted at different levels near the bases of the perianth-lobes; filaments free from each other; anthers dorsifixed. Ovary superior,

3-celled with numerous ovules. Style unbranched, about as long as the perianth-tube, with no stigmatic enlargement. Seeds spherical, 2 mm in diameter, black.

One species in the winter-rainfall area of South Africa, closely related to *Massonia*. It should be grown in a bulb-frame or greenhouse and dried off in summer.

Literature: Hall, H., Daubenya Lindley, *Journal of the Botanical Society of South Africa* **56**: 13–16, cover & frontispiece (1970); Jessop, J.P., Studies in the bulbous Liliaceae in South Africa, 6, *Journal of South African Botany* **42**: 401–37 (1976).

1. D. aurea Lindley (*D. fulva* Lindley). Illustration: *Edwards's Botanical Register* **21**: t. 1813 (1835), **25**: t. 53 (1839); *Botanical Magazine*, n.s., 700 (1975).

Outer flowers usually red or yellow but sometimes orange, inner yellowish. *South Africa (SW Cape Province)*. G1. Spring.

19. ALLIACEAE

Plants herbaceous, terrestrial, often with both bulbs and/or rhizomes, often smelling of onion or garlic. Leaves well developed. Flowers in umbels subtended by spathes, or solitary and subtended by spathes, usually actinomorphic. Perianth-segments 6, free or united. Stamens 6, filaments sometimes winged. Ovary of 3 united carpels, superior; placentation axile. Fruit a capsule.

There are 13 genera and 650 species. Five genera are native to Europe, 12 to North America, and many are cultivated as ornamentals, vegetables or flavourings. *Agapanthus* is sometimes separated off as Agapanthaceae.

1a. Plants with rhizomes or fleshy roots, bulbs absent 2
b. Plants with bulbs (which are sometimes borne on a rhizome) 3
2a. Perianth with 6 outgrowths forming a corona **2. Tulbaghia**
b. Perianth without outgrowths **1. Agapanthus**
3a. Perianth with free segments or with a very short tube 4
b. Perianth with a distinct tube at the base 10
4a. Plant smelling of onion or garlic 5
b. Plant not smelling of onion of garlic 6
5a. Each cell of the ovary with many ovules; flower-stalks swollen below the flowers; outer perianth-segments with 3–7 veins **4. Nectaroscordum**
b. Each cell of the ovary with 1–8, usually 2, ovules; flower-stalks not swollen below the flowers; outer perianth-segments 1-veined **3. Allium**
6a. Spathe 1 **9. Caloscordum**
b. Spathes 2 or more 7
7a. Leaves more or less terete **8. Muilla**
b. Leaves flat 8
8a. Filaments broadened at the base into a cup-like structure that bears 2 lateral teeth **6. Bloomeria**
b. Filaments not as above 9
9a. Flowers erect **5. Nothoscordum**
b. Flowers drooping **7. Bessera**
10a. Flower solitary (rarely 2 together), subtended by 2 spathes that are united into a tube below **11. Ipheion**
b. Flowers more than 2 together in umbels, subtended by 2 or 4 free spathes 11
11a. Stamens enclosed in the perianth-tube 12
b. Stamens projecting from the perianth-tube 13
12a. Perianth with outgrowths forming corona **2. Tulbaghia**
b. Perianth without corona **10. Leucocoryne**
13a. Spathers 4; flowers stalkless but appearing stalked because of the long perianth tube **15. Milla**
b. Spathes 2; flowers with distinct stalks 14
14a. Leaves not keeled beneath, veins distinct **12. Brodiaea**
b. Leaves keeled beneath, veins distinct 15
15a. Anthers versatile; anthers shortly stalked **11. Triteleia**
b. Anthers basifixed; ovary stalkless **14. Dichelostemma**

1. AGAPANTHUS L'Héritier

C.J. King

Perennial herbs with fleshy roots arising from a tuberous rootstock. Leaves 10–70 cm × 8–60 mm, usually basal, linear to strap-shaped with entire margins. Flowers few to many in an umbel on a usually stout scape 30–180 cm high. Perianth tubular to spreading, united towards the base, ranging from dark violet or deep blue to white. Stamens 6, inserted on the perianth-tube. Ovary of 3 cells, superior, developing into a many-seeded capsule. Seeds flat and winged, black.

All (except *A. africanus*) prefer a sunny position in a light, rich, well-drained soil, copiously watered in summer but kept fairly dry during the winter. The hardier forms may survive the winter in the open with a slight covering for the crowns, but, except in the mildest areas, the more tender forms are best placed outside in tubs for the summer and brought into a light but frost-free place in autumn. Propagation is by offsets, by divison, or by seed.

Leighton recognised 10 species and 15 subspecies in her monograph of *Agapanthus* (see below). She found that plants hybridise quite freely when grown near each other and that natural variation was so great that it would have been possible for the genus to be broken down into innumerable species or subspecies.

Subsequent research by other investigators tends to support this view. Indeed, Palmer (*Journal of the Royal Horticultural Society* **92**, 1967) was of the opinion that, as cross-pollination was so rife amongst cultivated plants, practically all garden forms of *Agapanthus* should be regarded as hybrids, and McNeil (*Journal of the Royal Horticultural Society* **97**, 1972) concluded that the whole *Agapanthus* population should be considered as one unstable mega-species.

In spite of the difficulties, a modification of Leighton's key and descriptions may enable a number of cultivated plants to be identified with some precision. However, the large numbers of individual clones that have been distinguished by cultivar names are often of complex parentage, and it may not be possible to assign these to a particular species or subspecies. Some of these cultivars were formerly included in the group of hardier plants raised or selected from existing garden forms by Palmer and known collectively as 'Headbourne Hybrids'.

Literature: Leighton, F.M., The Genus Agapanthus L'Héritier, *Journal of South African Botany*, Supplementary Volume **IV** (1965).

Leaves. Arising from an obvious stem: **5**. More than 4.5 cm broad: **1, 5, 6**.
Scape. More than 100 cm: **1, 5**.
Flower-stalks. More than **7** cm: **6**.
Perianth (see Figure 5, below). More than 5 cm: **2, 6**. Tubular: **1**; semi-open: **2, 3**. Thick in texture: **4**.

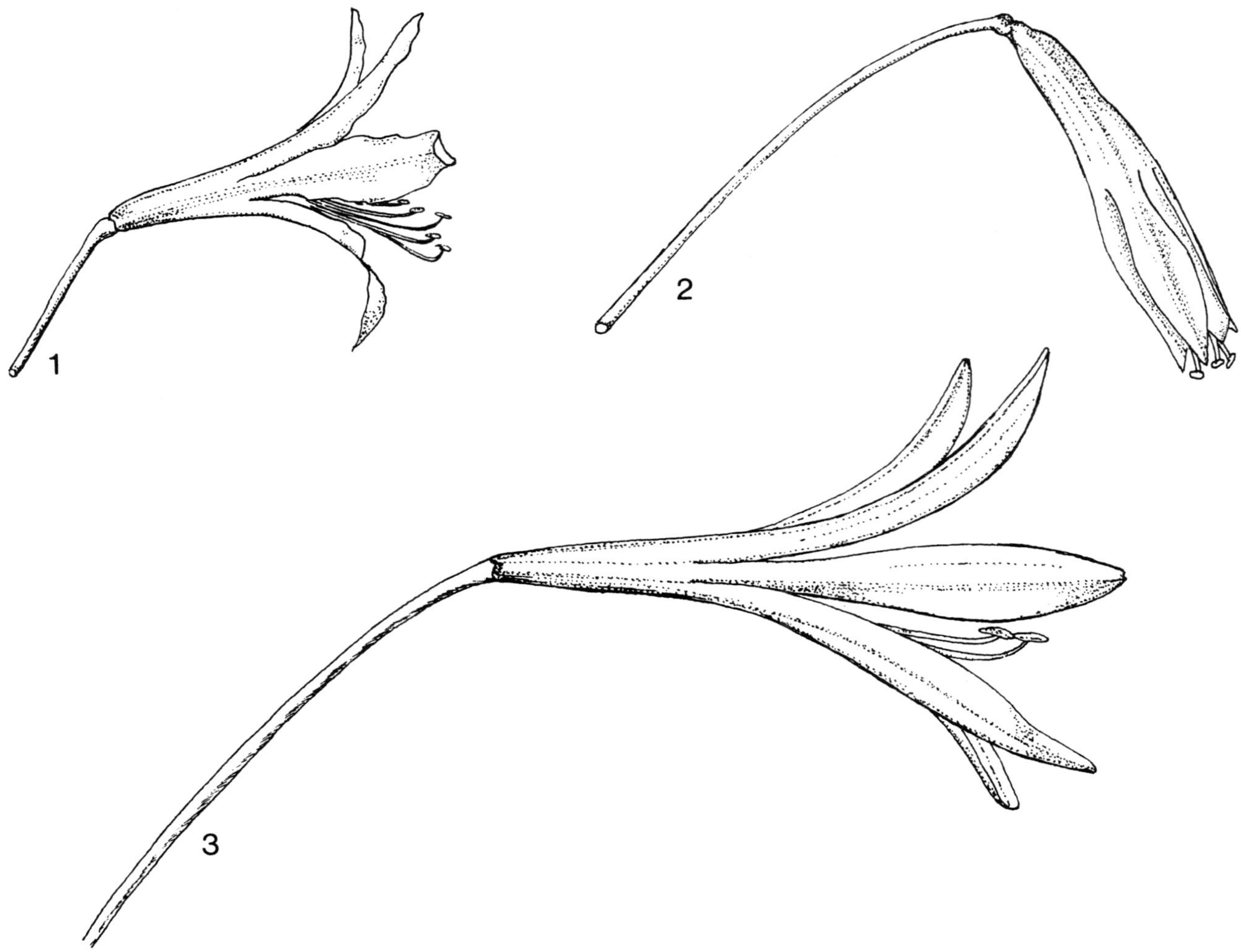

Figure 5. Flowers of Agapanthus species. 1, Open flower of *A. praecox* subsp. *minimus*. 2, Tubular flower of *A. inapertus* subsp. *inapertus*. 3, Semi-open flower of *A. nutans*.

Stamens. Projecting: **3, 6**.

1a. Perianth tubular (see Figure 5(2)) **1. inapertus**
b. Perianth open or semi-open (see Figure 5(1,3)) 2
2a. Perianth always semi-open, more than 3.5 cm **2. nutans**
b. Perianth open, or if semi-open, less than 3.5 cm 3
3a. Perianth semi-open, up to 3.5 cm **3. campanulatus**
b. Perianth open 4
4a. Perianth thick and waxy in texture; leaves not usually more than 35 × 2 cm **4. africanus**
b. Perianth thin in texture; leaves often longer and broader 5
5a. Perianth to 3.5 cm; leaves not usually more than 40 × 2.5 cm **3. campanulatus**
b. Perianth usually more than 3.5 cm; leaves often longer and broader 6
6a. Perianth not more than 5 cm; flower-stalks not more than 7 cm; stamens and style shorter than perianth **5. caulescens**
b. Perianth to 7 cm; flower-stalks to 12 cm; stamens and style usually as long as perianth **6. praecox**

1. A. inapertus Beauverd. Figure 5(2)
Illustration: *Botanical Magazine*, 9621 (1942); Letty, Wild flowers of the Transvaal, t. 13 (1962); Leighton, The genus Agapanthus, 42 & t. 12 (1965); *Flowering plants of Africa* **37**: t. 1479 & 1480 (1965–6)

Leaves to 70 × 6 cm. Scape to 1.8 m; flowers many. Flower-stalks 2–5.5 cm, erect at first, later nodding. Perianth 2.5–5 cm, tubular, bright blue to dark violet, sometimes white; tube at least as long as the lobes; lobes slightly spreading. Stamens about as long as perianth. Style usually slightly projecting when mature. *South Africa*. H4. Autumn.

Of the 5 subspecies, the 2 most frequently grown in European gardens are subsp. **inapertus**, with leaves *c.* 2.5 cm broad, and flowers deep blue or violet-blue, occasionally white, and subsp. **pendulus** (Bolus) Leighton (*A. pendulus* Bolus), which has leaves 5 cm or more broad, and very deep blue or dark violet flowers.

2. A. nutans Leighton. Figure 5(3).
Illustration: Leighton, The genus Agapanthus, 38 (1965)

Leaves 20–50 × 1–4 cm, often glaucous. Scape to 90 cm; flowers few to many. Flower-stalks spreading or nodding. Perianth 3.5–6 cm, nodding, semi-open, pale blue; tube 1.3–2.7 cm; lobes slightly spreading. Stamens shorter than perianth. Style finally almost as long as perianth. *South Africa.* H4. Late summer.

3. A. campanulatus Leighton. Illustration: *Botanical Magazine*, n.s., 380 (1962) – as *A. patens*; Leighton, The genus Agapanthus, 30, 32 (1965); *Flowering plants of Africa* **37**: t. 1478 (1965–6) – subsp. *patens*; Brickell (ed.), RHS A–Z encyclopedia of garden plants, 86 (2003).

Leaves usually 15–40 × 1–2.5 cm. Scape usually 40–70 cm, rarely to 100 cm; flowers few to many. Flower-stalks 2–7 cm, spreading. Perianth 2–3.5 cm, semi-open or open, pale to deep blue, sometimes white; tube 5–10 mm; lobes spreading. Stamens shorter than perianth, projecting. Style finally as long as stamens. *South Africa.* H4. Summer–early autumn.

Two subspecies are recognised: subsp. **campanulatus** with a bell-shaped perianth, and subsp. **patens** (Leighton) Leighton (*A. patens* Leighton), which has widely spreading or even reflexed perianth-lobes, and is generally a smaller, more slender, plant.

4. A. africanus (Linnaeus) Hoffmansegg (*A. umbellatus* L'Héritier; *A. minor* Loddiges). Illustration: *Loddiges' Botanical Cabinet* **1**: 42 (1817); Kidd, Wild flowers of the Cape peninsula, t. 3, f. 2 (1951); Hulme, Wild flowers of Natal, t. 21, f. 6 (1954).

Leaves 10–35 cm × 8–20 mm. Scape 30–60 cm; flowers few to many. Flower-stalks 1.5–5 cm, stiff. Perianth 2.5–5 cm, thick and waxy in texture, deep blue or blue-violet; tube 9–14 mm; lobes spreading. Stamens shorter than perianth. Style finally as long as stamens. *South Africa.* H5. Autumn.

The most difficult species to cultivate successfully since it requires watering in winter and to be kept moderately dry in summer.

5. A. caulescens Sprenger. Illustration: *Gartenflora* **50**: 281 (1901).

Leaves arising from an obvious stem, the basal sheathing leaves often only 5–15 cm, the upper 25–60 × 1.5–5 cm, green and glossy, acute or obtuse at apex. Stem 60–130 cm; umbel dense; flowers many. Flower-stalks 3–7 cm, spreading and somewhat drooping. Perianth 3–5 cm, bright to deep blue; tube 1–1.9 cm; lobes widely spreading, sometimes recurved. Stamens and style shorter than perianth. *South Africa.* H4. Late summer–early autumn.

Of the 3 subspecies, the one usually cultivated is subsp. **caulescens**, a more robust plant than the others, with broader, obtuse leaves.

6. A. praecox Willdenow. Figure 5(1), p. 131. Illustration: *Edwards's Botanical Register*: t. 699 (1823) –as *A. umbellatus* var. *minimus* – & t. 7 (1843) – as *A. umbellatus* var. *maximus*; Leighton, The genus Agapanthus, 20 & 22 (1965); Batten & Bokelmann, Wild flowers of the eastern Cape province, t. 9, f. 3 (1966) – subsp. *praecox*.

Leaves 20–70 × 1.5–5.5 cm. Scape 40–100 cm; flowers few to many. Flower-stalks 4–12 cm. Perianth 3–7 cm, bright to pale blue, sometimes white; tube 7–26 mm; lobes spreading. Stamens and style usually as long as perianth, sometimes shorter, projecting. *South Africa.* H5. Summer–autumn.

Three subspecies are recognised: subsp. **praecox**, a large, stout plant with flowers at least 5 cm in length; subsp. **orientalis** (Leighton) Leighton (*A. orientalis* Leighton) – illustration: Brickell (ed.), RHS A–Z encyclopedia of garden plants, 86 (2003) – a smaller plant forming dense clumps, having a stout scape bearing an often densely crowded umbel with flowers less than 5 cm; and subsp. **minimus** (Lindley) Leighton, an even smaller plant with leaves not more than 2.5 cm broad, a slender scape, and an uncrowded umbel with flowers less than 5 cm.

2. TULBAGHIA Linnaeus

V.A. Matthews

Rootstock a bulb or rhizome. Leaves 4–8, strap-shaped to linear, basal. Flowering stem solitary, erect, bearing an umbel of 6–40 flowers that is subtended by 2 bracts. Perianth united into a tube for about half its length; mouth of tube with a fleshy corona, which is cylindric or composed of 3 free scales. Anthers stalkless on the perianth-tube, in 2 whorls 1 above the other. Style short; stigma capitate. Capsule 3-celled, containing triangular black seeds.

A genus of 21 species, 5 of which are cultivated. *Tulbaghia* is native to southern temperate and tropical Africa. With the exception of *T. violacea*, the bulbs require protection from frost. They will grow in any soil and prefer a sunny position if planted outdoors.

Literature: Burbidge, R.B., A revision of the genus Tulbaghia (Liliaceae), *Notes from the Royal Botanic Garden, Edinburgh* **36**: 77–103 (1978); Burbidge, R.B., The genus Tulbaghia, *Lilies 1978/9 and other Liliaceae*, 78–84 (1978/9)

1a. Perianth-tube and lobes green or white **1. alliacea**
b. Perianth-tube and lobes purple 2
2a. Corona a cylinder split into 3 toothed lobes **2. fragrans**
b. Corona composed of 3 free scales **3. violacea**

1. T. alliacea Linnaeus. Illustration: *Flowering plants of South Africa* **17**: pl. 653 (1937); Rice & Compton, Wild flowers of the Cape of Good Hope, pl. 244 (1950).

Rootstock a rhizome. Leaves 15–25 cm × 3–5 mm, linear. Flowering stem 15–30 cm, bearing 6–10 flowers. Perianth-tube *c.* 6 mm, green; perianth-lobes 2–4 mm, green; corona 2–3 mm, orange-brown with 3 or 6 shallow lobes. *Southern Africa, Zimbabwe.* H5. Summer.

T. capensis Linnaeus. Illustration: *Botanical Magazine*, 806 (1805) – as *T. alliacea*; Mason, Western Cape sandveld flowers, pl. 8 (1972). Perianth olive-green and corona brownish purple, the latter up to 5 mm and 3-lobed with each lobe divided into 2. *South Africa (Cape Province).* H5–G1. Summer.

T. natalensis Baker. Illustration: *Flowering plants of South Africa* **25**: pl. 979 (1945). Perianth-tube 3–4 mm, white, sometimes tinged with purple, perianth-lobes 5–7 mm and corona 3–4.5 mm, yellowish orange or green with 3 toothed lobes. *South Africa (Natal).* H5. Summer.

2. T. fragrans Verdoorn (*T. daviesii* Grey; *T. pulchella misapplied*). Illustration: *Flowering plants of South Africa* **11**: pl. 438 (1931); Letty, Wild flowers of the Transvaal, pl. 11 (1962)

Rootstock a bulb. Leaves 30–60 × 1.5–2.5 cm, strap-shaped. Flowering stem to 60 cm, bearing 20–40 flowers. Perianth-tube 8–9 mm, light purple; perianth-lobes 6–8 mm, light purple, sometimes with ragged and inrolled margins; corona a 3–4 mm cylinder split into 3 toothed lobes, purple tinged with pink. *South Africa (Transvaal).* H5–G1. Late spring–summer.

T. fragrans differs from the other species in not, or scarcely, smelling of onion.

3. T. violacea Harvey. Illustration: *Botanical Magazine*, 3555 (1837); Perry, Flowers of the world, 19 (1972).

Rootstock a rhizome. Leaves 15–30 cm × 4–7 mm, linear. Flowering stem 30–65 cm, bearing 8–30 flowers. Perianth-tube 7–13 mm, purple; perianth-lobes 5–12 mm, purple; corona 1.5–3.5 mm, composed of 3 free fleshy scales, purple tinged with reddish, or white. *South Africa (Cape Province, Transvaal)*. H4. Early summer–late autumn.

'Silver Lace' is a large-flowered form.

3. ALLIUM Linnaeus

W.T. Stearn assisted by E. Campbell

Perennial herbs, mostly smelling of onion, usually with well-formed bulbs, arising in some species from a short rhizome. Leaves linear to elliptic, basal or sheathing the stem, flat to channelled, terete or semi-terete, solid or hollow. Flowers usually many (unless replaced by bulbils) in an umbel on a solid or hollow stem and enclosed at first within a spathe consisting of 1 or more bracts. Perianth star-shaped, cup-shaped or bell-shaped, with 6 usually free segments. Stamens 6; filaments free or united at base, simple or variously toothed; anthers opening by longitudinal slits. Ovary 3-celled, each cell with 1–8 ovules, usually 2; style slender; stigma usually simple but 3-lobed in a few species. Fruit a 3-celled capsule splitting between the septa. Seeds angled or rounded, blackish, in a few species with a spongy appendage.

A genus of about 690 species in Europe, Asia and America, most abundant in central Asia, but extending into tropical Africa, Sri Lanka and Mexico. Some species have wide ranges, notably *A. schoenoprasum;* some are very local. Nearly all are easy to cultivate, given good drainage and freedom from competition. Hybrids are almost unknown, the only well-authenticated cross being *A. cepa* × *fistulosum*. The diversity of bulb structure, particularly the nature of the outer covering (tunic), provides important characters for the distinction of sections and species, but these have been kept to a minimum in the following key owing to the natural reluctance of gardeners to dig up their plants when in flower. Since species are continually being introduced into cultivation that may not be included in the following account, reference should be made to Flora Europaea **5** (1980) for European species; to Flora Iranica **76** (1971) and Flora SSSR **4**: 112–280 (1935), translated and annotated in *Herbertia* **11**: 65–218 (1946), for central Asiatic species. Some species are of culinary importance, i.e. *A. cepa* (onion), *A. sativum* (garlic), *A. porrum* (leek), *A. fistulosum* (Japanese leek), *A. schoenoprasum* (chives), *A. tuberosum* (Chinese chives) and *A. chinense* (rakkyo or ch'iaot'ou).

Literature: Regel, E., Alliorum adhuc cognitorum monographia, *Acta Horti Petropolitani* **3**: 1–266 (1875); Stearn, W. T., Notes on the genus Allium in the Old World, *Herbertia* **11**: 11–34 (1946); Vvedensky, A.I., The genus Allium in the USSR, *Herbertia* **11**: 65–218 (1946); Moore, H.E., The cultivated Alliums, *Baileya* **2**: 103–23 (1954) & **3**: 137–49, 156–67 (1955); Ownbey, M. & Aase, H.C., The Allium canadense alliance, *Research studies of the State College of Washington, Monographic Supplement* **106** (1955); Stearn, W.T., Allium and Milula in the central and eastern Himalaya, *Bulletin of the British Museum (Natural History), Botany* **2**: 1159–91 (1960); Bothmer, R. von, Biosystematic studies in the Allium ampeloprasum complex, *Opera Botanica* **34**: 1–104 (1974); Wilde-Duyfjes, B.E.E. de, Revision of the genus Allium L. (Liliaceae) in Africa, *Belmontia* **7**: 1–237 (1977); Stearn, W.T., European species of Allium and allied genera of Alliaceae: a synonymic enumeration, *Annales Musei Goulandris* **4**: 83–198 (1978); Stearn, W.T., The genus Allium in the Balkan peninsula, *Botanische Jahrbücher* **102**: 201–13 (1981); Pastor, J. & Valdés, B., Revision del genero Allium en la Peninsula Iberica e Islas Baleares (1983); Yang, C.Y & Huang, J.H., A new variety of Allium from Xinjiang, *Bulletin of Botanical Research (China)* **21(2)**: 186 (2001); Seregin AP., Additions to Allium sect. Allium (Alliaceae) from North Africa, *Komarovia* **4**: 159–164 (2006); Khassanov F.O., Noroozi, J., & Akhani, H., Two new species of the genus Allium (Alliaceae) from Iran, *Rostaniha* **7**(suppl. 2): 119–129 (2006).

1a. Leaves completely withered or absent at flowering time 2
b. Leaves present although sometimes mostly withered at flowering time 3
2a. Stem completely covered with adpressed dry sheaths **49. callimischon**
b. Stem naked **23. tricoccum**
3a. Leaves clearly contracted into a stalk at base 4
b. Leaves scarcely or not stalked, in most species linear or thread-like 6
4a. Leaves sheathing the lower one-third to half of the stem; stamens longer than the pale yellow or greenish white perianth-segments **22. victorialis**
b. Leaves all basal; stamens shorter than the white perianth-segments 5
5a. Bulbs densely covered with netted fibres; leaves mostly withered at flowering time; stem terete **23. tricoccum**
b. Bulbs without netted fibres, only a few parallel fibres at base; leaves completely green at flowering time; stem angled **39. ursinum**
6a. Plant rarely producing flowering stems but forming clumps of hollow leaves, semi-circular in section, scarcely 1 cm wide but to 40 cm long **25. cepa** 'Perutile'
b. Plant producing flowering stems with flowers or bulbils or both 7
7a. Umbel with bulbils only, flowers absent or abortive 8
b. Umbel with flowers, sometimes also with bulbils 13
8a. Leaves basal, solitary, not sheathing the 3-angled stem **42. paradoxum**
b. Leaves sheathing lower part of the terete stem 9
9a. Leaves and stem inflated, 5–20 mm broad; bulbils large, 2–4 cm 10
b. Leaves and stem slender, not inflated; the leaves if flat and keeled to 3 cm broad, if hollow to 4 mm broad; bulbils less than 2 cm 11
10a. Leaves circular in section **26. fistulosum** 'Viviparum'
b. Leaves semi-circular in section **25. cepa** 'Viviparum'
11a. Leaves flat and keeled, 5–30 mm broad **59. sativum**
b. Leaves hollow, terete, to 4 mm broad 12
12a. Spathes 2, persistent, long **54. oleraceum**
b. Spathe 1, soon falling **63. vineale**
13a. Umbel almost stalkless in a clump or rosette of spreading, linear, ciliate leaves **38. chamaemoly**
b. Umbel on an evident stem 14
14a. Leaves hairy along the margin, sometimes also beneath 15
b. Leaves hairless 20
15a. Leaves 1–4 cm broad; umbel 10–15 cm or more across 16

b. Leaves to 2 cm broad; umbel 4–8 cm across 18
16a. Stem 15–40 cm; perianth-segments 1.2–2 cm, becoming rigid after flowering **69. christophii**
b. Stem 50–120 cm; perianth-segments 8–10 mm, reflexing and twisted (not rigid) after flowering 17
17a. Stem distinctly ribbed; bulbs blackish **72. stipitatum**
b. Stem lightly ribbed; bulbs grey **73. hirtifolium**
18a. Stamens about as long as or longer than perianth-segments **33. subvillosum**
b. Stamens one-third to two-thirds as long as perianth-segments 19
19a. Flower-stalks 3–5 times as long as perianth-segments **32. subhirsutum**
b. Flower-stalks half to 3 times the length of perianth-segments; anthers yellowish **34. trifoliatum**
20a. Stem and leaves inflated, hollow; stem 2–4 cm in diameter 21
b. Stem not inflated, usually slender and solid 23
21a. Leaves semi-circular in section, grooved above; perianth-segments 3–4.5 mm; stamens 4–6 mm **25. cepa**
b. Leaves circular in section; perianth-segments 6–9 mm; stamens 8–12 mm 22
22a. Flower-stalks 2–3 times as long as the perianth; bulb cylindric **26. fistulosum**
b. Flower-stalks slightly shorter than the perianth to 1 and a half times as long; bulb ovoid **27. altaicum**
23a. Filaments of outer and inner stamens markedly unlike, the outer simple or 3-toothed, the inner always with an expanded basal part bearing 3–5 teeth, the central tooth anther-bearing, the lateral teeth filamentous and longer 24
b. Filaments of all stamens simple or sometimes with short teeth at base very much shorter than the anther-bearing filament 29
24a. Leaves hollow, 1–4 mm broad 25
b. Leaves flat, keeled, 5–40 mm broad 27
25a. Spathe 1, soon falling **63. vineale**
b. Spathes 2, persistent 26
26a. Perianth-segments 3.5–5.5 mm; stamens projecting **62. sphaerocephalon**
b. Perianth-segments *c.* 10 mm; stamens included **64. heldreichii**
27a. Flowers mostly abortive and bulbils present; basal part of inner stamens shorter than or equalling the central anther-bearing tooth **59. sativum**
b. Flowers normal; basal part of inner stamens 2–3 times as long as the central anther-bearing tooth 28
28a. Bulblets (and bulbils if present) blackish; flower-stalks not more than 2 cm; stamens shorter than perianth **61. scorodoprasum**
b. Bulblets yellowish, not blackish; some flower-stalks more than 2 cm long; stamens equalling or slightly longer than perianth with anthers projecting **60. ampeloprasum**
29a. Leaves basal or almost so, sheathing not more than the lower one-quarter of the above-ground stem and parting from it at the same level or not much apart 30
b. Leaves sheathing more than the lower one-quarter of the above-ground stem and parting from it at different heights 96
30a. Perianth bright yellow 31
b. Perianth white, greenish, yellowish, pink, brownish, blue or purple 32
31a. Leaves 1.5–3.5 cm broad; ovary not prominently angled or crested **35. moly**
b. Leaves 3–7 mm broad; ovary prominently angled, 6-crested **36. scorzonerifolium**
32a. Stigma 3-lobed 33
b. Stigma entire 36
33a. Perianth white with longitudinal green stripes; stem triangular in section, flaccid after flowering 34
b. Perianth rose to purple-red; stem terete, rigid after flowering 67
34a. Leaf solitary **42. paradoxum**
b. Leaves 2–5 35
35a. Umbel 1-sided; flowers always pendent and bell-shaped; perianth-segments 1-veined **40. triquetrum**
b. Umbel loose; flowers spreading and star-shaped, later pendent and bell-shaped with perianth-segments converging around the capsule; segments 3-veined **41. pendulinum**
36a. Leaves 1–8 cm broad 37
b. Leaves 0.5–10 mm broad 57
37a. Flower-stalks markedly unequal, the longest at least 3 times as long as the shortest; umbel loose, *c.* 20–30 cm across 38
b. Flower-stalks almost equal; umbel dense or loose, less than 20 cm across 39
38a. Stem 30–60 cm; perianth purplish **78. schubertii**
b. Stem 10–25 cm; perianth pale brown **79. protensum**
39a. Perianth bell-shaped, the segments convergent and completely hiding the ovary; ovary with 6 well-marked apical swellings **15. macranthum**
b. Perianth cup-shaped to star-shaped , the ovary visible and without well-marked apical swellings 40
40a. Stamens distinctly longer than the perianth-segments 41
b. Stamens shorter than or equalling the perianth-segments 45
41a. Leaves ascending or erect; stem angled; umbel pendulous before opening **2. nutans**
b. Leaves spreading outwards; stem not angled; umbel erect before opening 42
42a. Stem 10–30 cm (rarely more), shorter than the leaves 43
b. Stem 30–200 cm, much longer than the leaves 44
43a. Leaves 3–6, linear **80. caspium**
b. Leaves 1–2, elliptic or narrowly ovate **81. mirum**
44a. Stem feebly ribbed; perianth-segments obtuse **75. giganteum**
b. Stem strongly ribbed; perianth-segments almost acute **76. macleanii**
45a. Perianth blue **13. sikkimense**
b. Perianth not blue 46
46a. Stem angled; leaves keeled 47
b. Stem terete; leaves not keeled 48
47a. Perianth-segments reddish purple, spreading then reflexing **14. wallichii**
b. Perianth-segments white, convergent after flowering **31. neapolitanum**
48a. Leaves 1–3, elliptic, broadly elliptic or lanceolate; stems 10–40 cm, often shorter than the leaves 49
b. Leaves 2–7, broadly linear; stem 40–100 cm, longer than the leaves 51
49a. Perianth-segments obovate **81. mirum**

b. Perianth-segments narrowly oblong to lanceolate 50
50a. Perianth-segments lanceolate, acute; stamens almost equalling the perianth-segments, the filaments thread-like except for broadened bases **77. karataviense**
b. Perianth-segments narrowly oblanceolate, obtuse; stamens half as long as the perianth-segments; filaments narrowly triangular almost to apex **68. akaka**
51a. Ovary on a short stalk visible after removal of perianth and stamens 52
b. Ovary stalkless 54
52a. Bulb-tunic elongated into a collar around leaves at base; perianth-segments 4–5 mm **70. suworowii**
b. Bulb-tunic not elongated; perianth-segments 6–10 mm 53
53a. Stem feebly ribbed **74. aflatunense**
b. Stem strongly ribbed **71. rosenbachianum**
54a. Perianth-segments and filaments of stamens dark purple **67. atropurpureum**
b. Perianth-segments and filaments white, pink or greenish 55
55a. Leaves sheathing the stem only to soil level; perianth-segments becoming reflexed or spreading; ovules 4–8 in each cell 56
b. Leaves sheathing the stem a little above soil level; perianth-segments erect and touching after flowering; ovules 2 in each cell 77
56a. Perianth-segments 1.5–3.5 mm broad, narrowly elliptic; filaments subulate above the base **65. nigrum**
b. Perianth-segments 1–1.5 mm broad, broadly linear; filaments narrowly triangular **66. cyrilli**
57a. Leaves thread-like, 0.5–2 mm broad 58
b. Leaves linear, flat, channelled or terete, 1–10 mm broad 65
58a. Stamens longer than perianth 59
b. Stamens shorter than perianth 63
59a. Perianth-segments white or yellowish 60
b. Perianth-segments blue or purplish 61
60a. Bulb-tunics ultimately breaking into fibres; flower-stalks smooth **4. ericetorum**
b. Bulb-tunics membranous, never fibrous; flower-stalks rough **5. albidum**
61a. Perianth-segments blue, obtuse; filaments of all stamens subulate, without teeth **11. cyaneum**
b. Perianth-segments purplish mauve or rose; filaments of outer stamens subulate, of inner stamens with a short tooth on each side 62
62a. Bulb-tunic with netted fibres; perianth-segments acute, purplish; filaments of inner stamens broadened in lower half **16. przewalskianum**
b. Bulb-tunic membranous; perianth-segments obtuse, rose; filaments of inner stamens not distinctly broadened **47. virgunculae**
63a. Spathes 2; tips of perianth-segments erect **43. moschatum**
b. Spathe 1; tips of perianth-segments recurving 64
64a. All flowers stiffly erect or ascending; perianth spotted **9. mairei**
b. Some flowers nodding; perianth deep rose **10. amabile**
65a. Outer and inner perianth-segments markedly dissimilar, the outer entire and recurved at the tips, the inner erect with minutely toothed upper margin **90. acuminatum**
b. Outer and inner perianth-segments not markedly dissimilar, the inner not minutely toothed 66
66a. Leaves cylindric, hollow **24. schoenoprasum**
b. Leaves flat or channelled or, if hollow, then semi-circular, flattened or 3–5 angled 67
67a. Perianth blue **13. sikkimense**
b. Perianth not blue 68
68a. Perianth 9–18 mm, reddish purple; stigma slightly or distinctly 3-lobed 69
b. Perianth 3–13 mm, usually white, lilac or pale purple; stigma entire 72
69a. Leaves 3–5; bulbs oblong, clustered on a short rhizome; perianth cylindric-bell-shaped; filaments of stamens subulate 70
b. Leaves 2; bulbs ovoid, without rhizome; perianth broadly bell-shaped or almost star-shaped; filaments of stamens narrowly or broadly triangular 71
70a. Bulb with persistent elongated fibrous tunic; umbel at first pendent, later erect **20. narcissiflorum**
b. Bulb without fibrous tunic or with only a few fibres at base; umbel always pendent **21. insubricum**
71a. Leaves 2–8 mm broad, longer than flower stem; stamens converging and covering ovary, inner stamens broadly triangular, outer narrowly triangular **37. oreophilum**
b. Leaves 1–3 mm broad, equalling or shorter than flower stem; all stamens very narrowly triangular, not converging 113
72a. Stamens distinctly longer than perianth-segments 73
b. Stamens shorter than or only slightly longer than perianth-segments 76
73a. Umbel nodding at flowering time, the stem curved just below the umbel; ovary with 6 crests **82. cernuum**
b. Umbel erect at flowering time; ovary rounded above, not crested 74
74a. Leaves hollow, 3–5 angled; perianth light violet; filaments 7.5–8 mm **46. chinense**
b. Leaves solid, flat; perianth white or yellowish; filaments 3–6 mm 75
75a. Bulb-tunics ultimately breaking into fibres; flower-stalks smooth **4. ericetorum**
b. Bulb-tunics membranous, never fibrous; flower-stalks rough **5. albidum**
76a. Perianth white, sometimes with a green, brown or reddish stripe down the middle of each perianth-segment 77
b. Perianth light purple, lilac, rose or reddish brown 81
77a. Perianth star-shaped; segments 4–7 mm, with faint green or brownish mid-line externally, reflexing or withering away from the capsule **18. tuberosum**
b. Perianth open, cup-shaped or bell shaped; segments 6–14 mm, with or without a reddish mid-line externally, erect after flowering 78
78a. Bulbs with membranous or crustaceous (not fibrous) outer tunic, ovoid or almost spherical; leaves solid 79
b. Bulbs with netted, fibrous outer tunic; if leaves solid, then ovary with apical projections (crests); if leaves hollow, then ovary without projections 114

79a. Spathes 2; perianth with purplish keel **30. massaessylum**
b. Spathe 1 but often becoming 3 or 4 lobed; perianth without purplish keel 80
80a. Bulb-tunic minutely and profusely pitted **28. roseum**
b. Bulb-tunic not pitted **29. zebdanense**
81a. Flower-stalks rough; perianth narrowly bell-shaped, red-purple, 6–8 mm; stamens united at base into a tube *c.* 3 mm long around the ovary **19. cyathophorum**
b. Flower-stalks smooth; perianth cup-shaped or star-shaped; stamens free or only shortly united at base 82
82a. Ovary rounded at the top without crests but sometimes with 6 small rounded knobs or 3 flat ridges 83
b. Ovary with 3 or 6 prominent crests 90
83a. Perianth star-shaped; segments purplish brown, contracted and not overlapping at base; bulb solitary but putting out slender rhizomes each terminated by a bulblet **96. glandulosum**
b. Perianth cup-shaped or star-shaped; segments not brown, slightly overlapping at base 84
84a. Bulbs cylindric or narrowly conical, clustered on a short persistent rhizome; stem 2-edged above 85
b. Bulbs narrowly ovoid to almost spherical, without a persistent rhizome; stem terete or only slightly angled above 86
85a. Leaves smooth and rounded beneath; stamens distinctly projecting **1. senescens**
b. Leaves sharply keeled beneath; stamens only slightly projecting **3. angulosum**
86a. Bulb-tunic crustaceous, minutely pitted; umbel often with bulbils; perianth-segments 7–12 mm **28. roseum**
b. Bulb-tunic fibrous, leathery or membranous, not pitted; umbel rarely or never with bulbils; perianth-segments 4–9 mm (rarely 10–15 mm as in *A. unifolium*) 87
87a. Leaves usually 1 or 2 111
b. Leaves 3 or more 88
88a. Bulb-tunics leathery; filaments of inner stamens 2–3 times as broad as outer ones **44. rubellum**
b. Bulb-tunics fibrous or membranous; filaments of inner and outer stamens equally broad 89
89a. Flower-stalks often less than twice as long as the perianth, nearly equal; ovary with 6 rounded knobs **84. geyeri**
b. Flower-stalks 1–3 times as long as the perianth; ovary without knobs but sometimes crested 90
90a. Stamens equalling or slightly longer than the perianth-segments **83. stellatum**
b. Stamens two-thirds to three-quarters as long as the perianth-segments 91
91a. Leaves much longer than the somewhat flattened stem, which is 5–10 cm 92
b. Leaves equalling or shorter than the terete stem, which is 20–30 cm 93
92a. Perianth-segments 8–15 cm, long-acuminate **88. A. falcifolium**
b. Perianth-segments *c.* 8 mm, acute **89. crenulatum**
93a. Perianth-segments 6–8 mm; filaments thread-like, almost equally slender from the apex to the slightly broadened base **91. campanulatum**
b. Perianth-segments 1–1.4 cm; filaments gradually broadened downwards 94
94a. Outer perianth-segments broadly elliptic, obtuse and mucronate, inner entire **92. dichlamydeum**
b. Outer perianth-segments lanceolate or narrowly ovate, acuminate, inner often finely toothed above 95
95a. Bulb-tunic with network of 4- or 6-sided meshes; flower-stalks 1–2 cm **90. acuminatum**
b. Bulb-tunic with close horizontal zigzag patterning; flower-stalks 2–3 cm **93. peninsulare**
96a. Spathes persistent, 2; 1 or both much longer than the umbel 97
b. Spathe often deciduous; if persistent, then 1; if 2, then both shorter than or just equalling the umbel 104
97a. Stamens long-projecting 98
b. Stamens included or with anthers just protruding 100
98a. Perianth purple (rarely white); ovary narrowly obovoid, much longer than broad **56. carinatum**
b. Perianth yellow or yellowish; ovary almost spherical, about as long as broad 99
99a. Flower-stalks unequal, 3–5 times as long as the perianth, 3–25 cm **57. flavum**
b. Flower-stalks almost equal, 1 and a half to 2 times as long as the perianth, to 8 mm **58. phrygium**
100a. Umbel loose, with unequal flower-stalks, the longest 1.5–7 cm 101
b. Umbel compact, the longest flower-stalks usually less than 2 cm 102
101a. Umbel with bulbils, few-flowered **54. oleraccum**
b. Umbel without bulbils, many-flowered **52. paniculatum**
102a. Perianth yellow **6. scabriscapum**
b. Perianth white or rose 103
103a. Leaves sheathing the lower five-sixths of stem; anthers whitish **55. tardans**
b. Leaves sheathing the lower one-third to half of stem; anthers yellow **53. pallens**
104a. Leaf-sheaths covering the stem almost up to the umbel; perianth-segments truncate **49. callimischon**
b. Leaves sheathing the lower one-quarter to two-thirds of the stem; perianth-segments acute or obtuse 105
105a. Umbel compressed, with erect or ascending flower-stalks; leaves thread-like, to 0.4 mm broad **48. cupani**
b. Umbel almost spherical or hemispherical or with pendulous flowers; leaves linear 106
106a. Leaves with sparse spreading hairs **45. chrysonemum**
b. Leaves hairless 107
107a. Perianth 1–1.7 cm, blue **12. beesianum**
b. Perianth 2.5–6 mm, not blue 108
108a. Perianth greenish yellow **6. scabriscapum & 8. obliquum**
b. Perianth blue or purplish pink 109
109a. Perianth purplish pink **7. hymenorhizum**
b. Perianth light blue 110
110a. Leaves 3-angled in section; filaments of inner stamens simple or with 2 short teeth near base **50. caeruleum**
b. Leaves semi-circular in section; filaments of inner stamens widened for lower two-thirds and with 2 short teeth **51. caesium**

111a. Bulb-tunic membranous with horizontal undulating markings; perianth-segments 1–1.5 cm **95. unifolium**
b. Bulb-tunic with netted fibres; perianth-segments 4–8 mm 112
112a. Ovary and capsule with short apical projections (crests) **86. textile**
b. Ovary and capsule without apical projections **87. drummondii**
113a. Inner perianth-segments with margin wavy or finely toothed in upper part **94. crispum**
b. Inner perianth-segments flat at margin **93. peninsulare**
114a. Leaves hollow; perianth-segments white with broad outer reddish stripe; ovary without apical projections (crests) **17. ramosum**
b. Leaves solid; perianth-segments without outer reddish stripe; ovary with apical projections 115
115a. Leaves several (up to 10); perianth-segments 1 cm **85. plummerae**
b. Leaves 2; perianth-segments 6–8 mm **86. textile**

1. A. senescens Linnaeus. Illustration: *Botanical Magazine*, 1150 (1808); Coste, Flore de la France **3**: 341 (1906).

Bulbs oblong to narrowly conical, *c.* 1 cm thick, clustered, on a short rhizome; outer tunics membranous. Leaves 4–9, 4–30 × 1 cm, basal, almost flat, not keeled beneath. Stem 7–60 cm, 2-angled above. Spathes 2 or 3 lobed, 5–8 mm. Umbel 2–5 cm, hemispherical; flower-stalks 8–20 mm. Perianth cup-shaped, lilac; segments 3.5–8 × 2–2.5 mm, the outer shorter. Stamens projecting, 4–8 mm, the inner with broader base. Capsule 4 mm. *Europe & Former USSR (Siberia).* H1. Summer–autumn.

This species comprises two main geographical races: (a) subsp. **senescens**, a robust plant, with leaves 5–10 mm broad and stems 3–60 cm, of northern Asia; (b) subsp. **montanum** (Fries) Holub (*A. montanum* Schmidt not Schrank; *A. lusitanicum* Lamarck; *A. fallax* Schultes & Schultes; *A. senescens* var. *calcareum* Wallroth) – illustration: Brickell (ed.), RHS A–Z encyclopedia of garden plants, 97 (2003) – with leaves 1–6 mm broad and stems 7–45 cm, from western Former USSR (Ukraine) to northern Portugal. There are numerous intermediate plants in gardens, some with very glaucous leaves (*A. glaucum* Schrader) and twisted leaves (*A. spirale* Willdenow).

2. A. nutuns Linnaeus. Illustration: Mathew, P.J. Redouté: Lilies and related flowers, 51 (1981).

Bulbs narrowly conical, to 2 cm thick, on a short rhizome; outer tunics membranous. Leaves 6–8, to 30 × 1.5 cm, basal, not keeled beneath. Stem 30–60 cm, 2-angled above. Spathe 2-lobed. Umbel *c.* 6 cm, almost spherical, nodding before flowering. Perianth cup-shaped, rose or lilac; segments 4–6 mm, the outer shorter. Stamens projecting, the inner nearly 3 times as broad as the outer and with a short tooth each side. *Former USSR (Siberia).* H1. Summer.

3. A. angulosum Linnaeus (*A. acutangulum* Schlechtendal). Illustration: Redouté, Liliacées **5**: 281 (1809); Reichenbach, Icones florae Germanicae et Helveticae **10**: 500 (1848); Coste, Flore de la France **3**: 341 (1906).

Bulbs cylindric or narrowly conical, 5 mm thick, clustered, on a short rhizome; outer tunics membranous. Leaves 4–6, 10–25 × 1.5–6 mm, basal, channelled, sharply keeled beneath. Stem 20–45 cm, 2-angled above. Spathe 2–5-lobed, *c.* 1 cm. Umbel 2.5–4.5 cm, hemispherical, many-flowered; flower-stalks 1–3 cm, almost equal. Perianth cup-shaped, pale purple; segments 4–6 × 1.5–2.5 mm, the outer shorter. Stamens just included, *c.* 6 mm, anthers becoming dark purple. Capsule 3.5 mm. *Europe to Siberia.* H1. Summer–autumn.

4. A. ericetorum Thore (*A. ochroleucum* Waldstein & Kitaibel). Illustration: Coste, Flore de la France **3**: 336 (1906).

Bulbs oblong to very narrowly conical, to 1.5 cm thick, clustered on a short rhizome; outer tunics breaking into parallel fibres. Leaves 3 or 4, to 25 cm × 3 mm, basal. Stem 10–40 cm, terete. Spathe 2-lobed, to 1 cm. Umbel 1–2.5 cm, hemispherical, many-flowered; flower-stalks 4–10 mm, almost equal, smooth. Perianth cup-shaped, white or yellowish; segments 3.5–4 × 1.5–2 mm. Stamens long-projecting, 4.5–6.5 mm; anthers brownish. Capsule 3.5–4 mm. *Yugoslavia and Carpathians to SW France & N Portugal.* H4. Summer.

5. A. albidum Bieberstein (*A. flavescens* Besser; *A. ammophilum* Heuffel). Illustration: Reichenbach, Icones florae Germanicae et Helveticae **10**: 499 (1848).

Bulbs oblong to very narrowly conical, to 1 cm thick, clustered on a short rhizome; outer tunics membranous. Leaves 5–9, to 14 cm × 2.5 mm, basal. Stem 10–30 cm, slightly ribbed. Spathe 2-lobed, to 1 cm. Umbel 1.5–2.5 cm, hemispherical or compressed, many-flowered; flower-stalks 8–15 mm, almost equal, rough. Perianth cup-shaped or star-shaped, white or yellowish; segments 3.5 × 1–1.5 mm. Stamens projecting, 3–4 mm; anthers yellow. Capsule 3 mm. *Bulgaria, Romania, Former European USSR.* H5. Summer.

6. A. scabriscapum Boissier & Kotschy.

Bulbs very narrowly conical, to 1 cm thick, clustered on a short rhizome; outer tunics with netted fibres. Leaves 4–6, shorter than the stem, 2–4 mm broad, sheathing the lower one-quarter to one-third of stem. Stem 15–50 cm, rough or smooth. Spathes 2-lobed, to 1.3 cm. Umbel hemispherical or almost spherical, many-flowered; flower-stalks 1.5–2.5 cm, almost equal. Perianth cup-shaped, yellow; segments 4–6 mm. Stamens slightly longer than perianth; the inner filaments twice as broad at base as the outer ones. *Turkey, Caucasus, C Asia, Iran, Iraq.* H4. Summer.

7. A. hymenorhizum Ledebour. Illustration: Ledebour, Icones **4**: 359 (1833).

Bulbs very narrowly conical, to 2 cm thick, clustered on a short rhizome; outer tunics breaking into leathery glossy strips. Leaves 4–6, to 30 cm × 6 mm, sheathing lower half of stem. Stem 30–90 cm. Spathe 2-lobed, not longer than flower-stalks. Umbel 2–3.5 cm, hemispherical to almost spherical, many-flowered, dense; flower-stalks 1–1.5 cm, almost equal. Perianth bell-shaped, purplish pink; segments 4–6 × 1.5–2 mm. Stamens long-projecting, *c.* 8 mm; anthers yellow. *Former USSR (Soviet Central Asia, W Siberia), Iran, Afghanistan.* H3. Summer.

8. A. obliquum Linnaeus. Illustration: *Botanical Magazine*, 1408 (1811); Redouté, Liliacées **7**: 364 (1812).

Bulbs oblong, to 2 cm thick, usually solitary, on a short rhizome; tunics membranous. Leaves 4–10, to 35 × 2 cm, sheathing lower half of stem. Stem 60–100 cm. Spathe 1, not longer than flower-stalks. Umbel 3.5–4 cm, hemispherical to almost spherical, many-flowered, dense; flower-stalks 1–2 cm, almost equal. Perianth cup-shaped or almost spherical, pale yellowish green;

segments 4–5 × 2.5 mm. Stamens projecting, *c.* 8–9 mm; anthers yellow. *Romania, Siberia, C Asia.* H3. Summer.

9. A. mairei Léveillé (*A. yunnanense* Diels). Illustration: Rix & Phillips, The bulb book, 146f (1981).

Bulbs very slender; outer tunics with netted fibres. Leaves 13–25 cm × 1–10 mm, thread-like, sheathing lower half of stem. Stem 10–40 cm, 2-angled. Spathe. Umbel with flowers stiffly erect; flower-stalks 1.5–3 cm, unequal, papillose. Perianth bell-shaped, light to bright pink or white with red spots; segments 8–12 × 1.5–3.2 mm, recurving above, the outer longer. Stamens included, 5–6 mm, united at base. *SW China (Xizang, Yunnan).* H5. Autumn.

10. A. amabile Stapf. Illustration: *Botanical Magazine*, 9257 (1931).

Bulbs very slender, on a short rhizome; outer tunics with netted fibres. Leaves 2–4, 10–18 cm × 1 mm, thread-like, sheathing lower half of stem, erect. Stem 10–20 cm, finely striped. Spathe 1, to 2 cm. Umbel 2–5-flowered, with some flowers pendent; flower-stalks 1.2–1.6 cm. Perianth bell-shaped, deep rose to magenta-crimson; segments 1.2–1.6 cm × 1.5–2 mm, keeled, recurving above. Stamens included, 5–6.5 mm, united at base. Capsule 4 mm. *SW China (Yunnan).* H5. Summer–autumn.

Doubtfully distinct as a species from *A. mairei.*

11. A. cyaneum Regel (*A. purdomii* W.W. Smith). Illustration: *Gartenflora*, t. 1317 (1890); *Botanical Magazine*, 9483 (1937); *Baileya* **3**: 161 (1955); Rix & Phillips, The bulb book, 147h (1981).

Bulbs cylindric, clustered, on a short rhizome; outer tunics fibrous but inconspicuously netted. Leaves 1–3, 15 cm × 1–2 mm, sheathing stem near base, semi-cylindric. Stem 10–45 cm. Spathe 1, 5–8 mm. Umbel 5–18-flowered, pendent; flower-stalks 4–10 mm. Perianth violet-blue to purplish; segments 4–6.5 × 2–2.5 mm, with darker blue or green mid-veins. Stamens projecting, 5–9 mm, united at base, the inner broader with a small tooth on either side, violet or blue. *W China.* H5. Summer.

12. A. beesianum W.W. Smith. Illustration: *Botanical Magazine*, 9331 (1933); Hay & Synge, Dictionary of garden plants in colour, 83 (1969).

Bulbs narrowly cylindric, 4–8 mm, clustered; outer tunics with netted fibres. Leaves 2–4, 15–20 cm × 4–10 mm, sheathing the lower half to one-third of stem, erect. Stem 25–50 cm. Spathe 1, to 1.2 cm. Umbel hemispherical, 6–12-flowered, more or less pendent; flower-stalks 7–10 mm. Perianth bell-shaped to tubular, bright blue to purplish-blue; segments 1–1.7 cm × 2.5–5 mm. Stamens included, 7–11 mm, blue. Capsule 3-angled. *W China.* H4. Summer.

13. A. sikkimense Baker (*A. kansuense* Regel; *A. tibeticum* Rendle). Illustration: *Gartenflora*, t. 1317 (1890); *Botanical Magazine*, 7290 (1893), 8858 (1920).

Bulbs cylindric, 3–5 mm, clustered on a short rhizome; outer tunics with netted fibres. Leaves 2–5, 7–30 cm × 2–5 mm, sheathing base of stem, flat. Stem 10–40 cm. Spathe 1, *c.* 2 cm. Umbel few to many-flowered, pendent; flower-stalks 2–6 mm, unequal. Perianth bell-shaped, blue-purple; segments 6–10 × 3–4 mm, acuminate, the outer shorter. Stamens included, the inner shortly toothed at base. *Himalaya to W China.* H4. Summer.

14. A. wallichii Kunth.

Bulbs hardly developed, cylindric to ovoid; outer tunics membranous. Leaves 60–90 cm × 8–20 mm, all sheathing the stem to the same level in the lower part, flat, keeled. Stem 30–75 cm, 3-angled. Spathes 2, 2.5–3.8 cm. Umbel 5–7.5 cm, many-flowered, loose; flower-stalks 2.5–3.8 cm. Perianth star-shaped, magenta to purple, papery; segments 7–11 mm, reflexed. Stamens included, 4–7 mm. Capsule 6 mm. *Nepal to W China.* H4.

15. A. macranthum Baker (*A. oviflorum* Regel). Illustration: *Gartenflora* **23**: t. 1134 (1883); *Botanical Magazine*, 6789 (1884).

Bulbs oblong to oblong-cylindric, 3–7 mm; outer tunics membranous. Leaves 3–many, 15–45 cm × 3 mm, channelled. Stems many, 20–30 cm, 3-angled. Spathe 2 cm. Umbel 7.5–10 cm, 3–12 (rarely to 50)-flowered, loose; flower-stalks 2.5–5 cm. Perianth bell-shaped, dark purple; segments 8–12 × 4 mm. Stamens included or just equalling perianth. Capsule 4 mm, 6-crested. *Himalaya to SW China.* H4. Summer.

16. A. przewalskianum Regel (*A. jacquemontii* Regel not Kunth).

Bulbs cylindric, 6–10 mm, clustered; outer tunics with netted fibres, reddish. Leaves 3 or 4, to 35 cm × 1 mm, almost basal. Stem 15–30 cm, terete. Spathe to 1 cm. Umbel 1.5–2.5 cm, hemispherical, many-flowered; flower-stalks 5–10 mm. Perianth star-shaped; segments 4–5 mm, mauve to lilac, spreading or reflexed. Stamens projecting, *c.* 7 mm, the inner broadened below the middle with a small tooth each side. *Himalaya to W China.* H4.

17. A. ramosum Linnaeus (*A. odorum* Linnaeus; *A. tataricum* Linnaeus filius). Illustration: Redouté, Liliacées **2**: 98 (1804); *Botanical Magazine*, 1142 (1808); Rix & Phillips, The bulb book, 145d (1981).

Bulbs cylindric, 1 cm, clustered, on a short rhizome; outer tunics with netted fibres. Leaves 4–9, to 35 cm, sheathing stem for one-third of its length, semi-terete, hollow. Stem 24–50 cm, terete to slightly angled. Spathes 1 or 2. Umbel 3–5 cm, usually compressed, few- to many-flowered; flower-stalks 1–3 cm. Perianth bell-shaped; segments 8–10 mm, white with dark red mid-veins, lanceolate-oblong. Stamens included, half the length of perianth-segments. Capsule 5–6 mm, broadest below middle. *C Asia.* H3. Summer.

18. A. tuberosum Sprengel (*A. uliginosum* G. Don; *A. tuberosum* Roxburgh). Illustration: *Herbertia* **11**: 231, 234, 235 (1946); *Botanical Magazine*, n.s., 386 (1962); Rix & Phillips, The bulb book, 144a (1981).

Bulbs cylindric, *c.* 1 cm broad, clustered on a short rhizome; outer tunics with netted fibres. Leaves 4–9, to 35 cm × 8 mm, sheathing the stem for about one-eighth of its length, solid, slightly keeled. Stem 25–50 cm, slightly angled. Spathe 1, usually becoming 2-lobed, shorter than the flower-stalks. Umbel 3–5 cm, hemispherical or compressed, many-flowered; flower-stalks 1–3 cm. Perianth star-shaped; segments 4–7 mm, white with a faint green or brown mid-line on the back. Stamens included, four-fifths of the length of perianth-segments. Capsule 4–5 mm, broadest above middle. *SE Asia (native or cultivated from Nepal to Japan).* H3. Late summer–autumn.

Although usually grown in Europe for its white, scented flowers and known as 'Chinese chives' or 'Kiu tsai', this is an important culinary herb in China and Japan.

19. A. cyathophorum Bureau & Franchet. Illustration: *Baileya* **2**: 119 (1954); *Botanical Magazine*, n.s., 252

(1955); Rix & Phillips, The bulb book, 147 (1981).

Bulbs cylindric, 3–5 mm, clustered; outer tunics breaking down into parallel fibres. Leaves 3–6, 18–24.5 cm × 1–5 mm, almost basal. Stem 19–41.5 cm, 3-angled. Spathe 1, entire or splitting into 2 or 3 lobes, 1 cm. Umbel 6–30-flowered, loose, without bulbils; flower-stalks 8–20 mm, unequal. Perianth bell-shaped, light reddish purple; segments 6–8 × 1.5–2 mm, the outer slightly narrower. Stamens included, 5 mm; filaments fused into a tube for half their length. Capsule 4 mm. *NW China.* H2. Summer.

The plants in cultivation belong to var. *farreri* (Stearn) Stearn (*A. farreri* Stearn).

20. A. narcissiflorum Villars (*A. grandiflorum* Lamarck; *A. pedemontanum* Willdenow). Illustration: Reichenbach, Icones florae Germanicae et Helveticae **10**: 504 (1848); Coste, Flore de la France **3**: 340 (1906); *Baileya* **2**: 119 (1954).

Bulbs oblong, 5 mm, clustered, on a short rhizome; outer tunic consisting of layers of parallel fibres. Leaves 3–5, 9–28 cm × 2–6 mm, sheathing stem all to about the same level, flat. Stem 15–35 cm, 2-edged. Spathe 1, often 2 or 3-lobed. Umbel 5–8-flowered, erect in flower; flower-stalks 8–18 mm. Perianth bell-shaped, purple; segments 1–1.5 cm, the outer narrower. Stamens included; filaments 5 mm. Capsule 5 mm. *SW Alps (France, Italy), N Portugal.* H4.

A. insubricum is often cultivated as *A. narcissiflorum.*

21. A. insubricum Boissier & Reuter (*A. narcissiflorum* var. *insubricum* (Boissier & Reuter) Fiori). Illustration: *Botanical Magazine*, 6182 (1875); Rix & Phillips, The bulb book, 147 (1981).

Bulbs oblong, 5 mm, clustered on a short rhizome; outer tunics membranous, sometimes with a few parallel fibres at the base. Leaves 3 or 4, 12–20 cm × 2–5 mm, all sheathing the stem to the same level in the lower part, flat. Stem 14–30 cm, 2-edged. Spathe 1, to 1.8 cm. Umbel 3–5-flowered, pendent; flower-stalks 8–20 mm. Perianth purple; segments *c.* 18 mm, the outer narrower. Stamens included, 6 mm. *N Italy.* H5.

22. A. victorialis Linnaeus. Illustration: Botanical Magazine, 1222 (1809); Redouté, Liliacées **5**: 265 (1809); Reichenbach, Icones florae Germanicae et Helveticae **10**: 508 (1848); Coste, Flore de la France **3**: 335 (1906).

Bulbs cylindric, 1–2 cm thick, clustered on a short rhizome; outer tunics densely fibrous. Leaves 2 or 3, 8–25 × 1.6–9 cm, sheathing the stem for up to half its length, narrowly lanceolate to broadly elliptic, narrowed at base into a short stalk. Stem 30–60 cm, 2-edged below. Spathes 1 or 2, shorter than flower-stalks, persistent. Umbel 3–5 cm, spherical or hemispherical, many-flowered; flower-stalks 1–3 cm, unequal. Perianth star-shaped, pale or greenish yellow; segments 4–5 × 2.25 mm, the outer narrower, all ultimately deflexed. Stamens simple, projecting. Ovary with distinct nectary pits. Capsule *c.* 4 mm. *S Europe, N Asia.* H3. Summer.

23. A. tricoccum Aiton. Illustration: *Baileya* **2**: 111 (1954); Rickett, Wild flowers of the United States **1**: t. 31 (1966).

Bulbs ovoid, clustered on a short rhizome; outer tunics with netted fibres. Leaves 2 or 3, 10–30 × 2.5–5 cm, basal, narrowed into long stalks, deciduous, withering before the flowers appear. Stem 10–40 cm, terete. Spathes 2. Umbel hemispherical, many-flowered, loose, without bulbils; flower-stalks 1–2 cm. Perianth white, star-shaped; segments 4–6 mm, obtuse. *E & C North America from Quebec to Virginia.* H2. Summer.

24. A. schoenoprasum Linnaeus (*A. sibiricum* Linnaeus; *A. purpurascens* Losa). Illustration: Reichenbach, Icones florae Germanicae et Helveticae **10**: 496 (1848); Smith & Sowerby, English Botany, Supplement **4**: 2934 (1849); Pratt, Grasses, sedges and ferns of Great Britain, edn 3, **5**: 225 (1873); Rix & Phillips, The bulb book, 143 (1981).

Bulbs very narrowly conical, clustered, borne on a short rhizome; outer tunics membranous. Leaves 1 or 2 (rarely more), up to 35 cm × 1–6 mm, sheathing up to one-third of stem, cylindric, hollow. Stem 5–50 cm, hollow. Spathe with 2 or 3 lobes, to 1.5 cm. Umbel 1.5–5 cm, 8–30-flowered, dense, without bulbils; flower-stalks 2–20 mm, more or less unequal. Perianth bell-shaped, lilac or pale purple, occasionally white; segments 7–15 × 2.5–4 mm, lanceolate. Stamens included, 5–7.2 mm. Capsule 4 mm. *Northern hemisphere.* H1.

A species with many local variants, some dwarf, some tall, long cultivated as a culinary herb (chives).

25. A. cepa Linnaeus. Illustration: *Botanical Magazine*, 1469 (1812); Reichenbach, Icones florae Germanicae et Helveticae **10**: 495 (1848); *Belmontia* **7**: 85 (1977).

Bulbs often flattened-spherical, to 10 cm thick; outer tunics membranous. Leaves up to 10, to 40 × 2 cm, basal or sheathing the lower part of stem, semi-circular in section. Stem to 100 cm, hollow, inflated in lower half, thick. Spathes often 3, shorter than umbel. Umbel 4–9 cm, dense, many-flowered, with bulbils, or with only bulbils ('Viviparum'); flower-stalks to 4 cm. Perianth star-shaped, white; segments 3–4.5 × 2–2.5 mm, white with green mid-veins, slightly unequal. Stamens projecting, *c.* 4–6 mm, the inner shortly toothed. Capsule 5 mm. *Not known in the wild.* H1. Summer.

Widely cultivated for its edible bulbs (onions); apparently derived from *A. oschaninii* B. Fedtschenko (*A. cepa* var. *sylvestre* Regel) of C Asia. 'Perutile' which is sometimes cultivated, rarely produces flowering stems.

26. A. fistulosum Linnaeus. Illustration: *Botanical Magazine*, 1230 (1809); Reichenbach, Icones florae Germanicae et Helveticae **10**: 495 (1848); *Belmontia* **7**: 88 (1977).

Bulbs cylindric, borne on a short rhizome; outer tunics membranous. Leaves 2–6, 6–30 cm × 5–15 mm, sheathing the lower one-quarter to one-third of stem, hollow, circular in section. Stem 12–70 cm, hollow, inflated in the middle. Spathes 1 or 2, almost equalling umbel. Umbel 1.5–5 cm, dense, many-flowered or with only bulbils ('Viviparum'); flower-stalks 3–20 mm, unequal. Perianth conically bell-shaped, yellowish white; segments unequal, outer 6–7 × 2 mm, inner 7–9 × 3 mm. Stamens long-projecting, 2–3 times as long as the perianth; filaments 8–12 mm. Capsule 4 mm. *Not known in the wild.* H1. Summer.

Widely cultivated, particularly in the Far East, where it originated, but unknown in a wild state.

27. A. altaicum Pallas.

Very like *A. fistulosum* but with more developed, narrowly ovoid bulbs to 4 cm thick, the membranous tunics reddish brown, the almost equal flower-stalks slightly shorter than or at most 1 and a half times as long as the yellowish, bell-shaped perianth. *Former USSR (Siberia), Mongolia.* H1. Summer.

28. A. roseum Linnaeus (*A. amoenum* G. Don; *A. incarnatum* Hornemann). Illustration: *Botanical Magazine*, 978 (1806); Redouté, Liliacées **4**: 213 (1808); Reichenbach, Icones florae Germanicae et Helveticae **10**: 486, 504 (1848); *Belmontia* **7**: 173–4 (1977); Rix & Phillips, The bulb book, 139 (1981).

Bulbs ovoid to spherical, 1.5 cm, with numerous bulblets; outer tunics crustaceous, with numerous minute pits or perforations. Leaves 2–7, 12–35 (rarely to 60) cm × 1–15 mm, sheathing the lower one-fifth of the stem. Stem 10–65 cm, solid. Spathe 1, deeply 3 or 4-lobed, 1.2–2.5 cm. Umbel 2–8 cm, 5–many-flowered, without bulbils (var. **roseum**), or with bulbils (var. **bulbiferum** de Candolle); flower-stalks 7–45 mm. Perianth bell-shaped to cup-shaped, pink or white, segments 7–12 × 3–5 mm. Stamens included; filaments 7 mm. Capsule 4 mm. *S Europe, N Africa, Turkey.* H4. Spring–summer.

29. A. zebdanense Boissier & Noë. Illustration: Rix & Phillips, The bulb book, 139 (1981).

Bulbs ovoid, 1–1.5 cm; outer tunics membranous, not pitted. Leaves 2, slightly shorter than stem, 3–6 mm broad, flat to channelled. Stem 25–40 cm. Spathe 1, 1–1.5 cm. Umbel 3–10-flowered; flower-stalks 8–13 mm, equal. Perianth bell-shaped, white; segments 9–13 mm, very obtuse. Stamens included, 4–7 mm. Capsule 4 mm. *Lebanon.* H5. Spring.

30. A. massaessylum Battandier & Trabut. Illustration: *Belmontia* **7**: 166 (1977); Pastor & Valdés, Revision del genero Allium, 135 (1983).

Bulbs ovoid or almost spherical, 1–2 cm; outer tunics papery, with sinuate patterning. Leaves 1–3, to 30 × 1 cm, basal, flat to channelled, hairless. Stem 15–40 cm. Spathes 2, persistent, 1–1.5 cm. Umbel 2–6 cm, compressed, many-flowered; flower-stalks 1–2.5 cm. Perianth cup-shaped; segments 1–1.4 cm, white, with purple mid-vein, somewhat acute. Stamens included, 5–6.5 mm. Capsule 3–5 mm. *Algeria, Morocco, Portugal & Spain.* H5. Spring.

31. A. neapolitanum Cyrillo (*A. cowanii* Lindley; *A. album* Santi; *A. lacteum* Smith). Illustration: *Edwards's Botanical Register* **9**: t. 758 (1823); *Botanical Magazine*, 3531 (1836); Reichenbach, Icones florae Germanicae et Helveticae **10**: 507 (1848); *Belmontia* **7**: 161 (1977); Rix & Phillips, The bulb book, 139, 143 (1981).

Bulbs almost spherical, 1–2 cm; outer tunics membranous or crustaceous, not pitted. Leaves usually 2, 8–35 cm × 5–20 mm, sheathing lower one-fifth to one-quarter of stem, keeled beneath, hairless. Stem 20–50 cm, slightly angled, solid. Spathe 1, 1.2–2.5 cm. Umbel 5–11 cm, many-flowered, without bulbils; flower-stalks 1.5–3.5 cm. Perianth cup-shaped or star-shaped, white; segments 7–12 × 4–6 mm. Stamens included, 6 mm. Capsule 5 mm, enclosed within shiny perianth-segments. *S Europe, N Africa.* H4. Spring.

32. A. subhirsutum Linnaeus (*A. hirsutum* Lamarck; *A. ciliatum* Cyrillo). Illustration: *Botanical Magazine*, 774 (1804); *Loddiges' Botanical Cabinet* **10**: 943 (1824); Reichenbach, Icones florae Germanicae et Helveticae **10**: 502 (1848); *Belmontia* **7**: 127–8 (1977); Rix & Phillips, The bulb book, 143 (1981).

Bulbs almost spherical, to 1.5 cm, outer tunics membranous, not pitted. Leaves 2 or 3 (rarely to 7), 8–50 cm × 1–20 mm, almost basal, flat, hairy at margin. Stem 7–30 cm, solid. Spathe 1, 8–25 mm. Umbel 2–8 cm, with few to many flowers, loose; *flower-stalks* to 4 cm, unequal, 3–5 times as long as perianth. Perianth star-shaped, white; segments 7–9 mm, unequal. Stamens included, 4–6 mm. Capsule 3 mm. *Mediterranean area.* H5. Spring.

33. A. subvillosum Schultes & Schultes (*A. subhirsutum* var. *subvillosum* (Schultes & Schultes) Ball). Illustration: Redouté, Liliacées **6**: t. 305 (1812); *Belmontia* **7**: 127 (1977); Pastor & Valdés, Revision del genero Allium, 119 (1983).

Bulbs almost spherical, to 2.5 cm; outer tunics membranous, not pitted. Leaves 2–5, 6–40 cm × 2–20 mm, almost basal, ciliate at margin. Stems 10–60 cm, terete. Spathe 1, becoming 3–4-lobed, persistent. Umbel 2.5–5 cm, hemispherical or compressed, many-flowered; flower-stalks to 2 cm. Perianth cup-shaped; segments 5–9 mm, white, acute or obtuse. Stamens slightly shorter to slightly longer than the perianth-segments; filaments 6–11 mm; anthers yellow. Capsule 3–4 mm. *W Mediterranean area, Portugal, Canary Islands.* H5. Spring.

34. A. trifoliatum Cyrillo (*A. subhirsutum* subsp. *trifoliatum* (Cyrillo) Arcangeli). Illustration: Bicknell, Flowering plants of the Riviera, t. 78 (1885).

Like *A. subhirsutum* but with flower-stalks 1.5–3 times as long as perianth-segments, which are tinged with pink or pink-veined and often redden on ageing, and anthers yellowish. *Mediterranean area.* H5.

35. A. moly Linnaeus. Illustration: *Botanical Magazine*, 499 (1800); Redouté, Liliacées **2**: 97 (1804); Reichenbach, Icones florae Germanicae et Helveticae **10**: 501 (1848); Rix & Phillips, The bulb book, 143 (1981).

Bulbs almost spherical, to 2.5 cm; outer tunics parchment-like. Leaves 1–3, 20–30 × 1.5–3.5 cm, almost basal, keeled beneath. Stem 12–35 cm. Spathes 2, less than 3.5 cm. Umbel 4–7 cm, many-flowered, rarely with bulbils (var. **bulbilliferum** Rouy); flower-stalks 1.5–3.5 cm, unequal. Perianth star-shaped; segments 9–12 × 4–5 mm, bright yellow, with greenish keels outside. Stamens included; filaments 5–6 mm. Capsule covered by the persistent perianth-segments. *E Spain & SW France.* H3. Summer.

36. A. scorzonerifolium de Candolle. Illustration: Redouté, Liliacées **2**: t. 99 (1804); Pastor & Valdés, Revision del genero Allium, 133 (1983).

Bulbs almost spherical, 1.5 cm; outer tunics parchment-like. Leaves 1–3, 18–40 cm × 3–7 mm, almost basal, keeled beneath. Stem 14–30 cm, angled. Spathe 1. Umbel 4–5 cm, few-flowered with bulbils (var. **scorzonerifolium**) or without bulbils and up to 18 flowers (var. **xericiense** (Perez Lara) Fernandes); flower-stalks 15–30 cm. Perianth star-shaped; segments 7–12 × 2–5 mm, yellow. Stamens included; filaments 4–7 mm. *Portugal & Spain.* H4. Summer.

37. A. oreophilum Meyer (*A. ostrowskianum* Regel). Illustration: *Gartenflora*, t. 775 (1873), 1089 (1882); *Botanical Magazine*, 7756 (1901); Rix & Phillips, The bulb book, 143 (1981).

Bulbs ovoid-spherical, 1–2 cm; outer tunics papery, grey. Leaves 2, longer than umbel, 2–8 mm broad, linear. Stem 5–20 cm. Spathe 8–15 mm. Umbel few-flowered, loose; flower-stalks 7–15 mm, almost equal. Perianth broadly bell-shaped; segments 8–11 mm, rose purple with darker mid-vein, ovate. Stamens included; filaments 3–5.5 mm, united into a ring for half their length. Stigma 3-lobed. Capsule 4 mm. Former USSR *(Caucasus, C Asia).* H4. Spring–summer.

38. A. chamaemoly Linnaeus. Illustration: *Botanical Magazine*, 1203 (1809); Redouté, Liliacées **6**: 325 (1812); Reichenbach, Icones florae Germanicae et Helveticae **10**: 501 (1848); *Belmontia* **7**: 155 (1977).

Bulbs ovoid, 5–10 mm, outer tunics parchment-like, pitted. Leaves 2–5 (rarely to 11), 3–27 cm × 3–10 mm, basal, usually spreading, flat, with margins ciliate, sometimes hairy on upper surface. Stem very short, solid. Spathe 1, 2–4-lobed. Umbel 1.5–2.5 cm, 2–20-flowered, almost stalkless on the rosette of leaves; flower-stalks 5–10 mm. Perianth star-shaped; segments white with green or purplish mid-veins. Stamens included, 5.5 mm. Capsule 4 mm. *S Europe, N Africa.* H5. Winter–spring.

39. A. ursinum Linnaeus. Illustration: Reichenbach, Icones florae Germanicae et Helveticae **10**: 507 (1848); Hogg & Johnson, Wild flowers of Great Britain **3**: 190 (1866); Ross-Craig, Drawings of British plants **29**: pl. 20 (1972); Rix & Phillips, The bulb book, 139 (1981).

Bulbs more or less cylindric; outer tunics papery with a few parallel fibres at the base. Leaves 2 or 3, 6–20 × 1.5–8 cm, basal, narrowed at base into a stalk. Stem 10–50 cm, 2 or 3-angled. Spathes 2, shorter than flower-stalks, persistent. Umbel 2.5–6 cm, 6–20-flowered; flower-stalks 1–4.5 cm, rough in subsp. **ursinum**, smooth in subsp. **ucrainicum** Kleopow & Oxner. Perianth star-shaped; segments 7–12 × 2–2.5 mm, white, lanceolate, acute. Stamens 5 mm. Capsule 3–4 mm. *Europe, Former USSR (Caucasus).* H1. Spring.

40. A. triquetrum Linnaeus. Illustration: *Botanical Magazine*, 869 (1805); Redouté, Liliacées **5**: 319 (1812); Reichenbach, Icones florae Germanicae et Helveticae **10**: 503 (1848); *Belmontia* **7**: 115–16 (1977); Rix & Phillips, The bulb book, 139 (1981).

Bulbs spherical to ovate, 1.2–2 cm; outer tunics membranous. Leaves 2–5, up to 50 × 1.7 cm, basal, keeled. Stems 1–4, 10–35 cm, 3-angled. Spathes 2, to 4 cm. Umbel 4–7 cm, 3–15-flowered, never with bulbils, usually 1-sided, loose; flower-stalks to 2.5 cm, 3-angled. Perianth bell-shaped, drooping; segments 1–1.9 cm × 2–5 mm, white with distinct green mid-veins, the outer broader. Stamens 6–7 mm, united at base. Stigma 3-lobed. Capsule 6–7 mm; seeds with an appendage. *W Mediterranean area, naturalised in Britain.* H4. Spring.

41. A. pendulinum Tenore (*A. triquetrum* var. *pendulinum* (Tenore) Regel). Illustration: *Loddiges' Botanical Cabinet* **11**: t. 1087 (1825); Reichenbach, Icones florae Germanicae et Helveticae **10**: t. 503 (1848); Coste, Flore de la France **3**: 340 (1906).

Bulbs almost spherical, *c.* 1 cm; outer tunics membranous. Leaves 2, to 25 cm × 7 mm, keeled. Stem 6–25 cm, 3-angled. Spathes 2. Umbel not 1-sided, loose; flower-stalks to 4 cm, ascending, later drooping. Perianth star-shaped; segments 3–5 × 1–1.5 mm, white with distinct green mid-vein. Stamens 4–5 mm. Stigma 3-lobed. Capsule 4–6 mm; seeds with an appendage. *C Mediterranean area.* H4. Spring.

42. A. paradoxum (Bieberstein) G. Don. Illustration: Flora Iranica **76**: 18 (1971); Rix & Phillips, The bulb book, 71 (1981).

Bulbs almost spherical, 1 cm; outer tunics membranous. Leaf 1, to 30 × 2.5 cm, keeled. Stem 15–30 cm, 3-angled. Spathes 2, 1–2.5 cm. Umbel with bulbils and usually 1-flowered, or without flowers (var. **paradoxum**), or without bulbils and up to 10 flowers (var. **normale** Stearn); flower-stalks 2–4.5 cm. Perianth bell-shaped; segments 8–12 × 6 mm, white with a faint green mid-vein. Stamens included, united at base. Stigma 3-lobed. Capsule 5 mm; seeds with a white appendage. *Former USSR (Caucasus), Iran.* H4. Spring.

This is usually represented in gardens by the weedy, bulbil-bearing var. *paradoxum*, introduced early in the nineteenth century and sometimes naturalised; the floriferous var. *normale* was introduced in the mid-twentieth century.

43. A. moschatum Linnaeus. Illustration: Reichenbach, Icones florae Germanicae et Helveticae **10**: t. 498 (1848); Pastor & Valdés, Revision del genero Allium, 97 (1983).

Bulbs narrowly ovoid, 1–1.5 cm, clustered; outer tunics breaking into fibres. Leaves 3–6, to 14 cm × 0.5 mm, thread-like, sheathing the lower one-quarter to one-third of the stem. Stem 10–35 cm. Spathes 2, short. Umbel compressed with 3–12 flowers, loose; flower-stalks 1–1.5 cm, almost equal. Perianth bell-shaped; segments 5–7 × 1.5–2 mm, whitish or pink with darker mid-vein, acute. Stamens included. *S Europe from E Spain to the Balkan Peninsula.* H4. Summer.

44. A. rubellum Bieberstein.

Bulbs ovoid, 7.5–15 mm, outer tunics leathery; bulblets yellow to brown. Leaves 3–5, to 30 cm × 1–5 mm, sheathing lower one-quarter of stem, hollow. Stem 10–60 cm. Spathe shorter than umbel. Umbel 2.5–4 cm, hemispherical, many-flowered; flower-stalks 5–15 mm, almost equal. Perianth bell-shaped; segments 5–6 × 1.5–2 mm, pink with darker mid-veins. Stamens included, 3–4 mm; inner filaments 1 and a half to twice as wide at base than the outer. Capsule 3 mm. *Former USSR (SE Russia, Caucasus), Afghanistan, W Himalaya.* H4.

45. A. chrysonemum Stearn. Illustration: *Annales Musei Goulandris* **4**: 150 (1978); Pastor & Valdés, Revision del genero Allium, 103 (1983).

Bulbs ovoid, *c.* 1.5 cm; outer tunics membranous. Leaves 3 or 4, to 15 cm × 2 mm, sheathing lower one-third of the stem, hairy. Stem 30–60 cm. Spathes 2. Umbel *c.* 4 cm, loose, many-flowered; flower-stalks 1–1.5 cm. Perianth cup-shaped; segments 4.5–5 × 2–2.4 mm, pale greenish yellow. Stamens slightly projecting; filaments yellow. Capsule 4 mm. *Spain.* H4. Summer.

46. A. chinense G. Don (*A. bakeri* Regel). Illustration: *Economic Botany* **14**: 70, 78 (1960).

Bulbs narrowly ovoid, 1–1.5 cm, clustered; outer tunics membranous. Leaves usually 3, to 30 cm × 1–3 mm, almost basal, hollow, 3–5-angled. Stem 28–30 cm. Spathes 2. Umbel hemispherical with up to 18 flowers, loose; flower-stalks 1–1.5 cm. Perianth cup-shaped; segments 5 × 3.5–4 mm, light violet, obtuse. Stamens projecting; inner 3 filaments with broadened bases and each with 2 short teeth. *China, cultivated in Japan as economic crop ('rakkyo').* H3. Autumn.

47. A. virgunculae Maekawa & Kitamura.

Bulbs 5–7 mm, clustered; outer tunics membranous. Leaves 3–5, 10–20 cm × 1 mm, thread-like, with numerous minute white dots. Stem 8–22 cm. Spathe 1. Umbel 2–12-flowered, loose; flower-stalks 1.2–1.5 cm. Perianth star-shaped; segments *c.* 5 × 3–4 mm, rose, obtuse. Stamens projecting, *c.* 6 mm, the inner 3 each with a tooth each side at base. Capsule *c.* 4 mm. *Japan (Kyushu).* H3.

48. A. cupani Rafinesque. Illustration: Fiori & Paoletti, Iconographia florae Italicae, 80 (1898); *Belmontia* **7**: 19 (1977).

Bulbs narrowly ovoid, 7–15 mm; outer tunics with netted fibres. Leaves 3–5, to 10 cm × 0.4 mm, thread-like, sheathing lower one-third to two-thirds of stem. Spathe 1, tubular at base. Umbel compressed, few-flowered; flower-stalks to 4 cm, very unequal. Perianth bell-shaped, segments 5.5–9 × 1.5–2 mm, whitish or pink with darker mid-vein. Stamens included, equal. Capsule 4 mm. *Mediterranean area.* H4. Summer.

49. A. callimischon Link. Illustration: Rix & Phillips, The bulb book, 177 (1981).

Bulbs narrowly ovoid, 1 cm; outer tunics leathery, breaking down into parallel fibres. Leaves 3–5, the lowest to 30 cm × 1 mm, sheathing stem almost up to umbel. Stem 9–38 cm. Spathe 1, 2–4 cm. Umbel 8–25-flowered, clustered; flower-stalks 7–25 mm, very unequal. Perianth cup-shaped; segments 5–7 × 1.5 mm, white with brown or reddish mid-veins, often becoming pink. *Greece, Crete & W Turkey.* H5.

This consists of two geographical variants: subsp. **callimischon,** with segments with a distinct mid-vein but unspotted, and subsp. **haemostictum** Stearn, with segments with dark red spots in upper part.

50. A. caeruleum Pallas (*A. azureum* Ledebour). Illustration: *Edwards's Botanical Register* **26**: t. 51 (1840); Rix & Phillips, The bulb book, 137 (1981); Brickell (ed.), RHS A–Z encyclopedia of garden plants, 95 (2003).

Bulbs almost spherical, 1–2 cm, outer tunics membranous. Leaves 2–4, 7 cm × 2–4 mm, sheathing lower one-quarter to one-third of stem, 3-angled. Stem 20–80 cm. Spathes 2, to 1.5 cm. Umbel 3–4 cm, almost spherical to hemispherical, many-flowered; flower-stalks 1–2 cm, almost equal. Perianth cup-shaped; segments 3.5–4.5 × 1.2–1.4 mm, blue with darker mid-vein. Stamens equalling perianth or slightly projecting, united at base; anthers blue. Capsule 3 mm. *N to C Asia.* H3. Summer.

51. A. caesium Schrenk.

Bulbs ovoid, outer tunics almost leathery. Leaves 2–4, 1–3 mm broad, sheathing up to half the stem, semi-cylindric, hollow. Stem 25–65 cm (rarely less). Spathes 2, to 1.5 cm. Umbel more than 1.5 cm, many-flowered, with some bulbils, or rarely all bulbils; flower-stalks 1.2–2 cm. Perianth-segments 4–5 × 2 mm, the outer slightly narrower, violet-blue with a brownish-green mid-vein, or more rarely white. Stamens included, united at base. Capsule 3–4 mm. *Former USSR (Siberia to C Asia).* H3. Summer.

52. A. paniculatum Linnaeus (*A. longispathum* Delaroche). Illustration: Redouté, Liliacées **6**: 316 (1812); Reichenbach, Icones florae Germanicae et Helveticae **10**: t. 487 (1848).

Bulbs ovoid, 1–2.5 cm; outer tunics membranous. Leaves 3–5, to 25 cm × 1.5 mm, sheathing lower one-third to half of stem, prominently ribbed beneath. Stem 30–70 cm. Spathes 2, unequal, each contracted into a long slender appendage very much longer than the flower-stalks. Umbel 3.5–7 cm, loose, without bulbils; flower-stalks 1–7 cm, very unequal. Perianth bell-shaped, white, pink, yellowish or greenish brown; segments 4.5–7 × 1.5–2.5 mm, obtuse. Stamens included or slightly projecting. Ovary ellipsoid, twice as long as broad. Capsule *c.* 5 mm. *Europe.* H1. Summer.

A very variable, widespread species divided into several subspecies, including subsp. **paniculatum** with white or pink flowers, and subsp. **fuscum** (Waldstein & Kitaibel) Arcangeli, with greenish, brownish or dirty white flowers.

53. A. pallens Linnaeus (*A. coppoleri* Tineo; *A. paniculatum* var. *pallens* (Linnaeus) Grenier & Godron). Illustration: Pastor & Valdés, Revision del genero Allium, 91 (1983).

Bulbs 1–1.5 cm; outer tunics membranous. Leaves 3 or 4, to 30 cm × 5 mm, thread-like to linear, sheathing lower one-third to half of the stem. Stem 10–30 cm, rarely more. Spathes 2, unequal, each contracted into a long slender appendage. Umbel 1.5–3 cm, compact, few- or many-flowered; flower-stalks 5–20 mm, almost equal. Perianth cup-shaped, white or rose-tinged; segments 3.5–5 × 1.5–2 mm, truncate or obtuse. Stamens included or with projecting anthers. Ovary ellipsoid, longer than broad. *S Europe.* H4. Summer.

54. A. oleraceum Linnaeus. Illustration: Smith & Sowerby, English Botany **7**: 488 (1798); Reichenbach, Icones florae Germanicae et Helveticae **10**: 487 (1848); Butcher, New illustrated British flora **2**: 652 (1961).

Bulbs 1–2 cm; outer tunics membranous. Leaves 2–4, to 25 cm × 6 mm, thread-like to linear, sheathing the lower half or more of the stem. Stem 25–100 cm. Spathes 2, unequal, each contracted into a slender appendage, the longest to 20 cm. Umbel with bulbils, loose; flower-stalks 1.5–6 cm, slightly flattened. Perianth cup-shaped, whitish but variably tinged with brown, green or pink; segments 5–7 × 2–3 mm, obtuse. Stamens included. Ovary narrowly obovoid or cylindric, truncate. *Most of Europe.* H1. Summer.

55. A. tardans Greuter & Zahariadi.

Bulbs 1–1.5 cm; outer tunics membranous, breaking into strips. Leaves 3 or 4, to 16 cm × 0.5 mm, thread-like, sheathing the lower five-sixths of stem. Stem 10–30 cm. Spathes 2, unequal, each contracted into a long slender appendage. Umbel 1.2–5 cm, compressed; flower-stalks 9–15 mm. Perianth cup-shaped; segments 5–6.5 × 1.8–2 mm, pale pink with green mid-veins, almost truncate. Stamens included or with anthers projecting. *Crete.* H5. Autumn.

56. A. carinatum Linnaeus. Illustration: Smith & Sowerby, English Botany **24**: 1658 (1807); Redouté, Liliacées **10**: 368 (1813); Reichenbach, Icones florae Germanicae et Helveticae **10**: 482 (1848); Bonnier, Flore complète **10**: 584 (1929).

Bulbs ovoid to oblong, *c.* 1 cm; outer tunics membranous, sometimes breaking into parallel fibres. Leaves 2–4, to 20 cm × 1–2.5 mm, sheathing stem for half to three-fifths of its length. Stem 30–60 cm. Spathes 2, unequal, each contracted into a slender appendage to 12 cm. Umbel *c.* 30-flowered, usually with bulbils; flower-stalks 1–2.5 cm, unequal, the outer curving downwards. Perianth cup-shaped, purple, segments 4–6 × 1.5–2 mm. Stamens long-projecting, 6.5–9 mm. Ovary oblong, much longer than broad. Capsule 5 mm, rarely produced. *W, C & S Europe, S Russia, Turkey.* H3. Summer.

Subsp. **carinatum** has numerous greenish bulbils in the umbel and few flowers and can become a weed. Subsp. **pulchellum** Bonnier & Layens (*A. pulchellum* G. Don) has no bulbils, but numerous flowers on purplish stalks and is an attractive garden plant.

57. A. flavum Linnaeus (*A. tauricum* Reichenbach). Illustration: Redouté, Liliacées **7**: 119 (1812); *Botanical Magazine*, 1330 (1810); Reichenbach, Icones florae Germanicae et Helveticae **10**: 485 (1848); Rix & Phillips, The bulb book, 147 (1981).

Bulbs ovoid, 1–1.5 cm, outer tunics membranous. Leaves 2 or 3, to 20 cm × 2 mm, sheathing lower one-third to half of stem. Stem 8–50 cm. Spathes 2, unequal, each contracted into a long slender beak to 11 cm. Umbel 9–60-flowered, hemispherical, never with bulbils; flower-stalks 3–25 mm, unequal, erect in fruit, the outer usually curving downwards. Perianth bell-shaped, lemon-yellow; segments 4.5–5 × 2 mm. Stamens projecting, 7–8 mm, yellow. Ovary spherical, as long as wide. Capsule 4 mm. *C Europe to W Asia.* H3. Summer.

58. A. phrygium Boissier.

Bulbs *c.* 1 cm; outer tunics membranous. Leaves sheathing half to two-thirds of stem, thread-like. Stem 23–30 cm. Spathes 2, unequal, each contracted into a long slender appendage. Umbel 2–2.5 cm, compact, spherical; flower-stalks 6–8 mm. Perianth cup-shaped, straw-coloured; segments 4 mm, acute. *Stamens projecting. Turkey.* H5.

59. A. sativum Linnaeus. Illustration: Reichenbach, Icones florae Germanicae et Helveticae **10**: 488 (1848); Bentley & Trimen, Medicinal plants **4**: 280 (1875); *Belmontia* **7**: 61 (1977).

Bulbs depressed ovoid, 2–7 mm thick, composed of 5–18 (rarely to 60) bulblets, outer tunics membranous to papery. Leaves 6–12, to 60 × 3 cm, sheathing lower half of stem, linear, flat or keeled. Stem 25–100 cm or more, solid. Spathes 4–20 cm, long-beaked. Umbel 2.5–5 cm, usually with few flowers (aborting in bud) and many bulbils; flower-stalks 1–2 cm, unequal. Perianth greenish-white, pink or rarely white or purple; segments 3–5 mm, the outer narrower. Stamens included or equalling perianth, the outer filaments simple or 3-toothed, the inner with the basal blade broad, one-third as long as central tooth, lateral teeth 2–4, much longer than central tooth. Capsules abortive, without seeds. *Not known in the wild.* H4.

Widely cultivated as a crop (garlic).

60. A. ampeloprasum Linnaeus. Illustration: *Botanical Magazine*, 1385 (1811); Redouté, Liliacées **7**: 385 (1813); Reichenbach, Icones florae Germanicae et Helveticae **10**: 489 (1848); *Belmontia* **7**: 67–8 (1977).

Bulbs 2–6 cm; outer tunics membranous, bulblets usually numerous, yellowish. Leaves 4–10, to 50 × 4 cm, sheathing lower one-third to half of stem, flat with rough margin. Stem 45–180 cm. Spathe 1, 4–11.5 cm, deciduous. Umbel 5–9 cm, dense, with up to 500 flowers, with or without bulbils; flower-stalks 4–5 cm, unequal. Perianth cup-shaped or bell-shaped, pink or dark red, with large papillae especially on keels; segments 4–5.5 × 1.3–2.4 mm. Stamens slightly to distinctly projecting; the outer filaments usually simple, 4–6 mm, the inner with basal blade as wide as perianth and 2 times as long as the central tooth. Capsule 4 mm. *S Europe, Former USSR (Caucasus), Iran, Turkey, N Africa.* H2. Summer.

Var. **babingtonii** (Borrer) Syme, found in Ireland, SW England and Channel Islands, has numerous bulbils and few flowers and is apparently a relic of former cultivation. **A. porrum** Linnaeus, the cultivated leek, is derived from the above species.

61. A. scorodoprasum Linnaeus. Illustration: Reichenbach, Icones florae Germanicae et Helveticae **10**: 490 (1848); Pratt, Grasses, sedges and ferns of Great Britain, edn 3, **5**: 224 (1873); Bonnier, Flore complète **10**: 2615 (1929); Rix & Phillips, The bulb book, 145 (1981).

Bulbs 1–2 cm; outer tunics membranous, sometimes breaking into parallel fibres, bulblets reddish black. Leaves 2–5, to 27 × 2 cm, linear, sheathing up to half of stem. Stem 25–90 cm. Spathe 1.5 cm, shortly beaked, deciduous. Umbel 1.5 cm, 0–many-flowered, usually with bulbils; flower-stalks to 2 cm, unequal. Perianth ovoid, lilac to dark purple; segments 4–7 × 1.5–3.5 mm. Stamens included; the outer filaments simple, 2.5–4.5 mm, the inner with basal blade more than 3 times length of central tooth, lateral teeth 2–3 times as long as central tooth. *C & E Europe, former USSR (Caucasus), Turkey, N Iran.* H3. Spring–summer.

Subsp. **scorodoprasum,** with purplish bulbils and few flowers, of northern and central Europe, is apparently a relic of former cultivation. It is evidently a bulbilliferous variant of subsp. **rotundum** (Linnaeus) Stearn, from southern Europe, which has an almost spherical or hemispherical umbel without bulbils.

62. A. sphaerocephalon Linnaeus (*A. descendens* Linnaeus). Illustration: Redouté, Liliacées **7**: 391 (1813); *Botanical Magazine*, 1764 (1815); Reichenbach, Icones florae Germanicae et Helveticae **10**: 492 (1848); *Belmontia* **7**: 48 (1977); Rix & Phillips, The bulb book, 141 (1981).

Bulbs spherical to ovoid, 1–3.5 cm thick; outer tunics membranous, sometimes breaking into parallel fibres. Leaves 2–6, 7–35 cm × 4–5 mm, sheathing stem for up to half, hollow, semi-cylindric. Stem 5–90 cm, hollow, terete. Spathes 2, to 2 cm, persistent. Umbel 1.6 cm, many-flowered, sometimes with bulbils; flower-stalks 2–30 mm, unequal. Perianth cylindric to narrowly ovoid, pink to dark reddish-brown; segments 3.5–5.5 × 1.2–2.5 mm, keeled, the outer broader. Stamens 4–7 mm, the outer simple (rarely 3-toothed), the inner 3-toothed with basal blade 1–2 times as long as central tooth, which is 1.5–3 mm, the lateral teeth shorter or longer than central tooth. Capsule 4 mm. *Europe, N Africa, W Asia.* H1. Summer.

Subsp. **sphaerocephalon** has red flowers, but subsp. **trachypus** (Boissier) Richter and subsp. **arvense** (Gussone) Arcangeli, have white, green-striped flowers.

63. A. vineale Linnaeus. Illustration: Smith & Sowerby, English Botany **28**: t. 1974 (1809); Reichenbach, Icones florae Germanicae et Helveticae **10**: t. 490 (1848); Ross-Craig, Drawings of British plants **29**: t. 19 (1972).

Bulbs ovoid, 1–2 cm; outer tunics breaking into parallel fibres. Leaves 2–4, 15–60 cm × 1.5–4 mm, sheathing lower one-third to two-thirds of stem, hollow. Stem 30–120 cm. Umbel with numerous greenish bulbils and few or no flowers, or 2–5 cm with numerous flowers and few or no bulbils; flower-stalks 5–30 mm. Spathe 1, deciduous. Perianth bell-shaped, greenish white, pink or dark red (var. **purpureum** H.P.G. Koch); segments 2–4.5 × 1.2–1.5 mm, somewhat acute or obtuse. Stamens slightly to distinctly projecting, the outer 3.5–4 mm, the inner 3-toothed, with basal blade slightly or much longer than central tooth, the lateral teeth much longer than central tooth. Capsule 3–3.5 mm. *Europe, N Africa, W Asia.* H1. Summer.

A pernicious weed which, in botanic gardens, may usurp the place of other species and be labelled accordingly.

64. A. heldreichii Boissier. Illustration: Strid, Wild flowers of Mount Olympus, t. 47 (1980).

Bulbs broadly ovoid; outer tunics membranous. Leaves 2–4, 5–30 cm × 0.5–3 mm, sheathing lower one-quarter to one-third of stem, hollow. Stem 20–60 cm. Spathes 2, 1–1.5 cm, persistent. Umbel 2.5–4.5 cm, spherical or hemispherical, many-flowered. Perianth bell-shaped, pink; segments about 10 × 2.5–3 mm, acute. Stamens included, the outer simple, *c.* 5.5 mm, the inner 3-toothed, with basal blade 4 times as long as central tooth, the lateral teeth 2–3 times as long as central tooth. Capsule 4–5 mm. *Northern Greece.* H5. Summer.

65. A. nigrum Linnaeus (*A. multibulbosum* Jacquin). Illustration: Redouté, Liliacées **2**: 102 (1805); *Botanical Magazine*, 1148 (1809); Reichenbach, Icones florae Germanicae et Helveticae **10**: 506 (1848); *Belmontia* **7**: 195 (1977); Rix & Phillips, The bulb book, 143 (1981).

Bulbs ovoid, 2.5–5 cm; outer tunics membranous. Leaves 3–6, to 50 × 8 cm, basal. Stem 40–90 cm. Spathe 1, becoming 2–4-lobed, to 3 cm. Umbel 5–10 cm, compressed or hemispherical, many-flowered, rarely with bulbils; flower-stalks 2.5–4.5 cm. Perianth star-shaped; segments 6–9 × 1.5–3.5 cm, later deflexed, white or pale lilac with greenish mid-veins. Stamens included, 5 mm, united at base into ring 1–1.5 mm broad; anthers yellow. Capsule 6–8 mm. *S Europe, N Africa, W Asia.* H4. Spring.

66. A. cyrilli Tenore.

Bulbs broadly ovoid, 1–2 cm; outer tunics membranous. Leaves 3–5, 20–30 × 1–3.5 cm, basal. Stem 50–60 cm. Spathes 2 or 3, to 2 cm. Umbel 4–7 cm, compressed or hemispherical, many-flowered; flower-stalks 1.5–2.5 cm, almost equal. Perianth cup-shaped; segments 6–7 × 1–1.5 mm, later deflexed, white with broad green mid-veins. Stamens just included, united at base into a ring 1.5–2 mm broad. *Italy to Turkey.* H4.

67. A. atropurpureum Waldstein & Kitaibel. Illustration: Reichenbach, Icones florae Germanicae et Helveticae **10**: 505 (1848); Rix & Phillips, The bulb book, 143 (1981).

Bulbs almost spherical, 1.5–3 cm; outer tunics membranous. Leaves 3–7, 15–35 × 1–4 cm, basal. Stem 40–100 cm. Spathes 2, to 12 cm. Umbel 3–7 cm, compressed; flower-stalks 2–4 cm. Perianth star-shaped, dark purple; segments 7–9 × 1 mm, later deflexed. Stamens 4–5 mm, united at base; anthers purple. Capsule 6–8 mm. *S Europe.* H4. Summer.

68. A. akaka Gmelin (*A. latifolium* Jaubert & Spach).

Bulbs almost spherical, 1.5–3 cm; outer tunics membranous. Leaves 1–3, to 20 × 6 cm, basal, elliptic-oblong. Stem 5–15 cm. Spathe 3 or 4-lobed. Umbel hemispherical, many-flowered, almost stalkless in the leaves; flower-stalks 8–30 mm. Perianth almost star-shaped; segments lilac, very narrowly elliptic, almost linear, rigid after flowering. Stamens erect, half as long as segments. Capsule 4–5 mm. *Turkey, Former USSR (Caucasus), Iran.* H5. Summer.

69. A. christophii Trautvetter (*A. albopilosum* C.H. Wright). Illustration: *Botanical Magazine*, 798 (1805); *Baileya* **3**: 143 (1955); Rix & Phillips, The bulb book, 145 (1981).

Bulbs spherical, 2–3 cm; outer tunics papery. Leaves 2–7, 15–40 × 1–3.8 cm, erect, somewhat glaucous with stiff spreading hairs beneath and on margins. Stem 15–40 cm, buried in ground, ribbed. Spathe 2.5 cm. Umbel many-flowered, loose; flower-stalks to 11 cm, almost equal. Perianth star-shaped, purple-violet; segments 1–1.8 cm, erect, rigid after flowering due to their thickened mid-veins. Stamens included, united at base; filaments dark purple. Capsule 5–8 mm. *Iran to Turkey.* H5. Summer.

70. A. suworowii Regel. Illustration: *Gartenflora* **30**: t. 1062 (1881); *Botanical Magazine*, 6994 (1888).

Bulbs almost spherical, 2–3 cm; outer tunics leathery, elongated above. Leaves 2–6, to 3 cm broad, basal, rough at margin. Stem 30–100 cm, indistinctly ribbed. Spathe 3- or 4-lobed. Umbel hemispherical, many-flowered; flower-stalks 1–4 cm, almost equal. Perianth star-shaped, rose-violet; segments 4–5.5 mm, almost linear, reflexed and twisted after flowering. Stamens slightly shorter to slightly longer than segments. Ovary almost stalkless, smooth. *C Asia, Afghanistan.* H4.

71. A. rosenbachianum Regel. Illustration: Flora Iranica **76**: 27 (1971).

Bulbs almost spherical, 1.5–2.5 cm; outer tunics papery. Leaves 2–4, much shorter than stem, 1–5 mm broad, basal. Stem 30–100 cm, ribbed. Spathe shorter than umbel. Umbel many-flowered, loose; flower-stalks 2–6 cm, unequal. Perianth star-shaped; segments 7–10 mm, later reflexed and twisted, dark violet, rarely white, with darker mid-veins. Stamens 7–10 mm, united at base; anthers violet. Ovary on a short stalk, papillose. Capsule 5 mm. *SW Asia.* H4. Spring.

A. jesdianum Boissier, from Iraq and Iran, is sometimes cultivated as the above but it differs in having larger bulbs with netted fibres, shorter stamens with a wider ring, and a larger capsule.

72. A. stipitatum Regel. Illustration: *Gartenflora* **30**: t. 1062 (1881).

Bulbs almost spherical, 3–6 cm; outer tunics papery, blackish. Leaves 4–6, 2–4 cm broad, hairy beneath. Stem 60–150 cm, distinctly ribbed. Spathes 2, *c.* 3 cm. Umbel hemispherical to almost spherical, many-flowered; flower-stalks 2.5–5 cm, almost equal. Perianth star-shaped, lilac-purple, rarely white; segments 8–9 mm, tapered for most of their length, later becoming reflexed and twisted. Stamens equalling perianth-segments, united at base. Ovary on a short stalk, papillose. Capsule 5 mm. *C Asia.* H4. Summer.

73. A. hirtifolium Boissier.

Bulbs almost spherical, 2.5–4 cm; outer tunics papery, breaking into strips and fibres. Leaves 4 or 5, to 5 cm broad, basal, hairy towards the base. Stem 80–120 cm, slightly ribbed. Spathe 2-lobed. Umbel spherical, many-flowered; flower-stalks 3.5–5 cm. Perianth star-shaped, purple, rarely white; segments 8–9 mm, almost linear, reflexed and twisted after flowering. Stamens almost equalling segments, united at base. Ovary on a short stalk, papillose. *Iran.* H5. Summer.

74. A. aflatunense B. Fedtschenko. Illustration: Hay & Synge, Dictionary of garden plants in colour, 658 (1969); Reader's Digest encyclopaedia of garden plants and flowers, 28 (1972).

Bulbs ovoid, 2–6 cm; outer tunics papery. Leaves 6–8, much shorter than stem, 2–10 cm broad, basal, strap-shaped. Stem 80–150 cm, feebly ribbed. Spathes 4. Umbel many-flowered, spherical;

flower-stalks 1.4–3.2 cm, almost equal. Perianth star-shaped; segments 7–8 mm, later reflexed and twisted, light violet with darker mid-veins. Stamens slightly projecting, not united at base. Ovary with a short stalk, papillose. Capsule 5 mm. *C Asia.* H4. Spring–summer.

75. A. giganteum Regel. Illustration: *Gartenflora*, t. 1113 (1883); *Botanical Magazine*, 6828 (1885); Flora Iranica **76**: 26 (1971); Rix & Phillips, The bulb book, 153 (1981).

Bulbs ovoid, 4–6 cm; outer tunics leathery, splitting into parallel fibres. Leaves 30–100 × 5–10 cm, basal. Stem 80–200 cm, slightly ribbed. Spathe half as long as umbel. Umbel spherical, many-flowered, dense; flower-stalks 6 cm, almost equal. Perianth star-shaped, purple-violet, rarely white; segments 5–7 mm, elliptic, unchanged after flowering. Stamens projecting, up to 1 and a half times as long as perianth, united at base; the inner filaments broader. Ovary stalkless. Capsule 4 mm. *C. Asia.* H4. Spring.

76. A. macleanii Baker (*A. elatum* Regel). Illustration: *Botanical Magazine*, 6707 (1883); *Gartenflora*, t. 1251 (1887); Rix & Phillips, The bulb book, 153 (1981).

Bulbs ovoid-spherical, 2–6 cm; outer tunics papery. Leaves 2–4, shorter than stem, 2–8 cm broad, basal. Stem 60–100 cm, prominently ribbed. Spathe one-third shorter than umbel. Umbel spherical, many-flowered, dense; flower-stalks 1.8–5.6 cm. Perianth star-shaped, bright violet; segments 6–8 mm, linear-lanceolate, unchanged after flowering. Stamens slightly projecting, united at base. Ovary stalkless. Capsule 5 mm. *SW Asia.* H4. Spring–summer.

77. A. karataviense Regel. Illustration: *Botanical Magazine*, 6451 (1879), 7294 (1893); Rix & Phillips, The bulb book, 141, 153 (1981); Brickell (ed.), RHS A–Z encyclopedia of garden plants, 96 (2003).

Bulbs depressed-spherical, *c.* 5 cm; outer tunics membranous. Leaves 2 or 3, 15–23 × 3–15 cm, basal, oblong to elliptic. Stem 10–25 cm, sometimes buried in ground to half its length. Spathes 2, less than 3.5 cm. Umbel 7.5–10 cm, many-flowered, spherical, dense; flower-stalks 1.5–4 cm, equal. Perianth star-shaped; segments 5–8 mm, later reflexed and twisted, light purple-pink with purple mid-veins. Stamens just projecting, with triangular bases. Capsule large, reversed heart-shaped. *C Asia.* H4. Spring.

78. A. schubertii Zuccarini. Illustration: *Botanical Magazine*, 7587 (1898); *Belmontia* **7**: 191 (1977); Rix & Phillips, The bulb book, 141 (1981).

Bulbs spherical to ovoid, 3–4 cm; outer tunics leathery, sometimes breaking down into parallel fibres. Leaves 4–8, 20–45 × 6 cm. Stem 30–60 cm, hollow, not distinctly tapering towards base. Spathes 2 or 3, 2–3 cm. Umbel 2–4 cm, many-flowered; flower-stalks 2–20 cm, very unequal. Perianth star-shaped; segments 6–10 × 2–2.5 mm, the outer longer, all becoming more or less rigid after flowering, white to pink or violet, with purplish mid-veins. Stamens included, 5.5–7.5 mm. Capsule 6–8 mm. *E Mediterranean area to C Asia.* H4. Spring.

79. A. protensum Wendelbo. Illustration: Flora Iranica **76**: 27 (1971); Rix & Phillips, The bulb book, 141 (1981).

Bulbs spherical, 2–3 cm; outer tunics leathery, splitting into fibres. Leaves 2–5, longer than stem, 1–5 cm broad, flat. Stem 10–30 cm, tapering towards base. Spathes 3, *c.* 2 cm. Umbel many-flowered, loose; flower-stalks 6–16 cm, very unequal, the longer ones with sterile flowers. Perianth broadly bell-shaped; segments 7–8 mm, pale brown with darkish mid-veins, erect after flowering. Stamens included, united at base. Ovary of fertile flowers on a short stalk, papillose. Capsule 4 mm. *Afghanistan, Iran, C Asia.* H4.

80. A. caspium (Pallas) Bieberstein. Illustration: *Botanical Magazine*, 4598 (1851).

Bulbs 2–4.5 cm, almost spherical; outer tunics membranous. Leaves 3–6, to 25 × 3.5 cm, basal. Spathes 2. Umbel 5–20 cm, many-flowered, loose; flower-stalks 3–15 cm, elongating in fruit, unequal. Perianth broadly bell-shaped, lilac; segments 5–11 × 2–3 mm, green. Stamens projecting, 1 and a half to 2 times as long as segments, united at base. Ovary on a short stalk. *SE Russia, Caucasus, C Asia.* H4. Summer.

81. A. mirum Wendelbo. Illustration: Flora Iranica **76**: t. 28 (1971).

Bulbs almost spherical, to 4 cm; outer tunics papery. Leaves 2, to 30 × 8 cm, elliptic or narrowly ovate, glaucous. Stem 8–40 cm. Umbel 5–9 cm, spherical, many-flowered; flower-stalks 1.5–3 cm. Perianth broadly bell-shaped; segments 8–12 mm, pale brownish purple to almost white with dark veins, narrowly obovate to elliptic, remaining erect and unchanged after flowering. Stamens about the length of the segments, united at base. *Afghanistan.* H4.

82. A. cernuum Roth. Illustration: Botanical Magazine, 1324 (1810); Redouté, Liliacées **6**: 345 (1812); Cronquist et al., Intermountain flora **6**: 513 (1977); Rix & Phillips, The bulb book, 145 (1981).

Bulbs narrowly ovoid, 1–2 cm, clustered, often on a short rhizome; outer tunics membranous. Leaves 4–6, 10–20 cm × 5–7 mm (rarely more), basal, flat. Stem 30–70 cm, terete or angled, abruptly curved at tip. Spathes 2, to 1 cm. Umbel 30–40-flowered, nodding; flower-stalks 8–18 mm, in fruit abruptly bent upwards at attachment point. Perianth cup-shaped, white to pink; segments 4–6 mm. Stamens long-projecting. Ovary and capsule with 6 conspicuous crests. *America from Canada to Mexico.* H2. Summer.

A very deep rose variant has been distinguished as **A. allegheniense** Small and an almost white variant as **A. oxyphilum** Wherry.

83. A. stellatum Ker Gawler. Illustration: *Botanical Magazine*, 1576 (1813); Gleason, Illustrated flora of the north-eastern United States and adjacent Canada **1**: 415 (1952).

Bulbs ovoid; outer tunics membranous. Leaves 3–6, 1–2 mm broad, basal. Stem 30–70 cm. Spathes 2. Umbel hemispherical, many-flowered, erect; flower-stalks 1–2 cm. Perianth cup-shaped, pink; segments 4–7 mm, ovate. Stamens equalling or longer than segments. Ovary and capsule with 6 conspicuous crests. *N America.* H2. Summer.

84. A. geyeri S. Watson. Illustration: Cronquist et al., Intermountain flora **6**: 511 (1977).

Bulbs ovoid or elongate, 1.5–2.5 cm, usually clustered; outer tunics with netted fibres. Leaves 3 or more, 10–20 cm × 2–4 mm. Stem 15–50 cm, terete to angled. Spathes 2 or 3, 1.5 cm. Umbel 10–25-flowered, sometimes with bulbils; flower-stalks 3–4 mm, becoming rigid and stiffly spreading in fruit. Perianth pink to white; segments 4–10 mm, often obscurely toothed. Stamens included. Capsule inconspicuously crested, enclosed within the persistent perianth-segments.

USA (Texas, New Mexico, Washington, Oregon, Nevada, Arizona). H3. Spring.

85. A. plummerae S. Watson.

Bulb narrowly ovoid, 2 cm, borne on a short rhizome; outer tunics becoming netted-fibrous. Leaves numerous (up to 10), to 27 cm × 6 mm, almost basal, channelled, linear, solid. Stem 30–50 cm. Umbel several-flowered, erect. Perianth white or pink, star-shaped; segments 10 mm, lanceolate, acute. Ovary with apical projection. *USA (Arizona), N Mexico.* H4. Summer.

86. A. textile Nelson & Macbride. Illustration: Hitchcock et al., Vascular Plants of the Pacific northwest **1**: 259 (1969); Cronquist et al., Intermountain flora **6**: 513 (1977).

Bulbs ovoid, 1–1.5 cm; outer tunics netted-fibrous. Leaves 1–3, usually 2, up to 30 cm × 5 mm, almost basal. Stem 5–30 cm. Spathes 2. Umbel hemispherical, many-flowered, loose; flower-stalks to 2 cm, unequal. Perianth bell-shaped, usually white, sometimes pink; segments 5–8 mm, acute or acuminate. Stamens included. Ovary inconspicuously crested. *USA (central states).* H2. Summer.

87. A. drummondii Regel (*A. nuttallii* S. Watson).

Bulbs ovoid, 1.5 cm; outer tunics with netted fibres. Leaves 10–15 cm × 1–4 mm, basal, flat or channelled. Stem 10–20 cm (rarely more). Spathes 2 or 3, 1–1.5 cm. Umbel several-flowered, erect. Perianth pink to white; segments 4–7 mm, ovate to ovate-lanceolate, becoming rigid in fruit. Stamens included. Capsule not crested. *N America (Texas to New Mexico and N Mexico to Nebraska).* H4. Spring.

88. A. falcifolium Hooker & Arnott. Illustration: Jepson, Flora of California **6**: 273 (1922); Abrams, Illustrated flora of the Pacific states **1**: 385 (1923).

Bulbs ovoid to spherical, 1.5 cm; outer tunics membranous. Leaves 1 or 2, to 25 cm × 9 mm, recurving, sickle-shaped. Stem 5–12 cm, flattened. Spathes 2, 1.2–1.9 cm. Umbel many-flowered, compact; flower-stalks 6–16 mm. Perianth bell-shaped, deep rose; segments 8–15 mm, often glandular. Stamens included, 4–9 mm. Ovary and capsule 3-crested. *W USA (California, Oregon).* H4. Summer.

89. A. crenulatum Weigand. Illustration: Abrams, Illustrated flora of the Pacific states **1**: 389 (1923).

Bulbs oblique, narrowly ovoid; outer tunics membranous. Leaves 1 or 2, 3–10 cm × 3–6 mm, basal, recurved with margins toothed. Stem 5–8 cm, 2-angled, with edges toothed. Spathes 2, 8–10 mm. Umbel hemispherical, few-flowered; flower-stalks 8–15 mm. Perianth pink, segments 6–12 mm, lanceolate, acute. Stamens included, 4 mm. Ovary and capsule 6-crested. *Western N America (British Columbia, Washington, Oregon).* H4. Summer.

90. A. acuminatum Hooker (*A. murrayanum* Regel). Illustration: *Gartenflora*, t. 770 (1873); *Botanical Magazine*, n.s., 213 (1953); Cronquist et al., Intermountain flora **6**: 517 (1977).

Bulbs ovoid to almost spherical, 1–1.5 cm; outer tunics membranous, netted, the meshes almost square or hexagonal. Leaves 2–4, shorter than stem, *c.* 1–3 mm broad, channelled. Stem 10–30 cm. Spathes 2, 1–3 cm. Umbel 12–30-flowered, loose; flower-stalks 1.2–2.4 cm, subequal. Perianth deep rose to white; segments 8–15 mm, acuminate, the outer with recurved tips and entire, the inner shorter, narrower and with slightly toothed tips, becoming rigid and keeled in fruit. Stamens included. Ovary and capsule inconspicuously 3-crested. *Western N America (British Columbia to N California, east to Montana, Colorado & Arizona).* H2. Spring.

91. A. campanulatum S. Watson (*A. bidwelliae* Watson). Illustration: Abrams, Illustrated flora of the Pacific states **1**: 391 (1923); Hitchcock et al., Vascular plants of the Pacific northwest **6**: 746 (1969).

Bulbs ovoid, 1.5–2 cm; outer tunics membranous, faintly netted with oblong, sinuous meshes. Leaves 2–3, equalling or shorter than stem, often withered at flowering time, 3–5 mm broad. Stem 10–30 cm. Spathes 2. Umbel hemispherical, many-flowered; flower-stalks 1.5–3.5 cm. Perianth cup-shaped or almost star-shaped, pale rose or white; segments 6–8 mm, acute or acuminate, becoming rigid and keeled in fruit. Stamens included. Ovary and capsule conspicuously 6-crested. *USA (California and Nevada).* H4. Summer.

92. A. dichlamydeum Greene. Illustration: Abrams, Illustrated flora of the Pacific states **1**: 394 (1923).

Bulbs ovoid, *c.* 1.5 cm; outer tunics membranous, with close horizontally zigzag impressed surface. Leaves 1–3, equalling stem, 1–2 mm broad. Stem 10–30 cm. Spathes 2. Umbel hemispherical; flower-stalks 1–1.5 cm. Perianth openly bell-shaped, deep rose-purple; segments 9–11 mm, oblong, the outer somewhat spreading, the inner erect. Stamens included. Ovary and capsule minutely crested. *USA (California).* H4.

93. A. peninsulare Lemmon. Illustration: Abrams, Illustrated flora of the Pacific states **1**: 395 (1923).

Bulbs ovoid, about 1.5 cm; outer tunics membranous, with close horizontally zigzag impressed surface. Leaves 2–4, about equalling stem, 1–6 mm broad. Stem 20–40 cm. Spathes 2. Umbel hemispherical, loose; flower-stalks 1.5–3 cm. Perianth bell-shaped, deep rose-purple; segments 1–1.3 cm, acuminate, the outer entire, longer and broader than the entire and crisped inner segments. Stamens included. Ovary and capsule minutely crested. *USA (California).* H4. Summer.

94. **A. crispum** Greene (*A. peninsulare* var. *crispum* (Greene) Jepson). Illustration: Abrams, Illustrated flora of the Pacific states **1**: 395 (1923).

Bulbs ovoid, about 1 cm; outer tunics membranous, with close horizontally zigzag impressed surface. Leaves 2, mostly shorter than stem, about 1.5 mm broad, basal. Stem 10–30 cm. Spathes 2. Umbel hemispherical, many-flowered, loose; flower-stalks 1–3 cm. Perianth bell-shaped; segments 9–12 mm, acute, the outer entire, the inner narrower with distinctly crisped margin. Stamens included. Ovary and capsule slightly crested. *USA (California).* H4. Summer.

95. A. unifolium Kellogg. Illustration: *Botanical Magazine*, 6320 (1877); Rix & Phillips, The bulb book, 147 (1981).

Bulbs obscure, with 2 bulblets, on a short rhizome; outer tunics cellular-netted. Leaf 1, shorter than stem, 7 mm broad, sickle-shaped, flat to slightly channelled. Stem 30 cm or more, terete. Spathes 2. Umbel 5 cm, 15–20-flowered; flower-stalks 2.5–3.8 cm, thickened at tips. Perianth widely bell-shaped, pink to lavender pink; segments 1–1.4 cm. Stamens included, 7–9 mm. *USA (Oregon, California).* H4. Spring–summer.

96. A. glandulosum Link & Otto. Illustration: *Edward's Botanical Register* **12**: t. 1034 (1826).

Bulbs *c.* 1.5 cm, almost spherical, ending a long slender rhizome; outer tunics membranous. Leaves 2–5, to 30 cm × 6 mm, equalling or longer than stem, 1–7 mm broad, almost basal, strongly veined. Stem 20–55 cm, 2-edged. Spathe 1, becoming 2-lobed. Umbel 4–9 cm, compressed, loose; flower-stalks unequal, 1.5–4 cm. Perianth star-shaped, brownish purple, segments 6–9 mm, lanceolate, acute. Stamens outspread, half the length of segments, united at base. Ovary without crests. *Mexico, USA (Texas & Arizona).* H4. Autumn.

4. NECTAROSCORDUM Lindley
Differs from *Allium* (p. 133) in that the outer perianth-segments have 3–7 veins, the flower-stalks are swollen below the flowers and the ovary contains many ovules.

A genus of 2–3 species from S Europe and W Asia, only 1 of which is cultivated; it is closely related to *Allium* and included in it by some botanists. It will grow in sun or shade and has no soil preferences. Propagation is by seed.

1. N. siculum (Ucria) Lindley (*Allium siculum* Ucria). Illustration: *Edwards's Botanical Register* **22**: t. 1913 (1836); Synge, Collins guide to bulbs, edn 2, pl. 1, f. 2 (1971); Mathew, The larger bulbs, 30, f. 4 (1978); Rix & Phillips, The bulb book, 152e (1981).

Stem to 120 cm. Leaves 3 or 4, basal, with a sharp keel beneath. Flowers up to 30, in a loose terminal umbel subtended by a deciduous spathe; flower-stalks unequal, drooping at first, becoming erect as the fruit develops. Perianth-segments 1.4–2.5 cm, whitish or cream, tinged with green and often pale pink outside. *W Mediterranean region.* H3. Late spring–summer.

The above description refers to subsp. *siculum*.

Subsp. **bulgaricum** (Janka) Stearn (*N. bulgaricum* Janka; *N. dioscoridis* (Sibthorp & Smith) Zahariadi; *Allium bulgaricum* (Janka) Prodan). Illustration: *Botanical Magazine*, n.s., 257 (1955); Rix & Phillips, The bulb book, 152d (1981); Mathew & Baytop, The bulbous plants of Turkey, f. 104 (1984). The greenish perianth-segments are strongly flushed with purple and edged with white. *SE Europe, NW Turkey, Former USSR (Crimea).*

5. NOTHOSCORDUM Kunth
V.A. Matthews
Perennial herbs with bulbs. Leaves linear, basal. Scape erect, bearing an umbel of 4–15 flowers, which is subtended by 2 papery bracts (spathe valves). Flower-stalks unequal. Perianth shortly united at the base. Ovary superior, 3-celled. Capsule 3-celled, containing angular black seeds.

A genus of 20 species, similar to *Allium* but with a terminal style and no smell of onion. Native to N and S America.

Cultivation and propagation as for *Allium* (p. 133).

1. N. gracile (Aiton) Stearn (*N. fragrans* (Ventenat) Kunth; *N. inodorum* misapplied). Illustration: Rix & Phillips, The bulb book, 143 (1981).

Bulbs ovoid with membranous outer tunics. Leaves 6–8, 20–40 cm, linear. Scape 30–75 cm, bearing 8–15 fragrant flowers. Perianth-lobes 9–15 mm, white or lilac with brownish or pink streaks and a pink or mauve central vein. Stamens shorter than perianth-lobes. Capsule 6–10 mm, obovoid. *Subtropical S America, Mexico.* H5. Spring–summer.

N. bivalve (Linnaeus) Britton. Illustration: Rickett, Wild flowers of the United States **2**: 45 (1967); Braun, The Vascular flora of Ohio **1**: 344 (1967). Leaves 3 or 4, umbels with 4–8 flowers, perianth-lobes white or yellowish with a green mid-vein and capsule more or less spherical. *SC & SE USA.* H5. Summer.

6. BESSERA Schultes
Differs from *Nothoscordum* (see above) in having drooping flowers with the stamens projecting, and a staminal tube that is toothed at the top. The perianth-lobes are 3–4 cm, each scarlet with a green mid-vein outside, and creamy-white inside except for the margins and midrib, which are scarlet.

A Mexican genus of 1, or possibly 2, species. Cultivation as for *Brodiaea* (see p. 148).

1. B. elegans Schultes. Illustration: *Edwards's Botanical Register* **25**: t. 38 (1839); *Botanical Magazine*, n.s., 270 (1956).

SW & SC Mexico. H5–G1. Summer.

7. BLOOMERIA Kellogg
Differs from *Nothoscordum* (see above) in having the flower-stalks jointed at the top, and the filaments broadened at the base into a cup-like structure with 2 short teeth at the top.

A genus of 3 species from SW USA and N Mexico. Cultivation as for *Brodiaea* (p. 148).

Literature: Ingram, J., A monograph of the genera Bloomeria and Muilla (Liliaceae), *Madroño* **12**: 19–27 (1953).

1. B. crocea (Torrey) Coville (*B. aurea* Kellogg). Illustration: Parsons, The wild flowers of California, 155 (1897); Abrams, Illustrated flora of the Pacific states **1**: 398 (1923); Rix, Growing bulbs, 71 (1983).

Perianth-segments 8–12 mm, spreading, each with a darker central stripe. *USA (S California), Mexico (Baja California).* H5. Spring.

8. MUILLA Watson
Differs from *Nothoscordum* (see above) in having almost terete leaves, the umbel subtended by more than one spathe and the ovary with many ovules.

The genus contains 5 species native to SW USA and N Mexico. Cultivation as for *Brodiaea* (p. 148).

Literature: see *Bloomeria* (above).

1. M. maritima (Torrey) Watson. Illustration: Abrams, Illustrated flora of the Pacific states **1**: 397 (1923).

Perianth-segments 3–6 mm, greenish-white with a brownish mid-vein. *USA (S California), Mexico (Baja California).* H5. Summer.

9. CALOSCORDUM Herbert
V.A. Matthews
Similar to *Nothoscordum* but umbel subtended by only one bract. Bulb with whitish papery tunic. Leaves 2–6, very narrowly linear, channelled above, shorter than flowering stem, often somewhat withered at flowering time. Stem 10–25 cm, bearing an umbel of 10–20 flowers that is subtended by a single bract. Flower-stalks unequal. Perianth bright pink, 6–8 mm, joined in the lower one-third, lobes spreading-reflexed, each with a darker central vein, perianth 5–7 mm across the lobes. Stamens borne on the perianth-tube. Ovary superior.

A genus of 1 species. Cultivation and propagation as for *Allium* (p. 133).

1. C. neriniflorum Herbert (*Nothoscordum neriniflorum* (Herbert) Bentham & Hooker; *Allium neriniflorum* (Herbert) Baker). Illustration: *Edwards's Botanical Register* **33**: t. 5 (1847); Rix & Phillips, The bulb book, 145 (1981). *N China, Former USSR (Pamir Mountains, SE Siberia), ?Mongolia.* H3. Late summer.

10. LEUCOCORYNE Lindley
S.G. Knees
Herbaceous perennials. Bulbs to 2 cm, spherical to ovoid, tunic dark brown. Leaves 2–5, deciduous, basal, narrowly linear, usually channelled, 15–25 cm × 1–3 mm, appearing after the flowers. Scapes to 50 cm, bearing umbels of 2–12 flowers subtended by 2 spathes. Perianth funnel-shaped to very widely funnel-shaped, white to purple or bluish violet. Stamens 3 with dorsifixed anthers. Staminodes 3. Stigmas capitate. Ovary superior, 3-celled. Ovules many.

A genus of 5 species, most smelling strongly of garlic. All are native to Chile. Protection from extreme conditions in a cool glasshouse is recommended and cultivation requirements are similar to *Freesia* (p. 291) or *Ixia* (p. 189).

Literature: Pizarro, C.M., Flores Silvestres de Chile (1966).

1a. Perianth purple-violet; scapes with 2–6 flowers **1. purpurea**
b. Perianth lilac-blue fading to white in the centre; scapes with 6–9 flowers **2. ixioides**

1. L. purpurea Gay (*L. ixioides* (Hooker) Lindley var. *purpurea* (Gay) Baker). Illustration: Pizarro, Flores silvestres de Chile, t. 9 (1966).

Scapes to 50 cm, with 2–6 flowers. Perianth-segments 1.5–2 cm, deep purple-violet throughout, broadly lanceolate with prominent mid-veins. Bracts channelled, 5–6 cm. Seeds with black testa. *Chile.* H5. Spring.

2. L. ixioides (Hooker) Lindley. Illustration: *Botanical Magazine*, 9457 (1936); Hunt (ed.), Marshall Cavendish encyclopedia of gardening **39**: 1085 (1968); Brickell (ed.), RHS A–Z encyclopedia of garden plants, 622 (2003).

Scapes to 45 cm with 6–9 flowers. Perianth-segments 1.2–1.5 cm, lilac-blue at apex, fading to white at throat, diamond-shaped to lanceolate. Stamens and staminodes yellow. *Chile.* H5. Spring.

L. odorata Lindley. Similar to *L. ixioides*, but generally with paler, very fragrant flowers. *Chile.* H5. Spring.

11. IPHEION Rafinesque
D.A. Webb
Perennials with small bulbs and fleshy roots. Leaves basal, linear. Flowers solitary (rarely in umbels of 2), with 2 partly united bracts on the scape some distance below the flower. Perianth with a narrowly bell-shaped tube and spreading lobes. Stamens within the perianth-tube; anthers dorsifixed, opening inwards. Ovary superior. Stigma small, obscurely 3-lobed. Fruit a many-seeded capsule.

A genus of 10 or more species from temperate S America. It has, from time to time been included in many other genera, giving rise to many synonyms, but is now generally recognised as distinct.

There is only 1 species in cultivation. It should be planted in fairly dry, gritty soil in a sunny place, but even in districts where it is hardy it forms an attractive pot-plant.

Literature: Traub, H.P. & Moldenke, H. N., The genus Ipheion: diagnosis, key to species and synonymy, *Plant Life* **11**: 125–30 (1955).

1. I. uniflorum (Graham) Rafinesque (*Triteleia uniflora* Lindley; *Milla uniflora* Graham; *Brodiaea uniflora* (Lindley) Engler). Illustration: *Botanical Magazine*, n.s., 185 (1952); Perry, Flowers of the world, 173 (1972); Rix & Phillips, The bulb book, 111 (1981).

Leaves 20–25 cm × 4–7 mm, somewhat fleshy, smelling of onion when bruised. Scape as long as leaves or slightly shorter; bracts brownish, scarious, forming a sheathing tube with 2 acute lobes. Flowers white or violet-blue, 4 cm in diameter, the lobes rather longer than the tube. Anthers and stigma just visible at the mouth of the tube. *Uruguay and warm-temperate Argentina.* H3–4. Spring.

The colour of the flowers varies from pure white to clear violet-blue, but the variants are scarcely worth naming. 'Froyle Mill', however, with flowers of a deep, dusky violet, is very distinct.

12. BRODIAEA J.E. Smith
S.J.M. Droop
Perennial herbs arising from corms that are more or less spherical and have dark brown, fibrous coats. Leaves linear, crescent-shaped in section, without a keel, with vein impressions weak or lacking. Flowers in an umbel subtended by scarious bracts. Perianth united into a tube below, lobes of the outer whorl narrower than those of the inner. Stamens 3, opposite the inner perianth-lobes, usually alternating with 3 staminodes; filaments joined to the perianth-tube for most of their lengths, lacking appendages; anthers basifixed. Ovary superior, stalkless, 3-celled with several ovules in each cell; stigma with 3 spreading, winged lobes. Seeds flattened, angled.

A genus of about 15 species from western N America, about 5 of which are found in cultivation. In the past, *Brodiaea* has included *Triteleia* (p. 149), *Dichelostemma* (p. 148) and *Ipheion* (above). They may be grown in open borders in suitable areas, or in pots, in a soil that is light and well-drained. Propagation is by offsets or by seed.

Literature: Hoover, R.F., A revision of the genus Brodiaea, *American Midland Naturalist* **22**: 551–74 (1939); Hoover, R. F., Further observations on Brodiaea and some related genera, *Herbertia* **11**: 13–23 (1955); Niehaus, T.F., A biosystematic study of the genus Brodiaea (Amaryllidaceae), *University of California Publications, Botany* **60**: 1–66 (1971); Niehaus, T.F., The Brodiaea complex, *The Four Seasons* **6(1)**: 11–21 (1980).

Perianth. To 2.4 cm: **1, 5**; 2.4 cm or more: **2, 3, 4**.
Perianth-tube. Splitting in fruit: **2**; not splitting in fruit: **1, 3, 4, 5**.
Perianth-lobes. Three times as long as tube: **2**; less than 2 and a half times as long as tube: **1, 3, 4, 5**. Widely spreading: **1**; ascending and recurved: **2, 3, 4, 5**.
Staminodes. Shorter than anthers, flat, acute: **3**. Apex notched: **1, 4**; apex 3-fid: **5**. Borne close to the anthers: **1, 2, 4**; borne distant from anthers: **3, 5**.

1a. Perianth-tube tightly constricted above the ovary; lobes widely spreading **1. minor**
b. Perianth-tube not or slightly constricted above ovary; lobes ascending and recurved 2
2a. Perianth-lobes about 3 times the length of the tube; tube thin, transparent and splitting in fruit **2. californica**
b. Perianth-lobes less than 2 and a half times as long as tube; tube thick, opaque, not splitting in fruit 3
3a. Anthers exceeding staminodes and stigma; perianth-tube funnel-shaped; staminodes flat, acute **3. elegans**
b. Anthers and stigma exceeded by staminodes; perianth-tube ovoid to narrowly bell-shaped; margins of staminodes more or less rolled inwards, apex notched or 3-fid 4
4a. Scape 5–25 cm; staminodes with margins conspicuously rolled inwards; apex notched **4. coronaria**

b. Scape 5–70 mm; staminodes with margins slightly rolled inwards, apex unequally 3-fid **5. terrestris**

1. B. minor (Bentham) Watson (*B. grandiflora* Smith var. *minor* Bentham). Illustration: Rickett, Wild flowers of the United States **5**: t. 18 (1971).

Scape 2–30 cm. Flower-stalks 1–3 cm. Perianth 1.7–2.4 cm, the lobes widely spreading, less than 2 and a half times times as long as tube, pink to purple; tube strongly constricted above the ovary, thick, opaque, not splitting in fruit. Staminodes with the margins rolled inwards, apex usually notched, slightly exceeding the anthers and stigma. Anthers 1–6 mm, filaments 0.5–4 mm. Ovary 3–5 mm, style 4–9 mm. *USA (California).* H5. Summer.

2. B. californica Lindley (*B. grandiflora* Smith var. *elatior* Bentham). Illustration: Rickett, Wild flowers of the United States **5**: t. 16 (1971).

Scape 20–70 cm. Flower-stalks 2–13 cm. Perianth 3–4.5 cm, lobes about 3 times as long as tube, pale to deep purple or pink, ascending and recurved; tube not constricted above the ovary, thin, transparent and splitting in fruit. Staminodes 1.6–2.7 cm, erect, linear, with margins wavy or rolled inwards, reaching the same level as the anthers and stigma. Anthers 9–13 mm, filaments 7–10 mm. Ovary 7–11 mm, style 1.8–2.2 cm. *USA (N California).* H5. Summer.

3. B. elegans Hoover (*B. coronaria* (Salisbury) Engler var. *mundula* Jepson). Illustration: Rickett, Wild flowers of the United States **5**: t. 17 (1971).

Scape 10–50 cm. Flower-stalks 5–100 mm. Perianth 3–4.5 cm, lobes less than 2 and a half times times as long as tube, deep purple, ascending and recurved; tube funnel-shaped, thick, opaque and not splitting in fruit. Staminodes 6–9 mm, erect and flat, acute, about the same level as the stigma, but shorter than the anthers. Anthers 4–10 mm, filaments 4–6 mm. Ovary 9–15 mm, style 7–15 mm. *USA (California, Oregon).* H5. Summer.

4. B. coronaria (Salisbury) Engler (*B. grandiflora* Smith; *B. howellii* Eastwood). Illustration: *Botanical Magazine*, 2877 (1829); Rickett, Wild flowers of the United States **5**: t. 17 (1971).

Scape 5–25 cm. Flower-stalks 1–5 cm. Perianth 2.4–4 cm, lobes less than 2 and a half times times as long as tube, pale to deep purple, ascending and recurved; tube often slightly constricted above the ovary, ovoid to narrowly bell-shaped, thick, opaque and not splitting in fruit. Staminodes close to anthers, with the margins conspicuously rolled inwards and apex notched, exceeding anthers and stigma. Anthers 4.5–7 mm, filaments 2–4.5 mm. Ovary 8–10 mm, style 7–11 mm. *USA (California, Oregon, Washington).* H5. Summer.

5. B. terrestris Kellogg (*B. grandiflora* Smith var. *macropoda* Torrey; *B. coronaria* (Salisbury) Engler var. *macropoda* (Torrey) Hoover).

Scape 5–70 mm. Flower-stalks 3–15 cm. Perianth 2–2.4 cm, lobes less than 2 and a half times times as long as tube, purplish pink, ascending and recurved; tube usually not constricted above the ovary, ovoid to narrowly bell-shaped, thick, opaque and not splitting in fruit. Staminodes distant from anthers, with the margins slightly rolled inwards, the apex unequally 3-fid with the central tooth smaller than the laterals, considerably exceeding anthers and stigma. Anthers 3–5 mm, filaments 1.5–3 mm. Ovary 4–5 mm, style 6–9 mm. *USA (California, S Oregon).* H5. Summer.

Subsp. *terrestris* is described above; other subspecies exist in the wild but are not generally in cultivation. Distinction between *B. terrestris* and *B. coronaria* is difficult in wild populations, and this may be reflected in gardens.

13. TRITELEIA Lindley

S.J.M. Droop

Like *Brodiaea*, but corms flattened and with pale brown coats, leaves distinctly keeled beneath and with evident veins; staminodes 0, fertile stamens 6, anthers versatile; ovary shortly stalked, seeds more or less spherical, angled.

A genus of 17 species from western N America, mostly from northern California and southern Oregon. Seven species and 1 garden hybrid are available in cultivation. The genus was formerly included in *Brodiaea* (Subgenus *Triteleia*). In the account below measurements of the perianth are from the base of the tube to the tip of the lobe when fully extended. Cultivation as for *Brodiaea* (p. 148).

Literature: Hoover, R.F., A systematic study of Triteleia, *American Midland Naturalist* **25**: 73–100 (1941); Lenz, L.W., Triteleia tubergenii, an amphidiploid of garden origin, *Aliso* **7**: 157–60 (1970); Lenz, L.W., A biosystematic study of Triteleia (Liliaceae), *Aliso* **8**: 221–58 (1975); Niehaus, T.F., The Brodiaea complex, *The Four Seasons* **6(1)**: 11–21 (1980).

Flower-stalks. More than 1 cm: **2**.
Perianth-tube. Longer than lobes: **1, 3, 7**; equalling lobes to *c.* half as long: **2, 4, 6**; less than half as long as lobes: **5**. Tapered at base: **3, 6, 7**; rounded at base: **1**; spreading from the base: **2, 4, 5**.
Stamens. Inserted on to perianth-tube at 2 levels at least 2 mm apart: **1, 2, 3**; inserted at more or less the same level: **4, 5, 6, 7**.
Filaments. Thread-like: **3, 6**; narrowly expanded at base: **1, 2**; broadly expanded at base: **5, 7**; broad, with parallel sides and apical appendages: **4**.
Ovary. Yellow: **2**. Shorter than its stalk: **3, 6, 7**.

1a. Stamens inserted on the perianth-tube at 2 levels that are more than 2 mm apart 2
b. Stamens inserted on the perianth-tube at more or less the same level 4
2a. Perianth-tube rounded at the base **1. grandiflora**
b .Perianth-tube acute, tapered or spreading at the base 3
3a. Flower-stalks usually more than 4 times as long as flowers; perianth-tube acute at base; ovary yellow **2. peduncularis**
b. Flower-stalks rarely more than 2 times as long as flowers; perianth-tube tapered at base; ovary green, blue or white **3. laxa**
4a. Stamens alternately long and short; filaments broad with forked apical appendages **4. ixioides**
b. Stamens equal or nearly so; filaments thread-like or broad only at the base, without appendages 5
5a. Perianth bowl-shaped, broadly spreading from the base, to 1.9 cm; stalk of ovary very short; perianth-lobes *c.* 3 times longer than tube **5. hyacinthina**
b. Perianth more or less funnel-shaped, tapered at the base, more than 1.9 cm; ovary-stalk considerably longer than ovary; perianth-lobes at most 2 times longer than tube 6
6a. Perianth yellow; lobes 1.5–2 times as long as tube **6. hendersonii**
b. Perianth blue or pale purple; lobes usually shorter than tube **7. bridgesii**

1. T. grandiflora Lindley (*Brodiaea grandiflora* Macbride not J.E. Smith; *B. douglasii* Piper). Illustration: *Botanical Magazine*, 6907 (1886).

Scape 20–70 cm. Flower-stalks 1–4 cm, usually shorter than the flowers, but elongating in fruit. Perianth 1.5–3 cm, bright blue or rarely white; tube as long as or longer than lobes, rounded at base. Stamens inserted in 2 whorls separated by more than 2 mm, filaments slightly dilated at base; anthers 2–4 mm. Ovary green or blue, usually longer than its stalk. *Western N America (British Columbia, Oregon & N Utah)*. H5. Summer.

2. T. peduncularis Lindley (*Brodiaea peduncularis* (Lindley) Watson). Illustration: Rickett, Wild flowers of the United States **5**: t. 17 (1971).

Scape 10–40 cm. Flower-stalks 2–18 cm. Perianth 1.5–2.8 cm, white or pale purple, broadly funnel-shaped; the tube equalling or shorter than the lobes, broadly acute at base. Stamens inserted in 2 whorls separated by more than 2 mm, filaments slightly dilated at base; anthers 2–4 mm. Ovary yellow, as long as or longer than its stalk. *USA (California)*. H5. Summer.

3. T. laxa Bentham (*Brodiaea laxa* (Bentham) Watson; *B. candida* Baker; *T. candida* (Baker) Greene). Illustration: *Edwards's Botanical Register* **20**: t. 1685 (1834); Rickett, Wild flowers of the United States **4**: t. 8 (1970) & **5**: t. 16 (1971); Brickell (ed.), RHS A–Z encyclopedia of garden plants, 1049 (2003).

Scape 10–70 cm. Flower-stalks 1–9 cm. Perianth 1.8–4.7 cm, blue or rarely white; tube usually slightly longer than lobes, narrowly funnel-shaped, tapered at base. Stamens inserted in 2 whorls separated by more than 2 mm, filaments 3–6 mm, thread-like; anthers 2–5 mm. Ovary green, blue or white, its stalk 2–3 times longer than the ovary at flowering, relatively shorter in fruit. USA *(California & S Oregon)*. H5. Summer.

T. × tubergenii Lenz (*Brodiaea × tubergenii* Anon.) is a fertile hybrid between *T. peduncularis* and *T. laxa*. It occurred spontaneously in a garden, and in most characters is intermediate between its parents.

4. T. ixioides (Aiton) Greene (*Brodiaea ixioides* (Aiton) Watson; *B. lutea* (Lindley) Morton; *Calliprora lutea* Lindley). Illustration: *Botanical Magazine*, 3588 (1837); Rickett, Wild flowers of the United States **4**: t. 8 (1970) & **5**: t. 16, 17 (1971).

Scape 6–60 cm. Flower-stalks 1–7 cm. Perianth 1–2.2 cm, yellow, with a dark line down each lobe; tube about half as long as lobes, acute at base. Stamens inserted in more or less 1 whorl near the top of the perianth-tube, each filament broad and continued behind the anther as forked projections; anthers 1.5–2 mm. Ovary green or yellowish, longer than its stalk. *USA (California)*. H5.

A variable species divided into several subspecies, more than one of which are in cultivation; their precise identification is, however, uncertain.

5. T. hyacinthina (Lindley) Greene (*Brodiaea hyacinthina* (Lindley) Baker; *B. lactea* Watson; *T. lutea* Davidson & Moseley). Illustration: *Journal of the Royal Horticultural Society* **89**: t. 66 (1964); Rickett, Wild flowers of the United States **5**: t. 16 (1971); Brickell (ed.), RHS A–Z encyclopedia of garden plants, 1049 (2003).

Scape 15–70 cm. Flower-stalks 5–50 mm. Perianth 8–18 mm, bowl-shaped, white or rarely blue or pale purple; the tube much shorter than the lobes, spreading away from the base. Stamens all inserted at the same level, filaments broadly dilated at base; anthers 1–2 mm. Ovary green, white or rarely pale blue or purple, its stalk very short at flowering, extending later. *Western N America (British Columbia to Idaho and California)*. H5. Summer.

6. T. hendersonii Greene (*Brodiaea hendersonii* (Greene) Watson). Illustration: Rickett, Wild flowers of the United States **5**: t. 16 (1971).

Scape 12–35 cm. Flower-stalks 1.5–4 cm. Perianth 2–2.3 cm, yellow often fading to bluish, each lobe with a dark line; lobes 1.5–2 times longer than the tube; tube narrowly funnel-shaped, tapered at base. Stamens more or less inserted at 1 level, filaments thread-like; anthers 1.5–2 mm. Ovary green or yellowish, its stalk about twice as long. *USA (SW Oregon)*. H5. Summer.

7. T. bridgesii (Watson) Greene (*Brodiaea bridgesii* Watson). Illustration: Rickett, Wild flowers of the United States **5**: t. 17 (1971).

Scape 10–50 cm. Flower-stalks 1.5–9 cm. Perianth 2.6–4 cm, blue or pale purple; the tube usually longer than the lobes, tapered at base. Stamens inserted at more or less 1 level, filaments dilated at base; anthers 3.5–4.5 mm. Ovary green, blue or pale purple. Ovary-stalk 3–4 times as long as ovary at flowering, relatively shorter in fruit. *USA (SW Oregon, California)*. H5. Summer.

14. DICHELOSTEMMA Kunth

S.J.M. Droop

Like *Brodiaea* but leaves distinctly keeled beneath and with obvious veins; bracts coloured; fertile stamens sometimes **6**; stigma with 3 short lobes.

A genus of 7 species from western N America, of which 4 are in cultivation. In the account that follows, measurements of the perianth extend from the base of the flower to the tip of the lobe when pulled straight. The genus was formerly part of *Brodiaea* (Subgenus *Dichelostemma*). Cultivation as for *Brodiaea* (p. 148).

Literature: Hoover, R.F., The genus Dichelostemma, *American Midland Naturalist* **24**: 463–76 (1940); Lenz, L.W., The nature of the floral appendages in four species of Dichelostemma (Liliaceae), *Aliso* **8**: 383–9 (1976); Niehaus, T.F., The Brodiaea complex, *The Four Seasons* **6(1)**: 11–21 (1980).

1a. Fertile stamens 6 **1. pulchellum**
b. Fertile stamens 3 2
2a. Perianth-tube red, lobes green, much shorter than the tube **2. ida-maia**
b. Perianth entirely blue or purple, lobes as long as or longer than tube 3
3a. Flowers in umbels; staminodes entire; perianth-tube much constricted above ovary **3. multiflorum**
b. Flowers in short racemes; staminodes deeply bifid; perianth-tube not much constricted above ovary **4. congestum**

1. D. pulchellum (Salisbury) Heller (*Brodiaea capitata* Bentham; *B. pulchella* Greene). Illustration: *Botanical Magazine*, 5912 (1871); Rickett, Wild flowers of the United States **4**: t. 8 (1970) & **5**: t. 16 (1971).

Scape 30–60 cm. Flowers in umbels. Bracts ovate, purple; flower-stalks 1–15 mm. Perianth 1.2–1.8 cm, white to purple; the lobes equalling or exceeding the tube, not widely spreading. Stamens 6 in 2 whorls, those opposite the outer perianth-lobes smaller than the others; filaments united below to form a short tube, connective of the larger anthers extending upwards as 2 triangular appendages. Style 4–6 mm. *USA (Oregon to California, east to Utah).* H5. Summer.

2. D. ida-maia (Wood) Greene (*Brodiaea ida-maia* (Wood) Greene; *B. coccinea* Gray; *Brevoortia ida-maia* Wood). Illustration: *Botanical Magazine*, 5857 (1870); Rickett, Wild flowers of the United States **5**: t. 18 (1971); Brickell (ed.), RHS A–Z encyclopedia of garden plants, 374 (2003).

Scape 30–90 cm. Flowers in an umbel. Perianth-tube 2–2.5 cm, bright red; lobes much shorter than the tube, greenish, recurved. Staminodes 3, each broader than long. Stamens 3, anthers 6–8 mm. Style 1.5–2 cm. *USA (Oregon, California).* H5. Summer.

3. D. multiflorum (Bentham) Heller (*Brodiaea multiflora* Bentham). Illustration: *Botanical Magazine*, 5989 (1872); Rickett, Wild flowers of the United States **5**: t. 18 (1971).

Scape 20–80 cm. Flowers in an umbel. Flower-stalks 3–15 mm. Perianth 1.6–2.1 cm, pale to dark purple; lobes about equalling the tube, widely spreading; tube much constricted above the ovary. Staminodes 3, each longer than broad, apex entire. Stamens 3, anthers 4–5 mm. Style 7–8 mm. *USA (Oregon, California).* H5. Summer.

4. D. congestum (J.E. Smith) Kunth (*Brodiaea congesta* J.E. Smith).

Scape 30–90 cm. Flowers in a short raceme. Perianth 1.4–2 cm, purple; lobes about equalling tube, not widely spreading; tube not much constricted above ovary. Staminodes 3, bifid. Stamens 3, anthers 4–5 mm. Style 5–6 mm. *USA (Washington to California).* H5. Summer.

15. MILLA Cavanilles

V.A. Matthews

Perennial herbs with bulbs. Leaves linear, flat to almost terete. Flowering stem erect bearing an umbel of erect flowers, subtended by 4 chaffy spathes that do not enclose the flower buds. Perianth united to above the middle, forming a long tube. Stamens borne at the mouth of the perianth-tube, shortly projecting. Ovary superior, borne on a long stalk.

A genus of 6 species native to the southern USA and northern C America. Only 1 species is cultivated and is kept in a cool greenhouse in colder areas. In warmer regions it can be planted outdoors in a sunny place in the spring, and the bulbs can be lifted and stored inside over the winter. Propagation is by offsets and seed.

Literature: Moore, H.E., The genus Milla and its allies, *Gentes Herbarum* **8**: 263–94 (1953).

1. M. biflora Cavanilles. Illustration: *Edwards's Botanical Register* **18**: t. 1555 (1832); *Flore des serres* **4**: t. 1459 (1861); *Baileya* **1**: 8 (1953); Rickett, Wild flowers of the United States **3**: 27 (1969).

Leaves 2–7, narrowly linear, channelled above and rounded below, to almost terete, 10–50 cm. Flowering stem 5–45 cm, with 1–6, occasionally to 8, fragrant, stalkless flowers that appear stalked because the perianth-lobes are united into a long 10–20 cm tube. Perianth-lobes 1.5–3.5 cm, spreading flat, white or occasionally flushed with lilac, with 3, 5 or 7 green central veins, or 1 broad central stripe. Anthers longer than the filaments. *Guatemala, Mexico and adjoining USA.* H5–G1. Late summer–autumn.

---

20. CONVALLARIACEAE

Plants terrestrial, entirely herbaceous, usually with rhizomes. Foliage leaves on flowering stem, with well developed blades. Flowers in racemes or clusters, without spathes when solitary. Perianth of 4 or 6 united segments, tubular to urceolate. Stamens 4 or 6. Ovary of 2 or 3 united carpels, superior or partly inferior; placentation axile; nectar secreted from glands on the ovary. Fruit a berry.

Seventeen genera and 210 species. Five genera native to Europe, 10 to North America, several cultivated for ornament.

1a. Ovary partly or fully inferior 2
b. Ovary superior 3
2a. Leaves with conspicuous stalk and blade; perianth united into a tube below **16. Peliosanthes**
b. Leaves without distinct stalks and blades; perianth-segments free to the base **14. Ophiopogon**
3a. Styles usually 3 (rarely 4), separate and distinct, or stigmas individually borne directly on the ovary 4
b. Style 1, sometimes 3-branched at the top, or style absent, stigmas united, borne directly on the ovary 5
4a. Leaves in 2 whorls on each flowering stem **13. Medeola**
b. Leaves borne singly on the flowering stems **10. Disporum**
5a. Perianth-segments free at the base, or united for less than one-tenth of their total length 6
b. Perianth-segments united at the base for more than one-tenth of their total length 13
6a. Flowering stem leafy over most of its length below the inflorescence 7
b. Flowering stem with only 2 or 3 leaves along its length, or with several leaves concentrated towards the base 9
7a. Flowers solitary or in pairs, their stalks fused to the stem for some distance so that they appear to be borne well above their bract, the stalks with a downward bend at about the middle of the free part **9. Streptopus**
b. Flowers and flower-stalks not as above 8
8a. Leafy part of the stem unbranched; perianth-segments neither swollen nor slightly inflated above the base **7. Smilacina**
b. Leafy part of the stem branched; perianth-segments often swollen or slightly inflated above the base **10. Disporum**
9a. Flowering stems leafless (scapes) 10
b. Flowering stems with at least a few leaves 12
10a. Filaments and style bent upwards in the middle: perianth bell-shaped, blue, purple or white. **15. Liniope**
b. Combination of characters not as above.
11a. Perianth-segments hairy outside **6. Clintonia**
b. Perianth-segments not hairy outside **2. Speirantha**
12a. Flowering stem bearing 2 or 3 heart-shaped leaves; flowers with parts in 4s **8. Maianthemum**
b. Flowering stems with a few leaves, which are not heart-shaped; flowers with parts in 3s **2. Speirantha**
13a. Flowering stems leafy 14
b. Flowering stems leafless, all leaves basal 16
14a. Leaves distinctly stalked; perianth spherical to bell-shaped **1. Convallaria**
b. Leaves not or shortly stalked; perianth not as above 15
15a. Anthers borne on a corona **12. Disporopsis**
b. Anthers borne individually on filaments **11. Polygonatum**

16a. Leaves evergreen, not stalked, thick and leathery **4. Rohdea**
b. Leaves deciduous, if evergreen then stalked, not thick and leathery 17
17a. Leaves stalked; flowers solitary, borne at ground level or almost so **5. Aspidistra**
b. Leaves not stalked; flowers on a scape in a terminal spike **3. Reineckia**

1. CONVALLARIA Linnaeus
A.C. Leslie
Hairless perennial herb, with creeping rhizomes. Stems erect, with several green or violet scales below and 1–4 leaves above. Leaf-blades ovate-lanceolate to elliptic, 3–23 cm × 5–110 mm, acute to acuminate; stalks 1–24 cm, the sheathing bases simulating an elongation of the stem. Scapes solitary, sharply angled, arising from scale axils, shorter than the leaves. Inflorescence a 1-sided raceme. Flowers 5–13, nodding; flower-stalks usually curved downwards and exceeding the ovate-lanceolate bracts. Perianth spherical to bell-shaped, 5–11 × 5–11 mm, white, united for half to two-thirds of its length, lobes 6, reflexed at tips. Ovary superior, 3-celled, with 4–8 ovules in each cell. Style simple; stigma capitate. Fruit a red berry.

One very variable species distributed throughout the north temperate regions. Widely cultivated for its long lasting, highly scented flowers. Once established, it can become an aggressive weed, although it is effective as ground cover under trees and shrubs. It grows best in semi-shade, in a humus-rich, moisture-retentive soil and, whilst tolerating deep shade, will not flower well in such conditions. Propagation is by division.

1. C. majalis Linnaeus. Illustration: Polunin, Flowers of Europe, t. 171 (1969); Keble Martin, The concise British flora in colour, t. 84 (1965); Ross-Craig, Drawings of British Plants **29**: 16 (1972); Brickell (ed.), RHS A–Z encyclopedia of garden plants, 301 (2003).

North temperate regions. H1. Spring.

There are variants in cultivation with pink flowers (var. **rosea** Reichenbach), with purple spots at the base of the filaments (forma **picta** Wilczek) and with either double white ('Flore Pleno') or double pink flowers. 'Prolificans' is a variant with a paniculate inflorescence, in which the flowers are often slightly malformed. There are also several larger-flowered or more vigorous forms as well as a number with variegated leaves, of which 'Variegata', with longitudinal creamy white stripes, is best known.

2. SPEIRANTHA Baker
J. Cullen
Stemless perennial herbs with thick rhizomes and numerous stolons. Leaves to 15 × 2.5–3 cm, erect, 6–8 in a rosette, oblanceolate, tapered slightly to the base but not stalked, apex acute. Scape with a few, small, sheathing, scale-leaves, to 15 cm. Flowers in a loose raceme, with bracts. Perianth-segments 6, free, spreading, white, 4–5 mm. Stamens 6, free. Ovary superior, 3-celled, with 3 or 4 ovules in each cell. Style 1; stigma 3-grooved. Fruit a berry.

A genus of a single species, not very frequently grown. Cultivation as for *Convallaria.*

1. S. convallarioides Baker. Illustration: *Botanical Magazine*, 4842 (1855) – as *Albuca ? gardeni*; *Journal of the Linnaean Society, Botany* **14**: t. 17 (1875).

China (Jiangxi). H5. Spring–summer.

J.D. Hooker (*Botanical Magazine*, cited above) thought, incorrectly, that the plant originated in South Africa; hence his doubtful placing of it in the genus *Albuca* (see p. 118).

3. REINECKEA Kunth
P.G. Barnes
Perennial herbs. Leaves persistent at the apex of prostrate, rhizome-like stems, linear-lanceolate, 15–35 cm, somewhat narrowed to the base. Flowers stalkless in a terminal spike on a scape 6–10 cm tall; bracteoles 4–7 mm, triangular-ovate. Flowers pale to deep pink, perianth 8–12 mm, becoming reflexed, united at the base to form a tube 3–5 mm long. Stamens and style equal, 8–12 mm. Ovary superior. Fruit a spherical red berry, seldom formed in cultivation.

A genus of a single species that is shade-tolerant, forming mats of evergreen foliage, growing freely in acid to neutral soils but sparse-flowering in cooler areas. It is propagated by division, or seed when available.

1. R. carnea (Andrews) Kunth. Illustration: *Botanical Magazine*, 739 (1804); *Illustration Horticole* **9**: t. 323 (1862) – 'Variegata'.

China, Japan. H4. Early summer.

The cultivar 'Variegata', with leaves striped with cream, is also cultivated.

4. ROHDEA Roth
J. Cullen
Perennial herbs with rhizomes and fibrous roots. Leaves in a basal rosette, leathery, thick, variable in shape, to 45 cm. Flowers in short, dense spikes borne among the leaves and shorter than them. Perianth greenish yellow, bell-shaped, united into a tube for most of its length. Stamens 6, borne on the perianth-tube; filaments very short or absent. Ovary superior, 3-celled, with 1 ovule in each cell. Stigma peltate; style absent. Fruit a spherical, red or yellow, usually 1-seeded berry.

A genus of a single species from SE China and Japan. It has been extensively cultivated in Japan as a foliage plant, and many cultivars have been selected there, including variants with variegated or twisted and curled leaves. The plant has not proved as popular in Europe, though cultivars have been imported from time to time. It is easily grown as a house plant or in a border, but some variants require protection during the winter. Propagation is by seed.

1. R. japonica (Thunberg) Roth. Illustration: *Botanical Magazine*, 898 (1806); Kitamura et al., Coloured illustrations of herbaceous plants of Japan **3**: t. 27 (1964).

SW China, Japan. H3.

5. ASPIDISTRA Ker Gawler
S.G. Knees & J.C.M. Alexander
Evergreen, stemless, perennial herbs with thick rhizomes and roots. Leaves all basal, arising singly, or in groups, at intervals along the rhizome. Leaf-blade leathery and glossy, lanceolate to elliptic, wedge-shaped at base, acute, with sunken midrib. Leaf-stalk winged, one-third to half as long as blade, subtended by a papery sheath. Flowers bisexual, with bracts, borne individually at ground level on short stalks, rather inconspicuous and usually obscured by the foliage. Perianth united to form a spherical, bell-shaped or urceolate flower, lobes 6–8, usually dull greenish brown, often with purple spotting. Stamens 6 or 8, included in perianth-tube. Ovary superior, 3 or 4-celled. Stigma peltate. Fruit a berry.

Eight species in E Asia, from the Himalaya to Japan. As these are normally forest-floor plants they are well-adapted to cultivation as pot-plants in cool, shady positions indoors, where they often thrive with minimum attention. Propagation is by division of the rhizomes, preferably in spring. **A. caespitosa** has recently been introduced into cultivation and may become more widely available in the future.

1a. Leaves with 10 or less veins 2
b. Leaves with 15 or more veins **1. elatior**
2a. Leaf-blades 20–40 cm; leaf stalks 10–25 cm **3. typica**
b. Leaf blades 15–20 cm; leaf-stalks 5–15 cm **2. lurida**

1. A. elatior Blume. Illustration: Davidson, Illustrated Dictionary of house plants, 39 (1983).

Leaves solitary. Leaf-blades ovate to lanceolate, 50–60 × 7–10 cm, with 16–20 veins, glossy dark green, sheathed at the base by a papery bract; leaf-stalks 1–5 cm. Flowers borne at soil level on short (5 cm) stalks arising directly from the rhizome. Perianth urceolate, 2.5–3 cm diameter, with 8 triangular, leathery lobes, recurved at apex, cream with purple spotting on outer face, wine-red within. Stamens 8, borne at the base of the perianth-lobes. Stigma a peltate disc, obscuring mouth of perianth. Fruit rarely produced in cultivation. *China.* H5. Spring.

'Variegata' (Illustration: Everett, The New York Botanical Gardens encyclopedia of horticulture, 278, 1980; Brickell (ed.), RHS A–Z encyclopedia of garden plants, 147, 2003) has leaves with longitudinal, cream variegation.

2. A. lurida Ker Gawler. Illustration: *Botanical Magazine*, 2499 (1824); *Edwards's Botanical Register* **8**: 628 (1822).

Leaves lanceolate, 15–20 cm, with 5–9 main veins, 2 or 3 per node, surrounded by a papery basal sheath of 8–12 cm; leaf-stalks 5–15 cm. Flowers almost stalkless on the rhizome, subtended by papery bracts. Perianth bell-shaped, xx cm in diameter, with 8, deep reddish purple, spreading lobes. Stamens 8, attached at base of perianth-tube. Stigma peltate, not completely covering perianth throat. *China.* H5. Spring.

Plants offered under this name are often *A. elatior.*

3. A. typica Baillon. Illustration: *Botanical Magazine*, 7484 (1896).

Rhizomes stout. Leaf-blades elliptic to lanceolate, 20–40 cm, with 7–9 main veins; leaf-stalks 10–25 cm. Perianth spherical, *c.* 2 cm in diameter, leathery, with 6 acute lobes, green with reddish purple spotting on outer face, purplish within. Stamens 6, attached at base of perianth-tube. Stigma peltate, 6-lobed. *China.* H5. Spring.

6. CLINTONIA Rafinesque

V.A. Matthews

Rootstock a short rhizome. Leaves basal, sheathing at the base, entire. Flowering stem erect, unbranched, leafless. Flowers borne in an umbel or raceme, occasionally solitary, bell-shaped to star-shaped. Perianth-segments free, downy outside at least at the base. Stamens borne at the base of the perianth-segments. Ovary superior. Stigma with 2 or 3 lobes. Fruit a berry with 2 or 3 cells.

A genus of 5 species from the Himalaya, E Asia and N America, all of which are in cultivation. They grow best in semi-shade.

1a. Flowers pinkish purple **4. andrewsiana**
b. Flowers white, greenish yellow or lilac 2
2a. Flowers 1 or 2; leaves usually hairy beneath **5. uniflora**
b. Flowers 3–30; leaves hairless except at the margins 3
3a. Filaments downy at the base **2. borealis**
b. Filaments hairless 4
4a. Flowers in a dense umbel; fruit black, with 2 seeds in each cell **3. umbellulata**
b. Flowers in a loose raceme; fruit blue or purplish black with 2–6 seeds in each cell **1. udensis**

1. C. udensis Trautvetter & Meyer (*C. alpina* Baker).

Leaves 2–8, 8–35 cm, oblanceolate to obovate, hairless except for the hairy margins. Flowering stem 10–85 cm, hairy, bearing a loose raceme of 2–10 more or less erect flowers. Perianth-segments 7–15 mm, yellowish green, white or lilac. Filaments hairless. Fruit dark blue to purplish black with 2–6 seeds in each cell. *Himalaya, E Asia.* H4. Summer.

2. C. borealis (Aiton) Rafinesque. Illustration: Rickett, Wild flowers of the United States **1**: 33 (1966).

Leaves 2–5, 10–30 cm, elliptic to obovate, hairless except for the hairy margins. Flowering stem to 30 cm, somewhat hairy, bearing a loose umbel of 2–8 nodding flowers. Perianth-segments 1.2–2 cm, greenish yellow. Filaments downy at the base. Fruit blue. *Eastern N America.* H4. Spring–early summer.

3. C. umbellulata (Michaux) Morong (*C. umbellata* Torrey). Illustration: Rickett, Wild flowers of the United States **1**: 33 (1966).

Leaves 2–5, 8–30 cm, oblanceolate to obovate, hairless except for the hairy margins. Flowering stem 15–30 cm, sparsely hairy, bearing a dense umbel of 5–30 erect or scarcely nodding flowers. Perianth-segments 5–10 mm, white, usually spotted with purple and green. Filaments hairless. Fruit black with 2 seeds in each cell. *Eastern N America*, H4. Late spring–summer.

4. C. andrewsiana Torrey. Illustration: Rickett, Wild flowers of the United States **5**: 47 (1971); Brickell (ed.), RHS A–Z encyclopedia of garden plants, 289 (2003).

Leaves 5 or 6, 15–25 cm, elliptic to broadly ovate, hairless except for the sparsely hairy margins. Flowering stem 25–50 cm, hairless or sparsely hairy, bearing a terminal umbel and often lateral umbels below. Perianth-segments 1–1.8 cm, pinkish purple. Filaments downy at the base. Fruit blue with 8–10 seeds in each cell. *USA (California).* H4. Summer.

5. C. uniflora (Schultes) Kunth. Illustration: Rickett, Wild flowers of the United States **5**: 47 (1971).

Leaves 2 or 3, 7–15 cm, oblanceolate to obovate, usually hairy beneath. Flowering stem 10–20 cm, hairy, bearing 1 or 2 erect flowers. Perianth-segments 1.8–2.5 cm, white. Filaments downy. Fruit blue with 2 or 3 seeds in each cell. *Western N America.* H4. Summer.

7. SMILACINA Desfontaines

V.A. Matthews

Perennial herbs with creeping rhizomes. Stems unbranched, leafy. Leaves alternate, stalkless or with short stalks. Flowers borne in terminal racemes or panicles; flower-stalks jointed. Perianth-segments free. Stamens borne at base of perianth-segments. Ovary superior. Fruit a few-seeded berry.

A genus of 25 species from N America, E Asia and the Himalaya. The species grow best in semi-shade in a soil that does not dry out. Propagation is by division. Now thought by many authors to be better placed in *Maianthemum* (p.154).

1a. Leaves hairless beneath **3. paniculata**
b. Leaves with short hairs beneath 2
2a. Flowers borne in a raceme **2. stellata**
b. Flowers borne in a panicle **1. racemosa**

1. S. racemosa (Linnaeus) Desfontaines (*S. amplexicaulis* Nuttall). Illustration: Peterson & McKenny, Field guide to wildflowers of north-eastern and north-central North America, 67 (1968); Hay & Synge, Dictionary of garden plants in colour, f. 1380 (1969); Hay & Beckett (eds), Reader's Digest encyclopaedia of garden plants and flowers, 668 (1971); Rickett, Wild flowers of the United States **6**: 26 (1973).

Stem to 90 cm, downy. Leaves elliptic to narrowly ovate, clasping or with short 1–4 mm stalks (in lower leaves), with short hairs beneath. Flowers in many-flowered panicles to 15 cm; flower-stalks 0.5–2 mm, downy. Perianth-segments 1–3 mm, white or greenish, shorter than stamens. Berry greenish to red, often mottled with red or purple. *N America, Mexico.* H4. Summer.

S. japonica Gray. Illustration: Kitamura et al., Coloured illustrations of herbaceous plants of Japan (Monocotyledoneae), pl. 28, f. 185 (1978). Lower leaves with stalks 1–1.5 cm; flower-stalks 2–5 mm, and usually densely hairy. *Japan, China, Korea.* H4. Late spring–summer.

S. oleracea (Baker) Hooker. Illustration: *Botanical Magazine*, 6313 (1877); *Journal of the Royal Horticultural Society* **43**: 416 (1893). Perianth-segments 5–7 mm, white, pink or purplish red; stamens shorter than perianth-segments; berry pinkish purple with dark spots. *Himalaya, N Burma.* H4. Summer.

2. S. stellata (Linnaeus) Desfontaines (*S. sessilifolia* (Baker) Watson). Illustration: Braun, Vascular Flora of Ohio **1**: 366 (1967); Peterson & McKenny, Field guide to wildflowers of north-eastern and north-central North America, 67 (1968); Weber, Rocky Mountain flora, 386 (1972).

Stem 20–60 cm, downy. Leaves lanceolate to oblong-lanceolate, stalkless or slightly clasping, minutely hairy beneath. Flowers in almost stalkless racemes with 6–20 flowers borne on minutely hairy stalks. Perianth-segments 3–6 mm, white or greenish, longer than stamens. Berry at first green with blue-black stripes, later turning dark red or sometimes dark blue. *N America, Mexico.* H4. Summer.

3. S. paniculata Martens & Galeotti. Illustration: *Botanical Magazine*, 8539 (1914); RHS dictionary of gardening **4**: 1966 (1951).

Stem 60–100 cm, hairless. Leaves narrowly ovate, with short stalks, hairless. Flowers in a many-flowered panicle to 15 cm. Perianth-segments 3–4 mm, white or pinkish. Berry at first green, spotted with purple, later turning red. *S Mexico to Panama.* G1. Spring.

S. trifolia (Linnaeus) Desfontaines. Illustration: Rickett, Wild flowers of the United States **1**: 37 (1966); Braun, Vascular Flora of Ohio **1**: 366 (1967); Peterson & McKenny, Field guide to wildflowers of north-eastern and north-central North America, 67 (1968). Stems to 30 cm. Leaves stalkless. Racemes 4–6 cm with 4–12 flowers. *Northern N America, Former USSR (Siberia).* H3. Summer.

8. MAIANTHEMUM Weber

A.C. Leslie

Perennial herbs, with slender creeping rhizomes, bearing solitary, long-stalked leaves. Flowering stems erect, with scales at base and 2 or 3 alternate leaves above. Flowers white, in terminal racemes. Perianth-segments 4, free, spreading or reflexed. Flower-stalks 1 to several at a node, slender, with bracts. Stamens 4, shorter than perianth. Ovary superior, 2-celled, with 2 ovules in each cell. Style simple, short; stigma slightly 2-lobed. Fruit a 1- or 2-seeded berry.

A genus of 28 species (including *Smilacina*, see above), extending around the whole north temperate region. Several are in general cultivation. Although invasive when grown amongst other herbaceous plants, they can provide effective ground cover under trees and shrubs, performing best in a humus-rich, moisture-retentive soil. Propagation is by division.

Literature: Ingram, J., Notes on Cultivated Liliaceae: 3. Maianthemum, *Baileya* **14**: 50–9 (1966); LaFrankie, J.V., Transfer of the species of Smilacina to Maianthemum (Liliaceae), *Taxon* **35**(3): 584–589 (1986).

1. M. bifolium (Linnaeus) Schmidt. Illustration: Keble Martin, The concise British flora in colour, t. 84 (1965); *Baileya* **14**: 53 (1966); Ross-Craig, Drawings of British plants **29**: pl. 15 (1972); Brickell (ed.), RHS A–Z encyclopedia of garden plants, 667 (2003).

Stems 5–20 cm, hairy above. Leaves 3–8 × 2.5–5 cm, broadly ovate, deeply cordate at base, with a broad sinus, acute to acuminate at apex, hairy beneath; at least the lower stem-leaves distinctly stalked. Racemes 1–5 cm, with 8–20 flowers. Perianth-segments 1–3 mm. Fruit 5–6 mm, pale green and spotted at first, eventually red. *W Europe to Japan.* H1. Spring.

Subsp. **kamtschaticum** (S.G. Gmelin) A.E. Murray (*M. dilatatum* (Wood) Nelson & Macbride). Illustration: *Baileya* **14**: 55 (1966); Hitchcock et al., Vascular plants of the Pacific northwest **1**: 798 (1969). Stems 1–3.5 cm, hairless. Leaves to 20 × 10 cm, hairless; at least the lower stem-leaves distinctly stalked. Racemes 2.5–7.5 cm. *Western N America, E Asia.* H1. Spring.

M. canadense Desfontaines. Illustration: Gleason, Illustrated flora of the northeastern United States and adjacent Canada **1**: 427 (1952); *Baileya* **14**: 50 (1966). Stems hairless or hairy. Leaves ovate to ovate-oblong, the base shallowly cordate, with a narrow sinus, hairy or hairless beneath; stem-leaves not or very shortly stalked. Fruit pale red. *Eastern N America.* H1. Spring.

Also recently introduced into cultivation are **M. henryi** (Baker) J.V. LaFrankie and **M. tatsiense** (Franchet) J.V. LaFrankie.

9. STREPTOPUS Richard

V.A. Matthews

Herbaceous perennials with rhizomes. Stems erect, leafy. Leaves alternate, stalkless or clasping. Flowers solitary or in pairs, nodding, borne on a stalk that is fused to the stem for some distance. Flower-stalks with a downward bend at about the middle. Perianth bell-shaped, joined at the extreme base. Stamens borne at base of perianth-lobes. Ovary superior. Style usually with 3 branches. Fruit a many-seeded berry.

7–10 species from the temperate northern hemisphere. They are best planted in a shady position in soil with plenty of humus. Propagation is by division or seed.

1a. Flowers white to greenish white; leaves clasping **1. amplexifolius**
b. Flowers pink or pinkish purple, at least at the base; leaves stalkless but not clasping **2. roseus**

1. S. amplexifolius (Linnaeus) de Candolle. Illustration: Weber, Rocky Mountain flora, 387 (1972); Grey-Wilson & Mathew, Bulbs, pl. 20 (1981).

Stem 40–100 cm, branched above, hairy or hairless. Leaves narrowly to broadly heart-shaped, clasping, hairless. Flowers in pairs on 5 cm stalks. Perianth-lobes *c.* 1 cm, greenish white, spreading widely or recurved from the middle. Anthers without horns, longer than the filaments.

Style very shortly divided into 3 at the apex. Berry orange-red to blackish crimson. *C & S Europe, N Asia, Japan, N America.* H4. Summer.

S. simplex Don. Illustration: *RHS Lily Year Book* **31**: f. 6 (1967). Flowers white, more often solitary than in pairs; style divided almost to the middle into 3 branches. *N India, Nepal, Burma, China.* G1. Summer.

2. S. roseus Michaux

Stem 25–60 cm, usually branched and often shortly hairy above. Leaves lanceolate to narrowly ovate, stalkless, margin ciliate. Flowers solitary on 1–2.5 cm stalks that may be hairy or not. Perianth-lobes 6–12 mm, pink or pinkish purple, the tips eventually recurved. Anthers 2-horned, shorter than filaments. Stigma 3-branched at the apex. Berry red. *North-eastern N America.* H3.

S. streptopoides (Ledebour) Frye & Rigg is a smaller plant, 10–30 cm, with 3–4 mm perianth-lobes that are pinkish or reddish at the base and yellowish green at the recurved tips. The stigma is not divided at the apex, but has 3 faces. The berry varies from red to maroon or black. *Northern N America, Former USSR (E Siberia), N Japan, Korea, China.* H3.

10. DISPORUM Don

V.A. Matthews

Perennial herbs with creeping rhizomes. Stems erect, leafy, branched. Leaves alternate, stalkless or shortly stalked. Flowers terminal, solitary or in a few-flowered cluster, usually drooping. Perianth-segments free, often slightly inflated or swollen at the base. Stamens borne at the base of perianth-segments; anthers opening by slits; filaments hairless. Ovary superior, 3-celled. Style 1, with an entire or 3-branched stigma, or styles 3, free. Fruit a berry.

Approximately 10 species from N America, Himalaya, E Asia and Malaysia. They grow best in semi-shade in a moist, humus-rich soil. Propagation is by seed, or division in the spring.

1a. Berry blue or black **3. sessile**
b. Berry orange or dark red 2
2a. Stamens equal to or longer than perianth **1. hookeri**
b. Stamens shorter than perianth 3
3a. Flowers yellowish green **2. lanuginosum**
b. Flowers white, greenish white or purplish 4
4a. Filaments stout; style 3-branched to halfway or more **4. cantoniense**
b. Filaments very slender; style entire or with 3 short branches **1. hookeri**

1. D. hookeri (Torrey) Nicholson.

Stem 30–100 cm. Leaves 3–14 cm, lanceolate to ovate, usually somewhat clasping, rough or shortly hairy beneath, margin with forward-pointing hairs. Flowers in clusters of 1–3, greenish white. Perianth-segments 9–20 mm, slightly inflated at the base. Stamens longer or shorter than perianth-segments; anthers hairless or hairy, shorter than filaments. Style hairless or with short hairs in the lower half. Berry red, usually with 4–6 seeds. *North-west N America.* H3. Summer.

Var. **hookeri.** Illustration: Rickett, Wild flowers of the United States **6**: t. 27 (1973); Niehaus & Ripper, Field guide to Pacific States wildflowers, 5 (1976). Leaves rough beneath; stamens usually shorter than perianth-segments; style usually hairless.

Var. **oreganum** (Watson) Jones (*D. oreganum* (Watson) Howell). Illustration: Clark, Wild flowers of British Columbia, 27 (1973). Leaves shortly hairy beneath; stamens usually longer than perianth-segments; style usually with short hairs in the lower half.

D. smithii (Hooker) Piper. Illustration: Rickett, Wild flowers of the United States **5**: 27 (1971); Niehaus & Ripper, Field guide to Pacific States wildflowers, 5 (1976). Flowers in clusters of 2–6, perianth-segments 1–3 cm; stamens shorter than perianth-segments; style shortly hairy for its full length; berry orange to red with 5–9 seeds. *Western coastal N America.* H4. Spring–summer.

D. trachycarpum (Watson) Bentham & Hooker. Illustration: Rickett, Wild flowers of the United States **6**: 27 (1973); Spellenberg, Audubon Society field guide to North American wildflowers – western region, f. 81 (1979). Leaves hairy beneath at first, losing their hairs with age, not clasping at the base; stamens equal to or slightly longer than perianth-segments; berry orange to red with 6–12 seeds. *Western N America.* H4. Summer.

2. D. lanuginosum (Michaux) Nicholson. Illustration: Rickett, Wild flowers of the United States **1**: 27 (1966); Peterson & McKenny, Field guide to wildflowers of north-eastern and north-central North America, 371 (1968).

Stem 30–90 cm. Leaves 3–12 cm, lanceolate to narrowly ovate, hairy beneath with forward-pointing marginal hairs. Flowers in clusters of 1–3, yellowish green. Perianth-segments 1.2–2 cm, slightly inflated at the base. Stamens shorter than perianth-segments; anthers shorter than filaments. Style hairless. Berry red, usually with 2 seeds, occasionally with 4–6. *Eastern central N America.* H3. Early summer.

3. D. sessile (Thunberg) Don. Illustration: Kitamura et al., Coloured illustrations of herbaceous plants of Japan (Monocotyledoneae), pl. 28, f. 187 (1978).

Stem 30–60 cm. Leaves 5–15 cm, oblong to narrowly oblong, hairless. Flowers in clusters of 1–3, white, greenish towards tip. Perianth-segments 2.5–3 cm, slightly inflated at the base. Stamens more or less equal to perianth-segments; anthers shorter than filaments. Style hairless, with 3 branches. Berry blue-black. *Japan.* H4. Late spring–early summer.

'Variegatum' has white-variegated leaves.

D. leschenaultianum Don. Illustration: *Botanical Magazine*, 6935 (1887). Flowers in clusters of 2–5, up to 1.8 cm. Stamens shorter than perianth-segments and anthers more or less equal to filaments. *S India, Sri Lanka.* G1. Early summer.

D. smilacina Gray. Illustration: Kitamura et al., Coloured illustrations of herbaceous plants of Japan (Monocotyledoneae), pl. 28, f. 188 (1978). Flowers usually solitary, erect, 1.5–1.8 cm, not inflated at the base. Berries black, drooping. *Former USSR (E Siberia), Japan.* H3. Spring.

4. D. cantoniense (Loureiro) Merrill (*D. pullum* invalid). Illustration: *Botanical Magazine*, 916 (1806) – as *Uvularia chinensis*; RHS dictionary of gardening **2**: 698 (1951).

Stem up to 90 cm, hairless. Leaves 5–15 cm, lanceolate, hairless. Flowers in clusters of 3–6, white to purplish. Perianth-segments *c.* 2.5 cm. Stamens shorter than perianth-segments; anthers shorter than filaments. Style 3-branched to halfway or more. Berry red. *Himalaya, China, E Indies.* H4. Early summer.

11. POLYGONATUM Miller

V.A. Matthews

Perennial herbs with horizontal creeping rhizomes. Stems unbranched, solitary, erect at least in lower part, leafy in upper part.

Leaves all borne on the stem, stalkless, alternate, opposite or whorled, often glaucous beneath. Flowers solitary or in clusters, borne in the leaf-axils. Bracts persistent or deciduous, usually small and membranous, rarely large and leafy. Perianth united into a tube, usually for more than half its length; corona absent. Stamens borne on the tube; anthers included, 2-lobed at the base. Ovary superior, 3-celled. Style slender, included. Stigma 3-lobed. Fruit a spherical berry, red, orange or blue-black when ripe.

About 50 species found in north temperate regions. They grow best in a fairly rich soil in a semi-shady or shady position. Propagation is by division.

Literature: Ownbey, R.P., The Liliaceous genus Polygonatum in North America, *Annals of the Missouri Botanical Garden* **31**: 373–413 (1944); Jeffrey, C., The genus Polygonatum (Liliaceae) in Eastern Asia, *Kew Bulletin* **34**: 435–71 (1980).

Upper leaves. Opposite or whorled: **1**, **2**; alternate: **2**, **3**, **4**, **5**, **6**, **7**, **8**, **9**. Hairless beneath (sometimes rough on veins): **1**, **2**, **3**, **5**, **6**, **7**, **8**, **9**; hairy on veins beneath: **4**, **6**.
Flowers. Pink or purplish: **1**, **2**, **3**; white, greenish or yellowish: **1**, **3**, **4**, **5**, **6**, **7**, **8**, **9**. Erect: **2**, **3**.
Stamens. Borne about halfway up tube: **1**, **3**, **4**, **8**, **9**; borne towards mouth of tube: **1**, **5**, **6**, **7**.
Filaments. Hairless or warty: **1**, **3**, **4**, **6**, **7**, **8**, **9**; hairy: **4**, **5**, **8**.

1a. Flowers erect 2
b. Flowers drooping 3
2a. Stem to 10 cm; leaves 1.5–4 cm; perianth-lobes 4–5 mm **3. hookeri**
b. Stem to 70 cm; leaves 7–15 cm; perianth-lobes *c.* 1.5 mm **2. roseum**
3a. Upper leaves opposite or whorled **1. verticillatum**
b. Upper leaves alternate 4
4a. Stamens borne about halfway up the perianth-tube 5
b. Stamens borne towards the mouth of the perianth-tube 7
5a. Leaves hairy on the veins beneath **4. hirtum**
b. Leaves hairless or minutely rough on the veins beneath 6
6a. Leaves sickle-shaped; fruit 3–4 mm in diameter **8. falcatum**
b. Leaves not sickle-shaped; fruit 8–12 mm in diameter **9. biflorum**
7a. Filaments downy **5. multiflorum**
b. Filaments hairless or warty 8
8a. Flowers 8–20 mm **6. odoratum**
b. Flowers 2.1–3.6 cm **7. stenanthum**

1. P. verticillatum (Linnaeus) Allioni (*P. macrophyllum* Sweet). Illustration: Ross-Craig, Drawings of British plants **29**: pl. 12 (1972); Fitter et al., Wild flowers of Britain and northern Europe, 269 (1974); Grey-Wilson & Mathew, Bulbs, pl. 20 (1981).

Stem 20–80 cm, angled, hairless or sometimes sparsely downy. Leaves opposite or whorled (lowermost alternate), 6.5–15 cm, lanceolate to narrowly ovate, minutely rough on veins beneath. Flowers drooping, solitary or in clusters of 2 or 3. Perianth 5–10 mm, white, cylindric although somewhat contracted in the middle. Stamens borne about halfway up the tube; filaments hairless. *Europe, temperate Asia.* H2. Spring–summer.

P. stewartianum Diels. Illustration: Brickell (ed.), RHS A–Z encyclopedia of garden plants, 837 (2003). Leaves (except lowermost) always in whorls, 5–10 cm and with a tendril at the apex. Perianth purplish pink; stamens borne near the mouth of the tube. *China.* H3. Summer.

P. sibiricum Delaroche. Illustration: Flora SSSR **4**: pl. 28 (1935). Stems 40–100 cm, round in section, hairless. Upper leaves in whorls, 6.5–25 cm, each with a tendril at the apex. Flowers solitary or in clusters of up to 30. Stamens borne near the mouth of the tube. *China, Mongolia, Former SE USSR.* H2. Summer.

2. P. roseum (Ledebour) Kunth. Illustration: *Botanical Magazine*, 5049 (1858).

Stem to 70 cm, round in section, hairless. Leaves alternate, opposite or in whorls of three, 7–15 cm, linear to narrowly lanceolate, often minutely rough on veins beneath. Flowers erect, solitary or in pairs. Perianth 1–1.2 cm, pinkish, cylindric with the lobes slightly reflexed, c. 1–1.5 mm. Stamens borne about halfway up the tube; filaments hairless. China, *Former USSR (Soviet Central Asia, W Siberia).* H3. Summer.

3. P. hookeri Baker. Illustration: *Hooker's Icones Plantarum* **23**: pl. 2218 (1892); *Bulletin of the Alpine Garden Society* **36**: 313 (1968); Hay & Beckett (eds), Reader's Digest encyclopaedia of garden plants and flowers, 541 (1971); Brickell (ed.), RHS A–Z encyclopedia of garden plants, 837 (2003).

Stem to 10 cm, hairless. Leaves alternate, clustered towards stem apex, 1.5–4 cm, linear to narrowly elliptic, hairless beneath. Flowers erect, solitary. Perianth 1.1–1.2 cm, pinkish, purplish, or occasionally greenish yellow, free lobes 4–5 mm, spreading. Stamens borne about halfway up the tube; filaments hairless. *E Himalaya, China.* H3.

4. P. hirtum (Poiret) Pursh (*P. latifolium* (Jacquin) Desfontaines). Illustration: Reichenbach, Icones florae Germanicae et Helveticae **7**: t. 965 (1845); Grey-Wilson & Mathew, Bulbs, pl. 20 (1981) – as *P. jacquinii*; Brickell (ed.), RHS A–Z encyclopedia of garden plants, 836 (2003).

Stem 20–120 cm, angled, sparsely downy. Leaves alternate, 7–12.5 cm, lanceolate to ovate, sparsely downy on veins beneath. Flowers drooping, solitary or in clusters of 2–5. Perianth 1–2.5 cm, white, tipped with green, cylindric. Stamens borne about halfway up the tube; filaments hairless or sparsely downy. *EC & SE Europe, Former European USSR, NW Turkey.* H3. Spring–summer.

P. humile Maximowicz (*P. officinale* '*Allioni*' var. *humile* (Maximowicz) Baker). Illustration: Flora SSSR **4**: pl. 28 (1935). Leaves not sickle-shaped, hairy on the veins beneath. Flowers solitary or in pairs. Filaments 4–5 mm. *Former USSR (Siberia), China, Japan, Korea.* H3. Summer.

5. P. multiflorum (Linnaeus) Allioni. Illustration: Ross-Craig, Drawings of British plants **29**: pl. 13 (1972); Fitter et al., Wild flowers of Britain and northern Europe, 269 (1974).

Stem 30–90 cm, round in section, hairless. Leaves alternate, 5–15 cm, lanceolate to ovate, hairless beneath. Flowers drooping, in clusters of 2–6. Perianth 9–20 mm, white, usually tipped with green, cylindric, somewhat contracted in the middle. Stamens borne towards the mouth of the tube; filaments sparsely downy. *Europe, temperate Asia.* H2. Spring–summer.

'Variegatum' has the leaves striped with white.

A hybrid between *P. multiflorum* and *P. odoratum* that displays characters intermediate between those of the parents, is the most common *Polygonatum* in cultivation, and called **P. × hybridum** Brügger (Illustration: Hay & Beckett (eds), Reader's Digest encyclopaedia of garden plants and flowers, 541, 1971; Perry, Flowers of the world, 172, 1972).

6. P. odoratum (Miller) Druce (*P. vulgare* Desfontaines; *P. officinale* Allioni; *P. japonicum* Morren & Decaisne; *P. odoratum* (Miller) Druce var. *thunbergii* (Morren & Decaisne) Hara). Illustration: Ross-Craig, Drawings of British plants **29**: pl. 14 (1972).

Stem 30–85 cm, angled, hairless. Leaves alternate, 3–15 cm, lanceolate to ovate, hairless beneath. Flowers drooping, fragrant, solitary or in clusters of 2–4. Perianth 8–20 mm, white, tipped with green, cylindric or widening towards the mouth. Stamens borne towards the mouth of the tube; filaments hairless, minutely warty or not. *Europe, temperate Asia.* H2. Spring–summer.

'Variegatum' has cream-striped leaves.

For the hybrid between *P. odoratum* and *P. multiflorum*, which may key out under *P. odoratum*, see note under *P. multiflorum*.

P. pubescens (Willdenow) Pursh. Leaves finely downy on the veins beneath; flowers yellowish green, the perianth-tube contracted at the base of the spreading lobes and minutely warty inside; filaments densely warty. *NE & NC USA, SE Canada.* H2.

7. P. stenanthum Nakai (*P. macranthum* (Maximowicz) Koidzumi).

Stem to 120 cm, round in section, hairless. Leaves alternate, 8.5–17.5 cm, lanceolate to ovate, hairless beneath. Flowers drooping, solitary or in clusters of 2–4. Perianth 2.1–3.6 cm, white, cylindric, the free lobes 4–7 mm. Stamens borne towards the mouth of the tube; filaments warty. *Japan, Korea.* H4.

8. P. falcatum Gray.

Stem 50–85 cm, hairless. Leaves alternate, 6–23 cm, narrowly lanceolate to ovate-elliptic, sickle-shaped, sometimes minutely rough on veins beneath. Flowers drooping, in clusters of 2–5. Perianth 1.1–2.2 cm, white, cylindric. Stamens borne about halfway up the tube; filaments usually hairless, occasionally hairy or warty, 5–7 mm. Fruit 3–4 mm in diameter. *Japan, Korea.* H4.

Often offered by nurserymen as *P. pumilum.*

9. P. biflorum (Walter) Elliott (*P. commutatum* (Schultes) Dietrich; *P. giganteum* Dietrich). Illustration: Rickett, Wild flowers of the United States **1**: 35 (1966).

Stem 40–200 cm, hairless. Leaves alternate, 4–18 cm, narrowly lanceolate to broadly elliptic, hairless beneath. Flowers drooping, solitary or in clusters of up to 15. Perianth 1.1–2.3 cm, whitish to greenish yellow, cylindric, with 3 or 4 mm lobes, often tipped with green. Stamens borne about halfway up the tube; filaments hairless, sometimes minutely warty. Fruit 8–12lmm in diameter. *E USA, SC Canada.* H3. Summer.

12. DISPOROPSIS Hance

S.G. Knees

Perennial Rhizome horizontally creeping, sometimes beaded, fleshy. Stem usually arching, rarely erect, simple, hairless. Leaves borne on the stem, usually alternate, shortly stalked, hairless. Inflorescences axillary, each a solitary flower or cluster of 2 to several flowers; bracts usually absent, rarely present. Perianth campanulate; segments 6, overlapping, fleshy, forming a tube for up to half of their length. Corona attached near apex of perianth tube, fleshy or membranous; lobes 6, alternate to or opposite perianth segments, apex often 2-cleft, sometimes notched, rarely entire. Anthers 6, opposite perianth segments. Ovary superior 3-celled; ovules 4–6 per cell. Style short; stigma capitate to slightly 3-lobed. Fruit a berry, several seeded.

A genus of six species, closely allied to *Polygonatum*, found China, Laos, Philippines, Thailand, Vietnam; six species (four endemic) are found in China. One species is in general cultivation and others have been introduced but may not have persisted. Propagation as *Polygonatum* (p. 155).

1. D. pernyi (Hua) Diels (*Polygonatum cyrtonema* invalid; *Aulisconema pernyi* Hua; *Disporopsis arisanensis* Hayata; *D. leptophylla* Hayata; *D. taiwanensis* S.S. Ying; *Polygonatum bodinieri* H. Léveillé; *P. ensifolium* H. Léveillé; *P. ensifolium* var. *didymocarpum* H. Léveillé). Illustration: Brickell (ed.), RHS A–Z encyclopedia of garden plants, edn 1, 373 (1996); Flora of China Illustrations, f. 265; Brickell (ed.), RHS A–Z encyclopedia of garden plants, 383 (2003).

Rhizome terete, 4–10 mm thick, green. Stem purple spotted, 20–40 cm. Leaves alternate; stalk 0.5–1.5 cm; blade lanceolate to elliptic, 5–14 × 1.2–6 cm, base rounded to obtuse, apex acuminate. Flowers usually solitary, occasionally in clusters of 2 or 3; flower-stalks 1–2 cm. Perianth white, occasionally dark reddish, 1.2–1.6 cm; tube 3–7 mm, not constricted; lobes 8–11 × 3–5 mm. Corona lobes 3–4 mm, membranous, apex deeply 2-cleft. Anthers 1.5–3 mm. Ovary subglobose, 3–5 mm. Style 2.5–3.5 mm. Berries 4–10 mm, dark purple or brown-purple at maturity, 1–3-seeded. *China.* H5. Spring.

Similar species, including **D. aspersa** (Hua) Engler, with a greenish yellow outer perianth, and **D. fuscopicta** Hance, with a white and purple outer perianth, have also become commercially available in recent years.

13. MEDEOLA Linnaeus

V.A. Matthews

Perennial herbs with horizontal, white tubers to 8 cm, smelling of cucumber. Stem simple, 20–90 cm, hairy at first, eventually becoming hairless. Leaves in 2 whorls, the lower whorl about halfway up the stem, with 5–11 lanceolate to obovate, shortly acuminate leaves; the upper whorl just below the flowers, with 3, occasionally 4 or 5, smaller, ovate leaves. Flowers 3–9, nodding, in an umbel at the top of the stem; flower-stalks to 2.5 cm, spreading or recurved, becoming more or less erect at fruiting-time. Perianth of 6, free, similar segments, 7–8 mm, pale greenish yellow, recurved. Stamens 6; filaments longer than anthers. Ovary superior. Styles 3, linear, recurved. Fruit a spherical, dark purple or red berry with 3 cells, 4–8 mm across.

A genus of only one species from North America. Cultivation as for *Trillium* (p. 160). Propagation is by seed.

1. M. virginiana Linnaeus. Illustration: Rickett, Wild flowers of the United States **1**: 25 (1966); Peterson & McKenny, Field guide to wildflowers of north-eastern and north-central North America, 103 (1968); Niering & Olmstead, Audubon Society field guide to North American wildflowers – eastern region, f. 3 (1979); Dean et al., Wildflowers of Alabama, 17 (1983).

Eastern N America. H3. Early summer.

14. OPHIOPOGON Ker Gawler

E. Campbell

Herbaceous perennials with short rhizomes and sometimes stolons as well. Roots fine or thickened, sometimes swollen and tuberous. Leaves all basal, in tufts, linear, leathery. Flowers in a terminal raceme or spike on a 3-angled or flattened scape. Perianth bell-shaped, of 6 free, equal, drooping, pale purple to white segments. Stamens 6, the filaments very short, the anthers lanceolate, acute, more or less united about the style. Ovary partly inferior, 3-celled. Style simple, straight or slightly curved; stigma small. Seeds fleshy, berry-like, exposed early, spherical or oblong, usually blue and often persistent.

A genus of about 20 species from India, China and Japan south to West Malaysia, Borneo and the Philippines. Only 3 are in general cultivation, and can be used as ground cover. They grow well in a light, sandy soil, but may require protection from cold during the winter. *O. japonicus* is sometimes grown as a submerged aquatic; it does not flower under these conditions. Propagation is by division.

1a. Tufted, leaves 1–1.5 cm broad with 9–13 veins; flower-stalks 1–1.5 cm; flowers 7–8 mm **1. jaburan**
b. Stoloniferous, leaves less than 7 mm broad, with up to 9 veins; flower-stalks to 1 cm; flowers 4–7 mm 2
2a. Leaves to 4 mm wide; scape to 12 cm; flower-stalks 2–6 mm **2. japonicus**
b. Leaves more than 4 mm wide; scape more than 12 cm; flower-stalks 5–10 mm **3. planiscapus**

1. O. jaburan (Kunth) Loddiges (*Flueggea jaburan* Kunth; *Slateria jaburan* (Kunth) Siebold; *Convallaria japonica* Linnaeus var. *major* Thunberg; *Mondo jaburan* (Kunth) Bailey).

Plant without stolons. Roots cord-like and without tubers. Leaves 30–80 × 1–1.5 cm, linear, stiff, with 9–13 veins and with rough margins. Scape 30–50 cm, flattened. Flower-stalks 1–1.5 cm. Flowers 3–8 at each node, in a loose raceme. Perianth 7–8 mm, white to pale purple. Anthers 4–5 mm. Seeds blue. *Japan*. H5–G1, Summer.

2. O. japonicus (Linnaeus) Ker Gawler (*Convallaria japonica* Linnaeus; *C. japonica* var. *minor* Thunberg; *Mondo japonicus* (Linnaeus) Farwell). Illustration: *Botanical Magazine*, 1063 (1808).

Roots slightly thickened and tuberous. Plant with stolons. Leaves mostly 10–20 cm × 2–4 mm, mostly curved, with 3–5 veins. Scape 7–12 cm, flattened. Flower-stalks 2–6 mm. Flowers 1–3 at each node in a loose raceme. Perianth 4–5 mm, pale purple to nearly white. Anthers 2.5 mm. Seeds deep blue. *Japan, China, Korea*. G1. Summer.

3. O. planiscapus Nakai (*O. japonicus* var. *wallichianus* Maximowicz).

Roots thickened, tuberous. Plant often stoloniferous. Leaves erect, 30–50 cm × 4–6 mm, with several veins. Scape 15–50 cm, weakly 3-angled near the base. Flower-stalks 5–10 mm. Flowers 1–3 at each node, in a loose raceme. Perianth 6–7 mm, pale purple or white. Anthers 2.5–3 mm. Seeds dull blue-black. *Japan*. H5–G1. Spring–summer.

'Nigrescens' is a dark, blackish green variant that is grown for its foliage, and is probably the same as the plant cultivated in America as 'Arabica'.

15. LIRIOPE Loureiro

E. Campbell

Perennial herbs with rhizomes and sometimes stolons. Leaves basal, linear, tufted. Flowers borne on a terete or flattened scape in a terminal spike or raceme; bracts small or large. Perianth of 6 free segments that are equal, ascending and persistent, forming a bell-shaped flower, purple, lilac or white. Stamens 6, filaments bent upwards in the middle; anthers oblong, obtuse. Ovary superior, 3-celled, with 2 ovules in each cell. Style columnar, bent upwards in the middle; stigma small. Fruit a capsule. Seeds 1–few, spherical, rather fleshy, usually dark purple.

A genus of about 4 species from Japan, China and Korea, of which only 2 are well known in cultivation. The genus has long been confused with *Ophiopogon* (above), and most species have names in this genus. The classification of *Liriope* is confusing, in spite of its small size.

1a. Leaves 4–7 mm broad; spike bearing 5–9 whorls of flowers, rather open; flowers pale purple to almost white **1. spicata**
b. Leaves *c.* 2 cm broad; spike bearing more than 10 whorls of flowers, very dense; flowers purple to bluish **2. muscari**

1. L. spicata (Thunberg) Loureiro (*Convallaria spicata* Thunberg; *Ophiopogon* spicatus (Thunberg) Loddiges).

Root-system shallow, roots slender, borne on a jointed stock just beneath the soil surface; some small tubers present. Leaves flexuous and grass-like, to 42 cm × 4–7 mm. Scapes slender, not exceeding the leaves. Bracts short and scarious. Flowers 20–36, in a loose spike, borne in 5–9 whorls each with 2–4 flowers. Perianth 2–5 mm across, pale purple to white. *Japan, China, Korea*. G1. Summer.

2. L. muscari (Decaisne) Bailey (*L. graminifolia* (Linnaeus) Baker var. *densiflora* Baker; L. spicata var. *densiflora* (Baker) Wright). Illustration: *Botanical magazine*, 5348 (1862); Brickell (ed.), RHS A–Z encyclopedia of garden plants, 640 (2003).

Root-system rather deep, roots slender, matted, arising from an upright stock that also bears a few thick tubers. Leaves stiff and grass-like, to 60 × 2 cm. Scape thick, flattened, usually exceeding the leaves. Bracts conspicuous, more or less leaf-like. Flowers 40–70, in a dense spike, borne in more than 10 whorls of 4–7 each. Perianth 5–8 mm across, purple to bluish. *China*.

Much confusion surrounds the name *L. graminifolia*: it has certainly been applied to both of the species included here at one time or another, and it seems likely that the species (if it exists at all in the wild) is not to be found in cultivation.

16. PELIOSANTHES Andrews

J. Cullen

Herbaceous perennials with long or short, erect or creeping rhizomes. Leaves borne at intervals along the rhizome and also frequently in a terminal tuft, conspicuously stalked, linear to oblong-ovate or obovate. Inflorescence a raceme with several to many flowers, arising from the rhizome, subtended by scale-leaves at the base; flowers 1–6 to each bract. Perianth united into a tube below, green, bluish, violet or purple, lobes 6, erect or spreading. Stamens 6, borne at the mouth of the perianth-tube, their filaments united and arching over and more or less covering the ovary; the anthers borne at the apex of the tube, slightly on the underside. Ovary usually at least partly inferior, 3-celled, with 2–4 basal ovules in each cell. Style columnar, often somewhat 3-lobed, exposed at the mouth of the staminal tube. Fruit a capsule that ruptures early, exposing the blue seeds.

A genus of a single species (see the paper by Jessop, cited below), formerly split into 2 genera (*Lourya* Baillon and *Peliosanthes* in the strict sense), with up to 12 species recognised within them. The

1 species is occasionally grown in warm glasshouses in a mixture of peat, sand and loam. Propagation is by division of the rhizomes.

Literature: Jessop, J.P., A revision of Peliosanthes (Liliaceae), *Blumea* **23**: 141–59 (1976).

1. P. teta Andrews.

Leaves 10–50 × 1–12 cm, stalks 4–50 cm. Inflorescence to 35 (rarely to 75) cm. Perianth-lobes to 8 mm. *Eastern Himalaya to SE Asia, Taiwan and China.* G2. Spring.

Jessop recognises 2 subspecies, both of which have been grown, and both of which have essentially the same distribution as the species:

Subsp. **teta.** Illustration: *Andrews' Botanical Repository* **10**: t. 605 (1810); *Botanical Magazine*, 1302 (1810). Flowers 2–6 in the axil of each bract, usually green, rarely blue.

Subsp. **humilis** (Andrews) Jessop (*P. humilis* Andrews; *P. albida* Baker; *P. violacea* Baker; *Lourya campanulata* Baillon). Illustration: *Andrews' Botanical Repository* **10**: t. 634 (1810); *Botanical Magazine*, 1532 (1813), 7110 (1890), 7482 (1896), 8276 (1909). Flowers solitary in each bract-axil, often white, bluish, violet or purple, more rarely green.

21. ASTELIACEAE

Plants terrestrial, herbaceous. Leaves well developed, borne on stems, often with a surface of scales that can be peeled off. Flowers in panicles, unisexual, whitish, actinomorphic. Perianth of 6 free segments. Stamens 6. Ovary of 3 united carpels, superior, ovules numerous, placentation axile; styles 3. Fruit a capsule.

There are 4 genera with 30 species; few are cultivated.

1. ASTELIA R. Brown

J. Cullen

Perennial herbs. Leaves evergreen, in rosettes, linear or narrowed just above the sheath, with a pronounced midrib and several other veins perceptible on the lower surface; covered with scales (at least beneath), which bear hair-like outgrowths at the base and may fuse to form a thin skin. Flowers functionally unisexual (plants dioecious), radially symmetric, borne in a panicle that may be stalked or held within the leaf rosette; primary bracts large. Perianth 6-lobed, membranous or fleshy, persistent. Stamens 6, anthers minute in female flowers. Ovary superior, 1- or 3-celled (visible in the pistillode in male flowers), ovules few to many, axile or parietal; styles 1 or 3. Fruit a berry.

A genus of about 25 species from Australia, New Zealand, Polynesia, Reunion, Mauritius and the Falkland Islands; all the cultivated species are from New Zealand, where several of them grow epiphytically. Their leaves are covered by remarkable scales whose structure is discussed and illustrated in the paper by Moore cited below.

They are grown mainly for their striking rosettes of leaves and may be cultivated in most garden soils. Propagation is by division in spring.

Literature: Skottsberg, C., Studies in the genus Astelia Banks & Solander, *Kunglinga Svenska Vetenskapsakademiens Handlingar*, ser. 3, **14**(2): 1–106 (1934); Wheeler, J.M., Cytotaxonomy of the large asteliads of the North Island of New Zealand, *New Zealand Journal of Botany* **4:** 95–113 (1966); Moore, L.B., Australasian asteliads, *New Zealand Journal of Botany* **4:** 201–40 (1966).

1a. Upper surface of mature leaf green, scaleless (at least when mature) 2
b. Upper surface of leaf covered by a skin of fused scales that obscures the green colour and peels in flakes or strips as the leaf ages 3
2a. Leaf 1–2 m, narrowed just above the sheath, drooping; lateral veins beneath 3 on either side of the midrib; ovary 1-celled **1. solandri**
b. Leaves to 1 m, arching, stiff and leathery; lateral veins beneath about 10 on either side of the midrib; ovary 3-celled **4. petriei**
3a. Leaf at most 50 cm, not densely felted beneath **5. nivicola**
b. Leaf 50–200 cm, densely felted beneath 4
4a. Leaves scarcely leathery, the upper half drooping; lateral veins beneath 3 or more on either side of the midrib **2. banksii**
b. Leaves leathery, arched stiffly outwards; lateral veins beneath 1 or 2 on either side of the midrib **3. nervosa**

1. A. solandri Cunningham (*A. cunninghamii* J.D. Hooker; *A. solandri* subsp. *hookeriana* (Kirk) L.B. Moore). Illustration: *Botanical Magazine*, 5175 (1860) & 5503 (1865).

Leaves scarcely leathery, drooping1–2 m × 2–3.5 cm, narrowed just above the sheath; upper surface green, scaleless, lower surface whitish-scaly or green dotted with whitish scale-bases; veins usually 3 on each side of the midrib beneath. Panicle 15–40 cm, borne on a stalk 30–100 cm. Perianth yellow, pink or dark red. Ovary 1-celled, placentation parietal. Berry greenish or brownish, surrounded at the base by the membranous remains of the perianth. *New Zealand.* H4–G1. Spring.

2. A. banksii Cunningham.

Leaves scarcely leathery, the upper half drooping, 1–2.5 m × 3–4.5 cm, narrow and tightly folded just above the sheath; upper surface covered by a skin of fused scales that peels off in long strips; lower surface felted, white; veins 3 or more on each side of the midrib. Panicle to 50 cm, borne on a stalk *c.* 40 cm. Perianth pale greenish cream. Ovary 3-celled, placentation axile. Berry white flushed with red-purple when ripe, the base surrounded by the brownish, membranous remains of the perianth. *New Zealand.* H4–G1. Spring.

3. A. nervosa J.D. Hooker (*A. montana* (Cheeseman) Cockayne; *A. cockaynei* Cheeseman). Illustration: Salmon, Field guide to the alpine plants of New Zealand, 207–209 (1968); Mark & Adams, New Zealand alpine plants, t. 99 (1973).

Leaves leathery, firm, arched stiffly outwards, 50–200 × 2–6 cm, narrowed and folded just above the sheath; upper surface covered with a skin of fused scales that is smoothed or somewhat ruffled, lower surface with a layer of white or brown scales overlying a layer of hairs; veins 1 or 2 on each side of the midrib. Panicle to 40 cm, borne on a stalk of up to 40 cm. Perianth greenish brown to dark red. Ovary 3-celled, placentation axile. Berry orange to red, the perianth persisting as a fleshy cup beneath it. *New Zealand.* H4–G1. Spring.

4. A. petriei Cockayne. Illustration: Mark & Adams, New Zealand alpine plants, t. 98 (1973).

Leaves leathery, stiff but arching, 25–100 × 2–6.5 cm, not narrowed above the sheath; upper surface green, without scales when mature, lower surface white-scaly; veins about 10 on either side of the midrib. Panicles 3–20 cm, the females often small, stalk short. Perianth pale green, becoming darker with age. Ovary 3-celled, placentation axile. Berry orange-yellow, borne on the fleshy remains of the perianth. *New Zealand.* H4–G1. Spring.

5. A. nivicola Cheeseman. Illustration: Mark & Adams, New Zealand alpine plants, t. 100 (1973).

Leaves erect or arching, 10–50 × 1–3 cm; upper surface with a conspicuous skin of fused scales that flakes off in older leaves, lower surface thickly felted; veins 6–many on each side of the midrib. Panicle rather small and dense, shortly stalked. Perianth greenish. Ovary 3-celled, placentation axile. Berry reddish orange. *New Zealand.* H4–G1. Spring.

22. TRILLIACEAE

Plants with rhizomes. Leaves borne on the stems, broad, opposite or in a single whorl. Flowers bisexual, actinomorphic, solitary or in umbels without spathes. Perianth of 4 or 6 free segments, rarely in 2 distinct whorls (3+3). Stamens 4 or 6. Ovary of 3–5 united carpels, superior, placentation axile or parietal. Fruit a capsule.

There are 4 genera and about 70 species. One genus is native to Europe, 2 to North America, and species of 3 are cultivated as woodland ornamentals.

1a. Flowers in an umbel **1. Scoliopus**
b. Flowers solitary 2
2a. Leaves 4 or more in a whorl; parts of the flower in 4s **2. Paris**
b. Leaves in whorls of 3; parts of the flower in 3s **3. Trillium**

1. SCOLIOPUS Torrey
V.A. Matthews

Perennial, hairless herbs with a short, slender rootstock. Leaves usually 2. Flowers borne in an umbel at the top of the short, underground stem, each flower thus appearing to be solitary. Perianth-segments 6, free, in 2 dissimilar whorls. Sepals (outer segments) broad, spreading to reflexed. Petals (inner segments) narrow, erect. Stamens 3, attached to the base of the sepals. Ovary superior. Style 1, very short. Stigmas 3, linear. Ovary 3-angled, 1-celled. Fruit a capsule.

A genus of 2 species from W USA; both are in cultivation. Cultivation as for *Trillium* (see below). Propagation is by seed.

1. S. bigelovii Torrey. Illustration: *Bulletin of the Alpine Garden Society* **35**: 190 (1967); Rickett, Wild flowers of the United States **5**: 45 (1971); Spellenberg, Audubon Society field guide to North American wildflowers – western region, f. 377 (1979).

Leaves 8–10 cm at flowering-time, subsequently growing to 20 cm, broadly elliptic, usually blotched with dark brown or purple. Flower-stalks 10–15 cm, 3-angled, tinged with red, curving down towards the ground at fruiting-time. Flowers with a strong, unpleasant smell. Sepals 1.5–2 cm, spreading or usually reflexed, pale green with longitudinal stripes of purple-brown. Petals linear, erect, purple. Stamens 5–6 mm. Stigmas 5–6 mm, becoming recurved. Capsule 1.5–1.8 cm, ellipsoid, bearing the persistent style and stigmas at the top. *W USA (California, SW Oregon).* H4. Late winter–spring.

S. hallii Watson. Illustration: Abrams, Illustrated flora of the Pacific States **1**: 450 (1923). Flower-stalks to 8 cm, petals 8–10 mm, stamens *c.* 4 mm, stigmas *c.* 2 mm. *W USA (Oregon).* H4. Spring.

2. PARIS Linnaeus
V.A. Matthews

Perennials with creeping rhizomes. Leaves 4 or more in a whorl near the top of the unbranched stem. Flowers solitary, terminal. Outer perianth-segments 4–6, green; inner perianth-segments 4–6, yellow, very narrow. Stamens 4–10 with short, flat filaments and basifixed anthers. Ovary superior. Styles free, usually 4. Fruit a fleshy, berry-like capsule.

There are about 20 species in the genus, native to temperate Eurasia. They grow best in a shady position.

1. P. quadrifolia Linnaeus. Illustration: Felsko, A book of wild flowers, pl. 150 (1956); Polunin, Flowers of Europe, f. 170 (1969); Fitter et al., Wild flowers of Britain and northern Europe, 269 (1974); Rose, The wild flower key, 407 (1981).

Stem 10–40 cm, hairless. Leaves usually 4, ovate to obovate, 5–15 cm, with short stalks. Perianth-segments usually 4, 2–3.5 cm. Fruit bluish black, spherical. *Europe, Former USSR (Caucasus, Siberia).* H3. Spring–summer.

P. polyphylla Smith. Illustration: Hutchinson, Families of flowering plants, edn 3, f. 381 (1973); Rix & Phillips, The bulb book, 105 (1981). Stem to 90 cm. Leaves 5–9, linear to oblong-lanceolate, stalks 2–3 cm. Perianth-segments 4–6, the inner 2–11 cm. Fruit greenish brown, splitting to reveal red seeds. *Himalaya, Burma, China, Thailand.* H4. Spring–summer.

3. TRILLIUM Linnaeus
V.A. Matthews

Perennial herbs with short, thick rhizomes. Stem simple, usually erect and hairless. Leaves 3, in a whorl at the top of the stem, with netted veins. Flower solitary, stalked or not. Perianth of 2 dissimilar whorls of 6 free segments. Sepals (outer whorl) 3, usually green. Petals (inner whorl) 3, varying in colour, rarely green. Stamens 6, with basifixed anthers. Styles 3. Ovary superior with 3 or 6 angles or wings. Fruit a 3-celled berry, usually spherical.

A genus of about 30 species occurring mainly in N America, with a few species from the W Himalaya to NE Asia. They require a rich, moist soil and partial shade. Propagation is by division and by seed.

Plants with aberrant flowers (e.g. with 4 sepals and 4 petals, or with stamens changed into petals or vice versa) occur in most species, and are deliberately cultivated in *T. grandiflorum* and *T. ovatum.*

Literature: Samejima, J. & K., Studies on the eastern Asiatic Trillium (Liliaceae), *Acta Horti Gotoburgensis* **25**: 157–259 (1962); Mitchell, R.J., The genus Trillium, *Journal of the Scottish Rock Garden Club* **11**: 271–83 (1969), **12**: 16–20 (1970), **12**: 115–23 (1970) & **13**: 13–19 (1972); Freeman, J.D., Revision of Trillium subgenus Phyllantherum (Liliaceae), *Brittonia* **27**: 1–62 (1975); Christian, P., Trilliums, *Bulletin of the Alpine Garden Society* **48**: 61–6, 139–43, 189–202, 339–42 (1980).

Stem. Lying along ground: **3**. Velvety or downy: **3, 8**.
Leaves. Mottled: **1, 2, 3, 4, 5, 6, 7, 8, 10, 11, 12**; not mottled: **3, 5, 7, 8, 9, 10, 12, 13, 14, 15, 16, 17, 18, 19, 20, 21, 22, 23, 24, 25, 26, 27**. Stalkless or nearly so: **2, 3, 4, 5, 6, 7, 8, 10, 11, 12, 13, 15, 18, 19, 20, 21, 23, 24, 25, 26, 27**; with a stalk 5 mm or more: **1, 9, 14, 16, 17, 22**.
Flowers. Stalkless, erect: **1, 2, 3, 4, 5, 6, 7, 8, 9, 10, 11, 12**; stalked with the flower held erect or horizontally: **16, 17, 18, 19, 20, 21, 22, 23, 24, 25, 26, 27**; stalked with the flower nodding: **13, 14, 15**.
Petals. Erect: **1, 2, 3, 4, 5, 6, 7, 8, 9, 10, 11, 12**; spreading: **16, 17, 18, 20, 21,**

22, 23, 24, 25, 26, 27; recurved or reflexed: **13, 14, 15, 19**. Less than 5 mm wide: **23**.

Basic colour of petals. White: **10, 11, 13, 16, 17, 18, 19, 20, 21, 22, 24, 25**; pink: **11, 13, 14, 16, 18, 19, 20, 21, 22, 24**; brownish, purplish or deep red: **1, 2, 3, 4, 6, 8, 9, 11, 12, 15, 20, 23, 26, 27**; yellow or greenish: **1, 2, 4, 5, 7, 8, 11, 18, 20, 23**.

1a. Flowers stalkless 2
b. Flowers with a stalk 15
2a. Leaves with stalks 5–200 mm 3
b. Leaves stalkless or nearly so 4
3a. Petals clawed at the base; sepals reflexed; stamens 1–1.5 cm **1. recurvatum**
b. Petals tapered towards the base but not clawed; sepals spreading, not reflexed; stamens 1.3–3 cm **9. petiolatum**
4a. Leaf apex acuminate 5
b. Leaf apex blunt to acute 7
5a. Petals purple or brownish purple **6. cuneatum**
b. Petals yellow to greenish yellow, sometimes purplish at the base 6
6a. Sepals 1.5–3.8 cm, apex acute **5. discolor**
b. Sepals 2.5–6.5 cm, apex blunt **7. luteum**
7a. Connectives prolonged up to 1.5 mm beyond anther sacs 8
b. Connectives prolonged 2–4 mm beyond anther sacs 14
8a. Petals white or pink 9
b. Petals red, brownish, purple, yellow or greenish yellow 10
9a. Anthers with lateral dehiscence, connectives green **10. albidum**
b. Anthers with introrse dehiscence, connectives purple **11. chloropetalum**
10a. Stigmas recurved, at least at the tips 11
b. Stigmas spreading to erect 12
11a. Leaves with minute white dots on upper surface; filaments 3–6 mm **8. viride**
b. Leaves lacking minute white dots; filaments 1.5–2.5 mm **4. underwoodii**
12a. Petals 2.5–5 cm, mainly pale yellow, never purple or red-brown **5. discolor**
b. Petals 5–10 cm, variously coloured, if less than 5 cm, then purple or red-brown 13
13a. Petals less than 1 cm wide; stamens only slightly longer than the ovary **12. angustipetalum**
b. Petals usually 1 cm or more wide; stamens about twice as long as the ovary **11. chloropetalum**
14a. Stem erect, hairless; stigmas erect **2. sessile**
b. Stem usually lying along the ground, with the leaves held more or less at ground level; stigmas spreading with recurved tips **3. decumbens**
15a. Petals less than 5 mm wide **23. govanianum**
b. Petals more than 5 mm wide 16
16a. Leaves stalkless or almost so 21
b. Leaves with stalks 17
17a. Flowers hanging, level with or below leaves 18
b. Flowers held erect above the leaves, at least at first 19
18a. Ovary deep red; petals *c.* 2 cm **13. cernuum**
b. Ovary pale green; petals 2.5–5 cm **14. catesbaei**
19a. Petals with bright or deep red marks at the base; anthers purple **16. undulatum**
b. Petals lacking red marks at the base; anthers yellow 20
20a. Flower-stalk 1.5–2.5 cm; petals white, 2.5–4.5 cm **17. nivale**
b. Flower-stalk 2.5–5 cm; petals white or pale pink, with pink-purple dots and lines, 2–2.5 cm **22. rivale**
21a. Sepals purple to purplish green, or reddish green 22
b. Sepals green, or if flushed with reddish purple, then petals not purple or purple-brown, or if petals pale pink, then flowers not smelling of wet dog 24
22a. Petals to 2 cm, ovate-circular to circular, blunt, or petals sometimes absent 23
b. Petals 3–8 cm, broadly lanceolate to ovate, acute **20. erectum**
23a. Flower-stalk 2–4 cm; sepals 1.5–2 cm **26. smallii**
b. Flower-stalk 5–7 cm; sepals 2.5–3 cm **27. amabile**
24a. Anthers red to purple 25
b. Anthers yellow or cream 26
25a. Flowers pleasantly scented, nodding; stigmas yellow **15. vaseyi**
b. Flowers smelling of wet dog, erect; stigmas purple **20. erectum**
26a. Petals overlapping and erect at the base, spreading above **18. grandiflorum**
b. Petals spreading from the base 27
27a. Flowers smelling of wet dog **20. erectum**
b. Flowers scentless or fragrant, sometimes unpleasant 28
28a. Flowers erect **25. kamtschaticum**
b. Flowers held in a more or less horizontal position 29
29a. Sepals reflexed **19. flexipes**
b. Sepals spreading 30
30a. Petals white fading to pink or red, or pink fading to whitish; stigmas erect or spreading **21. ovatum**
b. Petals white fading to pale reddish purple; stigmas short and thick, recurved **24. tschonoskii**

1. T. recurvatum Beck. Illustration: Rickett, Wild flowers of the United States **2**: 33 (1967); Peterson & McKenny, Field guide to wildflowers of northeastern and north-central North America, 241 (1968); Mohlenbrock, The illustrated flora of Illinois, flowering plants: lilies to orchids, 101 (1970); Courtenay & Zimmerman, Wildflowers and weeds, 5 (1972).

Stem 12–45 cm, hairless. Leaf-stalks 5–80 mm. Leaves lanceolate to broadly ovate or elliptic, mottled. Flower stalkless. Sepals 1.5–4 cm, lanceolate to ovate, reflexed. Petals 2–5 cm, erect, dark purple to reddish brown, or yellow, lanceolate to ovate with a distinct claw at the base. Stamens 1–1.5 cm with purple filaments; anthers purple, *c.* 1.5 mm. Ovary purple, 6-winged. Stigmas with recurved tips. *E USA.* H3. Spring.

2. T. sessile Linnaeus. Illustration: Hay & Synge, Dictionary of garden plants in colour, f. 889 (1969); *Horticulture* **50**(3): 39 (1972); Mathew, The larger bulbs, 81 (1978); Rix & Phillips, The bulb book, 103, 105 (1981).

Stem 10–30 cm, hairless. Leaves stalkless, 4–12 cm, elliptic to ovate or circular, blunt or acute, mottled. Flower stalkless, with a spicy scent. Sepals 1.3–4 cm, lanceolate to ovate, spreading, green, tinged with brown at the base. Petals 1.7–4.5 cm, erect, dark purple to greenish yellow (forma **viridiflorum** Beyer), narrowly elliptic to lanceolate. Stamens 9–24 mm, filaments shorter than the purple anthers; anther-connectives extending 2–3.5 mm beyond the anthers. Ovary purple, with 6 angles or wings. Stigmas erect. *NE USA.* H3. Spring–summer.

Many of the plants sold as *T. sessile* are *T. cuneatum.*

3. T. decumbens Harbison. Illustration: *Bulletin of the American Rock Garden Society* **39**: 112 (1981); Dean et al., Wildflowers of Alabama, 13 (1983).

Stem 5–20 cm, velvety, lying along the ground. Leaves stalkless, resting on the ground, 5–13 cm, ovate to almost circular, usually mottled with silvery-green in a V-shaped patch. Flower stalkless. Sepals 2–5 cm, lanceolate to ovate, spreading, green with a purplish flush. Petals 3.5–8 cm, erect, dark purple or reddish brown, sometimes fading to yellow-green above, oblanceolate to lanceolate, twisted. Stamens 1–2.5 cm, filaments shorter than the anthers; anther-connectives extending more than 2 mm beyond the anthers. Ovary with 6 angles or wings. Stigmas spreading with recurved tips. Fruit blackish. *SE USA (Alabama, Georgia).* H3. Spring.

4. T. underwoodii Small. Illustration: Dean et al., Wildflowers of Alabama, 15 (1983).

Stem 12–30 cm, hairless. Leaves stalkless, 5–15 cm, usually drooping, lanceolate to ovate-elliptic, acute, mottled. Flower stalkless, with a foetid smell. Sepals 2.5–5.5 cm, lanceolate to ovate, spreading. Petals 3.5–6.5 cm, erect, dark purple, maroon, brownish purple or greenish yellow, oblanceolate to narrowly elliptic. Stamens 8–20 mm, filaments 1.5–2.5 mm, shorter than the purple anthers; anther-connectives prolonged. Ovary purple, with 6 angles or wings. Stigmas recurved onto the ovary. *SE USA (Alabama, Georgia, N Florida).* H3. Spring.

5. T. discolor Hooker. Illustration: *Botanical Magazine*, 3097 (1831); Justice & Bell, Wild flowers of North Carolina, 18 (1968).

Stem 10–15 cm, sometimes more, hairless. Leaves stalkless, 5–14 cm, elliptic to almost circular, acute to acuminate, usually mottled with dark green. Flower stalkless, with a spicy scent. Sepals 1.5–3.8 cm, lanceolate to ovate, acute, spreading. Petals 2.5–5 cm, erect, pale yellow with a greenish or occasionally purplish base, oblanceolate, slightly twisted. Stamens 8–20 mm, filaments shorter than the anthers, purple. Ovary purple, with 6 ridges. Stigmas spreading to nearly erect. *SE USA (Georgia, North & South Carolina).* H3. Spring.

6. T. cuneatum Rafinesque *(T. hugeri* Small). Illustration: Rickett, Wild flowers of the United States **2**: 33 (1967); Justice & Bell, Wild flowers of North Carolina, 17 (1968); *Bulletin of the Alpine Garden Society* **48**:198 (1980).

Stem 30–60 cm, hairless. Leaves stalkless, ovate to circular, acuminate, mottled. Flower stalkless, with a spicy or musky scent. Sepals green, sometimes yellow-green or purple towards the apex. Petals *c.* 2.5 cm wide, erect, purple or brownish purple. *E USA.* H3. Spring.

Nurseries often offer plants of *T. cuneatum* under the name *T. sessile.*

7. T. luteum (Muhlenberg) Harbison (*T. viride* Beck var. *luteum* (Muhlenberg) Gleason; *T. sessile* Linnaeus var. *luteum* Muhlenberg). Illustration: Rickett, Wild flowers of the United States **2**: 33 (1967); Rix & Phillips, The bulb book, 103 (1981); Brickell (ed.), RHS A–Z encyclopedia of garden plants, 1048 (2003).

Stem 15–45 cm, hairless. Leaves stalkless, 5–15 cm, elliptic to-broadly ovate, acuminate, more or less mottled. Flower stalkless, smelling of freesias or lemon-oil. Sepals 2.5–6.5 cm, lanceolate to narrowly oblong-elliptic, blunt, spreading. Petals 3–9 cm, erect, yellow to greenish yellow, oblanceolate to obovate or narrowly elliptic. Stamens 8–25 mm, filaments shorter than the anthers. Ovary with 6 angles or ridges. *SE USA (Tennessee, Kentucky, North Carolina, N Georgia).* H3. Spring.

8. T. viride Beck. Illustration: Mohlenbrock, The illustrated flora of Illinois, flowering plants: lilies to orchids, 104 (1970); Rickett, Wild flowers of the United States **6**: 29 (1973).

Stem 20–40 cm, hairless or downy above. Leaves stalkless, 8–15 cm, narrowly to broadly elliptic, acute or blunt, slightly mottled or not, upper surface with tiny white dots. Flower stalkless, smelling of rotten fruit. Sepals 3–6 cm, lanceolate, acute, spreading to slightly deflexed. Petals 3–7 cm, erect, usually greenish or yellowish, purplish at the base, or sometimes purple or yellowish throughout, oblanceolate, sometimes twisted. Stamens 1.3–2.7 cm, filaments much shorter than the anthers. Ovary with 6 angles or wings. Stigmas with recurved tips. *North-central USA (Missouri, Illinois).* H3. Spring.

9. T. petiolatum Pursh. Illustration: Abrams, Illustrated flora of the Pacific States **1**: 451 (1923); Rickett, Wild flowers of the United States **6**: 29 (1973); Niehaus & Ripper, Field guide to Pacific states wildflowers, 257 (1976).

Stem 5–20 cm, hairless. Leaves stalked, 10–20 cm, elliptic to ovate, blunt. Flower stalkless. Sepals 2–5 cm, oblong-elliptic to oblanceolate, acute, spreading. Petals 3–5 cm, erect, incurved near tips, purple, brownish purple or greenish purple, linear to lanceolate, acute. Stamens 1.3–3 cm, filaments shorter than the anthers. Stigmas usually with recurved tips that extend between the stamens. *NW USA (Washington, Oregon, Idaho).* H3. Spring.

10. T. albidum Freeman. Illustration: Rickett, Wild flowers of the United States **5**: 37 (1971) – as *T. chloropetalum* with white petals; *Brittonia* **27**: 50 (1975).

Stem 20–60 cm, hairless. Leaves stalkless, 7–20 cm, ovate, blunt, mottled or not. Flower stalkless, sweetly scented. Sepals 2.7–7 cm, lanceolate, spreading. Petals 4–10 cm, erect, white, sometimes pink towards the base, oblanceolate to obovate. Stamens 1.5–3 cm, filaments much shorter than the anthers, pale green; anther-connectives green. Ovary green. *W USA (Washington, Oregon, N California).* H3. Spring.

11. T. chloropetalum (Torrey) Howell. Illustration: *Bulletin of the Alpine Garden Society* **42**: 21 (1974); Brickell (ed.), RHS A–Z encyclopedia of garden plants, 1048 (2003).

Stem 25–50 cm, hairless. Leaves stalkless, 8–20 cm, ovate to diamond-shaped, mottled. Flower stalkless, with a sweet or spicy scent. Sepals 3–6.5 cm, lanceolate, spreading. Petals 5–10 cm, erect, greenish yellow to purple, narrowly oblanceolate to obovate. Stamens 1.5–3.5 cm, purple, filaments much shorter than anthers, which have purple connectives. Ovary purple, more or less 6-angled. Stigmas straight, purple. Fruit reddish. *W USA (California).* H3. Spring.

Var. **giganteum** (Hooker & Arnott) Munz has petals that vary from white to red or red-purple. Illustration: Rickett, Wild flowers of the United States **5**: 37 (1971) – as *T. chloropetalum* with red petals.

12. T. angustipetalum (Torrey) Freeman (*T. sessile* Linnaeus var. *angustipetalum* Torrey).

Stem 20–65 cm, hairless. Leaves almost stalkless, 10–23 cm, ovate to broadly ovate, blunt, usually mottled. Flower stalkless, with a spicy or musty smell. Sepals 3–6 cm, narrowly lanceolate to narrowly oblong. Petals 4–5 cm, sometimes to 10 cm, to 1 cm wide, erect, dark purple or red-brown, linear. Stamens 1–2.2 cm, purple, filaments much shorter than the anthers. Ovary purple or red-brown, sharply 6-angled at the top. Stigmas dark purple, erect. *W USA (California)*. H4. Spring.

13. T. cernuum Linnaeus. Illustration: *Botanical Magazine*, 954 (1806); Rickett, Wild flowers of the United States **2**: 29 (1967); Courtenay & Zimmerman, Wildflowers and weeds, 5 (1972); *Bulletin of the Alpine Garden Society* **42**: 21 (1974).

Stem 25–60 cm, hairless. Leaves shortly stalked, diamond-shaped. Flower-stalk *c.* 2.5 cm, curving downwards below the leaves. Petals *c.* 2 cm, white or pale pink, margin wavy. Anthers dark purple. Ovary and fruit dark red. *E Canada, NE USA*. H3. Spring.

14. T. catesbaei Elliott (*T. nervosum* Elliott; *T. stylosum* Nuttall). Illustration: *RHS Lily Year Book* **25**: f. 10 (1961); Rickett, Wild flowers of the United States **2**: 31 (1967); Justice & Bell, Wild flowers of North Carolina, 19 (1968); *Journal of the Scottish Rock Garden Club* **13**: f. 2 (1972).

Stem 20–50 cm, hairless, red-tinged. Leaves shortly stalked, 5–7.5 cm, elliptic to ovate. Flower-stalk *c.* 5 cm, recurved, the flower borne level with or just below the leaves. Sepals recurved. Petals 2.5–5 cm, pale to deep pink, recurved. Ovary pale green. *SE USA*. H4. Spring–summer.

Much confusion surrounds the names *T. catesbaei*, *T. nervosum* and *T. stylosum*. Investigations presently being carried out may result in *T. nervosum (T. stylosum)* being separated from *T. catesbaei*.

15. T. vaseyi Harbison. Illustration: Rickett, Wild flowers of the United States **2**: 33 (1967); Justice & Bell, Wild flowers of North Carolina, 20 (1968).

Stem 30–60 cm, hairless. Leaves stalkless, 10–20 cm, broadly diamond-shaped. Flower stalked, fragrant, *c.* 10 cm in diameter, nodding. Petals maroon-red, overlapping, blunt. Stamens purple. Stigmas yellow. *SE USA (North & South Carolina, Georgia, Tennessee)*. H4. Spring–summer.

16. T. undulatum Willdenow (*T. erythrocarpum* Michaux). Illustration: Rickett, Wild flowers of the United States **2**: 29 (1967); Justice & Bell, Wild flowers of North Carolina, 20 (1968); *Bulletin of the Alpine Garden Society* **42**: 21 (1974) & **48**: 188 (1980); Rix & Phillips, The bulb book, 103 (1981).

Stem 10–30 cm, hairless, flushed with pink towards the base. Leaves stalked, to 15 cm or more, tapering to a sharp point. Flower borne on a short, erect stalk. Sepals *c.* 1 cm, green with red margins. Petals 2–3 cm, white or very pale pink, with a broad red band at the base that extends up the main veins, margin wavy. Filaments longer than the purple anthers. Ovary white. Fruit red. *Eastern N America*. H3. Spring–summer.

17. T. nivale Riddell. Illustration: *Botanical Magazine*, 6449 (1879); Rickett, Wild flowers of the United States **1**: 29 (1966); Courtenay & Zimmerman, Wildflowers and weeds, 5 (1972); Rix & Phillips, The bulb book, 37 (1981).

Stem 5–12 cm, hairless. Leaf-stalks 5–10 cm. Leaves 2.5–3.5 cm, broadly lanceolate, ovate or elliptic. Flower-stalk 1.5–2.5 cm, erect, reflexing with age. Sepals to 2 cm, green. Petals 2.5–4.5 cm, white. Filaments slightly shorter than the yellow anthers. Ovary green. *SE USA*. H4. Late winter–spring.

18. T. grandiflorum (Michaux) Salisbury. Illustration: *Horticulture* **35**(2): inside front cover (1957); Rickett, Wild flowers of the United States **2**: 29 (1967); *Bulletin of the Alpine Garden Society* **42**: 21 (1974) & **48**: 178 (1980); Rix & Phillips, The bulb book, 102 (1981).

Stem 20–45 cm, hairless. Leaves almost stalkless, to 30 cm, broadly ovate to almost circular. Flower-stalk *c.* 5 cm, erect. Sepals 3–5 cm, green. Petals 4–9 cm, white, gradually fading to pink, margin wavy, oblanceolate to almost circular, overlapping and erect at the base, then spreading above. Stamens 1.5–2.5 cm with greenish filaments and yellow anthers. Styles straight, almost erect. Fruit dark red. *Eastern N America*. H3. Spring–summer.

A number of forms of *T. grandiflorum* have been recognised: forma **flore pleno** Anon., has double flowers (illustration: *RHS Lily Year Book* **26**: f. 23, 1962); forma **parvum** Gates is a smaller plant than *T. grandiflorum* itself, with flowers which quickly turn pink and age to purple-pink; forma **roseum** Anon., has flowers which become pink within hours of opening and age almost to red; forma **variegatum** Smith has petals which vary from having a central green line to being wholly green with a narrow white edge (illustration: Everett, The New York Botanical Gardens illustrated encyclopedia of horticulture **10**: 3402, 1982).

19. T. flexipes Rafinesque. Illustration: Mohlenbrock, The illustrated flora of Illinois, flowering plants: lilies to orchids, 114 (1970); Rickett, Wild flowers of the United States **6**: 29 (1973).

Stem to 50 cm, hairless. Leaves stalkless, to 20 cm, acute to shortly acuminate. Flower-stalk holding the flower almost horizontally, flower usually fragrant. Sepals 2–5 cm, reflexed. Petals 2–5 cm, white or sometimes flushed with pink, spreading from the base, reflexed. Filaments shorter than the yellow or cream anthers. Ovary white or pinkish. Stigmas recurved. *Northern C USA (Kansas, Ohio, Illinois, Michigan, Missouri)*. H3. Spring.

Often sold under the name *T. cernuum*.

20. T. erectum Linnaeus. Illustration: Rickett, Wild flowers of the United States **2**: 29 (1967); *Journal of the Scottish Rock Garden Club* **12**: f. 36 (1970); *Bulletin of the Alpine Garden Society* **42**: 21 (1974); Rix & Phillips, The bulb book, 102 (1981).

Stem 15–60 cm, hairless. Leaves stalkless, broadly ovate to almost circular. Flower-stalks to 10 cm, erect, flowers smelling of wet dog. Sepals 3–5 cm, green, flushed with red, especially at the margin. Petals 3–8 cm, bright maroon-red, acute. Anthers 6–12 mm, red-purple. Ovary brown-purple. *Eastern N America*. H3. Spring-early summer.

The following variants are in cultivation: var. **album** Pursh (illustration: *Journal of the Scottish Rock Garden Club* **12**: f. 37, 1970), which has white or pale pink petals and yellow anthers; forma **albiflorum** Hoffmann, which has white petals tinged with green; forma **luteum** Louis-Marie with a stem to 12 cm and yellow petals lined with dark red.

21. T. ovatum Pursh. Illustration: *Journal of the Scottish Rock Garden Club* **12**: f. 5 (1970); *Bulletin of the Alpine Garden Society* **46**: 59 (1978); Spellenberg, Audubon Society field guide to North American wildflowers – western region, f. 49 (1979); Rix & Phillips, The bulb book, 102 (1981).

Stem 10–50 cm, tinged with red. Leaves stalkless, 5–15 cm, diamond-shaped with 5 obvious veins, acute or shortly acuminate. Flower-stalk 2–8 cm, erect. Flowers varying in smell from spicy to unpleasant. Sepals 1.5–6 cm. Petals 2.5–6.5 cm, white, fading to pink or red, ovate, spreading from the base. Anthers cream to yellow. Stigmas erect or spreading. *Western N America.* H3. Spring.

Double-flowered variants occur occasionally, as do variants with green-striped petals.

Forma **hibbersonii** Taylor & Szczawinski (*T. hibbersonii* invalid). Illustration: *Bulletin of the American Rock Garden Society* **31**: 46 (1973); *Syesis* **7**: 250 (1974); *Bulletin of the Alpine Garden Society* **42**: 21 (1974) & **48**: 138 (1980). Smaller in all its parts, with pink petals that eventually fade to whitish.

22. T. rivale Watson. Illustration: *Gardeners' Chronicle*, **54**: 43 (1913); *Botanical Magazine*, n.s., 444 (1964); *Bulletin of the Alpine Garden Society* **42**: 21 (1974) & **48**: 138 (1980).

Stem to 25 cm, usually much less, hairless. Leaf-stalks 5–20 mm. Leaves to 3 cm, narrowly ovate to ovate, acuminate. Flower-stalk 2.5–5 cm, erect. Sepals 1.2–1.6 cm, green. Petals 2–2.5 cm, white or pale pink, splashed and dotted with pink-purple, especially towards the base. Filaments white, shorter than the yellow anthers. Ovary cream. Stigmas recurved. *W USA (Oregon, California).* H4. Spring.

23. T. govanianum D. Don (*Trillidium govanianum* D. Don) Kunth. Illustration: Blatter, Beautiful flowers of Kashmir **2**: pl. 61 (1928).

Stem 10–30 cm, hairless. Leaves shortly stalked, 3.5–11 cm, ovate, acute. Flower stalked. Sepals 1–2 cm, lanceolate, green tinged with purple. Petals 1–2.5 cm, less than 5 mm wide, yellow to greenish brown or purplish. Filaments reddish, anthers yellow. Stigma reddish. Fruit purple. *Himalaya (Kashmir to Bhutan).* H3.

+24. **T. tschonoskii** Maximowicz. Illustration: *Acta Horti Gotoburgensis* **25**: pl. 3 (1962); *Journal of the Scottish Rock Garden Club* **11**: 276 (1969); Rix & Phillips, The bulb book, 105 (1981).

Stem *c.* 15 cm, hairless. Leaves stalkless, 7–17 cm, diamond-shaped to circular, acuminate. Flower stalked, held horizontally. Sepals 1–3.2 cm, broadly lanceolate to narrowly ovate, acute, greenish. Petals white, fading to pale reddish purple, slightly longer than the sepals, acute. Stamens 9–16 mm. Ovary pale green, sometimes sparsely dotted with purple. Stigmas short, thick, recurved. *Japan, Korea, SW China, Nepal.* H3. Spring–summer.

25. T. kamtschaticum Pallas. Illustration: *Acta Horti Gotoburgensis* **25**: pl. 2 (1962); *Journal of the Scottish Rock Garden Club* **11**: 274 (1969); *Bulletin of the Alpine Garden Society* **42**: 21 (1974).

Stem 25–30 cm, hairless. Leaves stalkless, 7–15 cm, diamond-shaped to ovate, acuminate. Flower stalked, erect. Sepals 2–5 cm, broadly lanceolate to ovate-oblong, green. Petals 2.5–4.5 cm, white, tinged with mauve as they age, acute. Filaments much shorter than the anthers. Fruit green, stained or spotted with purple at the top. *NE Asia.* H3. Late spring–summer.

26. T. smallii Maximowicz. Illustration: *Acta Horti Gotoburgensis* **25**: pl. 4 (1962).

Stem 20–40 cm, hairless. Leaves stalkless, 7–15 cm, diamond-shaped, acute. Flower-stalk 2–4 cm, flowers held horizontally at first, becoming erect with age. Sepals 1.5–2 cm, purplish green. Petals to 2 cm, thick, dark purple-brown, sometimes absent. Anthers 2.5–3.5 mm, brownish, often converted into petals. Ovary greenish or purple. Stigmas very short. *Japan.* H3. Spring.

27. T. amabile Miyabe & Tatewaki. Illustration: *Transactions of the Sapporo Natural History Society* **15**: 137 (1938).

Leaves stalkless, 12–15 cm, ovate to diamond-shaped. Flower-stalk 5–7 cm. Sepals 2.5–3 cm, broadly lanceolate, purple. Petals *c.* 2 cm, purple, broadly ovate to circular, sometimes absent. Anthers 6–7 mm, purple, often converted into petals or vice versa. Ovary pale yellowish, reddish at the top. *Japan.* H3. Spring.

The status of *T. amabile* is uncertain. Some botanists consider it to be an hybrid between *T. smallii* and *T. kamtschaticum*, or even the same as *T. smallii.*

23. ASPARAGACEAE

Plants herbaceous or woody. Leaves reduced to small scale-like spines on the stems, their function taken over by cladodes, which are flattened or much divided. Flowers borne among the cladodes, actinomorphic. Perianth of 6 free segments. Stamens 6. Ovary of 3 united carpels, superior; ovules many, axile. Fruit a berry.

One or possibly two genera, with about 300 species; one genus native to Europe.

1. ASPARAGUS Linnaeus

J. Cullen

Perennial herbs, shrubs or climbers, with rhizomes and usually with fusiform tubers. Leaves reduced to small scales at the nodes, their bases hardened and often projecting as spines. The function of the leaves is taken over by cladodes (leaf-like stems), which are either borne singly or in groups of 3–50 at the nodes in the axils of the scale-leaves; the cladodes may be flattened, leaf-like, awl-like or thread-like. Flowers bisexual or functionally unisexual, borne singly or in few- to many-flowered clusters among the cladodes, more rarely in terminal, umbel-like clusters, or in racemes borne on the older shoots; flower-stalks distinctly jointed. Perianth shortly united at the base, bell-shaped or with the 6 lobes widely spreading. Stamens 6. Ovary superior, 3-celled. Fruit usually a spherical berry containing 1–6 seeds.

A genus of 50–60 species from the Old World, absent from Australasia. Some 24 species are grown for the sake of their feathery shoots; some, wrongly called Smilax, are in demand as background material for bouquets, and 1 species is widely grown as a luxury vegetable. In using the account that follows, scale-leaves and their bases should be looked for on the main shoots.

Plants are relatively easy to grow in good garden soil, though the genus covers a wide range of hardiness; the plants grown in glasshouses should be given ample water and partial shade, though they require good ventilation. Propagation is mainly by seed, though cuttings can be used for some, and some can be divided at the roots.

Literature: Jessop, J.P., The genus Asparagus in southern Africa, *Bothalia* **9**: 31–96 (1966).

1a. Cladodes distinctly flattened 2
b. Cladodes rounded or angled in section, not flattened 10
2a. Cladodes 8–20 mm wide, leaf-like, with several parallel veins, borne singly **24. asparagoides**
b. Cladodes at most 7 mm wide, not as above, with a single vein or no visible veins, usually borne in clusters of 3–50 at the nodes 3
3a. Plants climbing or scrambling 4
b. Plants erect or trailing but not climbing 8
4a. Flowers borne singly or in clusters of 2 or 3 in the cladode-axils 5

b. Flowers in racemes of more than 2 or 3, borne on the older shoots 6
5a. Plant a shrub; cladodes 2.5–5 cm **8. lucidus**
b. Plant a herb; cladodes 5–15 mm **18. scandens**
6a. Cladodes 5–9 cm × 5–7 mm **21. falcatus**
b. Cladodes smaller 7
7a. Cladodes borne in a single plane on each main shoot; ridges of shoots and margins of cladodes roughened by papillae **23. drepanophyllus**
b. Cladodes borne in several planes on each main shoot; twigs and margins of cladodes smooth **22. aethiopicus**
8a. Cladodes 1–3 at each node, the cladodes of each shoot arranged in several planes; flowers in racemes **20. densiflorus**
b. Cladodes 3–5 at each node, all the cladodes on an individual shoot arranged in a single plane; flowers solitary or in pairs or in clusters of 1–4 at the ends of the twigs 9
9a. Flowers 1–2 at each node, borne among the cladodes; berry spherical **7. filicinus**
b. Flowers in clusters of 1–4 at the tips of the twigs; berry 3-lobed **17. madagascariensis**
10a. Flowers in racemes 11
b. Flowers in clusters of 1-many, borne among the cladodes at the nodes, or terminating the twigs 12
11a. Plant a climber **19. racemosus**
b. Plant erect, not climbing **20. densiflorus**
12a. Flowers in clusters of 1–4, terminating the twigs 13
b. Flowers in clusters of 1–15, borne at the nodes 14
13a. Cladodes 10 mm or more × 0.3–0.8 mm, 3- or 4-angled in section, the angles appearing as ridges which are papillose **11. umbellatus**
b. Cladodes to 8 × 0.1–0.2 mm, round in section, smooth **12. setaceus**
14a. Flowers 6-many in each cluster; spines on the older shoots conspicuous, projecting at about 90° to the shoot, more than 5 mm; old stems white **14. albus**
b. Flowers 1–3 in each cluster; spines smaller or absent, or, if large, then projecting obliquely downwards; old stems usually not white 15
15a. Young twigs white and deeply grooved 16
b. Young twigs not white or deeply grooved 17
16a. Ultimate branches bearing spines **15. laricinus**
b. Ultimate branches spineless **16. retrofractus**
17a. Cladodes 2–4-angled in section, the angles appearing on the surface as low ridges 18
b. Cladodes round in section, without ridges 21
18a. Cladodes in groups of 2 or 3 at each node **10. crispus**
b. Cladodes in groups of 5 or more at each node 19
19a. Cladodes 2–10 × 0.3–0.5 mm, glaucous, hard and sharp **6. acutifolius**
b. Cladodes 10–30 × 0.1–0.2 mm, green or dark green, soft and not sharp 20
20a. Perianth 6–8 mm; flower-stalk flexuous, exceeding the cladodes among which it is borne **1. tenuifolius**
b. Perianth 2–5 mm; flower-stalk straight, not or scarcely exceeding the cladodes among which it is borne **4. trichophyllus**
21a. Main stems conspicuously grooved and ridged, the ridges sometimes papillose 22
b. Main stems smooth, neither grooved and ridged nor papillose 24
22a. Scale-leaves not hardened at the base, the base continuing the scale and lying adpressed to the stem below the point of insertion of the leaf; cladodes in groups of 3–7 at each node **9. virgatus**
b. Scale-leaves hardened at the base, forming a tough, conical projection; cladodes in groups of 4–25 23
23a. Cladodes usually 1–6 cm × 0.5–1.2 mm, strongly 3-angled; main stem with smooth ridges **5. verticillatus**
b. Cladodes usually 5–20 × 0.2–0.5 mm, rather weakly 4-angled; main stem with papillose ridges **3. pseudoscaber**
24a. Plant erect, herbaceous; flowers 1–3 at each node **2. officinalis**
b. Plants climbing, woody; flowers 4 or more at each node **13. africanus**

1. A. tenuifolius Lamarck. Illustration: Bonnier, Flore compléte **10**: t. 596, f. 2666 (1929).

Stems to 1 m, herbaceous, erect; stem and branches smooth, the branches green. Scale-leaves with small, hardened bases that are not spine-like. Cladodes 1–3 cm × 0.1–0.2 mm, thread-like, 2–4-angled in section, soft and not sharp, green or dark green, in groups of 15–40 or more at each node. Flowers unisexual, borne in clusters of 1–3 among the cladodes; flower-stalks flexuous, exceeding the cladodes among which they are borne, jointed just below the flower. Perianth 6–8 mm. Berry 1–1.6 cm, red, containing 2–6 seeds. *Mediterranean area.* H5–G1. Summer.

2. A. officinalis Linnaeus. Illustration: Bonnier, Flore complète **10**: t. 596, f. 2667 (1929).

Stems to 1 m, herbaceous, erect; stem and branches smooth, the branches green. Scale-leaves with small, hardened, conical bases. Cladodes usually 1–2.5 cm, rarely smaller or larger, round in section, stiff, green or pale green, borne in groups of 4–25 at each node. Flowers usually unisexual, borne in clusters of 1–3 among the cladodes; flower-stalks arching downwards, mostly longer than the cladodes among which they are borne, jointed at or about the middle or just below. Perianth 4–6 mm. Berry 6–10 mm, red, containing usually 2–4 seeds. *Europe, N Africa, SW Asia; introduced elsewhere.* H1. Summer.

Widely cultivated as a luxury vegetable; variable, with a number of cultivars available.

3. A. pseudoscaber (Ascherson & Graebner) Grecescu (*A. officinalis* var. *pseudoscaber* Ascherson & Graebner).

Similar to *A. officinalis* but main stem grooved and ridged, the ridges papillose; cladodes usually 5–20 × 0.2–0.5 mm, rather weakly 4-angled in section, *Yugoslavia, Romania, Former W USSR (Ukraine).* H1. Summer.

4. A. trichophyllus Bunge.

Stems 2 m, woody, branched, erect; branches smooth, numerous, green. Scale-leaves on the main shoots with conspicuous, hardened, spine-like bases that point downwards. Cladodes 1.2–2.5 cm, 2–4-angled in section, appearing weakly ridged, stiff, pointed, pale green, borne in groups of 20–30 at the nodes. Flowers mostly unisexual, in groups of 1–3 in the axils of cladodes at the nodes; flower-stalks straight, shorter than the cladodes among which they are borne, jointed at the middle or just above. Perianth 2–3 mm. Berry 6–10 mm. *NE Asia, N China.* H2. Summer.

5. A. verticillatus Linnaeus.

Stem to 2.5 m, woody, branched, erect, conspicuously grooved and ridged, the ridges sometimes papillose. Scale-leaves with hardened bases that project downwards as small spines. Cladodes 1–6 cm × 0.5–1.2 mm, strongly 3-angled, hard, sharp, the angles papillose, borne in groups of 10–20 or more at each node. Flowers mostly unisexual, in groups of 1–3 at the nodes; flower-stalks deflexed, jointed above the middle. Perianth 2.5–4 mm. Berry 5–8 mm, black, containing 1–3 seeds. *SE Europe to Former USSR (W Siberia).* H2. Summer.

6. A. acutifolius Linnaeus. Illustration: Bonnier, Flore complète **10**: t. 595, f. 2664 (1929); Huxley & Taylor, Wild flowers of Greece and the Aegean, t. 378 (1977).

Stem to 2 m, woody, erect, white or grey, faintly ridged, the ridges papillose. Scale-leaves with the bases hardened to form downwardly pointing projections. Cladodes 2–10 × 0.3–0.5 mm, 2–4-angled in section, hard and sharp, glaucous, borne in groups of 10–30 or more at each node. Flowers mostly unisexual, borne in clusters of 1–4 among the cladodes; flower-stalks about as long as the cladodes, jointed below the middle. Perianth 3–4 mm. Berry 4–10 mm, black, containing 1 or 2 seeds. *Mediterranean area.* H5. Summer.

7. A. filicinus D. Don. Illustration: *Gardeners' Chronicle* **44**: 47, 48 (1908).

Stems to 2 m, herbaceous, erect; twigs, branches and cladodes borne in horizontal planes. Bases of scale-leaves hardened but scarcely projecting. Cladodes 6–20 × 1–2 mm, very regularly arranged, flattened (the branches looking superficially like a fern frond), dark green, in clusters of 3–5 at each node. Flowers usually unisexual, 1 or 2 (rarely more) at each node, borne among the cladodes; flower-stalks spreading, exceeding the cladodes, jointed in the upper half. Perianth 2–3 mm. Berry spherical, black. *Himalaya, SW China.* H4. Summer.

Very fern-like in appearance, and sometimes mistaken for a true fern.

8. A. lucidus Lindley.

Plant a woody climber to 3 m. Scale-leaves with hardened bases that project obliquely downwards as blunt spines. Cladodes 2.5–5 cm × 1–2 mm, flattened, bright green, borne in groups of 2–6 at each node, margins papillose. Flowers usually unisexual, in clusters of 1–3 among the cladodes; flower-stalks straight, spreading, shorter than the cladodes, jointed in the upper half. Perianth 2–2.5 mm. Berry 6–8 mm, brownish. *China, Korea, Japan, Taiwan.* H3. Summer.

9. A. virgatus Baker. Illustration: Saunders (ed.), Refugium botanicum, t. 214 (1870); Gibson, Wild flowers of Natal (coastal region), t. 11 (1975).

Stems to 1 m, woody, erect, straight, grooved and ridged. Scale-leaves not hardened at the base, the base continuing the scale and lying adpressed to the stem below the point of leaf insertion. Cladodes to 2.5 cm, round in section, borne in groups of 3–7 at the nodes. Flowers usually bisexual, in groups of up to 6 among the cladodes; flower-stalks 5–9 mm, jointed in the lower half. Perianth 3–4 mm. Berry red, containing a single seed. *Southern Africa.* G1.

10. A. crispus Lamarck. Illustration: Mason, Western Cape sandveld flowers, t. 9 (1972); *Gartenpraxis*, 53 (1983).

Stem to 1 m, herbaceous, erect or trailing, green. Scale-leaves with rather soft, spine-like projections at their bases. Cladodes 3–9 × 0.5–1 mm, 2–4-angled in section, borne in groups of 2 or 3 at the nodes. Flowers solitary, among the cladodes; flower-stalks 6–11 mm, jointed in the upper half. Perianth 4–5 mm. Berry 8–15 mm, ovoid, pale, containing 3–9 seeds. *South Africa (Cape Province).* G1.

11. A. umbellatus Link. Illustration: *Botanical Magazine*, 7733 (1900); *Lagascalia* **9**: 83 (1979).

Stems to 5 m, climbing, grooved and ridged, the ridges papillose. Scale-leaves with hardened bases that project obliquely downwards as small, blunt spines. Cladodes 10–30 × 0.3–0.8 mm, 3 or 4-angled in section, the angles appearing as papillose ridges, borne in groups of 10–30 (rarely fewer) at the nodes. Flowers usually bisexual, borne in umbel-like clusters of 1–4, terminating the twigs; flower-stalks 5–8 mm, jointed in the lower half. Perianth 5–7 mm. Berry 6–8 mm, yellow or orange, slightly 3-lobed, containing 1 or 2 seeds. *Canary Islands, Madeira.* G1. Summer.

Variable in its native habitat. A plant in cultivation under the name *A. myriocladus* misapplied (not *A. myriocladus* Baker, which is a synonym of *A. densiflorus*) is similar, but has thread-like, extremely fine cladodes. Its origin is unknown. For an illustration of the plant grown under this name, see *Gartenpraxis*, 55 (1983).

12. A. setaceus (Kunth) Jessop (*A. plumosus* Baker). Illustration: Gibson, Wild flowers of Natal (coastal region), t. 11 (1975).

Plant usually a woody climber or scrambler, often large, more rarely erect; stems smooth, green. Scale-leaves with hardened bases that project obliquely downwards as small, sharp spines. Cladodes to 10 mm, thread-like, green or dark green, borne in groups of 8–20 at the nodes. Flowers usually bisexual, solitary at each node, among the cladodes, and terminating the branches; flower-stalks 2–5 mm, jointed at about the middle. Perianth *c.* 3 mm. Berry *c.* 6 mm, red, containing 1–3 seeds. *Southern Africa.* G2. Autumn.

A commonly cultivated species, often used as background material in bouquets. It is variable, and several cultivars have been selected: the most commonly grown of these is 'Nanus' (*A. plumosus* var. *nanus* Anon.), which is of more compact, upright habit.

13. A. africanus Lamarck (*A. cooperi* Baker; *A. asiaticus* misapplied). Illustration: Gibson, Wild flowers of Natal (coastal region), t. 11 (1975); Clay & Hubbard, The Hawaii garden, tropical exotics, 91 (1977).

Plant a woody climber (rarely erect) to 3 m; stems smooth. Scale-leaves with hardened bases that project obliquely downwards as small spines. Cladodes 5–10 mm, round in section, thread-like, green or dark green, borne in groups of 8–20 at the nodes. Flowers usually bisexual, in clusters of 2 or 3, borne among the cladodes; flower-stalks straight, 5–8 mm, jointed in the lower half. Perianth 2.5–3.5 mm. Berry *c.* 6 mm, red, containing 1 seed. *Southern Africa.* G1. Spring.

14. A. albus Linnaeus.

Stem to 1 m, woody, erect, smooth or slightly ridged, white. Scale-leaves with conspicuous hardened bases that project outwards at more or less 90° from the shoot as strong, sharp spines. Cladodes 5–25 × 0.5–1.5 mm, more or less 3-angled in section, ridged, borne in groups of 10–20 at the nodes. Flowers usually bisexual, in clusters of 6–20, borne among the cladodes at the nodes; flower-stalks 3–7 mm, jointed in the lower half. Perianth 2–3 mm. Berry 4–7 mm, black, containing 1 or 2 seeds. *W Mediterranean area.* H5. Summer.

15. A. laricinus Burchell.

Stem to 2 m, woody, erect, grooved, white; young branches both grooved and white. Scale-leaves with bases hardened into sharp spines that project obliquely downwards; spines also borne on the ultimate branchlets. Cladodes 8–30 mm, round in section, rigid, borne in groups of 15–60 at the nodes. Flowers usually bisexual, in clusters of 3 or 4 among the cladodes at the nodes and sometimes in clusters of 7 or 8 terminating the branchlets; flower-stalks to 7 mm, jointed in the lower half. Perianth 2.5–3 mm. Berry *c.* 6 mm, red, containing 1–3 seeds. *South Africa, Botswana.* G1. Spring–early summer.

16. A. retrofractus Linnaeus.

Similar to *A. laricinus* but stems scrambling or weakly climbing, spines absent or few, always absent from the ultimate branchlets; berry orange, containing 1 seed. *South Africa, South West Africa.* G1. Spring–early summer.

17. A. madagascariensis Baker. Illustration: *Botanical Magazine*, 8046 (1905).

Stems to 60 cm or more, woody, erect, angled, green. Scale-leaves with hardened bases that project as downwardly hooked spines. Cladodes 1–1.7 cm × 5–7 mm, arranged in 1 plane on each shoot, flattened, lanceolate, acute, dark green, borne in groups of 3 at the nodes. Flowers usually bisexual, borne in terminal, umbel-like clusters of usually 4; flower-stalks erect, straight, jointed at or just below the middle. Perianth to 6 mm. Berry bright red, somewhat 3-lobed, containing 1–3 seeds. *Madagascar.* G2. Autumn.

18. A. scandens Thunberg. Illustration: *Botanical Magazine*, 7675 (1899); *Gartenpraxis*, 55 (1983). Brickell (ed.), RHS A–Z encyclopedia of garden plants, 146 (2003).

Plant herbaceous, scrambling or climbing; stems smooth or slightly ridged, green. Scale-leaves with small, hardened bases that scarcely project. Cladodes 5–15 × 0.8–1.6 mm, flattened, linear-lanceolate, borne in groups of 2 or 3, when 3, 1 of them longer than the other 2. Flowers usually bisexual, usually solitary, occasionally in groups of 2 or 3, among the cladodes; flower-stalks 8–12 mm, jointed in the upper half. Perianth 3–4 mm. Berry red, containing 1 seed. *South Africa (Cape Province).* G1. Summer.

19. A. racemosus Willdenow (*A. tetragonus* Bresler). Illustration: *Botanical Magazine*, 8288 (1909).

Plant a large, woody climber, to 7 m or more; main stems robust, finely ridged, papillose. Scale-leaves with hardened bases projecting as long sharp spines more or less at right angles to the shoot. Cladodes 1–2 cm × 1 mm, 3-angled in section, the angles appearing as ridges, greyish green, often strongly curved. Racemes with many flowers, borne on the older branches at nodes from which the cladodes have fallen. Perianth 2–4 mm. Berries to 6 mm, black, containing several seeds. *Himalaya, India, Sri Lanka and SW China to South Africa.* G1. Summer.

A variable species of very wide distribution.

20. A. densiflorus (Kunth) Jessop (*A. sprengeri* Regel; *A. meyeri* invalid; *A. myersii* invalid; *A. sarmentosus* misapplied). Illustration: *Botanical Magazine*, 8052 (1906); Hay et al., Dictionary of indoor plants in colour, t. 45 (1974); Gibson, Wild flowers of Natal (coastal region), t. 11 (1975); Clay & Hubbard, The Hawaii garden, tropical exotics, 93 (1977); *Gartenpraxis*, 53–4 (1983).

Erect or trailing shrubs to 1 m; stem finely ridged, green or brown. Scale-leaves with small, blunt, hardened bases that project almost vertically downwards. Cladodes 5–15 × 1–2 mm, usually flat, linear or linear-oblong, occasionally (in some cultivars) rather rounded in section, green, borne in groups of 1–3 at the nodes. Racemes with several to many flowers, borne among the cladodes at the nodes. Perianth 2–5 mm. Berry red, containing 1 seed. *South Africa.* G1. Summer.

A variable species, commonly grown, in which a number of cultivars has been recognised. 'Sprengeri', with a rather trailing, loose habit, is often seen, as is 'Myersii', in which the lateral branchlets are very regularly arranged and all more or less the same size, producing shoots that look like feathery green cylinders.

21. A. falcatus Linnaeus. Illustration: *Botanical Magazine*, 8751 (1918); Gibson, Wild flowers of Natal (coastal region), t. 11 (1975).

Stems climbing to 12 m or more, woody, smooth, grey-brown. Bases of scale-leaves hardened as large, sharp spines that project outwards and slightly downwards from the shoot. Cladodes flattened, 5–9 cm × 5–7 mm, linear-oblong, green, with a conspicuous midrib, borne in groups of 1–3 at the nodes. Flowers borne in racemes on the older shoots. Perianth 2.5–4 mm. Berry to 5 mm, containing 1 or 2 seeds. *Southern and E Africa to Sri Lanka.* G1. Summer.

22. A. aethiopicus Linnaeus (*A. ternifolius* J.D. Hooker). Illustration: *Botanical Magazine*, 7728 (1900).

Like *A. falcatus* but climbing to 7 m; cladodes 1–4 cm × 1–2 mm, borne in groups of 3–6; perianth to 3 mm; berry 5–7 mm. *C & southern Africa.* G2. Summer.

23. A. drepanophyllus Welwitsch.

Stems climbing, to 10 m, woody, faintly ridged, the ridges papillose. Scale-leaves with hardened bases that project as small, sharp spines. Cladodes flat, all borne in a single plane on each shoot, in groups of 3–5 at the nodes; the central cladode of each group 2–7 cm, the laterals somewhat smaller, all with papillose margins. Flowers in racemes borne on the older shoots. Perianth to 4 mm. Berry 1–1.2 cm, red, somewhat 3-lobed, containing 1 seed. *Tropical W&C Africa.* G2. Winter.

24. A. asparagoides (Linnaeus) Druce (*Medeola asparagoides* Linnaeus; *A. medeoloides* (Linnaeus filius) Thunberg). Illustration: *Botanical Magazine*, 5384 (1863); Mason, Western Cape sandveld flowers, t. 9 (1972); *Gartenpraxis*, 55 (1983).

Stem twining, to 1.5 m, smooth or slightly ridged. Scale-leaves with hardened but scarcely projecting bases. Cladodes flattened borne singly, 1.5–3.5 cm × 8–20 mm, leaf-like, with several parallel veins, ovate or ovate-lanceolate. Flowers solitary or in pairs in the axils of the cladodes; flower-stalks 5–8 mm, jointed in the upper half. Perianth-lobes 5–7 mm. Berry 6–8 mm, reddish, containing 1–4 seeds. *South Africa (Cape Province); naturalised in the Mediterranean area.* H5–G1. Spring.

24. RUSCACEAE

Plants woody. Leaves reduced to scales, their function taken over by flattened, leaf-like cladodes. Flowers often borne directly on cladodes in clusters, bisexual or unisexual, actinomorphic. Perianth of 6 free or united segments. Stamens 6 or 3, filaments united below. Ovary of 3 united carpels, superior, 1-celled. Fruit a berry.

Three genera with 8 species, one native to Europe.

1a. Plants climbing; flowers borne around the edges of the cladodes **2. Semele**
b. Plants not climbing; flowers not borne around the edges of the cladodes 2
2a. Flowers in clusters of 5–8 at the tips of the branches, separate from the cladodes **1. Danaë**
b. Flowers in clusters on the surfaces of the cladodes **3. Ruscus**

1. DANAË Medikus

P.F. Yeo

Evergreen shrubs with short rhizomes; stems 60–120 cm, branched. Leaves papery, those of main stems mostly about 1.5–2.5 cm, ovate-lanceolate, readily shed, those of axillary branches about 2 mm, persistent, 5 or 6 on each branch, subtending leaf-like stems (cladodes). Cladodes mostly 3–7 cm, ovate-lanceolate, asymmetric. Flowers bisexual, nearly spherical, in racemes of 5–8 at tips of branches. Perianth 2–3.5 mm, cream-coloured; segments joined for most of their length; mouth much narrower than the diameter of the flower. Stamens united into a tube enclosed within the perianth; anthers 6. Ovary superior; style slender, with 3 stigmatic lobes, reaching the mouth of the staminal tube. Fruit an orange-red berry with 1 or rarely 2 large seeds.

One species, easily distinguished from *Ruscus* in having the racemes separate from the cladodes. It forms a dense shrub with rich green, glossy foliage, but is not valuable for its berries, which are small and inconspicuous. It can be grown in sun or shade and is easily increased by division; raising it from seed is very slow. The shoots, preformed underground, emerge in spring and flower in the same season; once fully expanded their growth is ended.

1. D. racemosa (Linnaeus) Moench. (*D. laurus* Medikus). Figure 6 (1a–1f), p. 169. Illustration: Flora SSSR **4**: 445 (1935).

W. Asia. H2. Early summer.

2. SEMELE Kunth

P.F. Yeo

Evergreen climbers with short rhizomes. Stems 5–7 m, 1–2 cm thick, uniform in diameter, twining in upper part, green, bearing only scale-leaves, which soon become brown and chaffy, and axillary branches. Branches to about 90 cm, bearing ovate scale-leaves 4–8 mm long, up to about 40 in number, all or most subtending a cladode (a leaf-like stem). Cladodes mostly 2.5–7 × 1–5 cm and broadly ovate or sometimes heart-shaped, but the terminal ones ovate-lanceolate. Flowers unisexual, both sexes on the same plant, borne in clusters in notches on margins of cladodes; clusters 1-several per cladode. Perianth cream, united for about one-third of its length, lobes spreading; diameter of male flowers about 9 mm, of female about 6.5 mm. Stamens united into a tube with 6 anthers at the top, present also in female flowers but smaller, their anthers never opening. Ovary with a style topped by 3 stigmatic lobes emerging from the staminal tube. Fruit an orange-red berry containing 1 large seed.

The single species is a tall climber suitable for growing up a pillar; flowers moderately conspicuous but not colourful. The stems, unlike those of *Ruscus* and *Danaë*, are not completely preformed underground and unfold over a long period. Growth begins in spring, flowering takes place 12 months later and fruiting 9–12 months after that. Control of plant size should therefore be by removal of only a proportion of the shoots of each annual generation. Propagation is by division or by seed.

1. S. androgyna (Linnaeus) Kunth. Figure 6 (3a–3c), p. 169. Illustration: *Botanical Magazine*, 1898 (1816), 3029 (1830); RHS dictionary of gardening, **4**: 1935 (1951).

Canary Islands. G1. Early summer.

3. RUSCUS Linnaeus

P.F. Yeo

Evergreen herbs with short rhizomes and unbranched stems, or shrubs with branched stems. Leaves reduced to papery scales subtending branches or leaf-like stems (cladodes). Flowers unisexual, inconspicuous, borne in succession in clusters which are subtended by a bract and are usually solitary on the face of the cladode. Perianth-segments 6, free, the inner smaller than the outer, green, tinged or peppered with purple. Stamens united into a fleshy tube that is present in the flowers of both sexes; in the male flowers the tube is topped by stalkless anthers, represented in the female flower by minute, papery flanges. Ovary filling the space within the staminal tube, vestigial in male flowers; stigma more or less entire, protruding from the neck of the staminal tube. Fruit a red berry with thin flesh and 1–4 large seeds.

A genus of 6 species ranging from the Azores to the South Caspian region. *R. aculeatus* is suitable for tall ground-cover in dense shade; the dried shoots are frequently painted and used for decoration. The other species may be used for low ground-cover or as tub-plants. In species that are mainly dioecious, berries are occasionally produced on predominantly male plants. The plants are highly resistant to unfavourable conditions and neglect. They may be increased by division in autumn or spring; raising them from seed is very slow. The shoots, preformed underground, mostly emerge in spring and flower the following autumn or spring; once fully expanded, their growth is ended.

Literature: Yeo, P.F., A contribution to the taxonomy of the genus Ruscus, *Notes from the Royal Botanic Garden, Edinburgh* **28**: 237–64 (1968).

1a. Stem with 1 or more whorls of branches **1. aculeatus**
b. Stem unbranched or occasionally with 1 branch 2
2a. Inflorescence-bract 3.5–13 mm wide, with 5–15 veins **3. hypoglossum**
b. Inflorescence-bract not more than 3.5 mm wide, with 5 or fewer veins 3
3a. Cladodes twisted at the base so that the upper side faces downwards; staminal tube top-shaped or cup-shaped, striped and lobed **6. streptophyllum**
b. Cladodes not twisted at the base so as to invert the blade; staminal tube barrel-shaped, club-shaped or flask-shaped, without stripes or obvious lobes 4

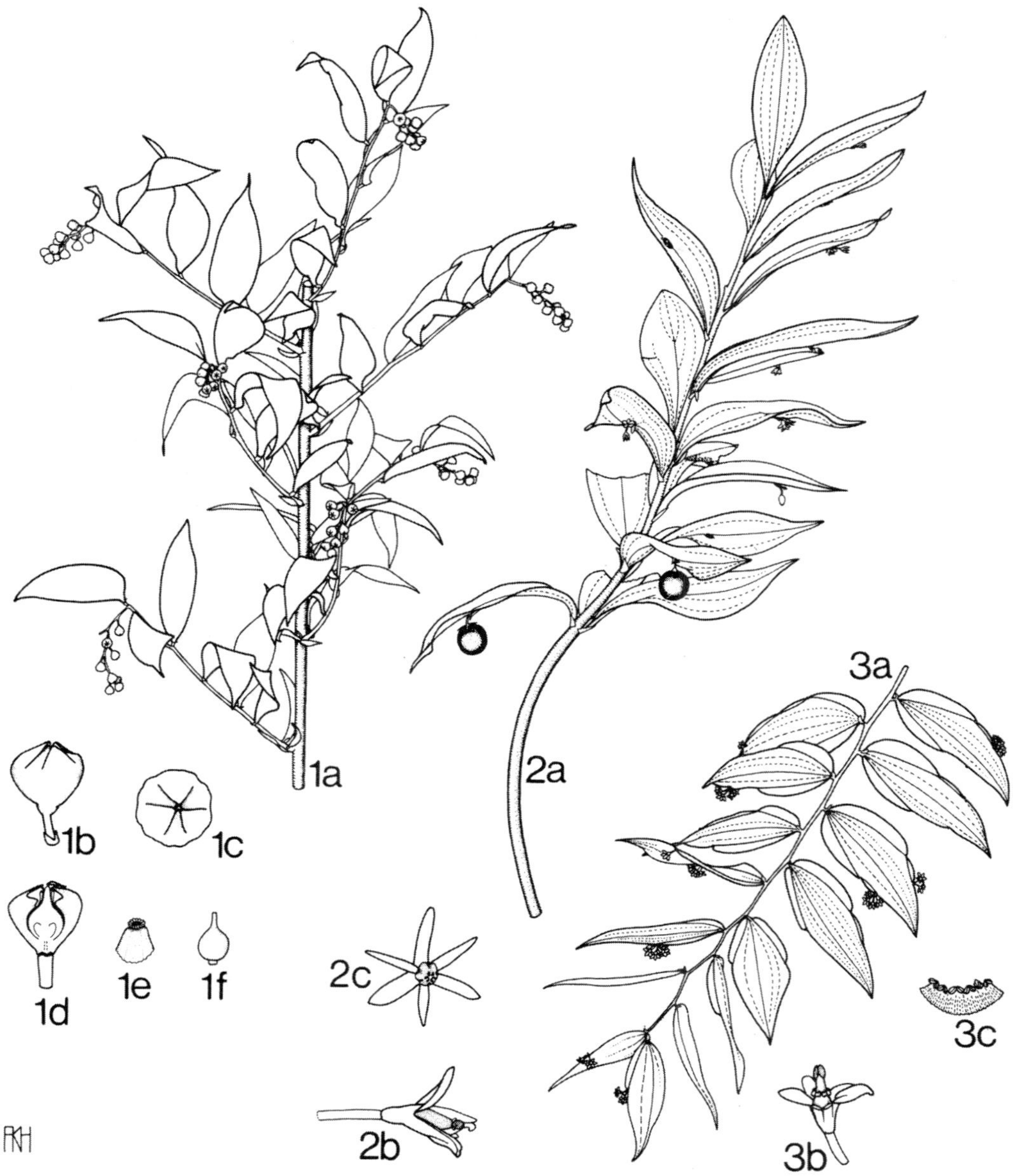

Figure 6. Shoots and flowers of *Danaë*, *Ruscus* and *Semele species*. 1, *Danaë racemosa* (a, flowering shoot; b, flower seen from the side; c, flower seen from above; d, flower in longitudinal section; e, staminal tube with anthers; f, ovary). 2, *Ruscus colchicus* (a, flowering and fruiting shoot; b, female flower, partly closed, seen from the side; c, male flower seen from above). 3, *Semele androgyna* (a, lateral branch in flower; b, functionally female flower with 2 perianth-lobes removed; c, staminal tube opened out, with sterile anthers).

4a. Cladodes with bract and flowers on underside (a cladode with these on the upper side occurs occasionally); bract with edges joined to the cladode at base; staminal tube 0.5–2.5 mm or occasionally to 3.5 mm **2. colchicus**

b. Cladodes with bract and flowers on the upper side or varying within the plant; edges of bract not evidently joined to cladode at base; staminal tube 2.75–3.5 mm 5

5a. Stems erect; cladodes usually ovate, dark green; male staminal tube thickest at the apex, female almost cylindric **4. hypophyllum**

b. Stems more or less oblique; cladodes usually obovate-lanceolate or oblanceolate, pale green; plants female; staminal tube thickest near the middle **5. × microglossum**

1. R. aculeatus Linnaeus. Illustration: Ross-Craig, Drawings of British plants **20**: t. 10 (1972); Graf, Exotica, edn 8, 1103 (1976).

Stems to 1.25 m, with several branches arising from more than 1 node. Cladodes mostly not more than 2.5 cm, very numerous, ovate, spine-tipped, rigid, twisted edge-on to stem. Plants essentially unisexual (except certain cultivated plants that produce self-fertilising, bisexual flowers giving rise to a rich crop of berries). Bract very small, on upper surface of cladode. Flowers on very short stalks. Perianth-segments spreading, outer about 2.5 mm.

Staminal tube about 2 mm, violet. Stigma spherical. *Mediterranean and Black Sea areas, extending to the Azores and N to Britain and Belgium.* H2. Autumn, spring; berries in winter.

Var. **angustifolius** Boissier is low-growing and slender and has cladodes 4–6 times as long as wide; cultivated plants are female.

2. R. colchicus Yeo. Figure 6 (2a–2c), p. 169. Illustration: *Notes from The Royal Botanic Garden, Edinburgh* **28**: 243, 244 (1968).

Stems to 60 m, unbranched, usually oblique or arching, with up to about 30 cladodes, the upper turned to one side. Cladodes 4–13.5 × 2.5–5.3 cm, broadly to narrowly ovate or elliptic to narrowly elliptic. Plants unisexual. Bract 4.5–8 × 2–3.5 mm, normally on underside of cladode, its edges joined to cladode at the base, green, with 3–5 veins. Flower-stalks 2–8 mm. Perianth-segments spreading or slightly recurved, the outer 2.5–4 mm. Staminal tube dark violet, sometimes with slightly paler stripes, in male flowers 0.5–2.5 or occasionally to 3.5 mm, top-shaped, cylindric or ovoid, in female flowers 1.25–2 mm, ovoid. Stigma spherical. *Black Sea coast of S Caucasus and NE Turkey.* H2. Autumn, spring.

3. R. hypoglossum Linnaeus. Illustration: Kerner & Oliver, Natural History of plants **1**: 333 (1894); Hegi, Illustrierte Flora von Mittel-Europa **2**: 266 (1909); *Notes from the Royal Botanic Garden, Edinburgh* **28**: 245 (1968); Brickell (ed.), RHS A–Z encyclopedia of garden plants, 942 (2003).

Stems to 40 cm, unbranched, arching, with up to 22 cladodes, mostly turned to one side. Cladodes 3–10 × 1–3.3 cm, obovate, obovate-lanceolate or broadly ovate, tapered at the base, thick, not dark in colour. Plants unisexual. Bract 1–3 cm × 3.5–13 mm, on upper side of cladode, its edges joined to cladode at base, green, with 5–15 veins. Flower-stalks 5–8 mm. Perianth-segments spreading, the outer 3–4 mm. Staminal tube violet, with faint paler bands, in male flowers 3–3.5 mm, nearly cylindric, in female about 2.5 mm, thickest at about the middle, not distinctly lobed. Stigma spherical. *Italy & Czechoslovakia to N Turkey.* H2. Autumn, spring.

4. R. hypophyllum Linnaeus. Illustration: Coste, Flore de 1a France **3**: 356 (1906); *Notes from the Royal Botanic Garden, Edinburgh* **28**: 247 (1968).

Stems to 70 cm, or, in the shade, to 1 m, erect, occasionally with one branch and with up to about 16 cladodes. Cladodes mostly 5–9 × 1.2–5.5 cm, usually ovate with a long acute apex, but sometimes other shapes, dark green. Plants bisexual. Bract usually 4.5–9 × 1–2 mm, on upper or lower surface of cladode (even on the same shoot) or sometimes in a notch at the edge, or one on each surface; its edges not joined to the cladode at the base, papery or green, with 1–3 or rarely 4 veins. Flower-stalks 5–10 mm. Perianth-segments spreading, the outer 4–4.5 mm. Staminal tube dull violet, sometimes with faint pale bands, in male flowers 3–4 mm, thickest at the tip, in female flowers 2.75–3 mm, nearly cylindric. Stigma spherical. *S Spain, N Africa, Sicily, SE France (Isle of Hyeres).* H4.

Flowers of both sexes may be produced on the same shoot, but at times shoots may bear many flowers of one sex exclusively.

5. R. × microglossum Bertoloni. Illustration: Bonnier, Flore complète **10**: t. 597 (1929); Graf, Exotica, edn 8, 1103 (1976); *Notes from the Royal Botanic Garden, Edinburgh* **28**: 246 (1968).

Stems to about 50 cm, unbranched, oblique or arching, with up to about 19 cladodes slightly turned to one side. Cladodes 4.5–10 × 1.4–3 cm, obovate-lanceolate, oblanceolate or, rarely, ovate-lanceolate, with a long acute apex, rather light green. Plants usually female. Bract 5.5–11.5 × 1.5–2.5 mm, on upper or lower surface of cladode but those of 1 shoot never all on the lower surface, its edges not or scarcely joined to cladode at base, green, with 3 or 4 veins. Flower-stalks 3–6 mm. Perianth-segments spreading, the outer about 4 mm. Staminal tube (female) 2.75–3.5 mm, rich, dark violet, broadest about the middle. Stigma spherical. *Garden origin.* H2. Flowering all year.

Evidently a hybrid between *R. hypoglossum* and *R. hypophyllum* and widely cultivated, usually under the name of one of the parents; sometimes naturalised. Most cultivated material appears to belong to a single clone.

6. R. streptophyllum Yeo. Illustration: *Botanical Magazine*, 2049 (1819); *Notes from The Royal Botanic Garden, Edinburgh* **28**: 248 (1968); Bean, Trees and shrubs hardy in the British Isles, edn 8, **4**: 242 (1980).

Stems to about 60 cm, unbranched, oblique and arching, with up to *c.* 18 cladodes in 2 rows. Cladodes 5–18 × 2.2–8 cm, broadly elliptic to ovate to narrowly elliptic or lanceolate, twisted at the base so that the upper surface faces downwards. Plants bisexual. Bract 4–11 × 1.5–2.5 mm, on upper surface of cladode, its edges not joined to the cladode at the base, papery, with 1 vein and sometimes 2 additional faint veins. Flower-stalks 6–23 mm. Outer perianth-segments about 3.5 mm, reflexed, inner divergent or spreading. Staminal tube 2.75–3.25 mm, whitish with 12 violet stripes, in male flowers top-shaped, in female cup-shaped in profile, in either sex with a 6-lobed rim and slightly hollowed apex, the anthers in male flowers occupying only the centre of the apex. Stigma mushroom-shaped. *Madeira.* G1.

25. PHILESIACEAE

Woody climbers, shrubs or subshrubs. Leaves on the stems well developed, leathery or parchment-like, not succulent, clearly stalked. Flowers conspicuous, usually large, actinomorphic. Perianth of 6 free or united segments. Stamens 6, sometimes attached to the base of the perianth. Ovary of 3 united carpels, superior, usually 1-celled; placentation usually parietal. Fruit a berry.

Seven genera and about 10 species, some cultivated for their large, attractive flowers. *Luzuriaga* is sometimes separated off as the *Luzuriagaceae*.

1a. Ovary 3-celled; flowers less than 4 cm — 2
b. Ovary 1-celled; flowers more than 4 cm — 4
2a. Inner perianth-segments densely fringed with long hairs; filaments united into a tube — **2. Eustrephus**
b. Inner perianth-segments not fringed; filaments free — 3
3a. Anthers opening by pores; flowers in cymes — **3. Geitonoplesium**
b. Anthers opening by slits; flowers solitary in the leaf-axils — **1. Luzuriaga**
4a. Plant a small upright shrub — **4. Philesia**

b. Plant climbing or twining 5
5a. Perianth-segments all equal in length **6. Lapageria**
b. Perianth-segments unequal, the outer somewhat shorter than the inner **5. × Philageria**

1. LUZURIAGA Ruiz & Pavon
P.F. Yeo
Slightly woody trailing plants with rooting stems, sometimes climbing trees, producing simple or branched lateral leafy shoots. Lowest leaves of lateral shoots often scale-leaves. Photosynthetic leaves jointed to the stems, very shortly stalked; blades with parallel veins and a few cross-veins, lower surface dark green but facing upwards, either because the shoot is pendent or because the leaf-stalk is twisted, upper surface (facing downwards) strongly glaucous except for green stripes following the veins. Flowers usually solitary, axillary, their stalks about as long as the perianth. Perianth-segments 6, free, white. Stamens 6; anthers basifixed or dorsifixed and versatile. Ovary superior, 3-celled, ovules few in each cell; style simple. Floral parts becoming spotted with orange-brown when dry. Fruit a berry.

Four species in Peru, Chile, the Falkland Islands and New Zealand. Slightly frost-tolerant plants with beautiful white flowers. They need cool, moist conditions.

Literature: Gardner, M.F., Philesia magellanica and notes on its temperate South American relatives, *New Plantsman* **2**(1): 38–42 (1995).

1. L. radicans Ruiz & Pavon. Illustration: *Botanical Magazine*, 6465 (1879); *New Plantsman* **2**(1): 41 (1995).

Aerial stems sparsely branched. Leaves 1–3.5 cm × 3–12 mm, linear-oblong to elliptic-ovate or broadly elliptic. Flowers few, solitary or rarely 2 or 3 together, nodding, fragrant. Perianth-segments *c.* 1.2–1.5 cm × 3–8 mm, narrowly elliptic, widely spreading. Stamens with thick filaments; anthers much longer than the filaments, basifixed, erect. Berry orange or red. *Chile.* H4. Summer.

L. polyphylla (W.J. Hooker) Macbride (*L. erecta* Kunth). Illustration: *Botanical Magazine*, 9192 (1860). Has numerous closely set branches and numerous closely set uniformly elliptic leaves, most of them subtending flowers; filaments longer than anthers; anthers dorsifixed, reflexed. *Chile.* H5. Summer.

2. EUSTREPHUS R. Brown
P.F. Yeo
Hairless, evergreen climbers to several metres, twining when young, growing from a small, tuberous rootstock. Foliage borne on branched lateral shoots arising from axils of scale-leaves on main stem. Leaves alternate, set in 2 rows, the lower ones on each branch being scale-leaves. Foliage-leaves jointed to the branch, very shortly stalked, with blades 5–10 cm, ovate to narrowly oblong, with upper surface facing upwards. Flowers bisexual, solitary, or several together in axils of upper leaves, nodding, on slender stalks *c.* 3 cm long, which are jointed about the middle. Perianth-segments 6, *c.* 6 mm, ovate or lanceolate, spreading, white, pink or pale purple, inner densely fringed with long hairs. Stamens 6, slightly shorter than perianth-segments; filaments very short, united into a tube; anthers erect in centre of flower, yellow. Ovary superior, 3-celled, with several ovules in each cell; style 1; stigma not or slightly lobed. Fruit to 1.3 cm in diameter, berry-like but eventually dehiscent, orange. Seeds black.

One species in Australasia. It requires protection from frost and, in summer, shade for part of the day and absence of very high temperatures. In hot exposed conditions it forms a bush. It is propagated by seed.

Literature: Schlittler, J., Die Gattung Eustrephus R. Br. ex Sims und Geitonoplesium (R. Br.) A. Cunn. Morphologischanatomische Studie mit besonderer Berucksichtigung der systematischen, nomenclatorischen und arealgeographischen Verhaltnissen, *Berichte der Schweizerischen botanische Gesellschaft* **61**: 175–239 (1951).

1. E. latifolius R. Brown. Illustration: *Botanical Magazine*, 1245 (1809); Baglin & Mullins, Australian wildflowers in colour, 105 (1969); Wrigley & Fagg, Australian native plants, 321 (1979), fruit; Morley & Toelken, Flowering plants in Australia, 339 (1983).

E Australia, New Guinea, New Caledonia. G1. Spring, summer.

3. GEITONOPLESIUM R. Brown
P.F. Yeo
Slender-stemmed, hairless, evergreen plants climbing to a height of several metres by twining. Foliage borne on branched lateral shoots that arise singly or in small groups from clusters of scale-leaves on the main stem. Leaves alternate, set in 2 rows, shortly stalked, jointed to the stem, lower surface darker than upper but facing upwards, either because the shoot is pendent or because the leaf-stalk is twisted. Flowers bisexual, in sparse cymes at tips of branches. Perianth-segments 6, entire. Stamens 6; filaments much shorter than the anthers, flattened and enlarged towards base; anthers attached just above their base, opening by 2 pores. Ovary 3-celled, superior, with a few ovules in each cell; style simple. Fruit berry-like, sometimes eventually dehiscent. Seeds black.

One very variable species, or perhaps 2, in SE Asia and Australia. A graceful, easily grown glasshouse climber requiring a minimum winter night temperature of 12 °C.

Literature: Schlittler, J., Die Gattung Eustrephus R. Br. ex Sims und Geitonoplesium (R. Br.) A. Cunn. Morphologischanatomische Studie mit besonderer Berucksichtigung der systematischen, nomenclatorischen und arealgeographischen Verhaltnissen, *Berichte der Schweizerischen botanischen Gesellschaft* **61**: 175–239 (1951).

1. G. cymosum (R. Brown) W.J. Hooker. Illustration: *Botanical Magazine*, 3131 (1832); Morley & Toelken, Flowering plants in Australia, 338 (1983).

Leaves commonly 2–8 cm × 4–8 mm and narrowly lanceolate, but very variable. Flowers nodding, bell-shaped. Perianth-segments *c.* 6 × 2 mm, green with whitish edges, sometimes tinged with purple or red, concealing the stamens. Fruit to 1.2 cm in diameter, black or dark blue. *SE Asia & Philippines to warm-temperate E Australia & Fiji.* G1.

4. PHILESIA Jussieu
P.F. Yeo
Shrub, 15–30 cm or, in nature, to 120 cm, with underground stolons. Main stem bearing scale-leaves and branches. Branches angled, with scale-leaves at base and shortly stalked photosynthetic leaves above. Leaf-blades pinnately veined, hard in texture, dark green above, glaucous beneath, recurved at the edges. Flowers solitary or few at the ends of the leafy branches, nodding, subtended by several pale green scale-leaves, overlapping each other and the base of the flower. Perianth of 3 outer sepal-like segments and 3 inner petal-like segments, the inner slightly

united at the base and each with a basal nectarial pouch. Stamens with filaments slightly united to perianth at base and united to each other for one-quarter to two-thirds of their length; anthers basifixed, opening by slits. Ovary 1-celled with 3 parietal placentas, ovules many; style club-shaped and slightly lobed at apex. Fruit a berry. Perianth becoming spotted with orange-brown when dry.

A single species from Chile. It requires peaty soil and, in dry districts, shade. It is propagated by division or layering, or from seed.

Literature: Gardner, M.F., Philesia magellanica and notes on its temperate South American relatives, *New Plantsman* **2**(1): 38–42 (1995).

1. P. magellanica Gmelin (*P. buxifolia* Willdenow). Illustration: *Botanical Magazine*, 4738 (1853); Bean, Trees and shrubs hardy in the British Isles, edn 4, **3**: 148 (1976); *New Plantsman* **2**(1): 39 (1995).

Leaf-blades 1.5–3.5 cm × 3–8 mm. Flowers narrowly bell-shaped. Outer perianth-segments 1.5–2.2 cm, narrowly ovate, obtuse, greenish or pinkish; inner 4.5–6.5 cm, purplish red. Stamens nearly as long as perianth. Style slightly longer than stamens. *Chile*. H5. Summer.

5. × PHILAGERIA Masters
P.F. Yeo
Similar to *Lapageria* in habit but much less vigorous. Leaf-blades 3-veined. Flowers arranged as in *Philesia*, similar to those of *Lapageria*, but outer perianth-segments not more than two-thirds the length of inner; stamens united to each other and to the perianth for *c.* 4 mm.

An artificial hybrid between *Lapageria* and *Philesia*. It requires peaty soil and cool, moist conditions and it resents disturbance. Propagation is by layering the new basal shoots; the layers have to be treated very carefully when being transplanted.

1. × P. veitchii Masters. Illustration: *Botanical Magazine*, n.s., 92 (1950).

Leaf-stalks 5–7 mm; leaf-blades 4–5.5 × 1–1.5 cm, lanceolate. Outer perianth-segments *c.* 3.8 × 1.6 cm, dark red with a bluish bloom; inner 6 × 3 cm, deep red with a slight bluish bloom. Stamens *c.* 4.3 cm. Style *c.* 4.8 cm, deep pink. *Known only in cultivation*. H5. Summer.

An artificial hybrid between *Lapageria rosea* and *Philesia magellanica*.

6. LAPAGERIA Ruiz & Pavon
P.F. Yeo
Woody climber, spreading strongly by underground stolons. Stems thick, twining, branched, to *c.* 10 m. Leaves alternate, with stalks to 1 cm; blades leathery, with parallel main veins and net-veins between them. Flowers 1–3 on short, scaly shoots in axils of upper leaves, pendent. Perianth-segments 6, entire, each with a pouched nectary at the base. Stamens 6, free from one another or very slightly joined at the base; filaments subulate; anthers basifixed, opening by slits. Ovary superior, 1-celled, with 3 parietal placentas, ovules many; style club-shaped and slightly lobed at apex. Fruit a berry. Seeds pale yellow or brownish.

A handsome climber for a cool greenhouse or walls and trellises outdoors. It needs cool, moist conditions and is better planted out than kept in pots.

Propagation is by seed, layering or pegging a detached and coiled shoot onto the surface of a sandy and peaty soil containing a little loam, and providing warm humid conditions. The berry is edible.

Literature: Gardner, M.F., Philesia magellanica and notes on its temperate South American relatives, *New Plantsman* **2**(1): 38–42 (1995).

1. L. rosea Ruiz & Pavon. Illustration: *Botanical Magazine*, 4447 (1849), 4892 (1856) – as 'Albiflora'; *Gartenflora*, t. 1445 (1897) – as 'Illsemannii'; Parey's Blumengärtnerei **1**: 323 (1958); *New Plantsman* **2**(1): 42 (1995).

Leaves 6–12 × 2–5 cm, ovate or ovate-lanceolate. Flowers bell-shaped, purplish red with small pale spots, varying to nearly white. Perianth-segments 6.5–9.5 cm, inner much wider at apex than outer. Stamens four-fifths as long as perianth. Style slightly longer than stamens. *Chile*. H5. Summer–winter.

26. SMILACACEAE

Dioecious, woody climbers or more rarely herbaceous plants. Leaves on the stems, leathery or parchment-like, not succulent, clearly stalked, stalks bearing 2 tendrils. Flowers inconspicuous, actinomoprhic, unisexual. Perianth of 6 free segments. Stamens 6. Ovary of 3 united carpels, superior; ovules few, axile. Fruit a berry.

Three genera with about 320 species; one genus native to both Europe and North America.

1. SMILAX Linnaeus
J.C.M. Alexander
Woody or herbaceous, evergreen or deciduous, perennial climbers or scramblers with rhizomes or tubers. Stems terete or angled, usually spiny at least below, branched. Leaves alternate, simple, sometimes shallowly lobed, papery to leathery, sometimes spiny on the margins, with 3–9 prominent veins interconnected by net-veins; lower leaves reduced to scales; leaf-stalks often very short, bearing a pair of tendrils (modified stipules) near the base, and sometimes a pair of stipule-like auricles. Flowers white to pale green, yellow or brown, lateral, in clusters, racemes, umbels, racemes of umbels, or solitary. Male and female flowers on separate plants. Perianth-segments free; male flowers with 6 free stamens borne at the base of the perianth-segments; female flowers with up to 6 staminodes and a superior ovary on which 1–3 stigmas are borne directly (styles absent). Fruit a spherical to ovoid, red, blue or black berry with 1–6 seeds.

About 200 species, mostly tropical but with some representatives in temperate Eurasia and N America. Several species yield sarsaparilla, formerly used medicinally but now only used for flavouring drinks and sweets. The Smilax of florists is *Asparagus asparagoides* (p. 167). The flowers are not beautiful, and most species are not reliably hardy, but can be effective if allowed to clamber through large trees. They are not particular as to soil`type and can be propagated by division, from seed or from half-ripe cuttings.

Literature: Killip, E.P. & Morton, C.V., A revision of the Mexican and Central American species of Smilax, *Carnegie Institute of Washington Publication* **461**, *Botany of the Maya area*: No. 12 (1936); Koyama, T., Materials towards a monograph of the genus Smilax, *Quarterly Journal of the Taiwan Museum* **13**: 1–61 (1960); Morton, C.V., Mexican Smilax, *Brittonia* **14**: 299–309 (1962); Duncan, W.H., The woody vines of the SE States, *Sida* **3**: 1–76 (1967).

1a. Herbaceous plants, lacking spines; flowers foetid **1. herbacea**

b. Woody plants, usually spiny at least at base; flowers not foetid 2
2a. Leaves glaucous beneath 3
b. Leaves not glaucous beneath 4
3a. Leaf-bases truncate to rounded; auricles shorter than leaf-stalks **2. glauca**
b. Leaf-bases usually cordate; auricles as long as leaf-stalks **3. discotis**
4a. Leaves strongly mottled with white or pale green **7. argyrea**
b. Leaves not strongly mottled 5
5a. Flowers in spikes or racemes, of umbels or clusters, or solitary 6
b. Flowers in a single umbel 8
6a. Mature leaves tapered to an acute or obtuse angle at base **4. smallii**
b. Mature leaves rounded, truncate, cordate, sagittate or hastate at base 7
7a. Flowers in clusters on axillary or terminal branches; leaves almost always spiny on margin **5. aspera**
b. Flowers in racemes of umbels or solitary; leaves not spiny on margin, sometimes spiny on veins beneath **6. regelii**
8a. Mature leaves tapered to an acute or obtuse angle at base 9
b. Mature leaves rounded, truncate or cordate at base 11
9a. Deciduous, rarely partially evergreen; berries red **8. walteri**
b. Evergreen, berries black 10
10a. Leaves papery **4. smallii**
b. Leaves thick and leathery **9. laurifolia**
11a. Leaf-tips rounded or shallowly notched **10. china**
b. Leaf-tips acute to obtuse 12
12a. Branches not spiny 13
b. Branches spiny 14
13a. Leaf-margins smooth; berries red or white **8. walteri**
b. Leaf-margins rough or finely toothed; berries bluish black **11. hispida**
14a. Stems terete with slightly raised lines; berries red **12. excelsa**
b. Stems 4-angled; berries blue to black **13. rotundifolia**

1. S. herbacea Linnaeus. Illustration: *Botanical Magazine*, 1920 (1817); Rickett, Wild flowers of the United States **1**: 45 (1966); Justice & Bell, Wild flowers of North Carolina, 16 (1968).

Herbaceous. Stem to 3 m, lacking spines, much branched, dying down each winter. Leaves 5–12 × 3–12 cm, triangular to ovate or lanceolate, rounded to shortly pointed at the tip, rounded to truncate or slightly cordate at the base, leathery, with 7–9 veins. Leaf-stalk to 8 cm. Flowers few to numerous, greenish, foetid; umbel-stalk 10–30 cm. Perianth-segments 3.5–6 mm. Berries almost spherical, *c.* 1 cm across, blue-black. *Eastern N America.* H2. Summer.

2. S. glauca Walter. Illustration: *Sida* **3**: 38 (1967).

Deciduous or semi-evergreen. Stems spiny, especially below, often glaucous; spines stiff and slender; young branches terete, green or brown. Leaves 4–13 × 3.5–10 cm, ovate to lanceolate, acute to obtuse at the tip, rounded to truncate or slightly cordate at the base, glaucous and papillose beneath, sometimes also above, with *c.* 7 veins, often variegated; auricles shorter than the leaf-stalks. Flowers 5–11 per umbel, yellow to brown; umbel-stalk 1–1.5 cm, arching, flattened, usually longer than the leaf-stalk. Berries spherical, 5–8 mm across, black, glaucous. *SE USA*, H2. Spring–summer.

Descriptions and illustrations in many European books and journals, e.g. *Botanical Magazine*, 1846 (1816) and RHS Dictionary of Gardening, differ in several respects from those in N American works. As *S. glauca* is native to N America, the American concept of the species has been followed here.

3. S. discotis Warburg.

Deciduous. Stems angled, with hooked spines to 4 mm. Leaves 4–10 × 2–5 cm, ovate, acute to obtuse at the tip, usually cordate at the base, glaucous beneath, with 3–5 veins. Leaf-stalk 2–4 cm; auricles large, equalling the leaf-stalks. Flowers greenish yellow; umbel-stalk to 4 cm. Berries blue-black. *China.*

4. S. smallii Morong (*S. lanceolate* misapplied). Illustration: *Sida* **3**: 39 (1967); Radford et al., Manual of the vascular flora of the Carolinas, edn 2, 288 (1968).

Evergreen. Stems to 3 m, terete and hairless, spiny below; spines 5–6 mm, upper stems spineless. Leaves to 15 × 7 cm, usually much smaller, lanceolate to ovate, abruptly acuminate at the tip, gradually tapered at the base, papery, hairless, with 5 veins, shiny dark green above, duller and paler beneath, not spiny. Leaf-stalks to 1.6 cm. Flowers few to numerous, in umbels, or spikes or racemes of umbels, green; umbel-stalks 1–10 mm, shorter than or equal to leaf-stalks. Perianth-segments: male 5–6 mm, female 3–4 mm. Styles 3. Berries spherical, 5–7 mm across, black. *SE USA, E Mexico & C America.* H5–G1. Summer.

5. S. aspera Linnaeus (*S. maculata* Roxburgh). Illustration: Schneider, Illustriertes Handbuch der Laubhölzkunde **2**: 863 (1912); Polunin & Smythies, Flowers of south-west Europe, 63 (1973); Ceballos et al., Plantas silvestres de la peninsula Iberica, 387 (1980).

Creeping, scrambling or climbing evergreen. Stems to 15 m, angled, sparsely to densely spiny, rarely spineless. Leaves very variable in shape and size, 4–11 × 2–10 cm, narrowly to broadly lanceolate, triangular, ovate, oblong or kidney-shaped, usually abruptly narrowed at the tip, cordate, hastate, sagittate or rarely truncate at the base, usually leathery, spiny on margins and on main veins (spines sometimes very few and not on every leaf), with 5–9 veins, shiny on both surfaces. Leaf-stalk to 2 cm, usually spiny. Flowers fragrant, 5–30 in clusters on axillary and terminal branches 2–15 cm long. Perianth-segments pale green, 2–4 mm. Berries *c.* 6 mm across, black or red. *Canary Islands, S Europe to Ethiopia and India.* H5–G1. Summer–autumn.

6. S. regelii Killip & Morton (*S. grandifolia* Regel; *S. utilis* Hemsley; *S. saluberrima* Gilg; *S. ornata* Hooker; *S. officinalis* Hanbury & Fluckinger). Illustration: *Botanical Magazine*, 7054 (1889).

Stem to 15 m, sharply 4-angled and spiny below, 4-angled or winged above; spines *c.* 1 cm. Lower leaves very variable, to 30 × 21 cm, ovate to oblong, rounded or acuminate at apex, cordate or hastate at base; upper leaves much smaller, lanceolate to oblong, gradually tapered at base; all papery, often spiny on the veins beneath, with 5–7 veins. Leaf-stalks to 7 cm. Male flowers solitary or in racemes of umbels to 6.5 cm long. Female flowers solitary, stalks to 10 cm though usually shorter. Berries to 1.3 cm across, black. *Northern C America.* G1.

7. S. argyrea Linden & Rodigas. Illustration: *Illustration Horticole* **39**: pl. 152 (1892).

Stems wiry, with short, stout spines. Leaves to 25 cm, lanceolate, acute or acuminate, 3-veined, dark green with white or pale green blotches. Leaf-stalks short. Flowers and fruit unknown. *Peru & Bolivia.* G1.

8. S. walteri Pursh. Illustration: Rickett, Wild flowers of the United States **2**: 41 (1967); *Sida* **3**: 39 (1967).

Deciduous, rarely partially evergreen. Stems slender, slightly angled, spiny near the base, yellowish or brownish; branches squarish in section, not spiny. Leaves 5–12 × 1.5–6.5 cm, ovate to lanceolate, obtuse or abruptly acute at the tip, broadly tapered, rounded or cordate at the base, with 5–7 veins. Flowers 6–15 in single umbels, yellow, green or brown, drying brownish orange; umbel-stalks 5–15 cm, flattened, usually shorter than leaf-stalks. Berries 8–12 mm across, red, rarely white. *E US A from New Jersey to Texas.* H5–G1. Spring–summer.

9. S. laurifolia Linnaeus. Illustration: Rickett, Wild flowers of the United States **2**: 41 (1967); *Sida* **3**: 38 (1967); Justice & Bell, Wild flowers of North Carolina, 16 (1968); Radford et al., Manual of the vascular flora of the Carolinas, edn 2, 288 (1968).

Evergreen. Stem terete, green, spiny below; branches angled, usually spineless, rarely very spiny. Leaves 5–20 × 1–7.5 cm, narrowly ovate to oblong-lanceolate, abruptly tapered at apex and base, margin often inrolled, thick and leathery, 3-veined, midrib much more prominent than lateral veins beneath. Flowers greenish, in single umbels; umbel-stalks shorter than or equal to leaf-stalks. Berries ovoid, 6–8 mm across, black. *SE USA.* H5–G1. Summer.

10. S. china de Candolle. Illustration: Makino, New illustrated flora of Japan, 858 (1963).

Deciduous. Stems to 5 m, climbing or scrambling, slightly spiny or spineless. Leaves to 8 cm, broadly ovate to circular, sometimes wider than long, rounded or notched at the tip, rounded, truncate or cordate at the base, leathery or papery, with 5–7 veins. Leaf-stalks 8–25 mm. Flowers yellowish green in single umbels; umbel stalks *c.* 2.5 cm. Berries *c.* 9 mm, spherical, red. *China, Korea & Japan.* H3. Spring.

11. S. hispida Muhlenberg (*S. pseudo-china* misapplied; *S. tamnoides* misapplied). Illustration: Radford et al., Manual of the vascular flora of the Carolinas, edn 2, 286 (1968).

Deciduous. Stems to 15 m, terete or slightly angled, densely spiny below; branches angled, lacking spines, hairless. Leaves 5–15 × 1.5–2 cm, lanceolate to ovate or almost circular, abruptly pointed at the tip, rounded to cordate at the base, with 5–9 veins, shiny, drying grey, margins rough or finely toothed. Leaf-stalks 6–18 mm. Flowers to 25 in single umbels; umbel-stalks 1.5–10 cm, 1 and a half times longer than leaf-stalks or more. Berries *c.* 6 mm, bluish black. *S & C USA.* H3. Spring–summer.

12. S. excelsa Linnaeus. Illustration: *Botanical Magazine*, 9067 (1924); RHS Dictionary of Gardening, edn 1, 1966 (1956).

Deciduous scrambler or climber. Stems to 20 m, terete with slightly raised lines, spines straight, to 7 mm; branches spiny. Leaves very variable, 5–13 cm, slightly less in width, broadly ovate to circular, acuminate at the tip, rounded, cordate or truncate at the base, membranous or slightly leathery, with 5–7 veins, margins rough or with fine teeth. Leaf-stalks to 1.2 cm. Flowers green, 4–12, in single umbels; umbel-stalks 7–20 mm. Perianth-segments: male 5–7 mm, female 4–5 mm. Berries 8–10 mm across, red. From *E Bulgaria through Turkey to USSR (S shores of the Caspian Sea).* H4. Summer.

13. S. rotundifolia Linnaeus. Illustration: *Sida* **3**: 38 (1967); Radford et al., Manual of the vascular flora of the Carolinas, edn 2, 286 (1968).

Deciduous or partly evergreen. Stems to 10 m, 4-angled, with spines to 8 mm on the angles; branches spiny. Leaves 5–15 cm, broadly ovate to circular, rounded to cordate at the base, thick and leathery, sometimes spiny on the margins, 5-veined, dark, shiny green. Leaf-stalks 6–12 mm. Flowers greenish yellow, in single umbels; umbel-stalks 5–15 mm, equal to slightly longer than leaf-stalks, flattened. Berries 5–8 mm across, blue to black. *Eastern N America.* H2. Spring–summer.

27. AGAVACEAE

Succulent herbs, often large or with woody bases. Leaves mostly in basal rosettes, succulent, ptyxis flat or revolute, each leaf terminating in a hard or soft spine. Flowers in racemes or panicles, often produced after prolonged vegetative growth, usually bisexual radially, or bilaterally symmetric less than 8 cm long. Perianth of 6 free or united segments. Stamens 6, usually borne near the top of the perianth-tube. Ovary of 3 united carpels, superior or inferior; ovules many, axile. Fruit a berry or dry and indehiscent.

There are 8 genera and perhaps 380 species. Seven genera are native to North America, and two are introduced into Europe. All are cultivated.

The genera in the family are relatively easily defined on floral characters, although, so defined, they are vegetatively very variable. Unfortunately, flowers are usually not available in cultivation – many of the species flower only at long intervals (Agaves are known colloquially as 'century plants'), or take many years to attain the flowering condition and then die; others, are often grown as house or office plants and generally die or are discarded before they attain sufficient size for flowering to take place. Thus, a key to the genera based on flowering material is not very helpful to the gardener. The formal, dichotomous key given here makes use of both floral and vegetative characters and shows the 'received' features used in separating the genera.

Literature: Bogler, D.J., Pires, J.C., & Francisco Ortega, J., Phylogeny of Agavaceae based on ndhF, rbcL, and its sequences: implications of molecular data for classification, *Aliso* **22**: 313–328 (2006).

1a. Ovary superior — 2
b. Ovary inferior — 3
2a. Perianth-lobes 2.5 cm or more; leaves without a conspicuous main vein — **1. Yucca**
b. Perianth-lobes much less than 2.5 cm; leaves with a conspicuous main vein from which the parallel lateral veins diverge at an acute angle — **8. Cordyline**
3a. Perianth bilaterally symmetric; plants herbaceous — **7. Polianthes**
b. Perianth radially symmetric; plants usually with woody stems or trunks or consisting of stemless rosettes, rarely herbaceous — 4
4a. Anthers clearly projecting from the perianth — 5
b. Anthers included in the perianth — 6
5a. Leaves dying off annually, without spines; perianth-lobes spreading or reflexed — **6. Manfreda**
b. Leaves persisting for several years, usually with spiny margins or at least ending in a spine; perianth-lobes erect — **5. Agave**

6a. Filaments and style swollen below the middle; apex of leaf narrowed to a persistent spine **4. Furcraea**
b. Filaments and style not swollen below; apex of leaf tapering but not narrowed to a persistent spine 7
7a. Leaves minutely toothed **3. Beschorneria**
b. Leaves not toothed but margins splitting off as fine threads **2. Hesperaloe**

1. YUCCA Linnaeus

J. Cullen

Plants stemless or with short stems or tall, woody trunks. Leaves borne in rosettes at the ends of the stems or branches, persistent, stiff or flexible, fleshy, margins entire, finely toothed or splitting off as threads, bases expanded and somewhat sheathing. Inflorescence a much-branched, erect or, rarely, hanging panicle; stalk long and conspicuous or short and hidden by the leaves. Flowers bisexual, subtended by bracts. Perianth bell-shaped, hemispherical or spherical, opening more widely at night, radially symmetric, of 6 lobes that are united at their bases for a short distance. Stamens 6, the filaments free from the perianth, or, more rarely, united to it for some distance, usually curved out towards their apices, which are swollen and broader than the anthers. Ovary superior, 3-celled, each cell with many ovules, usually tapering into the 3- or 6-lobed stigma (more rarely the style thread-like). Fruit a capsule or indehiscent and dry or fleshy. Seeds winged or wingless.

A genus of about 30 species from arid regions of the USA, Mexico, Guatemala and the West Indies. Its classification (particularly the recognition of sections) is based largely on characters of the fruit and seeds. In general, neither of these is available in cultivation (the complex adaptations to pollination by moth larvae render fertilisation away from the native habitats unlikely) and so the present account is based mainly on features of the stems, leaves, inflorescences and flowers and is therefore artificial to some extent. The existence of many hybrids (both wild and man-made) makes identification even more difficult, and the present account (particularly the keys) must be used with caution.

The species grow well in sandy or gravelly soils, generally in sites where frost is not too severe and rainfall is limited, or, if heavy, the soil is very well drained. They are readily propagated by seed (if it is available) or by stem or root (rhizome) cuttings.

Literature: Trelease, W., The Yucceae, *Report of the Missouri Botanical Garden* **13**: 27–133 (1902); Molon, G., Le Yucche (1914); MacKelvey, S.D., Yuccas of the south western United States **1** (1938), **2** (1947); Webber, J.M., Yuccas of the southwest, USDA Monograph **17** (1953); Matuda, E. & Lujan, I.P., Las Plantas Mexicanas del genero Yucca (1980).

Plant. Stemless: **11, 12, 13, 15, 16, 18**; with a distinct stem or trunk: **1, 2, 3, 4, 5, 6, 7, 8, 9, 10, 11, 14, 15, 17**.
Leaves. At most 2 cm wide: **1, 13, 14, 15, 17, 18**; more than 2 cm wide: **2, 3, 4, 5, 6, 7, 8, 9, 10, 11, 12, 16**. Margins entire or with a few obscure teeth: **7, 10**; margins finely but distinctly toothed: **1, 8, 9, 16, 17, 18**; margin splitting off as straight or curled threads: **2, 3, 4, 5, 6, 11, 12, 13, 14, 15**.
Inflorescence. Erect, borne essentially between the leaves (though overtopping them), stalk hidden: **1, 2, 6, 7, 8, 9, 10**; erect, borne on a distinct stalk which is clearly visible above the leaves: **3, 5, 11, 12, 13, 14, 15, 16, 17, 18**; hanging, but with a short but obvious stalk: **4**.
Filaments. Not arching out towards their apices: **18**.
Style. Thread-like, the ovary not tapering into it: **18**.
Fruit. Indehiscent, dry or fleshy: **1, 2, 3, 4, 5, 6, 7, 8, 9, 10**; a dehiscent capsule: **11, 12, 13, 14, 15, 16, 17, 18**.

1a. Leaf-margins without conspicuous threads (a few threads may be formed when the leaf is very old) 2
b. Leaf-margin with conspicuous threads 10
2a. Plant stemless 3
b. Plant with a distinct stem or trunk 5
3a. Style thread-like; filaments attached to the perianth at their bases, not arching outwards towards their apices **18. whipplei**
b. Style stout, the ovary tapering into it; filaments not or scarcely attached to the perianth at their bases, arching outwards towards their apices 4
4a. Panicle hairless; leaves with a yellowish brown, finely toothed margin **16. rupicola**
b. Panicle usually downy; leaves with entire, green margins **12. flaccida**
5a. Leaf-margins conspicuously though finely toothed 6
b. Leaf-margin entire or with a few scattered, inconspicuous teeth 9
6a. Leaves up to 1.7 cm wide, margins finely and evenly toothed 7
b. Leaves more than 2.5 cm wide, toothing small, coarse but uneven 8
7a. Branched tree to 15 m **1. brevifolia**
b. Shrub ultimately to 4.5 m, not or little branched **17. rostrata**
8a. Ovary stalked; flowers white, usually tinged with purple **8. aloifolia**
b. Ovary stalkless; flowers white or cream **9. elephantipes**
9a. Leaves channelled, blue-green, stiff and rigid, erect-spreading, surfaces rough **7. treculeana**
b. Leaves flat or pleated towards the apex, green, somewhat flexible, spreading or recurved, surfaces smooth **10. gloriosa**
10a. Plants without distinct stems, or with very short stems that are hidden by the leaves 11
b. Plants with distinct, often trunk-like stems 15
11a. Perianth-lobes almost entirely free **2. baccata**
b. Perianth-lobes united at the base into a distinct, though short, tube 12
12a. Style narrow at the base, swollen above though ultimately narrowing, green; leaves 5–13 mm wide **15. glauca**
b. Style oblong or tapering evenly to the apex, white or cream; leaves more than 1.5 cm wide 13
13a. Leaves tapering from base to apex, with straight or slightly curved threads on the margin **12. flaccida**
b. Leaves distinctly narrowed towards the base, with numerous, very curled threads on the margin 14
14a. Perianth-lobes 3–5 cm; leaves stiff and ascending **13. smalliana**
b. Perianth-lobes more than 5 cm; leaves usually flexible and recurved **11. filamentosa**
15a. Leaves at most 2.5 cm wide 16
b. Leaves more than 2.5 cm wide 18
16a. Style narrow at the base, then swollen above, ultimately narrowing, green **15. glauca**
b. Style oblong, not as above, white or cream 17

17a. Inflorescence hanging; leaves narrowed towards the base, 2–2.5 cm wide **4. filifera**
b. Inflorescence erect or spreading; leaves tapering evenly from the base, rarely more than 1.8 cm wide **14. elata**
18a. Ovary (including style) 4.5 cm or more 19
b. Ovary (including style) less than 4.5 cm 20
19a. Leaf with thick, generally curled threads on the margin; perianth-lobes united only at their extreme bases **2. baccata**
b. Leaf with few, thin, straight threads; perianth-lobes distinctly united into a short tube at their bases **3. arizonica**
20a. Leaves rigid, yellowish green, rough on both surfaces; style clearly distinguished from the ovary **6. torreyi**
b. Leaves flexible, blue-green, smooth on both surfaces; style not clearly distinguished from the ovary **5. schottii**

1. Y. brevifolia Engelmann (*Y. arborescens* (Torrey) Trelease; *Clistoyucca arborescens* (Torrey) Trelease).

Tree to 15 m, with rough bark, usually branched. Leaves in dense, elongate rosettes, spreading, 15–60 cm × 7–15 mm, lanceolate, somewhat narrowed towards the base, rigid, margins finely toothed, very sharply pointed. Inflorescence a panicle to 35 cm, erect, held just above the leaves, the stalk hidden; axis often roughly hairy. Flowers green or greenish yellow; perianth-lobes 2.5–5 cm, thick, incurved. Style oblong. Fruit indehiscent, dry and spongy, ovoid, 5–10 cm. Seeds unwinged. *USA (California to Utah).* H5–G1. Autumn.

2. Y. baccata Torrey. Illustration: *Illustration Horticole* **20**: t. 115 (1853); Rickett, Wild flowers of the United States **4**: t. 12 (1970); Matuda & Lujan, Las plantas mexicanas del genero Yucca, 66–7 (1980).

Stem distinct, though usually shorter than the leaves, thick. Leaves 30–70 × 3.5–5 cm, narrowly lanceolate, concave, rough on both surfaces, blue-green, margin splitting off as coarse, curled threads. Panicle shortly stalked, held between the leaves, to 60 cm. Flowers bell-shaped; perianth-lobes *c.* 7.5 cm, white or cream, often tinged with purple, united only at the extreme base. Ovary (including style) more than 4.5 cm. Fruit fleshy, indehiscent, spindle-shaped, 20–25 cm. Seeds unwinged. *USA (Colorado and Utah to New Mexico), N Mexico.* H5–G1. Autumn.

3. Y. arizonica MacKelvey. Illustration: Matuda & Lujan, Las plantas mexicanas del genero Yucca, 74 (1980).

Stem woody, simple, to 3 m. Leaves 60–80 × 3–4 cm, narrowly lanceolate, thick, straight, smooth, pale green or yellowish green, margins splitting off as fine, straight threads. Panicle 50–100 cm, long-stalked, flowers held well above the leaves. Flowers bell-shaped; perianth-lobes 5.5–12 cm, cream, often tinged with purple, forming a distinct, though short tube at the base. Ovary (including style) more than 4.5 cm. Fruit indehiscent, fleshy, 15–20 cm, with a long or short beak. Seeds unwinged. *USA (Arizona), Mexico.* H5–G1. Autumn.

4. Y. filifera Chabaud (*Y. australis* (Engelmann) Trelease). Illustration: *Botanical Magazine*, 7197 (1891), excluding habit drawing; Matuda & Lujan, Las plantas mexicanas del genero Yucca, 100 (1980).

Stem woody, trunk-like, to 10 m or more, ultimately branched, bark rough. Leaves to 55 × 2.5 cm, lanceolate, somewhat narrowed towards the base, rigid, margins splitting off as numerous fine, curled threads. Panicle hanging from a curved stalk, oblong, very dense. Flowers bell-shaped; perianth-lobes 3.8–5.2 cm, cream. Fruit indehiscent, oblong. Seeds wingless. *Mexico.* H5–G1. Autumn.

Y. valida Brandegee. Illustration: Matuda & Lujan, Las plantas mexicanas del genero Yucca, 97–8 (1980). Panicle conical, more or less erect or ascending, held well above the leaves. *Mexico.* H5–G1. Autumn.

5. Y. schottii Engelmann. Illustration: Rickett, Wild flowers of the United States **4**: t. 12 (1970); Matuda & Lujan, Las plantas mexicanas del genero Yucca, 84 (1980).

Stem woody, to 5 m, rarely branched. Leaves 50–80 × 3.5–5 cm, lanceolate, flexible, smooth, blue-green, the margins splitting off as rather sparse threads. Panicle erect, somewhat exceeding the leaves, the stalk mostly hidden, rather open; axes densely woolly. Flowers almost spherical, often wider than long; perianth-lobes 2.5–4 cm, white, the outer 3 with brownish apices. Ovary (including style) less than 4.5 cm, the ovary itself tapering gradually into the indistinct style. Fruit indehiscent, oblong. Seeds wingless. *SW USA, Mexico.* H5–G1. Autumn.

6. Y. torreyi Shafer. Illustration: Rickett, Wild flowers of the United States **4**: t. 12 (1970); Matuda & Lujan, Las plantas mexicanas del genero Yucca, 77 (1980).

Stem woody, ultimately to 4 m, usually simple. Leaves 30–100 × 3–5 cm, narrowly lanceolate, rigid, rough on both surfaces, yellowish green, the margins splitting off as tough threads that are curled at first. Panicle dense, oblong, held just above the leaves on a hidden stalk. Flowers spherical or bell-shaped, cream often tinged with purple; perianth-lobes 3.5–7.5 cm. Ovary (including style) less than 4.5 cm, the style more or less distinct from the ovary. Fruit indehiscent. Seeds unwinged. *USA (Texas, New Mexico), N Mexico.* H5–G1. Autumn.

7. Y. treculeana Carrière (*Y. canaliculata* Hooker). Illustration: *Botanical Magazine*, 5201 (1860); Rickett, Wild flowers of the United States **2**: t. 15 (1967); Matuda & Lujan, Las plantas mexicanas del genero Yucca, 76 (1980).

Stem woody, ultimately to 5 m, branched. Leaves 50–100 × 2.5–5 cm, very narrowly lanceolate, rigid, erect-spreading, channelled, rough on both surfaces, margins entire. Panicle ellipsoid, dense, borne just above the leaves, its short common stalk hidden. Flowers spherical to hemispherical, white or faintly tinged with purple; perianth-lobes 3–4.5 cm. Fruit indehiscent. Seeds unwinged. *USA (Texas), W Mexico.* H5–G1. Autumn.

8. Y. aloifolia Linnaeus. Illustration: Polunin & Everard, Trees and bushes of Europe, 181 (1976); Matuda & Lujan, Las plantas mexicanas del genero Yucca, 108 (1980); Brickell (ed.), RHS A–Z encyclopedia of garden plants, 1093 (2003).

Stem woody, simple or branched, to 8 m or more. Leaves 25–40 × 2.5–6 cm, narrowly lanceolate, rigid and flat, narrowed slightly towards the base, margin toothed, the teeth small and irregular. Panicle held just above the leaves, erect. Flowers bell-shaped, white often tinged with purple; perianth-lobes to 5.5 cm. Ovary shortly stalked. Fruit indehiscent, with purple pulp. Seeds unwinged. *SE USA, Mexico, West Indies.* H5–G1. Autumn.

A very variable species; variegated and coloured-leaved cultivars are grown; in var. **draconis** (Linnaeus) Engelmann the leaves are drooping.

9. Y. elephantipes Regel (*Y. guatemalensis* Baker). Illustration: *Botanical Magazine*, 7997 (1905); Chickering, Flowers of Guatemala, 123 (1973); Matuda & Lujan, Las plantas mexicanas del genero Yucca, 105–6 (1980).

Like *Y. aloifolia* but trunk to 10 m, leaves 50–100 × 5–7 cm, flowers white or cream, ovary stalkless, fruit with greenish pulp. *Mexico, Guatemala.* G1. Autumn.

10. Y. gloriosa Linnaeus (*Y. acuminata* Sweet). Figure 7(1), p. 178. Illustration: *Botanical Magazine*, 1260 (1810); Saunders et al., Refugium botanicum **5**: t. 316–20 (1872); Brickell (ed.), RHS A–Z encyclopedia of garden plants, 1093 (2003).

Stem woody, to 5 m, ultimately branched. Leaves to 50 × 5 cm, lanceolate, flat or pleated near the apex, somewhat narrowed towards the base, smooth, ascending, spreading or recurved, flexible, glaucous when young, green when older, margins entire or with a few inconspicuous teeth. Panicle large, erect, borne directly above the leaves on a hidden stalk. Flowers bell-shaped, greenish white, cream or reddish; perianth-lobes 5–7 cm. Fruit indehiscent, not fleshy. Seeds unwinged. *SE USA (North Carolina to Florida).* H5–G1. Autumn.

Probably the most widely grown of the species, and also the best-known of a complex of species that are difficult to distinguish. The following may be found in European gardens:

Y. recurvifolia Salisbury (*Y. recurva* Haworth). Illustration: Saunders et al., Refugium botanicum **5**: t. 321 (1872). Leaves less stiff, usually recurved in their upper half. *SE USA (Georgia to Missouri).* H5–G1. Autumn.

Y. desmetiana Baker. Leaves rigid, without an obvious spine at the apex; inflorescence, flowers and fruit apparently unknown. *N Mexico.* H5–G1.

Y. flexilis Carrière. Inflorescence usually with a distinct stalk, held well above the leaves. *Origin unknown.* H5–G1. Autumn.

11. Y. filamentosa Linnaeus (*Y. angustifolia* misapplied). Illustration: *Botanical Magazine*, 900 (1806).

Plant stemless or almost so, the stem, if present, usually short and hidden by the leaves. Leaves *c.* 50 × up to 2.5 cm, oblanceolate, very clearly narrowed towards the base, rather abruptly tapered to the apex, flexible and recurved, green or slightly glaucous, margin splitting off as stout, curled threads. Panicle clearly stalked, held well above the leaves. Flowers bell-shaped, white tinged with green, or cream; perianth-lobes 5–8 cm. Fruit a dehiscent capsule. Seeds winged. *E USA (New Jersey to Florida).* H5–G1. Autumn.

A variable species; several cultivars, including variegated forms, are available.

12. Y. flaccida Haworth (*Y. puberula* Haworth). Illustration: *Edwards's Botanical Register* **22**: t. 1895 (1836); Saunders et al., Refugium botanicum **5**: t. 323 (1872).

Like *Y. filamentosa* but leaves tapering from base to apex, margin splitting off as straight or slightly curled threads, panicle axes usually downy. *S USA (North Carolina to Alabama).* H5–G1. Autumn.

13. Y. smalliana Fernald (*Y. filamentosa* var. *smalliana* (Fernald) Ahles). Illustration: Rickett, Wild flowers of the United States **2**: t. 15 (1967).

Like *Y. filamentosa* but leaves stiff and ascending, perianth-lobes 3–5 cm. *SE USA (South Carolina to Florida).* H5–G1. Autumn.

14. Y. elata Engelmann (*Y. radiosa* (Engelmann) Trelease). Illustration: *Botanical Magazine*, 7650 (1899); Rickett, Wild flowers of the United States **4**: t. 12 (1970); Matuda & Lujan, Las plantas mexicanas del genero Yucca, 132 (1980).

Plants with an obvious woody stem to 3 m. Leaves very numerous, 40–70 cm × 8–15 mm, narrowly linear, tapering slightly from base to apex, flat, pale green, flexible, the margins splitting off as fine white threads. Panicle ellipsoid, borne on a distinct stalk, held well above the leaves. Flowers bell-shaped, cream-white, sometimes tinged with green or pink; perianth-lobes 3.5–5 cm. Fruit a dehiscent capsule. Seeds winged. *USA (Arizona), Mexico.* H5–G1.

15. Y. glauca Fraser (*Y. angustifolia* Pursh). Illustration: Saunders et al., Refugium botanicum **5**: t. 315 (1872); *Botanical Magazine*, 2236 (1821); Rickett, Wild flowers of the United States **4**: t. 13 (1970).

Plant stemless or with stem to 30 cm. Leaves 50–70 cm × 5–13 mm, linear or tapering slightly from the base, spreading, flexible, pale green, margins splitting off as fine threads. Panicle borne above the leaves on a long or short stalk. Flowers bell-shaped or almost spherical, greenish white, glossy; perianth-lobes thick, 4.5–6 cm. Style green, narrowed at the base, then broadening, ultimately narrowing again. Fruit a dehiscent capsule. Seeds with broad wings. *SE USA.* H5–G1. Autumn.

It is possible that **Y. stricta** Sims (illustration: *Botanical Magazine*, 2222, 1821) is the same species; if so, this name would have to be used.

16. Y. rupicola Scheele. Illustration: *Botanical Magazine*, 7172 (1891).

Plant stemless. Leaves 30–60 × 2–4 cm, narrowing somewhat towards the base, flexible, dark green, often wavy or twisted, margins finely toothed. Panicle held well above the leaves on a distinct stalk, hairless. Flowers bell-shaped, white or greenish white; perianth-lobes 4–7 cm. Fruit a dehiscent capsule, beaked at the apex. Seeds usually with narrow wings. *USA (Texas).* H5–G1. Autumn.

17. Y. rostrata Trelease. Illustration: Matuda & Lujan, Las plantas mexicanas del genero Yucca, 126 (1980).

Stems woody, ultimately to 4.5 m, not or little branched. Leaves 40–60 × 1.2–1.7 cm, linear or slightly broadened above the base, flat, thin and flexible, margins yellow or brown, finely toothed. Panicle borne well above the leaves on a distinct stalk. Flowers spherical to bell-shaped, white; perianth-lobes 4–5 cm. Fruit a dehiscent capsule with a long beak. Seeds more or less unwinged. *USA (Texas), N Mexico.* H5–G1. Autumn.

18. Y. whipplei Torrey (*Hesperoyucca whipplei* (Torrey) Baker). Illustration: *Botanical Magazine*, 7662 (1899); Rickett, Wild flowers of the United States **4**: t.12 (1970); Brickell (ed.), RHS A–Z encyclopedia of garden plants, 1094 (2003).

Plant stemless. Leaves 30–100 cm × *c.* 15 mm, linear, rigid, glaucous, margins yellowish, finely toothed. Panicle borne well above the leaves on a distinct stalk, oblong. Flowers cream-white, sometimes tinged purple, very fragrant; perianth hemispherical or bell-shaped, lobes 5–7 cm. Filaments clearly attached to the perianth at the base, their apices not turned outwards. Style thread-like, much narrower than the apex of the ovary, short; stigma 3-lobed. Capsule dehiscent, cylindric, with a short point at the apex. *USA (California) and adjacent Mexico.* H5–G1.

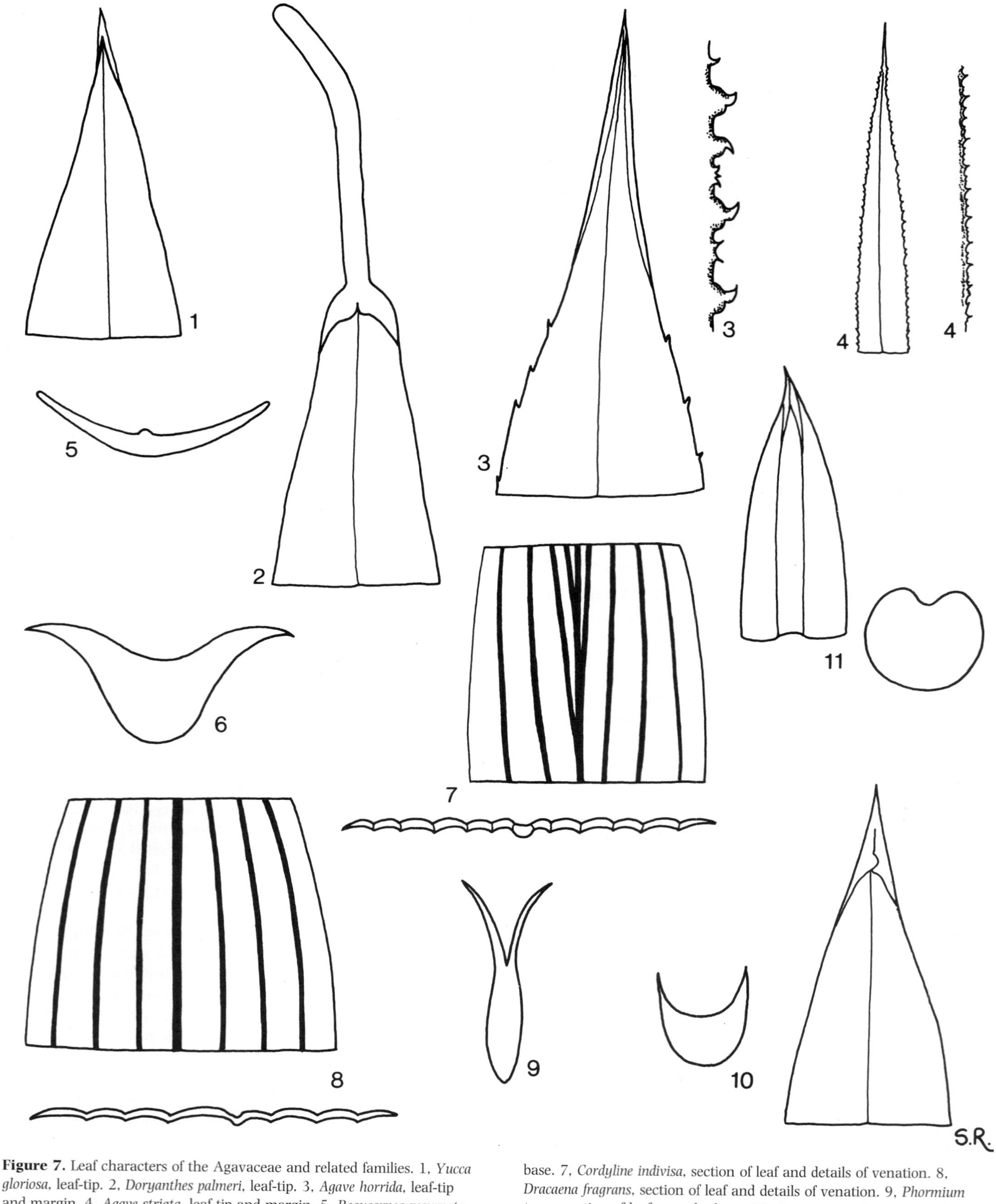

Figure 7. Leaf characters of the Agavaceae and related families. 1, *Yucca gloriosa*, leaf-tip. 2, *Doryanthes palmeri*, leaf-tip. 3, *Agave horrida*, leaf-tip and margin. 4, *Agave striata*, leaf-tip and margin. 5, *Beaucarnea recurvata*, section of leaf near the base. 6, *Agave attenuata*, section of leaf near the base. 7, *Cordyline indivisa*, section of leaf and details of venation. 8, *Dracaena fragrans*, section of leaf and details of venation. 9, *Phormium tenax*, section of leaf near the base. 10, *Sansevieria trifasciata*, section of leaf and leaf-tip. 11, *S. roxburghiana*, section of leaf and leaf-tip.

A variable species, divided into several subspecies by American botanists; it is not known which of these is in cultivation.

2. HESPERALOE Engelmann
J. Cullen
Stemless perennial herbs with swollen, bulb-like but fibrous bases. Leaves evergreen, forming clumps, linear but tapering, the margins splitting off as coarse or fine threads. Inflorescence usually a panicle, more rarely unbranched. Flowers bisexual, radially symmetric, cylindric bell-shaped; perianth-lobes 6, greenish or pink. Stamens 6, anthers included within the perianth. Ovary superior, 3-celled, ovules numerous in each cell. Fruit a capsule. Seeds numerous, thin and flat, black.

A genus of 3 species from Mexico and the adjacent USA. Cultivation as for *Yucca* (p. 175).

Literature: Trelease, W., The Yucceae, *Report of the Missouri Botanical Garden* **13**: 29–38 (1902).

1a. Flowers dark pink to orange-pink; marginal threads of the leaves very fine **1. parviflora**
b. Flowers green tinged with purple; marginal thread of the leaves coarse **2. funifera**

1. H. parviflora (Torrey) Coulter (*H. yuccifolia* Engelmann; *Yucca parviflora* Torrey). Illustration: *Botanical Magazine*, 7723 (1900); Rickett, Wild flowers of the United States **3**: t. 6 (1969); Brickell (ed.), RHS A–Z encyclopedia of garden plants, 530 (2003).

Leaves to 1.3 m × 3 cm near the base, with very fine marginal threads. Panicle to 1.2 m. Flowers pendent, dark pink to orange-pink, borne on red-pink stalks, perianth to 3 cm. *USA (Texas)*. H5–G1. Summer.

Plants with shorter, more bell-shaped, flowers have been called var. **engelmanii** (Krauskopf) Trelease (*H. engelmanii* Krauskopf).

2. H. funifera (Lemaire) Trelease (*Yucca funifera* Lemaire).

Leaves to 2 m × 5–6 cm near the base, marginal threads coarse. Panicle to nearly 3 m. Flowers green tinged with purple, borne on similarly coloured stalks; perianth to 2.5 cm. *N Mexico*. H5–G1. Summer.

3. BESCHORNERIA Kunth
C.J. Couper
Succulent perennial herbs with short, tuberous rootstocks. Leaves mostly in a basal rosette, persistent, ascending, lanceolate, glaucous, leathery and fleshy but flaccid, margins rough, minutely toothed; stem-leaves few, decreasing in size upwards, ultimately bract-like. Raceme or panicle erect or obliquely arched; bracts ovate, conspicuous, coloured and scarious. Flowers bisexual, clustered, stalked, pendent, their stalks jointed near the apex. Perianth tubular, tube very short and funnel-shaped; lobes 6, free almost to the base, radially symmetric, more or less equal, oblanceolate to lanceolate. Stamens 6, borne at the bases of the perianth-lobes, filaments thread-like, anthers dorsifixed. Ovary inferior, club-shaped, 3-celled with many ovules in each cell; style thickened at the base; stigma capitate. Capsule more or less spherical, 3-celled. Seeds flattened, brownish black.

A genus of about 12 species from Mexico. Cultivation as for *Agave* (p. 181).

1a. Leaves to 3 cm wide; flowers in an erect raceme **1. tubiflora**
b. Leaves more than 4 cm wide; flowers in panicles 2
2a. Leaves more than 60 cm; flowering stem 2.1–2.4 m **5. dekosteriana**
b. Leaves to 60 cm; flowering stem at most 2 m 3
3a. Flowering stem drooping, less than 1.2 m **2. yuccoides**
b. Flowering stem erect, more than 1.2 m 4
4a. Leaves keeled, stiff and fleshy; flowers downy **3. pubescens**
b. Leaves not keeled, thin and firm; flowers hairless **4. bracteata**

1. B. tubiflora Kunth. Illustration: *Botanical Magazine*, 4642 (1852); *Gartenflora* **24**: t. 851 (1875).

Stemless. Leaves 12 or more, 30–60 × 1.5–2.5 cm, linear, long-acuminate, rough on both surfaces, more or less recurved, thickened and triangular in section at the base, minutely striped. Raceme erect, axis 60–120 cm, brownish green; bracts ovate, violet-red. Flowers 2–4 in a cluster, *c.* 3.5 cm, drooping; stalks unequal and shorter than the bracts. Perianth-lobes *c.* 2.5 cm × 4 mm, reddish green, spathulate. Ovary *c.* 1.2 cm, 6-angled, purplish green. *Mexico*. G1.

2. B. yuccoides Hooker. Illustration: *Botanical Magazine*, 5203 (1860); *Gardeners' Chronicle* **46**: 138, 139, 309, 313 (1909); *Journal of the Royal Horticultural Society* **38**: 196 (1912); Hay & Synge, Dictionary of garden plants in colour, t. 1460 (1969).

Stemless. Leaves *c.* 20, 30–60 × *c.* 5 cm, lanceolate, narrowed to the base, rough on the lower surface, the apex sharp. Panicle with drooping, red branches on a stalk 90–120 cm; bracts lanceolate, acuminate, deep red-pink. Flowers 5–7.5 cm, dark green tinged with yellow, hairless. Perianth-lobes linear-oblong. Ovary cylindric. *Mexico*. G1.

3. B. pubescens Berger. Illustration: *Gardeners' Chronicle* **40**: t. 138 (1906).

Stemless. Leaves numerous, *c.* 60 × 5 cm, linear-lanceolate, thickly keeled, stiff and fleshy along the midrib, hardened at the apex, rough only near the tip. Panicle red, on a stalk 1.2–1.5 m; bracts triangular-lanceolate, bright red, hairless. Flowers 3 or 4 in a cluster, green, eventually becoming yellow, downy throughout; stalks 2–4 cm, hanging, hairless. Perianth-lobes linear, little spreading, hairless inside. Ovary *c.* 2.5 cm, conical-cylindric. *Mexico*. G1.

4. B. bracteata Jacobi. Illustration: *The Garden* **13**: 15 (1878); *Botanical Magazine*, 6641 (1882).

Stemless. Leaves 20–30, 30–60 × *c.* 5 cm, thin but firm, rough beneath, margins scarious or dry. Panicle borne on a terete, reddish brown stalk, 1.2–2 m; bracts ovate, pale red, scarious. Flowers 3.5–5 cm, deep red. Perianth-lobes *c.* 2.5 cm, oblanceolate, obtuse, incurved, greenish, becoming yellowish red outside, hairless. *Mexico*. G1.

5. B. dekosteriana Koch. Illustration: *Botanical Magazine*, 6768 (1884).

Stemless. Leaves 20 or more, 60–75 × 5–6 cm, oblanceolate, long-acuminate, base very thick and firm, smooth on the upper surface, rough on the margins and lower surface, dull green above, glaucous beneath. Panicle lax, borne on a yellow-brown stalk, 2.1–2.4 m; bracts ovate, white tinged with bright red. Flowers in clusters of 2 or 3, 3.5–6 cm; stalks to 2.5 cm. Perianth-lobes *c.* 3.5 cm, green, overlapping. Ovary green tinged with red, protruding a little into the perianth-tube. *Mexico*. G1.

4. FURCRAEA Ventenat
C.J. Couper
Monocarpic succulents. Leaves in a rosette that is basal in trunkless species, terminal

in those with a trunk, persistent, lanceolate, rigid, narrowed towards the base, margins finely toothed or spiny and with a spine at the apex. Panicle tall, loose, flowers often replaced by bulbils; bracts small, membranous; flower-stalks jointed. Flowers bisexual, radially symmetric. Perianth bell-shaped, of 6 lobes united into a short tube at the base, the lobes spreading horizontally. Stamens 6, attached at the apex of the perianth-tube, filaments erect, swollen basally; anthers oblong, versatile. Ovary inferior, oblong, 3-celled with many ovules in each cell. Style swollen basally with 3 prominent angles; stigma capitate. Fruit a capsule with many seeds.

A genus of about 20 species from tropical and subtropical America, separated from *Agave* because of its short-tubed flowers, short, thickened filaments and stoutly-based style. The genus is sometimes referred to incorrectly as *Fourcroya* Sprengel. Cultivation as for *Agave* (p. 181).

Literature: Michotte, F., Agaves et Fourcroyas, Culture et Exploitation, edn 3 (1931).

1a. Trunk present, at least 60 cm in height 2
b. Trunk absent, or if present, less than 40 cm in height 5
2a. Leaf-margins with hooked, dark brown spines along most of their length **4. selloa**
b. Leaf-margins entire or minutely toothed, without spines, or rarely with a few spines near the base 3
3a. Leaves 1.2–2.5 m, bright green **3. foetida**
b. Leaves to 1.2 m, blue-green or glaucous 4
4a. Trunk less than 2 m; leaves rigidly spreading, glaucous **2. bedinghausii**
b. Trunk more than 2.5 m; leaves flaccidly spreading, blue-green **1. roezlii**
5a. Leaves to 45 × 3.5–5 cm, dull or dark olive-green **9. undulata**
b. Leaves more than 50 × 5 cm, bright green 6
6a. Leaves distinctly keeled; marginal spines 2–3 mm **6. hexapetala**
b. Leaves not keeled; marginal spines more than 3 mm 7
7a. Trunk short but distinct, 15–20 cm; marginal spines of the leaves reddish brown **5. tuberosa**
b. Trunk absent or indistinct; marginal spines brown 8
8a. Leaves 60–130 × 6–7 cm; panicle-branches minutely downy **8. pubescens**
b. Leaves 1.5–2.3 m × 9–12.5 cm; panicle-branches hairless **7. elegans**

1. F. roezlii André (*Agave argyrophylla* Anon.; *Yucca argyrophylla* Nicholson; *Roezlia regia* Anon.). Illustration: *Revue Horticole*, 353 (1887); Jacobsen, Handbook of succulent plants, t. 610 (1960).

Trunk to 4 m × 30–35 cm. Leaves 1–1.2 m × 7.5–12 cm, gradually tapering, flaccidly leathery, fleshy, spreading, acute, concave or pleated, blue-green; margins thinner, upcurved, minutely toothed. Flowering axis 3–5 m, stout, downy; panicle large, pyramidal, freely branched, the branches spreading and drooping at their tips; bracts large, downy; flowers mostly in 3 s on short stalks. Perianth-lobes 5–6 cm, white, softly downy inside. Capsule ellipsoid. Elongate bulbils freely produced. *Mexico*. G1.

2. F. bedinghausii Koch (*Roezlia bulbifera* Siebert & Voss; *Yucca bulbifera* invalid; *Beschorneria multiflora* invalid). Illustration: *La Belgique Horticole* **13**: 327 (1863); *Botanical Magazine*, 7170 (1891); *The Garden* **40**: 143 (1891) & **52**: 197 (1897).

Trunk 90–180 × 15–22 cm. Leaves 50 or more, 60–120 × 5–10 cm, narrowed gradually to the apex and narrowed to *c.* 2.5 cm wide near the base, rigidly outcurving, stiff, flat, striped, glaucous on both surfaces, the upper smooth, the lower rough; margins minutely toothed. Flowering axis to 6 m, the panicle occupying most of it, the branches elongate and drooping; flower-stalks 6–8 mm. Perianth-lobes *c.* 2.5 cm, white tinged with green outside. Ovary *c.* 2.5 cm, downy. Capsule broadly oblong. Ovoid bulbils freely produced. *C Mexico*. G1.

3. F. foetida (Linnaeus) Haworth (*Agave foetida* Linnaeus; *F. gigantea* Ventenat). Illustration: Redouté, Liliacées **8**: t. 476 (1816); *Botanical Magazine*, 2250 (1821); *Gartenflora* **1**: t. 3 (1852); *Revue Horticole* **4**: 206, 207 (1857).

Trunk 90–120 cm. Leaves 40–50, 1.2–2.5 m × 10–20 cm, narrowed to 6–7.5 cm broad above the base, rigid, flat or channelled, the lower surface rough, all bright, glossy green, narrowed gradually to a sharp terminal spine; marginal spines absent or few and very small (to 3 mm). Flowering axis 6–12 m, the panicle occupying about half of its length, branches compound; flower-stalks 6–12 mm. Flowers strongly scented. Perianth-lobes *c.* 2.5 cm. Ovary *c.* 1.8 cm, hairless. Bulbils freely produced. *C & northern S America, naturalised and introduced elsewhere*. G1.

4. F. selloa Koch (*F. flavoviridis* Hooker). Illustration: *Botanical Magazine*, 6148 (1875); Jacobsen, Handbook of succulent plants, t. 611–13 (1960).

Trunk to 1.5 m × 10–12.5 cm. Leaves 30–40, 90–120 × 7.5–10 cm, narrowed to *c.* 3.5 cm wide at the base, spreading, very rigid, concave or pleated, the undersurface rough, bright or glossy dark green; terminal spine blunt, marginal spines 5–6 mm, hooked, brown, *c.* 3–4 cm apart. Flowering axis to 5 m, branches of the panicle mostly simple; bracts 7.5–10 cm, hairless; flower-stalks *c.* 1.2 cm. Perianth-lobes 2.5–4 cm. Ovary *c.* 2.5 cm. Bulbils freely produced. *Mexico, Guatemala & Colombia*. G1.

'Marginata', with pale leaf margins, is widely grown.

5. F. tuberosa (Miller) Aiton. Illustration: *Revue Horticole*, 234 (1877) & 394–5 (1880); *Report of the Missouri Botanical Garden* **18**: t. 1–4 (1907).

Trunk 15–20 × 5–7.5 cm. Leaves *c.* 30, 60–175 × 5–25 cm, firm, hairless, almost flat, smooth on the upper surface, rough on the lower; terminal spine blunt, margins upcurved towards the apex with horny, hooked, reddish brown spines, 5–9 mm long and 2–3 cm apart, sometimes absent towards the base and apex. Flowering axis to 9 m, branches compound. Flowers sweetly scented. Perianth-lobes 2.5–3 cm. Bulbils elongate, freely produced. *West Indies*. G1–2.

6. F. hexapetala (Jacquin) Urban (*F. cubensis* (Jacquin) Ventenat; *Agave hexapetala* Jacquin; *A. cubensis* Jacquin). Illustration: Nicholson, Illustrated dictionary of gardening **2**: 37 (1884–8); *Annales du Jardin Botanique de Buitenzorg*, Supplement **3**: t. 40 (1907).

Trunk very short, 7.5–10 cm thick. Leaves *c.* 30, 60–125 × 5–10 cm, narrowed to *c.* 3.5 cm wide above the base, spreading, rough and rather distinctly keeled on the lower surface, the upper surface bright, shining green; terminal spine blunt, brown, marginal spines 2–3 mm, triangular, brown, 1.6–2.5 cm apart. Flowering axis to 6 m; panicle loose, diamond-shaped, the lower branches slightly compound; flower-stalks very short. Perianth-lobes *c.* 2.5 cm, white, greenish outside. Ovary *c.* 1.8 cm. *Cuba, Haiti.* G1.

7. F. elegans Todaro. Illustration: Todaro, Hortus Botanicus Panormitanus **1**: t. 4 (1876–8); *Botanical Magazine*, 8461 (1912).

Trunk absent or very short. Leaves 40–50, 1.5–2.3 m × 9–12.5 cm, narrowed to *c.* 7.5 cm wide above the base, smooth above, rough beneath, bright green; terminal spine short, brown, marginal spines large, brown. Flowering axis 6–8 m, the panicle occupying most of its length, the branches very compound. Perianth-lobes *c.* 3.5 cm, greenish white. Ovary *c.* 2.5 cm. *Mexico.* G1.

8. F. pubescens Todaro. Illustration: *Botanical Magazine*, 7250 (1892); *Annales du Jardin Botanique de Buitenzorg*, Supplement **3**: t. 43 (1907).

Trunk absent. Leaves *c.* 30, 60–130 × 6–7 cm, rigid, very much thickened at the base on both surfaces, grooved towards the apex, green; terminal spine sharp, marginal spines *c.* 5 mm, triangular-hooked, brown, 1.6–3 cm apart. Flowering axis to 7 m; panicle *c.* 5.5m, branches ascending, curved, very compound, minutely downy. Perianth-lobes 2.5–3.5 cm. Ovary densely downy. Capsule large. Bulbils freely produced. *Mexico?.* G1.

9. F. undulata Jacobi (*F. pubescens* Baker *not* Todaro). Illustration: *Botanical Magazine*, 6160 (1875).

Trunk absent. Leaves 20–30, *c.* 45 × 3.5–5 cm, narrowed to 1.8–2.5 cm above the base, wavy, recurved, keel and midrib beneath rough, dull or dark green above; terminal spine blunt, marginal spines triangular-hooked, brown. Flowering axis *c.* 3.5 m; panicle 2–2.5 m, branches short, mostly simple, drooping, minutely downy. Perianth-lobes *c.* 3 cm, downy. Ovary hairless. *Mexico.* G1.

5. AGAVE Linnaeus

C.J. Couper & J. Cullen

Monocarpic or perennial leaf-succulents without fleshy underground stems. Stems absent or short, sometimes forming trunks. Leaves in a rosette, persisting for many years, small to very large, often very fleshy, usually hard and leathery, usually with spines or teeth on the margins (more rarely the margins unrolling as fine threads) and with a conspicuous terminal spine. Inflorescence a large spike or panicle with many bisexual, radially symmetric flowers. Perianth tubular, of 6 lobes, united for most of their length. Stamens 6, anthers projecting from the perianth. Ovary inferior, 3-celled, with many ovules in each cell. Fruit a capsule. Seeds flattened, black.

A difficult genus of over 100 species, mostly from C America and the West Indies, but extending both north and south from this centre; a few species are naturalised in other areas (e.g. the Mediterranean). Their large, fleshy leaves and tall inflorescences render them unsuitable for the making of good herbarium specimens, and their flowering pattern—monocarpic or flowering irregularly over long periods—means that much of their study must take place in the field.

Very many species are recorded as having been in cultivation, but their identification is dubious, and there is much uncertainty. The key to the species included here is based entirely on vegetative characters (flowering material being only irregularly available) and must be used with some caution. The leaves of most species bear spines – hardened, broadly based, usually distant and dark-coloured prickles that are sometimes borne on fleshy pads – along their margins; a few species have teeth rather than spines: these are small, narrowly based and evenly distributed; a few others have margins that are entire or finely roughened or saw-like; and, finally, a few have margins that split off as threads.

Agaves will grow well in relatively poor, sandy soils as long as they are well-drained. They will also grow well in pots, provided that these are large enough. Propagation is by seed or by suckers that have already rooted while still attached to the parent plant.

Literature: Jacobi, G.Z., Von Versuch zu einer systematischen Ordnung der Agaven (1864); Trelease, W., Species in Agave, *American Philosophical Society Proceedings* **49**: 232–237 (1910); Trelease, W., Agave in the West Indies, *Memoirs of the National Academy of Sciences* **11**: 1–55 (1913); Berger, A., Die Agaven (1915); Michotte, F., Agaves et Fourcroyas, Culture et Exploitation, edn 3 (1931); Gentry, H.S., The Agave family in Sonora, USDA agricultural handbook **399** (1972); Gentry, H.S., The Agaves of Baja California, *Occasional Papers of the California Academy of Sciences*, 130 (1978); Gentry, H.S., Agaves of continental North America (1982).

1a. Leaves with margins unrolling as hair-like threads 2
b. Leaves with smooth, toothed, spiny or horny margins that do not unroll as threads 4
2a. Mature rosettes to 25 cm in diameter **8. parviflora**
b. Mature rosettes more than 25 cm in diameter 3
3a. Leaves narrowed towards the base **2. filifera**
b. Leaves not narrowed towards the base **9. toumeyana**
4a. Leaves without spines or teeth on the margins, though the margins sometimes rough 5
b. Leaves with obvious spines or teeth on the margins 8
5a. Leaves conspicuously narrowed towards the base 6
b. Leaves not narrowed towards the base 7
6a. Leaves stiff, 90–130 × 9–12 cm, terminal spine hard **27. sisalana**
b. Leaves soft, to 75 × 12–16 cm, terminal spine soft **1. attenuata**
7a. Leaves oblong, thick, acutely 3-angled and keeled towards the apex, dark green with white lines on the surface and margins **6. victoriae-reginae**
b. Leaves linear, shortly tapering, flat, without white lines on the surface and margins **12. striata**
8a. Leaves soft and flexible, with teeth on the margins but without spines **10. albicans**
b. Leaves stiff and leathery, with spines on the margins 9
9a. Leaves conspicuously narrowed towards the base 10
b. Leaves not narrowed towards the base 21
10a. Rosette very large, more than 2 m in diameter when mature 11
b. Rosette less than 2 m in diameter when mature 14
11a. Terminal spine not decurrent on the leaf margins **14. americana**
b. Terminal spine decurrent on the leaf margins 12
12a. Leaves very dark blackish green; plant not producing suckers **19. atrovirens**
b. Leaves green, glaucous or whitish; plant producing suckers 13

13a. Leaves arching over, whitish green, 2–2.5 m **15. franzosinii**
b. Leaves ascending or erect, green or glaucous, 1–2 m **26. salmiana**
14a. Rosettes more than 65 cm in diameter when mature 15
b. Rosettes less than 65 cm in diameter when mature 17
15a. Mature plant producing suckers 16
b. Mature plant not producing suckers **21. marmorata**
16a. Leaves 12–18 cm broad, usually reddish in part, and with paler cross bands **17. colorata**
b. Leaves at most 10 cm broad, not coloured as above **18. palmeri**
17a. Leaves 7 cm or more wide 18
b. Leaves less than 7 cm wide 20
18a. Rosette to 25 cm in diameter; terminal spine of the leaf 3–4.5 cm **20. potatorum**
b. Rosette to 65 cm in diameter; terminal spine 1–3 cm 19
19a. Leaves 20–30 × 10–12 cm, glaucous or bloomed **22. parrasana**
b. Leaves 35–65 × 7–10 cm, green **11. polyacantha**
20a. Leaves 2–3 cm broad, glaucous or whitish; rosette producing suckers **25. macroacantha**
b. Leaves 4–7 cm broad, green; rosette not producing suckers **3. horrida**
21a. Rosette not producing suckers **7. xylonacantha**
b. Rosette producing suckers 22
22a. Rosette less than 65 cm in diameter **13. utahensis**
b. Rosette more than 65 cm in diameter 23
23a. Leaves more than 6 cm broad 24
b. Leaves less than 6 cm broad 26
24a. Leaves 25–40 × 8–12 cm **23. parryi**
b. Leaves 60–180 × 8–16 cm 25
25a. Leaf margins straight, spines 3–6 mm; trunk present **24. fourcroydes**
b. Leaf margins sinuous, spines 8–15 mm; trunk absent **16. scabra**
26a. Leaves 2.5–4 cm wide, margins straight, teeth generally directed towards the leaf base **4. lechuguilla**
b. Leaves 4–5 cm wide, margins sinuous, teeth directed outwards or upcurved **5. lophantha**

Subgenus **Littaea** (Tagliabue) Baker.
Flowers in a simple raceme or spike, usually in pairs (more rarely in 4 s or more) in the axils of each bract; axis of inflorescence narrower than the apex of the crown of the rosette.

1. A. attenuata Salm-Dyck. Figure 7(6), p. 178. Illustration: *Botanical Magazine*, 5333 (1862) – as *A. glaucescens*; *Gardeners' Chronicle* **17**: t. 63–4 (1896); Lamb & Lamb, Cacti and other succulents **2**: t. 267 (1959); Jacobsen, Handbook of succulent plants, t. 84 (1960).

Trunk to 1 m, prostrate, branching above ground, each branch ending in a rosette. Rosettes with 10–20 leaves. Leaves 50–75 × 12–16 cm, ovate-elliptic, narrowed towards the base, soft, spreading, glaucous; margins entire, pale; terminal spine needle-like but soft. Spike to 3.5 m, recurved. Flowers in 2s, 3.5–5 cm, greenish yellow. Spike with numerous plantlets after flowering. *Mexico*. H5–G1.

2. A. filifera Salm-Dyck. Illustration: *Gardeners' Chronicle* **7**: t. 49 (1877) & **4**: t. 135–7 (1881); Jacobsen, Handbook of succulent plants, t. 96 (1960); Brickell (ed.), RHS A–Z encyclopedia of garden plants, 88 (2003).

Rosettes trunkless, *c.* 65 cm in diameter with 60–100 leaves, producing suckers freely. Leaves 15–30 × 2–4 cm, lanceolate, long-tapering to the apex, narrowed towards the base, stiff, spreading and somewhat upcurved, shiny dark green with 2 or 3 white lines on the surface; margins pale, horny, splitting off as 5 or 6 long, thin threads; terminal spine 1–2 cm, brown, later grey, openly grooved. Spike to 2.5 m, flowers very dense. Flowers 3–3.5 cm, greenish at first, later reddish. *Origin uncertain, probably Mexico*. H5–G1.

A. schidigera Lemaire. Illustration: *Illustration Horticole* **9**: t. 330 (1862); *Botanical Magazine*, 5641 (1867). Rosette 90–100 cm in diameter; leaves 30–50 × 1.2–1.6 cm. *Mexico*. H5–G1.

3. A. horrida Jacobi. Figure 7(3), p. 178. Illustration: *Gardeners' Chronicle* **7**: t. 99 (1877); *Botanical Magazine*, 6511 (1880).

Rosette trunkless, to 60 cm in diameter with 50–100 leaves, not producing suckers. Leaves 20–35 × 4–7 cm, oblanceolate-oblong, narrowed towards the base, stiff and hard, glossy or yellowish green; margins grey, horny, sinuous, marginal spines 1–2 cm distant, curved in all directions and with accessory spines; terminal spine 2.5–4 cm, flattened above, red-brown becoming grey. Spike 3–4 m, dense. Flowers in 2s, 3.5–4 cm, pale yellowish green. *Mexico*. H5–G1.

Plants of this species are strictly monocarpic and die after flowering; propagation is by seed only.

4. A. lechuguilla Torrey. Illustration: *Report of the Missouri Botanical Garden* **7**: t. 31 (1896).

Rosettes *c.* 1 m in diameter with 40–60 leaves, producing suckers. Leaves 25–50 × 2.5–4 cm, triangular-oblong, curved or ascending, green or bluish, but not glaucous, pale-banded beneath; margins straight, narrow, horny, with slender spines; terminal spine 1.5–4 cm, slender, recurved, flattened, grey-brown. Spike 1–4 m. Flowers 3–4.5 cm. *USA (Texas), Mexico*. H5–G1.

5. A. lophantha Schiede. Illustration: Lamb & Lamb, Cacti and other succulents **2**: t. 270 (1959); Jacobsen, Handbook of succulent plants, t. 103 (1960).

Rosette *c.* 1 m in diameter with 30–40 leaves, producing suckers. Leaves 30–70 × 4–5 cm, lanceolate, long-tapering, stiff, hard, spreading, glossy green, the under surface with darker lines, the upper surface sometimes with a pale, longitudinal stripe; margin horny, marginal spines stout, triangular, straight or upcurved; terminal spine 1–2.5 cm, sharp, openly grooved. Spike to 4.5 cm, dense towards the apex. Flowers in 2s, 3.5–4.5 cm, pale greenish yellow. *Mexico*. H5–G1.

A very variable species, in which a number of varieties has been described. The most commonly grown is var. **univittata** (Haworth) Anon. (*A. univittata* Haworth), in which each leaf has a pale, longitudinal stripe along the upper surface.

6. A. victoriae-reginae Moore (*A. ferdinandi-regis* Berger). Illustration: *Gardeners' Chronicle* **18**: t. 148–9 (1882); Jacobsen, Handbook of succulent plants, t. 115–17 (1960); Rowley, The illustrated encyclopaedia of cacti and succulents, 169 (1978); Brickell (ed.), RHS A–Z encyclopedia of garden plants, 89 (2003). Rosette trunkless, to 60 cm in diameter with up to 200 leaves, not usually producing suckers. Leaves 10–20 × 4–6 cm, triangular-oblong, smooth, leathery, acutely 3-angled, keeled towards the apex, then narrowed into a blunt tip, dull, dark green marked with oblique, white lines; margins and keel without spines, smooth with white, horny bands; terminal spine 1–3 cm, brown, needle-like, often with additional, very short spines on either side. Spike to 5 m, loose below, dense above. Flowers usually in 3 s, 3–5 cm, pale green. *Mexico.* H5–G1.

7. A. xylonacantha Salm-Dyck. Illustration: *Botanical Magazine*, 5660 (1867).

Rosettes simple, borne on a short trunk, with *c.* 20 leaves, not forming suckers. Leaves 30–90 × 5–12 cm, oblong, tapering from the base, hard, rough, dull, glaucous or grey-green (rarely yellow-green), dark-lined beneath; margins broad, white, horny and with irregular spines, 8–15 mm long and 3–5 cm distant, borne on broad cushions; terminal spine 2.5–5 cm, straight, needle-like, grooved at the base, pale brown or grey. Spike to 6 m, sinuous. Flowers 3–8 together, 4–5 cm, greenish. *Mexico.* H5–G1.

8. A. parviflora Torrey. Illustration: *Report of the Missouri Botanical Garden* **7**: t. 30 (1896); Lamb & Lamb, Cacti and other succulents **2**: t. 271 (1959); Jacobsen, Handbook of succulent plants, t. 108–9 (1960); Gentry, The Agave family in Sonora, USDA agricultural handbook **399**: t. 10 (1972).

Rosette trunkless, 15–25 cm in diameter, producing suckers very rarely. Leaves 6–10 cm × 8–10 mm, oblong-linear, narrowed towards the base, stiff and hard, green, the margin splitting off as a number of white threads, minutely toothed towards the base; terminal spine weak, 5–8 mm, brown or greyish white. Spike 1–1.8 m, slender, loose, axis reddish. Flowers in 2s, 3s or 4s, 1.3–1.5 cm, pale yellow. *USA (Arizona), Mexico.* H5–G1.

9. A. toumeyana Trelease. Illustration: Lamb & Lamb, Cacti and other succulents **2**: t. 272 (1959).

Rosette trunkless, *c.* 30 cm in diameter, not producing suckers. Leaves 20–30 × 1.5–2 cm, linear, tapering slightly, dark green with some white markings on the upper surface; margins splitting off, at least in the upper part, into thin, white threads; terminal spine 1–2 cm. Spike dense. Flowers to 2.5 cm, greenish or pale yellow. *USA (Arizona).* H5–G1.

10. A. albicans Jacobi. Illustration: *Gardeners' Chronicle* **8**: t. 138 (1877); *Botanical Magazine*, 7207 (1891).

Trunk very short. Rosette with 20–30 leaves, not producing suckers. Leaves 30–60 × 7–13 cm, oblanceolate, narrowed towards the base, shortly tapering to the apex, soft, fleshy, fragile, pale green, glaucous or whitish; margins with weak teeth, at first colourless, later brown; terminal spine 1–2 cm, brown, soft, bristle-like, narrowly grooved. Spike *c.* 1 m, rather few-flowered above the middle. Flowers 4–6 cm, in pairs, reddish green, shortly stalked. *Mexico.* H5–G1.

Gentry (1982) treats this as *A. celsii* Hooker var. *albicans* (Jacobi) Gentry.

11. A. polyacantha Jacobi. Illustration: *Botanical Magazine*, 5006 (1857) – as *A. densiflora*; Jacobsen, Handbook of succulent plants, t. 96 (1960).

Rosette trunkless, to 65 cm in diameter with *c.* 30 leaves, at first simple, later forming mats, not producing suckers. Leaves 35–65 × 7–10 cm, oblanceolate, narrowed towards the base, tapering above, leathery, stiff and rigid, green (rarely glaucous?); margins sinuous with spines 2–3 mm, black-brown, triangular, confluent in the upper part into a brown, horny line; terminal spine 1–2.5 cm, grooved, black-brown. Spike to 2.5 m, dense. Flowers 5–7 cm, in pairs, light green or reddish. *Mexico.* H5–G1.

12. A. striata Zuccarini. Figure 7(4), p. 178. Illustration: *Botanical Magazine*, 4950 (1856); *Gardeners' Chronicle* **8**: t. 109 (1877); *Gartenflora* **29**: 24 (1880); Jacobsen, Handbook of succulent plants, t. 112 (1960).

Trunk short, but branched when old. Rosettes *c.* 1 m or more in diameter with 150–200 leaves. Leaves 25–60 cm × 5–10 mm, linear, tapering abruptly at the apex, 3-angled at the base, stiffly spreading, glaucous or grey-green with darker stripes; margins entire or roughened and saw-like; terminal spine 1–5 cm, needle-like, brown. Spike 3–3.6 m. Flowers closely packed, *c.* 3–4 cm, greenish yellow. *N Mexico.* H5–G1.

A. stricta Salm-Dyck (*A. striata* var. stricta (Salm-Dyck) Baker). Illustration: *Gartenflora* **31**: 56 (1882); *Cactus Journal* **1**: 139 (1898); Rowley, The illustrated encyclopaedia of succulents, 169 (1978); Brickell (ed.), RHS A–Z encyclopedia of garden plants, 88 (2003). Very similar to *A. striata* but leaves 25–50 cm × 8–10 mm, terminal spine 1–2 cm, 3 or 4-angled, spike to 2.5 m, dense. *S Mexico.* H5–G1.

Differs from *A. striata* in details of its flowers as well as in the characters mentioned above.

13. A. utahensis Engelmann. Illustration: *Report of the Missouri Botanical Garden* **7**: t. 32 (1896); Brickell (ed.), RHS A–Z encyclopedia of garden plants, 89 (2003).

Rosette trunkless, to 40 cm in diameter, usually producing suckers. Leaves 12–30 × 1.5–3 cm, tapering from the base, stiff, erect-spreading, glaucous; margins straight or sinuous, without a horny band towards the base, teeth 2–4 mm, 1–2.5 cm apart, hooked, triangular; terminal spine 2–4 cm, grey, grooved beneath, tip brown. Spike 1.5–4.5 m, sometimes slightly branched. Flowers in 2s or 4s, 2.5–3 cm, yellow. *SW USA (California, Arizona, Nevada, Utah).* H5–G1.

Subgenus **Agave.**
Inflorescence a panicle, the flowers borne in clusters at the ends of the branches; axis of inflorescence as broad as the crown of the rosette.

14. A. americana Linnaeus. Illustration: *Revue Horticole*, t. 77 (1904); Lamb & Lamb, Cacti and other succulents **2**: t. 266 (1959); Jacobsen, Handbook of succulent plants, t. 80–1 (1960).

Rosettes trunkless, 2–3 m in diameter with 20–60 leaves, producing suckers. Leaves 1–2 m × 15–30 cm, oblanceolate-spathulate, narrowed towards the base, rigidly spreading or reflexed towards the apex, smooth, leathery, usually glaucous; margins sinuous-toothed, marginal spines *c.* 8 mm, curved, blackish brown, later grey; terminal spine 3–5 cm, thick, obliquely flattened, rough at the base, not decurrent on the leaf margins. Panicle 5–8 m, slender, with 25–30 branches. Flowers 7–10 cm, pale yellow. *Mexico, widely naturalised in the Mediterranean area and elsewhere.* H5–G1.

A very variable species; some variants have variegated leaves.

15. A. franzosinii Baker. Illustration: *Gardeners' Chronicle* **12**: f. 31 (1892); *Botanical Magazine*, 8317 (1910); Jacobsen, Handbook of succulent plants, t. 97 (1960).

Rosette trunkless, to 4.5 m in diameter with 40–50 leaves, producing suckers. Leaves 2–2.5 m × 30–40 cm, narrowed towards the base, arching outwards, very rough, whitish grey variegated with green; margins sinuous-toothed, marginal spines *c.* 6 mm, triangular, hooked, black-brown, those at the base not confluent, the median 1.5–2 cm apart, the upper 5–6 cm apart; terminal spine 3–6 cm, needle-like, narrow, grooved, decurrent for *c.* 20 cm. Panicle to 11.5 m, elongate or widely cylindric-ovate. Flowers *c.* 8 cm, yellow, on stalks 1–2 cm. *Mexico.* H5–G1.

16. A. scabra Salm-Dyck. Illustration: *Report of the Missouri Botanical Garden* **22**: t. 75–9 (1911).

Rosette trunkless with *c.* 30 leaves, producing suckers. Leaves 60–110 × 12–16 cm, ovate, tapering, leathery, smooth, grey; margins sinuous-toothed, spines 8–15 mm, 1.5–2 cm distant, stout, straight or hooked; terminal spine 3.5–6 cm, widely grooved in the centre, decurrent. Panicle to 4 m. Flowers 6–8 cm, yellowish, shortly stalked. *Mexico.* H5–G1.

17. A. colorata Gentry. Illustration: Gentry, The Agave family in Sonora, USDA agricultural handbook **399**: t. 44 (1972).

Trunk short. Rosette with few leaves, producing suckers while young. Leaves 25–60 × 12–18 cm, broadly lanceolate, narrowed towards the base, thick, firm, glaucous, often red-tinted and cross-banded; margins conspicuously sinuous-toothed, marginal spines 5–10 mm, slightly flexed or straight, brown to greyish; terminal spine 3–5 cm, needle-like, grooved to the lower half. Panicle 2–3 m, narrow, with dense lateral branches in the upper half. Flowers 5–7 cm, yellowish green or coppery red. *Mexico (Sonora).* H5–G1.

18. A. palmeri Engelmann. Illustration: *Report from Missouri Botanical Garden* **7**: t. 48–52 (1896); *Cactus Journal* **2**: (1899); Gentry, The Agave family in Sonora, USDA agricultural handbook **399**: t. 35 (1972).

Rosette trunkless, 1–1.2 m in diameter with *c.* 30 leaves, ultimately producing suckers. Leaves to 80 × 7–10 cm, oblanceolate, narrowed towards the base, rigid, pale green, glaucous or reddish; margins sinuous-toothed, marginal spines of various sizes, usually alternating larger and smaller, purple or grey; terminal spine 3–6 cm, sharp, needle-like, grooved, long-decurrent as horny strips on the margins, brown or grey. Panicle 3–5 m, elongate. Flowers 4.5–5.5 cm, greenish yellow or waxy white, reddish in bud. *USA (New Mexico, Arizona), Mexico.* H5–G1.

19. A. atrovirens Salm-Dyck (*A. latissima* Jacobi; *A. coccinea* misapplied). Illustration: *The Garden* **76**: 251, 575 (1912); Jacobsen, Handbook of succulent plants, t. 83 (1960).

Rosette trunkless with 20–30 leaves, not producing suckers. Leaves 1.5–2 m × 25–40 cm, oblanceolate, narrowed towards the base, tapering from the upper one-third above, erect-spreading, very fleshy, dull dark green or blackish green; margins sinuous-toothed, upper spines distant, the lower confluent to a horny line; terminal spine *c.* 5 cm, narrowly grooved above, decurrent on the margins for up to 65 cm. Panicle 6–10 m, pyramidal, bearing 20–30 lateral branches. Flowers *c.* 10 cm, reddish to greenish yellow, borne on stalks to 1 cm. *Mexico.* H5–G1.

Material grown under this name may prove to be *A. salmiana* (p. 178).

20. A. potatorum Zuccarini. Illustration: Jacobsen, Handbook of succulent plants, t. 111 (1960).

Rosette trunkless, 20–25 cm in diameter with up to 80 leaves, not producing suckers. Leaves 25–40 × 9–18 cm, obovate-spathulate, narrowed towards the base, glaucous or green; margins usually sinuous-toothed, spines 5–10 mm, 1.2–3 cm distant, triangular, sharply pointed, at first yellow, brown or blackish, later grey, sometimes of 2 sizes alternating; terminal spine 3–4.5 cm, broadly furrowed above, little decurrent on the margins, brown. Panicle *c.* 3.5 m, pyramidal. Flowers 5.5–8 cm, yellow-green, on stalks 1–1.5 cm. *Mexico.* H5–G1.

Var. **verschaffeltii** (Lemaire) Berger is frequently cultivated: it has greyish green leaves and red- or yellow-brown spines.

21. A. marmorata Roezl. Illustration: *Botanical Magazine*, 8442 (1912).

Rosette more or less trunkless, to 2 m in diameter, with 30–50 leaves, not producing suckers. Leaves 1–1.3 m × 20–30 cm, oblanceolate-spathulate, narrowed towards the base, spreading, very rough, light grey, bluish or almost white; margins rough, sinuous between the spines; spines 6–12 mm, 4–5 cm distant, with smaller spines between the larger ones, all rusty brown; terminal spine 1.5–3 cm, needle-like, recurved, narrowly grooved, grey-brown. Panicle to 3.7 m, elongate. Flowers 4–5 cm, golden yellow. *Mexico.* H5–G1.

22. A. parrasana Berger (*A. wislizenii* Engelmann). Illustration: *Report of the Missouri Botanical Garden* **22**: t. 80–1 (1911); Jacobsen, Handbook of succulent plants, t. 107 (1960); Lamb & Lamb, Cacti and other succulents **1**: t. 435 (1960).

Rosette to 60 cm in diameter, not forming suckers. Leaves 20–30 × 10–12 cm, obovate, narrowed towards the base, shortly tapered to the apex, stiff, thick, smooth, dull glaucous green or bloomed; margins sinuous-toothed, the upper half with stout curved spines which are 5–10 mm; the upper 4–8 spines are confluent at their bases, all yellowish brown, later red, ultimately grey; terminal spine 2–3 cm, brown, decurrent to the upper, confluent, marginal spines. Panicle to 3.5 m. Flowers 5–6 cm, yellow. *Mexico.* H5–G1.

23. A. parryi Engelmann. Illustration: *Gardeners' Chronicle* **12**: t. 39 (1879); *Report of the Missouri botanical garden* **22**: t. 91–3 (1911); Brickell (ed.), RHS A–Z encyclopedia of garden plants, 88 (2003).

Rosette trunkless, with 20–30 leaves, producing suckers. Leaves 25–40 × 8–12 cm, broadly oblong, acute or at least tapering, grey or glaucous; margins sinuous, marginal spines 3–7 mm, 1–2.5 cm apart, narrowly triangular, straight or recurved, brown later grey, bases of the uppermost spines confluent; terminal spine 1.5–3 cm, more or less linear, openly grooved, flat to the tip on the upper side, brown later grey, long-decurrent. Panicle 3–5 m, branches curved upwards. Flowers 6–7.5 cm, stalked, red in bud, cream-yellow when open. *SW USA, Mexico.* H5–G1.

24. A. fourcroydes Lemaire. Illustration: *Memoirs of the National Academy of Sciences* **11**: t. 110–12 (1913); *Botanical Magazine*, 8746 (1918); Parey's Blumengärtnerei **1**: 355 (1958).

Trunk to 1 m. Rosette 2–2.5 m in diameter, with about 100 leaves, producing suckers. Leaves 1.2–1.8 m × 8–12 cm, linear or oblanceolate, rigid, smooth, glaucous or grey; margins straight, upcurved, marginal spines 3–6 mm, regularly spaced at 2–3 cm intervals, slender, dark brown; terminal spine 2–3 cm, narrowly conical, openly and shortly grooved above the base, dark brown. Panicle 5–6 m, elongate, pyramidal. Flowers 6–7 cm, greenish yellow. *Mexico.* H5–G1.

25. A. macroacantha Zuccarini. Illustration: *Botanical Magazine*, 5940 (1871) – as A. besseriana; *Gardeners' Chronicle* **8**: t. 27 (1877); *Report of the Missouri Botanical Garden* **18**: t. 18–26 (1907).

Trunk very short or absent. Rosette with 30–50 leaves, producing suckers. Leaves

25–35 × 2.5–3 cm, oblanceolate, narrowed towards the base, tapering towards the apex, erect-spreading, very rigid and stiff, glaucous to whitish grey; margins finely rough and cartilaginous between the spines, spines few, 3–4 mm, hooked and curved forwards, all similar; terminal spine 3–3.5 cm, straight, 3-angled, decurrent. Panicle to 3 m, slender, with 10–15 branches. Flowers 5–5.5 cm, numerous, greenish flushed with red, shortly stalked. *Mexico.* H5–G1.

26. A. salmiana Salm-Dyck (*A. atrovirens* var. *salmiana* (Salm-Dyck) Trelease). Illustration: *Gardeners' Chronicle* **29**: t. 51 (1886); *The Garden* **29**: 495 (1886); *Revue Horticole*, t. 40–1 (1872); Parey's Blumengärtnerei **1**: 357 (1958).

Rosette large, loosely leaved, forming suckers. Leaves 1–2 m × 20–35 cm, oblanceolate, narrowed towards the base, tapered above, firm and smooth, rigid, grey-green; margins sinuous, marginal spines 1–2 cm (including the fleshy cushions on which they stand), confluent at the base of the leaf, the median spines 2–3 cm distant, the upper 7–8 cm distant; terminal spine 5–10 cm, black-brown, long-decurrent. Panicle 9–19 m. Flowers 8–11 cm, greenish yellow. *Mexico.* H5–G1.

Var. **ferox** (Koch) Gentry (*A. ferox* Koch). Illustration: *Gardeners' Chronicle* **20**: t. 94 (1896) & **43**: t. 173 (1908); Jacobsen, Handbook of succulent plants, t. 95 (1960). Leaves shining green. *Mexico.* H5–G1.

27. A. sisalana Perrine. Illustration: *Die Tropenpflanze* **2**: 212 (1898) & **4**: 7 (1900); *Memoirs of the National Academy of Sciences* **11**: t. 113–15 (1913).

Trunk to 1 m. Rosette dense, stiff and spreading in all directions, producing suckers. Leaves 90–130 × 9–12 cm, narrowly oblanceolate, narrowed towards the base, grey-green or bright green; margins without spines (when mature) but with a cartilaginous edge; terminal spine to 2.5 cm, hard, conical, obliquely flattened on the upper surface, shallowly grooved, black-brown. Panicle 6–7 m. Flowers 5–6.5 cm, green. *Mexico, widely cultivated for fibres elsewhere.* H5–G1.

6. MANFREDA Salisbury

C.J. Couper

Perennial herbs with succulent roots, fleshy underground stems or bulbous rootstocks. Stems absent or very short. Leaves in a basal rosette, dying off annually, small, pliable, without spines, margins often wavy, green, often with brown, red or purple blotches. Spike loose with several bisexual flowers. Perianth radially symmetric, lobes 6, united below and forming a usually narrow and deep cylindric tube, the lobes equal or almost so, spreading or reflexed, green, yellow or reddish purple. Stamens 6, filaments borne on the perianth-tube, long-projecting, anthers versatile. Ovary inferior, 3-celled with numerous ovules in 2 rows in each cell; style long-projecting, stigma club-shaped, 3-lobed. Fruit an erect, apically dehiscent capsule. Seeds numerous, black.

A small genus related to *Agave* and requiring similar cultivation conditions (p. 181). Propagation is by seed.

1a. Leaves oblong-spathulate, reddish-striped; filaments borne at the base of the perianth-tube **1. virginica**
b. Leaves narrowly linear or linear, mottled with brown; filaments borne near the top of the perianth-tube 2
2a. Leaves to 2 cm broad; stamens about as long as the lobes of the perianth; capsules about as broad as long **3. maculosa**
b. Leaves more than 2 cm broad; stamens much longer than the lobes of the perianth; capsules much longer than broad **2. variegata**

1. M. virginica (Linnaeus) Salisbury (*Agave virginica* Linnaeus; *Polianthes virginica* (Linnaeus) Shinners). Illustration: *Botanical Magazine*, 1157 (1808); *Report of the Missouri Botanical Garden* **7**: t. 26–7 (1896); Gleason, New illustrated flora of the northern USA and Canada, edn 3, **1**: 445 (1952).

Rosette with 6–15 leaves. Leaves to 60 × 2.5–5 cm, oblong-spathulate, gradually tapered to the base, sharply tapered at the apex, somewhat grooved, thick, flaccid, dark green striped with red; margins wavy. Scape 80–180 cm, slender; spike 30–50 cm with *c.* 30 flowers. Flowers 2.5–5 cm, the lower on stalks 6–8 mm, greenish or brownish yellow, scented; perianth-lobes linear-oblong. Filaments borne at the base of the perianth-tube. Anthers *c.* 1.2 cm. Capsules 1.5–2 cm, longer than broad. *USA (Maryland to Missouri, Florida and Texas).* H5–G1. Summer.

2. M. variegata (Jacobi) Rose (*Agave variegata* Jacobi; *Polianthes variegata* (Jacobi) Shinners). Illustration: Saunders et al., Refugium botanicum **5**: t. 326 (1872).

Rosette with few leaves. Leaves 20–45 × 2–4 cm, lanceolate, narrowed slightly to the base and gradually upwards to a blunt point, deeply grooved, fleshy, glaucous and with brown markings; margins curved upwards. Scape 90–130 cm. Flowers *c.* 4 cm, almost stalkless, brownish green, fragrant; lobes narrow, about as long as the tube or longer. Stamens to 5 cm, filaments borne at the top of the perianth-tube. Anthers *c.* 8 mm. Capsules much longer than broad, 1.5–2.2 cm. *USA (S Texas), Mexico.* H5–G1. Spring.

3. M. maculosa (Hooker) Rose (*Agave maculosa* Hooker; *Polianthes maculosa* (Hooker) Shinners). Illustration: *Botanical Magazine*, 5122 (1859); *Gardeners' Chronicle*, 1194 (1872); Bailey, Standard cyclopaedia of horticulture **1**: t. 2319 (1916).

Rosette with few leaves. Leaves 15–30 × 1–2 cm, linear-lanceolate, recurved, concave, thick, fleshy, glaucous with brown or greenish markings; margins transparent with small, unequal teeth. Scape 90–120 cm; spike 20–30 cm, loosely flowered. Flowers 10–18, 4–5 cm, more or less stalkless, perianth greenish white but turning pink with age, fragrant; lobes oblong, shorter than the tube. Stamens slightly projecting, filaments borne at the top of the perianth-tube, anthers 9–16 mm. Capsules about as broad as long, 2–2.5 cm. *USA (S Texas), N Mexico.* H5–G1. Spring–summer.

7. POLIANTHES Linnaeus

C.J. Couper

Small perennial herbs with short, tuberous rootstocks. Leaves succulent, deciduous, without spines, mostly basal, those on the stem small and decreasing in size upwards. Flowers in loose racemes or spikes, bisexual, bilaterally symmetric, white or orange-red. Perianth-lobes 6, united below into a cylindric or funnel-shaped tube that is bent near the base and expanded near the middle, much longer than the lobes. Stamens 6, filaments inserted on the perianth-tube. Ovary inferior, 3-celled, with numerous ovules and a 3-lobed stigma. Fruit an ovoid capsule with the perianth persistent at its apex. Seeds flat.

A genus of 2 species from Mexico. They are easily grown in well-drained soils, though, if grown outdoors in areas prone to frost, the tubers should be lifted in the autumn and stored in a cool (but not cold), dry atmosphere. Propagation is mainly by suckers from the tubers.

1a. Flowers waxy white; perianth-tube *c.* 3 times longer than the lobes; each flower pair subtended by 1 bract **1. tuberosa**
b. Flowers pale red or orange; perianth-tube more than 3 times longer than the lobes; each flower pair subtended by 1 bract and 2 bracteoles **2. geminiflora**

1. P. tuberosa Linnaeus. Illustration: Saunders et al, Refugium botanicum **1**: t. 63 (1815); *Botanical Magazine*, 1817 (1816); Hay & Synge, Dictionary of garden plants, t. 630 (1969).

Tuber conical. Basal leaves 6–9, 30–45 × 1–2 cm, linear-lanceolate, smooth, entire, bright green spotted with brown on the back, reddish near the base; stem-leaves long-acuminate with clasping bases. Spike 50–100 cm, rigid, bracts leaf-like; each flower pair subtended by 1 bract only. Flowers 3.5–6 cm, the tube 2.5–3.5 cm, the lobes 1–2 cm, all waxy white, strongly fragrant. *Mexico*. H5–G1. Summer.

Normally seen in cultivation as a double-flowered variant.

2. P. geminiflora (Llave & Lexarza) Rose (*Bravoa geminiflora* Llave & Lexarza). Illustration: *Botanical Magazine*, 4741 (1853); *Contributions from the US National herbarium* **8**: t. 4 (1903).

Tuber shallow. Basal leaves 30–50 × 1–1.3 cm, linear, subulate, green; stem-leaves lanceolate, erect and adpressed to the stem. Spike 8–40 cm; each flower pair subtended by a bract and 2 bracteoles. Flowers *c.* 2 cm, strongly bent downwards near the base, pale red or orange, the perianth-tube more than 3 times longer than the lobes, fragrant. *Mexico*. H5–G1. Summer.

8. CORDYLINE R. Brown

J. Cullen

Plants woody, tufted or tree-like, branching usually falsely dichotomous. Leaves crowded in rosettes at the ends of the branches, persistent, usually lanceolate or sword-shaped, stalked, the lateral veins diverging from the conspicuous midrib at an acute angle. Flowers in dense, terminal panicles, which often appear lateral due to the growth of a new vegetative shoot from a nearby leaf axil. Flowers bisexual, bracteate (often with 2 bracts to each flower-stalk), stalked, radially symmetric. Perianth of 6 lobes united into a short tube at the base, the lobes usually reflexed. Stamens 6, anthers versatile, filaments flattened, attached to the top of the perianth-tube. Ovary superior, 3-celled, with several ovules in each cell. Fruit a small berry that dries with age; seeds several, curved, black.

A genus of about 20 species from SE Asia, Australasia, Polynesia and Hawaii, cultivated for their striking overall appearance. The genus is difficult to distinguish as a whole from *Dracaena* (p. 189), and the number of ovules in each cell of the ovary (several in *Cordyline*, one in *Dracaena*) is the only certain character, though the leaf venation (all veins parallel and entering the leaf from the base in *Dracaena*, parallel but branching from the midrib at a very acute angle in *Cordyline*) is usually characteristic; see Figure 7(7 & 8), p. 178. Most species have names under *Dracaena* and these names are often found in catalogues.

The plants are easily cultivated in a good, well-drained soil. A few are grown as pot plants in glasshouses and rooms, and require similar soil and a position with good light. Propagation as for *Dracaena* (p. 189).

1a. Leaves 10–20 cm, 2–3 times longer than broad **7. haageana**
b. Leaves 30–200 cm, more than 3 times longer than broad 2
2a. Mature leaves 1 m or more, 4–15 cm broad 3
b. Mature leaves to 1 m, 1–10 cm broad 4
3a. Leaves 4–8 cm broad, not glaucous beneath **5. banksii**
b. Leaves 10–15 cm broad, glaucous beneath **6. indivisa**
4a. Leaves less than 3 cm broad 5
b. Leaves more than 3 cm broad 6
5a. Perianth-lobes all equal, white or pinkish, 4–5 mm; flowers very widely spaced **3. pumilio**
b. Perianth-lobes unequal, the outer 3 lobes shorter than the inner 3, all pale purple; flowers very dense **4. stricta**
6a. Plant a tree to 20 m, usually with an obvious trunk; flowers scented **1. australis**
b. Plant a shrub to 4 m, without an obvious trunk; flowers not scented **2. fruticosa**

1. C. australis (Forster) Endlicher (*Dracaena australis* Forster). Illustration: *Botanical Magazine*, 5636 (1867).

Tree to 20 m, the trunk ultimately massive, to 1.5 m in diameter, usually little branched below, richly branched above, bark rough and fissured. Leaves 30–100 × 3–6 cm, lanceolate, light green on both surfaces, midrib and veins indistinct. Panicle to 1.5 m × 60 cm, much branched, the branches more or less at right angles and well spaced, the axes almost hidden by the flowers. Flowers almost stalkless, sweetly scented; perianth 5–6 mm, white, lobes equal, free almost to the base, slightly keeled, reflexed. Berry *c.* 4 mm in diameter, spherical, whitish. *New Zealand*. H5–G1.

C. baueri Hooker. Trunk slender, to 3 m, leaves 15–60 × 5–6.5 cm. *New Zealand (Norfolk Island)*. H5–G1.

2. C. fruticosa (Linnaeus) Chevalier (*C. terminalis* (Linnaeus) Kunth; *Dracaena glauca* Anon.). Illustration: Graf, Exotica, edn 8, 1070–1 (1973).

Shrub to 4 m, stems erect, slender and clustered. Leaves 30–60 × 5–10 cm, conspicuously stalked, the stalk grooved, the leaf green or variably tinged with red or purple. Panicles to 60 cm, richly branched, broad, with many flowers, the axis clearly visible between them. Flowers almost stalkless, *c.* 6 mm, white, yellowish or reddish, the 3 outer lobes somewhat shorter than the 3 inner, all reflexed. Fruit *c.* 8 mm in diameter, spherical, red. *Tropical SE Asia, Australia, Hawaii*. H5–G1.

Very variable in leaf coloration; numerous cultivars have been selected on the basis of this character.

C. rubra Kunth. Shrub to 5 m, leaves 30–40 × 4–5 cm; flowers *c.* 7 mm, pale purple. *Origin unknown*. H5–G1.

3. C. pumilio J.D. Hooker.

Shrub to 2 m, often flowering when smaller. Stem to 1.5 cm wide, arching. Leaves 30–100 × 1–2 cm, green, midrib conspicuous at least beneath, narrowed to a channelled stalk. Panicle to 60 cm, very open and with slender branches. Flowers almost stalkless, widely spaced; perianth 4–5 mm, all the lobes similar in size, recurved, white or sometimes pink outside. Fruit 4–5 mm in diameter, spherical, bluish or flecked with blue spots. *New Zealand*. H5–G1.

4. C. stricta (Sims) Endlicher (*Dracaena stricta* Sims; *Dracaena congesta* Anon.).

Illustration: Morley & Everard, Wild flowers of the world, t. 142 (1970).

Shrub to 4 m. Leaves 30–60 × 1.5–2.5 cm, linear-lanceolate, green. Panicle to 60 cm, pyramidal, richly branched. Flowers almost stalkless; perianth *c.* 7 mm, pale purple, the lobes reflexed, the outer 3 conspicuously shorter than the inner 3. *Australia.* H5–G1.

5. C. banksii Hooker (*Dracaena beuckelarii* Koch).

Shrub to 4 m, stems several, 10–15 cm in diameter. Leaves 1–2 m × 4–8 cm, linear-lanceolate, green, stalked, midrib conspicuous above, as are several lateral veins. Panicle to 2 m, richly branched and with many flowers, axes visible between the flowers, which are very fragrant. Perianth *c.* 1 cm, all lobes equal, spreading, white. Fruit 4–5 mm in diameter, spherical, white or bluish. *New Zealand.* H5–G1.

6. C. indivisa (Forster) Steudel (*C. hookeri* Kirk; *Dracaena indivisa* Forster). Figure 7(7), p. 179. Illustration: *Illustration Horticole* **7**: t. 264 (1860); Brickell (ed.), RHS A–Z encyclopedia of garden plants, 304 (2003).

Shrub to 8 m with massive, sparingly branched stems. Leaves 1–2 m × 10–15 cm, narrowly lanceolate, green above, glaucous beneath, the midrib broad and conspicuous, often red, sometimes the stronger lateral veins also red. Panicle to 1.6 m, compact, slightly branched, the axes hidden by the crowded flowers, which are borne on stalks 2–3 mm long. Perianth 7–8 mm, lobes all equal, strongly recurved, whitish inside, pale purple outside. Fruit *c.* 6 mm in diameter, spherical, bluish. *New Zealand.* H5–G1.

7. C. haageana Koch. Illustration: *Gartenflora* **20**: t. 675 (1871) & **27**: 115 (1878).

Small shrub to 1 m or more. Leaves 10–20 × *c.* 6 cm, broadly lanceolate, abruptly contracted into a long stalk, green. Panicle *c.* 30 cm, with few branches and with numerous flowers that are almost stalkless. Perianth *c.* 7 mm, lobes all equal, reflexed, white inside, purplish outside. Fruit *c.* 3 mm in diameter, spherical. *Australia (Queensland).* G1–2.

28. DORYANTHACEAE

Plant forming a loose to dense rosette, stems not obvious. Leaves leathery, blade-shaped, each leaf terminating in a cylindrical body. Flowers 10–15 cm long, bisexual, actinomorphic to weakly zygomorphic, in panicles. Perianth of 6 spreading segments. Stamens 6. Ovary of 3 united carpels, inferior; ovules many, axile. Fruit a capsule.

A single genus with 2 species, occasionally cultivated.

1. DORYANTHES Correa

C.J. Couper

Large, monocarpic or perennial leaf-succulents, without fleshy underground stems. Trunk absent. Leaves in a dense basal rosette, persistent, sword-shaped, the apex solid and cylindric but drying and soon falling. Inflorescence to 6 m, a terminal, spherical panicle of short spikes, the axis bearing much reduced, basally sheathing leaves. Flowers bisexual, bright red, often replaced by bulbils. Perianth radially symmetric, formed from 6 lobes that are united at their extreme bases into a short tube, the lobes sickle-shaped. Stamens 6, inserted at the bases of the perianth-lobes; anthers erect, basifixed, not projecting from the perianth. Ovary inferior, 3-celled, with many ovules in vertical rows; style elongate; , stigma capitate. Capsule with numerous, flattened, brown, kidney-shaped seeds.

A small genus from Australia. Cultivation is in general as for *Agave* (p. 181), but the plants prefer a richer soil. Propagation is by seed, by bulbils, or by suckers, which are sparingly produced after flowering.

1a. Leaves 1.8–2.4 m, ribbed, the cylindric point at the apex 10–15 cm **1. palmeri**
b. Leaves 1.2–1.8 m, smooth, the cylindric point at the apex 5–7.5 cm **2. excelsa**

1. D. palmeri Hill. Figure 7(2), p. 179. Illustration: *Flore des Serres* **10**: t. 2097–8 (1874); *Botanical Magazine*, 6665 (1883); *Revue Horticole*, 548 (1891); *Gartenflora* **61**: 324 (1912); Brickell (ed.), RHS A–Z encyclopedia of garden plants, 387 (2003).

Leaves 1.8–2.4 m × 10–15 cm, strongly ribbed, narrowed downwards to a winged, stalk-like base that is *c.* 2.5 cm broad; the cylindric apex of the leaf 10–15 cm. Flowering axis to 4 m, panicle 1.8–3 m, bearing numerous, short, ascending leaves. Bracts *c.* 5 cm, ovate, bright red. Perianth-lobes lanceolate, not longer than the ovary, bright orange-red outside, white inside. Stamens a little shorter than the perianth-lobes; anthers *c.* 1.2 cm, linear-oblong. *Australia (Queensland).* G1.

2. D. excelsa Correa. Illustration: *Botanical Magazine*, 1685 (1814); *Gartenflora* **42**: t. 421 (1864); *Revue Horticole*, 465–6 (1865), 548 (1891); *Flore des serres* **8**: t. 1912–13 (1869).

Leaves 1.2–1.8 m × 7.5–10 cm, smooth, narrowed gradually to a winged base that is *c.* 2.5 cm broad; the cylindric apex of the leaf 5–7.5 cm. Flowering axis to 6 m, bearing numerous ascending leaves. Bracts *c.* 7.5 cm, bright red. Perianth-lobes 6–7.5 cm, linear, longer than the ovary. Stamens almost as long as perianth-lobes; anthers *c.* 2.5 cm, linear. Ovary 3-angled in cross-section. *Australia (Queensland, New South Wales).* G1.

29. NOLINACEAE

Shrubs or trees, base of the trunk often swollen. Leaves not stalked, leathery or parchment-like. Flowers in panicles, mostly unisexual, actinomorphic; flower-stalks jointed near their bases. Perianth of 6 free or united segments. Stamens 6. Ovary of 3 united carpels, inferior; ovules 2 in each cell of the ovary, axile. Fruit dry, indehiscent.

Three genera with about 50 species. One is native to North America, and all are cultivated.

1a. Ovary 3-celled with 2 ovules in each cell **1. Nolina**
b. Ovary 1-celled with 2 or 3 ovules altogether 2
2a. Trunk conspicuously inflated at the extreme base **3. Beaucarnea**
b. Trunk not conspicuously inflated at base **2. Dasylirion**

1. NOLINA Michaux

C.J. Couper

Dioecious trees with woody trunks. Leaves many, persistent, borne in a rosette at the end of the trunk, narrowly linear, stiff, usually erect, more rarely ascending or drooping, often with rough margins. Panicle compound, at the apex of an almost leafless scape, its main branches subtended by tapering, triangular bracts. Flower-stalks jointed near their bases. Flowers mostly unisexual (occasionally a few bisexual), radially symmetric. Perianth of 6 segments that are whitish, persistent, papery, 1-nerved and small. Stamens 6, the filaments short and slender but dilated at their bases. Ovary superior, deeply 3-lobed, 3-celled; ovules 2 in each cell. Fruit a thin-walled, often inflated capsule with 1 seed per cell, often only 1 cell developing. Seeds spherical to oblong, wrinkled.

A genus of about 25 species from the southern USA and Mexico. Cultivation as for *Yucca* (000).

Literature: Trelease, W.T., The Desert Group Nolineae, *Proceedings of the American Philosophical Society* **50**: 404–43 (1911).

1a. Leaves 90–100 cm, tip entire, margins splitting off as threads **1. bigelovii**
b. Leaves more than 2 m, tip frayed, margins rough **2. longifolia**

1. N. bigelovii (Torrey) Watson (*Dasylirion bigelovii* Torrey; *Beaucarnea bigelovii* (Torrey) Baker). Illustration: *Contributions from the US National Herbarium* **16**: t. 109 (1916); Jacobsen, Handbook of succulent plants, 697 (1960); Wiggins, Flora of Baja California, 833 (1980).

Trunk to 3 m, its base club-shaped, to 70 cm in diameter, rough-barked, much narrowed towards the apex, with few, short branches and leaves in rosettes at their ends. Leaves 90–100 × 2–2.5 cm, bases triangular, the rest linear, slightly tapering, not keeled on the backs; margins splitting off as brown fibres, the tip entire, not frayed and brush-like. Perianth-segments of female flowers 2.5–3 mm, the outer 3 strongly reflexed. Fruit-stalks half as long as the capsules. *USA (Arizona) to Mexico (Baja California, Sonora).* G1. Spring–summer.

2. N. longifolia (Schultes) Hemsley (*Yucca longifolia* Schultes; *Dasylirion longifolium* (Schultes) Zuccarini; *Beaucarnea longifolia* (Schultes) Baker). Illustration: *Gardeners' Chronicle* **16**: 67 (1894); *Revue Horticole* **11**: 206 (1911); Jacobsen, Handbook of succulent plants, 689 (1960).

Trunk more than 2 m, rough-barked, swollen at the base with a thick, corky ring, narrowing towards the apex, with a few short branches and a dense crown of leaves. Leaves more than 2 m × 2.5 cm, long-tapering, thin, firm; margins minutely rough, the apex frayed and brush-like. Flowering axis to 2 m. Perianth-segments of female flowers *c.* 1.5 mm, white. Fruit-stalks as long as the inflated capsules. *Mexico.* G1. Spring–summer.

2. DASYLIRION Zuccarini

C.J. Couper

Dioecious shrubs or trees. Trunk short, stout, or absent. Leaves in a dense, terminal rosette, persistent, narrowly linear, hard and striped, broad and concave at the base, margins with hooked spines, apex entire or fibrous. Flowering axis tall, bearing dry, lanceolate, bract-like leaves. Panicle narrow, dense; bracts membranous, persistent, flower-stalks short, jointed near the apex. Flowers unisexual, bell-shaped, radially symmetric. Perianth-segments 6, oblong-ovate, whitish, all equal, finely toothed. Stamens 6, projecting. Ovary superior (rudimentary in male flowers), 1-celled with 2 or 3 ovules; style short, erect; stigmas 3. Fruit dry, indehiscent, 3-angled, the angles often winged, with 1 obtusely triangular seed.

A genus of about 15 species from southern USA to Mexico, easily grown in well-drained soil. Propagation is by seed.

Literature: Trelease, W.T., The desert group Nolineae, *Proceedings of the American Philosophical Society* **50**: 404–43 (1911).

1a. Trunk 80 cm or more 2
b. Trunk less than 80 cm, or absent altogether 4
2a. Leaves to 6 mm broad, apex entire **3. longissimum**
b. Leaves more than 8 mm broad, apex fibrous 3
3a. Leaves more than 2.5 cm broad, glaucous **1. wheeleri**
b. Leaves less than 1.8 cm broad, pale green, scarcely glaucous **2. acrotrichum**
4a. Leaf-apex entire **7. glaucophyllum**
b. Leaf-apex fibrous 5
5a. Leaves less than 1.5 cm broad, bright green **6. graminifolium**
b. Leaves more than 1.5 cm broad, glaucous 6
6a. Leaves rough; marginal spines 1–3 mm, deep yellow **5. serratifolium**
b. Leaves smooth; marginal spines 3–4 mm, orange or reddish **4. leiophyllum**

1. D. wheeleri Watson. Illustration: *Plant World* **10**: 255 (1907); de Wildeman, Icones selectae Horti Thenensis **6**: t. 225 (1908); *USDA Bulletin* **728**, t. 5 (1918).

Trunk at least 1 m. Leaves *c.* 100 × 2.5–4 cm, glaucous, margins finely toothed and with hooked, reddish brown spines that are 1.5–3 mm long and 1–2 cm apart. Flowering axis 2.5–5 m. Panicle 1.5–2.5 m, the branches flexuous and hanging. Bracts cup-shaped, irregularly toothed, 2–3 mm. Perianth *c.* 2 mm. Fruit 7–9 × 4–8 mm, narrowly reversed heart-shaped, deeply notched at the apex. Seeds 3–4 mm, 3-angled. *USA (Arizona to Texas), N Mexico.* G1. Summer.

2. D. acrotrichum Zuccarini (*D. gracile* Planchon). Illustration: *Botanical Magazine*, 5030 (1858); *Flore des serres*, **4**: t. 1448 (1861); *Gartenflora* **30**: 24 (1881); Brickell (ed.), RHS A–Z encyclopedia of garden plants, 358 (2003).

Trunk 90–150 cm, robust. Leaves 60–100 cm × 9–18 mm, the outer recurved, all pale green, margins distinctly and finely toothed between the hooked, pale yellow, brown-tipped spines that are 1–2 mm long; apex divided into 20–30 fibres. Flowering axis 2.5–4 m. Panicle 1.2–1.5 m, dense and cylindric. Bracts *c.* 2 mm, ovate, entire. Perianth *c.* 2 mm. Fruit *c.* 8 × 6–8 mm, obovoid, shallowly notched, wings broadening upwards, free from the style. *Mexico.* G1. Summer.

3. D. longissimum Lemaire (*D. quadrangulatum* Watson). Illustration: Parey's Blumengärtnerei **1**: 307 (1958).

Trunk 90–200 cm. Leaves very numerous, 1.2–1.8 m × *c.* 6 mm, spreading in all directions, 4-angled (the upper and lower surfaces raised to low keels), margin smooth, apex entire. Flowering axis 1.5–5 m. Fruit 7–9 mm, scarcely notched. *Mexico.* G1. Summer.

4. D. leiophyllum Engelmann.

Trunk less than 80 cm. Leaves *c.* 100 × 2.5–3 cm, smooth, glossy green or somewhat glaucous, margin with spines 3–4 mm long and 1–1.5 cm apart, recurved, becoming orange or reddish; apex fibrous. Panicle to 3 m. Perianth greenish, to 2 mm. Fruit 6–9 × 2–6 mm, broadly ellipsoid, openly and deeply notched at the apex, style thick. Seeds *c.* 3 mm. *USA (Texas, New Mexico), Mexico.* G1. Summer.

5. D. serratifolium (Schultes) Zuccarini (*Yucca serratifolia* Schultes; *D. laxiflorum* Baker). Illustration: *Report of the Missouri Botanical Garden* **14**: t. 12 (1903).

Plant trunkless or almost so. Leaves 60–100 × 1.5–3 cm, glaucous, rough, margins finely toothed and with hooked, deep yellow spines 1–3 mm long and

2–3 cm apart; apex fibrous. Panicle loose. Perianth *c.* 2 mm. Fruit 6–8 × 6–8 mm, more or less spherical, broadly winged, apex notched, style as long as the notch. *SE Mexico.* G1. Summer.

6. D. graminifolium (Zuccarini) Zuccarini (*Yucca graminifolia* Zuccarini).

Trunk short, less than 80 cm. Leaves 90–120 × 1.2–1.4 cm, linear, long-acuminate, bright green, glossy, margin finely toothed between the spines, which are 1–2 mm long, horny and yellowish; apex of 6–8 spreading fibres. Flowering axis 2.4–2.7 m. Panicle narrow. Perianth *c.* 2 mm. Fruit ellipsoid, 6–9 mm. *Mexico.* G1. Summer.

7. D. glaucophyllum Hooker. Illustration: *Botanical Magazine*, 5041 (1858); *Revue Horticole*, **11**: 87 (1911).

Trunk at most 30 cm. Leaves 60–120 × 1.2–1.8 cm, linear, acuminate, intensely glaucous, marginal spines 1–2 mm, hooked, deep yellow, the margin finely toothed between them; apex entire. Panicle to 1.2 m, dense. Perianth *c.* 2 mm, greenish white, lobes red at their tips. Fruit 6–8 mm broad, style half as long as the notch at the apex. *E Mexico.* G1. Summer.

3. BEAUCARNEA Lamarck

C.J. Couper

Dioecious trees. Trunk elongate, irregularly scarred by the remains of leaf-bases, its base strongly inflated, more deeply scarred than the part above. Leaves in rosettes, persistent, broadly linear, stiff, acuminate, grooved, the bases dilated and stem-clasping, completely hairless, margins smooth or slightly rough. Flowers unisexual (rarely a few bisexual), radially symmetric, numerous, very small, in a wide, terminal panicle. Each flower shortly stalked, white, slightly fragrant, subtended by a bracteole, the male flowers falling quickly. Perianth-segments 6. Ovary superior, 1-celled, with 2–3 ovules. Fruit a 3-winged capsule.

A genus of about 6 species from the southern USA and Mexico, formerly included in *Nolina*. Cultivation as for *Yucca* (p. 175).

1a. Leaves 90–180 cm, green, margins smooth **1. recurvata**
b. Leaves less than 90 cm, glaucous or grey, margins rough 2
2a. Leaves 60–90 cm × 6–12 mm, margins yellowish, slightly rough; capsule *c.* 8 × 6 mm **2. stricta**
b. Leaves 45–50 cm × 3–5 mm, at least the margins glaucous, rough; capsule *c.* 9 × 9 mm **3. gracilis**

1. B. recurvata Lemaire (*Nolina recurvata* (Lemaire) Hemsley; *B. tuberculata* Roezl; *N. tuberculata* invalid). Figure 7(5), p. 178. Illustration: *Illustration Horticole* **8**: 58 (1961); *Flore des serres* **18**: 26 (1869); *Gardeners' Chronicle* **46**: 4 (1909); Jacobsen, Handbook of succulent plants, 698 (1960).

Plant to 9 m. Trunk 4–6 m, 50 cm or more in diameter at the base, with few branches above. Leaves 90–180 × *c.* 2 cm, linear, tapering slightly, curved, thin, flat, though grooved, green, margins smooth. Flowering axis bearing flowering branches almost to the base. Perianth *c.* 1.5 mm. *Mexico.* G1.

2. B. stricta Lemaire (*Nolina stricta* (Lemaire) Ciferri & Giacomini; *B. glauca* Roezl).

Plant to 10 m. Leaves 60–90 cm × 6–12 mm, stiffly spreading, keeled, pale or glaucous grey, margins yellow, slightly rough, the surface with papillose grooves. Flowering axis bearing flowering branches almost to the base. Perianth *c.* 1.5 mm. Fruit 8 × 6 mm, seeds 3–4.5 mm. *Mexico.* G1.

3. B. gracilis Lemaire (*Nolina gracilis* (Lemaire) Ciferri & Giacomini; *B. oedipus* Rose).

Like *B. stricta* but leaves 45–50 cm × 3–6 mm, glaucous, very rough on the margins. Fruit *c.* 9 × 9 mm, seeds *c.* 3 mm. *Mexico.* G1.

30. DRACAENACEAE

Plants mostly woody. Leaves erect and blade-like, forming a tuft on a woody rhizome, or in terminal rosettes, not stalked, parchment-like or leathery. Flowers all bisexual, actinomorphic to weakly zygomorphic, in racemes or panicles, flower stalks not jointed. Perianth of 6 free segments. Stamens 6. Ovary of 3 united carpels, inferior; ovules 1–many, axile. Fruit a berry.

There are 2 genera, both commonly cultivated. *Cordyline*, currently included in the *Agavaceae* (000), may perhaps be better placed here.

1a. Plant tree-like with fleshy leaves, or shrubby with thin, woody stems and thin leaves, or herbaceous with thin leaves; rhizomes absent (our species) **1. Dracaena**
b. Plant not tree-like, usually stemless, leaves always fleshy; creeping or subterranean rhizomes present **2. Sansevieria**

1. DRACAENA Linnaeus

J.J. Bos & J. Cullen

Plants herbaceous or woody, sometimes forming trees. Roots usually orange. Leaves usually spirally arranged, occasionally in false whorls, stalked or stalkless, green or frequently variegated with white, silver or yellow, sometimes with red margins; venation always parallel. Inflorescence a panicle of varying form, with elongate branches or the branches short and condensed, the whole sometimes head-like; major bracts sometimes large, flowers 2 or more to each floral bract. Flowers bisexual, usually small, stalkless or with short stalks, usually fragrant and opening at night. Perianth radially symmetric, with 6 lobes that are united into a long or short tube at the base, rarely all almost free, the lobes spreading or recurving, white, greenish or with red midribs to each lobe. Stamens 6, inserted at the top of the perianth-tube, filaments slender or thickened. Ovary 3-celled, each cell with 1 ovule. Fruit a spherical or lobed, red, orange or yellow berry containing 1–3 seeds.

A genus of about 60 species, mostly from Africa. They are mainly grown for their patterned or variegated leaves rather than for their flowers, and tend to be retained only while in their juvenile stages (many which are popular as house plants are capable of becoming much too large for most sites); for several species apparently nothing is recorded of their flowers and fruits. Hence, the identification of cultivated material has to be based, as far as possible, on the leaves; this is not entirely satisfactory. Some plants are grown merely under cultivar names (e.g. 'Song of India') and are, perhaps, of hybrid origin. Plants grow well in glasshouses or in heated domestic rooms or offices; they require rather large pots if they are to reach a reasonable size. A compost of peat, sand and loam is suitable, and the plants should be placed in a position where they receive good light. They are readily propagated by cutting short pieces of stem and laying them on compost in a plant house with bottom heat.

Literature: Bos, J.J., Dracaena in West Africa, *Belmontia* **17**(80): 1–126 (1985).

1a. Leaves sword-shaped, with clasping, flaring bases; trees or large shrubs 2
b. Leaves more or less elliptic, with long or short stalks; small shrubs 9
2a. Freely branching shrub; most leaves less than 40 cm 3
b. Single-stemmed trees or small trees, generally not branching before flowering; most leaves more than 40 cm 4
3a. Leaves with distinct red margins **9. marginata**
b. Leaves without red margins **7. reflexa**
4a. Leaves scarcely narrowed towards the flaring base, glaucous; leaf-scars usually with patches of dark red resin **1. draco**
b. Leaves gradually narrowed towards the flaring base, green or variously variegated; resin absent 5
5a. Leaves to 3 cm wide with distinctly wavy margins towards the base **3. umbraculifera**
b. Leaves wider, margins not wavy 6
6a. Leaves with a distinct colourless (or rarely red) margin **6. aletriformis**
b. Leaves without distinct colourless or red margin 7
7a. Flowers on distinct, long-persisting stalks in groups of 3–5 towards the ends of the branches; leaves not variegated **2. arborea**
b. Flowers almost stalkless in well-spaced, many-flowered clusters; leaves variegated or not 8
8a. Leaves dark green, most less than 5 cm wide **5. deremensis**
b. Leaves light or bright green, most 5 cm or more wide **4. fragrans**
9a. Leaves variegated 10
b. Leaves not variegated 15
10a. Variegation of leaves consisting of irregular transverse bands of grey-white **11. goldieana**
b. Variegation of leaves not as above 11
11a. Variegation consisting of longitudinal bands of white or pale yellow **8. sanderiana**
b. Variegation consisting of separate or merging dots or rings 12
12a. Leaf-stalk almost as long as to longer than the blade; variegation consisting of few to many light green, transverse, oval blotches **15. phrynioides**
b. Leaf-stalk much shorter than the blade; variegation not as above 13
13a. Inflorescence unbranched **12. surculosa**
b. Inflorescence branched 14
14a. Inflorescence branches well developed, loosely or densely covered with clusters of flowers; leaf variegation of rounded, yellow dots **10. elliptica**
b. Inflorescence branches short, bearing distant, many-flowered clusters; variegation of yellow or whitish dots **13. × masseffiana**
15a. Leaf-stalk not more than one-sixth of the blade length 16
b. Leaf-stalk more than one-sixth of the blade length 17
16a. Inflorescence branched **10. elliptica**
b. Inflorescence not branched **12. surculosa**
17a. Leaf-stalk narrow, terete **14. aubryana**
b. Leaf-stalk strap-shaped, gradually merging into the leaf sheath **8. sanderiana**

1. D. draco (Linnaeus) Linnaeus. Illustration: *Botanical Magazine*, 4571 (1881); Graf, Exotica, edn 8, 1073, 1078, (1976); Bramwell & Bramwell, Wild flowers of the Canary Islands, t. 322 (1974); *Gartenpraxis*, 560 (1976).

Tree to 18 m or more with a thick trunk, unbranched till the first flowering, later with radiating branches that repeat the same pattern. Leaves stalkless, arching or reflexed, in dense rosettes at the ends of the branches, 30–60 × 2.5–4 cm, sword-shaped, base flaring and clasping, apex acute, glaucous, with no obvious midrib; leaf-scars with patches of dark red resin. Inflorescence a large, erect, terminal, much-branched panicle with flowers in groups of 4 or 5. Perianth *c.* 8 mm, lobed almost to the base, greenish white outside, white inside. Berries orange-yellow. *Atlantic islands.* H5–G1.

A striking tree, particularly when old.

2. D. arborea (Willdenow) Link. Illustration: *Gartenflora* **46**: t. 1438 (1897); *Belmontia* **17**(80): 24 (1985).

Tree to 20 m with a distinct trunk, ultimately branching above. Leaves stalkless, clustered at the ends of the branches, spreading or recurved, 50–150 × 4–10 cm, sword-shaped, with a prominent midrib on the lower surface. Panicles hanging, shortly branched, up to 2 m, flowers in 3 s, conspicuously stalked. Perianth *c.* 2 cm, creamy white, with a distinct tube at the base. Berry orange or red. *Tropical W Africa south to Angola.* G1–2.

3. D. umbraculifera Jacquin. Illustration: *Loddiges' Botanical Cabinet* **3**: t. 289 (1818).

Plant tree-like, 1–10 m, trunk erect and unbranched. Leaves stalkless, borne in a dense, umbrella-like rosette at the top of the trunk, spreading and recurving, 60–100 × up to 3 cm, linear-lanceolate, gradually narrowed to the flaring and clasping base, with a distinct midrib on both surfaces, margins distinctly wavy towards the base. Raceme terminal, head-like, borne immediately above the sheaths of the leaves, flowers in clusters at the ends of the branches. Perianth 5–6 cm, white tinged with red, tube longer than the lobes. *?Mauritius.* G2.

Introduced into Europe from a plant cultivated in Mauritius; its wild origin is uncertain.

4. D. fragrans (Linnaeus) Ker-Gawler. Figure 7(8), p. 178. Illustration: *Botanical Magazine*, 1081 (1808); Graf, Exotica, edn 8, 1072 (1976); Hay et al., Dictionary of indoor plants in colour, t. 196 (1974); Clay & Hubbard, The Hawaii garden, tropical exotics, 107 (1977).

Plant shrubby, erect, to 15 m but generally much smaller in cultivation, stem ultimately branched. Leaves borne in the upper part of the stem, stalkless, spreading and recurving, 50–150 × 5–10 cm, lanceolate, gradually narrowed to the flaring base, bright, pale green, the midrib conspicuous only beneath. Panicle terminal, erect, arching or hanging, its axis conspicuously zig-zag, the short branches bearing spherical clusters of flowers at their ends and often a further cluster near the base. Flowers very fragrant. Perianth 1.5–2.5 cm, white. Berry orange-red. *Tropical Africa, from Gambia to Ethiopia, south to Angola.* G2.

Variegated cultivars (e.g. 'Lindenii', 'Victoria' and 'Massangeana') are most frequently grown.

5. D. deremensis Engler. Illustration: Perry, Flowers of the world, 17 (1972); Graf, Exotica, edn 8, 1072, 1076–7 (1976); Hay et al., Dictionary of indoor plants in colour, t. 195 (1974).

Plant shrubby, erect, to 5 m, but often much smaller in cultivation. Leaves spreading, to 45 × 5 cm, tapering towards the flaring base, lanceolate, dark green, usually variegated in cultivated plants. Inflorescence a large, erect, branched panicle to 50 cm, the flowers in clusters of 20–30 together. Perianth *c.* 2 cm, tube *c.* half the total length, dark red outside, white within, strongly scented. *E tropical Africa.* 1

'Warneckii', which has leaves variegated with 2 longitudinal white stripes, and 'Bausei' with a central white band, are the most commonly grown variants; other cultivars are available.

6. D. aletriformis (Haworth) Bos (*D. rumphii* (Hooker) Regel; *D. latifolia* Regel). Illustration: Kaier, Indoor plants in colour, 52 (1961); Graf, Exotica, edn 8, 1073, 1079 (1973).

Plant shrubby, erect, to 2 m or more, often much smaller in cultivation. Leaves spreading and recurved, 60–75 × 5–10 cm, lanceolate, tapering towards the flaring base, midrib inconspicuous on the upper surface, more prominent beneath; margins narrow and white-translucent or rarely reddish. Inflorescence a large, terminal panicle with elongate branches, the flowers borne in clusters of 2 or 3. Perianth 2.5–3 cm, greenish white, the tube up to half the total length. Filaments expanded. Berry orange. *SE Africa.* G1–2.

Variable in the breadth of the leaves and the colour of their edges. A few cultivars are available (e.g. 'Latifolia', 'Variegata'). Plants in cultivation as *D. concinna* probably belong to this species.

7. D. reflexa Lamarck. Illustration: *Botanical Magazine*, 6327 (1877); Graf, Exotica, edn 8, 1078, 1104 (1976) – as *Pleomele reflexa*; Clay & Hubbard, The Hawaii garden, tropical exotics, 111 (1977).

Plant shrubby, erect, to 5 m in the wild, usually much less in cultivation, often branched. Leaves very variable, borne towards the top of the stem, spreading and recurved, to 20 × 5 cm, narrowly lanceolate to elliptic, tapering towards the flaring base, midrib inconspicuous above, convex beneath. Inflorescence a terminal panicle, its branches racemose. Perianth 1.2–2 cm, greenish outside, white inside, the tube to one-quarter of the total length. Berry orange-red. *Madagascar, Mauritius.* G2.

8. D. sanderiana Masters. Illustration: *Illustration Horticole* **40**: t. 175 (1893) – as *D. thalioides* var. *foliis variegatis*; Graf, Exotica, edn 8, 1075, 1077 (1976); Clay & Hubbard, The Hawaii garden, tropical exotics, 113 (1977); Brickell (ed.), RHS A–Z encyclopedia of garden plants, 389 (2003).

Plant shrubby, erect, to 1.5 m. Leaves spreading and recurved, 18–25 × 1–4 cm, lanceolate, tapering into a strap-shaped stalk, glossy green broadly variegated with longitudinal, silver-white stripes, rarely not variegated, midrib inconspicuous above. *Cameroun.* G2.

An obscure species; the variant with variegated leaves was only once collected in the wild, and it has not flowered in cultivation.

9. D. marginata Lamarck. Illustration: Graf, Exotica, edn 8, 1077 (1976); Hay et al, Dictionary of indoor plants in colour, t. 198 (1974); Brickell (ed.), RHS A–Z encyclopedia of garden plants, 388 (2003).

Plant shrubby, erect and branched, to 2 m. Leaves spreading and recurved, 30–40 × 1–2 cm, linear-lanceolate, tapering towards the flared and clasping base, red-margined, midrib inconspicuous above, convex beneath. Panicle terminal. Flowers white. Berries yellow, *Île de Réunion.* G2.

A doubtful species, apparently only once collected in the wild. A variant with white-variegated as well as red-margined leaves ('Tricolor') is sometimes grown.

10. D. elliptica Thunberg. Illustration: *Botanical Magazine*, 4787 (1854).

Plant shrubby, to 1 m, erect but with spreading branches. Leaves spreading and recurved, 7–20 × 4–6 cm, ovate or ovate-lanceolate, acute, abruptly rounded into short stalks, green marked with irregular yellow spots. Inflorescence a terminal panicle with elongate branches, the flowers borne in groups of 2 or 3. Perianth 1.2–2.5 cm, yellowish green, the tube forming up to half the total length. Berries red. *India, Indonesia (Java, Sumatra).* G2.

The variant with leaves spotted with yellow ('Maculata') is most commonly grown.

11. D. goldieana Lindley. Illustration: *Botanical Magazine*, 6630 (1882); Graf, Exotica, edn 8, 1075, 1079 (1976); Clay & Hubbard, The Hawaii garden, tropical exotics, 109 (1977).

Shrub to 2 m, stem erect. Leaves 15–30 × 5–13 cm, ovate, ovate-lanceolate or ovate-elliptic, spreading, tapering into a distinct stalk, which is up to one-quarter of the length of the blade, dark green variegated with silvery blotches forming transverse bars. Inflorescence a congested, head-like, terminal panicle borne between the sheaths of the upper leaves so that it appears to be without a common stalk (more rarely an elongate panicle). Perianth *c.* 2.5 cm, the tube longer than the spreading lobes, white. *Nigeria to Gabon.* G2.

12. D. surculosa Lindley (*D. godseffiana* Masters). Illustration: *Edwards's Botanical Register* **14**: t. 1169 (1828); *Botanical Magazine*, 7584 (1898); Graf, Exotica, edn 8, 1074 (1976); Hay et al., Dictionary of indoor plants in colour, t. 197 (1974).

Shrub *c.* 1 m, the branches erect or arching. Leaves spreading, borne in false whorls of 2–4, 5–20 × 2–6 cm, usually elliptic, acute, abruptly tapered to short but distinct stalks, dark green spotted with small, round, yellow or white dots. Inflorescence a terminal raceme, the flowers borne in groups of 1–5, or all flowers in a terminal cluster. Perianth 1.5–2.5 cm, white, the tube forming about half of the total length. Berries orange-scarlet. *Tropical Africa.* G2.

There are 2 varieties in cultivation: var. **surculosa** with distinct, pure white spots is usually found under the name *D. godseffiana* (when very densely spotted, as 'Punctulata' or 'Florida Beauty') and var. **maculata** J.D. Hooker, with very pale green spots or rings.

13. D. × maseffiana Bos. Illustration: *Gardeners' Chronicle* **121**: f. 111 (1947); *Notes from the Royal Botanic Garden, Edinburgh* **40**: 536 (1983).

Shrub to 2 m or more, branched. Leaves elliptic, tapering towards the base to short stalks that are stem-clasping, deep green covered with small, white or yellow, rounded spots. Inflorescence a panicle with short side branches, bearing distant, many-flowered clusters. Flowers white, the lobes with green midribs. *Garden origin.* G2.

A hybrid between *D. surculosa* and *D. fragrans,* raised in Puerto Rico but currently available in Europe. All the cultivated material belongs to 'Pennock'.

14. D. aubryana Morren (*D. humilis* Baker; *D. thalioides* Regel). Illustration: Graf, Exotica, edn 8, 1076 – as *D. kindtiana*, 1078 & 1104 – as *Pleomele thalioides* (1973); Clay & Hubbard, The Hawaii garden, tropical exotics, 115 (1977).

Unbranched shrub. Stems erect, short, covered by the overlapping leaf-sheaths. Leaves 15–60 or more × 5–12 cm, linear-lanceolate, arching, abruptly and sometimes asymmetrically rounded into a distinct, terete stalk that is almost as long as the blade. Inflorescence terminal, erect, a 15–50 cm spike-like raceme or sparsely branched panicle with the flowers in groups of 1–3. Perianth 1–5 cm, the tube shorter than the lobes, white with red lines outside, white within. *Tropical W Africa, south to Angola.* G2.

15. D. phrynioides Hooker. Illustration: *Botanical Magazine*, 5352 (1862); Graf, Exotica, edn 8, 1076 (1976).

Unbranched shrub. Stems short, covered by overlapping leaf sheaths. Leaves 10–75 × 3–9 cm, ovate, acuminate, abruptly narrowed into an erect stalk that is almost as long as the blade, holding the leaves well above the inflorescence. Inflorescence terminal, often nodding, borne at the top of the stem, between the sheaths of the upper leaves. Bracts ovate, overlapping, often tinged with red, sometimes abruptly tapered into long 'tails'. Perianth white, *c.* 2 cm, the tube distinctly longer than the lobes. Berries scarlet, distinctly horned. *Tropical W Africa.* G2.

This is a rather difficult species to grow well.

2. SANSEVIERIA Thunberg

C.J. Couper

Herbaceous, perennial leaf-succulents with creeping rhizomes, usually without obvious aerial stems. Leaves persistent, fleshy or leathery, flat, channelled, half-cylindric, cylindric or laterally compressed, rigid or flexible, margins entire and smooth or cartilaginous. Raceme or panicle basal, the axis with scale-like leaves or sheaths at the base. Flowers bisexual, solitary or in clusters, fragrant, borne on stalks that are jointed near the middle or towards the apex. Perianth of 6 lobes united below into a distinct tube, lobes narrow, radially symmetric. Stamens 6, projecting from the perianth. Ovary superior, 3-celled with 1 ovule in each cell. Style slender, thread-like. Fruit a berry containing 1–3 bony seeds.

A genus of over 50 species from S & E Africa, Arabia and India. They are easily grown, several of them being popular house plants that can stand a surprising amount of neglect. Propagation is by division of the rhizome or by leaf cuttings.

Literature: Brown, N.E., Sansevieria, a monograph of all the known species, *Kew Bulletin* (1915); Morgenstern, K.D., Sansevierias (1979).

1a. Plants with distinct, leafy stems which are 30 cm or more **1. arborescens**
b. Plants without stems or stems inconspicuous, much less than 30 cm 2
2a. Leaves terete in section, without a distinct groove running from near the apex almost to the sheath 3
b. Leaves flat, or if terete, then with a distinct groove running from near the apex almost to the sheath 4
3a. Stem short, but branching above ground; leaves 8–12 to a rosette **2. gracilis**
b. Stems absent; leaves 3 or 4 to a rosette **4. cylindrica**
4a. Leaves terete but distinctly grooved and with flattened sides **3. ehrenbergii**
b. Leaves fleshy but flat, not as above 5
5a. Leaves with a distinct red or white, horny margin 6
b. Leaves without a margin distinctly different from the substance of the leaf 8
6a. Rhizome very extensive (to 1 m or more), whitish or pale green above ground; leaves elliptic, oblong or broadly lanceolate, glaucous and marked with inconspicuous bars of pale green on both surfaces **11. grandis**
b. Rhizomes not extensive, brownish when above ground; leaves lanceolate, dull, dark green marked with paler bands or blotches that disappear with age on both surfaces 7
7a. Leaves to 45 cm, apex with a withered, whitish point that is *c.* 1.6 cm **10. hyacinthoides**
b. Leaves 45 cm or more, apex with a green, needle-like point that is 3–6 mm **9. metallica**
8a. Leaves not tapering towards the base **5. roxburghiana**
b. Leaves tapering towards the base 9
9a. Stem 2.5–12 cm; perianth-tube pale pinkish white, lobes tinged with mauve within, dull mauve or purplish outside **7. parva**
b. Stem absent or completely hidden by leaf bases; flowers dull pink outside, whitish inside, or pale greenish both outside and in 10
10a. Leaves 6–20 to a rosette; flowers dull pink or purplish outside **6. dooneri**
b. Leaves 1–5, rarely to 6, to a rosette; flowers greenish white outside **8. trifasciata**

1. S. arborescens Gerome & Labroy. Illustration: *Kew Bulletin*, 201 (1915); Morgenstern, Sansevierias, t. 6 (1979).

Stem 90–120 cm, erect, terete, leafy throughout. Leaves 20–45 × 2–4.5 cm, narrowly lanceolate or linear-lanceolate, spreading or recurved, smooth, flat or concave, narrowing slightly to the sheathing base, grass-green with no markings; margins recurved or wavy, hardened, whitish or reddish; tip 8–25 mm, stout, needle-like, green and hardened below, sharp and pale brown at the apex. Flowers unknown. *Tropical E Africa.* G1–2.

2. S. gracilis N.E. Brown. Illustration: *Kew Bulletin*, 204 (1915); Lamb & Lamb, Illustrated reference on cacti and other succulents **4**: t. 698 (1975); Morgenstern, Sansevierias, t. 34–5 (1979).

Stems 2.5–7.5 cm, branching above ground. Leaves 8–12 to a rosette, 22–75 cm, ascending or spreading, firmly flexible, smooth, channelled for 5–12 mm at the base, the rest cylindric, terete, gradually tapering, deep green with narrow, darker bands on the younger leaves; tip 2–6 mm, very acute, white or brownish. Raceme 6–7.5 cm, loose, spike-like. Flowers in pairs, white, the tube 2–2.5 cm, the lobes 1–1.2 cm, linear, obtuse. *Tropical E Africa.* G1–2.

3. S. ehrenbergii Baker. Illustration: *Hooker's Icones Plantarum* **23**: t. 2269 (1893); Morgenstern, Sansevierias, t. 26–9 (1979).

Stem to 25 cm. Leaves 5–9 to a rosette, 75–180 × 2.5–4 cm, terete, laterally compressed with flattened sides, rounded on the back, with a triangular channel down the face and 5–9 shallow grooves down the sides and back, tapering upwards, dark green with blackish green grooves, but without transverse markings; margins of channel acute, reddish brown with white membranous edges; tip 6–12 mm, hard and spine-like. Panicle to 2 m. Flowers 4–7 in a cluster, the tube

5–6 mm, the lobes *c.* 7 mm, linear, obtuse. *Tropical E Africa.* G1–2.

4. S. cylindrica Bojer. Illustration: *Botanical Magazine*, 5093 (1859); *Revue Horticole*, t. 109 (1861) & t. 70 (1901); Morgenstern, Sansevierias, t. 22 (1979).

Stems and aerial branches absent. Leaves 3 or 4 in a rosette, 75–150 cm, cylindric or compressed-cylindric, 2-ranked, stiffly erect, faintly rough, 2–3 cm thick at the base, gradually tapering, green or whitish green, with transverse bands of darker green, the banding becoming fainter with age, and with numerous longitudinal, dark green or blackish green lines that are impressed in older leaves; tip 4–6 mm, hardened, acute, whitish. Raceme spike-like, 35–75 cm on a stalk 60–90 cm. Flowers 5 or 6 in a cluster, white or pinkish, tube 1.5–2.5 cm, lobes 1.5–2 cm, linear, obtuse, margins rolled outwards. *Angola; South Africa?* G1–2.

5. S. roxburghiana Schultes & Schultes. Figure 7(11), p. 178. Illustration: Morgenstern, Sansevierias, t. 81 (1979).

Stemless. Leaves 6–24 in a rosette, the inner 20–60 × 1.2–2.5 cm, linear, ascending and slightly recurved, stiff, concave, channelled down the face, rounded on the back, green and transversely marked with regular, dark green bars on both surfaces, and with 6–11 longitudinal, dark green lines on the back; margins green becoming whitish; tip 6–50 mm, stout, subulate, soft, green. Raceme to 75 cm, spike-like. Flowers *c.* 4 in a cluster, the tube 6–7 mm, the lobes 8–9 mm, linear, obtuse. *India.* G1–2.

S. zeylanica (Linnaeus) Willdenow (*S. hyacinthoides* misapplied). Illustration: Redouté, Liliacées **5**: t. 290 (1810); *Edwards's Botanical Register* **2**: t. 160 (1816); Morgenstern, Sansevierias, t. 137 (1979). Very similar to *S. roxburghiana*, but leaves longer, thicker and more rigid, lighter green and with more numerous longitudinal, dark green lines on the back. *Sri Lanka.* G2.

6. S. dooneri N.E. Brown. Illustration: *Kew Bulletin*, 232 (1915); Morgenstern, Sansevierias, t. 25 (1979).

Stems absent or up to 5 cm, but hidden by the leaf bases. Leaves 6–20 in a rosette, 10–42 × 1.6–3 cm, strap-shaped or narrowly lanceolate, gradually narrowed to a somewhat stalk-like base, leathery but flexible, recurved-spreading, smooth, flat on the face, dark green with paler green, transverse bands on both surfaces, without longitudinal lines on the back; margins green; tip 5–60 mm, stout, soft, subulate, green. Raceme 30–40 cm, loose. Flowers 2 or 3 in a cluster, dull pink or pale purplish outside, whitish within, the tube *c.* 1.1 cm, the lobes 1.1–1.2 cm, linear. *Tropical E Africa.* G1–2.

7. S. parva N.E.Brown. Illustration: *Kew Bulletin*, 232 (1915); Lamb & Lamb, Illustrated reference on cacti and other succulents **4**: t. 700 (1975); Morgenstern, Sansevierias, t. 64–5 (1979).

Stems 2.5–12 cm. Leaves 6–14 in a rosette, 20–45 cm × 8–28 mm, linear, linear-lanceolate or lanceolate, ascending or recurved, spreading, smooth, leathery but not flexible, channelled on the face, rounded on the back, narrowed into a stalk *c.* 5 cm long; surfaces marked with distinct transverse, alternating bands of darker and paler green, the markings becoming paler with age; margins green; tip 3.5–7.5 cm, soft, subulate, green. Raceme *c.* 30 cm, loose and open. Flowers in pairs or borne individually, the tube 1–1.1 cm, the lobes 8–9 mm, linear, obtuse, dull mauve or purplish outside, white tinged with mauve inside. *Tropical E Africa.* G1–2.

8. S. trifasciata Prain (*S. guineensis* Gerome & Labroy not (Linnaeus) Willdenow). Figure 7(10), p. 178. Illustration: Parey's Blumengärtnerei **1**: 312 (1958); Lamb & Lamb, Illustrated reference on cacti and other succulents **4**: t. 517 (1971); Hay et al., Dictionary of indoor plants in colour, t. 441 (1974); Morgenstern, Sansevierias, t. 99–103 (1979).

Stemless. Leaves 30–120 × 2.5–7 cm, up to 5 or rarely to 6 in a rosette, linear-lanceolate or narrowly lanceolate, smooth, erect, stiff or firmly leathery, narrowing gradually to a concave-channelled stalk, more rarely (in some cultivars) smaller, stalkless and overlapping in a distinct, flat rosette; variously banded with pale and dark green, whitish or yellow, sometimes almost silver on both surfaces; margins of the same colour as the adjacent parts of the leaf; tip 3–36 mm, subulate, green. Raceme 30–75 cm. Flowers in clusters of 1–3, pale green or greenish white, tube 6–12 mm, the lobes 1.4–1.8 cm, linear, obtuse. *W tropical Africa.* G1–2.

Very variable. Numerous cultivars, based on the type of variegation and the form of the rosette, are available.

9. S. metallica Gerome & Labroy. Illustration: *Kew Bulletin*, 246 (1915); Morgenstern, Sansevierias, t. 60 (1979).

Stemless. Leaves 1–4 in a rosette, 45–150 × 5–12.5 cm, elongate-lanceolate or broadly strap-shaped, erect or spreading, smooth, tapering into a deeply channelled stalk, all dull dark green, obscurely marked above, more distinctly beneath with transverse bands or blotches of paler green that often disappear with age; margins whitish or pale reddish brown; tip 3–6 mm, soft, subulate, green. Raceme 45–120 cm, spike-like. Flowers 2–4 in a cluster, white, tube 1.2–1.6 cm, lobes 1.6–2.2 cm, linear-oblong, margins rolled outwards. *Tropical Africa.* G2.

10. S. hyacinthoides (Linnaeus) Druce (*S. thyrsiflora* (Petagna) Thunberg; *S. guineensis* (Linnaeus) Willdenow). Illustration: *Botanical Magazine*, 1179 (1809); Redouté, Liliacées **6**: t. 330 (1811); *Revue Horticole*, t. 174 (1915); Morgenstern, Sansevierias, t. 43 (1979).

Stemless but with a creeping rhizome that is brown when above ground. Leaves 2–4 to a rosette, 15–45 × 2.5–6 cm, lanceolate, erect, smooth, tapering into a concave, channelled stalk, surfaces dull green with numerous transverse, pale green bands that disappear with age; margins brownish red or whitish; tip *c.* 1 cm, usually whitish and withered. Raceme 45–75 cm, spike-like. Flowers 2–6 to a cluster, tube *c.* 1.8 cm, greenish white, lobes *c.* 1.8 cm, linear, margins rolled outwards, whitish. *S Africa.* G1.

11. S. grandis Hooker. Illustration: *Botanical Magazine*, 7877 (1903); Hay & Synge, Dictionary of garden plants in colour, t. 641 (1969); Morgenstern, Sansevierias, t. 38–41 (1979).

Rhizome long and creeping, to 1 m, whitish or pale green when above ground. Stem absent. Leaves 4 or 5 in a rosette, 30–60 × 8–15 cm, elliptic, oblong or broadly lanceolate, narrowed towards the base, stiffly leathery, flat with wavy, recurved margins, glaucous with regular, transverse bands of lighter green on both surfaces, though these more evident beneath; margins reddish brown, hard, with membranous white edges when young; tip 4–6 mm, acute, soft, white. Raceme *c.* 60 cm, compact, spike-like.

Flowers white, tube *c.* 1.8 cm, lobes *c.* 1.8 cm, linear, obtuse, margins rolled outwards. *Origin unknown, perhaps South Africa.* G1–2.

Perhaps no more than a variant of *S. hyacinthoides.*

31. PHORMIACEAE

Plants terrestrial, herbaceous. Leaves usually equitant, distichous, forming fans. Flowers in racemes, bisexual or occasionally male by abortion, actionomorphic or zygomorphic. Perianth of 6 united segments. Stamens 6. Ovary of 3 united carpels, usually 3-celled, placentation axile. Fruit a capsule or berry.

Seven genera with about 30 species, several cultivated.

1a. Perianth-lobes united into a tube at the base; anthers opening by slits **3. Phormium**
b. Perianth-segments entirely free; anthers opening by pores 2
2a. Filaments thickened but not hairy above; fruit a berry **1. Dianella**
b. Filaments hairy above; fruit a capsule **2. Stypandra**

1. DIANELLA Lamarck

E.H. Hamlet & V.A. Matthews

Rhizomatous perennials. Leaves in 2 ranks crowded at the base of the stem, linear or lanceolate, keeled beneath. Flowers borne in a panicle. Perianth-segments free, spreading, with 3–8 veins. Filaments thickened; anthers basifixed, opening by terminal pores. Ovary superior; style 1. Fruit a berry.

A genus of about 30 species found in tropical Asia, Australia, New Zealand and Polynesia. They need a sheltered position or cool greenhouse conditions in the colder parts of Europe. They are propagated by division or from seed.

1a. Leaves with smooth margins 2
b. Leaves with rough margins 3
2a. Anthers twice as long as filaments **4. laevis**
b. Anthers three or more times as long as filaments **5. revoluta**
3a. Flowers greenish white or purplish white **6. intermedia**
b. Flowers pure white or blue 4
4a. Anthers brownish yellow, shorter than or equal in length to the filaments **3. tasmanica**
b. Anthers yellow, longer than the filaments 5
5a. Flowering stem less than 60 cm; leaves 6–13 mm broad, rarely to 1.5 cm **2. caerulea**
b. Flowering stem more than 90 cm; leaves 1.3–3.3 cm broad **1. ensifolia**

1. D. ensifolia Redouté (*D. nemorosa* Lamarck). Illustration: *Botanical Magazine*, 1404 (1811); Yatabe, Iconographia florae Japonicae **1**(3): t. 58 (1893); Koorders, Exkursionsflora von Java **1**: 290 (1911).

Flowering stem 90–180 cm, leafy in lower part. Leaves up to 30 × 1.3–3.3 cm, lanceolate, margins and keel rough. Inflorescence a spreading panicle. Perianth-segments white or blue, 7.5–8 mm. Anthers yellow, longer than the filaments. Berry yellow when unripe, turning to bluish purple. *Tropics of the Old World.* G1–H5. Summer.

2. D. caerulea Sims. Illustration: *Botanical Magazine*, 506 (1801); *Addisonia* **18**: pl. 585 (1933).

Flowering stem 50–60 cm, sometimes branched. Leaves 10–60 cm × 6–13 mm, rarely to 1.5 cm, margins and keel rough. Inflorescence a loose panicle but sometimes clustered, 10–30 cm. Perianth-segments white or blue, *c.* 7.5 mm. Anthers pale yellow, 2–3 times as long as the filaments. Berry blue. *E & S Australia.* H5. Summer.

3. D. tasmanica Hooker. Illustration: Burbidge & Gray, Flora of the Australian Capital Territory, pl. 96 (1970); Hay et al., Dictionary of indoor plants in colour, pl. 182 (1974) – fruit; Costin et al., Kosciusko alpine flora, pl. 134–6 (1979); *Australian Plants* **12**: 156 (1983) – fruit.

Flowering stem to 150 cm. Leaves 30–120 × 1–3 cm, rigid, broadly lanceolate, keel rough, margins revolute and very sharply toothed. Panicle robust, much branched, loose, 30–60 cm. Perianth-segments pale blue. Anthers brownish yellow, shorter than, or equal in length to, the filaments. Berry blue to violet-blue, 1–2 cm. *SE Australia, Tasmania.* H4. Summer.

4. D. laevis Brown. Illustration: Black, Flora of South Australia, edn 3, **1**: f. 307 (1978).

Flowering stem to 90 cm. Leaves 20–60 cm × 6–10 mm, margins and keel smooth. Panicle loose and spreading, 20–50 cm. Perianth-segments blue. Anthers yellow, about twice as long as the filaments. Berry blue. *Australia (New South Wales).* H4. Summer.

5. D. revoluta Brown. Illustration: *Edwards's Botanical Register* **13**: t. 1120 (1828); Rotherham et al., Flowers and plants of New South Wales and South Queensland, pl. 137 (1975).

Flowering stem to 120 cm. Leaves 20–90 cm × 4–15 mm, linear, margins and keel smooth, margins revolute. Panicle loose, 20–60 cm. Perianth-segments blue, 7–10 mm. Anthers yellow, brown or black, three or more times as long as filaments. *S & E Australia, Tasmania.* H4. Summer.

6. D. intermedia Endlicher.

Leaves 30–150 × 1.2–1.8 cm, linear to lanceolate, margins and keel rough. Panicle 15–60 cm. Perianth-segments greenish white or purplish white, more than 5 mm. Anthers yellow. Berry blue, 8–15 mm. *Norfolk Island, Fiji.* G1. Summer.

D. nigra Colenso (*D. intermedia* Endlicher var. *norfolkensis* Brown). Illustration: Moore & Irwin, The Oxford book of New Zealand plants, 183 (1978). Differs from *D. intermedia* in having a more slender flowering stem and perianth-segments less than 5 mm. *New Zealand.* H4.

2. STYPANDRA R. Brown

E.H. Hamlet & V.A. Matthews

Differs from *Dianella* mainly in the filaments being woolly at the top, and in having a capsule rather than a berry.

A genus of 6 species from temperate Australia; one is sometimes cultivated. A sunny position is best, as is a well-drained soil. Propagation is by seed or division of the rhizome.

1. S. caespitosa R. Brown. Illustration: Cochrane et al., Flowers and plants of Victoria, 30 (1973); Willis et al., Field guide to the flowers and plants of Victoria, 41 (1975); Galbraith, Collins field guide to the wild flowers of south-east Australia, pl. 2 f., 7 (1977).

Flowering stem 30–60 cm. Leaves 10–35 cm × 5–8 mm. Flowers blue or yellowish, occasionally white, erect on 2.5–5 cm stalks. Perianth-segments 8–15 mm. *SE Australia.* H4. Summer.

3. PHORMIUM Forster & Forster

C.J. Couper

Rhizomatous perennial plants, ultimately becoming woody at the base. Leaves persistent, linear, folded towards the base and equitant, keeled, marked with many fine, close, longitudinal stripes. Flowering axis long, bearing alternate, deciduous bracts that are scarious, the upper bracts subtending and entirely enclosing the short, alternately branched flowering branches. Flower-stalks jointed near the apex. Flowers bisexual, bilaterally symmetric. Perianth-lobes 6, more or less equal, erect, united into a tube at the base. Stamens 6, projecting from the perianth. Ovary superior, 3-celled, elongate, ovules many in each cell. Fruit a long, many-seeded capsule, the seeds flattened and almost winged, black and shiny.

A genus of 2 species from New Zealand. They are easily grown in rich soils. Propagation is by seed or by division.

1a. Leaves to 3 m, stiff; flowers predominantly dull red, tips of the inner perianth-lobes not or only slightly recurved **1. tenax**
b. Leaves usually less than 2 m, flexible; flowers greenish with tones of orange or yellow, tips of inner perianth-lobes usually markedly recurved **2. cookianum**

1. P. tenax Forster & Forster. Figure 7(9), p. 178. Illustration: *Botanical Magazine*, 3199 (1832); *Illustration Horticole* **13**: t. 481 (1866); Hay & Synge, Dictionary of garden plants, t. 1732 (1969); *Journal of the Royal Horticultural Society* **102**: 111 (1977).

Base of plant usually brightly coloured. Leaves 1–3 m × 5–12 cm, stiff and often erect, at least in the lower part. Flowering axis usually erect, to 5 m. Flowers 2.5–5 cm, predominantly dull red, the tips of the inner perianth-lobes not or only slightly recurved. Capsule to 10 cm, though often much less, erect, 3-sided, abruptly contracted at the apex, not twisted, remaining firm and dark in age. Seeds 9–10 × 4–5 mm, elliptic, plate-like, sometimes twisted. *New Zealand*. H5–G1. Summer.

2. P. cookianum Le Jolis (*P. colensoi* Hooker; *P. hookeri* Hooker). Illustration: *Revue Horticole*, ser. 3, **2**: 5 (1848); *Illustration Horticole* **19**: t. 93 (1872); *Botanical Magazine*, 6973 (1888).

Base of plant usually pale. Leaves mostly less than 2 m, flexible, drooping. Flowering axis to 2 m, often inclined or arching. Flowers 2.5–4 cm, usually predominantly greenish with tones of orange or yellow, tips of inner perianth-lobes markedly recurved, usually 1 more so than the other 2. Capsule often more than 10 cm, occasionally to 20 cm, hanging, cylindric, gradually narrowed to the apex, twisted and becoming pale, fibrous and spirally curved with age. Seeds 8–10 mm, elliptic, plate-like. *New Zealand*. H5–G1. Summer.

32. HAEMODORACEAE

Herbs, sap often orange. Leaves mostly basal, often equitant. Flowers in cymes, racemes or clusters, bisexual, actinomorphic or weakly zygomorphic. Perianth of 6 free or united segments, persistent, often densely hairy outside. Stamens 6 or 3. Ovary of 3 united carpels, superior or inferior; ovules many, axile. Fruit a capsule.

Twenty-two genera with about 160 species. Two genera are native to North America, and about 5 are cultivated.

1a. Perianth-tube split equally **2. Conostylis**
b. Perianth-tube split more deeply on 1 side 2
2a. Ovules solitary in each cell; flowers clothed with black hairs **3. Macropidia**
b. Ovules numerous; flowers clothed in yellow or red, but not black hairs **1. Anigozanthos**

1. ANIGOZANTHOS Labillardière

E.C. Nelson

Perennial herbs. Leaves sword-shaped, arising from a rhizome, to 40 cm long, hairless. Flowers in a 1-sided raceme, densely clothed with hairs. Stamens 6. Style 1. Ovules numerous in each cell of ovary.

A genus of about 12 species restricted to western Australia, inhabiting the siliceous sandplain. Only 1 species is widely cultivated but new cultivars are being produced in Australia that may eventually reach Europe. *Anigozanthos* should be grown in full sun, in very well-drained compost; propagation is by seed, division, or tissue culture.

Literature: Geerinck, G., Revision du genre Anigozanthos Labill (Haemodoraceae d'Australie), *Bulletin Jardin Botanique National de Belgique* **40:** 261–76 (1970); *Growing Native Plants* **2:** 30–1 (1972).

1a. Inflorescence a spike-like raceme **2. manglesii**
b. Inflorescence a panicle **1. flavidus**

1. A. flavidus de Candolle. Illustration: *Botanical Magazine*, 1151 (1809); Erickson, Flowers and plants of western Australia, 66 (1973); Gardner, Wildflowers of western Australia, 16 (1973).

Robust perennial to 2 m tall. Leaves strap-like. Flower-spike to 2 m; flowers often yellow-green, sometimes pink, orange, red or green. *W Australia*. H5–G1. Summer.

This is the hardiest species but it is frost-sensitive. Other species may be cultivated in southern Europe.

A. pulcherrima Hooker. Illustration: *Botanical Magazine*, 4180 (1845); *Growing Native Plants* **2**: 31 (1972); Gardner, Wildflowers of western Australia, 16 (1973). Like *A. flavidus* but with bright yellow flowers and only 1 m tall. *W Australia*. Summer.

A. rufa Labillardière. Illustration: *Growing Native Plants* **2**: 30 (1972). Like *A. pulcherrima* but with maroon flowers. *W Australia*. H5–G1. Summer.

2. A. manglesii D. Don. Illustration: Gardner, Wildflowers of western Australia, 15 (1973).

Perennial with short rhizome. Leaves sword-like. Flower-spike with rich red velvet on stems and flowers, to 1 m tall. Perianth bright green with red hairs. *W Australia*. H5–G1.

A spectacular plant but short-lived and susceptible to 'ink disease'. It is best raised from seed and treated as a biennial.

2. CONOSTYLIS R. Brown

E.C. Nelson

Perennial herbs. Leaves in tufts, linear or strap-shaped. Flowers in cymes or panicles, not one-sided. Perianth clothed in hairs, lobes united into a tube at the base, splitting equally. Stamens 6. Style 1.

There are about 20 species in this genus, restricted to western Australia. Only 1 species is in cultivation here and it is uncommon. These plants need full sun and well-drained, sandy soil and generally require greenhouse protection in Europe.

1. C. candicans Endlicher.

Leaves arranged in a fan or tuft, strap-shaped, grey-green, to 30 cm long. Flowering stems longer than leaves. Flowers white-yellow, about 20 in each cyme. *W Australia*. H5–G1.

3. MACROPIDIA Harvey
E.C. Nelson
Perennial herb with sword-shaped leaves to 50 cm tall. Flowers in a branched panicle. Perianth green and covered with dense black hairs; lobes only united into a tube for *c.* half their length, reflexed at apex and curled back. Stamens 6. Ovule 1 in each carpel.

A genus containing only 1 species, sometimes placed within *Anigozanthos* (as Section *Macropidia* (Harvey) Geerinck). *Macropidia* is confined to western Australia (north of Perth) where it inhabits the woodlands and sandplain heath. It is frost-sensitive, and requires full sun and well-drained, sandy soil, or a cool, frost-free glasshouse in well-drained compost. Propagation is by seed or tissue culture.

1. M. fuliginosa (Hooker) Druce (*Anigozanthos fuliginosus* Hooker). Illustration: *Botanical Magazine*, 4291 (1846); Erickson, Flowers and plants of western Australia, 96–7 (1973); Gardner, Wildflowers of western Australia, 14 (1973).
W Australia. H5–G1.

33. IXIOLIRIACEAE

Terrestrial plants, herbaceous, with bulbs. Leaves borne on the lower parts of the stems. Flowers in heads, umbels or solitary, with spathes, actinomorphic. Perianth of 6 united segments. Stamens 6. Ovary of 3 united carpels, totally inferior; ovules numerous, axile. Seeds black.

A single genus with 4 species.

1. IXIOLIRION (Fischer) Herbert
J. Cullen
Perennial herbs with bulbs. Leaves mostly in a basal rosette, usually winter-persistent, also present on the stems. Flowers in a terminal umbel with or without several racemosely arranged flowers below it. Perianth radially symmetric, either of 6 free segments or united into a tube below and 6-lobed. Stamens 6, attached to the perianth-tube (if present) or to the bases of the perianth-segments or -lobes. Ovary 3-celled; ovules numerous in each cell, placentation axile. Fruit a capsule. Seeds numerous, black.

A genus of about 4 species from SW & C Asia; sometimes split into 2 genera, *Ixiolirion* in the strict sense, and *Kolpakowskia* Regel. Only 2 species are generally grown. They require a light, rich, well-drained soil and a sunny position.

Literature: Traub, H.P., The Ixiolirion Tribe, *Herbertia* **9:** 53–9 (1942).

1a. Perianth-segments completely free, the filaments attached at their extreme bases **1. tataricum**
b. Perianth with a distinct tube below, the filaments attached to it **2. kolpakovskyanum**

1. I. tataricum (Pallas) Herbert (*I. pallasii* Fischer & Meyer; *I. montanum* (Labillardière) Herbert). Illustration: *Edwards's Botanical Register* **30**: t. 66 (1844); *Gartenflora* **22**: t. 775 (1873) & **26**: t. 910 (1877).

Bulb ovoid, to 2.5 cm in diameter. Leaves mostly winter-persistent, mostly in a basal rosette. Stem to 40 cm, bearing 2 or 3 leaves. Flowers in an umbel with 0–4 racemosely arranged flowers below it. Perianth openly funnel-shaped, of 6 blue or violet-blue segments, each segment 2–5 cm with usually 3 darker lines down the middle. Filaments attached to the segments only at the extreme base. *SW & C Asia, India (Kashmir).* H4. Spring–summer.

Very variable; several units have been recognised from time to time as species, subspecies or varieties, under the names *ledebourii* Fischer & Meyer and *montanum* (Labillardière) Herbert, but it is uncertain which of these remain in cultivation.

2. I. kolpakovskyanum Regel (*Kolpakowskya ixiolirioides* Regel).

Similar to *I. tataricum* but flowers usually restricted to an umbel, perianth 2–2.5 cm, united into a distinct tube below, whitish violet, the filaments attached to the perianth-tube. *C Asia.* H3. Spring–summer.

34. AMARYLLIDACEAE

Herbs with bulbs or rhizomes or corms. Leaves basal, often in 2 ranks, ptyxis flat, rarely supervolute. Flowers solitary or in umbels, the flower or umbel subtended by 1–several spathes. Flowers bisexual, actinomorphic or zygomorphic. Perianth of 6 free or united segments. Stamens 6, anthers occasionally opening by pores. Ovary of 3 united carpels, inferior; ovules 2–many per cell, axile. Fruit a capsule or berry. Seeds black.

There are about 70 genera with about 850 species. Seven genera are native to Europe, 9 to North America. Many are cultivated, especially from the genera *Hippeastrum, Galanthus* and *Narcissus.* The limits of the family are controversial, and parts of the *Agavaceae* (p. 174–187) and *Liliaceae* (p. 62) are sometimes included within it. The delimitation of genera within the family is also difficult, frequently based on the presence or absence of a corona (and, if present, its nature). Different authors define the corona in different ways; in this account the small appendages (perianth-scales) or tufts of hairs found at the attachment of stamens to perianth in some genera *(Hippeastrum, Zephyranthes)* are not regarded as forming a corona.

Literature: Baker, J.G., Handbook of the Amaryllideae (1888). The American journal *Herbertia* (later incorporated into *Plant Life*), 1933 and continuing, is largely devoted to the family.

1a. Corona present, joining the bases of the filaments so that they are apparently borne on a tube, cup or ridge 2
b. Corona absent, or, if present, the filaments free from it; perianth-scales or tufts of hairs may be present at the bases of the filaments 13
2a. Scape hollow **33. Phaedranassa**
b. Scape solid 3
3a. Anthers basifixed **23. Chlidanthus**
b. Anthers dorsifixed, often versatile 4
4a. Ovules borne in 1 or 2 vertical rows in each cell of the ovary 5
b. Ovules bore side by side in each cell of the ovary 10
5a. Leaves with broad blades and distinct stalks 6
b. Leaves linear or strap-shaped, not divided into blades and stalks 7
6a. Perianth-lobes red, orange, yellow or green, without any white **29. Urceolina**
b. Perianth-lobes white, either completely or for the most part **30. Eucharis**
7a. Perianth yellow, pink or reddish, if white then with green stripes 8
b. Perianth completely white, without green stripes 9
8a. Flowers 1 or 2, more or less erect, yellow or pale yellow; corona *c.* 8.5 cm **26. Paramongaia**

b. Flowers in few-flowered umbels, usually pendent, not usually yellow; corona 1 cm or less **28. Stenomesson**
9a. Stamens directed horizontally towards the centre of the flower; stigma weakly 3-lobed **27. Pamianthe**
b. Stamens spreading; stigma capitate **25. Pancratium**
10a. Ovules 2–8 per cell, basal **32. Hymenocallis**
b. Ovules 2 or 3 per cell, axile 11
11a. Leaves with a conspicuous median white band **24. Vagaria**
b. Leaves without a conspicuous median white band 12
12a. Spathes much larger than the bracts of the individual flowers; perianth-tube tapered, straight **31. Caliphruria**
b. Spathes only slightly larger than the bracts of the individual flowers; perianth-tube cylindric and curved below, expanded above **30. Eucharis**
13a. Anthers basifixed, opening by pores 14
b. Anthers dorsifixed, usually versatile, opening by slits 16
14a. Flowers erect; filaments almost as long as anthers **21. Lapiedra**
b. Flowers pendent on arching stalks; filaments much shorter than anthers 15
15a. Segments of the inner whorl of the perianth different from those of theouter whorl **20. Galanthus**
b. All perianth-segments similar **22. Leucojum**
16a. Spathes 1 or 2, united to form a tube around the flower-stalks 17
b. Spathes 1–several (usually 2), not united below to form a tube 22
17a. Corona distinct, trumpet-, funnel- or cup-shaped, more rarely a disc or low ridge, arising at the junction of the perianth-tube and lobes **12. Narcissus**
b. Corona absent; perianth-scales or tufts of hairs sometimes present at the bases of the filaments 18
18a. Perianth-lobes at most 1 cm **13. Tapeinanthus**
b. Perianth-lobes 2 cm or more 19
19a. Perianth strongly bilaterally symmetric, mostly crimson, the bases of the 3 lower segments rolled to form a slightly open tube around the bases of the filaments and styles **3. Sprekelia**
b. Perianth radially symmetric or almost so, not as above 20
20a. Stigma simple, capitate or minutely 3-toothed; perianth usually yellow inside and out, rarely white **4. Sternbergia**
b. Stigma distinctly 3-fid; perianth usually not yellow, occasionally yellow inside 21
21a. Stamens of 2 different lengths, deflexed, curved upwards towards their tips **2. Habranthus**
b. Stamens all of the same length, more or less straight and spreading **1. Zephyranthes**
22a. Scape hollow 23
b. Scape solid 26
23a. Perianth cylindric, very narrowly funnel-shaped or bell-shaped, the tube longer than the lobes **11. Cyrtanthus**
b. Perianth funnel-shaped, often widely so, lobes longer than the tube 24
24a. Leaves and scape borne on a long aerial neck, which is blackish; flowers lilac-blue **7. Worsleya**
b. Leaves and scape borne more or less at soil level, bulb without a neck or the neck very short; flowers not lilac-blue 25
25a. Perianth-lobes each with a cushion-like swelling at the base **10. Vallota**
b. Perianth-lobes without such swellings **8. Hippeastrum**
26a. Ovules 1–3 in each cell of the ovary 27
b. Ovules 4 or more in each cell of the ovary 30
27a. Spathes 2; perianth-lobes spreading, tube curved 28
b. Spathes several; perianth-lobes erect, tube straight 29
28a. Uppermost stamen erect, pressed against the uppermost perianth-lobe, the other 5 deflexed, curved upwards towards their tips **6. Griffinia**
b. All stamens deflexed, curved upwards towards their tips, or all widely preading **5. Lycoris**
29a. Leaves very obviously in 2 ranks, usually thick and fleshy, without distinct midribs; rhizomes absent **17. Haemanthus**
b. Leaves not obviously in 2 ranks, thin and with distinct midribs; rhizomes present **18. Scadoxus**
30a. Perianth-tube long, cylindric or somewhat widened towards the apex, the lobes spreading at a wide angle into a cup- or bell-shape **14. Crinum**
b. Perianth-tube very short or absent, the lobes not spreading from it at a wide angle 31
31a. Scape obviously compressed and 2-edged; fruit a berry **19. Clivia**
b. Scape not obviously compressed and 2-edged; fruit a capsule 32
32a. Umbel spherical with 20–60 flowers on stalks that are at least 10 cm **15. Brunsvigia**
b. Umbel not as above, flowers usually fewer than 20, stalks always less than 10 cm 33
33a. Perianth-tube at most 5 mm, the lobes curving away from it, separate and distinct, margins often wavy **16. Nerine**
b. Perianth-tube at least 1 cm, the lobes forming a funnel-shape, overlapping or almost so, margins not wavy **9. Amaryllis**

1. ZEPHYRANTHES Herbert
E. Campbell

Perennial herbs with bulbs. Leaves usually present with the flowers, narrow, linear. Flowers solitary, borne on a hollow scape, subtended by a tubular spathe that is usually bifid towards the apex. Perianth usually united into a tube below, 6-lobed, the 3 inner lobes slightly narrower than the 3 outer, the tube often minutely scaly in the throat, more rarely of 6 free segments, radially symmetric. Stamens 6, erect, attached near the throat of the perianth-tube, anthers versatile. Ovary 3-celled with many ovules. Style thread-like, erect; stigma 3-fid. Fruit a capsule splitting between the septa. Seeds black.

A genus of about 60 species from warm regions of the western hemisphere, of which 6 are common in cultivation as border or pot plants.

Literature: Sealy, J.R., Zephyranthes, Pyrolirion, Habranthus and Hippeastrum, *Journal of the Royal Horticultural Society* **62**: 195–209 (1937); Herklots, G.A.C., Wind flowers: part 1, Zephyranthes, *The Plantsman* **2**: 8–19(1980).

Bulb. More than 3 cm thick: **4**; less than 3 cm thick: **1, 2, 3, 5, 6**.
Scape. Up to 15 cm: **1, 3, 4, 6**; usually more than 15 cm: **1, 2, 5**.
Flowering time. Spring: **1, 5, 6**; summer: **1, 2, 4, 6**; autumn: **2, 3**.
Flowers. Stalked: **2, 3, 4, 5, 6**; stalkless: **1**.

Perianth-lobes. White, sometimes tinged with green or pink: **1, 2, 4, 5**; pink: **3, 6**. Up to 4 cm: **1, 2, 3**; more than 4 cm: **2, 4, 5, 6**.

1a. Flowers stalkless **1. verecunda**
b. Flowers stalked 2
2a. Style shorter than stamens **2. candida**
b. Style longer than stamens 3
3a. Flower-stalk longer than spathe 4
b. Flower-stalk shorter than spathe 5
4a. Spathes 1.3–2 cm; flowers pink; perianth-lobes 3–4 cm **3. rosea**
b. Spathes *c.* 2.7 cm; flowers white tinged with green at the base; perianth-lobes 4–5 cm **4. tubispatha**
5a. Spathes 2.5–3 cm; flowers pure white; perianth-lobes *c.* 7.5 × 1.2 cm **5. atamasco**
b. Spathes 4–5 cm; flowers purplish pink; perianth-lobes 6.3–7.5 × 1.3–2 cm **6. grandiflora**

1. Z. verecunda Herbert (*Z. sessilis* Herbert). Illustration: *Botanical Magazine*, 2583 (1825).

Bulb ovoid, 2–2.5 cm across. Leaves 15–23 cm × *c.* 6 mm, recurved or almost erect. Scapes 2–3, 7.5–23 cm. Spathe *c.* 2.5 cm, cylindric in the lower half, bifid above. Flower stalkless. Perianth white tinged with pink, 3.8–5 cm, tube 2–3 cm, lobes 2.5–3.7 cm × 6–8 mm, spreading. *N Mexico*. H5–G1. Spring–summer.

2. Z. candida (Lindley) Herbert (*Argyropsis candida* (Lindley) Roemer). Illustration: *Botanical Magazine*, 2607 (1825); *Loddiges' Botanical Cabinet* **14**: t. 1419 (1828); Rix & Phillips, The bulb book, 182 (1981).

Bulb ovoid, *c.* 2.5 cm across. Leaves to 38 cm × 3–5 mm, erect. Scape 15–23 cm. Spathe *c.* 2.5 cm, cylindric in the lower half, not bifid. Flower-stalk *c.* 1.5 cm. Perianth pure white or slightly tinged with red, 3.8–5 cm, tube absent, segments *c.* 1.3 cm broad, spreading. Style shorter than stamens. *Argentina, Uruguay*. H5–G1. Summer–autumn.

3. Z. rosea (Sprenger) Lindley. Illustration: *Edwards's Botanical Register* **10**: t. 821 (1824); *Botanical Magazine*, 2537 (1825).

Bulb spherical, *c.* 2 cm across. Leaves 10–20 cm × *c.* 5 mm, spreading. Scapes 1–3, 10–15 cm. Spathe 1.3–2 cm, cylindric, bifid at the tip only. Flower-stalk more than 2 cm. Perianth pink, 3–3.5 cm, tube *c.* 4 mm, lobes 6–8 mm broad. Style longer than stamens. *West Indies, Guatemala*. G1. Autumn.

4. Z. tubispatha (L'Héritier) Herbert (*Z. insularum* Hume). Illustration: *Botanical Magazine*, n.s., 232 (1954–5).

Bulb spherical, 3–3.5 cm across. Leaves 12–33 cm × 4–7 mm, erect. Scape 12–15 cm. Spathe *c.* 2.7 cm, cylindric in the lower half, bifid above. Flower-stalk *c.* 3.5 cm. Perianth white tinged with green, 4–5 cm, tube *c.* 4 mm, lobes *c.* 1.3 cm broad. *West Indies, Venezuela, Colombia*. G1–H5. Summer.

5. Z. atamasco (Linnaeus) Herbert. Illustration: *Loddiges' Botanical Cabinet* **19**: t. 1899 (1833).

Bulb ovoid, to 2.5 cm across. Leaves *c.* 30 cm × 6 mm. Scape 15–30 cm. Spathe 2.5–3 cm, cylindric in the lower half, bifid above. Flower-stalk less than 2.5 cm. Perianth pure white, *c.* 7.5 cm, tube *c.* 1.2 cm, lobes *c.* 1.2 cm broad. *USA (Virginia, Florida, Alabama, Mississippi)*. H5–G1. Spring.

6. Z. grandiflora Lindley (*Z. carinata* Herbert; *Z. tsouii* Hu). Illustration: *Botanical Magazine*, 2594 (1825); *Edwards's Botanical Register* **11**: t. 902 (1825); Brickell (ed.), RHS A–Z encyclopedia of garden plants, 1098 (2003).

Bulb ovoid, *c.* 2.5 cm across. Leaves 15–30 cm × 5–8 mm, spreading. Scape 13–25 cm. Spathe 4–5 cm, cylindric in the lower half, slit on one side in the upper half. Flower-stalk *c.* 2.5 cm. Perianth pale purplish pink with a white throat, 6.3–7.5 cm, tube 3.1–3.3 cm, lobes 1.3–1.8 cm broad. Style longer than stamens. *West Indies, Cuba, Mexico, Guatemala*. H5–G1. Spring–summer.

2. HABRANTHUS Herbert

J. Cullen

Perennial herbs with bulbs. Leaves linear. Flowers solitary, terminal (in a reduced, 1-flowered umbel), subtended by a spathe (a pair of fused bracts) that is tubular in its lower half and usually somewhat bifid above. Perianth slightly bilaterally symmetric, funnel-shaped, with a short or very short tube below. Corona absent. Stamens 6, deflexed, then curving upwards towards the apex, of 2 different lengths. Style deflexed downwards then curving upwards, exceeding the stamens, stigma 3-fid. Fruit a top-shaped capsule containing numerous seeds.

A genus of 10–20 species from temperate S America (but see 3. *tubispathus*), very similar to *Zephyranthes* (p. 197), differing only in a slightly bilaterally symmetric perianth, the unequal, deflexed stamens and deflexed style. Cultivation as for *Zephyranthes*.

Literature: Sealy, J.R., Zephyranthes, Pyrolirion, Habranthus and Hippeastrum, *Journal of the Royal Horticultural Society* **62**: 195–209 (1937).

1a. Perianth 2.5–3.5 cm, bright yellow inside, coppery or purplish red outside **3. tubispathus**
b. Perianth 6–10 cm, red, pinkish mauve or whitish flushed with pink inside and out 2
2a. Perianth 7.5–10 cm, red or pink, the tube deeper mauve-pink **1. brachyandrus**
b. Perianth *c.* 6 cm, white flushed with pink, the tube green **2. robustus**

1. H. brachyandrus (Baker) Sealy (*Hippeastrum brachyandrum* Baker). Illustration: *Botanical Magazine*, 7344 (1894); *The Garden* **49**: t. 1048 (1896); *Horticulture* **53**: 82 (1975).

Bulb ovoid, to 2.5 cm in diameter, dark brown, with a long neck. Leaves linear, to 25 cm. Scape to 30 cm, spathe to 5 cm, flower-stalks slightly longer than the spathe. Perianth ascending-erect, 7.5–10 cm, bright red or pinkish mauve inside and out, the tube deeper red or pinkish mauve. Stamens to 2.5 cm. *S Brazil, Paraguay*. H5. Summer.

2. H. robustus Herbert (*Zephyranthes robusta* (Herbert) Baker). Illustration: *Botanical Magazine*, 9126 (1927); Brickell (ed.), RHS A–Z encyclopedia of garden plants, 499 (2003).

Similar to *H. brachyandrus* but leaves developing after flowering, scape at most 15 cm, perianth to 6 cm, white flushed with pink, the tube green, stamens *c.* 3 cm. *S Brazil, ?Argentina*. H5. Summer.

3. H. tubispathus (L'Héritier) Traub (*Amaryllis tubispatha* L'Héritier; *H. andersonii* Herbert). Illustration: *Edwards's Botanical Register* **16**: t. 1345 (1840); *Addisonia* **22**: t. 727 (1945); *Plant Life* **26**: 100 (1970).

Bulb ovoid, to 2.5 cm in diameter, with a short neck. Leaves 13–15 cm, narrowly linear. Scape to 15 cm. Spathe 2.5–3 cm, exceeded by the flower-stalk. Perianth erect, bright yellow inside, coppery or purplish red outside, 2.5–3.5 cm, tube very short. Stamens 1.2–1.7 cm. *S Brazil, Uruguay, E Argentina & S Chile.* H5. Summer.

The nomenclature of this species has been elucidated by Ravenna, *Plant Life* **26**: 99–103 (1970). The variant with the flowers coppery outside has been distinguished as *H. andersonii* var. *cupreus* Anon., but is no more than a selection.

H. texanus (Herbert) Steudel (*H. andersonii* var. *texanus* Herbert; *Zephyranthes texana* (Herbert) Herbert). Illustration: *Botanical Magazine*, 3596 (1837). Very similar, but smaller, perianth to 2.5 cm. *Southern USA (Texas).* H5. Summer.

Scarcely distinguishable from *H. tubispathus* but notable on account of its distribution. Perhaps an early introduction of *H. tubispathus* into N America.

3. SPREKELIA Heister
J. Cullen
Perennial herbs with bulbs. Leaves linear, present at flowering. Scape about as long as the leaves, hollow. Flower solitary, terminal (a reduced, 1-flowered umbel), subtended by a spathe that is tubular in its lower half, bifid above. Flower-stalk shorter than the spathe, the flower horizontal or slightly drooping. Perianth of 6 free segments, strongly bilaterally symmetric, the uppermost segment broadest, the 2 upper lateral segments sickle-shaped, arching downwards, the 3 lower segments with their basal parts rolled to form a tube (slightly open above) surrounding the lower parts of the stamens and style. Stamens deflexed, then curved upwards. Style deflexed, then curved upwards, exceeding the stamens. Stigma 3-fid. Ovary 6-angled, 3-celled with numerous ovules in each cell. Capsule 3-angled, containing many seeds.

A genus of a single species from Mexico and Guatemala. Cultivation as for *Hippeastrum* (p. 201), though it tolerates generally cooler conditions than most of the tropical species of that genus.

1. S. formosissima (Linnaeus) Herbert (*Amaryllis formosissima* Linnaeus). Illustration: *Botanical Magazine*, 47 (1788); *Journal of the Royal Horticultural Society* **87**: f. 80 (1962); *Horticulture* 48 (**10**): 37 (1970); Moggi & Guignolini, Flori da balcone e da giardino, f. 329 (1982).

Bulb to 5 cm in diameter, brown. Leaves to 50 × 2 cm. Spathe to 5 cm. Perianth 7–11 cm, bright crimson, the base of the segments sometimes striped or blotched with yellow. *Mexico, Guatemala.* G1. Spring.

4. STERNBERGIA Waldstein & Kitaibel
B. Mathew & J.C.M. Alexander
Perennial herbs with bulbs; tunics membranous, brown or black. Leaves basal, linear or strap-shaped to narrowly lanceolate, flat or shallowly channelled above, sometimes keeled beneath, often twisted. Flower-stems 1–several, sometimes below ground at flowering (Nos. 5 & 6), elongating and often arching in fruit. Flowers solitary, usually yellow (white in No. 4) appearing in spring or autumn. Spathe membranous, tubular below, split above. Perianth radially symmetric; tube cylindric, narrow; lobes 6, in 2 similar whorls, oblanceolate to obovate (very narrow in No. 6). Corona absent. Stamens 6, in 2 unequal whorls, borne at top of perianth-tube, shorter than perianth-lobes. Style 1, bearing an entire or minutely 3-toothed capitate stigma. Fruit a cylindric to spherical capsule containing numerous large, dark seeds often with fleshy appendages.

A genus of 7 or 8 species centred on Turkey, extending west to Spain, east to Kashmir and south to Israel. They are most likely to flourish if grown in a bulb frame where they can be given a dry dormant period in the summer to encourage the formation of flowers. *S. lutea* and *S. sicula* will also perform well in the open ground in a sheltered, sunny, well-drained site. The spring-flowering species, *S. candida* and *S. fischeriana*, are not hardy, and should be planted quite deeply to discourage the bulbs from dividing too often, which produces a succession of small, non-flowering plants. Most species can also be grown in pots in an alpine house.

Literature: Mathew, B., A review of the genus Sternbergia, *The Plantsman* **5**: 1–16 (1983).

1a. Flowers produced in spring 2
b. Flowers produced in autumn 3
2a. Flowers yellow; spathe 2.5–4.5 cm; perianth-lobes 2–4 cm **3. fischeriana**
b. Flowers white; spathe 5–5.5 cm; perianth-lobes 4.3–5 cm **4. candida**
3a. Leaves appearing at or before flowering-time; perianth-tube 2 cm or less 4
b. Leaves absent at flowering-time; perianth-tube usually 2–6.5 cm 5
4a. Leaves 7–12 mm wide, usually bright shiny green; perianth-lobes usually 3–5.5 × 1–2 cm **1. lutea**
b. Leaves 2–5 mm wide, deep green with greyish stripe; perianth-lobes usually 2–4 cm × 3–12 mm **2. sicula**
5a. Leaves 8–16 mm wide; perianth-lobes 3.5–7.5 × 1 cm or more **5. clusiana**
b. Leaves 1–4 mm wide; perianth-lobes usually less than 3 cm × 5 mm **6. colchiciflora**

1. S. lutea (Linnaeus) Sprengel. Illustration: *Botanical Magazine*, 290 (1795); Hay & Synge, Dictionary of garden plants in colour, 111 (1969); Rix & Phillips, The bulb book, 176 & 184 (1981); *The Plantsman* **5**: 1 (1983).

Bulb 2–4 cm across. Leaves 4–6, appearing with or just before the flowers, 7–12 mm wide, narrowly lanceolate, slightly channelled above, keeled beneath, usually bright shiny green. Flower-stem 2.5–20 cm, longer in fruit; spathe 3–6 cm. Flower deep yellow; tube 5–20 mm; lobes 3–5.5 × 1–2 cm, oblanceolate to obovate. Seeds lacking an appendage. *From Spain to Iran & Former USSR (Soviet Central Asia); seldom truly wild.* H4–5. Autumn.

2. S. sicula Gussone (*S. lutea* subsp. *sicula* (Tineo) Webb; *S. lutea* var. *graeca* Reichenbach; *S. lutea* var. *angustifolia* Anon.). Illustration: Polunin & Huxley, Flowers of the Mediterranean, f. 254 (1965); Rix & Phillips, The bulb book, 185 (1981); *The Plantsman* **5**: 7 (1983).

Similar to *S. lutea*. Bulb 1–2.5 cm across. Leaves 2–5 mm wide, linear, dark green with central grey stripe. Flower-stem 3–70 mm; spathe 3–6 cm. Perianth-tube 4–10 mm; lobes 2–3.4 cm × 4–12 mm. *Italy, Sicily, Greece, Crete, Aegean Islands & W Turkey.* H4–5. Autumn.

3. S. fischeriana (Herbert) Ruprecht. Illustration: *Botanical Magazine*, 7441 (1895); *Journal of the Royal Horticultural Society* **98**: 442 (1973); Rix & Phillips, The bulb book, 29 (1981); *The Plantsman* **5**: 9 (1983).

Bulb 2.5–3.5 cm across. Leaves 4–7, appearing before the flowers, 6–12 mm wide, strap-shaped, flat, not or scarcely keeled, usually dark greyish green. Flower-stem 3–15 cm, longer in fruit; spathe 2.5–4.5 cm. Flower pale yellow; tube 3–5 mm; lobes 2–3.5 cm × 5–8 mm, oblanceolate. Seeds with fleshy appendage. *Former USSR (Azerbaijan) & S Turkey to*

Kashmir (not recorded from Afghanistan). H4–5. Spring.

4. S. candida Mathew & Baytop. Illustration: *Journal of the Royal Horticultural Society* **104**: 302 (1979); Rix & Phillips, The bulb book, 29 (1981); *The Plantsman* **5**: 9 (1983); Brickell (ed.), RHS A–Z encyclopedia of garden plants, 1007 (2003).

Similar to *S. fischeriana.* Flower-stem 12–20 cm; spathe 5–5.5 cm. Flower white, fragrant; lobes 4.3–5 cm × 9–18 mm. *SW Turkey.* H4–5. Winter–spring.

5. S. clusiana (Ker Gawler) Sprengel (*S. macrantha* Baker). Illustration: *Botanical Magazine*, 7459 (1896); Mathew, Dwarf bulbs, pl. 78 (1973); Rix & Phillips; The bulb book, 176 (1981); *The Plantsman* **5**: 12 (1983).

Bulb 2.5–4 cm across. Leaves 5–12, appearing long after the flowers, 8–16 mm wide, strap-shaped, flat, not keeled, often twisted, greyish green. Flower-stem very short and below ground at flowering, raising ovary above ground in fruit. Spathe 5–10 cm. Flower bright to greenish yellow; tube 3–6.5 cm; lobes 3.7–7.5 × 1.1–3.3 cm, obovate to oblanceolate. Seeds with large fleshy appendage. *Turkey & Iran south to Israel & Jordan.* H4–5. Autumn.

6. S. colchiciflora Waldstein & Kitaibel. Illustration: Polunin, Flowers of Europe, pl. 172 (1969); Mathew, Dwarf bulbs, pl. 79 (1973); Rix & Phillips, The bulb book, 29 & 184 (1981); *The Plantsman* **5**: 12 & 15 (1983).

Bulb 5–15 mm across. Leaves 3–6, appearing long after the flowers, 1–4 mm wide, narrowly linear, not or scarcely keeled, twisted, sometimes strongly, dark green, sometimes glaucous. Flower-stem very short and below ground at flowering, raising ovary above ground in fruit. Spathe 2.2–4 cm. Flowers yellow; tube 3–6.5 cm; lobes 2.3–3.3 cm × 2–5 mm, narrowly oblanceolate. Seeds with large fleshy appendage. *SE Spain & Italy eastwards to Israel & Iran.* H4–5. Autumn.

5. LYCORIS Herbert

E. Campbell

Herbaceous perennials with ovoid bulbs. Leaves basal, linear or strap-shaped, usually appearing after the inflorescences. Flowers in a terminal umbel borne on a solid scape, the umbel subtended by 2 free spathes. Perianth bilaterally or radially symmetric, with a usually very short tube at the base, lobes equal, spreading or strongly recurved; perianth-scales minute. Stamens 6, erect or ascending, borne on the throat of the perianth-tube, deflexed, anthers versatile. Ovary 3-celled, with 2 or 3 ovules per cell. Style thread-like, stigma minute. Capsule almost spherical to ovoid. Seeds black-brown.

A genus of 10–12 species from China and Japan, of which 5 are commonly found in cultivation as pot plants. In northern Europe they require the protection of a greenhouse. Cultivation as for *Hippeastrum* (p. 201).

Literature: Traub, H.P. & Moldenke, H.N., Amaryllidaceae: Tribe Amarylleae, 165–82 (1949).

1a. Leaves 1.8–2.5 cm wide; flowers fragrant, rose pink, sometimes with purple or blue markings — 2
b. Leaves to 1.8 cm wide; flowers not fragrant, yellow or red — 3
2a. Spathes ovate; flowers rose pink, without purple-blue markings; perianth-tube absent — **1. sprengeri**
b. Spathes lanceolate; flowers pale rose pink with purple or blue markings; perianth-tube present — **2. squamigera**
3a. Flowers golden yellow; leaves 1.2–1.8 cm wide; bulb 5–6 cm in diameter — **3. aurea**
b. Flowers red; leaves 6–12 mm wide; bulb to 4 cm in diameter — 4
4a. Leaves 6–8 mm wide; perianth rose red to deep red, lobes very reflexed and with wavy margins — **4. radiata**
b. Leaves 1–1.2 cm wide; perianth bright red, lobes slightly reflexed and without wavy margins — **5. sanguinea**

1. L. sprengeri Baker

Leaves 30–60 × *c.* 2 cm, linear. Scape *c.* 30 cm. Spathes short, ovate, acute. Flower-stalks 2–4 cm. Flowers many, erect, fragrant. Perianth 7–8 cm, rose-pink, funnel-shaped, without a tube, the lobes recurved at the tips, margins not wavy. Stamens not projecting. *Japan.* H4. Summer.

2. L. squamigera Maximowicz. Illustration: *Botanical Magazine*, 7547 (1897).

Bulb ovoid, 4–5 cm in diameter. Leaves *c.* 30 × 1.8–2.5 cm, strap-shaped, produced in spring. Scape 50–70 cm. Spathes lanceolate, 2–4 cm. Flower-stalks 1–3 cm. Flowers 6–8, almost erect, fragrant. Perianth 9–10 cm, pale rose pink with blue or purple veining, funnel-shaped; tube *c.* 2.5 cm, lobes recurved at their tips, margins slightly wavy. *Japan.* H4. Summer.

The name *L. incarnata* Sprenger, which seems to be invalid, probably applies to *L. squamigera* or a variety of it.

3. L. aurea (L'Héritier) Herbert. Illustration: *Botanical Magazine*, 409 (1798); *Edwards's Botanical Register*, t. 611 (1822); Brickell (ed.), RHS A–Z encyclopedia of garden plants, 656 (2003).

Bulb 5–6 cm in diameter, ovoid. Leaves to 60 × 1.2–1.8 cm, strap-shaped, fleshy, glaucous, produced after the flowers. Scape to 60 cm. Spathes 3–5 cm, lanceolate. Flower-stalks 8–15 mm. Flowers 5 or 6, erect. Perianth 9.5–10 cm, golden yellow, funnel-shaped; tube 1.5–2 cm, lobes recurved at their tips, the margins usually very wavy. Stamens slightly projecting. *China, Japan.* H4. Spring–summer.

4. L. radiata (L'Héritier) Herbert. Illustration: *Edwards's Botanical Register*, t. 596 (1822); Brickell (ed.), RHS A–Z encyclopedia of garden plants, 656 (2003).

Bulb 2.5–3.5 cm in diameter, broadly ellipsoid. Leaves 30–60 cm × 6–8 mm, strap-shaped, dark green. Scape 30–50 cm. Spathes 2–4 cm, linear-lanceolate. Flower-stalks 6–15 mm. Flowers 4–6, drooping. Perianth 4–5 cm, rose-red to deep red; tube 6–8 mm, lobes very reflexed, margins very wavy. Stamens strongly projecting. *Japan.* H5. Summer–autumn.

L. albiflora Koidzumi is very similar, and is probably merely a variety.

5. L. sanguinea Maximowicz.

Bulb 1.5–2 cm in diameter. Leaves 1–1.2 cm wide, linear, dark green. Scape 30–50 cm. Spathes 2–4 cm, lanceolate. Flower-stalks *c.* 5 cm. Flowers 4–6, erect. Perianth 5–6 cm, bright red, funnel-shaped; tube 1.2–1.5 cm; lobes slightly recurved, margins not wavy. Stamens not projecting. *China, Japan.* H4. Summer–autumn.

6. GRIFFINIA Ker Gawler

J. Cullen

Perennial herbs with bulbs. Leaves distinctly stalked (in our species) with broadly oblong-lanceolate blades. Scape solid. Umbel with several flowers; spathes 2, free. Perianth horizontal, openly funnel-shaped, distinctly bilaterally symmetric, with short tube at the base. Stamens 6, spreading. Ovary 3-celled, with 2 ovules in each cell. Style somewhat deflexed, stigma capitate. Fruit a capsule containing few seeds.

A genus of 6 species from Brazil, of which only 1 is in general cultivation. Conditions as for the tropical species of *Hippeastrum* (p. 201).

Literature: Traub, H.P. & Moldenke, H.N., Amaryllidaceae: Tribe Amarylleae, 153–7 (1949).

1. G. hyacinthina (Ker Gawler) Ker Gawler (*Amaryllis hyacinthina* Ker Gawler; *Lycoris hyacinthina* (Ker Gawler) Herbert). Illustration: *Edwards's Botanical Register* **2**: t. 163 (1816) & **6**: t. 444 (1820); Everett, Illustrated encyclopedia of gardening **5**: opposite p. 822 (1981).

Bulb 5–6 cm in diameter. Leaf-blade 15–22 × 5–7.6 cm, stalk almost as long as the blade. Umbel with 8–10 flowers, flower-stalks very short. Perianth purplish, violet-blue or white, 5–6.5 cm. *Brazil* G2. Summer.

7. WORSLEYA Traub

J. Cullen

Bulb large, dark brown with a very long, aerial, trunk-like neck. Leaves several, evergreen, distinctly borne in 2 ranks, strap-shaped, curved downwards in a sickle shape, green with translucent, cartilaginous margins. Scape hollow, 2-edged, shorter than the leaves. Umbel with 4–14 flowers; spathes 4, 2 of them shorter and narrower than the others, free, longer than the flower-stalks. Perianth horizontal, funnel-shaped, somewhat bilaterally symmetric, with a short tube at the base. Corona absent. Stamens 6, deflexed and then curved upwards towards the apex, shorter than the perianth. Ovary inferior, 3-celled, with many ovules in each cell. Style deflexed, then curved upwards, exceeding the stamens. Stigma capitate and obscurely 3-lobed. Capsule large, wider than long.

A genus of a single species from Brazil, notable for the extremely long neck of its bulb. Cultivation as for *Hippeastrum* (p. 201), but the plant is frequently short-lived.

Literature: Traub, H.P. & Moldenke, H.N., Amaryllidaceae: Tribe Amarylleae, 22–4 (1949); Martinelli, G., Nota sobre Worsleya rayneri (J.D. Hooker) Traube & Moldenke, espécie ameaçada de extinçal, *Rodriguesia* **36**: 65–71 (1984).

1. W. rayneri (J.D. Hooker) Traub & Moldenke (*Amaryllis rayneri* J.D. Hooker; *A. procera* Duchartre not Salisbury; *Hippeastrum procerum* (Duchartre) Lemaire; *Worsleya procera* (Duchartre) Traub). Illustration: *Botanical Magazine*, 5883 (1871); *The Garden* **45**: 350 (1894); Everard & Morley, Wild flowers of the world, t. 190 (1970).

Bulb 13–15 cm in diameter, neck to 1 m. Leaves 60–100 × 4.5–7.5 cm. Perianth lilac-blue, 14–16 cm, tube to 2 cm. *S Brazil (Organ mountains)*. G1–2. Winter–spring.

8. HIPPEASTRUM Herbert

J. Cullen

Perennial herbs with bulbs. Leaves few, usually not stalked. Scape slender or conspicuously stout, hollow. Flowers 2–10 in an umbel subtended by 2 large, equal, free spathes that enclose the whole umbel in bud and persist beneath it in flower. Flowers erect, horizontal or drooping. Perianth bilaterally symmetric, with a long or short tube. Corona absent; sometimes small scales or tufts of hairs are found at the junction of the filaments and perianth. Stamens 6, attached to the perianth-tube, deflexed and then curving upwards towards the apex. Ovary 3-celled, with few to many ovules in each cell. Style deflexed downwards then curving upwards; stigma capitate or 3-fid. Fruit a spherical or 3-lobed capsule.

In the interpretation followed here, a genus of about 80 species from C & S America. In using the name *Hippeastrum* for these plants and *Amaryllis* (p. 204) for a South African species, common European practice is followed. However, H.P. Traub (see references below) and members of The American Plant Life Society, regard all the species here included in *Hippeastrum* as belonging to the genera *Amaryllis* or *Rhodophiala*, and treat the South African plant as a species of *Brunsvigia* (p. 216). This whole problem turns on the interpretation of what Linnaeus intended when he described *Amaryllis belladonna.* No generally acceptable solution has yet been agreed, though Goldblatt (*Taxon* **33**: 511–16, 1984) has reviewed the arguments and has proposed the name *Amaryllis* for conservation in the sense in which it is used here.

The small-flowered species (Nos. 1–5) are (with their wild allies) sometimes regarded as forming the genus *Rhodophiala* Presl, but a broad concept of *Hippeastrum* seems appropriate here. The large-flowered species (Nos. 6–15) and their allies form a very striking group of plants, much cultivated by enthusiasts, and many hybrids have been raised, named and distributed. Lists of these names appear from time to time in the journal *Plant Life*, which should be consulted for further information.

Hippeastrums are relatively easily grown in good soil. All can be grown in pots, but many benefit from being planted in a border. Potted plants can be forced to produce flowers earlier than the seasons indicated here. Enthusiasts use many specialised techniques for growing show plants; the journal *Plant Life*, and Traub's *Amaryllis Manual*, cited below, should be consulted for details.

Literature: Traub, H.P. & Moldenke, H.N., Amaryllidaceae: Tribe Amarylleae (1949); Traub, H.P., The genera Rhodophiala and Phycella, *Plant Life* **12**: 67–76 (1956); Traub, H.P., The Amaryllis Manual (1958).

Leaves. Linear, less than 2 cm broad (often much less): **1, 2, 3, 4, 5**; oblong, more than 2 cm broad, somewhat narrowed to the base but not stalked: **6, 7, 8, 9, 10, 11, 12, 13, 14**; oblanceolate, more than 2 cm broad, stalked: **15**.

Perianth. To 6 cm long: **1, 2, 3, 4, 5**; more than 6 cm long: **6, 7, 8, 9, 10, 11, 12, 13, 14, 15**. Tube 10–13 cm: **6**; tube 1–6 cm: **7, 8, 9, 10, 11, 12, 13, 14, 15**; tube less than 1 cm: **1, 2, 3, 4, 5**. Lobes yellowish green spotted with red: **1, 14**; lobes conspicuously striped red and green: **12**; lobes mauve with a superimposed darker mauve network: **15**.

Perianth-scales. Present and incurved, closing the throat of the perianth-tube: **12, 13, 14**; present but not incurved: **1, 2, 3, 4, 5**; present and very obscure, or absent: **6, 7, 8, 9, 10, 11, 15**.

Stigma. 3-fid: **1, 2, 3, 4, 7, 8, 12, 13**; capitate: **5, 6, 9, 10, 11, 14, 15**.

1a. Perianth up to 6 cm long; leaves linear, at most 2 cm broad, usually less than 1.3 cm broad — 2
b. Perianth 7–25 cm long; leaves oblong or oblanceolate, mostly more than 2.5 cm broad — 6
2a. Stigma capitate — **5. pratense**
b. Stigma 3-fid — 3
3a. Umbel 1–2-flowered; spathes 2.5–3.8 cm — **4. roseum**
b. Umbel 3–10-flowered; spathes 3.8–7.5 cm, rarely less — 4
4a. Stamens almost as long as perianth; perianth yellow or yellow tinged or spotted with red — **1. bagnoldii**

b. Stamens at most two-thirds of the length of the perianth; perianth yellow or red 5
5a. Perianth-lobes oblong, acute; spathes 2.5–5 cm **2. advenum**
b. Perianth-lobes oblanceolate, clawed, obtuse; spathes 5–7.5 cm **3. bifidum**
6a. Perianth-tube 10–13 cm **6. solandriflorum**
b. Perianth-tube at most 2.5 cm 7
7a. Stigma 3-fid 8
b. Stigma capitate 11
8a. Scales closing the throat of the perianth-tube 9
b. Scales obscure or absent, not closing the throat of the perianth-tube 10
9a. Each perianth-lobe mostly green, with many red stripes and a red margin **12. psittacinum**
b. Each perianth-lobe mainly red and without stripes, though with a greenish midrib **13. aulicum**
10a. Perianth 10–15 cm, tube *c.* 2.5 cm; scales very obscure or absent **7. vittatum**
b. Perianth 7.5–10 cm, tube *c.* 2 cm; small bristles or scales present at the bases of the filaments **8. rutilum**
11a. Perianth-scales conspicuous, closing the throat of the perianth-tube; perianth yellowish, copiously spotted with red **14. pardinum**
b. Perianth-scales absent or inconspicuous, not closing the throat of the perianth-tube; perianth not as above 12
12a. Leaves narrowed to an almost terete stalk; perianth reddish mauve with a superimposed network of reddish lines; ovules 1–3 per cell **15. reticulatum**
b. Leaves not narrowed to a stalk; perianth not as above; ovules numerous per cell 13
13a. Perianth only very slightly bilaterally symmetric, the lobes broadly overlapping, each with a white mark at the base **9. leopoldii**
b. Perianth distinctly bilaterally symmetric, lobes not as above 14
14a. Perianth-tube 2.2–3 cm; base of each perianth-lobe yellowish green or white, this area not forming a star shape when the whole flower is viewed **11. puniceum**
b. Perianth-tube to 1.3 cm; each perianth-lobe with a narrowly triangular white mark at the base, forming a star shape when the whole flower is viewed **10. reginae**

1. H. bagnoldii (Herbert) Baker (*Habranthus bagnoldii* Herbert; *Amaryllis bagnoldii* (Herbert) Dietrich; *Rhodophiala bagnoldii* (Herbert) Traub). Illustration: *Edwards's Botanical Register* **17**: t. 1396 (1831).

Bulb to 5 cm in diameter, almost black. Leaves linear, to 30 cm × 6 mm, somewhat glaucous, obtuse. Scape slender, to 30 cm. Umbel with 4–8 flowers; spathes 4–5 cm, flower-stalks 2–7 cm. Perianth ascending, openly funnel-shaped, yellow or yellow tinged or spotted with red, 3–5.5 cm, tube *c.* 5 mm. Perianth-scales small, not closing the throat of the tube. Stamens almost as long as the perianth. Style about as long as the perianth, stigma 3-fid. *Chile.* H5–G1. Winter.

2. H. advenum (Ker Gawler) Herbert (*Amaryllis advena* Ker Gawler; *Rhodophiala advena* (Ker Gawler) Traub). Illustration: *Botanical Magazine*, 1125 (1808); *Edwards's Botanical Register* **10**: t. 849 (1824); Saunders et al., Refugium botanicum **5**: t. 332 (1872).

Bulb ovoid, with a short neck, to 4 cm in diameter, brown. Leaves linear, to 30 cm × 5 mm, somewhat glaucous, blunt. Scape slender, to 50 cm. Umbel with 2–10 flowers; spathes 2.5–5 cm, flower-stalks 2–7 cm. Perianth horizontal or almost so, openly funnel-shaped, pink, red or yellow, 3.5–5 cm, tube to 5 mm, lobes oblong, acute. Perianth-scales small, not closing the throat of the perianth-tube, margins scalloped. Stamens at most two-thirds of the length of the perianth. Style almost as long as the perianth, stigma 3-fid. *Chile.* H5–G1. Winter.

3. H. bifidum (Herbert) Baker (*Amaryllis bifida* (Herbert) Sprengel; *A. kermesina* Lindley; *Rhodophiala bifida* (Herbert) Traub). Illustration: *Botanical Magazine*, 2597 (1825) & 2639 (1826); *Edwards's Botanical Register* **14**: t. 1148 (1828) & t. 1638 (1834).

Bulb ovoid, usually with a conspicuous neck, to 4 cm in diameter, dark brown. Leaves linear, to 45 × 1.3 cm, slightly glaucous, blunt. Scape 10–40 cm. Umbel with 2–7 flowers; spathes 5–7.5 cm, flower-stalks 2.5–5 cm. Perianth ascending to erect, rather narrowly funnel-shaped, bright to deep red, 4–5 cm, tube *c.* 3 mm, lobes oblanceolate, clawed. Stamens at most two-thirds of the length of the perianth. Style about as long as perianth, stigma 3-fid. *Argentina, Uruguay.* H5–G1. Winter.

Traub recognises 3 varieties (under his *Amaryllis bifida*): the above (var. **bifida**); var. **pulchra** (Herbert) Traub & Moldenke, which is a small plant with rather smaller flowers; and var. **spathacea** (Herbert) Traub & Moldenke (*A. spathacea* (Herbert) Schultes), in which the bulb is darker, the spathes incompletely separated, and the stamens and style are shorter.

4. H. roseum (Sweet) Baker (*Habranthus roseus* Sweet; *Amaryllis rosea* (Sweet) Traub & Uphof not Lamarck; *A. barlowii* Traub & Moldenke; *Rhodophiala rosea* (Sweet) Traub). Illustration: Sweet, British flower garden **2**: t. 107 (1825–27).

Bulb ovoid, with a short neck, to 2.5 cm in diameter, dark brown. Leaves linear, to 30 cm × 5 mm, glaucous. Scape to 20 cm. Umbel with 1–2 flowers; spathes 2.5–3.5 cm, flower-stalks 2–3 cm. Perianth almost horizontal, funnel-shaped, bright red, *c.* 5 cm, tube very short, greenish. Stamens shorter than the perianth and style, stigma 3-fid. *Chile (Chiloe).* H5–G1. Winter.

5. H. pratense (Poeppig) Baker (*Amaryllis pratensis* Poeppig; *Rhodophiala pratensis* (Poeppig) Traub). Illustration: *Edwards's Botanical Register* **28**: t. 35 (1842); *Botanical Magazine*, 3961 (1842); *The Garden*, 51 (1878).

Bulb ovoid, with a short neck, 3–4 cm in diameter, dark brown. Leaves linear, 30–50 cm × 6–13 mm, bright green, obtuse. Scape to 60 cm. Umbel with 2–8 flowers; spathes 3–5 cm, flower-stalks 2.5–4 cm. Perianth horizontal, openly funnel-shaped, bright red or purplish, the lobes often marked with yellow lines towards the base, 6–7 cm, tube *c.* 6 mm. Perianth-scales small. Stamens shorter than the perianth. Style about as long as the perianth, stigma capitate. *Chile.* H5–G1. Spring.

6. H. solandriflorum (Lindley) Herbert (*Amaryllis solandriflora* Lindley; *?A. elegans* Sprengel). Illustration: Lindley, Collectanea botanica, t. 11 (1821); *Botanical Magazine*, 2573 (1825) & 3542 (1837) – as *H. ambiguum*.

Bulb ovoid, with a short neck, 7–10 cm in diameter. Leaves oblong, blunt, to 50 × 2–2.5 cm. Scape 40–60 cm. Umbel with 2–6 flowers; spathes 5–8 cm, flower-stalks to 6 cm. Perianth horizontal, funnel-shaped, 18–25 cm, tube 10–13 cm, greenish white or white, sometimes striped or tinged with red. Perianth-scales small.

Stamens shorter than the perianth. Style as long as the perianth, stigma capitate though slightly 3-lobed. *Northern S America, south to Peru.* G2. Winter (sometimes also summer).

Variable in the precise colouring of the flowers; several varieties have been recognised to cover this variation. The species is easily recognised by its large, long-tubed flowers.

7. H. vittatum (L'Héritier) Herbert (*Amaryllis vittata* L'Héritier). Illustration: Lindley, Collectanea botanica, t. 12 (1821); *Botanical Magazine*, 129 (1795).

Bulb more or less spherical, 5–8 cm in diameter. Leaves oblong, obtuse, 45–60 × 2.5–3.5 cm, bright green. Scape 60–100 cm. Umbel with 2–6 flowers; spathes to 7.5 cm, flower-stalks to 7.5 cm. Perianth drooping, funnel-shaped, white striped with red, 9–15 cm, tube to 2.5 cm. Stamens shorter than the perianth. Style about as long as the perianth, stigma 3-fid. *Peru, Brazil.* G2. Early spring.

This species was crossed in the 1790s with *H. regime* to produce a range of popular hybrids known as *H.* × *johnsonii.*

8. H. rutilum (Ker Gawler) Herbert (*Amaryllis rutila* Ker Gawler; *A. striata* in the sense of Traub & Moldenke and Traub). Illustration: *Edwards's Botanical Register* **1**: t. 38 (1815), **3**: t. 226 (1817) & **7**: t. 534 (1821); *Botanical Magazine*, 2475 (1824).

Bulb more or less spherical, with a short neck, 5–8 cm in diameter. Leaves oblong, blunt, to 30 cm × at least 2.5 cm, bright green. Scape to 30 cm. Umbel with 2–4 flowers; spathes 3–6.5 cm, flower-stalks 3–6 cm. Perianth horizontal or drooping, bright red, scarlet, orange or reddish orange, each lobe with a greenish white base or green stripe along the midrib, total length 7.5–11 cm, tube greenish, *c.* 2 cm. Stamens shorter than the perianth. Style about as long as the perianth, stigma 3-fid. *S & C Brazil.* G2. Winter.

Variable in flower colour: var. **rutilum** (described above) has the perianth-lobes bright crimson, each with an elongate green base; var. **fulgidum** Baker has the perianth scarlet to red-orange, each lobe with a greenish stripe towards the base; and var. **crocatum** (Ker Gawler) Anon., has the perianth-lobes orange-yellow.

9. H. leopoldii Dombrain (*Amaryllis leopoldii* (Dombrain) Moore). Illustration: *Floral Magazine* **9**: t. 475 (1870); *Gardeners' Chronicle*, 733 (1870).

Bulb more or less spherical, with a short neck, to 7.5 cm in diameter. Leaves oblong, 45–60 × 2.5–4 cm, green. Scape stout. Umbel with 2 flowers; flower-stalks 3.5–3.7 cm. Perianth almost radially symmetric (the lowermost lobe only slightly narrower than the others), horizontal, openly funnel-shaped, bright red; the 5 larger lobes are broad and overlapping, each with a narrowly triangular white stripe at the base, giving the appearance of a 5-pointed star when the flower is viewed from the front; total length of perianth *c.* 11 cm, tube *c.* 1 cm. Perianth-scales small, edged with whitish bristles. Stamens shorter than the perianth. Style about as long as the perianth, stigma capitate. *Bolivia.* G2. ?Winter.

This species was introduced into cultivation in 1869, and was thought to have come from Peru. It was soon lost, however, and was not refound until the 1970s, when it was rediscovered in Bolivia (see *Plant Life* **30**: 81–3, 1974).

10. H. reginae (Linnaeus) Herbert (*Amaryllis reginae* Linnaeus; *A. alberti* Lemaire). Illustration: *Botanical Magazine*, 453 (1799).

Bulb more or less spherical, with a neck, to 7.5 cm in diameter. Leaves oblong, blunt, somewhat narrowed towards the base, to 60 × 2.5–4 cm, green. Scape to 50 cm. Umbel with 2–4 flowers; spathes 5–7.5 cm, flower-stalks 5–7.5 cm. Perianth horizontal or drooping, funnel-shaped, bright red with a 5-pointed, greenish white star at the base, 8–12 cm, tube to 1.3 cm. Stamens slightly shorter than the perianth. Style about as long as the perianth, stigma capitate. *C & northern S America.* G2. Winter–spring.

Plants in cultivation under the name *Amaryllis miniata* Ruiz & Pavon probably belong to this species. The name *Amaryllis alberti* applies to a variant with double flowers.

H. machupijchense Vargas. Illustration: *Botanical Magazine*, n.s., 821 (1981). Similar, but the flowers more open and the perianth red, each lobe with a green base that extends along the lobe as green stripes. *Peru.* G2.

11. H. puniceum (Lamarck) Voss (*Amaryllis belladonna* in the sense of Traub; *A. equestris* Herbert; *Hippeastrum equestre* (Aiton) Herbert). Illustration: *Botanical Magazine*, 305 (1795); *Edwards's Botanical Register* **3**: t. 234 (1817); *Gartenflora* **44**: t. 1418 (1895).

Bulb more or less spherical, with a short neck, to 5 cm in diameter, brownish. Leaves oblong, slightly tapered to the base, abruptly tapered to a blunt apex, 25–50 × 2.5–4.5 cm, bright green. Scape to 60 cm. Umbel with 2–4 flowers; spathes to 6 cm, flower-stalks 5–8 cm. Perianth horizontal, funnel-shaped, 10–13 cm, the tube *c.* 2.5 cm; tube green, lobes red or orange-red (rarely white), the base of each greenish white, the whitish parts not forming a star shape when the flower is viewed from the front. Stamens shorter than the perianth. Style somewhat exceeding the stamens, stigma capitate. *C & northern S America, Caribbean.* G2. Winter–spring.

Variable: a number of varieties has been described, based on flower colour and size. This is the species that Traub and his followers interpret as being the true *Amaryllis belladonna* of Linnaeus.

12. H. psittacinum (Ker Gawler) Herbert (*Amaryllis psittacina* Ker Gawler). Illustration: *Edwards's Botanical Register* **3**: t. 199 (1817).

Bulb more or less spherical, with a short neck, to 10 cm in diameter. Leaves oblong, tapered abruptly to a blunt apex, 30–50 × 2.5–3.8 cm. Umbel with 2–4 flowers; spathes exceeding the flower-stalks. Perianth horizontal, openly funnel-shaped, 10–12 cm; tube greenish, very short, the lobes mainly green shading to yellowish green above, the green area with numerous red stripes that merge into the red margins. Perianth-scales rather prominent, incurved and closing the throat of the perianth. Stamens shorter than perianth. Style about as long as perianth, stigma 3-fid. *S Brazil.* G2. Spring.

13. H. aulicum (Ker Gawler) Herbert (*Amaryllis aulica* Ker Gawler; *A. robusta* Otto & Dietrich; *H. aulicum* var. *robustum* (Otto & Dietrich) Voss). Illustration: *Botanical Magazine*, 1038 (1827) & 3311 (1834); *Edwards's Botanical Register* **6**: t. 444 (1820); Hay et al., Dictionary of indoor plants in colour, 280 (1974).

Similar to *H. psittacina* but leaves broader, spathes and flower-stalks longer, flowers 13–15 cm, each perianth-lobe red with slightly darker red veining, the base of each with a large, variably shaped, green blotch. *C Brazil, Paraguay.* G2. Winter.

14. H. pardinum (J.D. Hooker) Lemaire (*Amaryllis pardina* J.D. Hooker). Illustration:

Botanical Magazine, 5645 (1867); *Pacific Horticulture* **38**(3): 36 (1977).

Bulb more or less spherical, with a short neck, to 8 cm in diameter. Leaves oblong, somewhat narrowed towards the base, abruptly tapered to the blunt apex, to 60 × 5 cm, bright green. Umbel with 2 flowers; spathes *c.* 5 cm, flower-stalks 3.5–5 cm. Perianth horizontal or somewhat ascending, openly funnel-shaped, to 10 cm; tube to 1.3 cm, lobes oblanceolate, greenish or yellowish green flushed and spotted with red. Perianth-scales rather conspicuous, incurved, closing the throat of the perianth. Stamens a little shorter than the perianth. Style as long as the perianth, stigma capitate. *Bolivia.* G2. Winter–spring.

Formerly thought to originate from Peru (see *Plant Life* **30**: 81–3, 1974).

15. H. reticulatum (L'Héritier) Herbert (*Amaryllis reticulata* L'Héritier). Illustration: *Botanical Magazine*, 657 (1803); *Andrews' Botanical Repository* **1**: t. 179 (1801); *Gartenflora* **38**: t. 1297 (1889).

Bulb more or less spherical, with a short neck. Leaves thin, oblanceolate, tapered at the base into an almost terete stalk, to 30 × 2.5 cm. Umbel with 3–5 flowers; spathes to 5 cm, flower-stalks 2.5–4.5 cm. Perianth drooping, funnel-shaped, red-mauve with a network of darker veins, 10–12.5 cm, tube 1.3–2.5 cm. Stamens shorter than the perianth. Style about as long as the perianth, stigma capitate, slightly 3-lobed. Ovules 1–3 in each cell. *S Brazil.* G2. Winter.

9. AMARYLLIS Linnaeus

J. Cullen

Perennial herbs with bulbs. Leaves appearing after the flowers, oblong, somewhat narrowed to the base. Scape stout, solid. Flowers 5–many in an umbel subtended by 2 large, equal, free spathes that enclose the whole umbel in bud. Flowers ascending to erect. Perianth somewhat bilaterally symmetric, with a short tube and 6 oblanceolate lobes that are somewhat clawed, acute, the 3 outer lobes each with a small, hairy, inwardly pointing appendage just below the apex. Stamens 6, deflexed and then curving upwards towards the apex. Ovary 3-celled, with few ovules in each cell. Style deflexed and then curving upwards, stigma capitate. Fruit a capsule, splitting irregularly. Seeds few.

A genus of a single species from South Africa. For details of the controversy over its nomenclature, see under *Hippeastrum* (p. 201). The 1 species has been hybridised with *Crinum moorei* (p. 216) to produce a range of hybrids bearing the name × *Amarcrinum* Courts (× *Crinodonna* Anon.), with species of *Brunsvigia* (p. 216), producing hybrids named × *Amarygia* Ciferri & Giacomini (× *Brunsdonna* Anon.) and with species of *Nerine* (p. 217). Cultivation as for *Hippeastrum* (p. 201); the bulbs should not be disturbed or transplanted.

1. A. belladonna Linnaeus (*A. rosea* Lamarck; *Coburgia belladonna* (Linnaeus) Herbert; *Callicore rosea* (Lamarck) Hannibal; *Brunsvigia rosea* (Lamarck) Hannibal). Illustration: *The Garden* **43**: t. 490 (1893); *Flowering plants of Africa* **30**: t. 1200 (1954–5); *Horticulture* **51**(8): 23 (1973); Eliovson, Wild flowers of southern Africa, t. 3 (1980).

Scape to 60 cm or more. Umbel with usually 6 or more flowers; spathes to 8 cm. Perianth 6–10 cm, purplish pink or pink, often whitish towards the base or rarely all white; tube *c.* 1.2 cm. *South Africa.* H5. Late summer–autumn.

Variable in flower colour; numerous varieties (cultivars) have been selected and named.

10. VALLOTA Herbert

P.F. Yeo

Perennial herbs with bulbs; remaining evergreen in cultivation. Leaves rather numerous, strap-shaped, obtuse or acute, somewhat sheathing at the base. Inflorescence-stalk as long as or longer than the leaves. Umbels with 2 spathes and several smaller bracts. Flowers several, bisexual, radially symmetric, widely funnel-shaped; tube gradually enlarged in its apical half, lobes broad, somewhat longer than the tube, with cushion-like swellings at the throat. Stamens attached at the throat, shorter than the perianth-lobes; filaments not enlarged at the base; anthers dorsifixed. Style slightly longer than perianth, 3-lobed. Ovary 3-celled, oblong, with numerous ovules. Seeds black, winged at the base.

One species from South Africa with showy red or occasionally pink or white flowers; sometimes included in *Cyrtanthus*, in which case it must be called *C. purpureus*. It is usually grown in pots and should be disturbed as little as possible. Re-potting should be done in spring or early summer before the flower-stems are sent up and great care is necessary not to damage the roots. It is relatively inactive in winter, when it should be kept at 4–8 °C, but not dried out. Propagation is by offsets or seed; seedlings take 3 years to reach maturity.

1. V. speciosa (Linnaeus filius) Durand & Schinz (*V. purpurea* (Aiton) Herbert; *Cyrtanthus purpureus* (Aiton) Herbert). Illustration: *Botanical Magazine*, 1430 (1811); Parey's Blumengärtnerei **1**: 332 (1958); Huxley & Gilbert, Success with house plants, 390 (1979).

Leaves 20–45 × 2–6 cm, bright green. Stem 30–60 cm, hollow. Flower-stalks 2–5 cm. Flowers 7–10 cm long and about as wide, horizontal to erect, scarlet with whitish marks in throat or a white centre, or entirely white ('Alba'); tube 3–5 cm; lobes *c.* 3.5–4.5 × 2–3 cm. *South Africa.* G1. Summer–autumn.

11. CYRTANTHUS Linnaeus

E. Campbell

Herbaceous perennials with bulbs that sometimes have a neck made up of the sheathing bases of the old leaves. Leaves linear or strap-shaped, produced with or after the flowers, channelled or flat, sometimes flexuous. Flowers in a terminal umbel borne on a hollow scape; umbel subtended by 2 or 4 spathes. Perianth with a long tube that is straight or curved; lobes more or less equal, the outer 3 each with an incurved point or tuft of hairs at the apex. Stamens 6, borne on the perianth-tube, the anthers versatile. Ovary 3-celled with many ovules. Style thread-like, distinctly or indistinctly 3-lobed at the apex. Capsule oblong, 3-celled. Seeds black.

A genus of about 40 species from southern Africa, of which 6 are commonly found in cultivation as pot plants. They are easily grown; annual re-potting is beneficial. Propagation is by seed or offsets.

Literature: Dyer, R.A., A Review of the genus Cyrtanthus, *Herbertia* **6**: 65–103 (1939).

Bulb. More than 7 cm in diameter: **1, 2**; less than 6 cm in diameter: **3, 4, 5, 6**.
Leaves. More than 3 cm broad: **1**; up to 2 cm broad: **2, 3, 4, 5, 6**.
Scape. Green: **2, 3, 4, 5, 6**; purple: **1, 5**.
Spathes. Four: **1**; two: **2, 3, 4, 5, 6**.
Spotted: **4**.

Perianth. Red: **1, 2, 5, 6**; yellow: **1, 3**; white or cream: **3, 4**. More than 2 cm wide at the throat: **2**. Fragrant: **4**.
Style. Projecting: **2, 3, 4, 5, 6**; not projecting: **1, 5**.

1a. Leaves 3–6 cm broad; spathes 4 **1. obliquus**
b. Leaves up to 2.5 cm broad; spathes 2 2
2a. Flowers white or yellow 3
b. Flowers red or orange-red 4
3a. Leaves 1–2 mm broad; spathes 2–2.5 cm, linear-lanceolate, without spots when young; flowers 2–5, pale yellow **3. ochroleucus**
b. Leaves 5–10 mm broad; spathes 2.5–3.8 cm, lanceolate, spotted with red-brown when young; flowers 4–10, white or cream **4. mackenii**
4a. Leaves 35 cm or more; scape green; perianth 5 cm or more 5
b. Leaves 15–30 cm; scape tinged with purple; perianth to 5 cm **5. macowanii**
5a. Spathes to 8 cm; flowers 1–2 (rarely 3); throat of perianth 2.5–3 cm wide **2. sanguineus**
b. Spathes 4–5 cm; flowers 4–10; throat of perianth 8–10 mm wide **6. angustifolius**

1. C. obliquus (Linnaeus) Aiton. Illustration: *Botanical Magazine*, 1133 (1808); *Loddiges' Botanical Cabinet* **10**: t. 947 (1825).

Bulb almost spherical, *c*. 10 cm in diameter. Leaves 20–60 × 3–6 cm, strap-shaped, glaucous, twisted towards the apex, obtuse, channelled. Scape 20–60 cm, green tinged with purple. Flower-stalks 5–30 mm, recurved. Spathes 4, *c*. 3 × 1 cm, ovate-oblong. Flowers 6–12, hanging. Perianth *c*. 7 cm, yellow and red, sometimes tinged with green; tube straight or curved at the base, *c*. 4.5 cm, to 2 cm wide at the throat; lobes *c*. 2.5 × 1.5 cm, not spreading. Style indistinctly 3-fid. *South Africa (Cape Province)*. H5–G1. Spring–summer.

2. C. sanguineus (Lindley) Hooker. Illustration: *Botanical Magazine*, 5218 (1860); *The Garden*, t. 344 (1890).

Bulb ovoid, to 8 cm in diameter. Leaves *c*. 40 × 1–2 cm, linear-lanceolate, green or glaucous, channelled. Scape *c*. 30 cm, green. Flower-stalks to 2 cm, erect. Spathes 2, to 8 cm, lanceolate. Flowers 1 or 2 (rarely 3), almost erect. Perianth 7.5–10.5 cm, bright red; tube straight or curved, 4.5–6.5 cm, 2.5–3 cm wide at the throat; lobes 3–4 × 1.3–1.8 cm, spreading or recurved. Style projecting, 3-lobed. *South Africa (Cape Province, Natal)*. H5–G1. Summer.

Nordal (*Norwegian Journal of Botany* **26**: 183–192, 1979) splits this variable species into several subspecies; which of these are in cultivation is not known.

3. C. ochroleucus (Herbert) Burchell (*C. lutescens* Herbert).

Bulb spherical, to 2–2.5 cm in diameter. Leaves 15–30 cm × 1–2 mm, produced after flowering has started, linear to thread-like, slightly channelled. Scape 15–30 cm, green. Flower-stalks 5–25 mm, erect. Spathes 2, 2–2.5 cm, linear-lanceolate. Flowers 2–5, more or less erect. Perianth 4–5 cm, pale yellow; tube slightly curved, 3–4 cm, 5–7 mm wide at the throat; lobes to 1 cm, spreading. Style projecting, shortly 3-fid. *South Africa (Cape Province)*. H5–G1. Winter.

4. C. mackenii Hooker. Illustration: *Botanical Magazine*, 5374 (1863).

Bulb ovoid, 2.5–4 cm in diameter. Leaves 2–6, 20–30 cm × 5–10 mm, linear. Scape usually exceeding 30 cm, slightly glaucous. Flower-stalks 1–3 cm, erect. Spathes 2, 2.5–3.8 cm, lanceolate, spotted with red-brown when young. Flowers 4–10, more or less erect, fragrant. Perianth *c*. 5 cm, pure white or cream; tube slightly curved, 2–3 cm, 5 mm wide at the throat; lobes 6–7 mm, recurved. Style projecting, 3-fid. *South Africa (Natal)*. H5–G1. Summer.

5. C. macowanii Baker. Illustration: *La Belgique Horticole* **197**: t. 15 (1885).

Bulb ovoid, to 2.5 cm in diameter. Leaves 1–3, 15–30 cm × 3–7 mm, linear, dark green. Scape *c*. 30 cm, green tinged with purple. Flower-stalks to 2 cm. Spathes 2, 2.5–3.7 cm, lanceolate. Flowers 6–8, hanging. Perianth 3.5–4 cm, bright orange-red; tube curved, 2.8–3.5 cm, 4–5 mm wide at the throat; lobes 5–7 mm, spreading. Style slightly projecting, 3-toothed. *South Africa (Cape Province)*. H5–G1. Summer.

C. obrienii Baker. Leaves bright green, flowers bright, pale orange-red, style not projecting. *South Africa (Cape Province, Natal)*. H5–G1. Summer.

6. C. angustifolius (Linnaeus) Linnaeus. Illustration: *Botanical Magazine*, 271 (1794); *Loddiges' Botanical Cabinet* **4**: t. 368 (1820).

Bulb ovoid, to 4.5 cm diameter. Leaves 2 or 3, *c*. 45 cm × 7–20 mm, linear to strap-shaped, slightly channelled. Scape to 45 cm, green, stout. Flower-stalks 1–5 cm, more or less erect. Spathes 2, 4–5 cm, lanceolate, acute. Flowers 4–10, hanging. Perianth 6–8 cm, red; tube curved, 5–7 cm, 8–10 mm wide at the throat; lobes to 1 cm, spreading. Style projecting, 3-lobed. *South Africa (Cape Province)*. H5–G1. Spring–summer.

12. NARCISSUS Linnaeus

J. Cullen

Perennial herbs with bulbs. Leaves 1–several from each bulb, basal, erect, spreading or prostrate, almost cylindric in section to flat and broad. Scape present. Inflorescence an umbel of 2–20 flowers or flowers solitary, subtended by a usually scarious bract (spathe). Perianth with a tube below, 6-lobed. Corona almost always present, usually conspicuous, free from the stamens. Stamens 6, usually in 2 whorls, more rarely in 1. Ovary 3-celled with many ovules. Capsule ellipsoid to almost spherical, many-seeded. Seeds sometimes with an appendage.

'This genus exhibits, perhaps more acutely than any other, the taxonomic difficulties that arise from long-established cultivation, hybridisation and selection, and subsequent escape and naturalisation. For many populations, especially in S France and NW Italy, it is impossible to say with any confidence whether they are native or naturalised; and nearly all the early binomials are based on cultivated plants of which the origin was only guessed at. This not only gives rise to nomenclatural problems but it means also that the geographical background for specific delimitation is confused and defective' (D.A. Webb, Flora Europaea **5**: 78–82, 1980).

The above quotation neatly indicates the problems with *Narcissus* in terms of its classification and identification. It is a genus of perhaps 50 species (some considered as subspecies or varieties by some authors), most of which have been in cultivation, some of them for a very long time. Many thousands of hybrids have been raised, and many variants have been given cultivar names (see the *International Register*, cited below). They are grown for ornament, for naturalisation on banks and slopes, and as an important cut-flower crop. Most originate in Europe (especially France, Italy, Spain and Portugal), though a few are natives of N Africa and the eastern Mediterranean area.

The account given here largely follows that of Webb, except that some groups treated by him as subspecies are retained here at specific level: this has been done purely for convenience in use, and users are referred to the *Flora Europaea* account for a biologically more appropriate arrangement.

Narcissi are extremely popular garden plants, and are easily grown in borders, on grassy banks where they can naturalise and spread, or in pots. Pot-grown plants can be forced for early flowering. A sunny, open or partially shaded situation is best, and the plants will spread by means of bulbs and seed. Bulbs can be lifted and dried during the winter: this should be done if the plants become overcrowded. Lifting should not be done until the leaves have yellowed and are dying off, and the lifted bulbs should be stored in a cool, dry place. Propagation is by bulbs and seed.

Literature: Burbidge, F.W., The Narcissus: its history and culture (1875); Pugsley, H.W., Narcissus poeticus and its allies, *Journal of Botany* **53**: supplement 2 (1915); Pugsley, H.W., A monograph of Narcissus subgenus Ajax, *Journal of the Royal Horticultural Society* **53**: 17–93 (1933); Bowles, E.A., A Handbook of Narcissus (1934); Fernandes, A., Sur la phylogenie des espèces du genre Narcissus L., *Boletim da Sociedade Broteriana*, ser. 2, **25**: 113–90 (1951); Meyer, F.G., Narcissus species and wild hybrids, *American Horticultural Magazine* **45**: 47–76 (1966); Fernandes, A., Key to the identification of native and naturalized taxa of the genus Narcissus L., *Daffodil and Tulip Yearbook, 1968*, **33**: 37–66 (1967); *RHS Classified list and International Register of Daffodil names* (1954, 1955, 1958, 1961, 1965, 1969).

1a. Plant flowering in autumn 2
b. Plant flowering in winter, spring or early summer 5
2a. Corona rudimentary, formed by a low, white ridge that is sometimes 6-lobed; filaments longer than anthers **1. broussonetii**
b. Corona well-developed though sometimes small, not as above; filaments shorter than anthers 3
3a. Perianth and corona green **29. viridiflorus**
b. Perianth and corona not green 4
4a. Leaves 2–4.5 mm wide, present at flowering; corona greenish yellow **15. elegans**
b. Leaves to 1 mm wide, not present at flowering; corona orange **2. serotinus**
5a. Perianth-tube obconical, i.e. with a narrow base, widening evenly to the insertion of the lobes, funnel-shaped (narrowly so in some hybrids); stamens in a single whorl, the filaments as long as or longer than the anthers 6
b. Perianth-tube cylindric, parallel-sided for most of its length or widening a little towards the insertion of the lobes; stamens in 2 whorls, filaments of lower shorter than the anthers 19
6a. Flowers bilaterally symmetric, the stamens deflected downwards, the filaments then curving upwards their apices; corona funnel-shaped 7
b. Flowers radially symmetric, the stamens not as above; corona flattish, bell-shaped or cylindric 11
7a. Flower white, or white faintly flushed with green **34. cantabricus**
b. Flower orange-yellow to pale yellow or cream, or if whitish then flushed with yellow 8
8a. Flower-stalk at least 9 mm (rarely shorter); stamens usually not projecting from the corona 9
b. Flower-stalk absent or at most 5 mm; stamens usually projecting from the corona 10
9a. Leaves mostly erect; corona not contracted towards its margin **31. bulbocodium**
b. Leaves lying on the ground or at least arching downwards; corona somewhat narrowed towards its margin **32. obesus**
10a. Perianth-lobes 3–4.5 mm broad at the base; flower-stalk absent; scape 5–8 cm, shorter than the leaves **35. hedraeanthus**
b. Perianth-lobes 1–2.5 mm broad at the base; flower-stalk 1–5 mm; scape 15–20 cm, longer than the leaves **33. romieuxii**
11a. Perianth-tube narrowly funnel-shaped, 3–4 times as long as wide 12
b. Perianth-tube broadly funnel-shaped, up to 1 and a half times as long as wide 13
12a. Perianth and corona of the same colour; flower-stalks 2–3 cm **38. × odorus**
b. Perianth-lobes paler yellow than the corona; flower-stalks 1.2–1.5 cm **39. × incomparabilis**
13a. Perianth and corona white or cream **37. pseudonarcissus**
b. At least the corona yellow or orange 14
14a. Perianth-lobes reflexed at an angle of almost 180° to the corona, hiding the ovary **43. cyclamineus**
b. Perianth-lobes spreading or erect-spreading, not hiding the ovary 15
15a. Perianth-lobes to 1.5 cm; corona constricted near the middle **42. asturiensis**
b. Perianth-lobes more than 1.5 cm; corona not constricted near the middle 16
16a. Spathe 6–10 cm; flower-stalks 4–9 cm **36. longispathus**
b. Spathe 2–6 cm; flower-stalks to 3.5 cm 17
17a. Perianth-tube 1.5–2.5 cm **37. pseudonarcissus**
b. Perianth-tube 8–15 mm 18
18a. Corona 1.6–2.5 cm **41. minor**
b. Corona 3.5–4 cm **40. bicolor**
19a. Leaves cylindric or semi-cylindric, usually not channelled down the inner face, 1–3 mm wide; scape terete 20
b. Leaves mostly flat and channelled down the inner face, 3–25 mm wide, scape usually compressed 29
20a. Leaves smooth and rounded on the outer surface, usually dark green 21
b. Leaves with 2 keels on the outer surface or 4-angled, usually glaucous 26
21a. Perianth-tube straight, 2–3 cm 22
b. Perianth-tube straight or curved, up to 2 cm 23
22a. Leaves narrowly linear, rather flattened; corona c. 5 cm high, deeper yellow than the perianth-lobes **20. × tenuior**
b. Leaves almost terete; corona 2–4 mm high, of the same colour as the perianth-lobes **19. jonquilla**
23a. Longer flower-stalks longer than the spathe and usually longer than the flower 24
b. Longer flower-stalks included in the spathe, shorter than the flower 25
24a. Perianth-tube 1.5–2.2 cm; corona 4–5 mm high, about half the length of the perianth-lobes **23. fernandesii**

b. Perianth-tube 1–1.5 cm; corona 5–6 mm high, more than half the length of the perianth-lobes **22. willkommii**
25a. Leaves usually shorter than the scape; perianth-tube straight; corona more than 9 mm in diameter **21. assoanus**
b. Leaves usually longer than scape; perianth-tube curved; corona less than 9 mm in diameter **24. gaditanus**
26a. Flower solitary (rarely 2 together); spathe tubular for three-quarters of its length; flower-stalks 0–3 mm 27
b. Flowers 1–5 in each umbel; spathe tubular for up to half of its length; flower-stalks 6 mm or more 28
27a. Flower yellow; corona usually 6-lobed, rarely sinuous or almost entire **25. rupicola**
b. Flower white or cream; corona with sinuous margins **26. watieri**
28a. Leaves 4-angled, usually lying on the ground and twisted, the angles minutely rough; flower 1.2–1.7 cm in diameter **28. scaberulus**
b. Leaves not 4-angled, erect, smooth; flower 1.7–2.5 cm in diameter **27. calcicola**
29a. Perianth-lobes reflexed, hiding the ovary; corona cup-shaped, longer than broad **30. triandrus**
b. Perianth-lobes not or little reflexed, not hiding the ovary 30
30a. Umbel with 4–20 flowers (rarely fewer); corona without a red or scarious margin 31
b. Umbel with 1–3 flowers; corona with a red or scarious margin 42
31a. Corona yellow or orange 32
b. Corona white 37
32a. Flowers to 4 cm in diameter 33
b. Flowers 4–5 cm in diameter 35
33a. Perianth-lobes white **3. tazetta**
b. Perianth-lobes yellow or pale yellow 34
34a. Umbel with 1 or 2 flowers; perianth-lobes pale yellow **6. corcyrensis**
b. Umbel with 3–6 flowers; perianth-lobes bright yellow **8. × intermedius**
35a. Perianth-lobes white **7. cypri**
b. Perianth-lobes yellow or pale yellow 36
36a. Perianth-lobes cream or pale yellow; corona yellow **4. italicus**
b. Perianth-lobes bright or golden yellow; corona deep yellow or orange **5. aureus**
37a. Flower 1.2–2 cm in diameter 38
b. Flower 2–4 cm in diameter 40
38a. Corona about half the length of the perianth-lobes, lobed at the margins **14. dubius**
b. Corona at most one-third of the length of the perianth-lobes, margin entire or almost so 39
39a. Perianth-lobes ovate-oblong, more or less obtuse, about half the length of the perianth-tube; bulb 5–7 cm in diameter **12. pachybolbus**
b. Perianth-lobes ovate-lanceolate, acute, less than half the length of the perianth-tube; bulb smaller **13. canariensis**
40a. Flower 2–2.5 cm in diameter **11. panizzianus**
b. Flower 2.5–4 cm in diameter 41
41a. Leaves glaucous; scape very compressed, 2-edged; corona usually finely scalloped **9. papyraceus**
b. Leaves green; scape less flattened, not 2-edged; corona entire **10. polyanthus**
42a. Flowers in umbels of 2–3; pollen sterile **18. × medioluteus**
b. Flower solitary; pollen fertile 43
43a. Stamens of the lower whorl not projecting from the perianth-tube; perianth-lobes overlapping **16. poeticus**
b. Stamens all projecting from the corona; perianth-lobes not overlapping **17. radiiflorus**

Section **Aurelia** (Gay) Baker. Leaves flat. Flowers in umbels. Perianth-tube cylindric at the base, widening somewhat above; lobes spreading. Corona rudimentary. Filaments longer than the anthers; anthers dorsifixed, projecting from the perianth-tube. Flowering in autumn.

1. N. broussonetii Lagasca. Illustration: *Botanical Magazine*, 7016 (1888); Maire, Flore de l'Afrique du nord **6**: 74 (1959).

Bulb 3–4 cm, brown. Leaves broadly linear, glaucous, to 45 × 1.3 cm, flat or scarcely channelled on the inner face, striate on the outer. Flowers 3–12 in an umbel, spathe 3–3.5 cm, whitish; flower-stalks 5–25 mm. Perianth-tube cylindric at base, widening somewhat above, straight, 1.8–2.8 cm; lobes 1.2–1.6 cm, elliptic, somewhat acute, white. Corona rudimentary, reduced to a low white ridge that is sometimes 6-lobed. Stamens and style projecting from the perianth-tube. *Morocco*. H5–G1. Autumn.

Section **Serotini** Parlatore. Leaves thread-like, absent at flowering. Flowers usually solitary. Perianth-tube cylindric. Corona short. Filaments straight, shorter than the anthers to slightly longer; anthers dorsifixed, not projecting. Flowering in autumn.

2. N. serotinus Linnaeus. Illustration: Mathews, Lilies of the field, 2 (1969); Huxley & Taylor, Flowers of Greece, t. 387 (1977).

Bulb 1.5–2 cm, dark. Leaves appearing in spring, cylindric and thread-like, glaucous, 10–20 cm × *c.* 1 mm. Scape 10–25 cm. Flowers solitary or occasionally in an umbel of 2 or 3, erect, fragrant; spathe 1.5–3.5 cm, tubular for about half its length. Flower-stalks 7–30 mm. Perianth-tube 1.2–2 cm, cylindric, slightly expanded above; lobes 1–1.6 cm × 3–7 mm, oblong-lanceolate, acute, spreading, white. Corona 1–1.5 mm high, 3–4 mm in diameter, 6-lobed, orange. *Mediterranean area*. H4. Autumn.

Section **Tazettae** de Candolle (Section *Hermione* (Salisbury) Sprengel). Leaves flat or channelled. Flowers in umbels. Perianth-tube cylindric. Corona short. Filaments much shorter than anthers; anthers dorsifixed, included or those of the upper whorl projecting. Flowering mainly in spring.

3. N. tazetta Linnaeus. Figure 8(3), p. 208. Illustration: *Botanical Magazine*, 948 & 1026 (1807); Polunin & Huxley, Flowers of the Mediterranean, t. 257 (1965); Huxley & Taylor, Flowers of Greece, t. 386 (1977).

Bulb to 5 cm, dark. Leaves flat, green, striate, 20–50 cm × 5–25 mm, obtuse. Scape robust, compressed, to 45 cm. Flowers to 4 cm in diameter, in umbels of 3–15 (rarely fewer), fragrant; flower-stalks unequal, the longest to 7.5 cm. Perianth-tube 1.2–2 cm; lobes 8–22 mm, spreading, usually ovate and somewhat overlapping at their bases, white. Corona 3–6 mm high, 6–11 mm in diameter, yellow. Stamens not projecting from the corona. *W Mediterranean area*. H3. Winter–spring.

Variable in size; very small variants have been called *N. patulus* Loiseleur (*N. tazetta* subsp. *patulus* (Loiseleur) Baker).

4. N. italicus Ker Gawler (*N. tazetta* subsp. *italicus* (Ker Gawler) Baker; *N. ochroleucus* Loiseleur). Illustration: *Botanical Magazine*, 1188 (1809).

Similar to *N. tazetta* but flowers to 5 cm in diameter, the perianth-lobes cream to pale yellow, not or scarcely overlapping at

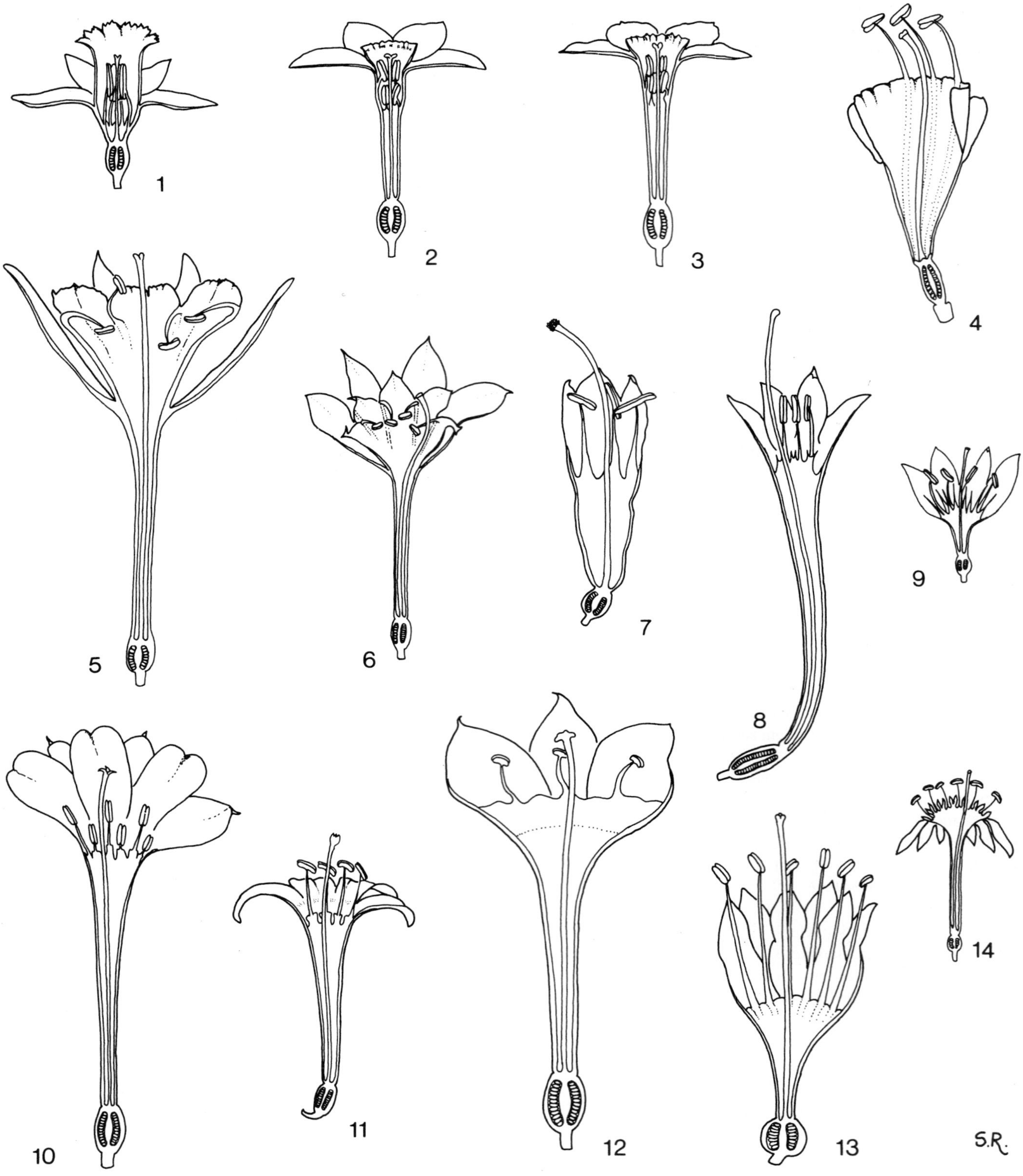

Figure 8. Floral sections of the Amaryllidaceae showing the coronas. 1, *Narcissus pseudonarcissus*. 2, *N. poeticus*. 3, *N. tazetta*. 4, *N. hedraeanthus*. 5, *Hymenocallis narcissiflora*. 6, *Pamianthe peruviana*. 7, *Phaedranassa carmioli*. 8, *Stenomesson variegatum*. 9, *Vagaria parviflora*. 10, *Chlidanthus fragrans*. 11, *Caliphruria subedentata*. 12, *Eucharis grandiflora*. 13, *Urceolina peruviana*. 14, *Pancratium maritimum*. Not to scale.

their bases; corona yellow. *N & E Mediterranean area.* H3. Spring.

Variable in the precise coloration of the perianth; some species (or subspecies or varieties) have been described on the basis of this variation.

5. N. aureus Loiseleur (*N. tazetta* subsp. *aureus* (Loiseleur) Baker; *N. bertolonii* Parlatore; *N. cupularis* (Salisbury) Schultes). Illustration: Grey-Wilson & Mathew, Bulbs, pl. 24 (1981).

Similar to *N. tazetta* but the perianth-lobes bright or golden yellow and the corona deep yellow or orange. *SE France, NW Italy, Sardinia; naturalised elsewhere.* H4. Winter–spring.

Also variable in size, colour of flower, and in sometimes having glaucous leaves.

6. N. corcyrensis (Herbert) Nyman (*N. tazetta* subsp. *corycyrensis* (Herbert) Baker).

Similar to *N. tazetta* but flowers sometimes reduced to 1 or 2 per umbel, perianth-lobes pale yellow, narrow, not overlapping at the base, sometimes somewhat reflexed. *S France, Italy, Balkan peninsula.* H4. Spring.

A rather doubtful species of uncertain distribution; it has been suggested that it is a natural hybrid of *N. serotinus* and *N. italicus.*

7. N. cypri Sweet. Illustration: Sweet, British Flower Garden **1**: t. 92 (1823–25).

Similar to *N. tazetta* but larger, scape compressed only above, flowers *c.* 5 cm in diameter, perianth-lobes white, ovate-oblong, overlapping at their bases, corona bright yellow. *Cyprus, ?Syria; introduced in Italy.* H5. Spring.

8. N. × intermedius Loiseleur. Illustration: Grey-Wilson & Mathew, Bulbs, pl. 26 (1981).

Leaves 30–45 cm × 5–8 mm, almost cylindric but channelled down the inner face. Scape to 40 cm, terete. Flowers in umbels of 3–6, fragrant. Perianth-tube 1.4–2 cm; lobes 1–1.8 cm, ovate, overlapping at their bases, bright yellow. Corona 3–4 mm, slightly deeper yellow than the perianth-lobes, its margin slightly lobed. *Spain, S France, Italy.* H4. Spring.

The hybrid between *N. tazetta* and *N. jonquilla.*

9. N. papyraceus Ker Gawler. Illustration: *Botanical Magazine,* 947 (1806); Polunin & Huxley, Flowers of the Mediterranean, t. 258 (1965); Ceballos et al., Plantas silvestres de la peninsula Iberica, 391 (1980).

Bulbs to 7 cm, dark. Leaves flat, slightly glaucous, 10–45 cm × 4–12 mm, obtuse. Scape 14–50 cm, compressed. Flowers 3–20 in an umbel, fragrant; flower-stalks of varying lengths, the longest to 5 cm. Flowers 2.5–4 cm in diameter. Perianth-tube 1.2–2.5 cm; lobes 1–1.8 cm, ovate or narrowly ovate, overlapping at their bases, spreading, white. Corona *c.* 3 mm, pure white, margin usually slightly and minutely scalloped. *SW Europe & Mediterranean area.* H4. Winter–spring.

10. N. polyanthus Loiseleur (*N. papyraceus* Ker Gawler subsp. *polyanthus* (Loiseleur) Ascherson & Graebner).

Similar to *N. papyraceus* but leaves green, scape terete, corona entire. *S France; naturalised in Spain & Italy.* H4. Winter–spring.

11. N. panizzianus Parlatore (*N. papyraceus* Ker Gawler subsp. *panizzianus* (Parlatore) Arcangeli; *N. barlae* Parlatore).

Similar to *N. papyraceus* and *N. polyanthus;* scape strongly compressed and 2-edged, flowers 2–2.5 cm in diameter. *Portugal, Spain, SE France, N Italy.* H4. Winter–spring.

12. N. pachybolbus Durieu. Illustration: *Botanical Magazine,* 6825 (1885).

Bulb large, to 7 cm, dark. Leaves flat, green, 30–50 cm × 6–12 mm. Scape to 40 cm, compressed, 2-edged and ridged. Flowers 1.2–2 cm in diameter, in an umbel of 6–20; flower-stalks somewhat unequal, the longest to 5 cm. Perianth-tube 1–1.5 cm; lobes 6–8 mm, ovate-oblong, obtuse, white. Corona *c.* 3 mm, white. *Morocco, Algeria.* H5. Winter.

13. N. canariensis Burbidge. Illustration: Burbidge, The Narcissus, t. 48 (1875).

Similar to *N. pachybolbus* but bulb smaller, perianth-lobes ovate-lanceolate, acute, less than 6 mm. *Canary Islands.* G1. Winter–spring.

14. N. dubius Gouan. Illustration: Grey-Wilson & Mathew, Bulbs, pl. 24 (1981).

Bulb 2–3 cm, dark. Leaves flat, glaucous, 12–25 cm × 3–5 mm. Scape to 20 cm, strongly compressed. Flowers in umbels of 2–6; flower-stalks unequal, the longest to 4 cm. Perianth-tube 1–1.4 cm, cylindric but widening slightly above; lobes 6–8 mm, broadly elliptic, overlapping at the base, white. Corona 3–4 mm, white, the margin more or less lobed. *S France, NE Spain.* H4. Spring.

Thought to be derived by allopolyploidy from a hybrid between *N. papyraceus* and *N. assoanus.*

15. N. elegans (Haworth) Spach. Illustration: Grey-Wilson & Mathew, Bulbs, pl. 24 (1981).

Bulb 2–3.5 cm, very dark. Leaves appearing before or with the flowers, glaucous, 8–25 cm × 2–4.5 mm, channelled. Scape to 35 cm. Flowers 2–3 cm in diameter, in umbels of 2–10, fragrant; flower-stalks unequal, the longest to 4.5 cm. Perianth-tube 1.1–2 cm, greenish; lobes 1.2–2 cm × 3–6 mm, narrowly oblong to elliptic, acute, not overlapping at the base, somewhat reflexed, white. Corona 1–2 mm high, 3–4 mm in diameter, yellowish green. Anthers not projecting from corona. *W Mediterranean area, Italy, Sicily.* H4. Autumn.

Variable in flower colour; a number of varieties has been recognised, based on this character.

N. × rogendorfii Battandier is the wild hybrid between *N. elegans* and *N. tazetta.* It is similar to *N. elegans* but has a larger corona, broader perianth-lobes and an umbel with up to 12 flowers. *Algeria.* H5. Autumn.

Section **Narcissus**. Leaves flat. Flower solitary. Perianth-tube cylindric. Corona small. Filaments much shorter than anthers; anthers dorsifixed, partly projecting or all included. Flowering in spring–summer.

16. N. poeticus Linnaeus. Figure 8(2), p. 208. Illustration: de Wit (ed.), Plants of the world **2**: t. 136 (1963–5); Grey-Wilson & Mathew, Bulbs, pl. 24 (1981).

Bulb to 4 cm, dark. Leaves flat, more or less glaucous, 20–40 cm × 6–10 or rarely to 12 mm. Scape to 50 cm, compressed. Flower-stalk 1–4.5 cm. Perianth-tube 2–3 cm; lobes mostly 2–2.5 cm, obovate to almost circular, overlapping at their bases without distinct claws, white or pale cream. Corona disc-like or cup-shaped, to 1.4 cm in diameter, yellow with red or scarious, minutely scalloped margin. Stamens in 2 whorls, the lower whorl included inside the perianth-tube, the upper whorl not projecting from the corona. *S Europe, from France to Former SW USSR.* H2. Spring–summer.

A very variable species in terms of flower size and the relative positions of the 2 whorls of stamens. Fernandes (1967) recognises 5 varieties:

Var. **poeticus**. Corona disc-like. *S France, Italy.*

Var. **verbanensis** Herbert (*N. verbanensis* (Herbert) Pugsley). Plant rather small. Corona cup-shaped, 2 mm high, 8–9 mm in diameter; flower 3.5–5 cm in diameter, perianth-lobes strongly mucronate. *Italy.*

Var. **hellenicus** (Pugsley) Fernandes (*N. hellenicus* Pugsley). Corona cup-shaped, *c.* 3 mm high, 1.2–1.4 cm in diameter; flower 3.5–5 cm in diameter, perianth-lobes obtuse though mucronate. *Greece.*

Var. **recurvus** (Haworth) Fernandes (*N. recurvus* Haworth). Flower 5–7 cm in diameter; corona cup-shaped, greenish yellow below the red margin. *Switzerland.*

Var. **majalis** (Curtis) Fernandes (*N. majalis* Curtis). Flower 5–7 cm in diameter; corona with a red margin and a whitish zone just below it. *France.*

As well as these (wild) varieties, many selections (including double-flowered variants) have been given cultivar names.

17. N. radiiflorus Salisbury (*N. angustifolius* Haworth; *N. poeticus* subsp. *radiiflorus* (Salisbury) Baker). Illustration: *Botanical Magazine*, 193 (1792); Grey-Wilson & Mathew, Bulbs, pl. 24 (1981).

Similar to *N. poeticus* but leaves 5–8 mm wide, perianth-lobes 2.2–3 cm, narrowly obovate, tapered to a conspicuous claw, not or scarcely overlapping at their bases; corona 2–2.5 mm high, 8–10 mm in diameter. *SC Europe & the western part of the Balkan peninsula.* H3. Spring–summer.

Not always clearly separable from *N. poeticus*, and, like it, very variable. Fernandes (1967) recognises the following varieties:

Var. **radiiflorus**. Corona cup-shaped, 2–2.5 mm high, 8–10 mm in diameter, bright yellow edged with red. *Switzerland, Austria, Yugoslavia.*

Var. **stellaris** (Haworth) Fernandes (*N. stellaris* Haworth). Corona cup-shaped like that of var. *radiiflorus*, but with a narrow, white zone within the red margin. *S France to Romania (? and farther east).*

Var. **poetarum** Burbidge & Baker. Corona flat and disc-like, wholly red. *Origin uncertain.*

Var. **exertus** (Haworth) Fernandas (*N. majalis* var. *exertus* Haworth; *N. exertus* (Haworth) Pugsley). Corona flat and disc-like, bright yellow with a scarlet margin. *Switzerland and adjacent France.*

18. N. × medioluteus Miller (*N. biflorus* Curtis). Illustration: Grey-Wilson & Mathew, Bulbs, pl. 26 (1981).

Bulb to 6 cm. Leaves flat, glaucous, 45–70 cm × 7–10 mm. Scape to 60 cm, compressed. Flowers usually 2 (more rarely 3 or 4) per umbel, fragrant; flower-stalks 2.5–3.5 cm. Perianth-tube 2–2.5 cm, widened towards the apex; lobes 1.8–2.2 cm, almost circular, white. Corona 3–5 mm high, 9–12 mm in diameter, bright yellow, with a red or scarious margin. *S France; widely naturalised elsewhere in Europe.* H3. Spring.

The natural hybrid between *N. poeticus* and *N. tazetta.*

Section **Jonquillae** de Candolle (Section *Apodanthi* Fernandes). Leaves narrow, usually not flat. Flowers solitary or in umbels. Perianth-tube cylindric or slightly widened above; lobes broad and overlapping. Corona small to large, its diameter greater than its height. Filaments much shorter than the anthers; anthers dorsifixed, included in the corona or partly projecting. Flowering mostly in spring.

19. N. jonquilla Linnaeus. Illustration: *Botanical Magazine*, 15 (1787); Genders, Bulbs, 273 (1973); Ceballos et al., Plantas silvestres de la peninsula Iberica, 392 (1980).

Bulb 1–2.5 cm, dark. Leaves erect, dark green, almost terete, 1–4 mm wide, channelled on the inner surface, slightly striate on the outer. Flowers in umbels of 2–5, very fragrant, yellow; flower-stalks unequal, the longest to 5 cm. Flowers 2–3.5 cm in diameter. Perianth-tube straight, 2–3 cm; lobes 1–1.5 cm, ovate to elliptic or obovate, mucronate. Corona 2–4 mm high, 9–15 mm in diameter, of the same colour as the perianth-lobes, margin finely scalloped. *Spain, Portugal; widely naturalised elsewhere in Europe.* H4. Winter–spring.

Cultivated on a commercial scale in S France for its perfume.

20. N. × tenuior Curtis. Illustration: *Botanical Magazine*, 379 (1797).

Like *N. jonquilla* but leaves narrowly linear and flat, corona flatter, to 5 mm high, of a deeper yellow than the perianth-lobes. *Garden origin.* H4. Spring.

The hybrid between *N. jonquilla* and *N. poeticus.*

21. N. assoanus Dufour (*N. juncifolius* of authors; *N. requienii* Roemer). Illustration: Rix & Phillips, The bulb book, 108 (1981).

Bulb 1–3 cm, dark. Leaves 1–2 mm wide, usually shorter than the scape, green. Scape to 25 cm. Flowers 1.6–2.2 cm in diameter, solitary or in pairs, fragrant, yellow; flower-stalks unequal, the longer to 2 cm, all included within the spathe. Perianth-tube 1.2–2 cm, straight; lobes 7–10 mm, spreading, obovate; corona 4–6 mm high, 1–1.7 cm in diameter. *S France, S & E Spain.* H5. Spring.

Fernandes (1967) recognised 2 varieties under *N. requienii*: var. **requienii** with yellow flowers and var. **pallens** (Willkomm) Fernandes (*N. pallens* Willkomm) with pale sulphur-yellow flowers.

22. N. willkommii (Sampaio) Fernandes (*N. jonquilloides* Willkomm).

Similar to *N. assoanus* but leaves 2–3 mm wide, flowers 1.4–1.9 cm in diameter, in umbels of 2 or 3, the longest flower-stalk exceeding the spathe, perianth-tube 1.1–1.5 cm, lobes 6–9 mm, corona 5–6 mm high, 7–11 mm in diameter. *S Portugal, SW Spain.* H5. Spring.

Hybrids with *N. gaditanus* occur in S Portugal.

23. N. fernandesii Pedro.

Similar to *N. assoanus* and *N. willkommii;* longer flower-stalk longer than the spathe, perianth-tube 1.5–2.2 cm, lobes 9–11 mm, corona 4–5 mm. *S Portugal.* H5. Spring.

A somewhat obscure species; in many respects intermediate between *N. willkommii* and *N. gaditanus.*

24. N. gaditanus Boissier & Reuter (*N. minutiflorus* Willkomm). Illustration: Taylor, Wild flowers of Spain and Portugal, 19 (1972); Rix & Phillips, The bulb book, 108 (1981).

Similar to *N. assoanus* but leaves usually exceeding the scape, the perianth-tube slightly to strongly curved, 8–17 mm, and the corona 4–8 mm in diameter. *S Portugal, S Spain.* H4. Winter–spring.

25. N. rupicola Dufour. Illustration: *Botanical Magazine*, 6473C (1880) & n.s., 577 (1970); Ceballos et al., Plantas silvestres de la peninsula Iberica, 394 (1980); Rix & Phillips, The bulb book, 108 (1981).

Bulb 1–2.5 cm, dark. Leaves more or less glaucous, to 15 cm × 1.5–3 mm, channelled on the inner face and with 2 keels on the outer, erect or prostrate, margins smooth. Scape exceeding the leaves, 1-flowered. Spathe conspicuous, tubular for three-quarters of its length; flower-stalk usually absent, rarely to 3 mm. Flowers yellow, not or faintly fragrant. Perianth-tube 1.3–2 cm; lobes 7–13 × 5–10 mm, obovate. Corona 2–5 mm high, 7–10 mm in diameter, usually 6-lobed, more rarely sinuous or almost entire. *N Portugal, C Spain, Morocco.* H4. Spring–summer.

Fernandes (1967) recognises a number of subspecies, of which the following are cultivated:

Subsp. **rupicola**. Whole plant glaucous; perianth-tube 1.3–2 cm. *C Spain, N Portugal.*

Subsp. **marvieri** (Jahandiez & Maire) Maire & Weiller (*N. marvieri* Jahandiez & Maire). Plant scarcely glaucous; perianth-tube 2–2.7 cm. *Morocco.*

26. N. watieri Maire (*N. rupicola* Dufour subsp. *watieri* (Maire) Maire & Weiller). Illustration: *Botanical Magazine*, 9443 (1936); Rix & Phillips, The bulb book, 108 (1981).

Similar to N. *rupicola* but flower-stalk to 1 mm, flower white or cream-white, the corona with sinuous margins. *Morocco.* H5. Spring.

N. atlanticus Stern. Illustration: *RHS Daffodil & Tulip Yearbook* **16**: f. 13 (1950). Similar to both *N. rupicola* and *N. watieri* but with larger, cream-white flowers that are strongly fragrant. *Morocco*. H5. Spring.

27. N. calcicola Mendonca. Illustration: *Boletim do Sociedade Broteriana*, ser. 2, **7**: 318 (1930); *Botanical Magazine*, n.s., 180 (1952); Grey-Wilson & Mathew, Bulbs, pl. 25 (1981).

Similar to *N. rupicola* but spathe tubular for up to half its length, subtending usually 2–5 flowers that are 1.7–2.5 cm in diameter; perianth-tube 1.3–1.8 cm, lobes 7–14 × 5–8 mm, corona 4–8 mm high, 6–9 mm in diameter, margins obscurely scalloped. *Portugal.* H4. Spring.

28. N. scaberulus Henriques.

Similar to *N. rupicola* but leaves 4-angled, often prostrate or arching over and twisted, and with rough angles; spathe tubular for up to one-quarter of its length, subtending 1–3 flowers. Flower-stalks 8–15 mm. Flowers 1.2–1.7 cm in diameter. Perianth-tube 1.2–1.7 cm; lobes 4.5–7 × 3–5 mm. Corona 2–5 mm high, 5–7 mm in diameter, margin entire or finely scalloped. *N Portugal.* H4. Spring.

29. N. viridiflorus Schousboe. Illustration: *Botanical Magazine*, 1687 (1815); Polunin & Smythies, Flowers of south-west Europe, t. 60 (1973); Grey-Wilson & Mathew, Bulbs, pl. 24 (1981).

Bulb 2–3 cm, dark or pale. Leaves appearing soon after the flowers, dark green, cylindric, hollow, to 30 cm × 4 mm. Flowers in umbels of 2–5, dull green with an unpleasant smell; flower-stalks very unequal, the longest to 7 cm. Perianth-tube 1–1.5 cm, narrow, widening somewhat towards the apex; lobes 1–1.6 cm, linear-oblong, acute, widely spreading or somewhat reflexed. Corona *c.* 1 mm high, 3–5 mm in diameter, 6-lobed. *Spain (Gibraltar area), Morocco.* H5. Autumn.

Hybrids with *N. serotinus* occur in the Gibraltar area.

Section **Ganymedes** Schultes & Schultes. Leaves flat to almost cylindric, narrow. Flowers solitary or in umbels. Perianth-tube cylindric; lobes narrow, reflexed. Corona more or less bell-shaped to almost cylindric, its diameter equalling or exceeding its height. Filaments very unequal, those of the lower whorl as long as the anthers, those of the upper whorl longer than the anthers; anthers dorsifixed. Flowering in spring.

30. N. triandrus Linnaeus. Illustration: *Botanical Magazine*, 48 (1793); Taylor, Wild flowers of Spain and Portugal, 31 (1972); Polunin & Smythies, Flowers of south-west Europe, t. 60 (1973); Rix & Phillips, The bulb book, 108 (1981).

Bulb 1–2 cm, dark. Leaves 15–30 cm × 1.5–3 mm, dark green, blunt. Scape about as long as to longer than the leaves, to 45 cm. Flowers solitary or up to 6 in an umbel, drooping, white to bright yellow. Perianth-tube 1–2 cm; lobes 1–2.5 cm (rarely to 3 cm), lanceolate to linear-oblong, strongly reflexed. Corona 5–25 mm high, 7–25 mm in diameter, margin sinuous. The upper 3 stamens and the style project from the corona. *Spain, Portugal, NW France.* H3. Spring.

A very variable species. Numerous attempts have been made to divide it up into species, subspecies or varieties, but none of these seems satisfactory. In general, plants with solitary, usually bright yellow, flowers are referred to **N. concolor** Link (*N. pallidulus* Graells; *N. triandrus* subsp. **pallidulus** (Graells) Webb). Illustration: Ceballos et al., Plantas silvestres de la peninsula Iberica, 394 (1980). Plants with 2 or 3 flowers that are pale yellow are referred to *N. triandrus* var. **cernuus** (Salisbury) Baker; plants with pure white flowers to *N. triandrus* var. **triandrus**; and plants with the perianth-lobes a deeper yellow than the corona are referred to **N. pulchellus** Salisbury, which may be a hybrid between *N. concolor* and *N. triandrus* var. *cernuus*. This scheme has simplicity to recommend it, but it is not adequate, and the group requires further study.

Section **Bulbocodii** de Candolle (genus *Corbularia* Haworth). Bulb small. Leaves narrow. Flowers solitary, bilaterally symmetric. Perianth-lobes narrow and spreading. Corona very conspicuous, funnel-shaped. Stamens with long filaments that are deflexed downwards and then curved upwards near their apices; anthers dorsifixed.

A complex of 5 very similar species, differing in size, flower colour and flowering time. There is little agreement in the literature as to what species should be recognised, what infraspecific units they should contain, and what their precise geographical distribution is. The account given here broadly follows Fernandes (1967) in including 5 species; these are often very difficult to distinguish. For 3 of them, an outline of the infraspecific units recognised by Fernandes is given; some of these variants are very similar and the keys can only be regarded as provisional. The whole group requires much further study.

31. N. bulbocodium Linnaeus (*Corbularia bulbocodium* (Linnaeus) Haworth). Illustration: *The Garden* **101**: 351 (1976); Ceballos et al., Plantas silvestres de la peninsula Iberica, 393 (1980) – var. *nivalis*; Rix & Phillips, The bulb book, 107 (1981); *Il Giardino Fiorito* **48**: 501 (1982).

Bulb 1.5–2 cm, whitish or pale brown. Leaves 2–several from each bulb, 1.5–4 mm broad, dark green, usually erect. Scape usually exceeding the leaves. Flowers horizontal or almost so, pale to deep yellow or orange-yellow, often tinged with green; flower-stalk at least 5 mm, usually more than 9 mm, at most 3 cm. Perianth-tube 4–25 mm; lobes 6–15 × 1.5–5 mm, linear to narrowly triangular, acute. Corona 7–25 × 9–35 mm, not at all narrowed at the margin. Stamens and style usually included in the corona, occasionally slightly projecting from it. *SW & W France, Spain, Portugal, N Africa.* H4. Winter–spring.

The most widely distributed of the species of the section. The following key provides a synopsis of those variants recognised by Fernandes that are thought to be in cultivation:

1a. Flower deep or golden yellow (subsp. **bulbocodium**) 2
b. Flower pale yellow 5
2a. Flower-stalk to 2 cm; perianth to 3.5 cm 3
b. Flower-stalk more than 2 cm; perianth 3.5 cm or more 4
3a. Leaves to 1.5 mm broad, the outer surface slightly striate; scape smooth (*W France, Spain, Portugal, Morocco*) var. **bulbocodium**
b. Leaves more than 1.5 mm broad, the outer surface conspicuously striate; scape slightly ridged (*Spain, Morocco*) var. nivalis (Graells) Baker
4a. Leaves to 2.5 mm broad, erect, about as long as the scape; flower 3–3.5 cm, the corona *c.* 2 cm in diameter (*W France and Spain*) var. **conspicuus** (Haworth) Baker
b. Leaves to 4 mm broad, recurved, longer than the scape; flower 3.5–5 cm, the corona *c.* 3 cm in diameter (*W Portugal*) var. **serotinus** (Haworth) Fernandes
5a. Flowers pale lemon yellow, 3.5–5 cm (subsp. **bulbocodium**. *Spain*) var. **citrinus** Baker
b. Flowers pale yellow, to 3.5 cm 6
6a. Flowering in spring; stamens usually projecting from the corona (subsp. **bulbocodium**. *Spain*) var. **graellsii** (Webb) Baker
b. Flowering in winter; stamens not projecting (subsp. **praecox**. *Morocco*) var. **praecox** Gattefosse & Weiller

32. N. obesus Salisbury (*N. bulbocodium* subsp. *obesus* (Salisbury) Maire). Illustration: *Botanical Magazine*, n.s., 650 (1973); Grey-Wilson & Mathew, Bulbs, pl. 25 (1981).

Very similar to *N. bulbocodium* but leaves very narrow, prostrate or arching and deflexed; perianth usually without any green tinge, the corona slightly narrowed at its margin. *Portugal, Spain.* H4. Spring.

33. N. romieuxii Braun-Blanquet & Maire (*N. bulbocodium* subsp. *romieuxii* (Braun-Blanquet & Maire) Emberger & Maire). Illustration: Rix & Phillips, The bulb book, 106 & 107 (1981); Brickell (ed.), RHS A–Z encyclopedia of garden plants, 714 (2003).

Bulb about 1 cm, dark brown. Leaves 2–several per bulb, as long as to somewhat shorter than the scape, striate on the outer side, dark green, more than 1 mm broad. Scape to 20 cm. Spathe whitish or dark-coloured. Flower-stalk 1–5 mm. Flowers pale yellow to almost white but with a yellow tinge or flush. Perianth-lobes 1–2.5 mm broad, acute, spreading. *Morocco.* H5. Winter–spring.

Very similar to both *N. bulbocodium* and *N. hedraeanthus*, but known only from Morocco. The following variants recognised by Fernandes may be grown:

1a. Flowers pale yellow (subsp. **romieuxii**) 2
b. Flowers whitish, flushed or tinged with yellow (subsp. **albidus** (Emberger & Maire) Fernandes) 3
2a. Spathe whitish, scarious; flower-stalk to 3 mm, rarely more var. **romieuxii**
b. Spathe dark-coloured; flower-stalk more than 3 mm var. **rifanus** (Emberger & Maire) Fernandes
3a. Perianth-lobes to 2.5 mm wide at the base, longer than the corona var. **albidus**
b. Perianth-lobes to 1.5 mm wide at the base, as long as, or shorter than the corona var. **zaianicus** (Maire et al.) Fernandes

The very early-flowering variant known as var. *mesatlanticus* Emberger & Maire, is probably the same as var. *romieuxii* above.

34. N. cantabricus de Candolle (*N. clusii* Dunal). Illustration: *Botanical Magazine*, 5831 (1870) – subsp. *monophyllus*; Ceballos et al., Plantas silvestres de la peninsula Iberica, 392 (1980) – subsp. *cantabricus*; Rix & Phillips, The bulb book, 107 (1981) – subsp. *monophyllus* & *cantabricus*; Brickell (ed.), RHS A–Z encyclopedia of garden plants, 710 (2003).

Bulb 1–2 cm, dark brown. Leaves 1 or more to each bulb, erect, 1–1.5 mm broad, dark green, as long as or longer than the scape. Scape erect, 10–25 cm. Spathe 2–3 cm, pale or dark. Flower-stalk 3–9 mm. Flowers horizontal, pure white or white flushed with green. Perianth-tube 1.2–2 cm; lobes 8–12 × 2–4.5 mm, linear-lanceolate. Corona 1.2–1.8 × 2–3.5 cm, margin expanded and frilled. Stamens and style not projecting from the corona. *S Spain, Morocco, Algeria.* H5. Winter.

Fernandes recognises 3 subspecies (as below), these further divided into varieties that are not included here, as it is uncertain which of them are in cultivation.

1a. Leaves mostly 1 to each bulb (*Spain, Morocco, Algeria*) subsp. **monophyllus** (Durieu) Fernandes
b. Leaves 2 or more to each bulb 2
2a. Leaves flexuous, 0.5–1.5 mm broad; stamens more or less equal (*Spain, Morocco*) subsp. **cantabricus**
b. Leaves stiff, 1–2.5 mm broad; stamens very unequal (*Morocco*) subsp. **tananicus** (Maire) Fernandes

35. N. hedraeanthus (Webb & Heldreich) Colmeiro (*Corbularia hedraeanthus* Webb & Heldreich). Figure 8(4), p. 208. Illustration: *Botanical Magazine*, n.s., 248 (1955); Rix & Phillips, The bulb book, 106 (1981); Grey-Wilson & Mathew, Bulbs, pl. 25 (1981).

Bulb *c.* 1 cm, dark brown. Leaves 1 per bulb, thread-like or terete, *c.* 1 mm broad, striate on the outer surface, dark green, usually exceeding the scape. Scape to 8 cm, emerging horizontally, later curved or erect. Spathe dark, 1 cm or more. Flower-stalk absent. Flowers horizontal, pale sulphur-yellow. Perianth-tube to 1.25 cm; lobes 3–4.5 mm broad, oblong, obtuse but with a fine, acute point, not spreading widely. Corona 8–13 mm. Stamens and style conspicuously projecting from the corona, the style exceeding the stamens. *Spain (Jaén, Ciudad Real).* H4. Winter–spring.

Section **Pseudonarcissi** de Candolle (Section *Ajax* (Haworth) Dumortier). Leaves flat, usually wide. Flowers usually solitary. Perianth-tube broadly funnel-shaped, the lobes spreading. Corona large, more or less cylindric, its height at least twice its diameter. Anthers usually basifixed, not projecting from the corona. Flowering in spring.

36. N. longispathus Pugsley. Illustration: *Botanical Magazine*, n.s., 246 (1955); Grey-Wilson & Mathew, Bulbs, pl. 26 (1981).

Bulb to 3 cm. Leaves at least 40 cm, often much longer, 1–1.5 cm wide. Scape exceeding the leaves, to 1.75 m. Spathe 6–10 cm, green at flowering. Flowers solitary or occasionally in pairs, ascending; stalks 4–9 cm. Perianth-tube 1–1.5 cm; lobes 2.5–3.2 cm, yellow, spreading, not twisted. Corona 2.5–3 cm, slightly deeper yellow than the lobes, its margin expanded and scalloped. *SE Spain.* H5. Spring.

37. N. pseudonarcissus Linnaeus. Figure 8(1), p. 208.

Bulb 2–5 cm, dark or rather pale. Leaves usually glaucous, 8–50 cm × 5–16 mm, erect or somewhat spreading, obtuse. Scape usually exceeding the leaves, to 90 cm. Spathe 2–6 cm, scarious. Perianth-tube 1.5–2.5 cm; lobes 1.5–5.5 cm, white to deep yellow, spreading, sometimes twisted. Corona 1.5–4.5 cm or more, white to deep yellow, the margin sometimes conspicuously expanded and then sometimes reflexed, minutely to deeply toothed or lobed. Stamens and style included in the corona, the anthers close and forming a ring beneath the stigma. *W Europe: long cultivated and naturalised in many places.* H2. Spring.

The most widespread of the species, and the most difficult in terms of its classification. It has been divided up into some 20 species (Pugsley, 1933), many of them further divided into varieties. But the existence of numerous intermediates (both wild and cultivated) makes this treatment unusable for many specimens. The most reasonable account seems to be that in *Flora Europaea*, and that is followed here with minor modifications.

1a. Corona white to pale yellow 2
b. Corona deep golden yellow 4
2a. Flower-stalk 1–2.5 cm; corona white to very pale yellow subsp. **moschatus**
b. Flower-stalk 3–10 mm; corona pale yellow 3
3a. Corona 3–4 cm, distinctly expanded and usually lobed at the margin subsp. **pallidiflorus**
b. Corona 2–3.5 cm, scarcely expanded at the margin and usually only slightly lobed subsp. **pseudonarcissus**
4a. Perianth-lobes to 2 cm 5
b. Perianth-lobes 2 cm or more 6
5a. Perianth-lobes twisted; corona expanded at the margin subsp. **major**
b. Perianth-lobes not twisted; corona not expanded at the margin subsp. **nevadensis**
6a. Perianth-lobes white or yellow, paler than the corona 7
b. Perianth-lobes deep golden yellow, of the same colour as the corona 8
7a. Flower-stalk 8–25 mm; corona conspicuously expanded at margin; flowers horizontal or ascending subsp. **nobilis**
b. Flower-stalk 3–12 mm; corona scarcely expanded at margin; flowers horizontal or drooping subsp. **pseudonarcissus**
8a. Leaves to 15 cm subsp. **portensis**
b. Leaves at least 20 cm 9
9a. Leaves 20–50 cm long, 5–15 mm broad subsp. **major**
b. Leaves 20–30 cm long, 6–10 mm broad subsp. **obvallaris**

Subsp. **pseudonarcissus**. Illustration: *Botanical Magazine*, n.s., 216 (1953); Grey-Wilson & Mathew, Bulbs, pl. 26 (1981). Leaves 12–35 cm × 6–12 mm. Flowers horizontal or drooping, stalks 3–12 mm. Perianth-lobes 2–3.5 cm, white to yellow, more or less twisted. Corona 2–3.5 cm, deeper yellow than the lobes, its margin scarcely expanded, not or little lobed. *W Europe except for Portugal and S Spain.* H2. Spring.

Subsp. **pallidiflorus** (Pugsley) Fernandes (*N. pallidiflorus* Pugsley; *N. macrolobus* (Jordan) Pugsley; *N. pallidus* misapplied). Illustration: Polunin & Smythies, Flowers of south-west Europe, t. 60 (1973); Grey-Wilson & Mathew, Bulbs, pl. 26 (1981). Leaves 15–30 cm × 5–12 mm. Flowers horizontal or drooping, stalks 3–10 mm. Perianth-lobes 3–4 cm, cream to pale yellow, more or less twisted. Corona 3–4 cm, deeper yellow than the lobes but still pale, margin expanded and usually reflexed. *France, Spain (Pyrenees and northern Spanish mountains).* H3. Spring.

Plants from the Pyrenees with large flowers are sometimes recognised as *N. macrolobus.*

Subsp **moschatus** (Linnaeus) Baker (*N. moschatus* Linnaeus; *N. tortuosus* Haworth; *N. albescens* Pugsley; *N. alpestris* Pugsley). Illustration: *Botanical Magazine*, 1300 (1810); Grey-Wilson & Mathew, Bulbs, pl. 26 (1981). Leaves 10–40 cm × 5–12 mm. Flowers horizontal or drooping. Perianth-lobes 2–3.5 cm, white or cream, twisted. Corona 2.5–4 cm, usually the same colour as the lobes, rarely pale yellow, its margins scarcely expanded. *France, Spain (Pyrenees and northern Spanish mountains).* H4. Spring.

Plants with the flowers completely white have been recognised as *N. alpestris* and plants with flowers at the larger end of the range have been treated as *N. tortuosus* (flower-stalks 1–1.8 cm) and *N. albescens* (flower-stalks 2–2.5 cm), but all these units intergrade.

Subsp. **nobilis** (Haworth) Fernandes (*N. nobilis* Haworth). Illustration: Taylor, Wild flowers of Spain and Portugal, 99 (1972); Ceballos et al., Plantas silvestres de la peninsula Iberica, 390 (1980); Grey-Wilson & Mathew, Bulbs, pl. 26 (1981). Leaves 15–50 cm × 8–15 mm. Flowers horizontal or ascending, stalks 8–25 mm. Perianth-lobes 3–3.5 cm, pale yellow, usually twisted. Corona 3–4.5 cm, deep golden yellow, margin expanded. *N Portugal, N Spain.* H3. Spring.

Plants with very large flowers which have the margin of the corona very conspicuously expanded have been recognised as *N. leonensis* Pugsley (*nobilis* var. *leonensis* (Pugsley) Fernandes). The name *N. gayii* Pugsley has been applied to similar plants with rather smaller, pale flowers; the origin of this variant is unknown.

Subsp. **major** (Curtis) Baker (*N. major* Curtis; *N. hispanicus* Gouan; *N. confusus* Pugsley). Leaves 20–50 cm × 5–15 mm. Flowers horizontal to almost erect, stalks 8–30 mm. Perianth-lobes 1.8–4 cm, yellow to deep yellow, twisted. Corona 2–4 cm, the same colour as the lobes, its margin more or less expanded. *S France, Spain, Portugal.* H4. Spring.

Plants with long flower-stalks that are conspicuously curved at the top have been recognised as *N. hispanicus;* and plants with shorter, more or less straight flower-stalks, as *N. confusus.*

Subsp. **portensis** (Pugsley) Fernandes (*N. portensis* Pugsley). Leaves 8–12 cm × 5–7 mm, much shorter than the scape. Flowers horizontal or slightly drooping, stalks 5–15 mm. Perianth-lobes 2–3 cm, not twisted, deep yellow. Corona 2.5–3.5 cm, the same colour as the perianth-lobes, rather flaring, but not conspicuously expanded at the margin. *N Portugal N & C Spain.* H3. Spring.

Subsp. **obvallaris** (Salisbury) Fernandes (*N. obvallaris* Salisbury). Leaves 20–30 cm × 8–10 mm. Flowers horizontal, stalks 1–1.5 cm. Perianth-lobes 2.5–3 cm, deep golden yellow. Corona 3–3.5 cm, the same colour as the lobes, its margins expanded and sometimes reflexed. *Origin uncertain; naturalised in Britain (Wales).* H2. Spring.

Subsp. **nevadensis** (Pugsley) Fernandes (*N. nevadensis* Pugsley). Illustration: Taylor, Wild flowers of Spain and Portugal, 34 (1972); Grey-Wilson & Mathew, Bulbs, pl. 26 (1981). Leaves 12–30 cm × 5–10 mm. Flowers horizontal to almost erect, often in umbels of 2–4; stalks 2–3.5 cm. Perianth-lobes 1.5–2 cm, pale yellow, not twisted. Corona 1.5–2.5 cm, bright yellow. *S Spain (Sierra Nevada).* H4. Spring.

38. N. × odorus Linnaeus. Illustration: *Botanical Magazine*, 934 (1806); Grey-Wilson & Mathew, Bulbs, pl. 26 (1981).

Bulb to 3 cm. Leaves bright green, 35–50 cm × 6–8 mm, keeled. Scape to 40 cm, more or less terete. Flowers in an umbel of 1–4, usually fragrant. Perianth-tube to 2 cm, narrowly funnel-shaped; lobes to 2.5 × 1.3 cm, bright yellow. Corona 1.3–1.8 cm, 1.7–2 cm in diameter, the margin lobed or almost entire. *Garden origin; naturalised in S Europe.* H3. Spring.

The garden hybrid between *N. jonquilla* and *N. pseudonarcissus.*

39. N. × incomparabilis Miller. Illustration: Grey-Wilson & Mathew, Bulbs, pl. 26 (1981).

Bulb to 3 cm. Leaves more or less flat, somewhat glaucous, 17–35 cm × 8–12 mm. Scape to 40 cm, compressed. Flower solitary, not or only slightly fragrant. Perianth-tube 2–2.5 cm, narrowly funnel-shaped; lobes 2.5–3 × 1.2–1.6 cm, spreading, pale yellow. Corona 1.3–2.2 cm, 1.7–2 cm in diameter, deep orange-yellow, its margin wavy and deeply lobed. *S & C France; naturalised elsewhere in Europe.* H3. Spring.

The natural hybrid between *N. poeticus* and *N. pseudonarcissus.*

40. N. bicolor Linnaeus (*N. abscissus* (Haworth) Roemer & Schultes). Illustration: *Botanical Magazine*, 1187 (1809); Polunin & Smythies, Flowers of south-west Europe, t. 60 (1973).

Bulb to 3.5 cm, dark. Leaves 30–35 × 1–2 cm. Scape to 40 cm. Flower solitary, horizontal; stalk 1.5–3.5 cm. Perianth-tube 6–12 mm; lobes 3.5–4 cm, cream or pale yellow, spreading, not twisted. Corona 3.5–4 cm, 2.5–3 cm in diameter, deep yellow, its margin scarcely expanded, entire or finely scalloped. *France & Spain (Pyrenees).* H3. Spring.

N. abscissus is the name that has been applied to wild plants with narrow leaves, narrower perianth-lobes and a corona 1.5–2 cm in diameter; these intergrade with *N. bicolor.*

41. N. minor Linnaeus (*N. provincialis* Pugsley; *N. pumilus* Salisbury; *N. nanus* Spach; *N. parviflorus* (Jordan) Pugsley). Illustration: *Botanical Magazine*, n.s., 410 (1962); Rix & Phillips, The bulb book, 107 (1981).

Bulb 2–3 cm, brown. Leaves glaucous, 8–25 cm × 3–14 mm. Scape to 25 cm. Flower solitary, horizontal or drooping; stalk 3–20 mm. Perianth-tube 9–15 mm; lobes 1.6–2.2 cm, pale or deep yellow, erect-spreading, sometimes twisted. Corona 1.6–2.5 cm, deep yellow, its margin toothed or lobed. *France, N Spain.* H3. Spring.

Variable in size and flower colour; this variation, which is not well correlated, has been used for the recognition of several species treated here as synonymous with *N. minor.*

42. N. asturiensis (Jordan) Pugsley (*N. minimus* misapplied). Illustration: *Botanical Magazine*, 9495 (1937); Polunin & Smythies, Flowers of south-west Europe, t. 60 (1973); *Gartenpraxis*, 21 (1976); Rix & Phillips, The bulb book, 107 (1981).

Bulb 1.5–2 cm, brown. Leaves glaucous, 5–15 cm × 2–8 mm. Scape usually to 10 cm, rarely more. Flower solitary, drooping; stalk 3–10 mm. Perianth-tube 5–9 mm; lobes 7–14 mm, erect-spreading, not twisted, yellow. Corona 8–16 mm, constricted at about the middle, inflated above and below the constriction, yellow, margin spreading or reflexed, variously lobed. *N Portugal, N & C Spain.* H4. Spring.

43. N. cyclamineus de Candolle. Illustration: *Botanical Magazine*, 6950 (1887); *Gartenpraxis*, 21 (1976); Rix & Phillips, The bulb book, 109 (1981); Brickell (ed.), RHS A–Z encyclopedia of garden plants, 710 (2003).

Bulb to 2 cm, brownish. Leaves bright green, 15–30 cm × 4–6 mm. Scape 15–30 cm. Flower solitary, drooping or pointing downwards almost vertically, spathe *c.* 2 cm, green at flowering, later scarious; flower-stalk 1.5–2.5 cm. Perianth-tube 2–3 cm; lobes to 2 cm, lanceolate, reflexed backwards at almost 180° to the corona, hiding the ovary, perianth-tube and part of the flower-stalk, deep yellow. Corona to 2 cm, cylindric but widening slightly above, its margin toothed or scalloped. *NW Portugal, NW Spain.* H4. Early spring.

A parent of many popular, early-flowering hybrids.

13. TAPEINANTHUS Herbert

J. Cullen

Very like *Narcissus.* Leaves absent at flowering. Flowers with perianth-tube extremely short, corona rudimentary or absent.

A genus of a single species from Spain and N Africa, very closely related to, and perhaps not distinct from, *Narcissus.* Cultivation as for *Narcissus* (p. 205).

1. T. humilis (Cavanilles) Herbert (*Narcissus humilis* (Cavanilles) Traub). Illustration: Stocken, Andalusian flowers and countryside, 93 (1969); Polunin & Smythies, Flowers of south-west Europe, t. 61 (1973).

Bulb to 1.5 cm. Leaves 1 or rarely 2 from each bulb, to 20 cm × 1 mm, appearing in spring and dying off before flowering. Scape 7–20 cm. Flower solitary (rarely 2 together), erect, yellow. Perianth-lobes to 1 cm × 2–3 mm. *SW Spain, N Africa.* H5. Autumn.

The validity of the generic name *Tapeinanthus* is uncertain, and this plant is sometimes found in the literature under the generic names *Braxireon* Rafinesque and *Carregnoa* Boissier.

14. CRINUM Linnaeus

E. Campbell edited by S.G. Knees

Herbaceous perennials with bulbs that usually have a distinct neck made up of the remains of sheathing leaf bases. Leaves linear or strap-shaped, channelled or flat, usually dying off in winter. Flowers in a terminal umbel subtended by 2 spathes, borne on a solid, compressed scape that arises laterally from the bulb. Perianth radially or bilaterally symmetric, with a long tube that may be straight or curved, the 6 lobes almost equal, spreading widely into a funnel- or bell-shape, white or white with various shadings of red or pink. Stamens 6, inserted in the throat of the perianth-tube, filaments arching or deflexed, anthers versatile. Ovary 3-celled, ovules numerous. Style ascending or deflexed. Fruit almost spherical, irregularly dehiscent. Seeds almost spherical.

A genus of about 120 species distributed throughout the tropics and warm temperate regions of the world. Some of the species will grow outdoors in a sunny position; these may require protection or lifting and drying during winter. All require a rich but well-drained soil. Propagation is by seeds or offsets.

Literature: Uphof, J.C., A Review of the species of Crinum, *Herbertia* **9**: 63–69 (1942); Nordal, I., Revision of the E African taxa of the genus Crinum, *Norwegian Journal of Botany* **24**: 177–94 (1977); Verdoorn, I.

C., The genus Crinum in southern Africa, *Bothalia* **11**: 27–50 (1973–5).

Bulb. Neck less than 15 cm: **1, 2, 6, 9, 12, 13, 14**; neck usually more than 15 cm: **3, 4, 5, 6, 7, 8, 9, 10, 11**.
Leaves. Up to 2.5 cm broad: **1, 5, 12**; more than 2.5 cm broad: **2, 3, 4, 5, 6, 7, 8, 9, 10, 11, 12, 13, 14**.
Scape. Green: **1, 3, 4, 6, 7, 8, 9, 11, 12**; green tinged with red or purple: **2, 5, 10, 12, 13, 14**.
Flowering time. Spring: **5, 8, 10, 11, 13**; summer: **1, 2, 3, 6, 7, 8, 9, 10, 11, 12, 13, 14**; autumn: **4, 8, 11, 13**; winter: **13**.
Flowers. 12 or fewer: **1, 2, 3, 4, 5, 6, 7, 12, 13, 14**; 15 or more: **2, 4, 5, 8, 9, 10, 11**. Fragrant: **2, 3, 4, 5, 6, 8, 9, 10, 11, 12, 13, 14**; not fragrant: **1, 7**.
Perianth-tube. Curved: **1, 2, 3, 4, 5, 6, 7, 12, 13, 14**; straight: **8, 9, 10, 11, 12, 13, 14**. Green: **2, 3, 5, 6, 7, 8, 9, 12, 14**; green tinged with red or purple: **1, 2, 4, 5, 7, 12, 13**; red or purple: **10, 11**.
Perianth-lobes. 1.5 cm or more wide: **1, 2, 3, 4, 5, 6, 7**; less than 1.5 cm wide: **8, 9, 10, 11, 12, 13, 14**.

1a. Perianth-tube distinctly curved; lobes 1.5 cm or more wide; stamens and style deflexed 2
b. Perianth-tube straight or very slightly curved; lobes less than 1.5 cm wide; stamens and style not deflexed 8
2a. Leaves folded to form a channel 3
b. Leaves flat 6
3a. Flowers usually 4–7 per umbel; perianth-lobes to 5 cm **1. campanulatum**
b. Flowers usually 6–20 per umbel; perianth-lobes 5 cm or more 4
4a. Bulb with neck less than 15 cm; leaf margins without teeth **2. latifolium**
b. Bulb with neck *c.* 30 cm; leaf margins toothed 5
5a. Leaf margins narrowly cartilaginous, not wavy; flowering in summer **3. bulbispermum**
b. Leaf margin rough, wavy; flowering in autumn **4. macowanii**
6a. Leaves erect, not curved or bent over; perianth-tube green tinged with purple; flowering in spring **5. zeylanicum**
b. Leaves recurved or bent over; perianth-tube green; flowering in summer 7
7a. Bulb without a neck or neck extremely short; flowers fragrant **6. jagus**
b. Bulb with neck 30–120 cm; flowers not fragrant **7. moorei**
8a. Umbels with 12 or more flowers; perianth-lobes linear, spreading or reflexed 9
b. Umbels with up to 12 flowers; perianth-lobes lanceolate, spreading or ascending 12
9a. Perianth-tube green; lobes less than 7.5 cm 10
b. Perianth-tube purple-red or green tinged with purple; lobes more than 7.5 cm 11
10a. Scape to 70 cm; flower-stalks very short; style as long as the filaments **8. asiaticum**
b. Scape more than 70 cm; flower-stalks 2.5–4 cm; style much shorter than filaments **9. pedunculatum**
11a. Scape green tinged with purple; flower-stalks 2.5 cm or more; leaves 60–110 cm **10. augustum**
b. Scape green; flowers stalkless; leaves 90–120 cm **11. amabile**
12a. Leaves linear; perianth-tube 7.5–10 cm **12. amoenum**
b. Leaves strap-shaped; perianth-tube 12.5–15 cm 13
13a. Leaves 2.5–4 cm wide, wavy; scape usually less than 30 cm **13. purpurascens**
b. Leaves 5–6.5 cm wide, not wavy; scape 45–60 cm **14. erubescens**

1. C. campanulatum Herbert (*C. aquaticum* Burchell). Illustration: *Flowering plants of Africa* **137**: t. 1455 (1965).

Bulb 4 cm in diameter with a neck 5 cm or more. Leaves 90–120 × 1.4–2.5 cm, linear, channelled, margins narrowly cartilaginous and toothed. Scape 30 cm or more, green. Flower-stalks 1–4.5 cm. Flowers 4–7, bell-shaped, not fragrant. Perianth-tube curved, to 5 cm, reddish green. Lobes 4–5 × 1.5–2 cm, white flushed with pink, the tips slightly recurved. Stamens deflexed. Style deflexed, usually longer than the stamens. *South Africa*. H5–G1. Summer.

2. C. latifolium Linnaeus (*C. speciosum* Herbert; *C. insigne* Sweet). Illustration: *Botanical Magazine*, 2217 (1821); *Edwards's Botanical Register* **15**: t. 1297 (1824).

Bulb to 15 cm in diameter, with a short neck. Leaves to 85 × 7.5–12 cm, strap-shaped, channelled, with a distinct midrib, margins rough. Scape 30–60 cm, green tinged with purple. Flowers stalkless, 10–20 in the umbel, bell-shaped, fragrant. Perianth-tube 7.5–12 cm, curved, green or green tinged with red. Lobes to 12 × 2.5 cm, white faintly tinged with red in their centres, the tips reflexed. Stamens deflexed. Style deflexed. *Tropical Asia*. G2. Summer.

3. C. bulbispermum (Burman) Milne-Redhead & Schweickerdt. Illustration: *Botanical Magazine*, 661 (1803); *Flowering plants of Africa* **29**: t. 1150 (1939).

Bulb 7.5–13 cm in diameter, neck to 30 cm. Leaves 60–90 × 7.5–11 cm, strap-shaped, channelled, reflexed, margins narrowly cartilaginous, toothed. Scape 50–90 cm, green. Flower-stalks 4.5–9 cm. Flowers 6–12 in the umbel, funnel-shaped, fragrant. Perianth-tube 7.5–10 cm, curved, green. Lobes 7.5–10 × *c.* 2.5 cm, white with a dark red streak, or entirely suffused with red. Stamens deflexed. Style deflexed. *South Africa*. H5–G1. Summer.

C. × powellii; *C. bulbispermum* × *C. moorei*. Intermediate between the parents. Large long-necked bulb. Broad, upright, channelled leaves to 1.5 m. Trumpet-shaped flowers, in clusters of 8–10 on tall stems. Perianth 8–10 cm, flared, pink or white. Garden origin. Late summer. H5–G1.

'Album'. Illustration: Brickell (ed.), RHS A–Z encyclopedia of garden plants, 323 (2003); Lord, Flora, 434 (2003).

4. C. macowanii Baker (*C. pedicellatum* Pax; *C. johnstonii* Baker). Illustration: *Botanical Magazine*, 6381 (1878) & 7812 (1902).

Bulb 6–25 cm in diameter with neck to 30 cm. Leaves 10–130 × 2–16 cm, linear or strap-shaped, bright green, channelled, margins wavy, rough and toothed. Scape 10–60 cm, green. Flower-stalks 1–4 cm. Flowers 4–20, bell-shaped, fragrant. Perianth-tube 7–15 cm, curved, green tinged with brownish red. Lobes 7–13 × 2–4 cm, white with a pink or red streak, tips reflexed. Stamens deflexed. Style deflexed, longer than the stamens. *South Africa*. H5–G1. Autumn.

5. C. zeylanicum Linnaeus (*C. ornatum* (Aiton) Bury; *C. yuccaeflorum* Salisbury; *C. scabrum* Herbert; *C. kirkii* Baker). Illustration: *Botanical Magazine*, 1171 (1809), 2180 (1820) & 6512 (1880).

Bulb 5–20 cm in diameter, neck to 30 cm. Leaves 75–120 × 2–12 cm, strap-shaped, green, with a distinct midrib, erect or spreading, margins often wavy, sometimes rough or toothed. Scape 25–70 cm, green tinged with purple. Flowers stalkless or almost so, 3–24, most often 4–8 in the umbel, bilaterally symmetric, bell-shaped, fragrant. Perianth-tube 9–16 cm, curved, green or tinged with purple. Lobes 7–13 × 2–4 cm, white with a broad purple or pale purple streak, the upper reflexed more than the lower. Stamens and style deflexed. *Tropical Asia, tropical E Africa.* G2. Spring.

6. C. jagus (Thompson) Dandy (*C. giganteum* Andrews; *C. podophyllum* Baker; *C. laurentii* Durand & de Wildeman; *C. rattrayi* Anon.). Illustration: *Botanical Magazine*, 5206 (1860) & 6483 (1880).

Bulb 2.5–20 cm in diameter, without a neck or neck very short. Leaves 20–90 × 4–12 cm, strap-shaped, bright or dark green, almost membranous, not channelled, with a distinct midrib, margins often wavy, sometimes rough. Scape 25–75 cm, green. Flowers stalkless or almost so, 2–12, bell-shaped, fragrant. Perianth-tube 10–20 cm, curved, green. Lobes 7–10 × 2.4 cm, pure white or with a pale green midrib, widely spreading and overlapping, tips reflexed. Stamens deflexed. Style deflexed, usually longer than the stamens. *Tropical Africa.* G2. Summer.

7. C. moorei Hooker. Illustration: *Botanical Magazine*, 6113 (1874); *Flowering plants of Africa* **34**: t. 1351 (1961).

Bulb 15–19 cm in diameter, neck 30–120 cm. Leaves 65–150 × 6–10 cm, strap-shaped, spreading, not channelled, with a distinct midrib, margins often wavy. Scape 45–70 cm, green. Flower-stalks 1–8 cm. Flowers 5–10, funnel-shaped, not fragrant. Perianth-tube 8–10 cm, curved, green or tinged with red. Lobes 8–10 × 2–4 cm, white usually suffused with pink, widely spreading. Stamens deflexed. Style much longer than the stamens, deflexed. *South Africa (Natal Cape Province).* H5–G1. Summer.

8. C. asiaticum Linnaeus. Illustration: *Botanical Magazine*, 1073 (1807); Brickell (ed.), RHS A–Z encyclopedia of garden plants, 322 (2003).

Bulb 10–12.5 cm in diameter with a neck 15–23 cm. Leaves 90–120 × 7.5–13 cm, strap-shaped, spreading, not channelled, margins sometimes rough. Scape 45–70 cm, green. Flower-stalks very short. Flowers *c.* 20, salver-shaped, fragrant. Perianth-tube straight, 7.5–10 cm, green. Lobes 6–7.5 cm × 6–8 mm, linear, white, spreading. Stamens not deflexed. Style ascending. *Tropical SE Asia.* G2. Spring–autumn.

9. C. pedunculatum Brown. Illustration: *Edwards's Botanical Register* **1**: t. 52 (1815).

Bulb to 10 cm in diameter, with a short neck. Leaves *c.* 120 × 10–12.5 cm, strap-shaped, glaucous, thick, recurved, channelled or not. Scape 60–90 cm, green. Flower-stalks 2.5–4 cm. Flowers 20–30, salver-shaped, fragrant. Perianth-tube straight, to 10 cm, green. Lobes 6 cm × 5–7 mm, linear, white, spreading. Stamens not deflexed. Style ascending, shorter than the stamens. *Australia (Queensland).* H5–G1. Summer.

10. C. augustum Roxburgh (*C. amabile* var. *augustum* (Roxburgh) Herbert). Illustration: *Botanical Magazine*, 2397 (1823); *Edwards's Botanical Register* **8**: t. 679 (1823).

Bulb *c.* 12.5 cm in diameter, neck to 30 cm. Leaves 60–110 × 7.5–10 cm, strap-shaped, bright green, spreading, broadly channelled, with a distinct midrib. Scape to 75 cm, green tinged with purple. Flower-stalks 2.5–3 cm. Flowers 20–30, salver-shaped, fragrant. Perianth-tube 7.5–10 cm, straight, purple-red. Lobes 7.5–10 × *c.* 1.2 cm, linear, white with a pink midrib, reddish purple with white margins on the back, spreading. Stamens ascending. Style ascending, shorter than the stamens. *Mauritius, Seychelles Islands.* G2. Spring–summer.

11. C. amabile Don. Illustration: *Botanical Magazine*, 1605 (1813).

Bulb 15–30 cm in diameter, neck *c.* 30 cm. Leaves 90–120 × 7.5–10 cm, strap-shaped, thick, spreading, not channelled, margins sometimes rough. Scape 60–90 cm, green. Flowers stalkless, 20–30, salver-shaped, fragrant. Perianth-tube 7.5–10 cm, straight, purple-red. Lobes 7.5–10 × *c.* 1.2 cm, linear, white suffused with purple, midrib dark red on the back. Stamens ascending. Style ascending. *Indonesia (Sumatra).* G2. Spring–autumn.

12. C. amoenum Roxburgh.

Bulb 5–10 cm in diameter, with a very short neck. Leaves 45–60 × 2.5–5 cm, linear, bright green, spreading, channelled or not, margins wavy. Scape 30–60 cm, green or tinged with purple. Flowers stalkless, 6–12, salver-shaped, fragrant. Perianth-tube 7.5–10 cm, straight or slightly curved, green or tinged with purple. Lobes 5–7.5 × *c.* 1 cm, lanceolate, white tinted on the back with pinkish red, spreading or ascending. Stamens ascending. Style ascending. *Burma.* G2. Summer.

13. C. purpurascens Herbert. Illustration: *Botanical Magazine*, 6525 (1880).

Bulb *c.* 5 cm in diameter, with a very short neck, producing stolons. Leaves 45–90 × 2.5–4 cm, strap-shaped, dark green, almost erect or recurved, channelled, margins wavy and rough. Scape usually less than 30 cm, green tinged with purple. Flowers stalkless, 6–10, salver-shaped, faintly fragrant. Perianth-tube 12.5–15 cm, straight or slightly curved, green tinged with purple. Lobes 6–7.5 cm × 8–10 mm, white tinged with purple, spreading or ascending. Stamens ascending. Style ascending, longer than the stamens. *Tropical W Africa.* G2. Can flower throughout the year, but mainly in autumn–winter.

14. C. erubescens Linnaeus. Illustration: *Botanical Magazine*, 1232 (1809); *Loddiges' Botanical Cabinet* **3**: t. 31 (1817).

Bulb small, neck short. Leaves 60–90 × 5–6.5 cm, strap-shaped, spreading and recurved, channelled or flat, margins toothed and rough. Scape 45–60 cm, green tinged with purple. Flowers almost stalkless, 6–12, salver-shaped, fragrant. Perianth-tube 12.5–15 cm, straight or slightly curved, green. Lobes 6–7.5 cm × 6–10 mm, lanceolate, white sometimes tinged with pinkish red on the back, spreading or ascending. Stamens ascending. Style ascending. *Tropical S America.* G2. Summer.

15. BRUNSVIGIA Heister

J. Cullen

Perennials with bulbs, the bulb large, subterranean or exposed. Leaves large, usually withering before flowering. Scape erect, somewhat compressed and angled. Umbel spherical, with 20–60 flowers borne on very long stalks; spathes 2, free. Perianth tubular-funnel-shaped, strongly bilaterally symmetric, with a short tube below, some or all of the lobes recurved, the whole bent just above the tube. Corona absent. Stamens 6, projecting from the perianth. Ovary 3-celled, with several ovules in each cell. Fruit a large capsule.

A genus of an uncertain number of species from South Africa. The 2 species included here are remarkable objects, with their large, many-flowered umbels. They are perhaps more striking than beautiful, and are relatively infrequently grown. They require a well-drained soil and a sunny position, and should have a period free from watering between the death of the umbel and the appearance of new leaves.

Literature: Dyer, R.A., Review of the genus Brunsvigia, *Plant Life* **6**: 63–83 (1950).

1a. Leaves downy above; bulb subterranean, the leaves lying flat on the ground; perianth 5.5–7 cm **1. orientalis**
b. Leaves not downy above; bulb mostly above ground, the leaves ascending to erect; perianth 7–8.5 cm **2. josephinae**

1. B. orientalis (Linnaeus) Ecklon (*Amaryllis orientalis* Linnaeus; *B. gigantea* Schultes). Illustration: *Flowering plants of Africa* **36**: t. 1440 (1964); Mason, Western Cape sandveld flowers, t. 17 (1972).

Bulb ovoid to spherical, mostly below ground, 10–15 cm in diameter. Leaves oblong-ovate, narrowed to the base, 30–45 × 7–12 cm or more, lying flat on the ground, downy above. Scape 30–50 cm. Umbel with 20–40 (rarely more) flowers; flower-stalks 10–20 cm. Perianth bright red or pink, 5.5–7 cm, tube *c.* 5 mm, lobes linear-lanceolate. Ovary 3-angled. Capsule strongly 3-angled. *South Africa (Cape Province).* G1. Autumn.

2. B. josephinae (Redouté) Ker Gawler. Illustration: *Edwards's Botanical Register* **3**: t. 192–3 (1817); *Botanical Magazine*, 2578 (1825); *Flowering plants of Africa* **31**: t. 1223 (1956); Eliovson, Wild flowers of southern Africa, frontispiece (1980).

Bulb ovoid, mostly above ground, to 30 cm in diameter, covered with matted, membranous tunics, Leaves ascending to erect, oblong, to 90 × 5–20 cm, hairless. Scape to 45 cm or more. Umbel with 20–30 (rarely to 60 in some variants) flowers; flower-stalks 15–20 cm, elongating in fruit. Perianth deep red with yellow spots, 7–8.5 cm, tube 1.25–1.5 cm, lobes linear. Ovary and capsule scarcely angled. *South Africa (Cape Province).* G1. Autumn.

16. NERINE Herbert

J. Cullen

Perennial herbs with bulbs, the bulb sometimes with an elongate neck. Leaves present or absent at flowering. Scape usually exceeding the leaves. Umbel with 4–20 flowers, subtended by 2 persistent spathes. Perianth usually bilaterally symmetric, more rarely radially symmetric, with a very short tube and 6 arching and spreading lobes that often have crisped margins. Stamens 6, attached to the perianth-tube at the base, usually deflexed, more rarely erect, sometimes with appendages at the bases of the filaments. Ovary inferior, 3-celled and -lobed, with usually 4 ovules in each cell. Fruit a membranous capsule.

A genus of about 30 species from South Africa, grown for the sake of their late, usually pink, flowers; most of the cultivated species with normally pink flowers have developed white-flowered variants. As well as the species, several hybrids, some of uncertain parentage, are grown. Nerines are somewhat tender in most of Europe, but can be grown outdoors or in pots in a greenhouse, in rather poor, well-drained soil in a sunny position; plants in shade do not produce flowers regularly. Propagation is by seed or bulbs.

Literature: Traub, H.P., Review of the genus Nerine Herb., *Plant Life* **23**: supplement (1967).

Leaves. Thread-like: **8**; 1.5–5 mm wide: **3, 4, 9, 10, 11**; 6–9 mm wide: **5, 6, 7**; more than 1 cm wide: **1, 2, 3, 5, 6**. Glaucous: **7**.
Scape and flower-stalks. Both hairless: **1, 2, 3, 5, 6, 7**; scape hairless, flower-stalks hairy: **9, 10**; both hairy: **4, 8, 11**.
Perianth. Radially symmetric: **1**.
Perianth-lobes. More than 4 cm: **2**. Usually with crisped margins: **1, 2, 3, 4, 5, 6, 8, 10, 11**; margins not crisped: **7, 9**.
Stamens and style. Deflexed: **2, 3, 4, 5, 6, 7, 8, 9, 10, 11**; not deflexed: **1**.
Filaments. Without appendages at the base: **1, 2, 3, 4, 5, 6, 7, 8**; with appendages: **9, 10, 11**.

1a. Filaments with appendages at the base 2
b. Filaments without appendages at the base 4
2a. Perianth pure white, lobes with smooth margins **9. pancratioides**
b. Perianth distinctly pink, if sometimes pale, lobes with crisped margins 3
3a. Scape hairy; filament-appendages small, lanceolate, entire **11. masonorum**
b. Scape hairless; filament-appendages to 1 cm, oblong, toothed **10. appendiculata**
4a. Stamens and style erect, not markedly deflexed downwards **1. sarniensis**
b. Stamens and style markedly deflexed downwards 5
5a. Leaves thread-like, terete **8. filifolia**
b. Leaves flat, 1.5–30 mm wide 6
6a. Leaves 1.5–3 cm wide 7
b. Leaves 3–13 mm wide 8
7a. Perianth-lobes 6–7.5 cm, deep pink with a deeper midrib **2. bowdenii**
b. Perianth-lobes 2.5–3 cm, pale pink throughout **3. flexuosa**
8a. Leaves glaucous; perianth white, margins of the lobes not crisped **7. pudica**
b. Leaves bright green; perianth pink or magenta, margins of the lobes crisped 9
9a. Leaves 3–8 mm wide; scape and flower-stalks hairy **4. angustifolia**
b. Leaves 6 mm or more wide; scape and flower-stalks hairless 10
10a. Perianth-lobes 1.7–2 cm, pale pink **5. undulata**
b. Perianth-lobes 2.5–3 cm, bright or rose-pink **6. humilis**

1. N. sarniensis (Linnaeus) Herbert. Illustration: *Botanical Magazine*, 294 (1801), 1089 & 1090 (1808) & 2124 (1820) – varieties; *Flowering plants of South Africa* **9**: t. 355 (1929); Eliovson, Wild flowers of southern Africa, t. 94 (1980).

Bulb more or less spherical, 2–3 cm in diameter. Leaves oblong, obtuse, bright green, 1–2 cm wide. Scape 25–50 cm, erect. Umbel with up to 20 flowers; flower-stalks unequal. Perianth radially symmetric, lobes to 3.5 cm, variously coloured, margins crisped. Filaments without appendages at their bases. Stamens and style erect, not deflexed. Ovary hairless. *South Africa (Cape Province).* H5. Autumn.

A variable species, particularly in the colour and size of the flowers. Traub recognises 8 varieties, some of them unknown in the wild. Var. **sarniensis** has bright crimson flowers; var. **plantii** Baker has dull or rosy crimson flowers and the perianth-lobes are somewhat clawed; var. **venusta** Baker has a scarlet perianth; var. corusca (Ker Gawler) Baker (*N. corusca* (Ker Gawler) Herbert) has large leaves and large, scarlet flowers; var. **rosea** Baker has rose-pink flowers; var. **alba** Traub has white flowers; var. **curvifolia** (Jacquin) Traub (*N. curvifolia* (Jacquin) Herbert) has rather thick, curved leaves and somewhat larger, bright scarlet flowers – forma **fothergillii** (Andrews) Traub (*N. fothergillii* Andrews) is a rather more robust variant; var. **moorei** (Leichtlin) Traub has a shorter scape and bright, scarlet flowers and may be a hybrid between var. *curvifolia* and *N. flexuosa.*

N. sarniensis has been naturalised on the island of Guernsey in the English channel for many years.

2. N. bowdenii Watson (*N. veitchii* Anon.). Illustration: *Botanical Magazine,* 8117 (1907); *Flowering plants of South Africa* **22**: t. 841 (1942); Eliovson, Wild flowers of southern Africa, t. 100 (1980); Rix & Phillips, The bulb book, 170 (1981).

Bulb somewhat elongate, with a conspicuous neck. Leaves 8–30 × 1.5–3 cm, tapering, obtuse. Scape to 45 cm. Umbel with 4–12 flowers; flower-stalks slightly 3-angled in section, to 5 cm. Perianth bilaterally symmetric, pink to deep pink or magenta, the lobes 5–7.5 cm, each with a darker pink stripe along the midrib, margins crisped in the upper half. Stamens and style deflexed. Filaments without appendages at their bases. Ovary hairless. *South Africa (Cape Province, Natal, Orange Free State).* H5. Autumn.

Somewhat variable in flower size and colour; the most widely grown of the species.

3. N. flexuosa (Jacquin) Herbert. Illustration: *Flowering plants of South Africa* **15**: t. 561 (1935).

Very similar to *N. bowdenii* but perianth pale pink or white, lobes 2.5–3 cm. South *Africa (Cape Province, Natal, Orange Free State).* H5. Autumn.

4. N. angustifolia (Baker) Watson, Illustration: *Flowering plants of South Africa* **17**: t. 658 (1937); *Botanical Magazine,* n.s., 244 (1954).

Bulb ovoid with a conspicuous neck. Leaves to 30 cm × 3–5 mm, slightly tapered, obtuse, bright green. Scape 40–70 cm, sparsely covered with short, spreading hairs. Umbel with up to 12 flowers; flower-stalks 1–7 cm, with a dense covering of spreading, gland-tipped hairs. Perianth magenta, bilaterally symmetric, lobes 3–5 cm, somewhat crisped, each with a darker midrib. Stamens and style deflexed. Filaments without appendages at their bases. Ovary covered with gland-tipped hairs. *South Africa.* H5. Autumn.

5. N. undulata (Linnaeus) Herbert. Illustration: *Botanical Magazine,* 369 (1797).

Bulb ovoid. Leaves shorter than the scape, 6–13 mm wide. Scape to 50 cm. Umbel with up to 12 flowers; flower-stalks 2–4 cm. Perianth bilaterally symmetric, pale pink, lobes 1.5–2 cm, with conspicuously crisped margins. Filaments without appendages. Ovary hairless. *South Africa.* H5. Autumn.

6. N. humilis (Jacquin) Herbert. Illustration: *Botanical Magazine,* 726 (1804); *Flowering plants of South Africa* **15**: t. 564 (1935).

Very similar to *N. undulata* but umbel with 10–20 bright or rose-pink flowers, perianth-lobes 2.5–3 cm, less conspicuously crisped. *South Africa (Cape Province).* H5. Autumn.

7. N. pudica J.D. Hooker. Illustration: *Botanical Magazine,* 5901 (1871).

Bulb spherical or elongate. Leaves 20–25 cm × 6–9 mm, glaucous. Scape to 50 cm. Umbel with 4–6 flowers; flower-stalks 2.5–4 cm. Flowers white, each segment with a pink keel; perianth bilaterally symmetric, lobes 2.5–3.5 cm, margins not crisped. Stamens and style deflexed. Filaments without appendages. Ovary hairless. *Origin uncertain, presumably South Africa.* H5. Autumn.

Described from a plant of uncertain origin which was grown at Kew; the species has not been found in the wild.

8. N. filifolia Baker. Illustration: *Botanical Magazine,* 6547 (1881); *Flowering plants of South Africa* **15**: t. 568 (1935); Eliovson, Wild flowers of southern Africa, t. 98 (1980); Brickell (ed.), RHS A–Z encyclopedia of garden plants, 722 (2003).

Bulb spherical with a short neck. Leaves bright green, thread-like, 15–20 cm. Scape to 30 cm, covered with short, spreading, glandular hairs. Umbel with 8–12 flowers; flower stalks 2.5–4 cm, glandular-hairy. Perianth bilaterally symmetric, pink, lobes 2.5–3 cm, margins crisped. Stamens and style deflexed. Ovary minutely hairy. *South Africa (Orange Free State).* H5. Autumn.

9. N. pancratioides Baker.

Bulb ovoid with a short neck. Leaves to 40 cm × *c.* 2 mm, linear, bright green. Scape to 60 cm. Umbel with 10–20 flowers; flower-stalks 2.5–4 cm, densely covered with short, spreading hairs. Perianth bilaterally symmetric, white, lobes 2–2.5 cm, their margins not crisped. Stamens and style deflexed. Filaments alternating with oblong, bifid appendages. Ovary hairy. *South Africa (Natal).* H5. Autumn.

10. N. appendiculata Baker.

Bulb ovoid. Leaves to 45 cm × 2.5–5 mm, channelled, linear. Scape to 80 cm or more. Umbel with 10–20 flowers; flower-stalks 3.5–5 cm, densely covered with short, spreading, gland-tipped hairs. Perianth bilaterally symmetric, pale to deep pink or magenta-pink, lobes 2.5–3 cm with crisped margins. Stamens and style deflexed. Each filament with 2 white, toothed appendages which are up to 1 cm long, at the base. Ovary hairy. *South Africa (Natal)* H5. Autumn.

11. N. masonorum Bolus. Illustration: *Flowering plants of South Africa* **15**: t. 570 (1935); Eliovson, Wild flowers of southern Africa, t. 99 (1980).

Bulb with a short neck. Leaves to 20 cm × 1.5 mm, somewhat thread-like. Scape to 30 cm, with a dense covering of short, spreading hairs. Umbel with 4–15 flowers; flower-stalks 1–3 cm, densely hairy. Perianth bilaterally symmetric, pink, lobes to 1.5 cm, each with a darker pink stripe along the midrib. Stamens and style deflexed. Filaments each with 2 lanceolate, entire appendages at the base. Ovary hairless. *South Africa (Cape Province).* H5. Autumn.

The name was originally, but incorrectly, spelled 'masoniorum'.

17. HAEMANTHUS Linnaeus

J.C.M. Alexander

Deciduous or evergreen perennial herbs. Bulbs ovoid to pear-shaped, often flattened, with fleshy, cream-coloured, pinkish or green scales that are either unequal and 2-ranked with rounded or irregular margins, or almost equal with horizontally truncate, even margins. Leaves usually 2, 4 or 6, basal thick, without a mid-rib, 2-ranked, erect, arched or prostrate, spathulate or strap-shaped to lanceolate, broadly elliptic, oblong or almost circular, hairy or hairless, usually ciliate, green, occasionally with darker green or purple bars or spots. Scape to *c.* 40 cm, appearing at about the same time as the current season's leaves or when leaves are absent, erect or curved, fleshy, often flattened, sometimes ridged or spotted, solid. Umbel with 4–150 flowers, inversely conical to hemispherical or spherical. Spathes 4–13, erect, spreading or bent downwards, stiff and fleshy or membranous. Flowers white to red; perianth-tube cylindric or narrowly bell-shaped; perianth-lobes 6, longer than tube, erect to slightly spreading, oblong to narrowly lanceolate. Corona absent. Stamens 6, borne at top of perianth-tube, usually projecting from flower. Ovary 3-celled, with 1 or 2 ovules per cell. Style slender, about equalling the stamens, minutely 3-lobed at apex. Fruit an ovoid to spherical, pulpy, white, yellow, pink or red berry; seeds fleshy, ovoid, red or green.

A genus now regarded as having 21, often rather variable, species from southern Africa, grown for their striking dense shaving-brush-like heads of white, pink or red flowers. Most species do best with a minimum temperature of *c.* 10 °C during the growing season, but can also be grown under cooler conditions or as house plants if given plenty of light. They should be planted in a mixture of peat and sandy loam with the bulbs completely covered (except Nos 3 & 4), and should only be repotted at infrequent intervals as they resent disturbance. Dilute liquid fertiliser can be applied when the young leaves are actively growing. Water should be given much more sparingly when the plants are resting, and the deciduous species (Nos 1, 2, 5–7) can be allowed to dry out completely. Propagation is easily effected from offsets.

Literature: Friis, I. & Nordal, I., Studies on the genus Haemanthus IV, *Norwegian Journal of Botany* **23**: 63–77 (1976); Snijman, D., A revision of the genus Haemanthus, *Journal of South African Botany*, Supplementary volume **12**: 1–139 (1984).

Plant. Deciduous: **1, 2, 5, 6, 7**; evergreen: **3, 4.**
Bulb-scales. Almost equal, margins horizontal: **1, 2, 3, 4**; unequal, 2-ranked, margins irregular or rounded: **5, 6, 7.** Becoming green in light: **3, 4.**
Leaves. Four or more: **3, 4.** Almost square or circular: **4, 5, 6**; broader than long: **6.** White-spotted: **3**; red-barred: **5.** Rough above: **6**; with spreading, white hairs above: **7.** Glaucous: **5, 7.** Margins red: **5, 7**; margins cartilaginous: **6.**
Scape. Appearing when leaves absent: **5, 6, 7**; appearing with or immediately before leaves: **1, 2, 3, 4.** Appearing beside new leaves (other leaves absent): **1, 2**; appearing between new leaves (other leaves present): **3**; appearing between old leaves (new leaves appear later): **4.** Very hairy: **1.** Furrowed: **6.** Six cm or less: **4**; 4–40 cm: **1, 2, 3, 5, 6, 7.**
Umbel. Dense: **1, 3, 4, 5, 6, 7**; loose: **2, 5, 6.**
Spathes. Four or 5: **7**; 4–9: **2, 3, 4, 5**; 7–10: **1**; 5–11: **6.** Membranous: **1, 2, 6**; stiff or fleshy: **3, 4, 5, 7.** Spreading or curved: **1, 2**; erect and overlapping: **4, 5, 7**; erect but separated: **3, 6, 7.** White: **3, 4**; pink, red or purple: **1, 2, 5, 6, 7.** 1.3–2 cm: **2**; 1.8–4 cm: **1, 3**; 2–7 cm: **5, 6**; 3–8 cm: **4, 7.**
Flowers. White: **1, 3, 4**; pink to red: **1, 2, 5, 6, 7**; red and white: **7.**
Perianth-tube. 3.5 mm or less: **2, 5, 6, 7**; 4 mm or more: **1, 3, 4, 5, 6.**
Perianth-lobes. Less than 2 cm: **2, 7**; more than 2 cm: **4, 5, 6, 7.**
Stamens. Not projecting from flower: **2.**

1a. Leaves present at flowering-time or immediately after, bulb ovoid to spherical, often green; scales almost equal, with horizontally truncate, even margins 2
b. Leaves not present at flowering-time; bulb ovoid to pear-shaped, never green; scales 2-ranked with rounded or irregular margins 5
2a. Evergreen; bulbs green if exposed to light; spathes white, firm and erect 3
b. Deciduous; bulbs not green if exposed to light; spathes pink, membranous, soon bent downwards 4
3a. Leaves floppy, 2.5–11.5 cm broad; scape 5–35 cm; spathes slightly separate **3. albiflos**
b. Leaves firm, 7–25 cm broad; scape less than 6 cm; spathes overlapping at least at base **4. deformis**
4a. Stamens not projecting from flower 5–15 mm; perianth-tube 2.5–3.5 mm **2. carneus**
b. Stamens projecting from flower; perianth-tube 5–12 mm **1. humilis**
5a. Spathes stiff and fleshy, erect and closely overlapping 6
b. Spathes firm but thin, erect to slightly spreading, overlapping at base only or slightly above 7
6a. Leaves often barred with reddish purple near base; scape usually spotted or streaked; spathes shorter than to slightly longer than flowers **5. coccineus**
b. Leaves unmmarked or slightly spotted at base; scape usually unmarked; spathes distinctly longer than flowers **7. pubescens**
7a. Scape unmarked, furrowed, often flattened; leaves rough, unmarked **6. sanguineus**
b. Scape usually spotted or streaked, not furrowed, at most slightly flattened; leaves smooth, usually barred with reddish purple near base **5. coccineus**

1. H. humilis Jacquin subsp. **hirsutus** (Baker) Snijman (*H. hirsutus* Baker; *H. candidus* Bull; *H. allisonii* Baker; *H. nelsonii* Baker; *H. carneus* misapplied). Illustration: *Botanical Magazine*, 9293 (1933); Pearse, Mountain splendour, 55 (1978); Eliovson, Wild flowers of southern Africa, 51 (1980); *Journal of South African Botany*, Supplementary volume **12**: 33 (1984).

Deciduous. Bulb to 8 cm across, ovoid to spherical; scales more or less equal, horizontally truncate. Leaves 2, rarely 3, soft, erect or arched, 15–30 × 5.5–13 cm, lanceolate to narrowly or broadly elliptic or oblong, light green, usually hairy beneath, sometimes above also, ciliate. Scape 15–30 cm, appearing with the new leaves, erect or slightly curved, pale green to reddish purple, usually densely hairy. Umbel dense, hemispherical, with 25–120 flowers; spathes 7–10, membranous, spreading or curved downwards, pink, 2–3 cm × 3–15 mm, narrowly to broadly triangular or lanceolate, acute. Flowers pale pink or white; perianth-tube 5–12 mm; lobes 1–1.7 cm, slightly spreading. Stamens projecting from flower 5–15 mm. Berries greenish white to cream or orange. *Southern Africa.* G1. Late summer.

H. humilis subsp. *humilis* with smaller umbels on slenderer almost hairless scapes is probably not in cultivation.

2. H. carneus Ker Gawler. Illustration: *Edwards's Botanical Register* **6**: t. 509 (1821); *Botanical Magazine*, 3373 (1834).

Similar to *H. humilis* subsp. *hirsutus.* Leaves prostrate, rarely arched. Scape 6–27 cm, appearing with or immediately before the new leaves, erect. Umbel loose, spherical; spathes 5–7, 13–20 × 2–5 mm, narrowly lanceolate. Flowers pink; perianth-tube 2.5–3.5 mm, slightly broader above; perianth-lobes 8.5–10 mm; stamens less than half perianth-lobe length. *Eastern South Africa.* G1. Late summer.

Often regarded as part of *H. humilis.*

3. H. albiflos Jacquin (*H. albomaculatus* Baker; *H. pubescens* misapplied). Illustration: *Botanical Magazine*, 1239 (1809); Batten

& Bokelmann, Wild flowers of the eastern Cape Province, pl. 17, 1 (1966); *Journal of the Royal Horticultural Society* **103**: 439 (1978); Everett (ed.), The New York Botanical Gardens illustrated encyclopedia of horticulture **5**: 1580 (1981).

Evergreen. Bulb to 8 cm across, sometimes flattened; scales more or less equal, horizontally truncate, turning green if exposed to light. Leaves 2, 4 or 6, erect, arched or prostrate, 9–40 × 2.5–11.5 cm, oblong to elliptic, sometimes white-spotted, ciliate, otherwise usually hairless. Scape 5–35 cm, arising between current season's leaves, erect or slightly curved, green, hairless or very slightly hairy. Umbel dense, ovoid or inversely conical, with 25–50 flowers; spathes 4–8, stiff, erect but separated, white with green veins, 1.8–4 cm × 6–30 mm, oblong to broadly ovate, usually ciliate. Flowers white; perianth-tube 4–7 mm; lobes 1–1.8 cm, slightly separated. Stamens projecting by up to 9 mm. Berries white, orange or red. *South Africa.* G1. Early summer.

Hairy-leaved forms known as var. *pubescens* Baker are no longer considered distinct.

4. H. deformis Hooker (*H. baurii* Baker). Illustration: *Botanical Magazine*, 5903 (1871); Gibson, Wild flowers of Natal (coastal region), pl. 13, 2 (1975); *Journal of South African Botany*, Supplementary volume **12**: 65 (1984).

Similar to *H. albiflos.* Leaves 2 or 4, strongly arched, 8–26 × 7–25 cm, very broadly oblong to almost square, unspotted. Scape to 6 cm, arising between previous season's leaves, flattened. Spathes 6 or 7, overlapping, 3–6 × 1.2–4.5 cm. Perianth-tube 7.5–9 mm; lobes 2.1–2.4 cm. Berries orange to red. *Eastern South Africa.* G1. Summer.

Sometimes considered an impoverished form of *H. albiflos*, though cultivation experiments show this not to be the case.

5. H. coccineus Linnaeus (*H. tigrinus* Jacquin; *H. coarctus* Jacquin; not *H. coccineus* Forsskal). Illustration: *Botanical Magazine*, 1075 (1808); Mason, Western Cape sandveld flowers, 58 (1972); *Journal of the Royal Horticultural Society* **100**: 582 (1975); *Journal of South African Botany*, Supplementary volume **12**: 99 & 101 (1984)

Deciduous. Bulb 3–15 cm across, ovoid to pear-shaped, flattened; scales in 2 overlapping rows, thick, with irregular or rounded margins. Leaves 2, rarely 3, firm, erect, arched or prostrate, 15–45 × 8.5–15 cm, strap-shaped to elliptic or almost circular, sometimes slightly glaucous, unmarked or barred with reddish purple, often hairy on backs, margins sometimes red, often ciliate. Scape 6–37 cm, appearing when leaves absent, erect, usually spotted or streaked, usually hairless. Umbel dense or spreading, inversely bell-shaped to conical, with 25–100 flowers; spathes 6–9, stiff and fleshy, rarely thin, erect or slightly spreading, overlapping, pink to scarlet, 2–6 × 1.5–4.8 cm, variously shaped. Flowers pink to scarlet; perianth-tube 2–5 mm, whitish; lobes 1.2–2.6 cm, narrow and spreading. Stamens projecting by up to 1 cm. Berries white to deep pink. *South & South West Africa.* G1. Late summer.

H. × clarkei Anon., the artificial hybrid between *H. coccineus* and *H. albiflos*, but more closely resembling the former, is sometimes grown.

6. H. sanguineus Jacquin (*H. rotundifolius* Ker Gawler; *H. incarnatus* Herbert). Illustration: Batten & Bokelmann, Wild flowers of the eastern Cape Province, pl. 18, 2 (1966); Jackson, Wild flowers of Table mountain, 107 (1977); *Journal of South African Botany*, Supplementary volume **12**: 69 (1984).

Similar to *H. coccineus.* Leaves 2, leathery, prostrate, 9–40 × 3.5–28 cm, often broader than long, dark green and usually rough above, paler, smooth, shiny and sometimes spotted below; margins cartilaginous. Scape to 27 cm, usually furrowed and compressed above, pink to dark red, unmarked and hairless. Spathes 5–11, erect but separated, thin, 20–70 × 7–50 mm, laterals keeled. Perianth-tube 2–9 mm; lobes 1–2.4 cm. *South Africa.* G1. Summer.

7. H. pubescens Linnaeus (*H. albiflos* Jacquin var. *pubescens* misapplied). Illustration: Mason, Western Cape sandveld flowers, 59 (1972); *Journal of South African Botany*, Supplementary volume **12**: 115, 119 & 121 (1984).

Deciduous. Bulb 4.5–8 cm across, ovoid to pear-shaped, flattened; scales in 2 overlapping rows, fleshy, with irregular or rounded margins, sometimes very loosely packed. Leaves 2, rarely 3, arched to prostrate, 10–20 × 1.5–4.5 cm, strap-shaped to slightly spathulate, dark green with white spreading hairs above, paler, less hairy and sometimes spotted below, sometimes glaucous and hairless all over; margin ciliate, sometimes red. Scape 4–28 cm, appearing when leaves absent, pink or red, usually unmarked and hairless. Umbel dense, inversely bell-shaped to conical, with 4–60 flowers. Spathes 4 or 5, erect, stiff and fleshy, red, rarely pink, 3–8 × 1–3 cm, abruptly pointed, sometimes ciliate. Flowers scarlet with white tube and tips, rarely pink; perianth-tube 2–5 mm, with swellings below lobes; lobes 1.3–2.6 cm, narrow and erect. Stamens projecting by up to 1.2 cm. Berries white to pale pink. *South & SW Africa.* G1. Summer.

There has been some confusion between non-flowering plants of *H. pubescens* and hairy-leaved forms of *H. albiflos*, previously separated as var. *pubescens* but no longer considered distinct.

18. SCADOXUS Rafinesque

J.C.M. Alexander

Perennial herbs. Bulb spherical (very narrow in No. 1), borne on a compact rhizome formed from old bulb-bases. Leaves 4–12, basal, spirally arranged (loosely 2-ranked in No. 1), erect or spreading. Leaf-stalks spotted, sheathing, forming a thick false-stem or neck above the bulb (except in No. 1), bearing a few scale-leaves. Leaf-blades elliptic to ovate, obtuse, acute or acuminate, tapered, rounded, truncate or slightly cordate at base, membranous, hairless, with a distinct midrib. Scape lateral in scale-leaf axil (apparently terminal in No. 1), arising with or before the leaves, terete below, angled above, solid, spotted. Spathes 4–many, erect and conspicuous at flowering or withering early. Flowering-head conical, hemispherical or spherical, bearing 10–200 erect (rarely pendulous) flowers. Perianth-tube cylindric; perianth-lobes 6, erect or spreading, linear to lanceolate; tips glandular-hairy. Corona absent. Stamens 6, borne at top of perianth-tube, projecting (except in No. 4); anthers yellow or red. Ovary spherical, 3-celled, with 1 or 2 ovules per cell; style longer than stamens; stigma entire or minutely 3-pointed. Berry fleshy, orange to red, with 1–3 pale seeds.

A genus of 9 species from tropical and southern Africa with 1 reaching Yemen and possibly Oman. Previously considered part of *Haemanthus*, the species are distinguished by their membranous leaves with a distinct midrib, whose bases usually form a thick false-stem. Having a more tropical distribution than *Haemanthus* (p. 218) they require warmer and moister conditions, but otherwise techniques for culture and propagation are similar.

Literature: Friis, I. & Nordal, I., Studies on the genus Haemanthus IV, *Norwegian Journal of Botany* **23**: 63–77 (1976).

1a. Leaf-stalks not sheathing, not forming a false-stem; leaves loosely 2-ranked; scape apparently terminal **1. cinnabarinus**
b. Leaf-stalks sheathing, forming a false-stem; leaves spirally arranged; scape lateral 2
2a. Spathes erect and conspicuous when flowers open **2. puniceus**
b. Spathes withered and drooping when flowers open 3
3a. Perianth-tube longer than 5 mm; perianth-lobes narrower than 5 mm; mature filaments longer than perianth-lobes **3. multiflorus**
b. Perianth-tube 5 mm or less; perianth-lobes broader than 5 mm; mature filaments shorter than perianth-lobes **4. pole-evansii**

1. S. cinnabarinus (Decaisne) Friis & Nordal (*Haemanthus lindenii* Brown). Illustration: *Gardeners' Chronicle* **8**: 437 (1890); de Wit, Plants of the world **2**: pl. 130 (1965); *Journal of the Royal Horticultural Society* **92**: 206 (1967); Maire, Flore d'Afrique central, Amaryllidaceae, 15 (1973).

Bulb 1–2 cm across, narrow and underdeveloped. Leaves 4–8, loosely 2-ranked; leaf-stalks 5–20 cm, not forming a false-stem; leaf-blades 5–20 × 4.5–14 cm, lanceolate to broadly ovate or elliptic, acute or obtuse, attenuate, truncate or almost cordate at base. Scape 15–45 cm, appearing with the leaves and apparently terminal. Spathes numerous, 2–4 cm × 7–11 mm, drooping and soon withering. Flowering-head conical to spherical, with 15–100 orange to pale red or scarlet flowers. Perianth-tube 3–15 mm; perianth-lobes 1–3 cm × 2.5–6 mm, with 5 (rarely 4–7) veins. Berries 1–2 cm, orange. *Sierra Leone to Angola & Uganda*. G2. Spring.

2. S. puniceus (Linnaeus) Friis & Nordal (*Haemanthus magnificus* (Herbert) Herbert; *H. insignis* Hooker; *H. magnificus* var. *insignis* (Hooker) Baker; *H. magnificus* var. *superbus* Baker; *H. natalensis* Poppe; *H. rouperi* Anon.). Illustration: *Botanical Magazine*, 4745 (1853) & 5378 (1863); *Flowering plants of Africa* **43**: pl. 1681 (1971); *Journal of the Royal Horticultural Society* **103**: 439 (1978).

Bulb *c.* 7 cm across. Leaves 7–12; false-stem to 50 cm; leaf-blades to 30 × 12 cm, oblong to elliptic, almost circular when young, often wavy-edged, acuminate, obtuse. Scape to *c.* 40 cm or more, appearing with the leaves. Spathes various; the outer 6 or 7 lanceolate, spathulate or rhombic, 3–10 × 1–5 cm, erect when flowers open, drooping later, reddish brown to green; the inner narrower and paler. Flowering-head conical, with 30–100 yellowish green to pink or scarlet flowers. Perianth-tube 3–12 mm; perianth-lobes linear, erect or almost so, 1.3–2.7 cm × 0.7–2 mm, with 3 (rarely 1–5) veins. Berries 5–10 mm, yellow. *E & southern Africa*. G2. Spring–summer.

'Konig Albert', with taller scapes, spreading perianth-lobes and erect green spathes, is a selection from the cross between *S. multiflorus* subsp. *katherinae* and *S. puniceus*. These hybrids were collectively known as *Haemanthus* × *hybridus* Wittmack or *H.* × *andromeda* Laplace.

3. S. multiflorus (Martyn) Rafinesque.

Bulb to 6.5 cm across. Leaves *c.* 5; false-stem to 60 cm; leaf-blades 11–32 × 4–20 cm, lanceolate to ovate, obtuse, acute or acuminate, tapered to almost cordate at base. Scape to 75 cm, appearing with or before the leaves. Spathes to 6 × 1.5 cm, linear to lanceolate, colourless or reddish, drooping when flowers open. Flowering-head hemispherical to spherical, with 10–200 scarlet flowers fading to pink. Perianth-tube 5–26 mm; perianth-lobes 1.2–3.2 cm × 0.5–5 mm, linear, spreading, with 3 (rarely 1–5) veins. Berries 5–10 mm, orange. *Tropical & southern Africa, Yemen, ?Oman*. G2. Spring–summer.

A widespread and very variable species that has been divided into 3 not very distinct subspecies:

Subsp. **multiflorus** (Haemanthus *sacculus* Phillips; *H. kalbreyeri* Baker; *H. tenuiflorus* Herbert; *H. lynesii* Stapf; *H. coccineus* Forsskal not Linnaeus). Illustration: *Botanical Magazine*, 3870 (1841) & 8975 (1923); *Journal of the Royal Horticultural Society* **94**: 22 (1969); Flora of tropical east Africa, Amaryllidaceae, 6 (1982). Very small to robust. Perianth-tube usually shorter than 1.5 cm; perianth-lobes usually narrower than 2.5 mm. *Distribution as for species.*

Subsp. **katherinae** (Baker) Friis & Nordal (*Haemanthus katherinae* Baker). Illustration: *Botanical Magazine*, 6778 (1884); Graf, Tropica, 63 (1978); Hay et al., Dictionary of indoor plants in colour, edn 2, pl. 262 (1985); Brickell (ed.), RHS A–Z encyclopedia of garden plants, 964 (2003). Plants robust, to 1.2 m. Leaves wavy. Perianth-tube longer than 1.6 cm; perianth-lobes 2.2–4 mm broad. *Eastern S Africa.*

Subsp. **longitubus** (Wright) Friis & Nordal (*Haemanthus longitubus* Wright; *H. mannii* Baker). Illustration: *Botanical Magazine*, 6364 (1878). Plants to 65 cm. Perianth-tube longer than 1.5 cm; perianth-lobes 1.4–3.5 mm broad. *W Africa*. G2.

4. S. pole-evansii Obermeyer. Illustration: *Flowering plants of Africa* **37**: pl. 1452 (1965); *Journal of the Royal Horticultural Society* **87**: 310 (1962); *Botanical Magazine*, n.s., 572 (1970).

False-stem to 85 cm; leaf-blades 24–52 × 3–17 cm. Scape to 1.2 m, bearing 30–70 pink or scarlet flowers. Perianth-tube 4–5 mm; perianth-lobes 2–2.5 cm × 5–10 mm, elliptic, spreading. Stamens 1.2–1.3 cm, shorter than perianth-lobes. Berries 1–2lcm, scarlet. *E Zimbabwe*. G2. Summer.

19. CLIVIA Lindley

A.C. Leslie

Perennials with short rhizomes and thick, fleshy roots. Leaves usually clustered at the end of the rhizome (at the apex of a distinct stem in one species), more or less in 2 ranks, strap-shaped, evergreen. Scapes solid, compressed, 2-edged, bearing umbels of numerous erect or pendulous flowers. Bracts several, the outer spathe-like. Flowers straight or curved, the outer perianth-lobes usually much narrower than the inner; perianth-tube very short. Stamens attached at the throat of tube, curved at base; anthers opening by slits. Stigma shortly 3-lobed. Fruit a red berry. Seeds few, spherical-angular.

A genus of 4 species from South Africa, one of which *(C. miniata)* is popular as a house or conservatory plant, valued for its bold evergreen foliage and umbels of striking orange flowers. They are sometimes listed as species of *Imantophyllum*.

Clivias will succeed in a good, well-drained soil and, once established, an annual top-dressing is all that is required.

During their resting period only a minimum of heat and water is necessary, but both should be increased at the onset of new growth. Flowering is reputed to be enhanced if the plants are underpotted. Propagation is effected by division after flowering or by seed.

The apparent distinctions between species Nos. 1–3 are not always clear-cut in cultivated specimens. Natural or artificial hybridisation may be involved. Species 2 and 3 are reputed not always to be distinguished readily in the wild.

Literature: Wittmack, L., Clivia (Imantophyllum) cyrtanthiflora van Houtte (Clivia nobilis and Clivia miniata), *Gartenflora* **53**: 225–8 (1904).

1a. Plants with a distinct stem **1. caulescens**
b. Plants usually lacking a distinct stem 2
2a. Flowers erect to spreading, broadly funnel-shaped **4. miniata**
b. Flowers spreading to pendulous, narrowly funnel-shaped 3
3a. Leaves gradually tapered at apex; umbels usually with 12–20 flowers **2. gardenii**
b. Leaves abruptly contracted at apex; umbels often with 40–60 flowers **3. nobilis**

1. C. caulescens Dyer. Illustration: *Flowering Plants of South Africa* **23**: t. 891 (1943).

Stems to 45 cm. Leaves 30–90 × 3–5 cm, margin entire, gradually narrowed at apex. Scapes *c.* 30 cm, with umbels of 15–20 pendulous flowers. Perianth narrowly funnel-shaped, straight or slightly curved; lobes 3.5–4 cm, red or pinkish red with yellowish margins and green tips. Stamens equalling or exceeding the perianth; anthers 2–2.5 mm. *South Africa.* G1. Spring.

2. C. gardenii J.D. Hooker. Illustration: *Botanical Magazine*, 4895 (1856).

Plant without stems. Leaves 45–75 × 2.5–4 cm, margin entire or slightly toothed, gradually tapered at apex. Scapes 30–60 cm, with umbels of usually 12–20 pendulous flowers. Perianth narrowly funnel-shaped, often strongly curved; lobes 3.5–5 cm, dull orange or brick red, with yellowish margins above and spreading green tips. Stamens equalling or exceeding the perianth; anthers 2.5–3 mm. *South Africa.* G1. Winter–spring.

3. C. nobilis Lindley (*Imantophyllum aitonii* J.D. Hooker). Illustration: *Edwards's Botanical Register*, t. 1182 (1828); *Botanical Magazine*, 2856 (1828); *Loddiges' Botanical Cabinet*, t. 1906 (1833); Brickell (ed.), RHS A–Z encyclopedia of garden plants, 290 (2003).

Plant without stems. Leaves 30–60 × 2.5–5 cm, margins toothed, abruptly contracted at apex, sometimes slightly notched. Scapes 30–50 cm, with umbels of often 40–60 pendulous flowers. Perianth narrowly funnel-shaped, straight or slightly curved; lobes 2–4 cm, red or yellowish red, with yellow margins and erect or slightly incurved green tips. Stamens equalling or exceeding perianth; anthers *c.* 2 mm. *South Africa.* G1. Spring–summer.

4. C. miniata (Lindley) Bosse. Illustration: *Botanical Magazine*, 4783 (1854); *Flowering Plants of South Africa* **1**: t. 13 (1921) & **11**: t. 411 (1931) – var. *flava*; Brickell (ed.), RHS A–Z encyclopedia of garden plants, 290 (2003).

Plant without stems. Leaves 40–60 × 3.5–6.5 cm, margin slightly toothed, gradually tapered at apex. Scapes 30–45 cm, with umbels of 12–20 erect to spreading flowers. Perianth broadly funnel-shaped, straight; lobes 5–7 cm, bright red or orange (rarely yellow: var. **flava** Phillips), with a yellow or white and yellow base, tips spreading. Stamens shorter than or equalling perianth; anthers 3.5–4.5 mm. *South Africa.* G1. Spring–summer.

There are several cultivars, selected on the basis of flower colour or improved flower size; plants with variegated foliage ('Striata') are also known.

C. × cyrtanthiflora (Van Houtte) Voss; *C. miniata* × *C. nobilis* has arisen in cultivation. It usually has pendulous flowers, ranging in colour from yellow to red, which are intermediate in size and shape between those of the parents.

20. GALANTHUS Linnaeus

C.D. Brickell revised by C.D. Brickell

Perennial herbs with bulbs. Leaves basal, 2, linear, strap-shaped or oblanceolate, enclosed in a tubular, membranous sheath at the base. Leaf vernation may be applanate (flat), explicative (plicate) or supervolute (conduplicate). Spathe of 2 fused bracts. Flowers pendent. Perianth-segments free, the outer 3 acute to more or less obtuse, spathulate or oblanceolate to narrowly obovate or linear, shortly clawed, erect-spreading; the inner half to two-thirds as long as the outer segments, usually notched (emarginate), oblong, spathulate or oblanceolate, tapered to the base, erect, with a green patch around the notch and sometimes also at the base. Considerable emphasis is placed on the shape and size of these green patches in the identification of closely related species. Stamens inserted at the base of the perianth, shorter than the inner segments. Filaments much shorter than the anthers. Anthers basifixed, opening by terminal pores. Style slender, exceeding the anthers; stigma capitate. Capsule ellipsoid, or almost spherical, opening by 3 flaps. Seeds light brown, each with an appendage.

A genus of some 20 species from Europe, Turkey, Iran and the Caucasus, most of which grow well in woodland conditions in alkaline or acid soils given a top-dressing of leaf-mould annually. Many hybrids have occurred in cultivation and numerous clones have been selected and named.

Literature: Beck, G., Die Schneeglöcken, eine monographische Skizze der Gattung Galanthus, *Wiener Illustrierte Garten Zeitung* **19**: 45–58 (1894); Gottlieb-Tannenhain, P. von, Studien über die Formen der Gattung Galanthus, *Abhandlungen der Zoologische-botanische Gesellschaft in Wien* **2**(4): 1–95 (1904); Traub, H.P. & Moldenke, H.N., The tribe Galantheae, *Herbertia* **14**: 85–114 (1947); Stern, F.C., Snowdrops and snowflakes (1956); Schwarz, O., Tentative keys to the wild species of Galanthus L., *Bulletin of the Alpine Garden Society* **31**: 131–6 (1963); Artjushenko, Z.T., Kritichekiy obzor roda Galanthus L., *Botanicheskii Zhurnal* **51**: 1437–51 (1966); Artjushenko, Z.T., Taxonomy of the genus Galanthus L., *RHS Daffodil & Tulip Year Book* **32**: 62–82 (1966); Artjushenko, Z.T., Galanthus L. (Amaryllidaceae) in Greece, *Annales Musei Goulandris* **2**: 9–21 (1974); Webb, D.A., The European species of Galanthus L., *Botanical Journal of the Linnaean Society* **76**: 307–13 (1978); Kamari, G., A biosystematic study of the genus Galanthus L. in Greece, Part I, *Botanische Jahrbücher* **103**: 107–35 (1982) & Part II, *Botanika Chronika* **1**: 60–98 (1981); Brickell, C.D., Galanthus L., in Davis, P.H. (ed.), Flora of Turkey **8**: 365–72 (1984); Davis, A.P., The genus Galanthus (1999); Davis, A.P., Galanthus L., in Guner, A., Ekim, T. & Baser, K.H.C. (eds), Flora of Turkey **11**: 265–71 (2001); Bishop M., Davis,

A.P. & Grimshaw, J., Snowdrops, a monograph of cultivated Galanthus (2001).

1a. Leaf vernation explicative **1. plicatus**
b. Leaf vernation applanate or supervolute 2
2a. Leaf vernation applanate 3
b. Leaf vernation supervolute 7
3a. Inner perianth segments with both apical and basal green patches **2. gracilis**
b. Inner perianth segments with only an apical green patch 4
4a. Upper leaf surface with a conspicuous grey median stripe **3. reginae-olgae**
b. Upper leaf surface without a conspicuous median stripe 5
5a. Leaves glaucescent to glaucous **4. nivalis**
b. Leaves glossy to matt green 6
6a. Upper leaf surface glossy green, rarely matt, lacking a median stripe **5. lagodechianus**
b. Upper leaf surface matt green, occasionally slightly glossy, with a pale median stripe **6. rizehensis**
7a. Leaves supervolute, glaucous 8
b. Leaves supervolute, deep to pale green 9
8a. Inner perianth segments with both apical and basal green patches, sometimes coalescing **7. elwesii**
b. Inner perianth segments with only an apical green patch **8. alpinus**
9a. Inner perianth segments with both apical and basal green patches **9. fosteri**
b. Inner perianth segments with only an apical green patch 10
10a. Leaves matt green with the apical green patch on the inner perianth segments at least half the length of the segment **10. ikariae**
b. Leaves glossy green, sometimes pale green, with the apical mark on the inner perianth segments less than half the length of the segment **11. woronowii**

1. G. plicatus Bieberstein.
Bulb almost spherical to ovoid, 2–4.5 × 1.7–2.8 cm. Leaves narrowly oblanceolate to strap-shaped, explicative (plicate), 5–21 × 1–1.5 cm at flowering, to 30 × 2 cm, upright to recurved at maturity, glaucous above and often with a paler central band on the upper surface, green to glaucescent on the underside, apex obtuse, hooded. Scape 9–22 cm. Outer perianth-segments convex, elliptic to obovate, 1.7–2.8 cm × 6–12 mm; inner segments flat, not flared at the apex, oblong-spathulate, 8–10 × 4–7 mm, with separate green patches at base and/or apex, these sometimes joined in the centre. Filaments 0.5–1 mm, anthers 6–7 mm. Capsule ellipsoid to almost spherical, to 1.6 × 1.2 cm. *N Turkey, E Romania, Former USSR (Crimea)*. H2. Early spring.

Two subspecies are recognised:

Subsp. **plicatus**. Illustration: Stern, Snowdrops and snowflakes, f. 10 (1956); Rix & Phillips, The bulb book, 12 (1981); Grey-Wilson & Mathew, Bulbs, pl. 22 (1981); Davis, The genus Galanthus, pl. 3, 26 & 27, (1999). Bishop et al. Snowdrops, pl. 14 (2001). Inner perianth-segments with apical green patch only. *N Turkey, E Romania, Former USSR (Crimea)*.

Subsp. **byzantinus** (Baker) Webb (*G. byzantinus* Baker). Illustration: Stern, Snowdrops and snowflakes, f. 11 (1956); Rix & Phillips, The bulb book, 15 (1981); Mathew & Baytop, Bulbous plants of Turkey, f. 1 (1984); Davis, The genus Galanthus, pl. 28 (1999); Bishop et al., Snowdrops, pl. 14 (2001). Inner perianth-segments with both apical and basal green patches, sometimes coalescing in the centre. *NW Turkey*.

2. G. gracilis Celakovsky (*G. bulgaricus* Velenovsky; *G.graecus* misapplied; *G. elwesii* J.D. Hooker subsp. *minor* Webb; ?*G. elwesii* var. *stenophyllus* Kamari). Illustration: Stern, Snowdrops and snowflakes, f. 9 (1956) – as *G. graecus*; Rix & Phillips, The bulb book, f. 13 (1981); Davis, The genus Galanthus, pl. 4, 30–2 (1999); Bishop et al., Snowdrops, pl. 17 (2001).

Bulb ovoid to almost spherical, 1.4–2.8 cm × 8–20 mm. Leaves linear, applanate, occasionally very narrowly oblanceolate, (3)5–15(21) cm × 3–9 mm at flowering, to 10–25 cm × 3–10 mm at maturity, frequently distinctly twisted (occasionally only slightly so), glaucous to glaucescent, apex obtuse. Scape 6–14 cm, glaucous. Outer perianth segments convex, narrowly to broadly obovate, 1.5–2.6 cm × 7–10(14) mm; inner segments flat, noticeably flared at apex, broadly obovate, 7–12 × 2–3.5 mm, with separate green patches at the apex and base, rarely joined in the centre. Filaments 1–2 mm, anthers 4–5 mm. Capsule ovoid to almost spherical, 1.1–1.6 × 1–1.4 cm. *Bulgaria, NE Greece, W Aegean Islands, Romania, S Ukraine, W Turkey*. H2. Early spring.

Two autumn–winter flowering species from Turkey, *G.cilicicus* and *G. peshmenii*, with glaucous or glaucescent, linear, applanate leaves, are closely related to *G. gracilis* (vide Davis, loc. cit. p. 131–2) although more frequently linked with *G. nivalis* and *G. reginae-olgae*:

G. cilicicus Baker (*G. nivalis* L. **subsp.** *cilicicus* (Baker) Gottlieb-Tannenheim). Illustration: *Gardeners' Chronicle* **23**: 79, f. 29 (1898); Stern, Snowdrops and snowflakes, 28, f. 5 (1956); Davis, The genus Galanthus, pl. 5 & 29 (1999). Differs from *G. gracilis* in flowering autumn to winter (November–January), with glaucous leaves 7–15 cm long at flowering, developing to 18–25 cm and recurving when mature, the inner perianth segments only having a single, apical, green patch. *SW Turkey*. H3–4.

G. peshmenii A.P. Davis & C.D. Brickell. Illustration: *The New Plantsman* **1**: 14, f. 1 (1994); Davis, The genus Galanthus, pl. 6, 33 (1999); Bishop et al., Snowdrops, pl. 18 (2001). Differs from *G. gracilis* in flowering from October to December, with the glaucescent to glaucous leaves absent or rarely to 5–6 cm at flowering, developing to *c.* 25 cm when mature; the inner perianth segments also only have a single, apical patch. It is distinguished from both *G. gracilis* and *G cilicicus* by the leaf anatomy, having a very well developed palisade cell layer in the leaves, lacking in both these species. *Greece (Kastelorrhizo), SW Turkey*. H3–4.

3. G. reginae-olgae Orphanides (*G. olgae* Boissier; *G. nivalis* Linnaeus subsp. *reginae-olgae* (Orphanides) Gottlieb-Tannenhain; *G. corcyrensis* (Beck) Stern). Illustration: Stern, Snowdrops and snowflakes, pl. 4 & f. 6, 7 (1956); Rix & Phillips, The bulb book, 183 (1981).

Bulb ovoid (rarely almost spherical), 1.5–3 × 1.5–2 cm. Leaves linear, applanate but with margins slightly revolute at the base as the leaves emerge, absent or to 11cm × 3–4 mm at flowering, to 30 × 3.5 cm and recurved at maturity, obtuse to acute at apex, upper surface dull green with a central, glaucous stripe, lower surface glaucous. Scape 5–22 cm. Outer perianth-segments convex, obovate to narrowly obovate, 1.5–2.5 cm × 5–12 mm; inner segments flat, not flared at the apex, cuneate or narrowly obovate, 9–12 × 4–7 mm, each with a green patch at the apex. Filaments 0.5–1 mm, anthers 3–4 mm. Capsule spherical to broadly ellipsoid, to 1.4 × 1.2–1.4 cm. H3. Autumn–early spring.

Two poorly defined subspecies are recognised, differing only in flowering period and leaf emergence:

Subsp. **reginae-olgae** (illustration: Davis, The genus Galanthus, pl. 2, 24 & 25 (1999); Bishop et al., Snowdrops, pl. 8, 2001) flowers October–November (December), the leaves absent or much shorter than the scape at flowering. *W Turkey, Greece, Corfu & Sicily.*

Subsp. **vernalis** Kamari (illustration: Davis, The genus Galanthus, pl. 22 & 23, 1999) is spring-flowering, the leaves always well-developed at flowering. In some seasons the flowering periods overlap. *Greece, Corfu, Sicily, Albania & Macedonia.*

4. G. nivalis Linnaeus. Illustrations: Polunin, Flowers of Europe, f. 173 (1969); Rix & Phillips, The bulb book, 13, f. h; Davis, The genus Galanthus, pl. 1, 20 & 21 (1999).

Bulb ovoid to almost spherical, 1.5–2.7 × 1–2.4 cm. Leaves linear, sometimes slightly broader in the upper third, applanate but occasionally slightly revolute at the base, 9–21 cm × 6–9 mm at flowering, to 30 × 1 cm at maturity, glaucescent, sometimes glaucous, apex obtuse to acute. Scape 11–18 cm. Outer perianth-segments convex, broadly obovate to narrowly obovate, 1.5–2.8 cm × 6–11 mm; inner perianth segments flat, not flared at the apex, narrowly obovate to cuneate, 7–12 × 4–6 mm, each with a green patch at the apex. Filaments 1–2 mm, anthers 5 mm. Capsule spherical or almost so, 1.4 × 1.2–1.4 cm. *Most of Europe (wild and naturalised).* H2. Early spring.

This species is extremely variable. Variants with large flowers from Italy have been described as subsp. **imperati** (Bertoloni) Baker; those with double flowers (forma *pleniflorus* Sell) are commonly naturalised and variants with separate spathes, sometimes elongated, have been described as var. **scharlockii** Caspary; other variants have yellow ovaries and yellow markings on the inner segments and have been called var. *flavescens* Mallett and also, by various authors, *G. nivalis* var. *lutescens* or *G. lutescens.* The earliest validly published epithet for these yellow-marked snowdrops is, however, var. **sandersii** Harpur-Crewe also referred to as Sandersii Group.

G. angustifolius Koss (*G. nivalis* subsp. *angustifolius* (Koss) Artjushenko). Illustration: Davis, The genus Galanthus, pl. 11, 39. (1999). Very similar to *G. nivalis* but leaves always glaucous, very narrow, 2–7 mm wide at maturity. The leaf anatomy indicates a close relationship with *G. alpinus* Sosn. *Northern Caucasus.* H2. Spring (March–May.)

G. trojanus A.P. Davis & N. Ozhatay. Illustration: Bishop et al., Snowdrops, pl. 28. (2001). Intermediate between *G. nivalis* and *G. rizehensis* but differing from both in the more robust habit and in the broader green, not glaucous or glaucescent, leaves, which are up to 24 cm long and to 1.8 cm wide. The green apical marks on the inner perianth segments also differ in shape and size from *G. nivalis and G. rizehensis.* NW Turkey. H2. Late winter–spring (March in the wild; January–February in cultivation.)

5. G. lagodechianus Kemularia-Nathadze (*G. cabardensis* Koss; *G. ketzkhovellii* Kemularia-Natadze; *G. kemulariae* Kuthaladze). Illustration: *Alpine Garden Society Bulletin* **30**: 2, 125. (1962); Davis, The genus Galanthus, pl. 17, 45, 46. (1999).

Bulb spherical to ovoid, 1–2.5 × 1.2–3 cm. Leaves linear, applanate, 8–18 × 0.5–1 cm at flowering, 15–30(40) × 0.6–1.3 cm at maturity, bright to dark green, matt to glossy, occasionally slightly glaucescent, apex acute. Scape 5–15 cm. Outer perianth segments convex, narrowly obovate, 1.5–3.0 × 1.0 × 1.5 cm; inner perianth segments flat, obovate, 0.8–1.2 cm × 3–6 mm, each with a green patch at the apex. Filaments 0.5 mm, anthers 4–5 mm. Capsule spherical to ellipsoid, 1–2 cm. *Caucasus, Transcaucasus (Armenia, Azerbaijan, Georgia, S Russia).* H3. Late winter–spring (January–April).

G. artjuschenkoae Gabrieljan. Illustration: *Flora, Vegetation and Plant Resources* **12**: 14 (1999). Closely related to *G. lagodechianus* but with matt, dark green, applanate leaves and slightly larger, more elongated flowers. It has been confused with *G. transcaucasicus* Fomin, which has supervolute, not applanate, leaves. *Armenia.* H3. February–March (April).

6. G. rizehensis Stern. Illustration: Stern, Snowdrops and snowflakes, f. 8 (1956); Rix & Phillips, The bulb book, 16 (1981); Mathew & Baytop, Bulbous plants of Turkey, f. 2 (1984); Davis, The genus Galanthus, pl. 16, 43 & 44 (1999).

Bulb ovoid to almost spherical, 1.5–3 × 0.8–1.5 cm. Leaves linear, applanate, 5–16 cm × 3–10 mm at flowering, 20–30 × 4–10(14) mm, erect to curved at maturity, deep green, matt, sometimes glossy, with a pale median stripe, apex obtuse. Scape 8–20 cm. Outer perianth segments slightly convex, narrowly obovate, 1.5–2 cm × 6–8(10) mm; inner perianth segments flat, not flared at the apex, obovate to cuneate, 0.8–1 cm × 4–6 mm, each with an apical green patch. Filaments *c.* 0.5 mm, anthers *c.* 5 mm. Capsule almost spherical, 0.7–8 × 1–1.5 cm. H3. *Western Transcaucasus (Georgia, S Russia), NE Turkey.* Late winter to spring (January–April).

7. G. elwesii J.D. Hooker (*G. graecus* Boissier; *G. maximus* Velenovsky; *G. elwesii* var. *whittallii* Moon). Illustration: *Botanical Magazine*, 6166 (1875); *The Garden* **57**: 45 (1900); Rix & Phillips, The bulb book, 12, 13 (1981); Davis, The genus Galanthus, pl.7, 8 & 33–35 (1999); Bishop et al., Snowdrops, pl. 10, 21 (2001).

Bulb almost spherical to ovoid, 2.1–3.5 cm. Leaves narrowly oblanceolate, supervolute, 5–32 × 0.6–3.5 cm at flowering, to 15–32(36) × 2.5–3.6 cm at maturity, erect, sometimes slightly twisted, glaucous, apex obtuse, hooded. Scape 12–20(28) cm. Outer perianth segments obovate to broadly obovate, 1.5–2.7 × 1–1.9 cm; inner perianth segments obovate to cuneate, 1–1.2 × 0.6–0.8 cm, with separate apical and basal green patches, sometimes coalescing at the centre. Filaments 1 mm, anthers 5–6 mm. Capsule almost spherical, 1–2 cm. *Bulgaria, NE Greece, Macedonia and Serbia, eastern Aegean Islands, Ukraine, Turkey.* H2. Spring.

Var. **monostictus** P.D. Sell (named from cultivated plants) has been applied to narrow-leaved variants with a single apical green patch on the inner perianth segments. These have long been known as *G. caucasicus* Baker, which was also named from cultivated plants. Wild populations in which all the plants have narrow leaves and single apical green marks on the inner perianth segments have been recorded from southern Georgia and near Antalya, Turkey. Illustration: Bishop et al., Snowdrops, pl. 22 (2001).

8. G. alpinus Sosn. (*G. caucasicus* (Baker) Grossheim; *G. schaoricus* Kemularia-Nathadze). Illustration: *RHS Daffodil and*

Tulip Year Book **30**: 73, f. 21 (1967); Davis, The genus Galanthus, pl. 9, 37 (1999).

Bulb ovoid to spherical, 2.1–3 × 1.5–2.0 cm, white or yellowish-white. Leaves narrowly oblanceolate, supervolute, 2.5–15 × 0.6–2.5 cm at flowering, 12–25 × 1–2.5 cm at maturity, glaucous, apex acute to obtuse, often cucullate. Scape 8–16 cm. Flowers globose-ellipsoid. Outer perianth segments broadly obovate, 1–2 × 0.8–1.2 cm; inner perianth segments obovate, 0.6–1 cm × 4–5 mm, each with a single, green, apical patch. Filaments 1 mm, anthers 4–5 mm. Capsule almost spherical, 0.7–1 cm. *Caucasus (Russia, Georgia, Armenia, Azerbaijan), NE Turkey.* H2. Spring (February–May).

Two varieties are recognized:

Var. **alpinus** is diploid and has whitish bulb-scales, broadly ellipsoid to globose flowers and produces fertile seed. It is widespread throughout the range of the species.

Var. **bortkewitschianus** (Koss) A.P. Davis (illustration: Davis, The genus Galanthus, pl. 9, 2001) is triploid and has yellowish bulb-scales, globose flowers and is sterile. *Only occurring in one site in the western Caucasus (Karbardino-Balcaria).*

G. koenenianus Lobin, C.D. Brickell & A.P.Davis. Illustration: Davis, The Genus Galanthus, pl. 10, 40 & 41 (1999). Closely related to *G. alpinus* var. *alpinus* but differing in having deeply furrowed abaxial leaf surfaces with usually faint yellowish-green marks at the bases of the inner perianth segments. *NE Turkey (Gumushane Province).* H2. Spring (February–March).

9. G. fosteri Baker (*G. latifolius* Ruprecht forma *fosteri* (Baker) Beck). Illustration: *Wiener Illustrierte Garten Zeitung* **19**: t. 17, f. 17 (1894); *Addisonia* **16**: pl. 513 (1931); Rix & Phillips, The bulb book, 16 (1981); Davis, The genus Galanthus, pl. 12, 42 (1999); Bishop et al., Snowdrops, pl. 6, 23 (2001).

Bulb almost spherical to ovoid, 2–2.5 × 1.5–2.2 cm. Leaves narrowly to broadly oblanceolate to strap-shaped, supervolute, (5)8–12(14) × 1–1.5 cm at flowering, to 25 × 2.2 cm and often recurved at maturity, bright, deep green, glossy or matt, apex obtuse, not or slightly hooded. Scape 8–23 cm. Outer perianth segments slightly convex, elliptic to narrowly obovate, 1.9–2.6 cm × 9–12(15) mm; inner perianth segments flat, slightly flared at the apex, obovate to cuneate, 8–14 × 3–8 mm, with separate green patches at the base and apex. Filaments *c.* 1 mm, anthers 4–5 mm. Capsule almost spherical to obovoid, 1–1.5 × 1 cm. *?Israel, Jordan, Lebanon, Syria & Turkey.* H5. Early spring (January–April).

Bulb-frame treatment or an open, well-drained, sunny site is required for successful cultivation.

10. G. ikariae Baker (*G. ikariae* subsp. *snogerupii* Kamari). Illustration: *Botanical Magazine*, 9474 (1937); Stern, Snowdrops and snowflakes, f. 12, 13 (1956); Rix & Phillips, The bulb book, 14, 15 (1981); Davis, The genus Galanthus, pl. 13, 50 & 51 (1999); Bishop et al., Snowdrops, pl. 24 (2001).

Bulb spherical to ovoid, 1.8–3.5 × 1.5–2.8 cm. Leaves strap-shaped to narrowly oblanceolate, supervolute, 5–16 cm × 5–20 mm at flowering, 25–30 cm × 5–30 mm at maturity, often recurved, bright or deep green, matt, rarely slightly glaucescent, apex obtuse, hooded. Scape 11–25 cm. Outer perianth segments convex, narrowly obovate, 1.7–2.6(3) × 0.8–1.6 cm; inner perianth segments flat, not flared at apex, narrowly obovate, 1–1.4 × 0.5–0.9 cm, each with an apical, green patch covering at least half the segment. Filaments 0.5–1.5 mm, anthers 5–6 mm. Capsule spherical or almost so, 1.5 × 1.4 cm. *Aegean Islands of Greece.* H3. Spring.

Much confused with *G. woronowii* Losinskaya and incorrectly known as *G. latifolius* Ruprecht (a synonym of *G. platyphyllus* Traub & Moldenke) and *G. ikariae* subsp. *latifolius*.

G. platyphyllus Traub & Moldenke (*G. latifolius* Ruprecht; *G.ikariae* Baker subsp. *latifolius* Stern *pro parte*). Illustration: Rix & Phillips, The bulb book, 15 (1981); Davis, The genus Galanthus, pl. 19, 55 & 56 (1999). Alpine species from the central Caucasus with supervolute, dark green leaves; distinctive in having blunt (not apiculate) anthers, obtuse inner perianth segments that are marginate (lacking an apical notch), and elongated, not more or less spherical, bulbs. *Caucasus (Georgia, Abkhazia & S Russia).* H2. Late spring to summer in the wild (April–July); spring (April–May) in cultivation.

G. krasnovii Khokhrjakov (*G. valentinae* Panjut). Illustration: Davis, The genus Galanthus, pl. 18, 52–4 (1999). Closely related to and very similar vegetatively to *G. platyphyllus*. The inner perianth segments are marginate as in that species but are elliptic to oblanceolate with an acute apex. The anther apex of *G. krasnovii* is apiculate not blunt as in *G. platyphyllus*. Georgia (Abkhazia, Adzhariya). H2. Spring–summer (March–May).

11. G. woronowii Losinskya (*G. ikariae* Baker subsp. *latifolius* Stern *pro parte*; *G. latifolius auct.* non Ruprecht; *G. ikariae auct.* non Baker *pro parte*). Illustration: *Botanical Magazine*, **164**: table 9669 (1946); *RHS Daffodil and Tulip Year Book* **30**: 70, f. 5 (1967); Davis, The genus Galanthus, pl. 14, 49; Bishop et al., Snowdrops, pl. 25 (2001).

Bulb almost spherical to ovoid, 2–2.5 × 1.5–1.7 cm. Leaves strap-shaped to narrowly oblanceolate, supervolute, 8–20 (25) × (0.8)1–2(3) cm at flowering, 12–30 × 1–3 cm when mature, glossy mid to dark green, sometimes light green, apex obtuse. Scape 6–20 cm. Outer perianth segments narrowly obovate, 1.5–2.5 × 0.7–1.4 cm; inner perianth segments 0.7–1.2 × 0.3–0.6 cm, each with a green apical patch. Filaments 1–1.5 mm, anthers 5–6 mm. Capsule spherical, 1–1.5 cm. *Caucasus, Transcaucasus (S Russia, Georgia) & northeastern Turkey.* H2. Winter–spring (January–April).

G. transcaucasicus Fomin (*G. nivalis* L. var. *caspius* Ruprecht). Illustration: Wendelbo, Tulips and Irises of Iran: f. 57 (1977); Davis, The genus Galanthus, pl. 15, 47 & 48. Similar to *G. woronowii* but with matt green, slightly glaucous, fairly wide leaves, 1–1.5 up to 2.5 cm. *Armenia, Azerbaijan & northern Iran.* H2. Winter–spring (December–April).

21. LAPIEDRA Lagasca

J. Cullen

Perennial herb with a bulb. Leaves few, basal. Flowers erect in an umbel of 4–9; spathe of 2 bracts. Perianth-segments all similar, free, spreading. Filaments as long as anthers; anthers basifixed, opening by pores. Capsule flattened-spherical, 3-lobed, surrounded by the persistent, withered perianth. Seeds few.

A genus of 2 species from the western Mediterranean area. Cultivation as for *Galanthus* (p. 222).

1. L. martinezii Lagasca. Illustration: Polunin & Smythies, Flowers of south-west Europe, 389 (1973).

Bulb to 5 cm in diameter. Leaves to 25 × 1 cm, appearing after the flowers, each with a pale band on the upper surface.

Perianth-segments 8–12 mm, white, each with a green stripe on the outside. *S Spain, N Africa.* H5. Late summer.

22. LEUCOJUM Linnaeus
D.A. Webb
Perennial herbs with bulbs. Flowers bell-shaped, nodding, white or pink, solitary or in umbels of up to 5 (rarely 7), subtended by a spathe of 1 or 2 bracts. Perianth-tube and corona absent. Perianth-segments free, all more or less alike. Anthers blunt at the tip, not pointed as in *Galanthus*, opening by pores. Capsule erect, more or less spherical; seeds numerous.

About 10 species in S, W & C Europe, NW Africa and SW Asia. *L. vernum* is easily grown in sun or semi-shade in a reasonably moisture-retentive soil. *L. aestivum* needs more moisture than do other species, and should be grown in semi-shade in a fairly heavy soil. The other species, though hardy in the warmest parts of Europe, are, on account of their small size and delicate habit, best cultivated in a cool greenhouse in pans of light, sandy soil.

Literature: Stern, F.C., Snowdrops and snowflakes (1956).

Flowering season. Late winter to early summer: **1, 2, 3, 4**; late summer or autumn: **5, 6**.
Leaves. 7–25 mm broad: **1, 2**; not more than 3 mm broad: **3, 4, 5, 6**.
Spathe. One bract: **1, 2, 5**; 2 bracts: **3, 4, 6**.

1a. Leaves at least 7 mm broad; scape stout, usually hollow 2
b. Leaves not more than 3 mm broad; scape very slender, solid 3
2a. Flowers in umbels of 2–7 (very rarely solitary); seeds black, without an appendage **1. aestivum**
b. Flowers solitary or in pairs; seeds white, with an appendage **2. vernum**
3a. Flowering in late summer or autumn; leaves appearing during or after the flowering period 4
b. Flowering in late winter or spring; leaves well developed before the flowers open 5
4a. Flowers usually solitary; flower-stalk not more than 5 mm; spathes 2 **6. roseum**
b. Flowers often in umbels of 2 or 3; longer flower-stalks 1.5–2.5 cm; spathe 1 **5. autumnale**
5a. Perianth-segments 1.3–2.5 cm **3. trichophyllum**
b. Perianth-segments 8–13 mm **4. nicaeense**

1. L. aestivum Linnaeus. Illustration: *Loddiges' Botanical Cabinet* **15**: 1478 (1829); Ross-Craig, Drawings of British plants **29**: t. 8 (1972); Rix & Phillips, The bulb book, 77 (1981).

Bulb to 4 cm. Leaves strap-shaped, *c.* 40 × 1.5 cm. Scape about as long as leaves, hollow, 2-winged. Spathe 1, 3–5 cm, green. Umbel with 3–5 (rarely 7) flowers; perianth-segments 1.5–2 cm, white tipped with green. Seeds black, without an appendage. *W, C & S Europe, eastwards to Turkey and the Caucasian region.* H2. Late spring–early summer.

The above description applies to the typical plant, subsp. *aestivum*. Subsp. **pulchellum** (Salisbury) Briquet, which is perhaps commoner in cultivation, is less robust and smaller in most of its parts, with only 2–4 flowers in the umbel and perianth-segments 1–1.4 cm. It is native to the W Mediterranean region, and flowers 3–4 weeks earlier than subsp. *aestivum*. A robust, large-flowered clone of subsp. *aestivum* is cultivated under the name of 'Gravetye'.

2. L. vernum Linnaeus. Illustration: *Botanical Magazine*, 1993 (1817); Javorka & Czapody, Iconographia florae hungaricae, f. 736 (1934); Rix & Phillips, The bulb book, 12 (1981).

Bulb *c.* 2.5 cm. Leaves to 25 cm × 8–25 mm, appearing with or just after the flowers. Scape 15–30 cm, slightly hollow, narrowly 2-winged, bearing usually a single flower (less often 2) and a spathe of 1 bract about equalling the flower-stalk. Perianth 1.5–2.5 cm, white, with a green or yellow spot near the tip of each segment. Seeds 7 mm, whitish, with an appendage. *Europe, from Belgium and Poland to the Pyrenees and Yugoslavia.* H2. Spring.

The plants with yellow spots on the perianth are sometimes distinguished as var. **carpathicum** Sweet. Var. **vagneri** Stapf is a robust variant with normally 2 flowers on each scape.

3. L. trichophyllum Schousboe (*L. grandiflorum* de Candolle). Illustration: *Botanical Magazine*, 9585 (1940); Maire, Flore de l'Afrique du nord **6**: 14 (1959); Rix & Phillips, The bulb book, 37 (1981).

Bulb *c.* 1.5 cm. Leaves to 30 cm × *c.* 1.5 mm. Scape slightly shorter, very slender, bearing 2–4 flowers. Flower-stalks to 5 cm, usually longer than the 2 spathes. Perianth 1.3–2.5 cm, white, variably tinged with pink, especially at the base, but sometimes pure white. Seeds 1.5 mm, black, without an appendage. Spain, *Portugal & Morocco.* H4–5. Spring.

Wild plants vary greatly in the size of their flowers, but the variation is continuous. Those in cultivation usually represent the large-flowered variant formerly distinguished as var. *grandiflorum* (de Candolle) Willkomm.

4. L. nicaeense Ardoino (*L. hiemale* de Candolle in part). Illustration: *Botanical Magazine*, 6711 (1883); Bicknell, Flowering plants of the Riviera, t. 70 (1885).

Bulb *c.* 2 cm. Leaves to 30 cm × 2 mm. Scape somewhat shorter, slender, bearing 1 flower (rarely 2 or 3). Flower-stalk *c.* 1.5 cm, equalling or shorter than the 2 spathes. Perianth 9–13 mm, white. Seeds 2–3 mm, black, with an appendage. *Coastal region of SE France, from Nice eastwards (almost extinct).* H5. Spring.

L. longifolium (Roemer) Grenier. Differs chiefly in its smaller bulb, longer scape with 1–4 flowers, flower-stalks usually exceeding the spathe, and seeds without an appendage. *Corsica.* H5.

5. L. autumnale Linnaeus. Illustration: Botanical Magazine, 960 (1806); *Gartenflora* **36**: t. 1261 (1886); Brickell (ed.), RHS A–Z encyclopedia of garden plants, 622 (2003).

Bulb *c.* 10 mm. Leaves to 15 cm, very slender, appearing during or just after the flowering period. Scape to 15 (rarely 25) cm, slender, bearing 1–4 flowers. Flower-stalks to 2.5 cm, usually exceeding the spathe. Perianth 9–15 mm, white, or tinged with pink, especially at the base. Seeds 1 mm, black, without an appendage. *W Mediterranean region.* H4–5. Autumn.

6. L. roseum Martin. Illustration: Stern, Snowdrops and snowflakes, t. 13 (1956).

Like *L. autumnale* but with spathes 2, shorter flower-stalks and smaller flowers, which are usually solitary and always pink. *Corsica, Sardinia.* H5. Autumn.

23. CHLIDANTHUS Herbert
S.G. Knees
Perennial herbs with bulbs and thick, fleshy roots; mature bulbs 3–4 cm in diameter. Leaves 5 or 6, linear, obtuse,

15–40 cm × 6–9 mm, appearing after flowers, surrounded by a papery sheath that is often up to one-third as long as the leaves. Scape solid, 12–30 cm, bracts 2, 7–9 cm, papery, soon withering, subtending an umbel of 3–5 stalked flowers, with stalks 9–12 mm. Flowers fragrant, bisexual, almost radially symmetric, perianth 4–7 cm with 6 yellow lobes, a tube at base, and corona small. Anthers basifixed. Ovary 3-celled, style 3-branched, projecting from the perianth. Fruit a capsule containing several black, flattened seeds.

A genus of 1 (possibly 3) species, native to southern Peru, tender in much of northern Europe; hardy in southern Europe. Bulbs should be potted in spring in a well-drained compost and then plunged in a frame. The pots should be lifted in autumn and kept dry in frost-free conditions until the following spring. Plants in permanent sites will benefit from a top dressing in spring. Propagation is by seeds or offsets.

Literature: Ravenna, P.F., Contributions to South American Amaryllidaceae VI, *Plant Life* **30**: 71–3 (1974).

1. C. fragrans Herbert. Figure 8(10), p. 208. Illustration: *Edwards's Botanical Register* **8**: t. 640 (1822); *Flore des serres* **4**: t. 326 (1848); Brickell (ed.), RHS A–Z encyclopedia of garden plants, 264 (2003). *Southern Peru.* H4. Spring.

24. VAGARIA Herbert

S.G. Knees

Perennial herbs with white fleshy roots and narrow, flask-shaped bulbs that are clustered together and covered to the neck with a papery brown tunic, often sheathing the scape and leaf bases for up to 15 cm. Leaves 4–7, linear, 10–20 cm × 8–12 mm, extending after flowering to 40–60 cm, obtuse and with a conspicuous median white band. Scape 15–25 cm, somewhat flattened and slightly 4-angled, solid. Bracts 2–4, papery, 3–5 cm long. Flowers bisexual, 6–9 per umbel, stalks 5–20 mm, apparently extending after flowering. Perianth-tube *c.* 1 cm, lobes 6, each narrowly elliptic to lanceolate with hooded apex, 2.5–3.5 cm, white with a conspicuous green midrib, somewhat keeled beneath. Corona arising at throat of perianth, consisting of 12 linear, acute teeth, occurring in pairs at the bases of the filaments. Anthers 6, versatile. Ovary 3-celled, ovules 2 or 3 per cell, placentation axile; stigma capitate. Fruit a capsule containing several 3-sided, glossy black seeds.

A genus of 4 species, 3 in N America and 1 in the Middle East, our species being the only one in cultivation. The tunicated bulbs and leaves suggest that the bulbs are often deeply buried in the wild, and it may be desirable to plant the bulbs well below soil level to help overwintering. They should be kept dry and frost-free in the winter and may either be repotted or given a suitable top-dressing in spring. Propagation is by offsets or seed (though this rarely ripens in northen Europe).

Literature: Herklots, G.A.C., Eurycles and Vagaria, *The Plantsman* **3**: 220–9 (1982).

1. V. parviflora Herbert. Figure 8(9), p. 208. Illustration: *Botanical Magazine*, 9406 (1935); *The Plantsman* **3**: 226, 227 (1982). *Syria & Israel.* H4. Summer.

25. PANCRATIUM Linnaeus

D.A. Webb

Perennial herbs with large bulbs and several broadly linear to strap-shaped basal leaves that are more or less in 2 ranks. Scape bearing an umbel of 3–15 flowers (rarely a solitary flower), subtended by usually 2 scarious spathes. Flowers large, white, fragrant. Perianth-tube present; perianth-lobes linear-lanceolate, spreading or almost erect; corona conspicuous, united to the lower part of the filaments, which thus appear to be inserted on its margin. Anthers dorsifixed. Stigma capitate. Ovary 3-celled with ovules in 1 or 2 vertical rows in each cell. Fruit a many-seeded capsule; seeds black, angular, dry.

A genus of about 16 species, extending from the Canary Islands through the Mediterranean region to tropical Asia, and southwards through W Africa to Namibia. It differs from *Hymenocallis* in its numerous seeds with a thin, dry, black testa. Many tropical species formerly included under *Pancratium* are now assigned to *Hymenocallis* (p. 230).

P. zeylanicum requires hot-house conditions. The other species can be cultivated outdoors in favoured regions of W & S Europe, but they require a very sunny position in well-drained soil. In the case of *P. maritimum*, the bulbs should be planted at a depth of about 30 cm and surrounded by sand.

1a. Free part of filament much longer than the anther 2
b. Free part of filament equalling or shorter than the anther 3
2a. Flowers in umbels; spathes 2 **1. illyricum**
b. Flower solitary; spathe 1 **4. zeylanicum**
3a. Flower-stalks 2–3 cm; perianth-tube 1.5–2.5 cm; corona one-third as long as perianth **2. canariense**
b. Flower-stalks 5–10 mm; perianth-tube to 7.5 cm; corona two-thirds as long as perianth **3. maritimum**

1. P. illyricum Linnaeus. Illustration: *Botanical Magazine*, 718 (1804); Coste, Flore de la France **3**: 382 (1906); Brickell (ed.), RHS A–Z encyclopedia of garden plants, 764 (2003).

Bulb-scales purplish black. Leaves *c.* 50 × 1.5–3 cm, glaucous, about equalling the scape. Flower-stalks 1–1.5 cm. Perianth-tube *c.* 2 cm, fairly stout. Perianth-lobes 5 cm, linear-oblong to narrowly elliptic. Corona much shorter than perianth, with pairs of long, narrow teeth alternating with the stamens. Free part of filament much longer than anther. *Islands of W Mediterranean (Corsica, Sardinia, Capraia).* H4–5. Late spring–early summer.

2. P. canariense Ker Gawler. Illustration: *Edwards's Botanical Register* **2**: t. 174 (1816); Bramwell & Bramwell, Wild flowers of the Canary Islands, f. 111 (1974).

Differs from *P. illyricum* chiefly in its somewhat broader leaves, much longer flower-stalks (up to 3 cm), shorter teeth to the corona, and free part of filament scarcely as long as the anther. *Canary Islands.* H5. Early autumn.

3. P. maritimum Linnaeus. Figure 8(14), p. 208. Illustration: Bonnier, Flore Complète **11**: t. 604 (1930); Maire, Flore de l'Afrique du nord **6**: 36 (1959); Goulandris & Goulimis, Wild flowers of Greece, t. 68 (1968).

Like *P. illyricum* but with pale bulb-scales, a very long neck to the bulb, and somewhat longer and narrower leaves. Flower-stalks 5–10 mm; perianth-tube to 7.5 cm, very slender. Corona two-thirds as long as perianth, with short, triangular teeth between the stamens. Free part of filament scarcely longer than the anther. *Mediterranean region & SW Europe.* H5. Summer.

4. P. zeylanicum Linnaeus. Illustration: *Botanical Magazine*, 2538 (1825).

Differs from all the above in having a solitary flower with a single spathe, and leaves of a fresh green, not glaucous. Flower-stalk very short; perianth-tube long and slender. Corona and filaments much as in *P. illyricum*. Tropical Asia. G2. Summer.

26. PARAMONGAIA Velarde

J.C.M. Alexander

Deciduous perennial herbs. Bulb to 6.5 cm across; leaf-bases not forming a false-stem or neck above the bulb. Leaves 6–8, all basal, bright green, sometimes glaucous, 2-ranked, to 75 × 5 cm, narrowly strap-shaped, stalkless, reaching full size after flowering. Scape erect, solid, to *c.* 75 cm, bearing 1 (rarely 2) long-lasting, yellow, fragrant flowers, to 18 cm across. Perianth-tube *c.* 10 cm; lobes to 9.5 × 4 cm, not thickened at base, spreading; corona *c.* 8.5 × 2 cm. Stamens 6, filaments borne well below corona margin, anthers versatile, not projecting. Style free from perianth-tube, not winged. Ovary 3-celled with ovules in 1 or 2 vertical rows in each cell. Fruit a capsule *c.* 4 cm long, with 50–75 seeds in rows (superposed).

A single species native to only 2 areas in coastal and montane Peru. This species is not easy to grow. The bulbs should be planted deeply in a mixture of loam, sand and peat with added leaf-mould and charcoal. In the wild they are as much as 30 cm below the surface. In the dormant season (summer) the bulbs should be baked at 21–4 °C, preferably in the pots in which they are being grown. When the leaves appear in about October, water should be freely given and similar temperatures maintained with a night minimum of about 18 °C. Plants do best in full sunshine and mature bulbs will produce one or two offsets per year.

1. P. weberbaueri Velarde. Illustration: *American Horticulturist* **54**(6): 30 (1975).

Peru. H2. Autumn.

27. PAMIANTHE Stapf

P.F. Yeo

Perennial herbs with bulbs. Leaves strap-shaped, evergreen or deciduous, sheathing one another at the base to form a false-stem. Flowers bisexual, radially symmetric, in umbels on an angled stalk; bracts conspicuous. Perianth-tube slender, straight, scarcely widened towards the throat; perianth-lobes obovate to oblanceolate, white. Stamens borne on a large white corona, attached in deep notches and directed horizontally towards the centre of the flower; anthers dorsifixed. Style weakly 3-lobed. Ovary 3-celled, with many ovules in 1 or 2 vertical rows in each cell. Seeds flattened, winged at the base.

Three species from tropical South America, of which *P. peruviana* is grown for its spectacular and exquisitely scented flowers; it is sometimes evergreen but can apparently be induced to become dormant in winter by drying it off. It may be grown in pots and requires a loam-based compost. Propagation is by offsets or, preferably, by seed, which takes 12–15 months to mature on the plant but then germinates rapidly if sown at once; seedlings grow quickly in humid conditions.

1. P. peruviana Stapf. Figure 8(6), p. 208. Illustration: *Botanical Magazine*, 9315 (1933); Hay et al., Dictionary of indoor plants in colour, 368 (1974); Everett, The New York Botanical Gardens illustrated encyclopedia of horticulture **7**: 2478 (1981); Brickell (ed.), RHS A–Z encyclopedia of garden plants, 764 (2003).

Plant 40–50 cm high. Pseudo-stem to 30 cm, composed of scale-leaves and leaf-bases; leaf-blades to 50 cm, 2–4 cm wide, spreading and drooping. Inflorescence-stalk with its visible part *c.* 10 cm. Flowers 2–4, *c.* 13 cm in diameter, shortly stalked, spreading. Perianth-tube 12–13 cm × 7–10 mm, green, solid at the base for *c.* 1 cm, divided into 3 by lengthwise partitions for the next 6–7 cm. Perianth-lobes to 12 × 3 cm, oblanceolate, divergent. Corona to 8 cm, bell-shaped, with pointed lobes; notches from which the stamens arise much deeper than those between the stamens. Style emerging on lower side of flower. *Peru.* G1. Spring–early summer.

28. STENOMESSON Herbert

S.G. Knees

Perennial herbs with bulbs, the brown tunic often exceeding the neck for up to 25 cm. Leaves 3–8, deciduous, linear to lanceolate, 20–35 × 2–3 cm at flowering time, often extending to 60 or 70 cm after flowering. Scape solid, sometimes 4-angled, 30–60 cm, bearing 2 broadly ovate spathes and 1–4 leaf-like bracts. Flowers shortly stalked, usually pendulous, 1–6 per umbel. Perianth tubular, with 6 ovate, hooded lobes, variously coloured and usually with a small, toothed corona arising just above the base of filaments. Anthers versatile, included in perianth, style projecting, stigma capitate. Ovary 3-angled, ovules many in 1 or 2 vertical rows in each cell. Fruit a capsule containing flattened black seeds.

A genus of about 20 species mostly native to the high Andes. Although several species were grown in the past there is little evidence of these now being widely cultivated, even in specialist collections, and most of what is grown is probably *S. variegatum*. This has been shown by Sealy (reference below) to be extremely variable and includes plants formerly grown as *S. incarnation* (Humboldt, Bonpland & Kunth) Baker, *S. luteoviride* Baker and *S. fuivum* Herbert. An open compost and a winter minimum of 5 °C are 2 major requirements for success with *Stenomesson*. Bulbs should never be allowed to dry out completely and liberal quantities of water are required during active growth. Propagation is normally by offsets.

Literature: Sealy, J.R., *Botanical Magazine*, n.s., 503 (1967); Ravenna, P.F., Contributions to the South American Amaryllidaceae IV & V, *Plant Life* **27**: 73–84 (1971).

1. S. variegatum (Ruiz & Pavon) Macbride. Figure 8(8), p. 208. Illustration: *Edwards's Botanical Register*, t. 1497 (1832); *Botanical Magazine*, n.s., 503 (1967); Hunt (ed.), Marshall Cavendish encyclopedia of gardening **7**: 2142 (1967); Rix, Growing bulbs, f. 2, 10 (1983).

Scape usually 4-angled. Flowers white, yellow, pink or reddish with broad green banding along the midrib near tips of the lobes. Perianth slender, tubular, 10–13 cm, with spreading lobes. *Bolivia, Ecuador, Peru.* H5. Spring–summer.

29. URCEOLINA Reichenbach

S.G. Knees

Perennial herbs with bulbs 4–7 × 1.5–5 cm and fibrous roots. Leaves 1 or 2, 35–50 × 1–15 cm, stalked for about one-quarter of their length, appearing with or after the flowers. Leaf-blades ovate to oblong, flattened, acute. Leaf-stalks *c.* 5 mm wide, almost winged. Scape solid, 12–25 cm, bearing umbels of 3–6 flowers subtended by 2 spathes. Perianth urceolate, with projecting style and stamens. Corona small. Anthers versatile. Ovary 3-celled with numerous ovules in vertical rows. Fruit a capsule.

A genus of 2 species, both native to S America, flowering more freely in pots than in the open ground. In northern Europe they require glasshouse protection and need to be kept dry during the dormant season. Propagation is by offsets, which are freely produced.

1a. Leaves linear, 1–5 cm wide, appearing after the flowers; perianth red or orange **1. peruviana**
b. Leaves ovate to oblong, 9–15 cm wide, appearing with the flowers; perianth yellow **2. urceolata**

1. U. peruviana (Presl) Macbride (*U. miniata* (Herbert) Bentham). Figure 8(13), p. 208. Illustration: *Edwards's Botanical Register*, t. 68 (1839); *Botanical Magazine*, n.s., 399 (1962); Mathew, The larger bulbs, 140 (1978).

Leaves flattened, strap-like, 40 × 1–5 cm, usually appearing after the flowers. Fused part of perianth 3–3.5 cm long, narrow at base for 1–1.3 cm, urceolate above, lobes spreading, 9–11 mm; bright red or orange, stamens projecting for *c.* 1.5 cm beyond the perianth. *Peru & Bolivia*. H5. Spring–summer.

2. U. urceolata (Ruiz & Pavon) M.L. Green (*U. pendula* (Herbert) Herbert). Illustration: *Botanical Magazine*, 5464 (1864); Everard & Morley, Wild flowers of the world, t. 190 (1970); Mathew, The larger bulbs, 140 (1978).

Leaf-blades ovate to oblong, 30–50 × 9–15 cm, usually appearing with the flowers. Perianth yellow, fused part 7.5–10 cm, narrow at the base for 2–3 cm, broadly urceolate above, lobes recurved, acute, green-tipped. *Peru*. H5. Spring–summer.

30. EUCHARIS Planchon & Linden

P.F. Yeo

Perennial herbs with bulbs. Leaves evergreen with a stalk and a distinct ovate or elliptic blade with numerous arching veins; stalks not sheathing one another. Flowers bisexual, radially symmetric, in an umbel on a stalk about as long as the leaves; bracts all similar in length. Perianth-tube long, slender, curved, somewhat widened above; perianth-lobes shortly oblong to broadly ovate (not linear), white. Stamens borne on a corona that usually bears a lobe on either side of each stamen, but may form a narrow, unlobed rim. Anthers dorsifixed. Style 3-lobed. Ovary 3-celled; ovules numerous, in vertical rows, more rarely 2 per cell. Seeds rounded, black or blue.

About 10 species from tropical S America, providing showy white flowers accompanied by glossy dark green leaves. The flowers of *E. amazonica* are suitable for cutting. Several hybrids have arisen in cultivation. The plants require shade, high humidity and high temperatures (15–20 ° C, rising to 25–30 °C in summer) and are often planted under benches and near heating pipes in hot-houses, but may also be grown in pots or tubs. They are propagated by careful removal of offsets or by seeds sown as soon as ripe.

Meerow considers that *Eucharis* and *Caliphruria* should be united, but for the purposes of this work it seems better to separate them.

Literature: Meerow, A., Two new species of pancratioid Amaryllidaceae from Peru and Ecuador, *Brittonia* **36**: 19–25 (1984); Meerow, A.W. & Dehgan, B., Re-establishment and lectotypification of Eucharis amazonica Linden ex Planchon (Amaryllidaceae), *Taxon* **33**: 416–22 (1984).

1a. Corona in the form of a narrow rim at throat of flower, without lobes or teeth **1. sanderi**
b. Corona 6 mm or more, lobed or toothed 2
2a. Flowering bulb with more than 1 leaf; perianth-lobes not recurved, the outer similar to the inner, broadly ovate **2. amazonica**
b. Flowering bulb with only 1 leaf; perianth-lobes slightly recurved, the outer about twice as long as wide, much wider than inner **3. candida**

1. E. sanderi Baker. Illustration: *Botanical Magazine*, 6676 (1883), 6831B (1885).

Plant to *c.* 45 cm. Leaves 2 on each flowering bulb. Leaf-stalks to 15 cm. Leaf-blades to 25 cm, with *c.* 6–10 veins on either side of the midrib. Inflorescence-stalk to *c.* 45 cm, with 2–6 flowers. Flower-stalks *c.* 5 mm. Flowers 4–6 cm in diameter, scentless, drooping by the curvature of the tube, which is *c.* 6 cm long and enlarged beneath the limb. Perianth-lobes broadly ovate, spreading or merely divergent, overlapping, inner and outer similar. Throat of flower cream-coloured within and striped with pale yellow. Corona a mere rim, not lobed, cream-coloured. Filaments 6–8 mm. Style much longer than the stamens. Ovules many in each cell of the ovary. *Colombia*. G2.

2. E. amazonica Planchon (*E. grandiflora* misapplied). Illustration: *Botanical Magazine*, 4971 (1857); *Flore des serres* **12**: 69 (1857); Hay et al., Dictionary of indoor plants in colour, 220 (1974); Everett, The New York Botanical Gardens illustrated encyclopedia of horticulture **4**: 1280 (1981).

Plant to *c.* 60 cm. Leaves several on each flowering bulb. Leaf-stalks 12–25 cm. Leaf-blades to *c.* 30 × 14 cm, with *c.* 20 veins on each side of midrib, dark glossy green. Inflorescence-stalk to *c.* 60 cm, with about 8 flowers. Flower-stalks *c.* 1.5–2 cm at first, lengthening to 3 cm. Flowers 10–12 cm in diameter, fragrant, drooping by the curvature of the tube, which is 6–8 cm long and gradually enlarged just beneath the limb; perianth-lobes *c.* 4 × 3 cm, broadly ovate, spreading, overlapping, inner and outer similar. Corona *c.* 12 mm × 2–2.5 cm, with an obtuse lobe on either side of each filament and a narrow sinus between adjoining lobes; white with a pale greenish yellow stripe in line with each filament. Filaments *c.* 6 mm, narrowly triangular. Style slightly exceeding the stamens. Ovules 9–12 in each cell of the ovary. *Peru, Ecuador*. G2. Summer, winter and occasionally at other times.

The closely related *E. grandiflora* Planchon & Linden (Figure 8(12), p. 208) from Colombia, is believed to have been lost from cultivation. It has long been confused with *E. amazonica*.

E. mastersii Baker. Illustration: *Botanical Magazine*, 6831A (1885). Similar to E. amazonica but corona half as long, with 12 equal broadly triangular teeth. *Colombia*. G2.

3. E. candida Planchon & Linden. Illustration: *Flore des serres* **8**: 107 (1852–3).

Similar to *E. amazonica* but with only one leaf per bulb at flowering time, with smaller scentless flowers, *c.* 7 cm in diameter, which have recurved perianth-lobes, the outer differing from the inner in being oblong, about twice as long as wide, with a narrower, less distinctly lobed, almost completely yellow corona cut nearly to the base, and with only 2 ovules in each cell of the ovary. *Colombia*. G2.

31. CALIPHRURIA Herbert

P.F. Yeo

Perennial herbs with bulbs. Leaves 2 or more to each bulb, evergreen with a stalk and a distinct elliptic to obovate blade; stalks not sheathing one another to form a pseudo-stem. Flowers bisexual, radially symmetric, in umbels on a stalk as long as or longer than the leaves, with 2 spathes concealing the floral bracts. Perianth-tube gradually widened from the base, straight, greenish. Perianth-lobes somewhat shorter than tube, approximately ovate, white. Stamens attached to throat of flower; filaments abruptly widened towards base, the enlargement on either side sometimes produced into a tooth; anthers dorsifixed. Style 3-lobed. Ovary 3-celled, with 2 or 3 ovules in each cell.

Three or perhaps more species from Colombia, sometimes included in *Eucharis*, but differing from it in the smaller flowers with a small corona. They are greenhouse subjects, which, if grown in pots or tubs, may be placed outdoors in summer. Propagation is by offsets.

1. C. subedentata Baker (*Eucharis subedentata* (Baker) Bentham & Hooker). Figure 8(11), p. 208. Illustration: *Botanical Magazine*, 6289 (1877); Nicholson, Illustrated dictionary of gardening **1**: 241 (1884).

Plant to *c.* 45 cm. Leaf-stalk *c.* 30 cm. Leaf-blade *c.* 20 cm, more than half as wide as long, with about 10 veins on either side of the pale midrib. Flower-stalks 1–1.5 cm. Flowers to *c.* 4 cm in diameter, horizontal or nodding, funnel-shaped; tube 3.5 cm; perianth-lobes *c.* 2 cm, slightly recurved at the tips. Stamens about half as long as perianth-lobes; filaments with the enlargement entire, toothed on one side or toothed on both, all conditions occurring in the same flower. Style not exceeding perianth-lobes. *Colombia.* G1.

C. hartwegiana Herbert (*Eucharis hartwegiana* (Herbert) Nicholson). Illustration: *Botanical Magazine*, 6259 (1876). Smaller in leaves and flowers than *C. subedentata* by about one-third. Leaf-blades thicker and firmer with inconspicuous veins. Leaf-stalks and inflorescence-stalks grey-green. Filaments with the basal enlargement produced on each side into a tooth about as long as the narrow part of the filament. Style exceeding perianth-lobes. *Colombia.* G1.

32. HYMENOCALLIS Salisbury

J.C.M. Alexander

Perennial herbs with spherical to ovoid bulbs. Leaves evergreen or deciduous, basal, usually 2-ranked, erect to arching, strap-shaped to sword-shaped or narrowly ovate, stalked or gradually narrowed to a stalkless base; bases sometimes forming a false-stem or neck above the bulb. Scape erect, 2-edged, solid, often bluish, bearing 1–several stalkless or shortly stalked, often fragrant, flowers, whose bases are enclosed in chaffy bracts. Perianth-tube narrow; perianth-lobes 6, linear to narrowly lanceolate, ascending, arched or spreading (adpressed to corona in No. 13). Corona funnel-shaped, remaining erect (subgenera *Hymenocallis & Ismene*) or becoming sharply bent (subgenus *Elisena*); margin erect or spreading, often complexly lobed or toothed. Lower parts of filaments fused to corona; free parts erect or hanging outwards (subgenera *Hymenocallis & Elisena*) or inturned (subgenus *Ismene*). Style long, equalling or projecting from the corona. Ovary 3-celled with 2 (rarely 3) or 4–8 basal ovules side by side in each cell. Fruit a fleshy capsule; seeds green, fleshy, often only 1 developing per cell or even per fruit.

A genus of 35–40 species, native to the tropics and subtropics of the New World, with a few species naturalised in E and W Africa and Sri Lanka. Many of the species were originally considered under *Pancratium*, a name now restricted to Old World species with many ovules in vertical rows, developing into black angular seeds. The genera *Ismene* Salisbury and *Elisena* Herbert are here included in *Hymenocallis* as subgenera. The differences between them are small and hybridisation is quite common.

All species perform well in light, fertile, well-drained soil, with plenty of moisture during the growing season. In general, the evergreen species require cool or warm greenhouse conditions and may need shading from strong sun. Many of the deciduous species can be grown outdoors in sheltered, sunny sites. In areas with frost it may be wise to lift the bulbs in autumn and keep them dry and frost-free before planting them out when the soil has warmed up. They are easily propagated from seed or offsets.

Leaves. Clearly stalked: **7–9**; stalkless, but narrower and parallel-sided towards the base: **10, 11, 12, 13, 14, 15**; stalkless, not narrow and parallel-sided towards the base: **1, 2, 3, 4, 5, 6, 11, 15**. Strap-shaped to sword-shaped, *c.* 5 times as long as broad or more: **1, 2, 3, 4, 5, 6, 10, 11, 12, 13, 14, 15**; elliptic to ovate, *c.* 3 times as long as broad or less: **7, 8, 9**.

Scape. 30 cm or less: **10**; 30–60 cm: **1, 2, 3, 4, 5, 7, 8, 9, 11, 12, 13, 14, 15**; 60 cm or more: **5, 6, 12, 14**.

Flowers. Yellow: **1**. Stalked: **1, 2, 3, 4, 5, 6, 8**; stalkless: **1, 7, 9, 10, 11, 12, 13, 14, 15**. Four or fewer per umbel: **1, 2, 3, 6, 10, 15**; 5–10 per umbel: **1, 2, 3, 4, 5, 7, 8, 9, 11, 12, 14**; 10 or more per umbel: **7, 8, 13, 14**.

Corona. Longer than perianth-tube: **3, 5, 6**; shorter than perianth-tube: **1, 2, 3, 4, 7, 8, 9, 10, 11, 12, 13, 14, 15**. Bending forward and becoming flattened: **5, 6**.

Stamens. Free part less than half corona length: **1, 2, 3, 4**; more than half corona length: **5, 6, 7, 8, 9, 10, 11, 12, 13, 14, 15**. Sharply inturned, not projecting from corona: **1, 2, 3**; projecting from corona: **3, 4, 5, 6, 7, 8, 9, 10, 11, 12, 13, 14, 15**; incurved over corona: **4**; parallel and more or less horizontal (upper 3 sometimes bending downwards across corona): **5, 6**; erect, spreading or hanging outwards: **7, 8, 9, 10, 11, 12, 13, 14, 15**.

Style. Not or scarcely projecting from corona: **1, 2, 3**; projecting 4 cm or more from corona: **4, 5, 6, 7, 8, 9, 10, 11, 12, 13, 14, 15**.

1a. Free parts of stamens less than half corona length, all sharply inturned and not projecting from corona, or all incurved and converging above corona (subgenus **Ismene**) 2
b. Free parts of stamens longer than, equal to or slightly shorter than corona, not all inturned or converging above corona 5
2a. Flowers yellow **1. amancaes**
b. Flowers white or greenish white 3
3a. Perianth-tube 3.8–5 cm **3. pedunculata**
b. Perianth-tube 7.6–10 cm 4
4a. Flowers shortly stalked; filaments *c.* 2.5 cm, converging above corona **4. × macrostephana**
b. Flowers stalkless; filaments *c.* 1.3 cm, sharply inturned, not projecting from corona **2. narcissiflora**
5a. Corona longer than perianth-tube, deeply cup-shaped, horizontal or downturned, bending downwards and becoming elliptic to oblong in cross-section before flower withers; stamens parallel or the upper 3 becoming deflexed (subgenus **Elisena**) 6

b. Corona shorter than perianth-tube, funnel-shaped, erect or ascending, remaining circular in cross-section; stamens erect, arched or hanging outwards (subgenus **Hymenocallis**) 7
6a. Perianth-tube *c.* 1 cm; filaments 6–7 cm, remaining parallel **5. longipetala**
b. Perianth-tube 4–5 cm; filaments 2.5–4 cm, the upper 3 becoming bent downwards **6. × festmlis**
7a. Leaves clearly divided into blade and stalk 8
b. Leaves stalkless, though often much narrower near base 10
8a. Perianth-tube 12.5–20 cm; ovules 4 or 5 per cell **7. tubiflora**
b. Perianth-tube 9 cm or less; ovules 2 per cell 9
9a. Perianth-tube 6.5–9 cm; leaf-blades at least 3 times longer than broad, 26–66 × 8–15.5 cm **8. speciosa**
b. Perianth-tube 4.5–5 cm; leaf-blades *c.* 2 times longer than broad, 10.5–30 × 4–15 cm **9. ovata**
10a. Corona 2 cm or less; leaves 3–5 **10. harrisiana**
b. Corona 2 cm or more; leaves usually 7 or more 11
11a. Free parts of stamens 2–3 cm; corona narrow and green at base; perianth-lobes spreading at right angles to tube **15. rotata**
b. Not this combination of characters 12
12a. Ovules 2 per cell; narrow part of mature leaf at least 2.5 cm wide 13
b. Ovules 4 or more per cell; narrow part of mature leaf 2.1 cm wide or less 14
13a. Perianth-tube 4–6.5 cm, shorter than lobes **11. caribaea**
b. Perianth-tube 10–15 cm, longer than lobes **12. latifolia**
14a. Broad part of mature leaf less than 3.8 cm wide; perianth-lobes adpressed to corona below, arched above; corona margin spreading **13. littoralis**
b. Leaf usually more than 5 cm wide; perianth-lobes ascending but not adpressed to corona; corona margin erect **14. pedalis**

1. H. amancaes (Ruiz & Pavon) Nicholson (*Ismene amancaes* (Ruiz & Pavon) Herbert). Illustration: *Botanical Magazine*, 1224 (1809).

Bulb spherical, 3.5–5 cm in diameter; neck 15–25 cm. Leaves 4 or 5, strap-shaped, bright green, erect to arching, 25–60 × 2.5–5 cm. Scape 11–33 cm, bearing 2–6, stalkless or shortly stalked, bright to deep yellow flowers. Perianth-tube 5–7.5 cm, slightly curved, greenish yellow. Perianth-lobes linear to narrowly lanceolate, 6–7.5 cm, greenish at tips. Corona 5–6 × 6–8.5 cm, funnel-shaped, irregularly 12-toothed. Free parts of filaments 1–2 cm, sharply inturned. Style equal to or slightly projecting from corona. Ovules 2 per cell. *Peru.* H5–G1. Summer.

2. H. narcissiflora (Jacquin) MacBride (*H. calathina* (Ker Gawler) Nicholson; *Ismene calathina* (Ker Gawler) Herbert; not I. *narcissiflora* Roemer). Figure 8(5), p. 208. Illustration: *Botanical Magazine*, 2685 (1826); RHS Dictionary of gardening, 1029 (1956).

Bulb spherical; neck 20–30 cm. Leaves 6–8, loosely 2-ranked, strap-shaped to sword-shaped, dark green, acute to obtuse, narrowed at base, 20–75 × 3.5–6 cm. Flowering stem 30–60 cm, bearing 1–5, fragrant flowers. Perianth-tube 7.5–10 cm, green. Perianth-lobes lanceolate, 7–9 × *c.* 1.3 cm, white. Corona 6–7.5 cm, bell-shaped to funnel-shaped with ragged, spreading lobes, white. Free parts of filaments 1–1.5 cm, sharply inturned. Style projecting. *Peru & Bolivia.* H5–G1. Summer.

3. H. pedunculata (Herbert) MacBride (*Ismene macleana* Herbert; *H. macleana* (Herbert) Nicholson; *Ismene virescens* Lindley). Illustration: *Botanical Magazine*, 3675 (1838).

Similar to *H. narcissiflora*. Neck *c.* 15 cm. Leaves 23–45 × 2–5.5 cm. Perianth-tube 3–5 cm; perianth-lobes 4–5.5 cm; corona 4–5 cm, green with white, spreading lobes. *Peru.* H5. Summer.

4. H. × macrostephana Baker. Illustration: *Botanical Magazine*, 6436 (1879); *The Garden* **89**: 209 (1925).

Bulb ovoid, *c.* 5 cm in diameter; neck long. Leaves 8 or 9 in a loose rosette, evergreen, oblanceolate, obtuse, bright green, 50–90 × 6–8 cm, *c.* 2.5 cm wide at base. Scape 30–45 × 2.5 cm, bearing 6–10, shortly stalked, fragrant flowers. Perianth-tube 6.5–8.5 cm, green below, white above. Perianth-lobes linear-lanceolate, 9–11 cm, white to pale greenish yellow, outer lobes with thick green tips. Corona 5.5–6 × 5–7.5 cm, funnel-shaped, obtusely 12-toothed, white. Free parts of filaments *c.* 2.5 cm; style projecting *c.* 4.5 cm from corona. *?Garden origin.* G2. Spring.

This is probably the result of a cross between *H. narcissiflora* and *H. speciosa*.

5. H. longipetala (Lindley) MacBride (*Elisena longipetala* Lindley). Illustration: *Botanical Magazine*, 3873 (1841).

Bulb 3.5–5 cm; neck to 50 cm. Leaves 6–8, 2-ranked, stiff, strap-shaped, acute, 45–55 × *c.* 4 cm. Scape 60–90 cm, bearing 5–10, white flowers on stalks *c.* 6 mm long. Perianth-tube *c.* 1 cm × 1 cm. Perianth-lobes linear, 7.5–10 cm, green at base, wavy. Corona *c.* 3 cm, becoming flattened and bent forward, marginal teeth bent inwards. Free parts of filaments 6–7 cm, parallel, upturned at tips. Style projecting 1–2 cm beyond stamens. *Ecuador & Peru.* H5–G1. Summer.

6. H. × festalis (Worsley) Schmarse (*Ismene festalis* Worsley). Illustration: *Gartenflora* **82**: 337 (1933); Brickell (ed.), RHS A–Z encyclopedia of garden plants, 556 (2003).

Neck *c.* 4.5 cm. Leaves *c.* 9, loosely 2-ranked, strap-shaped, 60–90 × 5–7 cm. Scape *c.* 120 cm, bearing *c.* 4, white, stalked, fragrant flowers. Perianth-tube 4–5 cm. Perianth-lobes 10–12 cm, the outer narrower and more deeply channelled. Corona *c.* 5 cm, becoming flattened and bent forward; marginal teeth bent. Free parts of stamens 2.5–4 cm, parallel or converging at first, the upper 3 become bent downwards. Style projecting *c.* 7.5 cm from corona. *Garden origin.*

The hybrid between *H. narcissiflora* and *H. longipetala*.

7. H. tubilora Salisbury (*H. undulata* (Humboldt et al.) Herbert; *H. moritziana* Kunth; *H. guianensis* (Ker Gawler) Herbert). Illustration: *Edwards's Botanical Register*, 265 (1818); *Gardeners' Chronicle* **27**: 89 (1900).

Bulb ovoid, 7.5–10 cm in diameter; neck short. Leaves 7–12, stalked, evergreen; blades elliptic to ovate, abruptly pointed, tapered to rounded at base, 20–38 cm × 7.5–15 cm; stalks 15–30 cm. Scape to 60 cm, bearing 5–20, stalkless, white flowers. Perianth-tube 12–20 cm, greenish. Perianth-lobes 8.5–14 cm × 4–6 mm. Corona 1.8–2.5 cm, funnel-shaped, margin erect or slightly spreading. Free parts of stamens 4–6.5 cm, ascending to erect. Style projecting beyond stamens, arching. *NE South America.* G2. Spring & autumn.

8. H. speciosa (Salisbury) Salisbury. Illustration: *Botanical Magazine*, 1453 (1812).

Bulb spherical, 7.5–10 cm in diameter. Leaves 12–20, stalked, evergreen; blades broadly elliptic to oblong, acuminate, tapering into the stalk, 25–65 × 8–16 cm; stalks 9–17 cm. Scape 30–45 cm, bearing 7–12, stalked, fragrant, white to greenish flowers; stalks to 1 cm. Perianth-tube 6.5–9 cm, greenish. Perianth-lobes 9.5–15 cm × 5–7 mm, white, recurved. Corona 2.5–5 cm, funnel-shaped, toothed. Free parts of stamens 3–5 cm, erect to hanging, filaments greenish white. Style bending downwards, almost as long as perianth-lobes. Ovules 2. *West Indies.* G1. Autumn–winter.

9. H. ovata (Miller) Sweet (*H. amoena* misapplied). Illustration: *Botanical Magazine*, 1467 (1812).

Similar to *H. speciosa.* Leaves 5–8. Flowers 5 or 6, stalkless; perianth-tube 4.5–6 cm; perianth-lobes 5.5–10 cm. Filaments green. *?West Indies.*

10. H. harrisiana Herbert. Illustration: *Botanical Magazine*, 6562 (1881).

Bulb spherical to ovoid, *c.* 4 cm in diameter. Leaves 3–5, deciduous, oblanceolate, much narrowed below though not with a true stalk, 20–34 cm overall, obtuse to acute; broad part 19–30 × 3–5.1 cm, oblanceolate to oblong; narrow part *c.* 7 cm. Scape 20–25 cm, bearing 1–3, stalkless, white flowers. Perianth-tube 8–13 cm, greenish below. Perianth-lobes 6–7.5 cm, linear. Corona 1.3–1.8 cm, funnel-shaped, margins spreading, 6-toothed. Free parts of stamens 2.5–3.5 cm, erect to hanging. Ovules 2. *Mexico.* G1. Spring–summer.

11. H. caribaea (Linnaeus) Herbert (*Pancratium caribaeum* Linnaeus, in part; *H. amoena* (Salisbury) Herbert). Illustration: *Botanical Magazine*, 826 (1805).

Bulb spherical, 7.5–10 cm in diameter. Leaves many, evergreen, shining, loosely 2-ranked, strap-shaped to sword-shaped, acute or obtuse, slightly narrowed below, stalkless, 30–90 × 5–10 cm. Scape 30–60 cm, bearing 8–10, stalkless, fragrant, white flowers. Perianth-tube 4–6.5 cm, greenish; perianth-lobes 8–12 cm, linear. Corona funnel-shaped, 2–3 cm, margin erect, 12-toothed. Free parts of stamens 3–5 cm, erect to hanging. Ovules 2 per cell. *West Indies.* G2. Summer–autumn.

12. H. latifolia (Miller) Roemer (*Pancratium caribaeum* Linnaeus, in part; *H. caribaea* misapplied; *H. caymanensis* Herbert; *H. keyensis* Small).

Very similar to *H. caribaea.* Scape 60–80 cm; perianth-tube 10–15 cm, always longer than the lobes. *USA (Florida) & West Indies.* G2. Summer.

13. H. littoralis (Jacquin) Salisbury (*H. senegambica* Kunth & Bouché, in part; *H. americana* (Miller) Roemer). Illustration: *Botanical Magazine*, 825 (1805) & 2621 (1826); Grey, Hardy Bulbs **2**: 47 (1938); Everett, The New York Botanical Gardens illustrated encyclopedia of horticulture **5**: 1758 (1981).

Bulb spherical, 7.5–10 cm in diameter. Leaves numerous, evergreen, 2-ranked, stalkless, 40–120 cm; upper part strap-shaped, 2–4 cm wide, acute, tapering to a narrower basal part, 1–2 cm wide. Scape 45–60 cm, bearing 6–11, white, stalkless, fragrant flowers. Perianth-tube 14–17 (rarely 10–20) cm, straight. Perianth-lobes 7.5–8 (rarely to 12.5) cm, adpressed to corona below, arched above. Corona 2.5–3 (rarely 2–3.5) cm, funnel-shaped below, widely flared above. Free parts of stamens 4–6 cm, erect to hanging. Ovules 4 or 5 (rarely 8) per cell. *Mexico, Guatemala & Northern S America; naturalised in W Africa & Sri Lanka.* G1. Spring.

14. H. pedalis Herbert (*H. senegambica* Kunth & Bouché in part). Illustration: *Edwards's Botanical Register*, t. 1641 (1834).

Similar to *H. littoralis*. Leaves 2.5–7 cm wide in upper part, tapering to 1.5–2 cm in lower part. Perianth-tube slightly curved. Perianth-lobes wavy-edged. Corona funnel-shaped, margins erect. *Colombia & Brazil; naturalised in W and possibly E Africa.* G2. Spring.

15. H. rotata (Ker Gawler) Herbert (*H. mexicana* misapplied; *H. lacera* Salisbury, in part). Illustration: Botanical Magazine, 827 (1805) & 1082 (1808).

Bulb ovoid, 3.5–5 cm in diameter, bearing stolons; neck *c.* 4 cm. Leaves 7 or 8, deciduous, linear to strap-shaped or sword-shaped, obtuse, 30–68 × 1.1–3.6 cm, tapered at base, or sometimes from near the middle to a narrower basal portion 7–17 mm wide. Scape 45–60 cm, bearing 2–4 (rarely to 8), stalkless, white, fragrant flowers. Perianth-tube 6–9.5 cm. Perianth-lobes 6.2–9.5 cm, spreading at right angles to tube. Corona 3–5 cm, narrow and green below, funnel-shaped and white above, margin spreading. Free parts of stamens 2–3 cm, erect. Ovary with 2 (rarely 4–6) ovules per cell. *Jamaica & southern USA.* H5–G1. Spring.

33. PHAEDRANASSA Herbert

S.G. Knees

Perennial herbs with bulbs. Leaves 1–4, stalked, appearing with, or a little after, the flowers. Leaf-blades lanceolate to elliptic, 15–30 × 3.5–7 cm, stalks 5–18 cm, somewhat channelled. Scape hollow, 2–70 cm, widest at the base, decreasing to less than half its diameter at the apex. Flowers pendulous, in umbellate clusters of 4–11. Perianth tubular, richly coloured, with combinations of green, yellow, crimson and scarlet. Corona a ring of small transparent teeth at the base of the filaments. Stamens projecting from the perianth, arising about 1 cm from its base, filaments unequal in length, anthers versatile. Ovary 3-celled, broadly triangular, style projecting from the perianth, stigma capitate. Fruit a capsule, deeply grooved and containing several black, flattened seeds.

A genus of about 6 species, all native to S America. None is widely cultivated and their morphological similarity has caused difficulty.

A good loamy soil and a position in a warm sunny site are both essential. If pot-grown, plants require a resting period once the leaves have died down, and should be kept dry until the following spring. Flowering periods may be rather erratic. Propagation is normally by offsets, although, occasionally, viable seed may be produced.

Literature: Ravenna, P.F., Contributions to South American Amaryllidaceae III, *Plant Life* **25**: 55–62 (1969).

1. P. carmioli Baker. Illustration: *Botanical Magazine*, n.s., 71 (1949); Graf, Exotica, edn 3, 102 (1963); Brickell (ed.), RHS A–Z encyclopedia of garden plants, 797 (2003).

Bulb 5 cm in diameter, with thin brown tunic. Leaves usually 2, 30–50 cm, elliptic with midrib almost keeled beneath. Scape 50 cm, flowers 7–9. Perianth tubular, 3.5–4.5 × 1 cm, green at base, tube scarlet, lobes green with yellow markings on margins, somewhat reflexed at the apex. Filaments white, anthers greenish yellow.

Ovary 3-angled, 6–8 × 4–5 mm. Capsule 1.5 cm × 5 mm. Seeds 1–1.2 cm × 4–6 mm. *Possibly native of Peru, but most collections originate from Costa Rica where it is widely grown.* H4. Spring–summer.

P. lehmannii Regel. Plants very similar to *P. carmioli* but differing in their shorter perianth-tube (3–3.5 cm) that is almost entirely red. The stamens project for 1.5–2 cm.

P. dubia (Humboldt, Bonpland & Kunth) Macbride. This species has a longer perianth-tube (4.5–5 cm) and stamens project for only 7 mm.

P. chloracra (Herbert) Herbert, *P. obtusa* (Herbert) Lindley, *P. ventricosa* Baker, and possibly also *P. schizantha* Baker are probably the same as *P. dubia*; *P. viridiflora* Baker may just be another colour form.

35. TECOPHILAEACEAE

Herbs with corms or tubers. Leaves linear or ovate, borne on the stems. Flowers usually solitary, bisexual, actinomorphic. Perianth of 6 united segments. Stamens 6, or 3+3 staminodes; anthers opening by pores. Ovary of 3 united carpels, half-inferior; ovules numerous, axile. Fruit a capsule.

There are 6 genera and about 20 species. *Tecophilaea cyanocrocus*, from Chile but now apparently extinct there, is widely grown in Europe for the sake of its blue flowers.

1. TECOPHILAEA Colla

E.C. Nelson

Corms with fibrous tunics. Leaves 2 or 3, linear, enclosed in a sheath when young, to 10 cm long, green, hairless. Perianth blue to purple with darker veins and white throat, 4 cm long.

A crocus-like plant with elegant blue flowers. The soil should be moist at flowering time but dry for the rest of the year. Propagation is by offsets. Two species have been described, but only 1 is in general cultivation.

1. T. cyanocrocus Leybold. Illustration: *Botanical Magazine*, 8987 (1923); Hay & Beckett (eds), Reader's Digest encyclopaedia of garden plants and flowers, 696 (1978); Brickell (ed.), RHS A–Z encyclopedia of garden plants, 1027 (2003).

Chile (Andes). H5.

Two cultivars are generally listed: 'Leichtlinii', with prominent white throat, and 'Violacea' with deep purple flowers.

36. HYPOXIDACEAE

Herbs with rhizomes or corms. Leaves usually basal, alternate, often pleated and hairy; ptyxis conduplicate or plicate. Flowers solitary or in racemes or heads, bisexual, actinomorphic. Perianth segments free, clearly 3+3, those of the outer series usually hairy outside. Stamens 6. Ovary inferior, of 3 united carpels; ovules many per cell, axile. Fruit a capsule or berry.

Often included within the *Amaryllidaceae* (p 198), there are 5 genera with about 150 species. A few are cultivated as ornamentals.

1a. Flowers in a head; leaves distinctly and conspicuously stalked **1. Curculigo**
b. Flowers solitary or in racemes; leaves not or obscurely stalked 2
2a. Flowers yellow or orange, rarely white when each perianth-lobe has a dark purple blotch at the base **2. Hypoxis**
b. Flowers red, pink or white, perianth-lobes without dark blotches at the base **3. Rhodohypoxis**

1. CURCULIGO Gaertner

J. Cullen

Rhizomatous herbaceous perennials. Leaves large, conspicuously stalked, the blades conspicuously pleated, hairy or not, borne in a tuft. Flowers in a dense, head-like raceme, each flower subtended by a large bract, the head borne at soil-level or on an elongate, aerial stalk that is shorter than the leaves. Perianth of 6 lobes united into a tube at the base (sometimes short). Stamens 6, borne on the perianth-tube. Ovary inferior, 3-celled, with many ovules in each cell. Fruit a berry containing several seeds.

A genus of 35 species from tropical Asia, Africa and America. Some of them are placed in the genus *Molineria* Colla by some authors (Geerinck and Hilliard & Burtt, reference below). Only 2 species are widely grown; they are easily cultivated, requiring warm, humid conditions and a peaty soil. Propagation is by offsets or division of the rhizome.

Literature: Geerinck, J. Genera des Haemodoraceae et des Hypoxidaceae, *Bulletin du Jardin Botanique National de Belgique* **69:** 70–2 (1969); Hilliard, O. & Burtt, B.L. Notes on some plants from Southern Africa: 7, *Notes from the Royal Botanic Garden, Edinburgh* **36:** 72–6, (1978).

1a. Head of flowers borne at soil level; bracts green, hairless **2. latifolia**
b. Head of flowers borne on an elongate, erect, aerial stalk; bracts brown, hairy **1. capitulata**

1. C. capitulata (Loureiro) Kuntze (*C. recurvata* Dryander; *Molineria recurvata* (Dryander) Herbert). Illustration: *Edwards's Botanical Register* **9**: t. 770 (1823); Parey's Blumengärtnerei **1**: 363 (1958); Iconographia Cormophytorum Sinicorum **5**: 547 (1976).

Rhizome oblique. Leaves large, erect then arching; stalk 30–60 cm, hairy; blade oblanceolate, acute, 60–90 × 5–15 cm, hairless or with some hairs on the veins. Scape 7–25 cm, hairy, bearing a deflexed head 2.5–7 cm long. Bracts brownish, hairy, 1.2–3 cm. Perianth with a very short tube at the base, yellow, lobes 6–8 mm. Berry spherical, 6–8 mm, seeds black. *Himalaya & SE Asia to Australia.* G2.

Plants with variegated leaves have been grown as 'Variegata' and 'Striata'.

2. C. latifolia Dryander. Illustration: *Edwards's Botanical Register* **9**: t. 754 (1823); *Botanical Magazine*, 2034 (1818); Keng, Orders and families of Malayan seed plants, f. 177 (1969).

Rhizome creeping. Leaves erect, then arching; stalk 7–30 cm; blade lanceolate or oblong-lanceolate, acuminate, 30–60 × 5–10 cm, hairless or somewhat hairy beneath. Scape absent, the flower-head borne at soil-level. Bracts green, hairless, 1.2–3.6 cm. Perianth with a distinct tube at the base, yellow, lobes 8–12 mm, densely hairy outside. Berry oblong, to 1.2 cm, seeds black and shining. *Burma, Malaysia, Indonesia & Borneo.* G2.

2. HYPOXIS Linnaeus

J. Cullen

Plants with corms and thick, fleshy roots. Leaves mostly basal, sometimes borne on the stem, ridged or not, often hairy, their bases sheathing and overlapping. Scape present, bearing a solitary flower or a 2–12-flowered raceme. Perianth-lobes 6, the outer 3 slightly broader than the inner and greenish and usually hairy outside. Stamens 6. Ovary inferior, 3-celled, with several ovules in each cell. Style present or very short; stigma 3-lobed. Capsule containing several seeds, opening by means of a lid formed from the ovary wall. Seeds black or brown, spherical or elongate.

A genus of perhaps 150 species from N America, Africa, tropical Asia and Australia. Only a small number of species is now grown, though many more were in cultivation in the late nineteenth century. The classification of the genus, and the identification of the species are difficult, and much further study is needed: species **1** included here (and the species mentioned under it) is often placed in the genus *Spiloxene* Salisbury, which differs from *Hypoxis* in the strict sense in being totally hairless, possessing linear anthers, and in having the 3 lobes of the stigma more distinct.

The plants are easily grown in good soils, and can be propagated by seeds or offsets.

Plant. Completely hairless: **1**; hairy in at least some part (leaves, scape, backs of perianth-lobes, ovary): **2, 3, 4, 5, 6**.
Leaves. Hairless: **1, 2**; hairy: **3, 4, 5, 6**. More than 3 cm broad: **2, 3**; to 1 cm broad: **1, 4, 5, 6**.
Flower. To 3 cm in diameter: **1, 2, 5, 6**; more than 3 cm in diameter: **1, 3, 4**.
Perianth-lobes. Each with a dark spot at the base: **1**. Outer lobes hairless or very sparsely hairy outside: **1, 5**; outer lobes densely hairy outside: **2, 3, 4, 6**.

1a. Whole plant completely hairless **1. capensis**
b. At least some part of the plant hairy 2
2a. Leaves 3–5 cm broad 3
b. Leaves to 1 cm broad 4
3a. Leaves completely hairless; flowers 2.5–3 cm in diameter; style very short, the stigma borne almost directly on the ovary **2. latifolia**
b. Leaves hairy, at least along the veins; flowers 3–3.6 cm in diameter; style evident **3. hemerocallidea**
4a. Flowers 3–3.6 cm in diameter; leaves rigidly leathery, deeply channelled, 30–45 cm **4. longifolia**
b. Flowers to 3 cm in diameter; leaves not as above, to 30 cm 5
5a. Leaves thread-like, *c.* 1 mm wide; outer perianth-lobes hairless or very sparsely hairy **5. hygrometrica**
b. Leaves flat, 2–10 mm wide; outer perianth-lobes very densely hairy **6. hirsuta**

1. H. capensis (Linnaeus) Druce (*H. stellata* Linnaeus filius; *Spiloxene capensis* (Linnaeus) Garside). Illustration: *Botanical Magazine*, 662 (1803), 1223 (1809); *Flowering plants of Africa* **39**: t. 1557A (1969).

Whole plant hairless. Corm 1–2 cm. Leaves mostly basal, 10–30 cm × 3–9 mm, the outer at least with pronounced midribs and V-shaped in section; stem leaves few, sheathing. Scape 5–25 cm, terminated by a solitary flower. Perianth-lobes widely spreading, 9–30 mm, yellow or white, each with an iridescent, purple spot at the base. Style short; stigma as long as the stamens. Seeds spherical. *South Africa (Cape Province).* G1. Spring–summer.

H. neocanaliculata Geerinck (*Spiloxene canaliculata* Garside, not *H. canaliculata* Baker). Illustration: *Flowering plants of Africa* **39**: t. 1557B (1969).
Outer leaves without midribs, U-shaped in section; perianth-lobes orange, each with a dark, non-iridescent spot at the base; seeds elongate and curved. *South Africa (Cape Province).* G1. Spring–summer.

2. H. latifolia J.D. Hooker. Illustration: *Botanical Magazine*, 4817 (1854).

Corm spherical, 6–8 cm in diameter. Leaves mostly borne on the stem, 30–60 × 3–5 cm, their bases sheathing and conspicuously overlapping, leathery, ribbed, hairless. Racemes borne in the leaf-axils, with 10–12 flowers. Flowers 2.5–3 cm in diameter. Perianth-lobes spreading, to 1.2 cm, yellow, the outer 3 densely hairy outside. Style very short; stigmas borne almost directly on the ovary. *South Africa (Natal).* G1. Spring–summer.

H. colchicifolia Baker is probably the same species; it is reputed to have a shorter, fewer-flowered, more corymbose raceme. It originates from the same general area.

3. H. hemerocallidea Fischer & Meyer (*H. elata* J.D. Hooker). Illustration: *Botanical Magazine*, 5690 (1868).

Corm spherical, 8–10 cm in diameter. Leaves mostly borne at the base of the scape, 45–60 × 3–5 cm, slightly leathery, ribbed, hairy along the veins and margins. Racemes borne in the leaf-axils, loose, with 6–12 flowers; axis hairy. Flowers 3–3.6 cm in diameter. Perianth-lobes spreading, to 1.8 cm, yellow, the outer 3 very densely hairy outside. Style evident. *South Africa.* G1. Spring–summer.

4. H. longifolia Baker. Illustration: *Botanical Magazine*, 6035 (1873).

Corm oblong, 4–5 cm in diameter. Leaves mostly borne at the base of the scape, 30–45 cm × 3–4 mm, leathery, deeply channelled, shortly hairy beneath. Raceme corymbose, with 2–4 flowers. Flowers 3–3.6 cm in diameter. Perianth-lobes spreading, 1.5–1.8 cm, yellow, the outer 3 densely hairy outside. Stigma evident. *South Africa.* G1. Spring–summer.

5. H. hygrometrica Labillardière.

Corm oblong, 6–10 mm. Leaves basal, thread-like, 6–15 cm × *c.* 1 mm, with sparse hairs on the outer side. Flowers solitary or 2–3 in a corymbose raceme; scape hairy towards the base. Perianth-lobes spreading, 1–1.5 cm, yellow, the outer 3 hairless or very sparsely hairy outside. Style elongate. *E Australia (Queensland to Tasmania).* G1. Spring–summer.

6. H. hirsuta (Linnaeus) Coville. Illustration: Rickett, Wild flowers of the United States **1**: t. 10 (1966).

Corm oblong, small. Leaves mostly basal, flat, to 30 cm × 2–10 mm, ribbed, hairy. Scapes borne between the leaves, densely hairy, bearing corymbose racemes of 3–7 flowers. Perianth-lobes 8–11 mm, yellow, the outer 3 densely hairy outside. Style evident. *Eastern N America.* H3. Spring–summer.

H. micrantha Pollard. Leaves 2–6 mm broad; raceme with 1–2 flowers. *SE USA.* G1. Spring–summer.

3. RHODOHYPOXIS Nel

J. Cullen edited by S.G. Knees

Perennial herbs with fleshy and fibrous roots. True stem absent. Leaves all basal, usually hairy. Scape present or absent (when absent the ovary of the flower is within the leaf-sheaths, at or below ground level). Flowers solitary or few. Perianth of 6 lobes united into a tube below, the lobes spreading, the 3 inner conspicuously clawed and each with a pronounced, inflexed curve at the junction of claw and blade, which closes the mouth of the tube. Stamens 6, borne on the perianth-tube at 2 levels. Ovary 3-celled, with numerous ovules in each cell; when borne at ground level, the ovary has a beak which raises the perianth into the air. Fruit a capsule opening by a lid, or, if borne at ground level, breaking up irregularly to release the seeds. Seeds black.

A genus of 6 species, all from southern Africa. Only 1 species is frequently grown, and numerous cultivars have been developed from it. It is usually easily grown in a light soil, with abundant moisture during the growing season (though care must be taken to prevent over-watering, as the long hairs on the leaves retain moisture). Mathew (reference below) gives a key to and descriptions of, other species that may appear in specialist collections. Propagation is by seed.

Literature: Hilliard, O.M. & Burtt, B.L., Notes on some plants from southern Africa: VII, *Notes from the Royal Botanic Garden*, Edinburgh **36:** 43–76 (1978); Mathew, B., Rhodohypoxis, *The Plantsman* **6:** 49–59 (1984).

1. R. baurii (Baker) Nel (*Hypoxis baurii* Baker).

Leaves mostly 4–10, 2.5–11 cm × 1.2–12 mm, linear-lanceolate to lanceolate, hairy, especially on the upper surface. Scape to 15 cm, hairy. Flowers solitary or in pairs, white, pink or red; perianth-tube to 2.5 mm, outer lobes 1–2 cm, somewhat hairy towards the base. *South Africa.*

Hilliard & Burtt recognise 3 varieties, of which 2 probably occur in cultivation:

Var. **baurii.** Illustration: *Botanical Magazine*, 9412, 1 (1935); Eliovson, Wild flowers of southern Africa, t. 104 (1980); Rix & Phillips, The bulb book, 151 (1981). Perianth-lobes red or deep, bright pink. *South Africa (Cape, Natal, Transkei) Lesotho.* H5–G1. Late spring.

Var. **platypetala** (Baker) Nel (*Hypoxis platypetala* Baker; *H. baurii var.* platypetala (Baker) Baker; *R. baurii* forma *platypetala* (Baker) Milne-Redhead; *R. platypetala* (Baker) Gray). Illustration: *Botanical Magazine*, 9412, 2 (1935); Gibson, *Wild flowers of Natal (inland region)*, t. 29 (1975); Eliovson, Wild flowers of southern Africa, t. 105 (1980); Rix & Phillips, The bulb book, 151 (1981). Perianth white or rarely pale pink. *South Africa (Natal, Cape, Transkei).* H5–G1. Late spring.

R. milloides (Baker) Hilliard & B.L. Burtt (*Rhodohypoxis palustris* Killick; *Hypoxis milloides* Bake). Similar, leaves 8–18 cm × 8–10 mm, erect, narrowly lanceolate, keeled. Scape slender 5–10 cm. Flowers erect, solitary, borne in succession, dark red-crimson, occasionally pink or white; perianth 2.5–3.5 cm across. *South Africa.* H5. Spring–late summer.

Numerous cultivars are available and taller plants such as 'Claret' may reach up to 12 cm.

R. rubella (Baker) Nel. Leaves thread-like, 1–4.5 cm × 1–1.5 mm, sparsely hairy; flowers solitary, the ovary subterranean and hidden by the leaf-sheaths, the perianth raised on a long ovary beak; fruits thin-walled, breaking up irregularly to release the seeds. *South Africa (Natal, Cape).* G1?.

Reputedly in cultivation, but more interesting than attractive.

37. VELLOZIACEAE

Shrubs with forked branches with persistent leaf-bases, or woody-based herbs. Leaves alternate, leathery; ptyxis conduplicate. Flowers solitary, terminal, actinomorphic, on naked stalks borne in terminal tufts of leaves. Perianth of 6 free or united segments, petal-like. Stamens 6 or more in 6 bundles. Ovary of 3 united carpels, inferior; ovules many, axile. Fruit a hard capsule, often with spines or glandular.

Two genera (sometimes further divided) and about 200 species. A few are cultivated as glasshouse ornamentals.

A family of 2 genera, *Vellozia* and *Barbacenia* (other segregate genera are recognised by some authors); unfortunately, there are two ways of circumscribing these genera, neither of which has won complete acceptance. If the genera are defined using the character of whether or not the perianth is united into a tube below, then those species with a perianth-tube are placed in *Barbacenia* and those without in *Vellozia* (including the species below, which is then correctly named *Vellozia elegans*). On this basis, *Vellozia* occurs in S America and in Africa and *Barbacenia* is restricted to S America. On the other hand, the genera can be separated using the number of stamens as the defining character. This means that those species with 6 stamens are placed in *Barbacenia* (including the species described below, which is then correctly named *Barbacenia elegans*). If so defined, the genus occurs in S America and Africa; those species with many stamens are then placed in the genus *Vellozia*, which, so defined, is restricted to S America. Further work on this problem is necessary; meanwhile, the 1 cultivated species can be referred to correctly as either *B. elegans* or *V. elegans*. In this account, the genus is defined as having 6 stamens and the name *Barbacenia* is used.

1. BARBACENIA Vandelli

J. Cullen

Perennial herbs or shrubs (ours). Leaves more or less linear, usually thinly leathery and evergreen. Perianth of 6 more or less equal, free segments. Stamens 6; anthers opening by longitudinal slits; filaments usually very short. Fruit a capsule.

A genus of about 90 species (see above) from Arabia, Africa and S America. Several species have been in cultivation from time to time, but only 1 seems to persist. It is easily grown in a freely draining compost in a warm place; in a glasshouse it can form dense mats and act as ground cover. Propagation is by seed or by division.

Literature: Creves, S., A revision of the Old World species of Vellozia, *Journal of Botany* **59**: 273–84 (1921).

1. B. elegans (J.D. Hooker) Pax (*Vellozia elegans* J.D. Hooker; *V. talboti* invalid; *Talbotia elegans* invalid). Illustration: *Botanical Magazine*, 5803 (1869); Gibson, Wild flowers of Natal (inland region), t. 17 (1975).

Plant slightly woody, forming dense mats to 30 cm high when well grown. Leaves in the upper part of the stem in 3 ranks, 9–20 cm × 6–20 mm, linear-oblong, tapered towards the apex where the margins are slightly toothed, widely V-shaped in section, keeled beneath. Buds pale purple, flowers white. Perianth-segments 1–1.8 cm, acute, somewhat spreading, enlarging and turning green after flowering. Ovary 3-winged. *South Africa (Natal).* H5–G1. Summer.

38. TACCACEAE

Herbs with scapes. Leaves all basal, broad, long-stalked. Flowers actinomorphic, in umbels, each umbel with an involucre of bracts, the inner of which are often dangling. Perianth of 6 united segments, more or less petal-like. Stamens 6. Ovary of 3 united carpels, inferior; ovules many, parietal. Fruit a berry or capsule.

A single genus with about 10 species, a few of which are cultivated.

Literature: Drenth, E., A revision of the family Taccaceae, *Blumea* **20**: 376–406 (1972).

1. TACCA Forster & Forster
V.A. Matthews
The only genus of the family; for description, see above. It is a pantropical genus with 10 species.

Taccas must be grown in a warm greenhouse and require a winter resting period during which water should be almost withheld. Propagation is by division of the rhizomes.

1a. Leaves lobed; involucral bracts 4–12 **1. leontopetaloides**
b. Leaves unlobed; involucral bracts 4 2
2a. Fruit *c.* 1 cm, pale yellow; floral bracts to 8 cm **3. plantaginea**
b. Fruit 2–5 cm, green, orange-red, purple or blackish; floral bracts to 25 cm 3
3a. Perianth-lobes usually reflexed, falling after flowering; floral bracts white or yellowish green, darker at the base **2. integrifolia**
b. Perianth-lobes spreading, persisting until fruiting; floral bracts pale green to violet-green **4. chantrieri**

1. T. leontopetaloides (Linnaeus) Kuntze (*T. pinnatifida* Forster & Forster; *T. artocarpifolia* Seeman). Illustration: *Botanical Magazine*, 6124 (1874) & 7299, 7300 (1893); *Hooker's Icones Plantarum* **26**: t. 2515, 2516 (1897) – as *T. viridis*.

Leaves 1–3, the blades to 70 × 120 cm, broadly obovate to ovate, 3-lobed with each lobe itself lobed or dissected; stalks 17–150 cm. Scape 20–170 cm. Umbel with 20–40 flowers. Involucral bracts 4–12, of different sizes. Floral bracts to 25 cm, dark purple to blackish brown. Flowers 6–17`mm, on drooping stalks to 6 cm, pale yellow, yellow-green or dark purplish green. Perianth-lobes persisting until fruiting. Fruit more or less spherical, 1.5–2.5 cm, green, becoming orange when ripe. *Old World tropics.* G2. Flowering at any time of the year.

2. T. integrifolia Ker Gawler (*T. aspera* Roxburgh; *T. laevis* Roxburgh; *T. cristata* Jack; *Ataccia cristata* (Jack) Kunth). Illustration: *Botanical Magazine*, 1488 (1812); Baillon, Histoire des plantes **13**: 167 (1894); Everard & Morley, Flowers of the world, t. 125 (1970).

Leaves 2–13, the blades 7–65 × 3–24 cm, very variable in shape, usually oblong or lanceolate but sometimes oblong-obovate or linear-lanceolate; stalks 4–40 cm. Umbel with up to 30 flowers. Involucral bracts 4. Floral bracts to 25 cm, white or yellowish green with a darker base. Flowers 1.4–2.7 cm, on stalks 5–40 cm, green, greenish violet or brownish purple, darkening with age. Perianth-lobes usually reflexed, falling after flowering. Fruit 2.5–5 cm, circular or triangular in cross-section, green to black tinged with purple. *NE India, Malaysia, Indonesia (Sumatra, Java), Borneo & Thailand.* G2. Mainly summer.

3. T. plantaginea (Hance) Drenth. Illustration: Das Pflanzenreich **92**: 12 (1928) – as *Schizocapsa plantaginea*.

Leaves 3–8, the blades 8–36 × 2–9 cm, oblong or lanceolate; stalks 5–30 cm. Scape 7–25 cm. Umbel with 6–20 flowers. Involucral bracts 4. Floral bracts to 8 cm. Flowers 9–17 mm, on stalks to 3 cm, white to green tinged with brownish purple. Perianth-lobes spreading, persisting until fruiting. Fruit *c.* 1 cm, triangular in cross-section, pale yellow. *S China, Thailand, N Vietnam & Laos.* G2.

4. T. chantrieri André. Illustration: Graf, Exotica, edn 8, 1468 (1976); Clay & Hubbard, The Hawaii garden, tropical exotics, 140 (1977); Brickell (ed.), RHS A–Z encyclopedia of garden plants, 378 (2003).

Leaves 3–12, the blades 17–55 × 4–22 cm, ovate, elliptic or lanceolate; stalks 11–45 cm. Scape 6–60 cm. Umbel with up to 25 flowers. Involucral bracts 4. Floral bracts to 20 cm, pale green or violet-green. Flowers 1–2.5 cm, on stalks 1–4 cm, greenish white when young, aging to red, purple or blackish. Perianth-lobes spreading, persisting until fruiting. Fruit 2–4 cm, triangular or circular in cross-section, green, orange-red or purple. *NE India & SE Asia.* G2.

39. DIOSCOREACEAE

Climbers with swollen rootstocks, sometimes with aerial stem-tubers. Leaves borne on the stems, usually alternate, stalked, often cordate or palmate; ptyxis flat or conduplicate. Flowers in axillary racemes, unisexual, actinomorphic, small. Perianth of 6 free or united segments, often greenish. Stamens 6 or 3, free or united. Ovary of 3 united carpels, inferior; ovules 2 per cell, axile. Fruit a capsule or berry.

Six genera and over 600 species. Three genera are native to Europe, 2 to North America. Many species are grown as vegetables (yams) in the warmer parts of the world, and a few species are grown as ornamental climbers.

The following terms are used for the direction of twining as seen when looking down on the plant from above: 'clockwise' indicates movement like the hands of a clock, as in the hop, *Humulus lupulus;* 'anticlockwise' indicates movement in the opposite direction, as in the runner bean, *Phaseolus coccineus.*

Literature: Knuth, R., Dioscoreaceae, in Engler, H.G.A., Das Pflanzenreich **87:** 1–387 (1924); Knuth, R., Dioscoreaceae, in Engler, H.G.A., & Prantl, K., Die natürlichen Pflanzenfamilien, edn 2, **15a:** 438–62 (1930).

1a. Fruit a berry **1. Tamus**
b. Fruit a capsule **2. Dioscorea**

1. TAMUS Linnaeus
C.J. King
Perennial, dioecious herbs with large, cylindric to ovoid tubers. Leaves alternate, heart-shaped. Flowers in axillary racemes. Perianth bell-shaped. Stamens 6, rudimentary in the female flower. Style with 3 bilobed stigmas. Fruit a berry with few, unwinged, spherical seeds.

A genus of 5 species from the Atlantic islands, Europe, N Africa and SW Asia, 1 of which is occasionally cultivated for the attractive but poisonous red berries. It thrives well in moist, well-drained soils in half-shade, and may become a pestilent weed. Propagation is by seed or by division of the rootstock.

1. T. communis Linnaeus. Illustration: Bonnier, Flore complète **10**: t. 597 (1929); Hegi, Illustrierte flora von Mittel-Europa **2**: t. 67 (1939); Ross-Craig, Drawings of British plants **29**: t. 9 (1972); Phillips, Wild flowers of Britain, 83(f) & 173(f) (1977).

Root-tuber ovoid, blackish, to 20 cm or more. Stem twining clockwise, to 4 m. Leaves 8–15 × 4–11 cm, deeply cordate at base, long-acuminate at apex, dark shining green, with 3–9 main veins diverging from the base. Flowers 3–6 mm in diameter, greenish yellow. Berry ovoid-ellipsoid, 1–1.2 cm, red, rarely yellowish, with 1–6, pale yellow, wrinkled seeds. *W & S Europe, N Africa, SW Asia.* H4. Summer.

2. DIOSCOREA Linnaeus

C.J. King

Dioecious climbers with twining stems arising from tuberous roots, sometimes bearing aerial tubers in the leaf axils. Perianth of male flowers bell-shaped. Perianth of female flowers deeply 6-lobed. Fruit a 3-angled or 3-winged capsule; seeds flat, usually winged.

About 600 species, mainly in tropical and subtropical regions. Several of the hardier species are grown in European gardens for their leafy, twining habit, and some of the tender ones, often with handsomely coloured leaves, are cultivated as house plants. The hardier species can be grown in any garden soil with good drainage, but they require a sunny position. The more tender species should be kept moderately dry during the winter resting period but need plenty of water during growth. Propagation is chiefly by division of the rootstock during the dormant season, or by aerial tubers where these are produced, but seed may also be used. Variegated-leaved kinds are grown from cuttings.

Root-tubers. Partially exposed: **3, 9**. Large: **3, 7, 9**.
Aerial tubers. Large: **1**.
Stems. Twining clockwise: **1, 4, 9**; twining anticlockwise: **2, 7, 8**. Angled: **3, 5, 7**; terete: **1, 4, 8**.
Leaves. Bicoloured or multicoloured above: **5, 6**. Always opposite: **7**. Narrow: **8**.

1a. Leaves uniformly green above 2
b. Leaves bicoloured or multicoloured above 8
2a. Root-tubers small or absent; aerial tubers large **1. bulbifera**
b. Root-tubers small or large; aerial tubers small or absent 3
3a. Root-tubers partially exposed 4
b. Root-tubers not exposed 5
4a. Leaves more than 5 cm broad **3. macrostachya**
b. Leaves less than 5 cm broad **9. elephantipes**
5a. Leaves opposite **7. batatas**
b. Leaves mostly alternate (occasionally some opposite in 2 and 4) 6
6a. Leaves mostly linear or linear-lanceolate, sometimes hastate **8. hastifolia**
b. Leaves ovate-lanceolate to broadly ovate 7
7a. Stem twining clockwise **4. balcanica**
b. Stem twining anticlockwise **2. cotinifolia**
8a. Leaves heart-shaped to ovate-lanceolate, 7-veined; male flowers stalkless **5. amarantoides**
b. Leaves heart-shaped to circular, 9–11-veined; male flowers stalked **6. discolor**

1. D. bulbifera Linnaeus. Illustration: Parey's Blumengärtnerei **1**: 366 (1958); Graf, Exotica, edn 8, 710 (1976); Graf, Tropica, 393 (1978).

Root-tubers small or absent. Stem terete, twining clockwise to 6 m, bearing aerial, axillary, almost spherical or angled, tubers to 10 cm across. Leaves alternate or opposite, ovate or heart-shaped, cuspidate, to 25 × 18 cm; leaf-stalks 6–14 cm. Male flowers in spikes 3–10 cm; female flowers in spikes 10–25 cm. Capsule to 2.5 × 1.5 cm. *Tropical Africa & Asia.* G2. Late summer.

2. D. cotinifolia Kunth (*D. malifolia* Baker). Illustration: Batten & Bokelmann, Wild flowers of the eastern Cape Province, t. 21, f. 3 (1966).

Root-tubers to 9 × 5 cm. Stem twining anticlockwise. Leaves alternate, sometimes opposite, broadly ovate, truncate or slightly cordate at base, to 8 × 5 cm; leaf-stalks 1.8–3 cm. Male flowers in racemes 5–7.5 cm; female flowers in few-flowered racemes 7.5–15 cm. Capsule to 3 cm. *South Africa.* G2. Summer.

3. D. macrostachya Bentham (*Testudinaria macrostachya* (Bentham) Rowley). Illustration: Graf, Tropica, 392, 393 (1978) – as *D. macrostachys*.

Root-tubers 20 cm or more across, partially exposed, the surface deeply grooved into irregular segments. Stem slightly angled, probably twining anticlockwise. Leaves alternate, ovate, acuminate at apex, cordate at base, to 20 × 18 cm, with long stalks. Male flowers in clusters of 2 or 3 along racemes 15–30 cm; female flowers solitary along racemes. Capsule to 2.5 cm. *C America, from SC Mexico to Panama.* G2.

4. D. balcanica Kosanin. Illustration: *Osterreichische Botanische Zeitschrift* **64**: t. 3 (1914); Jelitto & Schacht, Die Freiland Schmuckstauden **1**: 164 (1963).

Root-tubers to 2 cm in diameter. Stem terete, twining clockwise to 1.5 m or more. Leaves alternate or almost opposite, ovate or heart-shaped, shortly acuminate, to 7 × 6 cm, 9-veined, with long stalks. Fruiting spike to 7 cm, pendent. Capsule 2–2.5 cm across. *SW Yugoslavia & N Albania.* H4. Early summer.

5. D. amarantoides Presl.

Stem slightly angled. Leaves alternate, heart-shaped to ovate-lanceolate, to 10 × 5 cm, 7-veined. Male flowers stalkless, in dense spikes arranged in a panicle up to 40 cm. Capsule 2 × 1.3 cm. *Peru.* G2.

The typical plant is probably not in general cultivation, but a number of non-flowering cultivars, often placed under the name *D. multicolor*, appear to belong here. They have somewhat larger leaves (up to 13 × 8 cm), which are reddish purple beneath, but on the upper side are boldly patterned in various colours, usually with a metallic sheen (Illustration: *Illustration Horticole* **18:** 52, 1871; Graf, Exotica, edn 8, 710, 1976). Those most frequently grown are 'Argyraea', 'Chrysophylla', 'Eldorado', 'Melanoleuca' and 'Metallica'.

6. D. discolor Kunth. Illustration: Parey's Blumengärtnerei **1**: 367 (1958); Graf, Tropica, 392, 393 (1978); Brickell (ed.), RHS A–Z encyclopedia of garden plants, 380 (2003).

Root-tubers to 5–7 cm across. Leaves alternate, heart-shaped, cuspidate, 13–15 cm, dark green above with areas of lighter green, whitish along the midrib, reddish purple beneath, 9–11-veined. Male flowers in racemes; female flowers in spikes. *Tropical S America.* G2. Late summer.

D. discolor is very like *D. dodecaneura* Vellozo, native from Guyana to Paraguay, and may be only a form of that species.

7. D. batatas Decaisne (*D. opposita* misapplied). Illustration: *Flore des serres* **10**: 7 (1854–5).

Root-tubers to 90 cm. Stem slightly angled, twining anticlockwise to 3 m, and bearing small, aerial, axillary tubers. Leaves opposite, ovate, cordate at the base, 4–8 cm. Flowers in spikes, cinnamon-scented. *Temperate E Asia.* H2. Late summer.

8. D. hastifolia Nees (*Testudinaria hastifolia* invalid). Illustration: Erickson et al. (eds), Flowers and plants of Western Australia, 37 (1973).

Root-tubers to 12 × 3 cm. Stem terete, twining anticlockwise to 2 m. Leaves alternate, mostly linear or linear-lanceolate but sometimes hastate, to 8 cm. Male flowers in racemes 5–8 cm; female inflorescences very short with usually only 2 or 3 flowers. Capsule *c.* 2 cm across. *W Australia.* G2. Autumn.

9. D. elephantipes (L'Héritier) Engler (*Testudinaria elephantipes* (L'Héritier) Lindley). Illustration: *Botanical Magazine*, 1347 (1811); Graf, Tropica, 393 (1978); Brickell (ed.), RHS A–Z encyclopedia of garden plants, 380 (2003).

Root-tubers to 90 cm in diameter, partially exposed, the surface deeply grooved into irregular segments (the well-known 'elephant's foot'). Stem twining clockwise to 6 m. Leaves alternate, heart-shaped to kidney-shaped, 2–5 cm broad. Male flowers in racemes 5–7.5 cm; female flowers in racemes. Capsule *c.* 1.5 × 1.3 cm. *South Africa.* G1. Late summer.

40. PONTEDERIACEAE

Aquatic herbs. Leaves alternate, with sheathing bases, often stalked. Inflorescence racemose, borne in the axil of a spathe-like sheath. Flowers bisexual, actinomorphic or rarely zygomorphic. Perianth of 6 united segments, petal-like. Stamens usually 6. Ovary of 3 united carpels, superior; ovules 3–many, axile or parietal. Fruit a capsule.

Nine genera, 5 of which are native to North America, 4 of these introduced in Europe, and about 30 species of aquatic herbs. Species of about 5 genera are cultivated. *Eichhornia crassipes* has become a noxious weed in many tropical river systems.

1a. Stamens 3; flowers stalkless **1. Heteranthera**
b. Stamens 6; flowers stalked or not 2
2a. One stamen longer than the others **2. Monochoria**
b. All stamens equal 3
3a. Plants floating; leaf-stalks inflated at the base **3. Eichhornia**
b. Plants usually rooted in the substrate; leaf-stalks not inflated 4
4a. Perianth distinctly 2-lipped, hairy outside **4. Pontederia**
b. Perianth not 2-lipped, hairless outside **3. Eichhornia**

1. HETERANTHERA Ruiz & Pavon
E.C. Nelson

Annual or perennial herbs with submerged, floating or creeping stems. Leaves linear or with expanded blades and distinct stalks. Inflorescence a spike with many flowers. Perianth-lobes 6. Stamens 3, 1 usually much larger than the others. Fruit a capsule with numerous seeds.

A genus of about 10 species from tropical and subtropical America and Africa. They generally grow in shallow water or on mud. There are 3 species that are occasionally grown in aquaria, and 1 is naturalised in Europe. Propagation is by cuttings.

1a. Flowers yellow; all leaves submerged, linear **1. dubia**
b. Flowers blue; some leaves floating and not linear 2
2a. Floating leaves kidney-shaped **2. reniformis**
b. Floating leaves elliptic **3. zosterifolia**

1. H. dubia (Jacquin) MacMillan. Illustration: Britton & Brown, Illustrated flora of the northern United States and Canada, edn 2, **1**: 464 (1913).

Submerged perennial. Leaves stalkless, linear, to 15 cm. Flowers yellow, lying flat on the water surface. All stamens equal. *N & S America.* G1. Summer.

An easily cultivated plant for tropical aquaria. It does not attract algal growth, but is a poor oxygenator.

2. H. reniformis Ruiz & Pavon.

Submerged, creeping perennial. Floating leaves kidney-shaped, broader than long, to 3 × 3.5 cm; stalk to 10 cm. Flowers 2–5 in a spike, blue. Posterior stamen with a large, green anther, the other 2 stamens with smaller, yellow anthers. *N & S America.* H5–G1. Summer.

May be grown outdoors in the mildest areas, or in indoor aquaria in shallow water. It can spread rapidly, and can become a weed.

3. H. zosterifolia Martius. Illustration: Cook, Water plants of the world, 486 (1974).

Submerged, creeping perennial. Submerged leaves stalkless, linear, to 5 cm × 7 mm; floating leaves with stalks to 8 cm, blades elliptic, to 4 cm. Flowering spikes usually with 2 flowers. Perianth blue. All stamens alike, anthers pale blue. *Brazil, Bolivia.* G2. Summer.

This is a good oxygenator for tropical aquaria. In good light and warm water it will grow lushly and provide shade.

2. MONOCHORIA Presl
E.C. Nelson

Annual or perennial herbs, stems erect or creeping. Leaves stalked, emergent. Flowers blue, in elongate racemes. Perianth-lobes 6. Stamens 6, inserted on the perianth-tube. Anthers unequal, 5 small and yellow, the sixth larger and blue. Fruit a capsule with numerous seeds.

A genus of about 5 species from tropical habitats in Africa, Asia and Australia. Only 1 species is cultivated in Europe. Propagation is by division or by cuttings.

1. M. vaginalis (Burman filius) Kunth. Illustration: Cook, Water plants of the world, 487 (1974).

Stems creeping, rooted in the substrate. Leaves on long stalks, blades arrowhead-shaped. Raceme erect, to 60 cm. *Africa, India.* G2. Summer.

3. EICHHORNIA Kunth
E.C. Nelson

Annual or perennial herbs. Stems floating or creeping; stolons often present. Leaves submerged, floating or emergent; stalks sometimes inflated, blades expanded. Flowers in a spike or panicle. Perianth blue, tubular, with 6 lobes. Stamens 6, inserted on the perianth at 2 levels; anthers more or less equal. Styles 1 or 3. Fruit a capsule with numerous seeds.

A genus of 7 species, now widely dispersed through the tropics, though the majority were originally native in tropical America. One species is commonly grown in tropical aquaria and is now a serious weed in aquatic habitats in tropical regions. The plants should be grown in shallow water over a rich, loamy soil. Propagation is by division.

1a Base of leaf-stalk inflated, bulbous in appearance **1. crassipes**
b. Base of leaf-stalk not inflated **2. azurea**

1. E. crassipes (Martius) Solms-Laubach. Illustration: *Gartenflora* **37**: t. 1271 (1888); Cook, Water plants of the world, 484 (1974).

Plants usually floating; roots conspicuously clothed in fine, blue hairs.

Leaves with markedly swollen stalks, the swelling containing spongy, air-filled cells; blade ovate, to 8 cm. Spike to 15 cm with 8–12 flowers, 5–7 cm in diameter. Perianth blue, lobes 6. *Brazil, now widely naturalised in the tropics elsewhere.* G1. Summer.

Very easy to grow in warm water. Several cultivars are available, some of which have yellow or lilac-red flowers.

2. E. azurea (Swartz) Kunth (*Pontederia azurea* Swartz). Illustration: *Botanical Magazine*, 6487 (1880).

Like *E. crassipes*, but the leaf-stalks not swollen. Perianth-lobes joined at the base and apex of the tube, the tube in between often split. *Subtropical and tropical America.* G2. Summer.

4. PONTEDERIA Linnaeus

E.C. Nelson

Perennial herbs with submerged, floating or creeping stems. Leaves submerged, floating or emergent, linear or stalked. Flowers in a spike, blue. Perianth tubular, hairy outside, 2-lipped, each lip with 3 lobes. Stamens 6, anthers equal. Fruit a nutlet with 1 seed.

A genus of about 5 species from America, generally found growing in shallow water or on mud. Only 1 species is widely cultivated in Europe, where it is also naturalised; it is suitable for pond margins in very shallow water.

Literature: Lowden, R.M., Revision of the genus Pontederia L., *Rhodora* **75:** 426–87 (1973).

1. P. cordata Linnaeus. Illustration: *Botanical Magazine*, 1156 (1808), 2932 (1829) & 3753 (1839); Cook, Water plants of the world, 488 (1974).

A robust perennial to 1 m. Leaves erect; stalk to 25 cm, blades triangular to ovate, with a cordate base, to 20 cm broad. Flowers blue, densely arranged in a spike. Perianth to 8 mm in diameter. *N & S America.* H5. Summer.

One variant is reported to be taller, with a longer spike and lanceolate leaves, but is not as hardy as the species; it probably belongs to var. **lancifolia** (Muhlenberg) Torrey.

41. IRIDACEAE

Terrestrial herbs with rhizomes, corms or bulbs, rarely woody. Leaves alternate, often equitant and/or pleated; ptyxis usually conduplicate or plicate, rarely supervolute. Flowers solitary or in racemes or panicles, bisexual, actinomorphic or zygomorphic. Perianth of 6 free or united segments, petal-like, sometimes in 2 somewhat dissimilar whorls. Stamens 3, on the same radii as the outer perianth-lobes. Ovary of 3 united carpels, inferior; styles 3, often further divided; ovules few–many, usually axile. Fruit a capsule.

About 70 genera and 1,500 species. Nine genera (1 introduced) are found in Europe and 18 are native to North America. Species of many genera are cultivated for ornament. The genera are often difficult to distinguish, and care must be taken using the key below.

Literature: Goldblatt, P. & Manning, J.C., The iris family (2008).

1a. Perianth united into a tube at the base, which may be very short (some genera with extremely short perianth-tubes are keyed out under 1b as well as here) 2
b. Perianth-segments completely free (but including some genera with extremely short, inconspicuous, perianth-tubes) 38
2a. Leaves neither arranged in 2 ranks nor equitant 3
b. Leaves arranged in 2 ranks, usually equitant 7
3a. Leaves each with a white stripe on the upper surface **16. Crocus**
b. Leaves without such a stripe 4
4a. Styles petaloid **1. Iris**
b. Styles not petaloid 5
5a. Filaments united, at least at the base **18. Galaxia**
b. Filaments free 6
6a. Perianth-tube shorter than the lobes **19. Romulea**
b. Perianth-tube longer than the lobes **17. Syringodea**
7a. Perianth bilaterally symmetric (sometimes due only to the curved tube) 8
b. Perianth radially symmetric (tube always straight) 19
8a. Style branches 3, each bifid near the tip 9
b. Style unbranched or branches 3, each entire 12
9a. Inflorescence a spike, its axis bent, or at least curved, at about 90° at the level of the lowest flower, the flowers pointing upwards from the more or less horizontal axis **34. Freesia**
b. Inflorescence a spike, not as above, or a panicle 10
10a. Perianth-tube conspicuously widening upwards; stamens regularly arranged **37. Watsonia**
b. Perianth-tube cylindric, not or scarcely widening upwards; stamens usually asymmetrically arranged 11
11a. Basal leaves exceeding the stem-leaves; stem-leaves with undulate-crisped margins; corms with hard, woody tunics **35. Lapeirousia**
b. Basal leaves not exceeding the stem-leaves; stem-leaves not as above; corms with fibrous tunics **36. Anomatheca**
12a. Bilateral symmetry of the perianth mainly due to the curved tube, the lobes more or less radially symmetric 13
b. Bilateral symmetry of the flower due not only to the curved tube, but due also to the lobes, which are unequal in size and bilaterally symmetric 17
13a. Leaves conspicuously pleated; leaves and stems densely hairy **38. Babiana**
b. Leaves usually not pleated; leaves and stems usually not hairy (occasionally leaves only with a few fine hairs) 14
14a. Stamens symmetrically arranged **30. Hesperantha**
b. Stamens asymmetrically arranged, arching to the upper side of the flower 15
15a. Perianth-tube slender throughout or with a very slender basal part, abruptly expanded above **39. Crocosmia**
b. Perianth-tube gradually widening from the base 16
16a. Style exceeding the stamens **41. Tritonia**
b. Style about as long as the stamens **44. Gladiolus**
17a. The 3 lower lobes of the perianth very small and inconspicuous; bracts almost as long as the perianth-tubes **46. Anomalesia**
b. The 3 lower lobes of the perianth shorter than the others but still conspicuous; bracts much shorter than the perianth-tubes 18
18a. The 3 lower segments of the perianth not united to each other beyond the mouth of the perianth-tube; perianth red, orange or red and yellow **45. Chasmanthe**

b. The 3 lower lobes of the perianth united to each other beyond the mouth of the perianth-tube; perianth lavender to deep purple, with yellow stripes **43. Synnotia**
19a. Perianth-lobes distinctly in 2 whorls, the lobes of the outer whorl differing from those of the inner in shape and size 20
b. Perianth-lobes all similar, apparently in 1 whorl, or if obviously in 2 whorls then the lobes all similar in size and appearance 22
20a. Inner perianth-lobes differing from the outer in size, shape, posture and often in colour; style-branches petaloid 21
b. Inner perianth-lobes differing from the outer only in width; style-branches not petaloid **40. Melasphaerula**
21a. Rootstock consisting of finger-like tubers; inner perianth-lobes greenish yellow, blade of outer lobes dark purple-brown; ovary 1-celled **4. Hermodactylus**
b. Rootstock a bulb or rhizome; flowers never with the above colour combination; ovary 3-celled **1. Iris**
22a. Style-branches fringed **14. Ferraria**
b. Style-branches not fringed, or style unbranched 23
23a. Style-branches 3, each bifid at the tip **35. Lapeirousia**
b. Style-branches 3, each entire, or style unbranched 24
24a. Filaments united 25
b. Filaments free 28
25a. Style not branched, stigma entire or 3-toothed **26. Solenomelus**
b. Style 3-branched, sometimes slightly bifid, stigmas 3 26
26a. Flowers stalkless **33. Ixia**
b. Flowers with stalks (which may be concealed within bracts) 27
27a. Style-branches club-shaped, with rounded tips **20. Gelasine**
b. Style-branches narrowing from the base, tips acute or capitate **25. Sisyrinchium**
28a. Plant woody **28. Nivenia**
b. Plant not woody 29
29a. Leaves conspicuously pleated; stems and leaves densely hairy **38. Babiana**
b. Leaves not pleated; stems and leaves not hairy 30
30a. Stems arching over; bracts scarious, conspicuous; flowers pendent **32. Dierama**
b. Stems not arching over; bracts inconspicuous, green or scarious; flowers not pendent 31
31a. Plants with rhizomes 32
b. Plants with corms 34
32a. Flowers scarlet, crimson or pink **29. Schizostylis**
b. Flowers blue or purplish 33
33a. Perianth-lobes twisting spirally after flowering **27. Aristea**
b. Perianth-lobes not twisting spirally after flowering **24. Orthrosanthus**
34a. Corm-tunics hard and woody 35
b. Corm-tunics papery or fibrous 36
35a. Style longer than perianth-tube, branches usually short, recurved, exceeding the anthers **31. Geissorhiza**
b. Style shorter than or equal to perianth-tube, branches long, ascending, often not reaching the anther-tip **30. Hesperantha**
36a. Capsule thin-walled, the outlines of the seeds visible through it **33. Ixia**
b. Capsule with firm, membranous walls, the outlines of the seeds not visible 37
37a. Perianth-tube funnel-shaped; bracts entire or raggedly toothed near the tip **42. Sparaxis**
b. Perianth-tube cylindric; bracts notched near the tip **41. Tritonia**
38a. Fertile stamens 2 **21. Diplarrhena**
b. Fertile stamens 3 39
39a. Style-branches petaloid 40
b. Style-branches not petaloid 45
40a. Flower with an apparent perianth-tube that is, in fact, a beak on top of the ovary **3. Gynandriris**
b. Flower without a beak on top of the ovary 41
41a. Flowering stem with many long branches; claw of outer perianth-segments marked with transverse bands **2. Pardanthopsis**
b. Flowering stem unbranched or with a few short branches; claw of outer perianth-segments not marked with transverse bands 42
42a. Filaments united, at least at the base; rootstock a corm **5. Moraea**
b. Filaments usually free; rootstock a rhizome or bulb 43
43a. Leaves usually all basal; perianth-tube present, though sometimes extremely short **1. Iris**
b. Leaves basal and also borne on the stem; perianth-tube absent 44
44a. Leaves pleated **7. Cypella**
b. Leaves not pleated **6. Dietes**
45a. Filaments free 46
b. Filaments united at least at the base 54
46a. Perianth-segments distinctly in 2 whorls, the segments of the outer whorl differing from those of the inner in shape or size 47
b. Perianth-segments all similar, in 1 or 2 whorls 51
47a. Each perianth-segment with a band of hairs down the centre **13. Alophia**
b. Perianth-segments hairless, or if hairy, then not as above 48
48a. Flowers white to greenish yellow, to 1.8 cm in diameter **40. Melasphaerula**
b. Flowers bright yellow, blue or violet-blue, if white or pale yellow, then 2–10 cm in diameter 49
49a. Leaves pleated **7. Cypella**
b. Leaves not pleated 50
50a. Flowers 2–2.5 cm in diameter; outer perianth-segments yellow with spots at the base **8. Trimezia**
b. Flowers 3–10 cm in diameter; outer perianth-segments white, blue or violet-blue, if yellow then with bars at the base **9. Neomarica**
51a. Perianth-segments twisting spirally after flowering 52
b. Perianth-segments not twisting spirally after flowering 53
52a. Flowers blue **27. Aristea**
b. Flowers yellow or orange-red with red or purple spots **23. Belamcanda**
53a. Flowers stalkless; leaves not stiff, often curled **31. Geissorhiza**
b. Flowers shortly stalked; leaves stiff and leathery **24. Orthrosanthus**
54a. Style-branches fringed **14. Ferraria**
b. Style-branches not fringed 55
55a. Style-branches divided, bifid or notched 56
b. Style-branches entire 59
56a. Flowers 3–15 cm in diameter; stigmas on radii alternating with the anthers **10. Tigridia**
b. Flowers smaller; stigmas on the same radii as the anthers 57
57a. Perianth red, orange or yellow; segments coiling spirally after flowering **11. Rigidella**
b. Perianth purple or bluish; segments not coiling spirally after flowering 58
58a. Each perianth-segment with a band of hairs down the centre; anther-connective broad, fiddle-shaped **13. Alophia**

b. Perianth-segments hairless; anther-connective narrow, linear **12. Herbertia**

59a. Rootstock a corm 60

b. Rootstock a rhizome or just a cluster of fibrous roots 61

60a. Flowers blue; corm-tunics membranous **20. Gelasine**

b. Flowers not blue; corm-tunics netted **15. Homeria**

61a. Segments of the outer whorl of the perianth smaller than those of the inner whorl **22. Libertia**

b. Perianth-segments all of the same size **25. Sisyrinchium**

1. IRIS Linnaeus

V.A. Matthews & B. Mathew

Rhizomatous or bulbous perennial herbs. Leaves usually basal and in 2 ranks, flat, channelled, 4-angled or nearly cylindric. Flowers 1–several, borne within 2 spathes (Figure 9(1& 2), p. 242). Perianth radially symmetric, united at the base into a tube (hypanthial tube). Falls (outer lobes) narrowed towards the base into a claw (haft), the blade sometimes bearded. Standards (inner lobes) usually erect or arching, sometimes horizontal or deflexed, narrowed towards the base into a claw, rarely bearded or very reduced. Stamens 3, filaments free, borne at the base of the falls. Style-branches 3, each 2-lobed beyond the stigma, coloured and petaloid, each covering a stamen. Capsule cylindric to ellipsoid, more or less round to triangular in cross-section, often with 3 or 6 ribs; seeds numerous, sometimes bearing a fleshy appendage (aril). See Figure 10, p. 243.

A genus of about 250 species distributed throughout the northern hemisphere. Cultivation varies according to the species and is too complex to discuss here. Reference should be made to Mathew, B., The iris (1981) where the requirements of the species are discussed.

Literature: Dykes, W.R., The Genus Iris (1915), reprinted 1974; Lenz, L.W., A revision of the Pacific coast irises, *Aliso* **4**: 1–72 (1958); Cohen, V.A., A guide to the Pacific coast irises (1967); Grey-Wilson, C., The genus Iris, subsection Sibiricae (1971); Mathew, B., The iris (1981); Brearley, C., The crested irises of Asia, part 1, *The New Plantsman* **7**(2): 123–125 (2000); Brearley C., The crested irises of Asia, part 2, *The New Plantsman* **7**(3): 168–174 (2000); Mason, C.T., Iridaceae, iris family part 2: Iris L. and Nemastylis Nutt., *Journal of the Arizona–Nevada Academy of Science* **33**: 35–37 (2001).

1a. Plants with a bulb; leaves 4-angled or almost cylindric in cross-section or channeled 2

b. Plants usually with a rhizome, never bulbous; leaves flat, sometimes very narrow 19

2a. Leaves 4-angled or almost cylindric in cross-section, if channelled then standards erect and perianth-tube more than 5 cm; bulb-tunics netted and strongly fibrous 3

b. Leaves channelled; bulb-tunics papery to tough and leathery, never netted or fibrous 8

3a. Standards very reduced, bristle-like, 3–5 mm **92. danfordiae**

b. Standards not bristle-like, more than 2 cm 4

4a. Blade of falls pale yellow, spotted with green **93. winogradowii**

b. Blade of falls not pale yellow 5

5a. Leaves almost cylindric with 8 ribs; falls whitish with a dark violet apex **89. bakeriana**

b. Leaves 4-angled; falls not whitish with a dark violet apex 6

6a. Blade of falls pale to deep blue, sometimes violet-blue 7

b. Blade of falls whitish, greyish lilac, lilac-blue, purplish brown, violet-blue or purple, if blue then bracts green, not papery **88. reticulata**

7a. Leaves more or less absent at flowering time **90. histrioides**

b. Leaves present, often to 30 cm, at flowering time **91. histrio**

8a. Standards horizontal or deflexed, if erect then flowers smelling of cloves; dormant bulbs usually with thickened fleshy roots 9

b. Standards erect; bulbs with thin fibrous roots 15

9a. Stem to 15 cm 10

b. Stem 15 cm or more 12

10a. Blade of falls cream, yellow, greenish yellow or green, sometimes with pinkish brown staining **83. caucasica**

b. Blade of falls whitish, greyish, brownish, lilac, purple, reddish or blue 11

11a. Falls 2.5–5 cm **84. persica**

b. Falls 5–8 cm **81. planifolia**

12a. Blade of falls yellow or greenish **85. bucharica**

b. Blade of falls whitish, blue, or lilac to purple 13

13a. Claw of falls unwinged or with a narrow wing on each side **86. albomarginata**

b. Claw of falls with a broad wing on each side 14

14a. Internodes of stem visible at flowering time **87. magnifica**

b. Internodes of stem concealed by leaves at flowering time **82. aucheri**

15a. Flowers yellow; perianth-tube 3.5–5 cm **79. juncea**

b. Flowers blue to violet, reddish violet or bronze, if yellow or white then perianth-tube to 3 cm 16

16a. Plant leafless in winter, producing leaves in the spring **76. latifolia**

b. Plant producing leaves in autumn which persist through the winter 17

17a. Perianth-tube 1–10 mm **77. xiphium**

b. Perianth-tube 1–3 cm 18

18a. Flowers rich reddish violet; standards blunt at the apex **78. filifolia**

b. Flowers blue to violet-purple; standards pointed at the apex, or if blunt then falls with a sparse yellow beard **80. tingitana**

19a. Falls bearing an obvious beard of fairly long hairs 20

b. Falls lacking an obvious beard, but sometimes with cockscomb-like ridges on the blade or very short downy hairs on the claw 50

20a. Beard-hairs composed of several cells; stems branched or not; seed lacking a fleshy appendage 21

b. Beard-hairs composed of one cell only; stems unbranched; seeds with a fleshy appendage 37

21a. Stem almost absent, if present then unbranched, 5–30 cm, sometimes to 60 cm 22

b. Stem branched, usually 15–125 cm 28

22a. Bracts sharply keeled 23

b. Bracts rounded or sometimes the outer one with a slight keel 24

23a. Perianth-tube 1.5–2.5 cm **5. reichenbachii**

b. Perianth-tube 3–4.5 cm **6. suaveolens**

24a. Stem absent or up to 1 cm; plant leafless in winter 25

b. Stem at least 3 cm; leaves present throughout the winter 26

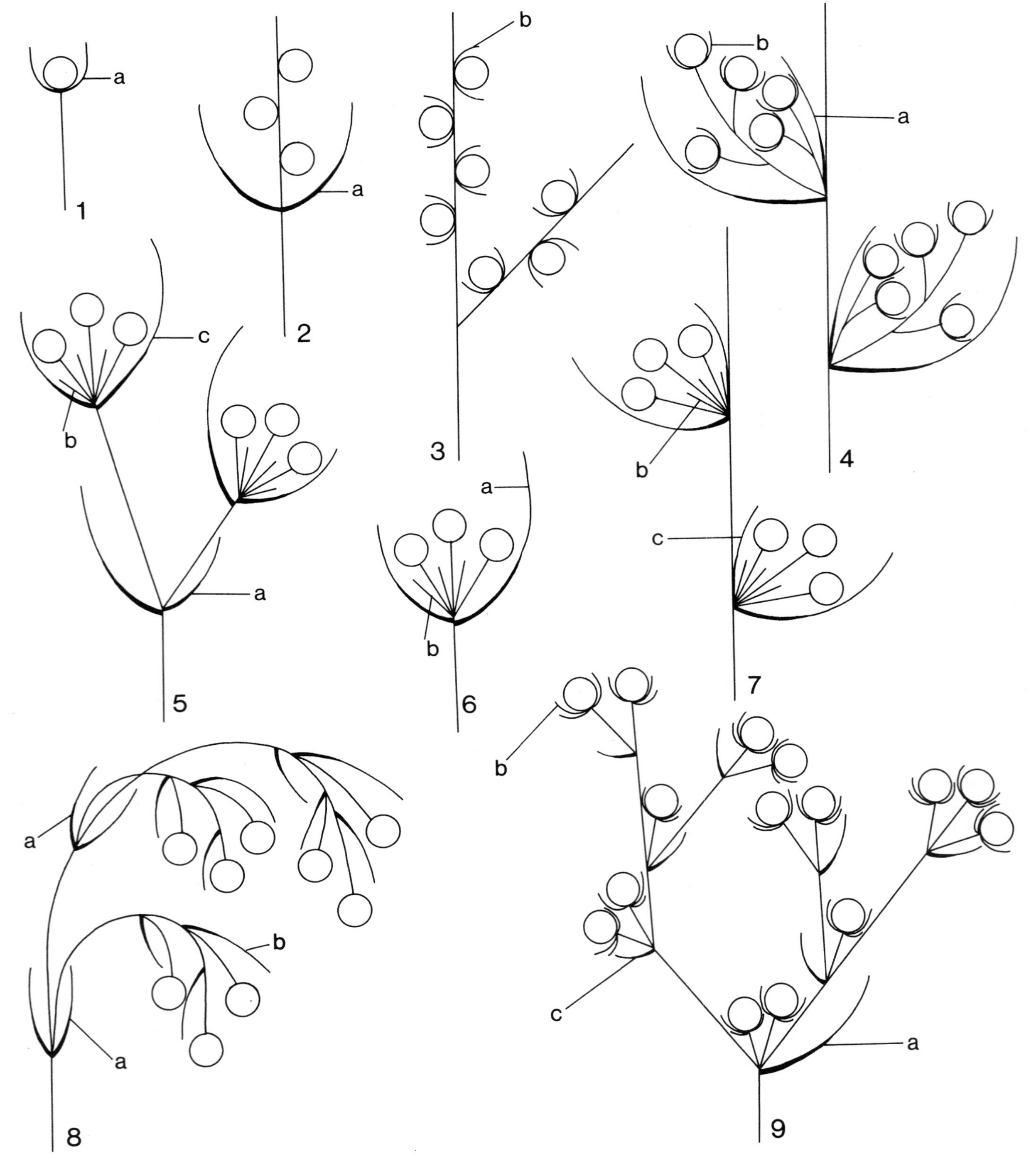

Figure 9. Inflorescences of the Iridaceae. 1, Solitary, terminal flower with 2 spathes (a), e.g. most *Iris* species. 2, Spike with 2 spathes (a), flowers without bracts, e.g. some *Iris* species. 3, Spike without spathes, each flower with 2 bracts (b), e.g. *Lapeirousia*. 4, Panicle, each group of flowers with 2 spathes (a), each flower with 2 bracts (b), e.g. *Orthrosanthus*. 5, Panicle with primary spathes (a), each group of flowers with secondary spathes (c), each flower with 1 bract (b), e.g. some *Sisyrinchium* species. 6, Terminal inflorescence with 2 spathes (a), each flower with 1 bract (b), e.g. some *Sisyrinchium* species. 7, Panicle without primary spathes, each group of flowers like those in No. 5 but stalkless with secondary spathes (c), each flower with 1 bract (b), e.g. some *Sisyrinchium* species. 8, Arching panicle, each group of flowers with 2 spathes (a), each flower with 1 bract (b) at the base of the stalk, e.g. *Dierama*. 9, Complex panicle with 1 primary spathe (a), each branch with 1 secondary spathe (c), each flower with 3 bracts (b), e.g. *Nivenia*.

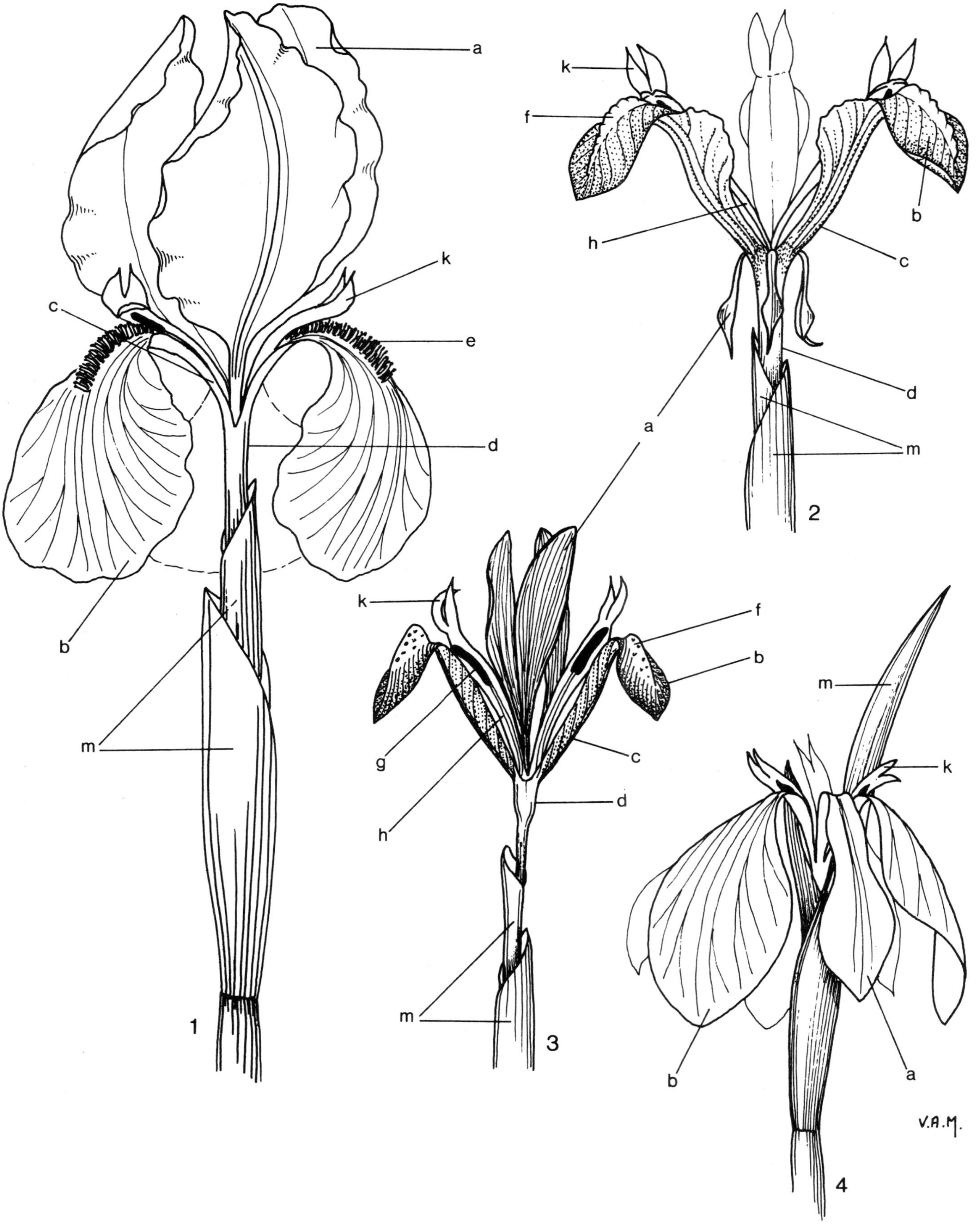

Figure 10. Flowers of Iris species. 1, Typical flower of Section Iris (p. 241). 2, Typical flower of Subgenus Scorpiris (p. 259). 3, Typical flower of Subgenus Hermodactyloides (p. 261). 4, *I. fulva* (p. 255) (a, standard; b, blade of fall; c, claw of fall; d, perianth-tube; e, beard; f, ridge; g, anther; h, style-branch; k, style-lobe; m, bract).

25a. Leaves strongly curved, less than 1 cm wide; flowers 3.5–4.5 cm in diameter **2. attica**
b. Leaves straight or slightly curved, 1–1.5 cm wide; flowers 5–6 cm in diameter **1. pumila**
26a. Flowers 8–9 cm in diameter; leaves abruptly narrowed to an incurved tip **10. albicans**
b. Flowers 4–8 cm in diameter; leaves not abruptly narrowed to an incurved tip 27
27a. Perianth-tube 5–7.5 cm **4. pseudopumila**
b. Perianth-tube 2–5 cm **3. lutescens**
28a. All bracts completely papery at flowering time **11. pallida**
b. Inner bracts green or purplish at least in part at flowering time 29
29a. Standards dirty yellow to clear yellow 30
b. Standards white, bluish, or lavender to purple 32
30a. Bracts green or purplish at least in the lower half, or completely greenish-membranous **8. variegata**
b. Bracts papery-translucent at the apex and margin 31
31a. Bracts sharply keeled; lobes of style-branches toothed **5. reichenbachii**
b. Bracts not keeled; lobes of style-branches not toothed **12. schachtii**
32a. Flowers 9–15 cm in diameter 33
b. Flowers 5–8 cm in diameter 34
33a. Standards with a small beard on the lower part; bracts to 11 cm, narrow **13. kashmiriana**
b. Standards beardless; bracts to 5.5 cm **9. germanica**
34a. Standards whitish with violet veins **8. variegata**
b. Standards lavender to violet or purple 35
35a. Bracts sharply keeled **5. reichenbachii**
b. Bracts not keeled 36
36a. Bracts transparent towards margin and apex **12. schachtii**
b. Bracts not transparent towards margin and apex **7. aphylla**
37a. Flowers lilac, purple or blue, mottled and blotched darker 38
b. Flowers of various colours, not mottled or blotched although sometimes veined or finely spotted 39
38a. Stem to 3 cm, rarely to 7 cm; perianth-tube 5–7.5 cm **25. kamaonensis**
b. Stem 5–15 cm; perianth-tube 2–3 cm **26. hookeriana**
39a. Both falls and standards bearded 40
b. Only the falls bearded 42
40a. Falls with a dark signal patch; beard of standards greenish or almost absent **22. korolkowii**
b. Falls lacking a dark signal patch; beard of standards yellow, lilac, purple or blue 41
41a. Falls lilac-blue; beard yellow **23. hoogiana**
b. Falls deep purplish-blue, brownish or purplish, sometimes shading to blue in the centre; beard yellow, lilac or blue; if falls lilac-blue then beard purple **24. stolonifera**
42a. Stem with 1–3 flowers; flowers yellow, 3–5 cm in diameter **14. bloudowii**
b. Stem bearing 1 flower; flowers of various colours but rarely yellow, at least 5 cm in diameter 43
43a. Ground colour of falls deep violet to blackish purple or purple-brown **20. atropurpurea**
b. Ground colour of falls white, cream, yellow, greenish, greyish, lilac-pink or bluish 44
44a. Beard of falls whitish, yellow, yellowish brown or reddish brown, sometimes with purple-tipped hairs, or purple hairs at edge of beard 45
b. Beard of falls purple or purplish brown 48
45a. Plant to 30 cm at flowering time 46
b. Plant more than 30 cm at flowering time, if less then leaves 1–1.5 cm wide or flowers yellow, spotted with brown 47
46a. Beard of falls with yellowish brown or reddish brown hairs, becoming purple at edge of beard **18. samariae**
b. Beard of falls with yellow hairs, sometimes whitish or tipped with purple **17. sari**
47a. Flowers 13–20 cm in diameter, the overall appearance greyish, brown or purple **15. gatesii**
b. Flowers 6–9 cm in diameter **16. lortetii**
48a. Flowering stem to 25 cm, sometimes to 30 cm 49
b. Flowering stem 30 cm or more **15. gatesii**
49a. Falls and standards circular, more or less equal in size **19. iberica**
b. Falls and standards differing in size, the falls usually smaller and narrower **21. acutiloba**
50a. Blade of falls with a central ridge or crest (occasionally with 3 ridges), which may be dissected or not 51
b. Blade of falls lacking a central ridge or crest 60
51a. Stem more or less absent, or if present then unbranched 52
b. Stem branched 54
52a. Leaves 2–5 mm wide **75. decora**
b. Leaves 5 mm or more wide 53
53a. Leaves 1–3 cm wide; falls with 3 ridges; perianth-tube 4–6 cm **67. cristata**
b. Leaves usually less than 1 cm wide; falls with 1 ridge; perianth-tube to 2 cm **68. lacustris**
54a. Flowers 3–5 cm in diameter 55
b. Flowers 5.5–10 cm in diameter 59
55a. Stem 10–35 cm; ridge on falls undissected, or if dissected then leaves only 2–5 mm wide; lobes of style-branches not fringed, or if fringed then bracts fused at the base 56
b. Stem 45–100 cm; ridge on falls dissected or frilly; lobes of style-branches fringed 58
56a. Bracts fused at the base; ridge on falls mainly white, yellow at the apex **71. gracilipes**
b. Bracts not fused at the base; ridge on falls yellow or orange-yellow, sometimes white or purple at the apex 57
57a. Leaves 2–5 mm wide; upper half of standards bent downwards **75. decora**
b. Leaves 1–1.5 cm wide; standards wholly erect **69. tennis**
58a. Falls with a fringed margin **72. japonica**
b. Falls with a scalloped-wavy margin **73. confusa**
59a. Flowering stem 25–35 cm, slightly branched; perianth-tube *c.* 2.5 cm **70. tectorum**
b. Flowering stem 30–200 cm, much-branched; perianth-tube *c.* 1.2 cm **74. wattii**
60a. Seeds scarlet **62. foetidissima**
b. Seeds not scarlet 61
61a. Style-branches bearing golden yellow glands on the upper surface towards the margin **66. unguicularis**
b. Style-branches lacking golden yellow glands 62

62a. Standards much reduced, very narrowly lanceolate **30. setosa**
b. Standards not reduced 63
63a. Flowers pale yellow or greenish yellow, the blade of the falls with an orange signal patch; bracts to 18 cm; rhizome usually vertical, covered with spiny, needle-like fibres **65. grant-duffii**
b. Flowers of various colours, if pale yellow then lacking an orange signal patch on the falls; bracts 3–12 cm; rhizome lacking spiny fibres 64
64a. Stem hollow, if solid then leaves glossy on 1 side, grey-green on the other 65
b. Stem solid 72
65a. Flowers yellow 66
b. Flowers violet, blue-violet or reddish purple, occasionally white 67
66a. Flowering stem 35–40 cm; leaves shorter than flowering stem; flowers 5–6 cm in diameter with erect standards **33. forrestii**
b. Flowering stem 60–75 cm; leaves more or less equal in height to flowering stem; flowers 6–8 cm in diameter, the standards held at an oblique angle **34. wilsonii**
67a. Stem solid **38. clarkei**
b. Stem hollow 68
68a. Bracts brown and papery at flowering time; leaves green on both sides **32. sibirica**
b. Bracts green or reddish at flowering time although sometimes brownish at the tip; leaves grey-green, or with 1 side glossy green 69
69a. Leaves equal in length to the flowering stem or more or less so 70
b. Leaves shorter than the flowering stem **35. delavayi**
70a. Leaves glossy green on one side, grey-green on the other **36. bulleyana**
b. Leaves grey-green on both sides 71
71a. Standards erect **31. sanguinea**
b. Standards held obliquely **37. chrysographes**
72a. Falls *c.* 1 cm **27. minutoaurea**
b. Falls 3 cm or more 73
73a. Falls basically white, lilac to purple, pinkish, orange-red or red 74
b. Falls basically cream or yellow 108
74a. Leaves 1–5 mm wide 75
b. Leaves 6 mm or more wide 85
75a. Blade of falls with an orange, yellow or whitish central patch or stripe 76
b. Blade of falls lacking such a patch or stripe 80
76a. Flowering stem 30–90 cm; perianth-tube 2–20 mm 77
b. Flowering stem 4–50 cm, if taller then perianth-tube 3.5–6.5 cm 78
77a. Flowers 8–15 cm in diameter; claw of falls yellow, the yellow spreading upwards to the base of the blade **50. ensata**
b. Flowers 5.5–7 cm in diameter; claw of falls greenish white with violet veins **54. prismatica**
78a. Perianth-tube 2–8 cm 79
b. Perianth-tube 6–15 mm **45. missouriensis**
79a. Blade of falls with a cream or yellowish central band, or band absent; apex of rhizome with remains of shiny brown leaf bases **63. tenuifolia**
b. Blade of falls with an orange stripe; old leaf bases not present at apex of rhizome **28. verna**
80a. Capsule rounded, not ribbed; bracts with pink margins **29. ruthenica**
b. Capsule ribbed; bracts lacking pink margins 81
81a. Capsule with 3 ribs 82
b. Capsule with 6 ribs 83
82a. Perianth-tube 5–30 mm **41. innominata**
b. Perianth-tube 2–3 mm **54. prismatica**
83a. Ribs of capsule evenly spaced 84
b. Ribs of capsule arranged in 3 pairs **57. sintenisii**
84a. Capsule with a beak; perianth-tube 2–3 mm with a slender beak at the base that is part of the ovary **64. lactea**
b. Capsule lacking a beak; perianth-tube *c.* 1 cm, lacking a beak at the base **45. missouriensis**
85a. Flowering stem to 6 cm; falls *c.* 3.5 cm **28. verna**
b. Flowering stem 10 cm or more; falls 4 cm or more, if less then blade lacking a central orange stripe 86
86a. Falls red, coppery-red or orange-red **52. fulva**
b. Falls basically white or shades of pink or lilac to purple 87
87a. Blade of falls with a central patch or stripe of orange or yellow 88
b. Blade of falls lacking such a patch or stripe, or patch, if present, whitish 101
88a. Flowering stem branched 89
b. Flowering stem unbranched 97
89a. Capsule with 3 ribs 90
b. Capsule with 6 ribs 94
90a. Stigma triangular **39. douglasiana**
b. Stigma 2-lobed 91
91a. Flowers 6–8 cm in diameter 92
b. Flowers 8–15 cm in diameter 93
92a. Blade of falls 3–4 cm wide, with a hairy yellow central patch **48. virginica**
b. Blade of falls 1.5–3 cm wide, central patch yellowish green, not hairy **49. versicolor**
93a. Leaves 4–12 mm wide with a prominent midrib **50. ensata**
b. Leaves 1.5–4 cm wide, lacking a prominent midrib **51. laevigata**
94a. Flowers white except for a yellow patch on the blade of the falls **59. orientalis**
b. Flowers blue to lilac or purple 95
95a. Leaves *c.* 2.5 cm wide **53. brevicaulis**
b. Leaves 3–12 mm wide 96
96a. Capsule with 6 evenly spaced ribs **45. missouriensis**
b. Capsule with 6 ribs arranged in 3 pairs **55. spuria**
97a. Capsule with 3 ribs 98
b. Capsule with 6 ribs 99
98a. Leaves 4–12 mm wide with a prominent midrib **50. ensata**
b. Leaves 1.5–4 cm wide, lacking a prominent midrib **51. laevigata**
99a. Capsule with 6 evenly spaced ribs 100
b. Capsule with 6 ribs arranged in 3 pairs **55. spuria**
100a. Flowers 2 or 3 in each set of bracts; leaves usually overtopping the flowering stem, which is usually branched **45. missouriensis**
b. Flowers 3–8 in each set of bracts; leaves equal in height to flowering stem or shorter; stem usually unbranched **46. longipetala**
101a. Flowering stem branched 102
b. Flowering stem unbranched 104
102a. Leaves 6–12 mm wide; capsule with 6 ribs arranged in 3 pairs **55. spuria**
b. Leaves 2–7 mm wide; capsule with 3 or 6 evenly spaced ribs 103
103a. Perianth-tube 2–3 mm; capsule with 3 ribs **54. prismatica**

b. Perianth-tube *c.* 10 mm; capsule with 6 ribs **45. missouriensis**

104a. Capsule with 3 ribs **43. macrosiphon**

b. Capsule with 6 ribs 105

105a. Ribs of capsule evenly spaced 106

b. Ribs of capsule in pairs at each of the 3 corners 107

106a. Flowers 2 or 3 in each set of bracts; leaves usually overtopping the flowering stem, which is usually branched **45. missouriensis**

b. Flowers 3–8 in each set of bracts; leaves equal in height to flowering stem or shorter; stem usually unbranched **46. longipetala**

107a. Flowering stem 20–40 cm, compressed or 2-winged **56. graminea**

b. Flowering stem 30–90 cm, not compressed or 2-winged **55. spuria**

108a. Flowers 12–18 cm in diameter 109

b. Flowers 10 cm in diameter or less 110

109a. Flowers lemon yellow; lobes of style-branches broadly triangular, 4–5 mm, strongly recurved **60. × monnieri**

b. Flowers golden yellow; lobes of style-branches narrowly triangular, *c.* 1 cm, not strongly recurved **61. crocea**

110a. Leaves 1–3 cm wide 111

b. Leaves up to 1 cm wide 112

111a. Capsule with 3 ribs; flowers 7–10 cm in diameter **47. pseudacorus**

b. Capsule with 6 ribs; flowers to 7 cm in diameter **55. spuria**

112a. Flowers 3–4 cm in diameter; capsule round in cross-section **29. ruthenica**

b. Flowers 4 cm or more in diameter; capsule ribbed, or if round then perianth-tube 5–8 cm 113

113a. Perianth-tube to 1 cm 114

b. Perianth-tube 1.5–12 cm 117

114a. Stigma triangular; capsule with 3 ribs 115

b. Stigma 2-lobed; capsule with 6 ribs 116

115a. Stem to 30 cm, unbranched; perianth-tube 5–10 mm **40. bracteata**

b. Stem 30–80 cm, usually branched; perianth-tube 2–3 mm **54. prismatica**

116a. Falls dingy yellow to golden yellow, usually with darker veining; flowers to 7 cm in diameter **55. spuria**

b. Falls cream to pale yellow, lacking darker veining; flowers 7–10 cm in diameter **58. kerneriana**

117a. Leaves persisting through the winter; capsule with 3 ribs 118

b. Leaves not persisting through the winter; capsule never with 3 ribs **63. tenuifolia**

118a. Leaves with pink, red or purplish coloration at the base 119

b. Leaves lacking pink, red or purplish coloration at the base 120

119a. Leaves 2–4 mm wide; perianth-tube 1.5–3 cm **41. innominata**

b. Leaves 7–9 mm wide; perianth-tube 3–6 cm **44. purdyi**

120a. Perianth-tube widened at the apex into a bowl-like shape; style-branches longer than the toothed lobes **43. macrosiphon**

b. Perianth-tube not widened into a bowl-like shape; style-branches more or less equal in length to the narrow entire lobes **42. chrysophylla**

Subgenus **Iris.**
Falls bearing a beard of fairly long hairs.

Section **Iris** (Section *Pogoniris* (Spach) Baker). Rhizomes stout, giving rise to fans of sword-shaped, usually fairly broad, leaves. Stems simple or branched, usually with 2 or more flowers. Falls and standards well developed, the falls bearing a prominent beard with each hair made up of several cells. Seeds lacking a fleshy appendage.

1. I. pumila Linnaeus. Illustration: Dykes, The genus Iris, pl. 32 (1915); Everard & Morley, Flowers of the world, pl. 42 (1970); Schauer & Caspari, Pflanzenführer, 119 (1978); Grey-Wilson & Mathew, Bulbs, pl. 27 (1981).

Stem more or less absent; plant 10–15 cm at flowering time. Leaves grey-green, 10–15 × 1–1.5 cm, almost straight to somewhat curved, absent in winter. Bracts 5–10 cm, the outer usually green and often slightly keeled, the inner rather membranous, not keeled. Flowers solitary, scented, 5–6 cm in diameter, yellow, blue, purple or a mixture of these colours. Perianth-tube 4–9 cm. Falls 3.5–6 cm; beard yellow or bluish. Standards 4–8 cm. *Eastern, central & SE Europe to the Urals.* H2. Spring.

I. pumila has been used frequently in the production of the often-grown dwarf bearded cultivars.

2. I. attica Boissier & Heldreich (*I. pumila* Linnaeus subsp. *attica* (Boissier & Heldreich) Hayek). Illustration: Huxley & Taylor, Flowers of Greece and the Aegean, f. 406–8 (1977); Mathew, The iris, f. 3 (1981).

Similar to *I. pumila* but smaller, only 5–10 cm at flowering time. Leaves 4–7 cm × 4–7 mm, strongly curved. Flowers 3.5–4.5 cm in diameter, whitish, yellow, blue, purple or bicoloured. *Greece, S Yugoslavia & NW Turkey.* H2. Spring.

Regarded by some botanists as a subspecies of *I. pumila.*

3. I. lutescens Lamarck (*I. chamaeiris* Bertoloni). Illustration: *Botanical Magazine*, 2861 (1828); Polunin, Flowers of Europe, f. 1692 (1969); Grey-Wilson & Mathew, Bulbs, pl. 27 (1981); Mathew, The iris, f. 5 (1981).

Plant 5–35 cm, stem unbranched. Leaves to 30 cm × 5–25 mm, straight or slightly curved, present throughout the winter. Bracts 3–5.5 cm, green or sometimes brownish and papery at the apex. Flowers 1 or 2, 6–7 cm in diameter, yellow, violet or a mixture of the two, occasionally white. Perianth-tube 2–3.5 cm. Falls 5–7.5 cm; beard yellow, at least the part on the blade of the falls. Standards 5.5–7.5 cm. *NE Spain, S France & Italy.* H2. Spring.

I. scariosa Link. Bracts dry and more or less transparent. Flowers 4–5 cm in diameter, reddish-violet, blue-violet, yellow or almost white. Falls to 4.5 cm. Standards to 3.5 cm. *Former USSR (Urals eastward to the Tien Shan).* H2. Spring.

I. subbiflora Brotero (*I. lutescens* Lamarck subsp. *subbiflora* (Brotero) Webb & Chater). Illustration: *Botanical Magazine*, 1130 (1808); Dykes, The genus Iris, pl. 33 (1913); Polunin & Smythies, Flowers of south-west Europe, pl. 62 (1973). Flowers 7–8 cm in diameter, deep violet with a perianth-tube 3.5–5 cm. Falls to 3.5 cm; beard white or violet, at least the part on the blade of the falls. *SW Spain, Portugal.* H2. Summer.

4. I. pseudopumila Tineo.

Stem *c.* 3 cm; plant 15–25 cm at flowering time. Leaves grey-green, to 20 × 1–1.5 cm, slightly curved, present throughout the winter. Bracts to 12 cm. Flowers solitary, 6–8 cm in diameter, white, yellow, purple or bicoloured. Perianth-tube 5–7.5 cm; beard of white hairs with yellow tips. *SE Italy, Sicily, Malta & W Yugoslavia.* H2. Spring.

5. I. reichenbachii Heuffel. Illustration: Dykes, The genus Iris, pl. 34 (1913); *Botanical Magazine*, 8812 (1919); Grey-Wilson & Mathew, Bulbs, pl. 27 (1981); Mathew, The iris, f. 3 (1981).

Plant 10–30 cm, stems unbranched or sometimes with a lateral branch. Leaves 8–35 × 1–1.5 cm. Bracts 3–5.5 cm, green, membranous on margin and at apex, sharply keeled. Flowers 1 or 2, 5–6.5 cm in diameter, greenish yellow, brownish purple or violet, often with darker veins. Perianth-tube 1.5–2.5 cm. Falls 4–6 cm; beard of yellow or bluish-purple hairs with white tips. Standards 4–6 cm. *SE Europe.* H2. Spring–early summer.

6. I. suaveolens Boissier & Reuter (*I. mellita* Janka; *I. rubromarginata* Baker).

Plant 8–15 cm, stems unbranched. Leaves sometimes purple-edged, to 22 cm × 4–10 mm, curved. Bracts 3.5–7 cm, sharply keeled. Flowers 1 or 2, 4.5–5.5 cm in diameter, yellow, purple, brownish purple or bicoloured. Perianth-tube 3–4.5 cm. Falls 3–5.5 cm; beard yellow or bluish. Standards 3.5–6 cm. *SE Europe, NW Turkey.* H2. Spring–early summer.

7. I. aphylla Linnaeus (*I. benacensis* Stapf; *I. nudicaulis* Lamarck). Illustration: *Botanical Magazine*, 5806 (1869); Dykes, Handbook of garden irises, pl. 21 (1924); Grey-Wilson & Mathew, Bulbs, pl. 27 (1981).

Plant 15–30 cm, stems branched usually from below the middle or from the base. Leaves 15–40 cm × 5–20 mm, curved. Bracts 3–6 cm, green, sometimes tinged with purple towards the tip, somewhat inflated. Flowers 1–5, 6–7 cm in diameter, dark purple to violet-blue. Perianth-tube 1.5–2.2 cm. Falls 4–6.5 cm; beard of white or bluish hairs, sometimes with yellow tips. Standards 4–6.5 cm, the claw often veined with brown. *C & E Europe, Former western USSR.* H2. Spring, occasionally again in autumn.

8. I. variegata Linnaeus (*I. regime* Horvat; *I. rudskyi* Horvat). Illustration: *Botanical Magazine*, 16 (1787); Grey-Wilson & Mathew, Bulbs, pl. 27 (1981); Mathew, The iris, 29 (1981); Mathew, P.J. Redouté: Lilies and related flowers, 189 (1982).

Plant 20–45 cm, stem branched in the upper half. Leaves deep green, 12–30 × 1–3 cm, slightly curved, ribbed. Bracts 3–5.5 cm, green, often purple-tinged, inflated. Flowers 3–6, 5–7 cm in diameter. Perianth-tube 1.8–2.5 cm. Falls 4.5–6 cm, ground colour whitish or pale yellow, heavily veined with reddish brown or violet, sometimes the veining running together into a complete brown-purple stain; beard yellow. Standards yellow, or sometimes whitish with violet veins. *C & SE Europe.* H1. Late spring–summer.

There is a group of cultivars with white standards and coloured falls that is known collectively as **I. amoena** de Candolle. It is likely that they arose by selection or hybridisation from *I. variegata*.

I. imbricata Lindley. Illustration: *Edwards's Botanical Register* **31**: t. 35 (1845); *Botanical Magazine*, 7701 (1900); *Journal of the Royal Horticultural Society* **90**: f. 28 (1965); Mathew, The iris, f. 2 (1981). Leaves grey-green, usually straight. Flowers 2 or 3, 7–9 cm in diameter, dull or pale yellow. Falls 5–6.5 cm, usually with brownish veins on the claw. *Iran, Former USSR (Transcaucasia).* H2. Spring–summer.

I. junonia,which is 50–65 cm tall with leaves 3–5 cm wide and flowers which vary from white to cream, yellow, pale blue or purple, may key out here (see under 9. *I. germanica*).

9. I. germanica Linnaeus. Illustration: *Botanical Magazine*, 670 (1805); Polunin, Flowers of Europe, f. 1693 (1969); Grey-Wilson & Mathew, Bulbs, pl. 27 (1981); Mathew, P.J. Redouté: Lilies and related flowers, 165 (1982).

Plant 50–120 cm, stems with 1 or 2 branches in the upper half. Leaves somewhat grey-green, 30–50 × 2–4.5 cm, almost straight, present throughout the winter. Bracts 3.5–5.5 cm, often tinged with purple in the lower half, membranous above. Flowers 3–5, scented, 9–10 cm in diameter, bluish violet. Perianth-tube 1.7–2.5 cm. Falls 5.5–9 cm; beard yellow. Standards 5.5–9 cm, often rather paler than falls. *Origin unknown, but probably native to the E Mediterranean area; widely naturalised.* H1. Late spring.

There are several named variants of *I. germanica* offered for sale, differing mainly in the colour of the flowers.

The following 5 similar plants have been described as species but may in fact be cultivars with a restricted distribution. It is doubtful whether they are truly wild, being usually recorded from cemeteries or near houses.

I. biliottii M. Foster. Stems 60–80 cm with 2 or 3 branches. Falls reddish purple with a beard of white hairs tipped with yellow. Standards bluish purple. *N Turkey.* H2. Early summer.

I. cypriana Baker & M. Foster. Outer bract brown and papery. Flowers *c.* 15 cm in diameter, lilac-blue, with falls that broaden towards the apex. Perianth tube *c.* 2.5 cm. *Cyprus.* H2. Late spring.

I. junonia Schott. Stems 50–65 cm with *c.* 4 branches. Flowers smaller than those of *I. germanica*, varying from white, cream or yellow to pale blue or purple. Claw of falls often whitish with brownish purple veins. *S Turkey.* H2. Early summer.

I. mesopotamica Dykes. Similar to *I. cypriana* but outer bract brown and papery in the upper third only. Perianth-tube 1.3–1.8 cm. Falls lavender-blue, the white claw with purple-brown veins; beard white and orange. *S Turkey, Syria & Israel.* H2. Late spring–early summer.

I. trojana Stapf. Illustration: Dykes, The genus Iris, pl. 37 (1913). Bracts rather narrow, flushed with purple at the base. Flowers *c.* 10 cm in diameter, with reddish purple falls and paler, bluer standards. Beard of white, yellow-tipped hairs. The veining on the claw of the falls is finer than in the rest of this group. *W Turkey.* H2. Early summer.

I. × sambucina Linnaeus (including *I. × lurida* Aiton, *I. × squalens* Linnaeus, *I. × neglecta* Hornemann). Illustration: *Botanical Magazine*, 669 (1803), 986 (1807); Mathew, P.J. Redouté: Lilies and related flowers, 175, 185 (1982). Stem *c.* 45 cm. Bracts green, flushed with purple. Falls reddish or brownish purple, the yellowish claw strongly veined; beard yellow or orange. Standards dull purple, often with a yellowish tinge. *N Italy, NW Yugoslavia; naturalised elsewhere.* H2. Spring.

All the variable hybrids between *I. pallida* and *I. variegata* are included under the name *I. × sambucina*.

I. × kochii Stapf grows to *c.* 45 cm and has red-purple or blue-purple flowers. The falls have a yellow beard and brown-veined claws. *N Italy.*

Thought to be a hybrid between *I. germanica* and *I. pallida* subsp. *cengialtii*.

10. I. albicans Lange. Illustration: Grey-Wilson & Mathew, Bulbs, pl. 27

(1981); Mathew, P.J. Redouté: Lilies and related flowers, 161 (1982).

Plant 30–60 cm, stems unbranched or with 1 stalkless lateral flower-head. Leaves grey-green, 1.5–2.5 cm wide, abruptly narrowed to an incurved tip, present throughout the winter. Bracts green or tinged with purple in lower part, papery-transparent towards the tip. Flowers 1–3, sweetly scented, 8–9 cm in diameter, white or blue. Falls with the claw flushed with greenish yellow; beard of white, yellow-tipped hairs. *Saudi Arabia, Yemen; naturalised elsewhere.* H2. Late spring.

The blue-flowered variant is known under the cultivar name 'Madonna'.

11. I. pallida Lamarck (*I. dalmatica* Anon.; I. pallida var. dalmatica Anon.). Illustration: *Botanical Magazine*, 685 (1803); Grey-Wilson & Mathew, Bulbs, pl. 27 (1981); Mathew, P.J. Redouté: Lilies and related flowers, 171 (1982).

Plant 15–120 cm, stems branched above the middle. Leaves grey-green, 20–60 × 1–4 cm, some usually persisting throughout the winter. Bracts 2–3.5 cm, silvery-papery. Perianth-tube 8–11 mm. Flowers 3–6, scented, 9–11 cm in diameter, lilac-blue. Falls 5–7.5 cm; beard yellow. Standards 5–7.5 cm. *W Yugoslavia.* H2. Late spring–early summer.

Subsp. **cengialtii** (Ambrosi) M. Foster (*I. cengialtii* Ambrosi). Illustration: Mathew, The iris, f. 1 (1981). Plant only to 45 cm. Leaves greener than subsp. *pallida*, usually withering in the winter. Bracts brownish-papery. Flowers usually 2, deep bluish purple. Beard of falls of white hairs tipped with yellow or orange. *NE Italy.* H2. Late spring–early summer.

I. pallida 'Variegata' (illustration: Hay & Synge, Dictionary of garden plants in colour, f. 1197, 1969) comes in two forms: 'Argentea' has the leaves striped with white and in 'Aurea' the leaves are yellow-striped.

I. florentina Linnaeus. Flowers white, slightly flushed with blue. *Origin unknown.* H2. Late spring.

This plant has been much confused with *I. albicans* from which it differs in the flowers being neither pure white nor blue, and the bracts being brown-papery. Lateral flowers, if present, are borne on short stalks. It is probably an albino cultivar close to *I. germanica*, and perhaps best regarded as *I. germanica* 'Florentina'.

12. I. schachtii Markgraf. Illustration: Mathew, The iris, f. 6, 7 (1981).

Plant 10–30 cm, stem with 1–3 branches. Leaves grey-green, to 1.5 cm wide. Bracts green, sometimes tinged with purple, transparent towards the margins and apex, slightly inflated. Flowers 5–6 cm in diameter, yellowish or purple. Perianth-tube 1.5–2 cm. Falls 5–6.5 cm. Standards 3.5–6.5 cm. *C Turkey.* H3. Spring.

I. albertii Regel. Illustration: Botanical Magazine, 7020 (1888); Dykes, The genus Iris, pl. 38 (1913). Plant 30–70 cm, with leaves 2–3 cm wide. Flowers 6–8 cm in diameter, lavender to violet. Claw of falls veined with brownish red. *Former USSR (Kazakstan).* H2. Late spring.

13. I. kashmiriana Baker. Illustration: *Botanical Magazine*, 6869 (1886), 9378 (1934); Mathew, The iris, 29 (1981).

Plant 75–125 cm, stems with 1 or branches. Leaves pale grey-green, to 60 × 2–3 cm, ribbed, straight. Bracts to 11 cm, green. Flowers 2 or 3 per branch, scented, 10–12 cm in diameter, white or pale lilac-blue. Perianth-tube 2.2–2.5 cm. Falls 6.5–9.5 cm; beard of dense white, yellow-tipped hairs. Standards 7–9 cm, with a similar but smaller beard on the lower part. *India (Kashmir).* H4. Early summer.

Section **Psammiris** (Spach) J. Taylor. Rhizomatous. Stems unbranched with 1–3 flowers. Flowers 3–5 cm in diameter, yellow; beard of falls made up of hairs composed of one cell only. Seeds with a fleshy appendage.

14. I. bloudowii Bunge.

Rhizome non-stoloniferous, with numerous brown fibres around the growing point. Plant 15–35 cm. Leaves to 30 cm × 7–13 mm, straight or slightly curved. Bracts very inflated. Flowers 2 or 3 per bract, *c.* 5 cm in diameter, yellow with a brownish or purple coloration on the lower part of the falls and standards. Beard yellow. *Former C & E USSR; ?China.* H2. Spring.

I. humilis Georgi (*I. arenaria* Waldstein & Kitaibel; *I. flavissima* Pallas). Illustration: *Edwards's Botanical Register*, t. 549 (1821); Grey-Wilson & Mathew, Bulbs, pl. 28 (1981). Plant 5–25 cm. Leaves 5–17 cm × 2–7 mm. Flowers 1–3, 3–4 cm in diameter. Falls yellow with purple veins and an orange beard, the blade spreading horizontally. Standards yellow or purple, veined with purple. *E Europe, Former USSR.* H2. Spring.

Section **Oncocylus** (Siemssen) Baker. Rhizomatous. Stems unbranched, 1-flowered. Flowers 5 cm or more in diameter, rarely yellow, beard of falls made up of hairs composed of 1 cell only. Seeds with a fleshy appendage almost as large as the seed itself and often whitish.

15. I. gatesii M. Foster. Illustration: *Botanical Magazine*, 7867 (1902); *Bulletin of the Alpine Garden Society* **39**: 287 (1971); Mathew, The iris, f. 18 (1981).

Rhizome compact, stout. Plant 45–60 cm. Leaves 5–7, grey-green, 5–9 mm wide, straight. Bracts 7.5–11.5 cm, pale green or pinkish brown. Flowers 13–20 cm in diameter, appearing greyish, brown or purplish depending on the colour and amount of veining and spotting. Perianth-tube 3–6 cm. Falls 8–10 cm, often with a small brown or blackish signal patch; beard yellow, brownish or purplish. Standards 8–10 cm. All perianth-lobes whitish or pale yellow, with fine veins and tiny spots of reddish brown or blackish. Style-branches 5–7 cm, lobes reflexed. *N Iraq, SE Turkey.* H4. Summer.

I. bismarckiana Damman & Sprenger (*I. nazarena* (Herbert) Dinsmore). Illustration: *Botanical Magazine*, 7986 (1904). Rhizome stoloniferous. Leaves 2–2.5 cm wide. Flowers 10–12 cm in diameter. Falls cream or yellowish, densely spotted and veined with reddish brown or purple, and with a large blackish purple signal patch; beard dark purple. Standards white, speckled and veined with purple or blue. *NE Israel, S Syria.* H4. Spring.

I. kirkwoodii Chaudhary. Illustration: *Botaniska Notiser* **125**: 498 (1972) & **128**: 387 (1975). Plant to 75 cm. Leaves 1–1.5 cm wide, curved and often drooping. Flowers 8–12 cm in diameter. Falls white, pale bluish or pale greenish, densely covered with deep purple veins and dots; beard with long brownish or purple hairs; signal patch dark purple. Standards similar but less heavily marked. *N Syria, S Turkey.* H4. Late spring.

I. westii Dinsmore (*I. susiana* Linnaeus forma *westii* (Dinsmore) Sealy). Illustration: *Botanical Magazine*, n.s., 550 (1969). Flowers 12–15 cm in diameter. Falls yellowish, heavily speckled and veined with deep brown or purplish; beard of long, sparse, purple hairs; signal patch deep velvety brown. Standards pale lilac, speckled and veined with darker lilac-blue. *Lebanon.* H4. Late spring.

I. sofarana M. Foster (*I. susiana* Linnaeus forma *sofarana* (M. Foster) Sealy). Illustration: *Botanical Magazine*, n.s., 550 (1969). Flowers 10–13 cm in diameter. Falls creamy-white, densely speckled and veined with brownish purple or violet; beard of sparse, dark purple hairs; signal patch blackish purple. Standards white, speckled and veined with violet or reddish purple. *Lebanon.* H4. Late spring.

I. susiana Linnaeus. Illustration: *Botanical Magazine*, 91 (1790). Flowers 10–12 cm in diameter. Falls greyish, heavily veined with deep purple; beard of sparse, deep purple hairs; signal patch velvety black. Standards greyish, heavily veined with deep purple. *Origin unknown, but probably Lebanon.* H4. Late spring.

I. haynei (Baker) Mallet. Illustration: *Gardeners' Chronicle* **35**: 266 (1904); *Journal of the Royal Horticultural Society* **29**: Proceedings f. 112 (1905). Plant to 40 cm. Flowers 10–12 cm in diameter, fragrant, with dark veins and spots on a pale ground giving an overall grey-lilac colour, with the standards more purplish. Perianth-tube *c.* 2.5 cm. Signal patch blackish brown. *N Israel.* H4. Spring.

16. I. lortetii Boissier. Illustration: *Botanical Magazine*, 7251 (1892); Dykes, The genus Iris, pl. 27 (1913); Everard & Morley, Flowers of the world, pl. 51 (1970); Mathew, The iris, f. 12, 13 (1981).

Plant 30–50 cm. Leaves grey-green, 1–1.5 cm wide, straight, abruptly narrowed at the apex. Flowers 8–9 cm in diameter. Perianth-tube to 5 cm. Falls to 7.5 cm, whitish or pale lilac with pink or maroon speckling and veining, the margin sometimes minutely toothed; beard reddish or yellowish, rather sparse; signal patch deep maroon. Standards similar in colour to falls, but veined rather than speckled. *S Lebanon.* H4. Late spring.

I. auranitica Dinsmore. Illustration: *Journal of the Royal Horticultural Society* **87**: f. 117 (1962); Warburton & Hamblen (eds), The world of irises, 18 (1978). Flowers yellow, spotted with brown, giving a bronze appearance. Falls with a maroon signal patch and a dense beard of yellow hairs that are tipped with purple. *Syria.* H4. Late spring.

17. I. sari Baker (*I. lupina* M. Foster). Illustration: *Botanical Magazine*, 7904 (1903); Dykes, The genus Iris, pl. 26 (1913); *Bulletin of the Alpine Garden Society* **45**: 113 (1977); Mathew, The iris, f. 16, 17 (1981).

Rhizome compact, stout. Plant 10–30 cm. Leaves 5–7, to 30 cm × 3–9 mm, curved or almost straight. Bracts 5–9.5 cm, green. Flowers 7–10 cm in diameter. Perianth-tube 2–3 cm. Falls 3.5–5.5 cm, cream, yellowish or greenish, veined with brownish purple or crimson, often minutely toothed and wavy on the margin; beard yellow; signal patch velvety brown or dark red. Standards 4–8.5 cm, similar in colour to falls or more bluish purple. Style-branches 3.5–5.5 cm, lobes scalloped. *C & S Turkey.* H4. Late spring.

I. barnumae Baker & M. Foster. Illustration: *Botanical Magazine*, 7050 (1889); Mathew, The iris, f. 6 (1981). Leaves to 5 mm wide. Flowers purple, rather indistinctly veined; beard yellow, the hairs sometimes purple-tipped; signal patch small and only slightly darker than the falls. *E Turkey, NE Iraq & adjacent Iran.* H4. Late spring.

Forma **urmiensis** (Hoog) Mathew & Wendelbo (*I. urmiensis* Hoog). Illustration: *Botanical Magazine*, 7784 (1901); Rechinger (ed.), Flora Iranica: Iridaceae, pl. 14 (1975). Flowers yellow. *SE Turkey, NW Iran.*

Subsp. **demavendica** (Bornmüller) Mathew & Wendelbo (*I. demavendica* (Bornmüller) Dykes). Illustration: *Botanical Magazine*, n.s., 448 (1964); Rechinger (ed.), Flora Iranica: Iridaceae, pl. 14 (1975). Leaves 4–7 mm wide. Flowers bluish violet; beard white or cream. *Iran (Elburz Mountains).*

I. meda Stapf. Illustration: *Botanical Magazine*, 7040 (1889); *Journal of the Royal Horticultural Society* **88**: f. 58 (1963); Mathew, The iris, f. 4 (1981). Leaves 1.5–4 mm wide. Flowers 5–7 cm in diameter, cream or whitish, suffused and veined with golden brown, especially towards the irregularly wavy edge of the segments. Beard yellow; signal patch purplish or brown. Style-branches 2–3 cm. *W Iran.* H4. Spring.

I. sprengeri Siehe (*I. elizabethae* Siehe). Illustration: *Gardeners' Chronicle*, ser. **3**, **36**: 50 (1904). Rhizome creeping, stoloniferous. Plant 5–15 cm. Leaves 3–5 mm wide. Perianth-tube 1–1.5 cm. Falls 5.5–6 cm, yellow, occasionally whitish, veined and spotted with purplish red; beard yellow; signal patch dark purple. Standards white with blackish and purplish red veins. Style-branches 2.5–3.5 cm. *C Turkey.* H4. Late spring.

I. paradoxa, which has standards 8–10 cm and much smaller falls almost entirely covered with a dense beard, may key out here (see under 20. *I. atropurpurea*).

18. I. samariae Dinsmore. Illustration: *Pacific Horticulture* **40**(3): front cover (1979).

Rhizome stout with many bristle-like fibres at the growing point. Plant 25–30 cm. Leaves to 35 × 1 cm, almost straight, clothing the stem throughout its length. Flowers 9–10 cm in diameter. Falls 4–5 cm, creamy white, spotted and veined with purple; beard of sparse reddish brown or yellowish brown hairs that become shorter and purple towards the beard edge; signal patch dark brown. Standards usually pinkish purple. Style-branches *c.* 5 cm. *NW Jordan.* H4. Late spring.

19. I. iberica Hoffmann.

Rhizome compact, not stoloniferous. Plant 15–20 cm. Leaves grey-green, strongly curved, 2–6 mm wide. Bracts 3–7 cm, greenish. Perianth-tube 2–3.5 cm. Falls 5.5–6.5 cm, whitish, heavily spotted and veined with brown, blade deflexed at an angle of *c.* 60° from the horizontal; beard purplish brown, signal patch brown or blackish. Standards 4.5–8.5 cm, white, cream or pale bluish, less veined than the falls, sometimes with sparse hairs at the base. Style-branches 3.5–7 cm, lobes scalloped. *SW Asia.* H4. Late spring.

Subsp. **elegantissima** (Sosnowsky) Fedorov & Takhtadjan (*I. elegantissima* Sosnowsky). Illustration: *Botanical Magazine*, 5847 (1870); *Bulletin of the Alpine Garden Society* **35**: 362 (1967); Rechinger (ed.), Flora Iranica: Iridaceae, pl. 12 (1975); Mathew, The iris, f. 14, 15 (1981). Plant usually 20–30 cm. Falls cream or pale yellow, heavily spotted and veined with brownish purple, blade sharply deflexed to an almost vertical position. Standards completely white or slightly veined at the base with brown. *NE Turkey, NW Iran, Former USSR (Armenia).* H4. Spring.

Subsp. **lycotis** (Woronow) Takhtadjan (*I. lycotis* Woronow). Illustration: *Botanical Magazine*, n.s., 580 (1970); Rechinger (ed.), Flora Iranica: Iridaceae, pl. 12 (1975); Mathew, The iris, f. 8 (1981). Falls and standards more or less equally heavily spotted and veined on a whitish ground, the falls with a concave blade standing out at an angle. *NE Iraq, SE Turkey, NW & W Iran, Former USSR (Armenia).* H4. Spring.

20. I. atropurpurea Baker. Illustration: *Gartenflora* t. 1361 (1891); Mathew, The iris, f. 7 (1981).

Rhizomes often stoloniferous. Plant 15–25 cm. Leaves somewhat grey-green, *c.* 15 cm, curved. Flowers *c.* 8 cm in diameter. Perianth-tube *c.* 3 cm. Falls to 5 cm, blackish purple, the blade veined with greenish yellow, the claw veined with reddish purple; beard yellow; signal patch blackish. Standards to 7.5 cm, reddish purple with faint blackish veins. *Israel.* H4. Spring.

I. atrofusca Baker. Illustration: *Botanical Magazine*, 7379 (1894); Warburton & Hamblen (eds), The world of irises, 193 (1978). Leaves pale green, straight. Perianth-tube 5–6.5 cm. Falls dark purple-brown, veined and dotted with red-black. Standards usually paler and more of a deep red colour. *Israel.* H4. Spring.

I. nigricans Dinsmore. Flowers 8–10 cm in diameter. Perianth-tube *c.* 6.5 cm. Falls dark brownish purple to deep purple, with darker veins and dots; beard dark, purple-tipped on a whitish ground. Standards whitish, very heavily veined with dark purple. *Jordan.* H4. Spring.

I. paradoxa Steven. Illustration: *Botanical Magazine*, 7081 (1889); *Journal of the Royal Horticultural Society* **90**: f. 21 (1965); Everard & Morley, Flowers of the world, pl. 51 (1970); Mathew, The iris, f. 5 (1981). Rhizome not stoloniferous. Flowers variable in colour; names have been given to the various forms that are listed below. Falls 2.5–4 cm, almost horizontal, oblong, very small in comparison with the large, rounded standards, almost entirely covered with a dense beard. Standards 7–10 cm, notched at the apex. Style-branches 2–3 cm, lobes minutely scalloped. *E Turkey, N Iran, Former USSR (Transcaucasia).* H4. Late spring.

Forma **paradoxa.** Falls blackish; standards deep violet.

Forma **atrata** Grossheim. Falls and standards blackish violet.

Forma **choschab** (Hoog) Mathew & Wendelbo. Falls blackish violet; standards white or pale lilac, veined with purple.

Forma **mirabilis** Gawrilenko. Falls golden yellow covered with a darker yellow to orange beard; standards pale yellow or pale blue.

I. barnumae,which has yellow, bluish violet or purple flowers with less obvious veining, may key out here (see under 17, *I. sari*).

21. I. acutiloba Meyer.

Rhizome slender, much-branched. Plant 8–25 cm. Leaves grey-green, 2–6 mm wide, curved. Bracts 5–8 cm, green or slightly pinkish. Flowers 5–7 cm in diameter, whitish, strongly veined with brown or grey, the segments rather pointed. Perianth-tube 1.5–2 cm. Falls 4–7 cm, with a sparse purplish or brownish beard, and 2 dark red-brown or blackish spots, 1 in the centre and 1 at the apex. Standards 4–8 cm. Style-branches 2–4 cm. *Former S USSR, Iran.* H3. Spring–early summer.

Subsp. **lineolata** (Trautvetter) Mathew & Wendelbo (*I. ewbankiana* M. Foster; I. helena (K. Koch) K. Koch). Illustration: *Botanical Magazine*, 9333 (1933); *Journal of the Royal Horticultural Society* **88**: f. 56 (1963) & **93**: f. 204 (1968); Hay & Synge, Dictionary of garden plants in colour, f. 791 (1969). Falls with only 1 dark spot. *Former USSR (Transcaucasia: S of Kura R), NW & NE Iran, adjacent Turkmenistan.*

I. mariae Barbey (*I. helenae* Barbey). Flowers 8–10 cm in diameter, lilac or pinkish veined with reddish brown. Claw of falls, beard and signal patch deep purple. *Israel.* H4. Late spring.

I. paradoxa, which has standards 8–10 cm and much smaller falls almost entirely covered with a dense beard, may key out here (see under 20. *I. atropurpurea*).

Section **Regelia** Lynch. Rhizomatous. Stems unbranched, usually bearing 2 flowers (1 in *I. afghanica*). Falls and standards both bearded, each hair of the beard composed of 1 cell only. Seeds with a fleshy appendage.

22. I. korolkowii Regel. Illustration: *Gartenflora* t. 766 (1873); *Botanical Magazine*, 7025 (1888); *RHS Iris Year Book 1967*, f. 6 (1968); Mathew, The iris, f. 10 (1981).

Rhizome thick, slightly stoloniferous, with fibrous remains of old leaves. Plant 40–60 cm. Leaves 5–10 mm wide, tinged with purple at the base. Bracts 9–10 cm. Flowers 2 or 3, 6–8 cm in diameter. Perianth-tube 2.5–3 cm. Falls 7–11 cm, creamy white veined with dark maroon, occasionally purplish with green or purple veins, blade deflexed, pointed; beard dark; signal patch blackish, brown or deep green. Standards 7–11 cm, similar in colour to falls, pointed; beard very sparse. Style-branches *c.* 4.5 cm with scalloped lobes. *Former USSR (Soviet Central Asia: Tien Shan & Pamir Alai), NE Afghanistan.* H4. Early summer.

I. korolkowii has been crossed with *I. stolonifera.*

I. afghanica Wendelbo. Illustration: *Notes from the Royal Botanic Garden, Edinburgh* **31**: 339 (1972); *Botanical Magazine*, n.s., 668 (1974); Mathew, The iris, f. 9 (1981); Heywood & Chant, Popular encyclopedia of plants, 181 (1982). Plant 15–35 cm. Leaves 2–5 mm wide. Flowers 1, sometimes 2, 8–9 cm in diameter. Falls cream or white, heavily veined with purplish brown; beard dark; signal patch purple. Standards 6–7.5 cm, pale yellow with a greenish beard. *NE Afghanistan.* H4. Spring.

23. I. hoogiana Dykes. Illustration: *Botanical Magazine*, 8844 (1920); Dykes, Handbook of garden irises, pl. 19 (1924); Mathew, The iris, f. 20 (1981); Brickell (ed.), RHS A–Z encyclopedia of garden plants, 576 (2003).

Rhizomes thick, stoloniferous. Plant 40–60 cm. Leaves to 50 × 1–1.5 cm, straight or slightly curved, tinged with purple. Flowers 2 or 3, scented, 7–10 cm in diameter, lilac-blue. Perianth-tube *c.* 2.5 cm. Falls and standards to 7.5 cm; beard yellow. *Former USSR (Soviet Central Asia: Pamir Alai).* H3. Late spring.

Several cultivars have been selected and named. *I. hoogiana* has been hybridised with *I. stolonifera.*

24. I. stolonifera Maximowicz (*I. leichtlinii* Regel; *I. vaga* M. Foster). Illustration: *Botanical Magazine*, 7861 (1902); Dykes, The genus Iris, pl. 29 (1913).

Rhizomes stoloniferous. Plant 30–60 cm. Leaves bluish green, to 60 cm × 5–15 mm, prominently veined. Flowers 2 or 3, 7–8 cm in diameter, a mixture of brown and purple. Perianth-lobes usually brownish at the wavy margins, shading to blue towards the centre; beard yellowish or blue. Falls to 7.5 cm. Standards to 6.5 cm. *Former USSR (Soviet Central Asia: Pamir Alai).* H3. Late spring.

A species that is very variable in stature and flower colour; some of the forms are commercially available and have been named.

I. heweri Grey-Wilson & Mathew. Illustration: *Journal of the Royal Horticultural Society* **99**: f. 87 (1974); Rechinger (ed.), Flora Iranica: Iridaceae, pl. 3, 15 (1975). Plant 10–15 cm, occasionally to 30 cm. Leaves green,

2–5 mm wide, strongly curved. Flowers *c.* 5 cm in diameter, deep purplish blue; beard lilac. Falls and standards 3.5–4.5 cm. *NE Afghanistan.* H4. Late spring.

I. darwasica Regel. Leaves grey-green, to 45 cm × 4–8 mm, straight. Flowers 5–6 cm in diameter, lilac with purple veins; beard purple. *Former USSR (Southern Soviet Central Asia).* H3. Spring.

Section **Pseudoregelia** Dykes. Rhizomatous. Stems unbranched with 1 or 2 flowers. Flowers lilac or purple, with darker blotches on the falls, beard of the falls made up of hairs composed of 1 cell only. Seeds with a fleshy appendage.

25. I. kemaonensis D. Don. Illustration: *Botanical Magazine*, 6957 (1887); Dykes, The genus Iris, pl. 30 (1913); Everard & Morley, Flowers of the world, pl. 107 (1970); Mathew, The iris, f. 11 (1981).

Rhizome thick and knobbly, producing rather fleshy roots. Plant to 7 cm. Leaves to 45 cm × 2–10 mm, overtopping the flowers. Flowers 1 or 2, stemless, scented, 4–5 cm in diameter. Perianth-tube 5–7.5 cm. Falls and standards 3.5–4 cm, lilac-purple, mottled and blotched darker, standards slightly paler than falls. Beard of white hairs with yellow tips. Style-branches *c.* 2.5 cm. *Himalaya from Kashmir to W China.* H3. Late spring.

The name is frequently misspelled *kumaonensis*.

26. I. hookeriana M. Foster. Illustration: *Botanical Magazine*, 7276 (1893); *Journal of the Royal Horticultural Society* **93**: f. 128 (1968).

Plant to 15 cm. Leaves to 30 × 1–2.5 cm. Flowers usually 2, scented, on stems 5–12 cm. Perianth-tube 2–3 cm. Falls and standards *c.* 5 cm, lilac, purple or bluish, sometimes white. *W Himalaya.* H3. Late spring.

I. tigridia Ledebour. Leaves to 10 cm × 1–4 mm. Flowers 1 or 2, on stems 10–15 cm, lilac to deep blue with purple mottling. Beard white, or hairs yellow-tipped, the area of the blade surrounding the beard being white, veined with purple. *Former SE USSR, Mongolia, NW China.* H3.

Subgenus **Limniris** (Tausch) Spach. Falls lacking a beard, but often with cockscomb-like ridges on the blade, or very short, downy hairs on the claw.

Section **Limniris.** Rhizomes often compact, clump-forming. Flower colour various. Falls with no crest-like ridge in the centre, occasionally with short, downy hairs on the claw.

Series **Chinensis** (Diels) Lawrence. Rhizomes thin, wiry, often far-creeping. Leaves with obvious ribs. Flowers yellow, flattish in appearance due to spreading falls and standards. Capsules triangular in cross-section.

27. I. minutoaurea Makino (*I. minuta* Franchet & Savatier). Illustration: *Botanical Magazine*, 8293 (1910); Mathew, The iris, f. 14 (1981).

Rhizomes thin, wiry. Plant 8–10 cm. Leaves 12–15 cm × 2–3 mm, elongating after flowering to *c.* 40 cm. Flowers solitary, 2–2.5 cm in diameter. Perianth-tube 2–2.5 cm. Falls *c.* 1 cm, yellow, often sparsely spotted with purple, spreading. Standards shorter than falls, paler yellow, spreading, the claw brownish. *Origin unknown: possibly Korea or China.* H2. Spring.

Series **Vernae** (Diels) Lawrence. Rhizomatous. Flowers white or lilac-blue. Capsule triangular in cross-section. Seeds spherical, with a fleshy appendage that dries out quickly after the seed is shed from the capsule.

28. I. verna Linnaeus. Illustration: Sweet, British flower garden, ser. 1, **1**: 68 (1824); Rickett, Wild flowers of the United States **1**: pl. 11 (1966); Hay & Synge, Dictionary of garden plants in colour, f. 95 (1969); Mathew, The iris, 82 (1981).

Plant 4–6 cm. Leaves green or grey-green, to 15 cm × 3–15 mm. Flowers 1 or 2, 3–5 cm in diameter. Perianth-tube 2–5 cm. Falls *c.* 3.5 cm, bright lilac-blue, sometimes white, the blade with a central orange stripe, claw orange or yellow. Standards lilac-blue, erect. *SE USA.* H3. Spring.

Series **Ruthenicae** (Diels) Lawrence. Rhizomatous. Flowers basically whitish. Capsule round in cross-section, the segments quickly opening and curling back. Seeds pear-shaped with a fleshy appendage.

29. I. ruthenica Ker Gawler. Illustration: *Botanical Magazine*, 1123 (1808), 1393 (1811); Dykes, The genus Iris, pl. 13 (1913); Grey-Wilson & Mathew, Bulbs, 161 (1981).

Rhizome shortly creeping. Plant 3–15 cm. Leaves green, to 30 cm × 2–5 mm, erect. Bracts 3–5 cm, membranous, greenish with pink margins. Flowers usually 1, occasionally 2, fragrant, 3–4 cm in diameter. Perianth-tube 8–10 mm. Falls 4.5–5 cm, whitish with bluish lavender or violet margins and veins, blade held horizontally. Standards 4–4.5 cm, almost erect, bluish lavender or violet. *E Europe, through C Asia to China & Korea.* H2. Late spring.

Series **Tripetalae** (Diels) Lawrence. Rhizomatous. Flowers basically blue to purple, standards narrowly lanceolate.

30. I. setosa Link. Illustration: *Edwards's Botanical Register*, t. 10 (1847); Dykes, The genus Iris, pl. 23 (1913); Clark, Wild flowers of the Pacific northwest, 86 (1976); Mathew, The iris, 85, f. 15 (1981).

Rhizomes fairly stout, often clad in the fibrous remains of old leaf bases. Plant 15–90 cm, stem usually branched. Leaves 20–50 × 1–2.5 cm, the bases often red-tinged. Bracts green, often with purple margins. Flowers 2 or 3, sometimes solitary, 6–9 cm in diameter. Perianth-tube 5–10 mm. Falls with a blue-purple to blue blade, which narrows abruptly into a narrow claw with blue or purple veins on a whitish or pale yellow ground. Standards blue-purple to blue, narrowly lanceolate, sharply pointed, sometimes with 2 lobes at the base of the blade. Style-branches with toothed lobes. *Northeast N America, Former E USSR, China, N Korea, Japan, Aleutian, Sakhalin & Kurile Islands to Alaska.* H1. Summer.

An extremely variable species in which several variants have been described. These are summarised below and are listed alphabetically, as the naming is somewhat chaotic and needs to be sorted out.

Forma **alpina** Komarov. A Siberian form with a very short stem.

Var. **arctica** (Eastwood) Dykes (*I. arctica* Eastwood). A dwarf Alaskan variety, the purple flowers variegated with white.

Subsp. **hondoensis** Honda. Stems to 75 cm bearing large purple flowers. *Japan.*

Subsp. **interior** (Anderson) Hultén. An Alaskan plant with narrow leaves and papery, violet bracts that are shorter than normal.

Var. **nasuensis** Hara. Stems to 1 m, bearing very large flowers and wide leaves. *Japan.*

Forma **platyrhyncha** Hultén. An Alaskan form whose solitary flowers have standards that are larger and wider than usual.

Forma **serotina** Komarov. A Siberian form with solitary, stemless flowers.

Forma **hookeri** G. Don (*I. setosa* subsp. canadensis (M. Foster) Hultén). A dwarf form from eastern N America, the stem unbranched, almost leafless and bearing 1, or rarely 2, lavender-blue flowers.

Series **Sibiricae** (Diels) Lawrence. Rhizomes stout. Stems hollow (except in *I. clarkei*). Leaves not persisting through the winter. Falls with 2 flanges at the base of the claw. Stigma triangular. Capsule almost round to triangular in cross-section. Seeds D-shaped to almost cubic.

31. I. sanguinea Donn (*I. orientalis* Thunberg). Illustration: Dykes, The genus Iris, pl. 1 (1913); Grey-Wilson, The genus Iris, subsection Sibiricae, 5 (1971).

Plant 30–75 cm, stems usually unbranched, hollow. Leaves slightly grey-green, 5–12 mm wide, more or less equalling stems, or slightly longer. Bracts green or reddish, sometimes brown and papery at the tip. Flowers 6–8 cm in diameter, borne on stalks that are almost equal in length. Blade of falls white with a network of reddish purple veins that merge into a solid band of colour near the margin, claw yellowish or orange, finely veined with purple. Standards smaller, erect. *Former SE USSR, Korea, Japan.* H2. Early summer.

'Alba' has white flowers, usually with some purple veins, and 'Violacea' has deep violet flowers with larger than normal standards. *I. sanguinea* has been hybridised with *I. sibirica*

32. I. sibirica Linnaeus. Illustration: Perry, Flowers of the world, 144 (1972); Grey-Wilson & Mathew, Bulbs, pl. 28 (1981); Mathew, The iris, 90 (1981); Mathew, P.J. Redouté: Lilies and related flowers, 179 (1982).

Plant 50–120 cm, stem branched, hollow. Leaves green, 25–80 cm × up to 4 mm, shorter than the flowering stems. Bracts 3–5 cm, brown and papery. Flowers 6–7 cm in diameter, borne on unequal stalks. Falls 3–6 cm, blue or blue-violet, the obovate or oblong blade with a white central area strongly veined with violet, blade narrowing abruptly to a paler claw with dark veining. Standards 2.5–5.5 cm. *C & E Europe, NE Turkey, Former USSR (Caucasus, European Russia to Lake Baikal).* H1. Late spring–summer.

Many cultivars are available, the flowers varying in colour from white to shades of blue or deep violet-blue.

33. I. forrestii Dykes. Illustration: Dykes, The genus Iris, pl. 3 (1913); Grey-Wilson, The genus Iris, subsection Sibiricae, 11 (1971); Mathew, The iris, 90 (1981); Brickell (ed.), RHS A–Z encyclopedia of garden plants, 575 (2003).

Plant 35–40 cm, stems unbranched, hollow. Leaves glossy on one side, grey-green on the other, much shorter than the flowering stem. Bracts green. Flowers 5–6 cm in diameter, scented. Falls yellow with brownish purple lines on the claw. Standards yellow, erect. *W China (Yunnan, Sichuan), N Burma.* H2. Summer.

34. I. wilsonii Wright. Illustration: *Botanical Magazine*, 8340 (1910); Dykes, The genus Iris, pl. 2 (1913); Grey-Wilson, The genus Iris, subsection Sibiricae, 13 (1971).

Plant 60–75 cm, stems unbranched, hollow. Leaves grey-green, more or less equalling the flowering stems and often rather drooping. Bracts green. Flowers 6–8 cm in diameter. Falls pale yellow, veined and dotted with purplish brown especially in the centre of the blade and on the claw. Standards pale yellow, held at an oblique angle, margins rather wavy. *W China.* H2. Summer.

35. I. delavayi Micheli. Illustration: *Botanical Magazine*, 7661 (1899); Hay & Synge, Dictionary of garden plants in colour, f. 1187 (1969); Grey-Wilson, The genus Iris, subsection Sibiricae, 15 (1971); Brickell (ed.), RHS A–Z encyclopedia of garden plants, 575 (2003).

Plant 90–150 cm, stems branched, hollow. Leaves grey-green, shorter than the flowering stems. Bracts green, papery at the tips. Flowers 7–9 cm in diameter. Falls pale or dark purple, the circular blade notched at the apex, with a large white patch in the centre. Standards similar in ground colour, held at an oblique angle. *W China (Yunnan, Sichuan).* H2. Summer.

I. delavayi and *I. wilsonii* have been crossed to produce plants with purple flowers whose falls have a yellow ground, veined with bluish purple.

36. I. bulleyana Dykes. Illustration: Dykes, The genus Iris, pl. 6 (1913); Grey-Wilson, The genus Iris, subsection Sibiricae, 17 (1971); Brickell (ed.), RHS A–Z encyclopedia of garden plants, 573 (2003).

Plant 35–45 cm, stems unbranched, hollow. Leaves glossy green on one side, grey-green on the other, more or less equal in length to the flowering stems. Bracts green, papery at the tips. Flowers 6–8 cm in diameter. Falls spreading to drooping, violet at the tips, the centre white or yellowish, dotted and streaked with deep violet, the claw greenish yellow. Standards violet, held at an oblique angle. *W China, N Burma.* H2. Summer.

Formerly considered to be a hybrid between *I. forrestii* and *I. chrysographes*, but recent Chinese work suggests it is probably a good species.

37. I. chrysographes Dykes. Illustration: Dykes, The genus Iris, pl. 4 (1913); Grey-Wilson, The genus Iris, subsection Sibiricae, 21 (1971); Brickell (ed.), RHS A–Z encyclopedia of garden plants, 574 (2003).

Plant 35–45 cm, stems unbranched, hollow. Leaves grey-green, 1–1.5 cm wide, more or less equal in length to the flowering stems. Bracts green. Flowers 6–7 cm in diameter, scented. Falls reddish violet, the blade drooping vertically, with or without gold streaks in the centre. Standards reddish violet, held at an oblique angle. *W China (Yunnan, Sichuan).* H2. Summer.

'Rubella' has very dark flowers.

I. chrysographes hybridises with both *I. forrestii* and *I. sibirica*.

I. dykesii Stapf. Illustration: Grey-Wilson, The genus Iris, subsection Sibiricae, 19 (1971). Similar to *I. chrysographes*, but the leaves sheath the stem for most of its length. Flowers larger, deep violet-purple, the centre of the blade of the falls with white and yellow veins. *Origin unknown.* H2. Summer.

A plant whose origin and status are shrouded in mystery. Probably it is Chinese. Grey-Wilson suggests that it may be a hybrid, but the parentage is so far unknown.

38. I. clarkei Baker. Illustration: *Botanical Magazine*, 8323 (1910); Dykes, The genus Iris, pl. 5 (1913); Grey-Wilson, The genus Iris, subsection Sibiricae, 23 (1971).

Plant to 60 cm, stems branched, solid. Leaves glossy green on one side, grey-green on the other, 1.3–2 cm wide, shorter than the flowering stems. Bracts green. Flowers 7–7.5 cm in diameter. Falls blue-violet to deep blue or reddish purple, the centre of the blade with a large, white area, veined with violet, claw somewhat yellowish. Standards spreading outwards. *E Himalaya.* H2. Early summer.

Series **Californicae** (Diels) Lawrence. Rhizomes tough, roots wiry. Stems usually unbranched, generally bearing 1 or 2 flowers. Falls usually with a horizontally-spreading blade. Stigma triangular (except in *I. purdyi*). Capsule triangular to rounded in cross-section, with 3 ribs.

39. I. douglasiana Herbert. Illustration: *Botanical Magazine*, 6083 (1874); Dykes, The genus Iris, pl. 8 (1913); Cohen, A guide to the Pacific coast irises, 32 (1967); Hay & Synge, Dictionary of garden plants in colour, f. 93 (1969).

Plant 15–70 cm, stems branched. Leaves to 1 m × 2 cm, stained with red at the base, persisting through the winter. Bracts 6–12 cm, held together or divergent. Flowers 2 or 3, 7–10 cm in diameter. Perianth-tube 1.5–2.8 cm. Falls 5–8.5 cm, lavender to purple, the blade with a yellowish central area and darker veins; sometimes the falls are nearly white with darker veins. Standards 4.5–7 cm. Style-branches 1.5–3.5 cm, the lobes with toothed margins. *W USA (S Oregon, California).* H2. Summer.

40. I. bracteata Watson. Illustration: *Botanical Magazine*, 8640 (1915); *Journal of the Royal Horticultural Society* **91**: f. 92 (1966); Cohen, A guide to the Pacific coast irises, 34 (1967); Rickett, Wild flowers of the United States **5**: pl. 20 (1971).

Plant 20–30 cm, stems unbranched. Leaves glossy green, often red at the base, to 1 cm wide, persisting through the winter; flowering stems bearing short bract-like leaves to the top, usually tinged with pink or purple-red. Bracts 5–9 cm, held closely around the flower-stalk and ovary. Flowers 2, 6–7.5 cm in diameter. Perianth-tube 5–10 mm. Falls 4.5–8 cm, cream to yellow, with brown or reddish purple veins, the blade with a central deeper yellow area. Standards 4.5–7.5 cm, erect, slightly less veined than the falls. Style-branches 2–3 cm. *W USA (S Oregon, N California).* H2. Early summer.

I. hartwegii Baker. Illustration: Dykes, The genus Iris, pl. 10 (1913); *Aliso* **4**(1): 32 (1958); Cohen, A guide to the Pacific coast irises, 10 (1967); Rickett, Wild flowers of the United States **4**: pl. 16 (1970). Leaves lacking red at the base, not persisting through the winter. Bracts divergent and separated from one another. Flowers pale yellow or lavender. *W USA (California).* H3. Summer.

I. tenax Lindley. Illustration: *Journal of the Royal Horticultural Society* **91**: f. 95 (1966); Rickett, Wild flowers of the United States **5**: pl. 19 (1971); Clark, Wild flowers of the Pacific northwest, 58 (1976); Mathew, The iris, 99 (1981). Leaves 3–5 mm wide, not persisting through the winter. Bracts divergent. Flowers 7–9 cm in diameter, white, cream, yellow or lavender to deep purple-blue. Falls, if purple, with a central patch of white or yellow. *NW USA (SW Washington, W Oregon).* H2. Early summer.

41. I. innominata Henderson. Illustration: *Journal of the Royal Horticultural Society* **91**: f. 93, 94 (1966); Hay & Synge, Dictionary of garden plants in colour, f. 1190 (1969); Rickett, Wild flowers of the United States **5**: pl. 21 (1971); Mathew, The iris, **97**, f. 16 (1981).

Plant 15–25 cm, stems unbranched. Leaves dark green, purplish at the base, 2–4 mm wide, persisting through the winter. Bracts 3.5–6 cm, held closely around the perianth-tube, margins papery. Flowers 1 or 2, 6.5–7.5 cm in diameter, cream, yellow, orange, lilac pink or pale bluish purple to dark purple. Perianth-tube 1.5–3 cm. Falls 4.5–6.5 cm, often veined, margin frilly. Standards slightly shorter, margin frilly. Style-branches 2–2.5 cm, lobes reflexed, toothed. *W USA (SW Oregon, NW California).* H2. Summer.

I. hartwegii, which has leaves that do not persist through the winter, divergent bracts and perianth-tube 5–10 mm, may key out here (see under 40. *I. bracteata*).

I. tenax, which has leaves that do not persist through the winter, flowers 7–9 cm in diameter with a perianth-tube 6–10 mm, may key out here (see under 40. *I. bracteata*).

42. I. chrysophylla Howell. Illustration: Cohen, A guide to the Pacific coast irises, 26 (1967); Rickett, Wild flowers of the United States **5**: pl. 21 (1971).

Plant almost stemless or to 20 cm. Leaves often slightly grey-green, 3–5 mm wide, longer than the flowering stems, persisting through the winter. Bracts 5–8.5 cm, held close together. Flowers usually 2, 6–7 cm in diameter, cream or pale yellow with deeper yellow or lilac veins. Perianth-tube 4.5–12 cm. Falls 4.5–6.5 cm, rather narrow, pointed. Standards 3–5.5 cm. Style-branches 1.5–2.5 cm, lobes more or less entire. *W USA (Oregon, N California).* H3. Late spring–summer.

I. tenuissima Dykes. Illustration: *Aliso* **4**(1): 69 (1958); *Journal of the Royal Horticultural Society* **91**: f. 90 (1966); Rickett, Wild flowers of the United States **5**: pl. 20 (1971). Plant 15–30 cm. Perianth-tube abruptly widened to form a conspicuous throat in the upper one-quarter to half. Flowers cream, with brown or purple veins on the falls. *W USA (N California).* H4. Early summer.

43. I. macrosiphon Torrey. Illustration: Dykes, The genus Iris, pl. 12 (1913); *Journal of the Royal Horticultural Society* **91**: f. 88 (1966); Cohen, A guide to the Pacific coast irises, **18**, f. 1 (1967); Rickett, Wild flowers of the United States **5**: pl. 20 (1971).

Plant almost stemless or to 25 cm, stem unbranched. Leaves grey-green, with no pink colour at the base, *c.* 6 mm wide, longer than the flowering stems, persisting through the winter. Bracts 4–4.5 cm, green, held closely together. Flowers 2, often fragrant, 5–6 cm in diameter, white, cream, yellow, lavender or purple, veined especially on the falls. Perianth-tube 3.5–8.5 cm, widened at the apex into a bowl-like shape. Falls 4–7 cm, the centre of the blade often whitish. Standards slightly shorter than falls. Style-branches 2–3.5 cm, with reflexed toothed lobes. *W USA (California).* H4. Summer.

I. munzii R.C. Foster. Illustration: *Aliso* **4**(1): 55 (1958); Cohen, A guide to the Pacific coast irises, 16 (1967). Plant to 75 cm. Leaves 1.5–2 cm wide. Bracts 6.5–11 cm, divergent, often separated from one another. Flowers 2–4, 6–7.5 cm in diameter, pale blue, or lavender to deep reddish purple, often with darker veins. Perianth-tube 7–10 mm. Falls and standards with wavy or frilly margins. *W USA (California).* H5. Summer.

I. hartwegii, which has leaves that do not persist through the winter, divergent bracts, flowers 6–8 cm in diameter and perianth-tube only 5–10 mm, may key out here (see under 40. *I. bracteata*).

44. I. purdyi Eastwood. Illustration: Dykes, The genus Iris, pl. 11 (1913); *Aliso* **4**(1): 59 (1958); Cohen, A guide to the Pacific coast irises, 30 (1967); Rickett, Wild flowers of the United States **5**: pl. 20 (1971).

Plant 20–35 cm. Leaves dark glossy green, tinged with pink or red at the base, to 8 mm wide; flowering stems bearing short bract-like leaves to the top. Bracts 5–7.5 cm, often flushed with red. Flowers

2, *c.* 8 cm in diameter, white or cream, the falls often tinged with lavender, and veined and spotted with purplish pink or brown. Perianth-tube 3.5 cm. Falls 5.5–8.5 cm, spreading. Standards 5–7 cm, spreading. Style-branches 2–3 cm. Stigma truncate to 2-lobed. *W USA (coastal N California).* H3. Summer.

I. fernaldii R.C. Foster. Illustration: *Aliso* **4**(1): 28, 29 (1958); *Journal of the Royal Horticultural Society* **91**: f. 89 (1966); Cohen, A guide to the Pacific coast irises, **24**: f. 2 (1967). Leaves grey-green, tinged with purple at the base. Flowers pale yellow, the falls with a deeper yellow central line, and sometimes with a pale purple tinge or veining. Stigma triangular. *W USA (C California).* H3. Summer.

Series **Longipetalae** (Diels) Lawrence. Rhizomes thick, clad in remains of old leaves. Stems persisting for 1 or more years after flowering. Flowers whitish to blue or lavender. Stigma 2-lobed. Capsule tapering at top and bottom, 6-ribbed.

45. I. missouriensis Nuttall (*I. tolmieana* Herbert). Illustration: *Botanical Magazine*, 6579 (1881); Rickett, Wild flowers of the United States **4**: pl. 16 (1970); Spellenberg, Audubon Society field guide to North American wildflowers – western region, f. 606 (1979); Brickell (ed.), RHS A–Z encyclopedia of garden plants, 578 (2003).

Plant to 50 cm, stem usually branched. Leaves pale green, 3–7 mm wide, usually longer than the flowering stems. Bracts 4–7 cm, green at the base, papery above. Flowers 2 or 3, pale blue or lilac to deep blue or lavender. Perianth-tube *c.* 1 cm. Falls *c.* 6 cm, strongly veined, blade often with a yellow signal patch. Standards shorter than falls. Style-branches to 2.5 cm, with toothed lobes. *Western N America.* H3. Late spring–summer.

I. pontica,which differs in growing only to 10 cm and having solitary violet flowers and completely membranous bracts, may key out here (see under 57. *I. sintenisii*).

I. tenax, which has leaves coloured pink at the base, and an unbranched stem carrying flowers that vary from pale lavender to deep blue-purple, through white to yellow, may key out here (see under 40. *I. bracteata*).

46. I. longipetala Herbert. Illustration: *Journal of the Royal Horticultural Society* **88**: f. 111 (1963); Rickett, Wild flowers of the United States **5**: pl. 20 (1971); Mathew, The iris, f. 17 (1981).

Leaves dark green, equal to or shorter than flowering stems. Bracts 7–12 cm. Flowers 3–8, whitish veined with lilac-purple. Falls 6–10 cm, often with a slightly yellowish signal patch. Standards notched. *W USA (coastal C California).* H3. Late spring–summer.

Series **Laevigatae** (Diels) Lawrence. Rhizomes stout. Flowers 6–15 cm in diameter. Stigma 2-lobed. Capsule triangular to nearly round in cross-section, 3-ribbed, thin-walled, breaking up irregularly rather than splitting into 3 cells. Seeds shiny.

47. I. pseudacorus Linnaeus (*I. pseudacorus* var. *acoriformis* (Boreau) Lynch). Illustration: Polunin, Flowers of Europe, f. 1690 (1969); Fitter et al., Wild flowers of Britain and northern Europe, 273 (1974); Grey-Wilson & Mathew, Bulbs, pl. 28 (1981); Mathew, P.J. Redouté: Lilies and related flowers, 173 (1982).

Rhizome stout, clad with fibrous remains of old leaf bases. Plant 73–160 cm, stems branched. Leaves grey-green, 50–90 × 1–3 cm, with a prominent midrib. Bracts 4–10 cm, green with a papery margin. Flowers 1–3 per branch, 7–10 cm in diameter, yellow. Perianth-tube 1–1.5 cm. Falls 5–7.5 cm, blade broadly ovate to almost circular, with a deeper yellow central area and usually with some violet or brown veining. Standards 2–3 cm, narrowly oblong. Style-branches 3.5–4.5 cm. *Europe to W Siberia, Caucasus, Turkey, Iran, N Africa.* H1. Summer.

'Alba' has cream flowers. 'Variegata' has the leaves striped with yellow. Var. **bastardii** (Boreau) Lynch has the blade of the falls lacking the darker yellow central patch.

48. I. virginica Linnaeus. Illustration: *Botanical Magazine*, 703 (1804); Rickett, Wild flowers of the United States **1**: pl. 11 (1966).

Plant 30–100 cm, stems often arching, branched. Leaves 1–3 cm wide, rather soft, drooping at the tips, as long as the flowering stems or longer. Flowers 1–4, 6–8 cm in diameter, blue, violet, lilac or pinkish lavender. Falls *c.* 7.5 cm, spreading, blade with a central yellow, hairy patch surrounded by a white, purple-veined area. Standards *c.* 5 cm, erect. *Southeastern USA.* H2. Summer.

Very similar to the next species: some botanists consider it to be the same. A hybrid between these two species is often offered for sale as **I.** × **robusta** E.S. Anderson, particularly as the cultivar, 'Gerald Darby' (illustration: Brickell (ed.), RHS A–Z encyclopedia of garden plants, 576, 2003), which has dark violet flowers and striking purple bases to the leaves and stems.

49. I. versicolor Linnaeus. Illustration: *Botanical Magazine*, 21 (1787); Rickett, Wild flowers of the United States **1**: pl. 11 (1966); Niering & Olmstead, Audubon Society field guide to North America wildflowers – eastern region, f. 620 (1979); Mathew, The iris, f. 19 (1981).

Plants 20–80 cm, stems branched. Leaves 35–60 × 1–2 cm, often pinkish at the base, as long as the flowering stems or shorter. Bracts 3–8 cm, green. Flowers 6–8 cm in diameter, violet, blue-purple, lavender or dull greyish purple. Falls 4–6 cm, widely spreading, blade often with a central greenish yellow blotch, surrounded by a white, purple-veined area that continues down the claw. Standards 2.5–4 cm. *Eastern N America.* H2. Summer.

'Kermesina' has reddish purple flowers.

50. I. ensata Thunberg (*I. kaempferi* Lemaire). Illustration: *Botanical Magazine*, 6132 (1874); Dykes, The genus Iris, pl. 19 (1913); Brickell (ed.), RHS A–Z encyclopedia of garden plants, 575 (2003).

Plant 60–90 cm, stems usually unbranched. Leaves 20–60 cm × 4–12 mm, shorter than the flowering stems, midrib prominent. Flowers 8–15 cm in diameter, purple or reddish purple. Perianth-tube 1–2 cm. Falls *c.* 7.5 cm, claw yellow, the yellow spreading upwards to the base of the ovate or elliptic blade. Standards *c.* 5lcm, erect. *Former E USSR, China, Japan.* H2. Summer.

Many named cultivars have been developed that may have larger flowers, double flowers, a more extensive colour range, and often flowers with spreading standards producing a flattish appearance (illustration: Perry, Flowers of the world, 145, 1972).

51. I. laevigata Fischer. Illustration: Perry, Flowers of the world, 145 (1972); Kitamura et al., Coloured illustrations of herbaceous plants of Japan (Monocotyledoneae), pl. 21, f. 143 (1964); Mathew, The iris, f. 18 (1981); Brickell (ed.), RHS A–Z encyclopedia of garden plants, 577 (2003).

Plant to 40 cm, stems often with 1 branch. Leaves 1.5–4 cm wide. Flowers 8–10 cm in diameter, white or bluish purple. Perianth-tube to 2 cm. Falls to 6.5 cm, claw yellow or whitish, this colour spreading upwards to the obovate or ovate-elliptic blade. Standards much smaller than falls, erect. *E Asia.* H2. Summer–autumn.

There are many named cultivars available, some of which extend the colour range. 'Variegata' has striped leaves.

Series **Hexagonae** (Diels) Lawrence. Rhizomatous. Flowers 5.5–12 cm in diameter. Capsules 6-ribbed. Seeds very large with a corky coat.

52. I. fulva Ker Gawler. Illustration: Dykes, The genus Iris, pl. 21 (1913); *Journal of the Royal Horticultural Society* **88**: f. 113 (1963); Rickett, Wild flowers of the United States **2**: pl. 22 (1967); Everard & Morley, Wild flowers of the world, pl. 166 (1970).

Plant 45–80 cm, stems straight or very slightly zig-zag. Leaves 1.5–2.5 cm wide, drooping at the tips. Flowers 5.5–6.5 cm in diameter, red, orange-red or coppery-red. Perianth-tube to 2.5 cm. Falls to 6.5 cm, drooping. Standards *c.* 5 cm, drooping, notched at the apex. Style-branches held at an oblique angle so they stand above the rest of the flower. *S USA (Mississippi valley).* H3. Summer.

I. × fulvala Dykes,with purple-red flowers, is occasionally grown. It is a hybrid between *I. fulva* and *I. brevicaulis.*

53. I. brevicaulis Rafinesque (*I. foliosa* Mackenzie & Bush). Illustration: Dykes, The genus Iris, pl. 20 (1913); Rickett, Wild flowers of the United States **1**: pl. 11 (1966); Hay & Synge, Dictionary of garden plants in colour, f. 1189 (1969).

Plant 30–50lcm, stems zig-zag. Stem leaves *c.* 2.5 cm wide, overtopping the flowers. Flowers terminal and borne in the leaf axils, 6–10 cm in diameter, bright blue-violet. Falls reflexed, blade broadly ovate, with a central yellow band usually surrounded by white, claw usually with greenish white veins. Standards much smaller than falls, widely spreading. *S USA (Mississippi valley).* H3. Summer.

I. hexagona Walter. Illustration: *Botanical Magazine*, 6787 (1884); Rickett, Wild flowers of the United States **3**: pl. 11 (1969). Stems not zig-zag. Flowers 10–12 cm in diameter, bluish purple or lavender. Blade of falls with a yellow signal patch. Standards erect. *SE USA.* H3. Summer.

Series **Prismaticae** (Diels) Lawrence. Rhizomes thin, far-creeping. Capsule triangular in cross-section with 1 rib at each corner. Seeds with smooth coats.

54. I. prismatica Ker Gawler. Illustration: *Botanical Magazine*, 1504 (1813); Dykes, The genus Iris, pl. 7 (1913); Rickett, Wild flowers of the United States **1**: pl. 11 (1966); Brickell (ed.), RHS A–Z encyclopedia of garden plants, 578 (2003).

Rhizomes far-creeping. Plant 30–80 cm, stems wiry, rarely straight, usually with 1 branch. Leaves bluish green, 50–70 cm × 2–7 mm. Bracts papery-brown. Flowers 2 or 3 at top of stem, 1 on the branch, 5.5–7 cm in diameter, borne on stalks to 4 cm. Perianth-tube 2–3 mm. Falls spreading widely, pale violet or violet-blue, blade whitish in the centre with violet veins, claw greenish white with violet veins. Standards violet, almost erect. Style-branches 2–3 cm, lobes recurved. *USA (E coast & S Appalachian mountains).* H2. Summer.

A variant with creamy white flowers is occasionally grown.

Series **Spuriae** (Diels) Lawrence. Rhizomes woody, roots wiry. Stems unbranched or closely branched. Outside of perianth-tube with 3 drops of nectar at the base of the perianth-lobes. Stigma 2-lobed. Capsule triangular in cross-section with 2 ribs at each corner, and an apical beak. Seeds hard, with a loose, shiny, papery coat.

55. I. spuria Linnaeus.

A variable species, of which the following subspecies are cultivated:

Subsp. **spuria.** Illustration: Grey-Wilson & Mathew, Bulbs, pl. 28 (1981); Mathew, P.J. Redouté: Lilies and related flowers, 183 (1982). Plant 50–80 cm. Bracts almost transparent at the tip. Flowers 6–8 cm in diameter, lilac or violet-blue veined with violet. Falls 4.5–6 cm, blade with a central yellow stripe, claw longer than the blade. *C Europe.* H2. Summer.

Subsp. **carthalinae** (Fomin) Mathew. Plant to 90 cm. Flowers white or sky blue, overtopping the leaves. *Former USSR (Georgia).* H2. Summer.

Subsp. **halophila** (Pallas) Mathew & Wendelbo (*I. halophila* Pallas). Plant 40–90 cm. Flowers 6–7 cm in diameter, dingy pale yellow to bright golden yellow, usually with darker veins. Falls 4–6 cm, the claw longer than the blade. *S Romania, Former USSR (Ukraine, Moldavia, Caucasus, W Siberia).* H2. Summer.

Subsp. **maritima** (Lamarck) Fournier. Plant 30–50 cm. Bracts green. Falls 3–4.5 cm, cream with purple veins, the blade dark purple, claw with a greenish stripe, claw longer than the blade. *SW Europe.* H2. Summer.

Subsp. **musulmanica** (Fomin) Takhtadjan (*I. klattii* Kemularia-Nathadze; *I. musulmanica* Fomin; *I. violacea* Klatt). Illustration: Rechinger (ed.), Flora Iranica: Iridaceae, pl. 12 (1975); Mathew, The iris, f. 20 (1981). Plant 40–90 cm. Flowers pale violet to deep lavender-violet with darker veins. Falls 5.5–8 cm, the blade with a central yellow stripe and sometimes a yellow suffusion towards the base, claw equal to or slightly longer than the blade. *E Turkey, N & NW Iran, Former USSR (S Caucasus).* H2. Summer.

Subsp. **notha** (Bieberstein) Ascherson & Graebner. Plant 70–90 cm. Flowers violet-blue to bright blue. Claw of falls with a broad yellow stripe, claw equal to or longer than the blade. *Former USSR (S Caucasus).* H2. Summer.

56. I. graminea Linnaeus. Illustration: Berrisford, Irises, pl. 29 (1961); Polunin, Flowers of Europe, f. 1686 (1969); Grey-Wilson & Mathew, Bulbs, pl. 28 (1981); Mathew, P.J. Redouté: Lilies and related flowers, 167 (1982).

Plant 20–40 cm, stem strongly flattened or 2 winged. Leaves 35–100 cm × 5–15 mm, the uppermost stem leaf overtopping the flowers. Bracts unequal, the lower larger and rather leaf-like. Flowers 1 or 2, fragrant, 7–8 cm in diameter. Falls 3–5 cm, violet, the centre of the blade with a white area veined with violet, claw conspicuously winged, often tinged with green or brown. Standards 2.5–4 cm, purple. Style-branches purple, often tinged with green or brown in the lower part. *NE Spain, through C Europe to Former W USSR, N & W Caucasus.* H2. Summer.

Var. **pseudocyperus** (Schur) Beck is a larger plant with unscented flowers. *Czechoslovakia, Romania.*

57. I. sintenisii Janka. Illustration: Cave, The iris, 81 (1959); Goulandris & Goulimis, Wild flowers of Greece, 185 (1968); Grey-Wilson & Mathew, Bulbs, pl. 28 (1981); Mathew, The iris, 111, f. 21 (1981).

Rhizome slender, clad in the brown fibrous remains of leaf bases. Plant

10–30 cm, stem unbranched, cylindric to slightly flattened, but not winged. Leaves 20–50 cm × 2–5 mm, usually longer than the flowering stems. Bracts 4–7 cm, papery, keeled. Flowers usually 1, sometimes 2, 5–6 cm in diameter with rather narrow perianth-lobes. Falls 3.8–4.5 cm, veined violet on a white ground giving a metallic violet-blue appearance. Standards 3–4 cm. Styles 2.5–3 cm, lobes recurved. *SE Europe, N Turkey, Former SW USSR*. H2. Summer.

I. pontica Zapalowicz. Plant *c.* 10 cm. Flowers solitary, violet, often veined darker. Falls 4.5–5.5 cm, blade almost circular, usually with a whitish or pale greenish yellow, violet-veined area in the centre, claw separated by a definite constriction. Standards 3.5–5 cm. *Romania, Former USSR (W Ukraine, Caucasus, ?C Asia)*. H2. Late spring.

58. I. kerneriana Baker. Illustration: Cave, The iris, 97 (1959); Thomas, Perennial garden plants, pl. III (1976); Mathew, The iris, 111 (1981); Brickell (ed.), RHS A–Z encyclopedia of garden plants, 577 (2003).

Rhizome slender, clad with the fibrous remains of the old leaf bases. Plant 20–30 cm, sometimes to 55 cm, stem unbranched. Leaves usually to 5 mm wide, sometimes to 1 cm, generally slightly shorter than the flowering stems. Bracts rather inflated, becoming chaffy as the fruit develops. Flowers 2–4, 7–10 cm in diameter, deep cream to pale yellow. Perianth-tube 7–10 mm. Falls 5–6.5 cm, blade with a deep yellow central blotch, strongly recurved so that the tip sometimes touches the stem. Standards 4.5–5.5 cm, erect, wavy-margined, notched at the tip. Style-branches following the curve of the falls. *N & C Turkey*. H2. Summer.

59. I. orientalis Miller (*I. ochroleuca* Linnaeus; *I. spuria* Linnaeus subsp. ochroleuca (Linnaeus) Dykes). Illustration: *Botanical Magazine*, 61 (1793); Goulandris & Goulimis, Wild flowers of Greece, 183 (1968); Hay & Synge, Dictionary of garden plants in colour, f. 1196 (1969); Perry, Flowers of the world, 145 (1972).

Plant 40–90 cm, stem usually with 1 branch, sometimes more. Leaves 1–2 cm wide. Bracts papery. Flowers 2 or 3, 8–10 cm in diameter, white. Falls with a circular blade 2.5–3.5 cm wide, with a large yellow signal patch. Standards to 8.5 cm, erect. *NE Greece, W Turkey*. H2. Summer.

60. I. × monnieri de Candolle.

Plant to 120 cm, stems stout. Leaves *c.* 60 cm. Flowers lemon-yellow, fragrant. Falls with a circular blade 3.5 cm wide and notched at the apex. Style-branches with strongly recurved, broadly triangular, 4–5 mm lobes. *Turkey*. H2. Summer.

Thought to be a hybrid between *I. orientalis* and *I. xanthospuria* Mathew & Baytop.

61. I. crocea R.C. Foster (*I. aurea* Lindley). Illustration: *Edwards's Botanical Register* **33**: t. 59 (1847); Dykes, The genus Iris, pl. 16 (1913).

Plant to 1.5 m, stems usually branched. Leaves to 75 × 1.5–2 cm. Flowers 2 or 3 per branch, 12–18 cm in diameter, deep golden yellow. Falls 7.5–8 cm, blade with a wavy margin and longer than the claw. Style-branches with narrowly triangular lobes *c.* 1 cm, not strongly recurved. *Origin unknown, possibly Kashmir*. H2. Summer.

Series **Foetidissimae** (Diels) Mathew. Rhizomatous. Leaves persisting through the winter. Seeds bright red, remaining attached to the capsule for a long time after it has split.

62. I. foetidissima Linnaeus. Illustration: Hay & Synge, Dictionary of garden plants in colour, f. 1188 (1969) – fruit only; Fitter et al., The wild flowers of Britain and northern Europe, 273 (1974); Grey-Wilson & Mathew, Bulbs, pl. 28 (1981); Mathew, P.J. Redouté: Lilies and related flowers, 163 (1982).

Plant 30–90 cm, stem branched, rather flattened. Leaves 30–70 × 1–2.5 cm, dark green. Bracts 5–10 cm, green. Flowers 1–5 per branch, 5–7 cm in diameter. Perianth-tube *c.* 1 cm. Falls 3–5 cm, blade lilac-grey veined with purple, fading to white in the centre, claw brownish, winged, sometimes as wide as the blade. Standards 2–4 cm, brownish, flushed with lilac. Style-branches brownish or yellowish. Seeds scarlet. *S & W Europe, N Africa, Atlantic islands*. H2. Summer.

Yellow-flowered variants are known mainly as var. **citrina** Syme and occur occasionally in cultivation. There is also a form with variegated leaves, and a very rare variant with white seeds.

Series **Tenuifoliae** (Diels) Lawrence. Rhizome small, almost vertical, apex with persistent shiny, brown remains of old leaf bases. Stigma 2-lobed. Capsule more or less rounded to triangular in cross-section, if 6-ribbed then with a rib at each corner and on each face. Seeds cubic to pear-shaped, wrinkled.

63. I. tenuifolia Pallas.

Plant 10–30 cm, stems with brown fibrous leaf remains around the base. Leaves to 40 cm × 1–3 mm. Bracts 7–10 cm, green. Flowers 1 or 2, scented, 4–6 cm in diameter. Perianth-tube 5–8 cm. Falls 4–6 cm, the blade with violet or lilac veins on a cream ground and with a cream or yellowish central stripe. Standards slightly shorter than falls, bluish violet or lilac. Style-branches with narrow, sharply pointed lobes. *Former SE USSR through C Asia to Mongolia & W China*. H2. Late spring.

I. songarica Schrenk. Illustration: Mathew, The iris, f. 22 (1981). Flowers 5–7 cm in diameter, up to 6 in each set of bracts, greyish lavender-blue, spotted and veined darker, the standards 6.5 cm or more. *C & E Asia*. H2.

Series **Ensatae** (Diels) Lawrence. Rhizomatous, clump-forming. Ovary with a slender apical beak beneath a very short perianth-tube, giving the perianth-lobes the appearance of being almost free from one another. Capsule 6-ribbed, beaked. Seeds spherical, shiny.

64. I. lactea Pallas (*I. biglumis* Vahl). Illustration: *Edwards's Botanical Register* **26**: t. 1 (1840); Mathew, The iris, 126 (1981).

Rhizomes tough, with wiry roots, clad with reddish or purple-brown fibrous remains of old leaf bases. Plant 6–40 cm. Leaves grey-green, 3–5 mm wide, equalling or longer than flowering stems, strongly ribbed. Flowers 2 or 3, scented, 4–6 cm in diameter, blue, blue-violet or purple. Perianth-tube 2–3 mm, with a slender beak at the base, which is part of the ovary. Falls rather narrow, claw whitish or yellowish with dark veins. *Former S USSR, China, Mongolia, Korea, Himalaya from Afghanistan to Xizang*. H2. Early summer.

Series **Syriacae** (Diels) Lawrence. Rhizome almost vertical, the rather swollen leaf bases giving a bulb-like appearance.

65. I. grant-duffii Baker. Illustration: *Botanical Magazine*, 7604 (1898); Cave, The iris, 32 (1959).

Rhizome clad in remains of leaf bases, which become spiny. Plant 25–30 cm, stems unbranched. Leaves grey-green, to 35 cm × 5–10 mm, prominently veined, margin white. Bracts to 18 cm, greenish, margin papery. Flowers solitary, slightly scented, *c.* 6 cm in diameter. Perianth-tube *c.* 7 mm. Falls 6–7 cm, sulphur-yellow or greenish yellow, blade with an orange signal patch and sometimes with a few black streaks, claw often with purplish veins. Standards *c.* 7 cm, similar in colour to the falls, or paler. Style-branches *c.* 4.5 cm, pale yellow. *Israel, Syria, ?Lebanon.* H4. Late spring.

Series **Unguiculares** (Diels) Lawrence. Rhizomes much-branched, forming large patches that eventually die in the middle to leave a circle of growth. Leaves persisting through the winter. Flowers stemless, but with a long perianth-tube. Style-branches joined at the base into a slender tube, and bearing golden yellow glands on the upper surface, towards the margin.

66. I. unguicularis Poiret (*I. stylosa* Desfontaines). Illustration: Hay & Synge, Dictionary of garden plants in colour, f. 1203 (1969); Perry, Flowers of the world, 146 (1972); Grey-Wilson & Mathew, Bulbs, pl. 28 (1981).

Stem absent or very short. Leaves 45–60 × to 1 cm, borne in erect, compact tufts. Bracts 6–13 cm, more or less green. Flowers solitary, fragrant, almost stemless, lavender. Perianth-tube 6–20 cm. Falls 7–8 cm, the blade with a yellow central band, usually surrounded by a white area veined with lavender, claw with dark veins. Standards 7–8 cm. Style-branches with golden yellow glands on the upper surface, towards the margin. *Algeria, Tunisia, S & W Turkey, W Syria, ?Israel.* H2. Winter–spring.

Several cultivars are available with white, deep violet, pinkish or variegated flowers. The following is the most distinct and often regarded as a separate species:

'Cretensis' (*I. cretensis* Janka; *I. cretica* Baker). Illustration: *Journal of the Royal Horticultural Society* **101**: 336 (1976); Huxley & Taylor, Flowers of Greece and the Aegean, f. 409, 410 (1977); *Gartenpraxis* **1980**: 113 (1980); Mathew, The iris, f. 23 (1981). Leaves 1–3 mm wide. Flowers with narrow segments less than 5.5 cm. Blade of falls violet or deep lavender at the apex, the rest of the blade and the claw white, veined with violet; centre of blade with an orange stripe. *S Greece, Crete.*

I. lazica Albov (*I. unguicularis* Poiret var. *lazica* (Albov) Dykes). Illustration: *Journal of the Royal Horticultural Society* **90**: f. 12 (1965). Leaves 15–30 cm × 8–15 mm, borne in distinct spreading fans. Falls lavender-blue, claw white-veined and spotted with darker lavender, the spots continuing onto the blade, which has a pale yellow central stripe. *NE Turkey, Former USSR (SW Caucasia).* H2. Autumn and spring.

Section **Lophiris** (Tausch) Tausch. Rhizomes often creeping with widely spreading stolons. Flowers white, lilac or blue. Falls with a ridge that is dissected (entire in *I. tenuis* and *I. speculatrix*), and sometimes has an extra ridge on each side.

I. tenuis and *I. speculatrix* do not fit well into this section, and probably do not belong here, but at the moment botanists are unsure of their position. In this account they are left, in the absence of further research, in Section *Lophiris.*

67. I. cristata Solander. Illustration: *Journal of the Royal Horticultural Society* **88**: f. 115 (1963); Rickett, Wild flowers of the United States **2**: pl. 22 (1967); *Green Scene* **3**(5): back cover (1975); Mathew, The iris, 72, f. 12 (1981).

Rhizomes much branched, small. Leaves to 15 × 1–3 cm at flowering time, often lengthening later. Flowers almost stemless, 3–4 cm in diameter, usually lilac-blue, sometimes white, lilac, purple or blue. Perianth-tube 4–6 cm. Blade of falls with a central white patch, along which run 3 crisped ridges that continue down the claw; on or around the ridges is a variable amount of yellow or orange-brown. Standards narrower than falls. *Eastern N America.* H2. Early summer.

68. I. lacustris Nuttall. Illustration: Rickett, Wild flowers of the United States **1**: pl. 11 (1966); Hay & Synge, Dictionary of garden plants in colour, f. 94 (1969); Brickell (ed.), RHS A–Z encyclopedia of garden plants, 577 (2003).

Very similar to *I. cristata* but smaller and with a stem to 5 cm. Leaves to 1 cm wide. Flowers sky-blue, occasionally white. Perianth-tube to 2 cm. Blade of falls with a central white patch and a yellow, central, minutely toothed ridge. *N America (Great Lakes region).* H1. Late spring–summer.

I. speculatrix Hance. Illustration: *Botanical Magazine*, 6306 (1877); Walden & Hu, Wild flowers of Hong Kong, pl. 35 (1977). Flowering stem 20–25 cm. Leaves to 1 m × 5–10 mm, with distinct cross-veining. Perianth-tube *c.* 1 cm, below which is a solid beak from the top of the ovary. Falls *c.* 2.5 cm, lilac, the blade with a central yellow ridge surrounded by a white area, speckled with purple, which in turn is surrounded by a dark purple area; claw violet-veined. Standards lilac, spreading obliquely. *SE China, Hong Kong.* H5. Spring–summer.

69. I. tenuis Watson. Illustration: Abrams, Illustrated flora of the Pacific states **1**: 463 (1923); *Aliso* **4**: 313 (1959); Rickett, Wild flowers of the United States **5**: pl. 19 (1971).

Rhizome far-creeping. Plant 30–35 cm, stem branched. Leaves *c.* 30 × 1–1.5 cm. Flowers 1 per branch, 3–4 cm in diameter. Falls to 3 cm, whitish, veined with blue-purple and blotched with yellow, central ridge yellow, undissected. Standards erect, notched at the tip. Style-lobes entire. *W USA (Oregon).* H2. Late spring.

70. I. tectorum Maximowicz. Illustration: *Botanical Magazine*, 6118 (1874); Dykes, The genus Iris, pl. 24 (1913); *Journal of the Royal Horticultural Society* **88**: f. 116 (1963); Hay & Synge, Dictionary of garden plants in colour, f. 1202 (1969).

Rhizome fat, greenish. Plant 25–35 cm, stems slightly branched. Leaves to 30 × 2.5–5 cm, ribbed. Flowers 2 or 3 per branch, 8–10 cm in diameter. Perianth-tube *c.* 2.5 cm. Falls to 5 cm, lilac with some darker veining and blotching, only slightly deflexed, blade wavy-margined with a white dissected ridge bearing a few dark spots. Standards spreading obliquely, margin wavy. Style-lobes irregularly toothed. *C & SW China, ?Burma; naturalised in Japan.* H2. Early summer.

'Alba' has flowers that are white except for a few yellow veins around the ridge and on the claw of the falls.

71. I. gracilipes A. Gray. Illustration: *Botanical Magazine*, 7926 (1903); Cave, The iris, 45 (1959); Kitamura et al., Coloured illustrations of herbaceous plants of Japan (Monocotyledoneae), pl. 20, f. 139 (1978); Mathew, The iris, 73 (1981).

Rhizome with short stolons. Plant 10–15 cm, stems branched. Leaves 5–10 mm wide. Bracts fused at the base, the fused part sheathing the ovary and the base of the perianth-tube. Flowers 1 per branch, 3–4 cm in diameter. Perianth-tube 1–1.5 cm. Falls *c.* 2.5 cm, lilac to lilac-blue, the centre of the blade with a white area veined with violet, and a white ridge that is yellow at the apex. *China, Japan.* H2. Early summer.

'Alba' has white flowers.

72. I. japonica Thunberg (*I. chinensis* Curtis). Illustration: *Botanical Magazine*, 373 (1797); Kitamura et al., Coloured illustrations of herbaceous plants of Japan (Monocotyledoneae), pl. 20, f. 138 (1978); Mathew, The iris, 70 (1981); Mathew, P.J. Redouté: Lilies and related flowers, 159 (1982).

Rhizome stoloniferous. Plant 45–80 cm, stems branched. Leaves green and glossy, to 45 cm. Flowers 3 or 4 per branch, 4–5 cm in diameter, appearing flattish because of the spreading perianth-lobes, white to pale bluish lavender. Perianth-tube to 2 cm. Falls with a fringed margin, the blade with an orange ridge surrounded by purple blotches. Standards to 3 cm, margin fringed. Style-branches with fringed lobes. *C China, Japan.* H3. Spring.

'Variegata' (illustration: Mathew, The iris, f. 13, 1981) has leaves striped with cream.

I. japonica has been crossed with *I. confusa* and *I. wattii.*

73. I. confusa Sealy. Illustration: Mathew, The iris, 70 (1981); Brickell (ed.), RHS A–Z encyclopedia of garden plants, 573 (2003).

Rhizomes with short stolons. Plant to 1 m, stems branched. Flowers 4–5 cm in diameter, flattish like those of *I. japonica*, white, the blade of the falls with a yellow ridge surrounded by yellow and sometimes purple spots. Falls and standards scalloped-wavy on the margin. Style-branches with fringed lobes. *W China (Yunnan).* H4. Late winter–spring.

74. I. wattii Baker. Illustration: Dykes, Handbook of garden irises, pl. 13 (1924); Everard & Morley, Wild flowers of the world, pl. 107 (1970).

Plant 1–2 m, stems branched. Leaves 60–90 × 3.5–7.5 cm, ribbed. Flowers 2 or 3 per branch, 5.5–8 cm in diameter. Falls to 5 cm, the blade deflexed to a vertical position, lilac-blue with a central whitish area spotted with darker lilac and deep yellow, and a central prominent whitish ridge with deep yellow spots, margin of blade crisped. Standards to 5 cm, spreading, lavender-blue, margins wavy. Style-branches usually with fringed lobes. *India (Manipur), W China (Yunnan).* H4. Spring–summer.

There is some doubt as to whether the plants in cultivation as described above are true *I. wattii;* they may be of hybrid origin (see *RHS Iris Year Book 1978*, 62–6, 1979). Further study of both wild and cultivated plants is necessary to clarify the confusion.

I. milesii M. Foster. Illustration: *Botanical Magazine*, 6889 (1886). Rhizomes fat, greenish. Plant 30–75 cm. Falls to 3 cm, the blade spreading or reflexed, pinkish lilac blotched with purple, central ridge toothed. Standards pinkish lilac. *Himalaya.* H3. Summer.

Subgenus **Nepalensis** (Dykes) Lawrence.
Rhizome tiny, consisting of a small growing point to which the plant dies back completely in winter; roots swollen at least in part. Flowers very short-lived, lasting less than a day. Blade of falls with a linear ridge.

75. I. decora Wallich (*I. nepalensis* D. Don). Illustration: Sweet, British flower garden, ser. 2, **1**: t. 11 (1829); Dykes, The genus Iris, pl. 39 (1913); *Journal of the Royal Horticultural Society* **94**: f. 100 (1969); Mathew, The iris, f. 24 (1981).

Rhizome covered in bristly fibres. Plant 10–30 cm, stem branched or not. Leaves 2–5 mm wide, longer than flowering stems, strongly ribbed. Flowers usually 2, slightly scented, 4–5 cm in diameter, pale bluish lavender to deep reddish purple. Perianth-tube 3.5–5 cm. Falls to 3.5 cm, spreading, the blade with an orange-yellow central ridge that becomes white or purple at the apex, claw whitish with purple veins. Standards spreading, the upper half bending downwards. Style-branches with broad lobes that have toothed or crisped margins. *Himalaya from Kashmir to China.* H3. Summer.

Subgenus **Xiphium** (Miller) Spach.
Bulbs with papery or leathery tunics; roots not swollen and fleshy. Leaves channelled. Stem unbranched with 1–3 flowers. Falls beardless (except in *I. boissieri*,which has a sparsely hairy band on the blade). Standards erect, well developed (but narrow in *I. serotina*).

76. I. latifolia (Miller) Voss (*I. anglica* Steudel; *I. xiphioides* Ehrhart). Illustration: Hay & Beckett (eds), Readers' Digest encyclopaedia of garden plants and flowers, 370 (1978); Perry, Flowers of the world, 143 (1972); Grey-Wilson & Mathew, Bulbs, pl. 29 (1981); Rix & Phillips, The bulb book, 153 (1981).

Plant to 60 cm. Leaves 25–60 cm × 5–8 mm, greyish white on upper surface, not persisting through the winter. Flowers usually 2 in each set of bracts, violet-blue. Perianth-tube *c.* 5 mm. Falls 6–7.5 cm, blade ovate-oblong, not sharply delimited from the claw and with a central yellow patch, claw widely winged. Standards 4–6 cm, oblanceolate. *NW Spain & Pyrenees.* H3. Summer.

Many cultivars are available with white, blue, violet or purple flowers.

77. I. xiphium Linnaeus (*I. hispanica* Steudel). Illustration: Polunin & Huxley, Flowers of the Mediterranean, f. 263 (1965); Hay & Synge, Dictionary of garden plants in colour, f. 785 (1969); Grey-Wilson & Mathew, Bulbs, pl. 29 (1981); Mathew, The iris, 139 (1981).

Plant 40–60 cm. Leaves greyish green, 20–70 cm × 3–5 mm, persisting through the winter. Bracts 6–10 cm, green. Flowers 1 or 2, usually blue or violet, occasionally white, yellow, bronze or bicoloured. Perianth-tube 1–3 mm. Falls 4.5–6.5 cm, blade almost circular with a central orange or yellow blotch, claw longer than blade, unwinged. Standards 4.5–6.5 cm, oblanceolate. *Spain, Portugal, SW France, S Italy, Corsica, Morocco, Algeria & Tunisia.* H3. Spring–early summer.

A variable plant, which has been introduced to gardens from all parts of its range. Some of the variants are available under the following names: 'Battandieri' (*I. battandieri* M. Foster), an African form with white flowers that have an orange ridge on the blades of the falls; 'Lusitanica' (*I. lusitanica* Ker Gawler) – illustration: *Botanical Magazine*, 679, 1803 – with yellow or bronze flowers, found in Portugal; 'Praecox' (var. *praecox* Dykes), an early-flowering variety from the Gibraltar area, with large flowers; 'Taitii' (*I. taitii* M. Foster), with pale blue flowers appearing in early summer.

I. xiphium is widely grown for the cut-flower trade.

I. serotina Willkomm. Illustration: *Botanical Magazine*, n.s., 733 (1977); Grey-Wilson & Mathew, Bulbs, pl. 29 (1981); Mathew, The iris, f. 25 (1981). Leaves dying away before flowering time. Flowers 2 or 3, violet-blue. Perianth-tube 5–10 mm. Falls 3.5–4.5 cm, blade with a yellow central line. Standards 7–14 mm, very narrow, almost bristle-like. *SE Spain, ? Morocco.* H3. Late summer.

78. I. filifolia Boissier. Illustration: *Botanical Magazine*, 5929 (1871); Dykes, The genus Iris, pl. 44 (1913); *Journal of the Royal Horticultural Society* **89**: f. 8 (1964);

Grey-Wilson & Mathew, Bulbs, pl. 29 (1981).

Plant 25–45 cm. Leaves 0.5–3 mm wide, appearing in autumn and persisting through the winter, the emerging young shoot being enclosed in a purple- and white-blotched sheath. Flowers 1 or 2, reddish-violet. Perianth-tube 1–3 cm. Blade of falls almost circular with a central orange patch. *S Spain, NW Africa.* H3. Summer.

79. I. juncea Poiret. Illustration: *Botanical Magazine*, 5890 (1871); Dykes, Handbook of garden irises, pl. 12 (1924); Grey-Wilson & Mathew, Bulbs, 164 (1981).

Bulb covered with tough leaf bases that split into fibrous points at the apex. Plant 30–40 cm. Leaves 0.5–3 mm wide, appearing in autumn and persisting through the winter. Flowers 2, fragrant, bright yellow. Perianth-tube 3.5–5 cm. *N Africa, S Spain, Sicily.* H3. Summer.

The following varieties have been in cultivation and may still be grown: var. **mermieri** Lynch with sulphur-yellow flowers; var. **numidica** Anon., with lemon-yellow flowers; and var. **pallida** Lynch with large, soft yellow flowers.

80. I. tingitana Boissier & Reuter. Illustration: *Botanical Magazine*, 6775 (1884); Hay & Synge, Dictionary of garden plants in colour, f. 784 (1969).

Plant stout, to 50 cm. Leaves to 45 cm, arching, silvery-green, appearing in autumn and persisting through the winter. Flowers 1–3 in each set of bracts, pale to deep blue. Perianth-tube 1–2.5 cm. Falls to 7.5 cm, blade obovate to almost circular with a central orange-yellow ridge. Standards to 10 cm. *NW Africa.* H3. Late winter–spring.

Var. **fontanesii** (Godron) Maire (*I. fontanesii* Godron) flowers later and is a more slender plant with flowers of deep violet-blue.

There are many garden hybrids between *I. xiphium* and *I. tingitana*, possibly also with influence from *I. latifolia*. These are sometimes grouped under the name **I. × hollandica** Anon., and known as Dutch iris. The flowers vary in colour from white, yellow and bronze to pale blue or purple (illustration: Hay & Synge, Dictionary of garden plants in colour, f. 788, 789, 790, 1969; Rix & Phillips, The bulb book, 152, 153, 1981).

I. boissieri Henriques. Illustration: *Botanical Magazine*, 7097 (1890); Grey-Wilson & Mathew, Bulbs, pl. 29 (1981).

Plant 30–40 cm. Flowers solitary, deep purple, with a yellow, sparsely bearded stripe in the centre of the blade of the falls. *N Portugal, NW Spain.* H3. Summer.

Subgenus **Scorpiris** Spach (Section *Juno* (Roemer & Schultes) Bentham). Bulbs with papery tunics; roots fleshy or thickened. Leaves channelled, held in 1 plane, often grey-green with a white margin. Falls beardless but with a ridge in the centre of the blade, which is often wavy or dissected. Standards much reduced, usually horizontal or deflexed (except in *I. cycloglossa*).

81. I. planifolia (Miller) Fiori & Paoletti (*I. alata* Poiret). Illustration: Hay & Synge, Dictionary of garden plants in colour, f. 779 (1969); Grey-Wilson & Mathew, Bulbs, pl. 29 (1981); Mathew, P.J. Redouté: Lilies and related flowers, 181 (1982); Rix & Phillips, The bulb book, 33 (1981).

Bulb large with fleshy, but not swollen, roots. Plant 10–15 cm. Leaves shiny green, 10–30 × 1–3 cm, concealing the stem. Bracts *c.* 10 cm, almost hidden by the leaves. Flowers 1–3, 6–7 cm in diameter. Perianth-tube 8–18 cm. Falls 5–8 cm, bluish violet, the blade with a yellow central ridge surrounded by violet veining, claw winged. Standards 2–2.3 cm, horizontal, oblanceolate, usually toothed. Stigma 2-lobed. *S Spain & Portugal, Sardinia, Sicily, Crete, N Africa.* H3. Winter.

There are variants that have white flowers, or flowers of dark blue with the blade of the falls edged with white.

I. palaestina (Baker) Boissier. Similar but smaller with greenish, whitish or bluish flowers. Standards 1–1.8 cm. Stigma not 2-lobed. *Coastal Israel & Lebanon, ?S Syria.* H4. Winter.

82. I. aucheri (Baker) Sealy (*I. fumosa* Boissier; I. sindjarensis Boissier). Illustration: Hay & Synge, Dictionary of garden plants in colour, f. 774 (1969); Mathew, Dwarf bulbs, 148 (1973); Mathew, The iris, f. 26 (1981); Rix & Phillips, The bulb book, 59 (1981).

Bulbs with fleshy, not very swollen, roots. Plants 15–40 cm. Leaves green, to 25 × 2.5–4.5 cm, concealing the stem at flowering time. Bracts 7–9 cm, long-pointed. Flowers 3–6, usually blue, rarely almost white. Perianth-tube 5–6.5 cm. Falls 4–6 cm, blade with a central yellow ridge and a wavy margin, claw winged. Standards 2–3.5 cm, horizontal or deflexed, obovate. *SE Turkey, N Iraq, N Syria, Jordan, NW Iran.* H4. Late winter–spring.

Usually offered in the trade under the name *I. sindjarensis*, and under this name it has been crossed with *I. persica* to produce the hybrid **I. × sindpers** Hoog (illustration: *Bulletin of the Alpine Garden Society* **18**: 148, 1950; *Botanical Magazine*, n.s., 419, 1963; Rix & Phillips, The bulb book, 53, 1981). The plants are only *c.* 25 cm and the flowers are blue or blue-mauve with a prominent orange or deep yellow ridge on the blade of the falls, which may have a darker apex. Another hybrid is 'Sindpur', in which the other parent is *I. galatica (I. purpurea)*; this resembles *I. aucheri* in general habit and size and shape of flower but is a translucent mauve colour.

I. graeberiana Sealy. Illustration: *Botanical Magazine*, n.s., 126 (1950). Leaves glossy green above, greyish below, margins whitish. Flowers blue, sometimes tinged with violet. Blade of falls with a white central ridge surrounded by a white area that is strongly veined. *Former USSR (Soviet Central Asia).* H4. Spring.

I. willmottiana M. Foster. Illustration: Rix & Phillips, The bulb book, 59 (1981). Flowers lavender or light purple, blotched with white, the blade of the falls marked with deeper purple, and with a white central ridge. Standards *c.* 1.5 cm, deflexed, diamond-shaped to 3-lobed. *Former USSR (Soviet central Asia: Pamir AM).* H3. Spring.

I. willmottiana is probably no longer in cultivation in Britain although it may still be grown in other parts of Europe. Plants grown under the name *I. willmottiana* 'Alba' are probably a form of *I. bucharica*.

I. pseudocaucasica, which differs in having leaves grey-green below, yellowish green or blue flowers and bracts 4–5 cm, may key out here (see under 83. *I. caucasica*).

83. I. caucasica Hoffmann. Illustration: Rechinger (ed.), Flora Iranica: Iridaceae, pl. 17 (1975); Mathew, The iris, f. 30 (1981); Rix & Phillips, The bulb book, 60, 61 (1981).

Bulb with only slightly thickened roots. Plant 10–15 cm. Leaves grey-green, 5–7, 10–12 × to 2 cm, concealing the stem at flowering time, margin white. Bracts 6–7.5 cm. Flowers 1–4, greenish or yellow-green. Perianth-tube 3–4.5 cm. Falls 3–4 cm, the blade with a central, yellow, scalloped ridge, claw winged. Standards 1.3–2.5 cm, horizontal, oblanceolate.

C & NE Turkey, NE Iraq, NW Iran, Former USSR (Caucasus). H3. Late winter–spring.

I. baldschuanica Fedtschenko. Illustration: Mathew, Dwarf bulbs, pl. 52 (1973). Roots of bulb very swollen. Perianth-tube 8–10 cm. Flowers cream, the blade of the falls with a central yellow ridge, the style-branches tinged with pinkish brown. Claw of falls with downturned margins. Standards horizontal to slightly deflexed, lanceolate. *Former USSR (S Tadjikistan), NW Afghanistan.* H4. Spring.

I. doabensis Mathew. Illustration: *Botanical Magazine*, n.s., 620 (1972); Rechinger (ed.), Flora Iranica: Iridaceae, pl. 23 (1975); Mathew, The iris, f. 29 (1981). Roots of bulb very swollen. Leaves shiny green. Flowers deep yellow. Perianth-tube 7–8 cm. Falls with a deeper yellow wavy ridge on the blade, often surrounded by purple veins, claw with downturned margins. Standards *c.* 8 cm, slightly reflexed. *NE Afghanistan.* H4. Spring.

I. fosteriana Aitchison & Baker. Illustration: *Journal of the Royal Horticultural Society* **90**: f. 211 (1965); Mathew, Dwarf bulbs, pl. 53 (1973); Rechinger (ed.), Flora Iranica: Iridaceae, pl. 19 (1975); Mathew, The iris, f. 26 (1981). Leaves green with white margins. Falls creamy yellow with a white band round the margin of the blade, which has a yellow central ridge surrounded by a darker yellow area veined with brown, claw unwinged. Standards purple, sharply reflexed. *Former USSR (S Turkmenistan), NE Iran, NW Afghanistan.* H4. Spring.

I. linifolia (Regel) Fedtschenko. Illustration: *Journal of the Royal Horticultural Society* **93**: f. 5 (1968). Bulb with very swollen roots. Plant 5–10 cm. Leaves 4–7 mm wide, revealing the stem at flowering time. Flowers pale yellow, the blade of the falls deeper with a central whitish or yellowish toothed ridge, claw unwinged. Standards *c.* 1 cm, horizontal to deflexed, more or less 3-lobed. *Former USSR (Central Asia: Tien Shan, Pamir).* H4. Spring.

I. pseudocaucasica Grossheim (*I. caucasica* misapplied). Illustration: *Botanical Magazine*, n.s., 405 (1962); Rix & Phillips, The bulb book, 33 (1981); Mathew, The iris, f. 32 (1981). Leaves 3–4. Bracts 4–5 cm. Flowers yellowish green or blue, the blade of the falls with a yellow wavy crest, the claw broadly winged, narrowing abruptly to the blade. *SE Turkey, NE Iraq, N & NW Iran, Former USSR (S Transcaucasia).* H4. Early summer.

I. palaestina, which has shiny green leaves, bracts *c.* 10 cm and greenish, whitish or bluish flowers with a perianth-tube 8–18 cm, may key out here (see under 81. *I. planifolia*).

I. drepanophylla, which differs in having tiny bristle-like standards only to *c.* 3 mm, may key out here (see under 85. *I. bucharica*).

84. I. persica Linnaeus. Illustration: *Botanical Magazine*, 1 (1787), 8059 (1906); Mathew, Dwarf bulbs, 148 (1973); Mathew, The iris, f. 29 (1981); Rix & Phillips, The bulb book, 58 (1981).

Bulb with rather thin roots. Plant to 10 cm. Leaves 3–4, green above, greyish beneath, margin whitish, to 10 cm × 5–15 mm. Outer bract 8–9 cm, rigid, green; inner bract shorter, whitish-papery. Flowers 1–4, translucent, silvery-grey, yellowish, brownish or grey-green. Perianth-tube 6–8 cm. Falls 3.5–4.5 cm, usually with a purplish or brownish blade, with a prominent central yellow ridge often brown- or purple-spotted, claw with upturned wings. Standards 1.5–2.5 cm, horizontal to deflexed, oblanceolate or more or less 3-lobed. Style-branches 3.5–4.5 cm. *S & SE Turkey, N Syria, NE Iraq.* H4. Winter–spring.

Several varieties have been described based on variation in flower colour, but may be no longer in cultivation. Recent work on the species does not recognise these varieties.

I. galatica Siehe (*I. purpurea* (Anon.) Siehe). Flowers 1 or 2, with a perianth-tube 4–6 cm, wholly reddish purple, or greenish yellow with a purple blade to the falls, or greenish purple with darker falls. Falls 2.5–5 cm, blade with a prominent yellow or orange central ridge, claw winged. Standards 1–2 cm, horizontal to slightly deflexed, spathulate to 3-lobed. Style-branches 2.5–3.5 cm. *Central N Turkey.* H4. Spring.

I. nicolai Vvedensky. Illustration: *Botanical Magazine*, n.s., 483 (1965); *Journal of the Royal Horticultural Society* **91**: f. 10 (1966); Mathew, The iris, f. 30 (1981). Bulb with swollen, radish-like roots. Plant 12–15 cm. Leaves scarcely developed at flowering time, the tips incurved, enclosing the stem in a tubular sheath, to 25 × 5–6 cm at fruiting time. Perianth-tube 8–11 cm. Falls whitish or lilac, the blade velvety deep violet at the apex, and with a central orange ridge with 2 deep violet veins on each side, claw margins downturned. Standards pale blue-lilac to whitish, more or less horizontal, obovate with inrolled margins. Style-branches pale blue-lilac to whitish, tinged with dull purple. *Former USSR (Central Asia: Pamir), NE Afghanistan.* H4. Spring.

I. rosenbachiana Regel. Illustration: *Botanical Magazine*, 7135 (1890); Mathew, Dwarf bulbs, 148 (1973). Similar to *I. nicolai* but leaves further developed at flowering time, the tips curving outwards, basic colour of flowers reddish purple. *Former USSR (Central Asia: Pamir).* H4. Spring.

I. rosenbachiana and *I. nicolai* are really only colour forms of one species: in Flora Iranica they have been united under the earlier name of *I. rosenbachiana.*

I. stenophylla Baker. Illustration: *Botanical Magazine*, 7734 (1900), 7793 (1901); Rix & Phillips, The bulb book, 58 (1981). Bulb with fleshy roots. Leaves very short at flowering time, lengthening eventually to 10–20 cm × 5–10 mm. Flowers usually solitary, violet-blue or lilac-blue. Falls usually with a darker blade, with a central yellow ridge surrounded by a violet-spotted whitish area. *S Turkey.* H4. Early spring.

Subsp. **allisonii** Mathew (illustration: Mathew, The iris, 166, 1981) has more leaves, which are 1.5–1.8 cm wide when fully grown, with wavy margins. The flowers are a bluer colour and along the claw of the falls is a line of hairs.

I. microglossa, which differs in having leaves with ciliate margins and pale lavender-blue to nearly white flowers, may key out here (see under 87. *I. magnifica*).

I. pseudocaucasica, which differs in having yellowish green or blue flowers with the perianth-tube 3–4 cm, may key out here (see under 83. *I. caucasica*).

85. I. bucharica M. Foster (*I. orchioides* misapplied). Illustration: Synge, Collins guide to bulbs, edn 1, pl. XI (1961); Hay & Synge, Dictionary of garden plants in colour, f. 776 (1969); Rechinger (ed.), Flora Iranica: Iridaceae, pl. 20, 21 (1975); Rix & Phillips, The bulb book, 60, 62 (1981).

Bulb with rather thin roots. Plant 20–40 cm. Leaves bright glossy green, curved, to 20 × 2.5–3.5 cm, with whitish margins. Bracts 7.5–9 cm, green. Flowers 2–6, golden yellow to almost white. Perianth-tube 4.5–5 cm. Falls *c.* 4 cm, blade yellow with blotches of green, brownish or dull violet on either side of the central yellow, scalloped ridge. Standards 1.5–2 cm, spreading to deflexed. Style-branches *c.* 3.5 cm. *Former USSR (Tadjikistan), NE Afghanistan.* H3. Spring.

I. aitchisonii (Baker) Boissier. Illustration: Rechinger (ed.), Flora Iranica: Iridaceae, pl. 17 (1975). Leaves grey-green, to 40 cm × 4–8 mm. Bracts 5–6 cm. Flowers 1–3, yellow, violet or bicoloured. Perianth-tube 2.5–3.5 cm. Claw of falls broadly winged. *E Afghanistan, Pakistan.* H3. Early spring–early summer.

I. drepanophylla Aitchison & Baker. Illustration: *RHS Iris Year Book 1967*, f. 9 (1968); *Journal of the Royal Horticultural Society* **93**: f. 206 (1968) & **99**: f. 85 (1974); Rix & Phillips, The bulb book, 62 (1981). Roots of bulb very swollen. Bracts 5–6.5 cm, membranous. Flowers 2–8, yellowish or greenish. Perianth-tube 3.5–4 cm. Falls with a yellow ridge on the blade, claw with downturned margins. Standards *c.* 3 mm, very narrow, almost bristle-like, slightly deflexed. *Former USSR (Turkmeniya), NE Iran, N & NW Afghanistan.* H4. Spring.

86. I. albomarginata R.C. Foster (*I. caerulea* Fedtschenko). Illustration: Mathew, The iris, f. 27 (1981).

Plant 20–30 cm. Leaves with white margin; internodes of stem visible. Flowers 2–5, blue. Perianth-tube *c.* 4 cm. Falls to 3.5 cm, blade with a white central ridge surrounded by a yellowish area, claw not broadly winged. Standards 2–2.5 cm, oblanceolate, deflexed. *Former USSR (Central Asia: Tien Shan, Alayskiy Khrebet).* H4. Spring.

I. warleyensis M. Foster. Illustration: *Botanical Magazine*, 7956 (1904); *Journal of the Royal Horticultural Society* **91**: f. 159 (1966); Mathew, Dwarf bulbs, pl. 54 (1973). Flowers purplish blue to deep violet, the blade of the falls darker, bordered with white, the central ridge whitish to yellow, toothed or crisped, surrounded by a yellow area. Standards 1–2 cm, narrowly linear to almost 3-lobed. *Former USSR (Central Asia: Pamir).* H3. Spring.

I. warleyensis has been crossed with *I. aucheri (I. sindjarensis)* to produce *I.* 'Warlsind', with yellow falls blotched with purple-brown on the blade, a yellow ridge, and standards and styles whitish (illustration: *Gartenpraxis* **9**: 438, 1978).

87. I. magnifica Vvedensky. Illustration: *Gartenpraxis* **9**: 437 (1978); Rix & Phillips, The bulb book, 60, 61 (1981); *Greenhouse & Garden* **9**(5): 26 (1984); Brickell (ed.), RHS A–Z encyclopedia of garden plants, 578 (2003).

Plant 30–60 cm. Leaves shiny green, 3–5 cm wide; internodes of stem visible. Flowers 3–7, to 7.5 cm in diameter, pale lilac or sometimes white. Perianth-tube 4.5–5 cm, blade with a white undissected ridge surrounded by an area of yellow, claw with broad wings. Standards 2–3 cm, obovate, horizontal to deflexed. *Former USSR (Central Asia).* H3. Spring.

I. cycloglossa Wendelbo. Illustration: Rechinger (ed.), Flora Iranica: Iridaceae, pl. 17 (1975); *Botanical Magazine*, n.s., 708 (1976); Rix & Phillips, The bulb book, 7 (1981). Flowers 1–3, 8–10 cm in diameter, clove-scented, blue-violet. Perianth-tube 3.5–4.5 cm. Falls 6–7 cm, white towards the centre of the blade and with a yellow blotch but no ridge. Standards 4–4.5 cm, at first erect, becoming horizontal as the flower ages. *NW Afghanistan.* H3. Spring.

I. microglossa Wendelbo. Illustration: *Journal of the Royal Horticultural Society* **91**: f. 8 (1966); Mathew, Dwarf bulbs, pl. 55 (1973); Rechinger (ed.), Flora Iranica: Iridaceae, pl. 18 (1975); Rix & Phillips, The bulb book, 62 (1981). Leaves grey-green, 1.5–2.5 cm wide, margins white, ciliate. Flowers 1–4, 4.5–5.5 cm in diameter, pale lavender-blue to almost white. Perianth-tube 3–4.5 cm. Falls 4–4.5 cm, central ridge of blade white or pale yellow. Standards 1.5–2 cm, oblanceolate. *NE Afghanistan.* H4. Spring.

I. aitchisonii, which is only 15–30 cm tall, with leaves 4–8 mm wide, and deep yellow, violet or sometimes bicoloured flowers with a perianth-tube 2.5–3.5 cm, may key out here (see under 85. *I. bucharica*).

Subgenus **Hermodactyloides** Spach
(Section *Reticulata* Dykes). Bulbs with fibrous, netted tunics, many bulblets often produced. Leaves usually 4-angled in cross-section (nearly cylindric in *I. bakeriana*, channelled in *I. kolpakowskiana*). Stems unbranched, flowers solitary, of typical iris type (except *I. danfordiae*, which has very reduced standards). Capsules more or less stalkless (except in *I. pamphylica*).

88. I. reticulata Bieberstein. Illustration: Synge, Collins guide to bulbs, edn 1, pl. 17 (1961); Hay & Synge, Dictionary of garden plants in colour, f. 780 (1969); Mathew, The iris, f. 34 (1981); Rix & Phillips, The bulb book, 52 (1981).

Plant 7–15 cm. Leaves 4-angled, very variable in their development at flowering time, hardly developed to overtopping the flowers. Bracts green. Flowers solitary, pale blue to deep violet-blue or reddish purple. Perianth-tube 4–7 cm. Falls 4–5 cm, blade usually with a central yellow ridge. Standards 3.5–5 cm. Style-branches 3–4 cm, lobes sometimes toothed. Stigma 2-lobed. *N & S Turkey, NE Iraq, N & W Iran, Former USSR (Caucasus).* H2. Early spring.

Many named cultivars of *I. reticulata* are available, differing mainly in the colour of their flowers (Illustration: Synge, Collins guide to bulbs, edn 1, pl. 17, 1961; Hay & Synge, Dictionary of garden plants in colour, f. 781, 782 & 783, 1969; Rix & Phillips, The bulb book, 54, 1981). Hybrids have been produced between *I. reticulata* and *I. bakeriana;* they can be identified by their leaves, which have *c.* 6 ribs.

I. kolpakowskiana Regel. Illustration: *Botanical Magazine*, 6489 (1880); Mathew, Dwarf bulbs, 148 (1973); Rix & Phillips, The bulb book, 55 (1981). Leaves channelled like those of Section *Scorpiris*, short at flowering time, lengthening later to 25 cm. Flowers pale lilac to pale purple, blade of falls dark reddish purple with an orange-yellow ridge. Stigma entire. *Former USSR (Soviet Central Asia: Tien Shan).* H3. Late winter.

I. pamphylica Hedge. Illustration: *Journal of the Royal Horticultural Society* **96**: f. 150 (1971); *Botanical Magazine*, n.s., 648 (1973); Rix & Phillips, The bulb book, 57 (1981); Mathew, The iris, f. 37 (1981). Plant 10–20 cm. Leaves 4-angled, 17–25 cm at flowering time, lengthening later to up to 55 cm. Falls 3.5–4 cm, blade brownish purple with a central slightly raised yellow area spotted with purple, claw greenish with purple veins. Standards *c.* 4 cm, pale blue, the lower part green spotted with purple. Capsules on a stalk, pendulous. *S Turkey.* H3. Late winter–spring.

I. vartanii M. Foster. Illustration: *Botanical Magazine*, 6942 (1887). Flowers usually greyish lilac to white, smelling of almonds. Blade of falls with darker veins and a pale yellow central ridge. Lobes of style-branches very long and narrow. *Israel, ?Syria.* H3. Autumn–winter.

89. I. bakeriana M. Foster. Illustration: Synge, Collins guide to bulbs, edn 1, pl. 17 (1961); Hay & Synge, Dictionary of garden plants in colour, f. 775 (1969); *Gartenpraxis* **9**: 437 (1978); Rix & Phillips, The bulb book, 56 (1981).

Similar to *I. reticulata* but the leaves are almost cylindric with 8 ribs. Flowers whitish, the falls with a dark violet apex, and violet spotting and veining around the cream-coloured ridge on the blade, and on the claw. Standards and style-branches bluish lilac. *SE Turkey, N Iraq, W Iran.* H2. Late winter–spring.

Very like *I. reticulata;* in the wild intermediate plants have been found. It is probably best regarded as a variety of *I. reticulata* as in Flora Iranica (1975), where it appears as var. *bakeriana* (M. Foster) Mathew & Wendelbo.

90. I. histrioides (G.F. Wilson) Arnott. Illustration: Dykes, The genus Iris, pl. 46, f. 1 (1913); *Botanical Magazine*, 9341 (1934); Mathew, The iris, f. 35 (1981).

Leaves 4-angled, more or less absent at flowering time, lengthening later to up to 50 cm. Bracts papery. Flowers 6–7 cm in diameter, blue. Blade of falls with a yellow ridge surrounded by a blue-spotted pale area, blade obviously distinct from claw. *Central N Turkey.* H2. Early spring.

Var. **sophenensis** (M. Foster) Dykes has deep violet-blue flowers with relatively little spotting, and narrower segments.

Several cultivars of *I. histrioides* are available, differing in the shade of blue or violet-blue. There is confusion surrounding 'Major', which varies in colour according to the nursery selling it.

I. histrioides has been hybridised with either *I. danfordiae* or *I. winogradowii* to produce *I.* 'Katharine Hodgkin', which has flowers of a curious grey-yellow-lilac mixture (illustration: Rix & Phillips, The bulb book, 53, 1981). 'Frank Elder' is a similar hybrid, in which *I. winogradowii* is a definite parent.

91. I. histrio Reichenbach. Illustration: Synge, Collins guide to bulbs, edn 1, pl. 17 (1961); *Journal of the Royal Horticultural Society* **91**: f. 187 (1966); Mathew, The iris, f. 36 (1981); Rix & Phillips, The bulb book, 52 (1981).

Leaves 4-angled, to 30 cm at flowering time, lengthing later to up to 60 cm. Bracts papery. Flowers 6–8 cm in diameter, pale blue. Blade of falls with a yellow ridge surrounded by a whitish blue-spotted area, blade not obviously distinct from claw. *S Turkey, Syria, Lebanon.* H2. Early spring.

Var. **aintabensis** G.P. Baker. Illustration: Synge, Collins guide to bulbs, edn 1, pl. 17 (1961); Mathew, Dwarf bulbs, pl. 51 (1973); Rix & Phillips, The bulb book, 52 (1981). A small variant whose leaves at flowering time are less advanced than those of *I. histrio* itself. Flower smaller, paler blue.

92. I. danfordiae (Baker) Boissier. Illustration: Synge, Collins guide to bulbs, edn 1, pl. 17 (1961); Hay & Beckett (eds), Readers' Digest encyclopaedia of garden plants and flowers, 360 (1978); Mathew, The iris, f. 38 (1981); Rix & Phillips, The bulb book, 54 (1981).

Plant 6–7 cm. Leaves 4-angled, varying at flowering time from 1 cm to overtopping flowers. Flowers bright yellow. Perianth-tube 3–6 cm. Falls 2.5–4 cm, blade with a deep yellow or orange central ridge surrounded by sparse green spots. Standards 3–5 mm, very reduced, bristle-like. Style-branches 2–3.5 cm with irregularly toothed lobes. *Turkey.* H2. Early spring.

93. I. winogradowii Fomin. Illustration: Synge, Collins guide to bulbs, edn 1, pl. 17 (1961); Hay & Beckett (eds), Readers' Digest encyclopaedia of garden plants and flowers, 370 (1978); Mathew, Dwarf bulbs, 148 (1973); Rix & Phillips, The bulb book, 53 (1981).

Plant to 20 cm. Leaves 4-angled, elongating to 40 cm. Flowers pale yellow. Perianth-tube *c.* 1.8 cm. Falls *c.* 5 cm, blade with an orange central ridge, surrounded by green spots that extend down the claw. Standards to 5 cm, oblanceolate. *Former USSR (Abkhazia).* H2. Early spring.

The following names appear in current commerical catalogues:

I. fimbriata. Plants offered under this name may be *I. japonica* or *I. tectorum*.

I. intermedia. A name of no botanical validity, sometimes used for bearded irises of hybrid origin that grow 35–65 cm tall.

2. PARDANTHOPSIS Lenz

V.A. Matthews & B. Mathew

Related to and resembling *Iris*. Stems much-branched with many small, short-lived flowers. Perianth-segments free, each fall with transverse bands on the claw. Perianth-segments twisting spirally after flowering.

The genus contains only 1 species, which, until recently, was included in *Iris*. Easily grown in a sunny position with plenty of moisture during the summer. Propagation is by seed, and seedlings should be raised to replace the parent plants, which are not long-lived.

1. P. dichotoma (Pallas) Lenz (*Iris dichotoma* Pallas). Illustration: *Edwards's Botanical Register*, t. 246 (1817); Sweet, British flower garden **1**: t. 96 (1823–25); *Botanical Magazine*, 6428 (1879).

Rhizome small with a mass of thick roots. Stems 40–100 cm, much-branched. Leaves 6–8, borne in a fan, to 30 × 2.5 cm. Spathes papery, 5- or 6-flowered. Flowers 3.5–4.5 cm in diameter, opening in the afternoon. Falls usually creamy white with purplish brown spots and bands in the centre and lower part, or deep reddish purple with whitish, purple-spotted area in the centre of the blade and with banding on the claw. Standards smaller than falls. Style-branches with 2 narrow lobes. *Former USSR (Siberia), N China, Mongolia.* H2. Summer.

P. dichotoma has been crossed with *Belamcanda chinensis* to produce × **Pardancanda norrisii** Lenz. These plants are variable, but generally resemble *P. dichotoma* in flower shape, with reduced style-branches. The flowers are in shades of salmon and apricot.

3. GYNANDRIRIS Parlatore

D.A. Webb & V.A. Matthews

Rootstock a corm, surrounded by fibrous tunics. Leaves 1 or 2, linear, channelled. Flowers in terminal and axillary cymes, short-lived, like those of *Iris* or *Moraea* in appearance with petaloid style-branches, and free perianth-segments differentiated into falls (outer) and standards (inner). Stamens partly united, closely adherent to the underside of the style-branches. Ovary extended upwards as a slender, sterile beak, on which the perianth is directly inserted; perianth-tube absent. Capsule with many seeds.

A genus of 9 species, from southern Africa and the Mediterranean area. The Mediterranean species need summer baking if they are to flower and do best in a bulb-frame in areas where the summer is not completely warm and dry. *G. setifolia* should be started into growth in the autumn and dried off the following spring. They grow best in a well-drained soil and can be propagated by seed or corm division.

Literature: Goldblatt, P., Systematics of Gynandriris (Iridaceae), a Mediterranean-southern African disjunct, *Botaniska Notiser* **133**: 239–60 (1980).

1. G. sisyrinchium (Linnaeus) Parlatore (*Iris sisyrinchium* Linnaeus). Illustration: Coste, Flore de la France **3**: 366 (1906); Polunin & Huxley, Flowers of the Mediterranean, f. 269 (1965); Taylor, Wild flowers of Spain and Portugal, 23 (1972); Reisigl & Danetsch, Flore Méditerranéenne, 111 (1979).

Stem 10–40 cm. Leaves usually 2, with a long, sheathing base; blade flexuous, usually lying on the ground. Stem bearing 1–4 compact cymes, each with 1–6 flowers. Bracts silvery or transparent. Flowers 3–4 cm in diameter, violet-blue or lavender. Falls 2–4 cm, the blade with a white and/or orange signal patch. Standards 2–3 cm, erect. Anthers 4.5–10 mm. Beak of ovary 2.5 cm. *From Portugal through the Mediterranean region and SW Asia to Pakistan.* H4. Spring.

G. monophylla Klatt. Illustration: *Botaniska Notiser* **133**: 255 (1980). Stem only to 5 cm and usually only 1 leaf present. Flowers 2–2.5 cm in diameter, pale slate-blue. Falls 1–2 cm, the blade with an orange signal patch surrounded by a white area. Anthers 2–4 mm. *E Mediterranean area,* H4. Spring.

G. setifolia (Linnaeus filius) R.C. Foster. Illustration: *Botaniska Notiser* **133**: 249 (1980). Stem 5–20 cm. Leaves 1 or 2. Flowers pale lavender-blue, the blade of the falls with a yellow signal patch. Standards reflexed. Anthers 2.5–4 mm. *South Africa (SW Cape).* H4. Winter.

4. HERMODACTYLUS Miller

D.A. Webb

Rootstock of 2–4 palmately branched tubers. Leaves in 2 ranks, narrowly linear, 4-angled, longer than the stem. Flowering stem to 40 cm. Flowers radially symmetric, terminal, solitary, partly enclosed by a pale green bract, yellowish green except for the blade of the falls which is very dark purple-brown to almost black. Falls (outer segments) 4–5 cm; standards (inner segments) 2–2.5 cm. Perianth-tube short, funnel-shaped. Ovary with a short, slender sterile beak at the apex.

A genus of 1 species, differing from *Iris* in its tuberous rootstock and 1-celled ovary with parietal placentation. It can be grown in any sunny, well-drained site, but in regions where the spring weather is likely to damage the flowers it is better under glass.

1. H. tuberosa (Linnaeus) Miller (*Iris tuberosus* Linnaeus). Illustration: Polunin, Flowers of Europe, f. 1683 (1969); Huxley & Taylor, Flowers of Greece and the Aegean, f. 411 (1977); Grey-Wilson & Mathew, Bulbs, pl. 30 (1981); Rix & Phillips, The bulb book, 57 (1981).

E & C Mediterranean region. H4. Spring.

5. MORAEA Miller

V.A. Matthews

Rootstock a corm with membranous, fibrous or netted tunics. Leaves 1 to several, usually linear. Stem with reduced bract-like sheathing leaves above. Flowers 1 to several, borne within bracts (spathes), which may be green or dry and papery. Flowers often short-lived. Perianth radially symmetric, segments free. Falls (outer segments) larger than standards (inner segments), reflexed or spreading, each bearing a conspicuous nectar guide. Standards reflexed to erect, entire or 3-lobed. Stamens opposite outer perianth-segments, adpressed to style-branches. Filaments completely joined, to joined at the base only. Style-branches petaloid with 2 lobes (crests). Fruit a capsule containing angled seeds.

A genus of about 100 species found in Africa south of the Sahara. They grow best in a well-drained soil and should be kept dry during cold weather. The species from tropical Africa are not hardy and require protection from frost; those from the SW Cape grow during the winter and flower in spring, hence they also need winter protection. Species from the E Cape are summer-flowering and better for outdoor cultivation in Europe. Propagation is by seed.

Literature: Goldblatt, P., Contributions to the knowledge of Moraea (Iridaceae) in the summer rainfall region of South Africa, *Annals of the Missouri Botanical Garden* **60**: 204–59 (1973); Goldblatt, P., The genus Moraea in the winter rainfall region of southern Africa, *Annals of the Missouri Botanical Garden* **63**: 657–786 (1976); Goldblatt, P., Systematics of Moraea (Iridaceae) in tropical Africa, *Annals of the Missouri Botanical Garden* **64**: 243–95 (1977).

Main colour of flowers. White: **2, 3, 4, 6, 7**; yellow: **1, 2, 4, 8**; lilac to purple, or brown: **1, 2, 3, 4, 7**; red or orange: **2, 5**.

Nectar guide. Yellow, brownish or bluish, sometimes spotted: **2, 3, 4, 8**; green or dark blue: **5**. Edged with a ring of a different colour: **2, 6, 7**.

Stem. Hairless: **1, 2, 3, 4, 6, 7, 8**; downy to long-hairy: **2, 5, 7**.

Standards. Three-lobed: **3, 4, 6, 7**; not 3-lobed: **1, 2, 3, 5, 8**. Reflexed: **1, 2, 3**; erect or spreading: **2, 3, 4, 5, 6, 7, 8**.

1a. Leaves absent or 1 — 2
b. Leaves 2 or more — 7
2a. Main colour of flowers orange to red — **5. neopavonia**
b. Main colour of flowers white, yellow, lilac, purple or brown — 3
3a. Standards 3-lobed — 4
b. Standards not 3-lobed — 6
4a. Stem finely downy to shaggily hairy; anthers 6–15 mm — **7. villosa**
b. Stem hairless; anthers 3–5 mm — 5
5a. Standards 1.5–2 cm — **6. aristata**
b. Standards to 1.2 cm — **4. tricuspidata**
6a. Main colour of flowers white or lilac, if yellow then stem with a long lower internode; falls 1.5–4 cm; standards 1.4–3.5 cm; anthers 4–8 mm — **2. fugax**
b. Main colour of flowers yellow; falls 3.5–7.5 cm; standards 3–6 cm; anthers 7–15 mm — **8. spathulata**
7a. Anthers *c.* 1 cm — **1. ramosissima**
b. Anthers 2.5–8 mm — 8
8a. Filaments joined only at the base, or united for up to half their length — **2. fugax**
b. Filaments joined for more than half their length or completely united — **3. bipartita**

1. M. ramosissima (Linnaeus filius) Druce (*M. ramosa* (Thunberg) Ker Gawler). Illustration: *Botanical Magazine*, 771 (1804); Courtenay-Latimer et al., The flowering plants of the Tsitsikama Forest and coastal national park, pl. 17 (1967); *Annals of the Missouri Botanical Garden* **63**: 690 (1976).

Leaves many, curved, to 30 × 1.5 cm. Flowering stems 50–120 cm, much branched, hairless. Flowers yellow, perianth-segments reflexed. Falls *c.* 3 cm with a brownish, spotted nectar guide, standards somewhat shorter. Filaments 1.2–1.5 cm, free or joined near the base; anthers *c.* 1 cm. Style-branches 2–2.5 cm. *South Africa (SW Cape).* H4. Spring.

M. polystachya (Thunberg) Ker Gawler. Illustration: Marloth, Flora of South Africa **4**: pl. 39A (1915); *Flowering plants of Africa* **35**: t. 1385 (1962); Batten &

Bokelmann, Wild flowers of the eastern Cape Province, pl. 34, f. 1 (1966); Brickell (ed.), RHS A–Z encyclopedia of garden plants, 699 (2003). Flowers lilac; falls to 5.5 cm with a yellow or orange nectar guide. Filaments to 1 cm, joined to the apex. Style branches *c.* 1 cm, lobes to 2 cm. *Southern Africa.* H5. Autumn.

2. M. fugax (de la Roche) Jacquin (*M. edulis* (Linnaeus filius) Ker Gawler; *M. longifolia* (Schneevoogt) Sweet; *M. odora* Salisbury). Illustration: *Botanical Magazine*, 613 (1803), 1238 (1809); Rice & Compton, Wild flowers of the Cape of Good Hope, pl. 188 (1951); Mason, Western Cape sandveld flowers, 77 (1972); *Annals of the Missouri Botanical Garden* **63**: 726 (1976).

Leaves usually 1, sometimes 2, trailing, usually much longer than stem. Flowering stems 12–40 cm, branched, with a conspicuous long, lower internode. Flowers white, violet or yellow, strongly scented. Falls 2.3–4 cm, the yellow ones with a darker yellow nectar guide, the white or violet ones with the yellow nectar guide surrounded by a bluish colour; claw downy. Standards 2–3.5 cm, erect to slightly reflexed. Filaments 6–10 mm, joined in the lower half; anthers 4–8 mm. Style-branches 1.5–2 cm, lobes 6–16 mm. *South Africa (SW Cape).* H4. Spring.

M. gawleri Sprengel (*M. sulphurea* Baker; *M. undulata* Ker Gawler). Illustration: *Botanical Magazine*, 7658 (1899); *Annals of the Missouri Botanical Garden* **63**: 690 (1976). Leaves 2, erect or spreading, somewhat coiled, margin flat or wavy. Flowers cream, yellow or brick red. Falls 1.2–2.8 cm with a dark yellow nectar guide spotted and lined with brown. Standards reflexed. Anthers 2–3 mm. *South Africa (SW Cape).* H4. Spring.

M. vegeta Linnaeus (*M. tristis* (Linnaeus filius) Ker Gawler). Illustration: Jacquin, Icones plantarum rariorum, t. 225 (1795); *Botanical Magazine*, 577 (1802); *Annals of the Missouri Botanical Garden* **63**: 694 (1976). Leaves several. Stem minutely downy. Flowers dull yellow to brownish, flushed with blue or purple. Falls 2–2.5 cm with a yellow, spotted nectar guide. Standards reflexed. Filaments 5–6 mm, joined in the lower one-third; anthers 3–4 mm. Style-branches 7–8 mm. *South Africa (SW Cape).* H4. Spring.

M. papilionacea (Linnaeus filius) Ker Gawler. Illustration: *Botanical Magazine*, 750 (1804); Marloth, Flora of South Africa **4**: pl. 39F (1915); Rice & Compton, Wild flowers of the Cape of Good Hope, pl. 187 (1951); *Annals of the Missouri Botanical Garden* **63**: 698 (1976). Leaves 2–4, usually downy on lower surface, margin ciliate. Flowering stem 10–15 cm, downy to long-hairy. Flowers yellow or brick red, scented. Falls *c.* 2.5 cm, the yellow nectar guides outlined with green or red. Standards reflexed. Filaments 3–6 mm, united at the base. Style-branches to 8 mm. *South Africa (SW Cape).* H4. Spring.

M. ciliata (Linnaeus filius) Ker Gawler. Illustration: *Annals of the Missouri Botanical Garden* **63**: 712 (1976); le Roux & Schelpe, Namaqualand & Clanwilliam: South African wild flower guide **1**: 47 (1981). Leaves 2–6, slightly downy, margin ciliate, often wavy. Flowers white, yellow, lilac or pale brownish, scented. Falls 2–3.5 cm, with a yellow nectar guide. Standards erect or spreading. Filaments united at the base. Style-branches 7–15 mm. *South Africa (SW Cape).* H4. Spring.

M. tricolor Andrews (*M. ciliata* (Linnaeus filius) Ker Gawler var. *barbigera* (Salisbury) Baker). Illustration: *Andrews' Botanical Repository*, t. 83 (1800); *Botanical Magazine*, 1012 (1807). Leaves usually 3, hairless, margin ciliate, flat or slightly wavy. Flowers red. Falls 2–2.5 cm, with a yellow, bearded nectar guide. Standards erect to spreading. Filaments 3–4 mm, joined at the base; anthers 2.5–3 mm. Style-branches to 5 mm. *South Africa (SW Cape).* H4. Spring.

M. thomsonii Baker. Illlustration: *Botanical Magazine*, 7976 (1904); Moriarty, Wild flowers of Malawi, pl. 5 (1975); *Annals of the Missouri Botanical Garden* **64**: 263 (1977); Cribb & Leedal, The mountain flowers of Tanzania, 180 (1982). Leaf 1, trailing, dry or absent at flowering time. Flowers pale lilac. Falls 1.5–2 cm with yellow to orange nectar guides, which are often brown-spotted. Standards erect, becoming spreading. Filaments 4–5 mm, joined in lower two-thirds. Style-branches *c.* 7 mm, lobes minute or to 5 mm. *E Africa from South Africa to Ethiopia.* G1. Summer.

M. stricta Baker. Illustration: Letty, Wild flowers of the Transvaal, pl. 36, f. 1 (1962); Pearse, Mountain splendour, 75, f. 3 (1978). Leaf 1, usually dry or absent at flowering time. Flowers pale lilac to mauve. Falls *c.* 2 cm, spotted with yellow on the blade. Standards erect. Filaments *c.* 3.5 mm, joined at the base. Style-branches 7–8 mm, lobes *c.* 3 mm. *South Africa (E Cape).* H4. Spring–summer.

3. M. bipartita L. Bolus (*M. polyanthos* misapplied). Illustration: *Annals of the Missouri Botanical Garden* **63**: 748 (1976) – as *M. polyanthos*.

Leaves usually 3, 2–6 mm wide. Flowering stem to 30 cm, branched. Flowers lilac, with reflexed perianth-segments. Falls 2–3 cm with a yellow nectar guide. Standards 1.5–2.5 cm, reflexed. Filaments 4–6 mm, joined in the lower two-thirds to three-quarters; anthers *c.* 5 mm. Style-branches *c.* 5 mm, lobes 5–8 mm. *South Africa (SW Cape).* H4. Late spring–summer.

M. polyanthos Linnaeus filius (*Homeria liladna* L. Bolus). Very similar but with filaments completely united and anthers *c.* 6 mm. Style-branches very narrow, without lobes. *South Africa.*

M. carsonii Baker. Illustration: *Annals of the Missouri Botanical Garden* **64**: 253 (1977); Tredgold & Biegel, Rhodesian wild flowers, pl. 7 (1979). Leaves 2, sometimes 3. Flowers mauve, with spreading perianth-segments. Falls 2–3 cm with yellow, purple-spotted nectar guides. Filaments 7–8 mm, joined in the lower two-thirds. Style-branches *c.* 8 mm, lobes 7–10 mm. *Zimbabwe, Zambia, Malawi & S Zaire.* G1. Spring–summer.

M. fergusoniae L. Bolus (*M. fimbriata* Klatt). Illustration: *Annals of the Missouri Botanical Garden* **63**: 698 (1976). Leaves 3–6, margin ciliate. Flowers white to lilac. Standards erect, lanceolate or 3-lobed. Filaments completely united or free in the upper quarter. *South Africa (SW Cape).* H4. Spring.

4. M. tricuspidata (Linnaeus filius) Lewis (*M. tricuspis* (Thunberg) Ker Gawler; *Iris tricuspis* Thunberg). Illustration: *Botanical Magazine*, 696 (1803); *Annals of the Missouri Botanical Garden* **63**: 768 (1976).

Leaf 1, often trailing, 4–7 mm wide. Flowering stem 25–60 cm, simple or branched. Flowers white. Falls 2.5–3 cm with a yellow, brown or bluish, speckled nectar guide, claw downy. Standards *c.* 1.2 cm, erect, 3-lobed. Filaments 6–7 mm, joined in the lower half; anthers *c.* 5 mm. Style-branches 6–7 mm, lobes 5–7 mm. *South Africa (SW Cape).* H4. Spring–early summer.

M. bellendenii (Sweet) N.E. Brown (*M. pavonia* (Linnaeus filius) Ker Gawler var. *lutea* (Ker Gawler) Baker). Illustration: *Botanical Magazine*, 772 (1804); Mason, Western Cape sandveld flowers, 77 (1972); *Annals of the Missouri Botanical Garden* **63**:

673, 768 (1976). Leaf to 1 cm wide. Flowering stem 50–100 cm, branched. Flowers yellow, the falls with a speckled nectar guide. Filaments 3–5 mm, joined except for the upper 1 mm. *South Africa (SW Cape).* H4. Spring.

M. unguiculata Ker Gawler. Illustration: *Botanical Magazine*, 593 (1802), 1047 (1807); *Annals of the Missouri Botanical Garden* **63**: 760 (1976). Flowers whitish, sometimes yellow, brown or violet. Falls 1–2.2 cm with yellow, speckled nectar guides. Filaments 4–6 mm, joined almost to the apex; anthers 3–4 mm. Style-lobes 2.5–4 mm. *South Africa (SW Cape).* H4. Spring–summer.

5. M. neopavonia R.C. Foster (*M. pavonia* (Linnaeus filius) Ker Gawler). Illustration: *Botanical Magazine*, 1247 (1809); *Journal of the Royal Horticultural Society* **75**: f. 154 (1950) – as *M. pavonia* var. *magnifica*; *Annals of the Missouri Botanical Garden* **63**: 770 (1976).

Leaf 1, basal, sparsely downy on undersurface, exceeding inflorescence. Flowering stems 30–60 cm, simple or with 1 branch, downy. Flowers orange to red. Falls 2.2–4 cm, spreading, nectar guide dark blue or green, often speckled. Standards 1.5–2.5 cm, lanceolate to 3-lobed, erect to spreading. Filaments *c.* 4 mm, joined except for the upper 1 mm; anthers *c.* 1 cm, longer than style-lobes. Style-branches *c.* 5 mm, lobes 1–2 mm. *South Africa (SW Cape).* H4. Spring–summer.

M. neopavonia has been hybridised with *M. villosa*.

6. M. aristata (de la Roche) Ascherson & Graebner (*M. glaucopsis* (de Candolle) Drapiez). Illustration: *Botanical Magazine*, 168 (1791); Redouté, Liliacées **1**: t. 42 (1803); Sweet, The British flower garden, **6**: t. 249 (1833–5); *Annals of the Missouri Botanical Garden* **63**: 776 (1976).

Leaf 1, basal, hairless, exceeding inflorescence. Flowering stem 25–35 cm, simple or sometimes with 1 branch, hairless. Flowers white. Falls 3–3.5 cm, spreading, nectar guide composed of concentric crescents of green, blue-violet or black, claw bearded. Standards 1.5–2 cm, 3-lobed, erect. Filaments 3–4 mm, joined for most of their length; anthers *c.* 5 mm. Style-branches *c.* 7 mm, lobes *c.* 7 mm. *South Africa (SW Cape).* H4. Late spring–summer.

7. M. villosa (Ker Gawler) Ker Gawler (*M. pavonia* (Linnaeus filius) Ker Gawler var. *villosa* (Ker Gawler) Baker). Illustration: *Botanical Magazine*, 571 (1802) & n.s., 112 (1950); *Journal of the Royal Horticultural Society* **75**: f. 154 (1950); Everard & Morley, Wild flowers of the world, pl. 87 (1970); *Annals of the Missouri Botanical Garden* **63**: 776 (1976).

Leaf 1, basal, downy on the undersurface, exceeding inflorescence. Flowering stem 20–40 cm, simple or branched, hairy. Flowers usually purple. Falls 3–4 cm, spreading, nectar guide yellow, surrounded by a dark contrasting colour, often blue or green, claw bearded. Standards 1.6–3 cm, spreading, 3-lobed. Filaments *c.* 5 mm, joined for most of their length; anthers 6–10 mm. Style-branches 5–7 mm, lobes 5–8 mm. *South Africa (SW Cape).* H4. Spring.

M. gigandra L. Bolus. Illustration: *Gardeners' Chronicle* **119**: f. 95, 99 (1946); *Botanical Magazine*, n.s., 188 (1952); *Annals of the Missouri Botanical Garden* **63**: 770 (1976); *Veld & Flora* **68**: 88 (1982). Flowers creamy-white or purple; creamy-white falls with an orange nectar guide spotted with green and surrounded by a band of blue; purple falls with a black basal blotch surrounded by a band of blue; claw indistinct, hairless. Standards creamy-yellow, greenish at the base with a purplish blue band. Anthers 1.3–1.5 cm, much longer than style-lobes. *South Africa (SW Cape).* H4. Spring.

8. M. spathulata (Linnaeus filius) Klatt (*M. spathacea* (Thunberg) Ker Gawler). Illustration: Batten & Bokelman, Wild flowers of the eastern Cape Province, pl. 28, f. 1 (1966); *Annals of the Missouri Botanical Garden* **64**: 267 (1977); Rix & Phillips, The bulb book, 152 (1981).

Leaf 1, basal, hairless, exceeding inflorescence. Flowering stem to 1 m, usually simple. Flowers pale to deep yellow, scented. Falls 3.5–7.5 cm, reflexed, nectar guides orange-yellow. Standards 3–6 cm, erect. Filaments 8–15 mm, joined in the lower two-thirds; anthers 7–15 mm, purple. Style-branches 1.2–1.8 cm, lobes to 1 cm. *Southern Africa.* H4. Spring–summer.

M. huttonii (Baker) Obermeyer. Illustration: *Botanical Magazine*, 6174 (1875); Batten & Bokelman, Wild flowers of the eastern Cape Province, pl. 34, f. 5 (1966); *Flowering Plants of Africa*, 1581 (1970). Nectar guide of falls edged with dark brown lines. Style-lobes 1–1.3 cm with a brown or purple basal blotch. *South Africa (E Cape), Lesotho.* H4. Summer.

M. moggii N.E. Brown. Illustration: Letty, Wild flowers of the Transvaal, pl. 36, f. 2 (1962); *Botanical Magazine*, n.s., 469 (1965). Nectar guide of falls with purple or brownish red lines of spots. Filaments joined in the lower half; anthers cream. Style-lobes 1–2 cm. *South Africa (E Cape).* H4. Summer.

6. DIETES Klatt

V.A. Matthews

Rhizome thick, creeping. Leaves several, in 2 ranks, leathery, linear to sword-shaped. Stem erect, hairless, branched above, bearing leaves at the lower nodes and bracts at the upper nodes. Flowers short-lived, in groups, each group subtended by a spathe-bract. Perianth radially symmetric, segments free, the outer larger than the inner, with an ascending claw and spreading blade and with a nectar guide at the base of the blade. Filaments usually free. Style short, divided into 3 large, flattened, petaloid branches that divide at the tips into paired style crests; stigma-lobes borne on the lower surface. Fruit a 3-celled capsule with many seeds.

A genus of 6 species from E, C and SE tropical Africa and coastal southern Africa, with 1 outlying species on Lord Howe Island. Cultivation as for *Moraea* (p. 263).

Literature: Goldblatt, P., Systematics, physiology and evolution of Dietes (Iridaceae), *Annals of the Missouri Botanical Garden* **68**: 132–53 (1981).

1a. Flowers yellow; leaves with a distinct, usually double, central vein **2. bicolor**

b. Flowers white; leaves lacking a distinct central vein 2

2a. Leaves 3–5 cm at widest point; claws of inner perianth-segments unmarked **1. robinsoniana**

b. Leaves up to 2 cm at widest point; claws of inner perianth-segments marked with dark brown towards the base or dotted with orange, if unmarked then outer perianth-segments 3.5 cm or less 3

3a. Outer perianth-segments 2.5–3.5 cm; stamens 8–15 mm; style-branches 7–9 mm; flowers lasting 1 day **3. iridioides**

b. Outer perianth-segments 4.5–6 cm; stamens 1.7–2.3 cm; style-branches 1.2–2 cm; flowers lasting 3 days **4. grandiflora**

1. D. robinsoniana (Mueller) Klatt (*Moraea robinsoniana* (Mueller) Bentham & Mueller). Illustration: *Botanical Magazine*, 7212 (1892); *Annals of the Missouri Botanical Garden* **68**: 142 (1981).

Stem 1–1.5 m. Leaves 3–5 cm at widest point. Flowers white. Outer perianth-segments *c.* 4.5 cm, claw *c.* 1.5 cm, blade *c.* 3 cm wide, with an orange nectar guide. Filaments to *c.* 1 cm; anthers 5–6 mm. Style *c.* 4 mm, branches *c.* 1 cm. *Australia (Lord Howe Island).* H5. Spring–summer.

2. D. bicolor (Steudel) Klatt (*Moraea bilcolor* Steudel). Illustration: *Flore des serres* **7**: t. 744 (1852); Batten & Bokelmann, Wild flowers of the eastern Cape Province, pl. 34, f. 6 (1966); *Flowering Plants of Africa* **39**: t. 1525 (1968); *Annals of the Missouri Botanical Garden* **68**: 144 (1981).

Stem 80–120 cm. Leaves 6–12 mm at widest point. Flowers yellow. Outer perianth-segments *c.* 3.5 cm, claw *c.* 1.2 cm, bearded and spotted with orange, blade to 2.3 cm wide, usually with a dark brown nectar guide. Filaments *c.* 6 mm; anthers 4–8 mm. Style *c.* 2 mm, branches 8–10 mm. *South Africa (E Cape).* H4. Spring–summer.

3. D. iridioides (Linnaeus) Klatt (*Moraea iridioides* Linnaeus; *M. catenulata* Lindley; *Dietes vegeta* misapplied). Illustration: *Edwards's Botanical Register*, t. 1074 (1827); Batten & Bokelmann, Wild flowers of the eastern Cape Province, pl. 31, f. 2 (1966); *Flowering Plants of Africa* **39**: t. 1524 (1968); *Annals of the Missouri Botanical Garden* **68**: 146 (1981).

Stem 15–60 cm. Leaves 6–20 mm at widest point, spreading. Flowers white. Outer perianth-segments 2.5–3.5 cm, claw *c.* 1.5 cm, hairy in centre and papillate, blade 1.2–1.6 cm wide, with a yellow nectar guide, claw often with orange dots. Filaments 5–9 mm; anthers 3–6 mm. Style 2–3 mm, branches 7–9 mm, blue or white, flushed with blue-violet. *Southern South Africa, north through eastern Africa to Kenya.* H5. Spring–summer.

'Johnsonii' has longer, erect leaves and larger flowers.

4. D. grandiflora N.E. Brown. Illustration: Batten & Bokelmann, Wild flowers of the eastern Cape Province, pl. 34, f. 3 (1966); Jeppe, Natal wild flowers, pl. 18f (1975); *Annals of the Missouri Botanical Garden* **68**: 151 (1981).

Stem 1–1.5 m. Leaves 1–2 cm at widest point. Flowers white. Outer perianth-segments 4.5–6 cm, claw 2–2.5 cm with a yellow beard, blade 2.5–3.5 cm wide, with a yellow nectar guide; claws of inner perianth-segments marked with dark brown towards the base. Filaments 1–1.3 cm, anthers 7–10 mm. Style *c.* 5 mm, branches 1.2–2 cm, flushed with mauve. *South Africa (E Cape through Transkei to Natal).* H5. Spring–summer.

7. CYPELLA Herbert

V.A. Matthews

Rootstock a long bulb with pleated tunics. Leaves pleated. Flowers short-lived, with the 3 outer perianth-segments large, spreading, and the 3 inner much smaller, usually erect and often hairy on the claw; segments free. Filaments free. Style longer than stamens, the branches sometimes petaloid. Fruit a capsule containing angular seeds.

A genus of about 15 species occurring from Mexico to Argentina. They are not generally hardy and so the bulbs should be lifted and stored in sand in a frost-free place during the winter. Propagation is by seed or offsets.

1a. Flowers yellow **1. herbertii**
b. Flowers blue or violet-blue **2. coelestis**

1. C. herbertii (Lindley) Herbert. Illustration: *Botanical Magazine*, 2599 (1825); Sweet, British flower garden, ser. 2, **4**: t. 33 (1829–31); *Flore des serres* **5**: t. 537 (1849); Rix & Phillips, The bulb book, 173 (1981).

Stem 30–50 cm. Leaves 15–30 × *c.* 2 cm, linear. Inflorescence usually much-branched. Flowers 4–6 cm across. Outer perianth-segments dull orange-yellow with a violet central stripe, claw concave; inner perianth-segments much smaller, yellow spotted with purple, rolled towards the tip. Stigma purple. *Brazil, Uruguay, Argentina; ? Paraguay.* H5–G1. Summer.

C. peruviana Baker. Illustration: *Botanical Magazine*, 6213 (1876). Differs in the inflorescence usually being unbranched, except in vigorous specimens. Flowers bright yellow, the lower part of each segment forming a cup banded with purple or brownish spots; inner segments very hairy on the central part. Stigma yellow, petaloid. *Peru, Bolivia.* G1. Winter.

2. C. coelestis (Lehmann) Diels (*C. plumbea* Lindley; *Phalocallis coelestis* (Lehmann) Ravenna). Illustration: *Botanical Magazine*, 3710 (1839); *The Garden* **60**: 10 (1901).

Stem 40–75 cm. Leaves basal, to 45 × 2.5 cm. Inflorescence unbranched. Flowers 5–8 cm across, dull blue with brownish shading and yellow-brown spots in the cup formed by the bases of the segments; inner segments with a yellow patch, and with hairs on the claw. Style bluish lilac. *Brazil, Uruguay, Argentina, ?Paraguay.* G1. Late summer.

C. herrerae R.C. Foster. Differs in having narrower basal leaves (7–8 mm) and short stem leaves bearing bulbils in the axils. Flowers deep blue to violet-blue, the central cup whitish with red-brown spots. *Peru.* G1. Winter.

8. TRIMEZIA Herbert

V.A. Matthews

Rhizome bulb-like, covered with the fibrous remains of old leaf bases. Leaves linear to linear-lanceolate. Perianth radially symmetric, segments free, in 2 dissimilar whorls, shortly clawed at the base. Stamens borne opposite the style-branches; filaments free. Fruit a dry capsule on a long stalk. Seeds black, angled.

A genus of 5 species from central and tropical South America and the West Indies. The plants grow best in a well-drained compost and need winter protection. Propagation is by seed.

1. T. martinicensis (Jacquin) Herbert. Illustration: *Botanical Magazine*, 416 (1797); Fournet, Flore illustrée des phanerogames de Guadeloupe et de Martinique, 246 (1978).

Stem to 20 cm, rarely more in cultivation. Leaves to 30 cm, linear, in a basal fan. Flowers short-lived, produced in succession from overlapping bracts, 2–2.5 cm across. Outer perianth-segments yellow with brownish or purplish spots at the base, to 2.5 cm, erect or incurved; inner much smaller, incurved, yellow. *S. America, West Indies; naturalised elsewhere in the Tropics.* H5. Mainly spring, spasmodically throughout the year.

9. NEOMARICA Sprague

V.A. Matthews

Rhizomes short, creeping. Leaves borne in basal fan, sword-shaped, with many veins. Flowering stems erect with a large, leaf-like bract. Flowers short-lived, borne in clusters, the stalks flat and winged. Perianth radially symmetric, segments free, in 2 dissimilar whorls, clawed at the base, outer spreading, inner smaller, usually erect, abruptly reflexed at the tip. Filaments free. Style-branches 2-fid or 3-fid, opposite the stamens. Fruit a capsule.

A genus of about 15 species from tropical America and West Africa. The illegitimate name *Marica* (Ker Gawler) Herbert was used in the past for this genus and still occasionally occurs in the literature. Ravenna has recently indicated (in Prance, G.T. & Elias, T.S. (eds), Extinction is forever, 257, 1977) that *Neomarica* and *Trimezia* overlap to such an extent that they should not be maintained as separate genera; however as his final work has not yet been published, these genera are retained in this Flora. The species are tender and require protection from frost. They thrive in any well-drained soil and should be kept fairly dry throughout the winter. Propagation is by seed or division of the rhizomes.

1a. Outer perianth-segments yellow **4. brachypus**
b. Outer perianth-segments white, blue or blue-lilac 2
2a. Leaves ribbed; outer perianth-segments white; style-branches 3-fid 3
b. Leaves not obviously ribbed; outer perianth-segments blue or blue-lilac; style-branches 2-fid **3. caerulea**
3a. Plant to 90 cm; flowers 6–10 cm across **1. northiana**
b. Plants to 60 cm; flowers to 5 cm across **2. gracilis**

1. N. northiana (Schneevoogt) Sprague. Illustration: *Botanical Magazine*, 654 (1803); Everett, The New York Botanical Gardens illustrated encyclopedia of horticulture **7**: 2302 (1981).

Stems to 90 cm. Leaves ribbed, to 60 × 5 cm. Flowers 6–10 cm across, fragrant. Outer perianth-segments white, inner segments violet towards the apex, both yellowish marked with red at the base. Style-branches 3-fid, the 2 lateral teeth erect, the central 1 often reflexed. *Brazil.* G1. Spring–summer.

This species may be viviparous, producing young plants on the old flowering stems.

2. N. gracilis (Hooker) Sprague. Illustration: *Botanical Magazine*, 3713 (1839).

Similar to *N. northiana* but shorter (to 60 cm) and with flowers less than 6 cm across. The style-branches have 3 erect teeth. *S Mexico to N Brazil.* G2. Summer.

3. N. caerulea (Ker Gawler) Sprague. Illustration: *Edwards's Botanical Register*, t. 713 (1823); Hooker, Exotic flora, **3**: 222 (1826); *Botanical Magazine*, 5612 (1866); Brickell (ed.), RHS A–Z encyclopedia of garden plants, 718 (2003).

Stems to 60 cm. Leaves not obviously ribbed, 90–180 × *c.* 3.5 cm. Flowers to 10 cm across. Outer perianth-segments blue or blue-lilac, inner deeper blue, both barred with yellow and brown at the base. Style-branches 2-fid, each with a horn-like appendage at the base. *Brazil.* G2. Summer.

4. N. brachypus (J.G. Baker) Sprague. Illustration: *Botanical Magazine*, 6380 (1878).

Leaves 40–50 × 2.5–4 cm. Flowers 7–8 cm across. Perianth-segments yellow, barred with reddish brown at the base. Style-branches 3-fid. *West Indies.* G1. Summer.

N. longifolia (Link & Otto) Sprague. Illustration: Everett, The New York Botanical Gardens illustrated encyclopedia of horticulture **7**: 2302 (1981).

Leaves *c.* 30 cm; flowers to 5 cm across. *Brazil.* G1. Summer.

10. TIGRIDIA Jussieu

J. Cullen

Perennial herbs with bulbs. Leaves mostly basal, equitant, in a fan. Spathes 2, enclosing 2–several flowers. Perianth of 6 free segments, the outer larger than the inner, with their bases overlapping to form a cup, the blades spreading or reflexed. Inner segments spreading, mostly ovate, the margins often upcurved near the middle so that the segment appears fiddle-shaped, with a nectary on the inner surface. Stamens 3, filaments united into a tube around the style. Stigmas 3, each bifid, on radii alternating with the anthers. Fruit a capsule.

A genus of 23 species from Mexico and Guatemala. In cool areas the bulbs should be planted in rich, well-drained soil in spring. In autumn, after flowering, the plants should be lifted and dried off, when the bulbs can be stored in a cool, frost-free place. In warmer areas the bulbs may be left in the ground. Propagation is by bulbs or by seed.

Literature: Molseed, E., The Genus Tigridia (Iridaceae) of Mexico and Central America, *University of California Publications, Botany* 54 (1970); Guglielmi, L., Tigridia, l'oceloxchitl degli Aztechi, *Il Giardino Fiorito* **49**: 380–1 (1983).

1a. Flower 10–15 cm in diameter; staminal tube 4–7 cm **1. pavonia**
b. Flower at most 6 cm in diameter; staminal tube 3–5 mm **2. violacea**

1. T. pavonia (Linnaeus filius) de Candolle (*T. pringlei* Watson). Illustration: *Botanical Magazine*, 532 (1801) & n.s., 544 (1969); Brickell (ed.), RHS A–Z encyclopedia of garden plants, 1038 (2003).

Bulb 3–5 × 1.5–4 cm. Stem to 1.5 m. Basal leaves 20–50 × 1.5–6 cm, lanceolate. Stem-leaves 1–several. Flowers 10–15 cm in diameter, pink, red, orange, yellow or white, often spotted or blotched with darker colours. Outer perianth-segments 7–10 cm, the blade spreading or reflexed. Inner segments much shorter, the blade spreading or reflexed. Staminal tube 4–7 cm, projecting conspicuously beyond the cup of the flower; anthers erect and incurving. Capsule 3–7 cm. *Mexico.* H5–G1. Summer.

This species originated in Mexico and has been cultivated there for a long time and is now naturalised in much of C & S America. It is very variable in flower colour and size, and many variants have been named and described in the past (some as 'varieties', others as species, e.g. *T. conchiflora* Sweet, *T. grandiflora* Salisbury, etc.).

2. T. violacea (Schlechtendahl) Otto & Dietrich. Illustration: *Botanical Magazine*, 7536 (1894).

Bulb 3–4 × 1.5–2 cm. Stems to 30 cm. Basal leaves short at flowering time, later to 45 cm, linear. Stem-leaf 1. Flowers 3–5 cm in diameter, purple to violet, the cup paler and spotted, its margins yellow. Outer perianth-segments with oblong, obtuse but shortly pointed, blades, spreading. Inner segments ovate, markedly pouched in the middle. Staminal tube 3–5 mm, not projecting from the cup; anthers spreading. Capsule 1.3–2 cm. *Mexico.* H5–G1. Summer.

11. RIGIDELLA Lindley

J. Cullen

Perennial herbs with bulbs. Leaves pleated, mostly basal. Flowers in cymes, mostly enclosed by 2 spathes. Flowers pendent or erect, fruits stiffly erect. Perianth of 6 free segments, the 3 outer large, overlapping and forming a cup at the base, the blades reflexed, all coiling spirally as the flower dies; the 3 inner segments smaller or much smaller, adpressed to the staminal column, each with a gland (nectary?) on the inner surface. Stamens 3, filaments united into a tube that projects conspicuously beyond the cup formed by the outer perianth-segments. Style enclosed within the staminal tube; stigmas 3, on the same radii as the anthers, each deeply bifid. Fruit a papery capsule.

A genus of 4 species from Mexico and Guatemala. They may be grown in pots, or, in favourable areas, in an open border. They require a rich, well-drained soil and much sunshine. Staking is generally necessary. The plants can be treated as half-hardy, the bulbs being lifted and dried off after growth ceases. The flowers each last only one day, the outer perianth-segments coiling spirally at the end of this, but are produced in succession over a fairly long period.

Literature: Cruden, R.W., The systematics of Rigidella, *Brittonia* **23**: 217–25 (1971).

1a. Flowers erect; inner perianth-segments 2.2–3 cm **3. orthantha**
b. Flowers pendent though fruits erect; inner perianth-segments 9–12 mm 2
2a. Flowers with dark spots or lines on the rim of the cup formed by the outer perianth-segments; basal leaves developed only after flowering **2. flammea**
b. Flowers without dark spots or lines; basal leaves present at flowering **1. immaculata**

1. R. immaculata Herbert. Illustration: *Edwards's Botanical Register* **27**: t. 68 (1841); *Flore des serres* **5**: t. 502 (1849) & ser. 2, **11**: t. 2215 (1875).

Bulb 4–5.5 × 2–3 cm. Stem to 1 m but often less. Basal leaves present at flowering, 60–90 × 2.5–5.5 cm. Stem-leaves usually 2. Flowers pendent. Outer perianth-segments 4.5–5.5 cm, the cup 1.3–1.5 cm, all scarlet. Inner segments 9–12 mm, flat, yellow with red tips. Style-branches 5–6 mm, without small projections between them. Capsule to 4 cm. *Mexico, Guatemala.* H5–G1. Spring–early summer.

2. R. flammea Lindley. Illustration: *Edwards's Botanical Register* **26**: t. 16 (1840).

Bulb to 5 × 2.5 cm. Stems to 1 m. Basal leaves developing after flowering, ultimately to 1 m or more. Stem-leaves usually few and reduced. Flowers pendent. Outer perianth-segments 5–6.5 cm, the cup 1.8–2 cm, scarlet and marked with dark spots or lines around the rim of the cup. Inner perianth-segments 9–11 mm, yellow with red tips, margins inrolled over the gland. Style-branches 3–4 mm, with short projections between each. Capsule 1.5–3.5 cm. *Mexico.* H5–G1. Spring–early summer.

3. R. orthantha Lemaire. Illustration: *Flore des serres* **1**: t. 107 (1845); *Botanical Magazine*, n.s., 667 (1974).

Bulb to 10 × 5 cm. Stems to 1 m. Basal leaves to 100 × 8.5 cm. Stem-leaves 1–2. Flowers erect. Outer perianth-segments 5–6 cm, scarlet, cup 1–1.2 cm. Inner segments 2.2–3 cm, flat, scarlet. Style-branches 5–7.5 cm, without projections between them. Capsule to 4 cm. *Mexico, Guatemala.* H5–G1. Spring–early summer.

This species has been hybridised with *Tigridia pavonia* (p. 267); the hybrid, which is apparently without a name, is sometimes grown.

12. HERBERTIA Sweet

J. Cullen

Perennial herbs with corms. Stems erect, simple or with few branches, the stem or the individual branches ending in many 1-flowered inflorescences. Leaves mostly basal, linear-lanceolate, tapered towards both ends, pleated. Spathes 2, completely hiding the flower-stalk and ovary, the inner longer than the outer. Perianth of 6 free segments, the 3 outer much larger than the 3 inner, all spreading or reflexed. Stamens 3, filaments united into a tube, the anthers spreading, connective narrow, linear. Stigmas 3, on the same radii as the stamens, each with a notched apex beyond which the anther projects. Fruit a many-seeded capsule.

A genus of perhaps 7 species, mostly from temperate S America. Its classification is confused, and information about the species in the wild is scattered. It has been much confused with *Alophia*, and, at one time, bore the name *Trifurcia* Herbert (see Goldblatt, P., *Brittonia* **27**: 373–85, 1976; *Annals of the Missouri Botanical Garden* **64**: 78–9, 1978). The flowers are very short-lived, lasting individually for only one day, but they are produced successively over a period of some weeks. The plants are uncommon in cultivation, and, in most areas, require cool greenhouse treatment, though the corms may be planted out in a sunny border in spring and lifted and dried off when the leaves begin to wither. In general, whether in pots or in a border, a well-drained soil or compost is necessary. Propagation is by seed.

1a. Stem 30–50 cm; leaves 18–20 cm × 5–8 mm **1. amatorum**
b. Stems 6–15 cm; leaves 4–10 cm × 3–5 mm 2
2a. Outer perianth-segments *c.* 2.8 × 1.2 cm, inner perianth-segments *c.* 8 × 2 mm **3. pulchella**
b. Outer perianth-segments *c.* 1.5 cm × 6–7 mm, inner perianth-segments *c.* 5.5 × 2 mm **2. lahue**

1. H. amatorum Wright (*Trifurcia amatorum* (Wright) Goldblatt). Illustration: *Botanical Magazine*, 8175 (1908).

Corm spherical, brown. Stems 30–50 cm, little branched. Leaves linear-lanceolate, 18–20 cm × 5–8 mm. Outer perianth-segments reflexed, obovate, *c.* 3.5 × 2–2.5 cm, purple with a white median stripe and a hairy, 2-lobed, yellow nectary at the base. Inner segments purple, spreading, *c.* 1.5 cm × 5 mm, the claw hairy. Anthers yellow, stigmas pale purple. *Southern S America (Uruguay only ?).* H5–G1. Spring.

2. H. lahue (Molina) Goldblatt (*Alophia lahue* (Molina) Espinosa; *Trifurcia lahue* (Molina) Goldblatt), Illustration: Correa, Flora Patagonica **2**: 170 (1969); Brickell (ed.), RHS A–Z encyclopedia of garden plants, 529 (2003).

Corm ovoid to spherical, brown. Stem 6–15 cm, simple or with 1 branch. Leaves linear-lanceolate, often sickle-shaped, 4–8 cm × 3–4 mm. Outer perianth-segments spreading, oblanceolate, violet marked with dark blue towards the base, *c.* 1.5 cm × 6–7 mm. Inner perianth-segments *c.* 5.5 × 2 mm, oblanceolate, violet-purple, glandular towards the base within. Anthers yellow, stigmas pale purple. *S Chile, S Argentina.* H5–G1. Spring.

The description and distribution given above are of the southern American plant, which is subsp. *lahue*. Subsp. **caerulea** (Herbert) Goldblatt (*Alophia caerulea* Herbert) occurs in the southern USA (Texas and Louisiana) and is taller, with longer leaves; it has been in cultivation.

3. H. pulchella Sweet (*Trifurcia pulchella* (Sweet) Goldblatt). Illustration: Sweet,

British flower garden **2**: t. 222 (1825–27); *Botanical Magazine*, 3862 (1841).

Corm ovoid, brown, Stems erect, 7.5–15 cm, not branched or with 1 branch. Leaves linear-lanceolate, *c.* 10 cm × 5 mm. Outer perianth-segments spreading or reflexed, blue or pinkish violet, sometimes with a median white stripe, dark violet or spotted with purplish blue at the base, *c.* 2.8 × 1.2 cm. Inner perianth-segments spreading, pale purple. Anthers yellow, stigmas reddish. *Southern S America as far north as S Brazil.* H5–G1. Spring.

13. ALOPHIA Herbert

J. Cullen

Perennial herbs with corms. Stems branched and bearing several leaves. Leaves linear-lanceolate, pleated. Inflorescences terminal, with few flowers, subtended by 2 spathes, the inner longer than the outer. Perianth of 6 free segments, the outer 3 a little longer than the inner 3, all ovate, each with a band of hairs down the centre. Stamens 3, filaments usually partly united, anthers spreading, each with a broad, fiddle-shaped connective. Stigmas 3, on the same radii as the stamens, each deeply divided into 2 ascending, thread-like branches. Fruit a many-seeded capsule.

A genus of a few species extending from the southern part of the USA through C America to S America. Only 1 species is cultivated, requiring conditions like those for *Herbertia.*

1. A. drummondil (Graham) Foster (*A. drummondiana* Herbert; *Gelasine punctata* Herbert). Illustration: *Botanical Magazine*, 3779 (1840).

Corm ovoid. Stems 15–45 cm. Leaves linear-lanceolate, tapered towards both ends, to 30 × 1.5 cm. Perianth-segments purple or purple-blue, spotted with brown near the centre; the outer 3, 2–2.5 cm; the inner 3, *c.* 1.5 cm with margins inrolled over the band of hairs. *S USA (Arkansas, Louisiana, Texas, Oklahoma), Mexico.* H5–G1. Early summer.

14. FERRARIA Burman

D.A. Webb

Rootstock a corm. Leaves in 2 ranks. Flowers in small cymes with numerous bracts. Perianth radially symmetric, tube short, lobes all similar. Stamens united for most of their length to form a tube around the style. Style-branches enlarged, fringed, somewhat petaloid, bifid. Capsule linear; seeds numerous.

A genus of about 10 species, from southern and tropical Africa.

Literature: de Vos, M.P., The African genus Ferraria, *Journal of South African Botany* **45**: 295–376 (1979).

1. F. crispa Burman (*F. undulata* Linnaeus). Illustration: Botanical Magazine, 144 (1791); Rice & Compton, Wild flowers of the Cape of Good Hope, t. 191 (1951); *Flowering plants of Africa* **34**: t. 1316 (1959).

Stem 20–50 cm. Lower leaves 15–30 cm, including the long sheathing base, the upper ones shorter; blade more or less erect, sword-shaped to sickle-shaped. Cymes numerous, mostly 2-flowered, in the axils of leaf-like bracts. Bracts *c.* 4 × 3 cm, boat-shaped, leaf-like. Perianth-lobes *c.* 2 cm, consisting of an oblong, erect claw and a slightly longer, spreading or slightly deflexed, strongly wavy blade, the whole dark brown with pale lines and spots, or pale yellow with brown lines and spots. Style-branches fringed with numerous long, hair-like processes. Flowers very short-lived, somewhat foetid. *South Africa (Cape Province).* H5–G1. Spring.

Hardy in favoured spots in S Europe, if the bulbs are planted deep, but usually grown under glass.

15. HOMERIA Ventenat

V.A. Matthews

Corm covered with netted tunics. Leaves solitary or several, linear. Stem erect, hairless, usually branched. Inflorescence with several flowers, each flower subtended by 2 leafy bracts. Perianth radially symmetric, segments free, almost equal or the inner ones slightly smaller, usually with a basal claw, the blade with a basal nectary that may extend down the claw. Stamens opposite the outer perianth-segments, filaments united. Style-branches 3, entire, flattened, with or without crests. Fruit a cylindric capsule, with a flat top or beaked.

A genus of 31 species from southern Africa, the cultivated ones coming from Cape Province in South Africa. Cultivation is as for *Ixia* (p. 289).

Literature: Goldblatt, P., Systematics and biology of Homeria (Iridaceae), *Annals of the Missouri Botanical Garden* **68**: 413–503 (1981).

1a. Leaves 2 or more **1. ochroleuca**
b. Leaves solitary 2

2a. Plant 15–35 cm; filaments 5–6 mm, downy at least at the base **3. collina**
b. Plant 30–75 cm; filaments 7–10 mm, hairless, or if sparsely downy in lower half then outer perianth-segments 3.5–4 cm 3

3a. Each perianth-segment with a diffuse nectary in the middle of the claw; anthers 4.5–8 mm; flowers with an unpleasant smell **1. ochroleuca**
b. Outer perianth-segments each with a distinct nectary at the base; anthers 8–11 mm; flowers not smelling unpleasant **2. flaccida**

1. H. ochroleuca Salisbury (*H. collina* (Thunberg) Salisbury var. *ochroleuca* (Salisbury) Baker). Illustration: *Botanical Magazine*, 1103, 1108 (1808); *Annals of the Missouri Botanical Garden* **68**: 454 (1981); Everett, The New York Botanical Gardens illustrated encyclopedia of horticulture **5**: 1698 (1981).

Stem 35–75 cm, once to several times branched. Leaves usually **1**, occasionally 2 or 3, 6–15 mm wide. Flowers yellow, sometimes orange in the centre, strongly and rather unpleasantly scented. Outer perianth-segments 3–4 cm, inner slightly smaller, nectaries diffuse. Filaments 7–10 mm, smooth and hairless; anthers 4.5–8 mm, divergent. Style divided just above the level of the bases of the anthers, branches 5–6.5 mm. *South Africa (SW Cape).* H4. Summer.

H. miniata (Andrews) Sweet (*H. lineata* Sweet). Illustration: *Andrews' Botanical Repository*, t. 404 (1804); Sweet, British flower garden **2**: pl. 152, 178 (1825–7); *Annals of the Missouri Botanical Garden* **68**: 486 (1981). Differs in having 2 or 3 leaves and flowers that may be reddish, pink, yellow or white. Outer perianth-segments 1.3–2.2 cm, with triangular nectaries dotted with green, the claw downy above the midline. Filaments downy at the base; anthers *c.* 2 mm, not diverging. *South Africa (SW Cape).* H4. Summer.

2. H. flaccida Sweet (*H. collina* (Thunberg) Salisbury var. *aurantiaca* (Zuccagni) Baker; *H. breyniana* (Linnaeus) Lewis var. *aurantiaca* (Zuccagni) Lewis). Illustration: *Botanical Magazine*, 1612 (1814); *Annals of the Missouri Botanical Garden* **68**: 465 (1981).

Stem 30–60 cm with several branches. Leaf 1, 9–13 mm wide. Flowers yellow to salmon-orange with deep yellow nectaries. Outer perianth-segments 3.5–4 cm, claws smooth or with tiny papillae in the lower part. Filaments 7–8 mm, smooth or with sparse papillae, downy below; anthers 8–11 mm, divergent towards the top. Style divided *c.* 3 mm above the level of the bases of the anthers, branches to 7 mm. *South Africa (SW Cape).* H4. Summer.

3. H. collina (Thunberg) Salisbury (*H. breyniana* (Linnaeus) Lewis). Illustration: *Annals of the Missouri Botanical Garden* **68**: 472 (1981); Brickell (ed.), RHS A–Z encyclopedia of garden plants, 540 (2003).

Stem 18–35 cm, unbranched or with few branches. Leaf 1, usually trailing, 4–10 mm wide. Flowers pale yellow or salmon-pink, fragrant, nectaries usually deep yellow edged with green. Outer perianth-segments 3–3.5 cm, claw often with papillae near the base. Filaments 5–6 mm, sparsely downy below; anthers 5–6 mm, divergent towards the top. Style divided slightly above the level of the bases of the anthers, branches 5–6 mm. *South Africa (SW Cape).* H4. Summer.

H. elegans (Jacquin) Sweet. Illustration: *Annals of the Missouri Botanical Garden* **68**: 463 (1981). Differs in having yellow flowers with a large green blotch in the central or upper one-third of each outer perianth-segment, which is usually shaded orange towards the top. Anthers 8–10 mm, divergent. *South Africa (SW Cape).* H4. Summer.

16. CROCUS Linnaeus

B. Mathew

Corm enclosed by several fibrous, papery or leathery tunics. Sheathing leaves up to 5, enclosing the aerial shoot. Leaves all basal, linear, not in 2 ranks, present at flowering time or absent and appearing later; upper surface green or grey-green with a central whitish stripe. Flowers 1 to several, produced in autumn, winter or spring, each carried on a short subterranean flower-stalk, which is sometimes subtended by a membranous spathe (prophyll – the 'basal spathe' of various authors). Bracts 1 or 2, membranous, subtending the ovary and sheathing the perianth-tube (the 'floral spathe' of various authors). Perianth radially symmetric with a long narrow tube; lobes 6, usually equal or inner whorl sometimes smaller. Style with 3 or more branches. Ovary subterranean. Capsule cylindric or ellipsoid, maturing above ground level by elongation of the flower-stalk; seeds numerous. See Figure 11, p. 271.

A genus of 80 species from Europe and W & C Asia, easily distinguished from all other genera by the funnel-shaped flowers with a very long perianth-tube. Most species can be cultivated in sunny situations in a wide range of soil types providing the drainage is good, allowing a dryish period during dormancy in summer. A few species tolerate or even prefer semi-shady conditions. Propagation is by seed or by natural vegetative corm division.

Literature: Maw, G., A monograph of the genus Crocus (1886); Bowles, E.A., Handbook of crocus and colchicum for gardeners, edn 2 (1952); Mathew, B., The crocus, a revision of the genus Crocus (Iridaceae) (1982).

The following key is constructed to identify plants that are representative of the species as they are generally known in cultivation; mutations such as albinos do not necessarily key out satisfactorily.

1a. Flowers produced from January to April 2
b. Flowers produced from September to December 22
2a. Flowers wholly pale creamy yellow to deep yellow or orange, sometimes striped or stained externally 3
b. Flowers white, blue, lilac, purple or violet, sometimes with a yellow throat, often striped externally 8
3a. Style with 6 or more branches **15. olivieri**
b. Style with 3 branches, these sometimes expanded and lobed at the apex 4
4a. Leaves usually numbering 10–20 from mature flowering-sized corms **16. korolkowii**
b. Leaves 8 or fewer per corm 5
5a. Leaves 2.5–4 mm wide, dark green; anthers markedly arrow-shaped **14. flavus**
b. Leaves very slender, usually 0.5–2 mm wide, usually grey-green; anthers not markedly arrow-shaped 6
6a. Perianth-lobes strongly striped, suffused or speckled with purplish brown on the exterior; corm-tunic netted-fibrous **10. angustifolius**
b. Perianth-lobes either unmarked on the exterior or, if striped or suffused with purplish brown, then corm-tunic not fibrous but papery or leathery with horizontal rings 7
7a. Corm-tunic fibrous, the fibres strongly netted; perianth-lobes obtuse to rounded **11. ancyrensis**
b. Corm-tunic membranous or papery, with rings at the base; perianth-lobes subacute to obtuse **12. chrysanthus**
8a. Anthers whitish **18. laevigatus**
b. Anthers yellow or rarely blackish 9
9a. Style divided into many slender orange to scarlet branches **17. fleischeri**
b. Style divided into 3 branches, sometimes each branch expanded and frilled or fringed at the apex 10
10a. Throat of perianth pale to deep yellow or orange 11
b. Throat of perianth with no yellowish coloration 16
11a. Corm-tunic papery, leathery or eggshell-like, with horizontal rings at the base **13. biflorus**
b. Corm-tunic fibrous or papery but with no rings at the base 12
12a. Leaves markedly greyish green above **6. versicolor**
b. Leaves deep green above 13
13a. Perianth-lobes creamy, buff, yellowish gold or silvery on the outside, usually with dark longitudinal stripes or veins 14
b. Perianth-lobes not creamy, buff, yellowish gold or silvery outside and usually with no conspicuous longitudinal stripes or veins **9. sieberi**
14a. Flowers mid to deep purple inside, with a deep yellow or orange, hairless throat; corm-tunic with parallel fibres **7. imperati**
b. Flowers mid to pale lilac or violet inside with a rather pale yellow, hairy throat; corm-tunic with netted fibres 15
15a. Bract solitary, clearly visible, sheathing the perianth-tube; leaves usually 2–6 mm wide at flowering time **3. etruscus**
b. Bracts 1 or 2, more or less equal and clearly visible, sheathing the perianth-tube; leaves usually 1–2 mm wide at flowering time **8. dalmaticus**
16a. Flowers white, sometimes flushed or strongly stained or striped with violet-blue 17
b. Flowers pale lilac to deep purple or reddish purple, sometimes with conspicuous narrow stripes outside 18
17a. Leaves 3–8, 0.5–3.5 mm wide, grey-green; bracts 2 **13. biflorus**
b. Leaves 2–4, usually 4–8 mm wide, green; bract solitary **1. vernus**

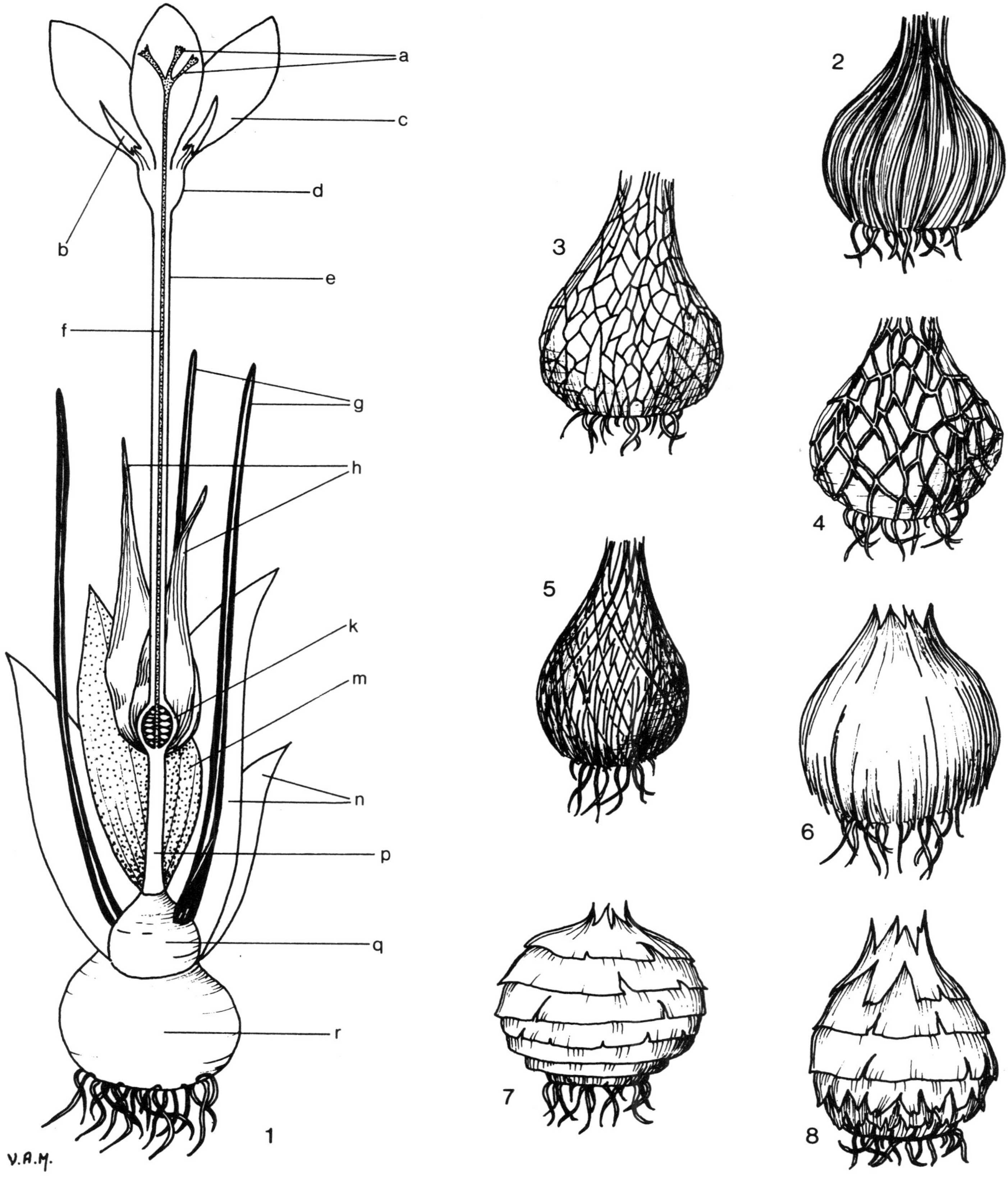

Figure 11. Diagnostic details of Crocus species. 1, Diagram showing the parts of a Crocus plant (a, style-branch; b, anther; c, perianth-lobe; d, throat; e, perianth-tube; f, style; g, leaf; h, bract; k, ovary; m, spathe; n, sheathing leaf; p, scape; q, new corm; r, old corm). 2–8, Types of corm-tunics. 2, Parallel fibres. 3, Finely netted fibres. 4, Coarsely netted fibres. 5, Interwoven fibres. 6, Papery with parallel fibres. 7, Non-fibrous, with rings at the base. 8, Non-fibrous splitting into teeth at the base.

18a. Flowers irregularly striped and veined on a paler ground **1. vernus**
b. Flowers uniformly striped and veined outside, sometimes only at the base, or flowers unstriped 19
19a. Flowers unstriped and more or less uniform in colour on the outside, or with only fine inconspicuous veining 20
b. Flowers with conspicuous stripes externally 21
20a. Flowers lilac-blue, purplish or reddish purple, usually with a white or very pale tube; leaves 2–3 mm wide **2. tommasinianus**
b. Flowers lilac, purple or violet, usually with the tube the same colour as or darker than the segments; leaves usually 4–8 mm wide **1. vernus**
21a. Corm-tunics with parallel fibres only; perianth-lobes usually 2–2.7 cm; tube usually 5–11 cm; style-branches yellow to orange **5. minimus**
b. Corm-tunics netted-fibrous at apex; perianth-lobes usually 2–3.5 cm; tube usually 3.5–7 cm; style-branches orange to red **4. corsicus**
22a. Inner and outer perianth-lobes very unequal in shape and size; style-branches lilac **35. banaticus**
b. Inner and outer perianth-lobes equal, or nearly so 23
23a. Anthers blackish maroon **13. biflorus**
b. Anthers yellow or creamy white 24
24a. Anthers creamy white 25
b. Anthers yellow 30
25a. Leaves absent at flowering time 26
b. Leaves present at flowering time, but sometimes only the tips visible 27
26a. Throat either marked with 2 yellow blotches at the base of each lobe (occasionally merging into a V-shape), or throat white **29. kotschyanus**
b. Throat with a continuous deep yellow zone **28. pulchellus**
27a. Style divided into 3 yellow branches; flowers creamy white with no conspicuous stripes on the outside **30. ochroleucus**
b. Style divided into many slender, yellow, orange or reddish branches 28
28a. Corm-tunics tough and splitting into triangular teeth at the base; flowers lilac or white usually striped or suffused with violet outside **18. laevigatus**
b. Corm-tunics papery, splitting into soft fibres at the base; flowers white or lilac-blue, only occasionally dark-veined 29
29a. Flowers usually lilac-blue, remaining open in dull weather and at night; filaments strongly hairy **20. tournefortii**
b. Flowers usually white, closing at night and in dull weather; filaments minutely hairy **19. boryi**
30a. Style clearly divided into more than 3 branches 31
b. Style divided into only 3 branches, these sometimes expanded at apex 36
31a. Leaves present at flowering time but sometimes with only the tips visible; throat usually pale to deep yellow 32
b. Leaves absent at flowering time and not visible for some time after the flowers have withered; throat either with no trace of yellow colour, or sometimes pale yellow 33
32a. Style divided into few (usually 3) orange to bright red branches, each shortly subdivided or lobed at the apex; throat deep yellow; flowers white or faintly lilac **22. niveus**
b. Style divided into 6 or more slender orange branches; throat white or pale yellow; flowers pale to deep lilac-blue or purple **26. serotinus**
33a. Flowers deep purple or rarely lilac-purple without conspicuous darker veining; corms stoloniferous with parallel-fibrous tunics **25. nudiflorus**
b. Flowers white, lilac-blue, purple-blue, violet-blue or deep purple, usually with conspicuous darker veining, especially towards the base of the lobes 34
34a. Perianth-lobes with dark longitudinal veins and conspicuous but finer cross-veining, usually speckled on the outside; corm-tunics non-fibrous, with rings at the base **27. speciosus**
b. Perianth-lobes with conspicuous longitudinal veins only, usually towards the base of the segments; corm-tunic coarsely netted-fibrous 35
35a. Flowers white to deep lilac-blue with a pale yellow throat **34. cancellatus**
b. Flowers deep purple or sometimes lilac with no yellow in the throat **24. medius**
36a. Throat yellow 37
b. Throat not yellow 40
37a. Bract silvery white; leaves grey-green, 0.5–1 mm wide **32. hadriaticus**
b. Bract greenish; leaves deep green, 1–3 mm wide 38
38a. Flowers mid to deep purple inside, prominently dark-veined and/or buff outside **23. longiflorus**
b. Flowers white to pale or pinkish lilac, not conspicuously dark-veined or buff outside 39
39a. Style-branches undivided at the apex; perianth-lobes 6–15 mm wide; corm-tunics papery, splitting lengthways **33. caspius**
b. Style-branches subdivided or lobed at the apex; perianth-lobes 1.5–3.5 cm wide; corm-tunics finely netted-fibrous **22. niveus**
40a. Flowers white, sometimes stained violet or brownish at the base **32. hadriaticus**
b. Flowers pale to deep lilac-purple 41
41a. Style-branches very long, deep red **31. sativus***
b. Style-branches short, white to orange **21. goulimyi**

1. C. vernus (Linnaeus) Hill.

Corm-tunic with fine parallel or slightly netted fibres. Leaves 2–4, usually 4–8 mm wide, often much shorter than the flower, green. Flowers white, purple or striped; throat white or purplish, hairy or hairless. Spathe present. Bract solitary. Anthers yellow; style divided into 3 deep yellow to orange-red, rarely whitish, branches, each much-expanded and fringed at the apex.

Subsp. **vernus** (*C. napolitanus* Mordant & Loiseleur; *C. purpureus* Weston). Illustration: Mathew, Dwarf bulbs, 88 (1973) – as *C. scepusiensis*; Rix & Phillips, The bulb book, 26a–e (1981); Mathew, The crocus, pl. 1a (1982). Flowers often pale to deep purple, or striped; perianth-lobes usually 3–5.5 cm × 9–20 mm. Style usually equalling or exceeding the stamens. *C, S & E Europe from Italy to W Russia.* H2. Spring.

The plants occurring at the eastern end of the range of distribution often have dark V-shaped markings at the tips of the perianth-lobes; they have been named *C. heuffelianus* Herbert and *C. scepusiensis* (Rehmann & Woloszczak) Borbas.

Subsp. **albiflorus** (Schultes) Ascherson & Graebner (*C. albiflorus* Schultes; *C. caeruleus* Weston). Illustration: Grey-Wilson & Mathew, Bulbs, 34 (1981) – subsp. *albiflorus*; Mathew, The crocus, pl. 1b (1982); Brickell (ed.), RHS A–Z encyclopedia of garden plants, 326 (2003). Flowers often white; perianth-lobes usually 1.5–3 cm × 4–10 mm. Style usually much shorter than the stamens. *W, C & S Europe from Czechoslovakia to the Pyrenees.* H2. Spring.

2. C. tommasinianus Herbert. Illustration: Synge, Collins guide to bulbs, edn 2, pl. 7 & 10 (1971); Hay & Beckett, Reader's Digest encyclopaedia of garden plants and flowers, 190 (1978); Mathew, Dwarf bulbs, 73 (1973); Rix & Phillips, The bulb book, 19 (1981).

Corm-tunic with fine parallel or slightly netted fibres. Leaves 3 or 4, 2–3 mm wide, equalling or exceeding the flowers, green. Flowers pale lilac to deep reddish purple, often silvery or creamy on the exterior, sometimes with darker tips to the lobes, rarely whitish-tipped; throat white, sparsely hairy. Spathe present. Bract solitary. Perianth-lobes 2.5–4.5 cm × 8–20 mm. Anthers yellow; style divided into 3 orange branches, each expanded and fringed at the apex. *S Yugoslavia to S Hungary.* H2. Spring.

3. C. etruscus Parlatore. Illustration: Synge, Collins guide to bulbs, edn 2, pl. 7 (1971); Grey-Wilson & Mathew, Bulbs, pl. 34 (1981); Rix & Phillips, The bulb book, 24c (1981); Mathew, The crocus, pl. 3 (1982).

Corm-tunic with distinctly netted fibres. Leaves 3 or 4, 2–6 mm wide, usually about equal to the flower, green. Flower with pale purple-blue interior, creamy or silvery and veined violet on the outside; throat pale yellow, hairy. Spathe present. Bract solitary. Perianth-lobes 3–4 cm × 9–13 mm. Anthers yellow; style divided into 3 orange branches, each expanded at the apex. *NW Italy.* H2–3. Spring.

4. C. corsicus Vanucci. Illustration: Synge, Collins guide to bulbs, edn 2, pl. 7 (1971); Grey-Wilson & Mathew, Bulbs, pl. 32 (1981); Rix & Phillips, The bulb book, 29i (1981); Mathew, The crocus, pl. 12 (1982).

Corm-tunic with fine fibres, netted in the upper part. Leaves 3 or 4, 1–1.5 mm wide, slightly shorter than or about equalling the flower, green. Flowers pale to deep purple on the interior, often yellowish with 1–3 violet stripes and finer veining on the exterior; throat white or pale purple, hairless. Spathe present. Bract solitary. Perianth-lobes usually 2–3.5 cm × 7–13 mm. Anthers yellow; style divided into 3 orange or red branches, each expanded or slightly subdivided at the apex. *C & N Corsica.* H3. Spring.

5. C. minimus de Candolle. Illustration: Synge, Collins guide to bulbs, edn 2, pl. 7 (1971); Grey-Wilson & Mathew, Bulbs, pl. 33 (1981); Rix & Phillips, The bulb book, 21i (1981); Mathew, The crocus, pl. 11 (1982).

Corm-tunic fibrous, with parallel fibres. Leaves 3–5, 0.5–1 mm wide, usually exceeding the flowers, green. Flowers mid to deep purple on the interior, usually yellowish with conspicuous violet veins, or stained violet, on the outside; throat white, hairless. Spathe present, bracts 2, very unequal, or rarely solitary. Perianth-lobes usually 2–2.7 cm × 4–8 mm. Anthers yellow; style divided into 3 yellow or orange branches, each obscurely lobed or expanded and frilled at the apex. *Sardinia & S Corsica.* H3. Spring.

6. C. versicolor Ker Gawler. Illustration: Synge, Collins guide to bulbs, edn 2, pl. 7 (1971); Rix & Phillips, The bulb book, 24a (1981); Mathew, The crocus, pl. 9 (1982).

Corm-tunics membranous, the outer ones becoming wholly fibrous with parallel fibres. Leaves 3–5, 1.5–3 mm wide, usually reaching the base of the flowers, markedly grey-green. Flowers white to mid-purple, usually conspicuously dark-striped on the exterior, sometimes on a yellowish ground; throat usually pale yellow, hairless. Spathe present. Bract solitary or accompanied by a narrow bracteole. Perianth-lobes 2.5–3.5 cm × 7–10 mm. Anthers yellow; style divided into 3 yellow or orange branches, each expanded and sometimes slightly lobed at the apex. *Mountains of SE France, NW Italy.* H3. Spring.

The most frequently cultivated variant of this species is a white-flowered, conspicuously striped cultivar known as 'Picturatus'.

7. C. imperati Tenore. Illustration: Hay & Beckett, Reader's Digest encyclopaedia of garden plants and flowers, 189 (1978); Grey-Wilson & Mathew, Bulbs, pl. 33 (1981); Rix & Phillips, The bulb book, 26d (1981).

Corm-tunic membranous, or the outer becoming fibrous, with parallel fibres. Leaves 3–6, 2–3 mm wide, equalling or exceeding the flowers, green. Flowers mid to deep purple on the interior, and yellowish with 1–5 conspicuous violet stripes on the outside; throat yellow to orange, hairless. Spathe present. Bracts 1 or 2. Perianth-lobes 2.3–4 cm × 7–18 mm. Anthers yellow; style divided into 3 deep orange or reddish branches, each expanded and shortly lobed.

Subsp. **imperati**. Bracts 2. Filaments 6–9 mm; anthers 1.2–2.1 cm. Style-branches deep orange or red. *W Italy (Naples southwards).* H3. Spring.

Subsp. **suaveolens** (Bertoloni) Mathew. Bract solitary. Filaments 3–5 mm; anthers 8–12 mm. Style-branches yellow or orange. *W Italy (N of Naples, to Rome).* H3. Spring.

Some of the commercial stocks of *C. imperati* are referable to subsp. *suaveolens*.

8. C. dalmaticus Visiani. Illustration: Synge, Collins guide to bulbs, edn 2, pl. 7 (1971); Grey-Wilson & Mathew, Bulbs, pl. 33 (1981); Rix & Phillips, The bulb book, 26c (1981); Mathew, The crocus, pl. 36 (1982).

Corm-tunic fibrous, the fibres markedly netted. Leaves 3–5, usually 1–2 mm wide, shorter than or equalling the flower, dark green. Flowers pale violet on the interior, usually yellowish or silvery with fine purple lines on the outside; throat yellow, sparsely hairy. Spathe absent. Bracts 2. Perianth-lobes 1.5–3.5 × 1–1.5 cm. Anthers yellow; style divided into 3 orange branches, each slightly expanded and fringed at the apex. *SW Yugoslavia, NW Albania.* H2. Spring.

The commercially available clone of *C. dalmaticus* has wider leaves than wild plants and may be of hybrid origin; stocks of it often show virus symptoms with the flowers spotted and streaked on the exterior.

9. C. sieberi Gay. Illustration: *Journal of Royal Horticultural Society* **74**: f. 6 – subsp. *atticus* – & f. 7b – forma *tricolor* (1949); Grey-Wilson & Mathew, Bulbs, pl. 33 (1981); Rix & Phillips, The bulb book, 18 g – subsp. *atticus* – & 24d – subsp. *sieberi* (1981); Mathew, The crocus, pl. 37a (1982).

Corm-tunic fibrous, the fibres markedly netted. Leaves 2–8, 1–6 mm wide, shorter than to equalling the flower, dark green. Flowers pale lilac to deep lilac-purple, or white with purple zones, or stripes, on the outside (subsp. *sieberi* from Crete); throat yellow to orange, hairless or hairy. Spathe absent. Bracts 2. Perianth-lobes 2–4 cm × 7–15 mm. Anthers yellow; style divided into 3 yellow to orange-red branches, each much-expanded and frilled or lobed at the apex. *S Albania, S Bulgaria, S Yugoslavia, Greece, Crete.* H2. Spring.

The most commonly cultivated variant of *C. sieberi* is subsp. **atticus** (Boissier & Orphanides) Mathew, which has lilac flowers and a rather coarsely netted corm-tunic; the white-flowered subsp. **sieberi** (*C. sieberi* var. *heterochromus* Halacsy; *C. sieberi* var. *versicolor* Boissier & Heldreich) from Crete is rarely seen in cultivation but a hybrid between it and subsp. *atticus*, called 'Hubert Edelsten', is often offered in catalogues; forma **tricolor** Burtt (a form of subsp. *sublimis* (Herbert) Mathew) has striking lilac-blue flowers with a white zone surrounding the yellow throat.

10. C. angustifolius Weston (*C. susianus* Ker Gawler). Illustration: Hay & Beckett, Reader's Digest encyclopaedia of garden plants and flowers, 191 (1978); Grey-Wilson & Mathew, Bulbs, pl. 34 (1981); Rix & Phillips, The bulb book, 24f (1981); Mathew, The crocus, pl. 43 (1982).

Corm-tunic fibrous, the fibres strongly netted. Leaves 3–6, 0.5–1.5 mm wide, usually equalling or exceeding the flower at flowering time, grey-green. Flowers yellow, heavily suffused or striped outside with purplish brown; throat yellow, hairless or minutely hairy. Spathe absent. Bracts 2. Perianth-lobes 1.7–3.4 cm × 6–13 mm. Anthers yellow; style divided into 3 yellow to deep orange-red branches. *SW Russia (Crimea, S Ukraine, Armenia).* H2. Spring.

A hybrid called **C. × stellaris** Haworth, somewhat similar in appearance to *C. angustifolius*, is sometimes cultivated. It has yellow flowers with dark stripes on the exterior; the probable parentage is *C. angustifolius* × *C. flavus.*

11. C. ancyrensis (Herbert) Maw. Illustration: Hay & Beckett, Reader's Digest encyclopaedia of garden plants and flowers, 187 (1978); Synge, Collins guide to bulbs, edn 2, pl. 7 (1971); Rix & Phillips, The bulb book, 25j (1981); Mathew, The crocus, pl. 44 (1982).

Corm-tunic fibrous, the fibres strongly netted. Leaves 3–6, usually 0.5–1 mm wide, shorter than to just exceeding the flower, greyish green. Flowers bright yellow or orange, sometimes with a purplish tube; throat yellow, hairless. Spathe absent. Bracts 2. Perianth-lobes 1.5–3 cm × 9–13 mm. Anthers yellow; style divided into 3 orange or orange-red branches. *C & N Turkey.* H2. Spring.

This is usually seen in cultivation as the selection known as 'Golden Bunch', which produces several flowers per corm.

12. C. chrysanthus (Herbert) Herbert. Illustration: Hay & Beckett, Reader's Digest encyclopaedia of garden plants and flowers, 187, 188 (1978); Synge, Collins guide to bulbs, edn 2, pl. 7 (1971); Grey-Wilson & Mathew, Bulbs, pl. 34 (1981); Rix & Phillips, The bulb book, 20b & c (1981).

Corm-tunic papery, leathery or eggshell-like, splitting into rings at the base. Leaves 3–7, 0.5–2.5 mm wide, shorter than to exceeding the flower, usually greyish-green. Flowers yellow to orange-yellow, sometimes striped or suffused with bronze or purple on the outside, occasionally creamy white; throat yellow, hairless. Spathe absent. Bracts 2. Perianth-lobes usually 1.5–3.5 cm × 5–11 mm. Anthers yellow, sometimes with blackish basal lobes; style divided into 3 yellow to orange branches. *Albania, Bulgaria, Greece, S Yugoslavia, E Roumania, W, C & S Turkey.* H2. Spring.

Variants of *C. chrysanthus* have been selected to give a larger flower size and a greater range of colour than in the wild forms. Some of the plants listed as '*C. chrysanthus* cultivars' are, however, forms of the related *C. biflorus* or hybrids between the two species.

13. C. biflorus Miller.

Corm-tunic papery, leathery or eggshell-like, splitting into rings at the base. Leaves usually 3–8, 0.5–3.5 mm wide, usually grey-green. Flowers with a white, lilac or blue ground colour, often conspicuously striped and veined with purple on the outside; throat yellow or white, hairless or finely hairy. Spathe absent. Bracts 2. Perianth-lobes 1.8–3 cm × 4–13 mm. Anthers yellow, sometimes with blackish basal lobes, or wholly blackish maroon; style divided into 3 yellow to orange branches. *S Europe & W Asia, from Sicily east to Crimea, Turkey, Caucasus and Iran.* H2. Spring or autumn (subsp. *melantherus* only).

A very variable and widespread species in which 14 subspecies are recognised but only a few of these are in general cultivation.

Subsp. **biflorus** (*C. pusillus* Tenore; *C. argenteus* Sabine; *C. praecox* Haworth; *C. biflorus* var. *parkinsonii* Sabine). Illustration: Rix & Phillips, The bulb book, 27h (1981); Mathew, The crocus, pl. 50a (1982). Flowers appearing in spring, white or lilac, usually conspicuously striped with purple or brownish on the outside; throat yellow. *Italy, Sicily, Rhodes, NW Turkey.*

Subsp. **weldenii** (Hoppe & Furnrohr) Mathew (*C. weldenii* Hoppe & Furnrohr). Illustration: Hay & Beckett, Reader's Digest encyclopaedia of garden plants and flowers, 188 (1978). Flowers appearing in spring, plain white, sometimes flushed with pale blue at the base or rarely all over the exterior of the lobes; throat not yellow. *W Yugoslavia, N Albania.*

Subsp. **alexandri** (Velenovsky) Mathew (*C. alexandri* Velenovsky). Illustration: Rix & Phillips, The bulb book, 24b (1981). Flowers appearing in spring, white, heavily stained with violet on the outside; throat not yellow. *SW Bulgaria, S Yugoslavia.*

The cultivar 'Lady Killer' is very similar to this subspecies.

Subsp. **melantherus** (Boissier & Orphanides) Mathew ('*C. crewei*' of many authors but not Hooker). Illustration: Rix & Phillips, The bulb book, 174e (1981); *Annales Musei Goulandris* **6**: 73 (1983). Flowers autumnal, white with conspicuous purple or greyish stripes or speckling on the outside; throat yellow. Anthers usually blackish maroon. *S Greece.*

Subsp. **pulchricolor** (Herbert) Mathew ('C. aerius'of many authors but not Herbert). Illustration: Hay & Beckett, Reader's Digest encyclopaedia of garden plants and flowers, 188 (1978); Mathew, The crocus, pl. 50 m (1982). Flowers appearing in spring, usually rich blue-violet, often darker towards the base of the lobes, normally unstriped on the outside; throat deep yellow. *NW Turkey.*

The cultivars 'Blue Bird' and 'Blue Pearl' almost certainly belong to this subspecies.

In addition to selected forms of *C. biflorus*, some hybrids have been raised between it and *C. chrysanthus*, for example 'Advance', which has a curious yellow and blue flower coloration.

14. C. flavus Weston (*C. aureus* Sibthorp & Smith; *C. lagenaeflorus* Salisbury; *C. luteus* Lamarck; *C. maesiacus* Ker Gawler; *C. lacteus* Sabine). Illustration: Bowles, Handbook of crocus and colchicum, edn 2, 98 (1952); Hay & Beckett, Reader's Digest encyclopaedia of garden plants and flowers, 187 (1978); Grey-Wilson & Mathew, Bulbs, pl. 33 (1981); Mathew, The crocus, pl. 61a (1982).

Corm-tunic papery, splitting at the base into parallel fibres. Leaves 4–8, 2.5–4 mm wide, shorter than to equalling the flower, green. Flowers yellow to orange-yellow; throat yellow, hairless or hairy. Spathe absent. Bracts 2. Perianth-lobes 2–3.5 cm × 6–12 mm. Anthers yellow; style divided into about 3 short, pale yellow to orange branches, usually shorter than the anthers. *S Yugoslavia, C & N Greece, Bulgaria, Roumania, NW Turkey.* H2. Spring.

The true species is not often cultivated but the most commonly seen yellow crocus in gardens (the large 'Dutch Yellow' or 'Golden Yellow') is very similar in appearance; this is a hybrid between *C. flavus* and *C angustifolius* of considerable antiquity. In the nineteenth century many variants and hybrids of *C. flavus* were known in gardens but these have now almost disappeared from cultivation.

15. C. olivieri Gay.

Corm-tunic membranous, splitting at the base into coarse triangular teeth. Leaves 1–4, usually 2–5 mm wide, slightly shorter than to just exceeding the flowers at flowering time, green. Flowers orange-yellow, sometimes stained or striped with purplish bronze on the outside; throat yellow, hairless or hairy. Spathe absent. Bracts 2. Perianth-lobes 1.5–3.5 cm × 4–12 mm. Anthers yellow; style divided into 6–15 slender yellow or orange branches.

Subsp. **olivieri** (*C. suterianus* Herbert). Illustration: Synge, Collins guide to bulbs, edn 2, pl. 7 (1971); Rix & Phillips, The bulb book, 20 g (1981); Mathew, The crocus, pl. 63a (1982). Flower usually unmarked on the outside; style divided into about 6 branches. *S Yugoslavia, SE Roumania, S Bulgaria, Albania, Greece, Turkey.* H2. Spring.

Subsp. **balansae** (Baker) Mathew (*C. balansae* Baker). Illustration: Mathew, The crocus, pl. 63b (1982). Flower usually striped or suffused with purplish brown or mahogany on the outside; style divided into 12–15 branches. *W Turkey, Aegean islands.* H2. Spring.

16. C. korolkowii Maw. Illustration: Synge, Collins guide to bulbs, edn 2, pl. 7 (1971); Mathew, Dwarf bulbs, 73 (1973); Rix & Phillips, The bulb book, 18b (1981); Mathew, The crocus, pl. 60 (1982).

Corm-tunic papery, splitting into many fine parallel fibres. Leaves usually 10–20, 1–2.5 mm wide, usually shorter than the flower at flowering time, green. Flowers yellow, usually conspicuously speckled or stained outside with brown, purple or black, sometimes on a buff ground; throat yellow, hairless. Spathe absent. Bracts 2. Perianth-lobes 2–3.5 cm × 6–12 mm. Anthers yellow; style divided into 3 orange branches. *N & E Afghanistan, N Pakistan, Former USSR (Soviet Central Asia).* H2. Spring.

17. C. fleischeri Gay. Illustration: Synge, Collins guide to bulbs, edn 2, pl. 7 (1971); Rix & Phillips, The bulb book, 20f (1981); Mathew, The crocus, pl. 73 (1982).

Corm-tunic fibrous, the fibres interwoven. Leaves 5–8, 0.5–1 mm wide, usually longer than the flower at flowering time, green or grey-green. Flowers white with a purplish or brownish base, sometimes with a median central stripe on the outside of each outer lobe; throat yellow, hairless. Spathe absent. Bracts 2. Perianth-lobes 1.7–3.1 cm × 4–6 mm. Anthers yellow; style divided into numerous slender orange to scarlet branches. *S & W Turkey, Aegean islands.* H3. Spring.

18. C. laevigatus Bory & Chaubard (*C. fontenayi* Reuter). Illustration: Synge, Collins guide to bulbs, edn 2, pl. 6 (1971); Rix & Phillips, The bulb book, 174 g & h (1981); Mathew, The crocus, pl. 76 (1982).

Corm-tunic smooth and tough, splitting at the base into long, narrowly triangular teeth. Leaves usually 3 or 4, 1–2.5 mm wide, usually about equalling the flower at flowering time, green. Flowers with a white or lilac ground colour, on the outside sometimes overlaid with silvery cream, buff or yellow and usually marked with 1–3 deep violet or purple stripes, rarely wholly suffused with dark purple outside; throat yellow, hairless. Spathe absent. Bracts 2. Perianth-lobes 1.3–3 cm × 4–18 mm. Anthers creamy white; style divided into many yellow to deep orange branches. *C & S Greece, Crete.* H3. Autumn or winter.

An extremely variable species, usually cultivated as the clone 'Fontenayi', which has conspicuously striped lilac flowers.

19. C. boryi Gay (*C. cretensis* Kornicke). Illustration: Grey-Wilson & Mathew, Bulbs, pl. 31 (1981); Rix & Phillips, The bulb book, 179 m (1981); Mathew, The crocus, pl. 77 (1982).

Corm-tunic papery, splitting at the base into many parallel fibres. Leaves 3–7, 1–3.5 mm wide, shorter than to much exceeding the flower at flowering time, dark green. Flowers creamy white, occasionally faintly veined purple on the outside; throat deep yellow, hairless or finely hairy. Spathe absent. Bracts 2. Perianth-lobes 2–5 cm × 7–23 mm. Anthers creamy white; style divided into many slender orange or reddish branches. *W & S Greece, SE Crete.* H3. Autumn.

20. C. tournefortii Gay. Illustration: Synge, Collins guide to bulbs, edn 2, pl. 6 (1971); Rix & Phillips, The bulb book, 1791 (1981); Mathew, The crocus, pl. 78 (1982).

Corm-tunic papery, splitting at the base into many parallel fibres. Leaves 5–10, 1–2.5 mm wide, shorter than or equalling the flower, dark green. Flowers remaining open in dull weather and at night, usually lilac-blue, sometimes veined darker towards the base, occasionally white (in Crete); throat usually deep yellow, hairless or hairy. Spathe absent. Bracts 2. Perianth-lobes 1.5–3.6 cm × 4–13 mm. Anthers creamy white; style divided into many slender orange or reddish branches, usually widely spreading and very conspicuous. *S Greece (Cyclades and Aegean islands), NE Crete.* H3. Autumn.

21. C. goulimyi Turrill. Illustration: *Botanical Magazine*, n.s., 354 (1960); Grey-Wilson & Mathew, Bulbs, pl. 3 (1981); Rix & Phillips, The bulb book, 182 h (1981); Mathew, The crocus, pl. 19 (1982).

Corm-tunic smooth and tough, splitting at the base into narrow triangular teeth. Leaves 4–6, 1–2.5 mm wide, often nearly as long as the flower at flowering time, green. Flowers pale to deep lilac-purple, the inner lobes often paler than the outer; throat white, hairy. Spathe present. Bract solitary. Perianth-lobes 1.6–3.8 × 1.1–1.8 cm. Anthers yellow; style divided into 3 white to orange branches, each slightly expanded or lobed at the apex. *S Greece (S Peloponnese).* H3. Autumn.

22. C. niveus Bowles. Illustration: *Botanical Magazine*, n.s., 146 (1951); Synge, Collins guide to bulbs, edn 2, pl. 6 (1971); Grey-Wilson & Mathew, Bulbs, pl. 31 (1981); Mathew, The crocus, pl. 18 (1982).

Corm-tunic fibrous, the fibres finely netted. Leaves 5–8, 1–2 mm wide, usually nearly as long as the flowers but sometimes with only the tips visible, dark green. Flowers usually white but sometimes pale lilac or the inner lobes white and the outer lilac; throat deep yellow, sparsely hairy. Spathe present. Bracts 2. Perianth-lobes 3–6 × 1.5–3.5 cm. Anthers deep yellow; style divided into 3 orange to bright red branches, each subdivided into many shorter branches or expanded and slightly lobed at the apex. *S Greece (S Peloponnese).* H3. Autumn.

23. C. longiflorus Rafinesque (*C. odorus* Bivona). Illustration: Synge, Collins guide to bulbs, edn 2, pl. 6 (1971); Rix & Phillips, The bulb book, 174f (1981); Grey-Wilson & Mathew, Bulbs, pl. 31 (1981); Mathew, The crocus, pl. 16 (1982).

Corm-tunic fibrous, the fibres netted, but sometimes rather obscurely so. Leaves 4–6, 1–3 mm wide, usually shorter than the flower at flowering time but sometimes only the tips visible, green. Flowers lilac to purple, usually dark-veined outside, sometimes on a yellowish external ground colour; throat yellow, hairless or sparsely hairy. Spathe present. Bract solitary. Perianth-lobes 2.2–4.3 cm × 7–16 mm. Anthers yellow; style divided into 3 orange-red or rarely yellow branches, each expanded and frilled or lobed at the apex. *SW Italy, Sicily, Malta.* H3. Autumn.

24. C. medius Balbis. Illustration: Synge, Collins guide to bulbs, edn 2, pl. 6 (1971); Hay & Beckett, Reader's Digest encyclopaedia of garden plants and flowers, 191 (1978); Rix & Phillips, The bulb book, 183r (1981); Mathew, The crocus, pl. 17 (1982).

Corm-tunic fibrous, the fibres strongly netted. Leaves 2 or 3, 2.5–4 mm wide, absent at flowering time and not appearing until long after, dark green. Flowers lilac to deep purple, veined darker towards the base; throat white, often veined with purple, hairless. Spathe present. Bract solitary. Perianth-lobes 2.5–5 × 1.2–1.7 cm. Anthers yellow; style divided into many slender deep orange to scarlet branches. *NW Italy, SE France.* H2. Autumn.

25. C. nudiflorus Smith. Illustration: Synge, Collins guide to bulbs, edn 2, pl. 6 (1971); Grey-Wilson & Mathew, Bulbs, pl. 32 (1981); Rix & Phillips, The bulb book, 175i (1981); Mathew, The crocus, pl. 14 (1982).

Corm usually producing stolons; tunics papery, splitting into parallel fibres and often becoming wholly fibrous. Leaves 3 or 4, 2–4 mm wide, absent at flowering time and not appearing until long after, dark green. Flowers deep purple or more rarely paler lilac-purple, not conspicuously veined; throat not yellow, usually hairless. Spathe present. Bract solitary. Perianth-lobes 3–6 cm × 9–20 mm. Anthers yellow; style divided into many slender orange branches. *SW France, N & E Spain.* H2. Autumn.

26. C. serotinus Salisbury.

Corm-tunic papery with parallel fibres or wholly fibrous with the fibres finely to coarsely netted. Leaves 3–7, 0.5–2 mm wide, usually shorter than the flower but sometimes equalling it or occasionally absent, but if so then developing immediately the flowers wither; green in colour. Flowers pale to deep lilac or occasionally purple, sometimes veined darker; throat white or pale yellow, hairy or hairless. Spathe present. Bract usually solitary but occasionally with a small bracteole. Perianth-lobes 2.5–3.8 cm × 8–13 mm. Anthers yellow; style divided into many slender orange branches.

Subsp. **serotinus**. Illustration: Grey-Wilson & Mathew, Bulbs, pl. 32 (1981); Mathew, The crocus, pl. 15a (1982). Corm-tunic coarsely netted. Leaves 3 or 4. Perianth-lobes 2.5–3.8 cm × 8–11 mm. *C & S Portugal.* H3. Autumn. This subspecies is rarely seen in cultivation.

Subsp. **clusii** (Gay) Mathew (*C. clusii* Gay). Corm-tunic finely netted. Leaves 4–7. Perianth-lobes usually 2.5–3.8 cm × 8–13 mm. *C & N Portugal, NW & SW Spain.* H3. Autumn.

Subsp. **salzmannii** (Gay) Mathew (*C. salzmannii* Gay; *C. asturicus* Herbert; *C. granatensis* Maw). Illustration: Synge, Collins guide to bulbs, edn 2, pl. 6 (1971) – subsp. *salzmannii*; Rix & Phillips, The bulb book, 175k (1981) – subsp. *salzmannii*; Mathew, The crocus, pl. 15c (1982). Corm-tunic papery, splitting into parallel fibres. Leaves 5–7. Perianth-lobes usually 3.5–5.5 × 1–1.7 cm. *N Africa, S, C & N Spain, Gibraltar.* H3. Autumn.

This is the most commonly cultivated variant of *C. serotinus*. In N Africa there occur very vigorous forms with leaves well developed at flowering time; the cultivar known as *C. salzmannii* 'Erectophyllus' is very similar to these. The deep-coloured cultivar *C. asturicus* 'Atropurpureus' is probably referable to subsp. *salzmannii*, although it may be of hybrid origin.

27. C. speciosus Bieberstein. Illustration: Synge, Collins guide to bulbs, edn 2, pl. 6 & pl. 10 (1971); Hay & Beckett, Reader's Digest encyclopaedia of garden plants and flowers, 190 (1978); Rix & Phillips, The bulb book, 175j (1981); Mathew, The crocus, pl. 74a (1982).

Corm-tunic leathery or eggshell-like, splitting into rings at the base. Leaves usually 4, 3–5 mm wide, absent at flowering time and not appearing until long after, deep green. Flowers lilac-blue to purple- or violet-blue, conspicuously veined darker and often speckled on the outside; throat whitish or sometimes faintly yellow, hairless. Spathe absent. Bracts 2. Perianth-lobes usually 3.7–6 × 1–2.2 cm. Anthers yellow; style divided into many slender yellow to deep orange branches. *Former USSR (Crimea, Caucasus), N Iran, N & C Turkey.* H2. Autumn.

There are several commonly cultivated clones, mainly differing in the depth of flower colour, and an albino, 'Albus'. The species occasionally hybridises with *C. pulchellus* to produce intermediates.

28. C. pulchellus Herbert. Illustration: Synge, Collins guide to bulbs, edn 2, pl. 6 (1971); Rix & Phillips, The bulb book, 182a (1981); Grey-Wilson & Mathew, Bulbs, pl. 32 (1981); Mathew, The crocus, pl. 75 (1982).

Corm-tunic papery or leathery, splitting into rings at the base. Leaves usually 4, 4–5 mm wide, absent at flowering time and not appearing until long after, dark green. Flowers clear pale to mid bluish-lilac with darker longitudinal veins; throat deep yellow, hairless or slightly hairy. Spathe absent. Bracts 2. Perianth-lobes 1.8–4 cm × 8–20 mm. Anthers creamy white; style divided into many yellow or orange branches. *S Yugoslavia, S Bulgaria, N Greece, NW Turkey.* H3. Autumn.

29. C. kotschyanus Koch (*C. zonatus* Gay). Illustration: Hay & Beckett, Reader's Digest encyclopaedia of garden plants and flowers, 189 (1978); Synge, Collins guide to bulbs, edn 2, pl. 6 (1971); Rix & Phillips, The bulb book, 176c (1981); Mathew, The crocus, pl. 23a (1982).

Corm often flattened and rather misshapen; tunic thinly papery with the fibres parallel near the base and weakly netted at the apex. Leaves 4–6, 1.5–4 mm wide, absent at flowering time and not appearing until long after, green. Flowers pale to mid bluish lilac with conspicuous darker veins; throat whitish, usually with 2 yellow blotches near the base of each lobe, these sometimes merging to form a larger V-shaped zone, or sometimes with no yellow in the throat. Spathe present. Bracts 2. Perianth-lobes 3–4.5 cm × 5–18 mm. Anthers creamy white; style divided into 3 creamy yellow to yellow branches, sometimes each slightly subdivided at the apex. *C & S Turkey, NW Syria, C & N Lebanon.* H3. Autumn.

The frequently cultivated var. **leucopharynx** Burtt lacks any yellow markings in the throat, which has a large clearly defined white zone; this clone is often erroneously referred to as *C. karduchorum* Maw, which is a related but distinct species, very rarely cultivated.

30. C. ochroleucus Boissier & Gaillardot. Illustration: Synge, Collins guide to bulbs, edn 2, pl. 6 (1971); Hay & Beckett, Reader's Digest encyclopaedia of garden plants and flowers, 190 (1978); Rix & Phillips, The bulb book, 174c (1981); Mathew, The crocus, pl. 26 (1982).

Corm-tunic thinly papery with the fibres parallel near the base and weakly netted at the apex. Leaves 3–6, 1–1.5 mm wide, usually visible but rather short at flowering time, green. Flowers creamy white; throat pale to deep yellow, hairy. Spathe present. Bracts 2. Perianth-lobes 2.3–3.7 cm × 7–10 mm. Anthers creamy white; style divided into 3 deep yellow branches, each shortly lobed or frilled at the apex. *SW Syria, Lebanon, N Israel.* H3. Autumn.

31. C. sativus Linnaeus (*C. orsinii* Parlatore; *C. sativus* var. *cashmirianus* Royle). Illustration: Synge, Collins guide to bulbs, edn 2, pl. 6 (1971); Rix & Phillips, The bulb book, 183n (1981); Mathew, The crocus, pl. 29a (1982).

Corm-tunic fibrous, the fibres finely netted. Leaves 7–12, 0.5–1.5 mm wide, usually equalling or exceeding the flower at flowering time, grey-green. Flowers lilac-purple, conspicuously veined darker and stained darker towards the base; throat lilac-purple, hairy. Spathe present. Bracts 2. Perianth-lobes 3.5–5 × 1–1.5 cm. Anthers yellow; style divided into 3 long red branches. *Unknown as a wild plant but occasionally persisting in semi-wild situations as a relic of cultivation.* H3. Autumn.

C. sativus is a sterile triploid of unknown origin. It was formerly widely cultivated in S Europe and Asia as the source of the spice saffron, prepared from its dried style-branches. It is still grown commercially to some extent in a few Mediterranean countries and in other parts of Europe as an ornamental.

C. cartwrightianus Herbert. Illustration: Mathew, The crocus, pl. 29 (1982). Differs mainly in having smaller flowers, which may be white or lilac-purple. *S Greece.* H3. Autumn.

32. C. hadriaticus Herbert. Illustration: Synge, Collins guide to bulbs, edn 2, pl. 6 (1971); Rix & Phillips, The bulb book, 182i (1981); Mathew, The crocus, pl. 33 (1982).

Corm-tunic fibrous, the fibres finely netted. Leaves 5–9, 0.5–1 mm wide, usually shorter than the flowers at flowering time but occasionally equalling them or sometimes very short with only the tips visible, grey-green. Flowers white, often stained brownish or violet at the base; throat deep yellow or rarely white, hairy. Spathe present. Bracts 2. Perianth-lobes 2–4.5 cm × 7–20 mm. Anthers yellow; style divided into 3 orange or red branches. *W & S Greece.* H3. Autumn.

Plants cultivated under the name *C. cartwrightianus* 'Albus' are usually referable to *C. hadriaticus.* The absence of yellow in the throat of the former is usually sufficient to distinguish between the two species. The name *C. hadriaticus* var. *chrysobelonicus* Herbert also appears in literature and catalogues but this is fairly typical of the species and does not merit a separate name.

33. C. caspius Fischer & Meyer. Illustration: Mathew, Dwarf bulbs, 73 (1973); Rix & Phillips, The bulb book, 179o (1981); Mathew, The crocus, pl. 57 (1982).

Corm-tunics papery, splitting longitudinally but not markedly fibrous. Leaves usually 4–6, 1–2 mm wide, equalling or exceeding the flowers, dark green. Flowers white to pinkish lilac; throat deep yellow, hairy. Spathe absent. Bracts 2. Perianth-lobes 2–4 cm × 6–15 mm. Anthers yellow; style divided into 3 orange branches, each expanded at the apex. *Former USSR, N Iran (Caspian region).* H3. Autumn.

34. C. cancellatus Herbert.

Corm-tunic fibrous, the fibres conspicuously netted. Leaves 4–7, 1–2 mm wide, normally absent at flowering time but rarely with the tips visible, grey-green. Flowers white or pale to deep lilac-blue, usually veined darker; throat pale yellow, hairy or hairless. Spathe absent. Bracts 2. Perianth-lobes 3–5.5 cm × 7–18 mm. Anthers yellow; style divided into many slender orange branches.

Two of the 5 subspecies are cultivated to a small extent:

Subsp. **cancellatus** (*C. cilicicus* Maw; *C. cancellatus* var. *cilicicus* (Maw) Maw). Illustration: Synge, Collins guide to bulbs, edn 2, pl. 6 (1971); Grey-Wilson & Mathew, Bulbs, pl. 32 (1981); Rix & Phillips, The bulb book, 182j (1981); Mathew, The crocus, pl. 39a (1982). Flowers pale to mid lilac-blue, usually veined with violet, rather slender. Perianth-tube 3–8 cm; lobes 3–4.2 cm × 6–11 mm, oblanceolate or narrowly elliptic, usually acute or almost so. *S Turkey, W Syria, Lebanon, N Israel.* H3. Autumn.

Subsp. **mazziaricus** (Herbert) Mathew. Illustration: Mathew, The crocus, pl. 39c (1982). Flowers creamy white to deep lilac-blue, often veined darker, usually rather goblet-shaped. Perianth-tube 5–15 cm; lobes usually 3–5.5 × 1–1.8 cm, obovate or oblanceolate, obtuse or almost acute. *S Yugoslavia, Greece, S & W Turkey.* H3. Autumn.

35. C. banaticus Gay (*C. byzantinus* Herbert; *C. iridiflorus* Reichenbach). Illustration: Synge, Collins guide to bulbs, edn 2, pl. 6 (1971); Mathew, Dwarf bulbs, 83 (1973); Rix & Phillips, The bulb book, 182, 1 (1981); Mathew, The crocus, pl. 79 (1982).

Corm-tunic finely fibrous, the fibres parallel at the base, netted at the apex. Leaves 1–3, 5–7 mm wide, absent at flowering time and not appearing until long after, green. Flowers lilac to purple, the inner lobes often paler than the outer and markedly smaller. Spathe present. Bract solitary. Outer perianth-lobes 3.7–5 × 1.3–2.5 cm, the inner 2.3–3 × 1.2–1.3 cm. Anthers yellow; style divided into many slender, widely spreading, lilac branches. *C, W & NW Roumania, NE Yugoslavia, SW Russia.* H2. Autumn.

17. SYRINGODEA Hooker

J.C.M. Alexander

Corm compressed or spherical with a pointed base; tunics hard and woody, the outer often cracked or split. Leaves 1–several, not 2-ranked, thread-like, terete or shallowly grooved on inner surface. Inflorescence-stalk short and subterranean. Bracts membranous, the inner smaller and bifid.

Flowers solitary, rarely 2, funnel-shaped. Perianth-tube long and slender, straight or slightly curved; lobes spreading, entire or notched. Filaments free; anthers erect. Style slender, 3-branched; branches entire or irregularly divided. Capsule spherical.

A genus of about 7 South African herbs, similar to *Crocus* and *Romulea*. Corms should be planted in May, about 5 cm deep, in a well-drained, rich, sandy loam and given plenty of water when in full growth. Plants should be given less water once growth ceases and protected in winter with a sheet of glass. The blooms last longer if the plants are grown in an alpine house.

Literature: de Vos, M.P., Die suid-Afrikaanse genus Syringodea, *Journal of South African Botany* **40**(3): 201–54 (1974).

1. S. pulchella Hooker. Illustration: *Botanical Magazine*, 6072 (1873); *Gardeners' Chronicle* **81**: 79 (1927); *Journal of South African Botany* **40**(3): 230 (1974).

Corm *c.* 1.2 cm in diameter, spherical to ovoid. Leaves 4–6, stiff, arched, 7.5–12 cm, hairless, bases enclosed in membranous sheaths. Flowers purple, *c.* 2.5 cm in diameter; perianth-tube 3–5 cm, cylindric; lobes *c.* 1.2 cm, narrowly triangular to obovate, deeply notched. Spathe-bracts 1.2–2 cm, lanceolate. Stigmas linear, undivided, barely exceeding the anthers. *South Africa.* H5–G1. Autumn.

18. GALAXIA Thunberg

V.A. Matthews

Corms with fibrous tunics that are often vertically ribbed. Stem underground or slightly above ground level, elongating in fruit. Leaves borne at the top of the stem, not 2-ranked. Perianth with a long tube that raises the flower above the leaves; lobes more or less equal, spreading. Filaments partly to completely joined, inserted at the throat of the perianth-tube. Stigmas entire or fringed. Ovary more or less stalkless, ovoid. Fruit a capsule containing many small, angled seeds.

A genus of 12 species from SW South Africa. The plants are best grown in a sandy soil, with the corms kept dry throughout the winter. Propagation is by seed or offsets.

Literature: Goldblatt, P., Biology and systematics of Galaxia (Iridaceae), *Journal of South African Botany* **45**: 385–423 (1979).

1a. Flowers completely yellow; stigmas fringed **1. ovata**
b. Flowers purple or pinkish usually with a yellow throat; stigmas entire **2. versicolor**

1. G. ovata Thunberg. Illustration: *Botanical Magazine*, 1208 (1809); *Journal of South African Botany* **45**: 413 (1979).

Leaves ovate, apex obtuse with a tiny point, margin thickened, minutely ciliate. Flowers yellow, perianth-tube 5–20 mm; perianth-lobes 1.6–2 cm. Filaments joined, 4–5 mm; anthers erect, 2–3 mm. Style extending well above anthers; stigmas fringed. *South Africa (SW Cape).* H4. Late spring–summer.

G. fugacissima (Linnaeus filius) Druce (*G. graminea* Thunberg). Illustration: *Botanical Magazine*, 1292 (1810); Marloth, Flora of South Africa **4**: pl. 45, f. F (1915); *Journal of South African Botany* **45**: 416 (1979). Differs in having linear leaves, filaments sometimes free in the upper third and diverging anthers. The flowers are scented. *SW South Africa.* H4. Summer.

2. G. versicolor Klatt (*G. ovata* Thunberg var. *purpurea* Ker Gawler). Illustration: *Botanical Magazine*, 1516 (1812); Marloth, Flora of South Africa **4**: pl. 45, f. E (1915); *Journal of South African Botany* **45**: 400 (1979).

Leaves lanceolate, often with a wavy margin that is minutely ciliate. Flowers purple or pinkish, usually with a yellow throat; perianth-tube 1–2.5 cm; perianth-lobes 1.5–2 cm. Filaments free at the tips, 5–6 mm; anthers curved, 1.5–3 mm. Style extending well above anthers; stigmas entire, often with a wavy margin. *South Africa (SW Cape).* H4. Summer.

19. ROMULEA Maratti

E.M. Rix

Perennial herbs with corms. Corms with hard, brown tunics, usually asymmetric at the base. Basal leaves usually 2, rarely 1 or up to 6, usually 4-grooved. Stem-leaves 1–6; all leaves linear, usually hairless and 4-grooved, not in 2 ranks. Flowering stems erect or recurved (especially in fruit), rarely absent. Flowers upright, surrounded by an outer bract and an inner bracteole. Perianth united into a short tube below, lobes all similar. Stamens 3, exceeding or shorter than the style, sometimes aborted. Style 1 with 3 bifid stigmas, the branches thread-like. Capsule borne on an elongate stem. Seeds spherical.

A genus with about 10 species in the Mediterranean area (extending northwards to SW England) and about 70 species in South Africa, mostly in Cape Province. The genus is similar to *Crocus* (p. 270), but is distinguished mainly by the asymmetric base of the corm and the long-stemmed flowers (absent in *R. macowanii* var. *alticola* —see p. 280).

The European and Asiatic species are mostly hardy and succeed well in very sandy soil, kept dry in summer. The African species generally need protection (at least in northern Europe), though the alpine *R. macowanii* and its varieties should be frost-hardy. Propagation is by seed or by offsets.

Literature: Beguinot, A., Revisione monografica del genere Romulea Maratti, *Malphighia* **21**: 49–122, 364–478 (1907), **22**: 377–469 (1908) & **23**: 55–117, 185–239, 257–96 (1909); de Vos, M., The genus Romulea in South Africa, *Journal of South African Botany*, Supplementary volume **9** (1972).

Leaves in section. Terete with 4 grooves: **1, 2, 3, 4, 5, 6, 8, 9, 10, 11, 13, 14**; terete with 8 sharp ridges: **7**; 4-winged: **12**.
Bracteole. Largely membranous, with green or brown keel: **1, 2, 3, 4, 5, 8, 13, 14**; largely green, with membranous margins: **6, 7, 9, 10, 11, 12**.
Flowers. Mostly more than 2 cm: **1, 2, 3, 4, 7, 9, 10, 11, 13, 14**; mostly less than 2 cm: **4, 5, 6, 8, 9**.
Perianth-lobes. Lilac or purple: **1, 2, 3, 4, 5, 6, 7, 8, 13**; white: **1, 8, 9, 13**; red, pinkish or magenta: **1, 6, 9, 10, 11, 13**; yellow: **1, 12, 13, 14**.
Throat of perianth. Yellow: **1, 6, 7, 8, 9, 10, 11, 12, 13, 14**; white or purple: **2, 3, 4, 5, 6, 8**; with dark blotches: **11**.
Anthers. Overtopping or equalling stigmas: **4, 5, 6, 8, 9, 10, 11, 12, 13, 14**; not reaching level of stigmas: **1, 2, 3, 7, 9**.

1a. Perianth more than 2 cm 2
b. Perianth less than 2 cm 16
2a. Stigmas overtopping anthers 3
b. Stigmas equalling or shorter than anthers 8
3a. Perianth-tube more than one-third of the total length of the perianth **2. tempskyana**
b. Perianth-tube less than one-third of the total length of the perianth 4
4a. Bracteole mostly green, with membranous margins 5
b. Bracteole mostly membranous, with green keel 6

5a. Leaves long and slender, 4-grooved **6. ramiflora**
b. Leaves stiff, erect and thick, 8-ridged **7. nivalis**
6a. Perianth deep violet, lobes obtuse; anthers reaching two-thirds or more of the length of the perianth **3. requienii**
b. Perianth violet, magenta, pink, yellow or white; anthers reaching at most two-thirds of the length of the perianth 7
7a. Basal leaves 2 **1. bulbocodium**
b. Basal leaves several **9. rosea**
8a. Perianth yellow 9
b. Perianth purple to pink or white 11
9a. Leaves winged, 2–5 mm wide, usually ciliate **12. hirta**
b. Leaves terete, 0.6–3 mm wide, hairless 10
10a. Stem longer than the flower **13. flava**
b. Stem shorter than the flower, ovary usually subterranean **14. macowanii**
11a. Flowering stem dichotomously branched **10. dichotoma**
b. Flowering stem simple or umbel-like 12
12a. Perianth scarlet or red, with a black blotch on each lobe **11. sabulosa**
b. Perianth-lobes without black blotches 13
13a. Perianth with yellowish throat 14
b. Perianth with white or purplish throat 15
14a. Corm rounded at the base **9. rosea**
b. Corm flattened at the base, with a crescent-shaped ridge **13. flava**
15a. Anthers reaching less than half the length of the perianth **4. ligustica**
b. Anthers reaching more than half the length of the perianth **5. linaresii**
16a. Corm rounded at the base **9. rosea**
b. Corm obliquely narrowed at the base 17
17a. Bracteole largely green **6. ramiflora**
b. Bracteole largely membranous 18
18a. Throat of perianth purple **5. linaresii**
b. Throat of perianth white or yellow **8. columnae**

1. R. bulbocodium (Linnaeus) Sebastiani & Mauri (*R. grandiflora* Todaro; *R. clusiana* (Lange) Nyman). Illustration: Rix & Phillips, The bulb book, 31 (1981); Brickell (ed.), RHS A–Z encyclopedia of garden plants, 909 (2003).

Leaves mostly recurved, 0.8–2 mm broad. Stem with 1–6 flowers. Bracts green with a narrow (*c.* 1 mm), membranous margin. Bracteole with only the central one-third green or purplish, the rest membranous. Perianth 2–3.5 (rarely to 5.5) cm, purplish, pinkish white or yellow; tube 3.5–8 mm; throat yellow, orange or white. Filaments hairy below; top of anthers not exceeding the white styles. *Mediterranean area, Portugal, NW Spain & S Bulgaria.* H3. Spring.

Several varieties occur; the 3 most common in cultivation are: var. **crocea** (Boissier & Heldreich) Baker, with yellow flowers; var. **leichtliniana** (Heldreich) Beguinot with white flowers that have a yellow throat; and var. **subpalustris** Baker with purplish flowers and a white throat.

2. R. tempskyana Freyn. Illustration: Rix & Phillips, The bulb book, 31 (1981).

Leaves mostly recurved, *c.* 1 mm broad. Stem 1-flowered. Bract green with narrow (1.5 mm), membranous margin. Bracteole membranous with a narrow, green, central strip. Perianth 2–3.7 cm, deep purple; tube narrow, 8.5–17 mm; lobes *c.* 5 mm wide, lanceolate, reflexed when the flower is open wide; throat purple. Filaments hairless; top of anthers reaching about halfway up the perianth-lobes, not overtopping the purple style. *Greece (Rhodes), SW Turkey, Cyprus, Israel.* H3. Spring.

3. R. requienii Parlatore. Illustration: *Botanical Magazine*, 9555 (1939).

Leaves recurved, *c.* 1 mm broad. Stem with 1–3 (rarely to 6) flowers. Bract green with narrow, membranous margins. Bracteole membranous with a narrow, green, central strip. Perianth 2–2.5 cm, deep purple; tube 5–8 mm; lobes 5–6 mm wide, obovate-oblong, rounded at apex, not reflexed; throat purple or whitish. Top of anthers reaching two-thirds or more up the perianth-lobes, not overtopping the style. *Corsica, Sardinia & Italy.* H4. Spring.

4. R. ligustica Parlatore. Illustration: Jordan & Fourreau, Icones ad floram Europae **2**: t. 342 (1903).

Leaves erect or recurved, 1–1.5 mm broad. Stem with 1 flower. Bract green with narrow membranous margin. Bracteole almost entirely membranous. Perianth 1.9–3.5 cm, lilac or violet; tube 5–7.5 mm; lobes obovate-elliptic, rounded to almost acute at the apex, not reflexed; throat not yellow. Filaments hairy at the base; top of anthers less than halfway up the perianth-lobes, overtopping the style. *Corsica, Sardinia & Italy.* H4. Spring.

5. R. linaresii Parlatore. Illustration: Jordan & Fourreau, Icones ad floram Europae **2**: t. 337 bis (1903); Mathew & Baytop, The bulbous plants of Turkey, t. 58 (1984).

Leaves mostly recurved, 1–1.5 mm broad. Stem with 1–2 flowers. Bract green with narrow (1 mm wide), membranous margins. Bracteole with soft greenish centre, otherwise membranous. Perianth 1.3–2.6 cm, violet-purple; tube 4–7 mm; lobes not reflexed; throat deep purple. Filaments hairy at the base; top of the anthers reaching more than halfway up the perianth, overtopping the style. *Sicily, Greece, W Turkey.* H4. Spring.

Two subspecies occur: subsp. **linaresii** from coastal sands in Sicily, has the perianth 2–2.6 cm with blunt lobes, whereas subsp. **graeca** Beguinot from further east has the perianth 1.3–2 cm with acute lobes.

6. R. ramiflora Tenore. Illustration: Jordan & Fourreau, Icones ad floram Europae **2**: t. 335 bis (1903).

Leaves erect or recurved, 0.75–1.5 mm. Stem with 1–4 flowers, elongating to 30 cm in fruit; flower-stalks elongating to 10 cm in fruit. Bract green, strongly veined. Bracteole green with a membranous margin. Perianth 1–2.2 cm, lilac or pink; tube 2.5–7 mm; lobes oblanceolate, acute; throat white or yellow. Filaments hairy or not; top of anthers reaching about two-thirds of the way up the perianth, at about the same level as the top of the style. *Mediterranean area.* H3. Spring.

Two subspecies occur: subsp. **ramiflora,** which has a perianth 1–1.8 cm, sometimes shorter than the bract and the style reaching to slightly below the tops of the anthers; and subsp. **gaditana** (Kunze) Marais, confined to the west and south of the Iberian peninsula, which has the perianth 2–3 cm, lilac or pink, green outside, and the style slightly exceeding the tops of the anthers.

7. R. nivalis (Boissier & Kotschy) Klatt. Illustration: Rix & Phillips, The bulb book, 31 (1981).

Leaves erect, stiff, *c.* 2 mm wide, with 8 equal, narrow ridges. Stem with 1–3 flowers, not elongating in fruit. Bract green with a narrow (*c.* 1 mm) membranous margin. Bracteole green with wide membranous margin. Perianth 2–2.5 cm, lilac; tube *c.* 3 mm; lobes lanceolate, blunt; throat yellow. Top of anthers about halfway up the perianth, not overtopping the style. *Lebanon.* H4. Spring.

8. R. columnae Sebastiani & Mauri. Illustration: Ross-Craig, Drawings of British plants **29**: t. 3 (1972).

Leaves recurved, 0.6–1 mm wide. Stem with 1–3 flowers; flower-stalks short. Bract green with narrow membranous margins. Bracteole almost entirely membranous. Perianth 1–1.5 cm, whitish to pale lilac with darker veins; tube 2.5–5.5 mm; lobes lanceolate or oblanceolate, acute; throat yellow to white. Filaments often hairless; top of anthers reaching about halfway up the perianth, overtopping the style. *W Europe, from SW England to the Mediterranean area.* H3. Spring.

9. R rosea (Linnaeus) Ecklon. Illustration: *Journal of South African Botany*, Supplementary volume **9**: f. 87 (1972).

Corm rounded at the base, the tunic splitting into bent teeth. Basal leaves several, erect or recurved, 1–2.5 mm wide. Stem short, with several flowers, not elongating in fruit; flower-stalks to 15 cm. Bract green with a very narrow membranous margin. Bracteole with wide membranous margins, with brown streaks. Perianth 1.5–4.5 cm, magenta, pinkish or white; tube 2–5 mm; lobes 3–10 mm wide; throat yellowish, sometimes with a bluish zone. Filaments hairy below; top of anthers reaching about halfway up the perianth, shorter than or exceeding the style. *South Africa, naturalised in S France and Guernsey and also in places in the southern hemisphere.* H4. Spring.

A very variable species; var. **rosea** has flowers 2.2–4.5 cm and anthers shorter than the style; var. **australis** (Ewart) De Vos has flowers 1.5–2.2 cm, usually pale pink, and anthers exceeding the style.

10. R. dichotoma (Thunberg) Baker. Illustration: *Journal of South African Botany*, Supplementary volume **9**: f. 57, 61 (1972).

Basal leaf 1 (rarely 2), erect or sometimes bent, 1–2 mm wide. Stem leaves 1–2. Stem 4–35 cm, erect, rigid, dichotomously branched near the top, with several flowers. Bract green with a very narrow membranous margin. Bracteole green with wide membranous margins, reddish towards the top. Perianth 2–4 cm, pink with deeper veins in the yellow throat; tube 3–5 mm; lobes 6–10 mm wide. Filaments minutely hairy below; top of anthers reaching less than halfway up the perianth, equalling or just overtopping the stigmas. *South Africa (Cape Province).* H4. Spring.

11. R. sabulosa Beguinot. Illustration: *Journal of South African Botany*, Supplementary volume **9**: f. 79, 93 (1972); Brickell (ed.), RHS A–Z encyclopedia of garden plants, 909 (2003).

Corm rounded at the base, the tunic splitting into bent teeth. Basal leaves almost erect to recurved, *c.* 1 mm wide. Stem with 1–4 flowers, short, hidden by the leaf bases; flower-stalks 4–14 cm. Bract stiff, green, with narrow, brown, membranous margin. Bracteole 2-keeled, green with membranous margin that is white below, brown above. Perianth 3–5 cm (rarely more); lobes 1.2–2 cm wide, acute, pointed or obtuse, shining scarlet or red, with a black, pale-edged blotch in the throat; tube 2–4 mm. Filaments hairless or slightly hairy below; top of anthers well below the middle of the perianth, overtopping the stigmas. *South Africa (Cape Province).* H4. Spring.

12. R. hirta Schlechter. Illustration: *Flowering plants of Africa* **29**: t. 1137 (1952).

Corm rounded at base, the tunic splitting into bent teeth. Basal leaves 3–6, 4-winged, wings ciliate or not, the whole leaf 2–5 mm wide. Stem with 1–4 flowers, short, hidden by the leaf bases; flower-stalks 4–20 cm. Bract green with very narrow membranous margin. Bracteole with wider membranous margins and tip. Perianth 1.8–3.5 cm; lobes 4–8 mm wide, more or less obtuse to acute, pale yellow, often with a brownish transverse band at about the middle; tube 4–5 mm. Filaments minutely hairy below; top of anthers below the middle of the perianth, overtopping the stigmas (sometimes only slightly). *South Africa (Cape Province).* H4. Spring.

13. R. flava (Lamarck) De Vos (*R. bulbocodioides* misapplied). Illustration: *Botanical Magazine*, 1392 (1811) – as *Trichonema caulescens.*

Basal leaf 1, minutely ciliate or hairless, 0.8–3 mm wide. Stem with 1–4 or more flowers, hidden by the leaf base or elongate and evident; flower-stalks 1.5–7 cm. Bract green with very narrow membranous margin. Bracteole membranous, sometimes with a brownish central strip. Perianth 2–4 cm; lobes 3–12 mm wide, acute to obtuse, usually yellow, sometimes white, rarely blue or pinkish; throat golden yellow; tube 3–7 mm. Filaments hairy below; top of anthers below the middle of the perianth, about level with the stigmas. *South Africa (Cape Province).* H4. Winter–spring.

14. R. macowanii Baker (*R. longituba* Bolus). Illustration: *Botanical Magazine*, n.s., 515 (1967).

Basal leaves 3–6, more or less erect or recurved, 0.6–1 mm wide. Stem with 1–3 flowers, hidden by the leaf bases; flower-stalks 1–8 cm, often below ground. Bract green, membranous below. Bracteole more membranous, with wide, white margin, tip green. Perianth 2.2–10.5 cm; lobes 5–15 mm wide, acute to obtuse, golden yellow; tube 2–6.5 cm; throat often orange-yellow. Filaments hairy below; tops of anthers about level with the middle of the perianth, level with the stigmas. *South Africa, Lesotho.* H4. Summer, rarely spring.

Three varieties occur: var. **alticola** (Burtt) De Vos, from over 2000 m in the Drakensberg in South Africa and Lesotho, has the perianth-tube 3.5–6.5 cm. Var. **macowanii** and var. **oreophila** De Vos have perianth-tubes to 3.3 cm; var. **macowanii** from the Karroo has larger flowers and wider perianth-lobes than var. **oreophila,** which is an alpine variant from 2400–2700 m in the northeast Cape and Lesotho.

20. GELASINE Herbert

J.C.M. Alexander

Corm *c.* 8 mm in diameter, almost spherical, with membranous tunics. Basal leaves 45–60 × 2.5 cm, 2-ranked, pleated, narrowly lanceolate, acuminate. Stem to 60 cm, bearing 3 or 4 reduced sheathing leaves and a terminal cluster of 2–4 funnel-shaped flowers. Spathe-bracts green, unequal. Flower-stalks short. Perianth radially symmetric, blue above, white with black spots at base; tube very short; lobes *c.* 1.8 cm, similar, obovate, each with a small point. Filament tube shorter than the triangular, blue, ascending anthers. Style short, divided into 3 short linear stigmas with rounded tips.

A genus of 1 species from eastern S America. *Gelasine* does well in light, well-drained soil in full sun, but will need protection from frost. If grown in a cool greenhouse, the foliage will continue to grow through the winter. It is easily propagated by cormlets or from seed that should be sown as soon as it is ripe.

1. G. azurea Herbert. Illustration: *Botanical Magazine*, 3779 (1840); *Journal of Horticulture* **26**: 488 (1874).

S Brasil & Uruguay. H5–G1. Spring.

21. DIPLARRHENA Labillardière

V.A. Matthews

Rhizome short. Leaves mostly basal, linear, in 2 ranks. Stem erect, hairless, simple or occasionally branched. Flowers short-lived, 2 or more within the terminal bracts. Perianth free, of 6 segments, the outer larger than the inner. Fertile stamens 2; filaments free. Style short, divided into 3 flattened branches. Fruit a 3-celled capsule containing flattened seeds.

A genus of 2 species from Australia and Tasmania, both of which are in cultivation. The plants should be cultivated as for *Libertia* but it should be remembered that they are slightly less hardy. Propagation is by seed, or division of the rhizome.

1a. Flowers 3–4 cm in diameter, usually 2 or 3 in each set of bracts **1. moraea**
b. Flowers 6–15 cm in diameter, 5 or 6 in each set of bracts **2. latifolia**

1. D. moraea Labillardière. Illustration: King & Burns, Wild flowers of Tasmania, 103 (1969); Rotherham et al, Flowers and plants of New South Wales and southern Queensland, f. 289 (1975); Brickell (ed.), RHS A–Z encyclopedia of garden plants, 381 (2003).

Stems 45–65 cm. Leaves 30–45 cm × 8–12 mm, shorter than the stems. Bracts 5–7 cm, each set enclosing 2 or 3 (occasionally more) flowers. Flowers 3–4 cm in diameter, fragrant. Outer perianth-segments reflexed-spreading, almost circular, one larger than the other two, *c.* 3 cm; inner segments erect, shorter, narrower, white flushed or veined with lilac or yellow towards the tips. *SE Australia, Tasmania*. H4. Summer.

2. D. latifolia Bentham. Illustration: Curtis, Endemic flora of Tasmania **6**: pl. 120 (1978).

More robust than *D. moraea*, with leaves to 2.5 cm wide. Flowers 6–15 cm in diameter, 5 or 6 in each set of bracts. *Tasmania*. H4. Summer.

22. LIBERTIA Sprengel

J.C.M. Alexander & D.A. Webb edited by P. Harrold

Perennials with short, creeping rhizomes and fibrous roots. Leaves numerous, mostly basal, equitant, linear, overlapping, flat and rigid (except no. 1). Flowering stems erect, bearing a few reduced leaves. Flowers radially symmetric, in loose clusters or panicles borne terminally and in the axils of the stem-leaves. Perianth-segments entirely free, spreading, the inner segments usually longer than the outer. Stamens 3; filaments slightly fused at base. Style with 3 entire, linear, spreading branches; fruit a many-seeded, 3-celled capsule.

A genus of up to 20 species from Australasia and the temperate or montane regions of S America, which does well in light peaty soil in a warm sunny border. Most of the species are hardy, though *L. caerulescens* (no. 5) requires protection from frost. Plants can be propagated from seed or by careful division in spring.

1a. Flowers blue **5. caerulescens**
b. Flowers white 2
2a. Flower-clusters dense; flower-stalks shorter than bracts **4. formosa**
b. Flower-clusters loose; flower-stalks longer than bracts 3
3a. Leaves soft, clear green; capsule spherical or broader than long **1. pulchella**
b. Leaves stiff, brownish green; capsule longer than broad 4
4a. Leaves mostly 20–30 cm × 4–6 mm; capsule 5–7 mm **2. ixioides**
b. Leaves mostly 30–50 cm × 7–12 mm; capsule 10–15 mm 5
5a. Leaves with conspicuous copper-orange mid-rib **6. peregrinans**
b. Leaves green **3. grandiflora**

1. L. pulchella Sprengel (*L. micrantha* Cunningham). Illustration: Galbraith, Collins field guide to the wild flowers of south-east Australia, pl. 51 (1977).

Basal leaves linear, soft, uniformly green, 5–15 cm × 2–4 mm. Stem 8–30 cm with a single reduced leaf in the lower half (rarely 1 or 2 lateral clusters also present). Flowers 2–8 in a stalkless terminal cluster. Spathe-bracts lanceolate, green, 6–12 mm. Flower-stalks 18–30 mm, slender, downy. Perianth-segments 4–6 mm, oblong to ovate, white, the outer almost as large as the inner. Stamens *c.* as long as perianth. Capsule almost spherical, 3–4 mm in diameter. *SE Australia, Tasmania, New Zealand and Papua New Guinea*. H4. Spring.

2. L. ixioides (Forster) Sprengel (not *L. ixioides* Gay nor *L. ixioides* Klatt). Illustration: Moore & Irwin, Oxford book of New Zealand plants, 187 (1978).

Basal leaves numerous, tufted, linear, rigid and leathery with cartilagenous margins, green, midrib pale, 20–30 cm or more × 4–6 mm. Stem 30–60 cm, branched in upper half. Inflorescence a broad panicle of many stalked umbels, each with 2–10 flowers. Bracts many, minute, green, lanceolate. Segments 6–9 mm, circular to oblong, white; the outer slightly shorter, oblong, white tinged with brown or green. Stamens *c.* 4 mm. Capsule conical to oblong, 5–7 mm. *New Zealand & Chatham Islands*. H4. Summer.

3. L. grandiflora (R. Brown) Sweet. Illustration: *Gardeners' Chronicle* **43**: 2 (1908); Kirk, British garden flora, 500 (1927); *Flora* **125**: 22 (1930).

Similar to *L. ixioides*. Leaves 30–75 cm × 7–12 mm. Umbels with 3–6 flowers, stalked. Inner perianth-segments 12–15 mm; the outer *c.* 6 mm, with a greenish brown keel. Capsule 10–15 mm, yellow. *New Zealand*. H4. Summer.

4. L. formosa Graham (*L. chilensis* Klotsch; *L. ixioides* Klatt). Illustration: *Botanical Magazine*, 3294 (1834); *The Garden* **70**: 175 (1906) & **75**: 545 (1911).

Basal leaves linear, rigid, dark green, 15–45 cm × 6–12 mm. Stem 50–120 cm, with 1 or 2 reduced sheathing leaves below the inflorescence. Flowers in many dense, stalkless, terminal umbels, on a simple or branched axis. Outer bract of each umbel large, obovate, membranous; inner bracts smaller, oblong. Flower-stalks shorter than outer bracts, 6–12 mm. Inner perianth-segments white or pale yellow, obovate to wedge-shaped, 12–18 mm; outer perianth-segments greenish brown, oblong, 6–9 mm. Stamens 6–10 mm. Capsule spherical, *c.* 6 mm in diameter. *Chile*. H3. Spring.

5. L. caerulescens Kunth. Illustration: *Gartenflora* **22**: t. 759 (1873).

Basal leaves linear, rigid, green, to 30 cm or more. Stem 30–60 cm, bearing 2–4 reduced leaves. Inflorescence 10–15 cm, of numerous many-flowered stalkless umbels on a single or sparsely branched axis. Outer bracts firm, ovate to lanceolate; inner bracts ovate, membranous. Flower-stalks very short. Inner perianth-segments pale blue, oblong, *c.* 6 mm; outer perianth-segments much smaller, greenish brown. Stamens as long as perianth. Capsule very small, spherical. *Chile.* H5–G1. Spring.

6. L. peregrinans Cockayne & Allan. Illustration: Dawson & Lucas, New Zealand coast & mountain plants, 40 (1996).

Leafy fans emerging from far-spreading rhizomes. Basal leaves stiff, linear, orange-green, with the middle veins thickened and forming a conspicuous copper-orange band, particularly if exposed to the full sun. Leaves 15–60 cm × 5–15 mm, with the outer margins also copper-orange. Stem short, not carrying the flowers or fruit above the leaves. Inflorescence a narrow panicle, upper bracts often brown. Flowers 1.5–2.0 cm in diameter, the white perianth segments spreading. The outer, smaller, oblong segments half the width and length of the inner, orbicular, segments. Filaments very shortly connate at base and the anthers *c.* 3–3.5 mm, dark red brown. Capsule *c.* 10 mm in diameter, more or less pyriform turning black and remaining undehisced until the next season's flowers are produced. *New Zealand.* H4. Spring.

23. BELAMCANDA Adanson
V.A. Matthews & B. Mathew
Rhizomatous perennial with branched stems and leaves borne in fans. Perianth-segments more or less alike, not differentiated into outer and inner, free, twisting spirally after flowering. Stamens free, not held against style. Styles 3, slender, not expanded and petaloid. Capsule splitting into 3 segments that reflex exposing many shiny black seeds attached to the central axis.

A genus of 1, possibly 2, species, best grown in any good soil, in sun or semi-shade with plenty of moisture during the growing period. It is not long-lived. Propagation is by seed or division.

1. B. chinensis (Linnaeus) de Candolle (*Pardanthus chinensis* (Linnaeus) Ker Gawler). Illustration: *Botanical Magazine*, 171 (1791); Bruggeman, Tropical plants, pl. 82 (1957); Everett, The New York Botanical Gardens illustrated encyclopedia of horticulture **2**: 395 (1981); Kitamura et al., Coloured illustrations of herbaceous plants of Japan (Monocotyledoneae), pl. 20, f. 136 (1981).

Stems 60–100 cm. Leaves to 25 × 1–2 cm. Flowers 3–12, 4–5 cm in diameter, yellowish or orange-red spotted with red or dark purple. *Former E USSR (Ussuri region), China, Japan, Taiwan, N India.* H2. Summer.

24. ORTHROSANTHUS Sweet
J.C.M. Alexander
Rhizome very short. Stem erect, bearing *c.* 12 stiff, leathery, acute, linear, equitant, 2-ranked leaves near the base and 1 or 2 reduced leaves higher up. Flowers in stalked or stalkless clusters forming a loose panicle, one per cluster open at a time, each lasting for only a few hours. Clusters each subtended by a pair of lanceolate to oblong spathe-bracts (see Figure 9(4), p. 242). Flowers each with a similar but smaller pair of bracteoles; flower-stalks very short. Perianth blue, radially symmetric; tube short; lobes oblong, spreading, similar. Filaments free, attached to top of perianth-tube; anthers linear. Style very short; branches entire, linear to oblong. Young capsule partly enclosed by spathe-bracts.

A genus of about 7 species, 2 from tropical America, the rest from Australia.

1a. Leaf-margin rough to touch; flowers purplish blue, on stalks 2.5–4 cm **1. chimboracensis**
b. Leaf-margin smooth; flowers pale blue on very short, barely exposed stalks **2. multiflorus**

1. O. chimboracensis (Humboldt, Bonpland & Kunth) Baker (*O. occisapungum* Ruiz). Illustration: *Botanical Magazine*, 8731 (1917); RHS Dictionary of gardening, edn 2, 1450 (1956 and reprints).

Rhizome short, stout, woody. Basal leaves to 40 cm × 6–10 mm, ribbed; margins finely toothed. Stem 30–60 cm, terete, sparsely branched. Flowers to 4 cm wide, purplish blue, in clusters of 3 or 4, forming a loose panicle. Outer spathe-bract *c.* 2.5 cm, lanceolate, acuminate, with a membranous margin and a sharp ciliate keel; the inner *c.* 1 cm. Flower-stalks 2.5–5 cm; bracteoles obtuse. Perianth-tube short; lobes *c.* 1.5 cm. Filaments *c.* 5 mm; anthers *c.* 5 mm. Style-branches *c.* 8 mm. Capsule 12–18 mm, 3-sided, dull brown. *Mexico to Peru.* H5–G1. Summer.

2. O. multiflorus Sweet (*Sisyrinchium cyaneum* Lindley). Illustration: *Edwards's Botanical Register*, 1090 (1821); RHS Dictionary of gardening, edn 2, 1450 (1956); Brickell (ed.), RHS A–Z encyclopedia of garden plants, 750 (2003).

Basal leaves 30–45 cm × 3–4 mm, linear, acute, rigid. Stem slender, equal to or longer than leaves. Flowers blue, in a narrow panicle of 5–8 clusters, some clusters stalkless. Spathe-bracts 12–15 mm, green with chaffy tips, enclosing several smaller bracteoles. Flower-stalks very short. Fully open flower 2.5–4 cm across; lobes all similar, obovate, obtuse to rounded, with a darker mid-vein. Filaments free; anthers projecting beyond flower and stigmas. Capsule *c.* 8 mm. *SW Australia.* H4. Spring–summer.

25. SISYRINCHIUM Linnaeus
J.C.M. Alexander
Rhizome absent or very short; roots fleshy or fibrous. Stem erect, simple or branched, terete or flattened, often winged. Leaves mostly basal, 2-ranked, linear to strap-shaped or very narrowly lanceolate, acute, pale to dark or bluish green, often blackening on drying. Stem-leaves broader and shorter. Spathe-bracts usually in pairs, lanceolate, acute, often with a white translucent margin. In branched species a lower (primary) pair of spathe-bracts gives rise to inflorescence-stalks, on the latter are further (secondary) pairs of spathe-bracts, from which the flower-stalks arise. In unbranched species, a single pair of spathe-bracts gives rise directly to the flower-stalks. Flower-stalks terete or flattened, often arched, lengthening in fruit. Flowers solitary or in clusters (see Figure 9(5–7), p. 242). Perianth radially symmetric, very shortly fused at base, lobes oblong, spreading, often notched with a short point in the notch. Stamens arising from base of perianth; filaments united into a cylindric or flask-shaped tube for some or all of their length. Ovary 3-celled, spherical to ovoid; ovules many. Style mostly hidden within filament-tube, with 3 acute or capitate, entire, stigmatic branches.

Variously estimated to contain between 70 and 200 hardy and half-hardy species from the New World. Some authors consider that 1 or 2 species may also be native to W Europe. Many have small flowers and are of little horticultural value. The cultivated species thrive on well-drained sandy soil in full sunlight and are easily propagated from seed. The generic names *Bermudiana, Marica* and *Olsynium* have been applied to species now considered under *Sisyrinchium*. The classification and naming of several of the species is very confused. Many names have been consistently misused, which makes use of references particularly hazardous. The situation is especially bad among the North American blue-flowered species. Parent (1980) considers that the widely used names *S. bermudiana* and *S. angustifolium* were unclear from the time of their original publication and should be abandoned. His suggestions have been followed in this account.

Literature: Parent, G.H., Le genre Sisyrinchium L. (Iridaceae) en Europe: un bilan provisoire, *Lejeunia* **99**: 1–40 (1980).

1a. Flowers bright yellow, sometimes tinged with green — 2
b. Flowers white, very pale yellow, off-white, blue, purple or violet — 3
2a. Stem unbranched, leafless — **1. californicum**
b. Stem usually branched and bearing leaves — **2. graminifolium**
3a. Flowers in distant clusters on an unbranched stem — **3. striatum**
b. Flower clusters 1 per stem or branch — 4
4a. Stems round in section — 5
b. Stems flattened or winged — 6
5a. Perianth-lobes less than 1.3 cm, white with pinkish purple veins, yellow at base; filament-tube flask-shaped; stigmas not protruding — **4. filifolium**
b. Perianth-lobes *c.* 1.8 cm, pinkish purple (rarely pure white); filament-tube cylindric; stigmas protruding — **5. douglasii**
6a. Stems unbranched — 7
b. Stems branched — 8
7a. Stems 4 mm or more in diameter — **7. montanum**
b. Stems 3 mm or less in diameter — **6. idahoense**
8a. Flowers white, off-white, very pale blue or very pale yellow — 9
b. Flowers blue or bluish purple — 10
9a. Stem 30–45 cm; perianth-lobes *c.* 12 mm; filament-tube inflated — **8. iridifolium**
b. Stem 7.5–20 cm; perianth-lobes 4–6 mm; filament-tube cylindric — **9. micranthum**
10a. Inflorescence-stalks thread-like, slender — **10. chilense**
b. Inflorescence-stalks winged or flattened — **11. graminoides**

1. S. californicum (Ker Gawler) Dryander (*Marica californica* Ker Gawler; *S. brachypus* (Bicknell) Henry; *S. borecde* (Bicknell) Henry; *S. convolution* misapplied). Illustration: *Botanical Magazine*, 983 (1807); Hay & Beckett, Reader's Digest encyclopaedia of garden plants and flowers, 667 (1978); Rickett, Wild flowers of the United States **5**: 73 (1971).

Stem to 60 cm, unbranched, flattened and winged, to 4 (rarely 6) mm in diameter, slightly glaucous, drying blackish. Leaves all basal, usually shorter than the stem, to 6 mm wide, linear, flat, obtuse. Spathe-bracts lanceolate, obtuse, margins translucent; the outer to 5 cm, the inner a little shorter. Flowers 3–7 in a single cluster; flower-stalks 1–4 cm, usually longer than the bracts. Perianth-lobes 1.2–1.8 cm, oblong, obtuse to acute, bright yellow with 5–7 darker veins. Stamens *c.* 7 mm; anthers orange to yellow, 3–5 mm. Mature capsule 7–12 × 6–8 mm, ellipsoid, 3-angled, blackish; seeds numerous, to 1.5 mm in diameter. *Western N America (Vancouver to California).* H3. Summer.

The name *S. brachypus* has been given to plants from the northern part of the range with short flower-stalks (1–2 cm) hidden in the spathe-bracts and seeds less than 1 cm in diameter. If this is accepted as a separate species then the name *S. brachypus* should be applied to many plants cultivated as *S. californicum.*

2. S. graminifolium Lindley (*S. majale* Link, Klotsch & Otto). Illustration: *Edwards's Botanical Register*, 1067 (1827).

Roots fleshy; stem to 45 cm, branched, flattened and winged, minutely hairy. Basal leaves linear, grass-like, shorter than stems; stem-leaves with broad, inflated sheathing-bases. Spathe-bracts oblong to lanceolate, with translucent margins, inflated; the outer 1.8–2.5 cm. Flower-clusters 3 or 4; flowers 4–8 per cluster; flower-stalks 1.5–2.5 cm. Perianth-lobes *c.* 1.3 cm, oblong to obovate, acute, yellow, often with a brownish purple spot near the base. Filaments united below into a cylindric tube. Capsule *c.* 8 mm. *Chile.* H5–G1. Spring.

Var. **maculatum** (Hooker) Baker (*S. maculatum* Hooker). Illustration: *Botanical Magazine*, 3197 (1832). Spots more pronounced; brownish purple patches also present in outer parts of alternate perianth-lobes.

Subsp. **nanum** (Philippi) Ravenna (*S. nanum* Philippi). Illustration: Smaller in stature with broader leaves.

S. convolutum Nocca (*S. graminifolium* misapplied; *S. alatum* misapplied; *S. iridifolium* misapplied). Illustration: Redouté, Liliacées **1**: pl. 47 (1803); Sanchez, Flora del valle de Mexico, f. 63 (1969). Densely tufted, roots fibrous. Basal leaves 15–20 cm × 4–8 mm. Stem broadly winged. Spathe-bracts 2.5–3 cm. Flower-stalks arched in fruit. Filaments fused only at base. *Mexico to Peru.* G2.

S. tenuifolium Humboldt & Bonpland (*S. hartwegii* Baker). Illustration: *Botanical Magazine*, 2117 (1819) & 2313 (1822); Sanchez, Flora del valle de Mexico, f. 63 (1969). Roots fleshy. Basal leaves linear to terete, 8–15 cm × 2–4 mm. Stem narrowly winged, 10–35 cm. Spathe-bracts 1.2–1.8 cm. Flowers 3 or 4 per cluster, on stalks protruding from the spathe-bracts and lengthening in fruit. Perianth-lobes 6–10 mm, yellow, striped with green, bending downwards when flower fully open. Capsule oblong, 6–9 mm, erect at maturity. *Mexico.* H4.

3. S. striatum Smith (*S. graminifolium, S. iridifolium, S. lutescens* and *S. pachyrhizum* misapplied). Illustration: *Botanical Magazine*, 701 (1803); Hay & Synge, Dictionary of garden plants, 175 (1969); Hay & Beckett, Reader's Digest encyclopaedia of garden plants and flowers, 667 (1978); Brickell (ed.), RHS A–Z encyclopedia of garden plants, 987 (2003).

Root fibres slender. Stem to 75 cm, erect, unbranched, flattened and narrowly winged, greyish green. Basal leaves 8–10, linear to narrowly lanceolate, 20–35 × 1.2–1.8 cm, greyish green; stem-leaves 1 or 2, narrowly lanceolate with sheathing bases, to 30 × 1.5 cm. Flowers pale greenish yellow, in several stalkless clusters spaced along stem; each cluster of 10–20 flowers enclosed in a pair of ovate, slightly chaffy spathe-bracts; flower-stalks equalling the bracts in length. Perianth-lobes 1.5–1.8 cm, oblanceolate, strongly veined with purplish brown. Filaments united only at base. Capsule spherical, *c.* 6 mm in diameter. *Chile & Argentina; naturalised in Isles of Scilly (SW Britain).* H5–G1. Summer.

A variegated form is available with yellow-striped leaves.

S. arenarium Poeppig (*S. cuspidatum* Poeppig; *S. striatum* misapplied). Illustration: Correa (ed.), Flora Patagonica **2**: 174 (1969). Roots fleshy. Stem scarcely flattened, hairy between flower-clusters. Basal leaves many, 10–40 cm × 3–6 mm. Flowers yellow with greyish brown veins; perianth-lobes 11–13 × *c.* 6 mm. Ovary very yellow-hairy. *Chile & Argentina.* H5–G1.

4. S. filifolium Gaudichaud-Beaupré (*S. junceum* Presl subsp. *filifolium* (Gaudichaud-Beaupré) Ravenna). Illustration: *Botanical Magazine*, 6829 (1885); Vallentin & Cotton, Plants of the Falkland Islands, pl. 55 (1921); Muñoz Pizarro, Sinopsis de la flora Chilena, t. 223 (1959).

Roots slender, slightly fleshy. Stem 15–35 cm, unbranched, terete. Basal leaves merging with stem-leaves, 3–6, linear, acute, 10–15 cm, 3-channelled; stem-leaves smaller. Spathe-bracts 5–12 cm. Flowers 2–8 in 1 or 2 terminal clusters, overtopped by final stem-leaf; clusters often hanging; flower-stalks *c.* 2.5 cm. Flowers *c.* 2.5 cm across, white with reddish purple veins, yellow at base. Perianth-lobes obovate, *c.* 1.2 cm. Filaments united at base. Capsule spherical to ovoid. *S Chile & Falkland Islands.* H2. Spring.

Subsp. **junceum** (Presl) Parent (*S. junceum* Presl; *S. juncifolium* Herbert). Illustration: Muñoz Pizarro, Sinopsis de la flora Chilena, t. 224 (1959); Correa (ed.), Flora Patagonica **2**: 181 (1969). Basal leaves usually 2, terete, rarely flattened. Spathe-bracts 2–2.5 cm. Flowers 13–29 mm wide; perianth-lobes *c.* 1 cm. Filament-tube narrowly flask-shaped. *S Peru, Bolivia, Chile & Argentina; naturalised in Canada (Ontario).*

5. S. douglasii Dietrich (*S. grandiflorum* Douglas; *S. inalatum* Nelson; *S. inflatum* (Suksdorf) St John; *S. angustifolium* misapplied). Illustration: *Botanical Magazine*, 3509 (1836); Hay & Beckett, Reader's Digest encyclopaedia of garden plants and flowers, 667 (1978); Rickett, Wild flowers of the United States **5**(1): 69 (1971) & **6**(1): 62 & 63 (1973).

Stem 10–30 cm, unbranched, terete. Basal leaves to 10 cm, linear with sheathing bases; stem-leaves few. Spathe-bracts very unequal; the outer to 12.5 cm; the inner shorter than the slender flower-stalks. Flowers usually 2 or 3 per cluster, 4–5 cm wide, purple, pink or white, sometimes with a basal green blotch, often slightly hanging. Perianth-lobes *c.* 1.8 cm. Filament-tube somewhat inflated. Style projecting, as long as perianth. *N America: British Columbia south to N California & Utah.* H2. Spring–summer.

Plants from the eastern part of the range with pink flowers and strongly inflated filament-tubes have been called *S. inflatum*.

6. S. idahoense Bicknell (*S. macounii* Bicknell; *S. anceps* misapplied; *S. bellum* misapplied; *S. montanum* misapplied). Illustration: Haskin, Wild flowers of the Pacific coast, 56 (1934); Rickett, Wild flowers of the United States **5**(1): 69 (1971).

Leaves all basal, 7–30 cm × *c.* 3 mm, linear to narrowly lanceolate, acute. Stem 20–50 cm, narrow, unbranched, bluish green, winged and often twisted. Spathe-bracts usually very unequal; the outer 3–6 cm, narrowly lanceolate, acute; the inner usually much smaller. Flower-stalks equal to or shorter than inner spathe-bract. Flowers 1–6 per cluster, dark purplish blue, yellow at base. Perianth-lobes *c.* 1.3 cm. *Western N America; naturalised in eastern Sweden.* H2. Summer.

Considered by some authors to be part of *S. graminoides*. Smaller plants (to 20 cm) with fewer, larger flowers and a large stem-bract have been called *S. sarmentosum* Greene (*S. birameum* Piper); if these are considered part of *S. idahoense*, then the name *S. sarmentosum* being earlier would apply to the whole species.

S. mucronatum Michaux (*S. bermudiana* Linnaeus var. *mucronatum* (Michaux) Gray; *S. gramineum* Lamarck; *S. junceum* misapplied). Illustration: *Botanical Magazine*, 464 (1799); Britton & Brown, Illustrated flora of the northern states & Canada, edn 2, 544 (1913). Leaves usually 2 mm wide or less. Stem stiffly erect, unbranched, almost thread-like. Flower-stalks longer than inner spathe-bract, spreading or arching. Spathe-bracts often reddish purple. Flowers violet. *Eastern N America.* H2. Spring–summer.

7. S. montanum Greene (*S. angustifolium* misapplied). Illustration: *Bulletin of the Torrey Botanical Club* **23**: pl. 265 (1896); Rickett, Wild flowers of the United States **6** (1): 62 & 63 (1973).

Stem to 60 cm, erect, unbranched, stiff, flattened and winged, pale green or bluish green. Leaves shorter than stem, to 4 mm wide, pale green, remaining green when dry. Spathe-bracts sometimes fused at base for *c.* 6 mm; the outer bract *c.* twice as long as the inner and very finely tapered. Flower-stalks shorter than inner spathe-bract, erect, making an angle with the stem of less than 30°. Flowers *c.* 1.3 cm across, deep purplish blue. Filaments almost completely fused. *N America from the Rocky Mountains east to the Atlantic Coast & Greenland; widely naturalised in Europe.* H2. Spring–summer.

Var. **crebrum** Fernald (*S. angustifolium* misapplied) is commonly grown and has dark green foliage often tinged with purple. *Eastern N America from Ontario to Virginia; Greenland.*

8. S. iridifolium Kunth (*S. chilense* misapplied; *S. iridioides* misapplied). Illustration: *Edwards's Botanical Register* **8**: t. 646 (1822); *Loddiges' Botanical Cabinet* **20**: t. 1979 (1833); Healy & Edgar, Flora of New Zealand **3**: f. 22 (1980).

Stem to 45 cm, branched, flattened and winged with bristly edges. Basal leaves shorter than stem, to 1 cm wide, linear to narrowly lanceolate, with bristly edges. Stem-leaves 2–4, similar but smaller. Spathe-bracts 2.5–3.8 cm, with bristly edges, similar in size and shape, or the inner a little smaller. Flowers in terminal clusters of 4–6, white to cream or pale yellow with purple veins merging into a dark purple centre. Flower-stalks *c.* equalling bract-length, becoming longer in fruit. Filaments more than half fused into a hairy, urn-shaped tube. Perianth-lobes *c.* 1.3 cm, oblong with a short, acute point, not overlapping when flower fully open. *Tropical S America (?Chile).* H5–G1. Summer.

S. laxum Sims. Illustration: *Botanical Magazine*, 2312 (1822). Similar, though not bristly; perianth-lobes *c.* 7 mm, ovate, overlapping when flower fully open. *Temperate S America.* H4. Summer.

9. S. micranthum Cavanilles. Illustration: *Botanical Magazine*, 2116 (1819); Burbidge & Gray, Flora of the Australian Capital Territory, 109 (1970).

Stem 7.5–20 cm, flattened and very narrowly winged, branched below, bearing several stem-leaves. Basal leaves less than half stem length, linear, acute. Flowers in small terminal clusters, occasionally some in lower axils, 2–6 per cluster. Spathe-bracts similar, lanceolate, 1.2–2 cm. Flower-stalks equal to or slightly shorter than spathe-bracts. Perianth-lobes off-white to pale yellow, reddish purple at base, 4–6 mm. Filaments fused into a very short, cylindric tube. *C American to NW Argentina; naturalised in Australia.* H3. Summer.

10. S. chilense Hooker (*S. geniculatum* misapplied). Illustration: *Botanical Magazine*, 2786 (1827).

Root fibres slender. Stem 15–45 cm, branched, flattened and narrowly winged, 2–3 mm wide; nodes prominent. Basal leaves linear to sword-shaped, 8–30 cm × 2–4 mm, veins prominent. Flowers in 4–6 clusters (rarely as many as 20), each cluster with 3–6 flowers. Inflorescence-stalks 10–12.5 cm, slender, thread-like, solitary in axils or in terminal groups of 4–6. Spathe-bracts similar, 1.8–2.5 cm, rigid, linear to lanceolate, acute. Flower-stalks very slender, 2.5–3.5 cm. Perianth-lobes *c.* 1.3 cm, purplish blue, yellow at base, oblong, notched at the top with a narrow tooth in the notch. Filaments almost totally united into a sparsely hairy cylindric tube. *Mexico to Uruguay & Chile; naturalised in Mauritius and possibly S France.* H4. Summer.

This species may not be in cultivation but is included here because its name has been widely misapplied to plants that are cultivated. Many other illustrations are available but they nearly all represent plants that do not fit Hooker's original description.

11. S. graminoides Bicknell (*S. angustifolium* Miller, ambiguous; *S. bermudiana* Linnaeus in part, ambiguous; *S. anceps* Cavanilles; *S. gramineum* Curtis, not Lamarck; *S. arenicola* Bicknell; *S. montanum* misapplied; *S. chilense* misapplied; *S. birameum* misapplied). Illustration: *Botanical Magazine*, 464 (1799); Rickett, Wild flowers of the United States **2**(1): 79 (1967); Fitter et al., Wild flowers of Britain and northern Europe, 273 (1974); Brickell (ed.), RHS A–Z encyclopedia of garden plants, 987 (2003).

Stem to 50 cm, almost always branched, flattened and broadly winged, flexuous, not stiff. Leaves equal to or shorter than the stem, to *c.* 6 mm wide, dark bluish green, turning brown when dry. Inflorescence stalks stout, flattened and winged. Spathe-bracts more or less equal, sometimes the outer longer. Flower-stalks longer than bracts, spreading, making an angle with the stem of at least 60°. Flowers 1.2–1.8 cm across, pale to dark purplish blue, yellow at base, greyish outside. Filaments fused into a slender cylindric tube. *Eastern N America (area imprecise through misidentification and misuse of names); ?Greenland; Ireland, SW Britain, Spain & France.* H2. Spring–summer.

Many authors consider the European population to be naturalised rather than native. Hybrids are recorded with *S. montanum* (no. 7).

Similar plants with slender, once-branched stems and dense tufts of old leaf-bases have been named *S. arenicola* Bicknell, a name sometimes found in horticultural literature and catalogues.

S. bellum Watson (*S. angustifolium* Miller var. *bellum* (Watson) Baker). Illustration: Rickett, Wild flowers of the United States **5**(1): 68 & 69 (1971). Stems narrowly winged; leaves *c.* half stem-length, *c.* 4 mm wide. Spathe-bracts very unequal. Flowers violet-blue; perianth-lobes *c.* 1.2 cm. *California & New Mexico.* H3.

Plants grown under this name are usually *S. idahoense* (no. 6).

S. atlanticum Bicknell. Leaves narrow, much shorter than stems, not darkening on drying. Stems slender, often branched more than once, nodes prominent. *Eastern N America.*

S. iridioides Curtis (*S. bermudiana* Linnaeus in part, ambiguous). Illustration: *Botanical Magazine*, 94 (1789). Flowers up to 2 times as large as those of *S. graminoides*, bright purplish blue. Spathe-bracts very broad, slightly unequal *Bermuda.*

26. SOLENOMELUS Miers

J.C.M. Alexander

Herbs with short rhizomes. Leaves 2-ranked, mostly basal or clustered near stem base, a few higher up. Flowers many, on short stalks, in clusters enclosed by a pair of spathe-bracts. Perianth yellow or blue, radially symmetric, with a slender cylindric tube, lobes spreading, similar. Filaments attached at throat, completely fused into a tube; anthers oblong. Style simple; stigma entire or with 3 very short teeth.

A genus of 2 species from temperate S America, closely related to *Sisyrinchium* but differing in its almost undivided or entire style. Cultivation as for *Sisyrinchium* (p. 263).

1a. Flowers yellow; perianth-tube *c.* 2.5 cm; leaves linear **1. pedunculatus**
b. Flowers blue; perianth-tube *c.* 5 mm; leaves almost terete **2. sisyrinchium**

1. S. pedunculatus (Hooker) Hochreutiner (*Sisyrinchium pedunculatum* Hooker; *Solenomelus chilensis* Miers; not *Sisyrinchium chilense* Hooker). Illustration: *Botanical Magazine*, 2965 (1830); *Transactions of the Linnaean Society* **19**: t. 8 (1845).

Basal leaves 15–30 cm, linear, grass-like, gradually tapered. Stem 30–45 cm, bearing a few reduced leaves and 1–3 long-stalked clusters of flowers in their axils. Spathe-bracts *c.* 2.5 cm, concave, ovate to oblong with white membranous margins. Flower-stalks very short. Perianth yellow; tube *c.* 2.5 cm; lobes *c.* 2.5 cm, oblanceolate to obovate, obtuse to rounded. Anthers much shorter than staminal tube. *Chile.* H3. Summer.

2. S. sisyrinchium (Grisebach) Diels (*Lechlera sisyrinchium* Grisebach; *S. lechleri* Baker). Illustration: Correa, Flora Patagonica **8**(2): 185 (1969).

Densely tufted. Basal leaves many, 10–30 cm × 1–1.5 mm, terete, acute, firm. Stem leaves fewer and smaller. Stem 10–30 cm, bearing 1 or 2 short-lived flowers. Spathe-bracts 2.5–3 cm, the outer lanceolate, rigid, green with white margins, the inner smaller. Flower-stalks to 1 mm. Perianth blue; tube *c.* 5 mm, very narrow; lobes 1.3–1.4 cm × 4 mm, oblanceolate. Anthers equal to or a little shorter than staminal tube. Stigma entire, forming a short tube at style apex. *Chile.* H3. Summer.

27. ARISTEA Aiton

J.C.M. Alexander

Herbs with creeping rhizomes bearing fibrous roots. Leaves mostly basal, 2-ranked, linear. Flowers usually blue, radially symmetric, in spikes, racemes or corymbs, on a long, erect, inflorescence-stalk that may bear a few reduced leaves. Spathe-bracts small. Perianth free almost to base; lobes similar, twisting spirally after flowering. Filaments free, inserted at top of the short perianth-tube. Style thread-like, with 3 small obovate stigmatic lobes. Fruit ovoid to cylindric, obtuse or acutely 3-angled.

A genus of about 50 species, 40 of which are in southern Africa, the rest further north in tropical Africa. Plants thrive in cold-house beds. Most of the cultivated species will tolerate a few degrees of frost but very rarely survive transplanting at anything beyond the seedling stage; it is useless to establish plants in a cool-house and then transplant them to an open bed. For outdoor culture, seedlings should be pricked out into gritty, well-drained soil in perishable pots and planted directly into sunny, sheltered positions. Seeds germinate freely.

1a. Leaves 3 mm wide or less, usually less than 15 cm long; flowers 1–4 per cluster **1. cyanea**
b. Leaves 6 mm wide or more, usually more than 15 cm long; flowers many per cluster 2

2a. Leaves 45 cm or less; stem to 45 cm, flattened **2. ecklonii**
b. Leaves 60 cm or more; stem to 120 cm, not flattened **3. confusa**

1. A. cyanea Aiton (*Ixia africana* Linnaeus; *A. africana* invalid; not *A. cyanea* de Wildeman or *A. africana* Hoffmannsegg). Illustration: *Botanical Magazine*, 458 (1799); Marloth, Flora of South Africa **4**: pl. 44 (1915); Jackson, Wild flowers of Table Mountain, 102 (1977).

Leaves 7.5–15 cm × 2–3 mm, linear, acute, somewhat stiff. Inflorescence-stalk 5–15 cm, bearing 1 or 2 reduced leaves and 1–4 clusters of short-stalked, blue flowers with yellow throats. Outer spathe-bract ovate, *c.* 8 mm, fringed; inner lanceolate. Perianth-lobes *c.* 1.2 cm, obovate, shallowly notched. Capsule oblong, acutely angled, *c.* 8 mm. *South Africa.* H5–G1. Spring and autumn.

2. A. ecklonii Baker. Illustration: Wood & Evans, Natal plants **1**: pl. 68 (1899); Macmillan, Tropical gardening & planting, edn 3, 190 (1925) & edn 5, 183 (1952); Eliovson, Wild flowers of southern Africa, edn 6, frontispiece (1980).

Leaves 15–45 cm × 6–12 mm, linear, acute, dark green, red at base, not rigid. Stem flattened, 30–60 cm, bearing several reduced leaves. Flowers in a loose panicle of many 3- or 4-flowered clusters; flower-stalks 6–12 mm. Outer spathe-bracts lanceolate, green, to 6 mm, with membranous margin. Perianth blue, 8–12 mm; outer lobes oblong, inner lobes ovate. Capsule cylindric, 12–18 mm. *Southern Africa.* H5–G1. Summer–autumn.

3. A. confusa Goldblatt (*A. capitata* misapplied and confused). Illustration: *Botanical Magazine*, 605 (1802); Marloth, Flora of South Africa **4**: pl. 44 (1915); Hamer, Wild flowers of the Cape, 36 (1926).

Plant robust, to 1.5 m; rhizome thick. Basal leaves 50–100 × 1.5 cm, rigid. Stem-leaves 3–5, to 45 cm. Flowers in a narrow, much branched panicle, fragrant; spathe-bracts ovate, brown and chaffy. Perianth deep blue; lobes *c.* 1.5 cm, ovate to wedge-shaped. Capsule 1.8–2.5 cm, oblong to cylindric, acutely angled. *SW South Africa.* H5–G1. Summer.

A. major Andrews (*A. thyrsiflora* misapplied and confused; *A. capitata* misapplied). Illustration: Rice & Compton, Wild flowers of the Cape of Good Hope, 192 (1951); *Journal of the Royal Horticultural Society* **93**: 471 (1968); Brickell (ed.), RHS A–Z encyclopedia of garden plants, 139 (2003).

Basal leaves to 3 cm wide, with brown margins. Spathe-bracts linear-lanceolate, to 1.2 cm, entire with white margin. *SW South Africa.* H5–G1. Spring–summer.

28. NIVENIA Ventenat

J.C.M. Alexander

Evergreen, perennial shrubs. Stems branching, rarely unbranched, woody below. Leaves linear, 2-ranked, acute, overlapping at base. Flowering-stem flat and 2-edged, ending in a spike, corymb or umbel subtended by a single spathe. Each flower or pair of flowers subtended by 3 bracts (see Figure 9(9), p. 242). Flowers radially symmetric, stalkless, falling entire on fading. Perianth with a distinct tube and 6 similar lobes. Flowers of 2 types (dimorphic): either with long stamens and short styles, or with long styles and short stamens. In some species both types occur, but on separate plants. Filaments free, borne at top of perianth-tube. Style-branches small, entire. Capsule spherical to cylindric.

A genus of 8 species from the South African Cape, unusual for the Iridaceae in their shrubby habit and dimorphic flowers; not to be confused with *Nivenia* Brown in the Proteaceae (correctly called *Paranomus* Salisbury). Plants perform well in full sun in light sandy soil, requiring plenty of moisture when actively growing. They can withstand some frost but not if conditions are damp. Easily propagated from seed.

Literature: Weimark, H., A revision of the genus Nivenia Vent., *Svensk Botanisk Tidskrift* **34**: 355–72 (1940).

1. N. corymbosa (Ker Gawler) Baker. Illustration: Botanical Magazine, 895 (1805); Grey, Hardy bulbs, **3**: 324 (1937); Everett, The New York Botanical Gardens illustrated encyclopedia of horticulture **7**: 2331 (1981).

Stem to 1.2 m, much branched; branches flattened. Bases of stem and branches densely leafy. Leaves linear, rigid, 5–20 cm × 2–5 mm, ascending at *c.* 45° to stem or branch. Flowering-stems 5–15 cm, flattened; flowers blue in dense corymbose panicles. Spathes 5–6 mm, oblong, brown. Perianth-tube 6–8.5 mm, emerging from bracts; lobes 6–10 × 3–6 mm, each with a small white basal spot. Stamens either concealed or protruding; style either concealed or 12.5–15 mm and protruding. *South Africa.* H5. Autumn.

N. fruticosa (Linnaeus filius) Baker. Illustration: *Svensk Botanisk Tidskrift* **34**: 360 (1940). Dwarf undershrub, rarely exceeding 20 cm. Upper parts of branches leafy. Leaves 3–9 cm × 1.5–2.5 mm. Flowering-stem to 2.5 cm; flowers blue in a single cluster; tube 1–1.3 cm; lobes 1–1.3 cm. Stamens projecting *c.* 6 mm from flower; style concealed in tube. *South Africa.* H5. Autumn.

29. SCHIZOSTYLIS Backhouse & Harvey

J.C.M. Alexander & D.A. Webb

Rootstock short, rhizome-like (rarely producing a corm); roots fibrous. Basal leaves 2–4, 2-ranked, 30–45 cm × 6–12 mm, linear, keeled; midrib distinct. Stem 30–90 cm, slender, terete, erect, bearing *c.* 3 reduced leaves. Flowers 4–14 in a 2-sided spike, radially symmetric, each with an unequal pair of green, lanceolate bracts. Perianth-tube 2.5–3 cm, narrowly conical, yellowish or reddish green; perianth-lobes *c.* 2.5 cm, similar, ovate to oblong, acute, overlapping, scarlet to crimson. Stamens 3; filaments slender, free, attached to top of perianth-tube; anthers bright yellow. Style equalling perianth-tube, divided into 3, entire, slender, red branches *c.* 1.8 cm long.

A genus from South Africa now generally thought to contain only 1 species; commonly grown for cut flowers. It is easily cultivated in wet peaty or sandy loam in full sun. An annual mulch of peat or compost helps to retain moisture and encourages growth. Propagation is mostly by division. In most modern treatments now included in *Hesperanthera* but maintained here because of its horticultural distinctness.

Literature: Wolff, K., Knees, S. & Cubey, S., Genetic relatedness and cultivar identification in a valuable garden species, Hesperanthera coccinea (Schizostylis coccinea), *Plant Genetic Resources* **8**: 1–10 (2009).

1. S. coccinea Backhouse & Harvey (*S. pauciflora* Klatt). Illustration: *Botanical Magazine*, 5422 (1864); Synge, Collins guide to bulbs, edn 1, pl. 32 (1961); Hay & Beckett, Reader's Digest encyclopaedia of garden plants and flowers, 652 (1978);

Brickell, RHS A–Z encyclopedia of garden plants, 966 (2003).

Eastern South Africa. H4. Summer–autumn.

The most commonly seen cultivars include 'Viscountess Byng', a vigorous plant with slender, pale pink flowers, and 'Mrs Hegarty' with rose pink flowers (illustration: Synge, Collins guide to bulbs, edn 1, pl. 32, 1961; Hay & Synge, Dictionary of garden plants in colour, 110, 1969). There is some confusion between these two cultivars and plants given the name *S. pauciflora*. Larger plants with deep brick-red flowers include 'Cindy Towe', 'Major' and 'Major Superba'.

30. HESPERANTHA Ker Gawler

J.C.M. Alexander

Corm spherical to ovoid, asymmetric; tunics woody, all at same level, or the older ones displaced upwards with bases irregularly split or evenly divided into overlapping segments. Basal leaves 2–4, in 2 ranks, linear or lanceolate, acute, hairless, often curved. Stem simple or branched, round in section, erect, bearing 1–3 leaves with inflated sheathing bases. Flowers few, in loose 1- or 2-sided spikes, each with a pair of leaf-like, chaffy or transparent bracts. Perianth-tube narrow, sometimes strongly curved; lobes 6, in 2 whorls, usually similar in size, shape and colour. Stamens symmetrically arranged, borne at top of perianth-tube; filaments free. Style shorter than or equal to perianth-tube; branches ascending, entire, long but not usually exceeding anthers.

A genus of about 55 species from Africa south of the Sahara, most strongly represented in South Africa. Most of the species have previously been included in *Ixia* or *Geissorhiza*. The flowers open in late afternoon and are deliciously scented. Cultivation conditions are similar to those for *Geissorhiza* (p. 288).

Literature: Foster, R.C., Studies in the Iridaceae V, *Contributions from the Gray Herbarium* **166**: 3–27 (1948); Goldblatt, P., Corm Morphology in Hesperantha, *Annals of the Missouri Botanical Garden* **69**: 370–8 (1982).

Basal Leaves. Linear: **1, 2, 4, 5, 6**; lanceolate: **3**.
Stem. Branched: **1, 2, 3, 6**; unbranched: **3, 4, 5**.
Flower-spike. One-sided: **4, 5**; 2-sided: **1, 2, 3, 6**.
Flower colour. Mostly white: **1, 2, 4, 5**; mostly yellow: **4, 6**; red, brown or purple on back of inner perianth-lobes: **2, 3, 5**; purple patch at base and apex of perianth-lobe: **6**.
Outer spathe-bract. Less than 1 cm: **3, 4**; 1–2 cm: **2, 5**; more than 2.5 cm: **1, 6**.
Perianth-tube. Curved: **1, 4, 5**; straight: **3, 6**. Less than 1 cm: **1, 3, 4, 6**; 1 cm or more: **2, 5**.
Perianth-lobes. Less than 1 cm: **4, 5**; 1–2 cm: **2, 3**; more than 2 cm: **1, 6**.

1a. Perianth-lobes more than 2 cm — 2
b. Perianth-lobes 2 cm or less — 3
2a. Perianth-tube *c.* 8 mm, strongly curved; flowers white, pointing downwards — **1. angusta**
b. Perianth-tube 3–5 mm, not strongly curved; flowers yellow or yellow and purple, not pointing downwards — **6. inflexa**
3a. Perianth-tube curved at base or above — 4
b. Perianth-tube straight — 5
4a. Basal leaves 4; stem 14–17 cm; perianth-tube 6–7 mm — **4. graminifolia**
b. Basal leaves 5–6; stem to 45 cm; perianth-tube to 1.2 cm — **5. radiata**
5a. Leaves lanceolate, 5–8 mm wide — **3. falcata**
b. Leaves linear, 4–5 mm wide — **2. buhrii**

1. H. angusta (Jacquin) Ker Gawler. Illustration: Batten & Bokelmann, Wild flowers of the eastern Cape Province, pl. 27, 5 (1966); le Roux & Schelpe, Namaqualand & Clanwilliam: South African wild flower guide , 46 (1981).

Basal leaves 2 or 3, to 32 cm × 3 mm, linear, acute, hairless, sheathing at base, slightly thickened at midrib and edges. Stem to 40 cm, branched; stem-leaves 1 or 2, to 22 cm. Main stem with 2–6 flowers in a loose 2-sided spike; side branches with 1 or 2 flowers. Flowers white. Outer bracts to 2.8 cm, membranous, notched at tip; inner to 2.5 cm, chaffy, 2-ridged, bifid. Perianth-tube *c.* 8 mm, strongly curved in upper part (causing flower to point downwards); lobes *c.* 2–4 cm × 7 mm, reflexed, ovate to oblong, acute or minutely notched. Stamens 4–7 mm. Style equal to perianth-tube; branches to 1.4 cm. *South Africa (Cape Province).* H5–G1. Spring.

Foster (1948) maintains that almost all the wild specimens collected as *H. angusta* should be called *H. bachmannii* Baker. This may also apply to plants grown as *H. angusta*.

2. H. buhrii L. Bolus.

Basal leaves 3, linear, acute, hairless, 14–21 cm × 4–5 mm. Stem 22–27 cm, branched; stem- and branch-leaves 9.5–11 cm, with inflated sheaths. Main stem with 6 or 7 flowers in a loose, 2-sided spike; branches each with 4–7 flowers. Outer bracts to 1.8 cm, leaf-like, membranous at tip, keeled; inner to 7 mm, membranous, 2-keeled. Perianth-tube 1–1.4 cm; lobes 1.5–2 cm × 6–9 mm, obovate, obtuse to acute, white; outer lobes flushed with pink or purple on backs. Stamens *c.* 1.3 cm. Ovary 3–5 mm. Style 7–8 mm; branches to 7 mm. *South Africa.* H5–G1. Spring–summer.

3. H. falcata Ker Gawler. Illustration: *Botanical Magazine*, 566 (1802); Marloth, Flora of South Africa **4**: pl. 42 (1915); Mason, Western Cape sandveld flowers, 68 (1972).

Basal leaves 2–4, lanceolate, hairless, to 8 cm. Stem simple or once branched, to 30 cm, with 1 or 2 small sheathing leaves. Flowers 2–10 in a loose, 2-sided spike. Outer bracts to 8 mm, oblong, leaf-like with chaffy edges. Perianth-tube straight, *c.* 8 mm; lobes to 1.2 cm, oblong, white; outer lobes red or purple on backs. Anthers *c.* 6 mm. Style-branches *c.* 4 mm. *SW South Africa.* H5–G1. Spring.

4. H. graminifolia Sweet. Illustration: *Botanical Magazine*, 1254 (1810); Hamer, Wild flowers of the Cape, 78 (1926).

Basal leaves 4, the lowest rudimentary, the others 7–14 cm × 2.5 mm, linear, curved, acute, hairless; midrib thick. Stem 14–17 cm, bearing 1 small sheathing leaf, 2–3.5 cm. Flowers 3–9 in a loose, 1-sided spike. Outer bracts to 8 mm, oblong to ovate, acute, leaf-like; inner to 1 cm, entire or bifid, 2-keeled. Perianth-tube 6–7 mm, curved from base; lobes 5–6 × 2 mm, obovate; the outer greenish, sometimes tinged with pink; the inner white to pale yellow. Stamens 4–5 mm. Ovary 2 mm, ovoid to oblong. Style *c.* 4 mm; branches *c.* 3 mm. *South Africa.* H5–G1. Autumn.

5. H. radiata Ker Gawler. Illustration: *Botanical Magazine*, 573 (1802); Trauseld, Wild flowers of the Natal Drakensberg, 41 (1969).

Basal leaves 5 or 6. Stem to 45 cm, unbranched; outer bracts to 1.8 cm. Perianth-tube to 1.2 cm, strongly curved; lobes to 8 mm; the inner white; the outer purplish red on backs. Style-branches *c.* 6 mm. *South Africa.* H5–G1. Summer–autumn.

6. H. vaginata (Sweet) Goldblatt (*Geissorhiza vaginata* Sweet; *H. inflexa* misapplied; *H. inflexa* (Delaroche) Foster var. *stanfordiae* (L. Bolus) Foster; *H. stanfordiae* L. Bolus). Illustration: Sweet, British flower garden, **2**: pl. 138 (1825–7).

Basal leaves 2, linear, curved, acute, hairless, to 15 cm × 8 mm, midrib slightly prominent. Stem to 32 cm, branched at base and once higher up, bearing 3 leaves 18–23 cm × 9–13 mm, with inflated sheathing bases. Flowers 2 or 3 in a loose, 2-sided spike. Outer bracts to 3 cm, oblong, keeled, membranous with translucent edges; inner to 2.5 cm, 2-keeled. Perianth-tube 3–5 mm, straight, abruptly widened at throat; lobes 3–3.5 × *c.* 1.5 cm, obovate to spathulate, notched, yellow with purple patches at base and tip, or pure yellow. Stamens *c.* 2 cm. Ovary 2–3 mm. Style 3–4 mm; branches 8–9 mm. *South Africa (Cape Province).* H5–G1. Spring.

Goldblatt (1970) has shown that the name *H. inflexa* does not apply to this species.

31. GEISSORHIZA Ker Gawler
J.C.M. Alexander

Corm ovoid, asymmetric, flat-based; tunics of hard, woody, overlapping layers, splitting longitudinally. Basal leaves lanceolate, 2-ranked, linear or thread-like, often curled, with margin or midvein often thickened. Stem simple or branched, erect, bearing a few reduced leaves. Flowers radially symmetric in a 1- or 2-sided spike, each with a pair of leaf-like bracts. Perianth-tube short, erect, widening above; lobes 6, in 2 similar whorls, spreading. Stamens borne at lobe-bases; filaments free. Style slender, longer than perianth-tube; branches short, entire, recurved.

A genus of 60–70 species from southern Africa, most of which were originally considered part of *Ixia.* It is similar to *Hesperantha,* from which it differs in its longer style with shorter branches. Most species are capable of withstanding a few degrees of frost, particularly if protected with sacking, straw, or glass. Plants do best in full sun in a well-drained sandy loam, and can be propagated by offsets or seed.

Literature: Foster, R., Studies in the Iridaceae II, A revision of Geissorhiza, *Contributions from the Gray Herbarium* **135**: 3–78 (1941).

Basal leaves. Less than 10 cm: **5**. Less than 3 mm wide: **1, 2**; 5–12 mm wide: **3, 4, 5**; more than 15 mm wide: **6**.
Stem-leaves. Bristly: **3**. Solitary: **1, 2, 4, 5**; 2 or more: **3, 6**. More than 15 cm: **3**.
Stem. Unbranched: **1, 2, 4, 5, 6**; once-branched: **4, 5**; much-branched: **3**.
Flowers. Yellow: **1, 6**; red, blue or purple: **2, 3, 4**; reddish purple and white: **5**.
Outer bract. Less than 1.5 cm: **1, 2, 3, 5**; 2–2.5 cm: **3**; *c.* 4 cm: **6**.
Perianth-tube. Less than 1 cm: **1, 2, 3, 4**; more than 1 cm: **5, 6**.
Perianth-lobes. Less than 2 cm: **1, 2, 3, 4, 5**; 2–3 cm: **4**; more than 4 cm: **6**.

1a. Veins of leaf all equally conspicuous 2
b. Midrib (and sometimes 2 other veins) more conspicuous than other veins 3
2a. Flowers yellow; stamens *c.* half perianth length **1. humilis**
b. Flowers purple; stamens almost as long as perianth **2. radians**
3a. Basal leaves 1.5 cm wide or more; perianth-lobes yellow with red midrib, longer than 2.5 cm **6. grandis**
b. Basal leaves 1 cm wide or less; perianth-lobes reddish or bluish purple (in part), shorter than 2.5 cm 4
4a. Stem-leaves 3 **3. aspera**
b. Stem-leaf 1 5
5a. Flower-spike 2-sided; inner perianth-lobes white **5. ovata**
b. Flower-spike 1-sided; all perianth-lobes reddish purple **4. erosa**

1. G. humilis (Thunberg) Ker Gawler. Illustration: *Botanical Magazine,* 1255 (1810).

Basal leaves 2, linear, acute, hairless, to 15 cm × 1 mm, thicker at margins and midrib. Stem-leaf 1, to 5 cm, with sheathing base 1–2 cm. Stem unbranched, 10–14 cm, round in section, hairless, purplish. Flowers yellow, *c.* 3 in a loose, usually 1-sided, spike. Outer bracts *c.* 1 cm, oblong to ovate, obtuse to truncate, leaf-like, white and chaffy near the apex. Inner bracts similar, slightly notched at tip. Perianth-tube to 4.5 mm; perianth-lobes *c.* 1.6 cm × 8 mm, obovate to elliptic, acute. Stamens *c.* 1 cm; anthers about equal to filaments. Ovary to 4 mm, conical. Style *c.* 1.3 cm. *South Africa (Cape Province).* H5. Summer.

2. G. radians (Thunberg) Goldblatt (*Ixia rochensis* invalid; *I. radians* Thunberg; *Geissorhiza rochensis* invalid). Illustration: *Botanical Magazine,* 598 (1802); Graf, Tropica, 525 (1978); Eliovson, Wild flowers of southern Africa, edn 6, 50 (1980).

Basal leaves 2, longer than inflorescence, *c.* 1.5 mm wide, thread-like, sometimes 4-angled, hairless. Stem-leaf 1, to 8.5 cm, sheath very inflated. Stem to *c.* 15 cm, unbranched, round in section, hairless. Flowers 1 or 2, purple, centres red, ringed with white. Outer bracts to 1.1 cm, ovate-lanceolate, leaf-like, with a membranous, 3-lobed apex; inner bracts a little smaller. Perianth-tube to 7 mm; perianth-lobes to 1.8 × 1.2 cm, obovate, obtuse. Filaments *c.* 8 mm; anthers 4–5 mm. Ovary to 3 mm. Style top level with anthers; the stigmas recurved, exceeding them. *South Africa (Cape Province).* H5. Spring.

3. G. aspera Goldblatt (*Ixia secunda* misapplied; not *I. secunda* Delaroche; *G. secunda* misapplied). Illustration: *Botanical Magazine,* 597 (1802); Marloth, Flora of South Africa **4**: pl. 43 (1915); Rice & Compton, Wild flowers of the Cape of Good Hope, pl. 197 (1951).

Basal leaves 2, to 28 cm × 6 mm, linear, acute, tapering towards the hase, hairless; margins, midrib and sometimes 2 other veins prominent. Stem-leaves 3, 17–25 cm, sparsely bristly, the lower 2 with sheathing bases. Stem to 40 cm or more, much branched, slender, bristly. Flowers bluish purple, 3–6 in a loose, 1-sided spike. Outer bracts to 1.2 cm, oblong to ovate, leaf-like, brownish above; inner bracts to 8 mm. Perianth-tube 1–2 mm; perianth-lobes to 1.2 cm × 4 mm, elliptic to obovate. Stamens *c.* 7 mm. Ovary 2–3 mm, ovoid to conical. Style 8 mm; stigmas 1.5 mm. *South Africa.* H5. Spring.

4. G. erosa (Salisbury) Foster (*Ixia erosa* Salisbury). Illustration: Eliovson, Wild flowers of southern Africa, edn 6, 50 (1980).

Basal leaves 2, linear, narrowed at base, to 23 cm × 7 mm; margins, midrib and sometimes 2 other veins prominent and hairy. Stem-leaf 1, to 14 cm; sheath inflated, hairy on veins. Stem to *c.* 3 5 cm, unbranched. Flowers 3–7 in a loose, 1-sided spike, reddish purple. Outer bracts to 2.3 cm, entire or 3-lobed, leafy below, brown and membranous above; inner bracts slightly shorter. Perianth-tube 2–4 mm; perianth-lobes to 2.4 × 1 cm, obovate, rounded. Stamens *c.* 1.5 cm. Ovary to 4 mm, conical. Style *c.* 1.4 cm; stigmas *c.* 3 mm. *South Africa.* H5. Spring.

Var. **kermesina** (Klatt) Foster (*G. hirta* (Thunberg) Ker Gawler) is more commonly grown and is usually called *G. hirta*. It is smaller, sometimes once-branched, with 2–4 reddish or bluish flowers per spike, and perianth-lobes to 15 × 8 mm. Illustration: Eliovson, Wild flowers of southern Africa, edn 6, 50 (1980).

5. G. ovata (Burman) Ascherson & Graebner (*Ixia ovata* Burman; *G. excisa* (Linnaeus) Ker Gawler). Illustration: Rice & Compton, Wild flowers of the Cape of Good Hope, pl. 194 (1951); Jackson, Wild flowers of Table Mountain, 94 (1977).

Basal leaves 2 or 3, to 7 × 1 cm, ovate to narrowly elliptic, acute, black-dotted, leathery with prominent margins and midrib. Stem-leaf 1, with sheathing base; free portion to 1.8 cm. Stem to 25 cm, simple, rarely once-branched. Flowers 1–7 in a loose, 2-sided spike, reddish purple and white. Outer bracts to 1.4 cm, oblong to ovate, acute or truncate, leaf-like, with reddish purple edges and tip; inner bracts slightly shorter. Perianth-tube to 2 cm, usually shorter; perianth-lobes to 18 × 6 mm, the outer reddish purple, the inner white. Stamens *c.* 6 mm. Ovary to 7 mm, ovoid to conical. Style to 2.6 cm; stigmas 2.5 mm. *South Africa.* H5. Spring.

6. G. grandis Hooker. Illustration: *Botanical Magazine*, 5877 (1870).

Basal leaves 2 or 3, to 20 × 2 cm, linear to narrowly elliptic; midrib prominent. Stem-leaves usually smaller. Stem to 30 cm, unbranched. Flowers *c.* 6 in a loose, 1-sided spike, yellow. Outer bracts *c.* 4 cm, ovate to elliptic, leaf-like; inner bracts to 2.5 cm. Perianth-tube to 2.5 cm, slender and curved; perianth-lobes to 4.5 × 1 cm, elliptic to obovate, obtuse, with red midribs. Stamens *c.* 1.5 cm. Ovary *c.* 8 mm, ovoid. Style *c.* 4 cm; stigmas *c.* 9 mm. *South Africa.* H5. Spring.

32. DIERAMA Koch

D.A. Webb edited by *P. Harrold*

Rootstock a corm. Leaves mostly basal, linear, stiff, 2-ranked. Stems exceeding the leaves, arching over towards the tip, each ending in a loose, 1-sided panicle with numerous, very slender primary branches, each of which terminates in a pendulous, crowded, spike-like raceme of 3–6 radially symmetric flowers (Figure 9(8), p. 242); bracts scarious. Perianth bell-shaped, its lobes *c.* twice as long as the tube. Filaments free. Stigma 3-lobed, lobes spathulate. Capsule spherical, with 6 seeds.

A genus of about 40 species, confined to southern and E tropical Africa; only about 5 are in cultivation. In older books and catalogues they may be found under *Sparaxis*. Although rather slow to become established, they flourish without much attention in any well-drained, sunny site. *D. pulcherrima* is particularly effective if planted at the edge of a pool where its arching stems overhang the water.

Literature: Hilliard, O. & Burtt, B.L., Dierama (1991).

1a. Perianth tube 5–8 mm in length 2
b. Perianth tube 9–14 mm in length 4
2a. Leaves narrow, 1–4 mm in breadth **5. pauciflora**
b. Leaves 3.5–6.5 mm in breadth 3
3a. Bracts broadly obovate; obtuse or shortly acute **3. dracomontanum**
b. Bracts lanceolate to oblong; acute to shortly acuminate **4. igneum**
4a. Leaves 8–10 mm wide; perianth usually deep purple **1. pulcherrima**
b. Leaves 6–8 mm wide; perianth usually pink **2. pendula**

1. D. pulcherrima (Hooker) Baker. Illustration: *Botanical Magazine*, 5555 (1866); Perry, Flowers of the world, 150 (1972); Hilliard, Burtt & Batten, Dierama, pl. 14 (1991); Brickell (ed.), RHS A–Z encyclopedia of garden plants, 376 (2003).

Densely tufted, eventually forming a large clump. Radical leaves several to many, 500–900 × 8–10 mm; reduced sheathing leaves 4–5. Flowering stems 0.9–1.8 m. Inflorescences pendulous, terminal spike 4–8 flowered; secondary inflorecences 4–8, each 2-5 flowered. Flowers somewhat crowded towards the tip of the peduncles. Bracts spathe-like, scarious, elliptic-lanceolate to elliptic-obovate, acute. Perianth usually deep purple, 35–54 mm long, the tube 10–14 mm. *South Africa.* H4. Summer.

2. D. pendula (Linnaeus) Baker. Illustration: *Journal of the Royal Horticultural Society* **39**: f. 9 (1913); Hilliard, Burtt & Batten, Dierama, pl. 16 (1991).

Consisting of single stems or a few tufted together. Radical leaves several to many, 600–900 × 4–8.5 mm; reduced sheathing leaves *c.* 5. Flowering stems 1–2 m in length. Infloresences pendulous, terminal spikes 5–9 flowered; secondary inflorescences 3–7, each 2–7 flowered. Flowers rather lax. Bracts spathe-like, narrowly obovate to elliptic. Perianth usually pink, 28–48 mm long, the tube 9–13 mm. *S & E tropical Africa, northwards to Ethiopia.* H4. Summer.

3. D. dracomontanum Hilliard. Illustration: Hilliard, Burtt & Batten, Dierama, pl. 35 (1991); Brickell (ed.), RHS A–Z encyclopedia of garden plants, 376 (2003).

Plants forming tufts. Radical leaves several, 300–650 × 3.5–6.5 mm; reduced sheathing leaves 3–4. Flowering stems 0.3–1.05 mm. Inflorescences pendulous or rarely suberect. Terminal spike 4–7 flowered; secondary inflorescences 2–5, each 2–5 flowered. Flowers somewhat crowded towards the tips of the peduncles. Bracts broadly obovate, obtuse to shortly acute. Perianth dark rose-pink or light coral-pink to red, 19–25 mm long, the tube 5–7 mm long. *South Africa.* H4. **Summer**

4. D. igneum Klatt. Illustration: Hilliard, Burtt & Batten, Dierama, pl. 17 (1991).

Plants usually a single stem, or forming small clumps. Radical leaves several to many, 300–900 × 2.5–6 mm; reduced sheathing leaves 3 or 4. Flowering stems 0.45–1.35 m. Inflorecences pendulous, terminal spike 4–11 flowers; secondary inflorecences 4–8, each 2–10 flowered. Flowers usually rather lax. Bracts lanceolate to oblong-obovate, acute to shortly acuminate. Perianth light to dark lilac or rosy-pink, 16–31 mm, tube 5–8 mm long. *South Africa.* H4. Summer.

5. D. pauciflora N.E. Brown. Illustration: Hilliard, Burtt & Batten, Dierama, pl. 7 (1991).

Plants usually forming dense clumps. Radical leaves several, 100–400 mm; reduced sheathing leaves **c.** 3. Flowering stems 0.3–0.6 m. Infloresences pendulous, sometimes reduced to a solitary, erect, 1–3 flowered spike, more commonly with up to 5 secondary inflorescences, each 1–3 flowered. Flowers rather lax. Bracts broadly obovate to broadly elliptic. Perianth bright pink or reddish, 20–28 mm long, the tube 5–7 mm. *South Africa.* H4. Summer.

33. IXIA Linnaeus

J. Cullen

Perennial herbs with corms with papery or fibrous tunics. Stems simple or little branched. Leaves few, mostly basal, 2-ranked, linear, lanceolate or thread-like, usually with a false midrib. Inflorescence a spike or a panicle of spikes with few, usually erect, branches; spikes with few to many flowers borne in 2 ranks or spirally arranged. Bracts short but exceeding the ovaries, usually 3-toothed at their.apices. Perianth of 6 equal or almost equal lobes, joined below into a short or long tube. Stamens attached at the throat of the tube or within it, filaments free or united into a short tube. Style thread-like, exceeding the stamens, stigma 3-fid. Fruit a thin-walled capsule opening by 3 flaps and containing many angular seeds.

A genus of about 45 species, all from the Cape Province of South Africa, where many are now rare. They are mostly grown in pots, though can be grown outdoors in sunny, frost-free areas. The corms should be planted in autumn in a well-drained compost; little water should be given during the winter, and the plants are best kept in a sunny, well-ventilated place. Watering can be increased in the spring, and should be continued after flowering until the leaves show signs of dying down.

Most of the plants grown today are hybrids of uncertain parentage (though *I. maculata* has made a significant contribution). The species are known to hybridise freely, and many different colour variants of many species were introduced to Europe (mainly through Holland) at the end of the eighteenth and the beginning of the nineteenth centuries. Much hybridisation took place then, and a wide range of flower colour is now available. Many variants were selected and named (see *Gardeners' Chronicle* 1873).

Literature: Lewis, G.J., South African Iridaceae: the genus Ixia L., *Journal of South African Botany* **28**: 45–195 (1962).

Leaves. To 2 mm wide: **1**.
Perianth. Predominant colour orange-yellow or yellow: **7**; cream, sometimes tinged with red: **8**; green: **2**; white, pale purple or bluish: **1, 3**; pink, red or purple: **4, 5**.
Perianth-tube. More than 4 cm: **8**; 1 cm or more: **2, 7**; 3–10 mm: **1, 2, 3, 4, 6**; less than 3 mm: **5**.
Perianth-lobes. Much shorter than tube: **8**; about as long as tube: **7**; 2–4 times as long as tube: **1, 2, 3, 4**; 6–8 times as long as tube: **5**.

1a. Perianth-tube 4–7 cm, much longer than the lobes **8. paniculata**
b. Perianth-tube to 1.8 cm, equalling lobes or shorter than the lobes 2
2a. Leaves thread-like, to 2 mm wide; stamens borne near the base of the perianth-tube **1. capillaris**
b. Leaves not thread-like, 2 mm or more wide; stamens borne in the throat of the perianth-tube 3
3a. Perianth-lobes green with a purple or blackish base **3. viridiflora**
b. Perianth-lobes variously coloured, not as above 4
4a. Filaments completely free from each other 5
b. Filaments united into a tube, at least at their base 7
5a. Perianth-tube 2–3 mm; leaves with the false midrib and margins somewhat thickened **5. campanulata**
b. Perianth-tube 4 mm or more; leaves with false midrib and margins not thickened 6
6a. Perianth-lobes to 2 times longer than the tube; perianth white, mauve or bluish; flowers faintly scented **2. polystachya**
b. Perianth-lobes 4 or more times longer than the tube; perianth pink, red or purple, rarely white; flowers not scented **4. patens**
7a. Flowers orange or yellow; perianth-lobes 1 and a half to 2 times longer than the tube **6. maculata**
b. Flowers variously coloured, rarely orange or yellow; the lobes of the perianth up to 1 and a half times longer than the tube **7. monadelpha**

1. I. capillaris Linnaeus filius. Illustration: *Botanical Magazine*, 570 (1803).

Corm 1–1.5 cm in diameter. Stems to 50 cm. Leaves usually 3, erect, thread-like, 6–18 cm × 0.5–2 mm, often spirally twisted. Spike with 1–4 white or pale blue flowers that are sometimes green at the throat. Perianth-tube 5–7 mm, slender below, swollen and cup-like above; lobes 1–1.5 cm × 4–6 mm. Stamens arising near the base of the perianth-tube, projecting, filaments free, 4–7 mm, anthers 3–5 mm. *South Africa (Cape Province)*. H5–G1. Spring–summer.

2. I. polystachya Linnaeus. Illustration: *Flowering plants of South Africa* **25**: t. 968 (1945–6).

Corm 8–25lmm in diameter. Stem 30–100lcm. Leaves 5–8, 15–50 cm × 2–8 mm (rarely broader). Spike or panicle loose or dense, with few to many flowers which are white, mauve or bluish with green, mauve or purple centres, sometimes tinged with yellow, faintly scented. Perianth-tube slender, 5–14 mm; lobes 1–2.8 cm × 5–9 mm. Stamens arising at the throat of the perianth-tube; filaments free, 3–4 mm; anthers 4–7 mm. *South Africa (Cape Province)*. H5–G1. Summer.

3. I. viridiflora Lamarck (*I. maculata* Linnaeus var. *viridis* Jacquin). Illustration: *Botanical Magazine*, 549 (1801); Eliovson, Wild flowers of southern Africa, edn 5, 51 (1955); Brickell (ed.), RHS A–Z encyclopedia of garden plants, 583 (2003).

Corm 1–3lcm in diameter. Stem 50–100lcm. Leaves 5–7, 40–55 cm × 2–4 mm, with 4–6 conspicuous veins. Spike with 12–many flowers, which are green with dark purple to almost blackish centres. Perianth-tube slender, 6–9 mm; lobes 1.6–2.5 cm × 7–11 mm. Stamens arising at the throat of the perianth-tube; filaments free, purple, 3–4 mm; anthers 9–13 mm, purple or yellow. *South Africa (Cape Province)*. H5–G1. Spring–summer.

4. I. patens Aiton. Illustration: *Botanical Magazine*, 522 (1801); *Gartenflora* **11**: t. 356 (1862).

Corms 1.5–2 cm in diameter. Stem 20–50 cm. Leaves 5–7, 10–35 cm × 6–12 mm, firm, with several prominent veins and cartilaginous margins. Spike loose with 5–15 flowers, which are various shades of pink, red or purple, rarely white, occasionally greenish in the centre. Perianth-tube slender, 4–6 mm; lobes with prominent veins, widely spreading, 1.6–2.4 cm × 5–12 mm. Stamens arising at the throat of the perianth-tube; filaments free, 4–6 mm; anthers 6–8 mm. *South Africa (Cape Province)*. H5–G1. Spring.

This species is perhaps threatened with extinction in its native habitat.

5. I. campanulata Houttuyn (*I. speciosa* Andrews; *I. crateroides* Ker Gawler). Illustration: *Botanical Magazine*, 594 (1803).

Corm 1–1.8 cm in diameter. Stems 10–15 cm. Leaves 5–10 cm × 2–5 mm, usually with 3–6 veins, the false midrib and margins somewhat thickened. Spike with usually 3–6 flowers, which are usually red, each lobe with a whitish band outside, occasionally entirely white or deep red. Perianth-tube slender, 2–3 mm; lobes 1.2–2.5 cm × 8–12 mm. Stamens arising at the throat of the perianth-tube; filaments free, 3–5 mm; anthers 6–8 mm. *South Africa (Cape Province).* H5–G1. Summer.

6. I. maculata Linnaeus (*I. conica* Salisbury). Illustration: *Botanical Magazine*, 539 (1801); *Flowering plants of South Africa* **9**: t. 329 (1929); *Journal of South African Botany* **28**: 139 (1962).

Corm 1–2lcm in diameter. Stem 15–50lcm. Leaves 10–35 cm × 2–7 mm, often spirally twisted, with several fine veins. Spike short, the 4–many flowers crowded, orange or yellow-orange with dark brown, purple or blackish centres. Perianth-tube slender, 5–8 mm; lobes 1.5–3 cm × 8–12 mm. Stamens arising at the throat of the perianth-tube; filaments 3–5 mm, united into a tube at the base; anthers 7–9 mm. *South Africa (Cape Province).* H5–G1. Spring–summer.

7. I. monadelpha Delaroche (*I. columnaris* Salisbury). Illustration: *Botanical Magazine*, 607 (1803); *Flowering plants of South Africa* **8**: t. 317 (1928).

Corms 1.3–2 cm in diameter. Stems 15–40 cm. Leaves 4–7, 8–30 cm × 3–10 mm, often spirally twisted, with several veins. Spike with 4–12 densely packed flowers, which are blue, violet, purple, pink or white, usually with a green or brown centre that is often outlined in another colour. Perianth-tube slender, 1–1.8 cm; lobes 1.3–2 cm × 6–11 mm. Stamens arising at the throat of the perianth-tube; filaments 3–6 mm, united for at least half of their length into a tube; anthers 7–8 mm, erect and almost touching each other. *South Africa (Cape Province).* H5–G1. Spring–summer.

8. I. paniculata Delaroche (*I. longiflora* Bergius). Illustration: *Botanical Magazine*, 256 (1794); Brickell (ed.), RHS A–Z encyclopedia of garden plants, 583 (2003).

Corm 1–1.5lcm in diameter. Stem 30–100lcm. Leaves 15–60 cm × 3–12 mm. Spike loose or dense, with 5–18 flowers, which are usually pale cream tinged or flushed with red, or occasionally pale yellowish. Perianth-tube slender, slightly and gradually widening above, 4–7 cm; lobes 1.5–2.5 cm × 3–8 mm. Stamens arising in the perianth-tube; filaments free, 5–6 mm; anthers 6–8 mm, within the perianth-tube or partly or wholly projecting from it. *South Africa (Cape Province).* H5–G1. Spring.

34. FREESIA Klatt

J. Cullen

Perennial herbs with corms with fibrous coats. Leaves in 2 ranks, mostly basal, with a few prominent veins and usually a false midrib. Stems usually unbranched or occasionally with 1 or 2 inflorescence-bearing branches. Inflorescence a spike, its axis bent sharply at an angle of about 90° at the level of the lowest flower, the flowers borne on the upper side of the axis. Bracts 2, membranous or green, just exceeding the ovary. Perianth consisting of a very narrow tube at the base, this gradually widening to a broader tube, ending in 6 lobes; the tube is curved. Stamens 3, free, not or scarcely projecting from the perianth-tube. Stigmas 3, each deeply divided into 2 lobes. Fruit a few-seeded capsule.

A genus of 19 species, all from the Cape Province of South Africa. Most of the freesias grown today, either as garden ornamentals or as an important cut-flower crop, are hybrids (*F.* × *hybrida* Bailey). Many of them are of uncertain parentage, due to the misapplication of names when the plants were introduced to Europe. All of the species included here may have been involved in hybridisation, but *F. lactea* (for size of flower and scent), *F. corymbosa* (for flower colour and scent) and *F. armstrongii* (for flower colour) have probably contributed most. There are surprisingly few good illustrations of the species.

Cultivation is generally as for most South African Iridaceae. The plants can be grown in pots in a cool greenhouse, or planted out in spring in a sunny border, the corms being lifted and dried off in autumn. For cultivation as a cut-flower crop, see the books by Mackenzie and Smith cited below.

Literature: Brown, N.E., Freesia Klatt and its History, *Journal of South African Botany* **1**: 1–31 (1935); Mackenzie, W.F., Freesias (1957); Smith, D., Freesias (1979); Kragtwijk, G., Freesia-sortiment, *Tuinbouwgids* **20**: 380–1 (1965).

1a. Bracts entirely or mostly brownish, membranous, often with the tips dark brown or almost black 2
b. Bracts green, sometimes with the margins minutely membranous, tips never dark brown or blackish 4
2a. Perianth-lobes entirely rose-pink **1. armstrongii**
b. Perianth-lobes yellow or greenish, some of them blotched with other colours 3
3a. Perianth pale to intense yellow, lower lobes each with an orange blotch **2. corymbosa**
b. Perianth brownish or greenish yellow, the backs of the lobes suffused with violet, the lower lobes each with a brownish blotch **3. refracta**
4a. Perianth 2.5–3.2 cm; bracts 4–5 mm **4. sparrmannii**
b. Perianth 3.4–6.5 cm; bracts 6–12 mm 5
5a. Flowers 5–6.5 cm, entirely white except for the yellowish base of the tube, without blotches on the lobes **6. lactea**
b. Flowers 3.6–4.6 cm, whitish or yellow, each of the 3 lower lobes, or at least the lowest, with a yellow or orange blotch **5. xanthospila**

1. F. armstrongii Watson.

Stems 20–40 cm. Leaves 5–13 cm × 6–16 mm, linear-lanceolate or linear-oblong. Bracts 4–6 mm, mostly membranous, the upper part purplish brown. Flowers not fragrant. Perianth 2.8–3.4 cm, the lobes 1–1.6 cm, the broad part of the tube 1.8–2.2 cm, the narrow basal part 6–8 mm; all rose-pink except for the basal part of the tube, which is yellow. *South Africa (Cape Province).* H5–G1. Spring.

2. F. corymbosa (Burman) N.E. Brown (*F. odorata* (Loddiges) Klatt; *F. refracta* var. *odorata* (Loddiges) Voss).

Stems 10–45 cm. Leaves 5–40 cm × 3–9 mm, at first flat on the ground, later erect, linear but gradually tapering. Bracts 4–6 mm, mostly membranous, tips dark brown to blackish. Flowers fragrant. Perianth 2.4–3.8 cm, the lobes 8–12 mm, the broad part of the tube 1.6–2.6 cm, the narrow basal part

6–14 mm; all yellow (entirely orange-yellow in var. **aurea** (Gumbleton) N.E. Brown) sometimes slightly flushed with purple, the throat and at least the lowermost lobe orange. *South Africa (Cape Province).* H5–G1. Spring.

3. F. refracta Klatt. Illustration: *Edwards's Botanical Register* **2**: t. 135 (1816).

Very like *F. corymbosa*, but flowers brownish or greenish yellow suffused with violet on the backs of the lobes, the lower lobes each with a brown or orange-brown blotch. *South Africa (Cape Province).* H5–G1. Spring.

Though the name *F. refracta* is widely used in the literature, the species to which it applies is one of the least attractive and least widely grown. The widespread use of the name is due to confusion with *F. corymbosa, F. sparrmannii* and *F. lactea.*

4. F. sparrmannii (Thunberg) N.E. Brown (*F. refracta var.* alba misapplied). Illustration: *Flowering Plants of South Africa* **1**: t. 11 (1920).

Stems 10–20 cm. Leaves ascending, 5–15 cm × 4–8 mm, linear but gradually tapering. Bracts 4–5 mm, green throughout. Perianth 2.4–3 cm, lobes *c.* 1 cm, broad part of tube *c.* 1.8 cm, narrow basal part to 1 cm; all white or creamy white (pale yellow in var. **flava** N.E. Brown). *South Africa (Cape Province).* H5–G1. Spring.

5. F. xanthospila Klatt.

Stems 15–25 cm. Leaves erect or ascending, 7.5–20 cm × 6–11 mm, linear or linear-lanceolate, gradually tapering. Bracts 7–8 mm, green throughout. Flower scented. Perianth 3.6–4.6 cm, the lobes 1.4–1.6 cm, the broad part of the tube 2–2.3 cm, the narrow basal part 8–10 mm; white with blotches of yellow on the lower lobes and sometimes on the base of the tube or (in var. **leichtlinii** (Klatt) N.E. Brown) the tube yellowish and the blotches orange. *South Africa (Cape Province).* H5–G1. Spring.

6. F. lactea Klatt (*F. alba* Foster; *F. refracta* var. *alba* (Foster) Anon.).

Plant 20–30 cm. Leaves ascending, 7–20 cm × 4–8 mm, linear, gradually tapering. Bracts 8–12 mm, green throughout. Flowers strongly and sweetly scented. Perianth 5–6.5 cm, lobes 1.2–1.4 cm, broad part of the tube 3–3.6 cm, narrow basal part 1.2–1.6 cm; all white or creamy white, except for the lower part of the tube, which is yellowish. *South Africa (Cape Province).* H5–G1. Spring.

This species, which has contributed its scent and large flower size to many hybrids, is usually found in the literature under the name *F. refracta* var. *alba.* This name is sometimes attributed to Baker or to other authors; it has also been used for *F. sparrmannii.*

35. LAPEIROUSIA Pourret

J. Cullen

Perennial herbs with corms covered by hard, woody coats. Stems erect, branched or not, usually bearing leaves. Basal leaves 1 or 2, usually longer than the stem-leaves; stem-leaves 2-ranked, usually with undulate-crisped margins, often bearing a keel or 2 keels on the lower surface, with branches in their axils. Flowers in a corymbose panicle, or in spikes at the ends of the stem and the branches; bracts 2, the outer larger than the inner (Figure 9(3), p. 242). Perianth with a long, narrow tube and 6 symmetrically or asymmetrically disposed lobes. Stamens 3, attached to the throat of the perianth-tube, placed to one side of the flower or symmetric. Stigmas 3, each deeply divided. Fruit an oblong-elliptic, membranous capsule with many angled or rounded seeds.

A genus of about 30 species from Africa, most of them from South Africa. Its name is frequently misspelled 'Lapeyrousia'. Several species traditionally included in it are treated here under *Anomatheca* (p. 293). Cultivation as for *Ixia* (p. 289).

Literature: Lawrence, G.H.M., Notes on Lapeirousia in cultivation, *Baileya* **3**: 131–6 (1955); Goldblatt, P., A revision of the genera Lapeirousia Pourret and Anomatheca Ker in the winter rainfall region of South Africa, *Contributions from the Bolus Herbarium*, 4 (1972).

1a. Inflorescence a corymbose panicle of spikes; flowers mostly blue, perianth-tube to 1 cm **1. corymbosa**
b. Inflorescence a simple spike (ending each branch); flowers red, violet, lavender, cream or white, perianth-tube 2 cm or more **2. pyramidalis**

1. L. corymbosa (Linnaeus) Ker. Illustration: *Botanical Magazine*, 595 (1801); Marloth, Flora of South Africa **4**: t. 23 (1915).

Corms *c.* 1 cm in diameter, dark brown to black, the coats woody, finally split longitudinally. Stems 5–45 cm, somewhat flattened and 2-winged. Basal leaves 8–10 cm, linear or linear-lanceolate, usually sickle-shaped. Inflorescence a corymbose panicle with many flowers, bracts 3–10 mm. Perianth more or less symmetric, the tube 4–10 mm, the lobes spreading, about as long as the tube, blue, each with an inverted V-shaped white mark near the base, the marks on the 6 lobes forming a star shape. *South Africa (Cape Province).* H5–G1. Spring.

Several subspecies exist in South Africa, differing in size and perianth colour, but only subsp. *corymbosa* (described above) appears to have been cultivated.

2. L. pyramidalis (Lamarck) Goldblatt (*L. fissifolia* (Jacquin) Ker; *L. bracteata* (Thunberg) Ker). Illustration: *Botanical Magazine*, 1246 (1810); *Flowering plants of South Africa* **6**: t. 239 (1926).

Corm bell-shaped, *c.* 1 cm in diameter. Stem at most 12 cm, terete, usually without branches. Basal leaf usually 1, linear to lanceolate; stem-leaves several, shorter and broader than the basal leaf, margins entire or undulate-crisped. Inflorescence a simple, few-flowered spike; bracts green and leaf-like. Perianth irregular, the tube 2–4 cm, the lobes unequal, the uppermost the largest, *c.* 8 mm, forming a hood over the anthers and stigma, all white, white flushed with purple or with reddish pink, the lower 3 lobes sometimes marked with a contrasting colour at the base. Anthers to the upper side of the flower, borne under the uppermost perianth-lobe. *South Africa (Cape Province).* H5–G1. Spring.

L. jacquinii N.E. Brown (*L. anceps* misapplied). Illustration: Sweet, British flower garden **2**: t. 143 (1825–7); le Roux & Schelpe, Namaqualand and Clanwilliam: South African wild flower guide **1**: 56 (1981). Plant small but often branched; stems 3-angled; stem leaves 2-keeled beneath; the angles of the stem and the keels of the leaves are undulate-crisped to irregularly toothed; perianth dark violet. *South Africa (Cape Province).* H5–G1. Spring.

L. fabricii (Delaroche) Ker (*L. anceps* misapplied; *Peyrousia aculeata* Sweet; *Lapeirousia aculeata* Sweet; *L. compressa* Pourret). Illustration: Sweet, British flower garden **1**: t. 39 (1823–5); le Roux & Schelpe, Namaqualand & Clanwilliam: South African wild flower guide **1**: 56 (1981). Plant small, usually branched; stems 2-angled and winged; stem leaves 1-keeled beneath; angles of stem and keels of leaves undulate-crisped to irregularly toothed; perianth cream or cream flushed with pink, sometimes with red spots on the lower lobes. *South Africa (Cape Province).* H5–G1. Spring.

The name *L. anceps* has been incorrectly used for both the above species. *L. anceps* (Linnaeus filius) Ker is a quite separate species, which is not now cultivated; it has stems to 30 cm, with many spreading, elongate branches and the perianth is white to pink.

36. ANOMATHECA Ker

J. Cullen

Herbaceous perennials with conical corms that have fibrous coats. Stems erect, branched or not. Basal leaves several, soft, with conspicuous false midribs, shorter than the 2-ranked stem-leaves. Inflorescence a simple spike with few flowers (in our species). Perianth with a tube and 6 lobes, of which the uppermost is somewhat larger than the others, rendering the flower slightly bilaterally symmetric. Stamens 3, asymmetrically disposed. Stigmas 3, each forked and recurved. Fruit a rough-surfaced capsule containing many rounded seeds.

A genus of 4 species from southern Africa, formerly included in *Lapeirousia* (p. 292), but sufficiently distinct in corm, leaf and habit characters. Cultivation as for Ixia (p. 289).

Literature: Lawrence, G.H.M., Notes on Lapeirousia in cultivation, *Baileya* **3**: 131–6 (1955); Goldblatt, P., A revision of the genera Lapeirousia Pourret and Anomatheca Ker in the winter rainfall region of South Africa, *Contributions from the Bolus Herbarium* 4 (1972).

1a. Stems flattened and winged; flowers green **4. viridis**
b. Stems terete; flowers not green 2
2a. Leaves to 10 cm, much shorter than the inflorescence, obtuse but with a small point **1.verrucosa**
b. Leaves 10–60 cm, about as long as the inflorescence, acute 3
3a. Perianth-lobes less than half as long as the tube, held at right angles to it **2. laxa**
b. Perianth-lobes more than half as long as the tube, curving outwards and forming a cup-shape **3. grandiflora**

1. A. verrucosa (Vogel) Goldblatt (*Lapeirousia juncea* (Linnaeus filius) Ker; *Anomatheca juncea* (Linnaeus filius) Ker). Illustration: *Botanical Magazine*, 606 (1802).

Corms conical. Stems terete, 8–20 cm. Basal leaves to 10 cm, much shorter than the inflorescence, linear-oblong, obtuse but with a small point and with a conspicuous false midrib. Inflorescence of several spikes. Perianth slightly irregular, pink, fragrant, tube 1–1.5 cm, the lobes *c.* 9 × 4 mm, the lower marked with red-purple spots. *South Africa (Cape Province).* H5–G1. Spring.

2. A. laxa (Thunberg) Goldblatt (*Lapeirousia laxa* (Thunberg) N.E. Brown; *A. cruenta* Lindley; *Lapeirousia cruenta* (Lindley) Baker). Illustration: *Edwards's Botanical Register* **16**: t. 1369 (1830); Eliovson, Wild flowers of southern Africa, edn 5, ii (1955); Hay & Synge, Dictionary of garden plants in colour, 100 (1969).

Corms conical, *c.* 1 cm in diameter. Stem terete, erect, 10–30 cm, not or little branched. Leaves mostly basal, equalling or exceeding the inflorescence, acute, with conspicuous false midribs. Inflorescence a simple spike with up to 6 flowers; bracts 6–8 mm. Perianth usually red, occasionally white or bluish, tube 2–3 cm, the lobes to 1.3 cm, slightly irregularly disposed, spreading at right angles to the tube, each of the 3 lower lobes with a dark red or purple spot near the base. Stamens borne towards the upper side of the flower, the 3 anthers touching. *South Africa, Moçambique.* H5–G1. Mostly spring.

3. A. grandiflora Baker (*Lapeirousia grandiflora* (Baker) Baker). Illustration: *Botanical Magazine*, 6924 (1887); Linley & Baker, Flowers of the veld, f. 18 (1972); Moriarty, Wild flowers of Malawi, t. 7 (1975).

Similar to *A. laxa* but stems and leaves to 60 cm, perianth brownish or purplish red, the 3 lower lobes each with a dark spot at the base, the tube 3–3.5 cm, the lobes 2.5–4 cm, curving outwards, forming a cup-shape. *Southern Africa north to Malawi.* H5–G1. Summer.

4. A. viridis (Aiton) Goldblatt (*Lapeirousia viridis* (Aiton) Bolus). Illustration: Mason, Western Cape sandveld flowers, t. 20 (1972).

Corm 1–2 cm in diameter. Stem to 35 cm, flattened and winged. Leaves much shorter than the inflorescence, margins smooth or undulate-crisped. Inflorescence a simple spike with up to 6 (rarely more) flowers; bracts *c.* 1 cm. Perianth-tube 2–3 cm, the lobes 1–2 cm, sharply reflexed from the apex of the tube, green. *South Africa (Cape Province).* H5–G1. Spring.

37. WATSONIA Miller

E. Campbell

Herbaceous perennials with more or less spherical corms. Leaves basal and borne on the stem, 2-ranked, usually rigid, sword-shaped to linear, each usually with a prominent main vein. Flowers in 2 ranks in a simple or branched spike. Perianth forming a curved tube at the base, the upper part gradually or abruptly widened, the lobes more or less equal or the 3 inner slightly wider than the 3 outer, all oblong or lanceolate. Stamens 3, arising below the throat of the perianth-tube. Ovary 3-celled, each cell with many ovules; style thread-like with 3 bifid stigmas. Fruit an oblong capsule.

A genus of about 60 species from South Africa; about 14 of them are in general cultivation today, though many more have been grown, and may still persist here and there. They are best grown in a well-drained soil in a sunny position, and require a good deal of water while growing. The corms of the spring-flowering species may be lifted and dried off for the winter. Propagation is by seeds or offsets.

Literature: Roux, J.P., Studies in the genus Watsonia Miller, *Journal of South African Botany* **46**: 365–78 (1980).

Leaves. Up to 90 cm: **1, 2, 3, 4, 5, 6, 7, 8, 9, 11**; 90 cm or more: **10**.
Stem. Up to 90 cm: **2, 5, 7, 8, 9**; 90 cm or more: **1, 3, 4, 5, 6, 10, 11, 12**.
Spike. Simple: **2, 7, 9**; branched: **1, 2, 3, 4, 5, 6, 8, 9, 10, 11, 12**.
Perianth. White: **7**; purplish red or lilac: **6, 11**; pink, red or orange-red: **1, 2, 3, 4, 5, 7, 8, 9, 10, 12**. Up to 6 cm: **2, 3, 4, 6, 7, 8**; 6 cm or more: **1, 4, 5, 8, 9, 10, 11, 12**.
Perianth-tube. Up to 4 cm: **2, 3, 4, 6, 7, 8**; 4 cm or more: **1, 5, 8, 9, 10, 11, 12**.

1a. Plant usually bearing bulbils in the axils of the leaves and bracts after flowering **1. bulbillifera**
b. Plant without bulbils 2
2a. Perianth to 4.5 cm, its tube to 2 cm 3
b. Perianth 4.5 cm or more, its tube 2 cm or more 4
3a. Perianth-tube 1.4–2 cm; bracts 2–3 cm **2. galpinii**
b. Perianth-tube to 1.3 cm; bracts 1–2 cm **3. marginata**

4a. Perianth-tube almost as long as the lobes **4. pyramidata**
b. Perianth-tube longer than the lobes 5
5a. Perianth-lobes up to 2.5 cm 6
b. Perianth-lobes more than 2.5 cm 10
6a. Perianth-tube 5 cm or more **5. beatricis**
b. Perianth-tube less than 5 cm 7
7a. Stem 1–1.5 m **6. wilmaniae**
b. Stem at most 90 cm 8
8a. Leaves 60–90 cm; spike compact **7. densiflora**
b. Leaves 15–60 cm; spike loose 9
9a. Leaves to 1.2 cm broad; spike branched **8. fulgens**
b. Leaves 1.8–2.5 cm broad; spike simple **9. meriana**
10a. Leaves more than 75 cm; perianth-tube to 4.3 cm **10. tabularis**
b. Leaves to 75 cm; perianth-tube 5 cm or more 11
11a. Basal leaves 4, stem leaves usually 5; perianth lilac-purple **11. wordsworthiana**
b. Basal leaves usually 8, stem leaf 1; perianth orange-red **12. fourcadei**

1. W. bulbillifera Mathews & Bolus. Illustration: *Flowering plants of South Africa* **19**: t. 726 (1939).

Basal leaves 5 or 6, to 60 × 6 cm. Bulbils present in the axils of the stem-leaves and bracts after flowering. Stem 1.2–1.5 m. Flowers few, in loose, branched spikes. Bracts to 2.2 cm, shorter than the internodes. Perianth brick red, pale outside, longitudinally striped with white inside; tube to 4.3 cm, lobes to 2.4 × 1.3 cm, the outer 3 oblong, the inner 3 obovate. *South Africa (Cape Province).* H5–G1. Spring–summer.

2. W. galpinii Bolus. Illustration: *Flowering plants of South Africa* **2**: t. 45 (1922).

Leaves 4–8, all basal, to 35 × 1.5 cm. Stem to 75 cm, occasionally branched. Flowers many in a fairly dense spike. Bracts 2–3 cm, longer than the internodes. Perianth bright red, tube to 2 cm, lobes to 1.5 × 1 cm, oblong to oblong-ovate. *South Africa (Cape Province).* H5–G1. Summer–autumn.

3. W. marginata (Ecklon) Ker Gawler. Illustration: *Botanical Magazine*, 608 (1802); *Flowering plants of South Africa* **19**: t. 748 (1939).

Basal leaves 3–4, 45–80 cm × 4–5 mm, stem-leaves 3. Stem 1.5–1.8 m. Flowers many, in a dense, branched spike. Bracts 1–2 cm, longer than the internodes. Perianth rose-pink, tube *c.* 1.3 cm, lobes 2–3 cm, ovate-oblong. *South Africa (Cape Province).* H5–G1. Spring–summer.

4. W. pyramidata (Andrews) Stapf (*W. rosea* Ker Gawler). Illustration: *Botanical Magazine*, 1072 (1807); *Flowering plants of South Africa* **25**: t. 974 (1945).

Basal leaves 5, to 73 × 2–2.5 cm, stem-leaves 6. Stem 1.2–1.6 m. Flowers many in a loose, branched spike. Bracts *c.* 1 cm, shorter than the internodes. Perianth rose-pink, tube 3–3.5 cm, lobes 3–3.5 cm, elliptic. *South Africa (Cape Province).* H5–G1. Summer.

5. W. beatricis Mathews & Bolus. Illustration: *Botanical Magazine*, 9139 (1928); *Flowering plants of South Africa* **19**: t. 744 (1939); *Journal of the Royal Horticultural Society* **75**: f. 229 (1950); Gledhill, Eastern Cape sandveld flowers, t. 17 (1977).

Basal leaves 3, to 75 × 3 cm, stem-leaves about 8. Stem to 90 cm, simple. Flowers many in a dense, branched spike. Bracts 1.5–3.5 cm, equal to or longer than the internodes. Perianth deep orange-red, tube *c.* 5 cm, lobes to 2.5 × 1–1.6 cm, oblong to ovate-oblong. *South Africa (Cape Province).* H5–G1. Summer–autumn.

6. W. wilmaniae Mathews & Bolus. Illustration: *Flowering plants of South Africa* **19**: t. 743 (1939).

Basal leaves 6, to 50 × 4 cm. Stem to 1.5 m. Flowers many in a fairly dense, 4-branched spike. Bracts to 2 cm, slightly longer than the internodes. Perianth purplish pink with darker stripes, tube to 4 cm, lobes to 2 × 1 cm, obovate. *South Africa (Cape Province).* H5–G1. Summer–autumn.

7. W. densiflora Baker. Illustration: *Botanical Magazine*, 6400 (1878).

Basal leaves 60–90 × 1.2–1.8 cm, stem-leaves present. Stem 60–90 cm. Flowers many in a dense, simple spike. Bracts 1.2–2.4 cm, longer than the internodes. Perianth rose-pink or sometimes whitish, tube *c.* 2.5 cm, lobes *c.* 1.8 cm, oblong. *South Africa (Transvaal, Natal, Orange Free State).* H5–G1. Summer–autumn.

8. W. fulgens (Andrews) Persoon (*W. angusta* Ker Gawler). Illustration: *Botanical Magazine*, 600 (1802).

Basal leaves several, 30–60 × 1.7–2.5 cm, stem-leaves present. Stem 60–90 cm, usually branched. Flowers many, in a loose, branched spike. Bracts to 2.5 cm, shorter than or sometimes slightly longer than the internodes. Perianth scarlet, tube 3.7–5 cm, sharply bent, lobes 1.8–2 cm, oblanceolate. *South Africa (Cape Province).* H5–G1. Summer.

9. W. meriana (Linnaeus) Miller (*W. ardernei* Sander). Illustration: Gledhill, Eastern Cape sandveld flowers, t. 17 (1977); Gibson, Wild flowers of Natal (inland region), t. 39 (1978).

Basal leaves 3–4, 30–60 cm × 5–12 mm, stem-leaves present. Stem 30–90 cm, simple or branched. Flowers few to several in a loose spike. Bracts *c.* 2.5 cm. Perianth bright rose-red, rarely scarlet or whitish, tube to 5 cm, lobes to 2.5 cm, oblong to oblong-ovate. *South Africa (Cape Province).* H5–G1. Summer.

W. coccinea Herbert. Leaves less than 30 cm, flowers 4–6, in a simple spike, perianth bright crimson. *South Africa (Cape Province).* H5–G1. Summer.

W. humilis Miller. Illustration: *Botanical Magazine*, 631 (1803). Leaves to 30 cm × 2.5–5 mm, flowers 4–6 in a simple spike, perianth bright pink. *South Africa (Cape Province, Natal).* H5–G1. Summer.

These 2 species may well be merely varieties of *W. meriana.*

10. W. tabularis Mathews & Bolus. Illustration: *Flowering plants of South Africa* **5**: 238 (1926).

Basal leaves 4–5, to 100 × 5 cm, stem-leaves 2–3. Stem 1.5–1.8 m. Flowers many on a 2–3-branched spike. Bracts 1–1.8 cm, shorter than the internodes. Perianth deep orange-red on the tube and the outsides of the outer lobes, paler salmon- or rose-pink within and on the outsides of the inner lobes; tube to 4.5 cm, lobes to 3.5 × 2.5 cm, oblong. *South Africa (Cape Province).* H5–G1. Summer.

11. W. wordsworthiana Mathews & Bolus. Illustration: *Flowering plants of South Africa* **19**: t. 742 (1939).

Basal leaves 4, to 60 × 2.5–3.5 cm, stem-leaves 5, to 55 cm. Stem to 1.6 m. Flowers many in a much-branched spike. Bracts 1.8–2.3 cm, shorter than the internodes. Perianth lilac-purple, tube *c.* 5 cm, lobes to 3 × 2 cm, obovate-oblong. *South Africa (Cape Province).* H5–G1. Summer.

12. W. fourcadei Mathews & Bolus. Illustration: *Flowering plants of South Africa* **5**: 235 (1926).

Basal leaves 8, to 60 × 4 cm, stem-leaf 1, to 15 cm. Stem 1–1.5 m. Flowers many in a much-branched spike. Bracts 1–1.5 cm, shorter than the internodes. Perianth orange-red, tube to 5.5 cm, lobes 2.5–3 cm, oblong-linear to obovate. *South Africa (Cape Province).* H5–G1. Summer.

38. BABIANA Ker Gawler

E. Campbell

Perennials with ovoid bulbs that nearly always extend upwards as a long neck. Leaves basal, 2-ranked, with sheathing stalks below ground and sometimes extending above, blades lanceolate, pleated, usually hairy. Flowers in a 2-ranked spike borne on an erect or ascending scape. Perianth bilaterally or radially symmetric, with a short or long tube. Stamens 3, free, arising near the throat of the perianth-tube. Ovary 3-celled with few to several ovules; style simple, stigmas short. Fruit a 3-celled oblong to spherical capsule.

A genus of about 60 species from South and southwest Africa, of which about 5 are in cultivation, usually as pot plants. They are easily grown in well-drained soils in a sunny position.

Literature: Lewis, G.J., The genus Babiana, *Journal of South African Botany*, Supplementary volume **3**: 1–149 (1960); Goldblatt, P. & Manning, J.C., A revision of the southern African genus Babiana Iridaceae: Crocoideae. *Strelitzia* **18**: 1–98 (2007).

1a. Perianth-tube more than 6 cm **1. tubulosa**
b. Perianth-tube less than 5.5 cm 2
2a. Flowers radially symmetric with equal or almost equal perianth-lobes; ovary hairy or rough 3
b. Perianth bilaterally symmetric, the 3 upper perianth-lobes longer than the others, or, if all lobes are more or less equal, then ovary hairless 4
3a. Perianth-lobes scarlet below, blue above, broad; stigmas flattened, 5 mm or more **2. rubro-cyanea**
b. Perianth-lobes various but not as above; stigmas not flattened, to 3 mm **3. stricta**
4a. Flowers 3.3–5 cm; ovary hairless **4. sambucina**
b. Flowers 5–8.5 cm; ovary hairy **5. plicata**

1. B. tubulosa (Burman) Ker Gawler (*B. tubata* (Jacquin) Sweet; *B. tubiflora* Linnaeus filius). Illustration: *Andrews' Botanical Repository* **1**: 5 (1797).

Leaves 6–8, blades 4–15 (rarely to 20) cm × 4–22 mm. Scape ascending, with 1–3 branches, below ground 7–12 cm, the aerial part short. Flowers 6–12 in a dense spike. Perianth cream with red markings on the lower 3 lobes, the tube flushed with purple outside and the outer lobes nearly always with a reddish median line outside; tube 6.5–9 cm; upper 3 lobes 3–3.5 cm × 8–14 mm, lower lobes 1.8–2.3 cm × 5–9 mm. *South Africa.* H5–G1. Spring.

2. B. rubro-cyanea (Jacquin) Ker Gawler (*B. stricta* (Aiton) Ker Gawler *var.* rubro-cyanea (Ker Gawler) Baker; *B. rubrocaerulea* Reichenbach). Illustration: *Botanical Magazine*, 410 (1798); Brickell (ed.), RHS A–Z encyclopedia of garden plants, 160 (2003).

Leaves 5–7, blades 6–15 cm × 6–20 mm. Scape ascending, 5–20 cm. Flowers 5–10 in a lax to fairly dense spike. Perianth-tube 1.5–2 cm; lobes red in the lower part, blue in the upper, 2–2.4 × 1–1.4 cm. Ovary downy or silky; stigmas 5 mm or more, flattened. *South Africa (Cape Province).* H5–G1. Spring.

3. B. stricta (Aiton) Ker Gawler. Illustration: *Botanical Magazine*, 621 (1802); Brickell (ed.), RHS A–Z encyclopedia of garden plants, 160 (2003).

Leaves 6–8, blade 4–12 cm × 6–12 mm. Scape erect, 10–20 cm or rarely more. Flowers 4–8 in loose spike, sometimes slightly fragrant. Perianth pale cream or dark purple, blue or mauve-blue, sometimes with dark, blood-red centre, rarely entirely sulphur yellow. Perianth-tube 1–1.8 cm; lobes 1.5–2.5 cm × 6–14 mm. Ovary densely hairy or silky; stigmas 2.5–3 mm, not flattened. *South Africa.* H5–G1. Spring.

A variable species divided into 5 varieties by Lewis (see above), differing in flower colour, perianth-tube length and characters of the stem and leaves; it is not known which of these are in cultivation.

4. B. sambucina (Jacquin) Ker Gawler (*B. spathacea* Ker Gawler; *B. stellata* Schlechter). Illustration: *Botanical Magazine*, 638 (1803), 1019 (1807); Hay et al., Dictionary of indoor plants in colour, edn 1, t. 51 (1974).

Leaves 5–6, 4–14 cm × 3–8 mm. Scape rarely extending more than 2 cm above ground, sometimes with 1 or 2 very short branches arising below ground. Flowers 2–6 in a compact spike, fragrant. Perianth pale blue-mauve, violet or purple, with a white mark near the middle on the lower lateral lobes; tube 3–5 cm; lobes all similar, 2–3.5 cm × 5–12 mm. Ovary hairless. *South Africa.* H5–G1. Summer.

A variable species, divided by Lewis into 3 varieties.

5. B. plicata Ker Gawler (*B. disticha* Ker Gawler). Illustration: *Botanical Magazine*, 576 (1802).

Leaves 6–8, 8–12 × 1–2 cm. Scape erect, 7–20 cm. Flowers 4–10 in a dense spike, fragrant. Perianth varying from violet to pale blue or rarely white, the lower half of the lateral lobes usually with a small or large yellow area marked with 2 purple dots at the base; tube 1.8–2.5 cm; the 3 upper lobes slightly larger than the others, 1.5–2.5 cm × 7–10 mm, the outer 3 lobes pointed. Ovary densely hairy or silky. *South Africa.* H5–G1. Summer.

39. CROCOSMIA Planchon

J. Cullen edited by S.G. Knees

Perennial herbs with small, flattened corms and creeping stolons that readily produce new corms. Leaves mostly basal, 2-ranked, conspicuously ribbed, sometimes pleated. Inflorescence usually a panicle of spikes, more rarely a simple spike, overtopping the leaves. Bracts small, scarcely exceeding the ovaries. Perianth with a curved tube, slender throughout or slender at the base and then abruptly expanded, the lobes spreading, all similar or the uppermost slightly longer than the others. Stamens 3, free, borne on the perianth-tube, arching to the upperside of the flower, the anthers either just projecting from the mouth of the perianth-tube, or with very long filaments, the anthers borne at the level of the tips of the perianth-lobes or beyond this. Style somewhat longer than the stamens, 3-fid at the apex. Fruit a capsule opening by 3 splits and containing many seeds.

A genus of about 7 species from southern Africa. Its limits are somewhat uncertain, and, as recognised at present, it contains species that have formerly been placed in *Antholyza, Curtonus, Montbretia* and *Tritonia.* They are relatively easily grown in a well-drained soil in a sunny position, and can be propagated by corms, stolons or seed. Lifting of the corms is not generally advisable. The one commonly grown hybrid (see below) can be invasive in some places.

Literature: Kostelijk, P., Crocosmia in gardens, *The Plantsman* **5**: 246–53 (1984); de Vos, M.P., The African genus Crocosmia Planchon, *Journal of South African Botany* **50**: 463–502 (1984); Goldblatt, P., Manning, J. & Dunlop, G., Crocosmia and Chasmanthe (2004).

1a. Leaves conspicuously pleated, 3–7 cm wide 2
b. Leaves ribbed but not pleated, up to 3 cm wide 3
2a. Ribs of the leaves bearing fine hairs; axis of the spike conspicuously zig-zag; perianth-tube 3–4 cm **1. paniculata**
b. Ribs of the leaves hairless; axis of the spike not zig-zag; perianth-tube less than 3 cm **2. masonorum**
3a. Flowers borne on 2 sides of the spike axis; perianth-tube slender, parallel-sided, the lobes 2–2.6 cm × 5–7 mm, spreading from its apex; stamens projecting beyond the perianth-tube **4. aurea**
b. Spikes 1-sided; perianth-tube slender at the base then abruptly expanded, the lobes 8–13 × 5–6 mm, not very widely spreading; only the anthers projecting from the mouth of the perianth-tube **3. pottsii**

1. C. paniculata (Klatt) Goldblatt (*Antholyza paniculata* Klatt; *Curtonus paniculatus* (Klatt) N.E. Brown). Illustration: Eliovson, Wild flowers of southern Africa, edn 5, xix (1955); Rix & Phillips, The bulb book, 169 (1981); Goldblatt et al., Crocosmia and Chasmanthe, pl. 9 (2004).

Plant to 1 m. Leaves to 90 × 6 cm, tapered to the base and apex, conspicuously pleated, with fine hairs on all the veins. Flowering stem erect. Inflorescence a panicle of spikes, the branches ascending to erect, each spike with its axis conspicuously zig-zag, especially towards the base. Flowers borne on 2 sides of the spike axis, facing outwards. Perianth 4.5–6 cm, orange, the tube curved, slender at the base then abruptly swollen, the lobes spreading, 1–1.5 cm. Stamens slightly projecting beyond the perianth-lobes. *South Africa.* H5–G1. Summer.

This species has been hybridised with *C. masonorum* and some of these hybrids are currently available. It has also been hybridised with *C. aurea* resulting in **C. × crocosmoides** (Leichtlin ex J.N. Gerard) Goldblatt (*Antholyza × crocosmoides* Leichtlin ex J.N. Gerard; *C. × latifolia* N.E. Brown). Despite the name being widely seen in catalogues Goldblatt believes this hybrid to be rare in cultivation.

2. C. masonorum (Bolus) N.E. Brown (*Tritonia masonorum* Bolus). Illustration: Synge, Collins guide to bulbs, edn 1, t. iv (1961); *Horticulture* **41**: 20 (1963); *The Garden* **101**: 463 (1976); Brickell (ed.), RHS A–Z encyclopedia of garden plants, 323 (2003); Goldblatt et al., Crocosmia and Chasmanthe, pl. 7 (2004).

Plant 1–1.25 m. Leaves to 1 m × 7 cm, tapered to the base and apex, conspicuously pleated, hairless. Flowering stem erect, inflorescence a simple spike or panicle of spikes, each of which arches to the horizontal; spike axis straight. Flowers borne in 2 rows on the upper side of the spike, so that they face upwards. Perianth 4–5 cm, reddish orange, the tube slender at the base then abruptly expanded, the lobes widely spreading. Stamens projecting beyond the perianth-lobes. *South Africa (Transkei).* H5–G1. Summer–autumn.

3. C. pottsii (Baker) N.E. Brown (*Tritonia pottsii* Baker; *Montbretia pottsii* (Baker) Baker). Illustration: *Botanical Magazine*, 6722 (1833); Goldblatt et al., Crocosmia and Chasmanthe, pl. 11 (2004).

Plant to 90 cm. Leaves to 90 cm × 8–16 mm, long-tapered to the apex, somewhat tapered to the base, ribbed but not pleated, hairless. Inflorescence a panicle of spikes (rarely a simple spike), the branches erect, not zig-zag. Flowers numerous, borne on 1 side of the spike, facing outwards or occasionally upwards. Perianth 2.5–3.5 cm, orange to scarlet, the tube slender at the base then abruptly expanded, the lobes not widely spreading, 8–13 × 5–6 mm. Anthers borne at the mouth of the perianth-tube or just beyond it. *South Africa.* H5–G1. Summer.

4. C. aurea (Hooker) Planchon (*Tritonia aurea* Hooker). Illustration: *Botanical Magazine*, 4335 (1847); Eliovson, Wild flowers of southern Africa, edn 5, xix (1955); Goldblatt et al., Crocosmia and Chasmanthe, pl. 5 (2004).

Plant to 1 m. Leaves to 90 × 1–2.2 cm, tapered to the apex, somewhat tapered to the base, ribbed but not pleated, hairless. Inflorescence a single, erect spike or a panicle of spikes with erect or ascending branches. Flowers borne on 2 sides of the straight spike axis, facing outwards. Perianth 3–5 cm, orange or yellowish, the tube slender and parallel-sided throughout, the lobes 2–2.6 cm × 5–7 mm, very widely spreading. Stamens projecting beyond the perianth-lobes. *South Africa.* H5–G1. Spring–early summer.

A variable species; a variant with a dark spot at the base of each perianth-lobe is sometimes grown as 'Maculata'.

C. × crocosmiiflora (Burbidge & Dean) N.E. Brown (*Montbretia crocosmiiflora* Lemoine). Illustration: *Flowering plants of South Africa* **4**: t. 152 (1924); Goldblatt et al., Crocosmia and Chasmanthe, pl. 12 (2004). This is the hybrid between *C. aurea* and *C. pottsii* and is more widely grown than either of its parents. It resembles *C. aurea* in general, but the flowers have a tube that is slightly widened above, the lobes do not spread as widely, and the stamens are about the same length as the perianth-lobes or shorter. The hybrid is very variable, and many cultivars have been selected, named and illustrated. They are often invasive, and require control.

40. MELASPHAERULA Ker Gawler

J.C.M. Alexander

Corm to 1.5 cm wide with hard black tunic. Basal leaves 6 or 7 in 2 ranks, 10–30 × *c.* 1.5 cm, linear to narrowly triangular, sheathing at base; midrib pale. Stem 30–75 cm, branched, slender, wiry, twisted and drooping. Flowers to 1.8 cm wide, stalkless in loose terminal and lateral spikes, 3–7 per spike. Bracts with chaffy margins. Perianth radially symmetric, with a very short tube; lobes lanceolate, acuminate, white to greenish yellow, sometimes streaked with purple; the inner 3 broader. Stamens arising at lobe-bases; filaments free. Style 3-branched, arched under inner perianth-lobes. Ovary sharply 3-angled.

A genus of a single species. Originally considered part of *Gladiolus*. Cultivation as for *Ixia* (p. 289).

1. M. ramosa (Linnaeus) N.E. Brown (*M. graminea* (Linnaeus filius) Ker Gawler; *M. parviflora* Sweet). Illustration: *Botanical Magazine*, 615 (1803); Rice & Compton, Wild flowers of the Cape of Good Hope, pl. 188 (1951).

South Africa. H5–G1. Spring.

41. TRITONIA Ker Gawler

J. Cullen

Perennial herbs with ovoid to spherical corms with papery or fibrous tunics. Leaves mostly basal, 2-ranked, linear, with false midribs. Inflorescence a simple spike. Bracts exceeding the ovaries. Perianth with a long or short tube gradually widening from the base, or cylindric, the lobes spreading or widely spreading, all more or less equal. Stamens 3, attached to the perianth-tube, filaments free, either curving variously so that the anthers are distant, or parallel, with the anthers close or touching and all lying on the lower side of the flower. Style exceeding the stamens, stigma 3-fid. Capsule ellipsoid, membranous, opening by 3 flaps and containing many seeds.

A South African genus of about 30 species. Cultivation as for *Gladiolus* (p. 298).

Literature: de Vos, M.P., The African genus Tritonia: part 1, *Journal of South African Botany* **48**: 105–63 (1982).

1a. Perianth orange or pinkish orange, the lobes considerably longer than the tube; stamens variously curved, the anthers distant **1. crocata**
b. Perianth pink, the lobes about as long as the tube; stamens with parallel filaments curved to the lower side of the flower, the anthers close or touching **2. disticha** subsp. **rubrolucens**

1. T. crocata (Linnaeus) Ker Gawler (*T. hyalina* (Linnaeus filius) Baker; *T. fenestrata* (Jacquin) Ker Gawler). Illustration: *Botanical Magazine*, 184 (1793), 704 (1804); Eliovson, Wild flowers of southern Africa, edn 6, 83 (1980); *Journal of South African Botany* **48**: 133 (1982).

Plant 25–50 cm. Corm ovoid or spherical, 1.5–3 cm. Leaves 5–30 cm × 4–15 mm, spreading, tapered to the apex and slightly so to the base. Spike with 6–10 flowers borne in 1 or 2 rows. Perianth usually bright orange, sometimes pinkish orange, the lower segments sometimes with a yellowish stripe at the base, tube 8–15 mm, lobes 2–2.8 cm, very rounded at the apex, the claws with translucent margins. Stamens curved and variously spreading, the anthers distant. *South Africa (Cape Province)*. H5–G1. Spring–early summer.

Variable in flower colour; there is, however, no reason for separating *T. hyalina* and *T. crocata*.

2. T. disticha (Klatt) Baker subsp. **rubrolucens** Foster (*T. rosea* Klatt). Illustration: *Botanical Magazine*, 7280 (1893).

Plant to 60 cm, though often less. Corm ovoid, *c.* 2.5 cm. Leaves to 55 cm × 8–15 mm, spreading, tapered to the apex, slightly so to the base. Spike with *c.* 8 flowers, somewhat 1-sided. Perianth pink, tube to *c.* 1.2 cm, lobes about the same length, obtusely pointed, the 3 lower with a yellow stripe at the base, claws without translucent margins. Stamens just projecting from the perianth-tube, the anthers close or touching, on the lower side of the flower. *South Africa (Cape Province, Natal)*. H5–G1. Summer–autumn.

42. SPARAXIS Ker

J. Cullen

Perennial herbs with corms with papery or fibrous tunics, the stems sometimes bearing small corms after flowering. Stems erect, rigid. Leaves mostly basal, in a fan, with several veins and a conspicuous false midrib. Spike with few flowers; bracts large, entire or raggedly toothed near the apex. Perianth radially symmetric; tube short; lobes equal. Stamens 3, symmetrically or asymmetrically arranged; filaments free, anthers sometimes coiled spirally. Style exceeding the stamens, the stigma 3-fid, the tips of the branches sometimes expanded. Capsule spherical, firm, containing several large, smooth and often shiny seeds.

A genus of 6 species, all from the Cape Province of South Africa. Following Goldblatt (reference below), *Streptanthera* Sweet is included in it. Cultivation as for *Ixia* (p. 289).

Literature: Goldblatt, P., The genus Sparaxis, *Journal of South African Botany* **35**: 219–52 (1969); The species of Sparaxis and their geography, *Veld & Flora* **65**: 7–9 (1978).

1a. Stamens symmetrically arranged, style straight; bracts entire, more than 1 cm 2
b. Stamens asymmetric, lying to the upper side of the flower, the style curved to the lower; bracts raggedly toothed, the solid part less than 1 cm 3
2a. Anthers coiled spirally, purple **1. elegans**
b. Anthers straight, yellow **2. tricolor**
3a. Stems producing many obvious, small corms after flowering, usually branched and bearing 1 leaf **4. bulbifera**
b. Stems not producing small corms, or, if so, these small and not obvious, usually unbranched and without leaves **3. grandiflora**

1. S. elegans (Sweet) Goldblatt (*Streptanthera elegans* Sweet; *Streptanthera cuprea* Sweet). Illustration: *Veld & Flora* **65** (1): cover (1978).

Corm 1–1.7 cm in diameter. Stem 10–30 cm, bearing small corms at the lowest nodes. Leaves tapered to the apex, 8–25 cm × 5–14 mm. Spike with 1–5 flowers; bracts to 2.2 cm, entire. Perianth-tube yellow, narrow at the base then widening, 6–8 mm; lobes ovate, obtuse, 1.8–2.2 × 1.4–1.7 cm, reddish orange fading to pink, rarely white, usually with a violet band at the base. Stamens symmetric, filaments yellow, 5–6 mm, anthers dark purple, coiled around the style, shorter than the filaments. Tips of the style-branches expanded. *South Africa (Cape Province)*. H5–G1. Spring.

2. S. tricolor (Schneevogt) Ker (*Ixia tricolor* Schneevogt). Illustration: *Botanical Magazine*, 381 (1797), 1482 (1812); Eliovson, Wild flowers of southern Africa, edn 5, 83 (1955); Brickell (ed.), RHS A–Z encyclopedia of garden plants, 997 (2003).

Corm 1–2lcm in diameter. Stem 10–40lcm. Leaves tapered to the apex, 5–30 × 1–2 cm. Spike loose with 2–5 flowers; bracts 2.5–3 cm, entire or slightly toothed at the tip. Perianth-tube *c.* 8 mm, yellow, narrow at the base, expanded above; lobes lanceolate, 2.5–3.5 × *c.* 1 cm, orange, red or purple, marked with black to dark red at the base. Stamens symmetric, filaments 6–7 mm, anthers 8–9 mm, yellow or white, straight. *South Africa (Cape Province)*. H5–G1. Spring.

A striking species, which has hybridised with others to produce a range of hybrids with variously coloured flowers, generally showing a 3-coloured effect.

3. S. bulbifera (Linnaeus) Ker (*Ixia bulbifera* Linnaeus). Illustration: *Flowering plants of South Africa* **9**: t. 338 (1929); *Veld & Flora* **65**(1): cover (1978).

Corms 9–16 mm in diameter. Stems 5–50 cm, usually with 1 branch and 1 leaf, bearing numerous, obvious corms in the leaf-axil after flowering. Leaves to 40 cm × 4–10 mm. Spike loose with 1–6 flowers; bracts deeply and raggedly toothed, the solid part less than 1 cm. Perianth-tube 1.4–1.6 cm, yellow, green outside at the base; lobes 2.5–2.8 cm, cream or white inside, the outside similar or with purple streaks, rarely entirely purple. Stamens asymmetric, filaments white, 7–8 mm, anthers white, curved,

7–8 mm. *South Africa (Cape Province).* H5–G1. Spring.

4. S. grandiflora (Delaroche) Ker (*Ixia grandiflora* Delaroche).

Corms 6–15 mm. Stem erect, 8–45 cm, bearing a few small corms after flowering. Leaves 3–30 cm × 4–13 mm, tapered above. Spike loose, with 1–6 flowers; bracts deeply and raggedly toothed, the solid part less than 1 cm. Perianth-tube 1–1.4 cm, yellow, purple or blackish; lobes variously coloured, 2.4–3 × 1.2–1.6 cm. Stamens asymmetric, filaments white or yellow, 7–9 mm, anthers white or yellow, a little longer than the filaments. *South Africa (Cape Province).* H5–G1. Spring.

A variable species; Goldblatt recognises 4 subspecies, 2 of which have been grown:

Subsp. **grandiflora**. Illustration: *Botanical Magazine*, 541 (1804); *Flowering plants of South Africa* **2**: t. 60 (1922). Perianth deep reddish purple, lobes often with a white base and whitish margins.

Subsp. **fimbriata** (Lamarck) Goldblatt (*Ixia fimbriata* Lamarck; *I. liliago* de Candolle; *S. grandiflora var.* liliago (de Candolle) Ker). Illustration: *Botanical Magazine*, 779 (1804); *Edwards's Botanical Register* **3**: t. 258 (1818); *Veld & Flora* **65** (1): cover (1978). Perianth-lobes narrower, the whole perianth creamy white or pale yellow, each lobe with a purple stripe outside and deep purple blotches at the base inside.

43. SYNNOTIA Sweet

J. Cullen

Perennial herbs with corms with fibrous coats. Stems erect, usually with 1–3 branches. Leaves 7–10, mostly basal, 2-ranked, each with several veins but without a conspicuous false midrib. Inflorescence a spike with few flowers; bracts membranous or brownish, irregularly toothed or torn at the apex, concealing the ovaries. Perianth bilaterally symmetric, the tube cylindric and narrow for most of its length, broadening above and curved where it broadens; lobes 6, the uppermost the largest, erect and hood-like over the stamens and stigmas; the 2 upper lateral lobes spreading; the 3 lower lobes are united to each other beyond the mouth of the tube, forming a lower lip, and these lobes are themselves deflexed. Stamens 3, free. Stigmas 3. Fruit a capsule with many spherical or angled seeds.

A genus of five species from the Cape Province of South Africa. The genus name was originally spelled *'Synnetia'* (Sweet, British flower garden **2**: t. 150, 1825–7), but Sweet himself corrected it to *Synnotia* at the end of the text to t. 154 of the same volume. Cultivation as for *Ixia* (p. 289).

Literature: Lewis, G.J., A revision of the genus Synnotia, *Annals of the South African Museum* **40**: 137–51 (1956).

1. S. variegata Sweet. Illustration: Sweet, British flower garden **2**: t. 150 (1825–7); Rice & Compton, Wild flowers of the Cape of Good Hope, t. 194 (1951); le Roux & Schelpe, Namaqualand and Clanwilliam: South African wild flower guide **1**: 55 (1981).

Corm ovoid or spherical, 1–2.2 cm in diameter. Stem 8–40 cm. Leaves 7–8, 2–15 cm × 5–20 mm, oblong, acute. Spike with 2–7 flowers; bracts 2–2.5 cm. Perianth lavender to deep purple with yellow stripes on the lower lip and in the tube (rarely these areas whitish), uppermost lobe 2–3 cm, lateral lobes 1.5–2 cm. *South Africa (Cape Province).* H5–G1. Spring.

44. GLADIOLUS Linnaeus

P.P. Yeo

Herbaceous, usually with a corm, often producing cormlets at base or on stolons. Lowest leaves reduced to sheaths, main leaves 2-ranked, usually equitant, linear to narrowly lanceolate, upper leaves much shorter, stem-clasping. Flowering stem terminating the leafy shoot or arising separately, simple or sparingly branched; flowers bisexual, stalkless, each with a bract and bracteole. Perianth in cultivated species bilaterally symmetric, with a tube at the base. Stamens arising on the tube, arching to the upper side of perianth. Style arching like the stamens and as long, 3-branched near apex.

About 180 species (including *Acidanthera* Hochstetter) in Africa, Madagascar and the Mediterranean Region, extending to N Europe. Many species have been cultivated but it is unlikely that more than 30 are to be found in Europe today. The Mediterranean and W Asian species are relatively hardy, being dormant in winter and flowering in early summer; they need warm conditions, and in N Europe some need to be grown in a bulb frame. The tropical African and South African summer rainfall species are mostly not hardy in most of Europe, and must be planted out in late spring for summer and early autumn flowering. The South African winter rainfall species are suited to the Mediterranean and N European Atlantic climate, but elsewhere must be grown in frost-free glasshouses.

There is a vast number of cultivars within *G.* × *hortulanus*, the florist's and border gladiolus. Being derived mainly from summer rainfall species, these are treated in the same way. There are also many hybrids (mostly not described here) less remotely derived from the South African species; their cultural treatment depends on the their parentage.

Literature: Lewis, G.J., et al., Gladiolus, a revision of the South African species, *Journal of South African Botany*, Supplementary volume **10** (1972); Hamilton, A.P., A history of the garden gladiolus, *The Garden* **101**: 424–8 (1976).

Length of flower includes the tube; recurved lobes must be straightened for measurement. The flower markings described in the informal key are not repeated in the descriptions of the species.

Dormancy and flowering. Dormant winter, flowering spring, early summer: **1, 2, 3, 4, 5**; dormant (usually kept in storage) winter, flowering spring, summer, autumn: **6, 7, 8, 9, 10, 11**; leafy in winter, flowering spring to summer: **12, 13, 14, 15, 16, 17**; leafy in winter, flowering autumn (without leaves): **12**.

Leaf width. Usually not more than 1 cm: **1, 3, 4, 13, 14, 15, 16, 17**; usually more than 1 cm: **2, 5, 6, 7, 8, 9, 10, 11, 12, 13**.

Upper middle perianth-lobe. Narrow and arched: **16**.

Perianth-lobes. Wavy: **6, 12, 17**.

Flower scent production. At night: **14, 17**; during day: **16**.

Anther length in relation to filament length. Longer: **1, 2**; equal or slightly shorter: **3, 4, 5, 11, 14, 17**; much shorter: **6, 7, 8, 9, 10, 12, 14, 15, 16**.

Principal colour of flowers. White: **6, 11, 12, 13, 14, 15, 16**; pale pink to violet: **6, 10, 12, 13, 15**; purplish pink to purplish red: **1, 2, 3, 4, 5, 6, 13**; red: **6, 7, 8, 9**; yellow: **6, 9, 10, 14, 17**; brownish: **9, 16**; greenish: **9, 12, 16**.

Flower markings. Three lower perianth-lobes, or 2 of them, with a fusiform white stripe bordered by a darker shade of the principal colour: **1, 2, 3, 4, 5, 13**.

Flower length, including tube. Not more than 5.5 cm: **1, 2, 3, 4, 5, 10, 16**.

1a. Corm more than 4 cm diameter; widest leaves at least 2 cm wide; flowers 12–20, crowded **6. × hortulanus**
b. Corm and leaves narrower; flowers fewer; if these characters do not apply the flowers are not crowded 2
2a. Anthers longer than filaments 3
b. Anthers equalling or shorter than filaments 4
3a. Flowers closely set in 1 row, dark dull violet; perianth-tube curved **1. atroviolaceus**
b. Flowers loosely set in 2 rows, purplish pink to purplish red; perianth-tube nearly straight **2. italicus**
4a. Flowers bright pinkish red to reddish purple; the 3 lower perianth-lobes with a dark-bordered cream-coloured central stripe; plant dormant in winter 5
b. Flowers differently coloured or, if as described above, plant growing in winter 7
5a. Flowers in 1 row **3. imbricatus**
b. Flowers in 2 rows 6
6a. Plant usually more than 50 cm; inflorescence often branched; flowers 10 or more **5. communis**
b. Plant usually less than 50 cm; inflorescence unbranched; flowers 10 or fewer **4. illyricus**
7a. Upper middle perianth-lobe narrow, nearly parallel-sided, arched; perianth-tube 1–1.5 cm **16. orchidiflorus**
b. Upper middle lobe broader, not nearly parallel-sided, not distinctly arched; perianth-tube more than 1.5 cm 8
8a. Main leaves 4-winged **14. tristis**
b. Main leaves flat or merely ribbed 9
9a. Main leaves to 3 mm wide (perhaps to 6 mm in cultivation) **17. liliaceus**
b. Main leaves mostly more than 5 mm wide 10
10a. Leaves 5–12, some of them 3–4 cm wide; flowers obliquely trumpet-shaped with strongly overlapping, abruptly pointed lobes, of which the lower have recurved tips **9. natalensis**
b. Leaves not more than 9, not more than 2.8 cm wide; flowers differently shaped 11
11a. Perianth-tube more than 4 cm 12
b. Perianth-tube not more than 4 cm 13
12a. Perianth-tube more than 7 cm, limb white with red marks in throat **11. callianthus**
b. Perianth-tube not more than 7 cm, limb white or pink with pale-centred purplish or red marks on lower lobes near the middle **12. undulatus**
13a. Perianth yellow or white to pink or violet 14
b. Perianth mainly brilliant red or intense purplish red 17
14a. Perianth-lobes abruptly spreading from the top of the tube **15. carneus**
b. Perianth-lobes gradually spreading 15
15a. Flowers 4–5.5 cm, limb bell-like with abruptly tapered lobes **10. papilio**
b. Flowers about 6.5–10 cm, trumpet-like with gradually tapered lobes 16
16a. Main leaves not emerging until flowering has begun, and then separately from the flowering stems **12. undulatus**
b. Main leaves emerging before flowering; flowering stems emerging from leafy shoots **13. × colvillei**
17a. Leaves 5–9; flowers pure bright red 18
b. Leaves usually 3 or 4; flowers purplish red **13. × colvillei**
18a. Perianth-lobes all about equally divergent, the lower 3 with an elongate central pale mark **7. cardinalis**
b. Upper middle perianth-lobes projecting forward as a hood, 3 or all of the others spreading widely to form a semicircular lip, the lower 3 with white basal area extending across most of their width **8. saundersii**

1. G. atroviolaceus Boissier. Illustration: Rix & Phillips, The bulb book, 150 (1981).

Plant 30–70 cm, growing in spring. Leaves 3, the lower 30–40 cm × 4–8 mm. Flowers 4–10, 4–4.5 cm, closely set in 1 row, dark dull violet; tube distinctly curved. Anthers longer than filaments. Seeds not winged. *Greece, Turkey, Iraq, Iran.* H2. Early summer.

2. G. italicus Miller (*G. segetum* Ker Gawler). Illustration: *Botanical Magazine*, 719 (1804); Grey-Wilson & Mathew, Bulbs, pl. 30 (1981).

Plant 50–110 cm, growing in spring. Leaves 3–5, the lower 18–50 cm × 5–17 mm. Flowers 6–16, 4–5 cm long, loosely spaced in 2 rows, purplish pink to purplish red; tube nearly straight; lobes scarcely overlapping. Anthers longer than filaments. Seeds not winged. *S Europe to Afghanistan, extending to NW Africa and Canary Isles.* H2. Early summer.

3. G. imbricatus Linnaeus. Illustration: Mathew, The larger bulbs, 65 (1978); Grey-Wilson & Mathew, Bulbs, pl. 30 (1981); Rix & Phillips, The bulb book, 151 (1981).

Plant 30–80 cm, growing in spring. Leaves 2 or 3, the lower 15–35 × 1–1.5 cm. Flowers 4–12, *c.* 3 cm, closely set in 1 row, pinkish red to reddish purple; tube strongly curved near throat; lobes usually overlapping. Anthers not longer than filaments. Seeds winged. *C & E Europe (N to the Baltic states of the Former USSR, naturalised in Finland north almost to the arctic circle), Turkey.* H2. Early summer.

G. palustris Gaudin, also with reddish purple flowers in 1 row, differs in its smaller size (to 50 cm) and loose spike of 2–6 flowers. *C, S & SE Europe.* H2. Early summer.

4. G. illyricus Koch. Illustration: Ross-Craig, Drawings of British plants **29**: pl. 5 (1972); Grey-Wilson & Mathew, Bulbs, pl. 30 (1981).

Plant 25–50 cm, growing in spring. Leaves 4 or 5, the lower 10–40 cm × 4–10 mm. Flowers 3–10, 3.5–5.5 cm, loosely spaced in 2 rows, reddish purple; tube slightly curved; lobes 6–16 mm wide, sometimes overlapping. Anthers not longer than filaments. Seeds winged. *S & W Europe.* H2. Early summer.

5. G. communis Linnaeus.

Plant 50–100 cm, growing in spring. Leaves 3–5, the lower 30–70 cm × 5–25 mm. Stem often with 1–3 branches. Flowers 10–20, 4–5.5 cm, in 2 rows, pink to deep purplish red; tube slightly curved; lobes more or less overlapping. Anthers not longer than the filaments. Seeds winged. Early summer.

Subsp. **communis**. Illustration: *Botanical Magazine*, 86 (1789); Bonnier, Flore complète **11**: pl. 601 (1931); Huxley & Taylor, Flowers of Greece and the Aegean, f. 413 (1977). Plant not more than 80 cm; lower leaves 30–50 cm × 1–15 mm. Flowers rather loosely spaced, usually pink;

lobes 10–20 mm wide. *C & S Europe, NW Africa.* H2.

Subsp. **byzantinus** (Miller) Hamilton (*G. byzantinus* Miller). Illustration: *Botanical Magazine*, 874 (1805); Grey-Wilson & Mathew, Bulbs, pl. 30 (1981). Lower leaves 30–70 × 1–2.5 cm. Flowers rather closely set, usually deep purplish red; lobes 1.5–2.5 cm wide. *S Spain, Sicily, NW Africa.* H3.

The most commonly cultivated European gladiolus; the cultivated variant is slightly different from the wild plant, is nearly sterile and has 90 chromosomes instead of 120.

6. G. × hortulanus Bailey. Illustration: Synge, Collins guide to bulbs, edn 1, pl. 15 (1961).

Corm usually more than 4 cm in diameter. Plant growing in summer. Largest leaves usually 2 cm or more wide. Flowers 12–20, closely set, mostly 6–15 cm wide, variously coloured, in the smaller cultivars more often with definite blotches on the lobes, in the larger more often uniform or bicoloured; lobes broad and strongly overlapping. *Garden origin.* H5. Summer.

This name covers the commonly grown garden and florist's hybrids; although selection has been mainly for large size, smaller cultivars have also been raised (Butterfly, Peacock and Primulinus gladioli); these make it difficult to separate *G. × hortulanus* from the species by means of a key. Nos 7–11, 14 and 15, among others, are parents of *G. × hortulanus*. The hybrid that served as the foundation for the development of the larger-flowered *G. × hortulanus* cultivars was *G. × gandavensis* Van Houtte (*G. natalensis × G. oppositiflorus* Herbert).

7. G. cardinalis Curtis. Illustration: *Botanical Magazine*, 135 (1790); *Flore des serres* **19**: 109 (1873); Bolus et al., A second book of South African flowers, 150 (1936); Lewis, Gladiolus, 24 (1972).

Corm to 3 cm in diameter. Plant 60–115 cm, growing in summer; stem often arched. Leaves 5–9, the lower 40–90 × 1.5–2.8 cm. Flowers *c.* 6, *c.* 8 cm long, closely spaced, indistinctly 2-rowed, bright red, the lower 3 lobes with a central whitish mark; tube 3–4 cm; lobes obtuse, upper slightly hooded, others widely spreading. Anthers much shorter than filaments. *South Africa (SW Cape Province).* H5. Summer.

When crossed with *G. carneus* this species gave rise to **G. × insignis** Paxton, from which was developed, probably by crossing with other species and hybrids, a group of cultivars known as Nanus Hybrids (not *G. nanus* Andrews, which is *Babiana nana* (Andrews) Sprengel).

8. G. saundersii Hooker. Illustration: *Botanical Magazine*, 5873 (1870); *Flowering plants of South Africa* **5**: t. 182 (1925); Lewis, Gladiolus, 24 (1972).

Corm 2.5–4 cm in diameter. Plant 40–90 cm, growing in summer. Leaves 7 or 8, the lower 25–60 cm × 6–26 mm. Flowers 3–8 or rarely up to 12, 6.5–11 cm long, red, the basal halves of lower 3 lobes whitish and speckled with red towards throat, midrib sometimes white to the tip; tube 3–3.8 cm, strongly curved, widened gradually; lobes with recurved tips, the upper hooded, the 3 lower and, usually, the 2 upper laterals, widely spreading and overlapping to form a semicircular lip. Anthers much shorter than filaments. *South Africa.* H5. Summer.

9. G. natalensis (Ecklon) J.D. Hooker (*G. psittacinus* J.D. Hooker; *G. quartinianus* Richard; *G. primulinus* Baker). Illustration: *Botanical Magazine*, 3032 (1830), 5884 (1871), 6202 (1875), 6739 (1884) & 8080 (1906); *Journal of the Royal Horticultural Society* **101**: 427 (1976).

Corms to 5 cm in diameter, sometimes producing stolons. Plant to 1.5 m, growing in spring and summer. Leaves 5–12, some to 60 × 1–3 cm, others to 30 × 1–4 cm. Flowers usually 8–12 cm, up to 25, rather loosely spaced, indistinctly 2-rowed, yellow, yellow streaked with orange-red, orange-red or red with central yellow areas on lobes, often with transitional red streaks or peppering, or coloured in similar combinations of green and brown; tube about half the length of the flower, curved, enlarging gradually; lobes acute, overlapping, diverging gradually, the 3 upper or the upper central forming a hood, the 3 lower with recurved tips, the middle one about twice as wide as the laterals. Anthers much shorter than filaments. *South Africa (E Cape Province) to Ethiopia and Arabia.* H5. Spring–autumn.

A late-flowering, red and yellow form with long stolons, 'Hookeri' (illustration: *Flowering plants of South Africa* **3**: pl. 116, 1923; Lewis, Gladiolus, 33, 1972), is widely cultivated in warm-temperate countries.

10. G. papilio J.D. Hooker (*G. purpureo-auratus* J.D. Hooker). Illustration: *Botanical Magazine*, 5565 (1866), 5944 (1872); *Flore des serres* **19**: 107 (1873); Lewis, Gladiolus, 48 (1972).

Corm 1–3 cm diameter, usually with stolons. Plant 50–90 cm, growing in summer. Leaves 7 or 8, the lower 20–100 cm × 5–20 mm or sometimes to 2.5 cm. Flowers 3–10, 4–5.5 cm, loosely arranged in 2 more or less distinct rows, bright or dull yellow, sometimes tinged with purple outside, or mainly pale pinkish violet; throat sometimes and lower lobes always blotched with brown or purplish red; tube 1.5–2.3 cm, bent near the throat; lobes more or less obtuse, overlapping, forming a usually drooping, bell-shaped limb. Anthers much shorter than filaments. *South Africa (Transkei to Transvaal), Lesotho, Swaziland.* H3. Summer.

11. G. callianthus Marais (*Acidanthera bicolor* Hochstetter; *A. murielae* invalid; *A. tubergenii* invalid). Illustration: *The Garden* **47**: 342 (1895); Synge, Collins guide to bulbs, edn 1, pl. 5, f. 1 (1961); Brickell (ed.), RHS A–Z encyclopedia of garden plants, 481 (2003).

Corms to 2.5 cm in diameter. Plant to 1.1 m, growing in spring and summer. Leaves slightly shorter than the stem, to 2.5 cm wide. Flowers up to 10, to 14 cm long, fragrant, loosely set, opening singly, white, all lobes except the upper with a purplish red rhomboidal mark near the base; tube to 10 cm, slender, curved; limb nodding, with nearly equal, divergent lobes. Anthers rather large. *E Africa (Ethiopia to Malawi).* H5. Autumn.

12. G. undulatus Linnaeus (*G. cuspidatus* Jacquin). Illustration: *Botanical Magazine*, 582, 591 (1802); Bolus et al., A second book of South African flowers, 154 (1936).

Corms to 30 cm in diameter. Plant 30–95 cm, growing in winter. Leaves about 5, 25–75 cm × 5–20 mm. Flowers 4–9, to *c.* 13 cm long, loosely set in 2 rows, greenish white, white, cream or pink, lower lobes with elongated or spade-shaped, pale-centred, red or purple marks; tube 5–7 cm, slender; lobes lanceolate, gradually divergent, with more or less wavy and recurved tips. Anthers much shorter than filaments. *South Africa (SW Cape Province).* H5. Spring–summer.

G. carmineus Wright. Illustration: *Botanical Magazine*, 8068 (1906); Bolus et al., A second book of South African flowers, 149 (1936). Growing in winter.

Main leaves appearing on shoots separate from the flowering shoots and shortly after them. Flowers erect, 6.5–10 cm; tube only 3–4.5 cm, widened upwards; lobes ovate to obovate, not wavy and recurved. *South Africa (SW Cape Province).* H5. Autumn.

13. G. × colvillei Sweet. Illustration: Sweet, British flower garden, **2**: t. 155 (1825–7); Rix & Phillips, The bulb book, 151 (1981).

Corm *c.* 2lcm in diameter. Plant *c.* 50 cm, growing in winter. Leaves 3 or 4, *c.* 25 × 1.2 cm. Flowers *c.* 8, *c.* 7 cm long, closely set in 2 rows, white to purplish red or pale pinkish violet, lower lobes with elongate pale marks, which may be dark-edged; perianth-lobes gradually divergent. Anthers shorter than filaments. *Garden origin.* H3. Early–late summer.

A cross between *G. cardinalis* and *G. tristis*, first raised in England in 1823; a number of cultivars has been named.

14. G. tristis Linnaeus. Illustration: *Botanical Magazine*, 272 (1794), 1098 (1808); *Flowering plants of South Africa* **5**: pl. 175 (1925); Bolus et al., A first book of South African flowers, 138 (1928); Lewis, Gladiolus, 161 (1972); *Journal of the Royal Horticultural Society* **101**: 426 (1976).

Corm to 3 cm diameter. Plant 40–150 cm, more usually 50–70 cm, growing in winter. Leaves 2–4, the lower about as long as the stem, 1.5–5 mm wide, twisted, 4-winged. Flowers 1–8 (vars. **tristis** and **concolor** (Salisbury) Baker), or 7–10, or at most 20 (var. **aestivalis** (Ingram) Lewis), 6–9.5 cm long, loosely set in 1 row, fragrant at night, white to buff or dull yellow (var. **concolor**) and flushed or peppered with violet to purplish brown outside, and sometimes along midribs of upper lobes inside; midribs often greenish (vars. **tristis** and **aestivalis**); tube 4–6 cm, curved above the middle, widened gradually above; limb gradually flared, lobes nearly equal, slightly twisted, gradually tapered to acute, recurved tips. Anthers shorter than filaments. *South Africa (S Cape Province).* H5. Spring–summer.

15. G. carneus Delaroche (*G. blandus* Aiton). Illustration: *Botanical Magazine*, 625, 645, 648 (1803), 1665 (1814) & 8923 (1938) – a small, slender form; Bolus et al., A second book of South African flowers, 153 (1936); Lewis, Gladiolus, 96, 99 (1972).

Corm to 3 cm in diameter. Plant variable, 20–100 cm tall, growing in winter. Leaves up to 5, usually to 30 cm × 5–10 mm but sometimes to 60 × 2 cm. Flowers usually 2–8, to 8 cm long, loosely spaced in 2 rows, white, cream, pale pink or pale pinkish violet, the three lower lobes usually with a reddish or yellowish patch, the throat sometimes blotched with red; tube 2–4 cm; lobes abruptly spreading. Anthers much shorter than filaments. *South Africa (SW Cape Province).* H5. Spring and early summer.

16. G. orchidiflorus Andrews. Illustration: *Botanical Magazine*, 688 (1803); Sweet, British flower garden, **2**: 156 (1825–7); *Flowering plants of South Africa* **4**: pl. 165 (1924); Lewis, Gladiolus, 144 (1972); Rice & Compton, Wild flowers of the Cape of Good Hope, pl. 203, f. 2 (1951); Mason, Western Cape sandveld flowers, 83, f. 4 (1972).

Corm to 2.5 cm in diameter. Plant to 40 cm, growing in winter. Leaves 3–8, the largest about as tall as stem and to 5 mm wide. Stem often branched, flexuous above. Flowers 5–15, 3–5 cm, loosely or closely spaced in 1 row or 2, fragrant, greyish white to greyish green or brownish cream, usually with a purplish sheen, the 3 upper with a brownish purple band, sometimes divided by a greenish stripe, the 3 lower with a basal yellow or yellow-green area bordered with brownish purple; tube 1–1.5 cm long; upper lobe narrow and arched, the others wide and spreading. Stamens lying close under perianth-lobe; filaments several times longer than anthers. *South Africa (W Cape Province to Orange Free State).* H5. Spring.

17. G. liliaceus Houttuyn (*G. grandis* Thunberg). Illustration: *Botanical Magazine*, 1042, lower figure (1807); Marloth, Flora of South Africa **4**: 157 (1915); Batten & Bokelmann, Wild flowers of the eastern Cape Province, pl. 29, f. 7 (1966).

Corm to 2.5 cm diameter. Plant to 60 cm or rarely to 90 cm, growing in winter. Leaves usually 3, the lowest about as tall as stem, 3 mm wide. Flowers 1–3 or rarely 5, 7–11.5 cm long, loosely set in 1 row, fragrant at night, dull yellow, densely streaked or mottled with pinkish brown or purple except on the basal halves of the lower lobes, sometimes red in throat, upper lobes sometimes translucent near base, the colour changing at night to dark purplish violet or violet; tube curved, 3.8–6 cm, enlarged upwards; lobes nearly equal, progressively spreading, tips recurved, wavy and tapered to an acute, obtuse or notched apex. Anthers slightly shorter than filaments. *South Africa (S Cape Province).* H5. Early spring–summer.

45. CHASMANTHE N.E. Brown

J. Cullen

Perennial herbs with large, flattened-spherical corms that have fibrous 'necks'. Leaves mostly basal, in 2 ranks, blade-like, with a false midrib or with 2 or 3 equal primary veins and no false midrib. Inflorescence usually a simple, erect spike, occasionally with 1 erect branch from the base, the flowers facing outwards. Bracts acute, longer than the ovaries. Perianth bilaterally symmetric, conspicuously curved, tubular below, the lower part of the tube very slender and sometimes twisted, broadening into the upper part; lobes 6, the upper much longer than the others and directed forwards, the 2 upper lateral lobes parallel to the uppermost, the 3 lower lobes deflexed. Stamens 3, attached to the perianth-tube where it widens, the anthers borne beneath the upper perianth-lobe. Style slightly longer than the stamens, stigma 3-fid. Capsule spherical, exceeding the bracts, opening by 3 splits and containing many ovoid, brightly coloured seeds.

There are 9 species from South and south-west Africa. Cultivation as for *Gladiolus* (p. 298).

Literature: Goldblatt, P., Manning, J. & Dunlop, G., Crocosmia and Chasmanthe (2004).

1a. Perianth-tube with the narrow base conspicuously twisted, broadening abruptly, the base of the broader part formed by 3 pouches — 2
b. Perianth-tube with the narrow base not twisted, broadening gradually, the base of the broader part without pouches — 3
2a. Spike with up to 7 flowers borne on 1 side — **1. aethiopica**
b. Spike with more than 7 flowers borne in 2 rows — **2. vittigera**
3a. Perianth bright red, the tube 1.8–2.5 cm — **3. caffra**
b. Perianth orange or scarlet and yellow, the tube 3.8–4 cm — **4. floribunda**

1. C. aethiopica (Linnaeus) N.E. Brown (*Antholyza aethiopica* Linnaeus). Illustration:

Goldblatt et al., Crocosmia and Chasmanthe, pl. 1 (2004).

Plant 20–70 cm. Leaves 20–55 × 1–2 cm, linear, tapering to the apex, narrowing slightly to the base, sometimes 1 side narrowing rather abruptly, with a false midrib. Spike usually simple, occasionally with 1 branch; flowers up to 7, borne on 1 side. Perianth scarlet and yellow, the tube *c.* 2.8 cm, its lower part narrow and conspicuously twisted, broadening abruptly into an upper part whose base is formed by 3 pouches; upper lobe *c.* 2.5 cm × 7 mm. *South Africa (Cape Province).* H5–G1. Spring–early summer.

2. C. vittigera (Salisbury) N.E. Brown (*Antholyza vittigera* Salisbury; *A. aethiopica var. vittigera* (Salisbury) Baker). Illustration: *Botanical Magazine*, 1172 (1809).

Very similar to *C. aethiopica* but leaves broader, spike with more than 7 flowers borne in 2 rows on the axis; perianth with dark stripes towards the base. *South Africa (Cape Province).* H5–G1. Spring–early summer.

3. C. caffra (Baker) N.E. Brown (*Antholyza caffra* Baker). Illustration: *Botanical Magazine*, 9470 (1937).

Plant to 50 cm. Leaves long-tapered to the apex, somewhat narrowed to the base, to 35 cm × 6 mm. without a false midrib but with 2 or 3 equally prominent primary veins. Spike usually simple with 12–20 flowers borne in 2 rows. Perianth bright red; tube 1.8–2.5 cm, broadening gradually from a basal part that is not twisted; upper lobe 2.5–3.2 cm. *South Africa (Cape Province, Natal).* H5–G1. Spring–early summer.

4. C. floribunda (Salisbury) N.E. Brown (*Antholyza floribunda* Salisbury; *A. praealta* Redouté). Illustration: *Botanical Magazine*, 561 (1801); Eliovson, Wild flowers of southern Africa, edn 5, xviii (1955); Goldblatt et al., Crocosmia and Chasmanthe, pl. 1 (2004).

Plant to 1 m or more. Leaves 20–65 × 2.5–5.5 cm, tapering to the apex and often abruptly and 1-sidedly to the base, with a conspicuous false midrib. Spike simple with 20–30 or more orange-red or scarlet and yellow flowers borne in 2 rows. Perianth-tube 3.8–4 cm, its basal part not twisted, broadening gradually above; upper lobe to 3.3 cm × 8–9 mm. *South Africa.* H5–G1. Summer.

46. ANOMALESIA N.E.Brown

J. Cullen

Similar to *Chasmanthe* but corms small, producing stolons that end in new corms; leaves with 2–3 equal primary veins (no false midrib); bracts almost as long as the perianth-tubes; perianth-tube with a slender base that is not twisted, broadening gradually above, the 6 lobes very unequal, the uppermost the longest, the 2 upper lateral lobes spreading upwards, above the upper lobe, the 3 lower lobes very small and inconspicuous. Stigmas 3. Fruit a capsule. Seeds flattened, winged.

A genus of 2 species from South Africa. Cultivation as for *Gladiolus* (p. 298).

1. A. cunonia (Linnaeus) N.E. Brown (*Antholyza cunonia* Linnaeus). Illustration: *Botanical Magazine*, 343 (1796).

Plant to 90 cm, though often much less. Leaves linear, long-tapering to the apex, to 30 × 1 cm. Spike with up to 7 flowers borne in 2 rows. Perianth scarlet to dark red, the lowermost lobe greenish, turned inwards, hidden by the others. *South Africa.* H5–G1. Spring.

A. splendens (Sweet) N.E.Brown (*Anisanthus splendens* Sweet; *Antholyza splendens* (Sweet) Steudel). Illustration: Sweet, British flower garden, **1**: t. 84 (1823–5). Leaves broader, perianth-tube greenish, the lobes mostly bright red, the lowermost deflexed downwards and outwards, bright red at the tip. *South Africa.* H5–G1. Spring.

42. JUNCACEAE

Herbs, often rhizomatous, rarely woody. Stems terete in section. Leaves usually basal, spirally arranged, sometimes reduced to sheaths; ptyxis supervolute. Flowers inconspicuous, in cymes, panicles, corymbs or heads, usually bisexual, actinomorphic. Perianth of 6 free segments. Stamens usually 6, pollen in tetrads. Ovary superior, of 3 united carpels; ovules 3–many, axile or parietal. Fruit a capsule (figure 12).

A family of 9 genera and 350 species. Two genera are native to both Europe and North America; species of both are occasionally grown.

1a Leaves hairless; capsule with numerous seeds **1. Juncus**
b . Leaves hairy; capsule with 3 seeds **2. Luzula**

1. JUNCUS Linnaeus

E.C. Nelson

Hairless perennial herbs with rhizomes. Stamens 6 or fewer, opposite the perianth segments and attached to their bases. Capsule with numerous seeds.

Only 1 species is widely cultivated in Europe, but others occur as weeds, especially in damp gardens and garden ponds in western regions. The genus contains about 225 species and is widely distributed. Propagation is by division of the rhizomes.

1. J. effusus Linnaeus. Illustration: Keble Martin, The concise British flora in colour, 87 (1965); Aeschimann et al., Flora alpina **2**:137.1.5 (2004).

Stems 50–150 cm. Leaves cylindric, as long as stems, with continuous white pith. Inflorescence dense or loose, with numerous pale brown flowers. Stamens 3, occasionally 6. *Cosmopolitan.* H1. Summer.

This species is represented in gardens only by cultivars, except when it is a weed. 'Spiralis' has spirally twisted leaves and stems; 'Vittatus' has leaves striped longitudinally with yellow. Both cultivars are suitable for pool margins and bog gardens.

2. LUZULA de Candolle

E.C. Nelson

Hairy perennial herbs with rhizomes and stolons. Leaves flat; leaf-sheath closed. Flowers in cymes, sometimes in dense heads. Capsule with 3 seeds.

Only 1 species is commonly cultivated in Europe, but 2 other native European species, *L. nivea* (Linnaeus) de Candolle and *L. luzuloides* (Lamarck) Dany, with decorative, whitish flowers, are sometimes grown. Propagation is by division of the rhizomes.

1. L. sylvatica (Hudson) Gaudin. Illustration: Grounds, Ornamental grasses, 192, 200 (1979); Aeschimann et al., Flora alpina **2**: 137.2.9 (2004).

Perennial, grass-like herb, with stolons. Leaves 5–30 mm wide, with white hairs. Perianth segments brown. Anthers up to 6 times as long as filaments. *Europe, SW Asia.* H1. Summer.

A plant suitable for growing in woodland gardens, where it will form good ground cover. A cultivar with white leaf margins ('Marginata') is available.

Figure 12. Diagram of some characters used in the identification of Juncaceae and Cyperaceae. 1, inflorescence of *Luzula*. 2, flower of *Luzula*. 3, flower of *Scirpus* (a, perianth bristles). 4, inflorescence of *Cyperus*. 5, spikelet of *Cyperus*. 6, flower of *Cyperus* (b, glume). 7, inflorescence of *Carex* (c, male spike; d, female spike). 8,9, portion of stem and leaf-base of *Carex* (e, leaf-sheath; f, ligule). 10, male flower of *Carex*. (g, glume). 11, female flower of *Carex*.

43. BROMELIACEAE

Herbs, often epiphytic. Leaves leathery, mainly basal, often spiny-margined and/or with elaborate scales covering the whole surface; ptyxis flat or supervolute. Flowers usually in terminal racemes or panicles, with conspicuous bracts, bisexual, usually actinomorphic. Sepals 3, usually small. Petals 3, free or united, usually larger than sepals. Stamens 6, anthers usually versatile. Ovary of 3 united carpels, usually inferior; ovules many, axile. Fruit a berry or capsule.

A family of 46 genera and 2,100 species of mainly epiphytic herbs. Ten genera are native to North America. About 80 species belonging to 17 genera are cultivated. *Ananas* provides the pineapple of commerce.

The inflorescences of the Bromeliaceae are very variable; they are basically spike-like racemes or panicles, but they may be reduced to a solitary flower, or condensed into a head-like structure which may be stalked or stalkless and borne between the leaf-sheaths. Panicles are described as 'bipinnate', when the axis regularly bears branches that are themselves regularly branched.

In cultivation the species flower at irregular intervals, so no flowering times are given in the account below.

Cultural requirements for these plants are of two kinds. On the one hand are the epiphytic species, which differ from most other plants in having specialised root-systems. For most of these, a well-drained medium is required, while the green-leaved species of *Tillandsia* and species of *Vriesea* and *Guzmania* require one that retains a considerable amount of water: *Osmunda* fibre, for example, is most useful. The terrestrial species, on the other hand, are more robust, and can easily be cultivated in general potting composts. However, species of *Aechmea*, *Billbergia* and *Dyckia* can be kept much drier, in a coarse gravel and sand mixture – as long as the tubular rosettes of those species that have them are regularly filled with water.

Literature; Smith, L.B. & Downs, R.J., Bromeliaceae, *Flora Neotropica* **14**: parts 1–3 (1974–9), from which most of the information presented here has been extracted; Rauh, W., Bromeliads for home, garden and greenhouse, English edn. (1979).

1a. Leaf margin entire, without spines or teeth 2
b. Leaf margin with spines or teeth 6
2a. Petals with their upper parts somewhat irregularly placed so that the corolla is not radially symmetric **3. Pitcairnia**
b. Petals with their upper parts regularly placed so that the corolla is radially symmetric 3
3a. Bases (claws) of the petals united into a tube that is longer than the sepals; flowers in many ranks **7. Guzmania**
b. Bases (claws) of the petals free or united into a tube that is shorter than the sepals; flowers usually in 2 ranks 4
4a. Petals with basal scales **6. Vriesea**
b. Petals without basal scales 5
5a. Inflorescence a loose panicle; flowers pendent, white **2. Fosterella**
b. Inflorescence spike-like or reduced to a solitary flower; flowers not pendent, usually coloured **5. Tillandsia**
6a. Scape absent or short and not immediately visible 7
b. Scape well-developed and evident 10
7a. Leaves rigid with stout marginal spines **12. Fascicularia**
b. Leaves flexible, margins toothed 8
8a. Inflorescence not sunk in the centre of the rosette, spherical, held above the leaf-sheaths **9. Cryptanthus**
b. Inflorescence sunk in the centre of the rosette, between the leaf-sheaths 9
9a. Inflorescence with the flowers stalkless, in flattened bundles covered by large bracts **10. Nidularium**
b. Inflorescence various, not as above, flowers stalked **8. Neoregelia**
10a. Inflorescence cone-like, usually crowned with a tuft of sterile, leaf-like bracts 11
b. Inflorescence not as above 12
11a. Inflorescence terminal, 10–25 cm when in fruit; scape bearing bracts **17. Ananas**
b. Inflorescence apparently lateral, 3–7 cm when in fruit; scape without bracts **11. Acanthostachys**
12a. Inflorescence with an involucre 13
b. Inflorescence without an involucre 14
13a. Petals without basal scales **10. Nidularium**
b. Petals with basal scales **13. Canistrum**
14a. Sepals each with a small apical spine, or at least sharply pointed **14. Aechmea**
b. Sepals without spines, not sharply pointed 15
15a. Sepals to 1 cm; petals to 3 cm 16
b. Sepals more than 1 cm; petals more than 3 cm 17
16a. Ovary completely superior **4. Dyckia**
b. Ovary completely inferior **14. Aechmea**
17a. Ovary completely superior to or, at most, half-inferior 18
b. Ovary completely inferior 19
18a. Petal blades twisted together spirally after flowering **1. Puya**
b. Petal blades not twisted together spirally after flowering **3. Pitcairnia**
19a. Petals erect or almost so, corolla radially symmetric **15. Quesnelia**
b. Petals irregularly placed, corolla bilaterally symmetric or petals each recurved in a spiral **16. Billbergia**

1. PUYA Molina

J. Cullen

Perennial, terrestrial herbs, usually with stems. Leaves very leathery, in dense rosettes, with distinct sheaths and spiny-margined blades that are not narrowed towards the base. Inflorescence usually a panicle. Sepals much shorter than the petals. Petals with distinct claw and usually broad blade, the blades coiled together spirally after flowering. Stamens usually somewhat shorter than the petals. Ovary superior or almost entirely so. Fruit a capsule.

A genus of about 170 species from S America. Only a small number is commercially available, and these are rarely seen outside specialist collections. Some of them are particularly prized for their metallic blue or blue-green petals.

1a. Branches of the inflorescence with flowers throughout (rarely flowerless towards the base) 2
b. Branches of the inflorescence flowerless in their upper halves or more, bearing reduced bracts there 5
2a. Branches of the inflorescence more than twice as long as the bracts that subtend them 3
b. Branches of the inflorescence at most one and a half times as long as the bracts which subtend them 4
3a. Petals twice as long as the sepals, with basal scales **2. coerulea**

b. Petals at most one and a half times as long as the sepals, without basal scales **1. spathacea**
4a. Inflorescence branches widely spreading; sepals covered with fine hairs when young, these falling quickly; scape less than 1 m **3. venusta**
b. Inflorescence branches erect; sepals persistently woolly; scape to 6 m **4. gigas**
5a. Leaves of the same colour on both surfaces **5. chilensis**
b. Leaves green above, white beneath **6. alpestris**

1. P. spathacea (Grisebach) Mez. Illustration: Flora Neotropica **14**: f. 20I–L (1974).

Stem erect, short; scape to 1 m. Leaves with spreading or recurved blades, 60–110 × 3–5 cm, grey-green above, pale-scaly beneath, margins with hooked spines to 4 mm. Panicle bipinnate with 10–20 lateral branches, covered with fine, white, stellate hairs; branches 7–30 cm, loosely flowered. Flower-stalks curved, *c.* 1.5 cm. Sepals 1.5–2.2 cm, pinkish red. Petals 2.5–3.3 cm, tongue-shaped, blue to dark green. *North-central Argentina.* H5–G1.

2. P. coerulea Lindley. Illustration: *Edwards's Botanical Register* **26**: t. 11 (1840); *Botanical Magazine*, 8194 (1908); Flora Neotropica **14**: f. 21C–G (1974).

Stem well developed, erect; scape to 2 m. Leaves in a rosette at the stem apex, 40–60 × 1–2 cm, sheaths brown, blades ashy white, margins with hooked, reddish brown spines to 5 mm. Inflorescence bipinnate or rarely simple; branches 5–30 cm, sometimes without flowers towards the base. Floral bracts broad, nearly as long as the sepals. Flower-stalks 1–2 cm. Sepals 1.2–2.4 cm. Petals to 5 cm, elliptic, obtuse, dark blue. *C Chile.* H5–G1.

A variable species; var. **violacea** (Brongniart) Smith & Looser (*P. violacea* (Brongniart) Mez), which has the floral bracts much shorter and narrower, is sometimes grown.

3. P. venusta Philippi. Illustration: Flora Neotropica **14**: f. 30A–C (1974); *The New Plantsman* **7**(2): cover (2000).

Stem to 40 cm; scape less than 1 m. Leaves to 30 × 3 cm, greyish on both surfaces, margins with stout, hooked spines to 5–7 mm. Inflorescence with numerous, widely spreading branches, each to 10 cm. Flower-stalks *c.*7 mm. Sepals 1.5–2 cm, covered with fine, stellate hairs when young, soon hairless. Petals to 3.5 cm, narrowly elliptic, obtuse, deep pinkish purple. *Chile.* H5–G1.

4. P. gigas André. Illustration: *Revue Horticole* **53**: f. 74 (1881); Flora Neotropica **14**: f. 33I (1974).

Scape to 6 m or more. Leaves 1–1.3 × *c.* 4 cm, hairless above, pale-scaly beneath, margin with hooked spines to 1.1 cm. Inflorescence dense, bipinnate, with more or less erect branches that are 13–15 cm, flowerless towards the base. Flower-stalks 1.5–2 cm. Sepals *c.* 2.6 cm, persistently woolly throughout. Petals to 5 cm. *Colombia.* G1.

5. P. chilensis Molina. Illustration: *Botanical Magazine*, 4715 (1863); Flora Neotropica **14**: f. 59 (1974); *The New Plantsman* **7**(2): 94 (2000); Brickell (ed.), RHS A–Z encyclopedia of garden plants, 869 (2003).

Stem woody, prostrate, simple or branched, to 5 m. Leaves in a dense rosette, *c.* 1 m × 5 cm, erect or stiffly spreading, soon hairless, margin with large (the lower to 1 cm), hooked spines. Inflorescence with many branches, the upper parts of these sterile and covered with reduced bracts. Flowers dense, their stalks *c.* 1.5 cm. Sepals to 3.5 cm, green. Petals *c.* 5 cm, narrowly elliptic, bright yellow or greenish yellow. *C Chile.* G1.

6. P. alpestris (Poeppig) Gay (*P. whytei* Hooker). Illustration: *Botanical Magazine*, 5732 (1868); Flora Neotropica **14**: f. 62A–D (1974).

Stem prostrate, branched. Leaves rather few in each rosette, to 60 × 2.5 cm, arching and recurved, green above, white beneath from a dense covering of stellate hairs, margins with hooked spines to 4 mm. Inflorescence with rather few branches, each with the upper half to two-thirds sterile and covered with bracts. Flower-stalks *c.* 1 cm. Sepals 2–2.5 cm. Petals *c.* 5 cm, elliptic, olive green. *S Chile.* H5–G1.

2. FOSTERELLA L.B. Smith

D. Philcox

Terrestrial herbs, stemless or nearly so. Leaves usually few in each rosette; sheaths distinct, blades thin, somewhat narrowed towards the base, entire. Scape well developed, slender. Inflorescence a loose panicle, hairless or finely scaly. Floral bracts inconspicuous; flower-stalks distinct, slender. Flowers small, pendent (in ours). Sepals free, shorter than the petals. Petals free, without basal scales, white. Ovary superior.

A genus of 12 or more species from central & western South America.

1. F. penduliflora (Wright) L.B. Smith (*Catopsis penduliflora* Wright; *Lindmania penduliflora* (Wright) Stapf). Illustration: *Botanical Magazine*, 9029 (1924); Flora Neotropica **14**: f. 73A–E (1974).

Leaves to 35 × 3.5 cm, linear-lanceolate, thin, with a soft apex, hairless above, scurfy beneath. Scape slender, hairless. Inflorescence to 20 cm, without an involucre, many-flowered. Flower-stalks 2–3 mm, slender. Flowers pendent. Sepals *c.* 3.5 mm, lanceolate, obtuse, green with white margins. Petals 8–9 mm, lanceolate-oblong, white. Stamens included in the corolla. *C Peru to NW Argentina.* G2.

3. PITCAIRNIA L'Héritier

J. Cullen

Terrestrial herbs. Leaves in many ranks, forming a dense rosette, mostly narrow, persistent or deciduous along a distinct, transverse line, rarely of 2 kinds. Scape usually present. Inflorescence a raceme or head-like spike (in ours). Flowers stalked or almost stalkless. Sepals free. Petals free, long and narrow, often with their upper parts rather irregularly arranged, with or without basal scales. Stamens about as long as petals. Ovary half-inferior to almost completely superior (in ours). Fruit a capsule.

A genus of 260 species, all but 1 from C & S America. Only 3 are widely grown, but more may be found in specialist collections.

1a. Leaves of 2 kinds, the outer reduced to spiny-margined spines, the inner linear, deciduous before flowering, along a distinct, transverse line; petals usually red **3. heterophylla**
b. Leaves all similar, persistent; petals yellow or orange 2
2a. Leaves to 1 m or more; sepals 1.5–2 cm, orange; petals 4.5–5 cm, pale yellow **1. xanthocalyx**
b. Leaves to 35 cm; sepals *c.* 2.3 cm, green; petals *c.* 6.5 cm, orange with yellow apices **2. andreana**

1. P. xanthocalyx Martius (*P. flavescens* Baker, not K. Koch). Illustration: *Botanical Magazine*, 6318 (1877); Flora Neotropica **14**: f. 116C–E (1974).

Stemless, scape to 1.3 m. Leaves all similar, persistent, 1 m or more, entire or margins minutely spiny. Inflorescence a raceme. Flower-stalks 1.5–2 cm. Sepals 1.5–2 cm, orange. Petals 4.5.–5 cm, pale yellow. Ovary half-inferior. *Mexico.* G2.

2. P. andreana Linden (*P. lepidota* Regel). Illustration: *Illustration Horticole* **20**: t. 189 (1873); *Botanical Magazine*, 6480 (1880); *Gartenflora* **22**: t. 772 (1873); Flora Neotropica **14**: f. 120I (1974).

Stemless, scape to 20 cm. Leaves all similar, persistent, to 35 × 3 cm, white beneath due to a dense covering of scales, entire. Inflorescence a raceme. Flower-stalks *c.* 1 cm. Sepals *c.* 2.3 cm, green. Petals *c.* 6.5 cm, bright orange with yellow apices. Ovary mostly superior. *Colombia.* G2.

3. P. heterophylla (Lindley) Beer. Illustration: Flora Neotropica **14**: f. 141A–F (1974); Brickell (ed.), RHS A–Z encyclopedia of garden plants, 824 (2003).

Stemless, scape to 20 cm. Leaves of 2 kinds, the outer reduced to spiny-margined spines, the inner green, linear, to 70 × 1.3 cm, deciduous before flowering, along a distinct, transverse line, entire above this line, spiny-margined below it. Inflorescence a condensed, head-like spike. Flower-stalks at most 3 mm. Sepals *c.* 3 cm. Petals to 5.5 cm, usually red. *C & S America, as far south as Peru.* G2.

A variable species; a white-flowered variant (var. **albiflora** Standley & Smith) is known in the wild, and may be occasionally grown.

4. DYCKIA Schultes
D. Philcox
Small to large herbs, terrestrial or growing on rocks. Leaves in dense rosettes, with fleshy sheaths and blades not narrowed at the base, usually spine-toothed, rigid. Scape conspicuous. Inflorescence simple or compound, without an involucre; primary bracts inconspicuous. Floral bracts variable in length, from shorter than the flower-stalks to longer than the flowers. Flowers relatively small, yellow to red. Sepals usually less than 1 cm, overlapping, usually free, shorter than the petals. Petals usually less than 3 cm, overlapping. Stamens longer or shorter than the corolla. Ovary superior.

A genus of over 100 species from Brazil, Paraguay, Uruguay and N Argentina. They are not widely acceptable as houseplants owing, in the main, to their rigid, inflexible leaves, which have vicious spines on the margins.

1a. Flowers in a compound panicle **1. frigida**
b. Flowers in a simple raceme 2
2a. Axes of the inflorescence densely and persistently scaly, scales sometimes felt-like **2. remotiflora**
b. Axes of the inflorescence hairless or sparsely scaly with the scales soon falling 3
3a. Leaf-blades very finely and regularly white-lined beneath, with scales in the deep furrows between the green nerves **3. brevifolia**
b. Leaf-blades not regularly white-lined beneath, the scales more or less covering the nerves **2. remotiflora**

1. D. frigida Hooker. Illustration: *Botanical Magazine*, 6294 (1877); Flora Neotropica **14**: f. 184E–I (1974).

Leaves *c.* 1 m, sheaths *c.* 9 cm wide, dark purplish brown; blades *c.* 3.5 cm wide, narrowly triangular, sharply pointed, closely adpressed-scaly beneath, spines 2–5 mm on outer leaves. Scape stout, ascending, scaly when young. Inflorescence paniculate with branches to 30 cm, many-flowered, more or less shortly brown-felted. Flower-stalks to 3 mm. Flowers 1.1–1.8 cm long. Sepals 5–8 mm, ovate, acute. Petals yellow or pale yellow. Stamens much shorter than the petals. *Brazil.* G1.

2. D. remotiflora Otto & Dietrich. Illustration: Flora Neotropica **14**: f. 187I–M (1974); Brickell (ed.), RHS A–Z encyclopedia of garden plants, 393 (2003).

Leaves 10–25 cm with large, ovate to circular sheaths and blades 8–12 mm wide, arching, narrowly triangular, sharply pointed, flat, dark green and without regular white lines beneath, with slender, curved spines 1–3 mm. Scape stout. Inflorescence 12–20 cm, simple, loose, densely to sparsely but persistently scaly. Flower-stalks very short, stout. Sepals 6–10 mm, ovate. Petals 1.1–2.3 cm, dark orange. Stamens not projecting from corolla. *Brazil, Uruguay & Argentina.* G1.

3. D. brevifolia Baker. Illustration: Flora Neotropica **14**: f. 192A–D (1974).

Leaves 10–20 cm, in a dense rosette with sheaths scarcely broader than the blades; blades 2.5–3.5 cm wide, lanceolate-triangular, acute, very thick, hairless above, minutely pale-scaly between the prominent nerves beneath, producing a finely and regularly white-lined appearance, with distant hooked spines to 2 mm. Scape stout. Inflorescence simple, loose or dense, many-flowered, scaly when young. Flower-stalks 2–4 mm, elongating in fruit. Flowers spreading then erect. Sepals to 8 mm, ovate, acute or obtuse. Petals *c.* 10 mm, bright yellow. Stamens not projecting from corolla. *Brazil.* G1.

5. TILLANDSIA Linnaeus
D. Philcox
Epiphytic herbs with or without stems. Leaves in rosettes, clustered or distributed along a stem, in 2 or many ranks, entire; blades strap-shaped to narrowly triangular or linear. Scape usually distinct. Inflorescence varied: usually a many-flowered spike with the flowers in 2 to many ranks, or the inflorescence reduced to a single flower. Floral bracts conspicuous to minute. Flowers mostly shortly stalked. Sepals usually symmetric, free or united to various degrees. Petals lacking basal scales, free. Ovary superior, hairless.

This is both the largest and most widespread genus of Bromeliaceae. It consists of about 400 species and covers the whole wide range of distribution throughout the Americas. Strangely though, only some 6 or 7 species are commonly found on the general market in Europe, although of recent date a few more are appearing ephemerally, attached to certain *objets d'art* as coffee table conversation pieces. Others not mentioned here are occasionally seen, having originated from the private, specialist collections in Europe. The identification of these should adequately be covered, if the need arises, by reference to Rauh's *Bromeliads.*

Literature: Rodriguez, A., Airplants: a selection of Tillandsia species to grow, *The New Plantsman* **2**(3): 162–181 (1995).

1a. Flowers borne all round the inflorescence axis; inflorescence unbranched 2
b. Flowers in 2 ranks on the inflorescence axis or solitary; inflorescence sometimes branched 4
2a. Plant stemless or almost so, leaves spreading in all directions; lower floral bracts gradually tapering into a narrow tail **1. stricta**
b. Plant with a distinct stem, leaves all curved to one side; lower floral bracts rounded, then abruptly narrowed into a tail 3

3a. Petals 2.8–3 cm, white **2. araujei**
b. Petals *c.* 2 cm, blue, pink or white **4. tenuifolia**
4a. Plant with a distinct stem bearing alternating bunches of leaves, forming a lichen-like festoon **3. usneoides**
b. Stem and plant not as above 5
5a. Leaf-sheaths dark, contrasting with the blades **5. tricolor**
b. Leaf-sheaths pale, of the same colour as the blades 6
6a. Floral bracts with prominent nerves; inflorescence-stalk elongate; petals dark blue, each with a small white spot at the base **6. lindenii**
b. Floral bracts without prominent nerves; inflorescence-stalk very short; petals entirely deep violet **7. cyanea**

1. T. stricta Solander. Illustration: *Botanical Magazine*, 1529 (1813); *Edwards's Botanical Register* **16**: t. 1338 (1830); Flora Neotropica **14**: f. 259A–D (1977); Brickell (ed.), RHS A–Z encyclopedia of garden plants, 1041 (2003).

Plant with a very short stem. Leaves 6–18 cm, many, in dense rosettes, covered with adpressed grey scales; blades very narrowly triangular, 4–11 mm wide at the base, attenuate. Scape erect to arching, slender; bracts overlapping, the lower leaf-like, the upper elliptic with linear blades, scaly. Inflorescence 2–7 cm long, 1.5–3.5 cm in diameter, simple, dense, with flowers borne all around the axis in several ranks. Lower floral bracts gradually tapering into narrow tails. Sepals 6–13 mm, lanceolate-ovate, hairless, united for up to 4 mm. Petals 1.5–2.5 cm, blue or purple. *Venezuela, Trinidad, Guyana, Surinam, Brazil to northern Argentina.* G2.

2. T. araujei Mez. Illustration: Flora Neotropica **14**: f. 260A–E (1977).

Plant with stem to 30 cm, simple or with few branches. Leaves 3–7 cm, in many ranks along the stem but all curved to one side; sheaths short, white, hairless at the base; blades stiff, attenuate to a sharply pointed apex, covered with fine, pale, brown-centred scales. Scape slender, ascending, hairless. Inflorescence 3–5 cm, simple, more or less loose, with 5–12 flowers borne all around the axis in several ranks. Lower floral bracts rounded, then abruptly narrowed into tails. Sepals 1.2–1.5 cm, lanceolate, acute, united for most of their length. Petals 2.8–3 cm, white, narrow. *Central E Brazil.* G2.

3. T. usneoides (Linnaeus) Linnaeus (*Renealmia usneoides* Linnaeus; *T. filiformis* invalid). Illustration: *Botanical Magazine*, 6309 (1877); Flora Neotropica **14**: f. 287A–C (1977); Brickell (ed.), RHS A–Z encyclopedia of garden plants, 1041 (2003).

Plant with long stems hanging in festoons up to 8 m; stems usually 1 mm or less in diameter, with the internodes 3–6 cm. Leaves 2.5–5 cm, in 2 ranks, densely grey-scaly, the sheaths to 8 mm, elliptic, inrolled; blades less than 1 mm wide, thread-like. Scape almost absent. Inflorescence reduced to a single flower. Flower almost stalkless. Sepals to 7 mm, narrowly ovate, acute, shortly united. Petals 9–11 mm, pale green or blue. *Southeastern USA to C Argentina & Chile.* G2.

4. T. tenuifolia Linnaeus. Illustration: Flora Neotropica **14**: f. 261B–E (1977).

Plant very variable with stem to 25 cm, often branching. Leaves 5–10 cm, dense and in many ranks along the stem, all curved to one side, minutely adpressed-scaly; sheaths barely distinct from the blades, which are 2–7 mm wide, narrowly triangular, channelled above. Scape erect, slender, largely concealed by the leaves. Inflorescence simple with 4–10 flowers borne all around the axis. Floral bracts rounded, then abruptly tapering into narrow tails. Sepals *c.* 10 mm, lanceolate, acute, hairless. Petals *c.* 2 cm, blue, white or rose. *West Indies to Bolivia & Argentina.* G2.

5. T. tricolor Schlechtendal & Chamisso (*Vriesea xiphostachys* Hooker). Illustration: Flora Neotropica **14**: f. 301D–F (1977).

Plant stemless. Leaves many, in dense rosettes, equalling or shorter than the inflorescence, the outer curving; sheaths elliptic-oblong, large, dark green; blades *c.* 1 cm wide at the base, linear-triangular, long-tapering, glaucous green. Scape slender, erect. Inflorescence simple or weakly branched, flowers in 2 ranks; spikes erect or diverging, 6–18 × 1.8–2.5 cm, dense, many-flowered, compressed with flat sides. Floral bracts erect, densely overlapping, ovate, acute, *c.* 3 × 1.8 cm, much longer than the sepals, incurved, leathery, smooth, hairless. Flowers stalkless. Sepals *c.* 2 cm, leathery. Petals *c.* 7 cm, erect, violet. Stamens projecting from corolla. *C America.* G2.

6. T. lindenii Regel (*T. lindeniana* Regel). Illustration: *Gartenflora* **18**: t. 619 (1869); Flora Neotropica **14**: f. 268A–C (1977); Brickell (ed.), RHS A–Z encyclopedia of garden plants, 1041 (2003).

Plant stemless. Leaves to 40 cm, many, arching, curved downwards, somewhat longitudinally striped with red-purple toward the base; sheaths small, elliptic; blades 1.2–1.8 cm wide, linear-triangular, more or less hairless above, minutely scaly beneath. Scape slender, erect, elongate; bracts overlapping. Inflorescence to 20 × 5 cm, simple, lanceolate, acute, with up to 20 flowers in 2 ranks, hairless. Floral bracts overlapping, 4–4.5 cm, slightly longer than the sepals, keeled, more or less leathery, prominently nerved, green to rose. Sepals *c.* 3.5 cm, elliptic, obtuse. Petals to 4 cm, deep blue, each with a small white spot at the base. *NW Peru.* G2.

7. T. cyanea K. Koch (*T. lindenii* Morren not Regel; *T. morreniana* Regel). Illustration: Flora Neotropica **14**: f. 269C–E (1977); Brickell (ed.), RHS A–Z encyclopedia of garden plants, 1040 (2003).

Plant stemless. Leaves to 35 cm, many, more or less erect, somewhat red-striped towards the base, finely scaly; sheaths *c.* 6 cm, distinct; blades 1–1.5 cm wide, linear-triangular. Scape erect or inclined, very short and almost hidden by the leaves; bracts overlapping. Inflorescence to 16 × 7 cm, simple, elliptic, obtuse or broadly acute, with up to 20 flowers. Floral bracts elliptic, keeled, without conspicuous nerves, exceeding the sepals, leathery, rose or red. Sepals *c.* 3.5 cm, free, obtuse or broadly acute. Petals 2–2.5 cm, deep violet. *Ecuador & Peru.*

This species is frequently confused with *T. lindenii*, but is easily distinguished by the apparent absence of a scape, giving the impression that the inflorescence is completely stalkless, while the inflorescence of *T. lindenii* is clearly stalked.

6. VRIESEA Lindley

D. Philcox

Stemless, perennial epiphytes. Leaves in rosettes, many-ranked, entire; blades usually tongue-shaped and sparsely scaly. Scape usually conspicuous. Inflorescence usually of spikes of flowers in 2 ranks, rarely of 1 or more spikes of many-ranked flowers. Floral bracts usually conspicuous. Flowers mostly shortly stalked. Sepals free or almost so, symmetric or almost so.

Petals free or united into a tube that is shorter than the sepals, each petal with 2 scales at the base. Ovary nearly or completely superior.

A genus of some 250 species from C & S America and the West Indies, of which only 6 are commonly cultivated in Europe. Many other species, however, have been or are currently grown in specialist collections, and many hybrids have been produced. The identification of the species in cultivation is therefore often very difficult.

1a. Inflorescence compound 2
b. Inflorescence simple 4
2a. Panicle to 40 × 4.5 cm; leaves to 5 cm wide **1. saundersii**
b. Panicle much larger, to about 1 m; leaves 6–10 cm wide 3
3a. Leaves wholly pale green, chequered; petals 4–5 cm; filaments dilated towards the apex **2. gigantea**
b. Leaves green with irregular, dark green or purple, transverse bands; petals *c.* 3.5 cm; filaments not dilated **3. hieroglyphica**
4a. Flowers all erect and overlapping; bracts keeled and incurved, longer than the sepals **4. splendens**
b. Flowers (at least the lowest) not overlapping, spreading and sometimes with evident space between them; bracts not keeled, not incurved, shorter than the sepals 5
5a. Stamens longer than the petals; petals bright yellow, sometimes green-tipped **5. psittacina**
b. Stamens shorter than the petals; petals greenish white **6. fenestralis**

1. V. saundersii (Carrière) Mez (*Encholirion saundersii* Carrière). Illustration: Flora Neotropica **14**: f. 390E–F (1977).

Leaves 20–30 cm, adpressed-scaly; sheaths broadly elliptic, covered with brown scales; blades 3.5–5 cm wide, broadly acute, grey-scaly, densely and finely brown-spotted beneath, the apex recurved. Scape erect, stout, hairless. Panicle to 40 × 4.5 cm, densely bipinnate, loose, with few flowers. Flowers yellow, stalks short, stout. Sepals *c.* 2 cm, obtuse. Petals *c.* 3.5 cm. Stamens projecting from corolla. *Brazil.* G2.

2. V. gigantea Gaudichaud (*V. tessellata* (Linden) Morren). Illustration: *Illustration Horticole* **14**: t. 516 (1867); Flora Neotropica **14**: f. 354D (1977).

Leaves 60–100 cm, many, in a large, broadly funnel-like rosette, minutely scaly on both sides; sheaths large, broadly elliptic; blades 6–9 cm wide, broadly acute, strap-shaped, wholly pale green, chequered. Scape erect, stout. Panicle loosely bipinnate, *c.* 100 × 40 cm, almost hairless, with many flowers in 2 ranks. Sepals to 3.7 cm, elliptic. Petals 4–5 cm long, obtuse, greenish yellow. Filaments dilated towards the apex. *S Brazil.* G2.

3. V. hieroglyphica (Carrière) Morren (*Massangea hieroglyphica* Carrière). Illustration: *Illustration Horticole* **31**: t. 514 (1884) & **42**: t.318 (1895); Flora Neotropica **14**: f. 372H–I (1977); Brickell (ed.), RHS A–Z encyclopedia of garden plants, 1083 (2003).

Leaves 50–80 cm, many, in a funnel-like rosette, minutely scaly, the inner erect, the outer recurving; sheaths slightly wider than the blades, dark brown beneath; blades 7–10 cm wide, strap-shaped, marked with dark green or purple transverse bands. Scape erect, stout. Panicle to 80 cm, bipinnate or further branched, with many flowers. Sepals *c.* 2.5 cm, narrowly elliptic. Petals *c.* 3.5 cm, yellowish. *Brazil.* G2.

4. V. splendens (Brongniart) Lemaire (*Tillandsia splendens* Brongniart). Illustration: *Flore des serres* **2**: t. 4 (1846); Flora Neotropica **14**: f. 396A–D (1977); Brickell (ed.), RHS A–Z encyclopedia of garden plants, 1083 (2003).

Leaves 40–80 cm, in dense, funnel-like rosettes, arching; sheaths indistinct, brown-scaly; blades 4–6 cm wide, strap-shaped, broadly acute or rounded, with broad, dark, irregular cross-bands, sparsely scaly. Scape erect; bracts clasping the scape, densely overlapping. Raceme to 55 × 6 cm, dense, lanceolate, strongly compressed, with many flowers. Floral bracts densely overlapping, bright red, exceeding the sepals, hairless. Flower-stalks *c.* 4 mm. Sepals *c.* 2.5 cm, wholly yellow, or red at the apex. Petals to 8 cm, yellow. Stamens projecting from corolla. *Venezuela to the Guianas.* G2.

There are several other varieties of this species, but only one is known so far to be cultivated, other than the one described above. This is var. **formosa** Witte (*V. longibracteata* (Baker) Mez), in which the leaf-blades are wholly green. *Venezuela, Guyana, Trinidad & Tobago.*

5. V. psittacina (Hooker) Lindley. Illustration: *Edwards's Botanical Register* **29**: t. 10 (1843); Flora Neotropica **14**: f. 400B–E (1977).

Leaves 30–50 cm, in a broad, funnel-like rosette; sheaths 7–9 cm, elliptic, green or pale brown, scaly; blades 2.2–2.8 cm wide, strap-shaped, somewhat narrowed toward the base, broadly acute, wholly green. Scape erect, slender; bracts overlapping, the lower green, the upper yellow and green. Raceme 17–23 cm, the flowers spreading and not overlapping. Floral bracts slightly shorter than the sepals, red with yellow apices. Flower-stalks to 5 mm, stout. Sepals 3.5–4 cm. Petals *c.* 6 cm, bright yellow or with green apices. Stamens projecting from corolla. *Brazil.* G2.

6. V. fenestralis Linden & André. Illustration: Flora Neotropica **14**: f. 360F–G (1977).

Leaves 35–50 cm, many, in a broad rosette, arching; sheaths large, 10–12 cm wide, marked with small red-brown spots; blades 6.5–8 cm wide, broadly rounded, pale green with narrow dark green lines, hairless above, minutely scaly beneath. Scape erect, green. Raceme to 50 cm, loose, with many spreading, non-overlapping flowers. Floral bracts much shorter than the sepals. Flower-stalks to 1 cm. Sepals *c.* 6.5 cm, green, spotted. Petals *c.* 4.5 cm, longer than the stamens, greenish white. *Brazil.* G2.

7. GUZMANIA Ruiz & Pavon

D. Philcox

Epiphytic herbs without (or rarely with) stems. Leaves entire, in many ranks. Scape usually conspicuous. Inflorescence bipinnate or simple, branches in many ranks, with many flowers. Sepals free to united for most of their length, symmetric or almost so. Petals with their claws united into a tube distinct from the blades, without scales near the base. Ovary superior. Fruit a capsule.

A genus of over 125 species mostly from C and tropical north-west S America.

1a. Sepals longer than the petals; leaves irregularly and transversely marked **3. musaica**
b. Sepals shorter than the petals; leaves not as above 2
2a. Inflorescence branched; leaf-blades gradually tapered **1. zahnii**
b. Inflorescence unbranched; leaf-blades broadly acute or rounded 3

3a. Inflorescence an elongate spike, not surrounded by bracts **2. monostachia**
b. Inflorescence corymbose, surrounded by large, brightly coloured bracts, forming a cup-like structure **4. lingulata**

1. G. zahnii (Hooker) Mez (*Caraguata zahnii* Hooker). Illustration: Flora Neotropica **14**: f. 436F–J (1977); Rauh, Bromeliads, t. 57 (1979).

Leaves to 60 cm, 20–30 in a rosette; sheaths obscurely brown-scaly, yellowish with dark red longitudinal stripes; blades *c.* 2.7 cm wide, purple-red throughout or green near the apex, marked with red-brown longitudinal stripes. Scape erect, hairless, red, with bright scarlet bracts. Panicle to 25 × 10 cm, bipinnate, pyramid-shaped, hairless, bright yellow throughout. Flowers almost stalkless. Sepals *c.* 1.8 cm, united for 2.5 mm. Petals *c.* 3 cm, united for 7 mm, elliptic, obtuse. *Costa Rica & Panama.* G2.

2. G. monostachia (Linnaeus) Mez (*Renealmia monostachia* Linnaeus). Illustration: Flora Neotropica **14**: f. 444C–E (1977); Rauh, Bromeliads, t. 9 (1979); Brickell (ed.), RHS A–Z encyclopedia of garden plants, 495 (2003).

Leaves to 40 cm, in dense rosettes, scaly, soon hairless; sheaths brownish; blades *c.* 2.5 cm wide, acute, yellow-green. Scape erect, hairless, with pale green bracts. Spike 8–15 × 2–3 cm, many-flowered, sterile toward the apex. Flowers 2.3–3 cm, white, erect. Sepals *c.* 1.8 cm, leathery, united for a quarter of their length. Petals united for most of their length. *USA (Florida), West Indies & Nicaragua to N Brazil & Peru.* G2.

3. G. musaica (Linden & André) Mez (*Tillandsia musaica* Linden & André). Illustration: Flora Neotropica **14**: f. 449G–J (1977); Rauh, Bromeliads, t. 56 (1979); Brickell (ed.), RHS A–Z encyclopedia of garden plants, 495 (2003).

Leaves to 70 cm, in a spreading rosette; sheaths purplish brown toward the base; blades 4–8 cm wide, broadly acute or rounded, marked with irregular, transverse lines, and with a short spine at the apex. Scape erect; bracts bright pinkish red. Spike more or less spherical, 12–25-flowered, hairless. Flowers more or less without stalks. Sepals 2.5–4.5 cm, united, leathery, yellowish. Petals shorter than the sepals and not projecting at flowering. *Panama & Colombia.* G2.

4. G. lingulata (Linnaeus) Mez (*Tillandsia lingulata* Linnaeus; *Caraguata latifolia* Beer). Illustration: Rauh, Bromeliads, t. 5, 6 (1979); Brickell (ed.), RHS A–Z encyclopedia of garden plants, 495 (2003).

Leaves 30–45 cm, in a dense rosette; sheaths conspicuous, more or less purplish brown at the base, occasionally marked beneath with fine, purple, longitudinal lines; blades to 4 cm wide, acute. Scape erect, stout; upper bracts usually reddish and forming an involucre. Raceme corymbose, to 7 cm wide, 10–50-flowered. Flowers erect, *c.* 4.5 cm; stalks short and stout. Sepals to 1.7 cm, free, hairless. Petals linear, hooded, white. *Belize, West Indies to Bolivia & Brazil.* G2.

Several varieties are found in cultivation in Europe, and generally are more commonly met with than the wild variety described above. The largest and most brilliant of these, var. **cardinalis** (André) Mez (*G. cardinalis* (André) Mez; *Caraguata cardinalis* (André) André) is very similar to the above but is distinguished by its showy, spreading, bright scarlet bracts. Var. **splendens** (Planchon) Mez, which until recently was commonly known in cultivation as *G. peacockii* (Morren) Mez, is a red-leafed plant in which the lower leaves are bright purple-red and the upper reddish green; the inflorescence terminates in a funnel-shaped, purplish red spike, with the small bracts in the centre yellowish and white-tipped. Var. **minor** (Mez) L.B. Smith (*G. minor* Mez) is a much smaller and more delicate plant, having thin, yellowish leaves *c.* 30 × 1.5 cm; the inflorescence is at the end of a short, stout scape, appearing as a raised cup of yellow or red bracts and whitish yellow flowers.

8. NEOREGELIA L.B. Smith

D. Philcox

Epiphytes with or without stems. Leaves dense, forming flat or tubular rosettes, usually broadly linear, flexible, spiny-margined, sheaths usually large and distinct, scaly or not, often variously spotted or banded, the inner often reddish or purplish, the apex rounded and with a small point. Inflorescence sunk centrally in the rosette, simple to densely umbellate, racemose or corymbose; bracts entire or toothed. Flowers stalked. Sepals united at least at the base. Petals violet, blue or white, rarely red, united at least at the base, without basal scales, blades spreading, acute to acuminate. Stamens attached to the petals, not projecting. Ovary inferior. Fruit a berry.

A genus, formerly known as *Aregelia* Mez, of more than 70 species largely natives of eastern Brazil, with a few species found in Amazonia and eastern Colombia and Peru. Very similar to *Nidularium* (p. 311); there is some confusion in the naming of cultivated material.

1a. Inner leaves of the rosette bright red at least at the base, unlike the outer 2
b. Inner leaves of the rosette like the outer, not bright red 5
2a. Leaf-blades sparsely and inconspicuously scaly beneath; sepals rounded or acute **2. carolinae**
b. Leaf-blades covered, at least beneath, with adpressed, grey scales; sepals acuminate 3
3a. Floral bracts minutely toothed; leaf-blades 1.5–2 cm wide **1. pineliana**
b. Floral bracts entire; leaf-blades 3–5 cm wide 4
4a. Leaves each with a bright red spot at the tip; petals wholly blue **8. spectabilis**
b. Leaves without red spots at the tips; petals blue above, white below **3. princeps**
5a. Leaf-sheaths forming a cylindric or ellipsoid tank distinctly higher than wide and often more or less constricted at the mouth 6
b. Leaf-sheaths forming a saucer- or funnel-shaped tank about as wide as high, or at least flaring above 7
6a. Leaves with spots or bands darker than the rest of the blade **4. ampullacea**
b. Leaves with spots or bands paler than the rest of the blade **6. chlorosticta**
7a. Sepals 1.2–1.6 cm; leaf-blades 2–3 cm wide **5. tristis**
b. Sepals 1.8–3 cm; leaf-blades 2–10 cm wide 8
8a. Leaves variegated with either light or dark spots **7. marmorata**
b. Leaves of one colour or banded, but not spotted 9
9a. Each leaf with a red spot at the apex **8. spectabilis**
b. Each leaf with a purple spot at the apex **9. concentrica**

1. N. pineliana (Lemaire) L.B. Smith (*Nidularium pinelianum* Lemaire; *Aregelia* *morreniana* (Antoine) Mez; *Neoregelia morreniana* (Antoine) L.B. Smith; *A. pineliana* (Lemaire) Mez). Illustration: Flora Neotropica **14**: f. 498A–D (1979).

Plant with a short stem or stemless, spreading by short runners. Leaves many, to 50 cm × 15 mm, distributed along the stem; sheaths purplish, densely scaly; blades 1.5–2 cm wide, linear, rounded and with a small point, green with greyish scales on both surfaces, inner leaves red. Inflorescence many-flowered, sunk between the sheaths of the inner leaves; floral bracts green, minutely toothed, hairless. Flowers 4–6 cm. Sepals 1.5–2 cm, acuminate, asymmetric, green, joined for 3–4 mm at the base. Petals more or less erect, to 4 cm, joined to above the middle, blades blue, tube white. *Known only from cultivation.* G2.

A variant in which the inflorescence is replaced by a tuft of enlarged, coloured, sterile bracts, forma **phyllanthidea** (Baker) L.B. Smith, is also known from cultivation.

2. N. carolinae (Beer) L.B. Smith (*Bromelia carolinae* Beer; *Nidularium carolinae* Baker; *Aregelia carolinae* (Beer) Mez).

Plant stemless. Leaves about 20, to 40–60 cm, forming a broad rosette; sheaths large, sparsely covered with minute, dark brown scales; blades 2.5–3.5 cm wide, somewhat densely toothed, glossy green, hairless above, obscurely scaly beneath, the outer leaves wholly green, the inner brilliant red at the base. Inflorescence many-flowered, sunk in the centre of the rosette; bracts entire or a few minutely toothed towards the apex. Flowers to 4 cm. Sepals 1.6–2 cm, asymmetric, each rounded and with a small point, united at the base for *c.* 5 mm. Petals 2.4–2.8 cm, joined to above the middle, dark blue. *Brazil.* G2.

A variant in which the leaf-blades are longitudinally striped white, green and reddish pink is known in cultivation as forma **tricolor** (Foster) L.B. Smith.

3. N. princeps (Baker) L.B. Smith (*Aregelia princeps* Baker; *Karatas meyendorfii* Antoine, in part; *A. marechalii* Mez, in part). Illustration: Flora Neotropica **14**: f. 499A–F (1979).

Stemless plant. Leaves 15–20, to 50 cm, forming a dense, spreading rosette; sheaths large, the outer circular in outline, green, densely scaly; blades 3–4.5 cm wide, with very short spines, the outer blades green, the inner bright red and smaller. Inflorescence sunk in the rosette, many-flowered; floral bracts entire. Flowers to 4 cm. Sepals *c.* 2.4 cm, asymmetric, acuminate, united for about 1.5 mm at the base, hairless, red. Petals *c.* 3.5 cm, linear, acuminate, united for much of their length, dark blue towards the apex, white below. *Brazil.* G2.

4. N. ampullacea (Morren) L.B. Smith (*Nidularium ampullaceum* Morren; *Aregelia ampullacea* (Morren) Mez). Illustration: *La Belgique Horticole* **35**: t. 14 (1885); Flora Neotropica **14**: f. 500 (1979).

Plant stemless, spreading by runners. Leaves few, 15–20 cm, in a dense rosette; sheaths large, forming an ellipsoid tank constricted at the mouth, densely scaly with brown to purple scales; blades to 1.6 cm wide, linear, rounded, spreading, green, hairless and with dark, narrow bands above, finely toothed. Inflorescence sunk deeply in the rosette, few-flowered. Flowers to 2.5 cm. Sepals *c.* 1.5 cm, green with white margins. Petals *c.* 2 cm, acute, united for three-quarters of their length, blue above, whitish below. *Brazil.* G2.

5. N. tristis (Beer) L.B. Smith (*Bromelia tristis* Beer; *Nidularium cyaneum* Linden & André; *Aregelia tristis* (Beer) Mez).

Plant stemless. Leaves 20–60 cm, in a dense, slender, funnel-like rosette; sheaths large, dark with pale spots, forming a saucer- or funnel-shaped tank; blades 2–3 cm wide, finely spiny with spines to 0.5 mm, green, hairless and with purple-brown spots above, sparsely scaly and broadly dark-banded beneath. Inflorescence sunk deeply in the rosette, many-flowered. Flowers to 3 cm. Sepals to 1.6 cm, asymmetric, united at the base for 3 mm. Petals to 2 cm, united at the base, violet-blue. *Brazil.* G2.

N. fosteriana L.B. Smith. Similar, but with coppery leaves tipped with red above, purplish beneath and greyish scaly on both surfaces. *Brazil.* G2.

6. N. chlorosticta (Baker) L.B. Smith (*Karatas chlorosticta* Baker; *Neoregelia sarmentosa* (Regel) L.B. Smith var. *chlorosticta* (Baker) L.B. Smith). Illustration: Rauh, Bromeliads, t. 106 (1979).

Plant stemless, forming a loose rosette. Leaves 20–30 cm, the inner purplish red with large, pale green spots; sheaths almost equalling blades, ovate to elliptic, sparsely brown-scaly, forming a cylindric or ellipsoid tank that is more or less constricted at the mouth; blades *c.* 2.5 cm wide, linear, rounded and with a small point, minutely toothed, scaly, especially beneath. Inflorescence many-flowered, sunk in the rosette. Sepals 1.2–1.8 cm, united at the base for up to 7 mm, asymmetric, shortly acuminate. Petals to 1.8 cm, blades elliptic, acute, blue-violet. *Brazil.* G2.

7. N. marmorata (Baker) L.B. Smith (*Karatas marmorata* Baker; *Aregelia marmorata* (Baker) Mez). Illustration: Rauh, Bromeliads, t. 105 (1979).

Plant stemless. Leaves 20–60 cm, about 15 in a dense, broad, funnel-like rosette; sheaths large, broadly elliptic, dark purple with pale green spots, more or less densely covered with dark brown scales, forming a saucer- or funnel-shaped tank; blades broadly rounded, slightly notched and with a small point, to 8 cm wide, toothed, green or spotted with purple, obscurely scaly. Inflorescence sunk deeply in the rosette, usually many-flowered. Sepals to 2.4 cm, asymmetric, green, more or less hairless, united at the extreme base. Petals 2.4–2.7 cm, acute, the lobes spreading with reflexed tips, pale violet to white. *Brazil.* G2.

8. N. spectabilis (Moore) L.B. Smith (*Nidularium spectabile* Moore; *Aregelia spectabilis* (Moore) Mez). Illustration: *Botanical Magazine*, 6024 (1873); Rauh, Bromeliads, t. 107 (1979).

Plant stemless. Leaves many, 40–45 cm, forming a broad, funnel-shaped rosette, with grey scales and pale-banded beneath; sheaths ovate or broadly elliptic; blades 4–5 cm wide, spreading, loosely toothed to often almost entire, outer leaves green above, lilac-purple with light cross-banding beneath, with a bright red spot at the tip, inner leaves red, or whitish towards the base. Inflorescence many-flowered, deeply sunk in the rosette. Flowers 4–4.5 cm. Sepals 1.8–2.4 cm, very asymmetric, more or less elliptic with a large semi-circular wing and long, linear, hooked blade, shortly united at the base. Petals to 3 cm, acuminate, blades spreading, blue. *Brazil.* G2.

9. N. concentrica (Vellozo) L.B. Smith (*Tillandsia concentrica* Vellozo; *Nidularium acanthocrater* Morren; *Nidularium laurentii* Regel; *Aregelia concentrica* (Vellozo) Mez). Illustration: *La Belgique Horticole* **34**: t. 9 (1884); Brickell (ed.), RHS A–Z encyclopedia of garden plants, 719 (2003).

Plant stemless, usually large. Leaves 20–40 cm, up to 30 in a dense, broad, funnel-like rosette; sheaths large, broadly elliptic, covered with pale brown, adpressed scales; blades 5–10 cm wide, broadly rounded and with a small point, wholly green except for a purple spot at the apex, or with dark purple spots above, or the inner leaves at times wholly purple, pale-scaly, especially beneath, densely spiny with stout, black spines to 4 mm long. Inflorescence many-flowered, sunk in the centre of the rosette. Flowers 4–5 cm. Sepals 2.2–3 cm, slightly asymmetric, green, free, hairless, hardened at the tip. Petals to 4 cm, narrow, acuminate, united to above the middle, white or pale blue. *Brazil.* G2.

9. CRYPTANTHUS Otto & Dietrich

D. Philcox

Small epiphytes with or without stems. Leaves in rosettes or many-ranked along the stem; sheaths inconspicuous; blades narrowly triangular or somewhat contracted into a more or less stalk-like base, flexible, spiny-margined, often variegated. Inflorescence spherical, either stalkless in the centre of the rosette or at the apex of the stem, but not sunk below the leaf-sheaths, compound. Sepals united. Petals white or yellow, united at the base. Stamens projecting from the corolla. Ovary completely inferior.

A genus of 20 species from Brazil, grown mainly for their attractive rosettes.

1a. Leaf-blades without spots or lines **1. acaulis**
b. Leaf-blades with spots or lines 2
2a. Leaf-markings consisting of irregular, dark, transverse bands **2. zonatus**
b. Leaf-markings consisting of pale longitudinal stripes **3. bivittatus**

1. C. acaulis (Lindley) Beer. Illustration: Flora Neotropica **14**: f. 518A (1979); Rauh, Bromeliads, t. 59 (1979).

Plant rarely stemless. Leaves 10–20 cm, narrowed between the sheath and the blade but not really stalked; sheaths short, blades 2–3 cm wide, narrowly lanceolate, wavy-margined, minutely toothed, covered beneath by a membrane of fused scales, scaly above, green. Inflorescence few-flowered. Flowers to 4 cm. Sepals *c.* 1.5 cm, united for more than half their length. Petals shortly united, spreading, white. *Brazil.* G2

The plants marketed under this name usually belong to 3 distinct varieties. In var. **ruber** Beer the leaves are tinged with red, while in var. **argenteus** Beer, the upper surface of the leaves is green and hairless; var. **acaulis** is described above.

2. C. zonatus (Visiani) Beer. Illustration: Flora Neotropica **14**: f. 518B–E (1979).

Plant stemless. Leaves to 20 cm, in a spreading rosette, narrowed between the sheath and the blade but not stalked; sheaths very short, toothed towards the apex; blades 2–4.5 cm wide, linear-lanceolate, wavy-margined, densely and minutely spiny with spines to 1 mm, dark-banded above, appearing white beneath. Inflorescence few-flowered. Flowers to 3.2 cm. Sepals to 1.9 cm, united for about three-quarters of their length. Petals united for more than half their length, white, slightly exceeding the stamens. *Brazil.* G2.

Two major variants of this species (other than the wild one described above) are known to exist in cultivation, although their availability on the European market is uncertain. They are forma **viridis** Mez, in which the leaves lack scales on the underside and appear green, and forma **fuscus** (Visiani) Mez, in which the leaf-blades are strongly tinged with red.

3. C. bivittatus (Hooker) Regel. Illustration: *Cryptanthus Society Journal*, **XIV**(2): 56–9 (1999); Brickell (ed.), RHS A–Z encyclopedia of garden plants, 328 (2003).

Plant stemless. Leaves 18–25 cm, the outer slightly narrowed between sheath and blade; sheaths short; blades to 4 cm wide, minutely toothed, wavy-margined, dark green with 2 paler, longitudinal stripes above, and with a membrane of pale brown scales beneath. Inflorescence with few flowers. Flowers *c.* 2.6 cm. Sepals united for more than half their length. Petals *c.* 2 cm, white. *Known only from cultivation, almost certainly from Brazil.* G2.

10. NIDULARIUM Lemaire

D. Philcox

Stemless perennial epiphytes. Leaves flexible, equalling to much exceeding the inflorescence; sheaths mostly forming funnel-like tanks; blades toothed or minutely toothed, spreading. Scape shorter than, or exceeding, the leaf-sheaths, mostly covered by bracts. Inflorescence compound with flowers in flattened bundles and covered by large bracts, axis usually very short. Floral bracts well developed but not visible. Flowers stalkless. Sepals more or less united at the base, symmetric or strongly asymmetric, with or without hardened points. Petals white to dark blue (in ours), without basal scales, united for half their length or more to form a tubular base, blades spreading or erect, rounded and somewhat hooded. Stamens not projecting from corolla. Ovary inferior.

A genus of more than 20 species, all natives of eastern Brazil. Very similar to *Neoregelia* (p. 309) but, unlike that genus, there has been little confusion in the naming of cultivated material.

1a. Petal-blades spreading, acute; inflorescence woolly, reddish brown; sepals 1–1.4 cm, united for 2–3 mm 2
b. Petal-blades erect, obtuse; inflorescence mostly adpressed-scaly to smooth; sepals generally larger and united for more than 3 mm 3
2a. Floral bracts minutely toothed; petals shortly united at the base; outer bracts of the inflorescence small and not forming an obvious involucre **1. burchellii**
b. Floral bracts entire; petals united for half their length; outer bracts of the inflorescence forming an evident involucre **2. microps**
3a. Scape without bracts below the inflorescence for most of its length **3. seidelii**
b. Scape, when visible, completely covered by bracts 4
4a. Leaf spines to 4 mm **4. fulgens**
b. Leaf spines to *c.* 1 mm **5. innocentii**

1. N. burchellii (Baker) Mez *(Aechmea burchellii* Baker; *Cryptanthus emergens* Lindman; *Aregelia burchellii* (Baker) Mez*)*. Illustration: Flora Neotropica **14**: f. 523A–E (1979).

Stemless plant with long, slender runners. Leaves 20–50 cm, few, in a more or less cylindric rosette, short and broad to long and narrow; sheaths entire or toothed toward the apex, scaly with brown, more or less adpressed scales; blades 2–5 cm wide, somewhat narrowed towards the base, acute with a small point, hairless to obscurely scaly, green, tinged with red beneath, minutely toothed. Scape short, about equalling the leaf-sheaths, reddish brown-scurfy, upper bracts ovate, acute, shorter than the inflorescence, not forming an obvious involucre. Inflorescence spherical, densely reddish brown-woolly. Floral bracts minutely toothed. Sepals *c.* 10 mm, acute, symmetric or almost so, united at the base for 2 mm. Petals *c.* 1.8 cm, united for 1–2 mm, white; blades spreading, flat, acute. *Brazil.* G2.

2. N. microps Mez (*Aregelia microps* (Mez) Mez). Illustration: Flora Neotropica **14**: f. 523F–G (1979).

Like *N. burchellii* but leaves 30–60 cm, more or less erect in a tubular rosette, outer bracts forming a distinct involucre, floral bracts entire, sepals 1–1.4 cm, united at the base for *c.* 3 mm, petals 2–2.5 cm, united for about half their length. *Brazil.* G2.

3. N. seidelii L.B. Smith & Reitz. Illustration: Flora Neotropica **14**: f. 525A–C (1979).

Plant stemless. Leaves 50–80 cm, many, in a broadly funnel-shaped rosette, much longer than the inflorescence; sheaths entire, densely covered with brown-centred, grey scales; blades to 4 cm wide, broadly acute with a fine point, narrowed towards the base, obscurely scaly. Scape erect, finely reddish brown-woolly, but becoming almost hairless, bearing 2 bracts. Inflorescence more or less cylindric, 11–15 × up to 8 cm. Flowers stalkless. Sepals *c.* 1.4 cm, slightly asymmetric, united at the base for 2.5 mm, spathulate. Petals united at the extreme base, blades 10 mm, obtuse, white. *Brazil.* G2.

4. N. fulgens Lemaire (*Karatas fulgens* (Lemaire) Antoine).

Plant stemless. Leaves 30–40 cm, many, in a broad, spreading rosette; sheaths much wider than the blades, entire, densely scaly with brown, adpressed scales; blades slightly narrowed toward the base, broadly acute, acuminate and ending in a short spine, 3–6 cm wide, sparsely spiny with spines to 4 mm, hairless above, obscurely spotted-scaly beneath, pale green. Scape very short. Inflorescence sunk in the centre of the rosette, many-flowered. Flowers stalkless. Sepals 2–2.4 cm, acute, shortly united, red. Petals *c.* 5 cm, united for about half their length or more, the tube white, the blades erect, dark blue with white margins, broadly rounded. *Brazil.* G2.

5. N. innocentii Lemaire (*Karatas innocentii* (Lemaire) Antoine). Illustration: *Illustration Horticole* **9**: t. 329 (1862); Flora Neotropica **14**: f. 526A–D (1979).

Plant stemless, propagating by runners. Leaves 20–60 cm, many, in a dense, spreading rosette, very variable in colour; sheaths entire or minutely toothed toward the apex, pale green or purplish, with small brown scales; blades 4–5.5 cm wide, acute to broadly rounded, hairless, shiny, with short spines. Inflorescence sunk in the centre of the rosette. Flowers to 6 cm. Sepals slightly asymmetric, 2.2–3 cm, acute, united for 3–9 mm, white or sometimes reddish. Petals *c.* 5 cm, united at the base, white, green at the base. *Brazil.* G2.

Several varieties are known in cultivation, all of which are very striking plants. Var. **innocentii** is clearly distinguished by the leaves being dark red beneath or on both surfaces, while the young bracts in the centre of the rosette are red, with or without a green apex – in all other varieties the leaf-blades are green. In var. **wittmackianum** (Harms) L.B. Smith the young bracts are wholly or mostly reddish purple, while var. **striatum** Wittmack is similar to the last but has the leaf-blades with longitudinal white lines. Very close to var. **striatum** is var. **lineatum** (Mez) L.B. Smith, which differs in the colour of the young bracts, which are green with red near the apex. This bract colouring also applies to var. **paxianum** (Mez) L.B. Smith, in which the leaves are green, each with a broad, white, longitudinal line.

11. ACANTHOSTACHYS Klotzsch

D. Philcox

Herbs, usually epiphytic, initially erect but often eventually pendent, spreading by short, stout, basal runners. Leaves few, to 1 m, not clearly contracted between the small sheath and the blade, covered by a membrane of fused scales, densely so beneath, less densely above; sheaths dark purplish brown, entire; blades linear, to 1.2 cm wide at base, much narrower above, strongly channelled with margins inwardly directed, olive green to reddish brown, sparsely spiny with spines *c.* 1.5 mm. Scape with pale scales, slender, with 2 bracts at the apex. Inflorescence appearing lateral, simple, dense, ovoid or cylindric, cone-like, 3–7 × 2–3.5 cm, with a small tuft of reduced, sterile bracts at the apex. Floral bracts broadly ovate with a spreading, triangular-acuminate apex, toothed or entire, bright orange to red, at first with pale scales, soon hairless. Flowers stalkless, 2.2–2.5 cm. Sepals free, nearly symmetric, 8–11 mm, acute, bright yellow. Petals *c.* 1.6 cm, erect, free, yellow, with 2 scales above the base. Stamens not projecting from the corolla. Ovary with just the upper part superior, the rest inferior.

A genus of 1 species, very popular for use in hanging baskets because of its pendent habit and long-lasting inflorescence.

1. A. strobilacea (Schultes) Klotzsch (*Hohenbergia strobilacea* Schultes). Illustration: Flora Neotropica **14**: f. 560A–G (1979); Brickell (ed.), RHS A–Z encyclopedia of garden plants, 62 (2003).

Brazil, Paraguay & NE Argentina. G2.

12. FASCICULARIA Mez

D. Philcox

Terrestrial, stemless or shortly stemmed epiphytes. Leaves forming very dense rosettes, but without a central cup; blades linear, rigid, strongly spiny-margined. Inflorescence simple, dense, sunk in the centre of the rosette. Flowers stalkless or shortly stalked. Sepals free, keeled, scaly, especially toward the apex. Petals free, obtuse, fleshy, blue, with 2 small scales near the base. Stamens not projecting from the corolla. Ovary completely inferior.

A genus of 3–5 species restricted in origin to Chile.

Literature: Nelson, E.C. & Zizka, G., Fascicularia (Bromeliaceae): which species are cultivated and naturalized in northwestern Europe, *The New Plantsman* **4**(4): 232–9 (1997).

1a. Bracts all shorter than the flowers; flowers 4 cm or more **1. pitcairniifolia**
b. Outer bracts clearly longer than the flowers; flowers 3.3–4 cm **2. bicolor**

1. F. pitcairniifolia (Verlot) Mez (*Hechtia pitcairniifolia* Verlot; *Rhodostachys pitcairniifolia* (Verlot) Baker). Illustration: *Botanical Magazine*, 8087 (1906); Flora Neotropica **14**: f. 571A–F (1979); Brickell (ed.), RHS A–Z encyclopedia of garden plants, 444 (2003).

Plant stemless. Leaves to 1 m, spreading; sheaths conspicuous, triangular-ovate, brownish beneath, white above, covered on both sides with a white membrane, minutely toothed; blades to 1.8 cm wide, linear, leathery, glaucous-green, more or less hairless at maturity above, sparsely white-scaly beneath, margin with spines 1.5 mm, those at the base spreading, the others forwardly directed. Inflorescence many-flowered, surrounded by the bright red inner leaves. Floral bracts shorter than the flowers. Sepals 1.7–2.2 cm, rounded, often slightly notched and incurved at the apex, with narrow, translucent margins, scaly towards the apex. Petals 2.2–2.4 cm, erect with recurved, spreading apices. *Chile.* H5–G1.

2. F. bicolor (Ruiz & Pavon) Mez (*Bromelia bicolor* Ruiz & Pavon; *Rhodostachys bicolor* (Ruiz & Pavon) Baker). Illustration: Rauh, Bromeliads, t. 92 (1979); *The New Plantsman* **4**(4): 232, 233 (1997).

Plant stemless or with a short, stout, erect stem, the rosette spreading to over 1 m in diameter when flowering. Leaves to 50 cm or more, very numerous; sheaths elongate, mostly covered with coarse, reddish brown scales; blades *c.* 1.5 cm wide, linear, attenuate, the apex with a strong spine, hairless above, brown-scaly beneath, glaucous-green becoming red, spiny with the basal spines recurved, the rest spreading or forwardly directed. Inflorescence 20–40-flowered, with the outer bracts forming an involucre, bracts longer than the flowers, ivory white. Sepals to 1.2 cm, obtuse and irregularly twisted at the apex, the margins more or less translucent, generally scaly but scales deciduous towards the base. Petals *c.* 2 cm, with slightly recurved or spreading apices. *Chile.* H5–G1.

13. CANISTRUM Morren

D. Philcox

Perennial, usually epiphytic herbs. Leaves in rosettes, variously spiny-toothed. Scape evident. Inflorescence compound, densely corymbose, bracts of involucre covering all but the petal apices. Flowers shortly stalked or almost stalkless. Sepals free or nearly so. Petals free, with appendages near the base. Stamens not projecting from corolla. Ovary completely inferior.

A small genus of some 7 species from Brazil.

1. C. lindenii (Regel) Mez (*Nidularium lindenii* Regel).

Leaves in a broad funnel-like rosette; sheaths large, entire, covered with adpressed, dark brown scales; blades to 10 cm wide, strap-shaped, broadly acute or rounded, green with spots of darker green, with broad spines to 3 mm. Scape stout, covered with dark brown wool. Inflorescence 7–8 cm in diameter without the bracts, with about 100 flowers, bracts yellowish white to almost white, faintly green at the apex. Flowers 3–3.5 cm, shortly stalked. Sepals 1.2–1.8 cm, free, whitish. Petals free, slightly longer than the sepals, white or pale green toward the apex, with 2 fringed scales at the base. *Brazil.* G2.

A very variable plant. In cultivation, the most common variety appears to be var. **roseum** (Morren) L.B. Smith (*C. roseum* Morren), which differs from the variety above in having a smaller inflorescence of some 50–90 flowers, surrounded by primary and outer bracts of rose to bright red. Illustration: Rauh, Bromeliads, t. 94 (1979); Brickell (ed.), RHS A–Z encyclopedia of garden plants, 227 (2003).

14. AECHMEA Ruiz & Pavon

D. Philcox

Epiphytes, usually stemless, of medium size, frequently spreading by short, basal runners. Leaves usually many, either in dense rosettes or grouped together with the sheaths often forming a tank, usually spiny-margined. Scape usually well developed. Inflorescence branched or simple, flowers stalkless or stalked, arranged in 2 opposite rows or in many rows. Sepals free or united, usually asymmetric, each sharply pointed or with an apical spine. Petals free, usually with obvious basal scales. Stamens shorter than the petals. Ovary inferior.

A genus of more than 170 species from C & S America and the West Indies. Many species not marketed have very striking inflorescences and would make ideal plants for cultivation, but their large size at flowering prevents this.

1a. Flowers stalked, stalks *c.* 6 mm **1. lueddemanniana**
b. Flowers stalkless or stalks much less than 6 mm 2
2a. Floral bracts attached to the axis by at least the lower parts of their sides as well as by their bases 3
b. Floral bracts attached to the axis only by their bases, sometimes very small or absent 5
3a. Flowers in many ranks; floral bracts finely toothed **4. fasciata**
b. Flowers in 2 ranks; floral bracts entire 4
4a. Floral bracts narrowly ovate, their margins attached to the axis only towards their bases **6. chantinii**
b. Floral bracts attached to the axis by most of their margins, forming cups **5. distichantha**
5a. Sepals acute but not spine-tipped 6
b. Sepals spine-tipped 8
6a. Inflorescence branched, at least in the lower half 7
b. Inflorescence unbranched **8. bromeliifolia**
7a. Inflorescence branched throughout **2. miniata**
b. Inflorescence branched only in the lower half **3. fulgens**
8a. Sepals free or almost so 9
b. Sepals joined for at least one-third of their length 10
9a. Inflorescence branched **7. purpureo-rosea**
b. Inflorescence completely unbranched **9. nudicaulis**
10a. Scape completely hidden by the leaf-sheaths **10. recurvata**
b. Scape visible above the leaf-sheaths 11
11a. Inflorescence branched, at least at the base 12
b. Inflorescence completely unbranched 14
12a. Petals yellow, drying purple **11. caudata**
b. Petals blue 13
13a. Scape persistently woolly **12. coelestis**
b. Scape hairy at first, soon hairless **13. gracilis**
14a. Petals yellow 15
b. Petals blue or purple 17
15a. Inflorescence loose, the axis visible **14. blumenavii**
b. Inflorescence dense, the axis mostly hidden by flowers and bracts 16
16a. Floral bracts broadly ovate, enfolding the base of the ovary; sepal blade *c.* 2 times as long as spine **15. lindenii**
b. Floral bracts narrowly triangular, not enfolding the base of the ovary; sepal spine and blade more or less equal **16. calyculata**
17a. Inflorescence 5–6 cm, with few (to 25) flowers; flowers 2–2.5 cm, petals 1.4–1.6 cm **13. gracilis**
b. Inflorescence 9–26 cm, with many (to 100) flowers; flowers *c.* 1.5 cm, petals *c.* 9 mm **17. gamosepala**

1. A. lueddemanniana (K. Koch) Mez (*Pironneava luddemanniana* K. Koch; *A. caerulea* Morren). Illustration: Flora Neotropica **14**: f. 597D–F (1979); Rauh, Bromeliads, t. 192 (1979).

Plant to 70 cm. Leaves 30–60 cm, forming a cup-shaped rosette, almost straight, densely pale-scaly, especially beneath; sheaths large; blades *c.* 4.5 cm wide, acute or rounded with a short strong point, margin with spines 1–2 mm. Scape slender, white-mealy. Inflorescence paniculate, cylindric, 12–30 × 5–10 cm, white-mealy. Flower-stalks *c.* 6 mm. Sepals *c.* 3.5 mm, each with a broad, lateral wing, mucronate, free. Petals *c.* 9 mm, pink and blue turning dark carmine-red. Fruit white becoming bright purple. *Mexico, Guatemala & Belize.* G2.

2. A. miniata (Beer) Baker (*Lamprococcus miniatus* Beer).

Leaves 25–54 cm, exceeding the inflorescence, in a dense, funnel-like rosette; sheath scaly with minute brown scales; blades 3–4 cm wide, narrowed toward the base, more or less acute or rounded and shortly pointed, hairless or finely scaly, green, more or less densely spiny with minute spines. Scape slender, hairless. Inflorescence 8–16 cm, branched to the top, ellipsoid or pyramidal, entirely red except for the petals; branches 2–10-flowered, spreading. Floral bracts minute. Flowers to 1.4 cm, stalkless. Sepals free, *c.* 4 mm, asymmetric, without spine-tips. Petals *c.* 10 mm, obtuse, blue. *Brazil.* G2.

A variety more commonly found in cultivation than that described above is var. **discolor** (Beer) Baker, which differs in having the leaves tinged with red, and shiny rather than scaly (illustration: Rauh, Bromeliads, t. 73, 1979).

3. A. fulgens Brongniart. Illustration: Brickell (ed.), RHS A–Z encyclopedia of garden plants, 81 (2003)

Leaves 30–40 cm, in a dense, funnel-like rosette; sheaths elliptic, minutely brown-scaly; blades 4–6.5 cm wide, broadly acute or rounded and shortly pointed, green, margins with minute spines. Scape red, hairless. Inflorescence 15–20 × 6–7 cm, narrowly pyramidal, branched at the base, but simple from the middle to the top, hairless. Floral bracts absent. Flowers to 2.2 cm, stalkless, spreading. Sepals to 5 mm, nearly free, asymmetric, without spine-tips, dark purple. Petals *c.* 1.2 cm, obtuse, blue or purple at first, soon turning red. *Brazil.* G2.

The commonest plant marketed is var. **discolor** (Morren) Brongniart, which has the under surface of the leaves reddish purple (illustration: Flora Neotropica **14**: f. 603A–E, 1979; Rauh, Bromeliads, t. 73, 1979).

4. A. fasciata (Lindley) Baker (*Billbergia fasciata* Lindley; *B. rhodocyanea* Lemaire; *A. rhodocyanea* invalid). Illustration; *Botanical Magazine*, 4883 (1855); Flora Neotropica **14**: f. 644C (1979); Rauh, Bromeliads, t. 68 (1979); Brickell (ed.), RHS A–Z encyclopedia of garden plants, 81 (2003).

Leaves 30–100 cm, forming a cylindric or funnel-like rosette, covered with pale, adpressed scales, pale-banded beneath; sheaths slightly wider than the blades, entire, sometimes purple-tinged; blades 3–8 cm wide, broadly rounded and short-tipped, margins with black spines that are 2–4 mm. Scape slender, white-woolly. Inflorescence 7–8 cm, simple or with a few branches at the base, densely pyramidal, white, shortly woolly. Floral bracts exceeding the sepals, pinkish red, the margins shortly attached to the axis and forming a pouch. Flowers 3–3.5 cm, in several ranks, stalkless. Sepals 1–1.2 cm, united for half their length, asymmetric, acute or with a short point. Petals 2.5–3 cm, blue or purple, red when dry. *Brazil.* G2.

This species has been sold for a number of years by many of the major suppliers under the invalid and incorrect name *A. rhodocyanea.* It is one of the best-known bromeliads, having been cultivated for over 150 years, and almost certainly the most popular.

5. A. distichantha Lemaire (*A. excavata* Baker; *A. platyphylla* Hassler). Illustration: *Botanical Magazine*, 5447 (1864); Rauh, Bromeliads, t. 71 (1979); Brickell (ed.), RHS A–Z encyclopedia of garden plants, 81 (2003).

Leaves 30–100 cm, forming a very dense rosette, covered with a membrane of fused scales; sheaths usually much wider than the blades, toothed toward the apex, entire elsewhere; blades 2.5–8 cm wide, rounded and short-tipped, margins with stout spines 3–5 mm long. Scape slender, white-woolly. Inflorescence always bipinnate, dense or loose, ovoid to cylindric, pink and white-woolly, except for the petals; spikes erect to spreading, the laterals with 2–12 stalkless flowers in 2 ranks, the terminal with more and many-ranked flowers. Floral bracts entire, attached to the axis by their margins for most of their length, forming a cup. Sepals 5–13 mm, free or shortly united, asymmetric. Petals purple or blue. *Brazil, Paraguay, Uruguay & Argentina.* G2.

6. A. chantinii (Carrière) Baker (*Billbergia chantinii* Carrière; *A. amazonica* Ule). Illustration: Rauh, Bromeliads, t. 66 (1979); Brickell (ed.), RHS A–Z encyclopedia of garden plants, 80 (2003).

Leaves 40–100 cm, densely scaly; sheaths large, chestnut-brown; blades 6–9 cm wide, acute or rounded and shortly spine-tipped, pale olive green with white banding beneath (though in some the leaf colour can range from almost black to red-tinged). Scape slender, white-woolly; bracts lanceolate, bright red. Inflorescence bipinnate; spikes long-stalked, with *c.* 12 flowers borne in 2 ranks. Floral bracts 1–1.3 cm, slightly longer than the ovary, scaly, attached to the axis by the bases of their margins. Flowers to 3.2 cm, stalkless. Sepals 1–1.2 cm, asymmetric, shortly united, without points. Petals *c.* 2 cm, obtuse, orange. *Colombia to Peru & Amazonian Brazil.* G2.

7. A. purpureo-rosea (Hooker) Wawra (*Billbergia purpureo-rosea* Hooker; *A. suaveolens* Knowles & Westcott). Illustration: *Botanical Magazine*, 3304 (1834).

Leaves 50–100 cm, in a dense tubular rosette; sheaths large, entire, covered with minute, adpressed, brown scales; blades of 2 types, the outer narrowly triangular and recurved, the inner strap-shaped, rounded, spine-tipped and erect, 3–6 cm wide, margins with black spines which are 2–5 mm, shiny green above, usually hairless. Scape slender, red, white-woolly. Inflorescence paniculate, many-flowered, 15–30 × 8–12 cm, white-woolly on the axis, elsewhere glandular-hairy; lower bracts shorter than the branches; branches spreading, each with up to 12 stalkless flowers. Floral bracts ovate, ending in long, slender spines. Sepals to 6 mm, asymmetric, ending in short spines, very shortly united, pinkish red. Petals to 1.5 cm, purple or blue. *Brazil.* G2.

8. A. bromeliifolia (Rudge) Baker (*Tillandsia bromeliaefolia* Rudge; *A. tinctoria* (Martius) Mez). Illustration: Flora Neotropica **14**: f. 661A–H (1979).

Leaves 60–100 cm, in a tubular rosette, covered with a membrane of white, united scales; sheaths usually much broader than the blades, entire except towards the top; blades 4–9 cm wide, acuminate to rounded with a short spine-tip or notched, green or rarely tinged with red, margins with spines to 1 cm. Scape stout, densely woolly. Inflorescence to 15 × 3–4 cm, simple, cylindric or ellipsoid, densely white-woolly with only the petals exposed at first. Floral bracts 2-keeled, leathery, shorter than the sepals. Flowers stalkless. Sepals *c.* 7 mm, leathery. Petals *c.* 1.5 cm, greenish yellow soon turning black. *C. America to Argentina.* G2.

9. A. nudicaulis (Linnaeus) Grisebach (*Bromelia nudicaulis* Linnaeus; *Billbergia pyramidata* Beer). Illustration: *Edwards's Botanical Register* **3**: t. 203 (1817); Reichenbach, Flora exotica **3**: t. 185 (1835); Rauh, Bromeliads, t. 74 (1979); Brickell (ed.), RHS A–Z encyclopedia of garden plants, 81 (2003).

Leaves 30–100 cm; sheaths large, forming a pitcher-shaped rosette, purple to purplish brown, densely and finely brown-scaly; blades 6–10 cm wide, broadly obtuse and shortly pointed, leathery, pale-scaly beneath, hairless above, margins coarsely spiny. Scape slender, erect to arching, white-woolly. Inflorescence simple, 5–25 cm, cylindric, pale woolly-scaly, soon becoming hairless. Floral bracts short and entire, or absent. Flowers *c.* 2.2 cm, stalkless. Sepals 5–10 mm, asymmetric, spine-tipped, free. Petals *c.* 1.2 cm, yellow. *Mexico & West Indies to northern S America & Brazil.* G2.

Several varieties of this species are known but it is uncertain how many of them have reached the European market. They are not totally dissimilar but var. **aureo-rosea** (Antoine) L.B. Smith is particularly outstanding by reason of its red petals and red-tinged calyx.

10. A. recurvata (Klotzsch) L.B. Smith (*Macrochordium recurvatum* Klotzsch; *A. legrelliana* (Baker) Baker). Illustration: Flora Neotropica **14**: f. 632H–K (1979); Rauh, Bromeliads, t. 80 (1979); Brickell (ed.), RHS A–Z encyclopedia of garden plants, 81 (2003).

Leaves 25–40 cm, forming compact, dense rosettes; sheaths large, sometimes longer than the blades, ovate and much broader than the blades, forming an ellipsoid tank, densely scaly; blades 1–2 cm wide, narrowly triangular, spreading abruptly or recurving just above the sheath, thick, green to reddish green, those in the centre becoming bright red during flowering, toothed. Scape short, completely hidden by leaf-sheaths. Inflorescence simple, few-flowered, ellipsoid or obovoid, scape white-woolly or hairless. Floral bracts equalling or slightly longer than the sepals, red. Flowers stalkless, 3.5–4.5 cm. Sepals 1.1–1.5 cm excluding the spine, which is *c.* 4 mm, slightly asymmetric, united from the base for more than one-third of their length. Petals to 3 cm, purple or pinkish red. *S Brazil, Paraguay, Uruguay & NE Argentina.* G2.

Other than the variety above, 2 others are met with in cultivation and both have their inflorescences within the leaf-sheaths. One of these, var. **ortgiesii** (Baker) Reitz (*A. ortgiesii* Baker) is distinguished by having strongly toothed leaves and bracts, while var. **benrathii** (Mez) Reitz (*A. benrathii* Mez) has leaves and bracts entire or nearly so.

11. A. caudata Lindman. Illustration: Flora Neotropica **14**: f. 634A–E (1979); Rauh, Bromeliads, t.182 (1979).

Leaves 50–100 cm, forming broad, funnel-like rosettes, minutely adpressed-scaly; sheaths ovate or elliptic, brown toward the base, entire; blades to 8 cm wide but often narrower, broadly rounded and sharply pointed, margins with spines to 1 mm. Scape white-woolly. Inflorescence paniculate, 10–25 × *c.* 11 cm, bipinnate to the middle or above, white-woolly. Flowers stalkless. Floral bracts 7–17 mm, red, ovate, each tipped by a long brown spine. Sepals 7–11 mm, including the long spine, united. Petals 1.2–1.5 cm, yellow, turning purplish on drying. *Brazil.* G2.

Var. **variegata** Foster is commonly seen on the market and is easily distinguished by its dark green leaves with longitudinal, creamy stripes and a pink tinge on new growth and near the base of older plants.

12. A. coelestis (K. Koch) Morren (*Hoplophytum coeleste* K. Koch). Illustration: Flora Neotropica **14**: f. 636A–B (1979); Rauh, Bromeliads, t. 183 (1979).

Leaves 30–100 cm, more or less erect in a dense rosette, scaly, especially beneath; sheaths large, scarcely wider than the blades, entire; blades 3–5 cm wide, broadly acute or rounded, margin with spines *c.* 0.5 mm, sometimes dark green-banded beneath. Scape slender, densely and persistently white-woolly. Inflorescence to 12 cm, paniculate, bipinnate at the base, simple above, densely white-woolly. Floral bracts ovate, red, each ending in a slender spine. Flowers stalkless, *c.* 2 cm. Sepals 3.5–6 mm excluding the long terminal spine, asymmetric, united for about one-third of their length. Petals *c.* 1.3 cm, blue. *Brazil.* G2.

Var. **albo-marginata** Foster is a very attractive plant that has dark green leaves banded, and with a regular and clearly defined, white margin.

13. A. gracilis Lindman. Illustration: Flora Neotropica **14**: f. 637A–F (1979).

Leaves 50–70 cm, in a more or less cylindric rosette; sheaths elliptic, purple-tinged; blades 3–4 cm wide, broadly rounded and shortly pointed, minutely toothed or almost entire, pale beneath and sometimes banded. Scape slender, woolly at first, the scales soon deciduous. Inflorescence 5–6 cm long, simple or with a few branches at the base, ovoid to pyramidal, hairless, with few flowers. Floral bracts ovate, slender-spined, red. Flowers 2–2.5 cm, stalkless, spreading. Sepals 6–7 mm (excluding the terminal spine, which is 3–5 mm), united for 3 mm. Petals 1.4–1.6 cm, pale blue with a darker apex. *Brazil.* G2.

14. A. blumenavii Reitz. Illustration: Flora Neotropica **14**: f. 638G–J (1979).

Leaves 40–70 cm, in a funnel-like rosette, curved, spreading; sheaths entire, scaly, tinged with violet above; blades *c.* 4 cm wide, broadly rounded or more or less truncate, shortly pointed, minutely toothed, often white-banded beneath. Scape slender, more or less hairless, dark violet. Inflorescence to 9 × 3.5 cm, simple, cylindric, with the flowers more or less whorled, loose, the axis visible. Floral bracts triangular, acuminate, with the lower quarter enfolding the ovary. Flowers *c.* 1.6 cm, stalkless. Sepals 4 mm (with a spine *c.* 3 mm), asymmetric with a large wing, united for half their length. Petals free, yellow, hooded at the apex. *Brazil.* G2.

15. A. lindenii (Morren) Baker (*Hoplophytum lindenii* Morren). Illustration: *Botanical Magazine*, 6565 (1881); Flora Neotropica **14**: f. 640A–E (1979).

Leaves 1–1.5 m in a rosette; sheaths broadly ovate, minutely brown-scaly, sometimes purplish above; blades to 5.5 cm wide, broadly rounded and shortly spine-tipped, green, covered with a whitish coat of more or less fused scales (especially beneath), margins with spines to 2.5 mm. Scape white-hairy. Inflorescence 7–8 × 4 cm, simple, cylindric or ellipsoid, with many flowers and bracts obscuring the axis. Floral bracts broadly ovate, enfolding the ovary, entire, red. Flowers *c.* 2 cm, stalkless. Sepals *c.* 5 mm (excluding the spine, which is *c.* 2 mm), asymmetric, shortly united, red. Petals yellow, turning brown. *Brazil.* G2.

A variety with the leaf-blades striped with yellow, var. **makoyana** Mez, is known from cultivation.

16. A. calyculata (Morren) Baker (*Hoplophytum calyculatum* Morren). Illustration: *La Belgique Horticole* **15**: 162 (1865); Flora Neotropica **14**: f. 640F–K (1979).

Leaves 30–100 cm, in a funnel-like rosette, minutely scaly; sheaths inconspicuous, slightly darker than the blades; blades 3–4 cm wide, broadly rounded, minutely spine-tipped, minutely toothed. Scape slender, white-woolly. Inflorescence 4–6.5 × 3–5 cm, usually simple, dense, the axis obscured by flowers and bracts, more or less white-woolly, except for the petals. Floral bracts narrowly triangular with a slender-spined tip, entire. Flowers stalkless. Sepals 3–5 mm (terminal spines slender, 3–4 mm), asymmetric, united for one-third to half their length, yellow. Petals *c.* 1.2 cm, yellow. *Brazil & N Argentina.* G2.

17. A. gamosepala Wittmack. Illustration: Flora Neotropica **14**: f. 642A–E (1979); Brickell (ed.), RHS A–Z encyclopedia of garden plants, 81 (2003).

Leaves 25–55 cm, in a dense, funnel-like rosette, pale-scaly especially beneath; sheaths narrow; blades 3–5 cm wide, nearly entire or with a few minute spines near the apex, broadly rounded, shortly spine-tipped. Scape slender, sparsely white-woolly, soon hairless. Inflorescence 9–26 cm, simple, cylindric, with numerous flowers. Floral bracts narrowly triangular, spine-tipped, entire, the lower exceeding the flower, the upper shorter than the ovary. Flowers *c.* 1.5 cm, stalkless. Sepals 4–5 mm (including the spine, which is 2.5–4 mm), asymmetric, united for half their length, hairless, red. Petals *c.* 9 mm, purple or blue. *Brazil.* G2.

15. QUESNELIA Gaudichaud

D. Philcox

Epiphytes, stemless, or rarely with a long stem. Leaves few or many, forming open, tubular or funnel-like rosettes; blades broadly linear, minutely toothed. Scape evident, well developed. Inflorescence simple or rarely with a few branches, the flowers in more than 2 ranks. Floral bracts usually showy. Sepals 1–3 cm, asymmetric or symmetric, free or shortly united. Petals free, 2.5 cm or more, usually erect, each with 2 scales near the base. Stamens shorter than the petals and not projecting from the corolla. Ovary completely inferior. Fruit a berry.

A genus of some 14 species from Brazil.

1a. Inflorescence dense **1. seideliana**
b. Inflorescence loose 2
2a. Leaves banded, in 2 ranks; sepals to 10 mm; floral bracts minute **2. marmorata**
b. Leaves not banded, not obviously in 2 ranks; sepals to 2.3 cm; floral bracts evident **3. liboniana**

1. Q. seideliana L.B. Smith & Reitz. Illustration: Flora Neotropica **14**: f. 681A–C (1979).

Leaves 35–40 cm, few in a tufted rosette, covered with white, closely adpressed, dark-centred scales; sheaths to 20 cm, tinged with dark purple; blades to 4.5 cm wide, rounded and spine-tipped, green, margins with dark spines to 1 mm. Scape slender, erect, finely white-woolly, becoming hairless; bracts few, remote, thin, enfolding the scape, entire. Inflorescence *c.* 3 cm, dense, simple, ellipsoid. Floral bracts *c.* 2 cm, broadly ovate, acute, about as long as the sepals. Flowers stalkless. Sepals *c.* 1.5 cm, white-scaly, united at the base for 3 mm. Petals *c.* 2.5 cm, bright sky-blue. *Brazil.* G2.

2. Q. marmorata (Lemaire) Read (*Billbergia marmorata* Lemaire; *Aechmea marmorata* (Lemaire) Mez). Illustration: *Illustration Horticole* **2**: t. 48 (1855); Flora Neotropica **14**: f. 682A–E (1979).

Leaves 40–60 cm, in a more or less cylindric rosette, clearly 2-ranked, irregularly banded with green and brown beneath; sheaths nearly as long as the blades, entire; blades 5–7 cm wide, oblong, broadly rounded or truncate, apically spined, distantly toothed, green. Scape short and erect, or longer, slender and curved downwards, hairless, red. Inflorescence 12–21 cm, loose, pyramidal, bipinnate at the base, hairless; bracts all large, some shorter than the branches, some longer, spreading or reflexed. Floral bracts minute, broadly triangular, entire. Flowers to 3 cm, stalkless. Sepals to 10 mm, strongly asymmetric, obtuse, unarmed, shortly united at the base, purple. Petals to 2.5 cm, purple or blue. *Brazil.* G2.

3. Q. liboniana (De Jonghe) Mez (*Billbergia liboniana* De Jonghe). Illustration: *Botanical Magazine*, 5090 (1858); Flora Neotropica **14**: f. 683A–E (1979); Brickell (ed.), RHS A–Z encyclopedia of garden plants, 876 (2003).

Leaves to 80 cm, in a slender, funnel-like rosette, not clearly in 2 ranks; sheaths elongate, covered with pale, brown-centred scales; blades 3–4.5 cm wide, broadly linear, acute, toothed, pale-scaly, sometimes banded beneath. Scape slender, erect. Inflorescence simple or sometimes with a 1–2-flowered branch at the base. Floral bracts narrowly triangular, the lower exceeding the ovary, the upper acute, much shorter. Flowers stalkless. Sepals to 2.3 cm, shortly united, red, hairless. Petals to 5 cm, dark blue. *Brazil.* G2.

16. BILLBERGIA Thunberg

D. Philcox

Epiphytes, without stems or nearly so. Leaves forming broadly funnel-like to long-tubular rosettes; blades mostly broadly linear with broad apices, often banded or spotted, variously toothed or very rarely entire. Scape erect or curved downwards; bracts large, thin, usually red. Inflorescence compound or simple. Flowers large, showy. Sepals 1 cm or more, free, hairless to densely mealy or woolly. Petals 2.5 cm or more, free, each with 2 scales near the base, blade long and narrow. Stamens projecting.

A genus of over 50 species mostly native to eastern Brazil but with some species found in Central America and south along the Atlantic coast to Argentina.

1a. Petals straight or loosely recurved, contorted as they die off; inflorescence often compound, axis often hairless 2
b. Petals coiled in a tight spiral, remaining so as they die off; inflorescence always simple, densely mealy 9
2a. Inflorescence completely hairless or rarely with a few scales on the sepal apices 3
b. Inflorescence scaly, at least on the bracts or sepals 8
3a. Flowers stalkless 4
b. Flowers stalked 6
4a. Floral bracts all large, equalling or exceeding the sepals **1. iridifolia**
b. Floral bracts much reduced toward the apex of the inflorescence, or all much shorter than the sepals 5
5a. Ovary truncate at the base; petals blue at the apex or wholly green; leaf-blades to 5 cm wide **2. distachia**
b. Ovary attenuate at the base; petals blue on the margins; leaf-blades 6–17 mm wide **3. nutans**

6a. Petals blue on the margins **3. nutans**
b. Petals blue at the apex only 7
7a. Sepals acute; leaves not spotted **4. lietzei**
b. Sepals broadly rounded and minutely spine-tipped; leaves spotted **5. leptopoda**
8a. Flower-stalks 5–10 mm; inflorescence more or less pendent; leaves spotted **6. chlorosticta**
b. Flowers stalkless or stalks less than 5 mm; inflorescence erect; leaves not spotted **7. pyramidalis**
9a. Ovary slightly or not at all furrowed or grooved, wholly white-mealy **8. decora**
b. Ovary furrowed or grooved with the ridges hairless, smooth and dark **9. zebrina**

1. B. iridifolia (Nees & Martius) Lindley (*Bromelia iridifolia* Nees & Martius). Illustration: *Edwards's Botanical Register* **13**: t. 1068 (1827); Flora Neotropica **14**: f. 688A–C (1979).

Leaves to 60 cm, few, in a short, tubular rosette, minutely pale-scaly beneath; sheaths large, entire; blades 2.5–4.5 cm wide, recurved, acute or slenderly acuminate, sparingly and minutely toothed to more or less entire. Scape slender, curved downwards, red, hairless; bracts large, acute, red, the upper overlapping. Inflorescence simple, pendent, loose and few-flowered, hairless. Floral bracts all exceeding the sepals. Flowers 5–6 cm, very shortly stalked. Sepals 2–2.5 cm, rounded, with a fine soft apex, more or less symmetric, usually dark blue toward the apex, red elsewhere. Petals 4–5 cm, blue at the apex, yellow elsewhere. *Brazil.* G2.

In var. **concolor** L.B. Smith, the petals are pale yellow throughout.

2. B. distachia (Vellozo) Mez (*Tillandsia distachia* Vellozo; *Billbergia caespitosa* Lindman). Illustration: Flora Neotropica **14**: f. 688D–E (1979).

Leaves 25–90 cm, few, in a slender, cylindric rosette; sheaths much wider than the blades; blades to 5 cm wide, narrowly triangular and attenuate, sometimes linear and rounded, sparingly and minutely toothed or entire, sparsely scaly beneath, not banded. Scape slender, curved downwards, hairless; bracts erect, overlapping, pinkish red, white-scaly beneath. Inflorescence with short, spreading branches, loose and few-flowered, hairless. Floral bracts much shorter than the ovary. Flowers stalkless. Sepals 1.7–2.5 cm, pale green, blue at the apex. Petals 4–5 cm, linear, obtuse, pale green, blue at the apex. *Brazil.* G2.

3. B. nutans Regel. Illustration: Flora Neotropica **14**: f. 688F–H (1979); Rauh, Bromeliads, t. 88 (1979); Brickell (ed.), RHS A–Z encyclopedia of garden plants, 182 (2003).

Leaves 30–70 cm, in clusters, rarely in rosettes; sheaths narrowly ovate, densely covered with small, white scales; blades linear or narrowly triangular, 6–17 mm wide, scaly beneath, green, margins with short, slender spines. Scape curved downwards, very slender, hairless; bracts erect, overlapping, pinkish red. Inflorescence with very short branches, usually few-flowered, hairless. Floral bracts much shorter than the ovaries. Flowers stalkless or shortly stalked, in 2 ranks. Sepals 1.5–2 cm, obtuse, pinkish red with dark blue margins. Petals 3.3–4.6 cm, linear, obtuse, pale green with dark blue margins, except for the extreme apex. *Brazil, Paraguay, Uruguay & Argentina.* G1.

Var. **schimperiana** (Baker) Mez has entire leaves and the petals blue at the extreme apex as well as on the margins.

The species has been crossed with *Cryptanthus bueckeri* and the resulting hybrid grown under the name × **Cryptbergia meadii** Wilson & Wilson.

4. B. lietzei Morren. Illustration: *La Belgique Horticole* **31**: t. 5–7 (1881); Flora Neotropica **14**: f. 689A–D (1979).

Leaves 35–55 cm, in a short, tubular rosette, whitish beneath but not banded; blades 2–3 cm wide, linear, acute and mucronate, margins with spines to 2 mm. Scape slender, red, hairless; bracts overlapping, papery, red, sometimes scaly towards the apex. Inflorescence simple, loose, 6–13-flowered. Floral bracts lanceolate, like the scape-bracts. Flower-stalks 3–5 mm, slender. Flowers 5–6 cm. Sepals 1.5–2 cm, red. Petals *c.* 4 cm, linear, blue toward the apex, the remainder greenish yellow or sometimes wholly green. *Brazil.* G2.

5. B. leptopoda L.B. Smith. Illustration: Flora Neotropica **14**: f. 690A–B (1979); Rauh, Bromeliads, t. 86 (1979).

Leaves 20–30 cm, few, in a short, tubular rosette, pale-scaly beneath, pale-spotted; sheaths large, pale violet; blades 2.5–3.5 cm wide, linear, acute, margins with spines to 2 mm. Scape erect, slender, hairless; bracts 4–5 cm, pinkish red, the upper overlapping. Inflorescence simple, few-flowered, hairless. Floral bracts exceeding the ovaries. Flower-stalks slender, 5–20 mm. Sepals 1.8–2.1 cm. Petals 3.5–4.5 cm, linear, green except for the dark blue apex. *Brazil.* G2.

6. B. chlorosticta Saunders (*B. saundersii* Dombrain). Illustration: Flora Neotropica **14**: f. 690A–B (1979).

Leaves 40–80 cm, usually 5–7, forming a tubular rosette; sheaths large; blades 3.5–4 cm wide, linear, broadly rounded or acute, shortly pointed, margins with spines to 1 mm. Scape slender, more or less curved downwards, white-mealy; bracts large, lanceolate, red. Inflorescence simple, more or less pendent, 7–17-flowered, white-mealy. Lowest floral bracts large, like the scape-bracts, the others minute, acuminate. Flower-stalks slender, 5–10 mm. Flowers 5–6 cm. Sepals *c.* 2 cm, asymmetric, red. Petals to 5.5 cm, linear, greenish yellow, dark blue towards the apex. *Brazil.* G2.

7. B. pyramidalis (Sims) Lindley (*Bromelia pyramidalis* Sims; *Billbergia thyrsoidea* Schultes). Illustration: *Botanical Magazine*, 1732 (1815); Flora Neotropica **14**: f. 701A–D (1979); Rauh, Bromeliads, t. 89 (1979); Brickell (ed.), RHS A–Z encyclopedia of garden plants, 182 (2003).

Leaves 40–100 cm, in a tubular rosette, often white-banded beneath; sheaths large, entire, somewhat purple-tinged, with a membrane of fused scales; blades 4–6 cm wide, broadly acute or more or less rounded, spine-tipped, minutely toothed. Scape erect, stout, white-mealy; bracts overlapping, with the upper massed below the inflorescence, rose. Inflorescence erect, to 15 cm, densely pyramidal or cylindric, densely mealy. Floral bracts minute, ovate, acute. Flowers with short stalks. Sepals 1.3–1.8 cm, slightly asymmetric, shortly united, pale red. Petals to 5 cm, red, blue toward the apex. *E Brazil* G2.

Also recorded from the West Indies, Venezuela, Paraguay and W Brazil, but doubtfully native from these areas. A variety having petals that are red throughout, var. **concolor** L.B. Smith, is known from cultivation.

8. B. decora Poeppig & Endlicher. Illustration: *Botanical Magazine*, 6937 (1887); Flora Neotropica **14**: f. 708 (1979).

Leaves to 70 cm, in a tubular or funnel-like rosette, white-banded beneath and sometimes yellow-spotted; sheaths large; blades to 6 cm wide, broadly acute or rounded and shortly spine-tipped. Scape curved downwards, white-mealy; bracts large, entire, bright pink, the upper ones massed below the inflorescence. Inflorescence simple, pendent, 10–15 cm, densely white-mealy. Floral bracts minute or lacking. Flowers 9–10 cm, stalkless or almost so. Sepals *c.* 1.2 cm, oblong to ovate. Petals to 7 cm, spirally recurved, green. Stamens green or greenish yellow. Ovary hardly or not at all grooved, without dark, hairless lines. *Amazonian Peru, Bolivia & Brazil.* G2.

9. B. zebrina (Herbert) Lindley (*Bromelia zebrina* Herbert). Illustration: *Botanical Magazine*, 2686 (1826); Flora Neotropica **14**: f. 710A–G (1979); Rauh, Bromeliads, t. 91 (1979).

Leaves to 1 m, about 6 in a tubular rosette, usually white-banded; sheaths large, entire; blades to 8 cm wide, broadly acute, adpressed-scaly, margins with spines to 4 mm. Scape curved downwards, pale-woolly; bracts large, entire, papery, pink, massed below the inflorescence. Inflorescence 15–40 cm, simple, pendent, many-flowered, white-mealy. Floral bracts minute, hidden by the meal. Flowers 7–8 cm. Sepals c. 8 mm, truncate. Petals to 6.5 cm, spirally recurved, linear, green becoming yellow. Ovary with thick, dark, longitudinal, hairless folds near the apex. *Brazil, Paraguay, Uruguay & NE Argentina.* G2.

17. ANANAS Miller
D. Philcox
Terrestrial herbs, not producing runners. Leaves scarcely enlarged at the base, forming a dense rosette. Scape evident, erect. Inflorescence terminal on the scape, densely cone-like, usually crowned with a tuft of sterile, leaf-like bracts. Flowers stalkless. Sepals free, obtuse, slightly asymmetric. Petals free, erect, violet or red, each bearing 2 slender, funnel-like scales toward the base. Stamens not projecting from the corolla. Ovaries combining with each other, the bracts and the axis to form a fleshy, compound fruit (syncarp).

A genus of 8 species from S America, of which only 2 are to be commonly found in cultivation, including the common edible pineapple.

1a. Floral bracts conspicuous, overlapping and covering the ovaries **1. bracteatus**
b. Floral bracts inconspicuous, not covering the ovaries at maturity **2. comosus**

1. A. bracteatus (Lindley) Schultes (*Ananassa bracteata* Lindley; *Ananas sativus* var. *bracteatus* (Lindley) Mez). Illustration: *Edwards's Botanical Register* **13**: t.1081 (1827); *Botanical Magazine*, 5025 (1858); Flora Neotropica **14**: f. 729A–B (1979).

Leaves to 1 m or more × 4 cm, margins coarsely spiny, spines all ascending and up to 10 mm. Scape to 50 cm. Floral bracts conspicuous, to 3.5 cm, coarsely toothed, overlapping and covering the ovaries, coloured at maturity. Flowers to 2.5 cm. Petals violet, becoming reddish. Fruit 15 × 12 cm or more, succulent, orange-red. *Brazil, Paraguay & Argentina.* G2.

A variety (var. **tricolor** (Bertoni) L.B. Smith), with the leaves variegated with longitudinal stripes, is sometimes grown.

2. A. comosus (Linnaeus) Merrill (*Bromelia comosa* Linnaeus; *A. sativus* Schultes). Illustration: Flora Neotropica **14**: f. 730 (1979).

Leaves to 1 m or more, scarcely more than 1.5 cm wide, bright green, channelled, margins with ascending spines to 2 mm. Scape 30–50 cm, stout. Floral bracts relatively inconspicuous, soon exposing the apices of the ovaries, weakly and minutely toothed or entire. Flowers *c.* 1.5 cm. Petals violet, to 1.2 cm. Fruit 20–25 × 12 cm at maturity, succulent, sweet and edible. *Origin almost certainly Brazil.*

This is a very variable species with many cultivars and forms. A variety having attractively variegated leaves in green, white and pink, is commonly cultivated as var. **variegatus** (Lowe) Moldenke. As with *A. bracteatus*, they can easily be grown by planting the top after it has been removed from the ripe fruit.

44. COMMELINACEAE

Terrestrial herbs. Leaves mostly borne on the stems, alternate, with closed basal sheaths; ptyxis involute or supervolute. Flowers in panicles or coiled cymes, rarely solitary, bisexual, actinomorphic or zygomorphic. Sepals 3, usually greenish. Petals 3 or rarely fewer. Stamens 3–6, anthers basifixed, filaments often hairy, staminodes 0–3. Ovary of 3 united carpels, superior; ovules few, axile. Fruit a capsule.

There are about 35 genera and 600 species. Thirteen genera are native to North America and species of 3 are naturalised in Europe. A few species are grown as ornamentals and house plants. The flowers, which open in strict succession along the cymes (cincinni), last a few hours only, the petals then deliquescing into a mass with the stamens and ovary, which makes dissection and study well-nigh impracticable. The cymes are also often so condensed and umbel-like as to make their construction difficult to discern. The diagnostic feature of *Tradescantia* and allied genera, however, is the fusion of the cymes in pairs to make a 2-sided unit, recognisable in *Tradescantia* itself by the accompanying paired bracts.

Literature: Clarke, C.B., Commelinaceae, in de Candolle, A., Monographiae phanerogamarum **3**: 115–324 (1881); Rohweder, O., Die Farinosae, in der Vegetation von El Salvador, C. Commelinaceae, *Abhandlungen aus dem Gebiet der Auslandskunde* **61**(ser. C, 18): 98–178 (1956).

Habit. Stemless or nearly so: **5, 6, 7**.
Leaves. In stout rosettes: **4, 10** (No. 2), **12** (No. 1). In 2 ranks: **1, 8, 9, 10, 11**. Striped or variegated: **1, 3, 5, 6, 10, 11, 12** (Nos. 1–4).
Cymes. Numerous, aggregated in compound cymes: **3, 4, 7, 10**. Fused in pairs: **10, 11, 12**.
Petals. United to form distinct tube: **7, 8, 12** (No. 2).

1a. Flowers bilaterally symmetric, the stamens of 2 or more types 2
b. Flowers radially symmetric, stamens all similar 7
2a. Inflorescence more or less enclosed by a folded bract **1. Commelina**
b. Inflorescence-bracts (if present) not folded 3
3a. Cymes condensed and fused in pairs, resembling small umbels; bracts none **11. Tripogandra**
b. Cymes not fused in pairs 4
4a. Petals hairless 5
b. Petals ciliate 6
5a. Cymes 1–2, rarely more in an umbel **2. Tinantia**
b. Cymes numerous, aggregated **3. Dichorisandra**
6a. Large, bromeliad-like epiphyte **4. Cochliostema**
b. Low, stemless or creeping plant with few leaves **5. Geogenanthus**

7a. Plants nearly stemless 8
b. Plants with evident stem, at least when flowering 9
8a. Petals free; leaves elliptic, stalked, clothed with brown hairs **6. Siderasis**
b. Petals united into a long tube; leaves linear, not conspicuously hairy **7. Weldenia**
9a. Cymes simple, stalked or stalkless 10
b. Cymes stalkless and associated or fused in pairs or clusters 11
10a. Cymes solitary, axillary and terminal, usually stalkless **8. Cyanotis**
b. Cymes stalked, borne in pairs or umbels on a terminal inflorescence-stalk **9. Gibasis**
11a. Leaves succulent; bracts inconspicuous, or leaves and bracts 2 cm or less **10. Callisia**
b. Leaves only slightly succulent; bracts conspicuous, paired **12. Tradescantia**

1. COMMELINA Linnaeus
D.R. Hunt

Usually perennial, often tuberous-rooted. Flowers variously coloured, bisexual or some unisexual, bilaterally symmetric, in single or paired cymes more or less enclosed by a folded, keeled, spathe-like bract, a few species also with subterranean, cleistogamous flowers. Sepals 3, free, unequal, the outermost hooded; petals 3, free, usually the upper 2 clawed, the lower reduced, more rarely all equal; fertile stamens 3, together on one side of the flower, often of 2 kinds, filaments naked; staminodes 3–2, with sterile anthers, filaments naked. Ovary 3–2-celled, with 1–2 ovules per cell.

About 200 species, mainly in the tropics, including some used for food, medicine or animal fodder. A few are grown in gardens for their bright, gentian-blue flowers.

1a. Bract 'closed', the back edges fused **1. erecta**
b. Bract 'open', the back edges free 2
2a. Plants prostrate or with very short stems 3
b. Plants erect, 30 cm or more 4
3a. Plant without tubers; lowermost petal reduced **2. communis**
b. Plant tuberous; petals equal **3. tuberosa**
4a. Leaves, at least the lower, almost flat, mostly 1–4 cm broad; bract not or shortly beaked **4. coelestis**
b. Leaves inrolled, recurved and sickle-shaped, 4–10 mm broad; bract often with a beak of 1.5–4 cm **5. dianthifolia**

1. C. erecta Linnaeus (*C. elegans* Kunth). Illustration: Rickett, Wild flowers of the United States **3**: t. 11 (1969).

Stems sprawling or erect, to 70 cm or more from slender, tuberous roots. Leaves lanceolate to ovate-lanceolate, acute or acuminate, rounded at base, 7–10 × 1–3 cm, thin, usually hairless. Inflorescence-stalk usually less than 1 cm; folded bract semi-ovate, shortly acuminate, 2–3.5 cm × 6–12 mm, hairy or hairless, back edges fused. Main cyme with 3–6 flowers. Flowers 2.5 cm in diameter; upper petals pale to deep blue, lower much reduced, white. *America, from southern USA to Argentina.* G1. Summer.

C. benghalensis Linnaeus. Leaves ovate, 5–7 × 2–4 cm, hairy. As well as normal flowers, cleistogamous, subterranean flowers are produced. *Old World tropics.* G2. Summer.

'Variegata', with cream-variegated leaves, is grown.

2. C. communis Linnaeus. Illustration: *Addisonia* **1**: t. 20 (1916); Aeschimann et al., Flora alpina **2**: 136.1.1 (2004).

Stems sprawling, to 70 cm, branching, sometimes rooting at the nodes, tubers absent. Leaves oblong-lanceolate, acute to acuminate, rounded and slightly stalked at base, mostly 6–10 × 1.5–3 cm, thin, nearly hairless. Inflorescence-stalks usually *c.* 2 cm, slightly thickened towards top; folded bract semi-ovate, shortly acute, rounded and open basally, 2–3 cm × 8–13 mm. Main cyme with 3–4 flowers. Flowers 2 cm in diameter; petals blue, the lower reduced. *China, Japan, naturalised in S Europe & USA.* H4. Summer–autumn.

C. diffusa Burmann (*C. nudiflora* misapplied). Similar but generally smaller. *Tropics.* G2. Summer–autumn.

3. C. tuberosa Linnaeus. Illustration: *Botanical Magazine*, 1695 (1814).

Prostrate, branching, roots tuberous. Leaves narrowly lanceolate, almost flat, 6–9 × 1–2 cm, hairy towards the translucent margins, sometimes slightly rough with short, stiff hairs. Inflorescence stalks erect, 3–6 cm, elongating to 10 cm or more in fruit. Folded bract ovate-lanceolate, more or less obtuse or shortly beaked, cordate and open towards the base, 2–3 × 1–1.5 cm, green, usually suffused or streaked with dark purple-blue, sparsely to densely hairy. Main cyme with 6–10 flowers. Flowers 3 cm in diameter, petals equal. *C & S America.* G1. Summer.

This, and *C. elliptica*, *C. coelestis* and *C. dianthifolia*, are the best-known elements of a large complex of intergrading forms often united as *C. tuberosa*, here treated separately in view of their horticulturally important differences in habit and leaves.

C. elliptica Kunth (*C. alpestris* Standley & Steyermark). Almost stemless. Leaves mostly basal, lanceolate, to 11 × 2.5 cm. Inflorescence-stalks scape-like, elongating to 20 cm or more in fruit. Folded bract to 4 × 3 cm. Flowers blue or white. *Mexico to Peru, Bolivia.* G1. Summer.

A readily recognised alpine species within the *C. tuberosa* complex.

4. C. coelestis Willdenow. Illustration: Hunt (ed.), Marshall Cavendish encyclopedia of gardening, 343 (1968); Everett, The New York Botanical Gardens illustrated encyclopedia of horticulture, 846 (1981); Brickell (ed.), RHS A–Z encyclopedia of garden plants, 299 (2003).

Erect, to 1 m or more, roots tuberous. Leaves oblong-lanceolate, the lower rounded to cordate at base, 8–18 × 1–4 cm, rough with fine, stiff hairs. Inflorescences crowded towards top of stem; stalks 1–2 cm, elongating to 5–8 cm. Folded bract semi-ovate, more or less obtuse or abruptly and shortly acuminate, almost cordate and open basally, suffused or streaked with dark purplish blue. Main cyme with 4–10 flowers. Flowers 2–3 cm in diameter, usually blue. *C & S America.* H5. Summer.

'Alba' is a white-flowered cultivar. A robust form has been distinguished as var. **bourgeaui** Clarke.

5. C. dianthifolia Delile. Illustration: Rickett, Wild flowers of the United States **3**: t. 11 (1969).

Erect, to 60 cm or more, branching at base from tuberous roots. Leaves distant, linear, inrolled and recurved, 8–12 cm × 4–10 mm, rough with short, stiff hairs. Inflorescence-stalks 2–4 cm, elongated to 10 cm or more. Folded bract 3–6 × 1–1.5 cm, usually with a long beak. Main cyme with 4–8 flowers. Flowers 2–3 cm in diameter; petals equal, blue or white. *Mexico, SW USA.* G1. Late summer.

2. TINANTIA Scheidweiler
D.R. Hunt
Annuals or short-lived perennials. Leaves elliptic to oblanceolate. Flowers bisexual or some unisexual, bilaterally symmetric, in short to elongate terminal cymes borne singly or aggregated in pairs, umbels or rarely whorls. Sepals 3, petals 3. Stamens 6, diverse, all fertile, free or the upper 3 united at base, anther-connective narrow. Ovary usually 3-celled with up to 5 ovules per cell.

An American genus of about 10 species. Besides the following, *T. anomala* (Torrey) Clarke (*Commelinantia anomala* (Torrey) Tharp) and *T. pringlei* (Watson) Rohweder (*C. pringlei* (Watson) Tharp) are occasionally cultivated.

1. T. erecta (Jacquin) Schlechtendal (*T. fugax* Jacquin). Illustration: *Botanical Magazine*, 1340 (1810).

Erect annual, to 1 m tall, with fleshy stems. Leaves elliptic, acuminate and narrowed to the base, 6–16 × 2–6 cm, thin, sparsely hairy. Inflorescence glandular-hairy; cymes in groups of 1–4, each with 3–20 flowers. Flowers *c.* 3 cm in diameter, stalks 1–2 cm, petals blue or purplish, *c.* 1.5 × 1.5 cm; stamens on the same radii as the sepals, bearded, the uppermost short; those on the same radii as the upper petals with appendages, the lowermost hairless. Capsule oblong, 7–11 × 4–5 mm, seeds 2–3 per cell, *c.* 3–3.5 mm, pale grey-brown, rough. *Tropical America.* G2. Late summer.

Usually treated as a half-hardy annual.

3. DICHORISANDRA Mikan
D.R. Hunt
Stout perennials, erect or rarely climbing, or almost without stems. Leaves spirally arranged or more or less in 2 ranks, relatively broad and thin. Flowers in terminal or rarely axillary compound cymes; sepals free, unequal, the upper hooded, often coloured like the petals; petals free or slightly united with the filament-bases, violet-blue, and often white at base, or white. Stamens 6–5, subequal or the 2 whorls more or less unequal; filaments hairless; anthers opening by apical pores (exceptionally by slits), connective narrow. Ovary 3-celled with several ovules per cell, stigma point-like. Fruit a capsule; seeds with fleshy, orange or red, false aril.

A genus of perhaps 30 species, mostly in tropical S America, one (*D. hexandra* (Aublet) Standley) extending to C America and the West Indies. Numerous species are listed in the horticultural reference books, but only 2 are found in cultivation today.

Dichorisandras require warm, humid greenhouse conditions with some shade in summer. They are usually propagated by cuttings or division. Species of the tropical African genus *Palisota* Endlicher are occasionally grown in botanical collections: they are similar but almost stemless, cymes densely congested; stamens 3, staminodes 3 or 2, bearded. Fruit an orange, pink or red berry.

1. D. thyrsiflora Mikan. Illustration: *Botanical Magazine*, n.s., 590 (1971); Graf, Tropica, 305 (1978).

Erect, to 2.5 m but flowering when much smaller, rootstock shortly rhizomatous. Leaves spirally arranged, elliptic-lanceolate, acuminate, wedge-shaped and almost stalked at base, 20–30 × 5–8 cm, hairless. Compound cyme *c.* 20 cm, dense; bracts subulate, cymes shortly stalked; flowers several, with very short stalks. Sepals 1 cm, violet with darker veins; petals almost circular, *c.* 1.2 × 1.1 cm, violet, white at base. Stamens 6, of two kinds: 4 erect, touching, with filaments 2 mm, anthers 5.5 mm; 2 spreading, with filaments 3.5 mm, anthers 4 mm. Capsule 1.2 cm; seed with a red false aril. *SE Brazil.* G2. Autumn.

D. reginae (Linden & Rodigas) H.E. Moore. Illustration: *Baileya* **5**: 120, 123 (1957); Brickell (ed.), RHS A–Z encyclopedia of garden plants, 374 (2003). Leaves in 2 ranks, often silver-striped and flecked. *Peru.* G2.

4. COCHLIOSTEMA Lemaire
D.R. Hunt
Robust, epiphytic perennials with bromeliad-like habit and large, oblong or lanceolate leaves. Flowers in axillary compound cymes; sepals free, more or less equal; petals free, more or less equal or 1 slightly larger, shortly clawed, violet-blue, long-ciliate. Fertile stamens 3, highly specialised, filaments united at the base, much dilated above, enveloping the anthers of the lateral stamens and produced into a tubular horn, anthers coiled; staminodes 3, the lateral well developed, hairy, the other rudimentary. Ovary 3-celled, hairless; ovules numerous, in 2 series in each cell. Capsule narrowly oblong, 3-valved, many-seeded.

A small genus of 2 species confined to Ecuador and Colombia and remarkable for the very curious, coiled anthers.

Literature: Read, R.W., Cochliostema velutinum (Commelinaceae), a new species from Colombia, *Baileya* **13**: 9–15 (1965).

1. C. odoratissimum Lemaire (*C. jacobianum* K. Koch & Linden). Illustration: *Illustration Horticole*, t. 217 (1859); Everard & Morley, Wild flowers of the world, t. 187 (1970).

Rosette very large, to 50–80 cm tall and 2 m in diameter. Leaves narrowly oblong-lanceolate, acute, dilated at base, 60–100 × 10–15 cm, leathery, hairless, midrib prominent beneath. Inflorescence-stalk 25–40 cm, flower-bearing part 15–25 cm. Flowers strongly scented, 3–6 cm in diameter; stalks 1.5–2.5 cm. Sepals 1.5–2.5 cm × 5–10 mm. Larger petal 1.5–3 cm × 5–12 mm, the other slightly shorter and broader; hooded stamens 1.2–2 cm overall, anthers 3–5 mm; lateral staminodes 5–10 mm, long-ciliate with pale violet hairs; ovary 2.5–4 mm, style 9–16 mm. *Colombia, Ecuador.* G2. Autumn.

5. GEOGENANTHUS Ule
D.R. Hunt
Stem short or creeping. Leaves few, at the stem-apex, very broad, stalked. Cymes with few flowers, borne near the stem-base. Flowers long-stalked, bisexual. Sepals 3, free; petals 3, free, equal or almost so. Stamens 5, free, unequal, 3 shorter, bearded, 2 longer, hairless; connectives narrow, cells dehiscing by slits. Ovary 3-celled, brown-hairy, with 4–5 ovules per cell. Capsule opening by 3 splits.

A small genus of 2 or 3 species in Amazonian Brazil, Colombia, Ecuador and Peru.

1. G. poeppigii (Miquel) Faden (*G. undatus* (K. Koch & Linden) Mildbraed & Strauss; *Dichorisandra musaica* var. *undata* (K. Koch & Linden) Miller). Illustration: Graf, Tropica, 305, 306 (1978).

Branched at base, to 15–25 cm. Leaves widely spreading, circular, almost heart-shaped, 6–12 × 4–10 cm, thin, transversely wavy, dark green above with several longitudinal silvery stripes, reddish beneath, finely hairy; stalk 1–1.5 cm, channelled. Inflorescence-stalk 8–15 mm, cyme with 2–5 flowers; flower-stalks 1.5–4 cm. Flowers *c.* 2.5 cm in diameter; sepals 8–10 mm, brown-hairy; petals violet-blue, 1.1–1.3 cm × 8–11 mm, ciliate at apex. Short stamens *c.* 6 mm with yellow-tipped hairs; long stamens 8–9 mm, white. Style 3–4 mm. *Brazil, Peru.* G2.

6. SIDERASIS Rafinesque
D.R. Hunt
Densely and softly reddish brown-hairy perennial, nearly stemless, shortly rhizomatous. Leaves in a rosette, elliptic to elliptic-obovate, abruptly acute, 15–25 × 5–9 cm, green with a silvery mid-stripe above, reddish beneath, conspicuously parallel-veined; stalk to 5 cm. Inflorescences axillary, stalks slender, 3–10 cm, cyme with 1–4 flowers. Flowers bisexual, 2.5–4 cm in diameter. Sepals free, 1–1.4 cm; petals violet to purplish pink, 1.2–2 × 1–1.8 cm. Stamens 6, equal, all fertile, filaments hairless, connectives narrow, anthers dehiscent by slits. Ovary 3-celled, with 4–5 ovules per cell. Capsule cells with 1–4 seeds or 1 cell empty.

A genus of 1 species.

Literature: Moore, H.E., Siderasis fuscata, *Baileya* **4**: 27–30 (1956).

1. S. fuscata (Loddiges) H.E. Moore, (*Pyrrheima fuscata* (Loddiges) Hasskarl). Illustration: *Botanical Magazine*, 2330 (1822); Graf, Tropica, 306 (1978).

E Brazil. G1. Summer.

7. WELDENIA Schultes
D.R. Hunt
Perennial, nearly stemless, roots tuberous. Leaves tufted or in rosettes, linear-lanceolate or broadly strap-shaped. Cymes stalkless, crowded into a dense head surrounded by the leaves. Flowers regular; calyx tubular, 3–4 cm, tips of lobes free; corolla white or very rarely blue, with a tube 4–6.5 cm, 1–2 mm in diameter, and ovate lobes 1–2 × 1 cm. Anthers projecting, the cells of each touching. Ovary cylindric, 3-celled with several ovules per cell. Capsule with many seeds.

A genus of a single species, distinctive on account of its long-tubed flowers.

Weldenia is a choice subject for the alpine house. It needs a deep pot and well-drained compost and must be kept dry and frost-free in winter. Propagation is by division of the rootstock or seeds.

1. W. candida Schultes (*Lampra volcanica* Bentham). Illustration: *Botanical Magazine*, 7405 (1895); Brickell (ed.), RHS A–Z encyclopedia of garden plants, 1087 (2003).

Mountains of Mexico & Guatemala. G1. Spring–summer.

8. CYANOTIS D.Don
D.R. Hunt
Usually perennial, often tuberous-rooted. Leaves narrow, often ciliate. Flowers violet-blue or purplish pink, bisexual, radially symmetric or nearly so, generally almost stalkless in dense, single, axillary cymes subtended by a leaf-like or reduced bract, and with conspicuous, sickle-shaped bracteoles. Sepals 3, free or joined below. Petals 3, free or joined with the filament-bases to form a short tube. Stamens 6, equal, fertile, filaments nearly always bearded, connectives narrow. Ovary 3-celled with 2 ovules per cell. Capsule dry, dehiscent.

A genus of about 30 species in the Old World tropics. Besides the 2 species described below, several others have been grown in botanical collections, including C. *barbata* D. Don (*C. hirsuta* Fischer & Meyer), *C. cristata* Schultes, *C. fasticulata* Schultes, *C. moluccana* (Roxburgh) Merrill and *C. speciosa* (Linnaeus) Hasskarl (*C. nodiflora* (Lamarck) Kunth).

1a. Leaves flat, brown-hairy; cymes open; petals free **1. kewensis**
b. Leaves channelled, hairs white; cymes congested; petals united at base **2. somaliensis**

1. C. kewensis (Hasskarl) Clarke (*Belosynapsis kewensis* Hasskarl). Illustration: *Botanical Magazine*, 6150 (1875); Graf, Exotica, edn 3, 602 (1963).

Prostrate perennial, creeping from its initial rosette, internodes and leaf-sheaths densely brown-hairy. Leaves of rosette lanceolate, acute, rounded at base, to 5 × 2 cm, those of side shoots overlapping in 2 ranks, often smaller, fleshy, dark green above, deep purple beneath, densely velvety. Cymes open, to 3 cm, with up to 8 flowers, the flowers often paired, *c.* 8 mm in diameter, bracteoles elliptic, acute, 5–8 × 1.5–4 mm; flower-stalks 3–4 mm. Sepals *c.* 3.5 × 1.5 mm, densely hairy. Petals free, ovate-elliptic, *c.* 5 × 2.5 mm, purplish pink. Stamens *c.* 6 mm, filaments bearded with violet hairs, anthers yellow with orange pollen. Ovary 1.5 mm, densely hairy; style *c.* 4 mm. *India.* G2. Flowers produced all year.

2. C. somaliensis Clarke. Illustration: Graf, Exotica, edn 3, 602 (1963); *Kakteen und andere Sukkulenten* **28**(4): 98 (1977); Brickell (ed.), RHS A–Z encyclopedia of garden plants, 333 (2003).

Stems of 2 kinds, some with non-flowering basal rosettes (often lacking in cultivated specimens), others stolon-like, creeping shoots, the latter potentially flowering or rooting and forming new rosettes. Leaves oblong-linear, acute and with a small point, stalkless, to 12 × 1.5 cm on rosettes, 1.5–4 cm × 5–10 mm on lateral shoots, these U-shaped in section, and often recurved, succulent, densely hairy below, margins long-ciliate, sheaths inflated, persistent, becoming papery. Bracts 1–1.5 cm; cymes with several flowers. Flowers *c.* 5 mm in diameter, purplish blue. *Somalia.* G2.

Close to *C. foecunda* Hasskarl, but distinct in the basal rosettes and overlapping, persistent leaf-sheaths. According to Faden & Suda (*Biological Journal of the Linnaean Society* **81**: 311, 1980), the identity of plants cultivated as *C. somaliensis* is uncertain.

9. GIBASIS Rafinesque
D.R. Hunt
Perennial (exceptionally annual), tufted from tuberous roots or creeping and rooting at nodes. Cymes individually shortly stalked, in pairs or umbels, fertile zone dense, bracteoles small, overlapping. Flowers regular; sepals and petals free. Stamens 6, equal, all fertile, bearded. Ovary 3-celled with 2 ovules per cell. Capsule dry, dehiscent.

A genus of about 11 species mostly confined to Mexico, one throughout tropical America.

Literature: Hunt, D.R., Revision of Gibasis, *Kew Bulletin* **41**(1): 107–129 (1986).

1. G. pellucida (Martens & Galeotti) Hunt (*G. schiedeana* (Kunth) Hunt). Illustration: *Botanical Magazine*, n.s., 636 (1973).

Creeping and soon leggy, rooting at nodes. Leaves in 2 ranks, obliquely lanceolate, acute to acuminate, rounded to cordate at the base, stalkless, 2.5–10 cm × 8–30 mm. Cymes paired. Flowers white, 7–14 mm in diameter. *Mexico.* G1. Flowering all year.

Formerly confused with *G. geniculata* and superficially similar to *Tripogandra multiflora*. More commonly grown in USA than in Europe. It is readily propagated by cuttings.

10. CALLISIA Linnaeus

D.R. Hunt

Perennial herbs with succulent leaves. Flowers usually radially symmetric, bisexual, stalked or stalkless. Cyme-pairs without conspicuous bracts, stalkless in the axils of the upper leaves, these normal or reduced almost to the sheath, or in stalked compound cymes. Sepals 3 (very rarely 2), usually translucent; petals 3 (very rarely 2), white or pink. Stamens 6 or 3–1, filaments typically hairless, rarely bearded, anther-connectives usually broad; anthers dorsifixed; stigma brush-like or papillose; ovary 3–2-celled, ovules 2 per cell. Capsule small, spherical, dehiscent.

As here understood in a broad sense, this is a genus of about 20 species native to tropical America. It includes species with the cymes fused in pairs (in 1 species associated in pairs or threes), regular flowers and the stamens all similar (though sometimes reduced in number), but lacking the conspicuous paired bracts beneath the paired cymes that characterise *Tradescantia.*

Literature: Hunt, D.R., Amplification of Callisia Loefl., *Kew Bulletin* **41**: 407–412 (1986).

1a. Leaves to 30 cm, in a rosette 2
b. Leaves 1–8 cm, spaced along the stem 3
2a. Inflorescence apparently axillary; flowers *c.* 1 cm in diameter; petals purplish pink; stamens 5 mm **1. warscewicziana**
b. Inflorescence terminal; flowers 5–8 mm in diameter; petals white; stamens 8–12 mm, projecting **2. fragrans**
3a. Leaves V-shaped in section; flowers purplish pink **3. navicularis**
b. Leaves flat; flowers white or colourless 4
4a. Fertile stamens 6 **4. elegans**
b. Fertile stamens 5 or fewer 5
5a. Paired cymes stalkless, axillary; flowers odourless; petals narrow **5. repens**
b. Paired cymes stalked, in delicate, many-flowered compound cymes; flowers scented; petals with a broad blade **6. multiflora**

1. C. warscewicziana (Kunth & Bouché) Hunt. Illustration: *Botanical Magazine*, 5188 (1860); *Baileya* **10**: 130 (1962).

Stem stout, eventually 1 m or more. Leaves narrowly oblong, acuminate, stalkless, to 30 × 6.5 cm, fleshy, hairless except for the sometimes ciliate, often purple, margin. Inflorescence apparently axillary, to 35 cm overall, branched below and sometimes viviparous. Cymes fused at base in pairs or threes, stalkless or almost so, cyme-axes to 2 cm, cymes with many flowers; bracteoles to 7 mm, stalks 1–1.3 cm. Flowers *c.* 1 cm in diameter. Sepals fleshy, persistent, 4–5 mm, hairless. Petals purplish pink, 6 × 6 mm. Stamens 6, *c.* 5 mm, usually hairless, connectives broad, yellow. Ovary glandular-hairy towards apex. Capsule 3.5 mm in diameter. *Guatemala.* G2. Spring–autumn.

2. C. fragrans (Lindley) Woodson. Illustration: *Edwards's Botanical Register* **26**: t. 47 (1840).

Main stem stout, flowering stems to 1.5 m, branching and producing stolons. Leaves elliptic-lanceolate, acute, stalkless, to 30 × 7 cm. Inflorescence terminal, branched, 50–100 cm overall. Paired cymes clustered, stalkless, chaffy, 6 mm. Flowers stalkless, 5–8 mm in diameter, scented. Sepals chaffy, scarcely exceeding the bracteoles. Petals white, lanceolate, 5 × 2.5 mm. Stamens 6, 8–12 mm, projecting; filaments hairless, anther-connective broad, thin, white. Ovary 3-celled, hairy on top, with 2 ovules per cell; stigma brush-like. *S Mexico.* G2. Winter–spring.

3. C. navicularis (Ortgies) Hunt (*Tradescantia navicularis* Ortgies). Illustration: Jacobsen, Handbook of succulent plants **2**: 893, 1134 (1960); Graf, Exotica, edn 3, 604 (1963).

Prostrate, with fleshy shoots of 2 intergrading types: short-shoots with tightly overlapping leaves, and 'stolons' with long internodes rooting and producing short-shoots or inflorescences at the nodes. Leaves ovate, acute, stalkless and boat-shaped, 2–3 × 1–2 cm, very fleshy, hairless except on the margins, often longitudinally lined, minutely green-dotted. Cyme-pairs terminal and axillary; bracts resembling the leaves. Bracteoles subulate, 7 mm, green. Flowers 1.5–2 cm in diameter, stalks 1–1.5 cm. Sepals 3, boat-shaped. Petals 3, purplish pink. Stamens 6; filaments bearded. *Mexico.* G1. Summer–autumn.

4. C. elegans Moore (*Setcreasea striata* invalid). Illustration: *Baileya* **6**: 141 (1958); Graf, Exotica, edn 3, 601, 602 (1963).

Creeping, to 60 cm, branching. Leaves in 2 ranks, oblong-ovate, acute, rounded at base, usually 5–7.5 × 2–3 cm, stiff, fleshy, dark green above with about 6 narrow whitish bands, often purplish beneath, velvety to the touch, and margins minutely ciliate. Inflorescences terminating all main shoots. Cyme-pairs terminal and axillary, stalkless; cymes with 4–5 flowers, bracteoles subulate, 6 mm. Flowers stalkless, 1 cm in diameter. Sepals 3.5–6.5 mm, ciliate on the keel. Petals white, clawed, blade 4.5–6.5 × 5–6.5 mm, claw 2.5–3 × 2–2.5 mm. Stamens 6, subequal, 6–8 mm, those opposite the petals slightly longer; filaments hairless, connective broadly triangular, acute, white or pale yellow. Ovary 3-celled with 2 ovules per cell, hairy at apex; stigma brush-like. *Mexico.* G2. Winter.

5. C. repens Linnaeus. Illustration: Graf, Tropica, 307 (1978); Brickell (ed.), RHS A–Z encyclopedia of garden plants, 209 (2003).

Slender-stemmed, prostrate, rooting at nodes and forming mats. Leaves ovate, acute, rounded to almost cordate at base, stalkless, usually 1–4 × 1–2 cm, fleshy, hairless except for the minutely ciliate margins. Flowering branches usually ascending, with progressively smaller leaves, the cyme-pairs scarcely projecting from the sheaths. Sepals 3, 2–3 mm. Petals 3, narrow, colourless, slightly shorter than the sepals. Stamens 3, or 6, *c.* 6–8 mm, with 1 or more staminodes projecting, connective broad, rounded, thin. Stigma brush-like. *Tropical America.* G1.

The common clone in cultivation, whose identity has been confirmed by study of its chromosomes, is very compact in habit and seems to be non-flowering, but the small, broad, pointed, fleshy leaves make confusion with other species unlikely.

6. C. multiflora (Martens & Galeotti) Standley. Illustration: *Botanical Magazine*, 4849 (1855); *Baileya* **9**: 16 (1961).

Creeping, to 80 cm, branching. Leaves oblong-lanceolate, acute or acuminate, rounded to almost cordate at base, stalkless, 3–9 × 1–2.5 cm, fleshy, rather pale green, velvety to the touch. Inflorescence much-branched. Cyme-pairs stalked; cymes with *c.* 3 flowers. Flowers smelling of violets, 1 cm in diameter; stalks 5 mm, glandular-hairy. Sepals 2–3 mm, green. Petals elliptic, 3–4 mm. Stamens 3, opposite the sepals; filaments hairless, connectives narrow. Ovary 3-celled, with 2 ovules per cell, hairless; style short; stigma papillose, somewhat 3-lobed.

Capsule 2 mm. *S Mexico, Guatemala.* G2. Winter.

11. TRIPOGANDRA Rafinesque

D.R. Hunt

Annual or perennial, without tubers. Habit and leaves diverse. Inflorescences terminal and axillary, cyme-pairs stalked, not subtended by leaf-like bracts. Flowers somewhat bilaterally symmetric; sepals and petals equal, free. Stamens 6, dimorphic, 3 with short filaments and 3 with long filaments. Ovary 3-celled with 2 ovules per cell, style short. Capsule dry, dehiscent.

A genus of 21 species in tropical America, from Mexico and the West Indies to Argentina.

Literature: Handles, W., The taxonomy of Tripogandra (Commelinaceae), *Rhodora* **77**: 213–333 (1975).

1. T. multiflora (Swartz) Rafinesque. Illustration: *Baileya* **8**: f. 27 (1960).

Creeping perennial, rooting at the nodes, flowering stems ascending, to 80 cm. Leaves ovate, acute, 3–9 × 1–3 cm, rather fleshy, hairless or hairy. Inflorescence-stalks to 3.5 cm; cymes with up to 16 flowers; flower-stalks 2–4 mm. Flowers 5–8 mm in diameter; sepals 2–4 mm; petals 2–4 × 1–2.2 mm, white or pink. Longer stamens 1.5–3 mm, bearded, shorter stamens 0.5–1.3 mm, hairless. Capsule 1.5–2.5 mm in diameter. *Jamaica, Trinidad, Costa Rica, W tropical S America to Argentina.* G2. Flowering all year.

One of several species grown in botanical collections. It is listed in some reference books as being in cultivation and alleged to be widely naturalised in the Azores, but these reports require confirmation. Some illustrations (e.g. in Graf, Exotica, edn 3, 604, 1963) depict *Gibasis pellucida* (p. 320), which is superficially similar but has radially symmetric flowers and the cymes simple, not fused in pairs.

12. TRADESCANTIA Linnaeus

D.R. Hunt

Perennials, habit diverse, roots fibrous or tuberous. Inflorescences mostly terminal, the cymes fused in pairs and subtended by paired, boat-shaped bracts similar to or differentiated from the leaves. Flowers bisexual, radially symmetric. Sepals 3, usually free. Petals 3, usually free sometimes united at base, blue-violet, purple, pink or white. Stamens 6, all fertile, similar and more or less equal; anthers versatile with broad connectives. Ovary 3-celled with 2 ovules (rarely 1) in each cell; stigma knob-like. Capsule dehiscent.

As here understood, *Tradescantia* is a genus of about 65 species native to N and S America. The winter-hardy *T. virginiana* group has long been popular in European gardens, and several non-hardy species (some of them often assigned to small segregate genera) are decorative foliage plants for the house or greenhouse.

Literature: Anderson, E. & Woodson, R.E., The species of Tradescantia indigenous to the United States, *Contributions from the Arnold Arboretum* **9**: 1–132 (1935); Hunt, D.R., Sections and series in Tradescantia (American Commelinaceae IX), *Kew Bulletin* **35**: 437–42 (1980); Hunt, D.R., Campelia, Rhoeo and Zebrina united with Tradescantia, *Kew Bulletin* **41**: 401–5 (1986).

Habit. Bromeliad-like: **1**; creeping, growth non-seasonal: **2**, **3**; tufted, growth seasonal: **5**, **6**.
Leaves. White-woolly: **6**; silver-striped above: **2**.
Corolla. Petals united into slender tube below: **2**.

1a. Plant bromeliad-like with short stem and numerous stiff, overlapping, linear-lanceolate leaves; bracts almost enclosing the flowers **1. spathacea**
b. Plant without the above combination of characters 2
2a. Creeping, growth usually non-seasonal; leaves lanceolate to broadly ovate, often longitudinally striped or variegated 3
b. Tufted or shortly creeping and ascending from rhizomatous base, growth seasonal; leaves linear to oblong-elliptic, rarely variegated 4
3a. Petals pink or violet-blue, united at base into a slender tube; leaves typically green, silver-striped above, purplish beneath **2. zebrina**
b. Petals white, free; leaves not silver-striped **3. fluminensis**
4a. Hardy species; leaves linear, green **5. virginiana**
b. Tender species; leaves elliptic or narrowly oblong 5
5a. Leaves densely white-woolly beneath **6. sillamontana**
b. Leaves hairless or hairy but not woolly beneath 6
6a. Flowers distinctly stalked, petals and stamens free; leaves elliptic-oblong to narrowly ovate **4. cerinthoides**
b. Flowers almost stalkless, petals clawed at base and shortly united with filament bases; leaves narrowly oblong **7. pallida**

1. T. spathacea Swartz (*Rhoeo spathacea* (Swartz) Stearn). Illustration: *Baileya* **5**: f. 57 (1957); Graf, Exotica, edn 3, 601, 604 (1963); Graf, Tropica, 307 (1978); Brickell (ed.), RHS A–Z encyclopedia of garden plants, 1045 (2003).

Erect, with short stem to about 20 cm. Leaves crowded in a rosette, linear-lanceolate, acuminate, scarcely narrowed above the sheath, 20–35 × 3.5–5 cm, fleshy, usually green above and purple beneath, hairless. Inflorescences axillary, stalk short, 2–4.5 cm; bracts deeply boat-shaped, 2–4.5 × 2.5–5 cm. Flowers numerous, scarcely projecting, 1–1.5 cm in diameter, white. Flower-stalks *c.* 1.5 cm, recurved in fruit. Sepals, petals and stamens free, equal. Ovary 3-celled with a single ovule per cell. *S Mexico, Belize, Guatemala; naturalised in West Indies.* G2. Flowering all year.

Widely cultivated; a variant with wholly green leaves is 'Concolor', and another with pale yellow stripes is 'Vittata'.

2. T. zebrina Bosse (*Zebrina pendula* Schnizlein). Illustration: Graf, Exotica, edn 3, 603 (1963); Everard & Morley, Wild flowers of the world, t. 187 (1970); Graf, Tropica, 305, 306 (1978); Brickell (ed.), RHS A–Z encyclopedia of garden plants, 1046 (2003).

Creeping or pendent, rooting at nodes. Leaves ovate-oblong to broadly ovate, acute, rounded at base and stalkless, 4–10 × 1.5–3 cm, slightly fleshy, usually bluish green and often silver-striped above, purplish beneath. Inflorescences terminal with stalks to *c.* 10 cm; bracts unequal, the outer 2–5 cm, the inner 1–2 cm; flower-stalks short. Flowers 1–1.5 cm in diameter, pink or violet-blue; calyx irregularly lobed, 5–8 mm, transparent; corolla-tube to 1 cm, slender, lobes ovate, 5–10 × 3–7 mm. Stamens borne on the petals, anther-connective white. *Mexico; widely naturalised in the tropics.* G2. Flowering all year.

Several clones are widely cultivated, including 'Quadricolor' and 'Purpusii'.

3. T. fluminensis Vellozo (*T. albiflora* Kunth). Illustration: Graf, Exotica, edn 3,

603 (1963); Graf, Tropica, 305, 306 (1978).

Creeping and rooting at the nodes. Leaves ovate to ovate-oblong, acute, asymmetric and sometimes slightly stalked at base, 2–10 × 1–3 cm, thin to fleshy, all green, or purplish beneath, or variegated with several cream or white longitudinal stripes, usually hairless except for the sheath. Paired cymes subtended by bracts similar to or somewhat smaller than the leaves. Flowers 1–1.5 cm in diameter; stalks 5–10 mm, recurved after flowering. Sepals to 9 mm; petals to 1.2 cm, white. *SE Brazil to N Argentina.* H5–G1. Flowering all year.

There are several clones in cultivation as house plants, the variegated, small-leaved variants being preferred. Some clones are hardier than others, and the base will overwinter out of doors in sheltered places as far north as SE England. Although *T. fluminensis* may prove divisible into more than one species, the distinctions between *T. fluminensis* and *T. albiflora* often reported in horticultural literature are false and misleading.

4. T. cerinthoides Kunth (*T. blossfeldiana* Mildbraed). Illustration: *Botanical Magazine*, n.s., 247 (1955); Graf, Exotica, edn 3, 603 (1963).

Creeping or ascending, branching, rooting at nodes. Leaves elliptic-oblong to narrowly ovate, acute, base constricted into the sheath, to 15 × 3.6 cm, glossy dark green and hairless above, purple and densely hairy beneath, ciliate. Inflorescences terminal and axillary from upper nodes. Inflorescence-stalk to 4 cm; bracts unequal, like the leaves but smaller; flower-stalks to 1.5 cm, hairy. Sepals free, *c.* 5.5 × 3 mm, purplish, hairy; petals free, 7–8 × 6 mm, pink above, white towards base. Stamens free, *c.* 4.5 mm, filaments white-bearded, anthers yellow; ovary hairless. Capsule 4 mm; seeds 2 mm. *?N Argentina.* G1.

A variegated form is also grown. The species was described from cultivated material and is closely allied to the following:

T. crassula Link & Otto. Illustration: *Botanical Magazine*, 2935 (1829); Graf, Exotica, edn 3, 604 (1963). Leaves green, usually hairless. Flowers white. *N Argentina, Uruguay.* G1. Flowering all year.

5. T. virginiana Linnaeus (*T.* × *andersoniana* Ludwig & Rohweder, invalid). Illustration: Hay & Synge, Dictionary of garden plants in colour, t. 1400 (1969); Perry, Flowers of the world, 77 (1972).

Tufted, stem-base shortly rhizomatous with slightly fleshy roots; flowering stems erect, 30–60 cm, withering after flowering. Leaves linear-lanceolate, 15–35 cm × 5–25 mm, slightly fleshy. Cymes many-flowered, subtended by bracts similar to the leaves but smaller. Flowers 2.5–3.5 cm in diameter with stalks 2–3 cm. Sepals 1–1.5 cm, hairy; petals 1.5–2 × 1–1.5 cm, violet, purplish or rarely white; stamens 1–1.5 cm, filaments very hairy.

The common tradescantias of gardens are mostly hybrids of *T. virginiana* with related species such as the hairless *T. canaliculata* Rafinesque. The name *T.* × *andersoniana* has been proposed for complex hybrids involving *T. ohiensis* Rafinesque, *T. subaspera* Ker Gawler and *T. virginiana*, but is invalid for lack of description. There are numerous named clones such as 'Isis'.

6. T. sillamontana Matuda (*T. pexata* H.E. Moore; *T. velutina* misapplied; *Cyanotis veldthoutiana* invalid). Illustration: *Baileya* **8**: 98, 101 (1960); Graf, Exotica, edn 3, 605 (1963); *Botanical Magazine*, n.s., 706 (1976); Brickell (ed.), RHS A–Z encyclopedia of garden plants, 1045 (2003).

Stems ascending and spreading from a persistent base and thick non-tuberous roots, to 20–30 cm, withering after flowering. Leaves more or less in 2 ranks, overlapping, elliptic or ovate, acute, rounded or sheathing at base, 4–6 × 1.5–2.5 cm, somewhat fleshy, sparsely hairy above, densely white-woolly beneath. Cyme-pairs terminal, with 6–10 flowers per cyme; bracts similar to the leaves; bracteoles thread-like, 5 mm, hairy; flower-stalks 1–1.5 cm, woolly, recurved after flowering. Flowers 2 cm in diameter. Sepals 6–7 mm; petals 9–14 × 9–12 mm, purple-pink; stamens 6–10 mm, filaments hairless. *N Mexico.* G1. Summer.

7. T. pallida (Rose) Hunt (*Setcreasea purpurea* Boom). Illustration: *Baileya* **5**: 150 (1957); Graf, Exotica, edn 3, 602 (1963); Everard & Morley, Wild flowers of the world, t. 187 (1970).

Stem branching and ascending from a persistent base, 20–40 cm. Leaves narrowly oblong, acute, trough-shaped, 8–15 × 2.5–3 cm when flat, slightly fleshy, pale glaucous green to deep purple, hairy or hairless. Cyme-pairs terminal, inflorescence-stalk to 10 cm; bracts 2, rarely 3, leaf-like, the outer longer and up to 4 × 2 cm, the inner smaller, to 1.8 × 2 cm; bracteoles broad, translucent. Flowers several, 1.5–2 cm in diameter; flower-stalks to 6 mm, hairy or hairless. Sepals 5.5 × 2 mm; petals pink, or pink and white, 1.5–1.8 cm × 8–12 mm, clawed at base and united with the filament bases for *c.* 1 mm; stamens-attached to base of petals, filaments bearded. *Mexico.* G1. Summer–autumn.

The widely cultivated form with dark purple stems and leaves ('Purpurea' or 'Purple Heart') was introduced from Mexico but is not known in the wild. It has become popular not only as a pot plant but also for summer bedding.

45. GRAMINEAE (Poaceae)

Originally compiled at the Royal Botanic Gardens, Kew; edited by S.G. Knees & N. Zantout

Annual or perennial herbs, rarely woody, often with rhizomes or stolons. Flowering stems (culms) usually cylindric, with hollow internodes and solid nodes, rarely solid throughout, with a growing point at the base of each internode. Leaves alternate, 2-ranked, consisting of sheath, ligule and blade. Sheaths surrounding the stem, with free, overlapping or united margins, sometimes with auricles at the mouth. Ligule situated at the junction of sheath and blade, usually membranous, sometimes a row of hairs or rarely absent. Blade linear to thread-like, rarely lanceolate to ovate, usually with more or less prominent, parallel ribs, generally constricted at the junction with the sheath, rarely shortly stalked. Flowers usually bisexual, consisting of 1–3 (rarely to 6 or more) stamens, and an ovary with 2 (rarely 3) styles; 2 (rarely 0, 3 or 6) small, translucent scales (lodicules) present near base of filaments; each flower enclosed in 2 bracts (a lower lemma and an upper palea), the whole structure forming a floret. Florets 1–many, inserted alternately on 2 sides of a slender, jointed axis (rachilla) and subtended by usually 2 (rarely 0,1 or 3) bracts (glumes), the whole forming a spikelet. Glumes sometimes represented by 1 or more bristles forming an involucre. Lemma often with a bristle (awn); palea usually membranous or transparent, 2-keeled, sometimes very small or absent. Lemma with a thickened, sometimes elongate and pointed base (callus). Ovary 1-celled, superior; ovule 1, attached on the inner side of the cell, to a point or line visible in fruit as the hilum. Fruit a grain (caryopsis) or rarely with a free, membranous pericarp. Figure 13, p. 326.

A large cosmopolitan family of about 660 genera and 9000 species. Over 500 species are mentioned in horticultural literature, but these are mainly employed in lawns, sports grounds and landscaping, or offered as novelties of special interest to flower arrangers. Relatively few of them can fairly lay claim to a place in the garden as ornamentals. Many are important cereal and fodder crops.

Following the order established in *Genera Graminum* by Clayton & Renvoize (1986), the bamboos (woody grasses) are placed here at the begining of the family with a separate general description and key to the genera. The editors are grateful to Dr T.A. Cope for helpful comments on the revison of this account.

Literature: The literature on the Gramineae is very extensive and most of it is not especially relevant to grasses cultivated for amenity. Good accounts of most of these (by C.E. Hubbard) can be found in the *Supplement to the Royal Horticultural Society's Dictionary of Gardening* (1969). Hubbard's *Grasses* (edn 1, 1954; edn 2, 1968) is an excellent source of good illustrations and accurate descriptions of the species native to and naturalised in the British Isles. More recently, Darke (1999 produced the well illustrated *Colour encyclopedia of ornamental grasses* which, as well as an aid to identification, includes useful chapters on propagation and cultivation.

Key To Groups

1a. Culms woody **Group A** (Bamboos)
b. Culms not woody **Group B** (Herbaceous grasses, p. 336)

Group A (Bamboos)
Woody, evergreen, perennial grasses. Rhizomes sometimes thick and short, sometimes thin and far-running. Culms emerging in spring, summer or autumn, as thick, pointed, asparagus-like shoots, growing rapidly; mature culms 20 cm to 46 m, 3 mm to 65 cm in circumference, usually hollow except at the nodes, which are marked by a more or less prominent line; below this there is usually another line which is the raised scar left by a sheathing leaf (culm sheath); these are either pushed off by the developing branches or are persistent; the culm sheaths are alternate, on 2 sides of the culm, smooth and shining inside, and usually have a stalkless, short-lived blade which often bears auricles and bristles at its junction with the sheath; a ligule is also present. Crowns sometimes pendent with the weight of the leaves and branches. Branches single at each node (except for *Chusquea*) but the branch itself is branched very near its base, so that many branches appear to arise at each node. Foliage leaves largest on new culms, with basal persistent sheaths, frequently with auricles and bristles at the sheath apex and always with a ligule; blades stalked on the sheaths, deciduous at a precise line or joint, with at least one margin with fine, forwardly pointing teeth, the veins parallel but joined by many transverse veinlets, the venation often showing as a distinct mosaic of tessellation. Inflorescence of panicles, spikes or racemes, sporadically produced in Europe. Spikelets various, with 0–4 (usually 2) glumes, and a varying number of florets (sometimes partly abortive in cultivation) which may be in part unisexual or otherwise imperfect. Lodicules usually 3. Stamens 3–6 in cultivated species. Style 1, stigmas 2–3.

Because bamboos flower and fruit uncertainly in Europe (see the paper by McClintock, cited below) the keys and descriptions given here are based almost entirely on vegetative characters. Identification using these is not entirely easy, and there is much confusion in the naming of the cultivated species. In order to use the account properly a hand lens ($\times$ 10 or $\times$ 15) is necessary, and the plants should be examined whole, if possible, at the most suitable time of the year, that is, soon after the culms have grown to their full height but while they retain their sheaths, which should be examined at the middle of the culm rather than above or below on mature plants. The bristles which occur on the culm and leaf-sheaths are important for identification: they may be absent, or present, when their colour and surface are important. The surface may be entirely smooth, or roughened with small projections which can be seen with the hand lens. In a few cases the bristles are rough near the base and smooth above. The presence or absence of vein tessellation in the leaves may be seen by holding them up to the light and examining them through the lens.

The size and development of many species varies considerably, depending on where they are being grown; a species which may do well to attain 5 m in northern Europe may well exceed 12 m further south; and similar differences in the degree of branching occur.

Though flowering is sporadic in Europe, and the flowers may be imperfect, whether or not and when and how a plant has flowered can be a useful pointer in identification. Despite constantly repeated statements to the contrary, flowering, in Europe at any rate, of any one species is by no means always synchronous, nor does the plant normally die as a result. The effect of flowering is to weaken or kill the culms that bear the flowers, which usually lose their leaves at the same time, but the rootstock can be expected to survive.

Bamboos form most of the subfamily Bambusoideae of the Gramineae; this subfamily is native over wide areas of E Asia, C and S Africa and America from Maryland south to Argentina. Because of uncertainties in classification, it is not possible to state the number of genera and species of bamboos, and there is much still to be learned about their classification.

Bamboos thrive best in sheltered environments which are not too dry; some species will form wind breaks. The safest time for transplanting is as growth is resumed in the spring and early summer, as the ground warms up. Establishment can take a considerable time, and demands watchful care, particularly with watering. Thereafter most will need no more than the old canes removing, but will gain from feeding and mulching. Propagation is usually by division. Pests and diseases are few and rare.

Literature: Munro, W., A monograph of the Bambusaceae, *Transactions of the Linnaean Society* **26**: 1–157 (1868); Rivière, A. & Rivière, C., Les Bambous (1878); Gamble, J.S., The Bambusaceae of British India, *Annals of the Royal Botanic Garden Calcutta* **7**: 1–133 (1894–96); Freeman-Mitford, A.B., The bamboo garden (1896); Satow, E., The cultivation of bamboos in Japan, *Transactions of the Asiatic Society of Japan* **27**: 1–127 (1899); Houzeau de Lehaie, J., Les Bambous, Nos. 1–10 (1906–08); Camus, E.G., Les Bambusees (1913); Lawson, A.H., Bamboos (1966); McClure, F.A., The bamboos (1966); Lin, Wei-chih, Studies on

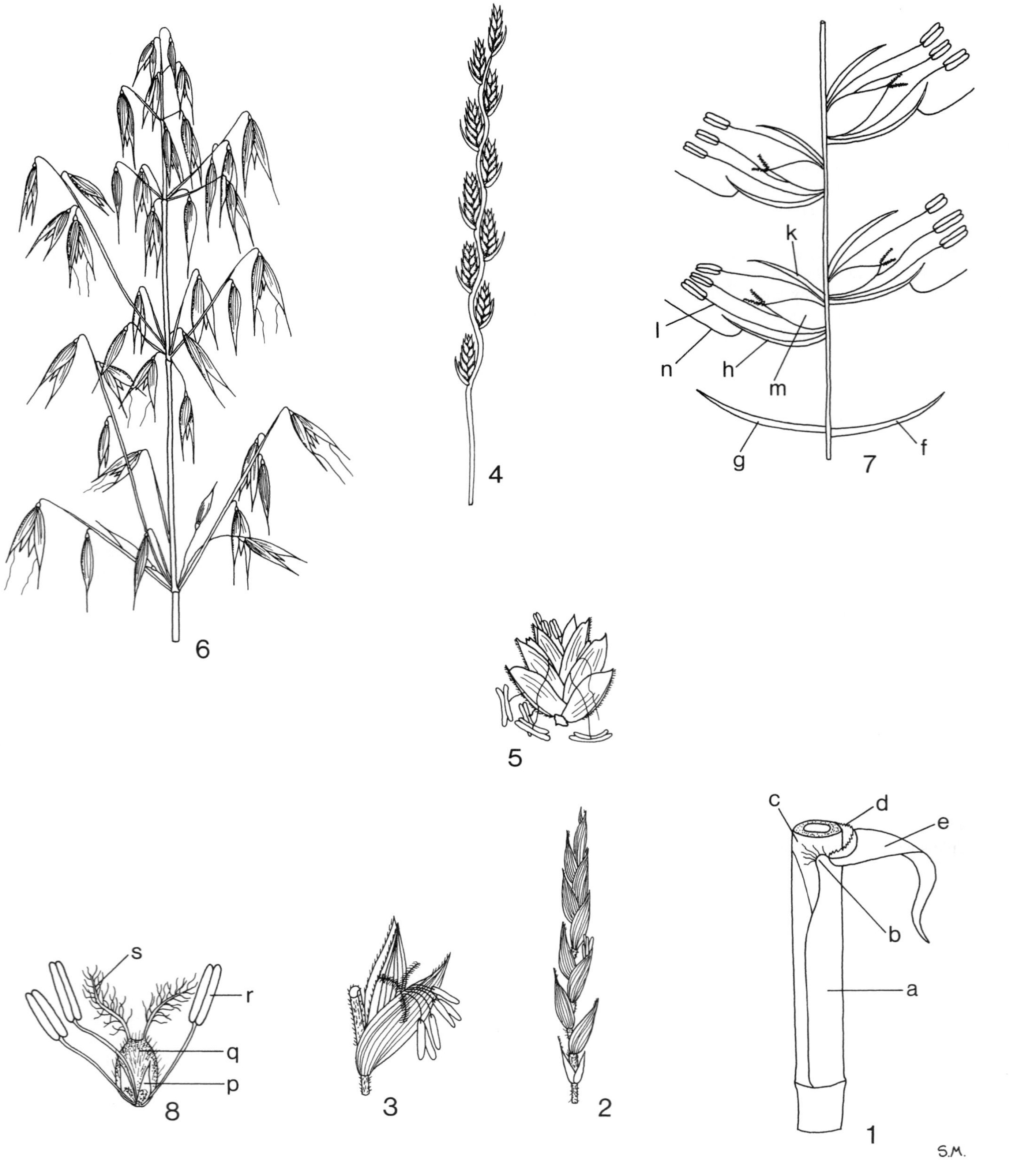

Figure 13. Diagram of some characters used in the identification of Gramineae. 1, portion of the culm of a bamboo (a, culm-sheath; b, auricle; c, bristles; d, ligule; e, blade of leaf). 2, spikelet of a bamboo. 3, flower of a bamboo. 4, a spike of stalkless spikelets. 5, a stalkless spikelet. 6, a panicle of stalked spikelets. 7, exploded diagram of a spikelet (f, lower glume; g, upper glume; h, lemma; k, palea; 1, stamen; m, ovary; n, awn). 8, flower of *Avena sativa* (p, lodicule; q, ovary; r, anther; s, stigma).

morphology of bamboo flowers, *Bulletin of the Taiwan Forestry Institute* 68 (1974); Suzuki, S., Intro to *Japanese Bambusaceae* (1978); McClintock, D., Bamboos, some facts and thoughts on their flowering, *The Plantsman* **1**: 31–50 (1979); Martin, F. & Demoly, J.-P., Bambous, *Bulletin de l'Association des Parcs Botaniques de France* **1**: 7–17 (1979), **2**: 13–25 (1979), **3**: 13–22 (1980), **4**: 25–31 (1981); Crouzet, Y., Les Bambous (1981). There is an American society for bamboo enthusiasts, which produces a journal, *The Journal of the American Bamboo Society* (1980 and continuing).

1a. Culms always solid when mature 2
b. Culms sooner or later hollow (except at the nodes) 4
2a. Culms D-shaped in section **13. Shibataea**
b. Culms terete in section 3
3a. Culms to 40 cm; branches 1–3 at each node **6. Pleioblastus**
b. Culms to 6 or rarely to 10 m; branches very numerous at each node **4. Chusquea**
4a. Culms not terete, at least above 5
b. Culms terete throughout 8
5a. Culms obtusely 4-cornered in section **10. Chimonobambusa**
b. Culms at least in part with an obvious groove 6
6a. Culms grooved throughout, or at least between those nodes where branches are borne **12. Phyllostachys**
b. Culms mainly terete below, if grooved then only above 7
7a. Culms with a large hollow, held stiffly erect; leaves 1.6 cm or more broad **11. Semiarundinaria**
b. Culms ultimately bending with the weight of the crown and with a small hollow; leaves at most 6 mm broad **2. Otatea**
8a. Branches always 1 at each node 9
b. Branches mostly more than 1 at each node 11
9a. Culms ascending or flopping; sheath bristles, if present, rough throughout 10
b. Culms erect; sheath bristles absent, or, if present, entirely smooth or rough only at the base **8. Pseudosasa**
10a. Culms persistently ascending; leaves 4–11 on each branch, at most 40 cm **7. Sasa**
b. Culms ultimately flopping; leaves 1–2 on each branch, to 60 cm **9. Indocalamus**
11a. Leaves without visible tessellation 12
b. Leaves with visible tessellation 15
12a. Culms 15 m or more in height, 20 cm or more in circumference **15. Dendrocalamus**
b. Culms to 15 m in height but usually much less, at most to 10 cm in circumference 13
13a. Leaves and sheaths silvery-grey, hairy **14. Bambusa**
b. Leaves and sheaths hairless or hairy but not as above 14
14a. New shoots appearing in late summer or autumn **10. Chimonobambusa**
b. New shoots appearing in spring **5. Arundinaria**
15a. Culm sheaths soon falling, though often hanging from a small attachment at the base for a time; branches forming on the culms from the base upward **11. Semiarundinaria**
b. Culm sheaths persistent, not hanging as above; branches forming on the culms from the top downwards 16
16a. Bristles rough **5. Arundinaria**
b. Bristles smooth 17
17a. Bristles and ligule dark in colour 18
b. Bristles and ligule whitish in colour 19
18a. Inflorescence simple, exserted, supported by narrow sheaths **1. Sinarundinaria**
b. Inflorescence compound, with many fascicles **3. Thamnocalamus**
19a. Culms finally with a small hollow, emerging in late summer; culm sheaths marbled, blades minute **10. Chimonobambusa**
b. Culms with a wide hollow, emerging in spring; culm sheaths not marbled, blade obvious, not minute **6. Pleioblastus**

1. SINARUNDINARIA Nakai
D. McClintock
Like *Arundinaria*, but with a much-branched rhizome, habit tufted, branches almost equal, arising horizontally, leaf bristles smooth and white. Inflorescence a small panicle. Spikelets with 2 glumes and 2–4 florets. Lodicules 3. Stamens 3. Stigmas 2–3.

1. S. nitida (Mitford) Nakai (*Arundinaria nitida* Mitford). Illustration: Lawson, Bamboos, 101 & t .14 (1966); Bean, Trees and shrubs hardy in the British Isles, edn 8, t. 16 (1970–80); Bartels, Gartengehölze, 416 (1981).

Culms to 6 m, *c.* 5 cm in circumference but often much less, variable in colour from dark purple to grey, arching above when mature. Sheaths soon shorter than the internodes, persistent for a year or longer in most cases, somewhat narrowed above, without auricles or bristles, usually purple, variably hairy or hairless. Nodes with white, waxy bloom beneath. Shoots unbranched till second year. Leaves usually *c.* 5 cm × 5 mm, sometimes to 10 × 1.3 cm, hairless, or hairy towards the base beneath. Sheaths hairless, ciliate, usually purple, sometimes with bristles; ligule 0.5–1 mm, often purple. *N China.* H2.

Various clones are cultivated; they are hardy if protected from wind. The species may be confused with *Arundinaria spathacea*, from which it may be distinguished by its smaller leaves, more tapered, less prominent culm sheaths and its general purple and grey coloration.

2. OTATEA (McClure & E.W.Smith) Calderon & Soderstrom
D. McClintock
Rhizome short or elongate. Young plants tufted, becoming more open with age. Culms slender, solid at first, later hollow, the upper internodes each with a shallow groove on one side, the whole culm ultimately arching with the weight of the crown. Sheaths soon deciduous, without bristles. Branches 3 at each node initially, later more numerous. Leaves without visible tessellation, 3–6 mm broad, the sheaths mostly without auricles or bristles. Inflorescence a panicle. Spikelets with 2 glumes and 3–7 florets. Lodicules 3. Stamens 3. Stigmas 2.

A genus of 2 species from C America (Mexico to Honduras), suitable only for very warm areas.

Literature: McClure, F.A., Genera of bamboos native to the New World, *Smithsonian Contributions to Botany* **9**: 116–19 (1973); Calderon, C.E. & Soderstrom, T.R., The genera of Bambusoideae (Poaceae) of the American continent, *Smithsonian Contributions to Botany* **44**: 21 (1980); Haubrich, R., The American bamboos, Part 1, *Journal of the American Bamboo Society* **1**: 29–30 (1980); Dunmire, J.R., Weeping Mexican bamboos, *Pacific Horticulture* **42**: 18 (1981).

1. O. acuminata (Munro) Calderon & Soderstrom (*Arundinaria longifolia* Fournier: *Arthrostylidium longifolium* (Fournier) Camus; *Bambusa longifolia* (Fournier) McClure; *Yushania aztecorum* McClure & E.W.Smith; *Otatea aztecorum* (McClure & E.W.Smith) Calderon & Soderstrom). Illustration: *Smithsonian Contributions to Botany* **9**: 117 (1973).

Culms flexuous, ultimately bending with the weight of the leaves, 2–6 m, 2–12 cm in circumference. Sheaths more or less hairy, sometimes the hairs deciduous, without auricles and bristles, or these poorly developed; blade elongate-triangular. Leaves 9–18 cm × 3–6 mm, numerous on non-flowering culms, very long-acuminate, often hairy towards the base beneath, hanging; sheaths downy; ligule 1–2 mm, downy. *Mexico.* H5.

3. THAMNOCALAMUS Munro
S.G. Knees
Rhizome culms in single clumps; mid-culm branch buds enclosed by single 2-keeled prophylls, the front of all culm buds open, with lateral branch axes always subtended by sheaths, first-year branches usually 5 at the mid-culm nodes. Leaf venation strongly tessellated. Inflorescence a dense contracted racemose panicle or raceme, branches usually subtended by substantial sheaths or occasionally by rings of hairs, and prophylls on the branches represented by single-keeled sheaths inserted near the point of branching or as the lower glumes of pedicellate spikelets.

Literature Stapleton, C.M.A. The bamboos of Nepal and Bhutan Part I: Bambusa, Dendrocalamus, Melocanna, Cephalostachyum, Teinostachyum, and Pseudostachyum (Gramineae: Poaceae, Bambusoideae). *Edinburgh Journal of Botany* **51** (1): 1–32 (1994); The Bamboos of Nepal and Bhutan Part II: Arundinaria, Thamnocalamus, Borinda, and Yushania (Gramineae: Poaceae, Bambusoideae). *Edinburgh Journal of Botany* **51** (2): 275—295 (1994).

1. T. spathiflorus (Trinius) Munro. Illustration: *Flora of Bhutan* **3** (2): 499, 509 (2000).

Rhizome neck 3.5–6 cm. Culms 3–5.5 m × 1–2 cm; internodes terete or slightly flattened above branches, 15–18 cm, grooved, ridged, initially densely white powdery, glabrous; wall 2–3 mm thick; nodes slightly to distinctly swollen; sheath scar prominent, with persistent remains of sheath base. Branches 3–6, erect or deflexed, densely white powdery, glabrous. Culm sheaths narrowly rounded–triangular, equal to or slightly longer than internodes, leathery, white powdery, glabrous or densely setose, margins yellow-brown ciliate, longitudinal ribs conspicuous; auricles absent or falcate, bristles absent or prominent; ligule arcuate, *c.* 1 mm, glabrous, margins fissured; blade erect, gray or gray-brown, triangular or linear-lanceolate, glabrous, margins usually rolled, serrulate. Leaves 2–5 per ultimate branch; sheath glabrous; auricles absent or prominent, britles absent, or many and readily deciduous, grey or reddish, 1–1.5 mm; ligule truncate, *c.* 1 mm, margins puberulous; blade narrowly lanceolate, 4.5–9 × 0.5–1 cm, glabrous, secondary veins 2-paired, transverse veins distinct, base broadly cuneate or nearly rounded, margins serrulate. *Himalayas (Bhutan, India, Nepal), W China.*

Represented in cultivation by two distinct varieties. Var **spathiflorus** with culm nodes only slightly swollen, culm sheaths densely hairy and with prominent auricles. Var. **crassinodus** (T. P. Yi) Stapleton has swollen culm nodes, hairless culm sheaths and auricles absent.

4. CHUSQUEA Kunth
D. McClintock
Rhizomes elongate, slender. Culms solid, terete, more than 1 cm in circumference. Sheaths persistent, without bristles. Branches numerous at each node. Leaves narrow, without bristles. Inflorescence usually a panicle. Spikelets with usually 4 glumes and with 1 (rarely 2) floret(s). Lodicules 3. Stamens 3. Stigmas usually 2, occasionally with 1 branched again.

A genus of over 90 species from Mexico to S Argentina. They are not easy to propagate vegetatively, and take a considerable time to establish. Once established, they can tolerate drier conditions than most bamboos.

Literature: Parodi, L.R., Sinopsis de las Graminas chilenas del Genero Chusquea, *Revista Universitaria, Universidad Catolica de Chile* **30** (1): 61–71 (1945).

1. C. culeou Desvaux (*C. andina* Philippi; *C. breviglumis* Philippi; *C. quila* misapplied). Illustration: Lawson, Bamboos, 149 (1966); Crouzet, Les Bambous, 55 (1981); Everett, Encyclopedia of horticulture **3**: 760 (1981); Brickell (ed.) RHS A–Z encyclopedia of garden plants, 275 (2003).

Plant loosely tufted with very short rhizomes. Culms 4–6 m, rarely to 10 m, *c.* 11–13 cm in circumference, yellowish, erect, tapering. Sheaths downy above, white, much longer than the internodes; blades narrow. Nodes hairy, quite prominent, with white, waxy bloom below. Branchlets developing in the second year and bursting through the sheaths, to 20 cm or more, bushy, for the most part unbranched. Leaves 5–10 × *c.* 1 cm, hairless, visibly tessellate, horny-edged and with the mid-vein prominent beneath; sheaths hairless, without bristles; ligule 0.5–1 mm, hairless. *Most of Andean South America.* H2.

There are 2 variants at least in cultivation: 1 as described above, the other (var. **tenuis** McClintock) with slenderer, shorter culms which are spreading.

C. cumingii Nees. Leaves 2–6 cm × 3–5 mm, midrib hardly apparent. *Chile.* H2. This may be the same species as *C. culeou;* if the 2 are combined, the correct name would be *C. cumingii.*

5. ARUNDINARIA Michaux
D. McClintock
Rhizomes slender and elongate, or compact and much branched. Culms erect, hollow, terete; new shoots emerging in spring. Sheaths persistent with well-developed blades and dark, rough bristles. Branches, several at each node, foliage developing from the top downwards. Leaves variably tessellate or not; sheaths bearing dark, tough bristles (rarely bristles absent); ligule often dark-coloured. Inflorescence racemose or paniculate. Spikelets with 2 or rarely 3 glumes and several florets of which the upper may be unisexual. Lodicules 3 or rarely 6. Stamens usually 3 or 6. Stigmas 3.

A genus of an uncertain number of species, mainly from the Himalaya, but also in America, China and South Africa. Species nos 4–7 are generally included in the genus *Thamnocalamus* Munro. This differs from *Arundinaria* essentially only in floral characters, so is not recognised here, though synonyms under it are cited.

1a. Leaves without visible tessellation **5. falconeri**
b. Leaves with visible tessellation 2
2a. Rhizomes far-running, culms spaced **3. jaunsarensis**
b. Rhizomes short, culms tufted 3
3a. Culms rough to touch **2. maling**
b. Culms smooth to touch 4

4a. Culm sheaths whitish, fairly persistent **7. tessellata**
b. Culm sheaths not whitish, deciduous 5
5a. Branches usually 1–3 at each node **1. amabilis**
b. Branches numerous at each node 6
6a. Culm sheaths with auricles and bristles **4. aristata**
b. Culm sheaths without auricles or bristles **6. spathacea**

1. A. amabilis McClure. Illustration: McClure, The bamboos, f. 67, 68 (1966).

Culms 3–12 m, 10–20 cm in circumference. Sheaths hairy. Nodes without white, waxy bloom below. Branches usually 1–3 at each node. Leaves *c.* 18 × 2 cm, occasionally to 36 × 4 cm, pale beneath, downy towards the base, with bristles and auricles; ligule *c.* 1 mm. *China, where it is much cultivated.* H2.

The species, which is seen occasionally in Europe, flowered in 1979–80.

2. A. maling Gamble (*A. racemosa* misapplied). Illustration: *Annals of the Royal Botanic Garden Calcutta* **7**: t. 8 (1896).

Culms to 5 m, *c.* 10 cm in circumference, glaucous, rough to the touch. Sheaths persistent, tapered and truncate, bristly hairy, downy inside, blades long and narrow; ligule fringed. Nodes with white, waxy bloom below. Leaves to 18 × 2 cm, sometimes downy towards the base beneath; sheaths ciliate above, with auricles and a few short bristles; ligule 1.5–2mm, minutely downy. *NE Himalaya.* H2.

3. A. jaunsarensis Gamble (*A. anceps* Mitford; *Yushania anceps* (Mitford) Lin). Illustration: *Annals of the Royal Botanic Garden Calcutta* **7**: t. 22 (1896); *Bulletin of the Taiwan Forestry Institute* **68**: t. 3 (1974); *The Plantsman* **1**: 37 (1979).

Rhizomes extensively creeping from the central tuft. Culms 3–6 m, arching when mature, 2–5 cm in circumference, with longitudinal lines, rough and with white, waxy bloom below the nodes. Sheaths shorter than the internodes, pale or purplish brown, not overwintering except sometimes on late shoots, ciliate and with auricles and bristles. Leaves usually *c.* 10 × 1 cm, sometimes to 16 × 2 cm, hairless, tessellate; sheaths hairless, usually with auricles and bristles; ligule *c.* 1 mm, downy. *C & NW Himalaya.* H2.

Flowering more or less freely, but by no means everywhere, since 1957. It is very invasive, and liable to suffer in hard winters. 'Pitt White' (*A. niitikayamensis* misapplied) has taller, stouter culms to 10 m, 12 cm in circumference; it has flowered since 1981.

4. A. aristata Gamble (*Thamnocalamus aristatus* (Gamble) Camus). Illustration: *Annals of the Royal Botanic Garden Calcutta* **7**: t. 17 (1896); *The Plantsman* **1**: 38 (1979).

Culms 5–8 m, rarely to 10 m, 4–6 cm in circumference, yellowish, soon smooth, speckled when old. Sheaths deciduous, tapering above, sparsely hairy, bristly towards the base, with auricles and bristles at the apex, blade short. Nodes with white, waxy bloom below. Branches reddish. Leaves *c.* 10 × 1 cm, downy beneath, with minute glands on the stalk, often twisted at the apex, visibly tessellate; sheaths dull purple, hairless but with a hairy callus at the apex, bearing bristles. Ligule to 3 mm, downy. *W Himalaya.* H3.

This species flowered in 1950–52, but is now uncommon in Europe.

A. spathiflora Trinius (*Thamnocalamus spathiflorus* (Trinius) Munro). Illustration: *Annals of the Royal Botanic Garden Calcutta* **7**: t. 16 (1896). Similar, but culms grey, white-bloomed at first, later pinkish in the sun, with dark bands below the nodes, somewhat zig-zag, branches 2–3, sheaths ciliate, with longer blades and without an apical callus, leaves greyer. *NW Himalaya.* H2.

5. A. falconeri (Munro) Rivière (*Thamnocalamus falconeri* Munro; *A. nobilis* Mitford; *A. pantlingii* misapplied). Illustration: *Annals of the Royal Botanic Garden Calcutta* **7**: t. 18 (1896); *Botanical Magazine*, 7947 (1904).

Culms 7–10 m, rarely to 20 m, 6–10 cm in circumference, the outer often spreading, greenish yellow, purple below the nodes, with no white waxy bloom. Sheaths deciduous, minutely downy, not ciliate, purplish, much longer than the internodes, without auricles or bristles; ligule broad, very shortly ciliate. Leaves to 16 × 1.6 cm but usually much smaller, hairless, pale green without visible tessellation; sheaths hairless, shortly ciliate, without bristles; ligule to 1.5 mm, downy. *NE Himalaya.* H2–3.

Very decorative, but suffering badly in harsh winters. The species flowered synchronously in 1929–31 and 1964–69, all the plants dying afterwards. It may be best placed in a separate genus together with some species currently in *Chimonobambusa.*

6. A. spathacea (Franchet) McClintock (*Fargesia spathacea* Franchet; *A. murielae* Gamble; *Sinarundinaria murielae* (Gamble) Nakai; *Thamnocalamus spathaceus* (Franchet) Soderstrom). Illustration: Hay & Synge, Dictionary of garden plants, t. 1901 (1969); *The Garden* **4**: 22, 24, 26 (1979).

Culms to *c.* 4 m, *c.* 4 cm in circumference, greenish with white, waxy bloom below the nodes, branching in the first year, arching when mature. Sheaths deciduous, rounded above and without auricles or bristles, downy and greenish purple when young, soon hairless and pale brown, noticeable for a short time while standing away from the culm before falling. Leaves to 15 × 2 cm, mostly much less, visibly tessellate, stalks purplish in the sun; sheaths often purplish, hairless, often without bristles; ligule 1–2 mm, downy. *China.* H2.

Much confused with *Sinarundinaria nitida*. Dwarf variants are grown in Europe. Some of these flowered in 1978–80 and then died.

7. A. tessellata (Nees) Munro (*Nastus tessellatus* Nees; *Thamnocalamus tessellatus* (Nees) Soderstrom & Ellis). Illustration: *Hooker's Icones Plantarum* **30**: t. 2930 (1911).

Culms tufted, 1.5–7 m, 6–8 cm in circumference, thick-walled, erect, smooth, with a dark purple ring below each node, unbranched and leafless during the first year. Sheaths fairly persistent, mostly longer than the internodes, thick, whitish, slightly downy, with auricles and bristles; blades long and narrow. Branches many, short, erect, purplish. Leaves 5–14 × 1–2 cm, visibly tessellate, long-acuminate, dark green; sheaths ciliate, with auricles and bristles; ligule 1.5 mm, fringed. *South Africa.* H2.

6. PLEIOBLASTUS Nakai

D. McClintock

Rhizomes mostly elongate, slender. Culms erect, almost always hollow but thick-walled, terete. New shoots arising in spring, leafing from the top downwards. Sheaths persistent, bearing smooth, whitish bristles. Branches usually 3–7 at each node or sometimes branches only 1–2, all in the lower part of the culm. Leaves with visible tessellation and with smooth, whitish bristles which are occasionally absent. Inflorescence a spike or raceme. Spikelets with 2 glumes and 5–13 florets. Lodicules 3, ciliate. Stamens 3. Stigmas 3.

A genus of about 20 species from China and Japan, formerly considered part of *Arundinaria*, mostly easy to grow, but some can form thickets.

Rhizomes. Markedly running: **1, 2, 4**.
Culms. More than 2 m: **1, 2, 5**; 1–2 m: **3, 6, 7**; less than 1 m: **4**.
Leaf-sheaths. Purple: **6, 7**.
Branches. 1–2, *rarely* 3: **3, 4, 6, 7**; 37: **1, 2, 5**.
Leaves. Variegated: **1, 3, 6, 7**. 12–20 times longer than broad: **2**. At least some with markedly oblique apices: **2, 5**. Ligule more than 3 mm: **2**.

1a. Leaves 12–20 times longer than broad; ligule 3 mm or more **2. gramineus**
b. Leaves 6–8 times longer than broad; ligule to 2 mm 2
2a. Upper margin of leaf-sheath oblique **5. simonii**
b. Upper margin of leaf-sheath horizontal 3
3a. Culms 3–4 m; branches numerous **1. chino**
b. Culms to 2 m; branches often fewer than 3 4
4a. Culms to 1 m 5
b. Culms 1–2 m 6
5a. Culms 60 cm or more; leaves variegated **6. variegatus**
b. Culms 50 cm or less; leaves not variegated **4. pygmaeus**
6a. Leaves entirely green **3. humilis**
b. Leaves variegated 7
7a. Leaves soon hairless **1. chino**
b. Leaves persistently downy **7. viridi-striatus**

1. P. chino (Franchet & Savatier) Makino (*Arundinaria chino* (Franchet & Savatier) Makino; *P. maximowiczii* (Rivière) Nakai). Illustration: Suzuki, Index to Japanese Bambusaceae, 304–9 (1978); Kitamura & Murata, Coloured illustrations of woody plants of Japan, t. 135 (1977).

Rhizome far-running. Culms 3–4 m, *c.* 4 cm in circumference, with white, waxy bloom below the nodes, purple when young. Sheaths very slightly downy when young, with white bristles. Branches from upper nodes spreading, rather short. Leaves 15–25 × 1.5–2.2 cm, green on both sides, slightly downy towards the base beneath; sheaths ciliate; ligule very short. *Japan.* H2.

A variable species fairly commonly grown; many variants have been described; they are difficult to distinguish and require further study. The most important are: forma **angustifolius** (Mitford) Muroi (*Bambusa angustifolia* Mitford; *Arundinaria angustifolia* (Mitford) Houzeau de Lehaie; *P. angustifolius* (Mitford) Nakai & Okamura), which is smaller (at most to 2 m), with narrower leaves which are usually striped with white (it has flowered since 1978); forma **humilis** (Makino) Suzuki, which is even smaller and confused with *P. humilis* (see below); forma **vaginatus** (Hackel) Muroi & Okamura (*Arundinaria vaginatus* Hackel; *P. vaginatus* (Hackel) Nakai; *P. chino* var. *vaginatus* (Hackel) Suzuki; *P. gracilis* (Makino) Nakai), which has culms which are somewhat purplish, mature leaves to 15 × 1.5 cm and even larger in young shoots (this is probably the most frequently seen variant, and is probably the one which has flowered sporadically since 1974). In addition to these there are 2 cultivars (probably clones) which are known only in cultivation: 'Argenteostriatus' (*P. argenteo-striatus* (Regel) Nakai; *Arundinaria chino* var. *argenteo-striata* (Regel) Makino), which is shorter (to 1 m), its leaves striped with cream, and is grown occasionally; 'Laydekeri' (*Arundinaria laydekeri* Bean; *P. chino* var. *laydekeri* (Bean) Nakai), which is also small and has leaves dully and irregularly mottled.

2. P. gramineus (Bean) Nakai (*Arundinaria graminea* (Bean) Makino). Illustration: Suzuki, Index to Japanese Bambusaceae, 23, 292–3 (1978); Kitamura & Murata, Coloured illustrations of woody plants of Japan, t. 134 (1979).

Rhizomes extensive. Culms 3–5 m, 3–6 cm in circumference, with white, waxy bloom below the nodes, hairless. Branches numerous, hanging. Leaves to 28 × 1.2 cm (rarely broader), long-tapered, somewhat drooping and twisted towards the apex; most apices oblique. Sheaths hairless, apex oblique; bristles, if present, long; ligule 3–4 mm, blunt, downy, *Japan.* H2.

This species is occasionally grown and has flowered frequently since 1948.

P. hindsii (Munro) Nakai (*Arundinaria hindsii* Munro; *Bambusa erecta* invalid; *B. gracilis* misapplied). Illustration: Suzuki, Index to Japanese Bambusaceae, 23, 288–9 (1978); Kitamura & Murata, Coloured illustrations of woody plants of Japan, t. 134 (1979). Similar, but rhizome shorter, and plant not so copiously leafy. Branches erect, leaves erect, darker, thicker, to 1.5 cm broad, ligule 3–4 mm, acute. *Hong Kong.*

This species flowered between 1905 and 1911, and is relatively uncommon.

3. P. humilis (Mitford) Nakai (*Arundinaria humilis* Mitford; *Sasa humilis* (Mitford) Camus). Illustration: Suzuki, Index to Japanese Bambusaceae, 322–3 (1978); *The Plantsman* **1**: 41 (1979).

Rhizome running. Culms to 1.5 m, *c.* 3 cm in circumference. Sheaths purple when young. Nodes usually hairless, with white, waxy bloom below. Branches 1–3, long, found low on the culm. Leaves to 20 × 1.8 cm, sometimes downy beneath towards the base; sheaths slightly downy, ciliate; ligule minute, downy. *Japan.* H2.

A variable species; several variants have been named, of which the following are the most important: forma **humilis**, with short, bushy branches well up the culm; a further variant, with no appropriate name in this species, but found as *Arundinaria pumila* Mitford, *A. gauntletti* invalid and *P. pumilus* (Mitford) Nakai, is more robust and compact and has conspicuously bearded bases to the culm sheaths, and fresh green leaves.

4. P. pygmaeus (Miquel) Nakai (*Arundinaria pygmaea* (Miquel) Mitford; *Sasa pygmaea* (Mitford) Camus). Illustration: Suzuki, Index to Japanese Bambusaceae, 310–13 (1978).

Rhizome running. Culms to 40 cm, *c.* 9 mm in circumference, usually solid and flattened above, hairless, with white, waxy bloom. Sheaths hairy. Nodes conspicuous, downy or bristly at least at first. Branches 1–2 at the lower nodes. Leaves 3–8 cm × 4–8 mm, slightly downy, like the sheaths. *Japan.* H2.

This is the smallest species of the genus and various clones are grown, as is var. **distichus** (Mitford) Nakai (*Bambusa disticha* Mitford; *Arundinaria disticha* (Mitford) Pfitzer; *Sasa disticha* (Mitford) Camus; *P. distichus* (Mitford) Muroi & Okamura), illustration: Brickell (ed.) RHS A–Z encyclopedia of garden plants, 829 (2003), in which the culms are always hollow, and up to 1 m tall, with usually hairless nodes, the leaves and their sheaths hairless or not, larger and more numerous, arranged in 2 rows on the culm. This variety flowered synchronously in 1967–70; it grades into the variant described above.

5. P. simonii (Carrière) Nakai (*Arundinaria simonii* (Carrière) Rivière; *Nipponocalamus simonii* (Carrière) Nakai). Illustration: Lawson, Bamboos, 109 & t. 16 (1966); Suzuki, Index to Japanese Bambusaceae, 294–5 (1978).

Rhizomes shortly running. Culms 4–8 m, *c.* 6 cm in circumference, hollow, the outer often arching. Sheaths conspicuous, long-persistent, bearded at the base when young. Nodes with white, waxy bloom below. Leaves to 25 × 3.5 cm, often smaller, minutely downy beneath, sometimes half green and half glaucous; sheath apex usually oblique; ligule *c.* 1 mm, downy. *Japan.* H2.

A commonly cultivated species, whose young leaves may be somewhat variegated at first, which flowered more or less generally in 1965–68 and 1981. 'Variegatus' (*Arundinaria simonii* var. *variegata* Hooker; *P. simonii* forma *variegata* (Hooker) Muroi; *A. simonii* var. *albo-striata* Bean; *P. simonii* var. *heterophyllus* (Makino) Nakai) is less tall and has its mature leaves variously broad or narrow, green or variegated. It is occasionally grown and has flowered since 1980—see McClintock, *Journal of the American Bamboo Society* (in press).

6. P. variegatus (Miquel) Makino (*Arundinaria variegata* (Miquel) Makino; *Sasa variegata* (Miquel) Camus; *P. fortunei* invalid; *A. fortunei* invalid). Illustration: *Flore des Serres* **15**: t. 1535 (1863); Reader's Digest encyclopaedia of garden plants and flowers **2**: 60 (1971); *The Plantsman* **1**: 45 (1979); Crouzet, Les Bambous, 34 (1981).

Rhizome shortly running. Culms to 75 cm, *c.* 2 cm in circumference. Sheaths minutely downy, ciliate. Nodes hairless, with white, waxy bloom below. Branches usually 1–2, mostly from low down on the culm, erect. Leaves *c.* 14 × 1.3 cm (rarely to 20 × 1.8 cm), tapered below, slightly downy on both sides, dark green with cream stripes; sheaths purplish, minutely downy, ciliate. *Japan.* H2.

This species flowered in about 1970 and has done so again since 1977. It is variable, and several named variants are grown, including dwarfs (less than half the height given above), which may be 'Pygmaeus' ('Nana'). Var. **viridis** (Makino) McClintock (*Arundinaria variegata* var. *viridis* Makino; *P. shibuyanus* Nakai) is the wild variant with leaves entirely green and the nodes sometimes downy. Seedlings from this variant are grown; their leaves are often broader and usually downy beneath; though some with hairless leaves are grown as forma **glaber** (Makino) McClintock.

7. P. viridi-striatus (André) Makino (*Arundinaria auricoma* Mitford; *A. viridistriata* (André) Makino; *A. fortunei* var. *aurea* invalid; *Bambusa argentea* invalid). Illustration: Hooker's Icones Plantarum **27**: t. 2613 (1901); Lawson, Bamboos, t. 9 (1966); Suzuki, Index to Japanese Bambusaceae, 24 (1978).

Rhizome shortly running. Culms to 1.5 m, *c.* 2 cm in circumference, purple. Sheaths markedly downy when young. Nodes hairy, with white, waxy bloom below. Branches 1–2, erect, from low down on the culm. Leaves *c.* 18 × 2.8 cm (rarely to 20 × 4 cm), rounded below, softly downy on both sides, especially beneath when young, brilliant yellow with green stripes of various breadths; sheaths densely downy, ciliate, purple. *Japan.* H2.

A widely grown species, which has a few spikelets in most years on mature culms. *P. kongosanensis* Makino is a name applied to the wild form, which has leaves entirely green.

7. SASA Makino & Shibata

D. McClintock

Rhizomes slender. Culms distant, ascending, hollow, terete, 2–3 m tall, 3–6 cm in circumference, unbranched below. Sheaths persistent, mostly shorter than the internodes, bristles spreading and rough or rarely absent. Nodes rather prominent, with white waxy bloom below. Branches 1 at each node. Leaves 4–11 per branch, thick and hairless, visibly tessellate; bristles rough, whitish, rigid, borne at right angles to the branch, sometimes absent; ligule downy. Inflorescence a panicle. Spikelets with 2 glumes and 4–10 flowers. Lodicules 3. Stamens 6. Stigmas 3.

A genus of probably 50 species including *Sasaella* and *Sasamorpha* treated separately in edition 1. They grow well in partial shade, but are not easy to control.

Literature: Haubrich, R., The Sasas, *Journal of the American Bamboo Society* **2**: 24–38 (1982).

1a. Rhizomes not far-running; culms with many branches above 2
b. Rhizomes far-running; culms with few-branches, all near the base 3
2a. Culms to 3 m; nodes hairless; bristles usually absent **1. kurilensis**
b. Culms to 1.6 m; nodes hairy below; bristles usually present on young leaf-sheaths **4. tsuboiana**
3a. Leaves and stalks often purplish beneath and towards base 4
b. Leaves and stalks green or yellowish but not purple 5
4a. Leaves with a broad, white margin **3. veitchii**
b. Leaves without a white margin **6. borealis**
5a. Leaves to 20 × 3 cm, midrib and blade green **5. ramosa**
b. Leaves to 40 × 9.5 cm, green with a yellow midrib **2. palmata**

1. S. kurilensis (Ruprecht) Makino & Shibata. Illustration: Suzuki, Index to Japanese Bambusaceae, 19, 118–19 (1978); Kitamura & Murata, Coloured illustrations of woody plants of Japan, t. 137 (1979).

Rhizome shortly running. Culms 0.5–3 m, 5–30 mm in circumference, with the central hollow about half the diameter, hairless, richly branched above. Sheaths over half the length of the internodes, almost hairless, ciliate, disintegrating irregularly. Nodes hairless. Leaves variable, 15–22 × 2.2–4.8 cm, shining above, sometimes toothed on one side only, veins prominent; sheaths minutely downy, ciliate; auricles and bristles (if any) very soon falling; ligule to 2 mm. *Japan, Korea.* H2.

Relatively uncommon in cultivation; it has flowered from 1978.

2. S. palmata (Burbidge) Camus (*Arundinaria palmata* (Burbidge) Bean; *Bambusa palmata* Burbidge; *S. senanensis* misapplied; *S. cernua* misapplied). Illustration; Suzuki, Index to Japanese Bambusaceae, 20, 170–3 (1978); Crouzet, Les Bambous, 61 (1981).

Rhizome extensive. Culms 2 m (rarely to 4 m), *c.* 3 cm in circumference, often leaning, often hairless, usually soon purple-streaked (forma *nebulosa* (Makino) Suzuki). Sheaths minutely downy towards the apex and the base; auricles and bristles frequently absent; ligule *c.* 1 mm, very broad. Leaves to 40 cm, acuminate, bright, shining green, paler beneath, the midrib yellow, the margin sometimes decaying in hard winters; stalks greenish yellow; sheaths usually without bristles; ligule *c.* 2 mm, downy. *Japan.* H2.

This species began flowering in 1962, but this phase of flowering is now dying out. There is a variant with the leaves striped longitudinally with yellow.

S. senanesis (Franchet & Savatier) Rehder. Illustration: Suzuki, Index to Japanese Bambusaceae, 174–7 (1978). Leaves narrower and densely downy beneath. *Japan.* H2.

3. S. veitchii (Carrière) Render (*Arundinaria veitchii* (Carrière) N.E. Brown; *S. albo-marginata* (Miquel) Makino). Illustration: Lawson, Bamboos, 146 (1966); Suzuki, Index to Japanese Bambusaceae, 190–2 (1978); *Horticulture* **7**: 26 (1979); Darke, Ornamental grasses, 275 (1999).

Rhizome extensive. Culms to 1.3 or rarely to 2 m, *c.* 2 cm in circumference, finally purplish, glaucous, hairy at first, hairs soon deciduous, with white, waxy bloom below the nodes. Lowest internodes short. Sheaths white-hairy at first, with auricles and bristles, sometimes decaying irregularly. Leaves to 25 × 6 cm, abruptly acuminate, rather dull green above, glaucous beneath, soon with a broad, white margin; stalks often purplish. Sheaths hairless, bristles soon falling; ligule to 1.6 mm, dark in colour. *Japan.* H2.

This species has not, so far as is known, flowered outside Japan. A small variant, with culms to 27 cm, leaves 3–14 cm × 8–32 mm, is grown; it is sometimes called forma *nana*, or *S. albo-marginata* forma *nana* Makino.

4. S. tsuboiana Makino. Illustration: Suzuki, Index to Japanese Bambusaceae, 20, 163 (1978).

Rhizomes very shortly running. Culms to 1.6 m, *c.* 1 cm in circumference, erect, with a very small central hollow. Sheaths hairy below at least when young, ciliate, not or hardly disintegrating untidily, less than half the length of the mature internodes. Leaves to 24 × 5 cm, usually with bristles, at least on the young shoots. *Japan.* H2.

5. S. ramosa (Makino) Makino & Shibata (*Bambusa ramosa* Makino; *Sasa vagans* misapplied; *Sasaella ramosa* Makino; *Arundinaria vagans* Gamble; *A. ramosa* (Makino) Makino; *Pleioblastus viridi-striatus* forma *vagans* (Gamble) Muroi; P. *kongosanensis* Makino 'Vagans'). Illustration: Makino, Illustrated flora of Japan, edn 1, t. 2633 (1948); Suzuki, Index to Japanese Bambusaceae, 22, 240–3 (1978); Kitamura & Murata, Coloured illustrations of woody plants of Japan, t. 136 (1979).

Rhizome extensive. Culms to 1.5 m, 1.5 cm in circumference, hairless. Nodes almost globular, hairless. Leaves to 20 × 3 cm, usually hairless above, distinctly downy beneath, the margins sometimes withering in harsh winters; sheath bristles few. *Japan.* H2.

This species first flowered outside Japan in 1981 (see McClintock, *Kew Bulletin* **38**: 191–5, 1983). There are dwarf variants, and another with variegated leaves.

6. S. borealis (Hackel) Makino (*Arundinaria borealis* (Hackel) Makino; *Sasamorpha borealis* (Hackel) Nakai; *Sasamorpha purpurascens* (Hackel) Nakai; *Sasamorpha purpurascens* var. *borealis* (Hackel) Nakai). Illustration: Suzuki, Index to Japanese Bambusaceae, 23, 270–1 (1978); Kitamura & Murata, Coloured illustrations of woody plants of Japan, t. 137 (1979).

Rhizome elongated. Culms to 2 m, often less, *c.* 6 cm in circumference, minutely downy, well branched above. Sheaths with long and short persistent hairs, tight, disintegrating irregularly. Leaves 2–3 per branch, 22–27 × 2.5–4 cm, long-acuminate, glaucous, slightly paler beneath, usually purplish towards the base and on the stalk and sheath; ligule *c.* 2.5 mm. *Japan, Korea.* H2.

Confused with *Pseudosasa japonica.* Has flowered since 1978.

S. masamuneana (Makino) Chao & Renvoize (*Arundinaria glabra* Nakai) is similar with leaves 12–20 × 2–2.8 cm. *Japan.* H4.

Often cultivated as 'Albostriata' with irregular creamy white stripes on the leaves.

8. PSEUDOSASA Nakai
D. McClintock
Rhizomes slender, usually clump-forming, occasionally running. Culms erect, to 6 m. Sheaths persistent, longer than the internodes, bristles usually absent, if present then smooth and white. Branches 1 at each node, the lower nodes without branches. Leaves 4–7 per branch, without auricles or bristles. Inflorescence a panicle. Spikelets with 2 glumes and 3–8 florets. Lodicules 3. Stamens 3 or rarely 4. Stigmas 3.

A genus of 6 species from E Asia.

1. P. japonica (Steudel) Makino (*Arundinaria japonica* Steudel; *Sasa japonica* (Steudel) Makino; *Yadakea japonica* (Steudel) Makino; *Bambusa metake* Miquel). Illustration: Lawson, Bamboos, t. 1–2 (1966); *Green Fingers* **8**: 58 (1976) & **10**: 32 (1978) – both as *Phyllostachys nigra* var. *henonis*; *The Plantsman* **1**: 42 (1979).

Culms to 6 m, *c.* 6 cm in circumference, thin-walled, richly branched above and arching when mature, unbranched in the first year. Sheaths densely and roughly hairy at first, closely persistent, blades rather narrow, entire. Nodes often oblique, with some white, waxy bloom below. Leaves up to 7 on each branch, to 36 × 3.5 cm, though usually less, rather thick, long-tapered, one-third green and two-thirds glaucous beneath; ligule 1.5–3 mm, minutely downy. *Japan, Korea.* H2.

Commonly cultivated, and probably in flower somewhere every year. Like *Sasamorpha borealis*, but larger, with white, waxy bloom below the nodes and the undersides of the leaves bicoloured.

9. INDOCALAMUS Nakai
D. McClintock
Differs from *Sasa* essentially in floral characters. The species described below has less prominent nodes without white, waxy bloom, short internodes which are exceeded by the sheaths and very large leaves (the largest of any bamboo hardy in Europe), which are borne 1–2 to a branch; bristles, when present, dark. Stamens 3. Styles 2.

A genus of about 20 species from Malaysia and China. The 1 cultivated species is easy to grow and will tolerate dense shade.

1. I. tessellatus (Munro) Keng (*Bambusa* tessellata Munro; *Sasa tessellata* (Munro) Makino & Shibata; *Sasamorpha tessellata* (Munro) Nakai; *B. ragamowskii* Nicholson; *Arundinaria ragamowskii* (Nicholson) Pfitzer). Illustration: Bailey, The standard cyclopedia of horticulture **1**: 445 (1942); *Arnoldia* **6**: t. 3 (1946).

Culms to 2 m, *c.* 4 cm in circumference, thick-walled, soon curved down with the weight of the leaves. Sheaths slightly downy but the hairs soon falling, rather loose. Internodes minutely downy. Leaves to 60 × 8.6 cm or more, sometimes with a line of hairs down one side of the midrib beneath; sheaths often purple towards the apex; ligule to 2 mm, dark in colour. *China.* H2.

This species is occasionally grown, and flowered in the USA between 1979 and 1982.

10. CHIMONOBAMBUSA Makino
D. McClintock
Rhizomes slender, mostly shortly running. Culms hollow, terete or 4-cornered; new shoots appearing in late summer or autumn. Sheaths without bristles, variably persistent. Branches 3–many, developing from the top downwards during the following year. Leaves visibly tessellate or not, with smooth bristles or bristles absent. Inflorescence a panicle. Spikelets with 0–2 glumes and several florets. Lodicules 3. Stamens 3. Stigmas 2.

A genus of several species from the Himalaya, China and Japan. Their late-emerging culms may not ripen sufficiently to overwinter, which renders them less reliable in cultivation than many others. Some of the species included here, together with *Arundinaria falconeri* (p. 329), should perhaps be transferred to a new genus.

1a. Culms obtusely 4-cornered **4. quadrangularis**
b. Culms terete 2
2a. Leaves visibly tessellate; branches usually 3 at each node; culm sheaths not ciliate **3. marmorea**
b. Leaves not visibly tessellate; branches many at each node; culm sheaths ciliate 3
3a. Culms 5–8 cm in circumference; sheaths soon deciduous; leaves usually less than 16 cm **1. falcata**
b. Culms 9 cm or more in circumference; culm sheaths fairly persistent; leaves 15–30 cm **2. hookeriana**

1. C. falcata (Nees) Nakai (*Arundinaria falcata* Nees; *Thamnocalamus falcatus* (Nees) Camus; *A. gracilis* invalid). Illustration: *Annals of the Royal Botanic Garden Calcutta* **7**: t. 11 (1896).

Culms 3–6 m, 5–8 cm in circumference, closely clumped, with a large hollow, terete, somewhat lined, glaucous at first, the inner erect, the outer spreading. Sheaths long-tapered, ciliate, not long-persistent, but emerging branches likely to push through them, as long as or longer than the internodes and dark red when young, later straw-coloured; ligule elongate, to 12 mm, toothed, blade narrow. Nodes hairy at first, sometimes with white, waxy bloom below. Branches numerous. Leaves usually *c.* 11 × 1.2 cm, rarely to 16 cm, not visibly tessellate, soon hairless beneath except for the midrib, the apex somewhat twisted; sheaths hairless but ciliate, usually without bristles; ligule 2.5–6.5 mm, downy, pointed. *NW Himalaya.* H3.

This species has flowered since 1978, mostly with specimens cultivated under glass.

2. C. hookeriana (Munro) Nakai (*Arundinaria hookeriana* Munro). Illustration: *Annals of the Royal Botanic Garden Calcutta* **7**: t. 15 (1896); Camus, Les Bambusees, t. 24b (1913).

Culms 5–10 m or more, 9 cm or more in circumference, with a large hollow, glaucous, usually striped with olive green and pink when young, sometimes with white, waxy bloom below the nodes. Sheaths almost hairless, often longer than internodes, fairly persistent, the broad base frequently remaining; ligule toothed. Nodes hairless, with numerous branches. Leaves 15–30 × 1–4 cm, minutely downy beneath, occasionally with thin, yellow stripes, not visibly tessellate; sheaths hairless, without bristles; ligule 2–3 mm, downy. *E Himalaya.* H3.

Flowered under glass in 1963.

3. C. marmorea (Mitford) Makino (*Bambusa marmorea* Mitford; *Arundinaria marmorea* (Mitford) Makino). Illustration: Suzuki, Index to Japanese Bambusaceae, 24, 334–5 (1978); Crouzet, Les Bambous, 54 (1981).

Culms 2–3 m, *c.* 2 cm in circumference, very thick-walled, terete, with hairless, purplish, rather short internodes. Sheaths thin, marbled purplish or pinkish brown at first, soon fading, almost hairless above, bushily hairy at the base when fresh, fairly persistent; blade minute. Nodes prominent. Branches usually 3, rather short and erect. Leaves to 16 × 1.2 cm, but often less (about half this size), visibly tessellate, minutely downy beneath; sheaths with bristles, downy; ligule minute. *Japan.* H2.

Flowers can be found in most years on some mature plants. 'Variegata' has white-striped leaves.

4. C. quadrangularis (Fenzi) Makino (*Bambusa quadrangularis* Fenzi; *Arundinaria. quadrangularis* (Fenzi) Makino; *B. angulata* invalid). Illustration: Lawson, Bamboos, 31 (1966); Suzuki, Index to Japanese Bambusaceae, 17, 98–9 (1978); Crouzet, Les Bambous, 64 (1981).

Rhizome far-running. Culms 3–7 m, rarely to 10 m, 6–12 cm in circumference, obtusely 4-cornered, slightly rough below the nodes. Sheaths hairless, thin, open, soon falling. Nodes very prominent, with a purple band below and with the bases of incipient aerial roots, which decrease in size up the culm. Branches 3–7 at each node, rather short, brittle. Leaves 14–25 × 1.4–2 cm, visibly tessellate, dark green; sheaths with bristles. *China.* H3.

The flowers of this species are unknown.

11. SEMIARUNDINARIA Nakai
D. McClintock
Culms usually tightly clumped (but capable of spreading), hollow, terete or with the upper internodes grooved, held stiffly erect. Sheaths soon falling, except on late shoots, but often hanging by their bases for some time, the inner surface highly polished, usually purplish red, without bristles. Branches erect, 3–8 at each node, developing from the base of the culm upwards, the lower nodes without branches. Leaves to 20 cm, visibly tessellate and with rough bristles. Inflorescence a number of spikes. Spikelets with glumes absent or rarely 1 present, with 3–6 flowers of which the upper 1–3 are male only. Lodicules 3, ciliate. Stamens 3. Stigmas 3.

A genus of perhaps 20 species from E Asia. They are easy to grow and excellent for tall hedging.

1. S. fastuosa (Mitford) Makino (*Bambusa fastuosa* Mitford; *Arundinaria narihita* Makino; *Phyllostachys fastuosa* (Mitford) Nicholson; *A. fastuosa* (Mitford) Makino; *A. narihira* Makino). Illustration: Suzuki, Index of Japanese Bambusaceae, 16, 84–7 (1978); *Horticulture* **7**: 27 (1979); Kitamura & Murata, Coloured illustrations of woody plants of Japan, t. 133 (1979); Crouzet, Les Bambous, 38 (1981).

Culms 3–6 m, 10–25 cm in circumference, held stiffly erect, and with a large hollow. Sheaths thick, hairless or slightly downy, auricles and bristles very small or absent; blades short, narrow. Leaves mostly 12–15 × 1.6–2 cm, broader on young shoots, thick, hairless, without auricles or bristles, often purple. *Japan.* H2.

Var. **yashadake** (Makino) Makino (*S. yashadake* (Makino) Makino). Illustration: Suzuki, Index of Japanese Bambusaceae, 16, 88–9 (1978); Kitamura & Murata, Coloured illustrations of woody plants of Japan, t. 132 (1979). Has leaves to 4 cm broad, leaves and sheaths with plentiful brown hairs at the base.

This species has flowered since 1957.

12. PHYLLOSTACHYS Siebold & Zuccarini

D. McClintock

Rhizomes short or elongate. Culms close or spaced, with hollow internodes which are grooved on alternate sides above each node (except sometimes for those below the lowest branches); nodes prominent. Branches developing in sequence from below upwards, 3 at each node, the outer unequal, the central small and depauperate, sometimes absent; branchlets numerous. Sheaths soon deciduous (those at the base sometimes more persistent), bristles rough or absent. Leaves 6–20 cm, visibly tessellate, hairless above, paler and sometimes downy near the base beneath; bristles rough but soon deciduous. Inflorescences spicate or racemose. Spikelets with usually 2 glumes and 5–13 florets. Lodicules 3, ciliate. Stamens 3. Stigmas 3.

A genus of perhaps 60 or more species, all originating in China, but very widely cultivated elsewhere.

Literature: McClure, F.A., Bamboos of the genus Phyllostachys under cultivation in the US, *USDA Handbook* 114 (1957); Haubrich, R., Handbook of bamboos cultivated in the United States, Part 1: the genus Phyllostachys, *Journal of the American Bamboo Society* **1**: 48–96 (1980); Wang Cheng-Ping and others, A taxonomical study of Phyllostachys, China, *Acta Phytotaxonomica Sinica* **18**: 15–19, 168–93 (1980).

Culms. Well over 10 m (at least in warmer areas); **3, 6, 9.** Downy: 7; hairy: **6**; slightly roughened: **2, 6, 7, 9.** Blackish: **5, 7**; blotched: **1, 3, 5, 7, 8**; bright yellow, often with green stripes: **6, 9**; only the groove yellowish: **2.** Lowest internodes very close and asymmetric: **1, 6.** Curved at the base: **6, 8.** Without white, waxy bloom below the nodes: **1, 3.**

Culm sheaths. Hairless: **1, 2, 3, 4, 5, 9**; hairy: **6, 7, 8.** Ciliate: **1, 2, 6, 7.** Without auricles or bristles: **1, 3, 4, 5, 7, 9.** Persistent near the base of the culm: **1, 6.**

Ligule. Downy: **1, 2, 4, 5, 7**; hairless: **3, 6, 8, 9.**

1a. Lowest nodes asymmetric and very close together 2
b. Lowest nodes symmetric and usually distant 3
2a. Most nodes with deep swollen bands immediately beneath **1. aurea**
b. Nodes all without swollen bands beneath **6. heterocycla**
3a. Culms with the groove of a different colour from the rest 4
b. Grooves the same colour as the rest of the culm 5
4a. Culms greyish green, grooves striped with yellow **2. aureosulcata**
b. Culms yellow, grooves green **3. bambusoides**
5a. Culms yellow with varying amounts of green striping 6
b. Culms green or greyish green, not yellow 8
6a. Nodes without white, waxy bloom beneath; green stripes few **3. bambusoides**
b. White, waxy bloom present below the nodes; green stripes many 7
7a. Culms smooth; sheaths hairy **6. heterocycla**
b. Culms initially slightly rough; sheaths hairless **9. viridis**
8a. Culms somewhat zig-zag; leaf-sheaths usually without auricles or bristles **5. flexuosa**
b. Culms straight; leaf-sheaths usually with auricles and bristles, at least when young 9
9a. Nodal ridges not prominent, single at lower, branchless nodes 10
b. Nodal ridges prominent, always 2 11
10a. Culms grey; sheaths hairy and with bristles; small auricles sometimes present; leaves 4–10 cm **6. heterocycla**
b. Culms green; sheaths hairless and without auricles or bristles; leaves 7–12 cm **9. viridis**
11a. Culm sheaths usually without spots; leaves often less than 5 cm 12
b. Culm sheaths spotted or streaked; leaves mostly more than 7 cm 13
12a. Culms grey-green, sheaths hairless, usually without auricles or bristles **4. congesta**
b. Culms black, blotched and spotted, or green; sheaths downy and usually with conspicuous auricles and bristles **7. nigra**
13a. Culms emerging in early summer, erect, without white, waxy bloom below the nodes; sheaths usually without auricles; leaves to 4.5 cm wide; sheaths with conspicuous bristles to 1.2 cm **3. bambusoides**
b. Culms emerging in spring, often curved at the base, with white, waxy bloom below the nodes; sheaths usually with auricles; leaves to 3.8 cm wide; sheaths with smaller deciduous bristles or bristles absent **8. viridi-glaucescens**

1. P. aurea (Carrière) Rivière (*P. formosana* Hayata). Illustration: Lawson, Bamboos, 117 (1966); Suzuki, Index to Japanese Bambusaceae, 11, 14, 72–73 (1978).

Culms 4–8 m, 9–12 cm in circumference, erect, smooth, hairless, greenish, greyish or brownish yellow, rarely blotched; each node mostly with a deep swollen band beneath; lowest internodes on many culms very short, and variously bulging and asymmetric. Sheaths (the lowermost often persisting on stouter culms) hairless except for a very narrow fringe of white hairs at the base, lightly spotted and streaked, without auricles or bristles, ciliate above. Leaves 6–15 × 1–2 cm, somewhat yellowish green, downy near the midrib beneath; sheaths usually with auricles and bristles which are often long and erect; ligule *c.* 1 mm, downy. *SE China.* H2.

This species, which is widely grown, especially in continental Europe, recently flowered in 1950 and 1964. A variant with leaves variegated with white ('Albovariegata') is sometimes found; 'Holochrysa', with orange-yellow culms is more frequent.

2. P. aureosulcata McClure (*P. nevinii* misapplied). Illustration: *USDA Handbook* **114**: 19, 20 (1957); Crouzet, Les Bambous, 68 (1981); Brickell (ed.) RHS A–Z encyclopedia of garden plants, 810 (2003).

Culms 3–6 m, 3–10 cm in circumference, sometimes zig-zag, greyish green, slightly rough and with white, waxy bloom below the nodes when young, grooves with green and yellow stripes which soon pale. Sheaths hairless, striped with pale yellow and olive green, those at mid-culm with auricles and bristles, blade small. Branches almost erect. Leaves to 15 × 2 cm, but usually much smaller, sheaths sometimes without auricles or bristles; ligule *c.* 2 mm, downy. *NE China.* H2.

3. P. bambusoides Siebold & Zuccarini (*P. quiloi* (Carrière) Rivière; *P. mazelii* Rivière; *P. reticulata* misapplied). Illustration; Polunin & Everard, Trees and bushes of Europe, t. 182d (1976) – as

P. nigra; Suzuki, Index to Japanese Bambusaceae, 11, 13, 74–5 (1978); Kitamura & Murata, Coloured illustrations of woody plants of Japan, t. 132 (1979); Crouzet, Les Bambous, 36 (1981).

Culms 6–20 m, 2–30 cm in circumference, hairless, green, stiffly erect. Sheaths almost hairless, obtuse above, with small auricles and kinked bristles, heavily marked and spotted; ligule short, blunt, blades mostly short. Nodes almost without waxy bloom. Leaves 6–20 × 1–4.5 cm, stalk to 8 mm; sheaths with auricles and prominent, persistent bristles to 1.2 cm; ligule to 2 mm, hairless. *China.* H2.

This species, which is quite commonly grown, has flowered synchronously since 1962. Shoots usually emerge later than those of the other species. It differs from the earlier emerging *P. viridis* 'Robert Young' in its smooth culms. For more detail see Soderstrom & Calderon, *Pacific Horticulture* **37**(3):7–14 (1976). It is variable and a number of clones have been given names. 'Castillonis' (*Bambusa castillonis* Carrière; *P. castillonis* (Carrière) Mitford) has culms which are yellow with a green groove; it flowered synchronously in 1965–66 and 1979–80 and may have died out. 'Allgold' ('*Holochrysa*'; '*Sulphurea*' misapplied), illustration: Brickell (ed.) RHS A–Z encyclopedia of garden plants, 811 (2003) is smaller with sparse green stripes on its yellow culms.

4. P. congesta Rendle. Illustration: *USDA Handbook* **114**: 28 (1957); Crouzet, Les Bambous, 73 (1981).

Culms 3–8 m, 6–16 cm in circumference, erect, tapered, grey-green; nodes with white, waxy bloom. Sheaths hairless, dark green suffused with pale purple, usually without auricles or bristles, blades short, somewhat wavy. Leaves to 15 × 1.5 cm but often much smaller, markedly downy near the base beneath; sheaths without auricles and with few or no bristles; ligule to 2 mm, minutely downy. *China.* H2.

Not commonly cultivated.

5. P. flexuosa (Carrière) Rivière. Illustration: Rivière and Rivière, Les Bambous, 269, 275 (1879); *USDA Handbook* **114**: 35, 36 (1957); Brickell (ed.) RHS A–Z encyclopedia of garden plants, 811 (2003).

Culms 3–7 m, 6–12 cm in circumference, erect or spreading, slightly ribbed, liable to darken with age to almost black, with white, waxy bloom below the nodes. Sheaths tapered above, hairless, smooth, with some dark markings, without auricles or bristles but with a tuft of hairs from the ligule. Leaves to 15 × 1.5 cm, usually smaller, often somewhat downy near the midrib beneath, sheaths without bristles or auricles (except in the case of fresh, young shoots which also have purple ligules at first); ligule *c.* 2 mm, downy. *China.* H2.

6. P. heterocycla (Carrière) Mitford (*P. pubescens* Houzeau de Lehaie; *P. edulis* invalid; *P. mitis* misapplied). Illustration: Suzuki, Index of Japanese Bambusaceae, 10, 12, 70–1 (1978); Crouzet, Les Bambous, 57 (1981).

Culms 5–20 m, 5–30 cm in circumference, tapered, distant, grey ultimately green or almost orange, curved near the base where the nodes are close together, terete and branchless in the lower part where the nodes have only 1 ridge; internodes velvety when young, soon becoming hairless, white-waxy. Sheaths to 1 m, thick, dark with many markings and covered with erect, brown hairs, the basal sometimes overwintering, auricles none or very small, bristles prominent; blades small, hairless. Leaves usually 4–10 cm × 4–10 mm; sheaths with no auricles or auricles poorly developed, and bristles; ligule *c.* 1 mm, hairless. *E China.* H3.

This species flowered in 1937. The variant described above, which is forma **pubescens** (Houzeau de Lehaie) McClintock, is commonly grown in warmer areas, but is difficult to propagate; the wild variant, often found in the literature as **P. pubescens** 'Heterocycla' or 'Kikkochiku' (illustration: *National Geographic Magazine*, October 1980:516, 520; Crouzet, Les Bambous, 36, 1981), has the lower internodes short, bulging on one side and grossly asymmetric on some culms (see Oshima, *Journal of the American Bamboo Society* **3**: 2–28, 1982). Forma **nabeshimana** (Muroi) Muroi (*P. pubescens* 'Bicolor') has the culms striped with yellow to varying degrees.

7. P. nigra (Loddiges) Munro. Illustration: *Botanical Magazine*, 7994 (1905); Lawson, Bamboos, 129 (1966); Suzuki, Index of Japanese Bambusaceae, 11, 15, 78–9 (1978); Darke, Ornamental grasses, 266 (1999).

Culms 3–5 m, 6–8 cm in circumference, slightly ribbed, olive green at first with white, waxy bloom, later usually shining black. Sheaths with rapidly deciduous hairs, unspotted except sometimes near the apex, usually with prominent, spreading auricles and bristles but sometimes these are absent; blades broad. Leaves 6–13 cm × 8–18 mm, with auricles and bristles absent or poorly developed or bristles erect; ligule to 1.5 mm, downy. *E China.* H2.

A commonly cultivated species, striking when in mature colour, which last flowered between 1932 and 1935. Several variants are grown. 'Punctata' (*P. nigra* var. *punctata* Bean) has culms finally heavily speckled with purplish brown. Var. **henonis** (Mitford) Stapf (*P. henonis* Mitford; 'Henonis'; *P. puberula* Makino). Illustration: *Hooker's Icones Plantarum* **28**: t. 2614 (1901). Culms to 10 m, unblotched and erect, downy and rough when young, broader, leafier and arching above. Widely grown. 'Boryana' (*P. boryana* Mitford) has culms finally marked with dark streaks.

8. P. viridi-glaucescens (Carrière) Rivière (*Bambusa viridi-glaucescens* Carrière). Illustration: *USDA Handbook* **114**: 61, 62 (1957); Polunin & Everard, Trees and bushes of Europe, t. 182e (1976).

Culms 4–7 m, 9–13 cm in circumference, often curved at the base, streaked, with white, waxy bloom below the nodes; sheaths downy, but the hairs soon falling, variable and usually rather lightly marked, tapered above, usually with auricles and bristles, ligule normally long. Leaves to 2.8 cm broad, bristles short and soon falling. *E China.* H2.

Flowered in 1979.

P. nidularia Munro. Illustration: Camus, Les Bambusees, t. 36A (1913); *USDA Handbook* **114**: 43, 44 (1957). Culm-sheaths with large blades and prominent nodal ridges which often bear thick, brownish hairs, sheaths with auricles and few or no bristles. *NE China.* Has flowered occasionally since 1968.

9. P. viridis (Young) McClure (*P. mitis* misapplied; *P. edulis* misapplied). Illustration: *USDA Handbook* **114**: 63 (1957); Crouzet, Les Bambous, 40, 58 (1981).

Similar to *P. heterocycla* but smaller, culms hairless, olive green, minutely pitted, often curved at the base and also above; sheaths ciliate, without other hairs, auricles or bristles; leaves fewer, 7–12 × 1.5–4 cm. *E China.* H3.

Variable. Somewhat similar to a large *P. heterocycla* and widely grown in warmer areas. 'Robert Young' ('*Sulphurea*' misapplied) has smaller culms which are yellowish green at first (later golden yellow), variably striped with green and with a dark band below each node. It differs from *P. bambusoides* 'Allgold' in its somewhat rough culms and plentiful green stripes.

13. SHIBATAEA Nakai
D. McClintock
Culms D-shaped in section, usually tufted (though rhizomes occasionally running), slender, at most 6 mm in circumference, solid, with prominent nodes. Sheaths fairly persistent, hairless, without auricles or bristles. Branches 2–5 at each node, short, all more or less equal. Leaves visibly tessellate, broadly lanceolate, without bristles. Inflorescence a panicle. Spikelets with 2–3 glumes and 1–2 florets. Lodicules 3. Stamens 3. Stigmas 3.

A genus of 5 species, 1 from Japan, the rest from China. They are attractive, distinct and easy to grow as long as the site is not too dry.

1. S. kumasasa (Zollinger) Nakai (*Bambusa ruscifolia* Munro; *Phyllostachys kumasasa* (Zollinger) Munro; *P. ruscifolia* (Munro) Satow). Illustration: Suzuki, Index of Japanese Bambusaceae, 100–1 (1978); *The Plantsman* **1**: 46 (1979); Kitamura & Murata, Coloured illustrations of woody plants of Japan, t. 132 (1979); Brickell (ed.) RHS A–Z encyclopedia of garden plants, 982 (2003).

Culms to 1.5 m, to 6 mm in circumference, somewhat zig-zag. Sheaths olive green with purplish ribs, hairless except for cilia, blade minute. Leaves 5–11 × 1.5–2.5 cm, often dead at apex, hairless above, hairy beneath when young, the hairs soon deciduous. *Japan.* H2.

Fairly frequently grown. Flowers from time to time.

14. BAMBUSA Schrader
D. McClintock
Rhizomes much-branched, short. Culms clump-forming to 15 m, but usually much less, to 10 cm in circumference, hollow, terete; new shoots emerging in late summer. Sheaths thick, deciduous, with rough bristles or bristles absent. Branches numerous at each node, of varying sizes. Leaves without visible tessellation, their sheaths with auricles and rough bristles. Inflorescence a number of spikes. Spikelets without glumes, or rarely 1 glume present, with 5–7 florets. Lodicules 3, ciliate. Stamens 3. Stigmas 3.

A genus of *c.* 100 or more species in subtropical Asia, Africa and America. At one time or another, most bamboos have been placed in this genus.

Literature: Holttum, R.E., On the identification of the common hedge bamboo of SE Asia, *Kew Bulletin* **11**: 207–11 (1956); Haubrich, R., Handbook of bamboos cultivated in the United States, Part II, *Journal of the American Bamboo Society* **2**: 2–20 (1981).

1. B. glaucescens (Willdenow) Holttum (*B. multiplex* (Loureiro) Steudel; *B. nana* Roxburgh; *Leleba multiplex* (Loureiro) Nakai). Illustration: *Annals of the Royal Botanic Garden Calcutta* **7**: t. 38 (1896); Suzuki, Index to Japanese Bambusaceae, 102–5 (1978); Kitamura & Murata, Coloured illustrations of woody plants of Japan, t. 134 (1979).

Culms 3–5 m, 6–9 cm in circumference, ascending, thin-walled, rough when young, somewhat arching above when mature, branched from the base. Sheaths soon hairless, unmarked, usually with conspicuous auricles and smooth bristles; ligule very short; blade triangular, decurrent on the sheath. Nodes rather prominent with white, waxy bloom beneath. Branches very numerous at each node, slender. Leaves variable in number and size, but generally *c.* 11 × 1.2 cm, occasionally to 16 × 2 cm, hairless above, silvery grey-hairy beneath; sheaths smooth, usually without auricles and with smooth bristles; ligule *c.* 0.5 mm. *China, widely cultivated elsewhere.* H5–G1.

A variable species with numerous cultivars. 'Alphonse Carr' has the mature culms yellow with green stripes, and 'Fernleaf' ('Gracillima') has very numerous, small leaves, markedly arranged in 2 ranks on each branch.

15. DENDROCALAMUS Nees
D. McClintock
Very similar to *Bambusa*, from which it is differentiated mainly by floral characters and its great size. Culms 15 m or more tall, 20 cm or more in circumference. Spikelets forming rounded structures in long spikes, with 2–3 glumes and up to 6 florets. Lodicules absent. Stamens 6. Stigma 1.
A genus of about 30 species, native to Asia.

Literature: Holttum, R.E., The bamboo of the Malay Peninsula, *Gardens' Bulletin Singapore* **16**: 89–104 (1958); Haubrich, R., Handbook of bamboos cultivated in the United States, Part II, *Journal of the American Bamboo Society* **2**: 2–20 (1981).

1. D. asper (Schultes) Heyne. Illustration: *Annals of the Royal Botanic Garden Calcutta* **7**: t. 76 (1896); Ochse & Bakhuizen van den Brink, Vegetables of the Dutch E Indies, 308, 309 (1931); *Sunset Magazine* (July 1980); Crouzet, Les Bambous, 37 (1981).

Culms 15–20 m, rarely to 30 m, 20–30 cm (rarely to 75 cm) in circumference, tufted, hairless, thin-walled, glaucous at first, covered at first with brown, closely adpressed hairs which persist near the base. Sheaths large, soon falling, pale green at first, thick and hard, ciliate and sparsely hairy, sometimes with small auricles and curved bristles; blade triangular, acute, adpressed-hairy on both sides. Nodes rather prominent, often with the base of the sheath still attached, hairy below and with white, waxy bloom; lower nodes with prominent root buds. Leaves 4–30 cm (rarely to 55 cm) × 1.5–4 cm (rarely to 11 cm), rough above, hairy towards the base or ultimately hairless beneath. Sheaths almost completely hairless, ciliate; auricles small or absent, bristles absent; ligule to 2.5 mm, downy. *Burma, China, much cultivated elsewhere.* H5–G1.

Group B (Herbaceous grasses)
Guide to the occurrence of some important characters in the genera of herbaceous grasses

Plants. Dioecious: **61**; monoecious: **92, 93, 94**.
Ligule. A line of hairs: **59, 60, 61, 62, 63, 65, 67, 69, 70, 71, 73, 74, 76, 78, 80, 81, 84**.
Inflorescence. A panicle: **16, 17, 18, 19, 20, 21, 24, 25, 26, 28, 29, 30, 31, 32, 36, 37, 38, 39, 42, 43, 45, 50, 59, 60, 61, 62, 63, 64, 65, 66, 67, 68, 72, 77, 78, 82, 83, 86, 87, 93**; a spike-like panicle: **22, 27, 34, 35, 40, 41, 44, 46, 47, 48, 49, 76, 80, 81, 85**; composed of racemose spikes or racemes: **70, 71, 73, 74, 75, 88**; composed of palmately arranged spikes or racemes: **69, 79, 84, 88, 89, 90, 91, 92**; of single spikes or racemes: **33, 51, 52, 53, 54, 54, 56, 57, 58, 71, 75, 94**.
Spikelets. Falling entire: **16, 23, 41, 47, 48, 54, 56, 57, 58, 66, 70, 72, 73, 75, 76, 78, 78, 80, 81, 82, 83, 84, 85, 86, 87, 88, 89, 90, 91, 92, 93,**

94; of two different kinds: **18, 22, 23, 57, 70, 86, 87, 89, 90, 91, 92, 93, 94**; with lowest floret male or sterile: **17, 31, 39, 40, 41, 59, 65, 72, 73, 74, 75, 76, 77, 78, 80, 81, 82, 83, 84, 85, 86, 87, 88, 89, 90, 91, 92, 93, 94**.

Glumes. The lower suppressed: **53, 74, 75, 77, 79, 80**; exceeding the lowest lemma: **18, 19, 20, 29, 30, 31, 32, 33, 34, 35, 36, 37, 38, 39, 41, 42, 43, 45, 46, 47, 48, 49, 56, 70, 82, 83, 84, 86, 87, 88, 89, 90**; awned: **38, 44, 47, 52, 54, 57, 58, 78, 94**.

Lemmas. With straight or flexuous awn: **51, 60, 61, 63, 68, 69, 73**.

1a. Spikelets unisexual, male spikelets different in appearance from female spikelets 2
b. All or some of the spikelets bisexual, rarely unisexual but then male spikelets similar to female spikelets 4
2a. Inflorescences bisexual, female below, male above, projecting **92. Tripsacum**
b. Inflorescences unisexual, the female enclosed 3
3a. Female inflorescence enclosed by leafy husks **93. Zea**
b. Female inflorescence enclosed by a hard globular jacket **94. Coix**
4a. Spikelets 2-flowered, falling entire at maturity, with the upper floret bisexual and the lower male or sterile; spikelets usually flattened from top to bottom 5
b. Spikelets with 1–many florets, breaking up at maturity above the more or less persistent glumes, or if falling entire then not with 2 florets with the upper floret bisexual and the lower male or sterile; spikelets usually flattened from side to side or terete 24
5a. Upper lemma firmer in texture than the lower glume; spikelets seldom paired 6
b. Upper lemma thinner in texture than the lower glume; spikelets paired 15
6a. Spikelets subtended by an involucre of 1 or more bristles 7
b. Spikelets not subtended by bristles 9
7a. Bristles retained on the branches after the spikelets have fallen **76. Setaria**
b. Bristles falling with the spikelet 8
8a. Involucral bristles free throughout, more or less thread-like **80. Pennisetum**
b. Involucral bristles flattened and united into a little disc below **81. Cenchrus**
9a. Inflorescence a panicle 10
b. Inflorescence composed of racemes 12
10a. Upper lemma leathery to hard and somewhat brittle **72. Panicum**
b. Upper lemma cartilaginous 11
11a. Upper floret flattened from side to side, the stigmas emerging laterally **78. Melinis**
b. Upper floret flattened from top to bottom, the stigmas emerging terminally **77. Tricholaena**
12a. Spikelet with a bead-like swelling at its base **74. Eriochloa**
b. Spikelet passing smoothly into its stalk without a bead-like swelling 13
13a. Upper lemma cartilaginous with flat, thin margins covering most of the palea and often overlapping **79. Digitaria**
b. Upper lemma leathery to hard and somewhat brittle with narrow, inrolled margins clasping only the edges of the palea 14
14a. Glumes 2 **73. Echinochloa**
b. Glume 1 **75. Paspalum**
15a. Spikelets in each pair alike 16
b. Spikelets in each pair different 19
16a. Raceme axis tough 17
b. Raceme axis fragile 18
17a. Inflorescence spike-like, cylindric **85. Imperata**
b. Inflorescence a group of elongate racemes **84. Miscanthus**
18a. Lower glume papery, convex, the veins raised **82. Spodiopogon**
b. Lower glume membranous to almost leathery, veins not raised **83. Saccharum**
19a. Internodes and stalk of the raceme with a translucent median line **88. Bothriochloa**
b. Internodes and stalk of the raceme not as above 20
20a. Racemes borne in a panicle 21
b. Racemes paired 22
21a. Spikelets flattened from side to side **87. Chrysopogon**
b. Spikelets flattened from top to bottom **86. Sorghastrum**
22a. Lower glume of stalkless spikelet rounded on back **91. Hyparrhenia**
b. Lower glume of stalkless spikelet laterally 2-keeled 23
23a. Leaf-blades not aromatic **89. Andropogon**
b. Leaf-blades aromatic **90. Cymbopogon**
24a. Inflorescence composed of spikes or racemes 25
b. Inflorescence a panicle 37
25a. Spikes or racemes several in a group 26
b. Spikes or racemes single 28
26a. Spikelets falling entire at maturity **70. Spartina**
b. Spikelets breaking up at maturity 27
27a. Racemes arranged along a central axis **71. Bouteloua**
b. Racemes all arising at the top of the stem **69. Chloris**
28a. Spikelets 1 at each node of the spike 29
b. Spikelets 2–4 at each node of the spike 32
29a. Inflorescence axis tough 30
b. Inflorescence axis fragile 31
30a. Spikelets 1-flowered **46. Mibora**
b. Spikelets 2- or more-flowered **51. Brachypodium**
31a. Spikelets with the side of the lemma towards the axis **58. Aegilops**
b. Spikelets with the back of the lemma towards the axis **33. Gaudinia**
32a. Spikelets with 1 floret **57. Hordeum**
b. Spikelets with 2 or more florets 33
33a. Inflorescence axis fragile **54. Sitanion**
b. Inflorescence axis tough 34
34a. Glumes absent or reduced to 2 short bristles **53. Hystrix**
b. Glumes well developed 35
35a. Spikelets with 1 fertile and 1 sterile floret **56. Taeniatherum**
b. Spikelets with 2–7 fertile florets 36
36a. Glumes 4—9-nerved **52. Elymus**
b. Glumes 3—5-nerved **55. Leymus**
37a. Spikelets with 1 floret 38
b. Spikelets with 2 or more florets, or of 2 kinds 49
38a. Glumes feathery; panicle spherical to oblong **44. Lagurus**
b. Glumes hairless to shortly hairy 39
39a. Lemma surrounded by fine white hairs from the base, the hairs one-third as-long-as to much longer than the lemma 40
b. Lemma hairless at base or with only a tuft of very short hairs there 41
40a. Lemma about half the length of the glumes **43. Calamagrostis**
b. Lemmas as long as the glumes **68. Muhlenbergia**

41a. Spikelets falling entire 42
b. Spikelets breaking up at maturity 44
42a. Glumes conspicuously awned **47. Polypogon**
b. Glumes not or shortly awned 43
43a. Lemma awned **48. Alopecurus**
b. Lemma awnless **16. Phaenosperma**
44a. Inflorescence a cylindric, spike-like panicle **49. Phleum**
b. Inflorescence an open or contracted panicle 45
45a. Lemma not strongly hardened and shiny when mature 46
b. Lemma strongly hardened and shiny when mature 47
46a. Lemma awnless or with a sharply bent awn **42. Agrostis**
b. Lemma with a long straight awn **45. Apera**
47a. Lemma awnless **20. Milium**
b. Lemma awned 48
48a. Lemma hairless; awn often falling off **19. Piptatherum**
b. Lemma hairy; awn persistent **18. Stipa**
49a. Spikelets of 2 different kinds, mostly in groups of 3–7 with 1 central bisexual spikelet and 2–6 male or sterile spikelets surrounding it 50
b. Spikelets not obviously of different kinds 52
50a. Male or sterile spikelets persistent **22. Cynosurus**
b. Sterile and fertile spikelets falling in groups 51
51a. Lemma of fertile spikelets awned **23. Lamarckia**
b. Lemma of fertile spikelets awnless **41. Phalaris**
52a. Inflorescence a dense, ovoid to spherical head of stalkless spikelets **27. Echinaria**
b. Inflorescence not a dense head of stalkless spikelets 53
53a. Ligule a row of hairs, sometimes surmounting a short membrane 54
b. Ligule membranous 61
54a. Lowest floret male or sterile 55
b. Lowest floret bisexual 56
55a. Inflorescence feathery **65. Phragmites**
b. Inflorescence not feathery **66. Uniola**
56a. Lemmas, at least of the female florets, hairy 57
b. Lemmas hairless 60
57a. Awn of lemma short, scarcely exceeding the teeth **62. Arundo**
b. Awn of lemma conspicuous 58
58a. Lemma with 7–9 veins **60. Chionochloa**
b. Lemma 3-veined 59
59a. Plant tufted **61. Cortaderia**
b. Plant with creeping rhizome **63. Hakonechloa**
60a. Lemma firm, somewhat cartilaginous **67. Eragrostis**
b. Lemma membranous **64. Molinia**
61a. Lowest floret male or sterile 62
b. Lowest floret bisexual 67
62a. Spikelets with 1 male or sterile floret below the fertile 63
b. Spikelets with 2 male or sterile florets, these often much reduced 64
63a. Spikelets awned **31. Arrhenatherum**
b. Spikelets awnless **59. Chasmanthium**
64a. Inflorescence an open or contracted panicle 65
b. Inflorescence a spike-like panicle 66
65a. Lemma of second floret from the base transversely wrinkled **17. Ehrharta**
b. Lemma of second floret from the base not transversely wrinkled **39. Hierochloe**
66a. Lemma awned **40. Anthoxanthum**
b. Lemma awnless **41. Phalaris**
67a. Lemma circular to broadly ovate, cordate **24. Briza**
b. Lemma usually lanceolate to ovate, not cordate 68
68a. Both glumes, or upper only, as long as or longer than the lowest lemma and usually as long as the whole spikelet and enclosing all the lemmas 69
b. Glumes usually shorter than lowest lemma; the other lemmas distinctly exceeding the glumes (except *Melica*) 75
69a. Lemmas awnless **34. Koeleria**
b. Lemmas awned 70
70a. Spikelets 1.1–3.2 cm; awns 1.2–5.5 cm 71
b. Spikelets 2–11 mm; awns to 1.7 cm 72
71a. Perennials **30. Helictotrichon**
b. Annuals **32. Avena**
72a. Spikelet falling entire at maturity **37. Holcus**
b. Spikelet breaking up at maturity 73
73a. Perennials **36. Deschampsia**
b. Annuals 74
74a. Awn arising on back of lemma **38. Aira**
b. Awn at apex of lemma **35. Rostraria**
75a. Lemmas keeled 76
b. Lemmas rounded on the back 78
76a. Spikelets borne in dense 1-sided clusters on the branches of a panicle **26. Dactylis**
b. Spikelets borne in loose or contracted panicles 77
77a. Lemmas 3-veined **67. Eragrostis**
b. Lemmas with 5–7 veins **25. Poa**
78a. Lemmas awned or with pointed tips 79
b. Lemmas awnless with blunt tips 80
79a. Ovary crowned by a fleshy hairy cap, the styles arising from beneath it **50. Bromus**
b. Ovary hairless or hairy, the styles arising from its tip **21. Festuca**
80a. Spikelets with 1–3 florets **29. Melica**
b. Spikelets with 3–20 florets **28. Glyceria**

16. PHAENOSPERMA Bentham
Tufted perennial with slender to stout stems, scarious ligules and flat, narrowly lanceolate leaf-blades which have numerous cross-veins. Inflorescence a loose panicle with clustered branches. Spikelets falling at maturity, elliptic-oblong, at length widely gaping, awnless, with 1 flower; glumes thinly membranous, the lower about one-half the length of the upper, which is as long as the spikelet; lemma as long as the spikelet, rounded on the back, becoming firmly papery, with 3–5 veins, hairless; grain spherical or almost so, with longitudinal grooves, and a free pericarp.

A genus containing a single species. Propagated from seed or by division. Hardy in sheltered places in the south and west of Europe, but succeeds well in cool greenhouses, in partial shade in ordinary garden soil. A remarkable grass on account of its unusual grains, and of ornamental value because of its graceful foliage and inflorescences.

1. P. globosa Bentham. Illustration: *Hooker's Icones Plantarum* **20**: t. 1991 (1891).

Stems 45–150 cm, erect. Ligules 5–20 mm. Leaf-blades finely pointed, gradually narrowed to the base, 30–60 × 1.2–3.6 cm, rough, green above, hairless beneath. Panicle lanceolate to narrowly ovate, 20–25 cm with spreading branches. Spikelets *c.* 4 mm, green; grain 2–3 mm long and broad. *NE India, Burma, China, Taiwan, Korea & Japan.* H5–G1. Summer.

17. EHRHARTA Thunberg

Annuals or perennials. Leaves flat or rolled; ligule membranous. Inflorescence a panicle or raceme. Spikelets stalked, flattened from side to side, with 3 florets, the 2 lowest reduced to the lemma, the uppermost complete; glumes persistent, membranous, shorter than the florets; lower lemmas tough, often hairy, awned or awnless; fertile lemma smaller and thinner.

A genus of 25–30 species in Southern Africa; a few naturalised in N America and Australia. The one cultivated species may be grown in the open in ordinary garden soil, the seed being sown in late spring.

1. E. erecta Lamarck. Illustration: Van Oudtshoorn, Guide to grasses of southern Africa, 119 (1999).

Loosely tufted perennial, with ascending, branched stems to 90 cm. Leaf-blades 5–15 cm × 3–9 mm; ligules to 3 mm. Inflorescence to 5–20 cm long, loose, pale green. Spikelets 3–4 mm, awnless; second lemma transversely wrinkled. *Southern Africa.* H5–G1. Autumn.

18. STIPA Linnaeus

Tufted perennials, rarely annuals. Leaves usually pleated or rolled, at least when dry, with prominent ribs on the upper surface. Inflorescence a panicle. Spikelets somewhat compressed from side to side, with 1 floret. Glumes usually more or less equal, translucent or membranous, much longer than the lemma, with 1–3 veins; lemma usually leathery, rolled, terete, entire or shortly bifid at apex, hairy, awned from apex or sinus; hairs usually confined to the veins and margins, those of the overlapping margins forming a single line. Callus bearded, usually long and pointed; awn usually with a thick twisted lower part (column), and a thinner straight upper part (limb), hairless or hairy, usually twice bent. Palea translucent, 2-veined, usually enclosed by the lemma. Rachilla not prolonged.

A genus of some 300 tropical and temperate species, mostly showing adaptations to dry conditions. It includes a number of useful ornamentals with attractive feathery awns, which are suitable for a dry, sunny border.

1a. Glumes 1–5 cm (rarely shorter); 2
b. Glumes not more than 9 mm; 6
2a. Ligules less than 1 mm 3
b. Ligules 1–15 mm 4
3a. Awn 7–12 cm, rough, with fine stiff projections **1. gigantea**
b. Awn *c.* 40 cm, with hairs 5–6 mm **2. tirsa**
4a. Awn 5–7 cm **8. tenuifolia**
b. Awn 10–30 cm 5
5a. Glumes *c.* 5 cm; lemma 1.3–2 cm **3. pennata**
b. Glumes 2.5–3.5 cm; lemma *c.* 8 mm **7. capillata**
6a. Ligule minute (up to 1 mm) 7
b. Ligule 1–13 mm 8
7a. Glumes 8–9 mm **4. calamagrostis**
b. Glumes 2.5–3.5 mm **6. arundinacea**
8a. Awn 6–12 mm **5. splendens**
b. Awn 5–9 cm **9. tenuissima**

1. S. gigantea Link. Illustration: Grounds, Ornamental grasses, t. 14 (1979); Darke, Ornamental grasses, 290, 291 (1999); Brickell (ed.), RHS A–Z encyclopedia of garden plants, 1008 (2003).

Stems to 2.5 m. Ligule very short. Panicle 20–50 cm, very loose. Glumes 2.5–3.5 cm, long-attenuate. Lemma 1.4–1.8 cm, membranous, with soft hairs 1–5 mm, 2-toothed at the tip; awn 7–12 cm, rough with minute projections throughout its length. *SW Europe.* H5. Summer.

2. S. tirsa Steven. Illustration: Aeschimann et al., Flora alpina **2**: 139.56.5 (2004).

Very robust. Leaves rolled, *c.* 0.5 mm in diameter, with a long, bristle-like tip, those at the base to 100 cm; ligule scarcely visible. Panicle dense; glumes long-attenuate; lemma 1.6–2.1 cm; awn *c.* 40 cm, the column hairless, the limb with hairs 5–6 mm long. *C, S & E Europe.* H2. Summer.

3. S. pennata Linnaeus. Illustration: Christiansen, Grasses, sedges and rushes in colour, f. 5 (1979); Aeschimann et al., Flora alpina **2**: 139.56.1 (2004).

Stems 25–40 cm. Leaves with a pointed tip; ligule *c.* 4 mm. Panicle dense *c.* 10 cm; glumes *c.* 5 cm; lemma 1.3–2 cm; awn 16–30 cm, usually 20–28 cm, the column pale, the limb hairy. *S & C Europe eastwards to Himalaya.* H4. Summer.

4. S. calamagrostis (Linnaeus) Wahlenberg. Illustration: Brickell (ed.) RHS A–Z encyclopedia of garden plants, 566 (2003).

Robust, tufted perennial, stems 60–120 cm. Leaves 2–3 mm wide; ligule minute. Panicle to 80 cm, rather loose. Spikelets 8–9 mm (rarely shorter), often purplish; glumes 8–9 mm, lanceolate, acute; lemma 3–4 mm, leathery, with hairs up to 4 mm long on the back; awn *c.* 10 mm, straight or curved. *S & SC Europe.* H3. Summer.

5. S. splendens Trinius.

Robust, tufted perennial, stems to 2.5 m. Ligule *c.* 6 mm. Panicle to 50 cm. Spikelets 5–6 mm; glumes 4–6 mm, upper glume three-quarters to four-fifths as long as the lower; lemma 5–6 mm; awn 6—12 mm. *E Europe to Siberia & C Asia.* H3. Summer.

6. S. arundinacea (Hooker) Bentham (*Anemanthele lessoniana* (Steudel) Veldkamp; *Oryzopsis lessoniana* (Steudel) Veldkamp. Illustration: Darke, Ornamental grasses, 140 (1999) – as *Anemanthele lessoniana*.

Tufted perennial, stems to 1 m; ligule to 1 mm. Leaves to 45 cm × 6 mm, soft and very fine, bright green, eventually gold or orange-tinted. Panicles open, rather loose, to 1 m; glumes 2.5–3.5 mm; lemma *c.* 2 mm; awn to 8 mm, the column thickened, twisted. *New Zealand.* H4. Summer.

7. S. capillata Linnaeus. Illustration: Darke, Ornamental grasses, 290 (1999).

Tufted perennial, stems to 1 m; ligules 3–15 mm. Leaves thread-like, 1–2 mm wide. Panicle linear, 10–25 cm; glumes 2.3–3.5 cm; lemma 1.1–1.5 cm; awn 10–20 cm. *C & S Europe to E Asia.* H3. Summer.

8. S. tenuifolia Steudel (*Austrostipa tenuifolia* (Steudel) Jacobs & Everette). Illustration: Jessop et al., Grasses of South Australia, 78 (2006).

Tufted perennial, stems 30–60 cm; ligule membranous, 1–4 mm. Leaves thread-like, 1–2 mm wide. Panicle open, 15–30 cm; spikelets solitary, *c.* 15 mm; glumes 1.2–1.5 cm; lemma *c.* 8 mm; awn 5–7 cm, hairless. *S Australia.* H4. Summer.

9. S. tenuissima Trinius (*Nassella tenuissima* (Trinius) Barkworth. Illustration: *Flora of Australia Poaceae 2* **44**A: 63 (2009).

Tufted perennial, stems erect or ascending, 25–70 cm; ligule membranous, 1.5–4 mm. Leaves 7–40 cm *c.* 0.5 mm, apex acute. Panicle open, linear, 10–25 cm, spikelets few, 5–9 mm; glumes 5–10 mm; lemma with a ring of hairs, 1.8–3.6 mm; awn 5–9 cm, column hairless. *Southern Africa, Australia and New Zealand, temperate N & S America.* H4. Summer.

19. PIPTATHERUM Palisot de Beauvois
Perennials. Leaves flat or rolled. Inflorescence a loose panicle. Spikelets somewhat compressed from side to side, with 1 floret. Glumes equal, the lower with 5 veins, the upper with 3–5 veins. Lemma obscurely 3–5-veined, leathery, with a terminal awn that is eventually shed. Palea about as long as lemma, leathery. Callus very small. Rachilla not prolonged.

A genus of about 25 species in the Old World subtropics. Propagation is by seed or by division of the rootstock, and the plants prefer a light, well-drained soil.

Literature: Freitag. H., The genus Piptatherum in SW Asia, *Notes from the Royal Botanic Garden, Edinburgh* **33**: 341–408 (1975).

1a. Spikelets 3–3.5 mm; awns 4–8 mm; panicle branches in clusters of 5–8 **1. miliaceum**
b. Spikelets 6–8 mm; awns 1–2 cm; panicle branches solitary or in clusters of 2–4 **2. paradoxum**

1. P. miliaceum (Linnaeus) Cosson. Illustration: Bonnier, Flore complète **12**: t. 662 (1934); Aeschimann et al., Flora alpina **2**: 139.55.1 (2004).

Loosely tufted with smooth, green, rigid stems, 45–150 cm. Leaf-blades finely pointed, to 30 cm × 3–10 mm; ligules very short. Panicles linear to oblong, loose, more or less nodding, 15–30 cm, the branches in clusters of 5–8. Spikelets stalked, very numerous, 3–3.5 mm, greenish or tinged with purple; lemmas smooth; awn 4–8 mm. *Mediterranean region.* H5. Summer–autumn.

2. P. paradoxum (Linnaeus) Palisot de Beauvois. Illustration: Bonnier, Flore complète **12**: t. 662 (1934); Aeschimann et al., Flora alpina **2**: 139.55.3 (2004).

Densely tufted perennial, with slender stems 45–120 cm. Leaf-blades broadly linear, pointed, to 30 × 1 cm, flat, smooth on both sides; ligules very short. Panicles loose, 15–30 cm; branches solitary or in clusters of 2–4. Spikelets elliptic-lanceolate, 6–8 mm, shortly stalked, green or tinged with purple; glumes lanceolate, acuminate; lemma narrowly elliptic, turning dark brown; awn 1–2 cm. *SW Europe, NW Africa, Atlantic Islands.* H5. Summer.

20. MILIUM Linnaeus
Annuals or perennials. Leaves flat. Inflorescence a panicle. Spikelets slightly flattened from top to bottom, with 1 floret. Glumes equal, membranous, longer than the floret, with 3 veins. Lemma with 5 veins, leathery and shiny in fruit. Palea as long as the lemma, leathery.

A genus of about 6 species in temperate parts of N America, Europe and Asia. Propagation by seed sown outside in spring, or by division of the rootstock.

1. M. effusum Linnaeus. Illustration: Hubbard, Grasses, edn 2, 276 (1968); Christiansen, Grasses, sedges and rushes in colour, f. 4 (1979); Darke, Ornamental grasses, 221 (1999); Aeschimann et al., Flora alpina **2**: 139.54.1 (2004).

Loosely tufted perennial, 45–180 cm. Leaf-blades 10–30 cm × 5–15 mm, hairless; ligules 3–10 mm. Panicles lanceolate to ovate or oblong, very loose, nodding, 10–30 × up to 20 cm, pale green, rarely purplish, the branches in clusters. Spikelets 3–4 mm. *Europe, Asia & northeastern N America.* H1. Spring–summer.

A variant with yellow leaves, 'Aureum', is widely grown.

21. FESTUCA Linnaeus
Tufted or rhizomatous perennials. Leaves folded, rolled or flat; sheaths open or closed. Inflorescence a panicle. Spikelets stalked, with 4 or more florets (rarely only 2), compressed from side to side. Glumes 2, the lower usually with 1 vein, the upper wider, usually with 3 veins. Lemma with rounded back, not keeled, with or without a terminal or almost terminal awn. Palea 2-keeled, scarious.

A large genus of about 300 species in temperate and subtropical regions throughout the world and on mountain tops in the tropics. The species are all hardy and can be sown outdoors in the spring in ordinary garden soil.

Literature: Hackel, E., Monographia Festucarum europeum (1882); Tzvelev, N.N., K sistematike i filogenii ovsyanits (Festuca L.) flory SSSR: I., *Botanicheskii Zhurnal* **56**: 1252–62 (1971).

1a. Leaf-blades blunt 2
b. Leaf-blades sharply pointed, very stiff and rigid 4
2a. Leaf-blades compressed from side to side and longitudinally grooved, especially when dry **1. valesiaca**
b. Leaf-blades cylindric 3
3a. Leaves green or greyish green **2. glacialis**
b. Leaves and usually the whole plant, bluish white, covered with a whitish wax **3. glauca**
4a. Ligule very short with 2 auricles; lemmas narrowly scarious **4. punctoria**
b. Ligule longer, without auricles; lemmas with broadly scarious tips **5. eskia**

1. F. valesiaca Gaudin. Illustration: Hegi, Illustrierte Flora von Mittel-Europa, **1**: f. 235 (1957); Aeschimann et al., Flora alpina **2**: 139.1.55.1 (2004).

Densely tufted, with slender stems, 20–50 cm. Leaf-blades blunt, hair-like, to 0.6 mm thick, flaccid or somewhat stiff, rough or very rough, glaucous, with 5 veins, grooved when dry. Panicles erect, oblong or ovate-oblong, rather dense, 5–10 cm. Spikelets with 3–8 florets, ovate or ovate-oblong, *c.* 6 mm, pale green or tinged with purple, more or less glaucous; lemma *c.* 4 mm, mostly hairless and smooth, tipped with an awn to 1.5 mm. *C & S Europe, Asia.* H3. Spring–summer.

2. F. glacialis (Hackel) Richter.
Densely tufted, with slender stems to 15 cm, coated at the base with the remains of old sheaths; blades thread-like, 0.5–0.6 mm in diameter, very blunt, slightly glaucous, smooth, with 5 veins. Panicles narrow, to 2.5 cm, the branches bearing 1, rarely 2 to 3, spikelets. Spikelets with 3–5 clustered florets, *c.* 6 mm; lemma 4 mm, pointed, awnless or very shortly awned. *Europe (Pyrenees and Alps).* H3. Summer.

3. F. glauca Lamarck. Illustration: Bonnier, Flore complète **12**: t. 681 (1934); Darke, Ornamental grasses, 198, 199 (1999); Brickell (ed.) RHS A–Z encyclopedia of garden plants, 448 (2003).

Densely tufted, bluish green, with numerous vegetative shoots and erect slender stems, 7.5–37.5 cm. Leaves smooth, the blades almost cylindric, thread-like, blunt, *c.* 0.8 mm in diameter, stiff, straight or curved, usually with 9 veins, ligules obscure. Panicles rather dense to very dense, ovate-oblong, erect, bluish green, 3–10 cm, with a smooth axis and short branches. Spikelets 5–8 mm, elliptic or oblong, with 4–7 flowers; lemma 4–5 mm, mucronate or shortly awned, hairless. *SE & C Europe.* H3. Spring–summer.

F. amethystina Linnaeus. Illustration: Darke, Ornamental grasses, 197 (1999), is similar in habit but with purple-red stems 50–80 cm; leaves slightly narrower; panicles slightly nodding, 10–15 cm;

spikelets *c.* 8 mm, lanceolate, with 3–7 flowers; lemma 4.5–5.2 mm, acuminate. *Europe—W Asia.* H3. Summer.

4. F. punctoria Sibthorp & Smith.

Densely tufted, with rigid stems, to 30 cm. Leaf-blades 5–7.5 cm, tightly inrolled, cylindric, 1–1.2 mm thick, sharply pointed, curved, very hard and rigid, smooth outside, bloomed, with 7–9 veins. Panicles narrowly oblong, dense 4–5 cm. Spikelets with 4–7 florets, oblong-ovate, 8–10 mm; lemma *c.* 5 mm, tipped with an awn to 2 mm, bloomed. *Turkey.* H5. Summer.

5. F. eskia de Candolle. Illustration: Bonnier, Flore complète **12**: t. 680 (1934).

Mat-forming, compact, with a creeping rootstock, and moderately slender stems 23–45 cm. Leaf-blades to 20 cm or more, *c.* 1 mm thick, very rigid, often curved, cylindric, very acute and sharp-pointed at the tip, smooth; ligules 3–8 mm. Panicles loose, nodding, ovate, 4–10 cm. spikelets ovate-lanceolate, *c.* 8 mm, more or less variegated with green, gold and purple, with 6–8 florets; lemma pointed, *c.* 6 mm, smooth or rough, with scarious apex, mucronate or with a very short awn. *Europe (Pyrenees).* H3. Summer.

22. CYNOSURUS Linnaeus

Perennials or annuals. Leaves flat. Inflorescence a more or less 1-sided, dense panicle. Spikelets of 2 different kinds, the fertile compressed from side to side, with 1–5 florets, the sterile below the fertile and consisting of narrow, rigid glumes and lemmas. Glumes of fertile spikelets thin, almost equal, acute. Lemma papery, with an apical awn. Palea 2-keeled, shortly bifid at apex, about as long as lemma.

A genus of 5 species mainly in the Mediterranean region, 1 widespread in Europe. Seed should be sown thinly in the spring, in light loamy soils.

1. C. echinatus Linnaeus. Illustration: Hubbard, Grasses, edn 2, 218 (1968); Aeschimann et al., Flora alpina **2**: 139.10.2 (2004).

Annual, with solitary or tufted, erect or ascending stems, 7–90 cm. Leaves to 20 cm × 3–10 mm; ligule blunt, to 10 mm. Panicles spike-like, dense, erect, 1-sided, bristly, 1–7.5 cm, green or purplish, shining. Spikelets in dense clusters, the fertile wedge-shaped, 8–12 mm, with 1–5 flowers; the lemmas bearing a fine straight awn, 6–15 mm, from near the tip; the sterile broadly obovate, 6–12 mm, consisting of up to 18 awned bracts. *Mediterranean region.* H5. Summer.

23. LAMARCKIA Moench

Annual. Leaves flat. Inflorescence a rather condensed 1-sided panicle. Spikelets of 2 kinds, all somewhat compressed from side to side, falling in groups of 3–5; fertile spikelets with 1 rudimentary and 1 fertile floret, surrounded by 2–4 sterile spikelets with many florets. Glumes thin, almost equal, acute. Lemma of fertile spikelets papery, awned, of the sterile spikelets membranous, unawned, Palea 2-keeled.

A genus with a single species. Seed should be sown in late spring in a light sandy soil; it can be sown in autumn, but the plants must then be overwintered under gl ass before planting out in spring.

1. L. aurea (Linnaeus) Moench. Illustration: Bonnier, Flore complète **12**: t. 677 (1934); Brickell (ed.) RHS A–Z encyclopedia of garden plants, 606 (2003).

Stems 10–30 cm. Leaf-blades finely pointed, 5–13 cm × 4–8 mm. Panicles oblong, 2.5–7.5 × 1.2–2.5 cm, straw-coloured, golden-yellow or sometimes purplish. Fertile spikelets *c.* 4 mm; sterile spikelets 4–8 mm, hiding the fertile. *Mediterranean region.* H5. Summer.

24. BRIZA Linnaeus

Almost hairless annuals or perennials. Inflorescence usually much-branched. Spikelets ovoid or broadly triangular, compressed from side to side; florets 4–20. Glumes ovate-circular, heart-shaped at the base, with 3–9 veins and with wide scarious margins. Lemma almost circular, heart-shaped at the base, with 7–9 veins, closely overlapping. Palea ovate to obovate, obtuse, shorter than to almost equalling the lemma, with 2 narrowly winged keels. Ovary hairless; styles terminal. Grain flattened on one side.

A genus of about 12 species in temperate regions mainly in the northern hemisphere. Most of the species are ornamentals growing well in any good humus-rich soil in the sun, the seeds being sown in autumn or spring. The panicles, once dried, are useful for decoration and should be gathered and dried as soon as fully grown.

1a. Perennial **1. media**
b. Annual 2
2a. Spikelets 1.2–2.5 cm, fewer than 12 **2. maxima**
b. Spikelets 3–5 mm, numerous **3. minor**

1. B. media Linnaeus. Illustration: Hubbard, Grasses, edn 2, 214 (1968); Christiansen, Grasses, sedges and rushes in colour, f. 37 (1979); Grounds, Ornamental grasses, t. 35 (1979); Darke, Ornamental grasses, 146, 147 (1999).

Perennial, tufted with shortly creeping rhizome and erect, slender, smooth stems 15–100 cm. Leaves to 15 cm × 2–7 mm, rough on the margins; ligule short, truncate. Panicle 4–18 cm, loose, erect at first. Spikelets broadly elliptic to broadly ovate, 4–8 mm long and wide, with 4–12 florets, usually purplish. *Europe, N & SW Asia.* H1. Summer.

2. B. maxima Linnaeus. Illustration; Hubbard, Grasses, edn 2, 212 (1968); Grounds, Ornamental grasses, t. 36 (1979); Brickell (ed.) RHS A–Z encyclopedia of garden plants, 194 (2003); Aeschimann et al., Flora alpina **2**: 139.15.2 (2004).

Annual with slender, smooth, solitary or tufted stems 10–60 cm. Leaves to 20 cm × 8 mm, rough on the margins; ligule 2–5 mm, acute. Panicles 2–10 cm, loose, nodding, with up to 12 large spikelets, these ovate or oblong, 1.2–2.5 cm × 8–15 mm, variously coloured, silvery, reddish brown, purplish or greenish, with 7–20 florets, hairy or hairless. *Mediterranean region.* H5. Spring–summer.

3. B. minor Linnaeus. Illustration: Hubbard, Grasses, edn 2, 210 (1968); Aeschimann et al., Flora alpina **2**: 139.15.3 (2004).

Annual with solitary or tufted, slender, smooth stems 10–60 cm. Leaves narrowly lanceolate, to 1.5 cm × 3–8 mm, rough; ligule lanceolate, 3–6 mm. Panicle 5–20 cm, broadly pyramidal, loose. Spikelets circular or triangular-ovate, 3–5 mm long and wide, greenish white or purplish, shining, with 4–8 florets. *W Europe and the Mediterranean region.* H4. Summer.

25. POA Linnaeus

Annuals or perennials. Inflorescence a panicle. Spikelets compressed from side to side, with 2–10 florets (rarely only 1). Glumes keeled, membranous, usually with 3 veins or the lower with 1 vein. Lemma with 5 veins, keeled, membranous, awnless or rarely with a short terminal awn. Palea 2-keeled, the keels with very small teeth or ciliate. Grain ellipsoid; hilum basal.

A genus of more than 250 species in temperate and cold regions, a few on mountains in the tropics. Many species are valuable pasture or hay grasses; others are important for lawns although they are little used for the purpose in Europe; a few are ornamental either on account of their foliage or their graceful panicles. They may be propagated from seed or by division, and generally are tolerant of a wide range of soils.

1a. Leaf-blades bright green, 5–10 mm wide; panicle loose **1. chaixii**
b. Leaf-blades bluish grey, 2–4 mm wide; panicle stiffly erect, loose or dense **2. glauca**

1. P. chaixii Villars. Illustration: Hubbard, Grasses, edn 2, 184 (1968); Aeschimann et al., Flora alpina **2**: 139.6.7 (2004).

Compactly tufted perennial, 60–120 cm, with moderately stout, erect stems. Leaves bright green; sheaths strongly compressed and keeled, rough; ligules to 1.5 mm; blades abruptly sharply pointed, to 45 cm long, folded or flat, 5–10 mm wide, rough. Panicles ovate to ovate-oblong, loose, 10–25 × 5–11 cm, green; branches in clusters of up to 7, spreading or the lower drooping. Spikelets ovate to oblong, 5–6 mm, with 3–4 florets; lemma lanceolate-oblong, pointed, 3.5–4 mm, minutely rough, hairless. *Europe & SW Asia.* H1. Spring–summer.

2. P. glauca Vahl. Illustration: Hubbard, Grasses, edn 2, 180 (1968); Aeschimann et al., Flora alpina **2**: 139.6.16 (2004).

Tufted, bluish grey perennial, 10–40 cm, with erect, slender, rather stiff stems, 2–3-noded below the middle. Leaves hairless; ligules 1–2.5 mm; blades abruptly pointed, 1.8–7.5 cm × 2–4 mm, stiff, smooth. Panicles stiffly erect, lanceolate to ovate, loose or dense, 1.8–10 × up to 3.8 cm, usually variegated with purple; branches mostly 2–3 together. Spikelets ovate to oblong, 4–6 mm, with 3–6 florets; glumes with 3 veins; lemma oblong, rather blunt, 3–4 mm, silkily hairy on the keels and veins up to middle. *N Eurasia, northern N America, and mountains further south.* H1. Summer.

26. DACTYLIS Linnaeus
Perennials. Leaves flat or rolled. Inflorescence a panicle with spikelets in dense clusters on the spreading or erect branches. Spikelets compressed from side to side with 2–5 florets. Glumes almost equal, keeled, somewhat curved, the lower with 1 vein, the upper with 3 veins. Lemma with 5 veins, papery, mucronate or shortly awned at apex. Palea about as long as lemma.

A genus of 4 or 5 species in Europe, the Mediterranean region and temperate Asia. The 1 cultivated species thrives on most good soils and is easily raised from seed sown outside in spring.

1. D. glomerata Linnaeus. Illustration: Hubbard, Grasses edn 2, 216 (1968); Christiansen, Grasses, sedges and rushes in colour, f. 36 (1979); Darke, Ornamental grasses, 183 (1999); Aeschimann et al., Flora alpina **2**: 139.9.1.1–1.4 (2004).

Densely tufted, 15–120 cm, with slender to stout, erect or spreading stems and compressed vegetative shoots. Leaves green or greyish green, sharply pointed, 10–45 cm × 2–12 mm, rough; ligule 2–12 mm. Panicles 1-sided, erect, oblong or ovate, very variable, 3–30 cm, of few branches, green, purplish or yellowish. Spikelets 5–10 mm, with 2–5 florets; lemmas stiffly ciliate or rough on the keels, tipped with an awn up to 1.5 mm long. *Temperate regions of the Old World.* H1. Summer–autumn.

27. ECHINARIA Desfontaines
Annual. Leaves usually flat. Inflorescence a dense, ovoid or spherical, prickly head. Spikelets almost stalkless, somewhat compressed from side to side, with 3–4 (rarely 1) florets. Glumes more or less equal, membranous, the lower with 2 (rarely 5) strong veins which run out at the margins, the upper with a midrib which runs out at the apex. Lemma leathery, with 5 (rarely 7) very strong veins prolonged as flattened awns, which are deflexed at maturity. Palea as long as the lemma, the 2 (rarely 5) veins prolonged as flattened awns.

A genus of 1 species. Seed should be sown, in the spring, in ordinary garden soil.

1. E. capitata (Linnaeus) Desfontaines. Illustration: Bonnier, Flore complète **11**: t. 650 (1931); Aeschimann et al., Flora alpina **2**:139.18.1 (2004).

Stems 5–20 cm, rigid, erect or spreading, leafy towards the base. Leaf-blades to 5 cm × 2 mm, green, downy. Flower-heads green, becoming hard and spiny, to 1.5 cm wide and long. *Mediterranean region.* H5. Summer.

28. GLYCERIA R.Brown
Hairless perennials. Leaves flat; sheaths often entire; ligule membranous. Inflorescence a panicle with a 3-sided main axis. Spikelets ovate to linear in outline, with 3–many florets. Glumes unequal, transparent, with 1 vein, shorter than the lowest lemma. Lemma awnless, ovate to oblong-lanceolate, rounded on the back, with 5–11 veins. Palea about as long as lemma.

A genus of about 20 species from wet places in the cooler parts of the northern hemisphere, a few in Australia, New Zealand and S America. Propagated by division of the rootstock or from seed. They are useful for planting on the banks of streams or lakes and may be used as border plants, except on dry soils.

1. G. maxima (Hartman) Holmboe (*G. aquatica* (Linnaeus) Wahlenberg; *Poa aquatica* Linnaeus). Illustration: Hubbard, Grasses, edn 2, 124 (1968); Christiansen, Grasses, sedges and rushes in colour, f. 52 (1979); Darke, Ornamental grasses, 201, 202 (1999); Brickell (ed.) RHS A–Z encyclopedia of garden plants, 566 (2003).

A stout leafy perennial with erect stems 1–2.5 m. Leaf-blades abruptly pointed, 30–60 cm × 8–20 mm, hairless. Panicle loose to rather dense, broadly ovate to oblong, 15–45 cm, much-branched, green to purplish. Spikelets oblong, 5–12 × 2–3 mm, with 4–10 florets; lemmas 3–4 mm. *Temperate Europe & Asia.* H1. Summer.

29. MELICA Linnaeus
Perennials. Leaves flat or rolled; sheaths entire. Inflorescence a usually rather loose, often simple, panicle. Spikelets slightly compressed from side to side, with 1–several bisexual florets, and 2–3 sterile lemmas forming a terminal, club-shaped structure. Glumes more or less unequal, firmly membranous, with 3–5 veins. Fertile lemma rounded on the back, leathery in fruit, the veins 5–9 (rarely to 13) but sometimes obscure. Palea membranous, 2-keeled, 2-toothed.

Literature: Hempel, W., Vorarbeiten zu einer Revision der Gattung Melica L., *Feddes Repertorium* **81**: 657–86 (1971), 84: 533–68 (1973).
A genus of 40–50 species in Europe, temperate Asia, temperate America and N and S Africa. They are propagated by

division or by seed sown outdoors in a light to medium loamy soil especially of a calcareous nature.

1a. Lemmas densely hairy towards and along the margins **1. ciliata**
b. Lemmas hairless 2
2a. Panicles dense, somewhat cylindric or 1-sided, 10–25 cm, with the branches bearing several spikelets 1–1.2 cm long **2. altissima**
b. Panicles loose, 1-sided, 4–20 cm, bearing few spikelets 4–8 mm long 3
3a. Ligules with a bristle-like outgrowth; panicles with spreading branches; spikelets with 1 fertile floret **3. uniflora**
b. Ligules truncate; inflorescence a raceme; spikelets with 2–3 fertile florets **4. nutans**

1. M. ciliata Linnaeus. Illustration: Christiansen, Grasses, sedges and rashes in colour, f. 40 (1979); Aeschimann et al., Flora alpina **2**: 139.19.5.1 (2004).

Densely tufted, greyish green; stems slender, erect, 20–45 cm. Ligules to 2 mm. Leaf-blades finely pointed, thread-like, 7.5–18 cm × 0.8–1 mm, or opening out and up to 3 mm wide, smooth beneath, closely ribbed above. Panicles spike-like, cylindric, dense, 5–15 × 1–1.2 cm, whitish or purplish. Spikelets shortly stalked, elliptic-oblong, 6–8 mm, falling entire at maturity; glumes ovate, pointed, 6–8 mm; lemmas pointed, *c.* 6 mm, with long white hairs along the margins. *Europe, N Africa & SW Asia.* H1. Summer.

2. M. altissima Linnaeus. Illustration: Darke, Ornamental grasses, 219 (1999); Brickell (ed.) RHS A–Z encyclopedia of garden plants, 684 (2003).

Stems tufted, leafy, 60–150 cm, with creeping rhizomes. Ligules 4–6 mm. Leaf-blades finely pointed, 10–23 cm × 5–12 mm, green, rough. Panicles erect, 1-sided, interrupted below, dense above 10–25 cm × 8–12 mm. Spikelets shortly stalked, oblong, 1–1.2 cm, with 2 fertile florets; glumes blunt, papery; fertile lemmas pointed, minutely rough. *C & E Europe, temperate Asia.* H2. Summer.

3. M. uniflora Retzius. Illustration: Hubbard, Grasses, edn 2, 226 (1968); Christiansen, Grasses, sedges and rushes in colour, f. 38 (1979); Aeschimann et al., Flora alpina **2**: 139.1.55.1 (2004).

Plant with slender creeping rhizomes, forming loose patches; stems 20–60 cm. Leaf-sheaths loosely hairy or hairless; ligule produced into a short bristle; blades finely pointed, 5–20 cm × 3–8 mm, bright green, shortly hairy above. Panicles very loose, sparingly branched, erect or nodding, 2.5–20 × up to 10 cm, with spreading branches. Spikelets 4–8 mm, elliptic-oblong, with 1 fertile floret, on fine stalks which are 2–5 mm, purplish or brownish; glumes equal or slightly unequal; fertile lemmas blunt, smooth. *Europe, SW Asia.* H1. Spring–summer.

4. M. nutans Linnaeus. Illustration: Hubbard, Grasses, edn 2, 224 (1968); Christiansen, Grasses, sedges and rushes in colour, f. 39 (1979); Aeschimann et al., Flora alpina **2**: 139.19.1 (2004).

Plants with slender rhizomes; stems loosely clustered or solitary, 20–60 cm, slender, angular. Ligules very short. Leaf-blades 5–20 cm × 2–6 mm, bright green, shortly haired above. Racemes loose, nodding, 1-sided, 4–15 cm. Spikelets 6–8 mm, elliptic-oblong, blunt, purplish or reddish purple, with 2 or 3 fertile florets on curved hair-like stalks which are 3–15 mm; glumes similar, papery-membranous; fertile lemmas blunt, minutely rough. *Europe, N & SW Asia.* H1. Spring–summer.

30. HELICTOTRICHON Besser

Tufted perennials. Leaves distinctly ribbed above; ligule of stem-leaves usually truncate or toothed; sheaths of basal leaves usually open to the base. Inflorescence a compound panicle. Spikelets usually numerous, erect or spreading, slightly compressed from side to side, with 2–4 fertile and 1 (or rarely 2) sterile florets; all fertile florets usually awned. Glumes lanceolate, unequal, membranous, acute, the lower with 1 vein, the upper with 3 veins. Lemma lanceolate, with 5–7 veins, hairless, slightly 2-toothed; awn sharply bent, arising at about the middle of the lemma, more or less terete in the lower part. Palea more or less bifid at apex, shortly ciliate on keels. Ovary densely hairy.

About 60 species in temperate regions of the northern hemisphere and S Africa, and on high mountains in the Old World tropics. Seeds should be sown outside in the spring in any good garden soil.

1a. Leaf-blades flat or folded, to 12 mm wide, green; lower sheaths minutely rough; spikelets with 4–7 florets, in stiff, erect, rather dense panicles **1. planiculme**
b. Leaf-blades usually rolled, to 1.6 mm in diameter, glaucous; lower sheaths smooth; spikelets with 2–3 florets, in loose, open panicles **2. sempervirens**

1. H. planiculme (Schrader) Pilger (*Avenula planiculmis* (Schrader) Sauer & Chmelitschek). Illustration: Hegi, Illustrierte Flora von Mittel-Europa, edn 2, **1**: f. 196 (1957).

Compactly tufted, with erect, moderately stout stems, 45–90 cm. Lower leaf-sheaths minutely rough; ligules to 1.2 cm, blades abruptly pointed, 7.5–40 cm × 5–12 mm, flat or folded, green, smooth. Panicles contracted, rather dense, 15–33 × 1.2–3.5 cm. Spikelets linear to oblong, 1.2–2.5 cm, with 4–7 florets, tinged with purple; glumes 8–15 mm; lemmas 1–1.5 cm, narrowly lanceolate-oblong, bearded at the base with hairs to 1.6 mm, awns 1.5–2.5 cm. *C & SE Europe & SW Asia.* H4. Summer.

2. H. sempervirens (Villars) Pilger. Illustration: Grounds, Ornamental grasses, t. 23 (1979); Darke, Ornamental grasses, 205 (1999); Brickell (ed.) RHS A–Z encyclopedia of garden plants, 518 (2003); Aeschimann et al., Flora alpina **2**: 139.32.6 (2004).

Densely tufted, with slender, erect, rigid stems, 30–120 cm. Leaf-sheaths hairless; ligules obscure, to 1 mm, blades finely pointed, 10–34 cm long, tightly inrolled and 0.6–1.6 mm thick, or opening out and up to 3 mm wide, glaucous, stiff. Panicles loose, open, 5–18 cm. Spikelets oblong, 10–15 mm, with 2–3 florets, yellowish or brownish, variegated with purple; glumes 1–1.2 cm; lemmas lanceolate, 8–12 mm, bearded at the base with hairs 3–4 mm; awns 1.5–2.5 cm. *SW Europe.* H4. Summer.

31. ARRHENATHERUM Palisot de Beauvois

Perennials; basal internodes often swollen and more or less spherical. Leaves flat, rolled when young, not strongly ribbed above. Inflorescence a panicle. Spikelets slightly compressed from side to side, usually with 3 florets, 2 fertile and 1 rudimentary. Lower floret male, with a sharply bent awn arising from the lower third of the back of the lemma, the upper female or bisexual, usually with a short, slender, straight bristle, rarely both florets bisexual and awned. Glumes unequal, translucent, the lower with 1–3 veins, the upper about as long as the florets, with

3 (rarely 5) veins. Lemma with 5–9 veins. Palea shorter than lemma, shortly 2-toothed at apex, ciliate on keels. Lodicules 2. Stamens 3. Florets falling together at maturity. Grain more or less terete, with hilum extending for one-half to two-thirds of its length.

A genus of about 6 species in Europe, N Africa and N & W Asia. Easily propagated by division.

1. A. elatius (Linnaeus) Presl (*A. avenaceum* Palisot de Beauvois). Illustration: Hubbard, Grasses, edn 2, 234 (1968); Christiansen, Grasses, sedges and rushes in colour, f. 26 (1979); Darke, Ornamental grasses, 141 (1999); Aeschimann et al., Flora alpina **2**: 139.33.1 (2004).

Loosely tufted. Stems 50–150 (rarely to 180) cm, smooth, shining. Leaves 10–40 cm × 10 mm, green, usually hairless or almost so, more or less rough with short, stiff projections; ligule 1–3 mm, obtuse, usually ciliate. Panicle 10–30 cm (rarely less), oblong or lanceolate, loose to rather dense. Spikelets 7–10 mm, oblong; rachilla between first and second florets not more than 0.6 mm. Glumes acute, the lower 4–6 mm, oblong-lanceolate, the upper 7–10 mm, ovate-lanceolate. Lemma 7–10 mm, oblanceolate, more or less acute, often shortly 2-toothed, hairless or with sparse hairs which are up to 1 mm. Awn of the lower floret 1–2 cm, arising in the lower third of the lemma. Anthers 4–5 mm. Grain 4–5 mm. *Most of Europe, but absent from much of the north east.* H2. Summer.

32. AVENA Linnaeus

Annuals. Leaves flat; ligule membranous. Inflorescence a panicle. Spikelets large, with 2–5 (rarely 1, or as many as 7) florets. Glumes lanceolate, acuminate, usually about equal, thinly papery, with a shiny scarious margin and several veins. Lemma usually leathery, 2-toothed or with 2 fine hairs (rarely almost entire), generally with an awn on the back; awn usually with a thick, twisted lower part, and a thinner, straight upper part, usually sharply bent. Palea tough, shorter than the lemma, with translucent margin and 2 ciliate keels.

A genus of about 10–15 species, cultivated in temperate regions or occurring as a weed in crops. Easily grown from seed sown in spring and autumn in ordinary garden soil.

Literature: Baum, B., Oats: wild and cultivated (1977).

1a. Spikelets persistent at maturity, the florets not separating from the glumes; lemmas hairless, awnless or with a straight awn **1. sativa**
b. Spikelets breaking up above the persistent glumes at maturity; lemmas hairy, especially at the base, awned, with the awn sharply bent and twisted below the bend 2
2a. Spikelets 3.3–4.3 cm; awns 5–7.5 cm **2. sterilis**
b. Spikelets and awns shorter **3. ludoviciana**

1. A. sativa Linnaeus. Illustration: Christiansen, Grasses, sedges and rushes in colour, f. 85 (1979); Aeschimann et al., Flora alpina **2**: 139.31.4 (2004).

Stems erect, slender or stout, 1–2 m. Leaves to 60 cm; ligules to 5 mm. Panicle loose or somewhat contracted, variable in size and shape. Spikelets 1.8–2.5 cm, with 1–3 florets, the lower 1–2-awned or all awnless. *Temperate regions.* H1. Summer.

2. A. sterilis Linnaeus. Illustration: Bonnier, Flore complète **12**: t. 665 (1934); Aeschimann et al., Flora alpina **2**: 139.31.5.1 (2004) – as *A. sterilis* subsp. *sterilis*.

Stems erect or spreading, solitary or tufted, slender to moderately stout, to 1 m. Leaves to 30 × 1.2 cm; ligules to 5 mm. Panicle loose, to 30 × 20 cm. Spikelets 3.3–4.3 cm with 3–5 florets; lower 2–3 lemmas awned, with the awns 5–7.5 cm. *Atlantic islands, Mediterranean region & SW Asia.* H5. Summer.

3. A. ludoviciana Durieu. Illustration: Hubbard, Grasses, edn 2, 240 (1968); Aeschimann et al., Flora alpina **2**: 139.31.5.2 (2004) – as *A. sterilis* subsp *ludoviciana*.

Stems solitary or tufted, erect, stout, to 2 m. Leaves to 60 cm × 6–15 mm, ligules to 8 mm. Panicle nodding, pyramidal, very loose, to 45 cm. Spikelets 1–3 cm with 2–3 florets; lower 2 lemmas awned, the awns 3–5 cm. *Europe & C Asia.* H1. Summer.

33. GAUDINIA Palisot de Beauvois

Annuals or biennials. Leaves flat. Inflorescence a spike with spikelets in 2 ranks; axis fragile, breaking up above the insertion of the spikelet. Spikelets compressed from side to side, stalkless, more or less adpressed to the concave axis, with 3–11 florets. Glumes unequal, shorter than or equalling the spikelet, the lower with 3 (rarely 5), the upper with 4–7 (rarely up to 11) strong veins. Lemma obscurely 7–9-veined, leathery, often with an awn on the back. Palea shorter than lemma.

Three species in southern Europe and the Mediterranean region. Seed should be sown thinly in the open in spring in light garden soils, and the spikes cut when young for use fresh or dried in interior decorations.

1. G. fragilis (Linnaeus) Palisot de Beauvois (*Avena fragills* Linnaeus). Illustration; Bonnier, Flore complète **12**: t. 694 (1934); Aeschimann et al., Flora alpina **2**: 139.34.1 (2004).

Annual. Stems 10–30 cm, erect or spreading. Leaf-blades finely pointed, 2.5–15 cm × 2–4 mm, softly hairy. Spikes 7.5–20 cm, green. Spikelets 1–2 cm; lower glume 3–5 mm, the upper 6–8 mm; lemmas 3–6 mm, lanceolate, bearing an awn 6–20 mm. *S Europe and the Mediterranean region, occasionally occurring elsewhere.* H5. Spring–summer.

34. KOELERIA Persoon

Tufted perennials. Leaves flat or rolled. Inflorescence a spike-like or loose panicle. Spikelets compressed from side to side, with 2–5 florets (rarely only 1). Glumes equal or unequal, the upper longer and usually equalling the first floret, the lower about two-thirds as long as the first floret, distinctly keeled, with 1–5 veins. Lemma lanceolate to ovate-lanceolate, keeled, mostly longer than the glumes, usually obtuse, with or without an awn. Palea as long as or shorter than the lemma, 2-keeled, bifid at the apex.

About 20–30 species in temperate regions and on mountains in tropical Africa. All the species are easy to grow and are usually propagated by division. Seed can be sown in the open in spring and will do well in most garden soils, especially those of a calcareous nature.

Literature: Domin, K., Monographic der Gattung Koeleria, *Bibliotheca Botanica*, 65 (1907).

1a. Leaves 2–5 cm, bluish green **2. glauca**
b. Leaves 5–20 cm, bright green **1. macrantha**

1. K. macrantha (Ledebour) Schultes (*K. cristata* (Linnaeus) Persoon; *K. gracilis* Persoon). Illustration: Christiansen,

Grasses, sedges and rushes in colour, f. 33 (1979); Darke, Ornamental grasses, 213 (1999); Aeschimann et al., Flora alpina **2**: 139.36.4 (2004).

Compactly tufted perennial, 10–50 cm, with stiff stems. Ligules to 1 mm. Leaf-blades with a fine blunt tip, to 20 cm long, rolled or flat, 1–2.5 mm wide, finely hairy or hairless. Panicle erect, very dense, narrowly oblong or tapering upwards, 2.5–10 cm × 5–20 mm, silvery, green or purplish. Spikelets with 2–3 florets, 4–6 mm, hairless or downy. *Europe, temperate Asia & N America.* H1. Summer.

2. K. glauca (Schrader) de Candolle. Illustration: Darke, Ornamental grasses, 213 (1999).

Tufted perennial, stems erect, 10–45 cm; Ligules *c.* 1 mm, membranous. Leaves 2–5 cm × 1.2 mm, stiff, glaucous, rough, hairy or hairless. Panicles 2–11 cm × 7–9 mm. Spikelets 4–5 mm, glumes 4–5 mm, lemmas 4–5 mm, membranous. *Europe to E. Asia.* H4. Summer.

35. ROSTRARIA Trinius

Annuals. Inflorescence a panicle. Spikelets laterally compressed, with 2–5 (rarely only 1, or as many as 10) florets. Glumes keeled, cartilaginous at the base. Lemma bifid, usually prominently 5-veined, keeled, with a straight or slightly curved awn inserted near the apex. Palea 2-keeled, bifid. Rachilla and callus hairless or with short hairs.

A genus of about 10 species in temperate Eurasia and N Africa. All are easy to grow, the seed being sown in spring outdoors in any good garden soil.

1. R. cristata (Linnaeus) Tzvelev (*Koeleria phleoides* (Villars) Persoon; *Lophochloa phleoides* (Villars) Reichenbach). Illustration: Bonnier, Flore complète **12**: t. 668 (1934); Aeschimann et al., Flora alpina **2**: 139.37.1 (2004).

Annual, with loosely tufted or solitary stems 5–45 cm. Ligule *c.* 1 mm. Leaf-blades finely pointed, 2.5–15 cm × 1.6–4 mm, flat, mostly hairy. Panicles linear to lanceolate or narrowly oblong, very dense, cylindric or lobed, 1–10 cm × 6–16 mm, pale green, spikelets 3–8 mm, with 3–8 flowers. *Mediterranean region & NE Africa.* H5. Summer.

36. DESCHAMPSIA Palisot de Beauvois

Perennials, usually tufted. Leaves flat or bristle-like. Inflorescence a panicle. Spikelets compressed from side to side, with 2 (rarely 3) bisexual florets. Glumes more or less equal, membranous, somewhat shorter than or equalling the florets, with 1–3 veins. Lemma obscurely 5-veined, truncate and irregularly toothed at apex, with short hairs at the base; awn arising on the back, straight or sharply bent. Palea as long as the lemma, with 2 rough keels. Rachilla prolonged.

A genus of about 40 species in temperate and cold regions, also on mountains in the tropics. The species below can be grown in most good soils and can either be grown from seed sown outside in autumn or spring or propagated by division of the rootstock.

1a. Leaves less than 1 mm wide, tightly inrolled; awn sharply bent **1. flexuosa**
b. Leaves more than 1 mm wide, flat or sometimes inrolled; awn straight **2. cespitosa**

1. D. flexuosa (Linnaeus) Trinius. Illustration: Hubbard, Grasses, edn 2, 250 (1968); Christiansen, Grasses, sedges and rushes in colour, f. 22 (1979); Darke, Ornamental grasses, 185 (1999); Brickell (ed.) RHS A–Z encyclopedia of garden plants, 365 (2003).

Loosely to densely tufted perennial, 8–90 cm, with smooth, slender, wiry stems. Leaves hairless; ligules blunt, to 3 mm; blades rather stiff, bristle-like, to 20 × 1 mm, rough towards the tips. Panicles open and very loose, 3.5–15 × 7.5 cm. Spikelets 4–6 mm, purplish, brownish or silvery, with awns 4–8 mm. *Eurasia, most of America, mountains in the Old World tropics.* H1. Summer.

2. D. cespitosa (Linnaeus) Palisot de Beauvois. Illustration: Hubbard, Grasses, edn 2, 252 (1968); Christiansen, Grasses, sedges and rushes in colour, f. 23 (1979); Darke, Ornamental grasses, 183–185 (1999); Brickell (ed.) RHS A–Z encyclopedia of garden plants, 365 (2003).

Densely tufted perennial, 30–150 cm, forming large tussocks, with erect, smooth stems. Leaves hairless; ligules to 1.5 cm; blades 10–60 cm × 2–5 mm, flat or inrolled, coarse, rough above on the ribs and margins. Panicles very loose and open, 10–45 cm long and up to 20 cm wide. Spikelets 4–6 mm, green, silvery, golden, purple or variegated, with awns to 4 mm. *Temperate regions, and mountains in tropical Africa & Asia.* H1. Summer.

37. HOLCUS Linnaeus

Perennials or annuals. Leaves flat, Inflorescence a panicle. Spikelets compressed from side to side, with 2–3 florets, the lower bisexual, the upper usually male. Glumes almost equal, membranous, longer than the florets, strongly keeled, the lower with 1 vein, the upper with 3 veins. Lemma obscurely 3–5-veined, leathery, shiny, the upper or both usually with an awn on the back. Palea membranous, slightly shorter than the lemma. Rachilla shortly prolonged.

A genus of about 8 species in Europe, temperate Asia and N and S Africa. They often invade neglected gardens and are difficult to eradicate.

1. H. mollis Linnaeus. Illustration: Hubbard, Grasses, edn 2, 264 (1968); Christiansen, Grasses, sedges and rushes in colour, f. 19 (1979); Aeschimann et al., Flora alpina **2**: 139.45.2 (2004).

Perennial, 15–45 cm, with tough creeping rhizomes and erect slender stems. Leaf-blades hairy; panicles pale. *Europe.* H1. Summer.

'Albovariegatus' is widely cultivated, illustration: Darke, Ornamental grasses, 207 (1999); Brickell (ed.) RHS A–Z encyclopedia of garden plants, 539 (2003).

38. AIRA Linnaeus

Annuals. Leaves often rolled. Inflorescence a panicle. Spikelets compressed from side to side, with 2 bisexual florets, the upper stalkless or almost so. Glumes about equal, membranous, equalling or longer than the florets, with 1–3 veins. Lemma with 5 veins in the basal half, bifid at apex, with an awn arising from below the middle on the back; sometimes the lower or both florets awnless. Palea shorter than lemma. Rachilla not prolonged. Grain more or less fusiform, longitudinally furrowed, hairless.

A genus of 9 species. Some are grown for their decorative value and for bouquets; they grow well sown in spring, in ordinary garden soil, especially the lighter kinds, and will thrive in semi-shade.

1a. Spikelets 2.5–3 mm **1. caryophyllea**
b. Spikelets 2–2.5 mm **2. elegantissima**

1. A. caryophyllea Linnaeus. Illustration: Hubbard, Grasses, edn 2, 258 (1968);

Christiansen, Grasses, sedges and rushes in colour, f. 22 (1979); Aeschimann et al., Flora alpina **2**:139.42.2 (2004).

Tufted annual or biennial, with slender stems, 8–30 cm. Leaf-blades to 5 cm × 1 mm, minutely rough; ligule to 5 mm. Panicles very loose, with widely spreading branches, to 10 cm; spikelet-stalks 1–10 mm. Spikelets 2.5–3 mm, silvery or purplish; glumes pointed; lemmas both awned, the awns projecting from the tips of the spikelets. *Europe, N&W Asia, N Africa, and mountains of tropical Africa.* H1. Spring–summer.

2. A. elegantissima Schur (*A. capillaris* Host; *A. elegans* invalid). Illustration: Hegi, Illustrierte Flora von Mittel-Europa, edn 2, **1**: f. 188 (1957); Aeschimann et al., Flora alpina **2**: 139.42.4 (2004).

Loosely tufted annual, to 30 cm, with very slender stems. Leaf-blades to 5 cm × 1 mm; ligule to 5 mm. Panicles very diffuse, 2.5–10 cm long and wide, with loosely divided, spreading branches; spikelet-stalks to 1 cm or longer, Spikelets 2–2.5 mm, silvery, scattered, 1–2-awned, the awns shortly projecting from the tips of the glumes. *Mediterranean region.* H4. Spring–summer.

39. HIEROCHLOE R. Brown

Perennials, smelling of coumarin (new-mown hay). Leaves flat or more or less rolled, those on the flowering stem with a short blade; ligule truncate, obtuse or acute. Inflorescence a panicle. Spikelets compressed from side to side, with 2 male florets and 1 terminal, bisexual floret. Glumes membranous, almost equal, ovate, about as long as florets, obscurely 3-veined. Lemmas of male florets membranous, obscurely 3–5-veined, obtuse, mucronate or awned. Palea shorter than lemma, membranous, with 2 veins. Lemma of bisexual floret obscurely 5-veined, hard and shiny in fruit, hairy towards apex. Palea shorter than lemma, membranous, with 1 vein. Stamens 3 in male florets, 2 in bisexual floret.

A genus of about 15 species in temperate and arctic regions. Propagated by division of the rootstock or from seed. Suitable for moist or damp places, although they will grow in most garden soils.

Literature: Weimarck, G., Variation and taxonomy of Hierochloe in the northern hemisphere, *Botanisker Notiser* **124**: 129–75 (1971).

1a. Panicles 4–10 cm; lower lemmas awnless; stems 20–45 cm **1. odorata**
b. Panicles 7.5–30 cm; lower lemmas awned; stems 45–135 cm or more **2. redolens**

1. H. odorata (Linnaeus) Palisot de Beauvois. Illustration: Hubbard, Grasses, edn 2, 266 (1968); Christiansen, Grasses, sedges and rushes in colour, f. 2 (1979); Aeschimann et al., Flora alpina **2**: 139.43.2 (2004).

Plant with slender creeping rhizomes, forming dense tufts or patches, with slender, erect stems 20–45 cm. Ligules 2–4 mm. Leaf-blades to 30 × 1 cm. Panicles ovate, 4–10 cm. Spikelets stalked, 3.5–5 mm, golden-brown, tinged with green or purple at the base; glumes ovate, blunt; male lemmas 3.5–5 mm, ciliate on the margins, minutely hairy, awnless or nearly awnless like the terminal lemma. *N America, Asia & Europe.* H1. Spring.

2. H. redolens (Vahl) Roemer & Schultes.

Plant with widely creeping rhizomes, forming large tufts, with slender to somewhat stout stems, 45–135 cm or more. Ligules 3–6 mm. Leaf-blades 30–75 cm × 5–12 mm, bright green. Panicles loose or dense, nodding, 7.5–30 cm. Spikelets shortly stalked, 6–10 mm, yellowish brown or golden-brown; glumes broadly lanceolate to ovate, pointed; male lemmas rough with very short, stiff hairs, ciliate on the margins and keel, 5–6 mm, with straight awns to 4 mm; terminal lemma awnless. *S America, SE Australia, New Zealand.* H4. Summer.

40. ANTHOXANTHUM Linnaeus

Annuals or perennials, often smelling of coumarin (new-mown hay). Leaves flat. Inflorescence a dense panicle. Spikelets compressed from side to side, of 2 sterile florets and a terminal, bisexual floret. Glumes membranous, very unequal, the lower with 1 vein, the upper with 3 veins. Lemma of sterile florets membranous, toothed or with 2 oblong, obtuse lobes at apex, with 3 veins and an awn on the back. Lemma of fertile floret somewhat hardened, shorter than the sterile ones, with 5–7 veins. Stamens 2. Lodicules absent.

A genus of about 15 species, native of Europe, N Asia, N & S America and the mountains of E Africa. Easily raised from seed sown in the open in ordinary garden soil.

Literature: Valdes, B., Revision de las Especies anuales del Genero Anthoxanthum, *Lagascalia* **3**: 99–141 (1973).

1. A. gracile Bivona

Slender, tufted annual, *c.* 30 cm, with stems ascending. Leaves linear, short, slender-pointed, ciliate, more or less hairy above; ligule short, much torn. Panicle simple, spike-like, ovate, with few florets. Spikelets awned, awn twice as long as glume. *C & E Mediterranean region.* H5. Spring.

41. PHALARIS Linnaeus

Annuals or perennials. Leaves flat. Inflorescence an ovoid or cylindric panicle, usually dense and unlobed. Spikelets strongly compressed from side to side, with 2–3 florets, the lower 1 or 2 reduced to lemmas, the uppermost bisexual. Glumes almost equal, papery, longer than the florets, with 3–5 veins, often winged on the keel. Lower lemmas small, linear to lanceolate; upper lemma with 5 veins, leathery. Palea leathery.

A genus of about 15 species, in temperate regions, mainly in the Mediterranean area. All are easily cultivated in ordinary garden soil, and are most useful for their ornamental inflorescences which may be used fresh, or dried for interior decoration. The annuals can be propagated from seed, sown thinly in spring in the open in a border, or in rows for cutting. The perennials may be increased by division of the rootstock, but are sometimes difficult to keep within bounds, the unsightly variegated form of *P. arundinacea* being a depressing feature of many an ill-tended border.

Literature: Anderson, D.E., Taxonomy and distribution of the genus Phalaris, *Iowa State Journal of Science* **36**: 1–96 (1961).

1a. Perennials with creeping rhizomes **1. arundinacea**
b. Annuals 2
2a. Spikelets of 2 kinds, clustered, 1 of each cluster fertile and surrounded by sterile spikelets, the clusters falling entire at maturity; glumes with toothed wings **2. paradoxa**
b. Spikelets all alike, breaking up at maturity and leaving the persistent glumes 3
3a. Fertile lemma 4–6 mm; sterile lemmas about half the length of the fertile **3. canariensis**

b. Fertile lemma *c.* 3 mm; sterile lemmas very dissimilar, the lower minute, the upper about one-third of the length of the fertile **4. minor**

1. P. arundinacea Linnaeus. Illustration: Hubbard, Grasses, edn 2, 274 (1968); Christiansen, Grasses, sedges and rushes in colour, f. 3 (1979); Darke, Ornamental grasses, 261–263 (1999); Brickell (ed.) RHS A–Z encyclopedia of garden plants, 798 (2003).

Perennial, 60–200 cm, with extensively creeping rhizomes. Stems stout, erect. Leaves hairless, green; ligules 6–10 mm; blades linear, finely pointed, 10–35 cm × 6–18 mm, firm, rough in the upper part. Panicles lanceolate to oblong, dense, lobed, 5–25 × up to 4 cm. Spikelets oblong, 5–6 mm, greenish, purplish, or whitish green; glumes narrowly lanceolate, pointed, minutely rough, almost wingless; fertile lemma narrowly ovate, 3–4 mm, finely hairy. *Europe, temperate Asia, N America, S Africa.* H1. Summer.

The variegated cultivar ('Variegata') is widely grown.

2. P. paradoxa Linnaeus. Illustration: Bonnier, Flore complète **11**: t. 646 (1931).

Annual, 10–75 cm, with tufted or solitary, erect or spreading, slender stems. Leaves green, hairless; upper sheaths inflated and often embracing the base of the panicle; ligules 3–10 mm; blades finely pointed, to 30 cm × 2–8 mm, rough. Panicles obovoid, ellipsoid or cylindric, mostly narrowed at the base, pale green or purplish, 2–9 cm × 8–20 mm. Spikelets in clusters of 6–7 falling together at maturity, with 1 in each cluster fertile and 5–8 mm, the rest sterile and smaller; glumes of fertile spikelet lanceolate, sharply pointed, winged on the keels in the upper part, and with a prominent tooth-like projection there, glumes of sterile spikelets smaller or more or less deformed; fertile lemma narrowly ovate, 3–4 mm. *W & C Europe, & Mediterranean region.* H4. Summer.

3. P. canariensis Linnaeus. Illustration: Hubbard, Grasses, edn 2, 272 (1968); Aeschimann et al., Flora alpina **2**: 139.53.2 (2004).

Annual, 20–120 cm, with slender to rather stout, erect, tufted, or solitary stems. Leaves hairless, green; ligules 3–8 mm; blades linear to linear-lanceolate, finely pointed, 5–25 cm × 4–12 mm, flat, rough. Panicles ovoid to ovoid-cylindric, very dense, erect, 1.5–6.5 × 1.2–2 cm wide, stiff, whitish except for the green nerves, rarely purplish. Spikelets obovate, 6–10 × up to 6 mm; glumes oblanceolate, abruptly pointed, broadly winged on the upper part of the keels, slightly hairy, minutely rough; fertile lemma narrowly ovate, 4–6 mm, hairy, becoming glossy and bright yellow. *W Mediterranean region.* H5. Summer.

4. P. minor Retzius. Illustration: Bonnier, Flore complète **11**: t. 646 (1931).

Annual, 10–90 cm, with slender, erect or ascending, tufted or solitary stems. Leaves hairless, green; ligules to 6 mm; blades linear, finely pointed, 5–15 cm × 3–6 mm, smooth. Panicles ovoid to cylindric, dense, 1.2–6.5 cm × 8–16 mm, rather soft, whitish green. Spikelets elliptic, 5–6 × 2.5–3 mm; glumes narrowly oblong, pointed, narrowly winged on the keels in the upper part, with the wings slightly toothed; fertile lemma *c.* 3 mm, narrowly ovate, greyish at maturity, finely hairy; sterile lemma to 1.5 mm. *Mediterranean region.* H5. Summer.

42. AGROSTIS Linnaeus

Perennials or annuals. Leaves flat or bristle-like. Inflorescence a panicle. Spikelets compressed from side to side, with 1 floret. Glumes more or less equal to somewhat unequal, membranous, longer than the floret, usually with 1 vein. Lemma membranous or scarious, truncate, with 3–5 veins, often with the lateral veins slightly projecting from the margin, sometimes with a sharply bent awn on the back. Palea shorter than the lemma, often very small. Rachilla prolonged or not. Callus usually hairy.

A genus of about 120 species. Some of the annual species with delicate open panicles are attractive and their inflorescences may be dried for ornamental purposes. Seeds of these should be sown thinly in the open in the spring; they may also be grown in pots for decorative purposes.

1. A. nebulosa Boissier & Reuter

Tufted annual to 35 cm, with flat, narrow leaf-blades; ligule blunt, to 5 mm. Panicles very diffuse and loose, to 15 cm long, with 10 or more basal branches. Spikelets *c.* 1.5 mm, awnless. Spain, *Portugal & Morocco.* H5. Summer.

43. CALAMAGROSTIS Adanson

Perennials. Leaves flat or rolled; ligule membranous. Spikelets narrow, with 1 floret, in dense to rather loose panicles. Lemma membranous, with 3–5 veins, shorter than the glumes, surrounded usually by long hairs from the base, with an awn arising from the back or apex. Rachilla separating above the glumes.

A genus of over 250 species in temperate regions throughout the world. Both species described below are grasses of open places in damp woods and fens and are therefore suitable for planting in such habitats in parks and wild gardens. Their inflorescences are also very decorative both in the living state and when dried. The plants are best propagated by division of the rootstock.

Literature: Wasiljew, W.N., Das System der Gattung Calamagrostis Roth, *Feddes Repertorium* **63**: 229–51 (1960).

1a. Leaf-blades hairy on the upper surface **1. canescens**
b. Leaf-blades hairless or with few hairs towards the base 2
2a. Leaf-blades entirely hairless; ligules 4–12 mm **2. epigejos**
b. Leaf-blades hairless except towards the base; ligules 2–4 mm **3. arundinacea**

1. C. canescens (Weber) Roth (*C. lanceolata* Roth). Illustration: Hubbard, Grasses, edn 2, 282 (1968); Christiansen, Grasses, sedges and rushes in colour, f. 16 (1979); Aeschimann et al., Flora alpina **2**: 139.50.4 (2004).

Stem slender, erect, forming loose tufts or patches, 60–120 cm high; leaves to 45 cm × 3–6 mm, shortly hairy above, rough; ligules 2–5 mm. Panicles lanceolate to oblong, moderately loose, wavy or nodding, 10–25 × 2.5–6 cm, purplish, greenish or yellowish. Spikelets lanceolate, 5–6 mm. *W, N & C Europe.* H3. Summer.

2. C. epigejos (Linnaeus) Roth. Illustration: Hubbard, Grasses, edn 2, 284 (1968); Christiansen, Grasses, sedges and rushes in colour, f. 15 (1979); Darke, Ornamental grasses, 152 (1999); Aeschimann et al., Flora alpina **2**: 139.50.1 (2004).

Stems erect, forming tufts or tussocks, 60–180 cm high; leaves to 60 cm × 4–10 mm, rough; ligules 4–12 mm. Panicles erect, lanceolate to oblong, dense, 15–30 × 3–5.5 cm, purplish, brownish or green. Spikelets 5–6 mm. *Throughout Europe & temperate Asia.* H1. Summer.

3. C. arundinacea (Linnaeus) Roth. Illustration: Flora of Bhutan **3**(2): 611 (2000).

Stems erect forming tufts 50–150 cm. Leaves to 50 cm × 2–10 mm, rough above, hairless except towards the base; ligules 2–4 mm. Panicles lanceolate, dense or loose, 8–18 × 1–4 cm, branches rough. Spikelets lanceolate, 4–6 mm. *Throughout Europe and temperate Asia.* H1. Summer–autumn.

Mostly known in cultivation as a parent of *C.* × *acutiflora* (Schrader) de Candolle (*C. arundinacea* × *C. epgejos*), of which there are many cultivars, including 'Karl Foerster', illustration: Darke, Ornamental grasses, 149, 150 (1999), forms erect vertical inflorescences to 2 m, with long-lasting, open, feathery panicles which persist through to winter.

44. LAGURUS Linnaeus
Annual. Leaves flat. Inflorescence a dense, ovoid or cylindric to almost spherical panicle. Spikelets compressed from side to side, with 1 floret. Glumes almost equal, linear, membranous, with 1 vein and a long, densely ciliate, bristle-like apex. Lemma with 5 veins, membranous, hairy at the base, with 2 long apical bristles and a sharply bent awn on the back. Palea somewhat shorter than lemma. Rachilla prolonged, ciliate.

A genus with a single species. Easily grown from seed in any good garden soil, but a light sandy one is best. Seeds sown in autumn should be protected from frost.

1. L. ovatus Linnaeus. Illustration: Hubbard, Grasses, edn 2, 314 (1968); Grounds, Ornamental grasses, t. 24 & 25 (1979); Brickell (ed.) RHS A–Z encyclopedia of garden plants, 606 (2003).

Softly hairy annual, 15–60 cm. Leaf-blades linear to narrowly lanceolate 1.2–20 cm × 1–12 mm, velvety. Panicles ovoid to oblong-cylindric or spherical, 5–60 × 5–20 mm, bristly, softly hairy, pale, rarely purplish. *Mediterranean region.* H4. Summer.

45. APERA Adanson
Annuals. Leaves flat or rolled. Inflorescence a panicle. Spikelets compressed from side to side, with 1 floret (rarely more). Glumes unequal, membranous, the lower with 1 vein, the upper with 3 veins and about as long as the lemma. Lemma obscurely 5-veined, papery, rounded on the back, with a long awn near the apex. Palea about equalling lemma. Rachilla shortly prolonged.

A genus of 3 species. They are easily raised from seed which may be sown in the open in spring. The inflorescences may be dried for decorative purposes.

1. A. spica-venti (Linnaeus) Palisot de Beauvois (*Agrostis spica-venti* Linnaeus). Illustration: Hubbard, Grasses, edn 2, 290 (1968); Christiansen, Grasses, sedges and rushes in colour, f. 11 (1979); Aeschimann et al., Flora alpina **2**: 139.12.1 (2004).

Annual, 25–30 cm; culms slender, mostly erect. Leaves hairless, to 10 mm wide; ligule oblong, 3–10 mm. Panicle ovate to oblong, 10–25 cm. Spikelets 2.5–3 mm, green or purple, with fine awns 5–10 mm. *Europe & N Asia.* H1. Summer.

46. MIBORA Adanson
Annuals. Leaves flat. Inflorescence spike-like, 1-sided. Spikelets compressed from side to side, with 1 floret. Glumes more or less equal, membranous, longer than the floret, with 1 vein, persistent. Lemma with 5 veins, truncate, densely hairy, thinner than the glumes. Palea as long as lemma, hairy.

A genus of 1 species. Seed should be sown in late summer or autumn in a light sandy soil.

1. M. minima (Linnaeus) Desveaux. Illustration: Hubbard, Grasses, edn 2, 338 (1968); Aeschimann et al., Flora alpina **2**: 139.14.1 (2004).

Compactly tufted, 2.5–15 cm, with numerous hair-like stems. Leaves hairless, blades to 2.5 cm long, flat or rolled and up to 0.5 mm wide. Racemes erect, 5–20 mm long, to 1 mm wide, reddish, purple, or green. Spikelets almost stalkless, adpressed to the axis, oblong, blunt, 2–3 mm; lemma densely hairy. *Mediterranean region, Britain.* H4. Winter–spring.

47. POLYPOGON Desfontaines
Annuals or perennials. Leaves flat. Inflorescence a dense panicle. Spikelets somewhat compressed from side to side, with 1 floret. Glumes more or less equal, with 1 vein, papery, rough with small, stiff hairs, longer than floret, usually with an apical awn. Lemma with 5 veins, transparent, truncate. Palea transparent. Rachilla not prolonged.

About 10 species, in warm temperate regions. The soft silky inflorescences are very attractive, and the plants are useful for the front of a border or bed. The panicles may also be used when young for interior decoration, but if left too long the spikelets are rapidly shed. Propagated from seed, sown thinly in the open in the spring in light loams.

1. P. monspeliensis (Linnaeus) Desfontaines. Illustration: Hubbard, Grasses, edn 2, 310 (1968); Aeschimann et al., Flora alpina **2**:139.49.1 (2004).

Annual, 7.5–60 cm. Stems tufted or solitary, slender. Leaves hairless; ligules 3–15 mm, blades pointed, 5–15 cm × 2–8 mm. Panicles very dense, narrowly ovate to oblong, cylindric or lobed, 1.8–15 × 1–3.5 cm, pale green or yellowish. Spikelets narrowly oblong, 2–3 mm; glumes blunt and slightly notched at the tip, minutely hairy on the margins, rough below; awns 4–8 mm. *Mediterranean region, NE & S Africa.* H4. Summer.

48. ALOPECURUS Linnaeus
Usually almost hairless annuals or perennials. Inflorescence a spike-like panicle. Spikelets compressed from side to side, with 1 floret. Glumes almost equal, often united below, with 3 veins. Lemma translucent, with 3 veins, usually awned from the back; margins often united below. Palea usually absent, or if present, small. Ovary hairless; styles usually united below. Rachilla breaking up below the glumes.

A genus of about 25 species, natives of temperate regions of the northern hemisphere. The variegated forms are useful in bedding, are hardy and not particular as to soil.

1. A. pratensis Linnaeus. Illustration: Hubbard, Grasses, edn 2, 334 (1968); Christiansen, Grasses, sedges and rushes in colour, f. 9 (1979); Darke, Ornamental grasses, 135 (1999); Aeschimann et al., Flora alpina **2**: 139.52.1 (2004).

A loosely to compactly tufted perennial with smooth erect stems, 30–120 cm. Leaves flat, rough to touch; sheath smooth; ligule membranous, to 2 mm, obtuse. Panicle 2.5–10 cm, dense, soft, pale green or purple. *Europe, N Asia.* H1. Spring–summer.

49. PHLEUM Linnaeus
Annuals or perennials. Leaves flat. Inflorescence a dense spike-like, ovoid or cylindric panicle. Spikelets strongly compressed from side to side with 1 floret. Glumes membranous, almost equal, longer than the lemma, keeled, shortly awned,

with 3 veins, with the margins overlapping along most of their length. Lemma membranous, truncate or obtuse, with 1–7 veins, unawned, hairless to densely ciliate. Palea equalling or almost equalling the lemma, ovate-lanceolate, obtuse, with 2 veins. Stamens 2–3.

A genus of about 15 species from temperate regions, mainly in Europe and the Mediterranean region. All species are hardy in NW Europe and are easy to grow in most garden soils. The annuals should be sown thinly in the open in the spring, the perennials may be propagated from seed or by division. The spike-like inflorescences, either fresh or dried, are useful for mixed bouquets and for vases.

1. P. pratense Linnaeus. Illustration: Hubbard, Grasses, edn 2, 322 (1968); Christiansen, Grasses, sedges and rushes in colour, f. 6 (1979); Aeschimann et al., Flora alpina **2**: 139.51.1 (2004).

Loosely to densely tufted perennial, with relatively stout, erect or ascending stems, 40–150 cm, usually swollen or bulbous at the base. Leaves hairless, green; ligules to 6 mm; blades finely pointed, to 45 cm × 3–8 mm, flat, usually minutely rough. Panicles dense, cylindric, 5–35 cm × 6–10 mm, greyish green or purplish, bristly. Spikelets oblong, 3–4 mm; glumes narrowly oblong, truncate, each bearing a rigid awn 1–2 mm, the keels fringed with stiff spreading hairs. *N, W & C Europe.* H3. Summer.

50. BROMUS Linnaeus

Annuals, biennials or perennials. Leaves flat or somewhat rolled; ligule membranous, often jagged. Inflorescence a panicle. Spikelets with 1–many florets; glumes with 1–9 veins, persistent at maturity, the upper usually the larger; lemma with few to many veins, awnless or with an almost terminal awn. Ovary crowned by a fleshy, hairy cap, the styles emerging from beneath it. Grain normally adherent to the palea and tightly enclosed within the lemma; hilum narrow, elongate.

A genus of over 100 species mainly in north temperate regions but also in S America, S Africa and on high mountains in the tropics. A few species are useful fodder grasses, and those described below all have ornamental panicles, which can be dried. They all thrive in ordinary garden soil and should be sown in autumn or spring.

Literature: Scholz, H., Zur Systematik der Gattung Bromus Subgenus Bromus, *Willdenowia* **6**: 139–59 (1970).

1a. Lemma with 3 awns **1. danthoniae**
b. Lemma with 1 awn or almost awnless 2
2a. Lower glume with 1 vein, upper glume with 3 veins; spikelets wedge-shaped, wider at the tip **2. madritensis**
b. Lower glume with 3–7 veins, upper glume with 5–9 veins; spikelets ovate or lanceolate, tapering towards the tip 3
3a. Lemma awnless or with a short awn to 1 mm **3. briziformis**
b. Lemma with an awn to 1.5 cm, flattened at the base 4
4a. Panicle compact, erect and rather dense; spikelet-stalks mostly equalling or shorter than spikelets **4. lanceolatus**
b. Panicle loose, nodding; spikelet-stalks longer than spikelets or panicle racemose 5
5a. Panicle usually 1-sided, racemose, with few spikelets **5. squarrosus**
b. Panicle usually compound with numerous spikelets **6. japonicus**

1. B. danthoniae Trinius

Annual with solitary or tufted, slender, erect stems to 45 cm. Leaves narrowly linear, to 10 cm, the blades and sheaths softly hairy. Panicle of 1 to few spikelets, narrow, erect. Spikelets lanceolate or oblong, to 5 cm, green or purplish, glistening, hairless or hairy; lemmas 1–1.3 cm long, *c.* 6 mm wide, with 3 awns from just below the tip, the central awn flattened in the lower half and becoming recurved and twisted at maturity, often purple or reddish purple. *SW & C Asia.* H4. Spring–summer.

2. B. madritensis Linnaeus. Illustration: Hubbard, Grasses, edn 2, 66 (1968); Aeschimann et al., Flora alpina **2**: 139.21.5 (2004).

Annual, 15–60 cm, with slender, solitary or tufted stems. Leaves hairy, especially on the lower sheaths; ligule to 4 mm. Panicle erect or slightly inclined, contracted and rather dense or somewhat loose, 3.5–15 cm long, to 5.5 cm wide, purple or green. Spikelets oblong, becoming wedge-shaped and gaping, 3–5.5 cm, compressed, with 6–13 florets; lemmas linear-lanceolate in side view, 1.2–2 cm, tipped with a fine, slightly diverging awn 1.2–2 cm. *Mediterranean region.* H4. Spring–summer.

3. B. briziformis Fischer & Meyer. Illustration: Britton & Brown, Illustrated flora of the northern United States and Canada, edn 2, **1**: 280 (1913).

Annual with solitary or tufted, slender, ascending stems, 30–60 cm. Leaf-sheaths softly hairy; leaves to 15 cm. Panicle loose, pyramidal, drooping, with the branches bearing 1 or rarely 2 spikelets, green or purplish. Spikelets ovate, 1.5–2.5 × 1–1.5 cm, with 10–15 florets; lemmas diamond-shaped or almost square, 8–15 × 4–8 mm, awnless or almost so. *SW & C Asia.* H4. Summer.

4. B. lanceolatus Roth. Illustration: Aeschimann et al., Flora alpina **2**: 139.21.19 (2004).

Annual with solitary or tufted, slender, erect stems, 20–60 cm. Leaves hairy; ligule short. Panicle erect, narrow, often compact or rather dense, of relatively few spikelets. Spikelets lanceolate to oblong, to 3 cm, with 10–20 florets, green or purplish; lemmas 1.2–1.5 cm, with a rather stout awn from just below the tip, the awn flattened and often purplish in the lower half, to 1.5 cm, becoming recurved and slightly twisted at maturity. *Mediterranean region.* H5. Summer.

5. B. squarrosus Linnaeus. Illustration: Hegi, Illustrierte Flora von Mittel-Europa, edn 2, **1**: f. 258 (1957); Aeschimann et al., Flora alpina **2**: 139.21.21 (2004).

Annual or biennial, with solitary or clustered, erect or ascending, slender stems, to 45 cm. Leaves hairy, the blades linear, to 6 mm wide. Panicle loose, of few spikelets, nodding, variable in size. Spikelets ovate to broadly oblong-elliptic, 1.8–2.5 cm long, up to 8 mm wide, with up to 20 florets, green or purplish; lemmas *c.* 1cm broad, with a spreading or recurved awn to 1 cm, from just below the tip. *Mediterranean region.* H5. Summer.

6. B. japonicus Thunberg. Illustration: Aeschimann et al., Flora alpina **2**: 139.21.20 (2004).

A variable annual or perennial with solitary or tufted, slender, erect or spreading stems to 60 cm. Leaves finely hairy, the blades to 10 cm × 6 mm; ligule to 4 mm. Panicle loose, nodding, of few to many spikelets, 5–15 cm long. Spikelets lanceolate to oblong, 1.8–2.5 cm, green or purplish, with 7–10 florets; lemmas 8–10 mm, awned from near the tip, with the awn 1–1.5 cm, often becoming slightly recurved. *Mediterranean region & temperate Asia.* H4. Summer.

51. BRACHYPODIUM Palisot de Beauvois

Perennials or annuals, rarely woody below, often with extensively branched rhizomes. Inflorescence a raceme of alternate, shortly stalked spikelets inserted in 2 ranks, with the back of the lemmas towards the axis. Spikelets 1 (rarely to 3) at each node, usually with numerous florets. Glumes unequal, shorter than the lowest floret. Glumes and lemmas acuminate, mucronate or with a straight or flexuous, apical awn. Palea equalling or a little shorter than the lemma, notched or truncate, the keels ciliate or rough with short, stiff hairs. Ovary hairy at the apex. Grain narrowly elliptic to oblanceolate; hilum linear.

A genus of about 16 species, mainly in temperate regions of the northern hemisphere. The *Brachypodium* described below is especially useful for establishing under trees, either by growing from seed or by dividing existing rootstocks.

Species of the genus *Lolium* Linnaeus are frequently grown as lawn, turf or forage grasses. They will key out to *Brachypodium* in the key to genera (p. 375), but can be distinguished by their spikelets, in which the lower glume is present, the upper absent.

1. B. sylvaticum (Hudson) Palisot de Beauvois. Illustration: Hubbard, Grasses, edn 2, 90 (1968); Christiansen, Grasses, sedges and rushes in colour, f. 67 (1979); Aeschimann et al., Flora alpina **2**: 139.22.1 (2004).

A tufted perennial with erect or spreading slender stems to 1m. Leaves linear-lanceolate, hairy, to 30 × 4–12 mm, yellow-green. Spikelets 18–45 mm, with 8–16 flowers; lemmas tipped with a fine awn which is up to 12 mm. *Most of Europe, temperate Asia & NW Africa.* H1. Summer.

52. ELYMUS Linnaeus

Tufted or rhizomatous perennials. Leaves flat or more or less rolled; ligule short, membranous. Inflorescence a spike, axis usually tough. Spikelets solitary or in groups of 2–3 at each node, almost stalkless, usually overlapping, with 2–11 florets; rachilla separating above the glumes and beneath each floret or sometimes the spikelets falling entire at maturity. Glumes with 3–9 veins, awnless or with short awns. Lemma lanceolate, with 5 veins. Palea 2-keeled.

A genus of about 150 species in north temperate regions. Propagated either by division of the rootstock or from seed sown outside in spring.

1a. Lemma awnless or with a short awn to 2.5 cm **1. virginicus**
b. Lemma awned, the awn to 4 cm, becoming curved or flexuous and spreading at maturity **2. canadensis**

1. E. virginicus Linnaeus. Illustration: Gould, The grasses of Texas, f. 84 (1975).

Stems tufted, erect, slender, 30–135 cm. Leaf-blades green or glaucous, finely pointed, to 25 cm × 5–15 mm, flat, rough with small, stiff hairs. Spikes erect, stiff, dense, 5–15 cm. Spikelets 1–1.5 cm paired, with 2–4 florets; glumes linear-lanceolate to narrowly lanceolate, 1–1.2 cm, smooth, yellowish, bowed out or arched at the base, strongly veined, rough with small, stiff hairs, firm and hardened above, tapering into straight awns of varying length; lemmas lanceolate, 1–1.2 cm, rough or hairy, tapering into straight awns to 2.5 cm, or awnless. *Canada & USA.* H2. Summer.

2. E. canadensis Linnaeus. Illustration: Gould, The grasses of Texas, f. 83 (1975); Darke, Ornamental grasses, 190 (1999).

Tufted, green or glaucous plant with erect, slender to stout stems, 75–180 cm. Leaf-blades very finely pointed, to 45 cm × 8–20 mm, flat or rolled, firm, rough or sparsely bristly above. Spikes dense, very bristly, nodding or drooping, 10–25 cm. Spikelets 1.2–1.6 cm (awns excluded) with 2–5 florets, in clusters of 2 to 4; glumes linear-lanceolate, stiff, rough with small, stiff hairs, awned from the tip with an awn to 1.6 cm; lemmas lanceolate, 1–1.6 cm, hairy, rough or hairless, awned from the tip with the awn 2.5–4 cm, becoming curved or flexuous. *Canada & USA.* H2. Summer.

E. magellanicus (Desvaux) A. Löve, illustration: Darke, Ornamental grasses, 190 (1999), is a striking blue-leaved species which is increasingly widely cultivated.

E. hispidus (Opiz) Melderis is also becoming more frequent in cultivation.

53. HYSTRIX Moench

Erect perennials, with flat leaf-blades. Inflorescence a loose, bristly spike. Spikelets with 2–4 florets, 1–4 at each node of a continuous, flattened axis, horizontally spreading or ascending at maturity. Glumes reduced to short or minute awns, the first usually rudimentary, both often absent from upper spikelets. Lemmas convex, rigid, tapering into long awns, with 5 veins, the veins obscure except towards the tip. Palea about as long as the body of the lemma.

A genus of 9 species in N America, N India, China and New Zealand. They may be propagated by division or from seed, and can be grown in an open border for their decorative spikes which may be cut and used fresh or dried.

1. H. patula Moench. Illustration: Darke, Ornamental grasses, 208 (1999).

Stems slender, 60–120 cm. Leaf-blades 8–15 mm wide. Spikes nodding, to 15 cm. Spikelets mostly in pairs, *c.* 1.5 cm, eventually spreading horizontally, the lemmas with awns to 4 cm. *N America.* H5. Summer.

54. SITANION Rafinesque

Tufted perennials with slender stems; ligules membranous. Leaf-blades flat or rolled. Inflorescence a very bristly spike. Spikelets with 2 to few florets, usually in pairs, alternating on opposite sides at each node of the fragile main axis of the spike, the axis breaking at the base of each joint at maturity; glumes bristle-like, with 1–2 veins, extending into 1–3 or more awns; lemmas rounded on the back, 5-veined, firm, slightly 2-toothed at the apex, the central vein extending into a long slender spreading awn, sometimes with 1 or more lateral awns.

A genus of 4 species in western N America. Propagated from seed. Useful in borders for their very ornamental bristly spikes; these must be cut when young if required for interior decoration as they soon break up.

1. S. hystrix (Nuttall) J.G. Smith

Stems rather stiff, 10–60 cm, erect or spreading. Leaves hairy or hairless; blades flat or rolled, 5–20 cm × 2–6 mm, rather stiff. Spikes erect, 5–20 cm. Glumes very narrow, extending into long awns which are covered with short, stiff hairs; lemmas hairless or short-hairy, smooth or rough, awned from the tip, the awns of glumes and lemmas finally widely spreading, 2.5–10 cm. *N America.* H2. Summer.

55. LEYMUS Hochstetter
Tufted perennials with elongated rhizomes (ours); ligules short, mebranous. Leaf-blades flat or rolled, hardened at apex, often to a sharp point. Stems stiff; inflorescence racemose; spikelets with 3–6 fertile florets, usually paired, the axis breaking at the base of each joint at maturity; spikelet callous hairless; glumes persistent, with 1 vein (ours); lemmas keeled; anthers 3.

A genus of about 40 species, all North temperate except for one in South America and Australasia. Previously included in *Elymus* (p. 350).

1. L. arenarius (Linneaus) Hochstetter. Illustration: Darke, Ornamental grasses, 214 (1999); Brickell (ed.) RHS A–Z encyclopedia of garden plants, 626 (2003).

Stems 60–180 cm, erect. Leaves 20–60 cm × 8–20 mm, metallic, blue-grey. Spikes 15–35 cm, erect. Lemmas 15–25 mm, leathery, 7-veined; anthers 7–8 mm. *North and South temperate regions.* H1. Summer.

56. TAENIATHERUM Nevski
Annuals. Leaves flat or rolled. Inflorescence a dense spike with spikelets in 2 ranks, separating at the nodes below the spikelets at maturity. Spikelets arranged in pairs at the nodes of the axis; florets 2, the lower bisexual, the upper rudimentary and sterile. Glumes joined at the base, narrowly subulate, rigid. Lemma lanceolate, with a flattened callus and a long, flexuous, rough awn.

A genus with probably only 2 species in S Europe and the Middle East. Seed should be sown in an open sunny border in well-drained soil. The spikes are suitable fresh or dried for interior decoration.

1. T. caput-medusae (Linnaeus) Nevski (*T. crinitum* (Schreber) Nevski).

Stems erect or ascending, 15–45 cm; ligules very short, truncate. Leaf-blades finely pointed, to 12.5 cm × 3 mm, flat or rolled, hairless or hairy. Spikes erect, 7.5–15 cm. Spikelets very long-awned; glumes very fine, rough, erect or spreading, 2–3 cm; lemmas smooth or rough on the back, 1–1.2 cm, narrowed into an erect or spreading, rough awn to 12 cm. *C & E Mediterranean region.* H5. Summer.

57. HORDEUM Linnaeus
Annuals or perennials. Leaves usually flat. Inflorescence a compressed, linear to oblong spike. Spikelets 3 at each node, dispersed together at maturity, the triplets arranged in 2 longitudinal rows, the fertile florets in 2, 4 or 6 longitudinal rows, each triplet with a central bisexual spikelet and 2 lateral bisexual, male or sterile spikelets. Florets 1 per spikelet, rarely 2. Glumes linear-subulate to lanceolate and awned, free to the base. Lemma 5-veined, ovate. Palea narrowly ovate, keeled. Rachilla usually prolonged in central spikelets.

A genus of about 25 species in temperate regions of the northern hemisphere and S America. They should be sown outdoors in spring in light garden soil. If grown for interior decoration, the spikes should be cut when young and hung upside down to dry, otherwise they will readily break up.

1a. Perennial; awns fine, hair-like, 3.5–9 cm; lateral spikelets mostly reduced to the awns **1. jubatum**
b. Annual; awns bristle-like, 1–2 cm; lateral spikelets with well-developed lemma **2. hystrix**

1. H. jubatum Linnaeus. Illustration: Britton & Brown, Illustrated flora of the northern United States and Canada, edn 2, **1**: 287 (1913); Darke, Ornamental grasses, 208 (1999); Aeschimann et al., Flora alpina **2**: 139.29.5 (2004).

Perennial, 23–60 cm; stems slender, erect or spreading, tufted or solitary. Leaf-blades 5–15 cm × 2–3 mm, flat, rough. Spikes very dense, brittle, very bristly, soft, nodding, 7.5–13 × 4–9 cm (including the awns), pale green or purplish. Lateral spikelets shortly stalked, sterile; glumes like those of the middle spikelet; lemma small when present. Middle spikelet complete: glumes bristle-like, very fine, spreading, minutely rough, to 7.5 cm; lemma lanceolate, *c.* 6 mm, with a very fine hair-like awn 3.5–9 cm. *N America, NE Asia.* H1. Summer–autumn.

2. H. hystrix Roth
Annual, 10–40 cm, with slender, ascending, solitary or tufted stems. Leaf-blades finely pointed, to 7.5 cm × 3.5 mm, flat, mostly softly hairy. Spikes oblong or ovate-oblong, very bristly, 2.5–6 × 1.2–1.8 cm, greyish green or tinged with purple. Lateral spikelets shortly stalked, sterile; glumes bristle-like, rigid, spreading, rough, 8–20 mm; lemma 8–12 mm. Glumes of middle spikelet bristle-like, 1–2.5 cm; lemma lanceolate, 5–10 mm, with a spreading awn 1–2 cm. Mediterranean region. *C Asia.* H4. Spring–summer.

58. AEGILOPS Linnaeus
Annuals. Leaves usually flat. Inflorescence a spike. Spikelets solitary at the nodes of the axis, all bisexual or the upper sterile and the lower vestigial. Florets 2–8, the upper usually male or vestigial. Glumes equal, leathery, truncate, often with 1 or more teeth or awns, usually rounded on the back. Lemma thin below, leathery, and strongly veined towards the toothed or awned apex. Palea with 2 keels.

A genus of about 20 species. The curious bristly inflorescence may be dried for decorative purposes. Seed should be sown in spring outdoors.

1. A. ovata Linnaeus. Illustration: Hegi, Illustrierte Flora von Mittel-Europa, edn 2, **1**: t. 39 (1957); Aeschimann et al., Flora alpina **2**: 139.25.3 (2004).

Stems slender, 10–30 cm. Leaves short, mostly hairy. Spikes short, ovoid or ellipsoid, 1–2.5 cm, of 2–4 spikelets, only the lower 2 fertile, the rest rudimentary. Glumes each with 3–5 long, spreading, rough awns. *Mediterranean region.* H5. Summer.

59. CHASMANTHIUM Link
Tufted to loosely rhizomatous perennials of diffuse habit. Leaf-blades linear to narrowly lanceolate, flat. Inflorescence an open or contracted panicle. Spikelets with few to many florets, strongly compressed from side to side. Glumes shorter than lemmas, 3–7-veined, acute to acuminate. Lemmas 5–15-veined, the keels toothed or ciliate, acuminate, entire or bifid, the lower 1–4 sterile.

A genus of five N American species. The panicles are very handsome, for use either in the open or in bouquets or vases. The species below may be propagated by seed or by division; it is suitable for partially shaded borders or open woods on good soil.

1. C. latifolium (Michaux) Yates (*Uniola latifolia* Michaux). Illustration: Gould, The grasses of Texas, f. 16 (1975); Darke, Ornamental grasses, 170 (1999); Brickell (ed.) RHS A-Z encyclopedia of garden plants, 260 (2003).

Loosely tufted plant with short, scaly rhizomes. Stems slender to stout, erect or spreading, 45–150 cm. Leaves smooth; ligules extremely short, minutely ciliate; blades 10–25 cm × 5–20 mm, spreading. Panicles loose, drooping, 10–30 cm, of relatively few spikelets. Spikelets oblong-lanceolate to broadly ovate, strongly flattened, with 6–18 florets, long-stalked, 1.5–4.5 × 1–2 cm, green or tawny; lemmas lanceolate, pointed, closely overlapping, 8–15 mm, the lowest sterile. *Eastern USA.* H2. Summer-autumn.

60. CHIONOCHLOA Zotov

Coarse perennials, forming tussocks from 20 cm to 2.5 m tall. Leaves mostly deeply grooved. Inflorescence a panicle. Spikelets with several florets. Glumes shorter than spikelets. Lemmas mostly with 7–9 veins, distinctly lobed, the awn a conspicuous prolongation of the middle vein, mostly bent or twisted at the base:

A genus of about 20 species from the mountains of New Zealand and southern Australia. It includes a number of handsome grasses, whose large, graceful panicles are very ornamental, particularly when the grass is planted in rock crevices so that it may be viewed against the sky. *C. conspicua* may be propagated by seed or by division of the rootstock.

1. C. conspicua (Forster) Zotov (*Arundo conspicua* Forster; *Cortaderia conspicua* (Forster) Stapf; *Danthonia cunninghamii* Hooker). Illustration; Buchanan, Indigenous grasses of New Zealand, t. 29 (1880); Brickell (ed.) RHS A–Z encyclopedia of garden plants, 263 (2003).

Densely tufted, forming large tussocks, with stout stems to 60–150 cm high. Leaves 45–60 cm × 6–8 mm, rigid, tough, flat or concave. Panicle erect or nodding, yellowish, 15–45 cm, loose or compact. Spikelets 8–12 mm, with 3–7 florets; lemmas silky-hairy at the base and on the lower half of the margins; awns straight, 1.8–2.2 cm. *New Zealand.* H4. Summer.

61. CORTADERIA Stapf

Large tufted perennials with separate male and female plants in some species and in others female and bisexual plants. Leaves flat; ligule a line of hairs. Inflorescence a large panicle. Spikelets compressed from side to side with 2–7 (rarely 1) florets. Glumes unequal, membranous, 1-veined. Lemma membranous, with shaggy hairs, 3-veined, with a terminal awn. Rachilla with shaggy hairs, breaking above the glumes and just above each floret.

A genus of about 15 species in south tropical and temperate S America and New Zealand. Since each plant is usually unisexual, uniformity of habit, flowering period, stature, and size and colour of the inflorescence can only be assured if the plants are propagated by division of the rootstock. Division and replanting is best carried out in April, and the site should be thoroughly prepared. A well-drained loam or deep sandy soil are the most suitable and protection from frost is advisable.

Literature: Stapf, O., The pampas grasses, Cortaderia Stapf, *Flora and Sylva* **3**: 171–6 (1905).

1a. Spikelets apparently all unisexual; male and female inflorescences on different plants, with 3–7 florets; lemma acuminate — 2
b. Spikelets either bisexual or female, borne on different plants, with 3–7 florets; lemma bifid, the lobes bristle-like — **1. richardii**
2a. Panicles erect or becoming 1-sided, moderately dense; glumes and lemmas 1.2–1.6 cm — **2. selloana**
b. Panicles loose and nodding; glumes *c.* 10 mm; lemmas to 1.2 cm — **3. jubata**

1. C. richardii (Endlicher) Zotov (*Arundo richardii* Endlicher). Illustration: Darke, Ornamental grasses, 173 (1999).

Plants either female or bisexual, densely tufted, with stout, erect, rigid stems 1.2–3 m high. Leaves numerous, 60–120 cm, narrow, curving, leathery. Panicles dense, much-branched, silvery white or yellowish white, 30–60 cm. *New Zealand.* H5. Summer.

2. C. selloana (Schultes & Schultes) Ascherson & Graebner. Illustration: Grounds, Ornamental grasses, t. 5–8 (1979)); Darke, Ornamental grasses, 174, 175, 176 (1999); Brickell (ed.) RHS A–Z encyclopedia of garden plants, 310 (2003).

Plants dioecious, densely tufted, with stout, erect stems 1.2–3 m high. Leaves numerous, crowded, narrow, 90–270 cm, glaucous, more or less arching, with very rough margins. Panicle 45–120 cm, oblong to pyramidal, silvery white or tinged with red or purple. *S America.* H5. Autumn.

Many cultivars are available, some with variegated leaves, some with coloured panicles, others rather dwarf.

3. C. jubata (Lemoine) Stapf. Illustration: *Botanical Magazine*, 7607 (1898); Darke, Ornamental grasses, 172 (1999).

Plants dioecious, densely tufted with stout, erect stems to 3 m high. Leaves numerous, crowded at the base, to 1.2 m × 1.2 cm, drooping, tough, rough with short, stiff hairs. Panicles 30–60 cm, with nodding or drooping branches, the lower to 35 cm, yellowish or tinged with mauve, red or purple. *Western tropical America.* H5. Autumn.

62. ARUNDO Linnaeus

Large rhizomatous perennials. Leaves flat. Inflorescence a loose panicle. Spikelets compressed from side to side, with few, mostly bisexual florets. Glumes almost equal, membranous, equalling the florets, with 3–5 veins, persistent. Lemma with 3–5 veins and with long soft hairs on the lower half of the back, not keeled. Rachilla hairless.

A genus of about 3 species of reed grasses distributed from the Mediterranean region to China and Japan. They should be grown on damp soils in a warm situation.

1. A. donax Linnaeus. Illustration: Hegi, Illustrierte Flora von Mittel-Europa, edn 2, **1**: f. 202 (1957); Darke, Ornamental grasses, 141 (1999); Aeschimann et al., Flora alpina **2**: 139.58.1 (2004).

Rootstock knotty, large. Stem 2–6 m, stout. Leaves alternate, regularly arranged, grey-green, almost smooth, arching, to 60 cm or more × 6 cm. Panicle much branched, erect and dense or somewhat drooping, at first reddish then white. Lemmas to 8 mm, bearing white hairs at base. *Mediterranean region.* H5. Autumn.

A variant with variegated leaves, 'Variegata', illustration: Darke, Ornamental grasses, 142 (1999), is available and is widely grown.

63. HAKONECHLOA Makino

Perennial with long scaly rhizomes, moderately slender stems, linear-lanceolate, finely pointed leaf-blades and minutely ciliate ligules. Inflorescence a loose panicle. Spikelets stalked, slightly compressed from side to side, with 3–5 florets; joints of the axis bearded at the tip; glumes persistent, pointed, membranous, with 3–5 veins, unequal, the lower shorter; lemmas much exceeding the glumes, overlapping, linear-lanceolate in side view, with a short straight awn from the tip, rounded on the back, membranous, with 3 veins, narrowed at the base into a bearded stalk.

A genus of 1 species of forest grass from the mountains of Japan. It is reasonably hardy, and flourishes in the partial shade of bamboos, in ordinary garden soil, spreading quickly by means of its rhizomes.

1. H. macra (Munro) Makino. Illustration: Grounds, Ornamental grasses, t. 38, 39

(1979); Darke, Ornamental grasses, 203 (1999); Brickell (ed.) RHS A–Z encyclopedia of garden plants, 500 (2003).

Stems leafy, ascending or spreading, 30–75 cm. Leaf-blades long and finely pointed, narrowed at the base, 4–25 cm × 5–15 mm, smooth. Panicle lanceolate to ovate, nodding, 9–17 cm; spikelets pale green, narrowly oblong to oblong, 1–2 cm; lemmas 5–9 mm, hairy on the margins, tipped with fine awns 3–4 mm long. *Japan.* H5. Autumn.

64. MOLINIA Schrank

Tufted perennials. Leaves flat; ligule a row of hairs. Inflorescence a panicle. Spikelets compressed from side to side, with 1–4 florets. Glumes almost equal, membranous, shorter than the florets, the lower with 1 vein or vein absent, the upper with 1–3 veins. Lemma membranous, with 3 veins, rounded on the back. Palea about equalling lemma.

A genus of 2 species in Europe, N and SW Asia. *M. caerulea* is well suited to damp acid soils, does well by the sides of streams and lakes, and is readily propagated by division of the rootstock.

1. M. caerulea (Linnaeus) Moench. Illustration: Hubbard, Grasses, edn 2, 350 (1968); Christiansen, Grasses, sedges and rushes in colour, f. 30 (1979); Darke, Ornamental grasses, 241–244 (1999); Aeschimann et al., Flora alpina **2**: 139.61.1 (2004).

Stems 15–120 cm, often forming large tussocks. Leaf-blades 10–45 cm × 3–10 mm, finely pointed, ligules densely ciliate. Panicles very variable, ranging from very dense and spike-like to open and very loose, light to dark purple, brownish or green. Spikelets 4–9 mm; lemmas pointed or blunt, 4–6 mm; stigmas purple. *Europe, N and SW Asia.* H1. Summer–autumn.

Many cultivars exist, among those especially widely grown are: 'Moorhexe', with very erect, narrow green leaves, 'Strahlenquelle' with arching inflorescences and 'Variegata' with striking yellowish cream variegation on the leaves.

65. PHRAGMITES Adanson

Rhizomatous perennials, with erect, stout or robust stems. Leaf-blades linear to narrowly lanceolate, flat, deciduous. Panicle large, dense, profusely branched, silkily hairy. Spikelets with 3–11 florets, the lowest male or sterile, the middle bisexual, the uppermost more or less reduced; rachilla bearded above with long silky hairs; glumes with 3–5 veins; lemmas narrow, rounded or slightly keeled on the back, hairless; lowest lemma much longer than the glumes, with 3–7 veins, more or less persistent; fertile lemmas acuminate, with 1–3 veins.

A genus of 3 to 4 aquatic species, widely distributed in temperate and tropical regions. *P. australis* is useful for the edges of lakes and large ponds on account of its ornamental habit and large plumose panicles.

1. P. australis (Cavanilles) Steudel (*P. communis* Trinius; *P. vulgaris* (Lamarck) Crépin). Illustration: Hubbard, Grasses, edn 2, 348 (1968); Christiansen, Grasses, sedges and rushes in colour, f. 29 (1979); Darke, Ornamental grasses, 264, 265 (1999); Aeschimann et al., Flora alpina **2**: 139.59.1 (2004).

Stems robust, 1.2–3 m, closely sheathed. Leaf-blades to 60 cm or more × 1–3 cm, tough, greyish green, long-tapering to a flexuous or curved tip, rough on the margins. Panicles erect or nodding, loose to dense, soft, silkily hairy, purplish or brownish, 15–45 cm. Spikelets 1–1.8 cm, with 3–6 florets; fertile lemmas surrounded by hairs 6–10 mm long. *Widespread in temperate regions and extending into the tropics.* H1. Autumn.

66. UNIOLA Linnaeus

Perennials, tufted or rhizomatous, with slender to stout stems. Leaf-blades linear to narrowly lanceolate, flat or rolled. Inflorescence a loose or contracted and dense panicle. Spikelets with few to many florets, strongly compressed from side to side, often large, falling entire. Glumes persistent, keeled, rigid, lemmas compressed and keeled, firm to tough, many-veined, the lower 1–6 sterile.

A genus of 4 species from N & S America which are useful as sand-binders and also very decorative on account of the beautifully formed spikelets. The species below is best sown in a sand-based seed compost.

1. U. paniculata Linnaeus. Illustration: Gould, The grasses of Texas, f. 199 (1975).

Tufted and with extensively creeping rhizomes. Stems stout, erect, 90–240 cm. Leaves smooth, the ligule a dense fringe of hairs; blades to 40 cm × 8 mm, stiff to rigid, tough. Panicles dense, erect or nodding, 16–40 cm long, to 10 cm wide. Spikelets oblong-lanceolate to ovate, much compressed, shortly stalked, with 8–24 florets, 2–3.7 × up to 1.2 cm, straw-coloured; lemmas lanceolate, mucronate, closely overlapping, 8–10 mm, the lower 4–6 sterile. *Eastern USA to the West Indies.* H2. Summer.

67. ERAGROSTIS Wolf

Annuals or perennials. Leaves flat or rolled, narrow; ligule a ring of hairs (rarely membranous). Inflorescence a panicle. Spikelets compressed from side to side, with 2–many florets. Glumes unequal to almost equal, membranous, shorter than the florets, 1-veined. Lemma 3-veined, membranous, usually, awnless, falling separately or with the palea and grain. Palea transparent, 2-keeled, often persistent. Stamens 3 or 2.

A genus of about 300 species in the tropics and subtropics. Annual species can be grown in light garden soil in the front of a border or bed, the seeds being sown in pots or in the open in late spring.

1a. Annual, without non-flowering shoots — 2
b. Perennial, with non-flowering shoots present — 3
2a. Lemmas to 0.8 mm — **1. japonica**
b. Lemmas 2–2.5 mm — **4. mexicana**
3a. Lemmas 3–3.5 mm; glumes about as long as the lowest lemmas — **2. trichodes**
b. Lemmas 2–2.5 mm; glumes shorter than the lowest lemmas — **3. curvula**

1. E. japonica (Thunberg) Trinius

Tufted annual with slender, erect stems, 15–60 cm. Leaf-blades pointed, flat, to 20 cm × 4 mm, hairless. Panicles usually open and loose, lanceolate to ovate-oblong, 7.5–25 cm, mostly over half the height of the plant, rather stiff, with spreading, profusely divided, whorled branches. Spikelets shortly stalked, loosely scattered, linear, with 4–10 florets, pale or purplish, to 4 × 1.3 mm; glumes to 0.8 mm long; lemmas lanceolate, oblong, very blunt, to 0.8 mm long, falling with the paleas at maturity. *Asia & Australia.* H3. Autumn.

2. E. trichodes (Nuttall) Wood. Illustration: Grounds, Ornamental grasses, t. 22 (1979); Darke, Ornamental grasses, 192 (1999).

Tufted perennial, with stiff, erect stems, 45–120 cm. Leaf-blades finely pointed, mostly flat, to 60 cm × 2–6 mm; leaf-sheaths bearded at the top. Panicles loose and open, oblong or elliptic, much-branched, 30–60 cm, about half the height of the plant. Spikelets 4–10 mm, long-stalked, lanceolate to lanceolate-oblong, with 3–9 florets, purple or green; glumes narrowly lanceolate, finely pointed, 3–3.5 mm; lemmas narrowly ovate or ovate-oblong, pointed, 3–3.5 mm. *USA.* H4. Autumn.

3. E. curvula (Schrader) Nees. Illustration: Chippindal & Cook, Grasses of southern Africa **1**: t. 17 (1976); Grounds, Ornamental grasses, t. 21 (1979); Darke, Ornamental grasses, 191 (1999).

Densely tufted perennial with erect, slender to stout stems, 30–120 cm. Leaf-blades with long, fine points, to 30 cm or more, rolled or opening out and 2–3 mm wide, rough above; sheaths usually hairy at the base. Panicles open and loose, or contracted, 7.5–30 cm, erect or nodding. Spikelets 4–12 × 1.6–2 mm, shortly stalked, narrowly oblong to oblong, dark olive-grey, with 3–18 florets; glumes to 2 mm; lemmas lanceolate-oblong, blunt, 2–2.5 mm. *S Africa.* H5. Autumn.

4. E. mexicana (Hornemann) Link

Loosely tufted annual, with slender erect or ascending, branched or simple stems, 15–90 cm high. Leaf-blades finely pointed, to 23 cm × 6 mm, flat, rough above. Panicles open and very loose, oblong to ovate, small to large, to 38 × 20 cm. Spikelets 5–8 mm, long-stalked, narrowly ovate to narrowly oblong, with 4–12 florets, purplish or greyish; glumes lanceolate, finely pointed, to 2 mm; lemmas ovate or ovate-oblong, pointed, 2–2.5 mm. *USA & Mexico.* H5. Summer–autumn.

68. MUHLENBERGIA Schreber

Annuals or perennials, with mostly slender stems. Leaf-blades thread-like or linear, flat or rolled; ligules membranous. Inflorescence a loose or contracted and dense panicle. Spikelets stalked, small, with 1 floret, awned or awnless. Glumes persistent, thin, with 1–3 veins, shorter than or as long as the lemma. Lemma narrow, membranous to firm, usually with 3 veins, awned or awnless, with the awn straight or flexuous.

A large genus of 120–130 species mainly in warm temperate and tropical America, a few in C and E Asia.

M. mexicana can be raised from seed sown outdoors in spring in light, well-drained soils in open borders.

1. M. mexicana (Linnaeus) Trinius. Illustration: Britton & Brown, Illustrated flora of the northern United States and Canada, edn 2, **1**: 185 (1913).

Tufted perennial, greyish green, 30–90 cm high, with short, stiff, curved, scaly rhizomes. Stems slender, erect or ascending, branched. Leaf-blades linear, finely pointed, 5–20 cm × 2–6 mm, flat. Panicles linear to narrowly lanceolate, erect or nodding, green or purplish, 4–15 cm, the branches densely covered with spikelets. Spikelets almost stalkless, 2–3 (rarely to 4) mm; glumes linear-lanceolate, finely pointed; lemma about as long as the glumes, shortly bearded at the base. *S Canada & USA.* H2. Summer–autumn.

69. CHLORIS Swartz

Annuals or perennials, tufted or with stolons, with slender stems. Leaves flat or folded; sheaths compressed and keeled; ligule a ciliate rim. Inflorescence of 1–many spikes, usually all arising from the apex of the stem, rarely racemose. Spikelets with 1 complete floret and 1 or more sterile florets above, stalkless, borne in 2 rows on 1 side of the spike-axis, compressed from side to side. Lemmas keeled, 3-veined, awned, usually ciliate on the keel and marginal veins.

A genus of about 50 species. Several species are ornamental and can be grown in the open border during summer, a light well-drained soil suiting them best, the seed being sown outside in May, or earlier in pots in a greenhouse.

1. C. virgata Swartz (*C. elegans* Kunth). Illustration: Chippindal & Cook, Grasses of southern Africa **1**: t. 4 (1973); Van Oudtshoorn, Guide to grasses of southern Africa, 231 (1999).

Annual, tufted, with ascending or horizontal stems 20–60 cm. Leaf-sheaths somewhat inflated; leaf-blades flat, 2–6 mm wide. Spikes 5–12, all arising from the stem-apex, silky or feathery, 2.5–7.5 cm, green or purplish. Spikelets bearded; lowest lemma *c.* 3 mm, with an awn to 10 mm. *Widespread in the tropics.* H5–G1. Autumn.

70. SPARTINA Schreber

Rhizomatous perennials. Leaves flat or rolled. Inflorescence a number of spikes arranged in a raceme. Spikelets strongly compressed from side to side, in 2 rows, closely adpressed to 1 face of the triangular spike-axis, each with 1 or rarely 2 florets. Glumes unequal, papery, the lower 1-veined, the upper with 1–3 (rarely to 6) veins, about as long as lemma. Lemma 1–6-veined, leathery, with a wide membranous margin. Palea slightly shorter than lemma.

A genus of about 14 species of saline or freshwater grasses from the coasts of W & S Europe, NW and S Africa, and inland in America and on the S Atlantic islands. All species of the genus have ornamental inflorescences of decorative value, but only 1 is suitable for cultivation in gardens. They are all propagated by division.

Literature: Saint-Yves, A., Monographia Spartinarum, *Candollea* **5**:19–100 (1932); Mobberley, D.G., Taxonomy and distribution of the genus Spartina, *Iowa State Journal of Science* **30**: 471–574 (1956).

1. S. pectinata Link. Illustration: Grounds, Ornamental grasses, t. 13 (1979); Darke, Ornamental grasses, 286, 287 (1999).

Densely tufted, 60–180 cm high, with stout, hard, scaly rhizomes. Stems stout. Leaf-blades 30–120 cm × 4–15 mm, flat, long and finely pointed, rough on the margins. Inflorescence erect, narrow; spikes 5–30, 2.5–10 cm, dense. Glumes toothed on the keels, the upper tapering into an awn 3–8 mm. *Canada & USA.* H2. Autumn.

71. BOUTELOUA Lagasca

Annuals or perennials, tufted or with stolons, with stiff, slender stems and narrow, flat or rolled leaves. Inflorescence of 1–many, dense, 1-sided spikes or racemes borne on a common axis. Spikelets stalkless or nearly so, in 2 rows on 1 side of the axis, with 1 bisexual floret, and with 1 or more rudimentary florets above it; lemmas 3-veined, the fertile one short-awned or mucronate, the rudimentary usually 3-awned.

A genus of 45–50 species, confined to warm temperate and tropical America, well represented in the USA. They may be propagated by division of the rootstock, or raised from seed sown in the open or in pots in loamy soil.

Literature: Gould, F.W., The genus Bouteloua, *Annals of the Missouri Botanical Garden* **66**: 348–416 (1979).

1a. Racemes persistent **1. gracilis**
b. Racemes deciduous **2. curtipendula**

1. B. gracilis (Kunth) Steudel (*B. oligostachya* Torrey; *Chondrosum gracile*

Kunth). Illustration: Gould, The grasses of Texas, f. 185 (1975); Darke, Ornamental grasses, 146 (1999); Brickell (ed.) RHS A–Z encyclopedia of garden plants, 188 (2003).

Densely tufted perennial, 25–55 cm, with slender, erect stems and linear, flat or inrolled leaves which are 1–2 mm wide. Inflorescence of usually 2, sometimes 1 or 3, dense, 1-sided, spreading or recurved spikes, these 2.5–5 cm long. Spikelets numerous, closely packed, *c.* 5 mm. *C & S USA & Mexico.* H5. Summer.

2. B. curtipendula (Michaux) Torrey. Illustration: Gould, The grasses of Texas, f. 178 (1975); Darke, Ornamental grasses, 145 (1999).

Perennial with scaly rhizomes; culms erect, tufted, 50–80 cm. Leaf-blades flat or almost rolled, 3–4 mm wide, rough with short, stiff hairs. Inflorescence 15–25 cm, of 35–50 spikes. Spikes each 1–2 cm long, purplish, spreading or pendent, mostly twisted to 1 side of the slender axis. Spikelets 5–8 per spike, adpressed or ascending, 6–10 mm; fertile lemma acute, mucronate; rudimentary lemma with 3 awns and more or less acute intermediate lobes, often reduced and inconspicuous. *Canada to Argentina.* H5. Summer.

72. PANICUM Linnaeus

Annuals or perennials. Inflorescence a panicle. Spikelets more or less flattened from top to bottom, unawned, with 2 florets; lower glume shorter than, upper glume as long as, the spikelet; lower floret male or sterile, its lemma resembling the upper glume, with or without a palea; upper floret bisexual, the lemma hard and rather brittle, clasping only the margins of the palea.

A genus of more than 400 species. It includes many valuable fodder grasses such as Guinea Grass, *P. maximum* Jacquin, a few cereals such as Common Millet, *P. miliaceum* Linneaus, and several species of ornamental value for gardens. The hardy species and the annuals flourish in ordinary garden soil, the annuals being sown in pots and transplanted to the open border when frosts have ceased, or sown outside in late spring. If required for interior decoration or bouquets, the panicles should be cut when first developed, otherwise their spikelets will be shed.

1a. Leaf-blades lanceolate to narrowly ovate, rounded at the base; spikelets minutely hairy; panicles terminal and axillary, the latter wholly or partly enclosed in the leaf-sheaths **1. clandestinum**

b. Leaf-blades linear to linear-lanceolate; spikelets hairless; axillary panicles, when present, never permanently enclosed in the sheaths 2

2a. Plants with long, creeping stolons; panicle of few, erect, spike-like racemes; spikelets blunt; lower glume almost as long as the spikelet **2. obtusum**

b. Plants tufted or rhizomatous; panicles generally open and loose, not of spike-like racemes; lower glume at most three-quarters of the length of the spikelet 3

3a. Perennial with stout, scaly rhizomes; leaves mostly hairless; lower glume two-thirds to three-quarters of the length of the spikelet **3. virgatum**

b. Annuals; leaves stiffly hairy; lower glume about half the length of the spikelet 4

4a. Panicles erect with widely spreading, much divided, stiff branches; spikelets 2–2.5 mm **4. capillare**

b. Panicles usually nodding; spikelets 5–6 mm **5. miliaceum**

1. P. clandestinum Linnaeus. Illustration: Britton & Brown, Illustrated flora of the northern United States and Canada, edn 2, **1**: 162 (1913).

Perennial, forming dense clumps. Stems slender to stout, erect or spreading, branched, 60–120 cm. Leaf-sheaths stiffly hairy to nearly hairless; blades lanceolate to narrowly ovate, rounded at the base, finely pointed, 7.5–20 cm × 12–30 mm, green, ciliate at the base. The terminal panicle open, loose, ovate, 7.5–15 × 3.5–10 cm, with smaller panicles wholly or partly enclosed in the leaf-sheaths. Spikelets elliptic, blunt, 3 mm, minutely hairy; lower glume up to one-third of the length of the spikelet; upper glume about equal to the lower lemma. *SE Canada & E USA.* H4. Autumn.

2. P. obtusum Kunth. Illustration: Gould, The grasses of Texas, f. 241 (1975).

Tufted perennials with long, widely creeping stolons which root at the densely hairy nodes. Stems wiry, erect or spreading, 20–75 cm. Leaf-blades linear, finely pointed, 3.5–30 cm × 2–8 mm, flat, hairless or slightly hairy. Panicles narrow, 3.5–12.5 cm long, to 12 mm wide, formed of few, erect, dense, 1-sided racemes. Spikelets shortly stalked, on 1 side of the raceme-axis, elliptic, blunt, 3–4 mm; lower glume slightly shorter than the spikelet, blunt; upper glume and lower lemma about equally long, blunt. *USA & Mexico.* H3. Autumn.

3. P. virgatum Linnaeus. Illustration: Gould, The grasses of Texas, f. 242 (1975); Darke, Ornamental grasses, 250–253 (1999).

Tufted perennial, 60–180 cm, purplish, glaucous or green, with stout, creeping, scaly rhizomes. Stems slender to stout. Leaves mostly hairless, blades linear, finely pointed, 15–45 cm × 3–15 mm, flat, erect, firm. Panicles very loose and open, 10–50 × 7.5–25 cm, with widely spreading, stiff, divided branches. Spikelets shortly stalked, elliptic-ovate, acuminate, 3–6 mm; lower glume two-thirds to three-quarters of the length of the spikelet, pointed; upper glume a little longer than the lower lemma, both acuminate. *S Canada & USA to C America.* H3. Autumn.

4. P. capillare Linnaeus. Illustration: Grounds, Ornamental grasses, t. 30 (1979); Brickell (ed.) RHS A–Z encyclopedia of garden plants, 765 (2003); Aeschimann et al., Flora alpina **2**: 139.59.1 (2004).

Tufted annual, 22–90 cm. Stems erect or spreading, often branched, slender to somewhat stout. Leaf-sheaths usually densely, stiffly rough from minute warts; leaf-blades linear or narrowly lanceolate, finely pointed, 10–30 cm × 5–15 mm, flat, stiffly hairy. Panicles very loose, open, one-third to half or more of the length of the plant, purple or green, with widely spreading, fine, much-divided branches. Spikelets stalked, elliptic, pointed, 2–2.5 mm; lower glume about half the length of the spikelet; upper glume and lower lemma about equal, acuminate. *S Canada & USA.* H3. Summer–autumn.

5. P. miliaceum Linnaeus. Illustration: Hegi, Illustrierte Flora von Mittel-Europa, edn 2, **1**: f. 169 (1957); Aeschimann et al., Flora alpina **2**: 139.71.1 (2004).

Annual, 30–120 cm. Stems solitary or tufted, stout, branched. Leaf-sheaths stiffly hairy, blades linear to linear-lanceolate, rounded at the base, finely pointed, 15–30 × up to 2 cm. Panicles loose or dense, usually nodding, green or purple, 10–30 cm. Spikelets stalked, ovate to elliptic, acuminate, 5–6 mm, prominently nerved; lower glume about half the length of the spikelet, pointed; upper glume and lower lemma equally long, acuminate; upper floret white, yellow to reddish or dark brown. *Warm regions of the world.* H4. Summer–autumn.

73. ECHINOCHLOA Palisot de Beauvois
Annuals or perennials. Ligule absent or a line of hairs. Inflorescence of 1-sided racemes arranged along a central axis. Spikelets in several rows, convex beneath, flattened above, toothed or awned, with 2 florets. Glumes unequal, the lower shorter than, the upper as long as, the spikelet. Lower floret male or sterile, the lemma resembling the upper glume; upper floret bisexual, the lemma hard and rather brittle, smooth, clasping only the margins of the palea.

A genus of 20–25 species in the tropics and warm temperate regions. Seed should be sown under glass and the plant protected from winter frosts.

Literature: Gould, F.W., Ali, M.A. & Fairbrothers, D.E., A revision of Echinochloa in the United States, *American Midland Naturalist* **87**: 36–59 (1972).

1. E. polystachya (Kunth) Hitchcock
Perennial, with coarse stems 90–180 cm; stem nodes densely bristly with adpressed, yellowish hairs; ligule a line of hairs. Leaf-blades to 2.5 cm wide, rough above. Panicle 10–30 cm, rather dense, with the lower racemes 4–7 cm. Spikelets *c.* 5 mm, awn to 10 mm. *USA (Texas), tropical America from Mexico & the West Indies to Argentina.* G1. Summer.

74. ERIOCHLOA Kunth
Annuals or perennials. Ligule a line of hairs. Inflorescence of 1-sided racemes arranged along a central axis. Spikelets flattened from top to bottom, with 2 florets; lower glume vestigial, attached to the swollen, bead-like, lowest rachilla internode; upper glume almost as long as the spikelet; lower floret male or empty, the lemma resembling the upper glume; upper floret bisexual, the lemma leathery, clasping only the margins of the palea.

A genus of about 30 species in the tropics. The 1 cultivated species needs a warm place in good garden soil.

1. E. villosa (Thunberg) Kunth
Loosely tufted annual, 10–60 cm. Inflorescence of 3–15 racemes each 1–2 cm; stalks with dense, long hairs. Spikelets 4.5–6 mm, elliptic; lower floret reduced to a lemma; upper lemma mucronate. *Asia.* H5. Summer.

75. PASPALUM Linnaeus
Hairless or sparsely hairy annuals or perennials. Leaves flat or rolled; ligule membranous; sheaths often bearded at the mouth. Inflorescence of 1-sided racemes arising like fingers from the apex of the stem or along a central axis, bearing adpressed spikelets. Spikelets circular to oblong, flattened from top to bottom, usually plano-convex with 2 florets; lower glume usually absent; upper glume as long as the spikelet; lower floret sterile, represented by a lemma as long as the spikelet; upper floret bisexual, the lemma more or less hard and brittle, its narrowly folded margins leaving the palea exposed.

A genus of more than 250 species. It includes several valuable pasture and fodder grasses. Many species have decorative inflorescences but most of these require warm greenhouse cultivation. They may be propagated from seed and increased by division, the half-hardy species being sown under glass and planted out after frosts have ceased.

1. P. ceresia (Kuntze) Chase
(*P. membranaceum* Lamarck; *P. ekgans* Roemer & Schultes). Illustration: Britton & Brown, Illustrated flora of the northern United States and Canada **1**: 106 (1896).

Loosely tufted perennial, hairy at the base, with slender ascending stems to 75 cm. Leaves glaucous; ligules very short; blades linear-lanceolate, finely pointed, 4–20 cm × 3–10 mm, flat, hairless or loosely hairy above. Inflorescence of usually 1–4 dense, ascending or arching racemes, these 2.5–8.5 cm; raceme-axis ribbon-like, purplish or bronze-green, with yellowish brown, translucent margins to 8 mm wide. Spikelets solitary, lanceolate or oblong, 3 mm, hidden by long, silvery, silky hairs; anthers bright yellow; stigmas purple. *Tropical S America.* H5–G1. Summer.

76. SETARIA Palisot de Beauvois
Annuals or perennials. Inflorescence a panicle, usually spike-like, the spikelets subtended by bristles which persist on the axis after the spikelets fall. Spikelets oblong to ovate; glumes unequal; lower floret male or sterile, as long as the spikelet; upper floret bisexual, its lemma hard and rather brittle, boat-shaped, often with a wrinkled surface.

A genus of about 100 species, in tropical and warm temperate regions. Several species are grown in warm, humid greenhouses, mainly for their foliage. They require rich loams and a plentiful supply of water. Many annual species have decorative spike-like panicles, with yellow, reddish or purplish bristles; when used for interior decoration the inflorescences should be cut when first developed, otherwise the spikelets are soon shed.

1a. Leaf-blades folded like a fan (pleated) between the veins, especially when young; bristles solitary **1. plicatilis**
b. Leaf-blades flat, not pleated; bristles usually several together 2
2a. Upper glume half to two-thirds of the length of the spikelet; upper lemma transversely wrinkled **2. pumila**
b. Upper glume as long as or nearly as long as the spikelet; upper lemma smooth or almost so 3
3a. Glumes and lower floret more or less persistent, the upper floret deciduous; panicle often lobed or interrupted, erect or nodding, purple or yellow **3. italica**
b. Spikelets falling entire at maturity; spikes very bristly, erect, cylindric or tapering upwards, green or purplish **4. viridis**

1. S. plicatilis (Hochstetter) Hackel
Loosely tufted perennial, with erect or ascending, slender to moderately stout stems, 45–150 cm. Leaf-sheaths compressed and keeled; ligule a dense fringe of hairs; leaf-blades linear to linear-lanceolate, 15–37.5 × 1–2.5 cm, narrowed to the base, finely pointed, closely pleated, green, rough. Panicles linear to lanceolate-oblong, loose or dense, erect or nodding, to 30 × 7 cm; bristles 6–16 mm. Spikelets narrowly ovate to elliptic-oblong, 3–3.5 mm; upper glumes half to three-quarters of the length of the spikelet; upper lemma smooth or nearly so. *E African highlands.* G2. Summer.

2. S. pumila (Poiret) Roemer & Schultes
(*S. glauca* misapplied; *S. pallidefusca* (Schumacher) Stapf & Hubbard). Illustration: Aeschimann et al., Flora alpina **2**: 139.76.1 (2004).

Annual, with slender, erect or ascending, tufted or solitary stems, 15–45 cm. Leaves green, sheaths compressed and keeled; ligule a dense fringe of hairs; blades linear, finely pointed, 5–30 cm × 2–10 mm, flat or folded, sparsely hairy or hairless. Panicles spike-like, very dense, erect or nodding, very bristly, 1.2–7.5 cm × 4–8 mm (excluding bristles); bristles several, 3–10 mm, yellowish to reddish yellow or orange-brown. Spikelets ovate-elliptic to broadly elliptic, blunt or slightly pointed, 2–3 mm; upper glume half to two-thirds of the length of the spikelet; upper lemma transversely wrinkled. *Warm temperate parts of the Old World.* G1–2. Summer–autumn.

3. S. italica (Linnaeus) Palisot de Beauvois. Illustration: Christiansen, Grasses, sedges and rushes in colour, f. 91 (1979); Aeschimann et al., Flora alpina **2**: 139.76.5 (2004).

Annual, with stout, erect or ascending stems to 150 cm. Leaf-blades linear to narrowly lanceolate, finely pointed, to 45 cm × 4–20 mm, flat, hairless, rough. Panicles erect or nodding, cylindric or lobed, dense, 2.5–30 cm × 8–30 mm; bristles to 15 mm. Spikelets broadly elliptic, blunt, 2–2.5 mm, persistent; upper glume two-thirds to four-fifths of the length of the spikelet; upper floret falling separately when ripe, smooth, variously coloured, white, cream, yellow, red, brown or black. *Warm temperate Asia.* H5–G1. Summer–autumn.

4. S. viridis (Linnaeus) Palisot de Beauvois. Illustration: Hubbard, Grasses, edn 2, 366 (1968); Christiansen, Grasses, sedges and rushes in colour, f. 90 (1979); Aeschimann et al., Flora alpina **2**: 139.71.4 (2004).

Loosely tufted annual with slender, erect stems, 10–60 cm. Leaves green; ligule a dense fringe of hairs; blades finely pointed, 3.5–30 cm × 4–10 mm, flat, hairless, minutely rough. Panicles spike-like, very bristly, cylindric or tapering upwards, very dense, erect, 2.5–10 cm × 4–10 mm (excluding bristles), greenish or purplish; bristles 1–3 together, to 10 mm. Spikelets elliptic-oblong, blunt, 2–3 mm; upper glumes as long as the spikelet; upper lemma very finely wrinkled. *Warm temperate Europe & Asia.* H5. Summer–autumn.

77. TRICHOLAENA Schrader

Perennials, rarely annuals. Inflorescence a panicle. Spikelets oblong in outline, awnless; lower glume very small or absent; upper glume as long as the spikelet, thinly membranous, indistinctly 5-veined, slightly notched to acute; lower floret male, its lemma similar to the upper glume; upper floret bisexual, the lemma thinly cartilaginous, clasping only the margins of the palea.

A genus of 4 species from Africa and the Mediterranean to India. They are best grown in light, well-drained loam in a greenhouse, to afford protection from rain and wind which would damage the panicles. They are propagated by seed.

1. T. teneriffae (Linnaeus) Link

Perennial, with slender, hard, rigid, ascending stems, 22–60 cm. Leaf-blades 2–15 cm × up to 5 mm, with a hard pointed tip, glaucous, downy above or hairless. Panicles linear, oblong or ovate, dense or loose, 2.5–12 cm. Spikelets ovate to oblong, *c.* 3 mm, pale green or purplish, silkily hairy with hairs exceeding the tip by up to 4 mm. *Mediterranean region & NE Africa.* G1. Summer–autumn.

78. MELINIS Palisot de Beauvois

Annuals or perennials. Leaf-blades flat or thread-like; ligule a ciliate rim. Inflorescence a panicle with slender branches. Spikelets more or less asymmetric in profile, compressed from side to side, often silky-hairy; lower glume small, sometimes distant from the upper; upper glume as long as the spikelet, firmly membranous to almost leathery, becoming thinner towards the tip, with 5 veins, more or less swollen on one side below the middle and sometimes tapering to a beak above, notched to 2-lobed (rarely entire), often awned from the sinus; lower floret usually male, its lemma resembling the upper glume or somewhat narrower and less swollen, palea usually ciliate; upper lemma smaller than the lower, often falling before the rest of the spikelet, cartilaginous, smooth, compressed from side to side.

A genus of about 22 species in tropical and S Africa, Madagascar, Arabia and India. Only 1 species is in cultivation, but there are many others suitable for greenhouse or open borders on account of their beautiful, colourful inflorescences.

1. M. repens (Willdenow) Zizka (*Rhynchelytrum repens* (Willdenow) Hubbard; *R. roseum* (Nees) Stapf & Hubbard). Illustration: Chippindal & Cook, Grasses of southern Africa **1**: t. 18 (1973); Darke, Ornamental grasses, 269 (1999).

Annual or short-lived perennial, with slender, erect or ascending stems 30–120 cm. Leaf-blades finely pointed, to 30 cm × 3–10 mm, flat. Panicle oblong to ovate, loose or dense, 5–20 × up to 10 cm. Spikelets 2.5–6 mm, densely silky-hairy with adpressed or spreading, purple, reddish or white hairs. *Tropical & S Africa.* H5–G1. Summer–autumn.

79. DIGITARIA Haller

Annuals or perennials. Leaves flat; ligule membranous. Inflorescence of 1-sided racemes arranged palmately or upon a short central axis and bearing the spikelets in adpressed groups of 1–5 or more. Spikelets flattened from top to bottom with 2 florets; lower glume small or absent; upper glume as long as or much shorter than the spikelet; lower floret sterile, represented only by a lemma which is usually as long as the spikelet; upper floret bisexual, the lemma cartilaginous, with its transparent margins enfolding and concealing most of the palea.

A genus of more than 250 species in tropical and warm temperate regions. The seeds of annual species should be sown outside in late spring in a light soil.

Literature: Henrard, J.T., A monograph of the genus Digitaria (1950).

1. D. sanguinalis (Linnaeus) Scopoli. Illustration: Hubbard, Grasses, edn 2, 372 (1968); Aeschimann et al., Flora alpina **2**: 139.74.1 (2004).

Green or purplish annual, 10–30 cm, with slender, ascending stems and mostly hairy leaves. Blades narrowly lanceolate, 2.5–10 cm × 3–8 mm, flat. Racemes 4–10, clustered at or near apex of stem, 4–15 cm, very slender, spreading. Spikelets in pairs, 2.5–3 mm, more or less elliptic, lower glume minute. *Temperate regions.* H5. Summer–autumn.

80. PENNISETUM Richard

Annuals or perennials. Inflorescence spike-like, each spikelet or cluster of spikelets enclosed by an involucre of slender bristles which are free throughout and fall with the spikelets. Spikelets lanceolate to oblong, with 2 florets; lower glume often minute; upper glume very small to as long as the spikelet; lower floret male or sterile, its lemma as long as the spikelet or much reduced; upper floret bisexual, its lemma membranous to thinly leathery.

A genus of about 120 species in tropical and warm temperate regions. Most of the wild species have decorative inflorescences, but few are cultivated. The species may be propagated by seed, or the perennials by division of the rootstock. Those suitable for cultivation in the open need a sunny position and a well-drained soil; the tender kinds require a warm greenhouse.

1a. Bristles, especially the inner, hairy in the lower part 2
b. Bristles hairless, rough with minute, stiff projections 4
2a. Spikelets 9–15 mm; bristles to 7.5 cm; inflorescences broadly cylindric to almost spherical, 2–9 cm **1. villosum**
b. Spikelets 4–6 mm; bristles to 4 cm; inflorescences cylindric, 5–30 cm 3
3a. Inflorescence loose, flexuous; inner bristles densely ciliate in the lower part, to 12 mm **2. orientale**
b. Inflorescences moderately dense, not flexuous; inner bristles loosely hairy in the lower part, to 4 cm **3. setaceum**
4a. Bristles to 3 cm, spreading; leaf-sheaths compressed; inflorescences 2.5–5 cm wide **4. alopecuroides**
b. Bristles to 1.5 cm; leaf-sheaths not compressed; inflorescences 8–20 mm wide **5. macrourum**

1. P. villosum Fresenius. Illustration: Grounds, Ornamental grasses, t. 31 (1979); Darke, Ornamental grasses, 261 (1999); Brickell (ed.) RHS A–Z encyclopedia of garden plants, 785 (2003); Aeschimann et al., Flora alpina **2**: 139.77.1 (2004).

Loosely tufted perennial with erect or ascending stems, 15–60 cm, loosely to densely hairy below the inflorescences. Leaf-blades finely pointed, 6.5–15 cm or more long, 2–6 mm wide, flat or folded. Inflorescence broadly cylindric to almost spherical, dense, erect or nodding, feathery, 2–9 × up to 5 cm, tawny or purplish; bristles numerous, spreading, loosely to densely hairy below the middle, the longer to 7.5 cm. Spikelets lanceolate, pointed, 9–15 mm. *Mountains of NE tropical Africa.* H5–G1. Summer–autumn.

2. P. orientale Richard. Illustration: Darke, Ornamental grasses, 259 (1999).

Tufted perennial, with short, stout rhizomes. Stems slender, erect or spreading, 15–90 cm, with bearded nodes. Leaf-blades to 10 cm or more × 4 mm, narrowly linear, with very finely pointed, straight or flexuous tips, rough with very small, stiff projections. Inflorescence loose, flexuous, pale or purplish, silky, 5–12.5 × up to 2.5 cm; axis hairy, beset with scattered minute stumps; bristles numerous, flexuous, 4–5 times as long as the spikelets, the inner densely ciliate in the lower half, the outer rough. Spikelets lanceolate, 5–6 mm, in clusters of 2 to 5. *SW Asia to C Asia & NW India.* H5. Summer–autumn.

3. P. setaceum (Forsskahl) Chiovenda. Illustration: Darke, Ornamental grasses, 259–261 (1999); Van Oudtshoorn, Guide to grasses of southern Africa, 83 (1999).

Densely tufted perennial, with erect, stiff, slender stems 23–90 cm, rough and sometimes hairy beneath the inflorescence. Leaf-blades finely pointed, to 30 cm × up to 3 mm, rolled or flat, erect, rigid, very rough. Inflorescence moderately dense, erect or slightly inclined, feathery, silky, pale or purplish, 10–30 × up to 4 cm; axis beset with minute stumps; bristles numerous, very slender, straight, loosely hairy in the lower part, one stouter than the rest and 1.6–4 cm, the others to 2.5 cm. Spikelets lanceolate, 4–6 mm, solitary or in clusters of 2 to 3. *N & NE tropical Africa, SW Asia, Arabia.* H5–G1. Summer–autumn.

4. P. alopecuroides (Linnaeus) Sprengel. Illustration: Darke, Ornamental grasses, 256, 257 (1999); Brickell (ed.) RHS A–Z encyclopedia of garden plants, 785 (2003).

Densely tufted perennial, with slender, erect stems 30–150 cm. Leaf-sheaths compressed; blades slender, finely pointed, 30–60 cm long, 4–8 mm wide when opened out, hairless or nearly so. Inflorescence narrowly oblong, cylindric, 5–20 × 2.5–5 cm, pale yellow or greenish to dark purple; axis closely studded with minute stumps; bristles numerous, straight or flexuous, very fine, rough, unequal, to 3 cm, spreading. Spikelets lanceolate, pointed, 6–8 mm, solitary or paired. *E Asia to E Australia.* H5–G1. Summer–autumn.

5. P. macrourum Trinius. Illustration: Darke, Ornamental grasses, 235 (1999); Van Oudtshoorn, Guide to grasses of southern Africa, 86 (1999).

Densely tufted perennial, with slender to stout, erect stems to 1.8 m. Leaves mostly basal, blades linear, very finely pointed, to 60 cm × 4–12 mm, flat or rolled, rough. Inflorescence cylindric, dense, erect or inclined, 7.5–30 cm × 8–20 mm, tawny to light brown or purplish; axis closely beset with minute stumps; bristles numerous, rough with very small, stiff hairs, one longer than the rest and to 1.5 cm long, the rest about as long as the spikelets. Spikelets lanceolate, pointed, 4–6 mm, solitary. *S Africa.* H5. Summer–autumn.

81. CENCHRUS Linnaeus

Annuals or perennials. Inflorescence spike-like, axis angular; each spikelet or cluster of spikelets enclosed by an involucre of more or less flattened bristles or spines which are united below to form a disc or cup which falls with the spikelets. Spikelets lanceolate or ovate, awnless, with 2 florets; lower glume up to half as long as spikelet, sometimes absent; upper glume as long as spikelet; lower floret male or sterile, its lemma as long as spikelet; upper floret bisexual, its lemma firmly membranous to leathery.

A genus of 25–30 species mainly in warm, dry parts of America, Africa and India. The species below can be raised from seed but needs protection from frost. It is an excellent forage grass and is also grown in gardens for its unusual spike-like inflorescence.

1. C. ciliaris Linnaeus (*Pennisetum cenchroides* Richard; *P. ciliare* (Linnaeus) Link). Illustration: Chippindal & Cook, Grasses of southern Africa **1**: t. 1 (1976).

Perennial, with a tough rootstock and slender, rigid stems, 15–90 cm, ascending from a sharply bent base. Leaves hairy or hairless, 7.5–30 cm × 2–4 mm; ligule a dense row of hairs. Inflorescence cylindric, moderately dense, 2.5–15 × 4–16 mm, pale or purplish. Spikelets solitary or in clusters of 2 to 3, surrounded by numerous bristles, the inner bristles densely ciliate, thickened and united at the base. *Africa eastwards to India.* H5–G1. Summer–autumn.

82. SPODIOPOGON Trinius

Perennial or annual, stems slender to stout; ligules membranous. Leaf-blades flat, linear to narrowly lanceolate. Inflorescence a loose or contracted panicle of spike-like racemes. Spikelets in pairs, 1 stalkless, the other stalked, in few-jointed, fragile racemes, all alike, falling entire at maturity, with 2 florets, the lower floret male and the upper bisexual; glumes firm, hairy; lemmas transparent, enclosed by the glumes, the upper awned.

A genus of 8 species in temperate and tropical Asia. Propagated from seed and increased by division. May be grown in ordinary garden soil, in a border, for its decorative, hairy inflorescence.

1. S. sibiricus Trinius. Illustration: Darke, Ornamental grasses, 287 (1999).

Perennial, with long, creeping, scaly rhizomes, and erect stems, 60–150 cm; ligules to 1.5 mm. Leaf-blades linear-lanceolate, finely pointed, 15–37.5 cm × 8–18 mm, green or purplish, mostly hairless. Panicles narrowly lanceolate, open or contracted, 10–20 cm; racemes to 1.8 cm, Spikelets 5 mm, narrowly ovate; glumes densely hairy with white hairs; awns 8–12 mm. *E Asia.* H3. Summer.

83. SACCHARUM Linnaeus
Perennials, often tall. Inflorescence a large, often plumose panicle with persistent branches bearing fragile racemes of paired, similar spikelets. Spikelets with 2 florets, lanceolate, flattened from top to bottom, enveloped in long hairs from the callus; glumes as long as spikelet; lower floret reduced to a transparent lemma; upper floret bisexual, the lemma transparent.

A genus of about 30 species indigenous to tropical and warm temperate regions, several of which are worthy of cultivation as border plants on account of the decorative value of their colourful panicles. They require sunny positions in borders or beds on well-drained soils, where they form dense stately tussocks.

1. S. ravennae (Linnaeus) Murray (*Erianthus ravennae* (Linnaeus) Palisot de Beauvois). Illustration: Coste, Flore de la France **3**: 559 (1906); Darke, Ornamental grasses, 274 (1999).

Stems usually robust, 1–2 m or more. Leaf-blades linear, to 1 m. Panicle 25–60 cm, narrowly ovate, plumose, lobed; racemes 1–2 cm, much shorter than the flexuous supporting branches. Spikelets 3.5–5 mm, the callus hairs about as long as the spikelet; lower glume membranous, hairy on the back or hairless, attenuate to an acute apex; upper lemma 2.5–3 mm, lanceolate, transparent, with a terminal awn 3–4 mm. Mediterranean region to *C Asia, NW India & Somalia.* H5–G1. Summer–autumn.

84. MISCANTHUS Andersson
Tufted or rhizomatous, usually with tall, stout, reed-like stems, short membranous ligules, and long, narrow, flat or folded leaf-blades. Inflorescence an oblong-elliptic or fan-shaped panicle of slender, hairy, spike-like racemes, the latter with a continuous axis. Spikelets surrounded with long hairs from the base, falling entire at maturity, paired, those of each pair unequally stalked, with 2 florets but with only the upper floret bisexual; glumes firm, awnless, enclosing the lemmas, the latter delicate, transparent, and the upper usually awned.

A genus of about 12 species in C, E and SE Asia and Polynesia, They are hardy and suitable for most types of soil, propagation being by division of the rootstock.

1a. Spikelets awnless **1. sacchariflorus**
b. Spikelets awned 2
2a. Leaves 45–75 cm; racemes 10–30 cm **2. sinensis**
b. Leaves 8–60 cm; racemes 7–15 cm **3. oligostachys**

1. M. sacchariflorus (Maximowicz) Hackel. Illustration: Darke, Ornamental grasses, 226 (1999).

Plant with long, creeping, stout, scaly rhizomes; stems erect, leafy, stout, 1.5–3 m. Leaf-blades linear, very finely pointed, 45–90 × 1–3 cm, flat, spreading, rigid, smooth except for the very rough margins. Panicles ovate or oblong, open and loose, or contracted, 20–40 × 7.5–13 cm, white, reddish or purplish, densely silky hairy; racemes numerous, 7.5–30 cm, very slender. Spikelets narrowly lanceolate, 4–5 mm, awnless, surrounded by soft, white or purplish hairs which are 6–12 mm. *E Asia.* H5–G1. Autumn.

2. M. sinensis Andersson. Illustration: Grounds, Ornamental grasses, t. 12 (1979); Darke, Ornamental grasses, 227—240 (1999); Brickell (ed.) RHS A–Z encyclopedia of garden plants, 695 (2003).

Plant tufted or clump-forming, leafy, with stout rhizomes, and erect, stout stems, 1–3 m. Leaf-blades linear, finely pointed, erect or spreading, 45–75 cm × 8–20 mm, hairy above or hairless, with very rough margins. Panicles dense to somewhat loose, fan-shaped, 25–40 cm long, to 15 cm wide, whitish, brownish or purplish, with a short main axis; racemes numerous, 10–30 cm, slender, erect, or slightly spreading. Spikelets lanceolate, 4–6 mm, awn 6–10 mm, surrounded by fine, white or purplish hairs up to 8 mm long. *China, Japan, Korea.* H5. Autumn.

Very variable. Numerous cultivars exist including: 'Ferner Osten' with pronounced red colouring in young inflorescences, illustration: Darke, Ornamental grasses, 228 (1999); the variegated, 'Puenktchen' illustration: Darke, Ornamental grasses, 234 (1999) and 'Strictus', illustration: Darke, Ornamental grasses, 236 (1999) and 'Silberfeder' (Silver Feather), illustration: Darke, Ornamental grasses, 235 (1999), with pale silvery gold inflorescences.

3. M. oligostachys Stapf. Illustration: Darke, Ornamental grasses, 224, 225 (1999).

Perennial, stems erect 80—180 cm. Leaves 8–60 × 0.6–2.5 cm, margins rough; ligules 1–3 mm, membranous with fine hairs. Racemes 2–12, 7–15 cm, spikelets paired, 6–8 mm; lemma 4–7 mm; awn 4–15 mm, column twisted. *Asia.* H4. Summer.

85. IMPERATA Cirillo
Rhizomatous perennials. Inflorescence a narrow, often spike-like panicle with tough branches bearing paired, similar spikelets, each borne upon a slender stalk. Spikelets with 2 florets, terete; glumes as long as spikelet, membranous; lower floret reduced to a transparent lemma; upper floret bisexual with unawned, transparent lemma and shorter palea; stamens 1 or 2.

A genus of 8 species in warm temperate and tropical regions. Propagated by seed or division of the rootstock.

1. I. cylindrica (Linnaeus) Rüuschel. Illustration: Bonnier, Flore complète **11**: t. 655 (1931); Darke, Ornamental grasses, 209 (1999); Brickell (ed.) RHS A–Z encyclopedia of garden plants, 566 (2003).

Stems slender to moderately stout, to 1.2 m. Leaf-sheaths usually hairless; ligule to 2 mm, blades narrowly linear, sharply pointed, tightly rolled or rarely flat, to 60 cm × 3–6 mm, rigid, glaucous. Panicles very dense, spike-like, cylindric, to 20 cm × 25 mm. Spikelets 4–5 mm, surrounded by white hairs which are 1.2–1.6 cm. *Mediterranean region.* H5–G1. Autumn.

86. SORGHASTRUM Nash
Annual or perennial; stems tufted, slender to stout, erect; ligules scarious. Leaf-blades narrow, flat or rolled. Inflorescence a loose or dense, narrow panicle, its branches terminating in few-jointed racemes. Spikelets solitary and stalkless at each raceme node, accompanied by the hairy stalk of a suppressed spikelet, falling entire at maturity, with 2 florets, the upper floret alone fertile; glumes leathery, the lower rounded or flattened on the back, the upper boat-shaped; lemmas transparent,

enclosed by the glumes, the upper with a bent and twisted awn.

A genus of about 20 species in tropical Africa, warm temperate and tropical N & S America. They are propagated by seed sown under glass, or planted out in the open in the spring. Some of the tropical species can be grown, but they require warm greenhouse treatment. All have beautiful, brownish, yellowish or reddish inflorescences, and may be cut and used for interior decoration, either fresh or dry.

1. S. nutans (Linnaeus) Nash. Illustration: Gould, The grasses of Texas, f. 303 (1975); Darke, Ornamental grasses, 284 (1999).

Stems 60–120 cm. Leaf-blades finely pointed, to 60 cm long, mostly 5–10 mm wide, rough. Panicle narrow, yellowish or brownish, contracted, rather dense, 7.5–35 cm; raceme-axis and spikelet-stalks greyish-hairy. Spikelets 6–8 mm, lanceolate, hairy; awn bent, 1–1.5 cm. *E & C USA*, H5. Summer–autumn.

87. CHRYSOPOGON Trinius

Perennial, mostly tufted. Leaf-blades linear, often harsh and glaucous. Inflorescence a terminal panicle with whorls of slender, persistent branches bearing terminal racemes, these reduced to a triad of 1 stalkless and 2 stalked spikelets. Stalkless spikelet compressed from side to side. Lower glume rounded on back. Lower floret reduced to a transparent lemma; upper lemma 2-toothed or entire, with a sharply bent awn.

A genus of 24 species from tropical and warm temperate regions of the world. They should be sown in late spring, when clear of frosts, or under glass.

1. C. fulvus (Sprengel) Chiovenda

Perennial, 30–150 cm. Panicle ovate, 3–15 cm, with fine, delicate branches which are brown-bearded at the tip. Stalkless spikelet narrowly oblong with a shortly pointed callus 1–1.5 mm; upper glume awned; upper lemma with an awn 2–3 cm. Stalked spikelet purplish or pallid, the lower glume awned. *India*. H5. Summer.

88. BOTHRIOCHLOA Kuntze

Annuals or perennials. Inflorescence of 1–many fragile racemes bearing paired, dissimilar spikelets, 1 stalkless, the other stalked; internodes of raceme–axis and spikelet–stalks with a translucent median line. Stalkless spikelet with 2 florets, flattened from top to bottom; lower floret reduced to a transparent lemma; upper floret bisexual with entire lemma passing into a hairless, sharply bent, awn. Stalked spikelet male or sterile, unawned.

A genus of 25–30 species in the tropical and warm temperate regions of both hemispheres. They should be sown in a warm place in spring.

1a. Inflorescence of racemes arising close together near the apex of the stem **1. ischaemum**
b. Inflorescence with a long central axis **2. saccharoides**

1. B. ischaemum (Linnaeus) Keng (*Andropogon ischaemum* Linnaeus). Illustration: Hegi, Illustrierte Flora von Mittel-Europa, edn 2, **1**: t. 22 (1957); Darke, Ornamental grasses, 144 (1999).

Tufted perennial; stems 15–100 cm, ascending. Inflorescence of 3–15 racemes arising close together near the apex of the stem, each 3–7 cm. Spikelets 3.5–4.6 mm, narrowly elliptic, papery, acute; awn 10–15 mm. *Europe northwards to N France & Poland* H2. Summer–autumn.

2. B. saccharoides (Swartz) Rydberg (*Andropogon saccharoides* Swartz; *A. argenteus* de Candolle; *A. laguroides* de Candolle).

Tufted perennial; stems 60–120 cm. Leaves often glaucous, hairless, 3–5 mm wide. Panicle oblong, dense, silvery white, silky hairy, to 15 cm; racemes numerous, erect, 1.8–5 cm. Spikelets 4 mm; awn 10–15 mm. *N America to Brazil.* H5(?). Summer–autumn.

89. ANDROPOGON Linnaeus

Annuals or perennials. Inflorescence of paired or palmately arranged, fragile racemes bearing paired, dissimilar spikelets, 1 stalked, the other stalkless. Stalkless spikelet with 2 florets, the callus inserted in the cup-like apex of the internode; lower glume with 2 keels; lower floret reduced to a transparent lemma; upper floret bisexual with a 2-toothed lemma passing into a hairless, sharply bent awn. Stalked spikelet male or sterile, awnless or with a very fine hair at the apex.

About 120 species, natives of warm temperate and tropical regions, few of which are of horticultural value, on account of their frequently coarse foliage. They can be grown in a warm border in deep fertile soil.

1. A. gerardii Vitman. Illustration: Gould, The grasses of Texas, f. 304 (1975); Darke, Ornamental grasses, 136,137 (1999).

A robust, glaucous perennial, often in large tufts, sometimes with short stolons; stems 1–2 m, usually sparingly branched towards the apex. Lower sheaths and blades sometimes with shaggy hairs, occasionally densely so, the blades flat, mostly 5–10 mm wide, the margins very rough. Racemes on the long-projecting terminal stalk, mostly 3–6, 5–10 cm long, usually purplish, sometimes yellowish. Stalkless spikelet 7–10 mm, the lower glume slightly furrowed, usually rough with short, stiff hairs, the awn sharply bent and tightly twisted below, 1–2 cm. Stalked spikelet not or only slightly reduced, male. *Canada to Mexico.* H5. Summer.

90. CYMBOPOGON Sprengel

Aromatic perennials, with tufted stems. Leaves linear to lanceolate; ligule membranous. Inflorescence usually profusely branched, with spathe-like bracts, each bearing in its axil a pair of slender, spike-like racemes. Spikelets in pairs on the raceme-axis, those of each pair dissimilar, 1 stalkless and bisexual, the other stalked and male, all 2-keeled; stalked spikelets always awnless.

A genus of 45 species in warm regions of the Old World. Many of the species produce aromatic oils used in perfumery. They can be grown in pots in a warm greenhouse and are propagated by division of the rootstock.

Literature: Soenarko, S., The genus Cymbopogon, *Reinwardtia* **9**: 225–375 (1977).

1. C. nardus (Linnaeus) Rendle

Densely tufted, leafy perennial, with erect, stout, smooth stems to 2.2 m. Leaves to 90 cm or more × 2 cm; ligule blunt, 2–4 mm. Inflorescence much-branched, dense, to 60 cm; spathes elliptic, 1.2–2 cm; racemes 1.2–1.5 cm, spreading, the axis ciliate. Spikelets awnless. *Tropical Asia; cultivated throughout the tropics.* G2.

The commercial source of oil of citronella; grown for its scented leaves.

91. HYPARRHENIA Fournier

Annuals or perennials. Inflorescence of pairs of fragile racemes bearing paired, dissimilar spikelets. Stalked spikelet with 2 florets, the callus oblique, lower glume rounded on the back, without keels; lower floret reduced to a transparent lemma;

upper floret bisexual, with a 2-toothed lemma passing into a hairy, sharply bent awn. Stalked spikelets male or sterile, usually longer than the stalkless, unawned or with a fine hair at the apex.

A genus of 60–70 species of tropical and warm temperate countries. They can be raised from seed and put into very large pots in good loamy soil in a tall, warm greenhouse.

Literature: Clayton, W.D., A revision of the genus Hyparrhenia, *Kew Bulletin Additional Series* II (1969).

1. H. cymbaria (Linnaeus) Stapf. Illustration: Die Natürlichen Pflanzenfamilien, edn 2, **14**c: 175 (1940); Van Oudtshoorn, Guide to grasses of southern Africa, 51 (1999).

Perennial, with slender to stout stems 1.8–6 m. Leaf-blades to 45 cm or more, 6–16 mm wide. Inflorescence often very large and dense, much-branched, to 60 cm or more; bracts subtending the racemes boat-shaped, pointed, horizontally spreading, thin, often red or purple, hairless; racemes 5–8 mm; awns 6–16 mm. *Tropical & S Africa*. G2. Summer–autumn.

92. TRIPSACUM Linnaeus

Perennial, often broad-leaved and robust, usually rhizomatous. Inflorescences terminal and axillary, of palmately arranged racemes, each raceme female below, with fragile, swollen internodes, male above, with tough, narrow internodes. Female spikelets single without a trace of pairing, deeply sunk in the internodes, the callus transversely truncate with central ridge or peg; lower glume hard and somewhat brittle, smooth, slightly winged at tip; lower floret sterile, without palea. Male spikelets paired, both stalkless or 1 raised on a free stalk; glumes papery; both florets male.

A genus of 13 species mainly in C America. They are propagated by seed or by division, and can tolerate quite severe frosts. They flourish in ordinary garden soil and are grown for their curious decorative inflorescences.

1. T. dactyloides (Linnaeus) Linnaeus. Illustration: Gould, The grasses of Texas, f. 330 (1975); Darke, Ornamental grasses, 295 (1999).

Forming large clumps, with short thick rhizomes. Stems erect, 1.2–3 m. Leaves green, hairless; blades finely pointed, to 60 × 1–3.5 cm, rough on the margins. Spikes usually 2–3 at the summit of the stem and solitary on the branches, 1.5–2.8 cm, hairless. Female spikelets 8–12 mm, with an ovate, rounded, polished lower glume. Male spikelets oblong, 6–8 mm. *USA to West Indies & tropical S America*. H5. Autumn.

93. ZEA Linnaeus

Annual. Leaves flat. Male inflorescence a terminal panicle of spike-like racemes; female inflorescence axillary, of numerous spikelets arranged in longitudinal rows on a thickened axis, the whole enclosed in leaf-sheaths. Male spikelets in pairs, 1 almost stalkless, the other stalked; florets 2; glumes equal, membranous; lemma and palea transparent; stamens 3, lodicules 2. Female spikelets with 2 florets, the lower sterile; glumes wider than long, fleshy below, transparent above; lemmas short, transparent; lodicules absent; style long, shortly bifid at apex, hairy throughout its length.

A genus of 4 species in C America. *Z. mays* is a staple cereal of all tropical regions and is also extensively grown for forage and silage and as a source of oil, syrup and alcohol. It can be grown from seed outdoors in countries with a warm summer, but without a reasonable sequence of warm, sunny days the grain will not ripen properly. In cooler countries it should be treated as half-hardy and raised under glass before planting out in May.

1. Z. mays Linnaeus. Illustration: Hegi, Illustrierte Flora von Mittel-Europa, edn 2, **1**: t. 22 (1957); Brickell (ed.) RHS A–Z encyclopedia of garden plants, 294 (2003); Aeschimann et al., Flora alpina **2**: 139.84.1 (2004).

Stems to 4 (rarely to 9) m, 2–6 cm in diameter, solid, erect, rooting at the lower nodes. Leaves 3–12 (rarely to 15) cm wide, wavy; ligule 3–5 mm, truncate. Male inflorescence to 20 × 20 cm, erect; spikelets 6–15 mm. Female inflorescence *c.* 20 cm long, stalk with short internodes; styles 15–25 (rarely to 40) cm long, projecting from the apex of the sheaths at and after flowering; grain 5–10 mm, compressed, usually wedge-shaped. *Unknown from the wild; first cultivated in Mexico*. H5–G1. Summer.

Many cultivars, with different ranges of hardiness, are available.

94. COIX Linnaeus

Annuals. Leaves flat. Inflorescence of 3 spikelets, 1 female and 2 sterile, enclosed in a bony jacket, and a terminal spike of male spikelets. Male spikelets in groups of 2 or 3, flattened from top to bottom; glumes papery; lemma and palea membranous; stamens 3. Female spikelets with almost circular, beaked glumes and membranous lemma and palea. Lodicules absent.

A genus of 3 or 4 species from tropical eastern Asia. *C. lacryma-jobi* may be cultivated as a curiosity in an open sunny border, the seed being sown under glass in heat in February or March, or in the open in May. The fruits can be strung and used for necklaces or curtains.

1. C. lacryma-jobi Linnaeus. Illustration: Hegi, Illustrierte Flora von Mittel-Europa, edn 2, **1**: f. 164 (1957); Brickell (ed.) RHS A–Z encyclopedia of garden plants, 294 (2003).

Annual with stout, erect, much-branched stems, 60–150 cm. Leaves narrowly lanceolate, 10–60 × 2.5–5 cm. Male racemes 1.2–5 cm. Spikelets 8–10 mm, fruit mostly ovoid-spherical, 6–12 mm, bony at maturity, shining, white, bluish white, grey, brown or black. *Widespread in the tropics*. H5–G1. Autumn.

46. PALMAE (Arecaceae)

Compiled at the Royal Botanic Gardens, Kew

Trees or shrubs or prickly woody scramblers. Leaves large and pleated, becoming pinnately or palmately lobed or divided; palmately lobed leaves bear an outgrowth (hastula) at the junction of stalk and blade ptyxis plicate. Inflorescence a fleshy panicle or spike, often the whole subtended by large, hard, spathe-like bracts. Flowers unisexual or bisexual, actinomorphic. Perianth of 6 fleshy, free or united segments. Stamens usually 6, rarely more. Ovary usually of 3 free or united carpels, 1–3-celled; ovules 3. Fruit a berry or drupe, sometimes very large.

A family of 212 genera and about 2,700 species of characteristic habit and appearance (palms), almost entirely tropical and subtropical in orgin. Two genera are native to Europe, 16 to North America. Species of about 30 genera are grown as ornamentals in Europe, particularly in the south.

It is difficult to ascertain which palms are grown in Europe, particularly in the south. The selection of species presented here is based on what might be expected to grow well in Europe, and is compiled with reference to the articles of Chabaud, Carvalho-e-Vasconcellos & do Amaral-Franco (see below) and on the basis of observations made in northern Italy. The account does not deal with palms found only in botanic gardens or in enthusiasts' collections, or with juveniles of many species that enter the pot-plant trade as ephemeral house plants with little chance of survival. The real palm enthusiast is recommended to read articles on palms in the journals *Gentes Herbarum* and *Principes* (journal of the Palm Society) and the entries by H.E. Moore in Hortus Third; several reference books such as McCurrach, Palms of the world, and its supplement by Langlois are of value, but contain many incorrectly named photographs.

Literature: Chabaud, B., Les palmiers de la Côte d'Azur (1915); Carvalho-e-Vasconcellos, J. de & do Amaral-Franco, E.J., Las palmeiras de Lisboa & Arredores, *Portugaliae Acta Biologica* (Serie B) **2**(4): 289–425 (1948); McCurrach, J.C., Palms of the world (1960); Moore, H.E., The major groups of palms and their distribution, *Gentes Herbarum* **11**(2): 27–140 (1973); Langlois, A.C., Supplement to palms of the world (1976); Uhl, N.W. & Dransfield, J., Genera Palmerum (1987); Ellison, D. & Ellison, A., Cultivated palms of the world (2001).

1a. Leaves bipinnate with fishtail segments **17. Caryota**
b. Leaves pinnate or palmate 2
2a. Leaves palmate or palmately ribbed or divided 3
b. Leaves pinnate or pinnately divided 16
3a. Leaf without a hastula **12. Nannorrhops**
b. Leaf with a hastula 4
4a. Leaf undivided except for very shallow marginal lobbing **7. Licuala**
b. Leaf divided into segments 5
5a. Leaf divided into segments along the upper ribs or between the ribs 6
b. Leaf divided into segments along the lower ribs 7
6a. Leaf divided to base along the upper ribs **7. Licuala**
b. Leaf divided not quite to base between ribs **5. Rhapis**
7a. Stem bearing spine-like roots; blade with a major split along the mid-line **4. Cryosophila**
b. Stem without spine-like roots; blade without such a major split 8
8a. Leaf-sheath disintegrating to form erect, needle-like spines **1. Rhapidophyllum**
b. Leaf-sheath without such spines 9
9a. Stem covered with a skirt of dead leaves; inflorescence bracts woody **11. Washingtonia**
b. Stem with or without persistent leaf-bases but not with a skirt; bracts thin or leathery 10
10a. Carpels free 11
b. Carpels joined together at tips or base or throughout 12
11a. Dwarf clustering palm; leaf-stalk with conspicuous spines **3. Chamaerops**
b. Usually erect solitary palms; leaf-stalk with inconspicuous spines **2. Trachycarpus**
12a. Carpels joined throughout **13. Sabal**
b. Carpels joined only in stylar region 13
13a. Seed with ruminate endosperm **10. Copernicia**
b. Seed with smooth endosperm 14
14a. Apparently stemless, clustering palm **8. Serenoa**
b. Erect, solitary (rarely clustering) palms 15
15a. Sepals free, overlapping **9. Brahea**
b. Sepals joined **6. Livistona**
16a. Leaflets V-shaped in cross-section at the base (induplicate) 17
b. Leaflets A-shaped in cross-section at the base (reduplicate) 18
17a. Leaflets acute, the lowermost modified as spines **14. Phoenix**
b. Leaflets with irregularly toothed apices **16. Arenga**
18a. Slender, climbing palm with barbed whips **15. Calamus**
b. Barbed whips absent 19
19a. Palms with crownshafts 20
b. Palms without crownshafts 25
20a. Leaflets with irregularly toothed apices **25. Ptychosperma**
b. Leaflets acute or bifid but not irregularly toothed 21
21a. Clustering palms 22
b. Solitary palms 23
22a. Crownshaft green or grey **19. Chrysalidocarpus**
b. Crownshaft orange to red **23. Cyrtostachys**
23a. Flowers in triads only near base of inflorescence (hence fruit only near the base); female flowers otherwise absent **20. Archontophoenix**
b. Flowers in triads (hence fruit) more or less throughout inflorescence 24
24a. Stamens 6 **22. Rhopalostylis**
b. Stamens 9–12 **21. Hedyscepe**
25a. Leaflets dark green on upper surface, chalky-white beneath **30. Microcoelum**
b. Leaflets more or less the same colour above and beneath 26
26a. Leaflets regularly arranged in one plane 27
b. Leaflets grouped and held in several planes 31
27a. Palm with very slender, reed-like stem less than 5 cm in diameter **18. Chamaedorea**
b. Trunk at least 5 cm in diameter 28
28a. Trunk not exceeding 10 cm in diameter; inflorescence spicate **24. Howea**
b. Trunk at least 15 cm in diameter; inflorescence branched 29
29a. Trunk very massive, at least 90 cm in diameter **28. Jubaea**
b. Trunk not exceeding 50 cm in diameter 30
30a. Leaf-stalks with marginal teeth **27. Butia**
b. Leaf-stalks not toothed **26. Cocos**
31a. Stem more or less absent; leaflets often splitting at the tips **31. Allagoptera**
b. Stem well developed, forming a trunk; leaflets not splitting at the tips 32
32a. Trunk smooth; leaf-stalk not toothed **29. Arecastrum**
b. Trunk covered in persistent leaf-bases; leaf-stalk toothed **32. Elaeis**

1. RHAPIDOPHYLLUM Wendland & Drude

Clustering, more or less stemless palm producing large clumps; plants mostly unisexual, or with a few flowers of the opposite sex. Leaf-sheaths disintegrating to form long, sharp, erect, needle-like spines. Leaf palmate, to *c.* 1 m, the stalk without teeth or spines, the blade divided into 5–12 or more spreading, stiff, 3-ribbed segments to 4 cm wide. Inflorescences short, buried in the leaf-sheaths and needles at the base of the leaf-stalks. Male and female flowers similar, wine-red, with 3 sepals, 3 petals, 6 stamens (fertile or sterile) and 3 separate carpels with short styles, or pistillodes.

Fruit ovoid, *c.* 2 × 1.5 cm, covered in woolly hairs. Seed with smooth endosperm.

One species native to the southern USA. Apparently the hardiest N American palm. It is slow growing and not widely cultivated. Propagation by seed.

Literature: Shuey, A.G. & Wunderlin, R.P., The needle palm: Rhapidophyllum hystrix, *Principes* **21**: 47–59 (1977).

1. R. hystrix (Pursh) Wendland & Drude (*Chamaerops hystrix* Pursh). Illustration: *Principes* **21**: 48, 51–3 (1977); Brickell (ed.), RHS A–Z encyclopedia of garden plants, 885 (2003).

Southern USA. H4–5.

2. TRACHYCARPUS Wendland

Small to moderate, solitary or clustering, dioecious palms, rarely monoecious or with unisexual and bisexual flowers on the same plant; trunks often covered with persistent, fibrous leaf-sheaths. Leaves palmate, divided into 1-ribbed segments. Inflorescences among the leaves, with several thin bracts on the inflorescence-stalk, branching to 3 orders. Flowers in clusters of 2–4, with 3 basally overlapping sepals and 3 overlapping petals. Male flowers with 6 stamens and 3 small pistillodes. Female flowers with 6 staminodes and 3 carpels. Bisexual flowers with 6 stamens and 3 carpels. Fruit spherical to kidney-shaped or oblong-ovoid with remains of stigma at apex. Seed with smooth endosperm and a lateral embryo.

About 6 species native to subtropical Asia, widely cultivated, and 1 species (*T. fortunei*) amongst the hardiest of palms. However, taxonomically the genus is still very poorly known; the differences between some of the species may be very slight, and little is known of their variation outside cultivation.

Literature: Kimnach, M., The species of Trachycarpus, *Principes* **21**: 155 (1977).

1a. Leaf-blade divided by splits of the same length; leaf-bases soon falling; fruit oblong-ovoid; seed with a longitudinal groove **1. martianus**

b. Leaf-blade divided by splits of varying lengths; leaf-bases persistent; fruit spherical to kidney-shaped; seed without a longitudinal groove 2

2a. Fibres on trunk loose and ruffled; leaf-base appendages ribbon-like, recurved, leaf-blades divided to more than half their length **2. fortunei**

b. Fibres on trunk closely adpressed; leaf-base appendages triangular, erect; leaf-blades divided to about half their length **3. takil**

1. T. martianus (Wallich) Wendland (*Chamaerops martiana* Wallich; *T. khasyanus* (Griffith) Wendland). Illustration: *Botanical Magazine*, 7128 (1890); *Principes* **21**: 156 (1977).

Trunk to 16 m, *c.* 10 cm in diameter, bare, ringed with leaf-scars. Leaves circular, very regularly divided to the middle into many segments, dark green above, usually glaucous beneath. Fruit glossy blue, oblong-ovoid, *c.* 12 × 10 mm. Seed with a longitudinal groove. *C & E Himalaya, Burma, N Thailand, probably also China*. H5–G1.

2. T. fortunei (W.J. Hooker) Wendland (*Chamaerops fortunei* W.J. Hooker; *C. excelsa* invalid; *T. excelsa* invalid). Illustration: Bean, Trees and shrubs hardy in the British Isles **4**: t. 86 (1980); McCurrach, Palms of the world, 250 (1960); Brickell (ed.), RHS A–Z encyclopedia of garden plants, 1044 (2003); Aeschimann et al., Flora alpina **2**: 133.1.1 (2004).

Trunk solitary, to 20 m, *c.* 10 cm in diameter, conspicuously covered with black, hair-like fibres from the old leaf-sheaths. Leaves circular, to 1.25 m across, dull green, sometimes paler beneath, variously divided to the middle or almost to the base into rather stiff, horizontal or drooping segments. Leaf-stalk to 1 m, finely toothed along its margins. Inflorescences not exceeding the leaves, with masses of golden yellow flowers. Fruit spherical to kidney-shaped, bluish, *c.* 1.2 cm across. Seed depressed in the centre. *Origin unknown, probably N Burma or C&E China; widely cultivated and becoming naturalised in China and Japan*. H4–5.

T. wagnerianus Roster. Illustration: *Principes* **21**: 160 (1977). Said to differ from *T. fortunei* by having smaller, more rigid leaves. It is known only in cultivation and probably represents only a form of *T. fortunei*.

T. fortunei var. **surculosa** Henry. This name was given to a clustering plant that still exists at Leonardslee, Sussex, England. It now consists of a few old stems and appears to have lost its suckering potential. In leaf form, this variety is similar to *T. wagnerianus*.

T. caespitosus Roster. Illustration: *Principes* **21**: 155 (1977). Named from plants cultivated in California. This may be the same as *T. fortunei* var. *surculosa*. Clustering *Trachycarpus* has been observed at Abbotsbury, Dorset, England, but it is not clear whether this is true clustering, or growth of seedlings at the base of the trunk.

3. T. takil Beccari. Illustration: *Principes* **21**: 159 (1977).

Very similar to *T. fortunei* but differing in the tendency of young plants to grow obliquely, the less divided leaf-blades, and the trunk fibres being closely adpressed to the trunk. *W Himalaya*. H5.

Named from plants cultivated in Italy from seed sent from Kumaon (India); the original plant still exists but is now very tall.

T. nanus Beccari (probable synonym *T. dracocephalus* Ching & Hsu), from Yunnan, China, is a very short-stemmed or apparently stemless *Trachycarpus*; it is not known in cultivation, but would obviously be well worth introducing into gardens.

3. CHAMAEROPS Linnaeus

Clustering, shrubby palms, mostly dioecious or rarely with a few flowers of the opposite sex in each inflorescence. Stems suckering and forming low clumps but occasionally producing an erect trunk to 5 m, covered in more or less persistent, fibrous leaf-sheaths. Leaves palmate; stalks with short spines pointing towards the apex; blade held stiffly and deeply divided into 1-ribbed segments. Inflorescences among the leaves, with a single large bract at the base and smaller bracts subtending the few branches. Flowers borne singly, but crowded together, golden yellow. Calyx a low, 3-toothed cup. Petals 3, overlapping. Stamens or staminodes 6. Carpels 3, free or reduced to pistillodes. Fruit rounded, reddish brown, fleshy, with a rancid smell. Seed with a slightly ruminate endosperm.

One species from S Europe and N Africa. One of the hardiest palms, widely cultivated throughout southern Europe. Propagation by seed or offsets.

1. C. humilis Linnaeus. Illustration: McCurrach, Palms of the world, 44 (1960); Brickell (ed.), RHS A–Z encyclopedia of garden plants, 259 (2003).

S Europe, N Africa. H4–5.

4. CRYOSOPHILA Blume

Small to medium, solitary, bisexual palms with trunks covered, at least at the base, with root-spines. Leaves palmate, divided more or less to the base along the mid-line and then into 1 to several, ribbed, acute segments; stalk with smooth margins. Inflorescence among the leaves, arched, with densely hairy, loosely sheathing bracts on the stalk and many crowded branches. Flowers solitary, creamy white to purplish. Sepals 3, shortly joined at the base. Petals 3, overlapping. Stamens 6, with filaments joined in their basal halves. Carpels 3, free, with narrow styles. Fruit yellowish or white. Seed spherical with smooth endosperm.

About 8 species in C & northern S America, some of considerable horticultural potential. One species is widely cultivated under glass. Propagation by seed.

1. C. warscewiczii (Wendland) Bartlett (*Acanthorrhiza warscewiczii* Wendland). Illustration; McCurrach, Palms of the world, 71, 72 (1960).

Trunk to 6 m or more, grey, with long root-spines near the base, becoming smooth near the crown. Leaves green on the upper surface, silvery beneath, divided into 50–60 segments that are up to 90 × 3.5 cm. Inflorescence to 60 cm or more. Fruit spherical to pear-shaped, white, to *c.* 2.5 cm. *Costa Rica to Panama.* G2.

5. RHAPIS Linnaeus

Clustering, slender, dioecious palms without teeth or spines, with reed-like or bamboo-like stems. Leaves palmate, divided nearly to the base into 2 to several, ribbed segments, shallowly or sharply toothed at the apex; sheaths fibrous, persistent; stalks well developed. Inflorescences among leaves; stalk with several empty bracts, then branching, the branches several, bearing yellow flowers. Male flowers stalkless with 3-toothed calyx, and 3-lobed corolla. Stamens 6. Female flowers similar but with calyx stalked; stamens absent. Carpels 3, separate. Fruit a small, 1-seeded berry. Seed 1, spherical, with smooth endosperm.

A genus of about 9 species from S China to Thailand, mostly very poorly known; 2 species have long been in cultivation. Several plants of as yet unknown identity have recently been introduced into cultivation in Thailand, whence they have spread to specialist palm collections. Seed of at least one has recently been offered for sale on a large scale. Propagation by seed, or by offsets removed with care. Established plants indoors are generally tolerant of neglect.

Literature: Bailey, L.H., Species of Rhapis in cultivation, *Gentes Herbarum* **4**: 199–208 (1939).

1a. Sheaths coarsely fibrous; at least some leaves with only 3–7 segments, others with up to 10, uniformly curving or drooping, 4–8 cm wide **1. excelsa**
b. Sheaths finely fibrous; leaves mostly with 9 or more segments, held rather stiffly, less than 3 cm wide **2. humilis**

1. R. excelsa (Thunberg) Henry (*R. flabelliformis* Aiton; *Chamaerops excelsa* Thunberg; *Trachycarpus excelsus* (Thunberg) Wendland). Illustration: *Gentes Herbarum* **4**: 200, 201, 204 (1939); Brickell (ed.), RHS A–Z encyclopedia of garden plants, 886 (2003).

Stems numerous, to 3 m or more tall, covered in coarse fibres. Some leaves with 3–7 segments, other with up to 10 segments, uniformly curving or drooping, broad at apex, 4–8 cm wide. *?S China.* G1.

2. R. humilis Blume.

Stems numerous, to 3 m, usually more slender than in *R. excelsa*, covered in very fine fibres. Leaves with 9 or more segments usually held rather stiffly, less than 3 cm wide, very narrow at the apex. *S China.* G1.

6. LIVISTONA R. Brown

Solitary, mostly very tall, bisexual palms. Leaves shortly but conspicuously palmate, the stalks usually with marginal spines, the blades deeply divided sometimes almost (but not quite) to the base, with 1-ribbed, bifid segments. Inflorescences among the leaves, with tubular, sheathing bracts subtending branches, branching to up to 4 orders. Flowers solitary or in clusters with the sepals joined into a low, 3-lobed cup, petals 3, free and edge-to-edge above, shortly joined at the base; stamens 6, joined into a ring; carpels 3, free except in the stylar region. Fruit spherical to ellipsoid. Seed with smooth endosperm and a lateral embryo.

About 29 species native to the tropics and subtropics of Asia and Australasia. Widely cultivated in the tropics, they also make very decorative pot-plants in the juvenile state. Propagation by seed.

Literature: Beccari, O., Asiatic Palms – Corypheae, *Annals of the Royal Botanic Gardens Calcutta* 13 (1935).

1a. Leaf-segments stiff, only slightly hanging; ripe fruit orange-red **1. rotundifolia**
b. Upper quarter or third of each leaf-segment conspicuously hanging; ripe fruit bluish green to black 2
2a. Trunk smooth, grey; leaf-segments very neatly and regularly hanging **2. chinensis**
b. Trunk brown, ridged with leaf-scars; leaf-segments more untidily hanging 3
3a. Leaf-stalk with strong spines throughout its length **3. australis**
b. Leaf-stalk with spines in the basal half only **4. decipiens**

1. L. rotundifolia (Lamarck) Martius. Illustration: McCurrach, Palms of the world, 121 (1960).

Trunk to 30 m, prominently ringed. Leaves more or less circular with large, entire central portion and many short, stiff, shortly 2-lobed segments. Inflorescence dividing at the very base into 3 equal branches, these branching to 3 orders. Flowers borne singly. Fruit orange-red, turning black on rotting. *Philippines, Sabah (offshore islands only), Celebes & Moluccas.* G2.

This species makes a beautiful pot-plant.

2. L. chinensis (Jacquin) Martius. Illustration: McCurrach, Palms of the world, 120 (1960); Brickell (ed.), RHS A–Z encyclopedia of garden plants, 643 (2003).

Trunk to 10 m, becoming smooth and grey. Leaf-stalk toothed near base; blade with large, entire central portion and numerous, deeply cleft, very neatly and regularly hanging segments. Inflorescence with main axis bearing several branches, each branching to 2–3 orders. Fruits bluish green, ellipsoid to almost spherical. *Japan (Ryukyu Islands).* G1. ?H5.

3. L. australis (R. Brown) Martius. Illustration: McCurrach, Palms of the world, 118 (1960); Brickell (ed.), RHS A–Z encyclopedia of garden plants, 643 (2003).

Trunk to 20 m, dull brown, conspicuously ridged. Leaf more or less circular in outline, to 1.75 m wide, divided to the middle into *c.* 70 soft, deeply 2-cleft segments with hanging tips; stalk spiny throughout its length. Inflorescence with hairy bracts and hairless, yellow branches, bearing flowers in 2s and 3s. Flowers hairless. Fruit purplish black, to 2.2 × 2 cm. *E Australia*. H5.

4. L. decipiens Beccari. Illustration: McCurrach, Palms of the world, 120 (1960).

Trunk smaller than L. *australis*. Leaves to *c.* 1.5 m, very deeply divided into *c.* 80 very narrow segments with hanging tips, stalk spiny near the base only. Inflorescence with more or less hairless bracts, and hairless branchlets bearing flowers in 2s and 3s. Fruit blackish, *c.* 1.3 cm in diameter. *Originally described from the French Riviera, but thought to have come from Australia.*

7. LICUALA Thunberg

Usually small, single-stemmed or clustered, bisexual palms. Leaves palmate or costapalmate; stalk usually with marginal spines and the blade sometimes undivided, or more often deeply divided into 1 to many, broad, wedge-shaped, ribbed segments with shallow apical lobing. Inflorescence among leaves, the stalk with tight, tubular bracts and ending in a spike; or with bracts subtending branches, themselves spicate or branched to 2 orders. Flowers solitary or in clusters of a few. Calyx cup-shaped with 3 apical lobes. Petals 3, joined in the basal half. Stamens 6, inserted at the mouth of the corolla tube, free, or joined into a ring. Carpels 3, free except for the joined styles, each with 1 ovule, usually only 1 developing. Fruit usually orange or red, with remains of stigma at apex. Seed with smooth endosperm and a lateral embryo.

About 108 species, from SE Asia extending through Malaysia to Australia and the New Hebrides. Many species are extremely decorative but they are often slow growing and most appear to require relatively high temperatures.

1a. Leaf-blade undivided **1. grandis**
b. Leaf-blade divided to the base into many segments **2. spinosa**

1. L. grandis Wendland. Illustration: *Botanical Magazine*, 6704 (1883); Brickell (ed.), RHS A–Z encyclopedia of garden plants, 627 (2003).

Stem solitary to *c.* 2 m. Leaves more or less circular, to *c.* 90 cm wide, undivided except for very shallow marginal lobing. Inflorescence to *c.* 1.5 m, branching to 2 orders. Flowers arranged singly. Staminal ring 3-lobed with 3 anthers at the tips of the lobes, 3 between. Fruit orange, *c.* 1.5 cm in diameter. *New Hebrides*. G2.

2. L. spinosa Thunberg. Illustration: McCurrach, Palms of the world, 117 (1960).

Stems clustered, to *c.* 5 m. Leaves more or less circular in outline, to *c.* 1.5 m across, divided into 18–19, several-ribbed segments. Inflorescence to 2 m, with about 10 primary branches, each of these divided in 3–10 hairy, flower-bearing branches. Flowers arranged in clusters or singly, finely hairy. Staminal ring with 6 equal teeth. Fruit *c.* 1 cm in diameter, bright orange-red. *SE Asia*. G2.

8. SERENOA J.D. Hooker

Small, colonial, apparently stemless or very shortly stemmed, bisexual palms, without teeth or spines except for the teeth on the leaf-stalk margins. Leaves palmate, to 75 cm across, divided for more than half their length into *c.* 20, stiff, 1-ribbed, 2-lobed segments, bluish green in colour. Inflorescences to *c.* 60 cm, with tubular bracts on the stalk subtending primary branches, these branching 1–2 times. Flowers creamy white, fragrant, arranged in pairs or singly. Sepals joined into a 3-lobed cup. Petals 3, joined in the basal half, edge-to-edge above. Stamens 6. Carpels 3, free except for the joined styles. Fruit black, 1-seeded, ellipsoid to almost spherical. Seed with smooth endosperm and a lateral embryo.

A genus of 1 species, much cultivated in the USA. Its status in Europe is not known, but it was once cultivated on the French Riviera. Propagation by seed.

1. S. repens (Bartram) Small (*S. serrulata* (Michaux) Nicholson). Illustration: McCurrach, Palms of the world, 234 (1960).

Southeast USA. H5–G1.

9. BRAHEA Martius

Dwarf to medium, solitary or rarely clustered, bisexual palms. Leaves shortly costapalmate; stalks smooth or with short spines; blade divided into 1-ribbed, bifid segments. Inflorescence elongate, among leaves, usually exceeding them in length, with tubular bracts subtending primary branches, to 2–3 orders. Flowers solitary or in clusters of 3. Sepals 3, free. Petals 3, free or shortly united. Stamens 6. Carpels 3, free except in the stylar region. Fruit ellipsoid to almost spherical with remains of stigma near apex and basal abortive carpels. Seed with smooth endosperm.

About 12 species confined to limestone in Mexico and C America. They are very decorative, and are often called *Erythea* Watson. Propagation by seed; they require full sun and a limey soil.

Literature: Bailey, L.H., Brahea and an Erythea, *Gentes Herbarum* **6**: 177–97 (1943).

1a. Leaves long-persistent, pale bluish green; inflorescences to 5 m, much longer than the leaves **1. armata**
b. Leaves not long-persistent, green or subglaucous; inflorescences to 2 m, not or only slightly longer than the leaves 2
2a. Leaf-blade paler beneath; inflorescence slightly longer than leaves **2. dulcis**
b. Leaf-blade green on both surfaces; inflorescence shorter than leaves **3. edulis**

1. B. armata Watson (*Erythea armata* (Watson) Watson). Illustration: *Gentes Herbarum* **4**: 106, 107 (1937); Brickell (ed.), RHS A–Z encyclopedia of garden plants, 192 (2003).

Trunk solitary, to 14 m, robust, covered in old leaves unless cut or burned. Leaves waxy blue, deeply divided into *c.* 50 segments, the stalk with stout, curved teeth. Inflorescence to 5 m, arching out of the crown, with hanging, creamy branches. Flowers in clusters of 3 on softly hairy branches. Fruit spherical, *c.* 2 cm in diameter, yellow, fleshy. *Mexico (Baja California)*. H5.

Slow growing but extremely decorative.

2. B. dulcis (Humboldt, Bonpland & Kunth) Martius (*B. calcarea* Liebmann). Illustration: *Gentes Herbarum* **4**: 120 (1937).

Trunk to 7 m or more, solitary and erect, prostrate at the base, occasionally suckering. Leaves not long-persistent; stalk with short hooked spines; blade stiff, to 1.7 m long, green above, paler beneath, with *c.* 60 segments. Inflorescence to 2 m, slightly longer than leaves. Flowers solitary, partly immersed in yellowish hairs. Fruit yellow, succulent, *c.* 1.5 cm in diameter. *W & C Mexico, Guatemala*. H5.

3. B. edulis Watson (*Erythea edulis* (Watson) Watson). Illustration: *Gentes Herbarum* **4**: 84, 86 (1937).

Trunk solitary to 10 m and 40 cm in diameter, becoming smooth. Leaves not long-persistent; stalks to 1.25 m × 6 cm, with short spines; blade green on both surfaces, to 2 m wide, soft, with *c.* 80 segments. Inflorescence not exceeding leaves, bearing flowers in clusters of 3 on softly hairy branches. Fruit black, *c.* 2.5 cm in diameter. *Mexico (Guadalupe Islands)*. H5.

Slow growing but very decorative.

10. COPERNICIA Martius

Moderately short to tall, solitary, bisexual palms with trunks frequently clothed with persistent leaf-bases, rarely with persistent leaves. Leaves palmate, with very short to long stalks that usually have marginal teeth; hastula conspicuous; blades of various shapes, shallowly divided into 1-ribbed segments. Inflorescences among, and scarcely to greatly exceeding the leaves, with tubular bracts on the stalk that subtend primary branches, these unbranched or up to 2 times branched. Flowers solitary or in clusters of 2–4, with sepals joined into 3-lobed cup, petals joined in the basal half, 6 stamens borne on the petals, and 3 carpels, which are more or less free except in the stylar region. Fruit ovoid to spherical, with abortive carpels and remains of stigma at apex. Seed 1, with deeply ruminate endosperm and a basal embryo.

About 29 species native to the West Indies and S America, with an astonishing range of species in Cuba. Many species are handsome ornamentals, but are generally slow growing. *C. prunifera* (Miller) H.E. Moore is the source of Carnauba wax. One species is said to be in cultivation in Europe. Propagation by seed.

Literature: Dahlgren, B.E. & Glassman, S.F., A revision of the genus Copernicia, *Gentes Herbarum* **9**: 1–232 (1961–63).

1. C. alba Morong. Illustration: *Gentes Herbarum* **9**: 18–19 (1961).

Stems to 30 m, covered (in young trees) with persistent leaf-bases. Leaf-stalk to 60 cm or more, toothed near the base, blade circular, *c.* 60 cm long, covered in white wax. Inflorescence with paired flowers. Fruit ovoid, *c.* 2 cm long. *Brazil to Argentina*. H5–G1.

11. WASHINGTONIA Wendland

Robust, solitary, bisexual palms, with trunks (in nature) clothed in persistent dead leaves. Leaves costapalmate with 1-ribbed segments; stalks with spines along margins. Inflorescences borne among leaves, exceeding them in length and bearing several empty bracts on the stalk, and woody, open, sword-shaped bracts each subtending a hanging inflorescence branch that bears slender, flower-bearing branches. Flowers cream, with tubular calyx; petals 3, deciduous; stamens 6; carpels 3, free, except at the very tip, where united into a common style. Fruit small, 1-seeded. Seed with smooth endosperm.

Two species native to southwest USA and NW Mexico. Widely cultivated in the drier parts of the subtropics, and areas with a Mediterranean climate. The 2 species are difficult to distinguish in the juvenile state. Propagation by seed.

Literature: Bailey, L.H., Washingtonia, *Gentes Herbarum* **4**: 53–82 (1936).

1a. Leaves grey-green, lacking a tawny patch around the hastula; mature leaves with hanging segments and filaments between them **1. filifera**
b. Leaves brilliant green with a conspicuous tawny patch around the hastula; mature leaves with stiff segments and usually without filaments between them **2. robusta**

1. W. filifera (Linden) Wendland. Illustration: *Gentes Herbarum* **4**: 55 (1936).

Robust palm to 25 m or more, with trunk not tapered, in nature covered in a dense, even skirt of old leaves. Crown open and broad. Leaves grey-green without a tawny patch around the hastula on the underside, with conspicuous hanging filaments and gracefully hanging egments. *Southwest USA (California and Arizona)*. H5.

2. W. robusta Wendland. Illustration: *Gentes Herbarum* **4**: 478 (1936); Brickell (ed.), RHS A–Z encyclopedia of garden plants, 1085 (2003).

Robust palm to 30 m or more, with trunk tapering from a swollen base, in nature covered in an untidy skirt of old leaves. Leaves brilliant green with a conspicuous tawny patch around the hastula on the underside, without conspicuous filaments and tending to have stiff segments. (Juvenile plants sometimes have leaf filaments.) *NW Mexico*. H5.

12. NANNORRHOPS Wendland

Small, bushy, clustering and stoloniferous, unarmed, bisexual palms, without teeth or spines; stems to 3 m or more, branching by equal forking (dichotomy). Leaves glaucous, shortly costapalmate, lacking a hastula; stalk well developed, smooth, segments 8–20 or more, to 1 m long, deeply 2-cleft to the middle, often becoming pendent. Inflorescence compound, ending the growth of the stem, with many lateral inflorescences borne in the axils of reduced leaves or bracts, branching to 3 orders. Flowers borne in clusters on the branches, white with dark purple anthers and pink carpels. Fruit spherical, orange, *c.* 2 cm in diameter, 1-seeded. Endosperm smooth.

One species in the semi-arid areas of Pakistan, Afghanistan, Iran and Arabia, rarely cultivated in S Europe. This relatively hardy palm deserves to be much more widely grown. Propagation by seed.

Literature: Moore, H.E., Flora Iranica **146**: 1–4 (1980).

1. N. ritchiana (Griffith) Aitchison (*Chamaerops ritchiana* Griffith). Illustration: Flora Iranica **146**: t. 1–4 (1980).

Pakistan, Afghanistan, Iran & Arabia. H5.

13. SABAL Adanson

Bisexual palms, apparently stemless or with well-developed trunks, without teeth or spines. Leaves costapalmate, blades either regularly divided for half to three-quarters of their length into 1-ribbed segments, or deeply and irregularly divided into less deeply divided pairs of segments; base of leaf-stalk split centrally, often long-persistent. Inflorescences among leaves, with tight, tubular bracts on the stalk subtending primary branches, these 1–3 times branched. Flowers creamy white, borne singly. Sepals united into a 3-lobed cup. Petals 3, overlapping, joined basally. Stamens 6, borne on the corolla. Carpels 3, joined. Fruit spherical to pear-shaped with remains of stigma at base. Seed more or less compressed, with smooth endosperm and a lateral embryo.

About 14 species in the tropics and subtropics of C & N America, and in northern S America. Several species are widely cultivated in the tropics and subtropics as ornamentals, but apparently only 2 species are cultivated in Europe, though there is no recent information available. Furthermore, *S. blackburniana* Glazebrook, described from cultivated material, may also be cultivated in S Europe but little is known of this species. Propagation by seed; germination rapid, but subsequent growth slow.

Literature: Bailey, L.H., Revision of the American palmettoes, *Gentes Herbarum* **6**: 367–459 (1944).

1a. Dwarf palm with subterranean stem **1. minor**
b. Palms with well-developed trunks **2. palmetto**

1. S. minor (Jacquin) Persoon (*S. adansonii* Guersent). Illustration: McCurrach, Palms of the world, 219 (1960); Brickell (ed.), RHS A–Z encyclopedia of garden plants, 943 (2003).

Stems subterranean or rarely erect, then very short. Leaves to 1 m, tinged with green, or bluish, stiff and nearly flat, shortly costapalmate, regularly divided to two-thirds or more of their length into 16–40 entire or very shortly lobed, 1-ribbed segments. Inflorescence erect, usually exceeding the leaves. Flowers whitish. Fruit glossy black, to 12 mm in diameter. *SE USA*. H5.

The hardiest species of *Sabal*.

2. S. palmetto (Walter) Loddiges. Illustration: McCurrach, Palms of the world, 223 (1960); Brickell (ed.), RHS A–Z encyclopedia of garden plants, 943 (2003).

Trunk to 30 m, rough and bare at maturity. Leaves to 2 m long, strongly costapalmate, regularly divided for about two-thirds of their length into 1-ribbed, deeply lobed segments, with filaments in the sinuses between them. Inflorescence usually exceeding the leaves. Flowers whitish. Fruit black, to 1.2 cm in diameter. *SE USA*. H5.

Besides these 2 species, several others are likely to be cultivated in the Mediterranean region; the interested gardener is referred to illustrated accounts of the genus in McCurrach, Palms of the world, and Stevenson, Palms of south Florida (1974).

14. PHOENIX Linnaeus

Solitary or clustering, dwarf to robust, dioecious palms. Leaves pinnate with acute segments that are V-shaped in section, the lowermost modified as spines. Inflorescence among the leaves, subtended by a single, usually deciduous bract, the stalk compressed and bearing single or clustered branches at the tip. Flowers creamy yellow, borne singly. Male flowers with sepals joined into a 3-lobed cup and 3 petals that are edge-to-edge, stamens usually 6. Female flowers with 3 sepals joined into a cup, and 3 overlapping petals; carpels 3, free. Fruit usually developing from only 1 carpel, cylindric to spherical with dry to fleshy mesocarp. Seed single, longitudinally grooved, with smooth endosperm and a basal or lateral embryo.

About 17 species native to drier parts of the Old World tropics and subtropics. Several species are cultivated as ornamentals. One species, *Phoenix dactylifera*, is the date palm, cultivated extensively as a staple and export crop in the Middle East and elsewhere. Despite the economic and horticultural potential of the genus, several of the species of *Phoenix* are poorly known; in the herbarium they are very difficult to distinguish, despite having distinctive features of habit.

Literature: Beccari, O., Rivista monografica delle specie del genere Phoenix Linn., *Malesia* **3**: 345–416 (1890).

1a. Dwarf or apparently stemless palms rarely exceeding 2m 2
b. Tall trees 3
2a. Leaflets delicate, rather limp, arranged very regularly in 1 plane **5. roebelenii**
b. Leaflets stiff, arranged regularly or in groups, in more than 1 plane **6. loureirii**
3a. Trunk *c.* 80–90 cm in diameter **2. canariensis**
b. Trunk not exceeding 50 cm in diameter 4
4a. Leaves glaucous, bluish green, with white powdery wax **1. dactylifera**
b. Leaves bright green or greyish green 5
5a. Stems clustered **4. reclinata**
b. Stem solitary **3. sylvestris**

1. P. dactylifera Linnaeus. Illustration: McCurrach, Palms of the world, 162–4 (1960).

Trunk to 30 m, rarely exceeding 30 cm in diameter, clustered at base, but suckers frequently removed. Leaf-stalk scars as high as or higher than wide. Leaves bluish green, covered in powdery wax, with leaflets to 45 cm, narrow, stiff, borne in pairs, clusters or more or less regularly arranged, diverging from the axis at an acute angle. Inflorescence elongate; petals of male flowers not acuminate; female flowers with calyx half as long as petals. Fruit cylindric or ellipsoid, 2.5–10 cm, with sweet edible flesh. Seed more or less pointed. *One of the oldest crops, probably originating in W Asia and N Africa*. H5.

P. theophrastii Greuter. Differs in having dry fruit. *Crete*.

This recently named species may be relatively hardy.

2. P. canariensis Chabaud. Illustration: *Principes* **18**: 85 (1974).

Solitary, trunk to 20 m, *c.* 80–90 cm in diameter, in age with leaf-stalk scars broader than high. Leaves very numerous, to 6 m long, sometimes twisted at an angle from the horizontal, with very many crowded, more or less regularly arranged, or paired, dark green leaflets in 1 plane. Inflorescence elongate. Petals of male flowers not acuminate. Female flowers with calyx nearly as long as petals. Fruit to 2 × 1.5 cm, cylindric to ellipsoid, yellow to reddish. *Canary Islands, widely cultivated elsewhere*. H5.

3. P. sylvestris (Linnaeus) Roxburgh. Illustration: McCurrach, Palms of the world, 169 (1960).

Trunk solitary to 16 m, *c.* 30 cm in diameter, at length with leaf-stalk scars higher than wide. Leaves numerous, appearing ragged, to 5 m long, greyish green but lacking white powdery wax; stalk short. Leaflets numerous, arranged in groups in 2–4 planes, to 45 × 2.5 cm. Inflorescence to 90 cm long. Male flowers with obtuse petals. Female flowers with calyx half as long as petals. Fruit to 3 cm, cylindric, orange-yellow. *India*. H5.

4. P. reclinata Jacquin (*P. spinosa* Schumacher). Illustration: McCurrach, Palms of the world, 165 (1960); Brickell (ed.), RHS A–Z encyclopedia of garden plants, 806 (2003).

Stems clustered, to 10 m or more, *c.* 15 cm in diameter. Leaves numerous, to 3 m long, graceful, arched, sometimes twisted. Leaflets *c.* 80 on each side, bright green, to 40 × 2.5 cm, arranged irregularly in several planes, rarely in 1 plane. Inflorescence to 1.2 m. Male flowers with acute or acuminate petals. Female flowers with calyx about half as long as petals. Fruit cylindric-ellipsoid, to 2 × 1 cm, orange-red to black. *Africa*. G2.

Very variable.

5. P. roebelenii O'Brien. Illustration: McCurrach, Palms of the world, 168 (1960).

Stem solitary (? rarely clustered) to 2 m, 5–10 cm in diameter, rough with remains of leaf-stalks. Leaves to 1.1 m, with *c.* 50 leaflets on each side, regularly arranged, to

20 cm × 5–7 mm, soft, bright green, borne in a single plane. Inflorescence to 45 cm. Male flowers with acuminate petals. Female flowers with calyx less than half as long as petals. Fruit blackish, *c.* 12 × 4 mm. *Laos*. G1.

A very elegant, dwarf date palm.

6. P. loureirii Kunth (*P. hanceana* Naudin; *P. humilis* Royle).

Variable, dwarf, apparently stemless, or with stem to 4 m, frequently clustering. Leaves more or less glaucous with leaflets in groups. Inflorescence-stalk to 90 cm in fruit. Male flowers with blunt petals. Female flowers with calyx half as long as corolla. Fruit red to black, 1.2–2 cm × 6–12 mm. *India to China*. G1.

A poorly understood species, which has been confused with *P. roebelenii.*

15. CALAMUS Linnaeus

Spiny, solitary or clustered, usually climbing, rarely apparently stemless or shrub-like, dioecious palms. Leaves pinnate, sometimes ending in a barbed whip, or with a similar whip borne on the leaf-sheath; leaflets always 1-ribbed, almost always acute, the whole leaf and sheath variously spiny. Inflorescence borne on the leaf-sheath, with tubular, spiny, persistent bracts, rarely the bracts splitting and expanding. Male flowers borne singly, with 3 joined sepals, 3 petals joined together at the base, and 6 stamens. Female flower borne in a pair with a sterile male flower, the ovary 3-celled. Fruit covered in neat vertical rows of reflexed scales. Seed usually 1 only, covered in a fleshy seedcoat, and with smooth or ruminate endosperm.

About 370 species, from W Africa, India and S China through SE Asia to Fiji and NE Australia. Of considerable economic importance as one of the genera producing 'rattan' canes used in furniture making. A few species may be cultivated as pot-plants.

1. C. ciliaris Blume. Illustration: McCurrach, Palms of the world, 33 (1960) – as *Calamus* species.

Slender, densely clustering; stems to 10 m, *c.* 1 cm in diameter, including sheaths. Leaves to 30 cm with numerous, very fine, regularly arranged leaflets. Sheaths, stalks and leaflets all densely covered with stiff, fine, greenish bristles. Fruit greenish white, rounded, *c.* 1 cm in diameter. *Indonesia (W Java, Sumatra)*. G2.

16. ARENGA Labillardière

Dwarf to very robust, solitary or clustered, monoecious or dioecious palms. Leaves pinnate or rarely undivided, the leaflets V-shaped in section, with jagged upper margins, often pale on the undersurface. Flowering usually beginning at the top of the stem and then downwards, followed by death of the stem. Inflorescences usually many-branched, rarely unbranched, 1–several at each node, with large scaly bracts and branches bearing triads (2 males, 1 female); inflorescence becoming unisexual by abortion. Female flowers with 3 overlapping sepals, 3 edge-to-edge petals, and ovary 3-celled. Fruit 1–3-seeded. Seeds with smooth endosperm.

About 17 species native to tropical and subtropical Asia. Several species are of economic importance, and, for this reason, may be cultivated in Europe in specialist collections; other species are cultivated as ornamentals in the tropics. Only 1 species is widely cultivated in Europe. Propagation by seed.

1. A. engleri Beccari. Illustration: McCurrach, Palms of the world, 16 (1960).

Stems clustered, often very short. Leaves to 1.75 m, with stalk to 80 cm. Leaflets linear, 28–32 on each side, held in 1 plane, to 3 cm wide, toothed only at the very tip. Inflorescences solitary. Male flowers maroon, very fragrant, with golden pollen. Female flowers green at first. Fruit spherical, red, *c.* 2 cm in diameter. Seeds 3. *Taiwan, Japan (Ryukyu Islands)*. H5–G1.

17. CARYOTA Linnaeus

Solitary or clustered, moderate to robust, monoecious palms with stems flowering from the top downwards and then dying. Leaf twice pinnate, with fishtail-like leaflets, with jagged teeth on the oblique apical margins. Inflorescences branching once, or very rarely unbranched, bearing triads of flowers throughout (2 male, 1 female). Male flowers with 3 overlapping sepals, 3 edge-to-edge petals, and stamens 6–100 or more. Female flowers with 3 overlapping sepals, 3 edge-to-edge petals joined at the base, and ovary with 3 cells. Fruit 1–2-seeded with irritant flesh. Seed with ruminate endosperm.

About 12 species native to Asia, poorly known, but of considerable local economic importance, and very decorative. Two species are widely cultivated. Propagation by seed.

1a. Stems clustering, rarely exceeding 10 m **1. mitis**
b. Stems solitary, to 15 m **2. urens**

1. C. mitis Loureiro. Illustration: McCurrach, Palms of the world, 36 (1960); Brickell (ed.), RHS A–Z encyclopedia of garden plants, 236 (2003).

Stems clustering, to *c.* 10 m tall and to *c.* 10 cm in diameter. Male flowers with *c.* 12–16 stamens. Fruit reddish, 1-seeded, *c.* 1.5 cm in diameter. *SE Asia*. G2.

2. C. urens Linnaeus. Illustration: McCurrach, Palms of the world, 37 (1960).

Stems solitary to *c.* 15 m and to 30 cm or more in diameter. Male flowers with 40–45 stamens. Fruit reddish, 1–2-seeded, to *c.* 2 cm in diameter. *India, Sri Lanka*. G2.

Apparently more cold-resistant than *C. mitis.*

18. CHAMAEDOREA Willdenow

Solitary or clustering, slender, dioecious palms, without teeth or spines. Leaves pinnate, or bifid with pinnate venation, the sheaths tubular but not forming a crownshaft; leaflets linear or asymmetric at base and apex. Inflorescences among or below the leaves, usually solitary but sometimes several at each node, branched or unbranched, with 3 or more bracts. Flowers solitary or rarely (in male plants) clustered or arranged in vertical rows, in the female rarely also in vertical rows. Branches of female inflorescence becoming orange or red in fruit. Male flowers with cup-like calyx. Petals variously free or united. Stamens 6. Pistillode prominent. Female flower with sepals and petals variously free or united. Staminodes 0–6. Ovary 3-celled. Fruit usually 1-seeded, red or black, with remains of stigma at base. Seed with smooth endosperm and a lateral embryo.

About 133 species, native to C & S America, whose classification is confused. Many species are highly ornamental and are cultivated on a wide scale in the USA and in specialist collections in Europe. Only 1 species is widely grown in Europe, but many more could be tried as pot-plants.

1. C. elegans Martius (*Collinia elegans* (Martius) Liebmann; *Neanthe bella* invalid). Illustration: McCurrach, Palms of the world, 57 (1960); Brickell (ed.), RHS A–Z encyclopedia of garden plants, 258 (2003).

Solitary to 2 m tall, but beginning to flower when still more or less stemless. Leaves dark green with 11–20 leaflets on each side, linear to lanceolate, to 20 × 2 cm. Inflorescence erect, with a long stalk, with 4–9 tubular bracts, and branching to 1 or rarely 2 orders, the male with more flexuous branches than the female. Male flowers yellow, globular; calyx lobed, petals joined except for a triangular opening, stamens minute. Female flower similar but with a flattened spherical ovary. Fruit spherical, black, *c.* 6 mm in diameter. *Mexico, Guatemala*. G1.

This is the commonest pot palm found for sale in chain stores throughout Europe; it is very tolerant of neglect, poor light conditions, and the dry atmosphere of rooms, but responds well to good cultivation. Fruit may be produced after hand pollination.

19. CHRYSALIDOCARPUS Wendland
Solitary or clustered, monoecious shrubs or trees, without teeth or spines. Stems usually conspicuously ringed. Leaf-sheaths open or tubular, sometimes forming a crownshaft. Leaves pinnate, usually with many slender, acute, 1-ribbed leaflets. Inflorescence among leaves, or below them, with 2 unequal, deciduous or more or less persistent bracts, and branching to 2 orders. Flowers borne in triads (2 males and 1 female) near the bases of the branches, with solitary or paired male flowers above. Male flowers with 3 overlapping sepals, 3 edge-to-edge petals, 6 stamens and a prominent pistillode. Female flowers with 3 overlapping sepals, 3 overlapping petals and an ovary with 3 cells and 1 ovule. Fruit with stigma scar at base. Endosperm smooth. Embryo lateral.

About 20 species in Madagascar, the Comoro Islands and Pemba Island. Several species are cultivated in the tropics, but only 1 species is cultivated in Europe as a pot-plant. Propagation by seed or offsets.

1. C. lutescens Wendland (*Areca lutescens* misapplied). Illustration: McCurrach, Palms of the world, 46 (1960); Brickell (ed.), RHS A–Z encyclopedia of garden plants, 266 (2003).

Stems many, densely clustering, sometimes aerially branching, to 10 m or more. Crownshaft present, tinged yellowish, covered in wax and black scales when young. Leaf-stalks and axis bright yellow or orange. Leaf ascending or curved. Leaflets narrow, regularly arranged, shining green, 40–60 on each side. Flowers golden yellow. Fruit cylindric-ellipsoid, violet-black, *c.* 1.5 cm long. *Madagascar*. G2.

Sucker shoots are freely produced, even under low light and dry conditions, but mature aerial stems are unlikely to be formed except under hot and humid conditions in specialist collections. Suckering juvenile plants are frequently sold as pot plants. *Areca lutescens* Bory is a synonym of *Hyophorbe indica* Gaertner, which is not generally cultivated.

20. ARCHONTOPHOENIX Wendland & Drude
Solitary, monoecious palms, without teeth or spines, with a smooth trunk ringed with leaf-scars and with a conspicuous crownshaft. Leaves pinnate, with acute, regularly arranged leaflets. Inflorescence below the leaves, subtended by 2 large, papery, deciduous bracts, and with a short stalk, and numerous branches, the lowermost further branched. Flowers in triads (2 male and 1 female) near the base, and solitary or paired (male) near the tips. Male flowers lopsided; sepals 3, small, acute and overlapping; petals 3, edge-to-edge, angled; stamens more than 6; pistillode as long as stamens. Female flowers symmetric with overlapping sepals and petals, and 1-celled ovary. Fruit with stigma scar at apex. Seed with ruminate endosperm and a basal embryo.

A genus of 2 or more species from Australia, both widely grown as ornamentals. Propagation by seed; rapidly germinating and fast growing.

Literature: Bailey, L.H., The king palms of Australia – Archontophoenix, *Gentes Herbarum* **3**: 391–409 (1935).

1a. Trunk enlarged at base; leaflets greyish white beneath **1. alexandrae**
b. Trunk scarcely swollen at base; leaflets green beneath **2. cunninghamiana**

1. A. alexandrae (Mueller) Wendland & Drude (*Ptychosperma alexandrae* Mueller). Illustration: *Gentes Herbarum* **3**: 398 (1935); Brickell (ed.), RHS A–Z encyclopedia of garden plants, 132 (2003).

Stem to 25 m tall, enlarged towards the base. Leaflets *c.* 5 cm wide, greyish white on the underside. Flowers white or cream-coloured. Fruit red, *c.* 1.25 cm. *Australia (N Queensland)*. G1.

Var. **beatricae** (Mueller) Bailey. Differs in having the trunk deeply ringed, and thus more or less stepped at the base.

2. A. cunninghamiana (Wendland) Wendland & Drude (*Ptychosperma cunninghamiana* Wendland; *Seaforthia elegans* misapplied). Illustration: *Gentes Herbarum* **3**: 393 (1935).

Stem scarcely swollen at the base. Leaflets *c.* 10 cm wide, green on the underside. Flowers lilac or purplish. Fruit red, more than 1.25 cm. *Australia (New South Wales, Queensland)*. H5–G1.

21. HEDYSCEPE Wendland & Drude
Solitary, monoecious palms without teeth or spines, with a smooth, green, conspicuously ringed trunk to 10 m tall, and conspicuous silvery blue crownshaft to 40 cm. Leaves pinnate with curved axis bearing regularly arranged leaflets, *c.* 40 on each side, stiff and ascending, to 55 × 2.5 cm. Inflorescence borne below the leaves, to 45 cm with a short stalk bearing 2 deciduous bracts, and branching to 2 orders, the branches hairless and bearing triads (2 male flowers, 1 female) more or less throughout. Male flowers asymmetric with 3 narrow, acute sepals, 3 angled, edge-to-edge petals, 9–12 stamens, and a pistillode as long as the stamens. Female flower with 3 overlapping sepals, 3 overlapping petals that are edge-to-edge at the very tip, 3 staminodes and a 1-celled ovary. Fruit broadly ellipsoid, to *c.* 4 × 2.5 cm, dull deep red, with remains of stigma near apex. Seed with smooth endosperm and a basal embryo.

A genus of 1 species; propagation by seed.

1. H. canterburyana (Moore & Mueller) Wendland & Drude (*Kentia canterburyana* Moore & Mueller). Illustration: *Principes* **10**: 16–18 (1966).

Lord Howe Island, off Australia. H5–G1.

A strikingly beautiful palm, but rather slow growing.

22. RHOPALOSTYLIS Wendland & Drude
Solitary, monoecious palms, without teeth or spines, with a smooth trunk and very conspicuous swollen crownshaft. Leaves pinnate, with leaflets acute or acuminate, regularly arranged. Inflorescence below the leaves with 2 deciduous bracts, the upper enclosed within the lower, and a short stalk, the axis branching to 2 orders. Flowers borne in triads (2 male and

1 female). Male flowers asymmetric, with 3 narrow, acute sepals, 3 edge-to-edge petals, 6 stamens, and a pistillode equalling the stamens. Female flowers symmetric with 3 overlapping sepals, 3 overlapping petals with edge-to-edge tips, and a 1-celled ovary. Fruit red with remains of stigma at apex. Seed with smooth endosperm and a basal embryo.

Three species, one each in New Zealand, Norfolk Island and Raoul Island. Propagation by seed. They are relatively hardy and very attractive palms best suited to the oceanic fringe of Europe.

Literature: Bailey, L.H., Rhopalostylis, *Gentes Herbarum* **3**: 429–35 (1935).

1a. Fruit spherical; fruiting perianth open and spreading **3. cheesemanii**
b. Fruit ovoid to ellipsoid; fruiting perianth cup-like 2
2a. Inflorescence to 60 cm; seed with adhering fibres, light in colour **1. sapida**
b. Inflorescence to *c.* 90 cm; seed dull brown **2. baueri**

1. R. sapida Wendland & Drude. Illustration: *Gentes Herbarum* **3**: 429, 430, 431 (1935).

Stem to 8 m tall, *c.* 23 cm in diameter, strongly and closely ringed. Leaves ascending, to 2.75 m long, with leaflets to 90 × 5 cm. Inflorescence to 60 cm long. Fruit ellipsoid, to 15 × 7 mm. Seed ellipsoid with tightly adhering fibres. *New Zealand.* H5.

2. R. baueri Wendland & Drude. Illustration: *Botanical Magazine*, 5735 (1868).

Stems to 12 m tall. Inflorescence to *c.* 90 cm long. Fruit ovoid to ellipsoid, to *c.* 17 × 13 mm. Seed spherical, dull brown. *Norfolk Island.* G1.

3. R. cheesemanii Beccari. Illustration: Cockayne, The vegetation of New Zealand, being Engler & Drude, Vegetation der Erde, edn 2, **14**: f. 94 (1928).

Stems to 12 m tall. Leaves to 3.4 m long, with leaflets to 100 × 5 cm. Inflorescence to 60 cm. Fruit spherical, *c.* 15 mm in diameter. Seed spherical, light brown. *Raoul Island.* G1.

Rare in cultivation.

23. CYRTOSTACHYS Blume

Clustering, monoecious palms, without teeth or spines. Leaves pinnate with sheaths forming a crownshaft; leaflets acute. Inflorescence below the leaves, with short stalk and 2 deciduous bracts, the axis branching to 2 orders. Flowers in triads (2 male and 1 female) in deep, open pits. Male flowers with 3 free, rounded, overlapping sepals, 3 edge-to-edge petals, and 6–15 stamens. Female flowers with 3 overlapping sepals, 3 overlapping petals and ovary 1-celled. Fruit ovoid to ellipsoid with remains of stigma at apex. Seed with smooth endosperm and basal embryo.

About 12 species from New Guinea and offshore islands; 1 species in Sumatra, Malaysia and Borneo. Propagation by seed.

1. C. renda Blume (*C. lakka* Beccari). Illustration: McCurrach, Palms of the world, 73 (1960).

Trunk to 10 m tall, *c.* 15 cm in diameter. Crownshaft brilliant orange to red. Leaves to 2 m long with axis and stalk bright orange, and with about 50 greyish green leaflets regularly arranged on each side, to 45 × 4 cm. Fruit blackish, borne on a dull crimson axis. *Sumatra, Malaysia, Borneo.* G2.

Extremely decorative but requires high temperatures; in Europe it is occasionally listed in nurserymen's catalogues.

24. HOWEA Beccari

Solitary, monoecious palms without teeth or spines, without crownshaft. Leaves pinnate; leaflets 1-ribbed, acute. Inflorescence with 1 or several spikes with long stalks subtended by a common bract in a leaf axil, each spike in bud enclosed by a thin bract at the apex of the stalk. Flowers in triads (2 male, 1 female), sunk in pits. Male flowers greenish brown with overlapping sepals, edge-to-edge petals, numerous stamens and a minute pistillode. Female flowers greenish brown with sepals and petals overlapping, ovary 1-celled. Fruit ellipsoid, *c.* 4 cm long, with remains of stigma at apex. Seed 1, with smooth endosperm and a basal embryo.

A genus of 2 species restricted to Lord Howe Island (off Australia), both widely cultivated as cool greenhouse subjects or houseplants. Propagation by seed; *H. forsteriana* is one of the commonest pot palms.

Literature: Bailey, L.H., Howea in cultivation, *Gentes Herbarum* **4**: 188–98 (1939).

1a. Tree to 20 m; leaves eventually more or less horizontal with well-developed stalks; leaflets horizontal or hanging; inflorescence with 3–8 spikes **1. forsteriana**
b. Tree to 7 m; leaves strongly arched with very short stalks; leaflets crowded, directed upwards; inflorescence with 1 spike **2. belmoreana**

1. H. forsteriana (Moore & Mueller) Beccari (*Kentia forsteriana* Moore & Mueller). Illustration: *Gentes Herbarum* **4**: 190–1 (1939); Brickell (ed.), RHS A–Z encyclopedia of garden plants, 549 (2003).

Tree to 20 m. Leaves to 3 m with long stalks, eventually held more or less horizontally and bearing rather distant, horizontal or hanging leaflets. Inflorescences to 1 m, with 3–8 spikes. Fruit narrowly ellipsoid. *Lord Howe Island.* G1.

2. H. belmoreana (Moore & Mueller) Beccari (*Kentia belmoreana* Moore & Mueller). Illustration: *Gentes Herbarum* **4**: 188 (1939).

Tree to 7 m. Leaves to 2 m, with very short stalks, strongly arched, and bearing crowded leaflets directed upwards from the axis. Inflorescence to 1 m, with 1 spike. Fruit broadly ellipsoid. *Lord Howe Island.* G1.

25. PTYCHOSPERMA Labillardière

Solitary or clustering, monoecious palms, without spines or teeth, with a smooth stem and conspicuous crownshaft. Leaves pinnate, leaflets slender to very broadly wedge-shaped, with conspicuously toothed apices. Inflorescence below the leaves, with a short stalk, and 2 large deciduous bracts, the inner enclosed within the outer; the axis branching to 2 orders. Flowers in triads (2 male flowers, 1 female) almost throughout the inflorescence, with single or paired male flowers confined to the tips. Male flower with 3 overlapping sepals, 3 edge-to-edge petals, numerous stamens and a pistillode usually as long as stamens. Female flower with 3 overlapping sepals, 3 overlapping petals, 1–6 tooth-like staminodes, and ovary 1-celled. Fruit with apical stigma and thin endocarp. Seed with 3–5 weak to very strong longitudinal grooves, smooth or ruminate endosperm, and a basal embryo.

About 30 species native to Australia, New Guinea, Solomon Islands and Micronesia. Propagation by seed.

Literature: Essig, F.G., A revision of the genus Ptychosperma Labill.

(Arecaceae), *Allertonia* **1**(7): 415–78 (1978).

1a. Stem solitary; leaflets to *c.* 80 × 8 cm; seed with ruminate endosperm **1. elegans**
b. Stems clustered; leaflets to *c.* 50 × 5 cm; seed with smooth endosperm **2. macarthurii**

1. P. elegans (R. Brown) Blume (*Seaforthia elegans* R. Brown). Illustration: McCurrach, Palms of the world, 193 (1960).

Stem solitary, to 8 m or more. Leaves to 2.5 m, with leaflets regularly arranged, *c.* 28 on each side, to 80 × 8 cm, with many long, twisted scales on the undersurface. Inflorescence green. Male flowers fragrant, green with white stamens. Fruit red, ovoid, *c.* 2 cm long. Seed with ruminate endosperm. *Australia.* G2–H5.

It is possible some records of this palm in outdoor cultivation may refer to *Archontophoenix cunninghamiana* (p. 369).

2. P. macarthurii (Wendland) Nicholson (*Actinophloeus macarthurii* (Wendland) Beccari). Illustration: McCurrach, Palms of the world, 194 (1960); Brickell (ed.), RHS A–Z encyclopedia of garden plants, 866 (2003).

Stems clustered, to 6 m or more. Leaves to *c.* 1.5 m, with *c.* 28 leaflets regularly arranged on each side, to 50 × 5 cm. Inflorescence green. Male flowers green with white stamens. Fruit red, *c.* 1.5 cm long. Seed with smooth endosperm. *New Guinea.* G2.

26. COCOS Linnaeus

Robust, solitary, monoecious tree, without spines or teeth. Leaves pinnate. Sheaths fibrous, open. Leaf-stalk deeply channelled on upper surface. Leaflets numerous, 1-ribbed, regularly arranged, acute. Inflorescence among leaves, with 1 order of branching and with 2 bracts, the lower inconspicuous, the upper woody, striped but not furrowed. Flowers cream, borne in triads (2 male, 1 female) at the bases of the branches and solitary or paired above. Male flowers with 3 overlapping, acute sepals, 3 edge-to-edge petals and 6 stamens. Female flower much larger than male, with 3 overlapping sepals and 3 overlapping petals. Ovary 3-celled. Fruit usually 1-seeded, with a thick fibrous husk and a thick bony endocarp with 3 pores near the base. Seed with smooth hollow endosperm partially filled with fluid.

Now considered a genus of 1 species. Origin uncertain, though there is evidence for an origin in the eastern Indian Ocean or western Pacific. The coconut is one of the most important tropical crops. Coconuts will not reach maturity in Europe outside specialist collections, but young seedlings still attached to the nut are frequently sold as pot plants. Many other palms were first included in *Cocos* but are now put in different genera. Some of them are still referred to by their old names in the horticultural trade.

1. C. nucifera Linnaeus. Illustration: McCurrach, Palms of the world, 53–5, 1960.

Origin uncertain; widely grown in the tropics. G2.

27. BUTIA (Beccari) Beccari

Solitary, monoecious palms. Leaves pinnate, leaflets acute, stalk and open sheath with stout, sharp teeth. Inflorescence among the leaves, with 2 bracts, the lower concealed among the leaves, the upper beaked, tubular in bud, splitting down one side, smooth or woolly but not grooved. Inflorescence axis branching once, with flowers in triads (2 male and 1 female) near the base, and solitary or paired male flowers above. Male flowers with 3 small, acute sepals joined at the base, 3 edge-to-edge petals and 6 stamens. Female flower with 3 free, overlapping sepals, 3 free, overlapping petals with short edge-to-edge tips, staminodes forming a short cup, and 3-celled ovary. Fruit 1-seeded with fleshy mesocarp, and bony endocarp pointed at both ends, slightly ridged and with pores near the base. Seed with smooth endosperm and basal embryo.

About 12 species native to tropical and subtropical S America, from Brazil to Argentina; generally very decorative. Propagation by seed.

Literature: Bailey, L.H., The genus Butia, *Gentes Herbarum* **4**: 16–50 (1936).

1a. Upper bract of inflorescence densely woolly **3. eriospatha**
b. Upper bract of inflorescence smooth 2
2a. Inflorescence branches straight; perianth short in fruit, not more than one-quarter of the length of the fruit **1. capitata**
b. Inflorescence branches flexuous; perianth longer in fruit, at least one-third of the length of the fruit **2. yatay**

1. B. capitata (Martius) Beccari (*Cocos capitata* Martius; *C. australis* misapplied). Illustration: *Gentes Herbarum* **4**: 27–34 (1936); Brickell (ed.), RHS A–Z encyclopedia of garden plants, 201 (2003).

Trunk to 6 m × 50 cm, covered in persistent leaf-bases. Leaves to 2.5 m, curving, with leaflets regularly arranged (rarely clustered) ascending from the axis, usually whitish beneath, or glaucous, to 75 × 2 cm. Upper bract to 90 cm, smooth. Inflorescence branches straight. Flowers yellow to reddish-tinged. Fruit ovoid, to 3 × 2.75 cm, bright yellow-orange; perianth in fruit not more than one-quarter of the fruit length. *Brazil.* H5.

2. B. yatay (Martius) Beccari (*Cocos yatay* Martius). Illustration: *Gentes Herbarum* **4**: 40 (1936).

Trunk to 10 m × 50 cm, eventually becoming bare in old age except below the crown. Leaflets more or less regularly arranged, glaucous beneath. Upper bract of inflorescence smooth, to 90 cm. Inflorescence branches flexuous. Flowers creamy yellow. Fruit ovoid, to 3 × 2.75 cm, dark yellow to orange or reddish-tinged; perianth in fruit at least one-third of the length of the fruit. *Argentina.* H5.

3. B. eriospatha (Drude) Beccari (*Cocos eriospatha* Drude). Illustration: *Gentes Herbarum* **4**: 43 (1936).

Trunk to 6 m, becoming bare. Leaves ascending then curving outwards, with hairy stalks. Leaflets more or less regularly arranged, green on upper surface, glaucous beneath, usually distinctly bifid at tips. Upper bract of inflorescence densely brown woolly-hairy. Fruit more or less flattened, to 20 × 22 mm, yellow, the perianth more or less obscured beneath the fruit. *Brazil.* H5.

28. JUBAEA Humboldt, Bonpland & Kunth

Solitary, monoecious tree palm, without teeth or spines, stem to 10 × 1 m. Leaves numerous, pinnate. Sheath open, fibrous; leaf-stalk short. Leaflets *c.* 120 on each side of axis, regularly arranged, or in groups of 2–5, stiff, green or silvery green, each with 2 apical lobes or a deep cleft. Inflorescence to 1 m or more, among the leaves, with long stalk and 2 major bracts, the lower concealed by the leaf-bases, the upper not deeply grooved externally, the axis branching once. Flowers reddish brown, borne in triads (2 males, 1 female) near the base and with solitary or paired male flowers above. Male flower maroon with yellow centre and many yellow stamens. Female flowers globular with overlapping sepals and petals, which are more or less of the same length, staminodes united into a ring and ovary with 3 cells, each with 1 ovule. Fruit ovoid, *c.* 4 × 2.5 cm, yellow, with thick endocarp and bearing 3 basal pores. Seed 1, with smooth, hollow endosperm.

One species in coastal C Chile, widely cultivated in areas with a Mediterranean climate. Propagation by seed. In Chile the species is used as a source of sap for the production of palm honey and is now very rare in the wild because of overexploitation.

1. J. chilensis (Molina) Baillon (*J. spectabilis* Humboldt, Bonpland & Kunth). Illustration: *Principes* **18**: 85 (1974).
C Chile. H5.

29. ARECASTRUM (Drude) Beccari
Solitary, monoecious tree palm, without teeth or spines, trunk smooth, grey-brown, to 15 m tall. Sheaths fibrous. Leaves pinnate, many borne in the crown, arching or drooping, to 5 m or more; stalk well developed. Leaflets to 100 × 3 cm, numerous, borne in groups of 1–5, and in several planes, giving the whole leaf a plumose effect. Inflorescence among the leaves, to 1 m long, with 2 bracts, the lower concealed, the upper woody and splitting down one side to become boat-shaped, deeply grooved on outside. Branches with creamy yellow flowers in triads (2 male, 1 female) almost throughout, with solitary or paired males near the tip. Male flowers with calyx 3-lobed, petals 3, edge-to-edge, 6 stamens and a minute pistillode. Female flowers with 3 overlapping sepals, and 3 petals that are overlapping except at the tips where they are edge-to-edge, staminodes in a low ring and a 3-celled ovary. Fruit yellow, ovoid, 2.5–4 cm long, with fleshy mesocarp and thick bony endocarp with 3 basal pores and a prominent intrusion into the single seed on one side. Endosperm smooth.

A genus of 1 species from subtropical S America. Propagation by seed; relatively short-lived.

1. A. romanzoffianum (Chamisso) Beccari (*Cocos romanzoffianum* Chamisso; *C. plumosa* W.J. Hooker; *Syagrus romanzoffiana* (Chamisso) Glassman). Illustration: McCurrach, Palms of the world, 13 (1960); Brickell (ed.), RHS A–Z encyclopedia of garden plants, 1014 (2003) – as *Syagrus romanzoffiana.*
Brazil to Argentina. H1.

Var. **australe** (Martius) Beccari, which has smaller fruit with the endocarp not narrowed at both ends, is also grown.

30. MICROCOELUM Burret & Potztal
Slender, solitary, monoecious palms, without teeth or spines. Leaves pinnate; stalks without teeth; leaflets 1-ribbed, obliquely 2-cleft or acuminate. Inflorescence among leaves, branched once, with a long stalk and 2 bracts, the lower inconspicuous, the upper woody and deeply furrowed. Flowers borne in triads (2 male and 1 female) near the base of the branches; upper part with solitary or paired male flowers. Male flower asymmetric, with 3 small sepals, 3 edge-to-edge petals and 6 stamens. Female flower with 3 broad, overlapping sepals and 3 broad, overlapping petals with short edge-to-edge tips; staminodes in a low ring; ovary with 3 cells, 1 ovule in each. Fruit 1-seeded, epicarp and mesocarp splitting into 3 parts exposing the smooth papery endocarp with basal pores. Seed with smooth hollow endosperm.

A genus of 2 species native to Brazil, 1 widely cultivated as a pot-plant. Propagation by seed.

1. M. weddellianum (Wendland) H.E. Moore (*Cocos weddelliana* Wendland; *Syagrus weddelliana* (Wendland) Beccari; *Microcoelum martianum* (Drude) Burret). Illustration: Langlois, Supplement to palms of the world, 171 (1976).

Eventually to *c.* 3 m. Leaves to 3.2 m with purplish scales along the stalk margin and the axis. Leaflets 50–60 on each side, regularly arranged, very narrow, dark green on upper surface, greyish beneath. Inflorescence to 1 m. Fruit *c.* 2.5 cm long. *Brazil*. G1.

31. ALLAGOPTERA Nees
Dwarf, apparently stemless, mostly clustering, monoecious palms, without spines or teeth. Leaves pinnate, with leaflets acute, but often splitting at the tip. Inflorescence among the leaves, erect, spike-like, with 2 bracts, the lower inconspicuous, the upper woody and grooved. Flowers fragrant, cream-coloured, in triads, 2 male and 1 female, very densely arranged. Male flowers lopsided, with 3 slender, elongate sepals joined at the base, 3 edge-to-edge petals, and 6 or more stamens. Female flower with 3 overlapping sepals, 3 overlapping petals, staminodes joined into a low ring and ovary 3-celled. Fruit 1-seeded with remains of stigma at apex, and endocarp with basal pores. Seed with smooth endosperm.

Five species native to Brazil and Paraguay. Propagation by seed.

1. A. arenaria (Gomes) Kuntze (*Diplothemium maritimum* Martius). Illustration: *Principes* **8**: 56, 57 (1964).

Leaves to 1.25 m with long stalks, leaflets *c.* 50 on each side, arranged in groups of 2–5, and borne in several planes, pale on undersurface. Inflorescence with stalk to 90 cm and spike to 15 cm. Male flowers with 10–16 stamens. Fruit covered in brown woolly scales. *Coastal Brazil (possibly hardy in warmest parts of Mediterranean coast)*. G2.

A coastal species, resistant to salt spray.

32. ELAEIS Jacquin
Robust, solitary, monoecious palms. Leaves pinnate, falling untidily leaving persistent bases. Leaf-stalk with fibrous, spine-like processes at the base, the midribs of the short lowermost leaflets also spine-like. Leaflets acute, 1-ribbed. Inflorescences unisexual among the leaves, with very short stalks and very congested branches with spine-like tips. Male flowers surrounded by united bracteoles, and with 3 free sepals, 3 edge-to-edge petals and 6 stamens. Female flower with 3 overlapping sepals, 3 overlapping petals and a 3-celled ovary. Fruit orange-red at base, usually blackish at the tip, with very oily mesocarp, and bony endocarp with 3 apical pores. Seed with smooth, hollow endosperm.

A genus of 2 species; 1 in S America, the other (the oil palm) native to tropical Africa, now very widespread as a tropical crop and becoming naturalised in some areas. Sometimes grown as an ornamental. Propagation by seed.

1. E. guineensis Jacquin. Illustration: McCurrach, Palms of the world, 83, 84 (1960).

Stem erect to *c.* 25 m, covered in persistent leaf-bases. Leaves to 5 m, with leaflets numerous, more or less grouped and held in several planes. Inflorescence branches to 15 cm, with sharp tips. Fruit ovoid, *c.* 3 cm long. *Tropical Africa*. G2.

The oil palm is the most important tropical source of vegetable oil. Seedlings grow rapidly and are sometimes sold in Europe as pot-plants.

47. ARACEAE

Herbs or woody climbers, sap often bitter, usually milky; rarely the plant is aquatic. Leaves usually stalked and broad, often lobed, veins forming a network; ptyxis supervolute. Flowers minute, stalkless on a spadix which is subtended by and often enclosed in a conspicuous spathe, unisexual or bisexual, actinomorphic. Perianth of 4—6 free or united segments, rarely absent. Stamens 2—8. Ovary 1—many-celled, superior or naked; ovules 1—many, axile, apical, basal or parietal. Fruit usually a berry.

A family of 115 genera and 2,000 species. Ten genera are native to Europe, 21 to North America. Species of about 40 genera are cultivated for ornament; several are food plants in the tropics.

Literature: Engler, A. (ed), Das Pflanzenreich, 11, 21, 37, 48, 55, 60, 64, 71, 73 (1905–20); Birdsey, M.R., The cultivated aroids (1951); Matuda, E., Las Araceas Mexicanas, *Anales del Instituto de Biologia, Mexico* **25**: 97–218 (1954); Bunting, G.S., Generic delimitation in Araceae, subfamily Monsteroideae, *Baileya* **10**: 21–31 (1962); Bunting, G.S., Commentary on Mexican Araceae, *Gentes Herbarum* **9**: 290–382 (1965); Bunting, G.S., Addenda to 'Commentary on Mexican Araceae', *Gentes Herbarum* **10**: 115–16 (1967); Madison, M., Notes on Caladium (Araceae) and its allies, *Selbyana* **5**: 342–77 (1981).

The following keys make provision for the identification of juvenile phases when these are prevalent in cultivation but it should not be assumed that all juvenile phases will be identifiable.

Key to genera (not requiring flowering material)

Specimens must first be assigned to one of the five slightly overlapping groups.

Group 1

Temperate and aquatic or marsh plants

Leaf-stalks. Without sheath: **9, 10**; with well-formed sheath: **22**; Widely channelled: **9**; narrowly channelled: **11**

Leaf-blades. Submerged or floating: **11**. Narrowly elliptic: **11, 22**; broadly ovate or oblong: **9, 10, 11**;

with sagittate base: **23**; with deeply cordate base: **22**.

Group 2

Tropical and subtropical aquatic and amphibious plants.

Habit. Floating rosettes of approximately truncate, hairy leaves: **41**; amphibious plants with pointed, hairless leaves: **39, 40**.

Group 3

Temperate tuberous plants grown in the open or under glass.

Plan of leaf-blade. Simple: **22, 32, 35, 36**; palmate: **37, 38**; pedate: **33, 34, 37, 38**.

Shape of leaf-blade when simple. Cordate or sagittate: **22, 32, 36**; elliptic lanceolate to linear: **22, 35**.

Leaf-blade venation. Mainly parallel: **22**.

Group 4

Tropical and subtropical tuberous plants usually with a dormant phase (young plants of Group 5, which may be stemless, are not included).

Plan of leaf-blade. With 3 or 5 pinnate or pinnately lobed divisions: **13**; pinnatisect: **26**; simple: **24, 25, 26, 27, 29, 30**.

Shape of leaf-blade when simple. Peltate: **24, 25, 26, 27**; not peltate but distinctly cordate or sagittate: **24, 25, 27, 30**; base weakly cordate or truncate: **24, 29**.

Posture of leaf-blade or its principal lobe. Drooping: **24, 25, 26, 27**; not drooping: **27, 29, 30**.

Leaf-stalk. Striped: **24, 29**; mottled: **13**; banded: **26, 27**; green: **24, 25, 26, 27, 30**; violet or purple? **24, 25, 26, 27**. Shorter than blade: **29**.

Group 5

Erect, trailing or climbing tropical plants without tubes (young plants of which may lack obvious above-ground stems, or tropical rosette plants).

1a. Sheath of leaf-stalk not embracing principal stem or the sheath next above (figure 14(3 & 4, p. 374); scale-leaves present among foliage leaves; sometimes rosette plants) 2
b. Sheath of leaf-stalk embracing principla stem and often the sheath next above (figure 14(1 & 2), p. 374); scale-leaves not usually present among foliage leaves except at base of plant or among inflorescences; not rosette plants 3
2a. Leaf-stalk with a pulvinus at the apex (figure 14(6), p. 374) – this is sometimes not obvious; several leaves should be examined **1. Anthurium**
b. Leaf-stalk without a pulvinus **18. Philodendron**
3a. Leaf-blade deeply lobed or compound, sometimes perforated 4
b. Leaf-blade simple, sometimes shallowly pinnately lobed 10
4a. Leaf-blade with 3 leaflets, or pedate **28. Syngonium**
b. Leaf-blade pinnatifid or pinnatisect 5
5a. Leaf-stalk with a pulvinus at the apex (figure 14(6), p. 374) – this is sometimes not obvious; several leaves should be examined 6
b. Leaf-stalk without a pulvinus 9
6a. Leaves pinnate with segments each tapered to a point **2. Rhaphidophora**
b. Leaves pinnate or pinnatifid with obliquely truncate segments (sometimes also perforated) or perforated or entire 7
7a. Leaves with pinprick-like holes near the midrib **3. Epipremnum**
b. Leaves with large perforations or none 8
8a. Leaves entire **2. Rhaphidophora**; **3. Epipremnum**
b. Leaves perforated or pinnatifid **2. Rhaphidophora**; **7. Monstera**
9a. Plant erect, without aerial roots **27. Alocasia**
b. Plant eventually climbing, with aerial roots **15. Rhektophyllum**
10a. Plant climbing or trailing on the ground 11
b. Plant erect 18
11a. Leaf-stalk with a pulvinus at the apex (figure 14(6), p. 374) – this is sometimes not obvious; several leaves should be examined 12
b. Leaf-stalk without a pulvinus 15
12a. Leaf-blade heart-shaped, strongly asymmetric **4. Scindapsus**
b. Leaf-blade tapered or rounded at the base 13
13a. Leaf-sheath not reaching nearly to pulvinus **19. Anubias**
b. Leaf-sheath reaching nearly to pulvinus 14
14a. Stem shortly creeping **8. Spathiphyllum**
b. Plant climbing **6. Rhodospatha**
15a. Leaf-blade sagittate **14. Nephthytis**; **28. Syngonium** (juvenile)
b. Leaf-blade not sagittate 16

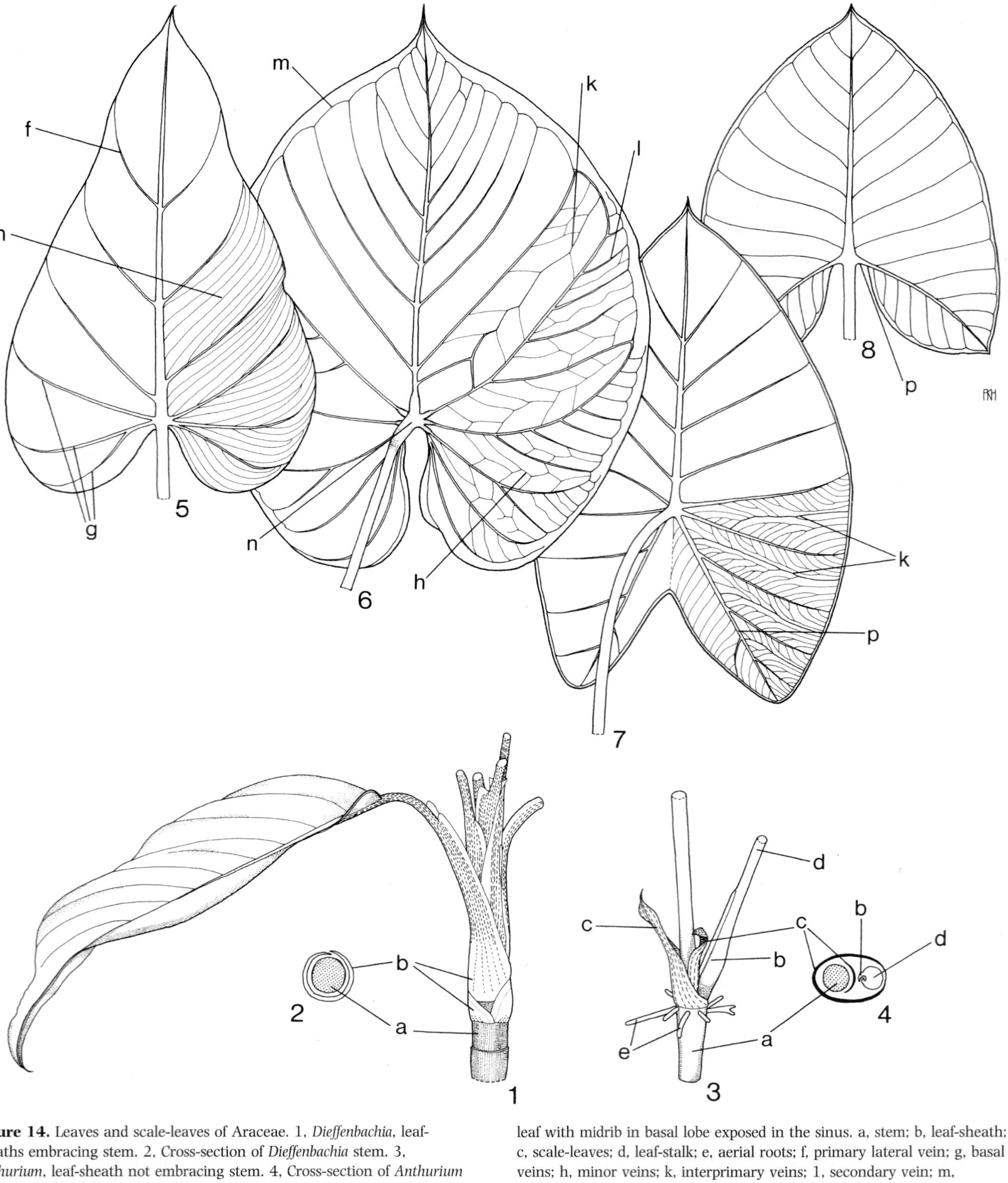

Figure 14. Leaves and scale-leaves of Araceae. 1, *Dieffenbachia*, leaf-sheaths embracing stem. 2, Cross-section of *Dieffenbachia* stem. 3, *Anthurium*, leaf-sheath not embracing stem. 4, Cross-section of *Anthurium* stem. 5, *Philodendron* leaf, minor veins almost parallel to primary lateral veins. 6, *Anthurium*, leaf with pulvinus at apex of stalk, minor veins netted. 7, *Colocasia*, peltate leaf with midribs in basal lobes. 8, *Xanthosoma*, leaf with midrib in basal lobe exposed in the sinus. a, stem; b, leaf-sheath; c, scale-leaves; d, leaf-stalk; e, aerial roots; f, primary lateral vein; g, basal veins; h, minor veins; k, interprimary veins; 1, secondary vein; m, marginal vein; n, pulvinus; p, midrib in basal lobe. (8 after *Das Pflanzenreich* **71**.)

16a. Base of leaf-stalk surrounded by 1 or 2 scale-leaves; leaf-blade net-veined and wrinkled **31. Callopsis**
b. Base of leaf-stalk not surrounded by scale-leaves; blade neither net-veined nor wrinkled 17
17a. Stem climbing; leaves shortly stalked, in 2 rows closely applied to the substrate **2. Rhaphidophora** (juvenile)
b. Stem trailing; leaves fairly long-stalked, clustered near upturned tips of shoots **20. Aglaonema**
18a. Leaf-stalk with a pulvinus at the apex (figure 14(6), p. 374) – this is sometimes not obvious; several leaves should be examined **5. Stenospermation**
b. Leaf-stalk without a pulvinus 19
19a. Leaf-blade with the subsidiary lateral veins mostly arising from the midrib and running parallel to the main laterals (figure 14(5), p. 374) **16. Homalomena**; **17. Schismatoglottis**; **20. Aglaonema**; **21. Dieffenbachia**
b. Leaf-blade with subsidiary lateral veins mostly arising from the primary laterals and forming a network, often joined to an interprimary vein (figure 14(6 & 7) p. 374) 20
20a. Leaf-blade less than 20 cm **28. Syngonium** (juvenile)
b. Leaf-blade normally exceeding 20 cm 21
21a. Leaf-blade peltate (figure 14(7), p. 374) **27. Alocasia**
b. Leaf-blade not peltate 22
22a. Main veins of basal lobe of leaf-blade exposed in the sinus for part of its length (figure 14(8), p. 374) **15. Rhektophyllum** (juvenile); **25. Xanthosoma**
b. Main vein of basal lobe of leaf-blade not exposed in the sinus **25. Xanthosoma**; **27. Alocasia**

Key to genera (requiring flowering material) (**4. Scindapsus** and **15. Rhektophyllum**) do not appear in this key as they do not flower in cultivation.)

1a. Floating aquatic plant with rosettes of more or less stalkless, downy leaves broadest at the apex **41. Pistia**
b. Plant otherwise 2
2a. Stipe several times longer than spadix, part of it wrapped in the sheathing part of the spathe 3
b. Stipe not longer than spadix 4
3a. Spathe reaching fertile part of spadix, its limb well developed, boat-shaped; marsh plant with large leaves **9. Lysichiton**
b. Spathe not reaching fertile part of spadix, its limb like a small bract; aquatic plants with submerged floating and emerging leaves **11. Orontium**
4a. Spadix united throughout its length to the mainly flat spathe **30. Spathicarpa**
b. Spadix free, or united to spathe for less than half its length 5
5a. Spathe in the form of an upwardly open funnel or chalice, white, yellow, pink or occasionally green; spacis without appendix **22. Zantedeschia**
b. Spathe not shaped as above; if upwardly open, then dark red, or spadix with an appendix, or both 6
6a. Whole spadix visible at flowering when viewed from the side, no part of it concealed by the spathe 7
b. At least the base of the spadix concealed or encircled by the spathe at flowering 19
7a. Spadix uniform, florets all or nearly all bisexual (figure 15(1–3), p. 377) 8
b. Spadix with zones bearing female, male and sometimes sterile florets exclusively, sometimes with 1 or more non-flowering zones (figure 15 (4–8), p. 377) 16
8a. Florets each with a perianth (figure 15(2 & 3), p. 377) 9
b. Florets without perianth 10
9a. Sheath of leaf-stalk not embracing principal stem or the sheath next above (figure 14(3 & 4), p. 374) **1. Anthurium**
b. Sheath of leaf-stalk embracing principal stem and often the sheath next above (figure 14(1 & 2), p. 374) **8. Spathiphyllum**
10a. Plant trailing or climbing, often by means of adhesive roots on the stem; leaf-blade simple in the juvenile phase, usually pinnately lobed, pinnatisect, pinnate or perforated in the adult phase 11
b. Plants not trailing or climbing; leaf-blades simple 15
11a. Leaf-blade, except in very young plants, with large perforations **7. Monstera**
b. Leaf-blade with pinprick-like perforations or not perforated 12
12a. Leaf-blade always variegated (in cultivation) 13
b. Leaf-blade not variegated 14
13a. Leaf-blade broadly ovate-cordate or broadly ovate-oblong, irregularly and extensively variegated with light green and yellow **3. Epipremnum**
b. Leaf-blade elliptic to elliptic-ovate, not cordate, variegated with scattered patches of light mottling between the veins **6. Rhodospatha**
14a. Leaf-blade in the adult phase with a row of pinprick-like holes on each side of the midrib; leaves of the juvenile phase with stalks nearly as long as blades, not arranged in 2 rows **3. Epipremnum**
b. Leaf-blade in the adult phase without pinprick-like holes; leaves of the juvenile phase with stalks much shorter than blades, arranged in 2 rows closely applied to the substrate **2. Rhaphidophora**
15a. Leaf-stalk without a pulvinus; leaf-blade about as wide as long **12. Calla**
b. Leaf-stalk with a pulvinus at the apex (figure 14(6), p. 374) – this is sometimes not obvious; several leaves should be examined **5. Stenospermation**
16a. Spathe spreading, flat but folded at the base, white **31. Callopsis**
b. Spathe erect, more or less boat-shaped or hooded, green or yellowish, at least outside 17
17a. Spadix with zone of female florets scarcely shorter than the zone of male florets **19. Anubias**
b. Spadix with zone of female florets much shorter than zone of male florets 18
18a. Leaf-blade with the base tapered, rounded or slightly cordate; leaf-stalk with a well-developed sheath **20. Aglaonema**
b. Leaf-blade sagittate; leaf-stalk with a very small sheath **14. Nephthytis**
19a. Lower part of spathe in the form of a closed tube formed by fusion of the margins 20
b. Lower part of spathe with free but overlapping margins 24
20a. Aquatic or amphibious plants with more or less vertical rhizomes and long stolons 21
b. Terrestrial and tuberous or shortly rhizomatous plants 22

21a. Each half of leaf-blade separately inrolled when young; female florets 25–50, spirally arranged **39. Lagenandra**
b. One half of leaf-blade rolled round the other when young; female flowers 4–7 in a single whorl **40. Cryptocoryne**
22a. Leaves pedate; spadix appendix *c.* 30 cm **34. Sauromatum**
b. Leaves simple; spadix appendix much less than 30 cm 23
23a. Spadix with the zones of male and female florets adjacent; leaf-blade heart-shaped or sagittate **36. Arisarum**
b. Spadix with a gap which sometimes bears sterile florets between the zones of male and female florets; leaf-blade tapered or truncate at base **35. Biarum**
24a. Florets all bisexual; spadix nearly as broad as long; temperate marsh plants flowering early in spring before their leaves appear **10. Symplocarpus**
b. Florets mostly unisexual; spadix not nearly as broad as long; plants otherwise 25
25a. Spadix with a sterile terminal appendix sometimes only a few millimetres long (it may be patterned as if with male florets (figure 15 (6–8), p. 377) 26
b. Spadix without sterile terminal appendix 34
26a. Spathe-limb falling off soon after flowering and while still fresh, the persistent remnant truncate **17. Schismatoglottis**
b. Spathe-limb not shed in this manner 27
27a. Female part of spadix completely joined to spathe **38. Pinellia**
b. Female part of spadix free from spathe 28
28a. Leaves compound or deeply lobed 29
b. Leaves simple 31
29a. Spadix unisexual or with male florets not touching or (usually) showing both of these conditions **37. Arisaema**
b. Spadix bisexual; male florets more or less touching 30
30a. Leaves palmate or with 3 or 5 pinnate or pinnately lobed divisions; only a single leaf present at a time **13. Amorphophallus**
b. Leaves pedate, several together **33. Dracunculus**
31a. Sterile florets between male and female zones of spadix each in the form of a thread or prong, often with a swollen base (figure 15(7), p. 377) **32. Arum**
b. Sterile florets each in the form of a flat plate, occasionally with a rudimentary ovary forming a boss in the middle, or sterile florets absent (figure 15(6), p. 377) 32
32a. Spadix without a constriction between male and female zones; sterile florets few and each with a central boss, or none; female florets each with a cup around the ovary (formed from fused staminodes) **23. Peltandra**
b. Spadix usually constricted in the zone of sterile florets (figure 15(6), p. 377); sterile florets without a central boss; female florets without staminodes 33
33a. Placentation parietal; ovules and seeds several–many **26. Colocasia**
b. Placentation basal; ovules and seeds 1 or few **27. Alocasia**
34a. Male florets consisting of groups of united stamens, the filaments of which form a stalk 6–7 mm long **29. Synandrospadix**
b. Anthers not borne on a long common stalk 35
35a. Leaf-sheath embracing the principal stem or the sheath next above (figure 14(3 & 4), p. 374), sometimes inconspicuous **18. Philodendron**
b. Leaf-sheath embracing the principal stem and often the sheath next above (figure 14(1 & 2), p. 374) 36
36a. Leaves compound (sometimes simple in the juvenile phase) 37
b. Leaves simple 38
37a. Stems loosely scrambling or closely applied to the substrate **28. Syngonium**
b. Plants essentially stemless **25. Xanthosoma**
38a. Leaf-blades with the areas between main lateral veins occupied mainly by veins diverging widely from the latter (figure 14(6 & 7), p. 374) 39
b. Leaf-blades with the areas between main lateral veins occupied mainly by veins running nearly parallel to the latter (figure 14(5), p. 374) 40
39a. Stigma undivided, resting on a disc formed by expansion of the style and which coheres with those of adjacent ovaries; ovary with 20–100 ovules **25. Xanthosoma**
b. Stigma divided by 3 or 4 grooves and placed directly on the ovary (style absent); ovary with 2–20 ovules **24. Caladium**
40a. Leaf-blade sagittate **23. Peltandra**
b. Leaf-blade not sagittate 41
41a. Leaf-stalk with a pulvinus at the apex (figure 14(6), p. 374)—this is sometimes not obvious; several leaves should be examined; plant creeping; female florets numerous; spadix with female zone longer than male **19. Anubias**
b. Leaf-stalk without a pulvinus; plant erect, or, if creeping, spadix with not more than 10 female florets; spadix with zone equal to or shorter than male 42
42a. Female florets with club-shaped staminodes (figure 15(5), p. 377); male florets with anthers united under a common, flat-topped connective (figure 15(4), p. 377), the pollen being extruded through pores on the sides **21. Dieffenbachia**
b. Female florets without staminodes; male florets with anthers free, arranged randomly or in groups of 2–4, the pollen being extruded from pores on their tops 43
43a. Ovary with 1 ovule; berry with 1 seed **20. Aglaonema**
b. Ovary with several–many ovules; berry with several–many seeds **16. Homalonema**

1. ANTHURIUM Schott
T.B. Croat

Plants usually epiphytic, erect, climbing or pendent, evergreen; stems non-tuberous. Leaf-stalks with a pulvinus at the apex (figure 14(6), p. 374); leaf-blades entire or palmately lobed, the major veins often joining to form a marginal vein; minor veins usually netted (figure 14(6), p. 374). Spathe flat or hooded, its edges not overlapping at base, erect, reflexed or spreading, often colourful. Spadix nearly cylindric (figure 15(1), p. 377), more or less tapered towards apex. Florets closely packed, bisexual (figure 15(2 & 3), p. 377); perianth segments 4; stamens 4, usually emerging just above the perianth segments. Ovary with 2 cells, usually with 1 ovule per cell, sometimes with 2 or more. Berries usually ovoid to spherical; colour various; flesh gelatinous. Seeds with sticky appendages at base or apex or both.

There are 700–900 species in the New World tropics. Approximately 28 species

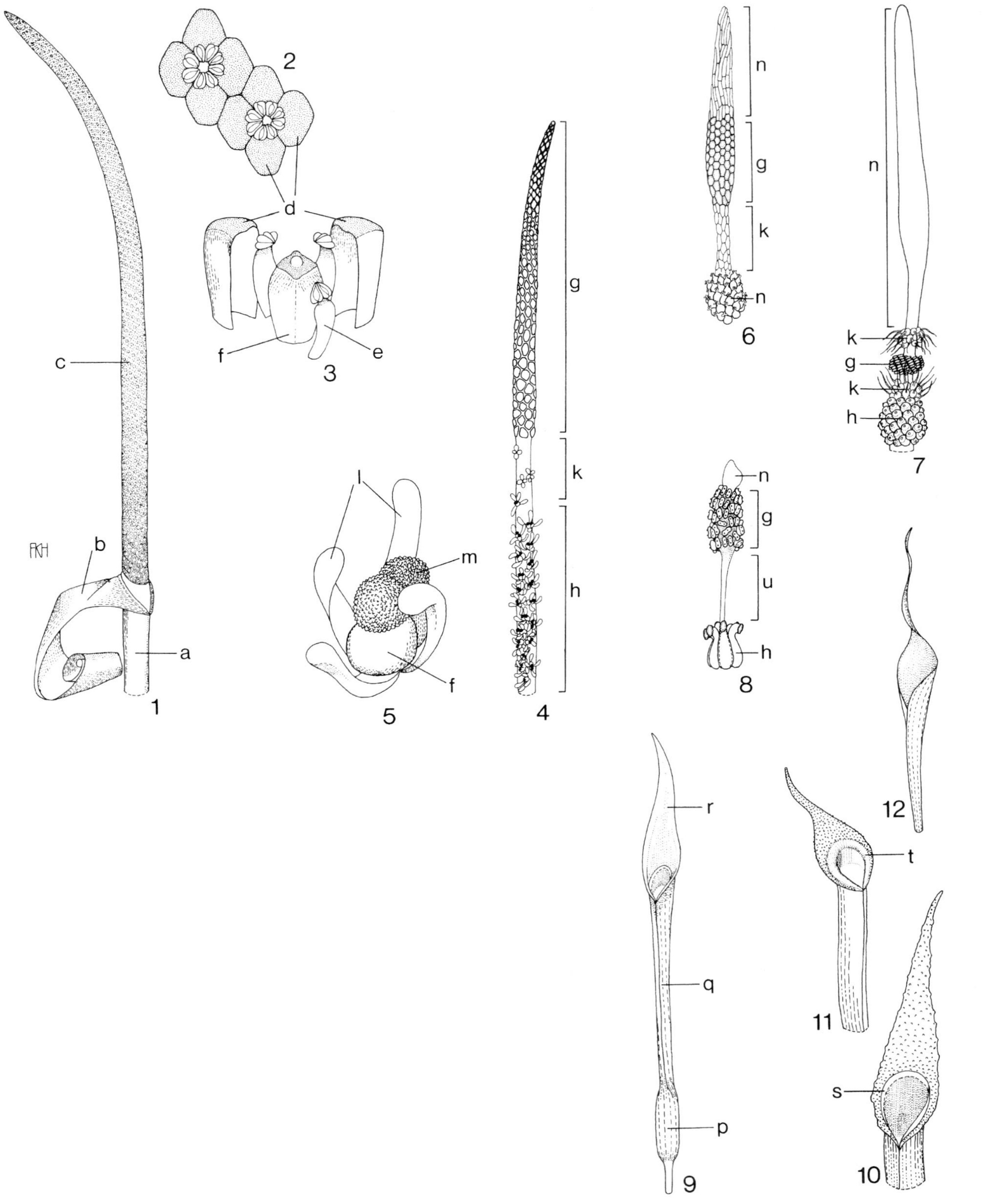

Figure 15. Inflorescences of Araceae. 1, *Anthurium*. 2, Two *Anthurium* florets seen from above. 3, *Anthurium*, bisexual floret 'exploded'. 4, *Dieffenbachia*, spadix. 5, *Dieffenbachia*, female floret. 6, *Alocasia*, spadix (sterile florets scale-like, sterile appendix scaly). 7,*Arum*, spadix (sterile florets thread-like with swollen bases, sterile appendix smooth). 8, *Cryptocoryne*, spadix. 9, *Cryptocoryne*, spathe, 10–12, Three different *Cryptocoryn* spathes. 10, with collar. 11, with collar zone. 12, without collar or collar zone. a, inflorescence-stalk; b, spathe; c, spadix; d, perianth segment; e, stamen; f, ovary; g, zone of male florets; h, zone of female florets; k, zone of sterile florets; 1, staminode; m, stigma; n, sterile appendix; p, floral chamber; q, tube; r, limb; s, collar; t, collar zone; u, sterile zone. (6 & 7 after *Das Pflanzenreich* **71** & **73**; 8–12 after de Wit, *Aquarienpflanzen*.)

are widely available. In species having deeply divided leaves, only those of the mature plant are described here; those of younger plants may be less divided. The basal veins are those which arise at the top of the leaf-stalk in leaf-blades which are heart-shaped or sagittate.

Most Anthuriums grow best under greenhouse conditions with high humidity and partial shade, but most of the species listed here adapt themselves well to other growing conditions and can survive considerable abuse. A loose, well-drained potting soil is essential and plants generally respond well to fertiliser.

Literature: Engler, A., Araceae, Pothoideae, Das Pflanzenreich **21**: 53–295 (1905); Madison, M., The species of Anthurium with palmately divided leaves, *Selbyana* **2**: 239–81 (1978); Croat, T.B. & Baker, R., Anthurium (Araceae) of Costa Rica, *Brenesia* **16**: 1–174 (1979); Bunting, G.S., Sinopsis de las Araceae de Venezuela, *Revista de 1a Facultad de Agronomia (Maracay)* **10**: 139–290 (1979); Mayo, S.J., Anthurium acaule (Jacq.) Schott & the Bird's Nest Anthuriums of the West Indies, *Kew Bulletin* **36**: 691–719 (1982); Croat, T.B., A revision of Anthurium for Mexico and Central America, Part I, Mexico and Middle America, *Annals of the Missouri Botanical Garden* (in press), Part II, Panama, *Annals of the Missouri Botanical Garden* (in press).

Habit. Forming a rosette: **21, 22, 24**; climbing: **4, 15, 25**.
Leaf-blade. Palmately lobed or divided: **1, 2, 3, 4, 5, 7**; entire or lobed at base: **8, 9, 10, 11, 12, 13, 14, 15, 16, 17, 18, 19, 20**; elongate, acute at base: **21, 22, 23, 24, 25, 26, 27**. Glandular-dotted: **21, 25, 26, 27, 28**. With velvety surface: **9, 10, 11, 12, 13**.
Inflorescence. Longer than leaf: **3, 4, 9, 10, 12, 13, 14** (rarely), **17, 18, 24, 27, 28**; shorter than leaf: **1, 2, 5, 6, 7, 8, 11, 14, 15, 16, 19, 21, 22, 23, 25, 26**; equal to leaf: **4, 8, 20, 21, 24, 26, 27**.
Spadix. White, creamy white, yellow: **9, 11, 14, 17, 18, 20, 26**; green: **3, 4, 8, 12, 13, 15, 16, 22, 24, 25**; pink, red, red-violet: **19, 20, 23, 27**; purplish: **1, 2, 3, 5, 7, 10, 16, 19, 20, 21, 22, 25**.

1a. Leaf-blade palmately lobed or divided 2
b. Leaf-blade simple, sometimes lobed 9
2a. Segments of leaf-blade all free to the base 3
b. Segments of leaf-blade united at the base 7
3a. Margins of leaf segments sinuous or deeply lobed **1. clavigerum**
b. Margins of leaf segments entire 4
4a. Leaf-blade with 11–15 segments, up to 1.5 cm wide **4. polyschistum**
b. Leaf-blades with 5–11 segments, these 4–10 cm wide 5
5a. Spathe deciduous after flowering 6
b. Spathe persisting even on the fruiting spadix **2. pentaphyllum** var. **pentaphyllum**
6a. Spadix 5–10 cm in flower, its stalk 2–7 cm **2. pentaphyllum** var. **bombacifolium**
b. Spadix 10–15 cm in flower, its stalk 5–12 cm **5. digitatum**
7a. Segments of leaf-blade commonly pinnatisect with 5–12 narrow divisions; inflorescence-stalk longer than the leaf-stalk **3. podophyllum**
b. Segments of leaf-blade not so deeply divided, sometimes wavy; inflorescence-stalk shorter than the leaf-stalk 8
8a. Central segment of leaf-blade wavy, twice as long as lateral segments; plant climbing; scale-leaves deciduous; berries deep purple **7. pedatum**
b. Central segment of leaf-blade not markedly wavy, not quite as long as lateral segments; plant with short stems; scale-leaves weathering and persisting; berries red **6. palmatum**
9a. Leaf-blade prominently or weakly lobed (cordate or sagittate) at base 10
b. Leaf-blade not lobed at base 22
10a. Leaf-blade either appearing conspicuously blistered or velvety with pale major veins 11
b. Leaf-blade neither blistered nor velvety 16
11a. Leaf-blade appearing conspicuously blistered; major veins usually not pale, all veins including minor veins (figure 14(6), p. 374) conspicuously sunken; marginal vein (figure 14(6), p. 374) arising from near base of blade **8. corrugatum**
b. Leaf-blade velvety, not blistered; major veins usually pale; the smaller veins not at all sunken; marginal vein not arising from base of blade 12
12a. Leaf-blade peltate **9. forgetii**
b. Leaf-blade not peltate (leaf-stalk borne at base of blade) 13
13a. Leaf-stalk with 4 wings or ridges **10. magnificum**
b. Leaf-stalk terete or grooved along upper side but not with 4 ridges or wings 14
14a. Leaf-blade 2 or more times as long as broad; berries red **11. warocqueanum**
b. Leaf-blade 1–1.5 times as long as broad; berries orange or white and purplish 15
15a. Leaf-blade to 25 cm, moderately thick, broadly ovate; major veins silver-grey; berries orange, spherical **12. clarinervium**
b. Leaf-blade to 40 cm, moderately thin, ovate; major veins usually pale green; berries ovoid, white at base, purplish above **13. crystallinum**
16a. Leaf-blade 2 or more times as long as broad; primary lateral veins reflexed towards the leaf-base near the midrib, then prominently arched as they approach the margin **14. veitchii**
b. Leaf-blade less than twice as long as broad; venation not as above 17
17a. Plant creeping, with branching stems to about 1 cm in diameter; leaf-blade less than 20 cm **15. radicans**
b. Plant erect, stems not branching, usually thicker; leaf-blade usually 25 cm or more 18
18a. Basal lobes of leaf-blade usually much longer than broad **16. watermaliense**
b. Basal lobes of leaf-blade about as broad as long, usually rounded 19
19a. Spathe narrowly lanceolate, thin, often twisted, green; leaf-blade dark green above with a satin-like sheen, the veins pale green; scale-leaves weathering to fibres **17. regale**
b. Spathe oblong-elliptic, oblong-lanceolate or wider, moderately thick, never twisted, usually white or reddish, sometimes tinged with green; leaf-blades medium green, the veins not markedly paler than the surface; scale-leaves persisting intact, reddish brown 20
20a. Spathe broadly ovate to almost rounded, bright red and puckered between the veins (though cultivars may vary in colour and are sometimes smooth); spadix white, tipped with yellow (cultivars various) **18. andraeanum**
b. Spathe oblong-elliptic, oblong-lanceolate, or ovate to narrowly obovate, rarely more or less rounded,

white or greenish, sometimes tinged with red, rarely pale red 21

21a. Inflorescence shorter than leaf; spathe whitish green; stems to 14 cm; berries orange **19. huixtlense**

b. Inflorescence as long as leaf; spathe white tinged with green to reddish; stems to 100 cm; berries reddish violet **20. nymphaeifolium**

22a. Plant forming a rosette; leaf-blade usually broadest above the middle, usually more than 60 × 20 cm; primary lateral veins mostly extending to the margin; the roots so dense as to be contiguous 23

b. Plant not forming a rosette; leaf-blade usually not broadest above the middle, usually less than 50 × 10 cm; primary lateral veins joining the marginal vein (figure 14 (6), p. 374) which arises at or near the base; stems long or short but the roots never contiguous 26

23a. Leaves black-dotted on both sides **21. hookeri**

b. Leaves not black-dotted 24

24a. Spathe moderately thin, often withering and deciduous, green; leaf-blade with the minor veins (figure 14(6), p. 374) prominently raised; leaf-stalk usually rounded on the back, sometimes prominently 3-ribbed 25

b. Spathe thick, usually persisting in fruit, usually tinged with purple in age; usually prominently reflexed; leaf-blade with minor veins (figure 14(6), p. 374) not prominently raised; leaf-stalk usually sharply 4-angled **22. schlechtendalii**

25a. Inflorescence usually as long as leaf, its stalk slender, usually spreading or pendent; spadix commonly 40 times as long as thick **24. crenatum**

b. Inflorescence usually only about half as long as leaf, its stalk stout, usually stiffly erect; spadix commonly 10–30 times (rarely to 35 times) as long as thick **23. crassinervium**

26a. Leaf peltate; surface velvety; major veins pale **9. forgetii**

b. Leaf not peltate; surface not at all velvety; major veins not pale 27

27a. Plant climbing; stem less than 6 mm in diameter; leaf-blade more or less elliptic, less than 16 cm **25. scandens**

b. Plant free-standing with short stems which are often 1 cm or more in diameter; leaf-blade oblong-elliptic to oblanceolate, never elliptic, usually more than 20 cm 28

28a. Spathe yellow-green; leaf-blade with marginal veins much more conspicuously sunken than the primary lateral veins; berries bright red **26. bakeri**

b. Spathe red, orange-red or violet-purple; leaf-blade with marginal veins no more conspicuously sunken than the primary lateral veins; berries white, red or orange 29

29a. Spathe red or orange-red, usually more than 6 cm **27. scherzerianum**

b. Spathe pale violet-purple, less than 4 cm **28. amnicola**

1. A. clavigerum Poeppig & Endlicher (*A. panduratum* Martius; *A. holtonianum* Schott; *A. kalbreyeri* N.E. Brown). Illustration: Graf, Exotica, edn 8, 132, 135 (1976); Graf, Tropica, 94, 95 (1978); *Brenesia* **16**: 120–1 (1979); Everett, Encyclopedia of horticulture **1**: 188 (1980).

Epiphytic creeper. Stem to 2 m, 3–5 cm in diameter. Leaf-blade 80 cm–2–m wide, palmatisect; segments 7–13 with margins sinuous to deeply lobed. Inflorescence pendent, its stalk to 90 cm, shorter than the leaf-stalk. Spathe 30–65 × 3–11 cm, violet-purple, lanceolate. Spadix 20–75 cm, pale purple. Berries obovoid, purple. *Nicaragua to the Guianas, Brazil and Bolivia.* G2.

2. A. pentaphyllum (Aublet) G. Don (*A. enneaphyllum* (Vellozo) Stellfeld; *A. undatum* Kunth). Illustration: *Selbyana* **2**: 255 (1978) – var. *pentaphyllum*.

Epiphytic creeper with short internodes. Leaf-stalk 19–44 cm, terete. Leaf-blade palmatisect; segments 5–9, 13–30 cm, oblanceolate to narrowly elliptic or ovate, the lateral unequal at base. Inflorescence erect, short, its stalk 1–13 cm. Spathe to 15 × 2.5 cm, pale green to purple, lanceolate to ovate, usually persisting in fruit or (var. **bombacifolium** (Schott) Madison (*A. aemulum* Schott)) falling soon after flowering. Spadix usually less than 10 cm long, 8–20 mm in diameter at base, pale violet-purple. Berries red to dark red-violet, more or less ellipsoid. *Trinidad and Atlantic coast of S America to Brazil* (var. *pentaphyllum*); *Mexico to Panama* (var. *bombacifolium*). G2.

3. A. podophyllum (Chamisso & Schlechtendal) Kunth. Illustration: Graf, Exotica, edn 8, 135, 137 (1976); *Selbyana* **2**: 253 (1978); *Annals of the Missouri Botanical Garden* **69**: f. 144, 156 (1982).

Terrestrial or on rocky cliffs. Stem short, 4–6 cm in diameter; internodes short. Leaf-stalk 40–80 cm, grooved above. Leaf-blade to 90 cm wide, palmatisect or palmate, thick; segments 7–11, stalkless or with stalk up to 2 cm long, each pinnatisect with 5–12 divisions or rarely with the margins merely wavy. Inflorescence erect-spreading, its stalk 50–100 cm, longer than the leaf-stalk. Spathe 5–8 cm, green, lanceolate. Spadix 6–15 cm, green or purplish, tapered to apex. Berries red. *Mexico (C Veracruz).* G2.

4. A. polyschistum Schultes & Idrobo (*A. putumayo* invalid; *A. pictamayo* invalid). Illustration: Graf, Exotica, edn 8, 1505 (1976); *Aroideana* **1**: 3 (1978); *Selbyana* **2**: 254 (1978); Everett, Encyclopedia of horticulture **1**: 189 (1980).

Epiphytic climber or terrestrial creeper. Stem with internodes 6–10 cm × 5–7 mm. Leaf-stalk 6–22 cm, nearly terete but weakly grooved along upper side. Leaf-blade palmate; segments 11–15, 8–15 × 0.5–1.5 cm, linear-lanceolate, thin, with the margins often crisped. Inflorescence-stalk 8–28 cm. Spathe 6–10 cm, green, narrowly lanceolate, reflexed. Spadix 6–14 cm long, 7 mm in diameter at base, pale green, erect. Berries obovoid, violet-purple to red-violet. *Colombia, Ecuador & northern Peru in the Amazon Basin.* G2.

5. A. digitatum (Jacquin) G. Don (*A. pentaphyllum* var. *digitatum* (Jacquin) G. Don). Illustration: Graf, Exotica, edn 8, 137 (1976); Everett, Encyclopedia of horticulture **1**: 188 (1980).

Epiphytic creeper. Stem to 2 cm in diameter; internodes short near the apex. Leaf-stalk 20–50 cm. Leaf-blade palmate; leaflets 7–13, 10–22 × 4–10 cm, nearly equal, stalkless or with stalk to 2 cm, thick, dark green. Inflorescence erect, much shorter than the leaf-stalk, its stalk 6–18 cm. Spathe 3–7 cm × 7–30 mm, green or lightly tinged purple, lanceolate to ovate, falling after flowering. Spadix 10–15 cm long, 8–20 mm wide at base, stalkless tapered towards apex, medium purple. Berries spherical, red to purple. *Venezuela.* G2.

Sometimes included in *A. pentaphyllum* (Aublet) G. Don.

6. A. palmatum (Linnaeus) G. Don (*A. fissum* Regel). Illustration: Graf, Exotica, edn 8, 1505 (1976); *Selbyana* **2**: 252 (1978).

Climbing epiphyte to 5 m. Stem with internodes 2–6 × 1.5–3 cm. Leaf-stalk 40–100 cm. Leaf-blade palmatisect; segments 5–11, 25–40 × 4–6 cm, oblanceolate. Inflorescence erect to spreading, its stalk 20–70 cm. Spathe 6–18 cm × 8–20 mm, lanceolate, green. Spadix 8–16 cm × 5–8 mm, tapered towards apex. Berries red, ovoid, rounded at apex. West Indies. *Lesser Antilles: St Lucia to Guadeloupe.* G2.

7. A. pedatum Humboldt, Bonpland & Kunth (*A. elegans* Engler; *A. fortunatum* Bunting). Illustration: Graf, Exotica, edn 8, 137, 140 (1976); *Selbyana* **2**: 253 (1978); *Aroideana* **4**: 19 (1981).

Terrestrial. Stem short, 2–6 cm in diameter; internodes short. Leaf-stalk 30–75 cm, terete. Leaf-blade palmatisect, leathery; segments 9–13, with wavy margins, unequal, the central one 20–45 × 5–8 cm, the lateral up to two-thirds as long as the central. Inflorescence-stalk half to five-sixths as long as the leaf-stalk. Spathe 6–10 × 2–4 cm, lanceolate, yellowish green, held horizontally. Spadix 7–12 cm × 5–10 mm, green to purplish, tapered, pendent. Berries deep purple. *Colombia (Western Andes).* G2.

8. A. corrugatum Sodiro (*A. papilionense* invalid; *A. papillosum* Markgraf). Illustration: Graf, Exotica, edn 8, 135 (1976); Graf, Tropica, 95 (1978).

Terrestrial. Stem somewhat climbing; internodes 6–12 × 2–3 cm. Leaf-stalk usually longer than blade, flattened and grooved along upper side. Leaf-blade 35–55 × 15–40 cm, ovate to more or less elliptic, sagittate, the surface appearing closely blistered or coarsely wrinkled with numerous deeply sunken primary lateral veins; the smaller veins sunken above, raised beneath; marginal vein (figure 14 (6), p. 374) arising from base of blade. Inflorescence erect, its stalk shorter than or equalling the leaf-stalk. Spathe to 20 × 3 cm, green, reflexed. Spadix about as long as spathe, 8–15 mm in diameter, stalkless, pale green, slightly tapered to the apex. Berries not known. *Colombia & Ecuador.* G2.

9. A. forgetii N.E. Brown. Illustration: *Aroideana* **1**: 18 (1978); Graf, Exotica, edn 8, 133 (1976).

Small epiphyte. Stem short with short internodes. Leaf-stalk 15–25 cm. Leaf-blade 20–30 × 15–22 cm, ovate, peltate, acuminate at apex, rounded or weakly notched at base, olive green and velvety on upper surface, with the major veins paler. Inflorescence 40–60 cm, erect, held about twice as high as the leaves. Spathe 8–15 cm, linear-lanceolate, green. Spadix about 15 cm, yellowish, gradually tapered. Berries narrowly ovoid, white, tinged with violet-purple. *Colombia.* G2.

The species readily hybridises with *A. crystallinum* and probably many of the plants sold as *A. forgetii* are such hybrids.

10. A. magnificum Linden. Illustration: Parey's Blumengärtnerei **1**: 169 (1958); Graf, Tropica, 93 (1978); *Aroideana* **1**: 18 (1978).

Probably terrestrial. Stem to 3 cm in diameter; internodes short. Leaf-stalk 25–40 cm, 4-sided and acutely ribbed. Leaf-blade 30–41 × 20–26 cm, ovate, deeply cordate, thick, olive green with greenish white major veins; upper surface matt, velvety. Inflorescence erect, its stalk 50–61 cm. Spathe 15–20 cm, lanceolate, recurved, green or sometimes tinged with red. Spadix 20–25 cm, tapered to apex, purplish. Berries not known. *Colombia.* G2.

11. A. warocqueanum J. Moore. Illustration: Das Pflanzenreich **21**: 197 (1905); Graf, Exotica, edn 8, 133 (1976); Graf, Tropica, 92, 95 (1978); Everett, Encyclopedia of horticulture **1**: 187 (1980).

Epiphytic creeper. Stem to about 1 m long and 1 cm in diameter; internodes short. Leaf-stalk 15–60 cm, terete, nearly erect. Leaf-blade mostly 50–90 × 11–25 cm, vertically pendent, narrowly ovate to oblong-ovate, somewhat leathery, dark green and velvety above; primary lateral veins pale green. Inflorescence nearly erect in flower, its stalk 6–30 cm. Spathe 8–10 × 1.5–2 cm, narrowly ovate-lanceolate, green. Spadix 6.5–30 cm × 5–12 mm, gradually tapered to apex, yellowish green. Berries red. *Colombia.* G2.

12. A. clarinervium Matuda. Illustration: Graf, Exotica, edn 8, 133 (1976); *Aroideana* **1**: 19, f. 1 (1978) – as *A. leuconeurum*; Graf, Tropica, 92 (1978).

Terrestrial, usually on limestone outcrops. Stem short, 1–2 cm in diameter. Leaf-stalk 7–16 cm, terete. Leaf-blade ovate, deeply cordate, dark green, velvety on upper surface, pale green along the major veins; lower surface pale, matt. Inflorescence erect, held well above the leaves, its stalk 16–38 cm. Spathe 3.7–6.5 × 1–1.5 cm, lanceolate, pale green. Spadix green, tinged with violet-purple. Berries bright orange, more or less spherical. *Mexico (C Chiapas).* G2.

13. A. crystallinum Linden & André. Illustration: *Illustration Horticole* **20**: 87 (1873); Graf, Exotica, edn 8, 133 (1976); *Aroideana* b18 (1978); Everett, Encyclopedia of horticulture **1**: 186 – also as *A. clarinervium* (1980).

Epiphytic. Stem to 25 cm long, 2–2.5 cm in diameter. Leaf-stalk terete. Leaf-blade 25–39 × 15–22 cm, reflexed more or less parallel to leaf-stalk, ovate to narrowly ovate, deeply cordate; upper surface dark green, velvety, with all major veins outlined in pale green. Inflorescence erect-spreading, its stalk 20–28 cm. Spathe 7–9 × 1.5–2 cm, oblong-lanceolate, tinged with red-violet. Spadix gradually tapered towards apex, green, turning yellow. Berries narrowly ovoid, white, tinged with violet-purple, beaked at apex. *Panama to Peru.* G2.

14. A. veitchii Masters. Illustration: Das Pflanzenreich **21**: 203 (1905); Parey's Blumengärtnerei **1**: 169 (1958); Graf, Exotica, edn 8, 130, 1505 (1976); Graf, Tropica, 92, 95 (1978).

Epiphytic. Stem 2–3 cm in diameter; internodes short. Leaf-stalk 20–70 cm, terete, nearly erect. Leaf-blade 25–100 × 12–30 cm, somewhat leathery, vertically pendent, oblong-ovate, cordate at the base with the sinus usually narrow or closed; primary lateral veins numerous, deeply sunken, each forming a broad arch and uniting with the marginal vein which arises from near the base. Inflorescence erect, shorter or rarely longer than the leaf, its stalk 13–50 cm. Spathe 7–13 cm, ovate to ovate-lanceolate, green, somewhat leathery, spreading. Spadix 5–9 cm long, to 1.6 cm in diameter, cream-coloured, cylindric, tapered slightly at both ends. Berries red. *Colombia.* G2.

15. A. radicans C. Koch. Illustration: Das Pflanzenreich **21**: 200 (1905); Graf, Exotica, edn 8, 130 (1976); *Aroideana* **1**: 37 (1978).

Decumbent creeper. Stem elongate with short internodes, to about 1 cm in diameter. Leaf-stalk 3–10 cm, grooved along upper side. Leaf-blade 10–16 × 6–9 cm, ovate, cordate, thick, often prominently arched along the midrib, appearing blistered above with the primary lateral and secondary veins deeply sunken. Inflorescence short, often hidden by the leaves, its stalk 2–4 cm, shorter than the leaf-stalk. Spathe 3.5–5 × 2–2.8 cm, thick, broadly ovate, cordate. Spadix 2–4 cm long and up to 1.5 cm in diameter, nearly cylindric but slightly tapered to apex, reddish green to brown. Berries ovoid, rounded at apex with an acute style. *Brazil (Bahia).* G2.

16. A. watermaliense Bailey. Illustration: Graf, Exotica, edn 8, 131, 134– also as A. 'Negrito', 135 – as *A. longilinguum* (1976); *Brenesia* **16**: 173 (1979); *Annals of the Missouri Botanical Garden* **69**: f. 224–5 (1982).

Terrestrial. Stem to 25 cm long and 1–2.5 cm in diameter; internodes short. Leaf-stalk 12–88 cm, erect-spreading, weakly flattened or grooved along upper side. Leaf-blade 30–60 × 19–36 cm, moderately thick, ovate-triangular, sagittate; basal lobes usually much longer than broad with a deep sinus between them; upper surface dark green. Inflorescence nearly erect, its stalk 12–66 cm, usually much shorter than the leaf. Spathe 15–21 × 2.3–8.5 cm, lanceolate-triangular, green to dark violet-purple or blackish. Spadix 7–10 cm, whitish to greenish, yellow or purple, prominently tapered to apex. Stamens unusually prominent, projecting well above perianth segments. Berries yellow to orange, ovoid to obovoid, beaked. *Costa Rica to Colombia.* G2.

17. A. regale Linden. Illustration: Graf, Exotica, edn 8, 130 (1976).

Probably epiphytic. Stem moderately short, to 2.5 cm in diameter; internodes short. Leaf-stalk 25–40 cm, grooved along upper side. Leaf-blade 30–57 × 19–36 cm, somewhat leathery, ovate, cordate, dark green on upper surface with a satin-like sheen and pale green veins; lower surface paler green. Inflorescence nearly erect, held above the leaves, its stalk 40–70 cm. Spathe 7–19 cm, green, lanceolate, reflexed. Spadix 10–25 cm, greenish white, prominently tapered. Berries not seen. *Peru (along Rio HualIaga).* G2.

18. A. andraeanum André. Illustration: Bailey, Manual of cultivated plants, edn 2, 178 (1949); Parey's Blumengärtnerei **1**: 32 (1956); Graf, Exotica, edn 8, 128, 1755 (1976); Graf, Tropica, 92 (1978).

Epiphytic. Stem usually less than 30 cm long, 1–2 cm in diameter; internodes short. Leaf-stalk 20–60 cm, nearly erect, terete. Leaf-blade 17–50 × 11–22 cm, leathery, narrowly ovate, sagittate, with prominent, sometimes overlapping, basal lobes; marginal vein (figure 14(6), p. 374) arising from the first basal vein. Inflorescence more or less erect, usually held above the leaves. Spathe 6–15 × 5–12.5 cm, bright red, broadly ovate to almost rounded, cordate, puckered, often with prominently raised veins. Spadix 4–9 cm × 6–8 mm, usually white, becoming yellowish at least towards the apex, slightly tapered to apex, usually curved slightly and directed downwards. Berries ovoid, reddish. *SW Colombia & NW Ecuador.* G2.

The above description refers to material of *A. andraeanum* collected in the wild. The species has been used in breeding programmes since it was introduced by André in 1876 and because few records were kept it is now difficult to say what the cultivated material bearing the name *A. andraeanum* really is.

No fewer than 26 hybrid names are listed in the Royal Horticultural Society's Dictionary of gardening. Almost half of these involve crosses with *A. nymphaeifolium* (here understood to include *A. ornatum* and *A. lindenianum*). Birdsey has proposed the name *A. × cultorum* for at least part of this material. Independently of hybridisation, *A. andraeanum* has undergone a great deal of spontaneous mutation, resulting in a wide array of flower forms (see 'Anthurium breeding in Hawaii' by H. Kamemoto, *Aroideana* **4**: 77–86, 1981).

Among the hybrids in cultivation which have *A. andraeanum* as one of the parents are:

A. × ferrierense Masters & T. Moore (*A. × carneum* Regel; *A. × roseum* Closon). Illustration: *Illustration Horticole* **32**: 175 (1885) & **33**: 31 (1886); *Revue de l'Horticulture Belge* **23**: 193 (1897) & **27**: 97 (1901); Graf, Exotica, edn 8, 128, 129 (1976); Graf, Tropica, 92, 94 – 'Rhodochlorum' (1978). The hybrid between *A. andraeanum* and *A. nymphaeifolium*, usually with longer leaf-stalks and broader leaf-blades than *A. andraeanum*, and with spathe less puckered, or smooth. Inflorescence variable in colour; spathe white, yellowish, pink or red; spadix white, pink or yellowish, or white tipped with pink or yellow.

A. × chelsiense N.E. Brown. Illustration: Graf, Exotica, edn 8, 128, 129, 131 (1976). This is most similar to *A. veitchii* but differs in having the spathe 13 × 9 cm, broadly heart-shaped, smooth, dark red, the spadix 8 cm long, and the leaf-blade with fewer, less arching primary lateral veins. It is the hybrid *A. andraeanum* × *A. veitchii*, with the former as the male parent.

19. A. huixtlense Matuda (*A. fraternum* misapplied). Illustration: *Annals of the Missouri Botanical Garden* **69**: f. 85, 89 (1982).

Terrestrial or epiphytic. Stem to 14 cm long and 3 cm in diameter. Leaf-stalk 15–70 cm long, terete, nearly erect. Leaf-blade 17–70 × 10.5–36 cm, more or less leathery, ovate to broadly ovate, sagittate, broadest just below point of attachment to leaf-stalk; sinus triangular to horseshoe-shaped; basal veins (figure 14(5), p. 374) joined at the base for 2–2.5 cm; marginal vein (figure 14(6), p. 374) arising from one of the lower primary lateral veins. Inflorescence erect-spreading, shorter than the leaf, its stalk 21–39 cm. Spathe 4.5–15 × 2.7–4.5 cm, pale whitish-green, oblong-lanceolate. Spadix 4.2–14 cm, dark pink to pale purple. Berries orange, spherical. *Mexico to Nicaragua.* G2.

Anthurium fraternum Schott, long confused with this species, is a distinct species from northern Colombia.

20. A. nymphaeifolium C. Koch & Bouché (*A. lindenianum* C. Koch & Augustin; *A. ornatum* Schott). Illustration: *Botanical Magazine*, 5848 (1870); Das Pflanzenreich **21**: 209 (1905); Graf, Exotica, edn 8, 131, not 141 (1976).

Terrestrial or rarely epiphytic. Stem 30–100 cm long and 2.5 cm in diameter; internodes short. Leaf-stalk 50–90 cm, terete, erect-spreading. Leaf-blade 31–70 × 19–40 cm, broadly ovate, deeply cordate, moderately thick; sinus open or closed; basal veins (figure 14(5), p. 374) united or free to the base; marginal vein usually arising from one of the primary lateral veins, sometimes from the uppermost basal vein (figure 14(6), p. 374). Inflorescence usually erect, equalling the leaves, its stalk 35–95 cm. Spathe 9–21 × 2–7.5 cm, oblong-elliptic to ovate or narrowly obovate, rarely more or less rounded and lobed at the base, white tinged with green to reddish throughout. Spadix 8–14 × 1–1.5 cm, white to pink, purplish or red, slightly tapered to the apex. Berries reddish. *Venezuela; ?Colombia (Cordillera Oriental).* G2.

Bunting's (1979) broad view of this species, including much variation, is followed here.

21. A. hookeri Kunth (*A. huegelii* Schott). Illustration: Graf, Exotica, edn 8, 138 (1976); Graf, Tropica, 95 (1978).

Epiphytic or on rocks, forming a rosette. Stem short, densely rooted. Leaf-stalk 2–14 × 1.5–2 cm, D-shaped to bluntly triangular in section. Leaf-blade 35–140 × 10–45 cm, usually obovate, rarely elliptic to ovate, black-dotted on both surfaces; primary lateral veins (figure 14(5), p. 374) mostly free to the margin, not linked by secondary lateral veins. Inflorescence more or less erect or spreading, its stalk 45–90 cm. Spathe 9–23 × 1.5–2.7 cm, oblong, green. Spadix 10–29 cm × 5–9 mm, gradually tapered, pale violet-purple. Berries obovate to cylindric, translucent, white, reddish to purplish at tip. *Lesser Antilles from St Kitts southward, Trinidad & Surinam.* G2.

This species has been confused with *A. crassinervium* but its leaves have glandular dots and their secondary veins run almost parallel between the primary laterals.

22. A. schlechtendalii Kunth (*A. crassinervium* misapplied; *A. tetragonum* (Hooker) Schott). Illustration: Graf, Exotica, edn 8, 138, 141 – as *A. tetragonum* (1976); *Annals of the Missouri Botanical Garden* **69**: f. 172–3 (1982).

Epiphytic or on rocks. Stem short, 2.5–5.5 cm in diameter, densely rooted. Leaf-stalk 12–23 cm × 5–20 mm, trapezoid in cross-section, the lower side sometimes ribbed, sometimes rounded. Leaf-blade 30–112 × 10–58 cm, obovate-elliptic, leathery; most primary lateral veins (figure 14(5), p. 374) extending to the margin. Inflorescence spreading, shorter than the leaf, its stalk 33–34 cm. Spathe 15–28 × 1.5–5 cm, thick, green to violet-purple, linear-lanceolate to lanceolate, strongly reflexed. Spadix 3–34 × 1.7–2 cm, green, often tinged with violet-purple, conspicuously tapered to apex. Berries 1–1.5 cm long, red, lanceolate-ellipsoid. *Mexico to Nicaragua.* G2.

23. A. crassinervium (Jacquin) Schott (*A. ellipticum* C. Koch & Bouché; *A. rugosum* Schott). Illustration: Graf, Exotica, edn 8, 138, 141 (1976).

Epiphytic, forming a rosette. Stem stout, moderately short, very densely rooted; internodes short. Leaf-stalk much shorter than blade, usually 5–20 cm, nearly quadrangular, broadly grooved along upper side, 3-ribbed beneath. Leaf-blade 45–110 × 9–35 cm, oblanceolate to obovate; primary lateral veins (figure 14 (5), p. 374) mostly extending to the margin; tertiary veins forming a network, prominently raised. Inflorescence more or less erect, its stalk to *c.* 60 cm, stout. Spathe mostly 10–15 × 1–2.5 cm at flowering time, oblong-lanceolate, moderately thin, spreading, green. Spadix 14–20 cm, dull red-violet at flowering time, gradually tapered to the apex. Berries red. *Venezuela (coastal region).* G2.

This species has long been confused with *A. schlechtendalii* and also with *A. hookeri.*

24. A. crenatum Kunth (*A. acaule* misapplied). Illustration: Graf, Exotica, edn 8, 138 (1976).

Epiphytic or on steep slopes or cliffs, with a bird's nest habit. Stem short, thick, densely rooted. Leaf-stalk 5–25 cm, broadly grooved along upper side. Leaf-blade 40–100 × 6–30 cm, oblong-lanceolate to oblanceolate, leathery; primary lateral veins (figure 14(5), p. 374) mostly extending to the margin. Inflorescence mostly spreading, its stalk to *c.* 1 m. Spathe 6–15 cm, linear-lanceolate, soon withering. Spadix 10–35 cm, greenish brown, long-tapered. Berries red, obovoid. *Hispaniola, Puerto Rico & Virgin Islands.* G2.

The species was long confused with the apparently rare *A. acaule* (Jacquin) Schott.

25. A. scandens (Aublet) Engler. Illustration: Graf, Exotica, edn 8, 139 only (1976); *Aroideana* **3**: 88 (1980); *Annals of the Missouri Botanical Garden* **69**: f. 169 (1982).

Climbing epiphyte. Stem elongate, slender, with internodes 5–40 × 4–6 mm. Leaf-stalk 1–6 cm, usually grooved along upper side, usually much shorter but occasionally longer than blade. Leaf-blade 4–16 × 1.5–7 cm, elliptic to ovate-lanceolate, densely black-dotted on lower surface; marginal vein (figure 14(6), p. 374) arising from the base. Inflorescence much shorter than the leaf, its stalk 2–8 cm, *c.* 1 mm in diameter. Spathe lanceolate to oblong-lanceolate, green, tightly reflexed after opening. Spadix 7–30 × 2–3 mm, green, sometimes becoming pale purple. Berries depressed-spherical, usually pale violet to almost white, with 2–10 or more seeds. *West Indies & S Mexico to S Brazil.* G2.

26. A. bakeri J.D. Hooker. Illustration: Graf, Exotica, edn 8, 115, 136, 139 (1976); Graf, Tropica, 92 (1978); *Brenesia* **16**: 112 (1979); Everett, Encyclopedia of horticulture **1**: 187 (1980).

Epiphytic. Stem less than 1.5 cm in diameter. Leaf-stalk to 17 cm, terete to sharply grooved above. Leaf-blade 19–55 × 2.8–9 cm, narrowly lanceolate-elliptic to narrowly oblanceolate, much paler and dotted with reddish brown on lower surface; marginal vein (figure 14(6), p. 374) arising from base of blade, conspicuously sunken and more prominent than the primary lateral veins. Inflorescence erect-spreading, its stalk 5.5–30 cm. Spathe pale yellow-green, oblong-lanceolate. Spadix 2–11 cm long, 5–15 mm wide at base, creamy white. Berries bright red, obovoid, pointed at apex. *Guatemala to Colombia.* G2.

27. A. scherzerianum Schott. Illustration: Graf, Exotica, edn 8, 129 (1976); Graf, Tropica, 92, 93 (1978); *Brenesia* **16**: 157 (1979); Everett, Encyclopedia of horticulture **1**: 186 (1980).

Epiphytic or terrestrial. Stem short, less than 1.5 cm in diameter. Leaf-stalk 4–20 cm, terete. Leaf-blade 5–26 × 1.5–6.5 cm, linear to elliptic or lanceolate, leathery, densely dotted on both surfaces. Inflorescence as long as or longer than the leaf; stalk 14–52 cm. Spathe 3.7–12 × 2.4–6 cm (often larger in hybrids), elliptic to ovate, bright red or orange-red. Spadix 2–8 cm, orange to red, tapered, usually coiled. Berries orange to red. *C Costa Rica (Cordillera Central & Cordillera Talamanca).* G2.

The group name *A.* × *hortulanum* Birdsey has been proposed for cultivars of *A. scherzerianum.* Two cultivars are worthy of special mention, 'Atrosanguineum', which differs from wild collections in having a much larger, dark red spathe, and 'Rothschildianum' (*A.* × *rothschildianum* Bergman & Veitch), with a large red spathe, heavily spotted with white, and a yellow spadix (Illustration: Graf, Tropica, 92, 1978).

28. A. amnicola Dressler (*A. lilacinum* Dressler not Bunting). Illustration: *Selbyana* **2**: 301 (1978); *Aroideana* **1**: 87 (1978).

Less than 20 cm tall. Leaf-stalk 3.5–9 cm, nearly terete, weakly grooved above. Leaf-blade lanceolate-elliptic to oblong-elliptic, sparsely and obscurely gland-dotted on upper surface; marginal vein arising from one of the basal veins (figure 14(6) p. 374). Inflorescence erect, its stalk 2–4 times as long as the leaf-stalk. Spathe pale violet-purple, narrowly to broadly ovate. Spadix 8–20 mm, dark violet-purple. Berries spherical, white. Seeds 1–2. *C Panama.* G2.

The species grows clinging to rocks in rapidly flowing streams.

2. RHAPHIDOPHORA Hasskarl
P.F. Yeo

Creeping or climbing evergreen plants with adhesive roots on the stems, sometimes with distinct juvenile and adult phases. Leaf-stalk with a pulvinus at the apex (figure 14(6), p. 374); sheath soon withering. Leaf-blade entire or pinnately lobed or divided, sometimes perforated; minor veins in leaf of adult parallel to the main lateral veins. Spathe boat-shaped, unfurling throughout its length at flowering time, deciduous. Florets without perianths, bisexual or a few female; stamens 4; ovary sometimes expanded at the top, 1-celled or incompletely divided into 2 or 3 cells, with several to many ovules in each cell. Seeds with a soft coat, straight.

A genus of about 60 species in tropical E Asia and the Pacific islands, a very few of which are cultivated. They need a well-drained but moist, rich soil and a support upon which to grow. Propagation is by dividing root-bearing parts of the stem. The genus name is often spelled *Raphidophora* (references to this issue are found in Flora of Java **3**:106 (1968) by Backer & Bakhuizen van den Brink).

Literature: Engler, A. & Krause, K., Monsteroideae & Calloideae, *Das Pflanzenriech* **37**: 17–53 (1908).

1a. Leaf-blade of adult phase irregularly pinnatifid or pinnatisect, the lobes obliquely truncate at the apex **1. celatocaulis**
b. Leaf-blade of adult phase regularly pinnatisect, the lobes gradually tapered to an acute point **2. decursiva**

1. R. celatocaulis (N.E. Brown) Knoll (*Pathos celatocaulis* N.E. Brown; *Monstera latevaginata* misapplied; *Marcgravia paradoxa* misapplied). Illustration: Birdsey, The cultivated aroids, 103 (1951); Graf, Exotica, edn 8, 226, 227 (1976).

Climber. Leaves of juvenile phase shortly stalked, arranged in 2 rows pressed to the substrate, with entire, elliptic-ovate blades 8.5–10 × 5.5–6.5 cm, overlapping their neighbours and concealing the stem. Leaves of adult phase with the stalks 10–15 cm, the blades 17–40 × 11–31 cm, entire, pinnatifid or pinnatisect, and sometimes perforated, with tips of lobes obliquely truncate. Inflorescence-stalk about as long as the leaf-stalk. Spathe 15 × 6 cm, very fleshy, yellow. Spadix 12 × 2.5 cm, pale yellow. *Borneo.* G2.

Usually seen in the juvenile phase.

2. R. decursiva (Wallich) Schott. Illustration: *Botanical Magazine*, 7282 (1893); Birdsey, The cultivated aroids, 105 (1951); Graf, Tropica, 110 (1978); Brickell (ed.), RHS A–Z encyclopedia of garden plants, 885 (2003).

Tall climber. Leaf-stalk 40–50 cm. Leaf-blade to 90 × 70 cm, regularly pinnatisect; segments 7–21 on each side, lanceolate with constricted base and tapered apex. Inflorescence-stalk much shorter than the leaf-stalk. Spathe *c.* 17.5 × 10 cm, fleshy, yellowish. Spadix *c.* 15 × 2–3 cm, yellowish, becoming grey-green. *India, Sri Lanka, N Burma.* G2.

Not usually seen in the juvenile state.

3. EPIPREMNUM Schott
M.T. Madison

Like *Rhaphidophora*, differing in the 1-celled ovary with usually 2 or 4 ovules, and in the hard-coated, kidney-shaped seeds.

A genus of about 12 species native to tropical SE Asia and the Pacific islands. The plants seldom flower in cultivation outside the tropics, as they require 5–10 m of host tree on which to climb before mature foliage is produced. In horticulture they are grown principally in hanging baskets or on poles for their attractive foliage which is tolerant of low light and low humidity. A moist, well-drained soil rich in humus is appropriate. The plants are propagated by cuttings. Plants growing out of doors in warmer regions will be damaged, but not killed, by a light frost.

Literature: Engler, A. & Krause, K., Monsteroideae & Calloideae, Das Pflanzenreich **37**: 54–67 (1908).

1a. Leaf-blade regularly pinnatisect and with tiny holes along the mid-rib, not variegated **1. pinnatum**
b. Leaf-blade entire or irregularly pinnatisect, variegated **2. aureum**

1. E. pinnatum (Linnaeus) Engler. Illustration: Das Pflanzenreich **37**: 61 (1908).

Climbing perennial, to 20 m tall, the stems green or brown, 1–4 cm thick. Leaf-blade dull green, pinnatisect with 8–20 divisions and with a row of pinprick-sized holes along each side of midrib. *Malaysia, Indonesia & New Guinea.* G1.

2. E. aureum (Linden & André) Bunting (*Pathos aureus* Linden & André; *Scindapsus aureus* (Linden & André) Engler; *Rhaphidophora aurea* (Linden & André) Birdsey). Illustration: *Illustration Horticole* **27**: 69 (1880); Birdsey, The cultivated aroids, 113 (1951); *Baileya* **10**: 156, 158 (1962); Graf, Exotica, edn 8, 228 (1976).

Climbing or trailing herb with broadly ovate, shiny leaves, more or less cordate at the base, variegated with golden yellow and green. In cultivation usually only the juvenile form with leaf-blades to 15 cm is found. Adult leaf-blade to 80 cm and irregularly pinnatisect. *Solomon Islands.* H5.

'Marble Queen' has cream and green variegation (Illustration: Graf, Exotica, edn 8, 138A, 1976).

4. SCINDAPSUS Schott
M.T. Madison

Plant evergreen, with tough stems climbing by means of adhesive, adventitious roots. Leaves entire, with parallel lateral veins. Spathe boat-shaped, deciduous. Spadix shorter than the spathe. Florets without perianths, each with 4 stamens and a 1-celled ovary with a solitary basal ovule. Seed more or less spherical, with a hard coat.

A genus of about 24 species from tropical Asia and the Pacific; only 1 species, *S. pictus*, is frequent in cultivation. It is slow growing and best grown in a hanging basket or clambering over rocks near a waterfall or pool. In pot-culture it requires a moist, well-drained soil rich in humus, and low to moderate light. Propagation is by cuttings.

Literature: Engler, A. & Krause, K., Monsteroideae & Calloideae, Das Pflanzenreich **37**: 67–80 (1908).

1. S. pictus Hasskarl var. **argyraeus** Engler. Illustration: *Gartenflora* **82**: 245 (1933); Graf, Exotica, edn 8, 228 (1976).

Slender, creeping climber. Stem 2–6 mm thick. Leaves 7–10 × 5–8 cm, ovate, cordate, silvery deep green marked with bright silver spots. Evidently a fixed juvenile form, of which flowers are unknown. *Java to Borneo.* G2.

5. STENOSPERMATION Schott
P.F. Yeo
Plants with horizontal rhizomes and erect, leafy stems. Leaf-stalk with a long sheath and a pulvinus at the apex (figure 14(6), p. 374); leaf-blade simple, with uniform, fine lateral veins. Inflorescence-stalk recurved at the top at flowering time. Spathe boat-shaped, deciduous. Spadix shorter than spathe, uniform. Florets bisexual, without perianths; stamens 4; ovary with an overhanging rim at the top, 2-celled, each cell with 4 or more ovules.

A genus of about 20 species from central and tropical S America. They are plants of slender habit with small but conspicuous drooping white spathes, and a few are occasionally grown in hothouse collections.

Literature: Engler, A. & Krause, K., Monsteroideae & Calloideae, *Das Pflanzenriech* **37**: 1–90 (1908).

1. S. popayanense Schott (*S. wallisii* Masters). Illustration: *Botanical Magazine*, 6334 (1877); Graf, Exotica, edn 8, 229 (1976).

Stem to 1 m. Leaf-stalk 15–17.5 cm. Leaf-blade 22–30 × 5–7 cm, mostly oblong-elliptic or oblong-lanceolate, rather dark green above with a narrow translucent margin; lateral veins uniform, very numerous. Spathe 8–12 cm long and nearly as wide, deeply hollowed, white, facing outwards from the shoot on its drooping stalk. Spadix 2.5–6 cm × 5–10 mm, yellowish white. *Colombia, Ecuador.* G2.

6. RHODOSPATHA Poeppig
P.F. Yeo
Evergreen climbing plants. Stem with leaves attached in 2 rows. Leaf-stalk with a pulvinus at the apex (figure 14(6), p. 374). Leaf-blade simple, with fine veins parallel to the main laterals and joined by small cross-veins. Inflorescence-stalk shorter than the leaf-stalk. Spathe boat-shaped, whitish or pink inside, soon shed. Spadix with a stipe, uniform. Florets without perianths, bisexual, or a few at the base female. Stamens 4, becoming twice as long as ovary. Ovary 2-celled, quadrangular in section, flat-topped. Ovules and seeds numerous.

About 12 species in the American tropics. They are suitable only for large hothouses, and should be treated like *Philodendron* (p. 388). Several have been in cultivation from time to time.

Literature: Engler, A. & Krause, K., Monsteroideae & Calloideae, *Das Pflanzenriech* **37**: 0–6 (1908).

1. R. picta Nicholson. Illustration: Graf, Exotica, edn 8, 226 (1976).

Leaf-stalk to 30 cm; sheaths reaching nearly to pulvinus, each to 1.2 cm wide at base; pulvinus *c.* 4 cm. Leaf-blade to 45 × 25 cm, elliptic or elliptic-ovate; upper surface deep green, with golden green mottling here and there between main lateral veins. Spathe 20–22 × 6.5–9 cm, pinkish. Spadix reaching tip of spathe, 1.6–1.8 cm thick, with a stipe *c.* 1.5 cm. *S America (origin not precisely known).* G2.

7. MONSTERA Adanson
M.T. Madison
Climbing or sprawling, evergreen plants with tough fibrous stems and large perforated leaves. Leaf-stalk with pulvinus at apex (figure 14(6), p. 374). Spathe boat-shaped, deciduous. Spadix shorter than the spathe. Florets without perianths, each with 4 stamens and a 2-celled ovary with 2 ovules in each cell. Seeds are soft.

A genus of about 25 species native to the New World tropics. Monsteras are much prized for their striking, perforated leaves and sweet, edible fruits. They are grown in pots, usually climbing on a pole, or rambling over stones near a pool or waterfall in a glasshouse. A moist, well-drained soil rich in humus is preferable. Though tolerant of low light and low atmospheric humidity, Monsteras will grow better in moderate light and with the moist air of a glasshouse. Propagation is by cuttings or seed.

Literature: Madison, M.T., A revision of Monstera (Araceae), *Contributions from the Gray Herbarium* **207**: 1–100 (1977).

1a. Leaf-blade thick, both perforated and pinnatifid **1. deliciosa**
b. Leaf-blade thin, perforated but not pinnatifid **2. adansonii**

1. M. deliciosa Liebmann (*Philodendron pertusum* Kunth & Bouché). Illustration: *Revue de l'Horticulture Belge* **30**: 125 (1904); Graf, Exotica, edn 8, 181 (1976); Brickell (ed.), RHS A–Z encyclopedia of garden plants, 699 (2003).

Climber to 20 m. Leaf-blade 25–90 × 25–75 cm, leathery, pinnatifid or pinnatisect, frequently perforated as well. Spadix 10–18 × 2.5–3 cm, swelling in fruit and becoming cream-coloured, the lower parts of the berries edible after the irritant upper portion has fallen away. *Mexico to Panama.* G1.

2. M. adansonii Schott var. **laniata** (Schott) Madison (*M. pertusa* (Linnaeus) de Vriese). Illustration: Graf, Exotica, edn 8, 177 (1976) – as *M. friedrichsthalii*.

Climber to 8 m. Leaf-blade 22–55 × 15–40 cm, ovate, thin, with numerous perforations. Spadix 8–13 × 1.5–2 cm, pale yellow. *Costa Rica to Brazil.* G1.

8. SPATHIPHYLLUM Schott
P.F. Yeo
Plants with rhizomes in or above the soil; with numerous leaves clustered near tips of rhizomes or on short, erect stems. Leaf-stalk nearly as long as blade or longer, with a long sheath embracing the stem at the base and with a pulvinus at the apex (figure 14(6), p. 374). Leaf-blade lanceolate or oblanceolate to oblong-ovate, dark green, often with the main lateral veins numerous and impressed. Inflorescence-stalk longer than the leaf-stalk. Spathe flat or slightly concave, white or greenish, its proportions similar to those of the leaf-blade. Spadix shorter than spathe, with a stipe which is often largely united with the spathe. Florets bisexual with 4–6 free or united, green or white perianth segments, 4–6 stamens with white or yellowish anthers, and a more or less conically tipped, white, incompletely 3-celled ovary; ovules 1–5 per cell.

One species in the Philippines and Indonesia and about 35 in tropical America. About 10 are grown as houseplants or in conservatories for their handsome foliage and inflorescences, which are often fragrant. They may also be grown for cut-flowers. They require a moist soil, rich in organic matter, and shade. Flowering is infrequent in the relatively dry atmosphere of occupied buildings. Dimensions of the plant are much affected by age and environment; those given below apply to well-grown plants. Many hybrids have been and continue to be raised in cultivation.

Literature: Bunting, G.S., The cultivated species of Spathiphyllum, *Baileya*. **9**: 109–17 (1960); Bunting, G.S., A revision of Spathiphyllum (Araceae), *Memoirs of the New York Botanical Garden* **10**(3): 1–54 (1960).

Leaf-stalk. With mottled sheath: **4**.
Leaf-blade. Less than 25 cm: **3**. Broadest above the middle: **1**. With 9 or fewer pairs of main veins: **3**.
Stipe. Long and largely joined to spathe: **4, 5, 6**; short and free from spathe: **1, 2, 3**.
Spathe. Spreading and recurved, not concave: **1, 2, 3**; erect or nearly so, concave: **4, 5, 6**. Mainly white on both sides: **2, 3, 5**; white with green back: **1**; yellow-green: **4, 6**.

1a. Tops of ovaries scarcely conical so that spadix is nearly smooth; spathe flat, spreading 2
b. Tops of ovaries conical, making the spadix rough; spathe concave, more or less erect 4
2a. Leaf-blade 13–20 cm; perianth segments green, so that spadix is patterned with green and white **3. floribundum**
b. Leaf-blade 25–50 cm; perianth segments white; spathe white or cream-coloured 3
3a. Leaf-blade about 3 times as long as broad; angle of divergence of lateral veins from midrib not more than *c.* 45° **1. cannifolium**
b. Leaf-blade usually about twice as long as broad; angle of divergence of lateral veins from midrib more than 45 **2. commutatum**
4a. Spathe white, becoming green in age, rounded at base **5. wallisii**
b. Spathe yellow-green, often acute at base 5
5a. Spathe 18–33 cm, leathery; ovules 6–16 **4. cochlearispathum**
b. Spathe 10–26 cm, papery; ovules 3–6 **6. phryniifolium**

1. S. cannifolium (Dryander) Schott. Illustration: *Botanical Magazine*, 603 (1802); Graf, Exotica, edn 8, 232 (1976); Everett, Encyclopedia of horticulture **9**: 3195 (1982).

Leaf-blade 25–45 × 8–16 cm, narrowly to broadly oblanceolate or narrowly elliptic, with tapered base; veins numerous, diverging from midrib at 45° or less and curving towards apex of blade, moderately impressed. Spathe 10–22 × 2.5–6.5 cm, oblong or elliptic, thick, spreading and recurved, its base clasping the stalk, upper surface white, lower green. Flowers fragrant. Spadix 4.5–12 cm × 5–20 mm, nearly smooth, cream-coloured; stipe 5–20 mm, free from spathe. Perianth segments united, white. Ovules usually 6–18. *Northern S America east of the Andes, Trinidad.* G2.

2. S. commutatum Schott. Illustration: *Gartenflora* **19**: 1 (1870); *Baileya* **9**: 115 (1961); Graf, Exotica, edn 8, 232 (1976).

Very similar to *S. cannifolium*, but with a relatively broader, elliptic, leaf-blade, 35–50 × 11–24 cm, its veins mostly diverging from midrib at more than 45°, spathe relatively broader, 14–24 × 5.5–8 cm, sometimes with the edges at the base continued down the stalk for up to 2 cm, creamy white with green edge; spadix slightly narrower. *Indonesia (N Celebes and some neighbouring islands), S & E Philippines.* G2.

3. S. floribundum (Linden & André) N.E. Brown. Illustration: *Illustration Horticole* **21**: 24 (1874); Birdsey, The cultivated aroids, 116 (1951); *Baileya* **9**: 116 (1961); Everett, Encyclopedia of horticulture **9**: 3194 (1982).

Leaf-blade 13–20 × 5.5–9 cm, elliptic to oblong or oblanceolate, with about 9 pairs of scarcely impressed veins diverging widely from the midrib; upper surface with a velvety lustre. Spathe 4–8 × 1.2–3 cm, lanceolate to oblong-elliptic, white, spreading and recurved, its base clasping the stalk. Spadix 2.5–5 cm × 7–8 mm, nearly smooth, white and green; stipe 3–8 mm. Perianth segments free or adherent in age, green. Ovules 4–6. *Colombia.* H2.

Valued for conservatory use but less suitable as a house plant.

S. patinii (Hogg) N.E. Brown (*S. candidum* (Bull) N.E. Brown). Illustration: *Illustration Horticole* **27**: 135 (1880); Graf, Exotica, edn 8, 231 (1976). Similar to *S. floribundum* but with much narrower leaf-blades (12–21 × 2.8–5 cm), with 4–6 pairs of main veins diverging from midrib at less than 45°, glossy, without velvety lustre on upper surface, with narrower spathe and smaller spadix. Back of spathe creamy white with green midrib. *Colombia.* H2.

4. S. cochlearispathum (Liebmann) Engler. Illustration: *Baileya* **9**: 110 (1961); Graf, Exotica, edn 8, 231 (1976).

Leaf-blade 45–65 × 15–23 cm, oblong, oblong-lanceolate or narrowly elliptic, with very numerous, widely diverging, impressed lateral veins; margin wavy; sheath of leaf-stalk and scale-leaves mottled with white and usually crisped. Spathe 18–33 × 6–10 cm, oblanceolate or elliptic, erect, concave, leathery, yellow-green. Spadix 5.5–11.5 cm, very rough from the long ovaries, white; stipe 4.5–8.5 cm, entirely united with spathe or the top 1 cm free. Perianth segments free. Ovules 6–16. *Mexico, near the Caribbean coast.* H2.

The tallest species, easily reaching 1.2 m.

5. S. wallisii Regel. Illustration: *Gartenflora* **26**: 323 (1877); Brickell (ed.), RHS A–Z encyclopedia of garden plants, 998 (2003).

Leaf-blade 24–36 × 5–10 cm, lanceolate-elliptic to oblong-elliptic, with *c.* 8–10 pairs of main veins arising from midrib at an angle of 45–50°; margin wavy. Spathe 7–17 × 2.5–7.5 cm, ovate to oblong-elliptic, concave, white, becoming green in age. Flowers fragrant. Spadix 1.5–8 cm, very rough from the long ovaries, white, with midrib green on the back; stipe 2–5 cm, united with spathe except in upper 5–10 mm. Perianth segments free, white. Ovules 6–12. *Columbia?* H2.

Rare in cultivation. The variant most frequently grown (often as *S. wallisii*) is 'Clevelandii' (*S. clevelandii* invalid; *S. kochii* misapplied; *S. candidum* misapplied). Illustration: Birdsey, The cultivated aroids, 118 (1951); Graf, Exotica, edn 8, 233 (1976); Everett, Encyclopedia of horticulture **9**: 3195 (1982). Leaf-blade 36–42 × 9–12 cm, relatively narrower than in *S. wallisii*, thin and drooping, glossy, with *c.* 12–15 pairs of impressed veins. Spathe *c.* 18 × 8.5 cm, erect or slightly spreading. Spadix 7–8.5 × 1.5–2 cm. 'Mauna Loa' (Illustration: Graf, Exotica, edn 8, 233, 1976; Huxley & Gilbert, Success with house-plants, 368, 1979) is probably descended in part from 'Clevelandii', from which it differs in its broader, less drooping leaf-blades and the abundance of erect, broadly elliptic spathes with the spadix centrally placed.

6. S. phryniifolium Schott (*S. friedrichsthalii* misapplied). Illustration: *Baileya* **9**: 112 (1961); Graf, Exotica, edn 8, 138A, 231, 233 (1976).

Leaf-blade 25–50 × 8–16 cm, lanceolate or sometimes oblong-elliptic, rounded at base, with numerous pairs of impressed veins arising from midrib at an angle of *c.* 60°. Spathe 10–26 × 3–9 cm, lanceolate to oblong-elliptic, papery, pale yellow-green, the edges at the base continued a little way down the stalk. Spadix 2.5–8 cm, rough from the long ovaries, white; stipe 2.5–6 cm, united with spathe except in upper 5–15 mm. Perianth segments free, white. Ovules 3–6. *Panama, Costa Rica.* G2.

9. LYSICHITON Schott
D.A. Webb
Robust, rhizomatous, stemless herbs, forming large clumps. Leaves ovate-oblong, truncate at the base, shortly stalked, without a sheath, appearing about the same time as the flowers. Spathe arising from ground level, withering before the fruit is ripe, the lower part consisting of a narrow sheath enclosing the very long stipe of the spadix, but with free margins, the upper part erect, elliptic, strongly concave. Spadix cylindric, on a long stipe, without appendix. Flowers crowded, bisexual; perianth-segments 4, often unequal; stamens 4; ovary 2-celled. Berries partly embedded in the spadix; seeds 2.

A genus of 2 closely related species from the North Pacific area. They may be cultivated on the banks of streams or the margins of ponds.

1. L. americanus Hultén & St John (*L. camtschatcensis* misapplied). Illustration: *Botanical Magazine*, 7937 (1904); Hay & Beckett, Reader's Digest encyclopaedia of garden plants and flowers, edn 2, 424 (1978); Brickell (ed.), RHS A–Z encyclopedia of garden plants, 657 (2003).

Rhizome short, thick, vertical. Leaf-blade 40–100 × 25–70 cm, rather fleshy; leaf-stalk very short, thick. Stipe lengthening during flowering, green above. Lower (sheathing) part of spathe white, wrinkled; upper part 15–35 cm, bright yellow. Spadix 5–21 cm, stout, greenish or yellowish. Berries green. *Western N America, from Alaska to C California.* H2. Spring.

L. camtschatcensis (Linnaeus) Schott. Illustration: Makino, Illustrated flora of Japan, 734 (1942); Kitamura et al., Coloured illustrations of herbaceous plants of Japan (Monocotyledoneae), t. 49, f. 317 (1964); Hay & Beckett, Reader's Digest encyclopaedia of garden plants and flowers, edn 2, 424 (1978). Very similar to L. *americanus*, but slightly smaller, with the spathe white instead of yellow and flowering rather later. *NE Asia, from S Kamchatka to C Japan.* H. Late spring.

10. SYMPLOCARPUS Salisbury
D.A. Webb
Fetid, stemless herbs with stout, vertical rhizome. Leaves numerous, forming large clumps; stalk channelled, without a sheath; blade ovate, cordate, acute, entire. Spathes borne at ground level, very strongly concave, somewhat suggestive of a shell. Spadix with a short stipe enclosed within the spathe, shortly cylindric or almost spherical. Florets bisexual; perianth-segments 4; stamens 4; ovary 1-celled. Fruits very deeply embedded in the spadix so as to form a compound fruit, succulent on the surface, dry and spongy within.

A genus of probably 2 species, from temperate E Asia and eastern N America. The species described below should be grown in damp, humus-rich soil in shade.

Literature: Krause, K., Monsteroideae & Calloideae, Das Pflanzenreich **37**: 150–2 (1908).

1. S. foetidus (Linnaeus) Nuttall. Illustration: *Botanical Magazine*, 3224 (1833); Parey's Blumengärtnerei **1**: 176 (1958); Everett, Encyclopedia of horticulture **10**: 3274, 3275 (1982).

Leaves to 80 × 40 cm, appearing after the flowers. Spathe yellowish green, dark brownish purple or spotted and striped with both colours. Spadix 2–3 cm across in flower; greatly enlarged in fruit. *Cool-temperate regions of E Asia; eastern N America, from Nova Scotia to Florida.* H2. Later winter or early spring.

11. ORONTIUM Linnaeus
D.A. Webb
Plant aquatic, rhizomatous. Leaves all basal, with long stalks, clustered, mixed with scale-leaves; blade ovate, entire, submerged, floating or emerging. Inflorescence-stalk a stipe, slender, about as long as leaves, white above, terminating in a slender yellow spadix. Spathe much reduced, consisting of a sheath surrounding the lower part of the very long stipe and bearing a small, bract-like green blade which withers early. Florets numerous, crowded, bisexual, with 4–6 perianth-segments, 4–6 stamens and a 1-celled ovary. Fruit green, deeply embedded in the spadix, containing a single seed.

A single species, cultivated in shallow water overlying a fair depth of loamy soil.

Literature: Krause, K., Monsteroideae & Calloideae, Das Pflanzenreich **37**: 152–3 (1908).

1. O. aquaticum Linnaeus. Illustration: Gleason, Illustrated flora of the northeastern United States and adjacent Canada **1**: 369 (1952); Mühlberg, Das grosse Buch der Wasserpflanzen, 368 (1980); Everett, Encyclopedia of horticulture **7**: 2444 (1981); Brickell (ed.), RHS A–Z encyclopedia of garden plants, 749 (2003).

E United States. H3. Late spring.

12. CALLA Linnaeus
D.A. Webb
Rhizomatous aquatic plants. Leaves all basal, mixed with scale-leaves, long-stalked, broadly ovate to kidney-shaped, shortly pointed, cordate; sheathing bases of the leaf-stalks persisting on the rhizome. Spathe similar to the leaf-blades in shape but smaller, erect, concave but scarcely enfolding the spadix even at its base, greenish outside, white on inner surface during flowering period, persistent in fruit. Two or 3 spathes per spadix are sometimes developed. Spadix green, shortly cylindric, with a short stipe and without appendix. Florets crowded, without perianths, mostly bisexual, but the uppermost usually male; stamens 6 or more; ovary 1-celled. Berry red, with 4–10 seeds.

A single species, easily cultivated in marshy ground or shallow water around the edge of a pond.

Literature: Krause, K., Monsteroideae & Calloideae, Das Pflanzenreich **37**: 154–5 (1908).

1. C. palustris Linnaeus. Illustration: *Botanical Magazine*, 1831 (1816); Hegi, Illustrierte Flora von Mittel-Europa, **2**: 165, 172 (1939) and edn 2, **2**(1): 321 (1967–80); Polunin, Flowers of Europe, t. 183 (1969); Aeschimann et al., Flora alpina **2**: 134.2.1 (2004).

Plants 15–30 cm high. Leaves 5–10 cm wide; inflorescence-stalk rising up to 30 cm above water; spathe 3–6 cm wide; spadix 2–3 cm. *Cool temperate and subarctic regions of North America and Eurasia.* H1. Summer.

13. AMORPHOPHALLUS Blume
P.F. Yeo
Plants each with an underground corm, usually hollowed on top, from which arise several scale-leaves, one large deciduous leaf with erect stalk, and one large inflorescence. Leaf-blade with 3 divisions (or by dichotomy 5 or more) which are once or twice pinnately lobed or cut. Spathe funnel-shaped or bell-shaped, often shorter than spadix. Spadix with a sterile appendix. Florets without perianths, unisexual; zones of male and female florets adjacent. Male florets with 1–6 stamens. Ovary with 1–4 cells, each with 1 ovule.

About 90 species in the Old World tropics. Grown mainly for the striking but evil-smelling inflorescence which emerges from the dry corm in spring, without need of soil or water. After flowering the corms are planted in a humus-rich soil and given supplementary fertiliser. The plants must be dried off in autumn and the corms stored at a temperature above 10 °C. The species treated here may be placed out of doors in summer and can be increased from cormlets.

Literature: Engler, A., Lasioideae, Das Pflanzenreich **48**: 61–109 (1911).

1a. Leaves not bearing cormlets; leaf-divisions beyond the dichotomies profusely bipinnatisect; spadix appendix narrowly conical **1. rivieri**
b. Leaves bearing cormlets; leaf-divisions beyond the dichotomies sparsely pinnate or bipinnatisect; spadix appendix ovoid **2. bulbifer**

1. A. rivieri Durieu. Illustration: *Flore des serres* **19**: 43, 45 (1873); Birdsey, The cultivated aroids, 29 (1951); Graf, Exotica, edn 8, 143 (1976).

Corm to 25 cm wide. Leaf-stalk to 100 cm, mottled. Leaf-blade to 1.3 m wide, its divisions once or twice dichotomous and then bipinnatisect. Inflorescence-stalk 50–70 cm; spathe 20–40 cm, dark brownish red inside, edges wavy; spadix nearly twice as long as spathe; appendix 20–55 cm, narrowly conical, dark reddish brown. *SE Asia.* G1. Spring.

The most commonly cultivated variant is 'Konjac', with the largest inflorescences and edible tubers, for which it is widely cultivated in the West Pacific region.

2. A. bulbifer (Curtis) Blume. Illustration: *Botanical Magazine*, 2072 (1819), 2508 (1824); Parey's Blumengärtnerei **1**: 177 (1958).

Corm to 7 cm wide. Leaves bearing cormlets. Leaf-stalk to 100 cm, mottled; leaf-blade to 60 cm wide; main divisions cut to the base into a few segments, rarely again divided. Inflorescence-stalk to 20 cm; spathe 15–20 cm, whitish with pale green or pink mottling outside and whitish, tinged with pink, inside; spadix about as long as spathe; appendix to 7 cm, ovoid, pinkish. *NE India, Burma; ?SW China.* G1. Spring.

14. NEPHTHYTIS Schott

P.F. Yeo

Plants evergreen. Stem a horizontal rhizome bearing leaves at short intervals. Leaf-stalk terete; sheath very small. Leaf-blade sagittate or hastate, the basal lobes about as long as the principal lobe; primary lateral veins joining the marginal vein (figure 14(6), p. 374), with netted veins between them. Inflorescence-stalk with several scale-leaves at base. Spathe open; hooded, flat, or reflexed. Florets without perianths, unisexual. Spadix with zones of male and female florets adjacent, the female zone much the shorter. Male florets of 1–4 stamens. Ovary containing 1 or 2 ovules.

Six species of West tropical Africa, sometimes cultivated in tropical collections. They require a moist soil high in organic matter, a hot humid atmosphere and some shade. The genus name is sometimes applied informally to the cultivated species of *Syngonium* (p. 401).

Literature: Engler, A., Lasioideae, Das Pflanzenreich **48**: 110–14 (1911); Bogner, J., On two Nephthytis species from Gabon and Ghana, *Aroideana* **3**: 75–85 (1980).

1. N. afzelii Schott. Illustration: Das Pflanzenreich **48**: 111 (1911); Graf, Exotica, edn 8, 224 (1976).

Leaf-stalk to 50 cm. Leaf-blade to 35 × 25 cm, sagittate; principal lobe triangular, slightly shorter than the ovate laterals; all lobes sharply pointed. Inflorescence-stalk slightly shorter than the leaf-stalk. Spathe 6–7 cm, hooded, green, unspotted. Spadix slightly shorter than spathe. Male florets with 4 stamens. Stigma disc-like. Berries few, to 9 mm, ovoid, orange. *West tropical Africa.* G2.

15. RHEKTOPHYLLUM N.E. Brown

P.F. Yeo

Evergreen, climbing plants with distinct juvenile and adult phases. Juvenile phase with thin stems and simple leaf-blades. Leaf-blades in juvenile phase hastate or sagittate, in adult phase ovate or oblong-ovate and with perforations and marginal incisions between wedge-shaped segments; main veins of basal lobes exposed in the sinus (figure 14(8), p. 374); minor veins netted. Inflorescence-stalk shorter than leaf-stalk. Spathe 9–10 cm, green outside and reddish purple inside. Florets without perianths, unisexual. Spadix with male zone 3 times as long as female. Male florets with 3 or 4 stamens. Ovary with 1 cell and 1 ovule.

A genus of 1 species from West Africa. Cultivated in the juvenile phase for its beautifully patterned leaves; it requires the same conditions as *Philodendron* (p. 388).

Literature: Engler, A., Lasioideae, Das Pflanzenreich **48**: 119–21 (1911).

1. R. mirabile N.E. Brown. Illustration: *Journal of Botany* **20**: 136, t. 230 (1882); Das Pflanzenreich **48**: 120 (1911); Birdsey, The cultivated aroids, 107 (1951); Graf, Exotica, edn 8, 224 (1976).

Leaf-blade of juvenile phase to 30 cm, hastate or sagittate with coarsely wavy margins; basal lobes broad with main veins exposed at the sinus for part of their length; principal lobe with 2 or 3 pairs of main lateral veins curving towards its apex; upper surface dark green, variegated with whitish between the veins in a pattern suggestive of fern fronds growing inwards from the margin. *Nigeria to Congo.* G2.

16. HOMALOMENA Schott

P.F. Yeo & D.J. Leedy

Plants evergreen, with a short stem above or below ground. Leaf-blade variable in shape; primary and subsidiary lateral veins parallel and joining a marginal vein (figure 14(6), p. 374), only the finest ones forming a network. Inflorescences several from one axil, sometimes accompanied by scale-leaves. Spathe margins overlapping, unrolling slightly above at flowering time. Spadix shorter or longer than spathe. Florets without perianths, unisexual. Zone of female florets shorter than the adjacent zone of male florets. Male florets usually with 2–4 stamens; lowest florets sometimes sterile. Female florets with or without 1 or more staminodes. Ovary incompletely divided into 2–4 cells, each with several to many ovules. Fruit is a berry.

About 140 species, mainly in S Asia and the SW Pacific region but a few in tropical S America. They are used and cultivated like *Dieffenbachia* (p. 393), but *H. wallisii* does not respond well to pot-culture.

Literature: Engler, A., Homalomeninae & Schismatoglottidinae, *Das Pflanzenriech* **55**: 25–81 (1912); Bogner, J., Homalomena lindenii (Araceae), *Baileya* **20**: 6–10 (1976).

1a. Leaf-stalk shorter than blade; leaf-blade elliptic-oblong, with blotchy variegation **1. wallisii**
b. Leaf-stalk longer than blade; leaf-blade triangular-ovate, with the main veins white to yellow **2. lindenii**

1. H. wallisii Regel. Illustration: *Gardeners' Chronicle* **7**: 108 (1877);

Birdsey, The cultivated aroids, 68 (1951); Graf, Exotica, edn 8, 176(1976).

Leaves crowded on a short, erect stem. Leaf-blade 13–20 × 6–8 cm, much longer than the stalk, more or less elliptic-oblong, rounded or cordate at base; upper surface smooth, dark green with a narrow white margin and bold irregular yellow blotches; lower surface rough, glaucous, tinged with red. Inflorescence longer than leaf-stalk. Spathe *c.* 8 cm, pale reddish purple, glossy within, constricted at the middle. Spadix as long as spathe. Female florets without staminodes. *Colombia.* G2.

Differs from *Aglaonema pictum* (p. 393) in the yellow rather than light green or silvery variegation, the white margin of the leaf-blade and the elongated spathe.

2. H. lindenii (Rodigas) Lindley (*Alocasia lindenii* Rodigas). Illustration: *Illustration Horticole* **33**: 111 (1886); Graf, Exotica, edn 8, 120, 125 (1976); *Baileya* **20**: 6, 8 (1976).

Stem to 25 cm, erect. Leaf-stalk whitish. Leaf-blade 20–25 × 12–18 cm, shorter than the stalk, triangular-ovate with a deeply cordate base, bright green and slightly glossy above; midrib and main veins impressed, white to yellow except at extremities. Inflorescence much shorter than leaf-stalk. Spathe 5–7 cm, pale green, scarcely constricted at the middle. Spadix as long as spathe. Female florets without staminodes. *New Guinea.* G2.

When damaged the plant smells of aniseed or chervil. The supposed intergeneric hybrid × *Homalocasia miamiensis* is thought by Bogner to be a form of *H. lindenii.*

17. SCHISMATOGLOTTIS Zollinger & Moritzi

P.F. Yeo

Plants evergreen. Stem rhizomatous or above ground and erect. Leaf-blade with minor veins nearly parallel to primary lateral veins (figure 14(5), p. 374); 1 or more marginal veins present. Inflorescence-stalk shorter than the leaf-stalk. Spathe margins overlapping, the upper part open at flowering time, dropping off while both parts are still fresh. Spadix sometimes with a sterile zone between male and female zones, bearing sterile florets sparsely or densely; usually with a club-shaped sterile appendix patterned with rudimentary florets (figure 15(6), p. 377). Florets without perianths, unisexual. Male florets with 2 or 3 stamens, filaments flattened. Female florets (in our species) with staminodes. Ovary 1-celled, ovules many.

About 100 species, mainly in tropical Asia, but a few in tropical America. Several have been brought into cultivation and are grown mainly for their variegated leaves. They grow best in a loose, humus-rich soil. The rhizomatous species are propagated by division and the erect-stemmed species by cuttings.

Literature: Engler, A., Homalomeninae & Schismatoglottidinae, Das Pflanzenreich **55**: 1–134 (1912).

1a. Stem erect, above ground; spadix with a sterile zone between male and female zones **1. concinna**
b. Stem an underground rhizome; spadix without a sterile intermediate zone 2
2a. Leaf-blade green, usually variegated irregularly with whitish yellow blotches **2. neoguineensis**
b. Leaf-blade with a bluish white band on either side, about half-way between midrib and margin **3. picta**

1. S. concinna Schott. Illustration: *Illustration Horticole* **28**: 71 (1881) – as *S. lavallei*, **29**: 173 (1882); Das Pflanzenreich **55**: 96 (1912); Graf, Exotica, edn 8, 230 (1976).

Stem erect, to *c.* 20 cm. Stem and leaf-stalk often reddish or pink. Leaf-stalk to *c.* 15 cm, not much longer than blade, slightly channelled towards apex. Leaf-blade to *c.* 15 × 4 cm, ovate-lanceolate with rounded or slightly cordate base; upper surface green, often with irregular grey-green blotches; lower surface green or purple. Spathe 6.5 cm; tube *c.* 2 cm. Spadix with female zone to 1.5 cm, intermediate sterile zone *c.* 1 cm, with scattered sterile florets, male zone *c.*1 cm and sterile appendix to 3 cm; male zone and appendix yellowish. *Borneo, Java, Sumatra.* G2.

'Immaculata' has the leaf-blade uniformly green above and purple beneath; 'Purpurea' has the leaf-blade variegated above and purple beneath.

2. S. neguineensis André. Illustration: *Illustration Horticole* **27**: 68 (1880); Birdsey, The cultivated aroids, 111 (1951); Graf, Exotica, edn 8, 230 (1976).

Stem an underground rhizome. Leaf-stalk flattened on the upper side towards apex, up to twice as long as blade, green or purplish. Leaf-blade 12–25 × 7–16 cm, ovate, cordate at base, thin; upper surface bright green with irregular blotches of whitish yellow, lower light green. Spathe *c.* 7 cm; tube 2.5–3 cm, green; limb white. Spadix with female zone *c.* 1.5 cm and male zone *c.* 1 cm, adjacent to female; appendix *c.* 1 cm, orange. *New Guinea.* G2.

3. S. picta Schott. Illustration: Graf, Exotica, edn 8, 229, 230 (1976).

Stem an underground rhizome. Leaf-stalk 20–30 cm, or sometimes to 60 cm, flattened on upper side towards apex, green. Leaf-blade 15–20 × 7.5–13 cm, or sometimes to 35 × 26 cm, narrowly ovate or oblong-ovate, cordate at base, thin; upper surface bright green and usually marked with a ragged, pale, glaucous band running along each side about half-way between midrib and margin. Spathe 5–6 cm; tube green; limb greenish yellow. Spadix with female zone 2 cm and male zone 1.5 cm, adjacent to female, pale yellow; appendix 5–7.5 mm. *Borneo, Celebes, Sumatra & Java.* G2.

18. PHILODENDRON Schott

G.S. Bunting

Plants usually climbing, evergreen, with stout stems bearing adventitious roots. Leaf-stalk commonly stout and about as long as the blade, with a well-developed sheath which does not embrace the stem (figure 14(3 & 4), p. 374). Leaf-blade often large, entire or variously lobed to pinnatifid or pedate, with pinnate venation (figure 14 (5), p. 374). Inflorescence-stalks 1 or more in leaf-axils. Spathe fleshy, rolled around the spadix, unfurling at flowering time in the upper half or more, sometimes nearly to the base, then hooded or boat-shaped, green or white or variously marked with red or purple, persistent until the fruits ripen. Spadix usually white or whitish. Florets without perianths, unisexual; zones of male and female florets touching; male florets of 2–5 or more stamens; lowest florets sterile; female florets without staminodes. Ovary with 2–many cells, each with 1–many ovules. Berries white to orange or red when ripe.

Perhaps 500 species of the American wet tropical forests, many of which are widely cultivated for their attractive foliage or spathes and durability in interior decorations, and also for landscaping in warm climates. Several striking hybrids are in cultivation, as well as some white-variegated forms.

Small plants in containers often have only juvenile foliage which may be distinct from the adult leaf form. Positive identification can be made only with an adult leaf, whose form is noted in the descriptions below.

Philodendrons are of easy culture in a continually humid, rich loam with good drainage, but some tolerate dry soils. Many species thrive in the low atmospheric humidity of heated dwellings, etc. and most tolerate remarkably low light, but some, such as *P. bipinnatifidum* (*P. selloum*), require high light intensity, thriving in full sun. Climbing species may be planted with a support to which their aerial roots can cling; if the support is moistened regularly and other conditions are adequate, the erect-growing stem will finally produce leaves of the adult form, while if left to grow pendent, the leaves (except in the case of *P. scandens*) will become very reduced and the plant of limited ornamental value. Propagation is by cuttings or layers. The so-called arborescent species such as *P. bipinnatifidum* are usually propagated by seeds and are often of mixed parentage.

Literature: Krause, K., Philodendrinae, Das Pflanzenreich **60**: 1–136 (1913); Bunting, G.S., Vegetative anatomy and taxonomy and the Philodendron scandens complex, *Gentes Herbarum* **10**: 136–68 (1968).

Habit. Prostrate: **5, 10, 12, 13**; erect or decumbent: **2, 5**; climbing: **1, 3, 4, 6, 7, 8, 9**.
Leaf-blade. Deeply lobed or pinnatisect: **1, 2, 3, 4**; sinuate (shallowly lobed): **1**; entire with base acute to truncate or notched and erect on leaf-stalk: **5**; entire with base cordate or sagittate and reflexed on leaf-stalk: **6, 7, 8, 9, 10, 11, 12, 13**. Green on upper surface: **1, 2, 3, 4, 5, 6, 7, 9, 10, 13**; brown or black-green on upper surface: **8, 9** (juvenile); grey-splotched: **11, 12**; strongly variegated with white and grey-green: **10**.
Leaf-stalk. More or less terete: **1, 3, 4, 7, 8, 9**; prominently flattened or channelled on upper surface: **2, 5, 9, 10**; narrowly winged near apex: **12**. Surface more or less rough or covered with fleshy scales: **3, 6, 8, 11**.
Spathe. Tube green outside, limb green or white (or pink) on both surfaces: **1, 2, 3, 4, 5, 6, 8, 9, 11, 13**; entire spathe green outside, limb red inside: **10**; tube purple outside, limb green or white on both surfaces: **1**; entire spathe brown-red, red, or purplish red outside, limb red or white inside: **2, 7, 12**. Covered with fleshy scales: **6**.

1a. Leaf-blade lobed or pinnatisect 2
b. Leaf-blade entire or margin only shallowly sinuate 5
2a. Each lateral segment of blade (discounting basal lobes) with a single principal vein arising from midrib; blade regularly pinnatifid or pinnatisect 3
b. Each lateral segment of blade (discounting basal lobes) with 2 or more principal veins arising from midrib; blade lobed or irregularly pinnatifid 4
3a. Leaf-blade once pinnatifid or pinnatisect; leaf-stalk more or less terete **1. angustisectum**
b. Leaf-blade twice pinnately cut or lobed; leaf-stalk flat on upper side with angular-ridged borders **2. bipinnatifidum**
4a. Leaf-blade with a prominent elliptic or obovate apical lobe, plus 2–5 or more deeply separated, oblong segments on each side of midrib, in addition to the pinnatifid basal lobes **3. pedatum**
b. Leaf-blade with a prominent elliptic or obovate apical lobe, the remainder shallowly 2-lobed on each side of midrib **4. bipennifolium**
5a. Leaf-blade narrowly obovate-oblong, erect on a much shorter, stout leaf-stalk whose upper surface is flat with angular-ridged borders **5. wendlandii**
b. Leaf-blade ovate or triangular in outline with prominent basal sinus, reflexed; leaf-stalk not as above, about as long as blade or longer 6
6a. Mature leaf-blade brown or black-green or suffused with red-purple on one or both surfaces 7
b. Mature leaf-blade more or less pure green on both surfaces, at least never brown or black-green or suffused with red-purple 10
7a. Leaf-stalks, scale-leaves, inflorescence-stalks and spathes covered with fleshy scales **6. verrucosum**
b. Leaf-stalks, scale-leaves, inflorescence-stalks and spathes never covered with fleshy scales 8
8a. New stems purple or red-purple **7. erubescens**
b. New stems green or brownish 9
9a. Stem marked with longitudinal white lines; leaf-blade narrowly ovate with sagittate base, usually black-green on upper surface; juvenile blade ovate and shallowly cordate with leaf-stalk winged in lower half **8. melanochrysum**
b. Stem not marked with white lines; leaf-blade ovate with cordate base, shiny and brownish to velvety dark green on upper surface; leaf-stalk never winged **9. scandens**
10a. Stem without persistent fibrous remains of old scale-leaves 11
b. Stem with persistent, fibrous remains of old scale-leaves 12
11a. Leaf-blade to *c.* 30 cm with only 3–4 principal veins arising from each side of midrib (discounting basal ribs) **9. scandens**
b. Leaf-blade to *c.* 60 cm with about 5–6 principal veins arising from each side of midrib (discounting basal ribs) **10. domesticum**
12a. Leaf-blade dark green with white midrib and principal veins **13. gloriosum**
b. Leaf-blade with veins not white 13
13a. Plant climbing; leaf-stalk rough **11. ornatum**
b. Plant prostrate; leaf-stalk crinkly-winged near apex **12. mamei**

1. P. angustisectum Engler (*P. elegans* Krause). Illustration: Das Pflanzenreich **60**: 116, 117 (1913); Birdsey, The cultivated aroids, 77 (1951); Graf, Exotica, edn 8, 212 (1976); Graf, Tropica, 113 (1978).

Climber of moderate size. Leaf-blade *c.* 60 × 45 cm, more or less reflexed, ovate in outline, pinnatisect with up to 16 linear, acute segments on each side; segments *c.* 2.5 cm wide, each with one vein; basal sinus triangular. Leaf-stalk terete, nearly as long as blade. Inflorescence-stalk short. Spathe *c.* 15 cm, green outside with pink margins, limb yellowish inside. *W Colombia*. G2.

P. lacerum (Jacquin) Schott. Illustration: Birdsey, The cultivated aroids, 85 (1951); Graf, Exotica, edn 8, 193, 221 (1976); Graf, Tropica, 115, 119 (1978). Plant tall, with long internodes. Leaf-blade to *c.* 75 cm, ovate-circular in outline, cordate at base, pinnately lobed less than halfway to midrib; lobes more or less wedge-shaped and obtuse; midrib and principal veins very prominent. Leaf-stalk to 90 cm, terete. Inflorescence-stalk to twice as long as spathe. Spathe *c.* 12 cm, with tube outside dull purple and limb greenish cream. *Cuba, Jamaica, Hispaniola.* G2.

P. × corsinianum Senoner. Illustration: *Botanical Magazine*, 8172 (1908); Birdsey, The cultivated aroids, 75 (1951); Parey's Blumengärtnerei **1**: 181 (1958); Graf, Tropica, 112 (1978). Plant tall with leaf-blade to 75 × 60 cm, ovate, cordate sagittate at base, shallowly pinnately lobed; marked beneath when new with metallic brownish purple between green veins. Leaf-stalk about as long as blade. Inflorescence-stalk short, with white lines. Spathe *c.* 19 cm, with tube purple outside and limb green. *Garden origin.* G2.

The first recorded hybrid of *Philodendron*, possibly involving *P. verrucosum*. Recommended for cultivation only in a warm, humid environment.

2. P. bipinnatifidum Endlicher (*P. selloum* C. Koch). Illustration: *Gartenflora* **29**: 353, t. 1029 (1880); Graf, Exotica, edn 8, 183, 185 – as *P.* × *barryi*, 187 (1976); Graf, Tropica, 112, 120, 121 (1978); Brickell (ed.), RHS A–Z encyclopedia of garden plants, 802 (2003).

Huge, tree-like plant, erect to *c.* 2 m, decumbent in age, with very stout aerial roots. Stem to 10 cm in diameter; internodes very short; leaf-scars prominent. Leaf-blade 1 m, more or less reflexed, glossy, ovate in outline, sagittate, regularly pinnatisect into many blunt-tipped segments, each often again pinnatifid or lobed, flat or wavy. Leaf-stalk about as long as blade, upper surface more or less flat between the 2 longitudinal ridges. Inflorescence-stalk *c.* 10 cm, solitary. Spathe *c.* 30 × 7 cm, green to dark purplish red on outside and red-edged, cream inside. *SE Brazil.* G2.

A handsome species whose adult phase is suitable only for a large area with bright light.

3. P. pedatum (W.J. Hooker) Kunth (*P. laciniatum* (Vellozo) Engler). Illustration: Das Pflanzenreich **60**: 112 (1913); Parey's Blumengärtnerei **1**: 182 (1958); Graf, Exotica, edn 8, 202, 212 (1976).

Tall climber of moderate size. Leaf-blade reflexed, glossy, ovate; the part along midrib *c.* 45 × 30 cm, irregularly pinnatifid into up to 5 oblong, obtuse segments on each side, each with 2 or more principal veins, in addition to the elliptic, obovate or rhombic terminal segment; the flaring basal lobes 27 × 17 cm. Leaf-stalk longer than blade, more or less terete. Spathe 14 cm, green outside with red edge. *S Venezuela to Surinam & SE Brazil.* G2.

The juvenile phase has 5-lobed leaves. This variable species develops long internodes and small leaves in adverse indoor conditions. Based on leaf variations, several varieties and cultivars have been named (e.g. *P.* 'Florida' and *P.* 'Florida Compacta'), and there are some hybrids involving this species.

P. squamiferum Poeppig. Illustration: *Illustration Horticole* **33**: 45 (1886); Birdsey, The cultivated aroids, 97 (1951); Graf, Exotica, edn 8, 202, 212 (1976). Climber of moderate size. Leaf-blade to 60 × 45 cm, reflexed, ovate in outline, deeply lobed or pinnatisect into 5 parts: a prominent, elliptic to rhombic terminal lobe, a pair of oblong-triangular and more or less sickle-shaped lateral segments, and prominent, elliptic basal lobes. Leaf-stalk terete, red near apex, covered with red or green fleshy scales. Inflorescence-stalk red, covered with fleshy scales. Spathe 10–15 cm, with tube red outside and blade cream. *Surinam to NE Brazil.* G2.

4. P. bipennifolium Schott (*P. panduraeforme* misapplied). Illustration: Martius, Flora Brasiliensis **3**(2): t. 34 (1878); Graf, Exotica, edn 8, 203, 221 (1976); Graf, Tropica, 118 (1978).

Tall climber of moderate size. Leaf-blade *c.* 45 × 15–25 cm, reflexed, glossy dark green, with a long, terminal, obovate lobe, the remainder much wider and formed by the sinuate lateral margins into a pair of blunt-angled lobes on each side of the midrib; basal sinus open. Leaf-stalk terete, somewhat shorter than blade. Inflorescence-stalk short. Spathe *c.* 11 cm, greenish cream outside. *SE Brazil.* G2.

An excellent species for indoor culture.

5. P. wendlandii Schott. Illustration: Das Pflanzenreich **60**: 27 (1913); Graf, Exotica, edn 8, 213 (1976); Graf, Tropica, 118 (1978).

Plant forming a rosette, erect to prostrate. Leaf-blade to 75 × 20 cm, erect on its stalk, narrowly obovate, approximately truncate or with small auricles at base; midrib 2.5 cm wide. Leaf-stalk *c.* 30 cm, stout, flat along upper side with angular borders. Spathe *c.* 18 cm, with tube light green suffused with purple outside, dark red inside, and limb cream. *S Nicaragua to Panama.* G2.

Several hybrids have been produced from this species.

P. cannifolium Kunth, not *P. cannifolium* (Rudge) Engler; (*P. martianum* Engler). Illustration: Martius, Flora Brasiliensis **3**(2): t. 31 (1878); Peyritsch, Aroideae Maximilianae, t. 38, 39 (1879); Graf, Exotica, edn 8, 219 – also as *P. warmingii* (1976); Graf, Tropica, 115, 117 (1978) – as *P. warmingii*. Stem prostrate; internodes short. Leaf-blade 45 × 15–20 cm, glossy, leathery, ovate, obtuse to truncate at base, erect on a stout, spongy leaf-stalk somewhat shorter than blade and wide-channelled above with angular margins. Inflorescence-stalks 11–17 cm. Spathe *c.* 17 cm and nearly cylindric, with tube dark red at base inside, otherwise green becoming cream above. *SE Brazil.* G2.

6. P. verrucosum Schott. Illustration: *Illustration Horticole* **18**: 192 (1871) & **20**: 12 (1873) – as *P. daguense*; *Gartenflora* **42**: 257 (1893); Birdsey, The cultivated aroids, 99 (1951); Graf, Tropica, 123 (1978).

Tall climber of moderate size. Leaf-blade to 60 × 40 cm, reflexed, ovate with sagittate-cordate base, very dark shimmering green on upper surface with light green zones along the furrowed midrib and principal veins, red-violet between veins beneath; margins wavy and shallowly sinuate. Leaf-stalk about as long as blade, more or less flattened on upper side and covered with red, green or white fleshy scales as are also the inflorescence-stalk, scale-leaf and spathe. Inflorescence-stalk *c.* 17 cm, marked with elevated, white, interrupted lines below. Spathe *c.* 21 cm, outside red-brown, becoming lime green above, inside pinkish red, becoming white above, or with white edge. *Costa Rica to Ecuador.* G2.

Thrives only under warm, humid conditions.

7. P. erubescens C. Koch & Augustin. Illustration: *Botanical Magazine*, 5071 (1858); *Flore des serres* **14**: 39 (1861); Birdsey, The cultivated aroids, 79 (1951); Graf, Tropica, 114, 116 (1978).

Tall climber of moderate size. New stems purple or red-purple, scale-leaves deep pink or pinkish brown. Leaf-blade to 40 cm, reflexed, slightly leathery, ovate-triangular, with shortly sagittate-cordate base, more or less glossy dark green on upper surface, suffused with purple beneath, especially on principal veins. Leaf-stalk about as long as blade, flattened above towards the blade, suffused with red-purple. Inflorescence-stalk and spathe purplish red outside, somewhat lighter coloured inside. *Colombia.* G2.

'Red Emerald' (illustration: Graf, Tropica, 118, 1978) is a selection from *P. erubescens*, which is more robust, with red-purple stems and leaf-stalks, and the principal veins red-purple on the undersurface of the leaf-blade.

8. P. melanochrysum Linden & André (*P. andreanum* Devansaye). Illustration: *Illustration Horticole* **20**: 198 (1873); *Revue Horticole* **58**: 36 (1886); Birdsey, The cultivated aroids, 73 (1951); Graf, Tropica, 115 (1978).

Tall climber of moderate size. Leaf-blade to 80 × 30 cm, reflexed-pendent, narrowly ovate with sagittate base; upper surface black-green with a velvety appearance and pale green ribs and principal veins, pinkish orange when new. Leaf-stalk to 50 cm, somewhat rough. Spathe *c.* 20 cm, with tube more or less green and limb white, slender-tipped. *Colombia.* G2.

A beautiful species when well grown, but demanding in cultural requirements. Recommended only for a warm, moist environment.

9. P. scandens C. Koch & Sello (*P. cordatum* misapplied). Illustration: Das Pflanzenreich **60**: 57 (1913)– in part as *P. oxycardium*; Birdsey, The cultivated aroids, 89 (1951) – as *P. oxycardium*; Graf, Exotica, edn 8, 203, 207 (1976) – *as P. oxycardium*; Graf, Tropica, 115, 121 (1978).

Tall climber of small size with terminal part of stem finally pendent and nearly reaching ground, where flowering occurs. Internodes elongate. Leaf-blade to 30 × 23 cm, reflexed, more or less glossy, heart-shaped with about 3 principal veins from each side of midrib. Leaf-stalk shorter than blade, slender, nearly terete, channelled on upper side. Inflorescence-stalk short. Spathe to 19 cm, green with more or less white limb. *Mexico & West Indies to Surinam, Peru & SE Brazil.* G2.

Perhaps the commonest philodendron in cultivation, thriving under adverse conditions in soil, water, or even when submerged in aquaria. Several variants occur, especially distinguishable in the juvenile growth phase. Subsp. **scandens** has juvenile leaves with a velvety or metallic appearance on upper surface; its forma **scandens** has leaves which are green above and green or red-purple beneath, while forma **micans** (C. Koch) Bunting (*P. micans* C. Koch) has leaves metallic brown (bronze) above and reddish to brown-red beneath, with more ample basal lobes that slightly overlap. Subsp. **oxycardium** (Schott) Bunting (*P. oxycardium* Schott) has its glossy juvenile leaves tinted brown when immature and green when mature.

P. imbe Endlicher. Illustration: Martius, Flora Brasiliensis **3**(2): t. 33 (1878); Peyritsch, Aroideae Maximilianae, t. 34 (1879); Graf, Tropica, 120 (1978). Climber. Leaf-blade *c.* 33 × 18 cm, reflexed, glossy, narrowly ovate, with only about 3 nearly horizontally spreading principal veins from each side of midrib; base sagittate with narrow sinus *c.* 7 cm long. Leaf-stalk terete, about as long as blade. Spathe *c.* 9 cm, the tube green outside and red inside, the limb cream. *SE Brazil.* G2.

10. P. domesticum Bunting (*P. hastatum* misapplied). Illustration: *Baileya* **14**: 90 (1966); Graf, Exotica, edn 8, 203, 204 (1976).

Climber of moderate size. Leaf-blade to 60 × 30 cm, reflexed, glossy, long-triangular, with sagittate base and 5–6 principal veins from each side of midrib. Leaf-stalk as long as blade, flattened on upper surface near the blade and with a longitudinal central ridge. Inflorescence-stalk nearly as long as spathe. Spathe *c.* 18 cm, green outside, deep pinkish red inside with green border. *Known only in cultivation.* G2.

P. 'Tuxla' is similar in appearance, but it is not certain from which species or hybrid it was developed.

P. cordatum (Vellozo) Kunth. Illustration: Vellozo, Flora Fluminensis **9**: t. 111 (1825). Tall climber. Leaf-blade to 45 × 25 cm or larger, ovate-triangular with sagittate base and more or less oblong, rounded-angular basal lobes separated by a narrow sinus to 15 cm long, or slightly overlapping. Spathe *c.* 15 cm, greenish. *SE Brazil.* G2.

Material of *P. scandens* has been offered for sale under this name.

P. ilsemannii Sander (*P. sagittifolium* misapplied). Illustration: *Möllers Deutsche Gärtnerei-Zeitung* **23**: 257 (1908); Parey's Blumengärtnerei **1**: 180 (1958); Graf, Exotica, edn 8, 201 (1976); Graf, Tropica, 114 (1978). Climber with narrowly ovate leaf-blade that is shallowly cordate at base, often with large white zones and few to numerous smaller white and grey-green irregular spots. Leaf-stalk winged in the lower part. *Origin unknown.* G2.

The juvenile phase of an undetermined species, strongly suggesting the variegated form of *P. cordatum.*

P. devansayeanum Linden. Illustration: *Illustration Horticole* **42**: 376 (1895).

Stem more or less prostrate with short internodes and erect, purplish red leaf-stalks to nearly twice as long as blade. Leaf-blade ovate, glossy; principal veins and midrib purplish red on under surface; new leaf orange-brown. *Peru (Andes).* G2.

11. P. ornatum Schott (*P. asperatum* C. Koch; *P. sodiroi* invalid). Illustration: Peyritsch, Aroideae Maximilianae, t. 40–2 (1879) – as *P. imperiale*; Birdsey, *The cultivated aroids*, 95 (1951); Graf, Exotica, edn 8, 208, 221, 222, 223 (1976); Graf, Tropica, 118, 123 (1978).

Climber of moderate to large size; upper part of stem hidden by persistent, fibrous remains of scale-leaves. Leaf-blade to 60 cm or more, reflexed, ovate with cordate base, shiny on upper surface; veins on under-surface early suffused with red. Leaf-stalk about as long as blade, flat above in the upper part, more or less rough, purplish when young. Inflorescence-stalk short, reddish. Spathe to 14 cm, ending in a thread-like structure to 2 cm long and having a green tube and white limb. *S Venezuela to Peru & SE Brazil.* G2.

Generally seen in the juvenile phase with leaf-blade blotched with silvery grey between the principal veins.

12. P. mamei André. Illustration: *Illustration Horticole* **43**: 293 (1896); Birdsey, *The cultivated aroids*, 87 (1951); Graf, Exotica, edn 8, 217, 223—centre row only (1976); Graf, Tropica, 114 (1978).

Stem prostrate; internodes very short. Scale leaves reddish, drying and persistent. Leaf-blade to 60 × 45 cm, reflexed, ovate, sagittate at base with more or less angular basal lobes, irregularly spotted with grey-green between the principal veins, these and lesser veins all furrowed on upper surface. Leaf-stalk nearly as long as blade, reddish at apex and base, flat on upperside and narrowly crinkly-winged near apex. Spathe *c.* 15 cm, brownish red outside. *Ecuador.* G2.

13. P. gloriosum André. Illustration: *Illustration Horticole* **23**: 194 (1876); Birdsey, The cultivated aroids, 81 (1951); Graf, Exotica, edn 8, 205 (1976).

Stem prostrate; internodes short. Dry scale-leaves persistent. Leaf-blade *c.* 40 × 33 cm, reflexed, broadly ovate with cordate-sagittate base; upper surface dark green with a velvety appearance and a white midrib and principal veins. Leaf-stalk much longer than blade, flattened above towards the top. Inflorescence-stalk *c.* 16 cm, reddish towards apex, marked with white lines. Spathe as long as inflorescence-stalk, with the tube green, flushed with pink, and the limb pink. *Colombia.* G2.

19. ANUBIAS Schott
P.F. Yeo
Plants evergreen. Stem a horizontal rhizome. Leaf-stalk long, with a pulvinus at the apex. Leaf-blade entire, elongate, variously shaped at base, with a marginal vein and with minor veins parallel to primary lateral veins. Inflorescence-stalk nearly as long as the leaf-stalk. Spathe open throughout or with margins overlapping in the lower half. Florets without perianths, unisexual. Zones of male and female florets adjacent. Male florets with 3–5 stamens united into a short column, the lowest sometimes sterile. Ovary containing numerous ovules.

About 15 species from tropical Africa. One species is sometimes grown at the borders of aquaria or in pots. It requires a wet soil, a humid atmosphere and some shade.

Literature: Engler, A., Anubiadeae etc., Das Pflanzenreich **64**: 2–9 (1915).

1. A. afzelii Schott (*A. lanceolata* misapplied). Illustration: Graf, Exotica, edn 8, 143 (1976); Mühlberg, Das grosse Buch der Wasserpflanzen, f. 170, 171 (1980).

Leaf-stalk to 40 cm, with a sheath more than half its length. Leaf-blade erect, to 55 × 17 cm, tapered at apex and base. Spathe 6–6.5 cm, green, closed at the base, contracted into a short point. Spadix longer than spathe, with a stipe; female zone 3 cm; male zone 3 cm, its basal half sterile. *West tropical Africa.* G2.

20. AGLAONEMA Schott
D. H. Nicolson
Plants evergreen, with erect or sometimes creeping stems. Leaf-stalk with well-developed sheath; leaf-blade entire, slightly asymmetric, sometimes variegated, the primary lateral veins with numerous fine minor veins parallel to them (figure 14(5), p. 374). Spathe with margins usually overlapping at base; upper part open and flat or inrolled. Spadix free from spathe or partially attached at base. Florets without perianths, unisexual. Zone of female florets adjacent to the much longer male zone, in which the stamens are not grouped into recognisable florets. Ovary with 1 cell, with 1 ovule. Differs from *Dieffenbachia* (p. 393) in having stamens free from each other and female florets which lack staminodes.

About 20 species in tropical Asia. The species are commonly grown for interior decoration, mostly for their variegated leaves. A minimum of 15 °C, plenty of water, and shade are required for best growth. They are easily propagated from short sections of stem half-buried in rooting medium or from fresh seeds. The species are variable in nature and the variants brought into cultivation have given rise, through hybridisation and selection, to many cultivars.

Literature: Nicolson, D.H., A revision of the genus Aglaonema (Araceae), *Smithsonian Contributions to Botany* **1**: 1–69 (1969).

Habit. Creeping: **1, 2**; erect: **3, 2, 3, 4, 5, 6, 7**.
Sheath-margin. Scarious: **6, 7**; membranous: **1, 2, 3, 4, 5**.
Spathe/spadix. Spathe shorter than spadix: **1, 2, 4**; spathe as long as or longer than spadix: **3, 5, 6, 7**.
Spadix. Broadest at end: **4**; broadest in middle: **1, 2**; elongate: **3, 5, 6, 7**.
Stipe. Attached to spathe: **3**; free from spathe: **1, 2, 4, 5, 6, 7**.

1a. Plants creeping; sheath of leaf-stalk not more than 1 cm 2
b. Plants erect; sheath much more than 1 cm 3
2a. Scale-leaves subtending foliage leaves persistent on drying; leaf-stalk equalling or longer than blade; leaf-base obtuse, more or less rounded **1. brevispathum**
b. Scale-leaves subtending foliage leaves immediately deliquescent; leaf-stalk equalling or shorter than blade; leaf-base slightly cordate to rounded **2. costatum**
3a. Stipe and major part of female zone of spadix united with spathe **3. modestum**
b. Stipe and female zone of spadix free from spathe 4
4a. Spadix club-shaped; spathe spherical **4. pictum**
b. Spadix cylindric; spathe oblong 5
5a. Sheath of leaf-stalk membranous (remaining green) **5. commutatum**
b. Sheath of leaf-stalk scarious (drying and cracking) 6
6a. Leaves elliptic, obtuse at base, with a broad silvery stripe running down each side of the upper surface **6. crispum**
b. Leaves narrowly elliptic, acute at base, with pale bars or blotches along lateral veins, or unvariegated **7. nitidum**

1. A. brevispathum (Engler) Engler. Illustration: Graf, Exotica, edn 8, 117 (1976)—as *A. hospitum*; Everett, Encyclopedia of horticulture **1**: 88 (1980).

Stem creeping. Scale-leaves persistent, like brown paper, clasping the leaf-stalks. Leaf-stalks as long as or longer than blade. Leaf-blade to 20 cm long, 2.5 times as long as broad or more, usually obtuse at base; surface with scattered blotches on both sides (forma **hospitum** (Williams) Nicolson), unvariegated (forma **brevispathum**), or with midrib only white (unnamed). Spathe to 3.5 cm, usually with a small point; stipe to 1 cm. Spadix as long as or longer than spathe. Female florets fewer than 10. *Continental SE Asia.* G2.

This species is very close to *A. costatum* and, upon further study, its distinguishing characters (persistence of scale-leaves, longer leaf-stalks, obtuse leaf-base) may prove insufficient to distinguish it as a species.

2. A. costatum N.E. Brown. Illustration: Birdsey, The cultivated aroids, 15 (1951); Graf, Exotica, edn 8, 114, 118 (1976) – as *A. costatum* and *A. costatum foxii*; Everett, Encyclopedia of horticulture 1: 88 (1980).

Stem creeping. Scale-leaves quickly deliquescent. Leaf-stalk half as long to as long as the blade. Leaf-blade to 20 cm long, up to 2.5 times as long as broad, usually slightly cordate at base; with white midrib and with scattered spots (forma **costatum**), or without spots (forma **immaculatum** (Ridley) Nicolson); or with green midrib and scattered spots (forma **virescens** (Engler) Nicolson), or without spots (forma **concolor** Nicolson). Spathe to 4 cm, abruptly acuminate; stipe to 1 cm. Spadix as long as or longer than spathe. Female florets fewer than 10. *Continental SE Asia.* G2.

Forma **immaculatum** and forma **virescens** are illustrated in Graf, Exotica, edn 8, 118 (1976).

3. A. modestum Engler (*A. acutispathum* N.E. Brown). Illustration: Addisonia **19**: 5

(1935); Birdsey, The cultivated aroids, 17 (1951); Graf, Exotica, edn 8, 119 (1976); Everett, Encyclopedia of horticulture **1**: 86 (1980).

Stem erect. Leaf-sheaths broad and membranous. Leaf-stalk half as long to as long as the blade. Leaf-blade 14–22 × 8–11 cm, not variegated (except in the rare 'Variegatum' and 'Shiningi'), ovate, the apex long-acuminate. Spathe exceeding spadix by 1–3 cm. Stipe and female zone of spadix joined to spathe. Male florets sterile, not shedding pollen (plant an apomictic triploid). Berry becoming orange. *SE Asia and S China.* G2.

One of the 2 most frequently grown species of the genus. It is considered to be a good-luck plant by the Chinese.

4. A. pictum (Roxburgh) Kunth. Illustration: Graf, Exotica, edn 8, 119 (1976); Everett, Encyclopedia of horticulture **1**: 87 (1980); Brickell (ed.), RHS A–Z encyclopedia of garden plants, 90 (2003).

Stem erect. Leaf-sheaths broad, translucent to scarious. Leaf-stalk much shorter than blade. Leaf-blade 10–16 × 3–6 cm, dull above, variegated, elliptic to ovate, apex often with a small point. Spathe spherical, to 3 cm long and 7 cm wide when unrolled. Stipe *c.* 1 cm. Spadix club-shaped, projecting from the spathe. *Sumatra.* G2.

The dull leaves of this species are distinctive. The variegation is quite irregular, usually light or silvery green on a dark green background.

5. A. commutatum Schott. Illustration (of most common aspect, *A. commutatum* var. *maculatum*): *Gartenflora* **14**: 130 (1865); *Botanical Magazine*, 5500 (1865); *Addisonia* **9**: 27 (1924); Graf, Exotica, edn 8, 119 (1976).

Stem erect. Leaf-sheaths usually membranous. Leaf-stalk usually the same length as the blade. Leaf-blade usually narrowly oblong-elliptic, 13–30 × 4–10 cm, the ashy variegation in irregular bars following the primary lateral veins. Spathe usually 1 cm longer than spadix. Stipe 5–10 mm. Male florets sterile, not shedding pollen (plants apomictic tetraploids and hexaploids). *Philippines.* G2.

Many cultivars, such as: 'Pseudobracteatum' (probably called 'Marmoratum' in France) with leaf-stalk white, blade heavily variegated in 3 shades (including yellow) in a somewhat marbled composition; 'Silver Queen' with leaf mostly silvery green, edged and streaked dark green ('Parrot Jungle' is probably the same); 'Treubii', similar but with very narrow leaves. These and others are illustrated by Graf (Exotica, edn 8, 114–19, 1976). 'Silver King' is patterned in 3 shades of green, and is blotchy rather than streaky. A recent cultivar, 'Manila', should probably be attributed to this species but, unlike others, produces pollen.

6. A. crispum (Pitcher & Manda) Nicolson (*A. roebelinii* Pitcher & Manda). Illustration: Birdsey, The cultivated aroids, 21 (1951); Graf, Exotica, edn 8, 117 (1976).

Stem erect. Leaf-sheaths with scarious margins. Leaf-stalk about half as long as blade. Leaf-blade 14–32 × 7–12 cm, elliptic, the silvery variegation covering most of each half of the leaf, leaving the midrib and marginal areas free. Spathe 1–2 cm longer than spadix. Stipe 3–7 mm. Philippines *(S Luzon).* G2.

Probably an apomictic triploid. 'Pewter' is the commonest cultivar, but one with a white leaf-stalk has also been raised.

7. A. nitidum (Jack) Kunth. Illustration: Birdsey, The cultivated aroids, 19 (1951); Graf, Exotica, edn 8, 117 (1976) – both as *A. oblongifolium curtisii.*

Stem erect. Leaf-sheaths with scarious margins. Leaf-stalk half as long to as long as the blade. Leaf-blade 20–45 × 7–16 cm, narrowly elliptic, not variegated or with bars or blotches along lateral veins. Spathe turning white with age, equalling spadix. Stipe 2–9 mm. Female florets 16–37, scattered. *Malaysia, Sumatra, Borneo.* G2.

This species is common in conservatories but rarely seen in houses.

21. DIEFFENBACHIA Schott

P.F. Yeo

Plants erect, evergreen, with stems becoming slightly woody. Leaf-stalk with a well-developed sheath; leaf-blade entire, slightly asymmetric, the primary lateral veins with numerous fine minor veins parallel to them (figure 14(5), p. 374). Spathe with margins overlapping at base; upper part open and hooded or flat, sometimes recurved at apex. Spadix nearly as long as spathe. Florets without perianths, unisexual. Zones of male and female florets about equal, adjacent or slightly separated, sometimes with a few sterile florets between them. Male florets with 4 or 5 stamens united into a short column. Female florets loosely spaced, each with 4 or 5 staminodes (figure 15(4 & 5), p. 377). Ovary with 2 or 3 cells, each with 1 ovule; stigmatic lobes 2 or 3. Differs from *Aglaonema* (p. 393) in the longer zone of female florets, which have well-developed staminodes.

25–30 species in tropical America. The genus is commonly grown for interior decoration and in glasshouses on account of its wide range of leaf variegations. A minimum temperature of 15 °C, plenty of water and some shade are required. Easily propagated from short sections of stem half-buried in a rooting medium. All parts of the plant are poisonous.

The leaf colour of some species is variable in nature and the variants brought into cultivation have also given rise to new cultivars by selection and hybridisation. Most cultivars belong to *D. seguine.* The leaf shape and colouring are somewhat influenced by the stage of growth of the plant and the degree of shading; the colouring is also influenced by the age of the individual leaf.

Literature: Engler, A., Anubiadeae, Aglaonemateae, Dieffenbachiae etc., Das Pflanzenreich **64**: 36–78 (1915).

Habit (inapplicable to young specimens). Branched, and with leaves all the way up the stem: **3, 5**; simple, with leaves only at the top: **1, 2, 4, 6.**

Leaf-blade variegation. None and midrib green: **2, 3, 4**; none and midrib white: **1, 2**; in 2 colours or shades: **4, 5**; in 3 colours; **4, 6.**

Leaf-blade shape. Ovate-lanceolate to lanceolate: **3, 4**; oblong, ovate, or broadly ovate: **1, 2, 4, 5, 6.**

1a. Leaf-blade green, except sometimes the midrib and its immediate borders 2
b. Leaf-blade variegated or spotted 6
2a. Midrib white or yellowish white 3
b. Midrib green 4
3a. Upper surface of leaf-blade with a sheen, leaf-stalk mottled **1. leopoldii**
b. Upper surface of leaf-blade without a sheen, matt; leaf-stalk not mottled **2. oerstedii** 'Variegata'
4a. Leaf-blade narrowly ovate to oblong-lanceolate, mostly more than 3 times as long as broad **3. humilis**
b. Leaf-blade broadly ovate or oblongovate, mostly not more than 3 times as long as broad 5
5a. Plant small, not more than about 1 m tall; leaf-blade to 25 cm **2. oerstedii**

b. Plant large, to 3 m tall; leaf-blade often to 40 cm or more **4. seguine**
6a. Leaf-blade variegated with irregular white or greenish white bands along the veins **4. seguine**
b. Leaf-blade not with pale bands along the veins 7
7a. Stem branched; leaf-blade to 75 cm, thin and drooping **5. bowmannii**
b. Stem unbranched; leaf-blade to 45 cm, firm 8
8a. Leaf-blade light yellowish green with the midrib dark green, and with dark green blotches and smaller white blotches **6. × bausei**
b. Leaf-blade differently variegated **4. seguine**

1. D. leopoldii Bull. Illustration: Graf, Exotica, edn 8, 172 (1976); Rochford & Gorer, The Rochford book of house plants, f. 12 (1963) – as *D. oerstedii.*

Leaf-stalk green or brownish with white mottling, mostly sheathed to the middle or above. Leaf-blade to 35 cm, 2–2.5 times as long as broad, ovate, rather shortly tapered at apex, rounded or weakly cordate at base; upper surface uniformly dark green with a satiny sheen, midrib white. Spathe to 17 cm, white; female florets with the staminodes 5 mm, weakly club-shaped; stigmas 2-lobed. *Costa Rica.* G2.

2. D. oerstedii Schott. Illustration: Graf, Exotica, edn 8, 168, 173 (1976).

Plant not more than 1 m. Leaf-stalk not mottled, sheathed to the middle or above. Leaf-blade to 25 cm, 2–2.5 times as long as broad, gradually tapered at apex, truncate or cordate at base; upper surface matt, rather dark green, sometimes with white midrib ('Variegata'). Spathe to 15 cm, whitish, becoming orange in fruit. *Mexico to Costa Rica.* G2.

3. D. humilis Poeppig. Illustration: *Baileya* **14**: 102 (1966).

Plant not more than 45 cm. Stem freely branched. Leaves crowded. Leaf-stalk sheathed nearly to the apex. Leaf-blade to 30 cm, mostly more than 3 times as long as broad, narrowly ovate to oblong-lanceolate, slightly glossy above. Spathe about 9 cm, increasing with age to 16 cm, green. Female florets with club-shaped staminodes. *Brazil, Peru.* G2.

4. D. seguine (Jacquin) Schott (*D. maculata* (Loddiges) G. Don; *D. picta* Schott). Illustration: see citations below under cultivars.

Plants usually attaining 1 m and sometimes 3 m. Leaf-stalk sheathed beyond the middle. Leaf-blade to about 40 cm, 2–2.5 times as long as broad, shortly tapered at apex, cordate at base. Spathe 15–18 cm. Female florets with club-shaped staminodes not exceeding the ovary. *Tropical America.* G2.

There are many cultivars, of which the commonest or most distinctive are described below; most do not exceed 1 m in height.

Leaf-blades unvariegated

'Viridis'. Illustration: Graf, Exotica, edn 8, 170 (1976). Leaf-blades irregularly white-variegated.

'Maculata'. Illustration: Rochford & Gorer, The Rochford book of house plants, f. 18 (1963); Graf, Exotica, edn 8, 168, 170 (1976). Upper surface freely spotted with white or yellowish white, the spots somewhat coalescent, not extending near the margin nor on to the midrib; leaf-stalk mottled.

'Superba' (*'Roehrs Superb'; 'Roehrs Superba'*). Illustration: Graf, Tropica, 106, 107 (1978). White areas much more extensive, though blotched with green and not affecting the midrib.

'Exotica' (*D. perfecta* invalid; *'Exotica Perfection'?*). Illustration: Hay et al., Dictionary of indoor plants in colour, t. 190 (1974); Graf, Exotica, edn 8, 168, 172 (1976); Graf, Tropica, 108 (1978). Similar, but midrib white or variegated.

'Baraquiniana'. Illustration: Birdsey, The cultivated aroids, 65 (1951). White spots very sparse, midrib white, leaf-stalk white and apparently longer than usual.

'Magnifica'. Illustration: Graf, Exotica, edn 8, 174 (1976). Spotted similarly to 'Baraquiniana' but leaf-stalks and midribs green.

'Lancifolia'. Illustration: Graf, Exotica, edn 8, 169 (1976). Leaf spotting like that of 'Maculata' but leaf-blade 5–6 times as long as broad, its veins forming a smaller angle with the midrib. Leaf-blades with yellowish white bands along the lateral veins.

'Amoena' (*D. amoena* misapplied). Illustration: Graf, Exotica, edn 8, 173 (1976). Plant to 2 m or more.

'Tropic Snow'. Illustration: Graf, Exotica, edn 8, 169 (1976); Graf, Tropica, 106, 107 (1978). Similar to 'Amoena' but less tall and with the bands of variegation coalescent and the leaf-blade wrinkled near the midrib.

'Jenmannii'. Illustration: Graf, Exotica, edn 8,168–70 (1976). To 70 cm with leaf-blades to 25 cm and 2.5–4 times as long as broad, drooping, on slender, shortly sheathed stalks. *Midrib pale, with a yellowish white, fringed band along either side.*

'Liturata' (*D. amoena 'Liturata'*). Plant to 3 m. Leaf-blades variegated in three colours or shades.

'Rudolph Roehrs' (*'Roehrsii'*). Illustration: Hay et al., Dictionary of indoor plants in colour, t. 191 (1974); Graf, Exotica, edn 8, 168, 169 (1976). White with green midrib and irregular green border, the white area becoming pale green by the darkening of the finest veins which appear as striations at close quarters.

'Memoria Corsii' (*'Memoria'*). Illustration: Birdsey, The cultivated aroids, 59 (1951); Graf, Exotica, edn 8, 170 (1976). Mainly grey-green, developing a darker colour with age by darkening of the finest veins and heavily blotched with dark green and sparsely spotted with white. Leaf-blades variegated with pale (not yellowish) green.

'Shuttleworthii' (*'Shuttleworthiana'*). Illustration: Graf, Exotica, edn 8, 174 (1976). With pale, scattered spots and a pale fringed border on either side of the midrib.

'Nobilis'. Illustration: Graf, Exotica, edn 8, 170 (1976). Plant to 3 m; leaf-blades with pale blotches.

D. × splendens Bull. Illustration: Birdsey, The cultivated aroids, 67 (1951); Graf, Exotica, edn 8, 172 (1976). Resembles *D. seguine* 'Baraquiniana' but has a shorter and broader leaf-blade and leaf-stalks mottled in light and dark green (*D. leopoldii* × *D. seguine* 'Maculata').

5. D. bowmannii Carrière (*D. reginae* invalid). Illustration: Graf, Exotica, edn 8, 173 (1976); Graf, Tropica, 105, 106 (1978).

Plant more than 1 m tall, with branched stem. Leaves persistent; leaf-blade to 75 cm, twice as long as broad or slightly more, shortly tapered at apex, rounded at base, thin and drooping, deep green with extensive paler green variegation and a few white spots, bluish green beneath. Spathe 15–18 cm, pale green. Female florets with staminodes 2 mm long. *Colombia, Brazil.* G2.

Descriptions of this species are inconsistent but the name is widely in use. The branched habit and large thin leaves are characteristic.

6. D. × bausei Masters & Moore. Illustration: Birdsey, The cultivated aroids,

55 (1951); Graf, Exotica, edn 8, 138A, 168, 171 (1976).

Plant to 1 m. Leaf-stalk mottled, fading to white at base where it contrasts with the dark green stem, sheathed to about the middle. Leaf-blade to about 30 cm, mostly 2.5–3 times as long as broad, ovate, tapering gradually from below the middle and ending in a long fine point; surface largely yellowish green, but with a dark green border and midrib and with sparse spots and blotches of both dark green and white. Spathe 12 cm, pale green. Female florets with club-shaped staminodes; stigmas with 2 or 3 lobes. *Garden origin.* G2.

The hybrid between D. *seguine* 'Maculata' and *D. weirii* Berkeley.

22. ZANTEDESCHIA Sprengel

P.F. Yeo

Plants with oblique, tuberous, underground rhizomes, which bear apical clusters of leaves or leaves and inflorescences. Scale-leaves absent. Leaf-stalk spongy. Leaf-blade wavy, lobed at the base or entire; main lateral veins joining a marginal vein, minor ones forming a network. Inflorescence as tall as the leaves or taller, fragrant; stalk spongy. Spathe normally white or brightly coloured, obliquely funnel-shaped, withering but persistent in fruit. Spadix much shorter than spathe, yellowish. Florets without perianths, unisexual. Zones of male and female florets adjacent. Male florets not individually recognisable. Female florets with or without staminodes. Ovary usually with 3 cells, each with 1–8 ovules. Style short; stigma entire. Fruit is a berry. Seed leathery.

Six species from southern Africa. They are valued for their showy, fragrant inflorescences and are extensively grown as cut-flowers. The range of variation has been increased by artificial hybridisation. All species require a rich, moist soil, and can be grown under glass with only protection from frost, but winter heating permits continued growth of *Z. aethiopica* and *Z. elliottiana. Z. aethiopica* is grown outdoors in W Europe as a waterside plant, with the tubers deeply planted to protect them from frost; some of its cultivars are hardy outdoors in southern England without this protection. Propagation is by separation of the lateral branches of the rhizomes.

Literature: Traub, H.P., The genus Zantedeschia, *Plant Life* **4**: 8–32 (1949); Shibuya, R., Intercrossing among pink Calla, white-spotted Calla and yellow Calla (1956); Letty, C., The genus Zantedeschia, *Bothalia* **11**: 5–26 (1973).

1a. Leaf-blade lanceolate, tapered at base **4. rehmannii**

b. Leaf-blade wider, cordate, sagittate or hastate 2

2a. Spathe bright yellow **3. elliottiana**

b. Spathe dull yellow, cream, white or occasionally pinkish or mainly green 3

3a. Staminodes present in female part of spadix; leaves usually unspotted; spathe not purple at base within **1. aethiopica**

b. Staminodes not present in female part of spadix; leaves usually white-spotted; spathe purple at base within **2. albo-maculata**

1. Z. aethiopica (Linnaeus) Sprengel (*Calla aethiopica* Linnaeus; *Richardia africana* Kunth). Illustration: *Botanical Magazine*, 832 (1805); Birdsey, The cultivated aroids, 130 (1951); Flowering Plants of Africa **30**: t. 1190 (1955); *Bothalia* **11**: 10, 11 (1973).

Plant to *c.* 1.75 m, evergreen in frost-free conditions. Leaf-blade to 45 × 25 cm, normally unspotted, broadly ovate, cordate with broad obtuse basal lobes which are bent upwards and sometimes slightly turned outwards. Spathe to 25 cm, green at base outside, otherwise white, widely open with a large limb, horizontal towards, and recurved at apex. Female part of spadix with conspicuous yellow staminodes. *South Africa (Cape Province, Natal, Transvaal, Lesotho). Widely naturalised in frost-free regions.* H3–G1. Spring–summer.

'Crowborough' is distinguished by being relatively hardy (H3) as are 'Green Goddess', which also has a mainly green spathe, and the miniature cultivar 'Little Gem'.

2. Z. albo-maculata (W.J. Hooker) Baillon (*Z. oculata* (Lindley) Engler). Illustration: *Botanical Magazine*, 5140 (1859), 5765 (1869); Birdsey, The cultivated aroids, 132 (1951); Letty, Wild flowers of the Transvaal, 10, 11 (1962); *Bothalia* **11**: 18–25 (1973).

Plant to *c.* 1.25 m, deciduous in winter. Leaf-blade to 45 × 23 cm, usually with numerous translucent white spots, hastate with main lobe oblong-triangular, or ovate-cordate (subsp. **valida**); basal lobes bent upwards. Spathe to 11.5 cm (or to 15 cm in subsp. **valida**), usually dark purple inside at base, elsewhere white, cream, straw-coloured, greenish yellow or, rarely, pinkish, moderately widely open (more so in subsp. **valida**), not recurved at apex. Female part of spadix without staminodes. *South Africa & E tropical Africa.* H5. Spring–summer.

Letty treats this as a variable species comprising 3 subspecies; unspotted leaves occur rarely in subsp. **albo-maculata**, sometimes in subsp. **macrocarpa** (Engler) Letty, and always in subsp. **valida** Letty.

3. Z. elliottiana (Watson) Engler (*Richardia elliottiana* Watson). Illustration: *Botanical Magazine*, 7577 (1898); Birdsey, The cultivated aroids, 134 (1951); Graf, Exotica, edn 8, 138A, 240 (1976).

Plant to *c.* 1.5 m, deciduous in winter. Leaf-blade to 28 × 25 cm, very broadly ovate, cordate with basal lobes bent upwards, with numerous translucent white spots. Spathe to 15 cm, bright golden yellow, duller outside, widely open above with narow spreading edge and recurved tip. Female part of spadix without staminodes. *?South Africa (unknown in nature).* H5. Spring–summer.

Letty suggests a hybrid origin for this plant. However, Shibuya found only trivial variation on selfing. R.E. Harrison, on the other hand (*Journal of the Royal Horticultural Society* **103**: 78, 1978), obtained plants with dark-centred spathes on crossing *Z. elliottiana* with *Z. rehmannii* and suggested that this indicated *Z. pentlandii* in the parentage of *Z. elliottiana.*

Z. jucunda Letty. Illustration: Letty, Wild flowers of the Transvaal, 15 (1962); *Bothalia* **11**: 14 & t. 3 (1973). Leaf-blade triangular-hastate, spotted. Spathe very like that of *Z. elliottiana* but with a dark purple area inside at base; inner surface slightly wrinkled. *South Africa (Transvaal).* H5. Summer.

Z. pentlandii (Watson) Wittmack (*Richardia pentlandii* Watson). Illustration: *Botanical Magazine*, 7397 (1895); *Gartenflora* **47**: 593 (1898); Letty, Wild flowers of the Transvaal, 14 (1962); *Bothalia* **11**: 16 & t. 4 (1973). Leaf-blade oblong-elliptic to oblong-lanceolate, sagittate or cordate, rarely spotted. Spathe similar to that of *Z. elliottiana* but with a dark purple area inside at base; apex not usually recurved. *South Africa (Transvaal).* H5. Summer.

The spathe is described and depicted in Botanical Magazine as wrinkled within but is shown as smooth in the plate by Letty.

4. Z. rehmannii Engler. Illustration: Birdsey, The cultivated aroids, 136 (1951); Letty, Wild flowers of the Transvaal, 7 (1962); *Bothalia* **11**: 12, 13 & t. 2 (1973); Graf, Exotica, edn 8, 240 (1976).

Plant usually to 60 cm, deciduous in winter. Leaf-blade to 40 × 7 cm, asymmetrically lanceolate, gradually tapered at both ends, rarely spotted. Spathe 7.5–12 cm, white to pink or dark purple, rather narrow; apex spreading or recurved. Female part of spadix without staminodes. *South Africa (Transvaal, Natal, Swaziland).* H5. Spring–summer.

Some cultivars are listed by Traub.

23. PELTANDRA Rafinesque

P.F. Yeo

Plants with horizontal rhizomes bearing rosettes of scale-leaves and long-stalked foliage leaves. Leaf-blade sagittate or hastate, with 2 or 3 marginal veins (figure 14(6), p. 374); lateral veins mostly parallel but the smallest joined by cross-veins; each basal lobe with a distinct midrib. Inflorescence-stalk as long as or slightly longer than leaf-stalks. Spathe with margins overlapping in the zone of female florets, open above; upper part decaying in fruit. Florets without perianths, unisexual. Male florets of 4 or 5 stamens united to form a flat plate with anthers around the edge, opening by pores at the top. Female florets with a cup-like structure round the ovary, derived from staminodes. Ovary 1-celled; ovules 1 or few; style present.

Four species of marsh-plants in E North America. They are grown out of doors as foliage plants in shallow water. Propagation is by division of the rhizomes, preferably in spring.

Literature: Engler, A., Philodendroideae, Anubiadeae-Peltandrae, Das Pflanzenreich **64**: 72–5 (1915).

1a. Female florets covering nearly half the length of the spadix; spathe-limb whitish, not wavy **1. sagittifolia**
b. Female florets covering one-third to one-fifth of the length of spadix; spathe-limb green with yellowish or whitish wavy edges **2. virginica**

1. P. sagittifolia (Michaux) Morong. Illustration: Das Pflanzenreich **64**: 73 (1915).

Leaf-blade to 15 cm, sagittate or slightly hastate. Spathe 7–10 cm, the tube green and the limb white, opening widely, with flat edges. Spadix covered with female florets for about half its length; female zone separated from male zone by scale-like sterile florets, each with a boss in the middle. Inflorescence-stalk erect in fruit. Berries red. *Southeastern USA.* H3. Early summer.

2. P. virginica (Linnaeus) Kunth. Illustration: Gleason, Illustrated flora of the northeastern United States and adjacent Canada **1**: 368 (1952); Rickett, Wild flowers of the United States **1**(1): 87 (1966); Graf, Tropica, 110 (1978); Everett, Encyclopedia of horticulture **8**: 2537 (1981).

Leaf-blade to 30 cm (larger after flowering time), more or less hastate. Spathe 10–20 cm, mainly green; limb with wavy yellowish or whitish margins, gaping only slightly. Spadix covered with female florets for one-fifth to one-third of its length, and with few or no sterile florets; sometimes with a small sterile appendix. Inflorescence-stalk recurved in fruit. *E & Southeastern USA.* H2. Late spring–summer.

24. CALADIUM Ventenat

M.T. Madison & P.F. Yeo

Plants with starchy, cylindric or spherical, subterranean tubers which are dormant in winter. Leaf-blade with base hastate to truncate or rounded, usually variegated with pink, red, white, and green patterns; main lateral veins with netted veinlets between them, joined to a marginal vein (figure 14(6), p. 374). Spathe with the margins overlapping below and the upper part open. Spadix shorter than the spathe, with female, sterile and male florets in adjacent zones. Florets without perianths. Stamens 2–5, joined into a column. Ovary with 2 or 3 cells, incompletely separated, each with several ovules; style none; stigma grooved. Berry whitish.

A genus of 7 or 8 species native to northern S America and naturalised throughout the tropics. Caladiums are cultivated in shady garden borders or as pot plants for their showy, variegated foliage. They grow well in moderate shade with a moist, well-drained soil rich in humus. Tubers should be lifted after the leaves die down in the autumn and stored in a cool dark spot until the following spring. Propagation is by division of tubers, or less often by seed.

Literature: Madison, M.T., Notes on Caladium (Araceae) and its allies, *Selbyana* **5**: 342–77 (1981).

1a. Leaf peltate 2
b. Leaf not peltate 3
2a. Leaf-blade not more than 10 cm **1. humboldtii**
b. Leaf-blade more than 15 cm **2. bicolor**
3a. Leaf-blade hastate **3. lindenii**
b. Leaf-blade with base weakly cordate, rounded or truncate **4. schomburgkii**

1. C. humboldtii Schott (*C. argyrites* Lemaire; *C. lilliputiense* Rodigas). Illustration: *Illustration Horticole* **5**: t. 185, no. 3 (1858); Graf, Tropica, 96, 98 (1976); Everett, Encyclopedia of horticulture **2**: 545 (1981); *Selbyana* **5**: 370 (1981).

To 25 cm. Tuber 1–2 cm in diameter, yellow within, suckering freely. Leaf-stalk 10–24 cm × 1–1.5 mm. Leaf-blade 5–10 × 2–4.5 cm, ovate, peltate, with sagittate base, dull deep green with white blotches and spots. Flowers unknown. *Brazil & Venezuela.* H5.

2. C. bicolor (Aiton) Ventenat (*C.* × *hortulanum* Birdsey; *C. marmoratum* Mathieu; *C. picturatum* C. Koch; *C. poecile* Schott). Illustration: *Botanical Magazine*, 2543 (1825); Graf, Exotica, edn 8, 149 (1976).

To 90 cm. Tuber 2.5–6 cm broad, flattened-spherical. Leaf-blade 18–46 × 12–25 cm, peltate, ovate to elliptic, with cordate or sagittate base, dark to medium green with or without white, pink or red spots, bands or blotches. Spathe 6–11 cm. Spadix with male part 2.5–6 cm, white. *Northern S America.* H5. Late spring–autumn.

Has produced an immense number of cultivars, including many in which the leaf-blade is largely white (see Graf, Exotica, edn 8, 147–62, 1976).

3. C. lindenii (André) Madison (*Xanthosoma lindenii* (André) Engler). Illustration: *Illustration Horticole* **19**: 3 (1872) – as *Phyllotaenium lindenii*; Birdsey, The cultivated aroids, 127 (1951); Graf, Tropica, 130 (1978); Everett, Encyclopedia of horticulture **10**: 3569 (1982).

To 80 cm. Tuber suckering freely. Leaf-stalk to 40 cm, covered with sparse, short, brown hairs, green with purple stripes. Leaf-blade 16–45 cm long, 10–20 cm wide across principal lobe, hastate, not peltate, hairy on the midrib; basal lobes with a pinnately branched midrib (figure 14(8), p. 374); upper surface green with midrib and main veins white or yellowish, the surface often paler green near midrib. Spathe 7.5–14 cm; tube whitish green; limb 7–9 cm, white. Spadix with male part 7–8 cm × 6–8 mm, white. *Colombia.* G1.

4. C. schomburgkii Schott. Illustration: *Illustration Horticole* **8**: t. 297, nos. 2 & 3 (1861).

To 45 cm. Tuber spherical. Leaf-stalk 16–23 cm. Leaf-blade about same length as stalk, triangular, ovate, broadly lanceolate or almost rhombic, with weakly cordate, truncate or rounded base, not peltate, usually variegated with pink or white, the margins often ruffled. Inflorescence-stalk 18–26 cm. Spathe *c.* 7 cm. *Guianas.* H5.

Represented by many cultivars (see Graf, Exotica, edn 8, 162–4, 1976).

25. XANTHOSOMA Schott

S.A. Thompson

Plants with starchy, cylindric or spherical, underground corms or with above-ground stems. Leaf-stalk with well-developed sheath; leaf-blade entire, sagittate, hastate, or pedately divided into 3–18 segments; main venation pinnate; minor veins netted. Spathe with margins overlapping at base and upper part open. Spadix shorter than spathe. Florets without perianths, unisexual. Zone of male florets 2 or more times as long as female zone, separated from it by sterile florets. Stamens 4–6, united into a column. Ovary of 2–4 cells with several to many ovules.

A genus of approximately 50 species native to the New World tropics. Several species of *Xanthosoma* are cultivated throughout the tropics for their starchy corms and edible leaves, as well as for ornament.

Xanthosoma species will succeed in most soils, but prefer rich, well-drained soil with abundant water. Propagation is by vegetative means; either the stem is cut into pieces as cuttings, or cormlets are planted directly. Flowering under cultivation is sporadic.

Literature: Engler, A., Colocasioideae, Das Pflanzenreich **71**: 41–62 (1920).

Habit. With stem: **1, 2**; stemless: **3, 4, 5, 6, 7**.

Leaves. Entire and sagittate: **1, 2, 3, 4**; entire and hastate: **5**; divided: **6, 7**.

1a. Plant with above-ground stem 2
b. Plant without stem 3
2a. Main veins of basal lobes of leaf-blade exposed in the sinus for part of their length (figure 14(8), p. 374) **1. undipes**
b. Main veins of basal lobes of leaf-blade not exposed in the sinus **2. sagittifolium**
3a. Leaves entire 4
b. Leaves divided 6
4a. Plant hairy **3. pilosum**
b. Plant hairless 5
5a. Leaf-blade sagittate **4. violaceum**
b. Leaf-blade hastate **5. brasiliense**
6a. Segments of leaf-blade 5–18, narrow, to 9 cm wide **6. helleborifolium**
b. Segments of leaf-blade 5–9, broad, to 53 cm wide **7. hoffmannii**

1. X. undipes (C. Koch) C. Koch (*X. jacquinii* Schott). Figure 14(8), p. 374. Illustration: Das Pflanzenreich **71**: 46 (1920).

Above-ground stem to 2 m tall and 20 cm thick. Leaf-stalk to 1 m; sheath wavy, green; leaf-blade 50–200 cm long and somewhat less wide at the base, broadly cordate-sagittate, main veins of basal lobes exposed in the sinus for 1–5 cm (figure 14(8), p. 374). Spathe-tube to 7 cm, ovoid, green outside, whitish yellow inside; spathe-limb 15–25 cm, pale yellowish green outside, whitish yellow inside. Female florets orange; sterile florets dull white; male florets yellow. *Tropical America from Mexico south to Peru; cultivated as a crop in the West Indies.* G2.

2. X. sagittifolium (Linnaeus) Schott (*Arum xanthorrhizon* Jacquin; *X. atrovirens* C. Koch). Illustration: *Botanical Magazine*, 4989 (1857).

Mature plant with above-ground stems 1 m long or more. Leaf-stalk to 1 m, bloomed; leaf-blade to 70 cm long and a little less wide, broadly sagittate-ovate, bloomed; main veins of basal lobes not exposed at the sinus. Spathe-tube to 7 cm, oblong-ovoid, green; spathe-limb to 15 cm, acuminate, greenish white. Spadix much shorter than spathe. *New World tropics, exact native distribution uncertain; widely cultivated in the tropics.* G2.

3. X. pilosum C. Koch & Augustin (*X. holtonianum* Schott; *Caladium puberulum* Engler). Illustration: Croat, Flora of Barro Colorado Island, 226 (1978).

Plant stemless, most parts covered with short fine hairs. Leaf-stalk to 40 cm, sheathed halfway or more; leaf-blade to 35 × 20 cm, sagittate-ovate to cordate-ovate, green, sometimes mottled with white spots; main veins of basal lobes shortly exposed in the sinus (figure 14(8), p. 374). Spathe-tube to 7 cm, oblong, green; spathe-limb to 10 cm, oblong-lanceolate, white. Sterile florets, violet. *Costa Rica to Colombia.* G2.

4. X. violaceum Schott. Illustration: Martius, Flora Brasiliensis **3**(2): t. 43 (1878); Birdsey, The cultivated aroids, 129 (1951); Graf, Exotica, edn 8, 238 (1976).

Plant stemless, robust, with large tuberous corm. Leaf-stalk 30–85 cm, dark brownish purple, bloomed; leaf-blade to 70 × 45 cm, sagittate-ovate, shortly acuminate at apex, bloomed at least when young, with midrib and main veins purple beneath and margin purple; main veins of basal lobes exposed for a short distance in the sinus in mature leaves (figure 14(8), p. 374). Spathe-tube to 10 cm, greenish violet outside, whitish yellow inside; spathe-limb to 20 cm, whitish yellow. Sterile florets violet to reddish brown. *New World tropics, exact native distribution uncertain; cultivated and naturalised throughout the tropics.* G2.

5. X. brasiliense (Desfontaines) Enagler (*X. hastifolium* C. Koch). Illustration: *Proceedings of the Florida State Horticultural Society* **85**: 90 (1972).

Plant stemless. Leaf-stalk 20–40 cm, narrowly sheathed to less than one-third of its length; leaf-blade of young plant ovate-sagittate, leaf-blade of adult plant to 40 × 15 cm, hastate, with outspread oblong basal lobes; main veins of basal lobes exposed for part of their length in the sinus (figure 14(8), p. 374). Spathe-tube to 5 cm, oblong-ovoid, green; spathe-limb to 12 cm, oblong-lanceolate, yellowish green. Female florets with yellow stigmas; sterile florets pink; male florets dull white. *New World tropics, exact native distribution uncertain; widely cultivated in the tropics, especially in the West Indies.* G2.

6. X. helleborifolium (Jacquin) Schott. Illustration: Croat, Flora of Barro Colorado Island, 226 (1978).

Plant stemless. Leaf-stalk to 85 cm, broadly sheathed in the lower part, mottled with purple and green; leaf-blade kidney-shaped in outline, deeply dissected into

5–18 oblong or lanceolate segments, the central to 36 × 9 cm. Inflorescence-stalks mottled like the leaf-stalks. Spathe-tube 3–6 cm, elliptic-ovoid, green; spathe-limb 6–10 cm, oblong-ovate, yellowish green to white. Spadix with a broad short stipe. *El Salvador to the Guianas & Peru; West Indies.* G2.

7. X. hoffmannii Schott (*X. barilletii* Carrière). Illustration: *Revue Horticole* **54**: 260 (1882) – as *X. barilletii.*

Plant essentially stemless. Leaf-stalk to 1 m, sometimes purple at base, broadly sheathed; leaf-blade kidney-shaped in outline, deeply dissected into 5–9 broadly ovate, acuminate segments, to 50 × 53 cm. Spathe-tube to 7 cm, oblong, green; spathe-limb to 12 cm, lanceolate, acuminate, white. Spadix on a short stipe, which is *c.* 5 mm long. *Mexico to Costa Rica; reported from Surinam.* G2.

The plant introduced to Europe in 1882 as *X. barilletii* has been tentatively assigned to *X. hoffmannii.*

26. COLOCASIA Schott

P.F. Yeo

Plants evergreen, in our species without an above-ground stem, sometimes with stolons, rhizomes or tubers. Leaf-blade peltate (figure 14(7), p. 374) and either ovate-cordate or sagittate, usually with 2 or 3 marginal veins (figure 14(6), p. 374); branches of primary lateral veins bent towards margin and joined by cross-veins to form rectangular compartments. Inflorescences 1 or several in each axil, much shorter than the leaf-stalk. Spathe constricted between the short overlapped part and the much longer, hooded or flattened limb, which is finally shed. Spadix shorter than spathe, with a short or long sterile appendix. Florets without perianths, unisexual. Zone of male florets much longer than zone of female florets, separated from it by a constriction bearing scale-like sterile florets (figure 15(6), p. 377). Male florets with 3–6 stamens joined to form a short column. Ovary 1-celled; placentation parietal; ovules several to many. Stigma stalkless or nearly so. Fruit is a berry.

Seven species in the Old World tropics. A few have variants with strikingly coloured foliage which are grown for ornament in hothouses or for planting out in summer in warm situations. Rich damp soil and warmth are necessary for vigorous growth. Propagation is by division of the rootstock or by seed.

Literature: Krause, K., Colocasioideae, Das Pflanzenreich **71**: 62–71 (1920); Holttum, R.E., Alocasia marshallii, *Gardeners' Chronicle* **150**: 217 (1961).

1a. Leaf-blade to *c.* 16 cm, broadly ovate, cordate **1. affinis**
b. Leaf-blade to *c.* 50 cm, ovate or elliptic, sagittate **2. esculenta**

1. C. affinis Schott (*Alocasia jenningsii* Veitch; *A. marshallii* Bull). Illustration: *Flore des serres* **17**: 183 (1868); *Illustration Horticole* **16**: t. 585 (1869); *Gardeners' Chronicle* **1872**: 801 (1872); Graf, Exotica, edn 8, 126 (1976).

Plant with a round tuber and many small rhizomes. Leaf-stalk 20–30 cm, transversely banded with purple. Leaf-blade 10–16 × 7–10 cm, broadly ovate, cordate, shortly acuminate at apex, with about 3 pairs of main lateral veins; main vein of each basal lobe without major lateral veins on the side towards the sinus; upper surface with large purplish blotches between the main veins, and a small to large central silvery patch. Spathe 6–9 cm, pale green. Appendix of spadix 3–4 cm, longer and much narrower than the remainder. *Tropical eastern Himalaya.* G2.

2. C. esculenta (Linnaeus) Schott (*C. antiquorum* Schott). Illustration: *Botanical Magazine*, 7364 (1894); *Baileya* **7**: 67 (1959).

Rootstock variable. Leaf-stalk to 1 m. Leaf-blade to 50 cm, ovate or elliptic, with rounded or angular basal lobes, acute apex and 4–8 pairs of main lateral veins; midrib of one or both basal lobes with a main lateral vein on the side towards the sinus (figure 14(7), p. 374). Spathe 15–35 cm; tube greenish; limb pale yellow. Appendix of spadix very variable in length, only slightly narrower than the remainder unless rudimentary. *Tropical E Asia.* H5.

Widely cultivated in warm-temperate to tropical climates for its edible tubers, 'taro' and 'dasheen'. These occur as numerous cultivars. For ornamental cultivation the 2 following are best known: 'Fontanesii' (Illustration: *Botanical Magazine*, 7732, 1900; Birdsey, The cultivated aroids, 45, 195; Brickell (ed.), RHS A–Z encyclopedia of garden plants, 296 (2003)) with the leaf-stalk, inflorescence-stalk and spathe-tube blackish violet and the upper surface of the leaf-blade dark green, with dark margin and veins; and 'Illustris' (Illustration: Birdsey, The cultivated aroids, 45, 1951) with the leaf-stalk brownish purple at the base, greener above, and the blade with a blackish violet blotch between each pair of main veins.

Distinctions from *Alocasia odora* are given under that species (p. 399).

27. ALOCASIA (Schott) G. Don

D.J. Leedy, T.B. Croat & P.F. Yeo

Plants normally evergreen, rhizomatous and tuberous, with erect or ascending, usually short, stems. Leaf-blade entire to pinnatisect, peltate, at least in the juvenile stage, and cordate or sagittate, with netted subsidiary veins, and with interprimary veins and a marginal vein (figure 14(6), p. 374). Leaf-stalk long, cylindric, with a long sheath. Inflorescence-stalk usually much shorter than the leaf-stalk. Spathe with margins overlapping below to form a cylindric or ovoid tube; limb usually boat-shaped, becoming reflexed, falling after flowering. Spadix shorter than spathe, with a sterile appendix. Florets unisexual, without perianths. Zone of female florets enclosed in spathe-tube, usually separated from male zone by sterile florets. Male florets of 3–8 stamens united into a short column. Sterile florets flattened, scale-like (figure 15(6), p. 377). Style present. Stigma with 3–5 lobes. Ovary with a single cell, sometimes with partitions in the upper part. Placentation basal. Berries with 1 or few seeds, ripening to orange or red. Differs from *Colocasia* in the basal, not parietal, placentation.

About 70 species have been named, though some are not now accepted. Species of *Alocasia* in their young stages may be grown as houseplants on account of their strikingly marked foliage. Mature stages of the larger species are suitable for glasshouse collections or conservatories.

Most species thrive in warm and humid conditions; some (*A. odora, A. cucullata, A. macrorrhiza*, etc.) can survive 2–3 °C in winter and many become dormant during the cooler weather. Plants should be under-potted (in small pots) and given a loose, well-drained potting medium. Sufficient crocking in the bottom of the container and up to 50% wood chips (bark or fern wood) in the mix provides the needed drainage, but necessitates frequent watering. Some growers recommend small additions of dolomite lime. The plants have moderate light requirements and thrive in dense shade. Propagation may be achieved from seeds, cuttings, or the corms which most species produce from the rhizomatous stem. The stems topple in old plants of some species.

Many interspecific hybrids of *Alocasia* have arisen in cultivation, only some of which are mentioned below. Their existence should be borne in mind if difficulties in identification are encountered.

Literature: Engler, A. & Krause, K., Colocasioideae, Das Pflanzenreich **71**: 71–115 (1920); Bunting, G.S., The genus Schizocasia, *Baileya* **10**: 113–20 (1962); Bunting, G.S. & Nicolson, D.H., The Alocasia plumbea confusion, *Baileya* **10**: 142–6 (1963); Dotort, F. & Thompson, T., Alocasias, *Aroideana* **2**: 35–51 (1979).

Plant size. 2 m or more: **2, 3, 4, 5**; 1 to 2 m: **6, 7, 8, 9**; less than 1 m: **1, 10, 11**.
Leaf-blade colour. All green (both surfaces): **1, 2, 4, 7**; purple beneath: **3, 6, 8, 9, 11**; lighter coloured midrib and lateral veins: **5, 6, 8, 9, 10**.
Leaf-blade shape. Lobed at the sides: **5, 6**.
Leaf-blade texture. Velvety: **10**.

1a. Leaf-blade with margins lobed or pinnatisect 2
b. Leaf-blade entire apart from the basal sinus, sometimes wavy 3
2a. Upper surface of leaf-blade with main veins and margin boldly marked in white; margin sinuous **6. sanderiana**
b. Upper surface of leaf-blade with the veins not or only slightly paler than background; margin deeply sinuous (in young plants) to pinnatifid **5. portei**
3a. Leaf-blade on both sides and leaf-stalk green, without variegation, or the blade with very irregular, more or less segmental variegation and the stalk striped with white 4
b. Leaf-blade on one or both sides purple, or with variegation following the veins; leaf-stalk sometimes variegated or marbled 6
4a. Leaf-blade to 12 cm, heart-shaped or triangular, hollow above attachment of leaf-stalk **1. cucullata**
b. Leaf-blade normally very much larger, ovate-sagittate, without a hollow 5
5a. Leaf of mature plant not peltate; main lobe of blade with 9–12 pairs of veins **2. macrorrhiza**
b. Leaf of mature plant peltate, the basal lobes being shortly united; main lobe of blade with 6–10 pairs of veins **4. odora**
6a. Leaf-blade with the veins curved towards base of blade as they leave midrib, arching towards its apex as they approach the margin, deeply sunken **11. cuprea**
b. Leaf-blade with the veins not curved towards base of blade near midrib 7
7a. Leaf-blade with at least the main veins pale, forming a conspicuous pattern 8
b. Leaf-blade without a conspicuous pale pattern following the veins 10
8a. Leaf not or scarcely peltate, lower surface green **10. micholitziana**
b. Leaf distinctly peltate, basal lobes joined for one-seventh of their length or more, lower surface purple 9
9a. Leaf-blade with the minor veins not pale **8. longiloba**
b. Leaf-blade with the minor as well as the major veins pale **9. veitchii**
10a. Leaf-blade to 35 cm, green on both surfaces **7. zebrina**
b. Leaf-blade to 75 cm, green or purple above, purple beneath **3. plumbea**

1. A. cucullata (Loureiro) G. Don. Illustration: Graf, Exotica, edn 8, 127 (1976); *Aroideana* **2**: 42 (1979); Everett, Encyclopedia of horticulture **1**: 117 (1980).

Plant less than 1 m. Stem to 40 cm or more with many suckers, branched and inclined. Leaf-blade 6–12 × 4–7 cm, heart-shaped or triangular, cordate, peltate with basal lobes united for less than half their length, somewhat cupped above attachment of stalk, glossy green; veins prominent, strongly curved towards apex of blade, crowded at the base. Spathe fleshy; tube 4–8 cm; limb 5–10 cm. *Sri Lanka to Burma.* H5 (can endure light frost with only leaf damage).

2. A. macrorrhiza (Linnaeus) G. Don (*A. indica* misapplied). Illustration: *Edwards's Botanical Register* **8**: 641 (1822); *Botanical Magazine*, 3935 (1842); Graf, Exotica, edn 8, 121 – and as A. odora, 122 – juvenile (1976); Brickell (ed.), RHS A–Z encyclopedia of garden plants, 99 (2003).

Plant to 3 m or more; vegetative parts bright green. Stem stout, to 1 m or more. Leaf-blade to 125 × 75 cm, erect or spreading, ovate, sagittate, not peltate in the mature plant; main lobe with 9–12 pairs of veins; upper surface smooth, shiny; uniformly green on both surfaces. Leaf-stalk to 1 m or more. Spathe-tube 6–8 cm, limb 12–16 cm. *Sri Lanka, widely naturalised elsewhere in warm regions.* H5.

'Variegata'. Illustration: Birdsey, The cultivated aroids, 25 (1951); Graf, Exotica, edn 8, 122 (1976). Leaf-blade with large or small irregular blotches of creamy white, grey-green and darker green. Leaf-stalk longitudinally striped with green and white.

This species, much cultivated in the tropics and subtropics for its edible (when cooked) rhizomes and shoots, is often confused with *A. odora*, which has peltate leaves in the mature state.

3. A. plumbea (C. Koch) Van Houtte (*A. indica* var. *metallica* misapplied; *A. macrorrhiza* var. *rubra* (Hasskarl) Furtado). Illustration: *Flore des serres* **21**: 93 (1875); Graf, Exotica, edn 8, 123 (1976); *Aroideana* **2**: 40 (1979).

Similar to *A. macrorrhiza*, but purplish in the colour of the vegetative parts. Leaf-blade 75 × 50 cm or more, with 6–8 pairs of veins in the main lobe, sometimes wavy; upper surface glossy purple or green, sometimes with purple veins or blotches; lower surface deep purple. Leaf-stalk purple or variously spotted or marbled. *Java.* G1.

Usually becoming dormant in winter, unless the temperature is maintained at or above 16 °C.

4. A. odora (Loddiges) Spach. Illustration: Graf, Exotica, edn 8, 125 (1976); *Aroideana* **2**: 43 (1979).

Very similar to (and confused with) *A. macrorrhiza*, differing mainly in the peltate leaf-blade of the mature plant, which also has only 6–10 pairs of veins in the main lobe. Spathe-tube 3–4 cm; limb 10–14 cm. *Tropical Asia.* H5.

Distinguishable from *Colocasia esculenta* (p. 398) by its erect or spreading, not drooping, leaf-blades, with pinnately veined basal lobes.

5. A. portei Schott (*Schizocasia portei* (Schott) Engler; *S. regnieri* Linden & Rodigas). Illustration: *Illustration Horticole* **34**: 17 (1887) – leaves of semi-mature plant; *Baileya* **10**: 114–17 (1962); Graf, Exotica, edn 8, 125 – juvenile, 230 (1976).

Plant to 2 m or more, with stem to 24 cm tall and 10 cm in diameter. Leaf-blade ovate-sagittate; that of juvenile plant with deeply sinuous margins; that of mature plant to *c.* 2 m, pinnatisect, its segments linear-lanceolate, obtuse, strongly wavy; surface dark metallic green with the main veins sometimes slightly lighter in colour. Leaf-stalk to 2 m or more, marbled with red-purple on green. Spathe tube 2–2.5 cm; limb 5–6 cm. *Philippines.* G2.

6. A. sanderiana Bull (*Schizocasia sanderiana* (Bull) Engler). Illustration: Birdsey, The cultivated aroids, 27 (1951); Graf, Exotica, edn 8, 127 (1976).

Plant to 2 m. Leaf-blade 30–40 × 12–20 cm, narrowly sagittate, thinly leathery, with deeply sinuous margin; upper surface deep glossy green, with major lateral veins and margin boldly marked in white; lower surface green or (in 'Gandavensis') purple. Leaf-stalk 25–40 cm or more, brownish green. Inflorescence-stalk nearly as long as the leaf-stalk. Spathe 11 cm, with green tube and cream-coloured limb. *Philippines.* G2.

Plants with shallowly sinuous leaf-blades are likely to be hybrids of this species, for example, with *A. lowii* var. *grandis* (*A.* 'Amazonica'. Illustration: *Aroideana* **2**: 48, 1979) or with *A. plumbea* (*A.* × *chantrieri* André. Illustration: *Illustration Horticole* **35**: 79, 1888; Graf, Exotica, edn 8, 125, 1976).

7. A. zebrina C. Koch & Veitch. Illustration: *Flore des serres* **15**: 81 (1862); Graf, Exotica, edn 8, 124 (1976).

Plant to 1 m or more. Leaf-blade to 35 × 20 cm, triangular, sagittate, scarcely peltate, green on both surfaces. Leaf-stalk to 1 m, transversely variegated with black and green. Inflorescence-stalk variegated. Spathe-tube 2.5–4 cm; limb 12–14 cm, whitish. Spadix appendix pink. *Philippines.* G2.

8. A. longiloba Miquel. Illustration: Graf, Exotica, edn 8, 124 (1976); *Aroideana* **2**: 46 (1979).

Stem 30 –75 cm. Leaf-blade 30–50 × 15 cm (or to *c.* 70 × 27 cm in 'Magnifica'), narrowly triangular-sagittate; basal lobes more than half as long as principal lobe, joined for one-seventh to one-fifth of their length, acute; main veins 6–7 pairs; midribs of basal lobes with no main lateral veins towards the sinus. Upper surface of blade rich deep glossy green with margin and main veins light greyish; lower surface glossy purple. Leaf-stalk 25–32 cm, slender, not patterned. Spathe to 10 cm; limb much longer than tube. *Malaysia, Borneo, Java.* G2.

This plant was described from the Malay peninsula; whether specimens from Borneo and Java are the same is uncertain. The group of species dealt with here under Nos. 8 and 9 is represented in nature by populations isolated by the sea and by their preference for limestone outcrops.

A. lowii W.J. Hooker. Illustration: *Botanical Magazine*, 5376 (1863). Leaf-blade to *c.* 50 × 18 cm; basal lobes one-third to half as long as principal lobe, joined for about one-quarter of their length, tapered to an obtuse tip. Main veins up to 13 pairs. Spathe *c.* 11 cm, greenish white; limb *c.* 3 times as long as tube. *Borneo.* G2.

Hybrids of this species with *A. cuprea* (*A.* × *sedenii* Veitch. Illustration: *Illustration Horticole* **24**: 154, 1877; Graf, Exotica, edn 8, 125, 1976) have the veins less distinctly coloured and show the impressed, curved veins and nearly united basal lobes of *A. cuprea.* For the plant known as *A. lowii* (var.) *grandis*, see under No. 9.

9. A. veitchii (Lindley) Schott (*A. lowii* W. J. Hooker var. *veitchii* (Lindley) Engler). Illustration: *Botanical Magazine*, 5497 (1865) – as *A. lowii* var. *picta*; Das Pflanzenreich **71**: 107 (1920).

Stem 30–75 cm. Leaf-blade to *c.* 45 × 20 cm, narrowly triangular-sagittate; basal lobes about half as long as principal lobe, joined for one-sixth to one-fifth of their length; main veins about 9 pairs; midribs of basal lobes with no main lateral veins towards the sinus. Interprimary veins slightly sinuous. Upper surface of blade deep green with margin, main veins (together with the surface adjacent to them) and minor veins light greyish or grey-green; lower surface purple. Leaf-stalk *c.* 45 cm, greenish to purplish with darker transverse bands. Spathe *c.* 11 cm; tube greenish; limb about 3 times as long as tube, pale yellow, tinged with red at tip. *Borneo.* G2.

See remarks under *A. longiloba.* The following all have pale minor veins:

A. watsoniana Masters. Illustration: *Gardeners' Chronicle* **13**: 569 (1893); Graf, Exotica edn 8, 123, 125, 127 (1976); *Aroideana* **2**: 41 (1979). Leaf-blade to 90 × 45 cm, elliptic-ovate, sagittate; basal lobes more than half as long as main lobe, joined for one-third to half of their length; main veins about 7–10 pairs; midrib of one of the basal lobes sometimes with 1 or 2 main lateral veins towards the sinus; main veins of principal lobe often with a conspicuous branch on the basal side, running out to the margin. Interprimary veins sharply zig-zag (figure 14(6), p. 374). Upper surface bluish green with margins, main veins and minor veins silvery. Leaf-stalk to 60 cm. Spathe with limb *c.* 4 times as long as tube. *Sumatra.* G2.

A. putzeysii N.E. Brown. Illustration: *Illustration Horticole* **29**: 11 (1882); *Gardeners' Chronicle* **19**: 501 (1883). Leaves smaller than in *A. watsoniana*, their blades more elliptic and with only *c.* 6 pairs of main veins; basal lobes joined for one-fifth to one-third of their length, their main veins without branches towards the sinus. Main veins of principal lobe without conspicuous branches; interprimary veins gently sinuous. *Java.* G2.

Plants presently cultivated as *A. putzeysii* may be incorrectly named. Another variant of this alliance in cultivation has the upper leaf surface with a greyish metallic sheen, the basal lobes elliptic and the sinus correspondingly abruptly narrowed towards the base; the midribs of the basal lobes have rather indistinct branches towards the sinus, and the blade is up to 80 cm long and has about 8 pairs of main veins; it is grown as *A. korthalsii* Schott and is perhaps illustrated by the figure labelled *A. lowii* var. *veitchii* in Graf, Exotica, edn 8, 123 (1976). (*A. thibautiana* Masters, illustration: *Illustration Horticole* **18**: 72, 1881, was stated by Engler and Krause to be synonymous with *A. korthalsii* but it is not the same as the plant just described.)

The plant known as *A. lowii* (var.) *grandis* (Illustration: Graf, Exotica, edn 8, 123, 1976) belongs here; it has broadly ovate-cordate leaf-blades with a narrow sinus, and rounded basal lobes joined for about one-third of their length. The midribs of the basal lobes give off 3 or 4 main veins towards the sinus. The subsidiary venation is very regular and the interprimary veins are nearly straight (figure 14(7), p. 374).

10. A. micholitziana Sander. Illustration: *Botanical Magazine*, 8522 (1913); *Aroideana* **2**: 37 (1979); Everett, Encyclopedia of horticulture **1**: 117 (1980).

Plant less than 1 m. Stem 30–35 cm. Leaf-blade to 25 cm or more, narrowly sagittate, sometimes peltate, with the basal lobes shortly united, wavy, rich green with white to cream midrib and lateral veins; upper surface with velvety texture. Leaf-stalk to 35 cm or more, paler green, with brown to purple marbling. Spathe-tube 2.5–3 cm, green, limb 9–10 cm, paler green than tube outside, whitish or yellowish inside. Spadix almost as long as spathe. *Philippines.* G2.

'Maxkowskii' has the leaf-blade less wavy and more cordate, darker green with whiter veining. Leaf-stalk clear green without marbling; a very popular cultivar known in commerce as 'African Mask'.

11. A. cuprea (C. Koch & Bouéch) C. Koch Illustration: *Botanical Magazine*, 5190 (1860); Birdsey, The cultivated aroids, 23 (1951); Graf, Exotica, edn 8, 124 (1976); *Aroideana* **2**: 39 (1979).

Plant less than 1 m. Leaf-blade usually 20–30 × 10–18 cm, oblong-ovate or elliptic, cordate, peltate, with basal lobes united for most of their length, shortly and abruptly acuminate at apex; upper surface greenish with purplish and pinkish orange metallic sheen (coppery) and darker main veins; lower surface intensely purple; major veins markedly sunken, curved towards the base of the blade as they leave the midrib and then towards its apex as they approach the margin. Leaf-stalk and inflorescence-stalk purple. Spathe 5–7 cm, purple; limb about same length as tube, greenish. *Malaysia, Borneo.* G2.

There is also a variant in which the leaf-blade is green on both surfaces. The hybrid with *A. lowii* is mentioned under that species (p. 400).

28. SYNGONIUM Schott

T.B. Croat

Plants evergreen, epiphytic or rooted in soil, the stems loose and climbing when young, adpressed to trees, and with short internodes when adult. Sap milky. Juvenile plants with simple, usually ovate to sagittate leaf-blades; pre-adult leaf-blade larger, commonly sagittate. Leaf-blade of adult plant (in cultivated species) with 3 or sometimes 5 leaflets, the lateral often unequal, sometimes cleft or divided (blade pedate). Usually several marginal veins. Inflorescence-stalk erect in flower, pendent in fruit. Lower portion of spathe with overlapping edges; limb of spathe usually boat-shaped, opening broadly, usually thin, falling after a few days. Spadix shorter than spathe. Florets unisexual, without perianths. Zone of female florets enclosed within spathe-tube; male and female zones separated by a few sterile florets. Male florets with 4 united stamens. Fruit compound, pulpy.

A genus of 33 species throughout the New World tropics. Relatively few species are commercially available and plants are usually sold in juvenile or pre-adult stages. Variation in leaf-markings has led to the naming of cultivars in species Nos. 2 and 3.

The plants are difficult to pollinate and the fruits are never seen in cultivation. They grow best under conditions of high humidity and partial shade but species currently in cultivation do remarkably well in average home or office conditions. The potting medium should be loose and well drained but plants should be watered frequently.

Literature: Croat, T.B., A revision of Syngonium (Araceae), *Annals of the Missouri Botanical Garden* **68**: 565–651 (1981).

1a. Each lateral division of the leaf-blade entire or with a subsidiary lobe or leaflet; inflorescences 3–5 per axil; spathe-tube nearly cylindric, frequently more than 5 cm; spathe-limb usually twice as long as visible part of spadix **1. auritum**
b. Each lateral division of the leaf-blade with 2–5 connected lobes or, more usually, leaflets; inflorescences 4–11 per axil; spathe-tube ellipsoid, less than 5 cm; spathe-limb only slightly longer than visible part of spadix 2
2a. Stem sometimes with numerous rough projections; leaf-stalk usually with a sharp rib along the upper side; middle segments of leaf-blade nearly equal in size and shape; male florets deeply hollowed at apex, showing no indication of fusion between anthers **2. angustatum**
b. Stem smooth, lacking any projections; leaf-stalk with at most an obtuse rib along the upper side; middle segments of leaf-blade usually somewhat unequal; male florets truncate at apex with an indication of the line of fusion between anthers **3. podophyllum**

1. S. auritum (Linnaeus) Schott (*Philodendron auritum* invalid, not Lindley; *P. trifoliatum* invalid). Illustration: Birdsey, The cultivated aroids, 123 (1951); Graf, Exotica, edn 8, 234, 237 (1976); *Annals of the Missouri Botanical Garden* **68**: 611 (1981).

Stem of adult plant slightly glaucous. Leaf-blade of juvenile plant ovate, sagittate to hastate at base. Leaf-blade of pre-adult sagittate to hastate. Leaf-stalk of adult 15–48 cm, sheathed usually for four-fifths of its length. Leaf-blade of adult with 3 or sometimes 5 leaflets, the laterals sometimes with an external subsidiary lobe or leaflet; central segment 20–30 × 6–20 cm, broadly elliptic. Inflorescences up to 5 per axil; stalks 7–13 cm. Spathe with tube 4–8.5 cm, nearly cylindric, dark green, tinged reddish within at base; limb 12–18 cm, mucronate at apex, cream outside, faintly tinged with violet-purple within. Spadix extending to half the height of the spathe-limb; female part 3.5–5 cm; male part 3.5–9 cm, creamy white. *Jamaica & Hispaniola.* G2.

2. S. angustatum Schott (*S. oerstedianum* Schott; *S. podophyllum* var. *oerstedianum* (Schott) Engler; *S. albolineatum* invalid; *Nephthytis triphylla* Bailey). Illustration: Birdsey, The cultivated aroids, 121 (1951); *Annals of the Missouri Botanical Garden* **68**: 607, 614, 619 (1981).

Stem not glaucous, usually sparsely warty in juvenile plant. Leaf-blade of juvenile plant heart-shaped, dark green, marked on upper surface along principal veins with grey-green. Leaf-blade of pre-adult sagittate or hastate. Leaf-stalk of adult 15–40 cm, sometimes bloomed, sheathed for half to four-fifths of its length. Leaf-blade of adult pedate; central segment 11–31 × 4–10 cm, elliptic to oblong-elliptic or oblanceolate; lateral divisions with 2–5 segments. Inflorescences about 7 per axil; stalk to 17 cm. Spathe 7–14 cm; tube 2–5 cm, ellipsoid, green and usually glaucous; limb cream. Spadix with female part 1.3–2.6 cm × 5–10 mm, greenish, male part 2.5–6.5 cm, pale yellow. *Mexico to Costa Rica.* G2.

3. S. podophyllum Schott (*S. gracile* (Miquel) Schott; *S. riedelianum* Schott; *S. vellozianum* Schott). Illustration: Birdsey, The cultivated aroids, 125 (1951); Graf, Exotica, edn 8, 234, 237 (1976); *Aroideana* **1**: 81 (1978) – as *S. auritum*; *Annals of the Missouri Botanical Garden* **68**: 632, 638, 641 (1981).

Stem sometimes glaucous. Leaves sometimes with grey-green markings. Leaf-blade of juvenile plant 7–14 cm, heart-shaped, of pre-adult sagittate or hastate. Leaf-stalk of adult 15–60 cm, sheathed for two-thirds of its length. Leaf-blade of adult pedate; central segment 16–38 × 6–17 cm, obovate to elliptic; lateral divisions with 2–5 united or free segments, the outermost variously auricled. Inflorescences 4–11 per axil; stalk usually less than 9 cm when in flower, sometimes glaucous. Spathe 9–11 cm; tube 3–4 cm, ovoid to ellipsoid, green, sometimes glaucous; limb greenish white to creamy white or rarely yellowish. Spadix with female part 1–2 cm × 6–9 mm, greenish white, male part 4–7 cm. *Mexico to the Guianas, Brazil & Bolivia.* G2.

29. SYNANDROSPADIX Engler
P.F. Yeo
Plant tuberous, with a resting period. Leaves few. Leaf-stalk to 30 cm. Leaf-blade 20–45 × 15–20 cm, ovate with cordate base, or sagittate. Inflorescences sometimes emerging before the leaves; stalk shorter than leaf-stalk, pale green with darker green stripes. Spathe to *c.* 15 cm, obliquely funnel-shaped, tailed, curved forwards so that open side is inclined downwards; margins recurved except at base where they overlap (some descriptions indicate that spathe unfurls completely); outer surface pale green with darker green stripes, inner brownish red with darker stripes of same colour. Spadix *c.* 12 cm, with a stipe; lower part united to spathe. Florets without perianths, mostly unisexual. Male florets with 4–5 united stamens; common filament 6–7 mm. Female florets with 3–5 staminodes in the form of ovate scales. Ovary with 3–5 cells, each with 1 ovule; style well developed.

One species from the eastern side of the Andes. It may be grown in a cool greenhouse for its striking inflorescences which are quite freely produced, but emit a powerful smell of decaying flesh.

1. S. vermitoxicus (Grisebach) Engler. Illustration: *Botanical Magazine*, 7242 (1892); Everett, Encyclopedia of horticulture **10**: 3276 (1982). *Argentina (Tucuman Province).* G1.

30. SPATHICARPA W.J. Hooker
P.F. Yeo
Small evergreen plants, producing leaves and inflorescences from a short tuberous rhizome. Leaf-stalk with a long sheath. Leaf-blade with netted veins. Inflorescence-stalk longer than the leaf-stalk or the leaf, slender. Spathe opening to expose all the florets, green. Spadix half-cylindric, united to spathe. Florets without perianths, unisexual. Male florets of 3–4 united stamens. Female florets each with 3 small staminodes. Ovary 1-celled; ovule 1.

About 6 species from tropical South America, one of which is occasionally cultivated in tropical collections for its curious inflorescences which are produced almost continuously. It needs damp, shady conditions in summer and lighter and drier conditions in winter. Propagation is by division or from seed, which, however, rapidly loses viability.

Literature: Engler, A., Aroideae & Pistioideae, Das Pflanzenreich **73**: 53–7 (1920).

1. S. sagittifolia Schott. Illustration: Das Pflanzenreich **73**: 55 (1920); Parey's Blumengärtnerei **1**: 193 (1958).

Leaf-stalk about twice as long as blade. Leaf-blade to 12 × 4 cm, sagittate or hastate, conspicuously net-veined, rich green. Inflorescence-stalk inclined. Spathe *c.* 5 cm, lanceolate, inclined so that inner surface faces upwards, recurved at apex, green. Spadix with florets in 4 or 5 rows, the 2 outer female, the others male. Male florets peltate. *Brazil, Paraguay, Argentina.* G2.

31. CALLOPSIS Engler
P.F. Yeo
Plants evergreen. Stem a horizontal rhizome with crowded, erect, lateral shoots, each consisting of 1–2 sheathing scale-leaves and a leaf or inflorescence-stalk. Leaf-stalk to 13 cm. Leaf-blade to 13 cm, entire, spreading, ovate with cordate base, wrinkled; main veins joined by cross-veins and joined to the marginal vein. Inflorescence borne above the leaves. Spathe 3–4 cm, white, flat but folded at the base. Spadix 1.5 cm, yellow, cylindric, joined to the spathe at the base. Florets without perianths, unisexual; zones of male and female florets adjacent. Female florets only 3–12, with prominent style and stigma turned to one side. Ovule 1.

One species. Sometimes cultivated in tropical collections. Requires a soil high in organic matter, shade from strong sun and a humid atmosphere.

Literature: Engler, A., Aroideae & Pistioideae, Das Pflanzenreich **73**: 58–9 (1920).

1. C. volkensii Engler. Illustration: *Botanical Magazine*, 8071 (1906); Das Pflanzenreich **73**: 59 (1920); Graf, Exotica, edn 8, 166 (1976). *Tanzania.* G2. Winter.

32. ARUM Linnaeus
D.A. Webb & P.F. Yeo
Tuberous herbs. Leaves all basal; stalk shortly sheathing at base, usually about twice as long as blade; blade hastate to sagittate, entire. Inflorescence-stalks shorter than leaf-stalks or nearly equalling them, usually visible above the ground. Lower part of spathe narrow, with overlapping margins, enclosing the floral region of the spadix and usually the lower part of its appendix; limb usually lanceolate, concave, longer than the spadix. Florets without perianths, unisexual. Spadix with zones of male and female florets usually separated by a zone of sterile florets; sterile florets usually present also above the male (figure 15(7), p. 377). Male florets with 3 or 4 stamens; ovary 1-celled. Berry red; seeds 1–6.

A genus of 12 species, ranging from the Canary Islands to Iran, and northwards to Sweden. *A. orientale*, *A. maculatum* and *A. italicum* are readily naturalised in shade or semi-shade and occur as native plants in gardens in some parts of Europe. The other species should be planted in brighter and warmer situations, though they may need some shade in the Mediterranean region. The floral odour is usually fetid, *A. dioscoridis* being particularly offensively scented, but in *A. creticum* the scent is sweet. Propagation is by division of the tubers.

Literature: Engler, A., Aroideae & Pistioideae, Das Pflanzenreich **73**: 67–99 (1920); Prime, C.T., Lords & ladies (1960).

Season of growth. Leaves appearing in autumn: **1, 2, 3, 8**; leaves appearing in spring: **4, 5, 6, 7**. Flowers appearing in autumn: **2**.
Tuber. Horizontal: **7, 8**; vertical: **1, 2, 3, 4, 5, 6**.
Spathe. Mainly white, yellow or green, though sometimes with purple spots on margin: **1, 3, 5, 6, 7, 8**; mainly purple inside: **2, 3, 4, 6**.
Appendix of spadix. With a stalk-like basal part much narrower than the upper part: **2, 4, 6, 7, 8**; base more or less uniform in diameter, not stalk-like: **1, 3, 5, 6**.

1a. Flowering in autumn **2. pictum**
b. Flowering in spring or early summer 2
2a. Female zone of spadix much shorter than male **1. creticum**
b. Female zone of spadix about as long as male or longer 3
3a. Spathe-limb uniformly dark purple inside and pale greenish outside; female zone of spadix about same length as male **4. palaestinum**

b. Spathe-limb with inner and outer surfaces not strongly contrasted or, if they are, part of the inner surface is boldly spotted; female zone of spadix much longer than male 4
4a. Spathe-limb marked in part with bold coalescent spots, usually of dark purple on a greenish or yellowish background but sometimes the reverse **3. dioscoridis**
b. Spathe-limb with scattered spots or none 5
5a. Leaves appearing in autumn **8. italicum**
b. Leaves appearing in spring 6
6a. Spadix with little or no space between male zone and the upper and lower zones of sterile florets; spathe 5–10 × 2–3 cm **5. hygrophilum**
b. Spadix with a clear space between male zone and the upper and lowerzones of sterile florets; spathe usually more than 10 × 3 cm 7
7a. Tuber horizontal; leaves often with blackish spots; fruiting spike 3–5 cm **7. maculatum**
b. Tuber vertical; leaves unspotted; fruiting spike 6–8 cm **6. orientale**

1. A. creticum Boissier & Heldreich. Illustration: *Botanical Magazine*, 101 (1950); Huxley & Taylor, Flowers of Greece and the Aegean, f. 418 (1977); Grey-Wilson & Mathew, Bulbs, pl. 35 (1981); Brickell (ed.), RHS A–Z encyclopedia of garden plants, 143 (2003).

Tuber vertical. Leaves appearing in autumn, hastate-sagittate, the blade 10–20 × 7–14 cm. Inflorescence-stalk usually exceeding the leaf-stalk. Spathe 15–25 cm, bright yellow, greenish yellow, white or rarely pale green. Zone of female florets much shorter than that of male; sterile florets few and small, sometimes absent. Appendix of spadix orange-yellow or purple, almost equalling the spathe. *Crete & Karpathos*. H4. Late spring.

2. A. pictum Linnaeus filius. Illustration: Coste, Flore de la France **3**: 433 (1906); Grey-Wilson & Mathew, Bulbs, pl. 35 (1981).

Tuber vertical. Leaves appearing in autumn, hastate; basal lobes short and rounded; stalk only slightly longer than blade. Inflorescence-stalk mainly below ground. Spathe 15–25 cm, dark reddish purple. Zone of female florets longer than that of male; sterile florets above the male but absent between male and female. Appendix of spadix dark purple, rather shorter than the spathe, the basal half forming a narrow stalk. *Islands of W Mediterranean*. H4. Autumn.

3. A. dioscoridis Sibthorp & Smith. Illustration: Das Pflanzenreich **73**: 70 (1920); Polunin & Huxley, Flowers of the Mediterranean, f. 224 (1965); Huxley & Taylor, Flowers of Greece and the Aegean, f. 419 (1977); Rix & Phillips, The bulb book, 157 (1981).

Tuber vertical. Leaves appearing in autumn, hastate-sagittate, acute; basal lobes triangular, acute. Inflorescence-stalk mainly below ground. Spathe 15–35 cm, usually for the most part dark purple, but sometimes pale green only slightly tinged with purple, always to some extent spotted with brown, yellow, black or dark purple, some of the spots coalescent. Zone of female florets twice as long as that of male. Sterile florets numerous, short. Appendix of spadix dark purple, not much shorter than the spathe. From *Turkey & Iraq to Israel*. H5. Spring.

4. A. palaestinum Boissier. Illustration: *Botanical Magazine*, 5509 (1865); Birdsey, The cultivated aroids, 41 (1951); Graf, Exotica, edn 8, 144 (1976); Rix & Phillips, The bulb book, 156 (1981).

Tuber vertical. Leaves appearing in spring, hastate; lobes broad, obtuse or acute. Spathe 15–20 cm, yellowish green at the base, the blade spreading, greenish outside, uniformly dark purple inside. Zones of male and female florets of about equal length. Appendix of spadix slender, nearly as long as the spathe, the lower part narrowed into a stalk. *Israel*. H4. Spring.

5. A. hygrophilum Boissier. Illustration: Das Pflanzenreich **73**: 71 (1920); Everett, Encyclopedia of horticulture **1**: 261 (1980).

Tuber vertical. Leaves appearing in spring, hastate, with obtuse basal lobes. Spathe 5–12.5 cm, mostly green, but usually with purple margin and sometimes inner surface suffused with purple; limb not much longer than tube. Zone of female florets twice as long as that of male; upper and lower zones of sterile florets adjacent or nearly so to male zone. Appendix of spadix dark purple, moderately or considerably shorter than spathe, of uniform diameter at base. *Cyprus to Syria & Israel; ?Morocco*. H4. Spring.

6. A. orientale Bieberstein. Illustration: Huxley & Taylor, Flowers of Greece and the Aegean, f. 418 (1977).

Tuber vertical. Leaves appearing in spring, sagittate. Spathe 7–15 cm, green, variably tinged with purple and sometimes almost entirely purple, but not spotted. Zone of female florets much longer than that of male; upper and lower zones of sterile florets clearly separated from the male zone. Appendix of spadix dark purple, with or without a stalk-like base, some what shorter than the spathe. Fruiting spike 6–8 cm. *C & E Europe, northwards to Sweden; SW Asia, eastwards to Iran*. H3. Early summer.

7. A. maculatum Linnaeus. Illustration: Prime, Lords & ladies, t. 1, 57 (1960); Proctor & Yeo, The pollination of flowers, 290 (1970); Ross-Craig, Drawings of British plants, **30**; 39 (1973); Grey-Wilson & Mathew, Bulbs, pl. 35 (1981).

Very like *A. orientale*, and only distinguished with certainty by the horizontal tuber. It often also differs, however, in having blackish spots on the leaves and spathe, which is otherwise greenish, in having the appendix to the spadix occasionally yellow, and always with a stalk-like base, and in the longer spathe (10–20 cm) and shorter fruiting spike (3–5 cm). *Europe*. H2. Late spring.

8. A. italicum Miller. Illustration: *Botanical Magazine*, 2432 (1823); Polunin & Huxley, Flowers of the Mediterranean, f. 220 (1965); Graf, Tropica, 95 (1978); Grey-Wilson & Mathew, Bulbs, pl. 35 (1981).

Tuber horizontal. Leaves appearing in autumn or early winter, hastate or sagittate, sometimes veined with white, rarely with dark spots. Spathe 15–40 cm, greenish yellow, rarely white, sometimes edged with purple. Zone of female florets much longer than that of male. Appendix of spadix yellow, with distinct, stalk-like base, much shorter than spathe. *From the Canary Islands to Cyprus, & northwards to England*. H3. Late spring.

33. DRACUNCULUS Schott
D.A. Webb & P.F. Yeo

Tuberous plants, *c.* 70–180 cm, with several large leaves. Leaves basal, their stalks with long, overlapping sheaths which together resemble a stem. Leaf-blade pedately divided or deeply pedately lobed. Lower part of spathe with overlapping margins enclosing lower part of spadix, limb large, widely open. Florets without perianths, unisexual. Spadix with zones of male and female florets adjacent or separated; appendix long. Male florets with 2–4 stamens. Ovary 1-celled. Berry green; seeds 6.

A genus of 3 species, all described below. They should be cultivated in fairly moist, humus-rich soil in a sheltered position in sun or semi-shade.

Literature: Engler, A., Aroideae & Pistioideae, Das Pflanzenreich **73**: 99–104 (1920) – as *Dracunculus* and *Helicodiceros;* Henderson, A., More on dragon plants, *The Garden* **107**: 498–9 (1982).

1a. Appendix of spadix and inner surface of spathe-limb covered with hairs or filaments **3. muscivorus**
b. Appendix of spadix and inner surface of spathe-limb without hairs or filaments 2
2a. Spathe within, and appendix of spadix dark purplish red **1. vulgaris**
b. Spathe white within; appendix of spadix yellow **2. canariensis**

1. D. vulgaris Schott (*Arum dracunculus* Linnaeus). Illustration: Fiori & Paoletti, Iconographia Florae Italicae, edn 3, 67 (1933); Polunin & Huxley, Flowers of the Mediterranean, f. 223 (1965); Huxley & Taylor, Flowers of Greece and the Aegean, f. 242 (1977); Rix & Phillips, The bulb book, 157 (1981).

Leaf-stalk marbled or spotted with dark purple; blade often spotted or striped with white, with 11–15 lanceolate segments, the largest to 20 cm long. Inflorescence-stalk longer than leaf-sheaths. Spathe 25–40 cm; tube 6 cm; limb 10–15 cm wide, tilted backwards, with wavy margins, uniformly dark reddish purple inside. Spadix with zones of male and female florets adjacent; sterile florets present above the male zone; appendix almost equalling the spathe, stout, blackish red. *Mediterranean region, from Corsica to S Turkey.* H3. Early summer.

2. D. canariensis Kunth. Illustration: Bramwell & Bramwell, Wild flowers of the Canary Islands, 108 (1974); *The Garden* **107**: 499 (1982) – lower figure.

Segments of leaf-blade 7–11, slightly smaller than in *D. vulgaris*, unspotted. Inflorescence-stalk purple-spotted. Spathe 25–35 cm; limb 5–6 cm wide, erect, with slightly inrolled and wavy margins, white inside. Spadix without sterile florets and with a slender yellow appendix, its male florets not enclosed in base of spathe. *Madeira & the Canary Islands.* H5. Early summer.

3. D. muscivorus (Linnaeus filius) Parlatore (*D. crinitus* Schott; *Helicodiceros muscivorus* (Linnaeus filius) Engler). Illustration: *Edwards's Botanical Register* **10**: 831 (1824); *Flore des serres* **5**: t. 445, 446 (1849); Fiori & Paoletti, Iconographia Florae Italicae, edn 3, 67 (1933); Graf, Exotica, edn 8, 145 (1976).

Leaf-stalk, inflorescence-stalk and outside of spathe-tube pale green with short darker green stripes. Leaf-blade deeply pedately cut into narrow lobes, the 3 central ones hastately arranged, the others disposed at various angles owing to twisting of each lateral axis. Spathe situated below the leaves; tube 15 cm; limb *c.* 25 × 20 cm, bent backwards horizontally, purplish brown with darker or lighter mottling, covered with coarse reddish hairs. Spadix with sterile florets above and below the male zone; appendix dark green or yellow, about half as long as spathe-limb, mainly horizontal, covered with bristle-like filaments. *Corsica, Sardinia & Balearic Islands.* H5. Early summer.

34. SAUROMATUM Schott
P.F. Yeo

Tuberous herbs, bearing one leaf at a time. Leaf-blade pedately cut or divided. Inflorescence appearing before the leaves; stalk short, so that the spathe may rest on the ground. Spathe with margins fused below to form a tube; limb lanceolate; both parts withering after flowering. Florets without perianths, unisexual. Spadix with zones of male and female florets about equal, with a long sterile zone between them which bears club-shaped sterile florets below, and a long curved appendix. Male florets indistinct. Ovary 1-celled, with 2–4 ovules.

A genus of 2 species from E and W Africa and S Asia, both cultivated. That described below is grown for its strikingly spotted leaf-stalks and spathes, which give it a sinister appearance. The tubers are sold as a curiosity for flowering indoors without soil or water, though the floral odour is sometimes offensive. Plants may be grown in the open or in greenhouses in a humus-rich well-watered soil. They are dormant in winter, when the tubers may be lifted if desired. Propagation is by offsets from the tubers.

Literature: Engler, A., Aroideae & Pistioideae, Das Pflanzenreich **73**: 122–7 (1920).

1. S. venosum (Aiton) Kunth (*S. guttatum* (Wallich) Schott; *Arum venosum* Aiton; *A. guttatum* Wallich; *S. nubicum* Schott). Illustration: *Edwards's Botanical Register* **12**: t. 1017 (1826); *Botanical Magazine*, 4465 (1849); *Flore des serres* **13**: 79 (1858); Everett, Encyclopedia of horticulture **9**: 3063 (1982).

Tuber to *c.* 15 cm in diameter. Leaf-stalk to 50 cm, green, often with darker, grey-green spots. Leaf-blade pedately cut or divided, with 7–15 lobes or leaflets. Inflorescence-stalk *c.* 5 cm. Spathe-tube 5–10 × 2–2.5 cm; limb 30–70 × 8–10 cm, channelled, becoming twisted, often with wavy margins; outer surface of limb dull purplish, inner surface yellowish or greenish, variously spotted with brownish purple. Spadix appendix to *c.* 30 cm, green, flushed with purple. *E & W Africa, Himalaya from Kashmir to N Burma, peninsular India.* H3–4. Spring–early summer.

Very variable in colour and extent of spotting; probably varying in hardiness according to origin.

35. BIARUM Schott
P.F. Yeo

Plants with small, flattened, spherical tubers producing roots above. Scale-leaves conspicuous around bases of foliage leaves and inflorescences. Leaf-blade entire, not lobed. Inflorescence-stalk concealed below ground or slightly exposed in fruit. Spathe with margins of lower part united to form a tube, withering before fruit is ripe. Florets unisexual, without perianths. Spadix with a wide space between zones of male and female florets on which are usually borne some slender sterile florets; appendix as long as remainder of spadix or longer. Male florets with 1 or 2 stamens. Ovary 1-celled, with 1 ovule. Fruiting spadix approximately spherical. Berry white or pale green, sometimes striped with purple.

A genus of about 21 species in the Mediterranean region and W Asia, of which a few are occasionally cultivated. The inflorescences have been described as looking like 'black holes in the ground with a tail protruding'. They are hardy and need hot conditions when dormant in summer to promote flowering. Identification in this genus is difficult, partly because leaves and flowers are not usually present at the same time.

Literature: Engler, A., Aroideae & Pistioideae, Das Pflanzenreich **73**: 133–43 (1920); Talavera, S., Revision de las especies españolas del genero Biarum Schott, *Lagascalia* **6**: 275–96 (1976); Boyce, P.C. A taxonomic revision of Biarum. *Botanical Magazine* **25**(1): 2–188 (2008).

1a. Spadix with sterile florets above and below male zone; spadix appendix to 4 mm in diameter **1. tenuifolium**
b. Spadix with sterile florets only below male zone; spadix appendix 6–8 mm in diameter **2. eximium**

1. B. tenuifolium (Linnaeus) Schott
Leaf usually 5–20 cm × 3–15 mm, narrowly lanceolate, often without a distinct stalk. Inflorescence appearing before the leaves. Spathe-tube to 6 cm; limb to 30 cm, purple inside and green outside. Spadix with zone of female florets 3 mm and zone of male florets 6–8 mm; sterile florets present above and below male zone; appendix to 4 mm in diameter, exceeding the spathe, much longer than remainder of spadix. *Mediterranean region from Greece westwards to Portugal.*

Very variable; some populations have been separated as distinct species or treated as varieties.

Subsp. **tenuifolium**. Illustration: *Botanical Magazine*, 2282 (1821); 19, 29, 605 (2008); Huxley & Taylor, Flowers of Greece, f. 422 (1977).

Plant relatively large. Spathe-limb many times longer than wide, erect, usually twisted. Spadix appendix 20–30 cm, erect. *C&E Mediterranean region.* H3. Autumn.

Subsp. **abbreviatum** (Schott) Engler. Illustration: Mathew, Dwarf bulbs, 50 (1973); Grey-Wilson & Mathew, Bulbs, pl. 35 (1981); Botanical Magazine, 35 (2008).

Plant smaller. Spathe-limb less than 10 cm, about half as wide as long, hooded. Spadix appendix *c.* 10 cm, emerging almost horizontally from spathe. *Italy, Yugoslavia, N Greece.* H3. Late summer.

2. B. eximium (Schott & Kotschy) Engler. Illustration: Das Pflanzenreich **73**: 138 (1920); Mathew, Dwarf bulbs, 50 (1973); *Botanical Magazine*, 52 (2008).

Leaf-stalk to 10 cm; leaf-blade to 9 × 4 cm. Inflorescence appearing before the leaves. Spathe *c.* 12 cm; limb 9 × 4 cm, nearly flat, velvety and blackish purple within, pale outside with dark brownish purple spots. Spadix about as long as spathe; female zone 5 mm, sterile zone and male zone each 1–1.5 cm; sterile florets present only below male zone; appendix 6–8 mm in diameter, not exceeding the spathe, about as long as remainder of spadix. *Turkey.* H3. Autumn.

36. ARISARUM Targioni-Tozzetti
D.A. Webb

Rootstock a tuber or rhizome. Leaves long-stalked, sagittate, entire. Spathe withering before fruit is ripe, the lower part with margins united so as to form a closed tube, the upper part open in front but curved forwards at the tip so as largely to conceal the spadix. Spadix shorter than or equalling the spathe, with a short floral zone at the base and a long appendix. Florets without perianths, unisexual, the female few, the male situated immediately above the female. Stamen 1; ovary 1-celled. Berry greenish, with about 6 seeds.

Two species from the Mediterranean region and the Atlantic islands. They are best cultivated in semi-shade in the rock garden or in a cool house in soil fairly rich in leaf-mould.

1a. Spathe acute to mucronate; appendix of spadix greenish **1. vulgare**
b. Spathe terminating in a long, thread-like tip; appendix of spadix whitish **2. proboscideum**

1. A. vulgare Targioni-Tozzetti (*A. simorrhinum* Durieu). Illustration: *Botanical Magazine*, 6023 (1873); Fiori & Paoletti, Iconografia della Flora Italiana, 62 (1895); Rix & Phillips, The bulb book, 156 (1981), Grey-Wilson & Mathew, Bulbs, pl. 35 (1981).

Rootstock an irregular tuber, from which arise an inflorescence and a single leaf. Leaves 5–12 × 3–10 cm; stalk to 30 cm, purple-spotted. Inflorescence-stalk 7–30 cm, slender. Spathe 4–6 × 1–1.5 cm, greenish, variably tinged or striped with dark purple. Appendix of spadix greenish. *Mediterranean region, Canaries & Azores.* H4–5. Late autumn–spring.

2. A. proboscideum (Linnaeus) Savi. Illustration: *Botanical Magazine*, 6634 (1882); Fiori and Paoletti, Iconografia della Flora Italiana, 62 (1895); Rix & Phillips. The bulb book, 159 (1971); Grey-Wilson & Mathew, Bulbs, pl. 35 (1981).

Rootstock a slender rhizome. Leaves 6–15 × 4–10 cm; stalk to 25 cm, not spotted. Inflorescence-stalk 8–15 cm, often inclined or prostrate. Spathe dark brown, paler towards the base, the hooded part and tube 2–4 cm, the thread-like tip 5–15 cm. Appendix of spadix whitish, concealed within the spathe. *C & S Italy, S Spain.* H4. Spring.

Cultivated mainly as a curiosity, the upper part of the spathe (which is usually all that protrudes from beneath the leaves) being strongly suggestive of the tail and hindquarters of a mouse.

37. ARISAEMA Martius
D.A. Webb

Plants with tubers (often with offsets or stolons), or rarely rhizomes. Leaves solitary or few, compound, long-stalked, all basal, though their sheathing base sometimes conceals the lower part of the inflorescence-stalk so as to make the leaves appear to be borne on a stem. Two or more whitish, papery, narrow-oblong sheathing scale-leaves are usually present, enfolding the bases of the leaf-stalks and the inflorescence-stalk. Lower part of spathe narrow, with overlapping margins, surrounding the floral part of the spadix and sometimes much of its appendix as well; upper part concave or nearly flat, erect or bent forwards at the tip, but seldom completely concealing the spadix. Spadix with a long appendix. Florets without perianths, unisexual, the male usually not touching; borne, in monoecious plants, immediately above the female. Stamens 1–5, united when more than 1. Ovary 1-celled; seeds few.

Most species are normally dioecious, including all those described below except *A. dracontium*. The condition is, however, variable in some species. Dioecious species occasionally bear some abortive male florets or filamentous sterile florets above the female.

A genus of 120–150 species, mainly from E Asia and the Himalayan region, but a few from N America and E Africa. They should be cultivated in semi-shade in damp, but not sodden, peaty soil.

Literature: Engler, A., Aroideae & Pistioideae, Das Pflanzenreich **73**: 149–220 (1920); Hara, H., A revision of the eastern Himalayan species of the genus Arisaema (Araceae), *Flora of Eastern Himalaya, Second Report (University Museum, Tokyo, Bulletin No. 2)*: 320–54 (1971); Ohashi, H. & Murata, J., Taxonomy of the Japanese Arisaema (Araceae), *Journal of the Faculty of Science of* *the University of Tokyo*, Section III, **12**: 281–336 (1980); Mayo, S.J., A survey of cultivated species of Arisaema, *The Plantsman* **3**: 193–209 (1982).

Tip of spathe. Very long and thread-like: **9**.
Tip of spadix. Very long and thread-like: **1, 2, 8**.
Leaves. Pedate: **7, 8**; palmate with **9** or more leaflets: **9**; with **3**, rarely **5**, leaflets: **1, 2, 3, 4, 5, 6**.
Sexuality of plant. Usually monoecious: **8**.

1a. Spathe terminating in a thread-like tip nearly as long as the rest of the spathe **9. consanguineum**
b. Spathe acute, acuminate, or with a slender tip much shorter than the rest of the spathe 2
2a. Spadix prolonged into a long, thread-like tip 3
b. Spadix obtuse or acute, but without a thread-like tip 5
3a. Leaves 2; limb of spathe broader than long, deeply 2-lobed, with a short filament arising from the sinus **1. griffithii**
b. Leaf solitary; limb of spathe longer than broad, not 2-lobed 4
4a. Leaf-blade pedate, with 7–15 leaflets; appendix of spadix whitish with green tip **8. dracontium**
b. Leaf-blade with 3 leaflets; appendix of spadix yellowish pink, with a dark purple tip **2. speciosum**
5a. Leaf-blade pedate, with 5–11 leaflets **7. serratum**
b. Leaf-blade with 3 (rarely 5) leaflets 6
6a. Upper part of spathe white, cream or pale pink **3. candidissimum**
b. Upper part of spathe mostly green or dark purple, though sometimes striped with white 7
7a. Spathe with conspicuous rounded auricles at base of blade **5. ringens**
b. Spathe without auricles 8
8a. Appendix of spadix white, much thicker at the tip than at the base **6. sikokianum**
b. Appendix of spadix brownish, more or less cylindric **4. triphyllum**

1. A. griffithii Schott. Illustration: *Botanical Magazine*, 6491 (1880); Rix & Phillips, The bulb book, 158 (1981); *The Plantsman* **3**: 201 (1982); Brickell (ed.), RHS A–Z encyclopedia of garden plants, 138 (2003).

Leaves 2; stalks stout, 12–20 cm, not spotted; leaflets 3, to 25 × 20 cm, broadly diamond-shaped to nearly circular. Inflorescence-stalk shorter than the leaf-stalks. Lower part of spathe *c.* 10 × 2.5 cm, dark purple with longitudinal white stripes; limb to 15 × 20 cm, dark purple with a network of green veins, deeply divided into 2 rounded lobes by a sinus 8–10 cm deep, from which protrudes a slender filament 5–10 cm long. Spadix with a stout floral zone surmounted by a dark purple appendix 5 cm long (excluding the thread-like tip), expanded into a wide disc near its base; thread-like tip 30 cm or more. *E Himalaya*. H4. Early summer.

2. A. speciosum (Wallich) Martius. Illustration: *Botanical Magazine*, 5964 (1872); Graf, Exotica, edn 8, 145 (1976); *The Plantsman* **3**: 207 (1982).

Leaf solitary; stalk 25–40 cm, stout, boldly spotted with purplish brown; leaflets 3, 20 × 12 cm, shortly stalked, triangular-ovate to lanceolate, acuminate, edged with reddish purple. Inflorescence-stalk much shorter than the leaf-stalk. Spathe 12–20 cm, the tubular base pale green, striped with dull red; limb ovate-lanceolate, long-acuminate, deep purple inside, striped with pale green, greenish outside. Appendix of spadix 5–7 cm, yellowish pink, terminating in a dark purple, thread-like tip 30–50 cm. *C & E Himalaya*. H5. Spring.

3. A. candidissimum Wright Smith. Illustration: *Botanical Magazine*, 9549 (1939); *Bulletin of the Alpine Garden Society* **38**: 409 (1970); Rix & Phillips, The bulb book, 158 (1981); Brickell (ed.), RHS A–Z encyclopedia of garden plants, 137 (2003).

Leaf solitary, appearing with the flower or just after; stalk 25–35 cm; leaflets 3, 10–20 × 7–18 cm, the middle one the largest. Inflorescence-stalk 12–20 cm. Spathe 8–15 cm, the lower part green, striped with white; limb ovate, acuminate, white, cream or pale pink. Male florets touching. Appendix of spadix 5–6 cm, tapered towards the tip, green. *S China (Yunnan and S Sichuan)*. H4. Early summer.

4. A. triphyllum (Linnaeus) Schott (*A. atrorubens* (Aiton) Blume). Illustration: *Botanical Magazine*, 950 (1806); Britton & Brown, Illustrated flora of the northern United States and Canada **1**: 361 (1896); Rix & Phillips, The bulb book, 154 (1981); Brickell (ed.), RHS A–Z encyclopedia of garden plants, 138 (2003).

Leaves usually 2, rarely 1 or 3; stalk 30–60 cm, slender, spotted; leaflets 3, 8–15 × 3–7 cm, ovate, acute. Inflorescence-stalk slightly shorter than leaves. Spathe 10–15 cm, green, dark purple, or green with purple stripes, the tip arching over the spadix but not concealing it. Appendix of spadix about 7 cm, brownish. *Eastern N America, from S Canada to Louisiana, and westwards to Kansas*. H2. Summer.

5. A. ringens (Thunberg) Schott (*A. praecox* Koch). Illustration: *Botanical Magazine*, 5267 (1861); Rix & Phillips, The bulb book, 155 (1981).

Similar to *A. triphyllum*, and formerly confused with it, differing chiefly in the long-acuminate or tailed leaflets and the spathe, which has 2 rounded auricles at the base of the blade, and whose upper part is curved sharply downwards so as to conceal the spadix. *S Japan, S Korea & E China*. H4. Spring.

6. A. sikokianum Franchet & Savatier (*A. sazensoo* misapplied). Illustration: *Botanical Magazine*, 9589 (1940); Graf, Tropica, 97 (1978); Rix & Phillips, The bulb book, 159 (1981); *The Plantsman* **3**: 205 (1982).

Leaves 2; stalk long, but colourless and sheathing in its lower part so that the leaves appear relatively short-stalked and borne on a stem; leaflets 3–5, to 15 × 10 cm, oblong-elliptic to broadly ovate, shortly acuminate. Inflorescence-stalk 10–30 cm. Spathe 15–20 cm, dark purple with green and white stripes; limb narrowly ovate, long-acuminate. Appendix of spadix 5 cm, slender below, thickened above and terminating in a spherical knob closing the throat of the spathe. *S Japan*. H4. Early summer.

7. A. serratum (Thunberg) Schott (*A. japonicum* Blume). Illustration: *Botanical Magazine*, 7910 (1903); Kitamura et al., Coloured illustrations of herbaceous plants of Japan (Monocotyledoneae), t. 53, f. 338 (1970).

Leaves 2; stalks 40–70 cm, sheathing for more than half their length; blade pedate with 5–11 lanceolate, more or less stalkless leaflets. Spathe 8–12 cm, the basal, tube-like portion constituting more than half the total length, shortly acuminate, green or purple with white stripes. Appendix of spadix 4–5 cm, slender, yellow. *C & S Japan*. H4. Spring.

8. A. dracontium (Linnaeus) Schott. Illustration: *Edwards's Botanical Register* **8**: 668 (1822); Gleason, Illustrated flora of the northeastern United States and adjacent Canada **1**: 367 (1952).

Leaf solitary; stalk to 80 cm, sheathing at the base; blade pedate, with 7–15 oblong-elliptic, shortly stalked segments. Spathe 5–7 cm, narrow, without any clear distinction between basal tubular part and blade, greenish. Spadix often with male and female florets; appendix white at the base, prolonged into a slender, green, thread-like tip about twice as long as the spathe. *Eastern USA & SE Canada.* H2. Late spring.

9. A. consanguineum Schott. Illustration: *Gardeners' Chronicle* **62**: 127 (1917); Das Pflanzenreich **73**: 175 (1920); *The Plantsman* **3**: 197 (1982); Brickell (ed.), RHS A–Z encyclopedia of garden plants, 137 (2003).

Leaf solitary; stalk to 80 cm, sheathing in its lower two-thirds, often spotted with dark red; blade palmate, with 11–21 linear-lanceolate, long-acuminate leaflets to 40 cm long. Inflorescence-stalk shorter than leaf-stalk. Spathe *c.* 20 cm, including the long, thread-like tip, which is nearly as long as the rest of the spathe, dark purple with fine white stripes. Spadix with a few sterile florets above the fertile; appendix 3 cm, slender, mostly concealed within the tube of the spathe. *From C China (Hubei) to NE India & NW Thailand.* H4. Early summer.

A. concinnum Schott. Illustration: Hara, Flora of eastern Himalaya, Second Report (University Museum, Tokyo, Bulletin No. 2), t. 19 (1971); Rix & Phillips, The bulb book, 158 (1981). Has the leaf-blade with 7–11 leaflets, which are broader (obovate-lanceolate) and thinner than in *A. consanguineum*, with wavy margins; spadix appendix thickened and roughened at the tip. *Himalaya & N Burma.* H5.

38. PINELLIA Tenore

D.A. Webb & P.F. Yeo

Slender plants growing from tubers. Leaves basal, their blades simple, lobed or compound. Inflorescence arising from tuber separately from the leaves. Spathe narrow, with margins overlapping below and with a flat or channelled limb above. Florets without perianths, unisexual. Spadix with zone of female florets entirely united with the spathe; male zone free; appendix slender. Male florets with 1 or 2 stamens. Ovary 1-celled. Berry with a single seed.

A genus of about 6 species from E Asia. They may be cultivated in well-drained, sandy soil, but *P. ternata* is apt to become a weed on account of the bulbils on the leaf-stalks.

Literature: Engler, A., Aroideae & Pistioideae, Das Pflanzenreich **73**: 220–5 (1920).

1a. Leaflets 3, not more than 12 cm; spadix appendix far exceeding spathe **1. ternata**
b. Leaflets 7–11, 15–18 cm; spadix appendix shorter than spathe **2. pedatisecta**

1. P. ternata (Thunberg) Breitenbach. Illustration: Das Pflanzenreich **73**: 223 (1920); Gleason, Illustrated flora of the northeastern United States and adjacent Canada **1**: 367 (1952); Makino, Illustrated flora of Japan, enlarged edn, 776 (1961); Kitamura et al., Coloured illustrations of herbaceous plants of Japan (Monocotyledoneae), t. 50, f. 323, 195 (1964).

Tuber 1–2 cm wide. Leaf often with a bulbil on the stalk or at base of blade. Leaf-blade simple in young plants, with 3 leaflets in adult plants. Leaflets 3–6 (rarely to 12) cm, ovate-elliptic to lanceolate. Inflorescence-stalk longer than leaf-stalk. Spathe 5–7 cm, green; limb two-thirds to twice as long as tube, downy inside, its apex clasping appendix of spadix. Spadix appendix 1–2 mm thick, much longer than the spathe, purple below, green above. *China, Korea & Japan.* H3. Summer.

2. P. pedatisecta Schott. Illustration: Das Pflanzenreich **73**: 223 (1920).

Tuber to 4 cm wide. Leaflets 7–11, the central ones 15–18 cm, ovate-lanceolate to lanceolate. Inflorescence-stalk shorter than leaf-stalk. Spathe 12–19 cm; limb 2–4 times as long as tube. Spadix appendix 1–2 mm thick, shorter than spathe, yellowish green. *N & W China.* H2.

39. LAGENANDRA Dalzell

N. Jacobsen

Plants evergreen, with creeping or erect rhizomes. Scale-leaves always present. Leaf-stalks shortly sheathed at the base. Leaf-blade with inrolled margins when young, narrowly elliptic to ovate, 10–50 cm long, with main veins running to the tip of the blade. Inflorescence-stalk much shorter than the leaf-stalk. Spathe of 2 parts, a basal floral compartment with united margins, and a limb with a narrow or broad opening, and usually with a tail-like appendage. Florets without perianths, unisexual. Spadix with zones of male and female florets separated, the sterile part slender. Female florets spirally arranged, each with a few ovules. Male florets each of 1 stamen which extrudes pollen through 2 horns. Fruiting spadix spherical; fruits somewhat fleshy, shed individually. Seeds large, ellipsoid, distinctly longitudinally ridged. The fruit takes about 6–9 months to mature, and the seeds germinate within a few days of being shed, without drying out.

About 10 species from Sri Lanka to southern India and tropical northeastern India, where they grow in and along shaded rivers. Several species are imported from Sri Lanka, but only 3 are commonly available. Generally, flowering specimens are necessary for a correct determination. Some species are poisonous.

In cultivation, Lagenandras require humid conditions and a rich soil, preferably without calcium. They are fine foliage plants for the aquarium, and the 3 numbered species can withstand being submerged for long periods. The rhizome should be placed at the surface of the soil. Flowering usually occurs from November to April. Propagated by cutting up the (frequently sprouting) rhizome.

Literature: De Wit, H.C.D., Aquarienpflanzen (1971); De Wit, H.C.D., Revisie van het Genus Lagenandra Dalzell (Araceae), *Mededelingen Landbouwhogeschool Wageningen* **13**: 1–18 (1978); De Wit, H.C. D., Aquariumplanten (1983).

1a. Spathe smooth or finely wrinkled outside **2. thwaitesii**
b. Spathe rough to warty outside 2
2a. Spathe more than 12 cm **1. praetermissa**
b. Spathe less than 10 cm **3. lancifolia**

1. L. praetermissa De Wit (*L. ovata* misapplied).

Leaf-blade 20–50 × 4–12 cm, narrowly ovate, smooth, fleshy, and shining green. Spathe rough, brownish to greenish outside, more or less red inside. *Sri Lanka.* G2.

L. ovata (Linnaeus) Thwaites (*Arum ovatum* Linnaeus; *Cryptocoryne ovata* (Linnaeus) Schott; *L. insignis* Trimen). Illustration: De Wit, Aquarienpflanzen, 233 (1971). Differs only in the large, inflated, roughly warted, black-purple spathe. *Sri Lanka to S India.* G2.

2. L. thwaitesii Engler. Illustration: De Wit, Aquarienpflanzen, 235 (1971).

Leaf-blade 10–20 × 1–4 cm, narrowly ovate, rough, dull dark green, mostly with a silvery margin. Leaf-stalk usually short. Spathe finely wrinkled. *Sri Lanka.* G2.

L. koenigii (Schott) Thwaites (*Cryptocoryne koenigii* Schott). Leaves 20–50 × 1–2 cm. Spathe completely smooth, often curved. *Sri Lanka.* G2.

3. L. lancifolia (Schott) Thwaites (*Cryptocoryne lancifolia* Schott). Illustration: De Wit, Aquarienpflanzen, 229 (1971).

Leaf-blade narrowly ovate, 6–15 × 2–8 cm, sometimes with leaf-stalk twice as long as blade or more. Spathe usually with a series of warts. *Sri Lanka.* G2.

40. CRYPTOCORYNE Wydler

N. Jacobsen

Plants evergreen, with short, stout, more or less vertical rhizomes and long stolons. Scale-leaves present in flowering specimens, sometimes also in sterile specimens. Leaves basal, in rosettes, with distinct stalks. Leaf-stalk shortly sheathed at the base. Leaf-blade with one half rolled round the other when young; main veins running to tip of blade; subsidiary veins connecting the main ones. Inflorescence-stalk much shorter than the leaf-stalk. Spathe of 3 parts, a basal floral compartment with fused margins, a tube with overlapping margins, and a limb (figure 15(9), p. 377). Floral compartment (containing spadix) separated from tube by a flap; mouth of tube usually marked by a collar or collar-zone (figure 15(10 & 11), p. 377). Florets without perianths, unisexual. Spadix with zones of male and female florets separated (figure 15(8), p. 377), the sterile part slender, with scent-producing bodies above the female zone. Female florets only 4–7, in a single whorl, united, each with many ovules. Male florets each of 1 stamen. Fruit compound, spherical, star-like when open. Seeds large, ovoid to kidney-shaped, germinating a few days after being shed, without drying out.

About 50 species in tropical areas from India and Sri Lanka to New Guinea. Aquatic to amphibious herbs growing in fast- or slow-running rivers in the rainforest. Many species are imported from Colombo, Bangkok, Singapore and Jakarta, but generally only a dozen are available at any one time. Flowering specimens are, in most cases, necessary for correct determination. Despite great variability, the leaves are characteristic for each species, but they are difficult to describe.

Cryptocorynes are cultivated in aquaria for their spectacular foliage, and they are among the more prized aquarium plants, even though many species are difficult to cultivate. Generally, they are not given adequate conditions (resulting in small, starved growth); they should have a rather rich, preferably calcium-free, mineral soil (but some species, as noted below, tolerate substantial amounts of calcium). Other species prefer more acid conditions and can be cultivated in almost pure sphagnum-peat. Water temperature should be 24–28 °C but the hardier species can tolerate 20 °C (e.g. *C. ciliata, C. undulata, C. wendtii, C. beckettii, C. lutea, C. parva, C.* × *willisii, C. pontederiifolia,* forms of *C. cordata, C. affinis, C. usteriana,* and forms of *C. crispatula*).

Propagation is by the usually plentiful long runners. Seed propagation is unimportant for commercial use. All species can be grown emerging from the water, which encourages flowering. Flowering occurs all the year but is most marked during winter.

Literature: De Wit, H.C.D., Aquarienpflanzen (1971); Rataj, K., Revision of the genus Cryptocoryne Fischer, *Studie Československé Akademie Véd, Praha,* 3 (1975); Jacobsen, N., Notes on Cryptocoryne of Sri Lanka (Ceylon), *Botaniska Notiser* **128**: 179–90 (1976); Rataj, K. & Horemann, T.J., Aquarium plants, tropical fish hobby (1977); Jacobsen, N., Chromosome numbers and taxonomy in Cryptocoryne (Araceae), *Botaniska Notiser* **130**: 71–87 (1977); Jacobsen, N., Akvarieplanter i farver, (1977), translated into: English (1979), Dutch (1979), German (1979) and Swedish (1980); Jacobsen, N., Cryptocoryner (1979), translated into: German (1982) and Dutch (1982); Jacobsen, N., The Cryptocoryne albida group of mainland Asia (Araceae), *Mededelingen Landbouwhogeschool Wageningen* **19**: 183–204 (1980); Arends, J.C., Bastmeijer, J. & Jacobsen, N., Chromosome numbers and taxonomy in Cryptocoryne (Araceae) II, *Nordic Journal of Botany* **2**: 453–63 (1982); De Wit, H.C.D., Aquariumplanten (1983).

Leaves. Green (sometimes brownish) without any purple: **1, 6, 7, 8, 13, 14, 16**; with a few purple markings: **2, 3, 4, 5, 8, 9, 10, 11, 12, 15**; with more or less dense purple markings: **2, 3, 4, 9, 10, 11, 12, 15**.

Scale-leaves. Usually present in non-flowering specimens: **8, 13, 15**.

Spathe. Tube not more than 1.5 cm: **7, 8, 9, 15**; tube at least 2 cm: **1, 2, 3, 4, 5, 6, 9, 10, 11, 12, 13, 14, 16**.

Limb of spathe. With a collar: **1, 2, 3, 4, 5, 6, 7, 8, 9, 12, 13**; with a collar-zone: **5, 8, 10, 11, 13, 14, 15**; without a collar-zone: **15, 16**. With a rough surface: **1, 6, 7, 9, 10, 13**.

1a. Scale-leaves usually present 2
b. Scale-leaves usually absent, except in flowering specimens 3
2a. Leaf-blade ovate (see also No. 15. *thwaitesii*) **8. pontederiifolia**
b. Leaf-blade narrowly ovate to linear **13. usteriana**
3a. Leaf-blade narrowly ovate to linear, more than 5 times as long as broad; limb of spathe more or less spirally twisted 4
b. Leaf-blade elliptic to ovate, usually not more than 4 times as long as broad; limb of spathe not spirally twisted 5
4a. Leaf narrowly ovate, with a distinct blade, green to dark green above, green with reddish veins or dark red beneath (see also No. 2. *undulata*); limb of spathe black-purple **12. affinis**
b. Leaf narrowly elliptic to linear, often without a distinct blade, green to rust-brown, smooth, wavy or blistered; limb of spathe whitish with purple lines **16. crispatula**
5a. Leaf-blade evenly green, without any purple 6
b. Leaf-blade purple, purple-marbled, purple-tinged, or with purple veins on the surface 9
6a. Plants usually more than 15 cm; leaf-blade narrowly ovate to linear, base cordate to truncate; limb of spathe ciliate **1. ciliata**
b. Plants usually smaller than 15 cm; leaf-blade narrowly ovate to obovate; limb of spathe not ciliate 7
7a. Leaf-blade spongy, obovate, 2–2.5 cm wide; limb of spathe with a long tail; collar absent **14. lingua**
b. Leaf-blade not spongy, narrowly ovate to ovate, 5–10 mm wide; limb of spathe short; collar present 8
8a. Plant 7–15 cm high; leaf-blade narrowly ovate to ovate; spathe

4–6 cm, with the limb (1–2 cm) upright **6. × willisii**
b. Plant 2–10 cm high; leaf-blade narrowly ovate to narrowly elliptic; spathe 2–3 cm, with the limb (3–8 mm) obliquely twisted **7. parva**
9a. Limb of spathe without a collar, but sometimes with a collar-zone 10
b. Limb of spathe with a pronounced collar 12
10a. Leaf-blade elliptic to ovate, with margin finely scalloped, appearing finely toothed; limb of spathe mostly with a long tail; collar-zone usually absent **15. thwaitesii**
b. Leaf-blade ovate to heart-shaped, margin entire; limb of spathe without a prolonged tail; collar-zone conspicuous 11
11a. Limb of spathe yellow, sometimes brown-tinged, surface more or less smooth **11. cordata**
b. Limb of spathe red, surface more or less rough **10. purpurea**
12a. Leaf-blade narrowly ovate to ovate; limb of spathe red-brown to purple, surface with small warts **9. minima**
b. Leaf-blade narrowly ovate, sometimes ovate; limb of spathe brownish to yellowish; surface more or less smooth 13
13a. Limb of spathe yellowish; collar of the same colour 14
b. Limb of spathe yellowish to dark brown; collar dark brown to black-purple 15
14a. Limb of spathe flat or recurved, slightly twisted; leaf-blade narrowly ovate to ovate, more or less stiff **5. walkeri**
b. Limb of spathe more or less obliquely twisted; leaf-blade narrowly ovate, soft **2. undulata**
15a. Limb of spathe flat or twisted once, not obliquely twisted; leaf-blade mostly brown to greenish, more or less stiff **4. beckettii**
b. Limb of spathe obliquely twisted; leaf-blade very variable, narrowly ovate to ovate, green to brown or red-brown, more or less flexible **3. wendtii**

1. C. ciliata (Roxburgh) Schott (*Ambrosina ciliata* Roxburgh; *C. elata* Griffith; *C. drymorrhiza* Zippelius). Illustration: De Wit, Aquarienpflanzen, 138–40 (1971).

To 1 m. Leaf-blade green, smooth, narrowly ovate to linear, with inconspicuous venation. Spathe with a long tube; limb with a characteristic ciliate margin. *India to New Guinea*, G2.

Rather variable in leaf form and size. A form with narrow leaves and long rooting runners (var. **ciliata**) is diploid, while a broader leaved form with short, fragile runners (var. **latifolia** Rataj) is triploid, but several other forms exist. It is a plant from the inner mangrove swamps and it grows well in a rich soil.

2. C. undulata Wendt (*C. willisii* Baum; *C. axelrodii* Rataj). Illustration: De Wit, Aquarienpflanzen, 205 (1971); *Botaniska Notiser* **129**: 186 (1976).

Submerged leaves 5–15 cm, narrowly ovate, green to brownish, with a distinct reddish venation and a truncate to cordate base. Emerged leaves shorter and broader, green with reddish veins, especially on the lower surface. (A triploid form has longer and broader leaves.) Limb of spathe twisted, dull yellowish to light brownish; collar of the same colour. *Sri Lanka*. G2.

Submerged, starved specimens often develop long internodes, 'stilt-roots', and narrow leaves, while plants under optimal conditions have larger and much more colourful leaves with a wavy margin.

3. C. wendtii De Wit. Illustration: De Wit, Aquarienpflanzen, 203 (1971); *Botaniska Notiser* **129**: 186 (1976).

Leaf-blade 5–25 × 1–3 cm, narrowly ovate to ovate, very variable in size and colour, green to brown or red-brown; base more or less cordate. Submerged leaves larger and more flexible. Limb of spathe more or less obliquely twisted, yellowish to brownish or reddish; collar black-purple. *Sri Lanka*. G2.

A most variable species with several different diploid and triploid forms. In some forms the leaves can change size and colour when transferred to new conditions of cultivation. One of the most undemanding species.

4. C. beckettii Trimen (*C. petchii* Alston). Illustration: De Wit, Aquarienpflanzen, 129 (1971); *Botaniska Notiser* **129**: 186 (1976).

Leaf-blade 5–10 cm, narrowly ovate to ovate, greenish to light or dark brown, with a truncate to slightly cordate base. Limb of spathe yellowish to brownish, more or less twisted; collar black-purple. *Sri Lanka*. G2.

Several leaf forms are found within both diploids and triploids.

5. C. walkeri Schott (*C. lutea* Alston). Illustration: De Wit, Aquarienpflanzen, 163, 169, 201 (1971); *Botaniska Notiser* **129**: 187 (1976).

Leaf-blade 5–10 cm, more or less ovate, rather stiff, more or less dark green; lower surface more brownish, often with reddish venation; base more or less cordate. Limb of spathe yellow (to greenish), slightly twisted or recurved; collar yellow. *Sri Lanka*. G2.

Submerged and emerged leaves rather similar, and stiffer than in the 3 preceding species.

6. C. × willisii Reitz (*C. lucens* De Wit; *C. nevillii* misapplied). Illustration: De Wit, Aquarienpflanzen, 167 (1971); *Botaniska Notiser* **129**: 186 (1976).

Leaf-blade 1–3 cm, narrowly ovate to ovate, green, inconspicuously veined; base truncate. Leaf-stalk sometimes with a reddish tinge. Limb of spathe somewhat twisted or recurved, more or less wrinkled, purple; collar yellowish to dark purple. *Sri Lanka*. G2.

Hybridisation experiments have shown that *C. × willisii* and *C. lucens* belong to a hybrid complex derived from *C. parva* on one side and at least *C. beckettii* and *C. walkeri* on the other. The hybrids are sterile.

7. C. parva De Wit. Illustration: De Wit, Aquarienpflanzen, 175 (1971); *Botaniska Notiser* **129**: 186 (1976).

Leaf-blade 5–15 mm, narrowly ovate to ovate, green, inconspicuously veined; base truncate. Limb of spathe more or less wrinkled, purple; collar dark purple, obliquely twisted. *Sri Lanka*. G2.

8. C. pontederiifolia Schott (*C. sulphurea* De Wit). Illustration: De Wit, *Aquarienpflanzen*, 179 (1971).

Leaf-blade 5–10 cm, ovate, cordate, green, sometimes slightly purple-marbled. Scale-leaves almost always present, even in non-flowering specimens. Limb of spathe upright, wrinkled, yellow; collar prominent and large; tube short. *Sumatra*. G2.

C. longicauda Engler (*C. caudata* N.E. Brown; *C. johorensis* Ridley). Illustration: De Wit, Aquarienpflanzen, 165 (1971). Leaf-blade 5–10 cm, ovate, cordate, green (to dull purple), sometimes strongly blistered. Limb of spathe red to dark purple, wrinkled, developed into a very long tail; collar present. *Borneo & Malaysia (Johore)*. G2.

Requires acid conditions.

9. C. minima Ridley. Illustration: De Wit, Aquarienpflanzen, 169 (1971).

Leaf-blade 2–4 cm, ovate, more or less cordate, green, sometimes with reddish marbling; lower surface reddish, often with reddish veins. Limb of spathe recurved, less than 1 cm broad, wrinkled, mostly reddish; collar present; tube short. *Malay peninsula.* G2.

C. griffithii Schott. Illustration: De Wit, Aquarienpflanzen, 159 (1971). Leaf-blade 4–8 cm, ovate; upper surface dark green to purplish; lower red. Limb of spathe recurved, more than 1 cm broad, strongly wrinkled, red to dark purple; collar present. *Malay peninsula.* G2.

10. C. purpurea Ridley (*C. hejnyi* Rataj; *C. griffithii* misapplied). Illustration: De Wit, Aquarienpflanzen, 181, 189 (1971).

Leaf-blade 4–8 cm, ovate, more or less cordate, green, sometimes reddish-marbled; lower surface reddish, often with reddish veins. Limb of spathe somewhat wrinkled, red; collar-zone prominent, sometimes yellowish. *Malay peninsula.* G2.

11. C. cordate Griffith (*C. siamensis* Gagnepain; *C. kerrii* Gagnepain; *C. blassii* De Wit; *C. evae* Rataj; *C. stonei* Rataj; *C. siamensis* Gagnepain var. *schneideri* Schöpfel). Illustration: De Wit, Aquarienpflanzen, 145, 189 (1971).

Leaf-blade 3–15 cm, narrowly ovate to ovate, more or less cordate, green to brown-purple or marbled, often more reddish beneath than above, sometimes somewhat blistered; texture varying from fragile to tough. Limb of spathe yellow, sometimes more or less brown-tinged, especially towards the margin; collar-zone conspicuous. *Malay peninsula.* G2.

Occurs in the diploid, tetraploid and hexaploid state. Leaves very polymorphic. The forms with a tougher leaf texture are generally the best in cultivation; the more fragile-leaved forms require acid conditions. Recently a form with variegated, whitish veins has come into cultivation, under the cultivar name 'Rosaefolia' (*C. siamensis* var. *schneideri*).

12. C. affinis N.E. Brown (*C. haerteliana* Milkuhn). Illustration: De Wit, Aquarienpflanzen, 125 (1971).

Leaf-blade 5–20 cm, narrowly ovate, more or less cordate, sometimes strongly blistered; upper surface green or tinged with red, with the veins often distinctly lighter coloured; lower surface red or green and tinged with red. Limb of spathe black-purple, spirally twisted; collar present. *Malay peninsula.* G2.

A fast-growing species that prefers somewhat calcareous conditions.

13. C. usteriana Engler (*C. aponogetifolia* Merrill). Illustration: De Wit, Aquarienpflanzen, 197 (1971).

Leaf-blade 20–100 × 1–4 cm, narrowly ovate to linear, green, blistered, with 2–6 strong veins. Scale-leaves always present. Limb of spathe wrinkled, dull violet, somewhat twisted; collar present. *Philippines.* G2.

A species that grows well under calcareous conditions.

14. C. lingua Engler (*C. spathulata* Engler). Illustration: De Wit, Aquarienpflanzen, 159 (1971).

Leaf-blade 1–3 cm wide, obovate, green and spongy. Limb of spathe yellowish with red spots and with a long tail; collar absent. *Borneo, introduced into Singapore.* G2.

15. C. thwaitesii Schott. Illustration: De Wit, Aquarienpflanzen, 193 (1971); *Botaniska Notiser* **129**: 187 (1976).

Leaf-blade elliptic to ovate, more or less cordate, green to purplish; margin often finely scalloped. Limb of spathe white, sometimes more or less spotted with red, developed into a shorter or longer tail that often bends forward; collar usually absent. *Sri Lanka.* G2.

16. C. crispatula Engler (*C. sinensis* Merrill; *C. balansae* Gagnepain; *C. tonkinensis* Gagnepain; *C. longispatha* Merrill; *C. bertelihansenii* Rataj; *C. yunnanensis* Li; *C. kwangsiensis* Li). Illustration: De Wit, Aquarienpflanzen, 129, 185 (1971).

Leaf-blade 10–70 cm × 2–40 mm, more or less linear, green to rust-brown; smooth to blistered; margin sometimes with fine, forward-pointing teeth. Limb of spathe spirally twisted, whitish with short or long purple lines; collar absent. *E India to Thailand, Vietnam & S China.* G2.

A very variable species, of which some of the narrow-leaved, smooth forms and some of the broad-leaved, blistered forms are well suited for cultivation, the latter preferring not too acid conditions. *C. crispatula* has sometimes been included in *C. retrospiralis* (Roxburgh) Kunth.

41. PISTIA Linnaeus

P.F. Yeo

Evergreen floating aquatics with a rosette of leaves and feathery roots below. Scale-leaves delicate, present on stolons which produce new rosettes. Leaves 20 × 7 cm, stalkless or nearly so, broadly wedge-shaped with truncate, sometimes notched, apex and 7–15 nearly parallel veins which are prominent beneath, but with no distinct midrib; both surfaces densely covered with minute hairs; upper surface light green. Spathe 1.5 cm, almost stalkless, hairy outside; margins overlapping below; upper half opening, acute. Spadix not evident; florets attached to back of spathe, without perianth, unisexual; two dissimilar scales present between male florets and female. Male florets few, combined into a single whorl on a short stalk, borne in the open part of the spathe. Female floret solitary. Ovary large, asymmetric, occupying almost the whole of the closed part of the spathe, to which it is extensively attached. Style present. Ovules numerous.

One rather variable species distributed throughout the tropics and extending into the subtropics, and characterised by the rosettes of distinctively shaped leaves. It is grown as an aquarium plant and requires a temperature of 21–26 °C. For overwintering, plants must be placed in shallow water and allowed to root into the soil. Alternatively, seed may be collected in summer and sown next spring in a fine wet soil at 30 °C.

Literature: Hooker, W.J. & Smith, J., Pistia stratiotes, *Botanical Magazine*, 4564 (1851); Engler, A., Aroideae & Pistioideae, Das Pflanzenreich **73**: *Araceae*, 250–62 (1920).

1. P. stratiotes Linnaeus. Illustration: *Botanical Magazine*, 4564 (1851); Das Pflanzenreich **73**: 251 (1920); Graf, Tropica, 129 (1978); Brickell (ed.), RHS A–Z encyclopedia of garden plants, 824 (2003). *Tropics and subtropics.* G2. Summer.

48. ACORACEAE

Herbs, aquatic or growing in damp places; all parts with a spicy aroma when crushed, without milky sap. Leaves all basal, narrow, parallel-veined. Inflorescence a spadix borne on a scape, the spathe appearing as a continuation of the scape so that the spadix projects laterally. Flowers bisexual, actinomorphic. Perianth of 6 segments. Stamens 6. Ovary of 2–3 united carpels, superior; style sessile. Fruit a berry.

A single genus, native to North America, introduced in Europe, with 2 species. One species is commonly cultivated.

1. ACORUS Linnaeus
D.A. Webb
Rhizomes branched, bearing linear leaves and stems that are leafless except for the spathe. Leaves equitant, without a distinct stalk. Spathe similar to the leaves (except for the basal part, which is narrow and stem-like), continuing the line of the inflorescence-stalk, so that the spadix appears lateral. Spadix stalkless, without appendix. Florets very numerous and densely packed, bisexual, with 6 obovate perianth-segments, 6 stamens, and a broadly ovoid, 2- or 3-celled ovary. Fruits not formed in Europe.

A genus of 2 species from S and E Asia and southern N America. They are easily grown in shallow water, marshy soil or well-watered pots. The tissues of both species (especially the rhizome) give off a spicy fragrance when bruised.

1a. Spadix at least 6 mm wide; leaves rather stiff, with a fairly prominent midrib **1. calamus**
b. Spadix not more than 5 mm wide; leaves soft and flexible, without a distinct midrib **2. gramineus**

1. A. calamus Linnaeus. Illustration: Boswell, English botany **9**: 11 (1883); Hegi, Illustrierte flora von Mittel-Europa, edn 3, **2**(1): 321 (1967–80); Graf, Tropica, 90 (1978); Aeschimann et al., Flora alpina **2**: 134.1.1 (2004).

Leaves to 125 × 2.5 cm, often transversely wrinkled in places, rather stiff, with a distinct and usually prominent midrib. Inflorescence-stalk 2- or 3-angled, narrower than the leaves and somewhat shorter. Spathe to 80 cm. Spadix 6–9 cm × 7–12 mm, pale yellow. *S&E Asia & southeastern USA; very widely naturalised elsewhere, so that native limits are uncertain.* H1. Summer.

A variant with leaves striped with cream and yellow ('Variegatus') is sometimes grown.

2. A. gramineus Aiton. Illustration: Makino, Illustrated flora of Japan, edn 2, 779 (1961); Graf, Exotica, edn 8, 143 (1976) – 'Pusillus' and 'Variegatus'; Graf, Tropica, 88 (1978) – 'Variegatus'

Like *A. calamus* but less robust, and smaller in all its parts, and with soft, flexible, grass-like leaves without a distinct midrib. Leaves to 50 cm × 8 mm. Spathe 7–15 cm. Spadix 5–10 cm × 3–4 mm. *NE India to Taiwan & C Japan.* H3. Summer.

Very variable in size. The smallest variants ('Pusillus') have leaves not more than 8 cm long and do not flower. 'Variegatus' has white-striped foliage.

49. LEMNACEAE

Small aquatic herbs, often floating, the plant body not differentiated into stem and leaves. Inflorescence minute, in a pocket on the margin of the plant body and consisting of 1 or more male flowers, each with 1–2 stamens, and a single female flower with a solitary ovary; perianth absent in flowers of both sexes.

There are 6 genera and about 30 species. Three genera are native to Europe, 4 to North America. A few species are grown in aquaria.

1a. Plant with 1 or more rootlets on its lower side 2
b. Plant rootless **Wolffia** (& **Wolffiella**)
2a. Plant with a single rootlet **Lemna**
b. Plant with 2 or more rootlets **Spirodela**

Literature: Landolt, E., Key to the determination of taxa within the family of Lemnaceae, *Veröffentlichungen des Geobotanischen Institutes der ETH, Stiftung Rübel, Zürich* **70**(1): 13–21 (1980). Several species of *Lemna* are illustrated in Aeschimann et al., Flora alpina **2**: 135.1.1–4 (2004).

50. PANDANACEAE

Dioecious trees or shrubs, often with stilt-roots. Leaves crowded, leathery, keeled, often with spiny margins; ptyxis conduplicate. Flowers in panicles or on spadices, unisexual. Perianth rudimentary or absent. Stamens many. Ovary 1-celled, superior or naked; ovules 1–many, basal or parietal. Fruit a syncarp, the woody or fleshy individual fruits fusing.

Three genera and 675 species of striking woody plants. Two genera are native to North America. Only a few species are grown as ornamentals or curiosities in Europe, generally in large glasshouses.

1. PANDANUS Parkinson
J.Cullen
Plants woody, dwarf or forming trees, usually dichotomously branched and with stilt-roots. Trunk with prominent, ring-like leaf-scars. Leaves basically arranged in 3 ranks, which are, however, twisted so that the leaves appear very conspicuously spirally arranged, in rosettes at the ends of the branches; margins and the underside of the midrib spiny. Stamens in elongate spikes. Female flowers aggregated into large, spherical or oblong heads. Fruit a syncarp of drupes, hard, spherical, conical or obconical.

A genus of 250 or more species from the Old World tropics. They are valued for their striking appearance, but most of them are too large for any but the largest greenhouses. The dwarfer species are occasionally grown as pot-plants, but these tend to remain in a juvenile phase, without flowers or fruits, and their identification (and relationship to the wild species) is difficult and imperfectly understood. The striking, corkscrew-like, spiral arrangement of the leaves distinguishes them easily from other woody monocotyledons.

They are relatively easy to cultivate, but require high temperatures and an abundance of water, particularly when actively growing. A rich compost is required, and pot-plants should be repotted each year before active growth begins. Good ventilation is essential, and light shade is required during very bright days.

Literature: There are numerous papers by H. St John, under the general title 'Revision of the genus Pandanus', covering species from various areas; most of these were published in the journal *Pacific Science* between 1960 and 1973.

1a. Leaves to 1.5 cm broad **4. pygmaeus**
b. Leaves at least 5 cm broad 2
2a. Plant tree-like; leaves uniformly glaucous green **1. utilis**
b. Plant dwarf; leaves variegated with white, silver-white or yellow 3
3a. Leaves with white or silver-white margins **2. veitchii**
b. Leaves with several longitudinal yellow stripes **3. sanderi**

1. P. utilis Bory. Illustration: Bailey, Standard cyclopedia of horticulture, Popular edn 3: f. 2743–5 (1935); Graf, Exotica, edn 8, 1372, 1373 (1976).

Tree to 20 m. Leaves 50–200 × 6–10 cm, glaucous, with reddish spines on the margins and midrib beneath. Male flowers fragrant. Female inflorescences long-stalked, solitary, pendent. Syncarp more or less spherical, 15–20 cm in diameter. *Madagascar or possibly Mauritius, but widely cultivated elsewhere in the tropics.* G2.

2. P. veitchii Masters & Moore. Illustration: Step, Favourite flowers of garden and greenhouse **4**: t. 293 (1897); Graf, Exotica, edn 8, 1373 (1976).

Plant to 60 cm high, but forming a large clump if given space. Leaves 50–100 × *c.* 7.5 cm, with spiny margins, dark green with white or silver-white bands along the margins. *?Polynesia.* G2.

A species of uncertain origin and status. It is usually grown as a pot–plant and rarely reaches its mature development. A very dwarf variant ('Compactus') is available.

3. P. sanderi Masters. Illustration: *Gardeners' Chronicle* **23**: 256 (1898); Graf, Exotica, edn 8, 1373 (1976).

Like *P. veitchii* but generally somewhat smaller, leaf margins minutely spiny, leaves dark green variegated with several longitudinal yellow stripes. *Malaysia, ?Indonesia.* G2.

Another species of uncertain origin and status. Several cultivars, differing in the degree and nature of the variegation, are available.

4. P. pygmaeus Thouars. Illustration: *Botanical Magazine*, 4736 (1853); Graf, Exotica, edn 8, 1373 (1976).

Shrub to 60 cm. Leaves 30–60 × at most 1.5 cm, green, minutely spiny on the margins and midrib beneath. *Madagascar.* G2.

51. SPARGANIACEAE

Emergent aquatic herbs. Leaves narrow, alternate. Flowers in unisexual, spherical heads. Perianth of a few scales. Stamens 3 or more. Ovary 1-celled, superior, more or less stalkless; ovule 1, apical. Fruit drupe-like.

A single genus with about 12 species; it is native to both Europe and North America. This family is often included in the next.

1. SPARGANIUM Linnaeus

D.M. Synnott

A genus of about 20 species from north temperate regions. They are easily grown in shallow water, e.g. at lake margins, where they provide useful cover for wild fowl. Propagation by seed.

Literature: Cook, C.D.K., Sparganium in Britain, *Watsonia* **5**: 1–10 (1961).

1a. Inflorescence branched, stems erect **2. erectum**
b. Inflorescence not branched, stems erect or floating 2
2a. Leaves triangular in cross-section; stems erect or floating **1. emersum**
b. Leaves not triangular in cross-section; stems floating **3. minimum**

1. S. emersum Rehmann (*S. simplex* Hudson). Illustration: Ross-Craig, Drawings of British plants **30**: t. 36 (1973); Keble Martin, The concise British flora in colour, 88 (1965).

Stems erect or floating, 20–60 cm. Leaves both erect and floating, triangular in cross-section. Inflorescence not branched, exceeded by the lowest leafy bract. Female capitula 3–6, the lower usually stalked. Male capitula 3–10, separated and stalkless on main axis. Fruit ellipsoid, 4–6 × 2–3 mm, stalked, with long, persistent style. *Europe, W & C Asia, N America.* H1. Summer.

2. S. erectum Linnaeus (*S. ramosum* Hudson). Illustration: Ross-Craig, Drawings of British plants **30**: t. 35 (1973); Keble Martin, The concise British flora in colour, 88 (1965); Aeschimann et al., Flora alpina **2**: 140.1.1.1 (2004) – as subsp *erectum*.

Stems erect, 50–100 cm. Leaves erect, triangular in cross-section, keeled below. Inflorescence branched. Fruits ellipsoid to pyramidal, 6–9 × 2–6 mm. *Europe, W&C Asia.* H1. Summer.

Sometimes divided into subspecies on the basis of the shape of the fruits.

3. S. minimum Wallroth. Illustration: Ross-Craig, Drawings of British plants **30**: t. 38 (1973); Keble Martin, The concise British flora in colour, 88 (1965).

Stems floating, 10–30 cm. Leaves floating or submerged, thin, flat, usually translucent. Inflorescence not branched; lowest leafy bract scarcely longer than inflorescence. Female capitula 2–3, stalkless and widely separated. Male capitulum 1. Fruit obovoid, stalkless, with short, beak-like, persistent style. *N temperate regions.* H1. Summer.

52. TYPHACEAE

Marsh herbs. Leaves narrow, alternate. Inflorescence of 2 unisexual, dense, superimposed spikes. Perianth-segments thread-like or scales. Stamens 2–5. Ovary 1-celled, superior on a hairy stalk; ovule 1, apical. Fruit dry.

A single genus native to both Europe and North America, with about 15 species, occasionally cultivated for ornament along rivers or streams.

1. TYPHA Linnaeus

E.C. Nelson

Hairless perennial herbs, with stout, creeping rhizomes. Stems erect. Leaves mostly basal, in 2 ranks, erect, with sheath closely enclosing stem.

A widely distributed genus of about 15 species. They are suitable for the margins of lakes and ponds, and may be propagated by division of the rhizomes or from seed.

1a. Leaves not more than 3 mm wide; female spike to 5 cm, oblong to cylindric **3. minima**
b. Leaves more than 3 mm wide; female spike more than 5 cm, not oblong 2
2a. Leaf-sheath usually with auricles; male and female spikes separated by 3–8 cm **1. angustifolia**
b. Leaf-sheath tapering into blade; male and female spikes touching or separated by less than 3 cm 3
3a. Female spike brown and mottled at maturity **2. latifolia**
b. Female spike silvery grey at maturity **4. shuttleworthii**

1. T. angustifolia Linnaeus. Illustration: Keble Martin, The concise British flora in colour, 88 (1965); Aeschimann et al., Flora alpina **2**: 141.1.1 (2004).

Perennial herb to 1.5 m. Leaves 3–6 mm wide. Male and female spikes separated by 3–8 cm; female spike 8–20 × 1.3–2.5 cm, dark or reddish brown at maturity. *Cosmopolitan.* H1. Summer.

More slender and less invasive than *T. latifolia;* suitable for small ponds.

2. T. latifolia Linnaeus. Illustration: Keble Martin, The concise British flora in colour, 88 (1965); Brickell (ed.), RHS A–Z encyclopedia of garden plants, 1061 (2003).

Robust perennial herb to 2 m or more. Leaves 8–20 mm wide. Male and female spikes touching or separated by not more than 2.5 cm; female spike 8–15 × 1.8–3 cm, dark brown but becoming mottled with age. *Almost worldwide but absent from the southern hemisphere S of 30° S.*

3. T. minima Funck. Illustration: Grounds, Ornamental grasses, 204 (1979); Brickell (ed.), RHS A–Z encyclopedia of garden plants, 1061 (2003); Aeschimann et al., Flora alpina **2**: 141.1.3 (2004).

Perennial herb 25–75 cm. Leaves not more than 3 mm wide. Male and female spikes touching or separated; female spike 1.5–5 cm long, oblong to cylindric, dark brown, sometimes subtended by a leaf-like bract. *C & S Europe, Asia eastwards to N China.* H1. Summer.

Suitable for small ponds.

4. T. shuttleworthii Koch & Sonder. Illustration: Coste, Flore de la France **3**: 437 (1906); Aeschimann et al., Flora alpina **2**: 141.1.6 (2004).

Similar to *T. latifolia* but not as tall or robust. Female spike becoming silvery grey with age. *S Europe.* H1. Summer.

53. CYPERACEAE

Herbs, sometimes large. Stems terete or 3-angled in section, usually solid. Leaves spirally arranged, with closed sheaths, ptyxis conduplicate. Flowers subtended by membranous bracts (glumes), spiral or in 2 ranks in spikes or spikelets, without empty glumes at the base, unisexual or bisexual. Perianth of scales, bristle-like hairs, or absent. Stamens 6 or rarely 3, with basifixed anthers. Ovary 1-celled, superior or naked, sometimes surrounded by a flask-shaped structure (utricle); styles 2–3; ovule 1, basal. Fruit nut-like.

A large family with 109 genera and about 5,500 species. Twelve genera are native to Europe, 26 to North America. Relatively few are cultivated for ornament, but interest in the group from horticulturists is increasing. In European Garden Flora, edn 1, there were just 4 genera but these have been revised in line with recent taxonomic work on the family and now there are 9. The descriptions are minimal and therefore are only appropriate for the taxa covered in this account.

Literature: Bruhl, J., Sedge genera of the world: Relationships and a new classification of the Cyperaceae, *Australian Systematic Botany* **8**: 125–305 (1995); Goetghebeur, P., Cyperaceae, in K. Kubitzki et al. (eds), The Families and Genera of Vascular Plants **4**: 141–90 (1998); Simpson, D.A. & Inglis, C.A., Cyperaceae of economic, ethnobotanical and horticultural importance: a checklist, *Kew Bulletin* **56**: 257–360 (2001); Govaerts, R. & Simpson, D.A., World checklist of Cyperaceae sedges (2007).

1a. Florets all unisexual; female florets enclosed in a sac (utricle), which persists round the fruit 2
b. Florets mostly bisexual; utricle absent 3
2a. Rachilla terminating in a hook; spikes simple **8. Uncinia**
b. Rachilla not terminating in a hook; spikes simple or compound **9. Carex**
3a. Inflorescence consisting of 2–8 primary branches terminating in spherical spikes **5. Scirpoides**
b. Inflorescence not as above 4
4a. Perianth bristles persistent, conspicuous and lengthening in fruit to form a white, cotton-like head **1. Eriophorum**
b. Perianth bristles absent, or, if present, then not conspicuous or lengthening in fruit 5
5a. Inflorescence with 1 terminal spikelet; lowest involucral bract glume-like, shorter than inflorescence **4. Eleocharis**
b. Inflorescence of 2 or more spikelets; lowest involucral bract leaf-like, exceeding inflorescence 6
6a. Spikelets distinctly flattened **7. Cyperus**
b. Spikelets rounded in section 7
7a. Inflorescence terminal, head-like or with 1–3 shortly stalked clusters of spikelets **2. Bolboshoenus**
b. Inflorescence lateral, lowermost involucral, bract stem-like 8
8a. Stems slender, to 1 mm wide; glumes to 2.5 mm **6. Isolepis**
b. Stems robust, 2 mm or more wide; glumes 2.5 mm or more **3. Schoenoplectus**

1. ERIOPHORUM Linneaus

S.G. Knees

Perennials. Leaves basal and along the stem. Stems rounded to trigonous near apex. Inflorescence terminal, spikelets solitary. Flowers bisexual or male-sterile with abortive stamens; perianth persistent, hairlike, smooth bristles, or very rarely 6 antrorsely barbed bristles, greatly elongate, essentially straight, usually obscuring most of scales in spikelet, much longer than achene; stamens 1–3; styles deciduous, linear, 3-fid. Achenes trigonous.

About 25 species from north temperate and Arctic areas. All species require full sun, plenty of moisture, acid soil and cool summer conditions to thrive. Propagate by seed or division in spring.

Literature: Jermy, A.C. et al., Sedges of the British Isles, edn 3 (2007).

1. E. angustifolium Honckeny. Illustration: Brickell (ed.), RHS A–Z encyclopedia of garden plants, 422 (2003); Jermy et al., Sedges of the British Isles, edn 3, 79 (2007).

Plants colonial from long-creeping rhizomes. Stems 15–75 cm. Leaves 2–60 cm × 2–7 mm, linear, channelled, flat to V-shaped in cross-section, narrowed to a long trigonous apex. Inflorescence with 2–9 spikelets, nodding in fruit; perianth bristles many, elongating to 4–5 cm in fruit, forming a white, cotton-like head; stamens 3, stigmas 3. Fruits 2–5 mm, obovoid, dull brown, smooth. *Europe.* H1. Spring.

2. BOLBOSCHOENUS (Ascherson) Palla

S.M. Walters & S.G. Knees

Rhizomatous perennials. Leaves keeled beneath. Inflorescence terminal, head-like or with 1–3 shortly stalked clusters of spikelets. Spikelets rounded in section. Nut flattened.

A genus of 6–15 species from Europe, N America and Australia.

Literature: Jermy, A.C. et al., Sedges of the British Isles, edn 3 (2007).

1. B. maritimus (*Scirpus maritimus* Linnaeus). Illustration: Clapham et al., Flora of the British Isles **4**: 53 (1952); Jermy et al., Sedges of the British Isles, edn 3, 103 (2007).

Robust, shortly rhizomatous, with 3-angled, leafy stems to 1 m. Leaves to 10 mm wide, keeled. Inflorescence dense, corymbose, with rather few, reddish brown, ovoid, stalkess spikes to 2 cm. Nut *c.* 3 mm, obovate, flattened. *Cosmopolitan, in saline habitats.* H2. Summer.

Grown by lakes and ponds, more often as 'Variegatus', with striped foliage.

3. SCHOENOPLECTUS (Reichenbach) Palla

S.M. Walters & S.G. Knees

Rhizomatous perennials. Inflorescence lateral, lowermost involucral bract stem-like. Nut flattened.

A genus of about 77 species with a cosmopolitan distribution. One of the most widely used rushes.

Literature: Jermy et al., Sedges of the British Isles, edn 3, (2007).

1. S. lacustris (Linnaeus) Palla *(Scirpus lacustris* Linnaeus; *Scirpus tabernaemontani* Gmelin). Illustration: Everett, The New York Botanical Gardens illustrated encyclopedia of horticulture **9**: 3090 (1982); Jermy et al., Sedges of the British Isles, edn 3, 103 (2007).

Rhizomatous, with smooth, cylindric, leafless stems to 2.5 m. Leaves, if present, long, slender, only produced under water. Inflorescence dense, with several ovoid, reddish brown spikes up to 8 mm. Nut *c.* 2 mm, broadly elliptic, flattened. *Cosmopolitan, though uncommon in the tropics.* H2. Summer.

Subsp. **lacustris**, with green stems and a smooth glume subtending the 3-angled nut, is occasionally planted (and often found) by lakes and in slow-moving streams. The commonly planted cultivars belong to subsp. **tabernaemontani** (Gmelin) Palla, a smaller plant with glaucous stems, papillose glumes and a biconvex nut, occurring mainly in coastal habitats in the wild; 'Albescens' has creamy white stems with narrow, green longitudinal stripes, whilst 'Zebrinus' has stems with alternate transverse bands of green and white.

4. ELEOCHARIS R. Brown

S.M. Walters

Tufted or rhizomatous, leafless herbs, with cylindric or angled stems sheathed at base and bearing solitary, bractless spikes. Florets as in *Scirpus.* Fruit a nut with persistent style-base.

A large cosmopolitan genus of aquatic and marsh plants, of very little horticultural value.

Literature: Jermy, A.C. et al., Sedges of the British Isles, edn 3 (2007).

1. E. acicularis (Linnaeus) Roemer & Schultes. Illustration: De Wit, Aquarium plants, English edn, 72–3 (1964); Aeschimann et al., Flora alpina **2**: 138.9.2 (2004); Jermy et al., Sedges of the British Isles, edn 3, 137 (2007).

Delicate, rhizomatous perennial with tufts of 4-angled, thread-like stems, to 30 cm if submerged, as is usual in cultivation. Spikelets (produced only when plant is out of water) 3–4 mm; nut *c.* 1 mm, with clearly separated style-base, and fine, deciduous bristles. *Subarctic and temperate regions of the N Hemisphere.* H3.

Commonly grown rooted in sand in aquaria, and tolerant of water temperatures up to 30 °C. Propagation is by division.

5. SCIRPOIDES Séguier

S.M. Walters & S.G. Knees

Tufted perennials. Stems cylindric. Inflorescence consisting of 2—8 primary branches terminating in spherical spikes, subtended and exceeded by a long, pointed bract. Nut 3-angled.

A genus of about 5 species from Eurasia, N America and possibly southern Africa.

Literature: Jermy, A.C. et al., Sedges of the British Isles, edn 3 (2007).

1. S. holoschoenus (Linnaeus) Soják. (*Scirpus holoschoenus* Linnaeus; *Holoschoenus vulgaris* Link). Illustration: Clapham et al., Flora of the British Isles **4**: 54 (1956); Jermy et al., Sedges of the British Isles, edn 3, 79 (2007).

Tufted, with smooth, cylindric stems to 1 m. Only the upper leaf-sheaths with blades, which are rigid and half-cylindric. Inflorescence consisting of an umbel of several globular spikes up to 10 mm. Nut *c.* 1 mm, more or less 3-angled. *From England & the Canary Islands, through S Europe & N Africa to C Asia.* H3. Summer.

A very distinctive sedge, easy to cultivate and tolerant of relatively dry soils, sometimes grown as 'Variegatus', with alternate rings of green and yellow tissue on leaf and stem.

6. ISOLEPIS R. Brown

S.M. Walters & S.G. Knees

Tufted annuals. Inflorescence lateral, lowermost involucral bract stem-like. Perianth bristles absent. Nut spherical.

A genus of 69 species with a wide dsitribution throughout the temperate world.

Literature: Jermy, A.C. et al., Sedges of the British Isles, edn 3 (2007).

1. I. cernua (Vahl) Roemer & Schultes (*Scirpus cernuus* Vahl; *S. gracilis* Koch; *S. savii* Sebastiani & Mauri). Illustration: Graf, Tropica, 390 (1978); Everett, The New York Botanical Gardens illustrated encyclopedia of horticulture **9**: 3091 (1982); Jermy et al., Sedges of the British Isles, edn 3, 153 (2007).

Delicate, tufted, often annual plant with very slender, thread-like, almost leafless stems to 30 cm. Inflorescence usually a solitary (sometimes 2 or 3) ovoid, greenish spike to 5 mm, apparently lateral with a short, pointed bract continuing the stem. Nut less than 1 mm, almost spherical, reddish brown. *Widespread, mainly in temperate regions.* H4. Summer.

Commonly grown in conservatories as a decorative pot-plant or for edging.

7. CYPERUS Linnaeus

S.M. Walters

Tufted or rhizomatous herbs with terminal, umbellate or capitate inflorescences. Spikes many-flowered, flattened; florets bisexual, in 2 ranks. Perianth absent; stamens usually 3. Fruit a nut.

A large, mainly tropical or subtropical, genus of about 500 species with few temperate representatives. Propagation is usually by division, but seed can also be used.

Literature: Kükenthal, G., Das Pflanzenreich **101** (1935–36); Jermy, A.C. et al., Sedges of the British Isles, edn 3, (2007).

Stems. Leafless: **1, 5**. More than 1.5 m: **1**.
Leaves. With 3 prominent, whitish veins: **6**.
Stolons. With terminal tubers: **3**.
Inflorescence. A compound umbel with many slender, drooping branches: **1**.
Spikelets. In dense, nearly spherical heads: **4**.

1a. Stems leafless or nearly so, with basal sheaths only — 2
b. Stems obviously leafy, at least at base — 3
2a. Inflorescence branches very slender and drooping, much exceeding the few bracts — **1. papyrus**
b. Inflorescence branches rather stout, more or less equalling the numerous bracts — **5. involucratus**
3a. Plant bearing slender underground stolons terminating in brown tubers — **3. esculentus**
b. Plant without tubers — 4
4a. Spikes in dense heads — **4. eragrostis**
b. Spikes solitary or loosely arranged — 5
5a. Leaves usually more than 10 mm wide; spikes narrowly lanceolate, often solitary — **6. albostriatus**
b. Leaves usually less than 10 mm wide; spikes linear, in a loose raceme — **2. longus**

1. C. papyrus Linnaeus. Illustration: *Gardeners' Chronicle* **3**: 81 (1875) – habit of Sicilian plant; Everett, The New York Botanical Gardens illustrated encyclopedia of horticulture **3**: 983 (1981); Brickell (ed.), RHS A–Z encyclopedia of garden plants, 341 (2003).

Giant rhizomatous, perennial 'reed' with bluntly 3-angled, leafless stems, often exceeding 2 m. Inflorescence a compound umbel with very many long, slender, drooping primary branches much exceeding the few bracts. Spikes slender, brownish, 20–30 together in larger spikes grouped at the ends of the branches. *S & tropical Africa, extending northwards to Egypt; naturalised in Sicily.* H5–G1.

Often cultivated as an ornamental aquatic in greenhouses; the classical source of Egyptian papyrus. *C. papyrus* 'Nanus' (*C. isocladus* Knuth; '*C. haspan viviparus*' misapplied) is a dwarf variant, sometimes treated as a distinct species, with stems to 80 cm; it is native to southern Africa, northwards to Zanzibar, and is grown as a pot plant in conservatories.

2. C. longus Linnaeus (*C. maximus* Anon.). Illustration: Roles, Illustrations of the British Flora **4**: 56 (1965); Aeschimann et al., Flora alpina **2**: 138.11.1 (2004).

Rhizomatous, with smooth, 3-angled, leafy stems to 1 m. Inflorescence a compound umbel, with several leafy bracts often exceeding the inflorescence, and unequal primary branches to 10 cm. Spikes *c.* 10 mm, more or less linear, reddish-brown, in loose racemes at the ends of the secondary branches. *Europe, SW & C Asia, N Africa.* H3. Summer.

The hardiest species, commonly grown in ponds, lakes and water gardens.

3. C. esculentus Linnaeus. Illustration: Coste, Flore de la France **3**: 463 (1906); Aeschimann et al., Flora alpina **2**: 138.11.3 (2004).

Rhizomatous, with slender stolons bearing terminal brown tubers, and leafy stems to 50 cm. Inflorescence a few-branched umbel with distant, grouped, yellowish brown spikes. *Native in the Mediterranean region and perhaps elsewhere; now widely naturalised in subtropical and warm temperate regions.* H5.

The cultivated plant, var. **sativus** Boeckeler, which yields the commercial edible tuber ('chufa' or 'tiger nut'), has numerous tubers crowded on short stolons, and rarely flowers. The wild plant is not cultivated in gardens, though it may occur as a weed.

4. C. eragrostis Lamarck (*C. vegetus* Willdenow). Illustration: Coste, Flore de la France **3**: 463 (1906); Brickell (ed.), RHS A–Z encyclopedia of garden plants, 341 (2003); Aeschimann et al., Flora alpina **2**: 138.11.5 (2004).

Shortly rhizomatous, with leafy stems to 60 (rarely to 90) cm. Leaves 4–8 mm wide, often equalling stems. Inflorescence a simple or compound umbel with 8–10 branches and 5–8 (rarely to 11) bracts exceeding the branches; spikes 8–15 mm, in dense, nearly spherical heads 1–2 cm in diameter, pale brown or straw-coloured. *Western USA and warm-temperate S America; naturalised in SW Europe.* H4. Summer.

5. C. involucratus Rottboell (*C. flabelliformis* Rottboell; *C. alternifolius* misapplied). Illustration: Graf, Tropica, 390 (1978); Mühlberg, Das grosse Buch der Wasserpflanzen, t. 183 (1980); Brickell (ed.), RHS A–Z encyclopedia of garden plants, 341 (2003).

Tufted, with short, erect, ridged, leafless stems to 1.5 m, clothed below with brown sheaths. Inflorescence a terminal umbel with an involucre of many long, spreading, leaf-like bracts; spikes 6–8 mm, mostly bunched at the ends of the branches, pale brown. *Africa; widely naturalised in the tropics and subtropics.* H5.

Commonly grown as a pot-plant standing in shallow water; in frost-free regions also found in water gardens. A dwarf variant, 'Gracilis', is usually attributed to this species: the stems are *c.* 40 cm and the bracts narrower than the type, and flowers are rarely produced. 'Variegatus' has stems and bracts variously striped and mottled with white.

6. C. albostriatus Schrader (*C. diffusus* misapplied; *C. elegans* misapplied). Illustration: Everett, The New York Botanical Gardens illustrated encyclopedia of horticulture **3**: 984 (1981).

Rhizomatous, with slender, leafy stems to 60 cm. Leaves and involucral bracts 8–16 mm wide, with 3 prominent, whitish nerves. Inflorescence a loose terminal umbel with thread-like branches and 6–9 bracts; spikes 6–13 mm, narrowly lanceolate, single or in groups of 2–4, pale brown. *South Africa.* G1.

Grown as a decorative pot-plant; 'Variegatus' has longitudinally striped leaves and bracts.

8. UNCINIA Persoon

S.G. Knees

Tufted or rhizomatous, evergreen, perennials. Leaves flat towards the base, channelled above; apex trigonous. Inflorescence a simple terminal spike with terminal male florets and lateral female florets; axis of spikelet extending beyond the utricle and terminating in a reflexed, hook-like glume.

A genus of 40–50 species, mostly in the southern hemisphere, except southern Africa. Propagate by seed or division. Plant in any well-drained fertile soil. Does not tolerate full sun on dry soils in warmer climates. One species is widely cultivated while *U. unciniata* (Linnaeus filius) Kükenthal and *U. egmontiana* Hamlin are increasingly available in European nurseries.

1. U. rubra Boott. Illustration: Darke, Ornamental Grasses, 298 (1999); Rice (ed.), RHS Encyclopedia of perennials, 466 (2006).

Plants coloured dark reddish brown throughout. Leaves 25–35lcm, glossy. Flowering stems 30–35lcm. Inflorescence a simple spike, 3–6lcm × 3–5lmm, dark brown; fruit 5–6lmm, beak 1–1.5lmm. *New Zealand.* H3–4. Late spring–summer.

9. CAREX Linnaeus

S.M. Walters revised by S.G. Knees

Rhizomatous or tufted perennials, with grass-like leaves in 3 rows, and stems often triangular in section. Inflorescence compound, terminal, containing (in all cultivated species) both male and female florets grouped in spikes. Perianth absent; male florets consisting of 2 or 3 stamens, female consisting of an ovary with 2 or 3 stigmas and enclosed in a sac or utricle. Fruit a small nut enclosed in the persistent utricle. In the key and descriptions 'fruit' denotes the utricle and the enclosed nut.

A very large genus with more than 1500 species, principally in the north temperate region, and mostly characteristic of wet habitats. The few species grown in gardens are mainly used on the edges of lakes, ponds and streams. Since several other species not included in this account are widespread and common native European plants, they may be present naturally in woodland, marsh, stream and lakeside habitats, and tolerated in such places within gardens. Propagation is by division of rootstock or rhizome.

Literature: Kükenthal, G., Das Pflanzenreich **38** (1909); Mohlenbrock, R. H., Sedges: Carex (1999); Jermy, A.C. et al., Sedges of the British Isles, edn 3 (2007).

1a. Fruit berry-like, red or purplish when ripe **1. baccans**
b. Fruit not berry-like, brownish, greenish or whitish 2
2a. Inflorescence a dense, whitish head 3
b. Inflorescence not capitate, usually greenish or brownish 4
3a. Inflorescence with 2–5 leaf-like bracts **2. baldensis**
b. Inflorescence without bracts **3. fraseri**
4a. Leaves half circular in cross-section, usually 1.5 mm wide or less 5
b. Leaves V-shaped, W-shaped or flat, 2 mm or more wide 9
5a. Stems and leaves green, yellowish green or bluish green, rarely reddish **4. comans**
b. Stems and leaves brown or reddish 6
6a. Plant forming high, arching clumps 50–100 cm **5. flagellifera**
b. Plants usually smaller, rarely exceeding 60 cm 7
7a. Diminutive plants, not exceeding 10 cm **6. berggrenii**
b. Plants significantly larger, 30–60 cm or more 8
8a. Female spikes 3 or 4, dark red; stigmas 3 **7. petriei**
b. Female spikes 5–8, pale brown; stigmas 2 **8. buchananii**
9a. Leaves 8–13 mm wide 10
b. Leaves 2–6 mm wide 13
10a. Plants not exceeding 45 cm; fruit not exceeding 3 mm 11
b. Plants stout, often to 1 m or more; fruit 4–8 mm 12
11a. Female spikes 2–4 on long stalks; leaf sheaths greenish brown **9. morrowii**
b. Female spikes 8–10; leaf sheaths dark purple **10. phyllocephala**
12a. Female spikes 3–5, c. 5 cm; fruit 4–6 mm **11. pseudocyperus**
b. Female spikes 4–6, 8–10 cm; fruit 6–8 mm **12. trifida**
13a. Fruit 6–9 mm 14
b. Fruit 2–4 mm 15
14a. Leaves c. *2 cm wide* **13. riparia**
b. Leaves 3–5 mm wide **14. muskingumensis**
15a. Leaves *c.* 2 cm wide **15. pendula**
b. Leaves 2–6 mm wide 16
16a. Dwarf 'alpine' plants, not exceeding 25 cm 17
b. Plants larger, 30–100 cm 19
17a. Leaves *c.* 4 mm wide, stiff, sharply pointed; fruit *c.* 4 mm **16. firma**
b. Leaves 1.5–3.5 mm wide, flat; fruit 2–3 mm 18
18a. Leaves 1.5–2 mm wide, mid-green; fruit yellowish green to brown **17. ornithopoda**
b. Leaves 2–3.5 mm wide, dark green; fruit purple-brown **18. conica**
19a. Leaves flat **19. oshimensis**
b. Leaves V-shaped in cross-section 20
20a. Leaves distinctly keeled beneath **20. elata**
b. Leaves V-shaped but not keeled **21. dipsacea**

1. C. baccans Nees. Illustration: *Botanical Magazine*, 7288 (1893); Darke, Ornamental Grasses, 154 (1999).

Tufted, with stems to 1 m, and long, curving leaves to 1.5 cm wide. Inflorescence a bracteate panicle of clustered spikes, bearing shiny, berry-like fruits at first pale red, later purplish. *Himalaya & Taiwan to Sri Lanka & Java.* H5–G1.

Cultivated as a decorative pot-plant in the cool greenhouse; can be grown outside in relatively frost-free areas.

2. C. baldensis Linnaeus. Illustration: Das Pflanzenreich **38**: 185 (1909); Aeschimann et al., Flora alpina **2**: 138.17.31 (2004).

Rhizomatous, with stems to 30 cm, and leaves 3–4 mm wide. Inflorescence a small, dense, whitish head of 3–9 spikes, subtended by 2–5 pale, spreading, linear bracts. Fruit 4–5lmm, pale. *E Alps.* H1. Summer.

3. C. fraseri Andrews (*Cymophyllus fraseri* (Andrews) Mackenzie). Illustration: *Botanical Magazine*, 1391 (1811) – as *C. fraseriana*; Das Pflanzenreich **38**: 95 (1909).

Shortly rhizomatous, with leafless flowering stems to 40 cm, and stiff, thick, evergreen leaves to 5 cm wide, appearing after flowering. Inflorescence a single, terminal, whitish, ovoid-spherical or oblong spike without any subtending bract. Fruit *c.* 6 mm, inflated. *Eastern USA, from Pennsylvania to South Carolina.* H4. Summer.

Suitable for damp, peaty, shady places; an unusually attractive species.

4. C. comans Berggren (*C. vilmorinii* Mottet). Illustration: Darke, Ornamental Grasses, 155 (1999).

Tufted, with slender stems to 40 cm and very narrow, drooping, pale green leaves, half-circular in section. Inflorescence brownish, with a terminal male spike and about 5 lateral female spikes. Fruit *c.* 2 mm, narrowly ovoid. *New Zealand.* H4. Summer.

Very common in New Zealand, in both reddish brown and green variants; only the latter seems to be in cultivation in Europe.

5. C. flagellifera Colenso. Illustration: Darke, Ornamental Grasses, 159 (1999).

Perennial forming high-arching clumps, 50–100 cm; evergreen in milder climates. Leaves reddish brown, 35–50 cm × 0.5–1 mm. Female spikes 3–5, distant, stalked, *c.* 3 cm × 4 mm, borne on stalks to 2 m. Stigmas 2. *New Zealand.* H3–4 Spring.

Grow in full sun or light shade in moist, fertile soil. 'Coca Cola' (illustration: Darke, Encyclopaedia of grasses, 265, 2007) has purple-brown foliage.

6. C. berggrenii Petrie. Illustration: Das Pflanzenreich **38**: 684 (1909).

Shortly rhizomatous, tufted plant with stems 1.5–3 cm long, glabrous, terete, distinctly flattened above, almost completely enclosed by light brown leaf-sheaths. Leaves 3–6 cm × 1–2 mm, linear, almost flat, veins very distinct. Spikes 3–5; terminal spike male, distinctly pedunculate; remaining spikes female, 5–8 mm long; leaf-like subtending bracts exceeding inflorescence. Fruit 2–3 × 1.5 mm, subtrigonous, red-brown to dark red-purple above, yellow below; beak minute. Stigmas 2 or 3. Nut *c.* 1.5 mm long, trigonous. *New Zealand.* H2–3. Spring.

7. C. petriei Cheesman.

Plants compact, erect, forming dense tufts to 30 cm. Leaves 20–30 cm × 0.5–1 mm, copper brown to bronze, often curled at the apex. Male spike terminal; female spikes 3 or 4, very dark red, *c.* 5 cm × 5 mm, lowest distant, on stiff stalks. Stigmas 3. Fruit *c.* 3 mm, brown. *New Zealand.* H2. Spring.

8. C. buchananii Berggren. Illustration: Darke, Ornamental Grasses, 155 (1999).

Tufted, with stems to 60 cm, and very narrow, hard, reddish brown leaves half-circular in section. Inflorescence pale brownish, with a terminal male spike and

5–8 lateral female spikes. Fruit *c.* 3 mm, flattened. *New Zealand.* H4. Summer.

Used as a coloured edging plant by water.

9. C. morrowii Boott. (*C. foliosissima* misapplied; *C. japonica* misapplied). Illustration: Graf, Tropica, 390 (1978).

Tufted, with stems to 30 cm, and long, stiff, long-pointed evergreen leaves to 8 mm wide. Inflorescence with 1 terminal male spike and 2–4 slender, long-stalked, cylindric female spikes to 2.5 cm. Fruit *c.* 3 mm, with long beak. *C & S Japan.* H5. Summer.

Grown as a pot-plant for the conservatory, usually as 'Variegata' (illustration: Darke, Ornamental Grasses, 161, 1999) or 'Albo-mediana', with longitudinal white stripes on the leaves. It can be used as a border plant in frost-free regions.

10. C. phyllocephala T. Koyama.

Stems erect, 35–45 cm. Leaf sheaths dark purple. Leaves 10–20 cm × 8—13 mm. Inflorescence capitate, spikes 8–10, with terminal male spike. Stigmas 3. Fruit *c.* 3 mm. *China, cultivated in Japan.* H3–4. Spring.

Most widely grown as 'Sparkler' (illustration: Darke, Ornamental Grasses, 165, 1999), with white or creamy white marginal variegation. H3–4.

11. C. pseudocyperus Linnaeus. Illustration: Jermy et al., Sedges of the British Isles, edn 2, 117 (1982); Aeschimann et al., Flora alpina **2**: 138.17.38 (2004).

Tufted, with stout, sharp-angled stems to 80 cm; leaves often exceeding stems and to 12 mm wide. Inflorescence with a single terminal male spike, and 3–5 broad, cylindric female spikes to 5 cm, the upper clustered, the lower nodding, all subtended by long, leafy bracts. Fruit 4–6 mm, ovoid, shining, with long beak. *Widespread in the temperate regions of both hemispheres; also in tropical America.* H1. Summer.

An attractive species sometimes grown by or in water; it prefers rather acid, peaty soils.

12. C. trifida Cavanilles. Illustration: Das Pflanzenreich **38**: 732 (1909); Moore & Edgar, Flora of New Zealand **2**: 236 (1976); Darke, Encyclopaedia of grasses, 275 (2007).

Very robust perennial forming dense tussocks to 1 m. Leaves 100–150 × 1–1.2 cm. Inflorescence with 6–10 sturdy, sessile, chestnut brown spikes to 10 × 1.5 cm; upper 2–4 male; lower 4–6 female. Stigmas 3. Fruit 6–8 mm. *New Zealand.* H3. Spring.

13. C. riparia Curtis. Illustration: Jermy et al., Sedges of the British Isles, edn 2, 115 (1982) & edn 3, 305 (2007); Aeschimann et al., Flora alpina **2**: 138.17.37 (2004).

Robust, somewhat tufted plant with short rhizomes, rough, sharply angled stems to 1.5 m, and long, keeled, greyish green leaves to 2 cm wide. Inflorescence with 2–6 large male spikes at top, and 1–5 more distant, lateral female spikes to 9 cm, almost erect or lower nodding, subtended by leafy bracts. Fruit *c.* 6 mm, ovoid with short beak. *Most of Europe, extending to N Africa & W Asia.* H1. Summer.

One of the commonest reed-swamp species by lakes and slow-flowing rivers in much of Europe; often present in suitable places in large gardens if specially planted, usually as 'Aurea' ('Bowles' Yellow') or 'Variegata' (illustration: Darke, Ornamental Grasses, 166, 1999), with golden or longitudinally striped leaves respectively.

C. acutiformis Ehrhart. Illustration: Aeschimann et al., Flora alpina **2**: 138.17.35 (2004). Very similar, but can be distinguished by its narrower leaves (not more than 1 cm wide), smaller fruits (*c.* 4 mm) and blunt (not pointed) male glumes. *Eurasia to southern Africa.* H1. Summer.

It is as likely as *C. riparia* to occur in reed-swamps in large gardens and parks.

14. C. muskingumensis Schweinitz. Illustration: Darke, Ornamental Grasses, 161 (1999); Rice (ed.), RHS Encyclopedia of perennials, 116 (2006); Darke, Encyclopaedia of grasses, 264 (2007).

Loosely tufted, spreading perennial, 40–100 cm. Leaves 12–25lcm × 3–5lmm, bright green, borne on erect stems. Inflorescence with 5–12 golden brown spikes; spikes 1–2.5lcm × 3–7lmm. Fruit 6–9lmm, cylindric. *North central North America.* H1. Spring.

Can be planted in water.

15. C. pendula Hudson. Illustration: Jermy et al., Sedges of the British Isles, edn 2, 127 (1982); Darke, Ornamental Grasses, 164 (1999); Brickell (ed.), RHS A–Z encyclopedia of garden plants, 232 (2003); Aeschimann et al., Flora alpina **2**: 138.17.41 (2004).

Robust, tufted, often forming large tussocks, with slender, graceful flowering stems to 1.5 m. Leaves shorter than stems, to 2 cm wide. Inflorescence with 1 terminal male spike and 4 or 5 distant, pendent, cylindric female spikes to 25 cm. Fruit *c.* 3 mm, ellipsoid, with short beak. *Europe from Denmark southwards, extending to N Africa & SW Asia.* H3. Summer.

Much the most familiar species in cultivation in W Europe, often planted in wet shady places, but surprisingly tolerant of drier ground.

16. C. firma Host. Illustration: Coste, Flore de 1a France **3**: 514 (1906); Aeschimann et al., Flora alpina **2**: 138.17.80 (2004).

Densely tufted or mat-forming plant, with stems to 20 cm and short, stiff, pointed leaves to 4 mm wide in rosettes. Inflorescence with 1 terminal male spike and 1–3 lateral, ovoid female spikes. Fruit *c.* 4 mm, oblong-ellipsoid. *Mountains of Europe, from the Pyrenees to the Carpathians; only on limestone.* H1. Summer.

The usual plant has blue-green leaves; in 'Variegata' these are longitudinally striped with white. Both are grown as 'alpines'.

17. C. ornithopoda Willdenow. Illustration: Reichenbach, Icones florae Germanicae et Helveticae **8**: pl. 240 (1846); Darke, Encyclopaedia of grasses, 265 (2007); Jermy et al., Sedges of the British Isles, edn 3, 415 (2007).

Shortly rhizomatous, stems tufted, of two kinds: apical leafy shoots with overwintering terminal buds and lateral short shoots 5–20 cm, terminating in an inflorescence. Leaves 5–25 cm × 1.5–2 mm, flat or slightly keeled. Inflorescence compact with 1 male spike and 2 or 3 female spikes. Fruit 2–3 mm, obovoid to pyriform, yellowish green to brown, pilose; beak absent. *Europe.* H1. Spring–early summer.

Most widely grown as 'Variegata'.

18. C. conica Boott. Illustration: Das Pflanzenreich **38**: 476 (1909).

Small tufted perennial, eventually forming mounds to 25 cm. Leaves to 15 cm × 2–3.5 mm, flat, rough. Male spike terminal, erect; female spikes 3 or 4,

1–2.5 cm, purple-brown. Fruit 2.5–3 mm. *Korea, Japan.* H2. Summer.

Most widely grown as 'Snowline' (illustration: Darke, Ornamental Grasses, 156, 1999), which has deep green leaves with conspicuous white margins.

19. C. oshimensis Nakai. Illustration: Darke, Encyclopaedia of grasses, 265 (2007).

Rhizomatous, densely tufted, evergreen perennial; stems 20–50 cm. Leaves 20–30 cm × 3–6 mm, flat. Spikes 3–5, erect, on stalks 1–5 cm, terminal male. Fruit *c.* 3mm. *Japan (Honshu).* H2–3. Spring.

Most widely grown as 'Evergold'. Illustration: Darke, Ornamental Grasses, 163 (1999); Brickell (ed.), RHS A–Z encyclopedia of garden plants, 232 (2003), which has broad median stripes of creamy white variegation, darkening to creamy yellow with age.

20. C. elata Allioni (*C. stricta* Goodenough; *C. hudsonii* Bennett). Illustration: Jermy et al., Sedges of the British Isles, edn 2, 205 (1982); Aeschimann et al., Flora alpina **2**: 138.17.97 (2004).

Robust, tufted, capable of forming large tussocks, with stems to 1 m, and long, sharply keeled leaves to 6 mm wide. Inflorescence of 1 or 2 (rarely 3) terminal male spikes, and 2–4 rather distant female spikes below, the latter each with a small, very narrow bract. Fruit *c.* 3 mm, flattened, with short, entire beak. *Europe, extending to N Africa & the Caucasus.* H5. Summer.

Usually seen in gardens as 'Aurea', with yellowish foliage.

21. C. dipsacea Berggren. Illustration: Rice (ed.), RHS Encyclopedia of perennials, 115 (2006).

Light green or reddish tinted tufts, 30–60 cm. Leaves 40–70 cm × *c.* 2 mm, V-shaped in cross-section. Female spikes 2–5, distant, sessile, 2–2.5 cm × 5 mm, reddish brown. Glumes obtuse, membranous, midrib often not reaching margin. Stigmas 2. Fruit 2–2.5 mm, margins scabrid. *New Zealand.* H3–4. Summer.

54. MUSACEAE

Large herbs or apparently small trees, with pseudostems made up of rolled leaf-stalks. Leaves with conspicuous main vein and parallel lateral veins, spirally arranged, as are the bracts. Flowers unisexual, zygomorphic, in spikes, many in the axils of each of the numerous bracts. Sepals 3, free or united, sometimes joined to the corolla or tubular and split down 1 side. Corolla 2-lipped. Stamens 5, sometimes with 1 staminode. Ovary of 3 carpels, 3-celled, inferior; ovules numerous. Fruit a banana.

Two genera and about 60 species, including the cultivated banana, *Musa sapientium*. A few species are cultivated for ornament.

1a. Plant solitary, not normally suckering, false stems distinctly swollen at the base; basal bracts persisting until the fruit is ripe; seeds usually more than 10 mm in diameter **1. Ensete**

b. Plant suckering freely; false stems cylindric, scarcely swollen at the base; basal bracts usually quickly deciduous; seeds rarely more than 6 mm in diameter **2. Musa**

1. ENSETE Horaninow

G. Argent

Solitary perennial herbs, flowering only once and then dying, not suckering from the base (unless damaged). Leaf-sheaths often spreading, rarely completely encircling the false stem. Inflorescence hanging, bracts usually persistent. Flowers many to each bract, in 2 rows. Upper lip often deeply lobed, the free segment 3-pointed or entire. Seeds usually more than 10 mm in diameter.

A genus of about 7 species from tropical Africa and Asia; they are rarely cultivated in Europe, but can make attractive architectural features.

Literature: Cheesman, E.E., Classification of the bananas I: the genus Ensete Horaninow, *Kew Bulletin* **2**: 97–106 (1947); Baker, R.E.D. & Simmonds, N.W., The genus Ensete in Africa, *Kew Bulletin* **8**: 405–16 (1953).

1a. False stem rarely more than 1 m; leaves entirely green; upper lip 3-lobed almost to the base, the free segment with toothed and pointed lateral lobes; seeds 1–1.2 cm in diameter, rarely less **2. superbum**

b. False stem to 5 m; leaves usually with reddish midribs; upper lip 3-lobed to less than halfway, the free segment with rounded lateral lobes; seeds 1.2–2.3 cm in diameter **1. ventricosum**

1. E. ventricosum (Welwitsch) Cheesman (*E. edule* Horaninow; *Musa ventricosa* Welwitsch; *M. ensete* Gmelin; *M. arnoldiana* De Wildeman). Illustration: *Botanical Magazine*, 5223, 5224 (1861); Brickell (ed.), RHS A–Z encyclopedia of garden plants, 407 (2003).

Swollen false stem to 5 m, only suckering if damaged. Leaf-blade to 6 × 1 m from a stalk that is up to 40 cm. Inflorescence more than 1 m, emerging horizontally, eventually drooping. Male bracts broadly lanceolate, tapering towards the apex for about two-thirds of their length. Basal flowers bisexual, the apical male. Upper lip 3-lobed to less than half way, the free segment with entire margins, rounded lateral lobes and a long, subulate central point. Stamens 6, 1 much smaller than the other 5. Fruit oblong, 6–10 cm. Seeds black, glossy, 1.2–2.3 cm in diameter. *Ethiopia to Angola.* H5–G1. Flowering irregular.

2. E. superbum (Roxburgh) Cheesman (*Musa superba* Roxburgh). Illustration: *Botanical Magazine*, 3849, 3850 (1841).

False stem very swollen, to 1 m. Leaf-blade *c.* 150 × 50 cm, stalk *c.* 50 cm. Inflorescence arching, later hanging vertically, rather short. Male bracts ovate, concave, tapering from about halfway to the broad apex, which may have a small, acute point. Basal flowers female with staminodes, the apical male. Upper lip 3-lobed almost to the base, the free segment with toothed margins, pointed lateral lobes and a long, subulate central point. Stamens 5. Fruit with numerous seeds, which are mostly 1–1.2 cm in diameter (rarely less). *S India.* H5–G1. Flowering irregular.

2. MUSA Linnaeus

G. Argent

Plants suckering and often forming large clumps. False stem cylindric, made up of completely encircling leaf-sheaths that are hardly swollen at the base. Bracts mostly deciduous. Flowers few to many per bract, in 1 or 2 rows. Upper lip with minute lobes, the free segment with a single point. Seeds rarely more than 5 mm in diameter.

A genus of about 40 species from SE Asia, commonly cultivated for fruit and ornament.

Literature: Moore, H.E., Musa and Ensete, the cultivated bananas, *Baileya* **5**: 167–94 (1957) & **6**: 69–70 (1958). Simmonds, N.W., The evolution of the bananas (1962); Simmonds, N.W., Bananas, edn 2 (1966).

Height. Less than 2 m: **1, 3, 5**; 3 m or more: **1, 2, 4**.
Inflorescence. Hanging: **1, 2**; erect: **3, 4, 5**.
Bracts. Yellow: **1, 2**; dark purple: **1**; scarlet: **5**; pink or purplish pink: **3, 4**.
Fruit. Hairy: **3**.

1a. False stem less than 2 m — 2
b. False stem more than 2 m — 4
2a. Inflorescence hanging; flowers in 2 rows per bract — **1. acuminata**
b. Inflorescence erect; flowers in 1 row per bract — 3
3a. Bracts pale purple; inflorescence-stalk and fruit hairy; leaves *c.* 2.5 times as long as broad with midrib purplish-flushed — **3. velutina**
b. Bracts red; inflorescence-stalk and fruit hairless; leaves *c.* 3.5 times as long as broad, completely green — **5. coccinea**
4a. Inflorescence erect; bracts purplish pink — **4. ornata**
b. Inflorescence hanging; bracts not purplish pink — 5
5a. Leaves and false stems without a glaucous waxy deposit; bracts dull yellow to brownish; leaf-stalks with broad, spreading, green, non-waxy margins — **2. basjoo**
b. Leaves and false stems with a glaucous, waxy deposit; bracts variable, usually dark purple; wings of leaf-stalk usually erect, if spreading then waxy and with a dead, brown margin — **1. acuminata**

1. M. acuminata Colla (*M. cavendishii* Paxton; *M. zebrina* Planchon). Illustration: Simmonds, Bananas, 20–1, 81 (1966); Graf, Tropica, 672 (1978).

Usually 4–6 m, mostly glaucous green. Upper edge of leaf-sheath with a brown, papery margin *c.* 1 cm wide. Inflorescence pendent, flowers in 2 rows per bract. Male bud pear-shaped, acutely pointed. Bracts variable in colour, usually dull purple outside, rolling back on opening to expose the flowers. Flowers white, cream or yellow, the free segment with a marked subapical wrinkle and the point turned inwards. Fruit variable, usually ripening yellow, with pale yellowish pulp. *SE Asia, from Thailand to N Australia.* H5–G2. Flowering irregular.

A very variable species that has given rise to most of the sweeter dessert bananas of commerce. The wild species is fully fertile with seeded fruits, but is unlikely to be found in cultivation. Various cultivars and hybrids with seedless fruits, which are formed without fertilisation, may be found in greenhouses and are difficult to name precisely. 'Dwarf Cavendish' (*M. cavendishii* Paxton) is one of the more distinct and common; it has a false stem *c.* 2 m high and leaf-blades about 2 times longer than wide and is sometimes known as the Canary or Chinese banana; it is cultivated on a moderate scale in Sicily and Crete, and is often planted in greenhouses because of its small size, resistance to cool conditions and sweet fruit.

M. × paradisiaca Linnaeus, with orange-yellow flowers and long, narrow, starchy, edible fruit, is a commonly used name, and strictly refers to the cultivar 'French Plantain', which is of hybrid origin (*M. acuminata* × *M. balbisiana* Colla).

M. × sapientum Linnaeus (*M. rosacea* Jacquin; *M. violacea* Baker), with white or pinkish flowers and short, stumpy, starchy, edible fruit, is another commonly grown variant; the name strictly applies to the cultivar 'Silk Fig', again thought to be of the same hybrid origin as *M.* × *paradisiaca*.

A variant of *M.* × *sapientum*, whose stem, leaves and fruit are tinged with purplish blue, is often sold as *M. violacea* Baker (an invalid name). The name *M. rosacea* correctly applies to this hybrid, but is commonly and incorrectly applied to *M. ornata*.

2. M. basjoo Siebold (*M. japonica* Thibaut & Keteleer). Illustration: *Botanical Magazine*, 7182 (1891); Noailles & Lancaster, Mediterranean plants and gardens, 101 (bottom right, without name) (1977); Brickell (ed.), RHS A–Z encyclopedia of garden plants, 703 (2003).

Plants 3.5–5 m, green, waxless. Leaf-sheath with a spreading, green, upper edge. Inflorescence pendent, flowers in 2 rows per bract. Male bud broadly ovoid, bluntly pointed, bracts dull yellowish to somewhat brownish, rolling back on opening. Flowers cream or pale yellow, the free segment reflexed backwards from a subapical wrinkle. Fruit green, seeded, inedible. *S Japan (Ryukyu Islands).* H5–G1. Summer.

3. M. velutina Wendland & Drude. Illustration: Clay & Hubbard, The Hawaii garden, tropical exotics, 149 (1977); Graf, Tropica, 671, 673 (1978).

Plant 1–1.5 m, purplish green, sparsely waxy. Leaves *c.* 2.5 times as long as broad with a flat, green or brownish edge to the leaf-sheath. Inflorescence erect, flowers in a single row per bract. Male bud slender, pointed, the bracts pale purple, margins slightly rolled under. Fruit pink, hairy, *c.* 9 × 3 cm, splitting when ripe to reveal white flesh in which are embedded numerous black seeds. *NE India.* G2. Flowering irregular.

4. M. ornata Roxburgh (*M. rosacea* misapplied; *M. rosea* misapplied). Illustration: *Edwards's Botanical Register* **9**: t. 706a & b (1823); Brickell (ed.), RHS A–Z encyclopedia of garden plants, 703 (2003).

False stems slender, 2–3 m, rather glaucous, waxy green. Leaf-blade 5–6 times as long as broad, to 2 m × 35 cm, upper edge of the sheath smooth, green and adpressed. Inflorescence erect, flowers in 1 row per bract. Male bud slender, acute, with pale purplish pink bracts that sometimes roll slightly before falling. Flowers deep orange-yellow with a paler free segment that is without a subapical wrinkle. Fruit 6–8 × 1.5–2 cm, greenish yellow when ripe, with white pulp and numerous black seeds. *Bangladesh (Chittagong hills).* G2. Flowering irregular.

5. M. coccinea Andrews (*M. uranoscopus* Loureiro). Illustration: *Botanical Magazine*, 1559 (1813); Graf, Tropica, 666, 673 (1978).

Plant *c.* 1 m, green, waxless. Leaves *c.* 3.5 times as long as broad and up to 1 m long, upper margin of the sheath with a flat green edge. Inflorescence erect, flowers in a single row per bract. Male bud rather cone-like, due to the persistence of the open, scarlet bracts that do not curl back. Fruit hairless, 4–5 × 2–2.5 cm, orange-yellow when ripe, pulp white with numerous black seeds. *S China, Vietnam, Laos, Kampuchea.* G2. Flowering irregular.

Names of doubtful application:

M. mannii Hooker. Illustration: *Botanical Magazine*, 7311 (1893). An imperfectly understood small species up to 1.3 m high, with purplish red bracts that do not curl back. *NE India.*

M. sanguinea Hooker. Illustration: *Botanical Magazine*, 5975 (1872). Similar to both *M. coccinea* and *M. mannii* but differing from both in having strongly rolled bracts and a hairy inflorescence-stalk. *NE India.*

55. STRELITZIACEAE

Herbs or shrubs with leaves and bracts in 2 ranks, leaves with a conspicuous mid-vein and parallel lateral veins. Flowers in coiled cymes in the axils of spathes, bisexual, zygomorphic. Sepals 3, free, very unequal. Corolla of 3 free or united lobes. Stamens usually 5+1 petaloid staminode. Ovary of 3 united carpels, inferior; ovules 1–many per cell, axile or apical; style 1. Fruit a capsule or indehiscent.

There are 5 genera and over 100 species. *Strelitzia reginae*, the bird-of-paradise flower, is widely cultivated, and is often used as a cut flower.

1a. Inflorescence of 1–5 spathes; stamens 5; aril reddish brown **1. Strelitzia**
b. Inflorescence of more than 5 spathes; stamens 6; aril ciliate, bluish **2. Ravenala**

1. STRELITZIA Aiton

G. Argent

Palm-like trees forming trunks, often suckering from the base, or herbs with short rhizomes. Leaves with open, partly sheathing bases, grooved stalks and mostly with broad, oblong or lanceolate blades. Inflorescence axillary, usually more or less horizontal, simple or compound. Flowers bilaterally symmetric, the calyx with 3 narrow segments, the corolla of 2, often arrow-shaped, paired parts (the 'tongue'), which enclose the stamens and style, and a shorter upper part. Stamens 5. Fruit a 3-celled capsule with numerous seeds surrounded by reddish brown, filamentous arils.

A genus of 4 species from southern Africa.

Literature: Moore, H.E. & Hyypio, P.A., Some comments on Strelitzia, *Baileya* **17**: 65–74 (1970).

1a. Herbaceous plants, without woody trunks; calyx orange, yellow or reddish **1. reginae**
b. Plant with a woody trunk; calyx white or purplish blue 2
2a. Inflorescence compound with more than 1 spathe; the unpaired, inner corolla segment to 2 cm **2. nicolai**
b. Inflorescence simple with only 1 sheathing spathe, although often with additional sterile bracts; the unpaired, inner corolla segment more than 2.5 cm 3
3a. Stem solitary; tongue forming a narrow, boat-shaped structure up to 1.25 cm broad, with smooth margins **3. alba**
b. Stem suckering; tongue forming a boat-shaped structure 2–3 cm wide, with wavy margins **4. caudata**

1. S. reginae Aiton (*S. parvifolia* Aiton; *S. juncea* Link). Illustration: *Botanical Magazine*, 119 (1797); Graf, Tropica, 669, 670 (1978); Brickell (ed.), RHS A–Z encyclopedia of garden plants, 1009 (2003).

Almost stemless and more or less herbaceous, 1–1.5 m. Leaf-blade oblong-lanceolate, 25–50 × 10–25 cm, acute to rounded or even notched at the apex, rounded or tapering and often wrinkled at the base, stalk oval in cross-section, not channelled, 25–100 cm. Inflorescence with flowers from a solitary green or purplish-flushed spathe *c.* 12 cm. Flowers opening in succession, usually with orange calyx and blue corolla, *c.* 10 cm. *South Africa.* H5–G1. Spring.

A variable species in flower colour, stature and size of the leaves; var. **parvifolia** (Aiton) Anon., has small, lanceolate leaf-blades, and var. **juncea** (Link) Ker Gawler has no blades at all.

2. S. nicolai Regel & Körnicke. Illustration: *Botanical Magazine*, 7038 (1889); *Flowering plants of Africa* **25**: t. 996 (1946); Graf, Tropica, 664 (1978); Brickell (ed.), RHS A–Z encyclopedia of garden plants, 1009 (2003).

Stems woody, in clumps, to 10 m. Leaf-blade oblong or ovate-oblong, *c.* 1.5 m × 60 cm, rounded or cordate at the base, stalk to 2 m. Inflorescences axillary with 3–5 purplish red spathes, which are 40–45 cm. Flowers *c.* 20 cm, the calyx white, the corolla white or light purplish blue, tongue boat-shaped, arrowhead-shaped at the base with the margins turned inwards and meeting in the middle. *South Africa (coastal regions of S Natal & NE Cape Province).* G1. Flowering intermittent.

3. S. alba (Linnaeus) Skeels (*S. augusta* Wright). Illustration: *Botanical Magazine*, 4167, 4168 (1845); *Flowering plants of Africa* **25**: t. 995 (1946); Graf, Tropica, 665, 669 (1978).

Stems woody, in clumps, to 10 m. Leaf-blade oblong-lanceolate, *c.* 2 m × 45–60 cm. Leaf-stalk *c.* 1 m. Inflorescence with 1 purplish red, glaucous spathe, 25–30 cm, and some sterile bracts. Flowers *c.* 20 cm, calyx and corolla white, tongue not arrow-shaped at the base but margins there turned inwards and overlapping. *South Africa (south coast of Cape Province).* G1. Flowering intermittent.

4. S. caudata Dyer. Illustration: *Flowering plants of Africa* **25**: t. 997 (1946).

Stems woody, suckering, to 6 m. Leaf-blade ovate to oblong, 1.5–1.75 m × 80–85 cm. Leaf-stalk 1.5–1.75 m. Inflorescence bracteate with a single, dark purplish red, somewhat glaucous spathe *c.* 30 cm. Flowers *c.* 20 cm, calyx white, corolla white or sometimes tinged with bluish purple at the base. Tongue forming a boat-shaped structure with obtuse basal lobes and a wavy margin. *South Africa (N Transvaal), Swaziland.* G2. Flowering intermittent.

2. RAVENALA Adanson

G. Argent

Palm-like tree with solitary, unbranched trunk to 25 m. Leaves like those of a banana, stalked, borne in a 2-ranked fan, *c.* 3 m. Inflorescence lateral, compound, composed of 6–12, 2-ranked spathes, in which are compact, cymose inflorescences. Flowers white with 3 calyx and 3 corolla segments, 2 of the corolla segments larger than the third. Stamens 6. Fruit a capsule with numerous seeds with bright blue arils.

A genus of 1 species from Madagascar.

1. R. madagascariensis Sonnerat. Illustration: Graf, Tropica, 672 (1978).

Madagascar. G2.

Plants of uncertain identity from SE Asia are often grown under this name; they are similar, except in producing occasional suckers and a terminal inflorescence. They may be a species of *Ravenalopsis* Nakai.

Species of the genus *Heliconia* Linnaeus are sometimes recorded as being in cultivation, but it is extremely doubtful if they are seen outside botanic gardens or specialist collections, although some are now becoming important in the cut-flower trade. The genus consists of herbaceous plants with false stems, which are made up of concentric circles of overlapping leaf-bases (a feature not seen in *Strelitzia* and *Ravenala*). Their classification is difficult, and it is not certain which species may be cultivated. They are often included in their own family, the Heliconiaceae.

56. ZINGIBERACEAE

Herbs with rhizomes. Leaves in 2 ranks, usually with open sheaths and ligules, aromatic; ptyxis supervolute. Flowers in racemes, heads or cymes, bisexual, zygomorphic. Calyx 3-lobed. Corolla of 3 free or united petals. Stamen 1, usually modified into a petal-like labellum, staminodes 3–5, usually petal-like. Ovary of 3 carpels, inferior, usually surmounted by epigynous glands, style often supported in a groove in the anther; ovules many, axile or parietal. Fruit a capsule.

There are about 40 genera and 1,000 species. Six genera are native to the southern parts of North America. Species of about 17 genera are grown as ornamentals, and species of several more are grown in the tropics as spice plants, e.g. ginger (*Zingiber*), cardamom (*Ellettaria*), turmeric (*Curcuma*), etc.

The Zingiberaceae are plants of the forest floor, requiring shade; they are very sensitive to high light intensity, particularly if grown in pots where the root-run is restricted. Damage resulting from this appears as yellowing of the foliage and browning of the leaf tips. Pest attack under glass is problematic with pot culture, particularly from red spider mite. Spraying has a limited effect, and biological control may be more suitable. A moist atmosphere, obtained by overhead spraying of the foliage and frequent damping down will also be helpful.

Literature: Schumann, K., Das Pflanzenreich **20** (1904); Holttum, R.E., The Zingiberaceae of the Malay Peninsula, *Gardens Bulletin of Singapore* **13**: 1–249 (1950); Smith, R.M., Synoptic keys to the genera of the tribes Zingibereae, Globbeae, Hedychieae and Alpineae in part, *Royal Botanic Garden Edinburgh, Departmental Publication Series* **2** (1981).

1a. Leafy shoots well developed, leaf-blades spirally arranged; plants not aromatic **1. Tapeinochilos**
b. Leafy shoots well developed, leaf-blades arranged in 2 ranks, or leafy shoots not well developed, leaves appearing tufted; plants aromatic 2
2a. Inflorescence borne separately from the leaves 3
b. Inflorescence terminal on the leafy shoot or from the centre of a tuft of leaves 9
3a. Inflorescence long, slender, prostrate; bracts well spaced **14. Elettaria**
b. Inflorescence congested; bracts often overlapping 5
4a. Anther crest wrapped around the projecting style **3. Zingiber**
b. Anther crest, if present, not as above 5
5a. Lateral staminodes large, petal-like 6
b. Lateral staminodes absent or reduced to small appendages 7
6a. Bracts joined to each other laterally in the lower part, forming pouches; anther versatile **10. Curcuma**
b. Bracts free to the base; anther not versatile **4. Kaempferia**
7a. Inflorescence held well above the ground, surrounded by an involucre of sterile bracts that are showy **13. Nicolaia**
b. Inflorescence often held at ground level, showy sterile bracts absent 8
8a. Lip large, almost circular, to *c.* 6 cm in diameter; fruit flask-shaped **12. Aframomum**
b. Lip elliptic, *c.* 1 cm wide; fruit spherical **11. Amomum**
9a. Filament arched; anther with lateral appendages; flowers often replaced by bulbils **2. Globba**
b. Filament not arched; anther without lateral appendages; bulbils rare 10
10a. Bracts joined to each other laterally in the lower part, forming pouches **10. Curcuma**
b. Bracts free to the base 11
11a. Lateral staminodes showy, petal-like 12
b. Lateral staminodes absent or small and tooth-like 17
12a. Anther versatile, almost always with conspicuous basal spurs 13
b. Anther never versatile, lacking conspicuous basal spurs 14
13a. Fruit elongate, slow to split; flowers usually purple, rarely yellow **8. Roscoea**
b. Fruit short, splitting readily; flowers yellow **9. Cautleya**
14a. Short-stemmed plants, often appearing tufted 15
b. Tall plants with distinct leafy stems 16
15a. Lip deeply 2-lobed; anther crest prominent **4. Kaempferia**
b. Lip entire or slightly lobed; anther crest not prominent **5. Boesenbergia**
16a. Lip showy; filament elongate **6. Hedychium**
b. Lip much reduced, at most 3–4 mm or entirely absent; filament not elongate **7. Brachychilum**
17a. Petals much larger than the lip; fruit elongate; flowers orange **15. Burbidgea**
b. Petals smaller than the lip; fruit round or oval; flowers rarely orange **16. Alpinia**

1. TAPEINOCHILOS Miquel

R.M. Smith

Plants sometimes branched. Leaves spirally arranged. Inflorescence terminal on a leafy shoot or borne separately, forming a dense head. Bracts overlapping, each subtending a single flower. Bracteoles absent. Flowers not, or scarcely exceeding the bracts. Lip inconspicuous. Lateral staminodes reduced to small appendages.

A genus of about 15 species from Indonesia, New Guinea and Australia.

1. T. ananassae Hasskarl (*T. pungens* Teysmann & Binnendijk). Illustration: *Journal of the Royal Horticultural Society* **83**: 71 (1958).

Plant to 2 m. Inflorescence resembling a red pineapple, borne separately from the leaves on a stalk 4–125 cm, the spike ovoid, to 12 cm. Bracts scarlet, strongly recurved, hard. Flowers yellow. *Indonesia.* G2.

2. GLOBBA Linnaeus

R.M. Smith

Plants slender, usually less than 1 m. Inflorescence terminal on the leafy shoot. Bracts often brightly coloured, each subtending 2–many flowers. Bracteoles open to the base. Flowers to 3 cm, the tube slender, much longer than the petals. Lip 2-lobed at apex. Lateral staminodes about the same size as the petals. Stamen with a long, curved filament, anther (in the cultivated species) with 2 small triangular appendages on either side. Flowers sometimes replaced by bulbils.

A genus of about 70 species, all from Asia. They require hot and humid conditions with a minimum night temperature of 18 °C. The plants must be dried out in the winter, allowing them to go dormant. This is most easily done with pot cultivation, but they may also be planted out in a border or, preferably, in a raised bed that can be kept dry during the winter, and whose soil can be renewed periodically.

1a. Bracts green; bulbils always copiously produced **1. marantina**

b. Bracts red-purple or red; bulbils very rarely produced 2
2a. Leaves with stalks to 10 cm; bracts red-purple **2. winitii**
b. Leaves without stalks; bracts red **3. atrosanguinea**

1. G. marantina Linnaeus (*G. bulbifera* Roxburgh; *G. schomburgkii* Hooker). Figure 16(1), p. 423. Illustration: *Botanical Magazine*, 6298 (1876).

Plant to 75 cm. Leaves more or less stalkless, elliptic, to 20 cm. Inflorescence to 10 cm, eventually hanging, bracts green, to 1.5 cm. Flowers yellow, often with orange markings at the base of the lip, sometimes entirely replaced by spherical, white bulbils. *SE Asia*. G2. Autumn.

2. G. winitii Wright. Illustration: *Botanical Magazine*, 9314 (1926); Brickell (ed.), RHS A–Z encyclopedia of garden plants, 488 (2003).

Plant to 1 m. Leaf-stalks to 10 cm, blade to 20 cm, lanceolate, cordate at the base. Inflorescence to 15 cm, hanging. Bracts red-purple, to 3.5 cm, reflexed, all usually fertile, each with a spike of 5–7 yellow flowers. *Thailand*. G2. Autumn–winter.

3. G. atrosanguinea Teysmann & Binnendijk. Illustration: *Botanical Magazine*, 6626 (1892).

Plant to *c.* 75 cm, stems purplish. Leaves more or less stalkless, to 20 cm, elliptic-lanceolate. Inflorescence erect, to 7 cm. Bracts bright red, all fertile, each subtending 6–7 flowers. *Borneo*. G2.

3. ZINGIBER Boehmer

R.M. Smith

Inflorescence cone-like, elliptic or cylindric, usually borne separately from the tall, leafy shoots. Flowers produced singly in the axils of broad, overlapping bracts. Bracteoles open to the base. Lip 3-lobed, the lateral lobes representing the lateral staminodes. Anther with an elongate crest that enfolds the upper part of the style. Fruit fleshy, held within the persistent bracts.

A genus of about 100 species, widely distributed in tropical *Asia*. *Z. officinale* is of economic importance as the source of commercial ginger. Cultivation as for *Globba* (p. 421).

Leaves. With 2-lobed ligule: **1, 2, 5**; with entire ligule: **3, 4**. Less than 1.5 cm wide: **2**.
Inflorescence. Stalkless: **1**. Cylindric: **3**.
Lip. Plain yellow: **1, 4, 5**; dark purple with yellow spots: **2, 3**.

1a. Inflorescence stalkless, arising at ground level **1. mioga**
b. Inflorescence held above ground on a distinct, erect stalk 2
2a. Lip dark purple with pale yellow spots; bracts not closely overlapping 3
b. Lip plain yellow; bracts closely overlapping 4
3a. Inflorescence robust, cylindric, 10–30 cm, the upper edges of the bracts stiffly incurved, forming pouches **2. spectabile**
b. Inflorescence ovoid, to 5 cm, margins of the bracts only slightly incurved **3. officinale**
4a. Ligule entire; leaves 5–8 cm wide **4. zerumbet**
b. Ligule 2-lobed; leaves 2–4 cm wide **5. purpureum**

1. Z. mioga (Thunberg) Roscoe. Illustration: *Botanical Magazine*, 8570 (1914).

Leafy shoots to 1 m. Leaves stalkless, to 30 cm, linear-lanceolate, ligule 2-lobed. Inflorescence more or less stalkless, ellipsoid, to *c.* 10 cm. Bracts ovate-oblong, green with reddish spots. Flowers yellow. *Japan*. G2.

2. Z. spectabile Griffith. Illustration: *Botanical Magazine*, 7967 (1904).

Leafy shoots to at least 2 m. Leaves stalkless, to 50 cm, oblong-lanceolate to elliptic, ligule entire. Inflorescence cylindric, to 30 cm, held on a robust stalk, 30–100 cm. Bracts yellowish, usually turning red with age, their tips incurved. Flowers pale yellow, the lip dark purple with pale yellow spots. *Malaysia*. G2.

3. Z. officinale Roscoe.

Leafy shoots to 1.5 m. Leaves stalkless, to 15 × 1.5 cm, linear-lanceolate, ligule slightly 2-lobed. Inflorescence to 5 cm, held on an erect stalk, 15–30 cm. Bracts green, broadly ovate. Flowers greenish yellow, lip dark purple with pale yellow spots. *Asia, precise area of origin unknown*. G2.

The rhizome is the source of commercial ginger and the plant is widely cultivated throughout the tropics.

4. Z. zerumbet (Linnaeus) J.E. Smith. Figure 16(2), p. 423. Illustration: *Botanical Magazine*, 2000 (1818).

Leafy shoots to 2 m. Leaves more or less stalkless, to 35 cm, lanceolate, ligule entire, prominent. Inflorescence cone-like, to 13 cm, held on an erect stalk, 20–45 cm. Bracts ovate, green becoming red. Flowers white to pale yellow. *India*. G2.

Widely cultivated as a medicinal plant. 'Darceyi', with variegated yellow-green leaves, is occasionally seen.

5. Z. purpureum Roscoe (*Z. cassumunar* Roxburgh). Illustration: *Botanical Magazine*, 1426 (1811).

Leafy shoots to 1.5 m. Leaves stalkless, to 35 cm, linear-lanceolate, ligule very small, 2-lobed. Inflorescence cone-like, to 12 cm, held on an erect stalk, 15–25 cm. Bracts ovate, slightly hairy, brownish purple. Flowers pale yellow. *India*. G2.

4. KAEMPFERIA Linnaeus

R.M. Smith

More or less stemless. Leaves often appearing tufted, sometimes variegated. Inflorescence terminal and appearing with the leaves, or borne on a separate leafless shoot, when the flowers usually appear before the leaves. Flowers single in the bract axils. Bracteoles open to the base, very thin, deeply split. Petals narrow. Lateral staminodes large and petaloid, often similar to the halves of the deeply 2-lobed, showy lip. Anther with a prominent, reflexed crest.

A genus of about 40 species widely distributed in tropical Asia. The African species formerly included in the genus have recently been transferred to *Siphonochilus* Wood & Franks, which is not in general cultivation.

Leaves. Lanceolate-oblong, held erect: **1, 2, 4, 5**; more or less circular, held horizontally: **3, 6, 7**. Red or purple beneath: **1, 6**; green beneath: **2, 3, 4, 7**; variegated green and creamy white on both surfaces: **5**.
Flowers. Borne on a leafless shoot: **1, 2**; terminal on a leafy shoot: **3, 4, 5, 6, 7**. Lilac: **3, 4**; white: **6**; white with some lilac on the lip: **1, 2, 5, 7**.

1a. Flowers borne on a separate leafless shoot, usually before the leaves 2
b. Flowers terminal on the short leafy shoot 3
2a. Leaves variegated light and dark green above, purple beneath; lip *c.* 5 cm **1. rotunda**
b. Leaves plain green on both surfaces; lip *c.* 2.5 cm **2. parishii**
3a. Flowers lilac 4
b. Flowers white, usually with some lilac on the lip 5

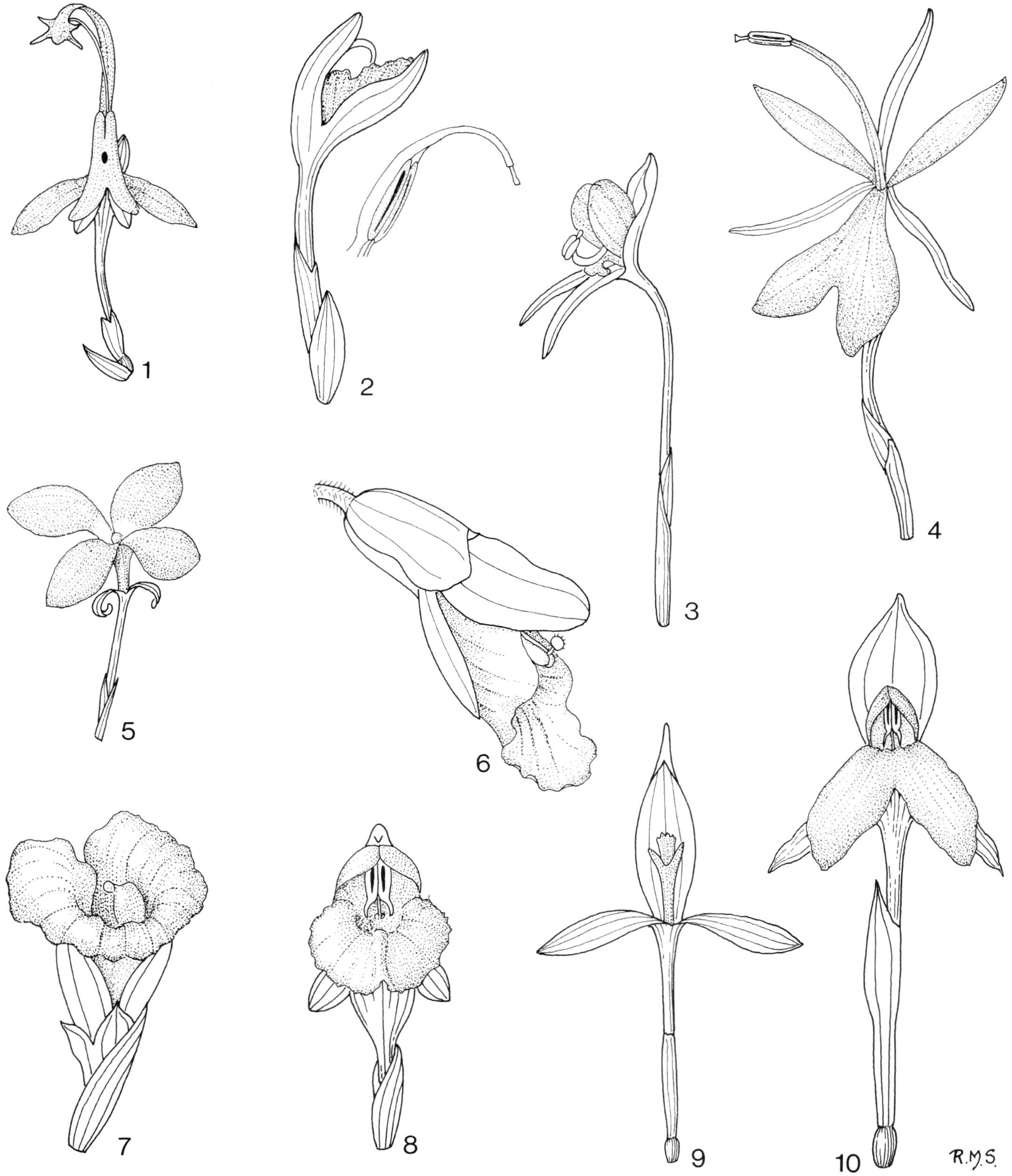

Figure 16. Flowers of the Zingiberaceae. 1, *Globba marantina*. 2, *Zingiber zerumbet* (with stamen enlarged). 3, *Brachychilum horsfieldii*. 4, *Hedychium coccineum*. 5, *Kaempferia elegans*. 6, *Alpinia zerumbet*. 7, *Costus afer*. 8, *Curcuma zedoaria*. 9, *Burbidgea schizocheila*. 10, *Roscoea purpurea*. Not to scale.

4a. Leaves held horizontally; stalk of inflorescence enclosed by the leaf-sheaths **3. pulchra**
b. Leaves more or less erect; stalk of inflorescence projecting beyond the leaf-sheaths **4. elegans**
5a. Leaves more or less erect, oblong-lanceolate, variegated creamy white and green **5. gilbertii**
b. Leaves held horizontally, more or less circular; if variegated then not as above 6
6a. Leaves green with paler green variegation above, reddish beneath; flowers pure white **6. roscoeana**
b. Leaves green and sometimes reddish on the margin above, paler green beneath; lip with lilac bands **7. galanga**

1. K. rotunda Linnaeus. Illustration: *Botanical Magazine*, 6054 (1873).

Leaves erect, short-stalked, to 45 cm, lanceolate, variegated green above, purple beneath. Inflorescence appearing separately from the leaves. Flowers white, lip lilac, deeply 2-lobed, lateral staminodes white. *SE Asia; widely cultivated, precise origin unknown.* G2. Summer.

2. K. parishii Hooker (*K. ovalifolia* Roscoe; *K. ovalifolia* Roxburgh misapplied). Illustration: *Botanical Magazine*, 5763 (1869).

Leaves erect, short-stalked, 15–20 cm, oblong-lanceolate. Inflorescence appearing separately from the leaves. Flowers white, lip bright purple, yellow in the throat, deeply 2-lobed. Lateral staminodes white. *Burma.* G2.

3. K. pulchra Ridley. Illustration: Brickell (ed.), RHS A–Z encyclopedia of garden plants, 593 (2003).

Leaves 2 or 3, held horizontally, to 14 cm, elliptic, dark green with paler variegation above. Inflorescence central to the leaf tuft and enclosed by the leaf-sheaths. Flowers lilac, lip deeply 2-lobed. *Thailand, Malay peninsula.* G2.

4. K. elegans (Wallich) Baker. Figure 16 (5), p. 423.

Leaves 1 or 2, more or less erect, to 15 cm, elliptic, sometimes with greyish spots above. Inflorescence central to the leaf tuft and shortly projecting from the leaf-sheaths. Flowers lilac. *Burma to Malay peninsula.* G2.

Very close to, and perhaps not distinct from, *K. pulchra.*

5. K. gilbertii Bull. Illustration: *Gartenflora* **32**: 215 (1883).

Leaves more or less erect, short-stalked, to 10 cm, oblong-lanceolate. Inflorescence central to the leaf tuft. Flowers white, the lip striped with violet, 2.5 cm, deeply 2-lobed. Lateral staminodes white. *India.* G2.

6. K. roscoeana Wallich. Illustration: *Botanical Magazine*, 1212 (1829).

Leaves usually 2, held horizontally, 9–11 × 8–9 cm, almost circular, deep green with pale green variegation above, reddish beneath. Inflorescence central to the leaf tuft. Flowers pure white, lip deeply lobed. *Burma.* G2. Autumn.

7. K. galanga Linnaeus. Illustration: *Botanical Magazine*, 850 (1805).

Leaves 2 or 3, held horizontally, to 15 × 10 cm, almost circular, green, sometimes with a red margin above, paler green beneath. Flowers white, lip with violet bands towards the base, deeply 2-lobed. *India.* G2. Summer–autumn.

5. BOESENBERGIA Kuntze

R.M. Smith

Short-stemmed plants with 1–several, usually long-stalked leaves. Inflorescence terminal. Flowers single in the axils of usually thin, lanceolate bracts. Bracteoles thin, non-tubular. Lateral staminodes petaloid. Lip often rather basin-shaped, entire or slightly lobed. Anther with or without a small crest.

A genus of about 30 species, all Asiatic, formerly known as *Gastrochilus* Wallich. Difficult in cultivation, both in pots and in borders. The plants are rarely vigorous and are particularly vulnerable to excessive watering during the winter. An open compost with a high proportion of peat appears to suit them best. Propagation is by division, but this is very slow.

1a. Leaves plain green; flowers pink **1. rotunda**
b. Leaves variegated; flowers white and yellow or yellow-orange 2
2a. Leaves dark green with a grey stripe along the mid-vein above, greyish green beneath; flowers white and yellow **2. vittata**
b. Leaves green with an irregular silver-white band along the mid-vein above, purple-brown beneath; flowers yellow and orange **3. ornata**

1. B. rotunda (Linnaeus) Mansfield (*B. pandurata* (Roxburgh) Schlechter; *Kaempferia pandurata* Roxburgh; *Gastrochilus panduratus* (Roxburgh) Ridley).

Leaves *c.* 4 per shoot, stalks 5–12 cm, blades to 28 × 10 cm, ovate-oblong, green. Flowers pink, lip *c.* 2.5 cm, slightly reflexed at the shortly 2-lobed tip. *Indonesia.* G2.

2. B. vittata (N.E. Brown) Loesener (*Kaempferia vittata* N.E. Brown; *Gastrochilus vittatus* (N.E. Brown) Valeton).

Leaves *c.* 4–6 per shoot, stalks 8–13 cm, blades 8–12 × 2–3 cm, elliptic, dark green and with a velvety sheen above, the mid-vein with a grey stripe along its length, grey-green beneath. Flowers white, lip *c.* 1.5 cm, basin-shaped, yellow in the centre. *Indonesia (Sumatra).* G2.

3. B. ornata (N.E. Brown) R.M. Smith (*Kaempferia ornata* N.E. Brown; *Gastrochilus ornatus* (N.E. Brown) Valeton). Illustration: *Illustration Horticole* **31**: t. 537 (1884).

Leaves several per shoot, stalks to 14 cm, blades 10–20 × 2.5–3 cm, lanceolate, upper surface with an irregular silver-white band along the mid-vein, purple-brown beneath. Flowers yellow, lip *c.* 1.8 cm, spotted with orange. *Borneo.* G2.

6. HEDYCHIUM Koenig

R.M. Smith

Leafy shoots to at least 2 m, with many stalkless or shortly stalked leaves. Inflorescence terminal, showy, with many flowers. Flowers 1–6, in the axils of overlapping or well-spaced bracts. Bracteoles usually tubular. Corolla tube slender, petals long and narrow, reflexed. Lip notched or deeply 2-lobed, narrowed at the base. Lateral staminodes petaloid. Filament usually long and slender. Anther without a crest.

A genus of about 40 species, all from Asia. *H. gardnerianum* may be grown outside in favoured localities: a sheltered border out of direct sunlight is ideal, with rich dressings of humus in the form of compost, peat or leaf-mould incorporated into the soil.

Literature: Schilling, T., A survey of cultivated Himalayan and Sino-Himalayan Hedychium species, *The Plantsman* **4**(3): 129–49 (1982).

Bracts. Broad and overlapping: **1, 2, 3, 4**; well spaced: **5, 6, 7, 8, 9, 10**.
Flowers. Deep orange or red: **3, 5, 6, 8, 9, 10**; white, often with a flush of another colour: **1, 2, 7**. One per bract: **5, 6**; 2–6 per bract: **1, 2, 3, 4, 7, 8, 9, 10**.

Calyx. Shorter than bract: **1, 2, 4, 5, 7, 9**; as long as, or longer than bract: **3, 6, 8, 10**.
Corolla tube. Shorter than to as long as bract: **4, 5**; longer than bract: **1, 2, 3, 6, 7, 8, 9, 10**.
Lip. Entire: **8**. About 5 cm across: **1**.
Stamen. Much exceeding lip: **4, 5, 8, 9, 10**; shorter than to scarcely longer than lip: **1, 2, 3, 5, 6, 7**. White: **1**.
Staminodes. Linear: **8**.

1a. Bracts broad and overlapping, hiding the main axis 2
b. Bracts narrow, well spaced, main axis clearly visible 5
2a. Flowers red; corolla tube hidden within the bract **4. greenei**
b. Flowers white, yellow or white flushed with yellow, orange or pink; corolla tube much longer than the bract 3
3a. Lip and staminodes yellow; calyx hairy; stamen distinctly longer than the lip **3. flavescens**
b. Lip and staminodes white, sometimes flushed with greenish yellow, yellow-orange or pink towards the base; calyx more or less hairless; stamen not, or scarcely longer than the lip 4
4a. Lip to 3.5 cm across, flushed with orange-yellow or pink **2. chrysoleucum**
b. Lip *c.* 5 cm across, yellow at base or the whole pure white **1. coronarium**
5a. Stamen shorter than or scarcely longer than lip 6
b. Stamen conspicuously longer than the lip 8
6a. Inflorescence narrowly cylindric, dense, at least 3 times as long as wide; corolla tube shorter than bract; flowers orange-red **5. densiflorum**
b. Inflorescence usually no more than 2 times as long as wide; corolla tube much longer than bract; flowers white or yellow-pink and white 7
7a. Flowers yellow-pink and white; bracts to 2.5 cm **6. spicatum**
b. Flowers white; bracts 4–5 cm **7. forrestii**
8a. Lip entire **8. speciosum**
b. Lip deeply 2-lobed 9
9a. Flowers white and yellow; leaves usually over 10 cm wide **9. gardnerianum**
b. Flowers red or varying shades of orange, red or purple; leaves to 4 cm wide **10. coccineum**

1. H. coronarium Koenig. Illustration: *Botanical Magazine*, 708 (1803).

Plant to 3 m. Leaves to 60 × 11 cm, lanceolate-acuminate. Inflorescence to 20 × 11 cm, bracts 4–6 cm, overlapping each other, each subtending 2–6 flowers. Flowers white. Calyx hidden by the bract. Corolla tube to 7 cm. Staminodes 3–5 cm, oblong-lanceolate. Lip about 5 cm across, usually yellow-green in the centre, deeply 2-lobed. Stamen white, shorter than the lip. Flowers very fragrant. *India to Indonesia.* G2. Spring.

Widely cultivated in the tropics. Var. **maximum** (Roscoe) Baker (illustration: *Edwards's Botanical Register* **12**: t. 1022, 1826) is also cultivated; it has broader leaves, bracts with conspicuously hairy margins, larger flowers and the filament tinged with pink. It originated in India.

2. H. chrysoleucum Hooker (*H. coronarium* Koenig var. *chrysoleucum* (Hooker) Baker). Illustration: *Botanical Magazine*, 4516 (1850).

Stems to 1.5 m. Leaves to 40 × 8 cm, oblong-lanceolate, acuminate. Inflorescence to 20 cm. Bracts *c.* 3 cm, ovate to lanceolate, overlapping each other, each subtending 2–6 flowers. Flowers white, calyx hidden by the bract. Corolla tube to 8 cm. Staminodes to 4 cm, flushed with orange-yellow or pink. Lip to 3.5 cm across, coloured as the staminodes, 2-lobed. Stamen as long as or scarcely longer than the lip, orange-pink. *India.* G2. Summer.

3. H. flavescens Roscoe (*H. flavum* Roxburgh misapplied; *H. coronarium* Koenig var. *flavescens* (Roscoe) Baker).

Stems to 3 m. Leaves to 60 × 8 cm, lanceolate, acuminate. Inflorescence to 20 × 8 cm. Bracts *c.* 3.5 cm, ovate, overlapping each other, each subtending at least 4 flowers. Flowers yellow, calyx about the same length as the bract. Corolla tube 7–8 cm. Staminodes lanceolate, 2.5–3 cm. Lip *c.* 3 cm across, 2-lobed. Stamen longer than the lip, yellow-orange. *Himalaya.* G2. Spring.

4. H. greenei W.W. Smith.

Stems to 2 m. Leaves 20–25 × *c.* 5 cm, oblong, acuminate. Inflorescence to *c.* 12 cm. Bracts 5–7 cm, ovate, overlapping each other, each subtending 2–3 flowers. Flowers red, calyx and corolla tube shorter than the bracts. Staminodes 1.5–2 cm, spathulate. Lip 3–4 cm across, darker than the rest, shortly 2-lobed. Stamen shorter than the lip, red. *Himalaya.* G2.

This species is said to form bulbils.

5. H. densiflorum Wallich. Illustration: *Botanical Magazine*, n.s., 325 (1958); Brickell (ed.), RHS A–Z encyclopedia of garden plants, 511 (2003).

Stems to 5 m. Leaves 30–40 × 5–6 cm, oblong-lanceolate. Inflorescence to 14 × 4 cm, dense, cylindric. Bracts *c.* 1.5 cm, oblong, not overlapping, each usually subtending a single flower. Flowers orange-red, calyx and corolla tube shorter than the bract. Staminodes elliptic, *c.* 1 cm. Lip to 7 mm across. Stamen red, just longer than the lip. *Himalaya.* G2.

6. H. spicatum J.E. Smith. Illustration: *Botanical Magazine*, 2300 (1822).

Stems to 1 m. Leaves 10–40 × 3–10 cm, oblong-lanceolate, acuminate. Inflorescence to 20 cm, loose. Bracts to 2.5 cm, lanceolate, well spaced, each subtending a single flower. Flowers yellow, pink and white, calyx longer than the bract. Corolla tube to 8 cm. Staminodes *c.* 1 cm, white and pink, spathulate. Lip 1.5–2 cm across, yellow, 2-lobed. Stamen shorter than the lip, red-orange. *Himalaya.* G2. Autumn.

7. H. forrestii Diels.

Stems to 1.25 m. Leaves 30–50 cm, narrowly lanceolate. Inflorescence to *c.* 25 cm, with many flowers. Bracts 4–5 cm, spreading, each subtending 2–3 flowers. Flowers white, sometimes flushed with pink, calyx shorter than the bract. Corolla tube 3–4.5 cm. Staminodes elliptic. Lip *c.* 3 cm across, 2-lobed. Stamen not exceeding the lip. *SW China (Yunnan).* G2.

8. H. speciosum Wallich. Illustration: Wallich, Plantae Asiaticae rariores **3**: t. 285 (1832).

Stems to 2.5 m. Leaves to 30 × 9 cm, oblong-lanceolate, acuminate. Inflorescence to 30 cm, moderately dense. Bracts 3–4 cm, oblong-ovate, well spaced, each subtending 2 flowers. Flowers yellow, calyx about as long as the bract. Corolla tube 4–5 cm. Staminodes linear, shorter than the petals. Lip *c.* 2.5 × 1.5 cm, oblong, entire. Stamen red, longer than the lip. *Himalaya.* G2. Summer.

9. H. gardnerianum Ker Gawler. Illustration: *Edwards's Botanical Register* **9**: t. 774 (1824); Brickell (ed.), RHS A–Z encyclopedia of garden plants, 512 (2003).

Stems to 2 m. Leaves 25–40 × 10–15 cm, lanceolate, acuminate. Inflorescence 25–35 cm, with many flowers. Bracts 3–5 cm, oblong, each subtending 1–2 flowers. Flowers white and yellow, calyx shorter than the bract. Corolla tube 5–6 cm. Staminodes 2.5–3 cm, lanceolate. Lip 1–2 cm across, 2-lobed. Stamen bright red, longer than the lip. *Himalaya*. H5–G1. Summer–autumn.

10. H. coccineum J.E. Smith. Figure 16 (4), p. 423. Illustration: *Edwards's Botanical Register* **14**: t. 1209 (1829).

Stems to 3 m. Leaves 30–50 × *c.* 3.4 cm, linear-lanceolate, acuminate. Inflorescence at least 20 cm, with many flowers. Bracts 1.5–2.5 cm, oblong, well spaced, each subtending 2–4 flowers. Flowers orange-red to purple, calyx and corolla tube about the same length as the bract. Staminodes *c.* 2 cm, oblong. Lip 2 cm across, deeply 2-lobed. Stamen much longer than lip. *Himalaya*. G2. Autumn.

Very variable in flower colour and leaf size. Var. **angustifolium** Roxburgh has very narrow leaves; var. **carneum** Roscoe has leaves to 6 cm wide, white to pink flowers and a spotted lip. The orange-flowered *H. aurantiacum* Roscoe is perhaps merely a variety of *H. coccineum*.

7. BRACHYCHILUM Petersen

R.M. Smith

Stems to 1 m. Leaves short-stalked, lanceolate to linear-lanceolate. Inflorescence terminal on the leafy shoot, rather loose. Bracts ovate, each subtending at least 2 flowers. Bracteoles tubular. Corolla tube long and slender. Petals hanging, twisted. Lip minute. Lateral staminodes elliptic. Stamen shorter than the petals.

A genus of only 1 species, closely related to *Hedychium*.

1. B. horsfieldii (R. Brown) Petersen. Figure 16(3), p. 423.

Leaves 18–30 cm. Flowers yellow, lip 3–4 mm, occasionally entirely suppressed. Stamen orange. Capsule orange, red within. *Indonesia (Java)*. G2.

8. ROSCOEA J.E. Smith

E.J. Cowley

Inflorescence terminal on the leafy stem, sometimes long-stalked. Bracts overlapping, each subtending a single flower. Bracteoles absent. Upper petal hooded. Lip entire or 2-lobed, sometimes each lobe notched. Staminodes petaloid. Anther versatile, spurred at the base. Fruit elongate, not splitting readily.

A genus of 17 species from China and the Himalaya. Plants are hardy in western Europe, and require a cool situation out of direct sunlight. They respond to rich feeding and compost, peat or leaf-mould should be incorporated into the soil when planting. Propagation by division in the spring, or by seed.

Literature: Cowley, E.J., A revision of the genus Roscoea, *Kew Bulletin* **36**: 747–77 (1982); Cowley, J., The genus Roscoea (2007).

Leaves. Conspicuously auricled at base: **2**.
Inflorescence. Usually stalked: **3, 6**.
Bracts. The lowest tubular: **3, 4**. Hairy at the top: **3, 5**.
Calyx lobes. Broad, rounded: **5**.
Corolla tube. Projecting from calyx: **5, 6**.
Petals. Pink: **3, 6**; yellow: **4**; purple and/or white: **1, 2, 5, 6**. The uppermost circular: **6**.
Anther spurs. With rounded ends: **4**.

1a. Bracts acute at apex 2
b. Bracts obtuse to truncate at apex 5
2a. Staminodes at least 2.5 cm, narrow with a long claw **1. purpurea**
b. Staminodes to 2 cm, relatively broad, with a short claw 3
3a. Leaves conspicuously auricled at the base **2. auriculata**
b. Leaves not auricled at the base 4
4a. Bracts equal to, or longer than the calyx, hairy at the top **3. scillifolia**
b. Bracts shorter than the calyx, hairless **4. cautleyoides**
5a. Upper petal broadest above the middle, 2.5 cm or more, mucronate at the apex **5. humeana**
b. Upper petal circular or elliptic, less than 2.2 cm, not mucronate at the apex 6
6a. Upper petal circular; first bract much shorter than the calyx **6. alpina**
b. Upper petal elliptic; first bract equal to or much longer than the calyx **3. scillifolia**

1. R. purpurea J.E. Smith (*R. procera* Wallich; *R. purpurea* var. *procera* (Wallich) Baker). Figure 16(10), p. 423. Illustration: *Edwards's Botanical Register* **26**: t. 61 (1840); Hooker, Exotic flora **2**: t. 144 (1825); Brickell (ed.), RHS A–Z encyclopedia of garden plants, 937 (2003).

Stems to 55 cm, sometimes flushed purple. Leaves 4–8, lanceolate to oblong-ovate, 7–25 cm. Inflorescence enclosed in the upper leaf-sheaths. Flowers pale purple or white with dark purple markings, 1 open at a time; upper petal narrow, 3.2–6 × 1–2.8 cm; lip 4.5–6.5 cm including the claw, which is 8–12 mm, entire or 2-lobed. Staminodes 2.5–4 cm, narrow. *Himalaya*. H5. Summer–autumn.

2. R. auriculata Schumann (*R. sikkimensis* invalid; *R. purpurea* misapplied). Illustration: *The Garden* **78**: 158 (1914).

Stems to 55 cm, sometimes flushed with purple. Leaves 3–10, linear to broadly lanceolate, auriculate at the base, 5–27 cm. Inflorescence enclosed in the upper leaf-sheaths. Flowers bright purple, 1 open at a time; upper petal broad, 2.8–3.7 × 1.5–2.8 cm. Lip 3.3–4.8 cm including the claw, which is 7–12 mm, deflexed. Staminodes white, 1.5–2 cm. *E Nepal, NE India (Sikkim)*. H5. Summer–autumn.

R. 'Beesiana', mentioned in nurserymen's catalogues, is a possible hybrid between *R. auriculata* and *R. cautleoides*, having stems up to 43 cm, with conspicuously auriculate leaf-bases and yellow flowers, sometimes with purple streaks. There is no known wild species that matches this description.

3. R. scillifolia (Gagnepain) Cowley (*R. capitata* var. *scillifolia* misapplied; *R. capitata* misapplied). Illustration: *New Flora and Silva* **11**: opposite p. 22 (1938).

Stems to 37 cm. Leaves 1–5, sometimes not fully developed at flowering, linear to linear-lanceolate, 6–12 cm. Inflorescence usually exposed on a stalk. First bract tubular, hairy at the top. Flowers pink, 1 open at a time. Upper petal elliptic, 1.4–2 cm × 6–10 mm. Lip narrow, entire or 2-lobed, 1.3–2 cm. *SW China (Yunnan)*. H5. Summer.

4. R. cautleyoides Gagnepain (*R. chamael* Gagnepain; *R. yunnanensis* Loesener; *R. sinopurpurea* Stapf invalid). Illustration: *Botanical Magazine*, 9084 (1926); Brickell (ed.), RHS A–Z encyclopedia of garden plants, 936 (2003).

Stems to 55 cm. Leaves 1–4, sometimes not fully developed at flowering, linear to lanceolate, 7.5–42 cm. Inflorescence exposed on a stalk. First bract tubular.

Flowers yellow, usually more than 1 open at a time; upper petal wider above the middle, mucronate, 2–3.8 × 1–2.7 cm. Lip deflexed, 2.6–4 cm (including the claw). Staminodes 1.2–1.8 cm. *SW China (Yunnan, Sichuan).* H5. Spring.

5. R. humeana Balfour & Smith. Illustration: *Botanical Magazine*, 9075 (1925); Brickell (ed.), RHS A–Z encyclopedia of garden plants, 936 (2003).

Stems to 34 cm. Leaves 1–2 (rarely 3), usually not fully developed at flowering, oblong to ovate, 2–22 cm. Inflorescence enclosed in the upper leaf-sheaths. Bracts hairy at the top, much shorter than the broadly lobed calyx. Flowers purple, many open at the same time. Corolla tube clearly projecting from the calyx. Upper petal wider above the middle, mucronate, 2.5–4.5 × 1.9–3 cm. Lip 2-lobed, deflexed, 2.2–4.5 × 2.8–3.5 cm including the claw. Staminodes small, shortly clawed. *SW China (Yunnan, Sichuan).* H5. Spring.

There is also a yellow-flowered form of this species.

6. R. alpina Royle (*R. longifolia* Baker; *R. intermedia* Gagnepain). Illustration: Royle, Illustrations of the botany of the Himalayan Mountains, t. 89 (1839).

Stems to 30 cm. Leaves 1–2, usually not fully developed at flowering, up to 4 after flowering. Inflorescence very short, not exposed. Flowers pink to purple, 1 open at a time. Corolla tube clearly projecting from the calyx. Upper petal circular, 1.3–2.2 × 1.1–2.4 cm. Lip narrow, 1.4–2.6 × 1–2 cm. Staminodes circular to elliptic, 6–12 × 4–8 mm. *Himalaya.* H5. Summer.

9. CAUTLEYA Hooker
R.M. Smith

Stems to 1 m. Inflorescence terminal on the leafy stem. Bracts each subtending a single yellow flower. Bracteoles absent. Lip deeply 2-lobed. Staminodes petaloid. Anther versatile, spurred at the base. Fruit splitting readily.

A genus of 2 or 3 species from the Himalaya. Cultivation as for *Roscoea*, but plants are less hardy and can be grown only in mild localities in sheltered borders.

1a. Bracts green, much shorter than the calyx; inflorescence with up to 12 flowers; plant very slender **1. gracilis**
b. Bracts green tinged with red, or entirely red, equalling or longer than the calyx; inflorescence with many flowers; plant robust **2. spicata**

1. C. gracilis (J.E. Smith) Dandy (*C. lutea* (Royle) Hooker). Illustration: *Botanical Magazine*, 6991 (1888).

Slender plant to 40 cm. Leaves 4–6, stalkless, up to 20 cm, ovate to oblong-lanceolate, usually purple beneath. Inflorescence axis dark red. *Himalaya.* H5–G1. Summer.

2. C. spicata (J.E. Smith) Baker. Illustration: Brickell (ed.), RHS A–Z encyclopedia of garden plants, 241 (2003)

Stems to 1 m. Leaves more or less stalkless or with stalks to *c.* 3 cm, the whole to 30 cm, lanceolate to oblong-lanceolate. Inflorescence robust, with many flowers, axis reddish green or dark red. *Himalaya.* H5–G1. Summer.

The plant often grown as *C. robusta* (a misapplied name) is probably no more than a larger form of *C. spicata.*

10. CURCUMA Linnaeus
R.M. Smith

Plants to 1 m. Inflorescence terminal on a usually short-stemmed leafy shoot or on a separate leafless shoot and sometimes appearing before the leaves. Bracts always of 2 kinds: flower-bearing (fertile), which are joined to each other for about a third to a half of their length forming pouches containing 2–7 flowers, and sterile (coma), which arise at the top of the inflorescence and are usually of a different colour. Bracteoles open to the base. Lip with a thickened central part. Lateral staminodes petaloid, folded under the hooded upper petal. Stamen with a versatile, usually spurred, anther.

A genus of perhaps 40 species from SE Asia, including the spice, turmeric (*C. longa*). This and *C. zedoaria* have been cultivated in SE Asia for a very long time, and are especially difficult for the botanist to classify. The modern Flora of Java Backer, C. A. & Bankhuizen van den Brink, R. C., Flora of Java **3**: 69, (1968) regards both as 'collective species', and a similar treatment has been followed here. More recent work in Indonesia suggests that many of these plants are sexually sterile triploids or pentaploids. Cultivation essentially as for *Globba* (p. 421).

Stem. Less than 45 cm: **2, 6**; larger: **1, 3, 4, 5, 7, 8**.
Leaves. Rounded or cordate at base: **4**. Downy beneath: **7**.
Bracts. All orange-red: **1**. Sterile bracts absent: **2**; sterile bracts whitish green: **3**; sterile bracts tinged with purple-pink: **4, 5, 6, 7, 8**.
Anther spurs. Absent: **1**.

1a. Bracts all orange-red; anther spurless **1. roscoeana**
b. Bracts variously coloured, never orange; anther spurred 2
2a. Inflorescence without sterile bracts at the top; plant to 30 cm **2. albiflora**
b. Inflorescence with sterile bracts at the top; plant larger 3
3a. Inflorescence terminal on the leafy shoot 4
b. Inflorescence borne on a leafless shoot 6
4a. Sterile bracts white or greenish white **3. longa**
b. Sterile bracts with at least some purple-pink markings 5
5a. Leaves rounded to cordate at the base **4. petiolata**
b. Leaves narrowed towards the base **5. amada**
6a. Plant slender, to 45 cm; rhizome small **6. angustifolia**
b. Plant much larger; rhizome robust 7
7a. Leaves downy beneath, often with light green variegation **7. aromatica**
b. Leaves hairless beneath, usually with a purple-brown 'cloud' down the centre, occasionally plain green **8. zedoaria**

1. C. roscoeana Wallich. Illustration: *Botanical Magazine*, 4667 (1852); Brickell (ed.), RHS A–Z encyclopedia of garden plants, 333 (2003).

Stems to *c.* 90 cm. Leaves long-stalked, blades to 30 cm. Inflorescence terminal, fertile and sterile bracts orange. Flowers yellow. Anther lacking basal spurs but with a prominent crest. *Burma.* G2. Summer–autumn.

2. C. albiflora Thwaites. Illustration: *Botanical Magazine*, 5909 (1871).

Stems to *c.* 30 cm. Leaves shortly stalked, blades to 15 cm. Inflorescence either terminal or lateral to the developing leaves. Sterile bracts absent, fertile bracts green. Flowers white, lip tinged with yellow. *Sri Lanka.* G2.

3. C. longa Linnaeus (*C. domestica* Valeton; *C. purpurascens* Blume).

Stems *c.* 1 m. Leaves long-stalked, blades to 50 cm. Inflorescence terminal. Sterile bracts white or greenish white, fertile bracts pale green. Flowers white and yellow. *Origin unknown, widely cultivated.* G2.

The turmeric of commerce.

4. C. petiolata Roxburgh. Illustration: *Botanical Magazine*, 5821 (1870).

Stems to 60 cm. Leaves long-stalked, blades to 25 cm, rounded or cordate at the base. Inflorescence terminal, sterile bracts purple, fertile bracts green. Flowers yellow and white. Anther spurs rather small, but distinct. *Burma.* G2.

5. C. amada Roxburgh. Illustration: Roscoe, Monandrian plants, t. 99 (1826).

Stems to 60 cm. Leaves long-stalked, blades to 40 cm. Inflorescence terminal. Sterile bracts pink-tinged, fertile bracts green. Flowers yellow and white. *India.* G2.

6. C. angustifolia Roxburgh. Illustration: *Asiatic Researches* **11**: t. 3 (1810).

Slender plant, stems to *c.* 45 cm. Leaves short-stalked, blades to 40 cm. Inflorescence borne separately from and before the leaves. Sterile bracts purple-pink, fertile bracts green. Flowers yellow. *Himalaya.* G2.

7. C. aromatica Salisbury. Illustration: *Botanical Magazine*, 1564 (1813).

Stems to 1 m. Leaves very long-stalked, blades to 60 cm, usually variegated light and dark green. Inflorescence borne separately from and before the leaves. Sterile bracts pink-tinged, fertile bracts pale greenish white. Flowers yellow-white, petals tinged with pink. *?India; widely naturalised.* G2.

8. C. zedoaria (Christmann) Roscoe (*C. heyneana* Valeton; *C. mangga* Valeton & van Zijp; *C. xanthorrhiza* Roxburgh). Figure 16(8), p. 423. Illustration: Roscoe, Monandrian plants, t. 109 (1807).

Stems to 1.5 m. Leaves long-stalked, blades to 80 cm, usually with a purple-brown coloration ('cloud') running down either side of the mid-vein. Inflorescence borne separately from and usually appearing before the leaves. Sterile bracts whitish, stained with pink or red, fertile bracts pale green with reddish margins. Flowers white and pale yellow usually tinged with pink. *SE Asia; precise origin unknown, widely cultivated and naturalised.* G2.

11. AMOMUM Roxburgh

R.M. Smith

Leafy shoots well developed. Inflorescence borne separately on a leafless stalk, usually cone-like. Bracts each subtending a single flower. Bracteoles usually tubular, rarely absent. Lip showy. Lateral staminodes reduced to very small appendages, or absent. Anther often crested.

A genus of about 90 species from Asia and Australasia.

1. A. compactum Maton (*A. cardamomum* Linnaeus misapplied; *A. kepulaga* Sprague & Burkill).

Stems to 3 m. Leaves stalkless, to 25 cm, narrowly lanceolate. Inflorescence congested, held on a stalk that is 5–10 cm. Flowers yellow. Lip *c.* 1 cm, elliptic. Anther crest 3-lobed. *Indonesia.* G2.

The round cardamom of Java.

12. AFRAMOMUM Schumann

R.M. Smith

Leafy shoots well developed. Inflorescence borne separately on a leafless stalk, congested. Bracts each subtending a single flower. Bracteoles open to the base. Lip (in the cultivated species) very showy, almost circular. Lateral staminodes reduced to small appendages. Anther crested.

A genus of perhaps 50 species from tropical Africa, the Mascarenes and Madagascar. Cultivation as for *Costus* (p. 430).

1. A. melegueta Schumann.

Stems to 2 m. Leaves more or less stalkless, to 20 cm, linear. Inflorescence fusiform, with few flowers. Flowers pale violet, the lip yellow in the centre. Fruit flask-shaped, bright red. *W Africa.* G2.

13. NICOLAIA Horaninow

R.M. Smith

Leafy shoots to 4 m. Inflorescence borne on a separate, erect, leafless stalk, surrounded by a conspicuous involucre of sterile bracts. Fertile bracts each with a single, tubular bracteole and a single flower. Lip short, erect, joined at the base to the lower part of the stamen and forming a distinct tube above the petals. Lateral staminodes absent. Anther without a crest.

A genus of 7 or 8 species from Malaysia, Indonesia and New Guinea, formerly called *Phaeomeria* Lindley.

1. N. elatior (Jack) Horaninow (*Phaeomeria speciosa* (Blume) Merrill; *P. magnifica* (Roscoe) Schumann; *N. imperialis* Horaninow; *N. speciosa* (Blume) Horaninow). Illustration: *Botanical Magazine*, 3192 (1832).

Inflorescence stalk to 1 m, head pyramidal, to 12 cm. Sterile bracts brownish red to pink, spreading. Flowers pink, lip deep brownish red with a white or yellow margin. Fruit spherical. *Malaysia.* G2.

14. ELETTARIA Maton

R.M. Smith

Leafy shoots to *c.* 4 m. Inflorescence borne separately from the leaves on a long, prostrate axis (rarely erect), often subterranean. Bracts well spaced, each subtending several flowers with tubular bracteoles. Flowers not showy, lateral staminodes reduced to small appendages. Anther with a small crest. Fruit spherical or oblong.

A genus of 3 or 4 species from tropical Asia. Cultivation as for *Boesenbergia* (p. 424).

1. E. cardamomum (Linnaeus) Maton.

Leaves to 1 m. Flowers white, lip streaked with violet. *India.* G2.

The fruits are the cardamoms of commerce.

15. BURBIDGEA Hooker

R.M. Smith

Leafy stems to 1 m. Inflorescence terminal on the leafy shoot, bracts absent or falling very quickly. Flowers borne singly (rarely in pairs), orange or yellow-orange, sometimes tinged with pink. Petals more conspicuous than the narrow, erect lip. Lateral staminodes absent. Anther with an elongate crest. Fruit elongate.

A genus of 5 or 6 species from Borneo.

Literature: Smith, R.M., The genus Burbidgea, *Notes from the Royal Botanic Garden, Edinburgh* **31**: 297–306 (1972).

1. B. schizocheila Hackett (*B. nitida* Hooker, misapplied). Figure 16(9), p. 423. Illustration: *Botanical Magazine*, 8009 (1905).

Stems to 50 cm. Leaves to 15 cm, elliptic, purplish beneath. Flowers orange. *Borneo.* G2.

16. ALPINIA Roxburgh

R.M. Smith

Leafy shoot well developed, simple or branched. Inflorescence terminal on the leafy shoot. Flowers arising singly, in pairs, or in spikes with few to many flowers; often in the axils of bracts, though these are sometimes absent. Bracteoles tubular, open to the base, or absent. Lip often showy. Lateral staminodes reduced to small appendages, occasionally absent. Fruit usually spherical. Some species were formerly placed in *Catimbium* Jussieu or *Languas* Small.

A very polymorphic genus of about 200 species from Asia and Australasia.

Leaves. Hairy all over the lower surface: **3, 5, 6**; hairless on lower surface: **1, 2, 4, 5, 7, 8, 9**. With small bristles on the margins: **4**.
Inflorescence. Erect: **1, 6, 7, 8, 9**; drooping or horizontal: **2, 3, 4, 5**.
Flowers. Borne singly: **6, 8**; borne 2 or more together: **1, 2, 3, 4, 5, 7, 9**.
Bract. Absent: **4, 5, 6, 7**; small, falling quickly: **4, 8, 9**; conspicuous and persistent: **1, 2, 3**. Bright red: **1**. Green and pink: **2**.
Bracteoles. Tubular: **1, 2, 3**; open to the base (often quickly deciduous): **4, 5, 6, 7, 8, 9**.

1a. Bracts bright red, persistent, inflorescence erect; flowers white **1. purpurata**
b. Bracts, if present, never bright red, if persistent then inflorescence arching over; flowers never pure white 2
2a. Leaves variegated; bracts persistent and conspicuous, green tinged pink **2. vittata**
b. Leaves plain green; bracts, if present, not conspicuous and persistent 3
3a. Inflorescence tightly congested; flowers orange **3. rafflesiana**
b. Inflorescence loose; flowers variously coloured, never orange 4
4a. Lip 2.5–6 cm, yellow heavily patterned with red 5
b. Lip to 2 cm, white lined with pink or red 8
5a. Leaves up to 2.5 cm wide, margins with well spaced bristles **4. calcarata**
b. Leaves 5–20 cm, margins sometimes hairy but not with bristles 6
6a. Inflorescence drooping **5. zerumbet**
b. Inflorescence erect 7
7a. Leaves softly hairy beneath; lip to 6 cm **6. malaccensis**
b. Leaves more or less hairless, except for the margin; lip to 3.5 cm **7. mutica**
8a. Inflorescence simple; flowers produced singly **8. officinarum**
b. Inflorescence often branched; flowers 3–5 **9. galanga**

1. A. purpurata (Vieillard) Schumann (*Guillainia purpurata* Vieillard). Illustration: Brickell (ed.), RHS A–Z encyclopedia of garden plants, 103 (2003).

Stems 3–4 m. Leaves stalked, to 80 cm, lanceolate. Inflorescence to 50 cm, erect, cylindric. Bracts bright red, each subtending 4–5 white flowers. Bracteoles tubular. Corolla tube slender, lip petaloid, usually 3-lobed. Lateral staminodes absent. Stamen with a small, almost circular crest. Capsule spherical. *Islands of SE Pacific.* G2.

Occasionally short leafy stems may replace the flowers in the axils of the lowermost bracts; these shoots may terminate in small inflorescences. In cultivation, completely sterile inflorescences can occur.

2. A. vittata Bull (*A. sanderae* Sander; *A. tricolor* Sander). Illustration: Graf, Exotica, edn 8, t. 268A, 1490 (1976).

Stems to *c.* 1 m. Leaves to *c.* 15 cm, narrowly lanceolate, green with white or cream variegation. Inflorescence nodding, to *c.* 15 cm. Bracts well spaced and persistent, green tinged with pink, each with 2–5 greenish flowers. Bracteoles tubular. Lip rather fleshy, not petaloid. *Islands of the SE Pacific.* G2.

Grown mainly for its attractive foliage.

3. A. rafflesiana Baker. Illustration: *Hooker's Icones Plantarum*, t. 1963 (1891).

Stems to *c.* 1.5 m. Leaves with stalks to 4 cm, blades to 60 cm, lanceolate, hairy on both surfaces. Inflorescence to 8 cm, tightly congested, held horizontally. Bracts 5–10 mm, each subtending 1–3 orange flowers. Bracteoles tubular. Lip 3-lobed. Lateral staminodes reduced to triangular appendages. Stamen with a minute tooth on either side at the top of the anther. Capsule spherical, green. *Malaysia.* G2.

4. A. calcarata Roscoe (*A. cernua* Sims). Illustration: *Botanical Magazine*, 1900 (1817).

Stems to 1 m, of rather slender habit. Leaves stalkless, to 30 cm, narrowly lanceolate, the margins with well-spaced, minute bristles. Inflorescence to 10 cm, held horizontally. Bracts absent, or very small and quickly deciduous. Bracteoles open to the base, quickly deciduous. Flowers usually arising in pairs. Lip to 3 cm, yellow, strongly patterned with red-purple lines. Lateral staminodes reduced to small appendages. Anther without a crest. *?China.* G2.

5. A. zerumbet (Persoon) Burtt & Smith (*A. speciosa* (Wendland) Schumann; *A. nutans* misapplied). Figure 16(6), p. 423. Illustration: *Botanical Magazine*, 1817 (1816); Brickell (ed.), RHS A–Z encyclopedia of garden plants, 103 (2003).

Stems to 3 m. Leaves more or less stalkless, to 60 cm, lanceolate, margin always hairy, the lower surface sometimes so. Inflorescence to 40 cm, drooping, main axis very hairy. Bracts absent. Flowers usually in pairs, each with a pink-tinged bracteole that enfolds the bud. Corolla white. Lip to 4 cm, yellow heavily marked with brownish red. Lateral staminodes reduced to small appendages, exceptionally up to 1 cm. Anther without a crest. Capsule spherical, red. *Tropical and subtropical Asia.* G2. Spring–summer.

6. A. malaccensis (Burmann) Roscoe. Illustration: *Edwards's Botanical Register* **4**: t. 328 (1818).

Very similar to *A. zerumbet* but leaves always softly hairy beneath, inflorescence erect and flowers borne singly. *India.* G2.

7. A. mutica Roxburgh. Illustration: *Botanical Magazine*, 8621 (1915).

Similar to *A. zerumbet* and *A. malaccensis* but less robust. Leaves to 50 cm, margins hairy. Inflorescence erect. Flowers arising in 2 s or 3 s. Bracteoles falling very quickly. Lip 3–3.5 cm. Lateral staminodes absent. *India, Malaysia.* G2.

8. A. officinarum Hance. Illustration: *Botanical Magazine*, 6995 (1888).

Stems to 1.5 m. Leaves more or less stalkless, to 30 cm, linear-lanceolate. Inflorescence erect. Flowers arising singly. Bracts minute, falling quickly. Bracteoles absent. Lip to 2 cm, white with red veins. Lateral staminodes reduced to small appendages. *China.* G2.

9. A. galanga (Linnaeus) Willdenow (*A. zingiberina* Hooker).

Stems to *c.* 2 m. Leaves very shortly stalked, to 50 cm, lanceolate. Inflorescence erect, usually branched; bracts small and falling quickly, each with 3–5 flowers. Bracteoles inconspicuous, open to the base. Corolla greenish white. Lip *c.* 2.5 cm, white with pinkish lines. Lateral staminodes reduced to small appendages. Capsule spherical, red. *Tropical and subtropical Asia.* G2.

57. COSTACEAE

Perennial herbs with rhizomes. Leaves spirally arranged, with closed sheaths (sometimes opened in later growth), not aromatic. Flowers as for *Zingiberaceae*, but staminodes 3, the laterals absent or reduced to short teeth.

A family of 5 genera and about 160 species, a few cultivated.

1. COSTUS Linnaeus
R.M. Smith
Plants to 3 m. Leaves spirally arranged. Inflorescence terminal or on a separate leafless shoot, forming a dense head of flowers. Bracts often with a leaf-like extension at tip, each subtending 1 or 2 flowers. Bracteoles tubular or open to the base. Flowers showy, lip more or less trumpet-shaped. Lateral staminodes absent, stamen petal-like, the anther usually attached in the middle.

A genus of about 90 species, the majority from tropical America and Africa, a few occurring in tropical Asia and Australia. The genus requires hot and humid conditions where a night minimum temperature of 18 °C can be maintained. Direct planting in a border, where the plants can enjoy a free root-run, is preferable to pot culture.

Regular mulching with peat and rich feeding are essential for healthy growth. Plants in pots suffer from root restriction and usually suffer heavy infestations of red spider mite. Propagation by division.

Leaves. With light and dark green bands: **2**. Sheaths with a ring of bristles near the top: **4**.
Bracts. Subtending 1 flower: **1, 2, 3**; subtending 2 flowers: **4, 5**. Tips leaf-like: **1**. Tips spiny: **3**.
Flowers. Plain orange: **1**.

1a. Plant to 50 cm; flowers plain orange **1. cuspidatus**
b. Plant to 1 m or more; flowers variously coloured, never plain orange 2
2a. Flowers golden yellow with orange-red venation; leaves densely downy, dark green with lighter bands between the veins above **2. malortieanus**
b. Flowers white, variously marked; leaves downy or not, not banded 3
3a. Bracts spine-tipped; flowers 1 per bract; lip white to pinkish white **3. speciosus**
b. Bracts never spine-tipped; flowers 2 per bract; lip white tinged with yellow, sometimes with purple stripes 4
4a. Leaf-sheaths with a row of bristles just below the top; calyx longer than the bracteole **4. lucanusianus**
b. Leaf-sheaths without bristles; calyx shorter than the bracteole **5. afer**

1. C. cuspidatus (Nees & Martius) Maas (*C. igneus* R. Brown). Illustration: *Botanical Magazine*, 6821 (1885).

Plant to 50 cm, stems usually purple-red. Leaves very short-stalked, 10–20 × 4–8 cm, elliptic, with a long, narrow tip. Inflorescence terminal, bracts with leaf-like tips. Flowers 1 per bract, orange, each with a tubular bracteole. Lip more or less circular, 5–7 cm across when spread out. *Brazil.* G2. Flowering throughout the year.

2. C. malortieanus Wendland. Illustration: *Botanical Magazine*, 5894 (1871); Brickell (ed.), RHS A–Z encyclopedia of garden plants, 314 (2003).

Plant to 1 m. Leaves short-stalked, to *c.* 20 × 18 cm, broadly elliptic, upper surface densely downy, banded light and dark green. Inflorescence terminal, bracts rounded at their tips. Flowers 1 per bract, yellow, each with a tubular bracteole. Lip broadly ovate, to 5 cm across when spread out, striped with red. *C America.* G2. Flowering throughout the year.

3. C. speciosus (Koenig) J.E. Smith.

Plant to 3 m. Leaves short-stalked, 12–25 × 3–6 cm, narrowly elliptic with pointed tips, softly hairy beneath. Inflorescence terminal, bracts with spine-like tips. Flowers 1 per bract, white or pinkish white, each with a non-tubular bracteole. Lip broadly obovate, to 10 cm across when spread out, yellow in the centre. *SE Asia to New Guinea; much cultivated in the tropics.* G2. Flowering throughout the year.

4. C. lucanusianus Braun & Schumann.

Plants 1–2 m. Leaf-sheaths with a conspicuous rim of bristles just below the top. Leaves 12–25 × 5–7 cm, lanceolate-oblong, downy beneath. Inflorescence terminal, bracts rounded at their tips. Flowers 2 per bract, white, each with a non-tubular bracteole, calyx tube longer than the bracteoles. Lip broadly obovate, 5 cm across when spread out, yellowish in the centre, striped with purple. *Tropical W Africa.* G2. Flowering throughout the year.

5. C. afer Ker Gawler. Figure 16(7), p. 423. Illustration: *Botanical Magazine*, 4979 (1857).

Like *C. lucanusianus* but without the bristles on the leaf-sheaths, the calyx tube shorter than the bracteoles. Flowers with or without purple markings. May produce terminal and separate inflorescences on the same plant. *Tropical W Africa.* Flowering throughout the year.

58. CANNACEAE

Herbs with rhizomes. Leaves spirally arranged without ligules; ptyxis supervolute. Inflorescences terminal with flowers in pairs. Calyx of 3 free sepals. Corolla of 3 united petals. Stamen 1, petal-like with a half-anther, staminodes several, petal-like. Ovary of 3 carpels, inferior; style petal-like; ovules many, axile. Fruit a warty capsule.

A single genus, native to southern North America, with about 50 species, several of which are grown as spectacular, half-hardy, bedding annuals in Europe.

1. CANNA Linnaeus
J.A. Ratter
Description as for family.

Several species of this genus are grown as ornamentals, and numerous hybrids have been produced. The spectacular flowers and foliage of the hybrid cannas make them important as summer bedding plants in the warmer parts of Europe. Most of them are complex hybrids, involving in their origin the species included here, as well as *C. coccinea* Aiton and various others, and are extremely variable in height, foliage colour and flower colour and form. They have been grouped as *C.* × *generalis* Bailey and *C.* × *orchioides* Bailey, the latter comprising the orchid-flowered cannas derived from hybrids of *C. flaccida*, and showing the distinctive broad, wavy-margined staminodes derived from that species.

Literature: Kränzlin, F., Das Pflanzenreich **56** (1912); Winkler, H., in Engler & Prantl, Die Natürlichen Pflanzenfamilien, edn 2, **15a**: 650–4 (1930). Good colour photographs of a number of named varieties of hybrid cannas are given in Graf, Tropica, 288 & 289 (1978).

1a. Flowers more or less hanging, 8–11 cm, long-tubed, waxy, reddish pink **4. iridiflora**
b. Flowers without the above combination of characters 2
2a. Petals reflexing; tube long, prominent; flower pale yellow, flaccid, reminiscent of that of a yellow flag iris **1. flaccida**
b. Petals remaining erect or ascending; tube very short or absent; flowers yellow or not 3
3a. Plant glaucous; staminodes yellow **2. glauca**
b. Plant not glaucous; staminodes red or orange **3. indica**

1. C. flaccida Salisbury. Illustration: Roscoe, Monandrian plants, t. 18 (1828); Das Pflanzenreich **56**: 51 (1912).

To 2 m. Leaves green, ovate-lanceolate, 20–50 × 10–13 cm. Racemes erect, loose and few-flowered. Flowers *c.* 12 cm, borne singly at each node of the inflorescence, yellow, soft or flaccid in texture, reminiscent of those of a yellow flag iris. Sepals *c.* 3 cm, reflexed. Petals forming a tube of *c.* 4 cm, and with strongly reflexed lobes of about the same length. Staminodes forming a tube considerably exceeding that of the petals, the 3 upper very broad, rounded and showy, with scalloped margins. *USA (South Carolina to Florida).* G1–2. Summer.

2. C. glauca Linnaeus. Illustration: Roscoe, Monandrian plants, t. 19 (1828); *Botanical Magazine*, 3437 (1835).

Plant glaucous, to 2 m. Leaves lanceolate, long-pointed, to 50 × 15 cm, hairless. Raceme usually simple. Flowers *c.* 9 cm, pale lemon yellow to darker yellow, more rarely with a tint or mottling of red or orange. Petals *c.* 8 cm, spoon-shaped but with pointed tips, united at the base into a tube *c.* 1.5 cm. Upper staminodes *c.* 8 cm, obtuse. *Tropical & subtropical America, from the West Indies to Bolivia & Argentina.* G1. Summer–autumn.

3. C. indica Linnaeus. Illustration: *Edwards's Botanical Register* **9**: t. 776 (1823) – colour atypical; Hooker, Exotic flora **1**: t. 53 (1823) – as var. *maculata*, with upper staminodes very speckled; Engler & Prantl, Die Natürlichen Pflanzenfamilien, edn 2, **15a**: 649 (1930).

Shoots slender, usually to 1.7 m. Leaves to 50 × 25 cm, ovate-lanceolate to oblong with an uneven, pointed tip, green, hairless. Inflorescence a raceme or somewhat branched panicle. Flowers to 6.5 cm. Sepals green. Petals *c.* 4 cm, spoon-shaped, pointed, united at the base into a tube *c.* 4 mm. Staminodes (including the lip) free almost to the base, the 3 upper straight, spreading, bright red or orange (rarely mottled or streaked), lip curled, yellow or orange with darker mottling or streaks; fertile stamen patterned like the lip. *Tropical and subtropical America.* G1. Autumn.

4. C. iridiflora Ruiz & Pavon. Illustration: Botanical Magazine, 1968 (1818); *Edwards's Botanical Register* **8**: t. 609 (1822); Graf, Tropica, 288 (1978).

Robust, shoots to 3 m. Leaves to 100 × 40 cm, broadly elliptic with a long, uneven, pointed tip, green, hairless. Inflorescence a drooping panicle. Flowers to 11 cm, waxy, reddish pink, more or less hanging. Petals *c.* 7 cm, united into a tube for about half their length, lobes narrow, pointed. Staminodes to 12 cm, forming a tube for about half their length, all reflexed above. *Peru.* G1. Autumn.

59. MARANTACEAE

Herbs with rhizomes. Leaves in 2 ranks, the stalk with a swollen band (pulvinus) at the apex; ptyxis supervolute. Inflorescence a panicle or spike with asymmetric flowers in zygomorphic pairs. Sepals 3. Corolla of 3 united petals. Stamen 1, with various petal-like staminodes. Ovary of 3 united carpels, inferior; ovules 1 per cell (occasionally 2 per cell, 1 aborting). Fruit a capsule, often fleshy.

There are 32 genera and about 500 species. Three genera are native to North America and species from 5 genera are grown in glasshouses in Europe for the sake of the frequently coloured or marked leaves. The flowers are individually asymmetric, but generally occur closely associated in pairs, so that the pair is zygomorphic about the vertical axis.

The species are mainly characteristic of forest floor habitats. Many are grown under glass, mainly for the decorative, coloured and patterned foliage, and some flower only rarely or never. They require shade, as exposure to bright sunlight will give rise to yellowing or blotching of the leaves, particularly if the plants are in pots.

Information on the wild habitat of many of the species in cultivation is scanty or altogether lacking; their geographical ranges can therefore be described only in very general terms.

Literature: Schumann, K., Das Pflanzenreich **11**: 1–184 (1902); Loesener, T., in Engle, A. & Prantl, K., Die Natürlichen Pflanzenfamilien, edn 2, **15a**: 654–93 (1930).

1a. Ovary 3-celled, with 3 ovules; flowers in compact heads or dense spikes (usually with spiral bracts); inflorescence often basal **1. Calathea**
b. Ovary 1-celled, with 1 ovule; inflorescences various, but never basal, if a raceme or spike, then bracts 2-ranked, never spiral 2
2a. Outer staminode 1; tall, erect, semi-aquatic plants **5. Thalia**
b. Outer staminodes 2, rarely absent; not erect, semi-aquatic plants 3
3a. Corolla tube elongate; inflorescence a raceme or spike with 2–4, 2-ranked bracts that are sometimes much exceeded by the branched flower axes **3. Maranta**
b. Corolla tube very short, wide; inflorescence a tight spike or raceme with many bracts, or a scarlet panicle 4
4a. Inflorescence a dense spike with persistent green, purple-tinged membranous bracts; flowers borne on 1 side; outer staminodes purple **2. Ctenanthe**
b. Inflorescence a panicle, raceme or spike with red, deciduous bracts; outer staminodes small or absent **4. Stromanthe**

1. CALATHEA Meyer
J.A. Ratter
Herbs with or without stems, with green or attractively marked and/or coloured leaves, often forming low clumps. Inflorescence usually a head or short spike with spiral bracts, more rarely a narrower spike with 2-ranked bracts. Outer staminode 1. Ovary 3-celled, each cell with 1 ovule. Capsule usually with 3 seeds.

A genus of about 300 species from tropical America. More than 90 specific names have been recorded in cultivation, but the accuracy of many of these is doubtful (as is the persistence of the plants). The 23 species included here cover those most likely to be found in general collections.

The plants require hot and humid conditions, and enjoy an open, loose compost, which allows root movement and free passage of air. Any peat in the mixture should be left in a rough state to prevent stagnation. They are easily propagated by division into single crowns. Shading is necessary for quick establishment, which will take place very rapidly in a case or growth chamber.

1a. Leaves not variegated 2
b. Leaves variegated 6
2a. Leaves linear-lanceolate, reddish-pilose along petiole and midrib **17. rufibarba**
b. Leaves without the above combination of characters 3
3a. Leaves red-purple beneath 4
b. Leaves greenish beneath 5
4a. Plant to 30 cm; leaves with pulvinus borne almost directly on the winged sheath, blade 10–14 cm; bracts and corolla orange **2. crocata**
b. Plant to 2 m; leaves with petioles *c.* 1 m, blade 20–60 cm; bracts yellowish; corolla white and violet **11. ornata**
5a. Plant to 2 m; leaf-blade to 80 cm, elliptic or oblong **3. cylindrica**
b. Plant to 30 cm; leaf-blade to 30 cm, nearly circular **16. rotundifolia**
6a. Lower surface of leaves purple, red or variously spotted or patterned in these colours 7
b. Lower surface of leaves green or green-variegated 22
7a. Upper surface of leaves with a distinct row of darker blotches or stripes on either side of the midrib 8
b. Leaves not as above 14
8a. Leaves with a row of distinct, usually elliptic blotches (often large and small alternating with each other) along each side of the midrib 9
b. Leaves not as above 11
9a. Leaves broadly ovate **8. makoyana**
b. Leaves linear-lanceolate to ovate-oblong 10
10a. Leaves linear-lanceolate, blade usually more than 20 cm, erect, conspicuously wavy **4. lancifolia**
b. Leaves ovate-oblong to oblong-lanceolate, blade 10–20 cm **22. wiotii**
11a. Leaf-blades less than 30cm 12
b. Leaf-blades 40–60 cm 13
12a. Upper surface of leaf with dark and pale green bands alternating to give a pinnate pattern; pulvinus *c.* 2 cm, minutely downy; inflorescence and its subtending foliage leaf borne on a long scape; bracts in 2 rows **6. lietzii**
b. Upper surface of leaf with a jagged pale pattern along the midrib and a sinuous, feathered, often interrupted, pale line running parallel to it along each side of the blade; pulvinus *c.* 8 cm, hairless; spike with bracts spirally arranged and a crown of sterile bracts above **20. veitchiana**
13a. Upper surface of leaf with a pinnate pattern of pale stripes running from midrib to margin; pulvinus hairless **23. zebrina**
b. Upper surface of leaf with a row of triangles of pale colour running out from the midrib but usually not reaching the margin; pulvinus downy **21. warscewiczii**
14a. Upper surface of leaf with a pinnate pattern of thin, pink or whitish stripes between the veins, often in groups; petiole to 1 m **11. ornata**
b. Leaf not as above 15
15a. Upper surface of the leaf with a dark green, feathered stripe along the midrib and margins and a pale yellowish green zone in between **14. princeps**
b. Upper surface of leaf with a pale or pinkish, often feathered, median stripe 16
16a. Upper surface of leaf with a distinct, narrow, often sinuous and feathered, pale whitish, silvery or pinkish arc on each side between the median stripe and margin 17
b. Such pale areas absent or somewhat indistinct 19
17a. Median stripe along midrib narrow, often rose-coloured; plant to 20 cm **15. roseo-picta**
b. Median stripe wider, whitish, silvery or pale green, feathered; plant usually taller 18
18a. Pulvinus *c.* 1.5 cm, minutely downy; median marking feathered but usually little lobed **13. picturata**
b. Pulvinus up to 8 cm, hairless; median marking feathered and deeply lobed **20. veitchiana**
19a. Midrib with deep yellow-green triangles on each side, which extend more than halfway to the margin **21. warscewiczii**
b. Midrib without yellow-green triangles on each side 20
20a. Median stripe broad, often occupying most of the leaf surface **13. picturata**
b. Median stripe narrower 21
21a. Upper surface of leaf dark green with a jagged median band that is whitish; blade *c.* 10 cm with petiole a little shorter; plant to 20 cm **18. undulata**
b. Upper surface of leaf with midrib feathered pale green and an indistinct band of the same colour close to the margin; blade to 40 cm with petiole about the same length; plant taller than 30 cm **7. lindeniana**
22a. Upper surface of leaf deep green with a white median band 23
b. Variegation patterns various but not as above 24
23a. Plant to 20 cm; leaf-blades to *c.* 13 × 4 cm; spike ellipsoid, *c.* 1.5 cm on a slender stalk to 12 cm; flowers white or pale violet **10. micans**
b. Plant to 60 cm; leaf-blades to 20 × 9 cm; spike almost spherical or ovoid, *c.* 5 cm on a stalk of the same length or longer; flowers violet **9. medio-picta**
24a. Leaf-blade *c.* 35 cm, oblong, acute, glossy dark green with a median band of black-green above; petiole very long **19. variegata**
b. Without the above combination of characters 25
25a. Leaf almost circular, with a pinnate pattern of alternating silvery white and green stripes **16. rotundifolia**
b. Leaves narrower, with patterns of dark lateral blotches or bands, but not as above 26
26a. Paired blotches in a row more or less midway between midrib and margin on each side; spike few-flowered, enclosed by the lowermost bract and borne on a long stalk **12. pavonii**
b. Blotches touching or almost touching the midrib, not in pairs on each side; spike usually on a stalk *c.* 3 cm 27
27a. Dwarf plant; leaf-blades 10–20 cm, pulvinus less than 1 cm **22. wiotii**
b. Plant taller; leaf-blades more than 20 cm, pulvinus less than 1 cm or not 28
28a. Background of upper surface of leaf bright pale green; pulvinus *c.* 5 mm **5. leopardina**

b. Background of upper surface of leaf grey-green; pulvinus *c.* 2 cm **1. bachemiana**

1. C. bachemiana Morren (*C. bella* (Bull) Regel; *Maranta bachemiana* invalid; *M. kegeliana* invalid). Figure 17(2), p. 434. Illustration: Engler & Prantl, Die Natürlichen Pflanzenfamilien, edn 2, **15a**: 679 (1930); Graf, Exotic plant manual, edn 4, 177 (1974); Graf, Exotica, edn 8, 1140, 1143 (1976); Graf, Tropica, 632 – as *C. kegeliana* – & 634 – as *C. bella* (1978).

Stemless, to 50 cm. Leaves with petiole *c.* 15 cm and pulvinus downy, *c.* 2 cm. Blade leathery, to *c.* 35 × 12 cm, ovate-lanceolate, narrowly acuminate, greenish silver-white above with dark green bands and often spots and sometimes a dark margin; green, rarely slightly purplish, beneath. Spike emerging directly from the soil, almost cylindric, *c.* 6 cm on a stalk *c.* 2.5 cm (sometimes to 10 cm). Flowers *c.* 5 cm (large for the genus). *Brazil.* G2.

2. C. crocata Morren & Jorisenne. Illustration: *Botanical Magazine*, 7820 (1902); Graf, Tropica, 631 (1978).

To 20 cm or taller. Leaves forming a clump, blade oblong or ovate-elliptic, *c.* 12 cm, green above, sometimes with a grey feather pattern, red-purple beneath, pulvinus borne almost directly on the top of the winged sheath. Inflorescence an ovoid or spherical head *c.* 4 cm, borne above the leaves on a scape. Bracts and corollas bright orange. *Brazil.* G2.

3. C. cylindrica (Roscoe) Schumann (*Phrynium cylindricum* Roscoe; *C. grandifolia* Lindley). Illustration: *Edwards's Botanical Register* **14**: t. 1210 (1827); Graf, Tropica, 634 (1978) – as *C. grandiflora*.

Robust plant with a stem to 2 m. Leaves with long petioles, pulvinus 2.5–5 cm, hairless. Blade to 80 × 33 cm, elliptic or oblong, shortly acuminate, rounded at the base, green on both sides. Inflorescence a cylindric spike. *c.* 15 cm. on a stalk to 30 cm, emerging from the central leaf-sheaths. Bracts green, spirally arranged. Flowers yellow or white. *S Brazil.* G2.

This species is sometimes misnamed *C. grandiflora* (Roscoe) Schumann, which is, in fact, a much smaller plant with a stalkless basal inflorescence bearing yellow flowers that are *c.* 6 cm. The confusion has no doubt arisen because of the similarity of the names 'grandiflora' and Lindley's synonym for the species, 'grandifolia'.

4. C. lancifolia Boom (*C. insignis* invalid, not *C. insignis* Petersen). Figure 17(7), p. 434. Illustration: Graf, Exotic plant manual, edn 4, 175 (1974); Graf, Exotica, edn 8, 1140 (1976); Graf, Tropica, 631, 632, 634 (1978).

Stemless, forming a compact clump to 60 cm. Petiole 15–40 cm; leaf-blade about the same length, linear-lanceolate, wavy, erect, upper surface with a row of alternating large and small, dark green, elliptic spots on either side of the midrib, background yellow-green, becoming darker at the margins, red-purple beneath. *Brazil.* G2.

5. C. leopardina (Bull) Regel. Illustration: *Gartenflora* **26**: t. 893 (1877); Graf, Exotic plant manual, edn 4, 177 (1974); Graf, Exotica, edn 8, 1143 (1976); Graf, Tropica, 633 (1978).

To 40 cm. Petiole about as long as the leaf-blade with pulvinus *c.* 5 mm, downy; blade to *c.* 20 × 9 cm, oblong or ovate-lanceolate, acuminate, bright pale green above with a row of dark green blotches along the midrib, pale green with deeper green marginal zones beneath, sometimes tinted bronze-purple. Inflorescence a more or less spherical, shortly stalked spike, *c.* 5 cm, borne amongst the leaf-bases. Bracts transparent and lined green. Flowers yellowish. *Brazil.* G2.

6. C. lietzii Morren. Figure 18(3), p. 435. Illustration: *La Belgique Horticole* **25**: t. 15–17 (1875); *Gartenflora* **27**: t. 935 (1878); Graf, Exotic plant manual, edn 4, 177 (1974); Graf, Exotica, edn 8, 1141 (1976); Graf, Tropica, 632 (1978).

Clump-forming plant to 60 cm with conspicuous, long, erect stalks bearing the inflorescence and subtending leaf, later growing on to produce stalked plantlets. Petiole short, the minutely downy pulvinus, *c.* 2 cm, often borne directly on the sheath; leaf-blade oblong-lanceolate, wavy, pointed at the apex, rounded to cordate at base, with somewhat ragged, dark and light green stripes alternating along the upper surface to give a pinnate pattern, red-purple beneath. Spikes *c.* 6 cm, narrow, subtended by a leaf that is usually solitary at flowering; stalk to 40 cm; bracts green, few, 2-ranked. Flowers white. *Brazil.* G2.

7. C. lindeniana Wallis. Illustration: *Gartenflora* **18**: t. 601 (1869); Graf, Exotic plant manual, edn 4, 176 (1974); Graf, Exotica, edn 8, 1141 (1976); Graf, Tropica, 632 (1978).

Clump-forming plant to 1 m. Petiole to 40 cm including the hairless pulvinus, which is 5 cm; leaf-blade to about 40 cm long, elliptic, somewhat pointed, upper surface dark green with a paler green, feathered median stripe and rather indistinct pale green zones running just inside the margins on each side, undersurface with red-purple zones corresponding to the dark green ones above. Spike ellipsoid, to 10 cm, almost stalkless, emerging directly from the soil. Bracts red-purple, spirally arranged. Flowers pale yellow. *NW Brazil, Peru.* G2.

8. C. makoyana Morren (*Maranta makoyana* Morren). Figure 17(4), p. 434. Illustration: *La Belgique Horticole* **22**: t. 24, 25 (1872); Parey's Blumengärtnerei **1**: 408 (1958); Graf, Exotic plant manual, edn 4, 175 (1974); Graf, Exotica, edn 8, 1143, 1186A (1976); Graf, Tropica, 632, 633, 635, 639 (1978).

Stemless clump-forming plant to 60 cm. Petiole shorter than or about as long as the leaf-blade; leaf-blade to 33 × 16 cm but usually much smaller, broadly ovate, pale green or cream above with a row of alternating large and small, dark green, more or less elliptic blotches on either side of the midrib, underside showing the same pattern in dark red-purple. Inflorescence emerging directly from the soil; spike ovoid, 2 cm, stalk *c.* 4 cm. Bracts green, spirally arranged. Flowers *c.* 2 cm, white, petals tinged with pink at the tips. *E Brazil.* G2.

9. C. medio-picta (Morren) Regel. Illustration: *Gartenflora* **28**: t. 934 (1878); Graf, Exotica, edn 8, 1136, 1139 (1976).

To 60 cm. Leaves with long petioles; blade to *c.* 20 × 9 cm, oblong to ovate-lanceolate with a short point, dark green above with the midrib feathered white, pale green beneath. Spike borne within the leaf-sheaths, more or less spherical or ovoid, *c.* 5 cm, stalk 5–10 cm. Bracts whitish or yellowish, spirally arranged. Flowers violet. *Brazil.* G2.

10. C. micans (Mathieu) Koernicke (*C. albicans* Schumann). Figure 18(5), p. 435. Illustration: Martius, Flora Brasiliensis **3**(3): t. 20 (1890); Graf, Exotic plant manual, edn 4, 176 (1974); Graf, Exotica, edn 8, 1141 (1976); Graf, Tropica, 632 (1978).

Small plant to 20 cm, but usually lower, with leaves close to the ground.

Figure 17. Leaves of *Calathea* species. 1, *C. zebrina*. 2, *C. bachemiana*. 3, *C. ornata*. 4, *C. makoyana*. 5, *C. pavonii*. 6, *C. warscewiczii*. 7, *C. lancifolia*.

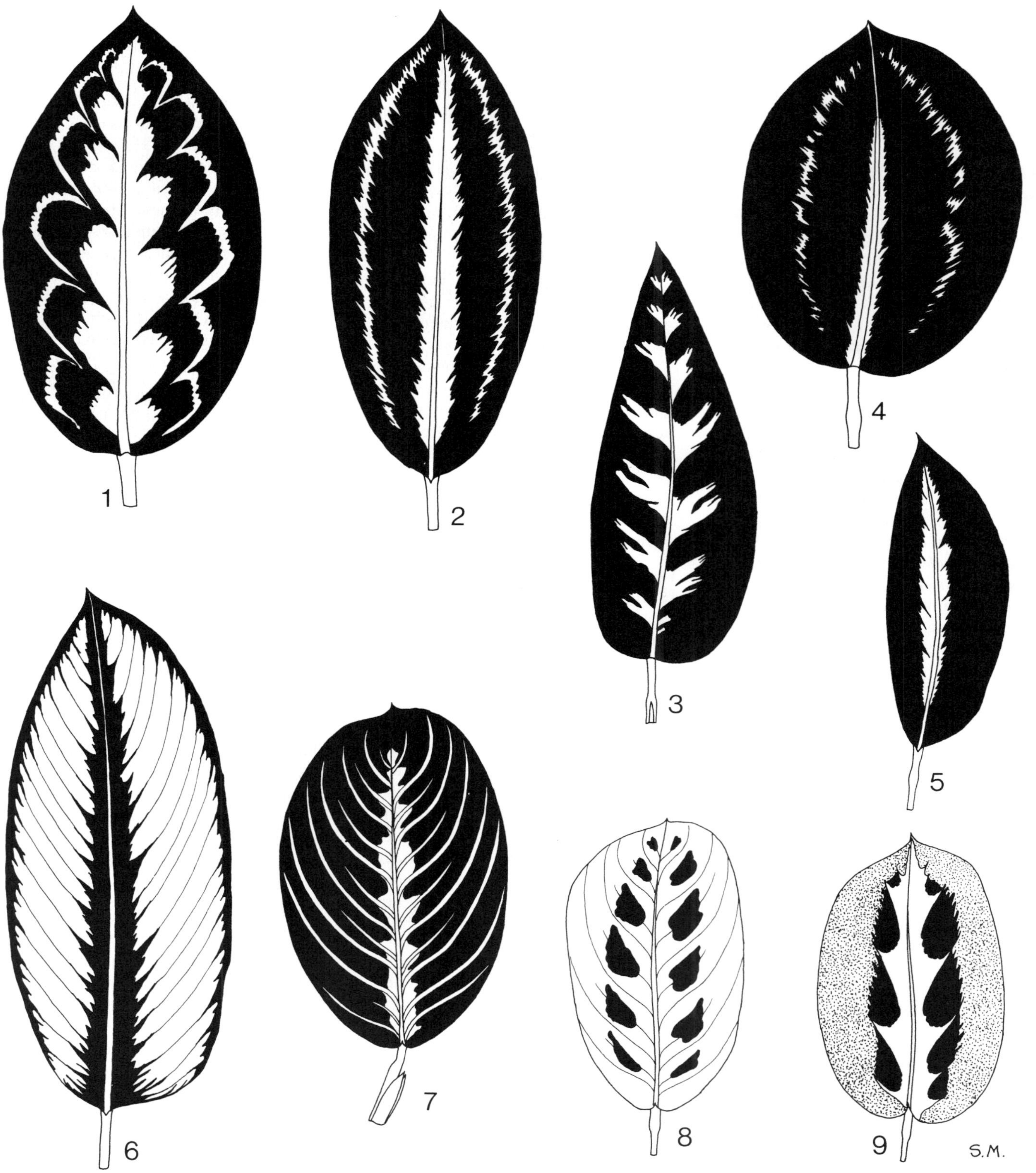

Figure 18. Leaves of *Calathea* and *Maranta* species. 1, *Calathea veitchiana.* 2, *C. picturata.* 3, *C. lietzii.* 4, *C. roseo-picta.* 5, *C. micans.* 6, *C. princeps.* 7, *Maranta leuconeura* var. *massangeana.* 8, *M. leuconeura* var. *kerchoviana.* 9, *M. bicolor.*

Petiole to 5 cm; leaf-blade to *c.* 13 × 4 cm, oblong-lanceolate to elliptic, dark green above with midrib feathered silvery white, grey-green beneath. Spike borne within the leaf-sheaths, ellipsoid, *c.* 1.5 cm, on a slender stalk to 12 cm. Flowers white or pale violet. *Mexico to Brazil.* G2.

11. C. ornata (Planchon & Linden) Koernicke (*C. sanderiana* invalid). Figure 17 (3), p. 434. Illustration: *Flore des serres* **4**: t. 413, 414 (1848); Graf, Exotic plant manual, edn 4, 177 (1974); Graf, Exotica, edn 8, 1136 (1976); Graf, Tropica, 636 (1978).

Robust stemless plant to 2 m. Leaves emerging from soil separately but usually forming a clump. Petioles over 1 m, pulvinus 5–10 cm; leaf-blade often erect, to 60 cm, ovate-lanceolate to broadly ovate, leathery, green above; usually with a pinnate pattern of thin, white or pinkish stripes, often in groups, starting well out from the midrib; red-purple beneath. Spike ovoid, *c.* 8 cm, borne on a stalk *c.* 30 cm. Bracts spirally arranged, yellowish. Flowers white and violet. *Tropical S America from Ecuador to Guyana.* G2.

Several variants of this species are recognised, e.g. 'Roseo-lineata' and 'Sanderiana'.

12. C. pavonii Koernicke (*C. tubispatha* Hooker). Figure 17(5), p. 434. Illustration: *Botanical Magazine*, 5542 (1865); Graf, Exotica, edn 8, 1148 (1976).

Plant 30–50 cm. Petiole very short, the pulvinus being borne almost directly on the winged sheath; leaf-blade to *c.* 25 × 14 cm, oblong-acute to ovate-elliptic, thin, pale green above with a row of dark brown paired blotches on each side between the midrib and margin, greenish beneath. Spike borne within the leaf-sheaths, cylindric, to 3 cm, inclined on a long, slender stalk. Bracts 2 or 3, green, the lowest forming an involucre around the others. Flowers yellow, few. *Peru.* G2.

13. C. picturata (Linden) Koch & Linden (*C. vandenheckei* (Lemaire) Regel). Figure 18 (2), p. 435. Illustration: Graf, Exotic plant manual, edn 4, 177 (1974); Graf, Exotica, edn 8, 1136, 1141 (1976); Graf, Tropica, 636 (1978).

Plant to 35 cm. Petiole to 30 cm with a pulvinus *c.* 1.5 cm; leaf-blade to *c.* 25 × 10 cm, elliptic, dark green with feathered whitish lines along the midrib and as arcs somewhat inside the margin (in 'Argentea' a single median whitish feathered area occupies most of the leaf surface), red-purple (sometimes very pale) or rarely green beneath. Spike narrowly cylindric, up to 10 cm, shorter than its stalk. Flowers white. *NW Brazil.* G2.

C. vandenheckei *is* a small, juvenile form of this species.

14. C. princeps (Linden) Regel. Figure 18 (6), p. 435. Illustration: Graf, Exotic plant manual, edn 4, 175 (1974); Graf, Exotica, edn 8, 1136, 1140 (1976); Graf, Tropica, 631, 633, 636 (1978).

Robust plant up to 2 m tall, but often smaller. Leaves often strikingly 2-ranked, reminiscent of the crown of *Ravenala* (p. 420). Petiole to *c.* 40 cm, pulvinus *c.* 5 cm; leaf-blade about as long as petioles, oblong-lanceolate, rounded at the base, upper surface dark green at the midrib and margins with a pale yellowish green zone between, red-purple beneath. *NW Brazil.* G2.

Graf (Tropica, 960, 1978) suggests that this species is a juvenile form of *C. altissima* (Poeppig & Endlicher) Koernicke.

15. C. roseo-picta (Linden) Regel. Figure 18(4) p. 435. Illustration: *Gartenflora* **18**: t. 610 (1869); Graf, Exotic plant manual, edn 4, 177 (1974); Graf, Exotica, edn 8, 1141, 1186A (1976); Graf, Tropica, 631, 636 (1978).

Plant scarcely reaching 20 cm. Petiole usually short but sometimes up to 12 cm, deep red; leaf-blade to 20 × 15 cm, broadly oval, obtuse, dark green above with a narrow pink or whitish stripe along the midrib and a thin, feathered, somewhat wavy line of the same colour (or whiter) running in an arc inside the margin, lower surface red-purple. Spikes *c.* 9 cm, more or less cylindric, stalk *c.* 9 cm or more. Bracts green, spirally arranged. Flowers white and violet. *NW Brazil.* G2.

16. C. rotundifolia (Koch) Koernicke (*C. fasciata* (Linden) Koernicke). Illustration: *Gartenflora* **13**: t. 452 (1864); Graf, Exotica, edn 8, 1143, 1146, 1186A (1976); Graf, Tropica, 633, 636, 640 (1978).

Low, stemless plant to 30 cm, forming clumps of leaves. Petiole up to 9 cm; leaf-blade to 30 × 27 cm, almost circular, often somewhat cordate at the base, leathery, green on both surfaces. Spike almost stalkless, nearly spherical, emerging directly from the soil. Bracts brown, spirally arranged. Flowers transparent white. *E Brazil.* G2.

The leaf of var. **fasciata** (Linden) Petersen has a pattern of silvery white, pinnately arranged bands and the under surface sometimes tinged purple.

17. C. rufibarba Fenzl. Illustration: *Botanical Magazine*, 7560 (1897); Graf, Exotica, edn 8, 1142, 1146 (1976).

Stemless, forming large dense clumps to 50 cm. Petiole and leaf-midrib (sometimes almost the whole plant) densely covered in reddish hairs. Petiole up to 20 cm; leaf-blade to 25 cm, linear-lanceolate to lanceolate, wavy, glossy green above, tinged with purple beneath. Spike basal, to 7 cm, borne on a stalk *c.* 5 cm, bracts spirally arranged, violet with reddish hairs. Flowers yellow, projecting. E Brazil. G2.

18. C. undulata Linden & André. Illustration: *Gartenflora* **25**: t. 852 (1876); Graf, Exotic plant manual, edn 4, 176 (1974); Graf, Exotica, edn 8, 1136, 1144 (1976); Graf, Tropica, 636 (1978).

Small, clumped plant to *c.* 20 cm. Petiole very short with pulvinus borne almost on the sheath; leaf-blade to *c.* 10 × 5 cm, oblong-elliptic, undulate, dark green above with a jagged whitish band along the midrib, purplish beneath. Spike *c.* 2 cm borne on a stalk that is *c.* 7.5 cm. Bracts spirally arranged, green edged and spotted with white. *Peru, Brazil.* G2.

19. C. variegata (Koch) Koernicke.

Robust plant. Petiole very long, with hairless pulvinus *c.* 6 cm; leaf-blade to 35 × 17 cm, oblong, acute, glossy deep green with a median band of black-green above, paler green with deeper bands beneath. Spike ellipsoid, *c.* 5 cm, borne on a stalk to 12 cm. Bracts spirally arranged. Flowers pale yellow. *Origin unknown.* G2.

20. C. veitchiana Hooker. Figure 18(1) p. 435. Illustration: *Botanical Magazine*, 5535 (1865); Graf, Exotic plant manual, edn 4, 175 (1974); Graf, Exotica, edn 8, 1139 (1976); Graf, Tropica, 636, 639 (1978).

Robust plant to 1 m. Petiole to 12 cm with hairless pulvinus, to 8 cm; leaf-blade to 30 × 15 cm, ovate-elliptic or oblong, obtuse, glossy green above with a median deeply lobed or jagged, pale stripe along the midrib and a pale sinuous stripe lying parallel to it on each side, pattern repeated in red-purple beneath. Spike top-shaped with a crown of sterile bracts above, *c.* 8 cm, on a stalk to 20 cm. Bracts spirally arranged, green. Flowers white and violet. *Peru.* G2.

21. C. warscewiczii (Mathieu) Koernicke. Figure 17(6), p 428. Illustration: *Gartenflora* **15**: t. 515 (1866); Graf, Exotic plant manual, edn 4, 175 (1974); Graf, Exotica, edn 8, 1139 (1976); Graf, Tropica, 635 (1978); *University of California Publications, Botany* **71**: t. 20 (1978).

Robust plant to 1 m. Petiole 10 cm or more, with the pulvinus downy, *c.* 3 cm; leaf-blade to *c.* 60 cm, oblong-lanceolate, dark velvety green above, feathered along the midrib with yellow-green triangles (like a row of shark fins on either side), red-purple beneath. Spikes ellipsoid, *c.* 6 cm, borne on long scapes bearing 2 leaves near their apices, the upper leaf subtending the spike. Bracts broad, spirally arranged, pale yellow outside. Flowers yellowish. *Costa Rica.* G2.

22. C. wiotti (Morren) Regel (*C. wiotiana* Morren). Illustration: *La Belgique Horticole* **25**: t. 15–17 & f. 7 (1875); Engler & Prantl, Die Natürlichen Pflanzenfamilien, edn 2, **15a**: f. 298a (1930); Graf, Exotic plant manual, edn 4, 177 (1974); Graf, Exotica, edn 8, 1140 (1976); Graf, Tropica, 632 (1978).

Dwarf plant, forming patches from a branching and spreading rhizome. Petiole *c.* 8 cm with a hairless pulvinus, *c.* 8 mm; leaf-blade 10–20 cm, ovate-oblong to oblong-lanceolate, rounded to almost cordate at the base, wavy, light green above with a row of darker green blotches (often alternating larger and smaller) along each side of the midrib, dull greyish green beneath with patches of yellow-green, or sometimes purple. *E Brazil.* G2.

23. C. zebrina (Sims) Lindley. Figure 17 (1), p. 434. Illustration: *Botanical Magazine*, 1926 (1817); Graf, Exotic plant manual, edn 4, 175 (1974); Graf, Exotica, edn 8, 1146, 1186A (1976); Graf, Tropica, 635, 636, 639, 640 (1978).

Robust plant up to 80 cm. Leaves with a hairless pulvinus, *c.* 4 cm, borne directly on the broadly winged sheath; leaf-blade to 65 × 25 cm, oblong-lanceolate, velvety dark green above with yellow stripes from the midrib to margins giving a pinnate effect, and shorter stripes bifurcating the ends of the green areas, red-purple beneath. Spike rather massive, *c.* 10 cm, almost spherical or ovoid, dense, with many flowers, borne on a stalk to 35 cm. Bracts broad, green tinted with violet. Flowers *c.* 4.5 cm, dark purple and white. *E. Brazil.* G2.

'Binotii' (var. *binotii* Bailey) is somewhat larger; it has narrower leaves to 90 cm, with the dark green areas much reduced, and the under surface greyish green.

Two other species, which are doubtfully in cultivation, will key out to *Calathea* in the generic key (p. 432). Both have 2 outer staminodes, whereas *Calathea* has only 1. They are:

Ataenidia conferta (Bentham) Milne-Redhead (*Phrynium confertum* (Bentham) Schumann), which has a usually simple stem up to 1 m, ovate-lanceolate leaves to 45 × 20 cm which have long petioles and are green on both sides, and the inflorescence consisting of stalkless or shortly stalked heads in the centre of the apical groups of leaves. Flowers are pale pink or purple. *W tropical Africa from Cameroun to Angola.*

Stachyphrynium jagorianum (Koch) Schumann, is a much smaller plant, up to 30 cm, with linear-oblong to oblong-lanceolate leaves, which have a pinnate pattern of blackish stripes from the midrib on the upper surface and a pale green lower surface. Flowers white on a nearly stalkless, cylindric spike. *Malaysia & Indonesia.*

2. CTENANTHE Eichler

J.A. Ratter revised by S.G. Knees

Perennial herbs, often with basal leaves with long petioles, and leaves on the flowering shoots with short petioles. Inflorescence of dense spikes or racemes with closely overlapping, persistent bracts; flowers usually borne on 1 side only. Corolla tube very short. Outer staminodes 2, petaloid. Ovary 1-celled, with 1 ovule.

A genus of about 20 species, 1 in Costa Rica, the rest in Brazil. Cultivation as for *Calathea*. In most species flowers are produced sporadically throughout the year.

Literature: Kennedy, H., Calathea burle-marxii and Ctenanthe burle-marxii (Marantaceae): two new cultivated species from Brazil, *Canadian Journal of Botany* **60**: 2365–2370 (1982).

1a. Leaves uniformly green or rarely with cream-coloured strips along the margins **4. setosa**
b. Leaves otherwise variegated 2
2a. Leaves pale yellow-green beneath **1. lubbersiana**
b. Leaves deep red or purple beneath 3
3a. Leaves 30–40 × 9–12 cm **2. oppenheimiana**
b. Leaves 15–20 × 4–7 cm **3. burle-marxii**

1. C. lubbersiana (Morren) Eichler. Illustration: Graf, Exotic plant manual, edn 4, 176 (1974); Graf, Exotica, edn 8, 1136, 1145 (1976); Graf, Tropica, 637 (1978).

Herb to *c.* 80 cm with widely branched shoots leaning under their own weight. Basal leaves with petioles to 10 cm. Lower internodes of shoots long; stem-leaves clustered above, with the green pulvinus borne almost directly on the broad, membranous-margined sheath. Leaf-blade 30 × 14 cm, oblong to oblong-lanceolate, abruptly contracted to a pointed apex, upper surface with pale yellow-green and dark green in an irregular speckling, paler beneath. Flowers white. *Brazil.* G2.

2. C. oppenheimiana (Morren) Schumann (*Stromanthe porteana* var. *oppenheimiana* invalid). Illustration: Graf, Exotic plant manual, edn 4, 175 (1974); Graf, Exotica, edn 8, 1140, 1186A (1976); Graf, Tropica, 637, 638, 639 (1978).

Vigorous herb to 1 m. Petiole hairless or downy, to *c.* 16 cm, pulvinus *c.* 3.5 cm. Leaf-blade leathery, 30–40 × 9–12 cm, ovate-lanceolate with a long, slender point, pinnately patterned with dark and pale green bands above, sometimes rather faint (and in 'Variegata', 'Tricolor', etc. with large cream areas reminiscent of variegated ivy), deep blood-red beneath. Spikes solitary or in pairs, to *c.* 5 cm. Bracts ovate, acute, hairless, membranous, pale green with pale purple along the margins. Flowers white. *E Brazil.* G2.

C. kummeriana (Morren) Eichler. Illustration: *Gartenflora* **79**: 378 (1930); Martius, Flora Brasiliensis **3**(3): t. 44 (1890). Similar but petioles of the basal leaves conspicuously hairy. *SE Brazil.*

3. C. burle-marxii Kennedy. Illustration: Brickell (ed.), RHS A–Z encyclopedia of garden plants, 330 (2003).

Herb with softly hairy, purple-tinged stems, 40–60 cm. Petiole and sheath green, purple-tinged; pulvinus olive-green. Leaf-blade 15–20 × 4–7 cm, ovate to obovate-oblong, pale green above, pinnately patterned with sickle-shaped dark green markings, purple beneath. Bracts light green. Flowers white; sepals transparent, corolla white, outer staminodes cream-white, callose staminode white above, yellow below, ovary white, capsule green. *SE Brazil.* G2.

4. C. setosa (Roscoe) Eichler. Illustration: Das Pflanzenreich **11**: 154 (1902).

Robust rhizomatous herb, to 1.5 m. Basal leaves with conspicuously hairy sheaths and usually long petioles up to 16 cm; stem leaves and some basal leaves with petioles as short as 2–3 cm; blades to *c.* 50 × 15–23 cm, oblong-lanceolate to oblong with a well-developed, crooked point, green on both surfaces (or rarely with cream-coloured strips along the margins). Inflorescence generally a cluster of spikes, each up to 7 cm, looking like a group of plump centipedes or woodlice. Bracts green, bristly, with long points. Flowers white and violet, borne on 1 side of the inflorescence only. *SE Brazil.* G2.

C. compressa (Dietrich) Eichler. Spikes solitary or paired and bracts ovate, acute *Brazil.*

3. MARANTA Linnaeus

J.A. Ratter

Erect or prostrate herbs, often with slender and much-branched stems (sometimes apparently dichotomous). Leaves often patterned. Inflorescence a spike, raceme or panicle with bracts in 2 rows, often with few nodes. Outer petaloid staminodes 2. Ovary 1-celled due to abortion of the other 2 cells, with 1 ovule. Fruit indehiscent, nut-like.

A genus of 30 species from tropical America. Cultivation as for *Calathea* (p. 531).

1a. Robust, erect plant up to 1 m or more; leaves ovate-lanceolate, acute at the apex, the basal with long petioles above the winged sheaths **1. arundinacea**
b. Small, low, often prostrate plants; leaves oblong, very obtuse at the apex; pulvinus borne almost directly on top of the sheath in all leaves (petiole at most 2 cm) 2
2a. Leaf with a broad, median, pale green stripe divided into a deep 'fishtail' pattern, and a dark green zone outside, sometimes divided by the tips of the fishtail into separate blotches, but these never clearly demarcated; bracts 4 or more **2. bicolor**
b. Leaf with a row of clearly demarcated brownish blotches on either side of the midrib and/or a herring-bone pattern picked out in white or red along lateral veins; bracts 2 **3. leuconeura**

1. M. arundinacea Linnaeus. Illustration: *Botanical Magazine*, 2307 (1822); Das Pflanzenreich **11**: 127 (1902); Graf, Exotic plant manual, edn 4, 176 (1974); Graf, Exotica, edn 8, 1145 (1976); Graf, Tropica, 637, 638, 640 (1978).

Erect, much-branched herb up to 1 m or more, with thick, tuberous, starch-filled rhizomes. Basal leaves with long petioles, stem-leaves with pulvinus borne more or less on the leaf-sheath; blade up to *c.* 30 × 8 cm, ovate-lanceolate, green on both sides or sometimes variegated with large creamy areas. Inflorescence often branched. Flowers white. *Tropical America; widely cultivated in the tropics.* G2.

The source of West Indian arrowroot, which comes from the rhizomes. 'Variegata' has leaf variegations in dark and light greens and greenish yellow.

2. M. bicolor Ker Gawler. Figure 18(9), p. 435. Illustration: *Edwards's Botanical Register* **10**: t. 786 (1824); Graf, Exotica, edn 8, 1147 (1976); Graf, Tropica, 637 (1978).

Low plant to *c.* 20 cm, with prostrate shoots. Leaves with pulvinus borne almost directly on the winged sheath, or more rarely with a short petiole up to 2 cm; leaf-blade more or less horizontal, to *c.* 16 × 12 cm, oblong to elliptic-oblong, upper surface with broad, median, pale green, fishtail patterned stripe and a dark green zone outside, the latter sometimes partially divided by the fishtails into blotches but these never with a clear demarcation; red-purple beneath. Spike solitary, bracts usually 4, sometimes more. Flower stalks not or little exceeding the bracts. Flowers white and violet. *Brazil & Guyana.* G2.

3. M. leuconeura Morren. Illustration: *La Belgique Horticole* **25**: 172, t. 9 (1875); Graf, Exotic plant manual, edn 4, 176 (1974); Graf, Exotica, edn 8, 1147, 1186A (1976); Graf, Tropica, 637, 638, 639, 640 (1978).

Low plant, often much branched at the base, with well-developed prostrate or sometimes semi-erect shoots. Leaves similar to *M. bicolor* but with a variety of colour patterns (see below) and usually rather smaller. Spike solitary, very slender, with only 2 rather distinct bracts, which are often considerably exceeded by the flower stalks. Flowers white or violet. *Brazil.* G2.

Several cultivated variants, based on differences in leaf pattern, are recognised:

Var. **kerchoviana** Morren. Figure 18 (8), p. 435. Illustration: *La Belgique Horticole* **29**: t. 5 (1879); Graf, Exotic plant manual, edn 4, 176 (1974); Graf, Tropica, 637 (1978); Brickell (ed.), RHS A–Z encyclopedia of garden plants, 676 (2003). Leaves vivid to pale greyish green above, with a row of clearly defined dark brown-green blotches along each side between midrib and margin; tinged with purple beneath, at least on the undersides of the blotches.

Var. **massangeana** Morren. Figure 18 (7), p. 435. Illustration: *La Belgique Horticole* **25**: t. 10 (1875); Graf, Exotic plant manual, edn 4, 176 (1974); Graf, Exotica, edn 8, 1147, 1186A (1976); Graf, Tropica, 637 (1978). Leaves usually with a pale green or whitish jagged stripe along the midrib and the lateral veins picked out to give a white herring-bone pattern on a dark, velvety green background, which often becomes paler towards the margins; small leaves often showing a row of blotches as in var. *kerchoviana;* underside red-purple, at least beneath the dark areas on the upper surface.

In the frequently cultivated variant known as 'Erythroneura' ('Erythrophylla') – illustration: Brickell (ed.), RHS A–Z encyclopedia of garden plants, 676 (2003) – the veins are picked out in red and the midrib is pale yellow-green.

4. STROMANTHE Sender

J.A. Ratter

Herbs with stems, basal leaves with long petioles. Stem-leaves in clumps on the flowering shoots. Inflorescence a raceme or fairly loose panicle with coloured, deciduous bracts. Outer staminodes small, 2 or absent. Ovary 1-celled due to abortion of the other cells, with 1 ovule. Fruit a capsule.

A genus of 13 species in tropical S America. Cultivation as for *Calathea* (p. 531).

1a. Inflorescence a raceme with sometimes a single side branch at the base; petals violet-purple **1. porteana**
b. Inflorescence a branching panicle; petals white **2. sanguinea**

1. S. porteana Grisebach. Illustration: Martius, Flora Brasiliensis **3**(3): t. 42 (1890).

Leaf-blades ovate-oblong or lanceolate, green above, red-purple beneath. Inflorescence a raceme, sometimes with a single side branch at the base. Bracts red. Petals violet-purple. *E Brazil.* G2.

2. S. sanguinea Sonder. Illustration: *Botanical Magazine*, 4646 (1852); Graf, Exotica, edn 8, 1147, 1148 (1976); Graf, Tropica, 637 (includes one photograph labelled '*Ctenanthe oppenheimiana* inflorescence'), 638 (1978).

Robust plants to 1.5 m, with basal, 2-ranked clumps of leaves producing long, rigid scapes on which are borne terminal groups of stem-leaves and inflorescences. Sheaths broad, membranous, pinkish or deep red, often hairy along the margins; petioles of some basal leaves to 25 cm, almost absent in others, and in the stem-leaves the pulvinus to *c.* 5 cm, borne almost directly on the sheath; blade to *c.* 55 × 12 cm, lanceolate to linear-lanceolate, dark green with slightly paler midrib above, red-purple beneath. Inflorescence paniculate. Bracts and sepals scarlet. Petals white. *SE Brazil.* G2.

5. THALIA Linnaeus

J.A. Ratter

Tall marsh plants with leaves basal, with long petioles. Inflorescence a rather loose panicle with many flowers, borne on a long scape. Bracts deciduous. Corolla tube short. Outer staminode 1, petaloid. Ovary 1-celled due to abortion of the other cells, with 1 ovule. Stigma 2-lipped. Fruit indehiscent.

A genus of about 12 species from tropical and warm-temperate America, 1 species extending into tropical Africa. The species will grow in a cool glasshouse, and will survive outside in warmer areas, provided they are planted *c.* 60 cm below the surface of the water. Compost containing raw humus or peat should be avoided, as these encourage anaerobic conditions under water. Frequent feeding is, however, necessary, and fertiliser mixed with puddled loam can easily be dropped at the base of the plants.

1a. Inflorescence, and sometimes the whole plant, usually covered in white meal; bracts bluish green, crowded, forming dense spikelets; flowers deep purple **1. dealbata**

b. Plant not mealy; panicle very diffuse with rather distant green bracts, not or hardly forming spikelets; flowers blue-violet **2. geniculata**

1. T. dealbata Fraser. Illustration: *Botanical Magazine*, 1690 (1815); Das Pflanzenreich **11**: t. 22 (1902); Brickell (ed.), RHS A–Z encyclopedia of garden plants, 1030 (2003).

Erect marsh plant to 2 m, usually covered in white meal. Leaves with hairless, green sheaths; petiole long, pulvinus *c.* 3 cm; blades to *c.* 50 × 25 cm, ovate-lanceolate with pointed apex, green on both surfaces. Inflorescence a panicle, up to 20 cm, subtended by a bladeless sheath. Bracts mealy, pale bluish green, crowded and overlapping to form spikelets. Flowers deep purple. *Southeastern USA.* H5–G1. Summer.

2. T. geniculata Linnaeus. Illustration: Roscoe, Monandrian plants, t. 49 (1828).

Differs from *T. dealbata* in the absence of meal; in the inflorescence, which is very diffuse (to *c.* 40 cm); the green bracts, up to 1 cm apart so that they do not form crowded spikelets; and in the blue-violet flowers. *Subtropical & tropical America from Florida to SE Brazil & N Argentina; also W Tropical Africa.* G2. Summer.

Ischnosiphon bambusaceus (Poeppig & Endlicher) Koernicke is said to be found sometimes in cultivation. It will key out to *Thalia* in the generic key, but is a tall climbing herb reaching 10 m or more, with a bamboo-like habit; the leaves are small, ovate-lanceolate, green above and bluish beneath. *Peru.*

60. ORCHIDACEAE

Terrestrial, epiphytic or saprophytic herbs. Leaves alternate or rarely opposite, often borne on succulent, swollen stems (pseudobulbs); ptyxis conduplicate or supervolute, rarely flat. Flowers solitary or in racemes or panicles, usually bisexual, zygomorphic, usually inverted by a twist in the flower-stalk or ovary. Calyx of 3 free sepals (rarely of 2 or 3 united sepals), all petal-like. Corolla of usually 3 free or united segments, the median usually modified into a lip (labellum) of varying complexity. Stamens usually 1, rarely, united to the style to form a column; pollen in masses (pollinia) of varying shapes and degrees of complexity, usually borne on a sticky pad (viscidium) or more rarely on 2 viscidia, to which they may be attached by pollinial tissue or by a structure of different texture and hardness (stipe). Ovary of 3 united carpels, inferior; ovules very numerous, parietal. Stigmas 3, the 2 lateral receptive, the median sterile and variously modified into a structure known as the rostellum. Fruit a capsule, seeds tiny, very numerous.

Probably the largest family of flowering plants (at least in terms of number of species), with about 800 genera and over 22,000 species. Thirty-five genera are native to Europe and 88 to North America. Many are cultivated, including numerous inter-generic hybrids (some involving up to 8 genera in their formation). It is extremely variable both vegetatively and florally, and the flowers show numerous adaptations to insect pollination. They vary from the extremely beautiful to the bizarre, and the plants hold a rather special place in gardening, with many enthusiasts growing them (and little else) on a large scale. The literature on the family is extensive, including many illustrated works, taxonomic accounts and horticultural guides, and there are several journals devoted solely to the family (see below).

Cultivation. The cultivation of orchids is a subject that has generated much discussion, controversy and publication. It is not possible to go into detail here, and the reader is referred to the books and journals cited below for further information. Most of the species included here are of tropical origin, and require greenhouse protection throughout the whole of Europe; the hardy species (of which there is a relatively small number, mainly in genera Nos. 1–28) can be grown outside.

Two different composts are used for the tropical species. The first is for species that are terrestrial in their native habitats (terrestrial compost), the second for those which are epiphytes (epiphytic compost).

Terrestrial compost is made up of 2 parts (by bulk) of fibrous loam, 2 parts of finely crushed bark or medium-grade peat, and 1 part of grit or sand; 50 grams of a slow-release general fertiliser should be added to each cubic metre of the mixture. Species with particularly vigorous root systems will benefit from an increase in the amount of bark or peat in the mixture.

Epiphytic compost consists of 3 parts of pine bark mixed with 1 part of rough peat or sphagnum moss; 50–70 grams of slow-release general fertiliser are added to each cubic metre of this mixture. Some epiphytic species grow best when attached to a piece of cork-oak bark, using the smallest possible amount of compost.

Control of pests and diseases in orchids is a rather complex and specialised matter, and the relevant literature should be consulted for precise information. Good hygiene and cultivation are, of course, essential.

Literature: As mentioned above, the literature on orchids is very extensive; only relatively recent publications are listed here, in 3 categories:

General taxonomic accounts: Schlechter, R., Die Orchideen (1914), edn 2, edited by E. Miethe (1927), edn 3, edited by F.G. Brieger, R. Maatsch & K. Senghas, appearing in parts and continuing (1972–present); Holttum, R.E., Revised flora of Malaya **1**: Orchids (1953); Hawkes, A.D., Encyclopaedia of cultivated orchids (1965); Sheehan, T. & Sheehan, M., Orchid genera illustrated (1979); Dietrich, H., Bibliographia Orchidacearum (1980); Bechtel, H., Cribb, P. & Launert, E., Manual of cultivated orchid species (German edn, 1980, entitled Orchideenatlas; English edn, 1981); Dressler, R.L., The orchids: natural history and classification (1981). Pridgeon, A.M., Cribb, P.J., Chase, M.W. & Rasmussen, F.N., Genera Orchidacearum **1** (1999), **2** (2001), **3** (2003), **4** (2005), and continuing.

Horticultural guides: Northen, R., Home orchid growing (1962); Richter, W., Orchideen (1969); Hartmann, W.I., Introduction to the cultivation of orchids (1971); Sander, D., Orchids and their cultivation, edn 9 (1979); Thompson, P.A., Orchids from seed (1979); Black, P.M., The complete book of orchid growing (1980); Williams, B.A. et al., Orchids for everyone (1980); Williams, B.S., The orchid grower's manual, edn 7 (1897, reprinted 1982).

Hybrids: Sander's list of orchid hybrids, first published in 1905; supplements are published as necessary by the Royal Horticultural Society (the registration authority for the names of hybrid orchids).

Several journals are devoted entirely to the family, containing articles of both taxonomic and horticultural interest; the most important of these are: *The Orchid Review,* 1893 and continuing (British); *Orchis,* 1906–20 (German); *The American Orchid Society Bulletin,* 1932–1996, from 1997 and continuing as *Orchids* (American); *Orchideen,* 1934 and continuing (Dutch); *Australian Orchid Review,* 1936 and continuing (Australian); *Orchid Digest,* 1937 and continuing (American); *Die Orchidee,* 1949 and continuing (German); *Orchid Journal,* 1952–55 (American); *The Orchadian,* 1963 and continuing (Australian); *Orchidata,* 1965 and continuing (American); *South African Orchid Journal,* 1970 and continuing (S African); *Orquidea,* 1971 and continuing (Mexican).

The identification of orchids can be difficult, particularly at the genus level, and because the plants are so variable and so different from those of most other plant families, a specialised terminology has been developed. The important terms used in this account are briefly defined below, and some indication is given of how the various characters are to be observed.

Growth habit. Orchids grow either monopodially (i.e. with a growing point that continues vegetative growth from year to year, the inflorescences being borne laterally), or sympodially (with a growing point that ceases growth after some time, generally by flowering, further growth being continued by a lateral bud formed on the older growth). This feature is often quite difficult to appreciate in individual specimens, and is not used, as such, in the key to genera (pp. 446–452). It is important, however, in that monopodial orchids do not produce pseudobulbs, whereas sympodial orchids may do so.

Pseudobulbs (see Figure 19, p. 441). A pseudobulb is a swollen fleshy stem; whether or not a particular plant has pseudobulbs or more normal stems is a very important feature in its generic identification. In a few species and genera there is no sharp dividing line between a 'normal' stem and a pseudobulb; the keys to the genera have been constructed, as far as possible, to allow for these to be identified whether they have been considered as having a stem or a pseudobulb. The pseudobulbs, if present, are borne on the rhizomes, and may consist of several internodes, when they will bear leaves or leaf-scars along their lengths; such pseudobulbs are described here as 'compound'. Alternatively, they may consist of a single internode, when they bear leaves or leaf-scars only at the apex; such pseudobulbs are described here as 'simple'. The pseudobulbs are borne on the rhizomes and may be close together or distant, or, in a few genera, they form chains built up of several pseudobulbs borne one on top of the other.

Leaves (see Figure 20, p. 442). The manner in which the leaves are packed when young is an important character in identification of the genus. The young leaves may be rolled, so that one margin overlaps the other and the back of the young leaf is rounded, or they may be folded once longitudinally. This character is generally easy to see in most plants when young leaves are present, and it is usually quite easy to see in mature leaves, as the rolling or folding persists at the base. In some genera that bear few leaves it may be difficult to decide on this character when the leaves are mature; the following features help in making a decision. Folded leaves generally retain a longitudinal line or small fold down the middle over most of their length, and usually have a definite keel on the outside towards the base; very hard, leathery leaves are almost all folded. Rolled leaves are often pleated or have prominent veins, do not usually have a single line or fold down the middle, and are rounded on their backs towards the base. In some genera and species the leaves are terete or cylindric and with a groove or grooves; such leaves cannot be said to be either rolled or folded when young, but they are keyed here with genera with folded leaves.

Inflorescences (see Figure 21, p. 443). These may be terminal on the stem or pseudobulb, or lateral; in the latter case the inflorescence can arise in a leaf axil or a leaf-scar axil on the pseudobulb or stem, or from the rhizome at the bases of the pseudobulbs. In a few genera in which the stems bear only 1 leaf (e.g. *Pleurothallis*), the racemes are considered here to be terminal, even though they might be interpreted as axillary to the leaf.

Flowers (see Figures 21 & 22, pp. 443 & 444). In all the keys and descriptions the flowers are considered to be resupinate (i.e. with their stalks and ovaries twisted through 180° so that the lip petal is lowermost with the column above it) unless the contrary is stated. In species with arching or hanging inflorescences the degree of resupination of the individual flowers may vary, those flowers towards the hanging apex being non-resupinate (they are effectively turned upside down by the hanging of the inflorescence), while those at the more erect base may be properly resupinate. However, in a few genera with hanging inflorescences (e.g. *Paphinia*) the flowers are all strictly resupinate.

In many genera the base of the column is prolonged at an angle to the rest, forming a column foot (figure 23, p. 445). In all genera with a foot the lip is attached at or near the end of it. In many of these genera the lateral sepals are also borne on the sides of the foot, which results in a humped, angular, chin-like projection being visible on the outside of the back of the flower. This projection is known as a

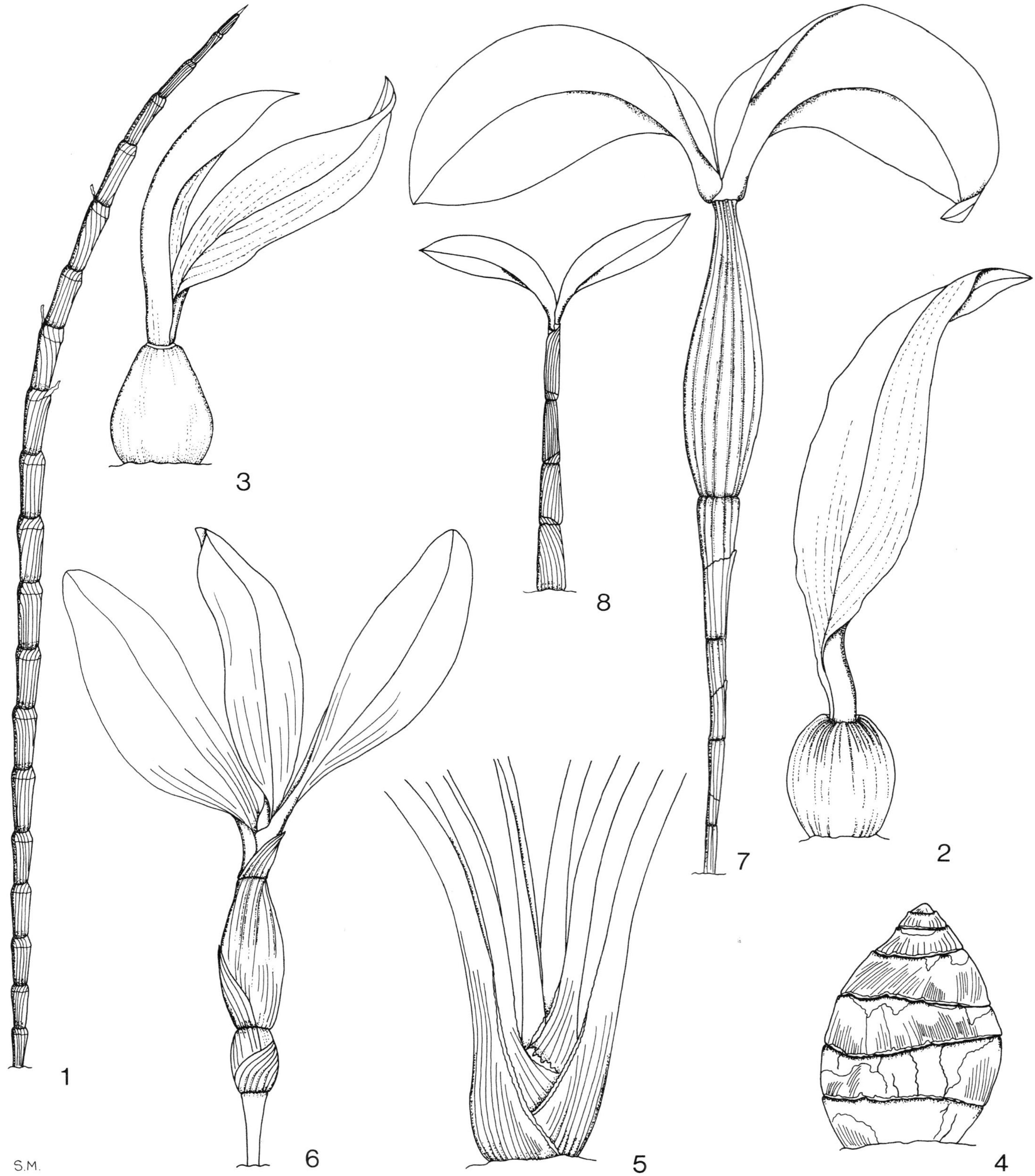

Figure 19. Diagram of some characters used in the identification of Orchidaceae. 1, a cane-like stem. 2, A simple pseudobulb bearing a single leaf. 3, a simple pseudobulb bearing 2 leaves. 4, a compound pseudobulb. 5, a compound pseudobulb enclosed in leaf-bases. 6, a fusiform compound pseudobulb. 7, a club-shaped compound pseudobulb. 8, a compound pseudobulb swollen at the base. (6, 7 & 8 are treated as pseudobulbs in the keys to the genera and species.)

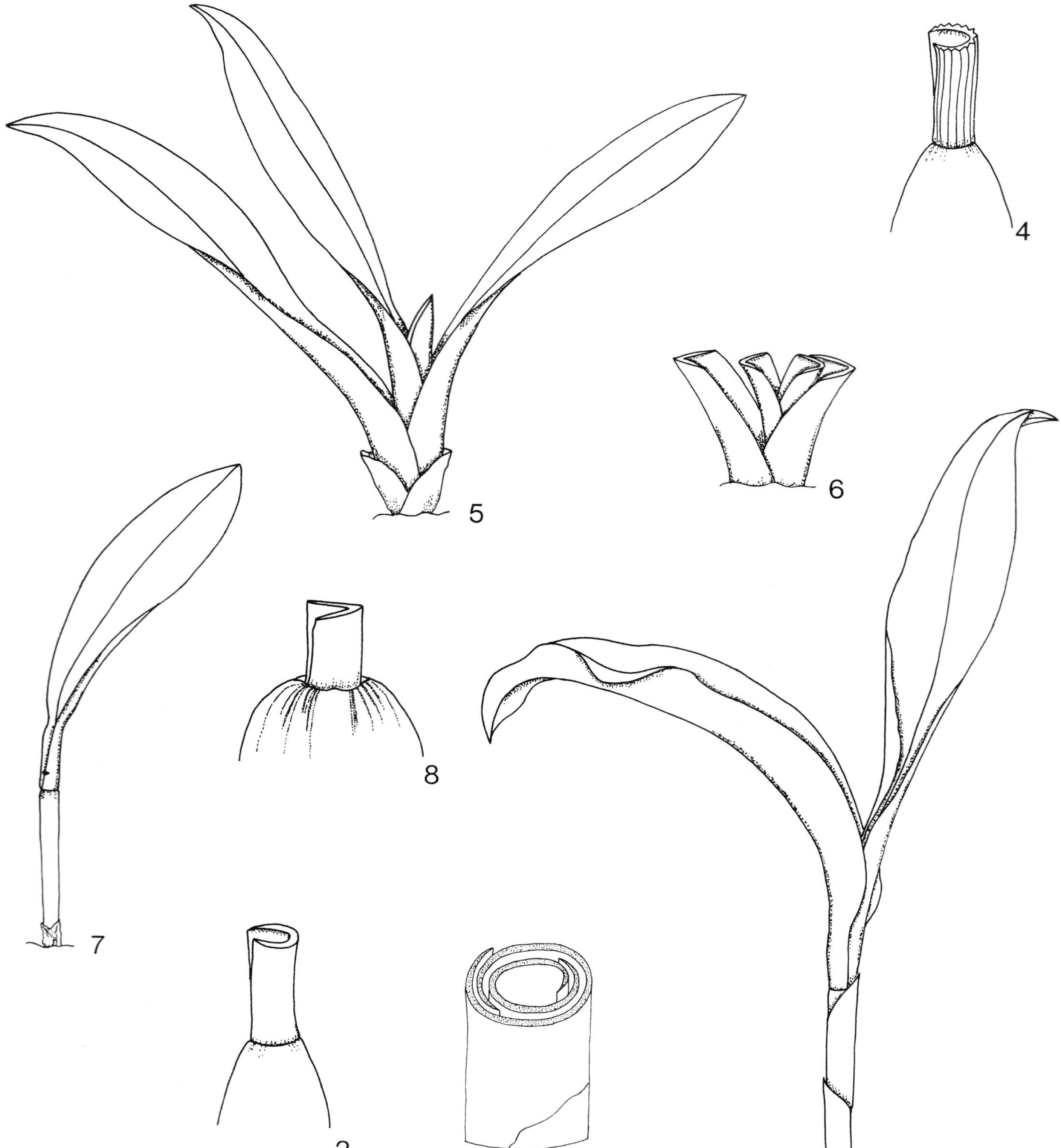

Figure 20. Diagram of some characters used in the identification of Orchidaceae. 1, a stem with rolled leaves. 2, cross-section of the bases of 2 rolled leaves. 3, a pseudobulb with the base of a rolled leaf. 4, a pseudobulb with the base of a rolled and pleated leaf. 5, a stem with folded leaves. 6, cross-section of the bases of folded leaves. 7, a stem of 1 internode with a single folded leaf. 8, a pseudobulb with the base of a folded leaf.

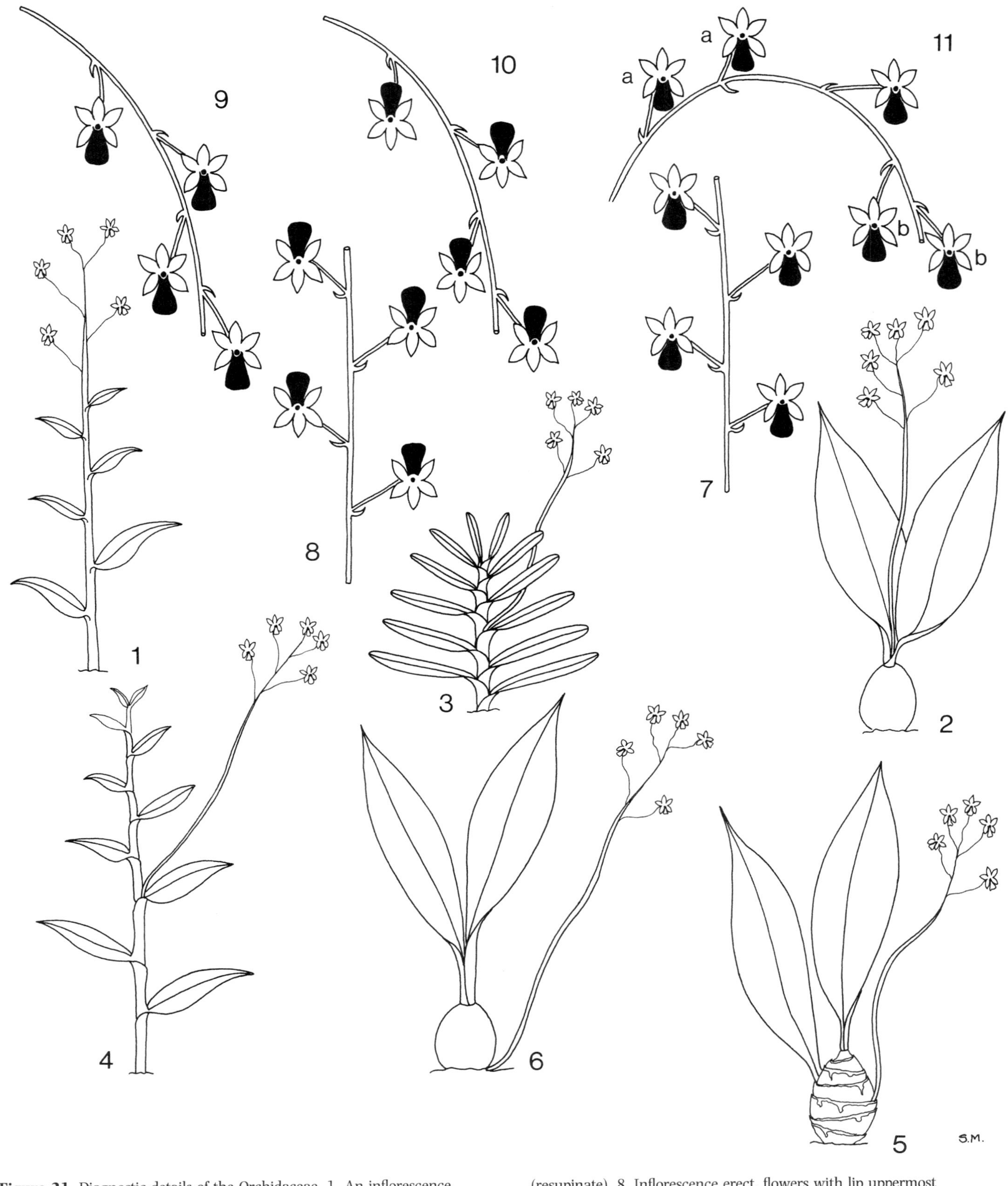

Figure 21. Diagnostic details of the Orchidaceae. 1, An inflorescence terminal on a stem. 2, An inflorescence terminal on a pseudobulb. 3, 4, Inflorescences lateral in the leaf axils on stems. 5, An inflorescence lateral to a leaf-scar on a compound pseudobulb. 6, An inflorescence basal to a pseudobulb. 7, Inflorescence erect, flowers with lip lowermost (resupinate). 8, Inflorescence erect, flowers with lip uppermost (non-resupinate). 9, Inflorescence hanging, flowers with lip lowermost (non-resupinate). 10, Inflorescence hanging, flowers with lip uppermost (resupinate). 11, Inflorescence arching, flowers with lip lowermost (a, resupinate; b, non-resupinate).

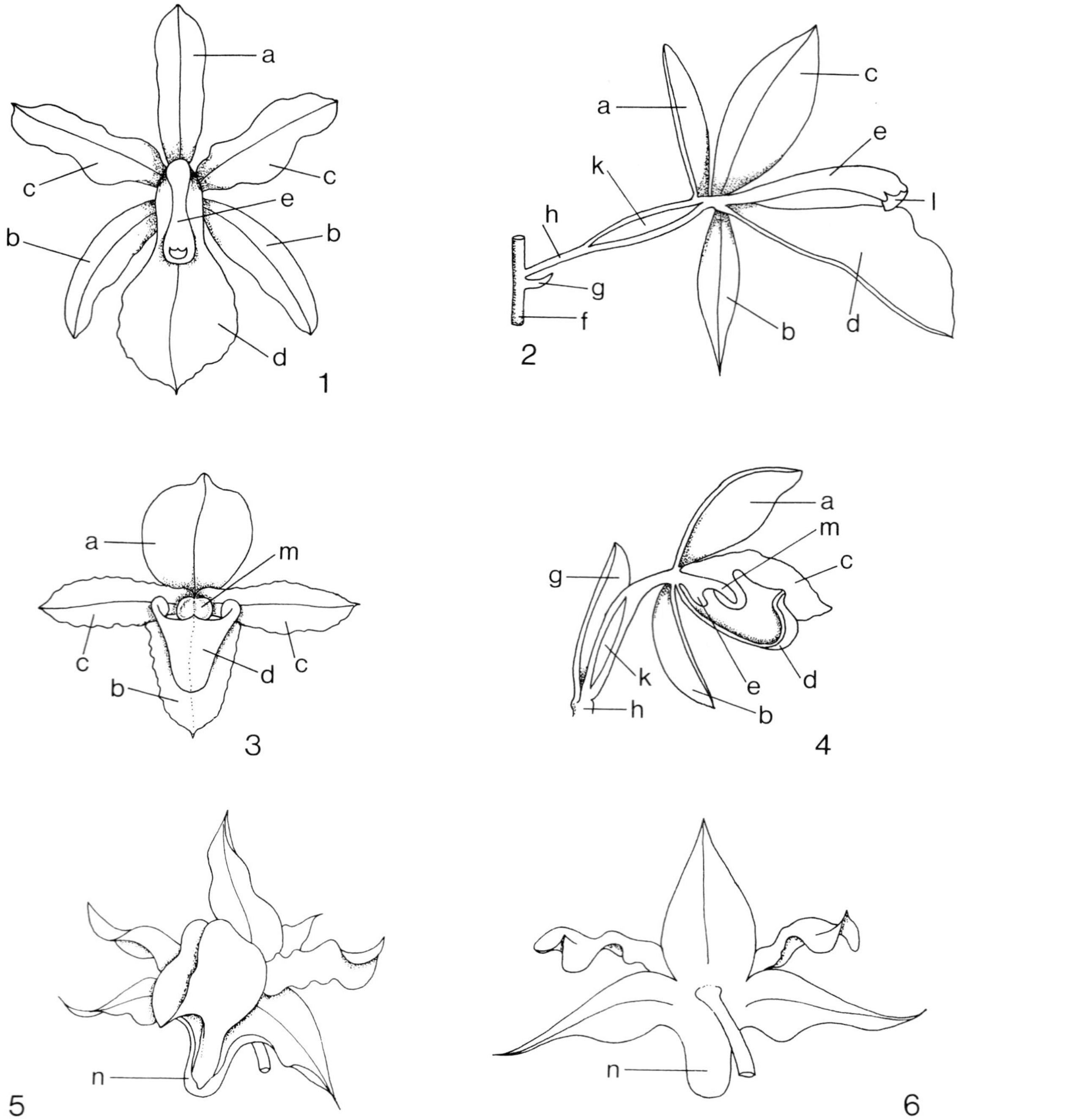

Figure 22. Diagnostic details of the Orchidaceae. 1, A generalised orchid flower (a, upper sepal; b, lateral sepal; c, petal; d, lip; e, column). 2, Section of a generalised orchid flower (a, upper sepal; b, lateral sepal; c, petal; d, lip; e, column; f, inflorescence-stalk; g, bract; h, flower-stalk; k, ovary; l, anther). 3, Flower of *Paphiopedilum* (a, upper sepal; b, lower sepal; c, petal; d, lip; m, staminode). 4, Section of flower of *Paphiopedilum* (a, upper sepal; b, lower sepal; c, petal; d, lip; e, column; g, bract; h, flower-stalk; k, ovary; m, staminode). 5, Flower of a *Dendrobium* from the front (n, mentum). 6, Flower of a *Dendrobium* from the back (n, mentum).

mentum (figure 22), and is of some importance in identification.

Pollinia (see figure 23, p. 445). Much of the classification of the family is based on the structure of the pollinia and their associated organs. In this account we have attempted to use these characters as little as possible, but the pollinia cannot be completely ignored, particularly their number. They can be extracted from the flower by sliding a needle or a match slowly upwards along the inner side of the column, when they will attach themselves to the needle, and can be examined through a hand-lens (a magnification of × 10 or × 15 is usually sufficient). The number of pollinia can usually easily be seen, though care must be taken, as the

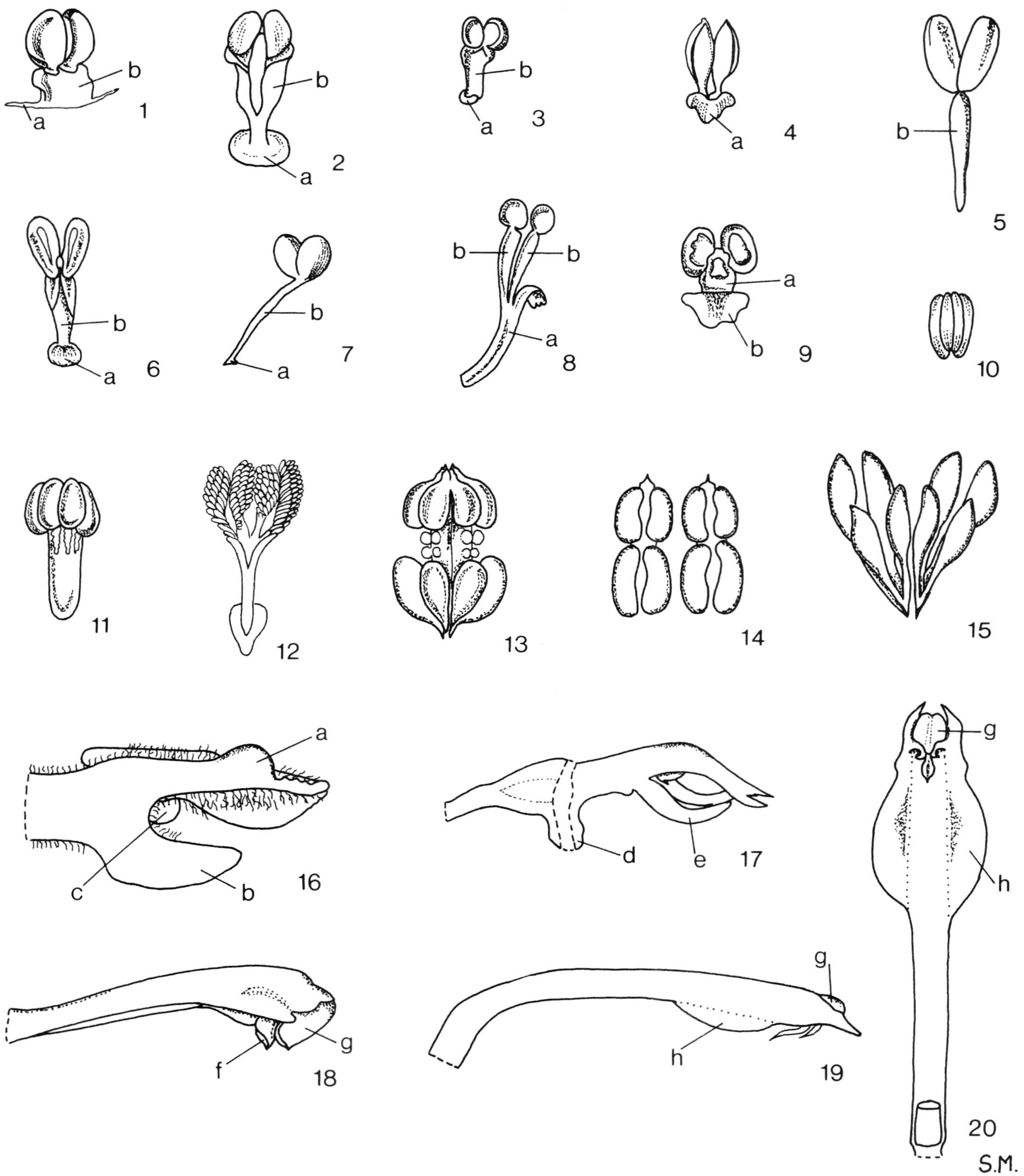

Figure 23. Characters used in the identification of the Orchidaceae. 1–15, Pollinia of various types. 1–9, Pollinia borne in groups of 2 (a, viscidium; b, stipe). 10, 11, Pollinia borne in groups of 4. 12, Pollinia borne in a group of 4, each made up of easily separable masses. 13–15, Pollinia in groups of 8. 16–20, Columns of various types. 16, Column and staminode of *Paphiopedilum* in section (a, staminode; b, column, c, pollinium). 17, Column with foot and lobes (d, foot; e, lobes). 18, Column with a beaked rostellum (f, rostellum; g, anther). 19, Side view of a winged column (g, anther; h, wing). 20, Lower view of a winged column (g, anther; h, wing).

pollinia are occasionally not all extracted at once. The pollinia are sometimes born jointly on a stalk known as the rostellum (see Figure 23, p. 445).

A few species have 2 deeply bilobed pollinia, which can look like 4; again, other species may have extra, infertile pollinia that are shrunken and colourless. In most of the tropical species in cultivation the pollinia are waxy and of well-defined shape. In some hardy species they are granular or powdery and in a few others they are each made up of several to many individual packets (known technically as massulae), which separate very easily from each other.

The column (see Figures 22 & 23, pp. 444 & 445). These are very variable in shape, size and details of structure; a selection is shown in the illustration.

The key to the genera that follows is divided into 5 group keys, with a key to the groups at the beginning. It makes use, as far as possible, of easily observed characters. Genera which are variable in some of the characters used in the key to the groups (e.g. whether or not they have pseudobulbs) have been deliberately keyed out in several of the groups. Many hybrids are cultivated in the Orchidaceae, including intergeneric hybrids to which 2 or more genera have contributed genetic materal. It is not, of course, possible to deal with these hybrids in this account. Reference should be made to Sander's list of orchid hybrids and its supplements for further information.

Key to groups

1a. Annual aerial growth dying off completely each autumn or winter, leaving no part above ground except, occasionally, a rosette of small leaves or a solitary leaf, formed after the current year's growth has died down **Group A**
b. Annual growth persisting throughout the year 2
2a. Leaves rolled when young (see Figure 31, p. 573) 3
b. Leaves folded once along the mid-line when young, or leaves terete or cylindric with a narrow groove (see Figure 31, p. 573) 4
3a. Plant with distinct pseudobulbs or stems swollen above or below **Group B** (p. 447)
b. Plant without pseudobulbs or swollen stems **Group C** (p. 448)
4a. Plant with distinct pseudobulbs or stems swollen above or below **Group D** (p. 448)
b. Plant without pseudobulbs or swollen stems **Group E** (p. 450)

Group A

1a. Leaves absent at flowering; lip with 3 ridges at the base, which become fleshy plates on the central lobe **40. Aplectrum**
b. Leaves present at flowering; lip not ridged as above 2
2a. Lip slipper-like 3
b. Lip not slipper-like 4
3a. Stem with 2 or more separated nodes and 2 or more leaves; lateral sepals pointing downwards, usually fused **1. Cypripedium**
b. Stem with 1 node and 1 leaf; lateral sepals pointing upwards, free **39. Calypso**
4a. A spur, or rarely 2 spurs, present, arising from the base of the lip or the base of the central sepal 5
b. Spur(s) absent 26
5a. Spurs 2 6
b. Spur 1 7
6a. Plant with 1 leaf which is held flat against the ground; flower 1 or rarely 2 **6. Corybas**
b. Leaves several, borne on erect stems; flowers numerous **5. Satyrium**
7a. Spur arising from the base of the concave or hooded central sepal **4. Disa**
b. Spur arising from the base of the lip 8
8a. Flower with the lip, which is similar to the petals, uppermost (i.e. flower not resupinate); spur 2 mm or less **16. Gymnadenia**
b. Flower with the lip, which is different from the petals, lowermost (i.e. flower resupinate); spur more than 2 mm 9
9a. Spur 2.5 cm or more 10
b. Spur less than 2.5 cm 16
10a. Lip strap-shaped, unlobed **18. Platanthera**
b. Lip not strap-shaped, variously lobed and dissected 11
11a. Uppermost leaf tightly sheathing the stem; column with 1 stigmatic area 15
b. Uppermost leaf not tightly sheathing the stem; column with 2 stigmatic areas 12
12a. Lip with a tooth at the entrance to the spur **21. Bonatea**
b. Lip without a tooth at the spur entrance 13
13a. Petals always entire; inflorescence usually glandular hairy; lip lobes not linear, laciniate or with thread-like ends **22. Cynorkis**
b. Petals often 2-lobed; inflorescence rarely glandular-hairy; lip lobes linear, usually laciniate or with thread-like tips 14
14a. Anthers widely spaced on broad connective; stigmas not stalked **20. Pecteilis**
b. Anthers close, connective narrow; stigmas stalked **19. Habenaria**
15a. Leaves often spotted, confined to basal rosette, not present on flowering stem **14. Orchis**
b. Leaves never spotted, present along flowering stem, not confined to basal rosette **13. Anacamptis**
16a. Sepals united at their bases to the column and lip **23. Stenoglottis**
b. Sepals free from column and lip 17
17a. Spur 10 mm or more, slender 18
b. Spur less than 10 mm, slender or not 21
18a. Leaves 1, rarely 2. **12. Ponerorchis**
b. Leaves several, lanceolate 19
19a. Spur cylindric, obtuse; flowers usually yellow **15. Dactylorhiza**
b. Spur thread-like, acute; flowers purple, pink or white 20
20a. Raceme conical; flowers with a foxy smell; lip deeply 3-lobed with 2 basal ridges **13. Anacamptis**
b. Raceme cylindric; flowers fragrant; lip shallowly 3-lobed without basal ridges **16. Gymnadenia**
21a. Lip spirally twisted or its lateral lobes with wavy margins 22
b. Lip not spirally twisted, margins not wavy 23
22a. Central lobe of lip 2 or more times longer than the lateral lobes, strap-shaped, spirally twisted; bracts shorter than the flowers **9. Himantoglossum**
b. Central lobe of lip less than 2 times longer than the lateral lobes, oblong, not spirally twisted; bracts longer than the flowers **10. Barlia**
23a. Lip strap-shaped, terminating in 3 short teeth; flowers greenish **17. Coeloglossum**
b. Lip not strap-shaped, variously lobed; flowers not greenish 24
24a. Leaves 1, rarely 2, often linear. **11. Amitostigma**
b. Leaves 3–10, lanceolate to oblong 25

25a. Root-tubers ovoid; leaves mostly in a basal rosette, the uppermost usually completely sheathing the stem; all bracts membranous, shorter than the flowers **14. Orchis**
b. Root-tubers lobed; leaves not in a basal rosette; upper leaves sheathing the stem at its base only, often transitional to the leaf-like bracts; bracts usually as long as, or longer than the flowers **15. Dactylorhiza**
26a. Stem with 2 more or less opposite leaves; lip strap-shaped, 2-lobed at the apex **82. Neottia**
b. Stem with more than 2 leaves, which are distributed along it or are in a basal rosette; lip not as above 27
27a. Flowers arranged in a tight spiral 28
b. Flowers not arranged in an obvious spiral 29
28a. Flowers white to pink; rostellum not ending in a sharp, rigid point **29. Spiranthes**
b. Flowers scarlet; rostellum ending in a sharp, rigid point **30. Stenorrhynchos**
29a. Lip divided into 2 parts by a constriction at the middle 30
b. Lip not divided into 2 parts 32
30a. Roots tuberous; leaves not pleated; bracts large, coloured like the sepals, sheathing the basal parts of the flowers **8. Serapias**
b. Roots fibrous, tubers absent; leaves pleated; bracts green and leaf-like, not sheathing the basal parts of the flowers 31
31a. Flowers shortly stalked, borne mostly on 1 side of the axis and horizontal or downwardly pointing; sepals and petals spreading **83. Epipactis**
b. Flowers not stalked, borne all round the axis, pointing upwards; sepals and petals not widely spreading **84. Cephalanthera**
32a. Lip small, usually sensitive to touch, hidden within a hood formed by the sepals and petals **28. Pterostylis**
b. Lip as large as or larger than, and not concealed by the sepals and petals, not sensitive to touch 33
33a. Lip velvety or covered with short hairs or with a geometric pattern or with a shining area or any combination of these, if lobed the lobes not long and slender **7. Ophrys**
b. Lip without any of the above characteristics, the lateral and central lobes slender, like the arms and legs of a man **14. Orchis**

Group B

1a. Pollinia 2 2
b. Pollinia 4 or 8 21
2a. Pseudobulbs compound 3
b. Pseudobulbs simple 10
3a. Flowers asymmetric, due to the twisting of the column to 1 side **165. Mormodes**
b. Flowers symmetric about the vertical plane 4
4a. Lip uppermost (i.e. flower not resupinate); flowers usually unisexual 5
b. Lip lowermost (i.e. flower resupinate); flowers bisexual 7
5a. Column thin but expanded conspicuously towards the apex, arching, without 2 backwardly directed, bristle-like outgrowths **168. Cycnoches**
b. Column equally thick throughout, straight, usually with 2 backwardly projecting, bristle-like outgrowths 6
6a. Flowers all bisexual; column without antennae **167. Clowesia**
b. Flowers mostly unisexual; column in the male flowers with 2 antennae **166. Catasetum**
7a. Raceme terminal; lip rolled around and partly concealing the column **160. Galeandra**
b. Raceme lateral; lip not rolled around and concealing the column 8
8a. Lip with a distinct, though short, spur **156. Eulophia**
b. Lip without a spur 9
9a. Lip with a distinct and conspicuous claw **159. Cyrtopodium**
b. Lip without a claw **158. Eulophiella**
10a. Sepals and petals directed forwards, incurved, the flower bell-shaped 11
b. Sepals and petals spreading or reflexed, the flower not bell-shaped 14
11a. Lip with a distinct claw, which bears the lateral lobes near its middle **169. Acineta**
b. Lip with a short claw or none, the lateral lobes borne at the extreme base 12
12a. Anterior part of the lip freely movable on the rest **170. Peristeria**
b. Anterior part of the lip not movable on the rest 13
13a. Pseudobulbs with 1–2 leaves; central lobe of lip not downy **171. Lacaena**
b. Pseudobulbs with 3–5 leaves; central lobe of lip downy **172. Lueddemannia**
14a. Column with a reflexed apex; lip very complex, bearing 2 nectar-secreting, peg-like glands near the base, which drip nectar into the bucket-like anterior part **179. Coryanthes**
b. Column without a reflexed apex; lip variously complex, but not as above 15
15a. Flowers in hanging racemes, the column (as viewed in the growing position) below the lip (i.e. flowers resupinate) 16
b. Flowers in hanging, arching or erect racemes, the lip (as viewed in the growing position) below the column (i.e. flowers not resupinate) 18
16a. Apex of the lip bearing white, club-shaped, hair-like processes **175. Paphinia**
b. Apex of the lip without such processes 17
17a. Upper sepal fused to the column for about half of the length of the latter; anther at the apex of the column **177. Gongora**
b. Upper sepal free from column; anther apparently on the back of the column **178. Cirrhaea**
18a. Basal part of the lip, or the whole lip, hollowed out and saccate **176. Stanhopea**
b. Lip not at all hollowed out 19
19a. Pseudobulbs with 2–4 leaves **108. Eriopsis**
b. Pseudobulbs with 1 leaf 20
20a. Lateral lobes of lip narrow, curving backwards and upwards and then forwards; axis of raceme without hairs **173. Houlletia**
b. Lateral lobes of lip oblong-triangular, curving upwards; axis of raceme downy **174. Polycycnis**
21a. Pollinia 8 22
b. Pollinia 4 32
22a. Racemes apparently terminal 23
b. Racemes obviously lateral 25
23a. Column with a distinct foot to which the lateral sepals are attached, forming a mentum 24
b. Column without a foot, mentum absent **33. Bletilla**
24a. Pseudobulbs formed from single internodes **85. Eria**

b. Pseudobulbs formed from several internodes **86. Pinalia**
25a. Column with a conspicuous foot to which the lateral sepals are attached, forming a mentum 26
b. Column without a foot, mentum absent 27
26a. Pseudobulbs simple **47. Coelia**
b. Pseudobulbs compound **46. Chysis**
27a. Lip with a projecting spur 28
b. Lip without a spur, though a pouch is sometimes present that does not project from the flower 29
28a. Leaves absent at flowering; axis of raceme, flower-stalks and backs of flowers hairy **41. Calanthe**
b. Leaves present at flowering; axis of raceme, flower-stalks and backs of flowers not hairy **43. Phaius**
29a. Central lobe of lip with a long, narrow claw **44. Spathoglottis**
b. Central lobe of lip without a long, narrow claw 30
30a. Inflorescence bracts broad, almost as long as the flower-stalks, often falling early 31
b. Inflorescence bracts narrow, much longer than the flower-stalks, persistent **45. Bletia**
31a. Lip not spurred, but saccate; callus raised and hairy **42. Gastrorchis**
b. Lip spurred, not saccate; callus ridged, not hairy **43. Phaius**
32a. Column with a distinct foot 33
b. Column without a foot 43
33a. Flower solitary 34
b. Flowers few to many, in racemes 36
34a. Perianth segments all incurved, flower almost spherical **107. Anguloa**
b. Perianth segments all spreading, flower widely open 35
35a. Side-lobes of lip entire, forming a wall along lower part of lip; flower facing forwards **105. Lycaste**
b. Side-lobes of lip not entire; flower inclined downwards **106. Ida**
36a. Lip freely movable on the column; flowers many, 2-ranked in slender racemes **36. Dendrochilum**
b. Lip not movable on the column; flowers various, not as above 37
37a. Callus of lip divided into finger-like projections **95. Acacallis**
b. Callus of lip not divided as above 38
38a. Callus of lip transverse, forming a curved frill across the base of the lip **94. Zygopetalum**
b. Callus of lip elongate or formed of a few longitudinal ridges or plates 39
39a. Pollinia 4, in 2 pairs, each pair borne on a separate stipe; column winged above 40
b. Pollinia 4, all borne on a common stipe (which may be very small); column not winged 41
40a. Pseudobulbs with 1 leaf **104. Bifrenaria**
b. Pseudobulbs with 2–3 leaves **102. Galeottia**
41a. Petals more or less equal in size to the upper sepal **108. Eriopsis**
b. Petals smaller than the upper sepal 42
42a. Leaves in 2 ranks; column somewhat swollen above **92. Warrea**
b. Leaves not in 2 ranks; column cylindric **103. Xylobium**
43a. Lip uppermost (i.e. flower not resupinate), with auricles that extend downwards below the column **80. Malaxis**
b. Lip lowermost (i.e. flowers resupinate), without auricles 44
44a. Lip pouched or sac-like towards the base; racemes dense, with many flowers **38. Pholidota**
b. Lip not pouched or sac-like towards the base; racemes with 1–4 flowers, or if more, then loose 45
45a. Lip bent sharply backwards at or about the middle; flowers at most 2.5 cm in diameter **81. Liparis**
b. Lip not bent sharply backwards; flowers more than 2.5 cm in diameter 46
46a. Leaves membranous, deciduous; lip unlobed or very shallowly 3-lobed, base rolled around and concealing the column **35. Pleione**
b. Leaves not membranous; lip clearly 3-lobed, lateral lobes erect or spreading, not concealing the column 47
47a. Central lobe of lip hairy, lateral lobes spreading **93. Pabstia**
b. Central lobe of lip hairless, lateral lobes erect **34. Coelogyne**

Group C

1a. Leaves with veins differently coloured from the rest of the leaf; pollinia made up of easily separable masses 2
b. Leaves with veins the same colour as the rest; pollinia granular or waxy, not made up of easily separable masses 5
2a. Lip uppermost (i.e. flower not resupinate); flowers asymmetric, due to twisting of the lip and column in opposite directions **25. Macodes**
b. Lip lowermost (i.e. flower resupinate); 3
3a. Flowers symmetric about the vertical plane 4
b. Flowers asymmetric, due to twisting of the lip and column in opposite directions **26. Ludisia**
4a. Lip not widening towards the apex, claw not fringed or toothed **24. Goodyera**
b. Lip abruptly widened to a transverse blade, claw fringed or toothed **27. Anoectochilus**
5a. Pollinia 2, borne on a triangular stipe **159. Cyrtopodium**
b. Pollinia 4 or 8, not borne on a stipe 6
6a. Axis of raceme, flower-stalks and the backs of the flowers hairy **41. Calanthe**
b. Plants not hairy as above 7
7a. Lip with an obtuse spur 8
b. Lip without a spur 9
8a. Inflorescence axillary, erect **43. Phaius**
b. Inflorescence terminal, hanging **37. Thunia**
9a. Lip saccate **42. Gastrorchis**
b. Lip not saccate 10
10a. Leaves many in 2 ranks; column without a foot, mentum absent **91. Sobralia**
b. Leaves 2 or 4 in opposite pairs; column with a distinct foot to which the lateral sepals are attached, forming a mentum **85. Eria**

Group D

1a. Column with a distinct foot, mentum often conspicuous 2
b. Column without a foot, mentum absent 17
2a. Pseudobulbs compound 3
b. Pseudobulbs simple (all nodes at, or very near, the apex), sometimes in chains 6
3a. Panicle large, terminal **161. Ansellia**
b. Racemes lateral (rarely apparently terminal) 4
4a. Petals less than one-third of the length of the upper sepal; raceme arising from the rhizome near the base of the pseudobulb **163. Grammangis**

b. Petals at least two-thirds of the length of the upper sepal; racemes usually borne on the pseudobulbs 5
5a. Pollinia 4 **88. Dendrobium**
b. Pollinia 8 **85. Eria**
6a. Pollinia 2 or 4 7
b. Pollinia 8 15
7a. Pseudobulbs borne one on top of another, forming erect chains **63. Scaphyglottis**
b. Pseudobulbs not as above 8
8a. Flowers or racemes terminal 9
b. Flowers or racemes lateral or basal 11
9a. Axis of raceme downy; flower with lip uppermost (i.e. not resupinate) **78. Polystachya**
b. Axis of raceme hairless; flower with lip uppermost (i.e. resupinate) 10
10a. Flowers more than 4.5 cm in diameter; lip 3-lobed, not pouched at apex **89. Epigeneium**
b. Flowers less than 2 cm in diameter; lip entire, pouched at apex **64. Domingoa**
11a. Lateral sepals much larger than upper sepal **90. Bulbophylllum**
b. All sepals more or less equal 12
12a. Bracts on the flowering scape clasping, overlapping or almost so, rarely with their apices spreading **110. Maxillaria**
b. Bracts on the flowering scape not clasping, distant, spreading 13
13a. Lip not 3-lobed, sometimes fringed **90. Bulbophyllum**
b. Lip distinctly 3-lobed 14
14a. Leaves terete **109. Scuticaria**
b. Leaves flat **96. Promenaea**
15a. Flower solitary **56. Meiracyllium**
b. Flowers in dense racemes 16
16a. Sepals fused at the base into a spherical or urn-shaped cup that conceals the petals **87. Cryptochilus**
b. Sepals almost free from each other but fused to the column foot; petals exposed **85. Eria**
17a. Flower with a spur, which may be formed by the bases of the sepals or by the base of the lip 18
b. Flower without a spur of any kind 23
18a. Spur formed by the base of the lip 19
b. Spur formed by the bases of the sepals 20
19a. Flowers 1 or 2, 2.5 cm or more in diameter **116. Trichocentrum**
b. Flowers 5 or more, up to 2.5 cm in diameter **157. Oeceoclades**
20a. Spur attached to the ovary **52. Broughtonia**
b. Spur free from the ovary 21
21a. Lip without appendages that project back into the sepal spur **117. Ionopsis**
b. Lip with 1 or 2 appendages that project back into the sepal spur 22
22a. Lip with 1 appendage **118. Rodriguezia**
b. Lip with 2 appendages **119. Comparettia**
23a. Sides of the lip attached to the column, at least at the base 24
b. Sides of lip entirely free from the column (though the point of lip attachment may be a little way up the column) 31
24a. Pollinia 2 25
b. Pollinia 4 or 8 27
25a. Pseudobulbs compound **162. Cymbidium**
b. Pseudobulbs simple 26
26a. Stigmatic surface divided into 2; sepals and petals free from the column **121. Cochlioda**
b. Stigmatic surface not divided; upper sepal and petals fused to the back of the column **125. Aspasia**
27a. Pollinia 4 28
b. Pollinia 8 30
28a. Sides of the lip attached to the column for almost the whole length of the latter **49. Epidendrum**
b. Sides of the lip attached to the column for at least the lower half of the latter 29
29a. Inflorescence a many-flowered raceme or panicle; column winged; capsule ellipsoidal **50. Encyclia**
b. Inflorescence 1–few flowered; column wingless; capsule trigonous **51. Prosthechea**
30a. Leaves ovate, elliptic or oblong, leathery **54. Cattleya**
b. Leaves linear, fleshy **61. Isabelia**
31a. Pollinia 2 32
b. Pollinia 4 or more (sometimes some sterile) 43
32a. Pseudobulbs compound 33
b. Pseudobulbs simple 34
33a. Stipe of pollinia divided; plants usually very large (raceme to 2 m) **164. Grammatophyllum**
b. Stipe of pollinia not divided; plant much smaller **162. Cymbidium**
34a. Sides of the lip rolled over the column **120. Trichopilia**
b. Sides of the lip not rolled over the column 35
35a. Sepals and petals directed forwards, incurved **126. Ada**
b. Sepals and petals spreading or reflexed 36
36a. Sepals and petals at least 6 times as long as broad **127. Brassia**
b. Sepals and petals relatively much broader 37
37a. Base of lip ascending more or less parallel with the column 38
b. Base of lip directed away from the column from its base, at an angle of *c.* 90° 40
38a. Lip with 2 basal auricles, which are fused to the base of the column **123. Helcia**
b. Lip without auricles 39
39a. Ridges on the base of the lip reaching up to and slightly clasping the column; upper sepal and petals directed upwards, lateral sepals directed downwards; flowers basically yellow **122. Gomesa**
b. Ridges on the base of the lip not reaching the column; sepals and petals usually evenly spreading; flowers usually not basically yellow **124. Odontoglossum**
40a. Lip with a distinct, parallel-sided claw **131. Sigmatostalix**
b. Lip scarcely clawed, or if with a claw then the claw not parallel-sided and conspicuous 41
41a. Lip usually simple, with 2 (rarely more) longitudinal ridges or lines of hairs near the base; flowers more or less flattened, not basically yellow; column not thickened towards the base 42
b. Lip usually lobed and bearing a complex callus of tubercles or ridges and tubercles; flowers not flattened, basically yellow; column thickened, with a fleshy plate below the stigma **130. Oncidium**
42a. Pseudobulbs and leaves grey-green; pseudobulbs usually with 1 leaf and some scales **128. Miltonia**
b. Pseudobulbs and leaves yellow-green; pseudobulbs usually with 2 leaves and some leaf bases **129. Miltoniopsis**
43a. Fertile pollinia 8 (sometimes some extra and sterile) 44
b. Fertile pollinia 4 (sometimes some extra and sterile) 47
44a. Column with narrow, acute teeth projecting beyond the anther but not adpressed to it; stigma irregularly lobed **60. Brassavola**

b. Column with small, blunt, apical teeth, which are adpressed to the anther; stigma triangular or semicircular 45
45a. Column slender with 3 blunt apical teeth; sepals and petals rarely wavy **57. Laelia**
b. Column stout with 1 blunt apical tooth; sepals and petals usually wavy 46
46a. Pseudobulbs solid **57. Laelia** (Schomburgkia)
b. Pseudobulbs hollow **58. Myrmecophila**
47a. Sepals united into a tube at the base **112. Trigonidium**
b. Sepals completely free from each other 48
48a. Lip freely movable on its attachment to the base of the column **111. Mormolyca**
b. Lip not freely movable on its attachment to the column 49
49a. Flowers to 2.5 cm in diameter; basal part of the lip not rolled around the column **81. Liparis**
b. Flowers 3 cm or more in diameter; basal part of the lip rolled around the column 50
50a. Leaves sharply toothed **52. Broughtonia**
b. Leaves entire 51
51a. Lip entire **55. Guarianthe**
b. Lip 3-lobed **54. Cattleya**

Group E

1a. A broad, flat staminode present on the column and extending beyond it; fertile anthers 2 2
b. Staminode absent; fertile anther 1 3
2a. Ovary 3-celled; sepals edge-to-edge in bud **2. Phragmipedium**
b. Ovary 1-celled; sepals overlapping in bud **3. Paphiopedilum**
3a. Inflorescence lateral, usually axillary 4
b. Inflorescence terminal, or borne directly on the rhizome 45
4a. Lip consisting mainly of a vertical pouch or sac, its blade relatively small, consisting of 3 lobes borne on the margins of the pouch 5
b. Lip with a conspicuous blade, which is lobed or not, spurred or not, but not as above 10
5a. Column without a foot; pollinia 2 6
b. Column with a foot; pollinia 4, often in 2 pairs 7
6a. Lateral lobes of lip fused to the column, central lobe hairy or fringed **147. Gastrochilus**
b. Lateral lobes of lip free from the column, central lobe neither hairy nor fringed **146. Ascocentrum**
7a. Plant leafless when flowering (though with small, green bracts on the scape) **133. Chiloschista**
b. Plant leafy when flowering 8
8a. Spur or pouch of lip without any calluses inside **134. Pteroceras**
b. Spur or pouch of lip with calluses inside 9
9a. Leaves in 2 ranks; central lobe of the lip smaller than the lateral lobes **132. Sarcochilus**
b. Leaves borne all around the stem, or 1-sidedly; central lobe of lip larger than the lateral lobes **149. Cleisostoma**
10a. Flower with a distinct spur projecting from the back 11
b. Flower without a spur, though the base of the lip sometimes slightly hollowed 24
11a. Lip distinctly and obviously 3-lobed 12
b. Lip entire or very slightly and inconspicuously 3-lobed 18
12a. Column without a foot 13
b. Column with a foot 16
13a. Sepals and petals 2–5 mm; rostellum linear, bifid, very conspicuous **148. Schoenorchis**
b. Sepals and petals at least 10 mm, often much larger; rostellum not as above 14
14a. Pollinia 4; flowers in panicles **145. Renanthera**
b. Pollinia 2; flowers in racemes 15
15a. Flower at most 3 cm in diameter, entirely white **152. Neofinetia**
b. Flower more than 4 cm in diameter, coloured, at least in part **141. Vanda**
16a. Lip freely movable on its attachment to the column foot; spur pointing forwards **137. Sedirea**
b. Lip not movable on its attachment to the column foot; spur pointing backwards or downwards 17
17a. Pollinia borne on a short, broad stipe; flower 8–10 cm in diameter **141. Vanda**
b. Pollinia borne on a long, slender stipe; flowers to 5 cm in diameter **136. Aerides**
18a. Column with a long foot, continued without a joint into the lip **135. Papilionanthe**
b. Column without a foot, or if with a short foot then lateral sepals not attached to form a mentum 19
19a. Rostellum small, rather inconspicuous, bifid **150. Angraecum**
b. Rostellum large, beaked, conspicuous, not bifid 20
20a. Sepals conspicuously broader than the petals; flower usually white, spotted or blotched with red or pinkish purple **138. Rhynchostylis**
b. Sepals and petals equally broad; flower white or greenish white, sometimes with the lip yellowish at the base but not spotted as above 21
21a. Pollinia 2, each borne on a separate stipe 22
b. Pollinia 2, both borne on a common stipe 23
22a. Lip about the same size as the sepals and petals, without a tooth-like callus in front of the mouth of the spur **153. Cyrtorchis**
b. Lip much broader than the sepals and petals, with a tooth-like callus in front of the mouth of the spur **154. Diaphananthe**
23a. Spur with a distinct, narrow mouth; lip equalling the sepals and petals in breadth or very slightly broader **151. Aerangis**
b. Spur with a wide, indistinct mouth; lip considerably broader than the sepals and petals **155. Eurychone**
24a. Column without a foot 25
b. Column with a distinct foot 33
25a. Pollinia 2 26
b. Pollinia 4 29
26a. Sepals and petals 7–15 mm; leaves strongly keeled, with overlapping bases **115. Lockhartia**
b. Sepals and petals much larger; leaves various, not strongly keeled, sheaths usually not overlapping 27
27a. Claw of lip fused to column; leaves very fleshy or leathery; plant a climber **31. Vanilla**
b. Claw of lip free from column; leaves usually not very fleshy or leathery; plant not a climber 28
28a. Leaves conspicuously veined, the apex acute or 2-lobed; lateral lobes of the lip erect and partially encircling the column **162. Cymbidium**

b. Leaves not conspicuously veined, the apex irregularly toothed; lateral lobes of lip not as above **141. Vanda**

29a. Flowers of 2 kinds in each raceme; raceme axis and the backs of the flowers woolly **144. Dimorphorchis**

b. Flowers all similar in each raceme; axis of raceme and the backs of the flowers not woolly 30

30a. Lip above the column (i.e. flower not resupinate); rostellum long-beaked, anther with a long, terminal appendage **114. Ornithocephalus**

b. Column above the lip (i.e. flowers resupinate); rostellum not long-beaked, anther without an appendage 31

31a. Flower solitary; lip usually without a callus or ridges, not hollowed out at the base **113. Dichaea**

b. Flowers in racemes; lip with a callus or longitudinal ridges, hollowed out near the base 32

32a. Lip freely movable on the column, with erect, oblong lateral lobes that are free from each other **143. Esmeralda**

b. Lip not movable on the column, the lateral lobes small, joined together by a fleshy plate above the ridge of the central lobe **142. Vandopsis**

33a. Pollinia 2 34

b. Pollinia 4 or 8 36

34a. Leaves flat **139. Phalaenopsis**

b. Leaves terete, or cylindric with a narrow, longitudinal groove 35

35a. Sepals and petals greenish yellow, brownish yellow or whitish, sometimes flushed with pink or violet **140. Paraphalaenopsis**

b. Sepals and petals yellow or greenish yellow, blotched with reddish brown **109. Scuticaria**

36a. Pollinia 8 **85. Eria**

b. Pollinia 4 37

37a. Racemes with few to many flowers 38

b. Flower solitary 39

38a. Lip 3-lobed, with 2 linear appendages on the claw; leaves clearly in 2 ranks **139. Phalaenopsis**

b. Lip 3-lobed, without appendages on the claw; leaves usually not obviously in 2 ranks **88. Dendrobium**

39a. Lip at most 1 cm **110. Maxillaria**

b. Lip at least 2 cm 40

40a. Callus of lip fringed **101. Huntleya**

b. Callus of lip not fringed 41

41a. Petals yellow with fringed or toothed margins **97. Chondroscaphe**

b. Petals basically white, cream, greenish, pink or purple, sometimes spotted with yellow at the base, margins entire 42

42a. Lip distinctly 3-lobed 43

b. Lip not 3-lobed 44

43a. Lateral lobes of the lip rolled over the column and overlapping; lip with little or no claw **98. Warszewiczella**

b. Lateral lobes of the lip spreading; lip with a conspicuous claw **100. Pescatoria**

44a. Column hooded, protruding over the lip callus **100. Pescatoria**

b. Column neither hooded nor protruding over the lip callus **99. Cochleanthes**

45a. Pollinia 2, 4 or 6, sometimes with some extra and sterile 45

b. Fertile pollinia 8 65

46a. Lateral sepals conspicuously fused at the base (the upper sepal may be fused to the laterals or free from them) 47

b. Lateral sepals free or very shortly and inconspicuously fused at the base 60

47a. Pollinia 4 48

b. Pollinia 2 49

48a. Stem with several leaves, which are up to 3 mm broad; upper sepal 8–9 mm **79. Isochilus**

b. Stems each with 1 leaf, which is at least 2.5 cm broad; upper sepal at least 2 cm **75. Restrepia**

49a. Upper sepal free, or almost so, from the fused lateral sepals 50

b. Upper sepal clearly fused at the base to the fused lateral sepals 54

50a. Stems with white sheaths **74. Specklinia**

b. Stems with grey or brown sheaths or sheaths absent 51

51a. Inflorescence dense **73. Acianthera**

b. Inflorescence lax 52

52a. Inflorescence zig-zag, or erect to spreading; flowers greenish to yellow, sometimes with red to purple markings **72. Stelis**

b. Inflorescence arched to pendant, flowers not as above 53

53a. Flowers dark purple **71. Pleurothallis**

b. Sepals greenish, petals orange to red **70. Lepanthes**

54a. Sepals fused at base and towards apex, leaving 3 ‘windows’ to the interior of the flower **76. Zootrophion**

b. Sepals fused only at the base 55

55a. Sepal tube forming a 3-angled cup slightly constricted at the mouth; column without a foot **64. Domingoa**

b. Sepal tube not as above; column with a foot (sometimes small) 56

56a. Lip with a distinct claw 57

b. Lip without a claw 58

57a. Leaves papillose on the upper surface; lip sensitive to touch; flowering stem densely hairy **68. Porroglossum**

b. Leaves not papillose; lip not sensitive; stem not densely hairy **67. Dracula**

58a. Fused lateral sepals forming an oblong, acute blade bearing 2 lateral tails **69. Trisetella**

b. Fused lateral sepals not as above, with a sinus or notch between the lateral lobes, which may be tailed or not 59

59a. Ovary and sepals covered with prickles **65. Diodonopsis**

b. Ovary and sepals without prickles **66. Masdevallia**

60a. Lip fused at least to the lower half of the column 61

b. Lip entirely free from the column 62

61a. Lip fused to the column for most of the length of the latter **49. Epidendrum**

b. Upper half of the column free from the lip **53. Barkeria**

62a. Pollinia 2; leaves, and racemes or panicles, borne directly on the rhizome (stem absent) **130. Oncidium**

b. Pollinia 4 or 6; stems present 63

63a. Pollinia 6, 4 large and 2 small; stems at most 4 cm **62. Leptotes**

b. Pollinia 4, all similar; stems more than 4 cm (often much more) 64

64a. Lip entire **55. Guarianthe**

b. Lip 3-lobed **54. Cattleya**

65a. Sepals to 2 cm; lip not rolled around column 66

b. Sepals 2.5 cm or more; lip rolled around, and often concealing the column 67

66a. Lip deeply concave or pouched; flowers pinkish purple, at least in part **48. Arpophyllum**

b. Lip flat; flowers white, yellow or greenish yellow **77. Octomeria**

67a. Lip conspicuously 3-lobed, without a claw, its lateral lobes rolled around the column **57. Laelia**
b. Lip entire or shallowly 3-lobed, sometimes with a conspicuous claw, its base rolled round the column 68
68a. Lateral sepals spreading **60. Brassavola**
b. Lateral sepals lying close together under the lip **32. Arundina**

1. CYPRIPEDIUM Linnaeus

J.C.M. Alexander

Terrestrial, aerial parts dying away in winter. Leaves deciduous, pleated (except No. 9), rolled when young, spirally arranged or in a single, almost opposite pair. Flowers persistent, solitary or up to 12 per raceme; bracts leaf-like. Sepals edge-to-edge in bud. Petals and upper sepal free, lateral sepals fused into a single lower sepal below the lip (free in No. 6). Lip extended into a sac-like pouch with an inturned rim (except No. 13), the opening partially blocked by a staminode. Stigmatic surface, 2 fertile anthers and staminode borne on the column. Rostellum absent. Pollen in exposed granular masses. Seeds fusiform in a single-celled ovary.

A genus of about 35 species, widely distributed throughout the northern hemisphere, mostly north of 35° N, though extending further south in Asia and Mexico. They can be cultivated outdoors on well-drained, humus-rich soils, performing best in partial shade. Though mostly hardy, they may need protection from severe frost, if not covered by snow. The name *Cypripedium* is still sometimes applied to species of *Paphiopedilum* and *Phragmipedium*, though the latter have long been considered distinct.

Literature: Franchet, A., Les Cypripedium de l'Asie Centrale et de l'Asie Orientale, *Journal de Botanique* **8**: 225–33, 249–56, 265–71 (1894); Luer, C.A., Native orchids of the United States and Canada (1975); Cribb, P. The genus Cypripedium (1997) – all species are well illustrated in this work; Bergel, G., Cypripedium – einige weniger bekannte Arten und Hybriden der Gattung, *Die Orchidee* **57(3)**: 46–50 (2006).

Leaves. Paired: **8, 9, 10, 11, 12, 13**; spirally arranged: **1, 2, 3, 4, 5, 6, 7**. Basal: **12**. Fan-shaped: **8**. 3-veined: **9**. Spotted: **10**. Membranous: **9**.
Lower stem. Hairless: **9, 10**; sparsely to moderately hairy: **1, 2, 3, 4, 5, 6, 7, 12, 13**; shaggy: **8, 11**.
Floral bract. Linear: **9**. Absent: **10**.
Flowers. Solitary: **2, 3, 6, 8, 9, 10, 11, 12, 13**; **1–3** per raceme: **1, 7, 11**; more than 4 per raceme: **4, 5, 11**. Clustered: **11**.
Lateral sepals. Free: **6**; fused into a rounded lower sepal: **7, 9**; fused into a pointed or bifid lower sepal: **1, 2, 3, 4, 5, 8, 10, 11, 12, 13**.
Petals. Longer than lip: **1, 2, 3, 6, 8, 9, 10, 11**; equal to or shorter than lip: **4, 5, 7, 12, 13**. Linear or narrowly lanceolate: **1, 2, 6**; spathulate: **13**; acute: **1, 2, 3, 4, 6, 8, 10, 11, 12**; rounded: **5, 7, 9, 13**. Twisted or curled: **1**.
Lip. Furrowed in front: **8, 9, 10, 11, 12**. With upstanding or turned-out rim: **13**. Spurred: **6**.

1a. Leaves 3 or more, spirally arranged 2
b. Leaves 2, almost opposite, basal or higher on the stem 8
2a. Lateral sepals free; lip with a blunt, conical spur **6. arietinum**
b. Lateral sepals fused; lip without a spur 3
3a. Petals linear to narrowly lanceolate, more than 2 cm, more than 3 times as long as broad 4
b. Petals oblong, elliptic or ovate (if lanceolate then less than 2 cm), less than 3 times as long as broad 5
4a. Petals twisted or curled; lower sepal entire, or bifid to less than 5 mm **1. calceolus**
b. Petals not twisted or curled; lower sepal usually bifid to more than 5 mm **2. cordigerum**
5a. Petals acute 6
b. Petals rounded 7
6a. Leaves 3 or 4; flowers usually solitary; lip red, purple, blackish or greenish brown **3. macranthos**
b. Leaves 5–16; flowers 1–8; lip yellow **4. irapeanum**
7a. Flowers 1–3 (rarely 4); petals more than 2 cm, upper sepal more than 2.5 cm **7. reginae**
b. Flowers 5–10; petals less than 2 cm; upper sepal less than 2.5 cm **5. californicum**
8a. Petals and sepals equal to or shorter than lip 9
b. Petals and sepals longer than lip 10
9a. Petals rounded, less than 2.5 cm; flower blotched; lip with out-turned rim **13. guttatum**
b. Petals acute, more than 3.5 cm; flower neither spotted nor blotched; lip with in-turned rim **12. acaule**
10a. Petals less than 3.5 cm; upper sepal less than 3 cm 11
b. Petals more than 3.5 cm; upper sepal more than 4 cm 12
11a. Lower stem hairless; leaves heart-shaped, membranous; petals rounded **9. debile**
b. Lower stem shaggy; leaves elliptic, pleated; petals acute **11. fasciculatum**
12a. Leaves spotted, broadly elliptic, obtuse; stem hairless below; floral bract absent **10. margaritaceum**
b. Leaves not spotted, fan-shaped; stem shaggy below; floral bract present **8. japonicum**

1. C. calceolus Linnaeus (*C. microsaccos* Kränzlin). Illustration: Sundermann, Europäische und mediterrane Orchideen, 216 (1975); Williams et al., Orchids of Britain and Europe, 25 (1978).

To 60 cm, sparsely hairy. Leaves 10–18 × 5–6 cm, ovate to elliptic, acute, spirally arranged. Flowers 1 or 2. Upper and lower sepals ovate, acuminate. Petals 3–7 cm, linear to narrowly lanceolate, longer than the lip, twisted or curled, acute. Sepals and petals purplish reddish brown. Lip dull to bright yellow, spotted inside with red or purple. Staminode oblong-ovate, broadest in upper half, white with red spots *From N & C Europe across Siberia to Korea.* H2. Spring–summer.

C. parviflorum var. **parviflorum** (Salisbury) Fernald (*C. parviflorum* Salisbury). Illustration: *Botanical Magazine*, 3024 (1830); *American Orchid Society Bulletin* **38**: 904 (1969); Luer, Native orchids of the United States and Canada, 51 (1975). Similar to *C. calceolus* though smaller in its floral parts, sparsely downy. Sepals and petals densely and minutely spotted with reddish brown, very twisted. Lip 1.5–3.4 cm. Staminode trowel-shaped to triangular-ovate, broadest in lower half, yellow with some red marks. *Northeast N America.* H2. Summer.

Plants with a glabrous upper sheathing bract and intensely sweet scent have been distinguished as var. *makasin* Farwell.

Var. **pubescens** (Willdenow) O.W. Knight (*C. pubescens* Willdenow; *C. veganum* Cockerell & Barker; *C. luteum* Aiton). Illustration: *Botanical Magazine*, 911 (1806); *American Orchid Society Bulletin* **38**: 905 (1969); Luer, Native orchids of the United States and Canada, 45, 47 (1975). Larger in its floral parts than var. *parviflorum*. Sepals and petals greenish yellow, unmarked or with with purple markings. Lip (2–)3–5.4 cm. *N America*. H2. Spring–summer.

Very small plants with yellow-green sepals and petals, and flat or scarcely twisted linear-lanceolate petals have been called var. *planipetalum* Fernald.

C. candidum Willdenow. Illustration: *Botanical Magazine*, 5855 (1870); *American Orchid Society Bulletin* **38**: 906 (1969); Luer, Native orchids of the United States and Canada, 53 (1975). Flowers similar in size to var. *parviflorum*. Petals and sepals green marked with purple, or almost completely purple; lip white, sometimes veined with purple, with red or purple spots inside, and with up to 5 leaves held very upright on each stem. *E & C USA, adjacent Canada*. H2. Spring–summer.

Hybrids with *C. parviflorum* have been named *C. × andrewsii* A.M. Fuller.

C. montanum Lindley. Illustration: *Botanical Magazine*, 7319 (1893); *American Orchid Society Bulletin* **38**: 904 (1969); Luer, Native orchids of the United States and Canada, 55 (1975). Lip to 6 leaves on each stem; flowers similar in size to var. *pubescens*. Petals and sepals uniformly brownish purple; lip compressed, white flushed with purple near the base. *Western N America east to Wyoming*. H2. Summer.

Forma *praeternictum* Sheviack produces pale yellow flowers with a white lip.

Forma *welchii* P.M. Brown has a crimson edge to the lip.

2. C. cordigerum Don. Illustration: *Botanical Magazine*, 9364 (1934); Pradhan, Indian Orchids **I**: 35 (1976); Senghas (ed.), *Proceedings of the 8th world orchid conference*, t. 4 (1976).

Differs from *C. calceolus* and its allies in its shorter, broader petals; lower sepal bifid to 1.8 cm; sepals and petals pale yellow, white or pale green; lip white with a few purple spots inside. *Himalaya*. H2. Spring.

3. C. macranthos Swartz (*C. speciosum* Rolfe; *C. franchetii* Rolfe). Illustration: *Botanical Magazine*, 2938 (1829); *American Orchid Society Bulletin* **44**: 769 (1975); Williams et al., Orchids of Britain and Europe, 27 (1978); *Orchids* **76**(11): 838 (2007).

To 50 cm, sparsely to moderately hairy. Leaves 3 or 4, 8–16 × 4–7 cm, ovate to elliptic, acute, spirally arranged. Flowers usually solitary. Upper and lower sepals broadly ovate, acuminate, violet to purplish pink or greenish brown, with darker veins. Petals 4–6.5 × 1.5–3 cm, broadly ovate, acute, a little longer than lip. Lip similar in colour, paler round the rim. *USSR, south of c. 60° N; Mongolia, N & W China, Korea, Japan*. H2. Spring–summer.

C. tibeticum Hemsley. (*C. froschii* Perner) Illustration: *Botanical Magazine*, 8070 (1906); Pradhan, Indian Orchids **I**: 35 (1976); Senghas (ed.), *Proceedings of the 8th world orchid conference*, t. 8 (1976); *Orchids* **76**(11): 847 (2007). Similar but distinguished by its larger flowers; petals and sepals pinkish yellow marked with a reddish purple network. Lip uniformly dark (sometimes almost black). *SW China*. H2.

C. himalaicum Hemsley. Illustration: *Botanical Magazine*, 8965 (1922); Morley & Everard, Wild flowers of the world, t. 109 (1970); *Orchids* **76**(11): 842 (2007). Distinguished by its slightly smaller flowers and narrower sepals and petals, which are greenish yellow, strongly marked with dark red lines; lip purple, paler behind and around the mouth. *Himalaya & W China*. H2.

C. fasciolatum Franchet (*C. wilsonii* Rolfe). Illustration: *Orchids* **76**(9): 679, 683 (2007). Described as having an almost spherical lip spotted with violet below and broadly striped above, as are the sepals and petals. *W. China*. H2.

C. × ventricosum Swartz. Illustration: Williams et al., Orchids of Britain and Europe, 26, (1978); *Orchids* **76(11)**: 838, (2007). Described as the hybrid between *C. macranthos* and *C. calceolus*, and reported from the wild and in cultivation. It can be separated from *C. macranthos* by its clawed staminode, and longer and narrower, often slightly twisted, petals.

4. C. irapeanum Llave & Lexarza. Illustration: Luer, Native orchids of the United States and Canada, 59 (1975); Senghas (ed.), *Proceedings of the 8th world orchid conference*, t. 4 (1976); *Orchid Digest* **41**: 209 (1977).

To 1.5 m, moderately hairy, especially below. Leaves 5–16, 5–15 × 3–9 cm, ovate to elliptic, acute, spirally arranged. Flowers 1–8. Upper sepal lanceolate, acuminate; lower sepal ovate, acuminate, entire. Petals 3–6.5 cm, oblong, acuminate, equal to or shorter than lip. Sepals and petals bright yellow. Lip bright yellow, flushed with red inside. *C & S Mexico, Guatemala*. G1. Summer.

5. C. californicum Gray. Illustration: *Botanical Magazine*, 7188 (1891); Luer, Native orchids of the United States and Canada, 63 (1975); *Orchid Digest* **43**: 112 (1979).

To 1.2 m, sparsely hairy. Leaves 5–10, 5–15 × 2–6 cm, elliptic to lanceolate, acute, spirally arranged. Flowers 5–10. Upper sepal oblong to elliptic, acuminate; lower sepal elliptic, shortly bifid. Petals 1.4–2 cm, oblong-lanceolate, blunt, equal to or shorter than lip. Sepals and petals yellowish green. Lip white. *USA (S Oregon, N California)*. H2. Spring–summer.

6. C. arietinum Brown (*C. plectrochilum* Franchet). Illustration: *Botanical Magazine*, 1569 (1813); Luer, Native orchids of the United States and Canada, 43 (1975); *American Orchid Society Bulletin* **44**: 384 (1977).

To 30 cm, sparsely hairy. Leaves 3–4, 1.5–3.5 cm, elliptic to lanceolate, acute, spirally arranged. Flower solitary. Upper sepal ovate-lanceolate, acuminate, green with purple veins; lateral sepals free, linear-lanceolate, greenish purple. Petals 1.3–2.2 cm, linear, acute, greenish purple, a little longer than lip. Lip with a blunt conical spur, white above with purplish veins coalescing below. *East & Central N America, from 42° to 52° N; C & SW China, N Burma*. H2. Spring–summer.

7. C. reginae Walter (*C. album* Aiton; *C. spectabile* Salisbury; *C. canadensis* Michaux). Illustration: *Botanical Magazine*, 216 (1793); Shuttleworth et al., Orchids, 17 (1970); Luer, Native orchids of the United States and Canada, 57 (1975).

To 1 m, densely hairy. Leaves 3–7, 10–25 × 6–16 cm, ovate to lanceolate, spirally arranged. Flowers 1–3. Upper sepal broadly elliptic to circular, acuminate; lower sepal similar, blunt. Petals 2.5–5 cm, oblong-elliptic, rounded. Sepals and petals white; petals shorter than lip. Lip pink streaked with white, or pure white. *Eastern N America*. H2. Spring–summer.

The name C. Reginae grex is also a synonym of *Paphiopedilum reginae*, the artificial hybrid *Paphiopedilum leeanum × fairrieanum*.

C. flavum Hunt & Summerhayes (*C. luteum* Franchet, not Aiton, not Rafinesque). Illustration: *Gardeners' Chronicle* **57**: 257 (1915); *Die Orchidee* **49** (4): 171 (1998). Similar but slightly smaller (to 60 cm), with less dense, brownish hair, and bright yellow flowers, which are faintly spotted or striped with purplish brown, particularly on lip. *W China*. H2.

C. Ulla Silkens grex (*C. reginae* × *flavum*) Illustration: *Die Orchidee* **49**(4): 172 (1998). Intermediate between parents, with white sepals and petals and a pale to rich pink lip. *Garden Origin.* H2.

8. C. japonicum Thunberg (*C. formosanum* Hayata). Illustration: *Botanical Magazine*, 9520 (1938); Shuttleworth et al., Orchids, 19 (1970); *American Orchid Society Bulletin* **44**: 768 (1975).

To 50 cm, shaggy below, sparsely hairy above. Leaves 2, in the upper half of the stem, 13–20 × 13–20 cm, broadly fan-shaped. Flower solitary. Sepals and petals green with purple spots at base; the acute lower sepal sometimes slightly bifid. Petals pointed, 4–6 cm, just longer than lip. Lip pale pink with purple spots, deeply furrowed in front. *Japan, Taiwan, S & C China.* H2.

9. C. debile Reichenbach. Illustration: *Botanical Magazine*, 8183 (1908); Maekawa, Wild orchids of Japan in colour, 85 (1971); Kitamura et al., Coloured illustrations of herbaceous plants of Japan (Monocotyledoneae), t. 1 (1978); *Orchids* **77**(2): 116–117 (2008).

To 25 cm, hairless. Leaves 2, about halfway up stem, 3–6 cm, heart-shaped, membranous, with 3–7 prominent veins. Bracts linear. Flower solitary, pendent. Sepals and petals pale green. Petals 1.3–2 cm, lanceolate, rounded, just longer than lip. Lip spherical, white marked with purple around the mouth. *Japan, W China.* H2.

10. C. margaritaceum Franchet (*C. ebracteatum* Rolfe). Illustration: *Gardeners' Chronicle* **46**: 419 (1909).

To 15 cm, hairless below. Leaves 2, about halfway up stem, 10–18 × 10–18 cm, broadly elliptic, spotted with purple. Bract absent. Flower solitary. Sepals and petals broadly ovate, acute, yellowish green spotted with purple. Petals 4–5 cm, longer than lip. Lip fleshy, angular, glandular-hairy, yellow with purple spots. *W China (Yunnan).* H2. Summer.

11. C. fasciculatum Watson (*C. knightae* Nelson). Illustration: *Botanical Magazine*, 7275 (1893); *American Orchid Society Bulletin* **38**: 906 (1969); Luer, Native orchids of the United States and Canada, 65 (1975).

To 30 cm, shaggy below. Leaves 2, in middle or upper half of stem, 5–12 × 3–8 cm, elliptic, obtuse. Flowers in clusters of up to 4, purplish brown to green. Sepals and petals acute, about twice as long as lip; petals 1.5–2.5 cm. *Western N America.* H2. Spring–summer.

C. elegans Reichenbach (illustration: *Orchids* **77**(2): 119, 2008) from the Himalaya is similar but differs by only producing solitary flowers.

12. C. acaule Aiton. Illustration: *Botanical Magazine*, 192 (1792); Shuttleworth et al., Orchids, 19 (1970); Luer, Native orchids of the United States and Canada, 41 (1975).

To 40 cm, sparsely hairy. Leaves 2, 10–28 × 5–15 cm, basal, elliptic, obtuse. Flower solitary. Sepals and petals shorter than lip, lanceolate, acute, yellowish green to purple; petals 4–6 cm. Lip pink or white, deeply furrowed in front. *Northeastern N America.* H2. Spring–summer.

13. C. guttatum Swartz (*C. yatabeanum* Makino). Illustration: *Botanical Magazine*, 7746 (1900); Luer, Native orchids of the United States and Canada, 67 (1975); Williams et al., Orchids of Britain and Europe, 29 (1978).

To 30 cm, sparsely hairy. Leaves 2, 5–12 × 2.5–7 cm, bluish green drying black, elliptic, acute. Flowers solitary, white blotched with purple or brown. Upper sepal elliptic, acuminate; lower sepal lanceolate, shortly bifid. Petals 1.3–2 cm, spathulate or with oblong, rounded tips, shorter than lip. Lip a deep, open pouch with upstanding or out-turned rim. *From European Russia across temperate Asia to Alaska & NW Canada.* H2. Summer.

2. PHRAGMIPEDIUM (Pfitzer) Rolfe

J.C.M. Alexander

Terrestrial, rarely epiphytic or on rocks. Leaves persistent, strap-shaped, leathery, 2-ranked, folded when young. Flowers in terminal racemes or panicles, soon falling. Sepals edge-to-edge in bud. Petals and upper sepal free, lateral sepals fused into a single lower sepal. Lip a sac-like pouch (rarely petal-like), sometimes with 2 triangular appendages on the rim; opening of lip partially blocked by a staminode. Stigmatic surface, 2 fertile anthers and staminode borne on the column. Rostellum absent. Pollen in exposed granular masses. Seeds fusiform in a 3-celled ovary.

A genus of 15–20 species from C & S America, most of which were originally placed in *Cypripedium* or *Selenipedium;* the latter name is sometimes used for them in the horticultural trade. True *Selenipedium* species are rare in cultivation. *Phragmipedium* species require cultivation conditions similar to those for the unmottled species of *Paphiopedilum* (p. 455).

Literature: Garay, L.A., The genus Phragmipedium, *Orchid Digest* **43**: 133–48 (1979).

Leaves. 30 cm or less: **1, 7, 8**; more than 30 cm: **2, 3, 4, 5, 6**. Less than 2 cm wide: **5**; 2–5 cm wide: **1, 2, 4, 6, 7, 8**; more than 5 cm wide: **3**. With yellow margins: **2**.

Flowers. 2–4 per inflorescence: **2, 3, 4, 7, 8**; 6 or more per inflorescence: **1, 2, 5, 6**. 8 cm wide or less: **1, 2, 3, 4, 5, 7**; more than 8 cm wide: **2, 3, 4, 5, 6, 8**.

Petals. 20 cm or more, more than 3 times longer than sepals: **3, 4**; 15 cm or less, less than 3 times longer than sepals: **1, 2, 5, 6**. Similarly shaped to sepals: **1, 7, 8**. Wavy: **2, 3, 4, 5, 6**; crisped: **6**; not wavy or crisped: **1, 7, 8**.

Staminode. Elliptic to circular: **1**; diamond-shaped to 3-lobed: **2, 3, 4, 7, 8**; triangular to semicircular: **5, 6**.

1a. Lip not pouched, similar to petals **4. lindenii**
b. Lip pouched 2
2a. Petals many times longer than sepals **3. caudatum**
b. Petals less than 3 times longer than sepals 3
3a. Flowers bright red, orange, yellow or purple 7
b. Flowers white and pink or brown and green shades 4
4a. Petals similar to sepals in shape, but larger **1. schlimii**
b. Petals longer and narrower than sepals 5
5a. Opening of lip without lobes 6
b. Opening of lip with 2 vertical, triangular lobes **6. longifolium**
6a. Petals not tapering; leaves more than 3 cm wide, with yellow margins **2. lindleyanum**
b. Petals narrow and tapering; leaves less than 2 cm wide, without yellow margins **5. caricinum**
7a. Flowers red, orange or yellow **7. besseae**
b. Flowers raspberry to purple, with yellow margin at pouch opening **8. kovachii**

1. P. schlimii (Linden & Reichenbach) Rolfe. Illustration: *Botanical Magazine*, 5614 (1866); *Orchid Digest* **43**: 134 (1979); Bechtel et al., Manual of cultivated orchid species, 343 (1981).

Leaves 4–8, 20–30 × 2.5–3 cm, strap-shaped, acute. Racemes or panicles to 50 cm, bearing 6–10 flowers; stems hairy. Flowers 5–6 cm wide. Sepals *c.* 2 cm, ovate, obtuse, white flushed with pink. Petals similar but larger, speckled with pink at base. Lip an ellipsoid pouch, mottled dark on pale pink. Staminode elliptic to circular, bright yellow. *Colombia.* G2. Summer.

2. P. lindleyanum (Lindley) Rolfe. Illustration: *American Orchid Society Bulletin* **44**: 235 (1975); *Orchid Digest* **43**: 134 (1979).

Leaves 4–7, 30–50 × 4–6.5 cm, lanceolate, acute. Racemes or panicles to 80 cm, bearing 3–7 flowers; stems reddish green, downy. Flowers 7–8 cm wide. Sepals elliptic, slightly hooded, yellowish green; upper sepal 3 × 2 cm; lower sepal a little broader. Petals *c.* 5 cm, oblong, rounded, with wavy margins, yellowish green at the base fading to greenish white at the tips; tips strongly purple-veined. Lip a broad, shallow pouch with incurved margins, yellow with brown veins. Staminode diamond-shaped to weakly 3-lobed, yellow. *Venezuela & Guyana.* G2.

P. sargentianum (Rolfe) Rolfe. Illustration: *Botanical Magazine*, 7446 (1895); *Orchid Digest* **43**: 134 (1979). Differs from *P. lindleyanum* in its narrower lip and shorter raceme (*c.* 40 cm), bearing 2–4 larger flowers (to 10 cm wide). *Brazil.* G2.

3. P. caudatum (Lindley) Rolfe (*P. warscewiczianum* (Reichenbach) Schlechter). Illustration: *Orchid Digest* **43**: 138 (1979); Bechtel et al., Manual of cultivated orchid species, 262 (1981); Shuttleworth et al., Orchids, 20 (1970).

Epiphytic or on rocks. Leaves 5–9, 40–60 × 6 cm, strap-shaped, light green; apex irregularly toothed. Racemes to 1 m, bearing 2–4 flowers; stems downy. Sepals lanceolate, tapering, off-white to yellowish green with darker green or orange veins; upper sepal 10–15 × 2–3 cm, curled forward over flower; lower sepal 7–10 × 3–4 cm. Petals to 60 × 1 cm, trailing, dark reddish to greenish brown, paler at the base. Lip a deep pinkish white pouch with pink or brown net-venation; rim turned out, yellow. Staminode diamond-shaped, off-white with dark purple corners. *Mexico through C America to Peru, Ecuador, Colombia & Venezuela.* G2. Spring–summer.

4. P. lindenii (Lindley) Dressler & Williams. Illustration: *Orchid Digest* **43**: 139 (1979); Bechtel et al., Manual of cultivated orchid species, 262 (1981).

Differs from *P. caudatum* in having a simple, unpouched lip similar in shape and colour to the petals, though a little broader. *Colombia, Ecuador & Peru.* G2. Summer.

5. P. caricinum (Lindley) Rolfe. Illustration: *Botanical Magazine*, 5466 (1864); *American Orchid Society Bulletin* **44**: 236 (1955); *Orchid Digest* **43**: 143 (1979).

Leaves 3–7, 35–50 cm × 5–15 mm, linear, stiff and sedge-like. Raceme or panicle 30–40 cm, bearing several flowers in succession; stems downy. Flowers 5–8 cm. Sepals 2–3 × 1–1.5 cm, ovate to lanceolate with wavy margins, yellowish or purplish green with a green base. Petals 5–10 cm × 5 mm, tapering, twisted, similar to sepals in colour. Lip a rounded greenish yellow pouch with brown veins. Staminode triangular, green. *Peru, Bolivia & Brazil.* G2. Spring–autumn.

6. P. longifolium (Warscewicz & Reichenbach) Rolfe. Illustration: *Botanical Magazine*, 5970 (1872); *American Orchid Society Bulletin* **44**: 225 (1975); *Orchid Digest* **43**: 146 (1979).

Leaves 5–8, 40–60 × 2–3 cm, strap-shaped, tapering to an acute tip. Raceme to 60 cm, bearing several flowers in succession; stems reddish purple, sparsely downy. Flowers to 18 cm wide. Sepals green or greenish yellow with dark green or greenish red veins; upper sepal ovate-lanceolate, 5 × 1.5 cm; lower sepal broadly ovate. Petals to 12 × 1 cm, tapering and twisted with wavy margins, greenish yellow with purple margins, solid purple at tips. Lip a deep pouch with 2 vertical triangular lobes on the rim, greenish yellow flushed with purple. Staminode broadly triangular to semicircular, greenish yellow; upper edge deep purple. *Costa Rica to Colombia.* G2. Winter.

P. boissierianum (Reichenbach) Rolfe. Illustration: *Orchid Digest* **43**: 146 (1979). Differs from *P. longifolium* in its shorter leaves and drooping petals with crisped margins. *Peru.* G2. Summer–autumn.

7. P. besseae C. Dodson & J. Kuhn. Illustration: *American Orchid Society Bulletin* **50**(11): 1308–10 (1981); *Orchid Digest* **71** (2): 78 (2007).

Leaves distichous, dark green, to 22 × 4 cm, narrowly elliptic. Flowers 1 to 4, bright red, 5–6 cm wide, produced in succession. Sepals ovate, obtuse, sparsely pubescent, *c.* 2.7 × 1.6 cm. Petals obliquely ovate, obtuse, *c.* 3 × 2 cm. Lip a pouch with involute margins forming a quadrate opening, *c.* 2.3 × 1.6 cm. Staminode obovate to almost pentagonal, with an emarginate apex. *Eastern Ecuador to Northern Peru.* G2.

Var. **flavum** Braem (*f. flavum* (Braem) O. Gruss & Roeth), a variety with bright yellow flowers much used in hybridisation.

Don Wimber grex (illustration: *Orchid Review* **106**: 264, 1998; *Die Orchidee* **55** (2): 219, 2004; *Orchids* **76**(2): 101, 2007) is the hybrid *P.* Eric Young grex (*besseae* × *longifolium*) × *P. dalessandroi.*

P. dalessandroi Dodson & Gruss. (*P. besseae* var. *dalessandroi* (Dodson & O. Gruss) A. Moon & P.J. Cribb). Illustration: *Die Orchidee* **47**(4): 216–17 (1996). Similar to *P. besseae* but with branched, villous inflorescence, bearing 9 or more orange flowers with oblique down-curved lateral petals, opening simultaneously. *Southern Ecuador.* G2.

Many hybrids listed with *P. besseae* as a parent were made using *P.* Jersey grex (illustration: *Die Orchidee* **47**(4): 218, 1996), which is *P. besseae* × *P. dalessandroi.*

8. P. kovachii Atwood, Dalstrom & Fernandez. (*P. peruvianum* Christenson; *P. moyobambii* Hort.) Illustration: *Orchids* **71**(7): 620–22 (2002) & **76**(11): 827 (2007).

Leaves 5–8, to 27 × 4 cm, linear, channelled, tapering to an acute apex. Raceme to 26 cm, bearing a spathiforme, lanceolate bract *c.* 4 cm. Flowers to 15 cm across, rich raspberry pink to purple with a white ring and yellow margin to the pouch opening. Upper sepal elliptic, to 5.5 × 3.5 cm, lower sepal rounded, *c.* 6.5 × 5 cm. Petals rounded, to 7 × 5 cm. Lip inflated forming a pouch *c.* 7 cm. Staminode shield-shaped. *Peru.* G2. Can be grown to flowering from seed in 18 months.

3. PAPHIOPEDILUM Pfitzer

J.C.M. Alexander

Epiphytic, terrestrial or on rocks. Stems very short, enclosed in leaf-bases. Leaves 4–several, ovate to strap-shaped, persistent, leathery, often mottled, in 2 opposite ranks, folded when young. Flowers solitary, or in terminal racemes bearing 2–8 flowers all open together, or many flowers in succession over several months. Sepals overlapping in bud. Petals and upper sepal free; lateral sepals united into a single lower sepal under the lip. Lip extended into a sac-like pouch, usually with side-lobes on the rim; opening of pouch partially blocked by a broad staminode borne on the column. Rostellum absent. Pollen in exposed granular masses. Seeds fusiform; ovary 1-celled. (Figure 22(3 & 4), p. 444.)

A genus of about 60 species from the Himalaya and S India through SE Asia to the Philippines, New Guinea and the Solomon Islands. Most of the species were originally placed in *Cypripedium*, which now contains only temperate species with rolled leaves. Many artificial hybrids have been produced that exploit the showiness of the upper sepal seen in the wild species. The petals are often strikingly marked with spots or warts and frequently bear marginal hairs. The flowers persist for several weeks if not pollinated.

Plants can be grown on bark or in well-drained pots in epiphytic compost. Lacking pseudobulbs, they must not be allowed to dry out, though less water should be given in winter. They need shading from direct summer sunlight. Species with mottled leaves do best in a warm, moist house; those with plain leaves generally prefer cooler conditions. Propagation is usually effected by division of large plants when repotting in early spring; some species have been successfully grown from seed, e.g. *P. delenatii.*

Literature: Asher, J.R., A checklist for the genus Paphiopedilum, *Orchid Digest* **44**: 175–84, 213–28 (1980) & **45**: 15–26, 57–65 (1981); Graham, R. & Roy, R., Slipper orchids (1982); Cribb, P. The genus Paphiopedilum, edn 2 (1998); Averyanov, L., Cribb, P., Phan Ke Loc, et al., Slipper orchids of Vietnam (2003).

1a. Lip with 2 side-lobes on rim 2
b. Lip without lobes on rim 32
2a. Flowers 2 or more, arising from different points on flowering stem 3
b. Flower solitary, rarely 2 or 3 arising from the same point on the flowering stem 5
3a. Petals spathulate, flat or slightly twisted **36. lowii**
b. Petals linear to oblong, twisted and wavy 4
4a. Flowers open 1 at a time; upper sepal almost circular; petals horizontal, hairy on margins **34. victoria-regina**
b. Flowers open all together; upper sepal elliptic; petals drooping, not hairy on margins **35. parishii**
5a. Leaves mottled on upper surface 6
b. Leaves not mottled on upper surface 25
6a. Lip with prominent tooth on outer edge of rim **26. acmodontum**
b. Lip without tooth 7
7a. Petals without spots or warts **16. dayanum**
b. Petals with spots or warts 8
8a. Each petal with 50 or more spots or warts, evenly distributed over at least the basal half of blade 9
b. Each petal with 40 or fewer spots or warts, usually concentrated near margins, mid-line or upper basal corner of blade; if more than 40, then never evenly distributed 14
9a. Spots or warts extending well into apical half of petal, though often absent from extreme tip 10
b. Spots or warts mostly in basal half of petal 13
10a. Petals distinctly broader in apical half **27. javanicum**
b. Petals oblong, elliptic or tapering from the base 11
11a. Petals acute, gradually tapering, almost flat, not curled, twisted or curved downwards **33. sukhakulii**
b. Petals obtuse or acute, curled, twisted or curved downwards 12
12a. Petals evenly marked with many small spots or warts **31. superbiens**
b. Petals irregularly marked with large spots, some with pale centres **32. argus**
13a. Petals elliptic, mostly purplish brown with dark spots or warts; upper sepal with *c.* 12 distinct purple and green veins **29. purpuratum**
b. Petals spathulate, mostly pinkish purple with faint pinkish purple spots; upper sepal with 15–20 indistinct veins **22. hookerae**
14a. Petal spots or warts almost all on or near margin 15
b. Some petal spots or warts away from margin, often on mid-line or in upper basal corner 20
15a. Upper sepal green to greenish yellow, often brown at base, indistinctlyveined; lip with bright green rim **20. bullenianum**
b. Upper sepal white, sometimes flushed with pink and green, distinctly veined with green or red and green; lip without green rim 16
16a. Upper sepal ovate, gradually tapering, white or off-white with green veins 17
b. Upper sepal almost circular, usually with a short abrupt tip, white usually flushed with pink and green, veined with purple or green 18
17a. Petals 4–5 cm, spathulate **25. venustum**
b. Petals to 10 cm, narrowly elliptic **16. dayanum**
18a. Upper and lower petal margins with 6–10 warts **19. lawrenceanum**
b. Upper petal margin with 4–8 warts (sometimes 1 or 2 on blade or lower margin) 19
19a. Petals almost horizontal, with dark purple veins; warts *c.* 1 mm in diameter **17. barbatum**
b. Petals curved downwards, with indistinct green veins; warts *c.* 3 mm in diameter **18. callosum**
20a. Petals with large, black, often coalescing spots; lip dull yellow to reddish brown, strongly veined with greenish brown **25. venustum**
b. Not the above combination of characters 21
21a. Petals green and bright pink or pinkish purple 22
b. Petals pale green, brown or brownish purple 24
22a. Spots or warts all in basal half of petal **21. appletonianum**
b. Spots or warts extending into apical half of petal 23
23a. Petals descending, broadly spathulate, with a prominent white patch **23. volonteanum**
b. Petals almost horizontal, narrowly spathulate, lacking a white patch **28. virens**
24a. Petals pale green with *c.* 10 large black spots on upper margin and mid-line; upper sepal with broad white margin **24. tonsum**
b. Petals reddish brown and dull green with many small spots in basal half, often concentrated in upper corner **30. mastersianum**
25a. Petal tips strongly recurved; upper sepal with bold interconnecting veins **15. fairrieanum**
b. Not the above combination of characters 26
26a. Upper sepal unspotted, finely or indistinctly speckled or with several indistinct stripes 27
b. Upper sepal with large, dark spots or 1 distinct central stripe 29

27a. Petals oblong, 4 cm; upper sepal *c.* 6 cm wide **9. charlesworthii**
b. Petals spathulate, 6 cm or more; upper sepal less than 4 cm wide 28
28a. Bract much shorter than ovary; petals with black or purple specks at base **10. hirsutissimum**
b. Bract almost equal to ovary; petals without specks **11. villosum**
29a. Upper sepal with dark central stripe **12. spicerianum**
b. Upper sepal with large, dark spots 30
30a. Flower-stalk and ovary with shaggy hairs **11. villosum**
b. Flower-stalk and ovary at most downy or felted 31
31a. Flowers 8 cm wide or less **13. exul**
b. Flowers 10 cm wide or more **14. insigne**
32a. Petals similar to sepals in shape and size; flowers solitary, rarely 2 or 3 arising from the same point on the flowering stem 33
b. Petals longer and thinner than sepals; flowers 2 or more, arising from different points on the flowering stem 37
33a. Lip cup-shaped to spherical; petals usually 4 cm or less; upper sepal unspotted 34
b. Lip deeply conical; petals more than 4 cm; upper sepal spotted or blotched 35
34a. Lip the same colour as petals and sepals; staminode white and yellow **4. niveum**
b. Lip pink; petals and sepals pink or white; staminode pink and yellow **5. delenatii**
35a. Flowers pale to deep yellow, finely flecked **1. concolor**
b. Flowers white or pale pink with large spots or blotches 36
36a. Flowering stem 3–10 cm; flower marked with coalescing blotches **3. godefroyae**
b. Flowering stem very short; flower marked with spots that coalesce only at bases of sepals and petals **2. bellatulum**
37a. Upper sepal unstriped or with 1–5 bold stripes **8. stonei**
b. Upper sepal with 10–20 stripes 38
38a. Basal part of petal evenly spotted with purple **6. rothschildianum**
b. Basal part of petal unspotted or with a few uneven spots **7. philippinense**

1. P. concolor (Lindley) Pfitzer. Illustration: *Botanical Magazine*, 5513 (1865); *Orchid Digest* **42**: 72, 74 & 75 (1978); Bechtel et al., Manual of cultivated orchid species, 253 (1981).

Leaves 4–5, 10–15 × *c.* 6 cm, elliptic to oblong, obtuse, mottled greyish green above, reddish purple below. Flower-stalk to 50 cm, purple, hairy, shorter than leaves, 1- or 2-flowered. Flower *c.* 5 cm wide, pale to deep yellow, finely flecked with reddish brown. Upper sepal circular, hooded; lower sepal ovate, acute. Petals elliptic to oblong, slightly curved downwards. Lip blunt, conical, equal to or shorter than the petals, without side-lobes. *Burma, Thailand, Cambodia, S Laos & S China.* G2. Winter–spring.

2. P. bellatulum (Reichenbach) Stein. Illustration: *Orchid Digest* **41**: 36 (1977); Graf, Tropica, 749 (1978); Bechtel et al., Manual of cultivated orchid species, 252 (1981).

Leaves 4–6, *c.* 25 × 8 cm, elliptic to oblong, mottled pale and dark green above, reddish below. Flower-stalk very short, 1- or 2-flowered. Flower to 8 cm wide, white or very pale yellow with bold reddish brown spots that coalesce only at the bases of the petals and sepals. Upper sepal almost circular, hooded. Petals almost circular, sometimes shallowly notched. Lip blunt, conical, without side-lobes. *Burma (Maymo plateau & Shan Highlands); adjacent Thailand.* G2. Spring.

3. P. godefroyae (Godefroy-Lebeuf) Stein (*P. leuchochilum* (Rolfe) Fowlie). Illustration: *Botanical Magazine*, 6876 (1886); *Orchid Digest* **41**: 33 (1977) & **44**: 178 (1980).

Similar to *P. bellatulum*. Leaves *c.* 4 cm wide, oblong. Flower-stalk 3–10 cm. Flower irregularly marked with coalescing spots and blotches. *Burma, Thailand (Kra isthmus & Birdsnest Islands) & S Vietnam.* G2. Spring.

Plants with unspotted, pure white lips have been called **P. leucochilum** (illustration: Rittershausen & Rittershausen, Orchids in colour, 94 (1979).

The name *P. ang-thong* Fowlie has been given to plants with fewer and finer spots on the flowers; these may be hybrids between *P. godefroyae* and *P. niveum.*

4. P. niveum (Reichenbach) Stein. Illustration: *Botanical Magazine*, 5922 (1871); *Orchid Digest* **41**: 33 (1977); Bechtel et al., Manual of cultivated orchid species, 255 (1981).

Leaves 6–8, 10–15 cm, elliptic to oblong, mottled grey and green above, deep purple below. Flower-stalk 12–15 cm, 1- or 2-flowered, deep purple with short hairs. Flower 6–8 cm wide, white with a fine stippling of reddish brown near the base of the petals. Upper sepal circular, hooded; lower sepal smaller, ovate, acute, bifid. Sepals streaked with purple near base. Petals broadly ovate to circular, scarcely pointed, slightly curled at the margins. Lip cup-shaped, with an incurved margin, without side-lobes. Staminode yellow and white. *Thailand, Malay peninsula & offshore islands.* G2. Spring–summer.

5. P. delenatii Guillaumin. Illustration: *Botanical Magazine*, 89 (1950); *Orchid Digest* **41**: 35 (1977); Bechtel et al., Manual of cultivated orchid species, 253 (1981); Cribb, The genus Paphiopedilum, edn 2, 77, 78, 80 (1998); Averyanov et al., Slipper orchids of Vietnam, 90, 94 (2003).

Similar to *P. niveum*. Leaves narrowly oblong to oblong, 8–11 × 3–4 cm. Flowers 7.5–10 cm across, sepals less than 4.5 cm, almost circular to broadly ovate, spreading. Petals and sepals uniformly light pink to pink-purple or almost white; lip deep pink, rarely white or yellow; staminode bright yellow centre and with a purple mark, flat with a broad apex. Stolons absent. *C Vietnam.* G2. Spring–summer.

Together with *P. delenatii* the 4 species that follow form a closely related group called Section Parvisepalum Averyanov & Cribb, in which the leaves are distinctly tesselated above and spotted or tinged with purple below and the rounded lip is longer than, or almost equalling, the petals.

P. vietnamense Gruss & Perner. Illustration: Averyanov et al., Slipper orchids of Vietnam, 97, 103 (2003); *Botanical Magazine*, n.s., 512 (2005). Leaves ovate to oblong-elliptic, 10–20 × 3–7 cm. Flowers 10–12 cm across, sepals more than 4.5 cm, broadly oblong to broadly ovate, sloping downwards. Petals and sepals uniformly light pink to pink-purple or almost white; lip deep pink, rarely white or yellow; staminode dull yellow to light green in centre, flat with a broad apex. Stolons absent. *Northern Vietnam (Thai Nguyen).* G2. Summer.

P. malipoense S.C. Chen & Z.H. Tsi. Illustration: *Botanical Magazine* n.s., 86 (1987); Cribb, The genus Paphiopedilum, edn 2, 83, 85–7 (1998); Averyanov et al., Slipper orchids of Vietnam, 106, 112 (2003).

Flowers 9–12 cm across, pale yellowish green to apple green, upper sepal and petals with brown stripes, upper sepal more than 4 cm, with an acuminate tip. Petals more than 4 cm, almost pointed at tip. Staminode more than 12.5 mm, broadly ovate, with a deep fold along the mid-vein near the top. Stolons present. *S. China, northern Vietnam, Laos.* G2. Summer.

P. micranthum T. Tang & F.T. Wang. Illustration: Cribb, The genus Paphiopedilum, edn 2, 92–8 (1998); Averyanov et al., Slipper orchids of Vietnam, 125–30 (2003). Flowers 6–8 cm across, pink to pink-purple, upper sepal and petals with purple stripes along veins, upper sepal less than 4 cm, blunt or shortly acute at tip. Petals less than 4 cm, rounded at tip. Staminode less than 12.5 mm, elliptic, with a fold along its length. Stolons present. *S.W. China, northern Vietnam.* G2. Summer.

P. armeniacum S.F. Chen & F.Y. Liu. Illustration: *Orchid Review* **106**(1223): 273 (1998); Cribb, The genus Paphiopedilum, edn 2, 72–5 (1998). Differs by all parts of the flower being bright yellow, except for the staminode, which is yellow with some red markings. *China (W.Yunnan).* G2. Summer.

6. P. rothschildianum (Reichenbach) Stein (*P. elliotianum* misapplied, not *P. elliotianum* (O'Brien) Pfitzer). Illustration: *Botanical Magazine*, 7102 (1890); *Orchid Digest* **40**: 180 (1980); Bechtel et al., Manual of cultivated orchid species, 256 (1981).

Leaves to 50 × 7 cm, elliptic to oblong, unmottled, shiny. Flowering stem longer than leaves, reddish, downy, with 2–6 flowers. Flowers *c.* 13 cm from top to bottom. Upper sepal *c.* 8 cm, ovate, acute, cream-coloured, with *c.* 15 black stripes, margins white. Petals to 15 cm, narrow, tapering, drooping, wavy with fine blunt tips and marginal hairs, cream with purple spots at base, yellowish green apically, veined with purple. Lip pointed, without side-lobes, purplish brown, yellow around rim; flattened sideways. *N Borneo (false records from New Guinea).* G2. Summer.

7. P. philippinense (Reichenbach) Stein (*Cypripedium laevigatum* Bateman). Illustration: *Botanical Magazine*, 5508 (1865); *Orchid Digest* **40**: 183 (1980); Bechtel et al., Manual of cultivated orchid species, 255 (1981).

Leaves to 30 × 4 cm, strap-shaped, unmottled, shiny. Flowering stem to 50 cm, hairy, bearing 2–6 flowers. Flowers *c.* 8 cm from top to bottom. Upper sepal *c.* 3.5 cm, ovate, acute, pale yellow with *c.* 12 brownish purple stripes. Petals to 15 cm, narrow, tapering, drooping, twisted and wavy, with marginal hairs, yellow at base with a few uneven spots, reddish purple to green at tips. Lip rounded, without side-lobes, yellow with green veins. *Philippines.* G2. Summer.

P. praestans (Reichenbach) Pfitzer (*P. glanduliferum* (Blume) Pfitzer). Illustration: *Gardeners' Chronicle* **26**: 776 (1886); Graf, Tropica, 746 (1978); *Orchid Digest* **40**: 182 (1980). Upper sepal 5 cm; petals to 12 cm, less drooping than in *P. philippinense. New Guinea.* G2. Summer.

8. P. stonei (Hooker) Stein. Illustration: *Botanical Magazine*, 5349 (1862); *Orchid Digest* **44**: 184 (1980); Bechtel et al., Manual of cultivated orchid species, 256 (1981).

Leaves 30–40 × 4 cm, strap-shaped, rounded, unmottled. Flowering stem to 60 cm, greenish purple, bearing 3–5 flowers. Flowers 14 cm from top to bottom. Upper sepal 6 cm, ovate, acuminate, white, unstriped or with 1–5 thick, dark stripes. Petals 12–15 cm, narrow, tapering, drooping, not twisted, dull yellow to green, with purple blotches, lacking hairs or warts. Lip rounded, without side-lobes, dull pink, yellowish green below, yellow around rim. *NW Borneo.* G2. Summer–autumn.

9. P. charlesworthii (Masters) Pfitzer. Illustration: *Botanical Magazine*, 7416 (1895); *Orchid Digest* **40**: 152 (1976); Bechtel et al., Manual of cultivated orchid species, 252 (1981).

Leaves 12.5–20 cm, linear to oblong, acute, hairless not mottled above, spotted with brownish purple beneath. Flower-stalk to 20 cm, shortly hairy, red-spotted. Flower 8–9 cm wide. Upper sepal *c.* 6 cm wide, almost circular, pink with a fine network of darker veins, white below, slightly reflexed above. Lower sepal much smaller, yellowish green. Petals *c.* 4 cm, oblong, obtuse-tipped, yellowish green with a fine network of brown veins. Lip greenish yellow to reddish brown. *Burma & NE India.* G2. Autumn.

10. P. hirsutissimum (Hooker) Stein. Illustration: *Botanical Magazine*, 4990 (1857); *Orchid Digest* **44**: 218 (1980); Bechtel et al., Manual of cultivated orchid species, 254 (1981).

Leaves 20–35 cm, oblong, not mottled above. Flower-stalk 15–30 cm, green with purple hairs. Bract shorter than ovary. Flower 10–12 cm wide, all parts shaggy on back. Upper sepal *c.* 4 × 3 cm, circular to ovate, wavy, green or greenish yellow with a central purplish brown patch, indistinctly speckled and streaked. Lower sepal smaller, pale green. Petals 6–7 cm, spathulate with hairy margins, base green to brown, finely spotted with black or purple, margins wavy, tip bright purple. Lip green to greenish brown. Staminode green with 2 basal warts. *NE India, Indo-Burmese border, Thailand & S China.* G1–2. Spring.

11. P. villosum (Lindley) Stein (*P. boxallii* (Reichenbach) Pfitzer). Illustration: *Orchid Review* **77**: 273 (1969); *Orchid Digest* **36**: 148 (1972); Bechtel et al., Manual of cultivated orchid species, 257 (1981).

Leaves 25–40 × 2–3 cm, oblong, uniformly dull green, purple-spotted underneath at the base. Flower-stalk 20–35 cm, green spotted with brown, shaggy. Bract almost equal to ovary. Flower glossy, 12–15 cm from top to bottom. Upper sepal to 5 cm, circular to ovate, reflexed at the sides, yellow or greenish yellow with a brownish purple base diffusing into indistinct stripes above. Petals 6–7 cm, spathulate, obtuse, sometimes notched, with a dark brownish purple mid-line separating an upper yellowish brown half from a lower greenish yellow half. Lip conical, yellow to green, flushed with brown. Staminode obovate, yellowish brown. *NE India, Burma, N Thailand & Laos.* G1–2. Winter–spring.

Variants from Burma with boldly mottled upper sepals and pink net-venation on the paler lip have been called **P. boxallii** (illustration: *Orchid Digest* **36**: 148, 1972).

12. P. spicerianum (Masters & Moore) Pfitzer. Illustration: *Botanical Magazine*, 6490 (1880); *Orchid Digest* **36**: 28 (1972); Bechtel et al., Manual of cultivated orchid species, 256 (1981).

Leaves 12.5–25 × 3 cm, oblong, uniformly dark green above, mottled beneath at the base. Flower-stalk to 30 cm, purple, slightly hairy. Bract much shorter than ovary. Flower 7–8 cm wide. Upper sepal broadly ovate, to 5 cm across, reflexed at the sides, bent forward, keeled above, white with a green base and a central dark purple stripe. Lower sepal narrower, greenish yellow with a purple stripe. Petals to 5 cm, oblong, obtuse, wavy, yellowish green with a dark purple stripe, flecked with purple at the base; margins wavy. Lip greyish brown. Staminode purple and white, circular with recurved margins. *NE India.* G2. Winter–spring.

P. druryi (Beddome) Stein. Illustration: *Orchid Digest* **38**: 35 (1974); *Botanical Magazine*, n.s., 764 (1978); Bechtel et al., Manual of cultivated orchid species, 253 (1981). Leaves light green with dark veins. Upper sepal yellow or yellowish green with a broad, purplish brown stripe. Petals elliptic to oblong, not wavy, hairy, yellow with dark central stripe. Lip pale yellow. *S India; possibly extinct in the wild.* G2. Spring–summer.

13. P. exul (Ridley) Kerchove. Illustration: *Botanical Magazine*, 7510 (1896) & 211 (1993); *Orchid Digest* **40**: 149 (1976); Bechtel et al. Manual of cultivated orchid species, 253 (1981).

Leaves 20–30 × 2–2.5 cm, oblong, obtuse, uniformly bright green. Flower-stalk 15–30 cm, with sparse red hairs. Bract nearly as long as ovary. Flowers to 8 cm wide. Upper sepal 3–4 × 2–3 cm, broadly ovate, greenish yellow, heavily spotted with brownish purple, with a broad, white, wavy margin. Petals *c.* 5 × 2 cm, oblong to elliptic with hairy margins, yellow with a faint central band of purplish brown. Lip yellow, suffused with brown. Staminode yellow, spotted with brown. *S Thailand.* G2. Spring–summer.

Very similar to *P. insigne* and perhaps a subspecies of it.

14. P. insigne (Lindley) Pfitzer. Illustration: *Botanical Magazine*, 3412 (1835); *American Orchid Society Bulletin* **42**: 809 (1973); Bechtel et al., Manual of cultivated orchid species, 254 (1981).

Leaves 20–30 cm, linear to lanceolate, uniformly yellowish green. Flower-stalk to 30 cm, purple, felted. Flower *c.* 10 cm wide. Upper sepal *c.* 5 × 4 cm, ovate, obtuse, wavy, mostly green or greenish yellow with brown spots; tip and some of margin white, unspotted. Lower sepal smaller, green with brown streaks. Petals 6–7 cm, wavy, oblong to narrowly spathulate, greenish yellow with a network of brown veins. Lip greenish or reddish brown, with a yellow rim. Staminode heart-shaped. *E. Himalaya.* G1–2. Autumn–spring.

Over 100 varieties have been described, of which the following are among the best known: var. **sanderae** Reichenbach (illustration: Graf, Tropica, 749, 1978; Bechtel et al., Manual of cultivated orchid species, 254, 1981), upper sepal primrose yellow with reddish brown spots, petals and lip yellow; var. **sanderianum** Sander is similar to var. *sanderae* but lacks spots.

P. × leeanum (Reichenbach) Kerchove. Illustration: Cogniaux & Goossens, Dictionnaire iconographique des orchidées, Cypripedium hybrids, t. 3 (1903). The hybrid between *P. insigne* and *P. spicerianum*. Upper sepal white with lines of purple spots; petals slightly drooping; lip greenish brown. *Himalaya & garden origin.* G2.

15. P. fairrieanum (Lindley) Stein. Illustration: *Botanical Magazine*, 5024 (1857); *Orchid Digest* **40**: 155 (1976); Bechtel et al., Manual of cultivated orchid species, 253 (1981); *Kew Magazine*, 47 (1985).

Leaves 9–15 cm, oblong, unmottled. Flower-stalk 9–15 cm, pale purple, sparsely hairy. Flower 6–8 cm from top to bottom. Upper sepal circular with a broad acuminate tip, pale green with *c.* 15 finely branched dark purple veins interconnecting near the wavy margin. Lower sepal smaller with unconnected veins. Petals 4–5 cm, oblong to lanceolate, curved downwards and strongly recurved at the tips, hairy and wavy on margins, green and white with green and purple veins. Lip inflated, greenish yellow, with a network of purplish brown veins. *Sikkim, Bhutan & NE India.* G2. Autumn.

Lindley in his original description misspelt the name 'fairieanum'; the man after whom he named this species was a Mr Fairrie.

16. P. dayanum (Lindley) Stein (*Cypripedium 'spectabile'* Reichenbach var. *dayanum* Lindley). Illustration: *Botanical Magazine*, n.s., 594 (1971); Williams et al., Orchids for everyone, 152 (1980); *Orchid Digest* **39**: 157 (1975) & **45**: 24 (1981).

Leaves 12–20 × 5 cm, oblong to lanceolate, strongly mottled pale and bluish green above. Flower-stalk 20–30 cm, downy, purple. Flower 10–12 cm wide. Upper sepal *c.* 6 × 3 cm, narrowly ovate, acuminate, white with *c.* 20 narrow green veins. Lower sepal similar but smaller and more pointed. Petals to 10 cm, narrowly elliptic to spathulate, with long marginal hairs, greenish brown at the base to pink at the tip, veined with green and purple, sometimes with small marginal warts. Lip purplish brown, conical. *N Borneo.* G2. Summer (may also flower in winter).

P. violascens Schlechter (*P. bougainvilleanum* Schoser). Illustration: *Orchid Digest* **42**: 100 (1978); Millar, Orchids of Papua New Guinea, 86 (1978); Bechtel et al., Manual of cultivated orchid species, 257 (1981). Upper sepal *c.* 3 × 3 cm, petals 4–4.5 cm, sickle-shaped, purple. Lip greenish brown. *Indonesia (West Irian).* G2. Autumn.

17. P. barbatum (Lindley) Stein. Illustration: *Botanical Magazine*, 4234 (1846); *Orchid Digest* **41**: 61 (1977); Bechtel et al., Manual of cultivated orchid species, 251 (1981).

Leaves 10–15 × 2–3 cm, oblong to lanceolate, mottled. Flower-stalk 20–30 cm. Bract ovate-lanceolate, a quarter of the ovary length; flower 8–10 cm wide. Upper sepal almost circular with a short abrupt tip, green at the base, white or pale pink above with *c.* 20 dark purple veins. Lower sepal shorter and narrower with greenish purple veins. Petals 5 cm, oblong, slightly wider at the tip, hairy on margins, held just below horizontal, reddish brown at the base, shading to purplish red at the tips, dark-veined, with 4–8 dark hairy warts about 1 mm in diameter on the upper margin. *S Thailand & Malay peninsula.* G2. Spring–summer.

18. P. callosum (Reichenbach) Stein. Illustration; *Botanical Magazine*, 9671 (1946); *Orchid Digest* **36**: 9 (1972); Bechtel et al., Manual of cultivated orchid species, 252 (1981).

Leaves 10–25 × 4–6 cm, oblong to elliptic, greyish green with dark mottling. Flower-stalk to 30 cm, brownish purple, densely downy; bract 2–3 cm, ovate, acute. Flower 7–9 cm wide. Upper sepal 6 cm, almost circular, with a short tip, white with *c.* 25 dark purple veins that are green below. Lower sepal shorter and narrower. Petals 5–7 cm, strongly curved downwards, indistinctly green-veined, pinkish yellow at the tips, with 5–8 black hairy warts *c.* 3 mm in diameter on the upper margin (occasionally 1 or 2 on the blade). Lip reddish brown tinged with green. *E Thailand, Cambodia & S Vietnam.* G2. Spring.

'Sanderae' (var. *sanderae* Anon.) – illustration: *Gardener's Chronicle* **15**: 663, 1894 – lacks red or yellow pigment and is white with brilliant green veins.

P. Maudiae grex. Illustration: *American Orchid Society Bulletin* **46**: 1089 (1977); Graf, Tropica, 746 (1978). Petals less deflexed than in *P. callosum*, with 1 or 2 warts on the lower margin as well as those on the upper. This cross was originally made between 'albino' varieties of *P. callosum* and *P. lawrenceanum* (lacking red or yellow pigment) resulting in an 'albino' hybrid. The name, however, strictly applies to all hybrids between these parents. *Garden origin.* G2.

P. callosum and *P. barbatum* probably represent the extreme forms of a single species, which runs from pure *P. barbatum* at the southern end of the Malay peninsula to *P. callosum* in N Thailand. Intermediates from the Kra isthmus have been referred to as *P.* × *siamense* Rolfe (*P. sublaeve* (Reichenbach) Fowlie).

19. P. lawrenceanum (Reichenbach) Pfitzer. Illustration: *Botanical Magazine*, 6432 (1879); *Orchid Digest* **41**: 61 (1977); Bechtel et al., Manual of cultivated orchid species, 254 (1981).

Leaves 15–30 × 5–7 cm, oblong to ovate with long sheathing bases, mottled dark and yellowish green. Flower-stalk 30–40 cm, purple, hairy. Bract ovate-oblong, much shorter than ovary. Flower 10–12 cm wide. Upper sepal 5 cm, circular, white flushed with pink; veins *c.* 20, purple above, green below. Petals 5–6 × 1.5 cm, horizontal, oblong to narrowly elliptic, obtuse, dull green with purplish veins; tips tinged with reddish purple. Each margin with 6–10 blackish purple hairy warts. Lip brownish purple. *N Borneo.* G2. Summer.

Var. **hyeanum** (Reichenbach) Pfitzer lacks red pigment and is green, white and yellow.

20. P. bullenianum (Reichenbach) Pfitzer (*P. linii* Schoser). Illustration: *Orchid Digest* **42**: 29 (1978); Bechtel et al., Manual of cultivated orchid species, 252 (1981).

Leaves 10–17.5 cm, faintly mottled with bluish green, often red beneath. Flower-stalk to 30 cm, hairy. Flower 7–8 cm wide. Upper sepal 3 cm, acute, reflexed above, green to greenish yellow, often brown at the base with *c.* 20 indistinct darker green veins. Petals spathulate, obtuse, more or less horizontal, apically pinkish purple with green margins, shading through brown to green at the base where there are a few dark shiny warts on marginal undulations. Lip slightly inflated, buff below, darker brown above with a green rim. Staminode with a narrow, parallel-sided sinus. *Tioman Isles off Malay peninsula, Borneo & Indonesia (Sulawesi & Ambon).* Summer–autumn.

21. P. appletonianum (Gower) Rolfe (*P. wolterianum* (Kraenzlin) Pfitzer). Illustration: Rittershausen & Rittershausen, Orchids in colour, 92 (1979); *Orchid Digest* **44**: 227 (1980); Bechtel et al., Manual of cultivated orchid species, 251 (1981).

Leaves *c.* 20 × 3–5 cm, elliptic, mottled. Flower-stalk to 50 cm, hairy. Flower *c.* 8 cm from top to bottom. Upper sepal *c.* 3.5 × 2.5 cm, ovate, green, paler above with darker green veins. Lower sepal smaller and narrower, acute. Petals 5–6 × 1–2 cm, elliptic to spathulate, obtuse to rounded, slightly wavy, drooping and half twisted; basal half bright green with brown spots concentrated near margins and mid-line; apical half pinkish purple. Lip 4–5 cm, pale greenish brown below, dark brown near the rim. Staminode sinus broad, diverging. *Laos, Thailand & Cambodia.* G2. Winter–spring.

Taller plants with broader, more drooping petals have often been treated separately as *P. wolterianum*. However, there are no consistent differences between the 2 species.

22. P. hookerae (Reichenbach) Stein. Illustration: *Botanical Magazine*, 5362 (1863); *Orchid Digest* **40**: 155 (1976) & **45**: 17 (1981).

Leaves 10–15 × 3–4 cm, ovate to oblong, mottled greyish on dark green. Flower-stalk to 20 cm, greyish purple, hairy. Flower to 10 cm wide. Upper sepal ovate, acute, yellowish with a green centre, indistinctly veined with darker green. Petals spathulate, acute, 4–5 cm, held below horizontal, bright pinkish purple at the tips shading to dull greenish yellow with many small pinkish spots in basal half. Lip yellowish green, tinged with reddish brown above. Sinus on staminode closed by overlapping lobes. *N Borneo.* G2. Spring–summer.

23. P. volonteanum (Masters) Pfitzer (*P. hookerae* (Reichenbach) Stein var. *volonteanum* (Masters) Hallier). Illustration: *Orchid Digest* **39**: 166 (1975) & **45**: 18 (1981).

Similar to *P. hookerae.* Petal spots purplish black, extending into outer half of petal but concentrated near margins and mid-line. Green and purple areas of petal separated by a patch of white or very pale green. *N Borneo.* G2.

24. P. tonsum (Reichenbach) Stein. Illustration: *Orchid Digest* **40**: 85 (1976); Bechtel et al., Manual of cultivated orchid species, 257 (1981); *Botanical Magazine*, n. s., 838 (1982).

Leaves 12–20 cm, oblong, ovate, mottled dark and light green. Flower-stalk 20–45 cm, purplish green. Flower 12 cm wide. Upper sepal 5–6 cm, broadly ovate, pale green, with 25–30 green and purple veins. Petals 6–7 cm, slightly below horizontal, narrowly spathulate, similar in colour and venation to the upper sepal but with a few large dark spots on the upper margin and in the mid-line; margins hairy near the tip. Lip greenish yellow flushed with pink. *Sumatra.* G2. Summer–autumn.

25. P. venustum (Sims) Pfitzer. Illustration: *Botanical Magazine*, 2129 (1820); *Orchid Digest* **36**: 177 (1972); Bechtel et al., Manual of cultivated orchid species, 257 (1981).

Leaves 15–25 cm, ovate-lanceolate, mottled dark on light green above, streaked and spotted with red beneath. Flower-stalk 20–30 cm, purple. Flower 8–10 cm wide. Upper sepal ovate, acute, white or off-white with *c.* 20 green veins. Petals 4–5 cm, horizontal or slightly descending, spathulate, hairy on margins, yellowish green or green at the base, reddish brown at the tip, with several irregular dark hairy spots on blade and margins. Lip dull yellow to reddish brown, strongly veined with greenish brown. *Nepal, Bhutan, Bangladesh & NE India.* G1–2. Winter–spring.

26. P. acmodontum Wood. Illustration: *Orchid Review* **84**: 350 (1976); *Orchid Digest* **41**: 60 (1977); Bechtel et al., Manual of cultivated orchid species, 251 (1981).

Leaves 15–20 × 4–5 cm, oblong to elliptic, mottled. Flower-stalk to 25 cm, sparsely hairy. Flowers 7–8 cm wide. Upper sepal *c.* 4 × 3 cm, broadly ovate, acuminate, off-white with *c.* 15 purple veins, flushed with pink at base. Petals 4–4.5 × 1.5–2 cm, obovate to spathulate, acute to acuminate, greenish yellow at base, reddish brown in outer half. Inner half irregularly spotted with purplish black. Lip *c.* 4 cm, yellowish green below, brown towards rim, with a prominent acute tooth on outer edge of rim. *Philippines.* G2. Spring.

27. P. javanicum (Lindley & Paxton) Pfitzer (*P. virens* misapplied; *P. purpurascens* Fowlie). Illustration: *Orchid Review* **85**: 159

(1977); *Orchid Digest* **45**: 24 (1981); Bechtel et al., Manual of cultivated orchid species, 251 (1981).

Leaves to 20 × 5 cm, oblong to elliptic, mottled. Flower-stalk to 25 cm. Flower *c.* 10 cm from top to bottom. Upper sepal *c.* 4.5 × 2.5 cm, ovate, acuminate, greenish yellow with 20–25 dark green veins; lower sepal similar but smaller. Petals *c.* 5 × 1.5 cm, narrowly obovate, broader in apical half, drooping, dull green, pink at tips, most of blade covered in small purplish black warts. Lip *c.* 4.5 cm, dull green to brownish green with darker veins. *Java & Borneo.* G2. Flowers intermittently.

28. P. virens (Reichenbach) Pfitzer (*P. javanicum* (Lindley & Paxton) Pfitzer var. *virens* (Reichenbach) Stein). Illustration: Reichenbach, Xenia Orchidacearum, **1**: 162 (1856); *Orchid Digest* **40**: 237 (1976) & **45**: 22 (1981).

Very similar to *P. javanicum* and often considered a subspecies of it. Petals narrowly spathulate, horizontal; basal two-thirds bright green, apical third pinkish purple; warts concentrated on margins and mid-line. Lip pale brown, darker near rim. *N Borneo.* G2.

Frequently confused with *P. javanicum* and thus mistakenly said to be native to Indonesia.

29. P. purpuratum (Lindley) Stein. Illustration: *Botanical Magazine*, 4901 (1856); *Orchid Review* **82**: 39 (1974); Bechtel et al., Manual of cultivated orchid species, 255 (1981).

Leaves 8–12 cm, oblong-elliptic, mottled dark and pale green. Flower-stalk 20–25 cm, deep purple, hairy. Flower 8–10 cm wide. Upper sepal circular, acuminate, white with about 12 alternating purple and green veins. Petals 4–5 cm, elliptic, hairy on margins, purplish brown with many small dark warts in basal half, paler at the tip. Lip purplish brown. *Hong Kong & adjacent China.* G2. Summer–winter.

30. P. mastersianum (Reichenbach) Stein. Illustration: *Botanical magazine*, 7629 (1898); *Orchid Digest* **45**: 19, 20 (1981); Bechtel et al., Manual of cultivated orchid species, 255 (1981).

Leaves 20–24 × 3–4 cm, oblong-elliptic, mottled dark and yellowish green. Flower-stalk 30–45 cm, deep purple, hairy. Flower to 10 cm wide, fleshy. Upper sepal circular, 5 cm broad, greenish yellow with about 20 dark green veins and a broad white or off-white margin. Petals 5–6 × 2 cm, horizontal, oblong to slightly spathulate, round-tipped, reddish brown in apical half, remainder dull green with many fine purplish warts above the mid-line. Lip dull reddish purple, yellow round the rim. *Indonesia (Borneo & Moluccas).* G2. Spring–summer.

31. P. superbiens (Reichenbach) Stein (*P. curtisii* (Reichenbach) Stein). Illustration: *Orchid Review* **83**: 394 (1975); *Orchid Digest* **42**: 177 (1978) & **45**: 60 (1981); Williams et al., Orchids for everyone, 152 (1980).

Leaves 15–20 cm, oblong-lanceolate, strongly mottled. Flower-stalk to 30 cm, purple, downy; bract ovate, a third of the ovary length; flower to 8 cm wide. Upper sepal broadly ovate, acuminate, pale green, purple at the base with about 25 dark green to purple veins. Lower sepal small, ovate-acuminate. Petals oblong, drooping, curled, hairy on margins, purple at the base shading to pale green, evenly marked with numerous small purplish black warts. Lip brownish purple. *Sumatra.* G2. Summer.

P. ciliolare (Reichenbach) Stein (*P. superbiens* (Reichenbach) Stein subsp. *ciliolare* (Reichenbach) Wood). Illustration: *Orchid Digest* **41**: 47 (1977); Rittershausen & Rittershausen, Orchids in colour, 90 (1979); Bechtel et al., Manual of cultivated orchid species, 252 (1981). Petals not curled, with dense marginal hairs; warts absent from outer third. *Philippines.* G2. Spring–summer.

32. P. argus (Reichenbach) Stein. Illustration: *Botanical Magazine*, 6175 (1875); *Orchid Digest* **41**: 61 (1977) & **45**: 57 (1981); Rittershausen & Rittershausen, Orchids in colour, 90 (1979).

Leaves 12–20 cm, oblong-lanceolate, pale green with darker mottling. Flower-stalk to 30 cm, purplish, glandular hairy. Bract pale green, half the ovary length. Flower 10 cm wide. Upper sepal broad, acuminate, off-white with about 20 dark green or purplish veins. Petals 7–8 cm, oblong, slightly tapering, hairy on margin, descending, off-white at the base, shading to pink at the tip, boldly marked all over with dark purple spots, some having pale centres. Lip yellow with dark green net-venation, red above, flushed with brown. *Philippines.* G2. Spring.

33. P. sukhakulii Schoser & Senghas. Illustration: *Orchid Review* **77**: 145 (1969); *Orchid Digest* **39**: 208 (1975); Bechtel et al., Manual of cultivated orchid species, 256 (1981).

Leaves 12–15 × *c.* 4.5 cm, narrowly elliptic, mottled. Flower-stalk to 25 cm. Flower *c.* 10 cm wide. Upper sepal 3.5–4 × 3 cm, broadly ovate, acuminate, yellowish green with 15–20 green veins. Petals 5–6 × 1.3–1.7 cm, oblong to narrowly elliptic, tapering to an acute point, almost horizontal, yellowish green, closely spotted with purplish black except at extreme tip. Lip *c.* 4.5 cm, conical, greenish brown. *Thailand.* G2. Autumn.

P. wardii Summerhayes. Illustration: *Botanical Magazine*, 9481 (1937); *Orchid Digest* **36**: 85 (1972) & **45**: 15 (1981); Pradhan, Indian orchids **1**: t. 19 (1976). Flowers usually browner than in *P. sukhakulii;* petals descending. *N Indo-Burmese border.* G2. Winter.

Very similar to *P. sukhakulii.* They should probably be considered as one species under *P. wardii,* which is the older name.

34. P. victoria-regina (Sander) Wood (*P. victoria-mariae* invalid). Illustration: *Botanical Magazine*, 7573 (1898); *Orchid Digest* **44**: 217 (1980).

Leaves oblong, to 40 × 6 cm, often faintly mottled. Flowering stem 30–60 cm or more, reddish brown, bearing several flowers in succession over several months. Flowers 7–8 cm wide. Upper sepal 3–4 cm wide, circular to broadly ovate, greenish yellow with 5–8 purple veins and a white to creamy yellow border. Petals 5 cm, narrowly oblong, horizontal, hairy on margins, twisted and wavy, greenish yellow, veined with reddish purple and with an unbroken border of reddish purple. Lip slightly compressed, purple, unspotted, with a green and yellow rim. *Sumatra.* G2. Intermittent flowering.

The subspecies described above is endemic to a small area in central Sumatra and is uncommon in cultivation. The following subspecies are more commonly grown:

Subsp. **chamberlainianum** (Sander) Wood (*P. chamberlainianum* (Sander) Stein). Illustration: *Botanical Magazine*, 7578 (1898); *Orchid Review* **84**: 140 (1976); Bechtel et al., Manual of cultivated orchid species, 252 (1981). Leaves to 10 cm broad; upper sepal yellow and purple; lip inflated, evenly spotted with purple. *Sumatra.* G2. Intermittent flowering.

Subsp. **glaucophyllum** (J.J. Smith) Wood (*P. glaucophyllum* J.J. Smith).

Illustration: *Orchid Review* **84**: 140 (1976); *Orchid Digest* **44**: 216 (1980); Bechtel et al., Manual of cultivated orchid species, 253 (1981). Leaves glaucous; petals slightly deflexed, marked with discontinuous purple spots on a white background. *E & C Java.* G2. Intermittent flowering.

35. P. parishii (Reichenbach) Stein. Illustration: *Botanical Magazine*, 5791 (1869); *Orchid Digest* **40**: 213 (1980); Bechtel et al., Manual of cultivated orchid species, 255 (1981).

Epiphytic. Leaves to 30 cm, strap-shaped, unmottled, bright green. Flowering stem to 60 cm, pale green, bearing 4–7 flowers. Flowers 8 cm wide. Upper sepal elliptic, pale yellowish green with faint veins. Petals to 15 cm, linear, gently tapering, drooping and twisted, green at base with marginal purple spots, purple at tips. Lip green to greenish purple. *Burma, NW Thailand & S China.* G2. Spring–summer.

36. P. lowii (Lindley) Stein. Illustration: *Gardeners' Chronicle*, 765 (1847); *Orchid Digest* **44**: 214 (1980); Williams et al., Orchids for everyone, 154 (1980).

Epiphytic or on cliffs. Leaves oblong, to 40×4 cm, unmottled, light green. Flowering stem to 1 m, arched, brownish purple, bearing 2–6 flowers. Flowers 15 cm across. Upper sepal 5 cm, broadly elliptic, short-pointed, concave above, with reflexed margins below, yellowish green flushed with purple, streaked with brown at base. Petals 8 cm, spathulate, spreading, slightly drooping at tips, slightly twisted, yellowish green with brown spots at base; tips bright purple. Lip blunt, oblong, pale brownish green with brown markings, flushed with purple near rim. *Malaysia & Indonesia.* G2. Spring–summer.

P. haynaldianum (Reichenbach) Stein. Illustration: *Botanical Magazine*, 6296 (1877); *Orchid Digest* **44**: 214 (1980); Bechtel et al., Manual of cultivated orchid species, 254 (1981). Upper sepal pale, marked with large irregular brown blotches. *Philippines.* G2. Spring.

Possibly better considered as a subspecies of *P. lowii.*

4. DISA Bergius

P.J.B. Woods & J.M.H. Shaw

Herbaceous, terrestrial perennials, occasionally evergreen, with tuberous roots, sometimes stoloniferous. Sterile shoots usually rosette-like. Leaves basal or on the stem, ovate to linear, mostly lanceolate. Inflorescence a dense or loose raceme or corymb. Flowers usually with lip lowermost (not obvious in single-flowered or corymbose inflorescences), less commonly with lip uppermost. Median sepal concave and usually hood-like, and with a basal spur, lateral sepals spreading. Petals smaller and narrower than sepals, often partially attached to the column. Lip small and narrow. Column short; pollinia 2, granular.

A genus of over 100 species from tropical and southern Africa and Madagascar. Cultivation can be difficult, and full details are given in the papers by Vogelpoel and Stoutamire cited below.

Literature: Linder, H.P., A revision of Disa excluding Sect. Micranthae, *Contributions from the Bolus Herbarium* **9**: 1–370 (1981); Vogelpoel, L., Disa uniflora – its propagation and cultivation, *American Orchid Society Bulletin* **49**: 961–72 (1980); Vogelpoel, L., Disa species and their hybrids, *American Orchid Society Bulletin* **49**: 1084–92 (1980); Stoutamire, W., Cultivated Disas in Ohio, *American Orchid Society Bulletin* **50**: 1195–200 (1981); Venter, J., The other Disas, *Orchids* (USA) **76**(7): 520–27 (2007).

1a. Lateral sepals about equalling upper sepal **1. uniflora**
b. Lateral sepals about one third longer than upper sepal **2. cardinalis**

1. D. uniflora Bergius (*D. grandiflora* Linnaeus). Illustration: *Botanical Magazine* 4073 (1844); *Orchid Review* **78**: 95 (1970), **85**: 229 (1977) & **86**: 127 (1978); Bechtel et al., Manual of cultivated orchid species, 204 (1981).

Stems to 60 cm. Leaves of sterile shoots in a loose basal rosette; leaves on flowering stems to 8, 15×1.25 cm, lanceolate, acuminate, the upper 3–4 smaller, sheathing. Inflorescence with 1–3 (rarely to 10) flowers, bracts longer than ovaries. Flowers 8–12 cm wide, lip lowermost, faintly scented, scarlet or carmine, orange to yellowish inside (pure white or yellow forms are occasionally recorded). Upper sepal *c.* 5×3 cm, ovate, the hooded basal part forming a narrow conical, laterally flattened spur, 1–1.5 cm. Lateral sepals broadly elliptic, acuminate, about equal to upper sepal in length. Petals *c.* 2 cm × 8 mm, narrowly obovate, erect. *South Africa (Cape Province).* G1. Winter.

2. D. cardinalis Linder. Illustration: *Flowering Plants of Africa* **46**(3): t. 1826 (1981); *Kew Magazine* **5**(1): t. 94 (1988).

Stems to 60 cm, from small tubers, with subterranean rhizomes forming clumps. Leaves 6–10 at the base of the stems, narrowly elliptic, 5–10 cm, acuminate, sheathing. Inflorescence with 8–40 flowers, bracts longer than ovaries. Flowers 2.5–3 cm wide, lip lowermost, bright red. Upper sepal *c.* 1.5×1 cm, boat-shaped, with a basal conical spur 4 mm. Lateral sepals elliptic, blunt or rounded, about one third longer than upper sepal. Petals *c.* 7×2 mm, narrowly oblong. *South Africa (Cape Province).* G1. Winter.

D. Kirstenbosch Pride grex is the garden hybrid *D. uniflora* × *cardinalis.* Illustration: *Kew Magazine* **5**(1): t. 94, 1988. It is intermediate in floral shape and size between its parents, but a richer colour.

5. SATYRIUM Swartz

P.J.B. Woods

Herbaceous, terrestrial perennials. Roots tuberous, ovoid. Stems unbranched, leafy. Aerial growth dying down in winter. Flowers several to many, in racemes or spikes, not resupinate, variously coloured. Lip uppermost, forming a hood. Spurs 2, attached to base of lip. Pollinia 2, granular.

A genus of perhaps 100 species, mainly from Africa, though 2 occur in Asia. Grassland (*Bulletin of the Alpine Garden Society* **46**: 301, 1978) recommends 'a compost providing humus and drainage …, that being used being one part each of loam, leaf-mould, peat and sand. A position in semi-shade where the plant can remain cool and moist throughout growth, and dry but not arid when dormant is most suitable'. Annual potting should be done before growth commences. Apparently plants may remain dormant over several years.

1. S. nepalensis Don. Illustration: *Annals of the Royal Botanic Garden Calcutta* **8**: t. 443, 444 (1898); *Orchid Review* **85**: 149 (1977); *Norwegian Journal of Botany* **26**: 286 (1979).

Tubers ovoid, produced singly during the flowering season at the ends of vigorous roots. Stem 10–75 cm, bearing 2–3 leaves. Leaves 7–25 × 3–9 cm, sheathing, oblong-ovate to lanceolate, obtuse, the upper shorter. Spike cylindric, 5–15 cm, flowers numerous, pink or white, scented, bracts longer than the flowers, deflexed. Sepals and petals to 8 mm, linear-oblong, pointed, slightly recurved, petals narrower than sepals (which have hairy margins in var. **ciliata** (Lindley) King & Pantling). Lip to 8 mm, as broad as or broader than long, concave, keeled outside, the base with 2 spurs, which are 8–12 mm. *N India to Sri Lanka & Burma.* H5–G1. Summer–autumn.

Plants offered under this name in the trade have occasionally proved to be *Herminium angustifolium, Habenaria arietina* or *Platanthera clavigera.*

6. CORYBAS Salisbury
P.J.B. Woods
Small, 1-leaved, terrestrial perennials. Aerial growth dying off in winter. Root tuberous, small and round. Leaves ovate to heart-shaped, held flat against the ground. Flower 1 (rarely 2 in a few species), large compared to the leaf size, shortly stalked (rarely not stalked), the stalk elongating in fruit. Upper sepal conspicuous, broad, concave or hooded, lateral sepals and petals thread-like, sometimes very small. Lip conspicuous, the basal part erect under the upper sepal, inrolled and tube-like around the column, the apical part expanded into a broad, often abruptly bent blade, the margin entire or fringed. Spurs 2, sometimes inconspicuous, placed on either side of the column. Pollinia 2, granular.

A genus of about 60 species from the Himalaya and E Asia, extending to Australasia. Cultivation as for *Pleione* (p. 482), or they can be grown in peaty compost, with high humidity when growing. When dormant, they should be kept cool and dry.

1. C. dilatatus (Rupp & Nicholls) Rupp. Illustration: *American Orchid Society Bulletin* **44**: 993 (1975), **46**: 989 (1977) & **47**: 1120, 1121 (1978); *Botanical Magazine*, n.s., 836 (1981).

Plant to 3 cm. Leaf to 3 cm wide, ovate to heart-shaped. Flower solitary, reddish purple. Upper sepal to 2.5 cm. Lip with the apical part white or white tinged with crimson, papillose towards the apex; margins crimson, translucent, coarsely toothed, blotched and purple-veined. *SE Australia, including Tasmania.* G1. Summer.

7. OPHRYS Linnaeus
J. Lamond & P.J.B. Woods
Perennial herbs. Tubers 2 (occasionally 3), spherical, fleshy. Leaves lanceolate, oblong or ovate, rolled when young, mostly in a basal rosette, sometimes spaced along the stem, when usually sheathing, usually appearing in autumn after the current season's growth has died back. Flowers few to *c.* 15 in a short or long, loose or rarely dense spike. Bracts leaf-like, often inrolled. Sepals spreading, usually greenish or pinkish, oblong or ovate, obtuse, all equal in length, the upper concave, erect or curved over the column. Petals usually smaller and narrower than the sepals, often hairy or velvety. Lip spurless, hairy, complex in structure, often like a bee, wasp or other insect, oblong, square, rounded or diamond-shaped, usually marked with a coloured, hairless, often complex, patterned area (the speculum), lobed or not, with or without humps or horn-like basal protuberances; apex often with a short, tooth-like, deflexed or upcurved appendage. Column erect or curved forwards; anther-connective beak-like, its apex obtuse or pointed. Pollinia 2, club-like, made up of small, easily separable masses.

A genus of 10 distinct species as demonstrated by molecular techniques. In the past, different workers have recognised between 16 and 251 species, however these entities represent populations that are the result of complex hybridisation. This phenomenon of past mass hybridisation also is displayed by several European genera in other families including *Rubus, Alchemilla* and *Taraxacum*, some of which have resorted to apomixis. *Ophrys* is distributed from NW Europe throughout the Mediterranean region to the Middle East. There is some degree of specificity between pollinators (mostly bees and wasps) and the species of *Ophrys*. The male insects are attracted and sexually stimulated by the scent, shape, colour and the tactile stimuli of the hairs on the lip of the orchid. Fuller descriptions of these complex flowers are given in Danesch & Danesch (cited below) and Wood, *Orchid Review* **89**: 298 (1981).

Literature: Nelson, E., Gestaltwandel und Abbildung erörtert am Beispiel der Orchidaceen Europas und der Mittelmeerländer mit einer Monographic und Ikonographie der Gattung Ophrys (1962) – all the species included in the following account are illustrated in this work, to which no further references are made; Danesch, E. & Danesch, O., Orchideen Europas, Ophrys-Hybriden (1972); Pedersen, H. A. & Faurholdt, N., Ophrys, the bee orchids of Europe (2007); Devey, D.S., Batemen, R.M., Fay, M. & Hawkins, J., Friends or relatives? Phylogenetics and species delimitation in the controversial European orchid genus Ophrys, *Annals of Botany* **101**(3): 385–402 (2008).

Sepals. Pinkish or whitish: **6, 7.**
Petals. Blackish or brownish purple: **1, 2**; greenish or yellowish: **3, 4, 5, 6, 7**; brownish: **4, 5**; pink or purplish: **6, 7.** Hairy or velvety: **1, 2, 6, 7**; hairless or papillose: **3, 4, 5.** Rounded: **2**; triangular: **6, 7**; oblong to lanceolate: **2, 3, 4, 5, 6, 7**; linear: **1, 7.**
Lip. 3-lobed: **1, 2, 3, 4, 6, 7**; entire: **5, 6.** Mostly yellow: **3.** Densely fringed: **2.** Speculum bright, shiny blue: **2, 4**; speculum H-shaped or of parallel lines: **5, 7.** With an apical appendage: **5, 6, 7**; without an apical appendage: **1, 2, 3, 4, 5.**
Anther-connective. Obtuse: **1, 2, 3, 4, 6**; acute: **5, 6, 7.**

1a. Lip with a tooth-like appendage at the apex 2
b. Lip without a tooth-like appendage at the apex 4
2a. Speculum of 2 more or less parallel lines, or if H-shaped then petals hairless **5. sphegodes**
b. Speculum various, not as above 3
3a. Lip distinctly 3-lobed **7. apifera**
b. Lip not 3-lobed, or lobes very indistinct **6. fuciflora**
4a. Lip 3-lobed, central lobe notched or not 5
b. Lip not lobed **5. sphegodes**
5a. Lip, at least at the margin, bright yellow and usually hairless **3. lutea**
b. Lip blue, violet, purple, reddish, brownish or black, lateral lobes occasionally yellow but then margins densely fringed 6
6a. Speculum bright, shining blue, covering almost the whole lip; margins of lip densely fringed with long hairs **2. speculum**
b. Speculum blue or not, restricted to the basal half of the lip; margins of lip not fringed 7
7a. Petals slender, antenna-like, blackish purple; lip sharply cleft at the apex **1. insectifera**
b. Petals oblong, greenish; lip notched at the apex **4. fusca**

1. O. insectifera Linnaeus (*O. muscifera* Hudson). Illustration: Danesch & Danesch, Orchideen Europas, Südeuropa, 128 (1969); Danesch & Danesch, Orchideen

Europas, Mittel-Europa, edn 3, 222, 223 (1972); Ross-Craig, Drawings of British plants **28**: t. 9 (1971); Mossberg & Nilsson, Orchids of northern Europe, 130, 131 (1979).

Plant to 60 cm. Leaves linear to oblong. Spike with up to 10 flowers. Sepals 6–8 mm, oblong-ovate, green. Petals 4–6 mm, linear, antenna-like, blackish purple, velvety. Lip 9–10 × 6–7 mm, oblong, 3-lobed, central lobe cleft at apex, blackish purple or dark violet, papillose, lateral lobes oblong. Speculum square or kidney-shaped, pale bluish violet. Anther-connective obtuse. *Most of Europe, uncommon in the extreme south and north and absent in the southeast,* H2. Spring–summer.

2. O. speculum Link (*O.vernixia* Brotero). Illustration: Danesch & Danesch, Orchideen Europas, Südeuropa, 126 (1969); Sundermann, Europäische und mediterrane Orchideen, 82 (1975); Landwehr, Wilde Orchideeën van Europa **2**: 377, 379 (1977); Grey-Wilson & Mathew, Bulbs, t. 37 (1981).

Plant 7–30 cm. Leaves oblong to lanceolate. Spike with up to 15 flowers. Sepals 6–8 mm, oblong-ovate, greenish or yellowish green, sometimes lined with brown. Petals one-third to half the length of the sepals, lanceolate, ovate-lanceolate or rounded, dark purple, hairy. Lip 1.3–1.5 cm, 3-lobed, margins densely fringed, hairs brownish purple to blackish purple, yellowish or brownish red, central lobe ovate, usually notched; lateral lobes oblong, yellowish. Speculum large, shining blue, with a yellow margin. Anther-connective obtuse. *Portugal, Mediterranean region.* H3. Spring.

3. O. lutea (Gouan) Cavanilles. Illustration: Danesch & Danesch, Orchideen Europas, Sudeuropa, 123, 124 (1969); Landwehr, Wilde Orchideeën van Europa **2**: 391 (1977); Williams et al., Orchids of Britain and Europe, 57 (1978); Grey-Wilson & Mathew, Bulbs, t. 37 (1981).

Plant 7–30 cm. Basal leaves ovate. Spike with up to 7 flowers. Sepals *c.* 1 cm, green. Petals one-third to half the length of the sepals, oblong, yellowish green, hairless or margins papillose. Lip 9–18 mm, oblong or rounded, 3-lobed, lateral lobes rounded, central lobe kidney-shaped, marginal zone broad, bright yellow, centre dark brown or blackish purple. Speculum oblong, 2-lobed, greyish or bluish grey. Anther-connective obtuse. *Portugal, Mediterranean region.* H3. Spring.

4. O. fusca Link. Illustration: Danesch & Danesch, Orchideen Europas, Südeuropa, 118 (1969); Landwehr, Wilde Orchideeën van Europa **2**: 383, 385, 387 (1977); Williams et al., Orchids of Britain and Europe, 55 (1978); Grey-Wilson & Mathew, Bulbs, t. 37 (1981).

Plant 10–40 cm. Leaves oblong or lanceolate. Spike with up to 8 flowers. Sepals 9–11 mm, greenish or yellowish green. Petals 6–8 mm, oblong, yellowish green or brownish green, indistinctly papillose. Lip 1.3–1.5 cm × 9–12 mm, oblong or obovate, purplish, purplish red or reddish brown, velvety, 3-lobed, lateral lobes not spreading, oblong, rounded, central lobe rounded, notched. Speculum oblong, usually divided down the middle, bluish, often with a paler or yellowish margin. Anther-connective obtuse. *Portugal, Mediterranean region, SW Romania.* H3. Spring.

Subsp. **iricolor** (Desfontaines) Swartz (*O. tricolor* Desfontaines). Illustration: Danesch & Danesch, Orchideen Europas, Südeuropa, 119 (1969); Sundermann, Europäische und mediterrane Orchideen, 84 (1975); Landwehr, Wilde Orchideeën van Europa **2**: 383 (1977); Williams et al., Orchids of Britain and Europe, 55 (1978). Spike with up to 4 flowers, petals often papillose, lip 2.5–3.1 cm, speculum bright blue. *C&E Mediterranean region.*

5. O. sphegodes Miller (*O. aranifera* Hudson). Illustration: Danesch & Danesch, Orchideen Europas, Südeuropa, 100–2 (1969); Ross-Craig, Drawings of British plants **28**: t. 8 (1971); Landwehr, Wilde Orchideeën van Europa **2**: 397 (1977); Mossberg & Nilsson, Orchids of northern Europe, 132, 133 (1979); *Orchid Review* **89**: 295 (1981).

Plant 10–45 cm. Basal leaves ovate-lanceolate. Spike with up to 10 flowers. Sepals 6–12 mm, greenish. Petals 4–8 mm, hairless, oblong-lanceolate, greenish or occasionally brownish, margins sometimes wavy. Lip 1–1.2 cm × 8–12 mm, entire, rounded, pale or dark brown or blackish brown, velvety, with or without basal protuberances. Speculum usually H-shaped, the cross-line sometimes absent, greyish or purplish blue. Anther-connective acute. *Europe.* H3. Spring–summer.

A variable species (see Wood, J.J., The subspecies of O. sphegodes in Cyprus and the eastern Mediterranean, *Orchid Review* **89**: 292–9, 1981). Subsp. **litigiosa** (Camus) Becherer (*O. litigiosa* Camus) has the lip 5–10 × 5–10 mm, pale or yellowish brown, usually without basal protuberances. Subsp. **mammosa** (Desfontaines) Soó (*O. mammosa* Desfontaines) has the lip 8–17 × 8–17 mm, blackish brown to blackish purple, usually with large basal protuberances. Both of these subspecies are from southern Europe. The specific name is sometimes misspelled 'sphecodes'.

6. O. fuciflora (Schmidt) Moench (*O. holoserica* (Burmann) Greuter). Illustration: Danesch & Danesch, Orchideen Europas, Südeuropa, 79a–c (1969); Danesch & Danesch, Orchideen Europas, Mitteleuropa, edn 3, 224–7 (1972); Ross-Craig, Drawings of British plants **28**: t. 7 (1971); Landwehr, Wilde Orchideeën van Europa **2**: 449, 461, 469 (1977); Mossberg & Nilsson, Orchids of northern Europe, 134, 135 (1979); Grey-Wilson & Mathew, Bulbs, t. 39 (1981).

Plant 15–55 cm. Basal leaves ovate-oblong. Spike with 2–6 flowers (occasionally more). Sepals 9–13 mm, pinkish, whitish or greenish. Petals about one-quarter to one-third of the length of the sepals, triangular to oblong-lanceolate, pink or purplish, hairy. Lip 9–13 mm, ovate or almost square, entire or rarely indistinctly 3-lobed, brownish or brownish purple, velvety; central part papillose, basal protuberances distinct, apical appendage upcurved, 3-toothed. Speculum purplish or bluish grey bordered by a bold pattern of vertical and horizontal bands of pale green or yellow. Anther-connective acute. *W, SW & C Europe.* H3. Spring–summer.

O. tenthredinifera Willdenow. Illustration: Danesch & Danesch, Orchideen Europas, Südeuropa, 92, 93 (1969); Sundermann, Europäische und mediterrane Orchideen, 98 (1975); Landwehr, Wilde Orchideeën van Europa **2**: 469, 471 (1977); Williams et al., Orchids of Britain and Europe, 43 (1978); Grey-Wilson & Mathew, Bulbs, t. 39 (1981). Similar, but petals broadly triangular, lip almost square with a notched apex, dark brown, marginal area usually yellowish; speculum violet or bluish with a pale greenish border, pattern usually simple; anther-connective obtuse. *Portugal, Mediterranean region.* H3. Spring.

7. O. apifera Hudson. Illustration: Danesch & Danesch, Orchideen Europas, Südeuropa, 94 (1969); Danesch &

Danesch, Orchideen Europas, Mitteleuropa, edn 3, 228–33 (1972); Ross-Craig, Drawings of British plants **28**: t. 6 (1971); *Quarterly Bulletin of the Alpine Garden Society* **44**: 59 (1976); Mossberg & Nilsson, Orchids of northern Europe, 136, 137 (1979).

Plants 15–50 cm. Basal leaves lanceolate to ovate. Spike with up to 10 flowers. Sepals 1–1.5 cm, pink or purplish, rarely whitish. Petals less than half as long as sepals, triangular to linear-lanceolate, greenish or rarely purplish, hairy. Lip 1–1.3 cm, ovate, 3-lobed, brownish or brownish purple; lateral lobes triangular, deflexed; basal protuberances *c.* 3 mm, apical appendage longer than wide, yellow, deflexed. Speculum H-shaped or shield- or W-shaped, violet-purple with a pale green margin, sometimes indistinct. Anther-connective pointed. *W, S & C Europe.* H2. Spring–summer.

O. scolopax Cavanilles. Illustration: Danesch & Danesch, Orchideen Europas, Südeuropa, 72–4 (1969); Landwehr, Wilde Orchideeën van Europa **2**: 437, 438 (1977); Williams et al., Orchids of Britain and Europe, 35 (1978); Grey-Wilson & Mathew, Bulbs, t. 39 (1981). Sepals 8–10 mm, the upper erect. Lip 8–12 mm, longer than wide, oblong-elliptic, with a narrow, hairless or papillose marginal zone, basal protuberances short, obtuse, apical appendage short, wider than long. Speculum often X-shaped. *Portugal, W & C Mediterranean region.* H3. Spring.

A very variable species; several of its subspecies and similar taxa are occasionally grown in specialist collections, for example: subsp. **cornuta** (Steven) Camus (*O. oestrifera* Bieberstein), which has a lip with a wide, hairless marginal zone and horn-like basal protuberances to 10 mm, and is found in the eastern Mediterranean region, the Caucasus and Iran; subsp. **heldreichii** (Schlechter) Nelson (*O. oestrifera* subsp. *heldreichii* (Schlechter) Soó), with lip 1.3–1.5 cm, also from the eastern Mediterranean region; and **O. umbilicata** Desfontaines (*O. scolopax* subsp. *orientalis* (Renz) (Nelson), with sepals pale greenish, whitish or rarely reddish pink, the central hooded over the column, from Cyprus, Turkey, Syria and the Lebanon.

8. SERAPIAS Linnaeus

J. Lamond & P.J.B. Woods

Terrestrial herbs. Roots tuberous, tubers 2–4, occasionally 5, stalkless and in some species also produced at the end of root-like stolons, fleshy, ovoid. Stems erect. Leaves lanceolate, mostly basal or spaced along the stem, rolled when young, later often folded. Aerial growth dying off during the current year. Spike loose- or dense-flowered; bracts very conspicuous, boat-shaped, usually purplish or glaucous. Sepals and petals forming a pointed hood, usually coloured like the bracts, veins distinct; sepals lanceolate; petals ovate at base, tapering to a long point. Lip 3-lobed, composed of a basal and an apical part, usually hairy on the upper surface, lateral lobes upturned, central lobe usually abruptly bent downwards, tongue-like, the basal part with a swollen area or a double ridge at the base. Pollinia 2, club-shaped, consisting of easily separable masses. Column apex beak-like.

A genus of about 6 species, between which the differences are not always clear, distributed from the Azores through the Mediterranean region to the Caucasus. Alpine house or cold-frame protection is necessary for plants grown in northern Europe. Pot culture and a compost incorporating coarse sand and limestone chips is recommended. New growth is produced during the autumn when water must be given sparingly and increased only when the plants are growing vigorously. As the growth dies down and the plants become dormant they require a warm, dry rest.

Literature: Nelson, E., Monographic und Ikonographie der Orchidaceen Gattungen Serapias, Aceras, Loroglossum, Barlia, 3–45 (1968); Christian, P., The genus Serapias in the wild and in cultivation, *Bulletin of the Alpine Garden Society* **43**: 188–98 (1975).

1a. Lip with a single, black, hump-like swelling at base **1. lingua**
b. Lip with 2 dark or pale ridges at base 2
2a. Apical part of lip about as broad as the basal part when flattened **2. cordigera**
b. Apical part of lip narrower than the basal part when flattened **3. vomeracea**

1. S. lingua Linnaeus. Illustration: Danesch & Danesch, Orchideen Europas, Südeuropa, 196, 197 (1969); Sundermann, Europäische und mediterrane Orchideen, 116 (1975); Landwehr, Wilde Orchideeën van Europa **2**: 355 (1977); Williams et al., Orchids of Britain and Europe, 67 (1978); Grey-Wilson & Mathew, Bulbs, t. 40 (1981).

Plant to *c.* 30 cm. Tubers 2–5, 1 stalkless. Leaves 3–6, to *c.* 15 cm × 5–12 mm, linear to lanceolate, sheathing bases usually unspotted. Spike with 2–5 (sometimes as many as 9) flowers, loose. Bracts *c.* 3–4 cm, about as long as the flowers. Hood *c.* 2 cm. Lateral lobes of lip purple; apical part to *c.* 1.9 cm × 7 mm, broadly lanceolate to narrowly ovate, hairless or shortly hairy, reddish, pinkish, violet or yellowish. *Portugal, N Africa, Mediterranean regions of S Europe east to Greece.* H5. Spring.

2. S. cordigera Linnaeus. Illustration: Danesch & Danesch, Orchideen Europas, Südeuropa, 192, 193 (1969); Landwehr, Wilde Orchideeën van Europa **2**: 369 (1977); Williams et al., Orchids of Britain and Europe, 65 (1978); Grey-Wilson & Mathew, Bulbs, t. 40 (1981).

Plant to 40 cm. Tubers 2–3, 1 stalkless. Leaves 3–8, to *c.* 15 × 1–1.5 cm, lanceolate, sheathing bases usually red-spotted. Spike 2–10-flowered, dense. Bracts *c.* 3.4–4.5 cm, shorter or scarcely longer than flowers. Hood *c.* 2.5 cm. Basal ridges of lip diverging, lateral lobes dark purple or reddish; apical part 2–3 × 1.6–2 cm, heart-shaped, very hairy, purplish to maroon. *Spain, Portugal, Mediterranean region to W Turkey.* H5. Spring.

S. neglecta de Notaris. Illustration: Danesch & Danesch, Orchideen Europas, Südeuropa, 189–91 (1969); Landwehr, Wilde Orchideeën van Europa **2**: 363 (1977); Williams et al., Orchids of Britain and Europe, 63 (1978); Grey-Wilson & Mathew, Bulbs, t. 40 (1981). Plant usually 15–25 cm, sturdy; sheathing leaf-bases unspotted; spike dense; bracts greenish. Basal ridges of lip parallel; apical part heart-shaped, pale yellowish or orange. *C Mediterranean regions of Europe.* H5. Spring.

3. S. vomeracea (Burman) Briquet (*S. longipetala* (Tenore) Pollini; *S. pseudocordigera* Moricand). Illustration: Danesch & Danesch, Orchideen Europas, Südeuropa, 185 (1969); Sundermann, Europäische und mediterrane Orchideen, 113 (1975); Landwehr, Wilde Orchideeën van Europa **2**: 357, 359 (1977); Grey-Wilson & Mathew, Bulbs, t. 40 (1981).

Plant 10–60 cm. Tubers 2, stalkless. Leaves 6–8, *c.* 6–11 cm × 5–12 mm, linear to lanceolate, sheathing bases green, unspotted. Spike to 10-flowered, loose. Bracts to 5 cm, longer than the flowers. Lip reddish or brownish, distinctly hairy, basal ridges parallel, lateral lobes often blackish: apical part *c.* 1.5–2 cm × 6–10 mm, lanceolate. *S Europe.* H5. Spring.

Subsp. **orientalis** Greuter (*S. orientalis* invalid). Illustration: Landwehr, Wilde Orchideeën van Europa **2**: 367 (1977); Williams et al., Orchids of Britain and Europe, 61 (1978). Plant 15–35 cm. Spike 5–7-flowered, dense. Bracts scarcely as long as or slightly longer than flowers, purplish. Apical part of lip broadly lanceolate. Although usually treated separately, we find it almost impossible to distinguish between this subspecies and *S. neglecta* but have followed recent accounts in placing it with *S. vomeracea.*

S. olbia Verguin (*S. gregaria* Godfery). Illustration: Danesch & Danesch, Orchideen Europas, Südeuropa, 195 (1969); Sundermann, Europäische und mediterrane Orchideen, 116 (1975); Landwehr, Wilde Orchideeën van Europa **2**: 365, t. 3, 373 (1977); Williams et al., Orchids of Britain and Europe, 67 (1978). Plant slender. Tubers 3–4, 1 stalkless. Spike few-flowered. *Southeast France.* H5. Spring.

Thought to be a natural hybrid between *S. lingua* and *S. parviflora.*

9. HIMANTOGLOSSUM Koch

P.J.B. Woods

Herbaceous perennials with 2 ovoid tubers. Aerial growth dying down in winter. Leaves in a basal rosette and arranged along the stem. Spike elongate, bracts equalling the flowers. Sepals and petals free, forming a hood over the column. Lip long, strap-shaped, 3-lobed, the central lobe bifid and spirally twisted, 2 or more times longer than the lateral lobes. A single spur present, arising from the base of the lip. Pollinia 2, club-shaped, composed of easily separable masses, attached to a single viscidium.

A genus of 4 species from Europe and SW Asia not often found in cultivation. Cultivation as for *Orchis.*

Literature: Wood, J.J., Notes on Himantoglossum, *Kew Bulletin* **38**: 75–78 (1983).

1. H. hircinum (Linnaeus) Sprengel. Illustration: Danesch & Danesch, Orchideen Europas, Mitteleuropa, edn 3, 130 (1972); Danesch & Danesch, Orchideen Europas, Südeuropa, 241 (1969); Williams et al., Orchids of Britain and Europe, 69 (1978); Mossberg & Nilsson, Orchids of northern Europe, 125 (1979).

Stem to 90 cm, purplish-blotched. Lower leaves 4–6, 5–51 × 3–5 cm, elliptic-oblong, obtuse, the upper leaves smaller, acute, their bases clasping the stem. Spike 10–50 cm, cylindric, with 15–18 loosely arranged flowers that emit a goat-like smell. Bracts shorter than the flowers. Sepals and petals pale green, sometimes purplish-lined, hooded, ovate, obtuse, the petals linear, 7–10 mm. Lip 3–5 cm × 2 mm, lateral lobes strap-shaped, central lobe ribbon-like, coiled in bud, spirally twisted when open, apex bifid, pale green suffused with purple, purple-spotted at base. Spur 2–2.5 mm, conical, downwardly pointing. *W, C & S Europe.* H4. Summer.

10. BARLIA Parlatore

P.J.B. Woods

Very similar to *Himantoglossum* but bracts as long as or longer than the flowers, the sepals erect, not hooded, and the lip much shorter, its central lobe not spirally twisted.

A genus of 2 species, from Mediterranean Europe and the Canary Isles, requiring cultural conditions similar to those for *Orchis.*

1. B. robertiana (Loiseleur) Greuter (*Himantolossum longibracteatum* (Bernardi) Schlechter). Illustration: Danesch & Danesch, Orchideen Europas, Südeuropa, 242, 243 (1969); Sundermann, Europäische und mediterrane Orchideen, 122 (1975); Landwehr, Wilde Orchideeën van Europa **2**: 345 (1977); Williams et al., Orchids of Britain and Europe, 97 (1978).

Large leaves in a rosette at the stem base. *Mediterranean region.* H5. Summer.

B. metlesicsiana Teshner. Illustration: Delforge, Orchids of Britain and Europe, 289 (1995). Leaves distributed along stem. *Canary Islands.* H5–G1. Spring–summer.

11. AMITOSTIGMA Schlechter

J.M.H. Shaw

Small perennial herbs with ovoid tubers. Stem with two basal scale-like bracts and one oblong to elliptic leaf. Inflorescence few-flowered, bracts shorter or equalling ovary. Flowers white, pink, mauve or yellow, with deep purple marking on lip. Sepals and petals free, blunt. Lip 3-lobed, mid-lobe further 2-lobed with spots or lines, base of lip wedge-shaped with a spur. Column short. Stigma under the anther with divergent side-lobes adnate to the lower margins of the lip. Anther cap small, low.

A genus of about 30 species distributed from Sikkim eastwards through China to Japan and Vietnam. The species are rarely seen, but a hybrid is occassionally available, requiring shaded frame or alpine house conditions.

1a. Leaves narrowly oblong to lanceolate; flowers purple, lip lobes ovate **1. keiskei**

b. Leaves linear; flowers white to pale pink, lip lobes lanceolate **2. kinoshitae**

1. A. keiskei (Maximowicz) Schlechter. Herb to 15 cm. Leaves narrowly oblong to lanceolate, 3–5 cm, obtuse. Flowers few in dense spike, purple; bracts 4–10 mm, broadly linear. Sepals obtuse, 1–3-nerved, to 4.5 mm; upper sepal elliptic, lateral sepals oblong. Petals equalling sepals, 2-nerved. Lip deeply 3-lobed, *c.*10 mm; lateral lobes ascending, ovate; terminal lobe longer, bifid. Spur to 2.5 mm. *Japan.* H5. Summer.

2. A. kinoshitae (Makino) Schlechter. Herb to 25 cm. Leaves linear, to 8 cm. Flowers 2–5, white to pale pink; bracts 3–8 mm, broadly lanceolate. Sepals obtuse, 3-nerved, to 4.5 mm; upper sepal elliptic, lateral sepals ovate. Petals about equalling sepals, ovate, 2-nerved. Lip 3-lobed, *c.* 10 mm; lateral lobes ascending, lanceolate; terminal lobe slightly longer, cuneate, shallowly bifid. Spur to 1.5 mm. *Japan.* H5. Summer.

A. enomoto-chidori grex (*A. keiskei* × *kinoshitae*; *A. Enomotoe* Hort.). A hybrid raised in Japan, which is intermediate between the parents, is sometimes grown in northern Europe. Flowers about 14, clustered at the stem apex. Upper sepal ovate, hooded, pink; lateral sepals spreading, pink, lanceolate-triangular, acute. Petals forming a pink, ovate hood with the dorsal sepal. Lip deeply 3-lobed, pale pink at extremities of lobes, white in centre, with dark red lines in throat, central lobes bifid. Spur short. *Garden Origin.* H5. Summer.

12. PONERORCHIS Reichenbach filius

J.M.H. Shaw

Small terrestrial herbs with spherical tubers. Leaves 1–3, basal. Inflorescence with lanceolate bracts, bearing pink, purple, or yellow, often spotted, flowers. Upper sepal often adnate to petals, forming a hood over the column. Lateral sepals spreading. Petals entire, smaller than upper sepal. Lip 3-lobed, with a small basal spur, which is about as long as the ovary. Column short, erect; anther bilocular. Anther cap triangular.

A genus of about 20 species distributed along the Himalaya through China to Japan, previously included in *Orchis*. The species cultivated require alpine house or frame conditions. Several hundred named cultivars and numerous groups exist in Japan.

Literature. Ebina, Y., A complete book of Ponerorchis graminifolia (in Japanese, 1987); Stewart, J., Ponerorchis, alpine gems from Japan, *The New Plantsman* **1**(1): 29–35 (1994); Cribb, P. & Perner, H., Ponerorchis, *The Alpine Gardener*, **70**: 290–291 (2002).

1a. Sepals 1-nerved; leaves linear **1. graminifolia**
b. Sepals 3-nerved; leaves lanceolate **2. joo-iokiana**

1. P. graminifolia Reichenbach (*Orchis graminifolia* (Reichenbach) Tang & Wang). Illustration: *The New Plantsman* **1**: 31 (1994).

Leaves linear, 7–12 cm × 3–8 mm, acute to acuminate. Flowers few, in one-sided racemes, pink, purple or other shades in cultivars; bracts lanceolate. Sepals 1-nerved, oblong, upper sepal forming a hood. Petals ovate, 1–2 nerved, forming a hood with the upper sepal. Lip deeply 3-lobed; lobes ovate. Spur 10–15 mm. *Japan*. H5. Summer.

A very large number of cultivars exists, as well as hybrids between different infra-specific taxa. Hybrids with several other species are grown in Japan, including a bigeneric hybrid, × *Ponerostigma* Wakikawa-chidori grex *(Amitostigma gracile* × *Ponerorchis kurokamiana)*.

2. P. joo-iokiana (Makino) Soó (*Orchis joo-iokiana* (Makino) Maekawa). Illustration: Kitamura et al., Coloured illustrations of herbaceous plants of Japan **17**: t. 2 (1964); Maekawa, Wild orchids of Japan in colour, t. 19 (1971).

Leaves usually 2, sometimes1–3, acute, 4–8 cm, not in a basal rosette. Flowers mostly on one side of the inflorescence, rose-purple; bracts longer than ovary. Sepals 3-nerved, lanceolate, lateral sepals oblique. Petals erect ovate, obtuse, with 2 faint nerves, shorter than sepals. Lip spreading, 13–15 mm, cuneate, longer than sepals, 3 lobed.Spur straight, to 17 mm. *Japan*. H5. Summer.

13. ANACAMPTIS Richard

P.J.B. Woods

Perennial herbs. Tubers ovoid. Stems to 75 cm with up to 8 leaves. Leaves always unspotted, lanceolate, upper leaves shorter, rolled when young, sheathing the flowering stem. Usually producing a rosette of leaves in autumn and winter-growing. Spike dense or lax, few to many-flowered. Bracts linear, slightly shorter than, equalling, or longer than ovary. Flowers pink, red or purple, occasionally white. Sepals and petals slightly concave, glabrous, upper sepal and petals incurved, forming a hood over the column, lateral sepals spreading. Lip shallowly or deeply 3-lobed, side-lobes rounded, oblong, trianular or linear, mid-lobe, if present, entire or bilobed, sometimes with 2 erect ridges leading towards the mouth of the spur. Spur present. Pollinia 2.

A genus of 12 species, in nature occurring in open grassland and calcareous dunes, moorland, open woodland and olive groves. Occurring naturally from northern to southern Europe, western Asia, (extending to Iran, Afghanistan and Yemen), and north Africa; suited for pot-culture as recommended for *Orchis*. Previously a monotypic genus, which now includes the species formely placed in *Orchis* with a chromosome count of 2n =32 or 36, and bearing sheathing leaves along the flowering stem.

Literature: Bateman, R.M., Pridgeon, A.M., & Chase, M.W., Phylogenetics of subtribe Orchidinae based on nuclear ITS sequences. 2. Infrageneric relationships and reclassification to achieve monophyly of Orchis sensu stricto, Lindleyana **12**: 113–41 (1997).

1a. Spur more than 11 mm, filiform **1. pyramidalis**
b. Spur 11 mm or less, not filiform 2
2a. Lip unlobed, ovate to fan-shaped **3. papilionacea**
b. Lip lobed 3
3a. Spur pointing upwards, or occasionally horizontal **2. morio**
b. Spur pointing downwards **4. coriophora**

1. A. pyramidalis Richard. Illustration: Danesch & Danesch, Orchideen Europas, Mitteleuropa, edn 3, 146, 147 (1972); Landwehr, Wilde Orchideeën van Europa **2**: 341, 343 (1977); Mossberg & Nilsson, Orchids of northern Europe, 123 (1979); Williams et al., Orchids of Britain and Europe, 81 (1978).

Perennial herbs. Tubers spherical or ovoid. Stems to 75 cm with up to 8 leaves. Leaves 15–25 × *c.* 2 cm, mostly basal, unspotted, lanceolate, upper leaves shorter, rolled when young. Aerial growth dying down in winter. Spike to 8 cm, conical, becoming shortly cylindric, dense. Bracts linear, slightly longer than ovary. Flowers pink, red or purple, occasionally white, emitting a faint fox-like smell. Sepals and petals 6–8 × *c.* 3 cm, upper sepal and petals incurved, lateral sepals spreading. Lip deeply 3-lobed, 6–9 × 8–10 mm, as broad as long or broader than long, lobes oblong, *c.* 4 mm, apex obtuse or truncate; base of lip with 2 erect ridges leading towards the mouth of the spur. Spur 1–1.4 cm, slender, downward pointing. Pollinia 2, club-like, made up of easily separated masses. *Europe, Mediterranean region, SW Asia*. H1. Summer.

2. A. morio (Linnaeus) R.M. Bateman, A.M. Pridgeon & M.W. Chase (*Orchis morio* Linnaeus). Illustration: Danesch & Danesch, Orchideen Europas, Südeuropa, 222 (1969); Landwehr, Wilde Orchideeën van Europa **1**: 245, 247 (1977); Williams et al., Orchids of Britain and Europe, 93 (1978).

Stem to 40 cm. Leaves to 10, basal, to 16 × *c.* 1.3 cm, lanceolate to broadly oblong; 1–2 sheath-like leaves on stem. Spike 3–13 cm, loose or dense, with few to many flowers; flowers reddish purple, mauve, or sometimes greenish or white. Bract about as long as ovary. Sepals 7–12 × 4–6 mm, oblong, apex rounded, often greenish- or purplish-veined, lateral sepals a little longer. Petals to 7 × 3 mm, oblong. Lip *c.* 10 × 10 mm, often broader than long, edges often folded back. Spur *c.* 8 mm, shorter than or as long as lip, cylindric, horizontal or upwardly curved. *Europe to W Iran*. H1. Spring.

A. longicornu (Poiret) R.M. Bateman, A.M. Pridgeon & M.W. Chase (*Orchis longicornu* Poiret). Illustration: Danesch & Danesch, Orchideen Europas, Südeuropa, 225 (1969); Landwehr, Wilde Orchideeën van Europa **1**: 251 (1977); Williams et al., Orchids of Britain and Europe, 95 (1978). Spur to 1.6 cm, much longer than lip. *W Mediterranean region*. H5. Spring.

3. A. papilionacea (Linnaeus) R.M. Bateman, A.M. Pridgeon & M.W. Chase (*Orchis papilionacea* Linnaeus). Illustration: Danesch & Danesch, Orchideen Europas, Südeuropa, 220 (1969); Sundermann, Europäische und mediterrane Orchideen,

138 (1975); Landwehr, Wilde Orchideeën van Europa **1**: 233 (1977).

Stem to 40 cm. Leaves to 10, to *c.* 10 × 1 cm, lanceolate, mostly basal. Spike ovoid with up to 15 flowers; flowers purplish, red or brownish. Bract longer than ovary. Sepals and petals 1–1.8 cm, narrowly ovate, strongly veined, upper sepal and petals forming a hood. Lip 1.2–2.5 × 1.2–2.5 cm, entire, fan-shaped or ovate, distinctly lined, margin shallowly toothed. Spur *c.* 1.2 cm, shorter than ovary, cylindric, downwardly curved. *S Europe to SW Asia.* H5. Spring.

4. A. coriophora (Linnaeus) R.M. Bateman, A.M. Pridgeon & M.W. Chase (*Orchis coriophora* Linnaeus). Illustration: Danesch & Danesch, Orchideen Europas, Südeuropa, 226 (1969); Landwehr, Wilde Orchideeën van Europa **2**: 301, 303 (1977); Williams et al., Orchids of Britain and Europe, 83 (1978).

Stem to *c.* 50 cm. Leaves to 7, *c.* 15 × 2 cm, linear to lanceolate, erect, on basal part of stem, often turning brown before completion of flowering. Spike 3–12 cm, ovoid to cylindric; flowers loose or dense, red, brownish or purplish, sometimes greenish. Bract a little longer than ovary. Sepals and petals hooded, tips free, sepals to 10 × 4 mm, ovate to lanceolate, the laterals wider than the upper; petals to 6 × 1–2 mm. Lip 5–11 mm, longer than broad, 3-lobed. Spur 5–9 mm, shorter than or equalling ovary, conical or cylindric, downwardly curved. *S, C & E Europe.* H2. Spring.

14. ORCHIS Linnaeus

P.J.B. Woods

Perennial herbs. Aerial growth dying down in winter. Tubers 2–3, spherical or ovoid. Leaves in a basal rosette, spotted or unspotted. Flowers in a short or long, dense- or loose-flowered spike, enclosed in a spathe-like sheathing basal leaf when emerging. Bracts membranous, never leafy, often coloured. Flowers in shades of red, purple, pink, yellow, often scented. Sepals and petals almost equal, free, all curved forward over the column, or the lateral sepals spreading and the upper sepal and petals incurved and hooded over the column. Lip mostly 3-lobed or apparently 4-lobed. Spur usually present, 8–25 mm, arising from the base of the lip. Pollinia 2, club-like.

A genus of about 33 species from Europe to Asia. In the past *Dactylorhiza* (p. 469) has been included under *Orchis* and occasionally nurserymen still offer plants under the latter name. Species formerly included in *Orchis* with leaves along the flowering stem are now treated under *Anacamptis*. Other species from eastern Asia are now treated under *Ponerorchis* and *Amitostigma*. *Aceras* is now included in *Orchis*.

Most species, apart from those with northerly distributions, are best suited for growing in pots in a cold-frame or unheated greenhouse. The compost should be extremely well drained and kept moist only while the plants are in active growth. The tubers should be repotted annually and kept almost dry while dormant.

Literature: Bateman, R.M., Pridgeon, A.M., & Chase, M.W., Phylogenetics of subtribe Orchidinae based on nuclear ITS sequences. 2. Infrageneric relationships and reclassification to achieve monophyly of Orchis sensu stricto, *Lindleyana* **12**: 113–41 (1997).

1a. Spur absent **5. anthropophora**
b. Spur present 2
2a. Lateral sepals spreading, only the petals and sometimes the upper sepal hooded over column 3
b. Sepals and petals all hooded or forwardly directed over column 4
3a. Spur to 1.2 cm, cylindric, not narrowed towards apex; or flowers yellow **1. mascula**
b. Spur 1.5 cm or more, slender, narrowed towards apex; flowers not yellow **2. anatolica**
4a. Sepals and petals much darker than lip; central lobe of lip divided into smaller lobes that are about as long as broad **3. purpurea**
b. Sepals and petals not darker than lip; central lobe of lip divided into smaller lobes that are twice or more longer than broad **4. militaris**

1. O. mascula Linnaeus. Illustration: Sundermann, Europäische und mediterrane Orchideen, 146 (1975); Landwehr, Wilde Orchideeën van Europa **1**: 261 (1977); Williams et al., Orchids of Britain and Europe, 101 (1978).

Stem to 60 cm. Leaves 3–11, 5–20 × 1.5–4.5 cm, lanceolate or oblanceolate to narrowly obovate, purplish-spotted or unspotted, mostly basal, upper leaves sheathing stem. Spikes to 32 cm, few to many, loosely or densely flowered; flowers purple, lilac, or sometimes white. Bract about as long as the ovary. Upper sepal to 10 mm, oblong, curved forwards; lateral sepals to 1.3 cm, obliquely oblong or ovate, erect or reflexed. Petals to 9 × 7 mm, obliquely ovate, hooded (with the upper sepal) over the column. Lip to 1.5 cm, 3-lobed or sometimes scarcely lobed, edges usually folded back, centre finely papillose, with purple lines or spots. Spur to 1.2 cm, about equal to ovary, cylindric, horizontal or erect. *Europe to W Iran.* H1. Spring.

O. provincialis Balbis. Illustration: Danesch & Danesch, Orchideen Europas, Südeuropa, 210, 211 (1969); Landwehr, Wilde Orchideeën van Europa **1**: 279 (1977); Williams et al., Orchids of Britain and Europe, 103 (1978). Flowers bright or pale yellow. *S Europe, N Africa to E Mediterranean region.* H5. Spring.

O. quadripunctata Tenore. Illustration: Danesch & Danesch, Orchideen Europas, Südeuropa, 206–8 (1969); Sundermann, Europäische und mediterrane Orchideen, 144 (1975); Landwehr, Wilde Orchideeën van Europa **2**: 297 (1977); Williams et al., Orchids of Britain and Europe, 99 (1978). Spur to 1.2 cm, curved downwards, uniformly narrow; lip base with 2–4 small purple spots. *E Mediterranean region.* H5. Spring.

2. O. anatolica Boissier. Illustration: Sundermann, Europäische und mediterrane Orchideen, 146 (1975); Landwehr, Wilde Orchideeën van Europa **1**: 283 (1977); Williams et al., Orchids of Britain and Europe, 99 (1978).

Stem to 40 cm. Leaves 3–8, to 13 × 2.5 cm, lanceolate to broadly lanceolate, unspotted, or blotched or spotted with purple, basal; stem with 1–2 sheath-like leaves. Spike to 15 cm, loose, with few to many flowers, flowers purplish to pale pink, sometimes white. Bract as long as or shorter than ovary. Sepals *c.* 10 × 4 mm, oblong to narrowly ovate; upper sepal slightly broader, hooded (with the slightly smaller petals) over the column. Lip to 1.4 cm, 3-lobed, edges often folded back, centre white with purple spots or lines. Spur 1.5–2.5 cm, longer than ovary, slender, wider at mouth, narrow at apex, upwardly curved or horizontal. *E Mediterranean region to W Iran.* H5. Summer.

3. O. purpurea Hudson. Illustration: Danesch & Danesch, Orchideen Europas, Mitteleuropa, edn 3, 114–17 (1972); Landwehr, Wilde Orchideeën van Europa **2**: 325 (1977); Mossberg & Nilsson, Orchids of northern Europe, 115 (1979).

Stem to 80 cm or more, robust. Leaves to 6 or more, to 23 × 6 cm, broadly lanceolate to elliptic, shining, unspotted, forming a basal rosette, 1–2 leaves sheathing the lower part of the stem. Spike to 17 × 6 cm, broadly cylindric, many-flowered; flowers brownish or reddish purple, rarely white, lip white or pinkish, its edges darker, with purple spots. Bract to 5 mm, much shorter than ovary. Sepals to 1.4 cm, ovate, acute. Petals shorter, linear-lanceolate. Lip to 1.5 cm, longer than broad, 4-lobed; lateral lobes oblong, central lobe divided into 2 smaller, oblong lobes that are about as long as broad and between which is a small tooth. Spur *c.* 5 mm, less than half the length of ovary, cylindric, downwardly curved. *Europe to N Africa.* H1. Spring–summer.

O. ustulata Linnaeus. Illustration: Danesch & Danesch, Orchideen Europas, Südeuropa, 230 (1969); Landwehr, Wilde Orchideeën van Europa **2**: 311 (1977); Mossberg & Nilsson, Orchids of northern Europe, 113 (1979). Stem to 25 cm. Sepals to 3.5 mm; lip with no tooth between the lobes of middle lobe; spur 2 mm. *Europe to the Caucasus.* H1. Spring–summer.

4. O. militaris Linnaeus. Illustration: Danesch & Danesch, Orchideen Europas, Mitteleuropa, edn. 3, 119–24 (1972); Landwehr, Wilde Orchideeën van Europa **2**: 319 (1977); Williams et al., Orchids of Britain and Europe, 89 (1978).

Stem to 45 cm, robust. Leaves 3–5, to 13 × 4 cm, broadly elliptic to oblanceolate, obtuse, unspotted, basal; 1–2 leaves sheathing lower part of stem. Spike to 10 × 4.5 cm, broadly cylindric, many-flowered; flowers whitish pink, rarely white, veins darker, lip purplish, white towards base. Bracts *c.* 3 mm, much shorter than ovary. Sepals to 1.5 cm, ovate-lanceolate, acute. Petals shorter, linear. Lip to 1.5 cm, longer than broad, 4-lobed; lateral lobes linear, central lobe divided into 2, almost transverse, oblong lobes that are about twice as long as broad and between which is a small tooth. Spur to 7 mm, about half the length of the ovary, cylindric, downwardly curved. *C Europe to the Caucasus.* H1. Spring–summer.

O. italica Poiret. Illustration: Sundermann, Europäische und mediterrane Orchideen, 134 (1975); Landwehr, Wilde Orchideeën van Europe **2**: 321 (1977); Williams et al., Orchids of Britain and Europe, 91 (1978). Leaves 5–8, margins wavy. Lobes of central lobe of lip linear, 4 or more times longer than broad. *Portugal & Mediterranean region.* H5. Spring.

5. O. anthroporphora (Linnaeus) Allioni (*Aceras anthropophorum* (Linnaeus) Aiton). Illustration: Danesch & Danesch, Orchideen Europas, Mitteleuropa, edn 3, 131 (1972); Danesch & Danesch, Orchideen Europas, Südeuropa, 240 (1969); Williams et al., Orchids of Britain and Europe, 67 (1978); Landwehr, Wilde Orchideeën van Europa **2**: t. 153 (1977).

Herbaceous perennial. Aerial growth usually dying down in winter or plant sometimes overwintering as a basal rosette. Stems to *c.* 50 cm arising from 2 ovoid tubers. Leaves unspotted, the lower crowded, spreading, 5–12 × *c.* 2.5 cm, lanceolate, obtuse, the upper smaller, erect, sheathing the stem. Spike to 20 cm, bearing 50 or more flowers. Sepals obovate-lanceolate, 6–7 mm, petals linear, slightly shorter, all yellowish or greenish yellow, hooded over the column. Lip *c.* 1.2 cm, shaped like a man, the lateral lobes being the arms and the divided central lobe the legs. Spur absent. Pollinia 2, club-shaped, composed of easily separable masses. *W Europe, Mediterranean region.* H5. Summer.

A species rarely seen in cultivation.

15. DACTYLORHIZA Nevski

J.C.M. Alexander

Terrestrial. Root-tubers palmately lobed or divided into finger-like sections. Aerial parts dying down in winter. Leaves 3–15, spirally arranged on flowering stem, not forming a basal rosette at flowering time, sometimes spotted or blotched with brownish purple; upper leaves often small and bract-like. Flowers several to many, in compact spherical, conical or cylindric racemes. Bracts leaf-like, green or purplish green. Sepals and petals free; lateral sepals spreading or erect; upper sepal and petals curved forwards forming a hood over the column. Lip simple or 3-lobed, spurred, usually marked with lines, dots or streaks. Column short; rostellum 3-lobed.

A temperate genus of 20–30 species from Europe, N Africa, Asia and N America. They are generally hardy and can be cultivated in full sun outdoors in a peat-bed. Cultivation conditions for the more southerly species are similar to those for *Serapias* and the less hardy species of *Ophrys* (see p. 463). The species comprising *Dactylorhiza* were long considered to be part of *Orchis*, and were separated as the genus *Dactylorchis* by Vermeulen in 1947, who was presumably unaware of Nevski's earlier name.

Hybridisation is rife in *Dactylorhiza* and the resulting crosses are often as fertile and plentiful in the wild as their parents. In addition, many frequently used characters are very variable. As a result the delimitation of species and other categories is not well defined and the classification is confused. The absence or presence of spots on the leaves is not a reliable character when attempting to identify wild populations but should suffice for the species in cultivation.

Literature: Nelson, E., Monographic und Ikonographie der Orchidaceen-Gattung Dactylorhiza (1976).

Height. Over 70 cm: **1, 2.**
Stem. Hollow: **1, 2, 3**; solid: **4.**
Leaves. Yellowish green: **1, 2**; mid to dark green: **1, 3, 4.** Spotted: **3, 4**; unspotted: **1, 2.** Hooded at tip: **2, 4**; not hooded at tip: **1, 3, 4.** Not bract-like near top of stem: **2.** Overtopping inflorescence: **2.**
Bracts. Longer than flowers: **1, 2, 3**; equal to or shorter than flowers: **1, 4.**
Inflorescence. More than 15 cm: **1.**
Flowers. White: **4**; pink: **1, 2, 4**; purplish pink: **1, 3**; crimson, brownish pink or brick-red: **2**; yellow: **2.**
Lip. More than 1.2 cm wide: **1.**
Spur. 1.2–1.5 cm: **1, 2.** Conical: **1, 2, 3**; cylindric: **2, 3, 4.**

1a. Leaves mid or dark green, spotted or blotched with brownish purple — 2
b. Leaves bright yellowish green, unspotted — 3
2a. Stem hollow; spur conical to cylindric — **3. majalis**
b. Stem solid; spur cylindric — **4. maculata**
3a. Lip more than 11 mm wide; leaves 8–14 — **1. foliosa**
b. Lip less than 10 mm wide; leaves 4–7 — **2. incarnata**

1. D. foliosa (Solander) Soó (*Orchis maderensis* Summerhayes). Illustration: *Botanical Magazine*, 5074 (1858); Landwehr, Wilde Orchideeën van Europa **2**: 219 (1977); Williams et al., Orchids of Britain and Europe, 119 (1978).

Stem to 70 cm, hollow. Leaves 8–14, to 20 cm, narrowly lanceolate to triangular, keeled, acute, stiff, ascending or arched, yellowish green, unspotted; upper leaves bract-like. Bracts equal to or shorter than the flowers and scarcely protruding from the inflorescence. Inflorescence 5–13 cm, cylindric, densely many-flowered. Flowers bright purplish pink. Sepals and petals 1–1.3 cm, lanceolate to ovate, rounded; lateral sepals a little broader. Lip 1–1.1 × 1.2–1.3 cm, always wider than long, transversely oblong or elliptic, shallowly 3-lobed, marked with lines and streaks; central lobe triangular. Spur 6–8 mm, narrowly conical, almost straight. *Madeira.* H5–G1. Spring–summer.

D. elata (Poiret) Soó. Illustration: Landwehr, Wilde Orchideeën van Europa **2**: 205–13 (1977); Williams et al., Orchids of Britain and Europe, 112 (1978); Grey-Wilson & Mathew, Bulbs, f. 44 (1981). To 110 cm. Bracts longer than flowers and protruding from inflorescence. Inflorescence 15–25 cm. Flowers pink to purple. Spur 1.2–1.5 cm. *SW Europe.* H5–G1. Spring–summer.

2. D. incarnata (Linnaeus) Soó. Illustration: Landwehr, Wilde Orchideeën van Europa **2**: 127–47 (1977); Williams et al., Orchids of Britain and Europe, 111 (1978); Grey-Wilson & Mathew, Bulbs, f. 44 (1981).

Stem to 80 cm, hollow. Leaves 4–7, 10–20 cm, narrowly triangular to lanceolate, keeled, stiff, hooded at tip, ascending or arched, bright yellowish green, unspotted, often overtopping the inflorescence. Upper 1 or 2 leaves bract-like. Lower bracts longer than flowers and usually protruding from inflorescence. Inflorescence 4–10 cm, conical to cylindric, densely many-flowered. Flowers pale to bright pink or brick-red. Sepals and petals 5–8 mm, lanceolate to ovate, acute or obtuse, usually erect or ascending. Lip 5–7 × 5–9 mm, diamond-shaped, sometimes 3-lobed, marked with hieroglyphic-like lines of darker pink. Spur 5–8 mm, broadly conical, slightly curved. *Europe & W Asia, uncommon in the south.* H2. Summer.

D. sambucina (Linnaeus) Soó. Illustration: Landwehr, Wilde Orchideeën van Europa **2**: 59–61 (1977); Williams et al., Orchids of Britain and Europe, 109 (1978); Grey-Wilson & Mathew, Bulbs, f. 43 (1981). To 30 cm. Leaves in lower half of the stem only, all much larger than bracts, never overtopping inflorescence. Flowers yellow, crimson or reddish brown, in small, spherical to ovoid heads. Spur 1.2–1.5 cm, equal to or longer than ovary, cylindric, obtuse. *Europe, N Africa & Turkey.* H5–G1. Summer.

3. D. majalis (Reichenbach) Hunt & Summerhayes (*D. latifolia* (Linnaeus) Soó – confused name). Illustration: Landwehr, Wilde Orchideeën van Europa **2**: 173–81 (1977); Williams et al., Orchids of Britain and Europe, 113 (1978); Grey-Wilson & Mathew, Bulbs, f. 44 (1981).

Stem to 60 cm, hollow. Leaves 4–8, to 15 × 2.5 cm, broadly lanceolate to elliptic or oblong, acute, not hooded at tip, mid to dark green with irregular brownish purple spots and blotches; sometimes 1 or 2 upper leaves bract-like. Lower bracts usually longer than flowers. Inflorescence 6–12 cm, cylindric, dense, many-flowered; flowers dark purplish pink. Sepals and petals 6–12 mm, lanceolate to oblong, obtuse, ascending or erect. Lip 6–9 × 9–12 mm, circular, diamond-shaped or transversely elliptic, simple or 3-lobed, with broad lateral lobes and a short, triangular central lobe, marked with lines and dots or dots alone. Spur 7–10 mm, conical to cylindric, equal to or shorter than ovary. *Europe (except extreme south), N Russia, Turkey & the Baltic.* H2. Summer.

4. D. maculata (Linnaeus) Soó (*Orchis ericetorum* (Linden) Marshall; *O. elodes* Grisebach; *D. maculata* (Linnaeus) Soó subsp. *ericetorum* (Linden) Hunt & Summerhayes). Illustration: Landwehr, Wilde Orchideeën van Europa **2**: 73–5, 95–7 (1977); Williams et al., Orchids of Britain and Europe, 121 (1978); Grey-Wilson & Mathew, Bulbs, f. 44 (1981).

Stem to 40 cm, solid. Leaves 5–12, to 20 × 3 cm, linear-lanceolate to oblong, acute to rounded, sometimes slightly hooded at tip, mid to dark green with irregular spots and blotches of brownish purple; some small upper leaves bract-like. Bracts about equal to flowers and not protruding from inflorescence. Inflorescence 5–10 cm, conical, ovoid or cylindric, dense, many-flowered; flowers white to pink. Sepals and petals 7–11 mm, narrowly lanceolate, obtuse, usually horizontal. Lip 7–8 × 10–11 mm, diamond-shaped to transversely elliptic, usually marked with darker pink dots and short lines, shallowly 3-lobed; central lobe shorter than lateral lobes. Spur 4–8 mm, cylindric, equal to or shorter than ovary. *Europe, N Africa & W Asia.* H2. Summer.

D. fuchsii (Druce) Soó (*D. maculata* subsp. *fuchsii* (Druce) Hylander). Illustration: Landwehr, Wilde Orchideeën van Europa **2**: 103–5, 107–9 (1977); Williams et al., Orchids of Britain and Europe, 121 (1978); Grey-Wilson & Mathew, Bulbs, f. 44 (1981). Very similar to *D. maculata* and often regarded as a subspecies of it. Lip more deeply 3-lobed, marked with hieroglyphic-like lines; central lobe narrow and usually longer than lateral lobes. *Europe (except extreme south) & W Asia.* H2. Summer.

16. GYMNADENIA R. Brown

P.J.B. Woods

Herbaceous, terrestrial perennials. Tubers palmately lobed. Leaves linear to lanceolate, arranged along lower part of stem, unspotted. Aerial growth dying down in winter. Flowers in a cylindric spike, sweetly scented. Upper sepal and petals incurved to form a hood over the column; lateral sepals spreading almost horizontally. Lip shallowly 3-lobed. Spur slender, shorter or longer than ovary. Pollinia 2, club-like, powdery. Ovary twisted or straight. Stigmas 2, lateral.

A genus of 16 species from Europe and temperate Asia. Cultivation as for *Orchis* (p. 468). Now includes the taxa formerly included in *Nigritella* with which it shares several unusual characters, such as the palmately lobed tubers.

1a. Ovary not twisted, lip uppermost **3. nigra**
b. Ovary twisted, lip lowermost 2
2a. Spur much longer than ovary; leaves oblong to lanceolate **1. conopsea**
b. Spur not longer than ovary; leaves linear **2. odoratissima**

1. G. conopsea (Linnaeus) R. Brown. Illustration: Landwehr, Wilde Orchideeën van Europa **1**: 221, 223 (1977); Williams et al., Orchids of Britain and Europe, 125 (1978); Mossberg & Nilsson, Orchids of northern Europe, 81 (1979).

Stems to 50 cm. Leaves to 7, arising along the lower third of the stem, to 25 cm × 6–35 mm, oblong to lanceolate. Spike to 25 cm, flowers crowded, rose or purple, sometimes white. Bracts lanceolate, as long as or longer than ovary. Sepals and petals 4–7 mm, upper sepal and petals elliptic to ovate, lateral sepals oblong, obtuse, spreading, margins slightly reflexed. Lip lowermost part of flower, 5–7 mm, 3-lobed, lobes almost equal. Spur to 2 cm, about twice as long as ovary, downwardly curved. Ovary twisted. *Eurasia.* H1. Spring–summer.

2. G. odoratissima (Linnaeus) Richard. Illustration: Danesch & Danesch, Orchideen Europas, Mitteleuropa, edn 3, 150, 151 (1972); Landwehr, Wilde Orchideeën van Europa **1**: 223, 225 (1977); Mossberg & Nilsson, Orchids of northern Europe, 83 (1979).

Stems to 40 cm. Leaves to 7, mostly basal, to 18 × 1.6 cm, linear, slightly infolded. Spike to *c.* 10 cm, flowers crowded, pale pink, purple, sometimes white. Bracts lanceolate, longer than ovary. Sepals and petals to 3 mm, upper sepal and petals elliptic to ovate, lateral sepals oblong, obtuse, spreading. Lip lowermost part of flower, *c.* 3 mm, 3-lobed, lateral lobes shorter than central lobe. Spur to 5 mm, as long as or shorter than the twisted ovary, horizontal or downwardly curved. *Sweden to N Spain, N Italy & W Russia.* H1. Spring–summer.

3. G. nigra (Linnaeus) Reichenbach (*Nigritella nigra* (Linnaeus) Reichenbach). Illustration: Sundermann, Europäische und mediterrane Orchideen, 178 (1975); Landwehr, Wilde Orchideeën van Europa **1**: 229, 231 (1977); Williams et al., Orchids of Britain and Europe, 123 (1978).

Herbaceous, terrestrial perennial. Tubers 2, palmately lobed. Stems to 28 cm. Leaves 6–11, crowded, to 11 cm × 1–9 mm, linear to narrowly lanceolate, channelled, mostly basal, unspotted. Aerial growth dying down in winter. Spike 1–3 cm × 7–25 mm, conical becoming ovoid or elongate, dense. Bracts slender, as long as or longer than flowers. Flowers *c.* 5 mm, blackish crimson, red, yellowish or rarely white, vanilla-scented; sepals and petals linear to lanceolate; tips acute. Lip uppermost part of flower, triangular to lanceolate-ovate, entire. Spur *c.* 2 mm, conical, shorter than ovary. Pollinia 2, granular. Ovary not twisted. *Scandinavia to N Spain & Greece.* H1. Summer.

17. COELOGLOSSUM Hartmann

P.J.B. Woods

Herbaceous, terrestrial perennials with 2 ovoid tubers. Leaves unspotted, the lower ovate or oblong, obtuse, the upper smaller, lanceolate and pointed. Aerial growth dying down in winter. Flowers slightly scented, 5–25 in a loose cylindric spike. Sepals and petals hooded over the column, greenish or yellow-green often tinged reddish, sepals ovate, petals relatively narrower, almost linear. Lip strap-shaped, the apex shortly 2-lobed with or without a short middle tooth, with a short, rounded spur. Pollinia 2, club-shaped, attached to separate viscidia.

A north temperate genus of 2 species. Cultivation as for *Dactylorhiza* (p. 469); in nature the 1 cultivated species occurs in calcareous grassland and at woodland edges. Some workers now include *Coeloglossum* in *Dactylorhiza* based on molecular evidence. However, further molecular work revealed that maintaning its separation is supported. The genus does hybridise with *Dactylorhiza*.

Literature: Bateman, R.M., Pridgeon, A.M., & Chase, M.W., Phylogenetics of subtribe Orchidinae based on nuclear ITS sequences. 2. Infrageneric relationships and reclassification to achieve monophyly of Orchis sensu stricto, *Lindleyana* **12**: 113–41 (1997); Devos, N., Raspé, O., Jacquemart, A-L. & Tyteca, D., On the monophyly of Dactlyorhiza Necker ex Nevski: is Coeloglossum viride (L.) Hartmann a Dactylorhiza? *Botanical Journal of the Linnean Society* **152**: 261–9 (2006).

1. C. viride (Linnaeus) Hartmann. Illustration: Danesch & Danesch, Orchideen Europas, Mitteleuropa, edn 3, 138 (1972); Landwehr, Wilde Orchideeën van Europa **1**: 227 (1977); Williams et al., Orchids of Britain and Europe, 127 (1978); Mossberg & Nilsson, Orchids of northern Europe, 74 (1979).

Stems 10–35 cm. Leaves 2–6, 3–10 cm. Spike 2–10 cm. Sepals and petals 4.5–6 mm. Lip 3.5–6 mm. *Eurasia, N America.* H1. Summer.

18. PLATANTHERA Richard

P.J.B. Woods

Herbaceous, terrestrial perennials. Tubers 2 or more, spindle-shaped, elongate. Stems erect, unbranched. Leaves basal or on the stem. Aerial growth dying down in winter. Raceme usually with many small or medium-sized, white or greenish flowers. Sepals and petals free, upper sepal and petals usually incurved over the column, lateral sepals spreading or recurved. Lip entire with the base extended into a spur. Anthers distinctly separated. Pollinia club-shaped, granular.

A genus of about 200 species mostly from temperate areas of Europe, Asia and N & S America. Cultivation as for *Dactylorhiza* (p. 469).

1a. Lobes of lip smooth — 2
b. Lobes of lip fringed, laciniate — 3
2a. Anther cells distinctly parallel, *c.* 2 mm apart at the base; spur 1.5–2.5 cm — **1. bifolia**
b. Anther cells distinctly divergent towards the base, where they are *c.* 4 mm apart; spur 1.5–3.5 cm — **2. chlorantha**
3a. Flowers orange — **3. ciliaris**
b. Flowers pink, rarely white — **4. psycodes**

1. P. bifolia (Linnaeus) Richard. Illustration: Sundermann, Europäische und mediterrane Orchideen, 182 (1975); Williams et al., Orchids of Britain and Europe, 129 (1978); Mossberg & Nilsson, Orchids of northern Europe, 68–70 (1979).

Stems to 50 cm. Tubers 2, ovoid, the tips narrow and elongate. Leaves 2 (rarely 3), basal, almost opposite, 7–25 × 2.5–8 cm, circular or elliptic or oblong-lanceolate, narrowing into a stalk; stem-leaves small and bract-like. Raceme 2.5–3.8 cm wide, cylindric, loose or dense, with many flowers; bracts lanceolate, as long as or slightly longer than the ovaries. Flowers 1.2–2 cm wide, white, scented. Lip 6–15 mm, strap-shaped, directed downwards, white, greenish towards the tip. Spur 1.5–2.5 cm, narrowly cylindric, slightly inflated, greenish towards the apex, often straight. Anther cells parallel, *c.* 2 mm apart. *Europe, N Africa, SW Asia.* H1. Summer.

2. P. chlorantha (Custer) Reichenbach. Illustration: Sundermann, Europäische und mediterrane Orchideen, 182 (1975); Landwehr, Wilde Orchideeën van Europa **1**: 31–3 (1977); Mossberg & Nilsson, Orchids of northern Europe, 70, 71 (1979); Williams et al., Orchids of Britain and Europe, 129 (1978).

Stems to 60 cm. Tubers 2, ovoid, the tips narrow, elongate. Leaves 2 (rarely 3), basal, almost opposite, 6–20 × 2.5–8 cm, oblong to oblong-lanceolate or obovate, narrowing into a stalk; stem-leaves small and bract-like. Raceme 3–4.5 cm wide, cylindric, loose or dense, with many flowers; bracts lanceolate, the lower somewhat longer than the ovaries. Flowers to 2.5 cm wide, greenish white, scented. Lip 8–18 mm, strap-shaped, directed downwards, greenish, the tip darker. Spur 1.5–3.5 cm, thickened towards the tip, greenish, darker at the tip. Anther cells divergent towards the base and *c.* 4 mm apart there. *Europe, SW Asia.* H1. Summer.

3. P. ciliaris (Linnaeus) Lindley (*Habenaria ciliaris* (Linnaeus) R. Brown; *Blephariglottis* *ciliaris* (Linnaeus) Linnaeus). Illustration: *Botanical Magazine*, 1668 (1814); Shuttleworth et al., Orchids, 25 (1970); Luer, Native orchids of the United States and Canada, 181 (1975); *Die Orchidee* **27**: 155 (1976) & **32**: 103 (1981).

Leaves lanceolate to oblong. Flowers orange; lip oblong, unlobed. *Eastern USA*. H1. Summer.

4. P. psycodes (Linnaeus) Lindley (*Habenaria psycodes* (Linnaeus) Sprengel). Illustration: *American Orchid Society Bulletin* **39**: 789 (1970); Luer, Native orchids of the United States and Canada, 196, 199 (1975); *Die Orchidee* **27**: 154, 155 (1976).

Leaves oblong to obovate. Flowers purplish, occasionally white; lip 3-lobed, lobes obovate or fan-shaped. *SE Canada, Northeastern USA*. H1. Summer.

19. HABENARIA Willdenow

P.J.B. Woods & J.M.H. Shaw

Terrestrial herbs. Roots tuberous, fleshy, elongate or ovoid. Stems erect. Leaves spaced along stem or sometimes basal, linear to lanceolate, sheathing at base, rolled when young. Aerial growth dying down in winter. Raceme few- to many-flowered. Upper sepal often forming a hood with the petals; lateral sepals spreading (in *H. bonatea* the basal parts of the sepals, petal lobes, column and lip are united). Lip variously lobed or fringed, base spurred, spur slender, long. Column short or long, slender or thick. Pollinia 2, club-like, breaking up into easily separable masses. Stigmas 2, borne on club-like arms, or stalkless.

The genus contains between 600 and 800 species distributed throughout temperate N America and east Asia and the subtropics and tropics of the Old and New World. Generic limits within this group are somewhat confused: Luer, for instance, place all the North American species under *Platanthera*. For the purpose of this Flora, *Bonatea* Willdenow and *Pecteilis* Rafinesque have been recognised as separate. Most species require greenhouse or alpine-house protection and are best grown in terrestrial compost or in one containing peat, leaf-mould, loam and sand in equal parts. Adequate water is necessary while the plants are actively growing, with cool, almost dry conditions when dormant.

Literature: Luer, C., Native orchids of the United States and Canada, 175–242 (1975); Seidenfaden, G., Orchid genera in Thailand V, *Dansk Botanisk Arkiv* **31**: 22–6, 65–137 (1977).

1a. Plant with tubers; lip not yellow 2
b. Plant with several swollen roots; lip bright yellow **3. xanthocheila**
2a. Leaves uniformly green without silvery patches 3
b. Leaves with rectangular silvery patches **4. erichmichelii**
3a. Margin of lip mid-lobe papillose-ciliate **1. carnea**
b. Margin of lip mid-lobe smooth **2. rhodocheila**

1. H. carnea R. Brown. Illustration: *The Garden* **47**: facing 182 (1895); Holttum, Flora of Malaya **1**: Orchids, edn 3, 86 (1964); *Dansk Botanisk Arkiv* **31**: 136 (1977); *Orchid Review* **115**: 286 (2007).

Plant 15–30 cm. Leaves 4–6, to 10 × 4 cm, lanceolate, mostly basal, olive green with pale spots. Raceme 4–15-flowered; flowers pink or reddish. Sepals and petals *c.* 1 cm, lateral sepals broader and longer than the upper; petals broad, ovate, veins several. Lip 3-lobed, *c.* 3 × 2.5 cm, sometimes as wide as long, lobes oblong or ovate, central lobe notched, surface and margins finely papillose-ciliate. Spur to 6 cm. *Malay Peninsula (Langkawi Islands), Thailand.* G2. Summer–autumn.

2. H. rhodocheila Hance (*H. militaris* Reichenbach). Illustration: *Botanical Magazine*, 7571 (1897); *American Orchid Society Bulletin* **48**: 366, 367 (1979); *Dansk Botanisk Arkiv* **31**: 135 (1977); Bechtel et al., Manual of cultivated orchid species, 216 (1981).

Tubers present. Sepals and petals green. Petals slender, oblong-spathulate, 1-veined. Lip orange to bright red, margins not papillose-ciliate. *China, Thailand, Malay Peninsula, Laos, Vietnam, Philippines.* G2. Summer–autumn.

3. H. xanthocheila Ridley. (*H. rhodocheila* var. *aureoflava* invalid, or 'Aureoflava'). Illustration: *Orchids* (USA) **71**(3): 223 (2002).

Like *H. rhodocheila* but differing by the production of horizontal coralloid roots instead of tubers. The flowers produce divergent ovate-triangular lateral sepals and a bright yellow lip. *Thailand, Malaysia (Penang).* G2. Summer.

4. H. erichmichelii Christenson. Illustration: *Orchids* (USA) **72**(2): 136 (2003) – as *H. erichmichaelii.*

Like *H. rhodocheila* but differing by short stems to 15 cm, leaves with transluscent, silvery, squarish windows between the primary veins, and a pastel pink lip. *Thailand.* G2. Summer.

A further related but undescribed species is sometimes cultivated as *H. rhodocheila*. It is distinguished by patterned leaves similar to *H. erichmichelii*, but has bracts almost equalling ovaries, a larger staminode, incurved column arms and an intense cerise lip.

The above four closely related species are all illustrated as *H. rhodocheila* in Vaddhanaphuti, N., Wild orchids of Thailand, 167 (2005).

20. PECTEILIS Rafinesque

J.M.H. Shaw

Terrestrial herbs with tubers. Stem erect with leaves or bracts. Leaves often 3 in a basal rosette, ovate to ovate-elliptic, fleshy, green. Flowers showy, large, white, cream or white with yellow lip. Sepals elliptic, forming a hood over the column. Petals linear, smaller than sepals. Lip 3-lobed, sometimes shallowly, lacking a callus, spurred, side-lobes spreading, often laciniate, mib-lobe smaller, entire. Column erect with widely spaced anthers. Pollinia 2. Stigmas stalkless, adhering to base of lip.

A small genus of about 8 species from southern and eastern Asia and Malaysia. Similar to *Habenaria* but distinguished by the column which displays widely spaced anthers and a stalkless stigma which is joined to the base of the lip. Cultivation as for *Habenaria*.

1a. Lip side lobes fimbriate to laciniate 2
b. Lip side lobes smooth **3. hawkesiana**
2a. Inflorescence of 1–2 flowers; spur 3–4 cm **1. radiata**
b. Inflorescence of 3–10 flowers; spur 9–14 cm **2. susannae**

1. P. radiata (Thunberg) Rafinesque (*Habenaria radiata* Thunberg). Illustration: *Australian Orchid Review* **33**: 5 (1968); *American Orchid Society Bulletin* **44**: 809 (1975); Maekawa, Wild orchids of Japan in colour, 90 (1971); Bechtel et al., Manual of cultivated orchid species, 257 (1981).

Plant *c.* 30 cm. Leaves 3–7, to 10 cm × *c.* 6 mm, linear, acute, the lowermost and uppermost shortest. Inflorescence 1–2-flowered, Sepals and petals *c.* 1 cm, ovate, margins of petals finely and irregularly toothed. Lip *c.* 1.5 × 2.5 cm or larger, 3-lobed; lateral lobes obovate, their outer margins laciniate; central lobe linear or oblong, shorter than lateral lobes, margins entire. Spur 3–4 cm. *Korea, Japan.* H5–G1. Summer.

2. P. susannae (Linnaeus) Rafinesque (*Habenaria susannae* (Linnaeaus) R. Brown). Illustration: Seidenfaden & Wood, Orchids of peninsular Malaysia and Singapore, 108, pl. 3D (1992); Pearce & Cribb, Orchids of Bhutan, 172 (2002);Vaddhanaphuti, Wild orchids of Thailand, 199 (2005).

Herb to1.3 m but usually less, growing from an ovoid tuber. Stem with leaves along it length, and tubular sheaths at the base. Leaves ovate-oblong, acute, 6–15 × 3–7 cm, sheathing the stem. Inflorescence lax, 3–10-flowered; floral bracts lanceolate, acute, 5 × 1 cm, with glandular hairs along the margins. Flowers sweet smelling, 4–10 cm wide, white. Upper sepal ovate to rhombic, obtuse, lateral sepals ovate. Petals narrow, to 1.3 cm. Lip 3-lobed, side-lobes with laciniate outer margins, mid-lobe linear to oblong, 3–4.5 × 0.5 cm. Spur 7–14 cm. *NE. India, China, Indochina, Thailand to Malaysia & Indonesia.* G1. Summer.

P. Kiat Tan grex (*P. susannae* × *hawkesiana*), produces pure white flowers, 5 cm across.

3. P. hawkesiana (King & Pantl) C.S. Kumar (*P. sagarikii* Seidenfaden). Illustration: Schuiteman & de Vogel, Orchid genera of Thailand, Laos, Cambodia and Vietnam, 107 (2000); Vaddhanaphuti, Wild orchids of Thailand 199 (2005); *Orchids* (USA) **76**: 750 (2007).

Herb to 25 cm, from an ovoid tuber. Leaves 2–3 in a basal rosette, ovate, 10–12 × 6–9 cm. Inflorescence 2–8-flowered; floral bracts lanceolate, 3 cm. Flowers slightly scented, white or cream with a bright yellow or cream lip. Upper sepal erect, ovate, 2.5 cm. Lip shallowly 3-lobed at base, 2.5 × 2 cm, side-lobes rounded, much smaller than mid-lobe. Spur 3–5 cm. *Burma (Myanmar), Thailand.* G1. Summer–autumn.

× **Pectabenaria Flamingo** grex (*Habenaria erichmichelii* × *P. hawkesiana*). Illustration: *Orchids* (USA) **76**: 751 (2007). It produces white to very pale pink flowers about 4 cm across. Lip shallowly 3-lobed, edges smooth. *Garden origin.* G1.

21. BONATEA Willdenow

J.M.H.Shaw

Terrestrial herbs with swollen storage roots and tubers. Stems with leaves, unbranched. Leaves distributed along the stem, green, often withered by flowering time. Inflorescence terminal with many flowers. Flowers in shades of green, yellow and white. Upper sepal usually hooded, lateral sepals joined to the base of the lip for part of their length. Petals 2-lobed for part of their length, the upper lobe usually pressed against the upper sepal and the lower lobe pressed against the stigma. Lip 3-lobed, with a spur at the base, the spur having a small tooth at the entrance. Column short.

A genus of 20 species from Africa and Arabia, closely related to *Habenaria* and sometimes included in it, but distinguishable by the lateral sepals and front lobe of the petals being joined to the lip, and a small tooth present in the mouth of the spur. Cultivation as for *Habenaria* (p. 472), but the plants do not like to be over-watered and appreciate a drier atmosphere.

Literature: La Croix, I. & E., African orchids in the wild and in cultivation, 153–7 (1997).

1a. Spur at least 9 cm — **4. steudneri**
b. Spur less than 7 cm — 2
2a. Upper petal lobe 15–25 mm; stolons absent — 3
b. Upper petal lobe less than 12 mm; stolons present — **3. bracteata**
3a. Leaves not withered at flowering time; lower petal lobe oblanceolate to linear — **1. speciosa**
b. Leaves withered at flowering time; lower petal lobe narrowly linear — **2. antennifera**

1. B. speciosa (Linnaeus) Willdenow (*Habenaria bonatea* Reichenbach). Illustration: *Botanical Magazine*, 2926 (1829); Bolus, Orchids of South Africa **2**: t. 44, 45 (1911); Schelpe, An introduction to the South African orchids, 75 (1966); *South African Orchid Journal*, cover (June, 1971); *American Orchid Society Bulletin* **47**: 994 (1978).

Plant 30–80 cm. Leaves many, often crowded along stem, to *c.* 15 × 5 cm, ovate-oblong or oblong, sometimes with purplish marks, obtuse. Raceme dense, bracts *c.* 3 cm, ovate-acuminate. Flowers to 15 or more, scented, greenish, the lip and stigmatic processes white tipped with green. Sepals ovate, 1.7–2.2 cm, lateral sepals longer than the upper. Petals 2-partite, the upper segment as long as the upper sepal, sickle-shaped; the lower segment as long as or longer than the lateral sepal, linear or filiform. Lip 3–3.5 cm, 3-lobed, the lobes thread-like, curved or bent; spur 3–3.5 cm, toothed at the mouth, thickened towards tip, greenish. *South Africa.* H5–G1. Spring–summer.

2. B. antennifera Rolfe (*Bonatea speciosa* var. *antennifera* (Rolfe) Sommerville). Illustration: *Flowering Plants of Africa* **36**: 1405 (1959).

Similar to *B. speciosa*, but differing mainly by more numerous flowers with narrower petal lobes, the lower petal lobe oblanceolate to linear, and longer lip lobes. Lip 40–45 cm, with a mid-lobe 25–35 mm, side-lobes 18–30 mm, slender and almost thread-like. Spur 3–5 cm. The leaves are usually withered by flowering time. In the wild it grows in drier areas and at higher altitudes. *Southern Africa.* G1. Spring–summer.

3. B. bracteata McDonald & McMurty (*Bonatea liparophylla* invalid). Illustration: La Croix, African orchids, pl. 28 (1997).

Herb 40–60 cm tall, with ovoid tubers. Stolons produced from stem base, at first arching over the ground, then tunnelling and producing terminal tubers. Leaves up to 9 × 3 cm, elliptic, acuminate. Inflorescence with 10–18 flowers, lax; floral bracts 1–4 cm. Sepals green, upper sepal 11 × 8 mm, ovate, acute, convex, lateral sepals 10–12 × 5 mm, ovate, with a prominent midrib. Petals with linear lobes, upper lobe 10 × 1 mm, joined to the upper sepal, lower lobe 10–16 × 1 mm, joined to the base of the lip for lower 2–4 mm. Lip 17 mm, 3-lobed from a short, 5 mm, undivided base; mid-lobe 10–12 × 1.5 mm, linear, acute. Spur 2–2.5 cm. *South Africa (Kwazulu-Natal).* G1. Spring–summer.

The only species known to produce stolons. The small tubers at the stolon tips must not be buried too deeply when repotting.

4. B. steudneri (Reichenbach) Durand & Schinz. Illustration: Pridgeon et al., Genera Orchidacearum **2**: 264 (2001); Demissew, Cribb & Rassmussen, Field guide to Ethiopian orchids 123 (2004).

Robust herb to 125 cm, with leaves all along the stem and tuberous roots. Leaves 10–20, oblong-lanceolate to ovate-elliptic, 7–19 × 3–6 cm, acute. Inflorescence loose, with 3–30 flowers, floral bracts similar to the uppermost leaves, 2–5 cm, shorter than flower stalk and ovary. Sepals to 3 × 1.5 cm, upper sepal elliptic-lanceolate, acute, lateral sepals bent downwards, lanceolate. Petals with upper lobe 20–30 × 1.5 mm, linear, pressed against upper sepal, lower lobe 3–7 cm × 1 mm, joined to base of lip for 6–20 mm, then curved out and downwards. Lip 3-lobed, with a narrow, undivided basal part, 1.5–3 cm; midlobe linear, bent back at the middle, 2–4 cm, side-lobes 2.5–8.5 cm. Spur 10–21 cm. *Southern and Eastern Africa, Arabia.* G1. Spring–summer.

22. CYNORKIS Thouars

J.M.H. Shaw

Terrestrial herbs with 1–3 small tubers, or rarely epiphytes with swollen storage roots. Stem, ovary and sepals often with glandular hairs. Leaves 1–4, with 1–2 sheaths below. Inflorescence a terminal raceme with 1–35 or more flowers. Flowers pink, orange-red to purple, white or yellow. Upper sepal usually forming a hood, and partly enclosing the petals. Lip entire or 3–5 lobed, with a basal spur. Column short and broad.

A genus of about 125 species with about 20 from Africa, and the remainder from Madagascar and the Mascarene Islands. Propagation is by division, and some species such as *C. fastigiata* will seed themselves about in collections.

Literature: Hermans, J. & Cribb, P., Cynorchis from Madagascar, *Orchid Review* **115**(1277): 248–55 (2007); Hermans, J. & C., Du Puy, D., Cribb, P. & Bosser, J. Orchids of Madagascar, edn 2, 137–70, 290–3 (2007).

1a. Lip with 3 lobes 2
b. Lip with 4 lobes 4
2a. Flowers 1(–2) **2. lowiana**
b. Flowers 3–35 or more 3
3a. Flowers lemon yellow or brick red-orange **6. gibbosa**
b. Flowers purple to deep pink **7. × mirabile**
4a. Flowers 1 or rarely 2 **8. uniflora**
b. Flowers 3 or more 5
5a. Lip with purple spots towards base **1. guttata**
b. Lip without purple spots at base 6
6a. Lip creamy-white with pink tones **4. fastigiata**
b. Lip purple to pink 7
7a. Floral bracts green, acute, less than half as long as flowers **5. purpurascens**
b. Floral bracts whitish-green with pink, long acuminate tips, almost as long as flowers **3. calanthoides**

1. C. guttata Hermans & Cribb. (*C. acuminata* invalid). Illustration: *Orchid Review* **115**: 250, 251 (2007) & **116**: 238 (2008).

Leaves solitary, elongated elliptic to strap-shaped, 10–50 × 3–8 cm, blunt at the tip. Inflorescence to 35 cm, with 10–35 or more flowers; floral bracts 18–22 × 2–3 mm, green with red tip. Flowers *c.* 2.5 cm across, creamy white or pink suffused with purple. Upper sepal oblong, 7–10 × 3–5 mm, concave, forming a hood over the column. Lateral sepals swept back, oblong, with a few glandular hairs at base. Petals narrowly oblong-obovate, 7–9 × 2–3 mm, held against the interior margins of the upper sepal. Lip 18 × 20–23 mm, 4-lobed, with purple spots towards the base, crenate along the edges. Spur slender, 26–32 mm. *Madagascar.* G1. Spring.

2. C. lowiana Reichenbach. Illustration: *Botanical Magazine* 7551 (1897); Hermans et al., Orchids of Madagascar, pl. 31B (2007); *Orchid Review* **115**: 253 (2007).

Plants with a single leaf, narrowly linear-lanceolate, 4.5–13 cm. Inflorescence 7–12 cm, 1–2 flowered. Flowers large; lip 3-lobed, *c.* 2 × 2.7 cm, reddish-purple with a dark spot on the disk. Spur thin, 2.5–4.5 cm. *Madagascar.* G1. Spring.

C. Kewensis grex (*C. lowiana* × *calanthoides*, this latter parent often cited in error as *C. purpurascens* or *purpurata*). Leaf solitary, lanceolate, to 30 cm. Lip with 4 lobes, lilac-purple with white and purple spots towards base, *c.* 3 cm. Spur *c.* 3 cm. *Garden origin.* G1. Spring.

3. C. calanthoides Kraenzlin (*C. uncinata* Perrier, *C. purpurascens* misapplied). Illustration: *Botanical Magazine* 7852 (1902); Hermans et al., Orchids of Madagascar, pl. 33A (2007); *Orchid Review* **115**: 250, 253 (2007); *Orchids* (USA) **77**: 419 (2008).

Epiphytic herb with 1–3 fleshy storage roots. Leaf solitary, ovate-oblong, large, to 40 × 10 cm, almost as long as inflorescence, apex acute. Inflorescence with 7–35 flowers; bracts white to pale green, broadly oval to lanceolate, almost as long as flowers, with a long acuminate pinkish tip. Flowers pink; upper sepal ovate, concave, lateral sepals arched; petals narrow, lanceolate-linear, blunt. Lip 4-lobed, mid-lobe divided into 2 to about half its length. Spur cylindrical. *Madagascar.* G1. Spring.

4. C. fastigiata Thouars. Illustration: Hermans et al., Orchids of Madagascar, pl. 28B (2007).

Plant with 1–2 leaves. Flowers *c.* 3 cm, creamy-white with red to purple tones. Lip 10–15 mm, unequally 4-lobed. Spur narrowly cylindrical, 1.8–3 cm. *Madagascar, Comoros, Seychelles, Mascarene Islands.* G1. Spring.

5. C. purpurascens Thouars. Illustration: Hermans et al., Orchids of Madagascar, pl. 32B (2007); *Orchid Review* **115**: 250 (2007); *Orchids* (USA) **77**: 419 (2008).

Plant 1-leaved. Floral bracts green, shorter than flower. Flowers mauve. Lip 4-lobed, 1.8–2.5 cm, narrowed into a channel towards the base. Spur to 4 cm. *Madagascar, Mascarene Islands.* G1. Spring.

6. C. gibbosa Ridley. Illustration: *Botanical Magazine* **13**: 296 (1996); Hermans et al., Orchids of Madagascar, pl. 29 (2007); *Orchid Review* **103**: 111 (1995) & **115**: 254 (2007).

Herb to 50 cm, bearing a single leaf, 10–30 × 3–8 cm, long-elliptic, which is dark spotted particularly when young. Inflorescence with 10–40 flowers forming a dense pyramidal head; main stem glandular hairy with 2–3 spotted bracts; floral bracts about half as long as the ovary. Flowers 2.5 cm across, brick red to salmon pink, with a lip 2–3 cm, 3-lobed, the central lobe notched or bilobed. Spur green, 2.3–2.7 cm. Column 12–14 mm, with 4 small lobes at the top. The pollinia have unusually long stalks, 12mm, specially designed for hawkmoth pollination. *Madagascar.* G1. Winter–spring.

Forma **aurea** Hermans. Illustration: Hermans et al., Orchids of Madagascar, pl. 30B (2007); *Orchid Review* **115**: 249 (2007). Differs by its pale lemon yellow flowers. *Madagascar.* G1. Winter–spring.

7. C. × mirabile Hermans & Cribb; *C. gibbosa* × *lowiana*. Illustration: *Orchid Review* **115**: 252 (2007).

Herb to 16 cm, with a single leaf 15 × 3.5 cm, dark green densely spotted with dark purple. Inflorescence to 15 cm, 3–4 flowered, with floral bracts 11 × 4 cm, green with a dark margin. Upper sepal 10 cm, hooded, glandular hairs present on base. Lip 19 × 10.5 cm, 3-lobed; mid-lobe with a raised ridge running from the disk to the base of the mid-lobe. Spur 2.5 cm. *Madagascar.* G1. Spring.

A natural hybrid differing from its parents by a 3-lobed lip, and distinctive purple spotted leaf and bracts.

8. C. uniflora Lindley. Illustrated: *Botanical Magazine*, 7564 (1897) – as *C. grandiflora*; *Orchids* (USA) **77**: 419 (2008).

Flowers 1(–2), large, with lip 4 × 2.5 cm, 4-lobed, violet-red with 2 paler spots on disk. *Madagascar*. G1. Spring.

C. gigas Schlechter. Illustration: Hermans et al., Orchids of Madagascar, pl. 30C (2007). Similar to *C. uniflora* but larger. Lip white with dark maroon blotches towards base. Spur 10 cm. *Madagascar*. G1. Spring.

23. STENOGLOTTIS Lindley

P.J.B. Woods

Herbaceous perennials, terrestrial, growing on rocks or epiphytic. Roots several, tuberous, oblong, fleshy. Leaves numerous, mostly basal and forming a rosette, membranous, rolled when young, usually dying back after flowering. Inflorescence erect, many-flowered, bearing several sheaths, upper ones smaller, not clasping stem. Basal part of sepals slightly joined to the lip and column. Petals often slightly broader than sepals, hooded over the column. Lip joined to the column, longer than sepals and petals, oblong, widening towards apex which is 3- to 5-, occasionally 7-lobed or toothed. Spur absent (in ours). Column short. Pollinia 2, club-like, attached to a short stalk, easily separable into distinct masses.

A genus of about 6 species confined to east and southern Africa. Cultivation in terrestrial compost. Plants should be given moist, shady conditions while growing vigorously.

Literature: Stewart, J., The genus Stenoglottis, *Kew Magazine* **6**(1): 9–22 (1989); McDonald, G., Notes on the genus Stenoglottis, *Plant Life* (South Africa) **35**: 10–15 (2006).

1a. Flowers to *c.* 20, arranged mostly to 1 side of inflorescence; lip 3-, occasionally 4-toothed **1. fimbriata**
b. Flowers to *c.* 100, evenly arranged in a cylindric inflorescence; lip 5-, occasionally 7-toothed **2. longifolia**

1. S. fimbriata Lindley. Illustration: *Botanical Magazine*, 5872 (1870); *Flowering Plants of South Africa* **15**: t. 585 (1935); *American Orchid Society Bulletin* **48**: 159 (1979); Stewart & Hennessy, Orchids of Africa, 51 (1981).

Leaves 5–12, to 12 × *c.* 2 cm, lanceolate or oblanceolate, arranged in a basal rosette, purplish-spotted or not, margins often wavy. Raceme loose, 1-sided, to 38 cm. Flowers few to *c.* 20, pale pink to mauve, lip paler with darker spots. Sepals 3–5 mm. Petals slightly broader than sepals. Lip 5–9 mm, its apical third 3- or occasionally 4-lobed or toothed; lobes almost equal, 2–4 mm. The ovary, sepals and petals are covered with small, crystal-like papillae. *Tropical E Africa to S Africa*. G1. Autumn.

2. S. longifolia Hooker. Illustration: *Botanical Magazine*, 7186 (1891); *Flowering Plants of South Africa* **24**: t. 933 (1944); *Orchid Review* **87**: 62 (1979); Bechtel et al., Manual of cultivated orchid species, 275 (1981).

Leaves many, to 18 × 2.5 cm, lanceolate, gradually narrowed at either end, tips slightly recurved, occasionally blackish-spotted, margins often wavy. Flowers to *c.* 100, crowded in a cylindric spike to 70 cm, pale pink, occasionally white, with scattered, dark marks or dots. Sepals *c.* 8 mm. Petals smaller than sepals. Lip to *c.* 1.2 cm, its apical part 5-, occasionally 7-toothed, teeth unequal, to *c.* 6 mm. *S Africa*. G1. Autumn.

24. GOODYERA R. Brown

P.J.B. Woods

Terrestrial herbs, rhizomes creeping, sometimes above ground, aerial growth persistent through the winter. Leaves membranous or fleshy, often asymmetric, mostly basal in rosettes, more rarely scattered on erect stems, often with differently coloured veins, rolled when young. Inflorescence erect, flowers few to many, arranged spirally or on 1 side, or uniformly in a cylindric spike or raceme. Sepals directed forwards or spreading. Petals directed forwards over the column. Lip saccate, the sac usually with hairs inside, entire, the tip usually pointed and reflexed. Column short, blunt or with 2 long or short teeth on the rostellum. Pollinia 2, club-shaped, each longitudinally divided and made up of easily separable masses.

A genus of about 80–100 species from the north temperate regions, and SE Asia and Australasia. As most of them are of creeping habit, they are best grown in wide pans, using terrestrial compost that should be kept moist throughout the year. Most of the species require shade. A periodic dressing of pine-needles is beneficial (see Christian, P., *Bulletin of the Alpine Garden Society* **43**: 322–4, 1975).

Literature: Coleman, R., Growing Goodyera oblongifolia, *American Orchid Society Bulletin* **64**(8): 864–68 (1995).

1a. Flowers in a cylindric spike, not spirally arranged 2
b. Flowers in a 1-sided spike, or spirally arranged 3
2a. Sepals with few hairs, reddish brown, tips white; leaves very dark, main veins red **1. colorata**
b. Sepals densely hairy, white with green mid-vein; leaves green with white veins **2. pubescens**
3a. Flowers 5 mm or more; leaves 3–11 cm, green, usually with a broad, white main vein, the rest net-veined **3. oblongifolia**
b. Flowers less than 5 mm; leaves 1–4 cm, usually green or dark, with or without net-venation **4. repens**

1. G. colorata (Blume) Blume. Illustration: Blume, Flore Javae **4**: t. 9b, f. 2 (1858); *La Belgique Horticole* **12**: 1, f. 4 (1862); Smith, Orchids of Java **2**: f. 93 (1909); Comber, Orchids of Sumatra, 57 (2001).

Stems *c.* 15 cm. Leaves *c.* 6, scattered on the lower part of the stem, *c.* 6 × 2.5 cm, blackish green with a red main vein, narrowly ovate, pointed, sheathing at base. Inflorescence *c.* 8 cm, hairy, with many flowers. Flowers *c.* 6 mm, sepals about 5 mm long, red towards the base, not spreading. Base of lip sac-like, hairy inside, the tip pointed and deflexed. *Indonesia (Java & Sumatra)*. G2.

G. schlechtendaliana Reichenbach. Illustration: Comber, Orchids of Sumatra, 54 (2001); Hirose & Yokoi, Variegated plants in color, 122 (1998); Huang, Flora of Taiwan **5**: 910 (2000). Stems 10–30 cm in flower. Leaves *c.* 8, along lower part of stem, shiny green with irregular pale green or yellow patches. Flowers mostly white, often with pink tinge, and a dark green to red spot at tip of lip. Upper sepal 8–11 mm, yellowish. Ovary glandular hairy, twisted. Lip hairy within. Column 5 mm. *India, Thailand, China, SE Asia, Japan, Sumatra*. G1–G2. Summer.

Many named cultivars exist in Japan, where it is known as Nishiki Ran, selected for patterns of variegation and leaf shape.

G. velutina Maximowicz. Illustration: Hirose & Yokoi, Variegated plants in color, 122 (1998); Huang, Flora of Taiwan **5**: 911 (2000). Sepals 9–11 mm, dark red-brown. Lip 7–8 mm. Column 7–8 mm. *Japan, China, Korea.* H5–G1. Summer.

Grown for variegated leaves, particularly in Japan.

2. G. pubescens (Willdenow) R. Brown. Illustration: *Botanical Magazine*, 2540 (1825); *Flore des serres* **15**: t. 1555 (1862–65); Luer, Native orchids of the United States and Canada, 141 (1975); *American Orchid Society Bulletin* **47**: 400 (1978) & **48**: 1115 (1979).

Rhizomes creeping, sometimes above ground. Stems to 50 cm, densely hairy. Leaves in rosettes, to 9 × 4 cm, oblong-elliptic, blunt, net-veined, veins white. Inflorescence cylindric, with up to 80 flowers, dense. Sepals *c.* 5 × 3 mm, ovate, bluntly pointed. Petals narrower, spathulate. Lip sac-like, the outer surface minutely warty, the tip pointed and recurved. Column with blunt rostellum. *Eastern USA.* H1. Summer.

3. G. oblongifolia Rafinesque (*G. decipiens* (Hooker) Hubbard; *G. menziesii* Lindley). Illustration: Luer, Native orchids of the United States and Canada, 143 (1975).

Rhizomes creeping, stoloniferous. Stems to 48 cm, densely hairy. Leaves to 11 × 4 cm, in rosettes, oblong-elliptic, blunt, mid-vein usually marked by a broad white line, the other veins forming a white, net-like pattern. Inflorescence 1-sided, flowers loose or dense, scented. Sepals 6–10 × 3–4 mm, oblong to narrowly ovate. Petals spathulate. Lip sac-like, its tip oblong, blunt, recurved. Column with pointed rostellum. *Western N America.* H2. Summer.

G. macrophylla Lowe. Illustration: Press & Short, Flora of Madeira, 527 (1994); Delforge, Orchids of Britain & Europe, 99 (1995). Leaves 7–20 × 3–7 cm in a rosette of 6–9, withering about flowering time, marked with white netted veins. Flowers 40–60, white, spirally arranged in a dense spike, 1.3 cm long with bracts longer than the ovaries. Upper sepal and petals forming a hood, back of sepals covered with glandular hairs. Lip fleshy, bluntly ovate, hairless. *Madeira.* H1–G1. Summer–autumn.

4. G. repens (Linnaeus) R. Brown. Illustration: Danesch & Danesch, Orchideen Europas, Mitteleuropa, edn 3, 201, 202 (1972); Landwehr, Wilde Orchideeën van Europa **2**: 541 (1977); Williams et al., Orchids of Britain and Europe, 137 (1978); Mossberg & Nilsson, Orchids of northern Europe, 52, 53 (1979).

Rhizome creeping, stoloniferous. Stems to 25 cm, glandular-hairy. Leaves 1.5–4 × 1–2 cm, ovate, blunt, forming a rosette, green or with a white, net-like pattern (var. **ophioides** Fernald). Inflorescence spirally arranged, or 1-sided, dense. Flowers glandular-hairy, cream-white, fragrant. Sepals 3–4.5 × 3 mm, ovate. Petals oblong-spathulate. Lip sac-like, its tip pointed, recurved. Column with a blunt rostellum. *Eurasia, N America.* H1. Summer.

G. hispida Lindley. Illustration: *Annals of the Royal Botanic Garden Calcutta* **8**: t. 375 (1898); Graf, Exotica, edn 8, 1264 (1978); Bechtel et al., Manual of cultivated orchid species, 215 (1981). Leaves reddish, elliptic-lanceolate, with a pink or white net-like pattern, rostellum pointed. *NE India, Malaysia, Thailand.* G2. Summer.

25. MACODES Lindley
P.J.B. Woods

Terrestrial. Stems short, fleshy, rooting at basal nodes. Leaves few, mostly basal, somewhat fleshy, attractively coloured and veined, rolled when young. Inflorescence erect, flowers few to about 40, asymmetric due to the twisting in opposite directions of the lip and column. Lip uppermost, 3-lobed, its base concave and containing 2 glands. Column with a single stigmatic area and with 2, thin, parallel wings on the front. Pollinia 4, club-like, made up of small, easily separable masses of pollen.

A genus of about 10 species, superficially similar to *Anoectochilus*, and distributed from Malaysia to New Guinea. Cultivation as for tropical terrestrial species with the substitution of beech or oak leaf-mould for peat and bark and the addition of perlite to improve drainage. High humidity and protection from slugs are essential.

1. M. petola (Blume) Lindley. Illustration: Blume, Flore Javae **4**: t. 31 (1858); *Orchid Digest* **38**: 29 (1974); Bechtel et al., Manual of cultivated orchid species, 224 (1981).

Leaves *c.* 6 × 4 cm, ovate, velvety, dark green, the main veins and cross-veins beautifully marked with bright golden yellow. Inflorescence *c.* 20 cm, downy, with 10–40 flowers. Flowers reddish brown, lip white. Sepals, petals and lip *c.* 5 mm. *Malaysia, Philippines to Indonesia (Sumatra).* G2. Autumn–spring.

26. LUDISIA Richard
J.M.H. Shaw

Terrestrial herbs with a thick succulent rhizome bearing a few internodes. Leaves with a short petiolate base that widens to sheath the stem, lamina oblong-lanceolate to elliptic, of variable colour but generally reddish-brown, velvety, with white, gold or red veins. Flowers inverted, twisted with the lip and column pointing in opposite directions, white with yellow anther. Sepals *c.* 10 × 5–7 mm, the dorsal forming a hood with the two petals. Lip slightly bilobed, 8 mm wide, attached to the lower third of column margins. Column twisted in opposite direction to lip, slightly widened towards the top. Ovary twisted, hairy.

A monotypic genus from Indochina and adjacent China, Malay peninsula. Cultivation as for *Macodes*, although *Ludisia* can withstand exposure to lower temperatures.

Literature: Ormerod, P., Taxonomic changes in Goodyerinae, *Lindleyana* **17**(4): 189–238 (2002).

1. L. discolor (Ker Gawler) Blume (*Haemaria discolor* (Ker Gawler) Lindley). Illustration: *Botanical Magazine* 2055 (1819); Seidenfaden & Wood, Orchids of peninsular Malaysia & Singapore, 64 & pl. 2a (1992).

G1. Summer.

The commonly cultivated plant has brownish red leaves with fine red to gold veination. Other selections include:

'Concolor' (var. *concolor* Ridley). Leaves deep olive green with a pink midrib.

'Dawsoniana' (var. *dawsoniana* (Low) Reichenbach; *Ludisia dawsoniana* (Low) Averyanov). Illustration: *Botanical Magazine* 7486 (1896). A larger more vigorous plant with leaves to 12 cm. Also sold as 'Red Velvet'.

'Nigrescens', (var. *nigrescens* invalid; var. *nigra* invalid**)**. Dark or blackish leaves.

'Ordinana' (var. *ordinana* Ridley). Leaves bright velvety green with pink veins.

'Otletae' (*Haemaria discolor* var. *otletae* (Rolfe) Ridley). Leaves narrowly lanceolate, finely netted.

27. ANOECTOCHILUS Blume
P.J.B. Woods

Terrestrial. Stems short, basal part rooting at nodes, upper part erect, fleshy. Leaves few, mostly basal, thin, fleshy, with distinctively coloured veins, rolled when young. Inflorescence erect, flowers few, symmetric. Lip lowermost, the apex widening abruptly to a transverse blade, the claw fringed, the base with a spur containing 2 glands. Column with 2 wings on the front and 2 distinct stigmatic areas. Pollinia 2, club-like, made up of small, easily separable masses of pollen on a common stipe.

A genus of about 30 species distributed from India to NE Australia. Cultivation as for *Macodes* (p. 476).

1. A. roxburghii (Wallich) Lindley. Illustration: *Annals of the Royal Botanic Garden Calcutta* **8**: t. 390 (1898); *Botanical Magazine*, 9529 (1938); *Orchid Digest* **39**: 132 (1975); *Dansk Botanisk Arkiv* **32**: 55 (1978).

Stems to 30 cm. Leaves 7 × 5.5 cm, ovate, velvety, brownish or pinkish green; veins marked in an attractive pink or yellowish net-like pattern, lower surface of leaf paler than upper. Inflorescence to 15-flowered, glandular-hairy. Sepals *c.* 1 cm, reddish or greenish red, reflexed; petals white or pinkish, forming a hood with the middle sepal, tips sickle-shaped. Lip *c.* 2 cm, inclined upwards. Spur bent abruptly away from ovary. *NE India to Vietnam.* G2. Autumn–spring.

A. setaceus (Blume) Lindley (*A. regalis* Blume). Illustration: *Edwards's Botanical Register* **23**: t. 2010 (1837); *Botanical Magazine*, 4123 (1844); *Flore de serres*, t. 6 (1846). Very similar, but lip downward pointing and spur parallel to ovary. *Sri Lanka, Java.* G2. Spring.

28. PTEROSTYLIS R. Brown

J. Lamond & P.J.B. Woods

Terrestrial herbs. Roots thread-like and fleshy with ovoid or spherical tubers. Stems erect. Leaves deciduous, mostly basal and forming a flat rosette, with 1–2 bract-like stem-leaves, or spaced along stem and linear to lanceolate; rolled when young. Flowers 1 or few, erect or pendent. Upper sepal and petals forming a hood, erect or curved forwards; lateral sepals joined along their basal inner margins, the tips tail-like or filament-like, diverging widely, erect or deflexed. Lip linear or oblong, apart from the tip, which is mostly hidden inside hood; base with a small, erect appendage variously divided at its tip. The lip is sensitive to touch so that a small insect alighting at the tip is projected upwards against the column, where it is trapped between column and lip; in squeezing its way out it detaches the pollinia. The lip gradually returns to its original position in 0.5–2 hours. Column slender, variously winged towards apex and extended into a short foot at base. Pollinia 4, powdery.

A genus of about 95 species distributed from Australia to New Zealand and extending to New Guinea and New Caledonia. A compost of loam, sphagnum moss, peat and sand in equal parts is adequate for all species; it should be kept dry and frost-free while the plants are dormant.

Literature: Jones, D.L. & Clements, M.A., A review of Pterostylis, *Australian Orchid Research* **4**: 1–168 (2002).

1a. Leaves linear or lanceolate, more than 5 times longer than broad, evenly spaced on the stem **1. banksii**
b. Leaves oblong to elliptic, less than 5 times longer than broad, mostly in a basal rosette **2. curta**

1. P. banksii Hooker. Illustration: *Botanical Magazine*, 3172 (1832); *Orchadian* **1**: 129 (1965); Salmon, New Zealand flowers and plants in colour, edn 2, 44 (1967); Cooper, A field guide to New Zealand native orchids, 60 (1981).

Plant robust, to 35 cm. Leaves to 6, to 25 × 1.5 cm, pointed, closely spaced along stem and extending above flower. Flower solitary, to 7 cm, erect, pale green with darker stripes; tips of sepals and petals pale orange to pink. Tips of lateral sepals filament-like, bent around hood and diverging with age; tips of upper sepal and petals curving forwards and downwards. Lip with greenish margins and reddish central part. *New Zealand.* H5–G1. Summer–winter.

2. P. curta R. Brown. Illustration: Rupp, Orchids of New South Wales, edn 2, 81 (1969); Nicholls, Orchids of Australia, t. 301 (1969); *American Orchid Society Bulletin* **41**: 804, 805 (1972); Pocock, Ground orchids of Australia, t. 113 (1972).

Plant robust, to 30 cm. Leaves usually 4, blade to 4 × 2 cm, ovate to elliptic, margins often wavy; stem-leaves 1–2, bract-like. Flowers usually solitary, *c.* 3 cm, erect, green flushed with brown. Tips of lateral sepals tailed, not extending beyond hood. Lip *c.* 2 cm, oblong-linear, characteristically twisted above the middle, reddish brown. *Australia, New Caledonia.* H5–G1. Summer–autumn.

P. pedunculata R. Brown. Illustration: *Australian Orchid Review* **34**: 142 (1969); Nicholls, Orchids of Australia, t. 309 (1969); Pocock, Ground orchids of Australia, t. 125 (1972). Flower to 2 cm, erect, green flushed with reddish brown or dark brown. Lateral sepals filament-like, extending beyond the hood. Lip *c.* 5 mm, ovate, not twisted, dark reddish brown. *Australia.* H5–G1. Winter–spring.

P. nutans R. Brown. Illustration: *Botanical Magazine*, 3085 (1831); *Australian Orchid Review* **34**: 142 (1969); Nicholls, Orchids of Australia, t. 292 (1969); Pocock, *Ground orchids of Australia*, t. 123 (1972); *American Orchid Society Bulletin* **44**: 992 (1975). Flower to 3 cm, conspicuously nodding, green, tips of sepals and petals pale brownish. *Australia & North Island of New Zealand.* H5–G1. Summer–autumn.

29. SPIRANTHES Richard

J. Lamond & P.J.B. Woods

Perennial herbs with slender or sturdy, fusiform tubers. Aerial growth dying off in winter. Leaves all basal in a rosette or evenly distributed on the stem, ovate-elliptic or lanceolate, rolled when young. Inflorescence elongate, twisted, with many flowers, which are close, spirally arranged and often scented. Sepals and petals almost equal, incurved, the tips free. Lip entire, its apical margins variously frilled or not. Spur absent. Pollinia 2, each made up of easily separable masses.

A genus of about 30 species mainly in temperate zones, the majority occurring in N America. They should be grown in pots containing a well-drained compost and protected in a cold-frame or greenhouse.

1a. Basal leaves ovate-elliptic, forming a flat rosette; inflorescence lateral **1. spiralis**
b. Basal leaves linear-lanceolate, not forming a rosette; inflorescence terminal **2. cernua**

1. S. spiralis (Linnaeus) Chevallier (*S. autumnalis* Richard). Illustration: Ross-Craig, Drawings of British plants **28**: 47 (1971); Landwehr, Wilde Orchideeën van Europa **2**: 534, 535 (1977); Grey-Wilson & Mathew, Bulbs, t. 47 (1981).

Tubers stout, 5–13 mm thick. Leaves 3–7, basal, 2–4 cm × 5–15 mm, ovate-elliptic, shortly and broadly stalked; upper leaves scale-like, sheathing the stem. Inflorescence to 20 cm, glandular-hairy above, produced laterally from below the leaves. Flowers 7–20, white, scented, in 1 spiral row. Bracts about as long as ovary. Sepals and petals 4–7 mm, the upper 3 and the lip forming a tube around the column. Lip apex truncate and scalloped. *Europe & N Africa to N Iran.* H2. Late summer–autumn.

S. lacera (Rafinesque) Rafinesque (*S. gracilis* (Bigelow) Beck). Illustration: Luer, Native orchids of the United States and Canada, 111 (1975). Tubers slender, 3–6 mm thick; inflorescence to 50 cm, bearing up to 40 flowers. *Eastern N America.* H1. Late summer.

2. S. cernua (Linnaeus) Richard (*S. odorata* (Nuttall) Lindley). Illustration: Luer, Native orchids of the United States and Canada, 119, 121 (1975).

Tubers slender, elongate. Basal leaves 2–6, 5–24 cm × 2–20 mm, linear-oblanceolate, the upper leaves small, sheath-like. Inflorescence to 50 cm, with up to 60 flowers; upper parts downy, hairs sometimes glandular. Flowers white with yellowish centres, arranged in 3–4 rows. Bract shorter or longer than ovary. Sepals and petals 1.1–1.2 cm, sepals forming a tube with the lip. Lip apex rounded, scalloped, recurved. *Eastern N America.* H4. Autumn.

S. aestivalis (Poiret) Richard. Illustration: Ross-Craig, Drawings of British plants **28**: 48 (1971); Landwehr, Wilde Orchideeën van Europa **2**: 537 (1977); Williams et al., Orchids of Britain and Europe, 139 (1978). Flowers 6–7 mm, arranged in 1 spiral row. *C & S Europe.* H2. Summer.

S. sinensis (Persoon) Ames (*S. australis* Lindley). Illustration: *Annals of the Royal Botanic Garden Calcutta* **8**: t. 369 (1898); *Dansk Botanisk Arkiv* **32**: 105 (1978); Williams et al., Orchids of Britain and Europe, 139 (1978); Pearce & Cribb, Orchids of Bhutan, 117 (2002). Flowers white or pink. Lip with 2 basal calluses. *Asia to Australasia.* G2. Summer.

30. STENORRHYNCHOS Richard

J.M.H. Shaw

Terrestrial herbs with fleshy storage roots. Leaves 6–12, in a basal rosette, green, sometimes with white spots, elliptic to lanceolate, acute. Inflorescence hairless, with numerous very showy, usually deep red, lanceolate bracts. Flowers unscented, tubular, with sepals in shades of pink to red, lip usually paler or white. Upper sepal erect, triangular to oblong-lanceolate, acute. Lip forming a rounded nectary at the base, the tip lanceolate and downwards pointing. Ovary hairless, slightly twisted.

A genus of about 4 species distributed from Mexico to NW South America, differing from *Spiranthes* by non-spiral inflorescence and features of the column. Cultivation is fairly easy in a free draining humus-rich compost. A dry resting period is not required.

Literature: Christenson, E.A., Drei neue Stenorrhynchos-Arten, *Journal für den Orchideenfreund* **12**(1): 5–25 (2005).

1. S. speciosum (Jacquin) Richard.

Herb to 60 cm, with a basal rosette of ovate-orbicular, acute, 4–20 × 2–6 cm leaves. Inflorescence with bracts longer than the flowers, lanceolate, acuminate, red, 2.5–4 cm. Sepals lanceolate, acuminate, 1–1.6 × 0.4 cm, recurved towards tip. Petals slightly narrower. Lip wedge-shaped to lanceolate, with an acute tip, shallowly 3-lobed, *c.* 1.5 × 0.5 cm. Column short, densely hairy, *c.* 5 mm. *Central to north-western South America.* G1. Summer.

Plants from Mexico and Costa Rica have been referred to *S. glicensteinii* Christenson.

S. albidomaculatum Christenson (*Spiranthes colorata* var. *maculata* N.E. Brown). Illustration: *Orchid Review* **115**: 174 (2007). A similar plant in which the leaves are patterned with white spots. *Colombia, Venezuela.* G1.

31. VANILLA Miller

J.C.M. Alexander

Perennial branched climbers with long green stems, initially terrestrial, often becoming epiphytic, bearing a leaf and 1 or 2 roots at each node. Leaves alternate, fleshy or leathery, not sheathing, sometimes reduced to scales, folded when young. Flowers in axillary spikes or racemes. Sepals and petals similar, fleshy, spreading; petals with distinct midrib. Lip with a narrow claw fused to the lower part of the column and an entire or obscurely lobed blade enclosing the upper part of the column. Pollinia 2, powdery-granular. Fruit a fleshy pod.

A genus of 107 species distributed throughout the tropics. *V. planifolia* is the source of natural vanilla essence and is widely cultivated in the range of the genus. The long climbing stems of *Vanilla* species must be firmly supported, and can be attached to bark at various points along their length. Much of their moisture is obtained from the atmosphere so the pots, containing epiphytic compost, can be quite small. The plants should be frequently syringed, even in winter. Propagation is effected by cutting up the stem into small sections that can be planted.

Literature: Portes, R., Le genre Vanilla et ses especies, in Bouriquet, G. (ed.), Le vanillier et la vanile dans le monde, (1954).

1. V. planifolia Andrews (*Myrobroma fragrans* Salisbury; *V. fragrans* Ames). Illustration: *Botanical Magazine*, 7167 (1891); Shuttleworth et al., Orchids, 30 (1970); *Orchid Digest* **41**: 181 (1977); Luer, Native orchids of Florida, 75 (1972).

Stem fleshy, terete. Leaves 15–20 × 3–5 cm, fleshy, ovate-elliptic to oblong, acute. Flowers 10–15 per spike, pale green or greenish yellow. Bracts 7 mm, triangular to ovate. Sepals and petals to 6 × 1 cm, linear-oblong to narrowly obovate. Lip to 4.5 × 2 cm; blade triangular with longitudinal papillose veins in the mid-line and a blunt, reflexed apical lobe. Column 3.5 cm, slender, downy. Ovary 4 cm, curved, maturing into a black fleshy cylindrical pod, 15–25 cm. *From C America and the West Indies south to Bolivia, Paraguay & N Argentina; naturalised in Madagascar, the Seychelles and elsewhere.* G2. Intermittent flowering.

Until recently, commercially cultivated plants in Africa, Madagascar and Asia, except the Philippines, were of a single clone propagated from the original type plant cultivated by the Marquess of Blandford. Since the 1990s a number of hybrids have been trialled in Madagascar, including Tsy Taitra grex (*V. planifolia* × *pompona*) and Manitra Ampotony grex (*V. planifolia* × Haapape grex).

V. pompona Schiede. Illustration: Bechtel et al., Manual of cultivated orchid species, 280 (1981). Similar to *V. planifolia* but generally more robust. Sepals and petals 8 cm, greenish yellow. Lip yellow or orange. Pod 3-sided. *From C America and the Lesser Antilles south through Peru & W Brazil to Bolivia.* G2.

V. grandiflora Lindley. Illustration: *Orchid Review* **115**: 263 (2007). Differs with larger flowers in racemes of 1–25. Pods strongly vanilla scented. *Peru.* G2.

V. aphylla Blume. Stem somewhat flattened. Leaves reduced to small green triangular scales. Flowers 3 or 4 together; sepals and petals pale green; lip white with pink hairs on the central lobe. *Burma (Tenasserim), N Malay peninsula & Java*. G2.

32. ARUNDINA Blume

J.C.M. Alexander

Terrestrial, to 3 m. Stems erect, slender and rigid, tufted, unbranched, many-leaved. Leaves 12–30 × 1.5–4 cm, linear to lanceolate, grass-like, in 2 opposite ranks, folded when young; sheaths overlapping at base. Flowers many, 6–7 cm wide, in simple or branched terminal spikes, fragrant; bracts prominent. Sepals and petals *c.* 4 cm, white to pale pinkish purple, acute; sepals *c.* 1 cm wide, lanceolate to elliptic; lateral sepals lying close together under lip; petals *c.* 2 cm wide, ovate to circular, wavy. Lip 4 × 3.5 cm, shallowly 3-lobed, oblong, rounded, base rolled around column, crisped, deeply notched at apex, 3-keeled; throat pale, usually with a yellow patch; blade white to bright pink. Column slender, without foot. Pollinia 8, flattened.

Formerly considered to contain several species, *Arundina* is now usually regarded as a single widespread and highly variable species. Distinctive forms from the extremes of the range appear to be interconnected by intermediates. Cultivation conditions are similar to those for *Sobralia* though a slightly higher minimum temperature is required.

Literature: Robbins, S., Bell, S. & Blakey, S., Arundina – the 'Cattleya' of the old world, *Orchid Review* **100**: 273–6 (1992).

1. A. graminifolia (Don) Hochreutiner (*Bletia graminifolia* Don; *A. speciosa* Blume; *A. chinensis* Blume; *A. bambusifolia* Lindley; *A. revoluta* Hooker). Illustration: *Botanical Magazine*, 7284 (1893); Shuttleworth et al., Orchids, 43 (1970); Graf, Tropica, 709 (1978).

Himalaya & S China, Sri Lanka, SE Asia & Pacific islands. G1. Flowers continuously.

33. BLETILLA Reichenbach

J.C.M. Alexander

Terrestrial, with corm-like pseudobulbs. Stems erect, produced laterally from the bases of old pseudobulbs. Leaves alternate, sheathing, thin and pleated, rolled in bud. Flowers several, in loose terminal racemes. Petals and sepals similar, free, pointing forward around lip. Lip 3-lobed; lateral lobes curled around column; central lobe with warts or ridges. Column free, long and narrow. Pollinia 8.

A genus of about 10 species from E Asia, named for its apparent similarity to the genus *Bletia* from C & S America, with which it is often confused. Species of *Bletilla* have sometimes been classified under *Limodorum, Arethusa* and *Cymbidium*.

Pseudobulbs should be planted in terrestrial compost and kept in a cold-frame. In mild areas, plants will flourish in the open. Very little water should be given in winter.

Literature: Cribb, P. & Thomas, S., Bletillas of western Sichuan, *Orchid Review* **106**(1220): 83–6 (1998); Weaver, R.E. & Stoutamire, W.P., Bletilla species, *Orchids* (USA) **77**(5): 358–65 (2008); Weaver, R. E. & Stoutamire, W.P., Bletillas in cultivation, *Orchids* (USA) **77**(5): 366–75 (2008).

1. B. striata (Thunberg) Reichenbach (*B. hyacinthina* (Smith) Pfitzer). Illustration: *Botanical Magazine*, 1492 (1812); Hay & Synge, Dictionary of garden plants in colour, 126 (1969); Bechtel et al, Manual of cultivated orchid species, 171 (1981).

Plant to 60 cm. Pseudobulbs horizontally flattened. Leaves 3–6, deciduous, 30–50 × 4–7 cm, elliptic, acute. Flowers 3–10 in a loose raceme, magenta. Sepals 2–3 cm × 5–10 mm, elliptic, acute; petals similar but broader. Lip 2.5–3 cm; central lobe longitudinally ridged, shallowly notched at apex. Column 2.5–3 cm, narrowly winged, purple at tip. *China & Japan*. H5–G1. Spring–summer.

B. sinensis (Rolfe) Schlechter. Differs in its smaller, rose-red flowers with darker tips to the sepals, petals and lip. *W China (Yunnan)*. H5–G1. Early summer.

34. COELOGYNE Lindley

J.C.M. Alexander & P.J.B. Woods

Epiphytic. Pseudobulbs cylindric, ovoid or fusiform, clustered or distant, with 1 or 2 leaves. Leaves linear to elliptic, leathery and pleated, rolled when young. Flowering stems erect to pendent, variously sheathed, terminal on pseudobulbs or basal and sometimes later developing into pseudobulbs. Flowers 1–many in loose racemes, often fragrant. Sepals and petals free; sepals often keeled; petals usually narrower, sometimes similar. Lip 3-lobed; lateral lobes erect; disc with entire, warty, toothed or fringed keels, which may extend onto central lobe. Column arched, exposed. Pollinia 4.

A genus of about 200 species from India, China, SE Asia and the Pacific islands, many of which have showy and long-lasting flowers. Himalayan species such as *C. cristata* and *C. barbata* perform best in cool conditions never exceeding 25 °C. They have a very definite resting period when no water should be given; this may last for several months. If the pseudobulbs shrivel during this time, light syringing will supply all the moisture required until growth begins again. Tropical species, e.g. *C. speciosa* and *C. dayana*, grow throughout the year and should be kept constantly moist with a minimum temperature of 15 °C. Good drainage is essential. The spreading habit of many of these species is best accommodated on bark or in hanging baskets. Well-established plants flower heavily and repotting should be avoided unless absolutely necessary.

Literature: Seidenfaden, G., Orchid genera in Thailand 3, Coelogyne, *Dansk Botanisk Arkiv* **29**(4) (1975); Clayton, D., The genus Coelogyne – a synopsis (2002).

Pseudobulbs. Clustered: **1, 2, 5, 6, 7, 9, 11, 14**; distant: **3, 4, 8, 10, 12, 13, 15**. One-leaved: **1**.

Leaves. 3 cm wide or less: **2, 3, 14**.

Flowering stem. Basal: **1, 5, 6, 7, 8, 9, 10, 11, 12, 13, 14, 15, 16**; on mature pseudobulbs: **2, 3, 4, 10**. Erect or arched (majority of flowers held at or above level of pseudobulbs): **1, 2, 3, 4, 10, 11, 12, 13, 14, 15**; pendent (majority of flowers held below level of pseudobulbs): **5, 6, 7, 8, 9, 15**. 5 cm or less: **12**; more than 50 cm: **4, 5, 6, 7, 10**; more than 100 cm: 7. Densely hairy: **6**; with a few scattered hairs: **6, 7**. Sheathed at base only: **1, 3, 5, 6, 8, 9, 10, 11, 12, 14, 15**; with a cluster of scales just below flowers: **4**; sheathed from base up to lowest flowers: **2, 7, 13**.

Flowers. Solitary: **1, 2, 3**; 10 or fewer per stem: **1, 3, 4, 5, 6, 8, 10, 11, 12, 14, 15**; more than 25 per stem: **5, 6, 7, 13**. About 2 cm wide: **2, 5**; 3–4.5 cm wide: **8, 9, 11, 14**, 56; more than 5 cm wide: **1, 2, 3, 4, 6, 7, 9, 10, 11, 12, 13, 14, 15**.

Bracts. Persistent: **5, 6, 7, 10, 12, 13**; deciduous: **1, 2, 3, 4, 8, 11, 14, 15**.

Petals. 2 cm or less: **3, 5, 6, 11, 12**; 5 cm or more: **10, 11, 12**. Narrower than sepals: **1, 2, 3, 4, 8, 9, 11, 13, 14, 15**; about as wide as sepals: **5, 6, 7, 10, 11, 12**.

Lip. 4.5 cm or more: **1, 2, 10**. Margin entire: **2, 5, 6, 7, 8, 9, 11, 13, 14**; margin toothed: **1, 4**; margin fringed: **3, 4, 12**; margin scalloped: **10**. With a distinct waist: **10**. With ringed patches:

11. With an even number of keels: **2, 3, 4, 7, 13**; with an odd number of keels: **1, 2, 3, 4, 5, 6, 8, 9, 11, 12, 14**. Keels entire: **3, 5, 11, 13**; keels wavy: **2, 8, 14, 15**; keels papillose: **1, 6, 10**; keels toothed: **7, 12**; keels fringed: **1, 2, 4, 9, 12**.

1a. Flowers solitary, if more, then not all open at the same time 2
b. Flowers more than 1, all open at the same time 5
2a. Lip margin not fringed 3
b. Lip margin fringed 4
3a. Pseudobulbs 1-leaved **1. speciosa**
b. Pseudobulbs 2-leaved **2. lawrenceana**
4a. Flowering stem short, with scales at base only **3. ovalis**
b. Flowering stem long, with a cluster of overlapping scales just below the flowers **4. barbata**
5a. Flowering stem pendent; majority of flowers held below level of pseudobulbs 6
b. Flowering stem erect or arched; majority of flowers held at or above level of pseudobulbs 10
6a. Flowers 2 cm wide or less; keels on lip not reaching base of central lobe **5. veitchii**
b. Flowers 3 cm wide or more; keels on lip reaching base of central lobe or further 7
7a. Inflorescence-stalk densely hairy **6. tomentosa**
b. Inflorescence-stalk hairless or with scattered hairs 8
8a. Main keels on lip 2; raceme to 100 cm with up to 100 flowers **7. pulverula**
b. Main keels on lip 3; raceme 20–60 cm with 9–30 flowers 9
9a. Flowers 10 or fewer, 4 cm wide or less **8. flaccida**
b. Flowers more than 10, 5 cm wide or more **9. swaniana**
10a. Flowers predominantly green **10. pandurata**
b. Flowers white, cream, yellow or brown 11
11a. Lip with ringed spots **11. nitida**
b. Lip without ringed spots 12
12a. Petals and lateral sepals more or less equal in width **12. cristata**
b. Petals distinctly narrower than lateral sepals 13
13a. Flowers 6–7 cm wide; flowering stem to 30 cm, with 10–18 flowers **13. asperata**
b. Flowers 5 cm wide or less; flowering stem 18 cm or less, with 10 flowers or fewer 14
14a. Leaves less than 2 cm wide **14. viscosa**
b. Leaves 3 cm wide or more **15. trinervis**

1. C. speciosa (Blume) Lindley. Illustration: *Botanical Magazine*, 4889 (1855); *Die Orchidee* **29**: 18 (1978); Bechtel et al., Manual of cultivated orchid species, 187 (1981).

Pseudobulbs 2–8 cm, ovoid, with 4 slightly concave sides, clustered, 1-leaved. Leaf-blade to 35 × 9 cm, narrowly elliptic, acuminate; leaf-stalk *c.* 5 cm. Flowering stem to 20 cm, basal, erect or arched, sheathed at base, bearing 1–3 flowers; bracts deciduous; flowers *c.* 6 cm wide, drooping. Sepals 5–7 × 1–1.8 cm, narrowly ovate, acuminate; petals 5–6 cm × 2–3 mm, linear; petals and sepals yellowish green. Lip 5 × 3–3.5 cm, 3-lobed, dark reddish brown with a white or yellow tip; keels 3, 2 long and 1 short, papillose, fringed; central lobe with a short broad claw. Column 3.5 cm, arched. *Java & Sumatra* (Malaysian records refer to *C. xyrekes* and those from Vietnam to *C. lawrenceana*). G2. Flowers continuously.

2. C. lawrenceana Rolfe. Illustration: *Botanical Magazine*, 8164 (1907); *Gardeners' Chronicle* **47**: 335 (1910); Bechtel et al., Manual of cultivated orchid species, 186 (1981).

Pseudobulbs 5–8 × 2.5–3 cm, ovoid to cylindric, clustered, 2-leaved. Leaves 20–30 × 3 cm, elliptic, acuminate, tapering into stalk. Flowering stem 15–20 cm, erect, on mature pseudobulb, sheathed throughout, 1-flowered; flower 5–7 cm wide. Sepals 5 × 1.5 cm, triangular to ovate, obtuse; petals 4.5 cm × 3–4 mm, linear, drooping; petals and sepals yellow. Lip 5 × 2.5–3 cm, 3-lobed, with 3 long, fringed keels and 2 shorter basal ones; disc and lateral lobes purplish red; central lobe white with a yellow stripe. Column 3–4 cm, club-shaped. *Vietnam*. G2. Spring.

C. holochila Hunt & Summerhayes (*C. elata* misapplied; *C. stricta* misapplied). Illustration: *Botanical Magazine*, 5001 (1857). Flowers 6–10 per raceme, *c.* 4.5 cm wide; sepals and petals white. Lip obscurely 3-lobed, yellow or orange with 2 sinuous, fringed keels. *NE India & Burma*. G2. Spring–summer.

C. miniata (Blume) Lindley (*Hologyne miniata* (Blume) Pfitzer). Illustration: Cogniaux & Goossens, Dictionnaire iconographique des orchidées, Coelogyne, t. 7 (1907); Van Steenis, Mountain flora of Java, t. 35 (1972). Pseudobulbs oblong to fusiform, distant, enclosed in scaly sheaths. Flowers several, *c.* 2 cm wide, bright red. Column with small foot. *Indonesia (Java & Bali; Sumatra)*. G2. Summer.

3. C. ovalis Lindley. Illustration: *Botanical Magazine*, 9255 (1931); *Orchid Review* **80**: 98 (1976); Bechtel et al., Manual of cultivated orchid species, 187 (1981).

Pseudobulbs 3–9 × 2 cm, fusiform to ovoid, sheathed below, distant, 2-leaved. Leaves 7–17 × 1.5–4.5 cm, lanceolate, acute; leaf-stalk short. Flowering stem to 12 cm, erect, on mature pseudobulb, 1–3 flowered, sheathed below; flowers to 6 cm wide; bracts deciduous. Sepals 3–3.5 × 1.2 cm, ovate, acute; petals 3 cm × 1 mm, linear, acute; sepals and petals pinkish brown. Lip 3–3.5 × 2–2.5 cm, oblong, 3-lobed, finely fringed, yellow to pinkish brown with purplish brown patches; keels 3, wavy, 2 long and 1 short. Column 1.5–1.8 cm, apically winged. *W Himalaya to Burma, China & Thailand*. G2. Summer.

C. fimbriata Lindley. Illustration: *Edwards's Botanical Register* **11**: t. 868 (1825); *Dansk Botanisk Arkiv* **29**(4): 18 (1975); Bechtel et al., Manual of cultivated orchid species, 186 (1981). Flowering stem to 5 cm; flowers 3–4 cm wide, pale yellow. Upper sepal 1.5–2.5 mm. Lip marked with purplish brown, with 2 extra, short keels on the central lobe. *NE India to China & Malaysia*. G2. Summer–autumn.

C. fuliginosa Lindley. Illustration: *Botanical Magazine*, 4440 (1849); *Dansk Botanisk Arkiv* **29**(4): 21 (1975); Bechtel et al., Manual of cultivated orchid species, 186 (1981). Flowers 5–6 cm wide; sepals and petals yellowish orange; lip dark brown; keels 4, slightly wavy, 2 long and 2 short. *Burma; ?Java*. G2. Winter.

4. C. barbata Griffith. Illustration: Veitch, Manual of orchidaceous plants: Eriae, 32 (1890); *Lindenia* **16**: 735 (1901); *Orchid World* **6**: 274 (1916).

Pseudobulbs to 9 × 2.5 cm, conical to ovoid, sheathed below when young, distant, 2-leaved. Leaves to 40 × 5 cm, oblong to lanceolate, acute or acuminate; leaf-stalk *c.* 5 cm. Flowering stem to 60 cm, erect, on mature pseudobulb, up to 10-flowered, with a cluster of overlapping scales just below the flowers; bracts deciduous. Sepals 3.5–4 × 1.5 cm, ovate to oblong, acute; petals 3.5–4 cm × 2–3 mm, linear, acute; sepals and petals white. Lip 3.5 × 3 cm, 3-lobed; central lobe laciniate; disc and central lobe dark brown to black, lateral lobes white; keels 3, fringed, middle one shorter. Column 2.5–3 cm. *Bhutan & NE India*. G2. Winter.

C. stricta (D. Don) Schlechter (*C. elata* Lindley). Illustration: *Orchid Review* **80**: 210 (1972). Lip less than 3 cm, keels 2, margin finely and shortly toothed. *Himalaya*. G2. Spring–summer.

The illustration under this name in *Botanical Magazine*, 5001 (1857) is not of this species.

5. C. veitchii Rolfe. Illustration: *Botanical Magazine*, 7764 (1901); *Orchid Review* **33**: 219 (1925).

Pseudobulbs 7.5–10 × 2–4 cm, fusiform, becoming grooved, at first almost enclosed in sheathing scales, clustered, 2-leaved. Leaf-blade to 30 × 8 cm, broadly elliptic, acute, distinctly 5-veined beneath; leaf-stalk *c.* 10 cm. Flowering stem 15–60 cm, basal, pendent, sheathed at base, bearing up to 30 flowers or more; bracts persistent; flowers *c.* 2 cm wide, white. Sepals 12–14 × 5 mm, oblong-lanceolate, acute, incurved and concave; petals 10–13 × 3 mm, lanceolate, acute, incurved and concave. Lip 1.5 × 1.3 cm, shallowly 3-lobed, with 3 short basal keels. Column 7 mm, widening apically with a winged slightly wavy margin. *New Guinea*. G2. Summer–autumn.

6. C. tomentosa Lindley (*C. massangeana* Reichenbach). Illustration: *Botanical Magazine*, 6979 (1888); *Dansk Botanisk Arkiv* **29**(4): 61, 62 (1975); Williams et al., Orchids for everyone, 75 (1980); Bechtel et al., Manual of cultivated orchid species, 187 (1981).

Pseudobulbs to 6 × 3 cm, fusiform to ovoid, clustered, 2-leaved. Leaf-blade to 50 × 10 cm, elliptic; leaf-stalk 5–10 cm. Flowering stem to 60 cm, basal, pendent, densely covered in short black hairs, base loosely sheathed, bearing up to 30 flowers; bracts persistent, hairy; flowers 4–7 cm wide, pale yellow to pale orange or salmon-coloured. Sepals 2.5–3.5 cm × 7–9 mm, lanceolate; petals slightly shorter and narrower. Lip 2–3 × 1.8 cm, 3-lobed, yellowish; lateral lobes brown with paler veins, keels 5, 3 long and 2 short, papillose at base, warty at apex. Column *c.* 6 mm. *Malaysia, Sumatra, Java*. G2. Spring.

7. C. pulverula Teijsmann & Binnendijk (*C. dayana* Reichenbach). Illustration: Schlechter, Die Orchideen, edn 2, 135 (1927); *Botanical Magazine*, n.s., 309 (1958); Graf, Tropica, 715 (1978); Bechtel et al., Manual of cultivated orchid species, 186 (1981).

Pseudobulbs 10–22 × 2–5 cm, fusiform to cylindric, clustered, 2-leaved. Leaf-blade 25–75 × 5–10 cm, elliptic, acute, with 7–9 prominent veins beneath; leaf-stalk *c.* 9 cm. Flowering stem to 100 cm, basal, pendent, with short scattered hairs, sheathed at base and above, bearing up to 50 flowers; bracts moderately persistent; flowers *c.* 7 cm wide, pale yellowish, brownish yellow or whitish. Sepals 3–3.5 cm × 5–7 mm, lanceolate, acute; petals a little smaller. Lip *c.* 3.5 × 2.5–3 cm, 3-lobed; lateral lobes brown with white veins; central lobe brown with a narrow white margin, keels 6, 2 long and 4 short. *Thailand, Malay peninsula, Sumatra, Borneo, Java*. G2. Winter–spring.

8. C. flaccida Lindley (*C. lactea* Reichenbach; *C. huettneriana* misapplied). Illustration: *Botanical Magazine*, 3318 (1834); Schlechter, Die Orchideen, edn 2, 136 (1927); *Orchid Review* **80**: 190 (1972); Bechtel et al., Manual of cultivated orchid species, 186 (1981).

Pseudobulbs to 12 × 2.5 cm, cylindric to fusiform, grooved, sheathed in shining purple scales below, 2-leaved. Leaf-blade to 16 × 3.6 cm or more, lanceolate, narrowing gradually to base, apex acuminate. Flowering stem 20–25 cm, basal, pendent, sheathed at base, bearing *c.* 8 flowers; bracts deciduous; flowers to 4 cm wide, white. Sepals 3–4 cm × 9 mm, lanceolate; petals 2.7–3.5 cm × 3 mm, linear. Lip 1.7 × 1.2 cm, 3-lobed, white; base of central lobe yellow; lateral lobes veined with reddish brown; disc with 3 wavy keels. Said to have an unpleasant smell. *E Himalaya*. G2. Winter–spring.

9. C. swaniana Rolfe. Illustration: *Botanical Magazine*, 7602 (1898); *Orchid Review* **83**: 150 (1975).

Pseudobulbs to 10 × 3.5 cm, fusiform, 4- to 6-angled, clustered, 2-leaved. Leaf-blade to 25 × 6 cm, elliptic; leaf-stalk 5–7.5 cm. Flowering stem to 40 cm, basal, pendent, with minute purple spots, loosely sheathed at base, bearing *c.* 20 flowers; bracts apparently deciduous; flowers *c.* 5 cm wide. Sepals 2.5 cm × 7–10 mm, lanceolate, white, or white tinged with pale brown; petals narrower, white. Lip 2.5 × 1.6 cm, 3-lobed, pale brown, darker at the margins and lobe tips; lateral lobes white-veined; keels 5, 3 long and 2 short. *Malay peninsula, Sumatra, Borneo, ? Philippines*. G2. Spring–summer.

10. C. pandurata Lindley. Illustration: *Botanical Magazine*, 5084 (1858); Shuttleworth et al., Orchids, 41 (1970); Skelsey, Orchids, 103 (1979).

Pseudobulbs 7–12 × 2–3 cm, cylindric to fusiform or ovoid, compressed, 2-leaved, distant. Leaf-blade 20–50 × 7 cm, lanceolate to elliptic, acute to acuminate; leaf-stalk *c.* 6 cm. Flowering stem to 50 cm, basal, erect or arched, sheathed below, with up to 15 flowers, bracts persistent; flowers *c.* 10 cm wide, fragrant, pale green. Sepals and petals 3.5–5.5 × 1.2–1.5 cm; sepals lanceolate to ovate, acute; petals lanceolate to slightly spathulate, acute. Lip 4.5 × 1.5–2.5 cm, 3-lobed with a long waist, mottled with purplish brown or black, 2-keeled at base; disc with warty ridges. *Malaysia, Sumatra & Borneo*. G2. Summer.

C. parishii Hooker. Illustration: *Botanical Magazine*, 5323 (1862); Graf, Tropica, 715 (1978); Bechtel et al., Manual of cultivated orchid species, 187 (1981). Flowering stem on mature pseudobulb, with 3–5 flowers; flowers 6–8 cm wide. Lip to 2.5 cm. *Burma*. G2. Spring–summer.

11. C. nitida (Wallich) Lindley (*C. ochracea* Lindley; not *C. nitida* misapplied, which is probably *C. punctulata* Lindley). Illustration: *Botanical Magazine*, 4661 (1852); Hay & Synge, Dictionary of garden plants in colour, 60 (1969); Bechtel et al., Manual of cultivated orchid species, 187 (1981).

Pseudobulbs 2–8 × 1–1.5 cm, cylindric to fusiform or narrowly ovoid, clustered, 2-leaved. Leaves 8–30 × 2–6 cm, lanceolate to elliptic, acute to acuminate; stalk short or absent. Flowering stem to 20 cm, erect or arched, basal, with 2–8 flowers, sheathed below, bracts deciduous; flowers 3–4 cm wide, fragrant. Sepals 2–3 cm × 7–9 mm, lanceolate, acuminate, upper sepal almost spathulate; petals 2–2.5 cm × 5 mm, lanceolate, acute; sepals and petals white. Lip 2 × 1–1.5 cm, 3-lobed, ovate, white with large yellow or orange patches ringed with reddish orange; keels 3, 2 long and 1 short. Column *c.* 1 cm. *W Himalaya to Burma, China, Thailand & Laos*. G2. Spring–summer.

C. corymbosa Lindley. Illustration: *Botanical Magazine*, 6955 (1887); Northen, Miniature orchids, 57 (1980); Bechtel et al., Manual of cultivated orchid species, 185 (1981). Pseudobulbs 2.5–4 × 1.5–2.5 cm, ovoid to fusiform. Flowers to 7 cm wide; lip 3 cm. *B Himalaya & NE India*. G2. Summer.

C. nervosa Richard (*C. corrugata* Wight). Illustration: Wight, Icones plantarum Indiae orientalis, t. 1639 (1851); *Botanical Magazine*, 5601 (1866). Pseudobulbs wrinkled. Sepals and petals more than 1 cm wide, elliptic. Lip with 3 equal wavy keels; central lobe longer than broad, mostly yellow with orange streaks. *S India*. G2. Autumn.

12. C. cristata Lindley. Illustration: *Botanical Magazine*, 8477 (1913); Williams et al., Orchids for everyone, 185 (1980); Bechtel et al., Manual of cultivated orchid species, 185 (1981).

Pseudobulbs 3.5–7.5 × 2–4 cm, cylindric to fusiform or ovoid, irregularly ridged, distant, 2-leaved. Leaves 12–30 × 2–3 cm, lanceolate, acute; stalk short or absent. Flowering stem 15–30 cm, erect or arched, basal, with 3–10 flowers, sheathed at base; bracts persistent; flowers 7–9 cm wide, white, very fragrant. Sepals and petals similar, 3.5–5.5 × 1.5–2 cm, oblong to lanceolate, acute or obtuse. Lip 3–4 × 3–3.5 cm, 3-lobed; keels 3, yellow and fringed, 2 long and 1 short. Column 2.5 cm. *E Himalaya*. G2. Winter–spring.

13. C. asperata Lindley. Illustration: Orchid Review **84**: 338 (1976); Millar, Orchids of Papua New Guinea, 74 (1978).

Differs from *C. pandurata* in its pale cream-coloured flowers, 6–7 cm wide; lip *c.* 3 cm, lined and speckled with brownish orange. *Malaysia, Sumatra to New Guinea*. G2. Spring–summer.

The illustration under this name in Sheehan & Sheehan, Orchid genera illustrated, 67 (1979) is probably of *C. pandurata*.

14. C. viscosa Reichenbach (*C. graminifolia* Parish & Reichenbach). Illustration: *Botanical Magazine*, 7006 (1885); *Dansk Botanisk Arkiv* **29**(4): 39 (1975).

Pseudobulbs 2.5–8 cm, ovoid, ribbed when mature, clustered, 2-leaved. Leaves 20–45 × 1–3 cm, linear. Flowering stem 10–15 cm, basal, erect, sheathed at base, bearing 2–7 flowers; bracts deciduous; flowers *c.* 5 cm wide, white, apparently fragrant. Sepals 2–3.5 cm × 6–9 mm, lanceolate; petals narrower. Lip 2 × 1–1.5 cm, 3-lobed, white; lateral lobes brownish-veined; central lobe dark yellow; keels 3, finely wavy, the middle one less distinct, purplish apically. *SW China, N India, Burma, Vietnam & Malaysia*. G2. Winter.

15. C. trinervis Lindley (*C. cinnamomea* Teijsmann & Binnendijk). Illustration: Smith, Orchideen von Java **2**: t. 109 (1905); *Dansk Botanisk Arkiv* **29**(4): 47 (1975).

Pseudobulbs to 6 × 3 cm, ovoid to narrowly ovoid, *c.* 2 cm apart, 2-leaved. Leaf-blade to 40 × 4.5 cm, lanceolate, acute, distinctly 3-veined beneath; leaf-stalk *c.* 7 cm. Flowering stem to 18 cm, basal, erect, sheathed at base, bearing 6–8 flowers; bracts deciduous; flowers *c.* 3 cm wide, cream or pale brown. Sepals 2.2 cm × 6 mm, lanceolate; petals 2.2 cm × 3 mm, linear. Lip to 1.8 × 1.4 cm, 3-lobed; keels 7, 3 long, 4 short. Column 1.4 cm. *Burma, Cambodia, Vietnam, Thailand, Malaysia & Java*. G2. Winter.

C. fuscescens Lindley. Illustration: *Annals of the Royal Botanical Garden Calcutta* **8**: t. 181 (1898); *Orchid Review* **80**: 188 (1972); *Dansk Botanisk Arkiv* **29** (4): 27 (1975). Flowers to 5 cm wide, pale brown. Lip *c.* 4 × 1.5 cm; lateral lobes forward pointing, less than 5 mm long. *NE India*. G2. Winter.

Var. **brunnea** (Lindley) Lindley. Illustration: *Botanical Magazine*, 5494 (1865); *Dansk Botanisk Arkiv* **29**(4): 29 (1975). Forward pointing lateral lobes more than 5 mm. Possibly not distinct. *Burma, Thailand, Laos & Vietnam*.

35. PLEIONE D. Don

J.C.M. Alexander

Epiphytic or terrestrial, often on rocks. Pseudobulbs clustered, spherical, ovoid or barrel-shaped, often sheathed, formed from bases of previous season's flowering stems. Leaves membranous, pleated, rolled when young, deciduous, 1 or 2 per flowering stem and produced with or soon after the flowers, or, in autumn-flowering species, several months later on mature pseudobulbs (Nos. 1 and 2). Flowering stems usually 1-flowered (occasionally 2- or 3-flowered in No. 5), sheathed below. Sepals and petals free, similar. Lip entire or very shallowly 3-lobed; lateral lobes or basal portion rolled or upturned around column, often concealing it; central lobe irregularly toothed, lacerate or fringed, with 2–9 longitudinal keels. Column slender, hooded, often winged at the apex. Pollinia 4.

A genus of about 12 Asian alpines, growing mostly between 1000 and 3500 m, most of which were originally described in *Coelogyne*. They should be grown in epiphytic compost in well-drained, shallow pots or wide pans in cool conditions with temperatures never exceeding 20–25 °C. In winter they should be protected from frost. Plants perform best if repotted annually; this should be done just after flowering has finished when new growth has just appeared. They should then be kept in shady, humid conditions, though no water should be given until the new roots are actively growing. Once growth ceases and the leaves drop, temperatures should be lowered and water withheld.

Literature: Hunt, P.P. & Vosa, C.G., The cytology and taxonomy of the genus Pleione, *Kew Bulletin* **25**: 423–32 (1971); Hunt, P.F. & Butterfield, I., The genus Pleione, *The Plantsman* **1**(2): 112–23 (1979); Cribb, P.J. & Butterfield, I., The genus Pleione, edn 2 (1999); Cribb, P.J. & Butterfield, I., Little known and confused species of Pleione, *The New Plantsman* **8**(3): 138–46 (2001); Bergel, G., Pleione – neues von der Gattung, *Die Orchidee* **57**(4): 445–53 (2006).

Pseudobulbs. Barrel-shaped: **1, 2**; spherical, slightly flattened: **6**; ovoid or pear-shaped: **3, 4, 5, 6, 7**. Lacking sheaths: **3**.

Leaves. Two on pseudobulbs: **1, 2**; one on flowering stems: **3, 4, 5, 6, 7**. Fully developed at flowering: **3**; fully developed soon after flowering: **4, 5, 6, 7**; fully developed several months later: **1, 2**.

Flowering stem. 4 cm or less: **4**; over 15 cm: **5, 6**.

Flowers. 5–7 cm wide: **1, 3, 4, 5, 6, 7**; 8 cm wide or more: **2, 4, 5**. Yellow: **7**.

Lip keel. Papillose: **1, 2, 6**; finely toothed: **6**; sinuous or toothed: **5**; finely barbed: **3, 4**; irregularly and bluntly toothed: **7**.

Lip margin. Irregular: **1**; irregularly toothed: **2**; finely fringed: **3, 4, 6, 7**; finely toothed: **5**

1a. Pseudobulbs barrel-shaped, narrowing abruptly above; leaves 2, on mature pseudobulbs — 2

b. Pseudobulbs spherical, ovoid or pear-shaped; leaves 1, on flowering stems — 3

2a. Flowers to 6 cm wide; keels on lip running almost to apex; flowering stem sheaths not warty **1. maculata**
b. Flowers to 8 cm wide; keels on lip two-thirds of lip length; flowering stem sheaths warty **2. praecox**
3a. Keels on lip barbed with fine, backwardly curving teeth 4
b. Keels on lip sinuous, irregularly cut or with upright teeth 5
4a. Leaves produced with flowers; flowering stems more than 3 times longer than pseudobulbs **3. hookeriana**
b. Leaves produced after flowers; flowering stem about 2 times longer than pseudobulbs **4. humilis**
5a. Leaves produced with flowers **5. bulbocodioides**
b. Leaves produced after flowers 6
6a. Flowers red, pink or purple **6. yunnanensis**
b. Flowers yellow to orange **7. × confusa**

1. P. maculata (Lindley) Lindley. Illustration: *Botanical Magazine*, 4691 (1853); *The Plantsman* **1**(2): 113 (1979); Bechtel et al., Manual of cultivated orchid species, 263 (1981).

On rocks. Pseudobulbs 3–4 cm, barrel-shaped, narrowing abruptly above, warty, 2-leaved. Leaves 15–30 × 2.5–3 cm, lanceolate to elliptic, acute. Flowering stems 5–10 cm, sheaths not warty, with a leafy bract above. Flowers 5–6 cm wide. Sepals and petals 3–3.5 cm × 5–7 mm, lanceolate, acute, white or pale cream; petals sometimes streaked with pink. Lip 2–3.5 × 2.5–3.5 cm, shallowly 3-lobed; lateral lobes triangular, curled around column; central lobe semicircular to oblong, wavy, margin irregular, slightly notched at tip, white with purple blotches and a central yellow patch. Keels 5–7, papillose, running almost to tip. *NE India, Bhutan, Burma & Thailand.* H5–G1. Winter.

2. P. praecox (J.E. Smith) D. Don (*Coelogyne wallichiana* Lindley; *P. lagenaria* Lindley; *P. reichenbachiana* (Moore & Veitch) Williams). Illustration: *Botanical Magazine*, 5370 (1863); *The Plantsman* **1**(2): 121 (1979); Bechtel et al., Manual of cultivated orchid species, 263 (1981).

Epiphytic or on rocks. Pseudobulbs 2–2.5 cm, barrel-shaped, narrowing abruptly above, purple with small greenish warts, 2-leaved. Leaves to 20 × 3.5 cm, lanceolate to narrowly elliptic, acute. Flowering stems to 15 cm, basal sheaths warty. Flowers *c.* 8 cm wide. Sepals and petals 4–6 cm × 5–15 mm, lanceolate, acute, pink to bluish purple. Lip 5–6 × 3 cm, shallowly 3-lobed; lateral lobes oblong to triangular, rolled around column; central lobe oblong, irregularly toothed, strongly notched, pinkish white with large marginal purple blotches and a central patch of yellow. Keels 5, papillose; lateral keels running about two-thirds of lip length. *Nepal, NE India, China & Burma.* H5–G1. Autumn.

3. P. hookeriana (Lindley) Williams. Illustration: *Botanical Magazine*, 6388 (1878); *The Plantsman* **1**(2): 114 (1979); Bechtel et al., Manual of cultivated orchid species, 263 (1981).

Epiphytic or on rocks. Pseudobulbs 2–3.5 cm, conical to ovoid, not sheathed. Leaves 1 per flowering stem, 3–15 × 3–4 cm, lanceolate to elliptic, acuminate. Flowering stems to 15 cm, bearing closely adpressed, smooth bracts below the leaf. Flowers 5–6.5 cm wide. Sepals and petals 2.5–3.5 cm, pink to pale purple, sepals-broader than petals, lateral sepals sickle-shaped. Lip 2–2.5 × 3–4 cm, shallowly 3-lobed; lateral lobes bluntly triangular, not concealing column, white to pale pink; central lobe semicircular to oblong, notched, finely fringed, pink with brown blotches and a central yellow patch. Keels 5–7, finely barbed. Column winged. *NE India, Nepal, Bhutan, Thailand & Laos.* H5–G1. Spring.

4. P. humilis (J.E. Smith) D. Don. Illustration: *Botanical Magazine*, 5674 (1867); *The Plantsman* **1**(2): 117 (1979).

Epiphytic or on rocks. Pseudobulbs 2.5–5 cm, pear-shaped, sheathed. Leaves 1 per flowering stem, 20–30 × 4–5 cm, produced after flowering, lanceolate. Flowering stem *c.* 4 cm, sheathed at base. Flowers 7–9 cm wide. Sepals and petals 3.5–4.5 cm, lanceolate to narrowly oblong, pale pinkish white, sepals broader than petals. Lip 3–4 × 2.5–3 cm, oblong, more or less unlobed, base rolled around column, apical portion finely fringed, pale pink strongly marked with crimson or orange. Keels 5–7, barbed. *NE India, Nepal & Burma.* H5–G1. Winter–spring.

5. P. bulbocodioides (Franchet) Rolfe (*Coelogyne pogonioides* Rolfe; *P. delavayi* (Rolfe) Rolfe; *P. formosana* Hayata; *P. pricei* Rolfe; *P. henryi* (Rolfe) Schlechter; *P. limprichtii* Schlechter). Illustration: *Botanical Magazine*, 8729 (1917) & n.s., 397 (1962), 421 (1963).

Epiphytic or on rocks. Pseudobulbs 2–4.5 cm, conical, ovoid or pear-shaped, pale or dark green to purple, sheathed. Leaves 1 per flowering stem, 10–50 × 3–5 cm, lanceolate to elliptic, acute, reaching full size after flowering. Flowering stems 10–20 cm, sheathed, with 1 (rarely 2 or 3) flowers. Flowers 5–12 cm wide. Sepals and petals 2.5–4 cm × 5–15 mm, lanceolate to asymmetrically elliptic, acute, pink to pale purple. Lip 2.5–4.5 × 3–4 cm, shallowly 3-lobed to bluntly diamond-shaped or almost circular, base rolled around column, apical portion finely toothed to lacerate, shallowly notched, white to pink, heavily spotted with pale brown or purplish pink or both. Keels 2–7, wavy or toothed. *W China, Burma & Taiwan.* H5–G1. Spring.

Variable. The synonyms quoted are frequently used in the horticultural trade to describe well-known forms that were formerly regarded as separate species; recent research suggests that some of them should be resurrected as valid species.

6. P. yunnanensis (Rolfe) Rolfe. Illustration: *Botanical Magazine*, 8106 (1906); *Orchid Review* **83**: 9 (1975).

On rocks. Pseudobulbs 2–3 cm, spherical, flattened, shiny, sheathed. Leaves 1 per flowering stem, 20–30 × 2–3 cm, lanceolate to elliptic, produced after flowering. Flowering stem 10–20 cm, sheathed at base. Flowers 5–7 cm wide. Sepals and petals 3–4 cm, oblong to lanceolate, obtuse, pinkish purple; sepals broader than petals. Lip oblong to circular; base rolled around column; apical portion finely fringed, purplish pink with purple blotches. Keels 5, finely toothed or papillose. *China (Yunnan).* H5–G1. Spring.

7. P. × confusa Cribb & Tang (*P. forrestii* misapplied). Illustration: *Botanical Magazine*, n.s., 501 (1967); *Orchid Review* **83**: 10 (1975).

On rocks. Pseudobulbs 2.5 cm, pear-shaped, sheathed. Leaves 1 per flowering stem, to 15 × 4 cm, lanceolate to elliptic, produced after flowering. Flowering stem 2–5 cm, sheathed at base. Flowers 5–6 cm wide. Sepals and petals *c.* 4 cm, lanceolate to elliptic, yellow or orange. Lip almost circular, obscurely 3-lobed, its base rolled around column; blade finely fringed, yellow with purple blotches; keels 5 or 6, irregularly and bluntly toothed. *Burma & China (Yunnan).* H5–G1. Spring.

Recent cytological and morphological studies have shown that this plant, generally known as *P. forrestii*, is a natural hybrid between *P. forrestii* Schlechter and *P. albiflora* Cribb & Tang. True *P. forrestii* has only recently been re-introduced to cultivation.

36. DENDROCHILUM Blume

J.C.M. Alexander

Epiphytic, rarely on rocks. Pseudobulbs fusiform, conical or ovoid, clustered or distant, 1- or 2-leaved. Leaves flat, narrow, leathery, stalked, tapered at both ends, rolled in bud. Flowering stems lateral, long and slender, arched, bearing many flowers in dense, 2-ranked racemes or spikes. Flowers small, fragrant. Sepals and petals spreading, similar; upper sepal often keeled; lateral sepals slightly fused to column foot. Lip simple or 3-lobed, fleshy below, loosely hinged to column foot, often with 2 or 3 keels. Column short, with narrow side-lobes and apical wings. Pollinia 4, pear-shaped.

A genus of 263 species from SE Asia. Most of the commonly cultivated species are in section *Platyclinis*, which has sometimes been recognised as a distinct genus. Cultivation conditions are similar to those for the tropical species of *Coelogyne* (p. 479).

Literature: Pedersen, H. Ae., The genus Dendrochilum in the Philippines – a taxonomic revision, *Opera Botanica* **131**: 1–205 (1997); Wood, J.J., Dendrochilum of Borneo (2001); Pedersen, H. Ae., Changes to Dendrochilum, *Orchid Review* **115** (1276): 220–226 (2007).

1a. Flowers less than 1 cm wide **1. filiforme**
b. Flowers more than 1 cm wide 2
2a. Lip distinctly shorter than petals and sepals; lateral lobes about half as long as central lobe; central lobe obtuse or rounded **2. glumaceum**
b. Lip about equal to petals and sepals; lateral lobes very small or absent; central lobe wedge- or fan-shaped **3. cobbianum**

1. D. filiforme Lindley (*Platyclinis filiformis* (Lindley) Bentham). Illustration: *Orchid Review* **77**: 154 (1969); Skelsey, Orchids, 109 (1979); Sander, Orchids and their cultivation, edn 9, t. 39 (1979).

Epiphytic. Pseudobulbs 2.5 × 1 cm, ovoid, densely clustered, 1- or 2-leaved. Leaves 12–20 × 1–2 cm, linear-lanceolate, acute. Flowering stem 20–45 cm, thread-like, arched, bearing up to 100 flowers or more. Flowers *c.* 6 mm wide in 2 ranks, fragrant, pale yellow or greenish yellow. Sepals and petals 2–3 × 1–1.5 mm, lanceolate to ovate or elliptic. Lip 2 × 2 mm, heart-shaped, shallowly notched, 2-ridged. Column *c.* 1 mm. *Philippines*. G2. Summer.

2. D. glumaceum Lindley (*Platyclinis glumacea* (Lindley) Bentham). Illustration: *Botanical Magazine*, 4853 (1855); Graf, Tropica, 724 (1978); Bechtel et al., Manual of cultivated orchid species, 203 (1981).

Epiphytic. Pseudobulbs 2–5 × 1–2 cm, ovoid, densely clustered, 1-leaved, fusiform, covered in red sheaths when young. Leaves 25–30 × 2–4 cm, narrowly lanceolate, acute, with a short stalk. Flowering stem to 40 cm, thread-like, arched, bearing many flowers; bracts spreading, papery, acute. Flowers 1.2–1.8 cm wide, 2-ranked, fragrant, off-white to pale yellow. Sepals 7–10 × 1–2 mm, linear-lanceolate, acute; petals 4–7 × 1 mm, linear-lanceolate. Lip 3–4 × 2 mm, 3-lobed, pale green, yellow or white, with 2 thick basal ridges; lateral lobes broadly triangular, rounded or obtuse, central lobe circular to broadly ovate, curved downwards. Column erect. *Philippines*. G2. Spring.

D. latifolium Lindley (*Platyclinis latifolia* (Lindley) Hemsley). Illustration: *Orchid Review* **77**: 155 (1969). Flowers greenish yellow. Lip striped with brown; lateral lobes linear to narrowly triangular, acute, deeply and finely toothed; central lobe obovate, obtuse. *Philippines*. G2. Spring.

3. D. cobbianum Reichenbach (*Platyclinis cobbiana* (Reichenbach) Hemsley). Illustration: Sheehan & Sheehan, Orchid genera illustrated, 83 (1979); Williams et al., Orchids for everyone, 187 (1980); Bechtel et al., Manual of cultivated orchid species, 203 (1981).

Epiphytic. Pseudobulbs 2.5–8 × 1–2 cm, narrowly conical, tapering above, 1-leaved. Leaves 6–35 × 2.5–6 cm, lanceolate to oblong, acuminate, softly leathery. Flowering stem 35–50 cm, slender, arched; bracts 4–7 mm, papery. Flowers 1.2–1.8 cm wide, in 2 twisted ranks, slightly spaced out, fragrant. Sepals and petals 7–10 × 2–3 mm, oblong to elliptic, obtuse, white. Lip 6–8 × 3 mm, wedge- or fan-shaped, yellow to orange; callus oblong. *Philippines*. G2. Autumn.

37. THUNIA Reichenbach

J.C.M. Alexander

Terrestrial. Stems tall, tufted, cane-like, sheathed below, densely leaved above. Leaves long, acuminate, slightly bluish, spreading, in 2 opposite ranks, rolled when young. Flowers densely clustered in pendent, terminal racemes; bracts large, boat-shaped, white to pale green. Sepals and petals free, similar. Lip unlobed, rolled around column, with 5–7 narrow fringed keels and a short obtuse spur. Column 2-winged at apex. Pollinia 8.

A genus of about 6 species from India, Burma and SE Asia, which could all be regarded as colour variants of a single species. They perform best if repotted annually on the appearance of new growth, and should then be kept shaded and slightly moist. Once the new growth is well established more water can be given. After leaf-fall plants should be kept cool and dry. Propagation by division of cuttings.

1. T. marshalliana Reichenbach. Illustration: Shuttleworth et al., Orchids, 43 (1970); Skelsey, Orchids, 140 (1979); Bechtel et al., Manual of cultivated orchid species, 276 (1981).

To 150 cm. Leaves 15–20 cm, lanceolate, acuminate, pale green. Flowers 5–12, to 12.5 cm wide, fragrant; bracts to 5 cm, white, boat-shaped. Sepals and petals 6.5–7.5 × 1.3–1.5 cm, lanceolate to ovate, acuminate, white. Lip 5–6 × 4.5 cm, broadly elliptic to oblong, yellow with a white border, margin irregular and fringed; blade with orange, branching, hairy ridges; spur *c.* 1 cm, shallowly 2-lobed. Column 2–2.5 cm. *Burma, Thailand & S China*. G2. Summer.

T. bracteata (Roxburgh) Schlechter (*T. alba* Reichenbach; *Phaius albus* Wallich). Illustration: *Botanical Magazine*, 3991 (1843); Graf, Tropica, 754 (1978). Robust, to 60 cm. Flowers 4–10, to 7.5 cm wide; lip less than 4 cm wide, oblong to ovate; ridges orange or purple. *N. India, Burma & Thailand*. G2. Summer.

T. bensoniae Hooker. Illustration: *Botanical Magazine*, 5694 (1868). Sepals and petals pale purple with darker tips. Lip white at base; blade dark purple with bright yellow, hairy ridges. *Burma*. G2. Summer.

38. PHOLIDOTA Hooker

J.C.M. Alexander

Epiphytic with creeping rhizomes. Pseudobulbs clustered, distant, or in branching chains, 1- or 2-leaved; leaves rolled when young. Flowering stems terminal, slender, erect, arched or pendent, several- to many-flowered; axis often zig-zag; bracts conspicuous, sometimes falling early. Flowers small, alternate in 2 ranks. Sepals and petals spreading or pointing forwards; sepals concave, lateral sepals often keeled; petals similar to sepals or narrower. Lip fused to column base, pouched below, with a small deflexed blade, often 2- or 3-lobed. Column short, winged around anther. Pollinia 4.

A genus of 31 species from India, China and SE Asia. Cultivation conditions are generally similar to those for the tropical species of *Coelogyne* (p. 479) though less water should be given from when the pseudobulbs are fully formed until new growth appears.

Literature: DeVogel, E.F., Revisions in Coelogyninae III. Thegenus Pholidota, *Orchid Monographs* **3**: 1–118 (1988).

1a. Pseudobulbs developed on top of older pseudobulbs, forming long, pendent chains **1. articulata**
b. Pseudobulbs developed side by side **2. pallida**

1. P. articulata Lindley. Illustration: Smith, Orchideen von Java **2**: t. 115 (1909); *Orchid Review* **85**: 300 (1977); Banerjee & Thapa, Orchids of Nepal, 122 (1978).

Pseudobulbs to 10 × 1.5 cm, cylindric, 2-leaved, developed on top of older pseudobulbs, forming long pendent, sometimes branched chains. Leaves 8–16 × 3–5 cm, elliptic, acute; leaf-stalk *c.* 1 cm. Flowering stem arched, *c.* 15 cm; bracts *c.* 1 cm, falling as flowers open. Flowers 10–20, *c.* 1.2 cm wide, fragrant, pink to dull brown. Sepals and petals similar, 7–8 mm, spreading, ovate to elliptic, acute or obtuse. Lip *c.* 5 mm; pouch with 5 yellow ridges, tapering to junction with blade; blade 3 × 5 mm, kidney-shaped, 2-lobed, orange at base. Column 3–4 mm. *Himalaya through SE Asia to Java*. G2. Spring–summer.

2. P. pallida Lindley (*P. imbricata* Lindley 1827, not 1824). Illustration: Shuttleworth et al., Orchids, 42 (1970); Sheehan & Sheehan, Orchid genera illustrated, 141 (1979); Bechtel et al., Manual of cultivated orchid species, 261 (1981).

Pseudobulbs 3–6 cm, conical, 1-leaved. Leaves 30 × 6 cm, lanceolate to elliptic, acute; leaf-stalk to 5 cm. Flowering stem to 30 cm, pendent, many-flowered; bracts 8–10 mm, persistent. Flowers 6–7 mm wide, off-white to pale pink. Upper sepal 5 × 3 mm; lateral sepals to 6 mm, fused below, keeled. Petals 4 × 1.5 mm, linear to oblong. Upper sepal and petals curved over column. Column almost circular, 3–4 mm. *Burma & S China to Australia*. G2. Spring–summer.

P. ventricosa (Blume) Reichenbach. Illustration: Smith, Orchideen von Java **2**: t. 114 (1909). Pseudobulbs 2-leaved. Leaf-stalks to 15 cm. Flowering stem erect; flowers 1–1.2 cm wide, greenish white. *Sumatra, Java & Borneo*. G2.

P. chinensis Lindley. Illustration: Sander, Orchids and their cultivation, edn 9, t. 35 (1979); Williams et al., Orchids for everyone, 192 (1980). Pseudobulbs spherical to oblong, 2-leaved. Flowering stem pendent; flowers 1.2–2 cm. Sepals and petals pinkish brown, petals linear; lip white. *China & Hong Kong*. G2. Spring–summer.

39. CALYPSO Salisbury

V.A. Matthews

Terrestrial, 5–20 cm. Rootstock a corm, producing a solitary leaf that dies down in winter. Leaf 3–10 cm, broadly ovate, pleated, rolled when young. Scape bearing 1 terminal flower. Sepals and petals 1.5–2 cm, lanceolate, spreading and ascending, often twisted, purplish pink. Lip 1.5–2.5 cm, inflated and saccate, with 2 small horns at the base, whitish to pale pink, marked with purple and with white or yellow hairs towards the base. Pollinia waxy, flat, in 2 pairs fixed to a detachable viscidium.

A genus with only 1 species, uncommon in cultivation. It succeeds best in a damp, semi-shaded position with plenty of leaf-mould.

1. C. bulbosa (Linnaeus) Oakes. Illustration: Luer, Native orchids of the United States and Canada, t. 95, 96 (1975); Williams et al., Orchids of Britain and Europe, 134 (1978).

N temperate regions. H2. Late spring–summer.

40. APLECTRUM (Nuttall) Torrey

V.A. Matthews

Terrestrial, to 55 cm. Rootstock a corm, producing a solitary leaf that dies down in winter. Leaf 10–20 cm, broadly elliptic, pleated, with silvery veins, rolled when young. Scape with 6–15 flowers. Sepals and petals 1–1.4 cm, oblanceolate, brown or yellowish brown, tinged with purple or brown. Lip 1–1.2 cm, 3-lobed, obovate, whitish with purple marks. Pollinia waxy, in 2 pairs, fixed to a detachable viscidium.

A genus with only 1 species, which is rather difficult in cultivation. It grows best in a shady, damp situation.

1. A. hyemale (Willdenow) Nuttall (*A. spicatum* (Walter) Britton et al.). Illustration: Luer, Native orchids of the United States and Canada, t. 88 (1975).

NE USA, SE Canada. H2. Late spring–early summer.

41. CALANTHE R. Brown

J. Cullen

Usually terrestrial. Pseudobulbs usually inconspicuous, more rarely large and grooved. Leaves large, thin, pleated, rolled when young, mostly evergreen, deciduous in some species with large pseudobulbs. Racemes arising from the leaf axils, each with a long axis, which is usually hairy. Bracts conspicuous, persistent or soon falling, usually hairy. Sepals and petals similar, spreading, usually directed upwards, often hairy outside. Lip simple or 3-lobed, usually spurred, with warts or ridges at the base. Column united to the lip for most of its length (except in No. 13). Pollinia 8, in 2 groups of 4, attached to a common base.

A genus of about 260 species with a distribution extending from Madagascar and the Mascarene Islands to the Himalaya, SE Asia generally, Japan and Australia. Many hybrids are cultivated and *Calanthes* were among the first orchids to be deliberately hybridised.

Two types of plant occur, deciduous and evergreen. Both should be grown in terrestrial compost. Deciduous species should be given some shade and abundant moisture when growing; after the fall of the leaves they should be kept cool and dry. The evergreen species require more shade, and should not be dried out.

Literature: Seidenfaden, G., Orchid genera in Thailand I: Calanthe, *Dansk Botanisk Arkiv* **29**(2): (1975); Cribb, J. & Thomas, S., The genus Calanthe in Mexico, *Orquidea* **13**: 227–32 (1993).

Pseudobulbs. Conspicuous: **1, 2, 3**; inconspicuous: **4, 5, 6, 7, 8, 9, 10, 11, 12, 13**.

Leaves. Fallen at flowering: **1**, **2**, **3**; present at flowering: **4**, **5**, **6**, **7**, **8**, **9**, **10**, **11**, **12**, **13**.
Raceme. Arising from pseudobulb: **1**, **2**, **3**; arising from a rosette of leaves: **4**, **5**, **6**, **7**, **8**, **9**, **10**, **11**, **12**; arising from the lower part of an elongate leafy stem: **13**. Corymbose: **8**.
Lip. Simple: **1**; 3-lobed: **2**, **3**, **4**, **5**, **6**, **7**, **8**, **9**, **10**, **11**, **12**, **13**. Without spur: **11**, **12**, **13**; spur 2–12 mm: **9**, **10**; spur 1.5–2.9 cm: **1**, **2**, **3**, **6**, **7**, **8**; spur 3–5 cm: **4**. With central lobe itself 2-lobed or notched: **2**, **3**, **4**, **5**, **6**, **8**, **9**, **10**, **12**, **13**; with central lobe entire: **7**, **8**, **9**, **10**, **11**. Free from column: **13**.

1a. Bracts falling as soon as the flower opens; lip free from the column; plant with conspicuous leafy stem **13. gracilis**
b. Bracts persistent; lip attached to column; leafy stem inconspicuous, leaves in a rosette or borne on pseudobulbs 2
2a. Plants with large pseudobulbs, leafless when flowering 3
b. Plants with inconspicuous pseudobulbs, leafy when flowering 5
3a. Lip undivided, without any lateral lobes **1. rosea**
b. Lip divided, with 2 distinct lateral lobes 4
4a. Sepals, petals and usually the lip white; sepals and petals 2 cm or more **2. vestita**
b. Sepals, petals and lip pink; sepals and petals to 1.5 cm **3. rubens**
5a. Spur absent 6
b. Spur present, long or short 7
6a. Sepals and petals spreading; upper surface of lip with 3 conspicuous ridges **11. tricarinata**
b. Sepals and petals reflexed; upper surface of lip without ridges **12. reflexa**
7a. Spur 2–12 mm 8
b. Spur 1.5–5 cm 9
8a. Spur to 3 mm, blunt, oblong, straight **10. brevicornu**
b. Spur 5–12 mm, pointed, tapering, curved **9. discolor**
9a. Lip with the central lobe itself unlobed **7. cardioglossa**
b. Lip with the central lobe itself lobed or notched 10
10a. Flowers all white, except for yellow warts on the base of the lip; raceme corymbose **8. triplicata**
b. Flowers with at least the sepals and petals green, yellow or violet; raceme not corymbose 11
11a. Lip white, the central lobe itself deeply lobed, its lobes oblong, diverging at a wide angle; sepals and petals greenish or yellow **6. herbacea**
b. Lip violet or greenish tinged with purple, the central lobe notched, its lobes rounded, scarcely diverging; sepals and petals usually violet, rarely greenish 12
12a. Spur 2–2.7 cm, broadening towards its apex **5. sylvatica**
b. Spur 3–5 cm, not broadening towards its apex **4. masuca**

1. C. rosea (Lindley) Bentham & Hooker. Figure 24(1), p. 487. Illustration: *Dansk Botanisk Arkiv* **29**(2): 41 (1975).

Pseudobulbs 10–15 cm, conspicuous, ovoid to oblong, grooved, narrowed just below the apex. Leaves lanceolate, fallen by flowering time. Raceme to 45 cm with 7–12 flowers. Bracts persistent, 1.5–2 cm. Flowers 4–5 cm in diameter, red-pink. Sepals and petals lanceolate, acute. Lip simple, larger than the petals, its base rolled around the column. Spur *c.* 1.5 cm, straight, oblong. *Thailand*. G1–2. Winter.

2. C. vestita Lindley. Figure 24(2), p. 487. Illustration: *Botanical Magazine*, 4671 (1852); *Dansk Botanisk Arkiv* **29**(2): 29 & t. 4 (1975); Bechtel et al., Manual of cultivated orchid species, 175 (1981).

Pseudobulbs to 8 cm, ovoid, grooved. Leaves lanceolate, fallen by flowering time. Raceme to 70 cm with 6–12 flowers. Bracts 2–3 cm, persistent. Flowers 6–6.5 cm in diameter, white or cream. Lip white blotched with yellow or rarely entirely pink, 3-lobed, the lateral lobes about the same size as the divergently notched central lobe. Spur 1.8–2.5 cm, tapering, curved. *SE Asia, from Thailand to Sulawesi*. G2. Winter–spring.

Variable in the precise coloration of the lip, a character formerly used in the recognition of varieties (cultivars). 'Regnieri' (*C. regnieri* Reichenbach), which has a pink lip, is frequently grown.

3. C. rubens Ridley. Figure 24(3), p. 487. Illustration: *Dansk Botanisk Arkiv* **29**(2): 33 & t. 5 (1975); Bechtel et al., Manual of cultivated orchid species, 175 (1981).

Similar to *C. vestita* but flowers *c.* 3 cm in diameter, pink. *Thailand, NE Malaysia*. G2.

4. C. masuca (Don) Lindley. Figure 24(4), p. 487. Illustration: Botanical Magazine, 4541 (1850); *Dansk Botanisk Arkiv* **29**(2): t. 2 (1975).

Pseudobulbs inconspicuous. Leaves 20–30 × 7–11 cm, narrowly elliptic. Raceme arising from the leaf rosette, to 70 cm, with many flowers. Bracts persistent, 1.3–2 cm. Flowers to 6 cm in diameter, pale reddish purple. Lip as long as the petals, 3-lobed, with oblong lateral lobes and a somewhat notched central lobe, deep reddish violet with yellow warts towards the base. Spur 3–5 cm, curved. *Himalaya, SW China, Thailand & Malaysia*. G1–2. Summer.

This species is now usually considered part of *C. sylvatica*. However, the distinctions between them, though imprecise, have some horticultural importance, so the 2 are maintained here separately.

5. C. sylvatica (Thouars) Lindley. Figure 24(5) p. 487. Illustration: Bechtel et al., Manual of cultivated orchid species, 175 (1981).

Pseudobulbs inconspicuous. Leaves 15–25 × 5–7 cm, narrowly elliptic. Raceme arising from the leaf rosette, 20–60 cm, with numerous flowers. Bracts persistent, 1.5–3 cm. Flowers 3–4 cm in diameter, usually reddish purple, more rarely greenish with the lip tinged with purple. Lip as long as the petals, 3-lobed, the lateral lobes small, oblong, the central reversed-heart-shaped, notched at the apex. Spur 2–2.7 cm, curved, broadening towards the apex. *N. Madagascar, Mascarene Islands*. G2. Spring–summer.

6. C. herbacea Lindley. Figure 24(6) p. 487. Illustration: *Annals of the Royal Botanic Garden Calcutta* **5**: t. 44 (1895).

Pseudobulbs inconspicuous. Leaves 20–25 × 5–8 cm, elliptic. Raceme arising from the leaf rosette, 30–40 cm, with numerous flowers. Bracts persistent, 1.8–2.5 cm. Sepals and petals greenish or yellowish. Lip longer than the petals, white with yellow warts at the base, 3-lobed, the lateral lobes oblong, the central deeply and divergently 2-lobed. Spur straight or S-shaped, 2–3 cm. *N India*. G1–2.

C. izu-insularis (Satomi) Ohwi. Illustration: Maekawa, Wild orchids of Japan, t. 142 (1979); *Botanical Magazine* **18**(2): 417 (2001). Similar but with pink sepals and petals, the lip white flushed with violet and the spur 1.6–2 cm. *Japan*.

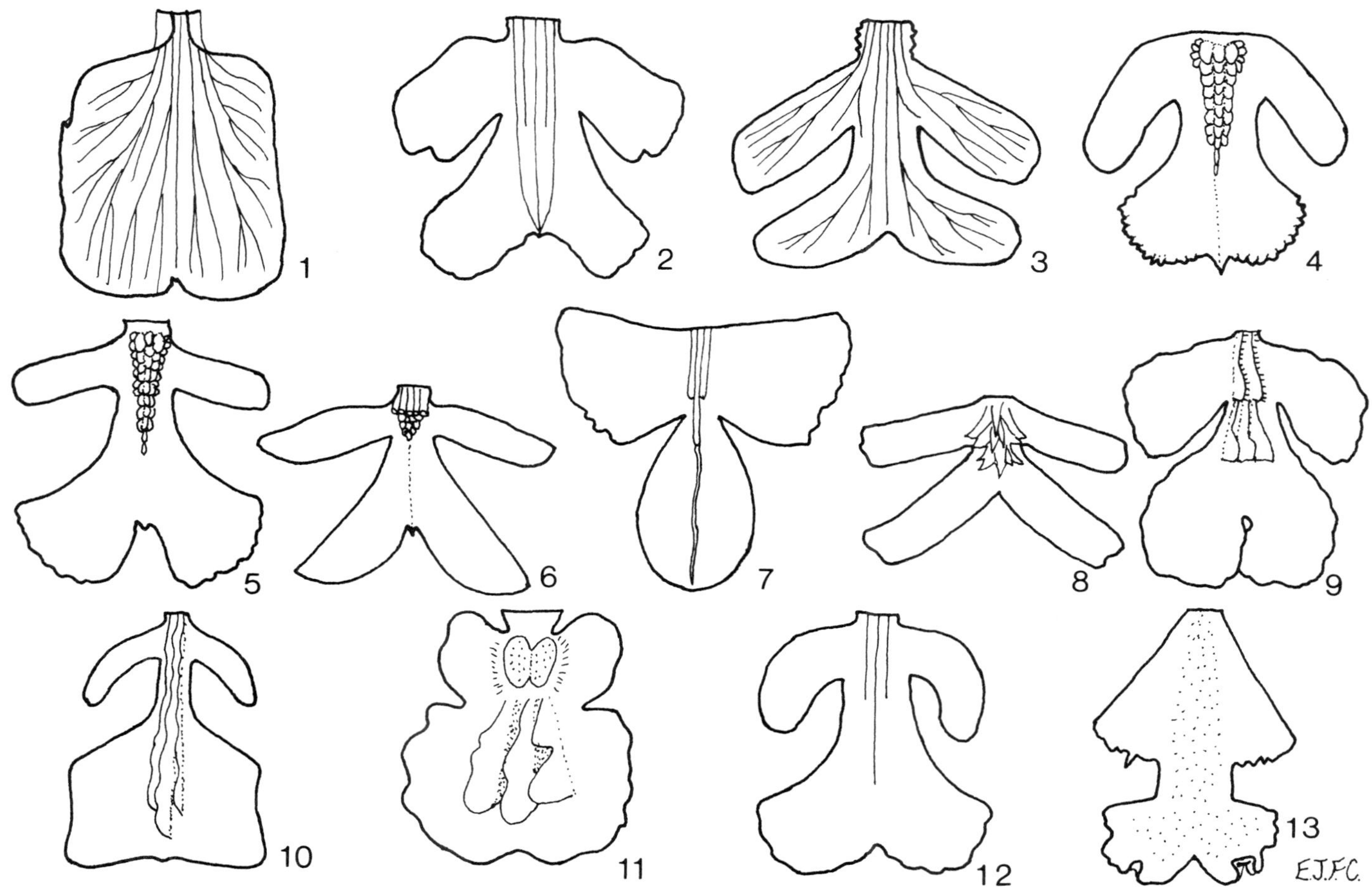

Figure 24. Lips of *Calanthe* species. 1, *C. rosea*. 2, *C. vestita*. 3, *C. rubens*. 4, *C. masuca*. 5, *C. sylvatica*. 6, *C. herbacea*. 7, *C. cardioglossa*. 8, *C. triplicata*. 9, *C. discolor*. 10, *C. brevicornu*. 11, *C. tricarinata*. 12, *C. reflexa*. 13, *C. gracilis*. (Compiled from various sources; not to scale.)

7. C. cardioglossa Schlechter. Figure 24 (7). Illustration: *Dansk Botanisk Arkiv* **29** (2): 37 & t. 6(1975).

Pseudobulbs inconspicuous. Leaves to 30 × 11 cm, lanceolate to elliptic. Raceme arising from the leaf rosette, to 70 cm, with 10–20 flowers. Bracts persistent, 1.5–2 cm. Flowers to 2 cm in diameter, pale reddish purple. Lip about as long as the petals, 3-lobed, with reversed-triangular, upcurving lateral lobes that are pale violet spotted darker, and an entire, diamond-shaped or elliptic central lobe. *Vietnam, Laos, Kampuchea & Thailand*. G2. Winter.

8. C. triplicata (Willemet) Ames (*C. veratrifolia* (Willdenow) R. Brown; *C. furcata* Lindley). Figure 24(8). Illustration: *Dansk Botanisk Arkiv* **29**(2): 16 & t. 1 (1975); Maekawa, Wild orchids of Japan in colour, t. 150 (1979); Bechtel et al, Manual of cultivated orchid species, 175 (1981).

Pseudobulbs inconspicuous. Leaves elliptic, to 40 × 13 cm. Raceme arising from the leaf rosette, to 1 m, the flowers corymbosely arranged at the top. Bracts persistent, 8–15 mm. Flowers 3–5 cm in diameter, white except for the yellow warts on the lip. Lip longer than the petals, 3-lobed, the lateral lobes oblong, the central deeply and divergently lobed. Spur 1.5–2.5 cm, curved, tapering. *SE Asia, Japan & E Australia*. G1–2. Spring.

9. C. discolor Lindley (*C. lurida* Decaisne; *C. amamiana* Fukuyama). Figure 24(9), p. 487. Illustration: *Botanical Magazine*, 7026 (1888); Lin, Native orchids of Taiwan, t. 46 (1975); Maekawa, Wild orchids of Japan in colour, t. 141, 143 (1975); Bechtel et al., Manual of cultivated orchid species, 174(1981).

Pseudobulbs inconspicuous. Leaves narrowly elliptic, 17–23 × 4–7 cm. Raceme arising from the leaf rosette, to 40 cm, with many flowers. Bracts persistent, 4–8 mm. Flowers 3–5 cm in diameter, reddish purple or yellow. Lip about as long as the petals, 3-lobed, the lateral lobes oblong or broadening towards their apices, the central reversed-heart-shaped, deeply notched, bearing 3 prominent ridges towards the base. Spur 5–12 mm, curved, tapering. *Japan, Korea & Taiwan*. G1–2. Spring.

C. striata R. Brown (*C. discolor* var. *flava* Yatabe; *Limodorum striatum* Banks 1791, not of Thunburg, 1784; *C. bicolor* Lindley; *C. kawakamii* Hayata; *C. sieboldii* Decaisne, and various combinations of these names at the level of variety or forma.) Illustration: Botanical Magazine **18**(2): 418 (2001). Raceme 40–80 cm. Flowers with yellow or yellow-brown sepals and petals, lip yellow. Central lobe of lip deeply 3-lobed with the central lobe bearing 2 pubescent ridges at the base, and 3–5 longitudinal ridges, the longest of which reach the apex. Spur 6–8 mm long. *Japan, Korea, Taiwan*. H5–G1.

A very vigorous selection of this species originating from Taiwan has been marketed as *C. 'kawakamiense'*. This name is evidently a corruption of the synonym *C. kawakamii*.

The name *C. bicolor* has been used erroneously for both the natural and garden hybrid of *C. discolor* × *striata*, for which the name *C.* Takane grex is available. It produces flowers that are smaller than *C. striata*, with a shorter, broader, bi-lobed mid-lobe to the lip, and distinctly bicoloured with both inner and outer surfaces of the sepals and petals brown, and the lip yellow.

10. C. brevicornu Lindley. Figure 24(10), p. 487. Illustration: *Annals of the Royal Botanic Garden Calcutta* **8**: t. 227 (1898); Pearce & Cribb, Orchids of Bhutan, t.10 (2002).

Pseudobulbs inconspicuous. Leaves 20–30 × 5–6 cm, narrowly elliptic. Raceme arising from the leaf rosette, to 50 cm, with many flowers. Bracts persistent, *c.* 1.5 cm. Flowers 2.5–4 cm in diameter, sepals and petals brown. Lip a little shorter than the petals, 3-lobed, the lateral lobes rounded, the central reversed-heart-shaped, deeply notched, purplish red with white margins, and with 3 yellow ridges near the base. Spur to 3 mm, oblong, blunt, straight. *Nepal, NE India (Sikkim)*. G1. Summer.

11. C. tricarinata Lindley. Figure 24(11), p. 487. Illustration: *Annals of the Royal Botanic Garden Calcutta* **8**: t. 223 (1898); *Botanical Magazine*, 8803 (1919); Lin, Native orchids of Taiwan, t. 49, 50 (1975); Maekawa, Wild orchids of Japan in colour, t. 138 (1979); *Botanical Magazine* n.s., 416 (2001).

Pseudobulbs inconspicuous. Leaves 18–30 × 5–9 cm, elliptic. Raceme arising from the leaf rosette, 30–40 cm, with several flowers. Bracts 8–20 mm, persistent. Flowers 2.5–4 cm in diameter, sepals and petals greenish yellow. Lip about as long as the petals, red-brown, 3-lobed, the lateral lobes small, rounded, the central reversed-heart-shaped, notched, margins wavy, bearing 3 very pronounced ridges. Spur absent. *Himalaya, China, Japan & Taiwan*. G1–2.

12. C. reflexa Maximowicz. Figure 24 (12), p. 487. Illustration: *Botanical Magazine*, 9648 (1943); Lin, Native orchids of Taiwan, t. 44, 45 (1975); Maekawa, Wild orchids of Japan in colour, t. 137 (1979).

Like *C. tricarinata* but generally smaller; sepals and petals reflexed, each narrowed into a bristle-like point, lip pale purple or white with oblong lateral lobes and an obovate, scarcely wavy, irregularly toothed but not notched central lobe that bears no ridges. *Japan, Korea (Quelpaert Islands) & Taiwan*. G1–2. Summer.

13. C. gracilis (Lindley) Lindley. Figure 24(13), p. 487. Illustration: *Botanical Magazine*, 4714 (1853); *Annals of the Royal Botanic Garden Calcutta* **8**: t. 222 (1898).

Pseudobulbs inconspicuous. Leafy stems to 40 cm. Leaves 15–20 × 3–4 cm, narrowly elliptic. Racemes arising laterally near the bases of the leafy stems and shorter than them, with 10–15 flowers. Flowers 2–3 cm in diameter, greenish yellow. Sepals and petals spreading. Lip free from the column, white with a yellow band, 3-lobed, the lateral lobes obliquely triangular, the central finely irregularly toothed and deeply notched. Spur absent. *Himalaya*. G1–2. Autumn.

Var. **venusta** (Schlechter) Maekawa (*C. venusta* Schlechter). Illustration: Maekawa, Wild orchids of Japan in colour, t. 136 (1979). Larger, with stems to 1 m, leaves 20–40 × 4–7 cm, sepals longer than the petals, lateral lobes of the lip ovate. *Japan*.

42. GASTRORCHIS Schlechter

J.M.H. Shaw

Terrestrial or rarely epiphytic herbs with pseudobulbs. Leaves pleated, not joined, petiolate. Inflorescence a raceme arising from the base of a pseudobulb. Sepals and petals free, spreading to hooded and incurved. Lip free from column, weakly 3-lobed, with the lateral lobes usually enfolding the base of the column. Spur absent.

A genus of 10 species confined to Madagascar, and included in *Phaius* by some authors, but differing in its basally saccate lip and raised and often hairy callus. Cultivation as for *Phaius* (see below).

Literature: Hermans, J. & C., Du Puy, D., Cribb, P. & Bosser, J., Orchids of Madagascar, edn 2, 187–90 (2007).

1. **G. humblotii** (Reichenbach) Schlechter (*Phaius humblotii* Reichenbach). Illustration: Cogniaux & Goossens, Dictionnaire iconographique des orchidées, Phaius, t. 1 (1898); Veitch, Manual of orchidaceous plants **6**: 12 (1890); Hermans et al., Orchids of Madagascar, edn 1, t. 27B (1999) & edn 2, t. 38D (2007).

Terrestrial. Pseudobulbs 4–6 × 3–4 cm, spherical to conical, ridged, few-leaved. Leaves 30–60 × 6–10 cm, broadly lanceolate, acuminate, membranous. Flowering stem robust, 30–80 cm, from base of pseudobulb, bearing 7–12 flowers; flowers 5–6 cm wide. Sepals and petals similar, 2.5–3 × 1.5–2 cm, obovate to elliptic, acuminate, pale pink with darker streaks of purple. Lip 3-lobed, lacking a spur, strongly crisped; lateral lobes oblong, erect to spreading, pinkish purple to reddish brown; central lobe semicircular to oblong, pink with white streaks; callus yellow, divided into 2 tooth-like ridges. Column club-shaped. *Madagascar*. G2. Summer.

G. tuberculosa (Thouars) Schlechter (*Phaius tuberculosus* (Thouars) Blume). Illustration: *Botanical Magazine*, 7307 (1893); *Proceedings of 8th world orchid conference*, t. 7 (1976); Bechtel et al., Manual of cultivated orchid species, 258 (1981). Pseudobulbs *c.* 2.5 cm × 5 mm. Sepals and petals white; lip with 3 yellow keels; lateral lobes yellow with red spots. *Madagascar*. G1.

43. PHAIUS Loureiro

J.C.M. Alexander

Terrestrial, rarely epiphytic, robust. Pseudobulbs spherical to ovoid (sometimes stem-like), bearing 3–10 leaves. Leaves large, lanceolate to elliptic, alternate, stalked, pleated, persistent, rolled when young. Flowering stems lateral, from pseudobulb bases or lower axils. Flowers few to many in loose or dense racemes. Sepals and petals similar, fleshy, free. Lip simple or 3-lobed with lateral lobes erect or curved around column, spurred, shortly fused to column at base. Column short. Pollinia 8, in 2 groups of 4.

A widespread genus of about 40 species ranging from tropical Africa eastwards to the Himalaya, N India, China & Japan, and south through SE Asia to N Australia and the Pacific islands. The spurless species are here treated as the separate genus *Gastrorchis* Schlechter. The terrestrial species should be grown in large, well-drained pots and kept moist and partially shaded during the growing season, though overhead spraying should be avoided. During the resting period small amounts of moisture can still be given and plants can be exposed to stronger light. The epiphytic species require more shading and are easily harmed by draughts.

1a. Pseudobulbs stem-like, narrowly fusiform to cylindric **1. mishmensis**
b. Pseudobulbs spherical, conical or ovoid 2
2a. Flowers 10 cm wide or more; sepals and petals lanceolate, more than 3 times as long as broad **2. tankervilleae**
b. Flowers 8 cm wide or less; sepals and petals obovate, elliptic or oblong, less than 3 times as long as broad **3. flavus**

1. P. mishmensis (Lindley) Reichenbach. Illustration: *Botanical Magazine*, 7497 (1896); Bose & Bhattacharjee, Orchids of India, 431 (1980); Bechtel et al., Manual of cultivated orchid species, 258 (1981).

Terrestrial. Stems to 140 cm, narrowly fusiform to cylindric, fleshy, bearing 6–8 leaves. Leaves 15–30 × 8–14 cm, lanceolate to ovate or oblong, acuminate. Flowering stems 1 or more from lower axils of pseudobulbs, to 60 cm, bearing 7–10 flowers; flowers 5–6 cm wide. Sepals and petals similar, pale pink to darker reddish brown; sepals to 3.5 × 1 cm, oblong, acuminate; petals somewhat smaller. Lip 3 × 2.5 cm, 3-lobed; lateral lobes oblong, rounded, pink speckled with red or purple, curved around column; central lobe square to oblong, shallowly lobed, with a central hairy ridge, pink speckled with red or purple, margins yellow. Spur 1.5–1.7 cm, tapering, curved. Column 1.2 cm, straight. *NE India, Burma, Thailand, Taiwan & Philippines*. G2. Summer–autumn.

2. P. tankervilleae (Banks) Blume (*P. bicolor* Lindley; *P. grandifolius* Loureiro; *P. blumei* Lindley; *P. wallichii* Hooker). Illustration: *Botanical Magazine*, 4078 (1844), 7023 (1888); Sheehan & Sheehan, Orchid genera illustrated, 133 (1979); Bechtel et al., Manual of cultivated orchid species, 258 (1981).

Terrestrial. Pseudobulbs 2.5–6 cm, conical to ovoid, few-leaved. Leaves 30–120 × 9–20 cm, elliptic to lanceolate, acuminate, membranous. Flowering stem lateral, from base of pseudobulb, to 2 m, bearing 10–20 flowers; flowers 10–12.5 cm wide. Sepals and petals similar, 4.5–6.5 × 1–1.3 cm, lanceolate, acuminate, reddish or yellowish brown, margin paler, greenish red on back. Lip 5 × 4.5 cm, 3-lobed; lateral lobes oblong, rounded, curved around column, pink to purplish red, yellow below; central lobe broadly ovate, acute or truncate, pink and yellow; throat yellow and purplish red; callus oblong; spur 1 cm, cylindric, curved. Column 2 cm, club-shaped. *C China, N India, Sri Lanka, through SE Asia to N Australia*. G2. Spring–summer.

3. P. flavus (Blume) Lindley (*P. maculatus* Lindley). Illustration: *Botanical Magazine*, 3960 (1842); Cogniaux & Goossens, Dictionnaire iconographique des orchidées, Phaius, t. 3 (1904); Schuiteman & de Vogel, Orchid genera of Thailand, Laos, Cambodia and Vietnam, t. 95 (2000).

Leaves with white spots. Sepals and petals yellow to greenish yellow, oblong, obtuse. Lip spurred, greenish yellow with reddish brown streaks in the heavily crisped margin. *SE Asia*. G2. Spring

44. SPATHOGLOTTIS Blume

J.C.M. Alexander

Terrestrial. Pseudobulbs conical to ovoid, sheathed, bearing 1–several leaves. Leaves narrowly lanceolate, tapering gradually to the stalk, pleated, prominently veined, rolled when young. Flowering stems robust, borne laterally from basal leaf axils, unbranched, basally sheathed, many-flowered. Sepals spreading, free; petals similar, often wider. Lip deeply 3-lobed; lateral lobes erect or spreading; central lobe lanceolate to spathulate, with a long narrow claw and a prominent basal callus and small basal teeth. Column slender, arched. Pollinia club-shaped, 8, in 2 groups of 4.

A genus of about 40 species distributed from N India through SE Asia and C China to Australia and the Pacific islands. *Spathoglottis* species perform best if repotted every spring in well-drained terrestrial compost. Plenty of water can be supplied through the summer but this should be decreased when the plants are dormant. Growing temperatures depend on the country of origin. Propagation is effected by dividing sections of rhizome with pseudobulbs.

Literature: Cribb, P.J. & Tang, C.Z., Spathoglottis in Australia and the Pacific Islands, *Kew Bulletin* **36**(4): 721–9 (1981). Spowart, R., Creating new forms of Spathoglottis, *Philippine Orchid Review* **12** (2): 2–4 (2003). Spowart, R., Color in Spathoglottis breeding, in *Proceedings of the Flora Filipina Conference* (2006), & *Philippine Horticultural Primer & Directory* **1** (1): 40–3 (2006).

1a. Flowers yellow **1. aurea**
b. Flowers purple **2. plicata**

1. S. aurea Lindley. Illustration: Holttum, Flora of Malaya **I**: Orchids, edn 3, 162 (1964); Shuttleworth et al., Orchids, 86 (1970).

Pseudobulbs to 20 cm, ovoid, with 2–4 leaves. Leaves to 100 × 5 cm. Flowering stem to 60 cm or more, bearing 4–10 flowers; flowers 6–7 cm wide, deep yellow; bracts 1–2 cm, spathulate, concave, spreading. Sepals and petals 2.5–3 × 1.2–1.5 cm, elliptic to ovate, obtuse, slightly concave. Lip 2.5–3 × 2.5–3 cm, deeply divided into 3 main lobes; lateral lobes oblong, rounded, broadest at base, erect; central lobe 3–4 mm wide, lanceolate, rounded, with 2 small triangular lobes or teeth at base; callus of 2 erect triangular teeth. Lateral lobes of lip, callus and base of central lobe spotted with crimson. Column 1–5 cm, narrowly club-shaped, arched. *Malay peninsula*. G2. Summer.

S. kimballiana Hooker. Illustration: *Botanical Magazine*, 7443 (1895); *Gardeners' Chronicle* **4**: 93 (1888).

Flowers to 7.5 cm wide, pale yellow. Lateral lobes of lip broadly spathulate. Sepals flecked with purple beneath. *Borneo*. G2. Spring.

S. fortunei Lindley. Illustration: *Edwards's Botanical Register* **31**: t. 19 (1845); Schlechter, Die Orchideen, edn 2, 308 (1927).

Leaves 15–20 × 1.5–2 cm. Bracts on flowering stem narrowly ovate, attenuate, not spreading. Flowers 3–4 cm wide. Lateral lobes of lip oblong, rounded, strongly sickle-shaped. *Hong Kong*. G2. Winter.

2. S. plicata Blume. Illustration: Holttum, Flora of Malaya **I**: Orchids, edn 3, 6 (1964); Shuttleworth et al., Orchids, 86 (1970); Bechtel et al., Manual of cultivated orchid species, 274 (1981).

Pseudobulbs to 7 × 5 cm, with 4–7 leaves. Leaves 30–120 × 2–7 cm. Flowering stem to 150 cm, bearing 5–25 flowers; bracts lanceolate, acuminate; flowers 3.5–4.5 cm wide, purple to pinkish purple. Sepals 2.3 × 1.2 cm, elliptic, acute; lateral sepals asymmetric; petals 2 × 14 cm, elliptic to ovate, obtuse. Lip 1.7 × 1.8 cm, 3-lobed; lateral lobes ovate to narrowly spathulate, rounded to truncate; central lobe spathulate, long-clawed, with 2 small teeth at base; callus triangular to heart-shaped, yellow. Column 1.2 cm, club-shaped, arched. *SE Asia from Sumatra to the Philippines (?India); naturalised in Florida, Hawaii and parts of tropical Africa.* G2. Flowers continuously.

S. vieillardii Reichenbach. Illustration: *Botanical Magazine*, 7013 (1888); Graf, Tropica, 754 (1978). Larger in all parts than *S. plicata;* flowers pale purple, to 6 cm wide. Lateral lobes of lip pinkish or brownish purple. *Malaysian & Indonesian islands.* G2. Autumn.

45. BLETIA Ruiz & Pavon

J.C.M. Alexander

Terrestrial, rarely epiphytic. Pseudobulbs almost spherical, slightly flattened, corm-like, bearing 1–several leaves. Leaves linear to lanceolate, pleated, rolled when young, deciduous. Flowers many, in loose racemes or panicles; flowering stems produced from lower nodes of pseudobulbs. Sepals and petals similar, free, spreading; lateral sepals closely pressed together at base but not fused. Lip free, simple or 3-lobed; lateral lobes erect or curved around column, central lobe broad, spreading, notched or more deeply 2-lobed, with 3–7 longitudinal toothed or scalloped keels. Column arched, apically winged. Pollinia 8, hard and compressed.

About 35 species ranging from Mexico, Florida and the West Indies south to Argentina. Several species now placed in other genera were originally described in *Bletia* and the name is still sometimes used for plants quite unlike the genus as currently recognised. *Bletia* species should be grown in terrestrial compost and repotted when new growth first appears; the pseudobulbs should be partially covered. Little water should be given once growth has ceased. Plants can be divided after flowering.

Literature: Dressler, R.L., Notes on Bletia, *Brittonia* **20**: 182–90 (1968).

1a. Lip 3.5 × 2–2.5 cm, central lobe deeply notched or 2-lobed **1. catenulata**

b. Lip 1.5–2 × 1.2–1.5 cm, central lobe almost square **2. purpurea**

1. B. catenulata Ruiz & Pavon (*B. sanguinea* Poeppig & Endlicher; *B. sherrattiana* Lindley; *B. watsonii* Hooker). Illustration: *Botanical Magazine*, 5646 (1867); Pabst & Dungs, Orchidaceae Brasilienses **1**: 229 (1975); Bechtel et al., Manual of cultivated orchid species, 170 (1981).

To 2 m. Pseudobulbs 2–6 cm, bearing 1–6 leaves in 2 opposite ranks. Leaves 20–100 × 1–10 cm, variable in shape, linear to lanceolate or elliptic, acuminate. Flowering stem to 2 m, bearing 3–15 flowers; flowers 5–6.5 cm wide. Sepals 3–3.5 cm × 7–10 mm, oblong, acute, red to pinkish purple. Petals 3–3.5 × 1.7–2 cm, ovate, rounded, darker than sepals. Lip 3.5 × 2–2.5 cm, circular to oblong, deeply 3-lobed; lateral lobes semicircular to oblong, erect to spreading, purple with darker veins; central lobe deeply notched to 2-lobed, rounded, pink or pinkish purple with 3–5 pale to bright yellow ridges. Column club-shaped. *Colombia, Ecuador, W Brazil & Bolivia.* G2. Summer.

2. B. purpurea (Lamarck) de Candolle (*B. verecunda* (Salisbury) R. Brown; *B. shepherdii* Hooker). Illustration: *Botanical Magazine*, 3319 (1834); Dunsterville & Garay, Venezuelan orchids illustrated **2**: 41 (1961); Luer, Native orchids of Florida, 227 (1972).

Terrestrial, rarely epiphytic, to 1.5 m. Pseudobulbs 2–4 cm, bearing 3–5 leaves. Leaves to 100 × 5 cm, linear-lanceolate, tapering, pale green. Flowering stem to 1.5 m, sometimes branched, bearing 3–80 flowers; flowers 2–2.5 cm wide, opening in succession. Sepals and petals 1.5–2.5 cm × 6–8 mm, pink to pinkish purple; upper sepal oblong to lanceolate; lateral sepals and petals ovate-lanceolate, acuminate, asymmetric; lateral sepals often with a central yellow stripe. Lip 1.5–2 × 1.2–1.5 cm, 3-lobed; lateral lobes triangular to diamond-shaped, scalloped, curved over column, purple; central lobe almost square, scalloped, purple, with 5–7 yellow scalloped ridges. Column white, 1–1.3 cm. *USA (Florida), Bahamas & Mexico, south through C America and the West Indies to Colombia, Venezuela & Guyana.* G2. Flowers throughout the year.

46. CHYSIS Lindley

J.C.M. Alexander

Epiphytic or on rocks. Pseudobulbs cylindric to fusiform or narrowly club-shaped, compound, fleshy, bearing several leaves in upper half, sheathed below. Leaves in 2 opposite rows, pleated, rolled when young, prominently veined, deciduous. Flowering stems lateral from lower leaf axils; flowers several in loose racemes. Sepals and petals free, spreading; lateral sepals arising obliquely from column foot forming a mentum. Lip 3-lobed, basally fused to column foot; lateral lobes broad, erect; central lobe spreading or reflexed, entire or notched; base of lip with keels or thickened veins. Column short and thick, 2-winged, with a prominent foot. Pollinia 8, often fused to column causing self-fertilisation.

A genus of 8–10 species from Mexico, south through Central America to Venezuela and Peru. They should be grown in epiphytic compost in pots or baskets. A warm, shady, humid environment gives best results when the plants are actively growing although lower temperatures and much less water should be applied once growth ceases and the leaves fall.

Literature: Allen, P.H., A revision of the genus Chysis, *American Orchid Society Bulletin* **24**: 664–66 (1955).

1. C. aurea Lindley. Illustration: *Botanical Magazine*, 3617 (1837); Skelsey, Orchids, 101 (1979); Shuttleworth et al., Orchids, 87 (1970); Bechtel et al., Manual of cultivated orchid species, 183 (1981).

Epiphytic. Pseudobulbs to 50 × 4 cm, cylindric to fusiform, pendent, bearing 4–15 leaves. Leaves to 45 × 7 cm, lanceolate, acuminate (when young) to attenuate. Flowering stem to 45 cm, bearing 3–12 flowers; bracts *c.* 2 cm, nearly equal to ovary; flowers to 6 cm wide, yellow with reddish lines and spots on petals and lip. Upper sepal 3–3.5 × 1.5 cm, oblong, bluntly tapering; lateral sepals 2.5 × 1.7 cm, triangular, asymmetric, obtuse. Petals 2.5–3 × 1–1.5 cm, spathulate to narrowly diamond-shaped, obtuse or rounded. Lip 2–2.5 × 2.5–3 cm, 3-lobed, very fleshy and crisped, lateral lobes oblong to ovate, sickle-shaped, erect; central lobe ovate to circular. Callus of 5 fleshy, hairy ridges. Column white, with a prominent foot. *Mexico to Venezuela, Colombia & Peru.* G2. Spring.

Var. **bractescens** (Lindley) Allen (*C. bractescens* Lindley). Illustration: *Botanical Magazine*, 5186 (1860); Williams et al., Orchids for everyone, 184 (1980); Bechtel et al., Manual of cultivated orchid species, 183 (1981). Bracts on flowering stem 2.5 cm or more, leaf-like. Flowers 6–8 cm wide; sepals and petals white with yellow streaks. *Mexico, Guatemala & Belize.* G2. Winter–spring.

C. limminghii Linden & Reichenbach. Illustration: *Botanical Magazine*, 5265 (1861); Cogniaux & Goossens, Dictionnaire iconographique des orchidées, Chysis, t. 3 (1901). Sepals and petals white with purple or purplish brown streaks; lip margin not crisped. *C Mexico*. G2. Summer.

C. laevis Lindley. Illustration: Cogniaux & Goossens, Dictionnaire iconographique des orchidées, Chysis, t. 2 (1901); Ospina & Dressler, Orquideas de las Americas, t. 42 (1974). Bracts about half the ovary length. Flowers yellow to orange with streaks of reddish purple. Lip pale yellow with pinkish purple spots and blotches; callus ridges hairless. *Mexico & Costa Rica*. G2. Summer.

47. COELIA Lindley

J.C.M. Alexander & J.M.H. Shaw

Epiphytic or terrestrial, often on rocks; rhizomes creeping. Pseudobulbs spherical to ovoid, smooth, simple, bearing several leaves. Leaves linear to lanceolate, acute, pleated, rolled when young. Flowers few to many, in robust lateral racemes or spikes, on sheathed stems. Upper sepal free; lateral sepals fused to column foot, forming a mentum. Petals free, lip sharply bent or pouched at junction of blade and claw; blade shallowly 3-lobed; central lobe triangular, acute. Column long and slender; column foot equalling it in length. Pollinia 8, waxy.

A genus of 5 species from C America and the West Indies. Sometimes 4 species are separated as *Bothriochilus* Lemaire, but recent anatomical and molecular studies have resulted in uniting the two genera. Cultivation requirements are similar to those for *Lycaste* (p. 546), though *C. machrostachya*, having fairly hard pseudobulbs, benefits from a longer dry period.

Literature: Williams, L.O., The orchid genera Coelia & Bothriochilus, *Botanical Museum Leaflets, Harvard University* **8**: 145–8 (1940); Pridgeon, A.M., Una revision de los genera Coelia y Bothriochilus, Orquidea **7**: 57–94 (1978).

1a. Pseudobulbs 5 cm or less; flowers few, about 5 cm long **1. bella**
b. Pseudobulbs 6 cm or more; flowers many, 1–1.5 cm long **2. macrostachya**

1. C. bella (Lemaire) Reichenbach (*Bothriochilus bellus* Lemaire). Illustration: *Botanical Magazine*, 6628 (1882); *Orchid Review* **82**: 314 (1974).

Terrestrial, to 1 m. Pseudobulbs 3–5 × 1.5–2.5 cm, spherical to ovoid, surmounted by a short stalk, bearing 3–5 leaves. Leaves 15–60 × 1–2 cm, linear-lanceolate, acuminate, with 3–5 veins, pale green. Flowering stems to 20 cm, heavily sheathed, bearing 2–6 fragrant flowers; flowers 5 × 3.5–5 cm, funnel-shaped. Sepals elliptic to oblong, obtuse, off-white with purple tips; upper sepal 3.5 × 1 cm; lateral sepals 5 × 1 cm, mentum 2 cm. Petals 3.5 × 1 cm, oblong to spathulate, obtuse or rounded, off-white with a central patch of purple. Lip pouched; blade 4.5 × 1.5 cm with small lateral lobes; central lobe narrowly triangular, inner face an orange, granular callus. Column 1.5 cm. *Mexico, Guatemala & Honduras*. G2. Summer.

2. C. macrostachya (Lindley (*Bothriochilus macrostachyus* (Lindley) Williams). Illustration: *Botanical Magazine*, 4712 (1853); Shuttleworth et al., Orchids, 71 (1970); Ospina & Dressler, Orquideas de las Americas, t. 40 (1974).

Epiphytic, to 1 m. Pseudobulbs 6–10 cm, spherical to ovoid, surmounted by a short stalk, bearing 3–6 leaves. Leaves to 100 × 2.5–4 cm, linear-lanceolate, acuminate, fleshy. Flowering stems to 60 cm, sheathed. Flowers many in a dense cylindric spike, 10–15 × 5–7 mm, pale to dark pink or purple. Sepals fleshy, warty on outer face; upper sepal 8–10 mm, ovate-lanceolate; outer part of lateral sepals 1.5 cm, semi-ovate to triangular, inner part forming a mentum. Petals 9 mm, oblong-lanceolate, asymmetric, membranous. Lip pouched; blade 12 mm, shallowly 3-lobed; central lobe narrowly triangular, acute. Column 1.5 cm. *Mexico to Panama*. G2. Summer.

48. ARPOPHYLLUM Llave & Lexarza

J.C.M. Alexander

Epiphytic or on rocks, with creeping, simple or branched rhizomes. Stems cane-like, sheathed, 1-leaved. Leaves fleshy or leathery, folded when young. Flowers many, small in dense terminal racemes; flowering stem with large basal sheath. Sepals spreading; lateral sepals inflated below and fused to column; petals smaller than sepals. Lip uppermost (flower non-resupinate), usually longer than sepals and petals, inflated and pouched below. Column slightly arched, with small foot. Pollinia 8, waxy, pear-shaped.

A small genus of 5 species from tropical America. Floral differences between the species are slight and specific distinctions are made largely on vegetative characters. Cultivation conditions are generally similar to those for *Cattleya* (p. 498), although a simpler epiphytic compost of equal parts of sphagnum and tree-fern fibre is adequate. Good ventilation is important.

Literature: Garay, L., Synopsis of the genus Arpophyllum, *Orquidea* **4**: 16–19 (1974).

1. A. spicatum Llave & Lexarza. Illustration: *Botanical Magazine*, 6022 (1873); Shuttleworth et al., Orchids, 70 (1970); Graf, Tropica, 710 (1978).

To 75 cm. Stem flattened, almost concealed in sheaths. Leaf to 50 × 4 cm, oblong, slightly sickle-shaped, obtuse, strongly folded along midrib, fleshy. Inflorescence to 30 × 3 cm, cylindric; stalks and ovaries with black warts; basal sheath to 15 × 2 cm. Flowers 8–10 mm wide, purplish pink. Upper sepal 4.5–5.5 × 1.5–3 mm, oblong to ovate, obtuse or with a small point; lateral sepals 5–6 × 2–3 mm, elliptic to oblong, acute to rounded, 3-veined, asymmetric, inflated at base. Petals 5–5.5 × 1.5–2 mm, linear to narrowly elliptic, obtuse or rounded, 1-veined. Lip 5.5–6 × 3.5 mm, constricted above pouch; blade ovate, curved over column, 7-veined. Column 3.5–4 mm. *Mexico*. G2. Spring.

A. giganteum Lindley. Illustration: Dunsterville & Garay, Venezuelan orchids illustrated **6**: 50 (1976); Bechtel et al., Manual of cultivated orchid species, 168 (1981). To 1 m. Leaf leathery, flattened, folded only in basal part, straight-sided. Inflorescence to 18 cm long. *Mexico to Costa Rica; Jamaica & Colombia*. G2. Spring.

49. EPIDENDRUM Linnaeus

J. Cullen

Terrestrial or epiphytic. Stems usually elongate and leafy, more rarely pseudobulb-like with 1–2 leaves, erect or hanging. Leaves acute or notched at the apex, folded when young. Flowers in racemes or panicles that are rarely umbel-like, bracts small or conspicuous. Flowers resupinate or not, variable in size. Sepals and petals similar, spreading. Lip joined to the column for most of the length of the latter, variable in shape, entire or 3-lobed. Pollinia 4, laterally flattened, parallel.

A genus of about 1500 species from the New World. Its relationships with other genera are controversial: here, *Epidendrum* is restricted to plants with the column united for all or most of its length to the lip. Species with the lip and column mostly free from each other are included in *Encyclia* (p. 493), *Prosthechea* (p. 496) and *Barkeria* (p. 498).

Plants should be potted in early spring in an epiphytic compost, using the smallest size of pot possible. Propagation is by division or by means of the young plantlets produced by some species after the flowering spikes have been cut back.

Literature: Ames, O., Hubbard, F.T. & Schweinfurth, C., The genus Epidendrum in the United States and Middle America (1936); Withner, C.L. & Harding, P., Cattleyas and their relatives. The debatable Epidendrums (2004).

Main axis. Stem-like: **2, 3, 4, 7**; pseudobulb-like: **1, 5, 6**.
Inflorescence. Panicle with many flowers: **1**; spherical raceme with many flowers: **2, 3**; elongate raceme with up to 5 flowers: **4, 5, 6, 7**.
Sepals and petals. To 3 cm: **1, 2, 3**; more than 3 cm: **4, 5, 6, 7**.
Lip. Not lobed: **7**; lobes more or less equal, the lip forming a cross-shaped structure at the top of the column: **2, 3**; lobes unequal, not as above: **1, 4, 5, 6**.

1a. Lip not lobed, though sometimes fringed or notched at the apex **7. medusae**
b. Lip distinctly 3-lobed 2
2a. Central lobe of lip narrowly triangular, narrowly oblanceolate or linear, projecting between and beyond the lateral lobes 3
b. Central lobe of lip various but not as above 5
3a. Stem elongate, bearing several leaves along its length **4. nocturnum**
b. Stem short, pseudobulb-like, bearing 1–2 leaves at its apex 4
4a. Lateral lobes of the lip deeply fringed along their outer margins **5. ciliare**
b. Lateral lobes of the lip entire or at most toothed along their outer margins **6. parkinsonianum**
5a. Inflorescence an almost spherical raceme borne on a long stalk covered with bladeless sheaths, which terminates the leafy shoot; lobes of the lip approximately equal in size, forming a cross at the apex of the column, often all fringed 6
b. Inflorescence usually a panicle, rarely an elongate raceme, borne directly on the leafy shoot or on an entirely leafless shoot; lobes of the lip various, usually unequal, rarely fringed, and not forming a cross at the top of the column **1. stamfordianum**
6a. Stem arching or scrambling, rooting from many of the nodes **2. ibaguense**
b. Stem erect, not rooting at all from the nodes **3. arachnoglossum**

1. E. stamfordianum Bateman. Illustration: *Fieldiana* **26**: 381 (1953); Bechtel et al., Manual of cultivated orchid species, 210 (1981).

Stems to 25 cm, erect, increasing in thickness upwards, then tapering, bearing 4–6 leaves near the apex. Leaves 13–24 × 3–7 cm, linear-oblong to linear-elliptic, slightly notched at the apex. Panicle (rarely raceme) arising on a leafless stalk from the base of the leafy stem or, more rarely, terminating a leafy shoot. Flowers greenish yellow spotted with red. Sepals and petals similar, 1.4–2.5 cm. Lip deeply 3-lobed, the lateral lobes oblong, recurved, the central deeply notched with divergent lobes, margins fringed. *Guatemala*. G2. Spring.

Two other species in cultivation will key out here:

E. paniculatum Ruiz & Pavon (*E. floribundum* Humboldt, Bonpland & Kunth). Illustration: *Botanical Magazine*, 3637 (1838). With tall, cane-like stems terminated by a panicle of flowers with greenish brown, reflexed sepals and very narrow, reflexed petals and a white lip with very divergently lobed central lobe. *C America, northern S America, West Indies*. G2.

E. myrianthum Lindley (*E. verrucosum* Swartz var. *myrianthum* (Lindley) Ames & Correll). Illustration: Bateman, Second century of orchidaceous plants, t. 163 (1867). Like *E. paniculatum* but with warty leaf-sheaths and lilac-purple sepals, petals and lip. *Guatemala*. G2.

2. E. ibaguense Humboldt, Bonpland & Kunth (*E. radicans* Ruiz & Pavon). Illustration: Hamer, Orquideas de El Salvador **1**: 211 (1974); Bechtel et al., Manual of cultivated orchid species, 209 (1981).

Stems to 1 m, scrambling or arching, producing roots from many of the nodes. Leaves ovate-oblong to oblong, notched at the apex. Raceme more or less spherical, borne on a long stalk covered with bladeless sheaths, which terminates the leafy shoot. Flowers variable in colour, orange, scarlet red, purplish red or yellow. Sepals and petals erect, 1–2.2 cm. Lip forming a cross at the top of the column, 3-lobed, the lobes all similar or the central larger, deeply fringed, the central 2-lobed. *C America, northern S America, West Indies*. G2. Flowers most of the year.

E. × obrienianum Rolfe; *E. ibaguense* × *E. evectum* Hooker Similar but has a longer inflorescence, the sepals and petals orange and the column tipped with yellow.

3. E. arachnoglossum Reichenbach. Illustration: *Revue Horticole* **1882**: 554 (1882).

Very similar to *E. ibaguense* but stems erect, not rooting, and flowers purplish pink throughout. *Colombia*. G2. Summer.

E. xanthinum Lindley. Illustration: *Botanical Magazine*, 7586 (1898). Similar, but flowers yellow and the apex of the column bright red. *Brazil*.

E. cnemidophorum Lindley. Illustration: *Botanical Magazine*, 5656 (1867). Also similar but with a more elongate inflorescence and the lip more obscurely 3-lobed, with entire lobes. *C America*.

4. E. nocturnum Jacquin. Illustration: *Fieldiana* **26**: 347 (1953); Bechtel et al., Manual of cultivated orchid species, 210 (1981).

Stem to 1 m, erect, leafy. Leaves oblong-lanceolate to elliptic, unequally notched at the apex. Raceme terminal with up to 5 flowers, borne directly on the leafy stems, with conspicuous, boat-shaped bracts. Sepals greenish white, petals white, all 3.5–9 cm, spreading. Lip white, 3-lobed, the lateral lobes reversed-triangular with finely toothed outer margins, the central lobe 2–2.5 cm, very narrowly triangular, projecting between and beyond the lateral lobes. *USA (Florida), C America, northern S America, West Indies*. G2. Summer–autumn.

5. E. ciliare Linnaeus. Illustration: *Botanical Magazine*, 463 (1799); Cogniaux & Goossens, Dictionnaire iconographique des orchidées, Epidendrum, t. 6 (1897); *Fieldiana* **26**: 319 (1953); Bechtel et al., Manual of cultivated orchid species, 209 (1981).

Stems short, pseudobulb-like, bearing 1–2 leaves at the apex. Leaves oblong-elliptic, somewhat unequally notched at the apex. Raceme terminal, with few flowers and conspicuous, boat-shaped bracts. Sepals and petals 4–9 cm, erect, pale green or yellowish. Lip white, 3-lobed, the lateral lobes narrowly reversed-triangular with deeply fringed outer margins, the central lobe 2.5–6 cm, linear, projecting between and beyond the lateral lobes. *C America, northern S America, West Indies*. G2. Winter.

6. E. parkinsonianum Hooker (*E. falcatum* Lindley). Illustration: *Botanical Magazine*, 3778 (1840).

Stems short, pseudobulb-like, usually hanging, bearing 1–2 or rarely more leaves. Leaves linear-lanceolate to lanceolate, acute. Raceme terminal with 1–3 flowers and conspicuous, boat-shaped bracts. Sepals and petals 5.5–8.5 cm, spreading, white, yellowish or yellowish green, becoming pale orange if not fertilised. Lip white or yellowish, deeply 3-lobed, the lateral lobes reversed-triangular with finely toothed outer margins, the central lobe 3.5–5.5 cm, projecting between and beyond the lateral lobes. *C America*. G2. Autumn.

E. oerstedtii Reichenbach (*E. costaricense* Reichenbach). Illustration: Cogniaux & Goossens, Dictionnaire iconographique des orchidées, Epidendrum, t. 3 (1897); *Botanical Magazine*, n.s., 375, (1962). Similar but with elliptic-oblong leaves and the central lobe of the lip narrowly oblanceolate. *Costa Rica*.

7. E. medusae (Reichenbach) Veitch (*Nanodes medusae* Reichenbach). Illustration: *Botanical Magazine*, 5723 (1868); Bechtel et al., Manual of cultivated orchid species, 210 (1981).

Stems to 20 cm, usually leafy, hanging. Leaves greyish green, lanceolate, tapering to an acute apex. Raceme with 1–3 large flowers. Sepals and petals 3–4 cm, spreading, yellowish green flushed with red. Lip broadly kidney-shaped, broader than long (to 6 cm broad), purplish brown, deeply fringed all round. *Ecuador*. G2. Summer.

There are several other rarely cultivated species, which, though unlike each other and *E. medusae*, will key out here because of their simple lips:

E. diffusum Swartz. Illustration: *Botanical Magazine*, 3565 (1837). Stems erect, leaves ovate, panicle wide, of small flowers (sepals and petals *c.* 8 mm), which are entirely yellowish green. *C America*.

E. eximium Williams. Illustration: *Orquidea* **2**: 245 (1972). Stems erect, cane-like, raceme of several flowers with yellow sepals and petals (*c.* 2.5 cm) and an oblong, entire, yellow, red-lined lip. *Mexico, El Salvador*.

E. latilabrum Lindley. Illustration: Martius, Flora Brasiliensis **3**(5): t. 31 (1898). Stems erect, zig-zag, cane-like, raceme few-flowered, umbel-like, sepals and petals 1.8–3.3 cm and the lip entire, broader than long, deeply notched. *Brazil*.

E. loefgrenii Cogniaux. Illustration: Martius, Flora Brasiliensis **3**(5): t. 49 (1898). Stems hanging with broad, overlapping leaves and a raceme of few, very small (*c.* 1 cm in diameter) greenish flowers with a broadly obovate, unfringed, notched lip. *Brazil*.

50. ENCYCLIA Hooker

J.C.M. Alexander

Epiphytic, with creeping rhizomes. Pseudobulbs erect, swollen, basally enclosed in chaffy sheaths. Leaves 1–4, folded when young, borne on top of the pseudobulb. Flowers in terminal racemes or panicles, often many flowered; basal spathiform bract absent. Sepals and petals free, often similar. Lip conspicuously 3-lobed, the lobes clasping the column. Claw of lip joined to the column for up to half the length of the latter, column usually terete, often apically winged, with small teeth at apex. Pollinia 4; rostellum at right angles to column. Fruit ellipsoid, not winged nor sharply angled.

A genus of about 150 species from C & S America, most of which were originally in *Epidendrum*. Cultivation as for *Epidendrum* (p. 491).

Literature: Dressler, R.L. & Pollard, G.E., The genus Encyclia in Mexico (1974); Withner, C.L., Cattleyas and their relatives **4**: 35–80 (1996), **5**: 85–136 (1998) & **6**: 1–194 (2000); Withner, C.L. & Harding, P., Cattleyas and their relatives. The debatable Epidendrums. (2004).

1a. Lip blade or central lobe more or less isodiametric, not tapering to a point 2
b. Lip blade or central lobe usually longer than broad, tapering to an obtuse or acute point **1. adenocaula**
2a. Lip (including claw) 3 cm or more **4. cordigera**
b. Lip (including claw) 2.5 cm or less 3
3a. Lip without lateral lobes on blade or claw **6. polybulbon**
b. Lip with lateral lobes on blade or claw 4
4a. Petals hooded at tip **5. selligera**
b. Petals not hooded at tip 5
5a. Lateral lobes of lip 12 × 6 mm or more, strongly constricted at base **2. alata**
b. Lateral lobes of lip 10 × 4 mm or less, triangular, lanceolate or narrowly spathulate **3. tampensis**

1. E. adenocaula (Llave & Lexarza) Schlechter (*Epidendrum nemorale* Lindley; *Epidendrum verrucosum* Lindley not Swartz). Figure 25(12), p. 494. Illustration: *Botanical Magazine*, 4606 (1851); *Orchid Digest* **37**: 45 (1973); Rittershausen & Rittershausen, Orchids in colour, 177 (1979).

Pseudobulbs 5–8 cm, conical to ovoid, clustered. Leaves 2 or 3 per pseudobulb, 11–35 cm × 5–30 mm, strap-shaped, acute or obtuse. Panicles 30–100 cm, many-flowered; stem, branches and ovaries covered in warts. Sepals and petals linear to lanceolate, acute, pale purplish pink; sepals 2.5–5.5 cm × 5–7 mm; petals 3–5 cm × 4–8 mm. Lip 3.5–4.5 cm, 3-lobed; lateral lobes lanceolate to oblong, *c.* 10 × 5 mm; central lobe circular to ovate, regularly scalloped, mucronate, with 1 or more streaks of dark red. Column 1.3–1.5 cm, club-shaped, apically winged; basal quarter fused to lip. *W Mexico*. G2. Spring–summer.

2. E. alata (Bateman) Schlechter. Figure 25(13), p. 494. Illustration: *Botanical Magazine*, 3898 (1842); Dressler & Pollard, The genus Encyclia in Mexico, t. 53 (1974); Hamer, Orquideas de El Salvador **1**: t. 14 (1974).

Pseudobulbs 6–12 cm, conical to ovoid, clustered. Leaves 2 or 3 per pseudobulb, 20–60 × 1.5–3.5 cm, strap-shaped to narrowly elliptic, acute. Panicles 30–200 cm, many-flowered. Sepals and petals 2.2–3 cm × 5–8 mm, greenish yellow with a central patch of reddish brown near the tip; margins curled back. Sepals lanceolate to elliptic, obtuse; petals elliptic to spathulate, obtuse or mucronate. Lip 1.8–2.3 cm, 3-lobed, pale greenish yellow, lateral lobes 1.2–1.6 cm, oblong to obovate, rounded, flared; central lobe circular to transversely diamond-shaped, fleshy, veined with dark red. Column 1–1.2 cm, apically winged; basal quarter

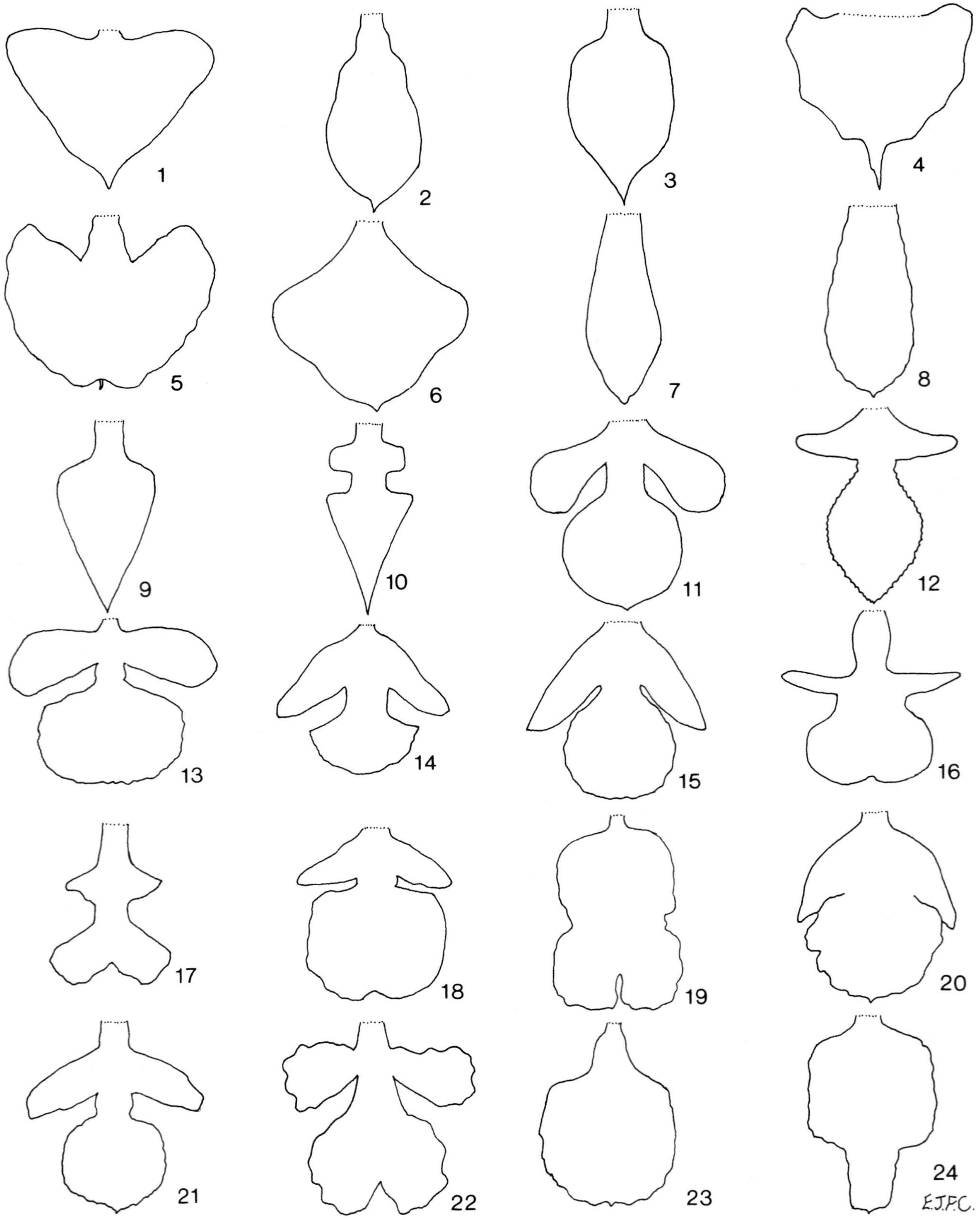

Figure 25. Lips of *Encyclia* and *Prosthechea* species. 1, *P. cochleata.* 2, *E. glumacea.* 3, *P. fragrans.* 4, *P. baculus.* 5, *P. radiata.* 6, *E. vespa.* 7, *P. vitellina.* 8, *E. boothiana.* 9, *P. brassavolae.* 10, *P. prismatocarpa.* 11, *E. patens.* 12, *E. adenocaula.* 13, *E. alata.* 14, *E tampensis.* 15, *E. bractescens.* 16, *E. michuacana.* 17, *E. varicosa.* 18, *E. cordigera.* 19, *P. mariae.* 20, *E. aromatica.* 21, *E. selligera.* 22, *E. advena.* 23, *E. polybulbon.* 24, *P. citrina.* (Compiled from various sources; not to scale.)

fused to lip. *S Mexico to Costa Rica*. G2. Summer.

3. E. tampensis (Lindley) Small. Figure 25(14), p. 494. Illustration: Shuttleworth et al., Orchids, 46 (1970); Luer, Native orchids of Florida, t. 57 & 58 (1972); *American Orchid Society Bulletin* **43**: 808 (1974).

Pseudobulbs 1–7 cm, ovoid, slightly separated. Leaves 1–3 per pseudobulb, 30–40 × 1.5–2 cm, linear to narrowly lanceolate. Panicles to 1 m, few- or many-flowered; flowers with a sweet honey scent. Sepals and petals 1.2–2.2 cm × 4–6.5 mm, lanceolate to obovate, greenish yellow to brown suffused with purple. Lip 1.5–2 cm, white, 3-lobed; lateral lobes triangular, upright or slightly flared; central lobe semicircular with a central patch of pink or pink streaks. Column 10 mm, apically winged; base fused to claw of lip. *USA (Florida)*. G2. Flowers intermittently.

E. aromatica (Bateman) Schlechter. Figure 25(20), p. 494. Illustration: Dressler & Pollard, The genus Encyclia in Mexico, t. 63 (1974); Hamer, Orquideas de El Salvador **1**: t. 15 (1974). Differs in its smaller (1–1.5 cm) cream or yellow lip with red or brown veins and an oblong to diamond-shaped, wavy-edged central lobe. *SE Mexico*. G2. Spring–summer.

E. bractescens (Lindley) Hoehne. Figure 25(15), p. 494. Illustration: *Botanical Magazine*, 4572 (1851); Dressler & Pollard, The genus Encyclia in Mexico, t. 46 (1974); Northen, Miniature orchids, C-31 (1980). Pseudobulbs 3 cm or less; leaves to 25 × 1 cm. Sepals and petals narrowly lanceolate, purplish brown with greenish yellow tips. Lip with long narrow lateral lobes and circular central lobe. *S Mexico*. G2. Spring–summer.

E. michuacana (Llave & Lexarza) Schlechter (*Epidendrum virgatum* Lindley). Figure 25(16), p. 494. Illustration: Dressler & Pollard, The genus Encyclia in Mexico, t. 20 (1974); Hamer, Orquideas de El Salvador **1**: t. 23 (1974). Lip 10–11 mm, with narrow lateral lobes arising from base of blade rather than from claw. Sepals and petals dark brown. Lip cream or pale yellow often spotted with purple. *S Mexico & Guatemala*. G2. Flowers intermittently.

E. varicosa (Lindley) Schlechter. Figure 25(17), p. 494. Illustration: Dressler & Pollard, The genus Encyclia in Mexico, t. 22a (1974); Hamer, Orquideas de El Salvador **1**: t. 23 (1974). Differs from *E. tampensis* in its papillose or warty notched lip with triangular lateral lobes arising from claw, and blade of 2 oblong-truncate lobes. Sepals and petals brown. Pseudobulbs with long slender necks. *S&W Mexico to Panama*. G2. Mainly winter.

4. E. cordigera (Humboldt, Bonpland & Kunth) Dressler (*Epidendrum atropurpureum* misapplied; *Epidendrum macrochilum* Hooker). Figure 25(18), p. 494. Illustration: *Botanical Magazine*, 3534 (1836); Dressler & Pollard, The genus Encyclia in Mexico, t. 57 (1974); Sheehan & Sheehan, Orchid genera illustrated, 89 (1979).

Pseudobulbs 3–11 cm, conical to ovoid, clustered. Leaves 2 or 3 per pseudobulb, 12.5–45 × 1.5–4.5 cm, narrowly elliptic. Racemes or panicles 15–75 cm with 3–15 flowers. Sepals and petals fleshy, 2.5–3.5 cm × 5–11 mm, green streaked with brown. Sepals obovate to elliptic, obtuse or mucronate; petals elliptic to spathulate. Lip 3.3–4 cm, cream or pink, centrally streaked with magenta, 3-lobed, lateral lobes oblong to lanceolate, central lobe broadly obovate. Column 1.5–1.7 cm; base fused to lip, white. *S Mexico to Colombia & Venezuela*. G2. Spring–summer.

5. E. selligera (Lindley) Schlechter (*E. oncidioides* misapplied). Figure 25(21), p. 494. Illustration: Reichenbach, Xenia Orchidacearum **3**: t. 233 (1892); Hamer, Orquideas de El Salvador **1**: t. 17 (1974); Dressler & Pollard, The genus Encyclia in Mexico, t. 59 (1974).

Pseudobulbs 4.5–8 cm, conical to ovoid, clustered. Leaves 2 per pseudobulb, 15–25 × 1.5–4 cm, elliptic to strap-shaped, obtuse. Panicles 25–80 cm, many-flowered. Sepals and petals pale green, strongly flushed with purplish brown. Sepals 1.8–2.4 cm × 5–9 mm, lanceolate to spathulate, obtuse. Petals 1.8–2.1 cm × 7–9 mm, spathulate, hooded, obtuse or mucronate. Lip 1.7–2.2 cm, 3-lobed, cream, pink or lavender; lateral lobes to 10 × 5 mm, broad-tipped, constricted at base; central lobe ovate to circular. Callus elliptic, running into 2 ridges on central lobe of lip. Column 1.1–1.3 cm; base fused to claw of lip. *S Mexico & Guatemala*. G2. Winter–spring.

Frequently confused with **E. oncidioides** (Lindley) Schlechter (illustration: Withner, Cattleyas and their relatives **6**: plate 47, 48, 2000; Coletanea de Orquideas Brasileiras, Encyclia, 17, 2003), which has smaller narrowly ellipsoid pseudobulbs and smaller yellow flowers flushed with brown. Dressler considers the latter to be very rare in cultivation. The following 2 species are also very similar:

E. hanburii (Lindley) Schlechter. Illustration: *Orchid Digest* **34**: 307 (1970); Dressler & Pollard, The genus Encyclia in Mexico, t. 60 (1974). Differs in its reddish purple sepals and petals and its pink, heavily veined lip. Callus ovate to rhombic, running into a single ridge on the central lobe of the lip. *S Mexico*. G2. Spring–summer.

E. advena (Reichenbach) Dressler (*Epidendrum capartianum* Lindley; *Epidendrum osmanthum* Barbosa-Rodrigues). Figure 25(22), p. 494. Illustration: *Botanical Magazine*, 7792 (1901); Coletanea de Orquideas Brasileiras, Encyclia, 7 (2003). Larger in all its parts than *E. selligera*. Pseudobulbs 7.5–13 cm. Petals spathulate to circular, about 1.5 cm wide, yellow streaked with pink. Lip more than 3 cm, wavy and scalloped, white with pink streaks. *Brazil*. G2. Autumn–winter.

E. patens Hooker (*Epidendrum odoratissimum* Lindley; not *Epidendrum patens* Swartz). Figure 25(11), p. 494. Illustration: *Botanical Magazine*, 3013 (1830); Pabst & Dungs, Orchidaceae Brasilienses **1**: 197, 298 (1975); Withner, Cattleyas and their relatives **6**: pl. 51 (2000); Coletanea de Orquideas Brasileiras, Encyclia, 18 (2003). Differs in its unspotted, greenish brown, smaller, elongated, narrowly spathulate petals and shorter (1–1.5 cm), cream-coloured lip, with oblong to obovate lateral lobes and an obovate central lobe. Flowers heavily scented. *SW Brazil* G2. Summer.

The *Botanical Magazine* plate shows the raceme growing from beside the pseudobulb, which would exclude this species from *Encyclia*. Subsequent authors make no reference to this; presumably the illustration is inaccurate.

6. E. polybulbon (Swartz) Dressler (*Dinema polybulbon* (Swartz) Lindley. Figure 25(23), p. 494. Illustration: *Botanical Magazine*, 4067 (1844); *Die Orchidee* **29**: lxxxvii (1978); Dressler & Pollard, The genus Encyclia in Mexico, t. 38 (1974).

Pseudobulbs 1–1.5 cm, ovoid, 5–40 mm apart. Leaves 2 or 3 per pseudobulb, 1–5 cm × 5–10 mm, elliptic to ovate, notched. Flowers solitary. Sepals 9–17 × 1.5–3 mm, linear-lanceolate, acute, brownish yellow. Petals similar, narrower with slightly hooked tips. Lip 1.5 cm, white, with a wavy, more or less circular, sometimes mucronate blade. Column *c.* 4 mm, fused to lip, winged, with two 3 mm apical processes. *S Mexico, Guatemala, Honduras, Cuba & Jamaica.* G2. Flowers intermittently.

51. PROSTHECHEA Hooker

J.M.H. Shaw

Epiphytic herbs with fusiform, often flattened, pseudobulbs. Leaves 1–5. Inflorescence racemose with 1–few flowers, often with a spathe like bract at the base. Flowers not resupinate. Lip almost entire to 3-lobed, adnate to lower part of column. Column usually swollen, without wings, with three teeth at the apex, the central tooth usually largest, erect and often covered by a fleshy knob; lateral teeth usually separated from the central tooth by deep narrow valleys. Pollinia 4. Capsule 3-winged or with 3 sharp angles, not ellipsoidal or globose.

A genus of about 100 species distributed from Florida and Mexico southwards throughout tropical America. Most species were previously included in *Epidendrum*, *Encyclia* or *Euchile*. Cultivation as for *Epidendrum*.

Literature: Higgins, W.E., A reconsideration of the genus Prosthechaea, *Phytologia* **82**: 370–83 (1997); Withner, C.L. & Harding, P., Cattleyas and their relatives. The debatable Epidendrums (2004).

1a. Flowers with lip at top (not resupinate) 2
b. Flowers with lip at bottom (resupinate) 5
2a. Pseudobulbs bearing a single leaf **2. fragrans**
b. Pseudobulbs bearing 2 or more leaves 3
3a. Sepals twisted or drooping, at least 5 times as long as broad 4
b. Sepals firm and erect, at most 3 times as long as broad **4. radiata**
4a. Petals at most 6 times as long as broad; flowers 2 or 3 per raceme; lip cream-coloured with radiating purple lines **3. baculus**
b. Petals at least 8 times as long as broad; flowers 4 or more per raceme; lip purplish brown, flushed with yellowish green **1. cochleata**
5a. Lip blade or central lobe more or less isodiametric, not tapering to a point 6
b. Lip blade or central lobe usually longer than broad, tapering to an obtuse or acute point 7
6a. Flowers green and white; lip notched **8. mariae**
b. Flowers yellow; lip truncate **9. citrina**
7a. Lip without lateral lobes on claw **7. prismatocarpa**
b. Lip with lateral lobes on claw 8
8a. Lip (including claw) 3 cm or more; petals and sepals linear, 3 cm or more, yellow or yellowish green **6. brassavolae**
b. Lip (including claw) 2 cm or less; petals and sepals elliptic or ovate, 2.5 cm or less, scarlet or orange **5. vitellina**

1. P. cochleata (Linnaeus) W.E. Higgins (*Encyclia cochleata* (Linnaeus) Dressler; *Hormidium cochleatum* (Linnaeus) Brieger; *Anacheilium cochleatum* (Linnaeus) Hoffmanns). Figure 25(1), p. 494. Illustration: *Botanical Magazine*, 572 (1801); Luer, Native orchids of Florida, t. 60 (1972); Dressler & Pollard, The genus Encyclia in Mexico, t. 7 (1974).

Pseudobulbs 5–26 cm, ellipsoid to pear-shaped, sometimes stalked, loosely clustered. Leaves 2–4 per pseudobulb, 20–33 × 3–4 cm, elliptic to lanceolate, acute. Raceme to 50 cm, many-flowered. Sepals and petals 3–7.5 cm × 3–7 mm, narrowly lanceolate, acute, twisted and drooping, pale green. Lip uppermost, 1–2.1 × 1.3–2.6 cm, triangular to heart-shaped, concave, deep purple flushed with yellowish green; base white with deep purple veins. Column 7–9 mm, green with purple spots; base fused to claw of lip. *USA (Florida); S & W Mexico to Colombia & Venezuela.* G2. Flowers continuously.

Var. **cochleata** occurs south of Florida, and bears one anther.

Var. **triandra** (Ames) W.R. Higgins is self-pollinating, and bears 3 anthers. It occurs in Florida.

P. glumacea (Lindley) W.E. Higgins (*Encyclia glumacea* (Lindley) Pabst; *Hormidium glumaceum* (Lindley) Brieger). Figure 25(2), p. 494. Illustration: *Edwards's Botanical Register* **26**: t. 6 (1840). Differs in its white flowers streaked with pink and its narrower acuminate lip, 1.8–2.5 cm × 8–12 mm. Base of raceme enclosed in long chaffy scales. *E Brazil & Ecuador.* G2.

2. P. fragrans (Swartz) W.E. Higgins (*Encyclia fragrans* (Swartz) Dressler; *Hormidium fragrans* (Swartz) Brieger; *Anacheilium fragrans* (Swartz) Acuña). Figure 25(3), p. 494. Illustration: *Botanical Magazine*, 1669 (1814); Dunsterville & Garay, Venezuelan orchids illustrated **4**: 87 (1966); Dressler & Pollard, The genus Encyclia in Mexico, t. 4 (1974).

Pseudobulbs 4.5–11 cm, narrowly ovoid to ellipsoid, 1–4 cm apart. Leaves 9–35 cm, solitary, elliptic to strap-shaped, obtuse or acute. Raceme 5–20 cm, with 2–6 white, cream or greenish, very fragrant flowers. Sepals 2–3.5 cm × 4–9 mm, lanceolate to elliptic, acuminate; petals 2–3 cm × 7–12 mm, elliptic to diamond-shaped, acuminate. Lip uppermost, 1.7–2.4 cm × 9–17 mm, ovate-acuminate with purple radial lines. Column 6–8 mm, basal half fused to lip. *Greater Antilles, S. Mexico to Brazil.* G2. Spring–summer.

3. P. baculus (Reichenbach) W.E. Higgins (*Encyclia baculus* (Reichenbach) Dressler & Pollard; *Epidendrum pentotis* Reichenbach; *Hormidium baculus* (Reichenbach) Brieger). Figure 25(4), p. 494. Illustration: *Botanical Magazine*, 702 (1975); Dressler & Pollard, The genus Encyclia in Mexico, t. 1 (1974); Graf, Tropica, 729 (1978).

Similar to *E. fragrans* but differing in its larger, gradually tapering pseudobulbs, 15–35 cm, bearing 2 or 3 leaves, and in its short 2- or 3-flowered racemes. Petals and sepals more than 3.5 cm. Lip blade with prominent basal auricles. *S Mexico to Brazil.* G2. Spring–summer.

4. P. radiata (Lindley) W.E. Higgins (*Encyclia radiata* (Lindley) Dressler; *Hormidium radiatum* (Lindley) Brieger; *Anacheilium radiatum* (Lindley) Pabst, Moutinho & Pinto). Figure 25(5), p. 494. Illustration: *Botanical Magazine*, n.s., 662 (1974); Dressler & Pollard, The genus Encyclia in Mexico, t. 8 (1974); Graf, Tropica, 729 (1978).

Pseudobulbs 7–11 cm, ellipsoid to ovoid, grooved, sometimes stalked, about 2.5 cm apart. Leaves 2–4 per pseudobulb, 14–35 × 1.5–3 cm, lanceolate to elliptic, obtuse or acute. Raceme 7–20 cm, with 4–12 cream or greenish white flowers. Sepals 1.5–2 cm × 5–7 mm, elliptic; petals 1.5–2 cm × 8–11 mm, broadly elliptic to diamond-shaped. Lip uppermost, 1–1.3 × 1.3–2 cm, semicircular to kidney-shaped, notched, with purple radial lines. Column 8–10 mm, basal half fused to lip. *C & S Mexico, Guatemala, Honduras & Costa Rica.* G2. Autumn–winter.

P. vespa (Vellozo) W.E. Higgins (*Encyclia vespa* (Vellozo) Dressler; *Epidendrum variegatum* Hooker (1832); *Hormidium variegatum* (Hooker) Brieger; *Anacheilium vespa* (Vellozo) Pabst, Moutinho & Pinto). Figure 25(6), p. 494. Illustration: *Botanical Magazine*, 3151 (1832); Pabst & Dungs, Orchidaceae Brasilienses **1**: 187 (1975); Graf, Tropica, 729 (1978). Differs in its small, pale yellowish green lip, scarcely longer than the column and its yellowish green petals and sepals densely spotted with blackish purple. Leaves faintly mottled. *West Indies, Venezuela, Guyana, Surinam, Ecuador, Peru & Brazil*. G2. Autumn–summer.

5. P. vitellina (Lindley) W.E. Higgins (*Encyclia vitellina* (Lindley) Dressler). Figure 25(7), p. 494. Illustration: *Botanical Magazine*, 4107 (1844); *Die Orchidee* **22**: 45 (1971); Dressler & Pollard, The genus Encyclia in Mexico, t. 31 (1974).

Pseudobulbs 2.5–5 cm, conical to ovoid, clustered. Leaves 1–3, 7–22 × 1–4 cm, lanceolate to elliptic, obtuse or acute. Raceme or panicle 12–30 cm, with 4–12 flowers. Sepals and petals lanceolate to elliptic, acute, orange to bright scarlet; sepals 1.5–2.5 cm × 3–8 mm; petals 1.5–2.5 cm × 6–11 mm. Lip 1.1–1.6 cm, yellow or orange with a red tip; blade elliptic to oblong, acute with deflexed sides. Column 6–7 mm; base fused to lip. *S Mexico & Guatemala*. G2. Spring–summer.

'**Majus**' (*Epidendrum vitellinum* var. *majus* Veitch) has shorter, denser racemes of larger, more brightly coloured flowers.

P. boothiana (Lindley) W.E. Higgins (*Encyclia boothiana* (Lindley) Dressler; *Diacrium bidentatum* (Lindley) Hemsley; *Epicladium boothianum* (Lindley) Small; *Hormidium boothianum* (Lindley) Brieger). Figure 25(8), p. 494. Illustration: Luer, Native orchids of Florida, t. 61 (1972); Dressler & Pollard, The genus Encyclia in Mexico, t. 29 (1974). Flowers smaller and fleshier. Sepals and petals 8–14 × 2–5 mm, rounded; yellowish green heavily spotted with purplish brown. Lip 7–10 mm, diamond-shaped or weakly 3-lobed, with down-curled margins, white to greenish yellow. *USA (Florida), S Mexico, Belize, Guatemala & West Indies*. G2. Winter.

Var. **boothiana** bears one anther and occurs south of Florida.

Var. **erythronioides** (Small) W.R. Higgins is self-pollinating, and bears 3 anthers. It occurs in Florida.

6. P. brassavolae (Reichenbach) W.E. Higgins (*Encyclia brassavolae* (Reichenbach) Dressler; *Hormidium brassavolae* (Reichenbach) Brieger). Figure 25(9), p. 494. Illustration: *Botanical Magazine*, 5664 (1867); Dressler & Pollard, The genus Encyclia in Mexico, t. 32 (1974); Rittershausen & Rittershausen, Orchids in colour, 177 (1979).

Pseudobulbs 9–18 cm, conical, ovoid or fusiform, clustered or up to 4.5 cm apart. Leaves 2 or 3 per pseudobulb, 15–30 × 3.5–5 cm, oblong to lanceolate, obtuse. Raceme 15–100 cm with 3–15 flowers. Sepals and petals linear-lanceolate, acuminate, pale yellowish green to brown. Sepals 3.5–5.5 cm × 3–6 mm; petals 3.2–4.7 cm × 2–4 mm. Lip 3–4.5 cm, cream-coloured with a purple tip; blade ovate to triangular, acuminate. Column 1.2–1.5 cm; basal half fused to lip. *S Mexico to W Panama*. G2. Summer–autumn.

7. P. prismatocarpa (Reichenbach) W.E. Higgins (*Encyclia prismatocarpa* (Reichenbach) Dressler; *Hormidium prismatocarpum* (Reichenbach) Brieger). Figure 25(10), p. 494. Illustration: *Botanical Magazine*, 5336 (1862); Shuttleworth et al., Orchids, 48 (1970); Graf, Tropica, 729 (1978).

Pseudobulbs to 30 cm, ovoid with a long 'neck'. Leaves 2–4 per pseudobulb, 25–40 × 3–6 cm, strap-shaped, obtuse. Raceme to 40 cm, many-flowered. Sepals and petals similar, 2–2.5 cm × 3–6 mm, linear-lanceolate to oblong, acute, pale greenish yellow with large purple blotches. Lip 2–2.5 cm, 3-lobed; lateral lobes *c.* 5 mm, obtuse and rounded, yellow or greenish yellow; central lobe narrowly triangular, acute, pale yellow or white with a large central patch of purple and a yellow tip. Column with basal third fused to lip, lobed and fringed at tip. *Costa Rica, Panama*. G2. Flowers intermittently.

8. P. mariae (Ames) W.E. Higgins (*Encyclia mariae* (Ames) Hoehne; *Hormidium mariae* (Ames) Brieger; *Euchile mariae* (Ames) Withner). Figure 25(19), p. 494. Illustration: *American Orchid Society Bulletin* **41**: 7 (1972); Dressler & Pollard, The genus Encyclia in Mexico, t. 36 (1974); Graf, Tropica, 729 (1978).

Flowers pendent, 2–4 per inflorescence. Petals and sepals green, spreading; lip white with green or yellowish green throat, unlobed or shallowly 3-lobed, apex notched. *E Mexico*. G2. Summer.

9. P. citrina (Llave & Lexarza) W.E. Higgins (*Encyclia citrina* (Llave & Lexarza) Dressler; *Cattleya citrina* (Llave & Lexarza) Lindley; *Epidendrum citrinum* (Llave & Lexarza) Reichenbach; *Euchile citrina* (Llave & Lexarza) Withner). Figure 25(24), p. 494. Illustration: *Botanical Magazine*, 3742 (1839); Dressler & Pollard, The genus Encyclia in Mexico, t. 37 (1974); Bechtel et al., Manual of cultivated orchid species, 207 (1981).

Flowers bright yellow, pendent, 1 or 2 per inflorescence. Lip shallowly 3-lobed, central lobe oblong, truncate. *Mexico*. G1. Spring.

52. BROUGHTONIA R. Brown

J.C.M. Alexander

Epiphytic herb, to 50 cm, with a creeping rhizome enclosed by overlapping sheaths. Pseudobulbs spherical to ovoid, cylindrical to slightly flattened, with 1 or 2 apical leaves. Flowering stems racemes or panicles, much longer than leaves, borne on top of pseudobulbs, with a small scale at each node, bearing few flowers. Sepals free, similar. Petals wider than sepals. Lip free, simple to weakly 3-lobed, margin entire to irregularly toothed; spur present or absent. Column free, winged towards base, exposed. Pollinia 4 similar, or 8, of which 4 rudimentary.

A genus of 6 species from the West Indies, which grows best on bark or rafts suspended near the roof. General conditions are similar to those for *Epidendrum* (p. 491); they should not be heavily shaded, nor should too much water be given in winter. Now includes species previously placed in *Laeliopsis* Lindley and *Cattleyopsis* Lemaire.

Literature: Sauleda, R.P. & Adams, R.M., A reappraisal of the orchid genera Broughtonia, Cattleyopsis and Laeliopsis, *Rhodora* **86**: 445–67 (1984); Withner, C.L., Cattleyas and their relatives **4**: 15–29, 81–84 (1996).

1a. Flowers red, yellow or white; spur present **1. sanguinea**
b. Flowers lilac; spur absent **2. domingensis**

1. B. sanguinea (Swartz) R. Brown (*B. coccinea* Lindley). Illustration: *Botanical Magazine*, 3536 (1836); Skelsey, Orchids, 96 (1979); Bechtel et al., Manual of cultivated orchid species, 173 (1981).

Epiphytic, to 50 cm. Pseudobulbs 3–6 cm, spherical to ovoid, slightly flattened, with 1 or 2 apical leaves. Leaves

7.5–15 × 2–4 cm, oblong, stiff and leathery, folded when young. Flowering stems to 50 cm, borne on top of pseudobulbs, with a small scale at each node, bearing 5–12 purplish red flowers, although cultivars with white or yellow flowers have been selected; flowers 2–3 cm wide. Sepals 1.5–2.5 cm × 5 mm, lanceolate, acute, extending into a long narrow spur that is fused to the ovary. Petals 1.5–2.5 × 1 cm, elliptic to ovate, rounded. Lip 2.5 × 2.5 cm, more or less circular, notched and irregularly toothed. Column free, exposed. Pollinia 4. *Jamaica.* G2. Intermittent flowering.

References to Cuba as a locality for this species are based on the similar *B. ortgesiana* (Reichenbach) Dressler (*Cattleyopsis ortgesiana* (Reichenbach) Cogniaux), which may be distinguished by projections on the column and 8 pollinia. The garden hybrid between the two is called Leona grex.

2. B. domingensis (Lindley) Rolfe (*Laeliopsis domingensis* (Lindley) Lindley; *B. lilacina* Henfrey). Illustration: Shuttleworth et al., Orchids, 52 (1970); *Orchid Digest* **40**: 212 (1976).

Pseudobulbs ovoid to fusiform, 5–8 cm. Leaves 1–3, apical, rigid, sharply toothed. Flowers 3–8, in terminal spikes; 3.5 cm wide, pale pinkish purple, spur absent. Lip rolled around column, yellow-throated, finely toothed. Column without foot, concealed. Pollinia 4. *Dominican Republic.* G2. Spring–summer.

53. BARKERIA Knowles & Westcott
J. Cullen

Epiphytes. Stems erect, elongate, cane-like, sometimes somewhat swollen at the base, leafy. Leaves leathery, folded when young. Raceme (rarely panicle) with few to many flowers. Flowers resupinate. Sepals and petals similar or the petals somewhat broader. Lip united to the column for up to half the length of the latter, entire. Column winged. Pollinia 4, laterally flattened, parallel.

A genus of 17 species from C America, closely related to and often included in *Epidendrum*, but differing in that at least half of the column is free from the lip; differing from *Encyclia* in the lack of true pseudobulbs. Cultivation as for *Epidendrum* (p. 491), but the plants are better grown on rafts or bark, as their elongate rhizomes are restricted in pots.

Literature: Halbinger, F., The genus Barkeria, *American Orchid Society Bulletin* **42**: 620–6 (1973); Withner, C.L., Cattleyas and their relatives **5**: 17–30 (1998).

1a. Lip joined to the column for almost half the length of the latter, its blade 1.3–1.8 cm, bearing 3 yellow ridges **3. skinneri**
b. Lip joined to the column for at most one-third of the length of the latter, its blade 2–3.5 cm, without ridges or the ridges not yellow 2
2a. Lip without ridges, though the main vein sometimes thickened **1. elegans**
b. Lip with 3–5 ridges **2. spectabilis**

1. B. elegans Knowles & Westcott (*Epidendrum elegans* (Knowles & Westcott) Reichenbach). Illustration: *Botanical Magazine*, 4784 (1854); *American Orchid Society Bulletin* **42**: 621 (1973).

Stems to 25 cm, erect, thick, leafy. Leaves linear-lanceolate to elliptic-lanceolate, acute. Raceme loose, with few flowers. Sepals and petals similar or the petals broader, all narrowly elliptic, acute, 2–3.5 cm. Lip 2–3.5 cm, obovate-elliptic to oblong, broadly rounded but with a small point, white with a large purple blotch and with a callus but without ridges. Column joined to the lip for up to one-third of its length. *Mexico, Guatemala.* G2. Spring.

2. B. spectabilis Lindley. Illustration: *Botanical Magazine*, 6098 (1874); *Fieldiana* **26**: 342 (1953); *American Orchid Society Bulletin* **42**: 619 (1973); Bechtel et al., Manual of cultivated orchid species, 170 (1981).

Very similar to *B. elegans* but often larger, raceme often with many flowers, flowers white, pink or deep purple, lip often spotted, bearing 3–5 ridges, which are the same colour as the rest. *C America.* G2. Winter.

Variable in flower colour and the precise shape of the lip. Halbinger recognises *B. lindleyana*, which is usually sunk in *B. spectabilis*, as a distinct species on the basis of lip shape and column length (15–18 mm in *B. spectabilis*, 6–14 mm in *B. lindleyana*).

3. B. skinneri (Lindley) Richard & Galeotti (*Epidendrum skinneri* Lindley). Illustration: *Botanical Magazine*, 3951 (1843); *American Orchid Society Bulletin* **42**: 624 (1973); Bechtel et al., Manual of cultivated orchid species, 170 (1981).

Very similar to *B. elegans* but smaller, inflorescence sometimes paniculate, flowers red to purple, lip 1.3–1.8 cm with 3 yellow ridges. Column joined to the lip for almost half its length. *Guatemala.* G2. Winter.

54. CATTLEYA Lindley
J.C.M. Alexander & J.M.H. Shaw

Epiphytic or terrestrial, often on rocks. Rhizomes creeping, branched, bearing erect, swollen or cane-like stems with several nodes. Leaves 1 or 2 (rarely 3) ovate, elliptic or oblong, borne on top of the stems, folded when young, leathery. Flowers solitary or in terminal racemes (except No. 7). Sepals and petals free, petals broader or narrower than sepals. Lip 3-lobed or entire, shortly fused to column at base in *Sophronitis* sensu stricto (Nos. 10, 11), but free from column in 'Brazilian Laelias' (Nos. 12–17); lateral lobes or basal part rolled around the column, often concealing it. Column free from lip, slightly flattened, often 2-winged. Pollinia 4 or 8 cohering in pairs.

A genus of about 100 species from C & S America; often used in bi- and trigeneric hybrids. The Central American *C. skinneri* group has been separated as the genus *Guarianthe*. *Cattleya* was recently expanded by Van den Berg (2008) to include species formerly placed in *Sophronitis* and *Laelia*. An earlier proposal to include the Brazilian Laelias in *Sophronitis* has not rested comfortably with many horticulturalists and there have been proposals to move them to other genera such as *Hoffmannseggella* H.G. Jones. *Cattleya* (*Sophronitis*) *coccinea* is often used to give bright red colours in intergeneric hybrids with *Cattleya, Guarianthe, Laelia, Brassavola, Epidendrum*, etc.

They require maximum light throughout the year, though new growth should be protected from strong sunlight. In the growing season water should be applied freely and the atmosphere kept moist with some air movement. Watering should be reduced during the winter but stems should not be allowed to shrink. Clusters of stems can be divided when repotting. Nos. 10 & 11 can be grown using tree-fern blocks or rafts in cool, shady conditions.

Literature: Withner, C., The genus Cattleya, *American Orchid Society Bulletin* **17**: 296–306, 363–9 (1948); Fowlie, J.A., Brazilian bifoliate Cattleyas and their color varieties (1977); Braem, G.J., Cattleya **1** (1984) & **2** (1986); Withner, C.L., Cattleyas and their relatives **1**: 1–147 (1988); Van den Berg, C., New combinations in the genus Cattleya Lindl., *Neodiversity* **3**: 3–12 (2008).

Stems. Cane-like: **3**, **4**, **5**, **6**; club-shaped: **12**, **13**, **14**; conical, ovoid or fusiform: **10**, **11**, **16**; swollen: **1**, **2**, **4**, **5**, **7**, **8**, **9**, **15**, **16**; swollen above: **1**, **2**, **4**, **5**, **8**, **9**; very slender throughout: **15**. Of 1 internode: **18**. 7 cm or less: **10**, **11**, **16**; 7 cm or more: **12**, **13**, **14**, **15**, **16**. Less than 15 cm: **1**, **3**, **7**; more than 15 cm: **1**, **2**, **3**, **4**, **5**, **6**, **8**, **9**.
Leaves. 1 per stem: **7**, **8**, **9**, **10**, **11**, **12**, **13**, **14**, **15**, **16**, **17**; 2 or 3 per stem: **1**, **2**, **3**, **4**, **5**, **6**, **7**, **15**. Spotted: **3**, **4**, **8**.
Flowering stems. Lateral: **7**.
Flowers. 1 or 2 per stem: **1**, **3**, **4**, **6**; 3–9 per stem: **1**, **2**, **4**, **5**, **6**, **7**, **8**, **9**; 10 or more per stem: **4**, **5**. Less than 6–7.5 cm wide: **3**, **4**, **5**, **10**, **11**, **15**; 8–10 cm wide: **1**, **2**, **3**, **4**, **5**, **6**, **7**, **8**; 11–13 cm wide: **1**, **2**, **7**, **9**; 14 cm wide or more: **9**.
Petals or sepals. Spotted: **2**, **3**, **4**, **5**; unspotted: **1**, **5**, **4**, **6**, **7**, **8**, **9**.
Lip. Unlobed: **6**; shallowly or obscurely lobed: **1**, **8**, **9**; 3-lobed: **2**, **3**, **4**, **5**, **7**; with very small basal lateral lobes: **6**. Broadest across lateral lobes: **1**, **2**, **4**, **5**, **8**, **9**; broadest across central lobe: **3**, **6**, **7**.
Lip lateral lobes. Semicircular: **3**. Overlapping central lobe: **1**, **2**.
Lip central lobe. Fan-shaped: **2**, **3**, **6**; kidney-shaped: **3**, **5**, **7**; fishtail-shaped: **2**, **4**, **5**.
Lip sinus. Broad and rounded: **7**; narrow and rounded: **2**; short and narrow: **1**; triangular: **3**, **4**, **5**.
Column. Fully exposed: **3**, **6**, **7**; partially or fully concealed: **1**, **2**, **4**, **5**, **8**, **9**.

1a. Fertile pollinia 4 — 2
b. Fertile pollinia 8 — 11
2a. Flowers lateral on slender leafless shoots — **7. walkeriana**
b. Flowers terminal on leafy stems — 3
3a. Stems bearing 2 or more leaves — 4
b. Stems bearing a single leaf — 10
4a. Lip simple, shallowly lobed, or with very small basal lateral lobes, never with deep sinuses — **6. bicolor**
b. Lip strongly 3- or 4-lobed — 5
5a. Claw of central lobe of lip at least one-quarter of lobe length — 9
b. Central lobe of lip shortly clawed or unclawed, separated from lateral lobes by narrow sinuses — 6
6a. Lateral lobes of lip strongly angled — **4. amethystoglossa**
b. Lateral lobes of lip rounded — 7
7a. Lip widest across central lobe; sinus broadly triangular or oblong — **3. aclandiae**
b. Lip widest across lateral lobes; sinus narrow or concealed by overlapping lobes — 8
8a. Petals narrower than sepals, usually 1 cm wide or less — **1. forbesii**
b. Petals equal to or broader than sepals, 1.5 cm wide or more — **2. intermedia**
9a. Column totally exposed; lateral lobes of lip absent or reduced to small basal projections — **6. bicolor**
b. Column partially or totally concealed; lateral lobes of lip well developed, triangular to oblong, strongly angled — **5. guttata**
10a. Flowers 10 cm wide or less — **8. lawrenceana**
b. Flowers more than 10 cm wide — **9. labiata**
11a. Lip free from column — 13
b. Lip fused to column at base — 12
12a. Leaves less than 3 cm wide; flowers 2–5, 2–3 cm wide; petals less than 1 cm wide — **10. cernua**
b. Leaves 3 cm wide or more; flowers solitary, 6–7 cm wide; petals 2 cm wide or more — **11. coccinea**
13a. Lip without internal keels or ridges — 14
b. Lip with internal keels or ridges — 16
14a. Column about half as long as lip or more — **12. perrinii**
b. Column distinctly less than half as long as lip — 15
15a. Petal length more than twice petal width — **13. grandis**
b. Petal length twice petal width or less — **14. purpurata**
16a. Flowering stem sheathed at base or with a small sheathing scale at each node — 17
b. Flowering stem without sheaths or scales — **17. pumila**
17a. Sepals and petals similar in shape — 18
b. Sepals distinctly narrower than petals — **17. pumila**
18a. Flowers 5 cm wide or more; sepals and petals orange to red — **15. cinnabarina**
b. Flowers less than 5 cm wide; sepals and petals white, yellow, pink or purple — **16. crispata**

1. C. forbesii Lindley (*C. pauper* (Vellozo) Stellfeld). Illustration: *Botanical Magazine*, 3265 (1833); Williams et al., Orchids for everyone, 105 (1980); Bechtel et al., Manual of cultivated orchid species, 179 (1981).

Epiphytic. Stems 10–20 × 1–1.5 cm, 3- or 4-noded, slightly swollen above, 2-leaved. Leaves 9–15 × 3–7 cm, oblong, leathery. Flowers 1–5, 9–11 cm wide. Sepals and petals 5–6 cm, pinkish yellow to pale brown; sepals to 1.5 cm wide; lateral sepals sickle-shaped; petals to 1 cm wide. Lip 4.5–5.5 × 3–3.5 cm, 3-lobed, off-white with pinkish brown veins and a central yellow stripe; lateral lobes semicircular to oblong, rolled around column, separated from the small central lobe by a short narrow sinus; central lobe 1.4–1.6 cm wide, white, almost circular. Column 4 cm. *S Brazil*. G2. Spring–summer.

C. dormaniana Reichenbach (x *Laeliocattleya dormaniana* (Reichenbach filius) Rolfe). Illustration: *American Orchid Society Bulletin* **45**: 978 (1976); Fowlie, Brazilian bifoliate Cattleyas, 17 (1977); Bechtel et al., Manual of cultivated orchid species, 178 (1981). Central lobe of lip larger (2.3–2.8 cm wide), dark purple, overlapped by lateral lobes when flattened out. Sometimes with 4 extra sterile pollinia. *S Brazil*. G2. Summer–autumn.

2. C. intermedia Graham. Illustration: *Botanical Magazine*, 2851 (1821); Fowlie, Brazilian bifoliate Cattleyas, 110 (1977); Bechtel et al., Manual of cultivated orchid species, 179 (1981).

Epiphytic or on rocks. Stems 15–40 × 1.5 cm, club-shaped, 2-leaved. Leaves 7–15 × 3.5–7 cm, elliptic to oblong, obtuse, fleshy. Flowers 3–10, 10–12.5 cm wide. Upper sepal 6–7 × 1.5 cm; lateral sepals shorter and broader. Petals 5.5–7 × 1.5–1.8 cm. Petals and sepals off-white to pale pink. Lip 5–6 × 4–4.6 cm, 3-lobed. Lateral lobes asymmetrically ovate, rolled around column, off-white to pale pink with irregularly toothed tips, separated from the central lobe by a narrow rounded sinus. Central lobe fishtail-shaped, purplish pink, minutely toothed. Column 2.5–3.5 cm, winged. *S Brazil*. G2. Autumn.

C. loddigesi Lindley. Illustration: Fowlie, Brazilian bifoliate Cattleyas, 123 (1977); Williams et al., Orchids for everyone, 106 (1980); Bechtel et al., Manual of cultivated orchid species, 180 (1981). Sepals and petals 2 cm wide or more, sparsely and finely spotted. Lateral lobes of lip broadly triangular, overlapping the semicircular central lobe. *S Brazil*. G2. Summer–winter.

C. harrisoniana Lindley (*C. loddigesii* var. *harrisoniana* Rolfe; *C. harrisoniae* Bateman). Illustration; *Botanical Magazine*, 4085 (1844); Fowlie, Brazilian bifoliate Cattleyas, 126 (1977). Stems slightly thicker above. Flowers deep pink, unspotted. Petals strongly sickle-shaped. Lateral lobes of lip rounded, pink with yellow veins; central lobe pale, wavy, slightly notched. *S Brazil*. G2. Spring & autumn.

3. C. aclandiae Lindley. Illustration: *Botanical Magazine*, 5039 (1858); Fowlie, Brazilian bifoliate Cattleyas, 32 (1977); Bechtel et al., Manual of cultivated orchid species, 178 (1981).

Epiphytic. Stems to 12 cm, 1- or 2-noded, slightly thicker above, 2-leaved. Leaves 4–10 × 2–3.5 cm, elliptic, fleshy, sparsely spotted with purple. Flowers 1 or 2 per raceme, 7–10 cm wide. Sepals and petals 4.5 × 1.5–2 cm, elliptic, yellowish green to brown, densely spotted with purplish brown; sepals acuminate; petals obtuse or rounded, wavy. Lip 4.5–5 × 3–3.5 cm, 3-lobed; lateral lobes semicircular, white to pale pink, not concealing column, separated from central lobe by a broad triangular sinus; central lobe kidney-shaped with a deep pointed notch, pink, minutely toothed, wavy. Column 2.5 cm, winged, deep pinkish purple. *E Brazil*. G2. Spring.

C. velutina Reichenbach. Illustration: Shuttleworth et al., Orchids, 62 (1970); Fowlie, Brazilian bifoliate Cattleyas, 20 (1977); Bechtel et al., Manual of cultivated orchid species, 182 (1981). Stems 20–30 cm, cane-like. Sepals and petals yellow with fine purple spots, wavy. Central lobe of lip fan-shaped, pale pink with darker veins and a broad yellow margin. *S Brazil*. G2. Summer.

4. C. amethystoglossa Linden & Reichenbach. Illustration: *Botanical Magazine*, 5683 (1868); Williams et al., Orchids for everyone, 105 (1980); Bechtel et al., Manual of cultivated orchid species, 178 (1981).

On rocks. Stems to 1 m, cane-like, slightly swollen above, with 4–6 nodes, 2-leaved. Leaves 16–23 × 5–7.5 cm, oblong to narrowly elliptic, obtuse, leathery, dark green. Flowers 6–12 per raceme, 7–10 cm wide, very pale pinkish purple spotted with dark purple; lip with darker lobes. Upper sepal 3.8–4.5 × 1.4–1.7 cm, ovate to oblong; lateral sepals shorter and broader, curved downwards. Petals 3.5–4.5 × 2–2.5 cm, lanceolate to obovate. Lip 3–4 × 2.7–3.5 cm, 3-lobed; lateral lobes ovate to triangular, with warty, slightly spathulate tips, rolled around column, separated from the central lobe by a narrow triangular sinus. Central lobe fishtail-shaped with warty ridges. Column 2 cm. *E Brazil*. G2. Spring–summer.

C. schilleriana Reichenbach. Illustration: Shuttleworth et al., Orchids, 61 (1970); Fowlie, Brazilian bifoliate Cattleyas, 36 (1977); Bechtel et al., Manual of cultivated orchid species, 181 (1981). Stems 15–20 cm, swollen. Leaves 6–10 cm, elliptic, spotted. Flowers 1 or 2 per raceme, to 9 cm wide. Sepals and petals yellow to brown, spotted. *E Brazil*. G2. Winter.

C. violacea (Humboldt, Bonpland & Kunth) Rolfe (*C. superba* Lindley). Illustration: *Botanical Magazine*, 4083 (1944); Fowlie, Brazilian bifoliate Cattleyas, 48 (1977); Bechtel et al., Manual of cultivated orchid species, 182 (1981). Flowers 1–5 per raceme. Sepals and petals pink, unspotted. *Venezuela, Guyana, Peru & Brazil*. G2. Spring–summer.

5. C. guttata Lindley (*C. elatior* Lindley). Illustration: *Botanical Magazine*, 3693 (1839); *American Orchid Society Bulletin* **45**: 979 (1976); Fowlie, Brazilian bifoliate Cattleyas, 102 (1977).

Epiphytic or on rocks. Stems 20–80 (rarely to 150) × 1–3 cm, slightly swollen above, with 3–8 nodes, 2- or 3-leaved. Leaves 15–25 × 5–7 cm, elliptic, leathery. Flowers 3–20 per raceme, 7.5–10 cm wide. Sepals and petals 3.5–4.5 cm × 8–10 mm, yellowish green to brown, boldly spotted with purplish red; lateral sepals shorter, sickle-shaped; petals slightly wavy. Lip 2.6–3.3 × 2.5–3 cm, 3-lobed; lateral lobes diamond-shaped, rounded at base, rolled around column, pale pink; central lobe fishtail-shaped, tapering gradually into a long claw, ridged, reddish purple. Column 2–2.5 cm. *S Brazil*. G2. Winter.

C. granulosa Lindley. Illustration: *Botanical Magazine*, 5048 (1858); Fowlie, Brazilian bifoliate Cattleyas, 71 (1977); Bechtel et al., Manual of cultivated orchid species, 179 (1981). Sepals and petals green, yellow or orange, finely spotted with red. Leaves 7–12 × 3.5–5 cm. Lip off-white to pale yellow; lateral lobes more or less triangular, not fully concealing column; central lobe kidney-shaped, tapering abruptly into a long narrow claw, pale pink, densely and finely spotted with purple. *E Brazil*. G2. Spring.

C. tigrina Richard (*C. leopoldii* Lemaire; *C. guttata* var. *leopoldii* (Lemaire) Rolfe). Illustration: *American Orchid Society Bulletin* **45**: 978 (1976); Fowlie, Brazilian bifoliate Cattleyas, 98 (1977); Bechtel et al., Manual of cultivated orchid species, 180 (1981). Flowering stem 10–25 cm. Central lobe of lip broad, fishtail-shaped, almost touching lateral lobes when flattened, abruptly contracted into claw, purple, ridged. *S Brazil*. G2. Summer.

C. elongata Barbosa Rodrigues. Illustration: *Botanical Magazine*, 7543 (1897); Fowlie, Brazilian bifoliate Cattleyas, 87 (1977). Very similar to *C. leopoldii*. Differs in its longer flowering stem (35–60 cm) and its unspotted, copper-coloured petals and sepals. *E Brazil*. G2. Autumn–winter.

6. C. bicolor Lindley. Illustration: *Botanical Magazine*, 4909 (1856); Fowlie, Brazilian bifoliate Cattleyas, 23 (1977); Bechtel et al., Manual of cultivated orchid species, 178 (1981).

Epiphytic. Stems 30–80 cm, slender, slightly thicker above, with 4–6 nodes, 2-leaved. Leaves 12–20 × 2–2.5 cm, oblong, leathery. Flowers 2–9 per raceme, 8–9 cm wide. Sepals 4.5–6 × 1–1.5 cm; lateral sepals broader, sickle-shaped. Petals 4.5–5 × 1.5–2 cm, wavy. Sepals and petals coppery to greenish brown. Lip 3.5 × 1.5 cm, without lobes, or with small basal flaps, long-clawed; blade fan-shaped, shallowly notched, wavy, minutely toothed, pink. Column 3 × 1.2 cm, pale pink, winged. *S Brazil*. G2. Summer–autumn.

7. C. walkeriana Gardner (*C. bulbosa* Lindley). Illustration: Shuttleworth et al., Orchids, 62 (1970); Fowlie, Brazilian bifoliate Cattleyas, 56 (1977); Bechtel et al., Manual of cultivated orchid species, 182 (1981).

Epiphytic or on rocks. Pseudobulbs 3–9 × 1–2 cm, 2- or 3-noded, bearing a single leaf (rarely 2-leaved). Leaves 5–10 × 3–5 cm, elliptic to ovate, rounded or notched. Flowers 1–3 per raceme, 8–9 cm wide, borne on slender, leafless, lateral shoots. Sepals and petals 4.5–5.5 cm, pinkish purple; sepals 1.2–1.5 cm wide, lanceolate, acuminate; petals 2.5–3.5 cm wide, diamond-shaped to obovate, rounded. Lip 3.5–4.5 × 2.5–3.5 cm, 3-lobed, broadest across central lobe. Lateral lobes triangular to obovate, pink, asymmetrically notched; sinus broad and rounded. Central lobe almost circular, notched, dark purplish red with a narrow central patch of yellow. Column 2.5–3 cm, pink, winged, exposed. *E Brazil*. G2. Spring.

C. nobilior Reichenbach. Illustration: *Orchid Digest* **36**: 33 (1972); Fowlie, Brazilian bifoliate Cattleyas, 68 (1977); Pabst & Dungs, Orchidaceae Brasilienses **1**: 192 (1975). Leaves 2 or 3. Flowers 10–12 cm wide. Petals and sepals pink; central lobe of lip pink with a large central yellow patch; lateral lobes of lip almost concealing column. *C Brazil*. G2. Summer.

8. C. lawrenceana Reichenbach. Illustration: *Botanical Magazine*, 7133 (1890); Bechtel et al., Manual of cultivated orchid species, 180 (1981).

Epiphytic. Stems to 25 cm, swollen above, 2- or 3-noded, bearing a single leaf. Leaves 8–20 × 4–5 cm, oblong, hard and leathery, sometimes purple-spotted. Flowers 3–8 per raceme, to 10 cm wide, pale to dark reddish purple, sometimes almost white; lip with a pale throat and a central reddish brown patch. Sepals 6.5–7 × 1.5–2 cm, elliptic to oblong; petals 6.5 × 2.5–4 cm, oblong to broadly elliptic, obtuse, wavy. Lip 5.5–6.3 × 4 cm, oblong to ovate, rolled around column. Column 2 cm. *Venezuela & Guyana*. G2. Spring–summer.

C. luteola Lindley. Illustration: *Botanical Magazine*, 5032 (1858); Bechtel et al., Manual of cultivated orchid species, 180 (1981). Flowers yellow, sepals and petals 5 cm or less. Lip to 3 cm, obscurely 3-lobed. *Ecuador, Peru, Brazil & Bolivia*. G2. Autumn.

9. C. labiata Lindley (*C. labiata* var. *labiata*). Illustration: *Botanical Magazine*, 3998 (1843); Bechtel et al., Manual of cultivated orchid species, 180 (1981).

Epiphytic. Stems to 30 cm, club-shaped, bearing a single leaf. Leaves 15–30 × 4–7 cm, oblong, obtuse, leathery. Flowers 3–5 per raceme, *c.* 15 cm wide. Sepals and petals 7–8 cm, pale pink; sepals 2–2.5 cm wide, lanceolate, acute; petals 5–7 cm wide, elliptic, wavy. Lip 7–9 × 4.5–7 cm, ovate, shallowly 3-lobed, rolled around column, bright reddish purple with a pink margin; throat white and yellow. Column 3–3.5 cm. *E Brazil*. G2. Autumn.

The *C. labiata* species complex consists of many different colour forms mainly from Colombia and Venezuela, most of which are in cultivation. They have been treated as varieties of *C. labiata* but are now usually regarded as separate species.

C. dowiana Bateman & Reichenbach filius (*C. labiata* var. *dowiana* (Bateman & Reichenbach filius) Veitch). Illustration: *Botanical Magazine*, 5618 (1867); Bechtel et al., Manual of cultivated orchid species, 178 (1981). Sepals and petals pale creamy yellow usually with red flecks; lip crimson to purple with gold veins, but without yellow patches towards the sides. *Costa Rica*. G2. Autumn.

Var. **aurea** (Linden) Williams & Moore (*C. aurea* Linden). Illustration: Withner, Cattleyas and their relatives **1**: 42, 43 (1988). Sepals and petals deep primrose yellow, usually without red flecks. Lip purple-red suffused with gold veins and yellow lateral zones. *Colombia*. G2. Autumn.

C. gaskelliana (N.E. Brown) Reichenbach filius (*C. labiata* var. *gaskelliana* N.E. Brown). Illustration: Bechtel et al., Manual of cultivated orchid species, 179 (1981). Stems fusiform. Flowers 15–18 cm wide. Sepals and petals white to pale pink; lip white to pale pink with a yellow throat and a small indistinct central patch of purplish pink separated from the yellow by an area of white or pale pink. *Colombia to Trinidad*. G2. Summer.

C. mendelii Linden & Rodigas (*C. labiata* var. *mendelii* (Linden & Rodigas) Reichenbach). Illustration: Bechtel et al., Manual of cultivated orchid species, 181 (1981); Orquideas nativas de Colombia **1**: 67 (1997). Differs from *C. gaskelliana* in its red-veined throat and larger central patch of purplish pink on the lip. *Colombia*. G2. Spring–summer.

C. mossiae C.S.P. Parish ex Hooker (*C. labiata* var. *mossiae* (Hooker) Lindley). Illustration: Bechtel et al., Manual of cultivated orchid species, 181 (1981). Very similar to *C. gaskelliana* and *C. mendelii*. Sepals and petals pink. Lip white; throat yellow with purple veins, which merge with a large central patch of purplish pink. *Northern Venezuela*. G2. Spring–summer.

C. schroederae Reichenbach filius (*C. labiata* var. *schroederae* (Reichenbach filius) Sander). Illustration: Bechtel et al., Manual of cultivated orchid species, 181 (1981); Orquideas nativas de Colombia **1**: 68 (1997). Very fragrant. Sepals and petals white or very pale pink. Lip pale pink with a purple tip. *NE Colombia*. G2. Winter–spring.

C. trianae Linden & Reichenbach filius (*C. labiata* var. *trianae* (Linden & Reichenbach filius) Duchartre). Illustration: Bechtel et al., Manual of cultivated orchid species, 182 (1981); Orquideas nativas de Colombia **1**: 61, 62 (1997). Differs from *C. schroederae* in its less fragrant flowers with a purple lip. *Colombia*. G2. Winter–spring.

C. candida (Kunth) F. Lehman (*C. quadricolor* Lindley; *C. chocoensis* Linden). Illustration: Orquideas nativas de Colombia **1**: 63 (1997). Regarded by some authors as a race of *C. trianae* to which it is very similar, differing by the lateral petals that are more triangular as opposed to ovate in *C. trianae*. *Colombia*. G2. Summer.

C. warneri Moore (*C. labiata* var. *warneri* (Moore) Veitch). Illustration: Bechtel et al., Manual of cultivated orchid species, 182 (1981). Flowers purplish pink; lip darker with a white and yellow throat. *SE Brazil*. G2. Summer.

C. warscewiczii Reichenbach filius (*C. labiata* var. *warscewiczii* (Reichenbach filius) Reichenbach filius; *C. gigas* Linden & André). Illustration: Bechtel et al., Manual of cultivated orchid species, 182 (1981); Orquideas nativas de Colombia **1**: 64–66 (1997). Flowers 17–20 cm wide. Lip mostly deep purplish crimson; throat greenish yellow, veined with crimson. *Colombia*. G2.

C. percivalliana (Reichenbach filius) O'Brien (*C. labiata* var. *percivalliana* Reichenbach filius). Illustration: Bechtel et al., Manual of cultivated orchid species, 181 (1981). Sepals and petals pinkish purple. Lip maroon, with a yellow throat, margin pink. *Colombia to W Venezuela*. G2. Winter–spring.

C. rex O'Brien. Illustration: Cogniaux & Goossens, Dictionnaire iconographique des orchidées, Cattleya, t. 22 (1899); *Botanical Magazine*, 8377 (1911). Sepals and petals white to very pale yellow. Lip purple-veined, yellow in basal half, apically pink with a white border. *Peru & Colombia*. G2. Summer.

C. maxima Lindley. Illustration: Bechtel et al., Manual of cultivated orchid species, 180 (1981). Petals about twice the width of the sepals. Flowers pale to deep pink. Lip strongly veined with purple, with a central yellow stripe. *Ecuador, Colombia & Peru*. G2.

10. C. cernua (Lindley) Van den Berg (*Cattleya cernua* (Lindley) Beer invalid; *Sophronitis cernua* Lindley; *Sophronitis pterocarpa* Lindley & Paxton). Illustration:

Botanical Magazine, 3677 (1839); Shuttleworth et al., Orchids, 69 (1970); Bechtel et al, Manual of cultivated orchid species, 274 (1981).

Pseudobulbs 1–2 cm, ovoid to cylindric, compressed. Leaves 2–2.8 × 1.5–2 cm, elliptic to oblong, slightly cordate at base, obtuse or abruptly pointed. Flowering stem to 6 cm, bearing 2–5 flowers; flowers 2–3 cm wide, reddish orange. Sepals and petals 1–1.5 cm × 5 mm; sepals ovate to lanceolate, acute; petals ovate to diamond-shaped, acute. Lip 1 cm × 8 mm, triangular to lanceolate, unlobed or shallowly 3-lobed, acute. Column 5 mm, club-shaped with 2 small, purple-tipped wings. *E Brazil*. G2. Winter.

11. C. coccinea Lindley (*Sophronitis coccinea* (Lindley) Reichenbach; *Sophronitis grandiflora* Lindley). Illustration: *Botanical Magazine*, 3709 (1839); Williams et al, Orchids for everyone, 196 (1980); Bechtel et al., Manual of cultivated orchid species, 274 (1981).

Pseudobulbs 1.5–3.5 cm, ovoid to fusiform. Leaves 3–6 × 1–1.5 cm, ovate to elliptic, fleshy, sometimes glaucous. Flowering stem to 8 cm, 1-flowered; flowers 6–7 cm wide, yellow to red or pinkish red. Sepals 1.5–2.5 × 1 cm, elliptic to oblong, acute; petals 2.5–3.5 × 2–3 cm, circular to broadly ovate, rounded. Lip 1.5–2 × 1.5–2 cm, 3-lobed; lateral lobes triangular, rolled around column; central lobe narrowly triangular. Column 5–10 mm, club-shaped, partly concealed. *E Brazil*. G2. Winter.

12. C. perrinii Lindley (*Sophronitis perrinii* (Lindley) Van den Berg & Chase; *Laelia perrinii* (Lindley) Bateman). Illustration: *Botanical Magazine*, 3711 (1840); *Orchid Digest* **41**:149 (1977); Bechtel et al., Manual of cultivated orchid species, 219 (1981).

Epiphytic. Stem 7–30 cm, ovoid to club-shaped, 1-leaved. Leaves 20–30 × 3–4.5 cm, oblong to strap-shaped, obtuse or slightly notched, leathery, often spotted underneath. Flowering stem 10–25 cm, robust, 2- or 3-flowered, basally sheathed; flowers 10–13 cm wide. Upper sepal 6.5 × 1.5 cm, lanceolate; lateral sepals smaller, sickle-shaped; petals 6.5 × 2 cm, ovate; sepals and petals pink. Lip 5 × 3.5 cm, 3-lobed; lateral lobes obtuse, white to pale pink, sometimes edged with purple; central lobe oblong to circular, crisped, with a broad purple margin and a pale yellow throat, lacking keels or ridges. Column 3–4 cm, white to pale pink, partially concealed. *SE Brazil*. G2. Autumn.

C. xanthina (Lindley) Van den Berg (*Sophronitis xanthina* (Lindley) Van den Berg & Chase; *Laelia xanthina* Lindley). Illustration: *Botanical Magazine*, 5144 (1859); Pabst & Dungs, Orchidaceae Brasilienses **1**: 210 (1975); Bechtel et al., Manual of cultivated orchid species, 220 (1981). Flowers 4 or 5 per flowering stem, less than 8 cm wide; sepals and petals yellow; petal margins recurved. Lip white and yellow, striped with pink. *E Brazil*. G2. Spring–summer.

C. crispa Lindley (*Sophronitis crispa* (Lindley) Van den Berg & Chase; *Laelia crispa* (Lindley) Reichenbach). Illustration: *Botanical Magazine*, 3910 (1842); *Orchid Digest* **41**: 144 (1977); Bechtel et al., Manual of cultivated orchid species, 218 (1981). Sepals and petals white or very pale purple, tinged with pink at base; petals crisped and wavy. Lip strongly crisped, lateral lobes white, central lobe purple with darker veins and a white margin, throat yellow. *S Brazil*. G2. Summer.

13. C. grandis (Lindley & Paxton) Chadwick (*Sophronitis grandis* (Lindley & Paxton) Van den Berg & Chase; *Laelia grandis* Lindley & Paxton). Illustration: *Botanical Magazine*, 5553 (1866); Graf, Tropica, 733 (1978); Bechtel et al., Manual of cultivated orchid species, 219 (1981).

Epiphytic. Stem 20–30 cm, club-shaped, 1-leaved. Leaves 20–35 × 3–4.5 cm, strap-shaped, obtuse. Flowering stem to 30 cm, basally sheathed, bearing 2–5 flowers; flowers 10–13 cm wide. Sepals 5–7 × 1–1.5 cm, lanceolate; petals 5–5.5 × 1.8–2 cm, diamond-shaped to elliptic, wavy. Petals and sepals greenish yellow to orange. Lip 5.5–6 × 3.5 cm, 3-lobed, white with purple veins; lateral lobes obtuse; central lobe oblong, obtuse, crisped and wavy, lacking keels or ridges. Column 2 cm, partially concealed. *S Brazil*. G2. Summer.

C. tenebrosa (Rolfe) Chadwick (*Sophronitis tenebrosa* (Rolfe) Van den Berg & Chase; *Laelia tenebrosa* Rolfe; *Laelia grandis* var. *tenebrosa* Gower). Illustration: Ospina & Dressler, Orquideas de las Americas, t. 64 (1974); *Orchid Digest* **41**: 144 (1977); Bechtel et al., Manual of cultivated orchid species, 220 (1981). Flowers 14–19 cm wide. Sepals 8–10 × 1.6–2 cm; petals 7–10 × 2.5–3.5 cm; petals and sepals greenish purple. Lip 6.5–8.5 × 4–6 cm, pale pinkish purple, darker inside with a deep purple throat. *E Brazil*. G2. Summer.

14. C. purpurata (Lindley & Paxton) Van den Berg (*Sophronitis purpurata* (Lindley & Paxton) Van den Berg & Chase; *Cattleya purpurata* (Lindley & Paxton) Beer invalid; *Laelia purpurata* Lindley). Illustration: Graf, Tropica, 737 (1978); Williams et al., Orchids for everyone, 107 (1980); Bechtel et al., Manual of cultivated orchid species, 220 (1981).

Epiphytic. Stem to 30 × 3 cm, fusiform to club-shaped, 1-leaved. Leaves 20–30 × 5 cm, oblong, rounded, leathery. Flowering stem to 30 cm, basally sheathed, bearing 2–7 flowers; flowers 15–20 cm wide. Sepals 7.5–11 × 2 cm, lanceolate to strap-shaped; petals 8–10 × 4–6 cm, oblong to ovate; petals and sepals white to very pale pink. Lip 7–10 × 6.5–8 cm, obscurely 3-lobed, wavy; lateral lobes curled around column; central lobe more or less circular; throat yellow, lateral lobes deep purple, margins paler, all veined with darker purple, lacking keels or ridges. Column 2.5–3 cm, pale green, concealed. *S Brazil*. G2. Early summer.

C. lobata Lindley (*Sophronitis lobata* (Lindley) Van den Berg & Chase; *Laelia lobata* (Lindley) Veitch; *Laelia boothiana* Reichenbach). Illustration: Pabst & Dungs, Orchidaceae Brasilienses **1**: 208 (1975); *Orchid Digest* **41**: 144 (1977). Flowers 13–15 cm wide, pink. Lip 6 × 5 cm with dark veins, throat deep pink. *Brazil*. G2. Spring.

15. C. cinnabarina (Bateman ex Lindley) Van den Berg (*Sophronitis cinnabarina* (Bateman ex Lindley) Van den Berg & Chase; *Laelia cinnabarina* Lindley; *Cattleya cinnabarina* (Bateman ex Lindley) Beer invalid; *Hoffmannseggella cinnabarina* (Bateman ex Lindley) H.G. Jones). Illustration: *Botanical Magazine*, 4302 (1847); Williams et al., Orchids for everyone, 191 (1980); Bechtel et al., Manual of cultivated orchid species, 218 (1981).

On rocks. Stems 10–25 cm, swollen below, 1- or 2-leaved. Leaves 10–25 × 3–3.5 cm, linear to oblong, acute, often purplish green. Flowering stems 35–50 cm, basally sheathed, bearing 5–15 flowers; flowers 5–6 cm wide, orange to red. Sepals and petals 3–3.5 cm × 5 mm, linear to lanceolate, acute. Lip

2–2.5 × 1 cm, 3-lobed, lateral lobes acute, curled around column; central lobe narrowly diamond-shaped to ovate, acute, with 3 ridges, crisped and reflexed. Column club-shaped, mostly concealed. *SE Brazil.* G2. Spring.

C. harpophylla (Reichenbach) Van den Berg (*Sophronitis harpophylla* (Reichenbach) Van den Berg & M.W. Chase; *Laelia harpophylla* Reichenbach; *Hoffmannseggella harpophylla* (Reichenbach) H.G. Jones). Illustration: Ospina & Dressler, Orquideas de las Americas, t. 63 (1974); Graf, Tropica, 733 (1978); Bechtel et al., Manual of cultivated orchid species, 219 (1981). Epiphytic; stems very slender. Flowering stem shorter than leaves, bearing 4–7 flowers. *SE Brazil.* G2. Winter–spring.

16. C. crispata (Thunburg) Van den Berg (*Sophronitis crispata* (Thunburg) Van den Berg & Chase; *Laelia crispata* (Thunberg) Garay; *Laelia crispilabia* Richard; *Laelia rupestris* Lindley; *Laelia flava* Lindley; *Cattleya flava* (Lindley) Beer invalid; *Hoffmannseggella flava* (Lindley) H.G. Jones). Illustration: *American Orchid Society Bulletin* **43**: 992 (1974); Pabst & Dungs, Orchidaceae Brasilienses **1**: 212 (1975); Northen, Miniature orchids, C-37 (1980).

On rocks. Stems 4–20 cm, not swollen, 1-leaved. Leaves 10–15 × 3–3.5 cm, strap-shaped to oblong, obtuse, leathery. Flowering stem 15–45 cm, with a single basal sheath, bearing 2–10 flowers; flowers 3–5 cm wide. Sepals and petals 2–2.5 cm × 8–10 mm, linear or oblong to elliptic, obtuse to rounded or acute, pink with pale centres. Lip 1.5–2 × 1.2 cm, 3-lobed; lateral lobes oblong, curled around column, pale pink; central lobe ovate to oblong, crisped and wavy, yellow with a pink margin, 2-ridged. Column concealed. *SE Brazil.* G2. Winter.

This very variable species has several colour forms which have been treated as species in the past.

C. longipes (Reichenbach) Van den Berg (*Sophronitis longipes* (Reichenbach) Van den Berg & Chase; *Laelia longipes* Reichenbach). Illustration: *Botanical Magazine*, 7541 (1897); *American Orchid Society Bulletin* **43**: 993 (1974); Bechtel et al., Manual of cultivated orchid species, 218 (1981). Stems conical to ovoid. Sepals and petals white, very pale pink or pale yellow. Lip pale yellow with a deep yellow throat. *SE Brazil.* G2. Summer.

17. C. pumila Hooker (*Sophronitis pumila* (Hooker) Van den Berg & Chase; *Laelia pumila* (Hooker) Reichenbach; *Cattleya marginata* Paxton; *Cattleya spectabilis* Pabst; *Cattleya pinellii* Lindley). Illustration: *Botanical Magazine*, 3656 (1839); Northen, Miniature orchids, C-40 (1980); Bechtel et al., Manual of cultivated orchid species, 220 (1981).

Epiphytic. Stems 2–3 cm, ovoid to fusiform, 1-leaved. Leaves 10–13 × 2–2.5 cm, linear to oblong, obtuse. Flowering stem 4–10 cm, 1(rarely 2)-flowered; flowers 8–12 cm wide. Sepals 4.5 × 1–1.5 cm, oblong to lanceolate, acute; petals 3.5–4.5 × 2–3 cm, elliptic, acute or obtuse, wavy; petals and sepals pinkish purple. Lip 4–4.5 × 3 cm, shallowly 3-lobed or unlobed, basal part pinkish purple, curled around column; central lobe oblong to circular, with 3–5 ridges, wavy, sometimes notched, dark purple with a white or pale yellowish green throat. Column 2–2.5 cm, concealed. *SE Brazil.* G2. Autumn.

C. jongheana (Reichenbach) Van den Berg (*Sophronitis jongheana* (Reichenbach) Van den Berg & Chase; *Laelia jongheana* Reichenbach). Illustration: *Botanical Magazine*, 6038 (1873); Northen, Miniature orchids, C-41 (1980); Bechtel et al., Manual of cultivated orchid species, 285 (1981). Leaves 3.5–4 cm wide. Sepals and petals 6–7 cm. Lip 5.5 × 3 cm; central lobe white, throat white and yellow with 7 wavy keels. *SE Brazil.* G2. Spring.

C. lundii (Reichenbach & Warming) Van den Berg (*Sophronitis lundii* (Reichenbach & Warming) Van den Berg & Chase; *Laelia lundii* Reichenbach & Warming; *Laelia regnellii* Rodrigues). Illustration: Northen, Miniature orchids, C-42 (1980); Skelsey, Orchids, 117 (1980); Bechtel et al., Manual of cultivated orchid species, 219 (1981). Leaves 2 per stem, 8–10 cm × 5 mm, linear, channelled and pointed. Sepals and petals 2–2.5 cm × 3–4 mm, white. Lip clearly 3-lobed, white with pink veins, wavy. *SE Brazil.* G2.

55. GUARIANTHE Dressler & W.E. Higgins

J.M.H. Shaw

Epiphytic herbaceous plants with swollen stems. Leaves 2, conduplicate, leathery. Inflorescence terminal, racemose, subtended by a papery, spathe-like bract. Flowers inverted with free sepals. Petals obovate, wider than sepals. Lip wrapped around the column, often with a wide spreading limb, not lobed. Column club-shaped, with a fleshy mid-tooth clasping the anther. Pollinia 4.

A genus of 4 species from Mexico, Central and northern South America, previously placed in Cattleya, from which it can be distinguished by the small slender column.

Literature: Fowlie, J.A., Observations on Cattleya skinneri and C. deckeri, *American Orchid Society Bulletin* **36**: 777–80 (1967); Hamer, F., Flora de Nicaragua **2**: 1640–2 (2001); Dressler, R.L. & Higgins, W.E., Guarianthe, a generic name for the 'Cattleya' skinneri complex, *Lankesteriana* **7**: 37–8 (2003).

1a. Flowers rich orange to yellow 2
b. Flowers purple to lilac, red or white 3
2a. Flowers orange, lip to 2.5 cm long, marked with red **1. aurantiaca**
b. Flowers yellow, lip over 2.5 cm long, often without red lines **4. × guatemalensis** var. **pachechoi**
3a. Lip 2.5–3 cm long, flower colour variable **4. × guatemalensis**
b. Lip 3.5 cm or longer, flower colour lilac to purple 4
4a. Flowers rosy-purple, lip 5 cm or more long **3. skinneri**
b. Flowers lilac to purple, lip 3.5 to 4 cm long **2. bowringiana**

1. G. aurantiaca (Lindley) Dressler & W.E. Higgins (*Cattleya aurantiaca* (Lindley) P.N. Don.) Illustration: Shuttleworth et al., Orchids, 58 (1970); Williams et al., Orchids for everyone, 105 (1980); Bechtel et al., Manual of cultivated orchid species, 178 (1981); Withner, Cattleyas and their relatives **1**: 41 (1988).

Epiphytic. Stems 20–35 cm, club-shaped, 2-leaved. Leaves 5–20 × 2.5–5.5 cm, ovate to elliptic, notched. Flowers several per raceme, 3–4 cm wide, orange to scarlet. Sepals 1.5–2.5 cm × 5 mm, lanceolate, acute. Petals 1.5–2.5 cm × 5 mm, ovate to lanceolate, acute. Lip 1.5–2.5 × 1 cm, ovate to elliptic; base rolled around column. Column 1 cm. *Mexico, Guatemala, Honduras & El Salvador.* G2. Summer.

2. G. bowringiana (O'Brien) Dressler & W.E. Higgins (*Cattleya bowringiana* O'Brien) Illustration: *Botanical Magazine*, n.s., 451 (1964); Shuttleworth et al., Orchids, 60 (1970); Williams et al., Orchids for everyone, 105 (1980).

Terrestrial, on rock or sand. Stems 20–35 cm × 5–10 mm, 4- or 5-noded, swollen at base, bearing 2 (rarely 3) leaves. Leaves 13–20 × 5–6 cm, oblong to elliptic, leathery, dark green. Flowers 5–20 (rarely to 45) per raceme, 6–7.5 cm wide, purple to magenta; lip darker, pale-throated. Sepals 3.8–5 × 1–1.4 cm, oblong to elliptic. Petals 4–5 × 2.5–3 cm, oblong to ovate. Lip 3.5–4 × 2.8–3.5 cm, oblong to ovate, base rolled around column. Column 1–1.2 cm, winged. *Belize & Guatemala.* G2. Autumn–winter.

3. G. skinneri (Bateman) Dressler & W.E. Higgins (*Cattleya skinneri* Bateman) Illustration: *Botanical Magazine*, 4270 (1846); Williams et al., Orchids for everyone, 106 (1980); Bechtel et al., Manual of cultivated orchid species, 181 (1981).

Epiphytic or terrestrial. Similar to *G. bowringiana* but stems pseudobulbous (2–3 cm in diameter) and flowers larger (7.5–10 cm wide), with a rounded lip. Flowers 4–12 per raceme. *Mexico to Costa Rica; ?Panama.* G2. Winter–spring.

G. deckeri (Klotzsch) Roeth. (*Cattleya deckeri* Klotzsch; *Cattleya skinneri* var. *autumnalis* L.O. Williams.) Similar to *G. skinneri* but differs in the acute, not rounded lip. *Mexico to Colombia, Venezuela.* G2. Autumn.

4. G. × guatemalensis (T. Moore) W. E. Higgins; *G. aurantiaca* × *skinneri* (*Cattleya* × *guatemalensis* T. Moore). Illustration: Withner, Cattleyas and their relatives **1**: 64–5 (1988).

Intermediate between the parents, and very variable morphologically, particularly in flower colour, which varies from deep red through pink to orange, and white. *Mexico to Nicaragua.* G2.

Var. **pachecoi** (Ames & Correll) J.M.H. Shaw (*Cattleya* × *pachecoi* Ames & Correll). Illustration: Withner, Cattleyas and their relatives **1**: 64, f. 24 (1988). Derived from *G. aurantiaca* × *guatemalensis*. Distiguished by yellow flowers with white column. *Guatemala, Nicaragua.* G2.

56. MEIRACYLLIUM Reichenbach

J.C.M. Alexander

Creeping rhizomatous epiphytes. Stems short, scaly, 1-leaved, slightly swollen, simple. Leaves stalkless, thick and leathery, folded when young. Flowers solitary or several, borne at junction of leaf and stem. Sepals similar, free except at the base where they are fused to the column foot, forming a short mentum. Petals narrower than sepals, spreading. Lip simple, deeply concave or pouched. Pollinia 8, in 2 groups of 4, narrowly club-shaped, waxy, unequal. Rostellum prominent.

A genus of 2 species from the Pacific coast of Mexico, Guatemala and El Salvador. Cultivation conditions are similar to those for *Cattleya* (p. 498), though night temperatures should be higher in winter.

Literature: Dressler, R.L., The relationships of Meiracyllium, *Brittonia* **12**: 222–5 (1960).

1. M. wendlandii Reichenbach (*M. gemma* Reichenbach). Illustration: *Orchid Digest* **41**: 69 (1977); *Die Orchidee* **30**: 76 (1979); Northen, Miniature orchids, C-57 (1980).

Stems to 1 cm, ascending. Leaves to 5 × 2.3 cm, oblong, obtuse to rounded. Flowering stems to *c.* 4 cm, 1-flowered. Flowers pinkish purple. Upper sepal 1–1.7 cm × 5 mm, ovate, acute or slightly acuminate; lateral sepals similar but narrower and slightly asymmetric. Petals 8–15 × 1–2.5 mm, oblong to lanceolate, acute, finely toothed. Lip 1–1.3 cm × 7 mm, ovate, concave; apex acuminate and curved downwards. Column *c.* 6 mm, acute, slightly narrowed at base. *Mexico, Guatemala & El Salvador.* G2. Summer.

M. trinasutum Reichenbach. Illustration: *Orchid Digest* **41**: 69 (1977); Northen, Miniature orchids, C-56 (1980); Bechtel et al., Manual of cultivated orchid species, 231 (1981).

Leaves 2.8–5 × 1.5–3.5 cm, elliptic to circular. Flowers several, pale pinkish purple, almost white below. Sepals 8–11 × 3.5–5 mm, oblong to elliptic, acuminate. Petals 7–10 × 3 mm, elliptic to spathulate, acute. Lip 7–9 × 4–5 mm, deeply pouched, acuminate. Column broad at base, round-tipped. *Mexico & Guatemala.* G2. Summer.

57. LAELIA Lindley

J.C.M. Alexander & J.M.H. Shaw

Epiphytic or terrestrial, often on rocks. Rhizomes creeping, bearing erect, solid, pseudobulbous or cane-like stems of several nodes. Leaves 1–3 per stem, folded when young. Flowers in terminal racemes, showy, resupinated or not. Sepals and petals free, rarely wavy. Lip 3-lobed, internally ridged or keeled; lateral lobes curved around column. Column free from lip; apical teeth lacking or 1–3, the middle one recurved and adpressed to anther cap. Pollinia 8. Ovary thickened towards the apex, 6-grooved, smooth or warty.

A genus of about 25 species extending from northern Mexico south to Brazil, Bolivia and the islands of Cuba and Jamaica. They are often used in multi-generic hybrids with *Cattleya, Guarianthe* and *Brassavola or Rhyncholaelia.* The Brazilian species formerly included in *Laelia* are now treated under *Sophronitis.* Most *Schomburgkia* species are also included here, except for *Schomburgkia* Section Chauno-schomburgkia Schlechter, which is transferred to *Myrmecophila* Rolfe. Many of the species were originally described in *Bletia* or *Cattleya* and can only reliably be distinguished from the latter on the number of pollinia. Cultural treatment should be as for *Cattleya* (p. 498), though the Mexican species need a longer and cooler resting period.

Literature: Jones, H.G., Review of sectional division in the genus Laelia of the Orchidaceae, *Botanischer Jahrbücher* **97**: 309–16 (1976); Halbinger, F. & Soto, M. Laelias of Mexico (1997).

Stem. Club-shaped: **6, 7**; conical, ovoid or fusiform: **1, 2, 3, 4, 5**. Ribbed: **1, 3**. 7 cm or less: **1, 2, 3, 4**; 7 cm or more: **1, 2**. 1-leaved: **1, 2, 4**; 2-leaved: **1, 2, 3**; 3-leaved: **1**.

Leaves. 20 cm or more: **1, 2**. 1.5–2.5 cm wide: **3**; 2.5–4.5 cm wide: **1, 2, 3, 4**; more than 4.5 cm wide: **5, 6, 7**.

Flowering stem. 10 cm or less: **4**; 10–25 cm: **3, 4**; 25–50 cm: **1, 2, 3, 7**; more than 50 cm: **5, 6**. Unsheathed: **4**; with several small sheathing scales: **1, 2, 3**.

Flowers. 1 per stem: **1, 4**; 2–7 per stem; **1, 2, 3, 4, 7**; 8–15 per stem: **1, 3, 5, 6, 7**. 6 cm wide or less: **2, 3, 7**; 7–12 cm wide: **1, 2, 5, 6**; 13 cm wide or more: **4**.

Sepals and petals. Pink to purple: **1, 2, 3, 4, 5, 7**; white: **3**; yellow, orange or red: **1, 6**. 4.5 cm or less: **3, 6, 7**; 5–7 cm: **1, 2, 5**; more than 7.5 cm: **4**.

Petals. Wider than sepals: **1, 2, 3, 4**; as wide as sepals: **5, 6, 7**. Less than 1 cm wide: **3**; more than 4 cm wide: **4**. Crisped: **7**.

Lip. Clearly 3-lobed: **1, 2, 3, 4**. Crisped: **5, 6**. With reflexed central lobe: **1, 3**. 3 × 2 cm or less: **3**; 4 × 2.5 cm or more: **1, 2, 4**. 1-ridged: **2, 4**; 2-ridged: **4**; 3-ridged: **3, 4**; 4 or 5 ridged: **5, 6, 7**.

Column. Concealed: **2, 3**; exposed or partly exposed: **1, 4, 5, 6, 7**.

1a. Column slender with 3 blunt apical teeth; sepals and retals rarely wavy 2

b. Column stout with 1 blunt apical tooth; sepals and petals usually wavy 5
2a. Flowering stems without sheaths or scales **4. speciosa**
b. Flowering stems sheathed at base or with small sheathing scale at each node 3
3a. Flowers 7 cm wide or more; lip 4 × 2.5 cm or more 4
b. Flowers 6 cm wide or less; lip 2.5 × 1.6 cm or less **3. rubescens**
4a. Lip with 3 longitudinal ridges; central lobe acute or acuminate **1. autumnalis**
b. Lip with 1 longitudinal ridge; central lobe blunt **2. anceps**
5a. Flowers more than 9 cm wide **5. superbiens**
b. Flowers less than 9 cm wide 6
6a. Flowers less than 6 cm wide; lip whitish, sometimes spotted **6. gloriosa**
b. Flowers 6 cm or more; lip pink-purple **7. undulata**

1. L. autumnalis Lindley (*L. gouldiana* Reichenbach). Illustration: *Botanical Magazine*, 3817 (1841); *Orchid Digest* **42**: 20 (1978); Bechtel et al., Manual of cultivated orchid species, 218 (1981).

Epiphytic or on rocks. Stems 10–15 cm, conical to pear-shaped, 2- or 3-leaved, ribbed. Leaves 12–20 × 3–4 cm, oblong to lanceolate, acute or obtuse, leathery. Flowering stems 30–100 cm, with several small sheathing scales, bearing 4–10 flowers; flowers 7.5–12 cm wide. Sepals and petals 5–6.5 cm, acuminate, pinkish purple with darker streaks; sepals 1.3 cm wide, lanceolate; petals 1.8 cm wide, ovate. Lip 4–5 × 2.5–3 cm, 3-lobed; lateral lobes rounded, partially enclosing column, white; central lobe oblong, acuminate, reflexed, purple with a white and yellow base, 3-ridged. Column 2.5–3 cm. *S Mexico*. G2. Winter.

L. furfuracea Lindley (*Laelia. anceps* var. *furfuracea* (Lindley) Rolfe). Illustration: *Botanical Magazine*, 3810 (1841); Northen, Miniature orchids, C-40 (1980). Stems 4–6 cm, ovoid, deeply grooved, 1-leaved. Flowers 1–3 per flowering stem; ovary mealy-glandular. *S Mexico*. G2. Winter.

2. L. anceps Lindley. Illustration: *Botanical Magazine*, 3804 (1841); Williams et al., Orchids for everyone, 107 (1980); Bechtel et al., Manual of cultivated orchid species, 218 (1981).

Similar to *L. autumnalis*. Stems 7–12 cm, flattened, 1- or 2-leaved. Flowering stem bearing 2–5 flowers. Petals narrowly elliptic to elliptic. Lip with 3 central longitudinal ridges, lateral lobes pinkish purple outside, yellow with purple veins and border inside; central lobe oblong, blunt, dark reddish purple, 1-ridged. *E & SE* Mexico, Guatemala to Honduras. G2. Winter.

Subspecies **dawsonii** (J. Anderson) Rolfe. Illustration: Halbinger & Soto, Laelias of Mexico, 62, 65 (1997). Distinguished from typical *L. anceps* by petals rhombic to elliptic-rhombic, lip with 3–5 central longitudinal ridges, and central lobe obovate to orbicular. The flowers are either white or pale pink with a dark purple lip. *Mexico (adjacent to Pacific)*. G2

A hybrid of garden origin known as **L. Autoceps** grex, derived from *L. anceps* × *autumnalis* (illustration: Halbinger & Soto, Laelias of Mexico, 123, t. 70, 1997) is occassionally grown.

L. aurea Halbinger & Salazar (*Laelia rubescens* var. *aurea* invalid) Illustration: Halbinger & Soto, Laelias of Mexico, 76, 78, 81 (1997). Stems often purple tinged, with longitudinal furrows and transverse pleats, with1 leaf. Inflorescence a compact raceme. Flowers yellow, fragrant, lip with red spot towards the base, and with short hairs on both surfaces. Middle of central lobe notched. *NW. Mexico*. G2. Autumn–winter.

3. L. rubescens Lindley (*L. acuminata* Lindley). Illustration: *Botanical Magazine*, 4099 (1844); *Orchid Digest* **42**: 20 (1978); Graf, Tropica, 733 (1978).

Epiphytic or on rocks. Stems 2.5–4.5 × 1.5–2.5 cm, clustered, ovoid, flattened, smooth, becoming furrowed, 1-leaved. Leaves 10–14 × 3–3.5 cm, oblong to lanceolate, obtuse, channelled, leathery. Flowering stem 25–50 cm, with several small sheathing scales, bearing 2–10 flowers; flowers 5–6 cm wide. Sepals 3.5 cm × 5 mm, lanceolate, acute; petals 2.5–3 cm × 8–10 mm, ovate to oblong, acute, wavy; sepals and petals white to pinkish purple. Lip 2.5–3 × 1.5 cm, 3-lobed; lateral lobes oblong, tips flared; central lobe oblong, acute, reflexed, wavy, white to pinkish purple; throat with dark purple patch surrounded by yellowish green. Column 8 mm, concealed. *S Mexico, Guatemala, Nicaragua & Costa Rica*. G2. Winter.

Forma **peduncularis** (Lindley) Halbinger (*L. peduncularis* Lindley) differs by its much longer flower stalks, larger flowers with entire margins to the petals and short recurved lip lobes. *Mexico (Guerrero, Oaxaca)*. G2. Autumn.

L. albida Lindley. Illustration: *Botanical Magazine*, 3957 (1842); Graf, Tropica, 733 (1978); Bechtel et al., Manual of cultivated orchid species, 217 (1981). Leaves 2 per stem, less than 1.5 cm wide, linear to lanceolate. Flowers white, faintly flushed with pink; lip 3-ridged. *Mexico*. G2. Winter.

4. L. speciosa (Kunth) Schlechter (*L. grandiflora* Lindley; *L. majalis* Lindley). Illustration: *Botanical Magazine*, 5667 (1867); *Die Orchidee* **22**: 239 (1974); *Orchid Digest* **42**: 21 (1978).

Epiphytic. Stems 3–5 cm, ovoid to fusiform, clustered, 1-leaved. Leaves 10–15 cm, oblong to lanceolate, acute or obtuse, stiff. Flowering stems 10–20 cm, lacking sheathing scales, bearing 1 or 2 flowers; flowers 15–20 cm wide. Sepals 10–12 × 1.5–2.5 cm, lanceolate, acute; petals 8–10 × 4.5–6 cm, ovate to diamond-shaped, obtuse, wavy; sepals and petals pale pinkish purple. Lip 7.5–9 × 4.5–5.5 cm, deeply 3-lobed; lateral lobes ovate to oblong, rounded, flared away from column, white to pale pinkish purple, edged with darker pink; central lobe oblong to circular, with 1–3 yellow keels, notched, white with pink streaks and a broad pink or pinkish purple border. Column exposed. *C Mexico*. G2. Spring–summer.

5. L. superbiens Lindley (*Schomburgkia superbiens* (Lindley) Rolfe). Illustration: *Botanical Magazine*, 4090 (1844); *Orchid Digest* **42**: 21 (1978); Bechtel et al., Manual of cultivated orchid species, 272 (1981).

Pseudobulbs to 1 m or more, fusiform, flattened, 1- or 2-leaved. Leaves to 35 × 6 cm, oblong, acute, leathery and stiff. Flowering stem to 1 m or more, robust, bearing 12–20 flowers; flowers 10–12 cm wide, fragrant; ovary plus flower-stalk 6–8 cm; bracts 5–7 cm. Sepals and petals 5–7 × 1.3–1.8 cm, oblong to narrowly lanceolate, acute, pink to pinkish purple, paler at base; petals slightly wavy. Lip 4–5.5 × 3–3.5 cm; lateral lobes rounded, crisped, greenish yellow with dark purple veins and a pink margin; central lobe ovate, notched, wavy, with 5 yellow ridges and a pink blade. Column 2.5–3 cm, club-shaped, curved. *Mexico, Guatemala & Honduras*. G2. Winter.

6. L. gloriosa (Reichenbach) Williams (*Schomburgkia gloriosa* Reichenbach; *S. crispa* Lindley 1844 not 1838; *S. crispa* misapplied, not *Laelia crispa* (Lindley) Reichenbach; *Schomburgkia fimbriata* Hoehne). Illustration: Pabst & Dungs, Orchidaceae Brasilienses **1**: 218 (1975); Dunsterville & Garay, Venezuelan orchids illustrated **6**: 398 (1976); Bechtel et al., Manual of cultivated orchid species, 383 (1981).

Epiphytic. Pseudobulbs to 30 × 5 cm, club-shaped with long tapering bases, ridged, 2- or 3-leaved. Leaves to 35 × 6 cm, oblong to lanceolate, acute or obtuse. Flowering stem to 75 cm, slender, bearing 8–15 flowers; flowers 7.5–8.5 cm wide; ovary plus flower-stalk to 9 cm; bracts to 6 cm. Sepals and petals 3.5–4 × 1 cm, oblong, obtuse (sepals each with a small point), wavy, yellowish brown with darker veins. Lip 2 × 2 cm, white to pale pink with yellowish brown margins and central lobe; lateral lobes kidney-shaped to ovate; central lobe broadly triangular, rounded, crisped, curled downwards, 3-ridged. Column 2 cm. *Venezuela, Guyana & Surinam*. G2. Winter.

L. lyonsii (Lindley) Williams (*Schomburgkia lyonsii* Lindley). Illustration: *Botanical Magazine*, 5172 (1860); *Gardeners' Chronicle* **26**: 203 (1899). Flowers *c.* 5 cm wide, cream-coloured with purple spots. *Jamaica*, G2. Summer–autumn.

7. L. undulata (Lindley) Williams (*Schomburgkia undulata* Lindley). Illustration: Dunsterville & Garay, Venezuelan orchids illustrated **2**: 256 (1961); Skelsey, Orchids, 38 (1979); Bechtel et al., Manual of cultivated orchid species, 272 (1981).

Epiphytic. Pseudobulbs to 25 cm, club-shaped with long tapering bases, ridged, 2-leaved. Leaves to 30 × 5 cm, lanceolate to oblong, obtuse, leathery. Flowering stem to 50 cm, slender, bearing 5–20 flowers; flowers 5–6 cm wide; ovary plus flower-stalk 5–7 cm, purplish green; bracts *c.* 5 cm. Sepals and petals 3–3.5 × 1.5 cm, oblong, acute, crisped and very wavy, purplish brown. Lip 3 × 1.5–2 cm, purple; lateral lobes broad and shallow, erect; central lobe oblong, shallowly notched, slightly curled downwards, wavy, 5-ridged. Column 1.5 cm. *Venezuela & Trinidad*. G2. Summer.

58. MYRMECOPHILA Rolfe
J.M.H. Shaw

Epiphytic or on rocks. Rhizomes branched, creeping, bearing erect, cylindric, 2–7 internoded pseudobulbs, often with cavities inhabited by ants in the wild (whence the generic name meaning 'ant-loving'). Leaves 1–5 per pseudobulb, narrowly elliptical to ovate with a broadly obtuse to acute apex. Flowers several, in tall terminal racemes or panicles. Flowering stems with up to 40 nodes, with a sheathing scale at each node. Flower-stalks with long or short, ovate-triangular bracts. Sepals and petals similar, free, or sepals fused at base, and enclosing nectary, spreading, wavy. Lip 3-lobed; lateral lobes loosely enclosing column, becoming erect or spreading; central lobe rounded or notched, often curled downwards, sometimes with keels or ridges, which are most prominent near the base of the column. Column stout, with 3 apical teeth. Pollinia 8. Ovary and capsule smooth, ellipsoidal.

A genus of 9–10 species from Mexico to northern Costa Rica, with disjunct species in Venezuela and the Cayman Islands, many of which were originally described in *Cattleya* or *Laelia*. For many years these species were included in *Schomburgkia* Section Chauno-schomburgkia Schlechter. Cultural requirements are similar to those for *Cattleya* (p. 498), though they perform best if repotted less frequently.

Literature: Kennedy, G.C., The genera Schomburgkia and Myrmecophila, *Orchid Digest* **43**: 203–11 (1979); Carnevali, G., Tapia, J.L., Williams, N.H. & Whitten, W.M., Sistematica, filogenia y biogeografia de Myrmecophila, *Lankesteriana* **7**: 29–32 (2003).

1. M. tibicinis (Lindley) Rolfe. (*Schomburgkia tibicinis* Lindley) Illustration: *Botanical Magazine*, 4476 (1849); Shuttleworth et al., Orchids, 66 (1970); Skelsey, Orchids, 137 (1979).

Epiphytic, very large and robust. Pseudobulbs to 60 × 8 cm or more, fusiform, hollow, 2- or 3-leaved. Leaves to 45 × 8 cm, elliptic to oblong, rounded or obtuse. Flowering stems to 5 m, branched, bearing 10–20 flowers; flowers 6–8 cm wide, opening in succession; ovary plus flower-stalk 3–5 cm; bracts 1 cm. Sepals and petals 4–5 × 1.5–2 cm, oblong to spathulate, obtuse or rounded, purplish blue flushed with brown, paler beneath. Lip 4–5 × 3–4 cm, 3-lobed, yellow, white and purple; lateral lobes semicircular to oblong; central lobe diamond-shaped, curled downwards. *Mexico to Panama*. G2. Spring.

59. RHYNCHOLAELIA Schlechter
J.M.H. Shaw

Low growing epiphytic herbs with thickened, club-shaped stems. Leaves 1, thick and fleshy, covered in white wax. Inflorescence with a single flower, emerging from a papery, spathiform bract. Flower resupinate, greenish white, fragrant. Sepals free, all similar. Petals broader than sepals, elliptic. Lip funnel-shaped, wrapped around the column, broad and spreading, and with a very fimbriate margin in *R. digbyana*. Column club-shaped. Pollinia 8, laterally flattened, in 4 pairs, sometimes with extra rudimentary pollinia between the larger ones. Ovary and capsule beaked.

A genus of 2 species from southern Mexico to Nicaragua, both of which have been extensively employed in multi-generic hybrids. Previously included in *Brassavola*, from which it may be distinguished by the presence of pseudobulbs, leaves covered with white wax, and not grooved.

Literature: Withner, C.L., The Cattleyas and their relatives **5**: 173–7 (1998).

1a. Lip margin entire; column less than 2 cm **1. glauca**
b. Lip margin deeply lacerate; column more than 2 cm **2. digbyana**

1. R. glauca (Lindley) Schlechter (*Brassavolia glauca* Lindley; *Laelia glauca* (Lindley) Bentham). Illustration: *Botanical Magazine*, 4033 (1843); Shuttleworth et al., Orchids, 68 (1970); Bechtel et al., Manual of cultivated orchid species, 172 (1981).

Epiphytic or terrestrial. Pseudobulbs 2–10 cm, fusiform to club-shaped, 1-leaved. Leaves 5–12.5 × 2.5–3.5 cm, oblong to elliptic, obtuse, leathery and glaucous. Flowering stems to 10 cm, terminal, 1-flowered; flower 8–10 cm wide, very fragrant. Sepals and petals similar, 5–6.5 cm, linear to lanceolate, obtuse, white to pale green or purple; sepals 1–1.5 cm wide, margins reflexed; petals 1.5–2 cm wide, margins wavy. Lip 5–5.5 × 4 cm, obscurely 3-lobed, wavy; lateral lobes rounded, bases rolled around column; central lobe oblong, abruptly pointed, throat with a narrow purple patch. Column 5 mm, partly exposed. *Mexico to Panama*. G2. Winter–spring.

2. R. digbyana (Lindley) Schlechter (*Brassavola digbyana* Lindley; *Laelia digbyana* (Lindley) Bentham). Illustration: *Botanical Magazine*, 4474 (1849); Williams et al., Orchids for everyone, 107 (1980); Bechtel et al., Manual of cultivated orchid species, 171 (1981).

Epiphytic. Pseudobulbs 12–20 cm, club-shaped, 3- or 4-noded, 1-leaved. Leaves 12–20 × 5–6 cm, elliptic, obtuse, leathery and slightly glaucous. Flowering stem to 15 cm, terminal, 1-flowered; flower 10–12 cm wide, fragrant. Sepals 8–9.5 × 2–2.5 cm, lanceolate, obtuse; petals somewhat wider; sepals and petals very pale green to greenish yellow, margins wavy. Lip 7–8 × 7.5 cm, obscurely 3-lobed, almost circular, margins deeply and finely lacerate; lateral lobes semicircular, bases rolled around but not concealing column; central lobe rectangular to oblong. Column 3–3.5 cm. *Mexico, Belize & Honduras*. G2. Summer.

60. BRASSAVOLA R. Brown

J.C.M. Alexander

Epiphytic or on rocks. Rhizomes creeping, bearing erect, narrow stems or pseudobulbs. Stems with several internodes, sheathed, 1-leaved. Leaves fleshy, mostly cylindric or linear, sometimes flattened and leathery, folded when young. Flowers 2–7, or rarely 12, with strong nocturnal fragrance, in terminal racemes. Sepals and petals free and spreading, similar, linear to narrowly lanceolate. Lip unlobed, or obscurely 3-lobed, base or claw rolled around and often concealing column, blade or central lobe elliptic, heart-shaped or ovate, often with a long acuminate point. Column free from lip, with acute projecting teeth at tip. Pollinia 8 (often a few extra abortive). Capsule ellipsoid, often beaked.

A genus of about 17 species extending from C America and the West Indies south to Brazil and northern Argentina, often used as a parent in bi- and trigeneric hybrids with *Cattleya* and *Laelia*. Cultural treatment should be as for *Epidendrum* (p. 491) though the *Brassavola* growth habit is more easily accommodated on blocks of tree fern or cork bark. They should have a definite winter rest.

Literature: Jones, H.G., Nomenclatural revision of the genus Brassavola of the Orchidaceae, *Annalen des Natur historischen Museums in Wien* **79**: 9–22 (1975); Withner, C.L., Cattleyas and their relatives **5**: 31–68 (1998).

Plant. Erect: **1, 2, 3**; pendent: **4**.
Leaves. Cylindric or linear: **1, 2, 3, 4**. Less than 12 cm long: **2**; 12–40 cm long: **1, 2, 3**; more than 40 cm: **2, 4**. Less than 1 cm wide: **1, 2, 3**; 1–2 cm wide: **4**; 2–4 cm wide: **1, 2**.
Flowers. Lateral: **4**; terminal: **1, 2, 3, 4**. 1–3 per raceme: **1, 2, 3, 4**; 4 or more per raceme: **2, 3**. Less than 7 cm wide: **2, 3**; more than 7.5 cm wide: **1, 3, 4**.
Petals and sepals. Spotted or warty: **2, 3**. Less than 5 cm long: **2**; 5 cm long or more: **1, 3, 4**. Less than 1 cm wide: **1, 2, 3, 4**; more than 1 cm wide: **1**.
Lip. Unclawed or with very short claw: **1, 2**; with claw about one-third of total lip length: **3, 4**. Lacerate: **1**. Long and gradually tapering: **1**. 3–4 cm long: **2**; 3.5–5.5 cm long: **3**; 6 cm long or more: **1, 4**. Less than 4.5 cm wide: **1, 2, 3**; more than 4.5 cm wide: **4**.

1a. Lip 6 cm or more, long and gradually tapering **1. cucullata**
b. Lip 5 cm or less, acute or obtuse, not gradually tapering 2
2a. Lip unclawed or with a very short claw **2. perrinii**
b. Lip claw about one-third of total lip length 3
3a. Plant erect; flowering stems terminal **3. nodosa**
b. Plant pendent; flowering stems lateral **4. acaulis**

1. B. cucullata (Linnaeus) R. Brown (*B. cuspidata* Hooker). Illustration: *Botanical Magazine*, 3722 (1839); Dunsterville & Garay, Venezuelan orchids illustrated **2**: 43 (1961); Bechtel et al., Manual of cultivated orchid species, 171 (1981).

Epiphytic. Stems to 25 cm, cylindric, thin, several-noded, 1-leaved. Leaves 20–40 cm × 5–10 mm, cylindric to linear, fleshy. Flowering stems short, bearing 1–3 flowers; flowers 8–10 cm wide, fragrant, white and greenish yellow tinged with pink or brown. Ovary plus flower-stalk 15–25 cm. Sepals and petals similar, 7–14 cm × 5–10 mm, lanceolate, tapering. Lip 6.5–10 × 1.5–2.5 cm, base circular to heart-shaped with lacerate margin; tip long and tapering. Column 1.5–2 cm, partly exposed. *Mexico & West Indies south to Ecuador*. G2. Autumn–winter.

2. B. perrinii Lindley. Illustration: *Botanical Magazine*, 3761 (1839); Cogniaux & Goossens, Dictionnaire iconographique des orchidées, Brassavola, t. 2 (1904).

Epiphytic or on rocks. Stems to 20 cm × 5 mm, cylindric or slightly angular, branched, 3- or 4-noded, 1-leaved. Leaves 10–25 cm × 5–80 mm, circular to triangular in cross-section, acute, deeply grooved, erect. Flowering stem 2–8 cm, terminal, 1- or 2-flowered; flowers 5–6 cm wide. Ovary plus flower-stalk 5–10 cm. Sepals and petals 3.5–4.5 cm × 4–6 mm, narrowly lanceolate, acute, greenish yellow. Lip 3–4 × 2–2.5 cm, ovate, acute, base rolled around column, white with a central patch of greenish yellow and green veins. Column 1 cm, partly exposed. *SE Brazil & Paraguay*. G2. Spring–summer.

B. tuberculata Hooker (*B. fragrans* Lemaire). Illustration: *Botanical Magazine*, 2878 (1829); Pabst & Dungs, Orchidaceae Brasilienses **1**: 219 (1975); Bechtel et al., Manual of cultivated orchid species, 172 (1981). Flowers 1–12; sepals cream-coloured, marked with reddish purple spots and warts. Lip ovate, blunt, white with a central patch of yellow. *Brazil*. G2. Summer–autumn.

B. flagellaris Rodrigues. Illustration: Bechtel et al., Manual of cultivated orchid species, 171 (1981). Flowers 5–15. Leaves 35–50 cm × 5 mm. Petals 2–3 mm wide. *E Brazil*. G2. Spring.

3. B. nodosa (Linnaeus) Lindley. Illustration: *Botanical Magazine*, 3229 (1833); Williams et al., Orchids for everyone, 182 (1980); Bechtel et al., Manual of cultivated orchid species, 172 (1981).

Epiphytic or on rocks. Stems 8–16 cm, cylindric, slightly thicker above, 4- or 5-noded, 1-leaved. Leaves 20–35 × 2.5–3.5 cm, linear, acute, fleshy and grooved. Flowering stem 8–22 cm, terminal, bearing 1-several flowers; flowers 7.5–8.5 cm wide, fragrant. Sepals and petals 6–9 cm × 5 mm, linear, acute, pale green to greenish yellow. Lip 3–5.5 × 2–4 cm, white; claw about one-third of total lip length, rolled around column; blade circular to ovate, acuminate. Column *c.* 1 cm, concealed. *Mexico & West Indies south to Colombia, Venezuela & Surinam*. G2. Winter.

B. subulifolia Lindley (*B. cordata* Lindley). Illustration: *Botanical Magazine*, 3782 (1840); Shuttleworth et al., Orchids, 68 (1970); Skelsey, Orchids, 45 (1979). Stems 2–8 cm, distinctly thicker above. Leaves to 20 × 1.5 cm. Flowers 3–5,

5–7 cm wide, spotted with red. Column about 5 mm. *Jamaica*. G2. Autumn.

4. B. acaulis Lindley & Paxton (*B. lineata* Hooker). Illustration: *Botanical Magazine*, 4734 (1853); Shuttleworth et al., Orchids, 67 (1970); Graf, Tropica, 708 (1978).

Similar to *B. nodosa*. Whole plant pendent. Leaves cylindric, grooved. Flowering stems lateral, 1- or 2-flowered. Flowers pale yellow to pale green. *Guatemala & Belize to Panama*. G2. Autumn.

61. ISABELIA Barbosa Rodrigues

J.C.M. Alexander & J.M.H. Shaw

Epiphytic with long creeping rhizomes. Pseudobulbs ovoid to fusiform, clustered or separated by long rhizomes, ridged, 1-leaved. Leaf linear, acute, channelled, folded when young, fleshy. Flowering stem 1- or 2-flowered; flowers white, pink or purple. Sepals and petals oblong to lanceolate, acute, free. Lip unlobed, lanceolate to ovate or diamond-shaped, fused to column at base. Column red or dark purple, club-shaped. Pollinia 4 or 8.

A genus of 3 species from southern Brazil, originally described in *Sophronitis*. The genus now includes *Sophronitella* Schlechter. Cultivation as for *Sophronitis*.

Literature: Cavestro, W. The genus Isabelia, *Orchid Review* **116**: 78–80, 152–3, 208–9 (2008).

1. I. violacea (Lindley) van den Berg & M.W. Chase (*Sophronitis violacea* Lindley, *Sophronitella violacea* (Lindley) Schlechter). Illustration: *Botanical Magazine*, 6880 (1886); Ospina & Dressler, Orquideas de las Americas, t. 71 (1974); Bechtel et al., Manual of cultivated orchid species, 274 (1981).

Pseudobulbs 1.5–3 cm × 3–10 mm, ovoid to fusiform, clustered, ridged, 1-leaved. Leaf 4–8 × 1 cm, linear, acute, channelled, folded when young, fleshy. Flowering stem to 5 cm, 1- or 2-flowered; flowers 2.5–4 cm wide, purple. Sepals and petals 1.7–2 cm × 3–5 mm, oblong to lanceolate, acute, free. Lip 1.7–2 cm × 5 mm, unlobed, lanceolate to ovate or diamond-shaped, abruptly acute, fused to column at base. *E Brazil*. G2. Winter.

62. LEPTOTES Lindley

J.C.M. Alexander

Dwarf epiphytes with creeping rhizomes. Stems short and cane-like, sheathed, 1-leaved, sometimes branched. Leaves cylindric to linear, acute, fleshy and channelled, folded when young. Flowers few or solitary, in terminal racemes. Sepals and petals free, similar, or petals broader than sepals. Lip free, 3-lobed; lateral lobes small, sometimes enfolding but not concealing the short fleshy column. Pollinia 6 (4 large and 2 small), waxy.

Five species from Brazil, Paraguay and Argentina. Cultivation requirements are similar to those for *Cattleya* (p. 498), though their dwarf habit is best accommodated on hanging rafts or baskets; the roots should be covered with moss.

1a. Flowers erect, 4–5 cm wide; sepals similar to petals; lip with a dark purple patch at base **1. bicolor**
b. Flower pendent, 3–4 cm wide; sepals wider than petals; lip uniformly pink **2. unicolor**

1. L. bicolor Lindley (*Tetramicra bicolor* (Lindley) Bentham). Illustration: *Botanical Magazine*, 3734 (1840); Shuttleworth et al., Orchids, 44 (1970); Pabst & Dungs, Orchidaceae Brasilienses **1**: 222 (1975); Skelsey, Orchids, 118 (1979).

Stem 1–4 cm, cylindric, fleshy. Leaves 5–15 cm × 5–10 mm. Flowering stem to 6 cm, bearing 2–4 flowers; flower-stalk plus ovary 4–5 cm; flower 4–5 cm wide, erect, white to very pale pink, yellow or green. Sepals and petals 2–2.5 cm × 3–5 mm, linear to narrowly lanceolate, acute with curled tips. Lip 1.5–2 cm × 5–8 mm, 3-lobed; lateral lobes triangular to circular, white or pale green, curved around column; central lobe oblong to narrowly ovate, acute, pale to deep purple with darker veins. Column 5 mm, greenish brown, exposed. *S Brazil & Paraguay*. G2. Winter–spring, autumn.

2. L. unicolor Rodrigues. Illustration: Pabst & Dungs, Orchidaceae Brasilienses **1**: 223 (1975); Northen, Miniature orchids, 100 (1980); Bechtel et al., Manual of cultivated orchid species, 221 (1981).

Stem 1–3 cm, cylindric. Leaves 2.5–6 cm × 4–8 mm, arched. Flowering stem 1–3 cm, 2-flowered; flowers 3–4 cm wide, pendent, white to pale pink or purple. Sepals 2–2.5 cm × 5 mm, lanceolate, acute; petals 1.5–2 cm × 3 mm, strap-shaped, acute. Lip 1.5–2 × 1–1.3 cm, 3-lobed; lateral lobes triangular, obtuse, flat, not enclosing column; central lobe diamond-shaped, acute. Column 5 mm, brown, exposed. *S Brazil & Argentina*. G2. Winter.

63. SCAPHYGLOTTIS Poeppig & Endlicher

J.C.M. Alexander

Epiphytic or on rocks. Pseudobulbs stipitate, in superposed chains with papery sheaths; leaves 1–3, all apical, folded when young. Flowers terminal or at stem nodes, in racemes or clusters, white or pale shades of blue, green, yellow or purple, bright orange or red. Lip articulated to column base or firmly fixed. Column base enlarged into a foot to which the lateral sepals are basally fused to form a mentum. Pollinia 4 or 6, waxy.

A tropical American genus of about 60 species. A group of 5 species from tropical America, lacking a movable lip, are sometimes separated as *Hexisea* Lindley, but recent molecular work supports its inclusion here. The generic name *Scaphyglottis* is conserved over *Hexisea*. Cultivation as for *Cattleya* (p. 498), though its straggling habit makes it more easily grown on rafts or bark suspended near the roof.

Literature: Dressler, R.L., The genus Hexisea, *Orquidea* **4**: 197–200 (1974); Withner, C.L., Cattleyas and their relatives **5**: 145–9 (1998); Dressler, R.L., Whitten, W.M. & Williams, N.H., Phylogenetic relationships of Scaphyglottis and related genera based on nrITS sequence data, *Brittonia* **56**: 58–66 (2004).

1a. Flowers white to pink, to 1.5 cm wide **1. amethystina**
b. Flowers red to orange, 2.5–3 cm wide **2. bidentata**

1. S. amethystina (Reichenbach) Schlechter. Illustration: Bechtel et al., Manual of cultivated orchid species, 272 (1981).

Leaves 1 or 2, to 15 × 1.5 cm, apical, linear, notched at apex. Flowers 1–1.5 cm wide, in terminal clusters, white to pale purplish pink. Lip 3-lobed, movable. Pollinia 4. *Guatemala to Panama*. G2. Spring.

2. S. bidentata (Lindley) (*Hexisia bidentata* Lindley). Illustration: *Botanical Magazine*, 7031 (1888); Sheehan & Sheehan, Orchid genera illustrated, 105 (1979); Bechtel et al., Manual of cultivated orchid species, 216 (1981).

Epiphytic herbs with sheathed stems composed of successive pseudobulbs, to 50 cm. Pseudobulbs to 10 × 1 cm, brownish purple, grooved. Leaves to 15 × 1.5 cm, strap-shaped, dark green, bifid or trifid at apex. Flowers 2.5–3 cm wide, bright reddish orange. Sepals 1.5–2 cm × 5 mm, lanceolate, acute or obtuse; petals similar but smaller. Lip to 1.5 cm × 5 mm, oblong, obtuse with a narrow claw fused to the column. Column 3–4 mm. *C America to Venezuela, Colombia & Peru.* G2. Flowering intermittent.

64. DOMINGOA Schlechter
J.C.M. Alexander & J.M.H. Shaw
Epiphytic or terrestrial. Rhizomes creeping, bearing simple, 1-leaved, club-shaped stems. Leaves fleshy, mottled with purple, folded when young. Flowers pinkish purple, terminal, in tall, slender, branched racemes or clusters. Sepals pointing forwards; lateral sepals fused basally to column foot forming a mentum. Lip unlobed, bent downwards, basally fused to column foot, with a small apical pouch. Column slender, slightly arched over upper part of lip. Pollinia 4, waxy.

A small genus of 5 species from Mexico to Nicaragua and the islands of Hispaneola, Cuba and Puerto Rico. *Nageliella* Williams, long known as *Hartwegia* Lindley, though this name had already been used for a genus in the Liliaceae, is now included in *Domingoa*. Cultivation requirements are similar to those for *Cattleya* (p. 498). Propagation is easily effected by division, though this should not happen frequently as the plants perform better when the pseudobulbs are densely clustered.

1. D. purpurea (Lindley) Van den Berg & Soto Arenas (*Nageliella purpurea* (Lindley) Williams; *Hartwegia purpurea* Lindley). Illustration: Hamer, Orquideas de El Salvador **2**: 139 (1974); *Orquidea* **4**: 115 (1974); Bechtel et al., Manual of cultivated orchid species, 234 (1981).

Stems 2–8 cm, narrowly fusiform to club-shaped, shallowly ridged. Leaves 3–12 × 1–2.5 cm, elliptic to ovate with sheathing bases. Flowering stems 10–5.0 cm with small, chaffy bracts, gradually lengthening over a few months. Flowers several, 1–1.5 cm wide, opening in succession. Sepals 7–10 × 3 mm, lanceolate to ovate, concave, acute or obtuse. Petals shorter and narrower than sepals, linear to lanceolate. Lip 10 × 4 mm, basally fused to column foot; middle portion linear, S-shaped; apex a small triangular pouch. Column *c.* 7 mm. *Mexico to Honduras.* G2. Summer.

65. DIODONOPSIS Pridgeon & M.W. Chase
J.M.H. Shaw
Small epiphytic herbs with thin stems. Leaves leathery, obovate to linear, blunt to acute, narrowed towards petiole. Inflorescence usually 1-flowered, with a bract at the base. Flowers inverted. Sepals ovate, fused at base forming a short cup, narrowed at the tip in to a thin projection. Petals oblong, apiculate, thickened along inner margin, with a projection at the base. Lip undivided, elliptical to oblong, blunt or pointed, hinged at the base. Column with a small extension at the foot. Pollinia 2. Ovary with grooves and ridges, papillose or with spiny hairs.

A genus of 5 species distributed from Costa Rica to Bolivia, and recently separated from *Masdevallia* following DNA studies, distinguished by keeled ovaries and petals with a projection at the base. Cultivation as for *Masdevallia* (p. 509).

1a. Plant erect; flowers more than 10 mm long **1. erinacea**
b. Plant creeping; flowers less than 5 mm long **2. pygmaea**

1. D. erinacea Reichenbach (*Masdevallia erinacea* Reichenbach; *Masdevallia horrida* Teuscher & Garay). Illustration: *American Orchid Society Bulletin* **29**: 23 (1960) & **37**: 520 (1968); *Die Orchidee* **30**: CXXV (1979); Northen, Miniature orchids, C-51 (1980); Luer, Icones Pleurothallidinarum **II**: 17 (1986).

Dwarf epiphyte. Leaves 3–6 cm × 1–3 mm, linear, acute. Flowers few per inflorescence, opening one at a time. Flowering stems 2.5–6 cm, longer than leaves; bract *c.* 5 mm. Sepals *c.* 1.5 cm, equally fused into a shallow cup, 4–8 mm, densely hairy on the veins; tails yellow, club-shaped, pendent, 3–5 mm. Upper sepal white to greenish yellow; lateral sepals densely spotted with maroon. *Coste Rica & Ecuador.* G2. Summer.

2. D. pygmaea (Kraenzlin) Pridgeon & M. W. Chase (*Masdevallia exigua* Ames & C. Schweinfurth; *Masdevallia pygmaea* Kraenzlin; *Masdevallia muriculata* Kraenzlin).

Rhizome creeping. Leaves narrowly oblanceolate, obtuse, to 2.5 cm. Inflorescence shorter than leaves, 1-flowered. Sepals white, 4 mm, united for about half their length, triangular-ovate for the basal 1 mm, the remainder a yellowish, narrow fleshy tail. Lip oblong. *Costa Rica to NW. Ecuador. G2. Summer.*

66. MASDEVALLIA Ruiz & Pavon
J.C.M. Alexander
Epiphytic or on rocks. Stems short and slender, erect, tufted, concealed in papery sheaths. Leaves 1, apical, linear to lanceolate, somewhat fleshy, leathery, minutely 3-toothed at tip, folded when young. Flowers 1–several, in terminal racemes borne at the leaf-base. Sepals fused below into a narrow or cup-shaped tube; lateral sepals usually further fused into a broad oblong blade; free portions of sepals spreading and usually long-tailed, separated by a sinus or notch. Petals and lip small, often concealed in the sepal-tube. Lip oblong, sometimes 3-lobed or fiddle-shaped. Column toothed at apex; column foot forming a short mentum with base of lateral sepals. Pollinia 2.

Over 500 species from Mexico, C and S America, mostly from the Andean cloud-forests, frequently grown for their large and brightly coloured sepals, whose tails often continue to grow after the flower has opened. Luer has recently published a scheme dividing *Masdevallia* into 20 genera. They perform best if grown in epiphytic compost or on tree-fern bark in cool, moist conditions with plenty of ventilation. Having no pseudobulbs, plants must never be allowed to dry out; they should be shaded from direct sunlight. Well-established plants can be divided in early spring.

Literature: Woolward, F.H., The genus Masdevallia (1890–6); Kraenzlin, F.W.L., Masdevallia, *Feddes Repertorium Beihefte* 34 (1925); Braas, L., The genus Masdevallia, *Orchid Review* **86**: 30–43 (1978); Luer, C.A., Icones Pleurothalidinarum IX, XXI, XXII, XXIII, XXV: The systematics of Masdevallia, Parts 1–5. Monographs in Systematic Botany **77** (2000), **82** (2000), **86** (2001), **87** (2002), **91** (2003); Thoerle, L. Masdevallia and affiliates (2007) – contains a useful key to all the species.

1a. Flowering stems winged or sharply angled 2
b. Flowering stems not winged or sharply angled 3
2a. Lateral sepals fused into a blade 2 cm or more wide **1. tovarensis**
b. Lateral sepals fused into a blade less than 2 cm wide **2. infracta**

3a. Two or more flowers open together on the same stem 4
b. Flowers open 1 at a time 5
4a. Flowering stems 20 cm or more; lateral sepals fused above sepal-tube; petals not toothed **8. schlimii**
b. Flowering stems less than 20 cm; sepals all equally fused; petals toothed **7. caloptera**
5a. Sepal-tube narrow and slightly curved 6
b. Sepal-tube broad and cup-shaped 9
6a. Lateral sepals not drawn out into fine tails **9. coccinea**
b. Lateral sepals drawn out into fine tails 7
7a. Flowering stems about equal to leaves; lip fringed **10. rosea**
b. Flowering stems longer than leaves; lip not fringed 8
8a. Free part of upper sepal (excluding tail) 2 cm or more **11. veitchiana**
b. Free part of upper sepal (excluding tail) 1 cm or less **12. amabilis**
9a. Free part of upper sepal (including tail) more than 7 cm **3. macrura**
b. Free part of upper sepal (including tail) less than 6 cm 10
10a. Apex of lip without papillae **4. rolfeana**
b. Apex of lip with papillae 11
11a. Upper sepal abruptly contracted into a thick fleshy tail; flowering stem equal to or slightly shorter than leaves **5. coriacea**
b. Upper sepal gradually tapered into a slender tail; flowering stem distinctly shorter than leaves **6. peristeria**

1. M. tovarensis Reichenbach. Illustration: *Botanical Magazine*, 5505 (1865); Skelsey, Orchids, 120 (1979); Bechtel et al., Manual of cultivated orchid species, 227 (1981).

Epiphytic. Leaves to 15 × 2 cm, lanceolate to obovate, obtuse or rounded. Flowering stem to 20 cm, angled, bearing up to 4 white flowers, all open together. Sepal-tube *c.* 6 mm; upper sepal 4–5 cm, finely tapered. Lateral sepals 3.5–4.5 cm, triangular to ovate, slender-tailed, fused for 2–2.5 cm. Petals *c.* 6 × 1.5 mm, linear, acute, asymmetric. Lip *c.* 5 × 2 mm, oblong to spathulate, rounded or acute. Column 4.5 mm, finely toothed. *Venezuela*. G2. Winter.

M. trochilus Linden & Andre (*M. ephippium* Reichenbach). Illustration: *Botanical Magazine*, 6208 (1876); Dunsterville & Garay, Venezuelan orchids illustrated **1**: 213 (1959); *American Orchid Society Bulletin* **45**: 104 (1976). Flowering stem 20–30 cm, 3-angled, bearing several flowers in succession. Flowers purplish brown with yellow tails. Upper sepal to 14 cm, blade small, circular, abruptly narrowed to a long tail. Lateral sepals fused into a deep cup; tails 7–10 cm. *Colombia & Ecuador*. G2. Spring–summer.

M. maculata Klotzsch & Karsten. Illustration: *Orchid Digest* **36**: 24 (1972); Bechtel et al., Manual of cultivated orchid species, 227 (1981). Flowering stem to 25 cm, bearing several flowers in succession. Sepals orange to purplish brown with yellow tails; sepal-tube *c.* 1.5 cm. Upper sepal 6–8 cm, narrowly triangular. Lateral sepals *c.* 8 cm, narrowly ovate, fused for *c.* 3 cm. Petals *c.* 6 mm, oblong, asymmetric, each with a small point. Lip *c.* 6 mm, spathulate to fiddle-shaped, papillose. Column *c.* 6 mm. *Venezuela, Colombia & Peru*. G2. Summer–autumn.

2. M. infracta Lindley. Illustration: Pabst & Dungs, Orchidaceae Brasilienses **l**: 225 (1975); Graf, Tropica, 737 (1978); Bechtel et al., Manual of cultivated orchid species, 225 (1981).

Epiphytic. Leaves 8–14 × 1.5–2.5 cm, lanceolate to oblong, obtuse. Flowering stem to 25 cm, angled, bearing several flowers in succession. Sepals pinkish purple with greenish yellow tails; sepal-tube 1–1.5 cm. Upper sepal 4–5 cm, triangular to ovate, slender-tailed. Lateral sepals 4–5 cm, oblong to ovate, fused for about 2 cm, slender-tailed. Petals 6–7 mm, linear, each with a small point, pink. Lip *c.* 6 mm, oblong, with a small point, pink and orange. Column *c.* 6 mm. *Peru & Brazil*. G2. Summer.

3. M. macrura Reichenbach. Illustration: *Botanical Magazine*, 7164 (1891); *Die Orchidee* **30**: 105 (1979); Bechtel et al., Manual of cultivated orchid species, 226 (1981).

On rocks. Leaves 20–30 × 5–7 cm, oblong to elliptic, obtuse, often notched. Flowering stem 20–30 cm, 1-flowered. Sepals red to brownish yellow with prominent veins and dark warts; tails long and slender, yellow; sepal-tube 1–1.5 cm, cup-shaped. Upper sepal 10–15 cm, lanceolate to narrowly triangular, tapered. Lateral sepals 9–13 cm, ovate to oblong, fused for 2–5 cm, tapered. Petals 8–10 mm, lanceolate to rhombic, rounded, yellow with dark stippling. Lip 6–8 mm, oblong, rounded, papillose, yellow with thick red margins. Column 6–8 mm, yellow. *Colombia & Ecuador*. G2. Winter–spring.

M. velifera Reichenbach. Illustration: *Gardeners' Chronicle* **1**: 745 (1887); *Journal of the Royal Horticultural Society* **33**: 393 (1908). Leaves 15–20 × 2.5 cm, lanceolate, obtuse. Flowering stem 6–8 cm. Sepal-tube *c.* 2 cm. Upper sepal 7–8 cm, ovate, slender-tailed, brownish yellow spotted with brown. Lateral sepals 7–8 cm, oblong, fused for *c.* 5 cm, shiny reddish brown with yellow tails. Petals *c.* 1.2 cm, oblong, greenish yellow. Lip *c.* 1.2 cm, oblong, dark purple, fringed. *Colombia*. G2. Winter.

4. M. rolfeana Kraenzlin. Illustration: Woolward, The genus Masdevallia, 55 (1893); *Orchid Digest* **32**: 188 (1968).

Epiphytic. Leaves 11–14 × 2 cm, obovate to elliptic, obtuse, keeled. Flowering stem *c.* 8 cm, 1-flowered. Sepals dark purple, tapered into slender green and yellow tails. Sepal-tube to 1 cm, cup-shaped. Upper sepal *c.* 4 cm, triangular to ovate. Lateral sepals *c.* 6 cm, oblong to ovate, fused for about 2 cm. Petals *c.* 6 mm, oblong to elliptic, obtuse, narrowed at base, purple. Lip 8–10 mm, oblong, apiculate, pink with darker spots. Column *c.* 8 mm, finely toothed at apex. *Costa Rica*. G2. Spring–summer.

M. estradae Reichenbach. Illustration: *Botanical Magazine*, 6171 (1875); *Journal of the Royal Horticultural Society* **26**: 146 (1901). Leaves 5–8 × 2 cm. Flowering stem *c.* 10 cm. Sepals equally fused into a cup-shaped tube, *c.* 6 mm. Upper sepal purple with a yellow margin, abruptly contracted into a slender, yellow tail. Lateral sepals oblong, purple below, pink above, tapered to a slender yellow tail. *Colombia*. G2. Spring.

5. M. coriacea Lindley. Illustration: *Gardeners' Chronicle* **21**: 95 (1897); *Orquideologia* **9**: 34 (1974); Bechtel et al., Manual of cultivated orchid species, 225 (1981).

Epiphytic or on rocks. Leaves 12.5–20 × 1–2 cm, linear to lanceolate, obtuse, keeled. Flowering stem 17–20 cm, 1(rarely 2)-flowered; flowering stem, bract and ovary flecked with purple, sepal-tube 1.5 cm, cup-shaped; sepals greenish white with lines of crimson spots. Upper sepal 3–4 cm, ovate, abruptly contracted into a broad, green, purple-flecked tail. Lateral sepals *c.* 4 cm, narrowly triangular, fused for 2–2.5 cm, without distinct tails. Petals 1.2–1.5 cm, oblong to spathulate, obtuse, white with a central purple stripe. Lip *c.* 1.2 cm, oblong, obtuse, papillose at tip, purple and green. Column *c.* 1 cm, finely toothed at tip, pale green. *Colombia*. G2. Spring–summer.

6. M. peristeria Reichenbach. Illustration: *Botanical Magazine*, 6159 (1875); *Orquideologia* **10**: 159 (1975); *Die Orchidee* **27**(6): 20 (1976).

Epiphytic. Leaves 10–15 × 2.5 cm, linear to lanceolate or oblong, obtuse. Flowering stem 5–8 cm, 1-flowered. Sepal-tube 1.5–2 cm, cup-shaped; sepals 4.5–6 cm, triangular to narrowly ovate, finely tapered, greenish yellow spotted with crimson; lateral sepals fused for *c.* 1.5 cm. Petals 1–1.2 cm, lanceolate to spathulate, obtuse, green. Lip 1.2–1.5 cm, oblong, narrowed at base, rounded to truncate, papillose, purple. Column *c.* 1.2 cm, finely toothed, greenish white. *Colombia*. G2. Spring–summer.

M. civilis Reichenbach & Warscewicz (*M. leontoglossa* misapplied). Illustration: *Botanical Magazine*, 5476 (1864); Dunsterville & Garay, Venezuelan orchids illustrated **4**: 128 (1966). Leaves 15–25 × 1–1.5 cm, linear, acute. Flowering stem 2.5–7 cm; sepals 3.5–4: cm, ovate, contracted into slender tails; outside of flower brownish purple below, green above, with crimson spots. Petals white with a central purple stripe. Lip oblong with a rounded, papillose tip, purple. *Peru*. G2. Spring–summer.

The name **M. leontoglossa** Reichenbach is frequently found in horticultural literature and catalogues. It has been treated as a synonym of *M. civilis* though the leaf shape is quite distinct in Reichenbach's original description. Illustrations in the *Botanical Magazine* (7245) and Woolward (The genus Masdevallia, 20) show strongly pendent flowering stems. This is not referred to by Reichenbach. *Colombia (Norte de Santander)*. G2. Summer.

M. corniculata Reichenbach. Illustration: *Botanical Magazine*, 7476 (1896); *Die Orchidee* **30**: 105 (1979). Leaves 20–25 × *c.* 4 cm, lanceolate. Flowering stem 7–10 cm, enclosed in an inflated sheath, which also conceals the leaf-base. Flowers yellow, spotted and blotched with orange. Sepal-tube *c.* 2 cm, cup-shaped. Lateral sepals fused for *c.* 3 cm. *Colombia*. G2. Summer–autumn.

7. M. caloptera Reichenbach (*M. biflora* Regel). Illustration: *Gartenflora* **40**: 1341 (1891); Bechtel et al., Manual of cultivated orchid species, 224 (1981).

Epiphytic. Leaves 6–8 × 1.4–2 cm, oblong to lanceolate, obtuse to rounded, older leaves tinged with red. Flowering stem 8–12.5 cm, bearing 2 or more flowers, 2 or more open together. Sepals *c.* 2 cm, equally fused for *c.* 6 mm, white with purple streaks, margins finely toothed; tails slender, yellow, Sepal-tube sharply indented above mentum. Upper sepal almost circular; lateral sepals triangular to oblong. Petals 4–5 mm, toothed, white with a central purple stripe. Lip *c.* 5 mm, oblong to shallowly 3-lobed, yellow and crimson. Column *c.* 4 mm, green and purple. *Colombia, Ecuador & Peru*. G2. Summer.

M. abbreviata Reichenbach (*M. polysticta* Reichenbach; *M. melanopus* misapplied). Illustration: *Botanical Magazine*, 6258 (1876) & 6368 (1878). Leaves 12.5–15 cm. Flowering stem 15–20 cm, bearing many flowers, several open together. Sepals *c.* 3 cm. Lip yellow. *Peru*. G2. Autumn–winter.

Very similar to *M. caloptera*. The name *M. melanopus*, which was given to a similar species with a single row of flowers all pointing in the same direction, has frequently been misapplied to *M. abbreviata*.

8. M. schlimii Lindley. Illustration: *Botanical Magazine*, 6740 (1884); Dunsterville & Garay, Venezuelan orchids illustrated **3**: 171 (1965); Bechtel et al., Manual of cultivated orchid species, 227 (1981).

Epiphytic. Leaves to 20 × 5 cm, obovate to elliptic, rounded or obtuse. Flowering stem to 3 5 cm, bearing up to 8 flowers, several open together. Sepals 5–6 cm, with long yellow tails; sepal-tube *c.* 6 mm, cup-shaped. Upper sepal triangular to ovate, finely tapered, greenish yellow, spotted with maroon below. Lateral sepals fused for 1.5–2 cm, oblong. Petals *c.* 6 mm, linear, pale yellow. Lip *c.* 6 mm, oblong to fiddle-shaped, acute, pink and yellow with purple spots and blotches. Column 6–8 mm. *Colombia & Venezuela*. G2. Summer.

M. racemosa Lindley. Illustration: *Die Orchidee* **23**: 141 (1972) & **30**: 105 (1979); *Orchid Review* **86**: 35 (1978); Bechtel et al., Manual of cultivated orchid species, 227 (1981). Creeping terrestrial, rarely epiphytic. Leaves 8–12.5 × 2 cm, oblong to lanceolate, obtuse. Sepal-tube *c.* 1.5 cm, straight and narrow; sepals bright scarlet. Upper sepal *c.* 1 cm, triangular, acuminate. Lateral sepals 3–3.5 cm, elliptic to circular, shortly apiculate, fused for *c.* 2.5 cm. *Colombia*. G2. Spring–autumn.

9. M. coccinea Lindley (*M. lindenii* André; *M. harryana* Reichenbach). Illustration: *Botanical Magazine*, 5990 (1872); Shuttleworth et al., Orchids, 37 (1970); Bechtel et al., Manual of cultivated orchid species, 225 (1981).

Terrestrial; stems *c.* 4 cm. Leaves 15–20 × 2–3 cm, oblong to lanceolate, obtuse to rounded. Flowering stem to 40 cm, 1-flowered, streaked with purple, flowers held well above leaves. Sepal-tube *c.* 2 cm, narrow and curved; upper sepal to 5 cm, narrowly triangular, finely tapered; tail erect or curved back. Lateral sepals *c.* 6 × 2 cm, ovate to oblong, acuminate, fused for 2–3 cm. Sepals purple, crimson, scarlet, orange, yellow or white; base of tube white, greenish yellow or yellow. Petals *c.* 1 cm, linear, notched, off-white. Lip *c.* 8 mm, oblong, shallowly fiddle-shaped, pinkish. Column *c.* 6 mm. *Colombia & Peru*. G2. Spring.

Var. **militaris** (Reichenbach) O. Gruss & M. Wolff (*M. militaris* Reichenbach; *M. ignea* Reichenbach). Illustration: *Botanical Magazine*, 5962 (1872). On rocks. Stems 5–8 cm. Flowering stem to 30 cm. Flowers held just above leaves. Very similar to *M. coccinea* but upper sepal tail drooping; sepals usually reddish brown to orange. Petals acuminate. *Colombia & Venezuela*. G2. Spring–summer.

10. M. rosea Lindley. Illustration: Graf, Tropica, 736 (1978); Northen, Miniature orchids, C-52 (1980).

Epiphytic. Leaves 12–20 × 2–3 cm, elliptic to obovate, acute. Flowering stem 12–20 cm, 1-flowered, arched. Sepal-tube 2.5–3.5 cm, narrow, slightly curved. Upper sepal 2.5–3.5 cm, tapering from the base into a long narrow tail. Lateral sepals 4–5 cm, oblong to narrowly triangular, fused for *c.* 2 cm. Sepals reddish purple and red. Petals 4–5 mm, oblong, off-white. Lip *c.* 5 mm, oblong, shallowly fiddle-shaped, pink and yellow at base; tip dark purplish brown, densely fringed. *Colombia & Ecuador*. G2. Spring–summer.

11. M. veitchiana Reichenbach. Illustration: *Botanical Magazine*, 5739 (1868); Graf, Tropica, 736 (1978); Bechtel et al., Manual of cultivated orchid species, 228 (1981).

On rocks. Leaves 15–25 × 2.5 cm, oblong to narrowly obovate, obtuse.

Flowering stem 30–45 cm, erect, 1-flowered. Sepal-tube *c.* 3 cm, cylindric. Upper sepal 5–7 × 2 cm, ovate to triangular, finely tapered. Lateral sepals 6–7.5 × 1.5–2.5 cm, triangular to ovate-oblong, fused for *c.* 3 cm, acuminate. Sepals bright orange inside, with short purplish hairs; yellowish with dark veins outside. Petals 1–1.5 cm, oblong, each with a small point, off-white. Lip 1–1.5 cm, oblong to shallowly 3-lobed, purple and white. *Peru.* G2. Spring–summer.

12. M. amabilis Reichenbach. Illustration: Veitch, Manual of orchidaceous plants **5**: 24 (1889); *American Orchid Society Bulletin* **42**: 240 (1973).

Leaves 12–18 × 1.5–2 cm, oblong to lanceolate, obtuse to rounded. Flowering stem 25–30 cm, erect, 1-flowered, tinged with pink. Sepal-tube 2–2.5 cm, narrow, slightly curved. Upper sepal *c.* 4 cm, triangular to ovate, acuminate, long-tailed, reddish orange with darker veins. Lateral sepals 4–4.5 cm, ovate to oblong, acuminate, fused for *c.* 1.5 cm, long-tailed, crimson. Petals *c.* 6 mm, oblong, each with a small point, yellow. Lip *c.* 6 mm, spathulate; base oblong. *Peru.* G2. Winter.

M. barleana Reichenbach. Illustration: *American Orchid Society Bulletin* **42**: 240 (1973); Bechtel et al., Manual of cultivated orchid species, 224 (1981). Leaves 8–12 × 2–2.5 cm. Flowering stem 15–25 cm. Sepal-tube 1.5–2 cm. Lateral sepals abruptly contracted into long tail. Sepals pink with darker veins. Petals with 3 almost equal teeth at apex. *Peru.* G2. Spring–summer.

67. DRACULA Luer

J.C.M. Alexander

Epiphytic, terrestrial or on rocks. Stems short, slender, erect, concealed in papery sheaths. Leaves 1, apical, linear to lanceolate, thinly leathery, sharply keeled, minutely 3-toothed at tip, folded when young. Flowers 1 or a few in succession on erect or arched racemes borne at leaf-base. Sepals fused below into a wide, shallow cup. Lateral sepals usually further fused into a broad blade; free parts of sepals ovate, long-tailed. Petals small, spathulate, split at tip into 2 bivalve-like flaps, often papillose. Lip hinged to column foot, clearly divided into a short claw and cup-shaped or pouched blade, 3-keeled below; marginally toothed. Column short; column foot slender, fused to base of lateral sepals forming a mentum. Pollinia 2.

More than 100 species, mostly from Colombia with a few representatives in C America, Ecuador and Peru. The species in this genus were originally described in *Masdevallia* Section Saccilabiatae Reichenbach, and more recently in Section Chimeroideae of the same genus. They are, however, quite distinct in floral morphology and appearance. They should be grown in conditions similar to those described for *Masdevallia* (p. 509) though, being native to slightly lower altitudes, they prefer higher temperatures.

Literature: Hawley, R.M., Masdevallia chimaera and the marvellous monsters, *American Orchid Society Bulletin* **46**: 600–9 (1977); Luer, C.A., Dracula, a new genus in the Pleurothallininae (Orchidaceae), *Selbyana* **2**: 190–8 (1978); Kennedy, G., The genus Dracula, *Orchid Digest* **43**: 31–8 (1979); Luer, C.A., Icones Pleurothalidinarium X: Systematics of Dracula. Monographs in Systematic Botany **46** (1993).

1a. Lip blade broader than long, kidney-shaped **1. bella**
b. Lip blade isodiametric or longer than broad **2. chimaera**

1. D. bella (Reichenbach) Luer (*Masdevallia bella* Reichenbach). Illustration: *Gardeners' Chronicle* **16**: 237 (1881); *Orchid Review* **77**: 207 (1969); *American Orchid Society Bulletin* **46**: 601 (1977).

Epiphytic or terrestrial. Leaves 12–20 × 2.5 cm, oblong to lanceolate, obtuse. Flowering stems to 18 cm, pendent, purple to purplish green, 1-flowered. Flowers to *c.* 25 cm long. Sepals hairless; sepal-tube 1–1.5 cm. Upper sepal 10–12 × 2–2.5 cm, triangular to ovate, finely tapered, greenish yellow, densely spotted with purplish brown; tail long and slender. Lateral sepals 8–10 × 1.5–2. cm, fused for *c.* 2.5 cm into an almost square blade; tails on outer corners, often crossed. Petals 6–8 mm, divided into a fan-shaped blade and a claw; blade unequally 2-flapped, yellow with brown markings, with papillae and teeth on the flaps. Lip-blade 1.2 × 2 cm, kidney-shaped, concave, white with radiating lines; claw *c.* 4 mm, pink and fleshy. Column *c.* 3 mm, stout, purplish brown. *Colombia.* G2. Winter–summer.

2. D. chimaera (Reichenbach) Luer (*Masdevallia chimaera* Reichenbach; *M. backhousiana* Reichenbach). Illustration: Shuttleworth et al., Orchids, 37 (1970); *American Orchid Society Bulletin* **46**: 600 (1977); Graf, Tropica, 738 (1978).

Terrestrial or on rocks, rarely epiphytic. Leaves 15–25 × 3–5 cm, elliptic to obovate, obtuse. Flowering stems to 60 cm, erect or ascending, green to purplish green, bearing 3–8 flowers. Sepal-tube 1–1.5 cm; lateral sepals fused for *c.* 2.5 cm; sepal blades 5–6 cm, triangular to ovate, tapered, yellow or greenish yellow, densely spotted and blotched with brownish purple or black, hairy and warty; tails 8–20 cm slender, purple. Petals *c.* 8 mm, spathulate, white with purple markings. Lip *c.* 2 cm, oblong to square or circular in surface view, deeply pouched, white to yellow or pinkish brown; claw pink and orange. Column *c.* 1 cm, minutely toothed at apex; column foot pink. *Colombia.* G2. Winter.

D. radiosa (Reichenbach) Luer (*Masdevallia radiosa* Reichenbach). Illustration: *Orchid Review* **38**: 247 (1930); *Orquideologia* **9**: 44 (1974) & **10**: 81 (1975); *American Orchid Society Bulletin* **46**: 604 (1977). Flowering stems erect or pendent, bearing 2–4 flowers. Sepals equally fused for *c.* 1.5 cm, brownish yellow with dark purplish brown papillae inside, pinkish or brownish yellow outside; free parts broadly ovate to rectangular, abruptly tapered; tails 7–10 cm, purplish brown. Petals 4–5 mm, yellow with purple spots; flaps dissimilar, papillose. Lip *c.* 6 mm in diameter, pinkish white. *Colombia.* G2. Spring–summer.

D. erythrochaete (Reichenbach) Luer (*Masdevallia erythrochaete* Reichenbach; *M. astuta* Reichenbach). Illustration; *Orquideologia* **9**: 43 (1974); Northen, Miniature orchids, C-30 (1980); Bechtel et al., Manual of cultivated orchid species, 225 (1981). Epiphytic. Leaves 15–25 × 1–1.5 cm. Flowering stem erect to pendent, bearing 1–3 flowers. Sepal-tube *c.* 6 mm; sepals *c.* 5 cm, ovate, abruptly tapered, pale yellow, spotted and veined with pinkish purple; tails purple. Petals *c.* 3 mm, pinkish. Lip 1–1.5 cm, pale pinkish yellow. *Costa Rica to Colombia.* G2. Autumn–winter.

D. vampira (Luer) Luer (*Masdevallia vampira* Luer). Illustration: *Orchid Digest* **43**: 34 (1979). Epiphytic. Flowering stem 20–40 cm or more, horizontal to pendent, bearing 5–7 flowers. Sepals hairless, very pale green with dense purplish black veins and tails. Petals *c.* 6 mm, white marked with purple. Lip 1.5–2.5 cm; blade concave, white with pink or yellow veins. *Ecuador.* G2. Summer.

68. PORROGLOSSUM Schlechter
J.C.M. Alexander
Similar to *Masdevallia*. Leaves obovate, lanceolate or elliptic, long-stalked. Flowers solitary or a few, open together or in succession, on a slender erect or arched stem. Sepals equally fused, forming a cup-shaped tube, free parts spreading. Petals small, meeting around column. Lip S-shaped in side profile, with a distinct claw, attached below apex of column foot, mobile and sensitive. Rostellum upright. Anther erect, on column apex. Pollinia 2, hard.

A genus of over 30 montane species from Venezuela to Peru and Bolivia, originally in *Masdevallia* but distinct in lip and column structure and in the sensitivity of the lip, which folds up suddenly against the column if stimulated by touch or shaking. Cultivation should be in cool, moist, shady conditions similar to those for *Masdevallia* (p. 509).

Literature: Sweet, H.R., The genus Porroglossum, *American Orchid Society Bulletin* **41**: 513–24 (1972); Luer, C.A., Icones Pleurothalidinarium IV: Systematics of Acostaea, Condylago and Poroglossum. Monographs in *Systematic Botany* 24. (1987).

1. P. echidnum (Reichenbach) Garay (*Masdevallia echidna* Reichenbach; *M. muscosa* Reichenbach; *P. muscosum* (Reichenbach) Schlechter). Illustration: *Botanical Magazine*, 7664 (1899); Northen, Miniature orchids, 149 (1980); Bechtel et al., Manual of cultivated orchid species, 267 (1981).

Leaves to 10 cm, elliptic, obtuse; upper surface papillose. Flowering stem to 15 cm, densely hairy, bearing 1 or 2 flowers. Sepal-tube 3–4 mm, free parts 2.5–3 cm × 6 mm, triangular to ovate; lateral sepals slightly asymmetric. Sepals yellow or yellowish brown. Petals 5–8 × 1 mm, pinkish. Lip 5–8 × 4 mm, triangular to diamond-shaped, long-clawed, front margin hairy. *Venezuela, Colombia & Ecuador*. G2. Flowering intermittent.

69. TRISETELLA Luer
J.C.M. Alexander
Similar to *Masdevallia*. Distinguished by its small, almost free, upper sepal and large, almost totally fused, lateral sepals with tails attached to side of blade, rather than arising at the apex as in *Masdevalia*. Lip arrow-shaped.

A genus of over 20 species distributed from Costa Rica to Bolivia, and Amazonian Brazil and Venezuela. Sometimes known incorrectly as *Triaristella* Brieger. Cultivation as for *Masdevallia* (p. 509).

Literature: Luer, C.A., Icones Pleurothalidinarium VI: Systematics of Pleurothallis sections Ancipita & Scopula and Trisetella. Monographs in Systematic Botany **31**. (1989).

1. T. reichenbachii Brieger (*Masdevallia triaristella* Reichenbach). Illustration: *Botanical Magazine*, 6268 (1876); Bechtel et al., Manual of cultivated orchid species, 228 (1981).

Leaves 2.5–5 cm × 5–8 mm, cylindric, channelled, rounded, tinged with purple. Flowering stems 7.5–10 cm, erect, bearing 1–3 flowers in succession. Upper sepal *c.* 2 cm, triangular to ovate, long-tailed; lateral sepal blade 15 × 2–4 mm, tails 1–1.5 cm. Sepals yellow and pinkish purple. Petals *c.* 4 mm, linear, each with a small point. Lip *c.* 6 mm, pinkish brown. *Costa Rica, Panama & Colombia*. G2. Summer–autumn.

70. LEPANTHES Swartz
J.M.H. Shaw
Small epiphytic or terrestrial herbs with clumping, trailing or hanging habit. Stems erect or hanging, bearing funnel-shaped sheaths. Leaves leathery, linear to ovate, or cylindric, pointed or blunt, stalked. Inflorescence a raceme, or bundle or a solitary flower, with tubular bracts. Flowers resupinate or non-inverted. Upper sepal elliptical to oblong, pointed or blunt, free or united with lateral sepals. Lateral sepals free or united, elliptic to ovate, often with a short pointed tip. Petals thin or fleshy, triangular or elongated or bi-lobed. Lip entire or 2- to 3-lobed, lateral lobes surrounding column, central lobe reduced to a minute appendix, often finely hairy, base joined to column base. Column cylindrical or almost so, hood and foot absent.

A genus of over 800 species ranging from the West Indies and southern Mexico to Bolivia and southern Brazil. Many species are cultivated in specialist collections. Cultivation as for *Pleurothallis* (see below).

1. L. calodictyon Hooker. Illustration: *Botanical Magazine* 5259 (1861); Bechtel, et al., Manual of cultivated orchid species, 3rd edn, 300 (1992); Pridgeon, Illustrated encyclopaedia of orchids 153 (1992).

Small epiphyte with clustered stems, 3–5 cm, bearing conical sheaths with a fringed margin. Leaves broadly ovate to almost circular, 1–2.5 × 1–2 cm, green with a brown netted pattern on upper surface, edges wavy. Inflorescences several together, shorter than leaf blade, usually with several flowers opening in succession. Flowers small with greenish sepals, which are spreading to bent back, elliptic-ovate, pointed, 2 × 1.5 mm, lateral sepals fused. Petals orange to red, bilobed, 2–3 × 1.5 mm, with each ovate lobe extended into a thin tip longer than the wider basal part of the lobe. Lip red to purple, spathulate, 1–1.5 mm, fleshy. Column 1–1.5 mm. *Western Colombia to western Ecuador*. G1. Summer.

Grown for its patterned leaves.

71. PLEUROTHALLIS R. Brown
J.C.M. Alexander
Epiphytic, rarely on rocks. Stems clustered, slender, with a few overlapping sheaths, 1-leaved. Leaves leathery, stalkless or shortly stalked, folded when young. Flowering stems 1–several, apical, erect or arched, bearing 1–many flowers in racemes. Flowers small, usually 2-ranked, not widely open, often pendent. Upper sepal free or very shortly fused to lateral sepals; lateral sepals partially to completely fused, concave, forming a small mentum with the column foot. Petals usually much smaller than sepals. Lip small, loosely hinged to column foot. Pollinia 2 (in our species).

A large genus of about 1000 species, widespread throughout the mountainous regions of the New World tropics and subtropics. Following molecular and anatomical studies many of the species have been transferred to *Stelis*. Cultivation conditions are similar to those for *Masdevallia* (p. 509).

Literature: Luer, C.A., Icones Pleurothalidinarum III: Systematics of Pleurtothallis. Monographs in Systematic Botany **20** (1986). Further parts covering the systematics of *Pleurothallis* are: **31** (1989); **52** (1994); **65** (1998); **72** (1998); **76** (1999); **79** (2000).

1. P. revoluta (Ruiz & Pavon) Garay (*P. macrophylla* Kunth; *P. roezlii* Reichenbach). Illustration: Cogniaux & Goossens, Dictionnaire iconographique des orchidées, Pleurothallis, 1.1 (1898); *Journal of the Royal Horticultural Society* **26**: 147 (1901); *Gartenflora* **50**: 272 (1901).

Stems 8–15 cm, with brown sheaths. Leaf 12–20 × 2.5–4 cm, oblong to lanceolate, acute, sometimes notched. Flowering stem to 30 cm, arched to pendent, bearing 5–10 flowers; bracts 1–1.5 cm. Flowers pendent, dark purple. Upper sepal 2.5–3 × 1–1.5 cm, elliptic, acute; lateral sepals fused into an elliptic, obtuse blade, 2.5–3 × 1.4–2 cm. Petals *c.* 1.5 cm, lanceolate, acute, fleshy, prominently 3-veined. Lip *c.* 1.5 cm, spathulate, fleshy, 2-ridged, papillose above. *Colombia.* G2. Winter.

P. maculatus N.E. Brown (*Cryptophoranthus maculatus* (N.E. Brown) Rolfe). Leaves 4–5.5 × 1.8–3.2 cm, spotted with purple above. Flowers many, *c.* 5 mm long, yellow *spotted* with purple. *Brazil.* G2. Summer.

72. STELIS Swartz

J.C.M. Alexander & J.M.H. Shaw

Tufted epiphytes. Stems slender, erect or ascending, sheathed, 1-leaved at apex. Leaves tapering to the base, fleshy or leathery, folded when young. Flowers many, in long, narrow, erect or arching racemes, borne at the junction of leaf and stem. Sepals equally fused for more than half their length, forming a 3-angled, inflated tube, slightly constricted at the mouth; free portions spreading. Petals and lip very small, concealed in sepal-tube. Lip 3-lobed or simple, fleshy, channelled. Column small, sometimes winged. Stigma 1–2-lobed. Pollinia 2, ovoid, waxy.

A genus of over 700 species from the New World tropics. Cultivation conditions are similar to those for *Masdevallia*.

Literature: Garay, L.A., Systematics of the genus Stelis Sw., *Botanical Museum Leaflets, Harvard University* **27**: 167–259 (1979).

1a. Upper sepal fused at the base to the fused lateral sepals forming a tube **1. tubatus**
b. Upper sepal free from the fused lateral sepals 2
2a. Lateral sepals fused for two-thirds of their length or more 3
b. Lateral sepals fused for half of their length or less 4
3a. Flowering stem shorter than leaves, if longer then lower part enclosed in a groove running about halfway up leaf-blade **2. immersa**
b. Flowering stem longer than leaves, lower part not enclosed in a groove running on leaf-blade **3. quadrifida**
4a. Stems 30 cm or more, flowering stem 1; sepals *c.* 3 cm **4. alta**
b. Stems usually less than 30 cm; flowering stems 1 or more; sepals 1 cm orless **5. gelida**

1. S. tubatus Loddiges (*Physosiphon tubatus* (Loddiges) Reichenbach; *P. loddigesii* Lindley; *P. guatemalensis* Rolfe). Illustration: *Botanical Magazine*, 4869 (1855); *Orchid Review* **79**: 220 (1971); Bechtel et al., Manual of cultivated orchid species, 262 (1981).

Stems 1.5–12 cm. Leaves 4–17 × 1.5–3 cm, elliptic to narrowly obovate. Flowering stem 8–45 cm; flowers 20–100, greenish yellow to brick-red. Sepals 6–22 mm; free portion oblong to lanceolate or elliptic. Petals 1.5–2.5 × 0.6–1 mm, obovate to spathulate, sometimes obscurely 3-lobed. Lip 2–3.2 × 1–2.5 mm, 3-lobed; lobes rounded. Column 2–3.5 mm. *Mexico & Guatemala.* G2. Summer.

Very variable in size and colour of flowers.

2. S. immersa (Linden & Reichenbach) Pridgeon & M.W. Chase (*Pleurothallis immersa* Linden & Reichenbach). Illustration: *Botanical Magazine*, 7189 (1891); Dunsterville & Garay, Venezuelan orchids illustrated **3**: 254 (1965); Ospina, *Proceedings of 7th world orchid conference*, 73 (1974).

Stems 2–7 cm, with brown sheaths. Leaf 7–20 × 2–4.5 cm, lanceolate to narrowly obovate, rounded to acute, with a central winged groove enclosing the lower part of flowering stem. Flowering stem 17–40 cm, erect, zig-zag, sometimes arched above, bearing 12–20 bell-shaped flowers. Flowers purplish brown or bright greenish yellow. Upper sepal 8–14 × 3–4.5 mm, lanceolate, obtuse; lateral sepals fused for *c.* two-thirds of their length into an elliptic blade, 8–13 × 5–7 mm. Sepals stiff, finely hairy. Petals 3–4 × 2–3 mm, spathulate, fleshy with thin margins. Lip *c.* 4 × 2 mm, triangular, clawed, 2-ridged, warty at tip. *Mexico to Colombia & Venezuela.* G2. Winter.

3. S. quadrifida (Lexarza) Solano & Soto Arenas (*Pleurothallis quadrifida* (Lexarza) Lindley; *Pleurothallis ghiesbrechtiana* Richard & Galeotti). Illustration: Dunsterville & Garay, Venezuelan orchids illustrated **2**: 299 (1961); Graf, Tropica, 752 (1978); Bechtel et al., Manual of cultivated orchid species, 265 (1981).

Stems to 15 cm, with grey sheaths. Leaf to 17 × 3 cm, oblong to narrowly elliptic, rounded, often notched, sometimes grooved along the back of the midrib. Flowering stem to 45 cm, usually longer than leaf, erect to spreading, bearing 10–30 flowers; flowers pendent, translucent, pale yellow to yellowish green. Upper sepal 7–10 × 1.5–2.5 mm, ovate, acuminate, concave, 3-veined; lateral sepals fused into an elliptic, rounded, minutely notched, 6-veined blade, 5–12 × 4–6 mm. Petals 6–10 × 2–4 mm, lanceolate, acute, 1-veined. Lip 4–6 × 1.5–3 mm, oblong, waisted; margin erect at base. *Mexico & West Indies south to Venezuela & Colombia.* G2. Winter.

4. S. alta Pridgeon & M.W. Chase (*Pleurothallis grandis* Rolfe; *Stelis grandis* (Rolfe) Pridgeon & M.W. Chase, invalid; *Dracontia grandis* (Rolfe) Luer). Illustration: *Botanical Magazine*, 8853 (1920); *Orchid Review* **30**: 99 (1922).

Stems 30–45 cm, with pale brown sheaths. Leaf 18–35 × 9–18 cm, ovate to broadly elliptic, acute to obtuse. Flowering stem to 50 cm, erect to spreading, loosely many-flowered. Flowers spreading to pendent, greenish brown, flushed and striped with brownish red. Upper sepal 2.5–3 cm, linear to lanceolate, acute; lateral sepals *c.* 3 cm, lanceolate, obtuse, fused for about half their length. Petals *c.* 1 cm, oblong, obtuse. Lip *c.* 2 cm, oblong to spathulate, with 2 fleshy incurved lobes at the tip. *Costa Rica.* G2. Summer–autumn.

5. S. gelida (Lindley) Pridgeon & M.W. Chase (*Pleurothallis gelida* Lindley; *Specklinia gelida* (Lindley) Luer). Illustration: Rickett, Wild flowers of the United States **2**: 125 (1967); Luer, Native orchids of Florida, 185 (1972); *American Orchid Society Bulletin* **45**: 194 (1976).

Stems 5–35 cm, with brown sheaths. Leaf to 25 × 6 cm, oblong to elliptic, rounded, shortly stalked. Flowering stems 1–several, 5–30 cm, erect, bearing 5–25 flowers; flowers pale yellowish green, translucent, softly hairy. Upper sepal *c.* 8 × 3 mm, lanceolate to ovate, obtuse, concave. Lateral sepals *c.* 8 × 3 mm, lanceolate, acute, fused for about half their length, pouched below. Petals *c.* 4 × 2 mm, oblong to spathulate, irregularly toothed at tip. Lip *c.* 2.5 mm, oblong, rounded or truncate. *S Mexico & Florida, south through C America & West Indies to Venezuela, Colombia, Ecuador & Peru.* G2. Flowers intermittently.

S. villosa (Knowles & Westcot) Pridgeon & M.W. Chase (*Pleurothallis schiedei* Reichenbach; *Pleurothallis ornata* Reichenbach). Illustration: Hamer, Las Orquideas de El Salvador **2**: 237 & t. 27 (1974); Northen, Miniature orchids, C-77 (1980). Stems to 1.5 cm. Leaf to 4 cm × 8 mm, lanceolate, spotted on back. Flowering stems 1–several, longer than leaf. Inflorescence loose, bearing 5–12 flowers. Sepals *c.* 7.5 mm, broadly elliptic to spathulate, obtuse to rounded, greenish yellow with purple warts and dangling white marginal appendages. *Mexico*. G2. Spring–summer.

Specklinia tribuloides (*Pleurothallis tribuloides*) may also key out here as there is some variation in the fusion of its lateral sepals.

73. ACIANTHERA Scheidweiler

J.M.H. Shaw

Epiphytic herbs, creeping or clumping, but rarely hanging, with smooth or angled stems with 2 or 3 ridges, enclosed by a few sheaths. Leaves leathery, round to elliptic and obovate, pointed or blunt. Inflorescence 1-flowered or a raceme borne at the stem apex or rarely from the rhizome, often emerging from a spathe. Flowers resupinate, with fleshy sepals bearing hairs on the outer surfce. Upper sepal triangular, obovate to elliptic, pointed or blunt. Lateral sepals ovate to elliptic, pointed or apiculate, united from the base to mid-point or sometimes almost the tips. Petals obovate, blunt or pointed, sometime toothed along the edges. Lip spathulate, elliptical, ovate or oblong, base often narrowed into a claw and bilobed, hinged to foot of column. Column usually cylindrical and smooth, hooded, sometimes winged, with a foot at the base.

A genus of about 200 species distributed from the Antilles and Mexico south to Argentina and Uruguay, with most species in Brazil, most of which were formerly incuded in *Pleurothallis*. Cultivation as for *Pleurothallis*.

1a. Leaves cone-shaped **1. pectinata**
b. Leaves flat, elliptical **2. saurocephala**

1. A. pectinata (Lindley) Pridgeon & M.W. Chase (*Pleurothallis pectinata* Lindley). Illustration: Schlechter, Die Orchideen, edn 2, 176 (1927); Pabst & Dungs, Orchidaceae Brasilienses **1**: 226 & 345 (1975).

Stems 10–15 cm, pendent. Leaf 7–10 cm, broadly elliptic, cone-shaped by fusion of lower margins, enclosing the base of the flowering stem. Flowering stem to 6 cm; flowers *c.* 15 in 2 dense ranks, yellow to green. *S Brazil*. G2. Spring–summer.

2. A. saurocephala (Loddiges) Pridgeon & M.W. Chase (*Pleurothallis saurocephala* Loddiges). Illustration: *Botanical Magazine*, 3030 (1830); Pabst & Dungs, Orchidaceae Brasilienses **1**: 351 (1975).

Stems to 15 cm, grooved. Leaf to 15 cm, elliptic. Flowering stems usually 2, equal to or slightly shorter than leaf; inflorescence dense, with fibrous chaffy bracts at the base. Sepals *c.* 1 cm, green outside, yellow spotted with brown inside. Petals minute. *S Brazil*. G2. Spring–autumn.

74. SPECKLINIA Lindley

J.M.H. Shaw

Small epiphytic or terrestrial herbs with a clumping or creeping habit. Stem erect, enclosed by tubular overlapping sheaths. Leaves leathery, orbicular to obovate, usually with a petiole. Inflorescence a raceme with tubular bracts. Flowers usually resupinate. Upper sepal triangular, elliptic to ovate, often with an elongated narrow tip, sometimes partly united with the lateral sepals to form a cup. Lateral sepals ovate, pointed, often drawn out to a long narrow tip, free or united to various degrees forming a synsepal. Petals elliptic to oblong or spathulate, often with fringed edges. Lip entire or 2- or 3-lobed, elliptic to pear-shaped, pointed to rounded at the tip, sometimes with a fringe, usually hinged at the base to the column foot. Column cylindrical to club-shaped, usually winged, with a basal foot.

A genus of about 200 species distributed from the Caribbean and Mexico to Bolivia and Brazil, many of which were formerly included in *Pleurothallis*, under which name they are often cultivated. Cultivation as for *Pleurothallis* (p. 513).

1a. Flowering stem to 1 cm; flowers red to purple **1. tribuloides**
b. Flowering stem to 15 cm; flowers yellow to green **2. grobyi**

1. S. tribuloides (Swartz) Pridgeon & M.W. Chase (*Pleurothallis tribuloides* Swartz). Illustration: Reichenbach, Xenia Orchidacearum **3**: 275 (1894).

Stem less than 1 cm, with white sheaths. Leaf to 7 × 1.5 cm, obovate. Flowering stem to 1 cm, bearing 1–3 pink to brownish red or purple flowers. Lateral sepals fused from one-third of their length to completely so. *C America, West Indies*. G2.

2. S. grobyi (Lindley) F. Barros (*Pleurothallis grobyi* Lindley). Illustration: *Botanical Magazine*, 3682 (1839); Shuttleworth et al., Orchids, 38 (1970); Skelsey, Orchids, 133 (1979).

Stems 1 cm or less, with white sheaths. Leaf to 7 × 1.1 cm, lanceolate to narrowly obovate, obtuse or rounded. Flowering stem to 15 cm, slender, erect, zig-zag, bearing 4–12 flowers in 2 opposite ranks; flowers translucent, pale yellow to green with scattered purplish veins. Upper sepal 4–10 × 1.5–3 mm, ovate, acuminate, rounded. Lateral sepals fused into an oblong to elliptic, slightly bifid blade, 5–14 × 4–7 mm. Petals 1.5–3 × 0.5–1 mm, lanceolate to spathulate. Lip 2–4 × 1 mm, tongue-shaped, arched, rounded. *Mexico & West Indies to Peru & Brazil*. G2. Summer.

75. RESTREPIA Kunth

J.C.M. Alexander

Dwarf, tufted epiphytes, rarely on rocks, occasionally creeping. Stems slender, 1-leaved at apex, concealed in flattened, papery, overlapping sheaths. Leaves stiff and leathery, folded when young. Flowers solitary on slender stalks, borne at the junction of leaf and stem. Upper sepal and petals lanceolate or narrowly triangular, drawn out into long slender tails with club-shaped tips. Lateral sepals almost totally fused into a broad, often bifid lower sepal. Lip oblong, often with fine lateral lobes near base; tip rounded, truncate or notched, papillose. Column slender, arched, club-shaped, with small column foot. Pollinia 4.

A genus of about 50 species from Mexico through C America to Argentina. Cultivation conditions are similar to those for *Masdevallia* (p. 509).

Literature: Luer, C.A., Icones Pleurothalidinarum XIII: Systematics of Restrepia. Monographs in Systematic Botany **59** (1996).

1a. Lower sepal marked with stripes or lines of spots separated by clear areas **3. antennifera**
b. Lower sepal marked with scattered spots without clear areas 2
2a. Leaves at least twice as long as wide **1. guttulata**
b. Leaves less than twice as long as wide **2. elegans**

1. R. guttulata Lindley (*R. maculata* Lindley). Illustration: Dunsterville & Garay, Venezuelan orchids illustrated **4**: 263 (1966); *American Orchid Society Bulletin* **47** (7): cover (1978); Bechtel et al., Manual of cultivated orchid species, 269 (1981).

Stems 1–15 cm. Leaves 5–8 × 2.5–3 cm, narrowly elliptic, obtuse, slightly keeled. Flowering stems to 7 cm, appearing behind the leaf through twisting of leaf-base. Upper sepal to 2.5 cm × 3 mm, lanceolate, pinkish purple with pale margins, finely tapered into a slender tail with a club-shaped tip. Lower sepal 2.5 × 1 cm, oblong to elliptic, slightly bifid, brownish green spotted with purple. Petals 13 × 1 mm, narrowly triangular, pale pink with a darker central stripe. Lip 10 × 2 mm, oblong with 2 narrow basal lobes, similar in colour to the lower sepal, papillose. Column *c.* 6 mm, club-shaped, pale with purple spots at base. *Venezuela, Colombia & Ecuador*. G2. Winter.

2. R. elegans Karsten. Illustration: *Botanical Magazine*, 5966 (1872); Dunsterville & Garay, Venezuelan orchids illustrated **1**: 375 (1959); Bechtel et al., Manual of cultivated orchid species, 269 (1981).

Stems 4–6 cm, erect or ascending. Leaves 4–6 × 3–4 cm, elliptic to circular, obtuse to rounded. Flowering stem to *c.* 4 cm, slender. Upper sepal 2.5 × 2–3 mm, lanceolate, attenuate, white with purple veins and tip. Lower sepal 2.4 cm × 9 mm, oblong, obtuse, yellow to brown, spotted with purple. Petals 14 × 1 mm, narrowly triangular, attenuate, pale with purple veins and tip. Lip 11 × 5 mm, 3-lobed; lateral lobes narrow and sickle-shaped, acute; central lobe fiddle-shaped, narrowing rapidly above the lateral lobes, then broadening above into an oblong notched blade. Column *c.* 6 mm, pale green. *Venezuela*. G2. Winter–spring.

3. R. antennifera Kunth. Illustration: Dunsterville & Garay, Venezuelan orchids illustrated **4**: 259 (1966); Shuttleworth et al., Orchids, 39 (1970); *American Orchid Society Bulletin* **43**: 621 (1974); *Botanical Magazine*, 6288 (1877).

Stems to *c.* 10 cm. Leaves to 5.5 × 3 cm, elliptic to ovate, obtuse or rounded. Flowering stems *c.* 2.5 cm, slender. Upper sepal 2.3 cm × 3 mm, lanceolate, attenuate, white with purple veins; tail green with pink tip. Lower sepal 2.3 × 1.2 cm, oblong to ovate, acute, bifid, pale yellowish brown with purple lines or dense lines of spots. Petals 1.5 cm × 1 mm, lanceolate to narrowly triangular, white with a purple mid-vein. Lip 10 × 3 mm, oblong to fiddle-shaped with very narrow, acute lateral lobes and a short, cylindric claw; tip truncate and papillose, pale brown, spotted with purple. Column *c.* 5 mm, club-shaped, pale brown. *Venezuela, Colombia & Ecuador*. G2. Winter–spring.

76. ZOOTROPHION Luer

J.C.M. Alexander & J.M.H. Shaw

Tufted epiphytes, rarely terrestrial. Stems erect or ascending, sheathed, 1-leaved at apex. Leaves erect, usually hard and leathery, folded when young. Flowers 1–many, clustered at junction of leaf and stem. Sepals fused at base and tip, leaving 2 lateral openings or windows, forming a box-like flower that resembles an animal's head. Lateral sepals forming a short mentum with column foot. Petals very small. Lip very small, free, shallowly 3-lobed. Column erect, winged, club-shaped, toothed at the apex. Pollinia 2, waxy.

About 16 species from the New World tropics, previously included in *Cryptophoranthus* Barbosa Rodrigues. Cultivation conditions are similar to those described for *Masdevallia* (p. 509). If grown in pots, plants should be well shaded and frequently repotted to avoid the development of stale, soggy compost.

1. Z. atropurpureus (Lindley) Luer (*Cryptophoranthus atropurpureus* (Lindley) Rolfe; *Pleurothallis atropurpurea* Lindley; *Masdevallia fenestrata* Hooker; not *Cryptophoranthus fenestratus* (Barbosa Rodrigues) Barbosa Rodrigues; *Z. schenckii* (Cogniaux) Luer). Illustration: *Botanical Magazine*, 4164 (1845).

Stem 3–5 cm. Leaves 3–9 × 1.5–3 cm, oblong to elliptic, acute to obtuse, shortly stalked. Flowers solitary, purplish crimson. Sepals *c.* 1.5 cm × 5 mm, acute, deflexed. Petals *c.* 4 × 2 mm, oblong, obtuse to truncate. Lip *c.* 4.5 × 1 mm, spear-shaped. Column *c.* 2 mm. *Cuba & Jamaica*. G2. Autumn.

Z. dayanum (Reichenbach) Luer (*Cryptophoranthus dayanus* (Reichenbach) Rolfe; *Masdevallia dayanum* Reichenbach). Illustration: *Botanical Magazine*, 8740 (1917); Ospina & Dressler, Orquideas de las Americas, f. 76 (1974); Vasquez & Ibisch, Orchids of Bolivia **1**: 438 (2000). Stem 10–15 cm. Leaves 6–9 × 3.5–5.5 cm, elliptic to circular. Flowers solitary or few, 3.5–4 cm long, white to pale yellow, spotted with pink, purple or crimson. *Colombia*. G2. Autumn.

77. OCTOMERIA R. Brown

J.C.M. Alexander

Creeping epiphytes, rarely on rocks. Stems slender, often densely tufted, erect, sheathed below, bearing a single leaf; sheaths sometimes frayed into chaffy strands. Leaves stalked or stalkless, folded when young, fleshy or leathery, flat to cylindric. Flowers 1–many, small, bell-shaped, in terminal clusters at the leaf-base. Sepals and petals free or slightly fused, lanceolate to narrowly ovate. Lip very small, simple or 3-lobed, sometimes shortly clawed, hinged to the base of the column. Column very short, with a small column foot. Pollinia 8.

A genus of about 150 species mostly from Brazil with representatives in the West Indies and other parts of C and S America. Cultivation conditions are generally similar to those for *Masdevallia* (p. 509). The plants should never be allowed to dry out, though less water should be given in winter.

1a. Leaves cylindric or semi-cylindric **3. juncifolia**
b. Leaves flat or slightly concave 2
2a. Stems densely tufted, 6–20 cm high **1. grandiflora**
b. Stems 1–4 cm apart, 3–7 cm high **2. graminifolia**

1. O. grandiflora Lindley. Illustration: Dunsterville & Garay, Venezuelan orchids illustrated **1**: 258 (1959) & **6**: 286 (1976); Pabst & Dungs, Orchidaceae Brasilienses **1**: 228 & 365 (1975); Bechtel et al., Manual of cultivated orchid species, 235 (1981).

Densely tufted epiphyte. Stems 6–20 cm, flattened above. Leaves 10–20 cm × 7–17 mm, linear to narrowly lanceolate, acute, stiff and leathery. Flowers several, yellow and greenish yellow or white. Sepals 1–1.2 cm × 3.5–5 mm, elliptic to ovate, obtuse. Petals 9–11 × 2.5–3.5 mm, lanceolate to ovate, acute or obtuse. Lip 6–8 × 4–6 mm, almost oblong, truncate or notched at tip, with small lateral lobes near base, white to pale yellow with 2 purple ridges and spots near base. Column 3–4 mm, curved, often with a pair of reddish spots near tip. *Venezuela, Trinidad, Surinam, Brazil, Bolivia & Paraguay*. G2. Autumn.

O. crassifolia Lindley. Illustration: Martius, Flora Brasiliensis **3**(4): t. 128

(1896); Pabst & Dungs, Orchidaceae Brasilienses **2**: 363 (1977). Stems 15–20 cm. Leaves 10–18 × 1.2–2 cm. Flowers densely clustered. Sepals and petals *c.* 6 × 2 mm. Lip 3.5–4 × 2.5–3 mm, oblong to elliptic, shallowly 3-lobed, acute to rounded at apex. Column 1.5–2 mm. *Brazil, Uruguay & Paraguay*. G2. Winter–summer.

2. O. graminifolia R. Brown. Illustration: *Botanical Magazine*, 2764 (1827).

Creeping epiphyte. Stems 3–7 cm, 1–4 cm apart. Leaves 6–11 cm × 5–10 mm, linear to narrowly lanceolate, acute. Flowers 1 or 2 on short individual stalks, pale yellow or greenish yellow, fragrant. Sepals 6–8 × 2.5–3 mm, ovate to lanceolate, acute. Petals 6–7.5 × 2–2.5 mm, lanceolate to elliptic, acute. Lip 4–6 × 2–3 mm, shallowly 3-lobed, pale yellow with purple ridges; central lobe oblong to circular. Column 2–2.5 mm, yellow. *West Indies & Brazil*. G2. Spring.

3. O. juncifolia Barbosa Rodrigues. Illustration: Martius, Flora Brasiliensis **3**(4): t. 132 (1896); Pabst & Dungs, Orchidaceae Brasilienses **1**: 228 & 372 (1975).

Tufted epiphyte. Stems 20–35 cm, very slender. Leaves 30–50 cm × 3–5 mm, very slender, cylindric with a shallow groove, arched. Flowers few (rarely numerous), yellow. Sepals and petals 8–10 × 2.5–3 mm, lanceolate to oblong, obtuse. Lip 4.5–5 × 3–3.5 mm, ovate to elliptic, centrally spotted with red. Column 2–2.5 mm. *S Brazil*. G2. Summer–autumn.

O. gracilis Lindley (not *O. gracilis* Barbosa Rodrigues). Illustration: Pabst & Dungs, Orchidaceae Brasilienses **1**: 372 (1975). Stems 3–9 cm. Leaves 6–12 × 1.5–2.5 mm, semi-cylindric, grooved. Flowers 1–8, pale yellow. Sepals and petals 3.5–4.5 × 1–1.5 mm, linear to oblong, obtuse. Lip 2.5–3 × 1.5 mm, circular to oblong, shallowly 3-lobed. Column *c.* 1.5 mm. *S Brazil*. G2. Spring.

78. POLYSTACHYA de Jussieu

J. Cullen

Epiphytic. Pseudobulbs variable, simple, ovoid, bearing 2 or more leaves. Leaves lanceolate, pleated, dark green, folded when young. Raceme or panicle with many flowers, its axis often finely downy, terminal. Flowers not resupinate (in ours). Sepals and petals similar in size and shape, arching forwards. Lip 3-lobed, not or scarcely exceeding the petals, often hairy or papillose. Column with a distinct foot, forming a conspicuous mentum. Pollinia 4, in 2 superposed pairs, almost spherical, borne on a short stipe with a circular viscidium.

A genus of about 200 species from the tropics, centred in Africa. Only 1 species is commonly available. Cultivation as for *Eulophia* (p. 596).

1. P. affinis Lindley (*P. bracteosa* Lindley). Illustration: *Botanical Magazine*, 4161 (1845); Bechtel et al., Manual of cultivated orchid species, 265 (1981).

Pseudobulbs 3–5 cm, compressed, circular in outline, each bearing 2 leaves. Leaves to 20 cm, thin, lanceolate, acute. Flowers in a raceme (rarely slightly branched), axis finely downy. Sepals and petals *c.* 1 cm, yellow with brown stripes. *W Africa*. G2. Summer.

A few other species may occur from time to time in specialist collections.

79. ISOCHILUS R. Brown

J.C.M. Alexander

Epiphytic, terrestrial or on rocks. Stems slender, erect and clustered or creeping. Leaves many, in 2 opposite ranks, linear to oblong, folded when young, with sheathing bases, often notched. Flowers terminal in 1- or 2-ranked racemes, sometimes solitary, white to pinkish purple. Upper sepal free; lateral sepals almost always at least half fused, inflated below, keeled or occasionally winged. Petals free, shorter and broader than sepals, short-clawed. Lip fused to base of column or column foot, short-clawed, narrow, often S-shaped. Column erect, sometimes with a small foot. Pollinia 4, waxy.

A genus of about 15 species from tropical America with rather variable floral characters interconnecting them. This variation has resulted in the use of many different names at species level. The degree of sepal fusion may vary considerably, even on a single plant. The reduction of all names in the genus to 2 species by Correll is not tenable. Cultivation conditions are generally similar to those for *Cattleya* (p. 498) though water should be given throughout the year.

Literature: Mirenda, T., Isochilus, *Orchids* **76**(10): 737 (2007).

1. I. linearis (Jacquin) R. Brown. Illustration: Dunsterville & Garay, Venezuelan orchids illustrated **1**: 178 (1959); Shuttleworth et al., Orchids, 71 (1970); Skelsey, Orchids, 115 (1979).

Epiphytic. Stems to 50 cm, erect, covered in grey leaf-sheaths. Leaves to 6 cm × 3 mm, linear, slightly tapering, asymmetrically notched, with prominent midrib. Flowers 1-several, in dense 1- or 2-sided racemes, *c.* 1 cm, pale pink to dark pinkish purple (rarely orange). Sepals 8–9 × 3 mm, triangular to narrowly ovate; lateral sepals fused to about half their length, inflated and forming a small mentum with column foot. Petals 6–8 × 3–4 mm, lanceolate, irregular, short-clawed. Lip 8 × 2 mm, linear to lanceolate, constricted in the middle, short-clawed. Column 3–4 mm, with a short foot. *Mexico to Peru & Brazil, West Indies*. G2. Summer.

80. MALAXIS Swartz

P.J.B. Woods

Perennial herbs, growing in the soil, on rocks, or as epiphytes. Stems fleshy, creeping, the tip leafy and erect, or crowded, or cylindric, or forming pseudobulbs. Leaves few to many, usually deciduous on older stems, broad, often coloured, usually pleated, often unequal at the sheathing base. Inflorescence terminal, erect; scape usually angled or shallowly winged, bracts deflexed. Flowers numerous, small, lip above the column (flowers non-resupinate). Sepals broader than petals, spreading. Petals linear, usually reflexed. Lip flat, entire, usually rounded or ovate, often with a hollow nectary at the base, which is auriculate and clasps the column; apical margins often toothed or lobed. Column short, winged. Pollinia 4, in 2 pairs, waxy.

A genus of about 300 species, widely distributed in temperate and tropical regions. Cultivation as for tropical, terrestrial species, with substitution of good leaf-mould (beech or oak) for crushed bark or peat.

1a. Inflorescence dense, the flowers crowded, their stalks less than 8 mm **2. discolor**

b. Inflorescence loose, flowers widely spaced, their stalks 8 mm or more **1. metallica**

1. M. metallica (Reichenbach) Kuntze. Illustration: *Botanical Magazine*, 6668 (1883); *La Belgique Horticole* **34**: plate opposite p. 281 (1884); Wood et al., The plants of Mount Kinabalu **2**: pl. 60B (1993).

Stems indistinct, cylindric, leaves up to 6, to 7.5 × 3.8 cm, elliptic to ovate, acute, dark metallic red-purple, with 3–5 veins, margins wavy. Flowers *c.* 1.5 cm in diameter, purplish; stalk and ovary paler, lip pale, apical margin irregularly toothed, auricles triangular. *Borneo.* G2. Late spring.

2. M. discolor (Lindley) Kuntze. Illustration: *Botanical Magazine,* 5403 (1863); Jayaweera, Flora of Ceylon **2**: 41 (1981).

Stems clustered, to *c.* 8 × 1 cm, oblong, leafless after flowering. Leaves 3–6, to 12 × 4 cm, ovate, acuminate, rich purple, pleated, main veins 5–7, margins wavy, sometimes green. Flowers *c.* 3 mm in diameter, at first yellow, changing to orange, then purplish. Lip crescent- or kidney-shaped, margins not toothed, auricles shallow. *Sri Lanka.* G2. Summer.

M. calophylla (Reichenbach) Kuntze (*Microstylis scottii* Ridley). Illustration: *Botanical Magazine,* 7268 (1892); *Annals of the Royal Botanic Garden Calcutta* **8**: t. 20 (1898); Seidenfaden & Woods, Orchids of peninsula Malaysia & Singapore 222 (1992).

Leaves bronze-green, margins paler, with spots, the spots sometimes distributed to the middle of the blade; flowers pink and cream, lip cleft to form 2 teeth, auricles long, lanceolate. *Himalaya to Malaysia and Borneo.* G2. Autumn.

81. LIPARIS Richard

J. Lamond & P.J.B. Woods

Perennial herbs, terrestrial, growing on rocks or epiphytic. Rhizome short. Stems fleshy, thicker at base, sometimes pseudobulbous. Leaves 1–7, membranous or leathery, jointed at the base or not, rolled or folded when young. Inflorescence racemose with few to many flowers. Sepals and petals oblong, reflexed, the petals usually narrower and a little shorter than the sepals. Lip broadly oblong to rounded, sharply bent backwards at or about the middle. Column long and slender, slightly winged. Pollinia 4, in 2 pairs, waxy.

A genus of about 320 species, distributed mostly in the tropics of the Old World. A few species also occur in temperate zones of the northern hemisphere. Many have little to recommend them as garden plants as their flowers are dull greenish or yellowish, but a few, not widespread in cultivation, are particularly attractive. *L. nervosa* should be grown in a compost recommended for terrestrial species whereas *L. viridiflora* should be grown in an epiphytic compost. *L. lilifolia* and *L. loeselii* are probably best grown under alpine-house or cold-frame conditions in a terrestrial compost containing one-third live sphagnum moss.

Literature: Seidenfaden, G., Orchid genera in Thailand IV, *Dansk Botanisk Arkiv* **31**: 1–105 (1976).

1a. Base of leaf unjointed, sheathing the stem; leaves membranous, broadly lanceolate to elliptic or ovate **1. nervosa**

b. Base of leaf jointed, not sheathing the stem; leaves leathery, oblong, lanceolate or oblanceolate **2. viridiflora**

1. L. nervosa (Thunberg) Lindley (*L. elata* Lindley). Illustration: *Edwards's Botanical Register* **14**: t. 1175 (1828); Maekawa, Wild orchids of Japan in colour, 325 (1971); Luer, Native orchids of Florida, 171 (1972); *Dansk Botanisk Arkiv* **31**: 33 (1976).

Pseudobulbs 3–7 × *c.* 3 cm, conical. Leaves 2–7, to 25 cm × 12 mm, elliptic to ovate or broadly lanceolate, pleated, acute or acuminate. Inflorescence 15–50 cm, loosely flowered, scape with 5 wings or grooves. Flowers 10–30, purplish. Sepals and petals to 8 × 2–5 mm, petals narrower than sepals. Lip to 8 × 7 mm, broadly oblong to obovate, apex notched, base with 2 short protruberances. *Tropics of America, Africa and Asia.* G2. Summer.

L. lilifolia (Linnaeus) Lindley. Illustration: *American Orchid Society Bulletin* **43**: 199 (1974) & **47**: 399 (1978); Luer, Native orchids of the United States and Canada, 309 (1975). Plant to 25 cm. Leaves 2, deciduous. Lip purplish, *c.* 10 × 8 mm. *Eastern N America.* H1–2. Spring–summer.

L. loeselii (Linnaeus) Richard. Illustration: Ross-Craig, Drawings of British plants **28**: t. 52 (1971); Danesch & Danesch, Orchideen Europas Mitteleuropa, edn 3, 207, 211 (1972); Mossberg & Nilsson, Orchids of northern Europe, 63 (1979). Plant to 25 cm. Leaves 2, deciduous. Lip yellowish green, *c.* 5 × 3 mm. *N & C Europe, NE USA.* H1. Spring–summer.

2. L. viridiflora Lindley (*L. longipes* Lindley). Illustration: *Annals of the Royal Botanic Garden Calcutta* **8**: t. 37 (1898); *Dansk Botanisk Arkiv* **31**: 85 (1976); Lin, Native orchids of Taiwan, 241 (1977).

Pseudobulbs 3–9 × 1–1.5 cm, cylindric, swollen towards the base. Leaves 2, to 28 × 2.5–3.5 cm, oblanceolate, apex shortly pointed. Inflorescence to 27 cm, densely flowered; scape flattened or narrowly winged. Flowers numerous, greenish, lip yellowish orange. Sepals and petals 2–3 × *c.* 1 mm, petals narrower than sepals. Lip to 3 × 2 mm, oblong, apex obtuse. *India, Sri Lanka & SE Asia.* G2. Winter.

82. NEOTTIA Guettard

P.J.B. Woods

Herbaceous perennials with short rhizomes. Aerial growth dying down in winter. Leaves 2 (rarely 3–4), ovate, almost opposite, arising at about the middle of the stem or below it. Flowers in a spike-like raceme. Sepals and petals more or less equal in size. Lip longer than the sepals and petals, strap-shaped, its apex 2-lobed. Spur absent. Pollinia 2, club-shaped, each longitudinally divided and made up of easily separable masses.

A genus of about 60 species distributed throughout the temperate regions with a few in tropical Asia, which now includes *Listera* R. Brown. Cultivation as for *Dactylorhiza* (p. 469).

1a. Plants to 20 cm, leaves to 2.5 cm; raceme with 4–12 flowers **1. cordata**

b. Plants to 60 cm, leaves 5–20 cm; raceme with many flowers **2. ovata**

1. N. cordata (Linnaeus) Richard (*Listera cordata* (Linnaeus) R. Brown). Illustration: Danesch & Danesch, Orchideen Europas, Mitteleuropa, edn 3, 135, 137 (1972); Landwehr, Wilde Orchideeën van Europa **2**: 533 (1977); Williams et al., Orchids of Britain and Europe, 141 (1978); Mossberg & Nilsson, Orchids of northern Europe, 49 (1979).

Rhizome slender, creeping. Stems slender, sometimes copper-coloured, 4.5–20 cm, upper part minutely downy. Leaves 2 (occasionally 4), ovate, shortly pointed, 1–2.5 cm. Raceme 1–6 cm, with 4–12 flowers. Bracts minute. Sepals greenish, oblong-elliptic, 2–2.5 mm, obtuse. Petals reddish. Lip purplish, linear, 3.5–4.5 mm, lateral lobes small and arising at the base, central lobe divided to about the middle, its segments diverging. *Europe, N America.* H2. Summer.

2. N. ovata (Linnaeus) Bluff & Fingerhuth (*Listera ovata* (Linnaeus) R. Brown). Illustration: Danesch & Danesch, Orchideen Europas, Mitteleuropa, edn 3, 134, 137 (1972); Landwehr, Wilde Orchideeën van

Europa **2**: 531 (1977); Williams et al., Orchids of Britain and Europe, 141 (1978); Mossberg & Nilsson, Orchids of northern Europe, 47 (1979).

Rhizome moderately thick. Stems 20–60 cm, hairless below, downy above. Leaves 2 (occasionally 3–4), ovate to broadly elliptic, distinctly 3–5-veined, obtuse, 5–20 cm. Raceme 7–25 cm, with numerous greenish or brownish green flowers. Bracts minute. Sepals incurved, ovate, obtuse, 4–5 mm. Petals about the same length as sepals but narrower. Lip yellowish green, strap-shaped, 7–15 mm, the apex divided for one-third to half its length, lateral lobes vestigial or absent. *Eurasia*. H1. Summer.

83. EPIPACTIS Zinn

P.J.B. Woods

Perennial herbs with creeping or vertical rhizomes and numerous fleshy but not tuberous roots. Aerial growth dying off in winter. Leaves spaced along the stem, spirally arranged or in 2 opposite rows, strongly nerved, pleated. Flowers held horizontally or hanging, often only on 1 side of the axis. Bracts green and leaf-like, not sheathing the basal parts of the flowers. Sepals and petals spreading or sometimes scarcely opening. Lip in 2 distinct parts, the basal cup-like, the apical flat and variously shaped. Spur absent. Pollinia rapidly breaking up.

A genus of about 20 species from north temperate regions. Cultivation as for *Dactylorhiza* (p. 469); some species, e.g. *E. gigantea* and *E. palustris*, prefer moister conditions.

Rhizome. Creeping: **4**, **5**; not creeping: **1**, **2**, **3**.
Leaves. Spirally arranged: **1**, **4**, **5**; in opposite rows: **2**, **3**.
Lip. Lateral lobes present: **4**, **5**; lateral lobes absent: **1**, **2**, **3**.

1a. Rhizome short, not creeping; lip without lateral lobes 2
b. Rhizome creeping, elongate; lip with lateral lobes 4
2a. Leaves spirally arranged **1. helleborine**
b. Leaves arranged in 2 opposite rows 3
3a. Axis, flower-stalks and ovary downy; flowers purple **2. atrorubens**
b. Axis, flower-stalks and ovary not downy; flowers yellowish green sometimes tinged with red or reddish violet **3. phyllanthes**
4a. Basal part of lip narrower than the ovate apical part **4. palustris**
b. Basal part of lip broader than the elongate apical part **5. gigantea**

1. E. helleborine (Linneaus) Crantz (*E. latifolia* (Linnaeus) Allioni). Illustration: Danesch & Danesch, Orchideen Europas, Mitteleuropa, edn 3, 190, 191 (1972); Sundermann, Europäische und mediterrane diterrane Orchideen, 206 (1975); Williams et al., Orchids of Britain and Europe, 15 (1978); Mossberg & Nilsson, Orchids of northern Europe, 35 (1979).

Rhizome short. Stems 1–3, arising close together, 35–90 cm, sparsely hairy. Leaves 3–10, 5–17 × 2.5–10 cm, spirally arranged, ovate or broadly elliptic, the upper lanceolate, acute or acuminate. Raceme 10–40 cm with 15–50 flowers, borne on 1 side of the axis, lowest bracts longer than flowers, the upper shorter. Sepals 1–1.3 cm, ovate, greenish; petals slightly smaller, pinkish or purplish. Apical part of the lip ovate, acute, recurved, with 2 basal bosses. *Europe, N Africa, SW Asia to Himalaya*. H2. Summer.

2. E. atrorubens (Hoffmann) Besser (*E. atropurpurea* Rafinesque). Illustration: Danesch & Danesch, Orchideen Europas, Mitteleuropa, edn 3, 186, 187, 194 (1972); Sundermann, Europäische und mediterrane Orchideen, 212 (1975); Williams et al., Orchids of Britain and Europe, 158 (1978); Mossberg & Nilsson, Orchids of northern Europe, 33 (1979).

Rhizome short. Stems 15–60 cm, downy, purplish-tinged. Leaves 5–10, 4–10 × 1.5–4.5 cm, arranged in opposite ranks along the stem, ovate to ovate-lanceolate, acute. Raceme to 25 cm with 8–18 flowers, lower bracts as long as flowers. Flowers wine-red, faintly fragrant. Sepals 6–7 mm, ovate, acuminate, petals slightly narrower. Lip 5.5–6.5 mm, basal part cup-shaped, apical part wider than long, with 3 rough basal bosses, apex acute or blunt, recurved. *Europe to C Asia*. H2. Summer.

3. E. phyllanthes G.E. Smith. Illustration: Landwehr, Wilde Orchideeën van Europa **2**: 493, 495 (1977); Williams et al., Orchids of Britain and Europe, 155 (1978); Mossberg & Nilsson, Orchids of northern Europe, 41 (1979).

Rhizome short. Stems 1 or occasionally 3, 20–45 cm, hairless or slightly hairy. Leaves 3–16, arranged in opposite rows on the stem, 3.5–7 × 3–5 cm, ovate or lanceolate, acute. Raceme to 15 cm with 15–35 flowers, bracts longer than flowers. Flowers greenish, hanging, opening only a little. Sepals 8–10 mm, petals sometimes slightly purplish-tinged. Basal and apical parts of the lip not always clearly distinct. *NW & C Europe*. H2. Summer.

4. E. palustris (Miller) Crantz. Illustration: Danesch & Danesch, Orchideen Europas, Mitteleuropa, edn 3, 183–5 (1972); Sundermann, Europäische und mediterrane Orchideen, 214 (1975); Williams et al., Orchids of Britain and Europe, 149 (1978); Mossberg & Nilsson, Orchids of northern Europe, 43 (1979).

Rhizome creeping. Stems to 50 cm, downy. Leaves 4–8, 5–15 × 2–4 cm, spirally arranged and often crowded, oblong to lanceolate, acuminate. Raceme to 15 cm, with up to 15 flowers, lower bracts equalling the flowers. Sepals 8–12 mm, ovate-lanceolate, greenish lined with red. Petals whitish flushed with pink. Lip 1–1.2 cm, basal part concave with erect lateral lobes, pinkish white, purple-lined and orange-dotted, narrower than the apical part, which is ovate, as broad as long, wavy-margined, white with a yellow bar across the base. *Eurasia, N Africa*. H2. Summer.

5. E. gigantea Hooker. Illustration: *Jahresberichte des Naturwissenschaftlichen Vereins in Wuppertal* **23**: t. 5 (1970); *American Orchid Society Bulletin* **71**: 240 (1971); Luer, Native orchids of the United States and Canada, 79 (1975); Rechinger (ed.), Flora Iranica **126**: t. 15, 16 (1978).

Rhizome creeping. Stems to 1 m, hairless. Leaves 4–12, 5–20 × 2–7 cm, spirally arranged, ovate to lanceolate, acuminate. Raceme to 30 cm with up to 15 flowers, bracts longer than flowers. Sepals 1.5–2 cm, ovate-lanceolate, greenish yellow, purple-veined, margins straight. Petals ovate, 1.3–1.5 cm, flushed with pink. Lip 1.4–1.5 cm, basal part concave with red warts, lateral lobes bluntly triangular, yellowish with reddish nerves, separated by a rigid fold from the apical part which is oblong-lanceolate with 2 basal ridges, orange or yellowish, the apex suffused with pink. *NW America*. H2. Summer.

E. royleana Lindley. Illustration: *Botanical Magazine, n.s.*, 214 (1993). Differs from *E. gigantea* by sepals with a broader base, margins slightly incurved. Petals wider with more reticulated veining. *Himalaya*. H1. Summer.

84. CEPHALANTHERA Richard
P.J.B. Woods
Perennial herbs with erect or short, creeping rhizomes. Roots fibrous. Aerial growth dying down in winter. Leaves evenly spaced along stem, pleated. Flowers stalkless or very shortly stalked in a loose terminal spike, pointing upwards; bracts green and leaf-like, not sheathing the basal parts of the flowers. Sepals and petals hooded, scarcely opening. Lip constricted at the middle into 2 distinct parts, the base concave and clasping the base of the column, the apex with several ridges. Spur absent. Pollinia 2, club-shaped, each longitudinally divided, powdery.

A genus of about 14 species from north temperate regions. Cultivation as for *Dactylorhiza* (p. 469).

1a. Flowers pinkish red; ovary hairy **3. rubra**
b. Flowers white; ovary not obviously hairy 2
2a. Bracts as long as or much longer than ovary; lip with 3–5 ridges **1. damasonium**
b. Bracts (except the lowermost) distinctly shorter than ovary; lip with 4–6 ridges **2. longifolia**

1. C. damasonium (Miller) Druce. Illustration: Landwehr, Wilde Orchideeën van Europa **2**: 513 (1977); Williams et al., Orchids of Britain and Europe, 145 (1978); Mossberg & Nilsson, Orchids of northern Europe, 31 (1979).

Stems 15–70 cm. Leaves 3–6, 5–8 × 2–3 cm, ovate-lanceolate to oblong-ovate, the upper narrowest. Spike with 3–5 or more flowers. Flowers *c.* 2 cm, loosely arranged, creamy white. Bracts as long as the ovary, lowermost much longer, ovate-lanceolate to lanceolate. Sepals and petals oblong to lanceolate, incurved, petals slightly shorter than sepals. Lip shorter than sepals and petals, yellowish orange at base; apical half heart-shaped, recurved, with 3–5 yellowish orange ridges. *Europe, N Africa, SW Asia.* H2. Spring–summer.

2. C. longifolia (Linnaeus) Fritsch. Illustration: Landwehr, Wilde Orchideeën van Europa **2**: 515 (1977); Williams et al., Orchids of Britain and Europe, 145 (1978); Mossberg & Nilsson, Orchids of northern Europe, 29 (1979).

Stems 10–50 cm. Leaves 4–12, 4–18 cm × 8–40 mm, linear-lanceolate to lanceolate. Spike with 10–20 or more flowers; flowers *c.* 1.8 cm, white, sepals and petals incurved but more open than in *C. damasonium*. Bracts, except for the lowermost, much shorter than ovary, mostly linear. Sepals lanceolate, petals broader and shorter. Lip shorter than sepals and petals, yellowish orange at base; apical half heart-shaped, recurved, with 4–6 yellowish orange ridges. *Europe, N Africa, to the Middle East & temperate Asia.* H2. Spring.

3. C. rubra (Linnaeus) Richard. Illustration: Danesch & Danesch, Orchideen Europas, Mitteleuropa, edn 3, 195 (1972); Landwehr, Wilde Orchideeën van Europa **2**: 517 (1977); Mossberg & Nilsson, Orchids of northern Europe, 27 (1979).

Stems 10–60 cm, upper part hairy. Leaves 2–8, 3–14 × 1–3 cm, oblong to lanceolate. Spike with 2–15 flowers, hairy; flowers 2–2.5 cm, opening fairly widely, pinkish red. Bracts mostly longer than ovary, but the upper much shorter, linear. Sepals and petals to 2.5 cm, ovate to lanceolate. Lip as long as sepals, white, with several brownish ridges at base; apical half lanceolate, recurved, red-violet with *c.* 10 yellowish brown ridges. *Europe, N Africa, SW Asia east to Iran.* H2. Spring–early summer.

85. ERIA Lindley
J.C.M. Alexander
Epiphytic or occasionally terrestrial, often densely hairy. Pseudobulbs ovoid, fusiform or cylindric, sometimes slender and cane-like, often densely sheathed. Leaves 1–several, folded when young (rolled in Nos. 1 & 2). Flowers in terminal or axillary racemes. Upper sepal free; lateral sepals united at base with column foot forming a pronounced mentum. Petals similar to sepals. Lip often 3-lobed, borne on column foot. Pollinia 8, waxy, pear-shaped or ovoid.

A very diverse genus of about 500 species from India, SE Asia and the Pacific islands. It is generally similar to *Dendrobium* but the flowers have 8 pollinia and are never spurred. Cultivation details are similar to those for the tropical species of *Dendrobium* (p. 521); the choice of pot, raft or bark will be governed by the stature and habit of the individual species. The resting period will depend on the fleshiness of the pseudobulb.

Literature: Seidenfaden, G., Orchid genera in Thailand X, *Opera Botanica* **62**: 1–157 (1982).

1a. Leaves rolled when young 2
b. Leaves folded when young 3
2a. Stems short, concealed; flowers 10–30 per raceme, hairy outside **1. javanica**
b. Stems long, exposed; flowers 2–6 per raceme, not hairy **2. coronaria**
3a. Flowers in dense racemes; flower-stalks hairy **4. spicata**
b. Flowers in loose racemes; flower-stalks not hairy **3. bractescens**

1. E. javanica (Swartz) Blume (*E. fragrans* Reichenbach; *E. stellata* Lindley). Illustration: *Botanical Magazine*, 3605 (1837); Williams et al., Orchids for everyone, 188 (1980); Bechtel et al., Manual of cultivated orchid species, 212 (1981).

Epiphytic. Pseudobulbs 5–8 × 1–2 cm, narrowly ovoid, of 1 internode, sheathed, bearing 1 or 2 leaves. Leaves 20–50 × 3.5–6 cm, lanceolate, acute, rolled when young. Flowering stem 30–60 cm, erect to arched, terminal, bearing up to 30 flowers. Flowers *c.* 4 cm wide, fragrant, white or cream-coloured. Flower-stalks and outside of sepals hairy. Sepals 2–2.5 cm, lanceolate to narrowly triangular, acute; petals smaller, slightly sickle-shaped. Lip 1–1.5 cm, oblong to lanceolate, 3- or 5-ridged, shallowly lobed, central lobe narrowly triangular, obtuse or acute; lateral lobes semi-ovate, erect. Column *c.* 8 mm; foot *c.* 5 mm. *Himalaya, Thailand, Malaysia, Indonesia & Philippines.* G2. Spring–summer.

2. E. coronaria (Lindley) Reichenbach (*Trichosma suavis* Lindley). Illustration: *Orchid Review* **82**: 283 (1974); Bechtel et al., Manual of cultivated orchid species, 212 (1981); *Opera Botanica* **62**: 41 & 151 (1982).

Terrestrial or epiphytic. Stems 7.5–25 × 5 mm, sheathed at base only, bearing 2 leaves near the apex. Leaves 10–25 × 3–5 cm, narrowly elliptic to ovate, leathery, acute, rolled when young. Flowering stem to *c.* 15 cm, erect, robust, terminal, bearing 2–6 flowers. Flowers 2.5–4 cm wide, white or pale yellow with crimson markings on the lip, fragrant. Sepals 1.5–2.2 cm × 5–8 mm, elliptic to ovate; petals *c.* 2 × 1 cm, oblong to elliptic, obtuse. Lip *c.* 2 × 1 cm, oblong to ovate, 3-lobed; lateral lobes oblong, rounded, erect; central lobe short, almost circular, with 5–7 wavy keels. Column longer than column foot. *E Himalaya to Thailand & Malay peninsula.* G2. Autumn–winter.

3. E. bractescens Lindley. Illustration: *Botanical Magazine*, 4163 (1845); Holttum, Flora of Malaya **I**: Orchids, 386 (1953); Millar, Orchids of Papua New Guinea, 62 (1978).

Pseudobulbs 6–15 cm, cylindric to fusiform, sheathed below, bearing 2 or 3 leaves. Leaves 12–20 cm, oblong to narrowly elliptic, obtuse. Flowering stems usually 2 per pseudobulb, 15–20 cm, erect to horizontal, many–flowered; bracts conspicuous, to 2.5 cm, oblong, rounded, cream-coloured to pale green. Flowers *c.* 2.5 cm wide and long, white to pale green, sometimes flushed with pink. Sepals narrowly triangular, acute; mentum pronounced. Petals lying close under upper sepal. Lip oblong, rounded, shallowly 3-lobed; lateral lobes oblong, rounded, pale purple; central lobe almost circular, with 3 or 5 pink, wavy keels. Column shorter than column foot. *E Himalaya, Andaman Islands, SE Asia, Indonesia & Philippines*. G2. Summer.

4. E. spicata (D. Don) Handel-Mazzetti (*E. convallarioides* Lindley). Illustration: Shuttleworth et al., Orchids, 72 (1970); Graf, Tropica, 732 (1978); Bechtel et al., Manual of cultivated orchid species, 213 (1981).

Pseudobulbs 5–15 × 1.5–3 cm, densely sheathed, bearing up to 4 leaves. Leaves narrowly elliptic, folded when young. Flowering stems shorter than leaves, densely many-flowered, erect. Flowers *c.* 1 cm wide, almost spherical, white to pale yellow. *E Himalaya & Burma*. G2. Spring.

86. PINALIA D. Don

J.M.H. Shaw

Epiphytic or terrestrial herbs with stems consisting of several enlarged internodes, each covered by a thin leaf sheath, and bearing leaves on the upper half or near the apex. Leaves linear, lanceolate, leathery with a distinct petiole. Inflorescence a raceme, leaving a concave scar on the stem. Sepals with hairs on the back. Upper sepal narrow, triangular or elliptic, lateral sepals triangular, attached to the column base to form a mentum. Petals similar to upper sepal. Lip 3-lobed, hinged at base to column foot, usually keeled and papillose. Column with a foot. Pollinia 8, club shaped.

A genus of about 160 species distributed from Himalayas through southern China, Indochina, Malaya to Australia and Pacific islands; sometimes included in *Eria*, but separated on the basis of molecular data and morphological characters such as the stem consisting of several internodes, as against one in *Eria*. Cultivation as for *Eria* (p. 520).

1. P. floribunda (Lindley) Kuntze (*Eria floribunda* Lindley). Illustration: *Edwards's Botanical Register* **30**: t. 20 (1844); *Orchid Review* **76**(9): cover (1968); Cootes, Orchids of the Philippines 116 (2001).

Stems to 30 × 1 cm, cylindric, slender, bearing 3–7 leaves in upper half. Flowering stems several, *c.* 15 cm, axillary, horizontal to pendent, densely many-flowered. Flowers *c.* 1 cm wide, white to pinkish red. Column dark purple. *Himalaya to Vietnam, Malay peninsula, Sumatra & Borneo*. G2. Summer–autumn.

87. CRYPTOCHILUS Wallich

J.C.M. Alexander

Epiphytic. Pseudobulbs cylindric to spherical, simple, crowded, developing from flowering stems. Leaves 1 or 2 per pseudobulb, leathery, usually shortly stalked, folded when young, appearing after flowering. Flowering stems sheathed below, with many flowers in dense 2-rowed spikes; flowers equal to or shorter than the persistent bracts. Sepals fused into a urn-shaped to spherical tube with free, spreading, triangular points. Petals linear to oblong, concealed in sepal-tube. Lip fused to column foot. Column short, straight, smooth at apex. Pollinia 8.

A genus of about 10 species distributed from the Himalaya to southern China, Thailand and Indochina. The affinities of *Cryptochilus* were uncertain and it was previously considered a member of the tribe *Vandae*; it is now thought to be allied to *Eria*, which has been confirmed by molecular studies. The flowers show adaptations to bird pollination. Cultivation conditions are similar to those for Himalayan species of *Dendrobium* (see below), allowing plants to become quite dry between waterings during the winter.

1. C. sanguinea Wallich. Illustration: *Orchid Review* **79**: 227 (1971); *American Orchid Society Bulletin* **46**: 244 (1977).

Pseudobulbs to *c.* 7 cm, ovoid, slightly compressed. Leaves 9–20 × 2–4 cm, oblong to lanceolate, acute. Flowers *c.* 2 cm long; bracts about equal to flowers in length. Sepals united into a scarlet, urn-shaped tube, swollen at base; tips of lobes acute. Petals small, oblong to obovate, obtuse, yellow, flushed with brown. Lip oblong, broader at tip, channelled, yellow. Column about half as long as lip, straight. Pollinia greenish blue. *Nepal & NE India (Sikkim, Meghalaya)*. G1. Summer.

C. lutea Lindley. Illustration: *Annals of the Royal Botanic Garden Calcutta* **8**: t. 221 (1898); Pearce & Cribb, Orchids of Bhutan, 366 (2002). Pseudobulbs cylindric. Leaves 10–15 × 1.5–2.5 cm, narrowly elliptic, acute. Sepal-tube *c.* 1 cm, spherical to cylindric, not swollen at base, yellow. Bracts longer than flowers. Petals diamond-shaped. Pollinia yellow. *NE India (Sikkim, Naga hills), Bhutan*. G1. Summer.

88. DENDROBIUM Swartz

P.J.B. Woods & J. Cullen

Sympodial epiphytes. Stems thin or more frequently fleshy and pseudobulbous, bearing several leaves or leaf-scars. Leaves variable, with or without a sheath enclosing the stem at the base, folded when young. Flowers 1–many in racemes borne on leafy or leafless stems. Upper sepal free, the laterals attached to the elongate column foot, forming a conspicuous, variously shaped mentum, which is sometimes spur-like. Petals larger or smaller than the sepals. Lip entire or more frequently 3-lobed, attached to the apex and sometimes to the sides of the column foot, very variable in shape, usually with conspicuous longitudinal ridges. Column short, though with an extended foot, frequently bearing 2 lateral horns. Pollinia 4 in 2 pairs.

A genus of about 900 species, widely distributed in Asia, in particular the Himalaya, Burma, Thailand, Malaysia, Indonesia and New Guinea, and extending to Australia. Its supraspecific classification is rather uncertain, both taxonomically and nomenclaturally. The account below follows the sections (and their order) proposed by Schlechter (*Feddes Repertorium Beihefte* **1**: 440–52, 1912); the sections themselves, however, are referred to as 'groups' because of uncertainties with regard to their correct nomenclature. The sectional names used by Schlechter and Kränzlin are cited as appropriate. In Schlechter (Brieger, Maatsch & Senghas (eds), Die Orchideen, edn 3, **1**(11 & 12): 636–752, 1981), Brieger divides the genus, as recognised here, into numerous segregate genera, formed largely by the raising of the Schlechter sections to generic level. This produces many new names, which are not cited here as synonyms. There is no doubt that the older, inclusive concept of the genus is appropriate here; whether the new genera will be widely accepted remains in doubt. The distinction between stem and pseudobulb is very blurred in this genus. We have used the latter term only those cases in which the stem is conspicuously swollen.

About 80 species are reputedly in cultivation, together with numerous hybrids. Because of the large size of the genus, the methods of cultivation differ somewhat, depending on the origins of the species. The Indian and the Burmese species require heat and moisture when growing, exposure to air and light during autumn, and a cooler, drier, rest period during the winter. The species from Australia and New Guinea require similar treatment, but need more frequent watering and a slightly higher winter temperature. All the species benefit from appropriate shading, especially when the new growth appears. Epiphytic compost should be used, and the containers should be as small as possible. Many species do well on bark, slabs or tree fern. Propagation is mainly by division of old plants, by small pseudobulbs, and by plantlets produced on the stems.

Literature: Kränzlin, F., Das Pflanzenreich **45**: 25–313 (1910) & 301–12; Kamemoto, H., Amore, T.D. & Kuehnle, A.R., Breeding Dendrobium orchids in Hawaii (1999); Lavrack, B., Harris, W. & Stocker, G., Dendrobium and its relatives (2000).

1a. Leaves without sheaths 2
b. Leaves with sheaths 17
2a. Stems wiry, creeping or hanging, branched, each branch terminating in a single, thick, hard-pointed leaf **1. pugioniforme**
b. Stems and leaves various, not as above 3
3a. Lip hairy on the upper surface, often fringed 4
b. Lip hairless on the upper surface, never fringed, sometimes wavy 9
4a. Stems creeping, bearing scattered pseudobulbs, each with 1 leaf **11. jenkinsii**
b. Pseudobulbs clustered, each with several leaves 5
5a. Leaves and pseudobulbs white-hairy **12. senile**
b. Leaves and pseudobulbs hairless 6
6a. Flowers entirely yellow 7
b. Lip yellow, sepals and petals white sometimes flushed with pink 8
7a. Pseudobulbs 4-angled; lip not fringed **14. densiflorum**
b. Pseudobulbs many-angled; lip fringed **13. chrysotoxum**
8a. Pseudobulbs 4-angled; sepals and petals white flushed with pink **16. farmeri**
b. Pseudobulbs not angled or with several slight ridges; sepals and petals white **15. thyrsiflorum**
9a. Basal part of the lip erect, ridges usually conspicuous, close to or contacting the column 10
b. Basal part of the lip diverging from the column, ridges usually shallow and not contacting the column 14
10a. Ovary and flower-stalk hairless 11
b. Ovary and flower-stalk roughly hairy 13
11a. Central lobe of lip obtuse and deeply notched; racemes hanging **10. woodsii**
b. Central lobe of lip acute, not notched; racemes erect 12
12a. Lateral lobes of lip large, spreading to form a cup-like structure around the column; flowers 7–9 cm in diameter **8. spectabile**
b. Lateral lobes of lip small, not as above; flowers *c.* 4 cm in diameter **9. atroviolaceum**
13a. Petals cream with purple spots and lines; lip with purple lines on all lobes; leaves to 30 × 10 cm **6. macrophyllum**
b. Petals cream, without spots or lines; lip cream with purple lines on the lateral lobes only; leaves 10–18 × 5–7 cm **7. forbesii**
14a. Stems drooping, very slender, wire-like in the lower half then dilated and 4-angled above; sepals very acuminate **5. tetragonum**
b. Stems erect, not as above; sepals not acuminate 15
15a. Stems tapered evenly from base to apex; lateral lobes of the lip acute (Figure 26(2), p. 523) **4. kingianum**
b. Stems uniformly thick or sometimes narrowed at the base; lateral lobes of the lip obtuse 16
16a. Central lobe of lip acute, upcurved; racemes 8–16 cm with 4–20 flowers; leaves thin **3. falcorostrum**
b. Central lobe of lip rounded and apiculate; racemes 10–60 cm with many flowers; leaves thick and leathery **2. speciosum**
17a. Stems and leaves strongly flattened, leaf-bases folded and overlapping **71. terminale**
b. Stems and leaves not as above 18
18a. Stems thickened only at the base, the rest thin and normally stem-like **70. crumenatum**
b. Stems not at all thickened, or thickened above the base 19
19a. Lip simple, unlobed, its margins attached to the column foot for some distance 20
b. Lip 3-lobed or simple, its margins not attached to the column foot 27
20a. Plants crystalline-papillose on leaves, sheaths, ovaries, flower-stalks and sometimes also on the flowers **49. cuthbertsonii**
b. Plants not at all crystalline-papillose, sometimes warty 21
21a. Lip hollowed, its apical part fringed, upcurved 22
b. Lip flat, its apical part neither fringed nor upcurved 23
22a. Leaf-sheaths and stem covered with fine warts; flowers orange or orange with yellow tips **48. subclausum**
b. Warts absent, except occasionally at the tops of the leaf-sheaths; flowers scarlet or purplish **47. lawesii**
23a. Leaves clustered at stem apex; ovary angled or winged; racemes with 2–4 flowers 24
b. Leaves evenly distributed along the length of the stem; ovary not angled or winged; racemes with many flowers 25
24a. Leaves linear, terete; ovary with 5 obscure wings (3 close together along the top of the ovary) **51. hellwigianum**
b. Leaves oblong to oblong-lanceolate, flat; ovary with 3 wings **50. vexillarius**
25a. Sepals and petals white (rarely bluish), lip apex pale purple; sepals 2–3 cm **44. amethystoglossum**
b. Flowers differently coloured; sepals less than 2 cm 26
26a. Flowers pink or reddish, rarely white; raceme conspicuously 1-sided **45. secundum**
b. Flowers orange; inflorescence not 1-sided **46. bullenianum**
27a. Mentum double, consisting of 2 swellings at right angles to each other 28
b. Mentum single, consisting of 1 swelling 29

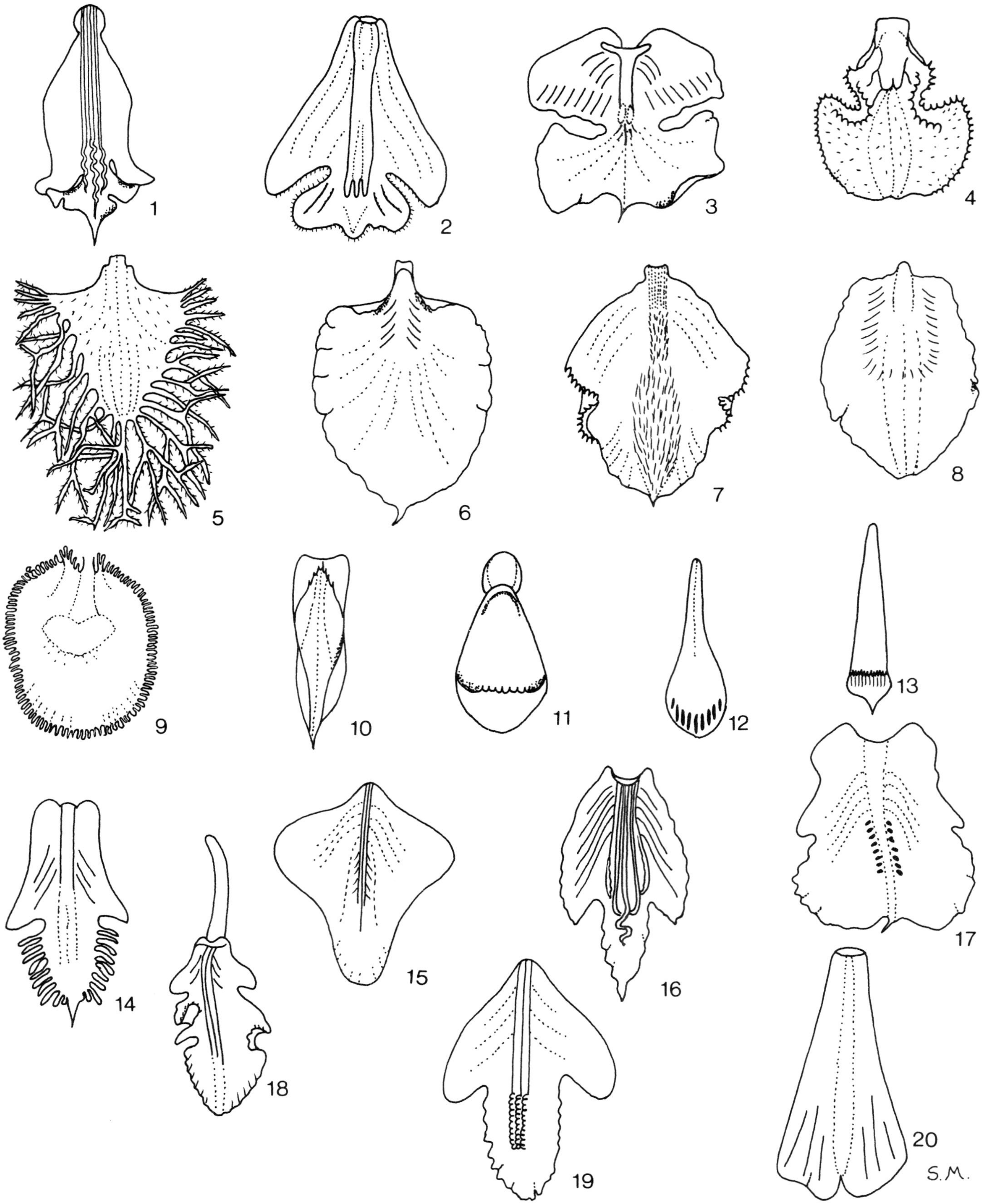

Figure 26. Lips of *Dendrobium* species. 1, *D. pugioniforme*. 2, *D. kingianum*. 3, *D. forbesii*. 4, *D. farmeri*. 5, *D. brymerianum*. 6, *D. findleyanum*. 7, *D. heterocarpum*. 8, *D. nobile*. 9, *D. fimbriatum*. 10, *D. secundum*. 11, *D. lawesii*. 12, *D. cuthbertsonii*. 13, *D. vexillarius*. 14, *D. delacourii*. 15, *D. phalaenopsis*. 16, *D. discolor*. 17, *D. formosum*. 18, *D. draconis*. 19, *D. crumenatum*. 20, *D. terminale*. (Compiled from various sources; not to scale.)

28a. Flowers 3–7 cm in diameter; petals broadly rounded, occasionally with a small point; calluses on lip numerous and dense **53. bigibbum**
b. Flowers 6–10 cm in diameter; petals acuminate; calluses on lip small and obscure **54. phalaenopsis**
29a. Leaf-sheaths hairy 30
b. Leaf-sheaths not hairy 36
30a. Stems no more than 8 cm; mentum broadly conical, blunt **69. bellatulum**
b. Stems taller; mentum spur-like 31
31a. Leaves hairy; petals as broad as or a little broader than sepals 32
b. Leaves hairless; petals much broader than sepals 33
32a. Lateral lobes of lip borne at its base; central lobe not fringed (Figure 26(18), p. 523) **67. draconis**
b. Lateral lobes of lip borne approximately halfway from base to apex; central lobe fringed **68. williamsonii**
33a. Ovary 3-keeled; base of lip green or purple-streaked 34
b. Ovary not keeled; base of lip red, yellow or orange 35
34a. Base of lip greenish **65. dearei**
b. Base of lip streaked with purple **66. sanderae**
35a. Flower *c.* 10 cm in diameter; lip scarcely 3-lobed, notched **63. formosum**
b. Flower up to 8 cm in diameter; lip clearly 3-lobed, irregularly fringed at the apex **64. infundibulum**
36a. Flowers not resupinate; lip orange with purple or brown markings **17. seidenfadenii**
b. Flowers resupinate; lip various but not as above 37
37a. Lip hairy on the upper surface; mentum relatively short for the size of the flower 38
b. Lip hairless on the upper surface; mentum relatively long for the size of the flower 64
38a. Raceme with 5 or more flowers and an axis longer than the individual flower-stalks 39
b. Flowers borne singly or in racemes of 2, 3 or rarely 4, axis absent or shorter than the individual flower-stalks 42
39a. Petals *c.* 3.5 cm broad, almost 2 times broader than the sepals; leaf-sheaths red-striped **35. pulchellum**
b. Petals at most 2–2.2 cm broad, less than 2 times broader than the sepals; leaf-sheaths not purple-striped 40
40a. Lip with the apical part and the sides curved upwards, forming a pouch, margin entire **18. moschatum**
b. Lip not as above, margin fringed 41
41a. Stems hanging; flowering stems usually leafy; fringing of the lip up to 4 mm deep **19. hookerianum**
b. Stems erect; flowering stems usually leafless; fringing of the lip at most to 2 mm deep **20. fimbriatum**
42a. Lip fringed to the naked eye 43
b. Lip not fringed to the naked eye, entire or finely toothed under a lens 48
43a. Fringe of lip greater in extent than the solid part; petals not broader than sepals (Figure 26(5), p. 523) **21. brymerianum**
b. Fringe of lip narrow; petals broader than sepals 44
44a. Sepals and petals yellow 45
b. Sepals and petals white, white tipped with pinkish lilac or entirely reddish purple or red 46
45a. Lip densely hairy beneath **23. ochreatum**
b. Lip more or less hairless beneath **22. chrysanthum**
46a. Sepals and petals 5–6 cm; lip mauve with 2 purple blotches but without any yellow **26. anosmum**
b. Sepals and petals up to 3 cm; lip variously coloured, always with some yellow 47
47a. Stem hanging or arching, 60–100 cm; upper surface of lip minutely hairy **37. devonianum**
b. Stem erect, to 15 cm; upper surface of lip densely velvety **27. loddigesii**
48a. Stem with the internodes thickening upwards conspicuously, thus appearing beaded 49
b. Stem with the internodes more or less equally thick throughout, not appearing beaded 52
49a. Lip without any yellow; racemes usually borne on leafy stems **28. moniliforme**
b. Lip with some yellow; racemes usually borne on leafless stems 50
50a. Leaves small, grass-like, 5–8 cm × 3–4 mm; centre of lip purple, outlined with pale yellow **38. falconeri**
b. Leaves not grass-like, much larger; centre of lip yellow 51
51a. Sepals 2–2.5 cm × 6–8 mm **29. findleyanum**
b. Sepals 3–4 cm × 9–12 mm **39. pendulum**
52a. Flowers mostly yellow, without any mauve or purple except occasionally on the lip, usually without any white or pink 53
b. Flowers basically white or purple or pink, without any yellow except occasionally on the lip 54
53a. Racemes borne on leafy stems **24. luteolum**
b. Racemes borne on leafless stems **25. heterocarpum**
54a. Sepals and petals either entirely white, white flushed with pink or purple, or purple, without tips of contrasting colour 55
b. Sepals and petals white with contrasting mauve, pink or purple tips 61
55a. Flowers without any yellow on the lip 56
b. Flowers with the lip mostly yellow (sometimes very pale) 60
56a. Flowers 5 cm or more in diameter 57
b. Flowers less than 5 cm in diameter 58
57a. Lip trumpet-shaped, its basal part more or less tubular, directed downwards, the apical part bell-shaped, directed forwards **30. lituiflorum**
b. Lip not trumpet-shaped, entirely directed obliquely forwards **36. nobile**
58a. Free part of lip very small, much smaller than mentum **43. aduncum**
b. Free part of lip much larger than the mentum 59
59a. Sepals white; lip white with 2 red blotches **31. transparens**
b. Sepals purple; lip mauve with 2 deeper purple blotches **32. parishii**
60a. Petals broader than sepals; lip with a line of small, papilla-like hairs down the middle **33. aphyllum**
b. Sepals and petals equally broad; lip with most of the upper surface velvety **34. primulinum**
61a. Flowers more than 5 cm in diameter; lip with no yellow blotch **36. nobile**
b. Flowers less than 5 cm in diameter; lip with a yellow blotch 62
62a. Sepals and petals 1.2–1.3 cm; lip not hairy beneath **40. crepidatum**
b. Sepals and petals 1.7–3.5 cm; lip hairy beneath 63

63a. Lip densely velvety above and beneath; petals entire **41. gratiosissimum**
b. Lip densely velvety above, sparsely hairy beneath; petals finely toothed **42. amoenum**
64a. Margin of lip fringed (Figure 26(14), p. 523) **52. delacourii**
b. Margin of lip not fringed 65
65a. Petals 1.5–2 times longer than lateral sepals 66
b. Petals as long as or a little longer than lateral sepals 69
66a. Raceme with 5 or more flowers; petals conspicuously twisted 67
b. Raceme with 2–3 flowers; petals not conspicuously twisted 68
67a. Flowers spotted with dark red or brown; lip mauve at base **55. helix**
b. Flowers with dark red venation; lip white at base **56. tangerinum**
68a. Central lobe of lip broader than long **57. minax**
b. Central lobe of lip longer than broad **58. stratiotes**
69a. Sepals greenish white, petals red **59. taurinum**
b. Sepals and petals pale yellow or dull yellow-brown 70
70a. Central lobe of the lip conspicuously undulate-crisped; sepals and petals cream or yellowish **60. crispilinguum**
b. Central lobe of the lip not undulate-crisped; sepals and petals dull yellow-brown or brown 71
71a. Sepals and petals very wavy; racemes with many flowers **61. discolor**
b. Sepals and petals not wavy; racemes with 4–12 flowers **62. mirbelianum**

GROUP A (Section **Rhizobium** Schlechter).

Leaves without sheaths; stems hanging, wiry, branched, each branch terminating in a single, hard-pointed leaf. This section of about 30 species and several natural hybrids is often treated at generic rank as *Dockrillia* Brieger. It is very popular in Australia and many hybrids have been produced.

Literature: Smee, G., Natural hybrids within the genus Dockrillia in Australia, *Orchadian* **12**(12): 532–45 (1999); Banks, D.P., The genus Dockrillia and its hybrids, *Orchadian* **13**(6): 347–56 (2001).

1. D. pugioniforme Cunningham (*Dockrillia pugioniformis* (Cunningham) Rauschert). Figure 26(1), p. 523. Illustration: Dockrill, Australian indigenous orchids, 369 (1969); Bechtel et al., Manual of cultivated orchid species, 201 (1981).

Stems thin, wiry, forming large hanging masses up to 2 m, each branch terminating in a single leaf. Leaves 1–7 cm × 5–20 mm, flat, ovate-acuminate, very sharply pointed. Raceme with 1–3 flowers. Sepals 8–12 mm, petals slightly shorter, all spreading, pale green. Lip 3-lobed with oblong lateral lobes and a triangular, acute, deflexed central lobe which is notched or wavy on either side of the apex, very pale green with purple or red markings on the lateral lobes and the base of the central. *Australia (New South Wales, Queensland)*. G2.

GROUP B (Subgenus **Dendrocoryne** Lindley): Section **Dendrocoryne** (Lindley) Schlechter (*Dendrocoryne* Brieger; *Thelychiton* Endlicher; *Tetrabaculum* Clements & Jones). Leaves without sheaths; lip shorter than sepals and petals, its basal part diverging from the column, hairless on the upper surface.

2. D. speciosum J.E. Smith (*Thelychiton speciosus* (Smith) Clements & Jones). Illustration: Cogniaux & Goossens, Dictionnaire iconographique des orchidées, Dendrobium, t. 24 (1900); Dockrill, Australian indigenous orchids, 385 (1969); Bechtel et al., Manual of cultivated orchid species, 202 (1981).

Stems 8–100 cm. Leaves 4–25 × 2–8 cm, usually ovate or oblong, thick and leathery. Racemes 10–60 cm with many flowers. Flowers variable, sepals and petals widely spreading or not. Sepals 2–4 cm, petals slightly shorter and much narrower, all white, cream or yellow. Lip shorter than the petals, white, cream or yellow with red or purple spots, lobes obtuse, the central lobe *c.* 2 times broader than long. *Australia (New South Wales, Queensland, Victoria)*. G1. Spring.

A very variable species, divided by Dockrill into 5 varieties; it is uncertain which of these are in cultivation.

D. jonesii Rendle (*D. ruppianum* Hawkes; *D. speciosum* var. *fusiforme* Bailey; *Thelychiton jonesii* (Rendle) Clements & Jones). Illustration: *Botanical Magazine*, n.s., 749 (1978). Similar, but with stems swollen at the base and then tapering and then broadening, central lobe of the lip 3 times broader than long. *Australia (Queensland), New Guinea*. G1.

3. D. falcorostrum Fitzgerald (*Thelychiton falcorostrus* (Fitzgerald) Clements & Jones). Illustration: Dockrill, Australian indigenous orchids, 407 & t. 11 (1969).

Stems 12–50 cm, erect, aggregated. Leaves 2–4, 6–14 × 1.5–3 cm, thin, narrowly obovate or rarely narrowly ovate. Racemes 8–16 cm with 4–20 flowers. Flowers 2.5–4 cm in diameter. Sepals 3–4 cm, oblong, petals slightly shorter and narrower, all white. Lip shorter than the petals, white with yellow and purple markings, the central lobe acute, upcurved, the lateral lobes obtuse. *Australia (New South Wales, Queensland)*. G2. Autumn.

4. D. kingianum Bidwill (*Thelychiton kingianus* (Bidwill) Clements & Jones). Figure 26(2) p. 523. Illustration: *Edwards's Botanical Register* **31**: t. 61 (1845); *Botanical Magazine*, 4527 (1850); Dockrill, Australian indigenous orchids, 415 (1969); Bechtel et al., Manual of cultivated orchid species, 198 (1981).

Stem 8–30 cm, broadest at the base, tapering upwards. Leaves 3–10 × 1–2 cm, narrowly ovate to narrowly obovate. Racemes with 2–10 flowers, which are 1–2.5 cm in diameter. Sepals 9–16 mm, triangular, the laterals erect and sickle-shaped; petals much narrower, all in various shades of pink or lilac, rarely white. Lip shorter than the petals to almost as long, pink with deep red stripes or blotches, the lateral lobes acute, the central transversely oblong, blunt or minutely apiculate. *Australia (New South Wales, Queensland)*. G1. Spring.

5. D. tetragonum Cunningham (*Tetrabaculum tetragonum* (Cunningham) Clements & Jones). Illustration: *Botanical Magazine*, 5956 (1872); Dockrill, Australian indigenous orchids, 421 (1969); Bechtel et al., Manual of cultivated orchid species, 202 (1981).

Stems hanging, aggregated, thin and wiry in the lower part, then dilated and 4-angled above. Leaves 2–5 near the stem apex, 3–8 × 1.5–2.5 cm, ovate, acuminate. Racemes short with few flowers; flowers 4–9 cm in diameter, very variable, very fragrant. Sepals 2–5 cm, very narrowly lanceolate, very acuminate; petals erect, shorter and narrower than the sepals, all greenish yellow with irregular brown or purple blotches. Lip much shorter than the petals, 3-lobed, very variable in shape, yellow with dark red markings. *Australia (New South Wales, Queensland)*. G1. Winter.

Var. **giganteum** Leaney (*Dendrobium cacatua* Clements & Jones; *Tetrabaculum cacatua* (Clements & Jones) Clements & Jones). Illustration: Jones, Native orchids of Australia 402 (2006). Differs by larger flowers, 6–8 × 4–6 cm, lacking darker colouration, lip white, midlobe narrower than side-lobes. *Australia (Queensland)* G1. Winter.

GROUP C (Section **Latouria** (Blume) Schlechter). Leaves without sheaths, mostly borne at the stem apex; lip hairless, as long as the petals, the basal part erect, close to the column, basal callus conspicuous.

Literature: Cribb, P.J., Key to the species of Dendrobium section Latouria with setose ovaries, *Botanical Magazine* **183**(1): 28 (1980); Cribb, P.J., A preliminary key to Dendrobium section Latouria, *The Orchadian* **6**: 277–83 (1981); Cribb, P.J., A revision of Dendrobium sect. Latouria, *Kew Bulletin* **38**(2): 229–306 (1983).

6. D. macrophyllum Richard. Illustration: *Botanical Magazine*, 5649 (1867); *Illustration Horticole* **35**: t. 57 (1888); *American Orchid Society Bulletin* **38**: 589 (1969).

Stems to 1 m × 2 cm, club-shaped. Leaves 2–3, to 30 × 10 cm, elliptic to obovate. Racemes to 30 cm with 8–10 (rarely more) flowers, erect. Sepals *c.* 3–3.5 cm, triangular, pale yellowish green, crimson-spotted, bristly outside. Petals shorter than the sepals, spathulate. Lip as long as petals, lateral lobes truncate, white, purple-veined inside, purple-spotted outside, the central lobe transversely elliptic, pale green with dark crimson spots outside. Basal callus 1.5 cm, white, with 3 ridges. *Java to New Guinea & Fiji*. G2. Summer–autumn.

7. D. forbesii Ridley. Figure 26(3), p. 523. Illustration: *Botanical Magazine*, 8141 (1907); Millar, Orchids of Papua New Guinea, 34 (1978).

Stems 15–45 × 2.5 cm, club-shaped. Leaves 2–3, 10–18 × 5–7 cm, elliptic. Racemes to 17 cm, with 5 or more flowers, erect. Sepals to 3.5 cm, triangular pale greenish yellow or cream, unspotted, bristly outside on the veins. Petals white, longer than sepals. Lip pale greenish cream, unspotted, bases of the lateral lobes purple-veined, the central lobe transversely elliptic. Callus obscurely 3-ridged. *New Guinea*. G2. Summer–spring.

8. D. spectabile (Blume) Miquel. Illustration: *Botanical Magazine*, 7747 (1900); Cogniaux & Goossens, Dictionnaire iconographique des orchidées, Dendrobium, t. 22 (1900); *Orchid Review* **74**: f. 59 (1966); Millar, Orchids of Papua New Guinea, (1978).

Stems to 60 cm, club-shaped. Leaves 2–5 near the apex of the stem, to 15 × 5 cm, oblong. Raceme to 30 cm, erect, with 5–25 slightly fragrant flowers. Sepals and petals *c.* 3.5 cm, narrowly triangular, pale yellow and marked with purple, margins wavy. Lip *c.* 5 cm, white, finely veined with purple, the lateral lobes large, forming a cup-shaped structure around the column, central lobe elongate, margins wavy. *New Guinea*. G2. Winter.

9. D. atroviolaceum Rolfe. Illustration: *Botanical Magazine*, 7371 (1894); *American Orchid Society Bulletin* **38**: 587, 589 (1969); Millar, Orchids of Papua New Guinea, 40 (1978); Bechtel et al., Manual of cultivated orchid species, 193 (1981).

Stems 15–30 cm, thickened upwards. Leaves 2–3, 7–15 × 2.5–5 cm, ovate-oblong. Raceme 10–20 cm, erect, bearing 3–8 fragrant flowers. Sepals *c.* 2.8 cm, triangular, petals spathulate, slightly longer than sepals, all cream or pale greenish white, purple-spotted outside. Lip pale green, the lateral lobes crimson-veined, margins violet, the central lobe ovate, purple-spotted outside. *New Guinea*. G2. Summer–spring.

D. rhodostictum Mueller & Kränzlin. Illustration: *Botanical Magazine*, 7900 (1903). Similar but flowers unspotted, the lip with green veins and a few purple spots on the margins of the lateral lobes. *New Guinea*. G2. Spring–summer.

D. johnsoniae Mueller. Illustration: *Journal of the Royal Horticultural Society* **87**: f. 23 (1962); *Botanical Magazine*, n.s., 560 (1969); Millar, Orchids of Papua New Guinea, 35 (1978). Also similar but with more slender stems; lip with longer, ovate central lobe. *New Guinea. Solomon Islands*. G2. Autumn–winter.

D. engae Reeve. Illustration: *The Orchadian* **6**: 123–4 (1979). Also broadly similar, but flowers sweetly scented, sepals without spots. *New Guinea*. G2. Summer–spring.

10. D. woodsii Cribb. Illustration: *Botanical Magazine*, n.s., 727 (1977) – as *D. fantasticum*; *The Orchadian* **6**: 283 (1981).

Stems to 20 (rarely to 40) cm, slender at the base, thickened upwards. Leaves 2 at the stem apex, *c.* 18 × 2–3.5 cm, elliptic to broadly lanceolate, gradually acuminate. Racemes to 10 cm, hanging, with up to 10 fragrant flowers. Sepals and petals 1–1.2 cm, whitish, bluntly triangular, the petals truncate. Mentum saccate. Lip white flushed with mauve, the lateral lobes flushed with green, the central lobe obtuse and deeply notched. *New Guinea*. G2. Autumn–winter.

D. aberrans Schlechter. Similar, but smaller in all its parts, the flowers white without any other colour. *New Guinea*. G2. Autumn-spring.

GROUP D (Section **Callista** (Loureiro) Schlechter).

Leaves without sheaths; lip 3-lobed, densely papillose-hairy on the upper surface.

Literature: Schelpe, E.A., Dendrobium section Callista, *Orchid Digest* **45**: 205–10 (1981).

11. D. jenkinsii Lindley. Illustration: *Edwards's Botanical Register* **20**: t. 1695 (1835); *Botanical Magazine*, 3643 (1838); Cogniaux & Goossens, Dictionnaire iconographique des orchidées, Dendrobium, t. 33 (1904).

Rhizome creeping. Pseudobulbs aggregated, oblong-ovoid, angled, each bearing 1 leaf. Leaves 5–8 × *c.* 2.5 cm, oblong. Racemes from the upper axils of the pseudobulb, arching, with 5–15 flowers *c.* 3 cm in diameter. Sepals and petals *c.* 1.5 cm, petals broader than the sepals, at first pale, later deep yellow. Lip longer than the petals, broadly circular, concave, yellow, which is deeper at the base, covered in short, white hairs. *Himalaya, Burma, SW China, Laos, Vietnam & Kampuchea*, G2. Spring.

12. D. senile Parish & Reichenbach. Illustration: *Botanical Magazine*, 5520 (1865); Seidenfaden & Smitinand, Orchids of Thailand **2**(2): t. 7 (1960).

Stem to 10 cm, club-shaped, bearing several leaves, covered, like the leaves, with long, white hairs. Leaves 5–8 cm, lanceolate. Racemes with 1–2 flowers on a short peduncle. Sepals and petals *c.* 2.5 cm, the petals broader than the sepals, all yellow. Lip ovate, as long as the petals, yellow, its surface densely downy. *Burma, Thailand & Laos*. G2. Spring–summer.

13. D. chrysotoxum Lindley. Illustration: *Botanical Magazine*, 5053 (1858); Cogniaux

& Goossens, Dictionnaire iconographique des orchidées, Dendrobium, t. 11, 11A (1898); Bechtel et al., Manual of cultivated orchid species, 195 (1981).

Pseudobulbs many-angled, bearing 2–5 leaves. Leaves 3–10 cm, oblong or linear-oblong. Racemes with 8–15 flowers, arching or hanging. Sepals and petals 2–2.2 cm, the petals a little broader than the sepals, all yellow. Lip as long as the petals, circular, the margin conspicuously fringed, orange-yellow, sometimes with a brown blotch at the base. *Himalaya, Burma, SW China (Yunnan), Laos & Thailand*. G2. Winter–spring.

Plants with the lip brown-blotched are sometimes called var. **suavissimum** (Reichenbach) Veitch (*D. suavissimum* Reichenbach).

14. D. densiflorum Lindley. Illustration: Botanical Magazine, 3418 (1835); Cogniaux & Goossens, Dictionnaire iconographique des orchidées, Dendrobium, t. 14 (1898); Bechtel et al., Manual of cultivated orchid species, 195 (1981).

Pseudobulbs 25–40 cm, 4-angled, bearing 3–5 leaves. Leaves 7.5–15 × 2–2.5 cm, lanceolate or elliptic-lanceolate. Racemes with many flowers, the scape arching over, the raceme completely pendent. Sepals and petals 2.5–2.8 cm, the petals somewhat broader than the sepals, all pale yellow. Lip as long as the petals, deep orange-yellow, margins somewhat finely toothed. *Himalaya, Burma, Vietnam*. G2. Spring.

Variants with white sepals and petals sometimes occur and have been cultivated: they are known as var. **schroederi** Anon. (lip orange-yellow) and var. **galliceanum** Linden (lip with white margin).

15. D. thyrsiflorum Reichenbach. Illustration: *The Garden* **30**: 544 (1886); Cogniaux & Goossens, Dictionnaire iconographique des orchidées, Dendrobium, t. 18 (1899); Seidenfaden & Smitinand, Orchids of Thailand **2**(2): t. 6 (1960).

Very similar to *D. densiflorum* but pseudobulbs not angled, or with many slight angles, sepals and petals white, the lip deep yellow. *Burma, Thailand*. G2. Spring–summer.

Often combined with *D. densiflorum*. The distinctions between the two are slight, and made more obscure by the existence of the varieties of *D. densiflorum* mentioned above. The precise distributions of both species are uncertain.

16. D. farmeri Paxton. Figure 26(4), p. 523. Illustration: *Annals of the Royal Botanic Garden Calcutta* **8**: t. 80 (1898); Cogniaux & Goossens, Dictionnaire iconographique des orchidées, Dendrobium, t. 30 (1903); Bechtel et al., Manual of cultivated orchid species, 196 (1981).

Pseudobulbs 20–30 cm, 4-angled, bearing 3–5 leaves. Leaves 8–18 × 3.5–5 cm, ovate-lanceolate. Raceme hanging, with arching scape, and with many flowers. Sepals and petals 2–2.5 cm, petals slightly broader than sepals, all white, often with pink tips. Lip as long as the petals, very densely papillose-hairy above, yellow with whitish margins and a pink apex. *Himalaya, Burma, Thailand & Malaysia*. G2. Spring.

GROUP E (Section **Eugenanthe** Schlechter).

Leaves with sheaths; racemes usually from leafless stems; lip not or very obscurely 3-lobed, its sides free from the column foot, often curved upwards around the column, hairy (usually densely so) above.

17. D. seidenfadenii Senghas & Bockemühl (*D. arachnites* Reichenbach not Thouars). Illustration: *Gardeners' Chronicle* **20**: 7 (1896); *Botanisk Tidsskrift* **65**: 335 (1970); *Die Orchidee* **29**: cii (1978).

Stem to 10 cm, swollen; flowering stems leafless. Leaves 4–6.5 cm, narrowly lanceolate. Racemes with 2–3 flowers, which are not resupinate. Flowers to 6 cm in diameter. Sepals and petals similar, 2–3 cm, orange-red. Lip as long as the petals, lanceolate, acute, orange marked with purple lines, and with ridges near the base only. *Burma*. G2. Spring–summer.

D. unicum Seidenfaden. Illustration: *Botanical Magazine*, n.s., 616 (1972); Bechtel et al., Manual of cultivated orchid species, 193 (1981); *Die Orchidee* **57**: 77 (2006). Very similar, but the stems are longer (to 25 cm) and the lip is oblong-elliptic, brownish orange, veined with brown, and has 3 ridges extending for most of its length. *Laos, Thailand*. G2.

Much material in cultivation as *D. arachnites* could well be *D. unicum*.

18. D. moschatum Swartz (*D. calceolaria* Carey; *D. cupreum* Herbert). Illustration: *Botanical Magazine*, 3837 (1840); *Annals of the Royal Botanic Garden Calcutta* **8**: t. 84 (1898); Seidenfaden & Smitinand, Orchids of Thailand **2**(2): t. 8 (1960); Bechtel et al., Manual of cultivated orchid species, 200 (1981).

Stems to 2 m, fleshy. Leaves 10–15 cm, lanceolate, acuminate. Racemes hanging, on leafless stems, with 10–15 flowers, axis elongate. Flowers 6–8 cm in diameter. Sepals and petals to 4 cm, pale orange, petals *c.* 2 cm broad, a little broader than the sepals. Lip shorter than the petals, concave, its margins erect, pale orange with 2 reddish brown blotches at the base. *Himalaya, Laos, Burma & Thailand*. G2. Spring–summer.

19. D. hookerianum Lindley. Illustration: *Botanical Magazine*, 6013 (1873); *Annals of the Royal Botanic Garden Calcutta* **8**: t. 83 (1898).

Stems up to 2.5 m, fleshy, pendent. Leaves 5–15 × 2.5–3 cm, lanceolate or oblong-lanceolate, acute. Racemes hanging, on leafy stems, with 5 or more flowers and an elongate axis. Flowers 7–10 cm in diameter. Sepals and petals to 4 cm, yellow, the petals finely toothed, broader than the sepals. Lip circular, yellow with 2 brown spots near the base, the margins fringed with the fringe *c.* 4 mm deep. *N India*. G2. Autumn.

20. D. fimbriatum Hooker. Figure 26(9), p. 523. Illustration: Cogniaux & Goossens, Dictionnaire iconographique des orchidées, Dendrobium, t. 9, 9A (1898); *Annals of the Royal Botanic Garden Calcutta* **8**: t. 82 (1898); Seidenfaden & Smitinand, Orchids of Thailand **2**(2): t. 8 (1960); Bechtel et al., Manual of cultivated orchid species, 196, 197 (1981).

Stems to 1.5 m, erect, fleshy. Leaves 8–15 × 2–3 cm, oblong-lanceolate. Racemes usually borne on leafless stems, with 5 or more flowers and an elongate axis. Flowers to 5.5 cm in diameter. Sepals and petals 2–3.5 cm, yellow-orange, the petals broader than the sepals. Lip orange, deeper orange or with 1 or 2 brown spots near the base, fringed, the fringe at most 2 mm deep. *Himalaya, Burma, Laos, Thailand, Vietnam & Malaysia*. G2. Spring.

Variable. Var. **oculatum** Hooker, with 1 or 2 brown spots on the lip, and var. **gibsoni** (Lindley) Gagnepain (*D. gibsoni* Lindley) with yellow, slightly smaller flowers, are both cultivated.

21. D. brymerianum Reichenbach. Figure 26(5), p. 523. Illustration: *Botanical Magazine*, 6383 (1878); *Lindenia* **4**: t. 183 (1888); Bechtel et al., Manual of cultivated orchid species, 195 (1981).

Stems to 50 cm, fleshy. Leaves 10–15 × 1.2–2 cm, lanceolate, acute. Racemes with 1–2 flowers, axis very short, borne almost terminally on leafless stems. Flowers 5–6 cm in diameter. Sepals and petals similar in size, 2.5–3.5 cm, yellow. Lip longer than the petals, mostly consisting of a fringe of branched threads, the central solid part small, all pale yellow. *Burma*. G2. Winter–spring.

Var. **histrionicum** Reichenbach (*D. histrionicum* (Reichenbach) Schlechter) with shorter stems, smaller flowers and the lip fringe less conspicuous, is also perhaps in cultivation.

22. D. chrysanthum Lindley. Illustration: *Edwards's Botanical Register* **15**: t. 1299 (1830); Cogniaux & Goossens, Dictionnaire iconographique des orchidées, Dendrobium, t. 2 (1897); *Annals of the Royal Botanic Garden Calcutta* **8**: t. 77 (1898); Bechtel et al., Manual of cultivated orchid species, 195 (1981).

Stems to 1.5 m, fleshy. Leaves 10–17 × 1–2.5 cm, lanceolate. Racemes with 2–3 flowers on a very short axis, borne on leafless stems. Sepals and petals 2–3 × 1.6–2 cm, petals slightly broader than sepals, all yellow, the sepals with reddish, warty ridges on their backs. Lip circular, yellow with 2 reddish brown blotches, shorter than the petals, the margin fringed, more or less hairless and shiny beneath. *Himalaya, Burma, Thailand*. G2. Autumn.

23. D. ochreatum Lindley (*D. cambridgeanum* Paxton). Illustration: *Botanical Magazine*, 4450 (1849); Cogniaux & Goossens, Dictionnaire Iiconographique des orchidées, Dendrobium, t. 16 (1899).

Very similar to *D. chrysanthum* but stems shorter, the petals little broader than the sepals and the lip densely hairy beneath. *E Himalaya, Thailand*. G2. Spring.

24. D. luteolum Bateman. Illustration: *Botanical Magazine*, 5441 (1864); Bateman, Second century of orchidaceous plants, t. 185 (1867).

Stems to 40 cm, fleshy. Leaves 6–10 × *c.* 3 cm, oblong-ovate or oblong-lanceolate, obtuse. Racemes with 2–3 flowers on a very short axis, borne on the leafy stems. Flowers 5–6 cm in diameter. Sepals and petals 2.5–2.8 cm, the petals a little broader than the sepals, all pale yellow. Lip oblong-ovate, almost as long as the petals, yellow with fine red lines towards the base. Mentum rather conspicuous. *Burma*. G2. Winter–spring.

25. D. heterocarpum Lindley (*D. aureum* Lindley). Figure 26(7), p. 523. Illustration: *Botanical Magazine*, 4708 (1853); *Annals of the Royal Botanic Garden Calcutta* **8**: t. 74 (1898); Cogniaux & Goossens, Dictionnaire iconographique des orchidées, Dendrobium, t. 10 (1898); Bechtel et al., Manual of cultivated orchid species, 198 (1981).

Stem to 40 cm, erect, fleshy. Leaves 10–20 × *c.* 2.5 cm, lanceolate or oblong-lanceolate, acute. Racemes with 2–3 flowers on a very short axis, borne on the leafless stems. Flowers 5–7 cm in diameter. Sepals and petals 2.5–3.5 cm, petals broader than sepals, all pale yellow. Lip oblong-ovate, yellow with red-orange lines and veins. *India, Sri Lanka, Burma, Thailand, Malaysia, Indonesia & Philippines*. G2. Spring–summer.

26. D. anosmum Lindley (*D. macrophyllum* Lindley not Richard; *D. superbum* Reichenbach). Illustration: Cogniaux & Goossens, Dictionnaire iconographique des orchidées, Dendrobium, t. 20 (1899); Bechtel et al., Manual of cultivated orchid species, 193 (1981).

Stems to 1.2 m, fleshy, hanging. Leaves 12–18 × *c.* 3 cm, oblong-lanceolate or ovate-oblong. Racemes with usually 2 flowers, axis short or absent, borne on leafless stems. Flowers to 10 cm in diameter. Sepals and petals 5–6 cm, petals much broader than sepals, all pinkish to purple. Lip broadly ovate, mauve with 2 deep purple blotches, fringed. *Philippines, Borneo, Moluccas*. G2. Winter–spring.

27. D. loddigesii Rolfe (*D. pulchellum* Loddiges not Lindley). Illustration: *Botanical Magazine*, 5037 (1858); Bechtel et al., Manual of cultivated orchid species, 199 (1981).

Stems 10–15 cm, thick. Leaves 4–7 × 1.25–2 cm, oblong-lanceolate. Racemes with 1–2 flowers, stalkless on leafless stems. Flowers *c.* 4 cm in diameter. Sepals and petals 2–2.5 cm, petals broader than sepals, all rose red. Lip as long as petals, circular, yellow with a purplish, conspicuously fringed margin. *SW China*. G2. Winter–spring.

28. D. moniliforme (Linnaeus) Swartz (*D. monile* (Thunberg) Kränzlin). Illustration: *Edwards's Botanical Register* **16**: t. 1314 (1830); *Botanical Magazine*, 4153 (1845); Bechtel et al., Manual of cultivated orchid species, 200 (1981).

Stem to 30 cm, with the internodes thickened upwards, somewhat beaded. Leaves to 7 cm, lanceolate, acute. Racemes with 2–3 flowers on a short axis, borne on leafy stems. Flowers *c.* 4 cm in diameter. Sepals and petals 2–2.5 cm, the petals somewhat broader than the sepals, all white or white flushed with pink. Lip shorter than the petals, elliptic, acute, white with red spots at the base and a red apex. *Japan, Korea; ?China*. G1. Spring.

29. D. findleyanum Parish & Reichenbach. Figure 26(6), p. 523. Illustration: *Botanical Magazine*, 6438 (1879); *The Garden* **49**: 496 (1896); Bechtel et al., Manual of cultivated orchid species, 197 (1981).

Stem to 30 cm, internodes thickened upwards, the stem appearing beaded. Leaves 7–8 × 1.3–3 cm, lanceolate. Racemes with 2 flowers on a short axis, usually borne on the leafless stems. Flowers 5–8 cm in diameter. Sepals and petals 2–2.5 cm, pale rose pink or mauve, rarely white suffused with these tones, the petals broader than the sepals. Lip ovate to almost circular, the centre deep yellow, the margin pale yellow or white, sometimes suffused reddish. *Burma, Thailand*. G2. Winter–spring.

30. D. lituiflorum Lindley. Illustration: *Botanical Magazine*, 6050 (1873); Bechtel et al., Manual of cultivated orchid species, 199 (1981).

Stem to 60 cm, fleshy. Leaves 7.5–10 × 1.5–2 cm, lanceolate to linear-lanceolate, acute. Racemes with 1–3 flowers on a short axis, borne on leafless stems. Flowers 5–7 cm, in diameter. Sepals and petals 2.5–3.8 cm, petals broader than sepals, all uniformly purple. Lip as long as petals, trumpet-shaped, its basal part more or less tubular, directed downwards, the apical part bell-shaped, directed forwards, purple margined with white, the inner part of the white margin sometimes yellowish. *NE India (Manipur), Burma*. G2. Spring.

31. D. transparens Lindley. Illustration: *Botanical Magazine*, 4663 (1852); Cogniaux & Goossens, Dictionnaire iconographique des orchidées, Dendrobium, t. 27 (1901); Bechtel et al., Manual of cultivated orchid species, 202 (1981).

Stems to 45 cm, rather thin but fleshy. Leaves 8–10 × *c.* 1.2 cm, lanceolate or linear-lanceolate. Racemes with 2–3 flowers and a very short axis, borne on leafless stems. Sepals and petals

1.8–2.5 cm, petals a little broader than the sepals, white or reddish. Lip ovate, shorter than petals, white with 2 red spots towards the base. *Himalaya*. G2. Spring–summer.

32. D. parishii Reichenbach. Illustration: *Botanical Magazine*, 5488 (1865); Bechtel et al., Manual of cultivated orchid species, 200 (1981).

Stems to 30 cm, hanging or prostrate, fleshy. Leaves to 10 cm, lanceolate to oblong-lanceolate. Racemes usually with 2 flowers on a short axis, borne on leafless stems. Flowers *c.* 5 cm in diameter. Sepals and petals 2.2–3.5 cm, the petals broader than the sepals, all uniformly lilac-purple. Lip shorter than petals, obovate, mauve-purple marked with 2 large dark red-purple spots towards the base. *Burma, SW China, Thailand, Laos & Kampuchea*. G2. Spring–summer.

33. D. aphyllum (Roxburgh) Fischer (*Limodorum aphyllum* Roxburgh; *D. pierardii* Roxburgh). Illustration: *Botanical Magazine*, 2584 (1825); *Annals of the Royal Botanic Garden Calcutta* **8**: t. 72 (1898); Cogniaux & Goossens, Dictionnaire iconographique des orchidées, Dendrobium, t. 26 (1901); Bechtel et al., Manual of cultivated orchid species, 193 (1981).

Stem to 1 m, slender. Leaves 6–12 × 2–3 cm, lanceolate to linear-lanceolate, acute. Racemes with 2–3 flowers on a very short axis, borne on leafless stems. Flowers 4.5–5.5 cm in diameter. Sepals and petals 2.5–3 cm, petals almost 2 times broader than the sepals, all pale pinkish lilac. Lip broadly ovate, as long as the petals, pale yellow, with a line of hairs down the middle. *Himalaya, SW China, Burma, Thailand, Vietnam, Laos, Kampuchea & Malaysia*. G2. Spring.

34. D. primulinum Lindley. Illustration: *Lindenia* **15**: t. 686 (1900); *Annals of the Royal Botanic Garden Calcutta* **9**: t. 98 (1906); Bechtel et al., Manual of cultivated orchid species, 201 (1981); *Die Orchidee* **32**: *Orchideenbewertung*, 9 (1982).

Similar to *D. aphyllum* but leaves unequally 2-lobed at apex, the petals little broader than the sepals and the upper surface of the lip velvety all over. *Himalaya, SW China, Burma, Thailand, Vietnam, Laos & Kampuchea*. G2. Spring.

35. D. pulchellum Lindley (*D. dalhousieanum* Wallich). Illustration: *Paxton's Magazine of Botany* **11**: t. 145 (1844); *Edwards's Botanical Register* **32**: t. 10 (1846); Cogniaux & Goossens, Dictionnaire iconographique des orchidées, Dendrobium, t. 7 (1897).

Stems to 1.2 m, thickened. Leaf-sheaths red-striped. Leaves 10–15 × *c.* 3 cm, linear-oblong, 2-lobed at apex. Racemes with 5–12 flowers on an elongate axis, borne on leafless stems. Flowers *c.* 7–8 cm in diameter. Petals *c.* 3.5 cm broad, almost twice as broad as the sepals, white. Lip shorter than the petals, slightly concave, white to pale yellow at the base, with 2 red-purple blotches. *Himalaya, Burma, Vietnam*. G2. Spring.

36. D. nobile Lindley. Figure 26(8), p. 523. Illustration: Cogniaux & Goossens, Dictionnaire iconographique des orchidées, Dendrobium, t. 1A, B, C (1896–97); *Annals of the Royal Botanic Garden Calcutta* **8**: t. 71 (1898); Bechtel et al., Manual of cultivated orchid species, 200 (1981).

Stems to 1 m, fleshy. Leaves 7–10 × 1–1.5 cm, lanceolate to ovate-lanceolate. Racemes with 1–2 or rarely 3 flowers on a short axis, borne on leafless stems. Flowers 7–10 cm in diameter. Sepals and petals 3–5 cm, the petals broader than the sepals, all white, white tipped with purple or entirely mauve. Lip almost circular, white or yellowish white, variously marked with pink or mauve. *Himalaya, China, Taiwan*. G2. Spring–summer.

The precise coloration of the flowers is very variable, and several named varieties, based on this feature, have been in cultivation.

37. D. devonianum Paxton. Illustration: *Botanical Magazine*, 4429 (1849); Cogniaux & Goossens, Dictionnaire iconographique des orchidées, Dendrobium, t. 23 (1900); Seidenfaden & Smitinand, Orchids of Thailand **2**(2): t. 9 (1960).

Stems to 1 m, thin but fleshy, hanging. Leaves 7–10 × *c.* 1.5 cm, lanceolate. Racemes usually with 2 flowers on a short axis, borne on leafless stems. Flowers *c.* 5 cm in diameter. Sepals and petals 2–2.3 cm, petals broader than the sepals, white tipped with pinkish purple. Lip circular, as long as the petals, white with 2 large yellow blotches towards the base, the apex pinkish purple, margin conspicuously fringed. *Himalaya, SW China, Burma, Thailand & Vietnam*. G2. Summer.

38. D. falconeri Hooker. Illustration: *Botanical Magazine*, 4944 (1856); Bateman, Second century of orchidaceous plants, t. 137 (1867).

Stems to 1 m, thin, hanging, the internodes thickened upwards, the stems appearing beaded. Leaves grass-like, 5–8 cm × 3–4 mm. Flowers usually borne singly on the leafless stems, occasionally 2–3 together on a short axis, 5.5–7 cm in diameter. Sepals and petals 3–4.5 cm, the petals broader than the sepals, white tipped with purple. Lip ovate, almost as long as the petals, white, the apex purple and the base with a large purple blotch surrounded by an orange-yellow zone. *NE India, Burma*. G2. Spring–summer.

39. D. pendulum Roxburgh (*D. crassinode* Benson & Reichenbach). Illustration: Cogniaux & Goossens, Dictionnaire iconographique des orchidées, Dendrobium, t. 19, 34 (1899); Bechtel et al., Manual of cultivated orchid species, 200 (1981).

Stems to 30 cm, internodes thickened upwards, the stems appearing beaded. Leaves 10–12 × 1.5–2 cm, lanceolate. Racemes with 2–3 flowers on a short axis, borne on the leafless stems. Flowers 5–8 cm in diameter. Sepals and petals 3–4 cm × 9–12 mm, the petals broader than the sepals, all white with red tips. Lip ovate to almost circular, the centre deep yellow, the margin white with a reddish apex. *Burma*. G2.

D. wardianum Warner. Illustration: Cogniaux & Goossens, Dictionnaire iconographique des orchidées, Dendrobium, t. 5, 5A (1897). Similar, perhaps the same species, but with less swollen internodes and larger flowers. *E. Himalaya, Burma*. G2.

40. D. crepidatum Lindley. Illustration: *Botanical Magazine*, 4993, 5011 (1857); Cogniaux & Goossens, Dictionnaire iconographique des orchidées, Dendrobium, t. 40 (1906); Bechtel et al., Manual of cultivated orchid species, 195 (1981).

Stems to 45 cm, fleshy. Leaves 5–10 × 1–1.25 cm, linear-lanceolate, acute. Racemes with 1–3 flowers on a very short axis, borne on leafless stems. Flowers 2.5–4.5 cm in diameter. Sepals and petals 1–2 cm, the petals a little broader than the sepals, white with pink tips. Lip about as long as the petals, ovate, yellow at the base, then white, the margin pink, sparsely hairy above, hairless beneath. *Himalaya, Burma, Thailand & Laos*. G2. Spring.

41. D. gratiosissimum Reichenbach. Illustration: Bechtel et al., Manual of cultivated orchid species, 197 (1981).

Stems to 1 m, fleshy, hanging. Leaves 7–12 × 1–1.5 cm, lanceolate, sheaths reddish. Racemes with 1–3 flowers on a very short axis, borne on leafless stems. Flowers 4–5 cm in diameter. Sepals and petals 1.8–3.5 cm, petals broader than sepals, all white tipped with pink. Lip broadly ovate, white, the base orange or yellow, the apex pink, densely velvety above and beneath. *Burma, Thailand.* G2. Spring.

42. D. amoenum Lindley. Illustration: *Botanical Magazine*, 6199 (1875); *Annals of the Royal Botanic Garden Calcutta* **8**: t. 69 (1898).

Stem to 60 cm, slender. Leaves to 10 cm, lanceolate or linear-lanceolate, acute. Racemes with 2–3 flowers on a very short axis, borne on leafless stems. Flowers 3–4 cm in diameter. Sepals and petals 1.7–2.5 cm, the petals finely toothed, sometimes irregularly so, broader than the sepals, all white tipped with purple. Lip as long as the petals, white with a yellow base and purple apex, densely velvety above, sparsely hairy beneath. *Himalaya.* G2. Spring.

Sometimes considered to be the same as *D. aphyllum* (p. 529).

43. D. aduncum Wallich. Illustration: *Botanical Magazine*, 6784 (1884); *Annals of the Royal Botanic Garden Calcutta* **8**: t. 67 (1898).

Stems hanging or ascending, red-spotted. Leaves to 8 cm, lanceolate or linear-lanceolate, somewhat red-lined beneath. Racemes with 1–2 flowers on a short axis, borne on leafless stems. Flowers 2.5–3 cm in diameter. Sepals and petals 1.6–1.8 cm, petals scarcely broader than sepals, all pale pinkish mauve. Lip with the free part very small, mostly pouch-like with an oblong projection at the base, whitish, much shorter than the mentum. *India (Sikkim), Bhutan, Burma.* G2. Summer–autumn.

GROUP F (Section **Pedilonum** (Blume) Lindley).

Leaves with sheaths, evenly spread along the stems; racemes with many flowers; lip mostly simple, its margins attached to the column foot for some distance, flat, blunt; ovary not angled or winged.

44. D. amethystoglossum Reichenbach. Illustration: *Botanical Magazine*, 5968 (1872); Davis & Steiner, Philippine orchids, 118 (1952).

Stems 60–90 cm. Leaves 5–10 cm. Racemes to 10 cm, drooping, borne on leafless stems and bearing 15–many flowers. Upper sepal *c.* 2 cm × 8 mm, lateral sepals to 3 cm, petals *c.* 2 cm, all white. Lip linear-spathulate, the basal part white with a backwardly pointing, fleshy, triangular ridge, the apical part amethyst-purple. *Philippines.* G2. Winter.

D. victoriae-reginae Loher. Illustration: Cogniaux & Goossens, Dictonnaire iconographique des orchidées, Dendrobium, t. 21 (1899); *Orchid Review* **20**: f. 6 (1912); *Botanical Magazine*, 9071 (1925). Similar but with fewer flowers in the raceme, the petals and sepals white in the lower part, violet-blue in the upper, the lip orange-yellow at the base, violet-blue above. *Philippines.* G2. Spring–summer.

D. glomeratum Rolfe. (*D. sulawesiense* Erfkamp & Gruss). Illustration: *Die Orchidee* **51**: 57–60 (2000). Plants similar to *D. mohlianum* in habit, with bright cerise to deep lavender flowers are marketed as *D. sulawesiense.*

45. D. secundum Lindley. Figure 26(10), p. 523. Illustration: *Edwards's Botanical Register* **15**: t. 1291 (1829); *Botanical Magazine*, 4352 (1848); Cogniaux & Goossens, Dictionnaire iconographique des orchidées, Dendrobium, t. 35 (1905); Bechtel et al., Manual of cultivated orchid species, 201 (1981).

Stems to 1 m. Leaves *c.* 10 × 3 cm, broadly oblong, mostly in the upper two-thirds of the stem. Racemes *c.* 12–15 cm, from the uppermost nodes, 1-sided. Flowers 1.5–1.8 cm in length, upper sepal *c.* 7 × 4 mm, petals narrow, all pink or reddish or rarely white. Lip simple, pink or reddish. *Burma, Laos, Vietnam, Kampuchea, Malaysia, Indonesia & Philippines.* G2. Winter.

46. D. bullenianum Reichenbach (*D. topaziacum* Ames). Illustration: Bechtel et al., Manual of cultivated orchid species, 194 (1981).

Stems to 60 cm, longitudinally grooved. Leaves 7–14 × 1.5–2.8 cm, oblong. Racemes short, borne on leafless stems, not 1-sided. Flowers *c.* 1.7 cm in length. Upper sepal lanceolate, *c.* 6 × 3 mm, lateral sepals obliquely triangular, *c.* 1.5 cm, petals *c.* 6 × 2 mm, oblong-spathulate, bluntly pointed, all orange. Lip *c.* 1.5 cm, narrowly elliptic or spathulate. *Philippines.* G2.

GROUP G (Section **Calyptrochilus** Schlechter; Section *Glomerata* Kränzlin).

Leaves with sheaths, uniformly spread along the fleshy stems; racemes short with few flowers; lip simple, its margins attached to the column foot for some distance, its apical part upcurved and fringed.

47. D. lawesii Mueller. Figure 26(11), p. 523. Illustration: *Botanical Magazine*, n.s., 524 (1969); Millar, Orchids of Papua New Guinea, 26 (1978); Bechtel et al., Manual of cultivated orchid species, 198 (1981).

Stems to 50 cm, hanging or erect, leafy for most of their length. Leaves 6–8 × *c.* 1.5 cm, clearly arranged in 2 ranks, lanceolate, gradually pointed. Racemes short with 1–5 (rarely to 10) flowers, occurring in the upper parts of leafless (occasionally leafy) stems. Upper sepal and petals *c.* 7 mm, lateral sepals *c.* 2.5 cm (from tip of sepal to tip of spur-like mentum), all scarlet or purplish. Lip mauve, boat-like, apex upcurved, its margin fringed. *New Guinea.* G2. Winter.

48. D. subclausum Rolfe. Illustration: *Feddes Repertorium Beihefte* **21**: t. 174, f. 651 & t. 175, f. 653 (1912); Millar, Orchids of Papua New Guinea, 26 (1978); Van Royen, Orchids of the high mountains of New Guinea, 344, 346, 353 (1980).

Habit very variable, stems mostly erect, 50–100 cm, often branched, finely warted on the leaf-sheaths. Leaves numerous, 2.5–7 cm × 2.5–10 mm, clearly arranged in 2 ranks and always twisted. Racemes very short, mostly on leafless stems, with up to 5 (rarely more) flowers. Flowers similar to those of *D. lawesii* but bright orange with pale orange or yellow tips. *New Guinea.* G2. Winter–spring.

D. dichroma Schlechter, *D. flammula* Schlechter, *D. nutriferum* J.J. Smith and *D. phlox* Schlechter, all from New Guinea, may well be the same species, but sufficient information is not yet available.

D. wentianum J.J. Smith. Illustration: Van Royen, Orchids of the high mountains of New Guinea, 338, 339, 351, 361 (1980). Similar but stems hanging, leaves to 2 cm, more or less ovate, often strongly reflexed, flowers much brighter orange. *New Guinea.* G1.

GROUP H (Section **Cuthbertsonia** Schlechter; Section *Leiotheca* Kränzlin in part).

Small tufted plants with crystalline-papillose leaves, sheaths and flower-stalks; racemes each with 1 non-resupinate flower; lip simple, obtuse, boat-shaped, its margins attached to the column foot for some distance.

49. D. cuthbertsonii Mueller (*D. sophronites* Schlechter). Figure 26(12), p. 523. Illustration: *Journal of the Royal Horticultural Society* **88**: f. 46 (1963); *Orchid Review* **76**: 248 (1968); *The Orchadian* **6**: 37, 38 (1978); Millar, Orchids of Papua New Guinea, 14, 15 (1978).

Plant compact, stems *c.* 1–2 cm (rarely to 5) × 4–7 mm, clustered, club-shaped or cylindric and narrowing abruptly upwards. Leaves almost apical, 2–5, linear-lanceolate to almost ovate, dark green and finely crystalline-papillose above, often purplish beneath. Flowers 2.5–4 cm long, 1.3–3.5 cm in diameter, terminal, borne singly on leafless stems, the colour extremely variable, commonly scarlet but frequently purplish, mauve, orange or white, sometimes bicoloured. Lip boat-shaped and strongly concave, with darker marginal markings. *New Guinea*. G1. Summer–spring.

GROUP I (Section **Oxyglossum** Schlechter).

Small tufted plants; leaves few, borne towards the upper parts of the stems, with sheaths; racemes terminal, with 2–4 flowers; lip tightly attached to the column and its foot, simple, sharply pointed; ovary winged or angled.

Literature: Reeve, T.M. & Woods, P.J.B., A preliminary key to the species of Dendrobium section Oxyglossum, *The Orchadian* **6**: 195–208 (1980); Reeve, T.M. & Woods, P.J.B., A revision of Dendrobium section Oxyglossum, *Notes from the Royal Botanic Garden Edinburgh* **46**(2): 161–305 (1989).

50. D. vexillarius J.J. Smith. Figure 26 (13), p. 523. Illustration: Millar, Orchids of Papua New Guinea, 35, 37 (1978) – under various specific names; Van Royen, Orchids of the high mountains of New Guinea, 376–93 (1980); *Australian Orchid Review* **46**: 106 (1981).

Stems 1–25 cm × 3–15 mm, clustered, cylindric, narrowing upwards. Leaves 2–4, 2–10 cm (rarely to 16) × 3–18 mm, oblong to oblong-lanceolate, blunt or shortly pointed, green above, often purplish beneath. Racemes stalkless from tips of leafy or leafless branches. Sepals and petals 2.5–5 × *c.* 3 cm, spreading, flat, variable in colour, purple, orange or yellow. Lip purplish, blackish or green, the sharp tip bright orange, reflexed. Ovary triangular in section. *New Guinea*. G1. Summer–spring.

A very variable plant both in habit and flower colour.

D. subacaule Lindley (*D. oreocharis* Schlechter; *D. tricostatum* Schlechter). Illustration: *Orchid Digest* **41**: 13 (1977); Northen, Miniature orchids, C-26 (1980). Similar but much smaller, stems and leaves to 1.5 cm, flowers bright orange, 9–18 mm. *New Guinea, Moluccas, Solomon Islands*. G1. Summer–spring.

51. D. hellwigianum Kränzlin (*D. cyananthum* Williams; *D. raphiotes* Schlechter).

Stems 1–8 cm × 2–6 mm, clustered, cylindric. Leaves 3–4, to 15 cm × 2.5 mm, round in section. Racemes with 2–4 flowers, stalkless on leafless stems. Flowers pinkish to purple or bluish, lip-tip acute, bright orange, not reflexed. Ovary 5-winged (the upper 3 wings so close as to give a triangular appearance). *New Guinea*. G1. Summer–spring.

D. violaceum Kränzlin. Illustration: Northen, Miniature orchids, C-21 (1980). Similar but the leaves not round in section, linear, to 25 cm × 3–6 mm, the flowers 3–4.5 cm, purplish, the lip-apex orange. *New Guinea*. G1. Summer–spring.

GROUP J (Section **Stachyobium** Schlechter)

Leaves with sheaths; racemes more or less terminal on the leafy stems; lip 3-lobed, fringed, its margin not attached to the column foot.

52. D. delacourii Guillaumin (*D. ciliatum* Hooker not Persoon). Figure 26(14), p. 523. Illustration: *Botanical Magazine*, 5430 (1864); Seidenfaden & Smitinand, Orchids of Thailand **2**(2): 231 (1960); Bechtel et al., Manual of cultivated orchid species, 195 (1981).

Stems to 45 cm, though often shorter, erect. Leaves 10–13 × 2–3 cm, oblong-elliptic, acute. Racemes erect, with several flowers. Sepals and petals 1–1.2 cm, the petals somewhat narrower than the sepals, greenish yellow or yellow. Lip obovate, 3-lobed, as long as the petals, the central lobe fringed with spathulate divisions, all greenish yellow with red veins. *Burma, Thailand, Laos, Vietnam & Kampuchea*. G2. Winter.

GROUP K (Section **Phalaenanthe** Schlechter).

Leaves with sheaths; racemes borne on leafless stems; petals much broader than sepals; mentum double, consisting of 2 swellings or humps at right angles to each other.

53. D. bigibbum Lindley. Illustration; *Botanical Magazine*, 4898 (1856); Bateman, Second century of orchidaceous plants, t. 169 (1867); *Lindenia* **7**: t. 317 (1891); Bechtel et al., Manual of cultivated orchid species, 194 (1981).

Stem to 50 cm, fleshy, spindle-shaped or cylindric. Leaves 5–12 × 2–3 cm, oblong-lanceolate or lanceolate. Raceme erect with many flowers, 3–7 cm in diameter. Sepals and petals 2–3.5 cm, petals broader than sepals, broadly rounded, occasionally with a small point, all usually reddish purple, rarely white. Lip shorter than the petals, 3-lobed, with a blunt or notched central lobe that bears many small calluses in the middle. Mentum bluntly conical. *Australia (N Queensland)*. G2.

Variable in flower colour and size. **D. superbiens** Reichenbach is the name applied to plants that are almost certainly natural hybrids between *D. bigibbum* and another Australian species, *D. discolor* (p. 532). These plants have been in cultivation (see *Lindenia* **7**: t. 294, 1891; Cogniaux & Goossens, Dictionnaire Iconographique des Orchidées, Dendrobium, t. 15, 1898) and differ from *D. bigibbum* in having the second hump of the mentum small and the petals relatively narrower (see Blake, S.T., *Proceedings of the Royal Society of Queensland* **74**: 29–44, 1964).

54. D. phalaenopsis Fitzgerald (*D. schroederanum* Gentil). Figure 26(15), p. 523. Illustration: Cogniaux & Goossens, Dictionnaire iconographique des orchidées, Dendrobium, t. 4 (1897); *Botanical Magazine*, 6817 (1885); *Lindenia* **6**: t. 280 (1890); Bechtel et al., Manual of cultivated orchid species, 200 (1981).

Similar to *D. bigibbum* but more robust, with stems to 60 cm; leaves to 15 cm. Flowers 6–10 cm in diameter. Sepals and petals 4–4.5 cm, petals much broader than sepals, acuminate, usually all red. Lip with a long, acute, central lobe which bears a few, small calluses, usually a deeper red than the petals and sepals. *Indonesia (Timor Laut)*. G2. Winter–spring.

Like *D. bigibbum*, this species is very variable in flower colour (shades of red, purple and white), and numerous varieties (cultivars) have been described on the basis of this variation. The origin of the species has long been obscure, but S.T. Blake (see reference under *D. bigibbum*) has shown conclusively that it is endemic to Timor Laut and does not occur as a native in Australia or New Guinea.

GROUP L (Section **Spatulata** Lindl.; Section *Ceratobium* Schlechter).

Leaves with sheaths; racemes from leafy or leafless stems; sepals and particularly petals often twisted, the petals usually very erect; lip with an entire central lobe.

Literature: Ossian, C.R., A review of the Antelope Dendrobiums (section Ceratobium), *American Orchid Society Bulletin* **50**: 1213, 1455 (1981) & **51**: 23, 362 (1982); Cribb, P., A revision of Dendrobium sect. Spatulata, *Kew Bulletin* **41**(3): 615–92 (1985).

55. D. helix Cribb. Illustration: *Orchid Review* **88**: 144 (1980); *The Orchadian* **6**: 175 (1980); *American Orchid Society Bulletin* **51**: 255, 258 (1981).

Stems to 1 m, cane-like. Leaves to 16 × 6.5 cm, elliptic or ovate-elliptic, obtuse and minutely 2-lobed at the apex. Racemes erect with 5 or more flowers. Sepals 2.6–3 cm × 8–12 mm, the upper spirally recurved; petals 3–4.5 cm, erect, spirally twisted 3 or 4 times; all yellow with red-brown blotches. Lip 3-lobed, shorter than the sepals, the central lobe recurved, base mauve, the rest yellow with red-brown spots. *New Guinea (New Britain)*. G2.

56. D. tangerinum Cribb (*Dendrobium* 'Tangerine'). Illustration: *Orchid Review* **88**: 145 (1980); *The Orchadian* **6**: 176 (1980); *American Orchid Society Bulletin* **50**: 1340, 1343 (1981).

Stems to 50 cm, erect, cane-like, clustered. Leaves to 9 × 2.5 cm, oblong-elliptic. Racemes spreading to erect, with *c.* 15 flowers. Upper sepal 2.3–2.5 cm × 6–7 mm, recurved, margins wavy; lateral sepals 1.6–2.3 × 1–1.2 cm; petals 3–3.4 cm × 4–5 mm, erect, spirally twisted 2 or 3 times; all orange-yellow, with dark red venation. Lip 3-lobed, shorter than the sepals, orange-yellow with 3 lilac ridges and a white base. *New Guinea*. G2.

57. D. minax Reichenbach. Illustration: Reichenbach, Xenia Orchidacearum **2**: 145 (1867); Das Pflanzenreich **45**: 143 (1910).

Stems to 30 cm. Leaves to 10 × 2 cm, oblong, unequally 2-lobed at the apex. Racemes with 2–3 flowers. Sepals 2.5–2.8 cm, petals to 5 cm, not twisted or with 1 slight twist, all greenish red, intensely red-streaked. Lip 3-lobed, the lateral lobes broadly triangular, the central transversely oblong, broader than long, with a small point, greenish red with purple lines. *Indonesia (Amboina, Moluccas, Sulawesi)*. G2. Summer.

58. D. stratiotes Reichenbach. Illustration: *Illustration Horticole* **34**: t. 602 (1886); Bechtel et al., Manual of cultivated orchid species, 202 (1981); *American Orchid Society Bulletin* **50**: 1337, 1343 (1981).

Stems to 60 cm or more. Leaves 8–12 × *c.* 2 cm, oblong, obtuse. Racemes with up to 5 flowers. Upper sepal, *c.* 3.5 cm, laterals *c.* 4.5 cm, petals to 6 cm, all white suffused with green. Lip with diamond-shaped lateral lobes and a central lobe that is broadly ovate or elliptic, longer than broad, acute, white with a greenish yellow base and red veins and spots. *Indonesia (Sunda Islands)*. G2. Summer.

59. D. taurinum Lindley. Illustration: *Edwards's Botanical Register* **29**: t. 28 (1843); *Lindenia* **13**: t. 621 (1897); *American Orchid Society Bulletin* **51**: 253, 258 (1982).

Stems to 1.5 m, erect, spindle-shaped. Leaves 10–15 × *c.* 6 cm, oblong or elliptic, rounded or slightly 2-lobed at the apex. Racemes erect with many flowers. Upper sepal *c.* 3 cm, laterals *c.* 4 cm, reflexed; petals 3–4 cm, reflexed; sepals white suffused with green, petals brownish red to pink, slightly twisted. Lip oblong-elliptic, pale brownish red to pink, about as long as the petals, with large lateral lobes and a small, somewhat wavy central lobe. *Philippines*. G2. Autumn–winter.

60. D. crispilinguum Cribb. Illustration: *Orchid Review* **88**: 146, 147 (1980); *The Orchadian* **6**: 177 (1980); *American Orchid Society Bulletin* **51**: 32 (1982).

Stems to 1 m or more, cane-like. Leaves to 12 × 1.5–3.5 cm, lanceolate or ovate-lanceolate, acute. Racemes with up to 20 flowers. Upper sepal recurved and somewhat twisted, 2.6–3 cm; lateral sepals reflexed, *c.* 2.6 cm; petals 2.5–3.3 cm, somewhat twisted; all cream or yellowish. Lip shorter than the petals with oblong lateral lobes and an ovate, very crisped central lobe, white suffused and veined red-purple. *New Guinea*. G2.

61. D. discolor Lindley (*D. undulatum* R. Brown not Persoon). Figure 26(16), p. 523. Illustration: *Edwards's Botanical Register* **27**: t. 52 (1841); Dockrill, Australian indigenous orchids, 469 (1969); Bechtel et al., Manual of cultivated orchid species, 196 (1981); *American Orchid Society Bulletin* **51**: 256–9 (1982).

Stems to 5 m, cane-like. Leaves 5–15 × 2–5 cm, ovate or elliptic, obtuse or slightly 2-lobed. Racemes erect with numerous flowers. Sepals and petals 2–5 cm, very wavy, somewhat twisted and reflexed, yellowish brown. Lip shorter than the petals, the central lobe ovate, bent downwards, not wavy, yellowish with 3 white or violet ridges. *Australia (Queensland), New Guinea*. G1. Winter–spring.

Variable in flower colour in the wild. See note under *D. bigibbum* (p. 531).

62. **D. mirbelianum** Gaudich (*D. wilkianum* Rupp). Illustration: Dockrill, Australian indigenous orchids, 475 (1969); *American Orchid Society Bulletin* **51**: 32, 147, 152 (1982).

Stems to 1 m, erect. Leaves 8–15 × 1.5–4 cm, ovate, usually suffused with red or purple. Racemes with 4–12 flowers. Sepals 1.8–3 cm, petals slightly longer, not or little twisted, all pale to dark brown. Lip shorter than the petals, 3-lobed, the central lobe oblong-ovate, acute, yellowish green with deep red veins and 4 ridges. *Australia (Queensland); ?New Guinea*. G2.

GROUP M (Section **Oxygenianthe** Schlechter; Section *Nigro-hirsutae* Kränzlin).

Leaves with sheaths which are conspicuously brown- or black-hairy; racemes borne on leafy stems; lip usually clearly 3-lobed, the lateral lobes usually arching upwards to the column, usually not hairy.

63. D. formosum (*D. infundibulum* Reichenbach not Lindley). Figure 26(17), p. 523. Illustration: Cogniaux & Goossens, Dictionnaire iconographique des orchidées, Dendrobium, t. 8 (1897); Seidenfaden & Smitinand, Orchids of Thailand **2**(2): t. 11 (1960); *Orchid Digest* **41**: 5 (1977); Bechtel et al., Manual of cultivated orchid species, 197 (1981).

Stems to 45 cm, thick. Leaves 10–12 × *c.* 3.5 cm, ovate-oblong, unequally 2-lobed at the apex. Racemes with 2 or more flowers, 9–11 cm in diameter. Sepals and petals 4–5 cm, petals much broader than the sepals, all white. Lip longer than petals, scarcely 3-lobed, obovate, wavy, notched, white with yellow markings. Mentum conical, acute. *Himalaya, Burma, Thailand.* G2. Winter–spring.

Variable in the size of the flowers and the precise coloration of the lip.

64. D. infundibulum Lindley. Illustration: *Botanical Magazine*, 5446 (1864); Cogniaux & Goossens, Dictionnaire iconographique des orchidées, Dendrobium, t. 6 (1897); *Orchid Digest* **41**: 5 (1977); Bechtel et al., Manual of cultivated orchid species, 198 (1981).

Similar to *D. formosum* but with flowers to 8 cm in diameter, sepals and petals at most 4 cm, lip distinctly 3-lobed, irregularly fringed. *Burma, Thailand.* G2. Spring–summer.

Variable; var. **jamesianum** (Reichenbach) Veitch (*D. jamesianum* Reichenbach), which has the lip papillose and marked with orange-yellow, is often grown.

65. D. dearei Reichenbach. Illustration: Cogniaux & Goossens, Dictionnaire iconographique des orchidées, Dendrobium, t. 36 (1905); Bechtel et al., Manual of cultivated orchid species, 195 (1981).

Stems to 1 m. Leaves to 6 cm, ovate-oblong, unequally 2-lobed at the apex, hairless. Raceme with 3–6 flowers. Sepals and petals 2.5–3.5 cm, petals broader than sepals, all white. Lip 3-lobed, about as long as petals, the central lobe somewhat deflexed and wavy, white, green at the base. Mentum conical, acute, *c.* 1.5 cm. Ovary 3-keeled. *Philippines.* G2. Spring–summer.

66. D. sanderae Rolfe. Illustration: *Botanical Magazine*, 8351 (1910).

Similar to *D. dearei* but the lip streaked with purple at the base and the sepals and petals to 4.5 cm. *Philippines.* G2. Autumn–winter.

Apparently sometimes confused with *D. sanderianum* Rolfe, a totally different plant not often seen in cultivation, and with the artificial hybrid *D.* × *ainsworthii* Moore (*D. heterocarpum* × *nobile*), which is sometimes found in older horticultural literature as *'D.* × *sanderae'* (see Hunt & Summerhayes, *Kew Bulletin* **20**: 55, 1966).

67. D. draconis Reichenbach (*D. eburneum* Reichenbach). Figure 26(18), p. 523. Illustration: Bateman, Second century of orchidaceous plants, t. 166 (1867); *Botanical Magazine*, 5459 (1864); *Orchid Digest* **41**: 5 (1977).

Stems to 45 cm, fleshy. Leaves 6–8 × 1–1.5 cm, lanceolate, apex unequally 2-lobed, with brown hairs on both surfaces. Sepals and petals 3–4 cm, the petals broader than the sepals, all creamy white. Lip longer than the petals, with 2 triangular lateral lobes borne at the base, the central lobe oblong-obovate, acute, somewhat wavy, entire, white with red and yellow lines towards the base. Mentum spur-like, 2–2.5 cm, narrow. *Burma, Thailand, Laos, Kampuchea & Vietnam.* G2. Spring–summer.

Cleistogamous variants of this species are known (see Pradhan, *Orchid Digest* **41**: 6, 1977).

68. D. williamsonii Day & Reichenbach. Illustration: *Annals of the Royal Botanic Garden Calcutta* **5**: t. 9 (1895); *Botanical Magazine*, 7974 (1904); Bechtel et al., Manual of cultivated orchid species, 203 (1981).

Stems to 25 cm. Leaves 5–10 × 1.5–2.5 cm, linear-oblong, with brown hairs on both surfaces. Racemes with 2 flowers. Sepals and petals 3–3.5 cm, equally broad, white inside, pale yellow outside. Lip longer than the petals with 2 rounded lateral lobes borne about half way along its length, the central lobe conspicuously fringed, creamy white to yellow with a large red-brown spot at the base. Mentum spur-like, *c.* 2.5 cm. *NE India.* G1. Spring.

69. D. bellatulum Rolfe. Illustration: *Botanical Magazine*, 7985 (1904); *Orchid Digest* **41**: 5 (1977).

Stems very short, usually less than 5 cm, clustered. Leaves lanceolate or narrowly ovate, apex unequally 2-lobed, with dark brown hairs on both surfaces. Racemes with 1–3 flowers. Sepals and petals 2–2.5 cm, equally broad, all white. Lip ovate, about as long as petals, with a notched, broadly oblong, deflexed central lobe and a thick, papillose ridge down most of its length, red at the base, the rest yellow. Mentum broadly conical, blunt. *SW China, Vietnam, Kampuchea, Laos & Thailand.* G1. Spring.

GROUP N (Section **Rhopalanthe** Schlechter).

Stems thickened only at the extreme base, tapering abruptly above this.

70. D. crumenatum Swartz. Figure 26 (19), p. 523. Illustration: *Botanical Magazine*, 4013 (1843); Holttum, Flora of Malaya **I**: Orchids, f. 3 (1953).

Stems to 1 m, the lower 4–5 internodes swollen, the rest thin. Leaves fleshy, obtuse or obscurely 2-lobed. Raceme leafless, with many fragrant flowers. Sepals and petals 2.5–3.5 cm, white or white suffused with pink. Lip as long as the petals, 3-lobed, the central lobe wavy and with a prominent crested callus, white with a yellow blotch. *Himalaya to Indonesia.* G2. Can flower at any time.

Flowering in this sparingly cultivated species follows a sudden, though often short, drop in temperature. The flowers last for only 1 day.

GROUP O (Section **Aporum** Schlechter).

Stems and leaves strongly laterally flattened, thick, leaves regularly alternating in 2 rows, jointed at the base; flowers produced singly from a cluster of chaffy bracts.

71. D. terminale Parish & Reichenbach. Figure 26(20), p. 523. Illustration: *Annals of the Royal Botanic Garden Calcutta* **8**: t. 55 (1898).

Stems 10–15 cm. Leaves to 2 × 0.5 cm. Flowers produced subterminally. Upper sepal and petals *c.* 5 × 3 mm, lateral sepals *c.* 13 × 6 mm. Lip simple, notched at the apex, *c.* 10 × 5 mm. *Burma.* G2.

89. EPIGENEIUM Gagnepain

J. Cullen

Epiphytes with long rhizomes bearing distant simple pseudobulbs that are 4-angled, at least when dry, and bear 2 leaves. Leaves leathery, folded when young, often unequally 2-lobed at the apex. Racemes erect with 1–many flowers, terminal on the pseudobulbs. Sepals and petals similar, the lateral sepals attached to the short column foot, forming a small mentum. Lip 3-lobed, attached to the base of the column foot and somewhat movable on it. Column short. Pollinia 4, parallel.

A genus of 35 species from Asia, formerly included in *Dendrobium* Swartz, *Sarcopodium* Lindley not Schlechtendahl or *Katherinea* Hawkes, of which 3 are occasionally grown. Cultivation in general as for the Indian and Burmese species of *Dendrobium* (p. 521), but these plants are better grown on rafts to accommodate their long rhizomes.

Literature: Summerhayes, V.S., Notes on Asiatic orchids II, *Kew Bulletin* 1957: 259–68; Seidenfaden, G., Orchid genera in Thailand IX, *Dansk Botanisk Arkiv* **34**: 68–82 (1980).

1a. Racemes with 10–15 or more flowers; sepals and petals white with green or yellow tips and maroon bases **1. lyonii**
b. Racemes with 1–2 flowers; sepals and petals brown or greenish white suffused with purple and/or spotted with brown 2
2a. Flowers 8–9 cm in diameter; central lobe of lip rhomboid, tapered towards its base **2. amplum**
b. Flowers 3.5–5 cm in diameter; central lobe of lip transversely oblong, not tapered towards its base **3. rotundatum**

1. E. lyonii (Ames) Summerhayes (*Dendrobium lyonii* Ames; *Sarcopodium lyonii* (Ames) Rolfe; *S. acuminatum* var. *lyonii* (Ames) Kränzlin; *Katherinea acuminata* var. *lyonii* (Ames) Hawkes). Illustration: Das Pflanzenreich **45**: 330 (1910); Davis & Steiner, Philippine orchids, 110 (1952).

Pseudobulbs 4–4.5 cm. Leaves 10–15 × 2–4 cm, oblong or elliptic. Racemes with 10–15 (or more) flowers. Sepals and petals to 4 cm, triangular, white with green or yellow tips and maroon bases. Lip to 3 cm, central lobe triangular, acute, whitish, bases of lateral lobes dark red. *Philippines*. G2.

2. E. amplum (Lindley) Summerhayes (*Dendrobium amplum* Lindley; *Sarcopodium amplum* (Lindley) Lindley; *Katherinea ampla* (Lindley) Hawkes). Illustration: *Annals of the Royal Botanic Garden Calcutta* **8**: t. 89 (1898); *Orchid Review* **78**: 117 (1970).

Pseudobulbs to 5 cm, ovoid or spindle-shaped. Leaves 10–15 × *c.* 5 cm, oblong or oblong-lanceolate, acute. Raceme with 1–2 flowers, 8–9 cm in diameter. Sepals lanceolate, petals linear-lanceolate, all 3–4 cm, greenish suffused brownish purple and/or with brown spots. Lip up to 3.5 cm, central lobe rhomboid, tapered to its base, brown, the lateral lobes coloured like the petals. *Nepal, N India*. G2.

E. coelogyne (Reichenbach) Summerhayes. Illustration: Bechtel et al., Manual of cultivated orchid species, 211 (1981). Similar, but with larger flowers that are more distinctly spotted and marked; lip *c.* 4.5 cm, dark purple. *Burma*. G2.

Not considered as a distinct species by Seidenfaden who treats it as a synonym of *E. amplum*, but well known in cultivation.

3. E. rotundatum (Lindley) Summerhayes (*Sarcopodium rotundatum* Lindley; *Dendrobium rotundatum* (Lindley) Hooker; *Katherinea rotundata* (Lindley) Hawkes). Illustration: *Annals of the Royal Botanic Garden Calcutta* **8**: t. 87 (1898).

Pseudobulbs 3–4 cm, ovoid. Leaves to 12 × 2.5 cm, lanceolate or oblong. Sepals and petals *c.* 2.5 cm, brown. Lip oblong *c.* 2.5 cm, the central lobe transversely oblong, not tapered to the base, yellowish and thickened in the middle. *N India, SW China*. G2.

90. BULBOPHYLLUM Thouars

J.C.M. Alexander & P.J.B. Woods

Epiphytic. Pseudobulbs simple, distant or clustered, 1- or 2-leaved. Leaves stalked or stalkless, folded when young, leathery. Flowering stems basal, erect or pendent (in our species), with distant, spreading, sheath-like scales. Flowers 1–many, in spikes, racemes or umbels. Sepals usually almost equal. Lateral sepals joined to column foot, forming a short mentum. Petals usually shorter than sepals, sometimes much reduced. Lip thick, not obviously lobed, sometimes fringed, loosely hinged to column foot. Column short, often with 2 or more apical arms; column foot long and narrow. Pollinia 4.

A genus of perhaps 1000 species from Asia, Australasia, Africa and tropical America. *Cirrhopetalum* Lindley is now treated as a section within *Bulbophyllum*. Being widely distributed, the species of *Bulbophyllum* show a range of cultural requirements. They can be grown in epiphytic compost in pots or on slabs of tree fern or bark. The deciduous species with hard pseudobulbs, mostly native to temperate regions, need a pronounced resting period once growth has ceased. The tropical species require a less marked rest and must not dry out completely. Propagation is by divison.

Literature: Seidenfaden. G., Orchid genera in Thailand VIII: Bulbophyllum, *Dansk Botanisk Arkiv* **33**(3) (1979); Vermeulen, J.J., Orchids of Borneo **2**, Bulbophyllum (1991); Siegerist, E.S., Bulbophyllums and their allies (2001).

Pseudobulbs. 2-leaved: **1**. Cylindric: **3**; spherical and flattened: **2**; ovoid: **4**; 3 cm or more and angled: **5, 9, 11**. Spherical in outline, 4-lobed: **8, 9**. Brownish: **5, 11**. Clustered: **2, 4, 11**.
Leaves. 7 cm or less: **6, 8, 9, 10**; 10 cm or more: **1, 3, 4**. 4 cm wide or more: **3, 4, 5, 6, 9**.
Flowering stem. More than 30 cm: **6, 7, 11**. Pendent: **2, 7, 8**. Arched above: **11**. With large scales: **5, 8**. With true leaf: **8**. Slender: **7**. Thickened or flattened: **1**.
Flowers. Solitary: **4, 9**; more than 10: **1, 2, 5, 10, 11**; many: 1.5 cm or less: **6, 7, 9, 10, 11**; 15 cm or more: **5, 8, 9**. In umbels: **3, 4**.
Bracts. Reaching base of sepals: **5, 7**; not reaching base of sepals: **6, 7, 8, 9, 10, 11**.
Sepals. More than 1.5 cm: **3, 4**; less than 1.5 cm: **1, 2**. Longer than petals: **1, 2, 3, 4**; almost equal to petals: **1, 4**. Drawn out into long points: **3**.
Upper sepal. Less than 1 cm: **6, 7, 10, 11**; more than 2 cm; **5**. Finely tapered: **5, 7, 8, 9**; abruptly tapered or with an apical hair or bristle: **6, 10, 11**. Almost circular: **6**; spathulate: **11**. With marginal hairs: **7, 8, 9, 10, 11**; with flat appendages: **8**.
Lateral sepals. 5 cm or less: **6, 7, 8, 9, 10, 11**; 10 cm or more: **5, 7, 8, 9**. Finely tapered: **5, 7, 8, 9**; oblong or linear: **6, 10, 11**. Warty near base: **9**. Upper sepal thickened: **1**.
Petals. Finely tapered: **5, 7**; oblong or linear: **6, 9**. Acute: **9, 11**. With apical hair or bristle: **6, 10**. With marginal hairs: **6, 7, 8, 9, 10, 11**; with concentration or tuft of hairs at tip: **9**; with flat appendages: **8**; with irregular margin: **7**.
Lip. Ovate to triangular: **3, 4**. Fringed towards base: **1**; with apical thread-like hairs: **2**.
With fleshy auricles: **6, 8**. 2-ridged: **9**.
Column. With 2 long projections: **5, 6**.

1a. Sepals all more or less equal 2
b. Lateral sepals much larger than upper sepal 5
2a. Pseudobulbs 2-leaved; flowering stem flat or swollen above **1. falcatum**
b. Pseudobulbs 1-leaved; flowering stem not flat or swollen 3
3a. Flowers 10 or more per raceme; upper sepal 1.5 cm or less **2. barbigerum**
b. Flowers 1–7 per raceme; upper sepal 2 cm or more 4
4a. Pseudobulbs 7 cm or more, cylindric to narrowly fusiform **3. ericssonii**
b. Pseudobulbs 5 cm or less, ovoid or narrowly ovoid **4. lobbii**
5a. Margin of upper sepal entire (apical bristle may be present) 6

b. Margin of upper sepal irregular, toothed or ciliate 7
6a. Petal margins entire, not hairy **5. medusae**
b. Petal with irregular or hairy margins **6. longiflorum**
7a. Lateral sepals with drawn out, tapered or thread-like tips 8
b. Lateral sepals not drawn out, if long then parallel-sided 10
8a. Petal margins evenly fringed or hairy **7. gracillimum**
b. Petals with small, flat, marginal appendages, or with hairs concentrated into an apical tuft, or denser near apex 9
9a. Petals with 15 or fewer small, flat marginal appendages **8. wendlandianum**
b. Petals with an apical tuft of many hairs **9. ornatissimum**
10a. Lateral sepals about 10 times longer than broad **10. makoyanum**
b. Lateral sepals less than 6 times longer than broad **11. auratum**

1. B. falcatum (Lindley) Reichenbach (*Megaclinium falcatum* Lindley). Illustration: *Edwards's Botanical Register* **12**: t. 989 (1826); Stewart & Campbell, Orchids of tropical Africa, 61 (1970); Bechtel et al., Manual of cultivated orchid species, 174 (1981).

Pseudobulbs 2–6 cm, ovoid or narrowly ovoid, 4-angled, distant, 2-leaved. Leaves to 16 cm × 6–23 mm, oblong to oblanceolate. Flowering stem to 35 cm, erect, forming a flattened green or reddish blade to 1.4 cm wide; flowers numerous in 2 ranks, 5–20 mm apart, *c.* 5 × 10 mm, green, reddish or purple. Upper sepal to 7 mm, linear; tip wider and thicker; lateral sepals shorter than upper sepal, 2.5–3.5 mm wide, sickle-shaped. Petals 1.7–3 mm, sickle-shaped, linear to oblong, obtuse and often thickened and yellow. Lip *c.* 2 mm, narrowly triangular, purplish. *W Africa to Uganda*. G2. Spring.

B. leucorrhachis (Rolfe) Schlechter (*Megaclinium leucorrhachis* Rolfe). Illustration: *Botanical Magazine*, 7811 (1901). Leaves to 18 × 3 cm. Flowering stem to 18 cm, swollen to *c.* 1.3 cm thick above; bracts triangular, reflexed; flowers *c.* 8 mm, yellow; lip small, tongue-shaped, fringed towards base. *W & C Africa*. G2. Spring.

2. B. barbigerum Lindley. Illustration: *Botanical Magazine*, 5288 (1861); Schlechter, Die Orchideen, edn 2, 320 (1927); *Orchid Review* **76**: 318 (1968); Bechtel et al., Manual of cultivated orchid species, 173 (1981).

Pseudobulbs to 3 × 1.5–2.5 cm, spherical, somewhat flattened, clustered, 1-leaved. Leaves to 10 × 2.8 cm, broadly oblong, obtuse. Flowering stem 10–20 cm, erect with *c.* 4 narrowly ovate-acuminate, sheath-like scales *c.* 9 mm long; floral bracts similar; flowers 8–20 mm long, reddish. Sepals 7–13 mm, greenish, narrowly lanceolate; petals minute. Lip *c.* 1 cm, yellowish, linear, with short marginal hairs; apex with a tuft of long, purplish, thread-like, apically swollen hairs; base hinged so that lip and hairs move in the slightest draught. *W & C Africa*. G2. Summer.

B. careyanum (Hooker) Sprengel. Illustration: *Botanical Magazine*, 5316 (1862); *Dansk Botanisk Arkiv* **33**: 146, 147 (1979); Pradhan, Indian orchids **2**: t. 130 (1979); Bechtel et al., Manual of cultivated orchid species, 173 (1981). Pseudobulbs distant. Flowering stem drooping; flowers many, crowded, yellow. Lateral sepals longer than upper sepal. Lip purple, downy, without long thread-like hairs. *E Himalaya, Burma & Thailand*. G2. Winter.

3. B. ericssonii Kränzlin. Illustration: *Botanical Magazine*, 8088 (1906).

Pseudobulbs *c.* 9 × 1.5 cm, oblong-cylindric, distant, 1-leaved. Leaf to 20 × 9 cm, broadly elliptic to ovate; apex shortly pointed; base rounded or shortly drawn out. Flowering stem 18–28 cm, erect, with 4 or 5 sheath-like scales *c.* 2 cm long; flowers 4–7, in umbels. Sepals to 11 × 1.5 cm, free, ovate, tapered to a long drawn out, often spirally curved, point, greenish with reddish spots. Petals to 4 × 1 cm, narrowly triangular, somewhat sickle-shaped, pale yellow, tapered to a long drawn out, often twisted, reddish point. Lip to 1.5 cm, triangular, strongly recurved, pale yellow with reddish apical markings. *Indonesia (Moluccas)*. G2. Winter.

4. B. lobbii Lindley. Illustration: *Botanical Magazine*, 4532 (1850); Schlechter, Die Orchideen, edn 2, 325 (1927); Shuttleworth et al., Orchids, 88 (1970); Bechtel et al., Manual of cultivated orchid species, 174 (1981).

Pseudobulbs 3–5 × 2–3 cm, broadly ovoid *c.* 8 cm apart, 1-leaved. Leaf to 35 × 9 cm, elliptic to broadly elliptic. Flowering stem 10 cm or more, 1-flowered; flowers to 7.5 cm wide, pale to brownish yellow, flushed with red or brown. Upper sepal to 5 × 1.7 cm, broadly lanceolate, acuminate; veins crimson-dotted; lateral sepals a little shorter and broader, downwardly curved, with purplish veins, apical half brownish. Petals to 3.5 × 1 cm, lanceolate, horizontal, with *c.* 9 purplish veins. Lip *c.* 1.3 × 1 cm, narrowly ovate, curved, basal margins upcurved, purplish with a bright orange circular callus. *NE India, through SE Asia to the Philippines*. G2. Flowers intermittently.

B. leopardinum (Wallich) Lindley. Illustration: *Botanical Magazine*, 9631 (1941); *Dansk Botanisk Arkiv* **33**: 29 (1979); Bechtel et al., Manual of cultivated orchid species, 174 (1981). Pseudobulbs clustered. Flowering stem 1–2 cm, with 1–3 flowers; flowers 2–3 cm wide, pale yellowish green, tinged or spotted with pink; petals about two-thirds of the sepal length; lip dark crimson, paler and spotted towards margins. *E Himalaya to Burma & N Thailand*. G2. Spring–summer.

5. B. medusae (Lindley) Reichenbach (*Cirrhopetalum medusae* Lindley). Figure 27 (4 & 11), p. 536. Illustration: *Botanical Magazine*, 4977 (1857); Shuttleworth et al., Orchids, 88 (1970); Skelsey, Orchids, 101 (1979).

Pseudobulbs 2–5 × 1–2 cm, ovoid, angled, brownish, *c.* 3 cm apart. Leaf 10–20 × 2–4.5 cm, oblong to elliptic, rounded, sometimes notched. Flowering stem 10–20 cm, erect, with large sheathing scales, bearing many small flowers in a dense umbel; bracts *c.* 2 cm; flowers 7.5–17 cm long, white to pale yellow with pink spots. Upper sepal 2.5–3 cm, linear, finely tapered; lateral sepals to 14 cm, narrow and drawn out, acute. Petals *c.* 5 mm, triangular, tapered, acute. Lip 1–2 mm. Column with 2 long apical projections. *Malaysia, Indonesia, Borneo; ?Philippines*. G2. Winter.

6. B. longiflorum Thouars not Ridley (*Bulbophyllum umbellatum* (Forster) J.J. Smith; *Cirrhopetalum umbellatum* (Forster) Hooker & Arnott; *C. thouarsii* Lindley; not *B. umbellatum* Lindley). Figure 27(3 & 10), p. 536. Illustration: *Botanical Magazine*, 4237 (1846); Stewart & Campbell, Orchids of tropical Africa, 69 (1970); Millar, Orchids of Papua New Guinea, 50 (1978).

Pseudobulbs 2.5–4.5 × 1–2 cm, conical to ovoid, angled, 2–4 cm apart. Leaves 8–18 × 3–4 cm, elliptic to oblong, rounded, notched. Flowering stem to 40 cm, erect or arched, with 3–7 flowers; bracts not reaching base of sepals; flowers 4–5 cm

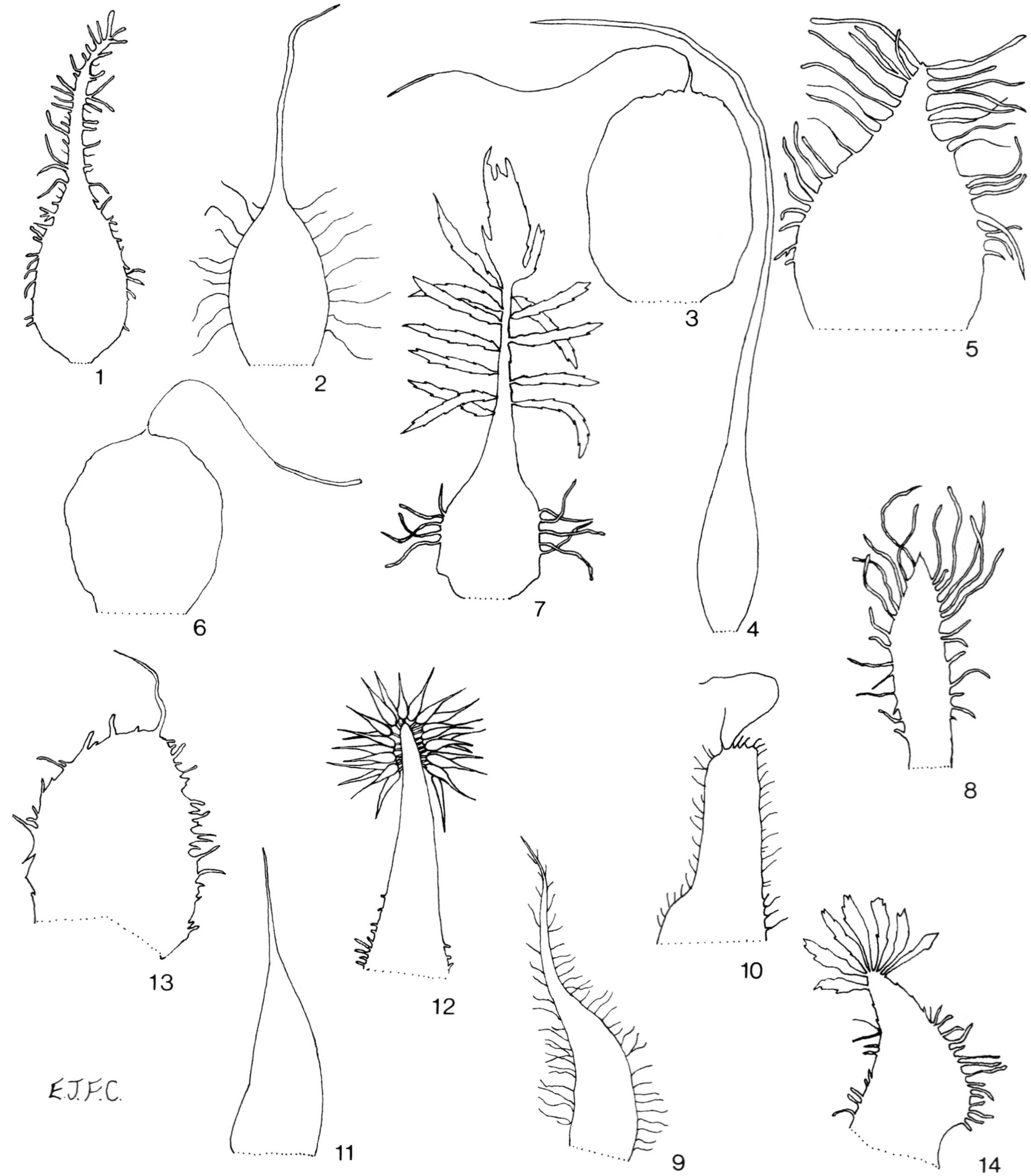

Figure 27. Upper sepals and petals of *Bulbophyllum* species. 1–7, upper sepals; 8–14, petals. 1 & 8, *B. putidum*. 2 & 9, *B. gracillimum*. 3 & 10, *B. longiflorum*. 4 & 11, *B. medusae*. 5 & 12, *B. ornatissimum*. 6 & 13, *B. picturatum*. 7 & 14, *B. wendlandianum*. (After Seidenfaden, *Dansk Botanisk Arkiv* **33**(3); not to scale.)

long, pale pinkish yellow. Upper sepal 7–10 mm, broadly ovate to circular, concave, purple-tipped, with a long apical bristle; lateral sepals 2–4 cm × 4–7 mm, oblong. Petals 4–10 mm, narrowly triangular, purple-tipped, with an apical bristle; margins hairy. Lip *c.* 7 mm, arched, purple. Column with 2 long apical projections. *E Africa, Indian Ocean islands, SE Asia, N Australia & Pacific islands*. G2. Summer.

B. picturatum (Loddiges) Reichenbach (*Cirrhopetalum picturatum* Loddiges). Figure 27(6 & 13), p. 536. Illustration: *Botanical Magazine*, 6802 (1885); *Orchid Review* **76**: 319 (1968); *American Orchid Society Bulletin* **42**(12): back cover (1973).

Petals broadly triangular. Lip with distinct fleshy auricles. Projections on column short, broad, 2-toothed. *NE India, Burma, Thailand & Vietnam*. G2. Spring–summer.

7. B. gracillimum (Rolfe) Rolfe (*Cirrhopetalum gracillimum* Rolfe; *Bulbophyllum psittacoides* (Ridley) Smith). Figure 27(2 & 9), p. 536. Illustration: *Dansk Botanisk Arkiv* **29**: 36 (1973); Bechtel et al., Manual of cultivated orchid species, 183 (1981).

Pseudobulbs 1–2 cm × 5 mm, ovoid, *c.* 2 cm apart. Leaf to 14 × 2 cm, oblong, rounded. Flowering stem to 35 cm, erect, very slender, bearing 8–10 flowers; bracts very small; flowers 2–2.5 cm long, purplish red. Upper sepal to 1 cm, ovate, concave, hairy on margin, finely tapered; lateral sepals variable, 2 cm or more, finely tapered; petals 5–10 mm, narrowly triangular, finely tapered, hairy on margin. Lip 1.5–2.5 mm, pinkish purple with a paler tip. *Malay peninsula, Indonesia, New Guinea, Solomon & Fiji Islands*. G2. Summer–autumn.

B. longissimum (Ridley) J.J. Smith (*Cirrhopetalum longissimum* Ridley). Illustration: *Botanical Magazine*, 8366 (1911); *Australian Orchid Review* **33**: 138 (1968); Bechtel et al., Manual of cultivated orchid species, 184 (1981). Pseudobulbs 3–7 cm apart. Flowering stem arched to pendent, bearing 3–7 flowers; bracts 1–2 cm. Lateral sepals 20–30 cm, finely tapered, pale pink with darker stripes. *Malay peninsula*. G2. Winter.

8. B. wendlandianum (Kränzlin) Dammer (*Cirrhopetalum wendlandianum* Kränzlin; *C. collettii* Hemsley). Figure 27 (7 & 14), p. 536. Illustration: *Botanical Magazine*, 7198 (1891); Sheehan & Sheehan, Orchid genera illustrated, 55 (1979); Northen, Miniature orchids, t. 16 (1980).

Pseudobulbs *c.* 2 × 2 cm, spherical but shallowly 4-lobed, 1.5–3 cm apart. Leaf 4–12 × 2–3 cm, elliptic, obtuse, sometimes notched. Flowering stem 10–15 cm, arched to pendent, bearing *c.* 5 flowers; base sheathed, 1-leaved, later developing into a pseudobulb. Flowers 12–15 cm long; bracts short. Upper sepal 1–2 cm, triangular, finely tapered, yellow with several small, flat, marginal appendages near the tip; margins hairy in lower half. Lateral sepals 8–12 cm, gradually tapered from the base, pink with darker red veins. Petals to 1 cm, triangular; colour, marginal hairs and appendages as in upper sepal. Lip *c.* 5 mm with swollen, bristly, pinkish purple auricles. *Burma & Thailand*. G2. Spring–summer.

9. B. ornatissimum (Reichenbach) J.J. Smith (*Cirrhopetalum ornatissimum* Reichenbach). Figure 27(5 & 12), p. 536. Illustration: *Botanical Magazine*, 7229 (1892); *Dansk Botanisk Arkiv* **29**: 28 (1973); Bechtel et al., Manual of cultivated orchid species, 184 (1981).

Pseudobulbs 2–5 × 1–2 cm, ovoid, angled, 2–5 cm apart. Leaf 7–16 × 3–6 cm, oblong to elliptic, obtuse or rounded. Flowering stem 10–20 cm, erect, bearing 3–5 flowers; flowers 4–9 cm long; bracts short. Upper sepal 1–1.5 cm, ovate, tapered, hairy on margins, yellowish with purple veins; lateral sepals 5–8 cm, linear-lanceolate, tapered, pinkish yellow with purple veins and spots. Petals 1–1.5 cm, lanceolate to narrowly triangular, with a dense apical tuft of bulbous-based hairs, yellow with purple veins. Lip to 1 cm, 2-ridged, rounded, bristly, purple. *E Himalaya & NE India*. G2. Autumn–winter.

B. putidum (Teijsmann & Binnendijk) J.J. Smith (*Cirrhopetalum putidum* Teijsmann & Binnendijk; *Mastigion putidum* (Teijsmann & Binnendijk) Garay, Hamer & Siegerist). Figure 27(1 & 8), p. 536. Illustration: Kupper & Linsenmaier, Orchids, 55 (1961); *American Orchid Society Bulletin* **43**: 886, 888 (1974); Seidenfaden & Wood, Orchids of peninsular Malaysia & Singapore, 438 (1992). Pseudobulbs almost spherical, obscurely 4-lobed, smooth. Leaves *c.* 5 × 1.5–3 cm, elliptic. Flowering stem *c.* 10 cm, bearing 1–3 flowers; flowers to 20 cm long, green and purple. Lateral sepals to 18 cm, finely tapered, warty near base. Lip tapered, acute, with 3 hairless keels. *E Himalaya, NE India, Laos, Malaysia, Indonesia & Philippines*. G2. Winter.

B. fascinator (Rolfe) Rolfe (*Cirrhopetalum fascinator* Rolfe; *Mastigion fascinator* (Rolfe) Garay, Hamer & Seigerist). Ilustration: *Botanical Magazine*, 8199 (1908). Similar to *B. putidum* and included by some authors, but said to differ by a lip with 2 hairy keels. *Laos, Malaysia*. G2. Winter.

10. B. makoyanum (Reichenbach) Ridley (*Cirrhopetalum makoyanum* Reichenbach). Illustration: *Botanical Magazine*, 7259 (1892); Shuttleworth et al., Orchids, 88 (1970); Graf, Tropica, 715 (1978).

Pseudobulbs 1–2 cm × 7–10 mm, broadly ovoid, 1–4 cm apart. Leaf 5–10 × 1.5–3 cm, oblong to elliptic, rounded, notched. Flowering stem 10–25 cm, erect, bearing 10–15 flowers; flowers 3–4 cm long, mostly pale yellow with red spots. Upper sepal *c.* 5 mm, triangular, acuminate, fringed with yellow hairs; lateral sepals *c.* 3.5 cm, linear, parallel-sided. Petals similar to upper sepal but narrower. Lip *c.* 2.5 mm, triangular. *Singapore, Borneo & Philippines*. G2. Winter.

11. B. auratum (Lindley) Reichenbach (*Cirrhopetalum auratum* Lindley; *Bulbophyllum campanulatum* Rolfe). Illustration: *Botanical Magazine*, 8281 (1909); *Dansk Botanisk Arkiv* **29**: 98 (1973).

Pseudobulbs 2–3 × 1–1.5 cm, fusiform to ovoid, angled, greenish brown, 1–2 cm apart. Leaf 10–15 × 1.5–2.5 cm, oblong, obtuse, notched. Flowering stem 10–15 cm, erect, arched above, bearing 8–12 flowers; flowers 2–2.5 cm long, mostly pale yellow with reddish veins. Upper sepal 6–7 mm, ovate-acuminate, with an apical hair, margins hairy. Lateral sepals 1.5–2 cm, oblong, sickle-shaped. Petals *c.* 5 mm, triangular, margins hairy. Lip triangular, acute. *Malaysia, Sumatra & Borneo*. G2. Winter.

B. mastersianum (Rolfe) J.J. Smith (*Cirrhopetalum mastersianum* Rolfe). Illustration: *Botanical Magazine*, 8531 (1913); *Dansk Botanisk Arkiv* **29**: 84 (1973); Bechtel et al., Manual of cultivated orchid species, 184 (1981). Pseudobulbs ovoid, clustered. Flowering stem 12–30 cm, bearing 6–8 flowers; flowers 4–5 cm long, yellow to orange. Lateral sepals 3–4 × *c.* 1 cm. *Indonesia (Moluccas) & Borneo*. G2. Winter.

91. SOBRALIA Ruiz & Pavon

J.C.M. Alexander

Terrestrial, rarely epiphytic. Stems erect, slender and unbranched, many-leaved. Leaves leathery, pleated, rolled when young, long-sheathed, in 2 opposite ranks. Flowers few or solitary, terminal or in the upper leaf axils. Bracts usually well developed, sometimes several to each flower. Sepals fused at base; petals free, broader than sepals. Lip slightly fused to base of column, simple or 2-lobed, rolled around column; blade expanded, wavy; small calluses and ridges often present. Column slender, without foot, 3-lobed at apex. Pollinia 8, in 2 groups of 4, granular.

A genus of about 100 species from C America and tropical S America. They perform best if repotted very infrequently and should be allowed to become well established in large pots or beds. When actively growing, plants should be kept well shaded and watered. Once growth ceases watering can be considerably reduced.

Literature: Rogers, B., Growing Sobralias, *Orchids* (USA) **76**(12): 912–19 (2007).

1a. Flowers less than 12 cm wide **1. sessilis**
b. Flowers more than 12 cm wide 2
2a. Bracts 2.5–5 cm; column *c.* 2.5 cm; lip 8 cm or less **2. leucoxantha**
b. Bracts 10 cm or more; column *c.* 3.5 cm; lip 8 cm or more 3
3a. Flowers stalked, purple or pinkish purple **3. macrantha**
b. Flowers almost stalkless, yellow **4. xantholeuca**

1. S. sessilis Lindley. Illustration: Dunsterville & Garay, Venezuelan orchids illustrated **1**: 397 (1959); *American Orchid Society Bulletin* **42**: 395 (1973).

To 90 cm. Leaves to 22 × 8 cm, stiff, narrowly elliptic to ovate, acute or obtuse. Flowers solitary, terminal, tapering smoothly into ovary and stalk; bracts several, dark brown. Sepals 4.5–5 × 1.5–2 cm, lanceolate, acute; petals 4–5 × 1–2 cm, lanceolate to ovate, acute or obtuse, strongly recurved; petals and sepals white. Lip 4.2–5 × 2.8–3.5 cm, ovate to circular, shallowly notched, slightly scalloped, rolled around column, white with yellow stripe in throat and patch on blade; callus of 2 small basal projections. Column 2.5 cm. *Brazil, Guyana & Venezuela*. G2. Summer.

S. decora Bateman. Illustration: Hamer, Orquideas de El Salvador, t. 34 (1974); Graf, Tropica, 759 (1978). Occasionally epiphytic. Leaves with warts and short stiff hairs beneath and on sheaths. Flowers to 10 cm wide, pale to dark purple. *Mexico to Honduras*. G2. Spring–summer.

S. cattleya Reichenbach. Illustration: *American Orchid Society Bulletin* **44**: 195 (1975). Flowers in several axillary panicles, each with 5–12 flowers, pinkish brown inside, white or cream outside. Lip maroon with yellow ridges. *Venezuela & Colombia*. G2.

2. S. leucoxantha Reichenbach. Illustration: *Botanical Magazine*, 7058 (1889); Graf, Tropica, 759 (1978); Bechtel et al, Manual of cultivated orchid species, 273 (1981).

Terrestrial. Stems to 150 cm, clustered. Leaves 20–30 × 3–4 cm, elliptic to lanceolate, acuminate, scurfy beneath; sheaths warty. Flowers solitary, terminal, 15–18 cm wide; bracts 2.5–5 cm. Sepals to 7 × 2.5 cm, linear to lanceolate; petals to 7 × 3–3.5 cm, lanceolate to oblong; petals and sepals white. Lip 6–8 × 5–6 cm, ovate, notched, wavy, rolled around column, white with a yellow to reddish orange throat. Column 2.5 cm. *Costa Rica & Panama*. G2. Summer–autumn.

3. S. macrantha Lindley. Illustration: *Botanical Magazine*, 4446 (1849); Skelsey, Orchids, 137 (1979); Bechtel et al., Manual of cultivated orchid species, 273 (1981).

Terrestrial or on rocks. Stems to 2 m, erect, clustered. Leaves 13–30 × 2–7.5 cm, lanceolate, acuminate, dark green above, paler beneath. Flowers solitary, 15–18 cm wide, fragrant; bracts to 11 cm, green. Sepals to 9 × 2.5 cm, lanceolate to elliptic, obtuse or acute, with curled tips; petals to 9 × 3–3.5 cm, elliptic, asymmetric, obtuse or rounded; sepals and petals pinkish purple. Lip to 10 × 7 cm, ovate, rolled around column, deeply scalloped and notched, slightly darker than petals, paler in throat, with 7 darker ridges; callus of 2 white, fleshy teeth near lip base. Column 3–4 cm, club-shaped. *Mexico to Costa Rica*. G2. Summer.

4. S. xantholeuca Reichenbach. Illustration: *Botanical Magazine*, 7332 (1894); Hamer, Orquideas de El Salvador, t. 34 (1974); Graf, Tropica, 759 (1978).

Epiphytic or on rocks. Stems to 180 cm, thick. Leaves 15–28 × 3–7 cm, lanceolate to oblong, acuminate; sheaths speckled with reddish brown. Flowers solitary, terminal, *c.* 15 cm wide, almost stalkless, bright yellow. Sepals 8–11 × 1.5–2 cm, lanceolate, acuminate; petals 8–10 × 2–2.5 cm, lanceolate to elliptic, acute or acuminate. Lip 8–11 × 6–7 cm, ovate to oblong, rolled around column, deeply notched, scalloped; throat streaked with deeper yellow. Column 2.5 cm. *Guatemala & El Salvador*. G2. Spring–summer.

92. WARREA Lindley

V.A. Matthews

Terrestrial. Pseudobulbs ovoid. Leaves few, in 2 ranks, usually lanceolate, pleated, rolled when young. Raceme loose, with few to several flowers, produced from the base of an immature pseudobulb. Sepals and petals spreading or curved inwards, the lateral sepals joined to the column foot to form a mentum. Petals similar to the upper sepal but smaller. Lip simple, usually with erect sides, disc bearing central ridges. Column stout, somewhat swollen above. Pollinia 4, in 2 pairs on a short, triangular stipe.

A genus of 3 species from C and northern S America. They grow best in a compost such as that used for *Phaius* (p. 488). Propagation by division.

1a. Sepals and petals white or pale yellow **1. warreana**
b. Sepals and petals reddish brown **2. costaricensis**

1. W. warreana (Lindley) Schweinfurth (*W. tricolor* Lindley; *W. bidentata* Lindley). Illustration: Lasser, Flora de Venezuela **15** (4): 31 (1970); Pabst & Dungs, Orchidaceae Brasilienses **2**: 230 (1977).

Plant to 1 m. Pseudobulbs 4–12 cm with 3–5 flowers. Leaves 30–70 cm, oblong-lanceolate. Raceme erect, usually longer than the leaves, with 6–15 flowers. Flowers somewhat hanging, white or pale yellow. Sepals and petals nearly circular, upper sepal to 3.5 cm. Lip 2–3.5 cm, white or reddish, marked with yellow and reddish purple, wavy at the edges and often with an apical notch. Disc with 3 ridges. *Northwest S America*. G2. Summer.

2. W. costaricensis Schlechter.

Plant to 75 cm. Leaves to 70 cm, lanceolate. Raceme erect, about as long as leaves. Flowers reddish brown. Sepals and petals oblong-ovate, *c.* 3 cm. Lip to 3 cm, nearly circular, entire, paler than sepals and petals, with reddish brown markings, disc narrow. *Costa Rica, Panama*. G2.

93. PABSTIA Garay

V.A. Matthews

Epiphytic. Pseudobulbs ovoid to oblong. Leaves 2–4 per pseudobulb, lanceolate, pleated, rolled when young. Scapes arising from the bases of the pseudobulbs, clothed in leafy bracts, with 1–4 flowers. Sepals and petals more or less equal or petals smaller. Lip 3-lobed. Column without a foot, hairy in front. Pollinia 4 on a short, linear or oblong stipe.

A genus of 5 species from Brazil usually found under the name *Colax* Lindley. Cultivation as for *Zygopetalum* (see below), *Maxillaria* (p. 546) and *Lycaste* (p. 546).

1a. Sepals white **1. jugosa**
b. Sepals green 2
2a. Sepals spotted **2. placanthera**
b. Sepals unspotted **3. viridis**

1. P. jugosa (Lindley) Garay (*Colax jugosus* (Lindley) Lindley). Illustration: *Botanical Magazine*, 5661 (1867); *Lindenia* **8**: t. 372 (1892); *Orchid Digest* **42**: 144 (1978); Bechtel et al., Manual of cultivated orchid species, 251 (1981).

Pseudobulbs with 2 leaves. Leaves 15–25 cm. Flowers 2–4 in each raceme, 5–7.5 cm in diameter, fragrant. Sepals white, ovate-oblong. Petals white spotted with brownish or reddish purple. Lip *c.* 2.5 cm, lateral lobes rounded and striped with bluish purple, central lobe rounded, striped and marked with bluish purple, velvety. Callus with ridges. *Brazil.* G2. Winter–spring.

The following 2 varieties have been in cultivation, and may still be available:

Var. **rufina** (Reichenbach) Garay (*Colax jugosus* var. *rufinus* Reichenbach). Flowers yellowish green spotted with dark purple and with brown on the lip.

Var. **viridis** (Godefroy-Leboeuf) Garay (*Colax jugosus* var. *punctatus* Reichenbach; *Colax puytdii* Linden & André). Illustration: *Illustration Horticole* **27**: t. 369 (1880). Flowers greenish yellow spotted with reddish purple.

2. P. placanthera (Hooker) Garay (*Colax placanthera* (Hooker) Lindley). Illustration: *Botanical Magazine*, 3173 (1832).

Pseudobulbs with 2–3 leaves. Flowers usually solitary. Petals smaller than sepals, yellowish green with brownish purple spots. Lateral lobes of lip pale green streaked with brown, central lobe with a whitish, purple-flushed central area, very shortly hairy. Callus smooth. *Brazil.* G2. Summer.

3. P. viridis (Lindley) Garay (*Colax viridis* (Lindley) Lindley). Illustration: Pabst & Dungs, Orchidaceae Brasilienses **2**: 226, 282 (1977).

Leaves 12–18 cm. Flowers 1–2 per scape. Sepals and petals directed forwards, green, petals with brownish spots that merge towards the centre. Lip with a pale purple, rhomboidal central lobe. *Brazil.* G2. Summer.

94. ZYGOPETALUM Hooker

V.A. Matthews

Epiphytic or terrestrial. Pseudobulbs ovoid, sheathed in basal leaves, bearing 2–5 leaves. Leaves lanceolate, pleated, with 5–9 conspicuous veins beneath, rolled when young. Racemes produced from the bases of the pseudobulbs. Sepals and petals spreading, lateral sepals attached to the column foot to form a short mentum. Petals similar to sepals or sometimes smaller. Lip 3-lobed, lateral lobes very small, erect or spreading, with a prominent callus, often ribbed. Pollinia 4, in 2 pairs, stipe variable in length.

A genus of 20 species from C and tropical S America, of which 5 are found in cultivation. They require water throughout the year, though the amount should be reduced in winter. Propagation is usually by division of large plants. All are very prone to attack by thrips, scale and red spider mite.

Petals. 2–2.5 cm: **4, 5**; 3–4 cm: **1, 2, 3**.
Lip. Hairy: **1, 2, 3**; hairless: **1, 4, 5**. Without spots, stripes or obvious veining: **5**.
Column. Yellow or green: **1, 3, 4**; white: **2**; purple: **5**.

1a. Lip neither spotted nor striped; callus shaped like a horse's hoof, more or less erect **5. maxillare**
b. Lip spotted or striped; callus not as above 2
2a. Lip hairy, sometimes minutely so 3
b. Lip hairless 5
3a. Petals shorter than upper sepal; column hairless **1. mackayi**
b. Petals equal to upper sepal; column hairy in front 4
4a. Central lobe of lip more or less ovate **2. crinitum**
b. Central lobe of lip kidney-shaped **3. intermedium**
5a. Leaves with 7–9 veins beneath; flowers *c.* 8 cm in diameter; lip 4–4.5 cm **1. mackayi**
b. Leaves with 5 veins beneath; flowers 5–6 cm in diameter; lip 2.5–3 cm **4. brachypetalum**

1. Z. mackayi Hooker. Illustration: Martius, Flora Brasiliensis **3**(5): t. 104 (1898); Pabst & Dungs, Orchidaceae Brasilienses **2**: 227, 283 (1977); *Gartenpraxis* **1980**: 272.

Pseudobulbs 4–7 cm with 2–3 leaves. Leaves 30–50 cm, with 7–9 veins beneath. Flowering stems to 75 cm, racemes usually with 5–7 flowers. Flowers fragrant, *c.* 8 cm in diameter. Sepals and petals greenish yellow with purplish or brownish blotches, sepals 4–4.5 cm, petals 3.5–4 cm. Lip 4–4.5 cm, central lobe ovate or fan-shaped, white with interrupted purple radiating lines, hairless or with tiny hairs, margins wavy, apex notched. Callus horseshoe-shaped, ridged, toothed along the front margin. Column *c.* 2 cm, hairless, yellowish green with reddish brown spots. *S & SC Brazil.* G2. Autumn–winter.

2. Z. crinitum Loddiges. Illustration: *Botanical Magazine*, 3402 (1835); *Orchid Digest* **36**: 128 (1972); Pabst & Dungs, Orchidaceae Brasilienses **2**: 226, 282 (1977); Bechtel et al., Manual of cultivated orchid species, 280 (1981).

Pseudobulbs 4–7 cm with 3–5 leaves. Leaves 25–40 cm with 7–9 veins beneath. Flowering stems 30–45 cm, raceme with 4–7 flowers. Flowers fragrant, 7–8 cm in diameter. Sepals and petals *c.* 4 cm, yellowish green with purplish brown blotches or bars. Lip 3–4 cm, central lobe broadly obovate, velvety, white or cream with densely hairy bluish purple or reddish veins. Callus yellow, more or less 2-lobed. Column 1.6–2 cm, hairy in front, white with purple stripes. *S Brazil.* G2. Winter.

Several varieties have been described, based mainly on colour variations of the veins of the lip.

3. Z. intermedium Lindley. Illustration: Pabst & Dungs, Orchidaceae Brasilienses **2**: 283 (1977); Bechtel et al., Manual of cultivated orchid species, 280 (1981).

Pseudobulbs 4–8 cm with 3–5 leaves. Leaves 25–50 cm with 7–9 veins beneath. Flowering stems 30–75 cm, raceme with 4–7 flowers. Flowers fragrant, 7–8 cm in diameter. Sepals and petals 3–4 cm, green or yellowish green with reddish brown or purplish markings. Lip 3–4 cm with a short claw, central lobe kidney-shaped, whitish spotted with purple and with purple veins, downy, margin wavy, apex notched. Callus horseshoe-shaped, ridged. Column 1.3–1.5 cm, velvety in front, pale green with purple stripes. *S Brazil, Peru, Bolivia.* G2. Autumn–spring.

4. Z. brachypetalum Lindley. Illustration: *Orchid Digest* **36**: 128 (1972); Pabst & Dungs, Orchidaceae Brasilienses **2**: 226 (1977).

Pseudobulbs 4–6 cm, somewhat compressed, with 2–3 leaves. Leaves 30–50 cm, with 5 veins beneath. Flowering stems 30–80 cm, raceme with 5–12 flowers. Flowers fragrant, 5–6 cm in diameter. Sepals 2.5–3 cm, petals 2–2.5 cm, all green with brown blotches. Lip 2.5–3 cm, central lobe more or less obovate, white with bluish violet veins and spots, hairless, apex notched. Callus white with purple stripes. Column 1–1.2 cm, yellowish with purple spots and stripes. *S Brazil.* G2. Autumn–winter.

5. Z. maxillare Loddiges. Illustration: *Botanical Magazine*, 3686 (1839); *Orchid Digest* **36**: 128 (1972); Pabst & Dungs, Orchidaceae Brasilienses **2**: 227, 283 (1977).

Pseudobulbs 4–8 cm with 2–3 leaves, borne on creeping, branched rhizomes. Leaves 20–40 cm with 5–7 veins beneath. Flowering stems 15–30 cm, racemes with 5–8 flowers, sometimes drooping. Flowers *c.* 5 cm in diameter. Sepals *c.* 2.5 cm, petals 2–2.3 cm, all green with dark brown blotches or bars. Lip with the central lobe more or less circular, bluish purple, hairless. Callus dark purple, more or less erect and shaped like a horse's hoof, ribbed. Column *c.* 1 cm, hairless, bluish purple. *S Brazil, Paraguay, N Argentina.* G2. Can flower at any time of year.

In the wild *Z. maxillare* grows only on tree ferns.

95. ACACALLIS Lindley

V.A. Matthews

Climbing epiphyte with a creeping rhizome bearing pseudobulbs at intervals of 2–5 cm. Pseudobulbs ovoid, 2–5 cm, with 1 or 2 leaves. Leaves 10–20 cm, narrowly elliptic, shortly stalked, rolled when young. Racemes with 3–7 flowers, longer than leaves. Flowers 2.5–5 cm in diameter. Sepals and petals 3.5–4 cm, oval, bluish mauve shading to white outside, flushed with pink inside. Lip 2.5–3 cm, clawed, kidney-shaped with a wavy margin, yellowish brown with a reddish or purplish centre and paler veins. Callus yellowish or pinkish, divided above into finger-like projections. Column 3-angled, with 2 wings and a long slender foot. Pollinia 4, in 2 pairs.

A genus of only 1 species. Best grown in moderate shade in a basket using an epiphytic compost that is kept moist.

1. A. cyanea Lindley (*Aganisia cyanea* (Lindley) Reichenbach). Illustration: *Botanical Magazine*, 8678 (1916); Dunsterville & Garay, Venezuelan orchids illustrated **2**: 33 (1961); Bechtel et al., Manual of cultivated orchid species, 161 (1981).

Brazil, Venezuela & Colombia. G2.

Forma **alba** Roeth & Gruss. Illustration: *Die Orchidee* **57**: 64 (2006). Differs in its pure white flowers.

96. PROMENAEA Lindley

V.A. Matthews

Epiphytic. Pseudobulbs more or less ovoid, compressed. Leaves 1–3 per pseudobulb, pleated, often rather glaucous, folded when young. Scapes produced from the bases of the pseudobulbs. Sepals and petals spreading. Lip 3-lobed with usually narrow, erect lateral lobes. Column with a short foot to which the lip is jointed. Disc of lip with a transverse callus. Pollinia 4, sessile on a broad, rounded viscidium.

A Brazilian genus containing 15 species. Cultivation as for *Zygopetalum* (p. 539). The pans or baskets are best hung near the glass.

1a. Sepals and petals spotted or marked **4. stapelioides**
b. Sepals and petals unspotted 2
2a. Sepals and petals yellow 3
b. Sepals and petals pale green or pale brown **3. microptera**
3a. Lip spotted with red towards the base **1. xanthina**
b. Lip whitish, spotted or barred with reddish purple **2. rollissonii**

1. P. xanthina (Lindley) Lindley (*P. citrina* Don). Illustration: *Botanical Magazine*, n.s., 697 (1975); Pabst & Dungs, Orchidaceae Brasilienses **2**: 214 (1977).

Pseudobulbs to 2 cm. Leaves to 7 cm. Scape usually with 1 flower, to 10 cm. Flowers *c.* 3.5 cm in diameter, fragrant. Sepals and petals bright yellow, petals slightly shorter than sepals. Lip bright yellow spotted with red on the lateral lobes and sometimes on the base of the central lobe, which is deflexed and more or less circular. Callus 3-lobed, the central lobe with a fleshy prominence and a toothed apex. *Brazil.* G2. Summer.

P. Crawshayana grex (*P. xanthine* × *P. stapelioides*). Illustration: *Orchids Australia* **15**: 19 (2003). It has pale yellow flowers spotted with reddish brown, especially so on the petals and lip. The central lobe of the lip is ovate and with a small point at the apex. The callus has a tubercled crest. *Garden origin.* G2. Summer

2. P. rollissonii (Lindley) Lindley. Illustration: *Edwards's Botanical Register* **24**: t. 40 (1838); Pabst & Dungs, Orchidaceae Brasilienses **2**: 276 (1977).

Pseudobulbs to 2.5 cm. Leaves 7.5–10 cm. Scape usually with 1 flower, arching or somewhat hanging. Flowers 3–5 cm in diameter. Sepals and petals pale yellow, petals shorter and broader. Lip whitish, spotted or barred with reddish purple, markings sometimes confined to the lateral lobes, central lobe slightly broader than long, and with a small point. *Brazil.* G2. Summer.

3. P. microptera Reichenbach. Illustration: *Botanical Magazine*, 8631 (1915); Pabst & Dungs, Orchidaceae Brasilienses **2**: 276 (1977).

Pseudobulbs 1.5–2.5 cm. Leaves 5–9 cm. Scapes with 1–3 flowers, 5–7 cm, arching. Flowers *c.* 5 cm in diameter. Sepals and petals pale green to pale brown. Lip white or yellowish white, central lobe elliptic-oblong with an acute, recurved apex, marked with purple. Callus spotted with purple and with a tubercled crest. *Brazil.* G2. Summer.

4. P. stapelioides (Link & Otto) Lindley. Illustration; *Edwards's Botanical Register* **25**: t. 17 (1839); *Botanical Magazine*, 3877 (1841); Pabst & Dungs, Orchidaceae Brasilienses **2**: 214, 276 (1977).

Pseudobulbs to 2.5 cm. Leaves 7.5–10 cm. Scapes often with 2 flowers, usually longer than the leaves, hanging. Flowers 3–4 cm in diameter. Sepals and petals greenish yellow outside, deep green inside, barred and spotted with dark purple. Lip dark purple, central lobe ovate to circular, with a small point, somewhat concave, paler on the margin. Callus with a 2-lobed crest and a tooth pointing to the base of the lip. *Brazil.* G2. Summer–autumn.

P. lentiginosa (Lindley) Lindley. The spotting of the flowers is reddish purple and the spots are more distinct, less coalesced into bars. The lip is similar in colour to the petals and the transverse crest of the callus has a central prominence, which is 3-toothed. *Brazil.* G2. Summer–autumn.

97. CHONDROSCAPHE (Dressler) Senghas & Gerlach
V.A. Matthews
Epiphytic. Pseudobulbs absent. Leaves stalked, jointed to the sheaths, oblong-lanceolate to strap-shaped, folded when young. Scape with 1 flower. Sepals almost equal, elliptic to lanceolate, joined to the column foot to form an acute mentum, lateral sepals not reflexed. Petals obovate-elliptic, often with fimbriate margins. Lip not 3-lobed, usually fimbriate. Callus in centre of disc, with 3 teeth, lower callus thickening present. Pollinia 4, almost linear, unequal.

A genus of about 14 species growing in C and north-western S America, formerly included in *Warscewiczella* Reichenbach or *Chondrorhyncha* Lindley. Cultivation as for *Zygopetalum* (p. 539).

1a. Petal margins with fine teeth **1. flaveola**
b. Petal margins with a long fringe **2. chestertonii**

1. C. flaveola (Linden & Reichenbach) Senghas & Gerlach (*Chondrorhyncha flaveola* (Linden & Reichenbach) Garay; *Chondrorhyncha fimbriata* (Linden & Reichenbach) Reichenbach). Illustration: Saunders, Refugium botanicum **2**: t. 107 (1872); Dunsterville & Garay, Venezuelan orchids illustrated **2**: 53 (1961); *American Orchid Society Bulletin* **44**: 1077 (1975).

Leaves to 45 cm, oblong, acute. Scapes erect or arching. Sepals *c.* 2.5 cm, elliptic, long-pointed, sometimes with wavy margins, pale yellow. Petals *c.* 3 cm, margins finely toothed, pale yellow. Lip 3–3.5 cm, fan-shaped, lobed, pale yellow, marked and spotted with brown or red at the base, the margin wavy and fringed. Callus hairless, surrounded by a lobed wall and numerous small calluses. *Colombia, Venezuela, Ecuador & Peru.* G2. Spring–autumn.

2. C. chestertonii (Reichenbach) Senghas & Gerlach. (*Chondrorhyncha chestertonii* Reichenbach). Illustration: *Lindenia* **9**: t. 405 (1893); Kupper & Linsenmaier, Orchids, 87 (1961); Escobar, Orquideas nativas de Colombia **1**: 79 (1997).

Very similar to *C. flaveola* but the sepals and petals are narrower and the petals have a long fringe on the margins. The front of the callus has 3 teeth. *Colombia.* G2. Summer.

98. WARSZEWICZELLA Reichenbach
V.A. Matthews
Epiphytic. Pseudobulbs absent. Stems short, covered by overlapping leaf-sheaths. Leaves in 2 ranks, borne in a fan, membranous, jointed to basal sheaths, folded when young. Scape shorter than leaves, with 1 flower. Sepals and petals usually spreading, the lateral sepals attached to the column foot, sometimes rolled and swept back to form spurs. Lip hinged to column foot, margins turned up towards base surrounding column, 3-lobed or not, often concave. Callus fleshy, with ridges radiating from the base. Column cylindric, not keeled, with a short foot. Pollinia 4, on a broad viscidium.

A genus of 10–12 species occuring in tropical America, previously included in *Cochleanthes* Rafinesque. Cultivation as for *Zygopetalum* (p. 539).

1a. Lip notched at the apex **2. amazonica**
b. Lip with an irregularly toothed or scalloped margin, or if entire then not notched at the apex 2
2a. Callus yellow with ridges of unequal length **1. discolor**
b. Callus white or violet with purple veins and with ridges of more or less equal length **3. wailesiana**

1. W. discolor (Lindley) Reichenbach (*Cochleanthes discolor* (Lindley) Schultes & Garay; *Zygopetalum discolor* (Lindley) Reichenbach). Illustration: *Botanical Magazine*, 4830 (1855); *Orchid Digest* **33**: 225 (1969); Sheehan & Sheehan, Orchid genera illustrated, 65 (1979); Bechtel et al., Manual of cultivated orchid species, 185 (1981).

Leaves to 30 cm, oblanceolate to strap-shaped. Scape to 15 cm. Flowers 5–7.5 cm in diameter. Sepals and petals 3–3.5 cm, white, sepals greenish or yellowish towards apex, petals often tinged with lilac towards the apex. Lip more or less 3-lobed, 2.5–4 cm, deep purple, often with a whitish margin, central lobe rounded, entire. Callus yellow, with ridges of unequal length. *Cuba, Honduras, Costa Rica, Panama & Venezuela.* G2. Spring–autumn.

2. W. amazonica Reichenbach & Warscewicz (*Cochleanthes amazonica* (Reichenbach & Warscewicz) Schultes & Garay; *Zygopetalum amazonicum* (Reichenbach & Warscewicz) Reichenbach). Illustration: *Lindenia* **8**: t. 337 (1892).

Leaves to 30 cm. Scape to 11 cm. Sepals and petals 2.5–3.5 cm, white, reflexed. Lateral sepals rolled and swept back to form spurs. Lip obscurely 3-lobed, *c.* 5 cm, white, veined with purple or reddish purple in the lower part, central lobe rounded, notched at the apex. Callus with a truncate apex with 3–7 teeth. *Peru, Brazil.* G2.

3. W. wailesiana (Lindley) Morren (*Cochleanthes wailesiana* (Lindley) Schultes & Garay). Illustration: Cogniaux & Goossens, Dictionnaire iconographique des orchidées, Warscewiczella, t. 2 (1897); *Orchid Digest* **33**: 225 (1969); Pabst & Dungs, Orchidaceae Brasilienses **2**: 232, 287 (1977).

Leaves 15–20 cm. Flowers *c.* 5 cm in diameter, scented. Sepals and petals white or cream. Lip to 2.2 cm, roundish, white, stained and striped with violet towards the centre, margin entire. Callus oblong, white striped with purple, with 5 ridges. *Brazil.* G2.

99. COCHLEANTHES Rafinesque
V.A. Matthews & J.M.H. Shaw
Epiphytic. Pseudobulbs absent. Stems short, covered by overlapping leaf-sheaths. Leaves in 2 ranks, borne in a fan, membranous, jointed to basal sheaths, folded when young. Scape shorter than leaves, with 1 flower. Sepals and petals usually spreading, the lateral sepals attached to the column foot. Lip flat, hinged to column foot but not enfolding foot, 3-lobed or not, often concave. Callus fleshy, with ridges radiating from the base. Column cylindric, distinctly keeled, with a short foot. Pollinia 4, on a broad viscidium.

A genus of about 2 species occuring in tropical America, which has included *Warscewiczella* Reichenbach. Cultivation as for *Zygopetalum* (p. 539).

1. C. flabelliformis (Swartz) Schultes & Garay (*Warscewiczella flabelliformis* (Swartz) Cogniaux; *W. cochlearis* (Lindley) Reichenbach). Illustration: Dunsterville & Garay, Venezuelan orchids illustrated **2**: 77, 241 (1961); Pabst & Dungs, Orchidaceae Brasilienses **2**: 232, 286 (1977); Bechtel et al., Manual of cultivated orchid species, 92 (1981).

Leaves to 30 cm, narrowly oblong to oblanceolate. Scape 5–15 cm. Flowers

5–6 cm in diameter, fragrant. Sepals and petals 2.5–3.5 cm, white or greenish white. Lip 2.5–3 cm, wider than long, broadly rounded, white marked with purple and with purple veins, margin irregularly toothed or scalloped. Callus white or violet with purple veins, semicircular with more than 11 ridges. *West Indies, Venezuela, Brazil.* G2. Spring–autumn.

C. aromatica (Reichenbach) Schultes & Garay. Similar to *C. flabelliformis*, with green sepals and petals. Lip white with blue to red lines, recurved. *Costa Rica, Panama.* G2. Spring–autumn.

100. PESCATORIA Reichenbach

V.A. Matthews

Epiphytic, without pseudobulbs. Leaves lanceolate, pleated, arranged in 2 ranks, folded when young. Scapes arising from the axils of basal bracts, each with 1 flower. Sepals and petals fleshy, rather concave, lateral sepals joined at the base and inserted on the column foot. Lip unlobed or 3-lobed, contracted at the base into a claw that merges with the column foot; central lobe convex. Callus semicircular, ribbed. Column with a short basal foot. Pollinia 4, in 2 pairs on a short, linear stipe.

A genus of about 28 species in southern C America and western tropical S America. The name is often misspelled 'Pescatorea'. *Bollea* Reichenbach is now included in *Pescatoria*, based on molecular and anatomical evidence. Cultivation as for *Zygopetalum* (p. 539).

Literature: Fowlie, J.A., A key and annotated check list to the genus Pescatorea, *Orchid Digest* **32**: 86–91 (1968).

Sepals and petals. Striped: **5**.
Lip. Bearing warts and bristles: **5**; bearing warts: **1**; without warts or bristles: **2, 3, 4, 6, 7**.
Callus. Yellow marked with brown: **1, 2, 6, 7**; reddish brown: **5**; purple or reddish purple: **3**; white, purple in front: **4**.

1a. Lip distinctly 3-lobed; column narrow 2
b. Lip not 3-lobed; column wide 7
2a. Sepals and petals striped; lip bearing bristles **5. lehmannii**
b. Sepals and petals without stripes; lip smooth or with warts 3
3a. Lip purple 4
b. Lip white or yellow 5
4a. Callus purple or reddish purple **3. dayana**
b. Callus white, purplish in front **4. wallisii**
5a. Callus purple or reddish purple **3. dayana**
b. Callus yellow 6
6a. Flowers 6–8 cm in diameter; lip bearing warts **1. cerina**
b. Flowers 5–6 cm in diameter; lip without warts **2. lamellosa**
7a. Lip violet **6. coelestis**
b. Lip yellow **7. lalindei**

1. P. cerina (Lindley) Reichenbach (*Zygopetalum cerinum* (Lindley) Reichenbach). Illustration: *Botanical Magazine*, 5598 (1866); *Flore des serres* **17**: t. 1815 (1868); *Orchid Digest* **32**: 87 (1968); Bechtel et al., Manual of cultivated orchid species, 258 (1981).

Leaves 20–60 cm. Scape 4–10 cm. Flowers 6–8 cm in diameter. Sepals white or cream, the laterals each with a greenish yellow blotch near the base. Petals similar to sepals. Lip 2– 3 cm, cream to yellow, central lobe oval, warty on the upper surface. Callus semicircular, ribbed, yellowish orange striped with reddish brown. Column white, often with a purple or reddish blotch near the base. *Costa Rica, Nicaragua, Panama.* G2.

2. P. lamellosa Reichenbach. Illustration: *Botanical Magazine*, 6240 (1876); *Die Orchidee* **33**(4): 142 (1982).

Differs from P. *cerina* in having slightly smaller flowers (5–6 cm in diameter), and a pale yellow lip with the central lobe more or less round and without warts. The sepals and petals are white to yellowish green and the large callus is yellowish orange striped with brown and sometimes with dark purple. *Colombia.* G2. Summer.

3. P. dayana Reichenbach.

Leaves 20–40 cm. Scape to 6 cm. Flowers 7–8 cm in diameter. Sepals and petals white with green tips. Lip white flushed with purple, with purple lines extending from the base of the purple or reddish purple callus. Column yellow with a reddish band near the base. *Panama, Colombia.* G2. Autumn–winter.

Three varieties are cultivated; some authors recognise them as subspecies, but it appears that they are no more than horticulturally desirable colour forms:

Var. **candidula** Reichenbach. Illustration: *Orchid Digest* **32**: 87 (1968). Sepals and petals completely white.

Var. **rhodacra** Reichenbach. Illustration: *Botanical Magazine*, 6214 (1876); Orchid Digest **32**: 87 (1968). Sepals and petals with reddish tips.

Var. **splendens** Reichenbach. Sepals and petals with purple tips, the lip deep purple.

4. P. wallisii Reichenbach. Illustration; *Flore des serres* **18**: t. 1828 (1869); *Handbook on Orchids, Brooklyn Botanic Garden Record* **23**(2): 40–1 (1967); *Orchid Digest* **32**: 87 (1968).

Flowers 7–8 cm in diameter. Sepals and petals white tipped with bluish violet. Lip deep violet, central lobe with a white margin and notched apex. Callus white, ribbed, purplish in front. Column white, pale violet below, hairy in front, *Ecuador.* G2.

5. P. lehmannii Reichenbach. Illustration: Cogniaux & Goossens, Dictionnaire iconographique des orchidées, Pescatorea, t. 2 (1898); *Orchid Digest* **32**: 90 (1968); *Die Orchidee* **33**(4): 140 (1982).

Leaves 30–45 cm. Scape to 15 cm. Flowers 6–9 cm in diameter. Sepals and petals white, yellowish towards the tips, striped longitudinally with reddish purple. Lip purple, central lobe oblong, apex notched, covered with warts and bristle-like whitish or purplish papillae. Callus reddish brown, ribbed. Column purple. *Colombia, Ecuador.* G2. Spring–autumn.

6. P. coelestis (Reichenbach) Dressler (*Bollea coelestis* (Reichenbach) Reichenbach; *Zygopetalum coeleste* Reichenbach). Illustration: *Botanical Magazine*, 6458 (1879); Ospina & Dressler, Orquideas de las Americas, t. 101 (1974); Escobar, Orquideas nativas de Colombia **1**: 45 (1997).

Leaves 30–60 cm. Flowers 7–10 cm in diameter, fragrant. Sepals and petals pale blue to deep violet blue, darker above the middle, yellowish at the tips, margin wavy. Lip heart-shaped at the base, with recurved tip and margin, dark violet, yellowish towards the base. Callus semicircular, pale brownish or yellow, with many ribs. *Colombia.* G2. Summer.

7. P. lalindei (Linden) Dressler (*Bollea lalindei* (Linden) Reichenbach; *Zygopetalum lalindei* Reichenbach; *Bollea patinii* Reichenbach). Illustration: *Botanical Magazine*, 6331 (1877); Escobar, Orquideas nativas de Colombia **1**: 46 (1997).

Flowers 6–8 cm in diameter, fragrant. Sepals and petals pale pink or purple, darker above the middle, upper sepal yellowish green at the tip, lateral sepals with yellowish lower margin; petals often with white margins. Lip ovate, narrowly heart-shaped at base, with recurved tip and margin, bright yellow. Callus semicircular, yellow, with many ribs. *Colombia*. G2. Summer–autumn.

'Patinii' has larger flowers (*c.* 10 cm in diameter) and petals without white margins with a flesh-coloured callus.

101. HUNTLEYA Lindley

V.A. Matthews

Epiphytic, without pseudobulbs. Leaves narrowly lanceolate, arranged in a fan, folded when young. Scapes arising from the leaf axils, shorter than the leaves, each with 1 flower. Sepals and petals more or less similar, spreading, glossy, the lateral sepals attached to the short column foot. Lip 3-lobed with small lateral lobes, attached to the column foot by a long, narrow claw, central lobe curving downwards. Callus semicircular, densely fringed. Column with broad, apical, lateral wings. Pollinia 4, in 2 pairs in which the members are unequal, on a short, reversed-triangular stipe.

A genus of about 13 species from C and northern S America. Cultivation as for *Zygopetalum* (p. 539).

1. H. meleagris Lindley (*H. burtii* (Endres & Reichenbach) Pfitzer). Illustration: Dunsterville & Garay, Venezuelan orchids illustrated **3**: 147 (1965); Hawkes, Encyclopaedia of cultivated orchids, 249 (1965); Bechtel et al., Manual of cultivated orchid species, 217 (1981).

Leaves 25–40 cm. Flowers 7.5–10 cm in diameter, fragrant. Sepals and petals lanceolate to ovate, pointed, whitish at the base shading into yellow, marked and flushed with reddish brown, usually with a median yellow cross-bar, and often appearing chequered. Lip 2.5–3.5 cm, white at the base, shading upwards to reddish brown, with a small point. Callus white or yellowish. Column 1.5–2 cm, creamy green, often striped with reddish brown, with a semicircular, down-pointing wing on each side and a hood covering the anther cap. *C America, north-western S America, Trinidad*. G2. Spring–autumn.

102. GALEOTTIA Richard

V.A. Matthews

Epiphytic or terrestrial. Leaves 2–3 per pseudobulb, pleated, rolled when young. Raceme with few flowers, arising laterally from the base of the new growth. Sepals and petals spreading, lateral sepals more or less swollen at the base and joined to the column foot. Lip 3-lobed. Callus formed from a series of longitudinal ribs or plates. Column with distinct wings. Pollinia 4, in 2 pairs on a short, fleshy stipe.

A genus of 11 species, most of which have at some time been included in *Zygopetalum* or *Batemannia* and which will be found in much of the literature under the incorrect name Mendoncella Hawkes. They are native to C and tropical S America, and are best grown in small pans, using an epiphytic compost with very free drainage to allow liberal watering during the growing season.

Literature: Hawkes, A.D., O genero Mendoncella, *Orquidea* **25**: 7–11 (1963); Christenson, E.A., Nomenclatural changes in neotropical Orchidaceae, *Lindleyana* **3** (4): 221–223 (1988).

1a. Flower *c.* 5 cm in diameter; central lobe of lip with entire margin; wings of column entire **3. burkei**
b. Flower 7–8 cm in diameter; central lobe of lip fringed; wings of column toothed 2
2a. Central lobe of lip hairy or warty between the veins; callus violet-streaked **1. fimbriata**
b. Central lobe of lip hairless or sparsely hairy; callus orange **2. grandiflora**

1. G. fimbriata (Linden & Reichenbach) Schlechter (*Mendoncella fimbriata* (Linden & Reichenbach) Garay). Illustration: *Gardeners' Chronicle* 660 (1856); *American Orchid Society Bulletin* **48**: 1223 (1979) & **50**: 37 (1981).

Epiphytic. Pseudobulbs 5–6.5 cm, ovoid, with 2 leaves. Leaves 15–30 cm. Raceme with 2–5 flowers. Flowers *c.* 8 cm in diameter. Sepals and petals greenish or brownish yellow striped with reddish brown, acuminate. Lip white, striped longitudinally with violet, lateral lobes erect, fringed, central lobe ovate, recurved, fringed, acuminate, hairy or warty between the veins. Callus violet-streaked, formed of a number of plates arranged like a fan. Column whitish yellow, wings toothed. *Colombia, Venezuela*. G2.

2. G. grandiflora Richard (*Mendoncella grandiflora* (Richard) Hawkes). Illustration: *Botanical Magazine*, 5567 (1866); *Annals of the Missouri Botanical Garden* **36**: 82 (1949).

Differs from *M. fimbriata* in having flowers 7–8 cm in diameter, the central lobe of the lip hairless or sparsely hairy and less sharply pointed at the apex, and an orange callus. *C America, Colombia*. G2. Spring–summer.

3. G. burkei (Reichenbach) Dressler & Christenson (*Mendoncella burkei* (Riechenbach) Garay; *Zygopetalum burkei* Reichenbach). Illustration: Dunsterville & Garay, Venezuelan orchids illustrated **3**: 333 (1965); *American Orchid Society Bulletin* **48**: 1223 (1979).

Terrestrial. Pseudobulbs 5–10 cm, narrowly oblong, with 2–3 leaves. Leaves 20–45 cm. Raceme with 5–6 flowers. Flowers *c.* 5 cm in diameter. Sepals and petals light green or light brown thickly striped or spotted with brown. Lip white, lateral lobes with entire margins, central lobe with recurved sides, margin entire. Callus reddish purple, ribbed. Column yellow or greenish, streaked in front with purple, wings entire. *Guyana, Venezuela*. G2.

103. XYLOBIUM Lindley

V.A. Matthews

Epiphytic or rarely terrestrial. Pseudobulbs ovoid or conical, more rarely slender and cylindric. Leaves 1–3 per pseudobulb, stalked, lanceolate to elliptic, pleated, with 3–6 conspicuous veins beneath, rolled when young. Raceme produced from the base of a mature pseudobulb. Bracts linear, usually exceeding the ovary. Sepals and petals spreading, the lateral sepals broader than the upper and joined to the column foot to form a mentum. Petals similar to the upper sepal but smaller. Lip curved upwards at the base, usually 3-lobed at the apex, the lateral lobes often standing more or less erect, with a central, usually elongate callus. Column cylindric. Pollinia 4, in 2 pairs, the pollinia of each pair unequal, borne on a short, triangular stipe.

A genus of 33 species from C and tropical S America and the West Indies. They should be grown in an epiphytic compost, and should be watered from the time that new leaves appear until November, then only occasionally. They should be freely ventilated during the summer.

Pseudobulbs. Ovoid, conical or shortly cylindric, to 10 cm: **2, 3, 4, 5, 6, 8, 9, 10**; cylindric and elongate, 15–27 cm: **1, 7**.

Sepals and petals. Spotted or with other markings: **2, 3, 6, 7**; lacking spots or other obvious markings: **1, 4, 5, 6, 8, 9, 10**.
Lip. With warts or papillae: **1, 2, 4, 5, 6, 7**; without warts or papillae: **3, 8, 9, 10**.

1a. Pseudobulbs with 1 leaf 2
b. Pseudobulbs with 2–3 leaves 6
2a. Pseudobulbs 15–20 cm, cylindric, elongate **1. pallidiflorum**
b. Pseudobulbs to 10 cm, ovoid to conical or shortly cylindric 3
3a. Flowers with spots 4
b. Flowers without spots 5
4a. Leaves with 3 conspicuous veins beneath; raceme hanging **3. colleyi**
b. Leaves with 5–6 conspicuous veins beneath; raceme more or less upright **2. leontoglossum**
5a. Raceme hanging **4. palmifolium**
b. Raceme upright **5. bractescens**
6a. Upper sepal 1–1.5 cm 7
b. Upper sepal 2–2.5 cm 9
7a. Racemes with 6–9 flowers **10. powellii**
b. Racemes with 10–22 flowers 8
8a. Lip creamy brown with pale brown veins **9. hyacinthinum**
b. Lip white, sometimes with pinkish stripes **8. foveatum**
9a. Pseudobulbs 12–27 cm, cylindric, elongate **7. elongatum**
b. Pseudobulbs to 9 cm, ovoid to shortly cylindric 10
10a. Leaves with 3 conspicuous veins beneath **3. colleyi**
b. Leaves with 5 conspicuous veins beneath **6. variegatum**

1. X. pallidiflorum (Hooker) Nicholson. Illustration: Dunsterville & Garay, Venezuelan orchids illustrated **1**: 439 (1959); *American Orchid Society Bulletin* **43**: 208 (1974).

Epiphytic or rarely terrestrial. Pseudobulbs 15–20 cm with 1 leaf. Leaves to 40 cm. Raceme upright with 3–9 flowers. Flowers white, pale green or yellowish green. Upper sepal *c.* 1.5 cm. Lip *c.* 5 cm, whitish grading to yellowish orange at the base, 3-lobed, the central lobe covered with white papillae or warts. *Northwest S America, West Indies.* G2.

2. X. leontoglossum (Reichenbach) Rolfe. Illustration: *Botanical Magazine*, 7085 (1889); *American Orchid Society Bulletin* **43**: 212 (1974); Dunsterville & Garay, Venezuelan orchids illustrated **6**: 443 (1976); Bechtel et al., Manual of cultivated orchid species, 280 (1981).

Epiphytic. Pseudobulbs 3.5–10 cm, ovoid to conical, with 1 leaf. Leaves to 35 cm, with 5–6 veins beneath. Flowers pale to bright yellow with reddish brown or pink spots. Upper sepal 1.8–2 cm. Lip 3-lobed, central lobe usually acute, densely covered with brown warts. *North-west S America.* G2.

3. X. colleyi (Lindley) Rolfe (*X. brachystachyum* Kränzlin). Illustration: Dunsterville & Garay, Venezuelan orchids illustrated **3**: 331 (1965); Pabst & Dungs, Orchidaceae Brasilienses **2**: 22, 279 (1977).

Epiphytic. Pseudobulbs to 4 cm, ovoid, with 1 or occasionally 2 leaves. Leaves to 50 cm, with 3 veins beneath. Raceme hanging, with 3–6 flowers. Flowers brown or reddish brown, spotted with purple and smelling of cucumber. Upper sepal 2–2.5 cm. Lip *c.* 2 cm, more or less oblong, smooth, dark purple to almost black, shining. *Venezuela, Guyana, Brazil & Trinidad.* G2.

4. X. palmifolium (Swartz) Fawcett (*X. decolor* (Lindley) Nicholson). Illustration: *Botanical Magazine*, 3981 (1843); Fawcett & Rendle, Flora of Jamaica **1**: t. 23 (1910).

Epiphytic. Pseudobulbs to 7.5 cm, narrowly ovoid, with 1 leaf. Leaves to 42 cm. Raceme somewhat hanging, loose, with few to many flowers. Flowers white or yellowish, scented (often unpleasantly so). Upper sepal *c.* 2 cm. Lip *c.* 1.5 cm, 3-lobed with small lateral lobes, the central lobe rounded or notched, warty, with a callus with 4–5 thickened ridges. *Cuba, Dominican Republic, Trinidad & Jamaica.* G2.

5. X. bractescens (Lindley) Kränzlin.

Pseudobulbs *c.* 3 cm, conical, with 1 leaf. Leaves to 27 cm, 3-veined beneath. Raceme flexuous and arching, loose with 5–7 flowers. Flowers dull yellow. Upper sepal *c.* 2 cm. Lip 3-lobed, reddish brown, recurved at the obtuse apex, bearing lines of warts, with the callus apically 3-lobed. *Ecuador, Peru, Brazil.* G2.

6. X. variegatum (Ruiz & Pavon) Garay & Dunsterville (*Maxillaria squalens* Lindley; *X. squalens* (Lindley) Lindley; *X. scabrilingue* Schlechter). Illustration: *Botanical Magazine*, 2955 (1829); Dunsterville & Garay, Venezuelan orchids illustrated **2**: 343 (1961); Bechtel et al., Manual of cultivated orchid species, 280 (1981).

Epiphytic or terrestrial. Pseudobulbs to 9 cm, ovoid to shortly cylindric, with 2–3 leaves. Leaves to 70 cm, 5-veined beneath. Raceme with 7–many flowers. Flowers white, pale yellow, pale pink or brownish, often marked or tinged with purple. Upper sepal *c.* 2.5 cm. Lip 1–2 cm, 3-lobed, the central lobe with rows of brown or purple papillae. *South C America and north-west S America.* G2.

7. X. elongatum (Lindley) Hemsley. Illustration: *American Orchid Society Bulletin* **43**: 208 (1974).

Epiphytic or terrestrial. Pseudobulbs 15–27 cm, cylindric, elongate, with 2 leaves. Leaves to 40 cm, with 3–5 veins beneath. Raceme arching or drooping, with 5–20 flowers. Flowers whitish, yellowish or pinkish with red, brown or violet markings. Upper sepal 2–2.5 cm. Lip *c.* 2 cm, usually purplish brown, 3-lobed, central lobe covered with dense papillae. *C America.* G2.

8. X. foveatum (Lindley) Nicholson (*X. stachyobiorum* (Reichenbach) Hemsley). Illustration: *American Orchid Society Bulletin* **43**: 209 (1974); Dunsterville & Garay, Venezuelan orchids illustrated **6**: 441 (1976).

Epiphytic. Pseudobulbs to 10 cm, more or less ovoid, with 2–3 leaves. Leaves to 40 cm, with 3 veins beneath. Raceme upright to arching, with 20–24 densely packed flowers. Flowers yellowish or sometimes white, fragrant. Upper sepal 1–1.5 cm, lateral sepals each with a dorsal keel projecting at the apex. Lip to 2 cm, 3-lobed, white with pinkish stripes in the basal part. *C America, northern S America, Jamaica.* G2.

9. X. hyacinthinum (Reichenbach) Schlechter. Illustration: Dunsterville & Garay, Venezuelan orchids illustrated **1**: 437 (1959); Lasser, Flora de Venezuela **15** (4): 333 (1970).

Epiphytic. Pseudobulbs to 6 cm, more or less ovoid, with 2–3 leaves. Leaves 25–50 cm. Raceme upright with 10–20 flowers. Flowers greenish cream, smelling of hyacinth. Upper sepal *c.* 1.3 cm, lateral sepals each with a dorsal keel projecting at the apex. Lip *c.* 1 cm, creamy brown with pale brown veins. *Venezuela.* G2.

10. X. powellii Schlechter. Illustration: Hawkes, Encyclopaedia of cultivated orchids, t. XIIB (1965).

Epiphytic. Pseudobulbs 4–5 cm, shortly cylindric, with 2 leaves. Leaves to 20 cm.

Racemes upright with 6–9 flowers. Flowers yellow or reddish brown, sometimes tinged with pale green, faintly scented. Upper sepal 1–1.5 cm. Lip *c.* 1.5 cm, 3-lobed, central lobe with 3 ridges, reddish brown. *Costa Rica, Panama.* G2.

104. BIFRENARIA Lindley

V.A. Matthews

Epiphytic. Pseudobulbs usually ovoid, often 4-angled. Leaves 1 per pseudobulb, stalked, rather leathery, linear-lanceolate to oblong, rolled when young. Racemes produced from the bases of the pseudobulbs. Sepals spreading, the laterals fused to the column foot, forming a short mentum. Petals similar to sepals or smaller, spreading or somewhat incurved. Lip joined to column foot, usually 3-lobed with the lateral lobes standing more or less erect, and with a central, usually elongate callus. Column winged above. Pollinia 4, in 2 pairs, each pair on a short, separate stipe attached to a broad, oblong, folded viscidium.

A genus of 10 species from tropical S America, nearly all of which are in cultivation. It is closely related to *Maxillaria*, which differs in producing only 1 flower per inflorescence, and to *Lycaste*, which has larger, more persistent leaves. Cultivation as for *Lycaste* (p. 546).

Flower diameter (when opened out flat). 2–3 cm: **1, 2, 3, 4**; 5–9 cm: **5, 6, 7, 8, 9**.

Sepals and petals. Basic colour orange or bright yellow: **1, 2, 3, 5**; basic colour red or purple: **7, 9**; basic colour white, pale yellow or green: **4, 5, 6, 8**. Spotted: **1**.

Lip. Length 1.5 cm: **1, 2, 4**; length 1.8–3.2 cm; **3, 8, 9**; length 3.5–5 cm: **5, 6, 7**. Entire: **4, 9**. Hairy: **5, 6, 7**.

1a. Flowers 2–3 cm in diameter when opened out flat — 2
b. Flowers 5–9 cm in diameter when opened out flat — 5
2a. Sepals and petals greyish cream shaded with pink — **4. racemosa**
b. Sepals and petals orange or bright yellow, spotted or not — 3
3a. Sepals and petals spotted — **1. aurantiaca**
b. Sepals and petals not spotted — 4
4a. Lip with a purple patch; sepals and petals ovate, *c.* 10 mm wide — **2. vitellina**
b. Lip lacking a purple patch; sepals and petals lanceolate, 5–7 mm wide — **3. aureo-fulva**
5a. Sepals and petals basically red or purple — 6
b Sepals and petals basically white, greenish or pale yellowish — 7
6a. Flowers to 9 cm in diameter when opened out flat; lip purple, 3-lobed — **7. tyrianthina**
b. Flowers to 5 cm in diameter when opened out flat; lip whitish or pink, entire — **9. atropurpurea**
7a. Lip hairy — 8
b. Lip not hairy — **8. tetragona**
8a. Sepals and petals basically green or yellowish green — **6. inodora**
b. Sepals and petals basically whitish or pale yellowish — **5. harrisoniae**

1. B. aurantiaca Lindley (*Rudolfiella aurantiaca* (Lindley) Schlechter). Illustration: *Botanical Magazine*, 3597 (1837); Pabst & Dungs, Orchidaceae Brasilienses **2**: 225, 281 (1977); Bechtel et al., Manual of cultivated orchid species, 271 (1981).

Pseudobulbs more or less ovoid, 3–6 cm, somewhat 4-angled and often flattened, bright or yellowish green, often red-spotted. Leaves 10–25 cm, often red-spotted beneath. Racemes drooping or erect, with 4–18 flowers. Flowers *c.* 2.5 cm in diameter, orange or deep yellow with brown or purplish red spots. Lip 1–1.5 cm, 3-lobed, the central lobe broadly triangular, often notched at apex, callus bright yellow. *Colombia, Venezuela, Guyana, Brazil & Trinidad.* G2.

This species is sometimes treated as a member of *Rudolfiella* Hoehne, a genus of about 9 species, no other members of which are in cultivation.

2. B. vitellina (Lindley) Lindley (*Stenocoryne vitellina* (Lindley) Kränzlin). Illustration: *Edwards's Botanical Register* **25**: t. 12 (1839); Pabst & Dungs, Orchidaceae Brasilienses **2**: 225, 281 (1977).

Pseudobulbs ovoid, to 4 cm, bluntly angular. Leaves to 30 cm. Racemes drooping or erect, with 5–8 flowers. Flowers orange-yellow, *c.* 2.5 cm in diameter, sepals and petals ovate, *c.* 10 mm wide. Lip 1.1–1.5 cm, 3-lobed, central lobe with a purple patch. *Brazil.* G2. Summer.

3. B. aureo-fulva (Hooker) Lindley (*Stenocoryne aureo-fulva* (Hooker) Kränzlin). Illustration: Pabst & Dungs, Orchidaceae Brasilienses **2**: 225 (1977).

Pseudobulbs ovoid, often wrinkled. Leaves to 15 cm. Racemes with 5–7 flowers. Flowers orange, 2–3 cm in diameter, sepals and petals lanceolate, 5–7 mm wide. Lip *c.* 2 cm, 3-lobed, central lobe lanceolate. *Brazil.* G2. Autumn.

B. aureo-fulva and *B. vitellina* are sometimes included in *Stenocoryne* Lindley, a genus differing in the detailed structure of the pollinia.

4. B. racemosa (Hooker) Lindley. Illustration: *Loddiges' Botanical Cabinet* **14**: t. 1318 (1828); Pabst & Dungs, Orchidaceae Brasilienses **2**: 225, 281 (1977).

Pseudobulbs ovoid. Racemes erect or drooping, with 4–10 flowers. Flowers greyish cream, shaded with pink, 2–3 cm in diameter. Lip 1.1–1.5 cm, entire, whitish, speckled with pink. *Brazil.* G2.

5. B. harrisoniae (Hooker) Reichenbach. Illustration: Pabst & Dungs, Orchidaceae Brasilienses **2**: 224, 280 (1977); Sheehan & Sheehan, Orchid genera illustrated, 45 (1979); Bechtel et al., Manual of cultivated orchid species, 170 (1981).

Pseudobulbs broadly ovoid, 5–9 cm, weakly or strongly 4-angled. Leaves to 30 cm. Racemes usually erect with 1–3 flowers. Flowers fragrant, 6–7 cm in diameter. Sepals and petals whitish or pale yellowish, often tinged with red. Lip 4–5 cm, 3-lobed, purple to dark red, usually with darker veins, central lobe with a wavy margin, notched at apex, hairy; callus yellow, hairy. *Brazil.* G2. Spring–summer.

Several varieties have been described, differing in flower size and colour.

6. B. inodora Lindley (*B. furstenbergiana* Schlechter). Illustration: *American Orchid Society Bulletin* **41**: 981 (1972); Pabst & Dungs, Orchidaceae Brasilienses **2**: 280 (1977).

Differs from *B. harrisoniae* in that the pseudobulbs usually have a dark band at the apex, the flowers are a little larger, fragrant or not, and the sepals and petals are green or yellowish green, sometimes tinged with purple. The lip may be white, yellow or pale pink, often suffused with a darker colour, and the callus is hairless. *Brazil.* G2. Early spring–summer.

7. B. tyrianthina (Loudon) Reichenbach. Illustration: *Botanical Magazine*, 7461 (1896); Pabst & Dungs, Orchidaceae Brasilienses **2**: 224, 280 (1977).

Differs from *B. harrisoniae* in being more robust, with flowers to 9 cm in diameter, purple. The lip is purple, whitish towards the base and densely hairy. *Brazil.* G2. Spring–summer.

8. B. tetragona (Lindley) Schlechter. Illustration: *Edwards's Botanical Register* **17**: t. 1428 (1831); Pabst & Dungs, Orchidaceae Brasilienses **2**: 224, 280 (1977).

Pseudobulbs to 9 cm, ovoid, 4-angled. Leaves to 45 cm. Racemes with 3–5 flowers. Flowers fragrant, *c.* 5 cm in diameter. Sepals and petals basically greenish yellow, streaked with brownish purple. Lip *c.* 3 cm, about as wide as long, 3-lobed, whitish, marked with purple towards the base, callus large and conspicuous. *Brazil.* G2. Early summer.

9. B. atropurpurea (Loddiges) Lindley. Illustration: *Loddiges' Botanical Cabinet* **19**: t. 1877 (1832); Pabst & Dungs, Orchidaceae Brasilienses **2**: 223, 280 (1977); Bechtel et al., Manual of cultivated orchid species, 170 (1981).

Pseudobulbs to 8 cm, ovoid, 4-angled. Leaves to 25 cm. Raceme usually borne horizontally, with 3–5 flowers. Flowers fragrant, *c.* 5 cm in diameter. Sepals and petals dark red flushed with yellow around the middle. Lip 1.8–2.2 cm, more or less entire, whitish or pink. *Brazil.* G2. Spring–summer.

105. LYCASTE Lindley

V.A. Matthews

Epiphytic or terrestrial. Pseudobulbs ovoid or ellipsoid, often laterally compressed, the base enclosed in papery or fibrous sheaths, of which the upper 1 or 2 are often leaf-bearing. Leaves usually 1–3 per pseudobulb, stalked, lanceolate to oblong-elliptic, pleated, rolled when young. Flowers solitary from the bases of the pseudobulbs, scapes shorter than the leaves, with several papery bracts. Lateral sepals broader than the upper, joined to the column foot to form a short mentum. Petals more or less equal to the upper sepal or shorter. Lip 3-lobed, the base continuous with or jointed to the column foot; lateral lobes erect; disc with a thickened callus. Pollinia 4, in 2 pairs, on a linear stipe.

A genus of 45 species from C and tropical S America, of which fewer than half are in cultivation. They require an epiphytic compost and should be grown in a cool to intermediate house with a definite resting period in winter. Watering may begin again when new growth emerges. Spraying should be delayed until after the new leaves have fully developed, to avoid unsightly black blotches.

Literature: Fowlie, J.A., The genus Lycaste (1970); Oakeley, H., Lycaste, Ida and Anguloa (2008).

Pseudobulbs. With spines at the apex: **1, 2, 3, 4, 5, 8, 12**; without spines, or spines vestigial (less than 3 mm): **6, 7, 9, 10, 11, 13**.

Lip. Length 2.5–3 cm: **1, 2, 5, 8, 9, 11, 12**; length 3–4 cm: **4, 6, 10**; length **4, 5**. 5 cm: **3, 7, 13**. Basic colour yellow or orange: **1, 2, 3, 4, 10**; basic colour cream or white: **5, 6, 7, 8, 9, 10, 11, 12**; basic colour reddish, brownish or purple: **10, 13**. Surface of central lobe hairless: **3, 4, 7, 8, 9, 10, 12, 13**; surface of central lobe hairy: **1, 2, 5, 6, 11**. Callus hairless: **1, 3, 4, 7, 8, 9, 10, 11, 12**; callus hairy: **2, 5, 13**.

1a. Pseudobulbs armed with apical spines more than 3 mm 2
b. Pseudobulbs unarmed or with vestigial spines less than 3 mm 8
2a. Central lobe of lip hairless 3
b. Central lobe of lip hairy, sometimes sparsely so 5
3a. Lip 4–5 cm, deep yellow with red spots **3. deppei**
b. Lip 3 cm or less, white, spotted or not 4
4a. Central lobe of lip narrowly ovate **12. xytriophora**
b. Central lobe of lip more or less circular **8. candida**
5a. Apex of callus a flap-like extension protruding over the base of the central lobe of the lip **1. aromatica**
b. Apex of callus not as above 6
6a. Central lobe of lip lanceolate **5. crinita**
b. Central lobe of lip ovate to more or less circular 7
7a. Callus hairy; lateral lobes of lip larger than the central lobe **2. cruenta**
b. Callus not hairy; lateral lobes of lip smaller than the central lobe **4. macrobulbon**
8a. Flowers to 10 cm in diameter 9
b. Flowers 12–18 cm in diameter 12
9a. Callus cream, yellow, pale orange, greenish or brownish 10
b. Callus orange-red or reddish purple, if yellow then spotted **10. macrophylla**
10a. Central lobe of lip narrowly oblong, densely hairy; upper sepal 6–6.5 cm **6. lasioglossa**
b. Central lobe of lip ovate, minutely downy or hairless; upper sepal 3–3.5 cm 11
11a. Lip whitish, heavily spotted with light pink, central lobe hairless **9. tricolor**
b. Lip creamy white flushed with brownish yellow, central lobe minutely downy **11. leucantha**
12a. Lateral sepals *c.* 6.5 cm; sinuses between lateral and central lobes of lip without a projection **13. skinneri**
b. Lateral sepals 9–9.5 cm; sinuses between lateral and central lobes of lip with a projection **7. schilleriana**

1. L. aromatica (Hooker) Lindley. Figure 28(1), p. 547. Illustration: *Botanical Magazine*, 9231 (1931); Fowlie, The genus Lycaste, 14 (1970).

Pseudobulbs 7–10 cm, strongly ribbed, apex with spines. Leaves 30–40 cm, developing after flowering. Scapes 11–15 cm. Flowers 4–6 cm in diameter, smelling of cinnamon. Sepals yellowish green, the upper 2.5–3 cm. Petals orange-yellow, more or less equal to sepals. Lip 2.5–3 cm, orange-yellow, spotted with dark orange, lateral lobes sickle-shaped, central lobe recurved with a finely toothed margin. Callus very shortly hairy with a flap-like extension protruding over the base of the central lobe. Column c. 2 cm, hairy below the stigma. *C America.* G2. Spring–summer.

2. L. cruenta (Lindley) Lindley. Figure 28 (2), p. 547. Illustration: *Edwards's Botanical Register* **28**: t. 13 (1842); Fowlie, The genus Lycaste, 26 (1970); *Orchid Digest* **42**: 209 (1978); Bechtel et al., Manual of cultivated orchid species, 223 (1981).

Pseudobulbs to 10 cm, apex with spines. Leaves 35–45 cm, deciduous at time of flowering. Scapes *c.* 15 cm. Flowers 5–7 cm in diameter, smelling faintly of cinnamon. Sepals greenish yellow, the upper 4–4.5 cm. Petals yellowish orange with red spots towards the base, 3.5–3.8 cm, recurved at apices. Lip *c.* 2.5 cm, orange with red elongate spots towards the base and with a red triangular patch at the base, lateral lobes larger than central lobe, central lobe more or less circular, recurved, margin minutely scalloped, upper surface shortly hairy. Callus ridged, hairy, red at the base. Column 1.5 cm, red, hairy below stigma. *Northern C America.* G2. Spring–summer.

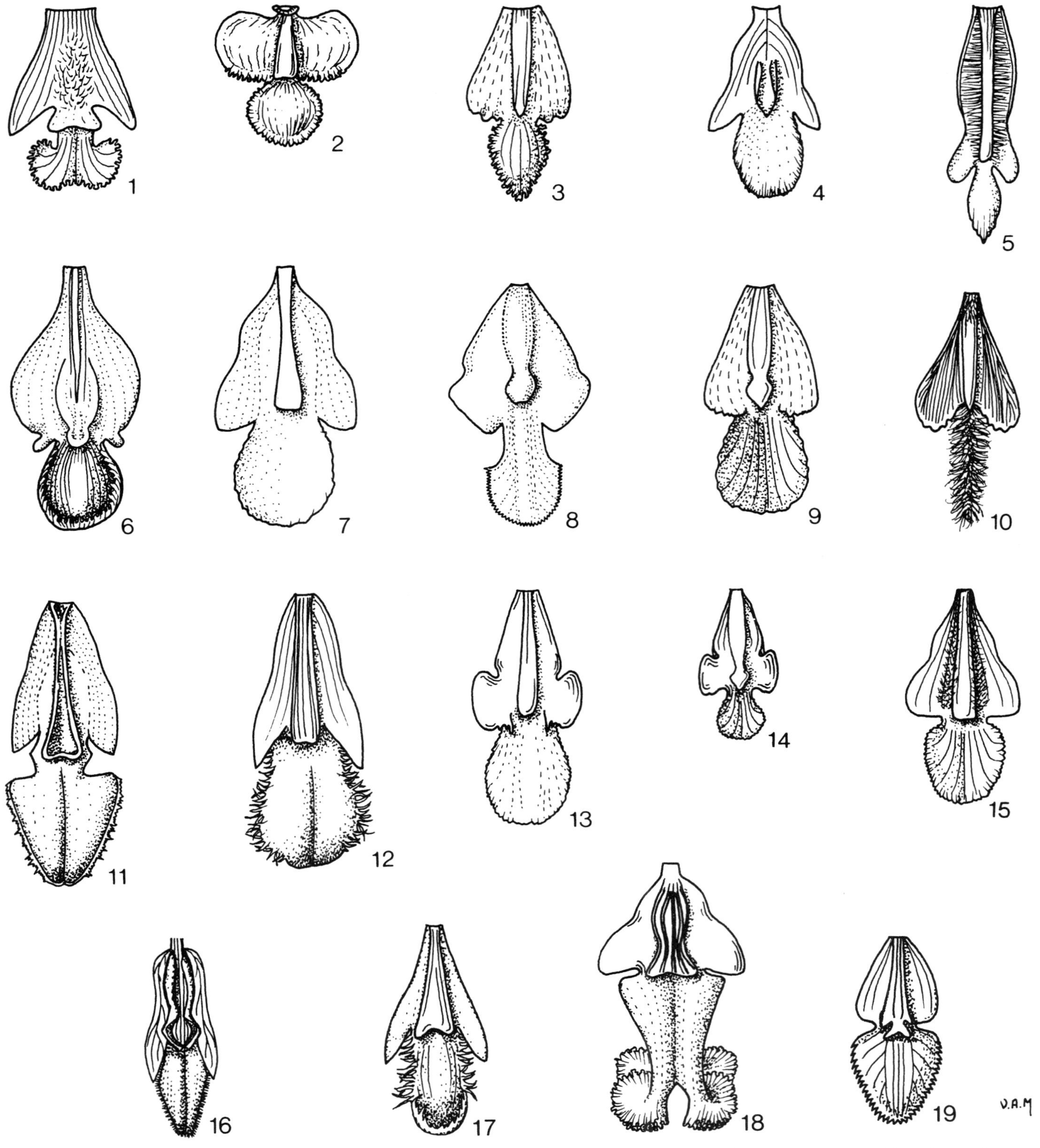

Figure 28. Lips of *Lycaste* and *Ida* species. 1, *L. aromatica.* 2, *L. cruenta.* 3, *L. deppei.* 4, *L. macrobulbon.* 5, *L. crinita.* 6, *L. schilleriana.* 7, *L. candida.* 8, *L. tricolor.* 9, *L. macrophylla.* 10, *L. lasioglossa.* 11, *I. heynderycxii.* 12, *I. fimbriata.* 13, *L. leucantha.* 14, *L. xytriophora.* 15, *L. skinneri.* 16, *I. locusta.* 17, *I. barringtoniae.* 18, *I. cinnabarina* (not opened out flat). 19, *I. dyeriana.* (Compiled from various sources; not to scale.)

3. L. deppei (Loddiges) Lindley. Figure 28 (3), p. 547. Illustration: *Loddiges' Botanical Cabinet* **17**: t. 1612 (1830); Fowlie, The genus Lycaste, 28 (1970); Bechtel et al., Manual of cultivated orchid species, 223 (1981).

Pseudobulbs to 8 cm, apex with spines. Leaves 30–50 cm. Scapes 12–20 cm. Flowers to 10 cm in diameter. Sepals green with reddish brown spots, the upper 4–7 cm. Petals creamy white with tiny reddish brown spots, shorter than sepals, recurved at apices. Lip 4–5 cm, deep yellow with red spots and pale red stripes towards the base, central lobe recurved, margin minutely scalloped. Callus cream to yellow with pale red markings, hairless. Column *c.* 2.5 cm, whitish spotted with red, hairy. *C America*. G2. Spring–summer.

4. L. macrobulbon (Hooker) Lindley. Figure 28(4), p. 547. Illustration: *Botanical Magazine*, 4228 (1846); Fowlie, The genus Lycaste, 32 (1970); *Orchid Digest* **42**: 209 (1978).

Pseudobulbs 5–8 cm, apex with spines. Leaves to 40 cm, developing after flowering. Scapes to 25 cm. Flowers 6–7 cm in diameter, fragrant. Sepals yellowish green, the upper 3.7–4 cm, gently recurved. Petals yellow with orange spots, 3–3.3 cm, recurved at apices. Lip 3–3.3 cm, deep yellow spotted with reddish brown, especially on the central lobe, which is recurved with a minutely toothed margin and hairy at the base. Callus with raised margins. Column *c.* 1.8 cm, pale yellow, spotted with red at base. *Colombia, Venezuela*. G2. Spring–summer.

5. L. crinita Lindley. Figure 28(5), p. 547. Illustration: Fowlie, The genus Lycaste, 24 (1970); *Orchid Digest* **42**: 207 (1978).

Pseudobulbs with apical spines. Leaves to 35 cm. Scapes 8–11 cm. Sepals light green, the upper 2.5–2.8 cm, somewhat recurved. Petals pale yellow, *c.* 2 cm, recurved at apices. Lip *c.* 2.5 cm, pale yellow spotted with orange-red, central lobe lanceolate, hairy. Callus densely hairy on margins. Column *c.* 1.5 cm, hairy at the base. *Mexico*. G2. Summer.

6. L. lasioglossa Reichenbach. Figure 28 (10), p. 547. Illustration: *Botanical Magazine*, 6251 (1876); Fowlie, The genus Lycaste, 30 (1970); Bechtel et al., Manual of cultivated orchid species, 223 (1981).

Pseudobulbs 5–10 cm, slightly ribbed, without apical spines. Leaves 40–50 cm, present at flowering. Scape to 25 cm. Flowers *c.* 10 cm in diameter. Sepals reddish brown, suffused with green outside and at apices, the upper 6–6.5 cm, the laterals with upswept apices. Petals pale yellow, 4–4.3 cm, recurved at apices. Lip 3–3.5 cm, pale yellow, suffused and spotted with red, central lobe recurved, narrowly oblong, densely hairy. Callus cream, tongue-shaped, grooved. Column 2.5–3 cm, hairy above the middle. *Guatemala*. G2. Winter–spring.

7. L. schilleriana Reichenbach (*L. hennisiana* Kränzlin; *L. longisepala* Schweinfurth). Figure 28(6), p. 547. Illustration: *Fieldiana* **30**(1): 647 (1960); Fowlie, The genus Lycaste, 43 (1970).

Pseudobulbs 7–10 cm, slightly ribbed, without apical spines. Leaves to 40 cm. Scapes 15–20 cm. Flowers 16–18 cm in diameter. Sepals pale green suffused with brown towards the base, upper sepal 9–10 cm, gently recurved. Petals 4–5 cm, white suffused with pale pink, recurved at apices. Lip 4.8–5.2 cm, white, suffused and spotted with pink, central lobe recurved, its margins finely scalloped, sinuses between lateral and central lobes with a tiny projection. Callus yellowish, striped pinkish at the base. Column *c.* 1.8 cm, shortly hairy. *Peru, Colombia & Surinam*. G2. Spring–summer.

8. L. candida Lindley. (*L. brevispatha* (Klotzsch) Lindley; *L. candida* Reichenbach). Figure 28(7), p. 547. Illustration: Reichenbach, Beiträge zu einer Orchideenkunde Central-Amerika's, t. 5 (1866); Bechtel et al., Manual of cultivated orchid species, 222 (1981); *Botanical Magazine* 215 (1993).

Pseudobulbs *c.* 6–7 cm, ribbed, apex with spines. Leaves to 50 cm. Scapes 6–10 cm. Flowers 4–5 cm in diameter, fragrant. Sepals brownish green, sometimes spotted with pink, recurved at apices, the upper *c.* 2.5 cm. Petals equal to sepals, paler, sometimes spotted or suffused with pink, apices recurved. Lip 2.5–3 cm, whitish with pink spots, concave towards the base, central lobe with wavy margins. Callus tongue-shaped with 2 ridges, yellowish grading to pink at the base. Column white with pink spots. *Southern C America*. G2. Winter–spring.

According to Oakeley (2008), the plant figured in Fowlie, The genus Lycaste, 36 (1970) as *L. brevispatha* is the hybrid *L.* × *daniloi* (*L. candida* × *L. angelae*), which shows a range of variation intermediate between the parents.

L. angelae Oakeley. Distinguished from *L. candida* by the lip broadly diamond-shaped, hardly lobed, with a callus that is flat, without grooves or ridges. *Costa Rica*. G2. Winter–spring.

9. L. tricolor (Klotzsch) Reichenbach. Figure 28(8), p. 547. Illustration: *Fieldiana* **26**(1): 555 (1952); Fowlie, The genus Lycaste, 39 (1970).

Pseudobulbs 5–9 cm, unarmed but with sharp apical ridges. Leaves 20–40 cm, deciduous at flowering. Scapes 8–12 cm. Flowers to 5.5 cm in diameter, never opening fully. Sepals greenish brown, sometimes with pink spots, recurved, the upper 3–4.5 cm. Petals shorter and paler than sepals, with pink spots. Lip 2.5–3 cm, whitish, heavily spotted with pink, central lobe with the apex strongly reflexed, margin finely wavy. Callus cream. Column 1.5–2 cm, sparsely hairy in front. *Southern C America*. G2. Spring–summer.

10. L. macrophylla (Poeppig & Endlicher) Lindley (*L. dowiana* Endres & Reichenbach). Figure 28(9), p. 547. Illustration; Fowlie, The genus Lycaste, 49, 53, 54 (1970); Pabst & Dungs, Orchidaceae Brasilienses **2**: 282 (1977).

Pseudobulbs 8–12 cm, ribbed, apex with vestigial spines *c.* 0.3 mm. Leaves to 65 cm. Scapes 8–50 cm. Flowers *c.* 9 cm in diameter, faintly scented or not. Sepals brownish green to dark reddish, recurved at apices, the upper 3–5 cm. Petals cream, suffused or not with yellow or pink, sometimes with pink spots, recurved at apices, shorter than sepals. Lip 3–4 cm, cream or yellowish, often spotted with pink to varying degrees, occasionally completely reddish. Callus tongue-shaped, grooved, yellow with red spots or pale reddish purple. Column *c.* 3 cm, cream, often with 2 purple spots at the base, hairy. *Southern C America & north-west S America*. G2. Late summer.

Related to *L. leucantha* and *L. xytriophora*. Several subspecies are recognised, some of which may be in cultivation.

11. L. leucantha (Klotzsch) Lindley. Figure 28(13), p. 547. Illustration: Paxton's flower garden **2**: 37–8 (1851–2); Fowlie, The genus Lycaste, 46 (1970); Bechtel et al., Manual of cultivated orchid species, 223 (1981).

Pseudobulbs to 6 cm, without apical spines. Leaves 40–50 cm. Scape 15–22 cm. Flowers *c.* 8 cm in diameter. Sepals pale green, sometimes brownish green, recurved at apices, the upper *c.* 4 cm. Petals creamy white, recurved towards apices 3.7–4 cm. Lip 2.7–3 cm, pale brownish yellow, central lobe ovate with a minutely wavy margin, minutely downy. Callus brownish yellow with a shallow groove. Column hairless. *Southern C America*. G2. Winter.

12. L. xytriophora Linden & Reichenbach. Figure 28(14), p. 547. Illustration: Reichenbach, Xenia Orchidacea **3**: t. 241 (1892); Fowlie, The genus Lycaste, 66 (1970); Bechtel et al., Manual of cultivated orchid species, 224 (1981).

Pseudobulbs 7–8 cm, apex with spines. Leaves 40–50 cm. Scapes 4–13 cm. Flowers 7–8 cm in diameter. Sepals pale green suffused with brown, the upper 3–4 cm, apices strongly reflexed. Petals cream suffused with green, shorter than sepals, recurved towards apices. Lip 2.5–3 cm, creamy white suffused with yellow and with mauve inside, central lobe narrowly ovate, sharply deflexed with minutely wavy margins. Callus linear, expanded at apex into an ovate projection, deep yellow grading to red at the base. Column 2.5–2.8 cm, white, hairy below stigma. *Panama, Ecuador*. G2. Summer.

13. L. skinneri (Lindley) Lindley (*L. virginalis* (Scheidweiler) Linden). Figure 28(15), p. 547. Illustration; Kupper & Linsenmaier, Orchids, 83 (1961); Fowlie, The genus Lycaste, 61 (1970); Bechtel et al., Manual of cultivated orchid species, 223 (1981).

Pseudobulbs to 10 cm, slightly ridged, without apical spines. Leaves 50–60 cm. Scapes 15–30 cm. Flowers 12–15 cm in diameter. Sepals cream, variously shaded with pink to lavender, recurved at apices, the upper 7–8 cm. Petals usually darker than sepals, reddish purple, recurved at apices, 4.5–7 cm. Lip 4–5.5 cm, pinkish, unmarked or mottled with purple, central lobe ovate, reflexed. Callus reddish, hairy. Column 3–4 cm, white, spotted with red towards the base, hairy in front. *C America*. G2. Winter–spring.

Probably the most commonly grown species. About 15 varieties, distinguished mainly by flower colour and size, have been in cultivation at various times, and some may still be available. *L. skinneri* has often been used in hybridisation.

L. guatemalensis Archila. (*L. skinneri* var. *ipala* invalid). Differs from *L. skinneri* by smaller size, sepals and petals narrow, lanceolate, white to pale pink. Lip narrow rose coloured. *Guatemala*. G2. Summer.

106. IDA Ryan & Oakeley

J.M.H. Shaw

Similar to *Lycaste*, but differing in the flowers nodding or pendulous, in shades of green, white, brown to orange, never spotted. Lateral sepals usually falcate. Lip lateral lobes reduced, triangular or rhomboidal, present only at the central part of the lip, mid-lobe fringed to toothed with a grooved to ridged central callus. Stalks on pollinia S-shaped.

A genus of about 36 species distributed in northern and central South America and the Caribbean, and previously included in *Lycaste* as Section Fimbriatae, recently separated on the basis of molecular work. Cultivation as for *Lycaste* (p. 546).

Literature: Ryan, A. & Oakeley, H.F., Ida, a new genus based on Lycaste section Fimbriatae Fowlie, *Orchid Digest* **67**(1): 8–10 (2003); Oakeley, H.F., Infrageneric classification of Ida, *Orchid Digest* **67**(1): 11–29 (2003); Oakeley, H.F., Lycaste, Ida and Anguloa (2008) – all the species are well illustrated.

1a. Pseudobulbs armed with apical spines more than 3 mm — 2
b. Pseudobulbs unarmed or with vestigial spines less than 3 mm — 3
2a. Lip green — **5. locusta**
b. Lip not green — **3. heynderycxii**
3a. Callus cream, yellow, pale orange, greenish or brownish — 4
b. Callus orange-red or reddish purple, if yellow then spotted — **4. cinnabarina**
4a. Lip 4–5.5 cm — 5
b. Lip 2.5–3.5 cm — **6. dyeriana**
5a. Lip yellowish orange with the apex of the central lobe inrolled; upper sepal 5–6 cm — **2. barringtoniae**
b. Lip creamy white or pale yellow with the apex of the central lobe not inrolled; upper sepal 8–10 cm — **1. fimbriata**

1. I. fimbriata (Poeppig & Endlicher) Ryan & Oakeley (*Lycaste fimbriata* (Poeppig & Endlicher) Cogniaux). Figure 28(12), p. 547. Illustration: *Orchid Digest* **38**: 144 (1974) & **67**(1): 15 (2003); *Botanical Magazine*, n.s., 767 (1978).

Pseudobulbs to 12 cm, ridged, without apical spines. Leaves 60–80 cm, present at flowering, then falling. Scapes 8–24 cm. Flowers 9–10 cm in diameter, slightly fragrant, appearing not to open fully. Sepals creamy white suffused with pale green, the upper 8–10 cm. Petals usually paler than sepals, 4.5–5 cm. Lip 4.5–5.5 cm, whitish or pale yellow with a slight green suffusion, central lobe oblong, recurved, lateral margins fringed. Callus with 5–7 ribs, yellowish to pale orange, apex rounded or notched and projecting over the base of the central lobe. Column white, hairy in front. *Northern S America*. G2. Late winter–summer.

Plants in cultivation and the literature are often called *Lycaste ciliata* (Ruiz & Pavon) Reichenbach, now *I. ciliata* (Ruiz & Pavon) Ryan & Oakeley. However this is not known to be in cultivation and is known only from the illustrations of the type collection made in Peru about 1785.

2. I. barringtoniae (Smith) Ryan & Oakeley. (*Lycaste barringtoniae* (Smith) Lindley). Figure 28(17), p. 547. Illustration: Fawcett & Rendle, Flora of Jamaica **1**: t. 23 (1910); Fowlie, The genus Lycaste, 74 (1970).

Pseudobulbs 6–8 cm, ribbed, without apical spines. Leaves 35–50 cm, falling at or just before flowering. Scapes 7–8 cm. Flowers 5–7 cm, fragrant, not opening fully. Sepals pale green suffused with brown, the upper 5–6 cm. Petals similar but *c.* 4.5 cm. Lip 4–4.5 cm, yellow-orange, sometimes with red markings, central lobe ovate, fringed, apex inrolled to form a shallow pocket. Callus yellow, with 5 ridges, the notched apex projecting forwards over the base of the central lobe. Column 3–5 cm, white. *Cuba, Jamaica*. G2. Spring–summer.

3. I. heynderycxii (Morren) Ryan & Oakeley. (*Maxillaria heynderycxii* Moren; *Lycaste gigantea* Lindley 1845, not Lindley 1843; *Lycaste longipetala* (Ruiz & Pavon) Garay misapplied). Figure 28(11), p. 547. Illustration: *Edwards's Botanical Register* **31**: t. 34 (1845); *Botanical Magazine*, 5616 (1866); Bechtel et al., Manual of cultivated orchid species, 223 (1981); *Orchid Digest* **67**(1): 22 (2003). Fowlie, The genus Lycaste, 83 (1970).

Pseudobulbs 7–12 cm with shallow furrows and apical spines. Leaves to 60 cm, present at flowering. Scapes 50–60 cm. Flowers faintly fragrant, appearing to be only half open. Sepals pale green suffused with brown, the upper 9.5–11 cm. Petals similar in colour, concave, *c.* 8 cm. Lip *c.* 6 cm, reddish brown, central lobe with a lighter orange border, recurved, with a deep central groove, apex folded upwards, lateral margins irregularly minutely toothed. Callus greenish brown with 2 ridges that merge into a notched plate. Column 2.5–3.5 cm, whitish, often orange at base, hairy below. *North-west S America.* G2. Spring.

I. gigantea (Lindley) Ryan & Oakeley. (*Lycaste gigantea* Lindley 1843). Illustration: Oakeley, Lycaste, Ida and Anguloa, 254 (2008). May be distinguished by the green flower with a green lip. *Colombia & Ecuador.* G2. Spring.

Much confused with the above.

4. I. cinnabarina (Rolfe) (*Lycaste denningiana* Reichenbach; *Lycaste cinnabarina* (Lindley) Rolfe). Figure 28(18), p. 547. Illustration: *Lindenia* **9**: t. 394 (1893); Dunsterville & Garay, Venezuelan orchids illustrated **3**: 167 (1965); Fowlie, The genus Lycaste, 56 (1970).

Pseudobulbs 7–10 cm, furrowed, without apical spines. Leaves 45–70 cm. Scapes to 16 cm. Flowers *c.* 10 cm in diameter. Sepals and petals cream, greenish or brownish yellow, the upper sepal 6–8 cm. Lip *c.* 4 cm, deep orange-red, central lobe ovate to circular, apex folded under and back, margin minutely toothed. Callus orange-red, with 5 ridges and a broadened truncate apex. Column 2–2.5 cm, white. *North-west S America.* G2.

5. I. locusta (Reichenbach) Ryan & Oakeley (*Lycaste locusta* Reichenbach). Figure 28(16), p. 547. Illustration: *Botanical Magazine*, 8020 (1905); Fowlie, The genus Lycaste, 76 (1970); *Orchid Digest* **67**(1): 8 (2003).

Pseudobulbs 6–10 cm, ribbed, apex with spines. Leaves 50–85 cm. Scapes to 30 cm. Flowers c. 8 cm in diameter. Sepals and petals pale greyish green, flushed or veined with darker green, sepals *c.* 5 cm, petals *c.* 4 cm with recurved apices. Lip 3.7–4 cm, dark green, lateral lobes lighter, central lobe oblong-ovate, recurved with a hairy border, apex notched. Callus dark green, spathulate with 2 ridges. Column white, hairy in front. *Peru.* G2. Spring–summer.

6. I. dyeriana (M.T. Masters) Ryan & Oakeley (*Lycaste dyeriana* M.T. Masters) Figure 28(19), p. 547. Illustration: *Botanical Magazine*, 8103 (1906); *Orchid Review* **89**: 327 (1981).

Plant hanging. Pseudobulbs 4–7.5 cm, angular, without apical spines. Leaves 18–30 cm. Scapes 7–13 cm, hanging. Flowers 4–5 cm in diameter, fragrant, not opening fully. Sepals pale green, the upper *c.* 5 cm. Petals darker green. Lip *c.* 3 cm, green, central lobes ovate-elliptic, reflexed, margin fringed. Callus green, grooved, 2-lobed on each side of the notched apex. Column *c.* 1.7 cm. *Peru, Ecuador.* G2. Summer–autumn.

In European Garden Flora, edn1, this epithet is attributed to Rolfe, but Masters published the name on 13 July 1895, while Rolfe published in August 1895.

107. ANGULOA Ruiz & Pavon

V.A. Matthews

Terrestrial or occasionally epiphytic. Pseudobulbs oblong, becoming ridged with age. Leaves usually 2–3 per pseudobulb, stalked, pleated, oblong, rolled when young. Scapes often several from the base of the pseudobulb, 15–30 cm, clothed in leafy bracts, bearing 1 flower. Flowers fragrant. Sepals and petals fleshy, concave and incurving, almost concealing the lip. Lip 3-lobed, hinged to the column foot so that it rocks, the lateral lobes erect. Callus linear-lanceolate with 2 apical lobes. Pollinia 4, on a linear stipe which has a slightly expanded viscidium.

A genus of 10 species from northern S America. Cultivation as for *Lycaste* (p. 546).

Literature: Oakeley, H.F., An annotated checklist of the genus Anguloa, *Orchid Digest Special Publication* (1999); Oakeley, H.F., Lycaste, Ida and Anguloa (2008).

Flower colour. Yellow: **3**; yellow flushed red or green: **1, 2, 4**; whitish or pink: **5, 6**.
Central lobe of lip. Surface hairless: **2, 4, 5, 6**; surface hairy: **1, 3**.

1a. Flowers whitish or pink 2
b. Flowers yellow, often flushed red or green 3
2a. Central lobe of lip wider than callus **5. uniflora**
b. Central lobe of lip narrower than callus **6. virginalis**
3a. Central lobe of lip with 2 tooth-like lobules, hairy 4
b. Central lobe of lip without lobules, hairless 5
4a. Flowers clear yellow **3. clowesii**
b. Flowers yellow flushed reddish and with red spots inside or occasionally entirely red inside **1. × ruckeri**
5a. Central lobe of lip oblong, fringed at the apex, not or scarcely recurved **2. brevilabris**
b. Central lobe of lip more or less 3-lobed, not fringed, recurved **4. cliftonii**

1. A. × ruckeri Lindley. Illustration: *Edwards's Botanical Register* **32**: t. 41 (1846); Kupper & Linsenmaier, Orchids, 81 (1961); *Orchid Digest Special Publication*, 5, 8 (1999).

Flowers dull yellowish flushed with red and with red spots or rarely entirely red inside. Lateral lobes of the lip more or less oblong, central lobe hairy with 2 tooth-like lobules. *Colombia-Venezuela.* G2. Spring–summer.

According to Oakeley (1999) *A.* × *ruckeri* is the hybrid *A. hohenlohii* × *clowesii*. The albino variant (var. **albiflora** Veitch) with pure white flowers may be in cultivation, and probably belongs under *A. eburnea*.

A. hohenlohii Morren. (*A. ruckeri* var. *sanguinea* Lindley) – illustration: *Botanical Magazine*, 5384, 1863; *Orchid Digest Special Publication*, 4–5, 8, 1999 – has flowers entirely or spotted orange-red inside.

2. A. brevilabris Rolfe. Illustration: *Botanical Magazine*, 9381 (1935); *Orchid Digest* **42**: 140 (1978); *Orchid Digest Special Publication*, 9 (1999).

Similar to *A. ruckeri* but with smaller flowers flushed with dull reddish green. The lateral lobes of the lip are very broad and the central lobe is oblong, hairless though fringed at the apex, and lacking lobules. *Colombia.* G2. Summer.

3. A. clowesii (*A. clowesii* var. *ruckeri* (Lindley) Foldats). Illustration: *Edwards's Botanical Register* **30**: t. 63 (1844); *Orchid Digest* **40**: 132 (1976); *Orchid Digest Special Publication*, 31 (1999).

Flowers yellow. Lip whitish to orange-yellow, lateral lobes more or less oblong, central lobe hairy, ovate, reflexed, with 2 tooth-like lobules. *Colombia, Venezuela.* G2. Spring–summer.

Very similar to *A. ruckeri*; in the opinion of some authors the 2 should be treated as varieties of the 1 species.

4. A. cliftonii Rolfe. Illustration: *Botanical Magazine*, 8700 (1917); *Orchid Digest* **40**: 132 (1976); Bechtel et al., Manual of cultivated orchid species, 167 (1981).

Flowers yellow splashed and marked with reddish purple. Lip pouched at the base, lateral lobes rounded, central lobe with a basal claw, more or less 3-lobed, hairless. *Colombia.* G2. Spring–summer.

5. A. uniflora Ruiz & Pavon. Illustration: *Edwards's Botanical Register* **30**: t. 60 (1844); *Orchid Digest* **40**: 132 (1976); Bechtel et al., Manual of cultivated orchid species, 168 (1981); *Orchid Digest Special Publication*, 13 (1999).

Flowers whitish, sometimes flushed with green, spotted and tinged inside with pink. Lip yellowish, spotted with pink, lateral lobes wide, rounded and inrolled, central lobe lanceolate, reflexed, wider than the callus. The inrolled lip tube slightly bulbous at the base. *Colombia, Venezuela, Peru*. G2. Spring–summer.

Several varieties, based on variation in flower colour, have been described.

6. A. virginalis Schlechter. Illustration: *Botanical Magazine*, 4807 (1854); *Orchid Digest* **40**: 132 (1976); Sheehan & Sheehan, Orchid genera illustrated, 37 (1979); *Orchid Digest Special Publication*, 16–17 (1999).

Very similar to *A. uniflora*, differing only in having the central lobe of the lip narrower than the callus and the lip kinked or bent at the base. *Colombia*. G2. Summer.

Often considered to be the same as *A. uniflora*. Pink-flowered variants have been found in the wild.

There are 2 other white or pale pink flowered Anguloas:

A. eburnea Williams. Illustration: *Orchid Digest Special Publication*, 16–17 (1999). Lip tube straight, like a test-tube, but wide, the lateral lobes short and reflexed at the tip to form a triangle. *Colombia, Ecuador, Peru*. G2. Summer.

A. tognettiae Oakeley. Illustration: *Orchid Digest Special Publication*, 20 (1999). Lip tube straight, but narrow, the lateral lobes long, square. *Venezuela, Colombia*. G2. Summer.

108. ERIOPSIS Lindley

V.A. Matthews

Epiphytic or terrestrial. Pseudobulbs elongate-cylindric or ovoid-oblong. Leaves 2–4 per pseudobulb, pleated, strongly veined, rolled when young. Raceme produced from the base of a pseudobulb, with many flowers. Sepals and petals more or less equal, sepals free or the laterals joined to the column foot to form a short mentum. Lip 3-lobed, lateral lobes erect, larger than the central lobe. Callus composed of 2 or more fleshy plates. Column curved, slightly swollen above. Pollinia 2 or 4, stipe very short, almost non-existent.

A genus of 6 species from southern C America and tropical S America. They should be grown in an epiphytic compost, with frequent watering when in growth. Light shading should be provided when young growth is present, but should be removed when the growth is mature; propagation by division.

1a. Callus of 2 erect fleshy plates with 2 separate lumps in front of each plate **1. biloba**
b. Callus of 2 erect, plate-like horns only **2. sceptrum**

1. E. biloba Lindley (*E. rutidobulbon* Hooker; *E. schomburgkii* (Reichenbach) Reichenbach; *E. wercklei* Schlechter). Illustration: *Botanical Magazine*, n.s., 611 (1972); Pabst & Dungs, Orchidaceae Brasilienses **2**: 214, 276 (1977).

Epiphytic or terrestrial. Pseudobulbs ranging from elongate-cylindric (to 45 × 3 cm) to ovoid-oblong (to 14 × 8 cm), surface smooth or finely wrinkled, greenish, brown or dark purplish brown, with 2–4 leaves at the apex. Leaves 20–50 cm, narrowly lanceolate to narrowly ovate. Scape 30–110 cm, erect to arching, raceme with 20–35 flowers. Flowers 2.5–5 cm in diameter. Sepals and petals 1.2–2.5 cm, yellow to orange-yellow or brownish yellow, reddish purple or brownish on margins or flushed with red to varying degrees. Lip 1.2–2.5 cm, lateral lobes yellowish with reddish purple spots, central lobe white with reddish purple spots, notched at the apex. Callus of 2 erect fleshy plates with 2 separate fleshy lumps in front of each. Column *c.* 1 cm, greenish yellow. *Tropical America from Costa Rica to Peru & Bolivia*. G2. Summer–autumn.

Extremely variable in flower colour, lip shape and colour of pseudobulbs.

2. E. sceptrum Reichenbach & Warscewicz (*E. sprucei* Reichenbach; *E. helenae* Kränzlin). Illustration: *Botanical Magazine*, 8462 (1912); Pabst & Dungs, Orchidaceae Brasilienses **2**: 214, 277 (1977).

Epiphytic. Similar to *E. biloba* but generally smaller with flowers *c.* 3.5 cm in diameter, sepals and petals 1.3–1.6 cm, and lip *c.* 1 cm. The callus consists of 2 erect plate-like horns, which are cream in colour. *Venezuela, Peru, Brazil*. G2. Summer.

109. SCUTICARIA Lindley

V.A. Matthews

Epiphytic. Pseudobulbs small, somewhat stem-like, borne on a stout, branching rhizome. Leaves narrowly cylindric with a longitudinal groove, continuous with the pseudobulbs, each of which bears 1 leaf. Raceme with 1–several flowers, arising laterally from a pseudobulb. Sepals and petals spreading, similar, the lateral sepals inserted on the column foot to form a mentum. Lip 3-lobed, joined to the column foot, lateral lobes erect, central lobe concave. Pollinia 2.

A genus of 5 species from tropical S America. They are best grown in epiphytic compost on rafts or blocks of tree fern suspended near the glass. They require ample moisture when growing, but should be kept dry when growth is completed, but not so much as to allow the foliage to shrivel.

1a. Leaves 60–150 cm; raceme with 1–3 flowers; central lobe of lip hairless **1. steelei**
b. Leaves 30–40 cm; raceme with 1 flower; central lobe of lip hairy **2. hadwenii**

1. S. steelei (Hooker) Lindley. Illustration: Pabst & Dungs, Orchidaceae Brasilienses **2**: 240, 296 (1977); *American Orchid Society Bulletin* **49**: 31 (1980).

Leaves hanging, 60–150 cm. Raceme usually hanging, with 1–3 flowers. Flowers 6–10 cm in diameter, fragrant. Sepals and petals yellow, blotched with reddish brown. Lip 3–4 cm, whitish to pale yellow marked with brownish pink especially on the lateral lobes, central lobe 2-lobed at apex, hairless. Callus yellowish orange with 3–5 apical teeth, hairy. Column whitish spotted with dark pink. *Brazil, Guyana, Colombia & Venezuela*. G2. Flowers throughout the year.

2. S. hadwenii (Lindley) Hooker. Illustration: *Botanical Magazine*, 4629 (1852); Pabst & Dungs, Orchidaceae Brasilienses **2**: 238 (1977).

Leaves hanging or sometimes more or less erect, 30–45 cm. Raceme with 1 flower, hanging or arching. Flowers 6–10 cm in diameter, fragrant. Sepals and petals yellowish green blotched with reddish brown. Lip pale yellow irregularly spotted with pinkish red, central lobe notched at apex, hairy. Callus with 3 apical teeth. Column whitish yellow flushed and spotted with red. *Brazil, Guyana*. G2. Spring–autumn.

S. strictifolia Hoehne. Illustration: *Orchid Digest* **41**: 235 (1977). Very similar to *S. hadwenii*, differing only in its erect foliage. The two may be the same, as plants of *S. hadwenii* growing in bright light may have the ability to produce erect leaves. *Brazil.* G2.

110. MAXILLARIA Ruiz & Pavon

V.A. Mathews

Epiphytic or terrestrial, with long or short rhizomes clothed in often papery sheaths and generally bearing pseudobulbs that are usually laterally compressed and bear 1–2 leaves. Leaves thin to leathery, with a prominent central vein, folded when young. Scapes produced in the axils of sheaths on the rhizome, solitary or in bundles, each bearing clasping, usually overlapping, bracts and 1 flower. Sepals free, the laterals usually oblique and joined to the column foot to form a mentum. Petals similar but smaller, usually joined to the upper sepal. Lip hinged to or continuous with the column foot, often erect and parallel to the column, entire or 3-lobed, usually shorter than the sepals. Column erect, somewhat curved, without wings. Pollinia 4, in 2 pairs, often 1 large and 1 small in each pair.

A genus of about 250 species from tropical and subtropical America. Recent molecular studies have resulted in *Maxillaria* being split into a number of smaller genera. As the taxonomy is still being published at the time of writing, the genus *Maxillaria* is retained in the broad sense, but synonyms and notes on the new genera have been added in the account below.

They should be grown in epiphytic compost; those with long rhizomes are best accommodated on rafts or blocks of tree fern. All should be kept moist throughout the year and should be shaded from bright sunlight. Maxillarias are prone to black spot on the foliage; this can be controlled by adequate ventilation and heating.

Literature: Christenson, E.A., Vue d'ensemble du genre Maxillaria, *Richardiana* **2**: 41–65 (2002); Blanco, M.A., Carnevali, G., Whitten, W.M., Singer, R.B., Koehler, S., Williams, N.H., Ojeda, I., Neubig, K.M. & Endara, L., Generic realignments in Maxillariinae, *Lankesteriana* **7**(3): 515–37 (2007).

Pseudobulbs. Absent: **26**. Bearing 1 leaf: **1, 2, 3, 4, 5, 6, 7, 8, 9, 10, 11, 13, 14, 15, 16, 17, 20, 21, 22, 23, 24, 25**; bearing 2 leaves: **2, 3, 16, 18, 19, 20, 21, 22**. Spaced at intervals along the rhizome: **2, 6, 12, 13, 14, 15, 16, 17, 18, 20, 21, 22, 25**; clustered: **1, 2, 3, 4, 5, 6, 7, 8, 9, 10, 11, 12, 13, 14, 15, 16, 19, 20, 21, 22, 23, 24**.
Leaves. Linear: **14, 15, 17, 18, 23**; relatively broader: **1, 2, 3, 4, 5, 6, 7, 8, 9, 10, 11, 12, 13, 16, 19, 20, 21, 22, 24, 25, 26**.
Sepals. Linear: **1, 2, 3, 4, 5, 6**; relatively broader: **7, 8, 9, 10, 11, 12, 13, 14, 15, 16, 17, 18, 19, 20, 21, 22, 23, 24, 25, 26**. Length to 3 cm: **13, 14, 15, 16, 17, 18, 19, 20, 21, 22, 23, 24, 25, 26**; length 3–8 cm: **1, 3, 4, 5, 6, 7, 8, 9, 12, 16, 18, 19, 21, 22**; length 10–15 cm: **2**.
Sepals and petals. Striped: **9**.
Central lobe of lip. Basic colour white, cream, yellow or orange: **2, 3, 4, 5, 6, 7, 8, 9, 10, 11, 12, 13, 14, 15, 16, 17, 18, 20, 21, 23, 25, 26**; basic colour red to maroon: **1, 12, 14, 16, 17, 21, 22, 23**. Spotted: **5, 7, 10, 12, 14, 17, 21, 22, 26**; not spotted: **1, 2, 3, 4, 6, 7, 8, 9, 11, 13, 14, 15, 16, 17, 18, 19, 20, 21, 22, 25, 26**. Entire: **1, 8, 11, 12, 14, 16, 17**; 3-lobed: **2, 3, 4, 5, 6, 7, 8, 9, 10, 13, 15,17, 18, 19, 20, 21, 22, 23, 24, 25, 26**.

1a. Sepals linear to narrowly strap-shaped, 3–15 cm — 2
b. Sepals ovate to oblong or broadly lanceolate, if narrower, then less than 3 cm — 8
2a. Sepals maroon, yellow at the base, apex maroon-orange — **1. nigrescens**
b. Sepals white, yellow or greenish, sometimes tinged with purple or brown, or red at the base — 3
3a. Sepals 10–15 cm, twisted — **2. fractiflexa**
b. Sepals 3–7.5 cm, rarely twisted — 4
4a. Sepals white on inner surface — **7. venusta**
b. Sepals pale yellow or greenish on inner surface, sometimes tinged with brown, or red at the base — 5
5a. Central lobe of lip white, recurved — **3. setigera**
b. Central lobe of lip deep cream to deep yellow, sometimes with purplish spots, not or scarcely recurved — 6
6a. Central lobe of lip obtuse at the apex, margin wavy, white — **4. luteoalba**
b. Central lobe of lip acute or mucronate, margin neither wavy nor white — 7
7a. Lip with some purplish spots; pseudobulbs 2.5–5 cm — **5. lepidota**
b. Lip lacking spots; pseudobulbs 5–8 cm — **6. ochroleuca**
8a. Sepals entirely white or cream — 9
b. Sepals not white or cream, or if so then marked with another colour — 12
9a. Flowers *c.* 2 cm in diameter; sepals 1.8–2.5 cm — 10
b. Flowers 9–15 cm in diameter; sepals 3.5–7 cm — 11
10a. Pseudobulbs borne at intervals along the rhizome; lip *c.* 1.2 cm with a downy central lobe — **13. alba**
b. Pseudobulbs clustered; lip 3.5–5 mm, central lobe hairless — **23. densa**
11a. Sepals lanceolate — **7. venusta**
b. Sepals ovate-oblong to triangular-ovate — **8. grandiflora**
12a. Leaves linear, grass-like — 13
b. Leaves lanceolate or broader — 17
13a. Pseudobulbs with 2 leaves — **18. chrysantha**
b. Pseudobulbs with 1 leaf — 14
14a. Lip marked with a number of reddish brown or purplish spots — **14. tenuifolia**
b. Lip unspotted — 15
15a. Pseudobulbs clustered; lip less than 6 mm — **23. densa**
b. Pseudobulbs borne at intervals along rhizome; lip to 1.3 cm — 16
16a. Apex of central lobe of lip yellowish — **17. variabilis**
b. Apex of central lobe of lip white — **15. sanguinea**
17a. Lateral sepals directed upwards, towards upper sepal — 18
b. Lateral sepals directed downwards, away from upper sepal, or spreading horizontally — 23
18a. Plant hanging, with leaves growing in a fan; pseudobulbs absent; sepals and petals pale greenish yellow — **26. valenzuelana**
b. Plant erect or creeping; pseudobulbs present; sepals and petals not greenish yellow, or if so, then striped — 19
19a. Sepals and petals bright pink to orange-red or red — 20
b. Sepals and petals not as above — 21
20a. Pseudobulbs 1–2 cm, borne at intervals along rhizome — **25. sophronitis**
b. Pseudobulbs 2.5–4 cm, clustered — **24. coccinea**
21a. Sepals and petals white, with purple blotches at the base **11. sanderiana**

b. Sepals and petals white, greenish yellow or purple, lacking purple blotches at the base 22
22a. Sepals with longitudinal stripes **9. striata**
b. Sepals with spots towards apex **10. fucata**
23a. Lip entire 24
b. Lip 3-lobed 26
24a. Scapes to 4 cm **17. variabilis**
b. Scapes 5–25 cm 25
25a. Scapes 4.5–9 cm **16. elatior**
b. Scapes to 25 cm **12. elegantula**
26a. Pseudobulbs with 2 leaves 27
b. Pseudobulbs with 1 leaf 30
27a. Central lobe of lip acute 28
b. Central lobe of lip obtuse or rounded 29
28a. Sepals and petals spotted **21. picta**
b. Sepals and petals not spotted **20. marginata**
29a. Flowers 2–3 cm in diameter **19. porphyrostele**
b. Flowers 4–4.5 cm in diameter **18. chrysantha**
30a. Callus hairless 32
b. Callus downy 31
31a. Side-lobes of lip upcurved and clasping column **22. cucullata**
b. Side-lobes of column erect **21. picta**
32a. Central lobe of lip bright pink, red, yellow or orange-yellow 33
b. Central lobe of lip pale yellow with a dark purple margin **20. marginata**
33a. Pseudobulbs borne at intervals along rhizome; scapes 1–2.5 cm 34
b. Pseudobulbs clustered; scapes 4–6 cm **24. coccinea**
34a. Pseudobulbs 1–2 cm; leaves 1–8 cm; central lobe of lip warty **25. sophronitis**
b. Pseudobulbs 1.5–6 cm; leaves 5–25 cm; central lobe of lip smooth **17. variabilis**

Maxilliaria in the strict sense (species 1–12)

Pseudobulbs usually present, usually 1-leaved, smooth, compressed, the stump of the old fallen leaf is usually projected above the pseudobulb as a persistent stalk. Inflorescences borne from the base of the newest pseudobulb, or from leaf axils at the top of the stem in the *M. exaltata* alliance. Floral bract longer or shorter than pedicel and ovary. Sepals and petals with abundant fibres. Column with obvious foot. Beaded hairs looking like pollen (pseudopollen) are produced on the lip surface, and is not present in any of the other related genera.

1. M. nigrescens Lindley. Illustration: Cogniaux & Goossens, Dictionnaire iconographique des orchidées, Maxillaria, t. 3 (1899); Dunsterville & Garay, Venezuelan orchids illustrated **3**: 189 (1965); *American Orchid Society Bulletin* **43**: 963 (1974).

Epiphytic. Pseudobulbs to 9 cm, clustered, ovoid, compressed, each with 1 leaf. Leaves strap-shaped, 20–35 cm. Scapes usually erect, to 12 cm. Flowers *c.* 10 cm in diameter. Sepals narrowly strap-shaped, 4.5–6 cm, maroon, yellow at the base and maroon-orange at the apex. Petals similar, but only 3.8–5.5 cm and usually darker than sepals. Lip entire, oblong, obtuse, 1.7–2 cm, deep maroon, flushed with brownish purple. *Colombia, Venezuela*. G2. Summer–autumn.

2. M. fractiflexa Reichenbach. Illustration: *Gardeners' Chronicle* **31**: 359 (1902).

Leaves strap-shaped, to 50 cm. Scapes erect. Flowers *c.* 15 cm in diameter. Sepals and petals linear, 10–15 cm, curved and twisted, yellowish tinged with brown and often purplish at the base and apex. Lip 3-lobed, white, part of the rounded lateral lobes and the central lobe red. Callus white. *Ecuador*. G2.

3. M. setigera Lindley. Illustration: *Botanical Magazine*, 4434 (1849); Dunsterville & Garay, Venezuelan orchids illustrated **3**: 192 (1965); *American Orchid Society Bulletin* **43**: 963 (1974).

Epiphytic. Pseudobulbs 3–6 cm, clustered, spherical, somewhat compressed, each with 1–2 leaves. Leaves narrow, 12–30 cm. Scapes 7–30 cm. Flowers fragrant, 10 cm or more across. Sepals narrowly strap-shaped, to 6.5 cm, white at base, greenish yellow towards apex. Petals similar, incurved, to 5 cm, with recurved margins. Lip 3-lobed, to 2.5 cm, lateral lobes white striped with purple, central lobe oblong, reflexed, white, margin toothed. Callus yellow, hairy. *Colombia, Venezuela, Guyana & Brazil*. G2. Autumn.

4. M. luteoalba Lindley (*M. luteo-alba* invalid). Illustration: Cogniaux & Goossens, Dictionnaire iconographique des orchidées, Maxillaria, t. 1 (1899); *American Orchid Society Bulletin* **43**: 962 (1974); Bechtel et al., Manual of cultivated orchid species, 229 (1981).

Epiphytic. Pseudobulbs 3–5 cm, clustered, oblong-ovoid, compressed, each with 1 leaf. Leaves linear-lanceolate, 25–45 cm. Scapes erect, 9–15 cm. Flowers very fragrant, 10 cm or more across. Sepals narrowly strap-shaped, 4–7.5 cm, whitish, or yellow at the base, becoming brownish orange towards the apex. Petals shorter, similar in colour although sometimes with purplish streaks. Lip 3-lobed, to 2.5 cm, downy inside, cream or yellow, lateral lobes purple-streaked, central lobe ovate, obtuse with a white wavy margin. Callus with 2 lateral ridges, downy towards base. *Panama, Colombia, Venezuela & Ecuador*. G2. Summer.

5. M. lepidota Lindley. Illustration: Dunsterville & Garay, Venezuelan orchids illustrated **6**: 253 (1976); Bechtel et al., Manual of cultivated orchid species, 229 (1981).

Epiphytic. Pseudobulbs 2.5–5 cm, clustered, oblong, each with 1 leaf. Leaves linear-lanceolate, 15–35 cm. Scapes to 15 cm, often with the flowers hanging. Sepals linear, 4–6 cm, yellow with red at the base and often brown at the tip. Petals similar to sepals but shorter. Lip 3-lobed, *c.* 2 cm, deep cream, spotted with dark purple or maroon towards the apex, reflexed, mealy on the upper surface. Callus grooved, hairy. *Colombia, Ecuador & Venezuela*. G2.

6. M. ochroleuca Lindley. Illustration; Pabst & Dungs, Orchidaceae Brasilienses **2**: 235, 291 (1977).

Epiphytic. Pseudobulbs 5–8 cm, compressed, each with 1 leaf. Leaves 25–40 cm. Scapes erect, 5–10 cm, many borne together in a bundle. Flowers fragrant. Sepals 3–3.5 cm, linear, slenderly pointed, pale yellow. Petals similar to sepals but shorter. Lip 3-lobed, pale yellow, central lobe oblong, obtuse, downy, deep yellow at apex. *Brazil*. G2.

7. M. venusta Linden & Reichenbach. Illustration: *Botanical Magazine*, 5296 (1862); Cogniaux & Goossens, Dictionnaire iconographique des orchidées, Maxillaria, t. 6 (1902).

Epiphytic. Pseudobulbs 4–7.5 cm, clustered, ovoid, much compressed, each with 1 leaf. Leaves strap-shaped, 30–38 cm. Scapes 10–25 cm. Flowers very fragrant, 12–15 cm in diameter, hanging. Sepals lanceolate, to 6.5 cm, white, usually with recurved margins. Petals similar, to 4.5 cm. Lip 3-lobed, to 2 cm, lateral lobes oblong, pale yellow with red margins, central lobe ovate, obtuse, reflexed, pale yellow with red margin and sometimes with red spots, mealy. Callus oblong, downy. *Colombia, Venezuela.* G2. Summer–autumn.

8. M. grandiflora (Humboldt, Bonpland & Kunth) Lindley. Illustration: Dunsterville & Garay, Venezuelan orchids illustrated **5**: 185 (1966); *Die Orchidee* **23**: 92 (1972).

Epiphytic. Pseudobulbs 3.5–7 cm, clustered, ovoid, compressed, each with 1 leaf. Leaves strap-shaped, 25–50 cm. Scapes erect but bent at apex, 10–25 cm. Flowers slightly scented, *c.* 10 cm across, drooping. Sepals ovate-oblong to triangular-ovate, 3.5–5.5 cm, white, sometimes with margins rolled under; lateral sepals with reflexed tips. Petals shorter and narrower than sepals, white, sometimes with pink streaks on the basal half, tips reflexed. Lip entire or very slightly 3-lobed, 2.5–3 cm, lateral lobes (if present) deep pink or purple, streaked paler, central lobe white or yellowish sometimes with a purple margin, wavy, reflexed, mealy. Callus mealy. *Northwestern S America.* G2. Spring–summer.

9. M. striata Rolfe. Illustration: Cogniaux & Goossens, Dictionnaire iconographique des orchidées, Maxillaria, t. 4 (1899); *Lindenia* **9**: t. 398 (1883–84).

Pseudobulbs 4.5–8 cm, clustered, ovoid, compressed, each with 1 leaf. Leaves broadly elliptic, to 25 cm. Scapes more or less erect or arching, to 30 cm. Flowers *c.* 12 cm in diameter. Sepals greenish yellow with reddish purple stripes, 4.5–7 cm, upper sepal oblong-lanceolate, lateral sepals triangular-lanceolate, forming a mentum with the column foot. Petals similar in colour, shorter, lanceolate, often with wavy margins. Lip erect, recurved, slightly 3-lobed towards apex, lateral lobes more or less obovate, white striped with reddish purple, central lobe ovate-lanceolate, white. Callus linear-oblong with many grooves. *Peru, Ecuador.* G2. Summer–autumn.

10. M. fucata Reichenbach. Illustration: *Botanical Magazine*, 9376 (1934); Shuttleworth et al., Orchids, 109 (1970).

Pseudobulbs 3.5–6 cm, clustered, narrowly ovoid, compressed, each with 1 leaf. Leaves narrowly oblong, 30–40 cm. Scapes erect, usually numerous, to 22 cm. Flowers *c.* 6 cm across, fragrant. Sepals 3–3.5 cm, ovate, reflexed at tips, white at the base, purplish in the middle and yellow at the apex, with reddish brown spots towards apex. Petals *c.* 2.5 cm, lanceolate-ovate, similar in colour to sepals but sometimes without spots. Lip 3-lobed, reflexed, lateral lobes yellowish, striped and edged with brown, central lobe ovate-oblong, notched at apex, yellowish, striped, spotted or suffused with brownish red. Callus yellow, strap-shaped, grooved and hairless. *?Ecuador.* G2. Late summer–autumn.

The name is often given erroneously as *M. fuscata* or *M. furcata*.

11. M. sanderiana Reichenbach. Illustration: Kupper & Linsenmaier, Orchids, t. 42 (1961); Bechtel et al., Manual of cultivated orchid species, 230 (1981).

Terrestrial or epiphytic. Pseudobulbs 2.5–5 cm, clustered, ovoid, compressed, each with 1 leaf. Leaves broadly oblong, 15–40 cm. Scapes ascending or horizontal, to 25 cm. Flowers 10–15 cm across, fragrant. Sepals white, blotched at base with reddish purple, upper sepal oblong, to 7.5 cm, lateral sepals triangular-ovate, slightly longer than upper sepal, forming a mentum with the column foot. Petals shorter than upper sepal, lanceolate-triangular, reflexed at tips, similar in colour to sepals. Lip entire, erect, recurved, elliptic with a crisped margin, white or yellow, streaked and stained with red or purple. Callus tongue-shaped, yellow, hairless. *Ecuador, Peru.* G2. Summer–early autumn.

12. M. elegantula Rolfe.

Pseudobulbs each with 1 leaf and scapes to 25 cm. The sepals and petals are white, yellow toward the tips, flushed and spotted with brownish purple or deep blue, margins revolute. Lip to 1.7 cm, elliptic, apex recurved and minutely toothed and crisped, usually purplish red on margin. *Ecuador, Peru.* G2.

May key out under *M. elatior.*

Maxillariella Blanco & Carnevali (species 13–17).

Pseudobulbs ovoid, separated by rhizome, 1- or 2-leaved. A few species have reduced pseudobulbs and monopodial growth. Flowers produced singly in axils of leaves or bracts. Floral bract shorter than stalk and ovary. Column foot very short. Lip simple or hardly 3-lobed, with a glossy callus.

Literature: Atwood, J.T., A review of the Maxillaria graminifolia suballiance, *Selbyana* **24**: 144–64 (2003).

13. M. alba (Hooker) Lindley (*Maxillariella alba* (Hooker) Blanco & Carnivali). Illustration: Hoehne, Flora Brasilica **12**(7): t. 91 (1953); Dunsterville & Garay, Venezuelan orchids illustrated **2**: 197 (1961).

Epiphytic. Pseudobulbs 2.5–5 cm, borne at short intervals along the rhizome, ovoid to narrowly ovoid, strongly compressed, each with 1 leaf. Leaves strap-shaped, 25–40 cm. Scapes 2.5–4 cm. Flowers slightly fragrant, *c.* 2 cm across. Sepals broadly lanceolate, 1.8–2.5 cm, creamy white. Petals similar, 1.6–1.8 cm. Lip 3-lobed towards the apex, hinged to the column foot, *c.* 1.2 cm, pale to bright yellow with an ovate or elliptic central lobe, which is downy. Callus strap-shaped, downy. *Most of C America & northern S America.* G2. Spring.

14. M. tenuifolia Lindley (*Maxillariella tenuifolia* (Lindley) Blanco & Carnivali). Illustration: Shuttleworth et al., Orchids, 110 (1970); *Die Orchidee* **23**: 92 (1972); Bechtel et al., Manual of cultivated orchid species, 230 (1981).

Terrestrial or epiphytic. Pseudobulbs 2–6 cm, borne at intervals of 2–4 cm along the rhizome, ovoid, somewhat compressed, each with 1 leaf. Leaves linear, 12–50 cm. Scapes 2–6 cm. Flowers coconut-scented, 4–5 cm in diameter. Sepals narrowly elliptic, 1.7–2.8 cm, dark red, heavily mottled with red or yellow. Petals similar, 1.5–2.4 cm, directed forwards. Lip 1.5–2.5 cm, entire, oblong, reflexed, dark red at base, yellow or whitish towards apex, spotted with purple or reddish brown. Callus downy, dark purple. *Mexico to Costa Rica.* G2. Summer–autumn.

15. M. sanguinea Rolfe (*Maxillariella sanguinea* (Rolfe) Blanco & Carnivali). Illustration: Hartmann, Introduction to the cultivation of orchids, 89 (1971).

Pseudobulbs to 3 cm, borne at intervals along the rhizome, ovoid, compressed, each with 1 leaf. Leaves linear, to 40 cm. Scapes to 3 cm. Sepals elliptic, to 2.5 cm, dark red or brownish, mottled with red or yellow. Petals similar, to 2 cm. Lip scarcely 3-lobed, to 1.3 cm, elliptic, dark red, pale yellow at base, white at apex. Callus dark red. *Costa Rica; ?Colombia.* G2. Summer.

16. M. elatior Reichenbach (*Maxillariella elatior* (Reichenbach) Blanco & Carnivali).

Illustration: *Botanical Magazine*, 9206 (1930).

Epiphytic. Pseudobulbs 4.5–9 cm, ovoid, compressed, scattered along the rhizome or often solitary at the end of the rhizome, each with 1–2 leaves. Leaves 12–40 cm, lanceolate. Scapes erect, 5–7.5 cm. Sepals 2.2–3.2 cm, oblong-elliptic, reddish yellow, often spotted or striped with red, upper sepal concave, lateral sepals spreading or recurved. Petals 1.7–2.5 cm, dull red. Lip entire, to 2 cm, ovate-oblong, recurved at apex, dark red. Callus *c.* 1 cm, blackish, hairless. *Mexico, Guatemala, Honduras & Costa Rica*. G2. Spring–summer.

17. M. variabilis Lindley (*Maxillariella variabilis* (Lindley) Blanco & Carnivali). Illustration: *Die Orchidee* **23**: 92 (1972).

Terrestrial or epiphytic. Pseudobulbs 1.5–6 cm, narrowly ellipsoid to ovoid, borne at 2–3 cm intervals along the rhizome, each with 1 leaf. Leaves 5–25 cm, very narrowly strap-shaped. Scapes 2–2.5 cm. Flowers 2–2.5 cm in diameter. Sepals 6–20 mm, ovate, white, greenish yellow, orange, deep red or purplish, often marked with red. Petals 5–14 mm, oblong-lanceolate, similar in colour to sepals. Lip entire or slightly 3-lobed near apex, up to 1.2 cm, hinged to column foot, erect, usually deep red and often with yellow blotch at apex. Callus strap-shaped, deep red, hairless. *C America, West Indies, Guyana*. G2.

M. curtipes Hooker (*Maxillariella curtipes* (Hooker) Blanco & Carnivali). Illustration: Cogniaux & Goossens, Dictionnaire iconographique des orchidées, Maxillaria, t. 2 (1899). Scapes 2–4 cm. Sepals and petals yellowish, overlaid with deep red inside and somewhat blotched towards the base, margins orange. Lip entire, oblong to broadly elliptic, reflexed, deep yellow, spotted and striped with reddish purple. Callus warty. *Mexico, Guatamela, El Salvador & Costa Rica*. G2. Spring.

M. houtteana Reichenbach (*Maxillariella houtteana* (Reichenbach) Blanco & Carnivali) was treated as a synonym of *M. curtipes* in European Garden Flora, edn 1, but is now accepted as distinct. Leaves shallowly 2-lobed at tip, not obviously narrowed into a petiole at the base. Peduncle exceeding the floral bract. Sepals and petals dull red. Sepals 2–2.5 cm long. Lip 2 cm long entire or shallowly 3-lobed. *Mexico to Venezuela*. G2. Summer.

Brasiliorchis Singer, Koehler & Carnevali (species 18–21).

Pseudobulbs with 2 leaves, with some sheaths below. Inflorescences several, simultaneously produced from the base of the most recent pseudobulb. Floral bract shorter than stalk and ovary. Sepals without fibres, and with dark spots outside. Lip 3-lobed.

Literature: Singer, R.B., Koehler, S. & Carnevali, G., Brasiliorchis: a new genus for the Maxillaria picta alliance, *Novon* **17**: 91–9 (2007).

18. M. chrysantha Rodriguez (*Brasiliorchis chrysantha* (Rodriguez) Singer, Koehler & Carnevali). Illustration: *Botanical Magazine*, 8979 (1923); Hoehne, Flora Brasilica **12**(7): t. 127 (1953).

Epiphytic. Pseudobulbs 1.5–2.5 cm, ovoid-conical, slightly compressed, produced at intervals of 2.5–4 cm along the rhizome, each with 2 leaves. Leaves 20–35 cm, linear. Scapes 7–10 cm. Flowers 4–4.5 cm in diameter. Sepals 2.8–3 cm, oblong, yellow with reddish or mauve margins. Petals 2.2–2.4 cm, similar in colour to sepals. Lip 3-lobed, recurved at apex, more or less circular, yellowish with violet blotches on the lateral lobes, margin of central lobe wavy. Callus strap-shaped, downy. *Surinam, Brazil*. G2.

19. M. porphyrostele Reichenbach (*Brasiliorchis porphyrostele* (Reichenbach) Singer, Koehler & Carnevali). Illustration: *Botanical Magazine*, 6477 (1880); Cogniaux & Goossens, Dictionnaire iconographique des orchidées, Maxillaria, t. 10 (1904); Pabst & Dungs, Orchidaceae Brasilienses **2**: 234, 290 (1977).

Pseudobulbs 2.5–3 cm, clustered, ovoid, compressed, each with 2 leaves. Leaves 12–20 cm, lanceolate. Scapes to 7.5 cm. Flowers *c.* 3 cm across, slightly fragrant. Sepals oblong, incurving, pale yellow. Petals shorter, incurving, pale yellow with purple stripes at the base. Lip 3-lobed, lateral lobes yellow striped with purple, central lobe oblong-circular, yellow. Callus yellow marked with purple. *Brazil*. G2. Winter–spring.

20. M. marginata (Lindley) Fenzl (*Brasiliorchis marginata* (Lindley) Singer, Koehler & Carnevali). Illustration: Hoehne, Flora Brasilica **12**(7): t. 141 (1953); Pabst & Dungs, Orchidaceae Brasilienses **2**: 290 (1977).

Pseudobulbs 4–6 cm, each with 1–2 leaves. Leaves 12–30 cm, lanceolate. Scapes 7.5–15 cm, erect or arching. Sepals 2–2.5 cm, narrowly oblong, yellow with a dark red margin and a dorsal red line. Petals 1.7–2.2 cm, similar in colour to sepals. Lip 3-lobed, lateral lobes pale yellow streaked with reddish purple, central lobe oblong, reflexed, acute, pale yellow with a scalloped purple margin. Callus strap-shaped, hairless. *Colombia, Ecuador, Brazil*. G2. Summer–autumn.

21. M. picta Hooker (*Brasiliorchis picta* (Hooker) Singer, Koehler & Carnevali). Illustration: *Botanical Magazine*, 3154 (1832); Bechtel et al., Manual of cultivated orchid species, 230 (1981).

Epiphytic. Pseudobulbs 4–7 cm, ovoid, somewhat compressed, clustered or borne at intervals along the rhizome, each with 1–2 leaves. Leaves to 30 cm, narrowly oblong. Scapes 12–16 cm. Flowers fragrant. Sepals *c.* 3 cm, oblong, incurved, whitish to yellow with purple or dark brown spots usually on the dorsal surface. Petals 2.5–2.8 cm, similar to sepals but with a red basal streak, very incurved. Lip 3-lobed, lateral lobes erect, white with purple marks, central lobe oblong, reflexed, acute, whitish with purple spots. Callus oblong, downy. *Colombia, Brazil*. G2. Spring–summer.

Camaridium Lindley (species 22–23).

Mostly with pseudobulbs separated by variable lengths of rhizome, some species with a juvenile sympodial phase, becoming monopodial at maturity. Floral bract longer than stalk and ovary, and overlapping base of upper sepal. Sepals and petals without fibres, sparkling surface texture.

Literature: Atwood, J.T., A revision of the Maxillaria neglecta complex in Mesoamerica, *Lindleyana* **8**: 25–31 (1993).

22. M. cucullata Lindley (*Camaridium cucullatum* (Lindley) Blanco). Illustration: Bechtel et al., Manual of cultivated orchid species, 229 (1981).

Sepals 2.5–5 cm, usually deep red, occasionally yellow, brownish, pinkish or almost black, variously striped and spotted with dark red. Petals shorter, similar in colour, with the tips meeting above the column. Lip 3-lobed, 1.5–3 cm long, hinged to column foot, yellow or whitish with reddish brown or purple spots, or occasionally wholly dark red, central lobe finely warty, side-lobes upcurved and clasping column. *Mexico to Panama*. G2. Summer–autumn.

Very variable in the size of the plant and the flowers and in the width of the sepals and petals.

M. meleagris Lindley (*Camaridium meleagris* (Lindley) Blanco; *Maxillaria lindeniana* Richard & Galeotti). Sepals and petals dirty yellow to salmon coloured, usually spotted with dark red. Sepals 1–3 cm. Lip dark red, 3-lobed, to 1.5 cm. *Mexico to Venezuela*. G2. Summer.

M. praestans Reichenbach (*Camaridium praestans* (Reichenbach) Blanco; *M. seymouriana* Atwood). Sepals, petals and lip yellow with red or purple markings. Sepals 3-5 cm. Lip about 3 cm, central lobe yellow. *Central America*. G2. Summer.

23. M. densa Lindley (*Camaridium densum* (Lindley) Blanco; *Ornithidium densum* (Lindley) Reichenbach). Illustration: Hamer, Orquideas de El Salvador **2**: 111 (1974); Bechtel et al., Manual of cultivated orchid species, 229 (1981).

Differs from *M. coccinea* in its very variable flower colour, which ranges from greenish white to yellow, purplish, maroon or brownish red. The pseudobulbs are often larger, the leaves linear and the flowers slightly smaller with a lip 3.5–5 mm, whose central lobe is more or less circular to broadly ovate, obtuse at the apex and with a small point. Callus plate-like, obtuse. *C America from Mexico to Honduras*. G2.

Ornithidium R. Brown (species 24–25).

Growth pattern either sympodial or monopodial. Some species are dimorphic, sympodial when juvenile, becoming monopodial with maturity. Pseudobulbs, when present, are ovoid with a shiny, minutely cracked surface, such as that of an old oil painting. Stems and leaves are olive green, which colour persists upon drying. Flowers are small, campanulate to globose, yellow, orange or red. Sepals and petals lack fibres.

24. M. coccinea (Jacquin) Hodge (*Ornithidium coccineum* (Jacquin) R. Brown). Illustration: *Loddiges' Botanical Cabinet* **4**: t. 301 (1819); Hooker, Exotic Flora **1**: t. 38 (1823).

Epiphytic. Pseudobulbs 2.5–4 cm, ovoid, more or less compressed, clustered, each with 1 leaf. Leaves 10–35 cm, lanceolate. Scapes 4–6 cm, borne in bundles. Flowers bright pink to red. Sepals 1.1–1.2 cm, concave, ovate to narrowly ovate. Petals 7–8 mm, similar. Lip 3-lobed, 6–8 mm, the central lobe curved downwards, narrowly ovate, apex acute. Callus fleshy, transverse. *West Indies, Colombia, Venezuela*. G2. Summer.

25. M. sophronitis (Reichenbach) Garay (*Ornithidium sophronitis* Reichenbach). Illustration: Dunsterville & Garay, Venezuelan orchids illustrated **1**: 237 (1959).

Epiphytic. Pseudobulbs 1–2 cm, borne at 2.5–6 cm intervals along the creeping rhizome, slightly compressed, each with 1 leaf. Leaves 1–8 cm, oblong, thick, apex often asymmetric. Flowers borne on scapes *c.* 1 cm. Sepals and petals ovate, bright orange-red; sepals 9–13 mm, petals smaller. Lip 3-lobed, *c.* 7 mm, yellow to orange-yellow, central lobe oblong, the upper surface warty and margin minutely toothed. Callus smooth, flat. *Venezuela*. G2.

Heterotaxis Lindley (species 26).

Flowers yellow, fleshy, campanulate. Sepals and petals with fibres. Column with a very short foot.

Literature: Ojeda, I., Carnevali, G. & Romero, G.A., New species and combinations in Heterotaxis Lindley, *Novon* **15**: 572–82 (2005).

26. M. valenzuelana (Richard) Nash (*Heterotaxis valenzuelana* (Richard) Ojeda & Carnivali). Illustration: Shuttleworth et al., Orchids, 111 (1970); Bechtel et al., Manual of cultivated orchid species, 231 (1981).

Hanging epiphyte without pseudobulbs. Leaves growing in a fan, 8–20 cm, lanceolate. Scapes to 5 cm, solitary or in bundles in the leaf axils. Sepals ovate, 1–1.4 cm, pale greenish yellow. Petals similar, 8–10 mm. Lip scarcely 3-lobed, *c.* 1 cm, pale brownish yellow with purple spots. Callus in 3 parts, glandular-downy. *C America, northern S America, West Indies*. G2. Summer.

111. MORMOLYCA Fenzl

V.A. Matthews

Epiphytic with creeping rhizomes. Pseudobulbs with a minutely warty surface, bearing a single leaf (up to 3 in *M. polyphylla*), subtended by several bracts. Leaves folded when young. Flowers solitary, arising from an axil of a pseudobulb bract. Sepals and petals without fibres; opening widely. Sepals spreading, more or less equal, free. Petals similar but smaller. Lip simple or 3-lobed. Column without a foot. Pollinia 4, in 2 pairs, waxy.

A genus of over 30 species native to C and tropical S America. Recently expanded to include *Maxillaria* Section Rufescens Christenson, and *Cyrtoglottis* Schlechter. There are many undescribed species from South America. Its name is often misspelled 'Mormolyce'. Cultivation as for *Maxillaria* (p. 552).

Literature: Garay, L.A. & Wirth, M., On the genera Mormolyca Fenzl and Cyrtoglottis Schlechter, *Canadian Journal of Botany* **37**: 479–90 (1959); Sweet, H.R., Orquideas Andinas poco conocidas V. Chrysocycnis Linden & Rchb.f., *Orquideologia* **6**: 3–10 (1971); Carnevali, G., Munoz, J.L.T. & Gomez-Juarez, M., A synopsis of the Maxillaria rufescens complex in Mexico, Central America and the Greater Antilles, *Brittonia* **53**: 454–65 (2001).

1. M. ringens (Lindley) Schlechter. Illustration: *Fieldiana* **26**: 591 (1952); *Canadian Journal of Botany* **37**: 487 (1959); Bechtel et al., Manual of cultivated orchid species, 234 (1981).

Pseudobulbs 2–5 cm, clustered or arising at intervals of 1–2 cm along the rhizome, rounded to ovoid, compressed, each bearing 1 leaf. Leaves 10–35 cm. Sepals 1.5–2 cm, yellowish green to light lavender, striped longitudinally with violet or dark red. Petals similar in colour. Lip *c.* 2 cm, pale purple to dark red, almost erect, jointed with the base of the column, central lobe broadly rounded to oblong-elliptic, curved downwards, hairless or minutely warty towards the margin. Callus triangular, apex entire or 3-toothed. *C America from Mexico to Costa Rica*. G2. Summer–autumn.

The flowers have a characteristic shape resulting from the upper sepal and petals being directed upwards and the lateral.

M. rufescens (Lindley) Blanco (*Maxillaria rufescens* Lindley). Illustration: *Edwards's Botanical Register* 1848 (1836); Dunsterville & Garay, Venezuelan orchids illustrated **2**: 219 (1961); *American Orchid Society Bulletin* **43**: 966 (1974).

Scapes 1–5 cm. Sepals 1–2.5 cm, greenish white, pinkish or yellow, usually flushed with brownish pink or maroon. Petals very slightly shorter, similar in colour, sometimes with deep red splashes on veins. Lip 3-lobed, hinged to column foot, lateral lobes pointed, somewhat sickle-shaped, central lobe oblong, slightly reflexed, usually pale or deep yellow, or reddish, spotted with maroon or brownish-red. Callus brownish yellow or deep red, warty and velvety. *Tropical America*. G2. Autumn.

An extremely variable species.

112. TRIGONIDIUM Lindley

V.A. Matthews

Epiphytic, with creeping rhizomes. Pseudobulbs with 1–4 leaves, ribbed, subtended by several fibrous bracts. Leaves strap-shaped, leathery. Flowers erect, bell-shaped, usually solitary, arising from the axils of bracts on the most recent pseudobulb. Sepals forming a basal tube around the petals and lip, recurved in the upper half. Petals much smaller than sepals with shiny thickenings at the tip. Lip 3-lobed, shorter than petals, almost erect, lateral lobes erect and parallel to column, central lobe fleshy. Callus usually strap-shaped. Column without a foot. Pollinia 4, on a short, laterally crescent-shaped stipe.

A genus of 14 species from C and tropical S America. Cultivation as for *Maxillaria* (p. 552).

1. T. egertonianum Lindley (*T. seemanni* Reichenbach). Illustration: Hawkes, Encyclopaedia of cultivated orchids, 475 (1965); Dunsterville & Garay, Venezuelan orchids illustrated **3**: 321 (1965).

Pseudobulbs 4–9 cm, more or less spherical to ellipsoid, compressed, each with 2 leaves. Leaves 20–60 cm. Scapes 15–35 cm. Sepals and petals yellowish green or pinkish with reddish brown or purple veins, upper sepal 2.5–4.5 cm, lateral sepals 2.5–4 cm, petals 1–2 cm with a purple or brown thickening just below the apex. Lip 5–10 mm, green or yellowish brown striped with dark purple or brownish red, central lobe reflexed, warty. *C America, Colombia, Venezuela*. G2. Spring.

113. DICHAEA Lindley

V.A. Matthews

Epiphytes without pseudobulbs. Leaves in 2 ranks, folded when young, the sheaths concealing the stem. Flowers solitary from the leaf axils. Sepals and petals free, similar, though petals usually smaller. Lip joined to the base of the column, clawed, usually 3-lobed, the lateral lobes often recurved, usually without a callus. Pollinia 4, ovoid.

A genus of 40 species occurring in tropical America and the West Indies. Only 4 species are generally cultivated. They are best grown on a block of tree fern or bark. Cultivation otherwise as for *Angraecum* (p. 593).

1a. Leaf-blade not jointed to the sheath, leaves persistent; lip with the narrow lateral lobes pointing backwards; capsule spiny 2
b. Leaf-blade jointed to the sheath, leaves deciduous; lip with backwardly pointing lateral lobes; capsule not spiny 3
2a. Sepals orange, lanceolate, warty on the outer surface **1. muricata**
b. Sepals pale yellow to pinkish, ovate, not warty **2. pendula**
3a. Lip 4.5–6 mm, with a callus; leaves green **3. graminoides**
b. Lip 6–7 mm, without a callus; leaves glaucous **4. glauca**

1. D. muricata (Swartz) Lindley. Illustration: Schultes, Native orchids of Trinidad and Tobago, 259 (1960); Lasser, Flora de Venezuela **15**(5): 461 (1970).

Stems hanging, 30–50 cm. Leaves 1.5–2 cm, ovate-oblong, acuminate, persistent, not jointed to the sheaths. Sepals orange, lanceolate, 7–10 mm; petals purplish-spotted, 5–8 mm; all warty on the outer surfaces. Lip 5–8 mm, pale purple with 2 recurved, narrow, lateral lobes. Capsule 1–2 cm, densely spiny. *Mexico to Brazil, West Indies*. G2. Winter.

2. D. pendula (Aublet) Cogniaux. Illustration; Lasser, Flora de Venezuela **15** (5): 466 (1970).

Stems hanging. Leaves 1–2.5 cm, elliptic-oblong to lanceolate, each with a small point, persistent, not jointed to the sheaths. Sepals 8–9 mm, pale yellow to pinkish, ovate. Petals *c.* 8 mm, pinkish-buff, sometimes with bluish purple marks near the tips, obovate. Lip 5–9 mm, white or pale purple with bluish purple blotches and bands, with 2 recurved, narrow side-lobes. Capsule 1–1.5 cm, spiny. *Southern C America to northern S America, West Indies*. G2. Summer–autumn.

3. D. graminoides (Swartz) Lindley.

Stem erect, 10–30 cm, strongly flattened. Leaves 3–4 cm, linear, jointed to the sheaths, eventually falling. Sepals and petals 5–8 mm, whitish, ovate to lanceolate. Lip 4.5–6 mm, whitish, heart-shaped at the base, shallowly 3-lobed, central lobe almost circular, with a small point. Callus present, linear. Capsule 1.5–2 cm, not spiny. *C and tropical S America, West Indies*. G2.

4. D. glauca (Swartz) Lindley.

Stems ascending or hanging, flattened. Leaves 3.5–7 cm, linear-oblong, glaucous, jointed to the sheaths, eventually falling. Flowers fragrant. Sepals and petals 7–8 mm, generally white but sometimes tinged and spotted with lilac and yellow. Lip 6–7 mm, whitish with a dark red spot at the base, claw broad, lateral lobes very short or absent, central lobe broadly ovate to kidney-shaped. Capsule 1–2 cm, not spiny. *Northern C America, Cuba, Hispaniola & Jamaica*. G2. Summer.

114. ORNITHOCEPHALUS Hooker

V.A. Matthews

Epiphytic, without pseudobulbs. Leaves fleshy, leathery, in fan-shaped clusters, densely overlapping and concealing the stem, jointed to sheaths, the part below the joint often thickened. Racemes erect or hanging, with several–many flowers, arising from the leaf axils. Flowers not resupinate. Sepals usually equal, free. Petals equal to or larger than sepals. Lip joined to the base of the column, simple or 3-lobed; callus present. Rostellum long-beaked. Anther with a long terminal appendage. Pollinia 4, on a slender stipe.

A genus of about 50 species from the American tropics. They are best grown in small shallow pans suspended near the glass, using an epiphytic compost. They require light but must be shaded from the direct rays of the sun. Moisture must be provided throughout the year, particularly during the summer.

1a. Leaves 7.5–15 cm; racemes 22–25 cm **3. grandiflorus**
b. Leaves 2.5–7.5 cm; racemes to 8 cm 2
2a. Sepals 2–2.5 mm; lip 4–5 mm, central lobe obtuse **1. iridifolius**
b. Sepals 3–4 mm; lip 5–6 mm, central lobe acute **2. gladiatus**

1. O. iridifolius Reichenbach.

Leaves 2.5–7.5 cm, linear-lanceolate. Racemes 4–8 cm, loose, with many flowers. Flowers white. Sepals 2–2.5 mm, almost circular to broadly elliptic, with a keel projecting beyond the tip. Petals 3–3.5 mm, fan-shaped. Lip 4–5 mm, deeply 3-lobed, lateral lobes almost circular, central lobe triangular-ovate. *Mexico, Guatemala*. G2. Summer.

2. O. gladiatus Hooker. Illustration: Hooker, Exotic flora **2**: t. 127 (1824);

Howard, Flora of the Lesser Antilles, 214 (1974).

Leaves 2.5–4 cm, lanceolate. Raceme to 5 cm, loose, with few flowers. Flowers greenish yellow to greenish white. Sepals 3–4 mm with a keel projecting beyond the tip. Petals 3–4 mm. Lip 5–6 mm, 3-lobed, central lobe linear-oblong with recurved margins, acute. *C. America, north-west S America, West Indies.* G2. Spring–summer.

3. O. grandiflorus Lindley (*Dipteranthus grandiflorus* (Lindley) Pabst). Illustration: Pabst & Dungs, Orchidaceae Brasilienses **2**: 263, 322 (1977).

Leaves 7.5–15 cm, narrowly oblong. Racemes 22–25 cm with many flowers. Sepals concave, white with a greenish mark at the base, petals similar but a little larger. Lip 3-lobed, white, greenish or yellowish at the base, concave, margins finely crisped. *Brazil.* G2. Summer.

115. LOCKHARTIA Hooker

J. Cullen

Epiphytic, without pseudobulbs. Stems erect, covered with 2-ranked, strongly keeled leaves that consist mainly of overlapping sheaths, the free blades small, folded when young. Flowers borne laterally, either singly or in open, few-flowered racemes, usually conspicuously bracteate with stem-clasping bracts. Sepals and petals similar, 7–15 mm, spreading or somewhat reflexed, yellow. Lip simple or more frequently complex with 2 lateral lobes and a central lobe that is often further divided and warty at its base. Column spreading, winged towards the apex, without a foot. Pollinia 2, attached directly to a small, circular viscidium.

Lockhartia is a genus of about 20 species from C & S America. Only 1 species is readily available in Europe, though others are grown in the USA (see Teuscher, H., The genus Lockhartia, *American Orchid Society Bulletin* **43**: 399–405, 1974). Cultivation as for *Oncidium* (p. 571).

1. L. oerstedtii Reichenbach (*L. verrucosa* Reichenbach; *L. robusta* Schlechter). Illustration: Saunders, Refugium botanicum **2**: t. 76 (1872); Bechtel et al., Manual of cultivated orchid species, 222 (1981).

Stems 15–30 cm. Leaves rounded or somewhat acute at the apex. Petals and sepals golden yellow. Lip with blunt, downwardly deflected, yellow and often red-spotted lateral lobes and a yellow central lobe which is often red-spotted at the base, oblong or reversed-heart-shaped, deeply notched at the apex and with 2 lateral, somewhat reflexed, lobe-like prolongations near its junction with the lateral lobes. *C America.* G2. Summer.

L. lunifera Reichenbach. Similar but with smaller flowers (sepals *c.* 7 mm), the petals slightly smaller than the sepals and the lateral lobes of the lip sickle-shaped. *N & E Brazil.*

116. TRICHOCENTRUM Poeppig & Endlicher

J. Cullen

Epiphytic. Pseudobulbs small, simple, to 1 cm, each with 1 leaf. Leaves thick, narrowly elliptic, acute, folded when young. Racemes spreading or arching, arising from the bases of the pseudobulbs, with 1–2 flowers, which are 2.5 cm or more in diameter. Sepals and petals similar in size and colour, spreading. Lip spurred, with a narrow basal part that often bears a number of ridges and a spreading, usually broad, notched blade. Column short, without a foot, conspicuously winged, the wings variable in shape, sometimes coarsely fringed. Anther downy. Pollinia 2, ovoid, stipe strap-shaped, longer and broader than the pollinia, viscidium small, circular.

A genus of about 20 species from C & S America. Cultivation as for *Paphinia* (p. 607), but good drainage is essential.

Literature: Pupulin, F., A revision of the genus Trichocentrum, *Lindleyana* **10**(3): 183–210 (1995).

1. T. albo-coccineum Linden (*T. albopurpureum* Reichenbach; *T. alboviolaceum* Schlechter). Illustration: *Botanical Magazine*, 5688 (1868); *Lindenia* **16**: t. 748 (1901); Hoehne, Flora Brasilica **12**(7): t. 165 (1953).

Leaves to 8 cm. Flowers to 5 cm in diameter. Sepals and petals oblong-obovate, olive green outside, brown fading to yellow at the tips inside. Lip notched at the apex, mostly white with 2 reddish violet patches at the sides near the base and with 4 yellow ridges. Spur about a quarter the length of the lip. Wings on column erect, acuminate. *E Peru & Brazil.* G1–2. Summer–autumn.

T. orthoplectron Reichenbach. Similar but with leaves 7–14 cm and the lip without ridges but with 5 violet lines. *Brazil. G1–2.*

T. pfavii Reichenbach. Illustration: *Gardeners' Chronicle* **17**: 117 (1882); *Illustration Horticole* **33**: t. 587 (1886); *Lindleyana* **10**: 198 (1995). Flowers slightly smaller with white, brown-spotted sepals and petals, the lip white with a red spot near the base, the spur very short, the wings of the column oblong, truncate. *Costa Rica.* G1–2. Autumn.

T. tigrinum Linden & Reichenbach. Illustration: *Illustration Horticole* **24**: t. 282 (1877); *Botanical Magazine*, 7380 (1894); Bechtel et al., Manual of cultivated orchid species, 277 (1981). Flowers *c.* 6 cm in diameter, sepals and petals yellow spotted with brown; the lip white with lateral violet spots, 3 yellow ridges and 2 small yellow subulate teeth on the margins near the ridges; the spur very short and the wings of the column coarsely fringed. *C America.* G1–2. Spring.

117. IONOPSIS Humboldt, Bonpland & Kunth

J. Cullen

Small rhizomatous epiphytes. Pseudobulbs simple, to 1.5 cm, each with 1 leaf; some leaves borne directly on the rhizomes. Leaves oblong-linear to narrowly elliptic, fleshy or not, folded when young. Flowers in panicles arising from the apices of the pseudobulbs. Upper sepal free, the lateral sepals united at their bases and forming a spur under the flower. Petals similar to the upper sepal. Lip with a narrow claw, which usually bears 2 ridges, and a reversed-heart-shaped blade. Column short, unwinged, without a foot. Pollinia 2, ellipsoid, stipe linear, slightly broadened above, longer than the pollinia, viscidium small, oblong, acute, forming an acute angle with the stipe.

A genus of 8 species widely distributed in tropical and subtropical America. Plants can be grown successfully on cork bark or blocks of tree fern hung near the glass in a warm house. Frequent spraying is necessary during the summer. Propagation by division.

1. I. utricularioides (Swartz) Lindley (*I. tenera* (Steudel) Lindley). Illustration: Ames, Orchidaceae **1**: t. 5 (1905); *Fieldiana* **26**: 598 (1953); Dunsterville & Garay, Venezuelan orchids illustrated **1**: 174–5 (1959); Bechtel et al., Manual of cultivated orchid species, 217 (1981).

Pseudobulbs cylindric. Leaves flat, not very fleshy, narrowly elliptic. Sepals and petals to 8 mm, pinkish white. Lip pinkish white with the claw intense purple, the blade deeply notched. *Tropical and subtropical America from USA (Florida) to Peru and Paraguay*. G2. Spring & autumn.

I. paniculata Humboldt, Bonpland & Kunth, from Brazil, is sometimes treated as a separate species; it is larger than *I. utricularioides*, with oblong, fleshy leaves, sepals and petals to 12 mm, and the blade of the lip less deeply notched.

118. RODRIGUEZIA Ruiz & Pavon

J. Cullen

Epiphytes with short or long rhizomes. Pseudobulbs distant or close, simple, each with 1–3 leaves. Leaves oblong, folded when young. Racemes with few to many flowers, arching or somewhat hanging, arising from the apices of the pseudobulbs. Upper sepal and petals similar, incurved, the lateral sepals fused at their bases, and forming a short spur. Lip directed forwards with a distinct claw and a notched or 2-lobed blade, bearing 2 ridges on the claw and a single appendage projecting back into the sepal spur. Column short with 2 large or small wings at the apex, without a foot. Pollinia 2, broadly ovoid, stipe very narrow, longer than pollinia, viscidium small, oblong, borne at an acute angle to the stipe.

A genus of about 30 species from C & S America, concentrated in Brazil. Cultivation as for *Paphinia* (p. 607), but *R. decora* should be grown on a raft or on a piece of bark. All are subject to attack by red spider mite and thrips.

- 1a. Rhizome long, pseudobulbs separated; lip with claw longer than blade and bearing 2 elongate, erect, toothed ridges; column wings elongate, ciliate at their ends **1. decora**
- b. Rhizome short, pseudobulbs close or touching; lip with claw shorter than blade, the ridges represented by 2 swellings; column wings minute, not ciliate 2
- 2a. Raceme 1-sided; flowers usually pink, occasionally almost orange or purple **2. secunda**
- b. Raceme not 1-sided; flowers white with a yellow blotch on the lip **3. venusta**

1. R. decora (Lindley) Reichenbach. Illustration: Hoehne, Iconografia de orchidaceas do Brasil, t. 239 (1949); Bechtel et al., Manual of cultivated orchid species, 270 (1981).

Rhizome long, pseudobulbs *c.* 10 cm apart, each compressed and with 1 leaf. Leaves 6–9 cm, oblong. Raceme with 6–10 flowers, arching, loose. Flowers *c.* 5 cm in diameter. Sepals and petals oblong, acute, yellow or white with red-purple spots. Lip much longer than sepals, the claw longer than the blade and bearing 2 toothed erect white- and red-spotted or rarely entirely red ridges; blade white, broader than long, deeply notched. Column wings elongate, ciliate at their ends. *S & C Brazil*. G2. Autumn.

2. R. secunda Humboldt, Bonpland & Kunth. Illustration: Dunsterville & Garay, Venezuelan orchids illustrated **1**: 376–7 (1959); Couret, Orquideas Venezolanas, 58–9 (1977); Sheehan & Sheehan, Orchid genera illustrated, 153 (1979); Bechtel et al., Manual of cultivated orchid species, 270 (1981).

Rhizomes short, pseudobulbs close or touching, each with 2–3 leaves. Leaves to 15 cm, oblong. Raceme arching, 1-sided, with many closely packed flowers. Sepals and petals ovate, pink, rarely orange or purple. Lip scarcely longer than the sepals, pink, with a short claw gradually broadening into the oblong blade which is slightly notched; ridges on the claw in the form of 2 swellings. Wings of the column minute, not ciliate. *Tropical America, from Panama to Ecuador, Trinidad & French Guiana*. G2. Spring–summer.

3. R. venusta (Lindley) Reichenbach (*R. fragrans* Lindley). Illustration: Hoehne, Iconografia de orchidaceas do Brasil, t. 240 (1949).

Similar to *R. decora* but with very fragrant flowers with narrower, acute, white sepals and petals, the lip white with a yellow centre, its blade almost circular, all borne in a raceme which is not 1-sided. *E Brazil*. G2. Autumn.

119. COMPARETTIA Poeppig & Endlicher

J. Cullen

Rhizomatous epiphytes. Pseudobulbs small, simple, close or touching, each with 1 leaf. Leaves narrowly elliptic, folded when young. Racemes arising from the bases of the pseudobulbs, arching or somewhat hanging, loose, with 6–10 flowers. Upper sepal and petals similar or the petals slightly broader, erect; lateral sepals united at the base and prolonged into a backwardly pointing spur. Lip much larger than the petals, with a short claw and broad blade, and with 2 appendages projecting back into the sepal spur. Column unwinged and without a foot. Pollinia 2, ovoid, stipe longer than pollinia, broadened above, viscidium small, circular or heart-shaped.

A genus of about 8 species from tropical America. Cultivation as for *Rodriguezia* (see above).

- 1a. Sepals and petals more than 1.5 cm; spur 3–5 cm **3. macroplectrum**
- b. Sepals and petals to 1.5 cm; spur 1–2 cm 2
- 2a. Sepals and petals rose-pink to violet **2. falcata**
- b. Sepals and petals bright shining scarlet throughout **1. coccinea**

1. C. coccinea Lindley. Illustration: *Edwards's Botanical Register* **24**: t. 68 (1838); Hoehne, Iconografia de orchidaceas do Brasil, t. 238 (1949); Bechtel et al., Manual of cultivated orchid species, 188 (1981).

Pseudobulbs 2–2.5 cm. Leaves to 10 cm, oblong, acute. Flowers bright shining scarlet. Sepals and petals *c.* 1.3 cm. Lip yellow outside. Spur to 1.8 cm. *S & C Brazil*. G2. Summer.

2. C. falcata Poeppig & Endlicher (*C. rosea* Lindley). Illustration: *Botanical Magazine*, 4980 (1857); *Fieldiana* **26**: 603 (1953); Dunsterville & Garay, Venezuelan orchids illustrated **1**: 86–7 (1959); Bechtel et al., Manual of cultivated orchid species, 188 (1981).

Vegetatively very like *C. coccinea*. Sepals and petals *c.* 1.3 cm, rose-pink to violet. Lip deeper violet than the petals inside, pink or violet outside. Spur to 2 cm. *Tropical America, from S Mexico & the West Indies to Bolivia*. G2. Winter.

3. C. macroplectrum Reichenbach & Triana. Illustration: *Botanical Magazine*, 6679 (1883); *Lindenia* **14**: t. 664 (1899); Sheehan & Sheehan, Orchid genera illustrated, 69 (1979); Bechtel et al., Manual of cultivated orchid species, 188 (1981).

Vegetatively like *C. coccinea*. Flowers clear violet with purple spots. Sepals and petals 1.5–2 cm. Lip violet outside. Spur 3–5 cm, very conspicuous. *Tropical S America from Ecuador to N Brazil*. G2. Autumn.

120. TRICHOPILIA Lindley

J. Cullen

Epiphytic. Pseudobulbs simple, small or large, each with 1 leaf. Leaves more or less elliptic or elliptic-oblong, acute, folded when young. Racemes arising from the bases of the pseudobulbs, arching or somewhat pendent with 1–5 flowers, which are often fragrant. Sepals and petals similar, spreading, narrow, sometimes twisted. Lip rolled around the column at its apex, forming a tube, obscurely 3-lobed, margins wavy. Column elongate, without a foot, with 3 fringed lobes at the apex. Pollinia 2, ovoid to almost spherical, stipe linear, slightly broadened above, longer than pollinia, viscidium small, forming an acute angle with the stipe.

A genus from C and northern S America of an uncertain number (15–30?) of doubtfully distinguished species. Various species, formerly included in the genus, have now been placed elsewhere (see Garay, *Orquideologia* **7**: 191–6, 1972). Five species of the reduced genus are found in cultivation.

They are grown in epiphytic compost, which must be well drained, in pots or tied on to sections of bark. Only occasional watering is required in the winter. Propagation by division.

1a. Petals and sepals conspicuously twisted **4. tortilis**
b. Petals and sepals not twisted 2
2a. Pseudobulbs almost circular in outline, 3–4 cm; sepals and petals 6–7 cm **1. suavis**
b. Pseudobulbs elongate in outline, 5–13 cm; sepals and petals 3–6 cm 3
3a. Lip red within **5. marginata**
b. Lip white with a yellow blotch and/ or red-brown spots within 4
4a. Sepals and petals 5.5–6 cm; lip white with a yellow blotch within **2. fragrans**
b. Sepals and petals to 5 cm; lip white with a faint yellow blotch and red-brown spots within **3. galeottiana**

1. T. suavis Lindley & Paxton. Illustration: *Botanical Magazine*, 4654 (1852); *Lindenia* **9**: t. 423 (1893); Sheehan & Sheehan, Orchid genera illustrated, 165 (1979); Bechtel et al., Manual of cultivated orchid species, 277 (1981).

Pseudobulbs 3–4 cm, almost circular in outline, compressed. Leaves to 20 cm, elliptic, broadened abruptly to almost cordate at the base. Raceme with 2–3 flowers. Sepals and petals 6–7 cm, white, sometimes faintly yellowish or greenish, rarely red-spotted. Lip white with a yellow blotch inside and red spots on the central lobe, or rarely (var. **alba** Anon.) without spots. *Costa Rica, Panama & Colombia*. G2. Spring.

2. T. fragrans (Lindley) Reichenbach. Illustration: Cogniaux & Goossens, Dictionnaire iconographique des orchidées, Trichopilia, t. 3 (1900); Dunsterville & Garay, Venezuelan orchids illustrated **3**: 316–17 (1965); Bechtel et al., Manual of cultivated orchid species, 277 (1981).

Pseudobulbs 10–13 cm, elongate in outline, compressed. Leaves to 17 cm, elliptic, tapered to the base. Racemes with 2–5 flowers. Sepals and petals 5.5–6 cm, white or greenish white. Lip white with a yellow blotch inside. *Tropical America from the West Indies to Bolivia*. G1–2. Winter.

3. T. galeottiana A. Richard (*T. picta* Lemaire). Illustration: *Illustration Horticole* **6**: t. 225 (1859); Cogniaux & Goossens, Dictionnaire iconographique des orchidées, Trichopilia, t. 5 (1900).

Like *T. fragrans*, but racemes with 2–3 flowers, sepals and petals up to 5 cm, greenish, sometimes with brown lines, lip white outside, yellowish or whitish with red-brown spots inside. *Mexico*. G2. Summer.

4. T. tortilis Lindley. Illustration: *Botanical Magazine*, 3739 (1839); Cogniaux & Goossens, Dictionnaire iconographique des orchidées, Trichopilia, t. 6 (1900); Bechtel et al., Manual of cultivated orchid species, 278 (1981).

Pseudobulbs 4–8 cm, elongate in outline, compressed. Leaves 10–15 cm, elliptic, tapered to the base. Raceme with 1–2 flowers. Sepals and petals up to 6 cm, conspicuously twisted, green, striped and spotted with red-brown. Lip white with red-brown spots inside. *C America from S Mexico to Honduras*. G2. Winter.

An albino variant, with pure white flowers, has been in cultivation.

5. T. marginata Henfrey (*T. coccinea* Lindley; *T. crispa* Lindley). Illustration: *Paxton's Flower Garden* **2**: t. 54 (1851–52); *Botanical Magazine*, 4857 (1855); Cogniaux & Goossens, Dictionnaire iconographique des orchidées, Trichopilia, t. 1, 2, 2A (1900).

Pseudobulbs 5–7 cm, elongate in outline, compressed. Leaves to 17 cm, elliptic, narrowed to the base. Raceme with 2–3 flowers. Flowers rather variable in colour. Sepals and petals to 6 cm, white or greenish white, variously tinged, spotted or lined with red. Lip white outside, mostly red within, the margin sometimes white. *C. America, Colombia*. G2. Spring.

121. COCHLIODA Lindley

J. Cullen

Epiphytic, with short rhizomes. Pseudobulbs simple, compressed, each with 1 or 2 leaves. Leaves linear to oblong, usually blunt, folded when young. Flowers in erect, arching or hanging racemes or rarely panicles, arising laterally. Sepals and petals spreading. Lip with an erect claw, which is fused to the column base, and a spreading and usually 3-lobed blade bearing 2 or 4 swellings near the base, the lateral lobes rounded and often reflexed, the central lobe pointing forwards, often laterally expanded from a narrow base, notched or 2-lobed. Column erect, curved, without a foot, with 2–3 lobes at the apex; stigmatic surface divided into 2. Pollinia 2, broadly ovoid, stipe linear, somewhat longer than the pollinia, viscidium oblong, almost as long as the stipe and forming an acute angle with it.

A genus of about 6 species from S America. Cultivation as for *Odontoglossum* (p. 561).

1a. Lateral sepals united; lip without obvious lateral lobes, tapering to its tip **4. sanguinea**
b. Lateral sepals free; lip with obvious lateral lobes, the central either oblong or enlarged laterally from a narrow base 2
2a. Bract equalling the ovary and its stalk; flower white or cream **2. densiflora**
b. Bract much shorter than the ovary and its stalk; flowers red to rose-purple 3
3a. Central lobe of lip oblong, not expanded laterally from a narrow base; column 3-lobed at apex **3. rosea**
b. Central lobe of lip expanded laterally from a narrow base; column finely toothed at apex **1. vulcanica**

1. C. vulcanica (Reichenbach) Veitch. Illustration: *Botanical Magazine*, 6001 (1872); Cogniaux & Goossens, Dictionnaire iconographique des orchidées, Cochlioda, t. 1 (1898).

Pseudobulbs 2.5–6 cm, clustered, compressed, each with 2 leaves. Leaves 7.5–15 × 1.5–3.5 cm, elliptic-oblong to oblong-linear. Raceme with 6–many flowers, almost erect. Bract shorter than the ovary. Flowers *c.* 4 cm in diameter, rosy purple. Sepals and petals *c.* 2.2 cm, lanceolate-oblong, acute. Lip fused to the lower half of the column, 3-lobed, the central lobe with a narrow base and laterally expanded, 2-lobed, apical part. Column red, finely toothed at the apex. *Ecuador, Peru.* G1. Autumn–winter.

C. noezliana Rolfe. Flowers scarlet with yellow blotches on the lip, the column 3-lobed at the apex. *Peru, Bolivia.*

2. C. densiflora Lindley.

Pseudobulbs to 5 cm, compressed, each with 1 leaf. Leaves to 14 × 1.8 cm, oblong. Racemes with *c.* 9 flowers, almost erect. Bract as long as the ovary and its stalk. Flowers *c.* 3 cm in diameter, white or cream. Sepals and petals *c.* 1.5 cm, the lateral sepals narrower than the rest. Lip fused to the lower half of the column, 3-lobed, the central lobe with a narrow base and laterally expanded, 2-lobed, apical part. Column with 2 auricles at the apex. *Peru.* G1.

3. C. rosea (Lindley) Bentham. Illustration: Cogniaux & Goossens, Dictionnaire iconographique des orchidées, Cochlioda, t. 3 (1898); Bechtel et al., Manual of cultivated orchid species, 185 (1981).

Pseudobulbs 3–5 cm, compressed, clustered, each with 1–2 leaves. Leaves 5.5–20 × 1–2.5 cm, elliptic-oblong to linear. Raceme with up to 20 flowers, erect or arched, rarely branched above. Bract much shorter than the ovary. Flowers 2–3.5 cm in diameter, rose-red. Sepals and petals *c.* 1.1 cm, narrowly elliptic to lanceolate. Lip fused to the lower third of the column, 3-lobed, the central lobe oblong. Column 3-lobed at the apex. *Ecuador, Peru.* G1. Spring.

The name *C. stricta* Cogniaux has been applied to plants resembling *C. rosea*, but thought to originate in Colombia. It is not certain what these plants are, or whether any are in cultivation now.

4. C. sanguinea (Reichenbach) Bentham.

Pseudobulbs 3–5 cm, ovoid, compressed, each with 2 leaves. Leaves to 25 × 1.5 cm, linear. Raceme with many flowers, almost erect, often branched above. Bract much shorter than the ovary. Flowers *c.* 3 cm in diameter, red, the apex of the column and the base of the lip white. Sepals and petals *c.* 1.5 cm, the lateral sepals united for about half their length. Lip small with scarcely perceptible lateral lobes and a tapering central lobe, fused to the lower half of the column. Column weakly 3-lobed at apex. *Ecuador, Peru.* G2. Autumn.

Sometimes placed in the genus *Symphyglossum*, where its correct name is *S. sanguinea* (Reichenbach) Schlechter.

122. GOMESA R. Brown

J. Cullen

Epiphytic, with short rhizomes. Pseudobulbs simple, clustered, oblong, compressed, each with 2 leaves. Leaves oblong or very narrowly elliptic, folded when young. Racemes arising from the bases of the pseudobulbs, arching, many-flowered. Sepals and petals spreading, greenish yellow, the upper sepal and the petals erect, the lateral sepals fused at the bases to some extent, spreading downwards. Lip greenish yellow, shorter than the sepals, with an erect basal part and a downwardly deflected, ovate or oblong-ovate blade that bears 2 large, finely toothed ridges from the base to the middle, which reach up to and slightly clasp the column. Column white, reddish around the rostellum, erect, without a foot. Pollinia 2, ovoid, stipe oblong, shorter than pollinia, viscidium small, making an acute or right angle with the stipe.

A genus of about 8 species from S America, mostly from Brazil. They are grown in a similar manner to *Odontoglossum* but require more heat in winter.

1a. Lateral sepals united only at the extreme base; blade of lip oblong-ovate **1. crispa**
b. Lateral sepals united for more than half their length; blade of lip ovate **2. planifolia**

1. G. crispa (Lindley) Klotzsch & Reichenbach. Illustration: Pabst & Dungs, Orchidaceae Brasilienses **2**: f. 1914 (1977); Sheehan & Sheehan, Orchid genera illustrated, 97 (1979); Bechtel et al., Manual of cultivated orchid species, 215 (1981).

Pseudobulbs 6–10 cm. Raceme many-flowered. Flowers *c.* 2 cm in diameter. Sepals and petals 9–10 mm, wavy, oblong, rounded at the apex, the lateral sepals united only at the base. Lip oblong-ovate. *Brazil.* G1. Spring.

2. G. planifolia (Lindley) Klotzsch & Reichenbach (*G. recurva* Loddiges, not R. Brown). Illustration: *Loddiges' Botanical Cabinet* **7**: t. 660 (1822); *Botanical Magazine*, 3504 (1836); Pabst & Dungs, *Orchidaceae Brasilienses* **2**: f. 1922 (1977).

Very like *G. crispa*, but sepals and petals scarcely wavy, rather broad, the lateral sepals united for about two-thirds of their length, the blade of the lip ovate, rounded. *Brazil, Paraguay, Argentina.* G1. Spring–summer.

The flowers of this species are said to be fragrant.

123. HELCIA Lindley

J. Cullen

A genus related to *Trichopilia* but keying near *Gomesa*, differing from the latter in its 1-leaved pseudobulbs, large flowers *c.* 7 cm in diameter, and the 3-lobed lip, whose small lateral lobes reach up to the column and whose central lobe is much larger, conspicuously wavy and without ridges extending to the column.

Cultivation as for *Odontoglossum*.

1. H. sanguinolenta Lindley (*Trichopilia sanguinolenta* (Lindley) Reichenbach). Illustration: *Botanical Magazine*, 7281 (1893); Bechtel et al., Manual of cultivated orchid species, 216 (1981).

Sepals and petals yellow with large brown spots, lip white with violet or purple spots. *Ecuador.* G2. Winter.

124. ODONTOGLOSSUM Humboldt, Bonpland & Kunth

J. Cullen

Rhizomatous epiphytes. Pseudobulbs simple, compressed, each bearing 1–3 leaves. Leaves often produced on the rhizome as well as on the pseudobulbs, variable in shape, folded when young. Flowers showy, in racemes or panicles, sometimes fragrant, bracts often conspicuous. Sepals spreading, free, or the lateral sepals united for a short distance at the base. Petals similar to, or shorter than the sepals. Lip complex, usually shorter than the sepals, with a short claw ascending parallel to the column (sometimes united to it) with or without small lateral lobes and with complex keels and swellings and a variably shaped blade, which is bent backwards near its base. Column long and slender, not thickened at the base, without a foot, often with auricles near the apex. Pollinia 2, ellipsoid, stipe linear or linear-oblong, somewhat longer than pollinia, viscidium almost as long as the stipe and forming an acute angle with it, narrowly oblong or triangular.

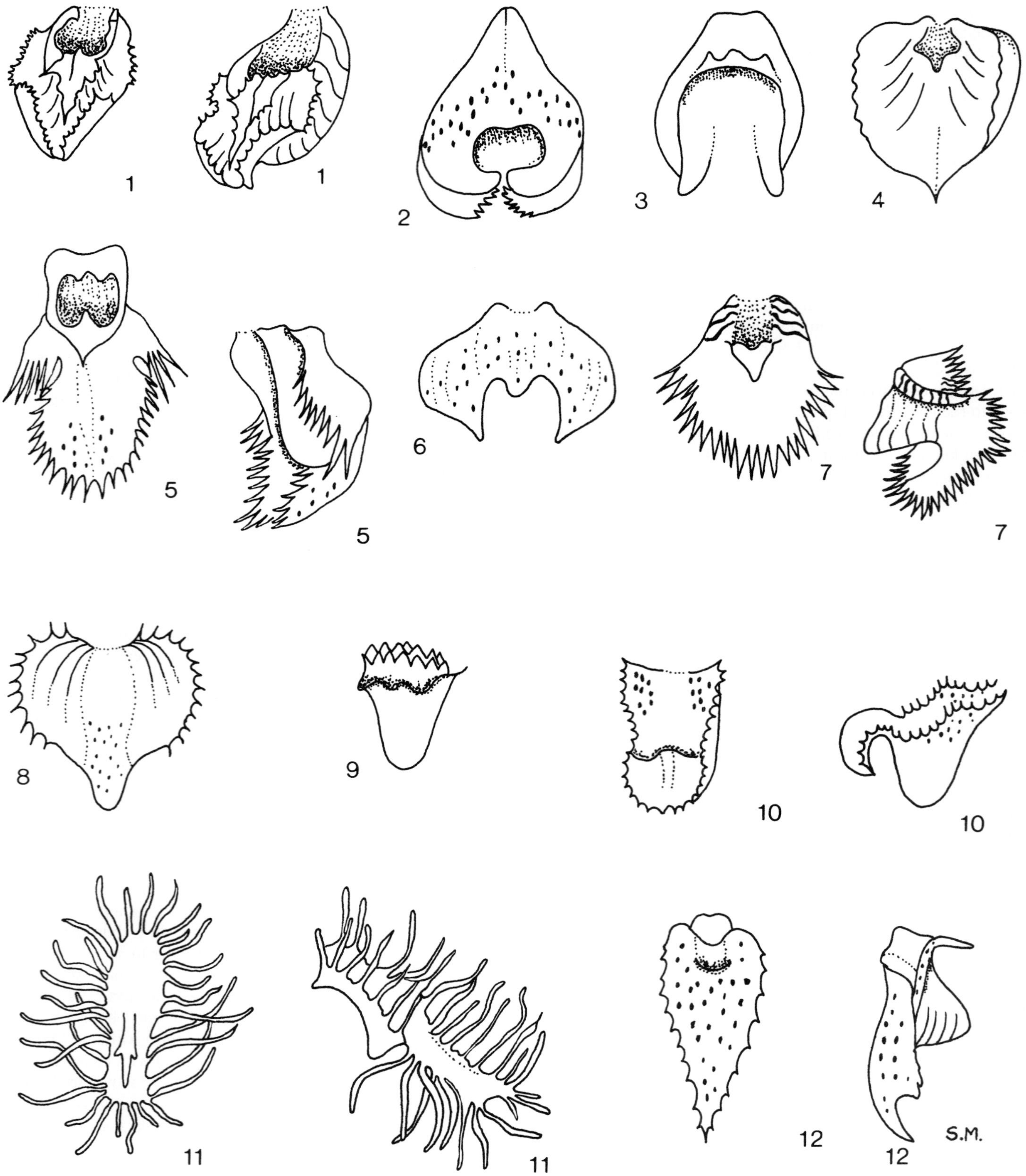

Figure 29. Lips of *Clowesia* and *Catasetum* species. 1, *Clowesia russelliana.* 2, *Catasetum integerrimum.* 3, *C. macrocarpum.* 4, *C. pileatum.* 5, *C. saccatum.* 6, *C. cernuum.* 7, *C. fimbriatum.* 8, *C. trulla.* 9, *C. microglossum.* 10, *C. atratum.* 11, *C. barbatum* (from 2 aspects). 12, *C. callosum.* (Compiled from various sources; not to scale.)

A large genus of about 200 species from C & S America, in need of further study. A very large number of species has been available in cultivation, but only about 20 are widely available today. Molecular work has suggested that *Odontoglossum* is better treated as part of *Oncidium*. Pending publication of the appropriate part of Genera Orchidacearum it is retained here, but synonyms in *Oncidium* are given where available in the species accounts below. Various treatments proposed advocate a number of other smaller genera; these are indicated in the synonymy. Many hybrids have been raised in the past, and many species are variable in flower colour, which has formed the basis for the recognition of numerous cultivars in some species. Species of *Odontoglossum* have been hybridised with species of several other genera.

Plants should be grown in an epiphytic compost. They require humid conditions combined with good ventilation, and may require shading during the hottest months. Watering should be reduced in winter. Propagation is by division, which is sometimes difficult, or by seed.

Literature: Escobar, R., El Genero Odontoglossum, *Orquideologia* **11**: 21–57, 119–160, 257–302 (1976); Halbinger, F., Odontoglossum y Generos Afines en Mexico y Centroamerica, *Orquidea* **8**: 155–282 (1982); Bockemühl, L., Odontoglossum, a monograph and iconograph (1989); Zelenko, H. & Chase, M.W. (ed.), Orchids: the pictorial encyclopaedia of Oncidium (2002).

Rhizome. Long, pseudobulbs distant: **7, 18**; short, pseudobulbs clustered: **1, 2, 3, 4, 5, 6, 8, 9, 10, 11, 12, 13, 14, 15, 16, 17, 19, 20**.
Pseudobulbs. Usually bearing 1 leaf: **1, 2, 3, 4, 5, 7, 8**; usually bearing 2–3 leaves: **6, 9, 20**.
Inflorescence. Panicle: **2, 9, 10, 11, 19**; raceme: **1, 3, 4, 5, 6, 7, 8, 11, 12, 13, 14, 15, 16, 17, 18, 20**.
Lip. (Figure 30, p. 564.) Blade waisted: **4, 7, 8, 9, 10, 11, 12, 13, 20**; blade not waisted: **1, 2, 3, 5, 6, 14, 15, 16, 17, 18, 19, 20**. Blade longer than broad: **1, 2, 3, 4, 5, 6, 7, 8, 9, 10, 11, 12, 13, 20**; blade broader than long: **14, 15, 16, 17, 18, 19**. Blade distinctly notched at the apex: **7, 15, 18**.
Column. Without auricles: **1, 3, 5, 6, 10, 18, 19**; with flap-like, entire auricles: **4, 13, 14, 16, 17**; with flap-like, toothed or finely toothed auricles: **7, 9, 11, 12, 15, 20**; with hair-like auricles: **2**; with auricles divided into a number of tooth-like processes: **8**.

1a. Pseudobulbs mostly bearing 1 leaf 2
b. Pseudobulbs mostly bearing 2–3 leaves 8
2a. Column with auricles near the apex 3
b. Column without auricles 6
3a. Sepals and petals conspicuously undulate-crisped; rhizome extensive, pseudobulbs distant **7. brevifolium**
b. Sepals and petals not undulate-crisped: rhizome short, pseudobulbs clustered 4
4a. Lip narrowing from just above its base (Figure 30(2), p. 564); column auricles formed by 2 hair-like processes **2. cirrhosum**
b. Lip broadening from just above the base; column auricles various, not as above 5
5a. Blade of lip toothed to lacerate, usually waisted in the middle, the base bearing a keel formed from several tooth-like processes (Figure 30(8), p. 564) **8. hallii**
b. Blade of lip entire, not waisted, the base bearing 2 entire keels (Figure 30(4), p. 564) **4. cervantesii**
6a. Blade of lip narrowly triangular to heart-shaped, acuminate (Figure 30(1), p. 564); raceme with more than 4 flowers **1. cordatum**
b. Blade broadly ovate or circular, rounded to the apex; raceme with 1–4 flowers 7
7a. Blade of lip irregularly toothed or lacerate (Figure 30(3), p. 564) **3. stellatum**
b. Blade entire though wavy (Figure 30(5), p. 564) **5. rossii**
8a. Column without auricles 9
b. Column with auricles 12
9a. Blade of lip distinctly waisted just below its middle (Figure 30(10), p. 564) **10. reichenheimii**
b. Blade various, not waisted as above 10
10a. Blade of lip longer than broad, tapering to the apex, not notched (Figure 30(6), p. 564) **6. nebulosum**
b. Blade broader than long, notched at the apex 11
11a. Blade of lip shining yellow barred with brown at the base, sometimes with additional brown spots on the surface **18. londesboroughianum**
b. Blade mostly white, purple at the base **19. cariniferum**
12a. Blade of the lip broader than long, usually notched at the apex 13
b. Blade longer than broad or as broad as long, tapering to the apex, not notched 14
13a. Sepals *c.* 2.5 cm, petals similar to sepals, white or pink, sometimes with a few reddish spots; inflorescence usually with more than 8 flowers **15. pendulum**
b. Sepals 5–8.5 cm, petals shorter and broader, all yellow, the sepals barred with brown, the petals brown in the lower half; raceme with 4–8 flowers **14. grande**
14a. Blade of the lip triangular to heart-shaped, tapering gradually to the apex 15
b. Blade of the lip oblong, waisted or not, tapering abruptly to the apex 16
15a. Blade of the lip 2.8 × 2.8 cm or more, pink, irregularly veined with white; petals ovate-elliptic **17. uroskinneri**
b. Blade up to 2 × 2.5 cm, violet, white or pink; petals elliptic-lanceolate to lanceolate **16. bictoniense**
16a. Flowers to 3 cm in diameter, pure white except for the yellow, red-spotted keels on the lip; leaves oblong-linear **20. pulchellum**
b. Combination of characters not as above 17
17a. Blade of the lip distinctly waisted 18
b. Blade oblong, not waisted 19
18a. Sepals and petals to 3.5 cm, yellow or green with brown bars or red spots; panicle many-flowered **9. laeve**
b. Sepals and petals at least 4 cm, dark brown with yellow veining; raceme with 4–12 flowers **11. harryanum**
19a. Sepals and petals white or pink, spotted with brown, elliptic, overlapping, the petals finely toothed **13. crispum**
b. Sepals and petals yellow with brown blotches, oblong-lanceolate or oblanceolate, not overlapping, petals wavy, not finely toothed **12. spectatissimum**

1. O. cordatum Lindley (*Rhynchostele cordata* (Lindley) Soto Arenas & Salazar; *Lemboglossum cordatum* (Lindley) Halbinger; *Odontoglossum maculatum* Hooker). Figure 30(1), p. 564. Illustration: *Lindenia* **9**: t. 430 (1893); Hamer, Orquideas de El

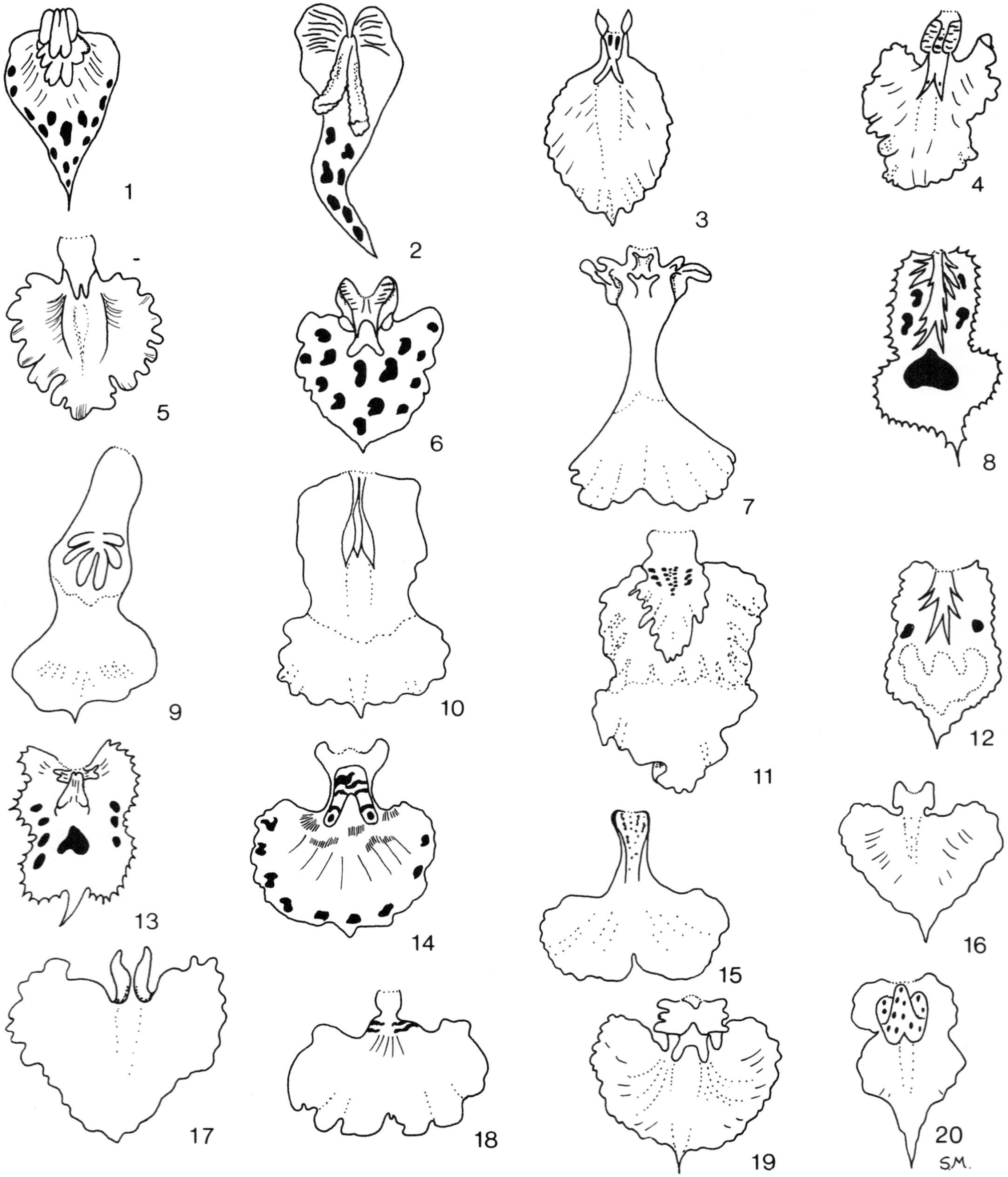

Figure 30. Lips of *Odontoglossum* species. 1, *O. cordatum*. 2, *O. cirrhosum*. 3, *O. stellatum*. 4, *O. cervantesii*. 5, *O. rossii*. 6, *O. nebulosum*. 7, *O. brevifolium*. 8, *O. hallii*. 9, *O. laeve*. 10, *O. reichenheimii*. z11, *O. harryanum*. 12, *O. spectatissimum*. 13, *O. crispum*. 14, *O. grande*. 15, *O. pendulum*. 16, *O. bictoniense*. 17, *O. uroskinneri*. 18, *O. londesboroughianum*. 19, *O. cariniferum*. 20, *O. pulchellum*. (Compiled from various sources; not to scale.)

Salvador **2**: 153 & t. 17 (1974); Williams et al., Orchids for everyone, 139 (1980); Bechtel et al., Manual of cultivated orchid species, 236 (1981).

Rhizome short, pseudobulbs clustered, 4.5–7.5 cm, each with 1 leaf. Leaves 9–30 × 3–4.5 cm, narrowly elliptic, acute. Raceme erect with few to many flowers. Sepals 3.5–5 cm × 6–12 mm, lanceolate, acuminate, petals 2.5–4 cm × 7–11 mm, ovate-lanceolate to elliptic-lanceolate, acuminate; all greenish, whitish or yellowish blotched with brown. Lip with a short claw with complex yellow keels that extend onto the blade; blade narrowly triangular to heart-shaped, acuminate, finely scalloped, white at the base with pink around the keels, the apex yellow spotted with brown. Column very shortly downy, without auricles. *C America, Venezuela*. G1. Summer.

2. O. cirrhosum Lindley (*Oncidium cirrhosum* (Lindley) Beer). Figure 30(2), p. 564. Illustration: *Botanical Magazine*, 6317 (1877); *Gartenflora* **41**: t. 1383 (1892); Cogniaux & Goossens, Dictionnaire iconographique des orchidées, Odontoglossum, t. 9 (1897); Bechtel et al., Manual of cultivated orchid species, 236 (1981).

Rhizome short, pseudobulbs clustered, 5–8 cm, each with 1 leaf. Leaves 10–30 × 2.5–3 cm, oblong-linear. Panicle arching, with many flowers. Sepals *c.* 4 cm × 7 mm, lanceolate, acuminate, terminating in a recurved point, petals broader and shorter; all white or cream with red-brown or red-purple blotches. Lip with a short claw bearing 2 small lateral lobes, yellow lined with red and with 2 forwardly projecting keels; blade white spotted reddish, narrowly heart-shaped to triangular, long-acuminate. Column bearing 2 hair-like auricles near the apex. *Ecuador, Peru*. G1. Spring.

3. O. stellatum Lindley (*Rhynchostele stellata* (Lindley) Soto Arenas & Salazar; *Lemboglossum stellatum* (Lindley) Halbinger; *Oncidium erosum* (Richard & Galeotti) Beer). Figure 30(3), p. 564. Illustration: Hamer, Orquideas de El Salvador **2**: 159 & t. 18 (1974).

Rhizome short, pseudobulbs clustered, 2–6 cm, each with 1 leaf. Leaves 6.5–15 cm × 8–25 mm, ovate to elliptic or oblanceolate. Raceme erect with 1–2 flowers. Sepals and petals similar, 2–2.8 cm × 3–5 mm, yellowish barred with brown, sometimes entirely brown, lanceolate, acute. Lip with a narrow claw bearing keels that extend onto the base of the blade as a short bifid plate; blade white or pink tinged with mauve, triangular-ovate, obtuse, margin toothed to lacerate. Column papillose, without auricles. *Mexico, Guatemala, El Salvador*. G1. Winter–spring.

4. O. cervantesii Lexarza (*Rhynchostele cervantesii* (Lexarza) Soto Arenas & Salazar; *Lemboglossum cervantesii* (Lexarza) Halbinger; *Oncidium cervantesii* (Lexarza) Beer). Figure 30(4), p. 564. Illustration: Cogniaux & Goossens, Dictionnaire iconographique des orchidées, Odontoglossum, t. 16 (1897); *Journal of the Orchid Society of Great Britain* **20**: 9 (1971); Williams et al., Orchids for everyone, 138 (1980); Bechtel et al., Manual of cultivated orchid species, 236 (1981).

Rhizome short, pseudobulbs clustered, 2–6.5 cm, each with 1 leaf. Leaves 4–30 × 1–3.5 cm, variable, ovate-lanceolate to elliptic-oblong. Raceme with 1–6 flowers, erect or arching, loose. Flowers fragrant. Sepals lanceolate, petals elliptic, otherwise similar, 1.8–3 cm × 6–20 mm, white to pink, marked in the lower third with concentrically arranged red spots and bars. Lip with a thick claw, white or pink, purple striped at the sides and with a yellow, papillose callus extending onto the blade as 2 points; blade heart-shaped, irregularly toothed, with a few spots. Column papillose with 2 circular auricles. *S Mexico, Guatemala*. G1. Winter–spring.

5. O. rossii Lindley (*Rhynchostele rossii* (Lindley) Soto Arenas & Salazar; *Lemboglossum rossii* (Lindley) Halbinger). Figure 30(5), p. 564. Illustration: *Edwards's Botanical Register* **25**: t. 48 (1839); Cogniaux & Goossens, Dictionnaire iconographique des orchidées, Odontoglossum, t. 6, 6A (1897).

Rhizome short, pseudobulbs loosely clumped, 3–6 cm, each with 1 leaf. Leaves 5–20 × 1.5–4 cm, elliptic or elliptic-lanceolate. Raceme with 2–4 flowers, erect or arched. Sepals 2.5–4.5 cm × 5–11 mm, oblong-elliptic to oblong-lanceolate, acute or acuminate, petals 2.5–3.8 cm × 8–19 mm, broadly elliptic to oblong-elliptic, obtuse to acute; all yellow, white or pink, the sepals and the bases of the petals spotted with brown. Lip with a narrow claw bearing a yellow, brown-spotted callus, which is notched at the apex with divergent tips; blade circular or nearly so to ovate, white or pink, wavy. Column without auricles. *C America, southwards to Nicaragua*. G1. Spring.

6. O. nebulosum Lindley (*O. apterum* Lexarza; *Oncidium apterum* (Lexarza) Beer; *Rhynchostele aptera* (Lexarza) Soto Arenas & Salazar; *Lemboglossum apterum* (Lexarza) Halbinger). Figure 30(6), p. 564. Illustration: *Lindenia* **8**: t. 350 (1892).

Rhizome short, pseudobulbs clustered, to 10 cm, each with 2 leaves. Leaves to 20 × 6 cm, elliptic or narrowly elliptic. Raceme with 5–7 flowers, loose, arching. Sepals and petals *c.* 3 cm, ovate, sepals acute, petals broader and more rounded at the apex, all white with red or greenish spots in the lower half. Lip shortly clawed, the claw bearing 2 yellow keels that project as divergent teeth over the base of the triangular (longer than broad), acute, slightly wavy, white, red- or yellow-spotted blade. Column without auricles. *Mexico*. G1. Summer.

7. O. brevifolium Lindley (*O. chiriquense* Reichenbach; *O. coronarium* Lindley; *Otoglossum brevifolium* (Lindley) Garay & Dunsterville). Figure 30(7), p. 564. Illustration: Cogniaux & Goossens, Dictionnaire iconographique des orchidées, Odontoglossum, t. 18 (1897); *Botanical Magazine*, 7687 (1899), 8725 (1917).

Rhizome creeping, pseudobulbs distant, 4–11 cm, each with 1 leaf. Leaves 10–30 × 6–9 cm, ovate to elliptic. Raceme with few to many flowers, erect or arching. Sepals and petals similar, 1.6–3 × 1.6–2 cm, obovate or obovate-oblong, obtuse, brown edged with yellow or yellow blotched with brown, margins conspicuously undulate-crisped. Lip with an indistinct claw bearing small lateral lobes and irregular keels, the blade oblong, usually waisted, whitish or yellow with transverse brown bars near the base. Column with conspicuous, variably developed auricles that are finely toothed. *Tropical America from Panama to Peru*. G1. Flowering irregular.

8. O. hallii Lindley (*Oncidium hallii* (Lindley) Beer). Figure 30(8), p. 564. Illustration: *Botanical Magazine*, 6237 (1876); *Lindenia* **4**: t. 158 (1888); Bechtel et al., Manual of cultivated orchid species, 237 (1981).

Rhizome short, pseudobulbs 5–10 cm, each with usually 1 leaf. Leaves 13.5–30 × 2–4.5 cm, elliptic-oblong. Raceme with many flowers, loose, arching.

Sepals 3.5–6 cm, ovate-lanceolate, acuminate, petals shorter and broader; all yellow with large purple-brown blotches. Lip with a short claw bearing a large crest of several yellow, tooth-like keels that extend onto the blade; blade white, variously spotted with red or purple, oblong, usually waisted, toothed to lacerate. Column with auricles divided into a number of tooth-like processes. *Colombia, Ecuador, Peru.* G1. Spring.

9. O. laeve Lindley (*Oncidium laeve* (Lindley) Beer; *Miltonia laevis* (Lindley) Rolfe; *Miltonioides laevis* (Lindley) Brieger & Lückel). Figure 30(9), p. 564. Illustration: *Botanical Magazine*, 6265 (1876); Bechtel et al., Manual of cultivated orchid species, 237 (1981).

Rhizomes short, pseudobulbs clustered, 5–12 cm, each with 2 leaves. Leaves 15–45 × 2.5–5.5 cm, oblong to narrowly elliptic, rounded to almost acute. Panicle with many flowers, arching. Sepals and petals 2.5–3.5 cm × 6–9 mm, linear-elliptic to linear-oblanceolate, yellowish blotched and barred with reddish brown. Lip with a short, thickened claw bearing 2–5 inconspicuous keels that run onto the blade, blade oblong, waisted, the basal part pale purple, the rest white, apex with a short point. Column without a ledge below the stigmatic surface and with finely toothed auricles. *S Mexico, Guatemala.* G1. Summer.

O. confusum Garay (*O. schroederianum* Reichenbach; *Miltonia schroederiana* (Reichenbach) Veitch; *Oncidium schroederianum* (O'Brien) Garay & Stacy). Illustration: Cogniaux & Goossens, Dictionnaire iconographique des orchidées, Miltonia, t. 3 (1897). Similar, but with green, red-spotted sepals and petals, the lip red in the lower half and the column with a transverse ledge below the stigmatic surface. *Costa Rica.* G1. Winter.

10. O. reichenheimii Linden & Reichenbach (*Oncidium reichenheimii* (Linden & Reichenbach) Garay & Stacy; *Miltonioides reichenheimii* (Linden & Reichenbach) Brieger & Lückel). Figure 30(10), p. 564. Illustration: Bechtel et al., Manual of cultivated orchid species, 239 (1981).

Very similar to *O. laeve* but the lip as in Figure 30(10) p. 564 and the column entirely without auricles. *Mexico.* G1. Summer.

For further information on the complex involving *O. laeve*, *O. reichenheimii* and *O. confusum*, see Garay, *Die Orchidee* **6**: 213–18 (1962), and Garay & Stacy, *Bradea* **1**(40): 393–424 (1974), in which all these species are considered to form part of the genus *Oncidium*.

11. O. harryanum Reichenbach. Figure 30(11), p. 564. Illustration: *Lindenia* **3**: t. 142 (1887); Cogniaux & Goossens, Dictionnaire iconographique des orchidées, Odontoglossum, t. 11 (1897); Williams et al., Orchids for everyone, 140 (1980).

Rhizome short, pseudobulbs clustered, 6–8 cm, each bearing 2 leaves. Leaves to 45 × 2–4 cm, oblong to oblong-elliptic, acute to obtuse. Raceme with 4–12 flowers. Sepals 4.5 × 1.3–2.5 cm, elliptic-oblong to elliptic, acute, petals somewhat smaller; all brown to dark brown, mottled darker, with yellow veins and with purple lines on a white ground at the bases of the petals. Lip with a short claw with a prominent yellow callus that is lacerate towards the front, blade oblong, waisted below the middle, yellowish at the base, the sides and the centre with purple lines on a white or yellow ground, the apical part white. Column with small, finely toothed auricles. *Colombia, Peru.* G1. Summer–autumn.

12. O. spectatissimum Lindley (*O. triumphans* Reichenbach). Figure 30(12), p. 564. Illustration: Cogniaux & Goossens, Dictionnaire iconographique des orchidées, Odontoglossum, t. 8, 8A, 8B (1897); Dunsterville & Garay, Venezuelan orchids illustrated **5**: 218–19 (1966).

Rhizome short, pseudobulbs clustered, 4–8 cm, each with 2 leaves. Leaves to 40 cm, linear-elliptic, acute. Raceme with 10–20 flowers, erect. Sepals and petals similar, to 4 cm, acute, margins undulate-crisped, yellow with numerous brown blotches. Lip with a short claw bearing a fan-shaped crest that is divided into processes, of which the inner 2 are the longest, yellow; blade oblong, acute, white or rarely yellow, the basal part pink or reddish, the apex brown, lacerate. Column with finely toothed auricles. *Colombia, Venezuela.* G1. Spring.

13. O. crispum Lindley (*O. alexandrae* Bateman). Figure 30(13), p. 564. Illustration; Cogniaux & Goossens, Dictionnaire iconographique des orchidées, Odontoglossum, t. 1 (1897); Schlechter, Die Orchideen, edn 2, t. 8 (1927); Williams et al., Orchids for everyone, 31 (1980); Bechtel et al., Manual of cultivated orchid species, 236 (1981).

Rhizome short, pseudobulbs clustered, 4–8 cm, each with 2 leaves. Leaves to 40 cm, linear-elliptic, acute. Raceme with 8–20 flowers, arched. Sepals and petals 3–4 cm, elliptic, acute, overlapping, margins wavy, petals toothed, white or pink spotted with brown. Lip with a short claw bearing a fan-shaped, laciniate, yellow keel with 2 longer projections in the middle, the blade oblong, wavy, finely toothed, acute, white, pink or pale purple, with brown spots. Column with 2 rounded, entire auricles. *Colombia.* G1. Winter.

This species is very variable and many selected variants (some of them perhaps accidental hybrids) were grown in the past. The species has also been used widely in hybridisation.

14. O. grande Lindley (*Rossioglossum grande* (Lindley) Garay & Kennedy). Figure 30(14), p. 564. Illustration: Schlechter, Die Orchideen, edn 2, t. 9 (1927); *Fieldiana* **26**: 615 (1953); *Die Orchidee* **31**: clxviii (1980); Williams et al., Orchids for everyone, 140 (1980).

Rhizome short, pseudobulbs 4–10 cm, each with 2 (rarely 1) leaves. Leaves 10–40 × 3–6.5 cm, elliptic to lanceolate, leathery, acute. Raceme with 4–8 flowers, erect. Sepals 5.5–8.5 × 1–2 cm, lanceolate, acuminate and recurved at the apex, yellow, heavily barred with brown, petals oblanceolate to oblong, rather blunt at the apex, brown in the lower half, yellow above, margins wavy. Lip more or less without a claw, with 2 small lateral lobes at the base, bearing a callus with variable horns and projections; blade broader than long, tapered to the base, not or scarcely notched at the apex, white with reddish, brownish or yellowish spots near the base and on the margins. Column very finely downy with 2 oblong or rounded auricles. *Mexico, Guatemala.* G1. Winter.

Rather variable in flower colour. This species and 5 others compose the genus *Rossioglossum* Garay & Kennedy (*Orchid Digest* **40**: 139, 1976 – illustration: *Die Orchidee* **31**: clxvi–clxx, 1980); the other species are found only in specialist collections.

15. O. pendulum (Lexarza) Bateman (*O. citrosmum* Lindley; *Cuitlauzinia pendula* Lexarza). Figure 30(15), p. 564. Illustration: Cogniaux & Goossens, Dictionnaire iconographique des orchidées, Odontoglossum, t. 19 (1897); Williams et al., Orchids for everyone, 139 (1980);

Bechtel et al., Manual of cultivated orchid species, 238 (1981).

Rhizome short, pseudobulbs clustered, 5–8 cm, each with 2 leaves. Leaves to 30 × 7 cm, oblong-elliptic, obtuse. Raceme with many lemon-scented flowers, dense, arching. Sepals and petals similar, *c.* 2.5 cm, ovate, shortly clawed, obtuse, white or pink, sometimes with a few reddish spots. Lip with an elongate claw bearing 2 yellow, red-spotted lateral lobes, without distinct keels; blade broader than long, heart-shaped at the base, notched at the apex, purple or pink. Column with 2 toothed auricles. *Mexico*. G1. Spring and autumn.

16. O. bictoniense (Bateman) Lindley (*Rhynchostele bictoniensis* (Bateman) Soto Arenas & Salazar; *Lemboglossum bictoniense* (Bateman) Halbinger). Figure 30(16), p. 564. Illustration: *Edwards's Botanical Register* **26**: t. 66 (1840); Hamer, Orquideas de El Salvador **2**: 151 & t. 18 (1974); Williams et al., Orchids for everyone, 138 (1980); Bechtel et al., Manual of cultivated orchid species, 235 (1981).

Rhizome short, pseudobulbs clustered, to 10 cm, each with 2–3 leaves. Leaves 10–45 × 1.5–5.5 cm, elliptic-oblong to linear. Raceme with many flowers, rarely branched, erect. Sepals and petals similar, 1.8–2.7 cm × 5–8 mm, elliptic-lanceolate to lanceolate, acute, pale yellowish green banded or spotted with brown. Lip with a short claw bearing a finely downy callus that clasps the column at the base and forms convex lobes at the apex, the blade 2 × 2.5 cm, heart-shaped, tapering to the apex, violet, white or pink, margin crisped. Column with 2 flap-like auricles. *Mexico, Guatemala, El Salvador*. G1. Winter–spring.

17. O. uroskinneri Lindley (*Rhynchostele uroskinneri* (Lindley) Soto Arenas & Salazar; *Lemboglossum uroskinneri* (Lindley) Halbinger). Figure 30(17), p. 564. Illustration: Cogniaux & Goossens, Dictionnaire iconographique des orchidées, Odontoglossum, t. 3, 3A (1897); Bechtel et al., Manual of cultivated orchid species, 239 (1981).

Similar to *O. bictoniense* but leaves lanceolate, sepals and petals 2.5–3 cm × 7–13 mm, dark red to greenish with brown bars and spots, blade of the lip 2.8 cm or more, pink, veined or spotted with white. *Guatemala, Honduras*. G1. Summer–autumn.

18. O. londesboroughianum Reichenbach (*Rhynchostele londesboroughiana* (Reichnbach) Soto Arenas & Salazar; *Mesoglossum londesboroughianum* (Reichenbach) Halbinger). Figure 30(18), p. 564. Illustration: *Illustration Horticole* **30**: t. 497 (1883).

Rhizome creeping, pseudobulbs distant, 6–8 cm, each bearing 2 (rarely 1) leaves. Leaves to 45 cm, lanceolate. Raceme with many flowers, occasionally branched, arching. Sepals and petals similar, 1.5–2 cm, ovate, obtuse, yellow with concentrically arranged red-brown spots towards the base. Lip with a short, keeled claw and a blade that is broader than long, heart-shaped at the base, notched at the apex, usually shining yellow, rarely spotted with brown, barred with brown at the base. Column without auricles. *Mexico*. G1. Winter.

19. O. cariniferum Reichenbach (*Oncidium cariniferum* (Reichenbach) Beer; *Miltonioides carinifera* (Reichenbach) Senghas & Luckel; *Collare-stuartense cariniferum* (Reichenbach) Senghas & Bockemühl). Figure 30(19), p. 564. Illustration: Williams et al., Orchids for everyone, 138 (1980); Bechtel et al., Manual of cultivated orchid species, 235 (1981).

Rhizome short, pseudobulbs close, 8–10 cm, each with 2 leaves. Leaves to 30 × 3–3.5 cm, oblong-elliptic, acute. Panicle with many flowers, arching. Sepals and petals similar, *c.* 3.5 cm, lanceolate, acute, purplish brown. Lip with a short claw bearing 2 small, oblong, purple lateral lobes and 2 divergent, purple keels; blade broader than long, heart-shaped at the base, notched and with a small point within the notch at the apex, white. Column without auricles. *Tropical America, from Costa Rica to Venezuela*. G1. Spring.

20. O. pulchellum Lindley (*Osmoglossum pulchellum* (Bateman) Schlechter). Figure 30(20), p. 564. Illustration: *Botanical Magazine*, 4104 (1844); Cogniaux & Goossens, Dictionnaire iconographique des orchidées, Odontoglossum, t. 18 (1897); Hamer, Orquideas de El Salvador **2**: 157 & t. 17 (1974); Bechtel et al., Manual of cultivated orchid species, 238 (1981).

Rhizome short, pseudobulbs clustered, to 10 cm, each with 2 leaves. Leaves 10–35 × 8–15 mm, oblong-linear, acute. Raceme with 6–10 flowers, loose, arching. Flowers very fragrant. Sepals 1–2 cm × 6–13 mm, obovate to elliptic, the laterals fused at their bases for *c.* 5 mm, white. Petals 1.3–2 cm × 7–15 mm, obovate, white. Lip complex, its claw ascending, the rest sharply curved outwards and then under, claw with a large, yellow, red-spotted callus extending as 2 blunt projections over the blade; blade waisted, acute, white, margins crisped. Column with 3 coarsely toothed, pinkish auricles. *Mexico, Guatemala, El Salvador*. G1. Winter.

125. ASPASIA Lindley

J. Cullen

Rhizomatous epiphytes. Pseudobulbs simple, elliptic to oblong, somewhat compressed, often furrowed, bearing 1–2 leaves. Leaves lanceolate to linear-lanceolate, acuminate, folded when young. Raceme lateral with 1–10 flowers. Sepals spreading, pointed, the upper united to the base of the column. Petals usually similar to the sepals but broader, both united to the base of the column. Lip united to the column by its lateral margins for about half the length of the column or more, then bent at right angles into a broad, 3-lobed blade, the lateral lobes small, the base bearing 2–4 keels. Column cylindric, short, without a foot. Pollinia 2, ellipsoid, stipe slightly longer than pollinia, narrowly club-shaped, viscidium oblong, slightly shorter than the stipe and forming an acute angle with it.

A genus of 5 species widely distributed in C and northern S America. Cultivation as for *Odontoglossum* (p. 561).

Literature: Williams, N.H., The taxonomy of Aspasia, *Brittonia* **26**: 333–46 (1974).

1a. Free part of column, above the junction with the lip, much shorter than the basal part; apex of lip entire, pointed, not curved upwards **3. lunata**

b. Free part of column as long as or longer than the basal part; apex of lip notched, often curved upwards 2

2a. Lip 1.7–2.1 cm wide, its apex curving upwards; anther cap not beaked **1. epidendroides**

b. Lip 1.3–1.5 cm wide, its apex not curving upwards; anther cap prominently beaked **2. variegata**

1. A. epidendroides Lindley. Illustration: *Botanical Magazine*, 3962 (1842); *Fieldiana* **26**: 625 (1953); Bechtel et al., Manual of cultivated orchid species, 169 (1981).

Pseudobulbs 5–12 cm, flattened on one side, convex on the other, bearing 2 leaves. Leaves 13–30 cm. Sepals and petals 1.5–2.5 cm, elliptic-obovate, yellow or yellow-green with transverse brown bars. Lip 1.7–2.1 cm wide, hairless at the base, white with purple markings towards the apex, yellow towards the base where there is a callus of 2–4 ridges, the apex notched, curving upwards. Lower half of the column united to the lip. Anther cap not beaked. *C America, Colombia.* G2. Spring.

2. A. variegata Lindley. Illustration: *Edwards's Botanical Register* **22**: t. 1907 (1836); *Botanical Magazine*, 3679 (1838); Pabst & Dungs, Orchidaceae Brasilienses **2**: t. 2082 (1977).

Similar to *A. epidendroides* but pseudobulbs and leaves a little smaller, lip 1.3–1.5 cm wide, downy at the base, its apex not curving upwards, and anther cap prominently beaked. *Tropical S America from Colombia to C Brazil.* G2. Spring.

3. A. lunata Lindley. Illustration: *Lindenia* **14**: t. 669 (1899); Pabst & Dungs, Orchidaceae Brasilienses **2**: t. 2080 (1977).

Pseudobulbs compressed. Leaves 14–25 cm, thin. Sepals 2.4–4.5 cm, linear-lanceolate, yellow-brown with dark red spots. Petals shorter and broader than the sepals. Lip 2–4 × 1.7–4 cm, oblong, conspicuously waisted, the apex entire, with a small point, the margin toothed, white on the margin and on the apical part, violet in the centre. Column united to the lip for most of its length, the free part very short. Anther cap not beaked. *C Brazil.* G2. Spring.

126. ADA Lindley

J. Cullen

Epiphytic, with short rhizomes. Pseudobulbs simple, ovoid, tapered, somewhat compressed, each with 1–3 leaves. Leaves lanceolate to narrowly elliptic, acute, folded when young. Racemes erect, arising from the bases of the pseudobulbs, with 8–13 shortly stalked, rather erect flowers. Sepals clawed, very narrowly elliptic, acuminate, orange; petals similar but shorter, sometimes with a dark red line or spots inside; all incurved, spreading only at their tips. Lip shorter than the petals, very narrowly elliptic, acuminate, with 1–2 elongate, raised ridges, the lip spreading away from the short column. Column without a foot. Pollinia 2, ovoid, stipe slightly longer than the pollinia, expanded above, viscidium shorter than the stipe and forming an acute angle with it.

A genus of 2 very similar species (often considered as 1), both of which are cultivated. Conditions as for *Odontoglossum* (p. 561).

1. A. aurantiaca Lindley. Illustration: *Botanical Magazine*, 5435 (1864); Cognaiux & Goossens, Dictionnaire iconographique des orchidées, Ada, t. 1 (1897); Bechtel et al., Manual of cultivated orchid species, 161 (1981).

Pseudobulbs 6–10 cm. Leaves 15–25 cm. Sepals 3–3.5 cm, petals 2–2.5 cm. Lip orange, bearing 2 elongate, irregularly toothed ridges. *Tropical S America, from Venezuela to Ecuador.* G1. Winter.

A. lehmannii Rolfe. Very similar (and perhaps not distinct), but with a white lip bearing an orange ridge. *Colombia.* G1. Autumn.

127. BRASSIA Lindley

J. Cullen

Epiphytic, with stout, creeping rhizomes, which give rise to leaves as well as pseudobulbs. Pseudobulbs large, simple, ovoid, compressed, each bearing 1–3 leaves. Leaves leathery, variable in shape, folded when young. Racemes arising from the bases of the pseudobulbs, arching, with 6–many flowers. Sepals spreading, linear-lanceolate, acuminate, the laterals often longer than the upper, all 6 or more times longer than broad. Petals similar to the sepals but shorter. Lip shorter than the sepals, stalkless at the base of the column, directed away from it at an obtuse angle, simple, bearing a keel that is grooved or composed of 2 ridges. Column without a foot. Pollinia 2, ovoid, attached independently to the small, oblong viscidium, stipe absent.

A genus of about 50 species from tropical and subtropical America, notable for its spider-like flowers with long, narrow petals and sepals. Six species, all belonging to section *Brassia* are available in Europe and are dealt with below; several other species, belonging to the section *Glumacea* Lindley are grown in the USA (see Teuscher, *Memoires du Jardin Botanique de Montreal* **55**: 15–21, 1962, reprinted in *American Orchid Society Bulletin* **42**: 1089–94, 1973); these have recently been placed in the genus *Ada*.

Plants should be grown in hanging baskets or on rafts or pieces of bark, in epiphytic compost. The atmosphere should be humid, though watering can be reduced during the winter. Propagation by division or by seed.

Literature: Kooser, R.G. & Kennedy, G.C., The genus Brassia, *Orchid Digest* **43**: 164–72 (1979).

1a. Blade of lip with green warts or excrescences **5. verrucosa**
b. Blade without warts or excrescences 2
2a. Blade of lip conspicuously broadened towards the apex so that its maximum breadth is at least twice the breadth at the base 3
b. Blade not or little broadened, its broadest part little broader than the base 4
3a. Blade of lip white or yellowish with purple spots; petals yellowish green with purple-brown spots towards the base **1. maculata**
b. Blade yellow with red-brown spots; petals yellow in the upper half, brown in the lower **2. gireoudiana**
4a. Ridges of the keel of the lip each ending in a separated tooth **4. caudata**
b. Ridges of the keel not ending in separated teeth **3. lanceana**

1. B. maculata R. Brown. Illustration: *Botanical Magazine*, 1691 (1814); Hamer, Orquideas de El Salvador **1**: 81 & t. 5 (1974); *Orchid Digest* **43**: 167 (1979).

Pseudobulbs 6–8 cm, each with 2 leaves. Leaves to 30 cm, oblong-linear. Raceme with 10–15 flowers. Sepals 6–7.5 cm, greenish yellow with a few brown-purple spots near the base; petals 4–5.5 cm, similar in colour to the sepals. Lip white or yellowish, purple-spotted, parallel-sided near the base then widely expanded towards the rounded or obtuse, though mucronate, apex. Ridges of the lip yellow, grooved, slightly downy. *C America, West Indies.* G2. Autumn and spring.

2. B. gireoudiana Reichenbach & Warscewicz. Illustration: Reichenbach, Xenia Orchidacearum **1**: t. 32 (1881); *Orchid Digest* **43**: 166 (1979); Bechtel et al., Manual of cultivated orchid species, 172 (1981).

Similar to *B. maculata* but sepals 10–13 cm, bright yellow with brown spots, petals *c.* 5 cm, bright yellow in the upper half, brown in the lower, lip yellow with red-brown spots and deep yellow, hairless ridges. *Costa Rica, Panama.* G2. Summer.

3. B. lanceana Lindley. Illustration: *Botanical Magazine*, 3577 (1837); Dunsterville & Garay, Venezuelan orchids illustrated **4**: 38–9 (1966); *Orchid Digest* **43**: 166 (1979).

Pseudobulbs 10–12 cm, each with 2 leaves. Leaves to 30 cm, lanceolate. Racemes loose, with 7–12 flowers. Sepals *c.* 6 cm, yellow with brown spots towards the base, petals *c.* 3 cm, similar in colour to the sepals. Lip oblong, acuminate, pale yellow-green with a few brown spots near the base. Ridges yellow, downy. *Tropical S America from Colombia to E Brazil.* G2. Autumn.

B. lawrenceana Lindley. Illustration: *Edwards's Botanical Register* **27**: t. 18 (1841); Dunsterville & Garay, Venezuelan orchids illustrated **2**: 48–9 (1961); *Orchid Digest* **43**: 166 (1979). Similar but with sepals *c.* 7.5 cm, yellow with red-brown spots, petals *c.* 4 cm, lip clear yellow or green, unspotted, sometimes a little widened towards the base. *Tropical S America, southwards to N Brazil and NE Peru.* G2. Summer.

4. B. caudata (Linnaeus) Lindley. Illustration: *Fieldiana* **26**: 625 (1953); Dunsterville & Garay, Venezuelan orchids illustrated **4**: 36–7 (1966); Luer, Native orchids of Florida, t. 76, f. 5, 6 (1972); Sheehan & Sheehan, Orchid genera illustrated, 51 (1979).

Pseudobulbs 6–15 cm, each with 2 leaves. Leaves to 35 cm, oblong-elliptic, lanceolate or oblong-lanceolate. Raceme loose, with 7–12 flowers. Sepals orange-yellow, spotted with red-brown, the upper 3.5–7.5 cm, the laterals 7.5–18 cm. Petals 1.5–3.5 cm, similar in colour to the sepals. Lip oblong, broadening slightly towards the apex, yellow or greenish with a few brown blotches towards the base. Ridges 2, hairless, each ending in a distinct, separated tooth. *Tropical and subtropical America from USA (Florida) to Bolivia and N Brazil.* G2. Autumn & spring.

5. B. verrucosa Lindley (*B. brachiata* Lindley). Illustration: *Edwards's Botanical Register* **33**: t. 29 (1847); Cogniaux & Goossens, Dictionnaire iconographique des orchidées, Brassia, t. 1 (1897); *Fieldiana* **26**: 629 (1953); *Orchid Digest* **43**: 169 (1979); Bechtel et al., Manual of cultivated orchid species, 173 (1981).

Pseudobulbs 6–10 cm, clustered, each with 2 leaves. Leaves to 45 cm, oblong-elliptic to oblanceolate. Raceme with 6–15 flowers, loose. Sepals yellowish green with dark brown spots towards the base, the upper 8–15 cm, the laterals 10–20 cm. Petals 4–8.5 cm, similar in colour to the sepals. Lip white, red-spotted, bearing green warts or excrescences, oblong at the base, widened towards the apex. Ridges ending in attached teeth. *Tropical America from S Mexico to Venezuela.* G2. Spring–summer.

128. MILTONIA Lindley

J. Cullen

Epiphytic with rhizomes. Pseudobulbs clustered or distant, grey-green or yellow-green, simple, usually compressed, each with 1–2 leaves. Leaves oblong, oblong-linear, or oblong-lanceolate, usually acute, folded when young. Racemes arising from the bases of the pseudobulbs, usually erect, with 1–several flowers; bracts sometimes conspicuous. Flowers often rather flat. Sepals and petals similar. Lip spreading away from the column base, usually simple, usually larger than the sepals, with 2 or more longitudinal ridges (or rarely a patch of hairs) at the base. Column short, with a narrow wing on each side at the base, without a foot, with auricles at the apex. Pollinia 2, ovoid to ellipsoid, stipe variable, shorter or longer than pollinia, viscidium shorter than the stipe and forming an acute angle with it.

A genus of about 12 species mainly from Brazil, with 1 in Peru. Much hybridisation has been done within the genus, and many hybrids are cultivated. Cultivation as for *Odontoglossum* (p. 561) but the plants require a higher winter temperature.

Literature: Sweet, H.R., The Miltonia complex in horticulture, *American Orchid Society Bulletin* **47**: 917–925 (1978).

Rhizome. Long, pseudobulbs distant: **1**; short, pseudobulbs clustered: **2, 3, 4, 5, 6, 7, 8, 9**.

Pseudobulbs. Greyish green, surrounded by the bases of a number of leaves: **1**; yellow-green, not surrounded by leaf-bases: **2, 3, 4, 5, 6, 7, 8, 9**.

Raceme. With 4 or more flowers: **4, 5, 6, 7, 8, 9**; with fewer than 4 flowers: **1, 2, 3**. Axis flattened: **3**.

Sepals. White, pink, red or purple: **1, 2, 4**; green to yellow, often spotted with brown: **3, 4, 5, 6, 7, 8, 9**.

Lip. Longer than sepals: **1, 2, 3, 4**; shorter than sepals: **5, 9**; about as long as sepals: **6, 7, 8**. Clearly waisted: **1, 5, 6, 7**; with oblong claw and abruptly widened blade: **3, 8**; widening gradually from the base and more or less flat: **2, 4**; widening gradually from the base, the sides curving up to the column: **9**.

1a. Pseudobulbs and leaves greyish green, the pseudobulbs surrounded by the bases of a number of leaves or leaf-sheaths; pseudobulbs usually with 1 leaf, usually distant **1. warszewiczii**

b. Pseudobulbs and leaves yellowish green, the pseudobulbs not surrounded by leaf-bases; pseudobulbs usually with 2 leaves and clustered 2

2a. Sepals and petals of a uniform colour, white, yellow, pink or purple, not spotted or blotched 3

b. Sepals and petals blotched or spotted with brown on a yellow or green ground 6

3a. Bract much shorter than the ovary and flower-stalk together; base of lip with 5–9 ridges **4. regnellii**

b. Bract very large and conspicuous, as long as the ovary and flower-stalk together, sheathing; base of the lip with 3 ridges or a patch of hairs 4

4a. Flowers 6–12 in a raceme; lip, sepals and petals yellow, lip with a patch of hairs at the base **5. flavescens**

b. Flowers solitary (rarely 2); lip, sepals and petals white, greenish yellow, purple or reddish, lip with 3 ridges 5

5a. Axis of raceme flattened; lip white with purple lines, oblong at the base, abruptly widened towards the apex **3. anceps**

b. Axis of raceme terete; lip purple or reddish, widening gradually from the base **2. spectabilis**

6a. Sides of the lip curving upwards to meet the column **9. candida**

b. Lip flat, or if somewhat concave, the sides not meeting the column 7

7a. Blade of the lip narrowly oblong at the base, abruptly widened to an apical part which is broader than long **8. cuneata**

b. Blade oblong, waisted towards the apex, the part beyond the waist longer than broad 8

8a. Sepals and petals little spreading, directed forwards; lip purple or pink with darker spots **6. russelliana**

b. Sepals and petals widely spreading; base of the lip purple with a median yellow stripe, apical part white **7. clowesii**

1. M. warszewiczii Reichenbach 1856 (*Odontoglossum weltonii* invalid; *Oncidium weltonii* W. Bull; *Miltonioides warszewiczii* (Reichenbach) Brieger & Lückel; *Chamaeleorchis warscewiczii* (Reichenbach) Senghas & Lückel; *Oncidium fuscatum* Reichenbach). Illustration: *Botanical Magazine*, 5843 (1870); *Lindenia* **8**: t. 384 (1892); Cogniaux & Goossens, Dictionnaire iconographique des orchidées, Miltonia, t. 8 (1900); Bechtel et al., Manual of cultivated orchid species, 253 (1981).

Similar to *M. phalaenopsis* but sepals and petals oblong or oblong-obovate, conspicuously undulate-crisped, mostly purple, the tips white, lip mostly purple, the margins white, usually with a yellow blotch or variegation in the centre. *Peru, Ecuador, Colombia*. G2. Spring.

Oncidium fuscatum is the prefered name for this species as demonstrated by molecular studies.

The names applied to this species are often confused with *Miltoniopsis warscewiczii* (Reichenbach) Garay & Dunsterville (below), to which it is not similar. Both species are illustrated in Betchel et al., and Zelencho & Chase, Orchids: the pictorial encyclopaedia of Oncidium,11, 89 (2002).

2. M. spectabilis Lindley. Illustration: *Edwards's Botanical Register* **23**: t. 1992 (1837); Cogniaux & Goossens, Dictionnaire iconographique des orchidées, Miltonia, t. 2, 2A (1900); Bechtel et al., Manual of cultivated orchid species, 232 (1981).

Pseudobulbs to 7 cm, clustered, ovoid, compressed, each with 2 leaves. Leaves to 15 cm, narrowly linear-oblong, green. Raceme with 1 or 2 flowers and a terete axis; bracts exceeding the flower-stalk and ovary together, sheathing. Sepals and petals to 4.5 cm, spreading, oblong, obtuse, uniformly white, light pink or reddish. Lip 3.4–6.1 × 2.1–4.5 cm, broadly obovate, widening gradually from the base, purple or reddish, with 3 yellow ridges at the base. *Venezuela, E Brazil*. G2. Summer.

M. moreliana Richard (*M. spectabilis* var. *moreliana* Henfrey; var. *atrorubens* O'Brien; var. *purpureo-violacea* Hooker; *M. rosea* Verschaffelt; *M. warneri* Nicholson). Illustration: *Botanical Magazine* 4425 (1849); Zelenko & Chase (ed.), Orchids: the pictorial encyclopaedia of Oncidium, 58 (2002). Differs from *M. spectabilis* by a usually larger flower size. Lip 4.6–6 × 4.2–61.5 cm. Petals and sepals dark purple. Once treated as a variety of *M. spectabilis* but now accepted at species level, see *Lindleyana* **17**(3): 122–129 (2002).

3. M. anceps Lindley (*Odontoglossum anceps* (Lindley) Klotzsch). Illustration: *Botanical Magazine*, 5572 (1866); Pabst & Dungs, Orchidaceae Brasilienses **2**: f. 2050 (1977).

Pseudobulbs 4.5–5.5 cm, each with 2 leaves, yellow-green, oblong, strongly compressed, clustered. Raceme with 1 flower, a flattened axis and the bract sheathing, exceeding the ovary and flower-stalk together. Sepals and petals 3.5–3.7 cm, spreading, oblong-lanceolate, obtuse, uniformly greenish yellow. Lip oblong, slightly waisted, exceeding the sepals, white with purple lines, and with 3 yellow ridges, each with a single red stripe at the base. *SE Brazil*. G2. Spring.

4. M. regnellii Reichenbach (*Oncidium regnellii* (Reichenbach) Reichenbach). Illustration; *Botanical Magazine*, 5436 (1864); Cogniaux & Goossens, Dictionnaire iconographique des orchidées, Miltonia, t. 7, 7A, 7B (1900); Hoehne, Iconografia de orchidaceas do Brasil, t. 267 (1949); Bechtel et al., Manual of cultivated orchid species, 232 (1981).

Pseudobulbs 5–8 cm, each with 2 leaves, yellow-green, clustered, ovoid. Leaves to 30 cm, linear-lanceolate, acute. Raceme with 4–7 flowers, erect, loose. Sepals and petals *c.* 3.5 cm, ovate-oblong, acute, white or yellow. Lip broadly obovate, larger than the sepals, obscurely 3-lobed, blunt at the apex, white or violet with white margins, with 7–9 low ridges at the base. *S Brazil*. G2. Summer.

5. M. flavescens (Lindley) Lindley (*Cyrtochilum flavescens* Lindley). Illustration: *Gartenflora* **39**: t. 1328 (1890); Hoehne, Iconografia de orchidaceas do Brasil, t. 263 (1949); Pabst & Dungs, Orchidaceae Brasilienses **2**: f. 2051 (1977).

Pseudobulbs 6–8 cm, each with 2 leaves, clustered, yellow-green, oblong. Leaves to 30 cm, linear-oblong, acute. Raceme with 6–12 flowers, erect, loose, bracts conspicuous, sheathing, longer than the ovary and flower-stalk together. Sepals and petals *c.* 3.5 cm, yellow, linear-oblong or linear-lanceolate, acute. Lip oblong, waisted, shorter than the sepals, pale yellow with reddish lines. *S & E Brazil, Paraguay, N Argentina*. G2. Summer–autumn.

6. M. russelliana Lindley. Illustration: *Edwards's Botanical Register* **22**: t. 1830 (1836); Pabst & Dungs, Orchidaceae Brasilienses **2**: f. 2058 (1977).

Pseudobulbs clustered, yellow-green. Leaves linear-oblong. Racemes several-flowered, erect, bracts inconspicuous. Sepals and petals 2.5–3 cm, oblong, acute, not spreading but directed forwards, green or greenish yellow heavily marbled or spotted with brown. Lip about as long as sepals, oblong, somewhat waisted towards the apex, purple or pink with darker spots, with many ridges at the base. *S Brazil*. G2. Winter–spring.

7. M. clowesii Lindley (*Odontoglossum clowesii* (Lindley) Lindley; *Brassia clowesii* (Lindley) Lindley). Illustration: *Botanical Magazine*, 4109 (1844); Cogniaux & Goossens, Dictionnaire iconographique des orchidées, Miltonia, t. 4 (1900); Bechtel et al., Manual of cultivated orchid species, 231 (1981).

Pseudobulbs 5–8 cm, clustered, yellow-green, elongate-ovoid, each with 2 leaves. Leaves to 30 cm, linear-lanceolate, acute. Raceme with 3–7 flowers, erect, loose, bracts inconspicuous. Sepals and petals 3–3.5 cm, spreading, oblong, acute, greenish yellow with brown spots. Lip about as long as sepals, waisted, the anterior part ovate, acute, violet in the lower part, white towards the apex, with several short ridges at the base. *E Brazil*. G2. Autumn.

8. M. cuneata Lindley (*M. speciosa* Klotzsch). Illustration: *Edwards's Botanical Register* **31**: t. 8 (1845); Cogniaux & Goossens, Dictionnaire iconographique des orchidées, Miltonia, t. 9 (1900); Pabst & Dungs, Orchidaceae Brasilienses **2**: f. 2055 (1977); Bechtel et al., Manual of cultivated orchid species, 232 (1981).

Pseudobulbs 5–8 cm, oblong-ovoid, yellow-green, clustered, each with 2 leaves. Leaves to 30 cm, oblong-lanceolate. Raceme with 5–8 flowers, erect. Sepals and petals 3–3.5 cm, oblong, acute, the margins somewhat wavy, yellow spotted with brown. Lip about as long as the sepals, narrowly oblong at the base, abruptly widened to an obovate, heart-shaped apical part that is broader than long, slightly notched at the apex, white, with 2 ridges at the base. *E Brazil*. G2. Spring.

9. M. candida Lindley (*Oncidium candidum* (Lindley) Reichenbach). Illustration:

Hoehne, Iconografia de orchidaceas do Brasil, t. 262 (1949); Pabst & Dungs, Orchidaceae Brasilienses **2**: f. 2053 (1977); Bechtel et al., Manual of cultivated orchid species, 231 (1981).

Pseudobulbs 5–8 cm, yellow-green, clustered, oblong-ovoid, each with 2 leaves. Leaves to 30 cm, linear-lanceolate, acute. Raceme with 3–7 flowers, erect, loose. Sepals and petals 4–4.5 cm, oblong, acute, greenish yellow with many brown spots. Lip shorter than the sepals, white or rarely pink, sometimes purplish in the centre, its sides rolled upwards to meet the column, somewhat wavy, with 3 ridges at the base. *E Brazil*. G2. Autumn.

129. MILTONIOPSIS Godefroy

J.M.H. Shaw

Resembling *Miltonia* but pseudobulbs and leaves greyish green, the pseudobulbs bearing 1 leaf, and surrounded by the bases of a number of leaves or leaf-sheaths. Flowers large showy and flattened. Lip broad and flat with a simple, flat callus, joined to the column foot by the central keel. Column short with a keel along the back, without a foot or wings. Pollinia 2.

A small genus of about 6 species ranging from Costa Rica to Peru. In the past they have been included in *Miltonia*, but molecular studies show the two groups are only distantly related. Cultivation is as for *Odontoglossum* (p. 561), in intermediate to warm conditions in shade.

1a. Lip with 2 backwardly pointed projections at the base, the blade not waisted 2
b. Lip without 2 backwardly pointed projections at the base, the blade waisted 3
2a. Sepals and petals white, acuminate, the petals usually with a deep purple blotch at the base; leaves striped with dark green lines beneath **2. roezlii**
b. Sepals and petals entirely pink, rounded to a blunt point; leaves without dark green lines beneath **1. vexillaria**
3a. Flowers *c.* 7 cm in diameter; lip with 2 small reddish or violet spots near the base, the rest white **3. endresii**
b. Flowers to 5.5 cm in diameter; lip mostly violet, the margins white **4. phalaenopsis**

1. M. vexillaria (Bentham) Godefroy-Lebeuf (*Miltonia vexillaria* (Bentham) Nicholson; *Odontoglossum vexillarium* Bentham). Illustration: Cogniaux & Goossens, Dictionnaire iconographique des orchidées, Miltonia, t. 1 (1900); Schlechter, Die Orchideen, edn 2, t. 10 (1927); Bechtel et al., Manual of cultivated orchid species, 233 (1981).

Pseudobulbs 5–6 cm, grey-green, distant, compressed, each with 1 leaf, subtended by 3–4 leaf-bases on each side. Leaves 15–25 cm, grey-green, linear-oblong. Raceme with 4–6 flowers, loose. Sepals and petals *c.* 4 cm, oblong-elliptic, rounded to a blunt point, pink. Lip much larger than the sepals, almost circular, with 2 backwardly directed projections and 3 ridges at the base, deeply notched at the apex, very variable in colour. *Colombia, N Ecuador*. G2. Spring–summer.

A very attractive species with the lip varying from pink to white, usually with red lines and a yellow blotch towards the base. Many cultivars exist and are widely grown.

2. M. roezlii (Reichenbach) Godefroy-Lebeuf (*Miltonia roezlii* (Reichenbach) Nicholson; *Odontoglossum roezlii* Reichenbach). Illustration: *Lindenia* **2**: t. 78 (1886); Cogniaux & Goossens, Dictionnaire iconographique des orchidées, Miltonia, t. 6 (1900); Sheehan & Sheehan, Orchid genera illustrated, 121 (1979); Bechtel et al., Manual of cultivated orchid species, 233 (1981).

Similar to *M. vexillaria* but leaves with dark green longitudinal lines beneath, sepals and petals acuminate, white, the petals with a purple blotch at the base, lip with 5 ridges at the base. *Colombia*. G2. Winter.

3. M. warscewiczii (Reichenbach) Garay & Dunsterville (*Miltonia endresii* Nicholson; *Odontoglossum warscewiczii* Reichenbach; *M. superba* Schlechter invalid). Illustration: Cogniaux & Goossens, Dictionnaire iconographique des orchidées, Miltonia, t. 10 (1900).

Pseudobulbs 5–6 cm, grey-green, subtended by 3–4 leaf-bases on each side, each 1-leaved, compressed, usually distant on the rhizome. Leaves 15–25 cm, grey-green, linear-oblong. Raceme with 4–6 flowers, loose. Flowers *c.* 7 cm in diameter. Sepals and petals 3–3.5 cm, ovate-elliptic, obtuse, white with red or purple stripes at the base. Lip larger than the sepals, broadly obovate, slightly waisted near the base, deeply notched and with a small point in the sinus at the apex, white with 2 small red or purple blotches at the base, and with 3 short papillose ridges. *Costa Rica*. G2. Winter.

If included with in *Miltonia* the correct name for this species is *M. endresii* Nicholson. If placed in *Miltoniopsis* the correct name is *Miltoniopsis warscewiczii*. The name *Miltonia warscewiczii* Reichenbach is a synonym of *Oncidium fuscatum* Reichenbach (see *Oncidium*, No. 5).

4. M. phalaenopsis (Linden & Reichenbach) Garay & Dunsterville (*Miltonia phalaenopsis* (Linden & Reichenbach) Nicholson; *Odontoglossum phalaenopsis* Linden & Reichenbach; *Miltonia pulchella* Linden). Illustration: *Lindenia* **7**: t. 334 (1891); Cogniaux & Goossens, Dictionnaire iconographique des orchidées, Miltonia, t. 11 (1900); Bechtel et al., Manual of cultivated orchid species, 233 (1981).

Pseudobulbs 2.5–4 cm, ovoid, grey-green, usually distant on the rhizome, subtended by 3–4 leaf-bases on each side, each with 1 leaf. Leaves 15–22 cm, green, linear. Raceme with 2–4 flowers, loose. Flowers to 5.5 cm in diameter. Sepals and petals ovate-elliptic, white. Lip larger than the sepals, broadly oblong, waisted towards the base, notched and with a small point within the sinus at the apex, white, mostly overlaid by red-purple in the form of streaks or lines that join together, covering most of the surface; callus with 3 short teeth at the base. *Colombia*. G2. Autumn.

130. ONCIDIUM Swartz

J. Cullen

Rhizomatous epiphytes. Pseudobulbs usually present, simple, usually bearing 1–3 leaves and subtended by a number of leaves or sheathing leaf-bases. Leaves variable, fleshy, leathery, sometimes terete, folded when young. Flowers usually numerous in racemes or panicles. Sepals and petals usually spreading, usually clawed, similar in size or the petals larger, the lateral sepals free or variably united at the base. Lip entire or 3-lobed, often waisted, very variable in shape and size, firmly united with the base of the column from which it diverges, with a complex tuberculate or ridged callus. Column with a fleshy plate below the stigma but without a foot, and with distinct, variably shaped, auricles or wings on either side. Rostellum short or beaked. Pollinia 2, ovoid to spherical, stipe linear, longer than pollinia, viscidium small, making an acute angle with the stipe.

A large genus (*c.* 450 species?) from subtropical and tropical America, of which over 100 species have been in cultivation. It is sometimes difficult to distinguish from both *Miltonia* and *Odontoglossum*, and it has been suggested that it should be broken up into a number of genera. Garay & Stacy (reference below), however, prefer to maintain it as a single genus. They divide the genus into many sections, some of which are interpolated into the present account with very short descriptions. More recently molecular evidence has suggested that there should be readjustments of many genera within this group, involving the recognition of several genera previously included within *Oncidium* and conversely the inclusion of *Cochlioda* and *Odontoglossum* within *Oncidium*. As the treatment of this group in Genera Orchidacearum has not yet been published traditional generic boundaries are retained here, with appropriate synonyms added.

The species are very variable as to their requirements in cultivation. They are usually grown in epiphytic compost in containers or on rafts. Some have a definite resting period while others do not, and treatment should vary accordingly. Propagation is by seed, division or the removal of small bulbs from mature plants.

Literature: Kränzlin, F., Das Pflanzenreich **80**: 25–290 (1922); Garay, L.A., A reappraisal of the genus Oncidium Sw., *Taxon* **19**: 443–67 (1970); Garay, L.A. & Stacy, J.E., Synopsis of the genus Oncidium, *Bradea* **1**(40): 393–424 (1974); Williams, N.H., Chase, M.W., Fulcher, T. & Whitten, W.M., Molecular systematics of the Oncidiinae based on evidence from four DNA sequence regions: expanded circumscriptions of Cyrtochilum, Erycina, Otoglossum, and Trichocentrum and a new genus [Zelenkoa], *Lindleyana* **16**(2): 113–139 (2001); Zelenko, H. & Chase, M.W. (ed.), Orchids: the pictorial encyclopaedia of Oncidium (2002).

1a. Sepals, petals and lip fleshy (Section *Cyrtochilum*) 2
b. Sepals fleshy or not; petals and lip not fleshy 4
2a. Lip oblong, tapering slightly to the apex (Figure 31(3), p. 573) **3. superbiens**
b. Lip broadly triangular at the base, abruptly narrowed to an acuminate apex 3
3a. Sepals and petals more than 4 cm, petals yellow **1. macranthum**
b. Sepals and petals less than 2 cm, petals reddish brown **2. microchilum**
4a. Pseudobulbs absent or rudimentary and less than 4 cm, rarely present and large, when the plant is very large (panicle up to 2 m) 5
b. Pseudobulbs present, usually more than 4 cm, always large and conspicuous for the size of the plant 13
5a. Sheaths enveloping the pseudobulbs all leaf-bearing (Section *Onusta*) **4. onustum**
b. Sheaths enveloping the pseudobulbs mostly without leaf-blades 6
6a. Leaves terete (Section *Cebolleta*) **5. cebolleta**
b. Leaves flat 7
7a. Leaves leathery; callus of lip hairless with many variable tubercles (Section Pluritaberculata) 8
b. Leaves fleshy; callus of lip downy, swollen (Section *Pulvinata*) 12
8a. Pseudobulbs absent or at most 1.5 cm 9
b. Pseudobulbs present and larger 11
9a. Sepals and petals unspotted, greenish yellow or yellow, bordered with yellow or crimson; flowers in racemes **10. bicallosum**
b. Sepals and petals yellow spotted with reddish brown or purple-brown; flowers in panicles 10
10a. Petals to 9 mm; lip 11 mm at its widest **8. nanum**
b. Petals more than 1.2 cm; lip more than 1.5 cm at its widest **9. cavendishianum**
11a. Panicle to 2 m or more with many branches and many flowers; sepals and petals to 2 cm **6. altissimum**
b. Panicle much smaller with few branches (rarely unbranched) and with 9–20 flowers; sepals and petals 2.5–3 cm **7. splendidum**
12a. Flowers to 1.8 cm in diameter; leaves grey-green, 7.5–15 cm **11. harrisonianum**
b. Flowers more than 2 cm in diameter; leaves green, 20–30 cm **12. divaricatum**
13a. Upper sepal and petals erect, antenna-like (Section *Glanduligera*) 14
b. Upper sepal and petals not as above 15
14a. Axis of raceme flattened above **13. papilio**
b. Axis of raceme not flattened above **14. kramerianum**
15a. Column downy with auricles in the form of forwardly projecting fleshy arms on the sides of the stigma (Section *Waluewa*) **15. pubes**
b. Column hairless with membranous auricles 16
16a. Sepals, petals and lip spread in 1 plane, the lip little more conspicuous than the rest (Section *Stellata*) 17
b. Sepals, petals and lip not spread in 1 plane, the lip much more conspicuous than the rest 18
17a. Lip white with red lines at the base, yellow towards the apex, lateral lobes triangular, pointing forwards (Figure 31(16), p. 573) **16. maculatum**
b. Lip mostly pink, lateral lobes oblong or almost square, directed outwards (Figure 31(17), p. 573) **17. hastatum**
18a. Rostellum elongate, proboscis-like (Section *Rostrata*) 19
b. Rostellum short, not projecting 20
19a. Lip, sepals and petals mostly pink except for the yellow ridges on the lip **18. ornithorrhynchum**
b. Lip, sepals and petals yellow except for the white ridges on the lip **19. cheirophorum**
20a. Callus of the lip with 2–4 parallel tubercles or ridges (Section *Concoloria*) **20. concolor**
b. Callus of lip with 3, 5 or 7 serial or scattered tubercles, rarely with accessory tubercles or papillae as well 21
21a. Lateral sepals longer than lip (Section *Barbata*) **21. longipes**
b. Lateral sepals shorter than the lip 22
22a. Lateral sepals united at the base to varying extents 23
b. Lateral sepals entirely free from each other 28
23a. Petals large, much more conspicuous than sepals (Section *Crispa*) 24
b. Petals and sepals similar in size (Section *Synsepala*) 27
24a. Wings of column sharply and conspicuously toothed **22. crispum**
b. Wings of column entire 25
25a. Sepals, petals and lip chestnut brown with yellow marbling on the margins; flowers usually in a raceme **23. forbesii**

Figure 31. Lips of *Oncidium* species. 1, *O. macranthum*. 2, *O. microchilum*. 3, *O. superbiens*. 4, *O. onustum*. 5, *O. cebolleta*. 6, *O. altissimum*. 7, *O. splendidum*. 8, *O. nanum*. 9, *O. cavendishianum*. 10, *O. bicallosum*. 11, *O. harrisonianum*. 12, *O. divaricatum*. 13, *O. papilio*. 14, *O. kramerianum*. 15, *O. pubes*, 16, *O. maculatum*. 17, *O. hastatum*. 18, *O. ornithorrhynchum*. 19, *O. cheirophorum*.

b. At least the lip mostly yellow, sometimes with red-brown spots at the base or on the margins; flowers in panicles 26
26a. Lip 3 × 4.5 cm; lateral sepals united for half their length **24. gardneri**
b. Lip *c.* 4 × 5 cm; lateral sepals united for a third of their length **25. marshallianum**
27a. Blade of lip 2.5–4 × 3.5–5.5 cm with 3 notches towards the apex, producing 4 lobes (Figure 32(26), p. 578) **26. varicosum**
b. Blade of lip 1.5–2 × 1.5–1.8 cm with 1 deep notch at the apex **27. flexuosum**
28a. Petals wider and more conspicuous than the sepals (Section *Excavata*) **28. excavatum**
b. Petals and sepals similar in size and colour 29
29a. Bracts small, scale-like, inconspicuous (Section *Oblongata*) 30
b. Bracts large and conspicuous (Section *Planifolia*) 32
30a. Lip white; sepals and petals to 1.7 cm **31. leucochilum**
b. Lip yellow; sepals and petals 2–7.5 cm 31
31a. Pseudobulbs with 1 leaf; sepals and petals 6–7.5 cm; auricles of the column toothed **29. ampliatum**
b. Pseudobulbs with 2 leaves; sepals and petals 2–3 cm; auricles of the column entire **30. tigrinum**
32a. Leaves 35–60 cm; lip wider across the central lobe than across the lateral lobes (Figure 32(32), p. 578) **32. sphacelatum**
b. Leaves 20–30 cm; lip wider across the lateral lobes than across the central lobe (Figure 32(33), p. 578) **33. wentworthianum**

Section **Cyrtochilum** (Kunth) Lindley. Flowers conspicuous, sepals, petals and lip all fleshy; column with fleshy auricles.

Literature: Dalström, S., A synopsis of the genus Cyrtochilum: taxonomic reevaluations and new combinations, *Lindleyana* **16**(2): 56–80, (2001), **16**(4): 255 (2001) & **17**(2): 91–2 (2002).

1. O. macranthum Lindley (*O. hastiferum* Reichenbach & Warscewicz; *Cyrtochilum macranthum* (Lindley) Kranzlin). Figure 31 (1), p. 573. Illustration: *Botanical Magazine*, 5743 (1868); *The Garden* **24**: t. 416 (1883); Cogniaux & Goossens, Dictionnaire iconographique des orchidées, Oncidium, t. 14 (1898); Bechtel et al., Manual of cultivated orchid species, 246 (1981).

Pseudobulbs 7–15 cm, oblong-conical, each with 2 leaves. Leaves 25–50 × 2.5–5 cm, oblanceolate to oblong, acute. Panicle to 3 m, each branch with 2–5 flowers. Sepals 2.5–3.5 cm, clawed, the laterals somewhat longer than the upper, dull yellow-brown. Petals circular to ovate, very shortly clawed, crisped-undulate, a little shorter than the sepals, golden yellow. Lip smaller than the petals, broadly triangular, abruptly narrowed to the recurved, acuminate apex, red or brownish purple towards the base, yellowish with red lines towards the apex, with a large white callus with 6–7 erect teeth. *Colombia, Ecuador, Peru*. G1. Summer.

2. O. microchilum Lindley (*Trichocentrum microchilum* (Lindley) M.W. Chase & Williams). Figure 31(2), p. 573. Illustration: *Fieldiana* **26**: 655 (1953); Bechtel et al., Manual of cultivated orchid species, 246 (1981).

Pseudobulbs 2.5–3.5 cm, spherical, compressed, each with 1 leaf. Leaves 12–25 × 3–6.5 cm, fleshy, elliptic, margins irregularly scalloped. Panicle to 1.5 m with many flowers. Sepals 1.2–1.4 cm, shortly clawed, circular to elliptic, reddish brown, the upper very concave. Petals almost clawless, oblong-elliptic, otherwise similar to sepals. Lip complex, very broadly triangular, much shorter than the petals, white spotted with red; callus dark red, covering most of the centre, with 5 or more blunt tubercles. Column with 2 slender, sickle-shaped wings. *Mexico, Guatemala*. G2. Summer.

3. O. superbiens Riechenbach (*Cyrtochilum halteratum* (Reichenbach) Krainzlin. Figure 31(3), p. 573. Illustration: *Botanical Magazine*, 5980 (1872); Cogniaux & Goossens, Dictionnaire iconographique des orchidées, Oncidium, t. 31 (1904); Bechtel et al., Manual of cultivated orchid species, 249 (1981).

Pseudobulbs 6–10 cm, clustered, each usually with 2 leaves. Leaves to 70 cm, linear-oblanceolate, acute. Panicle to 6 m, twining, with many flowers, Sepals clawed, ovate to almost circular, the upper 1.6–2 cm, the laterals somewhat longer, all greenish brown or brown, sometimes with yellow margins. Petals smaller than the sepals, oblong-ovate to almost circular, shortly clawed, margins crisped-undulate, yellow or white with reddish or brown bars in the lower half. Lip shorter than the petals, oblong, the apex recurved, purple, with a complex, toothed callus. Column with sickle-shaped wings. *Peru, Colombia, Ecuador & Venezuela*. G1. Winter.

Section **Onusta** Garay & Stacy. Pseudobulbs well developed, bearing leaves, or leafless with a terminal appendage; sheaths surrounding the pseudobulbs all leaf-bearing; leaves leathery or fleshy, finely toothed towards the apex.

4. O. onustum Lindley (*Zelenkoa onusta* (Lindley) M.W. Chase & Williams). Figure 31(4), p. 573. Illustration: *Lindenia* **11**: t. 498 (1895).

Pseudobulbs 2–4 cm, subtended by several leaves with blades, green mottled with brown, bearing 1 or 2 leaves or a terminal appendage. Leaves 5–12.5 × 1–1.8 cm, oblong or linear-oblong. Raceme (rarely branched) to 40 cm, often 1-sided, with many flowers. Sepals 7–8 mm, ovate or ovate-elliptic, yellow, the upper hooded. Petals *c.* 2 times as large as the sepals, almost circular, wavy, yellow. Lip larger than the petals, 3-lobed, the lateral lobes small, the central lobe almost circular, bearing a 3-lobed callus. Column with large, crescent-shaped auricles. *Panama, Colombia, Ecuador & Peru*. G2. Spring.

Section **Cebolleta** Lindley. Pseudobulbs small, each bearing 1 leaf; leaves terete in section.

5. O. cebolleta (Jacquin) Swartz (*Trichocentrum cebolleta* (Jacquin) M.W. Chase & Williams). Figure 31(5), p. 573. Illustration: *Botanical Magazine*, 3568 (1837); Dunsterville & Garay, Venezuelan orchids illustrated **2**: 246–7 (1961); Pabst & Dungs, Orchidaceae Brasilienses **2**: t. 1945 (1977); Bechtel et al., Manual of cultivated orchid species, 242 (1981).

Pseudobulbs 1.5–2 cm, conical to almost spherical, each with 1 leaf. Leaves 7–40 cm, fleshy, terete, slightly grooved, erect, often tinged or spotted with purple. Panicle up to 1.2 m with many flowers, its stalk spotted with purple. Sepals 6–10 mm, obovate, greenish yellow with red-brown spots. Petals similar, margins wavy. Lip 3-lobed, with lateral lobes rather large, oblong or obovate, the central lobe larger than the sepals, shortly clawed, kidney-shaped, deeply notched at the apex, yellow, callus consisting of a sharp projecting ridge surrounded by tubercles. Column with conspicuous, oblong, sometimes 2-lobed auricles. *Tropical America, from Mexico and the West Indies to N Argentina.* G2. Spring.

Section **Pluritubercuata** Lindley. Pseudobulbs absent or very small for the size of the plant; leaves leathery; callus of lip with many tubercles.

6. O. altissimum (Jacquin) Swartz. Figure 31(6), p. 573. Illustration: *Edwards's Botanical Register* **19**: t. 1651 (1834).

Pseudobulbs 5–7 cm, ovoid, compressed, each with 2 leaves. Leaves 25–30 cm, oblong or oblong-lanceolate, acute. Panicle to 2 m or more, with many flowers. Sepals and petals *c.* 2 cm, narrowly lanceolate or oblanceolate, somewhat wavy, greenish yellow with red or red-brown spots. Lip 3-lobed with small, oblong lateral lobes and a clawed, kidney-shaped central lobe that is slightly notched at the apex and bears a complex callus with about 10 tubercles, yellowish with a reddish band across the narrowest part. Column with small, triangular, entire auricles. *West Indies.* G2. Spring–summer.

For a discussion of the use of the name *O. altissimum*, see Garay & Stacy, *Bradea* **1** (40): 395–6 (1974).

7. O. splendidum Richard (*Trichocentrum splendidum* (Richard) M.W. Chase & Williams). Figure 31(7), p. 573. Illustration: Cogniaux & Goossens, Dictionnaire iconographique des orchidées, Oncidium, t. 7 (1897); Bechtel et al., Manual of cultivated orchid species, 248 (1981).

Pseudobulbs 4–5 cm, clustered, almost spherical, each bearing 1 leaf. Leaves to 100 × 3.5 cm, oblong-elliptic, usually purple-tinged. Panicle to 1.5 m with many flowers. Sepals *c.* 2.5 cm, shortly clawed, linear-elliptic to oblanceolate, wavy, yellow-green with broad, red-brown bands. Petals similar to sepals but up to 3 cm. Lip much larger than the sepals, 3-lobed, the lateral lobes very small, the central up to 4 cm wide, almost stalkless, kidney-shaped, deeply notched, somewhat wavy, yellow, with a callus composed of 3 conspicuous, bright yellow ridges. Column with small wings. *Guatemala.* G2. Winter–spring.

8. O. nanum Lindley (*Trichocentrum nanum* (Lindley) M.W. Chase & Williams). Figure 31(8), p. 573. Illustration: Dunsterville & Garay, Venezuelan orchids illustrated **1**: 272–3 (1959); Pabst & Dungs, Orchidaceae Brasilienses **2**: t. 1948 (1977); Bechtel et al., Manual of cultivated orchid species, 246 (1981).

Pseudobulbs to 7 mm, cylindric, each with 1 leaf. Leaves 7–24 × 2–7 cm, elliptic or oblong, acute. Panicle to 25 cm with many flowers. Sepals 7–9 mm, obovate, yellow spotted with orange, purple or brown. Petals similar but a little smaller. Lip slightly longer than the sepals, 3-lobed, the lateral lobes small, the central lobe transversely oblong to kidney-shaped, notched at the apex and with a complex, tuberculate callus at the base. Column with oblique, slightly toothed auricles. *Tropical S America from Colombia to Peru and C Brazil.* G2. Spring.

9. O. cavendishianum Bateman (*Trichocentrum cavendishianum* (Bateman) M.W. Chase & Williams). Figure 31(9), p. 573. Illustration: Cogniaux & Goossens, Dictionnaire iconographique des orchidées, Oncidium, t. 11 (1897); Bechtel et al., Manual of cultivated orchid species, 242 (1981).

Pseudobulbs absent or very small. Leaves 15–45 × 5–13 cm, elliptic to broadly lanceolate, acute or subobtuse. Panicle 60–150 cm, with many flowers. Flowers very fragrant. Sepals 1.2–1.7 cm, obovate, obtuse, margins undulate-crisped, yellow spotted with red. Petals similar but clawed. Lip larger than the sepals, deeply 3-lobed, the lateral lobes conspicuous, obovate, the central lobe large, transversely oblong to kidney-shaped, notched at the apex, yellow with red spots at the base; callus consisting of 5 tubercles. Column with sickle-shaped, red-spotted auricles. *S Mexico, Guatemala, Honduras.* G2. Spring.

10. O. bicallosum Lindley (*Trichocentrum bicallosum* (Lindley) M.W. Chase & Williams). Figure 31(10), p. 573. Illustration: *Edwards's Botanical Register* **29**: t. 12 (1843); *Botanical Magazine*, 4148 (1845); Bechtel et al., Manual of cultivated orchid species, 241 (1981).

Pseudobulbs very small or absent. Leaves 15–35 × 4–8.5 cm, oblong-elliptic, obtuse. Raceme (rarely branched) 20–65 cm with many flowers. Flowers fragrant. Sepals 1.5–2 cm, obovate to spathulate, obtuse, greenish yellow to deep yellow, rarely margined with red. Petals similar but with crisped margins. Lip larger than the sepals, deeply 3-lobed, the lateral lobes rather small, obovate, the central lobe transversely oblong to kidney-shaped, yellow, with a callus of 2 swellings and a variable number of tubercles, which are white with red spots. Column wings fleshy, sickle-shaped. *S Mexico, Guatemala, El Salvador.* G2. Autumn.

The following species, also belonging to this section, are sometimes grown:

O. carthagenense (Jacquin) Swartz (*Trichocentrum carthagenense* (Jacquin) M.W. Chase & Williams). Illustration: Bechtel et al., Manual of cultivated orchid species, 242 (1981). Leaves spotted with red, flowers white or pale yellow, densely spotted with red. *USA (Florida), West Indies, C America and northern S America.* G2. Summer.

O. pumilum Lindley (*Trichocentrum pumilum* (Lindley) M.W. Chase & Williams). Leaves 5–10 cm, panicle little longer than the leaves, flowers yellow, to 7 mm in diameter. *S Brazil, Paraguay.* G1. Spring.

O. stramineum Bateman (*Trichocentrum stramineum* (Lindley) M.W. Chase & Williams). Flowers whitish yellow, lip white, spotted with red. *Mexico.* G2. Spring.

Section **Pulvinata** Lindley. Pseudobulbs small, subtended by sheaths without blades; callus of the lip downy.

11. O. harrisonianum Lindley. Figure 31 (11), p. 573. Illustration: *Edwards's Botanical Register* **19**: t. 1569 (1833); Pabst & Dungs, Orchidaceae Brasilienses **2**: t. 1955 (1977); Bechtel et al., Manual of cultivated orchid species, 244 (1981).

Pseudobulbs to 2.5 cm, almost spherical, each bearing 1 leaf. Leaves 7.5–15 cm, greyish green. Panicle with 15–20 flowers, little branched, erect, loose, exceeding the leaves. Sepals and petals 6–9 mm, yellow with red longitudinal lines. Lip longer than the sepals, 3-lobed, the lateral lobes small, the central lobe clawed, transversely oblong or kidney-shaped, notched at the apex, yellow with a callus formed of 5 concave lobes. Wings of column sickle-shaped. *E Brazil.* G2. Autumn–winter.

12. O. divaricatum Lindley. Figure 31 (12), p. 573. Illustration: Pabst & Dungs, Orchidaceae Brasilienses **2**: t. 1951 (1977); Bechtel et al., Manual of cultivated orchid species, 243 (1981).

Pseudobulbs 2.5–4 cm, spherical, each bearing 1 leaf. Leaves 20–30 cm, oblong, obtuse. Panicle to 1.2 m, very branched, with many flowers. Sepals and petals 1–1.3 cm, clawed, obovate, brownish with yellow apices. Lip larger than the sepals, 3-lobed, with large lateral lobes and a rather small central lobe that is notched at the apex, all pale yellow, reddish at the base, with a callus consisting of 4 swellings. Wings of the column sickle-shaped, deflexed. *E Brazil.* G2. Autumn.

O. pulvinatum Lindley. Illustration: Cogniaux & Goossens, Dictionnaire iconographique des orchidées, Oncidium, t. 10 (1897). A much larger plant with the lateral lobes of the lip relatively smaller and undulate-crisped. *S & E Brazil.*

Section **Glanduligera** Lindley. Pseudobulbs conspicuous; upper sepal and petals linear, erect, antenna-like, lateral sepals curving downwards, much broader than the upper.

13. O. papilio Lindley (*Psychopsis papilio* (Lindley) H.G. Jones). Figure 31(13), p. 573. Illustration: *Edwards's Botanical Register* **11**: t. 910 (1825); *Botanical Magazine*, 3733 (1840); Cogniaux & Goossens, Dictionnaire iconographique des orchidées, Oncidium, t. 3 (1897); Dunsterville & Garay, Venezuelan orchids illustrated **2**: 254–5 (1961); Bechtel et al., Manual of cultivated orchid species, 247 (1981).

Pseudobulbs 3–5 cm, clustered, spherical, each bearing 1 leaf. Leaves 12–25 × 4–7 cm, ovate to elliptic, acute, green mottled with brown. Raceme or panicle to 1.2 m, erect, few-flowered, with only 1 or 2 flowers open at any one time, its axis conspicuously flattened towards the apex. Upper sepal and petals 8–13 cm, erect, linear, slightly expanded towards the apex, acute, with slightly wavy margins, purplish brown slightly mottled with greenish yellow. Lateral sepals oblong-lanceolate, crisped-undulate, brown with yellow transverse bands. Lip shorter than the lateral sepals, 3-lobed, the lateral lobes rather small, the central lobe clawed, almost circular to kidney-shaped, notched at the apex, mottled yellow and brown with a yellow centre, and with an obscurely 3-lobed callus. Wings of column oblong, lacerate, ending at the top in capitate, fleshy teeth. *Tropical S America from Trinidad to Peru.* G2. Flowering throughout the year.

14. O. kramerianum Reichenbach (*Psychopsis krameriana* (Reichenbach) H.G. Jones). Figure 31(14), p. 573. Illustration: *Lindenia* **6**: t. 246 (1890); Cogniaux & Goossens, Dictionnaire iconographique des orchidées, Oncidium, t. 26 (1900); Bechtel et al., Manual of cultivated orchid species, 245 (1981).

Very similar to *O. papilio*, but smaller, the axis of the panicle not flattened above, the central lobe of the lip rather shortly clawed. *Tropical America from Costa Rica to Ecuador*. G2. Autumn.

Section **Waluewa** (Regel) Schlechter. Pseudobulbs conspicuous; lateral sepals united for most of their length; column downy with curved, forwardly projecting, fleshy arms.

15. O. pubes Lindley. Figure 31(15), p. 573. Illustration: *Edwards's Botanical Register* **12**: t. 1007 (1826); Pabst & Dungs, Orchidaceae Brasilienses **2**: t. 1967 (1977); Bechtel et al., Manual of cultivated orchid species, 247 (1981).

Pseudobulbs 5–7 cm, cylindric, narrowed above, each bearing 1 leaf. Leaves 12–16 cm, lanceolate to elliptic, acute. Panicle 30–50 cm, erect, with many flowers. Sepals and petals 1–1.3 cm, obovate-oblong, brown with yellow transverse bars, the lateral sepals united for the whole of their length. Lip slightly shorter than the sepals, 3-lobed with small, spreading lateral lobes and a large, almost circular central lobe, yellow spotted with red-brown and bearing a large, tuberculate, downy callus. *Brazil.* G2. Summer.

Section **Stellata** Kränzlin. Pseudobulbs conspicuous; sepals and petals and lip in 1 plane, the lip not very different from the sepals and petals, giving the flower the appearance of a 6-pointed star.

16. O. maculatum (Lindley) Lindley (*Cyrtochilum maculatum* Lindley). Figure 31 (16), p. 573. Illustration: Cogniaux & Goossens, Dictionnaire iconographique des orchidées, Oncidium, t. 19 (1898).

Pseudobulbs 7–10 cm, ovoid, compressed, bearing 2 leaves. Leaves 18–25 × 2.5–5 cm, linear to oblong or oblong-elliptic, obtuse or acute. Raceme or panicle erect, up to 1 m. Flowers fragrant. Sepals and petals 1.5–3 cm, oblong-elliptic to oblong-lanceolate, pale yellow or yellow-green blotched with brown. Lip oblong, 3-lobed at about the middle, the lateral lobes directed forwards, triangular, the central lobe oblong, notched or rounded at the apex, white with red stripes at the base, yellow at the apex, with a callus of 4 united ridges, which end in upright teeth. Wings of column small and inconspicuous. *Mexico, Guatemala, Honduras*. G2. Summer.

17. O. hastatum (Bateman) Lindley (*Odontoglossum hastatum* Bateman). Figure 31(17), p. 573. Illustration: Bechtel et al., Manual of cultivated orchid species, 244 (1981).

Similar to *O. maculatum* but the lip mostly pink with oblong or almost square, outwardly directed lateral lobes, the central lobe ovate, acute. *S Mexico*. G2.

Section **Rostrata** Rolfe. Pseudobulbs conspicuous; rostellum elongate, proboscis-like.

18. O. ornithorrhynchum Humboldt, Bonpland & Kunth. Figure 31(18), p. 573. Illustration: *Edwards's Botanical Register* **26**: t. 10 (1840); *Botanical Magazine*, 3912 (1841); Cogniaux & Goossens, Dictionnaire iconographique des Orchidées, Oncidium, t. 22 (1899); Bechtel et al., Manual of cultivated orchid species, 247 (1981).

Pseudobulbs 2.5–9 cm, ovoid or ellipsoid, each bearing 2–3 leaves. Leaves 10–40 × 1–3 cm, linear-lanceolate to linear-elliptic, acute. Panicle up to 50 cm, erect. Sepals and petals 7–11 mm, shortly clawed, oblong-elliptic to oblanceolate-elliptic, pink or pinkish purple, obtuse. Lip 3-lobed, longer than the sepals, the lateral lobes small, rolled downwards and clasping the bases of the lateral sepals, the central lobe clawed, obovate, deeply notched at the apex, pink or pinkish purple with a small, reddish brown-spotted callus. Wings of the column broadly triangular from a narrow base, finely toothed. *C America, from S Mexico to Costa Rica*. G2. Autumn.

19. O. cheirophorum Reichenbach. Figure 31(19), p. 573. Illustration: *Botanical Magazine*, 6278 (1877); *Lindenia* **3**: t. 126 (1887); Bechtel et al., Manual of cultivated orchid species, 242 (1981).

Pseudobulbs 3–4 cm, ovoid, each with 1 leaf. Leaves to 15 cm, linear-oblong, acute. Panicle to 25 cm, erect or arching, with many fragrant flowers. Sepals and petals 6–13 mm, ovate, rounded at the apex, yellow. Lip 3-lobed, longer than the sepals, the lateral lobes spreading, rounded, the central lobe stalkless, oblong to almost circular, notched at the apex, yellow with a conspicuous white callus with 5 teeth. Wings of the column obovate, entire. *Costa Rica, Panama, Colombia*. G2. Winter.

Section **Concoloria** Kränzlin. Pseudobulbs conspicuous; lateral sepals united; lip entire and bearing a callus made up of 2 or 4 ridges.

20. O. concolor Hooker. Figure 32(20), p. 578. Illustration: *Botanical Magazine*, 3752 (1840); Cogniaux & Goossens, Dictionnaire iconographique des orchidées, Oncidium, t. 32 (1904); Pabst & Dungs, Orchidaceae Brasilienses **2**: t. 1989 (1977); Bechtel et al., Manual of cultivated orchid species, 242 (1981).

Pseudobulbs to 5 cm, ovoid, grooved, each bearing 2 leaves. Leaves to 15 cm, linear-lanceolate. Raceme with 6–12 flowers, arching. Sepals and petals 2.5–3 cm, oblong-elliptic, acute, yellow, the lateral sepals united for about half their length. Lip larger than the sepals, shortly clawed, diamond-shaped, slightly notched at the apex, yellow, bearing 2 red or orange ridges. Wings of the column oblong, entire, directed forwards. *SE Brazil, N Argentina*. G2. Spring.

Section **Barbata** Pfitzer. Pseudobulbs conspicuous; lateral sepals variously united, longer than the lip; callus of lip compound, composed of an uneven number of plates or tubercles.

21. O. longipes Lindley. Figure 32(21), p. 578. Illustration: *Botanical Magazine*, 5193 (1860); Bechtel et al., Manual of cultivated orchid species, 246 (1981).

Pseudobulbs 2–3 cm, elongate-ovoid, each with 2 leaves. Leaves 10–15 cm, oblong, acute. Raceme with 2–6 flowers, loose, scarcely exceeding the leaves. Sepals and petals 1.4–1.8 cm, oblong, rounded and finally pointed at the apex, yellow spotted with red-brown below or dark purple-brown throughout. Lip 3-lobed, the lateral lobes broadly obovate, attached to the broadly kidney-shaped central lobe by a broad, toothed claw, yellow with a central brown spot, bearing a large, yellow-spotted ridge with 3 teeth at its apex, the central tooth larger and blunter than the other 2. Wings of the column small, toothed. *SE Brazil*. G2. Spring.

Section **Crispa** Pfitzer. Pseudobulbs conspicuous; sepals and petals clawed, crisped, the lateral sepals variously united, the petals more conspicuous than the sepals; lip with a callus formed from an uneven number of tubercles.

22. O. crispum Loddiges. Figure 32(22), p. 578. Illustration: *Botanical Magazine*, 3499 (1836); *Edwards's Botanical Register* **23**: t. 1920 (1837); Cogniaux & Goossens, Dictionnaire iconographique des orchidées, Oncidium, t. 6, 6A (1897); Bechtel et al., Manual of cultivated orchid species, 242 (1981).

Pseudobulbs to 10 cm, ovoid, grooved, each bearing 2 leaves. Leaves to 20 cm, narrowly lanceolate acute. Panicle with many flowers, loose, arching, axis bluish green. Sepals 2.5–3 cm, oblong, crisped, brown with yellow spots towards the base. Petals similar but broader. Lip a little longer than the sepals, obscurely 3-lobed with very small lateral lobes and a large, kidney-shaped, strongly crisped central lobe that is brown with a yellow centre, bearing a prominent callus with an uneven number of teeth. Wings of the column oblong, finely toothed. *E Brazil*. G2. Autumn–spring.

Autumn-flowering material with rather small flowers may prove to be *O. enderianum* Masters (see Fowlie, *Orchid Digest* **43**: 190–5, 1979).

23. O. forbesii Hooker. Figure 32(23), p. 578. Illustration: *Botanical Magazine*, 3705 (1839); Cogniaux & Goossens, Dictionnaire iconographique des orchidées, Oncidium, t. 1 (1896); Pabst & Dungs, Orchidaceae Brasilienses **2**: t. 2024 (1977); Bechtel et al., Manual of cultivated orchid species, 244 (1981).

Very similar to *O. crispum* but the inflorescence unbranched, the sepals and petals chestnut brown with yellow marbling around the margins and the column with entire, red-spotted wings. *E. Brazil*. G2. Autumn.

24. O. gardneri Lindley. Figure 32(24), p. 578. Illustration: Pabst & Dungs, Orchidaceae Brasilienses **2**: t. 2013 (1977); Bechtel et al., Manual of cultivated orchid species, 244 (1981).

Very similar to *O. crispum* but sepals and petals oblanceolate, brown with yellow stripes on the margin, the lip with very small lateral lobes, the central lobe *c.* 3 × 4.5 cm, mostly yellow with a zone of red-brown spots around the margin, brown at the base, column with entire wings. *E. Brazil*. G2. Summer.

25. O. marshallianum Reichenbach. Figure 32(25), p. 578. Illustration: *Botanical Magazine*, 5725 (1868); Cogniaux & Goossens, Dictionnaire iconographique des orchidées, Oncidium, t. 8 (1897); Pabst & Dungs, Orchidaceae Brasilienses **2**: t. 2008 (1977).

Pseudobulbs 7–10 cm, ovoid, each with 2 leaves. Leaves 20–30 cm, oblong-lanceolate, acute. Panicle to 1.5 m with many flowers, erect. Sepals 2.3–2.5 cm oblong-obovate, greenish yellow with brown spots or transverse stripes. Petals to 3 cm, notched at the apex, golden yellow with dark brown spots in the centre. Lip 3-lobed with small, yellow and red-spotted lateral lobes and a large, broadly kidney-shaped, notched, yellow central lobe bearing a large, many-toothed callus. Wings of column oblong, entire. *E Brazil*. G2. Spring–summer.

Section **Synsepala** Pfitzer. Pseudobulbs conspicuous; sepals and petals similar, inconspicuous, the lateral sepals united for part of their length; lip with a callus made up of an uneven number of tubercles.

26. O. varicosum Lindley. Figure 32(26), p. 578. Illustration: Cogniaux & Goossens, Dictionnaire iconographique des orchidées, Oncidium, t. 18, 18A (1898); Pabst & Dungs, Orchidaceae Brasilienses **2**: t. 2029 (1977); *Orchids* (USA) **71**: 927 (2002); Zelenko & Chase (ed.), Orchids: the pictorial encyclopaedia of Oncidium 38, 39, 41 (2002).

Pseudobulbs 8–10 cm, elongate-ovoid, grooved, each bearing 2 leaves. Leaves 15–25 cm, oblong, acute. Panicle 80–150 cm, with many flowers, loosely branched. Flowers very variable in size. Sepals and petals 5–7 mm, oblong, lateral sepals united for about half their length, yellowish green spotted or barred with brown. Lip 3-lobed, much larger than the sepals, the lateral lobes small, rounded, the central lobe 2.5–4 × 3.5–5.5 cm, 3-notched towards the apex to produce 4 small lobes, golden yellow, bearing a callus of 2 trident-shaped flaps with several teeth in adjacent rows. Column with entire wings. *E & C Brazil*. G2. Winter.

O. insigne (Rolfe) Christenson (*O. varicosum* var. *insigne* Rolfe; var. *moortbeekiense* Linden; var. *baldin* invalid). Illustration: *Orchids* (USA) **71**: 925–7 (2002); Zelenko & Chase (eds), Orchids: the pictorial encyclopaedia of Oncidium 38, 39 (2002). Similar to *O. varicosum*, but differing by lip with a large deep red-brown irregular blotch and the sepals and petals uniformly coloured in a still darker shade of red-brown. Callus of 3 low parallel ridges terminating in a single, large, erect tooth without any adjacent rows of teeth. *Coastal E. Brazil*. G2. Winter.

Often grown under the names 'Baldwin', 'Baldim', 'Baldin', 'Baldin's Variety'. The parentage *O. varicosum* × *dasystyle* has been suggested, however this taxon is not a

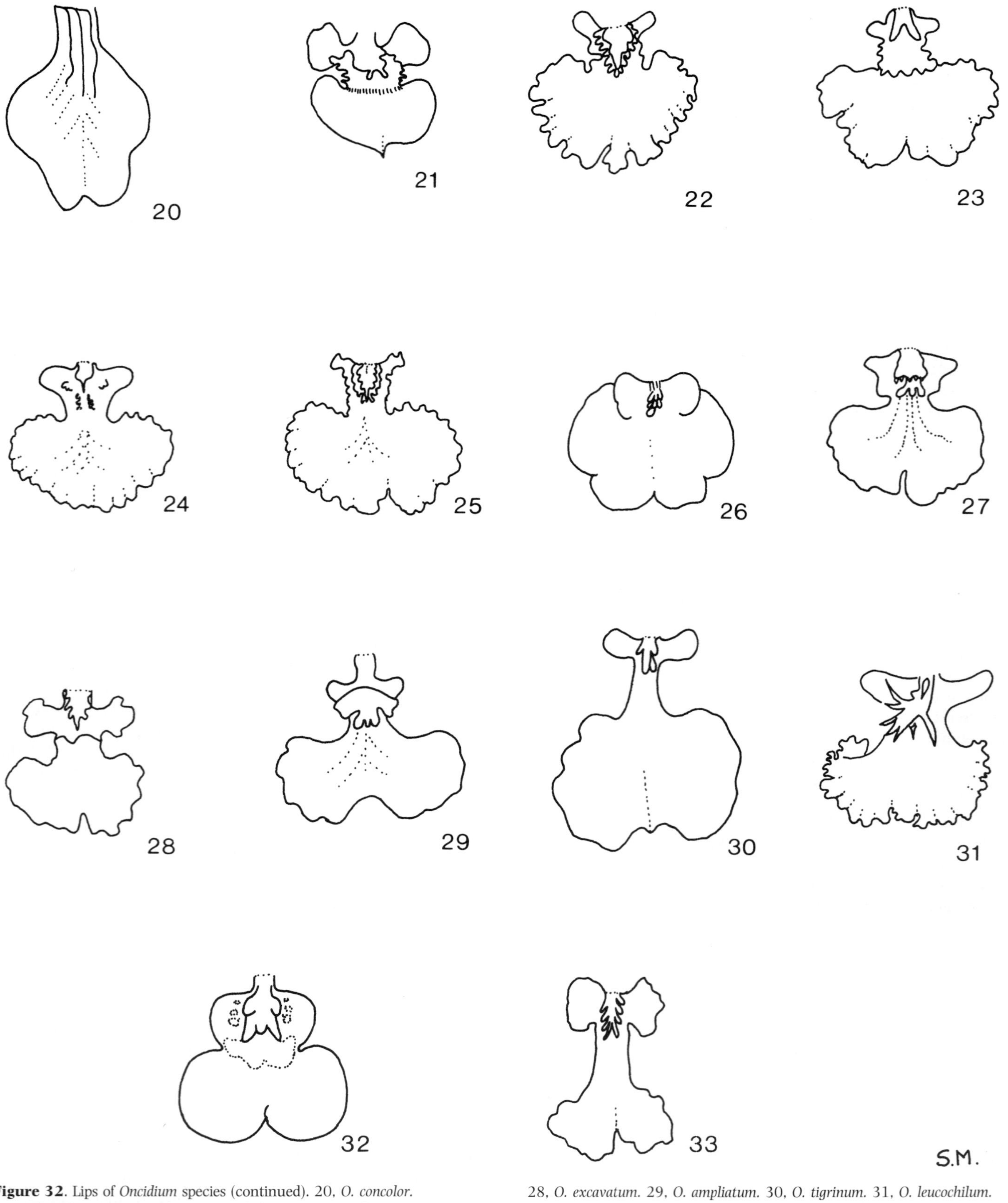

Figure 32. Lips of *Oncidium* species (continued). 20, *O. concolor.* 21, *O. longipes.* 22, *O. crispum.* 23, *O. forbesii.* 24, *O. gardneri.* 25, *O. marshallianum.* 26, *O. varicosum.* 27, *O. flexuosum.* 28, *O. excavatum.* 29, *O. ampliatum.* 30, *O. tigrinum.* 31, *O. leucochilum.* 32, *O. sphacelatum.* 33, *O. wentworthianum.* (Compiled from various sources; not to scale.)

hybrid, and breeds true from seed. The flower size is variable but the colouration is constant.

27. O. flexuosum Loddiges Figure 32(27), p. 578. Illustration: *Botanical Magazine*, 2203 (1821); Reichenbach, Flora Exotica **2**: t. 94 (1834); Pabst & Dungs, Orchidaceae Brasilienses **2**: t. 2024 (1977); Bechtel et al., Manual of cultivated orchid species, 243 (1981).

Pseudobulbs 3–4 cm, compressed, rather distant, each bearing 1–2 leaves. Leaves 10–20 cm, oblong, somewhat acute. Panicle to 80 cm, with many flowers. Sepals and petals oblong, acute, yellow with red-brown spots at the base, the lateral sepals united at the base. Lip 3-lobed, much longer than the sepals, the lateral lobes small, curved, erect, the central lobe 1.5–2 × 1.5–1.8 cm, clawed, kidney-shaped, notched at the apex, golden yellow, bearing a many-toothed callus. Wings of the column small, entire. *SE Brazil, Paraguay & Argentina*. G2. Autumn–winter.

Section **Excavata** Kränzlin. Pseudobulbs conspicuous; sepals and petals clawed, the lateral sepals free, the petals broader and more conspicuous than the sepals; callus of lip with an uneven number of tubercles.

28. O. excavatum Lindley. Figure 32(28), p. 578. Illustration: *Botanical Magazine*, 5293 (1862); Cogniaux & Goossens, Dictionnaire iconographique des orchidées, Oncidium, t. 20 (1899); Bechtel et al., Manual of cultivated orchid species, 243 (1981).

Pseudobulbs 7–12 cm, oblong-ovoid, somewhat compressed, each with 1–2 leaves. Leaves 30–50 × 2.5–4 cm, linear-oblong, acute. Panicle to 1.5 m, erect, with many flowers. Sepals 1.2–1.6 cm, oblanceolate-oblong, rounded and abruptly acute, the laterals narrower than the upper, all yellow spotted with red towards the base. Petals similar but slightly longer and much broader. Lip longer than the sepals, 3-lobed, the lateral lobes small, the central lobe 1.2–2.2 × 2 cm, clawed, yellow, red towards the base with a complex toothed and grooved callus. Wings of the column oblong, slightly notched. *Ecuador, Peru*. G2. Spring.

Section **Oblongata** Kränzlin. Pseudobulbs conspicuous; lateral sepals free; bracts small and inconspicuous, scale-like.

29. O. ampliatum Lindley. Figure 32(29), p. 578. Illustration: *Edwards's Botanical Register* **20**: t. 1699 (1835); *Fieldiana* **30**: 859 (1961); Dunsterville & Garay, Venezuelan orchids illustrated **2**: 242–3 (1961); Bechtel et al., Manual of cultivated orchid species, 240 (1981).

Pseudobulbs 3–7.5 cm, spherical or pear-shaped, compressed, each bearing 1 leaf. Leaves 8–50 × 2.5–10 cm, variable in shape, obtuse or broadly rounded. Panicle to 80 cm with many flowers. Upper sepal 7–9 mm, obovate, hooded, the lateral sepals longer and narrower, all yellow with small brown spots. Petals distinctly clawed, yellow with small brown spots. Lip longer than the sepals, 3-lobed, the lateral lobes small, the central shortly clawed, kidney-shaped, notched at the apex, yellow, with a small callus, which has 2 broad lateral lobes and 3 smaller lobes in front. Wings of the column toothed. *Tropical America from Guatemala to Peru*. G2. Spring.

30. O. tigrinum Llave & Lexarza. Figure 32(30), p. 578. Illustration: *Illustration Horticole* **22**: t. 221 (1875); Cogniaux & Goossens, Dictionnaire iconographique des orchidées, Oncidium, t. 4, 4A, 4B (1897); Bechtel et al., Manual of cultivated orchid species, 249 (1981).

Pseudobulbs 7–9 cm, spherical, with 2 leaves. Leaves 20–30 cm, narrowly oblong, acute, thinly leathery. Panicle to 80 cm, little branched, with many fragrant flowers. Sepals and petals similar, 2.5–3.5 cm, spreading, oblong to obovate, yellow with heavy brown spotting and marbling. Lip much larger than the sepals, 3-lobed, the lateral lobes small, the central lobe clawed, spherical to kidney-shaped, golden yellow, with a large, 3-lobed callus. Wings of the column oblong, entire. *Mexico*. G2. Winter.

31. O. leucochilum Bateman. Figure 32 (31), p. 578. Illustration: Cogniaux & Goossens, Dictionnaire iconographique des orchidées, Oncidium, t. 24 (1899).

Pseudobulbs 5–13 cm, ovoid, compressed, each with 1–2 leaves. Leaves 10–60 × 1.5–4.5 cm, oblong, obtuse to acute. Panicle to 3 m. Sepals and petals 1.3–2.3 cm, similar, oblong-elliptic to oblanceolate, bright green or greenish white blotched with red. Lip larger than the sepals, 3-lobed, the lateral lobes small, oblong to ovate, the central clawed, transversely oblong to kidney-shaped, notched at the apex, white, sometimes with red lines, with a purplish callus that ends in 5–9 slender teeth, some of them upcurved. Wings of the column oblong, irregularly scalloped. *Mexico, Guatemala, Honduras*. G2. Winter–spring.

The following species, also belonging to this section, is grown occasionally.

O. reflexum Lindley. Sepals and petals conspicuously reflexed, violet-brown tipped with yellow. *Mexico*. G2. Autumn.

Section **Planifolia** Bentham & Hooker. Pseudobulbs conspicuous; bracts conspicuous, as long as or little shorter than the stalked ovaries; lip with a callus composed of an uneven number of tubercles.

32. O. sphacelatum Lindley. Figure 32 (32), p. 578. Illustration: *Edwards's Botanical Register* **28**: t. 30 (1842); Dunsterville & Garay, Venezuelan orchids illustrated **4**: 196–7 (1966); Bechtel et al., Manual of cultivated orchid species, 248 (1981).

Pseudobulbs to 20 cm, ovoid-ellipsoid, tapering, each with 2 leaves. Leaves to 100 × 3.5 cm, linear-oblong or linear-lanceolate, acute. Panicle to 1.5 m, with many flowers, erect. Sepals 10–20 × 3–6.5 mm, clawed, elliptic to elliptic-obovate, yellow with reddish brown blotches. Petals similar but not clawed. Lip about as long as the sepals, 3-lobed, wider across the central lobe than across the lateral lobes, central lobe large, semicircular, notched at the apex and often at the sides as well, yellow, with a downy callus of 5–7 tubercles, some of which are divergent. Column with narrow, brown-bordered, slightly scalloped wings. *C America, Venezuela*. G2. Spring.

33. O. wentworthianum Lindley. Figure 32(33), p. 578. Illustration: Paxton's Flower Garden, f. 127 (1851–52); *Die Orchidee* **31**(5): clxxxv–clxxxvi (1980); Bechtel et al., Manual of cultivated orchid species, 249 (1981).

Pseudobulbs 7–10 cm, ovoid-ellipsoid, compressed, often mottled with brown, each bearing 2 leaves. Leaves 13–35 × 1.5–3 cm, linear-oblong to lanceolate. Panicle up to 1 m, with many flowers, arching. Sepals and petals 1.4–2.2 cm, spreading or somewhat reflexed, elliptic or elliptic-obovate, deep yellow irregularly blotched with brown. Lip 3-lobed, about as long as the sepals, wider across the lateral lobes than across the central lobe, the lateral lobes small, curved forwards, scalloped on their margins, the central lobe reversed-heart-shaped, deeply notched at the apex, yellow, with a fleshy reddish brown callus consisting of 3 teeth flanked by ridges. Wings of column triangular, finely scalloped, bordered with red-brown spots. *Mexico, Guatemala.* G2. Summer.

The following species, also belonging to this section, is occasionally grown.

O. incurvum Baker. Illustration: *Edwards's Botanical Register* **31**: t. 64 (1845); Cogniaux & Goossens, Dictionnaire iconographique des orchidées, Oncidium, t. 29 (1900); Bechtel et al., Manual of cultivated orchid species, 245 (1981). Sepals, petals and lip white with red spots. *Mexico.* Autumn.

131. SIGMATOSTALIX Reichenbach

J. Cullen

Small, rhizomatous epiphytes. Pseudobulbs simple, clustered or distant, each bearing 1–2 leaves. Leaves very narrow, grass-like, folded when young. Raceme (rarely branched) erect, arising from the bases of the pseudobulbs. Sepals and petals similar, widely spreading to reflexed. Lip conspicuously clawed, usually simple, with a callus near the base. Column very slender, arched, widened above, without a foot. Pollinia 2, ovoid, stipe oblong, expanded above with its sides upturned, longer than the pollinia and the small, almost circular viscidium.

A genus of probably more than 80 species from C & S America. Cultivation as for *Odontoglossum* (p. 561).

Literature: Kränzlin, F., Das Pflanzenreich **80**: 301–12 (1922).

1. S. graminea (Poeppig & Endlicher) Reichenbach. Illustration: Reichenbach, Xenia Orchidacearum **1**: t. 8 (1854).

Pseudobulbs 7–12 mm, clustered, oblong-cylindric to cylindric-ellipsoid. Leaves 3–5 cm × 2 mm, narrowly linear, acute or obtuse, obliquely notched at the apex. Raceme erect or arching. Sepals 2–2.5 mm, oblong or elliptic-lanceolate, acute. Petals slightly broader than sepals, both yellow with purple spots. Lip shortly and broadly clawed, blade ovate to almost square, notched, with rounded angles at the base. *Peru.* G2.

132. SARCOCHILUS R. Brown

V.A. Matthews

Epiphytic. Pseudobulbs absent. Stems covered with the remains of old sheathing leaf-bases. Leaves in 2 ranks, folded when young. Racemes lateral with 3–25 fragrant flowers. Sepals and petals similar, lateral sepals joined to the column foot. Lip hinged to column foot, shallowly pouched, 3-lobed, central lobe much smaller than lateral lobes, with a forward-pointing spur. Surface of lip with a large callus, and some calluses present on the inside walls of the spur. Pollinia 4, in 2 pairs.

A genus of about 12 species, mainly native to Australia. Though good drainage is essential to their culture, they should never be allowed to dry out. Treatment varies somewhat, depending on the origin of the species. Some are best grown in pots in an epiphytic compost, others on rafts or tree fern, close to the glass.

Literature: Upton, W.T., Sarcochilus orchids of Australia (1992).

1a. Sepals and petals white, pink or purple, without spots at the base; stems 5–12 cm, unbranched **1. falcatus**

b. Sepals and petals white spotted with red at the base; stems to 1 m, usually branched **2. hartmanni**

1. S. falcatus R. Brown. Illustration: Nicholls, Orchids of Australia, t. 452, 453 (1969); Dockrill, Australian indigenous orchids, t. 25 (1969).

Stems 5–7.5 cm. Leaves narrowly oblong, sickle-shaped, 5–14 cm, channelled. Racemes hanging, with 3–12 flowers. Flowers 2–3.5 cm in diameter. Sepals and petals oblong, 8–16 mm, white, sometimes with a purplish midrib on the back. Lip 3.5–6 mm, lateral lobes ovate-oblong, flushed with orange and with red stripes, central lobe broad, about a quarter of the length of the lateral lobes. Spur fleshy, sometimes marked with purple. Callus at opening of spur 2-lobed, spotted, some calluses on inside wall of spur. *Australia.* G2. Autumn–winter.

S. ceciliae Mueller. Illustration: Nicholls, Orchids of Australia, t. 456, 457 (1969). Stems to 12 cm, leaves often brown-spotted, racemes erect with 6–12 flowers, 4–6 mm in diameter, facing upwards, pale pink, mauve or purple, lip densely hairy, callus at opening of spur yellow, pronounced calluses present on inside walls of the spur. *Australia.* G2.

2. S. hartmannii Mueller. Illustration: Nicholls, Orchids of Australia, t. 455 (1969); *American Orchid Society Bulletin* **47**: 1122 (1978); Bechtel et al., Manual of cultivated orchid species, 271 (1981).

Stems to 1 m, usually branched. Leaves linear-oblong, 5–15 cm, channelled. Racemes arching or erect with 5–25 flowers. Flowers 1.5–3.5 cm in diameter. Sepals and petals white spotted with red at the base (rarely completely white or red). Lateral lobes of lip spotted and streaked with red, central lobe yellowish, spur white flushed with yellow. Callus at opening of spur 2-lobed, small calluses on inside walls of spur. *Australia.* G2.

S. fitzgeraldii Mueller. Illustration: *Botanical Magazine*, n.s., 201 (1953); Nicholls, Orchids of Australia, t. 454 (1969); Bechtel et al., Manual of cultivated orchid species, 271 (1981). Leaves sickle-shaped, racemes hanging, with up to 16 flowers, scape reddish, flowers 2.5–3.5 cm in diameter, lip *c.* 5 mm, callus at opening of spur yellowish with a smaller one behind. *Australia.* G2. Winter–spring.

133. CHILOSCHISTA Lindley

V.A. Matthews

Differs from *Sarcochilus* in that the plants are leafless when flowering, the petals are larger than the sepals and the 4 pollinia are in 2 pairs.

A genus of 3 species found in tropical Asia. Cultivation as for *Sarcochilus* (p. 580).

1. C. lunifera (Reichenbach) J.J. Smith (*Sarcochilus luniferus* (Reichenbach) Hooker). Illustration: *Botanical Magazine*, 7044 (1899).

Racemes drooping, with many flowers, scapes purple-spotted, shortly hairy. Flowers 1–1.5 cm in diameter. Sepals and petals yellow spotted with red or purple. Lip white, lateral lobes erect, central lobe minute, recurved, with a forward-pointing spur. Disc warty and hairy with 2 thick ridges. Anther cap with 3 spurs. *Sikkim, Nepal & Burma.* G2.

134. PTEROCERAS Hasskarl

V.A. Matthews

Differs from *Sarcochilus* in having the lip movably attached to the column foot, the spur of the lip lacking calluses on its inside walls, and the 4 pollinia in 2 pairs.

Thirty species found in NE India, SE Asia and Malaysia. Cultivation as for *Sarcochilus* (p. 574).

Literature: Holttum, R.E., The genera Sarcochilus R.Br. and Pteroceras Hassk., *Kew Bulletin* **14**: 263–76 (1960).

1. P. pallidum (Blume) Holttum (*Sarcochilus pallidus* (Blume) Reichenbach;

S. unguiculatus Lindley). Illustration: *Lindenia* **16**: t. 756 (1901).

Leaves oblong, fleshy, up to 20 cm. Raceme usually hanging, axis somewhat thickened. Flowers to 5 cm across, fragrant. Sepals and petals white to pale yellowish. Lip *c.* 6 mm, lateral lobes white streaked with red or purple, central lobe and spur yellowish, spotted with red or purple. *Malay peninsula, Java, Sumatra, Borneo & Philippines.* G2.

135. PAPILIONANTHE Schlechter

J.M.H. Shaw

Epiphytic plants with terete leaves and axillary few flowered inflorescences. Distinguished by a short, thick, non-pyramid shaped column, which at the base is developed into a long foot which merges imperceptably with the lip. Lip 3-lobed with lateral lobes either along side or wraped around the column. Pollinia 2.

A genus of 10 species from Ceylon (Sri Lanka) and the Himalaya to Indochina and Malaysia. Cultivation as for *Vanda* (p. 588).

Literature: Christenson, E.A., Sarcanthine genera 15: Papilionanthe., *American Orchid Society Bulletin* **63**(12): 1374–77 (1994).

1a. Lateral lobes of lip bi-lobed or forked 2
b. Lateral lobes of lip entire 3
2a. Leaves 7.5–10 cm; margins of sepals and petals not wavy; lateral lobes of lip with violet markings **1. subulata**
b. Leaves 10–20 cm; margins of sepals and petals wavy; lateral lobes of lip lacking violet markings **2. vandaraum**
3a. Lateral lobes of lip arched over column **3. teres**
b. Lateral lobes of lip spreading **4. hookeriana**

1. P. subulata (Koenig) Garay (*Aerides cylindricum* Lindley). Illustration: *Orchid Digest* **41**: 204 (1977).

Leaves terete, 7.5–10 cm. Raceme with 1 or 2 flowers. Flowers *c.* 2.5 cm in diameter. Sepals and petals white, sometimes flushed with pale pink, ovate-oblong. Lateral lobes of lip white with violet lines, central lobe white or yellowish with violet spots and 3 broad yellow ridges. Spur funnel-shaped, pinkish. *S India, Sri Lanka.* G2.

2. P. vandarum (Reichenbach) Garay (*Aerides vandarum* Reichenbach; *Aerides cylindricum* Hooker not Lindley). Illustration: *Botanical Magazine*, 4982 (1857); *Orchid Digest* **41**: 204 (1977); Bechtel et al., Manual of cultivated orchid species, 164 (1981).

Leaves terete, 10–20 cm. Raceme with 1–3 flowers. Flowers *c.* 5 cm in diameter. Sepals and petals white, somewhat reflexed, with wavy margins, petals twisted. Lip white flushed with yellow in the middle, central lobe clawed, expanding into 2 fringed lobes. Spur curved, 2–2.5 cm. *NE India, Burma.* G2. Spring.

3. P. teres (Roxburgh) Schlechter (*Vanda teres* (Roxburgh) Lindley). Illustration: *Botanical Magazine*, 4114 (1844); *Orchid Digest* **43**: 84 (1979); Bechtel et al., Manual of cultivated orchid species, 279 (1981).

Stem to 2 m. Leaves cylindric, terete, upright, rather blunt. Racemes to 30 cm with 3–5 flowers. Flowers 8–10 cm in diameter. Sepals and petals elliptic, lateral sepals white, petals and upper sepal violet or pink. Lip as long as sepals, lateral lobes large, arching over the column, reddish striped with yellow at the base, central lobe obovate, deeply 2-lobed, violet-pink to yellow with red spots towards the base. Spur short, conical. *E Himalaya to Thailand & Laos.* G2. Spring.

The hybrid *P. teres* × *P. hookeriana* is called Miss Joaquin grex, and is widely grown.

4. P. hookeriana (Reichenbach) Schlechter (*Vanda hookerianum* Reichenbach). Illustration: *The Garden* **23**: 10 (1883); Illustration: *Horticole* **30**: t. 484 (1883); Chan et al., Orchids of Borneo **1**: 214 (1994).

Similar to *P. teres* but with flowers to 5 cm in diameter, sepals and petals white, flushed and spotted with red, lip with spreading lateral lobes and a fan-shaped, white and red-spotted, shallowly 3-lobed central lobe. *Malaysia, Sumatra, Borneo.* G2. Autumn.

136. AERIDES Loureiro

V.A. Matthews

Epiphytic. Pseudobulbs absent. Stem producing long aerial roots, which either hang or attach themselves to a support. Leaves usually strap-shaped, thick, in 2 ranks, their bases sheathing the stem, folded when young, occasionally terete. Flowers usually fragrant, in hanging, many-flowered racemes arising from the leaf axils. Sepals and petals spreading, the 2 lateral sepals attached to the column foot. Lip simple or 3-lobed, continuous with the column foot, prolonged at the base into a hollow spur. Column short, usually with a long foot. Pollinia 2 unequally divided, on a single slender stipe.

A genus of some 40 species from tropical Asia. Cultivation as for *Vanda* (p. 588).

Literature: Seidenfaden, G., Contributions to the orchid flora of Thailand V, *Botanisk Tidsskrift* **68**: 68–80 (1973); Wood, J. & Kennedy, G.C., Some showy members of the genus Aerides, *Orchid Digest* **41**: 205–8 (1977).

Sepals and petals. Basic colour olive green: **4**; white, pale green or yellowish: **1, 2, 3**; pinkish or purplish: **2, 4, 5**. With defined spots at the tips: **1, 2, 3**.
Lip. Simple: **2, 4, 5**.
Margin of lip, or if lip 3-lobed, of central lobe. Entire: **1, 2, 5**; toothed or fringed: **1, 3, 4**.

1a. Basic colour of sepals and petals white, pale greenish or yellowish 2
b. Basic colour of sepals and petals olive green, pink or pinkish purple 4
2a. Spur curved upwards towards the rest of the flower **1. odorata**
b. Spur pointing forwards 3
3a. Lip simple with an entire margin **2. multiflorum**
b. Lip 3-lobed, the central lobes with a toothed or fringed margin **3. falcatum**
4a. Lip simple and fringed or central lobe of lip with a toothed margin **4. crispum**
b. Lip simple and entire or central lobe of lip entire though sometimes notched at apex 5
5a. Flowers 2–3 cm in diameter; lip simple, rounded or slightly notched at the apex **2. multiflorum**
b. Flowers 3–5 cm in diameter; lip 3-lobed or, if simple, then the apex pointed **5. crassifolium**

1. A. odoratum Loureiro (*A. suavissima* Lindley; *A. cornuta* Roxburgh; *A. ballantiana* Reichenbach). Illustration: *Botanical Magazine*, 4139 (1845); *Botanisk Tidsskrift* **68**: 72 (1973); Sheehan & Sheehan, Orchid genera illustrated, 31 (1979).

Stems to 50 cm, in old plants to 150 cm. Leaves 20–25 cm. Racemes dense, with many flowers. Flowers 2–3 cm in diameter. Sepals and petals creamy white tipped with pink or purple. Lateral lobes of lip white or occasionally yellowish, flushed and spotted with pink or purple, especially at the base, central lobe ovate to narrowly oblong, white with a broad central purplish band, margin entire or toothed. Spur curved upwards towards the flower, tipped with green or yellow. *SE Asia, from the Himalaya to Indonesia.* G2. Summer–autumn.

A. quinquevulnerum Lindley. Illustration: *Orchid Digest* **43**: 126 (1979). Similar, but sepals and petals greenish at the base, central lobe of the lip deep pinkish purple with a revolute margin, which is always toothed. *Philippines.* G2. Summer–autumn.

Often regarded as a variety of *A. odorata*.

A. lawrenceae Reichenbach. Illustration: *Lindenia* **9**: t. 401 (1893); Bechtel et al., Manual of cultivated orchid species, 164 (1981). Also similar, but sepals and petals greenish white becoming yellowish, lateral lobes of lip white with toothed margins, central lobe purple, or purple-tipped with lines of purple spots, margin toothed. *Philippines.* G2.

Possibly also merely a variety of *A. odoratum*.

2. A. multiflorum Roxburgh (*A. affine* Lindley). Illustration: *Botanical Magazine*, 4049 (1844); Cogniaux & Goossens, Dictionnaire iconographique des orchidées, Aerides, t. 2 (1898); *Gartenpraxis* **1980** (11): 512.

Leaves 18–25 cm. Racemes dense with many flowers, arching or hanging. Flowers 2–3 cm in diameter. Sepals and petals white, spotted, flushed and tipped with purplish pink, or occasionally completely pink. Lip pale purplish pink with a deeper central area, simple, oval, with the sides recurved and the apex rounded or slightly notched. Spur pointing forwards. *N India, Bhutan, SE Asia.* G2. Summer–autumn.

3. A. falcatum Lindley (*A. larpentae* Reichenbach). Illustration: *Botanisk Tidsskrift* **68**: 74 (1973); *Orchid Digest* **41**: 204 (1977).

Stems to 1.8 m. Leaves 15–35 cm. Raceme with many flowers, hanging. Sepals and petals white tipped and sometimes dotted with pink. Lateral lobes of lip white or mauve, sickle-shaped, central lobe pink or purple with a toothed margin. Spur short, greenish, pointing forwards. *NE India, SE Asia.* G2. Spring–summer.

A. houlletianum Reichenbach (*A. falcata* var. *houlletiana* (Reichenbach) Veitch). Illustration: Cogniaux & Goossens, Dictionnaire iconographique des orchidées, Aerides, t. 3 (1899); *Orchid Digest* **41**: 206 (1977). Sepals and petals yellow-buff tipped with reddish purple or brown, upper sepal and petals with finely toothed margins, lateral lobes of lip yellowish, marked with pinkish purple, central lobe pinkish purple, toothed and notched at the apex. *E Thailand, Vietnam, Laos & Kampuchea.* G2. Spring.

4. A. crispum Lindley (*A. brookei* Lindley). Illustration: *Orchid Digest* **43**: 126 (1979).

Stems to 1.5 m, often purplish. Leaves 15–25 cm. Raceme with many flowers, arching or hanging. Flowers *c.* 5 cm in diameter. Sepals and petals white flushed with pinkish purple. Lateral lobes of lip white striped with pinkish purple, central lobe broadly ovate, deep pinkish purple, margin toothed. Spur compressed, slightly curved. *Himalaya, India, Burma.* G2. Summer.

A. flabellatum Downie. Illustration: *Botanisk Tidsskrift* **68**: 77 (1973); *Orchid Digest* **41**: 206 (1977). Stems and leaves much shorter, raceme loose, sepals and petals olive green marked with reddish brown, lip simple, clawed, white blotched with red or purple, margin fringed, spur curved upwards. *Burma, Laos, Thailand.* G2. Summer.

A distinct species which may be more closely related to *Vanda*.

5. A. crassifolium Parish & Reichenbach (*A. expansa* Reichenbach). Illustration: Cogniaux & Goossens, Dictionnaire iconographique des orchidées, Aerides, t. 1 (1897); *Botanisk Tidsskrift* **68**: 76 (1973); *Orchid Digest* **41**: 204 (1977).

Leaves to 20 cm. Racemes with 10–50 flowers. Flowers 3–5 cm in diameter, fragrant. Sepals and petals pinkish purple, deeper towards the apex and white towards the base. Lateral lobes of lip crescent-shaped, pinkish purple, central lobe broadly ovate, often notched at the apex, darker in colour than sepals and petals. Spur with a greenish tip. *Burma, Laos, Thailand.* G2. Spring–summer.

A. roseum Lindley & Paxton (*A. fieldingii* Williams). Illustration: *La Belgique Horticole* **26**: facing p. 283 (1876); *Botanisk Tidsskrift* **68**: 78 (1973); *Orchid Digest* **41**: 206 (1977); Bechtel et al., Manual of cultivated orchid species, 163 (1981). Racemes branched, lip simple with a pointed tip, spur white. *N India, N Thailand, Laos.* G2. Spring–summer.

A. jarckianum Schlechter. Illustration: *Botanical Magazine*, 9274 (1932); *Orchid Digest* **43**: 126 (1979); Bechtel et al., Manual of cultivated orchid species, 163 (1981). Lateral lobes of lip very small, rounded, united nearly to their apex with the central lobe which is sharply curved upwards. *Philippines.* G2. Spring.

137. SEDIREA Garay & Sweet

V.A. Matthews

Epiphytic. Pseudobulbs absent. Stem short, *c.* 10 cm. Leaves in 2 ranks, linear-oblong, thick. Raceme axillary, loose, drooping, to 15 cm, with few flowers. Flowers *c.* 3 cm in diameter, fragrant. Sepals and petals white, cream or greenish, lateral sepals barred with purplish brown. Lip hinged to the column foot, 3-lobed, lateral lobes triangular, central lobe obovate, recurved, scalloped, white blotched with reddish purple. Spur funnel-shaped, obtuse, pointing forwards. Column *c.* 8 mm, curved, with a short foot. Pollinia 2, ovoid, each with a groove, on a narrowly triangular stipe.

Only 1 species, growing in Japan and Korea. Cultivation as for *Vanda* (p. 588).

Literature: Christenson, E.A., Sarcanthine genera 19: Sedirea, *American Orchid Society Bulletin* **64**(8): 860–862 (1995).

1. S. japonica (Linden & Reichenbach) Garay & Sweet (*Aerides japonica* Linden & Reichenbach). Illustration: *Orchid Digest* **41**: 207 (1977).

Japan, Korea. G1. Summer.

138. RHYNCHOSTYLIS Blume

V.A. Matthews

Epiphytic. Stems short, thick. Leaves fleshy, strap-shaped or linear, in 2 ranks, 2-lobed at the apex. Racemes lateral, erect or drooping, many-flowered. Sepals and petals spreading, petals narrower than sepals. Lip entire, spurred. Column short, dilated at base with a short foot; rostellum beaked. Pollinia 2.

A genus of about 15 species native to tropical Asia often found under the name *Anota* Schlechter. Cultivation as for *Gastrochilus* (p. 592).

Literature: Christenson, E.A., Sarcanthine genera 16: Rhynchostylis., *American Orchid Society Bulletin* **64**(2): 152–9 (1995).

1a. Apex of lip 3-lobed — 2
b. Apex of lip entire or notched — 3
2a. Lip pinkish purple, white towards the middle and base, and with 2 ridges near the base — **3. gigantea**

b. Lip completely purple, with 5 ridges near the base **4. violacea**

3a. Racemes erect; sepals and petals tipped with blue **1. coelestis**

b. Racemes drooping; sepals and petals with purple or pinkish markings **2. retusa**

1. R. coelestis (Reichenbach) Veitch. Illustration: *Lindenia* **7**: t. 300 (1891); *Handbook on Orchids, Brooklyn Botanic Garden Record* **23**(2): 40–1 (1967).

Racemes erect. Flowers *c.* 2 cm in diameter, fragrant. Sepals and petals white tipped with blue. Lip obovate-oblong, blue or violet-blue, white at the base. Spur slightly curved, compressed. *Thailand.* G2. Summer.

2. R. retusa (Linnaeus) Blume. Illustration: *Botanical Magazine*, 4108 (1844); *Orchid Digest* **39**: 8 (1975); Bechtel et al., Manual of cultivated orchid species, 269 (1981).

Racemes drooping. Flowers *c.* 2 cm in diameter, fragrant. Sepals and petals white or pinkish, marked with pinkish or purple, lateral sepals broader than upper sepal and petals. Lip variable in shape, usually lanceolate, tip rounded or sometimes notched, mauve. Spur truncate-conical, compressed. *From India and Sri Lanka to the Philippines.* G2. Summer.

'Alba' has completely white flowers and 'Holfordiana' has the sepals and petals spotted with crimson and a crimson lip.

3. R. gigantea (Lindley) Ridley (*R. densiflora* (Lindley) Williams; *Anota densiflora* (Lindley) Schlechter). Illustration: *Orchid Digest* **43**: 124 (1979); Bechtel et al., Manual of cultivated orchid species, 269 (1981).

Racemes drooping. Flowers 2.5–3 cm in diameter, fragrant. Sepals and petals white spotted with pinkish purple or pale purple. Lip oblong, 3-lobed at apex, pale purple or pinkish purple, white towards the middle and base and with 2 small ridges near the base. Spur very short, obtuse, compressed. *Burma, Thailand, Laos.* G2. Autumn–winter.

Var. **harrisoniana** (Hooker) Holttum has pure white flowers and longer racemes; var. *petotiana* Anon. is probably the same.

4. R. violacea (Lindley) Reichenbach (*Anota violacea* (Lindley) Schlechter; *Saccolabium violaceum* Lindley). Illustration: *Edwards's Botanical Register* **33**: t. 30 (1847).

Racemes arching or drooping. Flowers 2–2.5 cm in diameter, fragrant. Sepals and petals white spotted with pale violet. Lip oblong, 3-lobed at apex, purple, with 5 ridges near the base. Spur greenish, compressed. *Philippines.* G2. Winter–spring.

139. PHALAENOPSIS Blume

V.A. Matthews

Epiphytic. Stem short, pseudobulbs absent. Leaves in 2 ranks, usually persistent, fleshy, rarely with petioles. Raceme or panicle lateral, with 1–many flowers and with a straight or zig-zag axis. Sepals almost equal, free, usually spreading; petals similar or larger. Lip relatively small, often more brightly coloured than the sepals and petals, 3-lobed, continuous with the column foot or inserted at a right angle, lateral lobes erect with cushion-like swellings in the middle, central lobe fleshy, usually firmly attached to the lateral lobes. Between the lateral lobes and the base of the central lobe are various calluses and other ornamented processes. Column often constricted in the middle, usually wingless, produced into a short foot at the base. Pollinia 2, or 4 in *P. pulcherrima*, more or less spherical.

A genus of 40–55 species occurring in the Himalaya, SE Asia and N Australia. They are best grown in baskets in an epiphytic compost. They lack pseudobulbs and therefore they should not be allowed to dry out and the atmosphere should always be kept moist. Shading is essential in spring and early summer to prevent leaf scorch. Flowering can be encouraged by a drop in temperature.

Literature: Holttum, R.E., Cultivated species of the orchid genus Doritis Lindl., *Kew Bulletin* **19**: 207–12 (1965); Sweet, H.R., The genus Phalaenopsis (1980); Christenson, E.A., Phalenopsis, a monograph (2001); Chiba, M., Phalaenopsis species (2002) – a photographic album covering most taxa; Yukawa, T., Kita, K., Handa, T., Hidayat, T. & Ito, M., Molecular phylogenetics of Phalaenopsis and allied genera: Re-evaluation of generic concepts, *Acta Phytotaxonomica et Geobotanica* **56**(2): 141–61 (2005).

Leaves. Mottled or spotted above: **5, 6, 10, 11**.

Inflorescence. Erect or arching: **1, 2, 3, 4, 7, 8, 10, 11, 12, 13, 14, 15, 17, 18, 19, 20**; hanging: **2, 3, 4, 5, 6, 8, 9, 11, 12, 16, 17, 19**. With axis winged or flattened: **8, 13**.

Sepals and petals. Unmarked or with dots at the base only: **1, 2, 3, 4, 5, 6, 7, 10, 13, 16**; barred, blotched or spotted: **8, 9, 11, 12, 14, 16, 17, 18, 19, 20**; longitudinally striped towards the base or apex: **11, 15**.

Lip. Less than 2 cm: **1, 7, 8, 9, 10, 11, 12, 13, 15, 16, 17, 19, 20**; more than 2 cm: **2, 3, 4, 5, 6, 12, 13, 14, 15, 17, 18**. Central lobe hairless: **1, 2, 3, 4, 5, 6, 7, 8, 9, 10, 11, 12, 13, 14, 20**; with some hairs: **9, 12, 15, 16, 17, 18, 19, 20**. Warty: **9**. Mobile: **7**. Apex with 2 appendages: **2, 3, 4, 5, 6**; lacking 2 appendages: **7, 8, 9, 10, 11, 12, 13, 14, 15, 16, 17, 18, 19, 20**. Callus with some backwardly pointing warts or appendages: **12, 14**. Callus with 4 bristle-like appendages originating at the base: **7**.

1a. Pollinia 4; lip claw bearing 2 linear appendages **1. pulcherrima**

b. Pollinia 2; lip without linear appendages towads base 2

2a. Petals wider than sepals; apex of lip with 2 tendril-like or horn-like appendages 3

b. Petals as wide as or narrower than sepals, if wider then with lilac stripes at the base; apex of lip not as above 9

3a. Leaves purplish beneath, especially when young 4

b. Leaves not purplish beneath 8

4a. Scape purplish 5

b. Scape not purplish 6

5a. Mature leaves green above; petals white **3. aphrodite**

b. Mature leaves spotted above; petals white or pink **5. schilleriana**

6a. Petals pink **5. schilleriana**

b. Petals white 7

7a. Leaves spotted above; central lobe of lip white or yellowish **6. stuartiana**

b. Leaves green above; central lobe of lip deep pinkish purple **3. aphrodite**

8a. Scape purplish; callus horseshoe-shaped **4. sanderiana**

b. Scape green; callus more or less square **2. amabilis**

9a. Claw of lip fused to column foot; central lobe of lip mobile; 4 bristle-like appendages present above the callus **7. parishii**

b. Claw of lip not fused to column foot; central lobe of lip not mobile; bristle-like appendages absent 10

10a. Central lobe of lip anchor-shaped 11
b. Central lobe of lip not anchor-shaped 12
11a. Scape winged above; central lobe of lip without warts **8. cornu-cervi**
b. Scape not winged; central lobe of lip warty **9. mannii**
12a. Callus at junction of lobes of lip peltate **10. equestris**
b. Callus at junction of lobes of lip not peltate 13
13a. Central lobe of lip concave or flat with a high, fleshy central ridge or 5 parallel ridges, if ridge absent then lateral lobes rounded **11. fuscata**
b. Central lobe of lip convex with a thin central ridge, or ridge absent 14
14a. Petals broadly elliptic, length 2 times width or less **12. amboinensis**
b. Petals narrowly obovate, oblanceolate or elliptic, length more than 2 times width 15
15a. Central lobe of lip hairless 16
b. Central lobe of lip with at least some hairs towards the apex 18
16a. Callus with elongate, backwardly pointing warts **14. fasciata**
b. Appendages of callus all pointing forwards 17
17a. Sepals and petals without spots or transverse stripes; upper sepal 2–3.5 cm; lip 1.8–2.8 cm **13. violacea**
b. Sepals and petals with spots or transverse stripes; upper sepal 1.2–1.8 cm; lip 1–1.4 cm **20. maculata**
18a. Upper sepal 1.4–2 cm 19
b. Upper sepal 2–4 cm 20
19a. Sepals and petals white with green tips, sometimes with fine, transverse, reddish purple lines **16. fimbriata**
b. Sepals and petals white, cream or yellow with obvious, brown or brownish red bars and blotches **19. mariae**
20a. Sepals and petals with spots and circles **18. hieroglyphica**
b. Sepals and petals with transverse bars and/or blotches, or with longitudinal stripes 21
21a. Callus between lateral lobes of lip with warts **17. lueddemanniana**
b. Callus between lateral lobes of lip lacking warts 22
22a. Central lobe of lip with entire margin **15. sumatrana**
b. Central lobe of lip with margin toothed towards the apex **19. mariae**

1. P. pulcherrima (Lindley) J.J. Smith (*Doritis pulcherrima* Lindley; *P. esmeralda* Reichenbach). Illustration: Skelsey, Orchids, 112 (1979); Sheehan & Sheehan, Orchid genera illustrated, 84 (1979).

Epiphytic. Leaves to 20 cm, in 2 ranks, oblong, bright green, often purplish beneath. Flowers 2–2.5 cm wide, in erect racemes, which are 30–75 cm, with 3–20 flowers. Sepals and petals 1–1.2 cm, pale pink to dark purple, usually somewhat reflexed. Lip 3-lobed with a long claw bearing a linear appendage on each side, appendages pale pink sometimes flushed with yellow, or with purple spots. Lateral lobes erect, rounded, purple or striped with orange-red; central lobe oblong, acute, with a central ridge, purple or reddish. *Burma, Kampuchea, Thailand, Malay peninsula*. G2. Summer–autumn.

The flowers are variable in size and colour and are produced over a period of 2–3 months. A white-flowered variant, 'Alba' (forma *alba* Gruss & Roeth; var. *alba* invalid; var. *candidula* (Rolfe) Guillemin) is commercially available.

P. buyssoniana Reichenbach (*D. pulcherrima* var. *buyssoniana* invalid; *Doritis buyssoniana* (Reichenbach) J.M.H. Shaw). Plants larger in all their parts with the upper sepal to 2.2 cm long, and with the sepals and petals not reflexed. Tetraploid. *Thailand, Indochina*. G2. Summer–autumn.

2. P. amabilis (Linnaeus) Blume. Illustration: Kupper & Linsenmaier, Orchids, 111 (1961); *Orchid Digest* **38**: 192 (1974); Bechtel et al., Manual of cultivated orchid species, 259, 362 (1981); *Orchid Digest* **46**: 29 (1982).

Leaves elliptic to obovate, 15–50 cm, green above and beneath. Inflorescence simple or branched, often 1 m or more, loose with few to many flowers. Flowers fragrant, 6–10 cm in diameter. Sepals 3–4 cm, elliptic-ovate, white, often pinkish on the back. Petals to 4.5 cm, more or less circular, similar in colour to the sepals. Lip white, to 2.3 cm, lateral lobes oblanceolate with yellow margins and basal red markings, central lobe somewhat cross-shaped with triangular side-arms and 2 terminal, tendril-like appendages with yellow tips. Callus more or less square, yellow, spotted with red. *E Indies, Australia (NE Queensland)*. G2. Autumn–early spring.

There are 3 varieties in cultivation:

Var. **aurea** (Hort.) Rolfe. Most of the lateral lobes, the margin of the central lobe, and the terminal appendages are yellow. *Borneo.*

Var. **moluccana** Schlechter (subspecies *moluccana* (Schlechter) Christenson). The central lobe of the lip is linear-oblong with small lateral lobes. *Indonesia (Moluccas, Sulawesi).*

Var. **papuana** Schlechter (subspecies *rosenstromii* (Bailey) Christenson). The central lobe of the lip is narrowly triangular. *New Guinea, Australia (NE Queensland).*

3. P. aphrodite Reichenbach (*P. amabilis* var. *dayana* Warner & Williams). Illustration: *Orchid Digest* **38**: 192 (1974); *American Orchid Society Bulletin* **46**: 217 (1977); Bechtel et al., Manual of cultivated orchid species, 362 (1981).

Leaves broadly elliptic, 20–38 cm, green above, purplish beneath especially when young. Inflorescence simple or branched, arching or hanging, longer than the leaves, purplish. Flowers 7 cm or more in diameter. Sepals to 4 cm, elliptic, white, sometimes with tiny pink dots. Petals similar to the sepals but more or less circular. Lip white, *c.* 3 cm, lateral lobes ovate with yellow margins and dark pink dots and streaks towards the base, central lobe triangular with a dark pink flush at the base and yellow on the side-arms, apex with 2 tendril-like appendages. Callus with notched apex, each side with tooth-like projections, deep yellow with deep pink dots. *Philippines*. G2. Flowering all year.

Subspecies **formosana** Christenson (*P. amabilis* var. *formosa* Shimadzu). Flowers smaller, apple green. Inflorescence branching from basal nodes. *Taiwan*. G2.

P. × intermedia Lindley. Illustration; *Die Orchidee* **24**: 144 (1973); Rittershausen & Rittershausen, Orchids in colour, 111 (1979); Bechtel et al., Manual of cultivated orchid species, 260, 362 (1981). Leaves elliptic, to 30 cm, green above, brownish purple beneath. Inflorescence to 60 cm, simple or branched, with many flowers. Sepals 2.5–4 cm, ovate-elliptic, white with pink suffusion. Petals 2–3.5 cm, broadly elliptic, white with pink dots at the base. Lip pinkish purple, lateral lobes ovate, purplish spotted with deep pink, central lobe obovate, tapered towards the tip, deep pinkish purple, apex notched or with 2 horn-like or tendril-like appendages. Callus more or less square, yellow spotted with red. *Philippines*. G2. Spring.

A hybrid between *P. aphrodite* and *P. equestris.*

4. P. sanderiana Reichenbach. Illustration: *Orchid Digest* **38**: 221 (1974); Rittershausen & Rittershausen, Orchids in colour, 121 (1979); Sweet, The genus Phalaenopsis, 41 (1980); Bechtel et al., Manual of cultivated orchid species, 362 (1981).

Leaves 15–35 cm, elliptic, dark green above, often marked with silvery grey beneath. Inflorescence to 80 cm but usually less, purplish, simple or branched. Flowers 6–7.5 cm in diameter. Sepals to 3.5 cm, ovate-elliptic, pink, sometimes with white mottling or completely white. Petals similar to sepals but broader. Lip *c.* 3 cm, lateral lobes ovate, white spotted with pink, central lobe triangular with narrowly triangular side-arms, white streaked with brown or purple and sometimes yellow, apex with 2 tendril-like appendages. Callus horseshoe-shaped, white or yellow with brown, red or purple spots. *Philippines*. G2. Spring.

Very variable in flower colour and in the markings on the lip. On this basis several varieties have been recognised but it is generally accepted that they have no botanical standing.

5. P. schilleriana Reichenbach. Illustration: *Flore des serres* **5**: t. 1559 (1862–65); *Orchid Digest* **38**: 221 (1974); Hunt & Grierson, Country Life book of orchids, 61 (1978); Bechtel et al., Manual of cultivated orchid species, 261, 362 (1981).

Leaves to 45 cm, elliptic, dark green spotted with silvery grey above, purple beneath. Inflorescence to 90 cm and often bearing over 200 flowers, branched, hanging or arched, purplish. Flowers 5–7.5 cm in diameter. Sepals ovate-elliptic, to 3.5 cm, pink fading at margins to white, lateral sepals with deep pink spots towards the base. Petals more or less circular, to 4 cm, similar in colour to the sepals. Lip 2–2.3 cm, white to deep pink, lateral lobes elliptic, often broadly so, with reddish brown spots, central lobe ovate to rounded with 2 curved, horn-like appendages of varying size at the apex. Callus more or less square, yellow. *Philippines*. G2. Winter–spring, sometimes also late summer.

P. × leucorrhoda Reichenbach. Illustration: Cogniaux & Goossens, Dictionnaire iconographique des orchidées, Phalaenopsis hybrids, t. 1 (1902); Rittershausen & Rittershausen, Orchids in colour, 113 (1979). Leaves green above, spotted with silvery grey, purplish beneath. Inflorescence usually hanging, to 70 cm, with many flowers, purplish. Sepals 2.2–3.5 cm, elliptic to ovate. Petals 2.5–4 cm, kidney-shaped to almost circular. Sepals and petals white, often flushed with pink, or deep pink with white margins, lateral sepals sometimes with purple dots at the base. Lip 2.2–3 cm, lateral lobes spathulate, white or yellow with reddish purple spots, central lobe white with yellow and purple markings, sometimes entirely purple, variable in shape, tapering towards the apex which has 2 anchor-shaped or tendril-like appendages. Callus irregularly toothed, deep yellow, occasionally paler, densely spotted with dark red. *Philippines*. G2.

A hybrid between *P. aphrodite* and *P. schilleriana*.

6. P. stuartiana Reichenbach (*P. schilleriana* var. *alba* Roebelen; *P. schilleriana* var. *vestalis* Reichenbach). Illustration: Orchid Digest **38**: 221 (1974); *Der Palmengarten*, 187 (1979); Bechtel et al., Manual of cultivated orchid species, 261, 362 (1981).

Leaves elliptic-oblong, 15–35 cm, green blotched with grey above, purplish beneath. Inflorescence to 80 cm, with many flowers, branched, hanging. Flowers 5–6 cm in diameter, slightly fragrant. Sepals to 3.5 cm, upper sepal white, lateral sepals white with yellow on basal half, and with brownish red dots. Petals to 3.3 cm, somewhat square to circular, white, often with purplish brown dots at the base. Lip *c.* 2.5 cm, lateral lobes obovate, white at apex, yellow at base with reddish brown spots and with 2 white, horn-like appendages. Callus almost square with an apical projection on each side, orange, spotted. *Philippines*. G2. Winter.

In var. **punctatissima** Reichenbach (forma *punctatisima* (Reichenbach) Christenson) the dots on the flowers are pale purple. 'Larkin Valley' is the clone usually grown.

Forma **nobilis** (Reichenbach) Christenson produces sulphur yellow flowers.

7. P. parishii Reichenbach. Illustration: *Botanical Magazine*, 5815 (1870).

Leaves to 12 cm, elliptic to obovate. Inflorescence erect or arching, to 15 cm, with few or several flowers. Flowers opening simultaneously. Sepals white, the upper 6–8 mm, elliptic to almost circular, the laterals obovate to almost circular, 7–10 mm. Petals white, elliptic to obovate, 6–7 mm. Lip 10–15 mm, the short claw fused to the column foot, lateral lobes very small, triangular, yellow or whitish spotted with purple or brown, central lobe mobile, purple, triangular. Callus semicircular, fringed at margin. At the junction of the lobes there is a plate-like projection ending in 4 bristle-like appendages. Column white with purple spots. *E Himalaya, India*. G2.

P. lobbii (Reichenbach) Sweet (*P. parishii* var. *lobbii* Reichenbach). Illustration: *Orchid Digest* **37**: 168 (1973); *American Orchid Society Bulletin* **50**: 33–4 (1981); *Orchid Digest* **71**: 78 (2007).

Lip white or yellow with a pair of vertical brown stripes on each side, and the callus margin entire or only slightly toothed. *Eastern Himalaya, Burma, Vietnam*. G2.

P. parishii has only recently entered cultivation, hence most references to cultivated *P. parishii* refer to *P. lobbii*. Consequently most hybrids made with '*P. parishii*', up to about the year 2000, were made with *P. lobbii*.

8. P. cornu-cervi (Breda) Blume & Reichenbach. Illustration: *Botanical Magazine*, 5570 (1866); Rittershausen & Rittershausen, Orchids in colour, 118 (1979); *American Orchid Society Bulletin* **50**: 32 (1981); Bechtel et al., Manual of cultivated orchid species, 362 (1981).

Leaves to 25 cm, oblong. Inflorescence simple or branched, 10–40 cm, axis flattened. Flowers 3–5 cm in diameter. Sepals 1.8–2.3 cm, elliptic, keeled on the back, upper sepal with recurved margins, yellow or greenish with reddish brown blotches, lateral sepals often with blotches only on the upper half. Petals 7–18 mm, lanceolate, similar in colour to upper sepal. Lip *c.* 8 mm, lateral lobes almost square, with a callus below the truncate apex, whitish with reddish brown stripes, lateral lobes running into base of central lobe to form a flat, semicircular swelling, in front of which the central lobe is constricted and then expanded into a whitish, anchor-shaped portion with recurved lobes. Callus complex with 3 appendages in series, the first upcurved, yellow, the second forked, white, and the third lanceolate, purple; the second and third project over the central lobe. *SE Asia*. G2. Summer.

P. pantherina Reichenbach. Illustration: *American Orchid Society Bulletin* **38**: 510 (1969); Sweet, The genus Phalaenopsis, 61 (1980). Differs mainly in the larger (*c.* 1.5 cm), differently shaped lip. Below the swelling at the base of the central and lateral lobes, the central lobe has an isthmus that expands into a somewhat anchor-shaped portion with lobes at right angles. The callus appendages are complex. *Borneo.*

9. P. mannii Reichenbach (*P. boxallii* Reichenbach). Illustration: *Orchid Digest* **40**: 207 (1976); Bechtel et al., Manual of cultivated orchid species, 260, 362 (1981); *American Orchid Society Bulletin* **50**: 30–2 (1981).

Leaves to 35 cm, oblong. Inflorescence usually simple, hanging, with many flowers. Flowers *c.* 5 cm in diameter, opening in succession. Sepals obovate-lanceolate, yellow or green with brownish blotches, keeled on the back towards the apex, upper sepal 2–2.4 cm, margins rolled under, lateral sepals 2.2–2.5 cm. Petals lanceolate, margins rolled under, 1.7–2 cm, similar in colour to sepals. Lip white and purple, 9–11 mm, lateral lobes almost square with a fleshy callus, base of lateral lobes running into the central lobe to form a semicircular swelling from which the variable central lobe expands from a basal isthmus. Central lobe anchor-shaped, the lateral lobes toothed, upper surface warty and with an apical, often hairy callus. The basal callus appendages are complex. *Himalaya, Vietnam*. G2. Summer.

10. P. equestris (Schauer) Reichenbach (*P. rosea* Lindley). Illustration: *Flore des serres* **6**: t. 1645 (1865–67); Rittershausen & Rittershausen, Orchids in colour, 118 (1979); *Die Orchidee* **31**: 225 (1980); Bechtel et al., Manual of cultivated orchid species, 259, 362 (1981).

Leaves to 20 cm, oblong. Inflorescence simple or branched, almost erect to arching, purplish, with many flowers. Flowers 2.5–4 cm in diameter. Sepals oblong-elliptic, white to pink, the upper 1–1.7 cm, the laterals 1–1.5 cm. Petals elliptic, 8–15 mm, similar in colour to sepals. Lip deep pink or purplish, 1–1.4 cm, lateral lobes oblong, often with dark streaks, central lobe ovate with apex pointed, more or less concave in the middle. Callus peltate, 6–8-sided, yellow with red spots. *Philippines, Taiwan*. G2. Spring–autumn.

This species is the basis for most hybrids with a white flower and a contrasting strongly coloured lip.

P. lindenii Loher. Illustration: *Orchid Digest* **37**: 108 (1973); Bechtel et al., Manual of cultivated orchid species, 260 (1981); *Orchid Digest* **41**: 6 (1982). Leaves green mottled silvery white. Inflorescence simple, usually green. Lip 1.2–1.4 cm, lateral lobes obovate to strap-shaped, white with tiny orange or reddish dots at the base and 3 purplish lines above, central lobe more or less circular, with a small point, slightly concave in the middle, purplish pink towards the apex with 5–7 darker lines. *Philippines*. G2.

11. P. fuscata Reichenbach. Illustration: Sweet, The genus Phalaenopsis, 64 (1980).

Leaves to 30 cm, oblong. Inflorescence to 30 cm, simple or branched, with few flowers. Sepals and petals yellowish green blotched with brown, margins rolled under, sepals 1.4–1.8 cm, petals 1.2–1.5 cm. Lip yellowish with brown markings, 1.1–1.4 cm, lateral lobes almost square, twisted so that they touch, central lobe ovate to elliptic with a central ridge. Callus at junction of lobes with 2 forked appendages. *Malay peninsula, Borneo*. G2.

P. kunstleri Hooker. Illustration: *Botanical Magazine*, 7885 (1903); *Orchid Digest* **40**: 209 (1976). Similar but with larger flowers *c.* 5 cm in diameter. Sepals and petals yellow at base and apex, brown in the central part, sepals 1.8–2 cm, petals 1.5 cm. Lip yellow, 1–1.2 cm, lateral lobes almost square, central lobe almost circular with 1 or 2 brown stripes on each side of the central ridge. Callus at junction of lobes with 2 forked appendages. *Malaysia*. G2.

Considered to be the same as *R. fuscata* by some authorities.

P. cochlearis Holttum. Illustration: *Orchid Digest* **40**: 209 (1976). Inflorescence to 50 cm, branched. Sepals and petals white to pale yellow with 2 pale brown bars at the base, upper sepal 1.5–1.8 cm, lateral sepals 1.4–1.5 cm, petals 1.2–1.4 cm. Lip white or yellow, 9–12 mm, lateral lobes narrowly oblong, notched at apex with reddish streaks at the base, central lobe circular, concave, sometimes notched at apex, with reddish brown stripes and 5 shallow ridges. Callus with a pair of 2-lobed appendages. *Malay peninsula, Sarawak*. G2.

P. celebensis Sweet. Illustration: Sweet, The genus Phalaenopsis, 66 (1980). Leaves mottled with silvery white, sepals and petals white with lilac stripes at the base, sepals to 1.4 cm, petals to 9 mm, slightly wider than sepals. Lateral lobes of lip deep yellow, rounded, with a central fleshy ridge, central lobe up to 9 mm, concave, elliptic-ovate, with a small point at apex. Callus at junction of lobes more or less triangular. *Indonesia (Sulawesi)*. G2.

12. P. amboinensis J.J. Smith. Illustration: *Orchid Digest* **36**: 88 (1972).

Leaves to 25 cm, elliptic to obovate. Inflorescence arching to 45 cm, with few flowers; flowers opening in succession. Sepals and petals white or yellowish with greenish tips and brownish red bars, elliptic to ovate, sepals 2–3 cm, petals shorter. Lip white or yellowish, 2–2.2 cm, lateral lobes oblong, each with an orange spot and a callus, central lobe ovate, the central ridge with a toothed edge. Callus with a pair of 2-lobed appendages. *Indonesia (Amboina)*. G2. Summer.

P. gigantea J.J. Smith. Illustration: *Orchid Digest* **36**: 38 (1972); Bechtel et al., Manual of cultivated orchid species, 259, 362 (1981); *Orchid Digest* **41**: 29 (1982). Leaves to 90 cm, oblong-ovate. Flowers *c.* 5 cm in diameter, slightly fragrant, opening simultaneously, numerous, in hanging, dense inflorescences. Sepals and petals greenish or yellow, whitish at the base, blotched with reddish brown or purple. Lip *c.* 1.5 cm, lateral lobes triangular, curved, orange towards the apex, central lobe ovate with some tiny teeth on the lateral margins and an ovoid callus at the apex, striped and blotched with red or purple. Callus between lateral lobes with a pair of 2-toothed, deep yellow appendages. *Borneo*. G2.

P. javanica J.J. Smith (*P. latisepala* Reichenbach). Illustration: *Orchid Digest* **43**: 57 (1979). Leaves to 22 cm, elliptic. Flowers *c.* 3 cm in diameter. Sepals and petals broadly elliptic, white to yellow with brownish purple spots arranged in longitudinal lines. Lip distinctly clawed, 1–1.8 cm, lateral lobes linear-oblong, grooved in the middle, yellowish, apex with 2 teeth, central lobe elliptic, very fleshy, apex with a few hairs, purple, whitish towards the base. Callus between lobes consisting of a lanceolate, backwardly pointing appendage and a 2-toothed forwardly pointing appendage. *Java*. G2.

13. P. violacea Witte. Illustration: *Orchid Digest* **36**: 12 (1972); *Die Orchidee* **30**: 122–3 (1979); Bechtel et al., Manual of cultivated orchid species, 362 (1981).

Leaves 20–25 cm, generally less than 8 cm wide, elliptic to obovate. Inflorescence more or less erect or arching, simple, with a few, distant, flowers; axis flattened.

Flowers 5–7.5 cm in diameter, with a spicy fragrance. Sepals and petals bright purple with pale greenish tips. Sepals elliptic, not sickle-shaped, the tips of all 3 forming an equilateral triangle, 2–3.5 cm, keeled on the back towards the apex, petals elliptic, 2–3 cm, less than 7 mm wide. Lip reddish purple, 1.8–2.8 cm, lateral lobes linear-oblong, marked with yellow, central lobe ovate, abruptly pointed with a central ridge running into an apical callus. Callus between lateral lobes with warts and complex appendages, yellow. *Malay peninsula, Borneo, Sumatra*. G2. Summer.

Var. **alba** Teijsmann & Binnedijk. Illustration: *American Orchid Society Bulletin* **34**: 206 (1965). The sepals and petals are white with greenish tips. This can be difficult to separate from the white-flowered *P. bellina* f. *alba* Christenson.

P. bellina (Reichenbach) Christenson (*P. violacea* var. *bellina* Reichenbach; *P.* '*Borneo*', '*Borneo Form*'). Illustration: Chan et al., Orchids of Borneo **1**: 256 & pl. 17A (1994) – as *P. violacea*; Beaman et al., Orchids of Sarawak, pl. 26A (2001). Leaves generally more than 10 cm wide. Flowers with a lemony fragrance, white, greenish white, yellow or rarely orange, with a deep purple area on the inner half of the lateral sepals. The petals and sepals are never uniformly coloured as in *P. violacea*. Petals ovate, more than 1.3 cm wide. Lateral sepals sickle-shaped, the 3 sepal tips forming an isoceles triangle. *Borneo, Malay peninsula*. G2.

Adopted as the state flower of Sarawak. Prior to 2002 *P. bellina* was included in *P. violacea* when listing hybrids. Such hybrids can be identified by the lower half of the lateral sepals being more darkly coloured (usually dark pink) compared to the upper halves (usually light yellow to cream), a trait inherited from *P. bellina*.

14. P. fasciata Reichenbach. Illustration: *Orchid Digest* **40**: 207 (1976); Bechtel et al., Manual of cultivated orchid species, 259 (1981).

Leaves 14–20 cm, elliptic. Inflorescence more or less erect or arching, branched, longer than leaves. Flowers *c.* 4 cm in diameter. Sepals and petals yellow to yellowish green with transverse red-brown bars. Sepals ovate-elliptic, 2–3 cm, petals ovate to ovate-elliptic, 2.2–2.6 cm. Lip 2–2.7 cm, lateral lobes strap-shaped, erect, yellow, central lobe pale purple, oblong-ovate, convex with a central ridge running into an apical callus. Callus between lateral lobes orange, with elongate, backwardly pointing warts and a forwardly pointing forked appendage. *Phillippines*. G2.

In cultivation many of the plants offered under the name *P. lueddemanniana* var. *ochracea* are *P. fasciata*.

15. P. sumatrana Korthals & Reichenbach (*P. zebrina* Witte; *P. zebrina* Teijsmann & Binnedijk). Illustration: *American Orchid Society Bulletin* **37**: 1093 (1968); *Orchid Digest* **36**: 25 (1972); Bechtel et al., Manual of cultivated orchid species, 362 (1981).

Leaves 15–30 cm, oblong to obovate. Inflorescence erect or slightly arching, usually simple. Flowers *c.* 5 cm in diameter. Sepals and petals oblong-lanceolate, whitish to pale yellow with transverse, reddish brown or purplish bars. Sepals 2–4 cm, petals 2.2–3.5 cm. Lip to 2.5 cm, lateral lobes linear-oblong, apex 2-toothed with a third tooth on the inside, cream with yellow or brown margin and often orange spots, central lobe oblong-elliptic, very thick especially towards the hairy apex, convex with a central ridge, white with reddish or purple stripes on each side of the ridge. Callus between the lateral lobes a complex series of forked plates. Flowers with a slightly unpleasant scent. *Thailand, Malay peninsula, Java, Sumatra & Borneo*. G2.

P. corningiana Reichenbach. Illustration: *Orchid Digest* **40**: 105 (1976). Similar but sepals and petals greenish yellow towards the apex with longitudinal brown or purplish stripes merging into blotches towards the base, upper sepal obovate to oblanceolate, 3–4 cm, lateral sepals ovate, 2.5–3 cm, petals lanceolate 2.5–3.5 cm. Lip *c.* 2 cm, lateral lobes strap-shaped, whitish with a central orange-yellow callus, central lobe deep purplish pink, narrowly elliptic, convex with a central ridge running into a hairy apical callus. Callus between lateral lobes yellow, forked with a central groove. Below the callus is an appendage with finger-like projections. Flowers with a sugary fragrance. *Malaysia, Sarawak*. G2.

16. P. fimbriata J.J. Smith. Illustration: Sweet, The genus Phalaenopsis, 97 (1980); Bechtel et al., Manual of cultivated orchid species, 259, 362 (1981).

Leaves 14–23 cm, oblong-elliptic. Inflorescence to 30 cm, usually hanging, loose with many flowers. Flowers opening simultaneously. Sepals and petals white, shaded green towards the tips and sometimes with fine transverse reddish purple lines, sepals 1.5–2 cm, petals 1.5–1.8 cm, all elliptic. Lip 1.4–2 cm, lateral lobes oblong, white flushed with reddish purple, convex with a central toothed ridge running into an apical callus with white hairs, lateral margins of central lobe expanded and toothed towards the apex. Callus between the lateral lobes with 3 overlapping plates, each with 2 teeth. *Java, Sumatra*. G2.

17. P. lueddemanniana Reichenbach. Illustration: *Orchid Digest* **40**: 207 (1976); Bechtel et al., Manual of cultivated orchid species, 362 (1981).

Leaves 20–30 cm, oblong-elliptic. Inflorescence erect or hanging, simple or branched, longer than the leaves. Flowers 5–6 cm in diameter. Sepals and petals white with transverse brownish purple bars, sepals oblong-elliptic, 2–3 cm, petals slightly smaller. Lip pink to purple, 1.8–2.2 cm, lateral lobes oblong, central lobe oblong to ovate with a central ridge running into an apical callus with white hairs. Callus between lateral lobes warty and with a forked appendage. *Philippines*. G2. Summer.

Very variable in the size and colour of the flowers. Var. **delicata** Reichenbach (illustration: *Flore des serres* **6**: t. 1636, 1865–67) has white sepals and petals with narrow brown bars on the apical part and purple bars on the basal part; the lip is purple. Var. **ochracea** Reichenbach (illustration: *Orchid Digest* **40**: 207, 1976) has sepals and petals that are off-white with yellowish brown bars and a basal pale purple flush; this is often confused with yellow flowered *P. fasciata* (p. 587) and many plants cultivated under this name are, in fact, *P. fasciata*. According to Wallbrunn (*American Orchid Society Bulletin* **40**: 225, 1971), it may be better to consider *P. fasciata*, *P. hieroglyphica* and perhaps *P. pallens* as subspecies of a variable *P. lueddemanniana*.

18. P. hieroglyphica (Reichenbach) Sweet (*P. lueddemanniana* var. *hieroglyphica* Reichenbach). Illustration: *Orchid Digest* **40**: 207 (1976); Bechtel et al., Manual of cultivated orchid species, 260, 362 (1981).

Leaves to 30 cm, strap-shaped. Inflorescence more or less erect or arching, simple or branched, with many flowers. Sepals and petals ovate-elliptic, white, greenish or sometimes yellowish with brownish spots and circles. Sepals keeled on the back towards the apex, the upper

2.3–3.8 cm, the laterals 2.4–4.1 cm, petals 2.2–3.3 cm. Lip white or yellowish, 2–2.5 cm, lateral lobes oblong, apex notched, central lobe wedge-shaped with a central ridge running into an apical hairy callus. Callus between lateral lobes with elongate, forwardly pointing warts and 2 forked appendages. *Philippines*. G2. Summer.

19. P. mariae Warner & Williams. Illustration: *Botanical Magazine*, 6964 (1887); *Orchid Digest* **35**: 308 (1971); Bechtel et al., Manual of cultivated orchid species, 260 (1981).

Leaves 15–30 cm, strap-shaped. Inflorescence hanging, simple or branched. Flowers 2.5–5 cm in diameter. Sepals and petals oblong-elliptic, white or cream with transverse brownish red bars and blotches, sometimes with purplish spots at the base. Sepals 1.6–2.2 cm, petals 1.5–1.7 cm. Lip pale mauve or purple, lateral lobes broadly strap-shaped, toothed at the apex, central lobe ovate, the margins expanded and toothed towards apex, with a central ridge at the base and a hairy callus at the apex. Callus between lateral lobes with a series of 2-lobed appendages. *Philippines*. G2. Summer.

P. pallens (Lindley) Reichenbach (*P. foerstermanii* Reichenbach). Illustration: *Orchid Digest* **40**: 207 (1976); Sweet, The genus Phalaenopsis, 109 (1980); Bechtel et al., Manual of cultivated orchid species, 261, 362 (1981). Leaves elliptic to obovate, 12–18 cm. Inflorescence erect or arching, simple, with 1 or a few flowers. Flowers c. 5 cm in diameter. Sepals and petals oblong-elliptic, yellowish green with transverse brown bars, lines and spots. Sepals with a keel on the back, the upper 1.2–2.3 cm, the laterals 1.5–2.2 cm, petals 1.1–2 cm. Lip 1.3–1.7 cm, lateral lobes oblong, yellowish, central lobe white, narrowly ovate, margins usually expanded and toothed towards the apex, with a central ridge at the base and hairy callus towards the apex. Callus between lateral lobes with a pair of forked appendages. *Philippines*.

Var. **denticulata** (Reichenbach) Sweet has 2 or 3 reddish purple lines on either side of the central ridge of the central lobe of the lip.

20. P. maculata Reichenbach. Illustration: *Orchid Digest* **35**: 308 (1971) & **41**: 29 (1982).

Leaves 15–20 cm, strap-shaped. Inflorescence more or less erect to arching, usually simple, with few flowers. Flowers less than 2.5 cm in diameter. Sepals and petals oblong-lanceolate, cream to greenish white with a few purplish brown or red-brown blotches. Upper sepal 1.3–1.8 cm, lateral sepals 1.3–1.5 cm, petals c. 1.2 cm. Lip *c.* 1 cm, lateral lobes diamond-shaped, the apical part with a horseshoe-shaped callus, central lobe red to deep purple, elliptic with faint longitudinal grooves, very thick and fleshy. Callus between side-lobes 2-toothed; at the base of central lobe is a shallow, 2-toothed process. *Malay peninsula, Sarawak*. G2.

P. modesta J.J. Smith (*P. psilantha* Schlechter). Illustration; *Orchid Digest* **43**: 212 (1979). Leaves obovate, 11–15 cm. Inflorescence arched, usually simple, with few flowers. Sepals and petals elliptic, white or pale pink, with fine transverse purple stripes towards the base, sepals 1.2–1.5 cm, petals 1.1–1.5 cm. Lip 1.2–1.4 cm, lateral lobes oblong, narrowing towards the notched apex, yellow callus at middle, central lobe oblong, white, purple towards the apex, convex, with a basal ridge and an apical, usually hairless callus. Callus between the lateral lobes with a pair of forked appendages. *Borneo, Sulawesi*. G2.

Var. **bella** Gruss & Roellke. Leaves 30–40 cm. Flowers cupped, upper sepal elliptic-ovate, lateral sepals oblong-obovate, petals elliptic-oblanceolate. Said to be from Borneo, may be a hybrid between *P. javanica* and *P. modesta*.

140. PARAPHALAENOPSIS Hawkes

V.A. Matthews

Differs from *Phalaenopsis* in having terete leaves and often producing more than 1 raceme from the same point on the stem.

A genus of 4 species from W. Borneo (Kalimantan), all of which are in cultivation. Though the genus has been crossed with several other (e.g. *Vanda, Renanthera*), attempts to hybridise it with *Phalaenopsis* have so far failed. Cultivation as for *Phalaenopsis* (p. 583).

Literature: Sweet, H.R., The genus Phalaenopsis, 118–23 (1980); Chan, C.L., Lamb, A., Shim, P.S. & Wood, J.J., Orchids of Borneo **1**: 217–23 (1994) – all species are illustrated in detail; Yam, T.W., Breeding with Paraphalaenopsis, *American Orchid Society Bulletin* **63**(12): 1358–65 (1994).

1a. Sepals and petals greenish yellow to yellowish brown **1. denevei**

b. Sepals and petals white or cream, often flushed with pink or violet **2. serpentilingua**

1. P. denevei (J.J. Smith) Hawkes (*Phalaenopsis denevei* J.J. Smith). Illustration: *Orchid Digest* **37**: 12–13 (1973).

Leaves erect or hanging, to 70 cm. Raceme usually upright with 3–15 flowers, to 15 cm. Flowers *c.* 5 cm in diameter, fragrant. Sepals and petals pale greenish yellow to yellowish brown, with paler, wavy margins, *c.* 2.5 cm. Lateral lobes of lip triangular, yellow at base, purple above, with white margin, central lobe narrowly oblong, somewhat expanded and notched at apex, whitish, reddish or purplish towards apex. Callus more or less square with a slightly toothed margin, yellow with red lines. *W. Borneo*, G2. Spring–summer.

2. P. serpentilingua (J.J. Smith) Hawkes (*Phalaenopsis serpentilingua* J.J. Smith). Illustration: *Orchid Digest* **37**: 12 (1973).

Leaves as for *P. denevei*. Raceme usually upright, to 35 cm, with 7–many flowers. Flowers 3–4 cm in diameter, fragrant. Sepals and petals white or cream, sometimes tinged with violet. Lateral lobes of lip pale orange with reddish markings, central lobe narrowly oblong with 2 divergent apical lobes like a snake's tongue, white or pale yellow with transverse yellow and purple marks. Callus with a toothed margin, yellow with red marks. *W.* Borneo. G2.

P. laycockii (Henderson) Hawkes (*Phalaenopsis laycockii* Henderson). Illustration: *Orchid Digest* **37**: 12–13 (1973). Raceme with 9–15 flowers, 5–8 cm in diameter, sepals and petals 3.5–4.5 cm, whitish flushed with pink or pale purple, lateral lobes of lip white flushed with dark purple or red, central lobe narrowly oblong, margins reflexed with 2 short, divergent apical lobes, callus margins untoothed. *W. Borneo*. G2.

141. VANDA R. Brown

J. Cullen

Monopodial epiphytes with long or short stems, sometimes scrambling. Leaves in 2 ranks, often fleshy, flat or terete, often lobed or toothed at the apex, folded when young. Racemes axillary with few to many flowers. Sepals and petals similar or the lateral sepals larger than the rest, all more than 1.5 cm, usually spreading. Lip often complex, 3-lobed, the lateral lobes large or small, the central lobe often itself notched or lobed, bearing calluses of various forms, generally with 2 humped ridges at the entrance to the spur. Spur usually present, conical or more rarely cylindric. Column short and thick with or without a foot. Pollinia 2, usually borne on a short, broad stipe and with a large viscidium.

A genus of 50 or more species from the Himalaya and SE Asia. Its classification is confused, as are its relations with other genera. Species No. 1 is often referred to *Euanthe* Schlechter (recognised by its flat flowers with unequal sepals and the particular form of the spurless lip); No. 2 is referred by Garay to the genus *Holcoglossum* Schlechter (recognised by its narrow, cylindric spur, grooved leaves, footless, winged column and linear stipe). However, for the present purpose it seems sensible to maintain a wide circumscription of *Vanda*, citing synonyms as appropriate.

The number of species in cultivation is uncertain; 16 are included here, though only 8 are treated fully. A large number of hybrids is also available. Plants should be given as much light as possible, with only light shading in very sunny weather. They require an epiphytic compost and can be successfully grown in pots or on lengths of tree fern or bark. Propagation is by means of pieces of stem cut back to a well-developed aerial root.

Literature: Motes, M.R., Vandas, their botany, history and culture (1997).

Leaves. V-shaped in section: **3**, **4**; flat: **1**, **5**, **6**, **7**, **8**.
Sepals and petals. Spreading: **1**, **2**, **3**, **4**, **5**, **6**; curved forwards: **7**.
Lip. Central lobe with 2 divergent lobes at apex: **6**, **7**; with a downwardly projecting horn at apex: **7**. Shorter than sepals: **1**, **4**; as long as or longer than sepals: **3**, **4**, **6**, **7**.
Spur. Absent: **1**, **7**; cylindric, *c.* 2.5 cm: **2**; conical and much shorter: **4**, **5**, **6**, **7**.

1a. Leaves very narrowly V-shaped in section with a distinct groove above, at least near the base — 2
b. Leaves flat — 3
2a. Spur cylindric, acute, *c.* 2.5 cm; lateral sepals sickle-shaped, longer than the upper — **2. kimballiana**
b. Spur much shorter, blunt, conical; lateral sepals similar to the upper — **3. amesiana**
3a. Sepals and petals curved forwards; racemes with 2–5 flowers; plants small — **7. cristata**
b. Sepals and petals spreading; racemes with 6 or more flowers; plants often large — 4
4a. Flower flat with broad, overlapping sepals and petals; base of lip hollowed out, but spur absent — **1. sanderiana**
b. Flower not as above; lip with a distinct, if short, spur — 5
5a. Central lobe of lip expanded into a kidney-shaped apical part from an oblong base — **6. denisoniana**
b. Central lobe of lip oblong, sometimes waisted but then the apical part broadly rounded, not kidney-shaped — 6
6a. Flowers 7–10 cm in diameter; lip considerably smaller than the sepals — **4. caerulea**
b. Flowers 4–5 cm in diameter; lip about as long as the sepals — **5. tricolor**

1. V. sanderiana Reichenbach (*Euanthe sanderiana* (Reichenbach) Schlechter). Illustration: *Botanical Magazine*, 6983 (1888); *Lindenia* **12**: t. 547 (1896); Cogniaux & Goossens, Dictionnaire iconographique des orchidées, Vanda, t. 12 (1900); *Orchid Digest* **43**: 84 (1979).

Stem to 60 cm. Leaves to 45 × 5 cm, flat, oblong-linear, sharply truncate and 3-toothed at the apex. Raceme to 30 cm with 5–10 flowers. Flowers 9–10 cm in diameter, flat, sepals and petals overlapping, broadly elliptic; upper sepal to 5.5 cm, violet-blue like the petals, which are *c.* 4 cm; lateral sepals to 6 cm, yellow or greenish with red-brown marbling. Lip shorter than the petals, with a lower, hemispherical, saccate part bearing 2 semicircular lateral lobes, the central lobe broadly kidney-shaped with 3 low ridges. *Philippines*. G2. Autumn–winter.

2. V. kimballiana Reichenbach (*Holcoglossum kimballianum* (Reichenbach) Garay). Illustration: *Lindenia* **5**: t. 204 (1889); *Botanical Magazine*, 7112 (1890); Cogniaux & Goossens, Dictionnaire iconographique des orchidées, Vanda, t. 7 (1898).

Stem elongate. Leaves to 20 cm, fleshy, narrowly V-shaped in section, acute. Sepals and petals *c.* 2.5 cm, oblong-elliptic, white, wavy, the lateral sepals longer, sickle-shaped. Lip about as long as the sepals with small, acuminate, yellow, red-spotted lateral lobes and a large, nearly circular, toothed and notched, rose-red central lobe. Spur *c.* 2.5 cm, cylindric, curving downwards, acute. *Burma*. G2. Autumn.

3. V. amesiana Reichenbach. Illustration: *Botanical Magazine*, 7139 (1890); Cogniaux & Goossens, Dictionnaire iconographique des orchidées, Vanda, t. 1 (1897); *Orchid Digest* **43**: 44 (1979).

Stem short. Leaves to 20 cm, fleshy, narrowly V-shaped in section, acute. Raceme to 50 cm, with 15–30 flowers, axis red-spotted. Sepals and petals *c.* 2 cm, broadly oblong, white. Lip slightly longer than sepals, pink with white margin and stripes, lateral lobes erect, short, central lobe broadly ovate, wavy. Spur conical, very short and blunt. *Kampuchea, Burma, SW China*. G1. Winter.

4. V. caerulea Dudley. Illustration: Schlechter, Die Orchideen, edn 2, t. 13 (1927); *Orchid Digest* **43**: 44, 49 (1979); Bechtel et al., Manual of cultivated orchid species, 278 (1981).

Stem short, densely leafy. Leaves to 25 × 2.5 cm, oblong, blunt, flat. Racemes to 45 cm with 7–15 flowers. Flowers 7–10 cm in diameter. Sepals and petals oblong-obovate, clear blue, often with darker tessellation, petals often with twisted claws. Lip much shorter than the sepals, dark violet-blue with small whitish lateral lobes and an oblong, convex, 2–3-ridged central lobe. Spur conical, short, blunt. *E Himalaya, Burma, Thailand*. G2. Winter.

5. V. tricolor Lindley (*V. suavis* Reichenbach). Illustration: *Botanical Magazine*, 4432 (1894); Bechtel et al., Manual of cultivated orchid species, 279 (1981).

Stems to 1 m. Leaves to 45 × 5 cm, oblong. Racemes erect with 8–10 flowers. Flowers *c.* 5 cm in diameter, fragrant. Sepals and petals broadly ovate with broad, claw-like bases, wavy, yellow spotted with dark brown, their backs white. Lip as long as the sepals, lateral lobes small, erect, white, central lobe broadly oblong, waisted at the middle, the basal part broader than the apical, violet-red with purple stripes. Spur short, broad, laterally compressed. *Java*. G2. Winter.

V. lamellata Lindley (*V. boxallii* Reichenbach). Illustration: Cogniaux & Goossens, Dictionnaire iconographique des orchidées, Vanda, t. 9 (1898). Similar but with smaller, non-fragrant flowers, sepals and petals *c.* 2 cm, whitish (flushed red in 'Boxallii'), lip shorter than the sepals, the central lobe violet, oblong, notched, with 2–3 conspicuous ridges. *Philippines*. G2. Winter.

V. tesselata (Roxburgh) G. Don (*Epidendrum tesselatum* Roxburgh; *V. roxburghii* R. Brown). Illustration: *Botanical Magazine*, 2245 (1821); *Die Orchidee* **22**: 13 (1971); *Orchid Digest* **43**: 47 (1979); Bechtel et al., Manual of cultivated orchid species, 279 (1981). Also generally similar but with wavy sepals and petals that are tessellated yellow and brown, the central lobe of the lip oblong, slightly waisted, dark violet. *Tropical Asia from SE India to Malaysia*. G2. Winter.

6. V. denisoniana Benson & Reichenbach. Illustration: *Botanical Magazine*, 5811 (1869); Cogniaux & Goossens, Dictionnaire iconographique des orchidées, Vanda, t. 8 (1898); Seidenfaden & Smitinand, Orchids of Thailand **4**(1): t. 27 (1963).

Stem short. Leaves to 30 × 2 cm, linear, acutely 2-lobed at the apex. Racemes to 15 cm, with about 6 flowers, arching. Flowers *c.* 5 cm in diameter. Sepals and petals elliptic with broad claws, greenish to cream-white, the lateral sepals somewhat longer than the rest. Lip a little longer than the sepals, white with a yellow patch at the base, lateral lobes small, central lobe oblong at the base, bearing 4–5 ridges, expanding towards the apex to a kidney-shaped, notched portion. Spur oblong, blunt, laterally compressed. *Burma, Thailand.* G2. Spring.

V. caerulescens Griffith. Illustration: *Botanical Magazine*, 5834 (1870); *Orchid Digest* **43**: 44 (1979). Similar but with a longer raceme having many flowers, sepals and petals pale violet, the lip darker violet and the spur conical, curved downwards. *Burma, Thailand.* G2. Spring.

V. bensonii Bateman. Illustration: *Botanical Magazine*, 5611 (1866); Cogniaux & Goossens, Dictionnaire iconographique des orchidées, Vanda, t. 2 (1898).

Also similar but sepals and petals olive green with brown spots, the lip violet-pink, white at the base, the spur conical. *Burma, Thailand.* G2. Autumn.

V. insignis Blume. Illustration: *Botanical Magazine*, 5759 (1869); *Lindenia* **8**: t. 355 (1892); Cogniaux & Goossens, Dictionnaire iconographique des orchidées, Vanda, t. 3 (1897). Also broadly similar but with larger flowers in various shades of violet, yellow or white, and with the central lobe of the lip concave and wavy. *Indonesia (Moluccas)*. G2. Autumn.

7. V. cristata (Wallich) Lindley. Illustration: *Botanical Magazine*, 4304 (1847); *Gartenflora* **20**: t. 680 (1871); *Orchid Digest* **43**: 44 (1979); Bechtel et al., Manual of cultivated orchid species, 278 (1981).

Stems short, densely leafy. Leaves to 12 × 4 cm, linear, thick and fleshy, flat, 2-lobed at the apex. Racemes short with 2–5 flowers. Sepals and petals to 2.5 cm, curving forwards, oblong, greenish yellow. Lip with short, erect lateral lobes and with an oblong central lobe, which is 2-lobed at the apex with the lobes diverging, bearing a downwardly pointing horn-like projection between the lobes, the upper surface with 5 tuberculate ridges, all bright yellow to green with brown stripes. Spur short, blunt, conical. *E Himalaya*. G1. Winter.

V. alpina (Lindley) Lindley. Illustration: *Annals of the Royal Botanic Garden Calcutta* **8**: t. 289 (1898); Hara, Photo-album of plants of E Himalaya, t. 120 (1968). Similar but with hanging flowers, sepals and petals 1.3–1.8 cm, the lip with an oblong, truncate central lobe without a horn-like projection, the base saccate but without a spur. *Himalaya*. G1. Spring.

V. flavobrunnea Reichenbach (*V. pumila* Hooker). Illustration: *Annals of the Royal Botanic Garden Calcutta* **5**: t. 68 (1895) & **8**: t. 288 (1898); *Botanical Magazine*, 7968 (1904); *Orchid Review* **115**: 316–7 (2007). Also similar but sepals and petals *c.* 3 cm, greenish or yellowish white and the lip with an ovate, concave, red-striped central lobe. *Himalaya*. G1. Summer.

142. VANDOPSIS Pfitzer

J. Cullen

Large monopodial epiphytes. Leaves mainly at the base of the stem, in 2 ranks, large, flat or V-shaped in section, folded when young. Flowers in axillary racemes. Sepals and petals similar, fleshy, widely spreading. Lip 3-lobed, attached to the column by the base and by the small lateral lobes, which have a flap of fleshy tissue joining them above the ridge of the central lobe; central lobe shorter than the petals, often laterally flattened, saccate at the base and bearing a conspicuous ridge that is interrupted near the base. Column short with a projection at its base but without a foot. Pollinia 4, in 2 pairs, borne on a short, broad stipe with reflexed edges, the viscidium broad.

A genus of about 10 species from SE Asia. Cultivation as for *Vanda* (p. 588).

1a. Central lobe of lip as broad as long, not laterally flattened; leaves to 25 cm **3. parishii**

b. Central lobe of lip longer than broad, laterally flattened; leaves 30 cm or more 2

2a. Leaves recurved, flat in section; raceme 25–35 cm **1. gigantea**

b. Leaves straight, V-shaped in section; raceme 1–2 m **2. lissochiloides**

1. V. gigantea (Lindley) Pfitzer (*Vanda gigantea* Lindley; *Stauropsis gigantea* (Lindley) Bentham & Hooker). Illustration: *Botanical Magazine*, 5189 (1860); Bateman, Second century of orchidaceous plants, t. 142 (1865); Cogniaux & Goossens, Dictionnaire iconographique des orchidées, Stauropsis, t. 2 (1898); *Orchid Digest* **43**: 87 (1979).

Stems thick, fleshy. Leaves to 35 × 6 cm, fleshy, recurved, linear-oblong, flat, slightly unequally 2-lobed at the apex. Raceme 25–35 cm with up to 15 flowers. Sepals and petals 2.5–3 cm, oblong-obovate, yellow with red-brown blotches. Lip longer than broad, mostly yellow, lateral lobes suffused with purple. *Burma, Thailand, Malaysia*. G2. Spring–summer.

2. V. lissochiloides (Gaudichaud) Pfitzer (*Fieldia lissochiloides* Gaudichaud; *Vanda batemanii* Lindley; *Vanda lissochiloides* (Gaudichaud) Lindley; *Stauropsis lissochiloides* (Gaudichaud) Pfitzer). Illustration: Cogniaux & Goossens, Dictionnaire iconographique des orchidées, Stauropsis, t. 1 (1898); *Orchid Digest* **43**: 87 (1979); Bechtel et al., Manual of cultivated orchid species, 279 (1981); *Orchid Review* **116**: 87 (2008).

Very large plants. Leaves 30–50 cm, straight, V-shaped in section. Racemes 1–2 m with many flowers. Sepals and petals to 3.5 cm, bright purple or yellowish on the back, yellow with purple spots on the front. Lip longer than broad, short, yellow. *Philippines, Moluccas; ?Thailand*. G2. Summer.

3. V. parishii (Reichenbach) Schlechter (*Vanda parishii* Reichenbach). Illustration: Cogniaux & Goossens, Dictionnaire iconographique des orchidées, Vanda, t. 11, 11A (1898); Bechtel et al., Manual of cultivated orchid species, 279 (1981).

Stems short. Leaves to 25 × 6 cm, oblong-elliptic, unequally 2-lobed at the apex. Raceme to 35 cm with 5–6 flowers. Sepals and petals to 2.5 cm, obovate, greenish yellow spotted with brown (rose-pink, whitish at the base in var. **marriottiana** Reichenbach). Lip as broad as long, yellowish with red spots, and a conspicuous, vertical ridge. *Burma, Thailand*. G2. Summer.

Both var. *parishii* and var. *marriottiana* have been in cultivation; the latter is apparently more frequent. Garay (*Botanical Museum Leaflets, Harvard University* **23**: 374, 1974) suggests that this species should be transferred from *Vandopsis* to the genus *Hygrochilus*.

143. ESMERALDA Reichenbach

J. Cullen

Monopodial epiphytes with elongate, scrambling stems. Leaves in 2 ranks, unequally 2-lobed at their apices, folded when young, close, mainly at the bases of the stems. Flowers in axillary racemes. Sepals and petals similar, spreading. Lip freely movable on its attachment to the column, about as long as the petals, 3-lobed with small, erect, oblong lateral lobes and a larger, flat, ovate, auricled central lobe, which has a hollow sac concealed within the tissue at the base (perceptible as a hump on the back of the lip), the surface with 3–5 longitudinal ridges. Column large, fleshy, without a foot. Pollinia 4, in 2 pairs, stipe rapidly broadening to a broad, disc-like viscidium.

A genus of 2 species from the Himalaya and SE Asia. Cultivation as for *Vanda* (p. 588).

Literature: Tan, K., Taxonomy of Arachnis, Armodorum, Esmeralda and Dimorphorchis, *Selbyana* **1**: 1–15 (1975), 365–73 (1976).

1. E. cathcartii (Lindley) Reichenbach (*Vanda cathcartii* Lindley; *Arachnanthe cathcartii* (Lindley) Bentham & Hooker; *Arachnis cathcartii* (Lindley) J.J. Smith). Illustration: Cogniaux & Goossens, Dictionnaire iconographique des orchidées, Arachnanthe, t. 2 (1898); *Annals of the Royal Botanic Garden Calcutta* **8**: t. 278 (1898); *Orchid Digest* **43**: 89, 92 (1979).

Stem to 2 m, thick, hanging. Leaves to 15 × 3–4 cm. Racemes with 3–5 distant flowers. Sepals and petals to 4.5 cm, ovate to broadly elliptic, rounded at their apices, yellow with dense, red-brown, transverse stripes. Lip white with red stripes, the margin yellow, irregularly toothed and wavy. *E Himalaya*. G2. Spring–summer.

144. DIMORPHORCHIS Rolfe

J. Cullen

Large monopodial epiphytes with elongate stems. Leaves densely set in 2 ranks, folded when young. Flowers in long, loose, axillary racemes with densely woolly axis and flower-stalks, dimorphic, the first 1–4 flowers of each raceme orange-yellow with fine red spots and shorter and broader, more rounded sepals and petals than the rest, which are larger and have wavy sepals and petals that are greenish with large, confluent, red-brown spots; in each type of flower the sepals and petals are similar, spreading and woolly outside. Lip much shorter than the petals, hollowed out at the base and with a thin, high ridge in the centre, which ends in a fine, erect point. Column short, without a foot. Pollinia 4, in 2 pairs, borne almost horizontally, the stipe short and broad, broadening a little to the large, disc-like viscidium.

A genus from Borneo with 1 or 2 species. The 1 cultivated species is not common in cultivation, despite Holttum's remark (Flora of Malaya **I**, Orchids, 619, 1953): 'A large plant with a dozen inflorescences ranks with *Grammatophyllum* in full flower as one of the most remarkable objects in the Orchid world'. Cultivation as for *Vanda* (p. 588).

Literature: Tan, K., Taxonomy of Arachnis, Armodorum, Esmeralda and Dimorphorchis, *Selbyana* **1**: 1–15 (1975), 365–73 (1976); Soon, P.S., Notes on Sabah orchids, Part I, *Orchid Digest* **44**: 193–5 (1980).

1. D. lowii (Lindley) Rolfe (*Vanda lowii* Lindley; *Renanthera lowii* (Lindley) Reichenbach; *Arachnanthe lowii* (Lindley) Bentham & Hooker; *Vandopsis lowii* (Lindley) Schlechter; *Arachnis lowii* (Lindley) Reichenbach; *Renanthera rohaniana* Reichenbach). Illustration: Cogniaux & Goossens, Dictionnaire iconographique des orchidées, Arachnanthe, t. 4 (1898); *Orchid Digest* **44**: 194 (1980); Chan et al., Orchids of Borneo **1**: 136 (1994).

Stems to 2 m. Leaves to 70 × 5–6 cm, oblong, obtuse. Racemes to 2.5 m. Sepals and petals fleshy, 3.5–5 cm. Lip acute, whitish, violet within. *Borneo*. G2. Autumn.

Both var. *lowii* and var. *rohaniana* (Reichenbach) Tan, which has shorter racemes and slightly smaller, paler flowers, have been in cultivation.

145. RENANTHERA Loureiro

J. Cullen

Monopodial epiphytes with stout, branched stems. Leaves spirally arranged or in 2 ranks, well spaced, unequally 2-lobed at their apices, folded when young. Flowers in long, horizontal, axillary panicles. Upper sepal and petals similar, spreading upwards, the lateral sepals longer and broader, spreading downwards, often close together, all more than 1 cm. Lip much smaller than the petals, 3-lobed, with a conical spur and 2 calluses at the junctions of the central and lateral lobes. Column short, without a foot. Pollinia 4, borne on a short, linear-oblong, sometimes slightly waisted stipe, and with an oblong viscidium.

A genus of 10 or more species from E Asia. Cultivation as for *Vanda* (p. 588).

1a. Upper sepal 1.3–1.5 cm **4. pulchella**
b. Upper sepal 2 cm or more 2
2a. Central lobe of the lip with 5 yellow calluses at the base in addition to the 2 at the junctions of the lobes **3. imschootiana**
b. Central lobe of the lip with no additional calluses 3
3a. Lateral sepals red with deep crimson blotches; leaves 20–25 cm **2. storiei**
b. Lateral sepals without blotches; leaves 6–12 cm **1. coccinea**

1. R. coccinea Loureiro. Illustration: *Edwards's Botanical Register* **14**: t. 1131 (1828); *Botanical Magazine*, 2997, 2998 (1830).

Stems scrambling, to 2 m. Leaves 6–12 × 3–3.5 cm, unequally 2-lobed at the apex. Panicle to 70 cm. Upper sepal and petals oblong, obtuse, 2.5–2.8 cm, reddish spotted with scarlet. Lateral sepals oblong-spathulate, obtuse, somewhat wavy, 3.5–4.5 cm, clear scarlet. Lip with low, yellow, red-striped lateral lobes and a scarlet central lobe, which is yellowish towards the base. *Thailand, Laos, Vietnam, S China, possibly also in the Philippines, Java & Burma*. G2. Spring, autumn.

2. R. storiei (Storie) Reichenbach. Illustration: *Botanical Magazine*, 7537 (1897); Bechtel et al., Manual of cultivated orchid species, 268 (1981).

Very similar to *R. coccinea*, but stems shorter, leaves 20–25 × 3.5–4 cm, lateral sepals scarlet with crimson blotches. *Philippines*. G2. Summer.

3. R. imschootiana Rolfe. Illustration: *Botanical Magazine*, 7711 (1900); Bechtel et al., Manual of cultivated orchid species, 268 (1981).

Stem erect, scarcely scrambling. Leaves to 7 × 2 cm, oblong, shortly 2-lobed at their apices. Panicle to 45 cm. Upper sepal 2–2.2 cm, petals 3.7–4 cm, yellow spotted with red, lateral sepals clawed, wavy, elliptic, scarlet, or yellow flushed with red. Lip scarlet with 5 yellow calluses at the base of the central lobe. *NE India, Burma, Laos*. G2. Summer.

4. R. pulchella Rolfe.
Stem to 20 cm. Leaves 6.5–8 × 1.5–2 cm, in 2 ranks, narrowly oblong, unequally lobed at their apices. Panicle with few branches. Upper sepal 1.3–1.5 cm, lanceolate, obtuse; lateral sepals *c.* 1.7 cm, clawed, oblong-elliptic, obtuse, all yellow. Petals yellow below, red above. Lip with red lateral lobes, the central lobe yellow with 4 distinct teeth at the base. *Burma, perhaps also in Thailand*. G2. Summer.

146. ASCOCENTRUM Schlechter
J. Cullen
Monopodial epiphytes with short leafy stems. Leaves linear, unequally 2-lobed and toothed at the apex, folded when young. Flowers in axillary racemes. Sepals and petals similar, spreading. Lip small, consisting mainly of a vertical, cylindric, blunt pouch bearing at its apex 2 small, erect lateral lobes and a linear-oblong, acute central lobe that is much smaller than the petals. Column short, without a foot. Pollinia 2, spherical, borne on a short linear stipe and with a small, oblong viscidium.

A genus of about 5 species from E Asia (Himalaya to Borneo). Molecular studies have indicated that this genus belongs within *Vanda*. Cultivation as for *Gastrochilus* (see below).

Literature: Christenson, E.A., Sarcanthine genera 5: Ascocentrum., *American Orchid Society Bulletin* **55**(2): 104–111 (1986).

1a. Sepals and petals pink to violet-purple; leaves 1.7–2 cm broad **1. ampullaceum**
b. Sepals and petals orange or orange-yellow; leaves rarely exceeding 1 cm in breadth **2. miniatum**

1. A. ampullaceum (Lindley) Schlechter (*Saccolabium ampullaceum* Lindley). Illustration: *American Orchid Society Bulletin* **41**: 820 (1972); *Orchid Digest* **42**: 180 (1979).

Leaves stiff, 8–12 × 1.7–2 cm. Raceme with many flowers, erect, shortly stalked. Sepals and petals 1–1.2 cm, oblong, blunt, pink to violet-purple. Lip with pink to violet-purple pouch and central lobe (which is horizontal), the lateral lobes whitish. *NE India to Thailand*. G2. Spring.

2. A. miniatum (Lindley) Schlechter (*Saccolabium miniatum* Lindley). Illustration: Holttum, Flora of Malaya **I**: Orchids, 730 (1953); *Orchid Digest* **42**: 180 (1979); Bechtel et al., Manual of cultivated orchid species, 169 (1981).

Leaves *c.* 10–20 × 1 cm, fleshy. Raceme shortly stalked. Sepals and petals *c.* 6 mm, orange or orange-yellow. Lip similar in colour, the central lobe bent downwards. *E Himalaya to Borneo & Java*. G2. Spring.

147. GASTROCHILUS D. Don
V.A. Matthews & J. Cullen
Monopodial epiphytes with short stems. Leaves close together at the stem bases, in 2 ranks, fleshy, folded when young. Flowers fleshy, in axillary racemes, which are sometimes paired at each node. Sepals and petals similar, spreading. Lip consisting mainly of a vertical pouch with 2 lateral lobes joined to the column and a forwardly pointing, often hairy or fringed central lobe that is kidney- or crescent-shaped. Column short, without a foot. Anther ovate, projecting forwards on the column. Pollinia 2, on a linear stipe and with a small, oblong viscidium.

A genus of 15–20 species from the Himalaya and E Asia, including Japan. Only 3 species are commonly grown. Cultivation as for *Aerides* (p. 581) or *Vanda* (p. 588), but vegetative propagation is difficult.

Literature: Herklots, G.A.C., Nepalese and Indian orchids: Gastrochilus, *Orchid Review* **82**: 354–8 (1974); Christenson, E.A., Gastrochilus D. Don, with a synopsis of the genus, *American Orchid Society Bulletin* **54**(9): 1111–16 (1985).

1a. Central lobe of the lip with long white hairs on the upper surface **1. bellinus**
b. Central lobe of the lip without such hairs, though it may be papillose or fringed 2
2a. Leaves acute but not notched at the apex; stem 10–30 cm **3. acutifolius**
b. Leaves acute and 2-toothed at the apex; stem 1–3 cm **2. dasypogon**

1. G. bellinus (Reichenbach) Kunze. Illustration: *Die Orchidee* **23**: 8 (1972); Bechtel et al., Manual of cultivated orchid species, 150 (1981).

Stem usually less than 5 cm, stout. Leaves 15–30 cm, narrowly oblanceolate, curved, deeply notched at the apex. Flowers fragrant, in a cluster-like raceme of 4–7. Sepals and petals 9–14 mm, slightly incurved, greenish or pale yellow with brown or purple spots. Lip *c.* 1.3 cm, central lobe broadly elliptic with a toothed margin, white with purple or red spots, yellow towards the base, upper surface with long white hairs. *Burma, Thailand*. G2. Winter–spring.

2. G. dasypogon (Lindley) Kunze. Illustration: Sheehan & Sheehan, Orchid genera illustrated, 95 (1979); Bechtel et al., Manual of cultivated orchid species, 150 (1981).

Stem 1–3 cm. Leaves 10–15 cm, ovate-oblong, acute, notched at the apex. Raceme almost umbellate. Sepals and petals to 8 mm, obovate, pale yellow or greenish, sometimes spotted with purple. Lip 5–7 mm, central lobe kidney-shaped with a fringed margin, white or yellow with red or purple spots, not hairy. *Himalaya? Thailand & Sumatra*. G2. Autumn.

3. G. acutifolius (Lindley) Kunze (*Saccolabium acutifolium* Lindley). Illustration: *Annals of the Royal Botanic Garden Calcutta* **8**: t. 302 (1898); *Orchid Review* **82**: 358 (1974); *Die Orchidee* **27**: 127 (1976).

Stems 10–30 cm. Leaves to 12 × 2 cm, oblong, acute, not notched at the apex. Raceme with 10–20 flowers, compressed, umbel-like. Sepals and petals *c.* 1 cm, oblong, blunt, greenish yellow, usually with brown-purple spots. Lip white, the central lobe broadly triangular, papillose but not hairy on the upper surface, fringed. *Nepal, NE India*. G2. Winter.

148. SCHOENORCHIS Blume
V.A. Matthews
Differs from *Gastrochilus* in having leaves strap-shaped to terete and a simple or branched raceme with many small flowers; sepals and petals 2–5 mm; lip 3-lobed, spurred but not consisting mainly of a vertical pouch; pollinia 4, united in pairs attached to a long slender viscidium. Rostellum linear, bifid, very conspicuous.

Ten species ranging from the Himalaya to New Guinea. Cultivation as for *Vanda* (p. 588).

Literature: Christenson, E.A., Sarcanthine genera 2: Schoenorchis,

American Orchid Society Bulletin **54**(7): 850–54 (1985).

1. S. juncifolia Blume.

Stem to 30 cm, hanging, branched. Leaves to 16 cm, terete, often flushed with purple. Raceme to 10 cm, simple, hanging, dense, with many flowers. Sepals and petals oblong, shortly acuminate, bluish violet. Lip longer than sepals, with a more or less curved spur, the central lobe *c.* 2 mm, pale violet or white, recurved. *Sumatra, Java.* G2. Summer.

149. CLEISOSTOMA Blume

J. Cullen

Monopodial epiphytes with elongate stems. Leaves borne all around the stem, terete or flat, folded when young. Flowers numerous in axillary racemes or rarely panicles. Sepals and petals similar or the petals slightly smaller, all spreading, lateral sepals and petals not inserted on the column foot. Lip consisting mainly of a short, broad, blunt, vertical pouch, the blade 3-lobed with small lateral lobes and a larger, usually arrowhead-shaped, central lobe. The entrance to the pouch is largely blocked by a complex callus borne on the upper surface of its interior, and there is usually a longitudinal cross-wall from the lower part of its interior as well. Column short with a short foot. Pollinia 4, united in 2 round masses with the stipe linear, and the viscidium disc-like, or the stipe broadening rapidly from the apex to a saddle-shaped viscidium.

A genus of about 100 species from Asia, occurring from Nepal to New Guinea. The flowers are more interesting than attractive and few are grown. The genus is commonly found under the name *Sarcanthus* Lindley. Cultivation as for *Vanda* (p. 588).

Literature: Garay, L., On the systematics of the monopodial orchids I, *Botanical Museum Leaflets, Harvard University* **23**: 168–76 (1972); Seidenfaden, G., Orchid genera in Thailand II: Cleisostoma, *Dansk Botanisk Arkiv* **29**(3): 1–80 (1975).

1a. Leaves terete **1. appendiculatum**
b. Leaves flat **2. rostratum**

1. C. appendiculatum (Lindley) Jackson (*Sarcanthus appendiculatus* of various authors). Illustration: *Annals of the Royal Botanic Garden Calcutta* **5**: t. 76 (1895); *Hooker's Icones Plantarum* **22**: t. 2136 (1893).

Stem to 40 cm, hanging. Leaves 7–10 cm, terete, spirally arranged. Raceme with 10–15 flowers, stalked, hanging, longer than the leaves. Flowers 1.5–2 cm. Sepals and petals oblong, brownish yellow with red-violet veins. Lip with small, erect lateral lobes and a broadly triangular central lobe, rose-pink. Pouch cylindric, bright brownish yellow with purple veins. Stipe linear, viscidium disc-like. *E Himalaya, Burma.* G2. Summer.

C. filiforme (Lindley) Garay (*Sarcanthus filiformis* Lindley). Illustration: Bechtel et al., Manual of cultivated orchid species, 184 (1981). Similar, but with longer stems with leaves 20–25 cm borne on 1 side, the flowers *c.* 1.1 cm. *E Himalaya to Thailand.* G2. Summer–autumn.

C. simondii (Gagnepain) Seidenfaden (*Sarcanthus teretifolius* (Lindley) Lindley). Also similar, but with thick, short, often recurved leaves and a stipe that is narrow at the apex but rapidly broadens to a saddle-shaped viscidium. *Himalaya, Laos, Vietnam, Thailand & S China.* G2. Summer.

2. C. rostratum (Lindley) Garay (*Sarcanthus rostratus* Lindley). Illustration: Lindley, Collectanea botanica, t. 39 (1821); *Edwards's Botanical Register* **12**: t. 981 (1826); *Dansk Botanisk Arkiv* **29**(3): 29 (1975).

Stem to 25 cm, hanging. Leaves to 10 × 1–1.5 cm, flat. Raceme dense, with many flowers, usually shorter than the subtending leaves. Flowers to 1.3 cm. Sepals and petals yellowish green with purplish veins. Lip whitish with short lateral lobes and an acute, upwardly hooked central lobe, which is violet-pink. Pouch violet-pink. *S China, Vietnam, Laos & Thailand.* G2. Summer–autumn.

C. racemiferum (Lindley) Garay (*Sarcanthus racemifer* (Lindley) Reichenbach; *S. pallidus* Lindley). Illustration: Bechtel et al., Manual of cultivated orchid species, 184 (1981). Similar, but leaves unequally 2-lobed at the apex, flowers in erect panicles that exceed the subtending leaves. *Himalaya to Thailand.* G2. Spring–summer.

150. ANGRAECUM Bory

J. Cullen

Monopodial epiphytes with long or short stems. Leaves in 2 ranks, variable, overlapping and folded at least at the base, or distant and flat or terete, sometimes unequally 2-lobed at their apices, folded when young. Flowers variably resupinate or not resupinate, solitary in the leaf axils, or in axillary racemes. Sepals and petals similar, spreading. Lip stalkless, entire, with a flat or concave spreading blade extending backwards into a conspicuous spur. Column fleshy, short, without a foot, with 2 swellings, between which is the short, inconspicuous, bifid, tooth-like rostellum, beneath the anther. Pollinia 2, spherical, attached either to a short common stipe, or to individual stipes, viscidium single or double.

A genus of over 200 species centred mainly in tropical Africa but extending to Madagascar and Sri Lanka. Garay (reference below) divides it into 18 sections, but as only 6 or 7 species are found in cultivation, these are not described here. Species Nos. 1–3 belong to Section *Arachnangraecum* Schlechter; Nos. 4 & 5 to Section *Angraecum*; and No. 6 to Section *Dolabrifolia* (Pfitzer) Garay.

Cultivation as for *Phalaenopsis* (p. 583), but the smaller species should be suspended near the glass and the thick-leaved species should be given full air and light while avoiding a dry atmosphere. In winter the plants can be allowed to dry out to some extent, but not for long periods.

Literature: Garay, L., Systematics of the genus Angraecum Bory, *Kew Bulletin* **28**: 495–516 (1973).

Leaves. Terete: **3**; overlapping and folded at their bases only: **4, 5**; overlapping and folded for most of their length: **6**.
Flowers. Stalkless and single in the leaf axils: **6**; single and stalked, or in racemes in the leaf axils: **1, 2, 3, 4, 5**.
Lip. Narrowly ovate, longer than broad: **5**; broadly ovate, broader than long: **1, 2, 3, 4**; helmet-like: **6**.
Spur. Less than 1 cm: **6**; between 4 and 15 cm: **1, 2, 3, 4**; between 20 and 30 cm: **5**. Sharply bent near the middle: **1, 2**.

1a. Leaves overlapping and folded, at least at their bases 2
b. Leaves flat or terete, not overlapping 4
2a. Flowers stalkless; sepals and petals less than 1 cm **6. distichum**
b. Flowers stalked; sepals and petals more than 1 cm 3
3a. Lip longer than broad, narrowly ovate; spur 20–30 cm **5. sesquipedale**
b. Lip broader than long, broadly ovate; spur 6–8 cm **4. eburneum**

4a. Leaves terete **3. scottianum**
b. Leaves flat 5
5a. Spur 12–13 cm, tapering, curving forwards **2. infundibulare**
b. Spur *c.* 4.5 cm, sharply bent at the middle **1. eichlerianum**

1. A. eichlerianum Kränzlin. Illustration: *Botanical Magazine*, 7813 (1902); Bechtel et al., Manual of cultivated orchid species, 166 (1981).

Stems to 1 m, hanging, somewhat compressed. Leaves 8–12 × 4.5–5 cm, oblong-elliptic, flat, not overlapping. Flowers solitary or rarely in 2-flowered racemes in the leaf axils, stalk elongate, variably resupinate. Sepals and petals to 4.5 cm, spreading, oblong-lanceolate, acute, yellowish green. Lip broadly ovate, concave, broader than long, with a short point, white shading to greenish yellow at the base, with an elongate ridge. Spur *c.* 4.5 cm, funnel-shaped, sharply bent at the middle, obtuse. Pollinia with a common stipe. *W tropical Africa, from Nigeria to Angola*. G2. Summer.

2. A. infundibulare Lindley (*Mystacidium infundibulare* (Lindley) Rolfe). Illustration: *Botanical Magazine*, 8153 (1907); Stewart & Campbell, Orchids of tropical Africa, t. 9 (1970); Bechtel et al., Manual of cultivated orchid species, 166 (1981).

Very similar to *A. eichlerianum*, but leaves broader, the lip very broad and concave, the spur 12–13 cm, very wide-mouthed and funnel-shaped, yellowish, curved forwards in the lower half. *Tropical Africa from Zaire to Ethiopia*. G2. Winter.

3. A. scottianum Reichenbach. Illustration: *Botanical Magazine*, 6273 (1883); Cogniaux & Goossens, Dictionnaire iconographique des orchidées, Angraecum, t. 7 (1898); Stewart & Campbell, Orchids of tropical Africa, t. 12 (1970); Bechtel et al., Manual of cultivated orchid species, 167 (1981).

Stems *c.* 30 cm, thin, hanging. Leaves 8–10 cm, terete, not overlapping at the base. Flowers single in the leaf axils, on long stalks, variably resupinate. Sepals and petals 2.5–3 cm, spreading, oblong-linear, acute, greenish white. Lip broader than long, broadly ovate or transversely oblong, with a small point, white. Spur 8–10 cm, narrow, tapering, yellow. Pollinia with a common stipe. *Comoro Islands*. G2. Summer.

4. A. eburneum Bory. Illustration: *Botanical Magazine*, 5170 (1854); *Lindenia* **5**: t. 236 (1889); Cogniaux & Goossens, Dictionnaire iconographique des orchidées, Angraecum, t. 13 (1898); *Die Orchidee* **30**: cxxi (1979); Bechtel et al., Manual of cultivated orchid species, 166 (1981).

Stem to 1 m, erect. Leaves 40–50 cm, dense, their bases overlapping and folded, the free part oblong-lanceolate, unequally 2-lobed at the apex. Flowers not resupinate, 8–15 in axillary, 1-sided racemes, stalked, fragrant. Sepals and petals 2.5–3 cm, linear-oblong, acute, spreading, green or whitish green. Lip ovate, broader than long, concave, apiculate, white, often greenish at the base. Spur 6–8 cm, narrow, tapering, white with a green apex. Pollinia with a common stipe. *E tropical Africa, islands of the Indian Ocean*. G2. Autumn–winter.

A variable species. Subsp. **eburneum**, subsp. **superbum** (Thouars) Perrier de la Bathie (*A. superbum* Thouars), which has somewhat larger flowers, and subsp. **giryamae** (Rendle) Senghas & Cribb (*A. giryamae* Rendle), which has a smaller lip and a funnel-shaped spur that is 4–5 cm, are grown. This last subspecies occurs in Kenya and Tanzania.

5. A. sesquipedale Thouars. Illustration: *Botanical Magazine*, 5113 (1859); Cogniaux & Goossens, Dictionnaire iconographique des orchidées, Angraecum, t. 4 (1898); Stewart & Campbell, Orchids of tropical Africa, t. 13 (1970); Bechtel et al., Manual of cultivated orchid species, 167 (1981).

Stems to 1 m, thick, erect. Leaves dense, their bases overlapping and folded, the free part to 30 × 4–5 cm, oblong, unequally 2-lobed at the apex. Flowers usually 2–4 in axillary racemes, rarely single in the leaf axils, stalked, fragrant, resupinate. Sepals and petals to 6 cm, creamy white, spreading, lanceolate, acute, the petals somewhat shorter than the sepals. Lip narrowly ovate, longer than broad, tapering, its sides somewhat toothed, creamy white. Spur 20–30 cm, abruptly tapering from its junction with the lip. Pollinia with separate stipes. *Madagascar*. G2. Winter.

This is the orchid that led Darwin to predict the existence of a moth with a proboscis 20–30 cm long – a species unknown at the time of his prediction but since discovered in Madagascar.

6. A. distichum Lindley. Illustration: *Edwards's Botanical Register* **21**: t. 1781 (1835); *Botanical Magazine*, 4145 (1845); *Die Orchidee* **21**: 246 (1970); Bechtel et al., Manual of cultivated orchid species, 166 (1981).

Small plants, stems to 12 cm. Leaves 8–12 cm, fleshy, folded and overlapping for most of their length, the free part short, obtuse. Flowers solitary in the upper leaf axils, the ovary stalkless, not resupinate. Sepals and petals *c.* 5 mm, white, spreading, the petals slightly shorter and narrower than the sepals. Lip helmet-like, 3-lobed at the apex, the central lobe recurved. Spur *c.* 8 mm, cylindric. *Tropical Africa from Guinea to Uganda and Angola*. G2. Autumn.

151. AERANGIS Reichenbach

J. Cullen

Monopodial epiphytes with short, compressed stems. Leaves leathery, unequally 2-lobed at the apex, folded when young. Racemes with 7–many flowers, hanging from the leaf axils, the axis often conspicuously zig-zag. Bracts small, brownish. Flowers usually white, fragrant. Sepals and petals similar in shape and size (rarely the upper sepal smaller), spreading or somewhat reflexed. Lip smaller than the petals but often broader, projecting backwards into a short or long, curved or twisted and contorted spur, which has a distinct, narrow mouth. Column narrow, without a foot, rostellum large, usually beaked, usually extended at right angles to the column. Pollinia 2, each with a separate stipe or (in all species included here) attached to a single, linear stipe, viscidium oblong, sometimes slightly 2-lobed.

A genus of 35 species from tropical Africa and Madagascar. Cultivation as for *Angraecum* (p. 593).

Literature: Stewart, J.A., A revision of the African species of Aerangis, *Kew Bulletin* **34**: 239–319 (1979); Stewart, J.A., The genus Aerangis Reichenbach, *Orchid Digest* **70**(4): 196–225 (2006).

Habit. Plant very small (leaves 3–5 cm, sepals and petals 5–6 mm): **6**.
Sepals. Upper shorter and blunter than laterals: 5.2 cm or more: **1, 3, 4**.
Spur. Less than 1 cm: **6**; 2–4 cm: **4, 5**; 4–10 cm: **2**; more than 10 cm: **1, 3**. Contorted and twisted: **3**.

1a. Leaves 3–5 cm, spur 5.5–6 mm, not or only slightly longer than the ovary **6. hyaloides**
b. Leaves larger; spur much longer than the ovary 2

2a. Spur 2–3 cm; upper sepal 5–7 mm, blunt, shorter than the acute lateral sepals (which are 8–10 mm) **5. citrata**
b. Spur 4–27 cm; all sepals similar in size, 1.5–3 cm 3
3a. Spur 15–27 cm, twisted and contorted, the spurs of the individual flowers often twisted around each other **3. kotschyana**
b. Spur 4–17 cm, curved but not twisted or contorted 4
4a. Sepals and petals white tipped with pink; spur brownish; rostellum touching the lower side of the stigmatic cavity, which thus appears 2-lobed **4. biloba**
b. Sepals, petals and spur white; rostellum projecting outwards above the stigmatic cavity 5
5a. Raceme to 60 cm or more; leaves 20–25 × 5 cm **1. ellisii**
b. Raceme much shorter; leaves 6–12 × 1.5–3 cm **2. modesta**

1. A. ellisii (Reichenbach) Schlechter (*Angraecum ellisii* Reichenbach). Illustration: *Floral Magazine*, n.s., t. 191 (1875); *Lindenia* **2**: t. 92 (1886).

Stem short. Leaves 20–25 × 5 cm, oblong-obovate, dark green, unequally 2-lobed at the apex. Racemes 40–65 cm with 18–25 flowers. Sepals and petals 2.5–3 cm, white, spreading or reflexed, oblong-elliptic, acute. Lip similar to the petals but broader and with its margins reflexed. Spur 15–17 cm, curved, tapering. *Madagascar*. G2. Summer–autumn.

Introduced, apparently from Madagascar, in the nineteenth century, but not found there since.

A. articulata (Reichenbach) Schlechter (*Angraecum articulation* Reichenbach). Similar, but with a markedly zig-zag axis to its raceme. *Madagascar*. G2.

Both of these species are somewhat dubious. Perrier de la Bathie (Flore de Madagascar **49**(2): 108, 116, 1939) suggests that both are related to *A. stylosa* (see under No. 2 below), *A. ellisii* being perhaps a luxuriant variant of it and *A. articulata* a hybrid of it with some other unknown species.

2. A. modesta (Hooker) Schlechter (*Angraecum modestum* Hooker). Illustration: *Botanical Magazine*, 6693 (1883); Cogniaux & Goossens, Dictionnaire iconographique des orchidées, Angraecum, t. 2 (1898).

Stem to 15 cm. Leaves 6–12 × 1.5–3 cm, ovate-oblong, slightly unequally 2-lobed at the apex. Raceme to 30 cm at most, with 6–15 flowers, which are very distant. Sepals and petals *c.* 1.2 cm, lanceolate, more or less obtuse, the sepals with 5 nerves, white. Lip broader than the petals, acute. Spur 4–7 cm, curved, tapering. Column with a few short hairs on the upper side. *Madagascar & Comoro Islands*. G2. Spring.

A. stylosa (Rolfe) Schlechter (*Angraecum stylosum* Rolfe). Similar but with petals and sepals 1.5–2 cm, the sepals with 7 nerves and the column hairless. *Madagascar & Comoro Islands*. G2. Spring.

3. A. kotschyana (Reichenbach) Schlechter (*Angraecum kotschyanum* Reichenbach; *Angraecum kotschyi* Reichenbach; *Aerangis kotschyi* (Reichenbach) Reichenbach). Illustration: *Botanical Magazine*, 7442 (1895); Williamson, Orchids of southern central Africa, f. 108 & t. 175 (1977); Ball, Southern African epiphytic orchids, 33 (1978); *Kew Bulletin* **34**: 255 (1979).

Stem very short. Leaves 10–15 × 4.5–5 cm, oblong-elliptic to oblong-obovate, blunt or slightly 2-lobed at the apex, sometimes spotted with purple above. Raceme to 40 cm with 7–10 (rarely more) flowers. Sepals and petals elliptic-lanceolate, acute, to 2.5 cm, white tinged with pink towards their apices. Lip reversed-heart-shaped, waisted above. Spur 15–27 cm, twisted and contorted, those of the individual flowers often twisted together, slightly broadened towards the brownish apex. *Tropical Africa, from Guinea and Sudan to Zaire and South Africa (Transvaal)*. G2. Autumn.

4. A. biloba (Lindley) Schlechter (*Angraecum bilobum* Lindley). Illustration: *Edwards's Botanical Register* **27**: t. 35 (1841); *Kew Bulletin* **34**: 282 (1979).

Stem very short. Leaves to 15 cm, oblong-obovate, deeply and unequally 2-lobed at the apex. Raceme to 25 cm with 7–12 flowers. Sepals and petals *c.* 2 cm, lanceolate or narrowly elliptic, acute, white tipped with pink. Lip similar but broader. Spur *c.* 4 cm, brownish, curved. Anther conspicuously crested. Rostellum extending over the stigmatic cavity, touching its lower side, the cavity thus appearing 2-lobed. *Tropical W Africa, from Senegal to Cameroun*. G2. Summer.

5. A. citrata (Thouars) Schlechter (*Angraecum citratum* Thouars). Illustration: *Botanical Magazine*, 5624 (1867); *Lindenia* **5**: t. 238 (1889); Stewart & Campbell, Orchids of tropical Africa, t. 2 (1970).

Stem very short. Leaves 7–15 × 2–3.5 cm, oblong-obovate, very unequally 2-lobed at the apex. Raceme 10–30 cm with 15–60 flowers. Upper sepal 5–7 mm, obtuse, shorter than the lateral sepals, which are clawed, acute and 8–10 mm. Petals slightly shorter than the lateral sepals. All white with a faint yellow tinge. Lip clawed, obovate. Spur 2.5–3 cm, curved, tapering. *Madagascar*. G2. Winter–spring.

A. luteo-alba (Kränzlin) Schlechter var. **rhodosticta** (Kränzlin) Stewart (*Angraecum rhodostictum* Kränzlin; *Aerangis rhodosticta* (Kränzlin) Schlechter). Illustration: Hunt & Grierson, Orchidaceae, 110–11 (1973); *Kew Bulletin* **34**: 311 (1979). Will key out to *A. citrata* on its short spur, but sepals equal, 1–1.5 cm, column red. *Tropical Africa from Cameroun to Tanzania*. G2.

6. A. hyaloides (Reichenbach) Schlechter (*Angraecum hyaloides* Reichenbach). Illustration: Reichenbach, Xenia Orchidacearum **3**: t. 238 (1890).

Plant very small. Leaves 3–5 cm, oblong-obovate, shortly 2-lobed at the apex. Raceme 3–4 cm with 3–4 flowers. Sepals and petals 5–6 mm, whitish, translucent. Lip somewhat broader than petals. Spur 5.5–6 mm, not or scarcely exceeding the ovary, curved. *Madagascar*. G2. Spring.

152. NEOFINETIA Hu

J. Cullen

Very similar to *Aerangis*, but lip distinctly 3-lobed.

Literature: Craig, J.E., Neofinetia falcata, the Japanese 'furan', *American Orchid Society Bulletin* **40**: 114–119 (1971); Christenson, E.A., Sarcanthine genera 9: Neofinetia, *American Orchid Society Bulletin* **62**: 494–5 (1993); Rose, J., Neofinetia falcata hybrids, *American Orchid Society Bulletin* **63**: 384–91 (1994).

1. N. falcata (Thunberg) Hu (*Angraecum falcatum* (Thunberg) Lindley; *Angraecopsis falcata* (Thunberg) Schlechter). Illustration: Kitamura et al., Alpine plants of Japan **3**: f. 114 (1978); Hirose & Yokoi, Variegated plants in color, 175 (1998).

Stem very short. Leaves fleshy, to 7 cm. Raceme erect, shorter than the leaves, with 3–7 flowers. Flowers white, *c.* 3 cm in diameter. Lip with short lateral lobes and an oblong central lobe. Spur *c.* 4 cm. *Japan, southern Korea*. G1. Summer.

A large number of cultivars are grown in Japan.

N. richardsiana Christenson has been described from cultivated plants that were imported from China. It differs by leaves suffused purple and a shorter spur *c.* 1 cm long.

153. CYRTORCHIS Schlechter

J. Cullen

Monopodial epiphytes, stems short or elongate. Leaves in 2 ranks, thick, fleshy and leathery, unequally 2-lobed at their apices, folded when young. Racemes arching or hanging, arising from the axils of the older leaves (or leaf-bases), with several flowers and conspicuous, brownish bracts. Flowers fragrant, especially at night, white fading to dull orange if not pollinated. Sepals and petals similar in size and shape, usually recurved. Lip similar to the petals but prolonged backwards as a conspicuous, tapering spur. Column short, without a foot. Rostellum elongate, conspicuous, beak-like. Pollinia 2, each with a separate stipe, attached to an oblong or linear viscidium.

A taxonomically troublesome genus of 15 species from Africa, of which 2 are occasionally found in cultivation. Cultivation as for *Angraecum* (p. 593).

Literature: Summerhayes, V.S., African orchids XXVII, *Kew Bulletin* **14**: 143–56 (1960).

1a. Stipe considerably broadened upwards; viscidium oblong with a hardened upper part and a translucent lower part **1. arcuata**
b. Stipe scarcely broadened upwards; viscidium linear, entirely translucent **2. monteiroae**

1. C. arcuata (Lindley) Schlechter (*Angraecum arcuatum* Lindley; *Listrostachys arcuata* (Lindley) Reichenbach). Illustration: Stewart & Campbell, Orchids of tropical Africa, t. 25 (1970); *American Orchid Society Bulletin* **49**: 1230 (1980); Bechtel et al., Manual of cultivated orchid species, 192 (1981).

Stem 30–50 cm, thick. Leaves 7–15 × 1.5–2.5 cm, oblong. Racemes with 10–20 densely packed flowers. Sepals and petals 2–4 cm, lanceolate, acute, the petals sometimes a little shorter than the sepals. Spur 3–6 cm, curved, tapering. *S and tropical Africa northwards to Sierra Leone and Kenya.* G2. Spring–summer.

A very variable species, divided by Summerhayes into 5 subspecies; it is not known which of these are in cultivation.

2. C. monteiroae (Reichenbach) Schlechter (*Listrostachys monteiroae* Reichenbach). Illustration: *Botanical Magazine*, 8026 (1905).

Stem to 60 cm, leafy. Leaves 15–17 × 4.5–5.5 cm, oblong. Raceme with 10–15 rather distant flowers. Sepals and petals 1.5–2 cm, lanceolate. Spur 4–5 cm, curved, brownish yellow towards the apex. *Tropical Africa, from Sierra Leone to Angola and eastwards to Uganda.* G2. Spring.

154. DIAPHANANTHE Schlechter

J. Cullen

As for *Cyrtorchis* but sepals and petals erect, lip considerably larger than them and bearing a conspicuous tooth-like callus just in front of the narrow mouth of the spur.

1a. Leaves in a close tuft; lip fringed **1. pellucida**
b. Leaves distributed along the stem; lip not fringed **2. bidens**

1. D. pellucida (Lindley) Schlechter (*Angraecum pellucidum* Lindley; *Listrostachys pellucida* (Lindley) Reichenbach). Illustration: Cogniaux & Goossens, Dictionnaire iconographique des orchidées, Listrostachys, t. 1 (1906); Bechtel et al., Manual of cultivated orchid species, 204 (1981).

Stem compressed, leaves in a close tuft, each unequally 2-lobed at the apex. Flowers white or brownish pink. Lip triangular, conspicuously fringed. *Tropical Africa from Sierra Leone to Uganda.* G2. Winter.

2. D. bidens (Reichenbach) Schlechter (*Listrostachys bidens* Reichenbach). Illustration: *Botanical Magazine*, 8014 (1905); Bechtel et al., Manual of cultivated orchid species, 204 (1981).

Stems to 40 cm, leafy along their length. Leaves very conspicuously unequally 2-lobed. Flowers brownish pink. Lip triangular, notched at the apex, with a small point within the notch. *Guinea.* G2. Summer.

155. EURYCHONE Schlechter

J. Cullen

Monopodial epiphytes with very short stems. Leaves narrowly elliptic to oblanceolate, unequally 2-lobed at their apices, folded when young. Racemes lateral, with 5–10 flowers, arching, shorter than the leaves. Sepals and petals similar, thin, spreading. Lip about as long as the sepals but much broader, funnel-shaped, very obscurely 3-lobed, extended backwards as a wide-mouthed spur, which is thickened and recurved at its apex. Column without a foot. Rostellum beaked, conspicuous. Pollinia 2, almost spherical, borne on a narrow, linear stipe, viscidium large, oblong or oval.

A genus of 2 species from tropical Africa. Cultivation as for *Angraecum* (p. 593).

1. E. rothschildiana (O'Brien) Schlechter (*Angraecum rothschildianum* O'Brien). Illustration: *Gardeners' Chronicle* **34**: 131 (1903); Stewart & Campbell, Orchids of tropical Africa, t. 29 (1970).

Stem to 7 cm. Leaves 7–14 × 2–2.5 cm. Flowers fragrant. Sepals and petals 2–2.5 cm, each white with a pale green band. Lip to 2.5 cm wide, white with the centre bright green shading to brown or purple at the base, its sides curving upwards towards the column. Spur to 2.5 cm. *Tropical Africa, from Sierra Leone to Uganda.* G2. Summer.

156. EULOPHIA R. Brown

J. Cullen

Large, terrestrial plants. Pseudobulbs compound, sometimes stem-like or swollen. Leaves 2–several on each pseudobulb, pleated and with prominent veins, rolled when young. Racemes lateral with 5–many flowers. Sepals and petals similar to each other or the petals larger than the sepals, all spreading or the sepals sometimes reflexed. Lip pivoting freely on the base of the column, 3-lobed, shortly spurred or saccate at the base. Column without a foot. Pollinia 2, ovoid or laterally ovoid, stipe short and broad, viscidium almost circular.

A genus of about 300 species, mostly from tropical Africa. Several of those cultivated have at one time or other been placed in the genus *Lissochilus* R. Brown. Cultivation can be difficult, as most species are found in seasonally damp places. Abundant moisture is required during the growing season and drier conditions with maximum light while the plants are resting. Propagation by seed or by division.

1a. Petals much larger than the sepals, broadly obovate to almost circular 2
b. Petals similar in size to the sepals, relatively much narrower **1. guineensis**

2a. Petals yellowish or greenish yellow; flowers *c.* 3 cm in diameter; bracts lanceolate **4. streptopetala**
b. Petals white or white flushed with pink, or purple; flowers 4 cm or more in diameter; bracts ovate 3
3a. Sepals green; petals lilac-purple with darker veins; lateral lobes of the lip purple **2. gigantea**
b. Sepals purple; petals white flushed with pink; lateral lobes of the lip green striped with purple **3. horsfallii**

1. E. guineensis Lindley. Illustration: *Botanical Magazine*, 2467 (1824).

Pseudobulbs to 5 cm, each with 2–3 leaves. Leaves 30–45 cm, narrowly elliptic, narrowed to the base, acute. Racemes with 5–15 flowers. Flowers to 6 cm in diameter. Sepals and petals similar in size, narrow, acuminate, either whitish pink or greenish or brownish purple. Lip with the central lobe whitish pink to purple, streaked and spotted with darker purple. *Tropical Africa, from Gambia to Angola and Uganda*. G2. Autumn.

E. alta (Linnaeus) Fawcett & Rendle (*E. longifolia* Humboldt, Bonpland & Kunth; *E. woodfordii* (Lindley) Rolfe). Illustration: Hoehne, Flora Brasilica **12**(6): t. 1 (1942); Hawkes, Encyclopaedia of cultivated orchids, 208 (1965); Dunsterville & Garay, Venezuelan orchids illustrated **1**: 150 (1959); Ospina & Dressler, Orquideas de las Americas, f. 141 (1974). Leaves 100–120 cm, raceme with many flowers, which are 3–5.5 cm in diameter, greenish or brown, the central lobe of the lip purple. *Tropical Africa from Ghana and Sudan to Angola and Zimbabwe; tropical America from Mexico to Argentina*. G2. Autumn.

2. E. gigantea (Welwitsch) R. Brown (*Lissochilus giganteus* Welwitsch). Illustration: *Gardeners' Chronicle* **3**: 616–7 (1888).

To 5 m or more. Leaves to 1.2 m × 10 cm, bright green with prominent yellow midrib. Raceme with 20–100 flowers, bracts ovate, wrapped around the bases of the flower-stalks. Flowers 6–7 cm in diameter. Sepals green. Petals larger, lilac-purple with darker veins. Lip mostly purple throughout. *W tropical Africa from Congo to Angola*. G2. Spring–summer.

3. E. horsfallii (Bateman) Summerhayes (*Lissochilus horsfallii* Bateman; *L. porphyroglossa* Reichenbach; *E. porphyroglossa* (Reichenbach) Bolus). Illustration: *Botanical Magazine*, 5486 (1865).

Very like *E. gigantea* but to 2 m, raceme with 60–100 flowers, which are 4–6 cm in diameter, sepals purple, petals white or white flushed with pink, lateral lobes of the lip green striped with purple. *Tropical Africa from Sierra Leone to Mozambique*. G2. Winter.

Sometimes found under the name *Lissochilus roseus* Lindley, a name of doubtful application.

4. E. streptopetala Lindley (*Lissochilus krebsii* Reichenbach). Illustration: *Botanical Magazine*, 2931 (1829), 5861 (1870); Bolus, Orchids of S Africa **3**: t. 9–10 (1913); *Flowering plants of South Africa* **21**: t. 820 (1941); Gibson, Wild flowers of Natal (coastal region), t. 25 (1975).

Stem to 1.5 m. Leaves to 50 cm × 7 mm, lanceolate, bright green, with a prominent midrib. Raceme with up to 50 flowers, bracts lanceolate, not wrapped around the bases of the flower-stalks. Flowers *c.* 3 cm in diameter. Sepals green striped with purple. Petals yellow or greenish yellow. Lip yellow, the lateral lobes purplish or brownish. *South Africa, Zimbabwe*. G2. Winter.

157. OECEOCLADES Lindley

J. Cullen & J.M.H. Shaw

Plants very like *Eulophia* but pseudobulbs simple, each bearing 1–3 leathery leaves, which are folded when young. Pollinia 2, ovoid or pear-shaped, stipe very short or absent, viscidium semicircular or oblong.

A genus of about 30 species from the tropics, mostly in Africa and Madagascar, formerly known as *Eulophidium* Pfitzer. Cultivation as for *Eulophia* (p. 596).

Literature: Summerhayes, V.S., The genus Eulophidium Pfitzer, *Bulletin du Jardin Botanique de l'Etat, Bruxelles* **27**: 391–403 (1957); Garay, L.A. & Taylor, P., The genus Oeceoclades Lindl., *Botanical Museum Leaflets, Harvard University* **24**: 249–74 (1976); Hermans, J. & C., Du Puy, D., Cribb, P. & Bosser, J., Orchids of Madagascar, edn 2, 237–45 (2007).

1a. Each pseudobulb with 1 non-pleated, irregularly variegated leaf **1. maculata**
b. Each pseudobulb with 2–3 pleated, green leaves 2
2a. Inflorescence a raceme to 30 cm **2. saundersiana**
b. Inflorescence a panicle to 1.5 m **3. calcarata**

1. O. maculata (Lindley) Lindley (*Eulophia maculata* (Lindley) Reichenbach; *Eulophidium maculatum* (Lindley) Pfitzer; *Eulophidium ledienii* N.E. Brown). Illustration: *Gartenflora* **37**: t. 1288 (1888); Ospina & Dressler, Orquideas de las Americas, t. 141 (1974).

Pseudobulbs each bearing 1 leaf. Leaves 15–25 × 2.5–4 cm, not pleated, fleshy, leathery, irregularly variegated. Racemes 7–15 cm with several flowers. Sepals and petals 8–10 mm, pale brownish green. Lip 3-lobed with the central lobe itself 2-lobed, brownish green with 2 reddish or purple spots on the sides. *Tropical America, tropical Africa*. G2. Summer–autumn.

2. O. saundersiana (Reichenbach) Garay & Taylor (*Eulophidium saundersianum* (Reichenbach) Summerhayes). Illustration: Reichenbach, Xenia Orchidacearum **2**: t. 173 (1873).

Pseudobulbs each with 2–3 leaves, the leaves deciduous some distance above the pseudobulb, the persistent portion splitting and forming a stiff, fibrous crown. Leaves 10–18 × 5–7.5 cm, pleated, green. Raceme 10–30 cm with many flowers. Sepals and petals 1–1.2 cm, green, the petals marked with black lines. Lip 4-lobed, green with a few black lines. *Tropical Africa from Cameroun to Uganda*. G2. Summer–autumn.

This species grows in wet habitats and should not be kept as dry as the others.

3. O. calcarata (Schlechter) Garay & Taylor (*Eulophia paniculata* Rolfe; *Cymbidium calcaratum* Schlechter; *Eulophia calcarata* (Schlechter) Schlechter). Illustration: Hermans et al., Orchids of Madagascar, edn 2, pl. 49D (2007).

A large plant with conical pseudobulbs each bearing 2 leaves from the apex. Inflorescence to 1.5 m, paniculate. Lip 3-lobed with the central lobe 2-lobed, with 3 keels. *Madagascar*. G2. Summer–autumn.

Cultivated plants of this species are sometimes labelled as *O. paniculata*, however the name *O. paniculata* Lindley is a synonym of *Robiquettia succisa* (Lindley) Seidenfaden & Garay from eastern and south-eastern Asia.

158. EULOPHIELLA Rolfe

J. Cullen

Rhizomatous epiphytes. Pseudobulbs large, compund, each bearing 3–6 pleated, prominently veined leaves, rolled when young. Raceme basal or lateral with 7–25 flowers with prominent bracts. Flowers large, almost circular in outline. Sepals and petals all similar or the petals a little smaller, lateral sepals attached to the column foot. Lip 3-lobed, not spurred, attached directly to the column foot, without an obvious claw. Column with a conspicuous foot. Pollinia 2, ovoid or ellipsoid, stipe very short and broad, oval or conical.

A genus of 4 species from Madagascar, where they occur in seasonally damp places. Cultivation as for *Eulophia* (p. 596).

Literature: Bosser, J. & Morat, P., Contribution a l'étude des Orchidaceae de Madagascar IX: Les Genres Grammangis et Eulophiella, *Adansonia* **9**: 299–309 (1969); Kennedy, G.C., The genus Eulophiella, *Orchid Digest* **36**: 120–2 (1972).

1a. Leaves 100 × 8–10 cm or more; flower-stalk 4.5–5 cm; flowers 8–10 cm in diameter **1. roempleriana**
b. Leaves 60–80 × 3.5–5 cm; flower-stalk 2–3 cm; flowers 4–5 cm in diameter **2. elizabethae**

1. E. roempleriana (Reichenbach) Schlechter (*Grammatophyllum roemplerianum* Reichenbach; *E. peetersiana* Kränzlin; *E. hamelinii* Rolfe). Illustration: *Botanical Magazine*, 7612, 7613 (1898); *Gardeners' Chronicle* **23**: 200 (1898); *Orchid Digest* **36**: 120 (1972); Hermans et al., Orchids of Madagascar, edn 2, pl.38A, B (2007).

Pseudobulbs *c.* 20 cm, spindle-shaped, without fibrous leaf remains. Leaves 100 × 8–10 cm or more. Axis of raceme green, bracts brown. Flowers 8–10 cm in diameter. Sepals pink tinged purplish and with purple tips. Petals pink. Lip with large, pink lateral lobes that curve over the white column, the central lobe purple or white margined with purple and with 3–4 orange-yellow ridges. *Madagascar*. G2. Spring–summer.

2. E. elizabethae Linden & Rolfe. Illustration: *Botanical Magazine*, 7387 (1894) & n.s., 656 (1973); *Orchid Digest* **36**: 120 (1972); Hermans et al., Orchids of Madagascar, edn 2, pl.37 (2007).

Pseudobulbs *c.* 15 cm, spindle-shaped, covered with fibrous leaf remains. Leaves 60–80 × 3.5–5 cm. Axis of raceme and bracts red. Flowers 4–5 cm in diameter. Sepals red lined with white outside, white within. Lip with small lateral lobes, white, red at the base and with a large yellow blotch on the central lobe. *Madagascar*. G2. Spring.

159. CYRTOPODIUM R. Brown

J. Cullen

Terrestrial or epiphytic. Pseudobulbs often stem-like, compound, bearing several leaves, which are rolled when young and pleated, with conspicuous veins. Raceme or panicle with many flowers, lateral. Sepals and petals similar or the petals somewhat broader and shorter, the lateral sepals attached to the column foot. Lip 3-lobed, conspicuously clawed, the claw and lateral lobes spreading, the central lobe directed downwards. Column with a small foot. Pollinia 2, ovoid or ellipsoid, stipe triangular, short and broad.

A genus of about 30 species from C & S America. Cultivation as for *Eulophia* (p. 596).

1a. Sepals and petals equal, unspotted; lip yellow **1. andersonii**
b. Sepals narrower than petals, all with red-brown lines or spots; lateral lobes of lip dark red **2. punctatum**

1. C. andersonii (Lambert) R. Brown (*Cymbidium andersonii* Lambert). Illustration: *Botanical Magazine*, 1800 (1816); Bechtel et al., Manual of cultivated orchid species, 192 (1981).

Pseudobulbs stem-like to 1.5 m. Leaves to 50 cm, linear and tapering. Panicle with many flowers and ovate, acute, inconspicuous bracts. Flowers *c.* 5 cm in diameter. Sepals and petals spreading, more or less equal, yellow faintly flushed with green. Lip entirely yellow or orange-yellow, the central lobe spoon-shaped, wavy. *West Indies, Venezuela*. G2. Spring–summer.

2. C. punctatum (Linnaeus) Lindley (*Epidendrum punctatum* Linnaeus). Illustration: Hoehne, Iconografia de orchidaceas do Brasil, t. 153 (1949); Ospina & Dressler, Orquideas de las Americas, f. 139 (1974); Sheehan & Sheehan, Orchid genera illustrated, 77 (1979); Bechtel et al., Manual of cultivated orchid species, 192 (1981).

Like *C. andersonii* but with conspicuous bracts, sepals greenish yellow with brown stripes or spots, narrower than the yellow, red-spotted petals, and lip with dark red lateral lobes, the central lobe edged with red. *Tropical America from Costa Rica to C Brazil*. G2. Spring–summer.

C. virescens Reichenbach & Warming. Illustration: *Botanical Magazine*, 7396 (1895); Hoehne, Iconografia de orchidaceas do Brasil, t. 166 (1949). Pseudobulbs to 10 cm, each with 3–4 leaves, the flowers *c.* 3 cm in diameter, borne in a raceme, and the lip fleshy. *S Brazil and Paraguay*. G2. Spring–summer.

160. GALEANDRA Lindley

J. Cullen

Epiphytic. Pseudobulbs ovoid or stem-like, compound, bearing 2–7 leaves. Leaves rolled when young. Raceme terminal, open, with few to many flowers. Sepals and petals similar, spreading or reflexed. Lip somewhat cup-shaped, rolled around and partially concealing the column, with a short, downwardly directed or straight spur. Column without a foot. Pollinia 2, laterally ovoid, stipe more or less triangular, short and broad.

A genus of about 8 species from C & S America. They may be grown in epiphytic or terrestrial compost in pots. Cooler conditions are required during the resting season. Plants are very susceptible to attack by thrips and red spider mite. Propagation by division.

Literature: Rolfe, R.H., Galeandras, *Gardeners' Chronicle* **12**(2): 430–1 (1892); Teuscher, H., Die Gattung Galeandra, *Die Orchidee* **26**: 1–5 (1975).

1a. Pseudobulbs ovoid; flowers *c.* 5 cm in diameter **1. batemanii**
b. Pseudobulbs elongate, stem-like; flowers *c.* 10 cm in diameter **2. devoniana**

1. G. batemanii Rolfe (*G. baueri* misapplied). Illustration: *Edwards's Botanical Register* **26**: t. 49 (1840).

Pseudobulbs ovoid, each with 5–7 leaves. Leaves 15–20 cm, linear-lanceolate, acute. Racemes arching, with 12–many flowers, which are *c.* 5 cm in diameter. Sepals and petals reflexed, yellow or greenish, flushed with brown. Lip violet-red margined with white, deeply notched at the apex. Spur straight. *Mexico, Guatemala*. G2. Summer.

Genuine *G. baueri* Lindley, from French Guiana, is probably not in cultivation: it has fewer flowers and a yellow, purple-streaked or purple-flushed lip.

G. lacustris Barbosa-Rodriguez. Illustration: *Orchid Review* **83**: 165 (1975); *Die Orchidee* **26**: 3 (1975); Bechtel et al., Manual of cultivated orchid species, 214 (1981). Similar but with sepals and petals pale green suffused with purple, and the lip white with purple spots. *N & W Brazil, Peru*.

G. nivalis Masters. Also similar but with olive green sepals and petals and the lip white with a central violet blotch. *N&W Brazil.*

2. G. devoniana Schomburgk.
Illustration: *Botanical Magazine*, 4610 (1851); Dunsterville & Garay, Venezuelan orchids illustrated **2**: 157 (1961); *Die Orchidee* **26**: 2 (1975); Bechtel et al., Manual of cultivated orchid species, 214 (1981).

Pseudobulbs cylindric, stem-like, each with 2–5 leaves. Leaves to 20 × 1 cm, linear. Raceme arching, with few flowers. Flowers *c.* 10 cm in diameter. Sepals and petals linear, brown with greenish margins. Lip white, lined with violet, with 2 low ridges on the surface. Spur lightly curved. *Tropical S America, from Venezuela to N Brazil.* G2. Summer.

161. ANSELLIA Lindley
J. Cullen
Large epiphytes. Pseudobulbs compound, elongate and stem-like, bearing several leaves. Leaves folded when young. Raceme or panicle with 20–many flowers, terminal. Sepals and petals similar or the petals somewhat broader than the sepals, all spreading, the lateral sepals attached to the column foot, forming a small mentum. Lip without a spur, 3-lobed, the lateral lobes wrapped around the column, the central lobe with 2–3 ridges. Pollinia 2, each 2-lobed, appearing as 4, ovoid, stipe very short and broad, narrowly triangular.

A genus of 1 species from the southern half of Africa. Plants require careful potting (using terrestrial compost) and establishment, as they suffer from root disturbance. They should not be allowed to dry out completely during the resting season.

Literature: Summerhayes, V.S., African orchids IX, *Kew Bulletin* **1937**: 461–3 (1938); Khayota, B., The genus Ansellia, *Botanical Magazine* **10**(1): 27–35 (1993); Purver, C., Investigating Ansellia africana 'Mont Millais', *Orchid Review* 116(1282): 238–239 (2008).

1. A. africana Lindley (*A. gigantea* Reichenbach var. *nilotica* (N.E. Brown) Summerhayes). Illustration: *Botanical Magazine*, 4965 – excluding f. 3 – (1854) & 216 (1993); Sheehan & Sheehan, Orchid genera illustrated, 39 (1979); Bechtel et al., Manual of cultivated orchid species, 168 (1981).

Pseudobulbs to 80 cm. Leaves 4–7, 15–30 × 3–4 cm, linear to narrowly elliptic, prominently veined. Panicle to 75 cm with 20–many flowers. Flowers 2–2.5 cm in diameter. Petals broader than sepals or as broad as them, all greenish to yellow, usually with brownish or purplish spots. Lip 3-lobed, the lateral lobes greenish striped with brown, the central lobe yellow, oblong to circular with 2 bright yellow ridges. *Tropical and southern Africa.* G2. Winter.

162. CYMBIDIUM Swartz
J. Cullen
Epiphytic or terrestrial. Pseudobulbs obvious or inconspicuous, compound, with 2–several leaves. Leaves leathery, conspicuously veined, folded when young. Raceme lateral with 2–many flowers. Sepals and petals usually similar, spreading or curving forwards. Lip free or somewhat united to the base of the column, scarcely to obviously 3-lobed. Column narrow, without a foot. Pollinia 2, laterally ovoid, stipe very short and broad, triangular.

A genus of about 52 species from Asia. Most of the cultivated cymbidiums are hybrids, many of them complex. Species Nos. 9–10 are sometimes separated into the genus *Cyperorchis* Blume.

They should be grown in pots in an epiphytic compost and in cool conditions with a marked drop in temperature at night, which initiates flower-bud formation. They should be shaded from hot sunshine in the summer and in general should be well ventilated. Plants should be repotted after flowering. Propagation by division and commercially by meristem culture.

Literature: Hunt, P.F., Notes on Asiatic orchids V, Kew Bulletin **24**: 75–99 (1970); Pradhan, G.M., The Cyperorchis species of northern India, *Orchid Digest* **40**: 115–17 (1976); Du Puy, D. & Cribb, P.C., The genus Cymbidium, edn 1 (1988); Liu, Z.J., Chen, S.C. & Ru, Z.Z., The genus Cymbidium in China (2006); Du Puy, D. & Cribb, P.C., The genus Cymbidium, edn 2 (2007) – excellent illustrations of all species.

Flowers. Eight or more in each inflorescence: **1, 2, 3, 5, 6, 11**; inflorescence with up to 7 flowers: **7, 8, 9, 10**.
Sepals and petals. Directed forwards, not widely spreading: **10, 11**; widely spreading, flower up to 5 cm in diameter: **1, 2, 3, 4**; widely spreading, flower more than 5 cm in diameter: **5, 6, 7, 8, 9**. White or cream, sometimes flushed with pink: **6, 7, 9, 10**; yellow or greenish, rarely brown, often with dark lines or blotched: **1, 2, 3, 4, 8, 11**.
Raceme. Arching or hanging: **1, 2, 11**; erect or spreading: **3, 4, 5, 6, 7, 8, 9, 10**.

1a. Flowers 8 or more in each well-developed inflorescence 2
b. Flowers 1–7 in each well-developed inflorescence 7
2a. Flowers to 5 cm in diameter 3
b. Flowers more than 5 cm in diameter 6
3a. Sepals and petals directed forwards, the flower not opening widely **11. elegans**
b. Sepals and petals spreading, the flower opening widely 4
4a. Racemes hanging or arching 5
b. Racemes erect or spreading **3. ensifolium**
5a. Leaves distinctly stalked; lip diamond-shaped, scarcely 3-lobed, usually pink to purple **1. devonianum**
b. Leaves not stalked; lip clearly 3-lobed, yellowish or reddish **2. aloifolium**
6a. Sepals and petals white or white flushed with pink **6. insigne**
b. Sepals and petals green or yellowish, sometimes marked with red, purple or brown lines and spots **5. iridoides**
7a. Flowers solitary (rarely 2); not white **4. goeringii**
b. Flowers usually more than 1; if solitary then white 8
8a. Sepals and petals olive green to yellow, faintly spotted and lined with red or purple; leaves to 15 cm **8. tigrinum**
b. Sepals and petals white or ivory white sometimes flushed with pink; leaves 30–70 cm 9
9a. Sepals and petals erect, forwardly directed, the flower not opening widely **10. mastersii**
b. Sepals and petals spreading, the flower widely open 10
10a. Column bright crimson; lateral lobes of the lip large, more or less folding over the column **9. erythrostylum**
b. Column not crimson; lateral lobes of the lip not extending over the column **7. eburneum**

1. C. devonianum Paxton. Illustration: *Botanical Magazine*, 9327 (1933); Hawkes, Encyclopaedia of cultivated orchids, 192 (1965); *Orchid Digest* **41**: 8 (1977); Bechtel et al., Manual of cultivated orchid species, 190 (1981).

Pseudobulbs with 2–5 leaves. Leaves 30–40 × *c.* 3 cm, distinctly stalked, oblanceolate. Racemes equalling or exceeding the leaves, arching, with many flowers. Flowers 2.5–4 cm in diameter. Sepals and petals yellowish green or brown with pale purple streaks, spreading. Lip diamond-shaped, scarcely 3-lobed, variable in colour, usually pink to purplish, often with darker spots. *NE India*. G1. Spring–summer.

2. C. aloifolium (Linnaeus) Swartz (*C. simulans* Rolfe; *C. pendulum* (Roxburg) Swartz). Illustration: *Orchid Digest* **42**: 126 (1978); Bechtel et al., Manual of cultivated orchid species, 189 (1981).

Pseudobulbs scarcely obvious. Leaves 30–60 × 1.5–3 cm, not stalked, apex unequally notched. Racemes arching with many flowers, which are 3.5–4.5 cm in diameter. Sepals and petals brownish yellow, each with a purple median stripe, spreading, lip distinctly 3-lobed, yellowish or reddish with red lines and with 2 curved and interrupted ridges, finely downy. *From E Himalaya to S China and Malaysia*. G1. Spring.

C. bicolor Lindley (*C. aloifolium* misapplied). Leaves 30–70 × 1.2–3 cm, blunt at the apex, sepals and petals pale yellow to cream with a broad mid-stripe of maroon-brown and lip white or cream with a pale yellow patch at the base, with 2 straight to S-shaped, continuous or broken in the middle, ridges. *India, Indochina*. G1. Spring.

Subsp. **bicolor.** Sepals 4–6.5 mm longer than petals.

Subsp. **pubescens** (Lindley) Du Puy & Cribb. Sepals equalling or to 2 mm longer than petals. Lip densely hairy, with some glandular hairs towards base. Side-lobes acute.

Subsp. **obtusum** Du Puy & Cribb. Sepals equalling or to 2 mm longer than petals. Lip papillose or slightly hairy. Side-lobes blunt.

3. C. ensifolium (Linnaeus) Swartz. Illustration: *Botanical Magazine*, 1751 (1815).

Pseudobulbs scarcely obvious. Leaves 60–100 × 2–4 cm cm, linear to sword-like, acute. Raceme erect or spreading, with 8–12 flowers, exceeding the leaves. Flowers to 4.5 cm in diameter, fragrant. Sepals and petals greenish yellow with reddish brown lines, spreading. Lip clearly 3-lobed, white or yellowish with irregular, red-brown spots. *India, China, Japan and probably elsewhere in SE Asia*. G1. Summer.

C. floribundum Lindley (*C. pumilum* Rolfe). Smaller, with leaves 30–55 × 1.5–2 cm, the raceme shorter than the leaves, the sepals and petals sometimes brown and the lip white to pink, dotted with red and tipped with yellow. *W China*. G1. Summer.

C. kanran Makino (*C. sinokanran* T.C. Yen). Illustration: Du Puy & Cribb, The genus Cymbidium, edn 2, 291, 292 (2007). Terrestrial herb with narrowly ovoid pseudobulbs. Leaves 3–4, *c.* 30 × 1 cm, linear, acuminate. Inflorescence 25–60 cm, with 5–12 flowers and 3–6 sheaths up to 5 cm long. Flowers 5–7 cm across, scented. Sepals 3–4.5 cm × 4–7 mm, linear-elliptic, acuminate, dark to light green, often with a maroon stripe on the mid-vein towards the base. Petals 20–30 cm × 5–10 mm, ovate, acuminate, usually arching over the column. Lip 2–3 cm, side-lobes erect, flattened, mid-lobe 1.2–1.6 cm × 8 mm, oblong-ovate to triangular-ovate, blunt, with a callus of 2 ridges. *Southern China, Korea, Taiwan, Japan*. G1. Spring.

C. × nishiuchianum J.M.H. Shaw (*C. harukanran* invalid). Illustration: Masamune, Native orchids of Nippon **1**: 68–9 (1984); Maekawa, Wild orchids of Japan in colour, pl.168 (1971). Leaves 14–36 cm × 7–14 mm. Peduncle *c.* 16 cm, with 2–4 flowers. Floral bracts linear-lanceolate, 4 cm, acute. Flowers green to yellowish white. Sepals linear-oblong, 25–28 × 7 mm, petals shorter than sepals. Lip 2–2.5 × 1 cm, 3-lobed, oblong, lateral lobes erect, rounded. Column 11–13 mm. A natural hybrid, *C. kanran × goeringii*, known from Japan and Taiwan, and also produced in cultivation in Korea as *C. Meeark*.

4. C. goeringii (Reichenbach) Reichenbach (*C. virens* Reichenbach; *C. formosanum* Hayata; *C. forrestii* Rolfe; *C. yunnanense* Schlechter). Illustration: *Die Orchidee* **53**: 722–33 (2002); Du Puy & Cribb, The genus Cymbidium, edn 2, 284–6 (2007).

Terrestrial herb with pseudobulbs. Leaves 5–7, 25–40 × 0.5–1 cm, linear-elliptic, acute, with a serrated margin. Inflorescence bearing 1 or rarely 2–4 flowers, with 4–8 sheaths. Flower 4–5 cm across, nodding, sometimes scented. Sepals obovate to elliptic, blunt, 2.5–4 × *c.* 1 cm, apple-green to red-brown. Petals 2–3 × *c.* 1 cm, oblong-ovate to oblong-elliptic, blunt, usually covering the column. Lip 1.7–2.6 × 1–1.6 cm, unlobed to 3-lobed, with a broadly ovate to oblong mid-lobe, callus with 2 ridges, cream to pale yellow with red spots. Column 1.3–2 cm. *Southern China, Japan, Korea*. G1. Spring.

5. C. iridoides D. Don (*C. giganteum* Wallich). Illustration: *Botanical Magazine*, 4844 (1855); Bechtel et al., Manual of cultivated orchid species, 190 (1981).

Pseudobulbs ovoid, several-leaved. Leaves linear, tapering to the acute apex, 60–75 × 2–4 cm. Raceme erect or spreading, with 8–15 flowers, 8–10 cm in diameter. Sepals and petals green or yellowish green with red or brown stripes. Lip yellow, spotted with reddish brown, hairy. *Himalaya, W China*. G1. Winter.

C. hookerianum Reichenbach (*C. grandiflorum* Griffith). Illustration: *Botanical Magazine*, 5574 (1866). Flowers 9–13 cm in diameter (occasionally as much as 16 cm?), sepals and petals olive green, unstriped, lip yellow blotched with purple. *Himalaya, W China*. G1. Winter.

C. lowianum (Reichenbach) Reichenbach. Illustration: *Orchid Digest* **42**: 124 (1978); Bechtel et al., Manual of cultivated orchid species, 191 (1981). Flowers 9–11.5 cm in diameter, sepals and petals greenish yellow with faint red-brown stripes, lip variable, yellow, white or orange, usually with a V-shaped scarlet mark. *Burma*. G1. Winter.

C. tracyanum Rolfe. Illustration: *Botanical Magazine*, n.s., 56 (1949); Bechtel et al., Manual of cultivated orchid species, 191 (1981). Flowers fragrant, 10–15 cm in diameter, sepals and petals yellowish suffused and lined with red-brown, lip cream to yellow lined with purple or brown. *Burma, Thailand, Vietnam*. G1. Winter.

C. zaleskianum Linden is sometimes considered to be a hybrid between *C. giganteum* and *C. grandiflorum*, but may be a form of *C. tracyanum* with a well-marked lip.

C. wilsonii (Cooke) Rolfe. Illustration: *Botanical Magazine*, n.s., 704 (1976). Flowers *c.* 9 cm in diameter, sepals and petals green lined with reddish brown, lip yellowish white spotted with red-brown on the ridges and near the apex, lateral lobes lined with red-brown. *W China*.

Taylor & Woods consider this to be a cultivar of *C. iridoides* Don (in the notes to the illustration cited as *C. giganteum*).

6. C. insigne Rolfe (*C. sanderi* invalid). Illustration: *Botanical Magazine*, 8312 (1910).

Pseudobulbs spherical. Leaves 50–100 × 1.5–2 cm, linear, tapering to the acute apex. Raceme erect, exceeding the leaves, with 10–15 flowers. Flowers 7.5–10 cm in diameter. Sepals and petals white or white flushed with pink, their bases red-spotted. Lip white, flushed, lined and spotted with pink and with 2 yellow ridges. *Vietnam, Laos*. G1. Spring–summer.

7. C. eburneum Lindley. Illustration: *Botanical Magazine*, 5126 (1859); *Orchid Digest* **42**: 126 (1978); Bechtel et al., Manual of cultivated orchid species, 191 (1981).

Pseudobulbs scarcely evident. Leaves 45–60 × *c.* 2 cm, linear, tapering to the acute apex. Raceme erect with 1–2 flowers. Flowers usually 7.5–10 cm in diameter, fragrant. Sepals and petals white, spreading. Lip usually white with yellow ridges or a yellow centre, occasionally spotted with red or purple, the lateral lobes extending up to the column but not overlapping it, downy. *Himalaya, Burma, Vietnam*. G1. Spring.

Variable in the colour of the lip (see Pradhan, *Orchid Digest* **40**: 69–71, 1976).

8. C. tigrinum Hooker. Illustration: *Botanical Magazine*, 5457 (1864); *Orchid Digest* **42**: **129** (1978); Bechtel et al., Manual of cultivated orchid species, 191 (1981).

Pseudobulbs almost spherical, with 3–4 leaves. Leaves to 15 × 2.5 cm, broadly linear, acute. Racemes erect, exceeding the leaves, with 3–6 flowers. Flowers usually 5–7 cm in diameter (rarely larger). Sepals and petals olive green to yellow, faintly spotted and lined with red or purple, spreading. Lip mostly white, lined and spotted with purple, the margins often mostly purple. *Burma, Thailand*. G1. Summer–autumn.

9. C. erythrostylum Rolfe. Illustration: *Botanical Magazine*, 8131 (1907); *Orchid Digest* **42**: 129 (1978).

Pseudobulbs scarcely obvious. Leaves 30–45 × 1–1.5 cm, linear, tapering to the acute apex. Raceme erect, loose, with 4–6 flowers. Flowers 7.5–11 cm in diameter. Sepals and petals white. Lip white shaded with yellow and with red lines, the lateral lobes large, curved over the column and more or less overlapping. Column bright crimson. *Vietnam*. G1. Winter.

10. C. mastersii Lindley (*Cyperorchis mastersii* (Lindley) Bentham). Illustration: *Edwards's Botanical Register* **31**: t. 50 (1845); *Orchid Digest* **41**: 116 (1976) & **42**: 127 (1978); Bechtel et al., Manual of cultivated orchid species, 191 (1981).

Pseudobulbs scarcely obvious. Leaves 45–70 × 1.5 cm, linear, tapering to the acute apex. Racemes erect with 4–6 (rarely 7) flowers. Flowers fragrant, *c.* 6 cm, the sepals and petals forwardly directed, white or ivory white, sometimes flushed with pink. Lip white, sometimes spotted with pink, with a yellow centre, downy. *E Himalaya*. G1. Winter.

11. C. elegans (Blume) Lindley (*Cyperorchis elegans* Blume; *C. longifolium* D. Don). Illustration: *Botanical Magazine*, 7007 (1888); *Orchid Digest* **42**: 127 (1978).

Pseudobulbs scarcely obvious. Leaves 40–70 × 1–2.5 cm, numerous, linear, tapering to the acute apex. Raceme arching, with many flowers. Flowers 4–5 cm long, the sepals and petals forwardly directed, little spreading, pale to tawny yellow. Lip yellow marked with 2 orange lines, sometimes spotted with red, downy and sparsely ciliate. *Himalaya, W China*. G1. Winter.

163. GRAMMANGIS Reichenbach

J. Cullen

Like *Cymbidium* but the lateral sepals fused at the base and attached to the column foot forming a small but distinct mentum. Petals much smaller than the sepals and differently coloured.

1. G. ellisii Reichenbach. Illustration: *Botanical Magazine*, 5179 (1860); Bechtel et al., Manual of cultivated orchid species, 215 (1981); Hermans et al., Orchids of Madagascar, edn 2, pl.40D (2007).

Sepals *c.* 4 × 2 cm, yellow barred and blotched with brown. Petals white with pink tips. Lip white with reddish lines, central lobe small. *Madagascar*. G2. Summer.

164. GRAMMATOPHYLLUM Blume

J. Cullen

Large epiphytes. Pseudobulbs compound, distinct, or elongate and stem-like, with 2-several leaves. Leaves rolled when young. Raceme large, lateral, with many flowers; lower flowers sometimes sterile and abortive. Sepals and petals similar, spreading. Lip small, clawed, 3-lobed, the lateral lobes large, extending up to the column, the central lobe smaller, bearing several ridges. Column without a foot. Pollinia 2, laterally ovoid to almost spherical, stipe short and broad, divided into 2 branches above.

A genus of about 12 species from SE Asia. They are strong-growing plants requiring epiphytic compost and careful watering. *G. scriptum* benefits from cooler conditions during the resting period. Propagation by division in early spring.

Literature: Cootes, J., Orchids of the Philippines, 128–31 (2001).

1a. Pseudobulbs to 14 cm, distinct, with a few leaves at the apex: flowers to 8 cm in diameter **1. scriptum**

b. Pseudobulbs elongate, stem-like, bearing several leaves along their length; flowers to 10 cm in diameter **2. speciosum**

1. G. scriptum Blume (*G. fenzlianum* Reichenbach; *G. measuresianum* Weather; *G. multiflorum* Lindley; *G. rumphianum* Miquel). Illustration: *Botanical Magazine*, 7507 (1896); Sheehan & Sheehan, Orchid genera illustrated, 101 (1979).

Pseudobulbs to 14 cm, ridged, bearing 2–3 leaves at the apex. Leaves very narrowly elliptic, 30–35 × *c.* 8 cm. Raceme very long, with many flowers that are all fertile and up to 8 cm in diameter. Sepals and petals pale green or greenish yellow, suffused or spotted with brown or purplish brown. Lip small, hairy. *Borneo, Philippines, Indonesia (Moluccas)*. G2. Summer.

Variable as regards the colour of the lip, which varies from white to yellow and is striped with brown or purple.

2. G. speciosum Blume. Illustration: *Botanical Magazine*, 5157 (1860).

Pseudobulbs elongate, stem-like, leafy along their length. Leaves 45–60 × 3 cm, broadly linear, acute. Raceme 2 m or more with many flowers, the lower of which are abortive and sterile with 2 sepals, 2 petals, no lip and an abortive column. Fertile flowers *c.* 10 cm in diameter, sepals and petals pale greenish yellow with dull orange-brown spots. Lip small, hairy, yellow striped with red-brown. *Malaysia, Indonesia, Philippines*. G2. Autumn–winter.

165. MORMODES Lindley

J. Cullen

Epiphytic. Pseudobulbs conspicuous, compound, broadly spindle-shaped, each bearing 2–several leaves. Leaves strongly veined, rolled when young. Raceme lateral, from the central nodes of the pseudobulb, erect or arching, with inconspicuous bracts. Flowers variously coloured, often fragrant. Sepals and petals more or less equal, the upper 3 often converging to form a hood over the column. Lip clawed, not spurred, entire or 3-lobed, often with the margins rolled under. Column cylindric, always twisted to 1 side of the flower, which is thus asymmetric. Pollinia 2, ovoid, the stipe wineglass-shaped, longer than the pollinia and broader than them at the apex, the viscidium almost circular.

A genus of about 60 species from C & S America, which is taxonomically very confused, owing to the wide and little-understood variation in flower colour, shape and size. A few species are grown as curiosities. Cultivation as for *Catasetum* (see below).

Literature: Pabst, G.F.J., An illustrated key to the species of the genus Mormodes, *Selbyana* **2**: 149–55 (1978).

1a. Lip clearly 3-lobed (lateral lobes may be small) **1. maculata**
b. Lip entire 2
2a. Lip hairy above; flower usually uniformly dark purple **2. hookeri**
b. Lip hairless; flower usually of some other colour or combination of colours 3
3a. Lip flat or with the sides curved upwards; pseudobulbs 8–10 cm **3. aromatica**
b. Lip with the sides distinctly rolled under; pseudobulbs to 30 cm 4
4a. Lip ovate, tapering to the apex **4. colossus**
b. Lip broader than long to circular, rounded to notched (though with a small point) at the apex **5. buccinator**

1. M. maculata (Kotzsch) Williams (*Cyclosia maculata* Klotzsch; *M. pardinum* Bateman). Illustration: *Botanical Magazine*, 3879 (1842) & 3900 (1843); Bechtel et al., Manual of cultivated orchid species, 234 (1981).

Pseudobulbs 8–10 cm, with 4–5 leaves. Flowering stem to 40 cm, raceme with many flowers. Flowers *c.* 6 cm in diameter, yellow or pale yellow, with or without red-brown spots. Sepals and petals lanceolate, acute. Lip *c.* 5 cm with the margins rolled under, 3-lobed from the middle, the lateral lobes conspicuous. *Mexico*. G2. Autumn.

M. lineatum Lindley. Illustration: *Die Orchidee* **22**: 45 (1971). Similar but with slightly smaller flowers (*c.* 5 cm in diameter) and the lip *c.* 3 cm, with very small, incurved, tooth-like lateral lobes. *Guatemala*. G2. Autumn.

M. luxatum Lindley. Also similar but larger, with flowers *c.* 7 cm in diameter, the 2 upper sepals and upper petal converging to form a hood over the column, flowers sometimes white. *Mexico*. G2. Autumn.

2. M. hookeri Lemaire. Illustration: *Botanical Magazine*, 4577 (1851).

Pseudobulbs oblong-ovoid. Flowering stem *c.* 30 cm, raceme with 4–5 distant flowers. Sepals and petals reflexed, dark purple, their sides themselves reflexed. Lip pointing forwards, entire, its sides rolled under, hairy on the upper surface, purple but paler than sepals and petals. *Panama*. G2. Winter.

3. M. aromatica Lindley. Illustration: *Edwards's Botanical Register* **29**: t. 56 (1843); Hamer, Las Orquideas de El Salvador **2**: 124–5 & t. 14 (1974).

Pseudobulbs 8–10 cm with 4–5 leaves. Flowering stem weak, to 15 cm, with 5–8 rather distant flowers. Flowers *c.* 3 cm in diameter, yellowish deeply flushed with violet-brown and with darker spots. Sepals and petals similar, elliptic, acute. Lip entire, flat or with the sides curved upwards, with a triangular apex. *Mexico, El Salvador*. G2. Autumn.

4. M. colossus Reichenbach. Illustration: *Botanical Magazine*, 5840 (1870); Bechtel et al., Manual of cultivated orchid species, 233 (1981).

Pseudobulbs to 30 cm. Flowering stem to 70 cm, with 6–10 distant flowers, to 15 cm in diameter. Sepals and petals lanceolate, acuminate, reflexed, bright yellow, red towards the base. Lip ovate, acuminate, entire, the sides rolled under, bright yellow marked with red at the base. *Costa Rica*. G2. Spring.

5. M. buccinator Lindley (*M. lentiginosa* Hooker). Illustration: *Botanical Magazine*, 4455 (1849) & 8041 (1905); Couret, Orquideas Venezolanas, 39–42 (1977).

Pseudobulbs to 30 cm. Flowering stems to 25 cm with 7–10 flowers. Flowers extremely variable in colour, bright yellow to dark red-brown, variably spotted and lined. Sepals and petals narrowly elliptic, acute, spreading or reflexed at first, later curving forwards. Lip 2–2.5 cm wide, entire, broader than long, its sides very markedly rolled under, forming a trumpet shape. *Venezuela*. G2. Spring.

M. igneum Lindley & Paxton. Similarly variable in colour, but the lip 3–3.5 cm wide. *C America and northern S America*.

166. CATASETUM Richard

J. Cullen

Epiphytes with large, compound pseudobulbs bearing several leaves. Leaves large, pleated, rolled when young. Racemes arising from the bases of the pseudobulbs, erect or arching, with few to many flowers. Flowers usually not resupinate, monomorphic and bisexual, or, more usually, dimorphic and unisexual, very rarely trimorphic (male, female and bisexual). Sepals and petals similar, free, spreading or incurved. Lip variable in size, saccate or not. Column large, without a foot, usually acute in male and bisexual flowers, with (in all species treated below except No. 1) 2 processes (antennae) arising from near the apex and directed backwards onto the lip in the male flowers. Pollinia 2, ovoid or ellipsoid, the stipe elongate, linear or strap-shaped, its sides usually folded upwards, viscidium circular or broadly oblong.

The descriptions given here (with the exception of that for No. 1) refer to the male flowers; the female flowers are less well known and their identification is difficult. Plants bearing female flowers were at one time referred to the genus *Monachanthus* Lindley. The production of male and female flowers appears to be controlled by light and nutrition (see Dodson, *Annals of the Missouri Botanical Garden* **49**: 35–56, 1962). The antennae borne by the male flowers are sensitive, and touching of them by an insect causes the pollinia to be forcibly ejected. For more details on pollination mechanisms in the genus see Hills, Williams & Dodson, *Biotropica* **4**: 61–9 (1972). The genus is divided into 3 genera, *Clowesia*, *Dressleria* and *Catasetum*, by Dodson, *Selbyana* **1**: 130–8 (1975). Under this arrangement, species No. 1 covered here belongs to *Clowesia*, the rest to *Catasetum*.

Catasetum in the broad sense is a genus of about 70 species from C & S America; about a dozen of these are grown for the sake of their bizarre male flowers. The plants require sharp drainage and full light, combined with a humid atmosphere. They should be rested when the pseudobulbs are ripened. Propagation by seed or by division.

Literature: Mansfeld, R., Die Gattung Catasetum, *Feddes Repertorium* **30**: 257–75 (1932) & **31**: 99–125 (1933); Holst, A.W., The world of Catasetum (1999).

Flowers. Always bisexual: see **Clowesia**; usually unisexual: **1, 2, 3, 4, 5, 6, 7, 8, 9, 10, 11**.
Lip. (figure. 29, p. 562). Flat, not hollowed or saccate: **5, 7**; hollowed or saccate but with a flat blade: **3, 4, 6, 8, 9, 10, 11**; entirely hollowed, with no flat blade: **1, 2**. With conspicuous fringed or ciliate margins: **4, 6, 7, 8, 9, 10**; with more or less entire margins: **1, 2, 3, 5, 11**.
Antennae. Asymmetric: **1, 2, 3, 4**; symmetric: **5, 6, 7, 8, 9, 10, 11**.

1a. Flowers all bisexual; column without antennae **Clowesia** (p. 604)
b. Flowers mostly unisexual; column in the male flowers with 2 antennae 2
2a. Lip not at all saccate or hollowed 3
b. Lip distinctly saccate, the position of the sac and the degree of saccation variable 4
3a. Lip broadly obovate with 3 teeth at the apex, the laterals narrowly triangular, incurved, the central shorter, straight; margins of the rest of the lip entire (Figure 29(6), p. 562) **5. cernuum**
b. Lip triangular with rounded corners, the apex entire; margins lacerate-ciliate (Figure 29(8), p. 562) **7. trulla**
4a. Antennae not symmetric 5
b. Antennae symmetric 8
5a. Lip entirely helmet-shaped, hollowed out completely, without a flat blade 6
b. Lip variably saccate, but with a flat blade 7
6a. Basal angles of the lip projecting inwards, touching each other or almost so, toothed (Figure 29(2), p. 562) **1. integerrimum**
b. Basal angles of the lip not projecting inwards, entire (Figure 29(3), p. 562) **2. macrocarpum**
7a. Lip semicircular, truncate at the base, not lobed (Figure 29(4), p. 562) **3. pileatum**
b. Lip oblong, 3-lobed, the central lobe narrow, not truncate at the base (Figure 29(5), p. 562) **4. saccatum**
8a. Lip broadly obovate, fan-shaped, bearing a large humped callus in front of the mouth of the sac (Figure 29(7), p. 562) **6. fimbriatum**
b. Lip variously shaped, not as above, either without a callus or with the callus behind the mouth of the sac 9
9a. Lip small, deeper than long, containing a callus composed of hair-like teeth that almost entirely fill the sac (Figure 29(9), p. 562) **9. microglossum**
b. Lip various, longer than deep, callus not as above, or absent 10
10a. Lip narrowly ovate, its margin distantly and shortly toothed, green with red spots (Figure 29(12), p. 562) **11. callosum**
b. Lip oblong, its margin, at least in part, deeply, closely and coarsely toothed, not green with red spots 11
11a. Lip margin almost entirely divided into fleshy, blunt, hair-like teeth, its terminal portion narrowly triangular with a few teeth, with a small saccate area in the middle (Figure 29(11), p. 562) **10. barbatum**
b. Lip margin coarsely and sharply toothed except for the entire, semicircular terminal part; saccation extending from the base for more than half of the lip (Figure 29(10), p. 562) **8. atratum**

1. C. integerrimum Hooker (*C. maculatum* Bateman not Knuth). Figure 29(2), p. 562. Illustration: *Edwards's Botanical Register* **26**: t. 42 (1840); *Botanical Magazine*, 3823 (1840).

Pseudobulbs to 15 cm, spindle-shaped. Leaves to 40 cm, elliptic-lanceolate. Racemes erect, to 35 cm, with 6–12 non-resupinate flowers. Sepals 4.5–4.8 cm, elliptic or oblong-elliptic, acute, green outside, green suffused or spotted with red inside. Petals similar but slightly smaller, Lip helmet-shaped, wholly hollowed, the basal angles projecting inwards and touching each other or almost so, toothed, the whole greenish yellow or yellow, sometimes with red spots outside, greenish with brown spots inside. Antennae asymmetric. *Mexico, Guatemala, Honduras & Venezuela*. G2. Autumn.

2. C. macrocarpum Knuth (*C. tridentatum* Hooker). Figure 29(3), p. 562. Illustration: *Edwards's Botanical Register* **10**: t. 840 (1824); *Botanical Magazine*, 2559 (1825) & 3329 (1834); Hoehne, Iconografia de orchidaceas do Brasil, t. 170 (1949); Dunsterville & Garay, Venezuelan orchids illustrated **2**: 60–1 (1961).

Similar to *C. integerrimum* but raceme to 45 cm with 4–10 flowers, lip with its basal angles not projecting inwards, and not toothed except for the 3-toothed apex, entirely yellow within and without. *Tropical S America from Trinidad to Brazil*. G2. Autumn–winter.

3. C. pileatum Reichenbach (*C. bungerothii* N.E. Brown). Figure 29(4), p. 562. Illustration: *Botanical Magazine*, 6998 (1888); Cogniaux & Goossens, Dictionnaire iconographique des orchidées, Catasetum, t. 1, 1A (1897); Hoehne, Iconografia de orchidaceas do Brasil, t. 163 (1949); Dunsterville & Garay, Venezuelan orchids illustrated **1**: 70–1 (1959).

Pseudobulbs 15–25 cm, spindle-shaped. Leaves large, lanceolate. Raceme to 30 cm, arching, with 4–10 resupinate(?) flowers. Upper sepal and petals to 5 cm, erect, lateral sepals spreading, oblong-lanceolate, acuminate, white, pale yellow or white spotted with red. Lip semicircular in outline, truncate at the base, unlobed, saccate with a small sac whose mouth is just below the lip attachment, white, pale yellow or rarely red, the interior of the sac orange-red. Antennae asymmetric. *Tropical S America, south to Ecuador and N Brazil*. G2. Autumn.

4. C. saccatum Lindley (*C. christyanum* Reichenbach). Figure 29(5), p. 562. Illustration: *Botanical Magazine*, 8007 (1905); Hoehne, Iconografia de orchidaceas do Brasil, t. 172 (1949).

Pseudobulbs to 20 cm, fusiform. Leaves to 30 cm or more, lanceolate. Raceme arching, with up to 18 non-resupinate flowers. Upper sepal and petals close, erect, lateral sepals spreading, all 5–5.5 cm, narrowly lanceolate, acuminate, reddish green with darker red spots. Lip oblong, 3-lobed, its margins rolled downwards, lacerate-ciliate, the central lobe oblong, its basal part saccate, all reddish brown with a white callus in front of the mouth of the sac. Antennae asymmetric. *Tropical S America from E Peru to Guyana*. G2. Winter.

The most commonly cultivated variant is var. *christyanum* (Reichenbach) Mansfeld, which is described above.

C. fergusonii W.E. Higgins & Dalstrom. Illustration: *Selbyana* **28**: 107 (2008). Differs from *C. saccatum* by the lip profile, which is saddle shaped, with the centre white surrounded with apple-green and a maroon margin, saccate, with a sparsely fringed margin, 3-lobed, triangular, 2.3–2.7 × 0.9–1.3 cm. *Peru.* G2.

A plant that has been fairly widely cultivated and awarded since the 1980s, but was described only in 2008.

5. C. cernuum (Lindley) Reichenbach (*C. trifidum* Hooker). Figure 29(6), p. 562.

Pseudobulbs large. Leaves elliptic. Raceme to 30 cm, arching, bearing 10–15 non-resupinate flowers, themselves downwardly directed. Sepals and petals incurved, up to 3 cm, the petals usually a little shorter than the sepals, all oblong, acute, green with dark brown spots. Lip flat, greenish spotted with brown, broadly obovate with 3 teeth at the apex, the laterals elongate-triangular, incurved, the central shorter, straight. Antennae symmetric. *S Brazil.* G2. Spring–summer.

6. C. fimbriatum (Morren) Lindley. Figure 29(7), p. 562. Illustration: *Botanical Magazine*, 7158 (1891); Dunsterville & Garay, Venezuelan orchids illustrated **4**: 42–3 (1966).

Pseudobulbs to 20 cm, ellipsoid. Leaves 20 cm or more, elliptic. Raceme to 40 cm, arching, with 7–15 non-resupinate flowers. Upper sepal and petals close, erect, lateral sepals pointing downwards, all to 3.5 cm, lanceolate, acuminate, with margins rolled downwards, yellow or greenish, spotted or striped with purple. Lip broadly obovate, fan-shaped, saccate near the base with a large humped callus in front of the mouth of the sac, margins lacerate-ciliate, yellow or greenish spotted with purple towards the base. Antennae symmetric. *Tropical and subtropical S America.* G2. Spring–summer.

A very variable species.

7. C. trulla Lindley. Figure 29(8), p. 562. Illustration: *Edwards's Botanical Register* **27**: t. 34 (1841); Hoehne, Flora Brasilica **12**(6): t. 63 (1942).

Pseudobulbs 12–20 cm, variable in shape. Leaves 35–40 cm, oblong-linear. Raceme arching, with many non-resupinate flowers. Sepals and petals spreading, the upper sepal and petals erect, the lateral sepals outwardly directed, all to 2.5 cm, oblong-elliptic, green. Lip flat, triangular with rounded corners, the margin lacerate-ciliate except towards the apex, the whole whitish green tinged with red-brown in the centre, the apical part blunt, brown. Antennae symmetric. *S Brazil.* G2. Autumn.

8. C. atratum Lindley. Figure 29(10), p. 562. Illustration: *Edwards's Botanical Register* **26**: t. 63 (1838); *Botanical Magazine*, 5202 (1860).

Pseudobulbs 10–13 cm, spindle-shaped. Leaves to 30 cm, narrowly elliptic or lanceolate. Raceme to 30 cm, arching, bearing 12–15 resupinate(?) flowers. Sepals and petals to 2.5 cm, all more or less directed downwards, oblong, acute, green, densely marked with small brown transverse stripes. Lip oblong, saccate for two-thirds of its length from the base, margins lacerate-ciliate or lacerate-toothed, yellowish green with some brown spots and a white apical part. Antennae symmetric. *S & E Brazil.* G2. Spring.

9. C. microglossum Rolfe. Figure 29(9), p. 562. Illustration: *Botanical Magazine*, 8514 (1913).

Pseudobulbs to 10 cm, spindle-shaped. Leaves to 30 cm, oblong-elliptic to oblanceolate. Raceme to 65 cm, arching, with many non-resupinate flowers. Upper sepal and petals erect, lateral sepals directed downwards, all to 2.5 cm, oblong-lanceolate, acute, purplish. Lip small, consisting mostly of an obconic sac, which is deeper than long, with lacerate margins and containing a callus that terminates in numerous teeth or hair-like processes, yellow. Antennae symmetric. *Borders of Peru and Colombia.* G2.

10. C. barbatum (Lindley) Lindley (*Myanthum barbatum* Lindley; *M. spinosum* Hooker). Figure 29(11), p. 562. Illustration; *Botanical Magazine*, 3802 (1841); Dunsterville & Garay, Venezuelan orchids illustrated **3**: 54–5 (1965).

Pseudobulbs 10–12 cm, narrowly conical to spindle-shaped. Leaves lanceolate, wavy. Raceme to 45 cm, arching or erect(?), bearing numerous non-resupinate flowers. Upper sepal and petals erect, lateral sepals spreading, all to 3 cm, narrowly oblong, acute, green with small, dark purple, transverse bars. Lip oblong, white or pink, with a small sac at about the middle, the margin deeply divided into fleshy, blunt, hair-like teeth, with a narrowly triangular apical part and a callus divided usually into 3 fleshy teeth situated behind the mouth of the sac. Antennae symmetric. *Northern S America.* G2. Spring.

The variant most commonly found in cultivation is var. **spinosum** (Hooker) Rolfe, with the callus as described above.

11. C. callosum Lindley. Figure 29(12), p. 562. Illustration: *Botanical Magazine*, 6648 (1882).

Pseudobulbs to 8 cm, spindle-shaped. Leaves 15–30 cm, oblanceolate, greyish. Raceme to 35 cm, arching or erect, with 10–15 resupinate(?) or non-resupinate flowers. Upper sepal and petals erect, lateral sepals directed downwards, all 4–4.5 cm, narrowly lanceolate, acute, brownish. Lip narrowly ovate, green with red spots, its margin finely and distantly toothed, with a small sac near the base whose mouth is sited just in front of a humped callus. Antennae symmetric. *Venezuela, Guyana, Peru.* G2. Winter.

167. CLOWESIA Lindley

J.M.H. Shaw

Similar to *Catasetum* but differing by the production of bisexual flowers, in which the lip is free from the column and the pollinia are discharged by lifting the apex of the anther cap, rather than by pressure on the projections on the edge of the column alongside the anther as in *Catasetum*. The inflorescence is pendant.

A genus of about 7 species distributed from Mexico to Brazil, formerly placed in *Catasetum*.

Literature: Dodson, C.H., Dressleria and Clowisia: a new genus and an old one revived in the Catasetinae, *Selbyana* **1**(2): 130–7 (1975); Kennedy, G.C., The genus Clowesia, *Orchid Digest* **42**(3): 111–16 (1978).

1. C. russelliana (Hooker) Dodson (*Catasetum russellianum* Hooker). Figure 29 (1), p. 562). Illustration: *Botanical Magazine*, 3777 (1840); Dunsterville & Garay, Venezuelan orchids illustrated **4**: 41–2 (1966).

Pseudobulbs ovoid-ellipsoid. Leaves to 36 × 11 cm, broadly lanceolate. Raceme arching with many fragrant, non-resupinate flowers. Sepals and petals 3–4 cm, somewhat incurved, ovate-oblong to elliptic, white with green stripes. Lip oblong, deeply saccate for about one-third of its length, the sac projecting forwards under the ovate blade; margins lacerate-toothed, the blade bearing an elongate, double ridge with toothed margins; the whole white with green lines. *C America, Venezuela.* G2. Autumn.

C. thylaciochilum (Lemaire) Dodson (*Catasetum thylaciochilum* Lemaire). Similar, but with the sac of the lip curved backwards. *Mexico.* G2. Autumn.

168. CYCNOCHES Lindley
J. Cullen
Epiphytic. Pseudobulbs cylindric, tapering slightly towards the apex, compound. Leaves pleated, rolled when young. Racemes with several flowers, lateral from the axils of the upper leaves. Flowers not resupinate, usually unisexual. Sepals and petals similar, smaller in the female flowers than in the male. Lip fleshy and undivided (all female flowers and male flowers of No. 1) or variously divided or conspicuously appendaged (male flowers of Nos. 2 and 3), with a basal callus. Column short and fleshy in the female flowers, elongate, broadened above and arched in male flowers. Pollinia 2, ovoid, stipe narrowly ovoid, longer and narrower than the pollinia, viscidium oblong, 2-lobed.

A genus of 23 species from C & S America, whose classification is complicated and based almost entirely on the male flowers, which are more frequently seen than the female. Gregg (*Selbyana* **2**: 217, 1978) reports that 'plants growing in full sunlight with adequate moisture and embedded in a suitable substrate are robust and generally produce female flowers. Less robust plants often growing in shade or lacking adequate nutrition tend to produce male flowers.' Cultivation as for *Catasetum* (p. 602).

Literature: Allen, P.H., The swan orchids: a revision of the genus Cycnoches, *Orchid Journal* **1**: 173–84, 225–30, 349–54 & 397–403 (1952).

1a. Male inflorescence with up to 9 flowers: lip of male flowers simple and entire **1. ventricosum**
b. Male inflorescence with many flowers: lip of male flowers lobed or with marginal teeth or appendages 2
2a. Lip of male flower linear or lanceolate, flat or wavy with a forwardly pointing process on the claw and 2 linear, rounded teeth on each side **3. pentadactylon**
b. Lip of male flower clawed, ovate to circular, the margin drawn out into several processes, which are club-shaped or forked at their apices **2. egertonianum**

1. C. ventricosum Bateman.
Male flowers 9–15 cm in diameter. Sepals and petals green or yellowish green, spreading. Lip 6–6.5 cm, simple, fleshy, stalkless or with an elongate claw, which may be narrowly winged, white marked with dark green at the base. Column 2.8–3.2 cm. *Tropical America, from S Mexico to French Guiana*. G2. Summer.

A variable species, divided by Allen into several varieties; the most commonly grown is var. **chlorochilon** (Klotzsch) Allen (*C. chlorochilon* Klotzsch), which has large flowers with a stalkless lip. Illustration: Hoehne, Flora Brasilica **12**(6): t. 101, 1942; Dunsterville & Garay, Venezuelan orchids illustrated **2**: 81–2, 1961; Couret, Orquideas Venezolanas, 36, 1977; Sheehan & Sheehan, Orchid genera illustrated, 71, 1979; Bechtel et al., Manual of cultivated orchid species, 189, 1981.

C. loddigesii Lindley. Illustration: Hoehne, Flora Brasilica **12**(6): t. 100 (1942); Couret, Orquideas Venezolanas, 37 (1977). Similar, but with larger flowers with greenish brown sepals and petals and the lip with a conspicuously winged claw. *Venezuela & NE Brazil*. G2. Summer.

2. C. egertonianum Bateman.
Illustration: Hoehne, Flora Brasilica **12**(6): t. 109 (1942); Bechtel et al., Manual of cultivated orchid species, 189 (1981).

Flowers 5–6 cm in diameter, Sepals and petals greenish or greenish brown, spotted with brown or purple, sometimes yellow or almost white (var. **aureum**), more rarely pink. Lip of male flowers stalked, green or white, the apical part concave, ovate or circular, the margin with several elongate, club-shaped teeth, which may be forked at their apices. Column *c.* 20 mm. *C & S America*. G2. Autumn.

Variable in flower colour; both var. **egertonianum**, with greenish or greenish brown, darker spotted perianth, and var. **aureum** (Lindley) Allen (*C. aureum* Lindley), with yellow to almost white perianth, are grown.

C. maculatum Lindley. Similar but with the apical part of the lip drawn out into a strap-shaped lobe, and the marginal teeth not forked at their apices. *Venezuela*.

3. C. pentadactylon Lindley. Illustration: Flora Brasilica **12**(6): t. 104 (1942); Bechtel et al., Manual of cultivated orchid species, 189 (1981).

Flowers. 7–10 cm in diameter, sepals and petals green blotched with dark brown. Lip fleshy and rigid, linear-lanceolate, flat or wavy, clawed and with a forwardly pointing process on the upper surface of the claw, the blade narrow, with 2 large, linear, rounded teeth on each side. *E & C Brazil*. G2. Spring.

169. ACINETA Lindley
J. Cullen
Epiphytic with distinct, simple, ovoid or spindle-shaped, grooved pseudobulbs, each bearing 2–4 leaves. Leaves large, pleated, rolled when young. Flowers not resupinate, fleshy, bell-shaped, in lateral, hanging racemes that arise from the bases of the pseudobulbs. Sepals and petals similar in size or the sepals somewhat larger, incurved. Lip clawed, 3-lobed, the lateral lobes erect and borne near the middle of the claw, reaching up to the column, the central lobe relatively small, with a prominent callus between the lateral lobes. Column finely downy on the back, with a short foot to which the lateral sepals are attached. Pollinia 2, ellipsoid, stipe broadly linear, almost as long as the pollinia, the viscidium with a truncate or rounded basal part.

A genus of 10–15 species from C & S America. They should be grown in suspended baskets or rafts so that the hanging racemes can develop properly. An epiphytic compost should be used and abundant moisture provided when in growth. A lower temperature is required during the resting season. Propagation is by seed or division.

Literature: Schlechter, R., Die Gattung Acineta, *Orchis* **11**: 21–47 (1917); Christenson, E.A., The genus Acineta, *Orchid Digest* **70**(4): 232–51 (2006).

1a. Sepals and petals red, or pale brown and red-spotted; flowering in spring **3. superba**
b. Sepals and petals yellow, with or without red spots; flowering in autumn 2
2a. Flowers 4–5 cm in diameter; pseudobulbs with 2 leaves; racemes with 12–15 flowers **1. barkeri**
b. Flowers *c.* 6 cm in diameter; pseudobulbs with 3–4 leaves; racemes usually with more than 15 flowers **2. chrysantha**

1. A. barkeri Lindley. Illustration: *Botanical Magazine*, 4203 (1846); *Orchid Digest* **70**: 235 (2006).

Pseudobulbs *c.* 10 cm, with 2 leaves, ovoid. Leaves to 50 × 10 cm. Raceme with 12–15 flowers, which are 4–5 cm in diameter. Sepals and petals ovate, golden yellow. Lip yellow with a purple-red spot or spots near the base. *S Mexico*. G2. Autumn.

2. A. chrysantha (Morren) Lindley & Paxton (*A. densa* Lindley & Paxton). Illustration: *Botanical Magazine*, 7143 (1890); Bechtel et al., Manual of cultivated orchid species, 161 (1981); *Orchid Digest* **70**: 236 (2006).

Like *A. barkeri* but larger; pseudobulbs with 3–4 leaves, racemes usually with more than 15 flowers, petals red-spotted, lip with very prominent callus. *Tropical America from Mexico to Colombia*. G2. Autumn.

3. A. superba (Humboldt, Bonpland & Kunth) Reichenbach (*A. humboldtii* Lindley). Illustration: Dunsterville & Garay, Venezuelan orchids illustrated **1**: 40–1 (1959); Bechtel et al., Manual of cultivated orchid species, 161 (1981); *Orchid Digest* **70**: 248 (2006).

Pseudobulbs with 3 leaves. Racemes loose, with 6–12 flowers, which are *c.* 6 cm in diameter. Sepals and petals red, or pale brown and red-spotted. Lip yellow, reddish towards the base, and with dark purple spots. *Tropical S America from Venezuela to Ecuador*. G2. Spring.

The variant with brownish, red-spotted flowers is referred to as var. **fulva** (Hooker) Schlechter (illustration: *Botanical Magazine*, 4156, 1845).

170. PERISTERIA Hooker

J. Cullen

Epiphytes with simple, ovoid or spindle-shaped pseudobulbs, which are grooved and bear 3–5 leaves. Leaves stalked, often large, pleated, rolled when young. Racemes arising from the bases of the pseudobulbs, erect or hanging, with few to many flowers. Flowers spherical to bell-shaped, not widely open, often fragrant, resupinate or not. Sepals and petals similar or the petals somewhat smaller, fleshy, ovate to circular, incurved. Lip complex, joined to the base of the column by a basal part, which bears 2 lateral lobes that stand upright on either side of the column, and an erect or outwardly curving anterior part, distinguished from the basal part by a waist that is sharply bent or folded; the anterior part is freely movable on the rest. Column thick and fleshy, sometimes with a small wing on either side, the rostellum prominently beaked. Pollinia 2, ellipsoid, attached independently to the triangular viscidium, stipe absent.

A genus of about 10 species from C & S America. All can be grown like *Acineta*, but *P. elata*, which has erect racemes, is suitable for pot-culture, and will tolerate a terrestrial compost.

1a. Raceme erect, stiff, with 12–25 flowers **1. elata**
b. Raceme hanging, with 4–13 flowers 2

2a. Column with a narrow, downwardly pointing wing on either side; basal part of the lip with a crested surface **3. pendula**
b. Column without wings; basal part of lip smooth **2. cerina**

1. P. elata Hooker. Illustration: *Botanical Magazine*, 3116 (1832); Dunsterville & Garay, Venezuelan orchids illustrated **1**: 238–9 (1959); Ospina & Dressler, Orquideas de las Americas, t. 159 (1974); Sheehan & Sheehan, Orchid genera illustrated, 131 (1979).

Pseudobulbs to 13 × 10 cm. Leaves to 95 cm, narrowly elliptic, acute. Raceme to 1.5 m, held stiffly erect, with 12–25 flowers. Sepals and petals waxy, white, flower 4.5–5 cm in diameter. Lip with anterior part entire, curving outwards. Column unwinged. *Tropical America from Costa Rica to Venezuela and Colombia*. G2. Summer.

Known as the 'dove orchid' because of a resemblance between the column and the lateral lobes of the lip and a dove with outstretched wings.

2. P. cerina Lindley. Illustration: *Edwards's Botanical Register* **23**: t. 1953 (1837); Hoehne, Flora Brasilica **12**(6): t. 119 (1942).

Pseudobulbs to 11 × 5 cm. Leaves to 45 cm, narrowly elliptic, acute. Racemes hanging, 10–15 cm, with 7–13 flowers, which are *c.* 3 cm in diameter. Sepals and petals yellow. Lip with the basal part smooth, the anterior part erect, somewhat hooded, the margins orange-yellow and crisped. Column unwinged. *Trinidad, N Brazil*. G2. Summer.

3. P. pendula Hooker. Illustration: *Botanical Magazine*, 3479 (1836).

Vegetatively similar to *P. cerina*. Racemes hanging, with 4–7 flowers, which are 4.5–5 cm in diameter. Sepals and petals white or pale yellow, spotted with red. Column with a narrow, downwardly directed wing from the apex on either side. *Tropical S America from Trinidad to N Brazil and N Peru*. G2. Spring.

171. LACAENA Lindley

J. Cullen

Epiphytes with simple, ovoid or spindle-shaped, grooved pseudobulbs, each bearing 2 leaves. Leaves stalked, elliptic, pleated, rolled when young. Racemes hanging, arising from the bases of the pseudobulbs, with 8–30 flowers. Flowers openly bell-shaped, not resupinate. Sepals and petals similar, incurved. Lip attached to the column foot, complex in shape, 3-lobed, with a callus between the short, erect, basal, lateral lobes; the central lobe hairless, clawed, ovate or oblong, its anterior part not freely movable on the rest. Pollinia 2, ellipsoid, stipe linear, longer than the pollinia, somewhat expanded above, the viscidium with an acute base.

A genus of 2 species from C America; both are grown in conditions the same as those for *Acineta* (p. 605).

Literature: Jenny, R., Die Gattung Lacaena Lindley, *Die Orchidee* **30**: 55–61 (1979).

1a. Flowers 4.5–5 cm in diameter, lip with a hairy callus and ovate central lobe **1. bicolor**
b. Flowers 2.5–3 cm in diameter, lip with a hairless callus and oblong central lobe **2. spectabilis**

1. L. bicolor Lindley. Illustration: *Botanical Magazine*, n.s., 330 (1959); Ospina & Dressler, Orquideas de las Americas, f. 155 (1974).

Pseudobulbs 6–8 cm, ovate. Leaves 20–25 cm, elliptic. Racemes with 10–30 flowers, fragrant, 4.5–5 cm in diameter. Sepals and petals greenish white, greenish yellow or pinkish, scurfy outside. Lip white with a hairy, purple callus and purple spots on the shortly stalked, ovate central lobe. *Mexico, Guatemala, Honduras*. G2. Spring–summer.

2. **L. spectabilis** (Klotzsch) Reichenbach. Illustration: *Botanical Magazine*, 6516 (1880).

Like *L. bicolor* but racemes with 8–12 flowers, 2.5–3 cm in diameter; petals and sepals pinkish white, lip with hairless callus and an oblong, spade-like, conspicuously stalked, purple central lobe. *Mexico*. G1. Spring.

172. LUEDDEMANNIA Reichenbach

J. Cullen

Like *Lacaena* but pseudobulbs with 3–5 leaves, the sepals and petals somewhat more spreading and the lip simpler,

3-lobed, with curving, obliquely triangular lateral lobes and a narrowly triangular, downy central lobe, bearing a humped, downy callus.

Cultivation as for *Acineta* (p. 605).

1. L. pescatorei (Lindley) Linden & Reichenbach. Illustration: *Botanical Magazine*, 7123 (1890).

Sepals broader than petals, brown and scaly-hairy outside, yellow striped with red inside, petals yellow, lip yellow with red spots. *Colombia*. G2. Summer.

173. HOULLETIA Brongniart

J. Cullen

Usually epiphytic with simple pseudobulbs to 6 cm, each bearing 1 leaf. Leaves stalked, elliptic, acute, pleated, rolled when young. Raceme erect or arching, arising from the base of the pseudobulb, with 4–10 non-resupinate flowers that are often fragrant, axis hairless. Sepals and petals widely spreading, similar or the petals slightly smaller. Lip complex, attached to the column foot, divided by a narrow waist into a clawed basal part that is narrowly oblong and bears on each side a narrow, pointed lateral lobe, which curves backwards and upwards, then forwards; anterior part stalkless or shortly stalked, oblong or ovate, arrowhead-shaped at the base with pointed lobes. Pollinia 2, irregularly ellipsoid, stipe linear, longer than pollinia, slightly expanded above, viscidium elongate, acute. Rostellum conspicuously beaked.

A genus of about 10 species from C & S America. Cultivation as for *Acineta* (p. 605) except that Houlletias are best grown in pots, as the large leaves make suspension difficult.

1a. Sepals and petals oblong, of uniform colour, dark red or dark brown, not spotted or blotched **3. odoratissima**
b. Sepals and petals ovate to elliptic, with conspicuous brown blotches on a yellow ground 2
2a. Raceme arching; lip whitish with violet spots; petals each with a large tooth on the inner margin **1. lansbergii**
b. Raceme erect; lip yellow with red or brown spots; petals not toothed **2. brocklehurstiana**

1. H. lansbergii Linden & Reichenbach. Illustration: *Botanical Magazine*, 7362 (1894); *Fieldiana* **26**: 525 (1953).

Pseudobulbs *c.* 6 cm, ovate. Leaves to 40 cm, elliptic. Raceme arching, with 5–10 flowers, each *c.* 8.5 cm in diameter. Sepals and petals ovate to elliptic, yellow with conspicuous brown spots, each petal with a single tooth on the upper margin. Lip whitish with violet spots. *Costa Rica*. G2. Autumn.

2. H. brocklehurstiana Lindley. Illustration: *Botanical Magazine*, 4072 (1844); Hoehne, Flora Brasilica **12**(6): t. 136 (1942).

Vegetatively similar to *H. lansbergii*. Raceme erect with 5–10 flowers, each *c.* 7 cm in diameter. Sepals and petals ovate to elliptic, yellow with brown spots, petals untoothed. Basal part of the lip yellow with red spots, the anterior part purple. *E Brazil*. G2. Autumn–winter.

H. picta Linden & Reichenbach. Illustration: *Botanical Magazine*, 6305 (1877). Similar, but the bases of the sepals and petals entirely brown and the anterior part of the lip yellow with very dark brown spots. *Colombia*.

H. wallisii Linden & Reichenbach. Illustration: Ospina & Dressler, Orquideas de las Americas, f. 154 (1974). Also similar but with the lateral sepals united and the petals rather narrow. *Colombia*.

3. H. odoratissima Lindley & Paxton. Illustration: *Lindenia* **7**: t. 324 (1891).

Vegetatively similar to *H. lansbergii*. Raceme erect with 6–9 flowers, *c.* 7 cm in diameter. Sepals and petals oblong, dark brown or dark red, petals entire. Lip mostly white. *Colombia*. G2. Autumn.

A variant with red sepals and petals has been called var. **antioquiensis** Anon.

174. POLYCYCNIS Reichenbach

J. Cullen

Like *Houlletia* but plants not so large, the axis of the raceme downy, lip with the basal part somewhat clawed and bearing upcurving, oblong-triangular lateral lobes, the central lobe heart-shaped. Column slender, arching, abruptly widened at the apex.

Cultivation as for *Acineta* (p. 605).

1. P. barbata Reichenbach. Illustration: *Botanical Magazine*, 4479 (1849).

Raceme axis and flower-stalks purple, downy. Sepals and petals 2.5–2.7 cm, yellowish flushed with red and with brown-purple spots, lip similar in colour, prominently but sparsely hairy. *Costa Rica, Colombia*. G2. Winter.

175. PAPHINIA Lindley

J. Cullen

Epiphytes with rather small, simple pseudobulbs, each usually bearing 2 leaves. Leaves pleated, rolled when young. Racemes hanging, arising from the bases of the pseudobulbs, each with 1–5 flowers, which are resupinate and 8–14 cm in diameter. Sepals and petals widely spreading, the petals somewhat smaller than the sepals. Lip complex, 3-lobed, smaller than the petals, the claw and basal part ridged and crested, bearing the erect, forwardly directed lateral lobes; the central lobe movable on the rest, diamond-shaped, bearing a tuft of club-shaped, hair-like processes on the margin towards and at the apex. Column narrow with small wings or teeth towards the apex. Pollinia 2, ellipsoid, their surfaces roughened, stipe linear-oblong with the sides curved upwards near the top, viscidium very small. Rostellum conspicuously beaked.

A genus of about 7 species from C & S America. They require warm, humid conditions when growing and infrequent watering when resting. They are best grown in epiphytic compost in small pans, and given some shade during the summer.

Literature: Jenny, R., Die Gattung Paphinia Lindley, *Die Orchidee* **29**: 207–15 (1978).

1a. Flowers to 8 cm in diameter; lip red with white crests, ridges and hair-like processes **1. cristata**
b. Flowers 12–14 cm in diameter; lip dark purple, white towards the base, hair-like processes white **2. grandiflora**

1. P. cristata Lindley. Illustration: *Fieldiana* **26**: 527 (1952); Hoehne, Flora Brasilica **12**(7): t. 3 (1953).

Pseudobulbs *c.* 4 cm, somewhat compressed. Leaves 10–15 cm, lanceolate, acute. Flowers to 8 cm in diameter. Sepals and petals translucent white with brown or red stripes or spots. Lip red with white ridges, crests and hair-like processes. *Tropical S America, from Colombia and Trinidad to N Brazil*. G2. Autumn.

2. P. grandiflora Rodriguez. Illustration: Hoehne, Flora Brasilica **12**(7): t. 1 (1953).

Pseudobulbs to 5 cm. Leaves 20–25 cm, lanceolate to elliptic. Flowers 12–14 cm in diameter. Sepals and petals white, striped and spotted with red-purple, especially towards the apex. Lip dark purple, white towards the base and with white hair-like processes. *NW Brazil*. G2. Autumn.

The flowers are said to produce a strong, rather unpleasant smell.

176. STANHOPEA Hooker
J. Cullen

Epiphytic with small, simple, 1-leaved pseudobulbs. Leaves large, leathery, elliptic, stalked, pleated, rolled when young. Racemes arising from the bases of the pseudobulbs, hanging, with 2–10 flowers that have prominent, coloured bracts. Flowers usually fragrant, not resupinate. Sepals and petals similar, or petals sometimes a little smaller, all spreading or reflexed, the lateral sepals joined at the base. Lip complex, very fleshy, usually clearly divided into 3 parts, the basal part hollowed or saccate, the middle usually bearing a horn-like lateral lobe on either side, the interior part simple or 3-toothed or 3-lobed at the apex, jointed to the median part; more rarely the whole lip forming an irregular pouch. Column arched, slender, often winged towards the apex. Pollinia 2, ellipsoid, stipe linear, longer than pollinia, somewhat expanded above, its sides turned upwards, viscidium elongate.

A genus of from 25 to 50 species whose classification is chaotic. Most of the species are poorly known, variable and difficult to distinguish. About 10 species are reputedly in cultivation; they should be grown in the same manner as *Acineta* (p. 605).

Literature: Dodson, C.H. & Frymire, G.F., Preliminary studies in the genus Stanhopea, *Annals of the Missouri Botanical Garden* **48**: 137–72 (1961); Dodson, C.H., Clarification of some nomenclature in Stanhopea, *Selbyana* **1**: 46–55 (1975); Greer, B., The astonishing Stanhopeas (1998).

1a. Lip not divided into 3 parts, forming an irregularly margined pouch **1. ecornuta**
b. Lip clearly divided into 3 parts, not as above 2
2a. Lateral lobes (horns) borne on the basal part of lip **2. grandiflora**
b. Horns borne on the middle part of lip 3
3a. Horns less than 1 cm **3. lewisae**
b. Horns 2–3 cm 4
4a. Basal part of lip globular or saccate, its hollow centre linked to the middle part by a short groove; anterior part usually 3-toothed or 3-lobed **4. insignis**
b. Basal part of the lip oblong or boat-shaped, hollowed only near its base, the hollow linked to the middle part by an elongate groove; anterior part usually entire **5. oculata**

1. S. ecornuta Lemaire. Illustration: *Botanical Magazine*, 4885 (1855); *Fieldiana* **26**: 530 (1952); *Annals of the Missouri Botanical Garden* **48**: 145 (1961); *Die Orchidee* **27**: 173 (1976).

Leaves 35–40 cm. Racemes with 1–3 flowers. Sepals and petals creamy white, the petals usually spotted with purple towards the base, sepals 4.5–6.5 cm. Lip simple, mostly yellow, forming an irregularly margined pouch, without obvious lateral lobes. *C America, from Guatemala to Costa Rica*. G2. Summer.

2. S. grandiflora (Loddiges) Lindley not Reichenbach (*S. eburnea* Lindley). Illustration: *Botanical Magazine*, 3359 (1834); Hoehne, Flora Brasilica **12**(6): t. 117 (1942); *Annals of the Missouri Botanical Garden* **48**: 153 (1961); Couret, Orquideas Venezolanas, 28 (1977).

Leaves to 30 cm. Raceme with 2–4 flowers, stalks and ovaries with sparse, short, blackish hairs. Sepals *c.* 8 cm, white. Petals white, sometimes spotted with violet around their margins, violet at the base. Lip clearly 3-partite, the basal part purple-spotted, hollowed and bearing small, horn-like lateral lobes, the middle part solid, oblong, whitish spotted and streaked with purple, the anterior part ovate, acute, white. *Tropical S America from Trinidad to N Brazil*. G2. Summer.

3. S. lewisae Ames & Correll. Illustration: *Fieldiana* **26**: 533 (1952); *Annals of the Missouri Botanical Garden* **48**: 155 (1961).

Leaves 40–50 cm. Racemes with 3–5 flowers. Sepals and petals creamy white with reddish or purple spots, sepals 5–5.7 cm. Lip with a short, hollow, deep yellow, purple-spotted basal part, the middle part bearing horns that are less than 1 cm long on each side, the anterior part broadly ovate, obtuse, white densely spotted with red. *Guatemala*. G2. Summer.

A relatively recently described species whose distribution in cultivation is uncertain. It is possibly a naturally occurring hybrid between *S. ecornuta* and *S. jenishiana* (see Oestereich, *Orchidata* **7**: 209–10, 1969).

4. S. insignis Hooker. Illustration: *Botanical Magazine*, 2948, 2949 (1829); *Annals of the Missouri Botanical Garden* **48**: 158 (1961).

Leaves 20–30 cm. Racemes with 2–5 flowers. Sepals and petals yellowish white spotted with red or purple, the sepals *c.* 7 cm. Basal part of the lip spherical, saccate, purple, linked to the yellow, red-spotted, middle part by a very short groove; horns 2–3 cm, borne on the middle part; anterior part broadly ovate, weakly 3-lobed at the apex, yellowish white spotted with red. Column broadly winged. *SE Brazil*. G2. Autumn.

S. insignis is the best known of a complex of species that are very difficult to distinguish, one from another. Three other species of this complex are reputedly in cultivation:

S. hernandezii (Knuth) Schlechter (*S. devoniensis* Lindley). Sepals 5.5–6.5 cm, yellow or orange-yellow spotted with red; lip white spotted with purple, the anterior part broadly ovate with 3 equal, blunt teeth; column unwinged or very narrowly winged. *Guatemala*. G2. Autumn.

Easily recognised, looking like a smaller version of *S. tigrina*.

S. saccata Bateman. Illustration: *Fieldiana* **26**: 535 (1952); *Annals of the Missouri Botanical Garden* **48**: 161 (1961). Sepals 5–5.5 cm, greenish white to cream spotted with red or brown; lip orange-yellow spotted with red or brown, the anterior part 3-lobed at the apex, the central lobe shorter than the laterals; column slightly winged. *Guatemala, El Salvador*. G2. Autumn.

S. tigrina Lindley. Illustration: *Botanical Magazine*, 4197 (1845); *Annals of the Missouri Botanical Garden* **48**:159 (1961); *Die Orchidee* **31**: clxv (1980); Bechtel et al., Manual of cultivated orchid species, 275 (1981). Sepals *c.* 9 cm, yellowish white with confluent red-purple spots; lip yellow with violet spots, the anterior part broadly ovate with 3 equal, acute teeth; column very broadly winged. *S Mexico*. G2. Autumn.

Very distinct, with the largest flowers in the genus.

5. S. oculata (Loddiges) Lindley. Illustration: *Botanical Magazine*, 5300 (1862); Hoehne, Flora Brasilica **12**(6): t. 115 (1942); Sheehan & Sheehan, Orchid genera illustrated, 161 (1979); Bechtel et al., Manual of cultivated orchid species, 275 (1981).

Leaves 30–45 cm. Racemes with 4–10 flowers. Sepals 5.5–7 cm, sepals and petals variable in colour, white, yellow or orange, usually red-spotted. Basal part of the lip oblong when viewed from the side, hollowed only near its base, where the sides are notched, linked to the middle part by an elongate groove; middle part bearing horns that are 2–3 cm long; anterior part broadly ovate, acute or obtuse, not usually 3-lobed. Column winged. *Tropical America from S Mexico to Venezuela and N Peru.* G2. Summer–autumn.

As with *S. insignis*, this is the best known of a complex of species, of which 2 more (and perhaps others) are in cultivation:

S. jenishiana Reichenbach (*S. bucephalus* misapplied; *S. graveolens* misapplied; *S. grandiflora* Reichenbach, not (Loddiges) Lindley). Illustration: *Botanical Magazine*, 5278 (1861) & 8517 (1913); *Annals of the Missouri Botanical Garden* **48**: 165 (1961). Sepals and petals orange-yellow, sparsely purple-spotted, sepals *c.* 7 cm; basal part of lip boat-shaped when viewed from the side, its sides not notched; column winged. *Ecuador to C Peru.* G2. Summer.

S. wardii Loddiges. Illustration: *Botanical Magazine*, 5289 (1862); *Fieldiana* **26**:537 (1952) & **48**:167 (1961); Couret, Orquideas Venezolanas, 29–31 (1977). Sepals and petals orange-yellow with red-purple spots near the base, sepals *c.* 7 cm; basal part of the lip oblong when viewed from the side, its sides not notched; column very broadly winged. *Tropical America, from S Mexico to Venezuela & Peru.* G2. Summer.

177. GONGORA Ruiz & Pavon

J. Cullen

Epiphytic with simple, ovoid, grooved pseudobulbs, each usually with 2 leaves. Leaves elliptic to narrowly elliptic, stalked or not, pleated, rolled when young. Racemes hanging, arising from the bases of the pseudobulbs, with many flowers. Flowers often fragrant, resupinate. Sepals spreading, the upper attached to the column for about half its length or more. Petals smaller than the sepals, attached to the side of the column for part of their length. Lip borne on the column foot, complex, fleshy, strongly keeled, divided into 2 parts, the basal part often bearing narrow lateral lobes (horns), the anterior part forwardly directed, usually acuminate. Anther at the apex of the column. Pollinia 2, elongate-ovoid, stipe linear, longer than the pollinia, viscidium narrow. Rostellum prominently beaked.

A genus of about 25 species from Tropical America. A few of them are cultivated in the same manner as *Acineta* (p. 605) for the sake of their bizarre flowers.

1a. Lip without horns, its apex forming an upcurved hook; apex of petals truncate, the angles obtuse or prolonged into curved points **1. galeata**
b. Lip with distinct horns on the basal part, apex acute or acuminate, not upcurved; petals acute 2
2a. Petals very small, less than half the length of the column **2. truncata**
b. Petals more than half the length of the column 3
3a. Sepals yellow with red spots or mostly red; basal part of the lip with 2 spreading lobes at the base as well as 2 horns at its apex **3. quinquenervis**
b. Sepals mostly purple, made up of confluent purple spots on a white or greenish ground; basal part of lip with 2 horns at its apex but without lobes at the base **4. bufonia**

1. G. galeata (Lindley) Reichenbach (*Maxillaria galeata* Lindley; *Acropera loddigesii* Lindley). Illustration: *Botanical Magazine*, 3653 (1834); Bechtel et al., Manual of cultivated orchid species, 215 (1981).

Pseudobulbs *c.* 6 cm, ovoid. Leaves 25–35 cm. Sepals to 2.2 cm, ovate, concave, brownish yellow. Petals brown, joined to the column only at the base, truncate at the apex, the angles obtuse or prolonged into curved points. Lip yellow spotted with red, hornless, the apex prolonged into an upwardly curving, hook-like point. *Mexico.* G2. Summer.

2. G. truncata Lindley. Illustration: *Edwards's Botanical Register* **31**: t. 56 (1845); *Addisonia* **2**: t. 46 (1917).

Pseudobulbs 6 cm or more. Leaves to 30 cm. Sepals to 2.3 cm, ovate, pale yellow spotted with purple-brown. Petals very small, less than half the length of the column, acute. Lip yellow, unspotted, the basal part with 2 lobes at its base and 2 upcurving horns at its apex, the anterior part curved, abruptly acute. *S Mexico.* G2. Summer.

3. G. quinquenervis Ruiz & Pavon (*G. maculate* Lindley). Illustration: *Botanical Magazine*, 3687 (1835); *Lindenia* **5**: t. 208 (1889); *Die Orchidee* **30**: cxxiii–cxxiv (1979).

Pseudobulbs 5–12 cm. Leaves 20–60 cm. Sepals 1.5–2.4 cm, yellow with red spots or red, occasionally yellow with whitish spots, lanceolate, acuminate. Petals purplish, acute, attached to the column for about half their length. Lip mostly yellow with 2 spreading lobes near its base and 2 upcurving horns at the apex of the basal part, the anterior part very strongly keeled, acuminate. *C & S America.* G2. Summer.

4. G. bufonia Lindley. Illustration: *Edwards's Botanical Register* **26**: t. 2 (1841); Hoehne, Flora Brasilica **12**(6): t. 129 (1942).

Pseudobulbs 6 cm or more. Leaves to 30 cm, pale whitish green. Sepals *c.* 2.3 cm, mostly purple, the colour made up of confluent purple blotches on a white or greenish ground. Petals similar in colour to the sepals, attached to the column for about half their length, acute. Lip purple, the basal part with 2 upcurving horns at its apex, the anterior part keeled, acuminate. *S & E Brazil.* G2. Spring–summer.

178. CIRRHAEA Lindley

J. Cullen

Like *Gongora* but pseudobulbs each with 1 leaf, upper sepal and petals free from the back of the column, the lip 3-lobed, the lateral lobes directed backwards, the central lobe narrowly triangular, directed forwards, the whole with a narrow arrowhead shape. Anther on the back of the column.

Cultivation as for *Acineta* (p. 605).

1. C. dependens Reichenbach (*C. tristis* Lindley). Illustration: Hoehne, Iconografia de orchidaceas do Brasil, t. 191 (1949).

Sepals 2–2.5 cm, greenish with purple spots or flushed with purple, petals smaller, similar in colour. Lip violet. *Brazil.* G2. Summer.

179. CORYANTHES Hooker

J. Cullen

Epiphytic, with small, simple, grooved, ovoid or almost spindle-shaped pseudobulbs, each bearing 1–2 leaves. Leaves lanceolate, pleated, rolled when young. Racemes hanging, arising from the bases of the pseudobulbs, with 2–6 flowers. Flowers fragrant. Upper sepal reflexed, smaller than the 2 lateral sepals, which are spreading with their apices folded over backwards. Petals erect, smaller than the sepals, twisted or with wavy margins. Lip extremely complex, divided into 3 parts. The basal part, which consists of a stalk borne at an acute angle to the ovary terminating in a hollow, cup-like structure whose axis is at right angles to the stalk, is confluent with the column foot and bears 2 white, peg-like, nectaries near its base. The middle part, which is narrow and keeled, arises from the margin of the cup. The anterior part is bucket-like, hollowed out and with 3–5 lobes towards the apex, which lies close to the apex of the column. Column slender, its apex reflexed at right angles to the rest. Pollinia 2, ellipsoid, folded backwards along the surface of the curved stipe but arching away from it, viscidium narrowly triangular, almost as long as the stipe.

A genus of about 12 species from C America and the northern part of S America with striking, if bizarre, flowers that are difficult to describe; study of the illustrations cited is helpful in understanding the structure of the lip and column and their spatial relations. Nectar drips from the glands on the basal part of the lip and fills up the bucket-like anterior part, from where it overflows via the apical lobes, which form a spout.

Cultivation as for *Acineta* (p. 605); the plants should not be allowed to dry out completely during the resting period.

Literature: Kennedy, G., Some members of the genus Coryanthes, *Orchid Digest* **42**: 31–7 (1978).

1a. The sides of the middle part of the lip corrugated and ridged; lateral sepals 12–13 cm **1. macrantha**
b. The sides of the middle part of the lip smooth; lateral sepals to 9 cm 2
2a. Basal and middle parts of the lip finely downy outside; anterior part of the lip yellowish brown inside and out **2. speciosa**
b. Basal and middle parts of the lip not downy outside; anterior part of the lip yellow marbled with purple, at least within **3. maculata**

1. C. macrantha (Hooker) Hooker. Illustration: *Botanical Magazine*, 7692 (1900); Dunsterville & Garay, Venezuelan orchids illustrated **3**: 63 (1965); Couret, Orquideas Venezolanas, 35 (1977); *Orchid Digest* **42**:35 (1978).

Pseudobulbs to 12 cm, narrow. Leaves to 30 cm. Raceme usually with 2 flowers. Lateral sepals 12–13 cm, yellow with many elongate red spots. Petals with wavy margins, reddish yellow, red-spotted at the base. Basal part of lip with a purplish stalk, *c.* 2.5 cm, the cup orange-red; middle part yellow with red spots or entirely red, its sides ridged and corrugated; anterior part yellow, densely red-spotted. *Tropical S America, from E Peru to Trinidad.* G2. Spring.

2. C. speciosa Hooker. Illustration: Hoehne, Flora Brasilica **12**(6): t. 121, 122 (1942); *Fieldiana* **26**: 543 (1953); Couret, Orquideas Venezolanas, 34 (1977); *Orchid Digest* **42**: 35 (1978).

Pseudobulbs 7–15 cm. Leaves 35–55 cm. Raceme with 2–5 flowers. Sepals yellow, the laterals 6–8 cm. Petals yellow, somewhat twisted and with wavy margins. Basal part of the lip with a brownish stalk 1–1.2 cm, and a brownish cup, downy outside; middle part brown, its sides smooth, downy; anterior part pale yellowish brown. *Tropical America, from Guatemala to Peru & E Brazil.* G2. Spring?

3. C. maculata Hooker. Illustration: *Botanical Magazine*, 3102 (1831) & 3747 (1845); *Orchid Digest* **42**:32 (1978).

Pseudobulbs to 12 cm. Leaves to 30 cm. Raceme with 3–6 flowers. Sepals *c.* 9 cm, yellowish. Petals yellowish, margins wavy. Basal part of the lip with a yellowish or purplish stalk *c.* 1 cm, the cup yellow, hairless outside; middle part yellow or purplish, its sides smooth, hairless; anterior part yellow outside, yellow marbled with purple or rarely entirely purple within. *Guyana.* G2. Spring–summer.

GLOSSARY

abscission-zone. A predetermined layer at which leaves or fern-fronds break off.

achene. A small, dry, indehiscent, 1-seeded fruit, in which the fruit-wall is of membranous consistency and free from the seed.

aciculus. see *Nomocharis* Vol. 1, p. 102.

actinomorphic. Radially symmetric

acuminate. With a long, slender point.

adnate. Attached to an organ of a different kind (e.g. stamens adnate to corolla-tube)

adpressed. Closely applied to a leaf or stem and lying parallel to its surface, but not adherent to it.

adventitious. (1) Of roots: arising from a stem or leaf, not from the primary root derived from the radicle of the seedling. (2) Of buds: arising somewhere other than in the axil of a leaf.

aggregate fruit. A collection of small fruits, each derived from a single free carpel, closely associated on a common receptacle, but not united. *Ranunculus* and *Rubus* provide familiar examples.

allopolyploidy. Polyploidy in a plant of hybrid origin.

alternate. Arising singly, 1 at each node; not opposite or whorled (figure G1(2), p. 612).

anastomosing. Describes veins of leaves which rejoin after branching from each other or from the main vein or midrib.

anatropous. Describes an ovule which turns through 180° in the course of development, so that the micropyle is near the base of the funicle (figure G4(2), p. 615).

annual. A plant which completes its life-cycle from seed to seed in less than 1 year.

anther. The uppermost part of a stamen, containing the pollen (figure G3(3 & 4), p. 614).

apetalous. Describes a flower without a corolla (petals).

apical. Describes the attachment of an ovule to the apex of a 1-celled ovary (figure G4 (8), p. 615).

apiculate. With a small point.

apomictic. Reproducing by asexual means, though often by the agency of seeds, which are produced without the usual sexual nuclear fusion.

arachnoid. Describes hairs which are soft, long and entangled, suggestive of cobwebs.

areole. See Cactaceae, Vol. 2, p. 209.

aril. An outgrowth from the region of the hilum, which partly or wholly envelops the seed; it is usually fleshy.

ascending. Prostrate for a short distance at the base, but then curving upwards so that the remainder is more or less erect; sometimes used less precisely to mean pointing obliquely upwards.

attenuate. Drawn out to a fine point.

auricle. A lobe, normally 1 of a pair, at the base of the blade of a leaf, bract, sepal or petal.

awn. A slender but stiff bristle on a sepal or fruit.

axil. The upper angle between a leaf-base or leaf-stalk and the stem that bears it (figure G1(1), p. 612).

axile. A form of placentation in which the cavity of the ovary is divided by septa into 2 or more cells, the placentas being situated on the central axis (figure G4 (10), p. 615).

axillary. Situated in or arising from an axil (figure G1(1), p. 612).

back-cross. A cross between a hybrid and a plant similar to one of its parents.

basal. (1) Of leaves: arising from the stem at or very close to its base. (2) Of placentation: describes the attachment of an ovule to the base of a 1-celled ovary (figure G4(6 & 7), p. 615).

basifixed. Attached to its stalk or supporting organ by its base, not by its back (figure G3(3), p. 614).

berry. A fleshy fruit containing 1 or more seeds embedded in pulp, as in the genera *Berberis, Ribes* and *Phoenix.* Many fruits (such as those of *Ilex*) which look like berries and are usually so called in popular speech, are, in fact, drupes.

biennial. A plant which completes its lifecycle from seed to seed in a period of more than 1 year but less than 2.

bifid. Forked; divided into 2 lobes or points at the tip.

bilaterally symmetric. Capable of division into similar halves along 1 plane and 1 only (figure G3(9), p. 614).

bipinnate. Of a leaf: with the blade divided pinnately into separate leaflets which are themselves pinnately divided (figure G1 (18), p. 612).

blade. A broadened part, furthest from the base, of a petal, corolla or similar organ, which has a relatively narrow basal part – the claw or tube (figure G3(1), p. 614).

bract. A leaf-like or chaffy organ bearing flower in its axil or forming part of an inflorescence, differing from a foliage-leaf in size, shape, consistency or colour (figure G2(2), p. 613).

bracteole. A small, bract-like organ which occurs on the flower-stalk, above the bract, in some plants.

bulb. A seasonally dormant underground bud, usually fairly large, consisting of a number of fleshy leaves or leaf-bases.

bulbil A small bulb, especially one borne in a leaf-axil or in an inflorescence.

bulblet. A small bulb developing from a larger one.

caducous. Falling early.

caespitose. Forming a tuft.

calyptra. Applied to any cap-like covering.

calyx. The sepals; the outer whorl of a perianth (figure G3(1), p. 614).

campanulate. Bell-shaped; a broad tube terminating in a flared limbs or lobes.

campylotropous. Describes an ovule which becomes curved during development and lies with its long axis at right angles to the funicle (figure G4(4), p. 615).

capitate. Compact and approximately spherical, head-like.

capitellate. Grouped into very small, dense or compact clusters.

capitulum. An inflorescence consisting of small flowers (florets), usually numerous, closely grouped together so as to form a 'head', and often provided with an involucre.

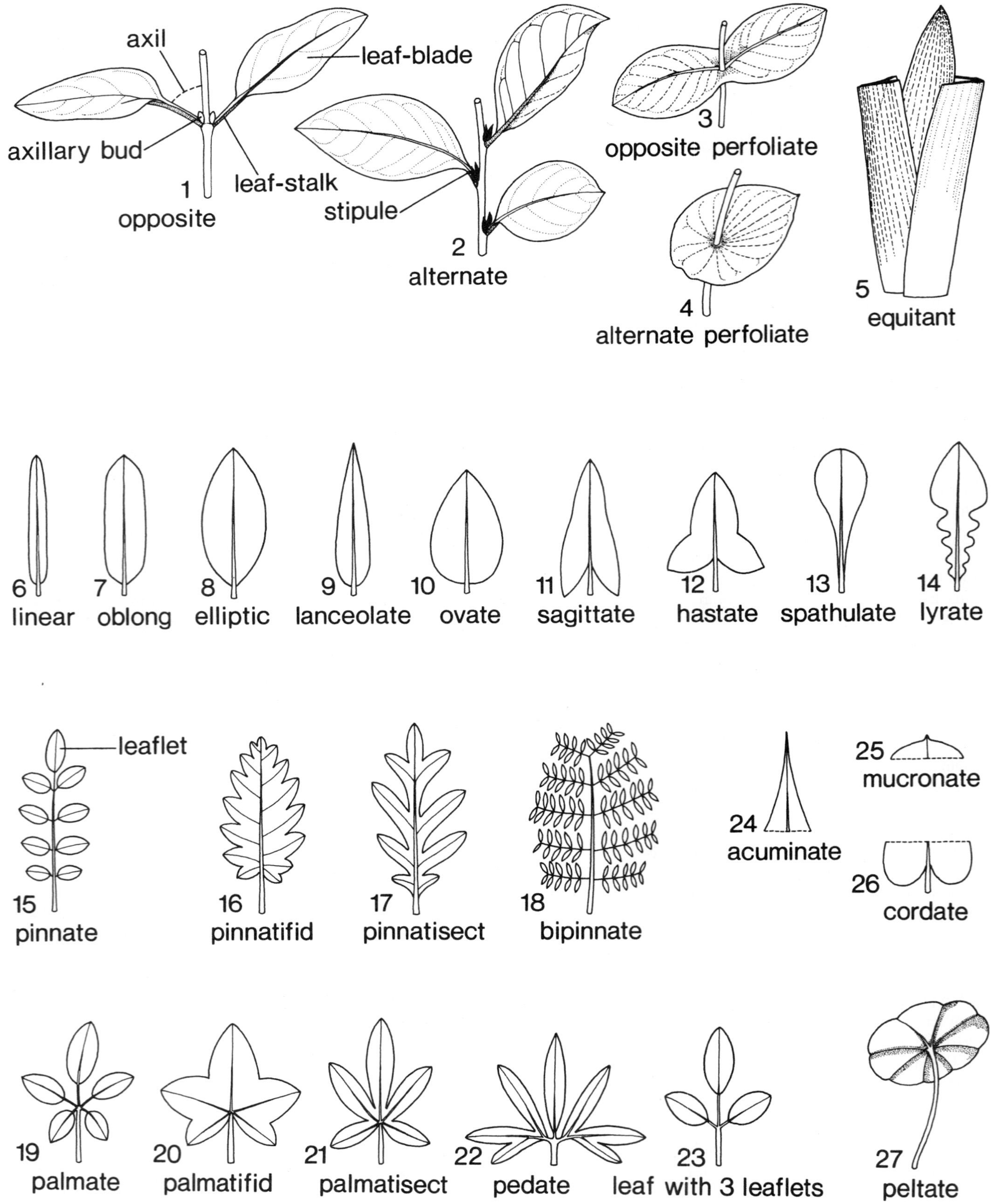

Figure G1. Leaves. 1–5, leaf insertion types. 6–14, leaf-blade outlines. 15–23, Leaf dissection types. 24–26, Leaf apex and base shapes. 27, Attachment of leaf-stalk to leaf-blade.

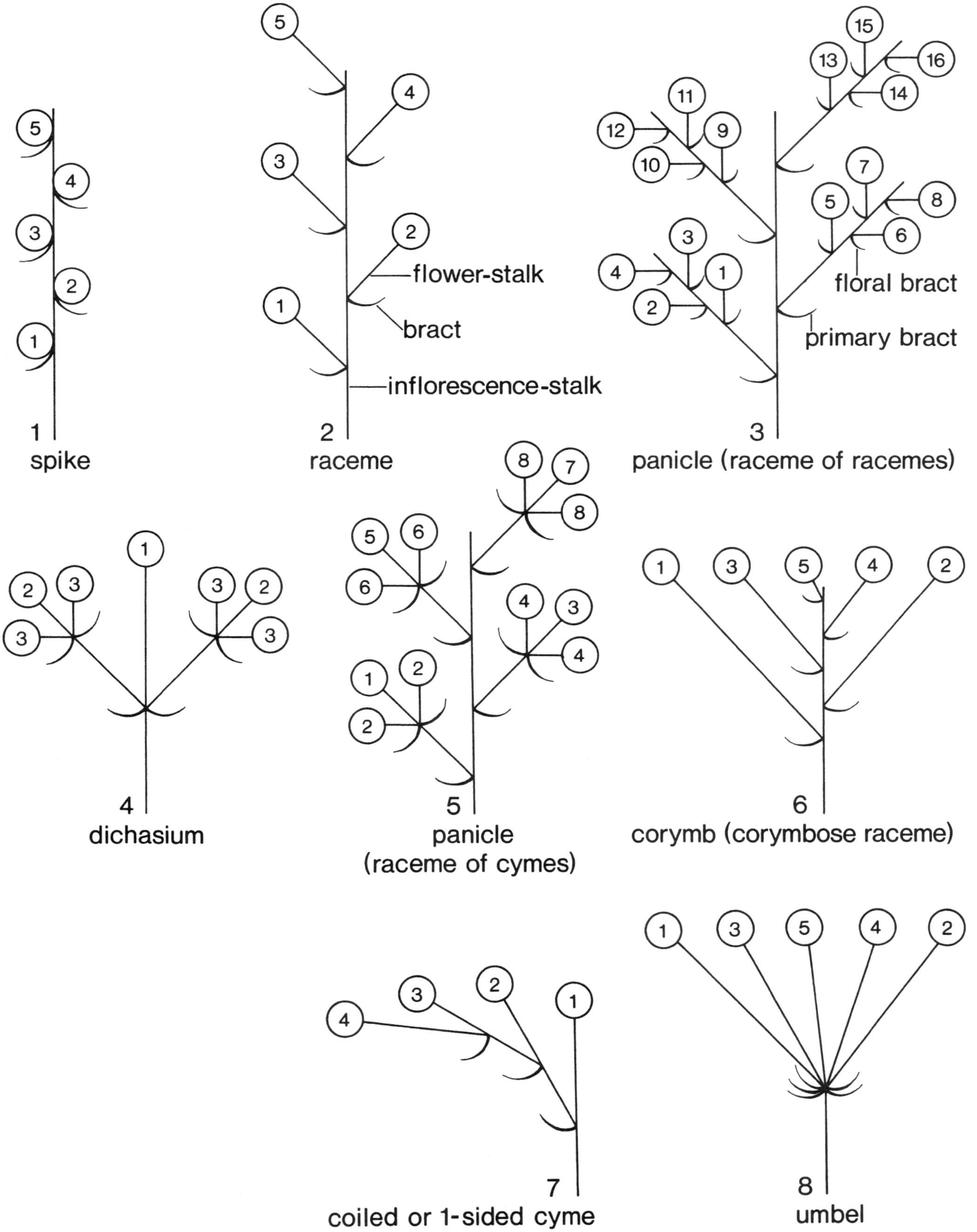

Figure G2. Inflorescences.

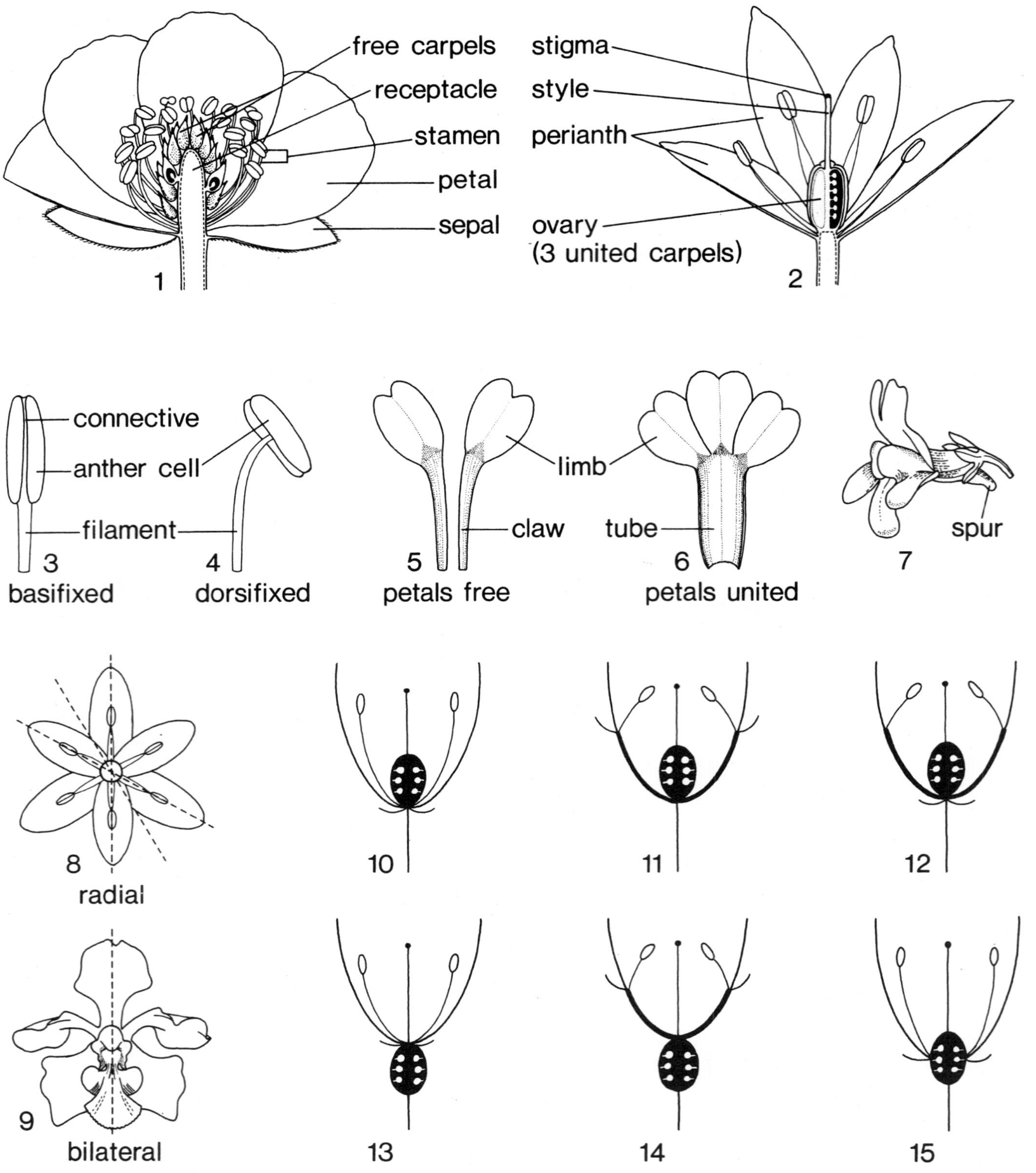

Figure G3. 1, 2, Two flowers illustrating floral parts. 3, 4, Two stamens showing alternative types of anther attachment. 5–7, Some terms relating to petals. 8, 9, Floral symmetry, planes of symmetry shown by broken lines. 10–15, Position of ovary. 10–12, Superior ovaries. 13, 14, Inferior ovaries. 15, Half-inferior ovary. 11, Perigynous zone bearing sepals, petals and stamens. 12, Perigynous zone bearing petals and stamens. 14, Epigynous zone bearing sepals, petals and stamens.

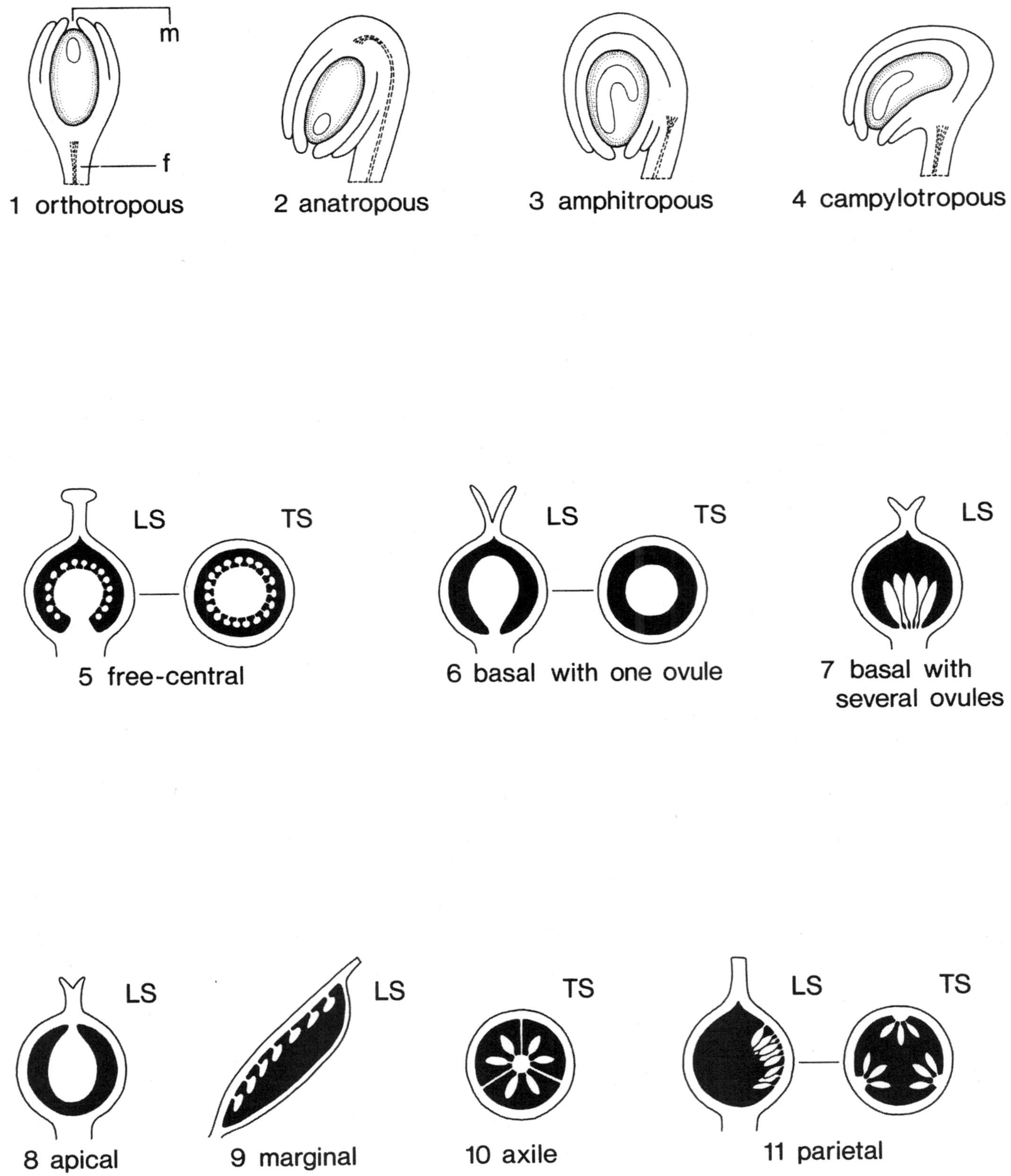

Figure G4. Ovules and placentation. 1–4, Ovule forms (f, funicle; m, micropyle). 5–11, Placentation types (LS, longitudinal section; TS, transverse section).

capsule. A dry, dehiscent fruit derived from 2 or more united carpels and usually containing numerous seeds.
carpel. One of the units (sometimes interpreted as modified leaves) situated in the centre of a flower and together constituting the gynaecium or female part of the flower. If more than 1, they may be free or united. They contain ovules and bear a stigma (figure G3(1 & 2), p. 614).
carpophore. See *Silene*, (Vol. 2, p. 184).
caruncle. A soft, usually oil-rich appendage attached to a seed near the hilum.
catkin. An inflorescence of unisexual flowers, made up of relatively conspicuous, usually overlapping bracts, each of which subtends a small apetalous flower or a group of such flowers; catkins are generally pendent, but some are erect.
cephalium. See Cactaceae, vol. 2, p. 209.
chromosome. One of the small, thread-like or rod-like bodies consisting of nucleic acid and containing the genes, which appear in a cell nucleus shortly before cell division.
cilate. Fringed on the margin with usually fine hairs (cilia).
circinate. Coiled at the tip, so as to resemble a crozier.
cladode. A branch which takes on the functions of a leaf (the leaves being usually vestigial). It may be flattened, as in *Ruscus* (vol. 1, p. 168), or needle-like, as in many species of *Asparagus* (vol. 1, p. 164).
clavate. Club-shaped, thickened towards the apex.
claw. The narrow base of a petal or sepal, which widens above into the limb or blade (figure G3(5), p. 614).
cleistogamous. Describes a flower with reduced corolla, which does not open but sets seed by self-pollination.
clone. The sum-total of the plants derived from the vegetative reproduction of an individual, all having the same genetic constitution.
column. (1) A solid structure in the centre of a flower, consisting of the style and stigmas united to the stamen or stamens. It is characteristic of the Orchidaceae (Vol 1. figures 22–23, p. 444, 445). (2) see Vol. 1, p. 339 (Graminae: *Stipa*).
compound. (1) Of a leaf: divided into separate leaflets. (2) Of an inflorescence: bearing secondary inflorescences in place of single flowers. (3) Of a fruit: derived from more than 1 flower.
compressed. Flattened from side to side.
comus. A tuft of infertile flowers or bracts sometimes found at the top of a dense inflorescence.
conduplicate. Of young leaves or leaflets in bud: folded upwards along the midrib.
cone. A compact, cylindrical or shortly conical inflorescence or fruiting inflorescence with closely overlapping bracts or scales.
connate. Attached to other organs of the same kind (e.g. petals connate).
connective. The tissue which separates the 2 lobes of an anther, and to which the filament is attached (figure G3(3), p. 614).
cordate. Describes the base of a leaf-blade which has a rounded lobe on either side of the central sinus (figure G1(26), p. 612).
corm. An underground, thickened stem-base, often surrounded by papery leaf-bases, and superficially resembling a bulb.
corolla. The petals; the inner whorl of the perianth (figure G3(1), p. 614).
corona. (1) A tubular or ring-like structure attached to the inside of the perianth (or perigynous or epigynous zone), either external to the stamens or united with their filaments; it is usually lobed or dissected (figure 163, Vol. 4, p. 495). (2) A ring-like structure or circlet of appendages on the outside of a tube formed by united filaments.
coronal scale. See *Silene*, vol. 2, p. 184
corymb. A broad, flat-topped inflorescence. In the strict sense the term indicates a raceme in which the lowest flowers have stalks long enough to bring them to the level of the upper ones (figure G2(6), p. 613), but the term *corymbose* is often used to indicate a flat-topped cymose inflorescence.
costapalmate. See Vol. 1, p. 361 (Palmae).
cotyledon. One of the leaves preformed in the seed.
crisped. (1) Of hairs: strongly curved, so that the tip lies near the point of attachment. (2) Of leaves, leaflets or petals: finely and complexly wavy.
cristate. With elevated, irregular ridges.
crownshaft. See Vol. 1, p. 361 (Palmae).
culm. See Vol. 1, p. 324 (Gramineae).
cultivar. A variant of horticultural interest or value, maintained in cultivation, and not conveniently equatable with an infraspecific category in botanical classification. A cultivar may arise in cultivation or may be brought in from the wild. Its distinguishing name can be Latin in form, e.g. 'Alba', but is more usually in a modern language, e.g. 'Madame Lemoine', 'Frühlingsgold', 'Beauty of Bath'.
cuneate. Wedge-shaped.
cupule. A group of bracts, united at least at the base, surrounding the base of a fruit or a group of fruits.
cyme. An inflorescence in which the terminal flower opens first, other flowers being borne on branches arising below it (figure G2(5 & 7), p. 613).
cystolith. A concretion of calcium carbonate found within the cells of the leaf in some plants; they can sometimesbe seen when the leaf is viewed against the light, or felt as tiny hard lumps when the leaf is drawn between finger and thumb.
decumbent. More or less horizontal for most of its length, but erect or semi-erect near the tip.
decurrent. Continued down the stem below the point of attachment as a ridge or ridges.
dehiscent. Splitting, when ripe, along 1 or more predetermined lines of weakness.
dendroid. Branched, tree-like.
dichasial. Resembling a dichasium.
dichasium. A form of cyme in which each node bears 2 equal lateral branches (figure G2(4), p. 613).
dichotomous. Dividing into 2 equal branches; regularly forked.
dimorphic. Of two forms.
dioecious. With male and female flowers or cones on separate plants.
diploid. Possessing in its normal vegetative cells 2 similar sets of chromosomes.
disc. A variously contoured, ring-shaped or circular area (sometimes lobed) within a flower, from which nectar is secreted.
dissected. Deeply divided into lobes or segments.
distichous. Alternate, but arranged on opposite sides of the stem, not spiral.
distylic. Having the flowers of different plants either with long styles and shorter stamens or with long stamens and shorter styles.
domatia. Pockets or dense tufts of hairs in the axils of the main veins on the underside of a leaf.
dorsifixed. Attached to its stalk or supporting organ by its back, usually near the middle (figure G3(4), p. 614).
double. Of flowers: with petals much more numerous than in the normal wild state.
drupe. An indehiscent fruit in which the outer part of the wall is soft and usually fleshy, but the inner part stony. A drupe

may be 1-seeded as in *Prunus* or *Juglans*, or may contain several seeds, as in *Ilex*. In the latter case each seed is enclosed in a separate stony endocarp and constitutes a pyrene.

drupelet. A miniature drupe forming part of an aggregate fruit.

ellipsoid. As *elliptic* but applied to a solid body.

elliptic. About twice as long as broad, tapering equally both to the tip and the base (figure G1(8), p. 612).

emarginate. Notched.

embryo. The part of a seed from which the new plant develops; it is distinct from the endosperm and seed-coat.

endocarp. The inner, often stony layer of a fruit-wall in those fruits in which the wall is distinctly 3-layered.

endosperm. A food-storage tissue found in many seeds, but not in all, distinct from the embryo and serving to nourish it and the young seedling during germination and establishment.

entire. With a smooth, uninterrupted margin; not lobed or toothed.

epicalyx. A group of bracts attached to the flower-stalk immediately below the calyx and sometimes partly united with it.

epigeal. The mode of germination in which the cotyledons appear above ground and carry on photosynthesis during the early stages of establishment.

epigynous. Describes a flower, or preferably the petals, sepals and stamens (or perianth and stamens) of a flower in which the ovary is inferior (figure G3(13 & 14), p. 614).

epigynous zone. A rim or cup of tissue on which the sepals, petals and stamens are borne in some flowers with inferior ovaries (figure G3(14), p. 614).

epiphyte. A plant which grows on another plant but does not derive any nutriment from it.

equitant. Used of leaves folded so that they are V-shaped in section at the base, the bases overlapping regularly, as in many Iridaceae (figure G1(5), p. 612).

exocarp. The outer, skin-like layer of a fruit-wall in those fruits in which the wall is distinctly 3-layered.

exstipulate. Without stipules.

fall. See Iridaceae (figure 10, Vol. 1, p. 243).

farina. The flour-like wax present on the stem and leaves of many species of Primula and of a few other plants.

fascicle. A bunch of leaves often enclosed at the base by a sheath.

fasciculate. A cluster or bundle, the unit of which are independent but appear to arise from a common point.

fastigiate. With all branches more or less erect, giving the plant a narrow tower-like outline.

filament. The stalk of a stamen, bearing the anther at its tip (figure G3(3 & 4) p. 614).

filius. Used with authority names to distinguish between parent and offspring when both have given names to species, e.g. Linnaeus (C. Linnaeus, 1707–1778), Linnaeus filius (C. Linnaeus, 1741–1783, son of the former).

fimbriate. Fringed.

flagellate. Bearing whip-like structures

floret. A small flower, aggregated with others into a compact inflorescence.

folioferous. Bearing leaf-like structures.

follicle. A dry dehiscent fruit derived from a single free carpel, and with a single line of dehiscence.

free. Not united to any other organ except by its basal attachment.

free-central. A form of placentation in which the ovules are attached to the central axis of a 1-celled ovary (figure G4(5), p. 615).

fruit. The structure into which the gynaecium is transformed during the ripening of the seeds; a *compound fruit* is derived from the gynaecia of more than one flower. The term 'fruit' is often extended to include structures which are derived in part from the receptacle *(Fragaria)*, epigynous zone *(Malus)* or inflorescence-stalk (Ficus) as well as from the gynaecium.

funicle. The stalk of an ovule (figure G4(1), p. 615).

fusiform. Spindle-shaped; cylindric, but tapered gradually at both ends.

gland-dotted. With minute patches of secretory tissue usually appearing as pits on the surface, as translucent dots when held up to the light, or both.

glandular. (1) Of a hair: bearing at the tip a usually spherical knob of secretory tissue. (2) Of a tooth: similarly knobbed or swollen at the tip.

glaucous. Green strongly tinged with bluish grey; with a greyish waxy bloom.

glochid. See Cactaceae, Vol. 2, p. 209.

glume. A small bract in the inflorescence of a grass (Gramineae) or sedge (Cyperaceae); also used in a narrower sense to denote the 2 small bracts at the base of a grass-spikelet (figure 13, Vol. 1, p. 326).

graft-hybrid. A plant which, as a consequence of grafting, contains a mixture of tissues from 2 different species. Normally the tissues of 1 species are enclosed in a 'eskin' of tissue from the other species.

gynaecium. The female organs (carpels) of a single flower, considered collectively, whether they are free or united.

gynophore. The stalk which is present at the base of some pvaries (and the fruits developed from them).

half-inferior. Of an ovary: with its lower part inferior and its upper part superior (figure G3(15), p. 614).

haploid. Possessing in its normal vegetative cells only a single set of chromosomes.

hastate. With 2 acute, divergent lobes at the base, as in a mediaeval halberd (figure G1 (12), p. 612).

hastula. See Vol. 1, p. 361 (Palmae).

haustorium. The organ with which a parasitic plant penetrates its host and draws nutriment from it.

herb. A plant in which the stems do not become woody, or, if somewhat woody at the base, do not persist from year to year.

herbaceous. Of a plant: possessing the qualities of a herb as defined above.

heterostylic. Having flowers in which the length of the style relative to that of the stamens varies from one plant to another.

hilum. The scar-like mark on a seed indicating the point at which it was attached to the funicle.

hybrid. A plant produced by the crossing of parents belonging to 2 different named groups (e.g. genera, species, subspecies etc.). An F1 hybrid is the primary product of such a cross. An F2 hybrid is a plant arising from a cross between 2 F1 hybrids (or from the self-pollination of an F1 hybrid).

hydathode. A water-secreting gland immersed in the tissue of a leaf near its margin.

hypocotyl. That part of the stem of a seedling which lies between the top of the radicle and the attachment of the cotyledon(s).

hypogeal. The mode of germination in which the cotyledons remain in the seed-coat and play no part in photosynthesis.

hypogynous. Describes a flower, or, preferably the petals, sepals and stamens (or perianth and stamens) of a flower in which the ovary is superior and the petals, sepals and stamens (or perianth and stamens) arise as individual whorls on the receptacle (figure G3(10), p. 614).

incised. With deep, narrow spaces between the teeth or lobes.
included. Not projecting beyond the organs which enclose it.
indefinite. More than 12 and possibly variable in number.
indehiscent. Without preformed lines of splitting; opening, if at all, irregularly by decay.
indumentum. Any hairy covering.
induplicate. See Vol. 1, p. 361 (Palmae).
inferior. Of an ovary: borne beneath the sepals, petals and stamens (or perianth and stamens) so that these appear to arise from its top (figure G3(13 & 14), p. 614).
inflorescence. A number of flowers which are sufficiently closely grouped together to form a structural unit (figure G2, p. 613).
infraspecific. Denotes any category below species level, such as subspecies, variety and form. To be distinguished from *subspecific*, which means relating to subspecies only.
integument. The covering of an ovule, later developing into the seed-coat. Some ovules have a single integument, others 2.
internode. The part of a stem between 2 successive nodes.
interprimary vein. See Vol. 1, p. 373 (Araceae).
involucel. A whorl of united bracteoles borne below the flower in some Dipsacaeae.
involucre. A compact cluster of bracts around the stalk at or near the base of some flowers or inflorescences or around the base of a capitulum; sometimes reduced to a hair or ring of hairs.
keel. A narrow ridge, suggestive of the keel of a boat, developed along the midrib (or rarely other veins) of a leaf, petal, sepal, or glume.
labellum. See Orchidaceae, vol. 1, p. 439.
lacinidate. With the margin deeply and irregularly divided into narrow and unequal teeth.
lamellate. Made up of thin plates.
lanceolate. 3–4 times as long as wide and tapering more gradually towards the tip (figure G1(9), p. 612).
layer. To propagate by pegging down on the ground a branch from near the base of a shrub or tree, so as to induce the formation of adventitious roots.
leaflet. One of the leaf-like components of a compound leaf (figure G1(15), p. 612).
legume. The typical fruit of the Fabaceae, formed from a single carpel and opening down both sutures; legumes are, however, variable, some indehiscent, others breaking into 1-seeded segments, etc.
lemma. The lower and usually stouter of the 2 horny or membranous bracts which enclose the floret of a grass (Gramineae) (figure 13), Vol. 1, p. 326).
lenticel. A small, slightly raised interruption of the surface of the bark (or of the outer corky layer of a fruit) through which air can penetrate to the inner tissues.
lignotuber. A woody tuber, generally underground.
ligule. (1) A small membranous flap (more rarely a line or hairs) at the base of a leaf-blade. (figure 13, Vol. 1, p. 326. (2) A whitish area, superficially somewhat similar to the ligule of a grass at the base of the upper surface of the leaf-blade in members of the Cyperaceae (figure 12, Vol. 1, p. 303).
limb. A broadened part, furthest from the base, of a petal, corolla or similar organ, which has a relatively narrow basal part – the *claw* or tube (figure G3(5 & 6), p. 614).
linear. Parallel-sided and many times longer than broad (figure G1(6), p. 612).
lip. (1) That petal in an orchid flower which differs from the other 2; it usually occupies the lowest position in the flower and serves as an alighting place for insects (figure 11, Vol. 1, p. 271). (2) A staminode or petal, suggestive of the lip of an orchid, in the flower of some other monocotyledonous families (e.g. Zingiberaceae). (3) A major division of the apical part of a bilaterally symmetric calyx or corolla in which the petals or sepals are united; there is normally an upper and a lower lip, but either may be missing (e.g. Labiatae).
lodicule. Small, scale-like structures in the floret of most grasses (Gramineae) which occure between the lemma and the reproductive organs. By their swelling they cause the lemma and palea to diverge at flowering time (figure 13(8), Vol. 1, p. 326).
lyrate. Pinnatifid or pinnatisect, with a large terminal and small lateral lobes.
marginal. Of placentation: describing the placentation found in a free carpel of which contains more than 1 ovule.
medifixed. Of a hair: lying parallel to the surface on which it is borne and attached to it by a stalk (usually short) at its mid-point.
membranous. Thin and semi-transparent.
mericarp. A carpel, usually 1-seeded, released by the break-up at maturity of a fruit formed from 2 or more joined carpels.
mesocarp. The central, often fleshy layer of a fruit-wall in those fruits in which the wall is distinctly 3-layered.
micropyle. A pore in the integument(s) of an ovule and the coat of a seed (figure G4 (1), p. 615)
monocarpic. Flowering and fruiting once and then dying.
monoecious. With separate male and female flowers or cones on the same plant; in flowering plants male flowers may contain non-functional carpels (and *vice versa*).
monopodial. A type of growth-pattern in which the terminal bud continues growth from year to year.
mucronate. Provided with a short narrow point at the apex (figure G1(25), p. 612).
mycorrhiza. A symbiotic association between the roots of a green plant and a fungus.
nectary. A nectar-secreting gland.
neuter. Without either functional male or female parts.
node. The point at which 1 or more leaves or flower parts are attached to an axis.
nut. A 1-seeded indehiscent fruit with a woody or bony wall.
nutlet. A small nut, often a component of an aggregate fruit.
obconical. Shaped like a cone, but attached at the narrow end.
obcordate. Inversely heart-shaped, the notch being apical.
oblanceolate. As *lanceolate*, but attached at the more gradually tapered end.
oblong. With more or less parallel sides and about 2–5 times as long as broad (figure G1(7), p. 612).
obovate. As *ovate*, but attached at the narrower end.
obovoid. As *ovoid*, but attached at the narrower end.
obsolete. Rudimentary: scarcely visible; reduced to insignificance.
ochrea. A sheath, made up of the stipules of each leaf-base, which surrounds the stem in many species of Polygonaceae; the ochrea is oftern scarious.
opposite. Describes 2 leaves, branches or flowers attached on opposite sides of the axis at the same node.
orthotropous. Describes an ovule which stands erect and straight (figure G4(1), p. 615).
ovary. The lower part of a carpel, containing the ovule(s) (i.e. excluding style and

stigma); the lower, ovule-containing part of a gynaecium in which the carpels are united (figure G3(2), p. 614).
ovate. With approximately the outline of a hen's egg (though not necessarily blunt-tipped) and attached at the broader end (figure G1(10), p. 612).
ovoid. As *ovate*, but applied to a solid body.
ovule. The small body from which a seed develops after pollination (figure G4, p. 615).
palea. The upper, and usually smaller and thinner, of the 2 bracts enclosing the floret of grass (Gramineae) (figure13, Vol. 1, p. 326).
palate. A swelling in the mouth of the corolla-tube.
palmate. Describes a compound leaf composed of more than 3 leaflets, all arising from the same point, as in the leaf of *Aesculus* (figure G1(19), p. 612).
palmatifid. Lobed in a palmate manner, with the incisions pointing to the place of attachment, but not reaching much more than halfway to it (figure G1(20), p. 612).
palmatisect. Deeply lobed in a palmate manner, with the incisions almost reaching the base (figure G1(21), p. 612).
panicle. A compound raceme, or any freely branched inflorescence of similar appearance (figure G2(3 & 5), p. 613).
papillose. Covered with small blunt protuberances (papillae).
parietal. A form of placentation in which the placentas are borne on the inner surface of the walls of a 1-celled ovary (figure G4 (11), p. 615).
pectinate. With leaves or leaflets in 2 opposite, regular, eye-lash-like rows.
pedate. With a terminal lobe or leaflet, and on either side of it an axis curving outwards and backwards, bearing lobes or leaflets on the outer side of the curve (figure G1(22), p. 612).
peltate. Describes a leaf or other flat structure with a stalk attached other than at the margin (figure G1(27), p. 612).
pepo. A gourd fruit; a 1-celled, many-seeded, inferior frui, with parietal placentas and pulpy interior.
perennial. Persisting for more than 2 years.
perfoliate. Describes a pair of stalkless opposite leaves of which the bases are united, or a single leaf in which the auricles are united, so that the stem appears to pass through the leaf or leaves (figure G1(3 & 4), p. 612).
perianth. The calyx and corolla considered collectively, used especially when there is no clear differentiation between calyx and corolla; used also to denote a calyx or corolla when the other is absent (figure G3(2), p. 614).
pericarpel. See Cactaceae, Vol. 2, p. 211.
perigynous. Describing a flower, or, preferably the petals, sepals and stamens (or perianth and stamens) of a flower in which the ovary is superior and the petals, sepals and stamens (or perianth and stamens) are borne on the margins of a rim or cup which itself is borne on the receptacle below the ovary (it often appears as though the sepals, petals and stamens, or perianth and stamens, are united at their bases) (figure G3(11 & 12), p. 614).
perigynous zone. The rim or cup of tissue on which the sepals, petals and stamens (or perianth and stamens) are borne in a perigynous flower (figure G3(11 & 12), p. 614).
petal. A member of the inner perianth-whorl (corolla) used mainly when this is clearly differentiated from the calyx. The petals usually function in display and often provide an alighting place for pollinators (figure G3(1), p. 614).
petaloid. Like a petal in texture and colour.
phyllode. A leaf-stalk taking on the function and, to a variable extent, the form of a leaf-blade.
pinna. The primary division of a pinnate frond or leaf; it may be simple or itself divided.
pinnate. Describes a compound leaf or frond in which distinct leaflets or pinnae are arranged on either side of the axis or rachis (figure G1(15), p. 612). If these leaflets are themselves of a similar compound structure the leaf or frond is termed bipinnate (similarly, tripinnate, etc.).
pinnatifid. Lobed in a pinnate manner, with the incision reaching not much more than halfway to the axis or rachis (figure G1(16), p. 612).
pinnatisect. Deeply lobed in a pinnate manner, with the incisions almost reaching the axis or rachis (figure G1 (17), p. 612).
pistillode. A sterile ovary in a male flower.
placenta. A part of the ovary, often in the form of a cushion or ridge, to which ovules are attached.
placentation. The manner of arrangement of the placentas.
plicate. Of young leaves in bud, folded along the main veins.
pollen-sac. One of the cavities in an anther in which the pollen is formed; in flowering plants each anther normally contains 4 pollen-sacs, 2 on either side of the connective and separated by a partition which shrivels at maturity.
pollinium. Regularly shaped masses of pollen formed by a large number of pollen grains cohering (Orchidaceae: figure 23, Vol. 1, p. 445).
polyploid. Possessing in its normal vegetative cells more than 2 sets of chromosomes.
proliferous. Giving rise to plantlets or additional flowers on stems, leaves or in the inflorescence.
protandrous. With anthers beginning to shed their pollen before the stigmas in the same flower are receptive.
protogynous. With stigmas becoming receptive before the anthers in the same flower shed their pollen.
pseudobulb. A swollen, above-ground internode, green (at least when young), (figure 19, Vol. 1, p. 441).
ptyxis. The way in which a leaf is rolled or folded (or not) in bud.
pulvinus. A swollen region at the base of a leaflet, leaf-blade or leaf-stalk.
pustulate. With low blister-like projections.
pyrene. A small nut-like body enclosing a seed, 1 or more of which, surrounded by fleshy tissue, make up the fruit of, for example, *Ilex*.
raceme. An inflorescence consisting of stalked flowers arranged on a single axis, the lower opening first (figure G2(2), p. 613).
rachilla. See Vol. 1, p. 324 (Gramineae).
rachis. Of a compound fern frond or pinna: the central axis.
radially symmetric. Capable of division into 2 similar halves along 2 or more planes of symmetry (figure G3(8), p. 614).
radiate. (1) Spreading outwards from a common centre. (2) Possessing ray flowers as in Compositae.
radicle. The root preformed in the seed and normally the first visible root of a seedling.
ramiform. Bearing branches.
raphe. A perceptible ridge or stripe, at one end of which is the hilum, on some seeds.
receptacle. The tip of an axis to which the floral parts or perigynous zone (when present) are attached (figure G3(1), p. 614).
reduplicate. See Vol. 1, p. 361 (Palmae).
reflexed. Bent sharply backwards from the base.

resupinate. See Vol. 1, p. 439 (Orchidaceae).
rhizome. A horizontal stem, situated underground or on the surface, serving the purpose of food storage or vegetative reproduction or both; roots or stems arise from some or all of its nodes.
rootstock. The compact mass of tissue from which arise the new shoots of a herbaceous perennial. It usually consists mainly of stem tissue, but is more compact than is generally understood by rhizome.
rostellum. See Vol. 1, p. 439 (Orchidaceae).
ruminate. See Vol. 1, p. 361 (Palmae).
runner. A slender above-ground stolon with very long internodes.
sagittate. With a backwardly directed basal lobe on either side, like an arrow-head (figure G1(11), p. 612).
samara. A winged, dry, indehiscent fruit or mericarp.
saprophytic. Dependent for its nutrition on soluble organic compounds in the soil. Saprophytic plants do not photosynthesise and lack chlorophyll; some plants, however, are *partially saprophytic* and combine the two modes of nutrition.
scabrid. Roughly hairy.
scale-leaf. A reduced leaf, usually not photosynthetic.
scape. A leafless flower-stalk or inflorescence-stalk arising usually from ground level.
scarious. Dry and papery, often translucent.
schizocarp. A fruit which, at maturity, splits into its constituent mericarps.
scion. A branch cut from one plant to be grafted on the rooted stock of another.
seed. A reproductive body adapted for dispersal, developed from an ovule and consisting of a protective covering (the seed-coat), an embryo and, usually, a food reserve.
semi-parasite. A plant which obtains only part of its nourishment by parasitism.
sepal. A member of the outer perianth whorl (calyx) when 2 whorls are clearly differentiated as calyx and corolla, or when comparison with related plants shows that the corolla is absent. The sepals most often function in protection and support of other floral parts (figure G3(1), p. 614).
septum. An internal partition.
sheath. The part of a leaf or leaf-stalk which surrounds the stem, being either tubular or with free but overlapping edges.
shrub. A woody plant with several stems or branches from near the base, and of smaller stature than a tree.
simple. Not divided into separate parts.
sinus. The gap or indentation between 2 lobes or teeth.
spadix. A spike with numerous florets borne on a usually fleshy axis which may project beyond the topmost floret, often wholly or partly enclosed by a large bract (*spathe*) (figure 15, Vol. 1, p. 377).
spathe. A large bract at the base of a flower or inflorescence and wholly or partly enclosing it. The term is used in some families to denote collectively 2 or 3 such bracts.
spathulate. With a narrow basal part, which towards the apex is gradually expanded into a broad blunt blade.
spicate. Similar to a spike.
spike. An inflorescence or subdivision of an inflorescence consisting of stalkless flowers arranged on a single axis (figure G2(1), p. 613).
spikelet. A small spike forming one of the units of a complex inflorescence (figure 13, Vol. 1, p. 326).
spur. An appendage or prolongation, more or less conical or cylindric, often at the base of an organ. The spur of a corolla or single petal or sepal is usually hollow and often contains nectar (figure G3(7), p. 614)
stamen. The male organ, producing pollen, generally consisting of an anther borne on a filament (figure G3(1), p. 614).
staminode. An infertile stamen, often reduced or rudimentary or with a changed function.
standard. See Iridaceae (figure 10, Vol. 1, p. 243).
stellate. Star-like, particularly of branched hairs.
stigma. The part of a style to which pollen adheres, normally differing in texture from the rest of the style (figure G3(2), p. 614).
stipule. An appendage, usually 1 of a pair, found beside the base of the leaf-stalk in many flowering plants, sometimes falling early, leaving scars. In some cases the 2 stipules are united; in others they are partly united to the leaf-stalk.
stock. A rooted plant, often with the upper part removed, on to which a scion may be grafted.
stolon. A far-creeping, more or less slender, above-ground or underground rhizome giving rise to a new plant at its tip and sometimes at intermediate nodes.
stoma. A microscopic ventilating pore in the surface of a leaf or other herbaceous part of a plant.
strophide. An oily appendage on a seed.
style. The usually slender upper part of a carpel or gynaecium, bearing the stigma (figure G3(2), p. 614).
subtend(ed). Used of any structure (e.g. a flower) which occurs in the axil of another organ (e.g. a bract); in this case the bract subtends the flower.
subulate. Narrowly cylindric, and somewhat tapered to the tip.
sucker. An erect shoot originating from a bud on a root or a rhizome, sometimes at some distance from the stem of the parent plant.
superior. Of an ovary: borne at the morphological apex of the flower so that the petals, sepals and stamens (or perianth and stamens) arise on the receptacle below the ovary (figure G3 (10–12), p. 614).
supervolute. Of young leaves in bud: rolled vertically so that one margin is inside and one outside.
suture. A line marking an apparent junction of neighbouring organs.
sympodial. A type of growth pattern in which the terminal bud ceases growth (often by the production of an inflorescence), further growth being carried on by lateral buds (see Vol. 1, p. 439 (Orchidaceae)).
syncarp. A multiple fruit made up of the products of several ovaries.
tendril. A thread-like structure which by its coiling growth can attach a shoot to something else for support.
terete. Approximately circular in cross-section; not necessarily perfectly cylindric, but without grooves or ridges.
tessellated. With a chequered pattern of light and dark squares.
tetraploid. Possessing in its normal vegetative cells 4 similar sets of chromosomes.
throat. The part of a calyx or corolla transitional between tube and limb or lobes.
trichome. See Cactaceae, Vol. 2, p. 209.
trigonous. 3-angled.
triploid. Possessing in its normal vegetative cells 3 similar sets of chromosomes. Most triploid plants are highly sterile.
tristylic. Having the flowers of different plants with long, short, or intermediate-length styles; the stamens of each flower are of 2 lengths which are not the same as the style-length of that flower.
truncate. As though with the tip or base cut off at right angles.
tuber. A swollen underground stem or root used for food-storage.

tubercle. A small, blunt, wart-like protuberance.

tunic. The dead covering of a bulb or corm.

turion. A specialised perennating bud in some aquatic plants, consisting of a short shoot covered in closely packed leaves, which persists through the winter at the bottom of the water.

umbel. An inflorescence in which the flower-stalks arise together from the top of an inflorescence-stalk. This is a *simple umbel* (figure G2(8), p. 613); in a *compound umbel* the several stalks arising from the top of the inflorescence-stalk terminate not in flowers but in secondary umbels.

undivided. Without major divisions or incisions, though not necessarily entire.

urceolate. Shaped like a pitcher or urn, hollow and contracted at or just below the mouth.

valve. The part of the fruit, covering the seeds, that falls from the septum during dehiscence (Cruciferae).

vascular bundle. A strand of conducting tissue, usually surrounded by softer tissue.

vein. A vascular strand, usually in leaves or floral parts and visible externally.

venation. The pattern formed by the veins in a leaf.

ventricose. Swollen or inflated on one side.

versatile. Of an anther: flexibly attached to the filament by its approximate mid-point so that a rocking motion is possible.

vessel. A microscopic water-conducting tube formed by a sequence of cells not separated by end-walls.

viscidium. See Vol. 1, p. 439 (Orchidaceae).

viviparous. Bearing young plants, bulbils or leafy buds which can take root; they can occur anywhere on the plant and may be interspersed with, or wholly replace, the flowers in an inflorescence.

whorl. A group of more than 2 leaves or floral organs inserted at the same node.

wing. A thin, flat extension of a fruit, seed, sepal or other organ.

xerophytic. Drought tolerant. Can also describe the environment in which drought-tolerant plants live.

zygomorphic. Bilaterally symmetric.

INDEX